22. $\displaystyle\int \frac{dx}{(a^2 + x^2)^2} = \frac{x}{2a^2(a^2 + x^2)} + \frac{1}{2a^3}\tan^{-1}\frac{x}{a} + C$

23. $\displaystyle\int \frac{dx}{(ax^2 + bx + c)^{n+1}} = \frac{2ax + b}{n(4ac - b^2)(ax^2 + bx + c)^n} + \frac{2(2n - 1)a}{n(4ac - b^2)}\int \frac{dx}{(ax^2 + bx + c)^n},$

$$\text{if } n > 0 \text{ and } b^2 \neq 4ac$$

24. $\displaystyle\int \frac{dx}{a^2 - x^2} = \frac{1}{2a}\ln\left|\frac{x + a}{x - a}\right| + C$

25. $\displaystyle\int \frac{dx}{(a^2 - x^2)^2} = \frac{x}{2a^2(a^2 - x^2)} + \frac{1}{2a^2}\int \frac{dx}{a^2 - x^2}$

26. $\displaystyle\int \sqrt{a^2 + x^2}\,dx = \frac{x}{2}\sqrt{a^2 + x^2} + \frac{a^2}{2}\ln(x + \sqrt{x^2 + a^2}) + C$

27. $\displaystyle\int x^2\sqrt{a^2 + x^2}\,dx = \frac{x(a^2 + 2x^2)\sqrt{a^2 + x^2}}{8} - \frac{a^4}{8}\ln(x + \sqrt{x^2 + a^2}) + C$

28. $\displaystyle\int \frac{\sqrt{a^2 + x^2}}{x}\,dx = \sqrt{a^2 + x^2} - a\ln\left(\frac{a + \sqrt{a^2 + x^2}}{x}\right) + C$

29. $\displaystyle\int \frac{\sqrt{a^2 + x^2}}{x^2}\,dx = -\frac{\sqrt{a^2 + x^2}}{x} + \ln(x + \sqrt{a^2 + x^2}) + C$

30. $\displaystyle\int \frac{dx}{\sqrt{a^2 + x^2}} = \ln(x + \sqrt{a^2 + x^2}) + C$

31. $\displaystyle\int \frac{x^2}{\sqrt{a^2 + x^2}}\,dx = \frac{x}{2}\sqrt{a^2 + x^2} - \frac{a^2}{2}\ln(x + \sqrt{a^2 + x^2}) + C$

32. $\displaystyle\int \frac{dx}{x\sqrt{a^2 + x^2}} = -\frac{1}{a}\ln\left|\frac{a + \sqrt{a^2 + x^2}}{x}\right| + C$

33. $\displaystyle\int \frac{dx}{x^2\sqrt{a^2 + x^2}} = -\frac{\sqrt{a^2 + x^2}}{a^2 x} + C$

34. $\displaystyle\int \sqrt{a^2 - x^2}\,dx = \frac{x}{2}\sqrt{a^2 - x^2} + \frac{a^2}{2}\sin^{-1}\frac{x}{a} + C$

35. $\displaystyle\int x^2\sqrt{a^2 - x^2}\,dx = \frac{a^4}{8}\sin^{-1}\frac{x}{a} - \frac{1}{8}x\sqrt{a^2 - x^2}(a^2 - 2x^2) + C$

36. $\displaystyle\int \frac{\sqrt{a^2 - x^2}}{x}\,dx = \sqrt{a^2 - x^2} - a\ln\left|\frac{a + \sqrt{a^2 - x^2}}{x}\right| + C$

37. $\displaystyle\int \frac{\sqrt{a^2 - x^2}}{x^2}\,dx = -\sin^{-1}\frac{x}{a} - \frac{\sqrt{a^2 - x^2}}{x} + C$

38. $\displaystyle\int \frac{dx}{\sqrt{a^2 - x^2}} = \sin^{-1}\frac{x}{a} + C$

39. $\displaystyle\int \frac{x^2}{\sqrt{a^2 - x^2}}\,dx = \frac{a^2}{2}\sin^{-1}\frac{x}{a} - \frac{1}{2}x\sqrt{a^2 - x^2} + C$

40. $\displaystyle\int \frac{dx}{x\sqrt{a^2 - x^2}} = -\frac{1}{a}\ln\left|\frac{a + \sqrt{a^2 - x^2}}{x}\right| + C$

41. $\displaystyle\int \frac{dx}{x^2\sqrt{a^2 - x^2}} = -\frac{\sqrt{a^2 - x^2}}{a^2 x} + C$

42. $\displaystyle\int \sqrt{x^2 - a^2}\,dx = \frac{x}{2}\sqrt{x^2 - a^2} - \frac{a^2}{2}\ln|x + \sqrt{x^2 - a^2}| + C$

43. $\displaystyle\int (\sqrt{x^2 - a^2})^n\,dx = \frac{x(\sqrt{x^2 - a^2})^n}{n + 1} - \frac{na^2}{n + 1}\int (\sqrt{x^2 - a^2})^{n-2}\,dx, \qquad n \neq -1$

John B. Fraleigh   University of Rhode Island

CALCULUS

ic
etry

dition

TES *by VICTOR KATZ*
University of District of Columbia

LEY PUBLISHING COMPANY

etts • Menlo Park, California • New York
• Wokingham, England • Amsterdam • Bonn
• Tokyo • Madrid • San Juan

1032543   3

WORLD STUDENT SERIES EDITION

Sponsoring Editor: David Pallai
Production Supervisor: Jack Casteel
Editorial and Production Services: Barbara G. Flanagan
Text Designer: Nancy Blodget/Books By Design
Illustrator: Illustrated Arts
Art Consultant: Loretta Bailey
Manufacturing Supervisor: Roy Logan
Cover Design: Marshall Henrichs

0201503638

PREFACE

This text is designed for a standard calculus sequence for students in the physical or social sciences. Such a sequence typically spans three semesters or four quarters. Students are expected to have a background of algebra and geometry, including some analytic geometry.

The changes in this third edition, involving topic presentation and order, are substantial. Several recommendations arising from the Tulane conference on calculus have been implemented, including

- a complete **dictionary of functions,**
- early treatment of the **trigonometric, exponential, and logarithm functions,**
- a **graphically oriented approach,**
- **calculator and computer exercises** with references to software.

Some Features Retained from the Previous Edition

- Informal expository style
- Intuitive explanations preceding formal definitions and theorems
- Standard notation and topic coverage
- Abundance of artwork, worked-out examples, and exercises
- Section summaries to emphasize main results and aid review
- Clearly marked exercises for a calculator or computer
- Answers to odd-numbered problems

Features New to This Third Edition

- The precalculus review in Chapter 1 closes with a complete **dictionary of functions.** For many students, the precalculus review now will not be necessary; occasional reference to the dictionary of functions will suffice.
- The study of **limits** in Chapter 2 has been streamlined. The formal ε, δ-definition has been delayed to optional Section 2.5, which also includes proofs of the sum, product, and quotient properties of limits.
- The **base e** for the exponential function is given a graphically oriented presentation in Chapter 3. The importance of the exponential function in some rate-of-growth situations is explained at once.

- Differentiation of **trigonometric, exponential, and logarithm functions** appears in Chapter 3, immediately following differentiation of polynomial functions and the product and quotient rules.

- Discussion of the **differential** is delayed to the final section of Chapter 3. The standard-analysis interpretation of differential notation, which students encounter in many other courses, is given an especially careful treatment here.

- The intuitive presentation of the **chain rule** in Section 3.5 is followed by a proof that is valid for many functions, but the complete proof is deferred to the final section of Chapter 3.

- An optional integral-based development of **ln** x and e^x is given in Section 7.2.

- **Taylor's theorem** is now treated before infinite series, providing better motivation for the study of series. Indeterminate forms also are studied before series.

- The treatment of **plane curves** and **polar coordinates** has been streamlined and is covered in a single chapter.

- The chapter on multiple integrals concludes with a new, optional section on **change of variables.**

- **Differential forms** and **line integrals** now appear at the start of the chapter containing divergence theorems, Green's theorem, and Stokes' theorem for a cohesive treatment.

- More **graphic-based exercises** have been included, emphasizing geometrically the derivative and the integral.

- In addition to clearly marked calculator and computer exercises, references to specific programs in the **Exploring Calculus software** are given at the end of appropriate sections.

- Some **historical notes** now appear.

Supplements for the Instructor

- **Complete Solutions Manual**   Contains the worked-out solutions for *all* the exercises in the text. Prepared by John B. Fraleigh, University of Rhode Island.

- **Complete Answer Book**   Contains the answers to all the computational exercises in the text.

- **Computerized Testing System—TestEdit**   Contains more than 3500 problems easily accessed by the computer in either multiple-choice or open-ended format. This testing system features problem editing, option to leave space for working exercises on the test, scrambling problems and answers, and printed answer keys. Available for the IBM PC.

- **Printed Test Bank**   Contains more than 3500 test questions to be used in creating quizzes or tests.

- **Transparency Masters**   Printed masters to create overhead projections of key theorems, definitions, proofs, tables, and figures.

Software-Related Supplements for the Instructor and the Student

- **Calculus Toolkit 2.0** Consists of thirty-three programs ranging from functions to vector fields. Enables the instructor and students to use the microcomputer as an "electronic chalkboard." Three-dimensional graphics are incorporated where appropriate. Available for both the Apple and the IBM PC.

- **Computer Explorations in Calculus** Activities and worksheets (using the Calculus Toolkit) explore concepts of calculus on the computer.

- **Student Edition of MathCAD** A very powerful free-form scratchpad. When the student inputs equations, MathCAD automatically calculates and displays the results as numbers or graphs. It also allows the student to plot results, to annotate the work with text, and to print the entire document. Available for the IBM PC.

- **Mathematical Modeling with MathCAD** Contains activities using MathCAD.

- **Exploring Calculus** Contains thirty-four programs that illustrate calculus principles by using twenty graphical displays, quizzes, demonstrations, tutorials, and games. Exercise manual. Prepared by John B. Fraleigh and Lewis I. Pakula, University of Rhode Island. Available for the IBM PC.

- **Master Grapher and 3D Grapher** A powerful interactive graphing utility for functions, polar equations, parametric equations, and other functions in two and three variables. Prepared by Franklin Demana and Bert Waits, Ohio State University. Available for MacIntosh, Apple, and the IBM PC.

- **Math PRO** An interactive problem generator that provides drill and practice exercises and includes help screens. Prepared by Joseph Mazur, Marlboro College. Available for the IBM PC.

This software is free to college instructors upon adoption.

Supplements for the Student

- **Student Solutions Manual** By John B. Fraleigh. Contains worked-out solutions to all *odd-numbered* exercises in the text and includes warnings about frequent errors.

- **Student Study Guide** By Maurice Weir, Naval Postgraduate School. Organized to correspond with the text, this workbook (in a semiprogrammed format) increases student proficiency.

Acknowledgments

Reviewers of text manuscripts perform a vital function in keeping authors in touch with reality. I wish to express my appreciation to all the reviewers of the manuscript for this edition, including the following:

Professor Raymond Beasley
Central State University (Oklahoma)

Professor Jacob Burbea
University of Pittsburgh

Professor Donald Cohen
SUNY–Agricultural and Technical College, Cobleskill

Professor Ben L. Cornelius
Oregon Institute of Technology

Professor Stuart Goldenberg
California Polytechnic State University

Professor Gary Grimes
Mount Hood Community College

Dr. William Keils
San Antonio College

Dr. Lawrence Maher
University of North Texas

Dr. Giles W. Maloof
Boise State University

Professor Sunny L. Norfleet
St. Petersburg Junior College

Professor Phillip Novinger
Florida State University

Professor G. Edgar Parker
James Madison University

Dr. George W. Schultz
St. Petersburg Junior College

Dr. James H. Yates
Central State University (Oklahoma)

I am very grateful to Victor Katz for providing the historical notes. I also wish to thank David Pallai and the rest of the mathematics staff at Addison-Wesley for their help and encouragement in the preparation of this edition.

Kingston, R.I.   J.B.F.
September 1989

A PREVIEW
OF CALCULUS
FOR THE STUDENT

We live in a changing world. When we walk from our living room to our kitchen, we contribute a change in the world. In a static world, everything would be frozen in just one constant condition and position. Calculus studies magnitudes that are changing rather than remaining constant. Calculus is a powerful tool for studying motion and other changing features of the universe.

Our study of calculus can be roughly divided into three parts:

1. *Differential calculus in the one-dimensional case* (Chapters 2–4 and parts of other chapters)

2. *Integral calculus in the one-dimensional case* (Chapters 5, 6, and 8 and parts of other chapters)

3. *Differential and integral calculus in higher-dimensional cases* (Chapters 14–16)

Differential calculus is used to study changing quantities in cases where we would compute a *quotient* if the quantities were held constant. For example, if an airplane travels in a straight line at *constant* speed, then the speed at any time is given by the *quotient*

$$\text{Speed} = \frac{\text{Distance traveled}}{\text{Elapsed time}}.$$

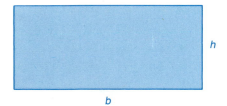

FIGURE 0.1 Rectangle of base width b and constant altitude h. Area $= b \cdot h$.

If the airplane's speed does not remain constant, then this quotient gives the *average speed* of the airplane for a given time period. When faced with nonconstant speed, we will learn to use calculus to find the *speed at any instant* if we know the total distance traveled at every instant of time. Differential calculus allows us to find such *instantaneous rates of change*.

Integral calculus is used to study changing quantities in cases where we would compute a *product* of the quantities if things remained constant. For example, if the speed remains *constant* for the airplane in the preceding paragraph, then the total distance traveled is given by the *product*

$$\text{Distance traveled} = \text{Speed} \times \text{Elapsed time}.$$

If the speed is not constant but is known at every instant of time, then integral calculus enables us to find the distance traveled. For another illustration, the area of the rectangle of base width b and constant altitude h, shown in Fig. 0.1, is the *product* $b \cdot h$. The shaded region in Fig. 0.2 has base width b but varying altitude, depending on where we are on the base. Integral calculus allows us to compute the area of this shaded region.

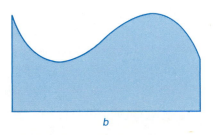

FIGURE 0.2 Region of base width b and varying altitude. Its area is found using integral calculus.

Division and multiplication are very closely related; they are almost different ways of saying the same thing, as in

$$3 = \frac{6}{2} \quad \text{and} \quad 2 \cdot 3 = 6.$$

Since differential calculus is concerned with quotients and integral calculus with products, it is not surprising to find that these two topics are also closely related.

Issac Newton (1642–1727) and Gottfried Leibniz (1646–1716) are generally credited with simultaneous but independent development of calculus as we recognize it today. However, it would be a mistake to suppose that those two great mathematicians invented calculus from scratch all by themselves. Archimedes (272?–212 B.C.) used principles at the heart of integral calculus in his work on the determination of areas of certain types of regions. Other mathematicians used the ideas behind calculus during the century preceding the work of Newton and Leibniz. Newton and Leibniz codified these ideas into roughly the form of calculus today. The independent work of Newton and Leibniz is a clear indication that calculus was then an area of mathematics ready to be born.

CONTENTS

PRECALCULUS REVIEW

1

This chapter provides a review of analytic geometry and functions to serve as a foundation for calculus. Much of the material in the chapter may be familiar to students. The dictionary of functions in Section 1.6 will suffice as reference for some. Those who start by examining this dictionary and have difficulty should refer to the appropriate sections (1.1–1.5) for a more detailed explanation.

1.1 COORDINATES AND DISTANCE

We are concerned primarily with real numbers in this text. A **real number** is a number that can be written as an unending decimal, positive or negative, or zero. For example,

$$3 = 3.000000 \ldots , \quad -\tfrac{2}{3} = -0.666666 \ldots , \quad \text{and} \quad \pi = 3.141592 \ldots$$

are real numbers. Certain types of real numbers have special names.

Integers Whole numbers: $\ldots -4, -3, -2, -1, 0, 1, 2, 3, 4 \ldots$

Rational numbers Fractions m/n, where m and n are integers and $n \neq 0$. Note that every integer can be expressed in this fashion, since $m = m/1$, so every integer is a rational number.

Irrational numbers Real numbers that are not rational. It can be shown that π and $\sqrt{2}$ are irrational numbers.

Pencil-and-paper arithmetic is easiest with integers and rational numbers, so we use them in many of the examples and exercises. Computers have made extensive approximate computations with irrational numbers feasible.

FIGURE 1.1   The number line.

It is visually helpful to identify real numbers with points on the **number line.** We can take a line extending infinitely in both directions and, using the real numbers, can make this line into an infinite ruler (see Fig. 1.1). We label any point on the line with 0 and any point to the right of 0 with 1; this fixes the *scale*. Each positive real number r corresponds to the point a distance r units to the right of 0, while a negative number $-s$ corresponds to the point a distance s units to the left of 0. The arrow on the line indicates the positive direction. The x to the right of the arrow indicates that we think of x as any real number on the line. In this context, x is known as a **real variable,** and the line is called the **x-axis.**

Inequalities

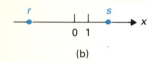

(a)

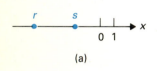

(b)

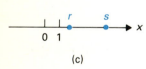

(c)

FIGURE 1.2 $r < s$

For real numbers r and s, the notation $r < s$ (read "r is less than s") means that r is to the left of s on the number line. The three parts of Fig. 1.2 illustrate that this criterion for $r < s$ is valid whether r and s are positive or negative.

EXAMPLE 1   Referring to the line shown in Fig. 1.1, we see that

$$2 < \pi \quad \text{and} \quad -2 < 3,$$

because 2 is to the left of π and -2 is to the left of 3. Note also that

$$-3 < -1,$$

although 3 is larger than 1, because the number -3 is to the left of -1 on the line. We also have

$$-\tfrac{2}{3} < \tfrac{1}{2}.$$

Any negative number is less than any positive number.   ■

This same relation $r < s$ between numbers r and s is sometimes written as $s > r$ (read "s is greater than r"). For example, if we want to consider all numbers greater than 2, we naturally say, "Consider all numbers x greater than 2," mentioning x before 2, and we abbreviate this as "all $x > 2$." The notation $r \leq s$ is read "r is less than or equal to s," while $s \geq r$ is read "s is greater than or equal to r."

Here are four arithmetic laws of inequalities that are often used.

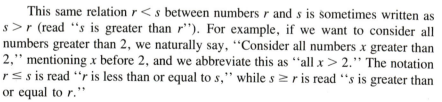

Arithmetic Laws for Inequalities

1. If $a \leq b$, then $a + c \leq b + c$.

2. If $a \leq b$ and $c \leq d$, then $a + c \leq b + d$.

3. If $a \leq b$ and $c > 0$, then $ac \leq bc$.

4. If $a \leq b$ and $c < 0$, then $bc \leq ac$.

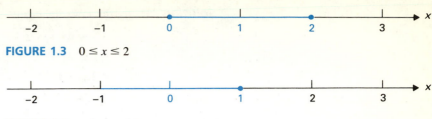

FIGURE 1.3 $0 \leq x \leq 2$

FIGURE 1.4 $-1 < x \leq 1$

EXAMPLE 2   Sketch on the line the points x that satisfy $0 \leq x \leq 2$.

Solution   The points are indicated by the colored line and points in Fig. 1.3. Both 0 and 2 satisfy the relation. ■

EXAMPLE 3   Sketch on the line the points x that satisfy $-1 < x \leq 1$.

Solution   The points are indicated by the colored line and point in Fig. 1.4. In this case -1 does not satisfy the relation, while 1 does. ■

Intervals

Some theorems of calculus deal with segments of the number line. It is customary to call such a segment an *interval* and to use the adjectives *closed* and *open* to describe whether the endpoints are considered part of the interval. Intervals are illustrated in Fig. 1.5.

> **Closed interval [a, b]**   all x such that $a \leq x \leq b$
>
> **Open interval (a, b)**   all x such that $a < x < b$
>
> **Half-open (or half-closed) intervals**
>
> > **[a, b)**   all x such that $a \leq x < b$
> >
> > **(a, b]**   all x such that $a < x \leq b$

EXAMPLE 4   Sketch the intervals $[0, 2]$ and $(-1, 1]$.

Solution   The closed interval $[0, 2]$ is sketched in Fig. 1.3, while Fig. 1.4 shows the half-open interval $(-1, 1]$. ■

EXAMPLE 5   Solutions of an inequality can sometimes be expressed in interval notation. Express the solutions of $-3 \leq 4x + 5 \leq 7$ in interval notation.

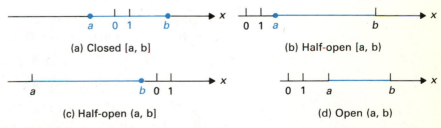

FIGURE 1.5   Intervals: (a) Closed $[a, b]$; (b) half-open $[a, b)$; (c) half-open $(a, b]$; (d) open (a, b).

Solution   Using the laws for inequalities listed before, we obtain

$$-3 \leq 4x + 5 \leq 7,$$
$$-8 \leq 4x \leq 2, \qquad \text{\textit{Adding} } -5$$
$$-2 \leq x \leq \tfrac{1}{2}, \qquad \text{\textit{Multiplying by} } \tfrac{1}{4}$$

which is the closed interval $[-2, \tfrac{1}{2}]$.   ■

Absolute Value

We all have an intuitive idea of the distance between points of the number line. Our notion of distance gives us insight into some other mathematical concepts. The *absolute value* of a number, denoted by $|r|$, is the distance between 0 and r on the number line. For example,

$$|5| = |-5| = 5$$

because both 5 and -5 are five units from 0. We can formally define the **absolute value** of r as

$$|r| = \begin{cases} r & \text{if } r \geq 0, \\ -r & \text{if } r < 0. \end{cases}$$

EXAMPLE 6   Simplify $\left|6 - |-3|\right| + |2 - 7|$.

Solution   We have $|-3| = 3$, so

$$\left|6 - |-3|\right| + |2 - 7| = |6 - 3| + |2 - 7|$$
$$= |3| + |-5| = 3 + 5 = 8.$$ ■

EXAMPLE 7   Find all real numbers x such that $|x|/x = 1$.

Solution   We find that

$$\frac{|x|}{x} = \begin{cases} \dfrac{x}{x} = 1 & \text{for } x > 0, \\[2mm] \dfrac{-x}{x} = -1 & \text{for } x < 0. \end{cases}$$

Thus the solutions of $|x|/x = 1$ consist of all $x > 0$.   ■

EXAMPLE 8   Solutions of an inequality involving absolute values can sometimes be expressed using interval notation. Express the solutions of $|x - 4| \leq 5$ using interval notation.

Solution   Since $|x - 4|$ is the distance from $x - 4$ to the origin, we see that $|x - 4| \leq 5$ means that we must have

$$-5 \leq x - 4 \leq 5.$$

The laws for inequalities show that

$$-1 \leq x \leq 9, \qquad \text{\textit{Adding} } 4$$

which is the closed interval $[-1, 9]$.   ■

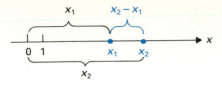

FIGURE 1.6   Distance between x_1 and x_2 if $x_1 \leq x_2$.

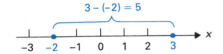

FIGURE 1.7   Distance between -2 and 3.

EXAMPLE 9   Express the solutions of $|(3 - 2x)/4| \leq 5$ using interval notation.

Solution   Since $|(3 - 2x)/4|$ is the distance from $(3 - 2x)/4$ to the origin, we see that $|(3 - 2x)/4| \leq 5$ means that we must have

$$-5 \leq \frac{3 - 2x}{4} \leq 5.$$

The laws for inequalities yield

$$-20 \leq 3 - 2x \leq 20, \qquad \textit{Multiplying by } 4$$

$$-23 \leq -2x \leq 17, \qquad \textit{Adding } -3$$

$$\frac{-23}{-2} \geq x \geq \frac{17}{-2}, \qquad \textit{Multiplying by } -\tfrac{1}{2}$$

$$-\frac{17}{2} \leq x \leq \frac{23}{2},$$

which is the closed interval $[-\frac{17}{2}, \frac{23}{2}]$. ∎

Computing Distance

Consider now the distance between any two points on the number line. The distance between the points x_1 and x_2, shown in Fig. 1.6, is surely $x_2 - x_1$. For any two points x_1 and x_2, where $x_1 \leq x_2$, the distance between them is $x_2 - x_1$.

EXAMPLE 10   Find the distance between -2 and 3 on the number line.

Solution   Since $-2 < 3$, the distance is $3 - (-2) = 5$, as indicated in Fig. 1.7. ∎

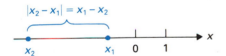

FIGURE 1.8   Distance between x_1 and x_2 if $x_1 \geq x_2$.

For *any* points x_1 and x_2, the distance between them is either $x_1 - x_2$ or $x_2 - x_1$, whichever is nonnegative. This nonnegative magnitude is the absolute value $|x_2 - x_1|$. Thus

$$\text{the distance between } x_1 \text{ and } x_2 \text{ is } |x_2 - x_1|.$$

See Figs. 1.6 and 1.8.

EXAMPLE 11   Use the absolute value formula to find the distance between -2 and 3.

Solution   For the points -2 and 3,

$$|3 - (-2)| = |5| = 5, \qquad \text{and also} \qquad |(-2) - 3| = |-5| = 5. \qquad ∎$$

Exercise 9 asks you to show that $(a + b)/2$ is the same distance from a as from b. This means that

$$\frac{a + b}{2} \text{ is the **midpoint** of } [a, b].$$

See Fig. 1.9.

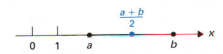

FIGURE 1.9   Midpoint $(a + b)/2$ of $[a, b]$.

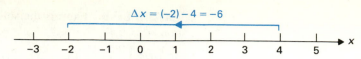

FIGURE 1.10 Directed distance from 4 to -2.

FIGURE 1.11 Directed distance from -8 to -5.

EXAMPLE 12 Find the midpoint of the interval $[-\frac{1}{2}, \frac{2}{3}]$.

Solution The midpoint is

$$\frac{-\frac{1}{2} + \frac{2}{3}}{2} = \frac{\frac{1}{6}}{2} = \frac{1}{12}.$$ ■

Often we need to know not only the distance from x_1 to x_2 but whether x_1 is to the left or right of x_2. The change $x_2 - x_1$ in x-value going from x_1 to x_2 (in that order) is positive if $x_1 < x_2$ and negative if $x_2 < x_1$. In calculus we will let Δx (read "delta x") be such a positive or negative change in x-value. Think, geometrically, of the directed distance

$$\Delta x = x_2 - x_1$$

as the *signed length* of the *directed line segment* from x_1 to x_2.

EXAMPLE 13 Find the directed distances from 4 to -2 and from -8 to -5.

Solution The directed distance from $x_1 = 4$ to $x_2 = -2$ is

$$\Delta x = x_2 - x_1 = -2 - 4 = -6.$$

To get to -2 from 4, we go six units to the *left*. See Fig. 1.10.
The directed distance from -8 to -5 is

$$\Delta x = -5 - (-8) = -5 + 8 = 3.$$

See Fig. 1.11. To get from -8 to -5, we go three units to the *right*. ■

Coordinates in the Plane

We take two number lines and place them perpendicular to each other in a plane, so that they intersect at the point 0 on each line (see Fig. 1.12). The number lines in Fig. 1.12 are called **coordinate axes.** The horizontal axis is the **x-axis** and the vertical axis is the **y-axis,** as shown by the labels at the arrows.

With each point in the plane, we associate an ordered pair (x_1, y_1) of numbers, as follows: The first number, x_1, gives the left–right position of the point according to the location of x_1 on the horizontal number line. Similarly, the second number, y_1, gives the up–down position of the point according to the

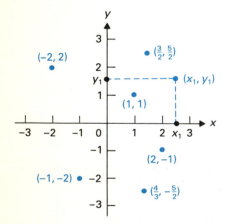

FIGURE 1.12 Coordinates in the plane.

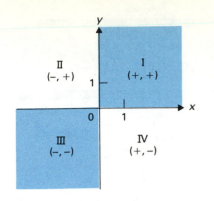

FIGURE 1.13 Quadrants of the plane.

location of y_1 on the vertical number line (see Fig. 1.12). Conversely, given any ordered pair of numbers, such as $(2, -1)$, there is a unique point in the plane associated with it.

EXAMPLE 14 Plot the points $(1, 1)$, $(-2, 2)$, $(-1, -2)$, and $(2, -1)$ in the plane.

Solution The positions of the points are shown in Fig. 1.12. ■

For the point (x_1, y_1), the number x_1 is the **x-coordinate** of the point, and y_1 is the **y-coordinate.** The coordinate axes divide the plane into four **quadrants,** according to the signs of the coordinates of the points. The quadrants are usually numbered as shown in Fig. 1.13. The point $(0, 0)$ is the **origin.** This introduction of coordinates allows us to use numbers and algebra as tools in studying geometry.

EXAMPLE 15 Sketch the portion of the plane consisting of the points (x, y) satisfying the relation $x \leq 1$.

Solution This portion of the plane is shown colored in Fig. 1.14. ■

EXAMPLE 16 Sketch the portion of the plane consisting of the points (x, y) satisfying *both* $-2 \leq x \leq 1$ and $1 \leq y \leq 2$.

Solution This portion of the plane is shown colored in Fig. 1.15. ■

The Distance Formula

We can compute the distance d between two points (x_1, y_1) and (x_2, y_2) in the plane. Look at the right triangle in Fig. 1.16. The lengths of the legs of this triangle are $|x_2 - x_1|$ and $|y_2 - y_1|$. By the Pythagorean theorem,

$$d^2 = |x_2 - x_1|^2 + |y_2 - y_1|^2. \tag{1}$$

Since the terms in Eq. (1) are squared, the absolute value symbols are not needed, and we obtain the following formula.

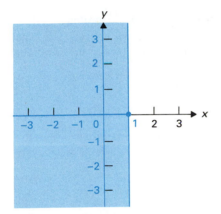

FIGURE 1.14 $x \leq 1$

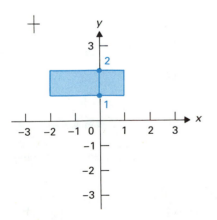

FIGURE 1.15 $-2 \leq x \leq 1$ and $1 \leq y \leq 2$

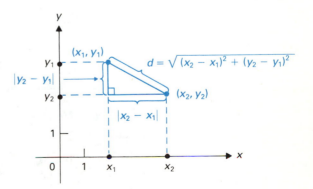

FIGURE 1.16 Distance between (x_1, y_1) and (x_2, y_2).

Distance Between (x_1, y_1) and (x_2, y_2)

$$d = \sqrt{(x_2 - x_1)^2 + (y_2 - y_1)^2}$$

EXAMPLE 17 Find the distance between $(2, -3)$ and $(-1, 1)$ in the plane.

Solution The distance is

$$\sqrt{(-1 - 2)^2 + [1 - (-3)]^2} = \sqrt{(-3)^2 + 4^2}$$
$$= \sqrt{9 + 16} = \sqrt{25} = 5.$$ ∎

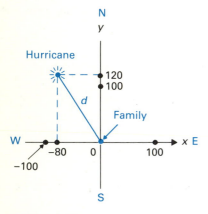

EXAMPLE 18 A hurricane with a destructive range of 150 miles is presently at a point 120 miles north and 80 miles west of a family. Determine whether the family is now within the destructive range of the storm.

Solution We consider the family to be at the origin of a coordinate system with x-axis pointing east and y-axis pointing north, as shown in Fig. 1.17. The storm is at the point $(-80, 120)$. The distance from this point to the family at $(0, 0)$ is then

$$\sqrt{(80 - 0)^2 + (120 - 0)^2} = \sqrt{80^2 + 120^2}$$
$$= \sqrt{6400 + 14{,}400} = \sqrt{20{,}800} \text{ miles.}$$

Since $150 = \sqrt{22{,}500}$, we see that $\sqrt{20{,}800} < 150$, so the family is within the destructive range of the hurricane. ∎

FIGURE 1.17 Is $d < 150$?

Circles

The **circle** with center (h, k) and radius $r \geq 0$ consists of all points (x, y) whose distance from (h, k) is r. See Fig. 1.18. Using the formula for the distance from (x, y) to (h, k), we see that this circle consists of all points (x, y) such that

$$\sqrt{(x - h)^2 + (y - k)^2} = r. \tag{2}$$

Squaring both sides of Eq. (2), we obtain the following equivalent relation.

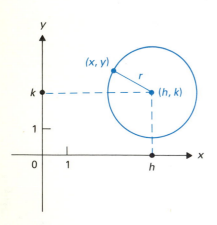

Equation of a Circle

$$(x - h)^2 + (y - k)^2 = r^2 \tag{3}$$

EXAMPLE 19 Find the equation of the circle with center $(-2, 4)$ and radius 5.

Solution From Eq. (3), we see that the circle has equation $[x - (-2)]^2 + (y - 4)^2 = 5^2$, or $(x + 2)^2 + (y - 4)^2 = 25$. The circle is sketched in Fig. 1.19. ∎

FIGURE 1.18 The circle $(x - h)^2 + (y - k)^2 = r^2$.

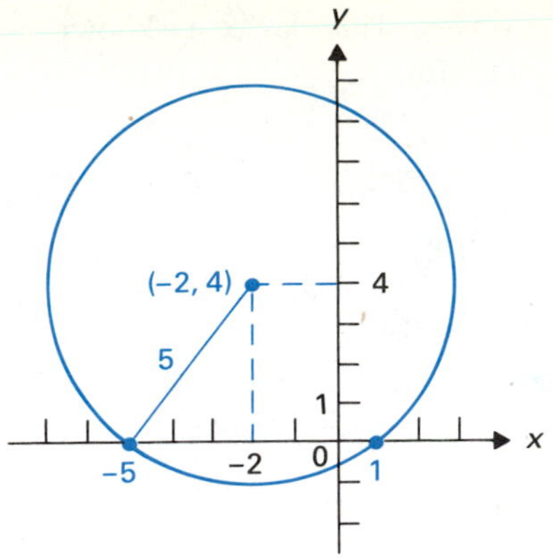

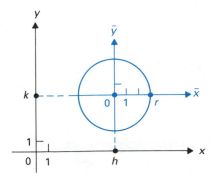

FIGURE 1.19 The circle $(x + 2)^2 + (y - 4)^2 = 25$.

EXAMPLE 20 Find the center and radius of the circle with equation $(x + 3)^2 + (y + 4)^2 = 18$.

Solution Rewriting the equation as

$$[x - (-3)]^2 + [y - (-4)]^2 = (\sqrt{18})^2,$$

we see from Eq. (3) that the circle has center $(-3, -4)$ and radius $\sqrt{18} = 3\sqrt{2}$. ∎

Sometimes an equation can be put in the form of Eq. (3) by an algebraic device known as *completing the square:*

$$x^2 + bx = \left(x^2 + bx + \frac{b^2}{4}\right) - \frac{b^2}{4} = \left(x + \frac{b}{2}\right)^2 - \frac{b^2}{4}.$$

An expression $x^2 + bx$ appearing in an equation can thus be changed to $[x + (b/2)]^2$ by adding $b^2/4$ to both sides of the equation. This device is used in the following example.

EXAMPLE 21 Show that $3x^2 + 3y^2 + 6x - 12y = 60$ describes a circle.

Solution We start by dividing by the common coefficient 3 of x^2 and y^2 and obtain

$$x^2 + y^2 + 2x - 4y = 20.$$

Now we complete the square on both the x-terms and the y-terms to get our equation in the form of Eq. (3). The steps are as follows:

$$(x^2 + 2x) + (y^2 - 4y) = 20,$$
$$(x^2 + 2x + 1) + (y^2 - 4y + 4) = 20 + 1 + 4,$$
$$(x + 1)^2 + (y - 2)^2 = 25.$$

Thus our equation describes a circle with center $(-1, 2)$ and radius 5. ∎

As illustrated by Example 21, every equation of the form $ax^2 + ay^2 + bx + cy = d$ and satisfied by at least one point (x_1, y_1) is the equation of a circle. We can put any such equation in the form of Eq. (3) to find the center and radius of the circle. However, there may be no points of the plane whose coordinates satisfy a particular equation of this type. For example, $x^2 + y^2 = -10$ is not satisfied for any point (x, y) in the plane because a sum of squares of real numbers cannot be negative.

Translation of Axes

There is another way of regarding Eq. (3) that is very useful. If we let $\bar{x} = x - h$ and $\bar{y} = y - k$, then Eq. (3) becomes

$$(\bar{x})^2 + (\bar{y})^2 = r^2. \qquad (4)$$

To interpret Eq. (4) geometrically, we take $\bar{x},\bar{y}$-axes parallel to the x,y-axes with $(x, y) = (h, k)$ as their origin, as shown in Fig. 1.20. Equation (4) is the equa-

FIGURE 1.20 $(\bar{x})^2 + (\bar{y})^2 + r^2$

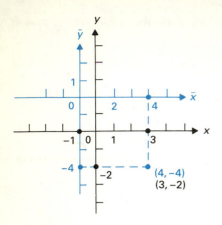

FIGURE 1.21 $(x, y) = (3, -2)$ has coordinates $(\bar{x}, \bar{y}) = (4, -4)$.

tion of the circle with respect to the new $\bar{x}, \bar{y}$-axes. This device is known as *translation of axes to (h, k)* and is often very useful.

Translation Equations

$$\bar{x} = x - h, \qquad \bar{y} = y - k.$$

These translation equations can be used to convert from x, y-coordinates to $\bar{x}, \bar{y}$-coordinates and vice versa, as illustrated in the next example.

EXAMPLE 22 Let translated $\bar{x}, \bar{y}$-axes be chosen with origin at the point $(h, k) = (-1, 2)$. Find $\bar{x}, \bar{y}$-coordinates of the point $(x, y) = (3, -2)$. See Fig. 1.21.

Solution Since $\bar{x} = x - h$ and $\bar{y} = y - k$, we see that the translated coordinates are $\bar{x} = 3 - (-1) = 4$ and $\bar{y} = -2 - 2 = -4$. Thus the point is $(\bar{x}, \bar{y}) = (4, -4)$. ∎

Using Calculators

Examples and exercises using calculators add substantially to an understanding and appreciation of calculus. We make no attempt to discuss in detail the pitfalls of using calculators. You may discover some of them by experimentation.

The calculator exercises that appear in this first chapter are designed to make sure that you know which buttons to push and understand the logic of your own calculator. The interesting use of calculators starts in Chapter 2.

EXAMPLE 23 Find the approximate distance between the two points $(\sqrt[3]{17}, \pi^{4/5})$ and $(-\sqrt{5}, \sqrt[4]{63})$.

Solution We must compute

$$\sqrt{(\sqrt[3]{17} + \sqrt{5})^2 + (\pi^{4/5} - \sqrt[4]{63})^2}.$$

The order for pushing the buttons depends on the logic used by the calculator. We compute $\sqrt[3]{17}$ as $17^{1/3}$, compute $\pi^{4/5}$ as $\pi^{0.8}$, and $\sqrt[4]{63}$ as $63^{0.25}$. Of course we always obtain an answer in decimal, or scientific, notation. Our calculator yields 4.81789405 for the answer. Another calculator's answer might have fewer or more digits or might differ in the final significant digit. But to two *decimal places*, the answer is 4.82, while to five *significant figures*, the answer is 4.8179. ∎

SUMMARY

1. The closed interval $[a, b]$ consists of all x such that $a \le x \le b$.

2. The distance from x_1 to x_2 on the real number line is $|x_2 - x_1|$.

3. The signed length of the directed line segment from x_1 to x_2 is

$$\Delta x = x_2 - x_1$$
$$= \text{(Number where you stop)} - \text{(Number where you start)}.$$

4. The midpoint of the interval $[a, b]$ is $(a + b)/2$.

5. The distance between (x_1, y_1) and (x_2, y_2) in the plane is

$$\sqrt{(x_2 - x_1)^2 + (y_2 - y_1)^2}.$$

6. The circle with center (h, k) and radius r has equation

$$(x - h)^2 + (y - k)^2 = r^2.$$

7. To find the center (h, k) and radius r of a circle $ax^2 + ay^2 + bx + cy = d$, complete the square on the x-terms and on the y-terms.

8. The relationship between x,y-coordinates and translated $\bar{x},\bar{y}$-coordinates with new origin at $(x, y) = (h, k)$ is given by

$$\bar{x} = x - h, \qquad \bar{y} = y - k.$$

EXERCISES 1.1

In Exercises 1 and 2, sketch as in Figs. 1.3 and 1.4 all points x (if there are any) that satisfy the given relation.

1. (a) $2 \leq x \leq 3$ (b) $x^2 = 4$ (c) $5 \leq x \leq -1$

2. (a) $x \leq 0$ (b) $x^2 < 4$ (c) $x^2 \leq 4$

3. Find the distance between the given points on the number line.
   (a) 2 and 5   (b) -1 and 4   (c) -3 and -6

4. Find the distance between the given points on the number line.
   (a) $-\frac{5}{2}$ and 12   (b) $-\frac{8}{3}$, and $-\frac{15}{3}$
   (c) $\sqrt{2}$ and $-2\sqrt{2}$ (d) $\sqrt{2}$ and π

5. Find the number described.
   (a) $|3 - 5|$ (b) $3 - |5|$

6. Find the number described.
   (a) $\big|4 - |2 - 7|\big|$ (b) $2/|-2|$

7. Find all x such that $|x + 2|/(x + 2) = 1$. (See Example 7.)

8. Find all x such that $|x - 3|/(x - 3) = -1$. (See Example 7.)

9. Show that, for any a and b on the number line, the distance from $(a + b)/2$ to a is the same as the distance from $(a + b)/2$ to b.

10. Find the midpoint of each of the following intervals.
    (a) $[-1, 1]$ (b) $[-1, 4]$ (c) $[-\frac{3}{2}, \frac{2}{3}]$

11. Proceed as in Exercise 10.
    (a) $[-6, -3]$ (b) $[-2\sqrt{2}, \sqrt{2}]$ (c) $[\sqrt{2}, \pi]$

12. Find the *signed* length Δx of the following directed line segments.
    (a) From 2 to 5   (b) From -8 to -1

13. Proceed as in Exercise 12.
    (a) From 3 to -7 (b) From 10 to 2

Each of the relations in Exercises 14–25 has as solution all points in a closed interval $[a, b]$. Find the interval in each case.

14. $-2 \leq x \leq 3$ 15. $6 \leq 3x \leq 12$

16. $4 \leq -4x \leq 8$ 17. $-3 \leq x + 2 \leq 4$

18. $8 \leq x - 3 \leq 15$ 19. $3 \leq 2x - 4 \leq 14$

20. $-7 \leq 8 - 3x \leq 29$ 21. $|x| \leq 4$

22. $|x - 1| \leq 4$ 23. $|x + 1| \leq 4$

24. $|2x - 3| \leq 7$ 25. $|8 - 3x| \leq 5$

26. Sketch the points (x, y) in the plane satisfying the indicated relations, as in Examples 15 and 16.
    (a) $x \leq y$ (b) $x = -y$
    (c) $y = 2x$ (d) $2x \geq y$

27. Proceed as in Exercise 26.
    (a) $x = 1$ (b) $-1 \leq x \leq 2$
    (c) $x = -1, -2 \leq y \leq 3$ (d) $x = y, -1 \leq x \leq 1$

28. Find the coordinates of the indicated point.
    (a) The point such that the line segment joining it to $(2, -1)$ has the x-axis as perpendicular bisector
    (b) The point such that the line segment joining it to $(-3, 2)$ has the y-axis as perpendicular bisector

29. Find the coordinates of the indicated point.
(a) The point such that the line segment joining it to $(-1, 3)$ has the origin as midpoint
(b) The point such that the line segment joining it to $(2, -4)$ has $(2, 1)$ as midpoint

30. Find the distance between the given points.
(a) $(-2, 5)$ and $(1, 1)$
(b) $(2\sqrt{2}, -3)$ and $(-\sqrt{2}, 2)$

31. Find the distance between the given points.
(a) $(2, -3)$ and $(-3, 5)$
(b) $(2\sqrt{3}, 5\sqrt{7})$ and $(-4\sqrt{3}, 2\sqrt{7})$

32. It can be shown that a triangle is a right triangle if and only if the sum of the squares of the lengths of two sides equals the square of the length of the third side. The right angle is then opposite the longest side of the triangle. Use this fact and the distance formula to show that $(1, 5)$, $(3, 2)$, and $(7, 9)$ are vertices of a right triangle, and find the area of the triangle.

33. Proceed as in Exercise 32 for $(-1, 3)$, $(4, 1)$, and $(1, 8)$.

34. To reach the Edwards' home from the center of town, you drive 2 miles due east on Route 37 and then 5 miles due north on Route 101. Assuming that the surface of the earth near town is approximately flat, find the distance, as the crow flies, from the center of town to the Edwards' home.

35. Refer to Exercise 34. Suppose you drive 6 miles due west on Route 37 and then 4 miles due south on Route 43 to reach the Hammonds' house from town. Find the distance from the Edwards' home to the Hammonds' as the crow flies.

36. See Exercise 34. Charlotte leaves the Edwards' home and jogs south on Route 101 at 7 mph. At the same time, John leaves town and jogs east on Route 37 at 5 mph. Find the distance s between them 15 minutes later. See Fig. 1.22.

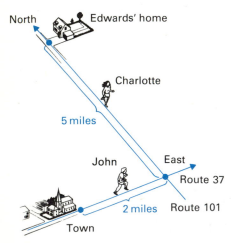

FIGURE 1.22

37. Refer to Exercises 34–36. Eleanor leaves the Hammonds' house at the same time as Charlotte and jogs north on Route 43 at 6 mph. Find the distance from Eleanor to Charlotte after 15 minutes.

In Exercises 38–41, find the equation of the circle with the given center and radius.

38. Center $(0, 0)$, radius 5

39. Center $(-1, 2)$, radius 3

40. Center $(3, -4)$, radius $\sqrt{30}$

41. Center $(-2, -3)$, radius $\sqrt{5}$

In Exercises 42–47, find the center and radius of the given circle.

42. $(x - 2)^2 + (y - 3)^2 = 36$ **43.** $(x + 3)^2 + y^2 = 49$

44. $(x + 1)^2 + (y + 4)^2 = 50$ **45.** $x^2 + y^2 - 4x + 6y = 3$

46. $x^2 + y^2 + 8x = 9$

47. $4x^2 + 4y^2 - 12x - 24y = -\frac{9}{2}$

48. Find the equation of the circle with center in the second quadrant, tangent to the coordinate axes, and with radius 4.

49. Find the equation of the circle having the line segment with endpoints $(-1, 2)$ and $(5, -6)$ as a diameter.

50. Find the equation of the circle with center $(2, -3)$ and passing through $(5, 4)$.

51. Find the point on the circle $x^2 + y^2 = 25$ that is diametrically opposite the point $(3, 4)$. [*Hint:* Draw a figure.]

52. Find the point on the circle $x^2 + y^2 - 4x + 6y = -12$ that is diametrically opposite the point $(1, -3)$. [*Hint:* Draw a figure.]

53. Find translated $\bar{x}, \bar{y}$-coordinates of the given points (x, y) with respect to a new origin at $(h, k) = (-3, 2)$.
(a) $(4, 6)$ (b) $(-1, 3)$ (c) $(0, 0)$

54. Proceed as in Exercise 53 for a new origin at $(h, k) = (-4, -1)$.
(a) $(-3, -2)$ (b) $(5, -3)$ (c) $(8, -1)$

Use a calculator for Exercises 55–61.

55. Estimate the midpoint of $[-2\sqrt{3}, 5\sqrt{7}]$.

56. Estimate the midpoint of $[\sqrt[5]{23}, -\sqrt[4]{50}]$.

57. Estimate the signed length of the directed line segment from $22\sqrt{2}$ to π^3.

58. Estimate the signed length of the directed line segment from $\sqrt[3]{17}$ to $-\sqrt[6]{43}$.

59. Estimate the distance between $(2, -3)$ and $(4, 1)$.

60. Estimate the distance between $(-3.7, 4.23)$ and $(8.61, 7.819)$.

61. Estimate the distance between $(\pi, -\sqrt{3})$ and $(8\sqrt{17}, -\sqrt[3]{\pi})$.

Use a calculator in Exercises 62–64 to estimate the center and radius of the given circle.

62. $(x - \pi)^2 + (y - \sqrt{\pi})^2 = 2.736$

63. $x^2 + y^2 + 3.1576x - 1.2354y = 3.33867$

64. $\sqrt{2}x^2 + \sqrt{2}y^2 - \pi^3 x + (\pi^2 + 3.4)y = \sqrt{17}$

 EXPLORING CALCULUS

A: points, lines, circles

1.2 LINES

Lines are critical to the study of calculus because a key idea of calculus is to approximate complicated, hard-to-handle formulas in x by very simple formulas of the form $mx + b$. We will see in this section that all points (x, y) in the plane satisfying $y = mx + b$ form a line.

The Slope of a Line

The **slope** m of a line is the number of units the line rises vertically for each unit of horizontal run from left to right. If a line rises 3 units for each unit it runs to the right, as in Fig. 1.23, the line has slope 3. We think of a fall as a negative rise, so that if a line falls 2 units for each unit it runs to the right, it has a rise of -2 per unit run, and its slope is -2. See Fig. 1.24. A horizontal line has a rise of 0 for any run and a slope of 0. A vertical line climbs straight up with a horizontal run of zero, so it is impossible to measure how much it climbs per unit horizontal run. Consequently, *the slope of a vertical line is undefined.*

We know that a line is determined by two points. We discover how to find the slope of a line through two given points in the next example.

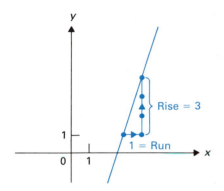

FIGURE 1.23 A line of slope 3.

EXAMPLE 1 Find the slope of the line through the points $(2, 4)$ and $(5, 16)$.

Solution As we go from a point (x_1, y_1) to a point (x_2, y_2) where $x_1 < x_2$, the horizontal run is the change $\Delta x = x_2 - x_1$ and the rise is the directed change $\Delta y = y_2 - y_1$. For our given points $(2, 4)$ and $(5, 16)$, we obtain

$$\Delta x = x_2 - x_1 = 5 - 2 = 3 \quad \text{and} \quad \Delta y = y_2 - y_1 = 16 - 4 = 12.$$

Thus the slope m of this line is given by

$$m = \frac{\text{Rise}}{\text{Run}} = \frac{\Delta y}{\Delta x} = \frac{12}{3} = 4. \qquad \blacksquare$$

As illustrated in Example 1, we can find the slope m of the line through (x_1, y_1) and (x_2, y_2) if $x_1 < x_2$ by finding Δx and Δy as we go from (x_1, y_1) to (x_2, y_2) and then taking the quotient:

$$m = \frac{\Delta y}{\Delta x} = \frac{y_2 - y_1}{x_2 - x_1}. \qquad (1)$$

If $x_2 < x_1$, then to go from left to right we should go from (x_2, y_2) to (x_1, y_1), and we obtain

$$m = \frac{\Delta y}{\Delta x} = \frac{y_1 - y_2}{x_1 - x_2} = \frac{y_2 - y_1}{x_2 - x_1}, \qquad (2)$$

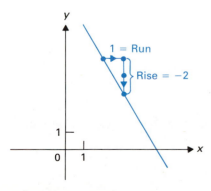

FIGURE 1.24 A line of slope -2.

which is the same formula as in Eq. (1).

The similar triangles in Fig. 1.25 show that the computation of the slope $(y_2 - y_1)/(x_2 - x_1)$ using points (x_1, y_1) and (x_2, y_2) on a line yields the same value as the ratio $(y_4 - y_3)/(x_4 - x_3)$ found using different points (x_3, y_3) and (x_4, y_4).

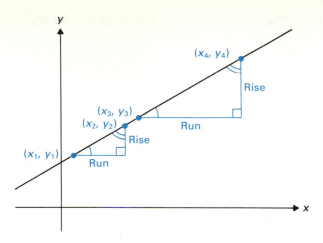

FIGURE 1.25   The slope ratios Rise/Run are equal for these two similar triangles.

Slope m of the Line Through (x_1, y_1) and (x_2, y_2)

1. $x_1 = x_2$ (vertical line): m is undefined.

2. $x_1 \neq x_2$: $m = \dfrac{\text{Rise}}{\text{Run}} = \dfrac{\Delta y}{\Delta x} = \dfrac{y_2 - y_1}{x_2 - x_1}$

EXAMPLE 2   Find the slope of the line through $(7, 5)$ and $(-2, 8)$.

Solution   From the slope formula in Eq. (2), we see that the line has slope

$$m = \frac{\Delta y}{\Delta x} = \frac{8 - 5}{-2 - 7} = \frac{3}{-9} = -\frac{1}{3}.$$ ∎

EXAMPLE 3   Determine the slope of the line through $(-3, 5)$ and $(-3, 11)$.

Solution   We have $\Delta x = -3 - (-3) = 0$, so the line is vertical. The slope is *undefined*.   ∎

The next example reinforces our understanding of slope as a measure of rise in practical situations.

EXAMPLE 4   A water-ski jump consists of a straight chute that rises 1.2 m above the water over a horizontal distance of 5 m. Taking an x-axis horizontal with origin at the bottom of the jump, find the slope of the jump. Give the physical interpretation of this slope.

Solution   As shown in Fig. 1.26, the jump goes from the origin $(0, 0)$ to the point $(5, 1.2)$. By Eq. (2), the slope of the jump is

$$\frac{\Delta y}{\Delta x} = \frac{1.2 - 0}{5 - 0} = 0.24.$$

The jump rises 0.24 m in height for every meter measured horizontally.   ∎

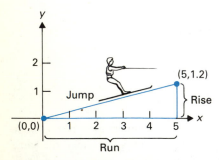

FIGURE 1.26   Water-ski jump rising 1.2 m over a distance of 5 m.

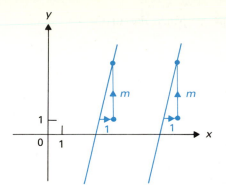

FIGURE 1.27 Parallel lines have the same slope m.

Parallel and Perpendicular Lines

Let two lines have slopes m_1 and m_2. Lines are parallel precisely when they have the same rise corresponding to equal runs, that is, when their slopes are the same. This is illustrated in Fig. 1.27.

Parallel Lines

Equal slopes: $m_1 = m_2$

EXAMPLE 5   Find the value c such that the line through $(-1, 4)$ and $(1, c)$ is parallel to the line through $(0, 3)$ and $(4, 6)$.

Solution   The line through the points $(-1, 4)$ and $(1, c)$ has slope $m_1 = (c - 4)/[1 - (-1)]$, and the line through $(0, 3)$ and $(4, 6)$ has slope $m_2 = (6 - 3)/(4 - 0) = \frac{3}{4}$. For the lines to be parallel, their slopes m_1 and m_2 must be equal, so

$$\frac{c - 4}{2} = \frac{3}{4},$$

$$c - 4 = \frac{6}{4} = \frac{3}{2},$$

$$c = 4 + \frac{3}{2} = \frac{11}{2}.$$ ∎

We now derive a condition for two lines to be perpendicular. By translation, we can assume that the point of intersection of the lines is the origin, as shown in Fig. 1.28. If the lines have slopes m_1 and m_2, then the points $(1, m_1)$ and $(1, m_2)$ lie on the lines as shown in the figure. For the angles shown in the figure, we see that

$$\alpha_1 + \beta_1 = 90°, \qquad \text{so} \quad \alpha_1 = 90° - \beta_1.$$

Thus $\alpha_1 + \beta_2 = 90°$ if and only if $(90° - \beta_1) + \beta_2 = 90°$ or if and only if $\beta_2 = \beta_1$. An analogous argument then shows that $\alpha_1 = \alpha_2$, so the triangles are similar. Corresponding sides are then proportional, and from Fig. 1.28, we see that

$$\frac{m_1}{1} = \frac{1}{-m_2}, \qquad \text{or} \qquad m_1 m_2 = -1. \qquad (3)$$

The relation (3) is the desired condition for the lines to be perpendicular.

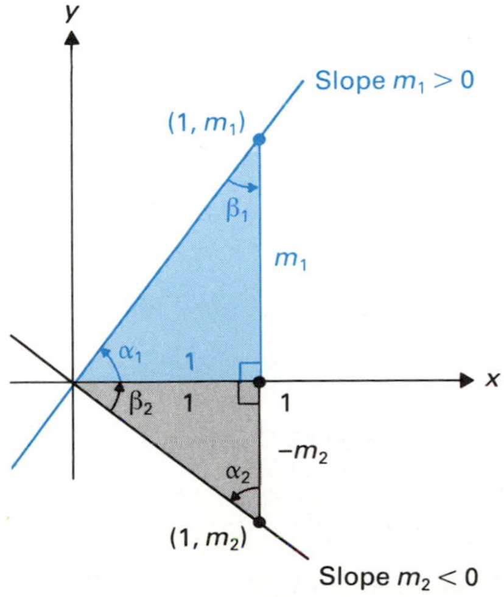

FIGURE 1.28 The lines are perpendicular when the two triangles are similar, or when $m_1/1 = 1/(-m_2)$.

Perpendicular Lines

Negative reciprocal slopes: $m_2 = -\dfrac{1}{m_1}$

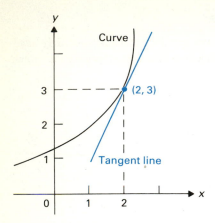

FIGURE 1.29 The tangent line to the curve at the point (2, 3).

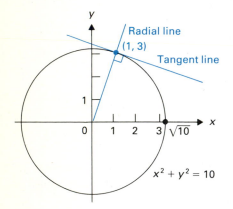

FIGURE 1.30 Tangent line and perpendicular radial line at (1, 3).

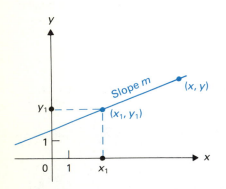

FIGURE 1.31 (x, y) on the line of slope m through (x_1, y_1).

EXAMPLE 6 Find the slope of a line perpendicular to the line through $(6, -5)$ and $(8, 3)$.

Solution The given line has slope

$$\frac{\Delta y}{\Delta x} = \frac{3 - (-5)}{8 - 6} = \frac{8}{2} = 4.$$

So a perpendicular line has slope $-\frac{1}{4}$. ∎

In geometric terms, differential calculus is concerned with finding the slope of a line that is "tangent to a curve." Figure 1.29 shows a tangent line to a curve at the point $(2, 3)$ on the curve. We cannot handle this problem in general yet, but we can find the slope of a line tangent to a circle at a point (x, y). The line from the center of the circle through a point (x, y) on the circle is the *radial line* through (x, y). The line *tangent* to the circle at (x, y) is the line perpendicular to the radial line there. Figure 1.30 shows the radial and tangent lines to the circle $x^2 + y^2 = 10$ at the point $(1, 3)$.

EXAMPLE 7 Find the slope of the line tangent to the circle $x^2 + y^2 = 10$ at the point $(1, 3)$ on the circle. See Fig. 1.30.

Solution We use the fact that the tangent line to a circle is perpendicular to the radial line at the point of tangency. The radial line goes through $(0, 0)$ and $(1, 3)$ and has slope $(3 - 0)/(1 - 0) = 3$. The tangent line thus has slope $-\frac{1}{3}$. ∎

The Equation of a Line

Let a given line have slope m and pass through the point (x_1, y_1) as shown in Fig. 1.31. We can find an algebraic condition for a point (x, y) to lie on the line. If the slope of the line that joins (x_1, y_1) and (x, y) is also m, then that line is parallel to the given line, because they have the same slope. But both lines go through (x_1, y_1), so they must coincide. Therefore a condition for (x, y) to lie on the given line is that

$$\frac{y - y_1}{x - x_1} = m, \tag{4}$$

or

$$y - y_1 = m(x - x_1). \tag{5}$$

EXAMPLE 8 Find the equation of the line through $(2, -3)$ with slope 7, shown in Fig. 1.32.

Solution The equation is $y - (-3) = 7(x - 2)$ or $y + 3 = 7(x - 2)$. This equation may be simplified to $y = 7x - 17$. The point $(3, 4)$ lies on this line, since $4 = 7 \cdot 3 - 17$. ∎

As indicated in Example 8, the *point-slope equation* (5) can be rewritten in the form

$$y = mx + b, \tag{6}$$

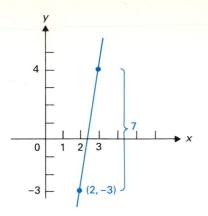

FIGURE 1.32 Line through $(2, -3)$ with slope 7.

where $b = y_1 - mx_1$. The constant b in Eq. (6) has a geometric interpretation. If we set $x = 0$ in Eq. (6), then $y = b$, so the point $(0, b)$ satisfies the equation and thus lies on the line. This point $(0, b)$ is on the y-axis, and b is the **y-intercept** of the line. See Fig. 1.33. If the line crosses the x-axis at $(a, 0)$, then a is the **x-intercept** of the line.

EXAMPLE 9 Find the intercepts of the line in Example 8.

Solution The equation is $y = 7x - 17$, so -17 is the y-intercept. To find the x-intercept, we set $y = 0$ and obtain $7x - 17 = 0$, so $x = \frac{17}{7}$. Thus the point $(\frac{17}{7}, 0)$ lies on the line, so $\frac{17}{7}$ is the x-intercept. ∎

The vertical line through $(a, 0)$ in Fig. 1.34 has undefined slope, so it does not have an equation of the form of Eq. (5) or (6). But surely a condition that (x, y) lie on the line is simply that $x = a$. Of course, $y = b$ is the horizontal line through $(0, b)$ shown in the figure. In any kind of coordinate system, it is important to know what sets of points are obtained by setting the coordinate variables equal to constants. We see that in our rectangular x,y-coordinate system, $x = a$ is a vertical line and $y = b$ is a horizontal line.

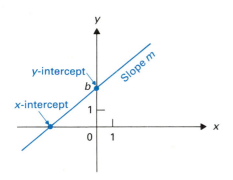

FIGURE 1.33 The line $y = mx + b$.

Vertical Line Through (*a*, *b*)

Equation: $x = a$

EXAMPLE 10 Find the equation of the line through the points $(-5, 3)$ and $(-5, 7)$.

Solution The slope of this line is undefined since $\Delta x = -5 - (-5) = 0$. Thus the line is vertical, with an equation of the form $x = a$, where a is the x-coordinate of every point on the line. In this case, the equation of the line is $x = -5$. ∎

We give a suggested outline to use to find the equation of a nonvertical line. (See Eq. 5.)

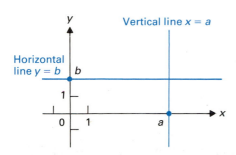

FIGURE 1.34 Equations of vertical and horizontal lines.

Finding the Equation of a Nonvertical Line

1. Find a *point* on the line: (x_1, y_1).
2. Find the *slope* of the line: m.
3. Write the *equation* of the line: $y - y_1 = m(x - x_1)$.

EXAMPLE 11 Find the equation of the line through $(-5, -3)$ and $(6, 1)$.

Solution We solve the problem according to the preceding outline.

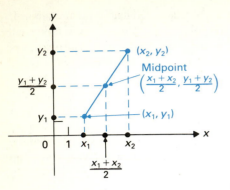

FIGURE 1.35 Midpoint of a line segment.

Point $(x_1, y_1) = (-5, -3)$

Slope $m = \dfrac{1 - (-3)}{6 - (-5)} = \dfrac{4}{11}$

Equation $y - y_1 = m(x - x_1),$

$$y - (-3) = \frac{4}{11}[x - (-5)],$$

$$y + 3 = \frac{4}{11}(x + 5)$$

The equation can be simplified to $11y + 33 = 4x + 20$ or $4x - 11y = 13.$ ■

Recall from Section 1.1 that a closed interval $[a, b]$ of the number line has as midpoint $(a + b)/2$. We see from Fig. 1.35 that, in the plane,

$$\left(\frac{x_1 + x_2}{2}, \frac{y_1 + y_2}{2}\right)$$

is the **midpoint** of the line segment joining (x_1, y_1) and (x_2, y_2).

EXAMPLE 12   Find the equation of the perpendicular bisector of the line segment joining $(-1, 4)$ and $(3, 8)$, shown in Fig. 1.36.

Solution   We need to find a point on the line and the slope of the line.

Point   The midpoint of the line segment joining $(-1, 4)$ and $(3, 8)$ lies on the bisecting line. From the midpoint formula just given, the midpoint is

$$\left(\frac{-1 + 3}{2}, \frac{4 + 8}{2}\right) = \left(\frac{2}{2}, \frac{12}{2}\right) = (1, 6).$$

Slope   The line is to be perpendicular to the line segment, which has slope

$$\frac{\Delta y}{\Delta x} = \frac{8 - 4}{3 - (-1)} = \frac{4}{4} = 1.$$

Thus the desired slope is $m = -1$.

Equation $y - y_1 = m(x - x_1),$

$$y - 6 = -1(x - 1),$$

$$y = -x + 7$$ ■

EXAMPLE 13   Find the equation of the line tangent to the circle $x^2 + y^2 = 10$ at the point $(1, 3)$.

Solution

Point $(1, 3)$

Slope   The line is perpendicular to the radial line through $(0, 0)$ and $(1, 3)$. This radial line has slope $\Delta y/\Delta x = \frac{3}{1} = 3$, so the tangent line has slope $-\frac{1}{3}$.

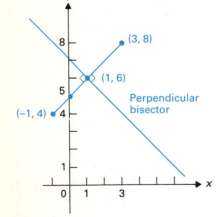

FIGURE 1.36 Perpendicular bisector of the line segment joining $(-1, 4)$ and $(3, 8)$.

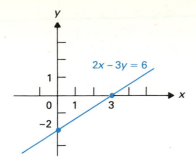

FIGURE 1.37 Sketch of $2x - 3y = 6$ using intercepts.

Equation   (We should no longer need to show substitutions as we did in Examples 11 and 12.)

$$y - 3 = -\tfrac{1}{3}(x - 1), \qquad \text{or} \quad 3y + x = 10 \qquad \blacksquare$$

Finally, observe that every equation $ax + by + c = 0$, where either $a \neq 0$ or $b \neq 0$, is the equation of a line. If $b = 0$, the equation becomes $x = -c/a$, which is a vertical line. If $b \neq 0$, the equation becomes $y = -(a/b)x - c/b$, which is a line with slope $m = -a/b$ and y-intercept $-c/b$.

EXAMPLE 14   Sketch all points (x, y) such that $2x - 3y = 6$.

Solution   We have seen students make a table of ten or more points satisfying such an equation, plot the points, and then draw a wobbly line through them. This is unnecessary. We know $2x - 3y = 6$ is the equation of a *line* and is thus determined by just two points. Setting $x = 0$, we obtain $(0, -2)$ as the y-intercept. Setting $y = 0$, we obtain $(3, 0)$ as the x-intercept. The line is shown in Fig. 1.37. $\blacksquare$

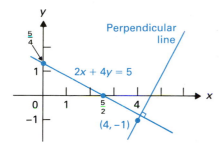

FIGURE 1.38 Line through $(4, -1)$ perpendicular to $2x + 4y = 5$.

EXAMPLE 15   Find the equation of the line through $(4, -1)$ and perpendicular to the line $2x + 4y = 5$, shown in Fig. 1.38.

Solution

Point $(4, -1)$

Slope   The given equation can be written $4y = -2x + 5$, $y = -\tfrac{1}{2}x + \tfrac{5}{4}$ in the form $y = mx + b$. Thus the given line has slope $-\tfrac{1}{2}$, so the desired perpendicular line has slope 2.

Equation $y + 1 = 2(x - 4), \qquad \text{or} \quad y = 2x - 9 \qquad \blacksquare$

EXAMPLE 16   Find the point of intersection of the lines $2x + y = 3$ and $x - 5y = 7$, shown in Fig. 1.39.

Solution   We must find (x, y) satisfying both equations at once. If we solve the equations simultaneously, we obtain

$$\begin{cases} 2x + y = 3 \\ x - 5y = 7 \end{cases} \qquad \begin{array}{r} 2x + y = 3 \\ -2x + 10y = -14 \\ \hline 11y = -11. \end{array} \quad \begin{array}{l} \textit{Multiplying by } -2 \\ \\ \textit{Adding} \end{array}$$

Thus $y = -1$, so $x = 5y + 7 = -5 + 7 = 2$, and the point of intersection is $(2, -1)$. $\blacksquare$

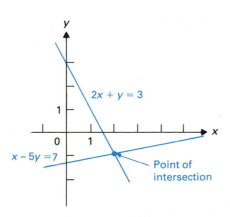

FIGURE 1.39 Point $(2, -1)$ of intersection of $2x + y = 3$ and $x - 5y = 7$.

Two quantities P and Q are *linearly related* if each can be computed from the other using a linear equation, such as $Q = mP + b$. Of course, this means geometrically that points (P, Q) having corresponding values of P and Q as coordinates lie on a line. For example, Fahrenheit and Celsius temperatures are linearly related. Exercise 72 asks you to find this relation. We give illustrations in our final two examples.

EXAMPLE 17   For a body immersed in a fluid, the fluid pressure P on the body is linearly related to the depth d of immersion. Suppose for a certain fluid that a depth of 24 cm corresponds to a pressure of 32 g/cm². Find the linear relation between P and d.

Solution   Since P must be zero when $d = 0$, the point $(d, P) = (0, 0)$ is one point satisfying the desired relation. We are given that another point is $(d, P) = (24, 32)$. Thus we have

Point $(0, 0)$

Slope $m = \dfrac{32 - 0}{24 - 0} = \dfrac{32}{24} = \dfrac{4}{3}$

Linear relation $P - 0 = \dfrac{4}{3}(d - 0)$,   or $3P = 4d$ ∎

EXAMPLE 18   Suppose a company's revenue remains constant and its costs also remain constant for any two equal periods of time. Let the company's profit over a period of t years after the company started be P. Then P and t are linearly related. If the company has accumulated profits, in thousands of dollars, of 500 after 10 years and of 640 after 12 years, find (a) the linear relations between P and t and (b) the initial cost to start the company.

Solution

(a) We are given that two points (t, P) satisfying the linear relation are $(10, 500)$ and $(12, 640)$.

Point $(10, 500)$

Slope $\dfrac{\Delta P}{\Delta t} = \dfrac{640 - 500}{12 - 10} = \dfrac{140}{2} = 70$

Linear relation $(P - 500) = 70(t - 10)$,   or $P = 70t - 200$

(b) From part (a), we find that when $t = 0$, the accumulated profit is -200. This means that the cost to start the company was $200,000. ∎

SUMMARY

1. A vertical line has undefined slope. If $x_1 \neq x_2$, the line through (x_1, y_1) and (x_2, y_2) has slope

$$m = \frac{\Delta y}{\Delta x} = \frac{y_2 - y_1}{x_2 - x_1}.$$

2. Lines of slopes m_1 and m_2 are

parallel if and only if $m_1 = m_2$,

perpendicular if and only if $m_1 m_2 = -1$,   or $m_2 = -\dfrac{1}{m_1}$.

3. A vertical line has equation $x = a$.

4. A horizontal line has equation $y = b$.

5. To find the equation of a line, find one point (x_1, y_1) on the line and the slope m of the line. The equation is then

$$y - y_1 = m(x - x_1).$$

6. The line $y = mx + b$ has slope m and y-intercept b.

EXERCISES 1.2

In Exercises 1–5, find the slope of the line through the indicated points, if the line is not vertical.

1. $(-3, 4)$ and $(2, 1)$ **2.** $(5, -2)$ and $(-6, -3)$

3. $(3, 5)$ and $(3, 8)$ **4.** $(0, 0)$ and $(5, 4)$

5. $(-7, 4)$ and $(9, 4)$

6. Find b so that the line through $(2, -3)$ and $(5, b)$ has slope -2.

7. Find a so that the line through $(a, -5)$ and $(3, 6)$ has slope 1.

8. Find the slope of a line perpendicular to the line through $(-3, 2)$ and $(4, -1)$.

9. Find b so that the line through $(8, 4)$ and $(4, -2)$ is parallel to the line through $(-1, 2)$ and $(2, b)$.

10. Find c such that the line through $(3, 1)$ and $(-2, c)$ is perpendicular to the line through $(4, -1)$ and $(3, 2)$.

11. Find the slope of the line through the centers of the circles $x^2 + y^2 + 4x - 2y = 8$ and $x^2 + y^2 - 8x + 6y = 0$.

In Exercises 12–15, use slopes to determine whether or not the four points are vertices of (a) a parallelogram, (b) a rectangle.

12. $A(2, 1)$, $B(3, 4)$, $C(-1, 2)$, $D(0, 5)$

13. $A(-1, 4)$, $B(3, 1)$, $C(1, 5)$, $D(1, 0)$

14. $A(3, 1)$, $B(2, -4)$, $C(6, 3)$, $D(5, -2)$

15. $A(4, 0)$, $B(7, 2)$, $C(2, 3)$, $D(5, 4)$

16. Let L be the line through $(-1, 4)$ with slope 5. Use slopes to find the y-coordinate of the point on L whose x-coordinate is (a) 2, (b) -3.

17. Let L be the line through $(2, -3)$ with slope -4. Use slopes to find the x-coordinate of the point on L with y-coordinate (a) 5, (b) 0.

In Exercises 18–21, use slopes to determine whether the points A, B, and C are collinear.

18. $A(3, -1)$, $B(5, 3)$, $C(2, -3)$

19. $A(0, 6)$, $B(3, 4)$, $C(6, 1)$

20. $A(1, 4)$, $B(6, 3)$, $C(8, 2)$

21. $A(-1, 3)$, $B(5, -6)$, $C(-3, 6)$

22. A fly moves in the direction of increasing x along the line L through $(0, 0)$ with slope 3. If the fly moves at a constant rate of 2 units/sec, find its position 5 sec after it is at the origin.

23. A bug moves in the direction of decreasing x along the line L through $(-1, 4)$ with slope $-\frac{4}{3}$. If the bug moves at a constant rate of 4 units/sec, find its position 5 sec after it is at $(-1, 4)$.

24. Find the slope of the line tangent to the circle $x^2 + y^2 - 2y = 4$ at the point $(1, -1)$ on the circle. [*Hint:* The tangent line is perpendicular to the radial line.]

25. Find all points on the circle $x^2 + y^2 + 2x - 4y = 15$ where the tangent line has slope 2. [*Hint:* The tangent line is perpendicular to the radial line.]

26. A toy train travels counterclockwise around the circle $x^2 + y^2 = 25$ at a constant rate of 1 unit/sec. The train flies off the circle at the point $(3, -4)$ and continues at the same speed in the direction tangent to the circle. Find the position of the engine 5 sec after it leaves the circle. (See Fig. 1.40.)

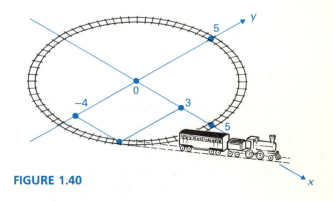

FIGURE 1.40

27. Show that the line joining the midpoints of two sides of a triangle is parallel to the third side. [*Hint:* Let the vertices of the triangle be $(0, 0)$, $(a, 0)$, and (b, c).]

28. Water freezes at 0° C and 32° F, and it boils at 100° C and 212° F. If points (C, F) are plotted in the plane, where F is the temperature in degrees Fahrenheit corresponding to a temperature of C degrees Celsius, then a straight line is obtained. Find the slope of the line. What does this slope represent in this situation?

29. The Easy Life Prefabricated Homes Company listed its super-deluxe ranch model for $50,000 in 1970. The company increased the price by the same amount each year and listed the same model for $110,000 in 1990. Find the slope of the segment drawn through points (Y, C) in the plane, where Y could be any year from 1970 to 1990 and C is the cost of this model ranch house in that year. What does this slope represent in this situation?

30. A house wall is 10 in. thick and is composed of four different types of material, as shown in Fig. 1.41. The graph in the figure shows the temperature in a cross section of the wall when the outside temperature is 5° F and the inside temperature is 65° F. Explain in terms of slopes which of the four types of materials is (a) the most efficient type of insulation, (b) the least efficient type of insulation.

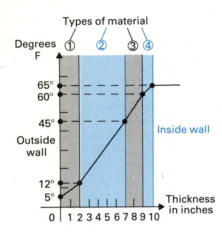

FIGURE 1.41
Temperature in a house wall.

In Exercises 31–50, find the equation of the indicated line.

31. Through $(-1, 4)$ with slope 5

32. Through $(2, -3)$ with slope -3

33. Through $(4, 2)$ with slope 0

34. Through $(-2, 1)$ with slope undefined

35. Through $(4, -5)$ and $(-1, 1)$

36. Through $(2, 5)$ and $(-3, 5)$

37. Through $(-3, 4)$ and $(-3, -1)$

38. Through $(-1, -2)$ and $(4, -3)$

39. Through $(-2, 1)$ parallel to $2x + 3y = 7$

40. Through $(4, -2)$ parallel to $4x - 2y = 5$

41. Through $(-3, -1)$ perpendicular to $x + 4y = 8$

42. Through $(4, -5)$ perpendicular to $3x + 2y = -6$

43. Through $(-2, 5)$ with x-intercept -3

44. Through $(1, -4)$ with y-intercept 6

45. The perpendicular bisector of the line segment from $(-1, 5)$ to $(3, 11)$

46. The perpendicular bisector of the line segment from $(2, -3)$ to $(8, 5)$.

47. Tangent to the circle $x^2 + y^2 = 25$ at $(-3, 4)$

48. Tangent to the circle $x^2 + y^2 + 2x - 4y = -4$ at $(-1, 3)$

49. Through the centers of the circles $x^2 + y^2 - 2x + 2y = 7$ and $x^2 + y^2 - 4x + 6y = 0$

50. Through the centers of the circles $x^2 + y^2 - x + 2y = 4$ and $x^2 + y^2 + 3x - 4y = 3$

In Exercises 51–54, determine whether the two lines are parallel, perpendicular, or neither.

51. $x + y = 6$, $2x - 2y = 7$

52. $3x + y = 8$, $12x + 4y = -5$

53. $4x - 3y = 6$, $3x - 4y = 8$

54. $6x - 2y = 7$, $x + 3y = 4$

In Exercises 55–58, find the slope, x-intercept, and y-intercept of the given line.

55. $x - y = 7$ **56.** $y = 11$

57. $x = 4$ **58.** $7x - 13y = 8$

In Exercises 59–64, sketch the line with the given equation.

59. $x + 2y = 6$ **60.** $2x - y = 4$

61. $4x + 3y = 12$ **62.** $4x = 9$

63. $8y = -11$ **64.** $3x - 2y = 0$

65. Find the point of intersection of the line $2x + 3y = 7$ and the line $3x + 4y = -8$.

66. Find the point of intersection of the line $2x - 5y = 4$ and the line $3x - 9y = 1$.

67. Find the distance from the point $(-2, 1)$ to the line $3x + 4y = 8$. [*Hint:* Find the point where the line through $(-2, 1)$ and perpendicular to $3x + 4y = 8$ meets $3x + 4y = 8$.]

68. Find the distance from the point $(-4, 3)$ to the line $x - y = 6$. (See the hint in Exercise 67.)

69. Show that the perpendicular bisectors of the sides of a triangle meet at a point. [*Hint:* Let the vertices of the triangle be $(-a, 0)$, $(a, 0)$, and (b, c).]

70. Find the equation of the circle through the points $(1, 5)$, $(2, 4)$, and $(-2, 6)$. [*Hint:* The center of the circle lies on the perpendicular bisector of each chord.]

71. Find the equation of the circle through the points $(0, 3)$, $(-2, 5)$, and $(-4, 5)$. (See the hint in Exercise 70.)

72. Referring to Exercise 28, find the linear relation giving the temperature F in degrees Fahrenheit corresponding to a temperature of C degrees Celsius.

73. A snowstorm starts at 3:00 A.M. and continues until 11:00 A.M. If there was 13 in. of old snow on the ground at the start of the storm and the new snow accumulates at a constant rate of $\frac{3}{2}$ in. per hour, find the depth d in inches at time of day t for $3 \le t \le 11$.

74. For a gas confined in a container of fixed volume, the pressure P (in g/cm^2) exerted by the gas on the container and the temperature T (in degrees Celsius) are linearly related. If the pressure is 50 g/cm^2 at a temperature of 16° C and 65 g/cm^2 at a temperature of 21° C, find the linear relation between P and T.

75. A spring is suspended vertically from a beam. The weight W (in kg) required to stretch a spring is linearly related to the distance d (in cm) the spring is stretched from its natural length, as long as the spring is not stretched too far. The spring is stretched 10 cm from its natural length when a weight is attached. An additional weight of 2 kg is required to stretch the spring an additional 12 cm. Find the linear relation between W and d.

76. The electric resistance R (in ohms) of a certain resistor is linearly related to the temperature T (in degrees Celsius). If

$R = 3$ ohms when $T = 10°$ C and $R = 3.003$ ohms when $T = 16°$ C, find (a) the linear relation between R and T, (b) the resistance R at 0° C.

77. Water drips from a leaking faucet at a constant rate. A cylindrical pail is set under the faucet. The depth d (in cm) of water in the pail is linearly related to the elapsed time t (in min). The depth at 5:00 P.M. is 2 cm and the depth at 6:00 P.M. is 5 cm. Find (a) the linear relation between d and t, (b) the time when the pail was set under the faucet.

 EXPLORING CALCULUS

B: slopezap game

1.3 FUNCTIONS AND THEIR GRAPHS

Functions

The area enclosed by a circle is a *function* of the radius of the circle, meaning that the area depends on and varies with the radius. If a numerical value for the radius is given, the area enclosed by the circle is determined. For example, if the radius is 3 units, then the area is 9π square units. Similarly, the area of a rectangular region is a *function* of both the length and the width of the rectangle; that is, the area depends on and varies with these quantities. If the length of a rectangle is 5 units and the width is 3 units, the rectangle encloses a region that has an area of 15 square units.

The study of how one quantity Q depends on and varies with other quantities is one of the major concerns of science. A rule that specifies Q for each possibility for the other quantities is exceedingly useful. Such rules are called *functions*.

DEFINITION 1.1

Function

A **function** f is a rule that assigns to each element x of some set X exactly one element y of a set Y. We write $y = f(x)$, read "y equals f of x."

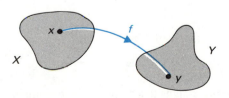

FIGURE 1.42 Schematic of a function f.

Figure 1.42 attempts to illustrate the definition. Previously in this text, we have used letters only to represent numbers. It is important to understand that a function f is *not* a number but rather should be regarded as a rule of assignment. We have attempted to denote this in Fig. 1.42 by labeling with f the colored arrow that indicates the function assigning to x in X the element y in Y.

Another intuitive picture of a function popular in elementary texts is a "magic box," as in Fig. 1.43. Drop in an element x, and an element y comes out at the end. A scientific calculator is such a magic box. For example, we punch in

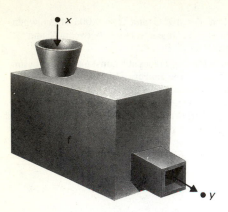

FIGURE 1.43   Another schematic of a function f.

a value for x and then press the sin x button to "perform the function," and an approximate y-value, where $y = f(x) = \sin x$, is shown as the display.

Let $y = f(x)$ for x in a set X and y in a set Y, as given in Definition 1.1. Since we think of y as depending on x, we call y the **dependent variable** and x the **independent variable.** The set X of all x on which f acts is the **domain** of the function f. The set of all y elements obtained by finding $f(x)$ for all x in X is the **range** of the function f. We will often speak of a function $f(x)$, but technically $f(x)$ is the *value* y of the function f at x, while f is the function.

From now until Chapter 14, we will use function notation and terminology only when both the domain and the range of the function are sets of real numbers. That is, when we write $y = f(x)$, both x and y will be real numbers. Such a function f is a **real-valued function of one real variable.**

When we are talking about some particular function f and writing $y = f(x)$, we have to describe the rule of assignment f that enables us to find the real number y starting with any real number x in the domain of f. Frequently, we will be interested in rules given by formulas. For example, we may discuss the

---

Squaring Function

$$y = f(x) = x^2 \qquad \text{for all } x. \tag{1}$$

The phrase "for all x" indicates that the domain of f consists of the set of *all* real numbers.

EXAMPLE 1   Let $y = f(x) = x^2 + 3$ for all x. Then

$$f(-2) = 7, \quad f(-1) = 4, \quad f(0) = 3, \quad f(1) = 4, \quad \text{and} \quad f(2) = 7.$$

Clearly for this function f, we have $f(-x) = f(x)$ for all x. The range of f consists of all $y \geq 3$. ∎

EXAMPLE 2   Let $y = f(x) = |x| - 4$ for all x. Then

$$f(-2) = -2, \quad f(-1) = -3, \quad f(-\tfrac{1}{2}) = -\tfrac{7}{2}, \quad f(0) = -4,$$
$$f(\tfrac{1}{2}) = -\tfrac{7}{2}, \quad f(1) = -3, \quad \text{and} \quad f(2) = -2.$$

Note that we can compute $f(\tfrac{9}{2})$ or $f(\pi)$, for that matter. Sometimes students feel that functions are defined only for integer values of x. This is *not* the case; the description

$$f(x) = |x| - 4 \qquad \text{for all } x$$

means that x can be *any* real number. The range of f consists of all $y \geq -4$. ∎

According to Definition 1.1, a function f assigns to a number x in its domain *exactly* one number y. This *uniqueness* of the number y assigned to x by f is a very important property of the function.

We sometimes have occasion to extract square roots, and we introduce the

---

Square Root Function

$$y = f(x) = \sqrt{x} \qquad \text{for all } x \geq 0. \qquad (2)$$

The domain of this function consists of all $x \geq 0$ since square roots of negative numbers are not real numbers. You may be accustomed to thinking of $\sqrt{4}$ as ± 2, since both 2^2 and $(-2)^2$ are equal to 4. However, for $\sqrt{x}$ to describe a *function*, we must consider that $\sqrt{4}$ yields just *one* of these possibilities. *So that $\sqrt{x}$ will denote a function, from now on we consider $\sqrt{x}$ to be only the* nonnegative *square root of* x. Thus we have $\sqrt{4} = 2$, $\sqrt{0} = 0$, $\sqrt{9} = 3$, and so on. If we want the negative square root, we will always use $-\sqrt{x}$.

Most of the time in this text, we describe a function f by just giving a formula for $f(x)$, such as $f(x) = \sqrt{x - 3}$, without specifying the domain of f. Here is the convention that everyone follows in this case.

Domain of a Function Given by a Formula

When $f(x)$ is defined by a formula in x with no domain specified, the domain consists of all values of x for which the formula can be evaluated and yields a *real* number. In particular:

1. division by zero is not allowed, and

2. square roots (or fourth roots, or any *even* roots) of negative numbers are not allowed.

EXAMPLE 3 Find the domain of the function f given by the formula $y = f(x) = \sqrt{x - 1}$.

Solution The domain consists of all x such that $x - 1 \geq 0$, that is, $x \geq 1$. The range of f consists of all $y \geq 0$. ∎

EXAMPLE 4 Find the domain of $f(x) = \dfrac{x^2 - 1}{x^2 - 9}$.

Solution Note that $f(2) = 3/(-5)$ and $f(5) = \frac{24}{16} = \frac{3}{2}$, but f is not defined at 3 since division by zero is not allowed. The domain of the function consists of all $x \neq \pm 3$. ∎

Sometimes we will want to discuss more than one function at the same time. In that case, we use different letters to represent different functions. The letters f, g, and h are commonly used to denote functions. We use g in the next two examples to become accustomed to letters other than f.

EXAMPLE 5   Find the domain of $g(x) = \dfrac{x}{\sqrt{x-3}}$.

Solution   We must be careful that both $x - 3 \geq 0$, so that $\sqrt{x-3}$ is real, and that $\sqrt{x-3} \neq 0$, so that we are not dividing by zero. Now $x - 3 \geq 0$ means $x \geq 3$, and $\sqrt{x-3} \neq 0$ means $x \neq 3$. Thus the domain consists of all $x > 3$. ∎

EXAMPLE 6   At the start of this section we said that the area A of a circle is a function of its radius r. Describe this function.

Solution   We will call this function g. Now r is the independent variable and A the dependent variable. We have

$$A = g(r) = \pi r^2 \qquad \text{for } r \geq 0.$$

The domain constraint $r \geq 0$ must be stated because we can't have a circle of negative radius. Of course, we could compute πr^2 for negative values of r. This time, the domain restriction is due to the geometric origin of the function. ∎

DEFINITION 1.2

Equal functions

Two functions f and g are **equal** if

1. their domains are identical and
2. $f(x) = g(x)$ for each number x in that common domain.

EXAMPLE 7   Show that the functions $f(x) = |x|$ and $g(x) = \sqrt{x^2}$ are equal.

Solution   The function f is defined for all x. Since $x^2 \geq 0$ and we can take the square root of any nonnegative number, we see that g has this same domain. For $x \geq 0$, we have $|x| = x = \sqrt{x^2}$, while for $x < 0$, both $|x|$ and $\sqrt{x^2}$ give the positive value $-x$. For example, $\sqrt{(-5)^2} = 5 = |-5|$. Thus f and g are the same function. ∎

The following example shows that in simplifying a formula describing a function, we have to be very careful not to change the domain of the function.

EXAMPLE 8   Determine whether the functions f and g given by

$$f(x) = \frac{x^2 - 1}{x - 1} \qquad \text{and} \qquad g(x) = x + 1$$

are equal.

Solution   Since

$$\frac{x^2 - 1}{x - 1} = \frac{(x-1)(x+1)}{x-1},$$

it is tempting to cancel the $(x - 1)$ factors and say that $f(x) = g(x)$, that is, that f and g are the same function. Note, however, that f is not defined at 1, since substituting 1 for x in $(x^2 - 1)/(x - 1)$ leads to division by zero. On the other hand, $g(1) = 1 + 1 = 2$. Thus f and g are not equal functions. Of course, $f(x) = g(x)$ for all $x \neq 1$. If we try to simplify the rule giving $f(x)$, we must write

$$f(x) = x + 1 \qquad \text{for } x \neq 1.$$ ∎

EXAMPLE 9   Let $f(x) = x^2 - 3x$. Find an expression in terms of Δx for $f(5 + \Delta x)$ and for $[f(5 + \Delta x) - f(5)]/\Delta x$. (We will soon encounter this problem in calculus.)

Solution   To compute $f(5 + \Delta x)$, we simply replace x by $5 + \Delta x$ in the formula $x^2 - 3x$ for $f(x)$. We obtain

$$
\begin{array}{ccccc}
f(x) & = & x^2 & - & 3x, \\
\downarrow & & \downarrow & & \downarrow \\
f(5 + \Delta x) & = & (5 + \Delta x)^2 & - & 3(5 + \Delta x)
\end{array}
$$
$$= [25 + 10(\Delta x) + (\Delta x)^2] - [15 + 3(\Delta x)]$$
$$= 10 + 7(\Delta x) + (\Delta x)^2.$$

Consequently,

$$\frac{f(5 + \Delta x) - f(5)}{\Delta x} = \frac{[10 + 7(\Delta x) + (\Delta x)^2] - (5^2 - 3 \cdot 5)}{\Delta x}$$

$$= \frac{10 + 7(\Delta x) + (\Delta x)^2 - 10}{\Delta x} = \frac{7(\Delta x) + (\Delta x)^2}{\Delta x}$$

$$= \frac{\Delta x(7 + \Delta x)}{\Delta x} = 7 + \Delta x \qquad \text{for } \Delta x \neq 0.$$ ∎

EXAMPLE 10   A function $y = f(x)$ does not have to be given by one single formula over its entire domain. Different formulas may be used for different parts of the domain. For example,

$$f(x) = \begin{cases} 2x - 4 & \text{for } x \geq 3, \\ |x| & \text{for } -5 < x < 3, \\ 1 + x & \text{for } x \leq -5 \end{cases}$$

describes a function with domain all real numbers x. Here

$$f(5) = 2 \cdot 5 - 4 = 6 \qquad \text{since } 5 \geq 3,$$
$$f(-2) = |-2| = 2 \qquad \text{since } -5 < -2 < 3,$$
$$f(-7) = 1 + (-7) = -6 \qquad \text{since } -7 \leq -5.$$

Any device that enables us to compute a single y-value for each x-value in some set X of numbers determines a function having the set X as domain. ∎

Graphs of Functions

Using the plane analytic geometry in Sections 1.1 and 1.2, we can draw helpful pictures of real-valued functions of one real variable. For such a function f, we may find the points (x, y) in the plane where $y = f(x)$. The points form the **graph** of the function.

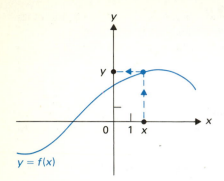

FIGURE 1.44 The graph of a function f, showing how to obtain the y-value from an x-value.

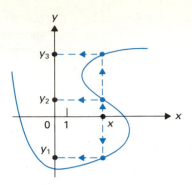

FIGURE 1.45 Not the graph of a function; one x-value must not give more than one y-value.

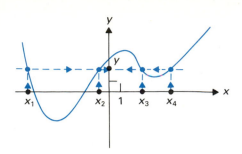

FIGURE 1.46 The graph of a function; different x-values may give the same y-value.

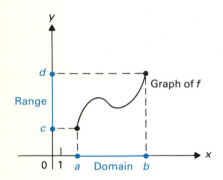

FIGURE 1.47 Domain of f is $[a, b]$. Range of f is $[c, d]$.

The graph of a function f is shown in Fig. 1.44. We can view the graph as giving geometrically the rule for computing the function at each x in the domain. That is, at x, go in a vertical direction until the graph is met, and then go horizontally to find the y such that $y = f(x)$. This geometric technique is indicated by the arrows in Fig. 1.44.

Remember that a function can assign *only one* y-value to each x-value. Thus the curve in Fig. 1.45 is *not* the graph of a function. For the x-value shown, there are three possible y-values, y_1, y_2, and y_3. A function cannot assign three y-values to one x-value. The following states the geometric condition that the graph of a function must satisfy.

Vertical Line Test

A vertical line can intersect the graph of a function in *at most one* point.

A horizontal line may intersect the graph of a function in many points, as shown in Fig. 1.46. Many different x-values are permitted to give the same y-value. Figure 1.47 illustrates graphically the *domain* and *range* of a function.

EXAMPLE 11 Sketch the graph of $y = f(x) = 2x - 4$.

Solution The graph of this function is just the graph of the equation $y = 2x - 4$, which is the line with slope 2 and y-intercept -4, shown in Fig. 1.48. ∎

EXAMPLE 12 Sketch the graph of the squaring function $s = g(t) = t^2$.

Solution The graph is shown in Fig. 1.49. Note the different letters on the axes, representing the letters used in defining the function. This graph will be discussed in more detail in the next section. ∎

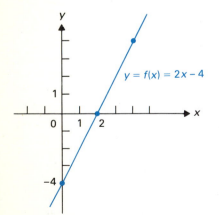

FIGURE 1.48 The graph of $f(x) = 2x - 4$.

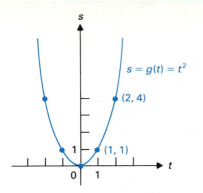

FIGURE 1.49 The graph of $s = g(t) = t^2$, the squaring function.

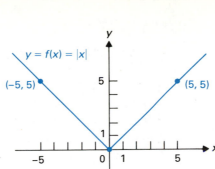

FIGURE 1.50 The graph of $f(x) = |x|$, the absolute value function.

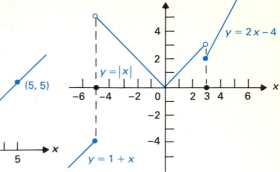

FIGURE 1.51 The graph of
$$f(x) = \begin{cases} 2x - 4 & \text{for } x \geq 3, \\ |x| & \text{for } -5 < x < 3, \\ 1 + x & \text{for } x \leq -5. \end{cases}$$

TABLE 1.1	
x	$y = x^4 - 2x^2 + 3$
-2	11
$-\frac{3}{2}$	3.5625
-1	2
$-\frac{1}{2}$	2.5625
0	3
$\frac{1}{2}$	2.5625
1	2
$\frac{3}{2}$	3.5625
2	11

EXAMPLE 13 Sketch the graph of $y = f(x) = |x|$.

Solution The graph of $y = f(x) = |x|$ is shown in Fig. 1.50. Note that $f(x)$ can also be described by

$$f(x) = \begin{cases} x, & \text{if } x \geq 0, \\ -x, & \text{if } x < 0. \end{cases}$$

This graph is one you should remember. ∎

EXAMPLE 14 Sketch the graph of the function f of Example 10.

Solution The function of Example 10 was defined by

$$f(x) = \begin{cases} 2x - 4 & \text{for } x \geq 3, \\ |x| & \text{for } -5 < x < 3, \\ 1 + x & \text{for } x \leq -5. \end{cases}$$

The graph of f is shown in Fig. 1.51. (The open dot ○ on the graph signifies no "endpoint" there for the line segment.) ∎

One way to sketch the graph $y = f(x)$ is to make a table of corresponding values of x and y, plot the points, and draw a curve through them. (Computing y-values can be tedious; techniques of calculus presented in Sections 4.3–4.5 will help you avoid such computations.) Exercises 51–56 provide drill on estimating function values using a calculator or a computer. A computer can easily make a table of approximate values for many important functions. A video terminal or plotter, used in conjunction with a computer and appropriate software, can provide graphs of functions.

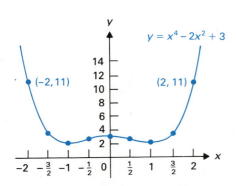

FIGURE 1.52 The graph of $f(x) = x^4 - 2x^2 + 3$.

EXAMPLE 15 Make a table of x- and y-values for $y = f(x) = x^4 - 2x^2 + 3$ with x-values every half-unit from $x = -2$ to $x = 2$. Then plot the graph.

Solution We obtain Table 1.1 and the graph shown in Fig. 1.52. Note that we chose different scales on the axes. ∎

SUMMARY

1. If $y = f(x)$, then x is the independent variable and y the dependent variable.

2. If $y = f(x)$, the domain of the function f consists of all allowable values of the variable x. The range of f consists of all values obtained for y as x goes through all values in the domain.

3. A function f assumes only *one* value $f(x)$ for each x in its domain. Thus, for instance, $\pm\sqrt{x}$ is *not* a function.

4. If $y = f(x)$ is described by a formula, the domain of f consists of all x where $f(x)$ can be computed and gives a real number. This often means just excluding x-values that would lead to division by zero or to taking even roots of negative numbers.

5. Two functions f and g are equal if their domains are identical and $f(x) = g(x)$ for each x in that common domain.

6. The graph of f consists of all points (x, y) such that $y = f(x)$.

7. Graphs can be sketched by making a table of x- and y-values and plotting the points (x, y). This task is ideally relegated to a computer.

EXERCISES 1.3

1. Express the volume V of a cube as a function of the length x of an edge of the cube.

2. Express the volume V of a sphere as a function of the radius r of the sphere.

3. Express the area A enclosed by a circle as a function of the perimeter s of the circle.

4. Express the area A of an equilateral triangle as a function of the length x of an edge of the triangle.

5. Express the volume V of a cube as a function of the length d of a diagonal of the cube. (A diagonal of a cube joins a vertex to the *opposite vertex,* which is the vertex farthest away.)

6. Bill starts at a point A at time $t = 0$ and walks in a straight line at a constant rate of 3 mi/hr toward point B. If the distance from A to B is 21 mi, express his distance s from B as a function of the time t, measured in hours.

7. Mary and Sue start from the same point on a level plain at time $t = 0$. Mary walks north at a constant rate of 3 mi/hr, while Sue jogs west at a constant rate of 5 mi/hr. Find the distance s between them as a function of the time t, measured in hours.

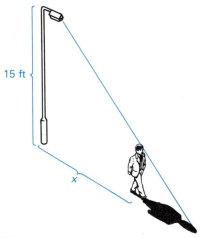

15 ft

x

FIGURE 1.53

8. Smith, who is 6 ft tall, starts at time $t = 0$ directly under a light 15 ft above the ground and walks away in a straight line at a constant rate of 4 ft/sec. See Fig. 1.53.
   (a) Express the length ℓ of Smith's shadow as a function of the distance x he has walked.

(b) Express the length ℓ of Smith's shadow as a function of the time t, measured in seconds.

(c) Express the distance x walked as a function of the length ℓ of Smith's shadow.

9. Let $f(x) = x^2 - 4x + 1$. Find the following.
(a) $f(0)$ (b) $f(-1)$ (c) $f(5)$

10. Let $g(t) = t/(1 - t)$. Find the following, if defined.
(a) $g(0)$ (b) $g(1)$ (c) $g(-1)$

11. Let $h(s) = (s^2 + 1)/(s - 1)$. Find the following.
(a) $h(-2)$ (b) $h(-1)$ (c) $h(3)$

12. Let $\phi(u) = \sqrt{4 + u^2}$. Find the following.
(a) $\phi(0)$ (b) $\phi(2)$ (c) $\phi(-\sqrt{12})$

13. Let
$$f(x) = \begin{cases} x^2 - x & \text{for } x < -3, \\ |x| & \text{for } -3 \le x \le 3, \\ 4x - 5 & x > 3. \end{cases}$$
Find the following.
(a) $f(1)$ (b) $f(4)$ (c) $f(-\tfrac{1}{2})$ (d) $f(-\tfrac{7}{2})$

14. Let
$$g(t) = \begin{cases} (t + 1)/t & \text{for } t > 0, \\ 4 & \text{for } t = 0, \\ (t + 3)/(t - 1) & \text{for } t < 0. \end{cases}$$
Find the following.
(a) $g(-1)$ (b) $g(0)$ (c) $g(-3)$ (d) $g(1)$

15. Let $f(x) = x^2$. Find an expression in terms of Δx for each of the following.
(a) $f(2 + \Delta x)$ (b) $f(2 + \Delta x) - f(2)$
(c) $\dfrac{f(2 + \Delta x) - f(2)}{\Delta x}$

16. Let $h(s) = s^2 - 3s + 2$. Find an expression in terms of Δs for each of the following.
(a) $h(1 + \Delta s)$ (b) $h(1 + \Delta s) - h(1)$
(c) $\dfrac{h(1 + \Delta s) - h(1)}{\Delta s}$

17. Let $g(t) = 1/t$. Find an expression in terms of Δt for each of the following.
(a) $g(-3 + \Delta t)$ (b) $g(-3 + \Delta t) - g(-3)$
(c) $\dfrac{g(-3 + \Delta t) - g(-3)}{\Delta t}$

18. Let $f(x) = \sqrt{x - 3}$. Find an expression in terms of Δx for each of the following.
(a) $f(7 + \Delta x)$ (b) $f(7 + \Delta x) - f(7)$
(c) $\dfrac{f(7 + \Delta x) - f(7)}{\Delta x}$

In Exercises 19–28, find the domain of the given function.

19. $f(x) = 1/x$

20. $f(x) = 1/(x^2 - 1)$

21. $f(x) = \dfrac{x}{x^2 - 3x + 2}$

22. $g(t) = \sqrt{t + 3}$

23. $f(u) = \sqrt{u^2 - 1}$

24. $g(t) = \dfrac{\sqrt{t - 2}}{t^2 - 16}$

25. $h(x) = \sqrt{x - 4}$

26. $k(v) = \dfrac{v^2}{\sqrt{9 - v^2}}$

27. $f(x) = \dfrac{x}{|x| - 1}$

28. $f(t) = \dfrac{t + 1}{\sqrt{|t - 3|}}$

In Exercises 29–36, give a simpler description of the function. Be sure you don't change the domain. Describe the domain explicitly where necessary. See Example 8.

29. $f(x) = \dfrac{2x}{x}$

30. $f(x) = \dfrac{x^2 + 3x}{x}$

31. $g(t) = \dfrac{t^2 - 4}{t + 2}$

32. $g(s) = \dfrac{s - 1}{s^2 - 1}$

33. $f(x) = \dfrac{|x|}{x}$

34. $f(x) = \dfrac{x^2 - 3x + 2}{x^2 - 1}$

35. $f(\Delta x) = \dfrac{(2 + \Delta x)^2 - 4}{\Delta x}$

36. $g(\Delta t) = \dfrac{(-3 + \Delta t)^2 + 2(-3 + \Delta t) - 3}{\Delta t}$

37. A portion of the graph of a function f is shown in Fig. 1.54. Estimate each of the following from the graph.
(a) $f(0)$ (b) $f(1)$ (c) $f(-1)$

38. Proceed as in Exercise 37 for the graph in Fig. 1.55.
(a) $f(0)$ (b) $f(2)$ (c) $f(-2)$ (d) $f(3)$ (e) $f(-3)$

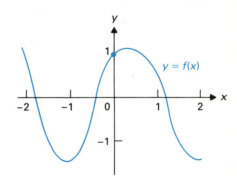

FIGURE 1.54
Graph for Exercise 37.

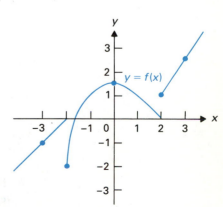

FIGURE 1.55
Graph for Exercise 38.

In Exercises 39–48, sketch the graph of the function over a convenient part of the domain near the origin.

39. $y = f(x) = x - 1$

40. $y = g(x) = -x^2$

41. $s = g(t) = t^2 - 4$

42. $y = f(x) = \sqrt{1 - x^2}$

43. $s = f(r) = -\sqrt{1 - r^2}$

44. $y = f(x) = 1/x$

45. $y = g(x) = \dfrac{1}{x - 2}$

46. $y = h(u) = \dfrac{-1}{u}$

47. $y = f(x) = |x - 3|$

48. $s = g(t) = \dfrac{1}{|t + 1|}$

49. Make a table of x-values and y-values for the function

$$y = f(x) = \frac{x + 1}{x - 1}$$

for $x = -1, -\frac{1}{2}, 0, \frac{1}{2}, \frac{3}{4}, \frac{7}{8}, \frac{9}{8}, \frac{5}{4}, \frac{3}{2}, 2, \frac{5}{2}$, and 3. Plot the points and draw the portion of the graph for all x in the domain such that $-1 \le x \le 3$.

50. Proceed as in Exercise 49 for $y = f(x) = x^3 - 3x^2 + 2$, computing y-values for x every half-unit, starting with $-\frac{3}{2}$ and ending at $\frac{7}{2}$.

Exercises 51–56 are designed to give practice in estimating function values using a calculator or a computer. If you are using a programmable calculator or a computer, key in the function f and find $f(x)$ for x given in parts (a), (b), and (c). If you are using a calculator that is not programmable, we suggest you find only one of these three values in each exercise. Use radian measure for all trigonometric functions. Round off your answers to five significant figures.

51. Estimate $f(x) = (x^4 - x)(2 - x^2)$ for $x =$ (a) 1.03, (b) 3.15, (c) −2.73.

52. Estimate $f(x) = x^2 \sin x$ for $x =$ (a) 5.26, (b) 6.28, (c) −1.27.

53. Estimate $f(x) = \dfrac{x^2 - \sqrt{\pi}}{x^3 - x^2}$ for $x =$ (a) 1.8, (b) 7.2, (c) 0.673.

54. Estimate $f(x) = 2^x \tan x$ for $x =$ (a) 2.35, (b) 4.21, (c) −1.23.

55. Estimate $f(x) = \dfrac{\sqrt{x^2 + 1.3^x}}{x + \cos x}$ for $x =$ (a) 4.17, (b) 0.13, (c) −1.28.

56. Estimate $f(x) = 2^{\sqrt{x^3 + 1}} \tan x$ for $x =$ (a) 0.83, (b) 2.57, (c) −0.71.

Use a calculator for Exercises 57 and 58.

57. Make a table of approximate values for the function

$$f(x) = \frac{x + 1}{\sqrt{x^3 + 1}}$$

using 11 equally spaced x-values starting with $x = 0$ and ending with $x = 10$. Use the data to draw the graph of the function over [0, 10]. [*Note:* Eleven values give ten intervals.]

58. Make a table of approximate values for the function $f(x) = \sin x^2$ using 13 equally spaced x-values starting with $x = 0$ and ending with $x = 3$. Use radian measure. Use the data to draw the graph of the function over [0, 3]. [*Note:* Thirteen values give twelve intervals.]

EXPLORING CALCULUS

C: graph 1 function

D: graph 1 to 8 functions (also tabulate)

1.4 ALGEBRA OF FUNCTIONS; SOME IMPORTANT GRAPHS

Algebra of Functions

We can perform familiar algebraic operations with functions. Given two functions f and g, we can form their

$$\text{sum } f + g, \quad \text{difference } f - g, \quad \text{product } fg, \quad \text{and} \quad \text{quotient } \frac{f}{g}.$$

If the functions are given by formulas, then these algebraic operations correspond to the same operations with the formulas. For example, if $f(x) = x^2$ and $g(x) = \sqrt{x}$, then $(f + g)(x) = f(x) + g(x) = x^2 + \sqrt{x}$.

We list here these familiar algebraic operations, giving their defining equations. We include one more operation, *composition* of the two functions, which is also useful in breaking a function into simple parts. For example, $(2x + 1)^5 = f(g(x))$, where $f(u) = u^5$ and $u = g(x) = 2x + 1$. For each operation in the list, the domain of the newly formed function consists of all x for which the new function can be computed to yield a real number.

Operations with Functions f and g

Addition $(f + g)(x) = f(x) + g(x)$

Subtraction $(f - g)(x) = f(x) - g(x)$

Multiplication $(fg)(x) = f(x)g(x)$

Division $\dfrac{f}{g}(x) = \dfrac{f(x)}{g(x)}$

Composition $(f \circ g)(x) = f(g(x))$

EXAMPLE 1   Describe the domain of each of the functions defined in the preceding list of operations.

Solution   Addition, subtraction, and multiplication can be computed for all x such that $f(x)$ and $g(x)$ are both defined. That is, the domain of $f + g, f - g$, and fg consists of all x in the *intersection* of the domain of f and the domain of g.

Division $f(x)/g(x)$ is defined for all x such that $f(x)$ and $g(x)$ are both defined and $g(x) \neq 0$. Thus the domain of f/g consists of all x in the intersection of the domain of f and the domain of g such that $g(x) \neq 0$.

For composition, $(f \circ g)(x) = f(g(x))$, we first have to be able to compute $g(x)$, that is, x must lie in the domain of g. Then we must be able to compute $f(g(x))$, that is, $g(x)$ must lie in the domain of f. Thus the domain of $f \circ g$ consists of all x in the domain of g such that $g(x)$ is in the domain of f.   ■

EXAMPLE 2   Let $f(x) = x^2$ and let $g(x) = \sqrt{x + 1}$. Evaluate $f + g, f - g$, $fg, f/g$, and $f \circ g$ at $x = 3$.

Solution   We find that $f(3) = 3^2 = 9$ and $g(3) = \sqrt{3 + 1} = 2$. Thus

$$(f + g)(3) = f(3) + g(3) = 9 + 2 = 11,$$
$$(f - g)(3) = f(3) - g(3) = 9 - 2 = 7,$$
$$(fg)(3) = f(3)g(3) = 9 \cdot 2 = 18,$$
$$\frac{f}{g}(3) = \frac{f(3)}{g(3)} = \frac{9}{2} = 4.5,$$
$$(f \circ g)(3) = f(g(3)) = f(2) = 4.$$ ■

It is very helpful to interpret the algebra of functions graphically.

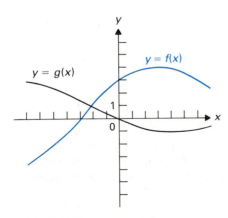

FIGURE 1.56   Graphs of f and g.

EXAMPLE 3   Figure 1.56 shows graphs of f and of g. Sketch the graph of $f + g$.

Solution   Since $(f + g)(x) = f(x) + g(x)$, we should add the y-values of $f(x)$ and $g(x)$ for each x. For example, at any point x_0 where the graph of f is above

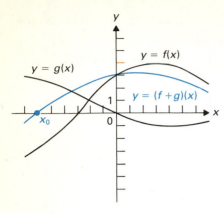

FIGURE 1.57   Graph of $f + g$.

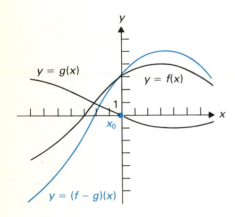

FIGURE 1.58   Graph of $f - g$.

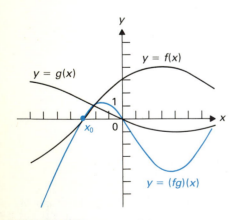

FIGURE 1.59   Graph of fg.

the x-axis and the graph of g is the same distance below the x-axis, we have $(f + g)(x_0) = 0$. The graph of this sum function is shown in color in Fig. 1.57.   ∎

EXAMPLE 4   Figure 1.56 shows graphs of f and of g. Sketch the graph of $f - g$.

Solution   Since $(f - g)(x) = f(x) - g(x)$, we should subtract the y-value $g(x)$ from the y-value $f(x)$ for each x. For example, for any value x_0 where $g(x_0) = 0$, we have $(f - g)(x_0) = f(x_0) - g(x_0) = f(x_0) - 0 = f(x_0)$. The graph of this difference function is shown in color in Fig. 1.58.   ∎

EXAMPLE 5   Figure 1.56 shows graphs of f and of g. Sketch the graph of fg.

Solution   Note that for a value x_0 such that either $f(x_0) = 0$ or $g(x_0) = 0$, we have $(fg)(x_0) = f(x_0)g(x_0) = 0$. The graph is sketched in Fig. 1.59.   ∎

EXAMPLE 6   Figure 1.56 shows graphs of f and of g. Sketch the graph of f/g.

Solution   Wherever $g(x) = 0$, the function $(f/g)(x) = f(x)/g(x)$ is undefined. For the graph of g in Fig. 1.56, we see that $g(x)$ is very close to zero near any point where $g(x) = 0$, and $f(x)$ is not near zero at those points. This means that the quotient $f(x)/g(x)$ is of very large magnitude and cannot be plotted as a y-value in our little figure near a value of x where $g(x) = 0$. Of course the sign of the y-values is determined by the signs of $f(x)$ and $g(x)$. The graph is sketched in Fig. 1.60.   ∎

Sketching the graph of $f \circ g$ from the graphs of f and g emphasizes the definition of function composition, as we see in the next example.

EXAMPLE 7   Sketch the graph of $f \circ g$ from the graphs of f and g shown in Fig. 1.56.

Solution   Recall that $(f \circ g)(x) = f(g(x))$. To sketch the graph, we simply choose some values of x, estimate $g(x)$ from the graph of g, and then estimate $f(g(x))$ from the graph of f. For example, we estimate that $g(3) \approx -1$ and that $f(-1) \approx 2.2$. Thus the graph of $f \circ g$ passes close to $(3, 2.2)$. The graph of $f \circ g$ is shown in Fig. 1.61.   ∎

Among the simplest functions are the *constant functions,* of the form $f(x) = c$ for all x, where c is some constant. Functions of the form

$$f(x) + c, \qquad cf(x), \qquad \text{and} \qquad \frac{c}{f(x)}$$

can be regarded as special cases of addition, multiplication, and division of functions. The exercises give you practice in sketching their graphs when the graph of f is given as well as more practice with problems like those solved in Examples 3–7.

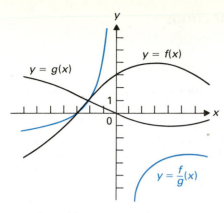

FIGURE 1.60 Graph of f/g.

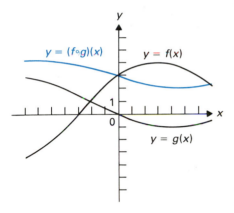

FIGURE 1.61 Graph of $f \circ g$.

Translation

Suppose we know the graph of $y = f(x)$. We can then sketch the graph of a function such as $y = g(x) = f(x - 2) + 1$ using translation of axes, as the next example shows.

EXAMPLE 8 The graph of $y = f(x)$ is shown in Fig. 1.62. Sketch the graph of $y = f(x - 2) + 1$.

Solution We write $y = f(x - 2) + 1$ as

$$y - 1 = f(x - 2). \tag{1}$$

We translate axes to the new origin $(2, 1)$ using the translation equations

$$\bar{x} = x - 2 \qquad \text{and} \qquad \bar{y} = y - 1,$$

as discussed in Section 1.1. These equations relate x,y-coordinates and $\bar{x},\bar{y}$-coordinates where the $\bar{x},\bar{y}$-origin is the point $(x, y) = (2, 1)$. Equation (1) becomes

$$\bar{y} = f(\bar{x}). \tag{2}$$

The $\bar{x},\bar{y}$-axes and the graph of $\bar{y} = f(\bar{x})$ are shown in Fig. 1.63. With reference to the x,y-axes, this is the desired graph of $y = f(x - 2) + 1$. Notice that it can be obtained by sliding the graph of $y = f(x)$ to the right two units and then up one unit. ■

As illustrated in Example 8, the graph of

$$y = f(x - h) + k$$

can be obtained by sliding the graph of $y = f(x)$ to the *right* or to the *left* the distance $|h|$, depending on whether h is *positive* or *negative,* respectively, and then sliding it *up* or *down* the distance $|k|$, depending on whether k is *positive* or *negative,* respectively.

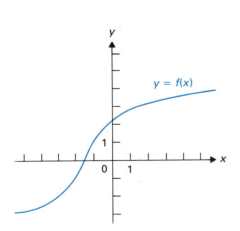

FIGURE 1.62 Graph of $y = f(x)$.

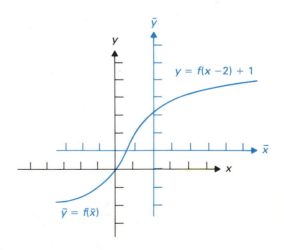

FIGURE 1.63 Graph of $y = f(x - 2) + 1$.

Graphs of Some Important Functions

Functions of the form

$$f(x) = a_n x^n + a_{n-1} x^{n-1} + \cdots + a_2 x^2 + a_1 x + a_0$$

are **polynomial functions.** Note that for a given x, the value $f(x)$ can be computed using only addition and multiplication of real numbers. This makes polynomial functions comparatively easy to work with. We know that the graph of a polynomial function of the form $ax + b$ is a line. We start our presentation of graphs with two other types of polynomial functions, monomial functions and quadratic functions.

Monomial Functions: The *monomial functions* are the *power functions*

$$x, \quad x^2, \quad x^3, \quad x^4, \quad x^5, \ldots, \quad x^n, \ldots$$

or constant multiples of them. When a function is given by a formula, we often refer to the formula as the function, to save writing. Thus we may refer to the function $4x^3$ rather than the function f where $f(x) = 4x^3$.

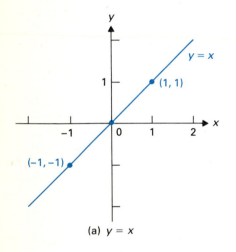

(a) $y = x$

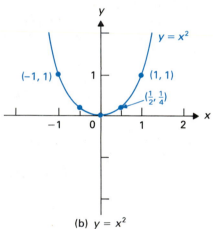

(b) $y = x^2$

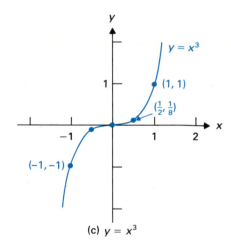

(c) $y = x^3$

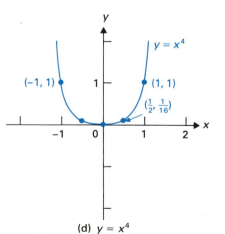

(d) $y = x^4$

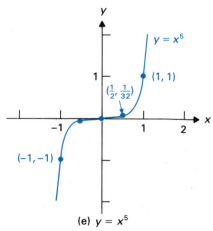

(e) $y = x^5$

FIGURE 1.64 Graphs of the first five monomial functions.

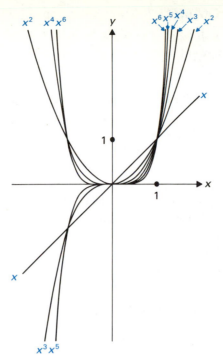

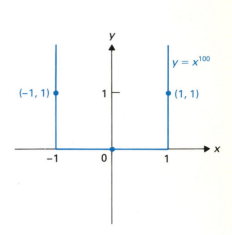

FIGURE 1.65   Computer-generated graphs of the first six monomial functions.

FIGURE 1.66   Attempt to draw the graph of x^{100}.

TABLE 1.2	
x	$y = x^{100}$
0.96	0.017
0.97	0.048
0.98	0.133
0.99	0.366
1.00	1.000
1.01	2.705
1.02	7.245
1.03	19.219
1.04	50.505

It is important to know the graphs of the monomial functions. We know that the graph of the function x is a straight line of slope 1 through the origin. The graph of the squaring function was shown in Fig. 1.49 on page 29. All the monomial graphs x^n go through the origin $(0, 0)$ and the point $(1, 1)$. If n is even, the graph of x^n goes through $(-1, 1)$, and if n is odd, the graph goes through $(-1, -1)$.

The different parts of Fig. 1.64 show the graphs of the first five monomial functions. Note in particular that the larger the value of n, the closer the graph is to the x-axis for $-1 < x < 1$. For example, $(\frac{1}{2})^4 < (\frac{1}{2})^2$, so the graph of x^4 is closer to the x-axis where $x = \frac{1}{2}$ than the graph of x^2 is. If $|x| > 1$, then the larger the value of n, the farther the graph of x^n is from the x-axis. For example, $2^5 > 2^3$, so x^5 is farther from the x-axis than x^3 where $x = 2$. Computer-generated Fig. 1.65 shows the graphs of the functions x, x^2, x^3, x^4, x^5, and x^6 on a single pair of axes. This figure indicates even more clearly that for high powers of n, the graph of x^n clings close to the x-axis for $-1 < x < 1$ and climbs very steeply for $x > 1$.

A calculator gives the approximate values in Table 1.2, which shows that it is hopeless to draw a true graph of x^{100} on a regular-sized piece of paper. An attempt, shown in Fig. 1.66, makes the graph appear as the x-axis for $-1 \leq x \leq 1$ and as upward vertical lines at $x = 1$ and $x = -1$.

Now that we know the graph of x^n, we can easily sketch the graphs of cx^n for any constant c and $y = (x - h)^n + k$ for any origin (h, k) of translated axes.

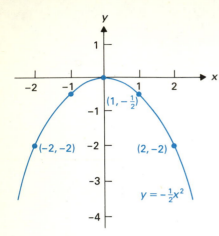

FIGURE 1.67 Graph of $y = -\frac{1}{2}x^2$.

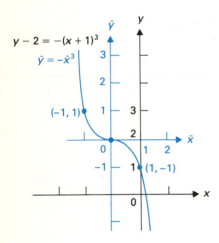

FIGURE 1.68 Graph of $y - 2 = -(x + 1)^3$, or $\bar{y} = -\bar{x}^3$.

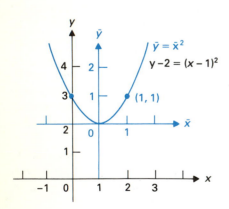

FIGURE 1.69 Graph of $y - 2 = (x - 1)^2$, or $\bar{y} = \bar{x}^2$.

EXAMPLE 9 Sketch the graph of $y = -\frac{1}{2}x^2$.

Solution The graph is shown in Fig. 1.67. It opens downward rather than upward. ∎

EXAMPLE 10 Sketch the graph of $y = 2 - (x + 1)^3$.

Solution Writing the equation as $y - 2 = -(x + 1)^3$, we let $\bar{x} = x + 1$ and $\bar{y} = y - 2$. The equation then becomes $\bar{y} = -\bar{x}^3$. We translate to $\bar{x},\bar{y}$-axes at $(h, k) = (-1, 2)$ and sketch the graph as shown in Fig. 1.68. ∎

Quadratic Functions: A quadratic function f is of the form $f(x) = ax^2 + bx + c$, where $a \neq 0$. Graphs of these functions are called *parabolas*. By completing the square and translating to $\bar{x},\bar{y}$-axes, we can put the equation $y = ax^2 + bx + c$ in the form $\bar{y} = d\bar{x}^2$ for some constant d. That is, the graph of a quadratic function is just a translation of the graph of a quadratic monomial function. The reason for this becomes clear in the following examples.

EXAMPLE 11 Sketch the graph of $y = x^2 - 2x + 3$.

Solution Completing the square on the x-terms, we have

$$y - 3 = x^2 - 2x,$$
$$y - 3 + 1 = x^2 - 2x + 1,$$
$$y - 2 = (x - 1)^2.$$

Setting $\bar{x} = x - 1$ and $\bar{y} = y - 2$, we have $\bar{y} = \bar{x}^2$. We translate to $\bar{x},\bar{y}$-axes at $(h, k) = (1, 2)$ and sketch the graph in Fig. 1.69. ∎

EXAMPLE 12 Sketch the graph of $y = 3x^2 + 12x + 11$.

Solution As the first step in completing the square on the x-terms, we factor the coefficient 3 of x^2 out of all the x-terms.

$$y - 11 = 3x^2 + 12x,$$
$$y - 11 = 3(x^2 + 4x), \qquad \textit{Factoring out 3}$$
$$y - 11 + 12 = 3(x^2 + 4x + 4), \qquad \textit{Adding 12 to both sides}$$
$$y + 1 = 3(x + 2)^2$$

We let $\bar{x} = x + 2$ and $\bar{y} = y + 1$, so the equation becomes $\bar{y} = 3\bar{x}^2$. The parabola is sketched in Fig. 1.70. ∎

Exponential Functions: For $a > 0$, the function $f(x) = a^x$, defined for all real numbers x, is the **exponential function with base** a. For example, if $f(x) = 2^x$, then $f(3) = 2^3 = 8$. We require $a > 0$ since $a^{1/2} = \sqrt{a}$ is not a real number if $a < 0$. Approximate values of an exponential function can be found using a computer or the y^x key on a scientific calculator. It is appropriate to emphasize here that all computations with a calculator or a computer are carried out using rational (fractional) approximations of real numbers. A calculator or computer cannot give you the *exact* value of 2^π. If six significant figures are used in

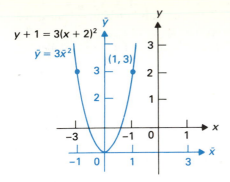

FIGURE 1.70 Graph of $y + 1 = 3(x + 2)^2$, or $\bar{y} = 3\bar{x}^2$.

calculation, it considers π to be

$$3.14159 = \frac{314159}{100000}.$$

Thus it considers 2^π to be

$$2^{314159/100000},$$

the 100,000th root of 2 raised to the power 314,159. It can't compute that number exactly either and gives as answer a rational approximation to it, presumably accurate to six significant figures. For the moment, it is best that you regard a^x for irrational values of x in the same way your calculator does, in terms of rational approximations. In Section 7.2, we will define a^x for all x in a precise way.

Figure 1.71 shows the graphs of the functions $(1.5)^x$, 2^x, and 3^x. These functions all rise very rapidly as x increases through positive values, and the larger the base, the faster they increase. Also, all the graphs go through the point $(0, 1)$. The graphs of $(\frac{1}{2})^x = 2^{-x}$ and $(\frac{1}{3})^x = 3^{-x}$ are sketched in Fig. 1.72. The graphs of 2^x and 2^{-x} are symmetric about the y-axis since each can be obtained from the other by replacing x by $-x$. This is true in general for the graphs of a^x and a^{-x}.

Recall these laws of exponents.

Laws of Exponents

1. $a^r a^s = a^{r+s}$ **2.** $(a^r)^s = a^{rs}$ **3.** $\dfrac{a^r}{a^s} = a^{r-s}$

We have $(4^3)(4^2) = (4 \cdot 4 \cdot 4)(4 \cdot 4) = 4^5 = 4^{3+2}$, $(4^3)^2 = (4 \cdot 4 \cdot 4)(4 \cdot 4 \cdot 4) = 4^6 = 4^{3 \cdot 2}$, and $4^3/4^2 = 4^1 = 4^{3-2}$.

Logarithm Functions: The graph in Fig. 1.73 shows a typical exponential function $y = a^x$ for $a > 0$ (in this case for $a = 2$). Given any positive y-value, we can find a *unique* value of x such that $a^x = y$. This *uniqueness* means that x

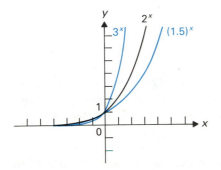

FIGURE 1.71 Exponential functions a^x for $a > 1$.

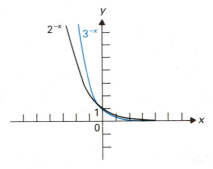

FIGURE 1.72 Exponential functions a^{-x} for $a > 1$.

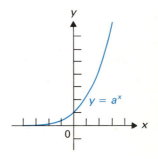

FIGURE 1.73 A typical exponential graph.

appears as a *function* of y with domain all $y > 0$. This function is called the **logarithm function with base** a and is written as $x = \log_a y$, read "x is the logarithm of y to the base a." Note that if we start with x and follow the labels on the arrows in

$$x \xrightarrow{\textit{find } y = a^x} y \xrightarrow{\textit{find } x = \log_a y} x,$$

we go from x on the x-axis up to y on the y-axis and then back to the same x again. That is,

$$\log_a(a^x) = x. \tag{3}$$

For example, $\log_3(3^4) = 4$. Thus the logarithm function with base a *undoes* the exponential function with base a; for this reason the logarithm function is called the **inverse** of the exponential function. If instead we start at y on the y-axis, go to x on the x-axis and then back to y according to the diagram

$$y \xrightarrow{\textit{find } x = \log_a y} x \xrightarrow{\textit{find } y = a^x} y,$$

we see that

$$a^{\log_a y} = y. \tag{4}$$

For example, $2^{\log_2 3} = 3$. This shows that the exponential function *undoes* the logarithm function with the same base; the exponential function is the **inverse** of the logarithm function. We summarize this relationship by saying that the functions a^x and $\log_a x$ are **inverse functions.** We discuss inverse functions in more detail in Section 7.1.

EXAMPLE 13   Simplify $4^{\log_2 3}$.

Solution   Motivated by relation (4), we try to express the exponential function in terms of base 2 rather than base 4 and apply the properties of exponents:

$$4^{\log_2 3} = (2^2)^{\log_2 3} = 2^{2(\log_2 3)} = (2^{\log_2 3})^2 = 3^2 = 9. \qquad \blacksquare$$

EXAMPLE 14   Simplify $\log_9(3^4)$.

Solution   Motivated by relation (3), we try to express the exponential function in terms of base 9 rather than base 3, using the properties of exponents:

$$\log_9(3^4) = \log_9((9^{1/2})^4) = \log_9(9^2) = 2. \qquad \blacksquare$$

On page 41 we list the properties of logarithm functions that correspond to the three properties of exponential functions on page 39.

By property 1, for example, $\log_2(32) = \log_2(2^5) = 5$ and $\log_2(4 \cdot 8) = \log_2 4 + \log_2 8 = \log_2(2^2) + \log_2(2^3) = 2 + 3 = 5$. To obtain property 1 for logarithm functions from property 1 for exponential functions, suppose that

$$c = \log_a r \qquad \text{and} \qquad d = \log_a s.$$

Then $a^c = r$ and $a^d = s$. By property (1) for exponential functions, we obtain

$$rs = a^c a^d = a^{c+d}.$$

This is precisely the assertion that $\log_a(rs) = c + d = \log_a r + \log_a s$.

Properties of Logarithm Functions

1. $\log_a(rs) = \log_a r + \log_a s$

2. $\log_a(r^s) = s(\log_a r)$

3. $\log_a\left(\dfrac{r}{s}\right) = \log_a r - \log_a s$

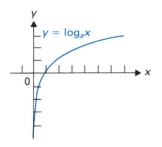

FIGURE 1.74 A typical logarithmic graph.

When we study a function f, it is customary to graph $y = f(x)$ rather than $x = f(y)$. We know that the graph of $x = \log_a y$ is shown in Fig. 1.73, because $y = a^x$ means the same as $x = \log_a y$. Interchanging x and y, we see that $y = \log_a x$ means the same as $x = a^y$. The graph of $x = a^y$ is easily sketched from Fig. 1.73 and is shown in Fig. 1.74. It is typical of the graph of $y = \log_a x$ for $a > 1$ (in this case for $a = 2$). An important feature of this graph is that y increases very slowly as x increases for large values of x.

SUMMARY

1. Operations with functions f and g include addition ($f + g$), subtraction ($f - g$), multiplication (fg), and division (f/g). The domain of each resulting function consists of all x for which the corresponding operation with $f(x)$ and $g(x)$ can be done.

2. Function composition is defined by $(f \circ g)(x) = f(g(x))$. The domain of $f \circ g$ consists of all x in the domain of g such that $g(x)$ is in the domain of f.

3. If the graph of $y = f(x)$ is given, then the graph of $y - k = f(x - h)$ can be found by taking translated $\bar{x}, \bar{y}$-axes with origin (h, k) and sketching $\bar{y} = f(\bar{x})$.

4. The graphs of monomial functions are illustrated in Figs. 1.64 and 1.65.

5. The graph of a quadratic function is a parabola. It can be sketched by completing the square and translating axes.

6. For $a > 0$, the function $f(x) = a^x$ is the exponential function with base a. Its domain consists of all real numbers.

7. For $a > 0$, the function $y = \log_a x$ is the logarithm function with base a. It is defined for all $x > 0$.

8. $y = a^x$ is equivalent to $x = \log_a y$.

9. $a^{\log_a x} = x$ and $\log_a(a^x) = x$, that is, the exponential and logarithm functions with base a are inverses of each other.

EXERCISES 1.4

In Exercises 1–10, sketch as much as you can of the graph of the indicated function from the graph $y = f(x)$ given in Fig. 1.75.

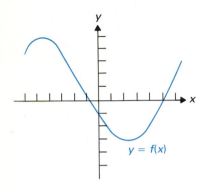

FIGURE 1.75

1. $y = 2f(x)$

2. $y = -f(x)$

3. $y = f(-x)$

4. $y = \dfrac{1}{f(x)}$

5. $y = f^2(x)$

6. $y = f(x - 1) + 2$

7. $y = f(x + 2) - 1$

8. $y = f(2x)$

9. $y = f\left(\dfrac{x}{2}\right)$

10. $y = f(2x - 4) - 2$

In Exercises 11–20, sketch as much as you can of the graph of the indicated function from the graphs $y = f(x)$ and $y = g(x)$ given in Fig. 1.76.

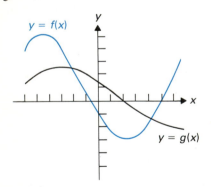

FIGURE 1.76

11. $y = (f + g)(x)$

12. $y = (f - g)(x)$

13. $y = (g - f)(x)$

14. $y = (fg)(x)$

15. $y = \dfrac{f}{g}(x)$

16. $y = \dfrac{g}{f}(x)$

17. $y = (f \circ g)(x)$

18. $y = (f + g)(-x)$

19. $y = (fg)(-x)$

20. $y = f(2x)g(x)$

21. Note that the graph of $y = |x|$ has a sharp point at the origin. Give a formula for a function $g(x)$ having a sharp point at $(x, y) = (-1, 3)$.

22. Sketch the graph of $f(x) = |x - 2| + |x + 2|$ using addition of functions.

23. Sketch the graph of $f(x) = |x + 1| - |x - 3|$ using subtraction of functions.

24. Sketch the graph of $f(x) = |x + 1||x - 1|$ using multiplication of functions.

In Exercises 25–45, sketch the graph of $y = f(x)$.

25. $y = -x^3$

26. $y = -x^4$

27. $y = \dfrac{x^2}{2}$

28. $y = 9x^2$

29. $y = \dfrac{1}{27}x^3$

30. $y = (x - 1)^2$

31. $y = (x + 2)^3$

32. $y = -(x + 1)^4$

33. $y = -4(x + 3)^2$

34. $y = x^2 + 3$

35. $y = x^3 - 1$

36. $y = 5 - x^4$

37. $y = 6 - \dfrac{1}{4}x^2$

38. $y = 4 + (x - 2)^2$

39. $y = (x + 1)^3 - 3$

40. $y = -1 - (x + 5)^4$

41. $y = 5 - 2(x + 3)^2$

42. $y = x^2 + 2x + 1$

43. $y = x^2 - 4x + 3$

44. $y = -x^2 - 6x + 5$

45. $y = -x^2 - 10x - 21$

In Exercises 46–59, use properties of logarithm and exponential functions to simplify the given expression.

46. $2^{\log_2 5}$

47. $3^{-\log_3 7}$

48. $3^{\log_5 1}$

49. $\log_4 64$

50. $\log_{25} 5$

51. $\log_3(3^4)$

52. $4^{\log_2 3}$

53. $2^{\log_4 2}$

54. $\log_2(4^3)$

55. $\log_9(3^8)$

56. $2^{3(\log_2 4)}$

57. $\log_{10}(2^{\log_2 10})$

58. $\log_3(2^{5(\log_2 3)})$

59. $2^{(\log_3 10)(\log_2 9)}$

In Exercises 60–65, sketch the graph of the indicated function.

60. $2^{x/2}$

61. 4^{-x}

62. $\log_3 x$

63. $\log_{0.5} x$

64. $\log_2(x + 2) - 4$

65. 2^{x+1}

◆ EXPLORING CALCULUS

E: function draw quiz

F: find equation quiz (options 1, 2, 7)

1.5 TRIGONOMETRY REVIEW

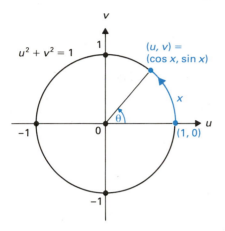

FIGURE 1.77 Point (u, v) on the unit circle corresponding to directed arc length x measured from $(1, 0)$.

The Sine and the Cosine

The circle $u^2 + v^2 = 1$ is shown in Fig. 1.77. It is important to note that the radius of this circle is 1; hence it is called a *unit circle*. The functions sin x and cos x may be evaluated as follows. If $x \geq 0$, we start at the point $(1, 0)$ on the circle and go *counterclockwise* along the circle until we have traveled the distance x on the arc. As shown in the figure, we will stop at some point (u, v) on the circle. Then

$$\sin x = v \quad \text{and} \quad \cos x = u. \qquad (1)$$

Equations (1) are also valid if $x < 0$, except that then we travel the distance $|x|$ *clockwise* around the circle from $(1, 0)$ to arrive at a point (u, v).

EXAMPLE 1 Find sin 0 and cos 0.

Solution If $x = 0$, we stop at the point $(1, 0)$ on the circle in Fig. 1.77, so $u = 1$ and $v = 0$. Thus sin $0 = 0$ and cos $0 = 1$. ∎

EXAMPLE 2 Find $\sin(-\pi/2)$ and $\cos(-\pi/2)$.

Solution With $x = -\pi/2$, we should travel *clockwise* a quarter of the way around the circle in Fig. 1.77, since the circle has circumference 2π. We arrive at the point $(u, v) = (0, -1)$, as shown in Fig. 1.78. Thus $\sin(-\pi/2) = -1$ and $\cos(-\pi/2) = 0$. ∎

The arc length x illustrated in Fig. 1.77 is the *radian measure* of the central angle θ shown in the figure. Now the radian measure of a central angle θ of a circle is

$$\text{Radian measure of } \theta = \frac{\text{Length of intercepted arc}}{\text{Radius}}.$$

Since the radius is 1 in the figure, the radian measure of θ is given by the length x of the arc. For example, a 360° angle has radian measure 2π since the length all the way around the circle is 2π. We see that

$$\text{Radian measure of } \theta = \left(\frac{\pi}{180}\right)(\text{Degree measure of } \theta),$$

because 180° corresponds to π radians.

EXAMPLE 3 Find the radian measure of an angle θ having degree measure 240°.

Solution Our conversion equation yields

$$\text{Radian measure of } \theta = \left(\frac{\pi}{180}\right)(240) = \frac{4}{3}\pi. \qquad ∎$$

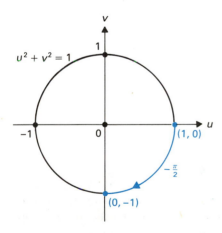

FIGURE 1.78 Point $(u, v) = (0, -1)$ on the unit circle corresponding to directed arc length $-\pi/2$ measured from $(1, 0)$.

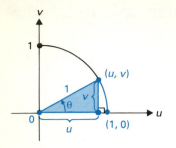

FIGURE 1.79 Right triangle of hypotenuse 1 and acute angle θ.

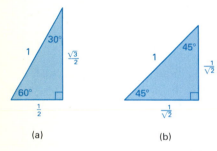

(a)   (b)

FIGURE 1.80

(a) $\sin 30° = \sin(\pi/6) = \frac{1}{2}$
$= \cos 60° = \cos(\pi/3)$;
$\sin 60° = \sin(\pi/3) = \sqrt{3}/2$
$= \cos 30° = \cos(\pi/6)$;
(b) $\sin 45° = \sin(\pi/4) = 1/\sqrt{2}$
$= \cos 45° = \cos(\pi/4)$.

TABLE 1.3

x	$\sin x$	$\cos x$
0	0	1
$\dfrac{\pi}{6}$	$\dfrac{1}{2}$	$\dfrac{\sqrt{3}}{2}$
$\dfrac{\pi}{4}$	$\dfrac{1}{\sqrt{2}}$	$\dfrac{1}{\sqrt{2}}$
$\dfrac{\pi}{3}$	$\dfrac{\sqrt{3}}{2}$	$\dfrac{1}{2}$
$\dfrac{\pi}{2}$	1	0

From Fig. 1.79, we see that if $0 < \theta < 90°$, then θ is an acute angle of the right triangle shown, and

$$\sin \theta = v = \frac{v}{1} = \frac{\text{Opposite side length}}{\text{Hypotenuse length}},$$

while

$$\cos \theta = u = \frac{u}{1} = \frac{\text{Adjacent side length}}{\text{Hypotenuse length}}.$$

You may have learned these definitions before. The advantage of using our definitions given in Eqs. (1) is that we can easily find $\sin x$ if $x > 90°$. For example, $\sin (3\pi/2) = \sin (270°) = -1$ since an arc length of $3\pi/2$ corresponds to the point $(0, -1)$ on the circle.

Recall from geometry the lengths of the sides of the right triangles shown in Fig. 1.80. Using these triangles, we obtain the values for $\sin x$ and $\cos x$ given in Table 1.3. Note that x is given in *radians* in Table 1.3. We will see in Section 3.5 that radian measure is the best to use in calculus.

> In the absence of a degree symbol, $\sin x$ and $\cos x$ always signify that x is in radians.

If we wish to use degree measure for some reason, we write $\sin x°$. Thus we have $\sin 30° = \frac{1}{2}$, but $\sin 30$ means the sine of 30 radians, which our calculator tells us is about -0.988.

EXAMPLE 4   Find $\sin(-2\pi/3)$ and $\cos(-2\pi/3)$.

Solution   Since $-2\pi/3$ radians corresponds to $-120° = -90° - 30°$, we see from Fig. 1.81 and triangle (a) in Fig. 1.80 that

$$\sin\left(-\frac{2\pi}{3}\right) = -\frac{\sqrt{3}}{2} \quad \text{and} \quad \cos\left(-\frac{2\pi}{3}\right) = -\frac{1}{2}. \quad \blacksquare$$

Tangent, Cotangent, Secant, and Cosecant

We define the remaining four basic trigonometric functions of x in terms of the point (u, v) in Fig. 1.77.

$$\tan x = \frac{v}{u} = \frac{\sin x}{\cos x}, \qquad \cot x = \frac{u}{v} = \frac{\cos x}{\sin x} = \frac{1}{\tan x},$$

$$\sec x = \frac{1}{u} = \frac{1}{\cos x}, \qquad \csc x = \frac{1}{v} = \frac{1}{\sin x}. \tag{2}$$

The trigonometric functions in Eqs. (2) are defined as quotients of other functions. Their domains thus exclude points where denominators may be zero. For example, $\cot x$ is not defined where $\sin x = 0$, which happens where $x = n\pi$ for any integer n.

EXAMPLE 5   Continuing Example 4, find the other four trigonometric functions of $-2\pi/3$.

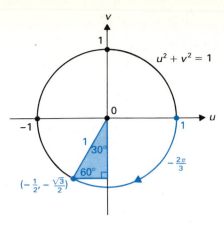

FIGURE 1.81 Computation of the trigonometric functions at $-2\pi/3$.

Solution From Fig. 1.81, we see that

$$\tan\left(-\frac{2\pi}{3}\right) = \frac{-\sqrt{3}/2}{-1/2} = \sqrt{3},$$

$$\cot\left(-\frac{2\pi}{3}\right) = \frac{1}{\tan(-2\pi/3)} = \frac{1}{\sqrt{3}},$$

$$\sec\left(-\frac{2\pi}{3}\right) = \frac{1}{-1/2} = -2,$$

$$\csc\left(-\frac{2\pi}{3}\right) = \frac{1}{-\sqrt{3}/2} = -\frac{2}{\sqrt{3}}.$$

■

EXAMPLE 6 Find the six trigonometric functions of $3\pi/4$.

Solution Using triangle (b) of Fig. 1.80, we see that an arc of length $3\pi/4$ starting from $(1, 0)$ brings us to the point $(u, v) = (-1/\sqrt{2}, 1/\sqrt{2})$ on the unit circle. This is shown in Fig. 1.82. Thus

$$\sin\frac{3\pi}{4} = \frac{1}{\sqrt{2}}, \qquad\qquad \cos\frac{3\pi}{4} = -\frac{1}{\sqrt{2}},$$

$$\tan\frac{3\pi}{4} = \frac{1/\sqrt{2}}{-1/\sqrt{2}} = -1, \qquad \cot\frac{3\pi}{4} = \frac{-1/\sqrt{2}}{1/\sqrt{2}} = -1,$$

$$\sec\frac{3\pi}{4} = \frac{1}{-1/\sqrt{2}} = -\sqrt{2}, \qquad \csc\frac{3\pi}{4} = \frac{1}{1/\sqrt{2}} = \sqrt{2}.$$

■

Identities

Referring to Eqs. (1) and Fig. 1.77, we have $\sin x = v$, $\cos x = u$, and $u^2 + v^2 = 1$. Therefore,

$$\sin^2 x + \cos^2 x = 1.$$

(Note that $\sin^2 x$ means $(\sin x)^2$.) This is a fundamental trigonometric identity. The trigonometric functions are often called the *circular functions* since by this identity, the point $(\cos t, \sin t)$ lies on the unit circle $x^2 + y^2 = 1$ for any value of t.

There are many other identities, some of which are readily obtained. For example, if we divide both sides of $\sin^2 x + \cos^2 x = 1$ by $\cos^2 x$, then

$$\frac{\sin^2 x}{\cos^2 x} + \frac{\cos^2 x}{\cos^2 x} = \frac{1}{\cos^2 x},$$

or

$$\tan^2 x + 1 = \sec^2 x.$$

Other identities that we will have occasion to use are

$$\cot^2 x + 1 = \csc^2 x, \qquad \sin 2x = 2 \sin x \cos x,$$

$$\cos 2x = \cos^2 x - \sin^2 x = 2 \cos^2 x - 1 = 1 - 2 \sin^2 x,$$

$$\sin(x + y) = \sin x \cos y + \cos x \sin y.$$

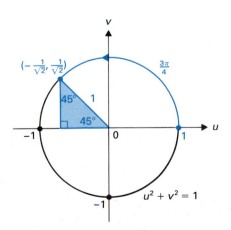

FIGURE 1.82 Computation of the trigonometric functions at $3\pi/4$.

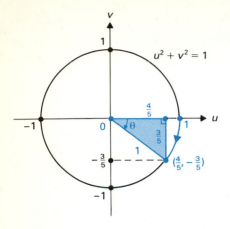

FIGURE 1.83 Computation of tan θ if $\sin \theta = -\frac{3}{5}$ and $-\pi/2 \le \theta \le \pi/2$.

EXAMPLE 7 If $-\pi/2 \le \theta \le \pi/2$ and $\sin \theta = -\frac{3}{5}$, find $\tan \theta$.

Solution To have $-\pi/2 \le \theta \le \pi/2$ and $\sin \theta < 0$, we must actually have $-\pi/2 < \theta < 0$. Thus $u = \cos \theta > 0$. From $\sin^2\theta + \cos^2\theta = 1$, we obtain

$$\cos \theta = \pm\sqrt{1 - \sin^2\theta} = \pm\sqrt{1 - \frac{9}{25}} = \pm\sqrt{\frac{16}{25}} = \pm\frac{4}{5}.$$

Since $\cos \theta > 0$, the value $\cos \theta = \frac{4}{5}$ is appropriate. See Fig. 1.83. Then

$$\tan \theta = \frac{\sin \theta}{\cos \theta} = \frac{-3/5}{4/5} = -\frac{3}{4}. \qquad \blacksquare$$

It is not our purpose to give a lot of drill in identities. The Dictionary of Functions (Section 1.6) lists a number of them for reference. The exercises ask you to prove a few identities that are easy to derive from our definition of $\sin x$ and $\cos x$. We illustrate the technique with an example.

EXAMPLE 8 Show that $\sin(x + \pi) = -\sin x$ and $\cos(x + \pi) = -\cos x$.

Solution Refer to Fig. 1.84. Let x correspond to a point (u, v) on the unit circle, so that $\cos x = u$ and $\sin x = v$. Then $x + \pi$ corresponds to the point $(-u, -v)$ on the circle. Consequently,

$$\sin(x + \pi) = -v = -\sin x$$

and

$$\cos(x + \pi) = -u = -\cos x. \qquad \blacksquare$$

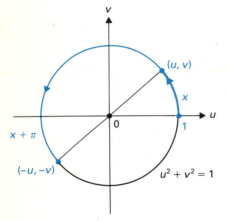

FIGURE 1.84 Arc x ends at (u, v); arc $x + \pi$ ends at $(-u, -v)$.

Graphs of the Six Basic Trigonometric Functions

The basic trigonometric functions are used so frequently that it is best to know their graphs without having to look them up, just as you know the graphs of x, x^2, x^3, x^4, and so on. It is feasible to learn the graphs of $\sin x$, $\cos x$, and $\tan x$ and then find the graphs of $\csc x$, $\sec x$, and $\cot x$ by taking reciprocals of the heights. That is, to find the graph of $y = \csc x$, we use the fact that

$$\csc x = \frac{1}{\sin x},$$

so at x where the graph of $\sin x$ has height $\frac{3}{10}$, the height to the graph of $\csc x$ is $\frac{10}{3}$.

We can easily check that the graph of $y = \sin x$ is that shown in Fig. 1.85. Note that the intercepts on the x-axis are multiples of π.

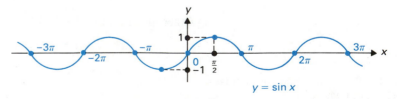

FIGURE 1.85 Graph of $y = \sin x$.

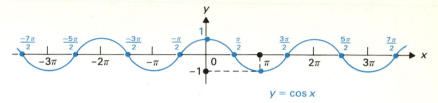

FIGURE 1.86   Graph of $y = \cos x$.

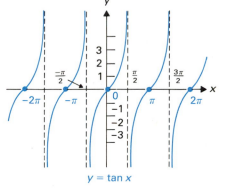

FIGURE 1.87   Graph of $y = \tan x$.

In view of the identity $\cos x = \sin(x + \pi/2)$, the graph of $\cos x$ can be obtained by translating the graph of $\sin x$ a distance $\pi/2$ to the left. The graph of $\cos x$ is shown in Fig. 1.86.

The graph of $\tan x$ has as height at each point $(\sin x)/(\cos x)$, if $\cos x \neq 0$. From the graphs of $\sin x$ and $\cos x$, we obtain the graph in Fig. 1.87 for $\tan x$. Note that the graph of $y = \tan x$ never touches the vertical lines $x = \pm\pi/2$, $x = \pm 3\pi/2$, $x = \pm 5\pi/2$, ...

Taking the reciprocals of the heights of the graph of $\tan x$, we obtain the graph of $\cot x = 1/(\tan x)$ in Fig. 1.88. Figures 1.89 and 1.90 show the graphs of $\sec x = 1/(\cos x)$ and $\csc x = 1/(\sin x)$.

The Graph of $y = a \cdot \sin[b(x - c)]$

We turn to the graph of $y = a \cdot \sin[b(x - c)]$ and first discuss the effect of the individual constants a, b, and c on the graph.

The graph of $\sin x$ oscillates in height between -1 and 1; the graph has *amplitude* 1. It is obvious that the graph of $a(\sin x)$ oscillates between $-a$ and a, that is, has **amplitude** $|a|$. The constant a controls the amplitude of the graph.

The graph of $\sin x$ repeats itself every 2π units on the x-axis. We say it has *period* 2π. The graph of $\sin bx$ will repeat as soon as bx changes by 2π, which is as soon as x increases by $|2\pi/b|$. Thus $\sin bx$ has **period** $|2\pi/b|$, and b controls the period of the graph.

Finally, we turn to the graph of $\sin(x - c)$. We have seen that the substitution $\bar{x} = x - c$, $\bar{y} = y - 0$ amounts to translating axes to the point $(c, 0)$. Thus the graph of $\sin(x - c)$ is the graph of $\sin x$ translated c units to the right for $c > 0$ and $|c|$ units to the left if $c < 0$. Note that $\sin(x - c)$ is zero when $x = c$, rather than when $x = 0$. The number c is a *phase angle* if $c > 0$.

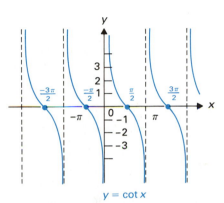

FIGURE 1.88   Graph of $y = \cot x$.

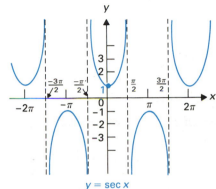

FIGURE 1.89   Graph of $y = \sec x$.

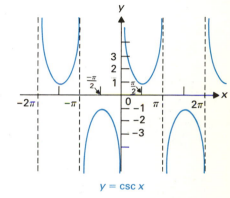

FIGURE 1.90   Graph of $y = \csc x$.

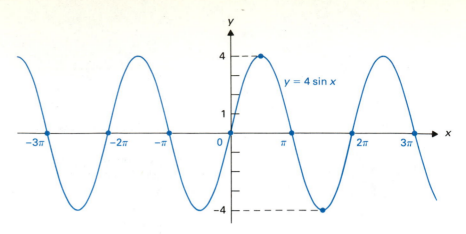

FIGURE 1.91   The graph of $y = 4 \sin x$ has amplitude 4.

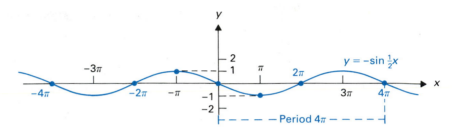

FIGURE 1.92   The graph of $y = -\sin \frac{1}{2}x$ has period $(2\pi/\frac{1}{2}) = 4\pi$.

EXAMPLE 9   Sketch the graph of $y = 4 \sin x$.

Solution   The graph is shown in Fig. 1.91. The amplitude is 4.   ■

EXAMPLE 10   Sketch the graph of $y = -\sin \frac{1}{2}x$.

Solution   The graph is shown in Fig. 1.92. Here $y = a \sin bx$, with $b = \frac{1}{2}$. The period is $2\pi/(\frac{1}{2}) = 4\pi$. Since $a = -1$, the sign of y is changed from $y = \sin \frac{1}{2}x$, so the graph starts down rather than up to the right of the origin.   ■

EXAMPLE 11   Sketch the graph of $y = \sin(x - \pi)$.

Solution   The graph is shown in Fig. 1.93. The phase angle is π. Setting $\bar{x} = x - \pi$ and $\bar{y} = y - 0$, the graph becomes $\bar{y} = \sin \bar{x}$ with respect to the $\bar{x},\bar{y}$-axes shown in the figure.   ■

EXAMPLE 12   Sketch the graph of $y = 3 \sin(2x + \pi)$.

Solution   First we rewrite:

$$y = 3 \sin(2x + \pi) = 3 \sin 2\left[x - \left(-\frac{\pi}{2}\right)\right].$$

The graph has amplitude 3 and period $2\pi/2 = \pi$. The graph is shown in Fig. 1.94. Think of moving the curve $y = \sin x$ a distance $\pi/2$ to the left and

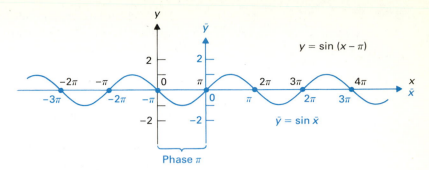

FIGURE 1.93   The graph of $y = \sin(x - \pi)$ has phase angle π.

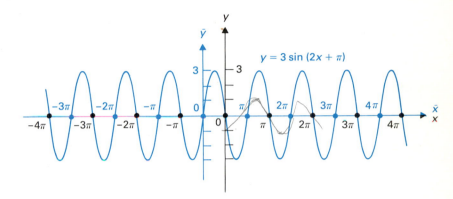

FIGURE 1.94   The graph of

$$y = 3 \sin(2x + \pi) = 3 \sin 2\left[x - \left(\frac{-\pi}{2}\right)\right]$$

has amplitude 3 and period $2\pi/2 = \pi$.

then having it oscillate three times as high (amplitude 3 rather than 1) and twice as fast (period π rather than 2π). ■

The effect of a, b, and c on the graph of $y = a \cdot \cos[b(x - c)]$ is precisely the same as for the graph of $a \cdot \sin[b(x - c)]$. Indeed, multiplying any of the six functions by a and replacing x by $b(x - c)$ has an analogous significance.

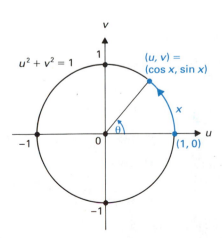

FIGURE 1.95   Point (u, v) on the unit circle corresponding to directed arc length x measured from $(1, 0)$.

SUMMARY

1. Referring to Fig. 1.95, we have

$$\sin x = v, \qquad\qquad \cos x = u,$$

$$\tan x = \frac{v}{u} = \frac{\sin x}{\cos x}, \qquad \cot x = \frac{u}{v} = \frac{\cos x}{\sin x} = \frac{1}{\tan x},$$

$$\sec x = \frac{1}{u} = \frac{1}{\cos x}, \qquad \csc x = \frac{1}{v} = \frac{1}{\sin x}.$$

2. A fundamental trigonometric identity, in addition to defining equations (1) and (2) given in the text, is

$$\sin^2 x + \cos^2 x = 1.$$

Other identities are listed on page 56 in Section 1.6, Dictionary of Functions.

3. The graphs of the six trigonometric functions are shown in Figs. 1.85–1.90.

4. In the graph of $y = a \cdot \sin[b(x - c)]$, the amplitude $|a|$ controls the height of the oscillation, the period (x-distance for repetition) is $|2\pi/b|$, and c is a phase angle if $b > 0$ and $c > 0$.

EXERCISES 1.5

In Exercises 1–30, find the indicated value of the function, if the function is defined there.

1. $\sin \dfrac{\pi}{3}$

2. $\cos \dfrac{3\pi}{2}$

3. $\tan \dfrac{5\pi}{6}$

4. $\sin \dfrac{4\pi}{3}$

5. $\sec \dfrac{5\pi}{4}$

6. $\tan \dfrac{3\pi}{4}$

7. $\csc \dfrac{\pi}{3}$

8. $\tan \dfrac{\pi}{2}$

9. $\cot \pi$

10. $\cot \dfrac{\pi}{2}$

11. $\sec \dfrac{5\pi}{2}$

12. $\csc 2\pi$

13. $\csc\left(-\dfrac{\pi}{4}\right)$

14. $\sin\left(-\dfrac{2\pi}{3}\right)$

15. $\sec\left(-\dfrac{3\pi}{4}\right)$

16. $\cot\left(-\dfrac{5\pi}{4}\right)$

17. $\csc \dfrac{7\pi}{6}$

18. $\cot\left(-\dfrac{2\pi}{3}\right)$

19. $\tan \pi$

20. $\cos 3\pi$

21. $\sec 2\pi$

22. $\sin 5\pi$

23. $\sin(-3\pi)$

24. $\cos(-3\pi)$

25. $\sin\left(-\dfrac{\pi}{2}\right)$

26. $\tan \dfrac{5\pi}{4}$

27. $\tan \dfrac{3\pi}{2}$

28. $\cot 5\pi$

29. $\sec \dfrac{9\pi}{4}$

30. $\csc \dfrac{23\pi}{6}$

31. If $-\pi/2 \leq \theta < \pi/2$ and $\sin \theta = -\frac{1}{3}$, find $\cos \theta$.

32. If $\pi/2 \leq \theta < 3\pi/2$ and $\tan \theta = 4$, find $\sec \theta$.

33. If $0 \leq \theta < \pi$ and $\cos \theta = -\frac{1}{5}$, find $\cot \theta$.

34. If $\pi \leq \theta < 2\pi$ and $\sec \theta = 3$, find $\tan \theta$.

35. If $\pi/2 \leq \theta < 3\pi/2$ and $\sin \theta = \frac{1}{4}$, find $\cot \theta$.

36. If $0 \leq \theta < \pi$ and $\cos \theta = \frac{1}{3}$, find $\sin 2\theta$.

37. If $-\pi/2 \leq \theta < \pi/2$ and $\sin \theta = -\frac{2}{3}$, find $\sin 2\theta$.

38. If $0 < \theta < \pi/2$ and $\tan \theta = 3$, find $\cos 2\theta$.

39. If $0 < \theta < \pi/2$ and $\sec \theta = 4$, find $\cos 2\theta$.

40. If $0 < \theta < \pi/2$ and $\cos \theta = \frac{1}{3}$, find $\sin 3\theta$.

41. As in Fig. 1.77, let the arc corresponding to x terminate at (u, v).
    (a) Where does the arc corresponding to $-x$ terminate?
    (b) Use the result in part (a) to verify identities $\sin(-x) = -\sin x$ and $\cos(-x) = \cos x$.

42. As in Fig. 1.77, let the arc corresponding to x terminate at (u, v).
    (a) Where does the arc corresponding to $x + \pi/2$ terminate?
    (b) Use the result in part (a) to verify identities
    $$\sin(x + \pi/2) = \cos x \quad \text{and} \quad \cos(x + \pi/2) = \sin x.$$

43. As in Fig. 1.77, let the arc corresponding to x terminate at (u, v).
    (a) Where does the arc corresponding to $x - \pi/2$ terminate?
    (b) Use the result in part (a) to derive identities similar to those in part (b) of Exercise 42.

In Exercises 44–52, use the identities on page 56 to verify each of the following identities.

44. $\tan(-x) = -\tan x$

45. $\sec(-x) = \sec x$

46. $\tan\left(x + \dfrac{\pi}{2}\right) = -\cot x$

47. $\sin\left(x - \dfrac{\pi}{2}\right) = -\cos x$

48. $\cos\left(x - \dfrac{\pi}{2}\right) = \sin x$

49. $\sec\left(x - \dfrac{\pi}{2}\right) = \csc x$

50. $\sin(x - y) = \sin x \cos y - \cos x \sin y$

51. $\cos 2x = 2\cos^2 x - 1 = 1 - 2\sin^2 x$

52. $\tan(x + y) = \dfrac{\tan x + \tan y}{1 - \tan x \tan y}$

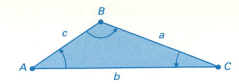

FIGURE 1.96
Triangle with angles A, B, C and sides a, b, c.

Let the angles of a triangle be A, B, and C with corresponding opposite sides of lengths a, b, and c, as shown in Fig. 1.96. In Exercises 53–58, find the desired quantity using the law of sines or the law of cosines given on page 56.

53. For $A = \pi/6$, $a = 5$, $c = 3$, find $\sin C$.
54. For $B = 3\pi/4$, $\sin C = 1/(2\sqrt{2})$, $b = 10$, find c.
55. For $c = 5$, $b = 7$, $A = \pi/4$, find a.
56. For $a = 6$, $c = 4$, $B = 3\pi/4$, find b.
57. For $a = 8$, $c = 4$, $b = 10$, find $\sin C$.
58. For $A = \pi/4$, $a = \sqrt{2}$, $c = \frac{4}{5}$, find $\tan C$.

In Exercises 59–74, find the amplitude and period and sketch the graph of the indicated function.

59. $y = 3 \sin x$
60. $y = 4 \cos x$
61. $y = -\frac{1}{2} \cos x$
62. $y = -\frac{1}{3} \sin x$
63. $y = \sin(-x)$
64. $y = \cos(-x)$

65. $y = 3 \sin 3x$
66. $y = 2 \cos\left(\dfrac{x}{2}\right)$
67. $y = -4 \cos 2x$
68. $y = -6 \sin\left(\dfrac{x}{3}\right)$
69. $y = -2 \sin\left(x - \dfrac{\pi}{2}\right)$
70. $y = -3 \cos(x + \pi)$
71. $y = 5 \cos\left(\dfrac{x}{2} - \dfrac{\pi}{4}\right)$
72. $y = 3 \sin(4x + \pi)$
73. $y = 5 \sin\left(\dfrac{x}{4} - \pi\right)$
74. $y = -2 \cos(2x + 5\pi)$

In Exercises 75–82, find the period (shortest x-distance for repetition) and sketch the graph of the indicated function.

75. $y = -\tan x$
76. $y = \cot 2x$
77. $y = 3 \sec x$
78. $y = \csc(2x - \pi)$
79. $y = \sin^2 x$
80. $y = 4 \cos^2 x$
81. $y = \tan^2 x$
82. $y = \sec^2 x$

EXPLORING CALCULUS

F: find equation quiz (option 6)

1.6 DICTIONARY OF FUNCTIONS

This section is designed for reference and review. For more complete information and examples, refer to the preceding three sections as indicated.

Terminology and Notation (Section 1.3)

Function A rule f that associates with each x in a set X exactly one element y of a set Y.

Notation $y = f(x)$

Independent variable The variable x in $y = f(x)$

Dependent variable The variable y in $y = f(x)$

Domain The set X

Range The set of all y-values $f(x)$ for all x in X

Real-valued function All y-values are real numbers.

Function of a real variable All x in X are real numbers.

Equal functions f and g The same domain X, and $f(x) = g(x)$ for all x in X

From this point on, all reference is to real-valued functions f and g of a real variable x.

When not explicitly specified, the **domain** of f consists of all x such that $f(x)$ can be computed. In particular, division by zero and even roots of negative numbers are not allowed.

A function need not be defined by one single rule or formula over its entire domain; it can be defined separately in different parts of its domain. For example, the **absolute value function** is defined by

$$|x| = \begin{cases} x & \text{for } x \geq 0, \\ -x & \text{for } x < 0. \end{cases}$$

Consequently,

$$\frac{|x|}{x} = \begin{cases} \dfrac{x}{x} & \text{for } x \geq 0, \\ \dfrac{-x}{x} & \text{for } x < 0, \end{cases} = \begin{cases} 1 & \text{for } x \geq 0, \\ -1 & \text{for } x < 0. \end{cases}$$

Graph of a function f   All points (x, y) in the plane such that $y = f(x)$.

Property   A vertical line can meet the graph of a function in at most one point.

Translation   The graph of $y - k = f(x - h)$ can be found by taking the origin of $\bar{x},\bar{y}$-axes at $(x, y) = (h, k)$ and drawing the graph of $\bar{y} = f(\bar{x})$.

Algebra of Functions   (Section 1.4)

Let f and g be functions with domains X_f and X_g. Let $X_f \cap X_g$ be the intersection of the domains, consisting of all real numbers x in both X_f and X_g.

	Definition	**Domain**
Sum	$(f + g)(x) = f(x) + g(x)$	x in $X_f \cap X_g$
Difference	$(f - g)(x) = f(x) - g(x)$	x in $X_f \cap X_g$
Product	$(fg)(x) = f(x)g(x)$	x in $X_f \cap X_g$
Quotient	$\dfrac{f}{g}(x) = \dfrac{f(x)}{g(x)}$	x in $X_f \cap X_g$, $g(x) \neq 0$
Composite	$(f \circ g)(x) = f(g(x))$	x in X_g and $g(x)$ in X_f

Polynomial Functions   (Section 1.4)

Polynomial function $f(x) = a_n x^n + a_{n-1} x^{n-1} + \cdots + a_1 x + a_0$

Monomial function $f(x) = cx^n$ for $c \neq 0$ and n a positive integer

Graphs of monomial functions

1. All pass through $(0, 0)$ and $(1, 1)$.

2. Symmetric about the y-axis if n is even and about the origin if n is odd.

3. The larger the value of n, the flatter the graph is near the origin, and the steeper the graph is near $x = \pm 1$ and for $|x| > 1$.

4. Fig. 1.97 shows the graphs of x, x^2, x^3, x^4, x^5, and x^6 on the same axes.

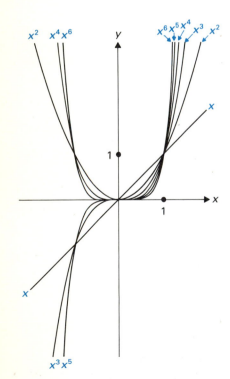

FIGURE 1.97   Computer-generated graphs of the first six monomial functions.

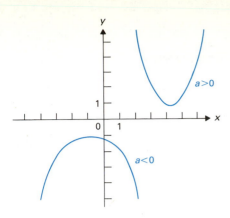

FIGURE 1.98 Graphs of parabolas $y = ax^2 + bx + c$.

Quadratic functions $f(x) = ax^2 + bx + c, a \neq 0$

Graph of a quadratic function   A parabola opening upward if $a > 0$ and downward if $a < 0$. See Fig. 1.98.

Rational Functions

Rational function $f(x)/g(x)$ where f and g are polynomial functions. Polynomial functions are the special case where $g(x) = 1$.

Domain   All x in the domains of both f and g such that $g(x) \neq 0$

Graphs of rational functions   To be discussed in Section 4.4.

Exponential and Logarithm Functions   (Section 1.4)

Exponential function with base a $f(x) = a^x, a > 0$

Domain of a^x   all real numbers x

Properties of a^x

1. $a^r a^s = a^{r+s}$ 2. $\dfrac{a^r}{a^s} = a^{r-s}$ 3. $(a^r)^s = a^{rs}$

4. $a^x > 0$ for all x 5. $a^{-x} = (1/a)^x$

Properties of the graph of a^x

1. Passes through $(0, 1)$
2. Rises from left to right if $a > 1$ (see Fig. 1.99a)
3. Falls from left to right if $0 < a < 1$ (see Fig. 1.99b)
4. Rises *very* rapidly for $a > 1$ and x large

Logarithm function with base a $f(x) = \log_a x, a > 0$. The value of $\log_a b$ is the power to which a must be raised to yield b.

Domain of $\log_a x$ All $x > 0$

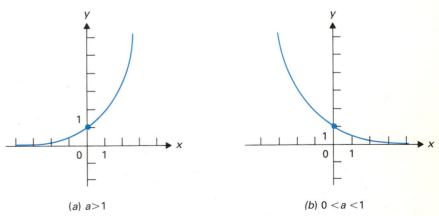

(a) $a > 1$ (b) $0 < a < 1$

FIGURE 1.99 Graphs of exponential functions $y = a^x$.

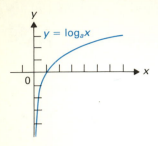

FIGURE 1.100 A typical logarithmic graph.

Properties of $\log_a x$

1. $\log_a(a^x) = x$
2. $a^{\log_a x} = x$ $\Big\}$ a^x and $\log_a x$ are inverse functions.
3. $\log_a(rs) = \log_a r + \log_a s$

4. $\log_a\left(\dfrac{r}{s}\right) = \log_a r - \log_a s$

5. $\log_a(r^s) = s(\log_a r)$

Properties of the graph of $\log_a x$

1. Passes through $(1, 0)$
2. Rises from left to right if $a > 1$ (see Fig. 1.100)
3. Falls from left to right if $0 < a < 1$ (graph not shown since this seldom occurs in practice)
4. Rises *very* slowly if $a > 1$ and x is large

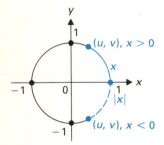

FIGURE 1.101 Points (u, v) the directed distance x from $(1, 0)$ along $u^2 + v^2 = 1$.

Trigonometric Functions   (Section 1.5)

Definition of the six basic functions   Let (u, v) be the point on the circle $u^2 + v^2 = 1$ the distance $|x|$ from $(1, 0)$ measured counterclockwise around the circle if $x > 0$ and clockwise around the circle if $x < 0$. See Fig. 1.101.

$$\sin x = v \qquad\qquad \cos x = u$$

$$\tan x = \frac{\sin x}{\cos x} = \frac{v}{u} \qquad \cot x = \frac{\cos x}{\sin x} = \frac{u}{v}$$

$$\sec x = \frac{1}{\cos x} = \frac{1}{u} \qquad \csc x = \frac{1}{\sin x} = \frac{1}{v}$$

Domains of the six basic functions   All x for $\sin x$ and $\cos x$, all x for which the denominators are nonzero for $\tan x$, $\cot x$, $\sec x$, and $\csc x$

Graphs of the six basic functions   See Figs. 1.102–1.107.

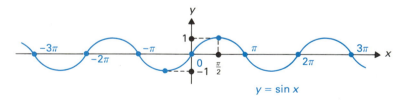

FIGURE 1.102 Graph of $y = \sin x$.

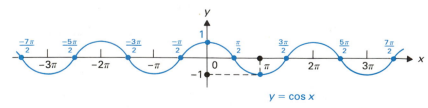

FIGURE 1.103 Graph of $y = \cos x$.

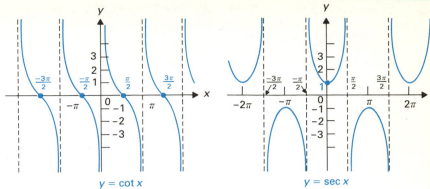

$y = \tan x$

FIGURE 1.104 Graph of $y = \tan x$.

$y = \cot x$

FIGURE 1.105 Graph of $y = \cot x$.

$y = \sec x$

FIGURE 1.106 Graph of $y = \sec x$.

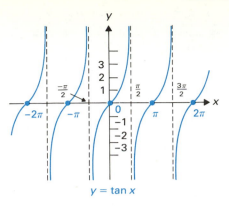

$y = \csc x$

FIGURE 1.107 Graph of $y = \csc x$.

Angle measurement   In applications, the independent variable for a trigonometric function often is the measure of an angle. The definition of the six basic functions used *radian measure,* which is given by the length of the arc intercepted on a unit circle by a central angle. The relationship between this radian measure x and degree measurement t (where one complete turn is 360°) is given by

$$x = \frac{\pi}{180}t.$$

radians   *degrees*

Radian measure is much more useful than degree measure in calculus, as we will see in Section 3.4.

Right-triangle trigonometry   For the right triangle with acute angle θ and sides with lengths labeled *Opposite, Adjacent,* and *Hypotenuse* in Fig. 1.108, we have

$$\sin \theta = \frac{\text{Opposite}}{\text{Hypotenuse}}, \qquad \cos \theta = \frac{\text{Adjacent}}{\text{Hypotenuse}},$$

$$\tan \theta = \frac{\text{Opposite}}{\text{Adjacent}}, \qquad \cot \theta = \frac{\text{Adjacent}}{\text{Opposite}},$$

$$\sec \theta = \frac{\text{Hypotenuse}}{\text{Adjacent}}, \qquad \csc \theta = \frac{\text{Hypotenuse}}{\text{Opposite}}.$$

Evaluation of trigonometric functions   Each of the six basic trigonometric functions satisfies the relation $f(x + 2n\pi) = f(x)$ for any integer n, so the values of the functions can be computed for any x if the values are known for x in the interval $0 \le x < 2\pi$. Values of these functions for any x can be estimated to several significant figures using a scientific calculator or a computer. For examples in this text, it is useful to know the values of these functions for all values x that are integer multiples of $\pi/6$ or $\pi/4$. Table 1.4 shows these values, which can be obtained without memorization using the familiar triangles in Figs. 1.109 and 1.110 and right-triangle trigonometry.

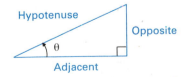

FIGURE 1.108 Sides of a right triangle.

Amplitude, period, phase $a \sin b(x - c)$ has *amplitude* $|a|$ and *period* $2\pi/|b|$, and c is a *phase angle* if $b > 0$ and $c > 0$.

TABLE 1.4

x	$\sin x$	$\cos x$	$\tan x$	$\cot x$	$\sec x$	$\csc x$
0	0	1	0	undefined	1	undefined
$\pi/6$	$1/2$	$\sqrt{3}/2$	$1/\sqrt{3}$	$\sqrt{3}$	$2/\sqrt{3}$	2
$\pi/4$	$1/\sqrt{2}$	$1/\sqrt{2}$	1	1	$\sqrt{2}$	$\sqrt{2}$
$\pi/3$	$\sqrt{3}/2$	$1/2$	$\sqrt{3}$	$1/\sqrt{3}$	2	$2/\sqrt{3}$
$\pi/2$	1	0	undefined	0	undefined	1
$2\pi/3$	$\sqrt{3}/2$	$-1/2$	$-\sqrt{3}$	$-1/\sqrt{3}$	-2	$2/\sqrt{3}$
$3\pi/4$	$1/\sqrt{2}$	$-1/\sqrt{2}$	-1	-1	$-\sqrt{2}$	$\sqrt{2}$
$5\pi/6$	$1/2$	$-\sqrt{3}/2$	$-1/\sqrt{3}$	$-\sqrt{3}$	$-2/\sqrt{3}$	2
π	0	-1	0	undefined	-1	undefined
$7\pi/6$	$-1/2$	$-\sqrt{3}/2$	$1/\sqrt{3}$	$\sqrt{3}$	$-2/\sqrt{3}$	-2
$5\pi/4$	$-1/\sqrt{2}$	$-1/\sqrt{2}$	1	1	$-\sqrt{2}$	$-\sqrt{2}$
$4\pi/3$	$-\sqrt{3}/2$	$-1/2$	$\sqrt{3}$	$1/\sqrt{3}$	-2	$-2/\sqrt{3}$
$3\pi/2$	-1	0	undefined	0	undefined	-1
$5\pi/3$	$-\sqrt{3}/2$	$1/2$	$-\sqrt{3}$	$-1/\sqrt{3}$	2	$-2/\sqrt{3}$
$7\pi/4$	$-1/\sqrt{2}$	$1/\sqrt{2}$	-1	-1	$\sqrt{2}$	$-\sqrt{2}$
$11\pi/6$	$-1/2$	$\sqrt{3}/2$	$-1/\sqrt{3}$	$-\sqrt{3}$	$2/\sqrt{3}$	-2

FIGURE 1.109   45°-45°-90° triangle.

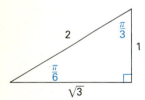

FIGURE 1.110   30°-60°-90° triangle.

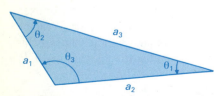

FIGURE 1.111   Angles and sides of a general triangle.

Some trigonometric identities

1. $\sin^2 x + \cos^2 x = 1$ *Fundamental identity*

2. $\tan^2 x + 1 = \sec^2 x$ **3.** $1 + \cot^2 x = \csc^2 x$

4. $\sin(x + 2n\pi) = \sin x$ **5.** $\cos(x + 2n\pi) = \cos x$

6. $\sin(-x) = -\sin x$ **7.** $\cos(-x) = \cos x$

8. $\sin\left(x + \dfrac{\pi}{2}\right) = \cos x$ **9.** $\cos\left(x + \dfrac{\pi}{2}\right) = -\sin x$

10. $\sin(x + \pi) = -\sin x$ **11.** $\cos(x + \pi) = -\cos x$

12. $\sin(x + y) = \sin x \cos y + \cos x \sin y$

13. $\cos(x + y) = \cos x \cos y - \sin x \sin y$

14. $\sin 2x = 2 \sin x \cos x$

15. $\cos 2x = \cos^2 x - \sin^2 x = 2\cos^2 x - 1 = 1 - 2\sin^2 x$

16. $\sin \dfrac{x}{2} = \pm\sqrt{\dfrac{1 - \cos x}{2}}$

17. $\cos \dfrac{x}{2} = \pm\sqrt{\dfrac{1 + \cos x}{2}}$

Two laws for the triangle in Fig. 1.111

Law of sines: $\dfrac{\sin \theta_1}{a_1} = \dfrac{\sin \theta_2}{a_2} = \dfrac{\sin \theta_3}{a_3}$

Law of cosines: $a_3{}^2 = a_1{}^2 + a_2{}^2 - 2a_1 a_2 \cos \theta_3$

EXERCISES 1.6

These exercises are arranged by topic and are similar to those in Sections 1.3–1.5. They are intended for the student who has skipped those sections and wishes to discover if further study of certain topics in them is necessary.

Terminology and Notation

In Exercises 1–5, find the domain of the function.

1. $f(x) = \sqrt{x - 3}$

2. $g(x) = \dfrac{1}{\sqrt{x + 5}}$

3. $f(t) = \dfrac{\sqrt{t + 4}}{t^2 - 1}$

4. $h(t) = (t^2 - 1)^{-2/3}$

5. $g(u) = \dfrac{u^2 - 3u}{\sqrt{4 - u}}$

6. Find the range of $f(x) = x^2$.

7. Find the range of $g(t) = \sqrt{t^2 - 4}$.

8. Find the range of $h(x) = 1/x$.

9. Are the functions $f(x) = (x^2 - 1)/(x + 1)$ and $g(x) = x - 1$ equal? Why or why not?

10. Are the functions

$$f(x) = (x^3 - x^2)/x^2 \quad \text{and} \quad g(x) = -(x - x^2)/x$$

equal? Why or why not?

Algebra of Functions

In Exercises 11–20, let $f(x) = x^2 - 6x$ and $g(x) = \sqrt{x + 3}$. Compute the indicated value, if it is defined.

11. $(f + g)(6)$

12. $(f/g)(0)$

13. $(g/f)(0)$

14. $(fg)(1)$

15. $(g - f)(1)$

16. $(f \circ g)(1)$

17. $(f \circ g)(5)$

18. $(g \circ f)(1)$

19. $(g \circ f)(0)$

20. $(f \circ g)(-3)$

Graphing

In Exercises 21–30, sketch the graph of the given functions.

21. $f(x) = x^2$

22. $f(x) = -\frac{1}{2}x^2$

23. $f(x) = -x^3$

24. $f(x) = x^3 - 4$

25. $f(x) = (x - 1)^2 - 2$

26. $f(x) = (x + 2)^2 - 3$

27. $f(x) = (x + 3)^3 - 1$

28. $f(x) = x^4 - 4$

29. $f(x) = x^2 + 2x + 2$

30. $f(x) = x^2 - 4x + 1$

Exponential and Logarithm Functions

31. Simplify $2^{\log_2 3}$.

32. Simplify $\log_3(3^4)$.

33. Simplify $\log_2(4^3)$.

34. Simplify $9^{\log_3 2}$.

35. Compute $\log_4 64$.

36. Compute $\log_8 2$.

37. Let $\log_a r = 2$ and $\log_a s = 3$. Find $\log_a(r^2/s)$.

38. For r and s in Exercise 37, find $\log_a(r^3 s^2)$.

39. Sketch the graph of $f(x) = 2^{x-1}$.

40. Sketch the graph of $f(x) = \log_3(x + 1)$.

Trigonometric Functions

41. Find the radian measure of an angle of $100°$.

42. Find the degree measure of an angle of -1.2 radians.

43. Find $\sin(2\pi/3)$ without using Table 1.4.

44. Find $\tan(5\pi/4)$ without using Table 1.4.

45. Find $\cos(17\pi/6)$.

46. Find $\csc(23\pi/4)$.

47. Sketch the graph of $\sin 2x$.

48. Sketch the graph of $-3 \cos x$.

49. Sketch the graph of $\tan(x/2)$.

50. Sketch the graph of $2 \sin(x + \pi/2)$.

SUPPLEMENTARY EXERCISES FOR CHAPTER 1

1. (a) Find the directed length Δx from -2 to 5.
(b) Sketch all points (x, y) in the plane that satisfy $x > y + 1$.

2. (a) Sketch on the number line all x such that $|x - 1| \le 2$.
(b) Find the midpoint of the interval $[-5.3, 2.1]$.

3. (a) Find the distance between $(2, -1)$ and $(-4, 7)$.
(b) Find the midpoint of the line segment joining $(-1, 3)$ and $(3, 9)$.

4. (a) Find the distance from $(-6, 3)$ to $(-1, -4)$.
(b) Sketch all points (x, y) in the plane such that $x \le y$ and $x \ge 1$.

5. (a) Find the equation of the circle with center $(2, -1)$ and passing through $(4, 6)$.
(b) Sketch all points (x, y) in the plane such that

$$(x - 1)^2 + (y + 2)^2 \le 4.$$

6. (a) Find the equation of the circle with $(-2, 4)$ and $(4, 6)$ as endpoints of a diameter.
(b) Find the center and radius of the circle $x^2 + y^2 - 6x + 8y = 11$.

7. (a) Find the slope of the line joining $(-1, 4)$ and $(3, 7)$.
(b) Find the slope of a line that is perpendicular to the line through $(4, -2)$ and $(-5, -3)$.

8. Find c such that the line through $(-1, c)$ and $(4, -6)$ is perpendicular to the line through $(-2, 3)$ and $(4, 7)$.

9. (a) Find the equation of the line through $(-4, 2)$ and $(-4, 5)$.
(b) Find the equation of the line through $(-1, 2)$ and parallel to the line $x - 3y = 7$.

10. (a) Find the equation of the line through $(-1, 4)$ and $(3, 5)$.
(b) Find the equation of the vertical line through $(3, -7)$.

11. (a) Find $\left|2 - |7 - 4|\right|$.
(b) Express the solutions of $|2x - 4| \le 5$ in interval notation.

12. (a) Sketch all points x on the number line such that
$$(x - 2)^2 < 9.$$
(b) Express the solutions of $|6 - 3x| \le 12$ in interval notation.

13. Find all values of c such that the point $(c, -3)$ is 8 units from the point $(4, 1)$.

14. Find the area of the right triangle with vertices $(-1, 4)$, $(2, 6)$, and $(-3, 7)$.

15. Find the equation of the smallest circle with center at $(7, 6)$ that is tangent to the circle $x^2 + y^2 - 2x + 4y = 11$.

16. Find the equation of the circle with center $(-1, 4)$ and tangent to the line $x = 5$.

17. Find the slope of the line tangent to the circle $x^2 + y^2 - 8x + 2y = -4$ at the point $(1, -3)$.

18. Use slopes to determine whether the points $(-1, 4)$, $(2, 8)$, and $(-7, -4)$ are collinear.

19. Find the equation of the line through $(-2, 4)$ and perpendicular to the line $3x - 4y = 5$.

20. (a) Find the equation of the line through $(-1, 3)$ with y-intercept 5.
(b) Find the equation of the line through $(2, 4)$ and parallel to the line through $(0, 5)$ and $(2, -3)$.

21. Let $f(x) = \dfrac{x^3 + 1}{(1 + |x|)^2}$. Find the following.

(a) $f(0)$ (b) $f(-2)$ (c) $f(3)$

22. Let $f(x) = \dfrac{x^2 - 3x + 2}{x^2 - 5x}$.

(a) Find the domain of f.   (b) Find $f(-2)$.

23. Let $f(x) = \sqrt{25 - x^2}$.
(a) Find the domain of f.   (b) Find $f(3)$.
(c) Sketch the graph of f.

24. Let $f(x) = \dfrac{x^2 - x}{x^2 + x - 2}$.

(a) Find the domain of f.
(b) Find a simpler description of $f(x)$.

25. Find the domain of the function g, where $g(t) = \dfrac{t^2 - 1}{t\sqrt{t + 3}}$.

26. Sketch the graph of the function $y = f(x) = 3 - (x + 4)^2$.

27. Sketch the graph of the function $y = f(x) = 2 + (x - 1)^4$.

28. Sketch the graph of the function $y = f(x) = 5 - x^2$.

29. Sketch the graph of the function $y = f(x) = 5(x + 2)^4$.

30. Sketch the graph of the function $y = f(x) = x^2 - 4x + 5$.

In Exercises 31–36, simplify the given expression.

31. $3^{\log_3 1}$ **32.** $\log_{25} 5$ **33.** $4^{\log_2 3}$

34. $\log_2(4^3)$ **35.** $\log_3(2^{5(\log_2 3)})$ **36.** $\log_9(3^9)$

37. If $\log_a r = 3$ and $\log_a s = \frac{1}{2}$, find the following.
(a) $\log_a(rs^2)$ (b) $\log_a(r/s^2)$ (c) $\log_{a^2}(r^3 s)$

38. Sketch the graphs of $y = 2^x$ and $y = 3^x$ on the same set of axes.

39. Sketch the graphs of $y = 3^{-x}$ and $y = 5^{-x}$ on the same set of axes.

40. Sketch the graph of $f(x) = \log_4 x$.

41. Find (a) $\tan \dfrac{5}{6}\pi$, (b) $\cos \dfrac{5}{4}\pi$.

42. Find (a) $\sin \dfrac{11}{6}\pi$, (b) $\cot \left(-\dfrac{4}{3}\pi\right)$.

43. Find (a) $\sin \dfrac{23}{2}\pi$, (b) $\csc \left(-\dfrac{7}{4}\pi\right)$.

44. Find (a) $\sin 17\pi$, (b) $\cos \dfrac{5}{6}\pi$.

45. If $0 \le \theta < \pi$ and $\sec \theta = -5$, find $\sin \theta$.

46. If $\pi/2 \le \theta < 3\pi/2$ and $\tan \theta = 2/3$, find $\sin \theta$.

47. If $\pi/2 \le \theta < 3\pi/2$ and $\tan \theta = \sqrt{3}$, find $\sin \theta$.

48. Find the amplitude and period and sketch the graph of $y = 3 \sin(2x - \pi)$.

49. Find the period and amplitude of $-\frac{1}{2} \cos(x/3)$.

50. Find the amplitude and period and sketch the graph of $y = -2 \sin 4x$.

51. Show that, for all real numbers a and b,
(a) $|a + b| \le |a| + |b|$,
(b) $|a - b| \ge |a| - |b|$.

52. Prove that, for any numbers a_1, a_2, b_1, and b_2, the following is true.
$$(a_1 a_2 + b_1 b_2)^2 \le (a_1^2 + b_1^2)(a_2^2 + b_2^2)$$

53. Prove algebraically from Exercise 52 and the formula for distance that the distance from (x_1, y_1) to (x_3, y_3) is less than or equal to the sum of the distance from (x_1, y_1) to (x_2, y_2) and the distance from (x_2, y_2) to (x_3, y_3). This is known as the *triangle inequality* in the plane. [*Hint:* Let
$$a_1 = x_2 - x_1, \qquad a_2 = x_3 - x_2,$$
$$b_1 = y_2 - y_1, \qquad b_2 = y_3 - y_2,$$
so that
$$x_3 - x_1 = a_2 + a_1 \quad \text{and} \quad y_3 - y_1 = b_2 + b_1.]$$

54. Show that if two circles
$$x^2 + y^2 + a_1 x + b_1 y = c_1$$
and
$$x^2 + y^2 + a_2 x + b_2 y = c_2$$

intersect in two points, the line through those points of intersection is

$$(a_2 - a_1)x + (b_2 - b_1)y = c_2 - c_1.$$

55. Use Exercise 54 to find the line through the points of intersection of the following pairs of circles, if they intersect.
(a) $x^2 + y^2 - 4x + 2y = 8$ and $x^2 + y^2 - 2x + 6y = -9$
(b) $x^2 + y^2 - 4x + 2y = 8$ and $x^2 + y^2 + 6x + 4y = 17$

56. Show that the distance from a point (x_1, y_1) to a line $ax + by - c = 0$ is

$$\left| \frac{ax_1 + by_1 - c}{\sqrt{a^2 + b^2}} \right|.$$

57. Use the formula in Exercise 56 to find the distance
(a) from $(-1, 2)$ to $4x - 3y = 10$,
(b) from $(-3, 4)$ to $5x - 12y = 2$.

58. Find the area of the triangle with vertices $(1, -2)$, $(5, 6)$, and $(-4, 3)$. [*Hint:* Find the point at the foot of the altitude from $(-4, 3)$ to the opposite side of the triangle.]

59. Solve the inequality $x^2 + 4x < 1$ for x.

60. Find the equation of the smaller circle tangent to both coordinate axes and passing through the point $(-3, 6)$.

61. Find the distance between the lines

$$x - 2y = 15 \qquad \text{and} \qquad x - 2y = -3.$$

62. Find the minimum distance between the circles

$$x^2 + y^2 - 2x + 4y = 139$$

and

$$x^2 + y^2 + 4x - 6y = 3.$$

63. Sketch the graph of the function $y = f(x) = \dfrac{|x^2 - 1|}{x - 1}$.

64. Sketch the graph of the function $y = f(x) = \dfrac{x^3 - 4x^2}{|x|}$.

65. If $f(x) = (2x - 7)/(x + 3)$, find a function g such that $g(f(x)) = x$ for all x in the domain of f.

LIMITS

2

A function $y = f(x)$ described as a quotient of two functions of x is not defined at a value of x where the denominator function becomes zero. For example,

$$y = f(x) = \frac{x^2 - 4}{x - 2}$$

is not defined where $x = 2$. Even though a function f is not defined at $x = a$, we often wish to know how function values $f(x)$ behave as x gets *very close to some value a*. We encounter this problem when we try to find how fast something is changing. Calculus is concerned with things that are changing.

2.1 THE TANGENT LINE PROBLEM AND LIMITS

The Problem

The function given by $y = mx + b$ has as graph a line, shown in Fig. 2.1. As x slides to the right at a uniform speed on the x-axis, the y-values on the graph change at a uniform rate. As we saw in Chapter 1, the slope m of this line measures the rise of this graph for each unit of run to the right on the x-axis.

The graph of $y = f(x)$ shown in Fig. 2.2 does not rise at a uniform rate as x runs with constant speed to the right on the x-axis. This graph, which is of an exponential character, rises faster and faster as x goes out the x-axis. Suppose that, when x reaches the point a as it runs out the x-axis at uniform speed, you could tell the point (x, y) that is staying above x on the graph, "Stop rising faster, and keep rising at the same rate that you are rising right now." If the point obeyed, it would leave the curve and start traveling to the right along the line *tangent* to the graph where $x = a$. This behavior is illustrated by the colored

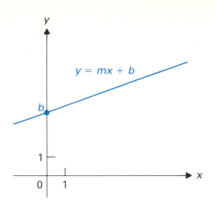

FIGURE 2.1 The line rises at a uniform rate.

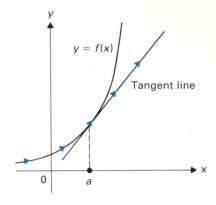

FIGURE 2.2 Tangent line to the graph of $y = f(x)$.

arrows on the graph and on the tangent line in Fig. 2.2. Since the slope of this tangent line measures how fast the line is rising per unit increase in x, this slope also measures how fast $y = f(x)$ is increasing *at the instant when $x = a$*. Differential calculus is concerned with such rates of change of function values $f(x)$ for specific values of x.

The last paragraph suggests that we tackle the problem of finding the slope $m_{\tan}$ of the tangent line to the graph of a function $y = f(x)$ when $x = a$. In doing this, we rely on our intuitive notion of the tangent line to the graph of a function at a point. Not all graphs have a tangent line at each point. Consider, for example, the graph of $y = g(x)$ shown in Fig. 2.3. This graph is formed by a line of slope $\frac{1}{2}$ for $x \leq a$ and a line of slope 2 for $x \geq a$. The graph rises at a uniform rate for $x < a$ and at a faster uniform rate for $x > a$. The rate of rise changes abruptly where $x = a$, and there is no (unique) tangent line to the graph where $x = a$. It is impossible to say how fast the graph is rising *at that point*. Very roughly speaking, a graph has a tangent line at a point where the graph is unbroken and does not have a "corner."

Finding $m_{\tan}$

We denote by $m_{\tan}$ the slope of the tangent line to a graph $y = f(x)$ where $x = a$. The four parts of Fig. 2.4 show both the *tangent* line where $x = a$ and a *secant* line through the points where $x = a$ and where $x = a + \Delta x$ for four different values of Δx. From the figure, it appears that for very small values of Δx, the secant line, with slope m_{sec}, almost coincides with the tangent line, so that the slope m_{sec} is almost equal to $m_{\tan}$. We express this by saying that

$$m_{\text{sec}} \text{ approaches } m_{\tan} \text{ as } \Delta x \text{ approaches zero}$$

or that

$$\text{the limit of } m_{\text{sec}} \text{ is } m_{\tan} \text{ as } \Delta x \text{ approaches zero}.$$

From Fig. 2.4(a), we see that the secant line goes through the points $(a, f(a))$ and $(a + \Delta x, f(a + \Delta x))$. Setting $\Delta y = f(a + \Delta x) - f(a)$, we obtain

$$m_{\text{sec}} = \frac{\Delta y}{\Delta x} = \frac{f(a + \Delta x) - f(a)}{\Delta x}. \tag{1}$$

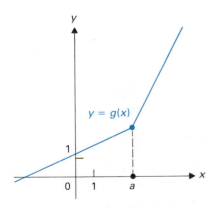

FIGURE 2.3 The graph has no tangent line where $x = a$.

We write

$$m_{\tan} = \lim_{\Delta x \to 0} \frac{\Delta y}{\Delta x} = \lim_{\Delta x \to 0} \frac{f(a + \Delta x) - f(a)}{\Delta x} \qquad (2)$$

to express the fact that the limiting value of $[f(a + \Delta x) - f(a)]/\Delta x$ as Δx approaches zero is $m_{\tan}$. It is time for numerical examples.

EXAMPLE 1 Find the slope $m_{\tan}$ of the line tangent to the graph of $y = x^2$ where $x = 1$.

Solution We take $f(x) = x^2$ and $a = 1$ in Eq. (2), and obtain

$$m_{\tan} = \lim_{\Delta x \to 0} \frac{f(1 + \Delta x) - f(1)}{\Delta x} = \lim_{\Delta x \to 0} \frac{(1 + \Delta x)^2 - 1^2}{\Delta x}$$

$$= \lim_{\Delta x \to 0} \frac{[1 + 2(\Delta x) + (\Delta x)^2] - 1}{\Delta x}$$

$$= \lim_{\Delta x \to 0} \frac{2(\Delta x) + (\Delta x)^2}{\Delta x} = \lim_{\Delta x \to 0} \frac{\Delta x(2 + \Delta x)}{\Delta x}$$

$$= \lim_{\Delta x \to 0} (2 + \Delta x) = 2. \qquad \blacksquare$$

In the preceding example, we "took the limit" in the last line of the computation. As Δx approaches zero, the value of $2 + \Delta x$ approaches 2. Note also that this limit cannot be found without this computation, for both numerator and denominator approach zero in the expression

$$\frac{(1 + \Delta x)^2 - 1^2}{\Delta x} \qquad (3)$$

as Δx approaches zero. It is necessary to simplify the numerator and obtain a factor Δx to cancel with the Δx in the denominator. This cancellation is permissi-

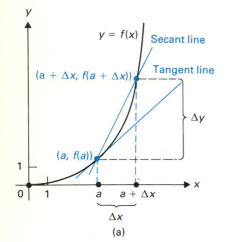

(a)

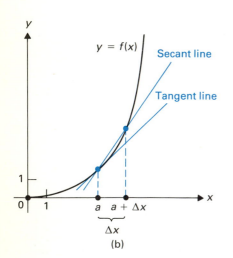

(b)

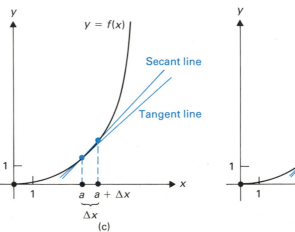

(c)

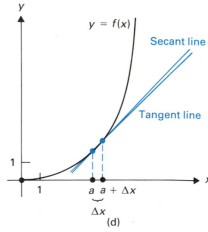

(d)

FIGURE 2.4 As Δx approaches zero, the secant lines approach the tangent line where $x = a$.

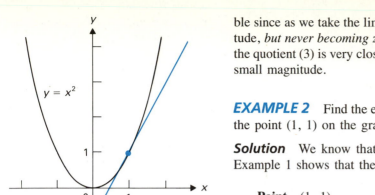

FIGURE 2.5 Tangent line where $x = 1$.

ble since as we take the limit, we think of Δx as becoming very small in magnitude, *but never becoming zero*. The limiting value 2 means that the numerator in the quotient (3) is very close to twice the denominator for Δx nonzero but of very small magnitude.

EXAMPLE 2   Find the equation of the line tangent to the graph of $f(x) = x^2$ at the point (1, 1) on the graph.

Solution   We know that the point (1, 1) is on the desired tangent line, and Example 1 shows that the slope of the line is 2. Thus we have

> **Point**   (1, 1),
>
> **Slope** $m_{\text{tan}} = 2$,
>
> **Equation** $y - 1 = 2(x - 1)$,   or $y = 2x - 1$.

The graph of $f(x) = x^2$ and the tangent line where $x = 1$ are shown in Fig. 2.5.

■

EXAMPLE 3   Find the slope m_{tan} of the line tangent to the graph of $y = g(t) = 1/t$ where $t = 2$.

Solution   In Eq. (2), we replace f by g, replace x by t, and take $a = 2$. We have

$$m_{\text{tan}} = \lim_{\Delta t \to 0} \frac{g(2 + \Delta t) - g(2)}{\Delta t} = \lim_{\Delta t \to 0} \frac{\dfrac{1}{2 + \Delta t} - \dfrac{1}{2}}{\Delta t}.$$

Again, both numerator and denominator approach zero, and we must simplify the numerator to get a factor Δt free to cancel with the Δt in the denominator. We have

$$m_{\text{tan}} = \lim_{\Delta t \to 0} \frac{\dfrac{1}{2 + \Delta t} - \dfrac{1}{2}}{\Delta t} = \lim_{\Delta t \to 0} \frac{\dfrac{2 - (2 + \Delta t)}{2(2 + \Delta t)}}{\Delta t}$$

$$= \lim_{\Delta t \to 0} \frac{\dfrac{-\Delta t}{2(2 + \Delta t)}}{\Delta t} = \lim_{\Delta t \to 0} \frac{-1}{2(2 + \Delta t)} = -\frac{1}{4},$$

because as $\Delta t \to 0$, we see that $2(2 + \Delta t)$ approaches $2(2) = 4$.   ■

A line is **normal** to a curve at a point if the line is perpendicular to the tangent line to the curve at that point.

EXAMPLE 4   Find the equation of the line normal to the graph of $y = g(t) = 1/t$ where $t = 2$.

Solution   Example 3 shows that the tangent line where $t = 2$ has slope $-\frac{1}{4}$. The normal line is perpendicular to the tangent line, so it has the negative reciprocal slope: $-1/(-\frac{1}{4}) = 4$. Thus in the t,y-plane, we obtain for the normal line

Point $(2, \frac{1}{2})$,

Slope 4,

Equation $y - \frac{1}{2} = 4(t - 2)$, or $2y = 8t - 15$.

The graph and normal line are shown in Fig. 2.6. ■

EXAMPLE 5 Let $f(x) = x^3 + 2x$. Find the approximation m_{sec} to the slope of the tangent line to the graph where $x = a = 2$, taking $\Delta x = 0.01$.

Solution Formula (1) for m_{sec} yields

$$m_{\text{sec}} = \frac{[(2.01)^3 + 2(2.01)] - (2^3 + 2 \cdot 2)}{0.01}$$

$$= \frac{12.140601 - 12}{0.01} = \frac{0.140601}{0.01} = 14.0601.$$

We will see in Chapter 3 that the slope m_{tan} is actually 14, so 14.0601 is not a bad approximation. A calculator is handy for computing m_{sec}. ■

EXAMPLE 6 Repeat Example 5, but take $\Delta x = 0.001$.

Solution Again we use formula (1) for m_{sec}. We expect a value closer still to 14. The computation was done using a calculator. We give the full display of the calculator:

$$m_{\text{sec}} = \frac{[(2.001)^3 + 2(2.001)] - (2^3 + 2 \cdot 2)}{0.001}$$

$$= \frac{12.014006 - 12}{0.001} = \frac{0.014006001}{0.001} = 14.00600099.$$

This is closer to the true value 14 of m_{tan} at $a = 2$ than the value 14.0601 obtained in Example 5 with the larger $\Delta x = 0.01$. ■

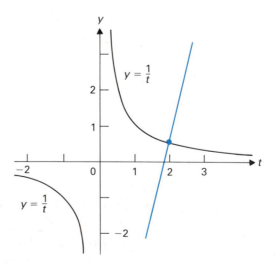

FIGURE 2.6 Normal line where $t = 2$.

EXAMPLE 7 Find the slope m_{tan} of the tangent line to the graph of $f(x) = 4x - 3x^2$ at the point where $x = a$.

Solution We use formula (2) but first we compute $m_{\text{sec}} = \Delta y / \Delta x$, simplifying as much as possible. We have, for $\Delta x \neq 0$,

$$
\begin{aligned}
m_{\text{sec}} &= \frac{f(a + \Delta x) - f(a)}{\Delta x} \\
&= \frac{4(a + \Delta x) - 3(a + \Delta x)^2 - (4a - 3a^2)}{\Delta x} \\
&= \frac{4a + 4(\Delta x) - 3a^2 - 6a(\Delta x) - 3(\Delta x)^2 - 4a + 3a^2}{\Delta x} \\
&= \frac{4(\Delta x) - 6a(\Delta x) - 3(\Delta x)^2}{\Delta x} \\
&= \frac{\Delta x[4 - 6a - 3(\Delta x)]}{\Delta x} = [4 - 6a - 3(\Delta x)].
\end{aligned}
$$

Then formula (2) yields

$$
m_{\text{tan}} = \lim_{\Delta x \to 0} \frac{\Delta y}{\Delta x} = \lim_{\Delta x \to 0} [4 - 6a - 3(\Delta x)] = 4 - 6a.
$$

The Notion of $\lim_{x \to a} f(x)$

We have seen that the tangent line problem appears when we try to find how fast the graph of $y = f(x)$ is rising and leads to the limit of a quotient whose numerator and denominator are both approaching zero. The independent variable in both the numerator and denominator is Δx. As we start our discussion of limits of quotients and other expressions, we change from Δx to x, t, or some other single letter for ease in writing.

While the limit of a quotient as both numerator and denominator approach zero is the most important situation for us at the moment, it is by no means the only type of limit that we will consider. Also, we wish to consider limits of expressions as the variable approaches some value a other than zero. We give a very intuitive definition of the limit of a function at a point in its domain and then proceed to illustrate with examples. A mathematically precise definition is given in Section 2.5.

DEFINITION 2.1

Limit of f at a (intuitive)

Let f be a function whose domain contains points that are arbitrarily close to a but different from a. The **limit of f at** a is L, written $\lim_{x \to a} f(x) = L$, if $f(x)$ can be made as *close to L* as we wish by taking any x *sufficiently close to a*, but different from a, in the domain of f.

This definition is too vague to be mathematically satisfactory, for what do *close* and *sufficiently close* mean? These ideas are explained precisely in Section 2.5. The definition here makes it clear that we must have points in the

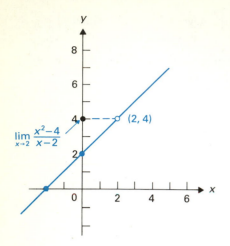

FIGURE 2.7   Graph of
$$y = (x^2 - 4)/(x - 2) = x + 2$$
for $x \neq 2$.

domain of f arbitrarily close to a to be able to even talk about $\lim_{x \to a} f(x)$; if there were no such points, it would make no sense to think of x approaching a without ever being equal to a. Note that the definition explicitly says that x must always be different from a. We never think of x attaining the value a, but just as being very close to a. We summarize the purpose of this limit concept and proceed with examples.

Purpose of a Limit

A limit is used to describe the behavior of a function *near* a point but not at the point. The function need not even be defined at the point. If it is defined there, the value of the function at the point does not affect the limit.

EXAMPLE 8   Let us try to find
$$\lim_{x \to 2} \frac{x^2 - 4}{x - 2}.$$
That is, can we discover whether $(x^2 - 4)/(x - 2)$ gets very close to some value L as x gets very close to 2?

Solution   Note that $(x^2 - 4)/(x - 2)$ is not defined if $x = 2$. For all $x \neq 2$,
$$\frac{x^2 - 4}{x - 2} = \frac{(x - 2)(x + 2)}{x - 2} = x + 2.$$
See Fig. 2.7. Thus
$$\lim_{x \to 2} \frac{x^2 - 4}{x - 2} = \lim_{x \to 2} (x + 2) = 4,$$
since, if x is very close to 2, then $x + 2$ is very close to 4.   ∎

EXAMPLE 9   Find
$$\lim_{x \to -1} \frac{x^2 - 1}{x^2 + 3x + 2}.$$

Solution   This function is not defined at -1. But we have
$$\frac{x^2 - 1}{x^2 + 3x + 2} = \frac{(x + 1)(x - 1)}{(x + 1)(x + 2)} = \frac{x - 1}{x + 2} \qquad \text{for } x \neq -1.$$
See Fig. 2.8. Hence
$$\lim_{x \to -1} \frac{x^2 - 1}{x^2 + 3x + 2} = \lim_{x \to -1} \frac{x - 1}{x + 2} = \frac{-2}{1} = -2.$$
That is, if x is very close to -1, then $x - 1$ is close to -2 and $x + 2$ is close to 1, so their quotient is close to -2.   ∎

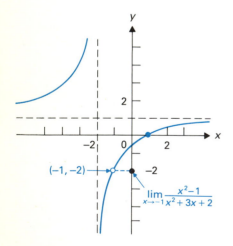

FIGURE 2.8   Graph of
$$y = (x^2 - 1)/(x^2 + 3x + 2)$$
$$= (x - 1)/(x + 2)$$
for $x \neq -1$.

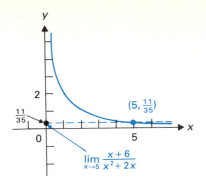

FIGURE 2.9 Graph of
$y = (x + 6)/(x^2 + 2x)$.

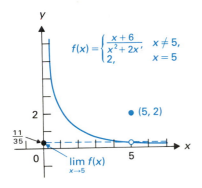

FIGURE 2.10 The value of $\lim_{x \to 5} f(x)$ does not depend on the value of $f(5)$.

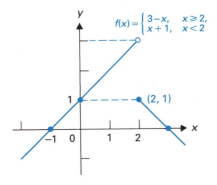

FIGURE 2.11 $\lim_{x \to 2} f(x)$ does not exist since the graph jumps at 2.

EXAMPLE 10 Find

$$\lim_{x \to 5} \frac{x + 6}{x^2 + 2x}.$$

Solution Here the function is defined at 5. As x becomes very close to 5, the numerator $x + 6$ is close to 11 while the denominator $x^2 + 2x$ is close to 35. Thus the quotient is close to $\frac{11}{35}$, so

$$\lim_{x \to 5} \frac{x + 6}{x^2 + 2x} = \frac{11}{35}.$$

See Fig. 2.9. In this case, the value of the limit is the value of the function at the point, but see the next example. ∎

EXAMPLE 11 Let

$$f(x) = \begin{cases} \dfrac{x + 6}{x^2 + 2x} & \text{for } x \neq 5, \\[2mm] 2 & \text{for } x = 5, \end{cases}$$

and find $\lim_{x \to 5} f(x)$.

Solution See Fig. 2.10; $\lim_{x \to 5} f(x) = \frac{11}{35}$, just as in Example 10. In computing $\lim_{x \to a} f(x)$, we think in terms of values $f(x)$ as x gets very close to a *but remains different from a*. The value $f(a)$ can be changed without having any effect on the limit. ∎

EXAMPLE 12 Discuss $\lim_{x \to 2} f(x)$ if

$$f(x) = \begin{cases} 3 - x & \text{for } x \geq 2, \\ x + 1 & \text{for } x < 2. \end{cases}$$

Solution The graph of f is shown in Fig. 2.11; $\lim_{x \to 2} f(x)$ does not exist, because if x is just slightly larger than 2, then $f(x)$ is close to $3 - 2 = 1$, while if x is slightly smaller than 2, $f(x)$ is close to $2 + 1 = 3$. There is no *single* value that $f(x)$ approaches for *all* x sufficiently close to but different from 2. We write "$\lim_{x \to 2} f(x)$ does not exist." ∎

EXAMPLE 13 Discuss $\lim_{x \to 2} f(x)$ if

$$f(x) = \begin{cases} 10 - 3x & \text{for } x > 2, \\ x^2 & \text{for } x < 2. \end{cases}$$

Solution The graph of f is shown in Fig. 2.12. This time, $\lim_{x \to 2} f(x) = 4$, because $10 - 3x$ is close to 4 for x just larger than 2, and x^2 is also close to 4 for x just smaller than 2. We did not define $f(2)$, but that does not matter. It could be given any value without changing the limit value. ∎

EXAMPLE 14 Discuss $\lim_{x \to 3} f(x)$ if $f(x) = \sqrt{x - 3}\sqrt{3 - x}$.

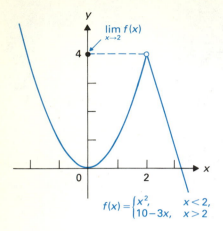

$$f(x) = \begin{cases} x^2, & x < 2, \\ 10 - 3x, & x > 2 \end{cases}$$

FIGURE 2.12 As $x \to 2$, $f(x) \to 4$. The graph does not jump at 2.

Solution Note that 3 is the *only* point in the domain of f. Thus it is meaningless to consider $\lim_{x \to 3} f(x)$ because we can't evaluate $f(x)$ for any point really close to 3 but not equal to 3. ■

We emphasize again that $\lim_{x \to a} f(x) = L$ is an assertion about values $f(x)$ only for x extremely close to a but different from a. This limit is completely determined if you know the values $f(x)$ for all $x \neq a$ in any very small interval having a as its midpoint.

EXAMPLE 15 Figure 2.13 shows the graphs of $f(x) = (\cos x) - 1$ and $g(x) = x^2$ for $-0.01 \leq x \leq 0.01$. (Note the scale on the y-axis also.) Estimate

$$\lim_{x \to 0} \frac{(\cos x) - 1}{x^2}$$

from the figure's magnification of the portions of the graphs of f and g near the origin.

Solution From the graphs, it appears that for x very close to zero, the graph of $g(x) = x^2$ is twice as far above the x-axis as the graph of $f(x) = (\cos x) - 1$ is below the x-axis. Thus we estimate that

$$\lim_{x \to 0} \frac{(\cos x) - 1}{x^2} = \frac{-1}{2} = -\frac{1}{2}.$$ ■

The exercises ask you to estimate more limits from such graphs.

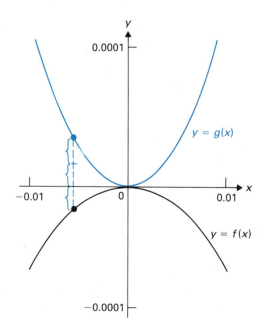

FIGURE 2.13

$$f(x) = (\cos x) - 1, \qquad g(x) = x^2 \qquad \text{for } -0.01 \leq x \leq 0.01$$

SUMMARY

1. The slope m_{sec} of the secant line to the graph of $y = f(x)$ through the points where $x = a$ and $x = a + \Delta x$ is

$$m_{sec} = \frac{\Delta y}{\Delta x} = \frac{f(a + \Delta x) - f(a)}{\Delta x}.$$

2. The slope m_{tan} of the tangent line to the graph of $y = f(x)$ at the point where $x = a$ is

$$m_{tan} = \lim_{\Delta x \to 0} \frac{\Delta y}{\Delta x} = \lim_{\Delta x \to 0} \frac{f(a + \Delta x) - f(a)}{\Delta x},$$

if this limit exists.

3. Limits are used to study the behavior of a function near a point a where the function may or may not be defined.

4. $\lim_{x \to a} f(x) = L$ means that $f(x)$ can be made as close to L as we wish by choosing any x in the domain of f that is sufficiently close to a but different from a.

EXERCISES 2.1

In Exercises 1–10, find the slope m_{sec} of the secant line to the graph of the function at the given point and for the given increment. Then find the slope m_{tan} of the tangent line to the graph at that point.

1. $f(x) = x^2$ where $x = 4$ for $\Delta x = 0.01$
2. $f(x) = 5x$ where $x = 2$ for $\Delta x = -0.005$
3. $f(t) = t^2 - 2t + 4$ where $t = -1$ for $\Delta t = -0.2$
4. $f(x) = 4x - 5x^2$ where $x = -2$ for $\Delta x = 0.01$
5. $g(s) = s^3 - 3s$ where $s = 1$ for $\Delta s = 0.1$
6. $g(t) = 1/t$ where $t = -2$ for $\Delta t = 0.1$
7. $h(u) = u + (1/u)$ where $u = -1$ for $\Delta u = -0.001$
8. $h(x) = 4/(x + 1)$ where $x = -2$ for $\Delta x = -0.01$
9. $g(v) = v^2 + 3v$ where $v = a$ for $\Delta v = 0.2$ [The answers will be in terms of a.]
10. $f(r) = r^2 - (1/r)$ where $r = c$ for $\Delta r = -0.1$ [The answers will be in terms of c.]
11. Use the answer to Exercise 1 to find the equation of the line tangent to the graph of $y = x^2$ where $x = 4$.
12. Use the answer to Exercise 3 to find the equation of the line tangent to the graph of $s = t^2 - 2t + 4$ where $t = -1$.
13. Use the answer to Exercise 5 to find the equation of the line normal to the graph of $x = s^3 - 3s$ where $s = 1$.
14. Use the answer to Exercise 7 to find the equation of the line normal to the graph of $v = u + (1/u)$ where $u = -1$.
15. Use the answer to Exercise 9 to find the equation of the line tangent to the graph of $z = v^2 + 3v$ where $v = -2$.

In Exercises 16–45, find the indicated limit if it exists.

16. $\lim\limits_{x \to 2} \dfrac{3x - 6}{x - 2}$

17. $\lim\limits_{t \to 0} \dfrac{4t^2 - 2t}{t}$

18. $\lim\limits_{u \to 1} \dfrac{|u - 1|}{u + 1}$

19. $\lim\limits_{u \to 2} \dfrac{|u + 2|}{u - 4}$

20. $\lim\limits_{x \to 1} \dfrac{x^2 - 1}{x^2 - x}$

21. $\lim\limits_{u \to 1} \dfrac{u^2 - 1}{u - u^2}$

22. $\lim\limits_{x \to 3} \dfrac{x^2 - 3x}{x^2 - 9}$

23. $\lim\limits_{t \to 2} \dfrac{t^2 - 4}{2t - t^2}$

24. $\lim\limits_{x \to 5} \dfrac{x^2 - 4x - 5}{x^3 - 5x^2}$

25. $\lim\limits_{u \to -2} \dfrac{u^2 - 4}{u^2 + 4}$

26. $\lim\limits_{t \to 0} \dfrac{t^3 + t^2 + 2}{t^3 + 1}$

27. $\lim\limits_{x \to 0} \dfrac{x^4 + 2x^2}{x^3 + x}$

28. $\lim_{s \to 0} \dfrac{s^3 - 2s^2}{s^4 + 3s^2}$

29. $\lim_{x \to 2} \dfrac{x}{x + 3}$

30. $\lim_{u \to 1} \dfrac{(u - 1)^2}{u - 1}$

31. $\lim_{x \to -1} \dfrac{x^2 + x}{x - 1}$

32. $\lim_{x \to 2} \dfrac{x^2 - 4}{x^2 - x - 2}$

33. $\lim_{\Delta x \to 0} (2 + \Delta x)$

34. $\lim_{\Delta t \to 0} \dfrac{4 + \Delta t}{2}$

35. $\lim_{\Delta x \to 0} [(2 + \Delta x)^2 - 4]$

36. $\lim_{\Delta x \to 0} \dfrac{(2 + \Delta x)^2 - 4}{\Delta x}$

37. $\lim_{\Delta t \to 0} \dfrac{[1/(3 + \Delta t)] - \frac{1}{3}}{\Delta t}$

38. $\lim_{\Delta x \to 0} \dfrac{|\Delta x|}{\Delta x}$

39. $f(x) = \begin{cases} x & \text{for } x < 0, \\ 1 & \text{for } x = 0, \\ x^2 & \text{for } x > 0 \end{cases}$

(a) $\lim_{x \to -2} f(x)$ (b) $\lim_{x \to 0} f(x)$ (c) $\lim_{x \to 3} f(x)$

40. $g(x) = \begin{cases} x + 1 & \text{for } x \geq 1, \\ 2x - 4 & \text{for } x < 1 \end{cases}$

(a) $\lim_{x \to -2} g(x)$ (b) $\lim_{x \to 1} g(x)$ (c) $\lim_{x \to 3} g(x)$

41. $g(t) = \begin{cases} \sqrt{t - 3} & \text{for } t > 3, \\ t^2 + 1 & \text{for } t < 3 \end{cases}$

(a) $\lim_{t \to -1} g(t)$ (b) $\lim_{t \to 3} g(t)$ (c) $\lim_{t \to 7} g(t)$

42. $g(u) = \begin{cases} |u|/u & \text{for } u > -2, \ u \neq 0, \\ u + 1 & \text{for } u < -2, \\ 3 & \text{for } u = -2 \end{cases}$

(a) $\lim_{u \to -3} g(u)$ (b) $\lim_{u \to -2} g(u)$
(c) $\lim_{u \to 0} g(u)$ (d) $\lim_{u \to 1} g(u)$

43. $f(x) = \begin{cases} -1/x^3 & \text{for } x < 0, \\ 10 & \text{for } x = 0, \\ 1/x & \text{for } x > 0 \end{cases}$

(a) $\lim_{x \to -2} f(x)$ (b) $\lim_{x \to 0} f(x)$ (c) $\lim_{x \to 3} f(x)$

44. $h(t) = \begin{cases} 1/(t - 1) & \text{for } t > 1, \\ 1/(t^2 - 1) & \text{for } -1 < t < 1, \\ 1/(t + 1) & \text{for } t < -1 \end{cases}$

(a) $\lim_{t \to -1} h(t)$ (b) $\lim_{t \to 0} h(t)$ (c) $\lim_{t \to 1} h(t)$

45. $f(x) = \begin{cases} 1/x & \text{for } x > 0, \\ 1/(x^3 + 2x^2) & \text{for } -2 < x < 0, \\ 1/(x^2 - 4) & \text{for } x < -2 \end{cases}$

(a) $\lim_{x \to 2} f(x)$ (b) $\lim_{x \to 0} f(x)$ (c) $\lim_{x \to -2} f(x)$

In Exercises 46–49, explain why it makes no sense to talk about the indicated limit.

46. $\lim_{x \to 2} \sqrt{x^2 - 9}$

47. $\lim_{x \to -3} \sqrt{-(x + 3)^2}$

48. $\lim_{x \to 4} \sqrt{8x - x^2 - 16}$

49. $\lim_{x \to 0} 1/\sqrt{x^2 - 9}$

In Exercises 50–55, estimate $\lim_{x \to 0} f(x)/g(x)$ from the indicated figure that shows the graphs of f and g very near zero.

50.

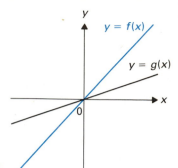

FIGURE 2.14

51.

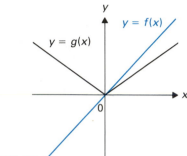

FIGURE 2.15

52.

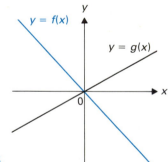

FIGURE 2.16

53.

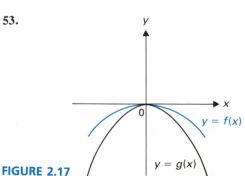

FIGURE 2.17

54.

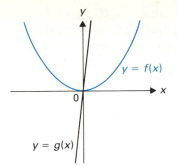

y = f(x)

y = g(x)

FIGURE 2.18

55.

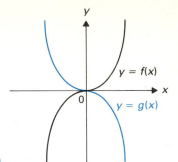

y = f(x)

y = g(x)

FIGURE 2.19

EXPLORING CALCULUS

G: limit demonstration

I: limit of m-secant

2.2 PROPERTIES OF LIMITS; TRIGONOMETRIC LIMITS

Properties of Limits

In the examples of Section 2.1, we often found the limit of a quotient $f(x)/g(x)$ as x approaches a by saying something like "$f(x)$ approaches 2 while $g(x)$ approaches 5, so the quotient approaches $\frac{2}{5}$. This intuitive argument can be placed on a firm mathematical foundation. We state a theorem that justifies such an argument. Proof of a portion of the theorem appears in Section 2.5.

THEOREM 2.1   Properties of limits

If $\lim_{x \to a} f(x) = L$ and $\lim_{x \to a} g(x) = M$, where the domains of f and g contain common points arbitrarily close to a but different from a, then

$$\lim_{x \to a} (f(x) + g(x)) = L + M, \qquad \textit{Sum property} \qquad \textbf{(1)}$$

$$\lim_{x \to a} (f(x) \cdot g(x)) = L \cdot M, \qquad \textit{Product property} \qquad \textbf{(2)}$$

$$\lim_{x \to a} \left(\frac{f(x)}{g(x)} \right) = \frac{L}{M} \qquad \text{if } M \neq 0, \qquad \textit{Quotient property} \qquad \textbf{(3)}$$

$$\lim_{x \to a} \sqrt{f(x)} = \sqrt{L} \qquad \text{if } L > 0. \qquad \textit{Root property} \qquad \textbf{(4)}$$

$\square$

To illustrate, if $f(x)$ is very near 2 and $g(x)$ is very near 5 when x is close to $a = -1$, then $f(x) + g(x)$ is very near $2 + 5 = 7$, and $f(x) \cdot g(x)$ is very near

$2 \cdot 5 = 10$ when x is close to -1. Similarly, $f(x)/g(x)$ is very near $\frac{2}{5}$ and $\sqrt{f(x)}$ is very near $\sqrt{2}$ for such x-values. Theorem 2.1 is certainly intuitively reasonable. We used property (3) of the theorem in several examples in Section 2.1.

Here is a sample of what we can do using properties (1), (2), and (3). Surely $\lim_{x \to a} x = a$. But then, from the product property (2),

$$\lim_{x \to a} x \cdot x = a \cdot a, \qquad \lim_{x \to a} x^3 = \lim_{x \to a} x^2 \cdot x = a^2 \cdot a = a^3.$$

From the sum property (1),

$$\lim_{x \to a} (x^3 + x^2) = a^3 + a^2.$$

Also, if $f(x) = 3$ for all x, then $\lim_{x \to a} f(x) = 3$. Using the product property,

$$\lim_{x \to a} 3 \cdot x^2 = 3a^2.$$

Similar arguments show that if $f(x)$ is any polynomial function, then $\lim_{x \to a} f(x) = f(a)$.

By the quotient property (3), if $g(x)$ is also a polynomial function and $g(a) \ne 0$, then $\lim_{x \to a} f(x)/g(x) = f(a)/g(a)$. Such a quotient of polynomial functions is a **rational function.** Thus the computation of the limit of a rational function as $x \to a$ amounts to evaluation of the function at the point a, provided that a does not make the denominator of the function zero. We state this as a corollary of Theorem 2.1.

COROLLARY   Limits of rational functions

The limit of any polynomial or rational function at a point a in the domain of the function is equal to the value of the function at that point.

The only "bad case" in computing the limit at a of a rational function $f(x)/g(x)$ is the case where $g(a) = 0$, so that a is not in the domain of $f(x)/g(x)$. When a denominator becomes zero at a point, we must try some algebraic trick, such as canceling a factor from both numerator and denominator, to try to find the limit at that point.

EXAMPLE 1   Find

$$\lim_{x \to 1} \sqrt{\frac{x + 3}{x + 35}}.$$

Solution   By property (3) and the preceding discussion, we know that

$$\lim_{x \to 1} \frac{x + 3}{x + 35} = \frac{1 + 3}{1 + 35}$$

$$= \frac{4}{36} = \frac{1}{9}.$$

By the root property (4), we see that

$$\lim_{x \to 1} \sqrt{\frac{x + 3}{x + 35}} = \sqrt{\frac{1}{9}} = \frac{1}{3}. \qquad \blacksquare$$

EXAMPLE 2   Find

$$\lim_{x \to 3} \frac{x^2 - 9}{x^2 - 4x + 3}.$$

Solution   Note that both the numerator and denominator become zero when $x = 3$. We cancel a common factor and use property (3), obtaining

$$\lim_{x \to 3} \frac{x^2 - 9}{x^2 - 4x + 3} = \lim_{x \to 3} \frac{(x - 3)(x + 3)}{(x - 3)(x - 1)} = \lim_{x \to 3} \frac{x + 3}{x - 1} = \frac{6}{2} = 3. \quad \blacksquare$$

EXAMPLE 3   Find

$$\lim_{x \to 5} \frac{x^2 - 25}{x + 4}.$$

Solution   We have $\lim_{x \to 5} [(x^2 - 25)/(x + 4)] = 0/9 = 0$. It is important to realize that a zero in a *numerator only* causes no problem. It is when the denominator becomes zero that we have to do some extra manipulation to find a limit.

$\blacksquare$

A Fundamental Trigonometric Limit: $\lim\limits_{x \to 0} \dfrac{\sin x}{x} = 1$

The fact that

$$\lim_{x \to 0} \frac{\sin x}{x} = 1 \tag{5}$$

is basic in the development of calculus for trigonometric functions. It is important to remember that we always use *radian measure* for the independent variable x in trigonometric functions, unless specifically stated otherwise. (See Exercise 26.) For a circle of radius 1, the (radian) measure of a central angle is equal to the length of the intercepted arc, as shown in Fig. 2.20. Note that the function $(\sin x)/x$ is not defined at 0 and that both numerator and denominator in Eq. (5) approach zero:

$$\lim_{x \to 0} (\sin x) = \lim_{x \to 0} x = 0.$$

Figure 2.20 shows part of the circle $u^2 + v^2 = 1$. A positive value of x is indicated by the central angle and by the length of the colored arc. The altitude of the small triangle in the figure is $\sin x$. Since the large and small triangles are similar, if d is the altitude of the large triangle, we have

$$\frac{d}{\sin x} = \frac{1}{\cos x},$$

so $d = \tan x$. We can see that the area of the small triangle in the figure is less than the area of the sector of the circle having arc of length x, which in turn is less than the area of the large triangle. The area of the sector of the circle is the fraction $x/2\pi$ of the area $\pi \cdot 1^2 = \pi$ of the whole circle, so we have

$$\frac{\sin x \cos x}{2} < \frac{x}{2\pi} \cdot \pi < \frac{\tan x}{2}. \tag{6}$$

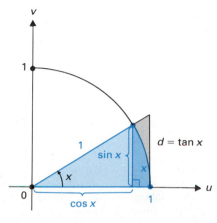

FIGURE 2.20   In terms of their area, Small triangle < Sector < Large triangle, or $\frac{1}{2} \sin x \cos x < (x/2\pi)\pi < \frac{1}{2} \tan x$.

Multiplying relation (6) by $2/(\sin x)$, we obtain

$$\cos x < \frac{x}{\sin x} < \frac{1}{\cos x}. \tag{7}$$

Relation (7) is valid also for x less than but near zero; this follows at once from the relations

$$\sin(-x) = -\sin x \quad \text{and} \quad \cos(-x) = \cos x.$$

As x approaches zero in Fig. 2.20, the length $\cos x$ of the base of the smaller triangle approaches 1. That is,

$$\lim_{x \to 0} (\cos x) = 1,$$

so

$$\lim_{x \to 0} \frac{1}{\cos x} = \frac{1}{1} = 1.$$

But from relation (7), we see that $x/(\sin x)$ is "trapped" between $\cos x$ and $1/(\cos x)$, both of which approach 1 as $x \to 0$, so we must have

$$\lim_{x \to 0} \frac{x}{\sin x} = 1.$$

Then

$$\lim_{x \to 0} \frac{\sin x}{x} = \lim_{x \to 0} \frac{1}{x/(\sin x)} = \frac{1}{1} = 1.$$

We have established the limit in Eq. (5).

EXAMPLE 4   Find $\lim_{x \to 0} [(\sin 5x)/x]$.

Solution   Let $u = 5x$, so $x = u/5$. As $x \to 0$, clearly $u \to 0$ also. Then

$$\lim_{x \to 0} \frac{\sin 5x}{x} = \lim_{u \to 0} \frac{\sin u}{u/5} = \lim_{u \to 0} \left(5 \cdot \frac{\sin u}{u} \right) = 5 \cdot 1 = 5.$$

We usually don't bother to write out the $u = 5x$ substitution. Just remember that

$$\lim_{u \to 0} \frac{\sin(u)}{u} = 1. \tag{8}$$

We can put the limit in this form by writing

$$\lim_{x \to 0} \frac{\sin 5x}{x} = \lim_{x \to 0} \left(5 \cdot \frac{\sin 5x}{5x} \right) = 5 \cdot 1 = 5. \quad \blacksquare$$

EXAMPLE 5   Find $\lim_{x \to 0} [(\sin 4x)/(\sin 3x)]$.

Solution   Putting the limit in the form of relation (8), as explained in Example 4, and using the product property (2) of limits in Theorem 2.1, we obtain

$$\lim_{x \to 0} \frac{\sin 4x}{\sin 3x} = \lim_{x \to 0} \left(\frac{4}{3} \cdot \frac{\sin 4x}{4x} \cdot \frac{3x}{\sin 3x} \right) = \frac{4}{3} \cdot 1 \cdot 1 = \frac{4}{3}. \quad \blacksquare$$

EXAMPLE 6   Find $\lim_{x\to 3} [\sin(x-3)]/(x^2 - x - 6)$.

Solution   Here we think of $x - 3$ as the u from Eq. (8). Then

$$\lim_{x\to 3} \frac{\sin(x-3)}{x^2 - x - 6} = \lim_{x\to 3} \frac{\sin(x-3)}{(x-3)(x+2)}$$

$$= \lim_{x\to 3} \left[\frac{\sin(x-3)}{x-3} \cdot \frac{1}{x+2}\right] = 1 \cdot \frac{1}{5} = \frac{1}{5}. \qquad \blacksquare$$

EXAMPLE 7   Use Eq. (5) and a trigonometric identity to show that

$$\lim_{x\to 0} \frac{(\cos x) - 1}{x} = 0. \tag{9}$$

[*Hint:* Multiply the numerator and denominator by $(\cos x) + 1$.]

Solution   Following the hint, we obtain

$$\frac{\cos x - 1}{x} = \frac{\cos x - 1}{x} \cdot \frac{\cos x + 1}{\cos x + 1}$$

$$= \frac{-\sin^2 x}{x(\cos x + 1)}$$

$$= -\frac{\sin x}{x} \cdot \frac{\sin x}{\cos x + 1}.$$

Since $\lim_{x\to 0} [(\sin x)/x] = 1$, we have

$$\lim_{x\to 0} \frac{\cos x - 1}{x} = \left[\lim_{x\to 0}\left(-\frac{\sin x}{x}\right)\right]\left(\lim_{x\to 0} \frac{\sin x}{\cos x + 1}\right) = -1 \cdot \frac{0}{2} = 0. \qquad \blacksquare$$

We will use both Eqs. (5) and (9) when we develop calculus of trigonometric functions.

Limits with Calculators

If $f(a + 0.01)$, $f(a - 0.005)$, $f(a + 0.002)$, and $f(a - 0.0001)$ have approximately the same value L, that is a good indication that $\lim_{x\to a} f(x) \approx L$. (The symbol $\approx$ means "approximately equals.") The numbers 0.01, -0.005, 0.002, and -0.0001 can of course be replaced by other values of appropriate, small magnitude, some positive and some negative. The appropriate magnitude of the numbers depends on the function and is best determined by trying a few. The next example illustrates this.

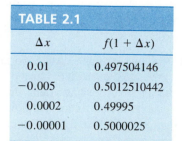

TABLE 2.1

Δx	$f(1 + \Delta x)$
0.01	0.497504146
−0.005	0.5012510442
0.0002	0.49995
−0.00001	0.5000025

EXAMPLE 8   Let

$$f(x) = \frac{\sin(x - 1)}{x^2 - 1},$$

where x is in radian measure. Use a calculator to estimate $\lim_{x\to 1} f(x)$.

Solution   In Table 2.1, we show values of $f(1 + \Delta x)$ for some values of Δx. It appears that $\lim_{x\to 1} f(x) \approx 0.5$. $\qquad \blacksquare$

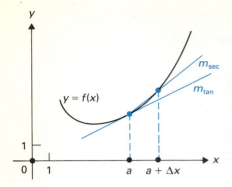

FIGURE 2.21   Approximation of the slope m_{tan} by the slope m_{sec}.

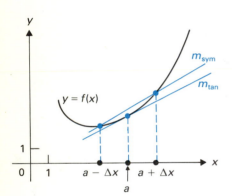

FIGURE 2.22   Better approximation of the slope m_{tan} by m_{sym}.

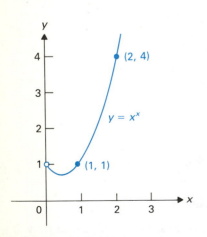

FIGURE 2.23   Graph of $y = x^x$.

Calculator Estimates for m_{tan}

Figure 2.21 shows the tangent line to a graph where $x = a$ and the secant line to the same graph through the points where $x = a$ and $x = a + \Delta x$. Figure 2.22 shows the same graph, the same tangent line where $x = a$, and the line through the points on the graph where $x = a - \Delta x$ and $x = a + \Delta x$ for the same Δx as in Fig. 2.21. These points $a - \Delta x$ and $a + \Delta x$ are symmetrically positioned on either side of a.

It appears from Figs. 2.21 and 2.22 that the symmetrically positioned line in Fig. 2.22 is likely to be more nearly parallel to the tangent line than is the secant line in Fig. 2.21. Let the slope of the symmetrically positioned line be m_{sym}. We expect m_{sym} to be a better estimate for m_{tan} than m_{sec} would be, using the same size Δx. Since the symmetric line goes through the points

$$(a - \Delta x, f(a - \Delta x)) \qquad \text{and} \qquad (a + \Delta x, f(a + \Delta x)),$$

its slope is

$$m_{sym} = \frac{f(a + \Delta x) - f(a - \Delta x)}{2 \cdot \Delta x}. \qquad (10)$$

When computing m_{sym}, we use only positive Δx values, because $-\Delta x$ and Δx lead to exactly the same symmetrically positioned line.

EXAMPLE 9   Use Eq. (10) to approximate m_{tan} for $f(x) = x^3 + 2x$, with $a = 2$ and $\Delta x = 0.01$, which we used in Example 5 in Section 2.1.

Solution   The computation is

$$\frac{f(a + \Delta x) - f(a - \Delta x)}{2 \cdot \Delta x} = \frac{[2.01^3 + 2(2.01)] - [1.99^3 + 2(1.99)]}{2 \cdot 0.01}$$

$$= \frac{12.140601 - 11.860599}{0.02}$$

$$= \frac{0.280002}{0.02} = 14.0001.$$

The correct answer is actually 14, so this approximation is significantly better than the approximation 14.0601, which we found in Example 5 in Section 2.1. And the computation is easy with a calculator.   ∎

Using a calculator and Eq. (10), we can estimate m_{tan} at points of the graphs of functions such as $\sin x$, 2^x, and x^x.

EXAMPLE 10   Estimate m_{tan} for $f(x) = x^x$ at $x = 3$.

Solution   Again we use m_{sym} in Eq. (10) for our estimate. After working Example 9, it seems reasonable to restrict ourselves to the smaller values of Δx and see if we have enough stabilization in the answer. We have done this in Table 2.2. It appears that the value of m_{tan} to six significant figures is 56.6625. The value to ten significant figures can be shown to be 56.66253179. The graph of $y = x^x$ is shown in Fig. 2.23.   ∎

TABLE 2.2

Δx	$m_{\text{sym}} = \dfrac{(3 + \Delta x)^{(3+\Delta x)} - (3 - \Delta x)^{(3-\Delta x)}}{2 \cdot \Delta x}$
0.005	56.6637952
0.001	56.66258235
0.0005	56.66254441
0.0001	56.6625325

You may wonder why we didn't compute m_{sym} in the preceding two examples for much smaller values of Δx, say $\Delta x = 0.0000001$. The reason is that for such Δx, we expect $f(a + \Delta x)$ and $f(a - \Delta x)$ to agree to many significant figures. If our calculator computes to twelve significant figures and $f(a + \Delta x)$ and $f(a - \Delta x)$ agree for ten significant figures, then $f(a + \Delta x) - f(a - \Delta x)$ has only two significant figures in the calculator. The quotient when divided by $2(\Delta x)$ can't be expected to be an accurate estimate of m_{tan}.

SUMMARY

1. If $\lim_{x \to a} f(x) = L$ and $\lim_{x \to a} g(x) = M$, then

$$\lim_{x \to a} (f(x) + g(x)) = L + M,$$

$$\lim_{x \to a} (f(x) \cdot g(x)) = L \cdot M,$$

$$\lim_{x \to a} \left(\frac{f(x)}{g(x)} \right) = \frac{L}{M}, \quad \text{if } M \neq 0,$$

$$\lim_{x \to a} \sqrt{f(x)} = \sqrt{L}, \quad \text{if } L > 0.$$

2. Limits as $x \to a$ of all polynomial and rational functions can be found by evaluating at a as long as a denominator does not become zero at a. If a denominator becomes zero at a, try to cancel a factor of the denominator with one in the numerator.

3. $\lim_{x \to 0} (\sin x)/x = 1$

4. $\lim_{x \to 0} [(\cos x) - 1]/x = 0$

5. When $y = f(x)$, the slope of the symmetrically positioned line through the graph where $x = a - \Delta x$ and $x = a + \Delta x$ is

$$m_{\text{sym}} = \frac{f(a + \Delta x) - f(a - \Delta x)}{2 \cdot \Delta x}.$$

In general, m_{sym} is likely to be a better approximation to m_{tan} than m_{sec} is for the same value of Δx.

EXERCISES 2.2

Theorem 2.1 presented four properties of limits. Exercises 1–10 examine further some properties of limits. Assume that all functions in these exercises have the set of all real numbers as domain.

1. Using Theorem 2.1, show that if $\lim_{x \to a} f(x) = L$ and $\lim_{x \to a} g(x) = M$, then $\lim_{x \to a} [f(x) - g(x)] = L - M$. [*Hint:* Use both property (1) and property (2) of Theorem 2.1.]

2. Using Theorem 2.1 or Exercise 1, show that if $\lim_{x \to a} f(x) = L$ and $\lim_{x \to a} [f(x) + g(x)] = M$, then $\lim_{x \to a} g(x) = M - L$.

In Exercises 3–10, give an example of functions $f(x)$ and $g(x)$ defined for all x and satisfying the given conditions, if such an example is possible. If it is impossible to give such an example, say why it is impossible.

3. Neither $\lim_{x \to 1} f(x)$ nor $\lim_{x \to 1} g(x)$ exists, but $\lim_{x \to 1} [f(x) + g(x)] = 2$.

4. $\lim_{x \to 0} f(x) = 2$ and $\lim_{x \to 0} g(x)$ does not exist, but $\lim_{x \to 0} [f(x) + g(x)] = 1$.

5. $\lim_{x \to 2} f(x)$ exists and $\lim_{x \to 2} g(x)$ does not exist, but $\lim_{x \to 2} [f(x)g(x)]$ exists.

6. $\lim_{x \to 1} f(x) = 2$ and $\lim_{x \to 1} g(x)$ does not exist, but $\lim_{x \to 1} [f(x)g(x)]$ exists.

7. Neither $\lim_{x \to 0} f(x)$ nor $\lim_{x \to 0} g(x)$ exists, but $\lim_{x \to 0} [f(x)g(x)]$ exists.

8. Both $\lim_{x \to 2} f(x)$ and $\lim_{x \to 2} g(x)$ exist, but $\lim_{x \to 2} [f(x)/g(x)]$ does not exist.

9. $\lim_{x \to 2} f(x)$ exists and $\lim_{x \to 2} g(x) = 4$, but $\lim_{x \to 2} [f(x)/g(x)]$ does not exist.

10. $\lim_{x \to 1} f(x)$ exists but $\lim_{x \to 1} \sqrt{f(x)}$ does not exist.

In Exercises 11–25, find the indicated limit, if it exists.

11. $\lim_{x \to 0} \dfrac{\sin x}{|x|}$

12. $\lim_{t \to 0} \dfrac{\sin 2t}{t}$

13. $\lim_{\theta \to 0} \dfrac{\sin \theta}{3\theta}$

14. $\lim_{t \to 0} \dfrac{\sin 2t}{\sin 3t}$

15. $\lim_{x \to 0} \dfrac{\cos 2x}{\cos 3x}$

16. $\lim_{x \to 0} \dfrac{\sin 2x}{\cos 3x}$

17. $\lim_{u \to 0} \sin u$

18. $\lim_{u \to 0} \sin(1/u)$

19. $\lim_{\theta \to 0} (\theta^2 \csc^2 \theta)$

20. $\lim_{x \to 0} (x \cot 2x)$

21. $\lim_{v \to 0} \dfrac{\tan v}{v}$

22. $\lim_{t \to 0} (t \sec t)$

23. $\lim_{x \to 2} \dfrac{\sin(x - 2)}{x^2 - 4}$

24. $\lim_{t \to -3} \dfrac{\sin(t + 3)}{t^3 + 3t^2}$

25. $\lim_{x \to 0} \dfrac{\cos^2 x - 1}{x^2}$

26. (a) Graph $y = x$ and $y = \sin x$ on the same axes for $-\pi/2 \le x \le \pi/2$.

    (b) Let $t°$ denote *degree measure* for trigonometric functions. Graph $y = t°$ and $y = \sin t°$ on the same $t°,y$-axes for $-90 \le t° \le 90$.

    (c) Find $\lim_{x \to 0} \dfrac{\sin(\pi x/180)}{x}$.

    (d) Using part (c), find $\lim_{t° \to 0} (\sin t°)/t°$ where $t°$ is as described in part (b). Interpret your answer in terms of the graph you drew in part (b).

27. Using a calculator or a computer, draw the graph of

$$y = \frac{\sin x}{x} \qquad \text{for } -4\pi \le x \le 4\pi, \ x \ne 0.$$

In Exercises 28–33, use a calculator to estimate the limit, if it exists. Use radian measure for all trigonometric functions.

28. $\lim_{x \to \sqrt{2}} \dfrac{x^2 + 2\sqrt{2}x - 6}{x^2 - 2}$

29. $\lim_{x \to 3} \dfrac{x - \sqrt{3x}}{27 - x^3}$

30. $\lim_{x \to 3} \dfrac{\sin(x - 3)}{x^2 - 9}$

31. $\lim_{x \to 0} (1 + x)^{1/x}$

32. $\lim_{x \to 0} \dfrac{\cos x - 1}{x^2}$

33. $\lim_{x \to 0} \dfrac{\sin x^2}{\cos^2 x - 1}$

In Exercises 34–41, use a calculator and Eq. (10) to approximate $m_{\tan}$ for the given function at the indicated x-value. You choose Δx. Use radian measure with trigonometric functions.

34. $\sin x$ at $x = 0$

35. $\tan x$ at $x = 1$

36. $\tan x$ at $x = \pi/4$

37. $\sqrt{x^2 + 2x - 3}$ at $x = 2$

38. $\dfrac{\sqrt{x^2 - 4x}}{x + 3}$ at $x = -2$

39. 3^x at $x = 2$

40. $2^{\sqrt{x+4}}$ at $x = 5$

41. x^x at $x = 1.5$

EXPLORING CALCULUS

G: limit demonstration

I: limit of m-secant

J: m-sec versus m-sym

2.3 ONE-SIDED LIMITS AND LIMITS AT INFINITY

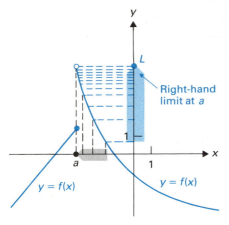

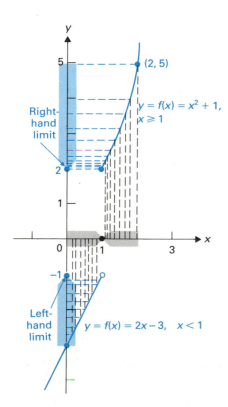

FIGURE 2.24 $\lim_{x \to a+} f(x) = L$

FIGURE 2.25 $\lim_{x \to 1+} f(x) = 2$;
$\lim_{x \to 1-} f(x) = -1$

$\lim_{x \to a+} f(x)$ and $\lim_{x \to a-} f(x)$

Recall that limits are used to describe the behavior of a function near a point, neglecting the value at the point itself. For the function f in Fig. 2.24, we see that $\lim_{x \to a} f(x)$ does not exist. However, as the figure shows, $f(x)$ approaches L if $x \to a$ *from the right-hand side only*. This illustration suggests that we introduce the notion of the limit from the right-hand side (or left-hand side) to help describe the behavior of some functions. We denote such a right-hand limit at a by

$$\lim_{x \to a+} f(x).$$

Thus for $f(x)$ in Fig. 2.24, we have $\lim_{x \to a+} f(x) = L$.

Example 1 involves the function $\sqrt{x}$ as $x \to 0$. We know that $\sqrt{x}$ is defined only for $x \geq 0$. While we could work with the regular limit, presented in Section 2.1, we often write it as a right-hand limit in such a case just to remind ourselves that we can approach zero only from the right-hand side in the domain of the function.

EXAMPLE 1 Find

$$\lim_{x \to 0+} \frac{x - \sqrt{x}}{\sqrt{x}}.$$

Solution Since this function involves $\sqrt{x}$, it is not defined for x negative, but it is defined if $x > 0$. We have

$$\lim_{x \to 0+} \frac{x - \sqrt{x}}{\sqrt{x}} = \lim_{x \to 0+} \frac{\sqrt{x}(\sqrt{x} - 1)}{\sqrt{x}}$$
$$= \lim_{x \to 0+} (\sqrt{x} - 1) = -1. \quad \blacksquare$$

EXAMPLE 2 Let

$$f(x) = \begin{cases} x^2 + 1 & \text{for } x \geq 1, \\ 2x - 3 & \text{for } x < 1. \end{cases}$$

Find $\lim_{x \to 1+} f(x)$.

Solution The graph of $f(x)$ is shown in Fig. 2.25. We are interested in values $f(x)$ for $x > 1$, so we use the formula $x^2 + 1$ to compute the function there. We have

$$\lim_{x \to 1+} f(x) = \lim_{x \to 1+} (x^2 + 1) = 1 + 1 = 2.$$

Note that $\lim_{x \to 1} f(x)$ does not exist. $\quad \blacksquare$

With Examples 1 and 2 as illustration, we give an intuitive definition of a right-hand limit.

DEFINITION 2.2

Right-hand limit (intuitive)

Suppose the domain of f contains points x arbitrarily close to but greater than a. Then $\lim_{x \to a+} f(x) = L$ if $f(x)$ can be made as close to L as we wish by taking any such x sufficiently close to a but greater than a.

Of course, we also have the notion of the left-hand limit at a of a function, $\lim_{x \to a-} f(x)$. Exercise 48 asks you to write out an intuitive definition like Definition 2.2. Only two words and one symbol in Definition 2.2 need to be changed.

EXAMPLE 3 Find $\lim_{x \to 1-} f(x)$ given in Example 2.

Solution Since we are interested in values $f(x)$ for $x < 1$, we use the formula $2x - 3$ to compute the function there. See Fig. 2.25. We have

$$\lim_{x \to 1-} f(x) = \lim_{x \to 1-} (2x - 3) = -1. \qquad \blacksquare$$

EXAMPLE 4 Find the left-hand and right-hand limits at $x = 1$ for

$$f(x) = \begin{cases} 2 - x & \text{for } x \geq 1, \\ 2x + 1 & \text{for } x < 1. \end{cases}$$

The graph is shown in Fig. 2.26.

Solution We have

$$\lim_{x \to 1-} f(x) = \lim_{x \to 1-} (2x + 1) = 3,$$

while

$$\lim_{x \to 1+} f(x) = \lim_{x \to 1+} (2 - x) = 1.$$

Of course, the actual value $f(1) = 1$ plays no role in the computation of a limit as $x \to 1$. Note that $\lim_{x \to 1} f(x)$ does not exist because $f(x)$ does not approach a *single* value as $x \to 1$. $\qquad \blacksquare$

Let $f(x)$ be defined on both sides of a for values of x arbitrarily close to a. As we might guess from Example 4, it is easy to see that

---

$\lim_{x \to a} f(x)$ exists if and only if $\lim_{x \to a+} f(x)$ and $\lim_{x \to a-} f(x)$ both exist and are equal.

We can use these ideas to show that $m_{\tan}$ does not exist for $f(x) = |x|$ at $a = 0$. Recall from page 62 that

$$m_{\tan} = \lim_{\Delta x \to 0} \frac{f(a + \Delta x) - f(a)}{\Delta x}$$

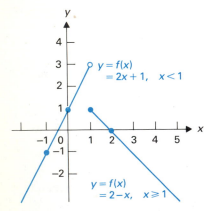

FIGURE 2.26 $\lim_{x \to 1-} f(x) = 3;$ $\lim_{x \to 1+} f(x) = 1$

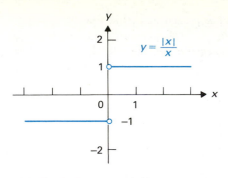

FIGURE 2.27 $\lim_{x \to 0}(|x|/x)$ does not exist. The graph jumps 2 units at $x = 0$.

if the limit exists. For $f(x) = |x|$ at $a = 0$, we obtain

$$m_{\tan} = \lim_{\Delta x \to 0} \frac{|0 + \Delta x| - |0|}{\Delta x} = \lim_{\Delta x \to 0} \frac{|\Delta x|}{\Delta x}.$$

Figure 2.27 shows that

$$\lim_{\Delta x \to 0+} \frac{|\Delta x|}{\Delta x} = 1, \quad \text{while} \quad \lim_{\Delta x \to 0-} \frac{|\Delta x|}{\Delta x} = -1.$$

Thus $\lim_{\Delta x \to 0} (|\Delta x|/\Delta x)$ does not exist, so $m_{\tan}$ does not exist for $|x|$ at $a = 0$.

The Notations ∞ and $-\infty$

The symbols ∞ (read "infinity") and $-\infty$ (read "negative infinity") can aid us in describing the behavior of functions. We can regard ∞ as being greater than every real number, and $-\infty$ as being less than every real number. However, ∞ and $-\infty$ are *not* real numbers and arithmetic should not be attempted with them. The notation $\lim_{x \to a} f(x) = \infty$ means that $f(x)$ can be made as large as we wish by taking x sufficiently close to a but different from a. Intuitively, we think of $f(x)$ as becoming as close to ∞ as we wish as x becomes closer and closer to a but never equal to a. In a similar fashion, we write $\lim_{x \to a} f(x) = -\infty$ if the values $y = f(x)$ can be forced as far down the y-axis as we wish by taking x sufficiently close to a but never equal to a.

EXAMPLE 5 Describe the behavior of $f(x) = 1/x^2$ as $x \to 0$.

Solution We note that x^2 is never negative, so $1/x^2$ is never negative. As x becomes closer and closer to zero, the values $1/x^2$ become larger and larger; indeed $1/x^2$ can be made just as large as we wish by taking x sufficiently close to zero. We express this by writing

$$\lim_{x \to 0} \frac{1}{x^2} = \infty. \qquad \blacksquare$$

The purpose in considering $\lim_{x \to a} f(x)$ is to describe the behavior of $f(x)$ near a. The expression $\lim_{x \to 0} (1/x^2) = \infty$ that we obtained in Example 5 describes this behavior near zero in a very concise way. We point out that $\lim_{x \to 0} (1/x^2)$ does not exist in the sense of Definition 2.1, for there is no *real number* that satisfies the condition for L described in the definition. The symbol ∞ tells us *why* the limit does not exist.

Of course, we can also use the symbols ∞ and $-\infty$ with one-sided limits. If one of these symbols is appropriate, the choice between them is just a question of the *sign* of $f(x)$ as $x \to a$ from the desired side. Use of ∞ and $-\infty$ is illustrated in the next two examples.

EXAMPLE 6 Find

$$\lim_{x \to 2-} \frac{x^2 - 5x}{x^2 - 4},$$

using the symbol ∞ or $-\infty$ if appropriate. The graph is shown in Fig. 2.28.

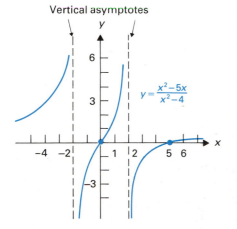

Vertical asymptotes

$y = \dfrac{x^2 - 5x}{x^2 - 4}$

FIGURE 2.28

$\lim_{x \to 2-} [(x^2 - 5x)/(x^2 - 4)] = \infty$

Vertical asymptote

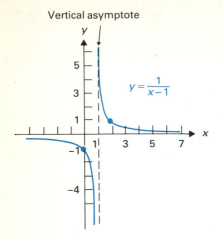

FIGURE 2.29 $\lim_{x\to1+}[1/(x-1)] = \infty$; $\lim_{x\to1-}[1/(x-1)] = -\infty$

Solution   Here the numerator does not approach zero, but the denominator does. Consequently, the limit will be ∞ or $-\infty$. It is just a question of *sign*. We have $\lim_{x\to2-}(x^2 - 5x) = -6$, which is negative. On the other hand, for x just a bit less than 2, $x^2 - 4$ is near zero but negative also. Thus the quotient becomes large but is positive, so the limit is ∞.   ■

The lines $x = -2$ and $x = 2$, which are dashed in Fig. 2.28 are *vertical asymptotes* of the graph.

EXAMPLE 7   Use right-hand and left-hand limits and the notations ∞ and $-\infty$ to describe the behavior of $f(x) = 1/(x - 1)$ as $x \to 1$.

Solution   Figures 2.29 and 2.30 show the graphs of $1/(x - 1)$ and $|1/(x - 1)|$. We see that

$$\lim_{x\to1+}\frac{1}{x-1} = \infty, \quad \lim_{x\to1-}\frac{1}{x-1} = -\infty, \quad \text{and} \quad \lim_{x\to1}\left|\frac{1}{x-1}\right| = \infty.$$ ■

Limits at Infinity or Negative Infinity

Sometimes we want to know the behavior of $f(x)$ for very large values of x. A natural way to phrase this problem is to discuss the behavior of $f(x)$ ''as x approaches ∞.'' The assertion $\lim_{x\to\infty} f(x) = L$ means that $f(x)$ will be as close to L as we wish if we take *any* sufficiently large x in the domain of f. As shown in Fig. 2.31, the graph of f then must be very close to the horizontal line $y = L$ for large x. This line $y = L$ is a *horizontal asymptote* of the graph.

EXAMPLE 8   Find $\lim_{x\to\infty}(1/x)$ and $\lim_{x\to-\infty}(1/x)$.

Solution   The graph of $f(x) = 1/x$ is shown in Fig. 2.32. We have both

$$\lim_{x\to\infty}\frac{1}{x} = 0 \qquad \text{and} \qquad \lim_{x\to-\infty}\frac{1}{x} = 0.$$ ■

EXAMPLE 9   Find $\displaystyle\lim_{x\to\infty}\frac{2x^2 - 3x}{3x^2 + 2}$.

Vertical asymptote

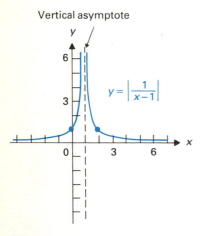

FIGURE 2.30 $\lim_{x\to1}|1/(x-1)| = \infty$

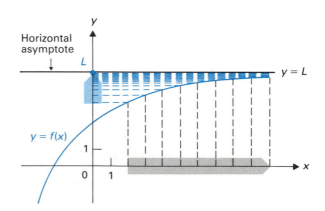

FIGURE 2.31 $\lim_{x\to\infty} f(x) = L$

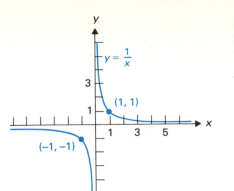

FIGURE 2.32 $\lim_{x\to\infty}(1/x) = 0$; $\lim_{x\to-\infty}(1/x) = 0$

Solution A nice technique is to factor the monomial terms of highest degree from both the numerator and the denominator. The factorization is not valid when $x = 0$, but we are interested in values of x far from 0. We have

$$\lim_{x\to\infty} \frac{2x^2 - 3x}{3x^2 + 2} = \lim_{x\to\infty} \left(\frac{2x^2}{3x^2} \cdot \frac{1 - \dfrac{3x}{2x^2}}{1 + \dfrac{2}{3x^2}} \right) = \lim_{x\to\infty} \left(\frac{2}{3} \cdot \frac{1 - \dfrac{3}{2x}}{1 + \dfrac{2}{3x^2}} \right)$$

$$= \frac{2}{3} \cdot \frac{1 - 0}{1 + 0} = \frac{2}{3}. \qquad \blacksquare$$

EXAMPLE 10 Repeat the technique of Example 9 to find

$$\lim_{x\to-\infty} \frac{2x^3 - 3x^2}{2x^2 + 4x - 7}.$$

Solution We have

$$\lim_{x\to-\infty} \frac{2x^3 - 3x^2}{2x^2 + 4x - 7} = \lim_{x\to-\infty} \left(\frac{2x^3}{2x^2} \cdot \frac{1 - \dfrac{3x^2}{2x^3}}{1 + \dfrac{4x}{2x^2} - \dfrac{7}{2x^2}} \right)$$

$$= \lim_{x\to-\infty} \left(\frac{x}{1} \cdot \frac{1 - \dfrac{3}{2x}}{1 + \dfrac{2}{x} - \dfrac{7}{2x^2}} \right) = -\infty,$$

because $x/1$ approaches $-\infty$ and the second factor approaches 1. $\qquad \blacksquare$

EXAMPLE 11 Use the technique of Example 9 to find

$$\lim_{x\to\infty} \frac{x^2 + 1}{2x - 3x^3}.$$

Solution We have

$$\lim_{x\to\infty} \frac{x^2 + 1}{2x - 3x^3} = \lim_{x\to\infty} \left(\frac{x^2}{-3x^3} \cdot \frac{1 + \dfrac{1}{x^2}}{\dfrac{2x}{-3x^3} + 1} \right) = \lim_{x\to\infty} \left(\frac{1}{-3x} \cdot \frac{1 + \dfrac{1}{x^2}}{\dfrac{-2}{3x^2} + 1} \right) = 0,$$

because the first factor approaches zero and the second approaches 1. $\qquad \blacksquare$

The preceding three examples make it clear that the limit of a rational function $f(x)$ as $x \to \infty$ is simply the limit of the quotient of the monomial term of highest degree in the numerator and the one of highest degree in the denominator. We regard these monomial terms of highest degree as *dominating* the other terms of the numerator and denominator as $x \to \infty$ or $x \to -\infty$. We can now simplify the computation of such limits.

EXAMPLE 12   Find
$$\lim_{x \to \infty} \frac{2x^5 - 3x^2 + 4x}{8x^3 - 10x^5}.$$

Solution   We have
$$\lim_{x \to \infty} \frac{2x^5 - 3x^2 + 4x}{8x^3 - 10x^5} = \lim_{x \to \infty} \frac{2x^5}{-10x^5} = \frac{2}{-10} = -\frac{1}{5}. \qquad \blacksquare$$

EXAMPLE 13   Find
$$\lim_{x \to \infty} \frac{4x^4 + 5}{8 - 3x^3}.$$

Solution   We have
$$\lim_{x \to \infty} \frac{4x^4 + 5}{8 - 3x^3} = \lim_{x \to \infty} \frac{4x^4}{-3x^3} = \lim_{x \to \infty} \frac{4x}{-3} = -\infty. \qquad \blacksquare$$

EXAMPLE 14   Find $\lim_{x \to \infty} \sin x$, if the limit exists.

Solution   For all integers n, we have $\sin(2\pi n) = 0$ and $\sin[(\pi/2) + 2\pi n] = 1$. This shows that $\sin x$ keeps attaining values zero and also 1 for arbitrarily large values of x. Consequently $\lim_{x \to \infty} \sin x$ does not exist. $\qquad \blacksquare$

EXAMPLE 15   Find $\lim_{x \to \infty} [(\sin x)/x]$, if the limit exists.

Solution   Since $|\sin x| \le 1$ for all x, we see that the numerator in $(\sin x)/x$ is never more than 1 in magnitude, while the denominator becomes huge as $x \to \infty$. Consequently, the quotient approaches zero, so
$$\lim_{x \to \infty} \frac{\sin x}{x} = 0. \qquad \blacksquare$$

SUMMARY

1. $\lim_{x \to a+}$ means the limit as x approaches a from the right-hand side and $\lim_{x \to a-}$ means the limit from the left-hand side.

2. The symbols ∞ (infinity) and $-\infty$ (negative infinity) can be useful in describing why some limits do not exist. Regard ∞ as being greater than every positive real number and $-\infty$ as being less than every negative real number.

3. The limit as $x \to \infty$ or as $x \to -\infty$ of a rational function is the limit of the quotient of the monomial term of highest degree in the numerator and the monomial term of highest degree in the denominator.

EXERCISES 2.3

In Exercises 1–47, find the indicated limit if it exists. Use the symbols ∞ and $-\infty$ to indicate the behavior near the point where appropriate.

1. $\lim\limits_{x \to 2} \dfrac{1}{2 - x}$

2. $\lim\limits_{x \to 2+} \dfrac{1}{2 - x}$

3. $\lim\limits_{x \to 2-} \dfrac{1}{2 - x}$

4. $\lim\limits_{t \to 2} \dfrac{1}{(2 - t)^2}$

5. $\lim\limits_{u \to 5+} \dfrac{u + 3}{u^2 - 25}$

6. $\lim\limits_{u \to -5+} \dfrac{u + 3}{u^2 - 25}$

7. $\lim\limits_{s \to 4+} \dfrac{s^2 + 3s + 5}{4s - s^2}$

8. $\lim\limits_{x \to 4+} \dfrac{x^2 - 3x - 5}{4x - x^2}$

9. $\lim\limits_{t \to 2-} \dfrac{t^2 - 7t + 4}{2 + t - t^2}$

10. $\lim\limits_{x \to 5-} \dfrac{x^2 - 25}{x^2 - 10x + 25}$

11. $\lim\limits_{u \to 4+} \dfrac{4u - u^2}{8u - u^2 - 16}$

12. $\lim\limits_{t \to 2+} \dfrac{t^2 - 3t + 2}{t^2 - 4t + 4}$

13. $\lim\limits_{x \to 2} \left| \dfrac{x^2 + 4}{x - 2} \right|$

14. $\lim\limits_{s \to 0+} \left(\dfrac{3}{s} - \dfrac{1}{s^2} \right)$
[*Hint:* Write $3/s - 1/s^2$ as a quotient of polynomials.]

15. $\lim\limits_{x \to 0} \left(\dfrac{1}{x^4} - \dfrac{1}{x} \right)$

16. $\lim\limits_{x \to 0-} \left(\dfrac{1}{x^3} - \dfrac{1}{x} \right)$

17. $\lim\limits_{t \to 0-} \left(\dfrac{t - 3}{t^4} + \dfrac{2}{t^2} \right)$

18. $\lim\limits_{x \to 0-} \left(\dfrac{5 + x}{x^2} - \dfrac{x^2 - 9}{x^3} \right)$

19. $\lim\limits_{x \to \infty} \dfrac{x + 1}{x}$

20. $\lim\limits_{x \to -\infty} \dfrac{3x^3 - 2x}{2x^3 + 3}$

21. $\lim\limits_{t \to -\infty} \dfrac{|t|}{t}$

22. $\lim\limits_{x \to \infty} \dfrac{x^3 + 2x}{x^2 - 3}$

23. $\lim\limits_{x \to \infty} \dfrac{x^2 - 2x + 1}{x^3 + 3x - 2}$

24. $\lim\limits_{x \to -\infty} \dfrac{x^2 - 2x + 1}{x^3 + 3x - 2}$

25. $\lim\limits_{x \to \infty} \dfrac{2x^3 + 4x + 2}{8x - 5x^2}$

26. $\lim\limits_{u \to -\infty} \dfrac{4u^5 - 8u^2}{3u^2 - 8u + 2}$

27. $\lim\limits_{t \to \infty} \dfrac{4t^3 - 8t^2 + 3}{7 - 2t^3}$

28. $\lim\limits_{x \to \infty} \dfrac{4x^{1/2} - 2x^{1/3} - 2}{3x^{1/3} - 5x^{3/4} + 5}$

29. $\lim\limits_{x \to -\infty} \dfrac{8x^{1/3} + 4x^{1/5} + 3}{5x^{1/9} - 7x^{1/5}}$

30. $\lim\limits_{x \to \infty} \dfrac{x^{5/4} - 3x^{4/3} + 2}{x^{7/6} - 3x^{10/9}}$

31. $\lim\limits_{u \to \infty} \dfrac{u^{2/3} + 4u^{5/6} - 3}{8u + 3u^{5/7}}$

32. $\lim\limits_{x \to -\infty} (x^2 + 3x)$

33. $\lim\limits_{x \to -\infty} (x^3 + 3x^2)$

34. $\lim\limits_{x \to \infty} (x^{1/2} - x^{1/3})$

35. $\lim\limits_{x \to -\infty} (x^{1/5} - x^{1/3})$

36. $\lim\limits_{x \to \infty} (x - \sqrt{x^2 + 1})$

37. $\lim\limits_{x \to 0-} \dfrac{|\sin x|}{x}$

38. $\lim\limits_{x \to 0-} \dfrac{\sin x}{x^2}$

39. $\lim\limits_{x \to 0-} \dfrac{\sin x^2}{x^3}$

40. $\lim\limits_{x \to 1+} \dfrac{x^2 - 1}{\sin(x - 1)^2}$

41. $\lim\limits_{x \to 1-} \dfrac{1 - x^2}{\sin(x - 1)^2}$

42. $\lim\limits_{x \to 0+} \dfrac{\cos x}{\sin x}$

43. $\lim\limits_{t \to 0-} \dfrac{\cos t}{\sin t^2}$

44. $\lim\limits_{u \to \infty} \dfrac{u}{\sin u}$

45. $\lim\limits_{t \to \infty} \dfrac{\sin t^2}{t^2}$

46. $\lim\limits_{x \to \infty} x \sin\left(\dfrac{1}{x} \right)$

47. $\lim\limits_{x \to \infty} \cos^2 x$

48. Write intuitive descriptions of the meaning of the assertion concerning a limit, as we did in Definition 2.2.
(a) $\lim\limits_{x \to a-} f(x) = L$ (b) $\lim\limits_{x \to a+} f(x) = -\infty$
(c) $\lim\limits_{x \to -\infty} f(x) = L$ (d) $\lim\limits_{x \to 0} f(x) = -\infty$

49. A rubber ball has the characteristic that, when dropped on a floor from height h, it rebounds to height $h/2$. See Fig. 2.33. It can be shown that if the ball is dropped from a height of h ft and allowed to bounce repeatedly, the total distance it has traveled when it hits the floor for the nth time is

$$h + 2h\left[1 - \left(\dfrac{1}{2} \right)^{n-1} \right] \text{ ft.}$$

Find the total distance the ball travels before it stops bouncing if it is dropped from a height of 4 ft.

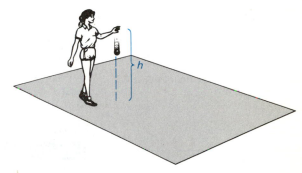

FIGURE 2.33

50. Taking 4000 miles as radius of the earth and 32 ft/sec² as gravitational acceleration at the surface of the earth, and neglecting air resistance, we can show that the velocity v with which a body must be fired upward from the surface of the earth to attain an altitude of s miles is given by the formula

$$v = \dfrac{8}{\sqrt{5280}} \sqrt{\dfrac{4000s}{s + (4000)(5280)}} \text{ mi/sec.}$$

Using this formula, find the velocity with which a body must be fired upward to escape the gravitational attraction of the earth.

In Exercises 51–56, decide whether the limit exists and estimate its value. Use a calculator to compute the function for at least four values very near the limit point (values of very large magnitude if $x \to \infty$ or $x \to -\infty$). Use radian measure with trigonometric functions.

51. $\lim_{x \to 0+} x^x$

52. $\lim_{x \to 0+} \left(\frac{1}{1-x} \right)^{-1/x^2}$

53. $\lim_{x \to 0+} (\cos x)^{1/\tan x}$

54. $\lim_{x \to -\infty} \left(1 + \frac{1}{x} \right)^{2x}$

55. $\lim_{x \to \infty} \left(1 + \frac{1}{x} \right)^x$

56. $\lim_{x \to \infty} \left(1 + \frac{1}{x} \right)^{x^2}$

 EXPLORING CALCULUS

G: limit demonstration

2.4 CONTINUITY

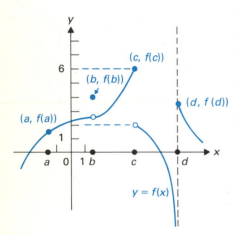

FIGURE 2.34 f is continuous at a but not continuous at b, c, or d.

The adjective *continuous* has a meaning in mathematics very close to its meaning in everyday speech. If we say a person talks continuously, we mean that the person talks with no break. Simplistically stated, a function is *continuous* on an interval if its graph is *unbroken* there and can be sketched without lifting the pencil off the page.

Continuous Functions

In simple terms, a function f is *continuous at a point a* in its domain if $f(x)$ remains close to $f(a)$ whenever x is close enough to a. That is, moving a small enough distance from a in the domain should result in only a small change in function value. Figure 2.34 shows the graph of a function that is continuous at the point a but not continuous at the points b, c, and d in its domain. For example, in Fig. 2.34 we have $f(c) = 6$. If we are at the point c on the x-axis and move any very small nonzero distance to the right, the function value drops at least four units. A graph must not jump or break at a if f is to be continuous at a.

DEFINITION 2.3

Continuity

A function f is **continuous at a point** a if

1. a is in the domain of f,

2. $\lim_{x \to a} f(x)$ exists, and

3. $\lim_{x \to a} f(x) = f(a)$.*

We have defined continuity of f at $x = a$ in terms of the limit of f at $x = a$. Note that condition (1) of Definition 2.3 requires that a lie in the domain of f,

---

*We also define f to be continuous at any point a *in its domain* that *does not* have other points of the domain arbitrarily close to it.

which is not required in the definition of $\lim_{x \to a} f(x)$. Condition (3) is another major conceptual change from the definition of $\lim_{x \to a} f(x)$; we emphasize this change:

> The value $f(a)$ plays an important role in the notion of continuity at a, while it plays no role at all in defining the limit at a.

EXAMPLE 1   Determine whether

$$f(x) = \begin{cases} \dfrac{x^2 - 9}{x + 3} & \text{for } x \neq -3, \\ 10 & \text{for } x = -3 \end{cases}$$

is continuous at $x = -3$.

Solution   We have

$$\lim_{x \to -3} f(x) = \lim_{x \to -3} \frac{(x - 3)(x + 3)}{x + 3} = -6.$$

But $f(-3) = 10$. Since $\lim_{x \to -3} f(x) \neq f(-3)$, this function is not continuous at -3.   ■

EXAMPLE 2   Determine whether

$$g(x) = \begin{cases} x^2 + 2 & \text{for } x > 1, \\ 5x - 1 & \text{for } x \leq 1 \end{cases}$$

is continuous at $x = 1$.

Solution   We have

$$\lim_{x \to 1+} g(x) = \lim_{x \to 1+} (x^2 + 2) = 3$$

while

$$\lim_{x \to 1-} g(x) = \lim_{x \to 1-} (5x - 1) = 4.$$

Therefore $\lim_{x \to 1} g(x)$ does not exist, so g is not continuous at $x = 1$.   ■

EXAMPLE 3   Determine whether

$$h(x) = \begin{cases} \dfrac{x^2 - x - 6}{x - 3} & \text{for } x \neq 3, \\ 5 & \text{for } x = 3 \end{cases}$$

is continuous at $x = 3$.

Solution   We have

$$\lim_{x \to 3} h(x) = \lim_{x \to 3} \frac{(x - 3)(x + 2)}{x - 3} = 5 = h(3).$$

Consequently h is continuous at $x = 3$.   ■

DEFINITION 2.4

Continuous function

A function f is **continuous** if it is continuous at every point in its domain.

Definition 2.3 gave the definition of continuity *at a point,* and Definition 2.4 defines the notion of a function continuous *in its entire domain.*

EXAMPLE 4   Determine whether each of the functions f, g, and h in Examples 1–3 is continuous.

Solution   The function f in Example 1 is not continuous, since that example shows that f is not continuous at $x = -3$, and -3 is in the domain of f.

The function g in Example 2 is not continuous, since that example shows that g is not continuous at $x = 1$, and 1 is in the domain of g.

We saw in Example 3 that h is continuous at the point 3 in its domain. At all points x other than 3, we have

$$h(x) = \frac{x^2 - x - 6}{x - 3}.$$

The corollary of Theorem 2.1 on page 72 shows that for any point $a \neq 3$,

$$\lim_{x \to a} h(x) = \lim_{x \to a} \frac{x^2 - x - 6}{x - 3} = \frac{a^2 - a - 6}{a - 3} = h(a),$$

since $(x^2 - x - 6)/(x - 3)$ is a rational function. Thus $h(x)$ is also continuous for all $x \neq 3$. Consequently, it is continuous at all points in its domain and therefore is a continuous function.   ■

EXAMPLE 5   Show that the function $f(x) = 1/x$ with graph shown in Fig. 2.35 is continuous at every point in its domain.

Solution   The domain of f consists of all $x \neq 0$. If $a \neq 0$, then the corollary of Theorem 2.1 shows that

$$\lim_{x \to a} f(x) = \lim_{x \to a} \frac{1}{x} = \frac{1}{a} = f(a).$$

By Definition 2.3, this means that $f(x) = 1/x$ is continuous at every point a in its domain.   ■

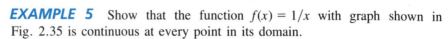

FIGURE 2.35   The function $f(x) = 1/x$ is continuous at every point in its domain. (Note that 0 is not in the domain.)

Looking at Example 5, where $f(x) = 1/x$, we see that it is impossible to define $f(0)$ in a way that would make f continuous at zero. This is the case for any rational function at a point where the denominator is zero, assuming the numerator and denominator have no common factor.

If f and g are both continuous at a, then using Theorem 2.1 we have

$$\lim_{x \to a} (f(x) + g(x)) = \lim_{x \to a} f(x) + \lim_{x \to a} g(x) = f(a) + g(a).$$

This shows that $f(x) + g(x)$ is continuous at a. A similar argument can be made for the function $f(x) \cdot g(x)$ and for $f(x)/g(x)$ if $g(a) \neq 0$. This gives us the following theorem.

THEOREM 2.2   Properties of continuous functions

Sums, products, and quotients of continuous functions are continuous. (Of course, quotients are not defined wherever denominators are zero.)   ☐

The corollary of Theorem 2.1 asserts that a limit of a polynomial or rational function at a point in its domain can be computed by simply evaluating the function at the point. This gives us at once a corollary to Theorem 2.2.

COROLLARY   Continuity of rational functions

Each polynomial function and each rational function is continuous at every point in its domain. (Rational functions are not defined where denominators are zero.)

EXAMPLE 6   Give an intuitive argument that the six basic trigonometric functions $\sin x$, $\cos x$, $\tan x$, $\cot x$, $\sec x$, and $\csc x$ are all continuous.

Solution   If we can show that $\sin x$ and $\cos x$ are continuous, then Theorem 2.2 will show that $\tan x$, $\cot x$, $\sec x$, and $\csc x$ are continuous.

Look again at Fig. 1.77, which shows how $\sin x$ and $\cos x$ are defined. We have $\sin x = u$ and $\cos x = v$ for the point (u, v) corresponding to the arc length x measured from $(1, 0)$ on the unit circle $u^2 + v^2 = 1$. As x approaches some number a, the points (u_x, v_x) on the circle corresponding to the values of x approach the point (u_a, v_a) corresponding to $x = a$. It is intuitively clear that $\lim_{x \to a} u_x = u_a$ and $\lim_{x \to a} v_x = v_a$. That is, $\lim_{x \to a} \cos x = \cos a$ and $\lim_{x \to a} \sin x = \sin a$. Thus $\sin x$ and $\cos x$ are continuous functions.   ■

Discontinuity

As we might expect, the term *discontinuous* means *not continuous*. Note that we make no classification of a function as continuous or discontinuous at a point not in its domain. The function is not concerned with such points. We now define precisely what we mean when we say that f is *discontinuous* at a point a.

DEFINITION 2.5

Discontinuity

A function f is **discontinuous at a point** a if

1. a is in the domain of f, and

2. f is not continuous at a.

Note that the function f of Example 1 is discontinuous at $x = -3$, and the function g of Example 2 is discontinuous at $x = 1$. These points are in the domains of the functions, and the functions are not continuous there. The function f in Fig. 2.34 is discontinuous at the points b, c, and d in its domain. In each case, the reason for the discontinuity is an abrupt jump, or break, in the graph.

Other types of discontinuities exist. Figure 2.36 shows the graph of a function f such that $f(x) = 0$ at $x = \pm 1, \pm \frac{1}{2}, \pm \frac{1}{3}, \pm \frac{1}{4}, \ldots$. The graph goes alternately up to 1 and down to -1 between the places where it is zero. If we define $f(0) = 0$, then 0 is in the domain of f, but f is not continuous at $x = 0$, since $\lim_{x \to 0} f(x)$ clearly does not exist. Thus f is discontinuous at 0. You may feel that a function such as the one shown in Fig. 2.36 is very contrived and could not be given by a simple formula. However, the function

$$f(x) = \begin{cases} \sin(\pi/x) & \text{for } x \neq 0, \\ 0 & \text{for } x = 0 \end{cases}$$

has the graph in Fig. 2.36.

Figure 2.37 shows the graph of the function

$$g(x) = \begin{cases} |x|\sin(\pi/x) & \text{for } x \neq 0, \\ 0 & \text{for } x = 0. \end{cases}$$

This function g is continuous at $x = 0$ since the oscillations of the graph decrease in height with a limiting height of zero as $x \to 0$.

Most functions corresponding to our physical life and activities are continuous. Our height is a continuous function of our age. Our speed, when driving safely, is a continuous function of time. Discontinuities in the physical world are apt to be destructive. For example, if we drive into a big tree at a speed of 40 mph, the speed of the leading edge of the front bumper is very nearly a discontinuous function of time at the instant of impact (the tree gives slightly). The speed drops at that instant from 40 mph to 0 mph. The speed of parts of the car farther to the rear decreases more slowly. Air bags attempt to make the driver's speed continuous in such an accident as well as to spread evenly over the upper body the force from the rapid drop in speed.

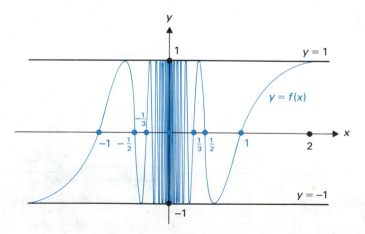

FIGURE 2.36 $f(0) = 0$. f is not continuous at $x = 0$ since $\lim_{x \to 0} f(x)$ does not exist. Neither the right-hand nor the left-hand limit exists at $x = 0$.

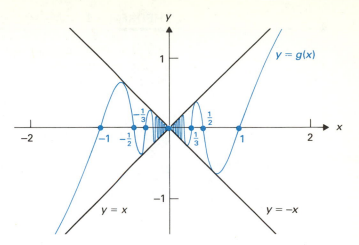

FIGURE 2.37 $g(0) = 0$. g is continuous at $x = 0$ because $\lim_{x \to 0} g(x) = 0$.

Two Properties of Continuous Functions

In Theorems 2.3 and 2.4, which follow, we give two mathematical properties of continuous functions that will be useful in Chapter 4. In Section 2.5, we restate Theorem 2.3 and prove it.

Let f be a function continuous at each point of a closed interval $[a, b]$. The graph of such a function f is shown in Fig. 2.38. *Intuitively, the continuity of f on $[a, b]$ means that the graph of f over $[a, b]$ can be sketched with a pencil without taking the pencil off the paper*. This geometric interpretation of *continuity of f* on $[a, b]$ makes Theorem 2.3 seem very reasonable.

Figure 2.39 shows the graph of a continuous function defined for all x in a closed interval $[a, b]$. The graph is higher at b than at a, that is, $f(a) < f(b)$. Choose any point c between $f(a)$ and $f(b)$ on the y-axis. As the figure indicates, there must be a point x_0 on the x-axis such that $f(x_0) = c$. Think of the horizontal line $y = c$ as a road. The point $(a, f(a))$ is on one side of the road, and $(b, f(b))$ is on the other side. We can't move continuously from one side of the road to the other without crossing the road somewhere, at some point (x_0, c).

HISTORICAL NOTE

THE INTERMEDIATE VALUE theorem was first given an analytic proof by Bernhard Bolzano (1781–1848) in 1817. Bolzano, a Bohemian theologian and mathematician, attempted to formulate clear concepts in mathematics, starting from indisputable first principles. He contributed to two important mathematical problems of the early nineteenth century—the study of Euclid's parallel postulate, which ultimately led to the discovery of non-Euclidean geometry, and the search for a solid foundation for analysis, which eliminated the use of infinitesimally small quantities. In connection with the latter problem Bolzano formulated a definition of continuity similar to that used today and a theorem asserting what is now known as the greatest lower bound property of the real numbers. From that property, which is closely related to the Dedekind cut principle discussed in Section 2.5, the intermediate value theorem follows readily.

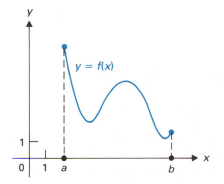

FIGURE 2.38 f is continuous on $[a, b]$, so its graph can be traced over $[a, b]$ without taking the pencil off the paper.

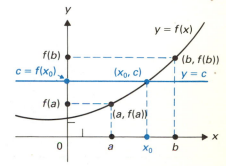

FIGURE 2.39   The graph of continuous f crosses the line $y = c$ to get from $(a, f(a))$ to $(b, f(b))$.

THEOREM 2.3   Weierstrass intermediate value theorem

Let f be a continuous function whose domain includes the closed interval $[a, b]$. If c is a number between $f(a)$ and $f(b)$, then there exists some point x_0 in $[a, b]$ such that $f(x_0) = c$.   □

Figure 2.40 shows that the conclusion of Theorem 2.3 need not hold if the function f is not continuous. The hypothesis of continuity is essential.

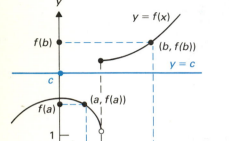

FIGURE 2.40 The noncontinuous graph need not intersect the line $y = c$ to get from $(a, f(a))$ to $(b, f(b))$.

EXAMPLE 7   A sapling 12 in. high is planted and grows into a tree 32 ft tall. Show that at some instant, the tree must be exactly 21.357 ft tall.

Solution   The height of the tree is a continuous function of time. The tree grows from 1 ft at some time $t = a$ to 32 ft at a later time $t = b$, and 21.357 is between 1 and 32. Theorem 2.3 shows that the tree is 21.357 ft tall at some time t_0.   ■

Example 7 reinforces our feeling that Theorem 2.3 is intuitively obvious. Example 8 is a bit more sophisticated, and still less obvious applications are given in the exercises.

EXAMPLE 8   A father has his minor son's car wired so that a loud buzzer is activated for 5 seconds each time the speed of the car is exactly 45 mph. The son has just 35 minutes to drive the car 28 miles to the airport to meet a friend's plane. Show that he can't get there in that time without being buzzed.

Solution   To drive 28 miles in 35 minutes, the son must average

$$28 / \tfrac{35}{60} = 28 \cdot \tfrac{12}{7} = 48 \text{ mph.}$$

He can't average 48 mph, starting with speed 0 mph at time $t = a$, without going a bit more than 48 mph at some time $t = b$. Since his speed is a continuous function of time, he will get buzzed at some time $t = c$ between $t = a$ and $t = b$, according to Theorem 2.3. Actually, he will be buzzed at least once more, between time $t = b$ and the time he stops at the airport.   ■

Our second property of continuous functions concerns the attainment of maximum and minimum values.

DEFINITION 2.6

Extrema

Let S be a subset of the domain of a function f. If x_1 is in S, and $f(x_1) \geq f(x)$ for all x in S, then $M = f(x_1)$ is the **maximum value attained** (or **assumed**) by f on S. Similarly, if x_2 is in S, and $f(x_2) \leq f(x)$ for all x in S, then $m = f(x_2)$ is the **minimum value attained** (or **assumed**) by f on S.

From Fig. 2.41, we see that 1 is the minimum value attained by the function $x^2 + 1$ on its entire domain. As shown in Fig. 2.42, the minimum value attained by x^2 on $[-2, 1]$ is 0, and the maximum value attained is 4.

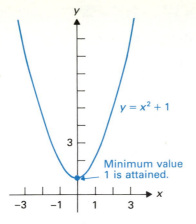

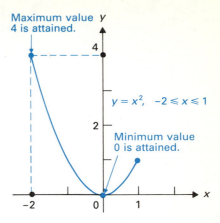

FIGURE 2.41 $f(x) = x^2 + 1$ attains the minimum value of 1 where $x = 0$.

FIGURE 2.42 $f(x) = x^2$ attains a minimum value of 0 where $x = 0$ and a maximum value of 4 where $x = -2$ over $[-2, 1]$.

THEOREM 2.4   Attainment of extreme values

A continuous function attains both a maximum value M and a minimum value m on any closed interval in its domain. □

   Figure 2.43 shows that the function f must be continuous or the conclusion of Theorem 2.4 need not be true. The only possible point where a maximum could be attained in Fig. 2.43 is at the "first point on the x-axis to the left of c." But there is no such point, for if d is to the left of c, then $(c + d)/2$ is halfway between c and d and thus closer to c than d is.
   Figure 2.44 shows that the interval must be closed or the conclusion of Theorem 2.4 need not hold. For the function f on the half-open interval $[a, b)$ in Fig. 2.43, the only conceivable place a maximum value could be attained would be at the "first point on the x-axis to the left of b." As explained in the preceding paragraph, there is no such point.

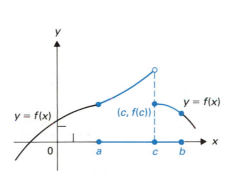

FIGURE 2.43 $f(x)$ attains no maximum value over $[a, b]$.

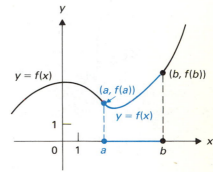

FIGURE 2.44 $f(x)$ attains no maximum value over half-open $[a, b)$.

SUMMARY

1. Let a be a point in the domain of a function f. The function f is continuous at a if $\lim_{x \to a} f(x)$ exists and is $f(a)$. If f is not continuous at a, then it is discontinuous at a. A continuous function is one that is continuous at every point in its domain.

2. We make no classification of a function being continuous or discontinuous at a point a unless a is in the domain of f.

3. Sums, products, and quotients of continuous functions are continuous functions. (Quotients are not defined where denominators are zero.) In particular, a rational function is continuous at every point in its domain.

4. Intermediate value theorem: A continuous function defined on a closed interval $[a, b]$ attains each value c between $f(a)$ and $f(b)$ at least once in the interval $[a, b]$.

5. A function continuous at each point of a closed interval in its domain attains a maximum value and also a minimum value on that interval.

EXERCISES 2.4

In Exercises 1–5, draw the graph of a function f, defined for all x, that satisfies the condition. (Many answers are possible.)

1. Continuous at all points but 2, and $\lim_{x \to 2} f(x) = 3$

2. Continuous at all points but 2, and $\lim_{x \to 2} f(x)$ undefined

3. Not continuous at -1 with $\lim_{x \to -1} f(x) = 1$, but continuous elsewhere with $\lim_{x \to 1} f(x) = 2$

4. Discontinuous at all integer values, but continuous elsewhere

5. Discontinuous at all positive integer values, but continuous elsewhere with $\lim_{x \to 0} f(x) = -\frac{3}{4}$

6. Is the function f defined by

$$f(x) = \begin{cases} \dfrac{x^2 - 9}{x - 3} & \text{for } x \neq 3, \\ 6 & \text{for } x = 3 \end{cases}$$

continuous? Why?

7. Is the function f defined by

$$f(x) = \begin{cases} \dfrac{4x^2 - 2x^3}{x - 2} & \text{for } x \neq 2, \\ 8 & \text{for } x = 2 \end{cases}$$

continuous? Why?

8. Is the function f defined by

$$f(x) = \begin{cases} x^2 & \text{for } x \geq 2, \\ 8 - 3x & \text{for } -1 \leq x < 2, \\ 12x + 1 & \text{for } x < -1 \end{cases}$$

(a) continuous at $a = 2$?
(b) continuous at $a = -1$?
(c) continuous? Why?

9. Is the function f defined by

$$f(x) = \begin{cases} 3x - x^2 & \text{for } x \geq 3, \\ x - 3 & \text{for } 1 < x < 3, \\ x^2 + 4x - 7 & \text{for } x \leq 1 \end{cases}$$

(a) continuous at $a = 3$?
(b) continuous at $a = 1$?
(c) continuous? Why?

10. Margaret was 20 in. long when she was born and grew to a height of 69 in. Use the intermediate value theorem to argue that at some time in Margaret's life, she was exactly 4 ft tall.

11. Give three more everyday applications of the intermediate value theorem, like the illustration in Exercise 10.

12. Use the intermediate value theorem to show that on August 4, at some point on the 37° meridian of the earth, there must be exactly 10 hours of daylight. We define daylight to mean that some portion of the sun is above the horizon. A meridian runs from the North Pole to the South Pole and is numbered according to the degrees of longitude, measured from the prime meridian of 0° through Greenwich, England.

13. Prove the following corollary of the intermediate value theorem.

COROLLARY Solution of $f(x) = 0$

Let f be continuous for all x in $[a, b]$ and suppose that $f(a)$ and $f(b)$ have opposite signs. Then the equation $f(x) = 0$ has at least one solution in $[a, b]$.

Exercises 14–22 work with the corollary of the intermediate value theorem stated in Exercise 13.

14. (a) Show that $x^3 - 5x^2 + 2x + 6 = 0$ has at least one solution in $[-1, 5]$.
    (b) Show that the equation in part (a) actually has three solutions in $[-1, 5]$.

15. (a) Show that $x^4 + 3x^3 + x + 4 = 0$ has at least one solution in $[-2, 0]$.
    (b) Does the equation in part (a) have any solution for $x \geq 0$? Explain.

16. Let $f(x) = x^2 - 4x + 3$.
    (a) Show that $f(0) > 0$ and $f(5) > 0$.
    (b) Show that $f(x) = 0$ has a solution in $[0, 5]$.
    (c) Why don't parts (a) and (b) constitute a contradiction of the corollary in Exercise 13?

17. Does $x^2 - 4x + 5 = 0$ have any solution in $[-4, 4]$? Why or why not?

18. Does $-x^2 + 6x - 9 = 0$ have any solution in $[0, 5]$? Why or why not?

19. Let $f(x)$ be continuous for all x. Mark each of the following true or false.
    _____ (a) If $f(a) = f(b)$, then $f(x) = 0$ has no solution in $[a, b]$.
    _____ (b) If $f(a)$ and $f(b)$ have opposite signs, then $f(x) = 0$ must have exactly one solution in $[a, b]$.
    _____ (c) If $f(a)$ and $f(b)$ have opposite signs, then $f(x) = 0$ has at least one solution in $[a, b]$.
    _____ (d) If $f(a)$ and $f(b)$ have the same sign, then $f(x) = 0$ must have either no solution in $[a, b]$ or an even number of solutions in $[a, b]$.
    _____ (e) If $f(a)$ and $f(b)$ have the same sign, then $f(x) = 0$ may have an infinite number of solutions in $[a, b]$.

20. Let f be a polynomial function of odd degree, so that $f(x) = a_n x^n + \cdots + a_1 x + a_0$, where $a_n \neq 0$ and n is an odd integer. Show that $f(x) = 0$ has at least one real solution. [*Hint:* Consider $\lim_{x \to \infty} f(x)$ and $\lim_{x \to -\infty} f(x)$, and then apply the corollary in Exercise 13 to a sufficiently large interval $[-C, C]$.]

21. A car on an oval racetrack passes the flag at the end of the third lap going exactly 96 mph. At the end of the fourth lap, the car is again going exactly 96 mph when passing the flag. Use the corollary in Exercise 13 to show that during the fourth lap, there were two diametrically opposite points of the oval where the car had equal speeds (not necessarily 96 mph, of course).

[*Hint:* Let S be the length of the oval track. For $0 \leq x \leq S/2$, let

$$f(x) = \text{Speed at } x - \text{Speed at } \left(x + \frac{S}{2}\right),$$

where x is the distance the car has traveled past the flag during the fourth lap. See Fig. 2.45.]

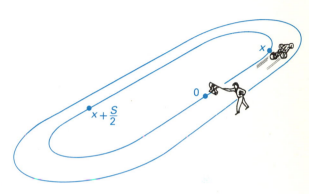

FIGURE 2.45

22. A square table with four legs of equal length teeters on diagonally opposite legs when placed on a warped floor. Using the corollary in Exercise 13, show that by rotating the table less than a quarter turn, you can make all four legs touch the floor, so that the table won't wobble anymore. [*Hint:* Number the legs 1, 2, 3, 4 in a counterclockwise order. Let $f(\theta)$ equal the sum of the distances from legs 1 and 3 to the floor minus the sum of the distances from legs 2 and 4 to the floor, when the table has been rotated counterclockwise through the angle θ degrees for $0 \leq \theta \leq 90$.]

23. Let $f(x) = a_n x^n + a_{n-1} x^{n-1} + \cdots + a_1 x + a_0$, where n is an even positive integer and $a_n > 0$.
    (a) Find $\lim_{x \to \infty} f(x)$ and $\lim_{x \to -\infty} f(x)$. (Remember how to find the limit at infinity of any rational function.)
    (b) Argue from part (a) that there exists $C > 0$ such that $f(x) > f(0)$ for $|x| > C$.
    (c) Apply Theorem 2.4 to $[-C, C]$ and show that $f(x)$ attains a minimum value over the entire x-axis.

24. Give the argument and conclusion analogous to Exercise 23 in the case that $a_n < 0$, assuming the other hypotheses remain the same.

 EXPLORING CALCULUS

O: zeros by bisection

2.5 PRECISE DEFINITIONS AND PROOFS (OPTIONAL)

This section shows how the intuitive definitions and arguments concerning limits and continuity given in the preceding sections can be made precise. We start by presenting some properties of the absolute value function. Then we give precise definitions of $\lim_{x \to a} f(x) = L$, of $\lim_{x \to a} f(x) = \infty$, and of $\lim_{x \to \infty} f(x) = L$. The exercises ask you to give similar definitions for other limit concepts that we have considered. We also show how to prove the properties of limits in Theorem 2.1 based on the definition of $\lim_{x \to a} f(x) = L$. Finally, we indicate how the intermediate value theorem can be proved.

Some Properties of the Absolute Value Function

The absolute value function is very useful in giving precise definitions of limits and in proving theorems involving limits. Observe that if $\delta > 0$, then $|x| < \delta$ if and only if $-\delta < x < \delta$. Replacing x by $x - a$, we see that $|x - a| < \delta$ if and only if $-\delta < x - a < \delta$, or, equivalently, $a - \delta < x < a + \delta$. Thus a convenient way of saying that the distance from x to a is less than a number $\delta > 0$ is to write $|x - a| < \delta$. If we also want to require that $x \neq a$, as we often do when working with limits, we can simply write $0 < |x - a| < \delta$.

Note that we always have $|a| = \sqrt{a^2}$, where we take the *positive* square root. In particular,

$$|ab| = \sqrt{(ab)^2} = \sqrt{a^2 b^2} = \sqrt{a^2}\sqrt{b^2} = |a||b|.$$

Since $x \leq |x|$ for all x, it follows that $ab \leq |a||b|$. Therefore,

$$|a + b| = \sqrt{(a + b)^2} = \sqrt{a^2 + 2ab + b^2}$$
$$\leq \sqrt{|a|^2 + 2|a||b| + |b|^2} = \sqrt{(|a| + |b|)^2}$$
$$= |a| + |b|.$$

We summarize these observations and some consequences of them that we will soon use. Let ε and δ be any positive numbers. Then

$$|f(x) - L| < \varepsilon \text{ if and only if } L - \varepsilon < f(x) < L + \varepsilon, \tag{1}$$
$$0 < |x - a| < \delta \text{ if and only if } a - \delta < x < a + \delta, \quad x \neq a, \tag{2}$$
$$|ab| = |a||b|, \qquad \textit{Multiplicativity} \tag{3}$$
$$|a + b| \leq |a| + |b|. \qquad \textit{Triangle inequality} \tag{4}$$

Definition of $\lim_{x \to a} f(x) = L$

Intuitively, $\lim_{x \to a} f(x) = L$ means that we can make $f(x)$ as *close to* L as we wish by taking any x sufficiently *close to* but different from a. This is a vague statement, for what does *close* mean? You may think that $f(x)$ is close to L if $L - 0.1 < f(x) < L + 0.1$, while a friend may say, "No, I want to have $L - 0.00001 < f(x) < L + 0.00001$." We can say that $\lim_{x \to a} f(x) = L$ only if *everyone* can be satisfied. So if someone demands to have $L - \varepsilon < f(x) < L + \varepsilon$ for some $\varepsilon > 0$, be it 0.1 or 0.00001, we must be sure that this will be true as long as x is within a certain distance, perhaps 0.05 or 0.003, of a but is not equal to a. That is, we must be able to find a $\delta > 0$ such that $L - \varepsilon < f(x) < L + \varepsilon$ will be true if $a - \delta < x < a + \delta$, but $x \neq a$. Figure 2.46 illustrates the choice

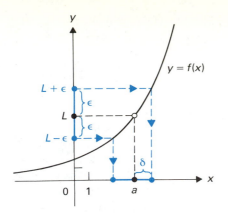

FIGURE 2.46 $\lim_{x \to a} f(x) = L$. Finding δ from a given ϵ.

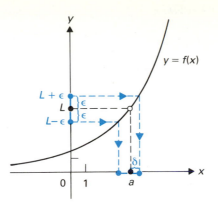

FIGURE 2.47   A smaller ϵ may require a smaller δ.

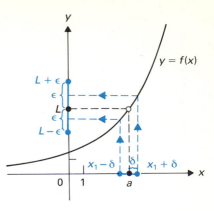

FIGURE 2.48   If one size for δ is satisfactory (see Fig. 2.46), then any smaller $\delta > 0$ is also satisfactory.

of δ from a given ϵ. We did not bother to define f at a in Fig. 2.46 because the value $f(a)$ plays no role in the discussion of $\lim_{x \to a} f(x) = L$. We mark $L - \epsilon$ and $L + \epsilon$ on the y-axis because this is where values $f(x)$ are plotted. For the function f in Fig. 2.46 we travel on horizontal lines from these points until we meet the graph and then down to the x-axis, as indicated by the arrows. As long as x is within the colored interval on the x-axis but is not a, we see that $f(x)$ lies between $L - \epsilon$ and $L + \epsilon$. We take as δ the *smaller* of the distances from the ends of this colored interval on the x-axis to a. Then the set of x-values where $a - \delta < x < a + \delta$, but $x \neq a$, surely lies within the interval we found on the x-axis starting from $L - \epsilon$ and $L + \epsilon$. Figure 2.47 illustrates that if the given ϵ is reduced in size, we expect that δ will be reduced in size also. Finally, Fig. 2.48 illustrates that for ϵ chosen as in Fig. 2.46, a still smaller δ than that in Fig. 2.46 will do. That is, we need to find only *some* $\delta > 0$ that will work, not necessarily the largest possible such δ.

With the preceding paragraph as motivation and using the equivalences (1) and (2) above, we give a mathematically precise definition of $\lim_{x \to a} f(x) = L$. When we say that the domain of f contains points *arbitrarily close to a but different from a*, we mean that for each $\delta > 0$ there is some point c in the domain of f such that $0 < |c - a| < \delta$.

DEFINITION 2.7

Limit of f at a

Let the domain of f contain points x arbitrarily close to a but different from a. Then $\lim_{x \to a} f(x) = L$ if for each $\epsilon > 0$ there exists $\delta > 0$ such that when $0 < |x - a| < \delta$ and x is in the domain of f, we have $|f(x) - L| < \epsilon$.

From the structure of this definition, we see that every demonstration that $\lim_{x \to a} f(x) = L$ must start "Let $\epsilon > 0$ be given." It is then up to us to demonstrate the existence of a number $\delta > 0$ such that $0 < |x - a| < \delta$ implies that $|f(x) - L| < \epsilon$ for x in the domain of f. As we mentioned before the definition, the size of the δ required usually depends on the size of the given ϵ. We illustrate with some simple examples, where we can use geometry as a guide.

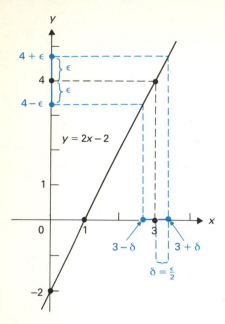

FIGURE 2.49 $\lim_{x \to 3}(2x - 2) = 4$. The graph has slope 2, so δ can be at most $\varepsilon/2$.

EXAMPLE 1   Use the ε, δ-characterization to demonstrate that

$$\lim_{x \to 3} (2x - 2) = 4.$$

This surely should be true; if x is close to 3, then $f(x) = 2x - 2$ is close to $6 - 2 = 4$.

Solution   Let $\varepsilon > 0$. We want to be sure that

$$4 - \varepsilon < f(x) < 4 + \varepsilon. \tag{5}$$

We mark $4 - \varepsilon$ and $4 + \varepsilon$ on the y-axis (see Fig. 2.49) because this is where $f(x)$ is plotted. Now we have to find $\delta > 0$ and mark $3 - \delta$ and $3 + \delta$ on the x-axis so that Eq. (5) is true if $3 - \delta < x < 3 + \delta$ but $x \neq 3$. Since the graph of $f(x) = 2x - 2$ is a line with slope 2, each unit of change on the x-axis produces two units of change of $f(x)$ on the y-axis. Thus a change of ε on the y-axis is produced by a change of only $\varepsilon/2$ on the x-axis, so we may let $\delta = \varepsilon/2$.

After these geometric considerations, we can give an algebraic argument that $\delta = \varepsilon/2$ works. Namely, if $0 < |x - 3| < \delta$, then we obtain in succession

$$3 - \delta < x < 3 + \delta, \qquad \textit{Using relation (1)}$$

$$3 - \frac{\varepsilon}{2} < x < 3 + \frac{\varepsilon}{2}, \qquad \delta = \varepsilon/2$$

$$6 - \varepsilon < 2x < 6 + \varepsilon, \qquad \textit{Multiplying by 2}$$

$$4 - \varepsilon < 2x - 2 < 4 + \varepsilon, \qquad \textit{Adding} -2$$

$$|(2x - 2) - 4| < \varepsilon. \qquad \textit{Using relation (1)}$$

Any smaller δ will work also.   ∎

A proof concerned with limits may contain a statement such as

"Let $\varepsilon > 0$ be given. Let $\delta = L^2\varepsilon/\sqrt{M}$."

Such a statement always means that the proof was devised starting with a given $\varepsilon > 0$ and computations were made to find a suitable δ for that ε, just as we did in the first paragraph of the solution of Example 1. Then the proof was written up by pulling this δ "out of the hat" and reversing the order of computations to show that it works, just as we did in the second paragraph of the solution. We will avoid this "out of the hat" technique. In the next example, we work to determine the necessary size of δ. Observe that the order of the computations we make is reversible.

EXAMPLE 2   Give an ε, δ-demonstration that $\lim_{x \to 1} (5 - 7x) = -2$.

Solution   Let $\varepsilon > 0$. We need

$$-2 - \varepsilon < 5 - 7x < -2 + \varepsilon. \tag{6}$$

Equation (6) is true if and only if

$$-7 - \varepsilon < -7x < -7 + \varepsilon, \qquad \text{or} \qquad 1 + \frac{\varepsilon}{7} > x > 1 - \frac{\varepsilon}{7}.$$

Thus we can take $\delta = \varepsilon/7$, because if x is within $\varepsilon/7$ of 1, then $5 - 7x$ is within ε of -2.   ∎

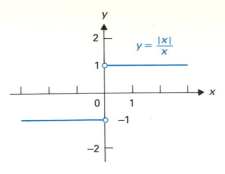

FIGURE 2.50 $\lim_{x\to 0} |x|/x$ does not exist. The graph jumps 2 units at $x = 0$.

EXAMPLE 3   Give an ε, δ-demonstration that $\lim_{x\to 0} |x|/x$ does not exist.

Solution   The graph of $|x|/x$ is shown in Fig. 2.50. For *every* $\delta > 0$,

$$\frac{|x|}{x} = 1 \quad \text{if } 0 < x < \delta, \qquad \text{and} \qquad \frac{|x|}{x} = -1 \quad \text{if } -\delta < x < 0.$$

The points 1 and -1 are two units apart, while for any possible limit L, the numbers $L - \varepsilon$ and $L + \varepsilon$ are 2ε units apart. Consequently, if $\varepsilon < 1$, it is impossible to have

$$L - \varepsilon < \frac{|x|}{x} < L + \varepsilon \qquad \text{for all } -\delta < x < \delta, \quad x \neq 0,$$

for *any* choice of L and $\delta > 0$. Thus given $\varepsilon = \frac{1}{2}$, there exists no $\delta > 0$ such that

$$L - \varepsilon < \frac{|x|}{x} < L + \varepsilon \qquad \text{for } -\delta < x < \delta, \quad x \neq 0,$$

no matter what L might be. This contradicts Definition 2.7, which asserts that for *each* $\varepsilon > 0$, in particular for $\varepsilon = \frac{1}{2}$, such $\delta > 0$ should exist. ∎

Proofs of the Limit Properties in Theorem 2.1

Theorem 2.1 gives the sum, product, quotient, and root properties for limits. Each property asserts that a limit exists and has a certain value. Thus each proof must start "Let $\varepsilon > 0$ be given" and proceed to demonstrate the existence of the required $\delta > 0$ in Definition 2.7. We find a suitable $\delta > 0$ using reversible steps, starting with what we want to show, as in Example 2. We now give a complete proof for the sum property and indicate the computations in reversible order for the other three properties, leaving complete written proofs of them to the exercises.

THEOREM 2.5   Limit of a sum of functions

Let $\lim_{x\to a} f(x) = L$, $\lim_{x\to a} g(x) = M$, and suppose that $f + g$ has points in its domain arbitrarily close to a but different from a. Then $\lim_{x\to a} (f + g)(x) = L + M$.

Proof: Let $\varepsilon > 0$ be given. We must find $\delta > 0$ such that $0 < |x - a| < \delta$ implies that

$$|(f + g)(x) - (L + M)| < \varepsilon \tag{7}$$

for x in the domain of $f + g$. Note that

$$|(f + g)(x) - (L + M)| = |f(x) + g(x) - L - M|$$
$$= |[f(x) - L] + [g(x) - M]|$$
$$\leq |f(x) - L| + |g(x) - M|. \qquad \textit{Using relation 4}$$

Thus we can achieve the required inequality (7) if we can determine $\delta > 0$ such that $0 < |x - a| < \delta$ implies both $|f(x) - L| < \varepsilon/2$ and $|g(x) - M| < \varepsilon/2$. Since $\lim_{x\to a} f(x) = L$, we can think of $\varepsilon/2$ as the "given epsilon" for this limit and find $\delta_1 > 0$ such that $0 < |x - a| < \delta_1$ implies that $|f(x) - L| < \varepsilon/2$. Also, we can find $\delta_2 > 0$ such that $0 < |x - a| < \delta_2$ implies that $|g(x) - M| < \varepsilon/2$.

Now we let δ be the *minimum* of the two values δ_1 and δ_2. Thus when $0 < |x - a| < \delta$, the conditions $0 < |x - a| < \delta_1$ and $0 < |x - a| < \delta_2$ hold *simultaneously*, so that *both* $|f(x) - L| < \varepsilon/2$ *and* $|g(x) - M| < \varepsilon/2$ are true. Reversing the order of our computations, we can see that

$$|(f + g)(x) - (L + M)| < \varepsilon/2 + \varepsilon/2 = \varepsilon.$$

This completes our proof.   □

Proofs of the product, quotient, and root properties have the same general structure as the proof for the sum property. Namely, compute the expression that needs to be made less than a given $\varepsilon > 0$. Then find numbers $\delta_i > 0$ such that $0 < |x - a| < \delta_i$ will make parts of the expression small enough so that if all the conditions hold simultaneously, the entire expression will be less than ε. Finally, select $\delta > 0$ to be the minimum of all the numbers δ_i.

To demonstrate the quotient property of limits, we need one other result, which we now prove.

THEOREM 2.6   Bounding away from 0

Let $\lim_{x \to a} f(x) = L \neq 0$. Then there exists $\delta > 0$ such that $0 < |x - a| < \delta$ and x in the domain of f implies that $|f(x)| > |L|/2$.

Proof: We need only take $\varepsilon = |L|/2$ and use a $\delta > 0$ whose existence is assured by Definition 2.7. Thus when $0 < |x - a| < \delta$, we have $|f(x) - L| < |L|/2$, so that

$$L - \frac{|L|}{2} < f(x) < L + \frac{|L|}{2}.$$

If $L > 0$, then $|L|/2 = L - |L|/2 < f(x)$, showing that $|f(x)| = f(x) > |L|/2$. If $L < 0$, then $f(x) < L + |L|/2 = L/2$, showing that

$$|f(x)| = -f(x) > \frac{-L}{2} = \frac{|L|}{2}.$$ □

We now proceed to show the computations necessary to prove the product, quotient, and root properties. We use the hypotheses in Theorem 2.1 concerning the domains of f and of g and the limits L of $f(x)$ at a and M of $g(x)$ at a. For the limit of the quotient $f(x)/g(x)$, we assume $M \neq 0$. For the root property, we assume $L > 0$.

Product Property: We have to make $|(fg)(x) - LM| < \varepsilon$. Note that

$$
\begin{aligned}
|(fg)(x) - LM| &= |f(x)g(x) - LM| \\
&= |f(x)g(x) - Lg(x) + Lg(x) - LM| \\
&= |g(x)[f(x) - L] + L[g(x) - M]| \\
&\leq |g(x)[f(x) - L]| + |L[g(x) - M]| \qquad \textit{Using relation (4)} \\
&= |g(x)||f(x) - L| + |L||g(x) - M|. \qquad \textit{Using relation (3)}
\end{aligned}
$$

We choose $\delta_i > 0$ such that for $0 < |x - a| < \delta_i$ and x in the domain of $f + g$ we have these consequences for $i = 1, 2,$ and 3:

$$\delta_1: \quad \text{makes } |g(x)| < |M| + 1, \qquad \delta_2: \quad \text{makes } |f(x) - L| < \frac{\varepsilon}{2(|M| + 1)},$$

$$\delta_3: \quad \text{makes } |g(x) - M| < \frac{\varepsilon}{2(|L| + 1)}.$$

In case $L = 0$

We then take δ to be the minimum of δ_1, δ_2, and δ_3.

Quotient Property: We need only show that $\lim_{x \to a} 1/g(x) = 1/M$ for $M \neq 0$ and then apply the product property to conclude that $\lim_{x \to a} f(x)/g(x) = \lim_{x \to a} [f(x)(1/g(x))] = L/M$. Thus we need to make $|1/g(x) - 1/M| < \varepsilon$. Note that

$$\left| \frac{1}{g(x)} - \frac{1}{M} \right| = \left| \frac{g(x) - M}{Mg(x)} \right| = \frac{1}{|M|} \cdot \frac{1}{|g(x)|} \cdot |g(x) - M|.$$

We choose $\delta_i > 0$ such that for $0 < |x - a| < \delta_i$ and x in the domain of g we have these consequences for $i = 1$ and 2:

$$\delta_1: \quad \text{makes } |g(x)| > |M|/2 \qquad \textit{(See Theorem 2.6)}$$

$$\delta_2: \quad \text{makes } |g(x) - M| < \frac{M^2 \varepsilon}{2}.$$

We then take δ to be the minimum of δ_1 and δ_2.

Root Property: We need to make $|\sqrt{f(x)} - \sqrt{L}| < \varepsilon$ if $L > 0$. Note that

$$|\sqrt{f(x)} - \sqrt{L}| = \left| \frac{\sqrt{f(x)} - \sqrt{L}}{1} \cdot \frac{\sqrt{f(x)} + \sqrt{L}}{\sqrt{f(x)} + \sqrt{L}} \right|$$

$$= \frac{1}{|\sqrt{f(x)} + \sqrt{L}|} \cdot |f(x) - L|.$$

We choose $\delta_i > 0$ such that for $0 < |x - a| < \delta_i$ and x in the domain of f we have these consequences for $i = 1$ and 2:

$$\delta_1: \quad \text{makes } f(x) > L/2 \qquad \textit{(See Theorem 2.6)}$$

$$\delta_2: \quad \text{makes } |f(x) - L| < \frac{(\sqrt{2} + 1)\sqrt{L}}{\sqrt{2}} \varepsilon.$$

We then take δ to be the minimum of δ_1 and δ_2.

Definitions of $\lim_{x \to a} f(x) = \infty$ and of $\lim_{x \to \infty} f(x) = L$

Intuitively, $\lim_{x \to a} f(x) = \infty$ if $f(x)$ can be made as large as we wish, that is, larger than any given number γ, provided that x is sufficiently close to a. An attempt to make this mathematically precise, analogous to the discussion preceding Definition 2.7, leads us to Definition 2.8.

DEFINITION 2.8

$\lim_{x \to a} f(x) = \infty$

Let the domain of f contain numbers arbitrarily close to a but different from a. Then $\lim_{x \to a} f(x) = \infty$ if for each real number γ there exists $\delta > 0$ such that when $0 < |x - a| < \delta$ and x is in the domain of f, we have $f(x) > \gamma$.

Intuitively, $\lim_{x \to \infty} f(x) = L$ means that $f(x)$ can be made as close to L as we wish by taking any x in the domain of f that is sufficiently large. An attempt to make this precise, analogous to the discussion preceding Definition 2.7, leads us to Definition 2.9. We say that the domain of f contains arbitrarily large numbers if for any number K we can always find numbers in the domain of f that are greater than K.

DEFINITION 2.9

$\lim_{x \to \infty} f(x) = L$

Let the domain of f contain arbitrarily large numbers. Then $\lim_{x \to \infty} f(x) = L$ if for each $\varepsilon > 0$ there exists γ such that when $x > \gamma$ and x is in the domain of f, we have $|f(x) - L| < \varepsilon$.

The exercises ask you to give more definitions of limits, such as $\lim_{x \to -\infty} f(x) = \infty$, using Definitions 2.7–2.9 as models.

Continuous Functions

Theorem 2.2 on the sums, products, and quotients of continuous functions follows at once from the corresponding properties for limits, since f is continuous at a point a in its domain if and only if $\lim_{x \to a} f(x) = f(a)$.

Proof of the intermediate value theorem depends on the following property of real numbers.

Dedekind Cut Principle: Let the closed interval $[a, b]$ be partitioned into two sets, A and B, such that each number in $[a, b]$ is in precisely one of these sets and such that each number in A is less than every number in B, as illustrated in Fig. 2.51. Then there is a number μ, the *cut number*, which is either the largest number in A or the smallest number in B.

We can think of constructing the decimal expansion of the cut number μ as follows. Let n_0 be the largest integer that is less than every number in B. Of the ten numbers

$$n_0, \ n_0 + 0.1, \ n_0 + 0.2, \ \ldots, \ n_0 + 0.9,$$

let n_1 be the largest one that is less than every number in B. Then of the ten numbers

$$n_1, \ n_1 + 0.01, \ n_1 + 0.02, \ \ldots, \ n_1 + 0.09,$$

let n_2 be the largest one that is less than every number in B. Continuing this process, we construct $n_3, n_4, \ldots$ to develop the decimal expansion of a number

FIGURE 2.51   A Dedekind cut of $[a, b]$.

μ. The construction shows that $\mu \leq \beta$ for every number β in B and that $\mu \geq \alpha$ for each number α in A. Since μ must be in one of these sets, it is either the largest element of A or the smallest element of B.

Using the Dedekind cut principle, we can prove the intermediate value theorem.

THEOREM 2.7   Intermediate value theorem

Let f be continuous on a closed interval $[a, b]$ and let c be a number between $f(a)$ and $f(b)$. Then there exists some number x_0 in $[a, b]$ such that $f(x_0) = c$.

Proof: We assume that $f(a) < f(b)$. (A similar argument can be made when $f(b) < f(a)$.) Let A be the set of all α in $[a, b]$ such that $f(x) \leq c$ for all x such that $a \leq x \leq \alpha$. (See Fig. 2.52.) Let B be the rest of $[a, b]$. If β is in B, then there is some $x \leq \beta$ in $[a, b]$ such that $f(x) > c$; otherwise, β would be in A. Each α in A must therefore be less than β. Thus we have a Dedekind cut partition of $[a, b]$. Let μ be the cut point of this partition. Now by continuity, $\lim_{x \to \mu} f(x) = f(\mu)$. If $f(\mu) < c$, we can find $\delta > 0$ such that $|f(x) - f(\mu)| < (c - f(\mu))/2$ for $|x - \mu| < \delta$. Thus

$$f(x) < f(\mu) + \frac{c - f(\mu)}{2} = \frac{f(\mu) + c}{2}$$

$$< \frac{c + c}{2} = c$$

for $|x - \mu| < \delta$, so $f(x) < c$ for all x in $[a, b]$ such that $x < \mu + (\delta/2)$. (See Fig. 2.53.) Thus $\mu + (\delta/2)$ is in A, contradicting the fact that μ is either the largest element of A or the smallest element of B. By a similar argument, if $f(x) > c$, then $\mu - (\delta/2)$ is in B for some $\delta > 0$, which is again a contradiction. Thus we must have $f(\mu) = c$. Taking $x_0 = \mu$, we have proved our theorem.
□

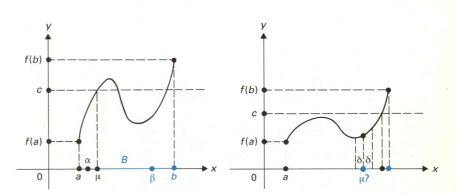

FIGURE 2.52   The Dedekind cut.   **FIGURE 2.53** $\mu + (\delta/2)$ is still in A.

A proof of Theorem 2.4 on attainment of maximum and minimum values for a continuous function in $[a, b]$ uses more theory than we wish to develop in this first course in calculus.

SUMMARY

See Definitions 2.7–2.9 for mathematically precise definitions of limits and Theorems 2.5 and 2.6 for illustrations of proofs using these definitions.

EXERCISES 2.5

In Exercises 1–6, if $\varepsilon > 0$ is given, find in terms of ε what size δ may be used in the ε, δ-characterization of a limit to establish the limit.

1. $\lim_{x \to a} x = a$

2. $\lim_{x \to a} c = c$, where c in $\lim_{x \to a} c$ is the constant function f defined by $f(x) = c$ for all x

3. $\lim_{x \to 4} 2x = 8$ 4. $\lim_{x \to -2} (-3x) = 6$

5. $\lim_{x \to -3} (14 - 5x) = 29$ 6. $\lim_{x \to -4} (2 - \frac{1}{3}x) = \frac{10}{3}$

7. Students often have trouble understanding the ε, δ-definition of a limit. We are convinced that this difficulty is rooted in logic. The definition uses both the *universal quantifier* ("for each") and the *existential quantifier* ("there exists"): *for each $\varepsilon > 0$, there exists $\delta > 0$.* . . . This exercise deals with this logical problem.
   (a) Negate the statement

    > For each $\varepsilon > 0$, there exists $\delta > 0$.

    That is, write a statement synonymous with "It is not true that for each $\varepsilon > 0$, there exists $\delta > 0$," without just saying, "It is not true that. . . ."
   (b) Negate the statement

    > For each apple blossom, there exists an apple.

   (c) Study your answer to part (a) in light of your answer to part (b). Decide whether your answer to (a) is correct.
   (d) Describe what one must show to demonstrate that $\lim_{x \to a} f(x) \neq c$.

8. Let f be a function. Classify each of the following "definitions" of the limit of f at a as either correct or incorrect. If it is incorrect, modify it to become correct.
   (a) The limit of f at a is c if for each number $\varepsilon > 0$ there exists $\delta > 0$ such that $|f(x) - c| < \varepsilon$, provided that $|x - a| < \delta$.
   (b) The limit of f at a is c if for some number $\varepsilon > 0$ there exists $\delta > 0$ such that $|f(x) - c| < \varepsilon$, provided that $0 < |x - a| < \delta$.
   (c) The limit of f at a is c if for each number $\varepsilon > 0$ there exists $\delta > 0$ such that $0 < |x - a| < \delta$ implies that $|f(x) - c| < \varepsilon$.
   (d) The limit of f at a is c if there exist some positive numbers ε and δ such that $0 < |x - a| < \delta$ implies that $|f(x) - c| < \varepsilon$.
   (e) The limit of f at a is c for $\delta > 0$, $0 < |x - a| < \delta$ implies $|f(x) - c| < \varepsilon$.
   (f) The limit of f at a is c if $|f(x) - c|$ can be made smaller than any preassigned $\varepsilon > 0$ by restricting x to elements different from a in some small interval $[a - \delta, a + \delta]$.
   (g) The limit of f at a is c if for each positive integer n there exists $\delta > 0$ such that

$$|f(x) - c| < \frac{1}{n},$$

provided that $0 < |x - a| < \delta$.

In Exercises 9–17, give a precise definition of the indicated limit.

9. $\lim_{x \to a+} f(x) = L$ 10. $\lim_{x \to a-} f(x) = L$

11. $\lim_{x \to a+} f(x) = -\infty$ 12. $\lim_{x \to \infty} f(x) = \infty$

13. $\lim_{x \to a-} f(x) = \infty$ 14. $\lim_{x \to a} f(x) = -\infty$

15. $\lim_{x \to -\infty} f(x) = L$ 16. $\lim_{x \to -\infty} f(x) = \infty$

17. $\lim_{x \to \infty} f(x) = -\infty$

18. Give an ε, δ-definition of continuity of a function f at a point a in its domain. Do not use the word *limit*.

19. Classify each of the following "definitions" as correct or incorrect. If it is incorrect, change it to be correct.
    (a) A function f is continuous at a if for each $\varepsilon > 0$ there exists $\delta > 0$ such that $|x - a| < \delta$ implies that $|f(x) - a| < \varepsilon$.
    (b) A function f is continuous at a if for each $\varepsilon > 0$ there exists $\delta > 0$ such that $|f(x) - c| < \varepsilon$, provided that $0 < |x - a| < \delta$.
    (c) A function f is continuous at a if for each $\varepsilon > 0$ there exists $\delta > 0$ such that
    $$|f(x) - f(a)| < \varepsilon,$$
    provided that $|x - a| < \delta$.
    (d) Let f be a function. The limit of f at a is $-\infty$ if for some γ there exists $\delta > 0$ such that $f(x) < \gamma$, provided that $0 < |x - a| < \delta$.
    (e) Let f be a function. The limit of f as x approaches a from the left is $-\infty$ if for each γ there exists $\delta > 0$ such that $f(x) < \gamma$, provided that $a - \delta < x \le a$.
    (f) A function f is continuous if for each a in its domain and each positive integer n there exists a positive integer m such that $|f(x) - f(a)| < 1/n$, provided that $|x - a| < 1/m$.

20. Give an ε, δ-proof that the function $f(x) = x$ is continuous.
21. Give an ε, δ-proof that a constant function $f(x) = c$ is continuous.
22. Give a detailed step-by-step proof that the function $f(x) = 2x^3 - x^2 + 4x - 7$ is continuous. Use Exercises 20 and 21 and the limit properties in Theorem 2.1, rather than epsilons and deltas.
23. Give a detailed statement and proof of the product property for limits stated in Theorem 2.1, $\lim_{x \to a} (fg)(x) = LM$.
24. Give a detailed statement and proof of the quotient property for limits stated in Theorem 2.1, $\lim_{x \to a} (f/g)(x) = L/M$.
25. Give a detailed statement and proof of the root property for limits stated in Theorem 2.1, $\lim_{x \to a} \sqrt{f(x)} = \sqrt{L}$.

> **EXPLORING CALCULUS**
>
> **H:** epsilon-delta demo
> **N:** calculus topics quiz (topic 5)

SUPPLEMENTARY EXERCISES FOR CHAPTER 2

1. Let $f(x) = 1/x$. Estimate the slope m_{tan} of the graph where $x = 1$ by finding the slope m_{sec} through the points where $x = 1$ and $x = 1 + \frac{1}{2}$.

2. Find an expression for m_{tan} at any point a if
$$f(x) = \frac{1}{(2x + 1)}.$$

3. Find m_{tan} at any point a if $f(x) = \sqrt{2x - 3}$.
4. Find m_{tan} for $f(x) = \sqrt{x} + x$ at any point $a > 0$.
5. Compute m_{tan} at $a = 2$, and then find the equation of the tangent line to the graph of $f(x) = x^2 - 3x$ at the point $(2, -2)$.

In Exercises 6–33, find the limit, using the notations ∞ and $-\infty$ where appropriate.

6. $\displaystyle\lim_{x \to 1} \frac{x^2 - 1}{x - 1}$

7. $\displaystyle\lim_{x \to -2} \frac{x + 1}{(x + 2)^2}$

8. $\displaystyle\lim_{x \to 4} \frac{x^2 - 3x - 4}{x^2 - 16}$

9. $\displaystyle\lim_{x \to -1} \frac{x^2 + 2x + 1}{x^2 - 2x}$

10. $\displaystyle\lim_{x \to 0} \left| \frac{x^2 - 3}{x} \right|$

11. $\displaystyle\lim_{x \to 2} \frac{x^2 - 3x + 2}{x^2 - 4}$

12. $\displaystyle\lim_{x \to 3} \frac{x^3 - 27}{x^2 + 9}$

13. $\displaystyle\lim_{x \to 2} \frac{x^2 - 4}{|x - 2|}$

14. $\displaystyle\lim_{x \to 2} \frac{x^2 - 4}{|x + 2|}$

15. $\displaystyle\lim_{x \to -3} \frac{x^2 + 3x}{x^2 + 2x - 3}$

16. $\displaystyle\lim_{x \to 1} \frac{x^2 - 3x + 2}{x + 1}$

17. $\displaystyle\lim_{x \to -2} \frac{x^3 + 2x^2}{x^2 - x - 6}$

18. $\displaystyle\lim_{x \to \infty} \frac{x^4 - 3x^2}{2x - 3x^4}$

19. $\displaystyle\lim_{x \to -\infty} \frac{x^4 + 100x^2}{14 - x}$

20. $\displaystyle\lim_{x \to 0} \frac{\sin x}{2x}$

21. $\displaystyle\lim_{x \to 5^-} \frac{|x - 5|}{x^2 - 25}$

22. $\displaystyle\lim_{x \to \infty} \frac{7 - 5x^2}{x^3 + 3x}$

23. $\displaystyle\lim_{x \to -\infty} \frac{14x^3 - 7x^2}{8x^3 + 4x}$

24. $\displaystyle\lim_{x \to 0} \frac{\sin x}{\cos x^2}$

25. $\displaystyle\lim_{x \to -1^+} \frac{x^2 + x}{x^2 + 2x + 1}$

26. $\displaystyle\lim_{x \to \infty} \frac{7 - 3x^2}{4x^2 + 3x - 2}$

27. $\displaystyle\lim_{x \to -\infty} \frac{8x + 4x^4}{3x^3 - 7}$

28. $\displaystyle\lim_{x \to \infty} x \sin(2/x)$

29. $\displaystyle\lim_{x \to 3^+} \frac{x^2 + x - 14}{x^2 - 8x + 15}$

30. $\displaystyle\lim_{x \to -2^-} \frac{4 + x^2}{4 - x^2}$

31. $\displaystyle\lim_{x \to \infty} \frac{4x^{2/3} - 7x}{x^{1/2} + x^{5/4} - 4}$

32. $\displaystyle\lim_{x \to 0} \frac{\sin(x - 2)}{x^2 - 4}$

33. $\displaystyle\lim_{x \to \infty} x \sin(1/x^2)$

34. Let

$$f(x) = \begin{cases} \dfrac{x^2 - 9}{x + 3} & \text{if } x \neq -3, \\ 6 & \text{if } x = -3. \end{cases}$$

Is f a continuous function? Why or why not?

35. State the intermediate value theorem.

36. Let

$$f(x) = \begin{cases} \dfrac{x^2 - 4x - 5}{x - 5} & \text{if } x > 5, \\ 2x - 4 & \text{if } x \leq 5. \end{cases}$$

(a) Find $\lim_{x \to 5-} f(x)$.
(b) Find $\lim_{x \to 5+} f(x)$.
(c) Is f continuous at $x = 5$? Why?
(d) Is f a continuous function? Why?

37. Show that $x^4 - 5x + 1 = 0$ has a solution in $[0, 1]$.

38. Is $f(x) = 1/x$ a continuous function? Why or why not?

39. State the theorem on the attainment of maximum and minimum values.

40. Let

$$f(x) = \begin{cases} \dfrac{|x| + x}{x + 1} & \text{for } x \neq -1, \\ 1 & \text{for } x = -1. \end{cases}$$

Is f continuous at $a = -1$? Why or why not?

41. Does $f(x) = x^2$ attain
(a) a maximum value on half-open $(0, 1]$? Why?
(b) a minimum value on half-open $(0, 1]$? Why?

42. Give an example of a function f discontinuous at $a = 1/n$ for every positive integer n but continuous at every other real number. [*Hint:* Be sure f is continuous at $a = 0$.]

43. Give an example of two functions defined for all x, neither of which is continuous at $x = 2$ but whose sum is continuous at $x = 2$.

44. Repeat Exercise 43 but for the *product* of the functions rather than the sum.

45. Give an example of a function defined for all real numbers x but not continuous at any point.

THE DERIVATIVE

3

This chapter marks the real beginning of our study of calculus and the use of calculus terminology and notation. The importance of calculus lies in its usefulness in studying *dynamic* situations, where quantities are changing, as opposed to *static* situations, where quantities remain constant.

In Section 3.1, we introduce the notions of average and instantaneous rates of change and some notation and terminology used in differential calculus. Much of the chapter is devoted to computation of derivatives. We develop formulas that often permit the computation of a derivative without actually working with limits. (However, the formulas themselves are derived by computing limits.)

3.1 THE DERIVATIVE

A Velocity Model

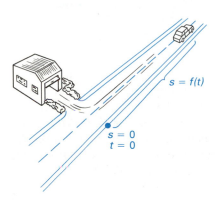

FIGURE 3.1 Car has traveled a distance $s = f(t)$ at time t.

Suppose a car traveling in one direction on a straight road goes a distance $s = f(t)$ at time t. See Fig. 3.1. We are not concerned now with the units for s and t; they might be yards and seconds, for example. If the car moves with a *constant velocity* of 2, then $s = 2t$ for $t \geq 0$. Consequently, the graph of this motion is the straight line of slope 2 shown in Fig. 3.2. If velocity remains constant with value m, then the total distance s traveled in elapsed time t is $s = mt$, which has as graph a line of slope m through the origin.

Now suppose that the graph of the distance function $s = f(t)$ is not a straight line. How could we find the velocity at a particular time $t = a$? We might first estimate the velocity at time $t = a$ by finding the average velocity over a very short time interval starting at time a. When we are driving a car, our velocity at any instant is close to our average velocity over the next hundredth of a second or

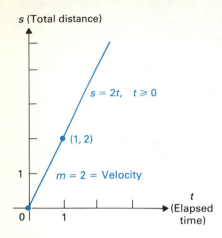

FIGURE 3.2 The graph of distance s against time t is a line when the velocity is constant.

our average velocity over the previous thousandth of a second. We find average velocity using the formula

$$\text{Average velocity} = \frac{\text{Change in distance}}{\text{Corresponding change in time}}. \quad (1)$$

EXAMPLE 1 The distance traveled by a car going north on a road is $s = f(t) = t^2$ at time $t \geq 0$. Estimate the velocity at time $t = 1$ by finding the average velocity from time $t = 1$ to time $t = 1.01$.

Solution At time $t = 1$, we have $s = 1^2 = 1$. At $t = 1.01$, $s = (1.01)^2 = 1.0201$. The change in distance is $1.0201 - 1 = 0.0201$. From formula (1),

$$\text{Average velocity} = \frac{0.0201}{0.01} = 2.01 \text{ velocity units.}$$

We expect the velocity at time $t = 1$ to be close to 2.01. ■

Figure 3.3 shows the graph of $s = f(t) = t^2$ for $t \geq 0$. Since the graph is not a straight line, the velocity does not remain constant; the increasing steepness of the graph as t increases indicates that the velocity increases as time increases. We are unable to illustrate the times $t = 1$ and $t = 1.01$ from Example 1 in Fig. 3.3 because we cannot accurately depict points so close together on the graph. Instead, we show a point where $t = a$ and a point where $t = a + \Delta t$. We should think of Δt as being small, in spite of its size in Fig. 3.3. We have from formula (1)

$$\text{Average velocity} = \frac{\Delta s}{\Delta t}.$$

But $\Delta s / \Delta t$ is the slope m_{sec} of the secant line through the points $(a, f(a))$ and $(a + \Delta t, f(a + \Delta t))$ on the graph.

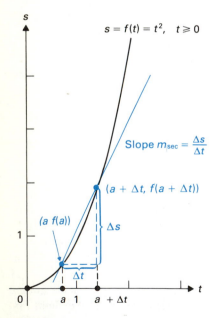

FIGURE 3.3 Average velocity = $\Delta s / \Delta t$, which is the slope m_{sec} of the secant line.

> ## Average Velocity from Time $t = a$ to Time $t = a + \Delta t$
>
> $$m_{\text{sec}} = \frac{f(a + \Delta t) - f(a)}{\Delta t}. \quad (2)$$

We are interested in the exact velocity at the instant $t = a$. Example 1 estimated it where $a = 1$ by finding the average velocity from 1 to $1 + \Delta t$, where $\Delta t = 0.01$. The smaller the size of Δt, the better we expect the approximation of the exact velocity to be. Example 2 illustrates that we can think of Δt as being negative as well as positive. Rather than finding the average velocity over the next 0.01 second, we find the average velocity over the previous 0.001 second, corresponding to taking $a = 1$ and $\Delta t = -0.001$ in formula (2).

EXAMPLE 2 Repeat Example 1 for the distance function $s = f(t) = t^2$ when $t = 1$, but find the average velocity corresponding to $\Delta t = -0.001$.

Solution Using formula (2), we have

$$\text{Average velocity} = \frac{f(1 - 0.001) - f(1)}{-0.001} = \frac{(0.999)^2 - 1^2}{-0.001}$$

$$= \frac{0.998001 - 1}{-0.001} = \frac{-0.001999}{-0.001} = 1.999.$$

We expect the average velocity 1.999 to be closer to the exact velocity at $t = 1$ than the average velocity 2.01 corresponding to $\Delta t = 0.01$, which we found in Example 1. ∎

If we find average velocities from a to $a + \Delta t$ for smaller and smaller Δt, perhaps running through values 0.01, 0.001, 0.0001, and so on, we should get closer and closer to the exact velocity at time a. Thus we have

Instantaneous Velocity at Time $t = a$

$$\lim_{\Delta t \to 0} \frac{\Delta s}{\Delta t} = \lim_{\Delta t \to 0} \frac{f(a + \Delta t) - f(a)}{\Delta t}. \qquad (3)$$

From Section 2.1, we see that this instantaneous velocity can be interpreted geometrically as the slope m_{tan} of the line tangent to the graph of $s = f(t)$ where $t = a$.

EXAMPLE 3 If distance s is given in terms of time t by $s = f(t) = t^2$, find the instantaneous velocity when $t = 1$.

Solution We first find the expression for average velocity $\Delta s/\Delta t$ in formula (2), taking $a = 1$. For $\Delta t \neq 0$, we have

$$\frac{\Delta s}{\Delta t} = \frac{f(1 + \Delta t) - f(1)}{\Delta t} = \frac{(1 + \Delta t)^2 - 1^2}{\Delta t}$$

$$= \frac{1 + 2(\Delta t) + (\Delta t)^2 - 1}{\Delta t} = \frac{2(\Delta t) + (\Delta t)^2}{\Delta t}$$

$$= \frac{\Delta t(2 + \Delta t)}{\Delta t} = 2 + \Delta t.$$

From formula (3),

$$\text{Instantaneous velocity} = \lim_{\Delta t \to 0} \frac{\Delta s}{\Delta t} = \lim_{\Delta t \to 0} (2 + \Delta t) = 2.$$

That is, as Δt becomes closer and closer to zero, the expression $2 + \Delta t$ becomes closer and closer to 2. Looking back at Examples 1 and 2, we see that the estimate 1.999 in Example 2, with $\Delta t = -0.001$, is indeed a better estimate of the exact velocity 2 than the estimate 2.01 in Example 1, with $\Delta t = 0.01$. ∎

Average and Instantaneous Rates of Change of $f(x)$

With the preceding velocity model as a guide, we now define average and instantaneous rates of change for a function. As usual, when describing a general mathematical formulation, we use x,y-notation. We assume that $y = f(x)$ and that the graph of f has a tangent line where $x = a$.

The **average rate of change** of y with respect to x from $x = a$ to $x = a + \Delta x$, where $\Delta x \neq 0$, is

$$m_{\text{sec}} = \frac{\Delta y}{\Delta x} = \frac{f(a + \Delta x) - f(a)}{\Delta x}. \tag{4}$$

The **instantaneous rate of change** of y with respect to x at $x = a$ is

$$m_{\text{tan}} = \lim_{\Delta x \to 0} \frac{\Delta y}{\Delta x} = \lim_{\Delta x \to 0} \frac{f(a + \Delta x) - f(a)}{\Delta x}. \tag{5}$$

These are very important formulas that you should remember. In Fig. 3.4, we show secant lines approaching the tangent line at a point where $x = a$ for a general function $y = f(x)$.

The average rate of change of $f(x)$ from a to b is also called the average rate of change of $f(x)$ *over the interval* $[a, b]$. From Eq. (4) we see that

$$\text{Average rate of change over } [a, b] = \frac{f(b) - f(a)}{b - a}. \tag{6}$$

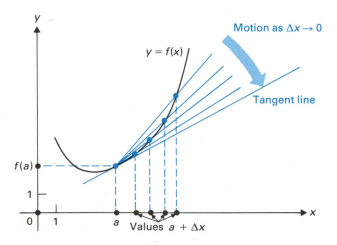

FIGURE 3.4   Secant lines approaching the tangent line where $x = a$.

EXAMPLE 4 Find the average rate of change of $f(x) = 1/(x - 1)$ over the interval $[2, 5]$.

Solution Using Eq. (6), we obtain

$$\frac{f(5) - f(2)}{5 - 2} = \frac{\frac{1}{4} - 1}{3} = \frac{-\frac{3}{4}}{3} = -\frac{1}{4}.$$ ∎

The Derivative of a Function

The derivative is the basic notion of differential calculus. Geometrically, the derivative of a function f at a point a in its domain is the slope m_{tan} of the tangent line to the graph of $y = f(x)$, if the tangent line exists and is not vertical. The preceding discussion shows that this slope is equal to the instantaneous rate of change of y with respect to x where $x = a$. We give two formal definitions for the concepts we have presented in this section.

DEFINITION 3.1

Difference quotient

The **difference quotient** of f from a to $a + \Delta x$ is

$$\frac{\Delta y}{\Delta x} = \frac{f(a + \Delta x) - f(a)}{\Delta x}. \tag{7}$$

It is the **average rate of change** of $f(x)$ with respect to x from a to $a + \Delta x$.

DEFINITION 3.2

Derivative; differentiable function

Let $f(x)$ be defined for $a - h < x < a + h$ for some $h > 0$. The **derivative** of f at a is

$$f'(a) = \lim_{\Delta x \to 0} \frac{f(a + \Delta x) - f(a)}{\Delta x}, \tag{8}$$

if this limit exists. It is the **instantaneous rate of change** of $f(x)$ with respect to x at a. If $f'(a)$ exists, then f is **differentiable** at a. A **differentiable function** is one that is differentiable at every point a in its domain.

The function f' in the notation $f'(a)$ is the *derived function,* and $f'(x)$ is the derivative of f at any point x where the derivative exists. Another notation for f' is Df. The notation y' is also used for the derivative; that is, when $y = f(x)$, we often write $y' = f'(x)$. Still another notation is

$$\frac{dy}{dx}.$$

At first glance, the notation dy/dx, due to the seventeenth-century mathematician Leibniz, seems to be a cumbersome way to denote a derivative, but it turns out to be very handy for remembering some formulas. It should be read "the derivative of y with respect to x." At the moment, regard dy/dx as a single symbol, not as a quotient. (A quotient interpretation will appear in Section 3.9.) Remember that $f'(a)$ is the slope $m_{\tan}$ of the line tangent to the graph $y = f(x)$ at $x = a$. It also has the interpretation of the instantaneous rate at which y is increasing compared to x at a. The notation dy/dx is suggestive of this rate of change of y with respect to x. The most important applications of differential calculus center on this rate-of-change interpretation suggested by dy/dx.

The computation of $f'(a)$ from Eq. (8) is exactly the same as the computation of $m_{\tan}$ given in Section 2.1. Here are two more illustrations, computing

$$Df(x) = y' = \frac{dy}{dx} = f'(x) = \lim_{\Delta x \to 0} \frac{f(x + \Delta x) - f(x)}{\Delta x}$$

at any point x.

EXAMPLE 5   If $y = f(x) = 1/(x - 1)$, find y'.

Solution

$$
\begin{aligned}
y' = f'(x) &= \lim_{\Delta x \to 0} \frac{f(x + \Delta x) - f(x)}{\Delta x} = \lim_{\Delta x \to 0} \frac{\dfrac{1}{x + \Delta x - 1} - \dfrac{1}{x - 1}}{\Delta x} \\[2ex]
&= \lim_{\Delta x \to 0} \frac{\dfrac{x - 1 - (x + \Delta x - 1)}{(x + \Delta x - 1)(x - 1)}}{\Delta x} \\[2ex]
&= \lim_{\Delta x \to 0} \frac{-\Delta x}{\Delta x (x + \Delta x - 1)(x - 1)} \\[2ex]
&= \lim_{\Delta x \to 0} \frac{-1}{(x + \Delta x - 1)(x - 1)} = \frac{-1}{(x - 1)^2}.
\end{aligned}
$$

■

EXAMPLE 6   If $g(x) = \sqrt{x + 3}$, find $g'(x)$.

Solution

$$
\begin{aligned}
g'(x) &= \lim_{\Delta x \to 0} \frac{g(x + \Delta x) - g(x)}{\Delta x} = \lim_{\Delta x \to 0} \frac{\sqrt{x + \Delta x + 3} - \sqrt{x + 3}}{\Delta x} \\[2ex]
&= \lim_{\Delta x \to 0} \frac{\sqrt{x + \Delta x + 3} - \sqrt{x + 3}}{\Delta x} \cdot \frac{\sqrt{x + \Delta x + 3} + \sqrt{x + 3}}{\sqrt{x + \Delta x + 3} + \sqrt{x + 3}} \\[2ex]
&= \lim_{\Delta x \to 0} \frac{(x + \Delta x + 3) - (x + 3)}{\Delta x (\sqrt{x + \Delta x + 3} + \sqrt{x + 3})} \\[2ex]
&= \lim_{\Delta x \to 0} \frac{\Delta x}{\Delta x (\sqrt{x + \Delta x + 3} + \sqrt{x + 3})} \\[2ex]
&= \lim_{\Delta x \to 0} \frac{1}{\sqrt{x + \Delta x + 3} + \sqrt{x + 3}} = \frac{1}{2\sqrt{x + 3}}.
\end{aligned}
$$

■

EXAMPLE 7   Example 4 shows that the average rate of change of $f(x) = 1/(x - 1)$ over the interval [2, 5] is $-\frac{1}{4}$. Using Example 5, compare the result of Example 4 with the instantaneous rate of change of $f(x)$ at the midpoint of the interval.

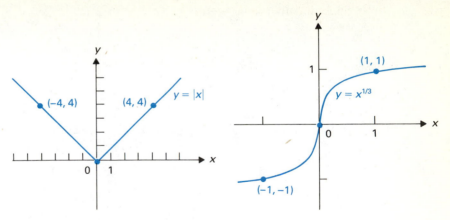

FIGURE 3.5 $|x|$ does not have a derivative at $x = 0$, where there is a sharp point on the graph.

FIGURE 3.6 $x^{1/3}$ does not have a derivative at $x = 0$, where the graph has a vertical tangent (the y-axis).

Solution   The instantaneous rate of change of $f(x)$ at the midpoint $\frac{7}{2}$ of the interval $[2, 5]$ is equal to the derivative $f'(\frac{7}{2})$ of f where $x = \frac{7}{2}$. Example 5 shows that $f'(x) = -1/(1 - x)^2$, so

$$f'\left(\frac{7}{2}\right) = \frac{-1}{(1 - \frac{7}{2})^2} = \frac{-1}{(-\frac{5}{2})^2} = \frac{-1}{\frac{25}{4}} = -\frac{4}{25}.$$

Example 4 shows that the average rate of change of the function over $[2, 5]$ is $-\frac{1}{4}$. The absolute difference between this average rate of change and the instantaneous rate of change at the midpoint is $|-0.16 - (-0.25)| = 0.09$. ■

EXAMPLE 8   Let $f(x) = |x|$. Show that $f'(0)$ does not exist.

Solution   The graph of f is shown in Fig. 3.5. Note that at $x = 0$,

$$\frac{\Delta y}{\Delta x} = \frac{|\Delta x|}{\Delta x} = \begin{cases} 1 & \text{if } \Delta x > 0, \\ -1 & \text{if } \Delta x < 0. \end{cases}$$

Thus $f'(0) = \lim_{\Delta x \to 0} (\Delta y / \Delta x)$ does not exist, because the right-hand limit is 1 while the left-hand limit is -1. This function $|x|$ is perhaps the simplest example of a function that is not differentiable at some point. Note that the graph, shown in Fig. 3.5, has a sharp point where $x = 0$ and no (unique) tangent line there. ■

EXAMPLE 9   Argue graphically that $f'(0)$ does not exist for the function $f(x) = x^{1/3}$.

Solution   The graph of $f(x) = x^{1/3}$ is shown in Fig. 3.6. The tangent line to the graph at the origin is vertical; that is, it is the y-axis. Since a vertical line has no slope, we see that $f'(0)$ does not exist for this function. We will see later that $f'(x) = 1/(3x^{2/3})$, and this expression is undefined when $x = 0$. ■

EXAMPLE 10   A graph $y = f(x)$ is shown for $-4 \le x \le 4$ in Fig. 3.7(a), using equal scales on the x- and y-axes. Sketch approximately the graph $y' = f'(x)$ over this interval.

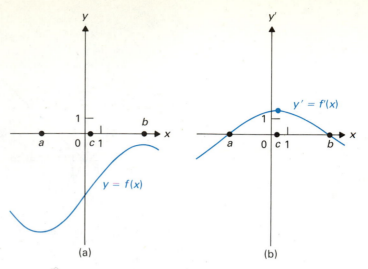

FIGURE 3.7 (a) Graph of f; (b) graph of f'.

Solution Since $f'(x)$ is the slope of the tangent line to the graph at $(x, f(x))$, we can plot some points on the graph $y' = f'(x)$ by estimating slopes of tangent lines at a few points on the graph of $y = f(x)$. In Fig. 3.7(a), the tangent lines to the graph at $x = a$ and $x = b$ are horizontal, so they have slope 0. Therefore, we mark the points $(a, 0)$ and $(b, 0)$ in Fig. 3.7(b) to be on the graph of $y' = f'(x)$. To get an idea how high and how low the graph $y' = f'(x)$ goes, we hunt for points where the graph $y = f(x)$ is steepest. The graph in Fig. 3.7(a) seems to have maximum steepness where $x = c$. We estimate the slope of this graph where $x = c$ to be $\frac{3}{2}$, so we plot the point $(c, \frac{3}{2})$ on the graph $y' = f'(x)$ in Fig. 3.7(b). Continuing in this fashion, we estimate the graph $y' = f'(x)$ to appear approximately as drawn in Fig. 3.7(b). ■

As shown in Example 8, the function $|x|$ is not differentiable at zero, but it is continuous there. *Thus continuity does not imply differentiability.* However, differentiability does imply continuity, as we proceed to show.

First note that

$$f'(a) = \lim_{x \to a} \frac{f(x) - f(a)}{x - a}, \tag{9}$$

if this limit exists. To see this we need only replace Δx by $x - a$ in the defining equation

$$f'(a) = \lim_{\Delta x \to 0} \frac{f(a + \Delta x) - f(a)}{\Delta x}$$

that we have given for the derivative. When $\Delta x = x - a$, the assertion $\Delta x \to 0$ is equivalent to $x \to a$.

THEOREM 3.1   Differentiable ⇒ continuous

If f is differentiable at $x = a$, then f is continuous at $x = a$.

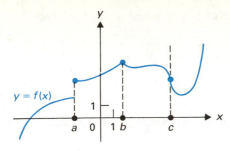

FIGURE 3.8 $f'(x)$ does not exist at a, which is a point of discontinuity; at b, where the graph has a sharp point; and at c, where there is a vertical tangent.

Proof: Let f be differentiable at a, so that $f'(a)$ exists. To show that f is continuous at a, we need only show that $\lim_{x \to a} f(x) = f(a)$. For $x \neq a$, we have

$$f(x) - f(a) = \frac{f(x) - f(a)}{x - a} \cdot (x - a). \tag{10}$$

From Eq. (10) and the product property in Theorem 2.1, we see that

$$\lim_{x \to a} [f(x) - f(a)] = \lim_{x \to a} \left[\frac{f(x) - f(a)}{x - a} \cdot (x - a) \right]$$

$$= \left[\lim_{x \to a} \frac{f(x) - f(a)}{x - a} \right] [\lim_{x \to a} (x - a)]$$

$$= f'(a) \cdot 0 = 0.$$

It follows that $\lim_{x \to a} f(x) = f(a)$, so f is continuous at a. □

In summary, functions are not differentiable at discontinuities or where their graphs have sharp points or vertical tangents. (Other features can also cause a function to fail to be differentiable.) Figure 3.8 shows this graphically.

Estimation of $f'(a)$ Using a Calculator

At the end of Section 2.2, we saw that m_{sym} is apt to be a better approximation than m_{sec} to m_{tan}. Since $m_{\text{tan}} = f'(a)$ by Eqs. (5) and (8), we see that for small Δx, the approximation

$$f'(a) \approx m_{\text{sym}} = \frac{f(a + \Delta x) - f(a - \Delta x)}{2 \cdot \Delta x} \tag{11}$$

(we read $\approx$ as "approximately equals") can be expected to be better than that given by the difference quotient Eq. (7) for the same value of Δx. The approximation Eq. (11) can be used with a computer or calculator to estimate the derivative of a function at any particular point. Of course, unless we know that f is differentiable at a, we should compute Eq. (11) with three or four different small values of Δx.

EXAMPLE 11 Use a calculator to attempt to estimate $f'(1)$ if $f(x) = 2^x$.

Solution Using approximation (11) and computing values of m_{sym} for four small values of Δx, we obtain the results shown in Table 3.1. It appears that $f'(1) \approx 1.38629$. ■

TABLE 3.1

Δx	$m_{\text{sym}} = \dfrac{2^{1+\Delta x} - 2^{1-\Delta x}}{2(\Delta x)}$
0.05	1.386571898
0.01	1.386305462
0.002	1.386294806
0.001	1.386294472

SUMMARY

1. The average rate of change of $f(x)$ over an interval $[a, b]$ in the domain of f is

$$\frac{f(b) - f(a)}{b - a}.$$

Let $y = f(x)$ be defined for all x between $a - h$ and $a + h$ for some $h > 0$.

2. The difference quotient of f from a to $a + \Delta x$ is

$$\frac{\Delta y}{\Delta x} = \frac{f(a + \Delta x) - f(a)}{\Delta x}.$$

It is the average rate of change of $y = f(x)$ with respect to x from a to $a + \Delta x$.

3. The derivative of f at a is

$$f'(a) = \lim_{\Delta x \to 0} \frac{f(a + \Delta x) - f(a)}{\Delta x},$$

if this limit exists. It is the instantaneous rate of change of $y = f(x)$ with respect to x at a.

4. A function f is differentiable if $f'(a)$ exists at every point a in the domain of f. Other notations for the derived function f' when $y = f(x)$ are Df, y', and dy/dx.

5. If $f'(a)$ exists, then f is continuous at a.

6. (Calculator) If $f'(a)$ exists, then

$$f'(a) \approx \frac{f(a + \Delta x) - f(a - \Delta x)}{2 \cdot \Delta x}$$

for small Δx.

EXERCISES 3.1

In Exercises 1–12, find the derivative of the given function at any point in its domain where the derivative exists.

1. $f(x) = x^2 - 3x$
2. $g(t) = 4t^2 + 7$
3. $f(u) = u^3 + u$
4. $h(s) = 2s^3 - 3s$
5. $f(x) = \dfrac{1}{2x + 3}$
6. $g(t) = \dfrac{t}{t + 1}$
7. $h(t) = t^2 + \dfrac{1}{t}$
8. $f(x) = \sqrt{x}$
9. $h(s) = \sqrt{2s - 1}$
10. $f(u) = \dfrac{1}{\sqrt{u}}$
11. $f(x) = |x|$
12. $g(t) = t + |t - 3|$

In Exercises 13–18, use the answers in the back of the text for Exercises 1, 3, 5, 7, 9, and 11, respectively, to find (a) the average rate of change of the function over the given interval, (b) the instantaneous rate of change of the function at the midpoint of the interval, and (c) the absolute value of the difference between the rates of change in parts (a) and (b).

13. $f(x) = x^2 - 3x$ on [0, 2]
14. $f(u) = u^3 + u$ on [0, 2]
15. $f(x) = \dfrac{1}{2x + 3}$ on [0, 1]
16. $h(t) = t^2 + \dfrac{1}{t}$ on [−2, −1]
17. $h(s) = \sqrt{2s - 1}$ on [1, 5]
18. $f(x) = |x|$ on [−2, 3]

In Exercises 19–26, sketch the graph of the derivative of the function having the graph in the indicated figure, as in Example 10 of the text.

19.

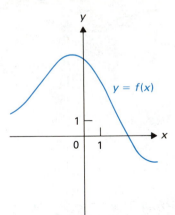

FIGURE 3.9

20.

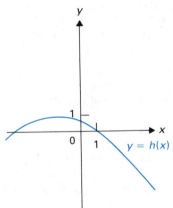

FIGURE 3.10

21.

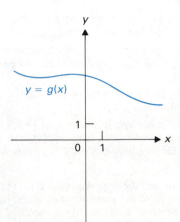

FIGURE 3.11

22.

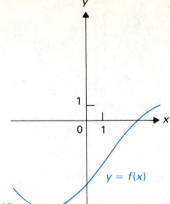

FIGURE 3.12

23.

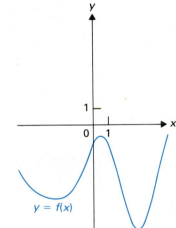

FIGURE 3.13

24.

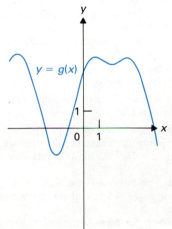

FIGURE 3.14

25.

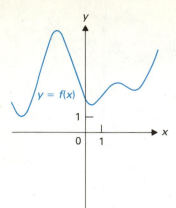

FIGURE 3.15

26.

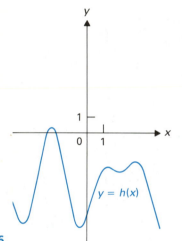

FIGURE 3.16

27. Suppose an object travels so that after t hours it has gone $s = f(t) = 3t^2 + 2t$ miles, for $t \geq 0$.
   (a) Find the average velocity of the object during the 2-hour time interval from $t = 3$ to $t = 5$.
   (b) Find the average velocity of the object during the 1-hour time interval from $t = 3$ to $t = 4$.
   (c) Find the average velocity of the object during the half-hour time interval from $t = 3$ to $t = \frac{7}{2}$.
   (d) On the basis of parts (a)–(c), guess the actual velocity of the object at time $t = 3$.

28. Referring to Exercise 27, find, from the expression for m_{tan} as a limit, the exact velocity at time $t = 3$.

29. To break open a clam, a seagull drops it from 32 ft above the ground. Neglecting air resistance, the distance it has fallen after t sec is $s = 16t^2$ ft, until it hits the ground. See Fig. 3.17.
   (a) For what time interval is the formula $s = 16t^2$ valid?
   (b) Find the *average velocity* of the clam during the time interval in part (a).
   (c) Find a formula for the instantaneous velocity of the clam at any time t in the interval found in part (a).

(d) At what time t is the instantaneous velocity of the clam equal to its average velocity over the time interval in part (a)?

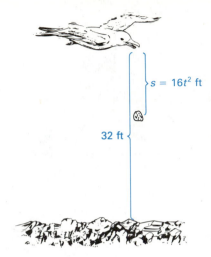

FIGURE 3.17

30. A gas is confined at constant temperature in a container of increasing volume. The pressure of the gas on the walls of the container at time t sec is given by

$$p = 24/(t + 2) \text{ g/cm}^2.$$

   (a) Find the average rate of change of pressure from time $t = 0$ to time $t = 2$ sec.
   (b) Find the instantaneous rate of change of pressure when $t = 0$.

31. Two positive electrical charges start 5 cm apart when $t = 0$ sec, and the distance between them is increased at a constant rate of 1 cm/sec. The force of repulsion between them is given by

$$F = 10/(t + 5)^2 \text{ dynes.}$$

   (a) Find the average rate of change of F from $t = 0$ to $t = 5$ sec.
   (b) Find the instantaneous rate of change of F when $t = 5$ sec.

32. If possible, give an example of a continuous function $f(x)$ such that $f'(5)$ does not exist but $f'(x)$ exists for all $x \neq 5$.

33. If possible, give an example of a continuous function $g(t)$ such that $g'(2)$ and $g'(-4)$ do not exist but $g'(t)$ exists for all $t \neq 2, -4$.

34. If possible, give an example of a differentiable function that is not continuous.

35. By sketching a graph in the interval $[-4.5, 4.5]$, indicate that there is a continuous function $f(x)$ such that $f'(n)$ does not exist for any integer n but $f'(x)$ exists for all x not an integer.

▦ Use a calculator and the approximation Eq. (11) in the text to find the derivative of the given function at the indicated point. Use radian measure with all trigonometric functions. You choose Δx.

36. $\sin 2x$ at $a = 0$

37. $[(x + 7)/(x^2 + 5)]^{1/3}$ at $a = 2.374$

38. x^x at $a = 2.36$

39. $\sin(\tan x)$ at $a = -1.3$

40. $(\sin x)^{\cos x}$ at $a = \pi/4$

41. $(x^2 - 3x)^{\sqrt{x}}$ at $a = 4$

42. Show that if $f'(a)$ exists, then

$$\lim_{\Delta x \to 0} \frac{f(a + \Delta x) - f(a - \Delta x)}{2 \cdot \Delta x} = f'(a).$$

[*Hint:* Use the fact that

$$\frac{f(a + \Delta x) - f(a - \Delta x)}{2 \cdot \Delta x}$$

$$= \frac{1}{2} \cdot \frac{f(a + \Delta x) - f(a)}{\Delta x} + \frac{1}{2} \cdot \frac{f(a + (-\Delta x)) - f(a)}{-\Delta x}$$

and Theorem 2.1.]

43. It is worth noting that

$$\lim_{\Delta x \to 0} \frac{f(a + \Delta x) - f(a - \Delta x)}{2 \cdot \Delta x}$$

may exist while $f'(a)$ does not. Give an example of a function $f(x)$ and a point a where this is true. [*Hint:* Consider Example 8.]

EXPLORING CALCULUS

I: limit of m-secant

J: m-sec versus m-sym

K: compute $f'(a)$, $f''(a)$

L: graph f, f', and f''

M: calculus draw quiz (option 1)

3.2 DIFFERENTIATION OF POLYNOMIAL FUNCTIONS

This section presents the first few of many formulas that can be used to quickly compute the derivatives of many functions. The process of finding a derivative is called *differentiation*. Mastering the technique of differentiation is as important for calculus as mastering the arithmetic operations was for all the mathematics you learned before.

We start by showing that the derivative of a constant function is zero at any point. Let $f(x) = c$, a constant, for all x. Then for any a,

$$f'(a) = \lim_{\Delta x \to 0} \frac{f(a + \Delta x) - f(a)}{\Delta x} = \lim_{\Delta x \to 0} \frac{c - c}{\Delta x} = \lim_{\Delta x \to 0} \frac{0}{\Delta x} = 0.$$

In Leibniz notation,

$$\frac{d(c)}{dx} = 0. \tag{1}$$

Now suppose $f(x) = x^n$ for a positive integer n. This time the computation of $f'(a)$ uses the *binomial theorem* of algebra to expand $(a + \Delta x)^n$. The binomial theorem gives an expanded formula for $(a + b)^n$ in terms of the products,

$$a^n, \quad a^{n-1}b, \quad a^{n-2}b^2, \quad a^{n-3}b^3, \quad \ldots, \quad ab^{n-1}, \quad b^n.$$

Namely,

$$(a + b)^n = a^n + na^{n-1}b + \frac{n(n-1)}{2}a^{n-2}b^2$$

$$+ \frac{n(n-1)(n-2)}{3 \cdot 2}a^{n-3}b^3 + \cdots + b^n. \qquad \textit{Binomial theorem}$$

Applying this formula with $b = \Delta x$, we obtain

$$f'(a) = \lim_{\Delta x \to 0} \frac{(a + \Delta x)^n - a^n}{\Delta x}$$

$$= \lim_{\Delta x \to 0} \frac{[a^n + na^{n-1}\Delta x + (n(n-1)/2)a^{n-2}(\Delta x)^2 + \cdots + (\Delta x)^n] - a^n}{\Delta x}$$

$$= \lim_{\Delta x \to 0} [na^{n-1} + (n(n-1)/2)a^{n-2}\Delta x + \cdots + (\Delta x)^{n-1}] = na^{n-1}.$$

Since a can be any point, this computation shows that

$$\frac{d(x^n)}{dx} = nx^{n-1}. \tag{2}$$

Next suppose that $u = f(x)$ and $v = g(x)$, and consider the sum function $h(x) = u + v = f(x) + g(x)$. A change Δx in x produces changes

$$\Delta u = f(x + \Delta x) - f(x) \qquad \text{and} \qquad \Delta v = g(x + \Delta x) - g(x)$$

in u and v. The change produced in $h(x) = u + v$ is

$$h(x + \Delta x) - h(x) = [(u + \Delta u) + (v + \Delta v)] - (u + v)$$

$$= \Delta u + \Delta v.$$

Suppose now that $f'(a)$ and $g'(a)$ both exist. Working at a and using Theorem 2.1 for the limit of a sum, we find

$$\lim_{\Delta x \to 0} \frac{\text{Change in } (u + v)}{\Delta x} = \lim_{\Delta x \to 0} \frac{\Delta u + \Delta v}{\Delta x}$$

$$= \lim_{\Delta x \to 0} \frac{\Delta u}{\Delta x} + \lim_{\Delta x \to 0} \frac{\Delta v}{\Delta x} = \frac{du}{dx} + \frac{dv}{dx}.$$

That is,

$$\frac{d(u + v)}{dx} = \frac{du}{dx} + \frac{dv}{dx} \qquad \textit{Sum rule} \tag{3}$$

at any point where the derivatives of both u and v exist. In function notation, Eq. (3) becomes

$$(f + g)' = f' + g'. \tag{4}$$

For the final result before we differentiate any polynomial function, let $u = f(x)$ and consider the function $c \cdot f(x)$ for a constant c. A change of Δx in x produces a change of Δu in u and therefore a change of $c \cdot \Delta u$ in $c \cdot f(x)$. This time we make use of Theorem 2.1 for the limit of a product. At any point x where $f'(x)$ exists,

$$\lim_{\Delta x \to 0} \frac{\text{Change in } c \cdot f(x)}{\Delta x} = \lim_{\Delta x \to 0} \frac{c \cdot \Delta u}{\Delta x}$$

$$= \left(\lim_{\Delta x \to 0} c \right) \left(\lim_{\Delta x \to 0} \frac{\Delta u}{\Delta x} \right) = c \cdot \frac{du}{dx},$$

so

$$\frac{d(c \cdot u)}{dx} = c \cdot \frac{du}{dx}. \tag{5}$$

In function notation, Eq. (5) becomes

$$(c \cdot f)' = c \cdot f'. \qquad (6)$$

Equations (4) and (6) are very important because they hold for any differentiable functions $u = f(x)$ and $v = g(x)$. We recommend learning them in words, as they are given in the Summary, independent of particular letters such as u and v. They deserve to be stated as a theorem.

THEOREM 3.2 Linearity property

If the functions $u = f(x)$ and $v = g(x)$ are both differentiable at the point x, then so are $u + v = f(x) + g(x)$ and $c \cdot u = c \cdot f(x)$ for any constant c. Furthermore, $(f + g)' = f' + g'$ and $(c \cdot f)' = c \cdot f'$, that is,

$$\frac{d(u + v)}{dx} = \frac{du}{dx} + \frac{dv}{dx} \qquad \text{and} \qquad \frac{d(c \cdot u)}{dx} = c \cdot \frac{du}{dx}. \qquad \square$$

EXAMPLE 1   Find the derivative of the function $4x^3 - 7x^2$.

Solution   Using Eqs. (4), (6), and then (2), we obtain

$$\frac{d(4x^3 - 7x^2)}{dx} = \frac{d(4x^3)}{dx} + \frac{d(-7x^2)}{dx} = 4\frac{d(x^3)}{dx} + (-7)\frac{d(x^2)}{dx}$$

$$= 4 \cdot 3x^2 + (-7)(2x) = 12x^2 - 14x \qquad \text{for all } x. \quad \blacksquare$$

Example 1 can be generalized to more than two summands to give a very nice formula for the derivative of any polynomial function. Namely,

$$\frac{d(a_n x^n + \cdots + a_2 x^2 + a_1 x + a_0)}{dx} = n a_n x^{n-1} + \cdots + 2a_2 x + a_1. \qquad (7)$$

EXAMPLE 2   Find the derivative of $g(t) = 4t^3 - 17t^2 + 3t - 2$.

Solution   We find that $g'(t) = 12t^2 - 34t + 3$. $\quad \blacksquare$

EXAMPLE 3   Find dy/dx if $y = (2x + 3)^2$.

Solution   We have $y = (2x + 3)^2 = 4x^2 + 12x + 9$. Thus $dy/dx = 4(2x) + 12(1) + 0 = 8x + 12$. We give a different method of solving this problem in Section 3.5. $\quad \blacksquare$

EXAMPLE 4   Find u' if $u = (s^2 + 4s^3)/5$.

Solution   We write $u = \frac{1}{5}(s^2 + 4s^3)$. By Theorem 3.2, we have

$$u' = \frac{1}{5} \cdot \frac{d(s^2 + 4s^3)}{ds} = \frac{1}{5}(2s + 12s^2). \quad \blacksquare$$

EXAMPLE 5   Find the average rate of change of $f(x) = x^3 - 2x^2$ over the interval $[-1, 3]$ and the instantaneous rate of change of $f(x)$ where $x = 2$.

Solution   We have

$$\text{Average rate of change} = \frac{\Delta y}{\Delta x} = \frac{f(3) - f(-1)}{3 - (-1)}$$

$$= \frac{9 - (-3)}{4} = \frac{12}{4} = 3.$$

The instantaneous rate of change when $x = 2$ is given by $f'(2)$. Now $f'(x) = 3x^2 - 4x$, so $f'(2) = 12 - 8 = 4$.   ■

Applications

We illustrate applications of the derivative to finding the equations of tangent lines and instantaneous rates of change. Note how easily we can solve such problems after only a few lessons in calculus! The problems would have seemed formidable only one week ago.

EXAMPLE 6   Find the equation of the line tangent to the graph of $y = f(x) = 3x^4 - 2x^2 + 3x - 7$ where $x = 1$.

Solution

Point $(1, f(1)) = (1, -3)$

Slope $f'(1) = (12x^3 - 4x + 3)|_{x=1} = 12 - 4 + 3 = 11.$ (The notation $|_{x=1}$ means "evaluated at $x = 1$.")

Equation $y + 3 = 11(x - 1)$,   or $y = 11x - 14$ ■

EXAMPLE 7   Find all points on the curve $y = 2x^3 - 3x^2 - 12x + 20$ where the tangent line is parallel to the x-axis.

Solution   The x-axis is the line $y = 0$ of slope zero. Thus we want to find all points where the tangent line to the given curve has slope zero. This slope is given by the derivative y', and we find that

$$y' = 6x^2 - 6x - 12 = 6(x^2 - x - 2) = 6(x - 2)(x + 1).$$

Thus, $y' = 0$ where $x = -1$ or $x = 2$. Computing the y-coordinates for those values of x, we find that the desired points are $(-1, 27)$ and $(2, 0)$.   ■

A car moves in one direction in a straight line. We consider the car to be traveling in the direction of increasing x on the x-axis. Let x be the position of the car on the x-axis at time t. The derivative dx/dt gives the velocity of the car and appears on the speedometer of the car at each time. The next example gives a specific illustration.

EXAMPLE 8   Find the velocity at time $t = 3$ if the position x of a car on the x-axis at time $t \geq 0$ is given by $x = t^2 + 2t$.

Solution   The velocity of the car when $t = 3$ is

$$\text{Velocity} = \frac{dx}{dt}\bigg|_{t=3} = (2t + 2)|_{t=3} = 6 + 2$$

$$= 8 \text{ (Units distance)/(Unit time)}.   ■$$

We now present a very simple economic model. Suppose a company is manufacturing some product in an ideal situation where there is no competition. This gives the company reasonable control over its own destiny. We further assume that the cost of producing an item, the revenue received from its sale, and the profit made are functions of the number x of units of the product manufactured per unit of time (a month, a year, and so on). This is a major assumption. It means that advertising decisions, transportation to market, and the like are all functions of this number x of units produced.

In many economic situations, we are interested only in *integer* values of a variable x. After all, a construction company is not going to build 31.347 houses! While some of our functions, such as cost, may be defined for only integer values of x, we will assume that there is some differentiable function $C(x)$ defined throughout a whole interval and giving the cost for all integer values of x in the interval. With this in mind, let

$C(x)$ = cost of producing x units, per unit time,

$R(x)$ = revenue received if x units are produced,

$P(x) = R(x) - C(x)$ = profit when x units are produced.

Economists are interested in the *marginal cost, marginal revenue,* and *marginal profit* when x units are produced. In a low-level economics course where calculus is not used, the marginal cost at x is defined to be the cost of producing one additional unit in the time period, so that

$$\text{Marginal cost} = C(x + 1) - C(x) = \frac{C(x + 1) - C(x)}{1}. \qquad (8)$$

From Eq. (8), we see that this marginal cost at x is approximately $C'(x)$, because Eq. (8) gives the approximation

$$C'(x) \approx \frac{C(x + \Delta x) - C(x)}{\Delta x},$$

where $\Delta x = 1$. In a higher-level course, the adjective *marginal* generally signifies a derivative. Thus the **marginal cost** is dC/dx, the **marginal revenue** is dR/dx, and the **marginal profit** is dP/dx.

EXAMPLE 9 A company manufactures a popular calculator. Its annual cost function in dollars is

$$C(x) = 90,000 + 500x + 0.01x^2,$$

where x is in hundreds of calculators produced per year. The 90,000 represents the annual capital outlay, in dollars, for the plant, insurance, and such fixed expenses. The coefficient 500 might represent the cost, in dollars, exclusive of fixed expenses, of producing 100 calculators if not too many are produced. The term $0.01x^2$ comes into play only as x becomes fairly large and might represent problems caused by crowding, storing excessive inventory, and an increase in the cost of materials if many calculators are produced, so that material becomes scarce and its cost increases. Find the marginal cost when $x = 100$ units.

Solution The marginal cost is given by the derivative

$$\frac{dC}{dx} = 500 + 2(0.01)x = 500 + (0.02)x.$$

When $x = 100$, the marginal cost is

$$500 + (0.02)(100) = 500 + 2 = 502.$$

Thus when 100 units (10,000 calculators) have been manufactured, we estimate the cost of manufacturing the next unit (100 calculators) to be \$502. ■

SUMMARY

1. The derivative of a constant function is the zero function; in symbols,

$$\frac{d(c)}{dx} = 0.$$

2. The derivative of a sum is the sum of the derivatives; in symbols,

$$(f + g)' = f' + g', \quad \text{or} \quad \frac{d(u + v)}{dx} = \frac{du}{dx} + \frac{dv}{dx}. \quad \textit{Sum rule}$$

3. The derivative of a constant times a function is the constant times the derivative of the function; in symbols,

$$(c \cdot f)' = c \cdot f', \quad \text{or} \quad \frac{d(c \cdot u)}{dx} = c \cdot \frac{du}{dx}.$$

4. $\dfrac{d(x^n)}{dx} = nx^{n-1}$ for any positive integer n

EXERCISES 3.2

In Exercises 1–13, find the derivative of the given function.

1. $3x - 2$
2. $8t^3 - 7t^2 + 4$
3. $2u^7 + 4u^2 - 3$
4. $15s^3 - 4s^6 + 2s^2 + 5$
5. $\dfrac{x^2 - 3x + 4}{2}$
6. $\dfrac{v^3 - 3v^2 + 2}{4}$
7. $(3t)^4 - (2t)^5$
8. $(w^2 - 2)(w + 1)$
9. $(s^2 + 2s)^2$
10. $t(3t + 2)(3t - 2)$
11. $(2z)^2(3z + 5)$
12. $8x^3 - 3(x + 1)^2 + 2$
13. $\dfrac{v(v - 1)(v + 1)}{3}$

In Exercises 14–19, find (a) the average rate of change of the function over the interval and (b) the instantaneous rate of change at the midpoint of the interval.

14. $f(x) = x^2 - 3x$ on [2, 4]

15. $g(t) = 2t^3 - 3t^2$ on $[-1, 1]$
16. $f(x) = x^4 - 16x^2 + 2x - 1$ on [0, 4]
17. $f(x) = x(3x - 5)^2$ on $[-1, 3]$
18. $f(s) = s^2(s - 1)$ on $[-2, 2]$
19. $f(x) = x(x - 2)(2x - 1)$ on [0, 1]

In Exercises 20–25, find the equations of (a) the tangent line and (b) the normal line (perpendicular to the tangent line) to the given curve at the indicated point.

20. $y = x^2 - 3x + 2$, where $x = 1$
21. $y = 3x^2 - 2x + 1$, where $x = -2$
22. $y = x^4 - 3x^2 - 3x$, where $x = 2$
23. $y = 2x^3 - 3x$, where $x = 2$
24. $y = (2x + 4)^2$, where $x = -2$
25. $y = x^2(2x + 5)$, where $x = -1$

26. (a) Compute $d(1/x)/dx$, assuming that the formula for $d(x^n)/dx$ in Eq. (2) also holds if $n = -1$.

(b) Verify your answer in part (a) by computing the limit of the appropriate difference quotient.

(c) Find

$$\frac{d}{dx}\left(\frac{3}{x} - 2x\right).$$

(d) Find

$$\frac{d}{dx}\left(\frac{1}{4x} - \frac{3}{x} + \frac{2}{5x}\right).$$

27. (a) Compute $d(\sqrt{x})/dx$, assuming that the formula for $d(x^n)/dx$ in Eq. (2) also holds if $n = \frac{1}{2}$.

(b) Verify that your answer in part (a) is correct by computing the limit of the appropriate difference quotient.

(c) Find

$$\frac{d}{dx}(3\sqrt{x} - 2x^2).$$

(d) Find

$$\frac{d}{dx}(\sqrt{5x} - \sqrt{7x}).$$

28. A spaceship travels in one direction on a straight path. After t hours, it has gone $x = f(t) = 4t^3 + 3t^2 + t$ miles for $t \geq 0$. Find the velocity of the spaceship as a function of the time t for $t \geq 0$.

29. If the length of an edge of a cube increases at a rate of 1 in./sec, find the (instantaneous) rate of increase of the volume when (a) the edge is 2 in. long, (b) the edge is 5 in. long.

30. Repeat Exercise 29, assuming that the edge of the cube is increasing at a rate of 4 in./sec. (Use Exercise 29 and common sense.)

31. Suppose that when a pebble is dropped into a large tank of fluid, a wave travels outward in a circular ring whose radius increases at a constant rate of 8 in./sec. See Fig. 3.18.

(a) Find the area of the circular disk enclosed by the wave 2 sec after the time the pebble hits the fluid.

(b) Find the (instantaneous) rate at which the area of the circular disk enclosed by the wave is increasing 2 sec after the time the pebble hits the fluid. (See Exercises 29 and 30.)

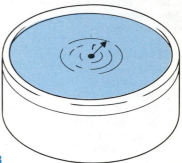

FIGURE 3.18

32. Referring to the company that manufactures calculators as described in Example 9, suppose the revenue function is

$$R(x) = 1000x - 0.05x^2.$$

Here $1000x$ appears because the first few calculators produced sell for $10 each. The term $-0.05x^2$ appears because if x is large there is a glut on the market, so the sales price falls. Find the marginal revenue after 100 units (10,000 calculators) have been manufactured.

33. Referring to Example 9 and Exercise 32, find the marginal profit after 100 units (10,000 calculators) have been manufactured.

3.3 DIFFERENTIATION OF PRODUCTS AND QUOTIENTS

The process of finding $f'(x)$ from $f(x)$ is known as *differentiation*. One nice feature of calculus is formulas that make differentiation feasible, at least for the functions used most often. We have already seen how quickly we can find the derivative of a polynomial function. This section gives formulas for finding the derivatives of a product $f(x) \cdot g(x)$ and a quotient $f(x)/g(x)$ in terms of the derivatives $f'(x)$ and $g'(x)$. The chain rule in Section 3.5 will complete our list of general differentiation formulas.

Let both f and g be differentiable functions, so that $f'(x)$ and $g'(x)$ exist. If we let $u = f(x)$ and $v = g(x)$, then a change Δx from x to $x + \Delta x$ produces changes Δu in u and Δv in v. Thus

$$f(x + \Delta x) = u + \Delta u \qquad \text{and} \qquad g(x + \Delta x) = v + \Delta v.$$

HISTORICAL NOTE

THE PRODUCT RULE first appears explicitly in a November 1675 manuscript of one of the co-inventors of the calculus, Gottfried Wilhelm Leibniz (1646–1716). After teaching himself Latin at an early age, Leibniz entered the University of Leipzig in 1661 and received a bachelor's degree in 1663, a master's degree in 1664, and a doctor of law degree in 1667.

One of Leibniz's lifelong goals was to develop an alphabet of human thought, a method of representing all fundamental concepts symbolically and, further, a way of combining such symbols to represent more complex thoughts. Though he never completed this project, his interest in developing symbols led him to the invention of the calculus notation used today.

Interestingly, his first attempt at the product rule was wrong. He conjectured that $d(xy)$ should be equal to $dx\,dy$, checked his conjecture on the example $x = cz + d$, $y = z^2 + bz$, made an error in his calculation, and concluded that his conjecture was correct. A few days later, however, he discovered his mistake and soon came up with the correct rule, which he wrote as $d(xy) = x\,dy + y\,dx$.

The change in $f(x) \cdot g(x)$ is therefore

$$f(x + \Delta x) \cdot g(x + \Delta x) - f(x) \cdot g(x) = (u + \Delta u)(v + \Delta v) - uv.$$

Therefore the difference quotient for $f(x) \cdot g(x)$ is

$$
\begin{aligned}
\frac{\text{Change in } u \cdot v}{\Delta x} &= \frac{(u + \Delta u) \cdot (v + \Delta v) - u \cdot v}{\Delta x} \\
&= \frac{uv + u \cdot \Delta v + v \cdot \Delta u + \Delta u \cdot \Delta v - uv}{\Delta x} \\
&= \frac{u \cdot \Delta v + v \cdot \Delta u + \Delta u \cdot \Delta v}{\Delta x} \\
&= u \cdot \frac{\Delta v}{\Delta x} + v \cdot \frac{\Delta u}{\Delta x} + \frac{\Delta u}{\Delta x} \cdot \Delta v.
\end{aligned}
\tag{1}
$$

We take the limit of Eq. (1) as $\Delta x \to 0$ to find the derivative of uv. Note that $v = g(x)$ is a continuous function since $g'(x)$ exists (Theorem 3.1). Thus $\lim_{\Delta x \to 0} \Delta v = 0$. Taking the limit, we have

$$
\begin{aligned}
\lim_{\Delta x \to 0} \frac{\text{Change in } u \cdot v}{\Delta x} &= \lim_{\Delta x \to 0} \left(u \cdot \frac{\Delta v}{\Delta x} + v \cdot \frac{\Delta u}{\Delta x} + \frac{\Delta u}{\Delta x} \cdot \Delta v \right) \\
&= u \cdot \frac{dv}{dx} + v \cdot \frac{du}{dx} + \frac{du}{dx} \cdot 0 = u \cdot \frac{dv}{dx} + v \cdot \frac{du}{dx}.
\end{aligned}
$$

This shows that

$$\frac{d(u \cdot v)}{dx} = u \cdot \frac{dv}{dx} + v \cdot \frac{du}{dx} \qquad \textit{Product rule} \tag{2}$$

at any point where du/dx and dv/dx both exist. In function notation, Eq. (2) becomes

$$(fg)' = f \cdot g' + g \cdot f'. \tag{3}$$

EXAMPLE 1 Find dy/dx in two ways if $y = (2x + 1)(3x - 2)$.

Solution

Method 1 From Eq. (2), we obtain

$$\frac{dy}{dx} = (2x + 1)\frac{d(3x - 2)}{dx} + (3x - 2)\frac{d(2x + 1)}{dx}$$

$$= (2x + 1)3 + (3x - 2)2 = 6x + 3 + 6x - 4 = 12x - 1.$$

Method 2 Note that if we multiply the factors $(2x + 1)(3x - 2)$ first, we obtain $y = 6x^2 - x - 2$, which again yields $dy/dx = 12x - 1$. ∎

EXAMPLE 2 Find s' if $s = f(t) = (t^2 + t)(t^3 - 7t^2 + 3t)$.

Solution Using Eq. (2) with $u = t^2 + t$ and $v = t^3 - 7t^2 + 3t$, we find that

$$s' = (t^2 + t)\frac{d(t^3 - 7t^2 + 3t)}{dt} + (t^3 - 7t^2 + 3t)\frac{d(t^2 + t)}{dt}$$

$$= (t^2 + t)(3t^2 - 14t + 3) + (t^3 - 7t^2 + 3t)(2t + 1).$$

This can be simplified algebraically, but such simplification is not necessary if we just want the derivative at one point. For example, we can find $s'(1) = f'(1)$ by substituting $t = 1$:

$$s'(1) = 2(-8) + (-3)3 = -16 - 9 = -25.$$ ∎

EXAMPLE 3 Suppose f is differentiable, with $f(3) = -2$ and $f'(3) = 5$. Find dy/dx where $x = 3$ if $y = (x^2 + 4x) \cdot f(x)$.

Solution We have

$$\frac{dy}{dx} = (x^2 + 4x) \cdot f'(x) + (2x + 4) \cdot f(x).$$

Thus

$$f'(3) = \frac{dy}{dx}\bigg|_{x=3} = 21 \cdot f'(3) + 10 \cdot f(3) = 21 \cdot 5 + 10(-2) = 85.$$ ∎

Now we turn to the differentiation of $f(x)/g(x)$ at $x = a$, under the assumptions that $f'(a)$ and $g'(a)$ exist and $g(a) \neq 0$. Again, let $u = f(x)$ and $v = g(x)$ have changes Δu and Δv produced by a change Δx in x at $x = a$. Then $u/v = f(x)/g(x)$ and

$$\frac{\text{Change in } u/v}{\Delta x} = \frac{\dfrac{u + \Delta u}{v + \Delta v} - \dfrac{u}{v}}{\Delta x}$$

$$= \frac{\dfrac{v(u + \Delta u) - u(v + \Delta v)}{v(v + \Delta v)}}{\Delta x}$$

$$= \frac{\dfrac{v \cdot \Delta u - u \cdot \Delta v}{v(v + \Delta v)}}{\Delta x} = \frac{v \cdot \dfrac{\Delta u}{\Delta x} - u \cdot \dfrac{\Delta v}{\Delta x}}{v(v + \Delta v)}.$$

We take the limit as $\Delta x \to 0$ to find the derivative of u/v. Note that $\lim_{\Delta x \to 0}(v + \Delta v) = v$, because $v = g(x)$ is continuous at $x = a$, since $g'(a)$ exists (Theorem 3.1). Taking the limit, we have

$$\lim_{\Delta x \to 0} \frac{\text{Change in } u/v}{\Delta x} = \lim_{\Delta x \to 0} \frac{v \cdot (\Delta u/\Delta x) - u \cdot (\Delta v/\Delta x)}{v(v + \Delta v)}$$

$$= \frac{v \cdot (du/dx) - u \cdot (dv/dx)}{v^2}.$$

This shows that

$$\frac{d(u/v)}{dx} = \frac{v \cdot (du/dx) - u \cdot (dv/dx)}{v^2} \qquad \textit{Quotient rule} \qquad \textbf{(4)}$$

at any point where du/dx and dv/dx exist and $v \neq 0$. In function notation, Eq. (4) becomes

$$\left(\frac{f}{g}\right)' = \frac{g \cdot f' - f \cdot g'}{g^2} \qquad\qquad \textbf{(5)}$$

EXAMPLE 4 Find y' if $y = (x^2 + 1)/(x^3 - 2x)$.

Solution Working with Eq. (4) and taking $u = x^2 + 1$ and $v = x^3 - 2x$, we find that

$$y' = \frac{(x^3 - 2x)[d(x^2 + 1)/dx] - (x^2 + 1)[d(x^3 - 2x)/dx]}{(x^3 - 2x)^2}$$

$$= \frac{(x^3 - 2x)(2x) - (x^2 + 1)(3x^2 - 2)}{(x^3 - 2x)^2} = \frac{-x^4 - 5x^2 + 2}{(x^3 - 2x)^2}.$$ ∎

EXAMPLE 5 Find y' if $y = (x^2 - 3x)/5$.

Solution If the denominator of a quotient is a constant, we can avoid using the quotient rule when differentiating. In this example, we may write

$$y = \tfrac{1}{5}(x^2 - 3x)$$

and obtain

$$y' = \tfrac{1}{5}(2x - 3).$$ ∎

EXAMPLE 6 Suppose $g(x)$ is differentiable with $g(1) = 4$ and $g'(1) = 2$. Find dy/dx at $x = 1$ if $y = (x^3 - 2x^2)/g(x)$.

Solution We have

$$\frac{dy}{dx} = \frac{g(x) \cdot (3x^2 - 4x) - (x^3 - 2x^2) \cdot g'(x)}{g(x)^2}.$$

Thus

$$\frac{dy}{dx}\Big|_{x=1} = \frac{g(1)(-1) - (-1)g'(1)}{g(1)^2} = \frac{4(-1) - (-1)(2)}{4^2}$$

$$= \frac{-2}{16} = -\frac{1}{8}.$$ ∎

We know that the formula

$$\frac{d(x^n)}{dx} = nx^{n-1} \tag{6}$$

holds if n is a positive integer. It also holds if n is a negative integer because then $-n$ is positive and, using Eqs. (4) and (6), we have

$$\frac{d(x^n)}{dx} = \frac{d(1/x^{-n})}{dx} = \frac{x^{-n} \cdot [d(1)/dx] - 1 \cdot [d(x^{-n})/dx]}{(x^{-n})^2}$$

$$= \frac{x^{-n} \cdot 0 - (-n)x^{-n-1}}{x^{-2n}}$$

$$= \frac{nx^{-n-1}}{x^{-2n}} = nx^{-n-1+2n} = nx^{n-1}.$$

Thus Eq. (6) holds for any integer n.

EXAMPLE 7 Compute s' if $s = 3/t^5$.

Solution We may compute s' by writing $s = 3 \cdot t^{-5}$ and differentiating this expression using Eq. (5). We obtain

$$s' = 3(-5)t^{-6} = \frac{-15}{t^6}.$$

This is easier than using the quotient rule (4).   ∎

In this section, we have seen the following theorem and corollary proved.

THEOREM 3.3   Product and quotient rules

If $u = f(x)$ and $v = g(x)$ are both differentiable functions, then so is $uv = f(x)g(x)$, and

$$\frac{d(uv)}{dx} = u\frac{dv}{dx} + v\frac{du}{dx}.$$

Also, $u/v = f(x)/g(x)$ is differentiable (of course points where $g(x) = 0$ are not allowed), and

$$\frac{d(u/v)}{dx} = \frac{v(du/dx) - u(dv/dx)}{v^2}.$$ □

COROLLARY   Derivative of x^n

We have $d(x^n)/dx = n \cdot x^{n-1}$ for every integer n.

It is best to learn the differentiation formulas in Theorem 3.3 in words rather than with particular letters u and v. Such verbal renditions are given in the Summary. We use "top" for numerator and "bottom" for denominator in this one instance of the quotient rule because these short words are easier to say and more suggestive. We recommend that you learn these formulas by repeating them over and over in words.

SUMMARY

1. *Product rule:* The derivative of a product is the first times the derivative of the second, plus the second times the derivative of the first. In symbols,

$$(fg)' = f \cdot g' + g \cdot f', \quad \text{or} \quad \frac{d(uv)}{dx} = u \cdot \frac{dv}{dx} + v \cdot \frac{du}{dx}.$$

2. *Quotient rule:* The derivative of a quotient is the bottom times the derivative of the top minus the top times the derivative of the bot-

tom, all divided by the bottom squared. In symbols,

$$\left(\frac{f}{g}\right)' = \frac{g \cdot f' - f \cdot g'}{g^2}, \quad \text{or} \quad \frac{d(u/v)}{dx} = \frac{v(du/dx) - u(dv/dx)}{v^2}.$$

3. $\dfrac{d(x^n)}{dx} = n \cdot x^{n-1}$ for every positive or negative integer n

EXERCISES 3.3

In Exercises 1–24, find the derivative of the indicated function. You need not simplify the answers.

1. $3s^2 + 17s - 5$

2. $20t^4 - \frac{3}{2}t^2 + 18$

3. $\dfrac{x^2 - 7}{3}$

4. $\dfrac{u^3 - 2u^2 + 4u}{4}$

5. $\dfrac{3}{y}$

6. $\dfrac{2}{v^3}$

7. $4u^3 - \dfrac{2}{u^2}$

8. $5x + 7 - \dfrac{1}{x^4}$

9. $(z^2 - 1)(z^2 + z + 2)$

10. $(3r^2 - 8r)(r^3 - 7r^2)$

11. $(x^2 + 1)[(x - 1)(x^3 + 3)]$

12. $[(v^2 - 5v)(2v + 3)](8 - 4v^2)$

13. $\left(\dfrac{1}{t^2} - \dfrac{4}{t^3}\right)(2t + 3)$

14. $(s^2 - 4s + 1)\left(\dfrac{1}{s^3} - \dfrac{4}{s}\right)$

15. $\dfrac{4r^2 - 3}{r}$

16. $\dfrac{8x^3 + 2x^2 + x}{x^2}$

17. $\dfrac{x^2 - 2}{x + 3}$

18. $\dfrac{4y^3 - 3y^2}{2y - 3}$

19. $\dfrac{(t^2 + 9)(t - 3)}{t^2 + 2}$

20. $\dfrac{(v^3 + 3v)(8v - 6)}{v^3 - 3v}$

21. $\dfrac{(2u + 3)(u^2 - 4)}{(u - 1)(4u^2 + 5)}$

22. $\dfrac{(8s - 6)(3s^2 - 2s)}{(2s + 1)(s^3 + 7)}$

23. $\left(\dfrac{x + 1}{2x + 3}\right)\left(\dfrac{1}{x} - \dfrac{1}{x^2}\right)$

24. $\left(\dfrac{4}{t^2} - \dfrac{1}{t^3}\right)\left(\dfrac{8t}{t^2 + 2}\right)$

The next section will show that

$$\frac{d(\sin x)}{dx} = \cos x \quad \text{and} \quad \frac{d(\cos x)}{dx} = -\sin x.$$

Use these facts and the formulas of this section to differentiate the functions $f(x)$ in Exercises 25–34.

25. $x(\sin x)$

26. $t^2(\cos t)$

27. $(\sin u)^2$

28. $\sin r \cos r$

29. $\tan x$

30. $\dfrac{\cos s}{\sin s}$

31. $\dfrac{v^3}{\sin v}$

32. $\dfrac{u^4}{\cos u}$

33. $\dfrac{\sin t}{t^2 - 4t}$

34. $\dfrac{x^3 - 3x^2}{\cos x}$

In Exercises 35–45, assume that the functions f and g are differentiable for all x.

35. If $f(1) = 3$, $f'(1) = -6$, and $y = x \cdot f(x)$, find $y'(1)$.

36. If $g(2) = -3$, $g'(2) = 4$, and $y = [g(x)]^2$, find $dy/dx|_{x=2}$.

37. If $f(-1) = 2$, $f'(-1) = 3$, and $y = (x^2 - 3x) \cdot f(x)$, find $dy/dx|_{x=-1}$.

38. If $g(3) = -1$, $g'(3) = 2$, and $s = t/g(t)$, find $s'(3)$.

39. If $f(5) = -1$, $f'(5) = -4$, and $y = x^2 f(x)/(2x + 3)$, find $dy/dx|_{x=5}$.

40. If $g(1) = 2$, $y = (x^3 - 2x^2)/g(x)$, and $dy/dx|_{x=1} = 4$, find $g'(1)$.

41. If $f(3) = -2$, $x = t \cdot f(t)/(t + 1)$, and $x'(3) = 5$, find $f'(3)$.

42. Let $u = x \cdot f(x)$ and $v = x^2 f(x)$. If $du/dx|_{x=1} = 5$ and $dv/dx|_{x=1} = 13$, find $f(1)$ and $f'(1)$.

43. Let $u = (x^2 - 3x)f(x)$ and $v = (2x - 3)f(x)$. If $du/dx|_{x=4} = 8$ and $dv/dx|_{x=4} = 3$, find $f(4)$ and $f'(4)$.

44. Let $u = x + f(x)$ and $v = 1/f(x)$. If $du/dx|_{x=2} = 5$ and $dv/dx|_{x=2} = -16$, find all possible values of $f(2)$ and $f'(2)$.

45. Let $u = f(x)g(x)$ and $v = f(x)/g(x)$. If $f(3) = 9$, $g(3) = 1$, $du/dx|_{x=3} = 13$, and $dv/dx|_{x=3} = -23$, find $f'(3)$ and $g'(3)$.

In Exercises 46–50, find the equations of the tangent line and the normal line to the graph of the given function at the indicated point.

46. $3x^2 - 2x$ at $(2, 8)$

47. $1/x$ at $(1, 1)$

48. $\left(\dfrac{3}{x} - \dfrac{2}{x^2}\right)$ at $(-1, -5)$

49. $\dfrac{2x + 3}{x - 1}$ at $(0, -3)$

50. $\dfrac{(x^2 - 3)(x + 2)}{(x + 1)}$ at $(0, -6)$

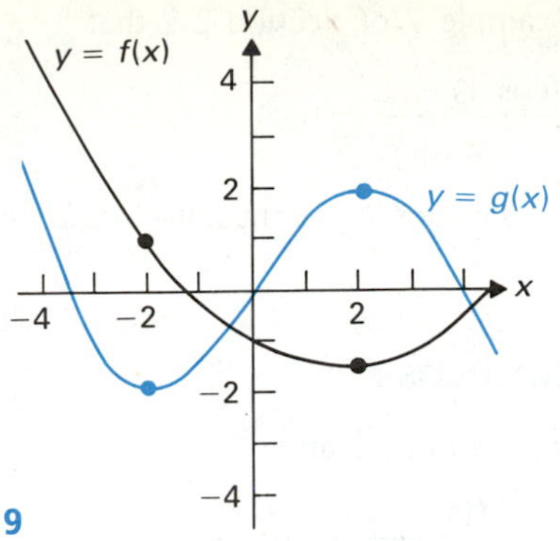

FIGURE 3.19

51. Fig. 3.19 shows the graphs $y = f(x)$ and $y = g(x)$ with equal scales on the x- and y-axes. Find the derivative of the product $f(x)g(x)$ at $x = 2$ by
   (a) sketching the graph $y = f(x)g(x)$ and estimating the slope of the tangent line where $x = 2$, and
   (b) estimating $f(2)$, $g(2)$, $f'(2)$, and $g'(2)$ from Fig. 3.19 and using the product rule.

52. Repeat Exercise 51 at $x = -2$.

53. Repeat Exercise 51 at $x = 0$.

54. Repeat Exercise 51 at $x = 2$, but this time find the derivative of the quotient $f(x)/g(x)$, modifying the instructions in parts (a) and (b) in the obvious way.

55. Repeat Exercise 54 at $x = 0$, but for the quotient $g(x)/f(x)$.

56. Repeat Exercise 54, but at $x = -2$.

The amount a family saves is a function $S(I)$ of its income I, and the amount it consumes is a function $C(I)$. The *marginal propensity to save* is dS/dI, and the *marginal propensity to consume* is dC/dI. Exercises 57–60 deal with this situation.

57. Explain the economic meaning of the marginal propensities to save and to consume.

58. Show that if the entire income is used to either save or consume, then

   Marginal propensity to consume
   $$= 1 - \text{Marginal propensity to save.}$$

59. The yearly savings of a family in terms of its annual income I is given by

   $$S(I) = \frac{2I^2}{5(I + 60,000)}.$$

   (a) Find the marginal propensity to save when $I = \$30,000$.
   (b) If the family consumes all the dollars it does not save, find the marginal propensity to consume when $I = \$30,000$.

60. Suppose the annual income of the family in Exercise 59 is extremely large. Find the limiting proportion of income saved as $I \to \infty$ by
   (a) computing the marginal propensity to save and examining it as $I \to \infty$,
   (b) examining the behavior of $S(I)$ as $I \to \infty$, as you learned to do in Chapter 2,
   (c) Would a problem that gave

   $$S(I) = \frac{2I^3}{5(I + 60,000)}$$

   be realistic?

61. The taxes T paid by a family in a certain country are given as a function of the income I by

   $$T(I) = \frac{3I^2}{80,000 + 5I}.$$

   Find the marginal tax rate when $I = \$10,000$.

62. In Exercise 61, find the limiting proportion of the income paid in taxes as $I \to \infty$ by
   (a) finding the marginal tax as $I \to \infty$,
   (b) examining the behavior of $T(I)$ as $I \to \infty$, as you learned to do in Chapter 2.

3.4 DIFFERENTIATION OF TRIGONOMETRIC FUNCTIONS

We are now ready to differentiate the trigonometric functions. Since they are not expressed in terms of functions whose derivatives we already know, we will have to go back to the definition of the *derivative as a limit*. When we do this with $\sin x$, we will run into the limits

$$\lim_{\Delta x \to 0} \frac{\sin \Delta x}{\Delta x} \quad \text{and} \quad \lim_{\Delta x \to 0} \frac{\cos(\Delta x) - 1}{\Delta x}.$$

In Section 2.2, we established the fundamental trigonometric limit

$$\lim_{x \to 0} \frac{\sin x}{x} = 1. \tag{1}$$

HISTORICAL NOTE

THE SINE as a line in a circle of a certain radius was used in ancient times, and the notion of the sine as a real-valued function began to appear in the late seventeenth century. The earliest occurrence of the derivative of the sine is found in a work of Roger Cotes (1682–1716). Cotes, a Cambridge graduate and a professor of astronomy, prepared the second edition (1713) of Isaac Newton's *Mathematical Principles of Natural Philosophy.*

Cotes's *On the Estimation of Errors,* published posthumously in 1722, dealt with the analysis of errors in astronomical observations. The particular lemma concerning the derivative of the sine reads, "The small variation of any arc of a circle is to the small variation of the sine of that arc as the radius to the sine of the complement [cosine]." Cotes's "small variations" can be thought of as differentials (see Section 3.9). If the radius is taken to be 1 and the arc to be x, his result can be written as $dx/d(\sin x) = 1/\cos x$. The proof uses similar triangles in the diagram.

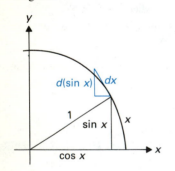

Using this limit, we showed in Example 7 of Section 2.2 that

$$\lim_{x \to 0} \frac{(\cos x) - 1}{x} = 0. \tag{2}$$

Remember that when interpreting x as an angle, we must use *radian measure* for these limits to be valid.

The Derivatives of sin *x* and cos *x*

We must go back to the definition of the derivative

$$f'(x) = \lim_{\Delta x \to 0} \frac{f(x + \Delta x) - f(x)}{\Delta x}$$

to find the derivative of $f(x) = \sin x$. Forming the difference quotient and using

$$\sin(x + y) = \sin x \cos y + \cos x \sin y,$$

trigonometric identity (12) in Section 1.6, we obtain

$$\frac{f(x + \Delta x) - f(x)}{\Delta x} = \frac{\sin(x + \Delta x) - \sin x}{\Delta x}$$

$$= \frac{\sin x \cos \Delta x + \cos x \sin \Delta x - \sin x}{\Delta x}$$

$$= \frac{\cos x \sin \Delta x + (\sin x)(\cos \Delta x - 1)}{\Delta x}$$

$$= \cos x \, \frac{\sin \Delta x}{\Delta x} + \sin x \frac{\cos \Delta x - 1}{\Delta x}.$$

Therefore, using the limits in relations (1) and (2), we have

$$f'(x) = \lim_{\Delta x \to 0} \frac{f(x + \Delta x) - f(x)}{\Delta x}$$

$$= (\cos x)\left(\lim_{\Delta x \to 0} \frac{\sin \Delta x}{\Delta x}\right) + (\sin x)\left(\lim_{\Delta x \to 0} \frac{\cos \Delta x - 1}{\Delta x}\right)$$

$$= (\cos x)(1) + (\sin x)(0) = \cos x.$$

Thus

$$\frac{d(\sin x)}{dx} = \cos x. \tag{3}$$

Note that when x measures an angle, x must be in *radians* for Eq. (3) to hold.

EXAMPLE 1   Let $y = x^2 \sin x$. Find y'.

Solution   Using the product rule and formula (3), we obtain

$$y' = x^2 \frac{d(\sin x)}{dx} + \frac{d(x^2)}{dx} \sin x$$

$$= x^2 \cos x + 2x \sin x. \qquad \blacksquare$$

EXAMPLE 2   Let $u = 3t \csc t$. Find du/dt.

Solution   Since $\csc t = 1/(\sin t)$, we have

$$u = 3t \csc t = \frac{3t}{\sin t}.$$

Using the quotient rule, we have

$$\frac{du}{dt} = \frac{(\sin t)(3) - (3t)(\cos t)}{\sin^2 t}$$

$$= 3\frac{\sin t - t \cos t}{\sin^2 t}. \qquad \blacksquare$$

We find the derivative of $g(x) = \cos x$ much as we found the derivative of $\sin x$. Trigonometric identity (13) of Section 1.6 is

$$\cos(x + y) = \cos x \cos y - \sin x \sin y.$$

Therefore,

$$\frac{g(x + \Delta x) - g(x)}{\Delta x} = \frac{\cos(x + \Delta x) - \cos x}{\Delta x}$$

$$= \frac{\cos x \cos \Delta x - \sin x \sin \Delta x - \cos x}{\Delta x}$$

$$= \frac{\cos \Delta x - 1}{\Delta x}(\cos x) - \frac{\sin \Delta x}{\Delta x}(\sin x).$$

Using the relations (1) and (2), we discover that

$$\lim_{\Delta x \to 0} \frac{g(x + \Delta x) - g(x)}{\Delta x} = 0(\cos x) - 1(\sin x) = -\sin x.$$

This shows that

$$\frac{d(\cos x)}{dx} = -\sin x. \qquad (4)$$

EXAMPLE 3   Find $f'(t)$ if

$$f(t) = \frac{\cos t}{3t}.$$

Solution   Using the quotient rule and formula (4), we have

$$f'(t) = \frac{(3t)(-\sin t) - (\cos t)(3)}{9t^2}$$

$$= -3\frac{t \sin t + \cos t}{9t^2}. \qquad \blacksquare$$

EXAMPLE 4   Use the identity $\sin 2x = 2 \sin x \cos x$ to find y' if $y = \sin 2x$.

Solution   Using the product rule with $y = 2 \sin x \cos x$, we see that

$$y' = 2[(\sin x)(-\sin x) + (\cos x)(\cos x)]$$
$$= 2(\cos^2 x - \sin^2 x).$$

By trigonometric identity (15) in Section 1.6, we have $\cos^2 x - \sin^2 x = \cos 2x$. Thus

$$\frac{d(\sin 2x)}{dx} = 2 \cos 2x.$$

In Section 3.5, we will see how to solve this problem without using identity (15).

■

Derivatives of the Other Basic Trigonometric Functions

Derivatives of the other four basic trigonometric functions are now readily found since they can be expressed as quotients involving the functions $\sin x$ and $\cos x$ and 1, all of which we can now differentiate. For example

$$\frac{d(\tan x)}{dx} = \frac{d[(\sin x)/(\cos x)]}{dx} = \frac{(\cos x)(\cos x) - (\sin x)(-\sin x)}{\cos^2 x}$$

$$= \frac{\cos^2 x + \sin^2 x}{\cos^2 x} = \frac{1}{\cos^2 x} = \sec^2 x.$$

Similarly, we find that

$$\frac{d(\cot x)}{dx} = -\csc^2 x, \qquad \frac{d(\sec x)}{dx} = \sec x \tan x,$$

and

$$\frac{d(\csc x)}{dx} = -\csc x \cot x.$$

We display these four formulas together.

$$\frac{d(\tan x)}{dx} = \sec^2 x \qquad \qquad (5)$$

$$\frac{d(\cot x)}{dx} = -\csc^2 x \qquad \qquad (6)$$

$$\frac{d(\sec x)}{dx} = \sec x \tan x \qquad \qquad (7)$$

$$\frac{d(\csc x)}{dx} = -\csc x \cot x \qquad \qquad (8)$$

We recommend that you memorize these formulas, together with formulas (3) and (4). If you know the derivatives of the three functions $\sin x$, $\tan x$, and $\sec x$, then the derivative of each of the corresponding cofunctions $\cos x$, $\cot x$, $\csc x$ is found by changing the derivative to the cofunction and changing sign. The formulas will take you only a few minutes to commit to memory. This text contains the formulas for the derivatives of these trigonometric functions, and computers and some calculators also have the capability to tell you that the derivative of $\sin x$ is $\cos x$. But people using calculus should not have to hunt through a text or

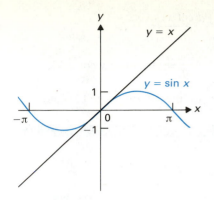

FIGURE 3.20 $y = x$ is tangent to $y = \sin x$ at the origin.

reach for a computer or calculator for such basic differentiation formulas. It simply takes too long.

EXAMPLE 5 Find the derivative of $f(x) = \sec^2 x$.

Solution We write $f(x) = \sec^2 x = (\sec x)(\sec x)$ and use the product rule, obtaining

$$f'(x) = (\sec x)(\sec x \tan x) + (\sec x \tan x)(\sec x) = 2 \sec^2 x \tan x. \quad \blacksquare$$

Graphical Interpretation of the Fundamental Limit

Since the derivative of $\sin x$ is $\cos x$, we see that the slope of the tangent line to the graph of $f(x) = \sin x$ at $x = 0$ is $\cos 0 = 1$. The equation of this tangent line, which passes through the origin, is thus $y - 0 = 1(x - 0)$, or $y = x$. Figure 3.20 shows the graphs of $y = x$ and $y = \sin x$ on the same axes. These graphs give us an intuitive understanding of the fundamental limit

$$\lim_{x \to 0} \frac{\sin x}{x} = 1.$$

The limit asserts that the ratio of the y-heights to the graphs of $\sin x$ and x gets as close to 1 as we wish if x is sufficiently close to zero. This happens precisely because $y = x$ is the tangent line to $y = \sin x$ at the origin. If we used a computer to draw the graphs of $y = x$ and $y = \sin x$ on the same axes with axis scales $-0.01 \le x \le 0.01$ and $-0.01 \le y \le 0.01$, we actually would not see it draw the second graph. The computer would put it right on top of the first graph. The graph of $\sin x$ is a bit closer to the x-axis than the line $y = x$ is, but the resolution of the computer screen is too coarse to show this difference.

EXAMPLE 6 Use the fundamental trigonometric limit to argue geometrically that the line tangent to the graph of $y = f(x) = \sin 2x$ at the origin is $y = 2x$, so that $f'(0) = 2$.

Solution From the fundamental limit, we know that

$$\lim_{x \to 0} \frac{\sin 2x}{cx} = 1$$

if and only if $c = 2$. Thus the line $y = 2x$ is the unique line through the origin such that the ratio of the y-value on the line to the y-value on the graph of $\sin 2x$ approaches 1. This indicates that the line $y = 2x$ is tangent to the graph of $y = \sin 2x$, so the derivative of $f(x) = \sin 2x$ at $x = 0$ must be 2. $\quad \blacksquare$

SUMMARY

Differentiation Formulas

1. $\dfrac{d(\sin x)}{dx} = \cos x$
2. $\dfrac{d(\cos x)}{dx} = -\sin x$

3. $\dfrac{d(\tan x)}{dx} = \sec^2 x$

4. $\dfrac{d(\cot x)}{dx} = -\csc^2 x$

5. $\dfrac{d(\sec x)}{dx} = \sec x \tan x$

6. $\dfrac{d(\csc x)}{dx} = -\csc x \cot x$

An angle x must be measured in *radians* for these formulas to be valid.

EXERCISES 3.4

In Exercises 1–20, find the derivative of the given function. You need not simplify your answers.

1. $x \cos x$

2. $x^2 \tan x$

3. $(t^2 + 3t)\sec t$

4. $\cos^2 u$

5. $\cot^2 v$

6. $\csc^2 s$

7. $\sin r \tan r$

8. $\dfrac{\csc t}{t}$

9. $\dfrac{\sin x}{1 - \cos x}$

10. $\dfrac{\tan y}{2 + \sec y}$

11. $\dfrac{\cot s}{3 + 2 \cos s}$

12. $\sin u \cot u$

13. $\cos x \tan x$

14. $\sin^2 r \cot r$

15. $\tan y \csc^2 y$

16. $\cot^2 t \sec^2 t$

17. $(1 + \sin u)^2$

18. $(1 - \cos s)^2$

19. $\dfrac{\tan v}{v^2 - 4v}$

20. $\dfrac{\csc x}{x + 3}$

21. Let $f(x) = \sin x$.
 (a) Find $f'(x)$.
 (b) Find the derivative of $f'(x)$, which is called the *second derivative* of $f(x)$ and is denoted by $f''(x)$.
 (c) Find the derivative of $f''(x)$, which is called the *third derivative* $f'''(x)$ of $f(x)$.
 (d) Find the derivative of $f'''(x)$, which is called the *fourth derivative* $f^{(4)}(x)$ of $f(x)$.
 (e) Using your answers to the previous parts, find $f^{(13)}(x)$.
 (f) Repeat part (e), but find $f^{(1034)}(x)$.

22. Repeat Exercise 21 for the function $f(x) = \cos x$.

23. Repeat Exercise 21 for the function $f(x) = x \sin x$.

24. Repeat Exercise 21 for the function $f(x) = x \cos x$.

Exercises 25–34 give practice in evaluating trigonometric functions as well as differentiating.

25. Find the equation of the line tangent to $y = \tan x$ where $x = \pi/4$.

26. Find the equation of the line tangent to $y = \csc x$ where $x = \pi/6$.

27. Find the equation of the line tangent to $y = \sin x$ where $x = 5\pi/3$.

28. Find the equation of the line tangent to $y = \sec x$ where $x = 3\pi/4$.

29. Find the equation of the line tangent to $y = \cot x$ where $x = 3\pi/4$.

30. Find the equation of the line tangent to $y = \cos x$ where $x = -7\pi/6$.

31. Find all x-values in the interval $-2\pi \le x \le 2\pi$ where the graph of $f(x) = \sin x$ has a horizontal tangent line.

32. Find all x-values in the interval $-2\pi \le x \le 2\pi$ where the graph of $f(x) = \sec x$ has a horizontal tangent line.

33. Find all x-values in the interval $-2\pi \le x \le 2\pi$ where the normal line to the graph of $f(x) = \cos x$ has slope -2.

34. Find all x-values in the interval $-2\pi \le x \le 2\pi$ where the tangent line to the graph of $f(x) = \sin x$ has slope 2.

35. Study Example 6 in the text and use a similar argument to find the equation of the tangent line to the graph of $f(x) = \sin(-3x)$ where $x = 0$. What is $f'(0)$?

36. Study Example 6 in the text and then find a polynomial function $p(x)$ such that $f(x) = \sin x^2$ and $p(x)$ have the same tangent line where $x = 0$. Find the equation of this tangent line. What is $f'(0)$?

37. Continuing the idea of Exercise 36, find a polynomial function $p(x)$ such that $f(x) = \sin(x^2 - 1)$ and $p(x)$ have the same tangent line where $x = 1$. Find the equation of this tangent line. What is $f'(1)$?

38. Repeat Exercise 37 where $x = -1$.

3.5 THE CHAIN RULE

The Chain Rule Formula

Consider the following question.

> *If a car is going three times as fast as a bicycle and the bicycle is going twice as fast as a runner, how many times as fast as the runner is the car going?*

Surely the car is going $3 \cdot 2 = 6$ times as fast as the runner. Recall that an interpretation of dy/dx is the *rate of change of y with respect to x.* Taking some liberties with the Leibniz notation, we can express the answer to the problem concerning the car, the bicycle, and the runner as follows:

$$\frac{d(\text{Car})}{d(\text{Runner})} = \frac{d(\text{Car})}{d(\text{Bicycle})} \cdot \frac{d(\text{Bicycle})}{d(\text{Runner})} = 3 \cdot 2 = 6.$$

Now we take a step toward more careful mathematics. Suppose $y = f(x)$ and $x = g(t)$, so that y appears as a **composite function** $f \circ g$ of t. Let $x = g(t)$ be differentiable at t_1 and let $y = f(x)$ be differentiable at $x_1 = g(t_1)$. The analogy between the following question and the one at the start of this section should be obvious.

> *If y is increasing three times as fast as x is at x_1, and x is increasing twice as fast as t is at t_1, how many times as fast as t is y increasing when $t = t_1$?*

Once again, it is clear that y must be increasing $3 \cdot 2 = 6$ times as fast as t is at t_1. In Leibniz notation, this becomes

$$\frac{dy}{dt} = \frac{dy}{dx} \cdot \frac{dx}{dt}. \qquad \textit{Chain rule} \qquad (1)$$

The chain rule illustrates the advantage of Leibniz notation in remembering formulas. We can remember Eq. (1) by pretending that the "dx's cancel." Using the composite function notation $(f \circ g)(t) = f(g(t))$, where $y = f(x)$ and $x = g(t)$, we can write Eq. (1) in function notation as

$$\begin{aligned} (f \circ g)'(t) &= f'(x) \cdot g'(t) \\ &= f'(g(t)) \cdot g'(t). \end{aligned} \qquad \textit{Chain rule} \qquad (2)$$

The preceding intuitive arguments do not constitute a proof of the chain rule, but they do help us to understand the meaning of the chain rule in terms of rates of change. A proof of the chain rule is given in Section 3.9. We can give a proof here that is valid for many functions f and g.

Suppose that $y = f(x)$ and $x = g(t)$ and that g is differentiable at $t = a$ and f is differentiable at $g(a)$. Suppose further that $\Delta x = g(a + \Delta t) - g(a)$ is nonzero whenever $|\Delta t|$ is sufficiently small but nonzero. Setting $\Delta y = f(g(a + \Delta t)) - f(g(a))$, we then have

$$\frac{\Delta y}{\Delta t} = \frac{\Delta y}{\Delta x} \cdot \frac{\Delta x}{\Delta t} \qquad (3)$$

whenever $|\Delta t|$ is very small but nonzero. Since $g'(a)$ exists, $x = g(t)$ is continuous where $t = a$ by Theorem 3.1, so as $\Delta t \to 0$ we must have $\Delta x \to 0$ also. We are assuming that Δx never becomes zero, so Eq. (3) holds whenever $|\Delta t|$ is very small but nonzero. We now take the limit of both sides of Eq. (3) as $\Delta t \to 0$ and remember the definition of the derivative. We obtain $dy/dt|_a$ on the left-hand side, and Theorem 2.1 on the limit of a product shows that we obtain $(dy/dx|_{g(a)}) \cdot (dx/dt|_a)$ on the right-hand side, which is Eq. (1). This proves the chain rule for such functions f and g. However, it is not difficult to construct functions $x = g(t)$ for which nonzero values Δt can be found of arbitrarily small magnitude that cause Δx to become zero.

We give a precise statement of the chain rule and proceed with examples.

THEOREM 3.4   Chain rule

Let $y = f(x)$ be differentiable at x_1 and $x = g(t)$ be differentiable at t_1, with $g(t_1) = x_1$. Then the composite function $y = f(g(t))$ is differentiable at t_1, and

$$\frac{dy}{dt}\bigg|_{t_1} = \frac{dy}{dx}\bigg|_{x_1} \cdot \frac{dx}{dt}\bigg|_{t_1}$$

If $y = f(x)$ and $x = g(t)$ are both differentiable functions, then for $y = f(g(t))$,

$$\frac{dy}{dt} = \frac{dy}{dx} \cdot \frac{dx}{dt}. \tag{4}$$

In the notation $f \circ g$ for the composite function, Eq. (4) becomes

$$(f \circ g)'(t_1) = f'(x_1) \cdot g'(t_1) = f'(g(t_1)) \cdot g'(t_1). \tag{5}$$

EXAMPLE 1   Let $f(x) = x^2 - x$ and $g(t) = t^3$. Find $(f \circ g)'(2)$.

Solution   Of course, $t_1 = 2$ yields $x_1 = g(2) = 2^3 = 8$. Using the chain rule, we have

$$(f \circ g)'(t) = f'(x) \cdot g'(t) = (2x - 1)(3t^2).$$

When $t = 2$ and $x = 8$, we obtain

$$(f \circ g)'(2) = 15 \cdot 12 = 180.$$

Alternatively, we could solve the problem by expressing y directly as a function only of t and differentiating, without using the chain rule. From $y = x^2 - x$ and $x = t^3$, we obtain $y = t^6 - t^3$. Then $dy/dt = 6t^5 - 3t^2$, so $dy/dt|_{t=2} = 6 \cdot 32 - 3 \cdot 4 = 192 - 12 = 180$. ∎

EXAMPLE 2   If the length of an edge of a cube increases at a rate of 4 in./sec, find the rate of increase of the volume per second at the instant when the edge is 2 in. long.

Solution   If an edge is of length x, then $V = x^3$. We wish to find the rate of change of V *with respect to time,* that is, dV/dt. Now

$$\frac{dV}{dt} = \frac{dV}{dx} \cdot \frac{dx}{dt} = (3x^2)\frac{dx}{dt}$$

by the chain rule. We are given that $dx/dt = 4$ in./sec. Thus when $x = 2$,

$$\frac{dV}{dt} = 12 \cdot 4 = 48 \text{ in}^3/\text{sec}.$$ ∎

Now we can finally see why radian measure of an angle is the convenient measure when doing calculus. Let x be radian measure and $t°$ be degree measure for an angle. Then

$$x = \frac{\pi}{180}t°$$

and

$$\frac{d(\sin x)}{dt°} = \frac{d(\sin x)}{dx} \cdot \frac{dx}{dt°} = (\cos x)\left(\frac{\pi}{180}\right).$$

If we used degree measure, we would have the nuisance factor $\pi/180$ in our differentiation formulas for trigonometric functions.

EXAMPLE 3   Let $y = \sin x^2$. Find dy/dx.

Solution   We let $u = x^2$, so that $y = \sin u$. By the chain rule,

$$\frac{dy}{du} = \frac{dy}{du} \cdot \frac{du}{dx}.$$

Our work in the preceding section shows that

$$\frac{d(\sin u)}{du} = \cos u,$$

and of course $du/dx = 2x$. Therefore,

$$\frac{dy}{dx} = \frac{dy}{du} \cdot \frac{du}{dx} = (\cos u)2x = 2x \cos x^2.$$ ∎

Example 3 really illustrates the tremendous power of the chain rule. We devoted the last section to differentiation of just $\sin x$, $\cos x$, $\tan x$, $\cot x$, $\sec x$, and $\csc x$. Now by imitating the procedure in Example 3, we can differentiate these trigonometric functions of u where u is any differentiable function of x that we know how to differentiate.

If u is a differentiable function of x, then

$$\frac{d(\sin u)}{dx} = (\cos u)\frac{du}{dx}, \qquad \frac{d(\cos u)}{dx} = (-\sin u)\frac{du}{dx},$$

$$\frac{d(\tan u)}{dx} = (\sec^2 u)\frac{du}{dx}, \qquad \frac{d(\cot u)}{dx} = (-\csc^2 u)\frac{du}{dx},$$

$$\frac{d(\sec u)}{dx} = (\sec u \tan u)\frac{du}{dx}, \qquad \frac{d(\csc u)}{dx} = (-\csc u \cot u)\frac{du}{dx}.$$

EXAMPLE 4   Let $y = \tan[3x/(x - 5)]$. Find dy/dx.

Solution   We simply use the formula

$$\frac{d(\tan u)}{dx} = (\sec^2 u)\frac{du}{dx}$$

with $u = 3x/(x - 5)$, and we use the quotient rule to find du/dx. We obtain

$$\frac{dy}{dx} = \sec^2\left(\frac{3x}{x - 5}\right) \cdot \frac{(x - 5)3 - 3x}{(x - 5)^2}$$

$$= \sec^2\left(\frac{3x}{x - 5}\right) \cdot \frac{-15}{(x - 5)^2}.$$ ∎

EXAMPLE 5   Let $y = \csc(\cos r)$. Find y'.

Solution   We *decompose* this function into $y = f(u) = \csc u$ and $u = g(r) = \cos r$. By the chain rule,

$$(f \circ g)'(r) = f'(u) \cdot g'(r),$$

so

$$y' = (f \circ g)'(r) = (-\csc u \cot u)(-\sin r)$$

$$= \csc(\cos r) \cot(\cos r) \sin r.$$ ∎

Differentiation of a Function to a Power

EXAMPLE 6   Find the derivative of the function $(4x^3 + 7x^2)^{10}$.

Solution   We could raise the polynomial $4x^3 + 7x^2$ to the tenth power and obtain one polynomial expression, but that would be hard work. Instead we let $y = (4x^3 + 7x^2)^{10}$ and let $u = 4x^3 + 7x^2$. Then $y = u^{10}$ and, by the chain rule, Eq. (1),

$$\frac{dy}{dx} = \frac{dy}{du} \cdot \frac{du}{dx} = 10u^9(12x^2 + 14x)$$

$$= 10(4x^3 + 7x^2)^9(12x^2 + 14x).$$

Our problem is solved.   ∎

The technique for differentiating a function to a power illustrated in Example 6 is so useful that we develop it into a formula. We know that if n is any integer, then

$$\frac{d(u^n)}{du} = nu^{n-1}. \tag{6}$$

If u is in turn a function of x, then the chain rule, Eq. (2), yields

$$\frac{d(u^n)}{dx} = \frac{d(u^n)}{du} \cdot \frac{du}{dx} = nu^{n-1} \cdot \frac{du}{dx}. \qquad \textit{Power rule} \tag{7}$$

Formula (7) is used very often. The Summary gives it in an easily memorized verbal form.

EXAMPLE 7   Find y' if $y = (x^3 - 2x)^5$.

Solution   We think of $x^3 - 2x$ as u in formula (7) and obtain

$$y' = 5(x^3 - 2x)^4 \cdot (3x^2 - 2).$$ ∎

We can extend formula (7) to rational exponents p/q, where p and q are integers and $q \neq 0$. Let u be a differentiable function of x. It can be shown that then $u^{p/q}$ is differentiable, except possibly at points where u becomes zero. Using formula (7) for integer exponents, we have

$$\frac{d[(u^{p/q})^q]}{dx} = q(u^{p/q})^{q-1} \cdot \frac{d(u^{p/q})}{dx} \tag{8}$$

and

$$\frac{d(u^p)}{dx} = pu^{p-1} \cdot \frac{du}{dx}. \tag{9}$$

Since $(u^{p/q})^q = u^p$, we obtain from Eqs. (8) and (9)

$$q(u^{p/q})^{q-1} \cdot \frac{d(u^{p/q})}{dx} = pu^{p-1} \cdot \frac{du}{dx},$$

so

$$\frac{d(u^{p/q})}{dx} = \frac{pu^{p-1}}{q(u^{p/q})^{q-1}} \cdot \frac{du}{dx} = \frac{p}{q} \cdot \frac{u^{p-1}}{u^{p-(p/q)}} \cdot \frac{du}{dx}$$

$$= \frac{p}{q} \cdot u^{p-1-p+(p/q)} \cdot \frac{du}{dx}.$$

Finally we obtain

$$\frac{d(u^{p/q})}{dx} = \frac{p}{q} u^{(p/q)-1} \cdot \frac{du}{dx}. \tag{10}$$

Formula (10) is precisely what we get from formula (7) with $n = p/q$.

EXAMPLE 8   Find the derivative of the function $\sqrt{1 + x^2}$.

Solution   From formula (10), we have

$$\frac{d(\sqrt{1 + x^2})}{dx} = \frac{d}{dx}[(1 + x^2)^{1/2}] = \frac{1}{2}(1 + x^2)^{-1/2} \cdot (2x)$$

$$= \frac{x}{\sqrt{1 + x^2}}.$$ ∎

EXAMPLE 9   Find y' if $y = \sin^2(5x^3)$.

Solution   Writing $y = (\sin 5x^3)^2$ and using formula (7) with $u = \sin 5x^3$, we obtain

$$y' = 2(\sin 5x^3) \cdot \frac{du}{dx}$$

$$= 2(\sin 5x^3) \cdot (\cos 5x^3)(15x^2) = 30x^2 \sin 5x^3 \cos 5x^3.$$ ∎

EXAMPLE 10 Find x' if $x = (s^2 - 2s)^{2/3} \cdot \sqrt{s^3 + 1}$.

Solution We use the product rule for differentiation and formula (10). The answer we obtain may not be elegant, but it can be found in one step. We have

$$y' = \left[(s^2 - 2s)^{2/3} \cdot \frac{1}{2}(s^3 + 1)^{-1/2} \cdot 3s^2 \right]$$
$$+ \left[\sqrt{s^3 + 1} \cdot \frac{2}{3}(s^2 - 2s)^{-1/3}(2s - 2) \right]. \qquad \blacksquare$$

The ability to differentiate a fairly complicated function like the one in Example 10 makes us really appreciate our differentiation formulas, especially the chain rule. Imagine what a problem we would have if we had to use the definition of the derivative and compute the limit of the difference quotient. This ability to differentiate makes calculus practical.

EXAMPLE 11 Let $y = g(u)$ where g is differentiable for all u. Suppose $u = 1/\sqrt{3x + 4}$ and $dy/dx|_{x=4} = \frac{3}{16}$. If possible, find $g'(4)$ and $g'(\frac{1}{4})$.

Solution By the chain rule,

$$\frac{dy}{dx} = \frac{dy}{du} \cdot \frac{du}{dx}.$$

When $x = 4$, we have $u = 1/\sqrt{3 \cdot 4 + 4} = 1/\sqrt{16} = \frac{1}{4}$. Knowing dy/dx and du/dx when $x = 4$, we can find $dy/du = g'(u)$ when $u = \frac{1}{4}$, but we have no way of finding $g'(4)$. Now $u = (3x + 4)^{-1/2}$, so by formula (7) we have

$$\frac{du}{dx} = -\frac{1}{2}(3x + 4)^{-3/2} \cdot 3$$
$$= \frac{-3}{2(3x + 4)^{3/2}}.$$

We then have

$$\frac{dy}{dx}\bigg|_{x=4} = \frac{dy}{du}\bigg|_{u=1/4} \cdot \frac{du}{dx}\bigg|_{x=4},$$
$$\frac{3}{16} = g'\left(\frac{1}{4}\right) \cdot \frac{-3}{2(16)^{3/2}} = g'\left(\frac{1}{4}\right) \cdot \frac{-3}{128}.$$

Thus

$$g'\left(\frac{1}{4}\right) = \frac{3}{16} \cdot \frac{128}{-3} = -8. \qquad \blacksquare$$

Example 11 of Section 3.8 will show that the formula

$$\frac{d(u^r)}{dx} = ru^{r-1} \cdot \frac{du}{dx} \qquad \text{Power rule} \qquad (11)$$

holds for any real exponent r, and we give a verbal rendition of formula (11) in the Summary.

SUMMARY

1. If $y = f(x)$ and $x = g(t)$ are both differentiable functions, then so is $y = f(g(t))$ and

$$\frac{dy}{dt} = \frac{dy}{dx} \cdot \frac{dx}{dt}. \qquad \textit{Chain rule}$$

In function notation, the formula is

$$[f(g(t))]' = f'(g(t)) \cdot g'(t).$$

2. The differentiation formulas for the trigonometric functions should be mastered in chain rule format, as for example

$$\frac{d(\tan u)}{dx} = (\sec^2 u)\frac{du}{dx}.$$

In words, the derivative of the tangent of something is the secant squared of that thing times the derivative of the thing.

3. The derivative of a function to a constant power is the power times the function to one less power, times the derivative of the function. In symbols,

$$\frac{d(u^r)}{dx} = ru^{r-1} \cdot \frac{du}{dx}. \qquad \textit{Power rule}$$

(We have demonstrated this formula only for rational $r = p/q$.)

EXERCISES 3.5

1. Given $y = f(x) = x^2 - 3x$ and $x = g(t) = t^3 + 1$, find $(f \circ g)'(1)$ by (a) using the chain rule, (b) expressing y as a function only of t and differentiating.

2. Repeat Exercise 1 for $f(x) = \sqrt{3x + 12}$ and $g(t) = t^2 - 2t$, where $t = 4$.

3. If $u = g(v) = (v^2 + 3v - 4)^{3/2}$ and $v = h(w) = w^3 - 3$, find $(g \circ h)'(2)$ by
   (a) using the chain rule,
   (b) expressing u directly as a function of w and differentiating.

4. Repeat Exercise 3 for $u = v^3 + 3v^2 - 2v$ and $v = w - 3$ at $w = 2$.

In Exercises 5–40, differentiate the given function. You need not simplify your answers.

5. $f(x) = (3x + 2)^4$

6. $g(t) = (8t^2 - 17t)^3$

7. $h(s) = (s^2 + 3s)^2(s^3 - 1)^3$

8. $g(x) = (x^3 - 2)^3(2x^2 + 4x)^2$

9. $g(u) = \dfrac{8u^2 - 2}{(4u^2 + 1)^2}$

10. $k(v) = \dfrac{(3v^3 - 2)^2}{4v^3 + 2}$

11. $f(x) = 4x^2 - 2x + 3x^{5/3}$

12. $h(t) = t^3 - 2t^2 - 3\sqrt{t}$

13. $g(w) = \dfrac{1}{\sqrt{w}}$

14. $h(y) = \dfrac{1}{\sqrt[3]{y}}$

15. $f(t) = t^{2/3} + t^{1/5}$

16. $h(z) = 4z^{1/4} + 9z^{7/3}$

17. $h(x) = \sqrt{2x + 1}$

18. $k(v) = (3v - 2)^{4/3}$

19. $f(x) = \dfrac{1}{\sqrt{5x^2 + 10x}}$

20. $g(u) = \dfrac{3}{(4u^3 - 5u)^{3/2}}$

21. $h(t) = \sqrt{t^2 + 1}$

22. $k(x) = (2x^3 - 1)^{2/3}$

23. $k(y) = \dfrac{\sqrt{y}}{y + 1}$

24. $h(w) = \dfrac{w}{\sqrt{w + 1}}$

25. $g(t) = \sqrt{3t + 4}\,(4t + 2)^{2/3}$

26. $f(u) = \dfrac{(u^2 + 4)^{1/3}}{\sqrt{8u - 7}}$

27. $f(x) = \sqrt{2x + 1}\dfrac{(4x^2 - 3x)^2}{2x + 5}$

28. $g(t) = \dfrac{\sqrt{t^2 + 4}}{3t - 8}(2t^3 + 1)^2$

29. $h(x) = \sin 2x$

30. $k(u) = \sin u^2$

31. $k(v) = \cos\sqrt{v}$

32. $h(t) = \sec(3t + 1)$

33. $g(w) = \cos^2(2 - 3w)$

34. $f(y) = \cot^2 y$

35. $f(z) = \sin^2 z \cos^2 z$

36. $h(s) = \tan s \sec 2s$

37. $g(s) = \sqrt{\cot^2 s + \csc^2 s}$

38. $k(w) = \sqrt{8w^2 + \cos^2 w}$

39. $f(x) = \sin(\tan 3x)$

40. $k(z) = \csc(z + \cos z^2)$

In Exercises 41–44, f and g are differentiable functions.

41. Let $y = [f(x)]^4$ and suppose $f'(1) = 5$ and $dy/dx|_{x=1} = -160$. Find $f(1)$.

42. Let $y = (f(x) + 3x^2)^3$ and suppose $f(-1) = -5$ and $dy/dx|_{x=-1} = 3$. Find $f'(-1)$.

43. Let $y = x/(x + 1)$ and suppose $x = g(t)$ and $g(5) = 1$. If $dy/dt|_{t=5} = 10$, find $g'(5)$.

44. Suppose $y = (f(x) + 3x)^2$ and $x = t^3 - 2t$. If $f(4) = 6$ and $dy/dt|_{t=2} = 180$, find $f'(4)$.

45. Find the equation of the line tangent to the circle $x^2 + y^2 = 25$ at the point $(3, 4)$.

46. Find the equation of the line normal to the graph of $1/x$ at the point $(2, \frac{1}{2})$.

47. Find the equation of the line tangent to the graph of $\sqrt{2x + 1}$ at the point $(4, 3)$.

3.6 DERIVATIVES OF EXPONENTIAL FUNCTIONS

One of the most exciting features of calculus is the rapidity with which we can learn to solve rate-of-change problems in our changing world. Our study of calculus proper started with this chapter. After only five lessons, we can already find the rate of change (derivatives) of all the functions listed in our dictionary of functions (Section 1.6) except for the exponential and logarithm functions. In this section, we learn to differentiate the exponential function and in Section 3.8, logarithm functions. The main emphasis of this chapter is the notion of the derivative and development of differentiation techniques. We will apply our differentiation techniques in Chapter 4.

The (Natural) Exponential Function

We turn to differentiation of the function $f(x) = a^x$ where $a > 0$. To differentiate this new type of function, we must go back to the definition of the derivative. We have

$$f'(x) = \lim_{\Delta x \to 0} \frac{f(x + \Delta x) - f(x)}{\Delta x} = \lim_{\Delta x \to 0} \frac{a^{x + \Delta x} - a^x}{\Delta x}$$

$$= \lim_{\Delta x \to 0} \frac{a^x \cdot a^{\Delta x} - a^x}{\Delta x} = \lim_{\Delta x \to 0} \left[\frac{a^{\Delta x} - 1}{\Delta x} \cdot a^x\right]$$

$$= \left[\lim_{\Delta x \to 0} \frac{a^{\Delta x} - 1}{\Delta x}\right] \cdot a^x.$$

Thus differentiation of a^x is reduced to the study of

$$\lim_{\Delta x \to 0} \frac{a^{\Delta x} - 1}{\Delta x}. \tag{1}$$

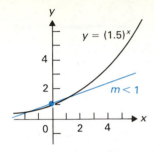

FIGURE 3.21 Tangent line to $y = (1.5)^x$ at $(0, 1)$.

But this limit (1) is precisely the derivative of a^x at $x = 0$ because

$$\frac{f(0 + \Delta x) - f(x)}{\Delta x} = \frac{a^{0+\Delta x} - a^0}{\Delta x} = \frac{a^{\Delta x} - 1}{\Delta x}.$$

Assuming that a^x is differentiable at $x = 0$, we see that

$$\frac{d(a^x)}{dx} = (\text{Derivative of } a^x \text{ at } x = 0) \cdot a^x. \qquad (2)$$

In this chapter, we will assume that a^x is indeed differentiable at $x = 0$, which certainly seems reasonable since the graphs in Fig. 1.99 look nice and smooth there, with no breaks or sharp corners. A computation of the derivative of a^x that does not depend on this assumption appears in Chapter 7.

From Eq. (2), it appears that one value for the base a of the exponential function a^x might be of special importance. If we can find a base a such that the tangent line to the graph of a^x where $x = 0$ has slope 1, then the expression in parentheses in Eq. (2) reduces to 1 and can be dropped from the formula. The exponential function to this base would then be unchanged upon differentiation! Figure 3.21 shows the graph of $y = (1.5)^x$. It is clear from the graph that the slope of its tangent line at $(0, 1)$ is less than 1. Figure 3.22 indicates that the slope of the tangent line to $y = 4^x$ at $(0, 1)$ is greater than 1.

We reach for our calculator and try to estimate the value a that will make the limit (1) equal to 1. Taking $\Delta x = 0.001$ as a value of Δx fairly close to zero, we proceed as shown in Table 3.2. We estimate that for some a where $2.7 < a < 2.8$, the limit (1) is equal to 1. If we use the more efficient method for numerically estimating the derivative, which appears at the end of Section 3.1, we can easily discover that the desired value of a satisfies $2.71 < a < 2.72$. Experimenting in a similar way with a computer, we can estimate the desired a even more accurately, obtaining $2.718281828 < a < 2.718281829$. The number a such that the limit (1) is equal to 1 is a very important number in mathematics and is always denoted by e. The value of e to thirteen significant figures is

$$e \approx 2.718281828459.$$

This number is fully as important as the number π. The function e^x is often referred to as the *exponential function*.

Equation (2) shows that we have

$$\frac{d(e^x)}{dx} = e^x.$$

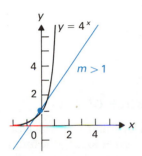

FIGURE 3.22 Tangent line to $y = 4^x$ at $(0, 1)$.

TABLE 3.2 Try to find a such that $(a^{0.001} - 1)/0.001$ is close to 1		
a	$\dfrac{a^{0.001} - 1}{0.001}$	*Classification of a*
2.0	0.6934	Too small
3.0	1.0992	Too big, but not much too big
2.8	1.0301	Still a bit too big
2.7	0.9937	A bit too small, but close

THEOREM 3.5   Derivative of e^x

There is a number e, which is approximately 2.71828, such that

$$\frac{d(e^x)}{dx} = e^x. \tag{3}$$

$\square$

In calculus, we use e almost exclusively as the base for exponential functions, just as we use the radian almost exclusively as the unit of measure for angles with trigonometric functions. It simplifies differentiation.

If we let $u = g(x)$ be any differentiable function of x and use the chain rule, we obtain

$$\frac{d(e^u)}{dx} = \frac{d(e^u)}{du} \cdot \frac{du}{dx} = e^u \frac{du}{dx}.$$

We emphasize this very important formula.

$$\frac{d(e^u)}{dx} = e^u \frac{du}{dx} \tag{4}$$

Be sure you understand the difference between formula (4) and the formula

$$\frac{d(u^n)}{dx} = nu^{n-1} \frac{du}{dx}. \tag{5}$$

Formula (4) gives the derivative of the *constant* e raised to the *variable* power u. Formula (5) gives the derivative of the *variable* u raised to the *constant* power n.

WRONG: $\dfrac{d(e^x)}{dx} = xe^{x-1}.$

EXAMPLE 1   Let $y = e^{3x}$. Find y'.

Solution   Formula (4) yields

$$y' = \frac{d(e^{3x})}{dx} = e^{3x} \frac{d(3x)}{dx} = 3e^{3x}.$$ ■

EXAMPLE 2   Let $y = e^{\sin x}$. Find y'.

Solution   Formula (4) yields

$$y' = \frac{d(e^{\sin x})}{dx} = e^{\sin x} \cdot \frac{d(\sin x)}{dx} = e^{\sin x} \cos x.$$ ■

The Natural Logarithm Function; Derivative of a^x

Recall that if $x = a^y$, then $y = \log_a x$, where $\log_a x$ is the logarithm of x with base a. The base e logarithm function, $\log_e x$, is called the **natural logarithm function** and is denoted by $\ln x$ (often read as "ell en of x"). Just as with exponential and logarithm functions using any base, we have

$$e^{\ln x} = x \tag{6}$$

and

$$\ln(e^x) = x. \tag{7}$$

Equations (6) and (7) show that e^x and $\ln x$ are *inverse functions*. On a scientific calculator without an e^x key, the two-key sequence INV $\ln x$ is used instead. We defer the differentiation of $\ln x$ to Section 3.8.

We now show how to differentiate a^x for any base $a > 0$. From Eq. (6) we see that $e^{\ln a} = a$, so we have

$$a^x = (e^{\ln a})^x = e^{(\ln a)x}. \tag{8}$$

Now $\ln a$ is a constant, so the derivative of $(\ln a)x$ is $\ln a$, just as the derivative of $3x$ is 3. From Eq. (8) and formula (4), we obtain

$$\frac{d(a^x)}{dx} = \frac{d[e^{(\ln a)x}]}{dx} = e^{(\ln a)x} \cdot \frac{d[(\ln a)x]}{dx}$$

$$= (e^{\ln a})^x \cdot (\ln a) = a^x \cdot (\ln a). \tag{9}$$

Combining this result with the chain rule, we obtain formula (10) for the derivative of a^u, where u is a differentiable function of x:

$$\frac{d(a^u)}{dx} = (\ln a)a^u \frac{du}{dx}. \tag{10}$$

The advantage of using base e for exponentials in calculus is now clear. Equation (7) shows that e can be characterized in terms of the natural logarithm as follows:

$\ln e = 1$, and e is the unique number with this property.

Figures 3.23 and 3.24 show the graphs of e^x and of $\ln x$. Since these functions are encountered so frequently, you should remember how their graphs look.

EXAMPLE 3   Let $y = 2^{3x+1}$. Find y'.

Solution   Formula (10) yields

$$y' = \frac{d(2^{3x+1})}{dx} = (\ln 2)2^{3x+1} \cdot \frac{d(3x+1)}{dx} = 3(\ln 2)2^{3x+1}. \quad\blacksquare$$

EXAMPLE 4   Let $y = x^2 3^{-x}$. Find y'.

Solution   We use formula (10) and the product rule, obtaining

$$y' = \frac{d(x^2 3^{-x})}{dx} = x^2 \cdot \frac{d(3^{-x})}{dx} + 3^{-x} \cdot \frac{d(x^2)}{dx} = x^2(\ln 3)3^{-x} \cdot \frac{d(-x)}{dx} + 3^{-x} \cdot 2x$$

$$= x^2(\ln 3)3^{-x}(-1) + 3^{-x}2x = 3^{-x}[-(\ln 3)x^2 + 2x]. \quad\blacksquare$$

EXAMPLE 5   Find all values of x where the graph of $y = 3^x - 2x$ has a horizontal tangent line.

Solution   For the graph to have a horizontal tangent line, the derivative of the function must be zero. Using formula (10), we find that

$$y' = (\ln 3)3^x - 2$$

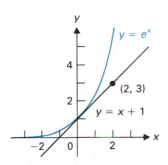

FIGURE 3.23   Graph of $y = e^x$ having tangent line of slope 1 at $x = 0$.

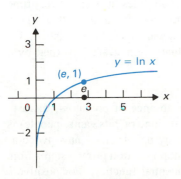

FIGURE 3.24   Graph of $y = \ln x$.

is zero only when

$$(\ln 3)3^x - 2 = 0, \qquad \text{so} \qquad 3^x = \frac{2}{\ln 3}.$$

To solve for a variable in an exponent, we take a logarithm, and we always prefer to work with base e since calculators have a $\ln x$ key. We have

$$\ln(3^x) = \ln\left(\frac{2}{\ln 3}\right),$$

so

$$x \ln 3 = \ln 2 - \ln(\ln 3).$$

Using a calculator, we obtain

$$x = \frac{\ln 2 - \ln(\ln 3)}{\ln 3} \approx 0.5453. \qquad \blacksquare$$

The Function $f(t) = Ae^{kt}$ in Applications

The primary purpose of this chapter is to introduce the notion of the derivative and develop differentiation techniques. As motivation, we consider the function $f(t) = Ae^{kt}$, which has many applications in everyday life.

Let us think of the variable t as measuring time, and let $f(t) = Ae^{kt}$, where A and k are constants. We see that

$$f(0) = Ae^0 = A \cdot 1 = A,$$

so A is the value of the function when $t = 0$. For this reason, we often use A_0 in place of A. Also,

$$f'(t) = A \frac{d(e^{kt})}{dt} = A(ke^{kt}) = k(Ae^{kt}) = k \cdot f(t).$$

The equation

$$f'(t) = k \cdot f(t) \tag{11}$$

shows that the rate of change of $f(t)$ is proportional to the value $f(t)$. For $k > 0$ and $f(t) > 0$, the increase of $f(t)$ is proportional to its size. We will show in Section 4.2 that there is a *unique* function f such that $f(0) = A_0$ and such that Eq. (11) is satisfied. This seems reasonable: If you know how big a thing is at time $t = 0$ and you know how fast it is changing at all times, its size at every time $t > 0$ should be determined uniquely. For the moment we assume that $f(t) = A_0 e^{kt}$ is the unique function f such that $f(0) = A_0$ and Eq. (11) is satisfied. We give three examples of such rate of growth situations that are governed by a function $f(t) = A_0 e^{kt}$.

EXAMPLE 6 *Population Growth*   In the absence of predators, disease, famine, and so on, the rate of growth of a population (rabbits, ants, people) is proportional to its size. If you have twice as many people, they have twice as many offspring. The concern scientists have about population growth stems from their understanding of the behavior of the exponential function. For positive k

(increasing population), the function $A_0 e^{kt}$ increases very rapidly once it acquires any physically significant size. The larger it is, the larger the derivative, so the faster it grows. It feeds on itself!   ■

EXAMPLE 7 *Decay of a Radioactive Substance*   A radioactive substance decays at a rate proportional to the amount of the substance present. When it is half gone, there is only half as much to contribute to decay over the next unit of time as at the beginning. In this decay situation, the constant k is negative and the most rapid rate of decay occurs for small positive values of t. Unfortunately, for some radioactive materials k is quite close to zero, so decay is very slow; the time required for half the substance to decay (the *half-life period*) seems quite long compared with our life span.   ■

EXAMPLE 8 *Continuously Compounded Savings*   A savings account in a bank that has interest *compounded continuously* grows at each instant at a rate proportional to the balance present. If the amount deposited at time $t = 0$ is A_0 dollars and the annual interest rate is $i\%$, then the balance in the account after t years is

$$A_0 e^{(i/100)t} \text{ dollars.}$$

Thus an interest rate of 8% compounded continuously corresponds to taking $k = 0.08$ in the function $A_0 e^{kt}$.   ■

We give two numerical illustrations of continuously compounded savings.

EXAMPLE 9   Referring to Example 8, suppose that $1000 is deposited in a savings account at 8.5% annual interest compounded continuously. Find the balance in the account after 20 years if both the deposit and the interest are left untouched.

Solution   By Example 8, the amount in the account after t years is given by $1000 e^{0.085t}$. When $t = 20$, the balance is therefore

$$1000 e^{(0.085)20} = 1000 e^{1.7} \approx 1000(5.47395) = \$5473.95.   ■$$

EXAMPLE 10   Find how long it takes the balance in a savings account to double if left untouched at 11% annual interest compounded continuously.

Solution   By Example 8, we must find t such that

$$A_0 e^{(0.11)t} = 2A_0.$$

Dividing by A_0 on each side of the equation, we must solve

$$e^{(0.11)t} = 2$$

for t. When solving for a variable in an exponent, take the logarithm of each side of the equation. We obtain

$$(0.011)t = \ln 2, \quad \text{so} \quad t = \frac{\ln 2}{0.11} \approx 6.3 \text{ years.}   ■$$

A Famous Limit

We work a bit more with Example 8 to arrive at a conjecture for another characterization of the number e. Suppose that interest on the original deposit of A_0 dollars is compounded annually at $i\%$. Then after one year, the balance will be

$$\underset{\textit{deposit}}{A_0} + \underset{\textit{interest}}{A_0 \cdot \frac{i}{100}} = A_0\left(1 + \frac{i}{100}\right).$$

That is, the balance in the account after one year can be found by multiplying the present balance by $1 + (i/100)$.

Continuing this multiplication for each successive year, we see that the balance after t years is

$$A_0\left(1 + \frac{i}{100}\right)^t.$$

If interest is compounded semiannually, that is, twice a year, the interest rate for each 6-month period is $(i/2)\%$, but there are $2t$ periods of 6 months in t years. In this case, the balance after t years is

$$A_0\left(1 + \frac{i/100}{2}\right)^{2t}.$$

Continuing this idea, we see that if interest is compounded n times a year, the balance in the account after n years is

$$A_0\left(1 + \frac{i/100}{n}\right)^{nt}. \tag{12}$$

Compounding continuously should amount to taking the *limit* of formula (12) as $n \to \infty$, so we should have

$$\lim_{n \to \infty} A_0\left(1 + \frac{i/100}{n}\right)^{nt} = A_0 e^{(i/100)t}. \tag{13}$$

From Eq. (13), we obtain

$$\lim_{n \to \infty}\left[\left(1 + \frac{i/100}{n}\right)^n\right]^t = e^{(i/100)t}.$$

In particular, taking $i = 100$ and $t = 1$, we expect that

$$\lim_{n \to \infty}\left(1 + \frac{1}{n}\right)^n = e,$$

which leads us to conjecture that

$$\lim_{x \to \infty}\left(1 + \frac{1}{x}\right)^x = e. \tag{14}$$

This conjecture is indeed true, as we show in Chapter 9.

Taking $x = 1,000,000$ in Eq. (14), a calculator tells us that

$$\left(1 + \frac{1}{1,000,000}\right)^{1,000,000} \approx 2.7182818,$$

which coincides with the decimal expansion of e for as many figures as the calculator can display.

We have now seen three characterizations of the number e.

Characterizations of e

1. e is the unique exponential base such that the tangent line to the graph of e^x where $x = 0$ has slope 1.

2. e is the unique number such that $\ln e = 1$.

3. $e = \lim_{x \to \infty} \left(1 + \frac{1}{x}\right)^x$.

SUMMARY

1. e is the unique exponential base such that the tangent line to the graph of the exponential function e^x where $x = 0$ has slope 1.

2. If u is a differentiable function of x, then

$$\frac{d(e^u)}{dx} = e^u \frac{du}{dx}.$$

3. $\ln x = \log_e x$ is called the natural logarithm function.

4. $e^{\ln x} = x$ and $\ln(e^x) = x$

5. e is the unique number such that $\ln e = 1$.

6. If u is a differentiable function of x, then

$$\frac{d(a^u)}{dx} = (\ln a)a^u \frac{du}{dx}.$$

7. The function $f(t) = A_0 e^{kt}$ has the properties $f(0) = A_0$ and $f'(t) = k \cdot f(t)$. It is the *unique* differentiable function with these properties. That is, every differentiable function f whose rate of change at every time t is proportional to $f(t)$ with the constant proportionality factor k has the form $f(t) = A_0 e^{kt}$.

8. $\lim_{x \to \infty} \left(1 + \frac{1}{x}\right)^x = e$

EXERCISES 3.6

In Exercises 1–25, find the derivative of the indicated function.

1. e^{2x}

2. e^{-s}

3. e^{4t+1}

4. e^{3-4x}

5. $e^{(x^2+1)}$

6. $e^{\sqrt{u}}$

7. te^t

8. $\dfrac{x}{e^x}$

9. $\dfrac{e^r + 1}{r}$

10. $e^{2u}\cos 3u$

11. $e^{3x}\tan 2x$

12. $\sin(e^x)$

13. $\sec(e^{2x})$

14. $\cot(e^{4t} + 1)$

15. $\sqrt{e^{2x} + 1}$

16. $\dfrac{e^{2s}}{e^s + 1}$

17. $\cos^2(e^t)$

18. 2^{3x-5}

19. $3^{\sin x}$

20. $2^{1/u}$

21. $5^{(t^2+t)}$

22. $2^{x-\sqrt{x}}$

23. $\sin(10^s)$

24. $e^u \cdot 2^{3u}$

25. $e^{3x} \cdot 10^{x+1}$

26. Find the equation of the line tangent to the graph of $y = e^x$ where $x = 0$.

27. Find the equation of the line normal to the graph of $y = e^x$ where $x = 1$.

28. Find all points where the line tangent to the graph of $y = e^x - x$ is horizontal.

29. Find all points where the line tangent to the graph of $y = e^x - ex + 7$ is horizontal.

For Exercises 30–34, refer to Examples 8–10 in the text.

30. A deposit of $10,000 is made to a Keogh savings account that grows at an annual rate of 10% compounded continuously. The deposit and interest are not touched. Find the balance in the account after 30 years.

31. Repeat Exercise 30 if the interest rate is 5% annually, compounded continuously.

32. Which provides more rapid growth in a savings account: an annual rate of 8% compounded annually or the same annual rate compounded continuously? Find the difference in the balances of accounts of each type after 15 years if each starts with a deposit of $10,000.

33. How long does it take the balance of a savings account to double at 18% annual interest, compounded continuously?

34. How long does it take the balance in a savings account to increase by a factor of 10 at 12% annual interest, compounded continuously?

3.7   HIGHER-ORDER DERIVATIVES AND MOTION

For an object moving on a straight line, the rate of change of its position with respect to time is called the *velocity*. The rate of change of its velocity with respect to time is called the *acceleration*. The rate of change of the acceleration with respect to time is called the *jerk*. Thus if the position of the object on the line (*x*-axis) is x, then

$$\text{Velocity} = \frac{dx}{dt}, \quad \text{Acceleration} = \frac{d(dx/dt)}{dt}, \quad \text{and} \quad \text{Jerk} = \frac{d\left(\dfrac{d(dx/dt)}{dt}\right)}{dt}.$$

We open this section by giving a more reasonable notation for such repeated derivatives. We then apply such derivatives to the study of motion on a line or in a plane.

Higher-Order Derivatives

If $y = f(x)$ is a differentiable function, then $y' = f'(x)$ is again a function of x, the *derived function f'*. We can attempt to find its derivative. The notations are

TABLE 3.3 Notations for derivatives of $y = f(x)$

Derivative	f'-notation	y'-notation	Leibniz notation	D-notation
1st	$f'(x)$	y'	dy/dx	Df
2nd	$f''(x)$	y''	d^2y/dx^2	D^2f
3rd	$f'''(x)$	y'''	d^3y/dx^3	D^3f
4th	$f^{(4)}(x)$	$y^{(4)}$	d^4y/dx^4	D^4f
5th	$f^{(5)}(x)$	$y^{(5)}$	d^5y/dx^5	D^5f
$\vdots$	$\vdots$	$\vdots$	$\vdots$	$\vdots$
nth	$f^{(n)}(x)$	$y^{(n)}$	d^ny/dx^n	D^nf

$$y'' = \frac{d(f'(x))}{dx} = \frac{d^2y}{dx^2} = f''(x).$$

The Leibniz notation d^2y/dx^2 is read ''the second derivative of y with respect to x.'' We can then find the derivative of $f''(x)$, which is the third derivative of $f(x)$. Table 3.3 is a summary of the various notations for derivatives.

EXAMPLE 1 Find the first six derivatives of $y = f(x) = x^4 - 3x^3 + 7x^2 - 11x + 5$, using Leibniz notation.

Solution We have

$$\frac{dy}{dx} = 4x^3 - 9x^2 + 14x - 11, \qquad \frac{d^2y}{dx^2} = 12x^2 - 18x + 14$$

$$\frac{d^3y}{dx^3} = 24x - 18, \qquad \frac{d^4y}{dx^4} = 24, \qquad \frac{d^5y}{dx^5} = 0, \qquad \frac{d^6y}{dx^6} = 0. \quad \blacksquare$$

EXAMPLE 2 If $f(t) = (3t + 2)^{5/3}$, find $f'''(2)$.

Solution We obtain

$$f'(t) = \frac{5}{3}(3t + 2)^{2/3} \cdot 3 = 5(3t + 2)^{2/3},$$

$$f''(t) = \frac{10}{3}(3t + 2)^{-1/3} \cdot 3 = 10(3t + 2)^{-1/3},$$

$$f'''(t) = -\frac{10}{3}(3t + 2)^{-4/3} \cdot 3 = -10(3t + 2)^{-4/3}.$$

Thus $f'''(2) = -10(8)^{-4/3} = -10 \cdot 2^{-4} = -\frac{10}{16} = -\frac{5}{8}.$ $\quad \blacksquare$

EXAMPLE 3 Find $f''(x)$ if $f(x) = e^{2x}\sin 3x$.

Solution Using the product rule and the formulas for differentiating $\sin u$ and e^u, we obtain

$$f'(x) = e^{2x}(\cos 3x) \cdot 3 + (\sin 3x)e^{2x} \cdot 2 = 3e^{2x}\cos 3x + 2e^{2x}\sin 3x.$$

Differentiating again, using the sum and product rules, we have

$$f''(x) = 3e^{2x}(-\sin 3x) \cdot 3 + (\cos 3x)(3e^{2x}) \cdot 2 + 2e^{2x}(\cos 3x) \cdot 3$$
$$+ (\sin 3x)(2e^{2x}) \cdot 2$$
$$= (-9 + 4)e^{2x}\sin 3x + (6 + 6)e^{2x}\cos 3x$$
$$= e^{2x}(-5 \sin 3x + 12 \cos 3x).$$ ■

Motion on a Line

FIGURE 3.25   The position x of a moving body is a function h of time t.

We now consider the motion of a body traveling on a line. We may as well assume that the line is the x-axis. The position of the body on the line at a time t corresponds to a point x on the line. Thus x appears as a function $h(t)$ of t. See Fig. 3.25. If x is a differentiable function of t, then dx/dt is the **velocity** v of the body at time t. Thus

$$v = \frac{dx}{dt}.$$

If $dx/dt > 0$, then a small positive increment Δt in t produces a positive increment Δx in x, so the body is moving in the positive x-direction; if $dx/dt < 0$, the increment Δx is negative so the body is moving in the negative x-direction. The magnitude of the velocity (without regard to sign) is the **speed,** so that

$$\text{Speed} = |v|.$$

Suppose v is also a differentiable function of t. Since $dx/dt = v$, we have

$$\frac{d^2x}{dt^2} = \frac{dv}{dt} = \text{Rate of change of velocity with respect to time.}$$

This derivative dv/dt of the velocity is the **acceleration** a of the body. Thus

$$a = \frac{dv}{dt} = \frac{d^2x}{dt^2}.$$

If $dv/dt > 0$, then a small positive increment Δt in t produces a positive increment Δv in v, so the velocity is increasing, while if $dv/dt < 0$, the increment Δv is negative, so the velocity is decreasing. For example, if $dx/dt < 0$, then the body is moving in the negative x-direction. If also $d^2x/dt^2 > 0$, then the velocity is increasing, perhaps from -2.1 ft/sec to -1.9 ft/sec as t increases a small amount Δt, so the speed will be decreasing from 2.1 ft/sec to 1.9 ft/sec. The body is moving in the negative x-direction but slowing down since the positive acceleration is opposite to the negative direction of motion.

EXAMPLE 4   Let the position x of a body on a straight line (x-axis) at time $t \geq 0$ be given by

$$x = 4 - \frac{1}{t + 1}.$$

Find the velocity and acceleration when $t = 3$.

Solution   We have $x = 4 - (t + 1)^{-1}$, so

$$v = \frac{dx}{dt} = (t + 1)^{-2} \qquad \text{and} \qquad a = \frac{d^2x}{dt^2} = -2(t + 1)^{-3}.$$

Thus

$$v|_{t=3} = \frac{1}{16} \quad \text{and} \quad a|_{t=3} = \frac{-2}{64} = -\frac{1}{32}.$$

The positive velocity indicates that the body is moving in the positive x-direction when $t = 3$, and the negative acceleration, with sign opposite to that of the velocity, indicates that the body is slowing down at $t = 3$. That is, the *speed* is decreasing.   ∎

EXAMPLE 5 If the position x of a body on a line (x-axis) at time t is given by $x = \sqrt{3t^2 + 4}$, find the velocity, acceleration, and speed when $t = 2$.

Solution We have $x = \sqrt{3t^2 + 4} = (3t^2 + 4)^{1/2}$, so

$$v = \frac{dx}{dt} = \frac{1}{2}(3t^2 + 4)^{-1/2}6t = \frac{3t}{\sqrt{3t^2 + 4}},$$

$$a = \frac{d^2x}{dt^2} = \frac{(\sqrt{3t^2 + 4})3 - 3t(\frac{1}{2})(3t^2 + 4)^{-1/2} \cdot 6t}{3t^2 + 4}$$

$$= \frac{(3t^2 + 4)3 - 9t^2}{(3t^2 + 4)^{3/2}}$$

$$= \frac{12}{(3t^2 + 4)^{3/2}}.$$

Thus

$$v|_{t=2} = \frac{6}{\sqrt{16}} = \frac{3}{2}, \qquad a|_{t=2} = \frac{12}{16^{3/2}} = \frac{12}{64} = \frac{3}{16}.$$

Now v is positive when $t = 2$, and we have

$$\text{Speed} = |v| = \frac{3}{2}.$$

Thus the body is moving to the right (assuming the x-axis is in the usual horizontal position), and the positive acceleration means that the velocity is increasing. Since the velocity is positive when $t = 2$, the speed is increasing also.   ∎

Consider an object moving near the surface of the earth on a vertical line (y-axis) perpendicular to the surface of the earth. We indicate such motion in Fig. 3.26. Note that we take the origin of the y-axis to be at the earth's surface and its positive direction to be upward. We neglect air resistance and assume that the only force on the object is due to gravity. Exercise 53 of Section 4.9 will show that the position of the object until it hits the earth is given by an equation of the form

$$y = y_0 + v_0t - \tfrac{1}{2}gt^2,$$

where y_0, v_0, and g are constants.

We can discover the physical meaning of these constants. When $t = 0$, we see that $y = y_0$, so y_0 is the *initial height* at time $t = 0$. Now

$$v = \frac{dy}{dt} = v_0 - gt.$$

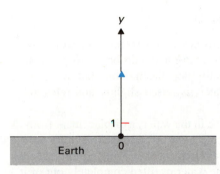

FIGURE 3.26 An object moving upward near the earth's surface.

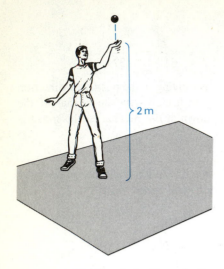

FIGURE 3.27   Sketch for Example 6.

When $t = 0$, we see that $v = v_0$, so v_0 is the *initial velocity* when $t = 0$. Finally,

$$a = \frac{d^2y}{dt^2} = -g,$$

so $-g$ is the acceleration of the body. For acceleration due to gravity near the earth's surface, $g \approx 32$ ft/sec^2 in the British system, and $g \approx 9.8$ m/sec^2 in the metric system.

EXAMPLE 6   A ball is thrown straight upward and is released 2 m above the surface of the earth with velocity 19.6 m/sec at time $t = 0$. See Fig. 3.27. Neglecting air resistance, find

(a) the height y above the surface at time t before the ball hits the earth,

(b) the velocity at time t,

(c) the maximum height the ball attains.

Solution

(a) The discussion before this example shows that

$$y = 2 + 19.6t - 4.9t^2 \text{ m.}$$

(b) We have $v = dy/dt = 19.6 - 9.8t$ m/sec.

(c) At the instant the ball has maximum height, it is stopped while its velocity is changing from positive to negative. By the intermediate value theorem, its velocity must be zero at that instant. This occurs when

$$19.6 - 9.8t = 0, \qquad \text{or} \qquad t = \frac{19.6}{9.8} = 2 \text{ sec.}$$

The height then is given by

$$y|_{t=2} = 2 + (19.6)2 - (4.9)4 = 2 + 39.2 - 19.6 = 21.6 \text{ m.} \quad \blacksquare$$

Simple Harmonic Motion

Consider a weight suspended vertically by a spring attached to a beam, as shown in Fig. 3.28. Suppose the weight is raised vertically a distance a, which still keeps the spring stretched a bit, and is then released. This weight bobs up and down. It can be shown that if the weight is released at time $t = 0$, then its position y in Fig. 3.28 at time t is given by

$$y = a \cos \omega t. \tag{1}$$

The constant ω in Eq. (1) depends on the stiffness of the spring. The weight bobs up and down with *amplitude a*, from a units above to a units below its position of rest. (Air resistance and friction eventually stop the motion, but we neglect them.) If a cube of wood, floating in water, is raised slightly and released, a similar motion results.

For the weight on the spring or the cube in the water, a displacement from its position of rest gives rise to a restoring force proportional to the magnitude of the displacement. (It is assumed that the elastic limit of the spring is not exceeded and that the cube is not pushed beneath the water or lifted completely out of it.) Motion governed by such a restoring force law is called *simple harmonic motion*. Our work in Section 17.3 will show that the position y in such a situation is

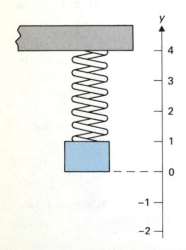

FIGURE 3.28   A weight hanging by a spring at position of rest $y = 0$.

always given by

$$y = a \cos \omega t + b \sin \omega t \qquad (2)$$

for some constants a and b. We assume this for now. Equation (1) is a special case of Eq. (2).

EXAMPLE 7 Let the position (in centimeters) for a body in simple harmonic motion be given by $y = 4 \cos 3t$ after t sec. Find

(a) the amplitude of the motion,

(b) the velocity when $t = \pi/2$ sec,

(c) the speed when $t = \pi/2$ sec,

(d) the acceleration when $t = \pi/2$ sec.

Solution

(a) The amplitude of the motion is 4 cm, the coefficient of $\cos 3t$.

(b) We have $v = dy/dt = -12 \sin 3t$, so when $t = \pi/2$,

$$v = -12 \sin 3t|_{t=\pi/2} = -12 \sin \frac{3\pi}{2} = -12(-1) = 12 \text{ cm/sec.}$$

(c) The speed is $|v| = 12$ cm/sec.

(d) The acceleration is $d^2y/dt^2 = -36 \cos 3t$, so when $t = \pi/2$, the acceleration is

$$-36 \cos 3t|_{t=\pi/2} = -36 \cos \frac{3\pi}{2} = 0 \text{ cm/sec}^2. \qquad \blacksquare$$

EXAMPLE 8 Let the position in simple harmonic motion be given by Eq. (2) with $\omega = 2$. If $y = 0$ cm and $v = -5$ cm/sec when $t = 0$ sec, find the equation for y.

Solution Putting $y = 0$ and $t = 0$ in Eq. (2), we obtain

$$0 = a \cos 0 + b \sin 0 = a(1) + b(0) = a,$$

so Eq. (2) now becomes

$$y = b \sin 2t.$$

Then

$$v = \frac{dy}{dt} = 2b \cos 2t.$$

Setting $v = -5$ and $t = 0$, we obtain

$$-5 = 2b(\cos 0) = 2b(1) = 2b, \qquad \text{so } b = -\tfrac{5}{2}.$$

Thus $y = -\tfrac{5}{2} \sin 2t$ cm at time t sec. $\blacksquare$

Exponential Decay of Velocity

Suppose you are coasting with constant velocity along an x-axis in the vacuum of outer space. Suppose further that you enter a medium, like air or water, that resists your motion with a force proportional to your velocity. We show that under these conditions, your velocity decays exponentially, becoming very small but never becoming mathematically zero.

Let's assume you start your stopwatch when you enter the medium, so that is the instant when $t = 0$. At that instant, you still have your coasting velocity, which we will call v_0. By Newton's second law of motion, $F = ma$, the force of resistance due to the medium is equal to the product of your mass and your acceleration. If we let v be your velocity, then for time $t \geq 0$ and some constant c, we have

$$F = -cv, \quad \text{and} \quad a = v'.$$

The minus sign appears in $F = -cv$ since the force F of resistance opposes the motion, and we think of c as a positive constant. Thus Newton's law $F = ma$ yields $-cv = mv'$, which we can write as

$$v' = -(c/m)v. \tag{3}$$

This is an equation of the form $f' = kf$ as in Eq. (11) of Section 3.6. The independent variable is t and the k in $f' = kf$ is $-c/m$ in the present situation. As explained after Eq. (11) of Section 3.6, we have

$$v = v_0 e^{-(c/m)t}$$

where v_0 is your velocity when $t = 0$. Thus for $t \geq 0$, your velocity is

$$v = v_0 e^{-(c/m)t}. \tag{4}$$

Equation (4) shows that your velocity decays exponentially, since $-c/m$ is negative, and the graph of $e^{-(c/m)t}$ has the shape shown in Fig. 3.29. Now $e^{-(c/m)t}$, or any exponential function, is never zero, so we see that, mathematically, your velocity never actually becomes zero.

EXAMPLE 9 You are coasting in empty space with constant velocity 1000 ft/sec. You and the space capsule in which you are enclosed have a combined mass of 20 slugs. At time $t = 0$, you enter a medium that resists your motion with a force proportional to your velocity. At time $t = 100$ sec, your velocity is 800 ft/sec. Find your velocity as a function of t for $t \geq 0$.

Solution Your velocity when $t = 0$ is $v_0 = 1000$ ft/sec, and your mass is $m = 20$ slugs. By Eq. (4), your velocity for $t \geq 0$ is

$$v = 1000 e^{-(c/20)t}$$

for some positive constant c. The corresponding value $v = 800$ ft/sec when $t = 100$ sec will allow us to find c. We have

$$800 = 1000 e^{-(c/20)100} \quad \text{or} \quad 0.8 = e^{-5c}.$$

Since our unknown c is in an exponent, we take the natural logarithm of both sides of this last equation, obtaining $\ln 0.8 = \ln(e^{-5c}) = -5c$, so $c = -(\ln 0.8)/5 = [\ln(1/0.8)]/5 = (\ln 1.25)/5$. Thus we have

$$v = 1000 e^{-[(\ln 1.25)/(5 \cdot 20)]t} = 1000 e^{-[(\ln 1.25)/100]t} = 1000(e^{\ln 1.25})^{-t/100}$$
$$= 1000(1.25)^{-t/100} \text{ ft/sec} \quad \text{for } t \geq 0. \quad \blacksquare$$

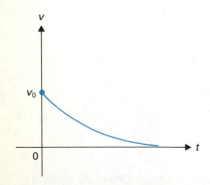

FIGURE 3.29 Exponential decay of velocity for $t \geq 0$.

Exercises 37–40, which are like the preceding example, give valuable practice in working with the logarithm and exponential functions.

Numerical Approximation of $f''(x_1)$ and $f'''(x_1)$ (Optional)

In Section 3.1, we saw that $f'(a)$ can be approximated numerically using

$$f'(a) \approx \frac{f(a + \Delta x) - f(a - \Delta x)}{2 \cdot \Delta x} \tag{5}$$

for small Δx. Now we will give formulas for the numerical approximation of $f''(a)$ and $f'''(a)$. First choose a small Δx and compute:

$$y_1 = f(a - 2 \cdot \Delta x), \qquad y_2 = f(a - \Delta x), \qquad y_3 = f(a),$$
$$y_4 = f(a + \Delta x), \qquad y_5 = f(a + 2 \cdot \Delta x).$$

It can be shown that

$$f''(a) \approx \frac{-y_1 + 16y_2 - 30y_3 + 16y_4 - y_5}{12(\Delta x)^2} \tag{6}$$

and

$$f'''(a) \approx \frac{-y_1 + 2y_2 - 2y_4 + y_5}{2(\Delta x)^3}. \tag{7}$$

These approximations are derived by finding the fourth-degree polynomial through the five points on the graph where x is equal to

$$a - 2 \cdot \Delta x, \quad a - \Delta x, \quad a, \quad a + \Delta x, \quad \text{and} \quad a + 2 \cdot \Delta x.$$

Formula (6) gives the second derivative of this polynomial where $x = a$, and formula (7) gives the third derivative where $x = a$. Note that formula (5) can be written as

$$f'(a) \approx \frac{-y_2 + y_4}{2 \cdot \Delta x}$$

in this notation. It can be shown that this is actually the derivative where $x = a$ of the second-degree polynomial through the points where $x = a - \Delta x$, a, and $a + \Delta x$.

EXAMPLE 10   Use formula (6) with $\Delta x = 1$ to approximate $f''(3)$ if $f(x) = 1/x$. Of course, $\Delta x = 1$ is large, but it makes a feasible computation for this noncalculator illustration.

Solution   We have $a = 3$ and $\Delta x = 1$, so

$$a - 2 \cdot \Delta x = 1, \qquad a - \Delta x = 2, \qquad a = 3,$$
$$a + \Delta x = 4, \qquad a + 2 \cdot \Delta x = 5,$$
$$y_1 = 1, \qquad y_2 = \tfrac{1}{2}, \qquad y_3 = \tfrac{1}{3}, \qquad y_4 = \tfrac{1}{4}, \qquad y_5 = \tfrac{1}{5}.$$

The approximation (6) becomes

$$f''(3) \approx \frac{-1 + 16 \cdot \tfrac{1}{2} - 30 \cdot \tfrac{1}{3} + 16 \cdot \tfrac{1}{4} - \tfrac{1}{5}}{12(1)^2}$$

$$= \frac{-1 + 8 - 10 + 4 - 0.2}{12} = \frac{0.8}{12} \approx 0.067.$$

Of course, if $y = x^{-1}$, then $y' = -1 \cdot x^{-2}$ and $y'' = 2x^{-3}$, so $f''(3) = 2/(3)^3 = \tfrac{2}{27} \approx 0.074$. Our error in approximation is about 0.007. ∎

SUMMARY

1. If $y = f(x)$, then the second derivative of y with respect to x is

$$\frac{d^2y}{dx^2} = \frac{d(dy/dx)}{dx} = f''(x) = y'' = D^2f,$$

and the nth derivative of y with respect to x is

$$\frac{d^ny}{dx^n} = f^{(n)}(x) = y^{(n)} = D^nf.$$

2. If x is the position of a body on a line (x-axis) at time t, then

$$\text{Velocity} = v = \frac{dx}{dt}, \qquad \text{Acceleration} = a = \frac{dv}{dt} = \frac{d^2x}{dt^2},$$

$$\text{Speed} = |v|.$$

3. The position of a body moving on a line (y-axis) in simple harmonic motion is given by an equation of the form

$$y = a \cos \omega t + b \sin \omega t,$$

where a, b, and ω are constants.

4. If the only force acting on motion on a line is a force of resistance proportional to the velocity, then the velocity decays exponentially but never becomes mathematically zero. (See Eq. 4.)

EXERCISES 3.7

In Exercises 1–8, find y', y'', and y'''. You need not simplify your answers.

1. $y = x^5 - 3x^4$

2. $y = \sqrt{x}$

3. $y = 1/\sqrt{5t}$

4. $y = u^{2/3}$

5. $y = \sqrt{x^2 + 1}$

6. $y = (3x - 2)^{3/4}$

7. $y = \dfrac{s}{s + 1}$

8. $y = x(x + 1)^4$

9. Find $f''(3)$ if $f(x) = \sqrt{2x + 4}$.

10. Find $g'''(-1)$ if $g(t) = 8t^3 - 3t^2 + 14t - 5$.

11. Find $f^{(4)}(2)$ if $f(s) = 1/s$.

12. Find $f^{(4)}(1)$ if $f(u) = u/(u + 1)$.

13. Find $f''(t)$ if $f(t) = t^2e^{2t}$.

14. Find $f'''(x)$ if $f(x) = x \sin 2x$.

15. Find $f''(s)$ if $f(s) = e^s \cos s$.

16. Find $f''(u)$ if $f(u) = \tan 3u$.

17. Find $f''(x)$ if $f(x) = \sec 2x$.

18. Find $f'''(r)$ if $f(r) = re^{-r}$.

19. Find a formula for $f^{(n)}(x)$ if $f(x) = xe^x$.

20. If the position x of a body on an x-axis at time t is given by $x = 3t^3 - 7t$, find the velocity and acceleration of the body when $t = 1$.

21. Repeat Exercise 20 if $x = \sqrt{t^2 + 7}$ and $t = 3$.

22. Let the position x on an x-axis of a body at time t be given by

$$x = 10 - \frac{20}{t^2 + 1} \qquad \text{for } t \geq 0.$$

(a) Show that, at any time $t \geq 0$, the body has traveled less than 20 units distance since time $t = 0$.

(b) Find the velocity of the body as a function of t.

(c) Interpret physically the fact that the velocity is positive for all $t > 0$.

23. Let $x = 10/(t + 1)$ be the position of a body on an x-axis for $t \geq 0$.
(a) Find the velocity as a function of t.
(b) Interpret physically the fact that the velocity is negative for all $t > 0$.
(c) Find the acceleration as a function of t.
(d) Interpret physically the fact that the acceleration is positive for all $t > 0$.
(e) Is the speed increasing or decreasing for $t > 0$?

24. A body is thrown straight up from the surface of the earth at time $t = 0$. Suppose that its height y in feet after t sec is $y = -16t^2 + 48t$.
(a) Find the velocity of the body at time t.
(b) Find the acceleration of the body at time t.
(c) Find the initial velocity v_0 with which the body was thrown upward.
(d) At what time does the body reach maximum height? [*Hint:* What is the velocity at maximum height?]
(e) Find the maximum height the body attains.
(f) From the physics of the problem, during what time interval is the equation $y = -16t^2 + 48t$ valid?

In Exercises 25–30, $y = k(t)$ is the position at time t of a body on the y-axis (positive direction upward). Fill in the blanks with the proper choice: *upward, downward, increasing, decreasing.*

25. If $dy/dt > 0$, the body is moving _____ as t increases.

26. If $d^2y/dt^2 < 0$, the velocity is _____ as t increases.

27. If $d^2y/dt^2 < 0$ and $dy/dt > 0$, the speed of the body is _____ as t increases.

28. If $d^2y/dt^2 < 0$ and $dy/dt < 0$, the speed of the body is _____ and the body is moving _____ as t increases.

29. If $d^2y/dt^2 > 0$ and $dy/dt < 0$, the speed of the body is _____ and the body is moving _____ as t increases.

30. If $d^2y/dt^2 > 0$ and $dy/dt > 0$, the speed of the body is _____ and the body is moving _____ as t increases.

31. A ball is tossed upward with a velocity of 7 m/sec. The ball is released 1.5 m above the ground at time $t = 0$. Neglecting air resistance, find
(a) the velocity when $t = 1$ sec,
(b) the speed of the ball when $t = 1$ sec,
(c) the maximum height the ball attains.

32. A boy standing on a balcony reaches out and tosses a ball straight upward with velocity 14 m/sec. The ball is released 15 m above the ground at time $t = 0$. See Fig. 3.30. Neglecting air resistance, find
(a) the velocity when $t = 1$,
(b) the speed when $t = 1$,
(c) the maximum height the ball attains,
(d) the elapsed time until the ball hits the ground.

33. Let the position of a body in simple harmonic motion be given by $y = -5 \sin \pi t$ cm at t sec. Find
(a) the amplitude,
(b) the velocity when $t = \frac{1}{3}$,
(c) the speed when $t = \frac{1}{3}$,
(d) the acceleration when $t = \frac{1}{3}$.

34. Let the position of a body in simple harmonic motion be given by $y = 10 \cos 2\pi t$ cm at time t sec. Find
(a) the velocity when $t = \frac{1}{4}$,

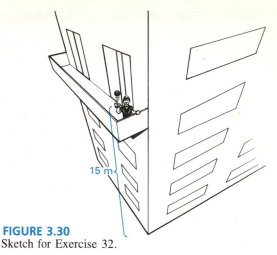

FIGURE 3.30
Sketch for Exercise 32.

(b) the speed when $t = \frac{1}{3}$,
(c) the acceleration when $t = \frac{1}{2}$.

35. Let the position of a body in simple harmonic motion be given by $y = a \cos \pi t + b \sin \pi t$. If $y = 0$ cm and $v = 3$ cm/sec when $t = 0$ sec, find a and b.

36. Let the position of a body in simple harmonic motion be given by $y = a \cos[(\pi/2)t] + b \sin[(\pi/2)t]$. If $y = -2$ cm and $v = -3$ cm/sec when $t = 0$ sec, find the acceleration when $t = \frac{1}{2}$.

37. Referring to Example 9, find the velocity v for time $t \geq 0$ if the data are changed so that $m = 10$ slugs, $v_0 = 5000$ ft/sec, and $v = 3000$ ft/sec when $t = 200$ sec.

38. Referring to Eq. (4), suppose that $c/m = 0.01$ in units corresponding to seconds for time and meters for distance. Find the time required to slow to half the initial velocity v_0. [*Hint:* Set $v = \frac{1}{2}v_0$ and solve for t.]

39. Suppose in Eq. (4) that $v = \frac{1}{2}v_0$ when $t = 500$. Find the value of t such that $v = \frac{1}{4}v_0$.

40. Repeat Exercise 39, but find the value of t such that $v = (1/\sqrt{2})v_0$.

In Exercises 41–46, use a calculator to find the second and third derivatives of the given functions at the indicated points, using approximations (6) and (7). You choose Δx.

41. $f(x) = \sqrt{x}(x^3 + 2x^2)$ at $x = 1$

42. $f(x) = \dfrac{\sin x}{x + 2}$ at $x = 3$ (radian measure)

43. $f(x) = 2^x$ at $x = 3$

44. $f(x) = x^x$ at $x = 1.5$

45. $f(x) = (x^2 + 3x)^{\sqrt{x}}$ at $x = 1$

46. $f(x) = x^{\sin x}$ at $x = 2$ (radian measure)

  EXPLORING CALCULUS

K: compute $f'(a)$, $f''(a)$

3.8   IMPLICIT DIFFERENTIATION; DERIVATIVE OF ln x

The graph of a continuous function f of one variable can be viewed as a curve in the plane. The curve has a tangent line at each point where f is differentiable.

Surely it is natural to consider the circle $x^2 + y^2 = 25$ in Fig. 3.31 to be a plane curve. However, this curve is not the graph of a function, since distinct points on the circle may have the same x-coordinate. For example, $(3, 4)$ and $(3, -4)$ are both on the circle. Let us restrict our attention to just a small part of the circle containing $(3, 4)$ and extending a short distance on both sides of $(3, 4)$. See the colored curve in Fig. 3.31. This small piece is the graph of a function. In fact, it is part of the graph of the function $y = f(x) = \sqrt{25 - x^2}$, which has as graph the whole upper semicircle. We would like to find the derivative $f'(3)$ of this function, that is, the slope of the tangent line to the circle at the point $(3, 4)$. We have

$$f'(x) = \frac{1}{2}(25 - x^2)^{-1/2}(-2x) = \frac{-x}{\sqrt{25 - x^2}},$$

so

$$f'(3) = \frac{-3}{\sqrt{16}} = -\frac{3}{4}.$$

We now show another way to find $f'(3)$, the slope of the tangent line to the circle $x^2 + y^2 = 25$ at $(3, 4)$. In this new method, we will not have to solve for y in terms of x. Thinking of y as some function of x near the point with which we are concerned, we differentiate both sides of the equation $x^2 + y^2 = 25$ with respect to x. Since we are differentiating *with respect to x* and since y is a function of x, the chain rule yields

$$\frac{d(y^2)}{dx} = 2y\frac{dy}{dx}.$$

Thus we obtain

$$x^2 + y^2 = 25, \qquad \textit{Given}$$

$$2x + 2y\frac{dy}{dx} = 0, \qquad \textit{Differentiating}$$

$$\frac{dy}{dx} = -\frac{2x}{2y} = -\frac{x}{y}, \qquad \textit{Solving for } dy/dx$$

$$\left.\frac{dy}{dx}\right|_{(3,\,4)} = -\frac{3}{4}. \qquad \textit{Evaluating at } (x, y) = (3, 4)$$

We obtained the same answer as in the preceding paragraph, but no radical appeared in this computation.

The computation we have just completed illustrates our topic in this section. Let a plane curve be given by an x,y-equation. The curve may not be the graph of a function, but a piece of the curve extending a short distance on both sides of a point (x_1, y_1) may be the graph of a function, as shown in Fig. 3.32. In this case, the x,y-equation defines y **implicitly** as a function of x near (x_1, y_1). An equation $y = f(x)$, on the other hand, defines y **explicitly** as a function of x. The terms

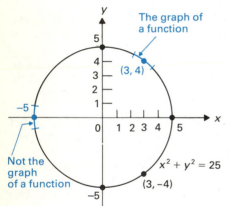

FIGURE 3.31   Some short arcs of $x^2 + y^2 = 25$ are graphs of functions; others are not.

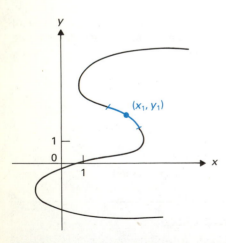

FIGURE 3.32   A portion of the curve extending a short distance both ways from (x_1, y_1) is the graph of a function.

implicit and *explicit* are used to distinguish between these cases. Turning back to our circle, we say that

$$x^2 + y^2 = 25 \text{ defines } y \text{ implicitly as a function of } x \text{ near } (3, 4),$$

while

$$y = \sqrt{25 - x^2} \text{ defines } y \text{ explicitly as a function of } x \text{ near } (3, 4).$$

For the circle $x^2 + y^2 = 25$, we can solve explicitly for y in terms of x and then find dy/dx. But often it is very difficult to solve an equation for y. For example, it is hard to solve for y if

$$y^5 + 3y^2 - 2x^2 = -4.$$

However, assuming that y is defined implicitly as a differentiable function of x near some point (x, y) on the curve, we can use the chain rule to find dy/dx without solving for y. This technique is *implicit differentiation* and was just illustrated. The labels on the steps of the computation there can be taken as an outline for the general technique. (See the highlighted steps at the bottom of this page.)

We give several more illustrations of the technique. (We use dy/dx rather than y' here, since the prime might be mistaken for an exponent in written work.)

EXAMPLE 1   Find dy/dx if $y^5 + 3y^2 - 2x^2 = -4$.

Solution   Viewing y as a function of x and using the chain rule, we have

$$\frac{d(y^5)}{dx} = \frac{d(y^5)}{dy} \cdot \frac{dy}{dx} = 5y^4 \frac{dy}{dx} \quad \text{and} \quad \frac{d(3y^2)}{dx} = 6y \frac{dy}{dx}.$$

Differentiating both sides of $y^5 + 3y^2 - 2x^2 = -4$ with respect to x, we obtain

$$5y^4 \frac{dy}{dx} + 6y \frac{dy}{dx} - 4x = 0.$$

We may now solve for dy/dx, obtaining

$$\frac{dy}{dx} = \frac{4x}{5y^4 + 6y}.$$

Finding *dy/dx* by Implicit Differentiation

STEP 1.   Write an equation in x and y.

STEP 2.   Differentiate both sides of the equation with respect to x, thinking of y as some function of x. (When differentiating any y-term, you must use the chain rule, so the derivative of that term will end with the factor dy/dx.)

STEP 3.   Solve algebraically for dy/dx. (Since dy/dx always appears to the first power only, this is always easy.)

STEP 4.   If you need the derivative corresponding to some point (x_1, y_1) on the curve, evaluate the expression for dy/dx in step 3 at that point.

This formula gives dy/dx at any point (x, y) on the curve where the denominator $5y^4 + 6y$ is nonzero. For example, we can easily see that $(2, 1)$ satisfies $y^5 + 3y^2 - 2x^2 = -4$ and therefore lies on the curve. Then

$$\left.\frac{dy}{dx}\right|_{(2,\ 1)} = \left.\frac{4x}{5y^4 + 6y}\right|_{(2,\ 1)} = \frac{8}{11}.$$ ■

EXAMPLE 2 Find ds/dt if $e^{st} + s \cos t = 4$.

Solution Differentiating implicitly with respect to t, we find that

$$e^{st}\left(s + t\frac{ds}{dt}\right) + s(-\sin t) + (\cos t)\frac{ds}{dt} = 0,$$

$$(te^{st} + \cos t)\frac{ds}{dt} = s(\sin t) - se^{st},$$

$$\frac{ds}{dt} = \frac{s(\sin t) - se^{st}}{te^{st} + \cos t}.$$ ■

As illustrated by the preceding examples, we obtain a quotient when finding a derivative by implicit differentiation. At certain points on the curve, the denominator may become zero. It can be shown that if, when computing dy/dx, this calamity does not occur at (x, y) and the curve is a sufficiently smooth one, then y is implicitly defined as a function of x near (x, y), and implicit differentiation does yield dy/dx.

EXAMPLE 3 Air is blown into a spherical balloon, increasing its volume V. Find the rate of increase of its surface area S with respect to V when V is 288π cm³.

Solution We wish to compute dS/dV when $V = 288\pi$ cm³. For a sphere of radius r, we have $V = \frac{4}{3}\pi r^3$ and $S = 4\pi r^2$. To obtain an equation involving just V and S, note that

$$r^3 = \frac{3V}{4\pi} \qquad \text{and} \qquad r^2 = \frac{S}{4\pi}.$$

Therefore,

$$r^6 = (r^3)^2 = \frac{9V^2}{16\pi^2} = (r^2)^3 = \frac{S^3}{64\pi^3}.$$

Then

$$\frac{S^3}{64\pi^3} = \frac{9V^2}{16\pi^2}, \qquad \text{so} \qquad S^3 = 36\pi V^2.$$

Differentiating this last equation with respect to V, we have

$$3S^2\frac{dS}{dV} = 72\pi V,$$

$$\frac{dS}{dV} = \frac{72\pi V}{3S^2} = 24\pi\frac{V}{S^2}.$$

When $V = \frac{4}{3}\pi r^3 = 288\pi$, we obtain

$$r^3 = \frac{3}{4}(288) = 3(72) = 216, \qquad \text{so} \qquad r = 6.$$

Then $S = 4\pi r^2 = 144\pi$. Our answer is therefore

$$24\pi\frac{V}{S^2}\bigg|_{\substack{S=144\pi \\ V=288\pi}} = \frac{24\pi \cdot 288\pi}{144^2\pi^2} = \frac{24}{72} = \frac{1}{3} \text{ cm}^2/\text{cm}^3. \qquad \blacksquare$$

Higher-order derivatives of implicit functions can be found by repeated implicit differentiation, as illustrated in the next example.

EXAMPLE 4   Find y'' if $x^2 + y^2 = 2$.

Solution   This time we use y' and y'' for the first and second derivatives in the implicit differentiation. For y defined implicitly as a function of x by $x^2 + y^2 = 2$, we obtain

$$2x + 2yy' = 0, \qquad \text{so} \quad y' = \frac{-x}{y}.$$

Differentiating this relation again implicitly with respect to x, we obtain

$$y'' = \frac{y(-1) - (-x)y'}{y^2} = \frac{-y + xy'}{y^2}.$$

Since $y' = -x/y$,

$$y'' = \frac{-y + x(-x/y)}{y^2} = \frac{-y^2 - x^2}{y^3}.$$

Remembering that $x^2 + y^2 = 2$, we have

$$y'' = \frac{-2}{y^3}. \qquad \blacksquare$$

Differentiation of Logarithm Functions

We can use implicit differentiation to find the derivative of $f(x) = \ln x$. Recall that $y = \ln x$ can be rewritten as

$$e^y = x.$$

Differentiating Eq. (1) implicitly, we obtain

$$\frac{d(e^y)}{dx} = 1,$$

or

$$e^y\frac{dy}{dx} = 1.$$

Then

$$\frac{dy}{dx} = \frac{1}{e^y} = \frac{1}{x},$$

that is,

$$\frac{d(\ln x)}{dx} = \frac{1}{x}, \qquad x > 0.$$

The chain rule gives us the general formula

$$\frac{d(\ln u)}{dx} = \frac{1}{u} \cdot \frac{du}{dx}, \qquad u > 0, \tag{2}$$

where u is a differentiable function of x.

EXAMPLE 5 Let $y = \ln(\sin x)$. Find dy/dx.

Solution Using formula (2) with $u = \sin x$, we obtain

$$\frac{dy}{dx} = \frac{1}{\sin x} \cdot \frac{d(\sin x)}{dx} = \frac{1}{\sin x}(\cos x) = \cot x, \qquad \sin x > 0. \quad \blacksquare$$

EXAMPLE 6 Let $y = \ln\left(\dfrac{x^2}{x^3 + 1}\right)$. Find dy/dx.

Solution If we use formula (2) immediately, we are faced with the task of finding du/dx where $u = x^2/(x^3 + 1)$. While we can do this using the quotient rule, it is best to first use the algebraic properties $\ln(a/b) = \ln a - \ln b$ and $\ln(a^n) = n(\ln a)$ of logarithm functions and then differentiate. We have

$$y = \ln(x^2) - \ln(x^3 + 1) = 2(\ln x) - \ln(x^3 + 1).$$

Now we differentiate and obtain

$$\frac{dy}{dx} = 2 \cdot \frac{1}{x} - \frac{1}{x^3 + 1} \cdot \frac{d(x^3 + 1)}{dx} = \frac{2}{x} - \frac{3x^2}{x^3 + 1}, \qquad x^3 + 1 > 0,$$

avoiding the quotient rule altogether. $\blacksquare$

Always see if you can use algebraic properties of logarithm functions to simplify before differentiating $\ln u$.

EXAMPLE 7 Differentiate $f(t) = \ln \dfrac{\sin^2 3t}{t^3}$.

Solution First we write

$$f(t) = \ln \frac{\sin^2 3t}{t^3} = 2 \ln(\sin 3t) - 3(\ln t)$$

Then we differentiate, obtaining

$$f'(t) = 2 \cdot \frac{1}{\sin 3t} \cdot \frac{d(\sin 3t)}{dt} - 3 \cdot \frac{1}{t} = \frac{2}{\sin 3t}(3 \cos 3t) - \frac{3}{t}$$

$$= 6(\cot 3t) - \frac{3}{t}, \qquad t > 0. \quad \blacksquare$$

Of course, we can also find $d(\log_a x)/dx$ by implicit differentiation. If $y = \log_a x$, then

$$a^y = x.$$

Implicit differentiation yields

$$(\ln a)a^y \cdot \frac{dy}{dx} = 1,$$

so

$$\frac{dy}{dx} = \frac{1}{\ln a} \cdot \frac{1}{a^y} = \frac{1}{\ln a} \cdot \frac{1}{x}, \qquad x > 0. \tag{3}$$

This time, the "nuisance factor" $\ln a$ arising from the base $a \neq e$ appears in the denominator. You see why we prefer to work with the *natural* exponential and logarithm functions as much as possible in calculus. The chain rule form of formula (3) is

$$\frac{d(\log_a u)}{dx} = \frac{1}{\ln a} \cdot \frac{1}{u} \cdot \frac{du}{dx}, \qquad u > 0. \tag{4}$$

EXAMPLE 8   Find dy/dx if $y = \log_2(x^3 + 1)$.

Solution   Formula (4) shows that

$$\frac{dy}{dx} = \frac{1}{\ln 2} \cdot \frac{1}{x^3 + 1} \cdot \frac{d(x^3 + 1)}{dx} = \frac{1}{\ln 2} \cdot \frac{3x^2}{x^3 + 1}, \qquad x^3 + 1 > 0. \quad ∎$$

EXAMPLE 9   Find ds/dt if $s = \log_{10}(\sin t \, \sec^3 t)$.

Solution   First we use algebraic properties of logarithm functions and simplify, obtaining

$$s = \log_{10} \sin t + 3(\log_{10} \sec t).$$

Formula (4) then yields

$$\frac{ds}{dt} = \frac{1}{\ln 10} \cdot \frac{1}{\sin t} \cdot \frac{d(\sin t)}{dt} + \frac{3}{\ln 10} \cdot \frac{1}{\sec t} \cdot \frac{d(\sec t)}{dt}$$

$$= \frac{1}{\ln 10} \cdot \frac{1}{\sin t}(\cos t) + \frac{3}{\ln 10} \cdot \frac{1}{\sec t}(\sec t \tan t)$$

$$= \frac{1}{\ln 10}(\cot t + 3 \tan t), \qquad \sin t \, \sec^3 t > 0. \quad ∎$$

Differentiating $f(x)^{g(x)}$

It can be shown that if $f(x)$ and $g(x)$ are differentiable functions, then $f(x)^{g(x)}$ is differentiable where $f(x) > 0$. Recalling that $e^{\ln u} = u$, we see that we have

$$f(x)^{g(x)} = (e^{\ln f(x)})^{g(x)} = e^{(\ln f(x))g(x)}. \tag{5}$$

This final expression is of the form $e^{k(x)}$, which we know how to differentiate.

EXAMPLE 10 Let $h(x) = x^{\tan x}$. Find $h'(x)$.

Solution First we rewrite $h(x)$ as in Eq. (5):

$$h(x) = e^{(\ln x)(\tan x)}.$$

Thus

$$h'(x) = e^{(\ln x)(\tan x)} \cdot \frac{d[(\ln x)(\tan x)]}{dx}$$

$$= x^{\tan x}\left[(\ln x)(\sec^2 x) + \frac{\tan x}{x}\right]. \qquad \blacksquare$$

EXAMPLE 11 Let $u = f(x)$ be a positive differentiable function. Derive the formula $d(u^r)/dx = ru^{r-1} \cdot du/dx$ where r is any real number, not necessarily rational.

Solution Let $y = u^r = e^{r(\ln u)}$. Then

$$\frac{dy}{dx} = e^{r(\ln u)} \cdot \frac{d[r(\ln u)]}{dx}$$

$$= e^{r(\ln u)} \cdot \frac{r}{u} \cdot \frac{du}{dx}$$

$$= u^r \cdot \frac{r}{u} \cdot \frac{du}{dx} = ru^{r-1} \cdot \frac{du}{dx}. \qquad \blacksquare$$

An alternative technique for differentiating these exponentials and also for converting differentiation of products and quotients to differentiation of sums and differences is *logarithmic differentiation*. Section 7.3 is devoted to that technique, but you already have the background to read Section 7.3 and learn logarithmic differentiation, if you wish.

We recommend that you commit formulas for differentiating logarithm functions to memory *before* you do any of the exercises.

SUMMARY

1. Let y be defined implicitly as a function of x. We can find dy/dx by following these steps.

   STEP 1. Write down an equation in x and y.

   STEP 2. Differentiate both sides of the equation with respect to x, thinking of y as some function of x and using the chain rule.

   STEP 3. Solve algebraically for dy/dx in terms of x and y.

2. If u is any differentiable function of x, then, for $u > 0$,

$$\frac{d(\ln u)}{dx} = \frac{1}{u} \cdot \frac{du}{dx} \qquad \text{and} \qquad \frac{d(\log_a x)}{dx} = \frac{1}{\ln a} \cdot \frac{1}{u} \cdot \frac{du}{dx}.$$

3. An exponential expression of the form $f(x)^{g(x)}$ can be differentiated by rewriting it as $e^{[\ln f(x)]g(x)}$.

EXERCISES 3.8

In Exercises 1–4, use the following two methods to find dy/dx when $x = x_1$ for y defined implicitly as a function of x near (x_1, y_1). (a) Find y explicitly as a function of x and differentiate, and (b) differentiate implicitly.

1. $x^2 + y^2 = 25$ and $(x_1, y_1) = (-3, 4)$

2. $x - y^2 = 3$ and $(x_1, y_1) = (7, 2)$

3. $y^2 - 2xy + 3x^2 = 1$ and $(x_1, y_1) = (0, -1)$ [*Hint:* Use the quadratic formula to solve $y^2 + (-2x)y + (3x^2 - 1) = 0$ for y.]

4. $2y^2 + 3x^2y - 4x = -2$ and $(x, y) = (1, -2)$ [See the hint to Exercise 3.]

In Exercises 5–20, find the indicated derivative at the given point by implicit differentiation.

5. $x^2 - y^2 = 16$; $\dfrac{dy}{dx}$ at $(x, y) = (5, -3)$

6. $t^3 + s^3 = 7$; $\dfrac{ds}{dt}$ at $(t, s) = (-1, 2)$

7. $uv = 12$; $\dfrac{dv}{du}$ at $(u, v) = (2, 6)$

8. $sr^2 = 12$; $\dfrac{dr}{ds}$ at $(s, r) = (3, -2)$

9. $xy^2 - 3x^2y + 4 = 0$; $\dfrac{dy}{dx}$ at $(x, y) = (-1, 1)$

10. $x^2e^t + t^2 \sin x = 1$; $\dfrac{dt}{dx}$ at $(x, t) = (1, 0)$

11. $(x - y)\sqrt{3x + 2y} = 8$; $\dfrac{dy}{dx}$ at $(x, y) = (4, 2)$

12. $3s^2t^3 + 4st^2 = 6 + \sqrt{t}$; $\dfrac{dt}{ds}$ at $(s, t) = (1, 1)$

13. $x/(x^2 + y^2) = -\frac{1}{5}$; dy/dx at $(x, y) = (-1, 2)$ [*Hint:* You can avoid using the quotient rule.]

14. $\ln y = \dfrac{xy}{x^2 + 1}$; dy/dx at $(x, y) = (0, 1)$

15. $e^y + y = x$; $\dfrac{dy}{dx}$ at $(x, y) = (1, 0)$

16. $t^2 + e^s + \sin s = 2$; $\dfrac{dt}{ds}$ at $(s, t) = (0, 1)$

17. $e^{tx} + 3t^2 = 4$; $\dfrac{dx}{dt}$ at $(t, x) = (1, 0)$

18. $uv + \cos(uv) = 1$; $\dfrac{dv}{du}$ at $(u, v) = (2, 0)$

19. $xy + \ln y = 4$; $\dfrac{dy}{dx}$ at $(x, y) = (4, 1)$

20. $e^t + \ln s = 1$; $\dfrac{ds}{dt}$ at $(t, s) = (0, 1)$

In Exercises 21–40, find the derivative of the given function. Use properties of logarithm functions first, where feasible, to simplify differentiation.

21. $\ln(3x + 2)$

22. $\ln\sqrt{t}$

23. $\ln x^3$

24. $\ln(\tan x)$

25. $\ln(\cos^2 s)$

26. $\ln\left(\dfrac{2x + 3}{x^2 + 4}\right)$

27. $(\ln x)^2$

28. $(\ln x)(\sin x)$

29. $\ln\sqrt{3x^3 - 4x}$

30. $\dfrac{\ln x}{x^2}$

31. $\ln(\sec u \tan u)$

32. $\ln[(t^2 + 4t)^2(3t - 2)^3]$

33. $\ln(\cos^2 x \sin^3 2x)$

34. $\ln(\ln t)$

35. $\log_{10} 2x$

36. $\log_2(r^2 + 1)$

37. $\log_2(\sin s)$

38. $\log_{10}\left(\dfrac{1}{x}\right)$

39. $\log_5 \dfrac{2x + 1}{x}$

40. $\log_3(s \sin s)$

41. If $x^2 - y^2 = 7$, find $\dfrac{d^2y}{dx^2}$ at the point $(4, 3)$.

42. Find $\dfrac{d^2y}{dx^2}$ at the point $(1, 2)$ if $y^2 - xy = 2$.

43. Find $\dfrac{d^2y}{dx^2}$ at the point $(2, -1)$ if $y^3 - xy = 1$.

44. Find $\dfrac{d^2y}{dx^2}$ at the point $(-2, 1)$ if $y^4 + x^2y = 5$.

In Exercises 45–53, use Eq. (5) and find the derivative of the given function.

45. x^x

46. x^{2x}

47. $t^{\sin t}$

48. $(\cos x)^x$

49. $(x^2 + 1)^{x+1}$

50. $(\sqrt{s})^s$

51. $u^{(e^u)}$

52. $x^{(2^x)}$

53. $(\sin t)^{\cos t}$

Two curves are **orthogonal** at a point of intersection if they have perpendicular tangent lines at that point.

54. Verify Fig. 3.33, showing that the two curves $y = x^2$ and $x^2 + 2y^2 = 3$ are orthogonal at the point $(1, 1)$ of intersection.

55. Show that the curves $2x^2 + y^2 = 24$ and $y^2 = 8x$ are orthogonal at the point $(2, 4)$ of intersection.

56. Show that for any values of $c > 0$ and k, the two curves $x^2 + 2y^2 = c$ and $y = kx^2$ are orthogonal at all points of intersection.

57. Show that for all nonzero values of c and k, the two curves $y^2 - x^2 = c$ and $xy = k$ are orthogonal at all points of intersection.

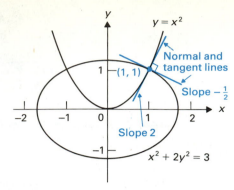

FIGURE 3.33
$y = x^2$ and $x^2 + 2y^2 = 3$ are orthogonal at $(1, 1)$.

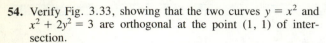

3.9 LINEAR APPROXIMATION AND DIFFERENTIAL NOTATION

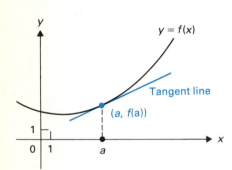

FIGURE 3.34   Tangent line to $y = f(x)$, where $x = a$.

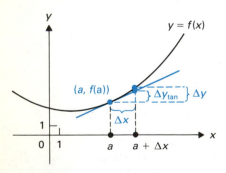

FIGURE 3.35 Δy = Change in height of the graph of f; Δy_{tan} = Change in height of the tangent line.

Scientists frequently encounter problems involving functions given by very complicated formulas that are difficult to work with. One frequently used technique is to *linearize* such a problem. Geometrically, this amounts to approximating the graph of a function near a point by the tangent line to the graph there. As indicated in Fig. 3.34, the tangent line to the graph of a function f at $(a, f(a))$ appears to be the line that clings closest to the graph near that point.

We will show that if $f'(a)$ exists, then the function f can be approximated very accurately, for x very close to a, by a linear function. This *approximation theorem* (Theorem 3.6) lies at the heart of calculus. Using this theorem, we will give a complete proof of the chain rule. After numerical illustrations of the theorem, we discuss the differential notations dy, dx, and dy/dx, which are used in engineering, physics, and other sciences that employ calculus.

Local Linear Approximation of a Function

Figure 3.35 shows the line tangent to the graph of a differentiable function f at a point $(a, f(a))$. We are interested in how closely the tangent line clings to the graph as x changes from a to $a + \Delta x$. As shown in the figure, this change Δx in x produces a change Δy in $f(x)$ and a change Δy_{tan} in the height of the tangent line. We assume that f is differentiable, so $f'(a)$ exists. From Fig. 3.35, we see that the slope of the tangent line at $(a, f(a))$ is $(\Delta y_{tan})/(\Delta x)$, so we have

$$f'(a) = \frac{\Delta y_{tan}}{\Delta x}, \qquad \text{or}$$

$$\Delta y_{tan} = f'(a) \cdot \Delta x. \qquad (1)$$

Figure 3.36 shows the difference $\Delta y - \Delta y_{tan}$, which can be viewed as the error at $x = a + \Delta x$ in approximating $f(x)$ by the line tangent to the graph where $x = a$. This error is a function of Δx, and we denote it by

$$E(\Delta x) = \Delta y - \Delta y_{tan}. \qquad (2)$$

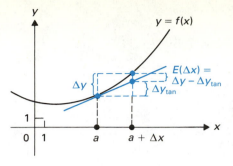

FIGURE 3.36 $E(\Delta x) = \Delta y - \Delta y_{\text{tan}}$

We claim that this error $E(\Delta x)$ is small compared with the size of Δx if Δx is itself sufficiently small. Namely, we now show that

$$\lim_{\Delta x \to 0} \frac{E(\Delta x)}{\Delta x} = 0. \qquad (3)$$

For the limit of a quotient to approach zero as the denominator approaches zero, the numerator must approach zero significantly faster than the denominator. Thus Eq. (3) tells us that $E(\Delta x)$ is of very small size compared with Δx, provided that Δx is itself sufficiently small. To demonstrate Eq. (3), we use the definition of $E(\Delta x)$ in Eq. (2) and also use Eq. (1):

$$\lim_{\Delta x \to 0} \frac{E(\Delta x)}{\Delta x} = \lim_{\Delta x \to 0} \frac{\Delta y - \Delta y_{\text{tan}}}{\Delta x} = \lim_{\Delta x \to 0} \frac{\Delta y - f'(a) \cdot \Delta x}{\Delta x}$$

$$= \lim_{\Delta x \to 0} \left[\frac{\Delta y}{\Delta x} - \frac{f'(a) \cdot \Delta x}{\Delta x} \right] = f'(a) - f'(a) = 0.$$

This establishes Eq. (3) and assures us that the approximation of Δy by Δy_{tan} has an error whose magnitude is only a very small percentage of the size of Δx if Δx is itself of sufficiently small magnitude.

We now state our result from Eq. (3) as a theorem. To use the standard notation, first let

$$\varepsilon(\Delta x) = \begin{cases} \dfrac{E(\Delta x)}{\Delta x} & \text{if } \Delta x \neq 0, \\ 0 & \text{if } \Delta x = 0. \end{cases}$$

Then $E(\Delta x) = \varepsilon(\Delta x) \cdot \Delta x$, and Eq. (3) becomes $\lim_{\Delta x \to 0} \varepsilon(\Delta x) = 0$. Since $\Delta y - \Delta y_{\text{tan}} = E(\Delta x)$, we may write $\Delta y = \Delta y_{\text{tan}} + E(\Delta x) = f'(a) \cdot \Delta x + \varepsilon(\Delta x) \cdot \Delta x$. Finally, it is conventional to write just ε in place of $\varepsilon(\Delta x)$ in the theorem to make the notation less cumbersome.

THEOREM 3.6   Approximation

Let $y = f(x)$ be differentiable at $x = a$, and let $\Delta y = f(a + \Delta x) - f(a)$. Then there exists a function ε of Δx, defined for small Δx, such that

$$\Delta y = f'(a) \cdot \Delta x + \varepsilon \cdot \Delta x,$$

where $\lim_{\Delta x \to 0} \varepsilon = 0$. $\qquad \square$

EXAMPLE 1   Illustrate Theorem 3.6 for $y = f(x) = 3x^2 + 2x$ by finding the function ε of Δx and showing directly that $\lim_{\Delta x \to 0} \varepsilon = 0$.

Solution   We work at a general point x rather than at $x = a$. We have

$$\begin{aligned} \Delta y &= f(x + \Delta x) - f(x) \\ &= [3(x + \Delta x)^2 + 2(x + \Delta x)] - (3x^2 + 2x) \\ &= 3x^2 + 6x(\Delta x) + 3(\Delta x)^2 + 2x + 2(\Delta x) - 3x^2 - 2x \\ &= 6x(\Delta x) + 3(\Delta x)^2 + 2(\Delta x) \end{aligned}$$

and

$$\Delta y_{\text{tan}} = f'(x)\Delta x = (6x + 2)\,\Delta x = 6x(\Delta x) + 2(\Delta x).$$

Consequently,

$$E(\Delta x) = \Delta y - \Delta y_{\tan} = 3(\Delta x)^2.$$

Thus

$$\lim_{\Delta x \to 0} \varepsilon(\Delta x) = \lim_{\Delta x \to 0} \frac{E(\Delta x)}{\Delta x}$$

$$= \lim_{\Delta x \to 0} \frac{3(\Delta x)^2}{\Delta x} = \lim_{\Delta x \to 0} 3(\Delta x) = 0. \qquad \blacksquare$$

Proof of the Chain Rule

We did not give a complete proof of the chain rule (Theorem 3.4) in Section 3.5. A proof can be made using Theorem 3.6. Let $y = f(x)$ be differentiable at x_1 and $x = g(t)$ be differentiable at t_1 with $g(t_1) = x_1$. Then an increment Δt in t produces a corresponding increment Δx in $x = g(t)$, and Δx in turn produces an increment Δy in $y = f(x)$. From Theorem 3.6 we know that

$$\Delta y = f'(x_1) \cdot \Delta x + \varepsilon \cdot \Delta x, \qquad \text{where } \lim_{\Delta x \to 0} \varepsilon = 0. \qquad (4)$$

Dividing Eq. (4) by Δt, we have

$$\frac{\Delta y}{\Delta t} = f'(x_1) \cdot \frac{\Delta x}{\Delta t} + \varepsilon \cdot \frac{\Delta x}{\Delta t}. \qquad (5)$$

Now $x = g(t)$ is continuous at t_1 since g is differentiable there, so as Δt approaches zero, Δx approaches zero also, and, consequently, ε approaches zero. Therefore Eq. (5) yields

$$\frac{dy}{dt}\Big|_{t_1} = \lim_{\Delta t \to 0} \left(f'(x_1) \cdot \frac{\Delta x}{\Delta t} + \varepsilon \cdot \frac{\Delta x}{\Delta t} \right) = f'(x_1) \cdot \frac{dx}{dt}\Big|_{t_1} + 0 \cdot \frac{dx}{dt}\Big|_{t_1}$$

$$= f'(x_1) \cdot \frac{dx}{dt}\Big|_{t_1} = \frac{dy}{dx}\Big|_{x_1} \cdot \frac{dx}{dt}\Big|_{t_1}.$$

This completes the proof.

The Linear Approximation Formula

Suppose that $f'(a)$ exists. Theorem 3.6 tells us that if Δx is of sufficiently small magnitude, we can approximate the actual change $\Delta y = f(a + \Delta x) - f(a)$ in the function by $\Delta y_{\tan} = f'(a) \cdot \Delta x$. The theorem also tells us that the error in this approximation is of small magnitude compared with the size of Δx, provided that Δx is itself sufficiently close to zero. Using the notation $\approx$ for approximation, we have, for Δx close to zero,

---

Approximate function change formula

$$\Delta y \approx \Delta y_{\tan} = f'(a) \cdot \Delta x, \qquad (6)$$

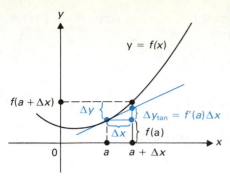

FIGURE 3.37 Approximation of $f(a + \Delta x)$.

Approximate function value formula
$$f(a + \Delta x) \approx f(a) + f'(a) \cdot \Delta x. \qquad (7)$$

Figure 3.37 illustrates these approximations graphically.

We give still another insight into the accuracy of approximation (6). Suppose that $f'(a) \neq 0$ so that $\Delta y \neq 0$ for Δx sufficiently close to zero. Then we have

$$\lim_{\Delta x \to 0} \frac{\Delta y_{\tan}}{\Delta y} = \lim_{\Delta x \to 0} \frac{f'(a) \cdot \Delta x}{\Delta y} = \lim_{\Delta x \to 0} \frac{f'(a)}{\dfrac{\Delta y}{\Delta x}} = \frac{f'(a)}{f'(a)} = 1.$$

The equation

$$\lim_{\Delta x \to 0} \frac{\Delta y_{\tan}}{\Delta y} = 1 \qquad (8)$$

means that for Δx sufficiently close to zero, *the approximation $\Delta y_{\tan}$ to Δy accounts for most of Δy.* That is, if Δx is close enough to zero, we can be sure that $\Delta y_{\tan}$ amounts to between 90% and 110% of Δy. By taking Δx even closer to zero, $\Delta y_{\tan}$ will amount to between 99% and 101% of Δy, and we obtain still greater accuracy as Δx gets closer and closer to zero.

We illustrate the use of approximation (7) to estimate the value of a function close to a point a where its value is known. Calculators give the true values of many functions at few, if any, points. Some of the techniques that calculators use to estimate function values depend on formulas (6) and (7).

EXAMPLE 2   Estimate $1/0.98^3$, using approximation (7).

Solution   Since we are interested in computing the reciprocal of a number cubed, we let $f(x) = 1/x^3 = x^{-3}$. Our estimate will be most accurate if we keep Δx as small as possible. Since we can compute $f(1)$ easily, we let $a = 1$ and $\Delta x = -0.02$. Now $f'(x) = -3x^{-4} = -3/x^4$. Then by approximation (7),

$$f(a + \Delta x) = f(0.98) \approx f(1) + f'(1)(-0.02)$$

$$= \frac{1}{1^3} - \frac{3}{1^4}(-0.02) = 1.06.$$

A calculator yields $f(0.98) \approx 1.0625$, so our estimate is quite good. The error of ≈ 0.0025 is much less than $|\Delta x| = 0.02$.   ∎

EXAMPLE 3   Estimate $\sqrt{25.01}$ using the approximation (7).

Solution   We have to choose $f(x)$, the value a, and the value Δx in approximation (7). Since we are trying to compute a square root, we let $f(x) = \sqrt{x}$. The approximation is best for small values Δx, so we choose a as close to 25.01 as we can and still compute $f(a)$ easily. Thus we take $a = 25$. This means that the value for Δx is $25.01 - 25 = 0.01$, which is acceptably small. Then $f'(x) = 1/(2\sqrt{x})$, and approximation (7) becomes

$$f(a + \Delta x) = \sqrt{25.01} \approx \sqrt{25} + \frac{1}{2\sqrt{25}}(0.01)$$

$$= 5 + \frac{1}{2 \cdot 5}(0.01) = 5.001.$$

Our calculator gives the estimate $\sqrt{25.01} \approx 5.0009999$, so our estimate is quite accurate. ∎

EXAMPLE 4   Use approximation (7) to estimate $\sin(29°)$.

Solution   We take $f(x) = \sin x$. Since we can compute $\sin(30°)$ easily, we choose a to correspond to $30°$. Recall that all our differentiation formulas for trigonometric functions are in terms of *radian measure,* so we must take $a = \pi/6$. Similarly, we must take $\Delta x = -\pi/180$, since $\pi/180$ is the radian measure corresponding to $1°$. Note that Δx is *negative* in this example, since we require that $a + \Delta x = 29\pi/180$.

Since $f'(x) = \cos x$, the approximation (7) becomes

$$\sin 29° = \sin\left(\frac{\pi}{6} + \frac{-\pi}{180}\right) \approx \sin\left(\frac{\pi}{6}\right) + \cos\left(\frac{\pi}{6}\right) \cdot \left(\frac{-\pi}{180}\right)$$

$$= \frac{1}{2} + \frac{\sqrt{3}}{2} \cdot \frac{-\pi}{180} = \frac{1}{2} - \frac{\pi\sqrt{3}}{360}$$

$$\approx 0.484885.$$

Our calculator gives 0.4848096 as its estimate for $\sin 29°$, so our error is about 0.000075. ∎

Estimates of the function values requested in Examples 2–4 can be found easily using a calculator or computer. For an application of formulas (6) and (7) that is not handled so immediately by a calculator, let us estimate the change in x required to produce a desired change Δy in the value $y = f(x)$. That is, we find Δx for which Δy_{tan} is equal to the desired change in y-value. Figure 3.38 illustrates this technique, and Example 7 indicates that it is very useful.

EXAMPLE 5   Estimate the value of x such that $x^3 = 8.05$.

Solution   Let $f(x) = x^3$. We have $f(2) = 8$. We set $a = 2$ and find Δx that will produce $\Delta y_{\text{tan}} = f'(a) \Delta x$ of 0.05, which is the difference $8.05 - 8$. We have $0.05 = f'(2)\Delta x$ so

$$\Delta x = \frac{0.05}{f'(2)} = \frac{0.05}{3 \cdot 2^2} = \frac{0.05}{12} \approx 0.00417.$$

Thus $2 + \Delta x = 2.00417$, and we expect that $f(2.00417) \approx 8.05$. A calculator yields $f(2.00417) \approx 8.0501$. ∎

EXAMPLE 6   A circle has radius 2 ft. By approximately how much should the radius be increased to produce an increase of 0.25 ft^2 in the area of the circle?

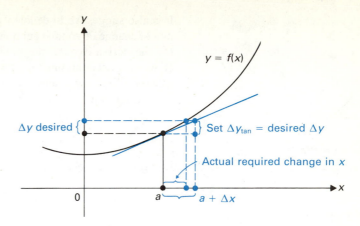

Estimated change Δx

FIGURE 3.38 Estimating the change in x required to produce $f(a) + \Delta y$.

Solution We have $A = \pi r^2$, so $\Delta A_{\tan} = (2\pi r)\Delta r$. We set $r = 2$ ft and find Δr that produces a value 0.25 ft² for $\Delta A_{\tan}$. Thus

$$0.25 = (2\pi \cdot 2)\Delta r,$$

so the radius should be increased by approximately

$$\Delta r = \frac{0.25}{4\pi} \text{ ft.} \qquad \blacksquare$$

EXAMPLE 7 Estimate a solution of the equation $x^3 - x^2 - 3 = 0$.

Solution Let $f(x) = x^3 - x^2 - 3$. Computing a few values, we find $f(-1) = -5, f(0) = -3, f(1) = -3, f(2) = 1,$ and $f(3) = 15$. The intermediate value theorem shows that the equation has a solution in the interval $[1, 2]$. Since $f(2) = 1$, we find Δx that will produce $\Delta y_{\tan}$ of -1 at $a = 2$ for $y = x^3 - x^2 - 3$. (See Fig. 3.39.) Since $\Delta y_{\tan} = (3x^2 - 2x)\Delta x$, we obtain

$$-1 = (3 \cdot 2^2 - 2 \cdot 2)\Delta x$$
$$= 8\Delta x,$$

so $\Delta x = -\frac{1}{8}$. Thus we expect $2 - \frac{1}{8} = \frac{15}{8}$ to be close to a solution of $x^3 - x^2 - 3 = 0$. A calculator shows that $f(\frac{15}{8}) \approx 0.076$. $\qquad \blacksquare$

The technique in Example 7 can be extended to Newton's method for solving an equation, which is developed in Section 4.7.

Differential Notation

We have seen how handy the Leibniz notation dy/dx for the derivative is in remembering the chain rule

$$\frac{dy}{dt} = \frac{dy}{dx} \cdot \frac{dx}{dt}.$$

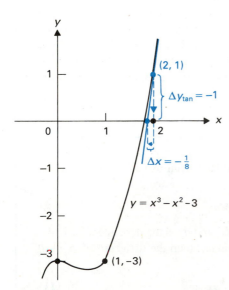

FIGURE 3.39 Finding Δx corresponding to $\Delta y_{\tan} = -1$ for the tangent line at $(2, 1)$.

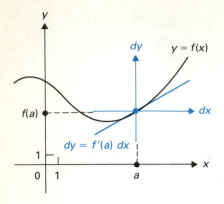

FIGURE 3.40   The tangent line is $dy = f'(a)\ dx$ with respect to local coordinate dx, dy-axes.

It is also suggestive to denote a rate of change of area per unit time by dA/dt, a rate of change of volume per unit radius by dV/dr, and so on. The notation dy/dx for the derivative and the notations dx and dy for *differential quantities* are commonly used throughout physics, engineering, and other sciences that employ calculus. We now discuss how dx and dy might be interpreted.

Recall that

$$f'(x) = \lim_{\Delta x \to 0} \frac{\Delta y}{\Delta x}.$$

Using Leibniz notation, we have

$$\lim_{\Delta x \to 0} \frac{\Delta y}{\Delta x} = \frac{dy}{dx}. \tag{9}$$

In Eq. (9), Leibniz viewed Δy as approaching an *infinitesimal quantity* dy and Δx as approaching an infinitesimal quantity dx as Δx approaches zero. He considered these infinitesimal quantities dx and dy to be nonzero where $f'(x) \neq 0$ but to be of *smaller magnitude than any nonzero real number*. The derivative thus became the ratio of these infinitesimal quantities dy and dx. Leibniz was roundly criticized for inventing such *infinitesimals* out of thin air and attempting to do mathematics with them. It is only within the last fifty years that mathematicians have put Leibniz's view on a firm mathematical foundation by showing that it is possible to augment the set of real numbers to include such infinitesimals. Using this augmented set with calculus is referred to as *nonstandard analysis*.

An interpretation of dx and dy in the set of real numbers is of heuristic value in studying calculus. Consider dx to mean the same as Δx, that is, to be an increment in x. If $dy/dx = f'(x)$ is to be interpreted as a quotient, we must define

$$dy = f'(x) \cdot dx.$$

When $x = a$ and $dx = \Delta x$, we then have $dy = f'(a) \cdot \Delta x = \Delta y_{\text{tan}}$. Thus if we consider dx to coincide with Δx, we do *not* identify dy with Δy but rather set $dy = \Delta y_{\text{tan}}$. In the equation $dy = f'(a) \cdot dx$, we regard dx as the independent variable and dy as the dependent variable. If we take a horizontal dx-axis and a vertical dy-axis at the point $(a, f(a))$ as shown in Fig. 3.40, we see that $dy = f'(a) \cdot dx$ is precisely the equation, with respect to these new coordinate axes, of the tangent line to the graph of f where $x = a$. The terminology in the following definition is commonly used.

DEFINITION 3.3

Differential

Let $y = f(x)$ be differentiable at a. The **differential of f at a** is the function of the single variable dx given by

$$dy = f'(a)\ dx. \tag{10}$$

In Eq. (10), the independent variable is dx, and dy is the dependent variable. If f is differentiable at all points in its domain, then the **differential dy or df** of $y = f(x)$ is

$$dy = f'(x)\ dx, \tag{11}$$

which associates with each point x the differential of f at that point.

EXAMPLE 8   Let $s = g(t) = t^3 - (1/t^2)$. Find the differential ds at $t = 2$.

Solution   Using Eq. (10) with g in place of f, t in place of x, and s in place of y, we have

$$ds = g'(2)\ dt.$$

We may write

$$s = g(t) = t^3 - \frac{1}{t^2} = t^3 - t^{-2},$$

so

$$\frac{ds}{dt} = g'(t) = 3t^2 - (-2)t^{-3} = 3t^2 + \frac{2}{t^3}.$$

Therefore

$$g'(2) = 3 \cdot 2^2 + \frac{2}{2^3} = 12 + \frac{1}{4} = \frac{49}{4},$$

so the differential at $t = 2$ is

$$ds = \frac{49}{4}\,dt.\qquad\blacksquare$$

EXAMPLE 9   If $y = f(x) = x^3 - e^{2x}$, then the differential dy is given by $dy = (3x^2 - 2e^{2x})\ dx$. $\blacksquare$

We like to regard the differential $dy = f'(a)\ dx$ of f at a as the best local linear approximation of f at a, as illustrated in Fig. 3.40. We regard the differential $df = f'(x)\ dx$ as the collection of *all* such local linear approximations of f, one for each x in the domain of the differentiable function f. The approximation formulas (6) and (7) can be written in differential notation as

$$\Delta y \approx dy = f'(a)\ dx \qquad (12)$$

and

$$f(a + dx) \approx f(a) + f'(a)\ dx. \qquad (13)$$

For this reason, approximation using formulas (6) and (7), or (12) and (13), is often called *approximation using differentials*. The only change from formula (6) to formula (12) and from formula (7) to formula (13) is one of notation. If formulas (12) and (13) were used in Examples 2–7, the only changes would be to replace Δx by dx and Δy_{tan} by dy wherever they occur.

To prepare you for your work in engineering, physics, and other sciences, we must use differential notation quite often in this text. However, we close this section by showing that the *standard analysis* interpretations of dy and dx that we have given must be regarded as only of heuristic and mnemonic value in calculus.

Let f and g be differentiable functions, and let $y = f(x)$ and $x = g(t)$. Let Δx be the change in x produced by a change Δt in t. Taking the standard analysis interpretations of the differentials of f and of g, we see than whenever $\Delta x \neq 0$, we have

$$\frac{dy}{dx} \cdot \frac{dx}{dt} = \frac{\Delta y_{\tan}}{\Delta x} \cdot \frac{\Delta x_{\tan}}{\Delta t}. \tag{14}$$

It is clear that in this interpretation of differentials, we cannot regard the chain rule as following by simply canceling dx from the numerator and denominator. In general, when $x = g(t)$, we have $\Delta x_{\tan} \neq \Delta x$. Also, $\Delta y_{\tan}$ for $y = f(g(t))$ given by the increment Δt is different from the numerator $\Delta y_{\tan}$ in Eq. (14) for $y = f(x)$ given by the increment Δx in $x = g(t)$ produced by the increment Δt.

SUMMARY

1. Let $f'(a)$ exist, let $\Delta y = f(a + \Delta x) - f(a)$, let $\Delta y_{\tan} = f'(a)\Delta x$, and let $E(\Delta x) = \Delta y - \Delta y_{\tan}$, which is the error in the approximation of Δy by $\Delta y_{\tan}$. Then

$$\lim_{\Delta x \to 0} \frac{E(\Delta x)}{\Delta x} = 0,$$

so this error becomes very small compared with the size of Δx for Δx sufficiently close to zero.

2. If $f'(a)$ exists and is nonzero, the approximation $f(a + \Delta x) \approx f(a) + f'(a) \Delta x$ is a good one for small values of Δx.

3. If $y = f(x)$, then the differential dy is $dy = f'(x) \, dx$.

EXERCISES 3.9

In Exercises 1–8, find the differential of the given function.

1. $y = f(x) = \dfrac{x}{x + 1}$

2. $s = g(t) = t^3 - 2t^2 + 4t$

3. $A = f(r) = \pi r^2$

4. $y = \sin 2x^2$

5. $y = e^{\tan 3x}$

6. $y = \ln(x \cos^2 5x)$

7. $s = 3 \tan^2 2t$

8. $z = \ln(\tan y)$

In Exercises 9–14, estimate the indicated quantity using formula (7) or, in differential notation, formula (13).

9. 0.999^{10}

10. $\dfrac{1}{10.05^5}$

11. $f(1.98)$ if $f(x) = \dfrac{x^3 + 4x}{2x - 1}$

12. $\cos 44°$

13. $\sin 62°$

14. $\ln(0.95)$

15. Estimate $\sqrt{101}$. [You may think that $\Delta x = 1$ is too large to give a good estimate, but the graph of $\sqrt{x}$ is turning so slowly at $x = 100$ that the tangent line is close to it for quite a distance.]

16. If $f(x) = x^4 - 3x^2$, then $f(2) = 4$. Use formula (6) or (12) to find the approximate value of x such that $f(x) = 3.98$.

17. If $f(x) = x^3/(x - 2)$, then $f(4) = 32$. Use formula (6) or (12) to find the approximate value of x such that $f(x) = 31.8$.

In Exercises 18–21, find the integer x where $|f(x)|$ is smallest and use formula (6) or (12) to estimate a solution of the equation $f(x) = 0$. (See Example 7.)

18. $f(x) = x^3 - 2$

19. $f(x) = x^3 - 7$

20. $f(x) = x^3 - 2x^2 + 18$

21. $f(x) = x^5 + x^3 - 42$

22. Estimate the change in volume of a cylindrical silo 20 ft high if the radius is increased from 3 ft to 3 ft 4 in.

FIGURE 3.41

23. Imagine the earth to be a ball of radius 4000 mi, and imagine a string tied around the equator of the earth. See Fig. 3.41. The string is cut, and six additional feet of string are inserted.
   (a) If the lengthened circle of string were lifted a uniform height above the equator all the way around the earth, estimate how high above the surface of the earth the string would be.
   (b) Discuss the accuracy of your estimate in part (a).

24. A ball has radius 4 ft, and a second ball has volume 1 ft^3 greater than the volume of the first ball. Estimate the difference in the surface area of these two balls. (The volume V and surface area A of a ball of radius r are given by $V = \frac{4}{3}\pi r^3$ and $A = 4\pi r^2$.)

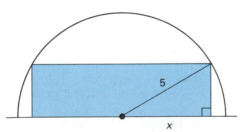

FIGURE 3.42

25. A rectangle is inscribed in a semicircle of radius 5 ft. See Fig. 3.42. Estimate the increase in the area of the rectangle if the length of its base (on the diameter) is increased from 6 ft to 6 ft 2 in.

26. A sphere has radius 4 ft. What change in radius will result in approximately 2 ft^3 increase in volume? (For a sphere, volume $V = \frac{4}{3}\pi r^3$.)

27. A silo consists of a cylinder of radius 4 ft and height 30 ft surmounted by a hemisphere. If the height of the cylinder remains 30 ft, approximately how much should the radius increase if the total volume is to be increased by 100 ft^3?

Scientists are sometimes interested in the *percent of error* in the measurement of a numerical quantity Q. If the error in computing Q is h, then the **percent of error** is

$$100\left|\frac{h}{Q}\right|.$$

Exercises 28–31 deal with estimating percent of error using differentials.

28. For a differentiable function f, let $Q = f(a) \neq 0$. Suppose Q is computed by "measuring a" and then computing $f(a) = Q$.
   (a) If a small error of Δx is made in the measurement of a, argue that the approximate resulting percent of error in the computed value for Q is $100|f'(a)\,\Delta x/f(a)|$.
   (b) If a small error of k percent of a is made in the measurement of a argue that the approximate resulting percent of error in the computed value for Q is

$$|ka\,f'(a)/f(a)|.$$

29. The radius of a sphere is found by measurement to be 2 ft $\pm$ 0.04 ft. Estimate the maximum percent of error in computing the volume of the sphere from this measurement of the radius. (For a sphere, volume $V = \frac{4}{3}\pi r^3$.)

30. One side of an equilateral triangle is found by measurement to be 8 ft with an error of at most 3%. Estimate the maximum percent of error in computing the area of the triangle from this measurement of the length of a side.

31. If it is desired to compute the area of a circle with at most 1% error by measuring its radius, estimate the allowable percent of error that may be made in measuring the radius.

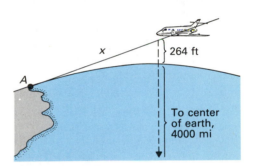

FIGURE 3.43

32. A plane flying over the ocean at night is headed straight toward a point A on the coast where a very strong light has been placed at the water line. Visibility is excellent, and the plane is flying at low altitude with the pilot's eyes 264 ft above the ocean. See Fig. 3.43. Assuming the radius of the earth is 4000 miles, use differentials to estimate the distance from the light to the plane when the light first becomes visible to the pilot. [*Hint:* If x is the distance from the light to the pilot and y is the distance from the pilot's eyes to the center of the earth, then

$$x^2 + 4000^2 = y^2.$$

(Draw a figure.) You will find that an attempt to estimate $x = \sqrt{y^2 - (4000)^2}$ by a differential for y near 4000 leads to difficulties. A successful technique is to estimate y^2 using a differential, then subtract 4000^2, and finally take the square root, possibly using another differential if the number is not a perfect square.]

33. Let $f(x) = x^2 - 2x$. Find $E(\Delta x)$ as a function of Δx and show directly that

$$\lim_{\Delta x \to 0} \frac{E(\Delta x)}{\Delta x} = 0.$$

(See Example 1.)

34. Repeat Exercise 33 for the function $f(x) = 1/x^2$.

In Exercises 35–40, use a calculator and the differential at the given value $x = a$ to find an approximate solution of the equation.

35. $e^x + x = 4$, $a = 1$

36. $2^x + x = 11.3$, $a = 3$

37. $4^{\sqrt{x}} - x = 13$, $a = 4$

38. $x^x = 4.15$, $a = 2$

39. $x^{2x} = 15.95$, $a = 2$

40. $x^{\cos x} = 6.3$, $a = 2\pi$

SUPPLEMENTARY EXERCISES FOR CHAPTER 3

1. Define the derivative $f'(a)$ of f at the point where $x = a$.

2. Use the definition of the derivative, not differentiation formulas, to find $f'(x)$ if $f(x) = x^2 - 3x$.

3. Use the definition of the derivative, not differentiation formulas, to find $f'(t)$ if $f(t) = 1/(2t + 1)$.

4. Find the equation of the tangent line to $y = x^3 - 3x^2 + 2$ where $x = 1$.

5. Find the equation of the normal line to $y = 4x^2 - 3x + 2$ at the point $(-1, 9)$.

6. Find the equation of the tangent line to the curve $y = x^4 - 3x^2 + 4x$ at the point $(1, 2)$.

7. Find the equation of the normal line to the curve $y = 3x^2 - 4x - 7$ at the point $(2, -3)$.

8. Find dy/dx if $y = (x^2 - 3x)(4x^3 - 2x + 17)$. You need not simplify your answer.

9. Find dx/dt if $x = (8t^2 - 2t)/(4t^3 + 3)$.

10. Find y' if $y = (x + 4)/(x^2 - 17)$.

11. Let f be differentiable and let $y = x^2 f(x)$. If $dy/dx|_{x=1} = 10$ and $f(1) = 5$, find $f'(1)$.

12. Let g be differentiable and let $s = (t^2 + 1)/g(t)$. If $ds/dt|_{t=3} = -4$ and $g(3) = 2$, find $g'(3)$.

13. Find y' if $y = (x^2 - 3x)/(2x^3 + 7)$.

14. Let $y = 3x^2 - 6x$ and let $x = g(t)$ be a differentiable function such that $g(14) = -2$ and $g'(14) = 8$. Find dy/dt when $t = 14$.

15. Find s' if $s = \sqrt{x^2 - 17x}$.

16. Find y' if $y = \dfrac{1}{\sqrt[3]{x^3 - 3x + 2}}$.

17. Let $y = \sqrt{x^2 - 3x}$ and $x = (t + 2)/(t - 1)$. Find dy/dt when $t = 2$.

18. Find s' if $s = (t^2 + 3)\sqrt{2t + 4}$.

19. Let f be a differentiable function and let $y = (f(x))^3 + 3\sqrt{f(x)} + 10$. If $f(3) = -1$ and $dy/dx|_{x=3} = 18$, find $f'(3)$.

20. Find y' if $y = \sin 2x \cos 3x$.

21. If $y = \cot x^2$, find dy/dx.

22. Find the equation of the line normal to the graph of $s = \sin 3t$ at the origin.

23. Find $f'(t)$ if $f(t) = te^{\sin t}$.

24. Find y' if $y = e^{\ln 4x}$.

25. Find v' if $v = u^3 e^{-u^2}$.

26. Find s' if $s = 3^{\cos 2t}$.

27. If \$5000 is deposited in an account that pays 9% annual interest compounded continuously, find the balance in the account after 12 years, assuming that the interest rate remains the same and no funds are deposited or withdrawn during that time.

28. Find d^3y/dx^3 if $y = 6(4x - 2)^{3/2}$.

29. Find d^3s/dt^3 if $s = \sqrt{3t + 4}$.

30. If the position x of a body on a line (x-axis) at time t is $x = t^2 - (t/3)$, find its velocity when $t = 2$.

31. Let the position x of a body on a line (x-axis) at time t be $x = t^3 + 2t$.
(a) Find the average velocity of the body from time $t = 1$ to time $t = 3$.
(b) Find the velocity of the body at time $t = 2$.

32. Let f be a twice-differentiable function and let $y = g(x) = x^3 f(x)$. If $g(-1) = 4$, $g'(-1) = 2$, and $g''(-1) = -3$, find $f''(-1)$.

33. If $y^3 + x^2 y - 4x^2 = 17$, find dy/dx.

34. Find ds/dt at $(4, -1)$ if $\sqrt{st^2} - 3st^3 = 14$.

35. Find all points where the curve $y^2 - xy + x^2 = 27$ has a horizontal tangent line.

36. Find dy/dx if $(x^2 - 3xy)/(x^3 + y^2) = y^3$.

37. Find dy/dx and d^2y/dx^2 at the point $(1, 2)$ if $y^2 - xy = 2$.

38. Find dx/dt and d^2x/dt^2 if $x = e^t(\ln t)$.

39. Find dy/dx if $y = \ln\left(\dfrac{x\sqrt{x^2 + 1}}{2x + 1}\right)$.

40. Find dr/ds if $r = \ln\left(\dfrac{\sqrt{s + 1}}{\sin 2s}\right)$.

41. Find dy/dt if $y = \log_2(3^t 4^{2t})$.

42. Find dy/dx if $y = \ln(\log_3 x)$.

43. Find s' if $s = (t + 1)^{2t}$.

44. Find y' if $y = (\ln x)^{e^x}$.

45. Find $dr/d\theta$ if $r = (\cos \theta)^{\tan \theta}$.

46. Estimate $\sqrt{63}$ using a differential.

47. Use a differential to estimate the percent increase in the area of a circle due to a 2% increase in the radius of the circle.

48. Use a differential to estimate the percent decrease in the radius of a ball required to produce a 5% decrease in its volume. (For a ball, $V = (\frac{4}{3})\pi r^3$.)

49. The radius of a ball is 4 ft. Use a differential to estimate the change in the surface area of the ball if its volume is increased by $\frac{1}{10}$ ft^3. (For a ball, volume $V = (\frac{4}{3})\pi r^3$, surface area $S = 4\pi r^2$.)

Exercises 50–55 indicate that some limits can be recognized as giving derivatives, or at least involving derivatives. Differentiation formulas may then aid in their computation.

50. If $f(x) = x^4 - 3x^2$, find
$$\lim_{\Delta x \to 0} \frac{f(2 + \Delta x) - f(2)}{\Delta x}.$$

51. If $f(x) = x/(x + 1)$, find
$$\lim_{\Delta x \to 0} \frac{f(3 + \Delta x) - f(3)}{2 \cdot \Delta x}.$$

52. If $g(x) = x^{10} - 7x^6 + 5x^4$, find
$$\lim_{\Delta x \to 0} \frac{f(1) - f(1 + 2 \cdot \Delta x)}{\Delta x}.$$

53. If $f(x) = 3x^{10} - 7x^8 + 5x^6 - 21x^3 + 3x^2 - 7$, find
$$\lim_{h \to 0} \frac{f(1 - h) - f(1)}{h^3 + 3h}.$$

54. Let $f(x) = \dfrac{x^3 - 2x^2}{x - 1}$. Find
$$\lim_{h \to 0} \frac{f(2 + 3h) - f(2 + h)}{h}.$$

55. Let $f(x) = \dfrac{x^2}{3x + 4}$. Find
$$\lim_{h \to 0} \frac{f(2h - 1) - f(-3h - 1)}{h}.$$

56. Let f and g be differentiable at $x = a$. Show that, if $f(a) = 0$ and $g(a) = 0$, then $(fg)'(a) = 0$.

57. Find a formula for the derivative of a product
$$f_1(x)f_2(x)f_3(x)\cdots f_n(x)$$
of n functions.

58. Let $p(x)$ be a polynomial function. Show that if $(x - a)^2$ is a factor of the polynomial expression, then both $p(a) = 0$ and $p'(a) = 0$.

59. Generalize the result in Exercise 58.

60. Find the equations of the lines through $(4, 10)$ that are tangent to the graph of $f(x) = (x^2/2) + 4$.

61. Let f be differentiable at $x = a$.
    (a) Find A, B, and C such that the polynomial function (parabola) $p(x) = Ax^2 + Bx + C$ passes through the three points on the graph $y = f(x)$ whose x-coordinates are $a - \Delta x$, a, and $a + \Delta x$. (For small Δx, the quadratic function $p(x)$ approximates $f(x)$ near a.)
    (b) Compute $p'(a)$ for $p(x)$ from part (a). What familiar expression do you obtain?

62. Let f and g be differentiable functions of x, and let $y = f(x)/g(x)$. Derive the formula for the derivative of the quotient $f(x)/g(x)$ using just implicit differentiation and the rule for differentiating a product.

63. The chain rule states that, under certain conditions,
$$\frac{d[f(g(t))]}{dt}\bigg|_{t=t_2} = f'(g(t_1)) \cdot g'(t_1).$$
Find a similar formula for
$$\frac{d[f(g(h(t)))]}{dt}\bigg|_{t=t_1}$$
for suitable functions f, g, and h.

64. Let f and g be differentiable functions satisfying
$$g'(a) \neq 0, \qquad g(a) = b, \qquad \text{and}$$
$$f(g(x)) = x.$$
Show that $f'(b) = 1/g'(a)$.

APPLICATIONS OF THE DERIVATIVE

4

This chapter introduces a few of the many applications of differential calculus to the sciences. It also shows how the derivative can be used in mathematics as an aid to studying functions and their graphs.

4.1 RELATED RATE PROBLEMS

Recall that the derivative dy/dx gives the instantaneous rate of change of y with respect to x. Leibniz notation is very useful in keeping track of just what rate of change we are working with. For example, imagine that a pebble is dropped into a calm pond and a circular wave spreads out from the point where the pebble was dropped. We might be interested in the following rates of change:

$$\frac{dr}{dt} = \text{Rate of increase of } radius \text{ per unit increase in } time,$$

$$\frac{dA}{dt} = \text{Rate of increase of } area \text{ per unit increase in } time,$$

$$\frac{dA}{dr} = \text{Rate of increase of } area \text{ per unit increase in } radius,$$

$$\frac{dC}{dt} = \text{Rate of increase of } circumference \text{ per unit increase in } time,$$

and so on. The Leibniz notation helps us remember which rate of change we want.

In many rate-of-change problems, we want to find the rate of change per unit of time of a quantity Q if we know such *time rates of change* of one or more

Outline for Related Rate Problems

Read the problem carefully. Draw a figure where appropriate and decide what letter variables to use. Then follow these steps.

STEP 1. Decide what rate of change is desired and express it in Leibniz notation:

$$\text{Find } \frac{dQ}{dt} \text{ when } \underline{\hspace{2em}}.$$

STEP 2. Decide what rates of change are given and express these data in Leibniz notation:

$$\text{Given } \frac{dr}{dt} = \underline{\hspace{2em}} \text{ and } \frac{ds}{dt} = \underline{\hspace{2em}} \text{ when } \underline{\hspace{2em}}.$$

STEP 3. Find an equation relating Q, r, and s. You may have to use a figure or some geometric formula.

STEP 4. Differentiate, with respect to time t, the relation in step 3 (often by implicit differentiation) to obtain a relation among dQ/dt, dr/dt, and ds/dt.

STEP 5. Put in values of r, s, and Q and of dr/dt and ds/dt corresponding to the instant when dQ/dt is desired, and solve for dQ/dt.

related quantities, say r and s. The outline at the top of this page suggests one convenient step-by-step procedure that you can follow in solving *related rate problems*. (The letters may be different from Q, r, and s in a problem, of course.)

The following examples illustrate the use of the outline.

EXAMPLE 1   If the radius r of a circular disk is increasing at the rate of 3 in./sec, find the rate of increase of its area when $r = 4$ in.

Solution   Let A be the area and r be the radius of the circular disk.

STEP 1.   Find dA/dt when $r = 4$ in.

STEP 2.   Given $dr/dt = 3$ in./sec.

STEP 3. $A = \pi r^2$

STEP 4. $dA/dt = 2\pi r \cdot dr/dt$

STEP 5.   When $r = 4$ and $dr/dt = 3$, $dA/dt = 2\pi \cdot 4 \cdot 3 = 24\pi$ in²/sec.   ∎

EXAMPLE 2   Ship A passes a buoy at 9:00 A.M. and continues on a northward course at a rate of 12 mph. Ship B, traveling at 18 mph, passes the same buoy on its eastward course at 10:00 A.M. the same day. Find the rate at which the distance between the ships is increasing at 11:00 A.M. that day.

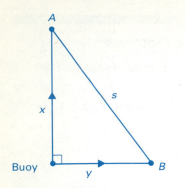

FIGURE 4.1 Positions of ships *A* and *B* relative to the buoy.

Solution We draw a figure and assign letter variables as shown in Fig. 4.1.

STEP 1. Find ds/dt at $t = 11:00$ A.M.

STEP 2. Given $dx/dt = 12$ mph and $dy/dt = 18$ mph.

STEP 3. $s^2 = x^2 + y^2$

STEP 4. $2s(ds/dt) = 2x(dx/dt) + 2y(dy/dt)$

STEP 5. At 11:00 A.M., ship A has traveled for 2 hours at 12 mph since passing the buoy, so it has gone $x = 24$ miles. Similarly, ship B has traveled for 1 hour at 18 mph, so it has gone $y = 18$ miles. Also, $dx/dt = 12$, $dy/dt = 18$, and

$$s = \sqrt{x^2 + y^2} = \sqrt{24^2 + 18^2} = 30.$$

Putting these values into the equation of step 4, we have

$$2 \cdot 30 \frac{ds}{dt} = 2 \cdot 24 \cdot 12 + 2 \cdot 18 \cdot 18 = 1224,$$

$$\frac{ds}{dt} = \frac{1224}{60} = 20.4 \text{ mph.} \qquad \blacksquare$$

EXAMPLE 3 Suppose a spherical balloon is being inflated in such a way that the volume of the sphere increases at a constant rate. Show that the rate of increase of the radius is inversely proportional to the surface area of the balloon.

Solution Let the spherical balloon have volume V, radius r, and surface area S at time t.

STEP 1. Find dr/dt and show that $dr/dt = k/S$ for some constant k.

STEP 2. Given $dV/dt = c$, where c is a constant.

STEP 3. The formula for the volume of a sphere is $V = \frac{4}{3}\pi r^3$.

STEP 4. $dV/dt = \frac{4}{3} \cdot 3\pi r^2 (dr/dt) = 4\pi r^2 (dr/dt)$

STEP 5. We know that $dV/dt = c$, and the formula for the surface area S of a sphere is $S = 4\pi r^2$. Thus

$$c = S \frac{dr}{dt}, \qquad \text{so} \qquad \frac{dr}{dt} = \frac{c}{S}. \qquad \blacksquare$$

EXAMPLE 4 Let a parallelogram have sides of lengths 8 and 12 in., and let a vertex angle θ be decreasing at a rate of 2°/min. Find the rate of change of the area of the parallelogram when $\theta = 30°$.

Solution Let A be the area of the parallelogram of altitude h, as shown in Fig. 4.2.

STEP 1. Find dA/dt when $\theta = 30° = \pi/6$ radians.

STEP 2. Given $d\theta/dt = -2°/\text{min} = -\pi/90$ radians/min. We must use radian measure because the differentiation formulas are in terms of radian measure. The negative sign appears because θ is decreasing.

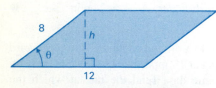

FIGURE 4.2 Parallelogram with sides 8 and 12, vertex angle θ, and altitude h.

STEP 3. We have $A = 12h$, and from Fig. 4.2 we see that $h = 8 \sin \theta$. Thus $A = 12(8 \sin \theta) = 96 \sin \theta$.

STEP 4. $dA/dt = 96(\cos \theta)(d\theta/dt)$

STEP 5. When $\theta = \pi/6$ and $d\theta/dt = -\pi/90$, we have

$$\frac{dA}{dt} = 96\frac{\sqrt{3}}{2} \cdot \frac{-\pi}{90} = -\frac{8\sqrt{3}\pi}{15} \text{ in}^2/\text{min.} \quad \blacksquare$$

EXAMPLE 5   A gas is confined in a container of variable volume that increases at a constant rate of 10 cm³/sec. The gas is kept at constant temperature. According to Boyle's law, the pressure of the gas on the walls of the container is inversely proportional to the volume. Suppose the pressure is 80 g/cm² when the volume is 400 cm³. Find the rate of change of the pressure when the volume is 600 cm³.

Solution   Let V be the volume of the container and P be the pressure of the gas.

STEP 1. Find dP/dt when $V = 600$ cm³.

STEP 2. We are given that $P = 80$ g/cm² when $V = 400$ cm³ and that $dV/dt = 10$ cm³/sec at all times.

STEP 3. By Boyle's law,

$$P = \frac{k}{V}$$

for some constant k. The data given in step 2 allow us to find k. Substituting in $P = k/V$, we have

$$80 = \frac{k}{400}, \quad \text{so} \quad k = 32,000.$$

Thus $P = 32,000/V = 32,000V^{-1}$.

STEP 4. $dP/dt = 32,000(-1)V^{-2} \cdot dV/dt = -(32,000/V^2)dV/dt$

STEP 5. Now $dV/dt = 10$ cm³/sec, so when $V = 600$ cm³, we have

$$\frac{dP}{dt} = -\frac{32,000}{600^2} \cdot 10 = -\frac{32}{36} = -\frac{8}{9} \text{ (g/cm}^2)/\text{sec.} \quad \blacksquare$$

EXAMPLE 6   A drinking glass is in the shape of a truncated cone with base of radius 3 cm, top of radius 5 cm, and altitude 10 cm, as shown in Fig. 4.3. A beverage is poured into the glass at a constant rate of 48 cm³/sec. Find the rate at which the level of the beverage is rising in the glass when it is at a depth of 5 cm.

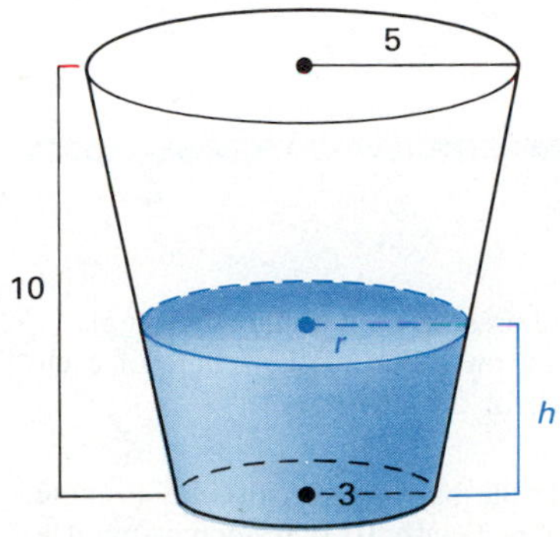

Solution   Let r be the radius at the top of a volume V of beverage in the glass when the depth is h, as shown in Fig. 4.3.

STEP 1. Find dh/dt when $h = 5$ cm.

STEP 2. Given $dV/dt = 48$ cm³/sec.

STEP 3. We need to find a relation between V and h. First, we express V in terms of r and h. The formula for the volume of a cone is

$$\tfrac{1}{3}\pi(\text{Radius})^2(\text{Altitude}).$$

FIGURE 4.3 Liquid of depth h and top radius r in a glass.

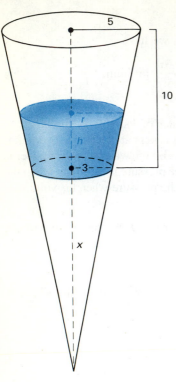

FIGURE 4.4   Continuation of the glass in Fig. 4.3 to a cone of altitude $10 + x$.

If the vertex of the cone before truncation was x cm below the base of the glass, then from similar triangles in Fig. 4.4, we see that

$$\frac{5}{3} = \frac{10 + x}{x}, \qquad 5x = 30 + 3x, \qquad 2x = 30, \qquad x = 15.$$

Then the volume of the beverage is

$$V = \tfrac{1}{3}\pi r^2(15 + h) - \tfrac{1}{3}\pi \cdot 3^2 \cdot 15.$$

Again, similar triangles in Fig. 4.4 show that

$$\frac{r}{3} = \frac{x + h}{x}.$$

Since $x = 15$, we have

$$\frac{r}{3} = \frac{15 + h}{15}, \qquad \text{so} \quad r = \frac{15 + h}{5} = 3 + \frac{h}{5}.$$

Thus

$$V = \frac{1}{3}\pi\left(3 + \frac{h}{5}\right)^2(15 + h) - \frac{1}{3}\pi \cdot 3^2 \cdot 15$$

$$= \frac{5}{3}\pi\left(3 + \frac{h}{5}\right)^3 - \frac{1}{3}\pi \cdot 3^2 \cdot 15.$$

STEP 4. $dV/dt = 5\pi(3 + h/5)^2 \cdot \tfrac{1}{5} \cdot dh/dt$

STEP 5.   Now $dV/dt = 48$ cm^3/sec when $h = 5$ cm, so

$$48 = 5\pi(3 + 1)^2 \cdot \frac{1}{5} \cdot \frac{dh}{dt} = 16\pi\frac{dh}{dt}.$$

Thus $dh/dt = 3/\pi$ cm/sec.   ■

SUMMARY

To solve related rate problems, follow the outline on page 183.

EXERCISES 4.1

1. Oil spilled from a tanker anchored in a harbor spreads out in a circular slick. When the radius of the slick is 100 ft, the area of the slick is increasing at 50 ft^2/sec. Find the rate of increase of the radius of the slick at that moment.

2. If the area of a circle is increasing at a constant rate of 200π in^2/sec, find the rate of increase of the circumference when the circumference is 100π inches.

3. Find the rate at which the area of an equilateral triangle is increasing when it is 10 in. on a side, if the length of each side is increasing at a rate of 2 in./min.

4. The vertex angle opposite the base of an isosceles triangle with equal sides of constant length 10 ft is increasing at a rate of $\tfrac{1}{2}$ radian/min.

(a) Find the rate at which the base of the triangle is increasing when the vertex angle is 60°.
(b) Find the rate at which the area of the triangle is increasing when the vertex angle is 60°.

5. A light tower is located 1 mile directly offshore from a point P on a straight coastline. See Fig. 4.5. The beam of light revolves at the rate of $\frac{1}{10}$ radian/sec. Find the rate at which the spot of light on the shore is moving along the coastline when (a) the spot is at P, and (b) the spot is 2 miles from P along the shore.

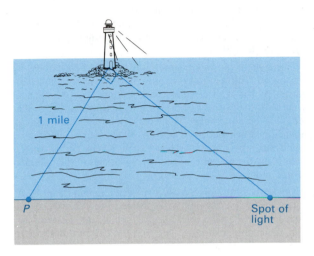

FIGURE 4.5

6. The length of a rectangle is increasing at 6 ft/min while its width is decreasing at a rate of 4 ft/min. Find the rate of change of the area when the length is 100 ft and the width is 30 ft.

7. A rectangle has diagonals of length 50 cm. If the acute angle θ between the diagonals is increasing at a constant rate of 3°/min while the diagonals remain the same length, find the rate of change of the area of the rectangle when $\theta = 60°$.

8. Repeat Exercise 7 for a parallelogram with diagonals of constant lengths 50 cm and 30 cm. The other data remain the same.

9. If V volts are measured across a resistance of R ohms in a circuit carrying a current of I amperes, then $V = IR$. At a certain instant, the resistance is 500 ohms and is decreasing at a rate of 2 ohms/min, while the current is 0.5 amp and is increasing at a rate of 0.01 amp/min. Find the rate of change of the voltage at that moment.

10. If a current of I amperes is flowing through a circuit of resistance R ohms, then W watts of power are used, where $W = I^2R$. If current flowing through a constant resistance of 5 ohms is increasing at a constant rate of 0.2 amp/min, find the rate of change of the power used when $I = 4$ amp.

11. An 18-ft ladder is leaning against a vertical wall. If the bottom of the ladder is pulled away from the wall at a constant rate of 6 in./sec, find the rate at which the top is sliding down the wall when the bottom is 8 ft from the wall.

12. Using the data in Exercise 11, find the rate of change of the area of the triangle formed by the ladder, the ground, and the wall when the bottom of the ladder is 8 ft from the wall.

13. A ship sails out of New York harbor at a rate of 20 ft/sec. The closest it comes to the Statue of Liberty is 1200 ft at a time t_0. Find the rate at which the distance s from the ship to the statue is increasing 25 sec later. (Assume the ship travels in a straight line during these 25 sec.) See Fig. 4.6.

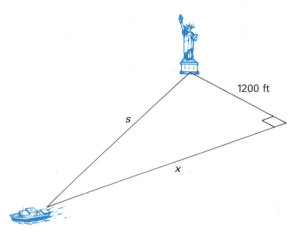

FIGURE 4.6

14. An airplane traveling in a straight line at 600 mph at a constant altitude of 4 miles passes directly over your head. Find the rate at which the distance from you to the plane is changing 48 sec later. (Neglect the curvature of the earth.)

15. A particle starts at the origin in the plane and travels on the curve $y = \sqrt{5x + 4} - 2$. If the x-coordinate of the particle increases at a uniform rate of 4 units/sec, find the rate of increase of the y-coordinate 7 sec after the particle starts from the origin.

16. A body travels on the circle $x^2 + y^2 = 25$. When the body is at the point $(3, -4)$, the x-component of the velocity is 2 units/sec. Find the y-component of the velocity at that point.

17. Jim is 6 ft tall and is walking at night straight toward a lighted street lamp at a rate of 5 ft/sec. If the lamp is 20 ft above the ground, find the rate at which his shadow is shortening when he is 30 ft from the lamppost.

18. A triangle has two sides of constant lengths 10 cm and 15 cm. The angle θ between the two sides increases at a constant rate of 9°/min. Find the rate of increase of the third side of the triangle when $\theta = 60°$.

19. The lengths of two sides of a triangle remain constant at 5 ft and 7 ft while the length of the third side decreases at a constant rate of 6 in./min. Let θ be the angle between the sides of constant lengths. Find the rate at which θ is changing when $\cos \theta = \frac{3}{5}$.

20. Sue is standing on a dock and pulling a boat in to the dock by means of a rope tied to a ring in the bow of the boat. If the ring is 2 ft above water level and her hands are 7 ft above water level, and if she is pulling in the rope at a

uniform rate of 2 ft/sec, find the speed with which the boat is approaching the dock when it is 12 ft from the dock.

21. A rocket fired straight up from the ground reaches an altitude of $5t^2$ ft after t sec. An observer whose eyes are 5 ft above the ground watches from 315 ft away. Find the rate at which the angle of elevation from the observer's eyes to the rocket is increasing when the angle is 45°. See Fig. 4.7.

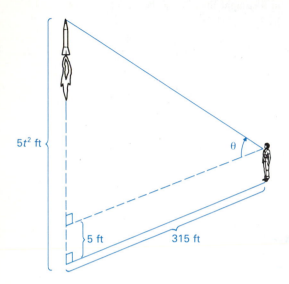

$5t^2$ ft

θ

5 ft   315 ft

FIGURE 4.7

22. A spherical balloon is being inflated so that its volume increases at a constant rate of 8 ft³/min. Find the rate of increase of the radius when the radius is 3 ft.

23. Assume that when a hard candy ball is dropped in a glass of water, it dissolves at a rate directly proportional to its surface area. Show that the radius of the ball decreases at a constant rate.

24. The trunk of a fir tree is in the shape of a cone. At a certain moment the diameter of the base of the tree is 1 ft and is

increasing at $\frac{1}{4}$ in./yr, while the height is 20 ft and is increasing at 6 in./yr. Find the rate of increase of the volume of wood in the trunk of the tree at that moment.

25. Water is being poured into an inverted cone (vertex down) of radius 4 in. and height 10 in. at a rate of 3 in³/sec. Find the rate at which the water is rising when its depth over the vertex is 5 in.

26. Sand is being poured at a rate of 2 ft³/min to form a conical pile whose height is always 3 times the radius. Find the rate at which the area of the base of the cone is increasing when the cone is 4 ft high.

27. The vertex angle of a right circular cone of constant slant height L cm is decreasing at a constant rate of $\frac{1}{8}$ radian/sec. Find the rate at which the volume is changing when the vertex angle is 90°.

28. In one type of prewalking exerciser, a child in a harness is suspended with feet just touching the floor by a cylinder of elastic rubber with circular cross section. We assume the volume of the cylinder of rubber remains constant while the length and radius change as the child bounces. If the child is bouncing downward at a rate of 30 cm/sec at an instant when the rubber cylinder has length 90 cm and diameter 3 cm, find the rate of change of the diameter of the cylinder at that moment.

29. A bridge goes straight across a river at a height of 60 ft. A car on the bridge traveling at 40 ft/sec passes directly over a boat traveling up the river at 15 ft/sec at time t_0. Find the rate at which the distance between the car and the boat is increasing 3 sec later.

30. In the operation of a tape recorder, a tape of constant thickness is wound from reel 1 onto reel 2. Let r_1 be the radius from the center of reel 1 to the outer edge of the tape on that reel at time t, and let r_2 be similarly defined for reel 2. Using calculus, show that at any time t,

$$\frac{dr_1/dt}{dr_2/dt} = -\frac{r_2}{r_1}.$$

[*Hint:* The volume of tape remains constant.]

4.2 THE MEAN VALUE THEOREM

We introduce the notions of local maxima and local minima of a function and present the mean value theorem for derivatives. The mean value theorem is used to determine bounds on function values from given bounds on the values of the derivative. The theorem is also useful for proving other theorems, as we show.

Local Maxima and Local Minima

In Section 2.4, we stated two important properties of continuous functions: the intermediate value theorem (Theorem 2.3) and the theorem on attainment of maximum and minimum values (Theorem 2.4). We restate Theorem 2.4 here for easy reference.

THEOREM 4.1   Attainment of extreme values

A continuous function assumes both a maximum value M and a minimum value m on any closed interval in its domain.   □

The function whose graph is shown in Fig. 4.8 assumes its maximum value in $[a, b]$ where $x = b$ and its minimum value where $x = a$. The value $f(x_1)$ is not a maximum for $f(x)$ on $[a, b]$ but is a maximum for $f(x)$ in the immediate vicinity of $x = x_1$. Similarly, $f(x_2)$ in Fig. 4.8 is a minimum value for $f(x)$ in the vicinity of x_2.

DEFINITION 4.1

Local maximum and minimum values

If $f(x_1)$ is a maximum value of $f(x)$ for $x_1 - h < x < x_1 + h$ for some $h > 0$, then $f(x_1)$ is a **relative maximum** or **local maximum** of $f(x)$. Similarly, $f(x_2)$ in Fig. 4.8 is a **relative minimum** or **local minimum** of $f(x)$.

It appears from Fig. 4.9 that if f is differentiable, then the graph of f has a horizontal tangent line at points corresponding to local maxima or minima. We prove this before we proceed to the main topic of this section.

THEOREM 4.2   Derivative at local extrema

If f is differentiable at x_1 and if $f(x_1)$ is a local maximum (or local minimum) of $f(x)$, then $f'(x_1) = 0$.

Proof: Recall that the definition of $f'(x_1)$ is

$$f'(x_1) = \lim_{\Delta x \to 0} \frac{f(x_1 + \Delta x) - f(x_1)}{\Delta x}. \qquad \textbf{(1)}$$

If $f(x_1)$ is a maximum, then for small Δx,

$$f(x_1 + \Delta x) \leq f(x_1).$$

Thus the numerator of the difference quotient in Eq. (1) is always negative or zero. If $\Delta x > 0$, the quotient is less than or equal to zero, while if $\Delta x < 0$, the quotient is greater than or equal to zero. The only way that $f'(x_1)$ can be simultaneously approached by numbers less than or equal to zero and numbers greater

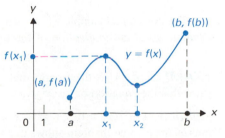

FIGURE 4.8 $f(x)$ has a local maximum at x_1 and a local minimum at x_2.

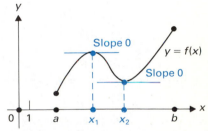

FIGURE 4.9 Tangent lines are horizontal at local maxima and minima.

than or equal to zero is if $f'(x_1) = 0$. An analogous argument shows that if $f(x_2)$ is minimum and $f'(x_2)$ exists, then $f'(x_2) = 0$.   □

The Mean Value Theorem

We are now ready to demonstrate Rolle's theorem, a special case of the mean value theorem. This special case serves as a lemma in the proof of the main theorem.

THEOREM 4.3   Rolle's theorem

If $f(x)$ is continuous in $[a, b]$ and differentiable for $a < x < b$ and if, furthermore, $f(a) = f(b)$, then there exists c where $a < c < b$ such that $f'(c) = 0$.

Proof: Figure 4.10 illustrates Rolle's theorem. By Theorem 4.1, we know that $f(x)$ attains a maximum value M and a minimum value m in $[a, b]$. If f is a constant function in $[a, b]$ so that $f(x) = f(a) = f(b)$ for all x in $[a, b]$, then $f'(c) = 0$ for any c where $a < c < b$. If f is not constant in $[a, b]$, then since $f(a) = f(b)$, either the maximum or the minimum of $f(x)$ must be attained at a point c where $a < c < b$. From Theorem 4.2, we then know that $f'(c) = 0$. This completes the demonstration of Rolle's theorem.   □

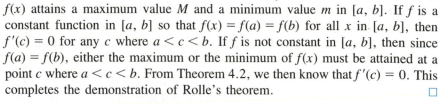

FIGURE 4.10   Position of c given by Rolle's theorem.

As indicated by Fig. 4.11, the value c described in Rolle's theorem need not be unique. Both c_1 and c_2 in Fig. 4.11 satisfy the requirement in the theorem.

EXAMPLE 1   Show that the hypotheses for Rolle's theorem are satisfied by $f(x) = x^3 - 3x + 4$ on $[-1, 2]$, and find a value c described by the theorem.

Solution   As a polynomial function, $f(x)$ is continuous and differentiable everywhere; therefore it is continuous and differentiable on $[-1, 2]$. We also have

$$f(-1) = 6 = f(2).$$

Thus the hypotheses of Rolle's theorem are satisfied.
We have $f'(x) = 3x^2 - 3$, which is zero where

$$3x^2 - 3 = 0, \qquad x^2 = 1, \qquad x = \pm 1.$$

Now -1 does not satisfy the relation $-1 < c < 2$ for c as required by Rolle's theorem. However, 1 does satisfy the relation. Thus $c = 1$, and the value for c is unique in this example.   ■

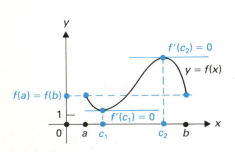

FIGURE 4.11   Both c_1 and c_2 satisfy the property described in Rolle's theorem.

The preceding example was a mere *illustration* of Rolle's theorem. The next example gives an *application* of the theorem.

EXAMPLE 2   Use the intermediate value theorem and Rolle's theorem to show that the equation $6x^3 + x^2 + x - 5 = 0$ has exactly one solution in $[0, 1]$.

Solution   Let $f(x) = 6x^3 + x^2 + x - 5$. Since $f(0) = -5$ and $f(1) = 3$, the intermediate value theorem shows that there is at least one value a in $[0, 1]$ such that $f(a) = 0$. Suppose there were also b in $[0, 1]$ such that $f(b) = 0$, and sup-

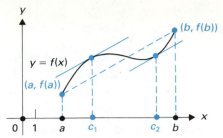

FIGURE 4.12 Mean value theorem; tangent lines where $x = c_1$ and $x = c_2$ are parallel to the line through $(a, f(a))$ and $(b, f(b))$.

pose $a < b$. Applying Rolle's theorem to $[a, b]$, we see that $f'(x) = 18x^2 + 2x + 1$ would be zero for some number c where $a < c < b$, and therefore $0 < c < 1$. But clearly $18x^2 + 2x + 1 > 0$ for all $x \geq 0$. Thus no such b can exist, and $x = a$ is the unique solution of $6x^3 + x^2 + x - 5 = 0$ in $[0, 1]$. ∎

The mean value theorem may be regarded as a generalization of Rolle's theorem. It is illustrated in Fig. 4.12. Both theorems assert that for f continuous in $[a, b]$ and differentiable for $a < x < b$, there exists at least one c where $a < c < b$ and where the tangent line to the graph of f is parallel to the line joining the points $(a, f(a))$ and $(b, f(b))$. The slope of this line is

$$\frac{f(b) - f(a)}{b - a},$$

so the conclusion of the mean value theorem will take the form

$$f'(c) = \frac{f(b) - f(a)}{b - a} \quad \text{or} \quad f(b) - f(a) = (b - a)f'(c)$$

for some c where $a < c < b$. There are two such points, c_1 and c_2, in Fig. 4.12.

THEOREM 4.4   Mean value theorem

Let $f(x)$ be continuous in $[a, b]$, and differentiable for $a < x < b$. Then there exists c where $a < c < b$ such that

$$f(b) - f(a) = (b - a)f'(c). \qquad (2)$$

Proof: To obtain this result from Rolle's theorem, we introduce the function g with domain $[a, b]$ whose value at x is indicated in Fig. 4.13. We can see from Fig. 4.13 that we should define g by

$$g(x) = f(x) - \left[f(a) + \frac{f(b) - f(a)}{b - a}(x - a) \right]$$

$$= [f(x) - f(a)] - \left[\frac{f(b) - f(a)}{b - a}(x - a) \right].$$

Since f is continuous in $[a, b]$ and differentiable for $a < x < b$, we see that g is also, and $g(a) = g(b) = 0$, so Rolle's theorem applies. Therefore for some c where $a < c < b$, we have $g'(c) = 0$. Now

$$g'(x) = f'(x) - \frac{f(b) - f(a)}{b - a},$$

so

$$0 = g'(c) = f'(c) - \frac{f(b) - f(a)}{b - a},$$

and $f(b) - f(a) = (b - a)f'(c)$. ☐

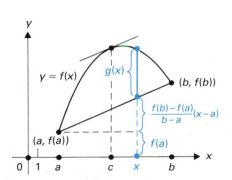

FIGURE 4.13 Proof of the mean value theorem:

$$g(x) = f(x) -$$
$$\left[f(a) + \frac{f(b) - f(a)}{b - a}(x - a) \right]$$

satisfies the hypotheses of Rolle's theorem.

EXAMPLE 3 Illustrate the mean value theorem with $f(x) = x^2$ in the interval $[0, 3]$.

Solution We must find c such that

$$f(3) - f(0) = (3 - 0)f'(c),$$

or such that $9 = 3 \cdot f'(c)$. Now $f'(x) = 2x$, and $9 = 3 \cdot 2c$ when $c = \frac{3}{2}$. Note that, as the mean value theorem states, we found a value c satisfying $0 < c < 3$. ∎

The adjective *mean* in the name of Theorem 4.4 is used in its mathematical sense signifying "average." We know that $(f(b) - f(a))/(b - a)$ is the average (mean) rate of change of $f(x)$ over $[a, b]$. *The mean value theorem asserts that this average rate of change is attained at some point.*

EXAMPLE 4 A car accelerates smoothly from 5 mph to 50 mph in 15 sec. Show that at some instant, the acceleration of the car is 3 mi/sec².

Solution *Smoothly* means that the velocity is a differentiable function of time. Suppose the velocity is 5 mph at time t_1 and 50 mph at time $t_2 = t_1 + 15$. Then

$$\frac{v(t_2) - v(t_1)}{t_2 - t_1} = \frac{50 - 5}{15} = \frac{45}{15} = 3 \text{ mi/sec}^2$$

is the average acceleration of the car during those 15 sec. By the mean value theorem, this value 3 mi/sec² is the exact acceleration at some instant c during that time interval. ∎

The mean value theorem is often used to give bounds on the size of $f(x)$ if we know bounds on the size of $f'(x)$. This application is much more significant than mere illustrations of the theorem, computing c as in Example 3.

EXAMPLE 5 Let $f(x)$ be differentiable. Suppose $f(2) = -1$ and $|f'(x)| \le 3$ for x in $[2, 6]$. Find bounds in terms of x on $f(x)$ in $[2, 6]$.

Solution For any $x \ne 2$ in $[2, 6]$, we have

$$\frac{f(x) - f(2)}{x - 2} = \frac{f(x) + 1}{x - 2} = f'(c)$$

for some c, $2 < c < x$, by the mean value theorem. But $|f'(c)| \le 3$ by assumption. Since $x - 2$ is positive, we then have

$$\left| \frac{f(x) + 1}{x - 2} \right| = \frac{|f(x) + 1|}{x - 2} \le 3.$$

Thus

$$|f(x) + 1| \le 3(x - 2),$$
$$-3(x - 2) \le f(x) + 1 \le 3(x - 2),$$
$$-1 - 3x + 6 \le f(x) \le -1 + 3x - 6,$$
$$-3x + 5 \le f(x) \le 3x - 7,$$

for x in $[2, 6]$. These are bounds on $f(x)$. See Fig. 4.14. For example, we have $-10 \le f(5) \le 8$. ∎

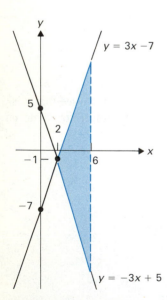

FIGURE 4.14 $(x, f(x))$ lies in the shaded triangle for $2 \le x \le 6$.

EXAMPLE 6 Let $f(x)$ be differentiable and suppose $f(0) = 2$ and $|f'(x)| \leq 4x$ for x in $[0, 5]$. Find bounds on $f(x)$ in the interval $[0, 5]$.

Solution As in Example 5, we have for x in $[0, 5]$, $x \neq 0$,

$$|f(x) - 2|/x \leq 4(x),$$
$$-4x(x) \leq f(x) - 2 \leq 4x(x),$$
$$-4x^2 + 2 \leq f(x) \leq 4x^2 + 2.$$

For example, we have $-34 < f(3) < 38$. ∎

Two Consequences of the Mean Value Theorem

The mean value theorem is very useful in proving other theorems. We give two illustrations.

THEOREM 4.5   Functions having zero derivative

Let f be differentiable for $a \leq x \leq b$ and suppose that $f'(x) = 0$ for all x in $[a, b]$. Then $f(x) = f(a)$ for all x in $[a, b]$; that is, $f(x)$ is constant throughout this interval.

Proof: We apply the mean value theorem to the interval $[a, x]$ where $a < x \leq b$, obtaining

$$\frac{f(x) - f(a)}{x - a} = f'(c)$$

for some c where $a < c < b$. But $f'(c) = 0$ by hypothesis, so

$$\frac{f(x) - f(a)}{x - a} = 0.$$

Consequently, $f(x) - f(a) = 0$, so $f(x) = f(a)$. □

In Section 3.6, we said that $f(t) = A_0 e^{kt}$ is the only function that is differentiable for all t and such that $f(0) = A_0$ and $f'(t) = k \cdot f(t)$. We can prove this using Theorem 4.5, which was in turn a consequence of the mean value theorem.

THEOREM 4.6   Unique solution of $f' = kf$, $f(0) = A_0$

The function $f(t) = A_0 e^{kt}$ is the *unique* function that is differentiable for all t and that satisfies

$$f(0) = A_0 \quad \text{and} \quad f'(t) = k \cdot f(t).$$

Proof: Suppose that $g(t)$ also satisfies the conditions given in Theorem 4.6. We consider the case where $A_0 \neq 0$. (Exercise 35 leads you through an argument if $A_0 = 0$.) Let

$$h(t) = \frac{g(t)}{f(t)}.$$

Since $f(t) = A_0 e^{kt} \neq 0$ for all t, we see that $h(t)$ is differentiable for all t. Now

$$h'(t) = \frac{f(t)g'(t) - g(t)f'(t)}{f(t)^2}.$$

Since $f'(t) = k \cdot f(t)$ and $g'(t) = k \cdot g(t)$, we have

$$h'(t) = \frac{f(t) \cdot k \cdot g(t) - g(t) \cdot k \cdot f(t)}{f(t)^2} = \frac{0}{f(t)^2} = 0$$

for all t. By Theorem 4.5, the function $h(t)$ must be a constant function. Since $h(0) = g(0)/f(0) = A_0/A_0 = 1$, we must have $h(t) = f(t)/g(t) = 1$ for all t, so $g(t) = f(t)$ for all t. $\qquad\square$

SUMMARY

1. If f is differentiable, then $f'(x) = 0$ at any point where the function has a local maximum or a local minimum.

2. *Rolle's theorem:* If $f(x)$ is continuous in $[a, b]$ and differentiable for $a < x < b$, and if $f(a) = f(b)$, then there is some number c where $a < c < b$ such that $f'(c) = 0$.

3. *Mean value theorem:* If $f(x)$ is continuous in $[a, b]$ and differentiable for $a < x < b$, then there is some number c where $a < c < b$ such that

$$\frac{f(b) - f(a)}{b - a} = f'(c).$$

4. If $f'(x) = 0$ throughout an interval, then $f(x)$ is constant in the interval.

EXERCISES 4.2

In Exercises 1–6, verify that $f(x)$ satisfies the hypotheses for Rolle's theorem on the given interval, and find all numbers c described by the theorem.

1. $f(x) = x^2 + x + 4$ on $[-3, 2]$

2. $f(x) = -3x^2 + 12x + 4$ on $[0, 4]$

3. $f(x) = \sin x$ on $[0, \pi]$

4. $f(x) = \cos x$ on $[0, 2\pi]$

5. $f(x) = x^3 - 6x^2 + 5x + 3$ on $[1, 5]$

6. $f(x) = (x^2 - 1)/(x + 2)$ on $[-1, 1]$

In Exercises 7–10, use the intermediate value theorem and Rolle's theorem to show that the given equation has exactly one solution in the given interval. (See Example 2.)

7. $x^2 - 3x + 1 = 0$ in $[2, 6]$

8. $10 - 5x - x^2 = 0$ in $[-1, 2]$

9. $\sin x = \frac{1}{4}x$ in $[\pi/2, \pi]$

10. $\cos x = x$ in $[0, \pi/2]$

11. Generalize Rolle's theorem to show that if f is a differentiable function and $f(x)$ is zero at r distinct points in an interval $[a, b]$, then $f'(x)$ is zero for at least $r - 1$ distinct points in $[a, b]$.

12. Let f be a differentiable function and suppose that $f'(x) = 0$ for at most three distinct values of x where $a < x < b$. What is the maximum possible number of distinct solutions of $f(x) = 10$ in $[a, b]$?

In Exercises 13–18, illustrate the mean value theorem for the given function over the given interval by finding all values of c described in the theorem.

13. $f(x) = 3x - 4$ on $[1, 4]$ 14. $f(x) = x^3$ on $[-1, 2]$

15. $f(x) = x - 1/x$ on $[1, 3]$ 16. $f(x) = x^{1/2}$ on $[9, 16]$

17. $f(x) = \sqrt{1 - x}$ on $[-3, 0]$

18. $f(x) = (x - 1)/(x + 3)$ on $[2, 4]$

19. Let f be a function satisfying the hypotheses of the mean value theorem on $[a, b]$.
    (a) Give the rate-of-change interpretation of
    $$\frac{f(b) - f(a)}{b - a}.$$
    (b) Give the interpretation of $f'(c)$ as a rate of change.
    (c) Restate the mean value theorem in terms of rates of change.

20. Two towns A and B are connected by a highway 10 miles long with a 60 mph speed limit. Mr. Smith is arrested on a speeding charge and admits having driven from A to B in 8 minutes. In a speeding offense, the court is allowed to impose a fine of $15 plus $2 for each mph in excess of the limit. Use the mean value theorem to show that the judge is justified in imposing a fine of $45 on Mr. Smith. [*Hint:* Use Exercise 19.]

21. For a quadratic function f given by $f(x) = ax^2 + bx + c$, show that the point between x_1 and x_2 where the tangent to the graph of f is parallel to the chord joining $(x_1, f(x_1))$ and $(x_2, f(x_2))$ is halfway between x_1 and x_2. What example in the text illustrates this?

22. Explain the fallacy in the following argument: The Elliot family spent $99.61 on food during one week, which was an average spending rate of $14.23 per day. Therefore, by the mean value theorem, the family must have spent exactly $14.23 for food on some day that week.

23. Let f be a differentiable function with domain containing $[a, b]$. Suppose that $m \leq f'(x) \leq M$ in $[a, b]$. Use the mean value theorem to show that
    $$f(a) + m(x - a) \leq f(x) \leq f(a) + M(x - a)$$
    for all x in $[a, b]$.

In Exercises 24–34, let f be differentiable for all x. Use the mean value theorem to find bounds on $f(x)$ from the given properties of f and f'. Be sure to state the values of x where the bounds are valid.

24. $f(0) = 3$, $|f'(x)| \leq 3$ for x in $[0, 8]$

25. $f(-2) = 1$, $|f'(x)| \leq 5$ for x in $[-2, 4]$

26. $f(1) = 7$, $|f'(x)| \leq 2\sqrt{x}$ for x in $[1, 3]$

27. $f(0) = 6$, $2 \leq f'(x) \leq 5$ for x in $[0, 6]$

28. $f(10) = -8$, $-1 \leq f'(x) \leq 3$ for x in $[10, 100]$

29. $f(2) = 4$, $f'(x) \leq 5$ for x in $[0, 2]$

30. $f(3) = 6$, $-2 \leq f'(x) \leq 5$ for x in $[-1, 3]$

31. $f(0) = -4$, $-1 \leq f'(x) \leq 4$ for $x \geq 0$

32. $f(0) = 5$, $2 \leq f'(x) \leq 3$ for $x \leq 0$

33. $f(0) = 1$, $3 \leq f'(x) \leq 7$ for all x

34. $f(2) = 5$, $-1 \leq f'(x) \leq 10$ for all x

35. To prove Theorem 4.6 for $A_0 = 0$, we need to show that if $g(0) = 0$ and g is differentiable with $g'(t) = k \cdot g(t)$ for all t, then $g(t) = 0$ for all t. Suppose $g(a) \neq 0$ and let $h(t) = g(t + a)$.
    (a) Compute $h(0)$.
    (b) Compute $h'(t)$ by the chain rule, and show that $h'(t) = k \cdot h(t)$.
    (c) Deduce from the proof of Theorem 4.6 in the text that $h(t) = g(a)e^{kt}$. Explain why this contradicts the hypothesis that $g(0) = 0$ and why our proof is now complete.

4.3 CRITICAL POINTS

Recall once more Theorem 4.1 of the preceding section:

Attainment of Extreme Values

A continuous function attains both a maximum value M and a minimum value m on any closed interval in its domain.

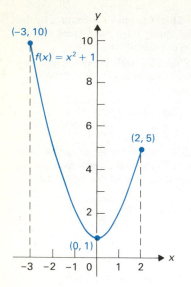

FIGURE 4.15 $f(x) = x^2 + 1$ has maximum value 10 and minimum value 1 on $[-3, 2]$.

The preceding section used this theorem to prove the mean value theorem. In this section we tackle the problem of computing M and m. First we illustrate the theorem.

EXAMPLE 1 Sketch the graph of $f(x) = x^2 + 1$ and use the graph to find the maximum and minimum values attained in $[-3, 2]$.

Solution The graph is sketched in Fig. 4.15. We see that the maximum value attained by $f(x) = x^2 + 1$ in $[-3, 2]$ is 10, which is attained at $x = -3$. The minimum value is 1, which is attained at $x = 0$. ∎

EXAMPLE 2 Repeat Example 1 for the function

$$f(x) = \sin x \quad \text{in} \quad [-2\pi, 2\pi].$$

Solution The graph is sketched in Fig. 4.16. The maximum value attained by $\sin x$ in $[-2\pi, 2\pi]$ is 1, which is attained at both $x = -3\pi/2$ and $x = \pi/2$. The minimum value is -1, which is attained at both $-\pi/2$ and $3\pi/2$. ∎

Suppose the extreme values M and m of $f(x)$ on $[a, b]$ do not occur at endpoints of the interval. Then M is also a local maximum of the function. If $f(x_1) = M$ and if f is differentiable at x_1, then Theorem 4.2 of the preceding section shows that $f'(x_1) = 0$. A similar analysis holds if f attains a local minimum at x_2 in the open interval (a, b). Thus if we find all points in the open interval (a, b) where either $f'(x)$ does not exist or where $f'(x) = 0$, we have all candidates for points in this open interval where f may attain a maximum or a minimum value.

DEFINITION 4.2

Critical point

A **critical point** of a function f is a point x_0 in its domain where either $f'(x_0) = 0$ or $f'(x_0)$ does not exist.

EXAMPLE 3 Find all critical points for $f(x) = (x - 1)^{2/3}$.

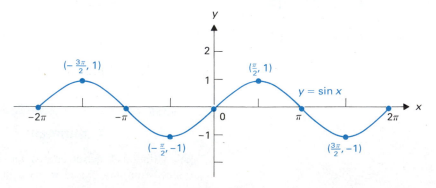

FIGURE 4.16 $f(x) = \sin x$ has maximum value 1 at $-3\pi/2$ and $\pi/2$ and has minimum value -1 at $-\pi/2$ and $3\pi/2$ on $[-2\pi, 2\pi]$.

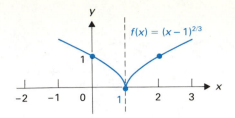

FIGURE 4.17 $f(x) = (x - 1)^{2/3}$ has minimum value 0 at $x = 1$, but $f'(1)$ does not exist since the tangent line there is vertical.

Solution Note that $f(x)$ is defined for all x. We have

$$f'(x) = \frac{2}{3}(x - 1)^{-1/3} = \frac{2}{3(x - 1)^{1/3}} \qquad \text{for } x \neq 1.$$

Thus $f'(x)$ is never zero, but $f'(1)$ does not exist. Consequently, 1 is a critical point, the only critical point. The graph of $y = f(x)$ is shown in Fig. 4.17. Clearly $f(x)$ attains a minimum value where $x = 1$, but $f'(1)$ does not exist. (The tanget line there is vertical.) This example illustrates that a derivative need not exist at a point where a function attains a local maximum or local minimum. ∎

EXAMPLE 4 Find all critical points of $f(x) = x/(x^2 + 1)$. Then graph the function to decide whether the critical points yield local maxima or local minima.

Solution Again, $f(x)$ is defined for all x. We have

$$f'(x) = \frac{(x^2 + 1)1 - x(2x)}{(x^2 + 1)^2} = \frac{-x^2 + 1}{(x^2 + 1)^2}.$$

We see that $f'(x)$ exists for all x. Now $f'(x) = 0$ when $-x^2 + 1 = 0$, or $x = \pm 1$. Thus the critical points are $x = \pm 1$.

Now $f(x) > 0$ for $x > 0$, and $f(x) < 0$ for $x < 0$. Also, $f(0) = 0$ and

$$\lim_{x \to \infty} \frac{x}{x^2 + 1} = \lim_{x \to -\infty} \frac{x}{x^2 + 1} = 0.$$

Therefore the graph of f must appear as in Fig. 4.18, so $(1, \frac{1}{2})$ corresponds to a local maximum and $(-1, -\frac{1}{2})$ to a local minimum. In this case, $f(1) = \frac{1}{2}$ is actually the maximum value and $f(-1) = -\frac{1}{2}$ the minimum value for the function over its entire domain. ∎

EXAMPLE 5 Sketch the graph of $f(x) = |\sin x|$ and from the graph find all critical points of f.

Solution The graph is sketched in Fig. 4.19. We see that local maxima equal to 1 occur at

$$x = \pm\frac{\pi}{2}, \ \pm\frac{3\pi}{2}, \ \pm\frac{5\pi}{2}, \ \ldots,$$

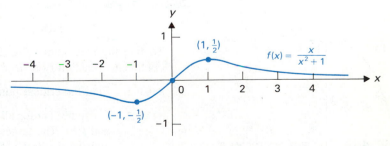

FIGURE 4.18 $f(x) = x/(x^2 + 1)$ has a maximum of $\frac{1}{2}$ at $x = 1$ and a minimum of $-\frac{1}{2}$ at $x = -1$.

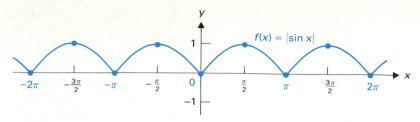

FIGURE 4.19 $f(x) = |\sin x|$ has maximum values of 1 where $f'(x) = 0$ and has minimum values of 0 where $f'(x)$ does not exist.

where $f'(x) = 0$, so all these points are critical points. It is clear from the graph that $f'(x)$ does not exist where

$$x = 0, \ \pm\pi, \ \pm2\pi, \ \pm3\pi, \ \ldots$$

because the graph has sharp points there. Thus these are also critical points of the function. They correspond to local minima equal to zero. ■

EXAMPLE 6   Find all critical points of $f(x) = x^3$, and determine any local maxima or minima by examination of the graph.

Solution   We have $f'(x) = 3x^2$ for all x. Of course, $f'(x) = 0$ only if $x = 0$, so $x = 0$ is the only critical point. The familiar graph of x^3 is shown in Fig. 4.20. We see that there is neither a local minimum nor a local maximum where $x = 0$. ■

To summarize our work, we have shown that

a local maximum or local minimum in an open interval must always occur at a critical point.

However, Example 6 shows that

a function might have some critical points where neither a local maximum nor a local minimum occurs.

We can now describe a method for attempting to find the maximum value M and minimum value m attained by a continuous function on a closed interval. This is the problem we posed at the start of this section.

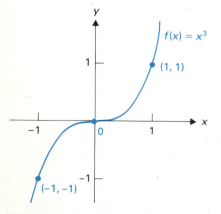

FIGURE 4.20 $f(x) = x^3$ has neither a local maximum nor a local minimum where $x = 0$, even though $f'(0) = 0$.

Finding Extrema on [a, b]

To find the maximum value M and minimum value m attained by a continuous function f on $[a, b]$:

STEP 1.   Find the critical points of f in the open interval (a, b).

STEP 2.   Compute $f(x)$ for all x found in step 1 and also compute $f(a)$ and $f(b)$. The largest value found is the maximum value M and the smallest value is the minimum value m.

This method may break down if f has an infinite number of critical points in $[a, b]$, but for many functions this problem does not occur.

EXAMPLE 7 Find the maximum and minimum values attained in $[-\frac{1}{2}, 2]$ by $f(x) = 3x^4 + 4x^3 - 12x^2 + 5$.

Solution

STEP 1. Since f is a polynomial function, it is differentiable, so critical points appear only where $f'(x) = 0$. Differentiating, we have

$$f'(x) = 12x^3 + 12x^2 - 24x$$
$$= 12x(x^2 + x - 2)$$
$$= 12x(x + 2)(x - 1).$$

Thus $f'(x) = 0$ for $x = -2, 0$, and 1. Of these three points, only 0 and 1 are in $[-\frac{1}{2}, 2]$.

STEP 2. Computation shows that

$$f(0) = 5, \qquad f(1) = 0, \qquad f(-\tfrac{1}{2}) = \tfrac{27}{16}, \qquad f(2) = 37.$$

Therefore 0 is the minimum value, attained at $x = 1$, and 37 is the maximum value, attained at $x = 2$. ∎

EXAMPLE 8 Find the maximum and minimum values attained by $f(x) = \sin x - \cos x$ on $[0, \pi]$.

Solution Again, all critical points occur where $f'(x) = 0$.

STEP 1. Differentiating, we obtain

$$f'(x) = \cos x + \sin x.$$

Thus $f'(x) = 0$ when $\cos x + \sin x = 0$, that is, when $\sin x = -\cos x$. This condition is equivalent to $\tan x = -1$. In $[0, \pi]$, we have $\tan x = -1$ only at $x = 3\pi/4$.

STEP 2. We have

$$f(0) = 0 - 1 = -1, \qquad f\left(\frac{3\pi}{4}\right) = \frac{1}{\sqrt{2}} - \frac{-1}{\sqrt{2}} = \frac{2}{\sqrt{2}} = \sqrt{2},$$
$$f(\pi) = 0 - (-1) = 1.$$

Thus $\sqrt{2}$ is the maximum value, attained at $x = 3\pi/4$, and -1 is the minimum value, attained at $x = 0$, over the interval $[0, \pi]$. ∎

SUMMARY

1. If $f(x)$ is continuous at each point in $[a, b]$, then $f(x)$ attains a maximum value M and also a minimum value m in $[a, b]$.

2. A critical point of f is a point x_0 in the domain of f where either $f'(x_0) = 0$ or $f'(x_0)$ does not exist.

3. If $f(x)$ is defined for $a < x < b$, then a local maximum or local minimum of f in this open interval can occur only at a critical point.

4. To find the maximum value M and minimum value m attained by a continuous function f on $[a, b]$, follow the two-step outline given on page 198.

EXERCISES 4.3

1. Mark each of the following true or false.
   _____ (a) If x_0 is a critical point of f, then it must be that $f'(x_0) = 0$.
   _____ (b) If $f'(x_0) = 0$, then x_0 must be a critical point of f.
   _____ (c) If f is differentiable in the open interval (a, b) and has a local maximum at x_0 in (a, b), then we must have $f'(x_0) = 0$.
   _____ (d) If the domain of f contains the open interval (a, b) and f has a local minimum at x_0 in (a, b), then x_0 must be a critical point of f.
   _____ (e) If the domain of f contains the open interval (a, b) and f has a local minimum at x_0 in (a, b), then we must have $f'(x_0) = 0$.

2. Mark each of the following true or false.
   _____ (a) Every function has either a local maximum or a local minimum at each critical point.
   _____ (b) Either the maximum or the minimum of a continuous function on a closed interval $[a, b]$ must be attained at one of the endpoints of the interval.
   _____ (c) It is possible for both the maximum value and the minimum value of a continuous function on a closed interval to be attained at the same point of the interval.
   _____ (d) If $f'(x_0) = 0$, then f must have either a local maximum or a local minimum at x_0.
   _____ (e) A point not in the domain of a function may be a critical point of the function.

3. If $f(x) = 1/x$, then $f'(x) = -1/x^2$. Thus $f'(0)$ does not exist. However, $x_0 = 0$ is not a critical point of f. Why not?

4. If $f(x) = \tan x$, then $f'(x) = \sec^2 x = 1/(\cos^2 x)$. Since $\cos(\pi/2) = 0$, we see that $f'(\pi/2)$ does not exist. Is $x_0 = \pi/2$ a critical point of f? Why or why not?

In Exercises 5–20, find all critical points of the given function. Use methods of the preceding chapters to sketch the graph of the function with enough accuracy to decide whether the function has a local maximum, a local minimum, or neither at each critical point.

5. $f(x) = 5 + 4x - 2x^2$ 6. $f(x) = x^2 - 4x + 3$

7. $f(x) = 4x^2 - 8x$ 8. $f(x) = -x^2 - 6x - 2$

9. $f(x) = 2x^3 - 4$ 10. $f(x) = x^3 - 3x^2$

11. $f(x) = x^4 - 1$ 12. $f(x) = 3 - x^5$

13. $f(x) = \dfrac{1}{x^2 - 1}$ [*Warning:* See Exercise 3.]

14. $f(x) = \dfrac{1}{x^2 + 1}$ 15. $f(x) = \dfrac{-x}{x^2 + 1}$

16. $f(x) = \dfrac{x}{1 - x^2}$ 17. $f(x) = \sec 2x$

18. $f(x) = \csc x$ 19. $f(x) = |\cos x|$

20. $f(x) = |\tan x|$

In Exercises 21–32, find the maximum and minimum values attained by the function in the given interval.

21. $f(x) = x^4$ in $[-2, 1]$ 22. $f(x) = \dfrac{1}{x}$ in $[2, 4]$

23. $f(x) = \dfrac{1}{x^2 + 1}$ in $[-1, 2]$

24. $f(x) = x^3 - 3x^2 + 1$ in
    (a) $[1, 3]$ (b) $[-1, 3]$ (c) $[-2, 4]$

25. $f(x) = x^2 + 4x - 3$ in
    (a) $[-5, -3]$ (b) $[-4, -1]$ (c) $[-4, 0]$

26. $f(x) = \dfrac{1}{x^2 - 1}$ in $[-3, -2]$

27. $f(x) = \dfrac{x^2 - x + 1}{x^2 + 1}$ in
    (a) $[-3, -2]$ (b) $[-2, 0]$ (c) $[0, 2]$ (d) $[-2, 2]$

28. $f(x) = \sin x$ in the intervals
    (a) $\left[0, \dfrac{\pi}{4}\right]$ (b) $\left[-\dfrac{\pi}{4}, \dfrac{\pi}{4}\right]$ (c) $\left[-\dfrac{3\pi}{4}, \dfrac{3\pi}{4}\right]$

29. $f(x) = \sin x + \cos x$ in the intervals
    (a) $\left[0, \dfrac{\pi}{2}\right]$ (b) $\left[-\dfrac{\pi}{2}, \dfrac{\pi}{2}\right]$ (c) $[0, \pi]$ (d) $\left[\dfrac{\pi}{2}, 2\pi\right]$

30. $f(x) = \sqrt{3} \sin x - \cos x$ in the intervals
    (a) $\left[0, \dfrac{\pi}{2}\right]$ (b) $[0, \pi]$ (c) $\left[0, \dfrac{3\pi}{2}\right]$ (d) $[0, 2\pi]$

31. $f(x) = e^x - x$ in $[-1, 1]$

32. $f(x) = x - \ln x$ in $[\tfrac{1}{2}, 2]$

In Exercises 33–36, the height y (in ft) of a body above the surface of the earth at time t sec is given. Find

(a) the maximum and minimum height of the body,
(b) the maximum and minimum velocity v of the body, and
(c) the maximum and minimum speed s of the body during the given time interval. Recall that $s = |v|$.

33. $y = 1600 - 16t^2,\ 0 \le t \le 5$

34. $y = 100 + 64t - 16t^2,\ 1 \le t \le 4$

35. $y = 1000 + 160t - 16t^2,\ 3 \le t \le 6$

36. $y = 100 + 320t + 16t^2,\ 2 \le t \le 4$

37. Let f be a polynomial function of even degree n such that the coefficient a_n of x^n is positive. Show that $f(x)$ attains a value that is minimum for all real numbers x. [*Hint:* Consider $\lim_{x \to \infty} f(x)$ and $\lim_{x \to -\infty} f(x)$, and apply Theorem 4.1 to a large interval $[-c, c]$.]

In Exercises 38–42, use the result of Exercise 37 and the two-step procedure outlined in the text to find the minimum value attained by the function for all real numbers x.

38. $3x^2 + 6x - 4$

39. $3x^4 - 8x^3 + 2$

40. $x^4 - 2x^2 + 7$

41. $3x^4 - 4x^3 - 12x^2 + 5$

42. $x^6 - 6x^4 - 8$

  EXPLORING CALCULUS

N: calculus topics quiz (topic 1)

4.4 GRAPHING USING LIMITS AND THE FIRST DERIVATIVE

This is the first of two sections devoted to using limits and derivatives in sketching graphs of functions. We start by using limits to examine the graph of a function f as $x \to \infty$ or $x \to -\infty$ and at points c where $|f(x)| \to \infty$ as $x \to c$. This technique is especially useful in graphing rational functions. Then we turn to the use of the sign (positive, negative, or zero) of values of the first derivative in sketching the graph of f.

Asymptotes

If $\lim_{x \to \infty} f(x) = c$ or $\lim_{x \to -\infty} f(x) = c$, then the horizontal line $y = c$ is a **horizontal asymptote** of the graph of f. This is illustrated in Fig. 4.21. If either the left-hand or right-hand limit of $f(x)$ at a is ∞ or $-\infty$, then the vertical line $x = a$ is a **vertical asymptote** of the graph. Figure 4.21 illustrates this also. We suggest that when graphing a function you first draw any horizontal or vertical

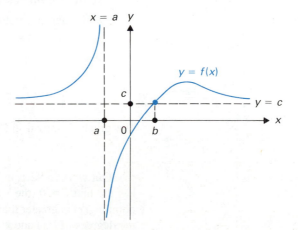

FIGURE 4.21 Horizontal asymptote $y = c$ intersects the graph where $x = b$. Vertical asymptote $x = a$ does not intersect the graph.

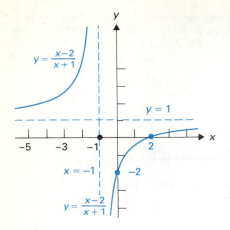

FIGURE 4.22 Vertical asymptote $x = -1$; horizontal asymptote $y = 1$; x-intercept 2; y-intercept -2.

asymptotes as dashed lines. It is possible for the graph of a continuous function to intersect a horizontal asymptote, as illustrated in Fig. 4.21, but the graph of such a function can never intersect a vertical asymptote. We illustrate further by graphing two functions of the form

$$f(x) = \frac{ax + b}{cx + d} \qquad \text{where} \quad c \neq 0. \tag{1}$$

EXAMPLE 1 Sketch $y = (x - 2)/(x + 1)$, finding the horizontal and vertical asymptotes and the x- and y-intercepts.

Solution The vertical asymptote is the line $x = -1$, since the denominator becomes zero at this value for x. This line is dashed in Fig. 4.22. Since

$$\lim_{x \to \infty} \frac{x - 2}{x + 1} = \lim_{x \to -\infty} \frac{x - 2}{x + 1} = 1,$$

the horizontal asymptote is $y = 1$, which is also dashed in Fig. 4.22. The *x-intercept* is 2, where the numerator is zero. The *y-intercept* is -2, obtained by setting x equal to zero.

To sketch the graph, we determine its behavior as it approaches the vertical asymptote by computing the left-hand and right-hand limits:

$$\lim_{x \to -1-} \frac{x - 2}{x + 1} = \infty \qquad \text{and} \qquad \lim_{x \to -1+} \frac{x - 2}{x + 1} = -\infty.$$

Thus the graph runs up along the asymptote on the left and down along the asymptote at the right. See Fig. 4.22. We can plot a few points to get a reasonably accurate graph. ■

EXAMPLE 2 Repeat Example 1 for $y = (-2x + 6)/(x + 3)$.

Solution Proceeding as in Example 1, we find that the graph has the following properties:

a vertical asymptote $x = -3$ where the denominator is zero,

a horizontal asymptote $y = -2$ since -2 is the limit at ∞ and $-\infty$,

an x-intercept 3 where the numerator is zero,

a y-intercept 2 where x is zero,

$$\lim_{x \to -3+} \frac{-2x + 6}{x + 3} = \infty, \qquad \text{and} \qquad \lim_{x \to -3-} \frac{-2x + 6}{x + 3} = -\infty.$$

The graph is shown in Fig. 4.23. ■

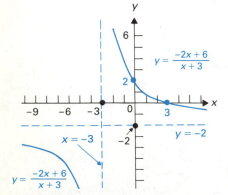

FIGURE 4.23 Vertical asymptote $x = -3$; horizontal asymptote $y = -2$; x-intercept 3; y-intercept 2.

Our work in Section 2.3 shows that the graph of a rational function $f(x)/g(x)$ has the line $y = 0$ (the x-axis) as horizontal asymptote if the degree of the polynomial $g(x)$ is greater than the degree of the polynomial $f(x)$. It also shows that if the degrees of $f(x)$ and $g(x)$ are the same, then the graph has a horizontal asymptote other than the x-axis. When the degree of $f(x)$ is greater than the degree of $g(x)$, we know that the y-coordinates of points on the graph approach ∞ or $-\infty$ as $x \to \infty$. However, in this case, we can obtain more detailed information by

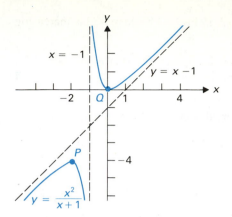

FIGURE 4.24 Graph of
$$f(x) = x^2/(x + 1).$$

executing polynomial long division and writing

$$\frac{f(x)}{g(x)} = q(x) + \frac{r(x)}{g(x)} \tag{2}$$

where the degree of the *remainder polynomial* $r(x)$ is less than the degree of $g(x)$. This degree relationship shows that

$$\lim_{x \to \infty} \frac{r(x)}{g(x)} = 0,$$

and we deduce from Eq. (2) that the graph of $f(x)/g(x)$ approaches the graph of $q(x)$ as $x \to \infty$ or $x \to -\infty$. We say that *the graph of $f(x)/g(x)$ is asymptotic to the graph of $q(x)$ for x of large magnitude.* We illustrate this technique in the next example.

EXAMPLE 3 Sketch the graph of $f(x) = x^2/(x + 1)$ for x of large magnitude and also near any vertical asymptotes.

Solution We know that $x = -1$ is a vertical asymptote, which we sketch as a dashed line in Fig. 4.24. We find that

$$\lim_{x \to -1^+} f(x) = \infty \quad \text{and} \quad \lim_{x \to -1^-} f(x) = -\infty.$$

Performing long division, we obtain

$$
\begin{array}{r}
x - 1 \\
x + 1 \overline{\smash{)}\ x^2 } \\
\underline{x^2 + x} \\
- x \\
\underline{- x - 1} \\
1
\end{array}
\qquad \text{so} \qquad \frac{x^2}{x + 1} = (x - 1) + \frac{1}{x + 1}.
$$

Thus the graph of $f(x)$ becomes asymptotic to the line $y = x - 1$ (an *oblique asymptote*) as $x \to \infty$ or $x \to -\infty$. This line is also shown dashed in Fig. 4.24. Since $1/(x + 1)$ is positive as $x \to \infty$ and negative as $x \to -\infty$, we see that the graph of $f(x)$ is just above this line as $x \to \infty$ and just below it as $x \to -\infty$. The entire graph is sketched in Fig. 4.24. Later in this section, we will see how to determine the high and low points labeled P and Q on this graph. ∎

Increasing and Decreasing Functions

We now explain how the first derivative can be helpful in graphing a function. Since the derivative gives the slope of the tangent line, it is geometrically plausible that the graph of $y = f(x)$ is rising where $f'(x) > 0$ and is falling where $f'(x) < 0$. In a moment, we will prove this using the mean value theorem. First we give a definition.

DEFINITION 4.3

Increasing and decreasing functions

Let $f(x)$ and $g(x)$ be defined in an interval. If $f(x_1) < f(x_2)$ when $x_1 < x_2$ in the interval, then $f(x)$ is **increasing** in the interval. If $g(x_1) > g(x_2)$ whenever $x_1 < x_2$ in the interval, then $g(x)$ is **decreasing** in the interval.

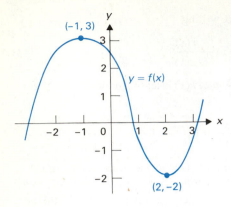

FIGURE 4.25 $f(x)$ is increasing for $x < -1$, $x > 2$; $f(x)$ is decreasing for $-1 < x < 2$.

EXAMPLE 4 Determine from the graph in Fig. 4.25 where $f(x)$ is increasing and where it is decreasing.

Solution From the graph in Fig. 4.25, it appears that $f(x)$ is increasing for $x < -1$, decreasing for $-1 < x < 2$, and increasing for $x > 2$. ■

Suppose $f'(x) > 0$ for all x in an interval. Let x_1 and x_2 be the interval, with $x_1 < x_2$. By the mean value theorem,

$$\frac{f(x_2) - f(x_1)}{x_2 - x_1} = f'(c) > 0, \tag{3}$$

where $x_1 < c < x_2$. In particular, Eq. (3) shows that $f(x_2) - f(x_1) > 0$, so $f(x_1) < f(x_2)$. Thus $f(x)$ is increasing in the interval. Similarly, if $f'(x) < 0$ for all x in an interval, then $f(x)$ is decreasing in the interval, because if $x_1 < x_2$ in the interval, an analogous argument using Eq. (3) shows that $f(x_1) > f(x_2)$. Conversely, if $f(x)$ is increasing, then both quotients

$$\frac{f(x + \Delta x) - f(x)}{\Delta x} = \frac{\text{Positive number}}{\text{Positive number}} \quad \text{for} \quad \Delta x > 0 \quad \text{and}$$

$$\frac{f(x + \Delta x) - f(x)}{\Delta x} = \frac{\text{Negative number}}{\text{Negative number}} \quad \text{for} \quad \Delta x < 0$$

are positive, and as $\Delta x \to 0$ these positive quotients must approach a nonnegative number, so $f'(x) \geq 0$. Similarly if $f(x)$ is decreasing, then $f'(x) \leq 0$. We summarize in a theorem.

THEOREM 4.7   Test for increasing or decreasing behavior

Let $f(x)$ be differentiable in an interval. If $f'(x) > 0$ throughout the interval, then $f(x)$ is increasing in the interval. If $f'(x) < 0$ throughout the interval, then $f(x)$ is decreasing in the interval. Conversely, if $f(x)$ is increasing in the interval, then $f'(x) \geq 0$ throughout the interval, while if $f(x)$ is decreasing, then $f'(x) \leq 0$. □

EXAMPLE 5 Let $f(x) = x^3 - 3x^2 + 2$. Find where $f(x)$ is increasing and where it is decreasing.

Solution We have $f'(x) = 3x^2 - 6x = 3x(x - 2)$, so $f'(x) = 0$ where $x = 0$ or $x = 2$. The points 0 and 2 separate the x-axis into three parts, and by the intermediate value theorem $f'(x)$ must have the same sign throughout each individual part. In Fig. 4.26 we show the sign of $f'(x)$ on each part of the x-axis. We can find the sign of $f'(x)$ on any part by simply evaluating at any point in that part. For example, since $f'(-1) = 3 + 6 = 9 > 0$, we see that $f'(x) > 0$ for $x < 0$. Since $f'(1) = 3 - 6 = -3 < 0$ and $f'(3) = 27 - 18 = 9 > 0$, we have $f'(x) < 0$ for $0 < x < 2$ and $f'(x) > 0$ for $x > 2$. Thus we see that $f(x)$ is increasing for $x < 0$ or $x > 2$, and $f(x)$ is decreasing for $0 < x < 2$. ■

FIGURE 4.26 The sign of $f'(x) = 3x(x - 2)$: $f(x)$ must decrease for $0 < x < 2$.

For $f'(x) = 3x(x - 2)$ in Example 5, we do not have to compute $f'(x)$ at points in all three of the intervals $x < 0$, $0 < x < 2$, and $x > 2$. We can argue that $f'(x) = 3x^2 - 6x$ must be positive for large x, since the term $3x^2$ dominates

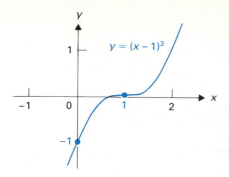

FIGURE 4.27 $f(x) = (x - 1)^3$ is increasing for all x, although $f'(1) = 0$.

for large x. Thus we can mark $+$ for $x > 2$. Now the factor $x - 2$ surely causes a sign change in $f'(x)$ where $x = 2$, so we can mark $-$ for $0 < x < 2$. The factor $3x$ causes a sign change again where $x = 0$, so we mark $+$ for $x < 0$. This enables us to simply zip across the x-axis, marking the intervals in Fig. 4.26 with the appropriate signs, once we have decided on the sign in *one* of the intervals. If we are going to use this technique, it is crucial to note that while a single linear factor $(x - a)$ changes sign as x goes from one side of a to the other on the x-axis, the square $(x - a)^2$ of this factor does not change sign. Note that $(x - a)^n$ changes sign at $x = a$ if n is an *odd* integer (either positive or negative) and does not change sign if n is *even*.

We must also take into account points not in the domain of f. For example, if $f(x) = x^2/(x + 1)$ as in Example 3, then $f'(x) = (x^2 + 2x)/(x + 1)^2$. We should mark -1 on the x-axis as an endpoint of intervals in which we are considering the sign of $f'(x)$. Since the exponent of $(x + 1)^2$ is even, $f'(x)$ does not change sign at $x = -1$. (See Fig. 4.24.)

EXAMPLE 6   Determine where $f(x) = (x - 1)^3$ is increasing and where it is decreasing.

Solution   We have $f'(x) = 3(x - 1)^2 = 0$ only where $x = 1$. Since the exponent 2 is even, we see that $f'(x)$ does not change sign as x increases through 1. Clearly $f'(x) \geq 0$ for all x, and $f(x)$ is increasing for all x. The graph is shown in Fig. 4.27.   ■

EXAMPLE 7   If the derivative of $f(x)$ is $f'(x) = x^2(x - 2)^3(x + 1)$, determine where $f(x)$ is increasing and where it is decreasing.

Solution   We draw an x-axis and record the intervals where $f'(x) < 0$ and where $f'(x) > 0$. See Fig. 4.28. Now $f'(x) = 0$ where $x = -1$, 0, or 2. Examination of the exponents shows that $f'(x)$ changes sign as x increases through -1 and through 2, but not as x increases through 0. If the factors were multiplied out, the expansion of $f'(x)$ would start with x^6, which must dominate the other terms as $x \to \infty$. Thus $f'(x)$ is positive for large x. We work systematically from right to left on the x-axis in Fig. 4.28, first labeling the points -1, 0, and 2. We have

$$f'(x) > 0 \qquad \text{for } x > 2,$$
$$f'(x) < 0 \qquad \text{for } 0 < x < 2 \text{ since the sign changes at 2,}$$
$$f'(x) < 0 \qquad \text{for } -1 < x < 0 \text{ since the sign does not change at 0,}$$
$$f'(x) > 0 \qquad \text{for } x < -1 \text{ since the sign changes at } -1.$$

Thus $f(x)$ is increasing for $x < -1$ and also for $x > 2$, and $f(x)$ is decreasing for $-1 < x < 2$.   ■

Local Maxima and Local Minima

FIGURE 4.28   The sign of $f'(x) = x^2(x - 2)^3(x + 1)$: $f(x)$ must increase for $x < -1$ and $x > 2$ and must decrease for $-1 < x < 2$.

Our graphical interpretation of the sign of $f'(x)$ helps to determine whether $f(x)$ has a local maximum, a local minimum, or neither at a critical point contained in an interval where the function is defined and continuous. Let x_1 be such a critical point, so that either $f'(x_1) = 0$ or $f'(x_1)$ does not exist. Then if $f'(x)$ changes sign as x increases through x_1, we have a local extremum there. If the sign of $f'(x)$

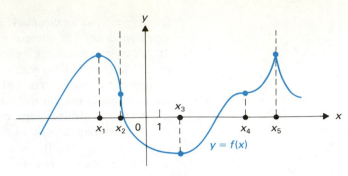

FIGURE 4.29 Critical points are x_1, x_2, x_3, x_4, x_5. $f'(x)$ changes sign at x_1, x_3, and x_5, where there are local extrema; $f'(x)$ does not change sign at x_2 or x_4.

changes from positive to negative as x increases, we have a local maximum as illustrated in Fig. 4.29 for the critical points x_1 and x_5. If the sign of $f'(x)$ changes from negative to positive as x increases through the critical point, we have a local minimum, as at the point x_3 in Fig. 4.29. If the sign of $f'(x)$ does not change, as at critical points x_2 and x_4 in Fig. 4.29, then there is no local extremum there. This analysis leads us to the steps described on page 207 for graphing a function using the sign of the first derivative.

EXAMPLE 8 Sketch the graph of $f(x) = x^3 - 3x^2 + 2$.

Solution

STEP 1. We have $f'(x) = 3x^2 - 6x = 3x(x - 2) = 0$ when $x = 0, 2$. These are the critical points. The corresponding points on the graph are $(0, 2)$ and $(2, -2)$.

STEP 2. As in Example 5, $f(x)$ is increasing for $x < 0$ and $x > 2$ and $f(x)$ is decreasing for $0 < x < 2$.

STEP 3. As x increases through $x = 0$, $f'(x)$ changes sign from positive to negative, so there is a local maximum at $x = 0$. Where $x = 2$, $f'(x)$ changes sign from negative to positive, so there is a local minimum there.

STEP 4. The graph is shown in Fig. 4.30. ∎

EXAMPLE 9 Sketch the graph of $f(x) = x^4 - 2x^3 + 1$.

Solution

STEP 1. We have $f'(x) = 4x^3 - 6x^2 = 2x^2(2x - 3) = 0$ when $x = 0, \frac{3}{2}$. These are critical points, corresponding to $(0, 1)$ and $(\frac{3}{2}, -\frac{11}{16})$ on the graph of f.

STEP 2. We see that $f'(x)$ changes sign as x increases through $\frac{3}{2}$, but not as x increases through 0. Since $\lim_{x \to \infty} f'(x) = \infty$, we see that $f'(x) > 0$ for $x > \frac{3}{2}$. Thus $f(x)$ is increasing for $x > \frac{3}{2}$. The sign change in $f'(x)$ at $\frac{3}{2}$ shows that $f(x)$ is decreasing for $0 < x < \frac{3}{2}$. The lack of sign change in $f'(x)$ at 0 shows that $f(x)$ is also decreasing for $x < 0$.

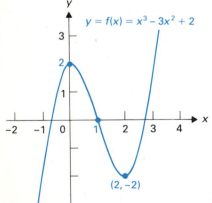

FIGURE 4.30 $f(x)$ increases for $x < 0$, $x > 2$; $f(x)$ decreases for $0 < x < 2$; local maximum where $x = 0$, local minimum where $x = 2$.

Graphing a Function $f(x)$ Using the Sign of the First Derivative

STEP 1. Compute $f'(x)$ and find the critical points of $f(x)$. Mark them on an x-axis and mark the corresponding points on the graph.

STEP 2. Determine where $f(x)$ is increasing ($f'(x) > 0$) and where it is decreasing ($f'(x) < 0$).

STEP 3. Determine at each critical point where f is continuous whether there is a

> local maximum: $f'(x)$ changes sign from positive to negative as x increases;
>
> local minimum: $f'(x)$ changes sign from negative to positive as x increases; or
>
> neither: $f'(x)$ does not change sign.

STEP 4. Sketch the graph using the information in steps 1–3, perhaps plotting a few more points.

STEP 3. Where $x = 0$, we know that $f'(x)$ does not change sign, so there is no local extremum there. Where $x = \frac{3}{2}$, $f'(x)$ changes sign from negative to positive as x increases, so there is a local minimum at $\frac{3}{2}$.

STEP 4. The graph is shown in Fig. 4.31. Since $f'(0) = 0$, we must make the graph flat, with a horizontal tangent line there. ∎

We conclude with an example illustrating that the sign of $f'(x)$ may change at a point not in the domain of f.

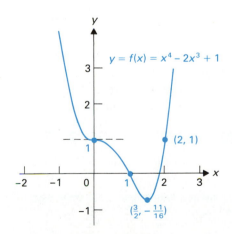

FIGURE 4.31 $f(x)$ increases for $x > \frac{3}{2}$; $f(x)$ decreases for $x < 0$, $0 < x < \frac{3}{2}$; no local extremum at critical point $x = 0$.

EXAMPLE 10 Sketch the graph of

$$f(x) = \frac{1}{x^2(x-1)} + 3.$$

Solution

STEP 1. Note that $f(x)$ is not defined at $x = 0$ or $x = 1$. We have

$$f'(x) = \frac{-1[x^2 + (x-1)2x]}{x^4(x-1)^2} = \frac{-3x^2 + 2x}{x^4(x-1)^2} = \frac{-3x + 2}{x^3(x-1)^2}.$$

Thus $f'(x) = 0$ where $3x = 2$, or $x = \frac{2}{3}$. This is the only critical point and corresponds to $(\frac{2}{3}, -\frac{15}{4})$ on the graph.

STEP 2. Our formula for $f'(x)$ shows that, by the odd/even exponent argument, we have sign changes as x increases through $\frac{2}{3}$ and 0 but not

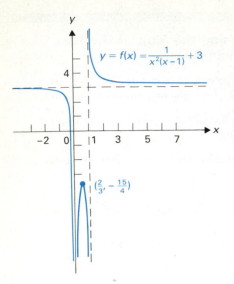

FIGURE 4.32 $f(x)$ increases for $0 < x < \frac{2}{3}$; $f(x)$ decreases for $x < 0$, $\frac{2}{3} < x < 1$, $x > 1$; local maximum where $x = \frac{2}{3}$; vertical asymptotes $x = 0$ and $x = 1$; horizontal asymptote $y = 3$.

through 1. Since $f'(x) < 0$ for large x, we see that

$$f(x) \text{ is decreasing for } x > 1,$$
$$f(x) \text{ is decreasing for } \tfrac{2}{3} < x < 1,$$
$$f(x) \text{ is increasing for } 0 < x < \tfrac{2}{3},$$
$$f(x) \text{ is decreasing for } x < 0.$$

STEP 3.   The critical point where $x = \frac{2}{3}$ corresponds to a local maximum because the sign of $f'(x)$ changes from positive to negative as x increases through $\frac{2}{3}$.

STEP 4.   The graph is sketched in Fig. 4.32. Note that there are vertical asymptotes $x = 0$ and $x = 1$ and that $y = 3$ is a horizontal asymptote.   ∎

SUMMARY

1. If $\lim_{x \to \infty} f(x) = \lim_{x \to -\infty} f(x) = c$, then $y = c$ is a horizontal asymptote of the graph of $y = f(x)$.

2. If $\lim_{x \to a} |f(x)| = \infty$, then $x = a$ is a vertical asymptote of the graph of $y = f(x)$.

Use of the first derivative in graphing

3. If $f'(x) > 0$ throughout an interval, then $f(x)$ is increasing in that interval.

4. If $f'(x) < 0$ throughout an interval, then $f(x)$ is decreasing in that interval.

5. A critical point x_1 where f is continuous is a candidate for a point where $f(x)$ has a local maximum or minimum.

   (a) If $f'(x)$ changes sign from negative to positive as x increases through x_1, then $f(x_1)$ is a local minimum.

   (b) If $f'(x)$ changes sign from positive to negative as x increases through x_1, then $f(x_1)$ is a local maximum.

   (c) If $f'(x)$ does not change sign as x increases through x_1, then $f(x_1)$ is neither a local maximum nor a local minimum.

EXERCISES 4.4

In Exercises 1–6, find the vertical and horizontal asymptotes and the x- and y-intercepts. Sketch the graphs.

1. $y = \dfrac{1}{x - 2}$

2. $y = \dfrac{2}{x + 3}$

3. $s = \dfrac{t - 1}{t}$

4. $x = \dfrac{5t + 10}{2t}$

5. $y = \dfrac{2x + 4}{x - 1}$

6. $s - \dfrac{r - 3}{2r + 4}$

In Exercises 7–10, find a polynomial function to which the graph of the given rational function is asymptotic as $x \to \infty$ or $x \to -\infty$. Sketch all line asymptotes and sketch the graph of the function where it is near them.

7. $y = \dfrac{x^2}{x - 1}$

8. $s = \dfrac{t^3}{t^2 - 1}$

9. $y = \dfrac{s^3 - 4}{s - 1}$

10. $y = \dfrac{x^4 + 2x}{x^3 + 1}$

In Exercises 11–28, find the intervals where the function is increasing and the intervals where it is decreasing.

11. $f(x) = x^2 - 6x + 2$

12. $f(x) = 4 - 3x - 2x^2$

13. $g(t) = t^3 + 3t^2 - 9t + 5$

14. $f(x) = x^3 - 6x^2 + 4$

15. $h(s) = 8 + 12s - s^3$

16. $f(x) = x^3 + 3x^2 + 3x - 5$

17. $f(x) = \dfrac{x}{x^2 + 1}$

18. $f(x) = \dfrac{x}{x^2 - 1}$

19. $f(x) = \dfrac{x^3}{x^2 - 4}$

20. $f(x) = (x^2 - 4)^{2/3}$

21. $f(x) = (x^2 - 8)^{1/3}$

22. $f(x) = e^x - x - 1$

23. $f(x) = e^{(x^2)} - ex^2 + 1$

24. $h(s) = s - 1 - \ln s$

25. $f(x) = 2(\ln x) - x^2 + 1$ **26.** $f(x) = x^x$

27. $g(t) = \sin t + \cos t$ for $-2\pi < t < 2\pi$

28. $h(u) = \sin u - \sqrt{3} \cos u$ for $-2\pi < u < 2\pi$

29. Find b such that the polynomial function $x^2 + bx - 7$ has a local minimum at 4.

30. Consider the polynomial function $ax^2 + 4x + 13$.
   (a) Find the value of a such that the function has either a local maximum or a local minimum at 1. Which is it, a maximum or a minimum?
   (b) Find the value of a such that the function has either a local maximum or a local minimum at -1. Which is it, a maximum or a minimum?

31. Consider the polynomial function f given by $ax^2 + bx + 24$.
   (a) Find the ratio b/a if f has a local minimum at 2.
   (b) Find a and b if f has a local minimum at 2 and $f(2) = 12$.
   (c) Can you determine a and b so that f has a local maximum at 2 and $f(2) = 12$?

32. Sketch the graph of a twice-differentiable function f such that $f(1) = 3$, $f(4) = 1$, f has a local minimum at 1, and f has a local maximum at 4.

33. It is possible for a twice-differentiable function f to satisfy the conditions of Exercise 32 and have no local maximum or minimum for $1 < x < 4$?

In Exercises 34–40, find
(a) the intervals where $f(x)$ is increasing,
(b) the intervals where $f(x)$ is decreasing,
(c) where $f'(x)$ exists and $f(x)$ has a local maximum, and
(d) where $f'(x)$ exists and $f(x)$ has a local minimum

from the given expression for $f'(x)$.

34. $f'(x) = x(x - 1)^2$

35. $f'(x) = (x - 3)^2(x + 1)^3$

36. $f'(x) = x^2(x + 4)^3(x - 1)^3$

37. $f'(x) = \dfrac{x^2}{(x - 1)(x + 2)}$

38. $f'(x) = \dfrac{(x - 1)(x + 2)}{(x + 3)^2}$

39. $f'(x) = \dfrac{x^2(x - 4)}{(x - 1)(x + 2)}$

40. $f'(x) = \dfrac{(x - 1)^3(x + 1)^2}{x^2(x + 4)}$

In Exercises 41–44, fill in the blanks.

41. A polynomial function of degree n can have at most _____ local extrema.

42. A polynomial function of even degree n with positive coefficient a_n of x^n can have at most _____ local maxima and at most _____ local minima.

43. A polynomial function of even degree n with negative coefficient a_n of x^n can have at most _____ local maxima and at most _____ local minima.

44. A polynomial function of odd degree n can have at most _____ local maxima and at most _____ local minima.

Exercises 45–48 refer to Figs. 4.33–4.36, which show graphs of *derivatives* of functions defined and continuous for all x. Mark the parts true or false.

45. If $f'(x)$ has the graph in Fig. 4.33, then
   _____ (a) $f(x)$ has a local minimum where $x = 2$,
   _____ (b) $f(x)$ has a local minimum where $x = 1$,
   _____ (c) $f(x)$ has a local minimum where $x = 3$,
   _____ (d) $f(x)$ has a local maximum where $x = 1$,
   _____ (e) $f(x)$ is increasing for $x > 2$.

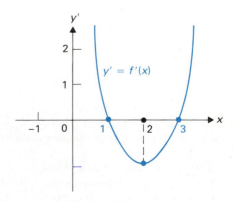

FIGURE 4.33
Graph of f'.

46. If $g'(x)$ has the graph in Fig. 4.34, then

_____ (a) $g(x)$ is decreasing for $x < 1$,

_____ (b) $g(x)$ is increasing for $x > 1$,

_____ (c) $g(x)$ is increasing for all x,

_____ (d) $g(x)$ has a local minimum where $x = 1$,

_____ (e) $g(x)$ has no local extremum where $x = 1$.

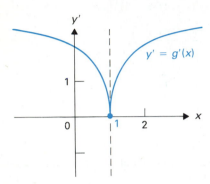

FIGURE 4.34
Graph of g'.

47. If $h'(x)$ has the graph in Fig. 4.35, then

_____ (a) $h(x)$ is decreasing for $0 < x < 4$,

_____ (b) $h(x)$ has a local maximum where $x = -2$,

_____ (c) $h(x)$ has a local minimum where $x = 4$,

_____ (d) $h(x)$ has a local minimum where $x = 0$,

_____ (e) $h(x)$ is increasing for $x > 2$.

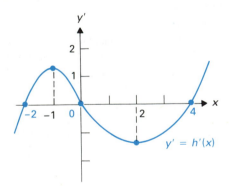

FIGURE 4.35
Graph of h'.

48. If $k'(x)$ has the graph in Fig. 4.36, then

_____ (a) the graph of $k(x)$ has a vertical tangent line where $x = 2$,

_____ (b) $k(x)$ has a local minimum where $x = 2$,

_____ (c) $k(x)$ has a local minimum where $x = -1$,

_____ (d) $k(x)$ is decreasing for $0 < x < 3$,

_____ (e) $k(x)$ has a local maximum where $x = 0$.

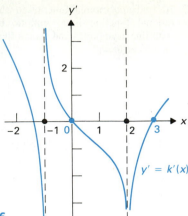

FIGURE 4.36
Graph of k'.

In Exercises 49–64, use the information given by the first derivative to sketch the graph of the function. The functions are multiples of some of those in Exercises 11 through 27 as noted. You may want to make use of your preceding work.

49. $f(x) = x^2 - 6x + 2$ (Exercise 11)

50. $f(x) = 4 - 3x - 2x^2$ (Exercise 12)

51. $y = g(t) = \frac{1}{8}(t^3 + 3t^2 - 9t + 5)$ (Exercise 13)

52. $y = h(s) = \frac{1}{8}(8 + 12s - s^3)$ (Exercise 15)

53. $f(x) = x^3 + 3x^2 + 3x - 5$ (Exercise 16)

54. $f(x) = \dfrac{x}{x^2 + 1}$ (Exercise 17)

55. $f(x) = \dfrac{x}{x^2 - 1}$ (Exercise 18)

56. $f(x) = \dfrac{1}{2} \cdot \dfrac{x^3}{x^2 - 4}$ (Exercise 19)

57. $f(x) = (x^2 - 4)^{2/3}$ (Exercise 20)

58. $f(x) = (x^2 - 8)^{1/3}$ (Exercise 21)

59. $f(x) = e^x - x - 1$ (Exercise 22)

60. $f(x) = e^{(x^2)} - ex^2 + 1$ (Exercise 23)

61. $h(s) = s - 1 - \ln s$ (Exercise 24)

62. $f(x) = 2(\ln x) - x^2 + 1$ (Exercise 25)

63. $f(x) = x^x$ (Exercise 26)

64. $x = g(t) = \sin t + \cos t$ for $-2\pi < t < 2\pi$ (Exercise 27)

65. A manufacturer has cost function $C(x)$ where x is the number of units manufactured. The *average cost* per unit is equal to $C(x)/x$. Show that maximum and minimum values of the average cost occur when the marginal cost is equal to the average cost.

EXPLORING CALCULUS

C: graph 1 function

F: find equation quiz (options 3, 4)

4.5 GRAPHING USING THE SECOND DERIVATIVE

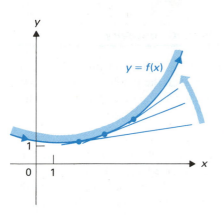

FIGURE 4.37 The tangent line turns counterclockwise as x increases; the graph is concave up.

The preceding section applied the first derivative as an aid in sketching the graph of a function. In this section, we discuss the application of the second derivative to sketching the graph of a function. The section concludes with an outline for sketching graphs using the information provided by both the first and the second derivatives.

Concavity

Let $f(x)$ be differentiable throughout an interval. Suppose that $f'(x)$ is increasing in that interval. Recall that $f'(x)$ is the slope of the tangent line to the graph at the point $(x, f(x))$. For the slopes of tangent lines to increase as x increases, the tangent lines must rotate *counterclockwise* as the graph is traveled in the direction of increasing x. This is the case in Fig. 4.37. In this situation, we say the curve is *concave up* (it holds water). By a similar argument, if $f'(x)$ is decreasing in the interval, the tangent lines must rotate *clockwise* as the graph is traveled in the direction of increasing x, as shown in Fig. 4.38. Then the curve is *concave down* (it spills water). Points where the concavity changes are *inflection points*, as illustrated in Fig. 4.39. We summarize this terminology in a definition.

DEFINITION 4.4

Concavity and inflection point

Let f be differentiable for $a < x < b$. The graph of f is **concave up** in this interval if $f'(x)$ is increasing there and is **concave down** in the interval if $f'(x)$ is decreasing there. An **inflection point** of the graph is a point where the concavity changes as x increases through that point.

Now suppose that $f''(x)$ exists for $a < x < b$. Since $f''(x)$ is the derivative of $f'(x)$, Theorem 4.7 shows that $f'(x)$ is increasing (so the graph of f is concave up)

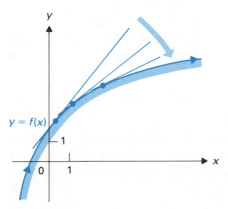

FIGURE 4.38 The tangent line turns clockwise as x increases; the graph is concave down.

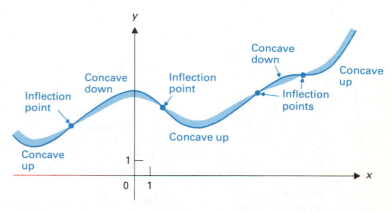

FIGURE 4.39 Inflection points occur where the concavity changes.

if $f''(x) > 0$. Similarly, if $f''(x) < 0$, then $f'(x)$ is decreasing, so the graph is concave down. The converse part of Theorem 4.7 shows that if the graph is concave up, then $f''(x) \geq 0$ and if the graph is concave down, then $f''(x) \leq 0$. We state this as a theorem.

THEOREM 4.8   The second derivative and concavity

Let f be twice differentiable for $a < x < b$. If $f''(x) > 0$ in this interval, then the graph is concave up, while if $f''(x) < 0$ there, then the graph is concave down. Conversely, if the graph is concave up, then $f''(x) \geq 0$, and if the graph is concave down, then $f''(x) \leq 0$.   □

Suppose the graph of $f(x)$ has an inflection point where $x = x_1$, and $f''(x)$ exists and is continuous near x_1. Then we must have $f''(x) > 0$ for x near x_1 on one side and $f''(x) < 0$ for x near x_1 on the other side. By the intermediate value theorem, we must then have $f''(x_1) = 0$. Thus for functions with continuous second derivatives, we can find all candidates for inflection points by finding where $f''(x) = 0$.

EXAMPLE 1   Discuss concavity and inflection points for

$$f(x) = x^3 - 3x^2 + 2.$$

Solution   We have $f'(x) = 3x^2 - 6x$ and $f''(x) = 6x - 6$. Thus $f''(x) = 0$ where $6x - 6 = 0$, or where $x = 1$. Since $6x - 6 < 0$ if $x < 1$, the graph is concave down if $x < 1$. The graph is concave up if $x > 1$ since $f''(x) = 6x - 6 > 0$ there. The concavity does change as x increases through 1, so the point $(1, 0)$ is an inflection point, as indicated in Fig. 4.40. We found the local maximum and local minimum in Example 8 of the preceding section.   ■

EXAMPLE 2   Discuss concavity and inflection points for $f(x) = x^4$.

Solution   If $f(x) = x^4$, then $f'(x) = 4x^3$ and $f''(x) = 12x^2$. Now $f''(0) = 0$, but $(0, 0)$ is not an inflection point since $f''(x) > 0$ on both sides of $x = 0$. Thus the graph in Fig. 4.41 is concave up for $x < 0$ and also for $x > 0$, and there are no inflection points. Actually, $(0, 0)$ is a local minimum since $f'(0) = 0$ and $f'(x) = 4x^3$ is less than zero if $x < 0$ and greater than zero if $x > 0$.   ■

EXAMPLE 3   Let $f''(x) = x^2(x - 1)^3(x + 2)$. Find the x-coordinates of all inflection points of the graph of f.

Solution   We see that $f''(x) = 0$ where $x = -2, 0, 1$. Using the odd/even exponent argument, we see that $f''(x)$ changes sign as x increases through -2 and 1 but not as x increases through 0. Thus inflection points occur where $x = -2$ and 1.   ■

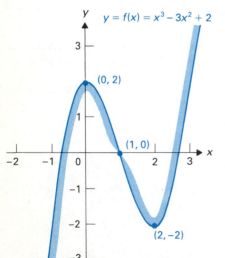

FIGURE 4.40 Concave down for $x < 1$, concave up for $x > 1$; inflection point $(1, 0)$.

EXAMPLE 4   Discuss concavity and inflection points of the graph of $f(x) = x^{1/3} + 1$.

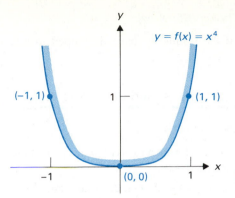

FIGURE 4.41 Concave up for $x < 0$ and $x > 0$; $f''(0) = 0$, but no inflection point where $x = 0$; $(0, 0)$ is a local minimum.

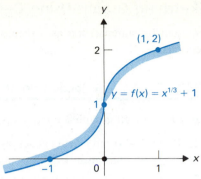

FIGURE 4.42 Concave up for $x < 0$, concave down for $x > 0$; $f''(0)$ does not exist, but $(0, 1)$ is an inflection point.

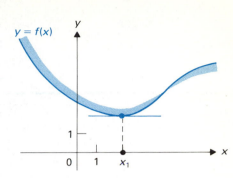

FIGURE 4.43 $f'(x_1) = 0$, $f''(x_1) > 0$; local minimum where $x = x_1$.

Solution We have

$$f'(x) = \frac{1}{3}x^{-2/3} \quad \text{and} \quad f''(x) = -\frac{2}{9}x^{-5/3} = \frac{-2}{9x^{5/3}}.$$

Thus $f''(x) \neq 0$ for any nonzero value of x, but $f''(0)$ does not exist. However, $f(x)$ is defined where $x = 0$, and $f(0) = 1$. Testing concavity, we have

$f''(x) > 0$ for $x < 0$, so the graph is concave up there,

$f''(x) < 0$ for $x > 0$, so the graph is concave down there.

Thus concavity does change as we go through $(0, 1)$ on the graph, and we consider $(0, 1)$ to be an inflection point of the graph. The graph is shown in Fig. 4.42. ∎

The Second Derivative Test for Local Extrema

Let f be a twice-differentiable function. Suppose that $f'(x_1) = 0$ and $f''(x_1) > 0$. Then the graph of f has a horizontal tangent where $x = x_1$, but it is concave up there, so $f(x)$ has a local minimum where $x = x_1$. See Fig. 4.43. On the other hand, if $f'(x_2) = 0$ but $f''(x_2) < 0$, then the graph has a horizontal tangent but is concave down where $x = x_2$, so $f(x)$ has a local maximum there. See Fig. 4.44. Thus if $f'(a) = 0$ but $f''(a) \neq 0$, the sign of $f''(a)$ can be used to determine whether there is a local maximum or local minimum where $x = a$. This method is called the *second derivative test*.

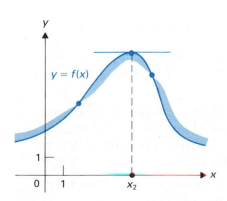

FIGURE 4.44 $f'(x_2) = 0$, $f''(x_2) < 0$; local maximum where $x = x_2$.

EXAMPLE 5 Use the second derivative test to determine whether $f(x) = x^3 - 3x^2 + 2$ has a local maximum or a local minimum at each point where $f'(x) = 0$.

Solution We have $f'(x) = 3x^2 - 6x = 3x(x - 2) = 0$ if $x = 0$ or $x = 2$. Also, $f''(x) = 6x - 6$. Then $f''(0) = -6 < 0$, so the graph is concave down at $(0, 2)$, which must be a local maximum. Finally, $f''(2) = 12 - 6 = 6 > 0$, so the graph is concave up at $(2, -2)$, which must be a local minimum. The graph was shown in Fig. 4.40. ∎

Sketching Graphs Using Derivatives

We start by summarizing as a theorem the results we have developed in this section.

THEOREM 4.9   Inflection points and the second derivative test

Let f be a function with a continuous second derivative.

1. If $f''(x_1) = 0$ and $f''(x)$ changes sign at x_1 as x increases through x_1, then the graph of f has an inflection point where $x = x_1$.

2. Second derivative test: If $f'(a) = 0$ and $f''(a) \neq 0$, then the graph has a local maximum where $x = a$ if $f''(a) < 0$, and the graph has a local minimum where $x = a$ if $f''(a) > 0$.   ☐

The steps listed at the bottom of this page provide an outline for sketching graphs using information given by both the first and the second derivatives.

EXAMPLE 6   Sketch the graph of $f(x) = x + \sin x$, finding all local maxima and minima and all inflection points.

Sketching the Graph of $f(x)$ Using Derivatives

STEP 1.   Compute $f'(x)$ and $f''(x)$.

STEP 2.   Find all critical points, that is, points where $f'(x) = 0$ or does not exist. Determine whether each critical point gives a local maximum, a local minimum, or neither, using the second derivative test or checking for a sign change of the first derivative.

STEP 3.   (Optional) Find intervals where the graph is increasing ($f'(x) > 0$) or decreasing ($f'(x) < 0$). If you have done an accurate job in step 2, these intervals will be evident from the graph when you sketch it.

STEP 4.   Find candidates for inflection points, where $f''(x) = 0$ or where it is not defined or is not continuous. Then check whether $f''(x)$ changes sign as x increases through each candidate point. If a sign change occurs, the point corresponds to an inflection point of the graph.

STEP 5.   (Optional) Find the intervals where the graph is concave up ($f''(x) > 0$) or concave down ($f''(x) < 0$). If you have done an accurate job in steps 1, 2, and 4, these intervals will be evident when you sketch the graph.

STEP 6.   Sketch the graph, using the information in steps 2 and 5. You may wish to plot a few additional points.

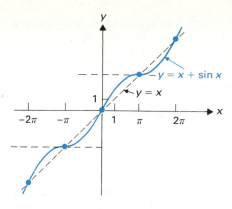

FIGURE 4.45 No local extrema; inflection points where $x = n\pi$; horizontal tangent lines at inflection points where $x = n\pi$ for n odd.

Solution

STEP 1. Differentiating, we obtain $f'(x) = 1 + \cos x$ and $f''(x) = -\sin x$.

STEP 2. Now $1 + \cos x = 0$ where $\cos x = -1$, which occurs when $x = (2n + 1)\pi$ for any integer n. But $f'(x) = 1 + \cos x$ cannot change sign as x increases through these points because $1 + \cos x \geq 0$ for all x. There are no local maxima or minima.

STEP 4. Skipping step 3 and turning to the second derivative, we find that $f''(x) = -\sin x$ is zero at $x = n\pi$ for all integers n, which include those points where $f'(x)$ is zero. At all these points, $f''(x) = -\sin x$ changes sign as x increases. Thus $x = n\pi$ corresponds to an inflection point for all integers n. If n is odd, then $f'(n\pi) = 0$, and there is a horizontal tangent at the inflection point.

STEP 5. The curve is concave up if $-\sin x > 0$, which occurs whenever $(2n - 1)\pi < x < 2n\pi$ for each integer n. It is concave down whenever $2n\pi < x < (2n + 1)\pi$.

STEP 6. The graph is shown in Fig. 4.45.   ■

EXAMPLE 7 Sketch the graph of

$$f(x) = \frac{x^4}{4} - \frac{4x^3}{3} + 2x^2 - 1,$$

finding all local maxima and minima and all inflection points.

Solution

STEP 1. Differentiating, we have

$$f'(x) = x^3 - 4x^2 + 4x \qquad \text{and} \qquad f''(x) = 3x^2 - 8x + 4$$
$$= x(x^2 - 4x + 4) \qquad\qquad\qquad = (3x - 2)(x - 2).$$
$$= x(x - 2)^2$$

STEP 2. From step 1, we see that $f'(x) = 0$ when $x = 0$ or $x = 2$. Since the root $x = 0$ comes from the single linear factor x, we see that $f'(x)$ does change sign, from negative to positive since $(x - 2)^2 > 0$, as x increases through 0. Alternatively, $f''(0) = 4 > 0$, so the graph is concave up at $x = 0$. This shows in two ways that $f(x)$ has a local minimum when $x = 0$, corresponding to $(0, -1)$ on the graph.

Although $f'(2) = 0$, the derivative $f'(x)$ does *not* change sign as x increases through 2, since the factor $(x - 2)^2$ has an *even* exponent. Thus there is no local extremum where $x = 2$.

STEP 4. From step 1, we see that candidates for inflection points occur where $x = 2$ and $x = \frac{2}{3}$. We see that $f''(x)$ does change sign where $x = 2$ because of the odd-powered factor $(x - 2)^1$, so $(2, f(2)) = (2, \frac{1}{3})$ is an inflection point. Another inflection point occurs where $x = \frac{2}{3}$, corresponding to the odd-powered factor $(3x - 2)^1$ in $f''(x)$.

STEP 6. The graph is sketched in Fig. 4.46.   ■

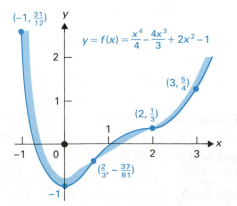

FIGURE 4.46 Local minimum of -1 where $x = 0$; inflection points $(\frac{2}{3}, -\frac{37}{81})$ and $(2, \frac{1}{3})$; horizontal tangent where $x = 0, 2$.

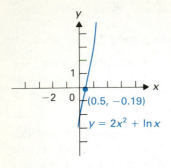

FIGURE 4.47 Inflection point at $(0.5, -0.19)$.

EXAMPLE 8 Sketch the graph of $f(x) = 2x^2 + \ln x$, finding all local maxima and minima and all inflection points.

Solution

STEP 1. Differentiating, we find that

$$f'(x) = 4x + \frac{1}{x} \qquad \text{and} \qquad f''(x) = 4 - \frac{1}{x^2}$$

$$= \frac{4x^2 + 1}{x}, \quad x > 0 \qquad \qquad = \frac{4x^2 - 1}{x^2}, \quad x > 0.$$

STEP 2. We see that $f'(x)$ is never zero since $4x^2 + 1 > 0$ for all x. Thus there are no local maxima or minima.

STEP 4. We find that $f''(x) = 0$ when $4x^2 - 1 = (2x - 1)(2x + 1) = 0$. This occurs when $x = \pm\frac{1}{2}$, but $-\frac{1}{2}$ is not in the domain of f since $\ln(-\frac{1}{2})$ is undefined. Thus the only candidate for an inflection point is where $x = \frac{1}{2}$. Now $f''(x)$ changes sign where $x = \frac{1}{2}$ since we have the single factor $(2x - 1)$. Thus we have an inflection point at

$$\left(\frac{1}{2}, \frac{1}{2} + \ln \frac{1}{2} \right) = (0.5, 0.5 - \ln 2) \approx (0.5, -0.19).$$

STEP 6. Since $f'(\frac{1}{2}) = 4$, we make the graph of $f(x)$ quite steep at the inflection point. We note that $\lim_{x \to \infty} f(x) = \infty$ because of the $2x^2$ term, and $\lim_{x \to 0+} f(x) = -\infty$ because of the $\ln x$ term. The graph is shown in Fig. 4.47. ∎

SUMMARY

Use of the second derivative in graphing

1. If $f''(x) > 0$ throughout an interval, the graph is concave up in that interval.

2. If $f''(x) < 0$ throughout an interval, the graph is concave down in that interval.

3. If $f''(x_2) = 0$, then $(x_2, f(x_2))$ is a candidate for an inflection point of the graph.

   (a) If $f''(x)$ changes sign as x increases through x_2, then $(x_2, f(x_2))$ is an inflection point.

   (b) If $f''(x)$ does not change sign as x increases through x_2, then $(x_2, f(x_2))$ is not an inflection point.

4. *Second derivative test:* If $f'(x_1) = 0$ and $f''(x_1) \neq 0$, then $f(x_1)$ is a local minimum if $f''(x_1) > 0$ and a local maximum if $f''(x_1) < 0$.

EXERCISES 4.5

1. Assume that f is twice differentiable and that f'' is continuous. Mark each of the following true or false.
_____ (a) If f has a local maximum at x_0, then $f'(x_0) = 0$.
_____ (b) If f has a local maximum at x_0, then $f''(x_0) < 0$.
_____ (c) If $f''(x_0) < 0$, then f has a local maximum at x_0.
_____ (d) If $f'(x_0) = 0$ and $f''(x_0) < 0$, then f has a local maximum at x_0.
_____ (e) If $f'(x_0) > 0$, then f is increasing near x_0.
_____ (f) If $f'(x_0) = 0$, then f cannot be increasing near x_0.
_____ (g) If f has an inflection point at x_0, then $f''(x_0) = 0$.
_____ (h) If $f''(x_0) = 0$, then f must have an inflection point at x_0.
_____ (i) If f has a local minimum at x_0, then the tangent line to the graph of f at $(x_0, f(x_0))$ is horizontal.
_____ (j) If f has an inflection point at x_0, then the tangent line to the graph of f at $(x_0, f(x_0))$ may possibly be horizontal.

2. Sketch the graph of a twice-differentiable function f such that $f(0) = 3$, $f'(0) = 0$, $f''(0) < 0$, $f(2) = 2$, $f'(2) = -1$, $f''(2) = 0$, $f(4) = 1$, $f'(4) = 0$, and $f''(4) > 0$.

3. Sketch the graph of a twice-differentiable function f such that $f''(x) < 0$ for $x < 1$, $f(1) = -1$, $f'(1) = 1$, $f''(1) = 0$, $f''(x) > 0$ for $x > 1$, and $f(3) = 4$.

4. Sketch the graph of a twice-differentiable function f such that $f'(x) > 0$ for $x > 2$, $f''(x) > 0$ for $x > 2$, $f''(2) = 0$, $f'(x) < 0$ for $x < 2$, and $f''(x) > 0$ for $x < 2$.

5. Do you think it is possible that, for a twice-differentiable function f, one can have $f(0) = 0$, $f'(0) = 1$, $f''(x) > 0$ for $x > 0$, and $f(1) = 1$? Why?

In Exercises 6–30, find
(a) the intervals where the graph of the function is concave up, and
(b) the intervals where the graph of the function is concave down from the given information.

6. $f(x) = 4 - x^2$

7. $g(t) = t^2 - 6t + 4$

8. $h(s) = \dfrac{1}{s+1}$

9. $f(x) = \dfrac{x^3}{3} + x^2 - 3x - 4$

10. $f(x) = \dfrac{x^3}{3} + x^2 + x - 6$

11. $g(u) = \dfrac{u^2}{u-1}$

12. $f(x) = x^4 - 4x + 1$

13. $f(x) = x^5 - 5x + 1$

14. $f(t) = 3t^4 - 4t^3$

15. $f(x) = x^4 - 2x^2 - 2$

16. $f(x) = x + 2\sin x$

17. $h(s) = s - 2\cos s$

18. $f(x) = x^2 - 8\ln x$

19. $f(x) = x^2 + 8\ln x$

20. $f(x) = xe^x$

21. $g(t) = e^t - t^2$

22. $f''(x) = (x-1)^3(x+2)^2$

23. $f''(x) = x(x+1)^2(x-1)$

24. $f''(x) = x^3(x+2)^2(x-4)$

25. $f''(x) = x(x-1)^3(x+3)^5$

26. $f''(x) = \dfrac{x^2-1}{x^2}$

27. $f''(x) = \dfrac{x^2-x}{x+3}$

28. $f''(x) = \dfrac{x(x-1)^3}{(x+2)^3}$

29. $f''(x) = \dfrac{x^3 - 2x^2}{(x+1)^2}$

30. $f''(x) = \dfrac{x^3}{(x-1)^2(x+4)}$

Exercises 31–34 refer to Figs. 4.48–4.51, which show the graphs of *derivatives* of functions defined and continuous for all x. Mark the parts true or false.

31. If $f'(x)$ has the graph in Fig. 4.48, then
_____ (a) the graph of f has an inflection point at $(0, 0)$,
_____ (b) $f(x)$ has a local maximum where $x = 0$,
_____ (c) $f(x)$ has a local maximum where $x = -1$,
_____ (d) the graph of f has an inflection point where $x = -1$,
_____ (e) the graph of f is concave down for $-2 < x < 0$.

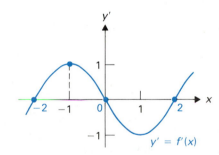

FIGURE 4.48
Graph of f'.

32. If $g'(x)$ has the graph in Fig. 4.49, then
_____ (a) the graph of g has an inflection point where $x = -1$,
_____ (b) the graph of g has an inflection point where $x = 1$,
_____ (c) the graph of g is concave down for $x < -1$,
_____ (d) the graph of g is concave down for $x > 1$,
_____ (e) $g(x)$ has a local minimum where $x = -2$.

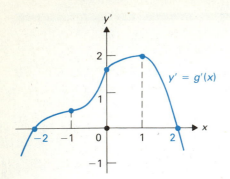

FIGURE 4.49
Graph of g'.

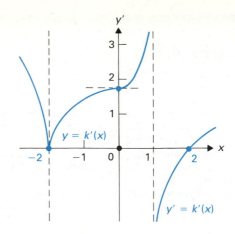

FIGURE 4.51
Graph of k'.

33. If $h''(x)$ has the graph in Fig. 4.50, then
_____ (a) the graph of h is concave up for $-1 < x < 1$,
_____ (b) the graph of h has an inflection point where $x = -1$,
_____ (c) the graph of h has an inflection point where $x = 1$,
_____ (d) $h'(x)$ has a local maximum where $x = 1$,
_____ (e) the graph of $h'(x)$ has an inflection point where $x = -1$.

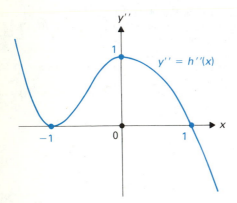

FIGURE 4.50
Graph of h''.

34. If $k'(x)$ has the graph in Fig. 4.51, then
_____ (a) $k(x)$ has a local minimum where $x = -2$,
_____ (b) the graph of k has an inflection point where $x = -2$,
_____ (c) the graph of k has an inflection point where $x = 1$,
_____ (d) $k(x)$ has a local maximum where $x = 1$,
_____ (e) the graph of k has an inflection point where $x = 0$.

In Exercises 35–49,

(a) find all local maxima,
(b) find all local minima,
(c) find all inflection points, and
(d) sketch the graph

of the given function. (Concavity was determined for these functions in Exercises 6–20, as noted.)

35. $f(x) = 4 - x^2$ (Exercise 6)
36. $g(t) = t^2 - 6t + 4$ (Exercise 7)
37. $h(s) = \dfrac{1}{s + 1}$ (Exercise 8)
38. $f(x) = \dfrac{x^3}{3} + x^2 - 3x - 4$ (Exercise 9)
39. $f(x) = \dfrac{x^3}{3} + x^2 + x - 6$ (Exercise 10)
40. $g(u) = \dfrac{u^2}{u - 1}$ (Exercise 11)
41. $f(x) = x^4 - 4x + 1$ (Exercise 12)
42. $f(x) = x^5 - 5x + 1$ (Exercise 13)
43. $f(t) = 3t^4 - 4t^3$ (Exercise 14)
44. $f(x) = x^4 - 2x^2 - 2$ (Exercise 15)
45. $f(x) = x + 2 \sin x$ (Exercise 16)
46. $h(s) = s - 2 \cos s$ (Exercise 17)
47. $f(x) = x^2 - 8 \ln x$ (Exercise 18)
48. $f(x) = x^2 + 8 \ln x$ (Exercise 19)
49. $f(x) = xe^x$ (Exercise 20)

In Exercises 50–53, sketch the curve with the given equation. [*Hint:* Regard the equation as giving x as a function g of y, and apply the theory in this section with x and y interchanged.]

50. $x = y^2 - 2y + 2$ **51.** $x = y^3 - 3y^2$
52. $x = \dfrac{1}{y^4 + 1}$ **53.** $x = \dfrac{y}{y^2 + 1}$

◖ EXPLORING CALCULUS

L: graph f, f', and f''
M: calculus draw quiz (topic 2)
N: calculus topics quiz (topic 2)

4.6 APPLIED MAXIMUM AND MINIMUM PROBLEMS

Many situations arise in which one wishes to maximize or minimize some quantity. For example, manufacturers want to maximize their profit. Builders may want to minimize their costs. Such extremal problems are clearly of great practical importance. We can often solve extremal problems with differential calculus, using the ideas developed in the preceding sections.

To find local maxima and minima of a function f, we may proceed as follows. Find all critical points x where either $f'(x) = 0$ or $f'(x)$ does not exist. At each critical point where $f'(x) = 0$, check if the derivative changes sign there or use the second derivative test to determine whether the point corresponds to a local maximum, a local minimum, or neither. Then examine the behavior of $f(x)$ near points where $f'(x)$ does not exist.

The steps listed at the bottom of this page give a suggested step-by-step procedure to use with applied maximum and minimum problems. We cannot overemphasize the importance of the second part of step 1 in the outline—deciding what you wish to maximize or minimize. If you don't know what you are trying to do, you won't have much success doing it. Step 1 sounds like a trivial directive, but many people have floundered aimlessly, writing equation after equation leading nowhere, without knowing where they were heading.

EXAMPLE 1   Find two numbers whose sum is 6 and whose product is as large as possible.

Solving an Applied Maximum/Minimum Problem

STEP 1.   Draw a figure where appropriate and assign letter variables. Decide what you want to maximize or minimize.

STEP 2.   Express the quantity that you want to maximize or minimize as a function f of *one* other quantity. (You may have to use some algebra.)

STEP 3.   Find all critical points, where $f'(x) = 0$ or $f'(x)$ does not exist.

STEP 4.   Decide whether the desired maximum or minimum occurs at one of the points you found in step 3. Frequently it will be clear that a maximum or minimum exists from the nature of the problem, and if step 3 gives only one candidate, that is the answer. If there is more than one candidate, or if there are points in the domain of f that are not *inside* an interval where f is differentiable, you may have to make a further examination.

STEP 5.   Put the answer in the requested form.

Solution

STEP 1. We want to maximize the product P of two numbers.

STEP 2. If the two numbers are x and y, then $P = xy$. Since $x + y = 6$, we have $y = 6 - x$, so $P = x(6 - x) = 6x - x^2$.

STEP 3. We have

$$\frac{dP}{dx} = 6 - 2x.$$

Thus $dP/dx = 0$ when $6 - 2x = 0$, or when $x = 3$.

STEP 4. Since $d^2P/dx^2 = -2 < 0$, we see that P has a maximum at $x = 3$.

STEP 5. We were asked to find the numbers x and y whose sum is 6 and whose product is maximum, not the maximum product. We know that $x = 3$, and consequently $y = 6 - x = 3$ also. Thus our answer is that both numbers must be *3*. ■

EXAMPLE 2 Use the methods of this section to find the point on the line $x + 3y = 6$ that is closest to the point $(-3, 1)$.

Solution

STEP 1. We want to minimize the distance s from a point (x, y) on the line to the point $(-3, 1)$, as shown in Fig. 4.52.

STEP 2. We have

$$s = \sqrt{(x + 3)^2 + (y - 1)^2}.$$

Since $x = 6 - 3y$, we obtain

$$s = \sqrt{(9 - 3y)^2 + (y - 1)^2}.$$

STEP 3. We really don't want to differentiate that radical expression, so we think for a moment and realize that s will be a minimum when s^2 is minimum, for positive s. Thus we decide to minimize

$$s^2 = (9 - 3y)^2 + (y - 1)^2.$$

We obtain

$$\frac{d(s^2)}{dy} = 2(9 - 3y)(-3) + 2(y - 1)$$

$$= 20y - 56$$

$$= 0 \quad \text{when} \quad y = \frac{56}{20} = \frac{14}{5}.$$

STEP 4. Since our problem obviously has a solution and we found only one candidate, we have the minimum distance when $y = \frac{14}{5}$. (Of course, $d^2(s^2)/dy^2 = 20 > 0$, which also shows that $y = \frac{14}{5}$ yields a local minimum.)

STEP 5. We were not asked for the minimum distance, but for the point (x, y) where the minimum distance is attained. Since $y = \frac{14}{5}$, we find that $x = 6 - 3y = 6 - \frac{42}{5} = -\frac{12}{5}$. Thus our answer is the point $(-\frac{12}{5}, \frac{14}{5})$. ■

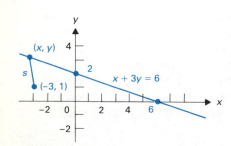

FIGURE 4.52 The distance s from $(-3, 1)$ to a point (x, y) on $x + 3y = 6$.

FIGURE 4.53 Cylindrical can of radius r and height h.

EXAMPLE 3   A manufacturer of dog food wishes to package the product in cylindrical metal cans, each of which is to contain a certain volume V_0 of food. Find the ratio of the height of the can to its radius to minimize the amount of metal, assuming that the ends and side of the can are made from metal of the same thickness.

Solution

STEP 1.   The manufacturer wishes to minimize the surface area S of the can shown in Fig. 4.53.

STEP 2.   The surface of the can consists of the two circular disks at the ends and the cylindrical side. If the radius of the can is r and the height is h, the top and bottom disks each have area πr^2 and the cylinder has area $2\pi rh$ (see Fig. 4.53). Thus

$$S = 2\pi r^2 + 2\pi rh.$$

We would like to find h in terms of r to express S as a function of the single quantity r. From $V_0 = \pi r^2 h$, we obtain $h = V_0/\pi r^2$. Thus

$$S = 2\pi r^2 + 2\pi r\frac{V_0}{\pi r^2} = 2\left(\pi r^2 + \frac{V_0}{r}\right).$$

STEP 3.   We easily find that

$$\frac{dS}{dr} = 2\left(2\pi r - \frac{V_0}{r^2}\right).$$

Thus $dS/dr = 0$ when $2\pi r - (V_0/r^2) = 0$, so $2\pi r^3 = V_0$, and

$$r^3 = \frac{V_0}{2\pi}.$$

STEP 4.   It is obvious that a minimum for S does exist from the nature of the problem. A can $\frac{1}{32}$ in. high and 6 ft across uses a lot of metal, as does one $\frac{1}{32}$ in. in radius and 1000 ft high. Somewhere between these ridiculous measurements there is a can of reasonable dimensions using the least metal. We found only one candidate in step 3, so we don't have to look further.

STEP 5.   We are interested in the ratio h/r. Now $V_0 = \pi r^2 h$, so

$$h = \frac{V_0}{\pi r^2} \qquad \text{and} \qquad \frac{h}{r} = \frac{V_0}{\pi r^3}.$$

From step 3, the least metal is used when $r^3 = V_0/2\pi$, so

$$\frac{h}{r} = \frac{V_0}{\pi V_0/(2\pi)} = 2. \qquad\blacksquare$$

Example 3 illustrates the practical importance of extremal problems. For a cylindrical can of minimal surface area containing a given volume, we should have $h = 2r$, so the height should equal the diameter. Thus few cylindrical cans on the supermarket shelves represent economical packaging, assuming that the ends and sides are of equally expensive material. The tuna fish cans are usually too short, and the soft drink and beer cans are too tall.

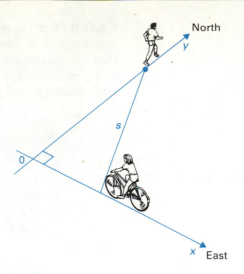

FIGURE 4.54   The distance *s* between Bill and Sue.

EXAMPLE 4   Two straight roads intersect at right angles, one running north and south and the other east and west. Bill jogs north through the intersection at a steady rate of 6 mph. Sue pedals her bicycle west at a steady rate of 12 mph and goes through the intersection 30 minutes after Bill. How close do Bill and Sue come to each other?

Solution

STEP 1.   We denote the roads by *x*,*y*-axes with the *y*-axis pointing north, as is usual on maps. See Fig. 4.54. We wish to minimize the distance *s* between Bill and Sue.

STEP 2.   Let time *t* be measured in hours, with $t = 0$ at the instant when Bill is at the intersection. Since Bill is jogging at a constant rate of 6 mph, he travels $6t$ miles in *t* hours, so his position at time *t* is $(0, 6t)$. Sue reaches the origin in the plane at time $t = \frac{1}{2}$ hour. She is riding west, corresponding to going to the *left* on the *x*-axis, at 12 mph. Thus at time *t*, she is at the point $-12(t - \frac{1}{2})$ on the *x*-axis, or at the point $(-12t + 6, 0)$ in the plane. Thus the distance between Sue and Bill is

$$s = \sqrt{(-12t + 6)^2 + 36t^2}.$$

STEP 3.   As in Example 2, we make our work easier by minimizing

$$s^2 = (-12t + 6)^2 + 36t^2,$$

which is minimum when *s* is minimum. Then

$$d(s^2)/dt = 2(-12t + 6)(-12) + 72t = 288t - 144 + 72t$$
$$= 360t - 144 = 0 \qquad \text{when} \quad t = \tfrac{144}{360} = \tfrac{12}{30} = \tfrac{2}{5}.$$

STEP 4.   Our problem obviously has a solution, and we have only one candidate, $t = \frac{2}{5}$, to give that solution. Thus the minimum distance *s* must occur 24 minutes after Bill goes through the intersection.

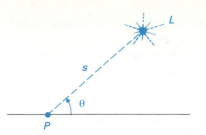

FIGURE 4.55 The distance s and angle of inclination θ from a point P on a level surface to a light L.

STEP 5. We were asked to find the minimum distance. Putting $t = \frac{2}{5}$ in the formula for s in step 2, we obtain

$$s = \sqrt{(\tfrac{6}{5})^2 + 36(\tfrac{2}{5})^2} = 6\sqrt{\tfrac{1}{25} + \tfrac{4}{25}} = 6/\sqrt{5} \text{ miles}$$

as the minimum distance. ∎

EXAMPLE 5 Let a point source of light be at a point L above a flat surface. The illumination that the light provides at a point P on the surface is inversely proportional to the square of the distance s from P to L and directly proportional to $\sin \theta$ where θ is the angle of elevation from P to L. See Fig. 4.55. How high above the sidewalk should a street light bulb be placed to provide maximum illumination on the sidewalk across the street at a point P, which is 24 ft from the base of the light pole?

Solution

STEP 1. In Fig. 4.56 the distance from P to the base B of the light pole is 24 ft. We wish to find $y \geq 0$ to maximize the illumination I.

STEP 2. Since I is directly proportional to $\sin \theta$ and inversely proportional to s^2, there exists a constant k such that

$$I = k \cdot \frac{1}{s^2} \cdot \sin \theta$$

$$= k \cdot \frac{1}{24^2 + y^2} \cdot \frac{y}{\sqrt{24^2 + y^2}}$$

$$= k \cdot \frac{y}{(576 + y^2)^{3/2}}.$$

STEP 3. Differentiating, we obtain

$$\frac{dI}{dy} = k \cdot \frac{(576 + y^2)^{3/2} - y(\tfrac{3}{2})(576 + y^2)^{1/2}(2y)}{(576 + y^2)^3}$$

$$= k \cdot \frac{\sqrt{576 + y^2}\,(576 + y^2 - 3y^2)}{(576 + y^2)^3}$$

$$= k \cdot \frac{576 - 2y^2}{(576 + y^2)^{5/2}}.$$

Thus $dI/dy = 0$ when

$$576 - 2y^2 = 0,$$

$$y^2 = \frac{576}{2} = 288 = 2(144),$$

$$y = \pm 12\sqrt{2}.$$

STEP 4. When $y = 0$, $\sin \theta = 0$, so the illumination I is zero. Also, as $y \to \infty$, the illumination $I \to 0$. Thus there is a maximum illumination I for some $y > 0$, and $y = 12\sqrt{2}$ ft is our only candidate.

STEP 5. Since we were asked to find the height y and not the maximum illumination, our answer is $y = 12\sqrt{2}$ ft. ∎

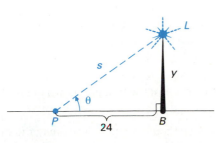

FIGURE 4.56 Light pole y ft high and 24 ft from P on a level surface.

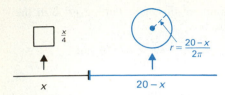

FIGURE 4.57 Line segments of lengths x and $20 - x$ bent to form a square of side $x/4$ and a circle of radius $(20 - x)/(2\pi)$.

EXAMPLE 6   A piece of wire 20 ft long is cut into two pieces of length x and $20 - x$. The piece of length x is bent into a square, and the piece of length $20 - x$ is bent into a circle, as shown in Fig. 4.57. Find the value of x such that the sum of the areas enclosed by the square and the circle is a maximum.

Solution

STEP 1.   We wish to maximize the sum A of the area A_1 of a square of perimeter x and the area A_2 of a circle of perimeter $20 - x$, subject to $0 \leq x \leq 20$.

STEP 2.   Since each side of the square has length $x/4$, we have $A_1 = x^2/16$. Since the radius of the circle is $(20 - x)/(2\pi)$, we see that $A_2 = \pi[(20 - x)/(2\pi)]^2$. Thus

$$A = A_1 + A_2 = \frac{x^2}{16} + \pi\frac{(20 - x)^2}{4\pi^2} = \frac{1}{16}x^2 + \frac{1}{4\pi}(20 - x)^2$$

for $0 \leq x \leq 20$.

STEP 3.   We obtain

$$\frac{dA}{dx} = \frac{1}{16} \cdot 2x + \frac{1}{4\pi} \cdot 2(20 - x)(-1) = \frac{1}{16}(2x) + \frac{1}{2\pi}(x - 20)$$

$$= \left(\frac{1}{8} + \frac{1}{2\pi}\right)x - \frac{10}{\pi} = \left(\frac{\pi + 4}{8\pi}\right)x - \frac{10}{\pi}$$

$$= 0 \quad \text{when} \quad x = \frac{10}{\pi} \cdot \frac{8\pi}{\pi + 4} = \frac{80}{\pi + 4} \approx 11.2.$$

STEP 4.   Since $d^2A/dx^2 = (\pi + 4)/(8\pi) > 0$, we see that the value $x = 80/(\pi + 4)$ corresponds to a local minimum, not a local maximum. Therefore the maximum of A for $0 \leq x \leq 20$ must be attained at an endpoint of the interval. When $x = 0$, $A = 20^2/(4\pi) = 100/\pi$. When $x = 20$, $A = 20^2/16 = 100/4$. Since $100/\pi > 100/4$, we see that the maximum area occurs when $x = 0$.

STEP 5.   We were asked to find the value of x for maximum total area, so $x = 0$ ft is our answer. The maximum total area is attained when the wire is not cut at all and the whole wire is bent into a circle. Our work shows that the minimum area is attained when $x \approx 11.2$ ft.   ∎

EXAMPLE 7   One vertex of a triangle lies at the center of a circle of radius a, and the other two vertices lie on the circle. Find the size of the vertex angle θ at the center of the circle if the area of the triangle is a maximum.

Solution

STEP 1.   We wish to find θ to maximize the area A of the triangle shown in Fig. 4.58.

STEP 2.   Let the altitude of the triangle be h and the base be b. The area of the triangle is then

$$A = \frac{1}{2}bh.$$

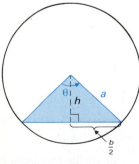

FIGURE 4.58 Triangle with vertex angle θ at the center of a circle of radius a and other vertices on the circle.

From Fig. 4.58, we see that $h = a \cos \theta/2$ and $b = 2a \sin \theta/2$. Thus

$$A = \frac{1}{2}\left(2a \sin \frac{\theta}{2}\right)\left(a \cos \frac{\theta}{2}\right) = a^2 \sin \frac{\theta}{2} \cos \frac{\theta}{2}.$$

STEP 3. We have

$$\frac{dA}{d\theta} = a^2\left[\left(\sin \frac{\theta}{2}\right)\left(-\sin \frac{\theta}{2}\right)\frac{1}{2} + \left(\cos \frac{\theta}{2}\right)\left(\cos \frac{\theta}{2}\right)\frac{1}{2}\right]$$

$$= \frac{a^2}{2}\left(-\sin^2 \frac{\theta}{2} + \cos^2 \frac{\theta}{2}\right) = 0,$$

where $\sin^2(\theta/2) = \cos^2(\theta/2)$, or $\tan(\theta/2) = 1$, or $\theta/2 = \pi/4$, or $\theta = \pi/2$.

STEP 4. The area is a maximum for some θ where $0 < \theta < \pi$, and $\theta = \pi/2$ is the only candidate.

STEP 5. We were asked to find the value of θ for maximum area, so $\theta = \pi/2$ is our answer. ∎

The final two examples relate to economic situations. We do not label the "steps" in the solutions, as we did in the preceding examples. You should analyze the solutions yourself and note that we are still following the procedure given in the outline.

EXAMPLE 8 A large home and auto store sells 9000 auto tires each year. It has *carrying costs* of 50¢ per year for each unsold tire stored (space, insurance, and so on.) The *reorder cost* for each lot ordered is 25¢ per tire plus $10 for the paperwork of the order. Find how many times per year the store should reorder and the size of the order, to minimize the sum of these *inventory costs*.

Solution Let x be the lot size of each order, that is, the number of tires ordered at a time. Then

$$\text{Number of orders per year} = \frac{9000}{x}.$$

The average number of unsold tires over the year is then $x/2$. Therefore

$$\text{Carrying costs per year} = 0.5\left(\frac{x}{2}\right) \text{ dollars}$$

and

$$\text{Reorder costs per year} = 10\left(\frac{9000}{x}\right) + 0.25(9000) \text{ dollars.}$$

Consequently, the total inventory costs per year are

$$C(x) = 0.5\left(\frac{x}{2}\right) + 10\left(\frac{9000}{x}\right) + 0.25(9000)$$

$$= \frac{1}{4}x + 90{,}000x^{-1} + 2250 \text{ dollars.}$$

Then

$$C'(x) = \frac{1}{4} - 90{,}000x^{-2} = \frac{1}{4} - \frac{90{,}000}{x^2}.$$

We easily see that $C'(x) = 0$ when $x^2 = 360{,}000$, or when $x = \sqrt{360{,}000} = 600$ tires. Since $C''(x) = 180{,}000x^{-3} > 0$, we do have a minimum. Therefore the tires should be ordered in lots of 600, and there should be $9000/600 = 15$ such orders each year. ∎

EXAMPLE 9 A fisherman has exclusive rights to a stretch of clam flats. Suppose that conditions are such that on his flats a population of p bushels of clams this year yields a population of $f(p) = 50p - \frac{1}{4}p^2$ bushels next year. (The term $-\frac{1}{4}p^2$ represents a decrease in population due to overcrowding if p is large.) How many bushels should the fisherman dig each year to sustain a maximum harvest year after year?

Solution The fisherman can dig (harvest) $h(p) = f(p) - p$ bushels of clams over the year without depleting the original population of p bushels. He wishes to maximize $h(p)$. We have

$$h(p) = f(p) - p = (50p - \tfrac{1}{4}p^2) - p$$
$$= 49p - \tfrac{1}{4}p^2.$$

Then

$$h'(p) = 49 - \tfrac{1}{2}p,$$

and $h'(p) = 0$ when $p = 98$. Since $h''(p) = -\frac{1}{2} < 0$, we do have a maximum. Thus the fisherman should first adjust the population of clams to 98 bushels, either by letting them increase without digging or by digging the excess population. Then he can dig

$$h(98) = 49(98) - \tfrac{1}{4}(98)^2 = 2401 \text{ bushels}$$

each year, or about 46 bushels per week, year after year. ∎

SUMMARY

Suggested outline for solving an applied maximum/minimum problem:

STEP 1. Decide what quantity you want to maximize or minimize, drawing a figure where appropriate.

STEP 2. Express the quantity as a function f of *one* variable.

STEP 3. Find all critical points of f.

STEP 4. Decide whether the desired maximum or minimum occurs at one of the points found in step 3.

STEP 5. Put the answer in the requested form.

EXERCISES 4.6

1. Find the maximum area a rectangle can have if the perimeter is 20 ft.

2. Generalize Exercise 1 to show that the rectangle of maximum area having a fixed perimeter is a square.

3. Find two positive numbers x and y such that $x + y = 6$ and xy^2 is as large as possible.

4. Find the positive number x such that the sum of x and its reciprocal is minimum.

5. Find two positive numbers such that their product is 36 and the sum of their cubes is a minimum.

6. Find the maximum distance, measured horizontally, between the graphs of $y = x$ and $y = x^2$ for $0 \leq x \leq 1$.

7. A rectangle has its base on the x-axis and its upper vertices on the parabola $y = 6 - x^2$, as shown in Fig. 4.59. Find the maximum possible area of the rectangle.

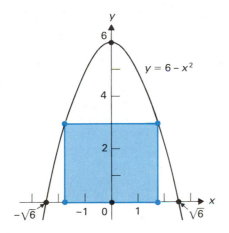

FIGURE 4.59

8. A trapezoid has endpoints of its lower base at $(2, 0)$ and $(-2, 0)$ on the x-axis, while its upper base is a horizontal chord of the parabola $y = 4 - x^2$. Find the maximum possible area of the trapezoid.

9. Find the minimum possible area of a right triangle with right-angle vertex at $(0, 0)$, legs on the positive x-axis and y-axis, and longest side passing through the point $(3, 7)$.

10. Find the point on the parabola $y = x^2$ that is closest to the point $(6, 3)$.

11. A cardboard box of 108 in³ volume with a square base and open top is to be constructed. Find the minimum area of cardboard needed. (Neglect waste in construction.)

12. An open box with a reinforced square bottom and volume 96 ft³ is to be constructed. If material for the bottom costs three times as much per square foot as material for the sides, find the dimensions of the box of minimum cost. (Neglect waste in construction.)

13. A parcel delivery service will accept parcels such that the sum of the length and girth is at most 96 in. (Girth is the perimeter of a cross section taken perpendicular to the longest dimension.) Find the maximum acceptable volume of (a) a rectangular parcel with square ends, (b) a cylindrical parcel.

14. A farmer has 1000 rods of fencing with which to fence three sides of a rectangular pasture; a straight river will form the fourth side of the pasture. Find the dimensions of the pasture of largest area that the farmer can fence.

15. A rancher has 1200 ft of fencing to enclose a double paddock with two rectangular regions of equal areas (Fig. 4.60). Find the maximum area that the rancher can enclose. (Neglect waste in construction and the need for gates.)

FIGURE 4.60
Paddock in Exercise 15.

16. A couple wish to fence a 6000 ft² rectangular plot of their property for their dogs. One side of the rectangle borders a straight road, and the ornamental fence for that side costs three times as much per foot as the rest of the fence. Find the dimensions of the plot for the most economical fencing.

17. An open gutter of rectangular cross section is to be formed from a long sheet of tin of width 8 in., by bending up the sides of the sheet. Find the dimensions of the cross section of the gutter so formed for maximum carrying capacity.

18. A gardener has some 8-in.-wide planks. He wants to build an irrigation trough with trapezoidal cross section to carry water to his garden. See Fig. 4.61. Find the value of θ in Fig. 4.61 so that the trapezoidal cross section has maximum area (which corresponds to maximum irrigation capacity for the trough).

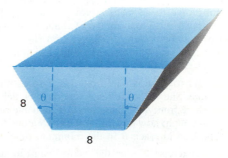

FIGURE 4.61
Trough in Exercise 18.

19. Find the dimensions of the rectangle of maximum area that can be inscribed in a semicircle of radius a. [*Hint:* You may find it easier to maximize the square of the area. Of course, the rectangle having maximum area is the one having the square of its area maximum.]

20. Find the volume of the largest right circular cylinder that can be inscribed in a right circular cone of radius a and altitude b. See Fig. 4.62.

FIGURE 4.62
Cylinder inscribed in a cone.

21. Find the altitude of the right circular cone of maximum volume that can be inscribed in a sphere of radius a.

22. Find the area of the largest isosceles triangle that can be inscribed in a circle of radius a.

23. Find the maximum possible area of a rectangle inscribed in an equilateral triangle of side a.

24. Of all sectors of circles where the perimeter of the sector is a fixed constant P_0, find the central angle θ of the sector of maximum possible area, subject to $0 \le \theta \le 2\pi$.

25. A rectangular cardboard poster is to contain 216 in^2 of printed matter with 2-in. margins at the sides and 3-in. margins at the top and bottom. Find the dimensions of the poster using the least cardboard.

26. The wreck of a plane in a desert is 15 mi from the nearest point A on a straight road. A rescue truck starts for the wreck at a point on the road that is 30 mi distant from A. If the truck can travel at 80 mph on the road and at 40 mph on a straight path in the desert, how far from A should the truck leave the road to reach the wreck in minimum time? See Fig. 4.63.

27. A woman in a rowboat is 1 mi from the nearest point A on a straight shore. She wishes to reach a point B, which is 2 mi along the shore from A. If she can row 4 mph and run along the shore at 6 mph, how far from A toward B along the shore should she land to reach B in the minimum time?

28. Suppose in Exercise 27 that the woman can run along the (possibly rocky) shore k times as fast as she can row. Describe, for all values $k > 0$, how far from A toward B she should land to reach B in the minimum time.

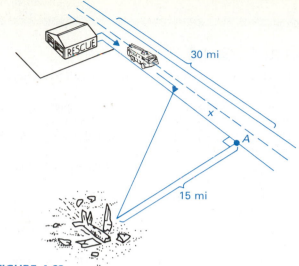

FIGURE 4.63

29. The illumination provided by a light at a point A is directly proportional to the intensity of the light and inversely proportional to the square of the distance from the point A to the light. Let light L_2 have c times the intensity of light L_1, and let the distance between the lights be a ft. Find the distance from L_1 on the line segment joining L_1 to L_2 where the illumination is minimal.

30. A piece of wire 100 cm long is cut into two pieces. One piece is bent to form a square, and the other is formed into an equilateral triangle. Find how the wire should be cut if the total area enclosed is to be (a) minimum, (b) maximum.

31. Answer Exercise 30 if both pieces of wire are bent into equilateral triangles.

32. A weight suspended from a beam by a spring is bouncing up and down. Its height above the floor at time t sec is $h = 48 + 12 \sin[(\pi/2)t]$ in. Find the maximum and minimum of
(a) the heights above the floor that the weight attains,
(b) the velocity of the weight,
(c) the speed of the weight,
(d) the acceleration of the weight.

33. A Norman window is to be built in the shape of a rectangle of clear glass surmounted by a semicircle of colored glass. The total perimeter of the window is to be 36 ft.
(a) Find the ratio of the height of the rectangle to the width that maximizes the area of the clear glass in the rectangular portion.
(b) If clear glass admits twice as much light per square foot as colored glass, find the ratio of the height of the rectangle to the width to admit the maximum amount of light.
(c) If colored glass costs four times as much per square foot as clear glass and the window is to be at least 4 ft wide, find the width of the window of minimum cost.

34. The strength of a beam of rectangular cross section is proportional to the width and the square of the depth. Find the dimensions of the strongest rectangular beam that can be cut from a circular log of radius 9 in.

35. Ship A travels due north and passes a buoy at 9:00 A.M. Ship B, traveling twice as fast due east, passes the same buoy at 11:00 A.M. the same day. At what time are the ships closest together?

36. A silo is to have the form of a cylinder capped with a hemisphere. If the material for the hemisphere is twice as expensive per square foot as the material for the cylinder, find the ratio of the height of the cylinder to its radius for the most economical structure of given volume. (Neglect waste in construction.)

37. An open box is to be formed from a rectangular piece of cardboard by cutting out equal squares at the corners and turning up the resulting flaps. Find the size of the corner squares that should be cut out from a sheet of cardboard of length a and width b to obtain a box of maximum volume.

38. The *probability p* of an event is a number in the interval $[0, 1]$ that measures how likely the event is to occur. For example, a sure event has probability 1, an impossible event has probability zero, and an event as likely to occur as not to occur has probability $\frac{1}{2}$.

Let p be the probability that a particular biased coin produces a head when flipped. We wish to estimate p. The coin is flipped n times, producing a certain sequence containing m heads and $n - m$ tails. It can be shown that if the probability of a head on one flip were x, then the probability of obtaining this particular ordered sequence of heads and tails would be $x^m(1 - x)^{n-m}$. Since this sequence of heads and tails did occur, it is natural to take as estimate for p the value of x in $[0, 1]$ that maximizes $x^m(1 - x)^{n-m}$. Find this estimate for p (the *maximum likelihood estimate*).

39. A fence a ft high is located b ft from the side of a house. Find the length of the shortest ladder that will reach over the fence to the house wall from the ground outside the fence.

40. A right circular cone has slant height 10 in. Find the vertex angle θ (see Fig. 4.64) for the cone of maximum volume.

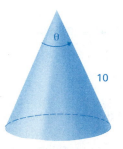

FIGURE 4.64

41. A statue 12 ft high stands on a pedestal 41 ft high. How far from the base of the pedestal on level ground should an observer stand so that the angle θ at his eye subtended by the statue is maximum if the eye of the observer is 5 ft from the ground? [*Hint:* The maximum value of θ occurs when $\tan \theta$ is maximum.]

42. According to Fermat's principle in optics, light travels the path for which the time of travel is minimum. Let light travel with velocity v_1 in medium 1 and velocity v_2 in medium 2, and let the boundary between the media form a plane, as shown in cross section in Fig. 4.65. Show that, according to Fermat's principle, light that travels from A to B in Fig. 4.65 crosses the boundary at a point P such that

$$\frac{\sin \theta_1}{\sin \theta_2} = \frac{v_1}{v_2}.$$

(This is the *law of refraction* or *Snell's law*.)

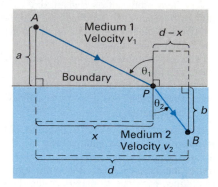

FIGURE 4.65
Refraction of light traveling from A to B.

Economics Exercises

43. A small company manufactures wood stoves. Its cost $C(x)$ and revenue $R(x)$ when x stoves per year are manufactured are given by

$$C(x) = 10{,}000 + 150x + 0.03x^2,$$
$$R(x) = 250x - 0.02x^2.$$

(a) Find the marginal profit when $x = 100$.
(b) How many stoves should the company make for maximum profit? What is that maximum profit?

44. For the company in Exercise 43, find the number of stoves manufactured for the minimum average cost and the number for the maximum average profit.

45. A supplier of stove wood finds that with x cords of wood stocked per year, the cost and revenue are

$$C(x) = 500 + 50x + 0.02x^2,$$
$$R(x) = 80x - 0.01x^2.$$

(a) Find the marginal revenue when $x = 100$ cords.
(b) Find the average revenue when $x = 100$ cords.
(c) Find the value of x that maximizes the profit.

46. For the supplier in Exercise 45, find the number of cords of wood for the maximum average profit.

47. Assuming the supplier in Exercise 45 sells only whole cords of wood, find the number of cords sold to produce the maximum average revenue by (a) using calculus blindly, (b) using common sense. Explain why the calculus method does not work.

48. The following is a simple economic model for the production of a single perishable item that must be sold on the day it is produced to avoid a loss from spoilage.

The producer has a basic plant overhead of a dollars per day. Each item produced costs b dollars for ingredients and labor. In addition, if the manufacturer produces x items per day, there is a daily cost of cx^2 dollars, resulting from crowded conditions and inefficient operation as more items are produced. (The value of c is usually quite small, so that cx^2 is of insignificant size until x becomes fairly large.) Each day, if a single item were produced, it could be sold that day for A dollars (the initial demand price). However, the price at which every item produced on a given day can be sold drops B dollars for each item produced on that day. (The number B reflects the degree of saturation of the market per item and is usually quite small.)

(a) Find an algebraic expression giving the daily profit if the manufacturer produces x items per day.

(b) Find, in terms of a, b, c, A, and B, the number x of items the manufacturer should produce each day to maximize the daily profit.

49. Suppose that in the economic model in Exercise 48 the government imposes a tax on the manufacturer of t dollars for each item manufactured.

(a) Determine the number x of items the manufacturer should produce each day to maximize the daily profit. [*Hint:* Calculus is not necessary if you worked Exercise 48. Just think how this changes the manufacturer's costs.]

(b) Find, in terms of a, b, c, A, and B, the value of t that will maximize the government's return, assuming that the manufacturer maximizes the profit as in part (a).

50. An appliance store sells 200 refrigerators each year. It has carrying costs of $10 per year for each unsold refrigerator stored. The reorder cost for each lot ordered is $5 per refrigerator plus $10 for the paperwork of the order. How many times per year should the store reorder, and at what lot size, to minimize its inventory costs?

51. A record store sells 100,000 records each year. It has carrying costs of 25¢ per year for each unsold record, reorder costs averaging 5¢ per record plus $5 for the paperwork of the order. How many times a year should the store reorder, and in what lot size, to minimize these inventory costs?

52. A college student plans to rent space in a shopping mall during the week before Christmas and sell 2000 Christmas trees. The supplier of trees can either deliver all 2000 trees at once or deliver them in equal lots, but there is a $40 charge over the cost of the trees for each delivery. The student figures it will cost $2 per tree on display to rent enough space in the mall for the enterprise. Find the lot size and the number of lots of trees the student should order to minimize these costs.

53. A population of p wild rabbits on a certain farm gives rise to a population of

$$f(p) = 6p - \left(\frac{1}{12}\right)p^2$$

next year if left undisturbed.

(a) What population will remain constant year after year if left undisturbed?

(b) What is the maximum sustainable number that can be trapped year after year? What is the corresponding population size?

54. A gardener finds that a population of p acceptable-sized Jerusalem artichoke roots in the patch will give rise to a population of $5p - (1/100)p^2$ acceptable-sized roots next year if left undisturbed.

(a) What population of acceptable-sized roots will remain constant year after year if left undisturbed?

(b) What is the maximum sustainable number of acceptable-sized roots that can be dug year after year? What is the corresponding population size?

4.7 NEWTON'S METHOD

Mathematics is often used to determine some numerical quantity by finding an equation that the quantity satisfies and then solving the equation. All of us have worked many problems of this type. An equation in an unknown x can always be expressed in the form $f(x) = 0$ by moving everything to the left-hand side. In algebra, we spent a lot of time learning to solve simple polynomial equations, such as $x^2 - x + 6 = 0$. However, we did not learn to solve $x^5 + 7x^3 - 20 = 0$, which we should regard as quite a simple equation. We will use Newton's method to approximate the only real solution of this equation in Example 4 of this section.

Suppose we wish to solve $f(x) = 0$ for a *differentiable* function f. This section describes *Newton's method* for finding successive approximations of a solution. As we will show, Newton's method really amounts to repeated approximation using differentials.

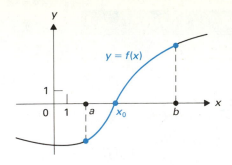

FIGURE 4.66 If f is continuous in $[a, b]$ and $f(a)$ and $f(b)$ have opposite sign, then $f(x_0) = 0$ for some x_0 between a and b.

First we find a number a_1, which we believe is close to a solution. We might, for example, substitute a few values of x to see where $f(x)$ is close to zero or use a computer to plot the graph. The following corollary of the intermediate value theorem is often useful. We asked you to prove this corollary to the theorem in Exercise 13 of Section 2.4. Figure 4.66 illustrates the corollary.

COROLLARY   To the intermediate value theorem

If f is continuous at every point in $[a, b]$ and if $f(a)$ and $f(b)$ have opposite sign, then $f(x) = 0$ has a solution x_0 where $a < x_0 < b$.

EXAMPLE 1   Use the corollary just stated to locate a solution of $x^3 + x - 1 = 0$ between two consecutive integers.

Solution   Let $f(x) = x^3 + x - 1$. Then $f(0) = -1$ and $f(1) = 1$, so by the corollary the equation $x^3 + x - 1 = 0$ has a solution where $0 < x < 1$. ■

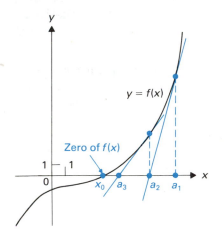

FIGURE 4.67 Newton's method for approximating a solution x_0 of $f(x) = 0$, starting with a_1 and finding a_2 and a_3.

Suppose now that we have found our approximate solution a_1 of the equation $f(x) = 0$. Look at the graph in Fig. 4.67. It appears that the tangent line to the graph of f at the point $(a_1, f(a_1))$ intersects the x-axis at a point a_2, which is a better approximation of a solution than a_1. Repeating this construction, starting with a_2, we expect to find a better approximation a_3, and so on.

Referring to Fig. 4.68, we can find a formula for the next approximation a_{i+1} if we know the approximation a_i. The tangent line to $y = f(x)$, where $x = a_i$, goes through the point $(a_i, f(a_i))$ and has slope $f'(a_i)$. Its equation is therefore

$$y - f(a_i) = f'(a_i)(x - a_i). \tag{1}$$

To find the point a_{i+1} where the line crosses the x-axis, set $y = 0$ in Eq. (1) and solve for x:

$$-f(a_i) = f'(a_i)(x - a_i),$$

$$x - a_i = -\frac{f(a_i)}{f'(a_i)}, \qquad x = a_i - \frac{f(a_i)}{f'(a_i)},$$

assuming $f'(a_i) \neq 0$. Thus we have the *Newton's method recursion formula:*

$$a_{i+1} = a_i - \frac{f(a_i)}{f'(a_i)}. \tag{2}$$

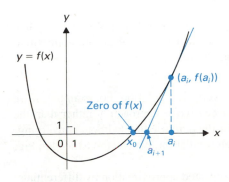

FIGURE 4.68 Newton's method: The tangent line at $(a_i, f(a_i))$ intersects the x-axis at a_{i+1}.

In mathematics, a *recursion formula* is one that allows us to compute the next in a sequence of values in terms of the values already found. In the case of Eq. (2), we compute the next approximation using only the last one found.

EXAMPLE 2   Use Newton's method to approximate $\sqrt{2}$ by approximating a solution of $f(x) = x^2 - 2 = 0$.

Solution   We have $f'(x) = 2x$. The recursion formula (2) becomes

$$a_{i+1} = a_i - \frac{a_i^2 - 2}{2a_i}.$$

TABLE 4.1

i	a_i	a_{i+1}
1	2	$2 - \dfrac{2}{4} = 1.5$
2	1.5	$1.5 - \dfrac{0.25}{3} = 1.416666$
3	1.416666	$1.416666 - \dfrac{0.0069425}{2.833332} = 1.414215$
4	1.414215	$1.414215 - \dfrac{0.000004}{2.828430} = 1.414214$
5	1.414214	

Table 4.1 shows successive approximations starting with $a_1 = 2$. These were easily found using a calculator. Only four iterations gave us at least six-significant-figure accuracy. ∎

A classical algorithm for computing the square root of a positive number is easily derived using Newton's method. Suppose we wish to compute $\sqrt{c}$ for $c > 0$. Let $f(x) = x^2 - c$. Then $f'(x) = 2x$. If a_i is an approximation for $\sqrt{c}$, then the recursion formula (2) becomes

$$a_{i+1} = a_i - \frac{a_i^2 - c}{2a_i} = a_i - \frac{a_i^2}{2a_i} + \frac{c}{2a_i} = a_i - \frac{a_i}{2} + \frac{c}{2a_i} = \frac{a_i}{2} + \frac{c}{2a_i}.$$

Thus

$$a_{i+1} = \frac{1}{2}\left(a_i + \frac{c}{a_i}\right). \tag{3}$$

It can be shown that the iterates given by Eq. (3) eventually approach $\sqrt{c}$ starting with *any* $a_1 > 0$.

EXAMPLE 3 Use a calculator and Eq. (3) to approximate $\sqrt{243}$, starting with $a_1 = 1$.

Solution This choice for a_1 is of course ridiculous. We are just illustrating that Eq. (3) yields $\sqrt{c}$ for any $a_1 > 0$. Results appear in Table 4.2. Our approximations have stabilized to seven significant figures. ∎

Newton's method will not always converge to a solution. For example, if we choose a_1 as shown in Fig. 4.69, the successive a_i approach ∞ rather than the solution of $f(x) = 0$. It is also possible for the a_i to oscillate without converging. Exercises 12 and 13 give examples of such oscillation, which is illustrated in Fig. 4.70.

Newton's method really amounts to repeated approximation by differentials. Recall that

$$dy = f'(a_i)\, dx.$$

TABLE 4.2

i	a_i	$a_{i+1} = \dfrac{1}{2}\left(a_i + \dfrac{243}{a_i}\right)$
1	1	122
2	122	61.99590
3	61.99590	32.95776
4	32.95776	20.16542
5	20.16542	16.10788
6	16.10788	15.59683
7	15.59683	15.58846
8	15.58846	15.58846
9	15.58846	

ISAAC NEWTON (1642–1727) was the foremost figure of the scientific revolution. One of the co-inventors of the calculus, he also developed a new theory of light, designed a reflecting telescope, and authored the *Principia,* the work that unified terrestrial and celestial physics under a few basic principles and defined the study of physics for the next two hundred years.

In grammar school Newton learned Latin and was introduced to mathematics, including geometry and trigonometry. At Trinity College, Cambridge, Newton spent most of his time studying mathematics on his own. When he received his bachelor's degree in 1665 he had read and mastered most of the mathematics known in his day and was well on his way to developing the basic ideas of the calculus. He wrote three papers on the subject from 1666 to 1671, but news of his discoveries spread slowly through the English mathematics community.

Newton's method of solving equations appears in these papers, not as an application of the derivative but as a preliminary to his more general method of solving polynomial equations of the form $p(x, y) = 0$ for y as a power series in x.

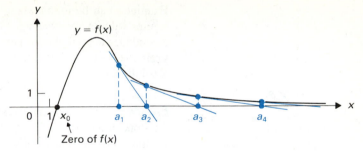

FIGURE 4.69   The iterates $a_1, a_2, a_3, a_4, \ldots$ approach ∞ rather than x_0.

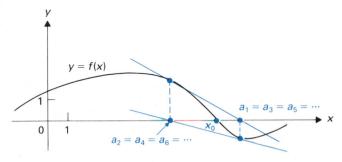

FIGURE 4.70   The iterates $a_1, a_2, a_3, a_4, \ldots$ oscillate and do not approach x_0.

If we set $dy = -f(a_i)$, which is the change we desire in y to make $f(x)$ zero, then

$$dx = \frac{-f(a_i)}{f'(a_i)}.$$

Thus we should change x from a_i to

$$a_i + dx = a_i - \frac{f(a_i)}{f'(a_i)},$$

which is precisely formula (2) for a_{i+1} in Newton's method.

It is only reasonable to use a calculator or computer these days to perform the computations involved in Newton's method. Just in case you don't have a calculator, Exercises 1–14 are designed to be feasible with pencil and paper. We now give three examples that really require a calculator for solution.

EXAMPLE 4   Using a calculator, find a root of $x^5 + 7x^3 - 20 = 0$. (A calculator that is programmable is handy, but by no means essential, for this computation.)

Solution   Let $f(x) = x^5 + 7x^3 - 20$. Clearly, $f(x) \le 0$ for $x \le 1$, while $f(x) \ge 0$ for $x \ge 2$. Since $f(1) = -12$ is closer to zero than $f(2) = 68$, we start with $a_1 = 1$. The recursion formula (2) becomes

$$a_{i+1} = a_i - \frac{a_i^5 + 7a_i^3 - 20}{5a_i^4 + 21a_i^2}.$$

Our calculator yields results given in Table 4.3. We have computed $f(a_i)$ also.

TABLE 4.3

i	a_i	$f(a_i)$
1	1	-12
2	1.461538462	8.522746187
3	1.335597387	0.9271519711
4	1.318225329	0.0155248157
5	1.317924405	0.0000045805
6	1.317924316	1×10^{-11}
7	1.317924316	1×10^{-11}

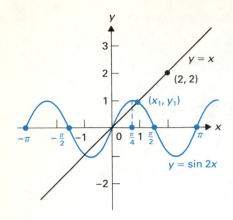

FIGURE 4.71 $\sin 2x = x$ for some $x_1 > 0$.

Examples can be found where the approximations stabilize to many significant figures without approaching a solution of $f(x) = 0$. (See Exercise 14.) Thus it is a good idea to compute $f(a_i)$ to check that we are indeed converging to a solution. From the data in Table 4.3, we are confident that 1.317924316 is indeed close to a solution. ∎

EXAMPLE 5   Note that the slope of the graph of $y = x$ is 1 at the origin, while the slope of the graph of $y = \sin 2x$ is 2 at the origin. Figure 4.71 then shows that we must have a solution of $\sin 2x = x$ for $x > 0$. Find this solution using Newton's method.

Solution   From the graph in Fig. 4.71, we see that $\sin 2x = x$ for some x where $\pi/4 < x < \pi/2$. We use Newton's method with $f(x) = \sin 2x - x$ and $a_1 = 1$. The recursion relation (2) becomes

$$a_{i+1} = a_i - \frac{\sin (2a_i) - a_i}{2 \cos(2a_i) - 1}.$$

Our calculator gave the data in Table 4.4. We are convinced by the table that $\sin 2x = x$ for $x \approx 0.9477471335$. ∎

EXAMPLE 6   Find a local extremum of $f(x) = x^6 + 3x^5 + 2x^4 + 7x^2 - 10x$ and classify it as a local maximum or local minimum.

Solution   We have

$$f'(x) = 6x^5 + 15x^4 + 8x^3 + 14x - 10$$

and

$$f''(x) = 30x^4 + 60x^3 + 24x^2 + 14.$$

We wish to find a solution of $f'(x) = 0$ and check the sign of $f''(x)$ at that solution. Since $f'(0) < 0$ and $f'(1) > 0$, we see that $f'(x) = 0$ for some value of x where $0 < x < 1$. We use Newton's method with $a_1 = 0$. The recursion relation (2) is

$$a_{i+1} = a_i - \frac{6a_i^5 + 15a_i^4 + 8a_i^3 + 14a_i - 10}{30a_i^4 + 60a_i^3 + 24a_i^2 + 14}.$$

Our programmable calculator provided the data in Table 4.5. Our calculator shows that $f''(a_7) \approx 31.90811905$, so we see that $f(x)$ must have a local minimum where $x \approx 0.5285768289$. ∎

In using Newton's method, we may compute $f'(a_i)$ numerically, using m_{sym}. Of course this can be done on a calculator. Since numerical differentiation can be very inaccurate, we should *always* check $f(a_i)$ to be sure we are indeed approaching a solution of $f(x) = 0$. We give one example that we worked out on our programmable calculator.

EXAMPLE 7   Using a calculator and Newton's method, find x such that $x^x = 100$.

Solution   Let $f(x) = x^x - 100$. Since $f(3) < 0$ and $f(4) > 0$, there must be a

TABLE 4.4

i	a_i	$f(a_i)$
1	1	-0.0907025732
2	0.9504977971	-0.0045200436
3	0.9477558227	-0.0000142334
4	0.9477471336	-0.0000000001
5	0.9477471335	-5×10^{-13}
6	0.9477471335	-2×10^{-13}

TABLE 4.5

i	a_i	$f'(a_i)$
1	0	-10
2	0.7142857143	7.935681561
3	0.5723744691	1.491926152
4	0.5313126087	0.0876442105
5	0.5285877659	0.0003489823
6	0.5285768291	0.0000000056
7	0.5285768289	0.0

TABLE 4.6

i	a_i	$f(a_i)$
1	4	156
2	3.744635441	40.3491215
3	3.620734188	5.500400569
4	3.597934159	0.1481299455
5	3.597285529	0.00011535
6	3.597285024	0.0000000004
7	3.597285024	0.0000000004

solution x such that $3 < x < 4$. We use Newton's method with

$$f'(a_i) \approx m_{\text{sym}} = \frac{f(a_i + \Delta x) - f(a_i - \Delta x)}{2(\Delta x)},$$

where $\Delta x = 0.00001$. Starting with $a_1 = 4$, we obtained the results in Table 4.6. Thus we see that $x^x = 100$ for $x \approx 3.597285024$. ■

SUMMARY

Newton's method to solve $f(x) = 0$: First decide on an approximate solution a_1 of $f(x) = 0$, and then determine successive approximations $a_2, a_3, a_4, \ldots$ of a solution using the recursion formula

$$a_{i+1} = a_i - \frac{f(a_i)}{f'(a_i)}.$$

EXERCISES 4.7

It is only reasonable to use a calculator these days in computations involved with Newton's method. However, Exercises 1–14 have been devised to be feasible with just pencil and paper.

In Exercises 1–4, use Newton's method to estimate the given square root, starting with the given value for a_1 and finding a_3. (See Eq. 3 in the text.)

1. $\sqrt{3}$, $a_1 = 2$

2. $\sqrt{8}$, $a_1 = 3$

3. $\sqrt{6}$, $a_1 = 3$

4. $\sqrt{12}$, $a_1 = 4$

5. Use Newton's method to estimate $\sqrt{7}$ until the difference between successive approximations is less than 0.0005.

6. Repeat Exercise 5, estimating $\sqrt{5}$.

7. Use Newton's method to estimate $\sqrt[3]{6}$, starting with $a_1 = 2$ and finding a_3.

8. Use Newton's method to estimate $\sqrt[3]{9}$, starting with $a_1 = 2$ and finding a_3.

9. Use Newton's method to estimate a solution of $x^3 + x - 1 = 0$, starting with $a_1 = 1$ and finding a_3.

10. Use Newton's method to estimate a solution of $x^3 + 2x - 1 = 0$, starting with $a_1 = 1$ and finding a_3.

11. Use Newton's method to estimate a solution of $x^3 + x^2 - 1 = 0$, starting with $a_1 = 1$ and finding a_3.

12. (a) Show that $x^4 + 4x^3 + 4x^2 - x - 1 = 0$ has a solution in $[-1, 0]$.
   (b) Attempt to find a solution of the equation in part (a) starting with $a_1 = 0$ and using Newton's method. What happens?

13. This exercise shows how oscillatory examples such as Exercise 12 can be constructed.
   (a) Show graphically that if $f(x)$ is differentiable with $f(0) = 1$, $f(1) = -1$, and $f'(0) = f'(1) = -1$, then $f(x) = 0$ has a solution in $[0, 1]$, but Newton's method starting with either $a_1 = 0$ or $a_1 = 1$ will lead to oscillation.
   (b) Construct a cubic $f(x) = ax^3 + bx^2 + cx + d$ having the properties in part (a). Follow these steps.
   (i)   Find the equation involving a, b, c, d corresponding to $f(0) = 1$.
   (ii) Repeat part (i) for the requirement $f(1) = -1$.
   (iii) Repeat part (i) for the requirement $f'(0) = -1$.
   (iv) Repeat part (i) for the requirement $f'(1) = -1$.
   (v)   Solve the system of four equations in a, b, c, d given by parts (i)–(iv) to find the desired cubic $f(x)$.

14. This exercise shows that it is possible to have the values a_1, $a_2, \ldots, a_n$ generated using Newton's method stabilize to many significant figures without approaching a solution of $f(x) = 0$. Checking values $f(a_i)$ as in Example 4 would guard against reaching a false conclusion in such a case.
   (a) Let $a_1 > a_2 > \cdots > a_n$ be any n distinct numbers. Let $c_1, c_2, \ldots, c_n$ be any n numbers all greater than 5, and let $d_1, d_2, \ldots, d_n$ be any n numbers. Argue graphically that there exists a differentiable function $f(x)$ such that $f(x) \geq 3$ for all x, while $f(a_i) = c_i$ and $f'(a_i) = d_i$ for $i = 1, 2, \ldots, n$.

(b) By part (a), there exists a differentiable function $f(x)$ where $f(x) \geq 3$ for all x and such that $f(1) = f(1.1) = f(1.01) = f(1.001) = \cdots = f(1.00000001) = 10$, while $f'(1.1) = 10/0.09$, $f'(1.01) = 10/0.009$, $f'(0.001) = 10/0.0009, \ldots, f'(1.0000001) = 10/0.00000009$. If Newton's method is used to solve $f(x) = 0$ starting with $a_1 = 1.1$, find $a_1, a_2, \ldots, a_8$. Are significant figures stabilizing? Are you approaching a solution of $f(x) = 0$?

Use a calculator in Exercises 15–20.

15. Use Newton's method to estimate a solution of $f(x) = x^3 + x + 16 = 0$ until the difference between successive approximations is less than 0.000005.

16. Use Newton's method to estimate $\sqrt{17}$, which is a solution of $x^2 - 17 = 0$, starting with $a_1 = 4$.

17. Use Newton's method to estimate $\sqrt[3]{25}$, which is a solution of $x^3 - 25 = 0$, starting with $a_1 = 3$.

18. Estimate a solution of $x - 2 \sin x = 0$, using Newton's method and starting with $a_1 = 2$.

19. Argue graphically that $\cos x - x = 0$ has a unique solution. Estimate it using Newton's method.

20. Use Newton's method to estimate the positive solution of $\cos x - x^2 = 0$.

In Exercises 21–26, use a calculator or computer and Newton's method to estimate a solution of the given equation.

21. $2^x + x = 9$

22. $3^x - 4x = 100$

23. $x^x = 5$

24. $x + \ln x = 400$

25. $2^x + 3^x = 50$

26. $5^x + x^x = 500$

EXPLORING CALCULUS

P: Newton method graphic

Q: Newton method compute

R: Newton method complex

4.8   ANTIDERIVATIVES AND DIFFERENTIAL EQUATIONS

Antiderivatives

We have spent a lot of time differentiating, that is, finding $f'(x)$ if $f(x)$ is known. Of equal importance is *antidifferentiating*, finding $f(x)$ if $f'(x)$ is known. We call $f(x)$ an **antiderivative** of $f'(x)$. Here is one indication of the importance of such a computation. We have seen that if we know the position x of a body on a line at time t, then $dx/dt = v$ is the velocity and $d^2x/dt^2 = a$ is the acceleration at time t. But, in practice, we often know the position and velocity only at the initial time $t = 0$, while we know the acceleration at every time $t \geq 0$. This is because we are often applying a controlled force F to produce the motion, perhaps by controlling the thrust of some motor. By Newton's second law of motion, $F = ma$, where m is the mass of the body. Thus if we know the force F, we know the acceleration $a = F/m$. Antidifferentiation of the acceleration will then find the velocity, and antidifferentiation of the velocity will find the position of the body. We will find the height of a freely falling body in a vacuum, subject only to gravitational acceleration, in Example 12.

We change notation a bit and let $f(x)$ always be the known function and $F(x)$ the desired antiderivative, so that

$$F'(x) = f(x).$$

EXAMPLE 1   Find an antiderivative $F(x)$ of $f(x) = x^2$.

Solution   Since differentiation drops the exponent of a monomial by 1, we naturally try x^3 as our antiderivative of x^2. But $d(x^3)/dx = 3x^2$. Thus we see that we should take $F(x) = \frac{1}{3}x^3$ as an antiderivative. Note that $(x^3/3) + 2$ is also an antiderivative of x^2, and indeed, if C is any constant, then

$$F(x) = \frac{x^3}{3} + C$$

is an antiderivative of x^2.

As we indicated in Example 1, if $F(x)$ is an antiderivative of $f(x)$, then $F(x) + C$ is also an antiderivative for any constant C. In this context, C is called an *arbitrary constant*. We can show that *all* antiderivatives of $f(x)$ are of the form $F(x) + C$; there are no others. Recall Theorem 4.5: If F is differentiable in $[a, b]$ and $F'(x) = 0$ for all x in $[a, b]$, then $F(x) = F(a)$ for all x in $[a, b]$. Suppose that $G'(x) = f(x)$ also for all x in $[a, b]$. Then

$$\frac{d(G(x) - F(x))}{dx} = G'(x) - F'(x) = 0.$$

By Theorem 4.5, $G(x) - F(x) = G(a) - F(a)$. If we let $C = G(a) - F(a)$, then $G(x) = F(x) + C$. The collection $F(x) + C$ of all of these antiderivatives is the **general antiderivative** of f. We summarize this work in a theorem.

THEOREM 4.10   Antiderivatives of *f(x)*

If $f(x)$ has an unbroken interval as domain and if $F'(x) = f(x)$, then all antiderivatives of $f(x)$ are of the form $F(x) + C$ for some constant C.   □

EXAMPLE 2   Find the general antiderivative of x^2.

Solution   The work in Example 1 shows that the general antiderivative of x^2 is $\frac{1}{3}x^3 + C$.

EXAMPLE 3   Find the general antiderivative of x^n for $n \neq -1$.

Solution   As in Example 1, our first attempt to find an antiderivative would be to try x^{n+1}. But $d(x^{n+1}) = (n + 1)x^n$. Thus we should take $x^{n+1}/(n + 1)$ as our trial antiderivative. We see at once that

$$\frac{d\left(\frac{1}{n + 1}x^{n+1}\right)}{dx} = x^n.$$

Thus the general antiderivative of x^n is

$$F(x) = \frac{1}{n + 1}x^{n+1} + C \qquad \text{for } n \neq -1.$$

The case $n = -1$, which causes the denominator $n + 1$ to become zero, will be treated in Example 8. There we will find an antiderivative of $x^{-1} = 1/x$.

Suppose that f and g are functions with the same domain. If F is an antiderivative of f and G is an antiderivative of g, then it is trivial to check that $F + G$ is an antiderivative of $f + g$ and that cF is an antiderivative of cf. We need only differentiate $F + G$ to obtain

$$\frac{d(F(x) + G(x))}{dx} = \frac{d(F(x))}{dx} + \frac{d(G(x))}{dx} = f(x) + g(x)$$

and differentiate cF to obtain

$$\frac{d(c \cdot F(x))}{dx} = c \cdot \frac{d(F(x))}{dx} = c \cdot f(x).$$

EXAMPLE 4   Use Example 3 and the remarks following the example to find the general antiderivative of a general polynomial function

$$f(x) = a_n x^n + \cdots + a_1 x + a_0.$$

Solution   We see at once that the general antiderivative is

$$F(x) = a_n \left(\frac{x^{n+1}}{n+1} \right) + \cdots + a_1 \frac{x^2}{2} + a_0 x + C.$$

For example, the general antiderivative of $3x^2 + 4x + 7$ is

$$3\left(\frac{x^3}{3}\right) + 4\left(\frac{x^2}{2}\right) + 7x + C = x^3 + 2x^2 + 7x + C. \qquad \blacksquare$$

EXAMPLE 5   Find the general antiderivative of $f(x) = 2/x^2 + 4/x^3$.

Solution   We write $f(x) = 2x^{-2} + 4x^{-3}$. Example 3 and the remarks following it show that the general antiderivative is

$$F(x) = 2\frac{x^{-1}}{-1} + 4\frac{x^{-2}}{-2} + C = \frac{-2}{x} - \frac{2}{x^2} + C. \qquad \blacksquare$$

EXAMPLE 6   Find the general antiderivatives of $f(x) = \sin ax$ and $g(x) = \cos ax$.

Solution   Since the derivative of $\cos x$ is $-\sin x$, we try $-\cos ax$ as an antiderivative of $\sin ax$. However, $d(-\cos ax)/dx = a \sin ax$. We see at once that the general antiderivative of $f(x) = \sin ax$ is

$$F(x) = -\frac{1}{a} \cos ax + C.$$

Similarly, we find the general antiderivative of $g(x) = \cos ax$ is

$$G(x) = \frac{1}{a} \sin ax + C. \qquad \blacksquare$$

EXAMPLE 7   Find the general antiderivative of $f(x) = e^{ax}$.

Solution   Since the derivative of e^{ax} is ae^{ax}, we see that an antiderivative of e^{ax} is $F(x) = (1/a)e^{ax}$. By Theorem 4.10, the general antiderivative is

$$F(x) + C = \frac{1}{a} e^{ax} + C. \qquad \blacksquare$$

EXAMPLE 8   Find the general antiderivative of $f(x) = 1/x$.

Solution   We know that the derivative of $\ln x$ is $1/x$, but this holds only for $x > 0$ since $\ln x$ is not defined for x negative. Now $\ln|x|$ is defined for all $x \neq 0$, so it has the same domain as $1/x$, and $\ln|x|$ coincides with $\ln x$ for $x > 0$. We conjecture that the derivative of $\ln|x|$ for $x < 0$ is also $1/x$. Let's check and see.

For $x < 0$, we know that $|x| = -x$, so $\ln|x| = \ln(-x)$. By the chain rule, we have

$$\frac{d(\ln(-x))}{dx} = \frac{1}{-x} \cdot \frac{d(-x)}{dx}$$

$$= \frac{1}{-x} \cdot (-1) = \frac{1}{x} \qquad \text{for} \quad x < 0.$$

Thus our conjecture is correct, and $\ln|x|$ is an antiderivative of $1/x$ for all $x \neq 0$.

The general antiderivative of $\dfrac{1}{x}$ is $\ln|x| + C$.

We will be using this fact frequently in the remainder of the text.   ■

Differential Equations

An equation involving a derivative of an "unknown" function that we want to find is a **differential equation.** Examples of differential equations, written in various notations, are

$$y' = \sin x, \qquad D^2 y - 2Dy + y = x^2,$$

$$\frac{d^3 y}{dx^3} + (\sin x)\left(\frac{dy}{dx}\right)^2 - y^3 = 0.$$

To solve such a differential equation means to determine all functions $y = f(x)$ that satisfy the equation. That is, y is regarded as an unknown function of x in the equation. Example 6 shows that $y = -\cos x + C$, where C may be any constant, describes all solutions of the first equation given above, because to solve $y' = \sin x$ clearly amounts to finding all antiderivatives of $\sin x$. In Chapter 17, we will see how to solve the second equation given above. We will not learn to solve the third equation above, but clearly one solution is $y = 0$. A differential equation can be expected to have an infinite number of solutions, as our solution $y = -\cos x + C$ of $y' = \sin x$ illustrates.

We want to find a function $y = F(x)$ that is a solution of the differential equation

$$y' = f(x).$$

If f has an antiderivative, that is, if the differential equation $y' = f(x)$ has a solution, then the general antiderivative of f is the **general solution of this differential equation.**

We can characterize the general solution $F(x) + C$ of the differential equation $y' = f(x)$ geometrically. If $G(x) = F(x) + k$, then the graph of G is simply the graph of F displaced k units "upward" if $k > 0$, or $|k|$ units "downward" if $k < 0$. Thus the set $F(x) + C$ of functions may be visualized geometrically as a collection of graphs with the property that any one of them is "congruent" to any other and may be transformed into the other by sliding up or down.

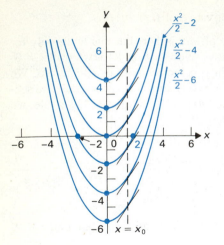

FIGURE 4.72 Some solutions $x^2/2 + C$ of the differential equation $y' = x$.

EXAMPLE 9 Sketch some graphs of solutions of the differential equation $y' = x$.

Solution Figure 4.72 shows some of the solutions. All the functions are antiderivatives of $f(x) = x$ and are of the form $(x^2/2) + C$. For antiderivatives F and G of f, the fact that $F'(x_0) = G'(x_0) = f(x_0)$ for each x_0 means that the graphs of F and G have the same slope (that is, parallel tangent lines) over x_0, as illustrated in Fig. 4.72. ∎

Consider an object moving on a line, which we assume to be the y-axis. As we mentioned at the start of this section, we might know the acceleration of the object, for the acceleration may be caused by some force that we can control. Such motion gives rise to a differential equation of the form

$$\frac{d^2y}{dt^2} = g(t),$$

where $g(t)$ is the acceleration at time t. We want to find the position y of our moving object as a function of t, that is, to find the solution of the differential equation giving the position for this object. A moment's thought shows that this should be physically determined if we know the position where the object started and the velocity with which it started. Suppose that at time $t = 0$, the *initial position* was y_0 and the *initial velocity* was v_0. We refer to these two requirements as *initial conditions* $y(0) = y_0$ and $y'(0) = v_0$ for the desired solution of $d^2y/dt^2 = g(t)$.

The adjective *initial* was undoubtedly first used to describe conditions at time $t = 0$. However, mathematicians currently use the term *initial conditions* with any differential equation to describe a specification of values of the unknown function or its derivatives at any *single value* of the independent variable. Thus we might ask for the solution of

$$\frac{d^2y}{dx^2} - \frac{dy}{dx} - 6y = x$$

that satisfies the *initial conditions* $y(3) = -4$ and $y'(3) = 5$. We will learn how to solve this problem in Chapter 17. Note that both these conditions describe behavior of the solution at the *same point, $x = 3$*. This is characteristic of initial conditions. Conditions that describe behavior at *different points* are referred to as *boundary conditions*.

If f is defined at x_0 and a solution F of $y' = f(x)$ exists, then for any y_0 we can find C such that the solution $G(x) = F(x) + C$ satisfies the initial condition expressed by

$$G(x_0) = y_0.$$

Namely, if we want to have

$$y_0 = G(x_0) = F(x_0) + C,$$

we simply take $C = y_0 - F(x_0)$.

EXAMPLE 10 Find the solution of $y' = x$ that satisfies the initial condition $y(1) = 3$.

Solution   We take $y = (x^2/2) + C$ and require that

$$3 = y(1) = \frac{1^2}{2} + C.$$

We find that $C = 3 - \frac{1}{2} = \frac{5}{2}$. Thus $y = (x^2/2) + \frac{5}{2}$. ∎

EXAMPLE 11   Find the solution of $y'' = \sqrt{x}$ that satisfies the initial conditions $y(1) = 5$ and $y'(1) = -2$.

Solution   From $y'' = \sqrt{x} = x^{1/2}$, we obtain, using Example 3,

$$y' = \frac{x^{3/2}}{\frac{3}{2}} + C_1 = \frac{2}{3}x^{3/2} + C_1.$$

Since $y'(1) = -2$, we have

$$-2 = \frac{2}{3}(1)^{3/2} + C_1 = \frac{2}{3} + C_1.$$

Thus $C_1 = -2 - \frac{2}{3} = -\frac{8}{3}$, so

$$y' = \frac{2}{3}x^{3/2} - \frac{8}{3}.$$

Example 3 then shows that

$$y = \frac{2}{3} \cdot \frac{x^{5/2}}{\frac{5}{2}} - \frac{8}{3}x + C_2 = \frac{4}{15}x^{5/2} - \frac{8}{3}x + C_2.$$

Since $y(1) = 5$, we have

$$5 = \frac{4}{15}(1)^{5/2} - \frac{8}{3} + C_2,$$

so

$$C_2 = 5 - \frac{4}{15} + \frac{8}{3} = 5 + \frac{36}{15} = 5 + \frac{12}{5} = \frac{37}{5}.$$

Thus

$$y = \frac{4}{15}x^{5/2} - \frac{8}{3}x + \frac{37}{5}.$$ ∎

We now give the application to a freely falling body in a vacuum, as promised at the start of this section.

EXAMPLE 12   If air resistance is neglected, the acceleration of a freely falling body near the surface of the earth is known to be about -32 ft/sec^2. (The negative sign is used since the acceleration is downward.) Find the height of a body above the surface of the earth at time t if the body has height 100 ft and (upward) velocity 30 ft/sec at time $t = 0$.

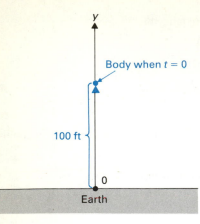

FIGURE 4.73 Freely falling body: When $t = 0$, then $y = 100$ ft and $v = dy/dt = 30$ ft/sec.

Solution We let y be the height of the body above the surface of the earth, as indicated in Fig. 4.73. We know

$$\frac{d^2y}{dt^2} = -32,$$

so

$$\frac{dy}{dt} = -32t + C_1.$$

When $t = 0$, $dy/dt = 30$, so $C_1 = 30$ and

$$\frac{dy}{dt} = -32t + 30.$$

Then

$$y = -32\frac{t^2}{2} + 30t + C_2$$

$$= -16t^2 + 30t + C_2.$$

When $t = 0$, $y = 100$, so $C_2 = 100$ and

$$y = -16t^2 + 30t + 100$$

is the equation giving the height of the body above the surface of the earth. Of course this equation is valid only when $y \geq 0$. We could find the time limits when $y = 0$ by solving the quadratic equation

$$-16t^2 + 30t + 100 = 0.$$

The quadratic formula and a calculator show that the time interval during which the formula is valid is about $-1.733 \leq t \leq 3.608$ sec. ∎

Exercise 55 asks you to obtain the familiar solution

$$y = y_0 + v_0 t - 16t^2$$

for $d^2y/dt^2 = -32$ with initial conditions $y(0) = y_0$ and $y'(0) = v_0$. This solution gives the position for an object falling freely in a vacuum near the surface of the earth.

SUMMARY

1. If $F'(x) = f(x)$, then $F(x) + C$ is the general antiderivative of $f(x)$, where C is an arbitrary constant.

2. The general antiderivative of x^n is $(x^{n+1})/(n + 1) + C$ if $n \neq -1$.

3. An antiderivative of a sum is the sum of antiderivatives, and an antiderivative of a constant times a function is the constant times an antiderivative of the function.

EXERCISES 4.8

In Exercises 1–34, find the general antiderivative of the given function.

1. 2

2. −10

3. $t^2 - 3t + 2$

4. $8v^3 - 2v^2 + 4$

5. $x^{1/2}$

6. $x^{-2/3}$

7. $4u + u^{1/2}$

8. $2s^{3/4} - 4s^{-1/2}$

9. $\dfrac{1}{r^2} + 3r + 1$

10. $\dfrac{2}{x^4} - \dfrac{3}{x^2} + x^2$

11. $\dfrac{x+1}{x^3}$

12. $\dfrac{y^2 - 3y + 2}{y^4}$

13. $\dfrac{3y^2 - 2y + 1}{\sqrt{y}}$

14. $(x^2 + 1)^2$

15. $t(t^2 + 1)$

16. $t^2(t^2 - t + 2)$

17. $\sqrt{x-1}$

18. $\sqrt{1-3x}$

19. $\dfrac{1}{\sqrt{u-1}}$

20. $\dfrac{2}{\sqrt{1-x}} + \sqrt{2x+5}$

21. $\sin x$

22. $\cos 3\alpha$

23. $5 \sin 8\theta$

24. $\sin(2x + 3)$

25. $\cos(2 - 4x)$

26. $\sec^2 3\theta$

27. $4 \csc^2 2\phi$

28. $\sec u \tan u$

29. $\csc 4t \cot 4t$

30. $\dfrac{3 \sin 2t}{\cos^2 2t}$

31. $e^{2x} + x^2$

32. $e^{-3u} - \sin 2u$

33. $\dfrac{2}{t} + \cos 3t$

34. $\dfrac{-3}{s} + \dfrac{1}{s^3}$

In Exercises 35–48, find the solution $y = F(x)$ of the differential equation that satisfies the given initial conditions.

35. $y' = 8$, $y(2) = -3$

36. $y' = 3x^2 + 2$, $y(0) = 1$

37. $y' = x + \sin x$, $y(0) = 3$

38. $y' = 3 \cos 2x$, $y\left(\dfrac{\pi}{4}\right) = -1$

39. $y' = x - \sqrt{x}$, $y(1) = \pi$

40. $y' = x + e^{-x}$, $y(0) = 2$

41. $y' = x^2 - \dfrac{4}{x}$, $y(1) = -3$

42. $y'' = 0$, $y'(0) = -1$, $y(0) = 3$

43. $y'' = 2$, $y'(0) = -2$, $y(0) = 1$

44. $y'' = 2x$, $y'(1) = -1$, $y(1) = 2$

45. $y'' = -\sin 2x$, $y'(0) = 2$, $y(0) = -1$

46. $y'' = \cos 3x$, $y'(0) = 4$, $y(0) = -2$

47. $y'' = e^{-3x}$, $y'(0) = -1$, $y(0) = 6$

48. $y'' = -\dfrac{2}{x^2}$, $y'(1) = 4$, $y(1) = -2$

49. Let $d^2y/dx^2 = f(x)$ have $y = F(x)$ as one solution for $a \le x \le b$. Argue that there exists a solution $y = G(x)$ of the differential equation on $[a, b]$ satisfying the *boundary conditions* $y(a) = \alpha$ and $y(b) = \beta$ for any choices of α and β.

50. Find the solution $y = G(x)$ of $d^2y/dx^2 = 0$ satisfying the boundary conditions $y(1) = -3$ and $y(2) = 0$. (See Exercise 49.)

51. Find the solution $y = G(x)$ of $d^2y/dx^2 = \sin x$ satisfying the boundary conditions $y(-\pi/2) = 0$ and $y(\pi/2) = 0$. (See Exercise 49.)

52. At time $t = 0$, a man standing on a balcony reaches over and tosses a ball upward at a speed of 32 ft/sec from a height of 48 ft above the ground. Neglecting air resistance,
(a) find the height of the ball as a function of t until the ball hits the ground,
(b) find the time when the ball hits the ground.

53. General equation for free fall in a vacuum: Suppose a body close to the surface of the earth is subject only to the force of gravity; that is, neglect air resistance. It is known that the acceleration of the body is then directed downward with magnitude $g = 32$ ft/sec^2. Thus if y measures the height of the body above the surface at time t, we have $d^2y/dt^2 = -32$. Suppose that at time $t = 0$, the body has *initial velocity* v_0 and *initial position* y_0.
(a) Find the velocity $v = dy/dt$ of the body as a function of t.
(b) Find the position y of the body as a function of t.

54. A body is attached to one end of a spring, whose other end is attached to a beam. The body hangs at rest in suspended position. See Fig. 4.74. If the body is pulled down a short distance and released, oscillatory motion results. Neglecting air resistance, the only force on the body is the spring's restoring force, which at each time t is proportional to the body's displacement s from its position at rest. A displacement of 1 unit results in a restoring force of k corresponding units.

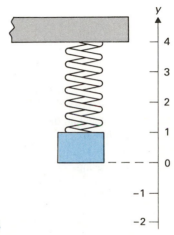

FIGURE 4.74

(a) Use Newton's law $F = ma$ to find the differential equation governing the motion of such a body of mass m. (Take as origin on a vertical s-axis the position of rest of the body.)

(b) If the body is pulled down a distance of 8 units, held for a moment, and then released at time $t = 0$, give the initial conditions that, together with your answer to part (a), would completely determine the motion of the body. (You are not expected to solve the differential equation.)

55. A body of mass m traveling through a medium (air, water, or the like) encounters a force of resistance from the medium proportional to the speed of the body. Suppose the magnitude $k > 0$ of this constant of proportionality is known, and suppose also that other forces on the body can be neglected. If the body enters the medium with a velocity of 80 ft/sec at time $t = 0$, find the differential equation and initial conditions which determine the distance s the body travels through the medium in time t.

56. A freely falling body of mass m near the surface of the earth encounters a force of air resistance proportional to its velocity in addition to the force produced by gravity. The body is dropped from a position of 5000 ft above the earth at time $t = 0$. Given that the air resistance proportionality constant is k (where units are in feet and seconds), find the differential equation and initial conditions that govern the altitude s of the body at time t until the body hits the ground.

57. Under suitable conditions (no predators, disease, and the like), the rate of growth of a population P is proportional to the size of P. If k is the constant of proportionality, give the differential equation that governs this situation.

SUPPLEMENTARY EXERCISES FOR CHAPTER 4

1. If the altitude of a right circular cone remains constant at 10 ft while the volume increases at a steady rate of 8 ft^3/min, find the rate of increase of the radius when the radius is 4 ft. (The volume V of a cone of base radius r and altitude h is $V = \frac{1}{3}\pi r^2 h$.)

2. If the lengths of two sides of a triangle remain constant at 10 ft and 15 ft while the angle θ between them is increasing at the rate of $\frac{1}{10}$ radian/min, find the rate of increase of the third side when $\theta = \pi/3$.

3. If W watts of power are being consumed, then $W = VI$ where the potential is V volts and the current is I amperes. Suppose that power consumption remains constant while, at a certain instant, the current is 15 amp and the potential of 120 volts is dropping at a rate of 0.5 volt/sec. Find the rate of change of the current at that instant.

4. State the mean value theorem without reference to the text.

5. If $f'(x)$ exists for all x and $f(4) = 12$, while $f'(x) \leq -3$ for x in $[-1, 4]$, use the mean value theorem to find the smallest possible value for $f(-1)$.

6. Let f be differentiable for all x. If $f(1) = -2$ and $f'(x) \geq 2$ for x in $[1, 6]$, use the mean value theorem to show that $f(6) \geq 8$.

7. Suppose $|f'(x)| < x + 2$ for x in $[1, 3]$. If $f(1) = 4$, find bounds in terms of x for $f(x)$ in $[1, 3]$.

8. If $f'(x) = (x - 2)^3(x + 1)^2$, find the intervals where $f(x)$ is (a) increasing, (b) decreasing.

9. Let $f(x) = 1/(x^2 - 4)$. Find the intervals where $f(x)$ is (a) increasing, (b) decreasing.

10. Suppose $f'(x) = (x + 1)^3(x - 5)^4$. Find all x-values where $f(x)$ has (a) a local maximum, (b) a local minimum.

11. If

$$f'(x) = \frac{(x + 1)(x - 4)^5}{(x - 2)^3},$$

find all x-values where $f(x)$ has (a) a local maximum, (b) a local minimum.

12. Find the maximum and minimum values attained by the function $x^4 - 8x^2 + 4$ in the interval $[-1, 3]$.

13. Find the maximum and minimum values attained by the function $x^3 - 3x + 1$ in the interval $[0, 3]$.

14. Find the maximum and minimum values attained by $f(x) = x/2 - \sin x$ in the interval $[0, 2\pi]$.

15. Find the maximum and minimum values attained by $f(x) = x + \sin x$ in the interval $[0, 2\pi]$.

16. If $f'(x)$ is as given in Exercise 8, find the intervals where the graph of f is (a) concave up, (b) concave down.

17. If

$$f''(x) = \frac{(x - 2)^3(x + 3)^5}{x^3},$$

find the intervals where the graph of f is (a) concave up, (b) concave down.

18. If $f(x) = x/(x - 2)$, find the intervals where the graph of f is (a) concave up, (b) concave down.

19. If $f'(x) = (x + 1)^3(x - 5)^4$, find all x-values of inflection points.

20. Sketch the graph of $y = x^3 - 3x^2 + 2$, finding and labeling all local maxima and minima and all inflection points.

21. Sketch the portion of the graph of $f(x) = x + \cos x$ for $0 \leq x \leq 2\pi$, finding and labeling all local maxima and minima and all inflection points for x in that interval.

22. Sketch the portion of the graph of $y = x/2 - \cos x$ for $0 \leq x \leq 2\pi$, and find and label all local maxima and minima and all inflection points for x in that interval.

23. Sketch the graph of $y = 2x^2 - x^4 + 6$, finding and labeling all relative maxima and minima and all inflection points.

24. A right circular cone has constant slant height 12. Find the radius of the base for which the cone has maximum volume. (The volume V of a cone of base radius r and altitude h is $V = \frac{1}{3}\pi r^2 h$.)

25. Find two positive numbers such that their sum is 6 and the product of one by the square of the other is maximum.

26. A rectangle has base on the x-axis and upper vertices on the parabola $y = 16 - x^2$ for $-4 \leq x \leq 4$. Find the maximum area the rectangle can have.

27. Find the area of the largest rectangle that can be inscribed in an equilateral triangle of side a ft.

28. At a certain instant, an equilateral triangle has sides of length 10 ft that are increasing at the rate of $\frac{1}{2}$ ft/min. Find the rate of increase of the area of the rectangle of maximum area that can be inscribed in the triangle. (See Exercise 27.)

29. A ski shop sells 1600 pairs of skis during the year. They have carrying costs of $5 per year for each pair of skis unsold and order costs averaging 50¢ per pair of skis plus $10 for the paperwork for each order. Find the lot size in which the skis should be ordered to minimize these inventory costs.

30. A family with a trout pond finds that a population of p trout one year produces a population of $51p - 0.1p^2$ five years later if left undisturbed. Find the maximum sustainable harvest of trout over a five-year period.

31. A solution of $f(x) = x^3 - 3x + 1$ is desired. Since $f(1) = -1$ and $f(2) = 3$, there is a solution between 1 and 2. Starting with 2 as a first approximation of the solution, find the next two approximations given by Newton's method.

32. Use Newton's method to estimate $\sqrt{30}$, starting with 3 as a first approximation and finding the next approximation.

33. Use a calculator and Newton's method to estimate the solution of $\cos x = x/2$ to six significant figures.

34. Find the general antiderivative of $f(x) = 3/x^3 - 4x + \cos(x/2)$.

35. A body moving on an s-axis has acceleration $a = 6t - 8$ at time t. If when $t = 1$ the body is at the point $s = 4$ and has velocity $v = -3$, find the body's position s as a function of time.

36. Find the solution of the differential equation $dy/dx = 3x^2 - 4x + 5$ that satisfies the initial condition $y = -2$ when $x = 1$.

37. Solve the initial value problem $d^2y/dx^2 = x - 3$ where $y = 5$ and $y' = -2$ when $x = 1$.

38. A company manufacturing x units of a certain product per year has a cost function $C(x) = 10,000 + 500x + 0.05x^2$ dollars and a marginal revenue of $MR(x) = 900 - 0.04x$ dollars per year. Find the profit if 100 units are manufactured in a year.

39. A farmer has 1000 ft of fencing. He wishes to use it to fence in one square and one circular enclosure, each having an area of at least 10,000 ft. What size enclosures should he fence to maximize the total area? (Use a calculator.)

40. If f is a twice-differentiable function for $a \leq x \leq b$ and $f(x)$ attains the same value at three distinct points in $[a, b]$, then show that $f''(c) = 0$ for some c where $a < c < b$.

41. State a generalization of the result in Exercise 40.

42. Suppose that f is differentiable for $0 \leq x \leq 10$ and $f(2) = 17$, while $|f'(x)| \leq 3$ for $0 \leq x \leq 10$.
(a) What is the maximum possible value for $f(x)$ for any x in $[0, 10]$?
(b) What is the minimum possible value for $f(x)$ for any x in $[0, 10]$?

43. Maureen is at the edge of a circular lake of radius a mi and wishes to get to the point directly across the lake. She has a boat that she can row at 4 mph, or she can jog along the shore at 8 mph, or she can row to some point and jog the rest of the way. How should she proceed to get across in the shortest time?

44. Town A is located 2 mi back from the bank of a straight river. Town B is located 3 mi from the bank of the same side at a point 15 mi farther downstream. The towns agree to build a pumping station on the bank of the river to supply water to both towns. How far downstream from the point on the bank closest to town A should they build the station to minimize the total length of the water mains to the two towns? (Solve this problem first by using calculus and then by "moving town A across the river" and using geometry. What have you learned?)

45. Show that there exists a polynomial function $f(x)$ of degree 5 such that Newton's method to solve $f(x) = 0$ starting with $x = 0$ yields the repetitive sequence 0, 1, 2, 0, 1, 2, 0, 1, 2, ...

46. Give an example of a function f and a number a_1 such that an attempt to solve $f(x) = 0$ using Newton's method and starting with a_1 leads to oscillation between positive and negative numbers of ever-increasing magnitude.

THE INTEGRAL

5

The first three sections of this chapter present the main ideas of integral calculus for functions of one variable, culminating in the fundamental theorem of calculus. Sections 5.4 through 5.6 are concerned with computational techniques and differential equations.

5.1 ESTIMATING AREA

In geometric terms, finding a derivative is equivalent to finding the slope of a line tangent to a curve. The main part of this section is devoted to explaining that integral calculus, viewed geometrically, is concerned with computing areas. As we started to study the derivative, we estimated the slope of a tangent line by computing m_{sec}, the slope of a secant line. Analogously, we will estimate areas by *Riemann sums*.

Summation Notation

In this chapter, we will often want to consider a sum of terms that are indexed by positive integers from 1 to n, such as

$$a_1 + a_2 + \cdots + a_n.$$

There is a very useful notation for writing such sums. The sum $a_1 + \cdots + a_n$ is denoted in *summation notation* by

$$\sum_{i=1}^{n} a_i,$$

which is read "the sum of the a_i as i runs from 1 to n." The Greek letter Σ (sigma) stands for *sum;* the value of i under the Σ is the **lower limit of the sum** and tells where the sum starts. The value above the Σ is the **upper limit of the sum** and tells where the sum stops. The letter i is the **summation index.** The choice of letter for summation index is of course not significant; the letters i, j, and k are commonly used. Thus

$$a_1 + \cdots + a_n = \sum_{i=1}^{n} a_i = \sum_{j=1}^{n} a_j = \sum_{k=1}^{n} a_k.$$

A typical sum that we will encounter in this chapter is

$$f(x_1) + f(x_2) + \cdots + f(x_n) = \sum_{i=1}^{n} f(x_i).$$

You will probably best grasp the use of the summation notation by studying some examples. In forming the sum, we simply replace the summation index successively by all integers from the lower limit to the upper limit and add the resulting quantities.

EXAMPLE 1   Write out $\sum_{i=1}^{2} a_i$, $\sum_{i=1}^{3} a_i$, and $\sum_{i=4}^{7} f(a_i)$.

Solution   We have

$$\sum_{i=1}^{2} a_i = a_1 + a_2,$$

$$\sum_{i=1}^{3} a_i = a_1 + a_2 + a_3,$$

$$\sum_{i=4}^{7} f(a_i) = f(a_4) + f(a_5) + f(a_6) + f(a_7). \qquad \blacksquare$$

EXAMPLE 2   Compute $\sum_{i=1}^{3} i^2$, $\sum_{j=2}^{4} (j-1)$, and $\sum_{k=0}^{2} (k^2+1)$.

Solution   We have

$$\sum_{i=1}^{3} i^2 = 1^2 + 2^2 + 3^2 = 14,$$

$$\sum_{j=2}^{4} (j-1) = (2-1) + (3-1) + (4-1) = 1 + 2 + 3 = 6,$$

$$\sum_{k=0}^{2} (k^2+1) = (0^2+1) + (1^2+1) + (2^2+1) = 1 + 2 + 5 = 8. \qquad \blacksquare$$

EXAMPLE 3   Show that $\sum_{i=1}^{n} (a_i + b_i) = \sum_{i=1}^{n} a_i + \sum_{i=1}^{n} b_i$.

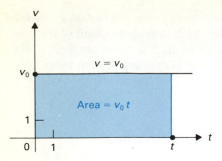

FIGURE 5.1   For constant velocity v_0, the distance traveled in time t is $v_0 t$, which numerically equals the area of the shaded rectangle.

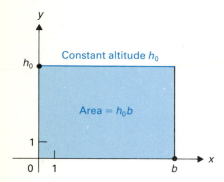

FIGURE 5.2   The area of a region of base width b and constant altitude h_0 is $A = h_0 b$.

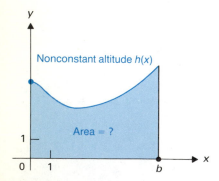

FIGURE 5.3   Find the area of the region of base width b and nonconstant altitude $h(x)$.

Solution   We have

$$\sum_{i=1}^{n} (a_i + b_i) = (a_1 + b_1) + \cdots + (a_n + b_n)$$

$$= (a_1 + \cdots + a_n) + (b_1 + \cdots + b_n)$$

$$= \sum_{i=1}^{n} a_i + \sum_{i=1}^{n} b_i. \qquad \blacksquare$$

The Integral Calculus Problem

Suppose you move in the positive direction on the x-axis, starting at $x = 0$ at time $t = 0$. If your velocity v remains constant and $t > 0$, then $v = x/t$, that is,

$$\text{Velocity} = \frac{\text{Distance}}{\text{Time}}. \qquad (1)$$

If your velocity does not remain constant, then this *quotient* gives the average velocity, and your velocity at any instant is found by differentiating:

$$v = \frac{dx}{dt}.$$

Calculus is concerned with the study of changing quantities, and this velocity model illustrates that differentiation is often appropriate when we would have computed a *quotient*, as in formula (1), if quantities had remained constant.

Integral calculus is frequently useful with changing quantities in cases where we would have computed a *product*, if quantities had remained constant. For your motion to the right on the x-axis with constant velocity, we see from formula (1) that the distance traveled is given for $t \geq 0$ by the *product*

$$\text{Distance} = \text{Velocity} \cdot \text{Time}. \qquad (2)$$

Figure 5.1 shows the graph where a constant velocity $v = v_0$ is plotted against time t. We see that the *distance $x = v_0 t$* traveled in time t is numerically equal to the area of the rectangle shaded in the figure.

The area A of a region of base width b and *constant* altitude h_0 is given by the *product $A = bh_0$*, as shown in Fig. 5.2. Figure 5.3 shows a region of base width b and altitude that *varies*, depending on where we are in the base. It is not so easy to find the area of the shaded region in Fig. 5.3. Integral calculus will solve the problem of finding the area of a region like that in Fig. 5.3.

Finding area is but one example in which a quantity (such as altitude) may vary and in which we must use integral calculus rather than just compute a single product. To put our study of integral calculus in perspective, we list a few other applications in which a product is appropriate if quantities are constant. We will discuss these applications later, using integral calculus, when one of the quantities varies as indicated.

1. The work done by a force acting on a body in the direction of motion is equal to the *product* of the magnitude of the force and the distance the body moves, but the magnitude of the force may vary as the body moves.

2. The total distance a body travels is the *product* of the magnitude of its velocity and the time it travels, but the magnitude of its velocity may vary with time.

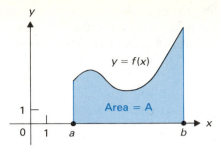

FIGURE 5.4 The region of area A under $y = f(x)$ over $[a, b]$.

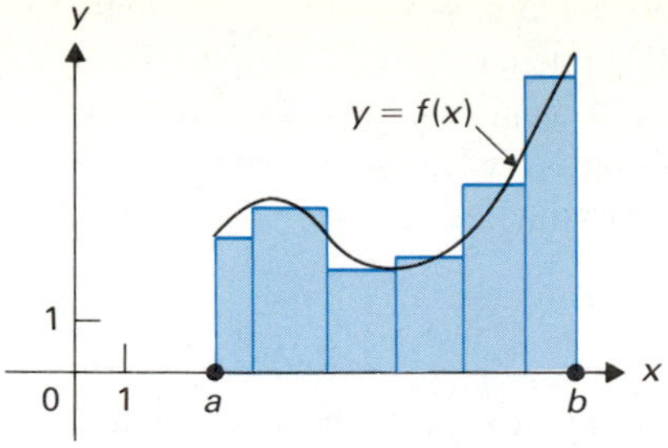

FIGURE 5.5 The sum of the areas of the shaded rectangles is an approximation to the area A in Fig. 5.4.

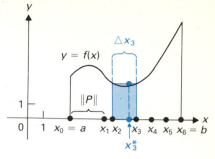

FIGURE 5.6 The contribution $f(x_3^*)\,\Delta x_3$ to a Riemann sum R_P for a partition P of $[a, b]$ into $n = 6$ subintervals.

3. The force exerted on an object by a fluid is the *product* of the pressure (force per unit area) exerted by the fluid and the area of the surface of the object, but the pressure may vary with the location on the surface. (The pressure near the bottom of a dam is greater than the pressure near the top.)
4. The volume of a solid is equal to the *product* of the area of the base and the altitude, but the altitude may vary with the position in the base.
5. The moment of a body about an axis is the *product* of its mass and its distance from the axis, but the distance from the axis may vary with the position in the body.

Summarizing, we will find that integral calculus is often appropriate when we would have computed a *product,* as in formula (2), if quantities had remained constant.

HISTORICAL NOTE

ARCHIMEDES (287–212 B.C.) made one of the earliest explicit calculations of an area bounded by curves. He showed that the area of a segment of a parabola is equal to $\frac{4}{3}$ of the area of the inscribed triangle. Using triangles inscribed in the parabola rather than rectangles to approximate this area, he added the area of the inscribed triangles using the rule for the sum of a geometric series.

Archimedes spent most of his life in Syracuse, Sicily. Generally regarded as the greatest mathematician of antiquity, he is also the earliest example of the dedicated scientist whose great powers of concentration enable him to solve difficult problems but who also neglects the more mundane aspects of life. Plutarch wrote that on many occasions Archimedes' concentration on mathematics ''made him forget his food and neglect his person, to that degree that when he was carried by absolute violence to bathe or have his body anointed, he used to trace geometrical figures in the ashes of the fire, and diagrams in the oil on his body.'' He died at the hands of a Roman soldier after the Roman siege of Syracuse because he refused to leave an interesting problem.

Estimating Areas Using Riemann Sums

With the preceding discussion for motivation, we turn to the estimation of the area A of a region like the one shaded in Fig. 5.3. Rather than always take the base of the region with its left-hand end at the origin as in Fig. 5.3, we will let the base be any interval $[a, b]$ on the x-axis.

Let f be a continuous function with domain containing $[a, b]$, and suppose that $f(x) \geq 0$ for all x in $[a, b]$. Since the graph of the *continuous* function f over $[a, b]$ is an unbroken curve, it would seem that the notion of the *area A of the region under the graph of f over* $[a, b]$, shown shaded in Fig. 5.4, should be meaningful.

We can find areas of triangles, rectangles, circles, trapezoids, and certain other familiar plane regions, but the region under the graph of f over $[a, b]$ may not be any of these. Since we can find areas of rectangles, we might approximate A by adding areas of rectangles with bases on the x-axis whose tops are horizontal line segments intersecting the graph of f, as shown in Fig. 5.5. The sum of the areas of these rectangles will give an estimate for A.

The bases of the rectangles in Fig. 5.5 partition $[a, b]$ into subintervals. As illustrated in Fig. 5.6 for $n = 6$, a **partition** P of $[a, b]$ into n subintervals is determined by selecting points x_i for $i = 0, 1, \ldots, n$ such that

$$x_0 = a < x_1 < x_2 < \cdots < x_{n-1} < x_n = b.$$

We let $\Delta x_i = x_i - x_{i-1}$, so that Δx_i is the length of the ith subinterval, as illustrated for $i = 3$ in Fig. 5.6. If x_i^* is a point of the ith subinterval, then $f(x_i^*)\,\Delta x_i$ is the area of the rectangle with the interval $[x_{i-1}, x_i]$ as base and with altitude $f(x_i^*)$, as illustrated for $i = 3$ in Fig. 5.6. The area of this rectangle approximates the area under the graph of f and over the interval $[x_{i-1}, x_i]$. The sum

$$R_P = f(x_1^*)\,\Delta x_1 + f(x_2^*)\,\Delta x_2 + \cdots + f(x_n^*)\,\Delta x_n \tag{3}$$

thus approximates the area A under the graph of f and over the interval $[a, b]$. We use the notation R_P for this sum in honor of the German mathematician Bernhard Riemann (1826–1866) who developed the general theory of such sums.

DEFINITION 5.1

Riemann sum

The sum R_P shown in Eq. (3) is a **Riemann sum** of order n for f over $[a, b]$.

For a partition P of $[a, b]$ into n subintervals, let the **norm** $\|P\|$ **of the partition** be the maximum of the lengths $\Delta x_1, \Delta x_2, \ldots, \Delta x_n$ of the subintervals. For example, in Fig. 5.6, we see that $\|P\| = \Delta x_1$. Our intuition tells us that in general, we obtain an accurate estimate for the area A under the graph of f and over $[a, b]$ provided that we make $\|P\|$ sufficiently small. (See Fig. 5.7.) Of course, making $\|P\|$ small requires that we make n large. Thus, for a Riemann sum R_P with $\|P\|$ sufficiently small, we expect to have

$$A \approx R_P = \sum_{i=1}^{n} f(x_i^*)\,\Delta x_i. \tag{4}$$

There are many ways to choose the points x_i^* in Eq. (4). One way is to select x_i^* so that $f(x_i^*)$ is the *maximum value* M_i of $f(x)$ for x in the ith subinterval (see Fig. 5.8). This gives a Riemann sum that we are sure is greater than or equal to A. Another way is to select x_i^* so that $f(x_i^*)$ is the *minimum value* m_i of $f(x)$ over the ith subinterval, which gives a Riemann sum that is less than or equal to A (see Fig. 5.9). Relations (5)–(9) summarize notation, terminology, and properties.

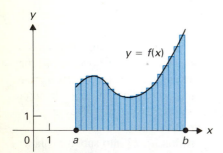

FIGURE 5.7 If $\|P\|$ is small, then R_P estimates the area A accurately.

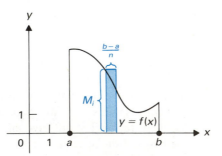

FIGURE 5.8 The maximum possible contribution to R_P from the ith subinterval.

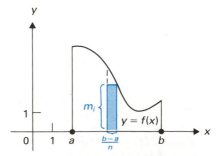

FIGURE 5.9 The minimum possible contribution to R_P from the ith subinterval.

HISTORICAL NOTE

GEORG FRIEDRICH BERNHARD RIEMANN (1826–1866) wrote a paper containing Riemann sums and the Riemann integral two years after qualifying to teach at the University of Göttingen. Such sums had in effect been considered from the very beginnings of calculus, but Riemann's precise definitions enabled him to extend the concept of an integral to various types of discontinuous functions.

Riemann had enrolled at Göttingen in 1846 to study theology but persuaded his father to allow him to devote himself to mathematics. Though he died of tuberculosis before his fortieth birthday, Riemann's many fundamental contributions to mathematics show the wisdom of his father's decision. Not only did he clarify the idea of a complex function and its associated Riemann surface, but he also introduced important ideas into the theory of Fourier series, the theory of surfaces, and the subject that was to become algebraic topology. Perhaps his most famous work relates to the zeta function

$$\zeta(s) = \sum_{n=1}^{\infty} n^{-s},$$

where s is a complex number and the sum is taken over all integers. Riemann noted that it was "probable" that all complex roots of ζ have the real part $\frac{1}{2}$. This "Riemann hypothesis" has inspired generations of mathematicians to important work, though to the present day it has been neither proved nor disproved.

Riemann Sums for a Partition P

$$R_{\max} = \text{Upper Riemann sum} = \sum_{i=1}^{n} M_i \, \Delta x_i \qquad (5)$$

$$R_{\min} = \text{Lower Riemann sum} = \sum_{i=1}^{n} m_i \, \Delta x_i \qquad (6)$$

$$R_P = \text{General Riemann sum} = \sum_{i=1}^{n} f(x_i^*) \, \Delta x_i \qquad (7)$$

$$R_{\min} \leq A \leq R_{\max} \qquad (8)$$

$$R_{\min} \leq R_P \leq R_{\max} \qquad (9)$$

Note that if $f(x)$ is *increasing* on $[a, b]$, then the maximum value M_i is attained at the right-hand end of the ith subinterval, while the minimum value m_i is attained at the left-hand end. If $f(x)$ is *decreasing* on $[a, b]$, then the reverse is true.

The remainder of this section is devoted to examples of Riemann sums. We use very small values of n in these examples and in most exercises, for we are just illustrating such sums rather than attempting to find really accurate estimates of areas. In our examples, and when writing a computer program to calculate Riemann sums, it is convenient to use **regular partitions** of $[a, b]$, where all subintervals have equal length. Since the length of the interval $[a, b]$ is $b - a$, we see that for a regular partition P of order n, we have

$$\Delta x_i = \frac{b - a}{n} \qquad \text{for} \qquad i = 1, 2, \dots, n. \qquad \textit{Regular partition}$$

A Riemann sum R_P then has the form

R_P for a Regular Partition P

$$R_P = \frac{b - a}{n} f(x_1^*) + \frac{b - a}{n} f(x_2^*) + \cdots + \frac{b - a}{n} f(x_n^*)$$

$$\qquad (10)$$

$$= \frac{b - a}{n} \sum_{i=1}^{n} f(x_i^*).$$

It is important to note that we can form Riemann sums even if $f(x)$ is negative for some x in $[a, b]$. The geometric interpretation of such a sum R_P for such a function is explained in the next section.

Of course, a Riemann sum R_P as in Eqs. (4) and (10) depends not only on the partition P but also on the function f, the interval $[a, b]$, and the choice of points x_i^* in $[x_{i-1}, x_i]$ for $i = 1, 2, \dots, n$. However, a notation that reflects all these dependencies would be unwieldy, so we simply use R_P.

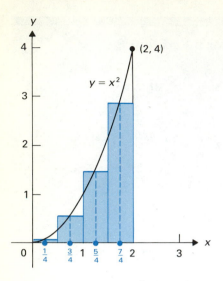

FIGURE 5.10 R_{mid} for x^2 over $[0, 2]$.

EXAMPLE 4   Use a regular partition with $n = 4$ to estimate the area A under the graph of $f(x) = x^2$ over $[0, 2]$ by a Riemann sum R_{mid}, taking for the points x_1^*, x_2^*, x_3^*, and x_4^* the midpoints of the first, second, third, and fourth intervals, respectively.

Solution   The estimation is illustrated in Fig. 5.10. We have

$$x_1 = \frac{1}{4}, \qquad x_2 = \frac{3}{4}, \qquad x_3 = \frac{5}{4}, \qquad x_4 = \frac{7}{4},$$

while

$$\frac{b - a}{n} = \frac{2 - 0}{4} = \frac{1}{2}.$$

The estimate R_{mid} for A given by Eq. (10) is then

$$\frac{1}{2}\left(\frac{1}{16} + \frac{9}{16} + \frac{25}{16} + \frac{49}{16}\right) = \frac{1}{2}\left(\frac{84}{16}\right) = \frac{21}{8} = 2.625.$$

We will see later that the exact value of A is $\frac{8}{3}$, so our estimate is not too bad.   ■

Midpoints of the subintervals are a natural choice for the points x_i^* in Eq. (10). We will continue to let

$$R_{\text{mid}} = \text{A midpoint Riemann sum}.$$

The exercises ask you to compute several more midpoint Riemann sums.

EXAMPLE 5   Using a regular partition, find the upper sum R_{max} and lower sum R_{min} with $n = 2$ for $f(x) = x^2$ over $[0, 1]$, estimating the area A under the graph.

Solution   Here $n = 2$, and

$$\frac{b - a}{n} = \frac{1 - 0}{2} = \frac{1}{2}.$$

It is easily seen from the *increasing* graph of x^2 in Fig. 5.11(a) that

$$m_1 = 0 \qquad \text{and} \qquad m_2 = \frac{1}{4},$$

while

$$M_1 = \frac{1}{4} \qquad \text{and} \qquad M_2 = 1,$$

as shown in Fig. 5.11(b). Thus

$$R_{\text{min}} = \left(\frac{1}{2}\right)\left(0 + \frac{1}{4}\right) \leq A \leq \left(\frac{1}{2}\right)\left(\frac{1}{4} + 1\right) = R_{\text{max}},$$

so

$$\frac{1}{8} \leq A \leq \frac{5}{8}.$$

Of course, these bounds on A are very crude.   ■

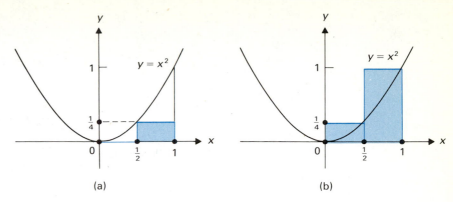

FIGURE 5.11 (a) The lower sum $R_{\min}$; (b) the upper sum $R_{\max}$.

EXAMPLE 6 Using a regular partition, find the upper sum $R_{\max}$ and the lower sum $R_{\min}$ with $n = 4$ for $1/x$ over $[1, 3]$, estimating the area A under the graph.

Solution Here $n = 4$, and

$$\frac{b - a}{n} = \frac{3 - 1}{4} = \frac{1}{2}.$$

It is easily seen from the *decreasing* graph of $1/x$ in Fig. 5.12(a) that

$$m_1 = \frac{1}{\frac{3}{2}} = \frac{2}{3}, \qquad m_2 = \frac{1}{2}, \qquad m_3 = \frac{1}{\frac{5}{2}} = \frac{2}{5}, \qquad m_4 = \frac{1}{3},$$

while

$$M_1 = \frac{1}{1} = 1, \qquad M_2 = \frac{1}{\frac{3}{2}} = \frac{2}{3}, \qquad M_3 = \frac{1}{2}, \qquad M_4 = \frac{1}{\frac{5}{2}} = \frac{2}{5},$$

as shown in Fig. 5.12(b). Then

$$R_{\min} = \frac{1}{2}\left(\frac{2}{3} + \frac{1}{2} + \frac{2}{5} + \frac{1}{3}\right) = \frac{1}{2}\left(1 + \frac{1}{2} + \frac{2}{5}\right) = \frac{1}{2} \cdot \frac{19}{10} = \frac{19}{20},$$

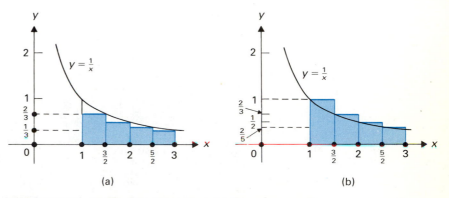

FIGURE 5.12 (a) The lower sum $R_{\min}$; (b) the upper sum $R_{\max}$.

and

$$R_{\max} = \frac{1}{2}\left(1 + \frac{2}{3} + \frac{1}{2} + \frac{2}{5}\right) = \frac{1}{2}\left(\frac{30 + 20 + 15 + 12}{30}\right) = \frac{77}{60}.$$

Therefore

$$\frac{19}{20} \le A \le \frac{77}{60}.$$

Now $\frac{19}{20} = 0.95$ and $\frac{77}{60} \approx 1.2833$. Section 5.3 will enable us to compute $A \approx 1.0986123$, which is indeed between the bounds we just found. ■

We can improve the bounds for A in Example 5 by taking a larger value of n.

EXAMPLE 7   Repeat Example 5 using $n = 4$.

Solution   As indicated in Fig. 5.13(a), we have

$$m_1 = 0, \qquad m_2 = \frac{1}{16}, \qquad m_3 = \frac{1}{4}, \qquad m_4 = \frac{9}{16},$$

while

$$M_1 = \frac{1}{16}, \qquad M_2 = \frac{1}{4}, \qquad M_3 = \frac{9}{16}, \qquad M_4 = 1,$$

as shown in Fig. 5.13(b). Of course,

$$\frac{b - a}{n} = \frac{1 - 0}{4} = \frac{1}{4}.$$

Thus

$$R_{\min} = \left(\frac{1}{4}\right)\left(0 + \frac{1}{16} + \frac{1}{4} + \frac{9}{16}\right)$$

$$\le A \le \left(\frac{1}{4}\right)\left(\frac{1}{16} + \frac{1}{4} + \frac{9}{16} + 1\right) = R_{\max}.$$

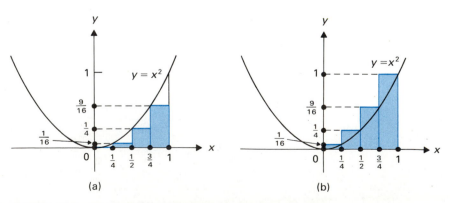

(a)   (b)

FIGURE 5.13   (a) The lower sum $R_{\min}$; (b) the upper sum $R_{\max}$.

The arithmetic works out to give

$$R_{\min} = \frac{7}{32} \le A \le \frac{15}{32} = R_{\max}.$$

This is an improvement over Example 5, which gave the bounds $\frac{4}{32} \le A \le \frac{20}{32}$.

We might estimate A by averaging $\frac{7}{32}$ and $\frac{15}{32}$, arriving at $\frac{11}{32}$. We will show in Section 5.3 that the exact value of A is $\frac{1}{3}$. Thus $\frac{11}{32}$ is not a bad estimate. ∎

As illustrated by Example 7, we expect a Riemann sum R_P for a regular partition to be a good estimate for A if n is large enough.

SUMMARY

1. To write out or compute a sum that is written symbolically using summation notation starting with

$$\sum_{i=k}^{n},$$

where k and n are integers and $k \le n$, replace the summation index i successively by all integers from the lower limit k to the upper limit n, and add the resulting quantities. For example,

$$\sum_{i=2}^{5} i^3 = 2^3 + 3^3 + 4^3 + 5^3.$$

2. Using Riemann sums as described in 3–8, we can estimate the area A over $[a, b]$ and under the graph of a positive continuous function f.

Let f be continuous on $[a, b]$, and let $[a, b]$ be partitioned into n subintervals. Let M_i be the maximum value and m_i the minimum value of $f(x)$ over the ith subinterval, and let x_i^* be any point of the ith subinterval of length Δx_i.

3. $R_{\max} = \displaystyle\sum_{i=1}^{n} M_i \, \Delta x_i$

4. $R_{\min} = \displaystyle\sum_{i=1}^{n} m_i \, \Delta x_i$

5. $R_P = \displaystyle\sum_{i=1}^{n} f(x_i^*) \, \Delta x_i$

6. $R_{\min} \le R_P \le R_{\max}$

7. For $f(x) > 0$, $R_{\min} \le A \le R_{\max}$

8. For a regular partition, $\Delta x_i = (b - a)/n$ for $i = 1, 2, \ldots, n$

EXERCISES 5.1

In Exercises 1–6, write out the sum.

1. $\displaystyle\sum_{i=0}^{3} a_i$ **2.** $\displaystyle\sum_{j=2}^{6} b_j^2$ **3.** $\displaystyle\sum_{i=1}^{4} a_{2i}$

4. $\displaystyle\sum_{k=4}^{6} (a_{2k} + b_k^2)$ **5.** $\displaystyle\sum_{i=1}^{5} c^i$ **6.** $\displaystyle\sum_{i=2}^{4} 2^{a_i}$

In Exercises 7–11, compute the sum.

7. $\displaystyle\sum_{i=0}^{3} (i + 1)^2$ **8.** $\displaystyle\sum_{j=2}^{4} 2^j$ **9.** $\displaystyle\sum_{i=1}^{3} (2i - 1)^2$

10. $\displaystyle\sum_{k=1}^{3} (2^k \cdot 3^{k-1})$ **11.** $\displaystyle\sum_{j=1}^{4} [(-1)^j \cdot j^3]$

In Exercises 12–17, express each sum in summation notation. (Many answers are possible.)

12. $a_1 b_1 + a_2 b_2 + a_3 b_3$

13. $a_1 b_2 + a_2 b_3 + a_3 b_4$

14. $a_1 + a_2^2 + a_3^3 + a_4^4$

15. $a_1^2 + a_2^3 + a_3^4$

16. $a_1 b_2^2 + a_2 b_4^2 + a_3 b_6^2$

17. $a_1^{b_3} + a_2^{b_6} + a_3^{b_9}$

18. Show, as in Example 3, that $\sum_{i=1}^{n} (c \cdot a_i) = c(\sum_{i=1}^{n} a_i)$.

19. Show, as in Example 3, that

$$\sum_{i=1}^{n} (a_i + b_i)^2 = \sum_{i=1}^{n} a_i^2 + 2\left(\sum_{i=1}^{n} a_i b_i\right) + \sum_{i=1}^{n} b_i^2.$$

20. Estimate the area under the graph of x^2 over $[0, 4]$ using the partition with $x_0 = 0$, $x_1 = 1$, $x_2 = 2$, and $x_3 = 4$ and the midpoint Riemann sum R_{mid}.

21. Estimate the area under the graph of $1/x$ over $[1, 5]$ by finding R_{mid} for the partition with $x_0 = 1$, $x_1 = \frac{3}{2}$, $x_2 = 2$, $x_3 = 3$, and $x_4 = 5$.

22. Estimate the area under the graph of x^2 over $[-1, 1]$ by finding R_{mid} for the regular partition with $n = 2$.

23. Find R_{max} and R_{min} for x^2 from 0 to 2 using the regular partition with $n = 2$. (The actual area under the graph over $[0, 2]$ is $\frac{8}{3}$.)

24. Find R_{max} and R_{min} for x^2 from 0 to 2 using the partition with $x_0 = 0$, $x_1 = \frac{1}{2}$, $x_2 = \frac{3}{4}$, $x_3 = 1$, $x_4 = \frac{3}{2}$, and $x_5 = 2$.

25. Estimate the area under the graph of $1/x$ over $[1, 2]$ by finding R_{max} and R_{min} for the regular partition with $n = 4$.

26. Find the exact value of the area under the graph of the constant function 4 over $[-2, 3]$.

27. Estimate the area under the graph of x^3 over $[1, 5]$ by finding R_{mid} for a regular partition with $n = 4$.

28. Estimate the area under the graph of $\sin x$ over $[0, \pi]$ by finding R_{max} and R_{min} for a regular partition with $n = 4$.

29. Estimate the area under the graph of $\sin^2 x$ over $[0, 2\pi]$ by finding R_{mid} for a regular partition with $n = 4$.

In Exercises 30–33, draw a sketch and use geometry to find the indicated area.

30. The area under the graph of $1 + x$ over $[0, 10]$

31. The area under the graph of $3 + (x/5)$ over $[0, 10]$

32. The area under the graph of $\sqrt{25 - x^2}$ over $[0, 5]$

33. The area under the graph of $4 + \sqrt{9 - x^2}$ over $[-3, 3]$ (See Fig. 5.14.)

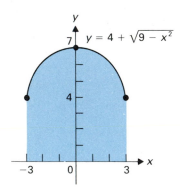

FIGURE 5.14

Estimation of Riemann sums is of practical importance. The estimates are more accurate for values of n larger than those in the preceding pencil-and-paper exercises. Use a calculator or computer in Exercises 34–44 to estimate the indicated area by computing R_{mid} using a regular partition and the given value of n.

34. The area under the graph of x^2 over $[0, 2]$, $n = 10$

35. The area under the graph of $\sqrt{x}$ over $[0, 2]$, $n = 10$

36. The area under the graph of $1/\sqrt{x}$ over $[1, 4]$, $n = 15$

37. The area under the graph of $\sin x$ over $[0, \pi]$, $n = 12$

38. The area under the graph of $1/x$ over $[1, 2]$, $n = 10$

39. The area under the graph of $\sin^2 x$ over $[0, 2\pi]$, $n = 10$

40. The area under the graph of $1/(1 + x^2)$ over $[0, 1]$, $n = 10$

41. The area under the graph of $\sqrt{25 - x^2}$ over $[0, 5]$, $n = 10$

42. The area under the graph of $\sin^2 x^2$ over $[0, 2\pi]$, $n = 12$

43. The area under the graph of $\ln x$ over $[1, 4]$, $n = 14$

44. The area under the graph of e^x over $[-1, 2]$, $n = 20$

EXPLORING CALCULUS

S: numerical integration graphic (option 1, midpoint Riemann sums)

5.2 THE DEFINITE INTEGRAL

The preceding section introduced Riemann sums for approximating the area A over an interval $[a, b]$ and under the graph of a positive continuous function f. Graphically, it seems that we obtain the best estimates for A by computing Riemann sums R_P where $\|P\|$ is small. This suggests that we might be able to find the exact value of A by taking the limit of Riemann sums R_P as $\|P\| \to 0$. Intuitively, the assertion $\lim_{\|P\| \to 0} R_P = A$ means that the Riemann sums R_P can be made as close to A as we please by taking $\|P\|$ sufficiently small but nonzero. (In the more precise language of Section 2.5, given any $\varepsilon > 0$, there must exist a number $\delta > 0$ such that $|R_P - A| < \varepsilon$ for all Riemann sums R_P where $\|P\| < \delta$.)

We have a lot of freedom in forming a Riemann sum

$$R_P = \sum_{i=1}^{n} f(x_i^*)\, \Delta x_i.$$

For example, x_i^* can be *any* point in the ith subinterval of the partition P of $[a, b]$. Assuming that the area A over $[a, b]$ and under the graph of a positive continuous function is a real number, we will give an intuitive argument that the limit of Riemann sums R_P as $\|P\| \to 0$ exists and is this area, no matter how the points x_i^* are chosen. Note that the requirement $\|P\| \to 0$ is not the same as $n \to \infty$. It is true that if $\|P\| \to 0$, then we must have $n \to \infty$. However, Fig. 5.15 indicates that we might have $n \to \infty$ without having $\|P\| \to 0$.

A careful treatment of these ideas is a topic for a course in advanced calculus. There the existence of the limit of the Riemann sums R_P as $\|P\| \to 0$ is demonstrated *without* assuming that A is a real number for all continuous functions f. Indeed, in advanced calculus the area is *defined* to be this limit.

The Definite Integral

Let f be continuous on $[a, b]$. We recall from the previous section the types of Riemann sums corresponding to one partition P of $[a, b]$ and relations that these sums satisfy:

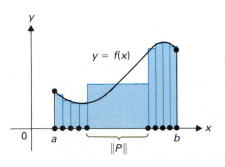

FIGURE 5.15 Riemann sums R_P where $\|P\|$ does not approach 0 may not approach A as $n \to \infty$.

For a Partition P

$$R_{\max} = \sum_{i=1}^{n} M_i\, \Delta x \qquad \textit{Upper Riemann sum}$$

$$R_{\min} = \sum_{i=1}^{n} m_i\, \Delta x_i \qquad \textit{Lower Riemann sum}$$

$$R_P = \sum_{i=1}^{n} f(x_i^*)\, \Delta x_i \qquad \textit{General Riemann sum}$$

$$R_{\min} \le A \le R_{\max} \qquad (1)$$
$$R_{\min} \le R_P \le R_{\max} \qquad (2)$$

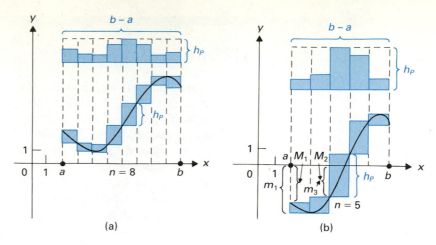

FIGURE 5.16 (a) $R_{\max} - R_{\min}$ for $f(x)$ nonnegative in $[a, b]$; (b) $R_{\max} - R_{\min}$ for $f(x)$ sometimes negative in $[a, b]$.

The error in the estimates $R_{\max}$ and $R_{\min}$ for the area A under the graph of f over $[a, b]$ can be at most $R_{\max} - R_{\min}$. Geometrically, $R_{\max} - R_{\min}$ corresponds to the sum of the areas of the small rectangles along the graph shown shaded in Fig. 5.16, where we have used regular partitions. (Figure 5.16a illustrates the case where $f(x) \geq 0$ for all x in $[a, b]$, and Fig. 5.16b illustrates the case where $f(x)$ is negative for some x in $[a, b]$.) Let h_P be the maximum height of these small rectangles; that is, h_P is the maximum of $M_i - m_i$. By lining up the shaded rectangles in a row, as shown at the top of Fig. 5.16, we find that the sum of the areas of the small rectangles is less than $h_P(b - a)$, so

$$R_{\max} - R_{\min} \leq h_P(b - a). \tag{3}$$

If f is a continuous function, then $f(x_i^*)$ is close to $f(x_i)$ for x_i^* sufficiently close to x_i. It is reasonable to expect that as $\|P\| \to 0$, so that the horizontal dimensions of the small rectangles shaded in Fig. 5.16 become very small, the vertical dimensions become small also. That is, we expect h_P to be close to zero for $\|P\|$ sufficiently small, so that

$$\lim_{\|P\| \to 0} h_P = 0. \tag{4}$$

In view of Eq. (3), $\lim_{\|P\| \to 0} h_P = 0$ implies that $R_{\max} - R_{\min}$ is close to zero for sufficiently small $\|P\|$. By Eq. (1), for $f(x) \geq 0$, both $R_{\max}$ and $R_{\min}$ therefore approach A as $\|P\| \to 0$. Equation (3) then shows that R_P, which is trapped between $R_{\max}$ and $R_{\min}$, must also approach A as $\|P\| \to 0$. It can be shown that the following theorem holds, even if $f(x)$ is not always nonnegative.

THEOREM 5.1   Common limit of Riemann sums

If f is continuous on $[a, b]$, then $\lim_{\|P\| \to 0} R_P$ exists. □

DEFINITION 5.2

Definite integral

Let f be continuous on $[a, b]$. The limit of Riemann sums R_P for f over $[a, b]$ as $\|P\| \to 0$ is the **definite integral of f over** $[a, b]$, written $\int_a^b f(x) \, dx$, so that

$$\int_a^b f(x) \, dx = \lim_{\|P\| \to 0} R_P. \qquad (5)$$

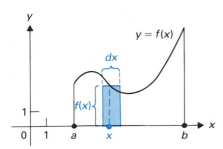

FIGURE 5.17 Heuristic meaning of $f(x) \, dx$ in the notation $\int_a^b f(x) \, dx$.

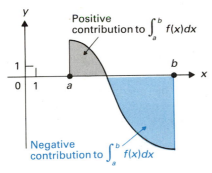

FIGURE 5.18 $\int_a^b f(x) \, dx = $ (Area of black region) − (Area of blue region).

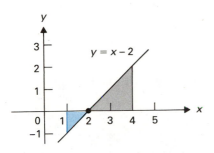

FIGURE 5.19

$$\int_1^4 (x - 2) \, dx = 2 - \frac{1}{2} = \frac{3}{2}.$$

There is a very useful heuristic interpretation of the notation $\int_a^b f(x) \, dx$ for the definite integral. We think of the *integral sign* $\int$ as an elongated letter S, standing for "sum." If we take the standard analysis interpretation $dx = \Delta x$ for the differential dx as explained in Section 3.9, then $f(x) \, dx$ is the area of the rectangle shown in Fig. 5.17. The notation for the integral thus suggests summing the area of such rectangles from $x = a$ to $x = b$; that is, the notation suggests a Riemann sum. This interpretation of the notation is very suggestive in applications and is widely used in the sciences that employ calculus.

We mention that $\int_a^b f(x) \, dx$ can also be defined if f is not continuous at some finite number of points in $[a, b]$ but satisfies $m \le f(x) \le M$ for x in $[a, b]$ and some numbers m and M. This is important for some applications. For example, the magnitude of the current flowing in an electric circuit can change abruptly several times over an interval of time.

We have drawn the graph of f in most of the figures for the case in which $f(x) \ge 0$ for x in $[a, b]$. If $f(x) < 0$ for x in the ith subinterval of a partition P of $[a, b]$, then both m_i and M_i are negative, so the contributions $m_i \, \Delta x_i$ to $R_{\min}$ and $M_i \, \Delta x_i$ to $R_{\max}$ are both negative. For a function f where $f(x)$ is sometimes negative and sometimes positive, $\int_a^b f(x) \, dx$ can be interpreted geometrically as the total area given by the portions of the graph above the x-axis minus the total area given by the portions of the graph below the x-axis. This is illustrated in Fig. 5.18.

EXAMPLE 1 Use geometry to find $\int_1^4 (x - 2) \, dx$.

Solution The graph $y = x - 2$ is shown in Fig. 5.19. Since the small shaded triangle is below the x-axis and the large one is above,

$$\int_1^4 (x - 2) \, dx = (\text{Area of large triangle}) - (\text{Area of small triangle})$$

$$= \frac{1}{2} \cdot 2 \cdot 2 - \frac{1}{2} \cdot 1 \cdot 1 = 2 - \frac{1}{2} = \frac{3}{2}. \qquad \blacksquare$$

EXAMPLE 2 Find $\int_{-5}^5 \sqrt{25 - x^2} \, dx$ graphically.

Solution Figure 5.20 indicates that the value of the integral is the area of the top half of the disk with center $(0, 0)$ and radius 5. Thus

$$\int_{-5}^5 \sqrt{25 - x^2} \, dx = \frac{1}{2} \pi \cdot 5^2 = \frac{25}{2} \pi. \qquad \blacksquare$$

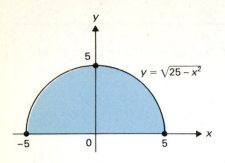

FIGURE 5.20 $\int_{-5}^{5} \sqrt{25 - x^2} \, dx =$
Area of the shaded half-disk.

Properties of the Definite Integral

Let $f(x)$ and $g(x)$ both be continuous in $[a, b]$. We state three properties of the definite integral in Eqs. (6), (7), and (8). Properties (6) and (7) parallel properties of the derivative. When the properties are stated in words, the parallel is striking. We can take our choice of "derivative" or "integral" in the following:

The $\begin{Bmatrix} derivative \\ integral \end{Bmatrix}$ of a sum is the sum of the $\begin{Bmatrix} derivatives \\ integrals \end{Bmatrix}$.

The $\begin{Bmatrix} derivative \\ integral \end{Bmatrix}$ of a constant times a function is the constant times

the $\begin{Bmatrix} derivative \\ integral \end{Bmatrix}$ of the function.

THEOREM 5.2   Properties of the integral

Three properties of the definite integral are

$$\int_a^b (f(x) + g(x)) \, dx = \int_a^b f(x) \, dx + \int_a^b g(x) \, dx, \tag{6}$$

$$\int_a^b c \cdot f(x) \, dx = c \cdot \int_a^b f(x) \, dx \qquad \text{for any constant } c, \tag{7}$$

$$\int_a^b f(x) \, dx = \int_a^h f(x) \, dx + \int_h^b f(x) \, dx \qquad \text{for any } h \text{ in } [a, b]. \tag{8}$$

$\square$

All these properties are suggested if we think of the interpretation in terms of areas. Figure 5.21 illustrates Eq. (8); surely the area under $f(x)$ from a to b is the area from a to h plus the area from h to b. Using the Riemann sums $R_P(f)$, $R_P(g)$, and $R_P(f + g)$ for f, g, and $f + g$, we have

$$R_P = \sum_{i=1}^{n} (f(x_i^*) + g(x_i^*)) \, \Delta x_i$$

$$= \sum_{i=1}^{n} f(x_i^*) \, \Delta x_i + \sum_{i=1}^{n} g(x_i^*) \, \Delta x_i = R_P(f) + R_P(g).$$

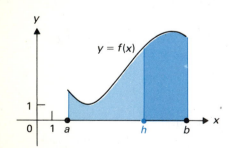

FIGURE 5.21

$$\int_a^b f(x) \, dx = \int_a^h f(x) \, dx + \int_h^b f(x) \, dx$$

Therefore,

$$\int_a^b (f(x) + g(x)) \, dx = \lim_{\|P\| \to 0} R_P(f + g) = \lim_{\|P\| \to 0} (R_P(f) + R_P(g))$$

$$= \lim_{\|P\| \to 0} R_P(f) + \lim_{\|P\| \to 0} R_P(g)$$

$$= \int_a^b f(x) \, dx + \int_a^b g(x) \, dx.$$

This establishes property (6). Property (7) is established in a similar way.

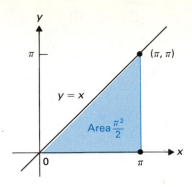

FIGURE 5.22

$$\int_0^\pi x \, dx = \left(\frac{1}{2}\pi\right)\pi = \frac{\pi^2}{2}$$

For another property, note that if $b = a$, then

$$\int_a^a f(x) \, dx = 0, \tag{9}$$

since the lengths of all subintervals in any partition are zero.

We would like $\int_a^b f(x) \, dx = \int_a^h f(x) \, dx + \int_h^b f(x) \, dx$ to hold whether or not h is between a and b. (Naturally we require that a, b, and h all lie in some interval where $f(x)$ is continuous.) Then if we set $b = a$, we would have

$$0 = \int_a^a f(x) \, dx = \int_a^h f(x) \, dx + \int_h^a f(x) \, dx,$$

which leads to

$$\int_h^a f(x) \, dx = -\int_a^h f(x) \, dx. \tag{10}$$

We are motivated by Eq. (10) to define $\int_b^a f(x) \, dx$ if $b > a$ by

$$\int_b^a f(x) \, dx = -\int_a^b f(x) \, dx. \tag{11}$$

EXAMPLE 3   Given that $\int_0^\pi \sin^2 x \, dx = \pi/2$, find $\int_0^\pi (x + \sin^2 x) \, dx$.

Solution   From property (6), we have

$$\int_0^\pi (x + \sin^2 x) \, dx = \int_0^\pi x \, dx + \int_0^\pi \sin^2 x \, dx.$$

From Fig. 5.22, we see graphically that $\int_0^\pi x \, dx = \pi^2/2$. Thus

$$\int_0^\pi (x + \sin^2 x) \, dx = \frac{\pi^2}{2} + \frac{\pi}{2} = \frac{\pi}{2}(\pi + 1). \qquad \blacksquare$$

EXAMPLE 4   If

$$\int_1^4 f(t) \, dt = -5 \quad \text{and} \quad \int_1^2 2 \cdot f(t) \, dt = -1,$$

compute $\int_2^4 f(t) \, dt$.

Solution   By property (8), we have

$$\int_1^4 f(t) \, dt = \int_1^2 f(t) \, dt + \int_2^4 f(t) \, dt.$$

From property (7), we see that $\int_1^2 2 \cdot f(t) \, dt = 2 \int_1^2 f(t) \, dt = -1$, so $\int_1^2 f(t) \, dt = -\frac{1}{2}$. Thus

$$-5 = -\frac{1}{2} + \int_2^4 f(t) \, dt,$$

so $\int_2^4 f(t) \, dt = -\frac{9}{2}$. $\blacksquare$

Computing Integrals by Limits of Sums (Optional)

The formulas

$$\sum_{i=1}^{n} i = \frac{n(n+1)}{2},$$ (12)

$$\sum_{i=1}^{n} i^2 = \frac{n(n+1)(2n+1)}{6},$$ (13)

and

$$\sum_{i=1}^{n} i^3 = \left[\frac{n(n+1)}{2} \right]^2$$ (14)

can be proved using mathematical induction. These formulas can be used to compute definite integrals of linear, quadratic, and cubic polynomial functions, as shown in Examples 5 and 6.

EXAMPLE 5   Find the integral $\int_0^4 (5x - 3)\, dx$ using Eq. (12).

Solution   Consider just regular partitions P of $[0, 4]$ with n subintervals of equal length $4/n$. If we take x_i as the right-hand endpoint of the ith subinterval, then

$$x_i = i \cdot \frac{4}{n} = \frac{4i}{n}$$

and

$$R_P = \frac{4}{n} \sum_{i=1}^{n} \left(5 \cdot \frac{4i}{n} - 3 \right)$$

$$= \frac{4}{n} \left[\frac{20}{n} \left(\sum_{i=1}^{n} i \right) - \sum_{i=1}^{n} 3 \right] = \frac{4}{n} \left[\frac{20}{n} \cdot \frac{n(n+1)}{2} - 3n \right]$$

$$= \frac{80(n+1)}{2n} - 12 = \frac{40(n+1)}{n} - 12.$$

Therefore

$$\int_0^4 (5x - 3)\, dx = \lim_{\|P\| \to 0} R_P = \lim_{n \to \infty} \left[\frac{40(n+1)}{n} \right] - 12$$

$$= 40 - 12 = 28. \qquad \blacksquare$$

EXAMPLE 6   Use Eqs. (12) and (13) to find $\int_2^4 (x^2 + 2x)\, dx$.

Solution   Consider just regular partitions P of $[2, 4]$ with n subintervals of equal length $2/n$, and let x_i be the right-hand endpoint of the ith subinterval. Then

$$x_i = 2 + i \cdot \frac{2}{n} = 2 + \frac{2i}{n},$$

and

$$R_P = \frac{2}{n} \cdot \sum_{i=1}^{n} \left[\left(2 + \frac{2i}{n} \right)^2 + 2\left(2 + \frac{2i}{n} \right) \right]$$

$$= \frac{2}{n} \cdot \sum_{i=1}^{n} \left(4 + \frac{8i}{n} + \frac{4i^2}{n^2} + 4 + \frac{4i}{n} \right)$$

$$= \frac{2}{n} \left(\sum_{i=1}^{n} 8 + \frac{12}{n} \cdot \sum_{i=1}^{n} i + \frac{4}{n^2} \cdot \sum_{i=1}^{n} i^2 \right)$$

$$= \frac{2}{n} \left[8n + \frac{12}{n} \cdot \frac{n(n+1)}{2} + \frac{4}{n^2} \cdot \frac{n(n+1)(2n+1)}{6} \right]$$

$$= 16 + \frac{12n(n+1)}{n^2} + \frac{4n(n+1)(2n+1)}{3n^3}.$$

Taking the limit as $\|P\| \to 0$ so that $n \to \infty$, we obtain

$$\int_{2}^{4} (x^2 + 2x) \, dx = \lim_{\|P\| \to 0} R_P = 16 + 12 + \frac{8}{3} = \frac{92}{3}. \qquad \blacksquare$$

SUMMARY

Let $f(x)$ and $g(x)$ be continuous in $[a, b]$.

1. $\displaystyle\int_{a}^{b} f(x) \, dx = \lim_{\|P\| \to 0} R_P$

2. $\displaystyle\int_{a}^{b} (f(x) + g(x)) \, dx = \int_{a}^{b} f(x) \, dx + \int_{a}^{b} g(x) \, dx$

3. $\displaystyle\int_{a}^{b} c \cdot f(x) \, dx = c \cdot \int_{a}^{b} f(x) \, dx \qquad$ for any constant c

4. $\displaystyle\int_{a}^{b} f(x) \, dx = \int_{a}^{h} f(x) \, dx \int_{h}^{b} f(x) \, dx \qquad$ for any h in $[a, b]$

5. $\displaystyle\int_{b}^{a} f(x) \, dx = -\int_{a}^{b} f(x) \, dx$

EXERCISES 5.2

1. Let $f(x) = 1 - x$ and consider regular partitions P of $[0, 2]$ with n subintervals of equal length.
   (a) Sketch the graph of f over this interval.
   (b) Find R_{max} for $n = 2$. (Note that $f(x)$ is negative for some values x in $[0, 2]$.)
   (c) Find R_{min} for $n = 2$.
   (d) What common value do both R_{max} and R_{min} approach as $n \to \infty$ so that $\|P\| \to 0$?

2. Consider regular partitions P of $[-1, 1]$ with n subintervals of equal length.
   (a) Find the upper sum R_{max} and the lower sum R_{min} for $n = 4$ and the function x^3.
   (b) From the graph of x^3, what common number do both R_{max} and R_{min} approach as $n \to \infty$?

In Exercises 3–12, sketch a region whose area is given by the integral. Determine the value of the integral by finding the area of the region.

3. $\displaystyle\int_0^2 x \, dx$

4. $\displaystyle\int_{-1}^2 x \, dx$

5. $\displaystyle\int_1^3 (2x + 3) \, dx$

6. $\displaystyle\int_{-3}^4 (2x - 1) \, dx$

7. $\displaystyle\int_{-1}^1 (x + 1) \, dx$

8. $\displaystyle\int_{-3}^1 (x - 1) \, dx$

9. $\displaystyle\int_{-3}^3 \sqrt{9 - x^2} \, dx$

10. $\displaystyle\int_0^4 \sqrt{16 - x^2} \, dx$

11. $\displaystyle\int_0^4 (3 + \sqrt{16 - x^2}) \, dx$

12. $\displaystyle\int_{-3}^3 (5 - \sqrt{9 - x^2}) \, dx$

13. Consider the regular partition P of $[0, 2]$ with $n = 2$ subintervals of equal length. Sketch the graph of a function f over the interval for which R_{max} is much closer to $\int_0^2 f(x) \, dx$ than the average of R_{max} and R_{min}.

14. Consider regular partitions P of $[0, 6]$ with n subintervals of equal length. Sketch the graph of a function f over $[0, 6]$ for which R_{max} for $n = 3$ is greater than R_{max} for $n = 2$. (This shows that while $\lim_{\|P\| \to 0} R_{max} = \int_0^6 f(x) \, dx$, it need not be true that $(R_{max}$ for $n + 1) \leq (R_{max}$ for $n)$.)

In Exercises 15–30, find the exact value of the integral using geometric arguments, properties of the integral, and the fact (which we will soon show) that $\int_0^\pi \sin x \, dx = 2$.

15. $\displaystyle\int_0^{\pi/2} \sin t \, dt$

16. $\displaystyle\int_{-\pi}^\pi \sin x \, dx$

17. $\displaystyle\int_0^{\pi/2} \cos x \, dx$

18. $\displaystyle\int_{-\pi/2}^{\pi/2} \cos u \, du$

19. $\displaystyle\int_{-2}^2 (y^3 - 3y) \, dy$

20. $\displaystyle\int_{-2}^2 (2x^3 + 4) \, dx$

21. $\displaystyle\int_{-3}^3 (v^5 - 2v^3 - 1) \, dv$

22. $\displaystyle\int_0^\pi \cos t \, dt$

23. $\displaystyle\int_0^{5\pi} \sin x \, dx$

24. $\displaystyle\int_0^{2\pi} |\sin x| \, dx$

25. $\displaystyle\int_{-\pi}^\pi |\cos s| \, ds$

26. $\displaystyle\int_{-1}^1 |2x| \, dx$

27. $\displaystyle\int_0^\pi (2 \sin x - 3 \cos x) \, dx$

28. $\displaystyle\int_0^{2\pi} (3|\sin r| + 2|\cos r|) \, dr$

29. $\displaystyle\int_0^{\pi/2} (2x - 3 \cos x) \, dx$

30. $\displaystyle\int_0^\pi (4x + 1 - 2 \sin x) \, dx$

In Exercises 31–40, assume that f and g are continuous for all x. Find the indicated integral from the given data.

31. $\displaystyle\int_0^1 (f(u) - g(u)) \, du = 3, \quad \int_0^1 g(u) \, du = -1,$

$\displaystyle\int_0^1 3 \cdot f(u) \, du = \underline{\quad\quad}$

32. $\displaystyle\int_0^1 (f(x) + 2 \cdot g(x)) \, dx = 8,$

$\displaystyle\int_0^1 (2 \cdot f(x) - g(x)) \, dx = -2, \quad \int_0^1 f(x) \, dx = \underline{\quad\quad}$

33. $\displaystyle\int_2^5 (f(x) - g(x)) \, dx = 1, \quad \int_2^5 2 \cdot g(x) \, dx = 4,$

$\displaystyle\int_2^5 f(x) \, dx = \underline{\quad\quad}$

34. $\displaystyle\int_1^2 f(s) \, ds = 4, \quad \int_2^4 3 \cdot f(s) \, ds = 5,$

$\displaystyle\int_1^4 2 \cdot f(s) \, ds = \underline{\quad\quad}$

35. $\displaystyle\int_0^2 g(t) \, dt = 3, \quad \int_1^2 2 \cdot g(t) \, dt = 8, \quad \int_0^1 g(t) \, dt = \underline{\quad\quad}$

36. $\displaystyle\int_0^4 g(x) \, dx = -3, \quad \int_4^2 3 \cdot g(x) \, dx = -2,$

$\displaystyle\int_0^2 2 \cdot g(x) \, dx = \underline{\quad\quad}$

37. $\displaystyle\int_1^{10} f(x) \, dx = 7, \quad \int_5^{10} 3 \cdot f(x) \, dx = 6,$

$\displaystyle\int_5^1 2 \cdot f(x) \, dx = \underline{\quad\quad}$

38. $\displaystyle\int_1^4 (f(x) - g(x)) \, dx = 10, \quad \int_4^1 (f(x) + g(x)) \, dx = 3,$

$\displaystyle\int_0^4 g(x) \, dx = 5, \quad \int_0^1 g(x) \, dx = \underline{\quad\quad}$

39. $\displaystyle\int_0^{10} f(v) \, dv = 5, \quad \int_0^6 f(v) \, dv = 3, \quad \int_{10}^4 2 \cdot f(v) \, dv = 6,$

$\displaystyle\int_6^4 f(v) \, dv = \underline{\quad\quad}$

40. $\displaystyle\int_1^8 g(w) \, dw = 4, \quad \int_5^1 2 \cdot g(w) \, dw = 6, \quad \int_2^8 g(w) \, dw = 5,$

$\displaystyle\int_2^5 g(w) \, dw = \underline{\quad\quad}$

In optional Exercises 41–48, use Eqs. (12), (13), and (14) and Definition 5.2 to find the value of the definite integral.

41. $\displaystyle\int_0^3 (x - 2)\, dx$

42. $\displaystyle\int_0^2 (4x^2 + 5)\, dx$

43. $\displaystyle\int_3^5 (7 - 2y)\, dy$

44. $\displaystyle\int_2^5 (x^2 - 4x + 2)\, dx$

45. $\displaystyle\int_0^1 x^3\, dx$

46. $\displaystyle\int_0^2 (2r^3 - r^2)\, dr$

47. $\displaystyle\int_1^2 (u - u^3)\, du$

48. $\displaystyle\int_1^3 (4x^3 - 3x^2 + 2x - 10)\, dx$

In Exercises 49–52, use a calculator or computer to estimate the given integral using R_P for a regular partition and the indicated value of n, where x_i^* is the midpoint of the ith subinterval.

49. $\displaystyle\int_1^2 x^x\, dx$, with $n = 8$

50. $\displaystyle\int_1^3 2^{\sin v}\, dv$, where $n = 10$

51. $\displaystyle\int_0^2 \cos t^2\, dt$, with $n = 16$

52. $\displaystyle\int_1^5 x^x\, dx$, with $n = 20$

> **EXPLORING CALCULUS**
>
> **S:** numerical integration graphic (option 1, midpoint Riemann sums)

5.3 THE FUNDAMENTAL THEOREM OF CALCULUS

Our development of the definite integral culminates in this section with the *fundamental theorem of calculus*. The theorem is as important as its name suggests, for it provides us with a technique for finding the exact value of many definite integrals. The theorem also shows that while not every continuous function is differentiable (recall that $|x|$ is not differentiable at $x = 0$), every function continuous on an interval has an antiderivative there.

Finding the Exact Value of $\int_a^b f(x)\, dx$

Let $f(x)$ be continuous in $[a, b]$. In the preceding section, we defined $\int_a^b f(x)\, dx$, and we have had some practice in estimating the integral using Riemann sums. The definite integral has a great variety of important applications, so it is highly desirable to have an easy way to compute $\int_a^b f(x)\, dx$. It turns out that the *exact* value of the integral can be found easily if we can find an *antiderivative F(x)* of $f(x)$. On page 248 of Section 5.1 we did a quotient-versus-product analysis of the derivative versus the integral. Since division and multiplication are opposite, or inverse, operations, it is not surprising that the opposite of differentiation, that is, antidifferentiation, should be involved in finding the integral.

We first state without proof a rule for computing $\int_a^b f(x)\, dx$ exactly for many functions f. Then we illustrate the rule. In the next subsection we state and prove the fundamental theorem of calculus, which includes this rule. Remember that we always assume that f is continuous in $[a, b]$.

Rule for Computing $\displaystyle\int_a^b f(x)\,dx$

Find $F(x)$ such that $F'(x) = f(x)$. Then

$$\int_a^b f(x)\,dx = F(b) - F(a). \tag{1}$$

EXAMPLE 1 Find the exact value of $\int_0^1 x^2 dx$, which we estimated in Section 5.1.

Solution An antiderivative of the function x^2 is $F(x) = x^3/3$. Thus, by rule (1),

$$\int_0^1 x^2\,dx = F(1) - F(0) = \frac{1}{3} - \frac{0}{3} = \frac{1}{3}.$$

After all our work trying to get a good estimate for this definite integral in Section 5.1, we should appreciate this easy computation. ∎

It is customary to denote $F(b) - F(a)$ by $F(x)]_a^b$. The *upper limit of integration* is b and the *lower limit of integration* is a. For example, we usually see $\int_0^1 x^2\,dx$ computed as follows:

$$\int_0^1 x^2\,dx = \frac{x^3}{3}\bigg]_0^1 = \frac{1}{3} - \frac{0}{3} = \frac{1}{3}.$$

EXAMPLE 2 Find $\int_1^3 (1/x)\,dx$, which we estimated as falling between 0.95 and 1.2833 in Example 6 of Section 5.1.

Solution In Example 8 of Section 4.8, we showed that the general antiderivative of $1/x$ is $\ln|x| + C$. Rule (1) then tells us that

$$\int_1^3 \frac{1}{x}\,dx = \ln|x|\bigg]_1^3 = (\ln 3) - (\ln 1)$$
$$= (\ln 3) - 0 = \ln 3.$$

Our calculator shows that $\ln 3 \approx 1.0986123$. ∎

EXAMPLE 3 Find the area of the region under one ''arch'' of the graph of $y = \sin x$, as shown in Fig. 5.23.

Solution Note that $-\cos x$ is an antiderivative of $\sin x$. The area is given by

$$\int_0^\pi \sin x\,dx = -\cos x\bigg]_0^\pi$$
$$= -\cos \pi - (-\cos 0)$$
$$= -(-1) - (-1) = 1 + 1 = 2. \quad ∎$$

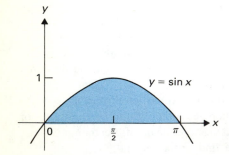

FIGURE 5.23 Region under one ''arch'' of $y = \sin x$.

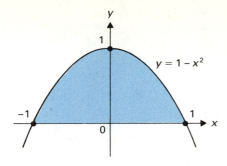

FIGURE 5.24 Region bounded by $y = 1 - x^2$ and the x-axis.

EXAMPLE 4   Sketch the region bounded by the graph of the polynomial function $1 - x^2$ and by the x-axis, and find the area of the region.

Solution   The function $1 - x^2$ has a maximum at $x = 0$ and has decreasing values as x gets farther from zero. The graph crosses the x-axis when $1 - x^2 = 0$, that is, when $x = \pm 1$. The graph is shown in Fig. 5.24, where we have shaded the region whose area we wish to find. The desired area is $\int_{-1}^{1} (1 - x^2)\, dx$ square units, and

$$\int_{-1}^{1} (1 - x^2)\, dx = \left(x - \frac{x^3}{3} \right) \Bigg]_{-1}^{1}$$

$$= \left(1 - \frac{1}{3} \right) - \left(-1 - \frac{-1}{3} \right) = \frac{2}{3} - \left(-\frac{2}{3} \right) = \frac{4}{3}. \qquad \blacksquare$$

EXAMPLE 5   Find $\int_{0}^{2} \sqrt{36 - 9x^2}\, dx$.

Solution   We have not yet learned how to find an antiderivative of $f(x) = \sqrt{36 - 9x^2}$. You might experiment for a bit, but unless you know more calculus than we have covered earlier in this text, you will not find an antiderivative of $\sqrt{36 - 9x^2}$.

Note that $\sqrt{36 - 9x^2} = 3\sqrt{4 - x^2}$, so our integral is equal to

$$3 \int_{0}^{2} \sqrt{4 - x^2}\, dx.$$

This is an integral we can evaluate geometrically, as discussed in the preceding section. As Fig. 5.25 indicates, $\int_{0}^{2} \sqrt{4 - x^2}\, dx$ is the area of a quarter-disk of radius 2. Thus

$$\int_{0}^{2} \sqrt{36 - 9x^2}\, dx = 3 \int_{0}^{2} \sqrt{4 - x^2}\, dx = 3 \cdot \frac{\pi \cdot 2^2}{4} = 3\pi. \qquad \blacksquare$$

It may seem now that Sections 5.1 and 5.2 on estimating $\int_{a}^{b} f(x)\, dx$ were a waste of time, in view of the powerful rule (1). However, a clear understanding of Sections 5.1 and 5.2 is needed to recognize applied problems that we can solve using an integral. Also, finding an antiderivative $F(x)$ of $f(x)$ may not be an easy matter, even if $f(x)$ is a fairly simple function. For example, it can be shown that while an antiderivative of $\sin x^2$ exists, it is impossible to express it as an *elementary function* (a function like $\ln(\cot x) + \sqrt{x + e^x}$ formed using a finite number of algebraic operations and rational, trigonometric, exponential, or logarithm functions). For integrals such as $\int_{0}^{1} \sin x^2\, dx$, we have to estimate the integral numerically. The preceding section showed one possible method.

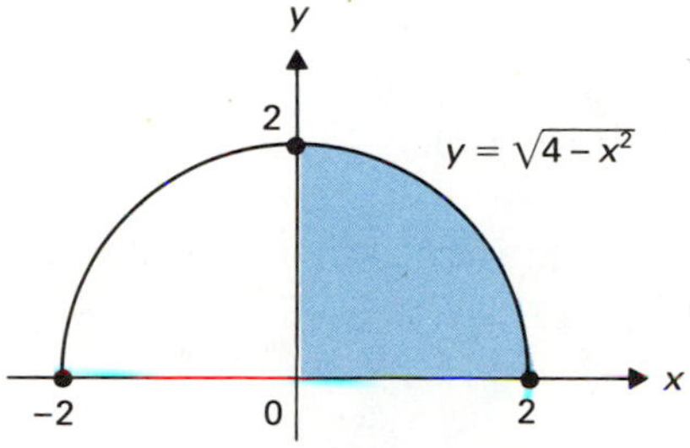

The Fundamental Theorem of Calculus

Now we show why antidifferentiation enables us to find the area under the graph of a function $f(x)$. The key is to consider the *integral function*

$$F(t) = \int_{a}^{t} f(x)\, dx \qquad \text{for } a \leq t \leq b, \tag{2}$$

FIGURE 5.25 Region with area $\int_{0}^{2} \sqrt{4 - x^2}\, dx$.

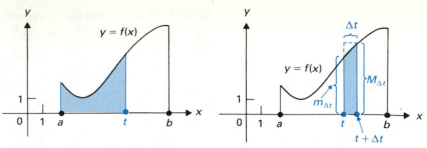

FIGURE 5.26 Region of area $F(t) =$ **FIGURE 5.27**
$\int_a^t f(x)\, dx$.

$$m_{\Delta t} \cdot \Delta t \leq \int_t^{t+\Delta t} f(x)\, dx \leq M_{\Delta t} \cdot \Delta t$$

so that if $f(x) \geq 0$, $F(t)$ is the area under the graph of f over $[a, t]$, as shown in Fig. 5.26. We will show that $F'(t) = f(t)$, and this will remove any remaining mystery.* If the derivative of the integral function F is f, then to find F we should find an antiderivative of f.

By a property of the integral,

$$\int_a^{t+\Delta t} f(x)\, dx = \int_a^t f(x)\, dx + \int_t^{t+\Delta t} f(x)\, dx.$$

Using the definition of the derivative, we have

$$\begin{aligned} F'(t) &= \lim_{\Delta t \to 0} \frac{F(t + \Delta t) - F(t)}{\Delta t} \\ &= \lim_{\Delta t \to 0} \frac{\int_a^{t+\Delta t} f(x)\, dx - \int_a^t f(x)\, dx}{\Delta t} = \lim_{\Delta t \to 0} \frac{\int_t^{t+\Delta t} f(x)\, dx}{\Delta t}. \end{aligned} \tag{3}$$

Referring to Fig. 5.27, we see that

$$m_{\Delta t} \cdot \Delta t \leq \int_t^{t+\Delta t} f(x)\, dx \leq M_{\Delta t} \cdot \Delta t, \tag{4}$$

where $m_{\Delta t}$ is the minimum and $M_{\Delta t}$ the maximum value of $f(x)$ in $[t, t + \Delta t]$. Therefore

$$m_{\Delta t} \leq \frac{\int_t^{t+\Delta t} f(x)\, dx}{\Delta t} \leq M_{\Delta t}. \tag{5}$$

Now $\lim_{\Delta t \to 0} m_{\Delta t} = \lim_{\Delta t \to 0} M_{\Delta t} = f(t)$ since f is continuous. We combine Eq. (3) with Eq. (5) to obtain

$$F'(t) = \lim_{\Delta t \to 0} \frac{\int_t^{t+\Delta t} f(x)\, dx}{\Delta t} = f(t). \tag{6}$$

This concludes our demonstration that $F'(t) = f(t)$, which is an important result in its own right.

---

*Note that $F(t)$ is defined for $a \leq t \leq b$. Derivatives of $F(t)$ at a and at b would have to be taken from one side only, a process we have not discussed. One forms the usual difference quotient but takes only positive Δt at a and only negative Δt at b. We will gloss over this point in what follows and refer to F as differentiable in $[a, b]$.

THEOREM 5.3   Fundamental theorem of calculus

Let f be continuous in $[a, b]$.

1. If $F(t) = \int_a^t f(x)\, dx$ for $a \le t \le b$, then F is differentiable and $F'(t) = f(t)$, that is,

$$\frac{d}{dt}\left(\int_a^t f(x)\, dx\right) = f(t). \tag{7}$$

2. If $F(x)$ is *any* antiderivative of $f(x)$, then

$$\int_a^b f(x)\, dx = F(b) - F(a). \tag{8}$$

$\square$

Note that part 1 of the fundamental theorem really asserts the *existence* of an antiderivative for a function f continuous on $[a, b]$. It even gives a formula for one such antiderivative F. We state this existence as a corollary.

COROLLARY

If f is continuous on $[a, b]$, then f has an antiderivative on this interval.

Part 1 of Theorem 5.3 was proved before the theorem was stated. Before we prove part 2, which is our rule for computing $\int_a^b f(x)\, dx$ given in rule (1), we give specific illustrations of part 1.

EXAMPLE 6   Find $\dfrac{d}{dt}\left(\displaystyle\int_0^t \sqrt{x^2 + 1}\, dx\right)$.

Solution   Part 1 of the fundamental theorem tells us immediately that

$$\frac{d}{dt}\left(\int_0^t \sqrt{x^2 + 1}\, dx\right) = \sqrt{t^2 + 1}. \qquad \blacksquare$$

EXAMPLE 7   Find $\dfrac{d}{dt}\left(\displaystyle\int_t^{-3} \sin^2 x\, dx\right)$.

Solution   Part 1 of the fundamental theorem concerns an integral with t as upper limit of integration. Here t is the lower limit. But we may write

$$\int_t^{-3} \sin^2 x\, dx = -\int_{-3}^t \sin^2 x\, dx.$$

Then

$$\frac{d}{dt}\left(\int_t^{-3} \sin^2 x\, dx\right) = -\frac{d}{dt}\left(\int_{-3}^t \sin^2 x\, dx\right) = -\sin^2 t. \qquad \blacksquare$$

EXAMPLE 8   Find $\dfrac{d}{dt}\left(\displaystyle\int_0^{t^2} \cos x^2\, dx\right)$.

Solution   We let $u = t^2$. Then by the chain rule and part 1 of the fundamental theorem,

$$\frac{d}{dt}\left(\int_0^{t^2} \cos x^2 \, dx\right) = \frac{d}{du}\left(\int_0^u \cos x^2 \, dx\right) \cdot \frac{du}{dt}$$

$$= (\cos u^2)2t = (\cos t^4)2t. \qquad \blacksquare$$

EXAMPLE 9   Find

$$\frac{d}{dt}\left(\int_{-2t}^t \frac{1}{1+x^2} \, dx\right).$$

Solution   Using properties of the integral, the chain rule as in Example 8, and part 1 of Theorem 5.3, we have

$$\frac{d}{dt}\left(\int_{-2t}^t \frac{1}{1+x^2}\, dx\right) = \frac{d}{dt}\left(\int_{-2t}^0 \frac{1}{1+x^2}\, dx + \int_0^t \frac{1}{1+x^2}\, dx\right) \quad \text{\textit{Property 8, page 260}}$$

$$= \frac{d}{dt}\left(-\int_0^{-2t} \frac{1}{1+x^2}\, dx\right) + \frac{1}{1+t^2} \quad \text{\textit{Property 11, page 261}}$$

$$= -\frac{1}{1+(-2t)^2} \cdot \frac{d(-2t)}{dt} + \frac{1}{1+t^2} \quad \text{\textit{Chain rule}}$$

$$= \frac{-1}{1+4t^2}(-2) + \frac{1}{1+t^2} = \frac{2}{1+4t^2} + \frac{1}{1+t^2}. \qquad \blacksquare$$

We now prove part 2 of the fundamental theorem, showing that Eq. (8) can be used to compute $\int_a^b f(x)\, dx$. Let $F(t) = \int_a^t f(x)\, dx$. Then

$$\int_a^b f(x)\, dx = F(b). \tag{9}$$

But $F(a) = \int_a^a f(x)\, dx = 0$, so we can write Eq. (9) as

$$\int_a^b f(x)\, dx = F(b) - F(a). \tag{10}$$

Theorem 5.3 asserts that Eq. (10) is true for *any* antiderivative of $f(x)$, not just our integral function F. Let $G(x)$ be any antiderivative, so $G'(x) = f(x)$. Since $F(x)$ and $G(x)$ have the same derivative $f(x)$, we know from Theorem 4.9 that $F(x) = G(x) + C$ for some constant C, and

$$\int_a^b f(x)\, dx = F(b) - F(a) = (G(b) + C) - (G(a) + C) = G(b) - G(a).$$

Hence $\int_a^b f(x)\, dx$ can be computed by finding *any* antiderivative of $f(x)$ and subtracting its value at a from its value at b, as stated in Eq. (8).

Interchanging a and b in Eq. (8) yields

$$\int_b^a f(x)\, dx = F(a) - F(b)$$

$$= -(F(b) - F(a)) = -\int_a^b f(x)\, dx,$$

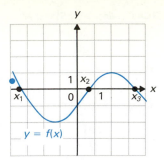

y

$y = f(x)$

FIGURE 5.28

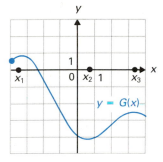

y

$y = G(x)$

FIGURE 5.29

which is consistent with our definition of $\int_b^a f(x)\, dx$ in the preceding section. We conclude with an example emphasizing that the *area function* in Eq. (2) is an antiderivative of f.

EXAMPLE 10 Figure 5.28 shows a graph $y = f(x)$ for x in the interval $[-4, 4]$. Sketch the approximate graph of an antiderivative G of the graph of f. Make the graph $y = G(x)$ start where $x = -4$ at the colored dot shown in the figure.

Solution Taking $a = -4$ in Eq. (2), we know that the *area function F* defined by

$$F(t) = \int_{-4}^{t} f(x)\, dx$$

for $-4 \le t \le 4$ is one function such that $F' = f$. However, $F(-4) = 0$, and we want the antiderivative of f having value approximately $\frac{1}{2}$ where $x = -4$. Thus we take

$$G(t) = \frac{1}{2} + \int_{-4}^{t} f(x)\, dx.$$

We can work from left to right and estimate values $G(t)$ by adding to $\frac{1}{2}$ the *positive* or *negative* contributions to the integral given by the areas of regions between the x-axis and the graph of f lying *over* or *under*, respectively, the x-axis. We estimate that the region over the interval $[-4, x_1]$ in Fig. 5.28 has area roughly 0.3, so that $G(x_1)$ is approximately $0.5 + 0.3 = 0.8$. We estimate the area of the region below the interval $[x_1, x_2]$ in Fig. 5.28 to be about 5.0, so that $G(x_2)$ is approximately $0.8 - 5.0 = -4.2$. The area of the region above the interval $[x_2, x_3]$ is approximately 1.6, so that we estimate $G(x_3)$ to be about $-4.2 + 1.6 = -2.6$. Finally, the area of the region below the interval $[x_3, 4]$ is about 0.2, so $G(4)$ is approximately $-2.6 - 0.2 = -2.8$.

Note also that f is the derived function of G, so that the graph of G is steepest where the magnitude of $f(x)$ is greatest. The graph of G is flattest where the magnitude of $f(x)$ is smallest. With this in mind and the points $(x_1, 0.8)$, $(x_2, -4.2)$, $(x_3, -2.6)$, and $(4, -2.8)$ we found, we sketch the graph of G in Fig. 5.29. ∎

SUMMARY

Fundamental Theorem of Calculus

Let f be continuous in $[a, b]$. Then

1. $\dfrac{d}{dt}\left(\displaystyle\int_a^t f(x)\, dx \right) = f(t);$

2. if $F'(x) = f(x)$, then $\displaystyle\int_a^b f(x)\, dx = F(b) - F(a).$

EXERCISES 5.3

1. State part 1 of the fundamental theorem of calculus without referring to the text.

2. State part 2 of the fundamental theorem of calculus without referring to the text.

In Exercises 3–50, use the fundamental theorem, properties of the definite integral, geometry, and, where necessary, the fact that $\int_0^\pi \sin^2 x \, dx = \pi/2$ to evaluate the definite integral.

3. $\displaystyle\int_0^1 x^3 \, dx$

4. $\displaystyle\int_{-1}^1 x^4 \, dx$

5. $\displaystyle\int_0^2 (t^2 + 3t - 1) \, dt$

6. $\displaystyle\int_4^9 \sqrt{s} \, ds$

7. $\displaystyle\int_1^8 x^{1/3} \, dx$

8. $\displaystyle\int_1^2 \frac{dt}{t^2}$

9. $\displaystyle\int_{-2}^{-1} \frac{du}{u^3}$

10. $\displaystyle\int_1^4 \left(\frac{2}{x^2} + \sqrt{x}\right) dx$

11. $\displaystyle\int_{-3}^{-1} \frac{x-2}{x^3} \, dx$

12. $\displaystyle\int_1^8 \frac{y^{1/3} + y^{1/2}}{y} \, dy$

13. $\displaystyle\int_0^\pi \sin u \, du$

14. $\displaystyle\int_0^{\pi/2} 4 \cos t \, dt$

15. $\displaystyle\int_0^{\pi/2} (\sin x + 2 \cos x) \, dx$

16. $\displaystyle\int_{\pi/4}^{\pi/2} (\sin x - \cos x) \, dx$

17. $\displaystyle\int_{-\pi/4}^{\pi/4} 3 \cos v \, dv$

18. $\displaystyle\int_0^{\pi/4} \sec^2 x \, dx$

19. $\displaystyle\int_{\pi/6}^{\pi/3} \csc^2 x \, dx$

20. $\displaystyle\int_{-\pi/6}^{\pi/3} \sec t \tan t \, dt$

21. $\displaystyle\int_{\pi/6}^{\pi/2} \csc x \cot x \, dx$

22. $\displaystyle\int_1^0 x^2 \, dx$

23. $\displaystyle\int_3^{-2} 4 \, dy$

24. $\displaystyle\int_3^{-2} (-4) \, dx$

25. $\displaystyle\int_{-1}^{-2} t \, dt$

26. $\displaystyle\int_4^4 \sqrt{u^3 + 1} \, du$

27. $\displaystyle\int_1^{-1} \sqrt{1 - x^2} \, dx$

28. $\displaystyle\int_0^1 (x + \sqrt{1 - x^2}) \, dx$

29. $\displaystyle\int_0^\pi (\sin x + \sin^2 x) \, dx$

30. $\displaystyle\int_{-\pi}^\pi \sin^2 x \, dx$

31. $\displaystyle\int_0^{2\pi} (\cos s + \sin^2 s) \, ds$

32. $\displaystyle\int_0^\pi 8 \sin^2 t \, dt$

33. $\displaystyle\int_0^\pi (2 \sin^2 x + 3\sqrt{\pi^2 - x^2}) \, dx$

34. $\displaystyle\int_{-\pi/2}^{\pi/2} \cos^2 x \, dx$

35. $\displaystyle\int_0^\pi (3s - 4 \cos^2 s) \, ds$

36. $\displaystyle\int_0^{\pi/2} (2 \sin^2 x - \cos^2 x) \, dx$

37. $\displaystyle\int_0^2 e^x \, dx$

38. $\displaystyle\int_0^{\ln 5} e^t \, dt$

39. $\displaystyle\int_0^{\ln 3} e^{2u} \, du$

40. $\displaystyle\int_{\ln 2}^{\ln 6} e^{-x} \, dx$

41. $\displaystyle\int_1^5 \frac{1}{x} \, dx$

42. $\displaystyle\int_1^e \frac{2}{y} \, dy$

43. $\displaystyle\int_2^{e^2} \left(s + \frac{1}{s}\right) ds$

44. $\displaystyle\int_1^3 \frac{x+2}{x} \, dx$

45. $\displaystyle\int_1^3 \left(\frac{d(x^2)}{dx}\right) dx$

46. $\displaystyle\int_0^1 \left(\frac{d(\sqrt{x^3 + 1})}{dx}\right) dx$

47. $\displaystyle\int_0^1 \left(\frac{d^2(\sqrt{x^2 + 1})}{dx^2}\right) dx$

48. $\displaystyle\int_0^2 \left(\frac{d^3(x^2 + 3x - 1)}{dx^3}\right) dx$

49. $\displaystyle\int_0^4 \left(\int_1^x \sqrt{t} \, dt\right) dx$

50. $\displaystyle\int_{-1}^3 \left(\int_x^2 4 \, dt\right) dx$

51. Sketch the region bounded by the curve $y = 9 - x^2$ and the x-axis, and find the area of the region.

52. Sketch the region bounded by the curve $y = 2x - x^2$ and the x-axis, and find the area of the region.

53. Sketch the region bounded by the curve $y = x^4 - 16$ and the x-axis, and find the area of the region.

In Exercises 54–64, use properties of the definite integral and Theorem 5.3 to find the indicated function of t.

54. $\displaystyle\frac{d}{dt}\left(\int_1^t x^2 \, dx\right)$

55. $\displaystyle\frac{d}{dt}\left(\int_{-50}^t \sqrt{x^2 + 1} \, dx\right)$

56. $\displaystyle\frac{d^2}{dt^2}\left(\int_2^t \sqrt{x^2 + 1} \, dx\right)$

57. $\displaystyle\frac{d}{dt}\left(\int_t^3 \frac{1}{1 + x^2} \, dx\right)$

58. $\displaystyle\frac{d}{dt}\left(\int_2^t \sqrt{x^2 + 4} \, dx + \int_t^{-1} \sqrt{x^2 + 4} \, dx\right)$

59. $\displaystyle\frac{d}{dt}\left(\int_{-t}^t \sqrt{3 + 4x^2} \, dx\right)$

60. $\displaystyle\frac{d^2}{dt^2}\left(\int_{-t}^t \sqrt{3 + 4x^2} \, dx\right)$

61. $\displaystyle\frac{d}{dt}\left(\int_1^{-t} \sqrt{x^2 + 1} \, dx\right)$

62. $\displaystyle\frac{d}{dt}\left(\int_1^{3t} \frac{1}{4 + x^2} \, dx\right)$

63. $\displaystyle\frac{d}{dt}\left(\int_0^{t^2 - 3} \sqrt{x^2 + 1} \, dx\right)$

64. $\displaystyle\frac{d}{dt}\left[\int_{t^2}^{t^3} \frac{1}{4 + 3x^2} \, dx\right]$

In Exercises 65–70, sketch the approximate graph of an antiderivative G of the graph of f shown in the indicated figure. Make the graph $y = G(x)$ start where $x = -4$ at the colored dot.

65.

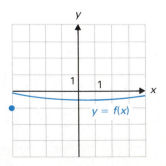

FIGURE 5.30

66.

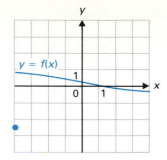

FIGURE 5.31

67.

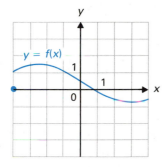

FIGURE 5.32

68.

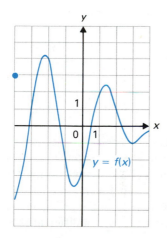

FIGURE 5.33

69.

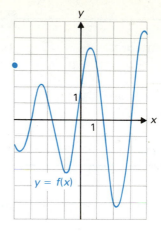

FIGURE 5.34

70.

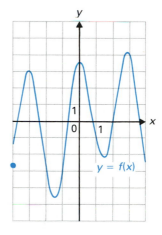

FIGURE 5.35

EXPLORING CALCULUS

M: calculus draw quiz (option 3)
N: calculus topics quiz (option 3)

5.4 THE INDEFINITE INTEGRAL

The fundamental theorem gives a powerful method for computing $\int_a^b f(x)\,dx$. The burden of the computation is placed squarely on finding an antiderivative $F(x)$ of $f(x)$.

DEFINITION 5.3

Indefinite integral

The general antiderivative of $f(x)$ is also called the **indefinite integral** of $f(x)$ and is written

$$\int f(x)\,dx$$

without any limits of integration. Computing an antiderivative is called **indefinite integration.**

EXAMPLE 1 Find $\int (x^3 + 2x^2 + 1)\,dx$.

Solution We have

$$\int (x^3 + 2x^2 + 1)\,dx = \frac{x^4}{4} + 2\frac{x^3}{3} + x + C,$$

where C is an arbitrary constant. ■

According to the preceding section, every function f that has a derivative in $[a, b]$ also has an antiderivative in $[a, b]$. To see this, note that if f is differentiable, it is continuous and, by the fundamental theorem, $F(t) = \int_a^t f(x)\,dx$ is an antiderivative of f in $[a, b]$. In practice, it is often more difficult to actually find an antiderivative of a function than to find its derivative. The reason is that there are no simple formulas to handle integration of products, quotients, or composite functions as there are for differentiation. Tables have been prepared that list indefinite integrals of a number of functions frequently encountered. A brief table of this type is found on the endpapers of this book. Such tables are often useful, but no table exists that gives the integral of every continuous function anyone will ever encounter. It can be proved that an antiderivative of an elementary function need not still be an elementary function. This is contrary to the situation for differentiation.

We start your training in integration now. This section presents a small number of integration formulas, which you are expected to remember and to be able to apply faster than you could find the desired formula in a table. For example, you should never have to look up $\int x^2\,dx$ in a table. The integration formulas we now give are those that arise from the differentiation formulas for basic elementary functions. For example, if u is a differentiable function of x, then, by the chain rule,

$$\frac{d(\tan u)}{dx} = (\sec^2 u) \cdot \frac{du}{dx},$$

so

$$\int (\sec^2 u) \cdot \frac{du}{dx}\,dx = \tan u + C.$$

Since $du = u'(x)\,dx = (du/dx)\,dx$, the preceding integration formula is usually written in the form

$$\int (\sec^2 u)\, du = \tan u + C.$$

The following list contains integration formulas that correspond to our differentiation formulas. Formulas 4–13 assume that u is a differentiable function of x.

1. $\displaystyle\int (f(x) + g(x))\, dx = \int f(x)\, dx + \int g(x)\, dx$

2. $\displaystyle\int c \cdot f(x)\, dx = c \cdot \int f(x)\, dx$

3. $\displaystyle\int x^n\, dx = \frac{x^{n+1}}{n+1} + C, \qquad n \neq -1$

4. $\displaystyle\int u^n\, du = \frac{u^{n+1}}{n+1} + C, \qquad n \neq -1$

5. $\displaystyle\int \sin u\, du = -\cos u + C$

6. $\displaystyle\int \cos u\, du = \sin u + C$

7. $\displaystyle\int \sec^2 u\, du = \tan u + C$

8. $\displaystyle\int \csc^2 u\, du = -\cot u + C$

9. $\displaystyle\int \sec u \tan u\, du = \sec u + C$

10. $\displaystyle\int \csc u \cot u\, du = -\csc u + C$

11. $\displaystyle\int \frac{1}{u}\, du = \ln|u| + C$

12. $\displaystyle\int e^u\, du = e^u + C$

13. $\displaystyle\int a^u\, du = \frac{1}{\ln a} a^u + C$

Formula (3) is a special case of formula (4), which is one of the most often used. The following example illustrates the use of formula (4).

EXAMPLE 2 Find $\int 2x\sqrt{x^2 + 1}\, dx$.

Solution If we let $u = x^2 + 1$, then $du = 2x \, dx$. From formula (4), we obtain

$$\int 2x\sqrt{x^2 + 1} \, dx = \int \underbrace{\sqrt{x^2 + 1}}_{u} \, \underbrace{(2x \, dx)}_{du}$$

$$= \int (u^{1/2}) \, du = \frac{u^{3/2}}{3/2} + C$$

$$= \frac{2}{3}(x^2 + 1)^{3/2} + C.$$

EXAMPLE 3 Find $\int x^2 \sin x^3 \, dx$.

Solution If we let $u = x^3$, then $du = 3x^2 \, dx$, so $x^2 \, dx = \frac{1}{3} \, du$. From formula (5), we obtain

$$\int x^2 \sin x^3 \, dx = \int (\sin \underbrace{x^3}_{u})(\underbrace{x^2 \, dx}_{\frac{1}{3} \, du}) = \int (\sin u)\left(\frac{1}{3} \, du\right) = \frac{1}{3}\int (\sin u) \, du$$

$$= -\frac{1}{3}\cos u + C = -\frac{1}{3}\cos x^3 + C. \ \blacksquare$$

In practice, we usually do not write out the substitution $u = g(x)$ in simple cases like those in Examples 2 and 3, but rather, realizing that the substitution is appropriate, we compute the integral in one step in our head. Suppose, for example, we wish to find $\int \cos 2x \, dx$. If we let $u = 2x$, then $du = 2 \, dx$, and

$$\int \cos 2x \, dx = \int \frac{1}{2} \cdot 2 \cos 2x \, dx = \frac{1}{2}\int (\cos \underbrace{2x}_{u})\underbrace{(2 \, dx)}_{du}.$$

We thus find from formula (6) that

$$\int \cos 2x \, dx = \frac{1}{2}\int (\cos \underbrace{2x}_{u})\underbrace{(2 \, dx)}_{du} = \frac{1}{2}\sin \underbrace{2x}_{u} + C.$$

We used the colored labels under the integrals to indicate what we were thinking, but we did not actually write u or du in the integral this time.

We illustrate this technique further with more examples.

EXAMPLE 4 Find $\int x^2 \sec^2 x^3 \, dx$.

Solution If $u = x^3$, then $du = 3x^2 \, dx$ and we would like to have an additional factor 3 in the integral so that formula (7) would apply. Of course, the supplied factor 3 for the "fix-up" must be balanced by a factor $\frac{1}{3}$, which we may write outside the integral by formula (2), since it is a constant. Using formula (7), we obtain

$$\int x^2 \sec^2 x^3 \, dx = \frac{1}{3}\int (\sec^2 \underbrace{x^3}_{u})\underbrace{(3x^2 \, dx)}_{du} = \frac{1}{3}\tan \underbrace{x^3}_{u} + C.$$

EXAMPLE 5 Find $\int \sin 3x \cos^2 3x \, dx$.

Solution If we let $u = \cos 3x$, then $du = -3 \sin 3x \, dx$, and we would like an additional factor -3 in our integral so that we can apply formula (4). We supply the factor -3 and the balancing factor $-\frac{1}{3}$ just as in Example 4. We obtain from formulas (2) and (4)

$$\int \sin 3x \cos^2 3x \, dx = -\frac{1}{3} \int \underbrace{(\cos^2 3x)}_{u^2}\underbrace{(-3 \sin 3x \, dx)}_{du}$$

$$= -\frac{1}{3}\frac{\cos^3 3x}{3} + C = -\frac{1}{9}\cos^3 3x + C. \qquad \blacksquare$$

As illustrated in the two preceding examples, formula (2) shows that we can "fix up an integral to supply a desired constant factor." *We warn you that a similar technique to supply a variable factor is not valid.* To illustrate,

$$\int x \, dx \neq \frac{1}{x} \int x \cdot x \, dx = \frac{1}{x} \int x^2 \, dx,$$

for $\int x \, dx = x^2/2 + C$, while

$$\frac{1}{x} \int x^2 \, dx = \frac{1}{x}\left(\frac{x^3}{3} + C\right) = \frac{x^2}{3} + \frac{C}{x} \qquad \text{for } x \neq 0.$$

For example, we can compute $\int x \sin x^2 \, dx$, because we can "fix up the desired constant 2" and use formula (5), but at the moment we can't compute $\int \sin x^2 \, dx$ because there is no way we can "fix up the variable factor $2x$" needed to apply formula (5).

EXAMPLE 6 Find $\int (\sin x)e^{\cos x} \, dx$.

Solution Formula (12) is appropriate here, taking $u = \cos x$. Then $du = -\sin x \, dx$. We introduce two minus signs and write the integral as

$$\int (\sin x)e^{\cos x} \, dx = -\int \overset{u}{\overbrace{e^{\cos x}}}\underbrace{(-\sin x)dx}_{du} = -e^{\cos x} + C. \qquad \blacksquare$$

EXAMPLE 7 Find $\int \tan x \, dx$.

Solution None of the thirteen formulas involves integrating the tangent function. However, if we note that $\tan x = (\sin x)/\cos x$ and that if $u = \cos x$, then $du = -\sin x$, we can use formula (11) to solve this problem. We obtain

$$\int \tan x \, dx = \int \frac{\sin x}{\cos x} \, dx = -\int \frac{\overbrace{-\sin x \, dx}^{du}}{\underset{u}{\underbrace{\cos x}}} = -\ln|\cos x| + C. \qquad \blacksquare$$

EXAMPLE 8   Find

$$\int \frac{\cos 3t}{(1 + \sin 3t)^5} \, dt.$$

Solution   We spot that $\cos 3t$ is the derivative of $1 + \sin 3t$, except for a constant factor. Thus we essentially want to integrate $du/u^5 = u^{-5} \, du$. We have

$$\int \frac{\cos 3t}{(1 + \sin 3t)^5} \, dt = \frac{1}{3} \int \underbrace{(1 + \sin 3t)}_{u}^{-5}\underbrace{(3 \cos 3t \, dt)}_{du}$$

$$= \frac{1}{3} \cdot \frac{(1 + \sin 3t)^{-4}}{-4} + C$$

$$= \frac{-1}{12(1 + \sin 3t)^4} + C. \blacksquare$$

EXAMPLE 9   Find $\int \sin^2 3x \, dx$.

Solution   This time none of the thirteen formulas will work. It is useless to try letting $u = \sin 3x$, for then $du = 3 \cos 3x$ and we do not have the function $\cos 3x$ present. A fast way to find this integral is to reach for a table of integrals. Formula 64 on the endpapers of this text tells us that

$$\int \sin^2 ax \, dx = \frac{x}{2} - \frac{\sin 2ax}{4a} + C.$$

We need only take $a = 3$ in this formula to obtain

$$\int \sin^2 3x \, dx = \frac{x}{2} - \frac{\sin 6x}{12} + C. \blacksquare$$

Example 9 illustrates how quickly some integrals can be found using a table of integrals. Very large computer programs have been written that are also quite successful at finding many indefinite integrals. Some pocket calculators are capable of limited indefinite integration, although displaying long answers is a problem. Tables are readily accessible and easy to carry around. You should familiarize yourself with the organization of the table on the endpapers of the text so you know where to look for a particular type of integral and what types of integrals appear in the table.

SUMMARY

1. The general antiderivative of $f(x)$ is also known as the indefinite integral, $\int f(x) \, dx$.

2. The following are thirteen important formulas for integrating.

$$\int (f(x) + g(x)) \, dx = \int f(x) \, dx + \int g(x) \, dx$$

$$\int c \cdot f(x) \, dx = c \cdot \int f(x) \, dx$$

$$\int x^n \, dx = \frac{x^{n+1}}{n+1} + C, \qquad n \neq -1$$

$$\int u^n \, du = \frac{u^{n+1}}{n+1} + C, \qquad n \neq -1$$

$$\int \sin u \, du = -\cos u + C$$

$$\int \cos u \, du = \sin u + C$$

$$\int \sec^2 u \, du = \tan u + C$$

$$\int \csc^2 u \, du = -\cot u + C$$

$$\int \sec u \tan u \, du = \sec u + C$$

$$\int \csc u \cot u \, du = -\csc u + C$$

$$\int \frac{1}{u} \, du = \ln|u| + C$$

$$\int e^u \, du = e^u + C$$

$$\int a^u \, du = \frac{1}{\ln a} a^u + C$$

EXERCISES 5.4

In Exercises 1–50, find the indicated integral without using tables.

1. $\displaystyle\int (x^3 + 4x^2) \, dx$

2. $\displaystyle\int \left(t + \frac{1}{t^2} \right) dt$

3. $\displaystyle\int (u + 1)^5 \, du$

4. $\displaystyle\int (2x + 1)^4 \, dx$

5. $\displaystyle\int \frac{dx}{(4x + 1)^2}$

6. $\displaystyle\int \frac{5}{(3 - 7s)^3} \, ds$

7. $\displaystyle\int \frac{t}{(3t^2 + 1)^2} \, dt$

8. $\displaystyle\int \frac{x^2}{(4x^3 - 1)^4} \, dx$

9. $\displaystyle\int \frac{x}{(4 - 3x^2)^3} \, dx$

10. $\displaystyle\int \frac{y}{\sqrt{y^2 + 4}} \, dy$

11. $\displaystyle\int \frac{4s + 2}{\sqrt{s^2 + s}} \, ds$

12. $\displaystyle\int \frac{2x^2}{(x^3 - 3)^{3/2}} \, dx$

13. $\displaystyle\int \frac{dx}{\sqrt{x}(\sqrt{x} + 1)^4}$

14. $\displaystyle\int \frac{dt}{\sqrt{t}(4 - 3\sqrt{t})^2}$

15. $\displaystyle\int x^2(x^3 + 4)^3 \, dx$

16. $\displaystyle\int (x^2 + 1)(x^3 + 1)^2 \, dx$

17. $\displaystyle\int \cos 3v \, dv$

18. $\displaystyle\int \sin (3x + 1) \, dx$

19. $\displaystyle\int x \cos (x^2 + 1) \, dx$

20. $\displaystyle\int \frac{\sin (\sqrt{s})}{\sqrt{s}} \, ds$

21. $\displaystyle\int \sin s \cos s \, ds$

22. $\displaystyle\int \cos x \sin^2 x \, dx$

23. $\int \sin 4x \cos^3 4x \, dx$ **24.** $\int \dfrac{t^2}{\sin^2 t^3} \, dt$

25. $\int x \sec^2 x^2 \, dx$ **26.** $\int \csc 2x \cot 2x \, dx$

27. $\int \tan t \sec^2 t \, dt$ **28.** $\int \sec^3 2v \tan 2v \, dv$

29. $\int x(\sec^2 x^2)(\tan^3 x^2) \, dx$ **30.** $\int \dfrac{x \cot x^2}{\sin x^2} \, dx$

31. $\int \dfrac{\sin s}{(1 + \cos s)^2} \, ds$ **32.** $\int \dfrac{\sec^2 2t}{(4 + \tan 2t)^3} \, dt$

33. $\int \dfrac{1}{\sec 4x} \, dx$ **34.** $\int \dfrac{\sin x}{\cos^2 x} \, dx$

35. $\int \dfrac{\tan 3v}{\cos^4 3v} \, dv$ **36.** $\int \dfrac{\sec^2 x}{\sqrt{1 + \tan x}} \, dx$

37. $\int (\tan 3t)\sqrt{\sec^2 3t + \sec^3 3t} \, dt$

38. $\int \dfrac{\cot 3x}{\sin 3x} \, dx$ **39.** $\int \csc^5 2x \cot 2x \, dx$

40. $\int \dfrac{\cos s}{(1 - \cos^2 s)^2} \, ds$ **41.** $\int \tan 2x \, dx$

42. $\int \cot 3t \, dt$ **43.** $\int \dfrac{x}{x^2 + 2} \, dx$

44. $\int e^{2s} \, ds$ **45.** $\int t^2 e^{t^3} \, dt$

46. $\int \dfrac{e^x}{e^x + 1} \, dx$ **47.** $\int (\cos y)3^{\sin y} \, dy$

48. $\int x \, 3^{x^2} \, dx$ **49.** $\int e^{2x}\sqrt{1 + 3e^{2x}} \, dx$

50. $\int \dfrac{2^t}{\sqrt{1 + 5 \cdot 2^t}} \, dt$

In Exercises 51–56, find the integral by using the tables on the endpapers of the text.

51. $\int \dfrac{3}{t^2 \sqrt{4 + t^2}} \, dt$

52. $\int \dfrac{1}{x^2 \sqrt{x^2 - 9}} \, dx$

53. $\int_0^{\pi} \cos^2 2x \, dx$ **54.** $\int \sin^4 v \, dv$

55. $\int \sin 2t \cos 3t \, dt$ **56.** $\int t^3 e^t \, dt$

5.5 DIFFERENTIAL EQUATIONS (VARIABLES SEPARABLE)

A **differential equation** is one containing a derivative or differential. In Section 4.8 we solved some differential equations of the form

$$\frac{dy}{dx} = f(x).$$

Many differential equations cannot be written in this form, where dy/dx equals a function of x alone. (Remember that, when differentiating implicitly, we frequently obtain an expression for dy/dx that involves both x and y.) An example of such a differential equation is

$$\frac{dy}{dx} = \frac{x}{y}. \tag{1}$$

We can solve this differential equation by the following device: *Rewrite the equation with all terms involving y (including dy) on the left and all terms involving x (including dx) on the right:*

$$y \cdot dy = x \cdot dx.$$

Since the differentials $y \cdot dy$ and $x \cdot dx$ are equal, it appears that we must have

$$\int y \cdot dy = \int x \cdot dx, \qquad \text{or} \qquad \frac{y^2}{2} + C_1 = \frac{x^2}{2} + C_2,$$

where C_1 and C_2 are arbitrary constants. This solution may be written as

$$y^2 + 2C_1 = x^2 + 2C_2, \quad \text{or} \quad y^2 - x^2 = 2C_1 - 2C_2.$$

Now $2C_1 - 2C_2$ may be any constant, so we can express it as a single arbitrary constant C, obtaining

$$y^2 - x^2 = C$$

as the solution of the differential equation. We consider an equation like this that defines y implicitly as a function of x to be an acceptable form for the solution. (Note that implicit differentiation of this relation indeed yields Eq. 1.)

Crucial to this technique is the ability to rewrite the differential equation with all terms involving y on one side and all terms involving x on the other. Such equations are *variables-separable* equations.

EXAMPLE 1 Solve

$$\frac{dy}{dx} = y^2 \sin x.$$

Solution This is a variables-separable equation. The steps of solution are

$$\frac{dy}{y^2} = \sin x \, dx, \qquad \int y^{-2} \, dy = \int \sin x \, dx,$$

$$\frac{y^{-1}}{-1} = -\cos x + C, \qquad \frac{1}{y} = \cos x + C,$$

where we used only one arbitrary constant C and replaced $-C$ by C again in the last step. ∎

EXAMPLE 2 Solve

$$\frac{dy}{dx} = \frac{\ln x}{xy^2}.$$

Solution The equation is variables-separable, and

$$y^2 \, dy = \frac{\ln x}{x} \, dx,$$

so

$$\int y^2 \, dy = \int \frac{\ln x}{x} \, dx,$$

and

$$\frac{1}{3} y^3 = \frac{(\ln x)^2}{2} + C$$

is the general solution. We may also express this as

$$2y^3 = 3(\ln x)^2 + 6C,$$

or, since $6C$ may again be any arbitrary constant, it is acceptable to simply write

$$2y^3 = 3(\ln x)^2 + C.$$

Such informality with arbitrary constants is conventional. ∎

EXAMPLE 3 Solve

$$\frac{dy}{dx} = \frac{\cos x}{\sin y}.$$

Solution Separation of variables leads to

$$\sin y \, dy = \cos x \, dx, \qquad \int \sin y \, dy = \int \cos x \, dx,$$

$$-\cos y = \sin x + C. \qquad \blacksquare$$

EXAMPLE 4 In Section 3.5, we discussed the equation

$$f'(t) = k \cdot f(t),$$

which asserts that the rate of growth of the function f is proportional to its size. We showed that $f(t) = Ae^{kt}$ is a solution of this equation. Derive this solution again, using the separation of variables technique.

Solution We set $y = f(t)$, and then the differential equation can be written as

$$\frac{dy}{dt} = ky.$$

Separating variables, we obtain

$$\frac{dy}{y} = k \, dt, \qquad \int \frac{dy}{y} = \int k \, dt, \qquad \ln|y| = kt + C,$$

$$e^{\ln|y|} = e^{kt+C}, \qquad |y| = e^{kt+C}, \qquad |y| = e^{C}e^{kt},$$

$$y = \pm e^{C}e^{kt}.$$

Now $\pm e^{C}$ may be any constant except zero. A check shows that $y = 0$ is a solution of $dy/dt = ky$, but we lost this solution when we divided by y to separate the variables. Therefore, if we let A be any constant, our solution can be written as

$$y = Ae^{kt},$$

which is what we wished to show. $\qquad \blacksquare$

The particular solution $y = h(x)$ of a differential equation $y' = F(x, y)$ such that $y_0 = h(x_0)$ is often called the solution of the initial value problem.

Initial Value Problem

$$y' = F(x, y), \qquad y(x_0) = y_0$$

EXAMPLE 5 Find the solution of the differential equation $dy/dx = x^2 y^3$ such that $y(1) = -1$.

Solution We have an initial value problem, so we must evaluate the constant in the general solution. First we separate variables and find the general solution:

$$\frac{dy}{y^3} = x^2 \, dx, \qquad \int y^{-3} \, dy = \int x^2 \, dx,$$

$$\frac{y^{-2}}{-2} = \frac{x^3}{3} + C, \qquad -\frac{1}{2y^2} = \frac{1}{3}x^3 + C.$$

Setting $x = 1$ and $y = -1$, we obtain $-\frac{1}{2} = \frac{1}{3} + C$, so $C = -\frac{1}{2} - \frac{1}{3} = -\frac{5}{6}$. Thus our desired solution is

$$-\frac{1}{2y^2} = \frac{1}{3}x^3 - \frac{5}{6}. \qquad \blacksquare$$

For an application, consider a body of mass m falling freely near the surface of the earth. The body is subject to a downward force mg due to gravity and an upward force due to air resistance. The body attains *terminal velocity* when these two forces are of equal magnitude, so the velocity no longer changes.

EXAMPLE 6 Suppose a body of mass m falling freely near the surface of the earth is subject to a force of air resistance of magnitude $k|v|$, where $k > 0$ and v is the velocity. Find v at time t if $v = 0$ when $t = 0$, and find the terminal velocity.

Solution We consider positive velocity v as directed downward, in the direction of motion of the falling body. By Newton's second law, the force acting on the body is

$$ma = m\frac{dv}{dt} = mg - kv,$$

where $a = dv/dt$ is the acceleration of the body. This is a variables-separable differential equation in v and t. We obtain

$$\frac{m \, dv}{mg - kv} = dt,$$

$$-\frac{m}{k}\ln|mg - kv| = t + C.$$

Setting $t = v = 0$, we see that $C = -(m/k) \cdot \ln(mg)$, so

$$-\frac{m}{k}\ln|mg - kv| = t - \frac{m}{k}\ln(mg),$$

$$\ln|mg - kv| = -\frac{k}{m}t + \ln(mg),$$

$$|mg - kv| = mge^{-kt/m}.$$

From the physical situation, we know that $mg - kv = ma \geq 0$. Thus we can drop the absolute value symbol, and we find that

$$kv = mg - mge^{-kt/m},$$

so

$$v = \frac{mg}{k}(1 - e^{-kt/m}). \tag{2}$$

Equation (2) is the desired expression for the velocity in terms of m, g, k, and t.

To find the terminal velocity, we could proceed in two ways. Physically, terminal velocity is achieved when the forces mg of gravity and kv of air resistance are of equal magnitude, or when $v = mg/k$. Alternatively, we can let $t \to \infty$ in Eq. (2), so that $e^{-kt/m} \to 0$, and see that v approaches mg/k. Equation (2) shows that the body never actually attains terminal velocity. However, it approaches it quite rapidly, since the exponential $e^{-kt/m}$ decreases rapidly as t increases. ∎

The problem of finding a family of curves orthogonal (perpendicular) to a given family of curves at all points of intersection is a problem in differential equations and can sometimes be solved by the techniques described in this section. We illustrate this problem of finding a family of *orthogonal trajectories* with an example.

EXAMPLE 7   Find the family of orthogonal trajectories to the hyperbolas $x^2 - y^2 = C$.

Solution   Differentiating the relation $x^2 - y^2 = C$ implicitly, we obtain

$$2x - 2y\frac{dy}{dx} = 0, \qquad \text{so} \qquad \frac{dy}{dx} = \frac{x}{y}.$$

The slope of a curve in the orthogonal family at (x, y) is the negative reciprocal, $-y/x$. Thus our orthogonal family of curves consists of the solutions of the differential equation

$$\frac{dy}{dx} = -\frac{y}{x}.$$

Then

$$\frac{dy}{y} + \frac{dx}{x} = 0,$$

so

$$\ln|x| + \ln|y| = K. \tag{3}$$

We may write Eq. (3) in the form

$$\ln|xy| = K.$$

Then $e^{\ln|xy|} = |xy| = e^K$, so $xy = \pm e^K$. As K runs through all constants, $\pm e^K$ runs through all constants except zero.

A special examination shows that the curves $x = 0$ and $y = 0$ are also orthogonal to the hyperbolas $x^2 - y^2 = C$; changing notation, we obtain, as the orthogonal family,

$$xy = K.$$

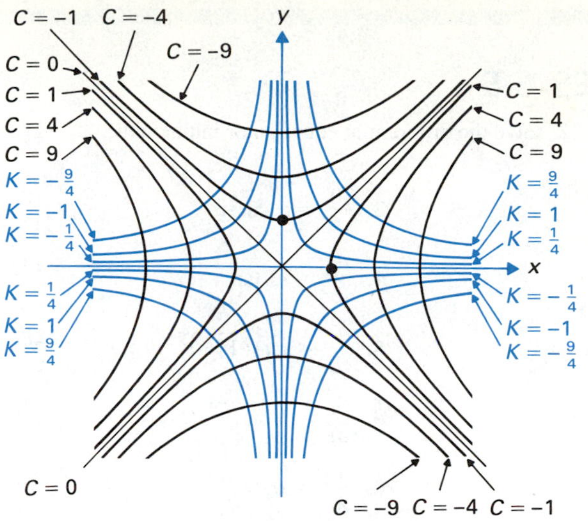

FIGURE 5.36 Orthogonal trajectories of hyperbolas $x^2 - y^2 = C$ are hyperbolas $xy = K$.

This is also a family of hyperbolas, with their axes tipped at 45° from the x,y-axes. Figure 5.36 shows some of the curves of these two orthogonal families. ∎

An example of a *non*-variables-separable equation is

$$\frac{dy}{dx} = x + y.$$

Chapter 17 discusses solutions of a few more types of differential equations.

SUMMARY

1. Suppose all the y terms (including dy) can be put on the left side of a differential equation and all the x terms (including dx) on the right, to become

$$g(y)\, dy = f(x)\, dx.$$

Then the solution of the differential equation is

$$\int g(y)\, dy = \int f(x)\, dx + C,$$

where C is an arbitrary constant. No further arbitrary constants need be included when computing $\int g(y)\, dy$ and $\int f(x)\, dx$.

EXERCISES 5.5

In Exercises 1–22, solve the differential equation or initial value problem.

1. $\dfrac{dy}{dx} = x^2y^2$

2. $\dfrac{dy}{dx} = x \cos^2 y$

3. $\dfrac{dy}{dx} = xy^2$

4. $\dfrac{dy}{dx} = e^x \tan y$

5. $\dfrac{dy}{dx} = \sin x \cos^2 y$

6. $\dfrac{ds}{dt} = s^2 t\sqrt{1 + t^2}$

7. $\dfrac{du}{dx} = x^3\sqrt{u}$

8. $\dfrac{dv}{dt} = \sqrt{5v - 7}(t \cos t^2)$

9. $\dfrac{dy}{dx} = xe^{2y}$

10. $\dfrac{dx}{dt} = xt$

11. $\dfrac{dy}{dx} = \dfrac{(1 + x^2)^2}{\sec^2 3y}$

12. $y' = 1 + x,\ y(1) = -1$

13. $y' = x \sin^2 y,\ y(2) = \dfrac{\pi}{2}$

14. $y' = \dfrac{x}{y},\ y(2) = -3$

15. $y' = x^2y + 2xy,\ y(2) = 1$

16. $y' = y^3e^{2x},\ y(0) = 2$

17. $\dfrac{dy}{dx} = xy^3,\ y = 1$ when $x = 0$

18. $\dfrac{dy}{dx} = \cos^2 y,\ y = \dfrac{\pi}{4}$ when $x = 1$

19. $\dfrac{dy}{dx} = \dfrac{y}{x},\ y = -4$ when $x = 3$

20. $\dfrac{dy}{dt} = y^2e^{2t},\ y = 2$ when $t = 0$

21. $\dfrac{dy}{dx} = \cos x \csc 2y,\ y = \dfrac{\pi}{2}$ when $x = 0$

22. $\dfrac{ds}{dt} = \sqrt{s + 1}\sqrt{3t + 1},\ s = 3$ when $t = 5$

In Exercises 23–27, find the family of orthogonal trajectories of the given family of curves. Sketch both families as in Fig. 5.36.

23. $y - x = C$

24. $y - x^2 = C$

25. $y^2 + x = C$

26. $x^2 + y^2 = C$

27. $y = C(x + 1)$

5.6 NUMERICAL INTEGRATION

If we can't find the indefinite integral $\int f(x)\,dx$, even in a table, then the fundamental theorem is of no help in computing $\int_a^b f(x)\,dx$. For example, it can be shown that the integral

$$\int \sqrt{4 - \sin^2 x}\,dx$$

cannot be expressed in terms of elementary functions such as polynomials, trigonometric functions, roots, exponentials, and logarithms. Numerical methods for computing a definite integral such as

$$\int_0^1 \sqrt{4 - \sin^2 x}\,dx$$

are therefore very important. With calculators and computers readily available, computation of such integrals is an easy matter if one is willing to settle for a good numerical approximation to the exact value.

The Rectangular Rule

One method of approximating $\int_a^b f(x)\,dx$ is to use a midpoint Riemann sum with a regular partition, taking midpoints of the subintervals for the points x_i^* where we compute $f(x_i)$. The integral is then computed as a sum of areas of rectangles, as indicated in Fig. 5.37. For obvious reasons, approximation by such Riemann sums is also known as the *rectangular rule*.

THEOREM 5.4 Rectangular rule

Let $[a, b]$ be divided into n subintervals of equal length and let x_i^* be the midpoint of the ith subinterval. Then, for f continuous in $[a, b]$,

$$\int_a^b f(x)\, dx \approx \frac{b - a}{n} [f(x_1^*) + f(x_2^*) + \cdots + f(x_n^*)] \tag{1}$$

for sufficiently large values of n. □

To find the midpoints x_i^* to use in the rectangular rule, we first compute the subinterval length $(b - a)/n$ and then find

$$x_1^* = a + \frac{1}{2} \cdot \frac{b - a}{n}.$$

We then find the remaining x_i^* by incrementing by $(b - a)/n$, that is,

$$x_2^* = x_1^* + \frac{b - a}{n}, \qquad x_3^* = x_2^* + \frac{b - a}{n},$$

$$x_4^* = x_3^* + \frac{b - a}{n}, \qquad \text{and so on.}$$

We know that

$$\int_1^5 x^4\, dx = \frac{x^5}{5}\bigg]_1^5 = \frac{5^5}{5} - \frac{1}{5} = 5^4 - \frac{1}{5} = 625 - \frac{1}{5} = 624.8.$$

In Example 1 we estimate $\int_1^5 x^4\, dx$ using $n = 4$ and the rectangular rule. The purpose is to compare this estimate with later estimates using two other rules.

EXAMPLE 1 Estimate $\int_1^5 x^4\, dx$ using the rectangular rule with $n = 4$.

Solution We have

$$\frac{b - a}{n} = \frac{5 - 1}{4} = 1.$$

Then

$$x_1^* = 1 + \frac{1}{2} \cdot 1 = \frac{3}{2}, \qquad x_2^* = \frac{5}{2}, \qquad x_3^* = \frac{7}{2}, \qquad x_4^* = \frac{9}{2}.$$

Since $f(x) = x^4$, the rectangular rule provides the estimate

$$\int_1^5 x^4\, dx \approx 1 \cdot \left[\left(\frac{3}{2}\right)^4 + \left(\frac{5}{2}\right)^4 + \left(\frac{7}{2}\right)^4 + \left(\frac{9}{2}\right)^4 \right]$$

$$= \frac{3^4 + 5^4 + 7^4 + 9^4}{16}$$

$$= \frac{9668}{16} = 604.25. \qquad ∎$$

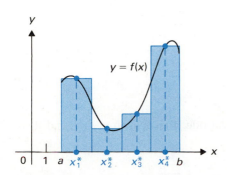

FIGURE 5.37

$$\int_a^b f(x)\, dx \approx [(b - a)/4][f(x_1^*)$$
$$+ f(x_2^*) + f(x_3^*) + f(x_4^*)]$$

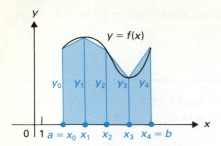

FIGURE 5.38 Approximation of $\int_a^b f(x)\,dx$ using areas of regions under four chords of the graph.

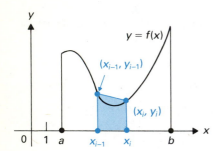

FIGURE 5.39 Trapezoid of area $(x_i - x_{i-1})[(y_{i-1} + y_i)/2]$.

It should not be necessary to give any more examples using the rectangular rule. We worked with such estimates in Section 5.1.

The Trapezoidal Rule

We again divide the interval $[a, b]$ into n subintervals of equal length, and we let the *endpoints* of the subintervals be

$$a = x_0, \qquad x_1, \qquad x_2, \qquad \ldots, \qquad x_n = b,$$

as illustrated in Fig. 5.38 for $n = 4$. Let the (signed) heights to the graph $y = f(x)$ at these endpoints be

$$y_0 = f(x_0), \qquad y_1 = f(x_1), \qquad y_2 = f(x_2), \qquad \ldots, \qquad y_n = f(x_n),$$

as in Fig. 5.38. Consider the chord of the curve $y = f(x)$ joining (x_{i-1}, y_{i-1}) and (x_i, y_i), as shown in Fig. 5.39. The shaded regions in Figs. 5.38 and 5.39 are trapezoids. The area of a trapezoid is the product of its altitude and the average length of the bases, so the area of the shaded trapezoid in Fig. 5.39 is

$$(x_i - x_{i-1})\frac{y_{i-1} + y_i}{2}.$$

Since we divide our interval $[a, b]$ into n subintervals of *equal* length,

$$x_i - x_{i-1} = \frac{b - a}{n}.$$

Adding the area of the trapezoids for $i = 1, \ldots, n$, we obtain

$$\frac{b - a}{n}\left(\frac{y_0 + y_1}{2}\right) + \frac{b - a}{n}\left(\frac{y_1 + y_2}{2}\right) + \frac{b - a}{n}\left(\frac{y_2 + y_3}{2}\right)$$

$$+ \cdots + \frac{b - a}{n}\left(\frac{y_{n-1} + y_n}{2}\right).$$

Factoring out $(b - a)/2n$, we obtain at once the trapezoidal rule.

THEOREM 5.5   **Trapezoidal rule**

For a continuous function f in $[a, b]$,

$$\int_a^b f(x)\,dx \approx \frac{b - a}{2n}(y_0 + 2y_1 + 2y_2 + \cdots + 2y_{n-1} + y_n) \tag{2}$$

for sufficiently large values of n. $\square$

To find the values y_i in the trapezoidal rule, first find $(b - a)/n$ and then compute x_i, where

$$x_0 = a, \qquad x_1 = x_0 + \frac{b - a}{n},$$

$$x_2 = x_1 + \frac{b - a}{n}, \qquad \ldots, \qquad x_n = b.$$

Then $y_i = f(x_i)$.

EXAMPLE 2 Use the trapezoidal rule with $n = 4$ to estimate

$$\int_1^3 \frac{1}{x}\, dx.$$

Solution We have

$$\frac{b-a}{n} = \frac{3-1}{4} = \frac{1}{2},$$

so

$$x_0 = 1, \qquad x_1 = \frac{3}{2}, \qquad x_2 = 2, \qquad x_3 = \frac{5}{2}, \qquad x_4 = 3.$$

Thus

$$y_0 = 1, \qquad y_1 = \frac{2}{3}, \qquad y_2 = \frac{1}{2}, \qquad y_3 = \frac{2}{5}, \qquad y_4 = \frac{1}{3}.$$

The trapezoidal rule yields

$$\int_1^3 \frac{1}{x}\, dx \approx \frac{3-1}{2 \cdot 4}\left(1 + 2 \cdot \frac{2}{3} + 2 \cdot \frac{1}{2} + 2 \cdot \frac{2}{5} + \frac{1}{3}\right)$$

$$= \frac{2}{8}\left(1 + \frac{4}{3} + 1 + \frac{4}{5} + \frac{1}{3}\right)$$

$$= \frac{1}{4}\left(\frac{15 + 20 + 15 + 12 + 5}{15}\right)$$

$$= \frac{1}{4}\left(\frac{67}{15}\right) = \frac{67}{60} \approx 1.1167.$$

Since the graph of $1/x$ is concave up over $[1, 3]$, it is clear that the approximation by chords is too large. The actual value of the integral to four decimal places is 1.0986. ∎

EXAMPLE 3 Estimate $\int_1^5 x^4\, dx$ using the trapezoidal rule with $n = 4$, and compare with Example 1.

Solution We have $(b - a)/n = 1$ as in Example 1. This time,

$$x_0 = 1, \qquad x_1 = 2, \qquad x_2 = 3, \qquad x_3 = 4, \qquad x_4 = 5.$$

Thus

$$y_0 = 1, \qquad y_1 = 2^4, \qquad y_2 = 3^4, \qquad y_3 = 4^4, \qquad y_4 = 5^4.$$

The trapezoidal rule yields

$$\int_1^5 x^4\, dx \approx \frac{1}{2}(1 + 2 \cdot 2^4 + 2 \cdot 3^4 + 2 \cdot 4^4 + 5^4)$$

$$= \frac{1332}{2} = 666.$$

The rectangular estimate 604.25 is closer to the true value 624.8 of the integral than is this estimate of 666. ∎

In real-world applications we do not take n as small as 4 when using the trapezoidal rule, and we would not use the trapezoidal rule to find definite integrals of either $1/x$ or x^4 since we can find antiderivatives of these functions. We chose small values of n to make Examples 2 and 3 short and easy to follow, and we selected the functions so that we could compare the estimates found with the true values.

When making numerical estimates, it is very important to know a *bound for the error* in the estimate. That is, we want to know some number $M > 0$ such that we are sure the magnitude of the error is less than M. It can be shown if f is continuous in $[a, b]$, then the error in the trapezoidal approximation (2) of $\int_a^b f(x)\,dx$ is

$$\frac{(b-a)^3}{12n^2}f''(c) \qquad \textit{Error in trapezoidal approximation} \qquad \textbf{(3)}$$

for some number c in $[a, b]$. To illustrate this formula for the approximation of $\int_1^3 (1/x)\,dx$ using $n = 4$ in Example 2, we note that for $f(x) = 1/x$ we have $f''(x) = 2/x^3$. Thus for some number c in $[1, 3]$, the magnitude of the error in the approximation in Example 2 is

$$\left|\frac{(3-1)^3}{12(4)^2} \cdot \frac{2}{c^3}\right| \le \frac{8}{12 \cdot 16} \cdot \frac{2}{1} = \frac{1}{12} \approx 0.0833. \qquad \textbf{(4)}$$

Since the actual value of this integral is $\ln 3 \approx 1.0986$, we see that the error in the approximate value 1.1167 obtained in Example 2 is about 0.0181. Thus the bound 0.0833 obtained from the error estimate indeed exceeds the magnitude of the error. Our bound 0.0833 is so large because we replaced c^3 in the denominator in relation (4) by the *smallest* value it can attain for c in $[1, 3]$, namely 1. The largest value that c^3 can attain in $[1, 3]$ is $3^3 = 27$. Replacing c^3 by 27 instead in relation (4), we obtain

$$\left|\frac{(3-1)^3}{12(4)^2} \cdot \frac{2}{c^3}\right| \ge \frac{8}{12 \cdot 16} \cdot \frac{2}{27} = \frac{1}{12 \cdot 27} \approx 0.0031,$$

and the error of approximately 0.0181 is indeed larger than this.

Simpson's (Parabolic) Rule

In the rectangular rule, the approximating rectangles have horizontal lines (constant functions $y = a$) as tops, and each line usually meets the graph at *one* point. In the trapezoidal rule, the top of a trapezoid may be any line (linear function $y = ax + b$), and each line usually meets the graph at *two* points. We can put a quadratic function $y = ax^2 + bx + c$ through *three* points of the graph. The graph of $y = ax^2 + bx + c$ is a *parabola*.

For Simpson's (parabolic) rule, we divide the interval $[a, b]$ into an *even number n* of subintervals of equal length having endpoints x_i for $i = 0, 1, \ldots, n$. We use the notation $y_i = f(x_i)$ as in the trapezoidal rule. We approximate the graph of f over the first *two* subintervals by a parabola through (x_0, y_0), (x_1, y_1), and (x_2, y_2), over the next two subintervals by a parabola through (x_2, y_2), (x_3, y_3), and (x_4, y_4), and so on (see Fig. 5.40). We then estimate the integral $\int_a^b f(x)\,dx$ by taking the areas under the parabolas.

For the case where $x_1 - x_0 = x_2 - x_1$, there is a convenient formula for the area under a parabola through (x_0, y_0), (x_1, y_1), and (x_2, y_2) in terms of the

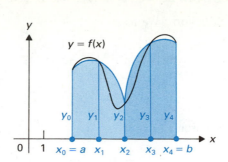
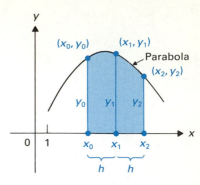

FIGURE 5.40 Estimating $\int_a^b f(x)\, dx$ by the areas of regions under two parabolas.

FIGURE 5.41 The shaded region has area $(h/3)(y_0 + 4y_1 + y_2)$.

heights y_0, y_1, and y_2. This formula allows us to estimate the integral without explicit computation of the equations of the parabolas approximating the graph of f. (Note that we did not have to find the equations of the linear approximations in the trapezoidal rule either.) It can be shown that if $y = ax^2 + bx + c$ is a parabola, as shown in Fig. 5.41, through the points (x_0, y_0), (x_1, y_1), and (x_2, y_2) where $x_1 - x_0 = x_2 - x_1 = h$, then

$$\int_{x_0}^{x_2} (ax^2 + bx + c)\, dx = \frac{h}{3}(y_0 + 4y_1 + y_2). \tag{5}$$

Soon we will show why Eq. (5) is true. The number h becomes $(b - a)/n$ in the approximation of $\int_a^b f(x)\, dx$. Thus Eq. (5) shows that the sum of the areas under the parabolas in approximating $\int_a^b f(x)\, dx$ is

$$\frac{b - a}{3n}(y_0 + 4y_1 + y_2) + \frac{b - a}{3n}(y_2 + 4y_3 + y_4) + \cdots$$

$$+ \frac{b - a}{3n}(y_{n-2} + 4y_{n-1} + y_n)$$

$$= \frac{b - a}{3n}(y_0 + 4y_1 + 2y_2 + 4y_3 + 2y_4 + \cdots + 2y_{n-2} + 4y_{n-1} + y_n).$$

This gives us the parabolic rule.

THEOREM 5.6   Simpson's parabolic rule

If f is continuous in $[a, b]$ and if n is even, then

$$\int_a^b f(x)\, dx \approx \frac{b - a}{3n}(y_0 + 4y_1 + 2y_2 + 4y_3 + 2y_4 + \cdots$$

$$+ 2y_{n-2} + 4y_{n-1} + y_n) \tag{6}$$

for n sufficiently large. $\square$

The computation of x_i and $y_i = f(x_i)$ for Simpson's rule is identical with that for the trapezoidal rule.

EXAMPLE 4   As in Example 2, take $n = 4$ and estimate $\int_1^3 (1/x)\, dx$.

Solution   We have

$$x_0 = 1, \qquad x_1 = \frac{3}{2}, \qquad x_2 = 2, \qquad x_3 = \frac{5}{2}, \qquad x_4 = 3,$$

$$y_0 = 1, \qquad y_1 = \frac{2}{3}, \qquad y_2 = \frac{1}{2}, \qquad y_3 = \frac{2}{5}, \qquad y_4 = \frac{1}{3}.$$

Simpson's rule yields

$$\int_1^3 \frac{1}{x}\, dx \approx \frac{3 - 1}{3 \cdot 4}\left(1 + 4 \cdot \frac{2}{3} + 2 \cdot \frac{1}{2} + 4 \cdot \frac{2}{5} + \frac{1}{3}\right)$$

$$= \frac{2}{12}\left(1 + \frac{8}{3} + 1 + \frac{8}{5} + \frac{1}{3}\right)$$

$$= \frac{1}{6}\left(\frac{15 + 40 + 15 + 24 + 5}{15}\right)$$

$$= \frac{1}{6}\left(\frac{99}{15}\right) = \frac{33}{30} = \frac{11}{10} = 1.1.$$

The correct answer to four decimal places is 1.0986, and comparison with Example 2 shows that Simpson's rule gives the more accurate estimate for the same value $n = 4$. ∎

It can be shown that if $f^{(4)}(x)$ is continuous in $[a, b]$, then the error in approximating $\int_a^b f(x)\, dx$ by Simpson's rule for a value n is given by

$$\frac{(b - a)^5}{180 n^4} f^{(4)}(c) \qquad \textit{Error in Simpson's rule} \qquad (7)$$

for some number c in $[a, b]$. For example, the magnitude of our error in estimating $\int_1^3 (1/x)\, dx$ by Simpson's rule with $n = 4$ in Example 4 is given by

$$\left|\frac{2^5}{180 \cdot n^4} \cdot \frac{24}{c^5}\right| \leq \frac{2^5}{180 \cdot 4^4} \cdot \frac{24}{1} = \frac{1}{60} \approx 0.0167.$$

The difference between our answer 1.1 in Example 4 and the actual value 1.0986 to four decimal places is 0.0014, so 0.0167 is indeed a bound for the magnitude of the error.

If $f(x)$ is a polynomial function of degree 3, then $f^{(4)}(x) = 0$ for all x. The preceding error formula then shows that for any estimate of $\int_a^b f(x)\, dx$ using Simpson's rule, the *exact value* for the integral is obtained, even if we use only $n = 2$. This seems surprising, since Simpson's rule involves approximation by only quadratic polynomials.

EXAMPLE 5   Estimate $\int_1^5 x^4\, dx$ using Simpson's rule with $n = 4$, and compare this estimate with the estimates in Examples 1 and 3.

Solution   We have, as in Example 3, $(b - a)/n = 1$ and

$$x_0 = 1, \qquad x_1 = 2, \qquad x_2 = 3, \qquad x_3 = 4, \qquad x_4 = 5,$$
$$y_0 = 1, \qquad y_1 = 2^4, \qquad y_2 = 3^4, \qquad y_3 = 4^4, \qquad y_4 = 5^4.$$

Simpson's rule then yields

$$\int_1^5 x^4\, dx \approx \frac{1}{3}(1 + 4 \cdot 2^4 + 2 \cdot 3^4 + 4 \cdot 4^4 + 5^4)$$

$$= \frac{1876}{3} \approx 625.3333.$$

This is much closer to the true answer 624.8 than is the rectangular estimate 604.25 in Example 1 or the trapezoidal estimate 666 in Example 3. ∎

In general, for a particular value of n, we may expect the rectangular and trapezoidal rules to be of roughly the same degree of accuracy. Simpson's rule is apt to give a significantly more accurate estimate for the same value of n, as our examples illustrated. (However, we can construct integrals where this is not true; see Exercise 17.) Since Simpson's rule is really no more work than the others, we recommend reaching for it when approximating a definite integral. These days, a calculator or computer is generally available to assist with the computations.

Derivation of Equation (5)

We now give the details of the derivation of Eq. (5). Recall that (x_0, y_0), (x_1, y_1), and (x_2, y_2) are points on the parabola $y = ax^2 + bx + c$. By translating axes if necessary, we can assume that $x_1 = 0$, so $x_0 = -x_2$ (see Fig. 5.42). Then

$$\int_{-x_2}^{x_2} (ax^2 + bx + c)\, dx = \left(\frac{ax^3}{3} + \frac{bx^2}{2} + cx\right)\Bigg]_{-x_2}^{x_2}$$

$$= \frac{ax_2^3}{3} + \frac{bx_2^2}{2} + cx_2 - \left(-\frac{ax_2^3}{3} + \frac{bx_2^2}{2} - cx_2\right) \quad (8)$$

$$= \frac{2}{3}ax_2^3 + 2cx_2.$$

On the other hand,

$$h = x_1 - x_0 = 0 - (-x_2) = x_2,$$

$$y_0 = f(x_0) = f(-x_2) = ax_2^2 - bx_2 + c,$$

$$y_1 = f(x_1) = f(0) = c,$$

$$y_2 = f(x_2) = ax_2^2 + bx_2 + c.$$

Thus

$$\frac{h}{3}(y_0 + 4y_1 + y_2) = \frac{x_2}{3}(ax_2^2 - bx_2 + c + 4c + ax_2^2 + bx_2 + c)$$

$$= \frac{x_2}{3}(2ax_2^2 + 6c) \quad (9)$$

$$= \frac{2}{3}ax_2^3 + 2cx_2.$$

Comparison of Eqs. (8) and (9) establishes Eq. (5).

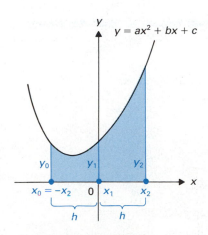

FIGURE 5.42 Region from x_0 to x_2 under a parabola translated so $x_1 = 0$, $x_0 = -x_2$.

SUMMARY

Let f be continuous in $[a, b]$ and let $[a, b]$ be subdivided into n subintervals of equal length $(b - a)/n$.

1. *Rectangular rule:* If x_i is the midpoint of the ith subinterval, then for sufficiently large n,

$$\int_a^b f(x)\, dx \approx \frac{b - a}{n}[f(x_1) + f(x_2) + \cdots + f(x_n)].$$

2. *Trapezoidal rule:* If $x_0 = a, x_1, x_2, \ldots, x_n = b$ are the endpoints of the subintervals and $y_0 = f(x_0), y_1 = f(x_1), y_2 = f(x_2), \ldots, y_n = f(x_n)$, then for sufficiently large n,

$$\int_a^b f(x)\, dx \approx \frac{b - a}{2n}(y_0 + 2y_1 + 2y_2 + \cdots + 2y_{n-1} + y_n).$$

3. *Simpson's rule:* If n is even and $y_0, y_1, y_2, \ldots, y_n$ are as described in the trapezoidal rule, then for sufficiently large n,

$$\int_a^b f(x)\, dx \approx \frac{b - a}{3n}(y_0 + 4y_1 + 2y_2 + 4y_3 + 2y_4 + \cdots$$
$$+ 2y_{n-2} + 4y_{n-1} + y_n).$$

4. In general, we expect Simpson's rule to be more accurate than either the rectangular rule or the trapezoidal rule for a given value of n.

5. Formulas (3) and (7) in the text give the error in the approximations in the trapezoidal rule and Simpson's rule.

EXERCISES 5.6

It is realistic to use a calculator or computer these days for approximating an integral. However, Exercises 1–15 are designed to be feasible with pencil and paper. Use the indicated rule and value of n to estimate the integral.

1. $\int_1^4 \frac{1}{x}\, dx$, rectangular rule, $n = 3$

2. $\int_1^4 \frac{1}{x}\, dx$, trapezoidal rule, $n = 3$

3. $\int_1^4 \frac{1}{x}\, dx$, Simpson's rule, $n = 2$

4. $\int_1^2 \frac{1}{x}\, dx$

    (a) trapezoidal rule, $n = 4$

    (b) Simpson's rule, $n = 4$

5. $\sqrt{3} = \int_0^{\pi/3} 2 \cos x\, dx$

    (a) rectangular rule, $n = 4$

    (b) Simpson's rule, $n = 4$

    (Use a table for cosine.)

6. $\pi = \int_0^1 \frac{4}{1 + x^2}\, dx$, trapezoidal rule, $n = 4$

    (Chapter 7 will show that this integral equals π.)

7. $\pi = \int_0^1 \frac{4}{1 + x^2}\, dx$, Simpson's rule, $n = 4$

8. $\int_0^\pi \dfrac{dx}{2 + \sin^2 x}$, trapezoidal rule, $n = 4$

9. $\int_0^\pi \dfrac{dx}{2 + \sin^2 x}$, Simpson's rule, $n = 4$

10. $\int_3^7 \dfrac{dx}{x - 1}$, Simpson's rule, $n = 4$

11. $\int_3^7 \dfrac{dx}{x - 1}$, trapezoidal rule, $n = 4$

12. $\int_1^2 \dfrac{x}{x + 1}\, dx$, Simpson's rule, $n = 2$

13. $\int_1^2 \dfrac{x}{x + 1}\, dx$, trapezoidal rule, $n = 2$

14. $\int_2^4 \dfrac{x^2}{x + 2}\, dx$, rectangular rule, $n = 4$

15. $\int_2^4 \dfrac{x^2}{x + 2}\, dx$, Simpson's rule, $n = 4$

16. Show that the bound for the error in Simpson's rule given in the text is consistent with the result obtained in Example 5.

17. Give an example of a definite integral and a value of n such that Simpson's rule gives a less accurate estimate than either the rectangular rule or the trapezoidal rule. [*Hint:* Consider $f(x) = |x|$.]

In Exercises 18–30, use a computer or calculator to estimate the integral, using the indicated rule and value for n.

18. $\int_0^\pi \sqrt{1 + \sin^2 x}\, dx$, Simpson's rule, $n = 10$

19. $\int_1^{21} \dfrac{1}{x}\, dx$, Simpson's rule, $n = 20$

20. $\pi = \int_0^1 \dfrac{4}{1 + x^2}\, dx$, Simpson's rule, $n = 10$

21. $\pi = \int_0^{1/2} \dfrac{6}{\sqrt{1 - x^2}}\, dx$, trapezoidal rule, $n = 10$

22. $\int_0^\pi \sqrt{4 - \sin^2 x}\, dx$, trapezoidal rule, $n = 10$

23. $\int_0^3 (1 + x)^x\, dx$, Simpson's rule, $n = 12$

24. $\int_0^4 \dfrac{dx}{\sqrt{6 - \cos^2 x}}$, Simpson's rule, $n = 8$

25. $\int_0^{\sqrt\pi} \sin x^2\, dx$, Simpson's rule, $n = 10$

26. $\int_0^1 \cos(x^2 + 1)\, dx$, Simpson's rule, $n = 10$

27. $\int_0^{\pi/4} \sec x\, dx$, trapezoidal rule, $n = 10$

28. $\int_0^1 2^x\, dx$, Simpson's rule, $n = 20$

29. $\int_0^4 \dfrac{x}{1 + \sin^2 x}\, dx$, Simpson's rule, $n = 10$

30. $\int_0^2 \dfrac{x}{4 - \cos^2 x}\, dx$, Simpson's rule, $n = 20$

EXPLORING CALCULUS

S: numerical integration graphic
T: numerical integration comparison
U: Simpson's rule compute

SUPPLEMENTARY EXERCISES FOR CHAPTER 5

1. Compute $\displaystyle\sum_{i=1}^4 \left(\dfrac{i}{2}\right)^2$.

2. Find the midpoint Riemann sum R_{mid} for $f(x) = x + x^2$ over $[0, 1]$, using the regular partition with $n = 4$.

3. Using the regular partition with $n = 4$, find the midpoint Riemann sum R_{mid} estimating $\int_0^{2\pi} \sin(x/2)\, dx$.

4. Find the upper sum R_{max} for $f(x) = 1/(x + 1)$ over $[2, 6]$, using the regular partition with $n = 4$.

5. Use geometry to find $\int_0^3 (2 + \sqrt{9 - x^2})\, dx$.

6. Use geometry to find $\int_{-5}^5 (5 - \sqrt{25 - x^2})\, dx$.

7. Find $\int_0^3 \sqrt{54 - 6x^2}\, dx$.

8. If $\int_{-1}^4 f(x)\, dx = 4$ and $\int_2^4 (3 - f(x))\, dx = 7$, find $\int_2^{-1} f(x)\, dx$.

9. If $\int_0^{10} (3 \cdot f(x) - g(x))\, dx = 5$ and $\int_{10}^0 (f(x) + g(x))\, dx = 2$, find $\int_0^{10} f(x)\, dx$.

10. If $\int_0^5 3 \cdot f(x)\, dx = 2$ and $\int_2^5 f(x)\, dx = -1$, find $\int_2^0 f(x)\, dx$.

11. If $\int_1^4 f(x)\, dx = 5$ and $\int_4^2 f(x)\, dx = -7$, find $\int_1^2 f(x)\, dx$.

12. Find $(d^2/dt^2)(\int_t^3 \cos x^2\, dx)$.

13. Find $\dfrac{d}{dt}\left(\int_1^{\sqrt t} \sqrt{2x^2 + x^4}\, dx\right)$.

14. Find $\dfrac{d}{dt}\left(\int_{-t}^{2t} \sin^2 x\, dx\right)$.

15. Find $\int_{-3}^{3} (x^2 - 1 + \sqrt{9 - x^2})\, dx$.

16. Find $\int_{1}^{3} (1/x^2)\, dx$.

17. Find $\int_{-1}^{1} x(x^2 + 1)^3\, dx$.

18. Find $\int_{0}^{1} x\sqrt{1 - x^2}\, dx$.

19. Find $\int_{1}^{2} \dfrac{x^2 + 1}{x^2}\, dx$.

20. Find $\int_{\pi/6}^{3\pi/4} \cot x\, dx$.

21. Find $\int_{0}^{\pi/2} \sin 2x\, dx$.

22. Find $\int_{0}^{\pi} \sin 3x\, dx$.

23. Find $\int_{0}^{\pi/6} \sec^2 2x\, dx$.

24. Find $\int_{\pi/6}^{\pi/3} \sin 2x\, dx$.

25. Find $\int_{0}^{\ln 3} e^{-2x}\, dx$.

26. Find the area of the region under the curve $y = x^3 + 2x$ and over the x-axis between $x = 1$ and $x = 2$.

27. Find the area of the region under one "arch" of $y = \sin 2x$ and over the x-axis.

28. Find the area of the region over the x-axis and under the curve $y = 16 - x^2$ between $x = -4$ and $x = 4$.

29. Find the area of the region bounded by $y = x^2 - 4$ and the x-axis.

30. Find $\int x^2(4x^3 + 2)^5\, dx$.

31. Find $\int \dfrac{x^2}{(x^3 + 1)^4}\, dx$.

32. Find $\int \sin 2x \cos 2x\, dx$.

33. Find $\int \dfrac{\cos 3x}{\sin^4 3x}\, dx$.

34. Find $\int \sec^3 x \tan x\, dx$.

35. Find $\int \sqrt{\csc x}\, \cot x\, dx$.

36. Find $\int \sin 3x \sec^2 3x\, dx$.

37. Find $\int \tan 3x\, dx$.

38. Find $\int \sqrt{\sec x}\, \sin x\, dx$.

39. Find $\int xe^{x^2}\, dx$.

40. Find $\int \dfrac{e^{-x}}{1 + e^{-x}}\, dx$.

41. Find $\int \dfrac{d^3(\sin x^2)}{dx^3}\, dx$.

42. Find the general solution of the differential equation $dy/dx = x^2\sqrt{y}$.

43. Find the general solution of the differential equation
$$\frac{dy}{dx} = x \cos^2 2y.$$

44. Find the general solution of the differential equation $dx/dt = 1/(x^2 t)$.

45. Find the solution of the differential equation $dy/dx = y^2 \cos 2x$ such that $y = -3$ when $x = \pi/4$.

46. Find the solution of the differential equation $dy/dx = e^{-x} \cot y$ passing through the origin.

47. Find the solution of the initial value problem $dy/dx = x^2 y^2$ such that $y = 2$ when $x = 1$.

48. Use the trapezoidal rule with $n = 4$ to estimate
$$\int_{-1}^{1} \frac{1}{x^2 + 1}\, dx.$$

49. Use Simpson's rule with $n = 4$ to estimate
$$\int_{0}^{2} \frac{1}{x + 2}\, dx.$$

50. Use a calculator and Simpson's rule with $n = 16$ to estimate
$$\int_{0}^{\pi/4} \tan x\, dx.$$

51. Use a calculator and Simpson's rule with $n = 12$ to estimate
$$\int_{1}^{3} \frac{x}{1 + x^2}\, dx.$$

Examples 5 and 6 of Section 5.2 showed how integrals can sometimes be found by evaluating the limit of a sum. Conversely, working backward through the examples, we see how a limit of a sum might be recognized as giving an integral and then evaluated using the fundamental theorem. In Exercises 52–57, use an integral to find each of the limits.

52. $\displaystyle\lim_{n\to\infty} \frac{4}{n} \cdot \sum_{i=1}^{n} \frac{4i^2}{n^2}$

53. $\displaystyle\lim_{n\to\infty} \frac{8}{n^4} \cdot \sum_{i=1}^{n} i^3$

54. $\displaystyle\lim_{n\to\infty} \frac{1}{n^{5/2}} \cdot \sum_{i=1}^{n} i\sqrt{i}$

55. $\displaystyle\lim_{n\to\infty} \frac{1}{n^5} \cdot \sum_{k=1}^{n} (1 + k)^3$

56. $\displaystyle\lim_{n\to\infty} \frac{1}{n^5} \cdot \sum_{k=1}^{n} (2k + 1)^4$

57. $\displaystyle\lim_{n\to\infty} \frac{1}{n^3} \cdot \sum_{k=1}^{n} (k - 100)^3$

APPLICATIONS OF THE INTEGRAL

6

At the start of Chapter 5, we indicated that integral calculus is related to the notion of a *product* in much the same way as differential calculus is related to a *quotient*. Calculus is the mathematics needed to study quantities of changing magnitude. The definite integral is potentially useful in any type of application where, in the case of quantities that do not vary, we would compute a *product*. (Since any quantity can be viewed as the product of itself with 1, this covers a very general situation.) We list here several things that may be given by a product of two quantities, some of which we mentioned before.

Area The area A of a plane region of base width b and constant altitude h (that is, a rectangle) is given by $A = bh$.

Volume For a solid with horizontal cross section of constant area A and with constant height h, the volume V is given by $V = Ah$.

Work If a body is moved a distance s by means of a constant force F acting in the direction of the motion, the work W done in moving the body is given by $W = Fs$.

Distance If a body travels in a straight line for a time t with a constant velocity v, the distance s traveled is given by $s = vt$.

Velocity If a body starting from rest travels in a straight line for a time t with a constant acceleration a, the velocity v attained is given by $v = at$.

Force If the pressure per square unit on a plane region of area A is a constant p throughout the region, then the total force F on the region due to this pressure is given by $F = pA$.

Moment The moment M about an axis of a body of mass m, all points of which are a constant (signed) distance s from the axis, is given by $M = ms$.

Moment of inertia The moment of inertia I about an axis of a body of mass m, all points of which are a constant distance s from the axis, is given by $I = ms^2$.

All these notions can be handled by integral calculus if the factors appearing in the products vary continuously. This chapter is devoted to such applications of the integral.

6.1 AREA AND AVERAGE VALUE

The Area of a Plane Region

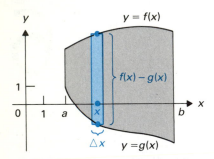

FIGURE 6.1 Thin vertical rectangle of area $(f(x) - g(x))\, \Delta x$.

This section shows how to find the area of a more general plane region than one under the graph of a continuous function from a to b. Consider, for example, the shaded region in Fig. 6.1 that lies between the graphs of continuous functions f and g from a to b. The area of this region can be estimated by partitioning $[a, b]$, taking thin rectangles like the one shown shaded in color in Fig. 6.1, and adding up the areas of such rectangles over the region. This rectangle is Δx units wide and $f(x) - g(x)$ units high, for x as shown in Fig. 6.1. (Note that $g(x)$ is itself negative. The vertical distance between the graphs of f and g over a point x is always $f(x) - g(x)$ if the graph of f is above that of g at x.) Thus the area of this rectangle is

$$(f(x) - g(x))\, \Delta x.$$

We wish to add up the areas of such rectangles and take the limit of the resulting sum as Δx grows smaller and the number of rectangles increases. As we know, the limit of such a sum will be $\int_a^b (f(x) - g(x))\, dx$.

In computing the area of a plane region, we must take care in finding the correct function to integrate, that is, in "setting up the integral." The area of the region bounded by the graphs of continuous functions f and g between a and b is not always $\int_a^b (f(x) - g(x))\, dx$. If $g(x) \le f(x)$ for all x in $[a, b]$, then the correct integral is indeed $\int_a^b (f(x) - g(x))\, dx$. However, if f and g have the graphs shown in Fig. 6.2, then $\int_a^b (f(x) - g(x))\, dx$ is numerically equal to the area of the first region minus the area of the second. The area of the total shaded region in Fig. 6.2 is best obtained by finding the areas of the first and second regions separately and then adding the areas; a correct expression is

$$\int_a^c (f(x) - g(x))\, dx + \int_c^b (g(x) - f(x))\, dx.$$

The area of the region between the graphs of f and g from a to b can always be expressed as

$$\int_a^b |f(x) - g(x)|\, dx.$$

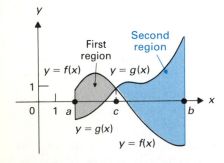

FIGURE 6.2 Total area is $\int_a^c (f(x) - g(x))\, dx$ $\quad + \int_c^b (g(x) - f(x))\, dx.$

Indeed, it is appropriate to *define* the area to be this integral. In practice, we do not work with the absolute value sign; rather, we look at a figure and set up one or more integrals to find the desired area. Setting up the integrals is the interesting part of the problem. Computation of the integrals should be regarded as a mechanical chore, much like adding a column of numbers, although most students haven't had as much experience in computing integrals.

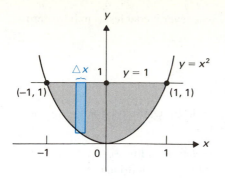

FIGURE 6.3 Thin vertical rectangle of area $(1 - x^2)\,\Delta x$.

Suggested steps for finding the area of a plane region are given at the bottom of this page. We illustrate with several examples.

EXAMPLE 1 Find the area of the region bounded by $y = 1$ and $y = x^2$ in two ways, once using Δx-increments and once using Δy-increments.

Solution 1 Δx-increments

STEP 1. Figure 6.3 shows the region bounded by $y = 1$ and $y = x^2$.

STEP 2. Figure 6.3 also shows a thin vertical rectangle of width Δx for this region.

STEP 3. The vertical height of the rectangle in Fig. 6.3 is the y-value at the top minus the y-value at the bottom, which we abbreviate by

$$\text{Vertical height} = y_{\text{top}} - y_{\text{bottom}}.$$

Since we are using Δx-increments, we must express this height as a function of x. We have $y_{\text{top}} = 1$ and $y_{\text{bottom}} = x^2$, so the height is $1 - x^2$. Thus the area of the rectangular strip is

$$A_{\text{strip}} = (y_{\text{top}} - y_{\text{bottom}})\,\Delta x = (1 - x^2)\,\Delta x.$$

STEP 4. Since we want to add up the areas of our rectangles over the region from $x = -1$ to $x = 1$, we have

$$A = \int_{-1}^{1} (1 - x^2)\, dx = \left(x - \frac{x^3}{3}\right)\Bigg]_{-1}^{1}$$

$$= \left(1 - \frac{1}{3}\right) - \left(-1 + \frac{1}{3}\right) = \frac{4}{3}.$$

Alternatively, we can make use of the obvious symmetry of our region about the y-axis and write

$$A = 2\int_{0}^{1} (1 - x^2)\, dx = 2\left(x - \frac{x^3}{3}\right)\Bigg]_{0}^{1} = 2\left(1 - \frac{1}{3}\right) - 0 = \frac{4}{3}.$$

Finding the Area of a Plane Region

STEP 1. Sketch the region, finding the points of intersection of bounding curves.

STEP 2. On the sketch, draw a typical thin rectangle arising from a partition P, vertical of width Δx or horizontal of width Δy.

STEP 3. *Looking at the sketch,* write down the area of this rectangle as a product of the width (Δx or Δy) and the length. Express this area entirely in terms of the variable (x or y) appearing in the Δ-increment.

STEP 4. Integrate between the appropriate (x or y) limits. (By definition of the integral, this amounts to adding the areas found in step 3 and taking the limit of the resulting sum as $\|P\| \to 0$.)

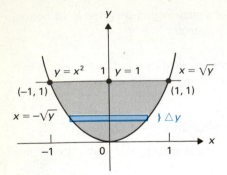

FIGURE 6.4 Thin horizontal rectangle of area $[\sqrt{y} - (-\sqrt{y})]\,\Delta y$.

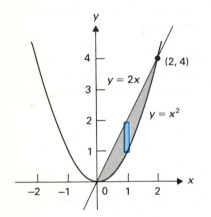

FIGURE 6.5 Thin vertical rectangle of area $(2x - x^2)\,\Delta x$.

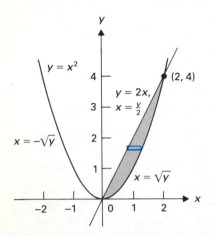

FIGURE 6.6 Thin horizontal rectangle of area $(\sqrt{y} - y/2)\,\Delta y$.

We prefer to use such symmetry whenever it enables us to have zero as one of the limits of integration.

Solution 2 Δy-increments

STEP 1. The region is sketched in Fig. 6.4.

STEP 2. We draw a thin horizontal rectangle of width Δy for the region in Fig. 6.4.

STEP 3. The horizontal dimension of the rectangle in Fig. 6.4 is the x-value at the right-hand end minus the x-value at the left-hand end, which we abbreviate to

$$\text{Horizontal length} = x_{\text{right}} - x_{\text{left}}.$$

Since we are using Δy-increments, we must express this horizontal length as a function of y. Both ends of the rectangle in Fig. 6.4 fall on the curve $y = x^2$, so we have $x_{\text{right}} = \sqrt{y}$ and $x_{\text{left}} = -\sqrt{y}$. Thus

$$(x_{\text{right}} - x_{\text{left}})\,\Delta y = \left[\sqrt{y} - (-\sqrt{y})\right]\Delta y = 2\sqrt{y}\,\Delta y.$$

STEP 4. Since we wish to add up the areas of our rectangles in the y-direction from $y = 0$ to $y = 1$, we have

$$A = \int_0^1 2\sqrt{y}\,dy = 2\int_0^1 y^{1/2}\,dy = 2 \cdot \frac{2}{3} y^{3/2}\Big]_0^1$$

$$= \frac{4}{3} \cdot 1^{3/2} - 0 = \frac{4}{3}. \quad \blacksquare$$

As indicated in Example 1, it is useful to remember that

$$A_{\text{strip}} = (y_{\text{top}} - y_{\text{bottom}})\,\Delta x \qquad \text{for a vertical strip,}$$
$$A_{\text{strip}} = (x_{\text{right}} - x_{\text{left}})\,\Delta y \qquad \text{for a horizontal strip.}$$

EXAMPLE 2 Find the area of the region bounded by $y = 2x$ and $y = x^2$ in two ways, as in Example 1.

Solution 1 Δx-increments

STEP 1. The region is sketched in Fig. 6.5.

STEP 2. Figure 6.5 shows a thin vertical rectangle of width Δx.

STEP 3. We need the vertical height of the rectangle in terms of x. Since the line $y = 2x$ is on top of the region and $y = x^2$ is at the bottom, we see that

$$A_{\text{strip}} = (y_{\text{top}} - y_{\text{bottom}})\,\Delta x = (2x - x^2)\,\Delta x.$$

STEP 4. We have

$$A = \int_0^2 (2x - x^2)\,dx = \left(x^2 - \frac{x^3}{3}\right)\Big]_0^2 = \left(4 - \frac{8}{3}\right) - 0 = \frac{4}{3}.$$

Solution 2 Δy-increments

STEP 1. The region is sketched in Fig. 6.6.

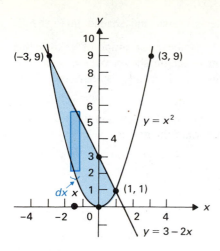

FIGURE 6.7 Thin vertical rectangle of area $dA = [(3 - 2x) - x^2]\, dx$.

STEP 2. Figure 6.6 shows a thin horizontal rectangle of width Δy.

STEP 3. We need the horizontal length of the rectangle in terms of y. Since $y = x^2$ is to the right of $y = 2x$ in the region, we have $x_{\text{right}} = \sqrt{y}$ and $x_{\text{left}} = y/2$. Thus

$$A_{\text{strip}} = (x_{\text{right}} - x_{\text{left}})\,\Delta y = \left(\sqrt{y} - \frac{y}{2}\right)\Delta y.$$

STEP 4. We wish to add the areas of the rectangles from $y = 0$ to $y = 4$, so

$$A = \int_0^4 \left(\sqrt{y} - \frac{y}{2}\right) dy = \int_0^4 \left(y^{1/2} - \frac{1}{2}y\right) dy = \left(\frac{2}{3}y^{3/2} - \frac{y^2}{4}\right)\Bigg]_0^4$$

$$= \left(\frac{2}{3}\cdot 8 - \frac{16}{4}\right) = \frac{16}{3} - 4 = \frac{4}{3}. \qquad \blacksquare$$

If we want to find the area of a region, we don't need to compute integrals for both Δx- and Δy-increments as we did in Examples 1 and 2. We set up and compute whichever of the integrals looks easier. In Examples 1 and 2, we would probably have used just the Δx-increments, making use of symmetry in Example 1, although the Δy-increments were equally simple in those examples. The next two examples show that one integral may be easier to use than the other. We will set up the integrals for those examples using differential notation, which we now discuss.

Differential Notation

Recall from Section 3.9 that if $y = f(x)$, then a heuristic standard analysis interpretation of dx is to consider dx to be the same as Δx, that is, to be an increment in x. This suggestive device is widely employed in applications. With this interpretation, the thin rectangle shaded in Fig. 6.7 is considered to have width dx, and the area of the thin rectangle is considered to be dA, so that

$$dA = (y_{\text{top}} - y_{\text{bottom}})\, dx.$$

We compute the area A by integrating dA between the appropriate limits a to b. In Section 5.3 we introduced the notion of the area function $F(x)$, which is equal to the integral of dA from a to x. We showed that the derivative of this area function is the integrand. Thus it is consistent with our work in Section 3.9 for us to consider dA to be the integrand multiplied by dx.

We work the next two examples in differential notation. It is important to become accustomed to this notation since it is widely used in physics and other sciences that employ calculus.

EXAMPLE 3 Find the area of the plane region bounded by the curves with equations $y = x^2$ and $y = 3 - 2x$, using dx-increments.

Solution

STEP 1. The curves are sketched and the region whose area we want to find is shown shaded in Fig. 6.7. The points $(-3, 9)$ and $(1, 1)$ of intersection of the bounding curves are found by solving the equations $y = x^2$ and $y = 3 - 2x$ simultaneously.

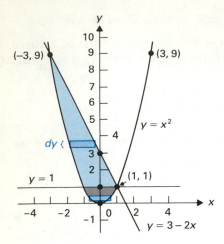

FIGURE 6.8 Upper rectangle has area

$$dA = [\tfrac{1}{2}(3 - y) - (-\sqrt{y})]\, dy;$$

lower rectangle has area

$$dA = [\sqrt{y} - (-\sqrt{y})]\, dy.$$

STEP 2.   A typical thin vertical rectangle is drawn and shaded in Fig. 6.7.

STEP 3.   The graph of the function $3 - 2x$ is above the graph of x^2 over this region, so the height of the thin rectangle shown over a point x is $(3 - 2x) - x^2$. The area of this thin rectangle is therefore $dA = [(3 - 2x) - x^2]\, dx$.

STEP 4.   Since we wish to add the areas of the thin rectangles in the x-direction from $x = -3$ to $x = 1$, the appropriate integral is

$$\int_{-3}^{1} [(3 - 2x) - x^2]\, dx = \left(3x - 2\frac{x^2}{2} - \frac{x^3}{3}\right)\Bigg]_{-3}^{1}$$

$$= 3 - 1 - \frac{1}{3} - (-9 - 9 + 9)$$

$$= \frac{32}{3}.$$

EXAMPLE 4   Solve Example 3 by taking the thin rectangles horizontally.

Solution

STEP 1.   The appropriate sketch is given in Fig. 6.8.

STEP 2.   Note that we can't take just one rectangle as typical for the whole region, for if the rectangle lies above the line given by $y = 1$, it is bounded on the right by the line with equation $y = 3 - 2x$, while a rectangle below the line given by $y = 1$ is bounded on both ends by the curve given by $y = x^2$. We split the region into two parts by the line with equation $y = 1$ and find the area of each part separately.

STEP 3.   To find the horizontal dimensions of the rectangles, we must solve for x in terms of y, obtaining $x = (3 - y)/2$ for the line and $x = \pm\sqrt{y}$ for the curve. The upper rectangle has area

$$dA = \left[\frac{1}{2}(3 - y) - (-\sqrt{y})\right] dy,$$

while the lower rectangle has area

$$dA = \left[\sqrt{y} - (-\sqrt{y})\right] dy.$$

STEP 4.   The appropriate integrals are

$$\int_{1}^{9} \left[\frac{1}{2}(3 - y) - (-\sqrt{y})\right] dy = \left(\frac{3}{2}y - \frac{y^2}{4} + \frac{2}{3}y^{3/2}\right)\Bigg]_{1}^{9}$$

$$= \left(\frac{27}{2} - \frac{81}{4} + \frac{2}{3} \cdot 27\right)$$

$$- \left(\frac{3}{2} - \frac{1}{4} + \frac{2}{3}\right)$$

$$= \frac{28}{3},$$

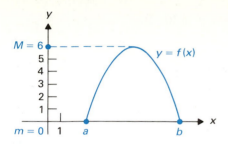

FIGURE 6.9 $f(a) = f(b) = 0$, but $f(x) > 0$ for $a < x < b$.

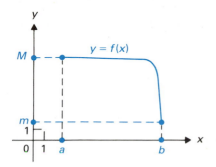

FIGURE 6.10 $f(x)$ is close to M over most of $[a, b]$.

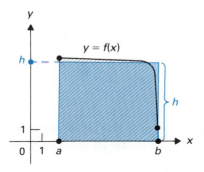

FIGURE 6.11 The area $h(b - a)$ of the rectangle equals the area $\int_a^b f(x)\, dx$ under the graph of f.

and

$$\int_0^1 \left[\sqrt{y} - (-\sqrt{y}) \right] dy = \int_0^1 2\sqrt{y}\, dy$$

$$= 2 \cdot \frac{2}{3} y^{3/2} \Big]_0^1 = \frac{4}{3} - 0 = \frac{4}{3}.$$

Thus the total area is $\frac{28}{3} + \frac{4}{3} = \frac{32}{3}$ square units. This computation is not as simple as that in Example 3. ∎

The Average Value of a Function

Let f be a continuous function with domain containing $[a, b]$, and consider the values $f(x)$ for x in $[a, b]$. These values can vary tremendously over an interval $[a, b]$, even if f is continuous. But we can try to develop some notion of the *average value of $f(x)$ over $[a, b]$*.

Perhaps the first natural attempt to find an average for $f(x)$ over $[a, b]$ is to form the average $(f(a) + f(b))/2$. After a little consideration, we see that we should reject such a definition of the average size for $f(x)$ over $[a, b]$ because $(f(a) + f(b))/2$ really reflects the size of $f(x)$ only at a and at b. For f with graph shown in Fig. 6.9, this average $(f(a) + f(b))/2$ is zero, even though $f(x) > 0$ for $a < x < b$.

Probably the next natural attempt to define an average for $f(x)$ over $[a, b]$ is to average the maximum size M and the minimum size m of all the $f(x)$ for x in $[a, b]$, if this maximum M and this minimum m exist. We know that M and m do exist if f is continuous in $[a, b]$. For the function f shown in Fig. 6.9, we have $M = 6$ and $m = 0$, so the average is $(M + m)/2 = \frac{6}{2} = 3$, which may seem like a reasonable value for the average of $f(x)$ over $[a, b]$. But for the function with graph shown in Fig. 6.10, the average of $f(x)$ over $[a, b]$ should surely be closer to M than to m, reflecting the fact that $f(x)$ is close to M over most of the interval $[a, b]$.

To take care of a situation like that shown in Fig. 6.10, we will regard the *average of $f(x)$ over $[a, b]$* as the height h that a rectangle with base $[a, b]$ must have for the area of the rectangle to be equal to the area of the region under the graph of f over $[a, b]$. This region under the graph of f over $[a, b]$ is shown shaded in Fig. 6.11, where a rectangle of equal area with height h is also indicated. The area of the rectangle is $h(b - a)$, while the area of the shaded region is $\int_a^b f(x)\, dx$, so if the areas are to be equal, we must have

$$h = \frac{1}{b - a} \int_a^b f(x)\, dx.$$

DEFINITION 6.1

Average value of a function

Let f be continuous in $[a, b]$. The **average** (or **mean**) **value** of $f(x)$ in $[a, b]$ is

$$\frac{1}{b - a} \int_a^b f(x)\, dx.$$

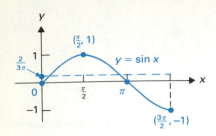

FIGURE 6.12 sin x has mean value $2/(3\pi)$ over $[0, 3\pi/2]$.

EXAMPLE 5   Find the average value of x^2 in $[0, 2]$.

Solution   We have

$$\int_0^2 x^2 \, dx = \frac{x^3}{3}\Big]_0^2 = \frac{8}{3}.$$

Thus the average value is

$$\frac{1}{2 - 0} \cdot \frac{8}{3} = \frac{4}{3}. \qquad \blacksquare$$

EXAMPLE 6   Find the mean value of sin x in $[0, 3\pi/2]$.

Solution   The function sin x takes on both positive and negative values in $[0, 3\pi/2]$, as shown in Fig. 6.12, but that does not affect the way we compute its mean value. We have

$$\int_0^{3\pi/2} \sin x \, dx = -\cos x\Big]_0^{3\pi/2} = 0 - (-1) = 1.$$

Thus the mean value is

$$\frac{1}{\dfrac{3\pi}{2} - 0} \cdot 1 = \frac{2}{3\pi}. \qquad \blacksquare$$

If m and M are the minimum value and maximum value for $f(x)$ over $[a, b]$, then we know that

$$m(b - a) \le \int_a^b f(x) \, dx \le M(b - a).$$

Dividing by $b - a$, we obtain

$$m \le \frac{1}{b - a} \int_a^b f(x) \, dx \le M,$$

which proves that the mean value of $f(x)$ in $[a, b]$ is between m and M. Since we are assuming that f is continuous in $[a, b]$, it follows by the intermediate value theorem that

$$\frac{1}{b - a} \int_a^b f(x) \, dx = f(c)$$

for some c where $a < c < b$. The existence of this c is known as the *mean value theorem for the integral*.

THEOREM 6.1   Mean value theorem for the integral

Let $f(x)$ be continuous in $[a, b]$. Then there exists a number c, where $a < c < b$, such that

$$\frac{1}{b - a} \int_a^b f(x) \, dx = f(c). \qquad \square$$

EXAMPLE 7 Illustrate Theorem 6.1 for $f(x) = x^2$ in $[0, 2]$.

Solution Example 5 showed that the average value of $f(x)$ in $[0, 2]$ is given by

$$\frac{1}{2-0} \int_0^2 x^2 \, dx = \frac{4}{3}.$$

Now $f(x) = x^2 = \frac{4}{3}$ when $x = \pm 2/\sqrt{3}$. The value $2/\sqrt{3}$ does indeed satisfy $0 < 2/\sqrt{3} < 2$, as required by Theorem 6.1, so we have $c = 2/\sqrt{3}$. ■

SUMMARY

1. The following is a suggested procedure (in differential notation) for finding the area of a region bounded by given curves. See page 299 for Δ-notation.

   STEP 1. Sketch the region, finding the points of intersection of the bounding curves.

   STEP 2. On the sketch, draw a typical thin rectangle either vertical and of width dx or horizontal and of width dy.

   STEP 3. Looking at the sketch, write down the area dA of this rectangle as a product of the width (dx or dy) and the length. Express dA entirely in terms of the variable (x or y) appearing in the differential:

   $$dA = (y_{\text{top}} - y_{\text{bottom}}) \, dx, \quad \text{or} \quad dA = (x_{\text{right}} - x_{\text{left}}) \, dy.$$

   STEP 4. Integrate dA between the appropriate (x or y) limits.

2. The mean or average value of $f(x)$ in $[a, b]$ is

   $$\frac{1}{b-a} \int_a^b f(x) \, dx.$$

3. *Mean value theorem for the integral:* If $f(x)$ is continuous in $[a, b]$, then there exists a number c, where $a < c < b$, such that

   $$\frac{1}{b-a} \int_a^b f(x) \, dx = f(c).$$

EXERCISES 6.1

In Exercises 1–31, find the total area of the region or regions bounded by the given curves.

1. $y = x^2$, $y = 4$
2. $y = x^4$, $y = 1$
3. $y = x$, $y = x^2$
4. $y = x$, $y = x^3$
5. $y = x^4$, $y = x^2$
6. $y = x^2$, $y = x^3$
7. $y = x^4 - 1$, $y = 1 - x^2$
8. $y = x^2 - 1$, $y = x + 1$
9. $x = y^2$, $y = x - 2$
10. $x = y^2 - 4$, $y = 2 - x$
11. $y = 1 - x^2$, $y = x - 1$
12. $y = \sqrt{x}$, $y = x^2$
13. $y = \sqrt{x}$, $y = x^4$
14. $x = 2y^2$, $x = 8$
15. $y = x^2 - 1$, $y = \sqrt{1 - x^2}$ [*Hint:* Use some known areas.]

16. $y = x^2$, $y = 10x - 9$

17. $y = x^2/4$, $y = 2x - 3$

18. $x = y^2 - 1$, $y = x/4 + 1$

19. $y = \sin x$, $y = 3 \sin x$ for $0 \le x \le \pi$

20. $y = \sin x$, $y = \cos x$ for $0 \le x \le \pi/4$, $x = 0$

21. $y = \sec^2 x$ for $-\pi/3 \le x \le \pi/3$, $y = 4$

22. $y = \csc^2 x$ for $\pi/3 \le x \le 2\pi/3$, $y = \frac{4}{3}$

23. $y = \sin 2x$, $y = \sec^2 x - 1$ for $0 \le x \le \pi/4$

24. $y = \sin x$, $y = 2x$, $x = \pi/2$

25. $y = \sin x$, $y = \sqrt{x}$, $x = \pi$

26. $y = x$, $y = 1/x$, $x = 2$

27. $y = 1/x$, $x + y = 5/2$

28. $y = \dfrac{9}{4x + 5}$, $x + y = 2$

29. $x + y = 1$, $y = \ln x$, $x = e$ (Use the table of integrals.)

30. $y = e^x - e$, $x = 0$, $y = 0$

31. $y = 4x/\pi$, $y = \tan x$

32. Express as an integral the area of the smaller region bounded by the curves $x^2 + y^2 = 4$ and $y = -1$. (See Fig. 6.13.)

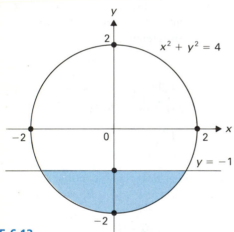

FIGURE 6.13

33. Express as an integral the area of the smaller region bounded by the curves $x = 2y^2$ and $x^2 + y^2 = 68$.

34. Express as one or more integrals the area of the region in the first quadrant bounded by the curves $x^2 + y^2 = 2$, $y = 0$, and $y = x^2$.

35. Express in terms of an integral the area of the larger region bounded by $x^2 + y^2 = 25$ and the line through $(-4, 3)$ and $(3, 4)$.

36. Express in terms of an integral the area of the larger region bounded by $x^2 + y^2 = 25$ and the line $x = -3$.

37. Find the value of c such that the triangle bounded by $y = 2x$, $y = 0$, and $x = 4$ is divided into two regions of equal areas by the line $y = c$. (See Fig. 6.14.)

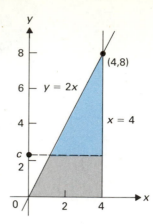

FIGURE 6.14

38. Find the value of c such that the region bounded by $y = x^2$ and $y = 4$ is divided into two regions of equal areas by $y = c$.

39. Find the average value of the function $1 - x^2$ in the interval $[-2, 2]$.

40. Find the average value of $\sin x$ in the interval $[0, \pi]$.

41. Find the value of c such that the average value of the function $x^4 - 1$ in the interval $[-c, c]$ is 0.

42. In one week, a small company makes a profit of \$150 on Monday, \$210 on both Tuesday and Wednesday, \$250 on Thursday, and \$75 on Friday.
    (a) Find the company's average daily profit that week.
    (b) Try to relate your answer in part (a) to the average value of some function $f(t)$ for $0 \le t \le 5$. Draw a sketch. [*Hint:* Your function will not be continuous, but there should be a natural idea of the area under the graph for $0 \le t \le 5$.]

43. Prove the mean value theorem for the integral by applying the mean value theorem for the derivative (Section 4.2) to $F(t) = \int_a^t f(x)\, dx$ on $[a, b]$.

Exercises 44–50 require a calculator or computer. Unless otherwise instructed, use Simpson's rule to estimate an integral giving the indicated area.

44. The area in Exercise 32

45. The area in Exercise 33

46. The area in Exercise 34

47. The area in Exercise 35

48. The area in Exercise 36

49. The area of the region bounded by $y = x^2$ and $y = \sin x$ (Use Newton's method to find one of the limits of integration. You will not need Simpson's rule.)

50. The area of the region bounded by $y = x^2$ and $y = \cos x$ [*Hint:* Proceed as suggested in Exercise 49.]

The remaining problems concern other applications of the integral. It is important to be able to recognize situations where an integral is appropriate.

51. The *length mass density* $\rho(x)$ of a rod at a distance x from one end of the rod is the mass a unit length of the rod would

have if it had the same composition and thickness everywhere as it has at the distance x. A tapered rod 3 ft long has mass density

$$\frac{1}{20} \cdot \frac{x + 5}{x + 1} \text{ slug/ft}$$

at the distance x ft from the thickest end of the rod. Find the mass of the rod. [*Hint:* Note that

$$\frac{x + 5}{x + 1} = 1 + \frac{4}{x + 1}.]$$

52. *Area mass density* (mass per unit area) is defined for a thin sheet of material in a manner analogous to the definition of length mass density in Exercise 51. A thin sheet of material covers the plane region bounded by $y = (x - 2)^2$ and $y = 1$. If units are in feet and the area mass density at (x, y) is $x/2$ slug/ft^2, find the mass of the sheet.

53. Repeat Exercise 52 if the area mass density at (x, y) is $y + 1$ slug/ft^2.

54. On a certain summer day, the temperature on the hour in degrees Fahrenheit from 6:00 A.M. to 6:00 P.M. is as follows.

Time	Temperature	Time	Temperature
6:00	55	1:00	85
7:00	60	2:00	82
8:00	65	3:00	80
9:00	67	4:00	87
10:00	70	5:00	70
11:00	78	6:00	65
12:00	83		

Use Simpson's rule to estimate the average temperature during the 12-hour period.

6.2 FINDING VOLUME BY SLICING

Volumes of Revolution

Let a plane region be revolved in the natural way about a given line (axis) lying in the plane. That is, each point P of the plane region describes a circular orbit, as shown in Fig. 6.15. This orbit bounds a disk having the given axis of revolution as perpendicular axis through its center. The three-dimensional region in Fig. 6.16 consisting of the points on all such orbits is the *solid of revolution* generated as the plane region is revolved about the axis. Using integral calculus, we can frequently find the volume of such a solid of revolution.

Consider the case in which a plane region under the graph of a continuous function f from a to b is revolved about the x-axis. Such a region is shown shaded in Fig. 6.17. The treatment for other regions is similar.

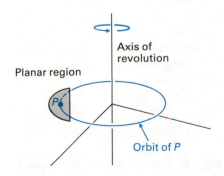

FIGURE 6.15 Orbit of a point P as a plane region is revolved about an axis.

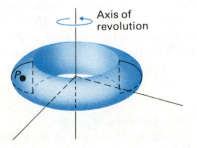

FIGURE 6.16 Solid of revolution.

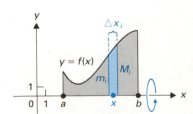

FIGURE 6.17 Vertical strip of width Δx_i with maximum height M_i and minimum height m_i.

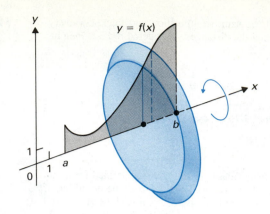

FIGURE 6.18 Slice generated by revolving the strip in Fig. 6.17 about the x-axis.

Let P be a partition of $[a, b]$ into n subintervals. Consider the contribution to the solid of revolution given by revolving the color-shaded strip in Fig. 6.17 about the x-axis. The strip sweeps out a slice, shown in Fig. 6.18, where the minimum radius is m_i and the maximum radius is M_i. The thickness of the slice is Δx_i.

The volume of a circular slice of radius r and thickness h is of course $\pi r^2 h$. Since the radius of the slice in Fig. 6.18 varies from m_i to M_i, its volume V_{slice} satisfies

$$\pi m_i^2 \, \Delta x_i \leq V_{\text{slice}} \leq \pi M_i^2 \, \Delta x_i.$$

Thus for the volume V of our whole solid of revolution, we have the relation

$$\sum_{i=1}^{n} \pi m_i^2 \, \Delta x_i \leq V \leq \sum_{i=1}^{n} \pi M_i^2 \, \Delta x_i. \tag{1}$$

Now for such partitions P, both extremes of the inequality (1) approach $\int_a^b \pi f(x)^2 \, dx$ as $\|P\| \to 0$, so our desired volume V of revolution is given by

$$V = \int_a^b \pi f(x)^2 \, dx. \tag{2}$$

The differential notation $\int_a^b \pi (f(x))^2 \, dx$ is very helpful and suggestive in this situation. For x as shown in Fig. 6.17, the radius of the slice in Fig. 6.18 is approximately $f(x)$, so the area of one face of the slice is roughly $\pi f(x)^2$. Since the thickness in our heuristic differential interpretation is $dx = \Delta x$, the approximate volume dV of the slice is $\pi f(x)^2 \, dx$. We then add all the little contributions to the volume and take the limit as $\|P\| \to 0$ by integrating. We don't memorize Eq. (2) but rather arrive at the correct integral

$$\int_a^b \pi f(x)^2 \, dx$$

by such a geometric argument.

The outline at the top of page 309 suggests steps to follow when using this slicing method to find the volume of such a solid of revolution. The steps are similar to those in the outline in Section 6.1 for finding the area of a plane region.

Finding the Volume of a Solid of Revolution by Slicing

STEP 1. Sketch the plane region that is to be revolved, finding the points of intersection of bounding curves.

STEP 2. On the sketch, draw a typical thin rectangle arising from a partition P and perpendicular to the axis of revolution, that is, either perpendicular to the x-axis and of width Δx or perpendicular to the y-axis and of width Δy.

STEP 3. *Looking at the sketch,* write down the volume V_{slice} of the slice swept out as the rectangle is revolved about the given axis. Express V_{slice} entirely in terms of the variable (x or y) appearing in the Δ-increment.

STEP 4. Integrate between the appropriate (x or y) limits. (Geometrically, this amounts to adding the volumes found in step 3 and taking the limit of the resulting sum as $\|P\| \to 0$.)

EXAMPLE 1 Derive the formula $V = \frac{4}{3}\pi a^3$ for the volume of a spherical ball of radius a from the formula $A = \pi a^2$ for the area of a circular disk of radius a.

Solution

STEP 1. A ball of radius a may be obtained by revolving the region bounded by $y = \sqrt{a^2 - x^2}$ and the x-axis about the x-axis. The region is shaded in Fig. 6.19(a), while Fig. 6.19(b) shows the ball.

STEP 2. A typical thin rectangle perpendicular to the x-axis and of width Δx is shaded in color in Fig. 6.19(a). When revolved about the x-axis, the rectangle sweeps out a circular slice.

STEP 3. The volume of the slice is the product of the area of the face of the disk and the thickness of the slice. The area of the face is πy^2 for the point (x, y) shown in Fig. 6.19(a). The thickness is Δx, so the volume is $V_{\text{slice}} = \pi y^2 \, \Delta x$. We must express this entirely in terms of x. Since $y = \sqrt{a^2 - x^2}$, we have $y^2 = a^2 - x^2$. Thus we obtain

$$V_{\text{slice}} = \pi (a^2 - x^2) \, \Delta x.$$

STEP 4. The appropriate integral is

$$V = \int_{-a}^{a} \pi (a^2 - x^2) \, dx = \pi \left(a^2 x - \frac{x^3}{3} \right) \bigg]_{-a}^{a}$$

$$= \pi \left(a^3 - \frac{a^3}{3} \right) - \pi \left(-a^3 + \frac{a^3}{3} \right)$$

$$= \frac{2}{3}\pi a^3 + \frac{2}{3}\pi a^3$$

$$= \frac{4}{3}\pi a^3.$$

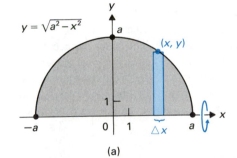

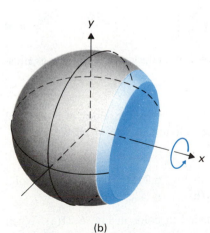

FIGURE 6.19 (a) Rectangle of width Δx and height $y = \sqrt{a^2 - x^2}$; (b) slice of the ball of revolution generated by the rectangle in (a).

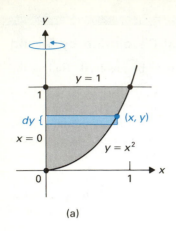

(a)

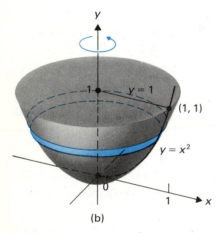

(b)

FIGURE 6.20 (a) Differential rectangle of width dy and length $x = \sqrt{y}$; (b) solid of revolution obtained from (a).

Alternatively, we could take advantage of the obvious symmetry of the region about the y-axis in Fig. 6.19(a) and write

$$V = 2 \int_0^a \pi(a^2 - x^2)\, dx = 2\pi\left(a^2 x - \frac{x^3}{3}\right)\Bigg]_0^a$$

$$= 2\pi\left(a^3 - \frac{a^3}{3}\right) - 0 = \frac{4}{3}\pi a^3.$$

We prefer to use the symmetry whenever it permits us to have zero as a limit in the integral. ■

EXAMPLE 2   Let the first-quadrant region bounded by the curves with equations $y = x^2$, $x = 0$, and $y = 1$ be revolved about the y-axis. Find the volume of the resulting solid. Use differential notation in forming the integral.

Solution

STEP 1.   The plane region to be revolved is shown shaded in Fig. 6.20(a). Revolving the region about the y-axis gives a solid bowl, as shown in Fig. 6.20(b).

STEP 2.   A typical rectangle of width dy is shown shaded in color in Fig. 6.20(a). When revolved about the y-axis, the rectangle sweeps out a thin, circular slice.

STEP 3.   The volume of such a slice is the product of the area of the circular face and the thickness (or height) of the slice. The area of a face is πx^2 for the point (x, y) shown in Fig. 6.20(a). The thickness of the slice is dy, so the volume of this slice is $dV = \pi x^2\, dy$. We must express the volume $\pi x^2\, dy$ entirely in terms of y. For the point (x, y) in Fig. 6.20(a), we have $x^2 = y$, so the volume of the slice becomes $dV = (\pi y)\, dy$.

STEP 4.   The appropriate integral is

$$V = \int_0^1 \pi y\, dy = \pi \frac{y^2}{2}\Bigg]_0^1 = \frac{\pi}{2} - 0 = \frac{\pi}{2}.$$ ■

The axis of revolution need not fall on a boundary of the region, as illustrated in the next example.

EXAMPLE 3   Find the volume generated when the plane region of Example 2 is revolved about the line $x = -1$. Use Δ-notation in setting up the integral.

Solution

STEP 1.   The appropriate sketch is given in Fig. 6.21(a). This time, the solid generated is shown in Fig. 6.21(b).

STEP 2.   A typical rectangle is shown shaded in color in Fig. 6.21(a). When revolved about the line $x = -1$, it sweeps out a thin, annular washer (a disk with a hole in it), as shown in Fig. 6.22.

STEP 3.   The volume of an annular washer is the volume of the disk slice minus the volume of the hole. The large radius of the whole disk slice is

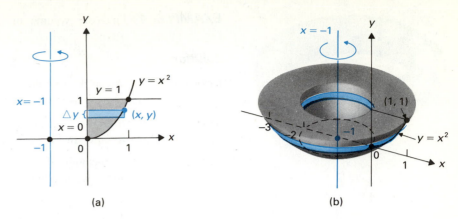

(a)   **(b)**

FIGURE 6.21 (a) Rectangle of width Δy to be revolved about the line $x = -1$; (b) solid of revolution obtained from (a).

$x - (-1) = x + 1$ for the point (x, y) shown in Fig. 6.22, so the disk slice has volume $\pi(x + 1)^2 \, \Delta y$. The hole has radius 1 and hence volume $\pi(1)^2 \, \Delta y$. Thus the volume of the annular washer is

$$V_{\text{slice}} = \pi(x + 1)^2 \, \Delta y - \pi(1)^2 \, \Delta y = \pi(x^2 + 2x) \, \Delta y.$$

We must express the volume $\pi(x^2 + 2x) \, \Delta y$ entirely in terms of y. For the point (x, y) in Fig. 6.22, we have $x^2 = y$, so the volume of the annular washer becomes $\pi(y + 2\sqrt{y}) \, \Delta y$.

STEP 4. The appropriate integral is

$$V = \int_0^1 \pi(y + 2\sqrt{y}) \, dy = \pi\left(\frac{y^2}{2} + \frac{4}{3}y^{3/2}\right)\Big]_0^1$$

$$= \pi\left(\frac{1}{2} + \frac{4}{3}\right) - \pi(0) = \frac{11}{6}\pi. \quad \blacksquare$$

We have seen that slices for volumes of revolution are of two types. If the thin rectangle reaches all the way to the axis of revolution, then we have a disk slice of radius r, as shown in Fig. 6.23(a). If the thin rectangle does not reach the axis of revolution, we have a washer slice of inner radius r and outer radius R, as shown in Fig. 6.23(b). The following formulas apply.

$$V_{\text{slice}} = \pi r^2 \, (\Delta x \text{ or } \Delta y) \qquad \text{\textit{Disk slice}}$$
$$V_{\text{slice}} = \pi(R^2 - r^2)(\Delta x \text{ or } \Delta y) \qquad \text{\textit{Washer slice}}$$

Of course, before integrating, we must express r and R in terms of the variable x or y that appears in Δx or Δy.

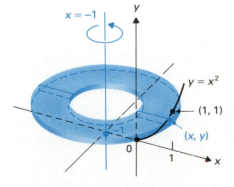

FIGURE 6.22 Washer slice obtained by revolving the rectangle in Fig. 6.21(a) about the line $x = -1$.

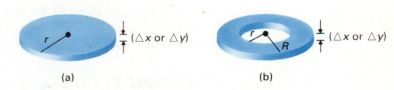

(a)   **(b)**

FIGURE 6.23 (a) Solid disk slice: Volume $= \pi r^2(\Delta x \text{ or } \Delta y)$; (b) washer slice: Volume $= \pi(R^2 - r^2)(\Delta x \text{ or } \Delta y)$.

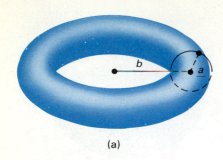

(a)

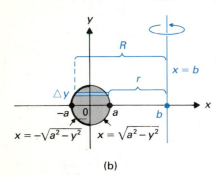

(b)

FIGURE 6.24 (a) A solid torus with circular cross sections of radius a; (b) rectangle of width Δy to be revolved about the line $x = b$.

EXAMPLE 4 Find the volume of the solid torus (doughnut) shown in Fig. 6.24(a).

Solution

STEP 1. The torus can be generated by revolving the disk $x^2 + y^2 \le a^2$ about the line $x = b$, as indicated in Fig. 6.24(b).

STEP 2. A typical thin rectangle of width Δy perpendicular to the line $x = b$ is shaded in color in Fig. 6.24(b).

STEP 3. Since our rectangle does not reach the axis of revolution $x = b$ in Fig. 6.24(b), we use the formula

$$V_{\text{slice}} = \pi(R^2 - r^2)\,\Delta y.$$

Since the right-hand semicircle has equation $x = \sqrt{a^2 - y^2}$ and the left-hand one has equation $x = -\sqrt{a^2 - y^2}$ in Fig. 6.24(b), we see that

$$r = b - \sqrt{a^2 - y^2},$$
$$R = b - (-\sqrt{a^2 - y^2}) = b + \sqrt{a^2 - y^2}.$$

Thus

$$\begin{aligned}
V_{\text{slice}} &= \pi\left[(b + \sqrt{a^2 - y^2})^2 - (b - \sqrt{a^2 - y^2})^2\right]\Delta y \\
&= \pi\left[(b^2 + 2b\sqrt{a^2 - y^2} + a^2 - y^2)\right.\\
&\qquad \left. -(b^2 - 2b\sqrt{a^2 - y^2} + a^2 - y^2)\right]\Delta y \\
&= 4\pi b\sqrt{a^2 - y^2}\,\Delta y.
\end{aligned}$$

STEP 4. Using the symmetry of the disk in Fig. 6.24(b) about the x-axis, we see that

$$V = 2\int_0^a 4\pi b\sqrt{a^2 - y^2}\,dy = 8\pi b\int_0^a \sqrt{a^2 - y^2}\,dy.$$

We cannot find an antiderivative of $\sqrt{a^2 - y^2}$ at the present time, but we recognize the integral $\int_0^a \sqrt{a^2 - y^2}\,dy$ as giving the area of the quarter-disk shaded in Fig. 6.25. Thus $\int_0^a \sqrt{a^2 - y^2}\,dy = \frac{1}{4}\pi a^2$, so

$$V = 8\pi b(\tfrac{1}{4}\pi a^2) = 2\pi^2 a^2 b. \qquad \blacksquare$$

More General Volumes by Slicing

The slicing method can often be used to find the volume of a solid that can be sliced up into parallel slabs whose faces have easily computed areas. A solid of revolution is an example of such a solid, for it can be sliced into parallel circular disks.

The general technique for finding volumes by slicing is perhaps best illustrated by examples. A step-by-step procedure for finding such volumes is found at the top of page 313.

FIGURE 6.25 $\int_0^a \sqrt{a^2 - y^2}\,dy$ is the area of the shaded region.

EXAMPLE 5 A 45° wedge of cheese has a semicircular base of diameter $2a$. If the cheese is cut perpendicular to the diameter of the semicircle, the cross section obtained is an isosceles right triangle with the right angle on the semicircle, as shown in Fig. 6.26(a). Find the volume of the cheese.

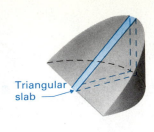

Triangular slab

(a)

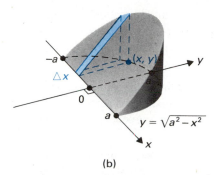

$y = \sqrt{a^2 - x^2}$

(b)

FIGURE 6.26 (a) Isosceles right triangular slice on a half-disk base; (b) triangular slice of volume $(y^2/2)\,\Delta x$, where $y = \sqrt{a^2 - x^2}$.

Finding the Volume of a Solid by the Slicing Method

STEP 1. Draw a figure; include an axis perpendicular to the cross section of known area (say, an x-axis).

STEP 2. Sketch a cross-sectional slice perpendicular to that x-axis.

STEP 3. Express the area $A(x)$ of the face of the cross-sectional slice in terms of its position x on the x-axis. The volume of the slice is then $A(x)\,\Delta x$, as in Fig. 6.28(c).

STEP 4. Integrate between appropriate x-limits.

Solution

STEP 1. We may take for the base of the cheese the half-disk bounded by the x-axis and the graph of $\sqrt{a^2 - x^2}$, as shown in Fig. 6.26(b).

STEP 2. The slice shown in Fig. 6.26(a) has thickness Δx, while its faces are isosceles triangles with legs of length y.

STEP 3. The area of such a triangle is $y^2/2$, so the volume of the slice is approximately $(y^2/2)\,\Delta x$. Since $y = \sqrt{a^2 - x^2}$, the slice has volume $[(a^2 - x^2)/2]\,\Delta x$.

STEP 4. The volumes of these slices should be added as x goes from $-a$ to a, so the appropriate integral is

$$\int_{-a}^{a} \frac{1}{2}(a^2 - x^2)\,dx = \frac{1}{2}\left(a^2 x - \frac{x^3}{3}\right)\Big]_{-a}^{a}$$

$$= \frac{1}{2}\left[\left(a^3 - \frac{a^3}{3}\right) - \left(-a^3 - \frac{(-a)^3}{3}\right)\right]$$

$$= \frac{1}{2}\left[2a^3 - \frac{2}{3}a^3\right] = \frac{2}{3}a^3. \qquad \blacksquare$$

EXAMPLE 6 A bookend has as base a half-disk of radius a. Sections of the solid perpendicular to the base and parallel to the diameter are squares. Find the volume of the bookend.

Solution

STEP 1. The bookend is shown in Fig. 6.27.

STEP 2. A square section is shaded in color.

STEP 3. For our choice of axes and the point (x, y) shown in the figure, the square slice has thickness Δx and sides of length $2y$. Thus the volume of the slice is

$$V_{\text{slice}} = (2y)^2\,\Delta x = 4y^2\,\Delta x.$$

STEP 4. We must express y in terms of x before integrating. We have $y = \sqrt{a^2 - x^2}$, so

$$V_{\text{slice}} = 4(a^2 - x^2)\,\Delta x.$$

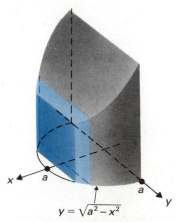

$y = \sqrt{a^2 - x^2}$

FIGURE 6.27 Cross section showing a square slice standing on a half-disk base: Slice volume $= (2y)^2\,\Delta x$, where $y = \sqrt{a^2 - x^2}$.

Then

$$V = \int_0^a 4(a^2 - x^2)\, dx = 4\left(a^2 x - \frac{x^3}{3}\right)\Big]_0^a$$

$$= 4\left(a^3 - \frac{a^3}{3}\right) = \frac{8}{3} a^3. \qquad \blacksquare$$

SUMMARY

1. A suggested step-by-step outline (in differential notation) for finding a volume of revolution by the slicing method is as follows.

   STEP 1. Sketch the plane region that is to be revolved, finding the points of intersection of bounding curves.

   STEP 2. On the sketch, draw a typical thin rectangle perpendicular to the axis of revolution, that is, either perpendicular to the x-axis and of width dx or perpendicular to the y-axis and of width dy.

   STEP 3. Looking at the sketch, write down the volume dV of the slice swept out as the rectangle is revolved about the given axis. This slice is either a circular disk of volume

   $$dV = \pi r^2\ (dx\ \text{or}\ dy)$$

   as in Fig. 6.28(a) or a circular washer of volume

   $$dV = \pi(R^2 - r^2)(dx\ \text{or}\ dy)$$

   as in Fig. 6.28(b). In either case, express dV entirely in terms of the variable (x or y) appearing in the differential (dx or dy).

   STEP 4. Integrate dV between the appropriate (x or y) limits.

2. A suggested step-by-step procedure (in differential notation) for finding a volume of a solid with parallel cross sections of known area is as follows.

   STEP 1. Draw a figure; include an axis perpendicular to the cross sections of known area (say, an x-axis).

   STEP 2. Sketch a cross-sectional slice perpendicular to that x-axis.

   STEP 3. Express the area $A(x)$ of the face of the cross-sectional slice in terms of its position x on the x-axis. The volume of the slice is then

   $$dV = A(x)\, dx,$$

   as in Fig. 6.28(c).

   STEP 4. Integrate the expression for dV between appropriate x-limits.

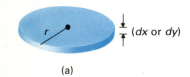

(a)

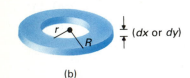

(b)

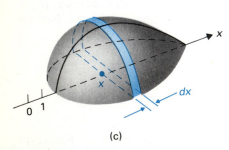

(c)

FIGURE 6.28   (a) Solid disk:

$$dV = \pi r^2(dx\ \text{or}\ dy);$$

(b) washer:

$$dV = \pi(R^2 - r^2)(dx\ \text{or}\ dy);$$

(c) Slice with face area $A(x)$ has volume $dV = A(x)\, dx$.

EXERCISES 6.2

In Exercises 1–22, use slicing to find the volume of the solid generated by revolving the region with the given boundary about the indicated axis.

1. Bounded by $y = 1 - x^2$, $y = 0$ about the x-axis.
2. Bounded by $y = x^2 + 1$, $y = 0$, $x = 1$, $x = -1$ about the x-axis.
3. Bounded by $x = \sqrt{3 - y^2}$, $x = 0$ about the y-axis.
4. Bounded by $y = \sqrt{x}$, $x = 4$, $y = 0$ about the x-axis.
5. Bounded by $y = \sqrt{x}$, $x = 4$, $y = 0$ about the y-axis.
6. Bounded by $y = \sqrt{x}$, $y = x^2$ about the x-axis.
7. Bounded by $y = x$, $y = x^2$ about the y-axis.
8. Bounded by $y = x^2$, $y = 3 - 2x$ about the x-axis.
9. Bounded by $y = \sec x$, $y = 0$, $x = \pi/4$, $x = -(\pi/4)$ about the x-axis.
10. Bounded by $y = \csc x$, $y = 0$, $x = \pi/6$, $x = 5\pi/6$ about the x-axis.
11. Bounded by $y = \sin x$ for $0 \le x \le \pi$, $y = 0$ about the x-axis (Use the table of integrals on the endpapers of this book.)
12. Bounded by $y = \sin x$ for $0 \le x \le \pi$, $y = 0$ about the line $y = 1$ (Use the table of integrals.)
13. Bounded by $y = \cos x$, $x = \pi/2$, $x = -(\pi/2)$, $y = -1$ about the line $y = -1$ (Use the table of integrals.)
14. Bounded by $y = \sin x$, $y = 2 \sin x$ for $0 \le x \le \pi$ about the x-axis (Use the table of integrals.)
15. Bounded by $y = e^x$, $y = 0$, $x = 0$, and $x = 1$ about the x-axis
16. The region in Exercise 15 about the line $y = -1$
17. Bounded by $y = 1/\sqrt{x + 1}$, $y = 0$, $x = 0$, and $x = 3$ about the x-axis
18. The region in Exercise 17 about the line $y = 1$
19. Bounded by $y = x^2$, $y = x$ about the line $x = 2$
20. Bounded by $y = x^2$, $y = 1$ about the line $y = 2$
21. Bounded by $y = x^2$, $y = 1$ about the line $x = 1$
22. Bounded by $y = 1/x$, $x = 1$, $x = 4$, $y = 0$ about the x-axis
23. Verify the formula $V = \frac{1}{3}\pi r^2 h$ for the volume of a right circular cone of height h with base of radius r. [*Hint:* Revolve the region bounded by the lines with equations $y = (r/h)x$, $y = 0$, and $x = h$ about the x-axis. See Fig. 6.29.]
24. Find the volume of the solid generated when the upper half ($y \ge 0$) of the elliptical disk $x^2/a^2 + y^2/b^2 \le 1$ is revolved about the x-axis.
25. A solid has as base the disk $x^2 + y^2 \le a^2$. Each plane section of the solid cut by a plane perpendicular to the x-axis is an equilateral triangle. Find the volume of the solid.
26. The base of a certain solid is an isosceles right triangle with hypotenuse of length a. Each plane section of the solid cut by a plane perpendicular to the hypotenuse is a square. Find the volume of the solid.
27. Find the volume of the smaller portion of a solid ball of

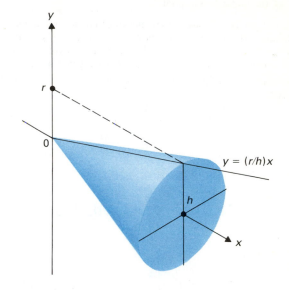

FIGURE 6.29

radius a cut off by a plane b units from the center of the ball, where $0 \le b \le a$.

28. A certain solid has as base the plane region bounded by $x = y^2$ and $x = 4$. Each plane section of the solid cut by a plane perpendicular to the x-axis is an isosceles right triangle with the right angle on the graph of $\sqrt{x}$. (See Fig. 6.30.) Find the volume of the solid.

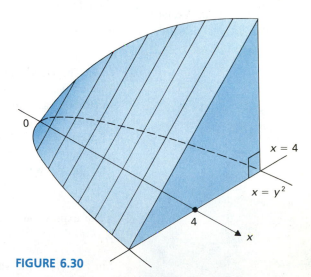

FIGURE 6.30

29. Let a solid have a flat base and altitude h. Suppose the area of a cross section parallel to the base and x units above the base is $c(h - x)^2$ for $0 \le x \le h$ for some constant c independent of the value of x. (This is true for tetrahedra, pyramids, and cones.) Show that the volume of the solid is $\frac{1}{3}h \cdot$ (Area of the base).

30. A rectangular swimming pool with vertical sides is 20 ft wide and 40 ft long. The depth of the water x ft from the shallow end of the pool is $2 + (x^2/200)$ ft. Find the volume of the water in the pool.

In Exercises 31–37, use a computer or calculator and numerical techniques (Simpson's rule with $n \geq 10$, Newton's method). Estimate the volume of the solid generated by revolving the region with the given boundary about the indicated axis.

31. Bounded by $1/(1 + x^2)$, $x = 0$, $x = 1$, $y = 0$ about the x-axis

32. Bounded by $1/(1 + x^2)$, $x = 0$, $x = 1$, $y = 0$ about the line $y = -1$

33. Bounded by $y = 2^x$, $x = 0$, $x = 2$, $y = 0$ about the x-axis

34. Bounded by $6/(1 + \sin x)$, $x = 0$, $x = \pi$, $y = 0$ about the line $y = -1$

35. Bounded by $y = x^2$, $y = \sin x$ about the x-axis

36. Bounded by $y = x^2$, $y = \cos x$ about the line $y = -1$

37. Bounded by $y = x$, $y = \sec x - 1$ for $-\pi/2 < x < \pi/2$ about the x-axis

6.3 FINDING VOLUME USING CYLINDRICAL SHELLS

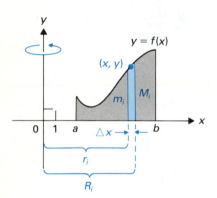

FIGURE 6.31 Strip to be revolved about the y-axis.

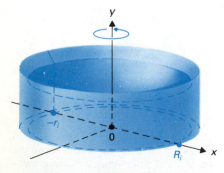

FIGURE 6.32 Cylindrical shell generated by the strip in Fig. 6.31.

We present another method for finding a volume of revolution. Suppose the region shown in Fig. 6.31 is revolved about the y-axis. This time the color-shaded strip in Fig. 6.31 sweeps out a cylindrical shell, as shown in Fig. 6.32. We can find an appropriate integral for the volume using these shells.

The volume of such a cylindrical shell should be approximately the product of the surface area of the cylinder and the thickness of the shell (the wall of the cylinder). The surface area in turn is the product of the perimeter of the circle and the height of the cylinder. For the point (x, y) shown in Fig. 6.31, the perimeter is $2\pi x$ and the height is y. Thus the volume V_{shell} of the cylindrical shell should be about $2\pi xy \, \Delta x$. Since $y = f(x)$, we would expect, adding the volumes of these shells and taking the limit of the resulting sum, that the volume of the whole solid of revolution would be

$$\int_a^b 2\pi x f(x) \, dx.$$

The preceding integral was set up using rough estimates, in the manner in which we always set up an integral for the shell method. A justification of the method can be made as follows. Let m_i and M_i be the minimum and maximum values of $f(x)$ over the ith interval of length Δx_i in a partition P of $[a, b]$. Referring to Fig. 6.31 and taking first the minimum radius r_i and minimum height m_i for the cylindrical shell and then the maximum radius R_i and maximum height M_i, we see that

$$2\pi r_i m_i \, \Delta x_i \leq V_{\text{shell}} \leq 2\pi R_i M_i \, \Delta x_i. \tag{1}$$

The expression

$$2\pi R_i M_i \, \Delta x_i$$

in Eq. (1) is the value of

$$2\pi x f(x) \, \Delta x_i$$

if we evaluate $2\pi x$ at $x_i^* = R_i$ and $f(x)$ at the (possibly different) point x_i', where $f(x)$ attains its maximum value M_i in this ith subinterval. A theorem known as Bliss's theorem, which we will not prove, shows that Riemann sums for $g(x) f(x)$

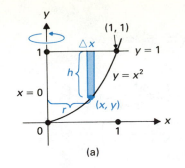

(a)

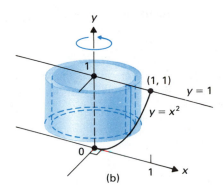

(b)

FIGURE 6.33 (a) Differential strip of width Δx to be revolved about the y-axis; (b) cylindrical shell generated by the strip in (a).

where the two functions f and g are evaluated at possibly *different* points x_i^* and x_i' in the ith interval still approach the integral. That is,

$$\int_a^b g(x)f(x)\,dx = \lim_{\|P\|\to 0} \sum_{i=1}^n f(x_i^*)g(x_i')\,\Delta x_i.$$

Using this fact and setting $g(x) = 2\pi x$, we see at once from Eq. (1) that

$$V = \int_a^b 2\pi x f(x)\,dx. \tag{2}$$

The outline of steps in Section 6.2 for finding the volume of revolution by the slicing method needs only slight modification to serve for computing volumes of solids of revolution by the shell method. See the bottom of this page.

EXAMPLE 1 In Example 2 of Section 6.2, we found the volume of the solid generated by revolving the first-quadrant region bounded by $y = x^2$, $x = 0$, and $y = 1$ about the y-axis. Repeat this computation, using the shell method.

Solution

STEP 1. The region is shaded in Fig. 6.33(a).

STEP 2. A thin rectangle parallel to the y-axis and of width Δx is shaded in color in Fig. 6.33(a). Figure 6.33(b) shows the shell generated by this rectangle.

STEP 3. We must compute $V_{\text{shell}} = 2\pi rh\,\Delta x$ in terms of x, for r and h shown in Fig. 6.33(a). Referring to the figure and the point (x, y) shown there, we have $r = x$, while

$$h = y_{\text{top}} - y_{\text{bottom}} = 1 - y = 1 - x^2.$$

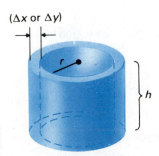

(Δx or Δy)

FIGURE 6.34 Cylindrical shell

$$V_{\text{shell}} = 2\pi rh(\Delta x \text{ or } \Delta y).$$

Finding the Volume of a Solid of Revolution by the Shell Method

STEP 1. Sketch the plane region that is to be revolved, finding the points of intersection of bounding curves.

STEP 2. On the sketch, draw a typical thin rectangle parallel to the axis of revolution, either horizontal with width Δy or vertical with width Δx.

STEP 3. *Looking at the sketch,* write down the volume

$$V_{\text{shell}} = 2\pi rh\,(\Delta x \text{ or } \Delta y)$$

of the shell swept out as the rectangle is revolved about the given axis. (See Fig. 6.34.) Express r and h in terms of the variable (x or y) in the increment.

STEP 4. Integrate between the appropriate (x or y) limits.

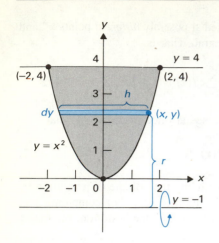

FIGURE 6.35   Differential strip of width dy to be revolved about the line $y = -1$.

Thus

$$V_{\text{shell}} = 2\pi x(1 - x^2)\,\Delta x = (2\pi x - 2\pi x^3)\,\Delta x.$$

STEP 4.   The volume is

$$V = \int_0^1 (2\pi x - 2\pi x^3)\,dx = \left(2\pi \frac{x^2}{2} - 2\pi \frac{x^4}{4}\right)\Big]_0^1$$

$$= \left(\frac{2\pi}{2} - \frac{2\pi}{4}\right) - 0 = \frac{\pi}{2}.$$ ■

You should also become accustomed to differential notation where V_{shell} becomes $dV = 2\pi rh\ (dx$ or $dy)$.

EXAMPLE 2   Using the shell method, find the volume when the region bounded by $y = 4$ and $y = x^2$ is revolved about the line $y = -1$. Use differential notation to set up the integral.

Solution

STEP 1.   The region is sketched in Fig. 6.35.

STEP 2.   A thin rectangle parallel to the line $y = -1$ and of width dy is shaded in color in Fig. 6.35.

STEP 3.   The volume of the shell is $dV = 2\pi rh\,dy$. We must express r and h in terms of y. In terms of the coordinates of the point $(x,\ y)$ shown on the curve $y = x^2$, we have $r = y - (-1) = y + 1$, while $h = 2x = 2\sqrt{y}$. Thus

$$dV = 2\pi(y + 1)2\sqrt{y}\,dy = 4\pi(y + 1)\sqrt{y}\,dy.$$

STEP 4.   The volume is

$$V = \int_0^4 4\pi(y + 1)\sqrt{y}\,dy = 4\pi\int_0^4 (y^{3/2} + y^{1/2})\,dy$$

$$= 4\pi\left(\frac{2}{5}y^{5/2} + \frac{2}{3}y^{3/2}\right)\Big]_0^4 = 4\pi\left(\frac{64}{5} + \frac{16}{3}\right) = 64\pi\left(\frac{4}{5} + \frac{1}{3}\right)$$

$$= 64\pi\frac{17}{15} = \frac{1088\pi}{15}.$$ ■

EXAMPLE 3   Find the volume of the solid generated when the region bounded by $y = \sin x$ for $0 \le x \le \pi$ and $y = 0$ is revolved about the y-axis. Set up the integral using differential notation.

Solution

STEP 1.   The region is shaded in Fig. 6.36.

STEP 2.   A thin rectangle parallel to the y-axis and of width dx is shaded in color in Fig. 6.36.

STEP 3.   We must compute $dV = 2\pi rh\,dx$ in terms of x. For the point $(x,\ y)$ shown in Fig. 6.36, we have $r = x$ and $h = y = \sin x$. Thus $dV = 2\pi x(\sin x)\,dx$.

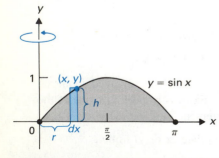

FIGURE 6.36   Differential strip to generate a shell of volume $dV = 2\pi rh\,dx$.

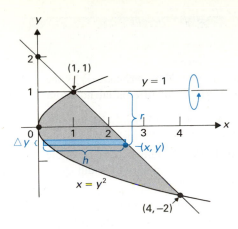

FIGURE 6.37 Horizontal strip to generate a shell of volume $2\pi rh\,\Delta y$.

STEP 4. The appropriate integral is

$$\int_0^\pi 2\pi x(\sin x)\,dx = 2\pi \int_0^\pi x(\sin x)\,dx.$$

The integral table on the endpapers of this book tells us that

$$\int x(\sin x)\,dx = -x\cos x + \sin x + C.$$

The volume is therefore

$$V = 2\pi \int_0^\pi x(\sin x)\,dx = 2\pi(-x\cos x + \sin x)\Big]_0^\pi$$
$$= 2\pi[-\pi(-1) + 0] - 2\pi(0) = 2\pi^2. \quad \blacksquare$$

We could not have found the volume in the preceding example by slicing because we do not yet know how to solve $y = \sin x$ for x in terms of y. The shell method was easy. The next example also illustrates that often one method is easier than the other.

EXAMPLE 4 Find the volume of the solid generated when the region bounded by $x = y^2$ and $x + y = 2$ is revolved about the line $y = 1$.

Solution

STEP 1. The points of intersection of $x = y^2$ and $x + y = 2$ are found by solving simultaneously:

$$y^2 + y = 2, \qquad y^2 + y - 2 = 0,$$
$$(y + 2)(y - 1) = 0, \qquad y = -2 \text{ or } 1,$$
$$(x, y) = (4, -2) \text{ or } (1, 1).$$

The region is shaded in Fig. 6.37.

STEP 2. Using the slicing method with vertical rectangles perpendicular to the line $y = 1$ would necessitate setting up two integrals, one integral for $0 \le x \le 1$ and one for $1 \le x \le 4$. Using the shell method, with the thin horizontal rectangle of width Δy shaded in color in Fig. 6.37, we need only one integral.

STEP 3. We must express $V_{\text{shell}} = 2\pi rh\,\Delta y$ in terms of y. For the point (x, y) in Fig. 6.37, we have

$$r = y_{\text{top}} - y_{\text{bottom}} = 1 - y, \qquad h = x_{\text{right}} - x_{\text{left}} = (2 - y) - y^2.$$

Thus

$$V_{\text{shell}} = 2\pi(1 - y)(2 - y - y^2)\,\Delta y = 2\pi(2 - 3y + y^3)\,\Delta y.$$

STEP 4. The volume is

$$V = \int_{-2}^{1} 2\pi(2 - 3y + y^3)\,dy = 2\pi\left(2y - \frac{3y^2}{2} + \frac{y^4}{4}\right)\Big]_{-2}^{1}$$
$$= 2\pi\left(2 - \frac{3}{2} + \frac{1}{4}\right) - 2\pi(-4 - 6 + 4)$$
$$= 2\pi \cdot \frac{3}{4} - 2\pi(-6) = \frac{27\pi}{2}. \quad \blacksquare$$

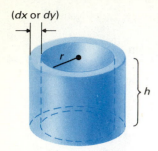

(dx or dy)

FIGURE 6.38 Cylindrical shell:
$$dV = 2\pi rh(dx \text{ or } dy).$$

SUMMARY

To find a volume of revolution about a horizontal or vertical axis by the shell method, use the following steps, given in differential notation.

STEP 1. Sketch the plane region that is to be revolved, finding the points of intersection of bounding curves.

STEP 2. On the sketch, draw a typical thin rectangle parallel to the axis of revolution, that is, either horizontal with width dy or vertical with width dx.

STEP 3. *Looking at the sketch,* write the differential volume $dV = 2\pi rh \cdot (dx \text{ or } dy)$ of the shell swept out as the rectangle is revolved about the given axis. (See Fig. 6.38.) Express r and h in terms of the variable (x or y) in the differential.

STEP 4. Integrate the differential volume dV in step 3 between the appropriate (x or y) limits.

EXERCISES 6.3

In Exercises 1–18, use the shell method to find the volume of the solid generated by revolving the region with the given boundary about the indicated axis.

1. Bounded by $y = x^2$, $y = 1$ about the line $y = 1$
2. Bounded by $y = x^2$, $y = 1$ about the line $x = 2$
3. Bounded by $y = x$, $y = x^2$ about the y-axis
4. Bounded by $y = x$, $y = x^2$ about the x-axis
5. Bounded by $y = \sqrt{x}$, $x = 4$, $y = 0$ about the y-axis
6. Bounded by $y = \sqrt{x}$, $x = 4$, $y = 0$ about the line $x = 5$
7. Bounded by $y = \sqrt{x}$, $y = x^2$ about the y-axis
8. Bounded by $y = \sqrt{x}$, $y = x^2$ about the x-axis
9. Bounded by $x = y^2$, $y = x - 2$ about the line $y = 2$
10. Bounded by $y = x^2$, $y = 3 - 2x$ about the line $x = 1$
11. Bounded by $y = \cos x$ for $\pi/2 \le x \le 3\pi/2$, $y = 0$ about the y-axis
12. Bounded by $y = \sin x$ for $0 \le x \le \pi$, $y = 0$ about the line $x = -1$
13. Bounded by $y = \sin x$ for $0 \le x \le \pi$, $y = 0$ about the line $x = 2\pi$
14. Bounded by $y = \cos x$ for $-\pi/2 \le x \le \pi/2$, $y = 0$ about the line $x = 2$
15. Bounded by $y = e^x$, $y = 0$, $x = 0$, and $x = 1$ about the y-axis (Use the integral table if necessary.)

16. The region in Exercise 15 about the line $x = -1$
17. Bounded by $y = 9/x^2$, $y = 0$, $x = 1$, and $x = 3$ about the y-axis
18. The region in Exercise 17 about the line $x = 4$
19. Use the method of cylindrical shells to derive the formula $V = \frac{1}{3}\pi r^2 h$ for the volume of a right circular cone with height h and base of radius r. [*Hint:* Revolve the region bounded by the lines $y = (r/h)x$, $y = 0$, and $x = h$ about the x-axis.]
20. Derive the formula $V = \frac{4}{3}\pi a^3$ for the volume of a sphere of radius a, using the method of cylindrical shells.
21. Use the method of cylindrical shells to find the volume of the doughnut (torus) generated by revolving the disk $x^2 + y^2 \le a^2$ about the line $x = b$ for $b > a$. (See Fig. 6.39.) [*Hint:* Part of the integral can be evaluated as the area of a familiar region.]
22. A fluid flows through a circular pipe of radius R cm. The velocity of the flow in cm/sec at the distance r cm from the center of the pipe, for $0 \le r \le R$, is given by $v(r) = k(R^2 - r^2)$, where k is a constant. Find the total rate of flow, in cm^3/sec, of the fluid through the pipe.

🔲 In Exercises 23–29, use the method of cylindrical shells with a calculator or computer and numerical techniques (Simpson's rule with $n \ge 10$, Newton's method). Find the volume of the solid

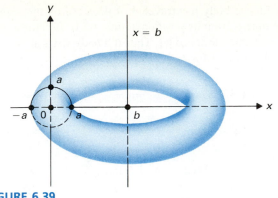

FIGURE 6.39
Torus in Exercise 21.

generated by revolving the region with the given boundary about the indicated axis.

23. Bounded by $y = \tan x$, $y = 0$, $x = \pi/4$ about the y-axis
24. Bounded by $y = \sec x - 1$ for $-(\pi/2) < x < \pi/2$, $y = 1$ about the line $x = 2$
25. Bounded by $y = \sin \sqrt{x}$ for $0 \le x \le \pi^2$, $y = 0$ about the y-axis
26. Bounded by $y = 3^x$, $x = 0$, $x = 2$, $y = 0$ about the line $x = 3$
27. Bounded by $x = 2^y$, $y = 0$, $y = 3$, $x = 0$ about the line $y = -1$
28. Bounded by $y = 10/(x^2 + 1) - 2$, $y = 0$ about the line $x = 3$
29. Bounded by $y = x^2$, $y = \sin x$ about the y-axis

6.4 DISTANCE

We will now apply integral calculus to find the distance traveled by a body moving on a straight line if we know the velocity of the body at any time t. The body may be moving back and forth on the line. In that case, we will have to distinguish clearly whether "distance" means the *displacement* from the point where the body starts at the initial time to the point where the body is at a later time or the *total distance* the body travels, as measured, for instance, by an odometer in a car.

If a body travels on a line with constant velocity v, then the distance s traveled after time t is given by the product $|v|t$. The absolute value sign is used since v is negative if the body is moving in the negative direction on the line. Thus $|v|$ is the *speed* of the body.

EXAMPLE 1 If the velocity at time t sec of a body traveling on a line is $2t + t^2$ ft/sec, find how far the body travels from $t = 2$ sec to $t = 4$ sec, using integral calculus.

Solution Since distance equals the product of speed and time when speed is constant, we see that at time t, the distance traveled over the next small time interval Δt sec is about $(2t + t^2) \Delta t$ ft. We wish to add these distances as t ranges from 2 to 4 and take the limit of the resulting sum as $\Delta t \to 0$. The appropriate integral is

$$\int_2^4 (2t + t^2)\, dt = \left(t^2 + \frac{t^3}{3} \right)\Bigg]_2^4 = \left(16 + \frac{64}{3} \right) - \left(4 + \frac{8}{3} \right) = \frac{92}{3}.$$

Thus the body travels $\frac{92}{3}$ ft. ∎

The velocity of a body moving on the x-axis is considered positive if it is moving to the right and negative if it is moving to the left. Thus the integral of the velocity v from t_1 and t_2 will give the total distance traveled toward the right minus the distance traveled toward the left between these times. This may not give a true picture of how far the body has traveled; it does tell us the *signed displacement* of the body at time t_2 from its position at time t_1. To find the actual distance traveled, we must integrate the speed $|v|$. Thus if a body starts at time t_1, stops at time t_2, and has velocity $v(t)$ for $t_1 \leq t \leq t_2$, then

$$\text{Displacement} = \int_{t_1}^{t_2} v(t)\, dt,$$

$$\text{Total distance traveled} = \int_{t_1}^{t_2} |v(t)|\, dt.$$

EXAMPLE 2 Suppose the velocity at time t of a body traveling on a line is $\cos(\pi t/2)$ ft/sec. Thus at time $t = 0$, the body is moving at the speed of 1 ft/sec in the positive direction, while at time $t = 2$, the body is moving in the negative direction at the speed of 1 ft/sec. Find the total distance the body travels from time $t = 0$ to time $t = 2$ and also the displacement.

Solution To find the total distance traveled, we need to compute

$$\int_0^2 \left| \cos \frac{\pi t}{2} \right| dt.$$

Now $\cos(\pi t/2)$ is positive for $0 \leq t < 1$ and negative for $1 < t \leq 2$. Thus we have

$$\int_0^2 \left| \cos \frac{\pi t}{2} \right| dt = \int_0^1 \left(\cos \frac{\pi t}{2} \right) dt + \int_1^2 \left(-\cos \frac{\pi t}{2} \right) dt$$

$$= \left(\frac{2}{\pi} \sin \frac{\pi t}{2} \right)\Big]_0^1 - \left(\frac{2}{\pi} \sin \frac{\pi t}{2} \right)\Big]_1^2$$

$$= \left(\frac{2}{\pi} - 0 \right) - \left(0 - \frac{2}{\pi} \right) = \frac{4}{\pi}.$$

Therefore the body travels $4/\pi$ ft from time $t = 0$ to time $t = 2$ sec.

The displacement from start to stop for this body is

$$\int_0^2 \left(\cos \frac{\pi t}{2} \right) dt = \left(\frac{2}{\pi} \sin \frac{\pi t}{2} \right)\Big]_0^2 = \frac{2}{\pi}(0 - 0) = 0.$$

Thus this body returns at $t = 2$ to the point where it started at $t = 0$. ∎

EXAMPLE 3 Let the acceleration $a = d^2x/dt^2$ of a body traveling on the x-axis be $6t$. If the body starts with initial velocity $v_0 = -3$ at time $t = 0$, find (a) the velocity v as a function of t, (b) the total distance the body travels from time $t = 0$ to $t = 4$.

Solution From $d^2x/dt^2 = 6t$, we obtain

$$v = \frac{dx}{dt} = 3t^2 + C.$$

Since $v = -3$ when $t = 0$, we have $C = -3$, so

$$v = 3t^2 - 3.$$

This answers part (a).

For part (b), note that v is negative for $0 \le t < 1$ and positive for $1 < t \le 4$. Thus

$$s = \int_0^4 |3t^2 - 3| \, dt = \int_0^1 -(3t^2 - 3) \, dt + \int_1^4 (3t^2 - 3) \, dt$$

$$= -(t^3 - 3t) \Big]_0^1 + (t^3 - 3t) \Big]_1^4$$

$$= -(-2 - 0) + [52 - (-2)]$$

$$= 2 + 54 = 56. \qquad \blacksquare$$

SUMMARY

1. If a body on a line has velocity $v(t)$ at time t and starts at time t_1 and stops at time t_2, then

$$\text{Displacement} = \int_{t_1}^{t_2} v(t) \, dt,$$

$$\text{Total distance traveled} = \int_{t_1}^{t_2} |v(t)| \, dt.$$

EXERCISES 6.4

In Exercises 1–14, the velocity v of a body on a line is given as a function of time t. Find

(a) the displacement from beginning point to ending point and
(b) the total distance traveled

in the indicated time interval.

1. $v = t - 4$, $0 \le t \le 2$
2. $v = t - 3$, $1 \le t \le 6$
3. $v = 2t - t^2$, $0 \le t \le 4$
4. $v = 3t^2 - 4$, $2 \le t \le 3$
5. $v = t^2 - 3t + 2$, $0 \le t \le 1$
6. $v = t^2 - 3t + 2$, $0 \le t \le 2$
7. $v = t^2 - 3t + 2$, $0 \le t \le 3$
8. $v = \sin t$, $0 \le t \le 3\pi$
9. $v = \cos t$, $0 \le t \le 5\pi$
10. $v = \sin(\pi t/2) + \cos(\pi t/2)$, $0 \le t \le 1$
11. $v = \sin(\pi t/2) + \cos(\pi t/2)$, $0 \le t \le 2$
12. $v = \sin(\pi t/2) + \cos(\pi t/2)$, $0 \le t \le 4$
13. $v = |t - 5|$, $0 \le t \le 5$
14. $v = |t - 5|$, $0 \le t \le 10$
15. Let the velocity v of a body on a line be $v = 3e^{-t}$ for $t \ge 0$.
    (a) Find the distance s the body travels as a function of the time t.
    (b) How far does the body travel throughout eternity?

In Exercises 16–21, the acceleration a as a function of time t and the initial velocity v_0 when $t = 0$ are given. Find

(a) the velocity of the body as a function of t and
(b) the total distance the body travels in the indicated time interval.

16. $a = 3$, $v_0 = 0$, $0 \le t \le 2$
17. $a = 2t - 4$, $v_0 = 3$, $0 \le t \le 3$
18. $a = \sin t$, $v_0 = 0$, $0 \le t \le 3\pi/2$
19. $a = -1/\sqrt{t + 1}$, $v_0 = 2$, $0 \le t \le 4$

20. $a = 6t - 1/(t + 1)^3$, $v_0 = 2$, $0 \le t \le 2$

21. $a = 6/(t + 1)^2$, $v_0 = -3$, $0 \le t \le 2$

22. Suppose that $v(s)$ is the velocity of a body when it has traveled a distance s moving in one direction on a line. Argue that the time required for the body to travel a distance c is $\int_0^c [1/v(s)]\, ds$. [*Hint:* If the velocity v is *constant,* how long does it take the body to travel a distance s?]

23. Referring to Exercise 22, suppose one could build a spaceship that could achieve a velocity v that increases exponentially as a function of the distance s traveled, so that $v = Ae^{ks}$ for some positive constants k and A. Show that such a spaceship could travel an infinite distance in a finite time.

24. Show that for the spaceship with velocity described in Exercise 24, its acceleration dv/dt is directly proportional to the square of its velocity. [*Hint:* Use the chain rule to compute dv/dt.]

The remaining problems concern other applications of the integral. It is important to be able to recognize situations where an integral is appropriate.

25. The population of a country is 200 million people and is growing continuously at a rate of 1% per year. Estimate the increase of the population of the country over the next 10 years. [*Hint:* See Example 4 of Section 5.5.]

26. The world demand for a natural resource in 1990 is 40 million units/year, and the demand is growing continuously at a rate of 5% per year. If this same exponential growth continues, find the number of units of the natural resource consumed over the next 10 years. [*Hint:* See Example 4 of Section 5.5.]

27. The marginal cost (in dollars per unit) of producing the nth unit of a product is approximately $0.0003n^2 + 0.1n + 25$. Suppose that 100 units have been produced. Use an integral to estimate the cost of producing an additional 20 units.

28. Referring to Exercise 27, suppose the marginal profit (in dollars per unit) on the nth unit produced is approximately $300 - 0.006n^2$. Use an integral to estimate the profit on the additional 20 units produced.

29. A company estimates that the rate of growth of its accumulated (total) sales of a product after t days during the first year of its manufacture will be $50e^{0.03t}$ units per day. Use an integral to estimate the accumulated sales during the 100-day period from day 101 through day 200.

30. Money flows into a savings account continuously at a rate of $5000 per year and the bank pays continuously compounded interest on the balance in the account at an annual rate of 10%. Show that if the account is started at time $t = 0$ years with a balance $B = 0$, then the change in the balance of the account between $t = T_1$ and $t = T_2$ years is given by $\int_{T_1}^{T_2} 5000e^{0.1t}\, dt$. (Economists call this the *amount of continuous money flow into the account* for $T_1 \le t \le T_2$.) [*Hint:* If $B(t)$ is the balance in the account after t years, argue that $dB/dt = 0.1B + 5000$ and solve this differential equation. Then differentiate to obtain dB/dt as a function of t.]

EXPLORING CALCULUS

N: calculus topics quiz (topic 7)

6.5 ARC LENGTH

The preceding three sections presented applications of integral calculus to finding area, volume, and distance. We now learn how to find the *length of a curve* (or *arc*). First we partition the curve into small pieces. Next, we approximate these pieces by line segments. Using *integral calculus,* we sum the lengths of the approximating segments and take the limit of the sum as we partition the curve into smaller and smaller pieces.

Length of a Graph

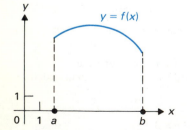

FIGURE 6.40 Find the length of the colored arc.

A function $y = f(x)$ is **continuously differentiable** if $f'(x)$ exists for all x in the domain of f and if f' is itself a continuous function. We wish to find the length of the portion of the graph of a continuously differentiable function $y = f(x)$ for x in $[a, b]$, as shown in Fig. 6.40.

We form a partition P of $[a, b]$ into n subintervals, as illustrated for the regular partition with $n = 4$ in Fig. 6.41. Let s be the length of the graph of f over $[a, b]$. We take as approximation to s the sum of the lengths of the chords

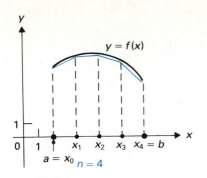

FIGURE 6.41 Approximating arc length using chords.

shown in Fig. 6.41, where the ith chord joins $(x_{i-1}, f(x_{i-1}))$ and $(x_i, f(x_i))$. The length of this chord is

$$\sqrt{(x_i - x_{i-1})^2 + (f(x_i) - f(x_{i-1}))^2} = (x_i - x_{i-1})\sqrt{1 + \left(\frac{f(x_i) - f(x_{i-1})}{x_i - x_{i-1}}\right)^2}$$

$$= (\Delta x_i)\sqrt{1 + \left(\frac{f(x_i) - f(x_{i-1})}{x_i - x_{i-1}}\right)^2}. \quad (1)$$

Our hypotheses on f enable us to apply the mean value theorem to f on $[x_{i-1}, x_i]$, and we have

$$\frac{f(x_i) - f(x_{i-1})}{x_i - x_{i-1}} = f'(x_i^*)$$

for some point x_i^* in $[x_{i-1}, x_i]$. From Eq. (1) we obtain, as length of the ith chord,

$$(\Delta x_i)\sqrt{1 + f'(x_i^*)^2}, \quad (2)$$

which yields the approximation

$$s \approx (\Delta x_1)\sqrt{1 + f'(x_1^*)^2} + (\Delta x_2)\sqrt{1 + f'(x_2^*)^2} + \cdots + (\Delta x_n)\sqrt{1 + f'(x_n^*)^2} \quad (3)$$

for the length of the arc. The right-hand side of approximation (3) is a Riemann sum for the function $\sqrt{1 + f'(x)^2}$ over $[a, b]$. As $\|P\| \to 0$, we expect the approximation to s to become more and more accurate, so we arrive at the integral formula

$$s = \int_a^b \sqrt{1 + f'(x)^2}\, dx \quad (4)$$

for arc length s of the graph of a continuously differentiable function f over $[a, b]$.

EXAMPLE 1 Find the length of the curve $y = f(x) = x^{3/2}$ for $0 \le x \le 1$.

Solution The arc whose length is desired is shown in Fig. 6.42. It is not really necessary to sketch the arc in a problem like this. The sketch in Fig. 6.42 will not be used to set up the appropriate integral. When finding areas and volumes, sketches were used for this purpose.

If $y = f(x) = x^{3/2}$, we have

$$f'(x) = \frac{3}{2}x^{1/2}.$$

By Eq. (4), the length of our curve is therefore

$$\int_0^1 \left(1 + \frac{9}{4}x\right)^{1/2} dx = \frac{4}{9} \cdot \frac{(1 + \frac{9}{4}x)^{3/2}}{\frac{3}{2}}\bigg]_0^1 = \frac{4}{9} \cdot \frac{2}{3}\left[\left(\frac{13}{4}\right)^{3/2} - 1\right]$$

$$= \frac{8}{27} \cdot \frac{13^{3/2}}{8} - \frac{8}{27} = \frac{13^{3/2} - 8}{27}. \quad \blacksquare$$

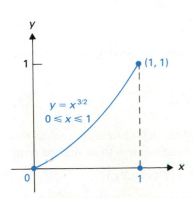

FIGURE 6.42 The length of the arc is $\int_0^1 \sqrt{1 + (9x/4)}\, dx$.

EXAMPLE 2 Find the arc length of the portion of the graph of $y = f(x) = x^2$ for $0 \le x \le 2$.

Solution   We have $f'(x) = 2x$, so

$$\sqrt{1 + f'(x)^2}\, dx = \sqrt{1 + 4x^2}\, dx.$$

Thus

$$s = \int_0^2 \sqrt{1 + 4x^2}\, dx.$$

Using formula 26 in the integral table on the endpapers, we find that

$$\int_0^2 \sqrt{1 + 4x^2}\, dx = \frac{1}{2}\int_0^2 \sqrt{1 + (2x)^2}\,(2\,dx)$$

$$= \frac{1}{2}\left[\frac{2x}{2}\sqrt{1 + 4x^2} + \frac{1}{2}\ln(2x + \sqrt{4x^2 + 1})\right]\Bigg|_0^2$$

$$= \frac{1}{2}\left[\frac{4}{2}\sqrt{17} + \frac{1}{2}\ln(4 + \sqrt{17})\right] - 0$$

$$= \sqrt{17} + \frac{1}{4}\ln(4 + \sqrt{17}) \approx 4.647. \qquad \blacksquare$$

The radical that appears in Eq. (4) frequently makes evaluation of the integral difficult; examples have to be chosen with care to enable us to compute the integral by taking an antiderivative of $\sqrt{1 + f'(x)^2}$. But we know there are numerical methods (such as Simpson's rule) that we can easily use if integration causes difficulty.

EXAMPLE 3   Find the arc length of one arch of the graph of $y = f(x) = \sin x$ over the interval $[0, \pi]$.

Solution   Since $f'(x) = \cos x$, the desired arc length is given by

$$s = \int_0^\pi \sqrt{1 + \cos^2 x}\, dx.$$

We do not find $\int \sqrt{1 + \cos^2 x}\, dx$ in the table on the endpapers of this text. Reaching for some computer software and using Simpson's rule with $n = 100$, we obtain the estimate 3.8202 for the desired arc length. $\qquad \blacksquare$

Differential Notation

Once again, differential notation is of heuristic value in viewing Eq. (4) for arc length. Suppose we approximate

$$s = \int_a^b \sqrt{1 + f'(x)^2}\, dx \qquad (5)$$

using a Riemann sum, taking for the points x_i^* the *left-hand endpoints* of the intervals $[x_{i-1}, x_i]$. Figure 6.43 and the Pythagorean theorem show that the term

$$(\Delta x_i)\sqrt{1 + f'(x_i^*)^2} = \sqrt{(\Delta x_i)^2 + [f'(x_i^*)(\Delta x_i)]^2}$$

is equal to the length s_i of a short segment of the tangent line to the curve, shown in the figure. Thus we can also regard arc length over $[a, b]$ as the limit as

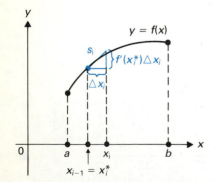

FIGURE 6.43   Length s_i of a tangent line segment.

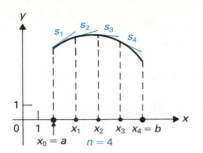

FIGURE 6.44

Arc length $\approx s_1 + s_2 + s_3 + s_4$

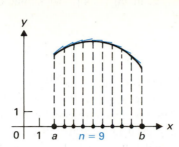

FIGURE 6.45 The approximation usually improves as n increases.

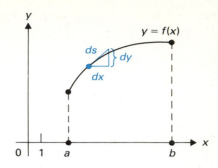

FIGURE 6.46 Differential right triangle at a point on the graph of f.

$\|P\| \to 0$ of a sum, for a partition P, of such approximating tangent line segments to the curve. This is illustrated in Fig. 6.44 for a regular partition with $n = 4$. Figure 6.45 is the analogous illustration with $n = 9$.

Recall from Section 3.9 that in our standard analysis interpretation of dx and dy, we regard dx as an increment Δx in x, while dy is the change Δy_{tan} of the y-coordinate on the tangent line corresponding to this increment in x. Figure 6.46 shows the same triangle as in Fig. 6.43, but with sides labeled using differential notation; in particular, the hypotenuse of this right triangle is labeled ds. We consider ds to be the *differential of arc length*, and we interpret ds geometrically, in our standard analysis approach, as the length of a small tangent line segment to the curve. Once again, we see that calculus deals with approximation of a function f by *linear functions* having as graphs tangent lines to the graph of f. Note that in differential notation, we find arc length s by integrating ds between appropriate limits.

We consider the triangle in Fig. 6.46 to be a *differential triangle* for the graph of f. From this triangle, we obtain the relation $ds = \sqrt{(dx)^2 + (dy)^2}$. We give a formal definition.

DEFINITION 6.2

Differential of arc length

Let f be a continuously differentiable function, and let $y = f(x)$. The **differential ds of arc length** for f is

$$ds = \sqrt{(dx)^2 + (dy)^2}. \tag{6}$$

Note that formal manipulation of Eq. (6) yields

$$ds = \sqrt{(dx)^2 + (dy)^2} = \sqrt{1 + \left(\frac{dy}{dx}\right)^2} \; dx$$

$$= \sqrt{\left(\frac{dx}{dy}\right)^2 + 1} \; dy. \tag{7}$$

From Eq. (7), we obtain the formula $\sqrt{1 + f'(x)^2}$ for the function to be integrated over $[a, b]$ to obtain the length of the graph of f over that interval when $y = f(x)$. Reversing the roles of the variables x and y, we see that if $x = g(y)$ for

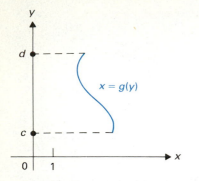

FIGURE 6.47 Arc length s of $x = g(y)$ for $c \le y \le d$ is
$$s = \int_c^d \sqrt{1 + g'(y)^2} \, dy.$$

HISTORICAL NOTE

ARC LENGTH was considered by René Descartes (1596–1650), one of the inventors of analytic geometry. In 1637 he wrote that the human mind could discover no rigorous and exact method of determining the ratio between curved and straight lines, that is, of determining exactly the length of a curve. Scarcely two decades later, several human minds showed that he was wrong. The most general procedure for finding the length of a curve was worked out in 1659 by Henrich van Heuraet (1634–1660).

Van Heuraet was born in the Netherlands and studied mathematics at the university in Leiden. Though he showed great promise in mathematics, he died at an early age.

In one of only two mathematical articles that he published, Van Heuraet developed essentially the modern procedure for rectifying a curve and illustrated it with the familiar example of the semicubical parabola $y^2 = x^3$, for which he found a numerical answer. He also noted that though the length of a parabolic arc can be found only by determining the area under a hyperbola, a problem that was at the time unsolvable, explicit solutions can be found for the length of the curves $y^4 = x^5$, $y^6 = x^7$, $y^8 = x^9$, and so on.

$c \le y \le d$ as in Fig. 6.47, the length of the graph of g over this interval is given by

$$s = \int_c^d \sqrt{1 + g'(y)^2} \, dy.$$

Since $g'(y)$ is dx/dy in differential notation, we see that the final formula for ds in Eq. (7) corresponds to this integral.

In summary, this differential notation is of considerable heuristic value. It is also widely used in scientific literature, so we should understand its use and practice using it to become familiar with it.

EXAMPLE 4 Find the length of the arc of the curve $x = 2y^3 + 1/(24y)$ for $1 \le y \le 3$, using differential notation in forming the integral.

Solution We use ds in the form

$$ds = \sqrt{1 + \left(\frac{dx}{dy}\right)^2} \, dy.$$

We easily compute

$$\frac{dx}{dy} = 6y^2 - \frac{1}{24y^2}.$$

Thus

$$ds = \sqrt{1 + \left(6y^2 - \frac{1}{24y^2}\right)^2} \, dy$$

$$= \sqrt{1 + (6y^2)^2 - \frac{1}{2} + \frac{1}{(24y^2)^2}} \, dy$$

$$= \sqrt{(6y^2)^2 + \frac{1}{2} + \frac{1}{(24y^2)^2}} \, dy$$

$$= \sqrt{\left(6y^2 + \frac{1}{24y^2}\right)^2} \, dy = \left(6y^2 + \frac{1}{24y^2}\right) dy.$$

Therefore we have

$$s = \int_1^3 \left(6y^2 + \frac{1}{24}y^{-2}\right) dy = \left(2y^3 - \frac{1}{24y}\right)\Bigg]_1^3$$

$$= \left(54 - \frac{1}{72}\right) - \left(2 - \frac{1}{24}\right) = 52 + \frac{2}{72} = 52 + \frac{1}{36} = \frac{1873}{36}. \quad \blacksquare$$

SUMMARY

1. If $f'(x)$ is continuous, then the length of arc of the graph $y = f(x)$ from $x = a$ to $x = b$ is

$$s = \int_a^b \sqrt{1 + f'(x)^2} \, dx.$$

2. The differential ds of arc length takes the following forms.

$$ds = \sqrt{(dx)^2 + (dy)^2}$$
$$= \sqrt{1 + \left(\frac{dy}{dx}\right)^2}\, dx$$
$$= \sqrt{\left(\frac{dx}{dy}\right)^2 + 1}\, dy$$

EXERCISES 6.5

In Exercises 1–10, find the length of the curve with the given equation.

1. $y^2 = x^3$ from $(0, 0)$ to $(4, 8)$
2. $y = \frac{1}{3}(x^2 - 2)^{3/2}$ from $x = 2$ to $x = 4$
3. $9x^2 = 16y^3$ from $y = 3$ to $y = 6$, $x > 0$
4. $9(x + 2)^2 = 4(y - 1)^3$ from $y = 1$ to $y = 4$, $x \geq -2$
5. $y = x^{2/3}$ from $x = 1$ to $x = 8$
6. $y = x^3 + 1/(12x)$ from $x = 1$ to $x = 2$
7. $y = x^3/3 + 1/(4x)$ from $x = 1$ to $x = 3$
8. $x = y^4 + 1/(32y^2)$ from $y = -2$ to $y = -1$ [*Hint:* Be sure to take the *positive* square root.]
9. $x = y^5/20 + 1/(3y^3)$ from $y = 1$ to $y = 2$
10. $x^{2/3} + y^{2/3} = a^{2/3}$

Consider the arc length function $s(x) = \int_a^x \sqrt{1 + f'(t)^2}\, dt$ where $y = f(x)$. A similar expression holds if $x = g(y)$. In Exercises 11–15, approximate the short length of curve indicated using the appropriate arc length function and using approximation by the differential ds.

11. The approximate arc length of the curve $y = \sin x$ from $x = 0$ to $x = 0.03$
12. The approximate arc length of the curve $y = \cos x$ from $x = \pi/6$ to $x = 8\pi/45$
13. The approximate arc length of the curve $y = x^2$ from $x = 3$ to $x = 3.01$
14. The approximate arc length of the curve $x = 2/y^2$ from $y = 1$ to $y = 1.05$
15. The approximate arc length of the curve $x = \ln y$ from $y = 1$ to $y = 1.05$

In Exercises 16–20, use a calculator or computer to compute the approximate arc length, using Simpson's rule with some $n \geq 10$.

16. The length of $y = x^3$ from $x = 1$ to $x = 5$
17. The length of $y = 1/x$ from $x = \frac{1}{2}$ to $x = 2$
18. The length of $x = \sin^2 y$ from $y = 0$ to $y = \pi$
19. The length of $y = e^{2x}$ for $-1 \leq x \leq 2$
20. The length of $y = 3 \cos 2x + 4 \sin (x/2)$ from $x = 0$ to $x = \pi$

6.6 AREA OF A SURFACE OF REVOLUTION

In Section 6.5, we found the length of an arc. In this section, we revolve such arcs about axes and find the *area of the surface* generated.

We assume again that f is a continuously differentiable function. If a portion of the graph of $y = f(x)$ for $a \leq x \leq b$ is revolved about the x-axis, it generates a *surface of revolution* as shown in Fig. 6.48. Using the suggestive differential notation, suppose we take a small tangent line segment of length ds at (x, y) on the curve and revolve this segment about the x-axis as shown in Fig. 6.49. The surface area generated by this segment would seem to be approximately that of a

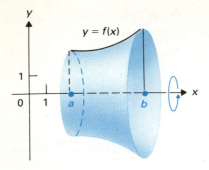

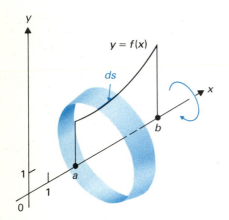

FIGURE 6.48 Surface generated by revolving $y = f(x)$ about the x-axis.

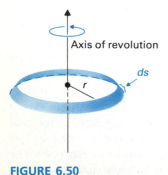

FIGURE 6.49 Strip of surface generated by the tangent line segment of length ds.

cylinder of radius $|y|$ and height ds, that is, approximately $2\pi|y|ds$. Since $y = f(x)$ and $ds = \sqrt{1 + (f'(x))^2}\, dx$, we arrive at the formula

$$S = \int_a^b 2\pi|f(x)|\sqrt{1 + (f'(x))^2}\, dx \qquad (1)$$

for the area S of the surface of revolution about the x-axis.

The preceding paragraph indicates how we can easily remember formula (1). A justification of the formula can be made along the following lines. Form a partition P of the interval $[a, b]$ into n subintervals. Consider the chords to the curve over the subintervals, as in the previous section. Formula (2) in the last section showed that the length of the ith chord of the curve can be written

$$(\Delta x_i)\sqrt{1 + f'(x_i^*)^2}$$

for some x_i^* in the ith subinterval. Let c_i and c_i' be chosen in the ith subinterval so that

$|f(c_i)|$ is the maximum value of $|f(x)|$ over the subinterval,
$|f(c_i')|$ is the minimum value of $|f(x)|$ over the subinterval.

Then the surface area element generated by revolving this ith chord satisfies

$$2\pi|f(c_i')|(\Delta x_i)\sqrt{1 + f'(x_i^*)^2} \leq \text{Area element}$$
$$\leq 2\pi|f(c_i)|(\Delta x_i)\sqrt{1 + f'(x_i^*)^2}.$$

By the theorem of Bliss referred to in Section 6.3, both

$$\sum_{i=1}^{n} \left(2\pi|f(c_i')|\sqrt{1 + f'(x_i^*)^2}\,\Delta x_i\right)$$

and

$$\sum_{i=1}^{n} \left(2\pi|f(c_i)|\sqrt{1 + f'(x_i^*)^2}\,\Delta x_i\right)$$

approach the integral in formula (1) as $\|P\| \to 0$. This justifies formula (1).

If an arc is revolved about an axis other than the x-axis, formula (1) is modified in the natural way. Namely, we "add by integrating" the contributions $2\pi r \cdot ds$ where ds is the length of a tangent line segment to the curve and r is the radius of the circle through which this tangent segment is revolved. See Fig. 6.50. The general formula in differential notation for surface area of revolution is thus

$$\text{Surface area} = S = \int_a^b 2\pi \cdot \text{Radius of revolution} \cdot ds, \qquad (2)$$

where a and b are appropriate limits and ds is the differential of arc length.

EXAMPLE 1 Find the area of the surface generated by revolving the arc of $y = f(x) = x^2$ from $(0, 0)$ to $(1, 1)$ about the y-axis.

Solution We have $f'(x) = 2x$, so

$$ds = \sqrt{1 + 4x^2}\, dx.$$

FIGURE 6.50

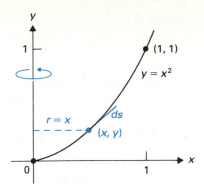

FIGURE 6.51 Tangent line segment of length ds is at distance $r = x$ from the x-axis of revolution.

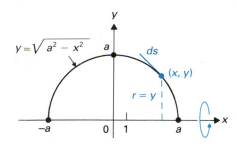

FIGURE 6.52 Surface area of revolution is $2\pi y\,ds$ integrated from $x = -a$ to $x = a$.

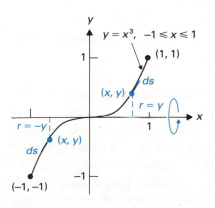

FIGURE 6.53 $r = y$ for ds above the x-axis, $r = -y$ for ds below the x-axis; thus $r = |y| = |x^3|$.

The radius of revolution is x, as indicated in Fig. 6.51. Then formula (2) becomes

$$S = \int_0^1 2\pi x\,ds = \int_0^1 2\pi x\sqrt{1 + 4x^2}\,dx = \frac{2\pi}{8}\int_0^1 8x(1 + 4x^2)^{1/2}\,dx$$

$$= \frac{\pi}{4}\cdot\frac{(1 + 4x^2)^{3/2}}{\frac{3}{2}}\Bigg]_0^1 = \frac{\pi}{6}(1 + 4x^2)^{3/2}\Bigg]_0^1 = \frac{\pi}{6}(5\sqrt{5} - 1). \qquad\blacksquare$$

EXAMPLE 2 Derive the formula $A = 4\pi a^2$ for the surface area of a sphere having radius a.

Solution We view the sphere as the surface generated by revolving the semicircle $y = f(x) = \sqrt{a^2 - x^2}$ for $-a \le x \le a$ about the x-axis, as shown in Fig. 6.52. Here formula (1) is appropriate. Differentiating, we have

$$f'(x) = \frac{1}{2}(a^2 - x^2)^{-1/2}(-2x) = \frac{-x}{\sqrt{a^2 - x^2}}.$$

We obtain as our integral

$$\int_{-a}^a 2\pi\sqrt{a^2 - x^2}\,\sqrt{1 + \frac{x^2}{a^2 - x^2}}\,dx = \int_{-a}^a 2\pi\sqrt{a^2 - x^2}\,\sqrt{\frac{a^2 - x^2 + x^2}{a^2 - x^2}}\,dx$$

$$= \int_{-a}^a 2\pi\sqrt{a^2}\,dx = 2\pi ax\Big]_{-a}^a$$

$$= 2\pi a^2 - (-2\pi a^2) = 4\pi a^2. \qquad\blacksquare$$

EXAMPLE 3 Find the area of the surface generated by revolving the arc of $y = x^3$ from $(-1, -1)$ to $(1, 1)$ about the x-axis.

Solution We have $dy/dx = 3x^2$, so

$$ds = \sqrt{1 + \left(\frac{dy}{dx}\right)^2}\,dx = \sqrt{1 + 9x^4}\,dx.$$

From Fig. 6.53, we see that $r = |y| = |x^3|$. The absolute value symbol is crucial here, since $x^3 < 0$ for $-1 \le x < 0$. Thus

$$S = \int_{-1}^1 2\pi|x^3|\sqrt{1 + 9x^4}\,dx.$$

Since $x^3 < 0$ for $-1 \le x < 0$, and $x^3 \ge 0$ for $0 \le x \le 1$, we have

$$S = \int_{-1}^0 2\pi(-x^3)\sqrt{1 + 9x^4}\,dx + \int_0^1 2\pi x^3\sqrt{1 + 9x^4}\,dx$$

$$= -\frac{2\pi}{36}\int_{-1}^0 (1 + 9x^4)^{1/2}(36x^3)\,dx + \frac{2\pi}{36}\int_0^1 (1 + 9x^4)^{1/2}(36x^3)\,dx$$

$$= -\frac{\pi}{18}\cdot\frac{2}{3}(1 + 9x^4)^{3/2}\Bigg]_{-1}^0 + \frac{\pi}{18}\cdot\frac{2}{3}(1 + 9x^4)^{3/2}\Bigg]_0^1$$

$$= -\frac{\pi}{27}(1 - 10\sqrt{10}) + \frac{\pi}{27}(10\sqrt{10} - 1) = \frac{2\pi}{27}(10\sqrt{10} - 1).$$

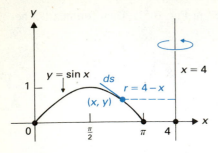

FIGURE 6.54

$$r = x_{\text{right}} - x_{\text{left}} = 4 - x$$

We would usually have used symmetry and computed S by

$$S = 2 \int_0^1 2\pi x^3 \sqrt{1 + 9x^4}\, dx,$$

but we wished to call attention to the need to be sure that r is nonnegative.   ■

EXAMPLE 4   Find the area of the surface generated by revolving the arc of $y = \sin x$ for $0 \le x \le \pi$ about the line $x = 4$.

Solution   We have $dy/dx = \cos x$, so $ds = \sqrt{1 + \cos^2 x}\, dx$. From Fig. 6.54, we see that $r = 4 - x$. Thus

$$S = \int_0^\pi 2\pi(4 - x)\sqrt{1 + \cos^2 x}\, dx.$$

This integral is one we have to compute numerically. Using Simpson's rule with $n = 20$, we find that

$$S = 2\pi \int_0^\pi (4 - x)\sqrt{1 + \cos^2 x}\, dx \approx 58.308.$$ ■

SUMMARY

If an arc of a smooth curve is revolved about an axis, then the area of the surface generated is

$$S = \int_a^b 2\pi \cdot \text{Radius of revolution} \cdot ds$$

for appropriate limits a and b.

EXERCISES 6.6

In Exercises 1–12, find the area of the surface obtained by revolving the given curve about the indicated axis.

1. $y = 4x$ from $(0, 0)$ to $(1, 4)$ about the y-axis
2. $x = 2y$ from $(2, 1)$ to $(6, 3)$ about the x-axis
3. $x = y - 2$ from $y = 2$ to $y = 5$ about the line $y = 3$
4. $y = x - 3$ from $x = 3$ to $x = 5$ about the line $x = -1$
5. $y = x^3$ from $x = -1$ to $x = 2$ about the x-axis
6. $y = \sqrt{x}$ from $x = 1$ to $x = 3$ about the x-axis
7. $y = \frac{2}{3}x^{3/2}$ from $x = 0$ to $x = 2$ about the y-axis (Use the table of integrals on the endpapers of this book.)

8. $y = |x|$ from $x = -1$ to $x = 1$ about the x-axis
9. $y = \frac{1}{3}(x^2 - 2)^{3/2}$ from $x = \sqrt{2}$ to $x = 4$ about the y-axis
10. $y = x^4/4 + 1/(8x^2)$ from $x = 1$ to $x = 2$ about the y-axis
11. $y = x^3 + 1/(12x)$ from $x = 1$ to $x = 2$ about the x-axis
12. $x^{2/3} + y^{2/3} = a^{2/3}$ for $y \ge 0$ about the x-axis

13. Assuming the surface of the earth to be a sphere of radius 4000 miles, find the surface area of the zone of the earth between the equator and 45° north latitude.

14. If a cone has base radius r and slant height s, its surface area is πrs. Derive this formula by revolving the segment of the

line $y = (r/h)x$ from $x = 0$ to $x = h$ about the x-axis. (See Fig. 6.55.)

15. Derive the formula for the surface area of a frustrum of a cone having base radius R, top radius r, and altitude h, using the methods of this section.

In Exercises 16–23, use a calculator or computer to find the area of the surface obtained by revolving the given curve about the indicated axis. Use Simpson's rule with $n \geq 10$.

16. $y = x^2$ from $x = 0$ to $x = 2$ about the x-axis

17. $y = \sin x$ from $x = 0$ to $x = \pi$ about the y-axis

18. $x = y^3 + 3y^2$ from $y = 0$ to $y = 4$ about the y-axis

19. $y = x^2$ from $x = 0$ to $x = 2$ about the line $y = -4$

20. $y = x^2$ from $x = 0$ to $x = 2$ about the line $x = -3$

21. $x = y^2 + 2y$ from $y = 1$ to $y = 5$ about the line $y = 10$

22. $x = \sin y$ from $y = 0$ to $y = \pi$ about the line $x = 2$

23. $y = \tan x$ from $x = -\pi/3$ to $x = \pi/3$ about the line $y = -\pi$

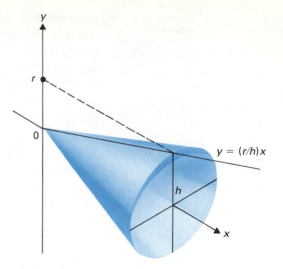

6.7   WORK AND FORCE EXERTED BY A FLUID

In the preceding sections, we described each application of the integral by taking a partition P of an interval and forming an appropriate Riemann sum of terms of the form $f(x_i^*)\Delta x_i$. We arrived at an integral by taking the limit of such Riemann sums as $\|P\| \to 0$. We also described how to form such integrals using the heuristic differential notation. By now, the Riemann sum argument should be very familiar to you. In this section, we emphasize recognizing an application of the integral and forming the appropriate integral using differential notation. You should be able to pass to the more careful partition and Riemann sum argument. Our presentation here is closer to that of typical physics or engineering texts and will prepare you for reading such texts.

Work

If a body is moved a distance s by means of a constant force F in the direction of the motion, then the work W done in moving the body is the product Fs. For example, if a body is being pushed along the x-axis by a constant force of 20 lb directed along the axis, then the work done by the force on the body in pushing from position $x = 0$ to position $x = 10$ ft is $W = 20 \cdot 10 = 200$ ft·lb.

If the force acting on the body does not remain constant but is a continuous function $F(x)$ of the position x of the body, then the work done over a short interval of length dx is approximately $F(x)\,dx$ for some position x in this interval. Adding all these contributions to the work from starting position $x = a$ to stopping position $x = b$, we obtain

$$W = \int_a^b F(x)\,dx. \tag{1}$$

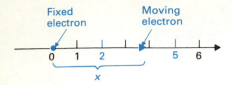

FIGURE 6.56 Repelling force on the moving electron is k/x^2.

EXAMPLE 1 By Hooke's law, the force F required to stretch (or compress) a coil spring is proportional to the distance x it is stretched (or compressed) from its natural length. That is, $F(x) = kx$ for some constant k, the *spring constant*. Suppose that a spring is such that the force required to stretch it 1 ft from its natural length is 4 lb. For this spring, $k = 4$. Find the work done in stretching the spring a distance of 4 ft from its natural length.

Solution Work is defined as the product of force and distance, if the force remains constant and is in the direction of the motion. Thus as our spring is stretched an additional small distance dx at a distance x from its natural length, the work done is approximately $F(x)\,dx = 4x\,dx$. We wish to add all these little pieces of work from $x = 0$ to $x = 4$ and take the limit of the resulting sum as increments $dx \to 0$. The appropriate integral is

$$\int_0^4 4x\,dx = 4\frac{x^2}{2}\bigg]_0^4 = \frac{64}{2} - 0 = 32.$$

Thus 32 ft·lb of work are done. ∎

EXAMPLE 2 Two electrons a distance s apart repel each other with a force k/s^2, where k is some constant. If one electron remains fixed at the origin on the x-axis, find the work done by the force of repulsion in moving another electron from the point 2 to the point 5 on the x-axis.

Solution We sketch the situation in Fig. 6.56. When the second electron is at a point x, the force of repulsion is k/x^2. The work done in moving the second electron an additional small distance dx is approximately $(k/x^2)\,dx$. Thus the total work done is

$$W = \int_2^5 \frac{k}{x^2}\,dx = -\frac{k}{x}\bigg]_2^5 = -\frac{k}{5} - \left(-\frac{k}{2}\right) = k\left(\frac{1}{2} - \frac{1}{5}\right) = \frac{3k}{10}\ \text{units.} \quad ∎$$

EXAMPLE 3 Suppose a long, heavy, flexible chain lies coiled on the floor. If the chain weighs 2 lb/ft, estimate the work W done in raising one end of the chain vertically 4 ft off the floor.

Solution Since the chain weighs 2 lb/ft, we lift with the force

$$F(y) = 2y \text{ lb}$$

when the end of the chain is y ft off the floor, as shown in Fig. 6.57. The work done as the chain is lifted an additional dy ft off the floor is therefore approximately $(2y)\,dy$. We add these small contributions to the work and take the usual limit of the sum using an integral. Since the end of the chain is lifted 4 ft, the appropriate integral is

$$\int_0^4 (2y)\,dy = y^2\bigg]_0^4 = 16 - 0 = 16 \text{ ft·lb.} \quad ∎$$

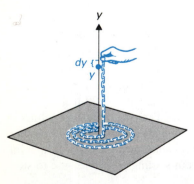

FIGURE 6.57 Chain weighing 2 lb/ft lifted by a force $2y$ when the end is y ft off the floor.

EXAMPLE 4 A cylindrical tank of radius 2 ft and height 8 ft is filled with water to a height of 6 ft. The tank is emptied by dropping in a hose, pumping all the water up to the top of the tank, and letting it spill over. If water weighs 62.4 lb/ft^3, find the work done to empty the tank.

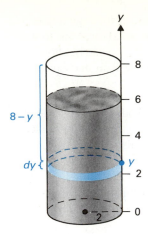

FIGURE 6.58 The circular slice of water weighs $62.4(4\pi) \Delta y$ lb and must be raised $8 - y$ ft to the top of the tank.

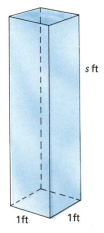

FIGURE 6.59 The column of water s ft high and 1 ft square weighs $62.4s$ lb.

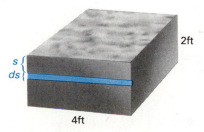

FIGURE 6.60
Pressure $p = 62.4s$;
Area of strip $= 4\ ds$;
Force on strip $= 62.4s(4)\ ds$

Solution We take a y-axis upward from the bottom of the tank, as shown in Fig. 6.58. The circular slice of water at height y and thickness dy shown in the figure has volume

$$V_{\text{slice}} = \pi 2^2\ dy = 4\pi\ dy \text{ ft}^3.$$

This slice then weighs $62.4(4\pi)\ dy$ lb. Therefore a force of $62.4(4\pi)\ dy$ lb is required to raise it, and it must be raised the distance $(8 - y)$ ft to the top of the tank. The work done on such slices from $y = 0$ to $y = 6$ ft is summed and the limit is taken in the usual fashion using an integral. The total work is then given by

$$W = \int_0^6 (8 - y)(62.4)(4\pi)\ dy = 249.6\pi \int_0^6 (8 - y)\ dy$$

$$= 249.6\pi\left(8y - \frac{y^2}{2}\right)\Bigg]_0^6 = 249.6\pi(48 - 18)$$

$$= 249.6\pi(30) = 7488\pi \text{ ft·lb}.$$

(Note that since we are using 62.4 lb as the approximate weight of a cubic foot of water, we can consider only the first three digits of 7488π to be significant. The same is true for the solutions in Examples 5 and 6.) ∎

Force Exerted by a Fluid

If the pressure per square unit on a plane region of area A is a constant p throughout the region, then the total force F on the region due to the pressure is the product pA. Suppose the pressure does not remain constant throughout the region. We then attempt to find the total force by integrating $p\ dA$ where dA is the area of a small piece of the region throughout which the pressure remains roughly constant.

The pressure exerted by the weight of a fluid at a depth s in the fluid acts equally in all directions. A cubic foot of water weighs approximately 62.4 lb, so a column of water s ft high and 1 ft square, such as the one shown in Fig. 6.59, weighs $62.4s$ lb. Thus the pressure in water at a depth of s ft is approximately $62.4s$ lb/ft^2.

EXAMPLE 5 One side of a rectangular tank measures 4 ft wide by 2 ft high. If the tank is filled with water, find the total force of the water on this side of the tank.

Solution Consider a small horizontal strip of width ds across the side of the tank, as shown in Fig. 6.60. The pressure is almost constant throughout this strip. Let s be the distance from the top of the tank down to this strip. The pressure at that depth is $p = 62.4s$ lb/ft^2. Since the area of the strip is $dA = 4\ ds$, the force dF due to water pressure on this strip is approximately $dF = p\ dA = 62.4s(4)\ ds$. To find the total force F on this side of the tank, we integrate, obtaining

$$F = \int_0^2 62.4s(4)\ ds = 4(62.4) \int_0^2 s\ ds = 4(62.4)\frac{s^2}{2}\Bigg]_0^2$$

$$= 4(62.4)(2 - 0) = 8(62.4) = 499.2 \text{ lb.} \quad ∎$$

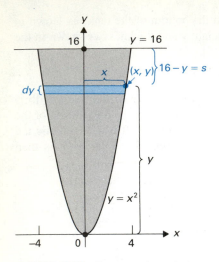

FIGURE 6.61 Force on the strip is

$62.4(16 - y)(2x) \, dy =$
$\qquad 62.4(16 - y)(2\sqrt{y}) \, dy.$

EXAMPLE 6 Suppose a dam 16 ft high has the shape of the region bounded by the curves with equations $y = x^2$ and $y = 16$. Find the total force due to water pressure on the dam when the water is at the top of the dam. (It can be shown that this force depends only on the height of the water at the dam and not on the distance the water is backed up behind the dam.)

Solution The region bounded by the curves $y = x^2$ and $y = 16$ is shaded in Fig. 6.61. For small dy, the pressure at a depth s on the strip heavily shaded in Fig. 6.61 is nearly constant at $62.4s$ lb/ft^2. The area of this strip is $2x \, dy$, so the force on it is approximately $62.4s(2x) \, dy$ lb. We want to add these quantities $62.4s(2x) \, dy$ over the region as y ranges from 0 to 16 and take the limit of the resulting sum as increments $dy \to 0$. We must express this quantity entirely in terms of y. For the point (x, y) in Fig. 6.61, we have $s = 16 - y$, and $y = x^2$, so $x = \sqrt{y}$. Thus

$$62.4s(2x) \, dy = 62.4(16 - y)2\sqrt{y} \, dy$$
$$= 124.8(16y^{1/2} - y^{3/2}) \, dy.$$

The appropriate integral is

$$\int_0^{16} 124.8(16y^{1/2} - y^{3/2}) \, dy = 124.8\left(16\left(\frac{2}{3}\right)y^{3/2} - \frac{2}{5}y^{5/2} \right)\Bigg]_0^{16}$$

$$= 124.8\left(16\left(\frac{2}{3}\right)64 - \left(\frac{2}{5}\right)1024 \right) - 0 = 124.8(1024)\left(\frac{2}{3} - \frac{2}{5}\right)$$

$$= 124.8(1024)\frac{4}{15} = 34{,}078.72.$$

Thus the force is about 34,100 lb. ∎

SUMMARY

1. If the force acting on a body in the direction of its motion along a line is the function $F(x)$ of the position x of the body, then the work done by the force in moving the body from position $x = a$ to position $x = b$ is

$$W = \int_a^b F(x) \, dx.$$

2. The pressure exerted by water at depth s is about $62.4s$ lb/ft^2. Let a submerged vertical plate have its top at a depth a and its bottom at depth b. If the area of a thin horizontal strip of the plate at depth s is dA, then

$$\text{Total force on one side of the plate} = \int_a^b 62.4s \, dA.$$

EXERCISES 6.7

1. Find the work done in stretching the spring of Example 1 from 2 ft longer than its natural length to 6 ft longer than its natural length.

2. If a spring has a natural length of 2 ft and a force of 10 lb is required to compress the spring 2 in. (to a length of 22 in.), find the work done in stretching the spring from a length of 26 in. to a length of 30 in.

3. If 36 ft·lb of work are done in stretching a spring 3 ft from its natural length, find the spring constant k.

4. A force of 10 lb stretches a spring 2 ft. The spring is stretched from its natural length until 90 ft·lb of work has been done. How far was the spring stretched?

5. A flexible chain 6 ft long and weighing 2 lb/ft lies coiled on the floor. Find the work done when one end of the chain is raised vertically (a) 3 ft, (b) 6 ft.

6. Answer Exercise 5 if the end of the chain is raised vertically 8 ft.

7. A cylindrical tank of radius 2 ft and height 10 ft is full of water and is emptied by dropping in a hose, pumping all the water up to the top of the tank, and letting it spill over. Find the work done to empty the tank.

8. Work Exercise 7 for a conical reservoir (with vertex down) of height 10 ft and top radius 2 ft.

9. A rectangular tank of water has a base measuring 4 ft by 8 ft and is 4 ft high. The tank is emptied by pumping the water to the top and letting it spill over the side. If 14,976 ft·lb of work are needed, how deep was the water in the tank?

10. A bucket of cement weighing 100 lb is winched to a height of 30 ft using a cable weighing 1 lb/ft. (See Fig. 6.62.) Find the work done.

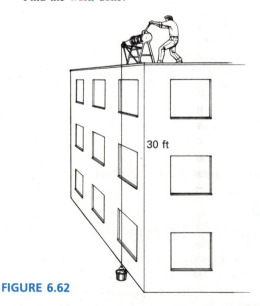

FIGURE 6.62

11. Work Exercise 10 if the bucket has a hole in it and cement falls out at a uniform rate of $\frac{1}{2}$ lb for each foot it is raised.

12. Two electrons a distance s apart repel each other with a force k/s^2 where k is some constant. If an electron is at the point 2 on the x-axis, find the work done by the force in moving another electron from the point 4 to the point 8.

13. Referring to Exercise 12, suppose one electron is fixed at the origin and another is fixed at the point 1 on the x-axis. Find the work done by the forces of repulsion in moving a third electron along the x-axis from the point 2 to the point 6.

14. According to Newton's law of gravitation, the force of attraction of two bodies of masses m_1 and m_2 is $G(m_1 m_2 / s^2)$, where s is the distance between the bodies and G is the gravitational constant. If the distance between two bodies of masses m_1 and m_2 is a, find the work done in moving the bodies twice as far apart.

15. Let a body of mass m move in the positive direction on the x-axis subject only to a force F acting in the positive direction along the axis. (The magnitude of the force may vary with the time t.) If the body has velocity v, the *kinetic energy* of the body is $\frac{1}{2}mv^2$. Show that the work done by the force from time t_1 to time t_2, for $t_1 < t_2$, is equal to the change in the kinetic energy of the body between these times. [*Hint:* By Newton's second law of motion,

$$F = ma = m\frac{dv}{dt} = m\frac{dv}{dx}\cdot\frac{dx}{dt} = m\frac{dv}{dx}v.$$

Consider both F and v as functions of the position x_t of the body at time t, and express the work

$$W = \int_{x_{t_1}}^{x_{t_2}} F(x)\,dx$$

as an integral involving v, using Newton's law.]

16. A rectangular plate measuring 1 ft by 2 ft is suspended vertically in water with the 2-ft edges horizontal and the upper one at a depth of 5 ft. Find the force of the water on one side of the plate. (Of course, the force on the other side is equal, but in the opposite direction, so the plate does not move.)

17. Work Exercise 16 for a vertically suspended circular plate of radius 2 ft if the top of the plate is at a depth of 4 ft.

18. The top of a rectangular swimming pool measures 20 ft by 40 ft. The pool is 1 ft deep at one end and slopes uniformly to a depth of 11 ft at the other end. Find the force of the water on the bottom of the pool when it is filled.

19. Find the force of the water on one side of the pool in Exercise 18.

20. A vertical dam is in the shape of a semicircle of radius 36 ft, with the diameter of the circle at the top of the dam. Find the force on the dam due to the water when the water level is at the top of the dam.

21. A vertical dam is in the shape of an isosceles trapezoid with the long 60-ft base at the top of the dam and the short 30-ft base at the bottom. The altitude of the trapezoid is 21 ft. Find the force of the water on the dam when the water is at the top of the dam.

22. A cylindrical drum of diameter 2 ft and length 4 ft is filled with water. Find the force due to the water on one end of the drum if the drum is lying on its side.

23. For the drum in Exercise 22, find the force due to the water on the cylindrical wall of the drum if the drum is standing on end.

24. A sphere of radius 2 ft is submerged in water so that its top is at a depth of 10 ft. Find the force of the water on the surface of the sphere. [*Hint:* Don't try to find some expression to integrate, but use symmetry and a calculus-type argument. For a small piece of surface of area dA on the upper hemisphere at depth $12 - x$, there is a symmetrically located piece of surface of equal area dA at depth $12 + x$ on the lower hemisphere. Find the total force on these two pieces, and then "sum" over the upper hemisphere in the integral sense.]

25. A cylindrical drum 3 ft in diameter and 5 ft long is submerged horizontally in water, with the axis of the cylinder parallel to the surface of the water, so that its top edge is at a depth of 4 ft. Use the method suggested in the hint of Exercise 24 to find the force of water on the cylindrical wall of the drum.

26. For the drum in Exercise 25, find the force of water on one end of the drum, using the method suggested by Exercise 24.

The remaining problems concern other applications of the integral. It is important to be able to recognize situations where an integral is appropriate.

27. Electric power consumption is measured in kilowatt hours, where consuming one kilowatt hour is equivalent to using 1000 watts of power for a duration of one hour. Also, $W = VI$ where W is in watts, V is in volts, and I is the current in amperes. Suppose the voltage remains constant at $120V$ and the current I after t hours is given by $I = 22 + 5\sqrt{t}$ amp for $0 \le t \le 4$. Find the power consumed (in kilowatt hours) during that time.

28. Referring to Exercise 27, find the average rate of power consumption during the time interval.

29. Referring to Exercise 27, suppose the voltage remains constant at $120V$, the current is measured at 5-minute intervals for an hour, starting at 6:00 P.M., and the following data are obtained.

Time	Current (amp)	Time	Current (amp)
6:00	20	6:35	32
6:05	25	6:40	25
6:10	25	6:45	23
6:15	20	6:50	20
6:20	22	6:55	18
6:25	30	7:00	20
6:30	35		

Use Simpson's rule to estimate the power consumption, in kilowatt hours, during that hour.

30. A sample of the pollen count per cubic centimeter of water at a depth of h cm in a pond yields the following data.

h	Pollen	h	Pollen
0	10	50	5
10	8	60	5
20	8	70	4
30	6	80	3
40	6	90	2
		100	0

Based on these data, use Simpson's rule to estimate the pollen count in a cube of water one meter on each edge with the top face of the cube horizontal at the surface of the pond.

SUPPLEMENTARY EXERCISES FOR CHAPTER 6

1. Find the area of the region bounded by the curves $y = \sqrt{2x}$ and $y = \frac{1}{2}x^2$.

2. Find the area of the region bounded by $x = 4 - y^2$ and $x = 3y$.

3. Find the area of the region bounded by $x = 0$, $y = \sin(x/4)$, and $y = \cos(x/4)$ for $0 \le x \le \pi$.

4. Find the area of the plane region bounded by $y = x^4$ and $y = 16$.

5. Find the area of the region in the first quadrant bounded by $y = x^2$, $y = x^2 + 9$, $x = 0$, and $y = 25$.

6. Find the volume of the solid generated by revolving the region bounded by $y = x^2$ and $y = x + 2$ about the line $x = -1$.

7. Find the volume of the solid generated when the region in Exercise 3 is revolved about the y-axis.

8. Find the volume of the solid generated when the region in Exercise 4 is revolved about the line $y = 16$.

9. Find the volume of the solid generated by revolving the region bounded by $y = 4 - x^2$ and $y = 0$ about the line $y = -1$.

10. A solid has as base the region bounded by $y = \cos(x/2)$ for $-\pi \le x \le \pi$ and the x-axis. Cross sections of the solid perpendicular to the x-axis are squares. Find the volume of the solid. (Use the table of integrals.)

11. A solid has as base the region bounded by $x = \sqrt{a^2 - y^2}$ and the y-axis. Cross sections of the solid perpendicular to the y-axis are equilateral triangles. Find the volume of the solid.

12. Use the method of cylindrical shells to find the volume of the solid generated by revolving the region bounded by $y = \sin 2x$ for $0 \le x \le \pi/2$ and $y = 0$ about the y-axis. (Use the table of integrals.)

13. Use the method of cylindrical shells to find the volume of the solid generated by revolving the region bounded by $y = 10/(4 + x^2)$, $x = 1$, $x = 3$, and $y = 0$ about the y-axis.

14. Find the total distance traveled on a line from $t = 1$ to $t = 4$ by a body with velocity $v = \sin \pi t$.

15. The velocity of a body traveling on a line is given by $v = 4 - t^2$. If the body starts at time $t = 0$, at what time has the body traveled a total distance of 16 units?

16. The velocity of a body traveling on a straight line is $v = (2t - 4)$ ft/sec. Find the total distance the body travels from time $t = 0$ to time $t = 10$ sec.

17. Let the acceleration of a body traveling on a line at time t be $(2t + 2)$ ft/sec^2, and let the initial velocity at time $t = 0$ be -8 ft/sec. Find the total distance the body travels from time $t = 0$ to time $t = 3$ sec.

18. Find the arc length of the curve $y = 2(\frac{5}{3} + x)^{3/2}$ from $x = 0$ to $x = \frac{11}{3}$.

19. Find the arc length of the curve $x = \frac{1}{5}y^{5/2} + (1/\sqrt{y})$ from $y = 1$ to $y = 4$.

20. (a) Express as an integral the length of the curve $y = \sqrt{x}$ from $x = 1$ to $x = 16$.
    (b) Use a calculator to evaluate approximately the integral in part (a).

21. Find the area of the surface of revolution generated by rotating the arc $y = x^3/\sqrt{3}$ from $x = 0$ to $x = 1$ about the x-axis.

22. Find the area of the surface of revolution generated by revolving the arc of $x = \sqrt{y}$ for $1 \le y \le 4$ about the y-axis.

23. (a) Express as an integral the surface area generated when the curve $y = 2 \sin x$ for $0 \le x \le \pi$ is revolved about the line $y = -1$.
    (b) Use a calculator to evaluate approximately the integral in part (a).

24. The work done in stretching a spring 1 ft from its natural length is 18 ft·lb. Find the work done in stretching it 4 ft from its natural length.

25. The force of attraction between two bodies is k/s^2 lb, where s is the distance between them measured in feet. If the bodies are 10 ft apart, find the work done in separating them an additional 10 ft.

26. Water flows from a pipe at time t minutes at a rate of $40/(t + 1)^2$ gal/min for $t \ge 0$. Find the amount of water that flows from the pipe during the first hour of flow.

27. The face of a dam is in the form of an isosceles triangle with 40-ft base at the top of the dam and equal 30-ft sides at the sides of the dam. Find the total force of the water on the dam when the water is at the top of the dam.

28. A square of side 2 ft is submerged in water with a diagonal vertical and the top corner of the square at the surface of the water. Find the force of the water on one side of the square.

29. Estimate with an integral
$$\sum_{k=1}^{800} \frac{k^{3/2}}{100,000}.$$

30. Estimate with an integral
$$\sum_{k=1}^{4000} \frac{1000}{(1000 + k)^2}.$$

31. A carpenter has a contract to hang 100 doors in a housing project. It takes him 1 hr to hang the first door. Using the expertise he continually attains, he finds that the time required to hang the nth door is $3\sqrt{n} - 1$ minutes less than 1 hr. Estimate the time required for him to hang all 100 doors.

32. A manufacturer finds that a newly hired employee is able to seal $40 + 2\sqrt[3]{n}$ cartons during the nth hour of work, except that, once a level of 60 cartons per hour is reached, there is no further increase. Estimate the number of cartons the employee can seal during the first 1500 hours of work.

OTHER ELEMENTARY FUNCTIONS

7

This chapter is devoted to further study and applications of the logarithm and exponential functions and to enlarging our stockpile of functions for calculus. The logarithm, exponential, inverse trigonometric, and hyperbolic functions have many important applications, often as solutions of differential equations that arise in the sciences.

7.1 INVERSE FUNCTIONS

This section studies inverse functions and their derivatives. It provides a foundation for a calculus-based development of the logarithm and exponential functions in Section 7.2 and the introduction of inverse trigonometric functions in Section 7.5.

Invertibility

Recall that when $y = f(x)$, the function f is a rule that assigns to each number x in the domain of f a single number y. The graph of a function gives us a geometric way to obtain y from x, as illustrated in Fig. 7.1: We follow the arrows. The curve in Fig. 7.2 is not the graph of a function because from a single x-value we can obtain more than one y-value by following the arrows.

A vertical line (parallel to the y-axis) cannot intersect the graph of a function in more than one point.

However, several different x-values may give rise to the same y-value, as illustrated in Fig. 7.3.

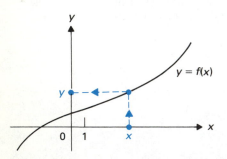

FIGURE 7.1 Follow the arrows to find $y = f(x)$ from x graphically.

340

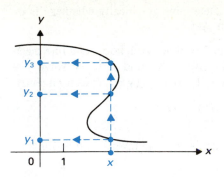

FIGURE 7.2 Not the graph of a function: Following arrows from x leads to three y-values.

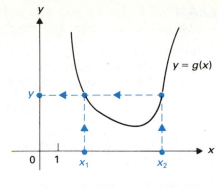

FIGURE 7.3 The graph of a function: Two x-values may give the same y-value.

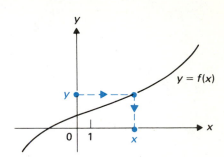

FIGURE 7.4 x is given as a function of y: Following arrows from a y-value gives just one x-value.

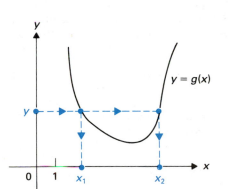

FIGURE 7.5 x is not given as a function of y: Following arrows from a y-value can give more than one x-value.

In this section, we are interested in those functions $y = f(x)$ whose graphs can be viewed as also defining x as a function of y. That is, we want to be able to reverse the directions of the arrows shown in Fig. 7.1 and still have a function. Starting with a y-value in the range of f, we want to obtain *just one* x-value. This can be done for the graph of f in Fig. 7.1, as shown in Fig. 7.4. However, for the graph in Fig. 7.3, one y-value may lead to two x-values, as shown in Fig. 7.5.

For the graph of $y = f(x)$ also to define x as a function of y, each horizontal line (parallel to the x-axis) must meet the graph in at most one point.

A function whose graph has this property is called *invertible*. In this case, the function that assigns to each y-value in the range of f the x-value obtained by following the arrows as in Fig. 7.4 is called the *inverse function* of f, denoted by f^{-1}.

Caution: The superscript -1 in f^{-1} must *not* be regarded as an exponent; that is, $f^{-1}(y)$ *is not* $1/f(y)$. However, $(f(y))^{-1} = 1/f(y)$.

For an invertible function f, we may then write both $y = f(x)$ and $x = f^{-1}(y)$; both functions are defined by the same graph. Let us summarize in a definition.

DEFINITION 7.1

Invertibility

A function f, where $y = f(x)$, is **invertible** if each horizontal line meets the graph of f in at most one point. The function that assigns to each y-value in the range of f the corresponding x-value in the domain of f is the **inverse function** of f and is denoted by f^{-1}.

For an invertible function f, we then have both

$$y = f(x) \qquad \text{and} \qquad x = f^{-1}(y). \tag{1}$$

The *domain* of f^{-1} is the range of f, and the *range* of f^{-1} is the domain of f. To find a rule for f^{-1} from a given f, we attempt to solve $y = f(x)$ for x in terms of y, as illustrated in Example 1.

EXAMPLE 1 Let $y = f(x) = 2x + 1$. Determine graphically whether f is invertible and, if so, express x as $x = f^{-1}(y)$.

Solution The graph of f is shown in Fig. 7.6. Since each horizontal line meets the graph only once, the function f is invertible. To find f^{-1}, we solve $y = 2x + 1$ for x in terms of y. Thus $2x = y - 1$, so $x = \frac{1}{2}(y - 1)$. Consequently,

$$x = f^{-1}(y) = \tfrac{1}{2}(y - 1).$$ ∎

EXAMPLE 2 Determine whether $y = h(x) = x^2$ is an invertible function and, if so, find $h^{-1}(y)$.

Solution The graph of h is shown in Fig. 7.7. Clearly, it is possible for a horizontal line to meet the graph in more than one point. For example, the line $y = 4$ meets the graph in two points, since $h(2) = 4$ and $h(-2) = 4$. Consequently, h is not invertible. ∎

EXAMPLE 3 Determine whether $y = g(x) = \sqrt{x}$ is invertible and, if it is, find $g^{-1}(y)$.

Solution The graph of g is shown in Fig. 7.8. Clearly, any horizontal line meets the graph in at most one point. Solving $y = \sqrt{x}$ for x in terms of y, we obtain $x = y^2$. Since the domain of g^{-1} must be the range of g, where $y \geq 0$, we have

$$x = g^{-1}(y) = y^2 \qquad \text{for } y \geq 0.$$ ∎

EXAMPLE 4 Let f be an invertible function, and suppose that $f(3) = -2$ and $f(4) = 5$. Find (a) $f^{-1}(-2)$, (b) $f(f^{-1}(5))$, and (c) $f^{-1}(f(3))$.

Solution

(a) Since $f(3) = -2$, we have $f^{-1}(-2) = 3$.

(b) Since $f(4) = 5$, we have $f(f^{-1}(5)) = f(4) = 5$.

(c) Since $f(3) = -2$, we have $f^{-1}(f(3)) = f^{-1}(-2) = 3$. ∎

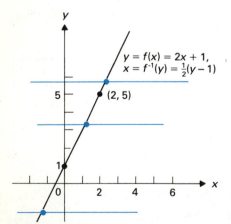

FIGURE 7.6 $f(x) = 2x + 1$ is invertible: Each horizontal line meets the graph at most once.

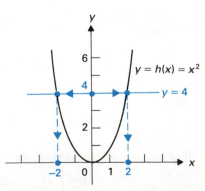

FIGURE 7.7 $h(x) = x^2$ is not invertible: A horizontal line may meet the graph in more than one point.

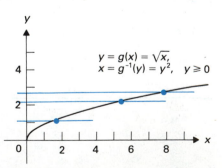

FIGURE 7.8 $g(x) = \sqrt{x}$ is invertible: $g^{-1}(y) = y^2$ for $y \geq 0$.

Example 4 illustrates that if f is invertible, then

$$f^{-1}(f(x)) = x \qquad \text{and} \qquad f(f^{-1}(y)) = y. \qquad (2)$$

We think of f^{-1} as "undoing" or reversing f and, similarly, f reverses f^{-1}.

Before we leave this subsection, we introduce one bit of standard terminology. We have seen that invertible functions are those where a single point of the domain gives rise to a single point of the range (definition of a function) and also a single point of the range gives rise to a single point of the domain (definition of invertibility). This single-point-to-single-point concept is abbreviated by the term *one to one*.

DEFINITION 7.2

One-to-one function

A function f is **one to one** if distinct points x_1 and x_2 in the domain are carried by f into distinct points y_1 and y_2 in the range. That is, if $x_1 \neq x_2$, then $f(x_1) \neq f(x_2)$.

Thus the invertible functions are precisely those that are one to one. Indeed, this is usually the way one sees an invertible function defined, rather than with the geometric Definition 7.1 that we gave. However, we want to develop as much geometric understanding as possible.

Let f be an increasing function. Then if $x_1 < x_2$, we have $f(x_1) \neq f(x_2)$, so f is one to one. Similarly, a decreasing function is one to one. This gives the following theorem.

THEOREM 7.1   Invertibility of increasing or decreasing functions

---

If f is either an increasing function or a decreasing function throughout its domain, then f is invertible. $\qquad \qquad \square$

Since a differentiable function f is increasing on an interval throughout which $f'(x)$ is positive and decreasing over an interval where $f'(x)$ is negative, we obtain the following corollary.

COROLLARY   The sign of $f'(x)$ and invertibility

Let f be differentiable and let the domain of f be either an interval or the set of all real numbers. If either $f'(x) > 0$ throughout the domain of f or $f'(x) < 0$ throughout the domain, then f is invertible.

Graphs of Inverse Functions

In a moment we will discuss derivatives of inverse functions. We have always discussed derivatives of functions $y = f(x)$, where we think of the *domain* as falling on the horizontal axis and the *range* as part of the vertical axis. The derivative is then the slope of the tangent line to the graph. To make use of this familiar geometric picture as we try to differentiate $x = f^{-1}(y)$, we want to turn the plane so that the y-axis is horizontal and the x-axis is vertical. This can be

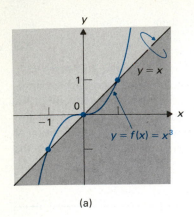

(a)

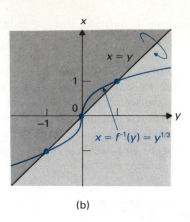

(b)

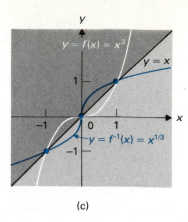

(c)

FIGURE 7.9 Obtaining the graph of $y = f^{-1}(x)$ from the graph of $y = f(x)$, illustrated for $y = f(x) = x^3$. To get from (a) to (b), flip the plane over about $y = x$; to get from (b) to (c), interchange the variable names x and y.

accomplished by flipping the plane over about the line $y = x$, as shown in Figs. 7.9(a) and (b). In those figures, we show the graph of $y = f(x) = x^3$ or, equivalently, $x = f^{-1}(y) = y^{1/3}$. In Fig. 7.9(c), we interchange the letters x and y on the axes, so we have the graph of $x = f(y) = y^3$ or, equivalently, $y = f^{-1}(x) = x^{1/3}$. To interpret the derivative of f^{-1} as the usual slope dy/dx of the tangent line, we should work with $y = f^{-1}(x)$.

EXAMPLE 5 Show that $y = f(x) = (x - 1)/(x + 2)$ is invertible, find $f^{-1}(x)$ and sketch the graphs of $y = f(x)$ and $y = f^{-1}(x)$.

Solution From $y = (x - 1)/(x + 2)$, we obtain

$$y(x + 2) = x - 1,$$
$$yx - x = -2y - 1,$$
$$x(y - 1) = -2y - 1$$
$$x = -\frac{2y + 1}{y - 1}.$$

Since each y-value gives at most one x-value, we see that f is indeed invertible. We sketched the graph of $f(x) = (x - 1)/(x + 2)$ in Fig. 7.10. The graph indicates again that f is invertible; each horizontal line meets the graph in at most one point. We see that

$$x = f^{-1}(y) = -\frac{2y + 1}{y - 1},$$

so

$$f^{-1}(x) = -\frac{2x + 1}{x - 1}.$$

To obtain the graph of $y = f^{-1}(x)$, we just flip the graph of $y = f(x)$ over about the line $y = x$, as shown in Fig. 7.11. The white curve in the figure shows where the graph of $y = f(x)$ was before it was flipped to become the graph of $y = f^{-1}(x)$. ∎

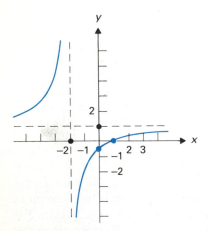

FIGURE 7.10 Graph of $y = f(x) = (x - 1)/(x + 2)$.

Derivatives of Inverse Functions

Let f be a differentiable and invertible function, and suppose that $f(a) = b$. Since f is differentiable, $f'(a)$ exists, and the graph of f has a tangent line where $x = a$. From Fig. 7.12, we see that the graph of f^{-1} then has a tangent line where $x = b$. If the tangent line to the graph of f at $(a,\ b)$ is not horizontal, then the tangent line to the graph of f^{-1} at $(b,\ a)$ will not be vertical. Under these circumstances, $(f^{-1})'(b)$ should exist. Computing the slopes of the tangent lines as indicated in Fig. 7.12, we should have

$$(f^{-1})'(b) = \frac{1}{f'(a)}. \tag{3}$$

We can also arrive at Eq. (3) by implicit differentiation. Let $y = f^{-1}(x)$, so $x = f(y)$. Assuming that dy/dx does exist and differentiating $x = f(y)$ implicitly, we obtain

$$1 = f'(y) \cdot \frac{dy}{dx}, \tag{4}$$

so

$$\frac{dy}{dx} = (f^{-1})'(x) = \frac{1}{f'(y)} \tag{5}$$

if $f'(y) \neq 0$. Since $f'(y) = dx/dy$, we may express Eq. (5) by

$$\frac{dy}{dx} = \frac{1}{dx/dy}. \tag{6}$$

Once more, the correct formula is easy to remember using differential notation.

While we don't claim to have proved the following theorem, we have made it at least plausible.

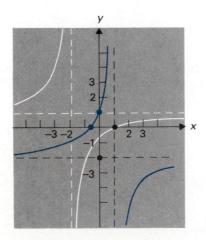

FIGURE 7.11 White curve:
$$y = f(x) = (x - 1)/(x + 2);$$
blue curve:
$$y = f^{-1}(x) = -(2x + 1)/(x - 1).$$

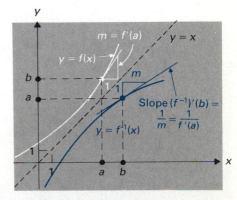

FIGURE 7.12 If the graph of f has slope $m = f'(a)$ where $x = a$ and $f(a) = b$, then the graph of f^{-1} has slope $1/m = 1/(f'(a))$ where $x = b$.

THEOREM 7.2   The derivative of f^{-1}

Let f be a differentiable and invertible function, and let $f(a) = b$. If $f'(a) \neq 0$, then f^{-1} is differentiable at b and, furthermore,

$$(f^{-1})'(b) = \frac{1}{f'(a)}.$$

EXAMPLE 6   Let $y = f(x) = x^3$. Find $(f^{-1})'(8)$ in two ways.

Solution 1   Now $f(2) = 8$, and $f'(2) = 3x^2|_2 = 12 \neq 0$. Theorem 7.2 shows that f^{-1} is differentiable at 8, and

$$(f^{-1})'(8) = \frac{1}{f'(2)} = \frac{1}{12}.$$

Solution 2   If $y = f(x) = x^3$, then $x = f^{-1}(y) = y^{1/3}$. Then

$$(f^{-1})'(8) = \frac{1}{3}y^{-2/3}\bigg|_8 = \frac{1}{3} \cdot 8^{-2/3} = \frac{1}{3} \cdot \frac{1}{8^{2/3}} = \frac{1}{3} \cdot \frac{1}{4} = \frac{1}{12}. \qquad \blacksquare$$

EXAMPLE 7   Let $y = f(x) = \sin x$ for $-\pi/2 \leq x \leq \pi/2$. Show graphically that f is an invertible function, and sketch the graph of $y = f^{-1}(x)$. Find a formula for $(f^{-1})'(x)$ in terms of x, wherever it exists.

Solution   The graph of f is shown in Fig. 7.13. Note that it is not the whole graph of the function $\sin x$, but just the portion for which $-\pi/2 \leq x \leq \pi/2$. Since a horizontal line meets this increasing graph in at most one point, we see that f is invertible. The graph of $y = f^{-1}(x)$ is obtained by flipping the graph of f over about the line $y = x$. The graph is shown in Fig. 7.14.

To find $(f^{-1})'(x)$, we let $y = f^{-1}(x)$. Then $x = f(y) = \sin y$. From Eq. (6), we have

$$\frac{dy}{dx} = (f^{-1})'(x) = \frac{1}{dx/dy} = \frac{1}{\cos y}.$$

We must express $\cos y$ in terms of x to complete the problem. Now $\sin y = x$, and $\cos^2 y + \sin^2 y = 1$. Consequently,

$$\cos y = \sqrt{1 - \sin^2 y} = \sqrt{1 - x^2}$$

or

$$\cos y = -\sqrt{1 - \sin^2 y} = -\sqrt{1 - x^2}.$$

Since $-\pi/2 \leq y \leq \pi/2$ in Fig. 7.14, we see that $\cos y \geq 0$, so

$$\cos y = \sqrt{1 - x^2}.$$

Thus

$$(f^{-1})'(x) = \frac{1}{\cos y} = \frac{1}{\sqrt{1 - x^2}}.$$

Note that $(f^{-1})'(x)$ is not defined where $x = \pm 1$, although this point is in the domain of f^{-1}. $\qquad \blacksquare$

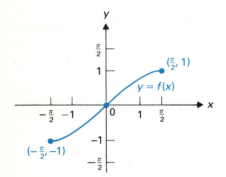

FIGURE 7.13   Graph of $f(x) = \sin x$ for $-\pi/2 \leq x \leq \pi/2$.

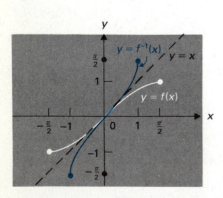

FIGURE 7.14

$f(x) = \sin x$, $-\pi/2 \leq x \leq \pi/2$,
$f^{-1}(x) = \sin^{-1}(x)$, $-1 \leq x \leq 1$

The function f^{-1} of Example 7 is called the *inverse sine function* and is denoted by $\sin^{-1}x$. In the example, we found its derivative. We showed that

$$\frac{d(\sin^{-1}x)}{dx} = \frac{1}{\sqrt{1 - x^2}}. \tag{8}$$

You may have noticed a fascinating feature of Eq. (8). The derivatives of the six elementary trigonometric functions are again trigonometric expressions. However, the derivative of the function $\sin^{-1}x$ is given by an *algebraic expression*. Could it be that trigonometry and algebra are really closely related? Chapter 10 will throw more light on this question.

SUMMARY

1. If $y = f(x)$ also defines x as a function of y, then f is invertible and x is the inverse function f^{-1} of y. That is, $x = f^{-1}(y)$.

2. A function $y = f(x)$ is one to one if it always carries two distinct points of the domain into two distinct points of the range, so that if $x_1 \neq x_2$, then $f(x_1) \neq f(x_2)$.

3. A function $y = f(x)$ is invertible if each horizontal line meets the graph in at most one point. Equivalently, f is invertible if it is one to one.

4. If f is invertible, then the domain of f^{-1} is the range of f and the range of f^{-1} is the domain of f. If a is in the domain of f and b is in the domain of f^{-1}, then $f^{-1}(f(a)) = a$ and $f(f^{-1}(b)) = b$.

5. The graph of $y = f^{-1}(x)$ may be obtained from the graph of $y = f(x)$ by flipping the plane over about the line $y = x$.

6. Let f be an invertible function and let $f(a) = b$. If $f'(a)$ exists and is nonzero, then $(f^{-1})'(b)$ exists, and

$$(f^{-1})'(b) = \frac{1}{f'(a)}.$$

In differential notation,

$$\frac{dx}{dy} = \frac{1}{dy/dx}.$$

EXERCISES 7.1

In Exercises 1–18, determine whether f is invertible. If it is, find $f^{-1}(x)$ and sketch the graph of both $f(x)$ and $f^{-1}(x)$.

1. $f(x) = x - 1$
2. $f(x) = 2x + 4$
3. $f(x) = 3 - x$
4. $f(x) = 8 - 2x$
5. $f(x) = 4$
6. $f(x) = -3$
7. $f(x) = x^2 - 3$
8. $f(x) = 4 - 3x^2$
9. $f(x) = x^3 + 1$
10. $f(x) = 1 - x^3$
11. $f(x) = \sqrt{x}$
12. $f(x) = 3 - 2\sqrt{x}$

13. $f(x) = \sqrt{x^2}$

14. $f(x) = |x| + 1$

15. $f(x) = \dfrac{x-1}{x+1}$

16. $f(x) = \dfrac{2x-1}{x+2}$

17. $f(x) = \dfrac{x^2}{x+1}$

18. $f(x) = \dfrac{x-1}{x^2}$

In Exercises 19–26, decide whether f is invertible and, if it is, sketch the graph of both $f(x)$ and $f^{-1}(x)$.

19. $f(x) = \cos x$

20. $f(x) = \tan x$

21. $f(x) = \cos x,\ 0 \le x \le \pi$

22. $f(x) = \tan x,\ -\pi/2 < x < \pi/2$

23. $f(x) = \sin x,\ 0 \le x \le \pi$

24. $f(x) = \sec x,\ -\pi/2 < x < \pi/2$

25. $f(x) = \sec x,\ 0 \le x \le \pi,\ x \ne \pi/2$

26. $f(x) = \cot x,\ 0 < x < \pi$

27. Mark each of the following true or false.

_____ (a) Every one-to-one function is invertible.

_____ (b) Every invertible function is one to one.

_____ (c) Every increasing function is invertible.

_____ (d) Every invertible function is either increasing or decreasing over its entire domain.

_____ (e) Every invertible function whose domain is an interval must be either increasing or decreasing over its entire domain.

_____ (f) Every continuous invertible function whose domain is an interval must be either increasing or decreasing over its entire domain.

_____ (g) No polynomial function of even degree is invertible.

_____ (h) Every polynomial function of odd degree is invertible.

_____ (i) Every monomial function of odd degree is invertible.

_____ (j) None of the six elementary trigonometric functions is invertible.

In Exercises 28–34, assume all functions are invertible and differentiable. Fill in the blanks.

28. If $f(2) = 3$, then $f^{-1}(3) =$ _____ .

29. If $g(3) = -4$, then $g(g^{-1}(-4)) =$ _____ .

30. If $h^{-1}(4) = 2$ and $h^{-1}(2) = 5$, then $h(h^{-1}(2)) =$ _____ .

31. If $f(3) = -5$ and $f'(3) = 2$, then $(f^{-1})'(-5) =$ _____ .

32. If $f(-2) = 4$ and $(f^{-1})'(4) = -3$, then $f'(-2) =$ _____ .

33. If $g(4) = 3$ and $g'(1) = 2$, then $(g(g^{-1}))'(3) =$ _____ .

34. If $f(4) = 3$ and $f'(3) = -2$, then $(f^{-1}(f))'(4) =$ _____ .

35. If $y = f(x) = 2x + 4$, find $(f^{-1})'(3)$ in two ways.

36. If $y = f(x) = x^3 + 1$, find $(f^{-1})'(28)$ in two ways.

37. If $y = g(x) = (x-1)/(x+2)$, find $(g^{-1})'(0)$ in two ways.

38. If $y = h(x) = (2x-3)/(x+1)$, find $(h^{-1})'(1)$ in two ways.

39. Let $f(x) = \cos x$ for $0 \le x \le \pi$.
(a) Show that f is invertible; $f^{-1}(x)$ is denoted by $\cos^{-1}x$.
(b) Find the domain of $\cos^{-1}x$.
(c) Find $d(\cos^{-1}x)/dx$.
(d) Find dy/dx if $y = \cos^{-1}(2x+1)$.

40. Let $f(x) = \tan x$ for $-\pi/2 < x < \pi/2$.
(a) Show that f is invertible; $f^{-1}(x)$ is denoted by $\tan^{-1}x$.
(b) Find the domain of $\tan^{-1}x$.
(c) Find $d(\tan^{-1}x)/dx$.
(d) Find dy/dx if $y = (\tan^{-1}x)^2$.

41. Let $g(x) = \cot x$ for $0 < x < \pi$.
(a) Show that g is invertible; $g^{-1}(x)$ is denoted by $\cot^{-1}x$.
(b) Find the domain of $\cot^{-1}x$.
(c) Find $d(\cot^{-1}x)/dx$.
(d) Find dy/dx if $y = \cot^{-1}(x^2)$.

42. Let $h(x) = \sec x$ for $0 \le x < \pi/2$ or $\pi \le x < 3\pi/2$.
(a) Show that h is invertible; $h^{-1}(x)$ is denoted by $\sec^{-1}x$.
(b) Find the domain of $\sec^{-1}x$.
(c) Find $d(\sec^{-1}x)/dx$.
(d) Find dy/dx if $y = \sec^{-1}(1/x)$.

43. Let $f(x) = \csc x$ for $0 < x \le \pi/2$ or $\pi < x \le 3\pi/2$.
(a) Show that f is invertible; $f^{-1}(x)$ is denoted by $\csc^{-1}x$.
(b) Find the domain of $\csc^{-1}x$.
(c) Find $d(\csc^{-1}x)/dx$.
(d) Find dy/dx if $y = \csc^{-1}(\sqrt{x})$.

 EXPLORING CALCULUS

N: calculus topics quiz (topic 4)

7.2 AN INTEGRAL-BASED DEVELOPMENT OF $\ln x$ AND e^x (OPTIONAL)

You have probably never seen a careful definition of an exponential function such as 2^x for *all* real numbers x. We can define 2^n for any positive integer n by

$$2^n = \underbrace{2 \cdot 2 \cdots 2}_{n \text{ factors}}.$$

Then we can define 2^0 to be 1 and 2^{-n} to be $1/2^n$. All these definitions of exponentiation seem well grounded.

Suppose that we try to define $2^{m/n}$ to be $(\sqrt[n]{2})^m$. This raises a problem. For example, when describing $2^{4/3}$, how do we know that there really *is* some number b such that $b^3 = 2$? That is, how do we know that $\sqrt[3]{2}$ exists? We try to make the existence of such a number plausible by saying that $1.2^3 = 1.728$ while $1.3^3 = 2.197$, so there should be some number between 1.2 and 1.3 whose cube is 2. For a skeptic, we can compute $1.25^3 = 1.953125$ and $1.26^3 = 2.000376$ and say, ''See how close these cubes are to 2 now; surely some number between them and just a bit less than 1.26 will have its cube equal to 2.'' We are actually approaching $\sqrt[3]{2}$ as a *limit*.

Integers and rational numbers (fractions) are much easier for us to understand than other real numbers. In terms of decimal notation, a general real number actually involves a *limit*. For example, π appears as a limit of a *sequence* of rational numbers, namely

$$3, \ 3.1, \ 3.14, \ 3.141, \ 3.1415, \ 3.14159, \ 3.141592, \ \ldots.$$

This is a much more sophisticated concept than an integer or a fraction. (We will discuss sequences in Chapter 10.)

Suppose now that we accept that $2^{m/n}$ makes sense for every rational number m/n. How do we describe what 2^π is? Again, we try to describe 2^π as a limit. That is, we say that π has a decimal expansion

$$\pi = 3.1415926 \ \ldots.$$

Since 3.14 is fairly close to π, we may take $2^{314/100}$ as an approximation to 2^π. Since 3.14159 is even closer to π, we may take $2^{314159/100000}$ as an even better approximation to π. What we are really doing is describing 2^π by

$$2^\pi = \lim_{m/n \to \pi} 2^{m/n}$$

for rational numbers m/n without having any good mathematical reason to believe that this limit exists. This type of presentation is unsatisfactory to mathematicians, who like to see everything placed on a firm foundation. We have described exponential functions in this intuitive but rather loose way so that you can use all real exponents in your work as soon as possible.

In Section 3.6, we extended this intuitive description to the calculus of logarithm and exponential functions. In that section, we based our definition of e and the formula for the derivative of e^x on some figures that made it plausible that for some base e, the graph of e^x has a tangent line of slope 1 where $x = 0$. We also discussed briefly the application of exponential functions to situations where a quantity is growing at a rate proportional to the magnitude of the quantity, such as a savings account with interest compounded continuously. We will discuss more applications in Section 7.4.

A little knowledge of limits and calculus does provide a more elegant and satisfactory way to define and develop exponential and logarithm functions, and that is the purpose of this section. However, we would be in error if we claim that we are now giving a firmly grounded presentation. Except for Section 2.5, our presentation of limits and the derivative was not mathematically rigorous; in particular, our argument for the existence of the definite integral was largely intuitive. Limits and the derivative and integral are treated more carefully in courses in advanced calculus or real analysis that are taken primarily by mathematics majors. However, one can make a case for saying that in those courses, the gaps in rigor are simply moved to other locations. To push everything always all the way back to the axioms of set theory that constitute the foundation of mathematics is really an exercise in inefficiency.

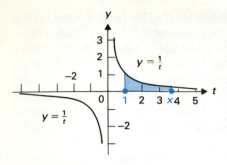

FIGURE 7.15 $\ln x = \int_1^x (1/t)\, dt =$ (Area of the shaded region).

This section consists of an outline for a calculus-based definition and development of exponential and logarithm functions. The in-text exercises ask you to fill in many of the details. This treatment of these functions is totally independent of the discussion of them in Chapter 3.

The Function ln x

The fundamental theorem of calculus (Theorem 5.3) tells us that $1/x$ for $x > 0$ does have an antiderivative, namely $F(x)$, given by

$$F(x) = \int_1^x \frac{1}{t}\, dt \qquad \text{for } x > 0.$$

The value $F(x)$ is equal to the area of the region shaded under the curve in Fig. 7.15.

DEFINITION 7.3

Natural logarithm

The function $\ln x$ defined by

$$\ln x = \int_1^x \frac{1}{t}\, dt \qquad \text{for } x > 0$$

is the **(natural) logarithm function.**

Calculus of ln x

EXERCISE 1   Show that $d(\ln x)/dx = 1/x$ for $x > 0$. [*Hint:* See Theorem 5.3.]

EXERCISE 2   Show that if u is a differentiable function of x, then

$$\frac{d(\ln u)}{dx} = \frac{1}{u} \cdot \frac{du}{dx}.$$

EXERCISE 3   Find dy/dx if $y = \ln(\sin x)$.

EXERCISE 4   Show that for $x < 0$, we have $d(\ln(-x))/dx = 1/x$.

EXERCISE 5   Show that for a continuous function $u = f(x)$, we have $\int (1/u)\, du = \ln|u| + C$.

EXERCISE 6   Find $\int \tan x\, dx$.

Properties of ln x

It is no surprise that the calculus of $\ln x$ falls out so readily since we defined $\ln x$ using an integral. What is surprising is that we can recover the algebraic proper-

LOGARITHMS WERE INVENTED by John Napier (1550–1617) by 1615 and developed into a computing tool through the tables of Henry Briggs (1561–1631) and others in the 1620s. It was not until the middle of the century, however, that the connection between logarithms and the area under the hyperbola was noted. The Belgian mathematician Gregory of St. Vincent (1584–1667) showed in 1647 that if x_1, x_2, x_3, and x_4 are four positive numbers such that $x_2:x_1 = x_4:x_3$, then the area under the hyperbola $xy = 1$ over $[x_1, x_2]$ is equal to that over $[x_3, x_4]$. His argument involved showing that appropriately chosen Riemann sums over each interval give the same approximation.

Two years later, the Belgian Jesuit Alfonso Antonio de Sarasa noticed that Gregory's result implied that the area $A(x)$ under the hyperbola from 1 to x had the basic logarithmic property $A(uv) = A(u) + A(v)$. If one could calculate the area under the hyperbola one could calculate natural logarithms. The search for the means of calculating this area ultimately led to the power series methods of Newton and others in the 1660s, methods that were instrumental in Newton's invention of the calculus.

ties of logarithm functions familiar from high school using this definition. (See Exercise 10.)

EXERCISE 7 Show that ln x is an increasing function.

EXERCISE 8 Show from Definition 7.3 and the preceding exercises that

(a) ln $1 = 0$,

(b) ln $x > 0$ if $x > 1$,

(c) ln $x < 0$ if $0 < x < 1$.

EXERCISE 9 Let $f(x)$ and $g(x)$ be two differentiable functions having an interval as common domain. Show that if $f'(x) = g'(x)$ and $f(c) = g(c)$ for some value c in their common domain, then $f(x) = g(x)$ for all x in their domain. [*Hint:* Use Theorem 4.10.]

EXERCISE 10 Let a be a positive constant. Use Exercises 9 and 8(a) to show that for all $x > 0$,

(a) ln $ax = $ ln $a + $ ln x,

(b) ln$(x/a) = $ ln $x - $ ln a,

(c) ln $x^n = n($ln $x)$ for every integer n. (We use an integer n rather than any real number r as exponent since we have not yet formally defined x^r for every real number r and all $x > 0$.)

EXERCISE 11 Use the properties of ln x in Exercise 10 to differentiate

$$f(x) = \ln \frac{(x^2 + 1)(x - 1)^3}{\cos^4 x}.$$

EXERCISE 12 Show that

(a) $\lim_{x \to \infty} \ln x = \infty$,

(b) $\lim_{x \to 0+} \ln x = -\infty$.

[*Hint:* Consider ln(2^n) and ln(2^{-n}) and use Exercises 10 and 7.]

EXERCISE 13 Show that the graph of ln x has slope 1 at $x = 1$ and is always concave down. Use this information and Exercises 7, 8, and 12 to sketch the graph of ln x.

The Exponential Function

EXERCISE 14 Show that ln x is a one-to-one function having as range the set of all real numbers. [*Hint:* Use Exercises 7 and 12 and the intermediate value theorem, Theorem 2.3.]

Natural exponential function and e

DEFINITION 7.4

The **(natural) exponential function** is the inverse of the natural logarithm function and is denoted by exp(x). The real number e is defined by $e = $ exp(1).

EXERCISE 15 Using Exercise 12, give the domain and the range of exp(x). Find ln(e). Sketch the graph of exp(x) from the graph of ln x using Section 7.1.

EXERCISE 16 Show that for every integer n, we have exp(n) = e^n. [*Hint:* Show that ln(exp(n)) = ln(e^n) and use Exercise 14.]

---

e^x

DEFINITION 7.5

For every real number x, the number e^x is defined by $e^x = $ exp(x). [We now drop the notation exp(x) in favor of e^x.]

EXERCISE 17 Show that

(a) $e^{\ln x} = x$ for all $x > 0$,

(b) $\ln(e^x) = x$ for all x.

EXERCISE 18 Show that

(a) $e^a \cdot e^b = e^{a+b}$,

(b) $\dfrac{e^a}{e^b} = e^{a-b}$,

(c) $(e^a)^n = e^{na}$ for every integer n.

[*Hint:* Take the natural logarithm of each side and use Exercise 10.]

Calculus of e^x

EXERCISE 19 Show that $d(e^x)/dx = e^x$.

EXERCISE 20 Show that if u is a differentiable function of x, then $d(e^u)/dx = e^u(du/dx)$.

EXERCISE 21 Let $y = e^{\sin 3x}$. Compute dy/dx.

EXERCISE 22 Show that if u is a differentiable function of x, then $\int e^u \, du = e^u + C$.

EXERCISE 23 Find $\int x e^{x^2} \, dx$.

Logarithm and Exponential Functions to Other Bases

We know that for $a > 0$, we have $a = e^{\ln a}$ and $(e^b)^n = e^{nb}$, so that $a^n = (e^{\ln a})^n = e^{n(\ln a)}$ at least for integers n. This motivates the following definition.

DEFINITION 7.6

Exponentials and logarithms with base a

Let $a > 0$. The **exponential function with base a** is defined by $a^x = e^{a(\ln x)}$. For positive $a \neq 1$, the inverse of the exponential function a^x is the **logarithm function with base a** and is denoted by $\log_a x$.

EXERCISE 24 Show that $\ln x^r = r \ln x$ and $(e^a)^r = e^{ra}$ for all real numbers a and r and all $x > 0$. (Note that we had shown this in Exercises 10 and 18 only in the case where $r = n$, an integer.)

EXERCISE 25 Show that for $a > 0$ and for all real numbers b and c, we have

(a) $a^b \cdot a^c = a^{b+c}$,

(b) $\dfrac{a^b}{a^c} = a^{b-c}$,

(c) $(a^b)^c = a^{bc}$.

EXERCISE 26 Show that if u is a differentiable function of x, then

$$\frac{d(a^u)}{dx} = (\ln a)a^u \cdot \frac{du}{dx}.$$

EXERCISE 27 If $y = x^2 3^{\cos 2x}$, find dy/dx.

EXERCISE 28 Show that if u is a continuous function of x, then

$$\int a^u \, du = \frac{1}{\ln a} a^u + C.$$

EXERCISE 29 Find $\int (\sin 2x) 10^{\cos 2x} \, dx$.

EXERCISE 30 Show that for a, b, and c all positive, we have

(a) $\log_a(bc) = \log_a b + \log_a c$,

(b) $\log_a(b/c) = \log_a b - \log_a c$,

(c) $\log_a(b^c) = c \log_a b$.

EXERCISE 31 Show that if u is a differentiable function of x, then

$$\frac{d(\log_a u)}{dx} = \frac{1}{\ln a} \cdot \frac{1}{u} \cdot \frac{du}{dx}.$$

EXERCISE 32   Let $y = \log_5[3x/(2x + 1)]$. Find dy/dx.

EXERCISE 33   Describe the domain of $h(x) = f(x)^{g(x)}$ in terms of the domains of f and g and the values of the functions in those domains.

EXERCISE 34   Let $y = x^{2x}$. Find dy/dx.

7.3 LOGARITHMIC DIFFERENTIATION

Suppose we want to differentiate an exponential expression of the form $f(x)^{g(x)}$. One way to do this is to recall the identity $e^{\ln a} = a$ and use it to write

$$f(x)^{g(x)} = (e^{\ln f(x)})^{g(x)} = e^{g(x)\,\ln(f(x))}.$$

If we can differentiate $g(x)$ and $f(x)$, we then have no difficulty differentiating $e^{g(x)\,\ln(f(x))}$. We used this method in Section 3.8.

EXAMPLE 1   Let $y = x^{\sin x}$. Find y'.

Solution   We write

$$y = x^{\sin x} = (e^{\ln x})^{\sin x} = e^{(\sin x)(\ln x)}$$

and differentiate to obtain

$$y' = e^{(\sin x)(\ln x)}\left[(\sin x)\frac{1}{x} + (\ln x)(\cos x)\right].$$

Since $e^{(\sin x)(\ln x)}$ is just another way of writing $x^{\sin x}$, we can write our answer as

$$x^{\sin x}\left[\frac{\sin x}{x} + (\ln x)(\cos x)\right]. \qquad \blacksquare$$

The purpose of this section is to present another technique, logarithmic differentiation, for handling the type of problem in Example 1. First we make a few general comments about the usefulness of the logarithm function. It has two important features:

1. Taking the logarithm of large positive numbers converts those numbers to much smaller numbers.

2. Taking a logarithm converts the operations of multiplication, division, and exponentiation to addition, subtraction, and multiplication respectively.

If we run into difficulty working with large numbers, we might try taking logarithms. For example, suppose we wish to see how big

$$500! = 500 \cdot 499 \cdot 498 \cdots 3 \cdot 2 \cdot 1$$

is and even to find a few significant figures for this number. If we reach for our computer to multiply these five hundred numbers, it will probably give us an *overflow* message, meaning that the numbers are too large for it to handle. Under these circumstances, we can take a logarithm and compute instead

$$\ln(500!) = \ln(500) + \ln(499) + \ln(498) + \cdots + \ln(3) + \ln(2) + \ln(1).$$

The computer has no trouble adding these small numbers. By then computing $\log_{10}(500!) = \ln(500!)/\ln(10)$, we find that $500! \approx 1.22013682599 \times 10^{1134}$. It is no wonder that our computer had trouble computing it directly!

The second feature of the logarithm function enables us to convert differentiation of products, quotients, and exponentials to differentiation of sums, differences, and products respectively. This very useful technique is known as *logarithmic differentiation*. Solving problems with this technique gives us valuable practice in implicit differentiation and in using properties of the logarithm function. To solve the problem in Example 1, differentiating $y = x^{\sin x}$, using this method, we take the logarithm of both sides of the equation and then differentiate implicitly. We obtain

$$\ln y = (\sin x)(\ln x),$$

$$\frac{1}{y} \cdot y' = (\sin x)\frac{1}{x} + (\ln x)(\cos x),$$

$$y' = y\left[(\sin x)\frac{1}{x} + (\ln x)(\cos x)\right],$$

$$y' = x^{\sin x}\left[\frac{\sin x}{x} + (\ln x)(\cos x)\right].$$

It would seem that this method of differentiating $y = f(x)$ is valid only where $f(x) > 0$, since we cannot take the logarithm of a negative number. However, if $x < 0$, then

$$\frac{d[\ln(-x)]}{dx} = \frac{1}{-x}(-1) = \frac{1}{x}. \tag{1}$$

For values of x for which $y = f(x)$ is negative, we can form the equation $-y = -f(x)$ where both sides are positive and then take the logarithm, obtaining $\ln(-y) = \ln(-f(x))$. Equation (1) and the chain rule show that we obtain

$$\frac{1}{y} \cdot y' = \frac{1}{f(x)} \cdot f'(x). \tag{2}$$

It is thus permissible to form Eq. (2) whether $f(x)$ is positive or negative.

The advantage of logarithmic differentiation is more striking when we have complicated products, quotients, or exponentials to differentiate.

EXAMPLE 2 Find y' if $y = 2^x(\sin x)^{\cos x}$.

Solution The technique of logarithmic differentiation yields

$$\ln y = \ln(2^x) + \ln((\sin x)^{\cos x}),$$

$$\ln y = x(\ln 2) + (\cos x) \cdot \ln(\sin x),$$

$$\frac{1}{y} \cdot y' = (\ln 2) + (\cos x) \cdot \frac{1}{\sin x}(\cos x) + (\ln(\sin x))(-\sin x),$$

$$y' = y\left[(\ln 2) + \frac{\cos^2 x}{\sin x} - (\sin x)(\ln(\sin x))\right],$$

$$y' = 2^x(\sin x)^{\cos x}[(\ln 2) + \cos x \cot x - (\sin x)(\ln(\sin x))]. \quad \blacksquare$$

EXAMPLE 3   Find dy/dx if $y = x^x/(x + 1)^2$.

Solution   Using logarithmic differentiation, we have

$$\ln y = \ln(x^x) - \ln(x + 1)^2, \qquad \ln y = x(\ln x) - 2 \ln(x + 1),$$

$$\frac{1}{y} \cdot \frac{dy}{dx} = x \cdot \frac{1}{x} + (\ln x) - \frac{2}{x + 1},$$

$$\frac{dy}{dx} = y\left(1 + (\ln x) - \frac{2}{x + 1}\right),$$

$$\frac{dy}{dx} = \frac{x^x}{(x + 1)^2}\left(1 + (\ln x) - \frac{2}{x + 1}\right). \qquad \blacksquare$$

You must not think that logarithmic differentiation can be used only with the exponential functions introduced in this chapter. We can use the technique to advantage with problems like those in Chapter 3.

EXAMPLE 4   Find dy/dx if

$$y = \frac{x^3 \sin^2 x}{(x + 1)(x - 2)^2}.$$

Solution   Using logarithmic differentiation, we have

$$\ln y = 3(\ln x) + 2 \ln(\sin x) - \ln(x + 1) - 2 \ln(x - 2),$$

$$\frac{1}{y} \cdot \frac{dy}{dx} = \frac{3}{x} + 2\frac{\cos x}{\sin x} - \frac{1}{x - 1} - \frac{2}{x - 2},$$

$$\frac{dy}{dx} = \frac{x^3 \sin^2 x}{(x + 1)(x - 1)^2}\left(\frac{3}{x} + 2 \cot x - \frac{1}{x - 1} - \frac{2}{x - 2}\right).$$

This solution looks quite different from the one we would have found in Chapter 3 using the product and quotient rules, but it is no less valid. $\blacksquare$

SUMMARY

1. Logarithmic differentiation may simplify the differentiation of products, quotients, and exponentials.

EXERCISES 7.3

In Exercises 1–20, find the derivative of the given function using logarithmic differentiation.

1. 2^{3x}

2. $10^{x^2 + 2x}$

3. $x^2 3^{2x-1}$

4. $4^{1/x}/x$

5. $2^x 3^{2x}$

6. $x(3^{\sin x})$

7. x^{2x}

8. $(\sin x)^x$

9. $x^{\tan x}$

10. $\cos (x^x)$

11. $(x - 1)^{x^2}$

12. $2^x x^{2x}$

13. $7^x \cdot 8^{-x^2} \cdot 100^x$

14. $5^{x^2}/7^x$

15. $\dfrac{5^{\sin x}}{2^{1/x}}$

16. $x^2 e^x \cos x$

17. $\dfrac{x^2 3^x}{\tan x}$

18. $(x^2 + 1)\sqrt{2x + 3}(x^3 - 2x)$

19. $\dfrac{x^{\sin x}}{(\cos x)^x}$

20. $\dfrac{x^{2x} \cos x}{\sqrt{2x + 1} \, (x^2 - 3)^3}$

21. Find the equation of the tangent line to the graph of $y = x^x$ where $x = 1$.

22. Find the equation of the normal line to the graph of

$$y = \frac{(x^2 - 3)(2x - 4)^4}{\sqrt{4x + 5} \, (2x - 1)^{3/2}}$$

where $x = 1$.

23. Use a differential to estimate $(2.01)^{2.01}$.

24. Use a differential to estimate $(5.02)^2/1.01^{2.02}$. [*Hint:* Note that $5.02 = 3 + 2.02 = 3 + 2(1.01)$.]

▣ Use a computer or calculator for Exercises 25–30.

25. Estimate all solutions of $x^{2x} - 5(x^x) - 6 = 0$, using Newton's method.

26. Repeat Exercise 25 for $x^x = \cos x$.

27. Repeat Exercise 25 for $x^x = 2^x + 4$.

28. Estimate $\displaystyle\int_1^3 x^x \, dx$.

29. Estimate $\displaystyle\int_1^{10} \log_{10} \sqrt{x} \, dx$.

30. Estimate $\displaystyle\int_\pi^{2\pi} x^{\sin x} \, dx$.

7.4 APPLICATIONS TO GROWTH AND DECAY

Growth Governed by the Equation $dy/dt = ky$

As we stated in Section 3.6, in many situations the rate at which a numerical quantity changes is proportional to that quantity. This section is devoted to examining several such situations. If the numerical quantity y is given by $y = f(t)$ at time t, then we must have

$$\frac{dy}{dy} = ky, \qquad \text{that is,} \qquad f' = k \cdot f \tag{1}$$

where k is the constant of proportionality. Here are descriptions of four situations where differential equations like Eq. (1) occur.

Radioactive decay It is known that the rate of decay of any particular radioactive element is proportional to the amount of the element present. This means that if $Q = f(t)$ gives the quantity of the element present at time t, then

$$\frac{dQ}{dt} = -cQ \tag{2}$$

where c is a positive constant of proportionality. (The negative sign occurs because the quantity Q present is *decreasing* as time increases.) Since Eq. (2) can be written in the form $f' = -c \cdot f$, we see that Eq. (2) is an equation of the form of Eq. (1), where $k = -c$ is a negative constant.

Retarded motion Suppose a body traveling through a medium (like air or water) is subject only to a force of retardation proportional to the velocity of the body through the medium. Let $v = f(t)$ be the velocity at time t. By Newton's second law of motion, the force F is ma, where m is the mass of the body and $a = dv/dt$ is the acceleration. Thus $m(dv/dt) = -cv$ where c is some positive

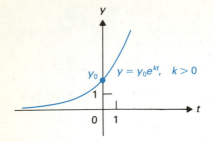

FIGURE 7.16 Exponential growth occurs if $k > 0$.

constant of proportionality, so

$$\frac{dv}{dt} = -\frac{c}{m}v. \tag{3}$$

Since $v = f(t)$, we may write Eq. (3) in the form $f' = (-c/m) \cdot f$, which is again an equation of the form of Eq. (1), with $k = -c/m$.

Population growth   Under "ideal" conditions with no overcrowding, predators, or disease, the rate of growth of a population P (whether it be people or bacteria) is proportional to the size of the population. This means that if $P = f(t)$ is the size of the population at time t, then

$$\frac{dP}{dt} = cP \tag{4}$$

for some constant c. Equation (4) can be written as $f' = c \cdot f$, which is again of the form of Eq. (1).

Bank interest   Banks often offer savings accounts in which the interest is "compounded continuously." If the amount in such a savings account at time t years is $S = f(t)$ and if the interest rate is c percent per year, then

$$\frac{dS}{dt} = \frac{c}{100}S.$$

This is again a differential equation of the form of Eq. (1), where $k = 0.01c$.

In Example 4 of Section 5.5, we saw that the solution of the differential equation $dy/dt = ky$ is $y = f(t) = Ae^{kt}$. Note that $f(0) = Ae^0 = A$, so A is the *initial value* of the function y when $t = 0$. For this reason, we will now write this constant A as y_0. The solution thus takes the form

$$y = y_0 e^{kt}. \tag{5}$$

The two constants y_0 and k can be determined if the value of $f(t)$ is known at two different times t_1 and t_2. The sign of k determines whether $f(t)$ is *increasing* (if $k > 0$) or *decreasing* (if $k < 0$) as the time t increases; this is indicated by Figs. 7.16 and 7.17.

Illustrations

We illustrate the preceding discussion with some specific examples.

EXAMPLE 1   Show that the time t_h required for half of an initial amount of a radioactive element to decay is independent of the initial amount of the element present. (This time t_h is the *half-life* of the element.)

Solution   If an amount Q of the element is present at time t and if Q_0 is the initial amount present at time $t = 0$, then we know by the radioactive decay equation that $dQ/dt = -cQ$ for some positive constant c. Our preceding work then shows that

$$Q = Q_0 e^{-ct}.$$

Since $Q = \frac{1}{2}Q_0$ when $t = t_h$, we have

$$\tfrac{1}{2}Q_0 = Q_0 e^{-ct_h},$$

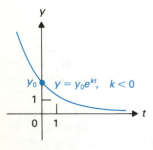

FIGURE 7.17 Exponential decay occurs if $k < 0$.

so

$$e^{ct_h} = 2.$$

Taking the (natural) logarithm of each side of this equation, we obtain

$$ct_h = \ln 2,$$

so

$$t_h = \frac{1}{c} (\ln 2).$$

Thus t_h depends only on the constant c and is independent of the initial amount Q_0 of the element present. ∎

EXAMPLE 2 A certain culture of bacteria grows at a rate proportional to its size. If the size doubles in 3 days, find the time required for the culture to increase to 100 times its original size.

Solution If we let $P = f(t)$ be the size of the culture after t days, then $dP/dt = kP$ for some constant k. The form of the solution (5) shows that

$$P = P_0 e^{kt}$$

where P_0 is the initial size of the culture. When $t = 3$ days, $P = 2P_0$, so

$$2P_0 = P_0 e^{3k},$$
$$2 = e^{3k},$$
$$\ln 2 = 3k,$$
$$k = \frac{1}{3} (\ln 2).$$

Thus we have

$$P = P_0 e^{(1/3)(\ln 2)t}.$$

We wish to find t when $P = 100P_0$. We have

$$100P_0 = P_0 e^{(1/3)(\ln 2)t},$$
$$100 = e^{(1/3)(\ln 2)t},$$
$$\ln 100 = \frac{1}{3} (\ln 2)t,$$
$$t = \frac{3(\ln 100)}{\ln 2} \approx 20 \text{ days.}$$ ∎

EXAMPLE 3 Suppose a body traveling through a medium is subject only to a force of retardation proportional to its velocity. If the body is traveling at 100 ft/sec at time $t = 0$ and at 20 ft/sec when $t = 3$ sec, find the velocity of the body at time 9 sec and also find the total distance the body travels through the medium.

Solution From the retarded motion equation, we know that $dv/dt = -kv$ for some positive constant k, so by the preceding section,

$$v = v_0 e^{-kt}.$$

From the data, the initial velocity v_0 is 100 ft/sec, so

$$v = 100e^{-kt}.$$

When $t = 3$, we have $v = 20$, so

$$20 = 100e^{-3k}. \tag{6}$$

We can use Eq. (6) to find k. We find that $e^{3k} = 5$, so taking logarithms we have

$$3k = \ln 5 \quad \text{and} \quad k = \frac{1}{3}(\ln 5).$$

Thus

$$v = 100e^{-(\ln 5)t/3} = 100 \cdot 5^{-t/3}.$$

When $t = 9$, we obtain

$$v\big|_{t=9} = 100 \cdot 5^{-9/3} = \frac{100}{125} = \frac{4}{5} \text{ ft/sec.}$$

If s is the distance traveled by the body from time $t = 0$, then

$$\frac{ds}{dt} = v = 100e^{-(\ln 5)t/3}.$$

Integrating, we find that

$$s = -\frac{300}{\ln 5}e^{-(\ln 5)t/3} + C$$

for some constant C. Now $s = 0$ when $t = 0$, so we have

$$0 = -\frac{300}{\ln 5} \cdot 1 + C$$

$$C = \frac{300}{\ln 5}.$$

Hence, the distance s as a function of t is given by

$$s = -\frac{300}{\ln 5} \cdot 5^{-t/3} + \frac{300}{\ln 5}.$$

As $t \to \infty$, $5^{-t/3} \to 0$ and $s \to 300/\ln 5$. Thus the body travels a total distance of $300/(\ln 5)$ ft through the medium. ∎

EXAMPLE 4   Savings with interest compounded continuously increase at a rate proportional to the amount in the savings account. If S is the amount in the account after t years and if interest is compounded continuously at $c\%$, then

$$\frac{dS}{dt} = \frac{c}{100}S.$$

Find how long it takes savings to triple at 5%, compounded continuously.

Solution   The differential equation becomes

$$\frac{dS}{dt} = \frac{5}{100}S = \frac{1}{20}S.$$

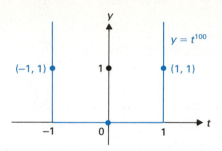

FIGURE 7.18 Attempt to draw the graph of t^{100}.

Therefore

$$S = S_0 e^{t/20}.$$

When $S = 3S_0$ so the original amount S_0 has been tripled, then

$$3S_0 = S_0 e^{t/20},$$
$$3 = e^{t/20},$$
$$\ln 3 = \frac{t}{20},$$
$$t = 20(\ln 3) \approx 22.$$

Thus money triples in about 22 years if compounded continuously at 5%. ∎

Unrestricted Exponential Growth

We hope you understand how rapidly the exponential function e^t increases as $t \to \infty$. In Fig. 1.66, we attempted to draw the graph of $y = x^{100}$ to demonstrate how rapidly it increases for $x \geq 1$. Figure 7.18 shows our best attempt to draw the graph of $f(t) = t^{100}$. It looks as though it increases faster than e^t, but the graph shows the behavior only near the origin. If t is large enough, then e^t is much larger than t^{100} (or than t^{1000}, for that matter). Suppose $t = 10,000$. Then $e^{10,000}$ is the product of 10,000 factors e, while $10,000^{100}$ is the product of only 100 factors, each of size 10,000. Our calculator shows that

$$e^{10,000} = (e^{100})^{100} \approx (2.6881171 \cdot 10^{43})^{100} \approx 8.8068 \cdot 10^{42} \cdot 10^{4300}.$$

Thus

$$e^{10,000} \approx 8.8068 \cdot 10^{4,342} \qquad \text{and} \qquad 10,000^{100} = 10^{400}.$$

Any possibility of unrestricted exponential growth e^t, corresponding to large values of t, is a cause of great concern.

EXAMPLE 5   According to the 1971 *World Book Encyclopedia*, the world population in 1971 was about 3,692,000,000 and was increasing at an annual rate of about 1.9%. Suppose the rate of increase of population is proportional to the population, which continues to increase at the continuous rate of 1.9% annually. If the surface of the earth is a sphere of radius 4000 mi, estimate the year when there will be one person for every square yard of surface area of the earth, including oceans as well as continents.

Solution   If the population t years after 1971 is $P = f(t)$, then

$$\frac{dP}{dt} = \frac{1.9}{100}P = 0.019P.$$

Solution (5) shows that then

$$P = 3,692,000,000 e^{0.019t}.$$

We want to find t when P is the number of square yards of surface area of the earth. Since the area of a sphere is $A = 4\pi r^2$, we would have at that time

$$P = 4\pi \left(4000 \cdot \frac{5280}{3}\right)^2 \text{ yd}^2.$$

Using a calculator to solve

$$4\pi\left(4000 \cdot \frac{5280}{3}\right)^2 = 3{,}692{,}000{,}000e^{0.019t},$$

we have

$$e^{0.019t} = \frac{4\pi(4000 \cdot 5280)^2}{9 \cdot 3{,}692{,}000{,}000},$$

$$0.019t = \ln\left[\frac{4\pi(4000 \cdot 5280)^2}{9 \cdot 3{,}692{,}000{,}000}\right]$$

$$\approx \ln 168{,}691$$

$$\approx 12.0358,$$

$$t \approx 633 \text{ years.}$$

This shows that there would be one person for every square yard of the earth's surface in the year $1971 + 633 = 2604$. Taken in the context of the history of the human race, the year 2604 might be like "next month" as we view our own lives. ■

The preceding example explains the great concern many people have about the "population explosion." Some denounce the concern, saying that we can build cities in space to avoid overcrowding on earth. Exercise 29 asks you to compute the year when there would be one person for every 3 cubic yards of space out as far as the moon, under the growth assumptions in Example 5. It takes a bit longer; the year would be 4029. In the context of the history of our race, that might be like "next season" to us.

We turn to an economic discussion. Government officials strive to achieve an "ideal," long-term, stable economic growth of about 3% annually, after correction for inflation. Such growth would be exponential and would lead to absurdities, as our next example illustrates.

EXAMPLE 6 In 1982, roughly 7,000,000 cars were manufactured in the United States. Suppose the number manufactured were to increase continuously at an annual rate of 3%. If the average weight of a car is 0.7 ton, find the year when the entire mass of the earth, which weighs about $6.6 \cdot 10^{20}$ tons, would have to be used to manufacture that year's cars in the United States.

Solution If $N = f(t)$ is the number of cars manufactured t years after 1982 in the United States, then we have

$$\frac{dN}{dt} = 0.03N,$$

so

$$N = 7{,}000{,}000e^{0.03t}.$$

We wish to find t when $N = 6.6 \cdot 10^{20}/0.7$. Solving

$$\frac{6.6 \cdot 10^{20}}{0.7} = (7 \cdot 10^6)e^{0.03t}$$

using a calculator, we have

$$e^{0.03t} = \frac{6.6 \cdot 10^{14}}{4.9},$$

$$0.03t = \ln\left(\frac{6.6 \cdot 10^{14}}{4.9}\right) \approx 32.534,$$

$$t \approx 1084.5 \text{ years.}$$

Thus in the year $1084 + 1982 = 3066$, all the earth would be used to make just that year's cars, leaving no earth on which to drive the cars! ∎

Example 6 illustrates that "boom" periods of 3% growth must lead inevitably to painful "bust" periods. We who study calculus can see the consequences of creating an economic condition governed by $dQ/dt = kQ$ for $k > 0$. However, zero growth, or even more realistic linear growth, such as having the gross national product increase by the equivalent of a flat \$500,000,000 each year, is not a popular idea. Until the population as a whole has studied $dQ/dt = kQ$, we will continue to suffer the "boom and bust."

SUMMARY

1. The general solution of the differential equation $dy/dt = ky$ is $y = Ae^{kt}$. The arbitrary constant A is the initial value y_0 of y, when $t = 0$.

2. There are several important situations in which the rate of growth (or decay) of some quantity is proportional to the amount $Q(t)$ present at time t. Examples are decay of a radioactive substance, growth of a population, and savings with interest compounded continuously. In all these cases, the differential equation $dQ/dt = kQ$ governs the growth (or decay).

3. Long-term exponential growth can lead to catastrophe.

EXERCISES 7.4

1. If the half-life of a radioactive element is 1600 yr, how long does it take for $\frac{1}{3}$ of the original amount to decay?

2. A certain culture of bacteria grows at a rate proportional to the size of the culture. If the culture triples in size in the first 2 days, find the factor by which the culture increases in size in the first 10 days.

3. A culture of bacteria increases at a rate proportional to the size of the culture. If 5000 bacteria are initially present and

the culture increases to 12,000 bacteria after 1 day, find how many days elapse before 10,000,000 bacteria are present.

4. Suppose half of a radioactive substance decomposes in 50 yr. Find how long it takes for $\frac{31}{32}$ of it to decompose by
   (a) reasoning that half decomposes in the first 50 years and then half of the remaining half, or $\frac{1}{4}$ more, decomposes during the next 50 years, and so on, and
   (b) working with the exponential equation describing the amount remaining at time t.

5. A population grows at a rate proportional to its size and doubles in 10 years. If the initial population is 50, find the number of years required for it to increase to 12,800 by
   (a) reasoning that after the first 10 years the population would be 100, after the next 10 it doubles again to 200, and so on, and
   (b) working with the exponential equation describing the population at time t.

6. A population grows at a rate proportional to the number present. If the population doubles in 20 days, find the factor by which the population increases after 10 days
   (a) without using the exponential equation giving the population at time t and
   (b) using the exponential equation.

7. Suppose the population in Exercise 6 triples in 20 days. Find the factor by which it increases in 5 days
   (a) without using the exponential equation giving the population at time t and
   (b) using the exponential equation.

8. If the half-life of a radioactive substance is 50 yr, how long does it take for 100 lb of the substance to completely decompose?

9. *Carbon dating* Radioactive carbon-14 is constantly created in the upper atmosphere by cosmic radiation and is absorbed by living organisms. The ratio of carbon-14 to regular carbon in a living organism is approximately the same as this ratio in its environment. Once the organism dies, it ceases to absorb carbon-14, so the ratio of carbon-14 to regular carbon in its remains starts to decrease as the carbon-14 decays. Assuming that the ratio in the environment has remained roughly constant for many thousands of years, we can calculate the approximate age of a fossil by determining the proportion of its original carbon-14 that has decayed. Taking 5700 years as the approximate half-life of carbon-14, estimate the age of a fossil in which 80% of its original carbon-14 has decayed.

10. Find how long it takes savings to double at 5.5% interest, compounded continuously.

11. Mr. and Mrs. Brice put $10,000 into a savings account in 1975. They plan to leave it there until 1995, when they will withdraw it and all the accrued interest to make a down payment on a new house. During that 20-year period, their savings will earn 5% interest, compounded continuously. How much will the bank pay the Brices in 1995?

12. When you worked Exercise 11, you discovered that the Brices will receive $27,182.82 in 1995. During the time the bank has the Brices' money, the bank loans it continually for home mortgages and receives what amounts to 9% interest, compounded continuously. If it costs the bank an average of $200 per year to handle its investments arising from the Brices' $10,000 deposit, how much profit will the bank have made after paying the Brices in 1995?

13. Answer Exercise 12 if the bank instead continually uses the money generated by the Brices' $10,000 for car loans at 12% interest, compounded continuously. Assume the same expenses as in Exercise 12.

14. What percent annual interest rate, compounded continuously, must a bank offer on a savings account if an undisturbed and untaxed balance doubles every 8 years?

15. A sum of money deposited in a savings account grows undisturbed at a continuously compounded annual rate of $c\%$. If the account has a balance of $10,000 after 5 years and a balance of $12,000 after 7 years, find
   (a) the value of c and
   (b) the initial amount S_0 deposited in the account.

16. *Rule of 72*
   (a) Show that the number of years required for the balance to double in an undisturbed savings account with interest compounded continuously at $k\%$ is approximately $69/k$.
   (b) The *rule of 72* takes $72/k$ as the time required for the balance to double in part (a). To see two possible reasons for using 72 in place of 69, do the following.
     (i) Compare the integer divisors of 69 with those of 72.
     (ii) Compute the time required for the balance to double at 5% interest *compounded annually* and compare that number with $69/5$ and with $72/5$. Repeat the computation at 8% interest *compounded annually*. [Recall that when a principal P is compounded annually at $k\%$ interest, the balance after t years is $P(1 + k/100)^t$.]

17. Suppose the average worker's annual salary in 1985 was $20,000. Suppose salaries increase continuously at a rate proportional to their size. If the rate of increase is kept constant at 5% annually, find the year in which the average worker's salary will be
   (a) $1,000,000,
   (b) $1,000,000,000.
   (c) What might be done to keep workers "happy" with raises but also prevent working with such cumbersome large numbers?

18. If you buy a $2000 IRA (individual retirement account) on your 20th birthday, and if it earns 9% annual interest, compounded continuously, find the cash value of the IRA on your 65th birthday by
   (a) assuming the 9% per year means per calendar year and
   (b) assuming the 9% per year means per banker's year, which is 360 days. Assume there are 11 leap years between your 20th and 65th birthdays, and neglect any calendar adjustment in the year 2000.

19. With reference to Exercise 18, suppose that inflation runs at 5% annually, compounded continuously, while you are between age 20 and 65 and that you would not have had to pay taxes on the original $2000 if you had not invested it in the IRA when you were 20.
   (a) How much money would you need at age 65 to equal the purchasing power of the $2000 when you were 20?
   (b) Assuming that the cash value of your IRA at age 65 is $122,000 and you then have to pay $48,000 deferred taxes on that amount, by what factor has the purchasing power of your original investment increased during the 45 years it was invested?

20. A tank initially contains a solution of 100 gal of brine with a salt concentration of 2 lb/gal. Fresh water is added at a rate of 4 gal/min, and the brine is drawn off at the bottom of the tank at the same rate. (See Fig. 7.19.) Assume that the concentration of salt in the solution is kept homogeneous by stirring. If $f(t)$ is the number of pounds of salt in solution in the tank at time t, show that f satisfies a differential equation of the form of Eq. (1).

FIGURE 7.19

21. With reference to Exercise 20, find the amount of salt in solution in the tank after 25 min.

22. With reference to Exercise 20, find the time when there is just 1 lb of salt left in the tank.

23. With reference to Exercise 20, find the amount of salt in the tank after 1 hr if fresh water flows in and brine is drawn off at the rate of 2 gal/min while the other data remain the same.

24. A body traveling in a medium is retarded by a force proportional to its velocity. If the velocity of the body after 4 sec is 80 ft/sec and the velocity after 6 sec is 60 ft/sec, find the initial velocity of the body.

25. In Exercise 24, find the total distance the body travels through the medium.

26. Newton's law of cooling states that a body placed in a colder medium will cool at a rate proportional to the difference between its temperature and the temperature of the surrounding medium. Assume that the temperature of the surrounding medium remains constant and let $f(t)$ be the difference between the temperature of the body and the temperature of the surrounding medium at time t. Show that f satisfies a differential equation of the form of Eq. (1).

27. With reference to Exercise 26, a body having a temperature of 90° is placed in a medium whose temperature is kept constant at 40°. If the temperature of the body after 15 min is 70°, find its temperature after 40 min.

28. Describe all solutions of the differential equation $d^2y/dt^2 = k(dy/dt)$ where k is a constant. [*Hint:* Let $u = dy/dt$.]

29. Referring to Example 5 in the text, suppose it were possible to have people live in space near the earth, say as far out as the moon. Use a calculator to estimate the year in which there would be one person for every 3 cubic yards of such space, assuming the moon is 250,000 mi from the earth.

7.5 THE INVERSE TRIGONOMETRIC FUNCTIONS

The Inverse Sine Function $\sin^{-1}x$

The graph of $y = \sin x$ is shown in Fig. 7.20. The function $\sin x$ does not have an inverse because, given a y-value c where $-1 \le c \le 1$, there are many values of x such that $\sin x = c$, as indicated in Fig. 7.20. In Fig. 7.21, we picked out by color an unbroken portion of the graph of $\sin x$ that covers the range $-1 \le y \le 1$ just once. We naturally picked out the unbroken arc closest to the origin. We use this colored arc in Fig. 7.21 to define the inverse sine function. That is, given a y-value where $-1 \le y \le 1$, there is a unique x-value where $-\pi/2 \le x \le \pi/2$ such that $\sin x = y$. We use the notation $x = \sin^{-1}y$ or $x = \arcsin y$ to denote this

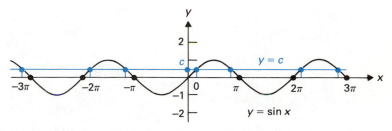

FIGURE 7.20   The function $\sin x$ has no inverse because one y-value c may correspond to many x-values.

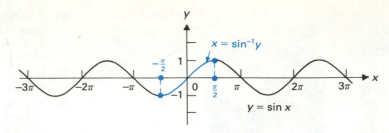

FIGURE 7.21   The arc of $y = \sin x$ where $-\pi/2 \le x \le \pi/2$ is used to define the inverse sine function.

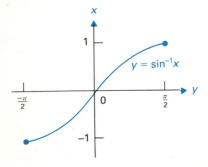

FIGURE 7.22   Changing axis labels from Fig. 7.21 gives the graph $y = \sin^{-1}x$.

inverse sine function. We will use the first notation, $x = \sin^{-1}y$, and we emphasize that the -1 *must not be considered as an exponent.*

The colored arc in Fig. 7.21 is the graph of $x = \sin^{-1}y$. To obtain the graph of $y = \sin^{-1}x$, we first interchange the x and y labels on the axes, as shown in Fig. 7.22. Then we flip the plane about the line $y = x$ to bring the axes into their usual position, with the x-axis horizontal, as shown in Fig. 7.23. From Fig. 7.23, we see that the domain of $\sin^{-1}x$ is $-1 \le x \le 1$; the range is $-\pi/2 \le y \le \pi/2$. We will see in a moment that it is very important to remember this range.

EXAMPLE 1   Find $\sin^{-1}(\tfrac{1}{2})$.

Solution   Since $\sin(\pi/6) = \tfrac{1}{2}$ and $\pi/6$ falls in the range $-\pi/2 \le y \le \pi/2$ of $\sin^{-1}x$, we see that $\sin^{-1}(\tfrac{1}{2}) = \pi/6$.   ∎

EXAMPLE 2   Find $\sin^{-1}(-\sqrt{3}/2)$.

Solution   We know that $\sin(4\pi/3) = -\sqrt{3}/2$, but $4\pi/3$ does not fall in the range $-\pi/2 \le y \le \pi/2$ of $\sin^{-1}x$. The angle falling within this range whose sine is $-\sqrt{3}/2$ is $-\pi/3$. Thus $\sin^{-1}(-\sqrt{3}/2) = -\pi/3$.   ∎

One way to find values of $\sin^{-1}x$ is to use a calculator. When radian mode is used, the calculator provides values in the correct range from $-\pi/2$ to $\pi/2$.

We develop calculus for $y = \sin^{-1}x$, and in the process we see why we are so fussy about the range of $\sin^{-1}x$. If $y = \sin^{-1}x$, then $x = \sin y$, and we find dy/dx by differentiating $x = \sin y$ implicitly:

$$x = \sin y,$$

$$1 = (\cos y)\,\frac{dy}{dx},$$

$$\frac{dy}{dx} = \frac{1}{\cos y}.$$

We would like to express dy/dx in terms of x rather than y. Since $\sin y = x$, we have

$$\cos y = \pm\sqrt{1 - \sin^2 y} = \pm\sqrt{1 - x^2}.$$

Since $-\pi/2 \le y \le \pi/2$, *we see that* $\cos y \ge 0$, *so the choice*

$$\cos y = \sqrt{1 - x^2}$$

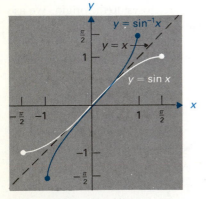

FIGURE 7.23   Flipping the plane in Fig. 7.22 about the line $y = x$ brings the axes into their usual position.

is appropriate. Our differentiation formula thus depends on our choice of range for $\sin^{-1}x$. We then obtain

$$\frac{dy}{dx} = \frac{1}{\sqrt{1-x^2}},$$

so that

$$\frac{d(\sin^{-1}x)}{dx} = \frac{1}{\sqrt{1-x^2}}.$$

If u is a differentiable function of x, then by the chain rule,

$$\frac{d(\sin^{-1}u)}{dx} = \frac{1}{\sqrt{1-u^2}} \cdot \frac{du}{dx}. \tag{1}$$

EXAMPLE 3 Find dy/dx if $y = \sin^{-1}2x$.

Solution From Eq. (1), we have

$$\frac{d(\sin^{-1}2x)}{dx} = \frac{1}{\sqrt{1-(2x)^2}} \cdot \frac{d(2x)}{dx} = \frac{1}{\sqrt{1-4x^2}} \cdot 2 = \frac{2}{\sqrt{1-4x^2}}. \quad \blacksquare$$

From Eq. (1), we obtain the important integration formula

$$\int \frac{du}{\sqrt{1-u^2}} = \sin^{-1}u + C. \tag{2}$$

We now know one reason to study the inverse sine and the other inverse trigonometric functions: We obtain new and important integration formulas.

Frequently, we encounter an integral of the form

$$\int \frac{dx}{\sqrt{a^2-x^2}}$$

where $a^2 \neq 1$. Rewriting this integral to take the form of Eq. (2), we have

$$\int \frac{dx}{\sqrt{a^2-x^2}} = \frac{1}{a} \int \frac{dx}{\sqrt{(a^2/a^2)-(x^2/a^2)}} = \int \frac{(1/a)\,dx}{\sqrt{1-(x/a)^2}}.$$

Taking $u = x/a$, so $du = (1/a)\,dx$ and using Eq. (2), we obtain

$$\int \frac{dx}{\sqrt{a^2-x^2}} = \sin^{-1}\left(\frac{x}{a}\right) + C.$$

If u is a differentiable function of x, the chain rule then gives the useful formula

$$\int \frac{du}{\sqrt{a^2-u^2}} = \sin^{-1}\left(\frac{u}{a}\right) + C. \tag{3}$$

EXAMPLE 4 Find

$$\int \frac{dx}{\sqrt{1-9x^2}}.$$

Solution   We "fix up" the integral to fit formula (3), with $a = 1$ and $u = 3x$, obtaining

$$\int \frac{dx}{\sqrt{1 - 9x^2}} = \int \frac{dx}{\sqrt{1 - (3x)^2}} = \frac{1}{3} \int \frac{\overbrace{3\,dx}^{du}}{\sqrt{1 - \underbrace{(3x)^2}_{u}}}$$

$$= \frac{1}{3} \sin^{-1}(3x) + C. \qquad \blacksquare$$

EXAMPLE 5   Find

$$\int_{-1.6}^{1.2} \frac{dx}{\sqrt{4 - x^2}}.$$

Solution   Formula (3) shows that

$$\int_{-1.6}^{1.2} \frac{dx}{\sqrt{4 - x^2}} = \sin^{-1}\left(\frac{x}{2}\right)\Bigg]_{-1.6}^{1.2}$$

$$= \sin^{-1}(0.6) - \sin^{-1}(-0.8)$$

$$\approx 0.6435 - (-0.9273) = 1.5708. \qquad \blacksquare$$

The Other Inverse Trigonometric Functions

We have developed the function $\sin^{-1}x$ in detail. We will now consider the other inverse trigonometric functions all at once and leave the derivation of the appropriate calculus formulas as exercises.

Figure 7.24 shows the graphs of all the elementary trigonometric functions. None of them has an inverse, in the sense that a horizontal line $y = c$ may meet any of the graphs in more than one point. The colored arcs in Fig. 7.24 show the portions of the graphs that are selected to define the inverse functions. Note that for the functions $\sec x$ and $\csc x$, it is impossible to pick one *unbroken* arc that covers both the values $y \geq 1$ and the values $y \leq -1$.

Figure 7.25 shows the graphs of the inverse functions of the independent variable x. Table 7.1 summarizes the basic information concerning them. We have discussed the information for $\sin^{-1}x$ in detail. The domains and ranges for the other inverse functions are apparent from Fig. 7.25. We ask you to establish the differentiation formulas in the exercises.

In Table 7.1, note that the derivatives of $\cos^{-1}x$, $\cot^{-1}x$, and $\csc^{-1}x$ are just the negatives of the derivatives of $\sin^{-1}x$, $\tan^{-1}x$, and $\sec^{-1}x$ respectively. This means that the "inverse cofunctions" give us no additional integration formulas. That is, we would normally write

$$\int \frac{-1}{\sqrt{1 - x^2}} = -\int \frac{dx}{\sqrt{1 - x^2}}$$

$$= -\sin^{-1}x + C$$

rather than

$$\int \frac{-1}{\sqrt{1 - x^2}} = \cos^{-1}x + C.$$

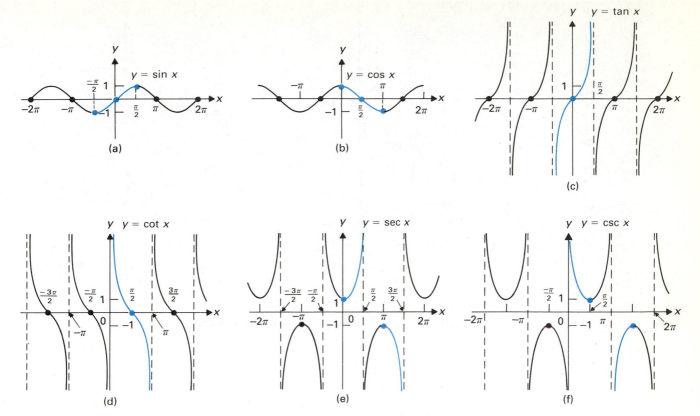

FIGURE 7.24 Portions of the graphs of trigonometric functions used to define their inverse functions.

The values of the inverse trigonometric functions falling in the y-ranges shown in Table 7.1 are called the **principal values** of those functions. There is no uniform agreement as to the principal values for $\sec^{-1}x$ and $\csc^{-1}x$. Some people prefer the second-quadrant range $\pi/2 < y \le \pi$ for principal values of $\sec^{-1}x$ where $x \le -1$. With that choice, we have $\sec^{-1}x = \cos^{-1}(1/x)$. Since calculators normally have keys to compute only $\sin^{-1}x$, $\cos^{-1}x$, and $\tan^{-1}x$, that choice might seem to be the proper one for today's technology. However, with that choice, we would have $d(\sec^{-1}x) = \pm 1/(x\sqrt{x^2-1})$, where the negative sign would occur for $x \le -1$. We ask you to show this in Exercise 63. Also, in Section 8.5 we will want to write $\sqrt{\sec^2 t - 1} = \sqrt{\tan^2 t} = \tan t$ for all t in the principal value range of $\sec^{-1}x$. This is true for our choice of principal values, where t is a first- or third-quadrant angle. However, it would not be true for t a second-quadrant angle because $\tan t$ is negative in the second quadrant. We chose what we feel is the lesser of two evils: more difficult evaluation of $\sec^{-1}x$ using a calculator.

We want to be able to compute $\sec^{-1}x$ using a calculator. Since $\sec\theta = 1/(\cos\theta)$, Fig. 7.26 indicates that we have

$$\sec^{-1}x = \begin{cases} \cos^{-1}(1/x) & \text{for } x \ge 1, \\ 2\pi - \cos^{-1}(1/x) & \text{for } x \le -1. \end{cases} \qquad (4)$$

We illustrate the use of Eq. (4) with a calculator in the next example.

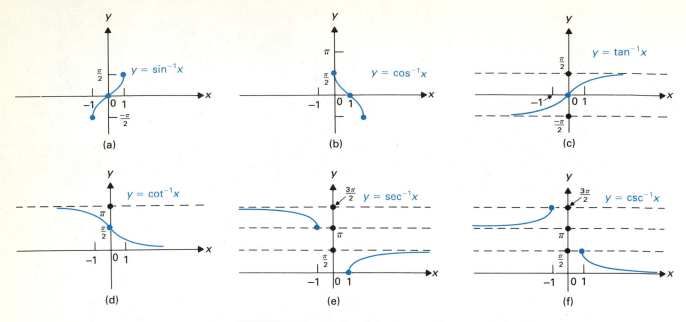

FIGURE 7.25 Graphs of the inverse trigonometric functions.

TABLE 7.1	Inverse trigonometric functions		
Function	*Domain*	*Range*	*Derivative*
$\sin^{-1}x$	$[-1, 1]$	$\left[-\dfrac{\pi}{2}, \dfrac{\pi}{2}\right]$	$\dfrac{1}{\sqrt{1 - x^2}}$
$\cos^{-1}x$	$[-1, 1]$	$[0, \pi]$	$\dfrac{-1}{\sqrt{1 - x^2}}$
$\tan^{-1}x$	All x	$-\dfrac{\pi}{2} < y < \dfrac{\pi}{2}$	$\dfrac{1}{1 + x^2}$
$\cot^{-1}x$	All x	$0 < y < \pi$	$\dfrac{-1}{1 + x^2}$
$\sec^{-1}x$	$\begin{cases} x \geq 1 \\ x \leq -1 \end{cases}$	$\begin{cases} 0 \leq y < \dfrac{\pi}{2} \\ \pi \leq y < \dfrac{3\pi}{2} \end{cases}$	$\dfrac{1}{x\sqrt{x^2 - 1}}$
$\csc^{-1}x$	$\begin{cases} x \geq 1 \\ x \leq -1 \end{cases}$	$\begin{cases} 0 < y \leq \dfrac{\pi}{2} \\ \pi < y \leq \dfrac{3\pi}{2} \end{cases}$	$\dfrac{-1}{x\sqrt{x^2 - 1}}$

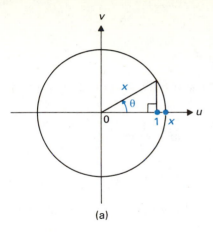

(a)

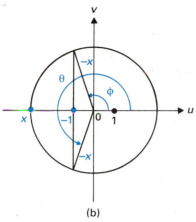

(b)

FIGURE 7.26 (a) $\theta = \sec^{-1}x = \cos^{-1}(1/x)$ for $x \geq 1$; (b) for $x \leq -1$, $\theta = \sec^{-1}x, \phi = \cos^{-1}(1/x), \theta = 2\pi = \phi$.

EXAMPLE 6   Find

$$\int_{-5}^{-2} \frac{1}{x\sqrt{x^2 - 1}}\, dx.$$

Solution   From Table 7.1 and Eq. (4), we have

$$\int_{-5}^{-2} \frac{1}{x\sqrt{x^2 - 1}}\, dx = \sec^{-1}x \Big]_{-5}^{-2} = \sec^{-1}(-2) - \sec^{-1}(-5)$$

$$= \left[2\pi - \cos^{-1}\left(-\frac{1}{2}\right)\right] - \left[2\pi - \cos^{-1}\left(-\frac{1}{5}\right)\right]$$

$$= \cos^{-1}\left(-\frac{1}{5}\right) - \cos^{-1}\left(-\frac{1}{2}\right)$$

$$\approx 1.7722 - 2.0944 = -0.3222. \qquad \blacksquare$$

Formulas 5–13 follow from Table 7.1 and the chain rule. The integration formulas provide a motivation for studying the calculus of the inverse trigonometric functions. Formulas (12) and (13) are obtained from formulas (7) and (9), just as we obtained formula (11) from formula (5) earlier in this section.

Differentiation Formulas

5. $\dfrac{d(\sin^{-1}u)}{dx} = \dfrac{1}{\sqrt{1 - u^2}} \cdot \dfrac{du}{dx}$ 6. $\dfrac{d(\cos^{-1}u)}{dx} = \dfrac{-1}{\sqrt{1 - u^2}} \cdot \dfrac{du}{dx}$

7. $\dfrac{d(\tan^{-1}u)}{dx} = \dfrac{1}{1 + u^2} \cdot \dfrac{du}{dx}$ 8. $\dfrac{d(\cot^{-1}u)}{dx} = \dfrac{-1}{1 + u^2} \cdot \dfrac{du}{dx}$

9. $\dfrac{d(\sec^{-1}u)}{dx} = \dfrac{1}{u\sqrt{u^2 - 1}} \cdot \dfrac{du}{dx}$ 10. $\dfrac{d(\csc^{-1}u)}{dx} = \dfrac{-1}{u\sqrt{u^2 - 1}} \cdot \dfrac{du}{dx}$

Integration Formulas

11. $\displaystyle\int \frac{du}{\sqrt{a^2 - u^2}} = \sin^{-1}\left(\frac{u}{a}\right) + C$ 12. $\displaystyle\int \frac{du}{a^2 + u^2} = \frac{1}{a}\tan^{-1}\left(\frac{u}{a}\right) + C$

13. $\displaystyle\int \frac{du}{u\sqrt{u^2 - a^2}} = \frac{1}{a}\sec^{-1}\left(\frac{u}{a}\right) + C$

EXAMPLE 7   Find

$$\int_{0}^{1/2} \frac{dx}{\sqrt{1 - x^2}}.$$

Solution   We have

$$\int_{0}^{1/2} \frac{1}{\sqrt{1 - x^2}}\, dx = \sin^{-1}x \Big]_{0}^{1/2} = \sin^{-1}\frac{1}{2} - \sin^{-1}0$$

$$= \frac{\pi}{6} - 0 = \frac{\pi}{6}. \qquad \blacksquare$$

EXAMPLE 8 Find

$$\int_0^3 \frac{1}{1+4x^2}\,dx.$$

Solution Using formula (12) with $a = 1$ and $u = 2x$, so $du = 2\,dx$, we find that

$$\int_0^3 \frac{1}{1+4x^2}\,dx = \frac{1}{2}\int_0^3 \frac{2}{1+(2x)^2}\,dx = \frac{1}{2}\tan^{-1}2x \Big]_0^3$$

$$= \frac{1}{2}\tan^{-1}6 - \frac{1}{2}\tan^{-1}0 = \frac{1}{2}\tan^{-1}6.$$

If we desire a decimal approximation for $\tan^{-1}6$, we use a calculator and find that $\tan^{-1}6 \approx 1.406$. ∎

EXAMPLE 9 Find $\int_0^1 \sqrt{4-x^2}\,dx$.

Solution Using formula 34 from the integral table on the endpapers of this book, we find that

$$\int_0^1 \sqrt{4-x^2}\,dx = \left(\frac{x}{2}\sqrt{4-x^2} + \frac{4}{2}\sin^{-1}\frac{x}{2}\right)\Big]_0^1$$

$$= \left(\frac{1}{2}\sqrt{3} + 2\sin^{-1}\frac{1}{2}\right) - 0$$

$$= \frac{\sqrt{3}}{2} + 2\frac{\pi}{6} = \frac{\sqrt{3}}{2} + \frac{\pi}{3}. \quad ∎$$

SUMMARY

1. A summary of the six inverse trigonometric functions is given in Table 7.1.

2. A summary of differentiation and integration formulas is given on page 371.

EXERCISES 7.5

In Exercises 1–18, find the given quantity.

1. $\sin^{-1}1$

2. $\cos^{-1}\left(\dfrac{-1}{\sqrt{2}}\right)$

3. $\tan^{-1}(-\sqrt{3})$

4. $\csc^{-1}2$

5. $\sec^{-1}\left(\dfrac{-2}{\sqrt{3}}\right)$

6. $\sin^{-1}\left(\dfrac{\sqrt{3}}{-2}\right)$

7. $\cot^{-1}1$

8. $\sec^{-1}1$

9. $\csc^{-1}(-1)$

10. $\cos^{-1}(-1)$

11. $\sec^{-1}(-1)$

12. $\csc^{-1}\left(\dfrac{-2}{\sqrt{3}}\right)$

13. $\tan^{-1}0$

14. $\sin^{-1}(-1/\sqrt{2})$

15. $\sec^{-1}(-\sqrt{2})$

16. $\tan^{-1}1$

17. $\tan^{-1}(-1)$

18. $\sin^{-1}(-1/2)$

19. Find $\lim_{x\to\infty}\tan^{-1}x$.

20. Find $\lim_{x\to-\infty}\tan^{-1}x$.

21. Find $\lim_{x\to\infty}\sec^{-1}x$.

22. Find $\lim_{x\to-\infty}\sec^{-1}x$.

23. Find $\lim_{x\to\infty}\cot^{-1}x$.

24. Find $\lim_{x\to-\infty}\cot^{-1}x$.

In Exercises 25–35, find the derivative of the given function.

25. $\sin^{-1}(2x)$

26. $\cos^{-1}(x^2)$

27. $\tan^{-1}(\sqrt{x})$

28. $x\,\sec^{-1}x$

29. $\csc^{-1}\left(\dfrac{1}{x}\right)$

30. $(\sin^{-1}x)(\cos^{-1}x)$

31. $(\tan^{-1}2x)^3$

32. $\sec^{-1}(x^2+1)$

33. $\dfrac{1}{\tan^{-1}x}$

34. $\sqrt{\csc^{-1}x}$

35. $(x+\sin^{-1}3x)^2$

36. A careless mathematics professor asked his calculus class on their final examination to find dy/dx if $y=\sin^{-1}(1+x^2)$. What is wrong with this problem?

In Exercises 37–41, find the integral without the use of tables.

37. $\displaystyle\int_{-1}^{1}\dfrac{1}{1+x^2}\,dx$

38. $\displaystyle\int\dfrac{1}{\sqrt{1-9x^2}}\,dx$

39. $\displaystyle\int_{-1}^{\sqrt{3}}\dfrac{1}{\sqrt{1-(x^2/4)}}\,dx$

40. $\displaystyle\int_{2/\sqrt{3}}^{2}\dfrac{1}{x\sqrt{x^2-1}}\,dx$

41. $\displaystyle\int\dfrac{1}{3x\sqrt{4x^2-1}}\,dx$

⊞ In Exercises 42–49, find the integral without using a table, but use your calculator or computer in the final step to obtain a decimal value for the integral, as in Examples 5 and 6 in the text.

42. $\displaystyle\int_{0}^{0.75}\dfrac{dx}{\sqrt{1-x^2}}$

43. $\displaystyle\int_{-0.2}^{0.3}\dfrac{dx}{1+x^2}$

44. $\displaystyle\int_{2}^{10}\dfrac{dx}{x\sqrt{x^2-1}}$

45. $\displaystyle\int_{-3}^{-8}\dfrac{dx}{x\sqrt{x^2-1}}$

46. $\displaystyle\int_{0}^{3}\dfrac{dx}{\sqrt{16-x^2}}$

47. $\displaystyle\int_{-2}^{4}\dfrac{dx}{4+x^2}$

48. $\displaystyle\int_{6}^{10}\dfrac{dx}{x\sqrt{x^2-4}}$

49. $\displaystyle\int_{0}^{2}\dfrac{\cos x}{\sqrt{4-\sin^2 x}}\,dx$

In Exercises 50–57, use tables if necessary to find the definite integral.

50. $\displaystyle\int_{0}^{1}\dfrac{1}{\sqrt{4-x^2}}\,dx$

51. $\displaystyle\int_{-1}^{1}\sqrt{1-x^2}\,dx$

52. $\displaystyle\int_{0}^{2}x^2\sqrt{16-x^2}\,dx$

53. $\displaystyle\int_{1}^{\sqrt{3}}\dfrac{\sqrt{4-x^2}}{x^2}\,dx$

54. $\displaystyle\int_{-4}^{-4/\sqrt{3}}\dfrac{\sqrt{x^2-4}}{x}\,dx$

55. $\displaystyle\int_{2}^{3}\dfrac{1}{\sqrt{4x-x^2}}\,dx$

56. $\displaystyle\int_{0}^{1}\sqrt{2x-x^2}\,dx$

57. $\displaystyle\int_{2}^{3}x\sqrt{4x-x^2}\,dx$

58. Derive the formula for $d(\cos^{-1}x)/dx$.

59. Derive the formula for $d(\tan^{-1}x)/dx$.

60. Derive the formula for $d(\cot^{-1}x)/dx$.

61. Derive the formula for $d(\sec^{-1}x)/dx$.

62. Derive the formula for $d(\csc^{-1}x)/dx$.

63. In defining $\sec^{-1}x$, some mathematicians prefer to choose branches of the graph of the secant so that the range (set of principal values) of $\sec^{-1}x$ is $0\le y<\pi/2$ or $\pi/2<y\le\pi$. With this definition, we see that the relation $\sec^{-1}x=\cos^{-1}(1/x)$ holds.

   (a) Show by an example that the relation $\sec^{-1}x=\cos^{-1}(1/x)$ does not hold for the definition of $\sec^{-1}x$ in Table 7.1.

   (b) Show that if $\sec^{-1}x$ is defined to have range described in this exercise, then the formula for the derivative of the inverse secant changes from that in Table 7.1 to
   $$\dfrac{d(\sec^{-1}x)}{dx}=\dfrac{1}{|x|\sqrt{x^2-1}}.$$

64. Find the area of the plane region bounded by $x=0$, $y=0$, $x=1$, and $y=1/(1+x^2)$.

65. The definite integral formed to solve Exercise 64 has value $\pi/4$. Make use of this fact and use Simpson's rule with a calculator and $n=20$, or with a computer and $n=200$, to estimate the value of π.

7.6 THE HYPERBOLIC FUNCTIONS

The hyperbolic functions are defined in terms of exponential functions, and their inverses can be expressed in terms of logarithm functions. (The inverse of the exponential function is the logarithm function, so this is not surprising.) This section gives just a brief summary of hyperbolic functions. Our goal is to give you enough familiarity with them so that they won't appear mysterious or difficult if you encounter them in your work.

The Functions

The hyperbolic functions have many points of similarity with the trigonometric functions. The reason for this similarity is shown in a course in *functions of a complex variable*. There are six basic hyperbolic functions, just as there are six basic trigonometric functions; the hyperbolic functions are the *hyperbolic sine*, denoted by sinh x, the *hyperbolic cosine*, or cosh x, the *hyperbolic tangent*, or tanh x, and so on. We usually pronounce "sinh x" as though it were spelled "cinch x."

We will define the hyperbolic functions in terms of exponential functions; this is very different from the way we defined the trigonometric functions. However, if you study *complex analysis* later, you will discover that the trigonometric functions can also be defined in terms of exponential functions of a "complex variable." Table 7.2 gives the definitions of the hyperbolic functions. We have to remember the definitions for sinh x and cosh x, and then we proceed exactly as for the trigonometric functions.

We can see that

$$\cosh^2 x - \sinh^2 x = 1 \tag{1}$$

for all x, namely,

$$\left(\frac{e^x + e^{-x}}{2}\right)^2 - \left(\frac{e^x - e^{-x}}{2}\right)^2 = \frac{e^{2x} + 2 + e^{-2x}}{4} - \frac{e^{2x} - 2 + e^{-2x}}{4}$$

$$= \frac{2}{4} + \frac{2}{4} = 1.$$

Equation (1) is the basic relation for hyperbolic functions and plays a role similar to the relation $\sin^2 x + \cos^2 x = 1$ for the trigonometric functions. For the trigonometric functions, the point $(\cos t, \sin t)$ lies on the circle with equation $x^2 + y^2 = 1$ for all t; for the hyperbolic functions, the point $(\cosh t, \sinh t)$ lies on the curve with equation $x^2 - y^2 = 1$, which is called a *hyperbola*. This explains the name "hyperbolic functions."

The graphs of the six hyperbolic functions are easily sketched. To sketch the graph of sinh x, we simply take half the difference in height between the graphs of e^x and e^{-x}, while for cosh x we average the heights of these two graphs. Since $\lim_{x \to \infty} e^{-x} = 0$, we see that the graphs of both sinh x and cosh x are close to the graph of $e^x/2$ for large x. The graphs of the six hyperbolic functions are sketched in Fig. 7.27.

TABLE 7.2   The six hyperbolic functions

Function	Definition	Function	Definition
sinh x	$\dfrac{e^x - e^{-x}}{2}$	cosh x	$\dfrac{e^x + e^{-x}}{2}$
tanh x	$\dfrac{\sinh x}{\cosh x}$	coth x	$\dfrac{\cosh x}{\sinh x}$
sech x	$\dfrac{1}{\cosh x}$	csch x	$\dfrac{1}{\sinh x}$

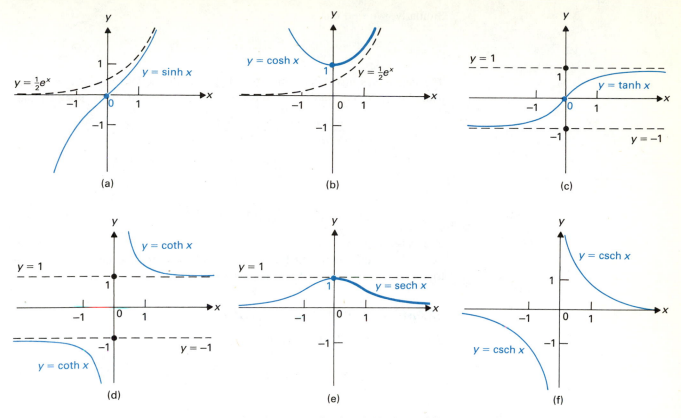

FIGURE 7.27 Graphs of the six hyperbolic functions.

The hyperbolic functions occur naturally in the physical sciences. For example, a flexible cable suspended from two supports of equal height hangs in a "catenary curve." It can be shown that with suitable choice of axes and scales, the curve has the equation $y = a \cosh(x/a)$ for some constant a.

Note from the graphs that $\sinh x$, $\tanh x$, $\coth x$, and $\operatorname{csch} x$ all yield a unique x for each y, so the inverse hyperbolic functions $\sinh^{-1}y$, $\tanh^{-1}y$, $\coth^{-1}y$, and $\operatorname{csch}^{-1}y$ exist. While $\cosh x$ and $\operatorname{sech} x$ do not satisfy this condition, we abuse terminology just as we did for trigonometric functions and pick the branches of their graphs indicated by the darker curves in Fig. 7.27 (b) and (e) to define the functions $\cosh^{-1}y$ and $\operatorname{sech}^{-1}y$.

Calculus of the Hyperbolic Functions

Developing calculus of the hyperbolic functions is straightforward, since the functions are defined in terms of the exponential function. Turning to the derivative of $\sinh x$, we find that

$$\frac{d(\sinh x)}{dx} = \frac{d}{dx}\left(\frac{e^x - e^{-x}}{2}\right)$$

$$= \frac{1}{2} \cdot \frac{d}{dx}(e^x - e^{-x}) = \frac{1}{2}(e^x + e^{-x}) = \cosh x.$$

Similarly, we find that

$$\frac{d(\cosh x)}{dx} = \sinh x.$$

The derivatives of the remaining four hyperbolic functions can then be found using Eq. (1) just as for the corresponding trigonometric functions. For example,

$$\frac{d(\tanh x)}{dx} = \frac{d}{dx}\left(\frac{\sinh x}{\cosh x}\right)$$

$$= \frac{\cosh x \cosh x - \sinh x \sinh x}{\cosh^2 x}$$

$$= \frac{1}{\cosh^2 x} = \operatorname{sech}^2 x.$$

The left-hand portion of Table 7.3 gives the derivatives of the hyperbolic functions. These differentiation formulas parallel those for the derivatives of the trigonometric functions, except for occasional differences in sign.

EXAMPLE 1   Find y' if $y = \sinh(x^2 + 1)$.

Solution   Using Table 7.3 and the chain rule, we have

$$y' = \cosh(x^2 + 1)\,\frac{d(x^2 + 1)}{dx} = 2x \cosh(x^2 + 1).$$ ∎

EXAMPLE 2   Find y' if $y = \sinh^2 3x \tanh^3 4x$.

Solution   Using Table 7.3, differentiation rules, and the chain rule, we have

$$y' = (\sinh^2 3x)(3 \tanh^2 4x \operatorname{sech}^2 4x)4 + (\tanh^3 4x)(2 \sinh 3x \cosh 3x)3$$

$$= 12 \sinh^2 3x \tanh^2 4x \operatorname{sech}^2 4x + 6 \sinh 3x \cosh 3x \tanh^3 4x.$$ ∎

TABLE 7.3   Derivatives of the hyperbolic functions

Function	Derivative	Function	Derivative
$\sinh x$	$\cosh x$	$\sinh^{-1} x$	$\dfrac{1}{\sqrt{1 + x^2}}$
$\cosh x$	$\sinh x$	$\cosh^{-1} x$	$\dfrac{1}{\sqrt{x^2 - 1}},\quad x > 1$
$\tanh x$	$\operatorname{sech}^2 x$	$\tanh^{-1} x$	$\dfrac{1}{1 - x^2},\quad |x| < 1$
$\coth x$	$-\operatorname{csch}^2 x$	$\coth^{-1} x$	$\dfrac{1}{1 - x^2},\quad |x| > 1$
$\operatorname{sech} x$	$-\operatorname{sech} x \tanh x$	$\operatorname{sech}^{-1} x$	$\dfrac{-1}{x\sqrt{1 - x^2}},\quad 0 < x < 1$
$\operatorname{csch} x$	$-\operatorname{csch} x \coth x$	$\operatorname{csch}^{-1} x$	$\dfrac{-1}{|x|\sqrt{1 + x^2}}$

If $y = f(x) = \sinh x$, then $x = \sinh^{-1}y$ and dx/dy can be obtained by differentiating $y = \sinh x$ implicitly with respect to y:

$$y = \sinh x,$$

$$1 = (\cosh x)\,\frac{dx}{dy},$$

$$\frac{dx}{dy} = \frac{1}{\cosh x} = \frac{1}{\sqrt{1 + \sinh^2 x}} = \frac{1}{\sqrt{1 + y^2}}.$$

Interchanging x and y so that x is the dependent variable, we obtain

$$\frac{d(\sinh^{-1}x)}{dx} = \frac{1}{\sqrt{1 + x^2}}.$$

The derivatives of the other inverse hyperbolic functions are found similarly and are given in the right-hand portion of Table 7.3. These differentiation formulas can be combined with the chain rule in the usual way.

EXAMPLE 3 Find the derivative of $\sinh^{-1}(\tan x)$.

Solution We have

$$\frac{d(\sinh^{-1}(\tan x))}{dx} = \frac{1}{\sqrt{1 + \tan^2 x}} \cdot \sec^2 x$$

$$= \frac{\sec^2 x}{|\sec x|} = |\sec x|.\qquad\blacksquare$$

In the preceding section, we studied the inverse trigonometric functions because they led to important integration formulas, such as

$$\int \frac{du}{\sqrt{1 - u^2}} = \sin^{-1}u + C. \tag{2}$$

The formula $d(\sinh^{-1}x)/dx = 1/\sqrt{1 + x^2}$ leads to the integration formula

$$\int \frac{du}{\sqrt{1 + u^2}} = \sinh^{-1}u + C. \tag{3}$$

This formula should surely be as important as formula (2). However, $\sinh^{-1}x$ can be expressed in terms of the logarithm function, and with a calculator or computer, values of the function $\ln x$ are readily found. Most calculators do not have a $\sinh^{-1}x$ key. Thus one generally consults an integral table and uses in place of formula (3)

$$\int \frac{du}{\sqrt{a^2 + u^2}} = \ln(u + \sqrt{a^2 + u^2}) + C. \tag{4}$$

We now show that we indeed have

$$\sinh^{-1}x = \ln(x + \sqrt{1 + x^2}),$$

as you would guess by comparing formulas (3) and (4). The exercises ask you to imitate the procedure in Example 4 to find the logarithmic expressions for the other inverse hyperbolic functions.

EXAMPLE 4 Show that $\sinh^{-1}x = \ln(x + \sqrt{1 + x^2})$.

Solution Let $y = \sinh^{-1}x$. Then

$$x = \sinh y = \frac{e^y - e^{-y}}{2},$$

so

$$e^y - e^{-y} = 2x.$$

Multiplying this equation by e^y, we obtain

$$e^{2y} - 1 = 2xe^y,$$
$$(e^y)^2 - 2x(e^y) - 1 = 0. \tag{5}$$

We regard Eq. (5) as a quadratic equation in the unknown e^y and solve it using the quadratic formula:

$$e^y = \frac{2x \pm \sqrt{4x^2 + 4}}{2} = x \pm \sqrt{x^2 + 1}.$$

Since $e^y > 0$ for all y, we see that the plus sign is appropriate, so

$$e^y = x + \sqrt{x^2 + 1}.$$

Then

$$\sinh^{-1}x = y = \ln(x + \sqrt{x^2 + 1}). \quad \blacksquare$$

EXAMPLE 5 Use formula (4) and a calculator to find a 4-decimal approximation of

$$\int_0^5 \frac{x}{\sqrt{1 + x^4}}\, dx.$$

Solution We can "fix up" our integral to fit formula (4) with $a = 1$, $u = x^2$, and $du = 2x$, as follows:

$$\int_0^5 \frac{x\, dx}{\sqrt{1 + x^4}} = \frac{1}{2}\int_0^5 \frac{\overbrace{2x\, dx}^{du}}{\sqrt{1 + \underbrace{(x^2)^2}_{u}}}$$

$$= \frac{1}{2}\ln(x^2 + \sqrt{1 + x^4})\Big]_0^5$$

$$= \frac{1}{2}\ln(25 + \sqrt{626}) - \frac{1}{2}\ln 1$$

$$\approx 1.9562 - 0 = 1.9562. \quad \blacksquare$$

From the formulas for the derivatives of the inverse hyperbolic functions found in Table 7.3, we obtain the integration formulas listed in the Summary. The Summary also gives the logarithmic form of the integration formulas. The reason for the split in the formula for $\int[1/(a^2 - u^2)]\, du$ is that $\tanh^{-1}x$ and

$\coth^{-1}x$ have algebraically identical derivatives, but the domain of $\tanh^{-1}x$ is $|x| < 1$ while the domain of $\coth^{-1}x$ is $|x| > 1$. Since

$$\frac{1}{1 - u^2} = \frac{\frac{1}{2}}{1 - u} + \frac{\frac{1}{2}}{1 + u},$$

we have, as the logarithmic alternative,

$$\int \frac{du}{1 - u^2} = -\frac{1}{2} \ln|1 - u| + \frac{1}{2} \ln|1 + u| + C$$

$$= \frac{1}{2} \ln\left|\frac{1 + u}{1 - u}\right| + C.$$

EXAMPLE 6 Find

$$\int \frac{1}{x\sqrt{4 - x^2}} \, dx.$$

Solution Using formula (9) in the Summary, we have

$$\int \frac{1}{x\sqrt{4 - x^2}} \, dx = -\frac{1}{2} \operatorname{sech}^{-1}\left|\frac{x}{2}\right| + C$$

$$= -\frac{1}{2} \ln\left|\frac{2 + \sqrt{4 - x^2}}{x}\right| + C. \qquad \blacksquare$$

SUMMARY

1. The hyperbolic functions are defined in Table 7.2.

2. The graphs of the hyperbolic functions are shown in Fig. 7.27.

3. Differentiation formulas are in Table 7.3.

4. Integration formulas given by the inverse hyperbolic functions (and also by logarithm functions) are as follows.

$$\int \frac{du}{\sqrt{a^2 + u^2}} = \sinh^{-1}\left(\frac{u}{a}\right) + C = \ln(u + \sqrt{a^2 + u^2}) + C \quad \textbf{(6)}$$

$$\int \frac{du}{\sqrt{u^2 - a^2}} = \cosh^{-1}\left(\frac{u}{a}\right) + C = \ln(u + \sqrt{u^2 - a^2}) + C \quad \textbf{(7)}$$

$$\int \frac{du}{a^2 - u^2} = \begin{cases} \dfrac{1}{a} \tanh^{-1}\left(\dfrac{u}{a}\right) + C, & |u| < a \\[2mm] \dfrac{1}{a} \coth^{-1}\left(\dfrac{u}{a}\right) + C, & |u| > a \end{cases} \quad \textbf{(8)}$$

$$= \frac{1}{2a} \ln\left|\frac{a + u}{a - u}\right| + C$$

$$\int \frac{du}{u\sqrt{a^2-u^2}} = -\frac{1}{a}\operatorname{sech}^{-1}\left|\frac{u}{a}\right| + C$$

$$= -\frac{1}{a}\ln\left|\frac{a+\sqrt{a^2-u^2}}{u}\right| + C \qquad (9)$$

$$\int \frac{du}{u\sqrt{a^2+u^2}} = -\frac{1}{a}\operatorname{csch}^{-1}\left|\frac{u}{a}\right| + C$$

$$= -\frac{1}{a}\ln\left|\frac{a+\sqrt{a^2+u^2}}{u}\right| + C \qquad (10)$$

EXERCISES 7.6

In Exercises 1–6, prove the given relation for hyperbolic functions.

1. $1 - \tanh^2 x = \operatorname{sech}^2 x$
2. $\coth^2 x - 1 = \operatorname{csch}^2 x$
3. $\sinh(-x) = -\sinh x$
4. $\cosh(-x) = \cosh x$
5. $\sinh(x + y) = \sinh x \cosh y + \cosh x \sinh y$
6. $\cosh(x + y) = \cosh x \cosh y + \sinh x \sinh y$
7. Use Exercises 5 and 6 to find "double-angle" formulas for $\sinh 2x$ and $\cosh 2x$.
8. If $\sinh a = -\frac{3}{4}$, find $\cosh a$, $\tanh a$, and $\operatorname{csch} a$.

In Exercises 9–17, derive the formula for the derivative of the given function. In each exercise, you may use the answer to any previous exercise.

9. $\cosh x$
10. $\coth x$
11. $\operatorname{sech} x$
12. $\operatorname{csch} x$
13. $\cosh^{-1} x$
14. $\tanh^{-1} x$
15. $\coth^{-1} x$
16. $\operatorname{sech}^{-1} x$
17. $\operatorname{csch}^{-1} x$

In Exercises 18–35, find the derivative of the given function.

18. $\sinh(2x - 3)$
19. $\cosh(x^2)$
20. $\tanh^2 3x$
21. $\operatorname{sech}(\sqrt{x})$
22. $\coth(e^{2x})$
23. $\operatorname{csch}(\ln x)$
24. $\operatorname{sech}^3 3x$
25. $\sinh^2 x \cosh^2 x$
26. $(e^{-x} + \tanh x)^2$
27. $\sinh^{-1}(2x)$
28. $\cosh^{-1}(\sec x)$
29. $\tanh^{-1}(\sin 2x)$
30. $\coth^{-1}(e^{2x+1})$
31. $\operatorname{sech}^{-1}(x^2)$
32. $\operatorname{csch}^{-1}(\tan x)$
33. $e^x \sinh^{-1}(e^x)$
34. $\cosh^{-1}(x^2 + 1)$
35. $\tanh^{-1}(\cot 3x)$

In Exercises 36–44, compute the given integral without using tables.

36. $\int \tanh x\, dx$
37. $\int \coth x\, dx$
38. $\int \sinh^2 x \cosh x\, dx$
39. $\int \sinh(3x + 2)\, dx$
40. $\int x \cosh(x^2)\, dx$
41. $\int \operatorname{sech}^2 3x\, dx$
42. $\int \frac{1}{(e^x + e^{-x})^2}\, dx$
43. $\int \frac{e^x + e^{-x}}{e^x - e^{-x}}\, dx$
44. $\int (\cosh 2x)(e^{\sinh 2x})\, dx$

In Exercises 45–59, use tables if necessary to compute the given integral.

45. $\int_0^{1/2} \frac{x}{1 - x^2}\, dx$
46. $\int_0^{1/4} \frac{1}{1 - 4x^2}\, dx$
47. $\int \frac{x}{\sqrt{1 + 4x^2}}\, dx$
48. $\int \frac{1}{\sqrt{4 + x^2}}\, dx$
49. $\int \frac{1}{\sqrt{1 - e^{2x}}}\, dx$
50. $\int \frac{\sin 2x}{(\cos 2x)\sqrt{1 + \cos^2 2x}}\, dx$
51. $\int_0^1 \sqrt{4 + x^2}\, dx$
52. $\int_0^4 x^2\sqrt{9 + x^2}\, dx$
53. $\int \frac{\sqrt{16 + x^2}}{x}\, dx$
54. $\int \frac{\sqrt{2 + x^2}}{3x^2}\, dx$
55. $\int \frac{\sin^2 x \cos x}{\sqrt{9 + \sin^2 x}}\, dx$
56. $\int_3^5 x^2\sqrt{x^2 - 9}\, dx$
57. $\int \frac{e^{3x}}{\sqrt{e^{2x} - 16}}\, dx$
58. $\int x \sinh 2x\, dx$
59. $\int \cosh^3 4x\, dx$

In Exercises 60–64, use the technique of Example 4 in the text to derive the logarithmic expression for the given inverse hyperbolic function.

60. $\cosh^{-1} x$
61. $\tanh^{-1} x$
62. $\coth^{-1} x$
63. $\operatorname{sech}^{-1} x$
64. $\operatorname{csch}^{-1} x$

SUPPLEMENTARY EXERCISES FOR CHAPTER 7

1. If $f(x) = (x - 1)/(x + 2)$, find $f^{-1}(x)$.

2. If $f(x)$ is invertible and $f(2) = 4$ while $f'(2) = 3$, find $(f^{-1})'(4)$.

3. (a) Is $\sin x$ an invertible function? Why?
(b) Is $\sin^{-1}x$ an invertible function? Why?

4. Show that $f(x) = x + \sin x$ is an invertible function.

5. What property must a graph of $y = f(x)$ satisfy if f is invertible?

6. Let f be an invertible function. Describe the domain and the range of f^{-1}.

7. Define what is meant by a one-to-one function.

8. Give an example of an invertible differentiable function whose inverse is not differentiable at some point.

9. (a) Give the definition of $\ln x$ as an integral.
(b) Sketch the graph of $\ln(x/2)$.

10. Show from the definition of $\ln x$ as an integral that, for each integer $n > 1$,

$$\frac{1}{2} + \frac{1}{3} + \cdots + \frac{1}{n} < \ln n < 1 + \frac{1}{2} + \frac{1}{3} + \cdots + \frac{1}{n-1}.$$

[*Hint:* Use the definition of $\ln n$ as an integral.]

11. (a) Define the number e in terms of an integral.
(b) Sketch the graph of e^{-x^2}.

12. Solve for x if $(2^x)(3^{x+1}) = 4^{-x}$ by changing all exponentials to base e.

13. Solve the equation in Exercise 12 by taking the natural logarithm of each side.

14. (a) Simplify $(1/\sqrt{e})^{\ln 5}$.
(b) Solve for x if $3^x/5^{x-1} = 20$.

15. (a) Solve for x if $\ln x - 2(\ln x^3) = \ln 10$.
(b) Simplify $25^{\log_5 7}$.

16. (a) Find dy/dx if $y = \ln[x^3(x+1)^2(\sin x)]$.
(b) Find $\int \cot 2x \, dx$.

17. (a) Find y' if $y = e^{\sin^{-1}x}$.
(b) Find $\int (\sec^2 3x) e^{\tan 3x} \, dx$.

18. Differentiate by changing to base e.
(a) 10^{x^2-x} (b) $(\cos x)^{\sin x}$

19. Differentiate by changing to base e.
(a) 2^{3x+4} (b) $x^{5x}, \quad x > 0$

20. Find the derivatives in Exercise 19 using logarithmic differentiation.

21. (a) Find y' if $y = 3^{\sin x}$ by changing to base e.
(b) Find y' if $y = (3^x \cdot 4^x)/5^x$ using logarithmic differentiation.

22. Using logarithmic differentiation, find y' if $y = (x \sin x^2)/(x + 1)^4$.

23. Let $y = e^{x \sin x}(\cos x)^x$. Find y' by logarithmic differentiation.

24. Use logarithmic differentiation to find dy/dx if $y = (\ln x)^x (x^{\sin x})$.

25. Find the equation of the tangent line to the graph of

$$y = \frac{(x^2 + 1)^3(x - 1)^4}{(3x - 1)^2\sqrt{5x + 1}}$$

at the point $(0, 1)$ using logarithmic differentiation.

26. If savings are compounded continuously at 5% interest, how long does it take for the savings to double?

27. The only force on a body traveling on a line is a force of resistance proportional to its velocity. If the velocity is 100 ft/sec at time $t = 0$ and 50 ft/sec at time $t = 10$, find
(a) the velocity of the body as a function of t,
(b) the distance traveled as a function of t.

28. Suppose $f'(t) = 3 \cdot f(t)$. Find $f(5)/f(2)$.

29. (a) Evaluate $\sin^{-1}(-1/2)$.
(b) Evaluate $\tan^{-1}(-\sqrt{3})$.

30. Evaluate
(a) $\cos^{-1}(\frac{1}{2})$,
(b) $\sec^{-1}(-2)$.

31. Evaluate
(a) $\tan^{-1}(-1)$,
(b) $\sin^{-1}(-1)$.

32. (a) Find $\cos^{-1}(-\sqrt{3}/2)$.
(b) Find $\lim_{x \to \infty} \tan^{-1}[(1 - x^2)/2]$.

33. Find dy/dx if $y = \tan^{-1}4x$.

34. Find dy/dx if $y = \sec^{-1}(x^2)$.

35. Find dy/dx if $y = \sin^{-1}(\sqrt{x})$.

36. Find dy/dx if $v = (\tan^{-1}2x)^3$

37. Find $\displaystyle\int_{-\sqrt{2}}^{\sqrt{2}} \frac{dx}{4 + x^2}$.

38. Find $\displaystyle\int_{\sqrt[4]{2}}^{\sqrt{2}} \frac{x \, dx}{x^2\sqrt{x^4 - 1}}$.

39. Evaluate $\displaystyle\int_{-1}^{\sqrt{3}} [1/(1 + x^2)]dx$.

40. Evaluate $\displaystyle\int_{-1/4}^{1/4} (1/\sqrt{1 - 4x^2}) \, dx$.

41. (a) Give the definition of $\sinh x$ in terms of the exponential function.
(b) Differentiate $\operatorname{sech}^3 2x$.

42. (a) Give the definition of $\cosh x$ in terms of the exponential function.
(b) Find $\displaystyle\int_0^1 \cosh x \, dx$ in terms of e.

43. (a) Sketch the graph of $y = \cosh(x + 1)$.
(b) Differentiate $\coth^3(2x + 1)$.

44. Find the expression for $\sinh^{-1}2x$ in terms of the logarithm function.

TECHNIQUES OF
INTEGRATION

8

We had some practice integrating in Chapters 5, 6, and 7, where we developed several integration formulas. To save turning pages, they are collected on page 383. We omit those involving the hyperbolic functions and their inverses. Formulas not mentioned before for $\int \sec x \, dx$ and $\int \csc x \, dx$ are included. These formulas are easily verified by differentiation.

One important method of integration is the use of a table of integrals. Proper use of a table of integrals is so important that we include a section on this topic at the end of this chapter. Very large computer programs have been developed to find formal derivatives and indefinite as well as definite integrals and to perform many other tasks of calculus. It may well be that such programs will be generally accessible in the near future and will largely replace the use of tables as an efficient and reliable way to find many indefinite integrals.

You should learn how to *transform* certain integrals that do not appear in a table into integrals that you know or that a table contains. The first three sections of this chapter give three such methods: integration by parts, partial fraction decomposition, and substitution techniques. Sections 8.4 and 8.5 describe how to integrate, without the use of tables, many frequently encountered functions. These sections continue to develop your facility in the general techniques presented in Sections 8.1, 8.2, and 8.3, especially the technique of substitution.

If you are unable to integrate a function to find a *definite* integral, it is easy to use Simpson's rule with calculators and computers. The importance of such a convenient numerical method of solution cannot be overemphasized.

8.1 INTEGRATION BY PARTS

For differentiation, certain rules and formulas enable us to compute derivatives of constant multiples, sums, products, and quotients of functions whose derivatives are known. While it is easy to compute antiderivatives of constant multiples and sums of functions with known antiderivatives, *there are no simple rules or formulas for integrating products or quotients of functions with known antiderivatives*. In this section, we present our only formula for integrating a product of functions. The technique is known as *integration by parts*. The formula is used not only as a tool for integration but also in some theoretical considerations.

Integration Formulas

1. $\displaystyle\int (u + v)\, dx = \int u\, dx + \int v\, dx + C$

2. $\displaystyle\int cu\, dx = c \int u\, dx + C$

3. $\displaystyle\int u^n\, du = \frac{u^{n+1}}{n+1} + C, \qquad n \neq -1$

4. $\displaystyle\int \frac{1}{u}\, du = \ln|u| + C$

5. $\displaystyle\int \frac{du}{\sqrt{a^2 - u^2}} = \sin^{-1}\left(\frac{u}{a}\right) + C$

6. $\displaystyle\int \frac{du}{a^2 + u^2} = \frac{1}{a}\tan^{-1}\left(\frac{u}{a}\right) + C$

7. $\displaystyle\int \frac{du}{u\sqrt{u^2 - a^2}} = \frac{1}{a}\sec^{-1}\left(\frac{u}{a}\right) + C$

8. $\displaystyle\int \sin u\, du = -\cos u + C$

9. $\displaystyle\int \cos u\, du = \sin u + C$

10. $\displaystyle\int \sec^2 u\, du = \tan u + C$

11. $\displaystyle\int \csc^2 u\, du = -\cot u + C$

12. $\displaystyle\int \sec u \tan u\, du = \sec u + C$

13. $\displaystyle\int \csc u \cot u\, du = -\csc u + C$

14. $\displaystyle\int \sec u\, du = \ln|\sec u + \tan u| + C$

15. $\displaystyle\int \csc u\, du = -\ln|\csc u + \cot u| + C$

16. $\displaystyle\int e^u\, du = e^u + C$

17. $\displaystyle\int a^u\, du = \frac{1}{\ln a}a^u + C$

The Formula

The formula for integration by parts is easily obtained from the formula for the derivative of a product. Recall that if f and g are differentiable functions, then their product is differentiable, and we have

$$\frac{d}{dx}[f(x)g(x)] = f(x)g'(x) + g(x)f'(x). \tag{1}$$

Taking the indefinite integral of both sides of Eq. (1), we obtain

$$\int \frac{d}{dx}[f(x)g(x)]\, dx = \int f(x)g'(x)\, dx + \int g(x)f'(x)\, dx,$$

so that

$$f(x)g(x) + C = \int f(x)g'(x)\, dx + \int g(x)f'(x)\, dx. \tag{2}$$

Thus

$$\int f(x)g'(x)\, dx = f(x)g(x) - \int g(x)f'(x)\, dx. \tag{3}$$

It is not necessary to include specifically an arbitrary constant C in Eq. (3) since both sides of the equation contain an indefinite integral. If we let $u = f(x)$ and $v = g(x)$ so that $du = f'(x)\, dx$ and $dv = g'(x)\, dx$, then Eq. (3) becomes

$$\int u\, dv = uv - \int v\, du, \tag{4}$$

which is the formula for *integration by parts*.

The Technique

Suppose we must integrate a product of two functions, that is, we must find

$$\int f(x) \cdot g(x)\, dx = \int \underbrace{g(x)}_{u} \cdot \underbrace{(f(x)\, dx)}_{dv} = \int \underbrace{f(x)}_{u} \cdot \underbrace{(g(x)\, dx)}_{dv}. \tag{5}$$

Suppose that F and G are antiderivatives of f and g, respectively, so that

$$F'(x) = f(x) \qquad \text{and} \qquad G'(x) = g(x).$$

Then formula (4) allows us to transform the integral (5) as follows.

Integration by Parts

The problem of finding $\int f(x) \cdot g(x)\, dx$ may be reduced to finding

$$\int F(x) \cdot g'(x)\, dx \qquad \text{or} \qquad \int f'(x) \cdot G(x)\, dx. \tag{6}$$

That is, we can change the problem of integrating a product to the problem of integrating the derivative of one factor times an antiderivative of the other.

EXAMPLE 1 Find $\int x \sin x \, dx$.

Solution The two factors are x and $\sin x$ and, according to integrals (6), we can reduce the problem to finding either

$$\int \frac{x^2}{2}(\cos x) \, dx \qquad \text{or} \qquad \int 1(-\cos x) \, dx.$$

The first of these integrals is more difficult to deal with than the original problem, but the second one is easy. Accordingly, we wish to differentiate x and integrate $\sin x$. Using formula (4), we write out the substitution as

$$u = x \qquad dv = \sin x \, dx$$
$$du = dx \qquad v = -\cos x.$$

Formula (4) tells us the following:

> The integral of the product $u \, dv$ in the top row of the array equals the product uv of the functions at the ends of the dashed diagonal minus the integral of the product $v \, du$ in the bottom row.

Thus we obtain

$$\int x \sin x \, dx = x(-\cos x) - \int (-\cos x) \, dx$$

$$= -x \cos x + \int \cos x \, dx = -x \cos x + \sin x + C. \qquad \blacksquare$$

EXAMPLE 2 Find $\int xe^{3x} \, dx$.

Solution We wish the first factor x were not present. It would be changed to a factor 1 by differentiation. Thus we make the substitution

$$u = x \qquad dv = e^{3x} \, dx$$
$$du = dx \qquad v = \frac{1}{3}e^{3x}.$$

Using formula (4), we find that

$$\int xe^{3x} \, dx = x \cdot \frac{1}{3}e^{3x} - \int \frac{1}{3}e^{3x} \, dx = \frac{1}{3}xe^{3x} - \frac{1}{9}e^{3x} + C. \qquad \blacksquare$$

EXAMPLE 3 Find $\int x^3 \sqrt{1 + x^2} \, dx$.

Solution We don't know how to integrate $\sqrt{1 + x^2}$, and if we try differentiating $\sqrt{1 + x^2}$ and integrating x^3 using formula (4), we only make things worse. We borrow one x-factor of x^3 to put with $\sqrt{1 + x^2}$, for we can integrate $x\sqrt{1 + x^2}$. This leads to the substitution

$$u = x^2 \qquad dv = x\sqrt{1 + x^2} \, dx$$
$$du = 2x \, dx \qquad v = \frac{1}{3}(1 + x^2)^{3/2}.$$

Formula (4) shows that

$$\int x^3\sqrt{1 + x^2}\, dx = x^2 \cdot \frac{1}{3}(1 + x^2)^{3/2} - \frac{1}{3}\int 2x(1 + x^2)^{3/2}\, dx$$

$$= \frac{x^2}{3}(1 + x^2)^{3/2} - \frac{1}{3} \cdot \frac{2}{5}(1 + x^2)^{5/2} + C$$

$$= \frac{x^2}{3}(1 + x^2)^{3/2} - \frac{2}{15}(1 + x^2)^{5/2} + C.$$ ∎

EXAMPLE 4 Compute $\int x^2 e^x\, dx$.

Solution Integration by parts of the product of x^2 and e^x leads to the computation of either $\int 2xe^x\, dx$ or $\int (x^3/3)e^x\, dx$. The former integral $\int 2xe^x\, dx$ is simpler than our original integral, and this new integral can obviously be further reduced to $\int 2e^x\, dx$ by an additional application of the formula for integration by parts. Thus we make the substitution

$$u = x^2 \qquad dv = e^x\, dx$$
$$du = 2x\, dx \qquad v = e^x$$

and obtain

$$\int x^2 e^x\, dx = x^2 e^x - \int 2xe^x\, dx. \tag{7}$$

To compute the remaining integral $\int 2xe^x\, dx$, we integrate by parts again. (Do not be confused by the use of u again for the function $2x$ rather than x^2; it is customary in this "looping" process.) Our substitution this time is

$$u = 2x \qquad dv = e^x\, dx$$
$$du = 2\, dx \qquad v = e^x$$

and Eq. (7) yields

$$\int x^2 e^x\, dx = x^2 e^x - \int 2xe^x\, dx = x^2 e^x - \left(2xe^x - \int 2e^x\, dx\right)$$

$$= x^2 e^x - 2xe^x + 2e^x + C.$$ ∎

Any function can be regarded as the product of itself and 1. Thus for $\int f(x)\, dx$ we can always try the substitution

$$u = f(x) \qquad dv = 1\, dx$$
$$du = f'(x)\, dx \qquad v = x.$$

EXAMPLE 5 Find $\int \ln x\, dx$ using integration by parts.

Solution The substitution

$$u = \ln x \qquad dv = 1\, dx$$
$$du = \frac{1}{x}\, dx \qquad v = x$$

yields

$$\int \ln x \; dx = x \ln x - \int x \cdot \frac{1}{x} \; dx$$

$$= x \ln x - \int 1 \; dx = x \ln x - x + C.$$ ∎

EXAMPLE 6 The inverse trigonometric functions can be integrated by parts. Find $\int \sin^{-1}x \; dx$.

Solution We set

$$u = \sin^{-1}x \qquad\qquad dv = 1 \; dx$$
$$du = \frac{1}{\sqrt{1-x^2}} \; dx \qquad v = x.$$

Then

$$\int \sin^{-1}x \; dx = x \sin^{-1}x - \int \frac{x}{\sqrt{1-x^2}} \; dx$$

$$= x \sin^{-1}x + \frac{1}{2} \int -2x(1-x^2)^{-1/2} \; dx$$

$$= x \sin^{-1}x + \frac{1}{2} \cdot \frac{(1-x^2)^{1/2}}{\frac{1}{2}} + C$$

$$= x \sin^{-1}x + \sqrt{1-x^2} + C.$$ ∎

EXAMPLE 7 Compute $\int e^x \sin x \; dx$.

Solution This time, both substitutions for integration by parts lead to $\int e^x \cos x \; dx$, which seems to be just as hard as the original problem. The following is an effective trick. We make the substitution

$$u = e^x \qquad\qquad dv = \sin x \; dx$$
$$du = e^x \; dx \qquad v = -\cos x$$

and obtain

$$\int e^x \sin x \; dx = -e^x \cos x + \int e^x \cos x \; dx.$$ (8)

We then integrate by parts again, *continuing to let u be the exponential term,* so that

$$u = e^x \qquad\qquad dv = \cos x \; dx$$
$$du = e^x \; dx \qquad v = \sin x.$$

We obtain from Eq. (8)

$$\int e^x \sin x \; dx = -e^x \cos x + \left(e^x \sin x - \int e^x \sin x \; dx \right)$$

$$= (e^x \sin x - e^x \cos x) - \int e^x \sin x \; dx.$$ (9)

We may now solve Eq. (9) for $\int e^x \sin x \, dx$, and we find that

$$\int e^x \sin x \, dx = \frac{1}{2}(e^x \sin x - e^x \cos x) + C.$$

(If we had let $u = \cos x$ and $dv = e^x \, dx$ for the second integration by parts, we would have obtained the identity $\int e^x \sin x \, dx = \int e^x \sin x \, dx$.) ∎

The technique illustrated in Example 7 involves finding an equation in which the integral is regarded as an unknown and then solving for the integral. We give another illustration of this technique.

EXAMPLE 8   Find $\int \sec^3 ax \, dx$.

Solution   Since we can integrate $\sec^2 ax$, we try the substitution

$$u = \sec ax \qquad\qquad dv = \sec^2 ax \, dx$$

$$du = a \sec ax \tan ax \, dx \qquad v = \frac{1}{a}\tan ax.$$

We obtain

$$\int \sec^3 ax \, dx = \frac{1}{a}\sec ax \tan ax - \int \sec ax \tan^2 ax \, dx.$$

Using the identity $\tan^2 ax = \sec^2 ax - 1$ and the integration formula (14) on page 383, we obtain

$$\int \sec^3 ax \, dx = \frac{1}{a}\sec ax \tan ax - \int (\sec ax)(\sec^2 ax - 1) \, dx$$

$$= \frac{1}{a}\sec ax \tan ax - \int \sec^3 ax \, dx + \int \sec ax \, dx$$

$$= \frac{1}{a}\sec ax \tan ax - \int \sec^3 ax \, dx + \frac{1}{a} \ln|\sec ax + \tan ax|.$$

Solving for $\int \sec^3 ax \, dx$, we have

$$2 \int \sec^3 ax \, dx = \frac{1}{a}\sec ax \tan ax + \frac{1}{a} \ln|\sec ax + \tan ax| + C,$$

so

$$\int \sec^3 ax \, dx = \frac{1}{2a}(\sec ax \tan ax + \ln|\sec ax + \tan ax|) + C.$$ ∎

EXAMPLE 9   Integration by parts is a basic technique for finding *reduction formulas* such as

$$\int \sin^n x \, dx = -\frac{1}{n}\sin^{n-1} x \cos x + \frac{n-1}{n} \int \sin^{n-2} x \, dx \qquad \text{for } n \geq 2,$$

which reduce an integral to a less complicated one. Derive this reduction formula.

Solution   We form the substitution

$$u = \sin^{n-1}x \qquad\qquad dv = \sin x \, dx$$
$$du = (n-1)\sin^{n-2}x \cos x \, dx \qquad v = -\cos x,$$

which yields

$$\int \sin^n x \, dx = -\sin^{n-1}x \cos x - \int (n-1)(\sin^{n-2}x)(-\cos^2 x) \, dx. \qquad \textbf{(10)}$$

From Eq. (10) and the identity $-\cos^2 x = \sin^2 x - 1$, we have

$$\int \sin^n x \, dx = -\sin^{n-1}x \cos x - \int (n-1)(\sin^{n-2}x)(\sin^2 x - 1) \, dx$$

$$= -\sin^{n-1}x \cos x - (n-1) \int \sin^n x \, dx$$

$$+ (n-1) \int \sin^{n-2}x \, dx. \qquad \textbf{(11)}$$

From Eq. (11), we obtain

$$n \int \sin^n x \, dx = -\sin^{n-1}x \cos x + (n-1) \int \sin^{n-2}x \, dx.$$

On division by n, we obtain the reduction formula for $\int \sin^n x \, dx$.   ∎

When integrating by parts, we have to decide what part of the integrand to call u and what part to call dv. Sometimes we wish to lower a power of x or to change an integrand involving a logarithm or inverse trigonometric function to an algebraic expression. The preceding examples indicate that we should then let u be the power of x or the logarithm or inverse trigonometric function.

SUMMARY

1. Integration by parts is used for finding $\int f(x) \cdot g(x) \, dx$ when the product of the derivative of one of the functions and an antiderivative of the other gives an easier integration problem.

2. If we make the substitution

$$u = f(x) \qquad\qquad dv = g(x) \, dx$$
$$du = f'(x) \, dx \qquad v = \int g(x) \, dx,$$

then $\int f(x) \cdot g(x) \, dx$ equals the product uv of the functions at the ends of the dashed diagonal minus the integral of the product $v \, du$ of the terms in the bottom row. In symbols,

$$\int u \, dv = uv - \int v \, du.$$

EXERCISES 8.1

In Exercises 1–46, find the indicated integral or formula without the use of tables.

1. $\int x \cos x \, dx$

2. $\int x \sin 5x \, dx$

3. $\int x e^{3x} \, dx$

4. $\int x^3 \cos x^2 \, dx$

5. $\int x^3 \sin 2x^2 \, dx$

6. $\int x^2 e^{-x^3} \, dx$

7. $\int x^2 \sin x \, dx$

8. $\int x^3 e^x \, dx$

9. $\int x^3 e^{x^2} \, dx$

10. $\int x^5 \sin x^3 \, dx$

11. $\int x \tan^2 x \, dx$

12. $\int x \csc^2 3x \, dx$

13. $\int x \sec^2 2x \, dx$

14. $\int x^3 \sqrt{4 - x^2} \, dx$

15. $\int \frac{x^3}{\sqrt{1 + 2x^2}} \, dx$

16. $\int \frac{x^3}{\sqrt[3]{9 - x^2}} \, dx$

17. $\int x^5 \sqrt{1 + x^2} \, dx$

18. $\int \frac{x^2}{(1 + x^2)^2} \, dx$

19. $\int \frac{x^3}{(1 + x^2)^2} \, dx$

20. $\int \ln x^2 \, dx$

21. $\int x(\ln x) \, dx$

22. $\int_1^e x^3(\ln x) \, dx$

23. $\int \ln(1 + x) \, dx$

24. $\int (\ln x)^2 \, dx$

25. $\int \ln(1 + x^2) \, dx$

26. $\int_0^{1/2} \sin^{-1}(2x) \, dx$

27. $\int \cos^{-1} x \, dx$

28. $\int \cot^{-1} 5x \, dx$

29. $\int \tan^{-1} 3x \, dx$

30. $\int x \sec^{-1} x \, dx$

31. $\int x \tan^{-1} x \, dx$

32. $\int x \cot^{-1} 3x \, dx$

33. $\int x^2 \sin^{-1} x \, dx$

34. $\int \sin(\ln x) \, dx$

35. $\int \cos(\ln x) \, dx$

36. $\int e^{ax} \cos bx \, dx$

37. $\int e^{ax} \sin bx \, dx$

38. A reduction formula for $\int (\ln ax)^n \, dx$

39. A formula for $\int \sqrt{ax + b}/x^2 \, dx$ in terms of
$$\int [1/(x\sqrt{ax + b})] \, dx$$

40. A reduction formula for $\int x^n e^{ax} \, dx$

41. A reduction formula for $\int x^n \sin ax \, dx$

42. A reduction formula for $\int x^n \cos ax \, dx$

43. A reduction formula for $\int \cos^n ax \, dx$

44. A reduction formula for $\int \sin^n ax \cos^m ax \, dx$, which reduces the power of $\sin ax$ by 2

45. A reduction formula for $\int \tan^n ax \, dx$

46. A reduction formula for $\int \sec^n ax \, dx$

Work Exercises 47–55 without using tables.

47. Find the area of the region bounded by $y = x$ and $y = x \sin x$ for $0 \le x \le \pi/2$.

48. Find the area of the region bounded by $y = \ln x^2$, $y = 0$, and $x = e$.

49. Find the area of the region bounded by $y = 1 - \cos \sqrt{x}$, $x = \pi^2$, and the x-axis for $0 \le x \le \pi^2$. [*Hint:* Note $dx = 2\sqrt{x} \, d(\sqrt{x})$.]

50. Find the volume of the solid generated by revolving the plane region bounded by $y = \sin x$ and $y = 0$ for $0 \le x \le \pi$ about the y-axis.

51. Find the volume of the solid generated by revolving the plane region bounded by $y = xe^x$, $y = 0$, and $x = 1$ about the y-axis.

52. Find the volume of the solid generated by revolving the plane region bounded by $y = \ln x$, $y = 0$, and $x = e$ about the x-axis.

53. The velocity at time t of a body moving on the x-axis is given by $v = t \cos t$. If the body is at $x = 3$ when $t = 0$, find its position on the x-axis when $t = 3\pi$.

54. The moment of inertia I of a particle of mass m about an axis is the product of m and the square of the distance from the particle to the axis. A thin plate of constant mass density (mass per unit area) k covers the region bounded by $y = \sin x$ and $y = 0$ for $0 \le x \le \pi$. Find the moment of inertia of the plate about the y-axis.

55. Repeat Exercise 54 if the area mass density of the plate at (x, y) is $k(1 + x)$.

8.2 INTEGRATION OF RATIONAL FUNCTIONS

We would not find a formula for

$$\int \frac{x^6 - 2}{x^4 + x^2} \, dx$$

in most tables. This section shows how this integral, and any other integral of a rational function (a quotient of polynomial functions), can be transformed into a sum of integrals that we know how to find or that are given in many tables. The complete technique involves four steps, which we state here in abbreviated form.

Technique for Integrating Rational Functions

STEP 1. Perform long division if necessary.

STEP 2. Factor the denominator into irreducible polynomials.

STEP 3. Find the partial fraction decomposition.

STEP 4. Integrate the result of step 3.

Each step is explained in detail in a separate subsection.

Step 1: Perform Long Division if Necessary

We wish to find

$$\int \frac{f(x)}{g(x)} \, dx,$$

where $f(x)$ and $g(x)$ are polynomial functions. If the polynomial $f(x)$ in the numerator has degree greater than or equal to the degree of the denominator polynomial $g(x)$, perform polynomial long division and write

$$\frac{f(x)}{g(x)} = q(x) + \frac{r(x)}{g(x)} \tag{1}$$

for polynomials $q(x)$ and $r(x)$, where the degree of $r(x)$ is less than the degree of $g(x)$. Since the polynomial function $q(x)$ is readily integrated, we have essentially reduced the problem of integrating $f(x)/g(x)$ to a problem of integrating $r(x)/g(x)$.

EXAMPLE 1 Do step 1 for

$$\int \frac{x^6 - 2}{x^4 + x^2} \, dx.$$

Solution Polynomial long division yields

$$\begin{array}{r}
x^2 - 1 \\
x^4 + x^2 \overline{\smash{\big)}\ x^6 \qquad\qquad -\ 2} \\
\underline{x^6 + x^4} \\
-\ x^4 \\
\underline{-\ x^4 - x^2} \\
x^2 - 2.
\end{array}$$

Thus

$$\int \frac{x^6 - 2}{x^4 + x^2}\, dx = \int \left(x^2 - 1 + \frac{x^2 - 2}{x^4 + x^2} \right) dx. \qquad \blacksquare$$

EXAMPLE 2 Do step 1 for

$$\int \frac{x^2 + x - 3}{(x - 2)(x^2 - 1)}\, dx.$$

Solution Since the degree 2 of the numerator polynomial is less than the degree 3 of the denominator polynomial, no long division is necessary. $\qquad \blacksquare$

Step 2: Factor the Denominator into Irreducible Factors

For step 2, we factor the denominator polynomial $g(x)$ in Eq. (1) into a product of (possibly repeated) first-degree and irreducible quadratic factors, so that the quadratic factors cannot be factored further into a product of real first-degree factors. *It is a theorem of algebra that such a factorization exists.*

EXAMPLE 3 Continuing with the integral in Example 1, do step 2 for

$$\int \left(x^2 - 1 + \frac{x^2 - 2}{x^4 + x^2} \right) dx.$$

Solution The factorization of the denominator $x^4 + x^2$ into irreducible first-degree and quadratic factors is given by $x^4 + x^2 = x \cdot x(x^2 + 1)$. Example 5 describes a test that can be used to establish that $x^2 + 1$ cannot be factored further, but you probably know this already. $\qquad \blacksquare$

EXAMPLE 4 Continuing with the integral in Example 2, do step 2 for

$$\int \frac{x^2 + x - 3}{(x - 1)(x^2 - 1)}\, dx.$$

Solution The factorization of $(x - 2)(x^2 - 1)$ into irreducible factors is
$$(x - 2)(x^2 - 1) = (x - 2)(x - 1)(x + 1). \qquad \blacksquare$$

EXAMPLE 5 Factor $x^3 + 1$ into irreducible factors.

Solution As you learned in algebra, if $x = a$ is a root of a polynomial equation $g(x) = 0$, then $x - a$ is a factor of $g(x)$. Since $x = -1$ is a root of $x^3 + 1 = 0$, we see that $x - (-1) = x + 1$ is a factor of $x^3 + 1$. The remaining factor can be found by long division:

$$
\begin{array}{r}
x^2 - x + 1 \\
x + 1 \overline{\smash{\big)}\ x^3 \qquad\qquad + 1} \\
\underline{x^3 + x^2} \\
- x^2 \\
\underline{- x^2 - x} \\
x + 1 \\
\underline{x + 1} \\
0
\end{array}
$$

Thus

$$x^3 + 1 = (x + 1)(x^2 - x + 1). \tag{2}$$

We now have to decide whether $x^2 - x + 1$ is irreducible. A quadratic polynomial $ax^2 + bx + c$ is irreducible precisely when $ax^2 + bx + c = 0$ has no real roots. The quadratic formula shows that this is the case precisely when $b^2 - 4ac < 0$. For $x^2 - x + 1$, we have

$$b^2 - 4ac = (-1)^2 - 4 \cdot 1 \cdot 1 = -3 < 0,$$

so $x^2 - x + 1$ is irreducible. Thus Eq. (2) is the factorization of $x^3 + 1$ into irreducible factors. ∎

Step 3:   Find the Partial Fraction Decomposition

It is a theorem of algebra (whose proof we will not attempt) that $r(x)/g(x)$ in Eq. (1) can be written as a sum of *partial fractions,* which arise from the factors of $g(x)$ as we now describe. A first-degree factor $ax + b$ of $g(x)$ of multiplicity 1 gives rise to a single term

$$\frac{A}{ax + b} \tag{3}$$

for some constant A, while such a factor with multiplicity n, say $(ax + b)^n$, gives rise to a sum of such terms, one for each power of the factor up to the nth power. For example, a multiple factor $(ax + b)^4$ gives rise to the sum

$$\frac{A}{(ax + b)^4} + \frac{B}{(ax + b)^3} + \frac{C}{(ax + b)^2} + \frac{D}{ax + b} \tag{4}$$

for some constants A, B, C, and D.

An irreducible quadratic factor $ax^2 + bx + c$ of $g(x)$ of multiplicity 1 gives rise to a single term

$$\frac{Ax + B}{ax^2 + bx + c} \tag{5}$$

for constants A and B, while such a factor of multiplicity n gives rise to a sum of such terms, one for each power of the factor up to the nth power. For example, a multiple factor $(ax^2 + bx + c)^3$ gives rise to the sum

$$\frac{Ax + B}{(ax^2 + bx + c)^3} + \frac{Cx + D}{(ax^2 + bx + c)^2} + \frac{Ex + F}{ax^2 + bx + c} \tag{6}$$

for some constants A, B, C, D, E, and F.

EXAMPLE 6   Continuing with the integral in Examples 1 and 3, do step 3 for

$$\int \left[x^2 - 1 + \frac{x^2 - 2}{x^2(x^2 + 1)} \right] dx.$$

Solution   As Eqs. (4) and (5) indicate, there should be constants A, B, C, and D such that

$$\frac{x^2 - 2}{x^2(x^2 + 1)} = \frac{A}{x^2} + \frac{B}{x} + \frac{Cx + D}{x^2 + 1}. \tag{7}$$

Adding the quotients on the right-hand side of Eq. (7), we see that

$$\frac{x^2 - 2}{x^2(x^2 + 1)} = \frac{A(x^2 + 1) + Bx(x^2 + 1) + (Cx + D)x^2}{x^2(x^2 + 1)}. \tag{8}$$

The two numerators in Eq. (8) must be equal, and we form the

Numerator Equation

$$A(x^2 + 1) + Bx(x^2 + 1) + (Cx + D)x^2 = x^2 - 2. \tag{9}$$

We must find constants A, B, C, and D so that the numerator equation (9) holds identically for all x. If a first-degree factor appears in our factorization of the denominator, then we can find one unknown constant easily by setting x in Eq. (9) equal to the value that makes that factor zero. Thus setting $x = 0$, which makes the first-degree factor x of the denominator in Eq. (8) equal to zero, we obtain from Eq. (9)

$$A = -2. \qquad \textit{Setting } x = 0$$

Since we have no more *first-degree* factors of the denominator, we must find the values for B, C, and D by another method. These remaining constants are found by equating coefficients of like powers of x on the two sides of the numerator equation (9). We proceed:

$$B = 0, \qquad\qquad\qquad \textit{Equating x-coefficients}$$
$$A + D = 1, \quad D = 1 - A = 1 + 2 = 3, \qquad \textit{Equating } x^2\textit{-coefficients}$$
$$B + C = 0, \quad C = -B = 0. \qquad\qquad \textit{Equating } x^3\textit{-coefficients}$$

Thus, using Eq. (7), we have

$$\int \left[x^2 - 1 + \frac{x^2 - 2}{x^2(x^2 + 1)} \right] dx = \int \left(x^2 - 1 + \frac{-2}{x^2} + \frac{0}{x} + \frac{0x + 3}{x^2 + 1} \right) dx$$

$$= \int \left(x^2 - 1 - \frac{2}{x^2} + \frac{3}{x^2 + 1} \right) dx. \qquad \blacksquare$$

EXAMPLE 7   Continuing with the integral in Examples 2 and 4, do step 3 for

$$\int \frac{x^2 + x - 3}{(x - 2)(x + 1)(x - 1)} dx.$$

Solution   From formula (3), we have

$$\frac{x^2 + x - 3}{(x - 2)(x + 1)(x - 1)} = \frac{A}{x - 2} + \frac{B}{x + 1} + \frac{C}{x - 1}$$

$$= \frac{A(x + 1)(x - 1) + B(x - 2)(x - 1) + C(x - 2)(x + 1)}{(x - 2)(x + 1)(x - 1)}$$

for some constants A, B, and C. Equating numerators yields the

---

Numerator equation

$$A(x + 1)(x - 1) + B(x - 2)(x - 1) + C(x - 2)(x + 1)$$
$$= x^2 + x - 3. \qquad (10)$$

If we set x equal to each value that makes a linear factor zero in Eq. (10), we obtain

$$6B = -3, \qquad B = -\tfrac{1}{2}, \qquad \textit{Setting } x = -1$$
$$-2C = -1, \qquad C = \tfrac{1}{2}, \qquad \textit{Setting } x = 1$$
$$3A = 3, \qquad A = 1. \qquad \textit{Setting } x = 2$$

Thus

$$\int \frac{x^2 + x - 3}{(x - 2)(x + 1)(x - 1)} \, dx = \int \left(\frac{1}{x - 2} - \frac{\tfrac{1}{2}}{x + 1} + \frac{\tfrac{1}{2}}{x - 1} \right) dx. \quad \blacksquare$$

Step 4:   Perform the Integration

EXAMPLE 8   To conclude the work in Examples 1, 3, and 6, we perform step 4 to find

$$\int \left(x^2 - 1 - \frac{2}{x^2} + \frac{3}{x^2 + 1} \right) dx.$$

Solution   The desired integral is

$$\frac{x^3}{3} - x + \frac{2}{x} + 3 \tan^{-1} x + C.$$

Thus, referring back to Example 1, we see that

$$\int \frac{x^6 - 2}{x^4 + x^2} \, dx = \frac{x^3}{3} - x + \frac{2}{x} + 3 \tan^{-1} x + C. \quad \blacksquare$$

EXAMPLE 9   To conclude the work in Examples 2, 4, and 7, we perform step 4 to find

$$\int \left(\frac{1}{x - 2} - \frac{\tfrac{1}{2}}{x + 1} + \frac{\tfrac{1}{2}}{x - 1} \right) dx.$$

Solution   The desired integral is

$$\ln|x - 2| - \frac{1}{2}\ln|x + 1| + \frac{1}{2}\ln|x - 1| + C$$

$$= \ln|x - 2| + \frac{1}{2}\ln\left|\frac{x - 1}{x + 1}\right| + C.$$

Referring back to Example 2, we see that

$$\int \frac{x^2 + x - 3}{(x - 2)(x^2 - 1)}\,dx = \ln|x - 2| + \frac{1}{2}\ln\left|\frac{x - 1}{x + 1}\right| + C. \qquad \blacksquare$$

In this final step of the process, we may have to find an integral of the form

$$\int \frac{A}{(ax + b)^n}\,dx$$

or

$$\int \frac{Bx + C}{(ax^2 + bx + c)^n}\,dx$$

for some integer value of $n \geq 1$. Here is a complete list of the integration formulas we may need for step 4.

Basic Formulas for Integration of Rational Functions

$$\int \frac{1}{ax + b}\,dx = \frac{1}{a}\ln|ax + b| + C \tag{11}$$

$$\int \frac{1}{(ax + b)^n}\,dx = \frac{-1}{a(n - 1)} \cdot \frac{1}{(ax + b)^{n-1}} + C, \qquad n \neq 1 \tag{12}$$

$$\int \frac{1}{ax^2 + bx + c}\,dx = \frac{2}{\sqrt{4ac - b^2}}\tan^{-1}\frac{2ax + b}{\sqrt{4ac - b^2}} + C, \tag{13}$$
$$b^2 < 4ac \qquad New$$

$$\int \frac{2ax + b}{ax^2 + bx + c}\,dx = \ln|ax^2 + bx + c| + C \tag{14}$$

$$\int \frac{1}{(ax^2 + bx + c)^{n+1}}\,dx = \frac{2ax + b}{n(4ac - b^2)(ax^2 + bx + c)^n}$$
$$+ \frac{2(2n - 1)a}{n(4ac - b^2)}\int \frac{1}{(ax^2 + bx + c)^n}\,dx, \tag{15}$$
$$4ac \neq b^2 \qquad New$$

$$\int \frac{2ax + b}{(ax^2 + bx + c)^n}\,dx = \frac{-1}{(n - 1)(ax^2 + bx + c)^{n-1}} + C, \tag{16}$$
$$n \neq 1$$

Formulas (11) and (14) are readily obtained from the familiar formula

$$\int \frac{du}{u} = \ln|u| + C, \tag{17}$$

and we should not need to look up formula (11) or (14). For example, using formula (17), we have

$$\int \frac{2}{3x + 5} \, dx = \frac{2}{3} \int \frac{3}{3x + 5} \, dx = \frac{2}{3} \ln|3x + 5| + C.$$

Formulas (12) and (16) are instances of the familiar formula

$$\int u^n \, du = \frac{u^{n+1}}{n + 1} + C \qquad \text{for } n \neq -1, \tag{18}$$

and we should not need to refer to tables for formula (12) or (16) either. This leaves just formula (13) and the reduction formula (15), which we may wish to look up when we need them. Example 10 illustrates the use of formulas (13) and (15).

EXAMPLE 10   Find

$$\int \frac{x - 2}{(x^2 - 5x + 7)^2} \, dx.$$

Solution   We first "fix up" the integral so that part of the numerator is the derivative $2x - 5$ of the factor $x^2 - 5x + 7$ in the denominator, and we can use formula (16). We have

$$\int \frac{x - 2}{(x^2 - 5x + 7)^2} \, dx = \frac{1}{2} \int \frac{2x - 4}{(x^2 - 5x + 7)^2} \, dx$$

$$= \frac{1}{2} \int \frac{(2x - 5) + 1}{(x^2 - 5x + 7)^2} \, dx$$

$$= \frac{1}{2} \int \frac{2x - 5}{(x^2 - 5x + 7)^2} \, dx$$

$$+ \frac{1}{2} \int \frac{1}{(x^2 - 5x + 7)^2} \, dx$$

$$= \frac{1}{2} \cdot \frac{-1}{(x^2 - 5x + 7)} + \frac{1}{2} \int \frac{1}{(x^2 - 5x + 7)^2} \, dx.$$

Using formulas (15) and (13), we have

$$\frac{1}{2} \int \frac{1}{(x^2 - 5x + 7)^2} \, dx = \frac{1}{2} \cdot \frac{2x - 5}{1 \cdot 3(x^2 - 5x + 7)}$$

$$+ \frac{1}{2} \cdot \frac{2 \cdot 1 \cdot 1}{1 \cdot 3} \int \frac{1}{x^2 - 5x + 7} \, dx$$

$$= \frac{2x - 5}{6x^2 - 30x + 42}$$

$$+ \frac{1}{3} \cdot \frac{2}{\sqrt{3}} \tan^{-1}\left(\frac{2x - 5}{\sqrt{3}}\right) + C.$$

Putting everything together, we have

$$\int \frac{x - 2}{(x^2 - 5x + 7)^2} \, dx = \frac{-1}{2x^2 - 10x + 14} + \frac{2x - 5}{6x^2 - 30x + 42}$$

$$+ \frac{2}{3\sqrt{3}} \tan^{-1}\left(\frac{2x - 5}{\sqrt{3}}\right) + C. \qquad \blacksquare$$

Comments and Additional Examples

We should mention that the actual *execution* of the technique described in this section may break down at step 2, factoring the denominator. While our algebraic theory assures us that the denominator polynomial $g(x)$ in Eq. (1) does have a factorization into first-degree and irreducible quadratic factors, *finding* such a factorization may be a very tough job. Roots of $g(x) = 0$, and hence *first-degree* factors of $g(x)$, can be approximated by Newton's method. But factorization is the only problem we might encounter in using this technique; all the other steps are mechanical chores.

EXAMPLE 11   Find

$$\int \frac{13 - 7x}{(x + 2)(x - 1)^3} \, dx.$$

Solution   Since the degree of the numerator is less than that of the denominator, we do not need to perform long division as in step 1, and the denominator is already factored for step 2. For the partial fraction decomposition (step 3), we let

$$\frac{13 - 7x}{(x + 2)(x - 1)^3} = \frac{A}{x + 2} + \frac{B}{(x - 1)^3} + \frac{C}{(x - 1)^2} + \frac{D}{x - 1}.$$

The numerator equation is

$$A(x - 1)^3 + B(x + 2) + C(x + 2)(x - 1) + D(x + 2)(x - 1)^2 \qquad \text{(19)}$$
$$= 13 - 7x.$$

We find as many of the unknown constants as possible using the zeros, -2 and 1, of the linear factors in the denominator. We have, from Eq. (19),

$$-27A = 27, \quad A = -1, \qquad \textit{Setting } x = -2$$
$$3B = 6, \quad B = 2. \qquad \textit{Setting } x = 1$$

To find C and D, we need two equations containing them found by equating coefficients. The easiest equations to find are often those corresponding to the terms of highest degree and to the constant terms. Computing the constant terms amounts to setting $x = 0$. We obtain, from Eq. (19),

$$A + D = 0, \quad D = -A = 1, \qquad \textit{Equating } x^3\textit{-coefficients}$$
$$-A + 2B - 2C + 2D = 13, \qquad \textit{Setting } x = 0$$
$$2C = -A + 2B + 2D - 13$$
$$= 1 + 4 + 2 - 13$$
$$= -6,$$
$$C = -3.$$

Hence

$$\int \frac{13 - 7x}{(x + 2)(x - 1)^3} \, dx = \int \left[\frac{-1}{x + 2} + \frac{2}{(x - 1)^3} + \frac{-3}{(x - 1)^2} + \frac{1}{x - 1} \right] dx$$

$$= -\ln|x + 2| - \frac{1}{(x - 1)^2} + \frac{3}{x - 1} + \ln|x - 1| + C. \quad \blacksquare$$

EXAMPLE 12 Find

$$\int \frac{3x^4 + 2x^3 + 8x^2 + x + 2}{x^5 + 2x^3 + x} \, dx.$$

Solution Long division is unnecessary, and factoring the denominator yields

$$x^5 + 2x^3 + x = x(x^2 + 1)^2.$$

For the partial fraction decomposition, we let

$$\frac{3x^4 + 2x^3 + 8x^2 + x + 2}{x^5 + 2x^3 + x} = \frac{A}{x} + \frac{Bx + C}{(x^2 + 1)^2} + \frac{Dx + E}{x^2 + 1}.$$

The numerator equation is

$$A(x^2 + 1)^2 + (Bx + C)x + (Dx + E)(x^3 + x)$$
$$= 3x^4 + 2x^3 + 8x^2 + x + 2. \quad \textbf{(20)}$$

Working to find the constants, we have, from Eq. (20),

$$A = 2, \qquad\qquad\qquad \textit{Setting } x = 0$$
$$A + D = 3, \quad D = 3 - A = 3 - 2 = 1, \qquad \textit{Equating } x^4\textit{-coefficients}$$
$$E = 2 \qquad\qquad\qquad \textit{Equating } x^3\textit{-coefficients}$$
$$2A + B + D = 8, \quad B = 8 - 2A - D$$
$$= 8 - 4 - 1 = 3, \qquad \textit{Equating } x^2\textit{-coefficients}$$
$$C + E = 1, \quad C = 1 - E = 1 - 2 = -1. \qquad \textit{Equating } x\textit{-coefficients}$$

Thus we have

$$\int \frac{3x^4 + 2x^3 + 8x^2 + x + 2}{x(x^2 + 1)^2} \, dx = \int \left[\frac{2}{x} + \frac{3x - 1}{(x^2 + 1)^2} + \frac{x + 2}{x^2 + 1} \right] dx. \quad \textbf{(21)}$$

Using formulas (16), (15), and (13), we obtain

$$\int \frac{3x - 1}{(x^2 + 1)^2} \, dx = \frac{3}{2} \int \frac{2x}{(x^2 + 1)^2} \, dx - \int \frac{1}{(x^2 + 1)^2} \, dx$$

$$= \frac{3}{2} \cdot \frac{-1}{x^2 + 1} - \left[\frac{2x}{1(4)(x^2 + 1)} + \frac{2 \cdot 1}{1 \cdot 4} \int \frac{1}{x^2 + 1} \, dx \right] \quad \textbf{(22)}$$

$$= \frac{-3}{2(x^2 + 1)} - \frac{x}{2(x^2 + 1)} - \frac{1}{2} \tan^{-1}x + C.$$

We also find that

$$\int \frac{x + 2}{x^2 + 1} \, dx = \frac{1}{2} \int \frac{2x}{x^2 + 1} \, dx + 2 \int \frac{1}{x^2 + 1} \, dx$$

$$= \frac{1}{2} \ln|x^2 + 1| + 2 \tan^{-1}x + C. \quad \textbf{(23)}$$

Then Eqs. (21), (22), and (23) yield

$$\int \frac{3x^4 + 2x^3 + 8x^2 + x + 2}{x(x^2 + 1)^2} \, dx$$

$$= 2 \ln|x| - \frac{3}{2(x^2 + 1)} - \frac{x}{2(x^2 + 1)} - \frac{1}{2} \tan^{-1}x$$

$$+ \frac{1}{2} \ln|x^2 + 1| + 2 \tan^{-1}x + C$$

$$= \frac{1}{2} \ln|x^6 + x^4| - \frac{x + 3}{2(x^2 + 1)} + \frac{3}{2} \tan^{-1}x + C. \quad \blacksquare$$

SUMMARY

To integrate a quotient of polynomials, the following steps are suggested.

STEP 1. Perform polynomial long division, if necessary, to reduce the problem to integrating a rational function whose numerator has degree less than the degree of the denominator.

STEP 2. Factor the denominator into linear and irreducible quadratic factors.

STEP 3. Obtain the partial fraction decomposition of the rational function.

STEP 4. Integrate the summands in the resulting decomposition of the original rational function, using formulas (11)–(16) in this section.

EXERCISES 8.2

Use the methods described in this section to find the integral.

1. $\int \frac{x - 1}{x^2} \, dx$

2. $\int \frac{x^5 - 4x^3 + 2x^2 - 7}{x^3} \, dx$

3. $\int \frac{x^3 - 2x}{x + 1} \, dx$

4. $\int \frac{x^4 + 3x^2}{2x - 3} \, dx$

5. $\int \frac{x^3 + 2x + 1}{x^2 + 1} \, dx$

6. $\int \frac{x^4 - x}{x^2 + 4} \, dx$

7. $\int \frac{1}{x^2 - 1} \, dx$

8. $\int \frac{1}{x^2 - x} \, dx$

9. $\int \frac{x^3 + 4}{x^2 - 4} \, dx$

10. $\int \frac{x^3 + x^2 - x + 8}{x^2 + 2x} \, dx$

11. $\int \frac{x^4 - 2x^2 + 6}{x^2 - 3x + 2} \, dx$

12. $\int \frac{x^4 - x^2 - 4x - 2}{x^3 - x} \, dx$

13. $\int \frac{3x + 4}{x^3 - 2x^2 - 3x} \, dx$

14. $\int \frac{2x^4 + 6x^3 - 10x^2 - 3x + 2}{x^3 + 3x^2 - 6x - 8} \, dx$

15. $\int \frac{x^3 + 4x^2 - 5x - 4}{x^3 + 3x^2 - x - 3} \, dx$

16. $\int \dfrac{2x + 1}{x^3 - x^2} \, dx$

17. $\int \dfrac{x - 1}{x^3 + 2x^2 + x} \, dx$

18. $\int \dfrac{x^4 + 1}{x^4 + 2x^3} \, dx$

19. $\int \dfrac{x^2 - 3}{(x - 2)(x + 1)^3} \, dx$

20. $\int \dfrac{2x^2 + 2x - 2}{x^3 + 2x} \, dx$

21. $\int \dfrac{5x^3 - x^2 + 4x - 12}{x^4 - 4x^2} \, dx$

22. $\int \dfrac{2x^2 - x + 3}{x^3 - x^2 + x - 1} \, dx$

23. $\int \dfrac{6x^2 + 3x + 1}{x^3 - x^2 + x - 1} \, dx$

24. $\int \dfrac{3x^3 + 3x^2 - 5x + 7}{x^4 - 1} \, dx$

25. $\int \dfrac{2x^3 + 5x^2 - 2x + 16}{x^4 - 2x^2 - 8} \, dx$

26. $\int \dfrac{x + 3}{x^2 - 5x + 7} \, dx$

27. $\int \dfrac{5x - 4}{3x^2 - 4x + 2} \, dx$

28. $\int \dfrac{x + 4}{(x^2 - x + 1)^2} \, dx$

29. $\int \dfrac{3x - 1}{(x^2 - 2x + 2)^2} \, dx$

30. $\int \dfrac{x^3 + 2x^2 + 3x + 1}{x^4 + 2x^2 + 1} \, dx$

31. $\int \dfrac{x^3 - x^2 + 2x + 1}{(x^2 - x + 1)^2} \, dx$

32. $\int \dfrac{-x^3 + 10x^2 - 19x + 22}{(x - 1)^2(x^2 + 3)} \, dx$

33. $\int \dfrac{3x^6 + 9x^4 - 4x^3 + 7x^2 + 3}{x^2(x^2 + 1)^3} \, dx$

Work Exercises 34–38 without using tables.

34. Find the area of the plane region bounded by $y = 12/(4 - x^2)$ and $y = 4$.

35. Find the volume of the solid generated when the region in Exercise 34 is revolved about the line $x = 1$.

36. The velocity at time t of a body moving on the x-axis is $(t + 1)/(t^2 + 5t + 6)$. Find the total distance the body travels for $0 \le t \le 4$.

37. Find the solution of the initial value problem $y' = x(y^2 - 1)$, $y(0) = 3$.

38. Find the general solution of the differential equation $dy/dx = (y^2 - y - 2)/(x^2 - x)$.

8.3 SUBSTITUTION

Various substitution techniques for integration are presented in this section and in Sections 8.4 and 8.5. This section deals primarily with *algebraic substitutions*, which are often made to change an integral to a more familiar form or to attempt to simplify a radical in an integral. Trigonometric substitutions are studied in Sections 8.4 and 8.5.

Chapter 5 introduced the type of substitution technique illustrated in the following example.

EXAMPLE 1 Find $\int x(x^2 + 3)^4 \, dx$.

Solution We note that $x \, dx$ is the differential of $x^2 + 3$, except for a constant factor. We let $u = x^2 + 3$, so $du = 2x \, dx$ and $x \, dx = \frac{1}{2} \, du$. Then

$$\int x(x^2 + 3)^4 \, dx = \int u^4 \frac{1}{2} \, du = \frac{1}{2} \cdot \frac{u^5}{5} + C = \frac{1}{10}(x^2 + 3)^5 + C. \quad \blacksquare$$

Success in this technique for finding $\int f(x) \, dx$ depends on spotting a factor of the integrand to serve as u' for some expression u that already appears in the integral; then $u' \, dx = du$. If $f(x)$ can be expressed as $g(u)u'$ and if $\int g(u) \, du =$

$G(u) + C$, then by the chain rule

$$\frac{d(G(u))}{dx} = \frac{d(G(u))}{du} \cdot \frac{du}{dx}$$

$$= g(u)u' = f(x),$$

so $G(u)$ is indeed an antiderivative of $f(x)$. The validity of such substitution thus follows at once from the chain rule for differentiation.

Here is another substitution example.

EXAMPLE 2   Find $\int x\sqrt{x + 1}\, dx$ by substitution.

Solution   To eliminate the radical, we let $u = \sqrt{x + 1}$. Then $u^2 = x + 1$, so

$$x = u^2 - 1 \quad \text{and} \quad dx = 2u\, du.$$

(*This substitution for dx must be computed. A common error is just to replace dx by du.*) Therefore

$$\int x\sqrt{x + 1}\, dx = \int (u^2 - 1)u \cdot 2u\, du = 2\int (u^4 - u^2)\, du$$

$$= 2\left(\frac{u^5}{5} - \frac{u^3}{3}\right) + C$$

$$= 2\left[\frac{(\sqrt{x + 1})^5}{5} - \frac{(\sqrt{x + 1})^3}{3}\right] + C. \qquad \blacksquare$$

Often there is more than one substitution that will handle an integral. We solve the problem in Example 2 again using a substitution that simplifies the radical rather than eliminating it altogether.

EXAMPLE 3   Find $\int x\sqrt{x + 1}\, dx$ by substitution.

Solution   This time we let $u = x + 1$ so that $x = u - 1$ and $dx = du$. The integral becomes

$$\int x\sqrt{x + 1}\, dx = \int (u - 1)\sqrt{u}\, du = \int (u^{3/2} - u^{1/2})\, du$$

$$= \tfrac{2}{5}u^{5/2} - \tfrac{2}{3}u^{3/2} + C$$

$$= \tfrac{2}{5}(x + 1)^{5/2} - \tfrac{2}{3}(x + 1)^{3/2} + C.$$

This is equivalent to the answer we obtained in Example 2. $\qquad \blacksquare$

In the type of substitution illustrated in Examples 2 and 3, we let $x = h(u)$, so $dx = h'(u)\, du$. We then write

$$\int f(x)\, dx = \int f(h(u)) \cdot h'(u)\, du = G(u) + C. \qquad (1)$$

Again, the chain rule shows that

$$\frac{d(G(u))}{dx} = \frac{d(G(u))}{du} \cdot \frac{du}{dx} = f(h(u)) \cdot h'(u) \cdot \frac{du}{dx} = f(h(u)) \cdot \frac{dx}{du} \cdot \frac{du}{dx}$$

$$= f(h(u)) = f(x),$$

so $G(u) = G(h^{-1}(x))$ is indeed an antiderivative of $f(x)$. After obtaining the indefinite integral $G(u) + C$ in Eq. (1), we want to express $G(u)$ in terms of the original variable x. To do this, we solve the substitution $x = h(u)$ for u in terms of x and substitute the resulting expression for u in $G(u)$.*

As mentioned in Example 2, a common error when substituting $x = h(u)$ is to replace dx by du instead of replacing dx by $h'(u)\, du$. *Don't forget to compute dx when substituting in $\int f(x)\, dx$.*

An *algebraic substitution* $x = h(u)$ is one where $h(u)$ is a function involving only arithmetic operations and roots (no trigonometric functions, for example).

EXAMPLE 4 Use an algebraic substitution to find the integral

$$\int x^3 \sqrt{4 - x^2}\, dx.$$

Solution To eliminate the radical, we let $u = \sqrt{4 - x^2}$ so that $u^2 = 4 - x^2$. Implicit differentiation yields $2u\, du = -2x\, dx$, so

$$x\, dx = -u\, du.$$

We thus have

$$\int x^3 \sqrt{4 - x^2}\, dx = \int x^2 \sqrt{4 - x^2}\, x\, dx = \int (4 - u^2)u(-u\, du)$$

$$= \int (-4u^2 + u^4)\, du = -\frac{4u^3}{3} + \frac{u^5}{5} + C.$$

Since $u = \sqrt{4 - x^2}$, we obtain

$$\int x^3 \sqrt{4 - x^2}\, dx = -\frac{4(4 - x^2)^{3/2}}{3} + \frac{(4 - x^2)^{5/2}}{5} + C. \qquad \blacksquare$$

EXAMPLE 5 Find

$$\int \frac{x^3}{(x^2 - 1)^{3/2}}\, dx.$$

Solution We let $u = \sqrt{x^2 - 1}$, so $u^2 = x^2 - 1$. Then $2u\, du = 2x\, dx$, so $u\, du = x\, dx$. We then obtain

$$\int \frac{x^3}{(x^2 - 1)^{3/2}}\, dx = \int \frac{x^2(x\, dx)}{(x^2 - 1)^{3/2}} = \int \frac{(u^2 + 1)(u\, du)}{u^3}$$

$$= \int \frac{u^3 + u}{u^3}\, du = \int \left(1 + \frac{1}{u^2}\right) du$$

$$= u - \frac{1}{u} + C = \sqrt{x^2 + 1} - \frac{1}{\sqrt{x^2 - 1}} + C. \qquad \blacksquare$$

*To solve $x = h(u)$ for u in terms of x, the function h must have an inverse. Thus we should restrict our substitutions to invertible functions with continuous derivatives. Strictly speaking, we really should think of our substitution as $u = h^{-1}(x)$, defining what u is to be in terms of x, since x is given. We did this in Examples 1 and 2. We then go to $x = h(u)$ to compute dx. However, we will downplay these theoretical considerations and feel free to substitute $x = h(u)$ in those cases where it seems most natural.

We should not think that all integrals with radicals will yield to algebraic substitution. Look at the following example.

EXAMPLE 6   Find $\int x^4 \sqrt{9 - x^2}\, dx$.

Attempted Solution   It is natural to try the substitution $u = \sqrt{9 - x^2}$. Then $u^2 = 9 - x^2$, so $2u\, du = -2x\, dx$, or $x\, dx = -u\, du$. We obtain

$$\int x^4 \sqrt{9 - x^2}\, dx = \int x^3 \sqrt{9 - x^2}\, x\, dx = \int x^3 u(-u\, du).$$

But since $u^2 = 9 - x^2$, we have $x^2 = 9 - u^2$, so

$$x^3 = x^2 \cdot x = (9 - u^2)\sqrt{9 - u^2}$$

and our integral becomes

$$\int x^3 u(-u\, du) = \int (9 - u^2)\sqrt{9 - u^2}\, u(-u\, du)$$

$$= -\int (9u^2 - u^4)\sqrt{9 - u^2}\, du.$$

We have not eliminated the radical and clearly have just as tough an integral. We will see how to find this type of integral in Section 8.5.   ∎

Sometimes a substitution can be used to simultaneously eliminate different fractional powers from an integrand.

EXAMPLE 7   Find

$$\int \frac{x^{1/2}}{1 + x^{1/3}}\, dx.$$

Solution   To eliminate the fractional powers $x^{1/3}$ and $x^{1/2}$ at the same time, we let $u = x^{1/6}$. Then $x = u^6$, so that $dx = 6u^5\, du$ and

$$\int \frac{x^{1/2}}{1 + x^{1/3}}\, dx = \int \frac{u^3}{1 + u^2}6u^5\, du.$$

Application of the method of partial fractions yields

$$6\int \frac{u^8}{u^2 + 1}\, du = 6\int \left(u^6 - u^4 + u^2 - 1 + \frac{1}{u^2 + 1} \right) du$$

$$= 6\left(\frac{u^7}{7} - \frac{u^5}{5} + \frac{u^3}{3} - u + \tan^{-1}u \right) + C.$$

Since $u = x^{1/6}$, we have

$$\int \frac{x^{1/2}}{1 + x^{1/3}}\, dx = 6\left(\frac{x^{7/6}}{7} - \frac{x^{5/6}}{5} + \frac{x^{1/2}}{3} - x^{1/6} + \tan^{-1}x^{1/6} \right) + C.$$ ∎

Suppose now that a substitution $x = h(u)$ is involved in the computation of a *definite* integral $\int_a^b f(x)\, dx$. Then we may choose to change the x-limits, a and b, to u-limits and evaluate the indefinite integral as soon as it is found in terms of

u, using these u-limits. The u-interval for integration is generally different from the x-interval $[a, b]$. Alternatively, we can find an indefinite integral in terms of x as in the preceding examples and then compute the definite integral using the x-limits. We illustrate both approaches in the next example.

EXAMPLE 8 Find

$$\int_0^3 \frac{x}{\sqrt{x+1}} \, dx.$$

Solution 1 We let $u = \sqrt{x+1}$, so $x + 1 = u^2$. Then $dx = 2u \, du$. When $x = 0$, we see that $u = \sqrt{0+1} = 1$. Similarly, when $x = 3$, we have $u = \sqrt{3+1} = 2$. Then

$$\int_0^3 \frac{x}{\sqrt{x+1}} \, dx = \int_{u=1}^2 \frac{(u^2 - 1)(2u \, du)}{u} = \int_{u=1}^2 (2u^2 - 2) \, du$$

$$= \left(2\frac{u^3}{3} - 2u \right)\Big]_1^2 = \left(\frac{16}{3} - 4 \right) - \left(\frac{2}{3} - 2 \right)$$

$$= \frac{14}{3} - 2 = \frac{8}{3}.$$

Solution 2 Using the same substitution and computation of an indefinite integral as in Solution 1, we have

$$\int_0^3 \frac{x}{\sqrt{x+1}} \, dx = \int_{x=0}^3 (2u^2 - 2) \, du = \left(2\frac{u^3}{3} - 2u \right)\Big]_{x=0}^2$$

$$= \left(2\frac{(x+1)^{3/2}}{3} - 2\sqrt{x+1} \right)\Big]_0^3$$

$$= \left(2 \cdot \frac{8}{3} - 2 \cdot 2 \right) - \left(\frac{2}{3} - 2 \right)$$

$$= \frac{14}{3} - 2 = \frac{8}{3}.$$

While we have used only algebraic substitutions so far in this section, the substitution technique is very general. We deal with other substitutions in the following sections. We present one example involving a nonalgebraic substitution.

EXAMPLE 9 Find

$$\int e^{\sin^{-1}x} \, dx.$$

Solution This integral looks pretty hopeless, but we can try to simplify it by the substitution

$$u = \sin^{-1}x.$$

Then $x = \sin u$, and $dx = \cos u \, du$. The integral becomes

$$\int e^{\sin^{-1}x} \, dx = \int e^u \cos u \, du.$$

A table (or integration by parts) yields

$$\int e^u \cos u \, du = \frac{e^u}{2}(\cos u + \sin u) + C.$$

Since $u = \sin^{-1}x$, we know that $-\pi/2 \le u \le \pi/2$, so $\cos u \ge 0$. We then have

$$\sin u = x \quad \text{and} \quad \cos u = \sqrt{1 - \sin^2 u} = \sqrt{1 - x^2}.$$

Thus we have

$$\int e^{\sin^{-1}x} \, dx = \frac{e^{\sin^{-1}x}}{2}(x + \sqrt{1 - x^2}) + C. \quad\blacksquare$$

SUMMARY

1. If $x = h(u)$, then $\int f(x) \, dx$ becomes $\int f(h(u)) \cdot h'(u) \, du$. If the latter integral can be computed as $G(u) + C$, then solve the original substitution for u in terms of x and substitute in $G(u) + C$ to obtain $\int f(x) \, dx$ in terms of x.

2. Suppose a substitution $x = h(u)$ is used in finding a definite integral $\int_a^b f(x) \, dx$. We may wish to find the u-limits corresponding to the x-limits, a and b, and compute the definite integral using those u-limits as soon as we find an indefinite integral in terms of u.

3. Integration of expressions involving fractional powers may yield to an algebraic substitution designed to eliminate all the fractional powers.

EXERCISES 8.3

1. Integrals that yield easily to one technique often yield easily to another. Show that $\int x\sqrt{x + 1} \, dx$ of Example 2 can be found using integration by parts.

2. Repeat Exercise 1 for $\int x^3\sqrt{4 - x^2} \, dx$ of Example 4.

In Exercises 3–37, find the integral. For the definite integrals, change to limits for the substitution variable, as illustrated in Example 8, Solution 1.

3. $\int \dfrac{x}{\sqrt{1 + x}} \, dx$

4. $\int \dfrac{\sqrt{x - 1}}{x} \, dx$

5. $\int x^2\sqrt{4 + x} \, dx$

6. $\int x\sqrt{3 - 2x} \, dx$

7. $\int \dfrac{\sqrt{x}}{\sqrt{x} + 1} \, dx$

8. $\int \dfrac{\sqrt{x} - 1}{\sqrt{x}} \, dx$

9. $\int \dfrac{\sqrt{x + 2} - 1}{\sqrt{x + 2} + 1} \, dx$

10. $\int \dfrac{x^2}{\sqrt{x - 1}} \, dx$

11. $\int \dfrac{3x + 2}{\sqrt{x + 1}} \, dx$

12. $\int_2^5 \dfrac{x}{\sqrt{x - 1}} \, dx$

13. $\int_{-3}^0 x\sqrt{1 - x} \, dx$

14. $\int_1^4 \dfrac{\sqrt{x}}{1 + \sqrt{x}} \, dx$

15. $\int \dfrac{x}{\sqrt{1 + x^2}} \, dx$

16. $\int \dfrac{x - 3}{\sqrt{4 - x^2}} \, dx$

17. $\int \dfrac{x^3}{\sqrt{1 + x^2}} \, dx$

18. $\int \dfrac{\cos x}{\sqrt{4 - \sin^2 x}} \, dx$

19. $\int x^3(x^2 + 1)^{3/2} \, dx$

20. $\int \dfrac{x^3 + 2x}{\sqrt{x^2 - 2}} \, dx$

21. $\int \dfrac{x^3 - x}{\sqrt{x^2 + 8}}\, dx$

22. $\int \dfrac{x^5}{\sqrt{1 - x^3}}\, dx$

23. $\int \dfrac{x^3}{(x^2 - 1)^{5/2}}\, dx$

24. $\int x^5\sqrt{9 + x^3}\, dx$

25. $\int_3^5 x\sqrt{25 - x^2}\, dx$

26. $\int_0^4 x^3\sqrt{9 + x^2}\, dx$

27. $\int_0^3 \dfrac{x^3}{\sqrt{x^2 + 16}}\, dx$

28. $\int \dfrac{\sqrt{x}}{1 + x^{1/4}}\, dx$

29. $\int \dfrac{x + \sqrt{x}}{x^{1/4} - 1}\, dx$

30. $\int \dfrac{x^{1/2}}{4 + x^{1/3}}\, dx$

31. $\int \dfrac{x^{2/3}}{1 + x^{1/2}}\, dx$

32. $\int \dfrac{dx}{x^{1/2} + x^{1/3}}$

33. $\int e^{\cos^{-1}x}\, dx$

34. $\int x e^{\cos^{-1}x}\, dx$

35. $\int x e^{\sin^{-1}x}\, dx$

36. $\int_0^1 e^{\sin^{-1}x}\, dx$

37. $\int_0^1 e^{\cos^{-1}x}\, dx$

Work Exercises 38–41 without using tables.

38. Find the area of the region bounded by $y = x\sqrt{2 - x}$ and the x-axis for $0 \le x \le 2$.

39. Find the volume of the solid generated when the region in Exercise 38 is revolved about the y-axis.

40. A body is moved along a straight line by a force of $(\sqrt{s}+2)/(\sqrt{s}+1)$ lb acting in the direction of the motion, where s is the distance the body has moved from its starting position. Find the work done by the force in moving the body 16 ft from its starting position.

41. Find the solution of the initial value problem $y' = (x\sqrt{y} + 1)/(y\sqrt{x} + 2)$, $y(2) = 8$.

8.4 INTEGRATING POWERS OF TRIGONOMETRIC FUNCTIONS

Integration of Odd Powers of sin x and cos x

An integral of the form

$$\int \sin^m x \cos^n x\, dx,$$

where either m or n is an odd positive integer, can be computed by the following device. If m is odd, "save" one factor $\sin x$ to serve as u' for $u = \cos x$. Then change all other factors of $\sin^2 x$ to $1 - \cos^2 x$. The result is an integral of the form

$$\int f(\cos x) \sin x\, dx,$$

where $f(\cos x)$ is a polynomial in $\cos x$. This integral can readily be found. If n is odd, a similar procedure can be followed, "saving" one factor $\cos x$ and changing all other factors $\cos^2 x$ to $1 - \sin^2 x$.

EXAMPLE 1 Find $\int \sin^3 x \cos^2 x\, dx$.

Solution We have

$$\int \sin^3 x \cos^2 x\, dx = \int \sin x(1 - \cos^2 x) \cos^2 x\, dx$$

$$= -\int (\cos^2 x - \cos^4 x)(-\sin x)\, dx$$

$$= -\tfrac{1}{3}\cos^3 x + \tfrac{1}{5}\cos^5 x + C. \qquad \blacksquare$$

EXAMPLE 2   Find $\int \cos^5 x \, dx$.

Solution   We have

$$\int \cos^5 x \, dx = \int (1 - \sin^2 x)^2 \cos x \, dx$$

$$= \int (1 - 2 \sin^2 x + \sin^4 x) \cos x \, dx$$

$$= \sin x - \tfrac{2}{3} \sin^3 x + \tfrac{1}{5} \sin^5 x + C.$$ ∎

An integral $\int \sin^m x \cos^n x \, dx$ can be found using this technique if either m or n is not an integer, as long as the other is an odd integer.

EXAMPLE 3   Find

$$\int \frac{\sin^3 2x}{(\cos 2x)^{3/2}} \, dx.$$

Solution   We have

$$\int \frac{\sin^3 2x}{(\cos 2x)^{3/2}} \, dx = \int \frac{(1 - \cos^2 2x)}{(\cos 2x)^{3/2}} (\sin 2x \, dx)$$

$$= -\frac{1}{2} \int [(\cos 2x)^{-3/2} - (\cos 2x)^{1/2}](-2 \sin 2x \, dx)$$

$$= -\frac{1}{2} [-2(\cos 2x)^{-1/2} - \frac{2}{3}(\cos 2x)^{3/2}] + C$$

$$= \frac{1}{\sqrt{\cos 2x}} + \frac{1}{3}(\cos 2x)^{3/2} + C.$$ ∎

Integration of Even Powers of sin x and cos x

An integral of the form

$$\int \sin^m x \cos^n x \, dx,$$

where both m and n are nonnegative *even* integers, is more tedious to find than the case where either m or n is odd. This time, we use the trigonometric identities

$$\sin^2 ax = \frac{1 - \cos 2ax}{2},$$

$$\cos^2 bx = \frac{1 + \cos 2bx}{2},$$ **(1)**

which are derived from the more familiar relation $\cos 2x = \cos^2 x - \sin^2 x$. (See Exercise 1.) We use identities (1) repeatedly until we obtain just constants and odd powers of cosine functions. The technique is best illustrated by examples.

EXAMPLE 4   Find $\int \sin^4 x \, dx$.

Solution   Using identities (1) repeatedly, we have

$$\int \sin^4 x \, dx = \int (\sin^2 x)^2 \, dx$$

$$= \int \left(\frac{1 - \cos 2x}{2} \right)^2 dx$$

$$= \frac{1}{4} \int (1 - 2 \cos 2x + \cos^2 2x) \, dx$$

$$= \frac{1}{4} \int \left(1 - 2 \cos 2x + \frac{1 + \cos 4x}{2} \right) dx$$

$$= \frac{1}{8} \int (3 - 4 \cos 2x + \cos 4x) \, dx$$

$$= \frac{3}{8} x - \frac{1}{4} \sin 2x + \frac{1}{32} \sin 4x + C. \quad \blacksquare$$

EXAMPLE 5   Find

$$\int \sin^2 x \cos^2 x \, dx.$$

Solution   We use the relation $\sin x \cos x = \frac{1}{2} \sin 2x$ followed by identities (1) and obtain

$$\int \sin^2 x \cos^2 x \, dx = \frac{1}{4} \int \sin^2 2x \, dx = \frac{1}{4} \int \frac{1 - \cos 4x}{2} \, dx$$

$$= \frac{1}{8} x - \frac{1}{32} \sin 4x + C. \quad \blacksquare$$

Integration of Expressions such as sin *mx* cos *nx* for $m \neq n$

The relations

$$\sin mx \sin nx = \tfrac{1}{2}[\cos(m - n)x - \cos(m + n)x], \qquad (2)$$

$$\sin mx \cos nx = \tfrac{1}{2}[\sin(m - n)x + \sin(m + n)x], \qquad (3)$$

$$\cos mx \cos nx = \tfrac{1}{2}[\cos(m - n)x + \cos(m + n)x] \qquad (4)$$

enable us to integrate $\sin mx \sin nx$, $\sin mx \cos nx$, and $\cos mx \cos nx$. Formulas (2), (3), and (4) are easily derived from the more familiar formulas

$$\sin(mx + nx) = \sin mx \cos nx + \cos mx \sin nx, \qquad (5)$$

$$\cos(mx + nx) = \cos mx \cos nx - \sin mx \sin nx. \qquad (6)$$

See Exercise 2.

EXAMPLE 6 Find

$$\int \sin 2x \cos 3x \, dx.$$

Solution We have, using formula (3),

$$\int \sin 2x \cos 3x \, dx = \frac{1}{2} \int [\sin(-x) + \sin 5x] \, dx$$

$$= \frac{1}{2} \int (-\sin x + \sin 5x) \, dx$$

$$= \frac{1}{2} \cos x - \frac{1}{10} \cos 5x + C. \quad \blacksquare$$

Integration of Powers of tan x and cot x and Even Powers of sec x and csc x

We make use of the identities

$$1 + \tan^2 x = \sec^2 x \quad \text{or} \quad \tan^2 x = \sec^2 x - 1 \qquad (7)$$

and

$$1 + \cot^2 x = \csc^2 x \quad \text{or} \quad \cot^2 x = \csc^2 x - 1. \qquad (8)$$

To integrate an even power of sec x, we "save" one factor $\sec^2 x$ to serve in du and change all the other powers of $\sec^2 x$ to powers of $u = \tan x$.

EXAMPLE 7 Find

$$\int \sec^6 x \, dx.$$

Solution We obtain

$$\int \sec^6 x \, dx = \int (1 + \tan^2 x)^2 \sec^2 x \, dx$$

$$= \int (1 + 2 \tan^2 x + \tan^4 x) \sec^2 x \, dx$$

$$= \int \sec^2 x \, dx + 2 \int \tan^2 x \sec^2 x \, dx + \int \tan^4 x \sec^2 x \, dx$$

$$= \tan x + 2 \frac{\tan^3 x}{3} + \frac{\tan^5 x}{5} + C. \quad \blacksquare$$

We now illustrate how to integrate a power of tan x using identity (7).

EXAMPLE 8 Find $\int \tan^5 x \, dx$.

Solution We obtain

$$\int \tan^5 x \, dx = \int \tan^3 x \, (\sec^2 x - 1) \, dx$$

$$= \int \tan^3 x \, \sec^2 x \, dx - \int \tan^3 x \, dx$$

$$= \frac{\tan^4 x}{4} - \int \tan x \, (\sec^2 x - 1) \, dx$$

$$= \frac{\tan^4 x}{4} - \int \tan x \, \sec^2 x \, dx + \int \tan x \, dx$$

$$= \frac{\tan^4 x}{4} - \frac{\tan^2 x}{2} + \int \frac{\sin x}{\cos x} \, dx$$

$$= \frac{\tan^4 x}{4} - \frac{\tan^2 x}{2} - \ln|\cos x| + C. \qquad ■$$

Powers of $\cot x$ and even powers of $\csc x$ can be integrated in a similar fashion using identity (8) and $d(\cot x) = -\csc^2 x \, dx$.

Reduction Formulas for Powers of tan x, cot x, sec x, and csc x

Odd powers of $\sec x$ and $\csc x$ are tough to integrate. They can be done using integration by parts. Example 8 in Section 8.1 showed how to find $\int \sec^3 ax \, dx$. We ask you to find $\int \csc^3 x \, dx$ in Exercise 33. Of course, an alternative way to integrate any power of $\tan x$, $\cot x$, $\sec x$, or $\csc x$ is to use a table and make repeated use of the appropriate reduction formula:

$$\int \tan^n ax \, dx = \frac{\tan^{n-1} ax}{a(n-1)} - \int \tan^{n-2} ax \, dx, \qquad n \neq 1, \tag{9}$$

$$\int \cot^n ax \, dx = -\frac{\cot^{n-1} ax}{a(n-1)} - \int \cot^{n-2} ax \, dx, \qquad n \neq 1, \tag{10}$$

$$\int \sec^n ax \, dx = \frac{\sec^{n-2} ax \, \tan ax}{a(n-1)}$$
$$+ \frac{n-2}{n-1} \int \sec^{n-2} ax \, dx, \qquad n \neq 1, \tag{11}$$

$$\int \csc^n ax \, dx = -\frac{\csc^{n-2} ax \, \cot ax}{a(n-1)}$$
$$+ \frac{n-2}{n-1} \int \csc^{n-2} ax \, dx, \qquad n \neq 1. \tag{12}$$

This procedure will reduce the problem to the integration of the function to the first power (or to integration of a constant). Then we can use

$$\int \tan x \, dx = -\ln|\cos x| + C, \tag{13}$$

$$\int \cot x \, dx = \ln|\sin x| + C, \tag{14}$$

$$\int \sec x \, dx = \ln|\sec x + \tan x| + C, \tag{15}$$

$$\int \csc x \, dx = -\ln|\csc x + \cot x| + C, \tag{16}$$

which we have already seen. You are asked to derive the reduction formulas (9)–(12) in the exercises, using identities (7) and (8) and integration by parts.

EXAMPLE 9 Find $\int \sec^5 2x \, dx$.

Solution We have, using formula (11),

$$\int \sec^5 2x \, dx = \frac{\sec^3 2x \tan 2x}{2 \cdot 4} + \frac{3}{4} \int \sec^3 2x \, dx$$

$$= \frac{\sec^3 2x \tan 2x}{8} + \frac{3}{4}\left(\frac{\sec 2x \tan 2x}{2 \cdot 2} + \frac{1}{2} \int \sec 2x \, dx \right)$$

$$= \frac{\sec^3 2x \tan 2x}{8} + \frac{3}{16} \sec 2x \tan 2x$$

$$+ \frac{3}{16}\ln|\sec 2x + \tan 2x| + C. \qquad \blacksquare$$

Finally, remember always to see whether there is a simple way to find an integral before blindly reaching for a "rule." Consider, for example, three methods for finding $\int \tan x \sec^2 x \, dx$.

Method 1

$$\int \tan x \sec^2 x \, dx = \int \frac{\sin x}{\cos x} \cdot \frac{1}{\cos^2 x} \, dx = \int \frac{\sin x}{\cos^3 x} \, dx$$

can be done by the $t = \tan(x/2)$ substitution described in Exercise 47. However, that is a lot of work.

Method 2

$$\int \tan x \sec^2 x \, dx = \int \tan x(1 + \tan^2 x) \, dx = \int (\tan x + \tan^3 x) \, dx$$

can be done using formulas (9) and (13). This is an improvement over method 1.

Method 3 Since $d(\tan x) = \sec^2 x \, dx$, then

$$\int \tan x \sec^2 x \, dx = \frac{\tan^2 x}{2} + C.$$

This is the easiest method of all.

We illustrate that if a calculator or computer program is at hand, Simpson's rule is often a faster way to find a *definite* integral, even if the necessary integration formulas are contained in a convenient table.

EXAMPLE 10 Find $\int_0^\pi \sec^7(x/4)\,dx$.

Solution We could use a reduction formula and find $\int \sec^7(x/4)\,dx$ and then evaluate it from zero to π to find the definite integral. However, our calculator contains a built-in program for Simpson's rule. Timing ourselves from the moment we turned the calculator on, we found that

$$\int_0^\pi \sec^7\left(\frac{x}{4}\right) dx \approx 8.998128946$$

in 89 seconds flat. We couldn't do it that fast using a table. We used $n = 20$ in Simpson's rule. Taking a larger value of n for comparison, we find that the $n = 20$ approximation was probably accurate to four significant figures. ■

SUMMARY

1. All the indefinite integrals discussed in this section can easily be found using an integral table. Definite integrals that would require much work with tables are often computed faster using a computer and Simpson's rule.

2. Odd powers of $\sin x$ can be integrated by "saving" one factor $\sin x$ to serve in du and changing all other factors $\sin^2 x$ to powers of $u = \cos x$ using $\sin^2 x = 1 - \cos^2 x$. Odd powers of $\cos x$ are integrated similarly.

3. Even powers of $\sin x$ or $\cos x$ can be integrated using

$$\sin^2 ax = \frac{1 - \cos 2ax}{2}$$

and

$$\cos^2 bx = \frac{1 + \cos 2bx}{2}$$

repeatedly.

4. Formulas (2), (3), and (4) given in this section can be used to integrate expressions involving $\sin mx \sin nx$, $\cos mx \cos nx$, and $\sin mx \cos nx$.

5. Powers of $\tan x$ and $\cot x$ and even powers of $\sec x$ and $\csc x$ can be integrated using

$$1 + \tan^2 x = \sec^2 x$$

and

$$1 + \cot^2 x = \csc^2 x.$$

6. All powers of tan x, cot x, sec x, and csc x can be integrated using the reduction formulas (9)–(12) and formulas (13)–(16) of this section.

EXERCISES 8.4

1. Derive identities (1) from the relation $\cos 2x = \cos^2 x - \sin^2 x$ and other trigonometric relations.

2. Derive formulas (2), (3), and (4) from formulas (5) and (6) and other well-known trigonometric relations.

In Exercises 3–35, compute the indicated integrals without the use of tables.

3. $\int \sin^2 x \cos^3 x\, dx$

4. $\int \sin^3 x \cos^4 x\, dx$

5. $\int \sin^4 x \cos^5 x\, dx$

6. $\int \dfrac{\sin^3 x}{\cos^2 x}\, dx$

7. $\int \sin^3 2x \sqrt{\cos 2x}\, dx$

8. $\int \dfrac{\cos^3 x}{\sqrt{\sin x}}\, dx$

9. $\int \dfrac{\cos^3 x}{(\sin x)^{3/2}}\, dx$

10. $\int \sin^2 3x\, dx$

11. $\int \cos^2 5x\, dx$

12. $\int \sin^2 3x \cos^2 3x\, dx$

13. $\int \cos^4 2x\, dx$

14. $\int \sin^4 x \cos^2 x\, dx$

15. $\int \sin^6 x\, dx$

16. $\int \cos^6 x\, dx$

17. $\int \dfrac{1}{\sin^2 x}\, dx$

18. $\int \dfrac{1}{\cos^2 x}\, dx$

19. $\int \sec^4 x\, dx$

20. $\int \csc^4 x\, dx$

21. $\int \sec^2 x \tan^2 x\, dx$

22. $\int \sec^4 x \tan^2 x\, dx$

23. $\int \sec^2 x \tan^4 x\, dx$

24. $\int \dfrac{\cos^2 x}{\sin^4 x}\, dx$

25. $\int \dfrac{\sin 3x}{\cos^3 3x}\, dx$

26. $\int \tan^3 x\, dx$

27. $\int \tan^4 x\, dx$

28. $\int \dfrac{dx}{\cos^6 x}$

29. $\int \tan^7 2x\, dx$

30. $\int \tan^3 x \sec^2 x\, dx$

31. $\int \csc^4 x \cot x\, dx$

32. $\int \csc^4 5x \cot^3 5x\, dx$

33. $\int \csc^3 x\, dx$ [*Hint:* Use integration by parts.]

34. $\int \csc^5 x\, dx$ [*Hint:* Use integration by parts.]

35. $\int \sec^5 x\, dx$ [*Hint:* Use integration by parts.]

In Exercise 36–41, derive the indicated reduction formula.

36. Formula (9) of this section

37. Formula (10) of this section

38. Formula (11) of this section

39. Formula (12) of this section

40. $\int \sin^n ax\, dx = \dfrac{-\sin^{n-1} ax \cos ax}{na} + \dfrac{n-1}{n} \int \sin^{n-2} ax\, dx$

41. $\int \cos^n ax\, dx = \dfrac{\cos^{n-1} ax \sin ax}{na} + \dfrac{n-1}{n} \int \cos^{n-2} ax\, dx$

Work Exercises 42–46 without using tables.

42. Find the volume of the solid generated by revolving the region bounded by $y = \sin x$ and $y = 0$ for $0 \le x \le \pi$ about the x-axis.

43. Find the volume of the solid generated by revolving the region bounded by $y = \tan x$, $y = 0$, and $x = \pi/4$ about the x-axis.

44. Find the length of the curve with equation $y = \ln(\cos x)$ from $x = -\pi/3$ to $x = \pi/3$.

45. The velocity at time t of a body moving on a coordinate axis is $v = \sin^3 t$.
(a) Find the displacement of the body when $t = 3\pi$ from its position when $t = 0$.
(b) Find the total distance the body has traveled during the interval $0 \le t \le 3\pi$.

46. The acceleration at time t of a body moving on a coordinate axis is $a = \cos^2 3t \sin^3 3t$. Find the change in the velocity of the body from time $t = 0$ to time $t = \pi$.

The remaining exercises are devoted to a substitution that enables us to find the integral of any quotient of polynomial expressions in $\sin x$ and $\cos x$. The substitution converts such an integral into the integral of a rational function of a variable t. In theory, the rational function of t can be integrated by the method of partial fractions. Suppose that

$$t = \tan(x/2) \qquad \textbf{(17)}$$

so that

$$x = 2 \tan^{-1} t. \qquad \textbf{(18)}$$

47. (a) Starting with the identity $\sin x = 2 \sin(x/2) \cos(x/2)$, express $\sin x$ in terms of $\tan(x/2)$ and then use substitution (17) to express $\sin x$ in terms of t.
(b) Starting with the identity $\cos x = 2 \cos^2(x/2) - 1$, express $\cos x$ in terms of $\tan(x/2)$ and then use substitution (17) to express $\cos x$ in terms of t.
(c) Using substitution (18), find dx in terms of t and dt.
(d) Using parts (a)–(c), find the integral in terms of t ob-

tained by using substitution (17) with the integral

$$\int \frac{dx}{1 + \sin x + \cos x}.$$

48. Use substitution (17) and the text answer for Exercise 47 to find

$$\int \frac{dx}{\sin x + \tan x}.$$

49. Use substitution (17) and the text answer for Exercise 47 to find

$$\int \frac{dx}{3 \sin x + 4 \cos x}.$$

8.5 TRIGONOMETRIC SUBSTITUTION

Integrals Involving $\sqrt{a^2 \pm x^2}$ and $\sqrt{x^2 - a^2}$

The familiar trigonometric identities

$$a^2 - a^2 \sin^2 t = a^2 \cos^2 t, \qquad \textbf{(1)}$$

$$a^2 + a^2 \tan^2 t = a^2 \sec^2 t, \qquad \textbf{(2)}$$

$$a^2 \sec^2 t - a^2 = a^2 \tan^2 t \qquad \textbf{(3)}$$

are sometimes useful in eliminating radicals from integrals involving $\sqrt{a^2 - x^2}$, $\sqrt{a^2 + x^2}$, or $\sqrt{x^2 - a^2}$. Suppose, for illustration, we wish to eliminate a radical expression $\sqrt{x^2 - a^2}$ from an integral. A substitution where $x = a \sec t$ would result in

$$\sqrt{x^2 - a^2} = \sqrt{a^2 \sec^2 t - a^2} = \sqrt{a^2 \tan^2 t}, \qquad \textbf{(4)}$$

according to identity (3). We would like to replace $\sqrt{a^2 \tan^2 t}$ by $a(\tan t)$, but we must be careful. While $\sqrt{a^2 \tan^2 t}$ is positive, $a(\tan t)$ might be positive or negative. We can assume $a > 0$ and focus on the sign of $\tan t$. We should think of the substitution that yields $x = a \sec t$ as really being defined by

$$t = \sec^{-1}(x/a), \qquad \textbf{(5)}$$

since x was the given variable and t is to be defined. By our definition of the principal values of the inverse secant function, we have

$$0 \le t < \pi/2 \qquad \text{or} \qquad \pi \le t < 3\pi/2. \qquad \textbf{(6)}$$

For these values of t in Eq. (6), the function $\tan t$ is nonnegative. Thus for the substitution $t = \sec^{-1}(x/a)$, we indeed have

$$x = a \sec t, \qquad dx = a \sec t \tan t \, dt,$$

$$\sqrt{x^2 - a^2} = \sqrt{a^2 \tan^2 t} = a(\tan t),$$

and we have eliminated the radical.

			Radical	
Radical	Substitution	x-replacement	replacement	dx-replacement
$\sqrt{a^2 - x^2}$	$t = \sin^{-1}(x/a)$	$x = a\sin t$	$a\cos t$	$dx = a\cos t\, dt$
$\sqrt{a^2 + x^2}$	$t = \tan^{-1}(x/a)$	$x = a\tan t$	$a\sec t$	$dx = a\sec^2 t\, dt$
$\sqrt{x^2 - a^2}$	$t = \sec^{-1}(x/a)$	$x = a\sec t$	$a\tan t$	$dx = a\sec t\tan t\, dt$

TABLE 8.1

Table 8.1 shows the trigonometric substitutions suggested by identities (1), (2), and (3). Although the second column of the table gives the actual definition of the substitution, we always think in terms of the *x*-replacement column when deciding which substitution to choose to eliminate the radical. We checked that the last row of table entries is correct in the preceding paragraph. In Exercises 1 and 2, we ask you to check that the first two rows in the radical replacement column are correct.

Substitutions using Table 8.1 result in integrals of trigonometric functions. These integrals are sometimes readily found or can often be found using methods of the preceding section.

EXAMPLE 1 Find

$$\int x^3\sqrt{4 - x^2}\, dx,$$

which is not given in the brief integral table on the endpapers of this book.

Solution If we use the substitution $x = 2\sin t$, then $dx = 2\cos t\, dt$, and we obtain

$$\int x^3\sqrt{4 - x^2}\, dx = \int 8\sin^3 t\,(\sqrt{4 - 4\sin^2 t})(2\cos t)\, dt$$

$$= \int 8\sin^3 t\,(4\cos^2 t)\, dt$$

$$= 32\int \sin t\,(1 - \cos^2 t)(\cos^2 t)\, dt$$

$$= 32\int (\cos^2 t - \cos^4 t)\sin t\, dt$$

$$= -\frac{32\cos^3 t}{3} + \frac{32\cos^5 t}{5} + C.$$

From $x = 2\sin t$, we obtain

$$\cos t = \sqrt{1 - \sin^2 t}$$

$$= \sqrt{1 - \left(\frac{x}{2}\right)^2} = \frac{1}{2}\sqrt{4 - x^2},$$

so we obtain for our integral

$$\int x^3\sqrt{4 - x^2}\, dx = -\frac{4(4 - x^2)^{3/2}}{3} + \frac{(4 - x^2)^{5/2}}{5} + C. \qquad \blacksquare$$

HISTORICAL NOTE

A RELATION BETWEEN trigonometric and logarithm functions is suggested by the table of integrals. Formulas 26 and 34, or 30 and 38, exhibit cases involving the square of the variable in which a change of sign means the difference between the integral's being a logarithm and its being an inverse trigonometric function. This curious relationship was noted by Roger Cotes (1682–1716) in the table of integrals he worked out and led to Cotes's result that, in modern notation,

$$i\theta = \ln(\cos\theta + i\sin\theta).$$

(Euler later rediscovered this result in the form

$$e^{i\theta} = \cos\theta + i\sin\theta.)$$

The relationship between the integrals was extensively analyzed by Nicholas Saunderson (1682–1739), who published a commentary on Cotes's work that was evidently used as a text at Cambridge.

Another mathematician who noted that a change of sign in an integrand may change the integral from a logarithm to an inverse trigonometric function was Thomas Simpson (1710–1761), who published a text on calculus in 1737. Simpson is today best remembered for Simpson's rule (Section 5.6), which he published in 1743 but which had actually been discovered by James Gregory many years earlier.

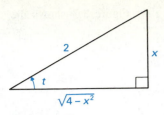

FIGURE 8.1   Triangle for $x = 2 \sin t$.

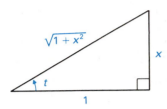

FIGURE 8.2   Triangle for $x = \tan t$.

When we make a substitution such as $x = 2 \sin t$ in Example 1 and then integrate, the answer obtained is in terms of trigonometric functions of t rather than a function of x. The triangle shown in Fig. 8.1 is helpful in figuring out the trigonometric functions of t in terms of x. If $x = 2 \sin t$, then $\sin t = x/2$, so we label the side opposite the angle t with x and the hypotenuse with 2. The remaining side is then found using the Pythagorean theorem.

EXAMPLE 2   Compute

$$\int x^2 \sqrt{1 + x^2} \; dx.$$

Solution   This time, we use the substitution $x = \tan t$, so that $dx = \sec^2 t \; dt$. The triangle in Fig. 8.2 shows the relation between t and x. We obtain

$$\int x^2 \sqrt{1 + x^2} \; dx = \int \tan^2 t \; \sqrt{1 + \tan^2 t} \; \sec^2 t \; dt$$

$$= \int \tan^2 t \; \sec^3 t \; dt$$

$$= \int (\sec^2 t - 1) \; \sec^3 t \; dt$$

$$= \int (\sec^5 t - \sec^3 t) \; dt.$$

We use reduction formula 98 for powers of secant from the table on the end-papers and obtain

$$\int \sec^5 t \; dt = \frac{\sec^3 t \tan t}{4} + \frac{3}{4} \int \sec^3 t \; dt.$$

Our integral thus takes the form

$$\int (\sec^5 t - \sec^3 t) \; dt = \frac{\sec^3 t \tan t}{4} - \frac{1}{4} \int \sec^3 t \; dt$$

$$= \frac{\sec^3 t \tan t}{4} - \frac{1}{4} \left(\frac{\sec t \tan t}{2} + \frac{1}{2} \int \sec t \; dt \right)$$

$$= \frac{\sec^3 t \tan t}{4} - \frac{1}{8} \sec t \tan t$$

$$- \frac{1}{8} \ln|\sec t + \tan t| + C.$$

From Fig. 8.2, we obtain $\sec t = \sqrt{1 + x^2}$. Thus we have

$$\int x^2 \sqrt{1 + x^2} \; dx = \frac{x(1 + x^2)^{3/2}}{4} - \frac{1}{8} x \sqrt{1 + x^2}$$

$$- \frac{1}{8} \ln|\sqrt{1 + x^2} + x| + C. \quad \blacksquare$$

EXAMPLE 3   Find $\int \sqrt{a^2 - x^2} \; dx$ without the use of tables.

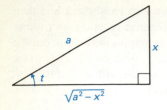

FIGURE 8.3 Triangle for $x = a \sin t$.

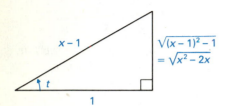

FIGURE 8.4 Triangle for $x - 1 = \sec t$.

Solution We let $x = a \sin t$, so that $dx = a \cos t \, dt$. Figure 8.3 shows the triangle. Then

$$\int \sqrt{a^2 - x^2} \, dx = \int \sqrt{a^2 - a^2 \sin^2 t} \, (a \cos t) \, dt$$

$$= a^2 \int \cos^2 t \, dt = a^2 \int \frac{1 + \cos 2t}{2} \, dt$$

$$= \frac{a^2}{2} \left(t + \frac{1}{2} \sin 2t \right) + C = \frac{a^2}{2} (t + \sin t \cos t) + C$$

$$= \frac{a^2}{2} \left(\sin^{-1} \frac{x}{a} + \frac{x}{a} \frac{\sqrt{a^2 - x^2}}{a} \right) + C$$

$$= \frac{1}{2} a^2 \sin^{-1} \frac{x}{a} + \frac{x}{2} \sqrt{a^2 - x^2} + C. \qquad \blacksquare$$

EXAMPLE 4 Find $\int_0^2 \sqrt{16 - x^2} \, dx$.

Solution The indefinite integral $\int \sqrt{16 - x^2} \, dx$ is the special case of Example 3 where $a = 4$. Thus we use $x = 4 \sin t$, corresponding to the substitution $t = \sin^{-1}(x/4)$. The x-limit zero then corresponds to the t-limit $\sin^{-1} 0 = 0$, and the x-limit 2 corresponds to the t-limit $\sin^{-1}(\frac{1}{2}) = \pi/6$. Using the third line of the computation in Example 3, we see that

$$\int_0^2 \sqrt{16 - x^2} \, dx = \frac{16}{2} \left(t + \frac{1}{2} \sin 2t \right) \Big]_{t=0}^{\pi/6}$$

$$= 8 \left(\frac{\pi}{6} + \frac{1}{2} \cdot \frac{\sqrt{3}}{2} \right) - 0 = \frac{4\pi}{3} + 2\sqrt{3}. \qquad \blacksquare$$

Integrals Involving $\sqrt{ax^2 + bx + c}$, Where $a \neq 0$

By completing the square, we may be able to handle the integral of an expression involving $\sqrt{ax^2 + bx + c}$ by the methods of this section. We illustrate with two examples.

EXAMPLE 5 Find $\int \sqrt{x^2 - 2x} \, dx$ without using tables.

Solution We have

$$\int \sqrt{x^2 - 2x} \, dx = \int \sqrt{(x - 1)^2 - 1} \, dx.$$

We let $x - 1 = \sec t$, so $dx = \sec t \tan t \, dt$. Figure 8.4 shows the triangle. Then

$$\int \sqrt{(x - 1)^2 - 1} \, dx = \int \sqrt{\sec^2 t - 1} \, \sec t \tan t \, dt$$

$$= \int \tan^2 t \, \sec t \, dt = \int (\sec^2 t - 1) \sec t \, dt$$

$$= \int (\sec^3 t - \sec t) \, dt.$$

We should know $\int \sec t\, dt$. Example 8 in Section 8.1 shows that

$$\int \sec^3 t\, dt = \frac{1}{2} \sec t \tan t + \frac{1}{2} \ln|\sec t + \tan t| + C.$$

We now obtain for our original integral

$$\int \sqrt{x^2 - 2x}\, dx = \int \sec^3 t\, dt - \int \sec t\, dt$$

$$= \tfrac{1}{2} \sec t \tan t + \tfrac{1}{2} \ln|\sec t + \tan t| - \ln|\sec t + \tan t| + C$$

$$= \tfrac{1}{2} \sec t \tan t - \tfrac{1}{2} \ln|\sec t + \tan t| + C$$

$$= \tfrac{1}{2}(x - 1)\sqrt{(x - 1)^2 - 1}$$

$$\qquad - \tfrac{1}{2} \ln|(x - 1) + \sqrt{(x - 1)^2 - 1}| + C$$

$$= \tfrac{1}{2}[(x - 1)\sqrt{x^2 - 2x} - \ln|x - 1 + \sqrt{x^2 - 2x}|] + C. \quad ■$$

EXAMPLE 6   Find $\int (1/\sqrt{1 + 4x - x^2})\, dx$.

Solution   Completing the square, we have

$$\int \frac{dx}{\sqrt{1 + 4x - x^2}} = \int \frac{dx}{\sqrt{5 - (x - 2)^2}}.$$

We use $x - 2 = \sqrt{5} \sin t$, so that $dx = \sqrt{5} \cos t\, dt$. Then

$$\int \frac{dx}{\sqrt{5 - (x - 2)^2}} = \int \frac{\sqrt{5} \cos t}{\sqrt{5 - 5 \sin^2 t}}\, dt = \int \frac{\sqrt{5} \cos t}{\sqrt{5} \cos t}\, dt$$

$$= \int 1 \cdot dt = t + C = \sin^{-1}\frac{x - 2}{\sqrt{5}} + C. \quad ■$$

SUMMARY

1. If $x = a \sin t$, then $\sqrt{a^2 - x^2} = a \cos t$.
2. If $x = a \tan t$, then $\sqrt{a^2 + x^2} = a \sec t$.
3. If $x = a \sec t$, then $\sqrt{x^2 - a^2} = a \tan t$.
4. To eliminate the radical from $\sqrt{ax^2 + bx + c}$ for $a \neq 0$, complete the square and then use the appropriate substitution.

EXERCISES 8.5

1. Verify that the radical replacement given in Table 8.1 for the substitution $t = \sin^{-1}(x/a)$ is correct. (Check the sign of the radical, as we did following Eq. 4.)

2. Repeat Exercise 1 for the substitution $t = \tan^{-1}(x/a)$ in Table 8.1.

It is our experience that students often have as much trouble choosing the appropriate trigonometric substitution and then transforming the integral as they have in finally performing the integration. In Exercises 3–24, give the appropriate substitution to eliminate the radical and then find the transformed integral. Do not perform the integration.

3. $\int \sqrt{9 - x^2}\, dx$

4. $\int \dfrac{x^2}{\sqrt{4 + x^2}}\, dx$

5. $\int \dfrac{dx}{\sqrt{x^2 - 16}}$

6. $\int (25 - x^2)^{3/2}\, dx$

7. $\int \dfrac{x + 2}{\sqrt{x^2 - 1}}\, dx$

8. $\int \dfrac{dx}{x\sqrt{4 - x^2}}$

9. $\int \sqrt{1 + 4x^2}\, dx$

10. $\int x^2\sqrt{9x^2 - 1}\, dx$

11. $\int \dfrac{x^4}{\sqrt{1 - 16x^2}}\, dx$

12. $\int \sqrt{16 - 4x^2}\, dx$

13. $\int \dfrac{dx}{(9 + 5x^2)^{3/2}}$

14. $\int (8x^2 - 4)^{5/2}\, dx$

15. $\int \sqrt{2 + 3x^2}\, dx$

16. $\int \sqrt{x^2 - 2x + 5}\, dx$

17. $\int \sqrt{x^2 + 4x + 3}\, dx$

18. $\int \dfrac{x\, dx}{\sqrt{3 - 2x - x^2}}$

19. $\int \dfrac{x^2\, dx}{\sqrt{-2x - x^2}}$

20. $\int \sqrt{4x^2 + 6x + 5}\, dx$

21. $\int \dfrac{dx}{\sqrt{8 - 4x - 3x^2}}$

22. $\int e^x\sqrt{e^{2x} + 4e^x + 5}\, dx$

23. $\int \dfrac{e^x\, dx}{\sqrt{15 + 2e^x - e^{2x}}}$

24. $\int e^{2x}\sqrt{e^{2x} + 6e^x + 5}\, dx$

In Exercises 25–46, find the integral without using tables.

25. $\int \dfrac{x}{\sqrt{1 + x^2}}\, dx$

26. $\int \dfrac{x - 3}{\sqrt{4 - x^2}}\, dx$

27. $\int_0^{\sqrt{3}} \dfrac{x^3}{\sqrt{1 + x^2}}\, dx$

28. $\int (16 - x^2)^{3/2}\, dx$

29. $\int_1^{\sqrt{5}} x^3\sqrt{x^2 - 1}\, dx$

30. $\int \sqrt{4 - x^2}\, dx$

31. $\int \dfrac{x}{\sqrt{5 + x^2}}\, dx$

32. $\int_0^1 \dfrac{dx}{\sqrt{x^2 + 1}}$

33. $\int \dfrac{x - 1}{\sqrt{x^2 - 16}}\, dx$

34. $\int \dfrac{dx}{x\sqrt{x^2 - 4}}$

35. $\int_0^{\sqrt{6}} \sqrt{1 + 4x^2}\, dx$

36. $\int \dfrac{dx}{(4 - x^2)^{3/2}}$

37. $\int \dfrac{x^3\, dx}{(9 + x^2)^{5/2}}$

38. $\int_4^5 x^3(x^2 - 16)^{3/2}\, dx$

39. $\int e^{4x}\sqrt{1 - e^{2x}}\, dx$

40. $\int \dfrac{e^{4x}}{\sqrt{1 + e^{2x}}}\, dx$

41. $\int_{-1}^0 \sqrt{x^2 + 2x + 2}\, dx$

42. $\int \sqrt{x^2 - 6x + 8}\, dx$

43. $\int_{1/2}^1 \dfrac{dx}{\sqrt{2x - x^2}}$

44. $\int \dfrac{dx}{4x^2 + 8x + 12}$

45. $\int \dfrac{x}{\sqrt{x^2 - 4x + 5}}\, dx$

46. $\int_{-1}^0 \dfrac{3x - 2}{x^2 + 2x + 2}\, dx$

47. (a) By integration, derive the formula $A = \pi r^2$ for the area of the circle $x^2 + y^2 = r^2$ of radius r.
   (b) We claim that your work in part (a) does not give a valid proof, based on our work in this text, that the formula $A = \pi r^2$ for the area of the circle is correct. Why not?

Work Exercises 48–53 without using tables.

48. Find the area of the portion of the disk $x^2 + y^2 \le 16$ that lies above the line $y = 2$.

49. Find the arc length of the parabola $y = x^2$ from $x = 0$ to $x = 1$.

50. Find the arc length of the curve $y = e^x$ from $x = 0$ to $x = \ln 4$.

51. Find the area of the surface generated by revolving the portion of the parabola $y = x^2$ for $0 \le x \le 1$ about the x-axis.

52. The velocity at time t of a body moving on a coordinate axis is $v = \sqrt{9 + t^2}/t^2$ for $t \ge 1$. Find the distance the body travels for $1 \le t \le 4$.

53. Find the general solution of the differential equation $y' = y\sqrt{1 - x^2}/\sqrt{y^2 + 1}$.

8.6   INTEGRATION USING TABLES

Integration using tables is a very important technique that you should practice. In the table of integrals on the endpapers of this book, the formulas are given using the variable x since that is the way they are stated in most tables. For a differentiable function u, it follows, from the chain rule for differentiation and $du = u'(x)\, dx$, that the formulas hold if x is replaced by u and dx by du.

Before starting to find integrals using a table, you should determine how the table is organized. Formulas do not appear in random order in a table of integrals. For example, in the table on the endpapers of this text, we start with integrals of algebraic functions and then continue with integrals involving trigonometric functions, then those involving exponential and logarithm functions, and finally those involving hyperbolic functions. These individual categories are also organized. For example, the integrals of algebraic functions start with those involving linear functions $ax + b$ and then proceed to quadratics. Within those involving linear functions, the ones not involving radicals appear first. Such organization is a big help when trying to find an integral using a table.

EXAMPLE 1   Compute

$$\int_1^{12} \frac{dx}{x\sqrt{5x + 4}}.$$

Solution   While we could solve this problem without tables using the substitution $u = \sqrt{5x + 4}$, it is faster to use a table, and probably there is less chance of error as well.

The integral involves a square root of a linear expression, and we quickly find that the appropriate formula in the table is formula 19(b).

$$\int \frac{dx}{x\sqrt{ax + b}} = \frac{1}{\sqrt{b}} \ln\left|\frac{\sqrt{ax + b} - \sqrt{b}}{\sqrt{ax + b} + \sqrt{b}}\right| + C, \quad \text{if } b > 0$$

Table formula 19(b)

Using this formula, we obtain

$$\int_1^{12} \frac{dx}{x\sqrt{5x + 4}} = \frac{1}{2} \ln\left|\frac{\sqrt{5x + 4} - 2}{\sqrt{5x + 4} + 2}\right|\Bigg]_1^{12}$$

$$= \frac{1}{2} \ln\left|\frac{8 - 2}{8 + 2}\right| - \frac{1}{2} \ln\left|\frac{3 - 2}{3 + 2}\right|$$

$$= \frac{1}{2}\left[\ln\left(\frac{3}{5}\right) - \ln\left(\frac{1}{5}\right)\right] = \frac{1}{2} \ln 3. \quad \blacksquare$$

If you have access to a computer and a program for Simpson's rule, you can use them as a fast way to estimate the integral in Example 1 to six or more significant figures.

EXAMPLE 2   Find $\int \sin^5 2x \, dx$.

Solution   The table integration formula

$$\int \sin^n ax \, dx$$

$$= \frac{-\sin^{n-1} ax \cos ax}{na} + \frac{n - 1}{n} \int \sin^{n-2} ax \, dx \qquad \textit{Table formula } 66$$

is known as a *reduction formula;* the problem of integrating $\sin^n ax$ is *reduced* to integrating $\sin^{n-2} ax$. We apply this formula repeatedly until the problem is reduced to integrating either $\sin ax$ or $\sin^0 ax = 1$. We have

$$\int \sin^5 2x \, dx = \frac{-\sin^4 2x \cos 2x}{5 \cdot 2} + \frac{4}{5} \int \sin^3 2x \, dx$$

$$= \frac{-\sin^4 2x \cos 2x}{10} + \frac{4}{5}\left(\frac{-\sin^2 2x \cos 2x}{3 \cdot 2} + \frac{2}{3} \int \sin 2x \, dx \right)$$

$$= -\frac{\sin^4 2x \cos 2x}{10} - \frac{2 \sin^2 2x \cos 2x}{15} - \frac{4}{15} \cos 2x + C. \quad \blacksquare$$

We emphasize again that each integration formula in x given in the table can be modified, using the chain rule, to give a more general result. For example, the table formula

$$\int \frac{dx}{x^2\sqrt{a^2 + x^2}} = -\frac{\sqrt{a^2 + x^2}}{a^2 x} + C, \qquad \textit{Table formula 33}$$

together with the chain rule for differentiation, yields the formula

$$\int \frac{du}{u^2\sqrt{a^2 + u^2}} = -\frac{\sqrt{a^2 + u^2}}{a^2 u} + C \qquad \text{(1)}$$

for any differentiable function u.

EXAMPLE 3   Find

$$\int \frac{\sec^2 x \, dx}{(\tan^2 x)\sqrt{3 + \tan^2 x}}.$$

Solution   It is unlikely that we would find this integral with $\tan^2 x$ and $\sec^2 x$ in *any* table. We should notice at once that $\sec^2 x$, which appears as a factor in the numerator, is the derivative of $\tan x$, which appears elsewhere in the integral. This suggests that we might be able to simplify things if we let $u = \tan x$. Then $du = \sec^2 x \, dx$, and the integral becomes

$$\int \frac{du}{u^2\sqrt{3 + u^2}}.$$

Taking $a^2 = 3$ in formula (1) and remembering that $u = \tan x$, we obtain

$$\int \frac{\sec^2 x \, dx}{(\tan^2 x)\sqrt{3 + \tan^2 x}} = -\frac{\sqrt{3 + \tan^2 x}}{3 \tan x} + C. \quad \blacksquare$$

As in the previous example, *you should train yourself to spot when the integrand contains, as a factor, a function that is the derivative of some other function u in the integrand.* Even if a table of integrals has the formula you need, discovering which formula it is sometimes requires much ingenuity.

EXAMPLE 4   Find

$$\int \frac{x \, dx}{1 + \sin(3x^2)}.$$

Solution   Our table does not contain this integral in its present form. However, we should note at once that x is the derivative of $3x^2$, except for the constant factor 6, which we can "fix up." Thus if we let $u = 3x^2$, we have $du = 6x\,dx$, so $x\,dx = \frac{1}{6}\,du$. Our integral becomes

$$\frac{1}{6} \int \frac{du}{1 + \sin u}.$$

We search the table for an appropriate formula and find

$$\int \frac{dx}{1 + \sin ax} = -\frac{1}{a} \tan\left(\frac{\pi}{4} - \frac{ax}{2}\right) + C. \qquad \textit{Table formula } 78$$

With $a = 1$ and u in place of x, this formula becomes

$$\int \frac{du}{1 + \sin u} = -\tan\left(\frac{\pi}{4} - \frac{u}{2}\right) + C.$$

Thus

$$\int \frac{x\,dx}{1 + \sin(3x^2)} = -\frac{1}{6} \tan\left(\frac{\pi}{4} - \frac{3x^2}{2}\right) + C. \qquad ∎$$

EXAMPLE 5   Find $\int \cos 7x \cos 3x\,dx$.

Solution   This time we find the formula

$$\int \cos ax \cos bx\,dx = \frac{\sin(a - b)x}{2(a - b)} + \frac{\sin(a + b)x}{2(a + b)} + C,$$

$$\textit{Table formula } 68(c)$$

which we may use directly with our integral. Taking $a = 7$ and $b = 3$, we obtain

$$\int \cos 7x \cos 3x\,dx = \frac{\sin 4x}{8} + \frac{\sin 10x}{20} + C. \qquad ∎$$

EXAMPLE 6   Find

$$\int \frac{x^3}{\sqrt{4 + 5x^2}}\,dx.$$

Solution   The table does not contain a formula that we can use directly. However, it does contain formulas with $\sqrt{ax + b}$ in the denominator, and we can think of $\sqrt{ax + b}$ as $\sqrt{au + b}$. Thus we try $u = x^2$ to see if the integral can be made to fit such a formula in the table. With $u = x^2$, we have $du = 2x\,dx$, so

$$\int \frac{x^3}{\sqrt{4 + 5x^2}}\,dx = \frac{1}{2} \int \frac{x^2 \cdot 2x}{\sqrt{4 + 5x^2}}\,dx = \frac{1}{2} \int \frac{u}{\sqrt{4 + 5u}}\,du.$$

We search for something like this with x in place of u in the table and find

$$\int \frac{x\,dx}{\sqrt{ax + b}} = \frac{2(ax - 2b)}{3a^2} \sqrt{ax + b} + C. \qquad \textit{Table formula } 17$$

We then use this formula with x replaced by $u = x^2$, $a = 5$, and $b = 4$ and obtain

$$\int \frac{x^3}{\sqrt{4 + 5x^2}}\, dx = \frac{1}{2} \cdot \frac{2(5x^2 - 8)}{75}\sqrt{5x^2 + 4} + C$$

$$= \frac{5x^2 - 8}{75}\sqrt{5x^2 + 4} + C.$$ ∎

SUMMARY

1. All formulas for $\int f(x)\, dx$ in the tables should be regarded as formulas for $\int f(u)\, du$, where u may be any differentiable function of x.

2. When faced with a complicated integral that you can't find easily in a table, try to see whether some factor of the integrand can be viewed as the derivative of some other portion u. Then try to write the integral in the form $\int f(u)\, du$ and see if you can find $\int f(x)\, dx$ in the table. See Examples 3, 4, and 6.

EXERCISES 8.6

In Exercises 1–50, find the given integral the fastest way you can, including the use of tables (but excluding looking up the answer in the back of the text).

1. $\int_2^4 x^2\, dx$

2. $\int \frac{3}{x^2\sqrt{4 + x^2}}\, dx$

3. $\int \frac{3x}{(\sqrt{4 + x^2})^3}\, dx$

4. $\int \frac{1}{x^2\sqrt{x^2 - 9}}\, dx$

5. $\int \frac{5}{x\sqrt{10x - x^2}}\, dx$

6. $\int 3x\sqrt{2x + 5}\, dx$

7. $\int \frac{-2x}{\sqrt{3 - 7x}}\, dx$

8. $\int 4x(2x + 3)^{10}\, dx$

9. $\int x^3\sqrt{4x^2 + 1}\, dx$ [*Hint:* Use table formula 13.]

10. $\int \frac{x^5}{\sqrt{2x^2 + 3}}\, dx$ [*Hint:* Use table formula 18.]

11. $\int x^3(4x^2 - 1)^7\, dx$ [*Hint:* Use table formula 8.]

12. $\int \frac{x^3}{\sqrt{4x^2 - 1}}\, dx$

13. $\int \frac{dx}{x^3\sqrt{16 - x^4}}$

[*Hint:* Multiply numerator and denominator by x.]

14. $\int \frac{dx}{x^3\sqrt{x^4 - 1}}$

[*Hint:* Multiply numerator and denominator by x.]

15. $\int \frac{dx}{x\sqrt{4x^2 - x^4}}$

[*Hint:* Multiply numerator and denominator by x.]

16. $\int \frac{1}{x^3\sqrt{4 + x^4}}\, dx$

17. $\int_0^\pi \cos^2 2x\, dx$

18. $\int_0^{\pi/4} \sin^2 x\, dx$

19. $\int \sin^4 x\, dx$

20. $\int \cos^6 2x\, dx$

21. $\int \sin 2x \cos 3x\, dx$

22. $\int_0^\pi \sin 2x \sin 4x\, dx$

23. $\int \sin 4x \cos 4x\, dx$

24. $\int \cos 2x \cos 5x\, dx$

25. $\int \sin^4 x \cos^2 x\, dx$

26. $\displaystyle\int \frac{3}{1 + \cos x}\, dx$

27. $\displaystyle\int \frac{x}{1 - \cos x^2}\, dx$

42. $\displaystyle\int \frac{\sin 3x}{\cos^2 3x \sqrt{\cos^2 3x - 25}}\, dx$

28. $\displaystyle\int x \sin 3x\, dx$

29. $\displaystyle\int x \cos 4x\, dx$

43. $\displaystyle\int \frac{\cot 2x}{\sin 2x \sqrt{\sin^2 2x + 9}}\, dx$

30. $\displaystyle\int x^2 \sin 2x\, dx$

31. $\displaystyle\int x^3 \cos x\, dx$

44. $\displaystyle\int \frac{\tan^2 x \sec^2 x}{\sqrt{3 \tan x + 4}}\, dx$

32. $\displaystyle\int \sec^2 4x\, dx$

33. $\displaystyle\int \tan^2 3x\, dx$

45. $\displaystyle\int (\sec^2 2x \tan 2x)\sqrt{1 - 4 \sec 2x}\, dx$

34. $\displaystyle\int x \cot^2 x^2\, dx$

35. $\displaystyle\int \tan^4 2x\, dx$

46. $\displaystyle\int \frac{\sec^2 x}{\tan^2 x \sqrt{25 - \tan^2 x}}\, dx$

36. $\displaystyle\int x \sec^4 x^2\, dx$

37. $\displaystyle\int x^2 \cos^2 x^3\, dx$

47. $\displaystyle\int \frac{\sec^2 x}{\tan^2 x \sqrt{25 + 4 \tan^2 x}}\, dx$

38. $\displaystyle\int x(\sin^2 x^2)(\cos x^2)\, dx$

48. $\displaystyle\int \frac{\cos x}{\sin^2 x \sqrt{4 \sin^2 x - 1}}\, dx$

39. $\displaystyle\int (\sin x \cos x)(2 \sin x + 4)^8\, dx$

49. $\displaystyle\int \frac{\tan 3x\, dx}{\sqrt{4 \sec 3x - \sec^2 3x}}$

[*Hint:* Multiply numerator and denominator by sec 3x.]

40. $\displaystyle\int \frac{\tan x \sec^2 x}{\sqrt{2 \tan x + 3}}\, dx$

50. $\displaystyle\int \frac{\tan 3x}{\sec 3x \sqrt{25 - 4 \sec^2 3x}}\, dx$

41. $\displaystyle\int \frac{\cos x}{\sin^2 x \sqrt{4 - \sin^2 x}}\, dx$

[*Hint:* Multiply numerator and denominator by sec 3x.]

SUPPLEMENTARY EXERCISES FOR CHAPTER 8

In Exercises 1–39, find the given integral without using tables.

1. $\displaystyle\int x^2 \cos 3x\, dx$

2. $\displaystyle\int \sin^{-1} 3x\, dx$

17. $\displaystyle\int \frac{dx}{x\sqrt{9 - x^2}}$

18. $\displaystyle\int \frac{x^{1/3}}{x^{1/6} + 1}\, dx$

3. $\displaystyle\int x \ln x\, dx$

4. $\displaystyle\int e^x \sin 2x\, dx$

19. $\displaystyle\int \frac{(x - 1)^{1/2} - 7}{3 - (x - 1)^{1/4}}\, dx$

20. $\displaystyle\int \frac{\sin^3 x}{\cos^2 x}\, dx$

5. $\displaystyle\int x \sin^2 x\, dx$

6. $\displaystyle\int x^5 e^{x^2}\, dx$

21. $\displaystyle\int \sin^2 4x\, dx$

22. $\displaystyle\int \tan^3 x \sec^2 x\, dx$

7. $\displaystyle\int x \cos^3 x\, dx$

8. $\displaystyle\int \tan^{-1} 2x\, dx$

23. $\displaystyle\int \sec^3 2x\, dx$

24. $\displaystyle\int x \cos^2 3x\, dx$

9. $\displaystyle\int \frac{x^4 + 1}{x^3 + x}\, dx$

25. $\displaystyle\int \cos^3 x \sin^2 x\, dx$

26. $\displaystyle\int \cos^4 2x\, dx$

10. $\displaystyle\int \frac{8x^2 + 3x - 7}{4x^3 - 12x^2 + x - 3}\, dx$

27. $\displaystyle\int \cot^2 x \csc^4 x\, dx$

28. $\displaystyle\int \cot^4 x\, dx$

11. $\displaystyle\int \frac{5x + 40}{2x^2 - 7x - 15}\, dx$

12. $\displaystyle\int \frac{x^3 - 3x + 2}{x^2 - 1}\, dx$

29. $\displaystyle\int \sin^5 2x\, dx$

30. $\displaystyle\int \cos^2 x \sin^2 x\, dx$

13. $\displaystyle\int \frac{\sqrt{x - 1}}{1 + \sqrt{x - 1}}\, dx$

14. $\displaystyle\int \frac{(x - 1)^{3/2}}{x}\, dx$

31. $\displaystyle\int \sec^4 3x \tan^2 3x\, dx$

32. $\displaystyle\int \cot^3 2x\, dx$

15. $\displaystyle\int \frac{x^3}{\sqrt{x^2 + 4}}\, dx$

16. $\displaystyle\int \frac{x}{\sqrt{x + 3}}\, dx$

33. $\displaystyle\int \sec^4 x \tan x\, dx$

34. $\displaystyle\int \frac{\cos^3 x}{\sin^5 x}$

35. $\displaystyle\int \frac{dx}{\sqrt{9 - 4x^2}}$

36. $\displaystyle\int \sqrt{16 - 9x^2}\, dx$

37. $\displaystyle\int \frac{x^2}{\sqrt{4 - x^2}}\, dx$

38. $\displaystyle\int \sqrt{4 + x^2}\, dx$

39. $\displaystyle\int \frac{dx}{\sqrt{8 - 2x - x^2}}$

40. Use the formula

$$\int \frac{dx}{x^2\sqrt{a^2 - x^2}} = -\frac{\sqrt{a^2 - x^2}}{a^2 x} + C$$

to find

$$\int \frac{\cos x}{\sin^2 x\sqrt{9 - 4\sin^2 x}}\, dx.$$

41. Use the formula

$$\int \frac{dx}{x^2\sqrt{a^2 + x^2}} = -\frac{\sqrt{a^2 + x^2}}{a^2 x} + C$$

to find

$$\int \frac{dx}{x^{5/2}\sqrt{9 + x^3}}.$$

42. Use the formula

$$\int \frac{x\,dx}{\sqrt{ax + b}} = \frac{2(ax - 2b)}{3a^2}\sqrt{ax + b} + C$$

to find

$$\int_0^{\pi/4} \frac{\sin 4x}{\sqrt{3 + \cos 2x}}\, dx.$$

43. Use the reduction formula

$$\int \cos^n ax\, dx = \frac{\cos^{n-1} ax \sin ax}{na} + \frac{n - 1}{n}\int \cos^{n-2} ax\, dx$$

to find $\int_0^{\pi/8} \cos^4 2x\, dx$.

LOCAL BEHAVIOR
OF A FUNCTION

9

It is too bad that not all functions are polynomials. Calculus of polynomial functions, even integration, is readily handled, and polynomial function values are easy for a computer to calculate. However, many functions we encounter are not polynomial functions. For example, $\sin x$ is not a polynomial function because $\sin x$ is always between -1 and 1, while every nonconstant polynomial takes on values of arbitrarily large magnitude as $x \to \infty$.

This chapter is concerned with local behavior of functions, that is, with the behavior of $f(x)$ for x very close to some point a. We first encountered this problem in Chapter 2 where we studied $\lim_{x \to a} f(x)$. In Sections 9.1 and 9.2, we assume that $f(x)$ has derivatives of orders less than or equal to n at a, and we study approximation of $f(x)$ by a polynomial function of degree n near $x = a$. As we just mentioned, polynomial functions are very convenient functions to work with. We will see that if a function has derivatives of sufficiently high order n, it is often possible to approximate $f(x)$ accurately by a polynomial of degree n for x in a fairly wide interval about a. Chapter 10 considers the limit of this process as $n \to \infty$. Section 9.3 is concerned with $\lim_{x \to a} f(x)$ as in Chapter 2, but here we use techniques of calculus and Section 9.2 to find such limits.

Sections 9.4 and 9.5 deal with the existence of $\int_a^b f(x)\, dx$ in cases where the domain of f includes the interval $a < x \le b$ but may not include a. Again, the existence of the integral for a continuous function f is determined by the local behavior of $f(x)$ for x near a. It is possible for this integral to make sense even if $\lim_{x \to a+} f(x) = \infty$, as long as $f(x)$ does not approach infinity too fast.

9.1 TAYLOR POLYNOMIALS

In this section, we try to find a good polynomial approximation of a function $f(x)$ near some point in the domain of the function. That is, we want a polynomial function g whose graph clings close to the graph of f near a designated point. We let x_0 be this point (we will be using a_i for coefficients of the approximating polynomial, so we prefer not to use a to denote this point in the domain of f). We start by observing that it is easiest to find derivatives at $x = x_0$ of a polynomial function if the polynomial is written in powers of $x - x_0$ rather than in powers of x, since $c(x - x_0)^i$ has value 0 when $x = x_0$.

If $f(x) = 4 - x + 2x^2 + x^3$, then $f(0) = 4$ is the simplest value of $f(x)$ to compute. Also, the simplest value to compute of $f'(x) = -1 + 4x + 3x^2$ is $f'(0) = -1$, and $f''(0) = 4$ is the simplest value to compute of $f''(x) = 4 + 6x$. We think of $f(x)$ as being expressed by a polynomial formula **centered at** zero.

On the other hand, if $g(x) = 2 + 3(x + 1) - 2(x + 1)^2 - 4(x + 1)^3$, then $g(-1) = 2$ is the simplest value of $g(x)$ to find. Also, $g'(-1) = 3$ is the simplest value to find for $g'(x) = 3 - 4(x + 1) - 12(x + 1)^2$, and $g''(-1) = -4$ is the simplest value to find for $g''(x) = -4 - 24(x + 1)$. Here, $g(x)$ is expressed by a polynomial formula centered at -1.

The next example indicates that any polynomial function can be expressed by a polynomial formula centered at x_0 for any real number x_0.

EXAMPLE 1 Express $f(x) = 4 - x + 2x^2 + x^3$ by a polynomial formula centered at $x_0 = 2$.

Solution We simply replace x in our given formula by $(x - 2) + 2$ and expand using the binomial theorem. We have

$$
\begin{aligned}
f(x) &= 4 - x + 2x^2 + x^3 \\
&= 4 - [(x - 2) + 2] + 2[(x - 2) + 2]^2 + [(x - 2) + 2]^3 \\
&= 4 - [(x - 2) + 2] + 2[(x - 2)^2 + 4(x - 2) + 4] \\
&\quad + [(x - 2)^3 + 6(x - 2)^2 + 12(x - 2) + 8] \\
&= (4 - 2 + 8 + 8) + (-1 + 8 + 12)(x - 2) \\
&\quad + (2 + 6)(x - 2)^2 + (x - 2)^3 \\
&= 18 + 19(x - 2) + 8(x - 2)^2 + (x - 2)^3.
\end{aligned}
$$
∎

We now proceed to try to find a good polynomial approximation of $f(x)$ near x_0. The preceding computations suggest that it is natural to work with polynomials centered at x_0. We will run into products of the form

$$2 \cdot 1, \qquad 3 \cdot 2 \cdot 1, \qquad 4 \cdot 3 \cdot 2 \cdot 1.$$

Recall that n *factorial,* written $n!$, is given by

$$n! = (n)(n - 1)(n - 2) \cdots (3)(2)(1).$$

It is convenient to define $0! = 1$. We will frequently use the relation

$$(n + 1)! = (n + 1)(n)(n - 1)(n - 2) \cdots (2)(1) = (n + 1)n!$$

In Chapter 3, we approximated a differentiable function f near a point in its domain by a linear function, namely, by the function whose graph is the tangent line to the graph of f at that point. (See Fig. 3.34 in Section 3.9, for example.) If f is differentiable at x_0, then the tangent line to the graph of f at the point $(x_0, f(x_0))$ has equation $y - f(x_0) = f'(x_0)(x - x_0)$. Thus the corresponding linear approximation to $f(x)$ is given by

$$f(x) \approx f(x_0) + f'(x_0)(x - x_0).$$

We proceed to try to find even better approximations using polynomial functions of degree greater than 1, if f has higher-order derivatives.

Let $f(x)$ be a function defined in a neighborhood of x_0, that is, for $x_0 - h < x < x_0 + h$ for some $h > 0$, and let $f^{(n)}(x_0)$ exist. We want to find the polynomial

$$g(x) = a_0 + a_1(x - x_0) + a_2(x - x_0)^2 + \cdots + a_n(x - x_0)^n \qquad \textbf{(1)}$$

of degree n centered at x_0, which approximates $f(x)$ near x_0. Surely we want to require that $g(x_0) = f(x_0)$ and that $g'(x_0) = f'(x_0)$, so that the functions have the same slope at x_0. If we also require that $g''(x_0) = f''(x_0)$, then the rates of change of slope will be the same; the graphs will bend or curve at the same rate at x_0. It seems reasonable to require that $g(x)$ have as many as possible of the same derivative values at x_0 as $f(x)$ has. We determine the coefficients $a_0, a_1, a_2, \ldots, a_n$ to make this true. Computing, we find that

$$g(x) = a_0 + a_1(x - x_0) + a_2(x - x_0)^2 + \cdots + a_n(x - x_0)^n,$$
$$g'(x) = a_1 + 2a_2(x - x_0) + 3a_3(x - x_0)^2 + \cdots + na_n(x - x_0)^{n-1},$$
$$g''(x) = 2a_2 + 3 \cdot 2a_3(x - x_0) + \cdots + n(n - 1)a_n(x - x_0)^{n-2},$$
$$\vdots \qquad\qquad \vdots$$
$$g^{(n)}(x) = n(n - 1)(n - 2) \cdots 3 \cdot 2 \cdot 1 \cdot a_n = n!a_n.$$

Then

$$g(x_0) = a_0,$$
$$g'(x_0) = a_1,$$
$$g''(x_0) = 2a_2,$$
$$g'''(x_0) = 3 \cdot 2a_3 = 3!a_3,$$
$$\vdots \qquad\qquad \vdots$$
$$g^{(n)}(x_0) = n!a_n.$$

Setting these equal to the corresponding derivatives of $f(x)$ at x_0, we have

$$f(x_0) = a_0, \qquad\qquad\qquad a_0 = f(x_0),$$
$$f'(x_0) = a_1, \qquad\qquad\qquad a_1 = f'(x_0),$$
$$f''(x_0) = 2a_2, \qquad\qquad\qquad a_2 = \frac{1}{2!}f''(x_0),$$

which lead to

$$f'''(x_0) = 3!a_3, \qquad\qquad\qquad a_3 = \frac{1}{3!}f'''(x_0),$$
$$\vdots \qquad \vdots \qquad\qquad\qquad \vdots \qquad \vdots$$
$$f^{(n)}(x_0) = n!a_n, \qquad\qquad\qquad a_n = \frac{1}{n!}f^{(n)}(x_0).$$

It is important to note that the coefficients $a_0, a_1, \ldots, a_n$ of the polynomial function $g(x)$ in Eq. (1) are *uniquely* determined by the conditions that $g^{(i)}(x_0) = f^{(i)}(x_0)$ for $i = 0, 1, \ldots, n$. (We define $f^{(0)}(x)$ to be $f(x)$.)

DEFINITION 9.1

Taylor polynomials

The polynomial function

$$g(x) = T_n(x) = f(x_0) + f'(x_0)(x - x_0) + \frac{f''(x_0)}{2!}(x - x_0)^2$$

$$+ \frac{f'''(x_0)}{3!}(x - x_0)^3 + \cdots + \frac{f^{(n)}(x_0)}{n!}(x - x_0)^n \quad (2)$$

$$= \sum_{i=0}^{n} \frac{f^{(i)}(x_0)}{i!}(x - x_0)^i$$

is the nth **Taylor polynomial for** $f(x)$ **at** x_0. (We define $f^{(0)}(x) = f(x)$ and $0! = 1$ so that we may include the constant term $f(x_0)$ in our formal sum.)

In Definition 9.1, we said "the nth Taylor polynomial" rather than "the Taylor polynomial of degree n" since it is possible that $f^{(n)}(x_0) = 0$ so that the degree of $T_n(x)$ is less than n. These polynomials are named in honor of the English mathematician Brook Taylor (1685–1731).

EXAMPLE 2 Find the Taylor polynomial $T_3(x)$ at $x_0 = 2$ for the polynomial function $f(x) = 4 - x + 2x^2 + x^3$ of Example 1.

Solution Computing derivatives, we obtain

$$f(2) = (4 - x + 2x^2 + x^3)|_2 = 4 - 2 + 8 + 8 = 18,$$
$$f'(2) = (-1 + 4x + 3x^2)|_2 = -1 + 8 + 12 = 19,$$
$$f''(2) = (4 + 6x)|_2 = 4 + 12 = 16,$$
$$f'''(2) = 6|_2 = 6.$$

Formula (2) then yields

$$T_3(x) = 18 + 19(x - 2) + \frac{16}{2!}(x - 2)^2 + \frac{6}{3!}(x - 2)^3$$

$$= 18 + 19(x - 2) + 8(x - 2)^2 + (x - 2)^3.$$

Note that this is the same polynomial expression centered at $x_0 = 2$ that we obtained in Example 1 for $f(x)$. The polynomial formula centered at x_0 for a polynomial function must yield the same derivatives at every point as the expression centered at zero. The comment on uniqueness before Definition 9.1 shows that if $f(x)$ is a polynomial function of degree n, then the polynomial expression for $f(x)$ centered at x_0 must be $T_n(x)$ at x_0, as given by formula 2. ∎

EXAMPLE 3 Find the Taylor polynomial $T_7(x)$ for the function $\sin x$ at $x_0 = 0$.

Solution We must compute $\sin 0$ and the derivatives of orders less than or equal to 7 of $\sin x$ at $x_0 = 0$. For $f(x) = \sin x$, we obtain $f(0) = \sin 0 = 0$, while

$$f'(0) = \cos 0 = 1, \qquad f''(0) = -\sin 0 = 0,$$
$$f'''(0) = -\cos 0 = -1, \qquad f^{(4)}(0) = \sin 0 = 0.$$

At this point, we note that since $f^{(4)}(x) = f(x) = \sin x$, the derivatives will start repeating. Thus the derivatives of $\sin x$ at 0, starting with the first derivative, are

$$1, \quad 0, \quad -1, \quad 0, \quad 1, \quad 0, \quad -1, \quad 0, \quad 1, \quad 0, \quad -1, \quad 0, \quad \ldots$$

for as far as we wish to take them. Since $x_0 = 0$, we have $x - x_0 = x$, and the seventh Taylor polynomial is

$$T_7(x) = 0 + 1 \cdot x + \frac{0}{2!}x^2 + \frac{-1}{3!}x^3 + \frac{0}{4!}x^4 + \frac{1}{5!}x^5 + \frac{0}{6!}x^6 + \frac{-1}{7!}x^7$$

$$= x - \frac{x^3}{3!} + \frac{x^5}{5!} - \frac{x^7}{7!}. \qquad\blacksquare$$

Figures 9.1–9.4 show the approximation to $\sin x$ by $T_n(x)$ at $x_0 = 0$ for $n = 1, 3, 9,$ and 19. The graphs of the approximations seem to cling to the graph of $\sin x$ farther and farther out as n increases. Think what might happen in the next chapter when we let n approach infinity!

EXAMPLE 4 Estimate $\sin 3°$ using the Taylor polynomial $T_3(x)$ for $\sin x$ at $x_0 = 0$. Compare with the value for $\sin 3°$ given by a calculator.

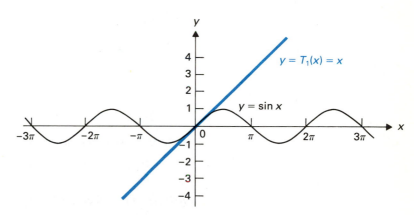

FIGURE 9.1 Approximation of $\sin x$ near zero by $T_1(x) = x$.

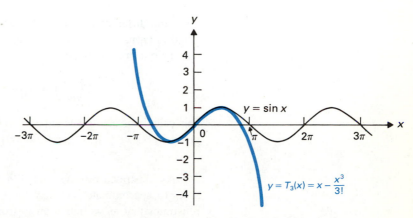

FIGURE 9.2 Approximation of $\sin x$ near zero by $T_3(x) = x - x^3/3!$.

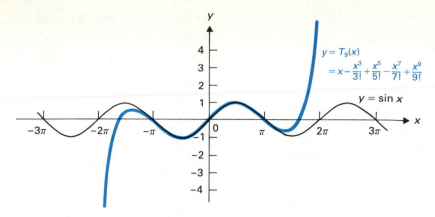

FIGURE 9.3 Approximation of $\sin x$ near zero by

$$T_9(x) = x - x^3/3! + x^5/5! - x^7/7! + x^9/9!.$$

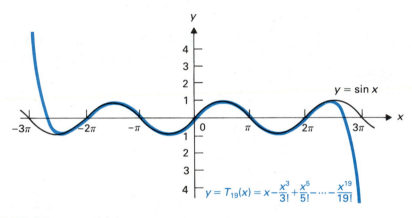

FIGURE 9.4 Approximation of $\sin x$ near zero by

$$T_{19}(x) = x - x^3/3! + x^5/5! - x^7/7! + \cdots - x^{19}/19!.$$

Solution From the computations in Example 3, we see that

$$T_3(x) = x - \frac{x^3}{3!}.$$

Since our differentiation formula for $\sin x$ and this Taylor polynomial were computed in terms of x *radians*, we must change $3°$ to $3\pi/180 = \pi/60$ radians and take as our approximation

$$T\left(\frac{\pi}{60}\right) = \frac{\pi}{60} - \frac{1}{6}\left(\frac{\pi}{60}\right)^3 \approx 0.052336.$$

Our calculator also gives $\sin 3° \approx 0.052336$, so the error in the approximation given by $T_3(x)$ is less than 0.000001. ∎

As illustrated by Figs. 9.1–9.4, we expect the Taylor polynomial for $f(x)$ at x_0 to approximate $f(x)$ best for x very close to x_0. The following example illustrates that as we move away from x_0, the error in approximation by $T_n(x)$ often increases.

EXAMPLE 5 Repeat Example 4, but this time approximate sin 30° using $T_3(x)$ for sin x at $x_0 = 0$. Compare the error in the approximation of sin 30° with the error in the approximation of sin 3° found in Example 4.

Solution Since 30° corresponds to $x = \pi/6$ radians, our approximation this time is

$$T\left(\frac{\pi}{6}\right) = \frac{\pi}{6} - \frac{1}{6}\left(\frac{\pi}{6}\right)^3$$
$$\approx 0.4996742.$$

We know that sin 30° = 0.5, so our error this time is about 0.00033, which is many times larger than the error in approximating sin 3° in Example 4. ∎

Figures 9.1–9.4 also suggest that the approximation of a function by $T_n(x)$ may improve as n becomes larger. Example 6 illustrates this.

EXAMPLE 6 Estimate sin 30° using the Taylor polynomial $T_5(x)$ for sin x at $x_0 = 0$. Compare the error in the approximation with the error using $T_3(x)$ in Example 5.

Solution The computations in Example 3 show that

$$T_5(x) = x - \frac{x^3}{3!} + \frac{x^5}{5!}.$$

Since 30° corresponds to $\pi/6$ radians, our calculator shows that

$$T_5\left(\frac{\pi}{6}\right) = \frac{\pi}{6} - \frac{1}{6}\left(\frac{\pi}{6}\right)^3 + \frac{1}{120}\left(\frac{\pi}{6}\right)^5$$
$$\approx 0.5000021.$$

Since sin 30° = 0.5, the error this time is about 0.0000021, which is much smaller than the error 0.00033 obtained using $T_3(x)$. ∎

We conclude this section with two more examples that compute Taylor polynomials. Formula (2) giving the Taylor polynomial of degree n is very important, and you should memorize it. We warn you about two mistakes that students often make:

1. Be sure to *evaluate* the derivatives at the point x_0. We have seen students use the unevaluated formulas for the derivatives of the functions in formula (2).
2. Don't forget the factorials in the denominators in formula (2).

EXAMPLE 7 Find the Taylor polynomial $T_8(x)$ for the function cos x at $x_0 = \pi$.

Solution As in Example 3, we find that cos $\pi = -1$ and that the derivatives of cos x at π, starting with the first derivative, are

$$0, \quad 1, \quad 0, \quad -1, \quad 0, \quad 1, \quad 0, \quad -1, \quad 0, \quad 1, \quad 0, \quad -1, \quad \ldots$$

for as far as we wish to take them. Dropping the terms with coefficients zero, we obtain the polynomial

$$T_8(x) = -1 + \frac{(x - \pi)^2}{2!} - \frac{(x - \pi)^4}{4!} + \frac{(x - \pi)^6}{6!} - \frac{(x - \pi)^8}{8!}. \quad \blacksquare$$

EXAMPLE 8   Find the fifth Taylor polynomial $T_5(x)$ at $x_0 = 0$ for f, where $f(x) = 1/(1 - x) = (1 - x)^{-1}$.

Solution   Computing derivatives, we obtain

$$f'(x) = (1 - x)^{-2},$$
$$f''(x) = 2(1 - x)^{-3},$$
$$f'''(x) = 3 \cdot 2(1 - x)^{-4} = 3!(1 - x)^{-4},$$
$$f^{(4)}(x) = 4!(1 - x)^{-5},$$
$$f^{(5)}(x) = 5!(1 - x)^{-6}, \ldots.$$

Thus the derivatives of f at zero, starting with the first derivative, are

$$1!, \quad 2!, \quad 3!, \quad 4!, \quad 5!, \quad 6!, \quad 7!, \quad 8!, \quad \ldots$$

for as far as we wish to go. Since $f(0) = 1$, we obtain

$$T_5(x) = 1 + x + \frac{2!}{2!}x^2 + \frac{3!}{3!}x^3 + \frac{4!}{4!}x^4 + \frac{5!}{5!}x^5$$
$$= 1 + x + x^2 + x^3 + x^4 + x^5$$

as the fifth Taylor polynomial. $\quad \blacksquare$

SUMMARY

1. If $f(x)$ has derivatives of orders less than or equal to n at x_0, then the nth Taylor polynomial for $f(x)$ at x_0 is

$$T_n(x) = f(x_0) + f'(x_0)(x - x_0) + \frac{f''(x_0)}{2!}(x - x_0)^2$$

$$+ \frac{f'''(x_0)}{3!}(x - x_0)^3 + \cdots + \frac{f^{(n)}(x_0)}{n!}(x - x_0)^n$$

$$= \sum_{i=0}^{n} \frac{f^{(i)}(x_0)}{i!}(x - x_0)^i.$$

The polynomial $T_n(x)$ and the function $f(x)$ have the same derivatives of orders less than or equal to n at x_0.

2. A polynomial function $f(x)$ of degree n expressed in powers of x can also be expressed as a polynomial centered at x_0, that is, in powers of $x - x_0$. Such an expression can be found either by replacing x by $(x - x_0) + x_0$ or by computing $T_n(x)$ at x_0.

EXERCISES 9.1

1. Approximation of $f(x)$ near x_0 by

$$T_1(x) = f(x_0) + f'(x_0)(x - x_0) \qquad (3)$$

has been studied before in this text. Find where this occurred, and specify the change in notation that makes the formula for $T_1(x)$ in Eq. (3) become the formula for the approximation given earlier in the text.

In Exercises 2–13, find the polynomial formula centered at the indicated point x_0 for the given polynomial function. Use the method in either Example 1 or Example 2 where indicated.

2. $x - 2$ at $x_0 = 3$, Example 1

3. $2x + 4$ at $x_0 = -1$, Example 1

4. $\frac{1}{2}x + 5$ at $x_0 = 3$, Example 2

5. $-5x + 2$ at $x_0 = -2$, Example 2

6. $3x^2 + 2x - 4$ at $x_0 = 1$, Example 1

7. $\frac{1}{2}x^2 - 3x + 1$ at $x_0 = -2$, Example 1

8. $3x^2 - x - 1$ at $x_0 = 5$, Example 2

9. $-4x^2 + 3x - 2$ at $x_0 = -3$, Example 2

10. $2x^3 - x^2 + 4x - 2$ at $x_0 = 2$

11. $x^4 - 3x$ at $x_0 = -1$

12. $3x^4 - 4x^3 + 2x^2 - x + 1$ at $x_0 = 1$

13. $-5x^4 + 2x^3 - 3x^2 + 2x - 1$ at $x_0 = -2$

14. Show that if f is a polynomial function of degree r, then $T_n(x) = T_r(x)$ at any point x_0 for $n \geq r$.

In Exercises 15–26, find the indicated Taylor polynomial for the function at the given point x_0. If an estimate is requested, use the Taylor polynomial you found and a calculator to make the estimate. Then find the value your calculator gives for the quantity estimated, and give the approximate error in the estimate.

15. $T_6(x)$ for e^x at $x_0 = 0$, estimate e

16. $T_5(x)$ for $\dfrac{1}{1 + x}$ at $x_0 = 0$, estimate $\dfrac{1}{1.1}$

17. $T_4(x)$ for $\sin x$ at $x_0 = \pi/2$, estimate $\sin 88°$

18. $T_2(x)$ for $\tan x$ at $x_0 = 0$, estimate $\tan 5°$

19. $T_6(x)$ for $\dfrac{1}{x}$ at $x_0 = 1$, estimate $\dfrac{1}{0.8}$

20. $T_3(x)$ for $\sqrt{x}$ at $x_0 = 4$, estimate $\sqrt{4.05}$

21. $T_4(x)$ for $x^2 + x - \cos x$ at $x_0 = 0$

22. $T_5(x)$ for $3x + 2\cos x$ at $x_0 = \pi/2$

23. $T_8(x)$ for $x^3 + e^x$ at $x_0 = 0$

24. $T_5(x)$ for $3x - e^x$ at $x_0 = -1$

25. $T_4(x)$ for $e^x + \ln(1 + x)$ at $x_0 = 0$

26. $T_{10}(x)$ for $x^5 + e^{-x}$ at $x_0 = 0$

27. (a) Find the Taylor polynomial $T_7(x)$ for $\sin 2x$ at $x_0 = 0$.
    (b) Compare your answer in part (a) with the answer in Example 3. What do you notice?

28. (a) Find the Taylor polynomial $T_3(x)$ for $\sin x^2$ at $x_0 = 0$.
    (b) Compare your answer in part (a) with the answer in Example 3, as in Exercise 27(b).
    (c) Guess the Taylor polynomial $T_{10}(x)$ for $\sin x^2$ at $x_0 = 0$.

29. (a) Find the Taylor polynomial $T_3(x)$ at $x_0 = 0$ for $1/(1 - x^2)$.
    (b) Compare your answer in part (a) with the answer in Example 8. What do you notice?
    (c) Guess the Taylor polynomial $T_8(x)$ at $x_0 = 0$ for $1/(1 - x^2)$.

30. Let f and g be functions that have derivatives of orders less than or equal to n at x_0. Show that the nth Taylor polynomial for $f + g$ at x is the sum of the Taylor polynomials for f and g at x_0.

31. Show that if $T_n(x)$ is the nth Taylor polynomial for $f(x)$ at $x_0 = 0$, then $T_n(rx)$ is the nth Taylor polynomial for $g(x) = f(rx)$ at $x_0 = 0$.

9.2 TAYLOR'S THEOREM

The preceding section presented the Taylor polynomials with the goal of approximating a function $f(x)$ near a point x_0. It is very dangerous to work with approximate quantities unless you know that the approximations you make are good enough to produce the accuracy you need. In this section we discuss the accuracy in approximating $f(x)$ by $T_n(x)$. We will discover that we can often actually compute a number $B > 0$ such that the error in this approximation has magnitude at most B.

Let $f(x)$ have derivatives of orders less than or equal to n at x_0, so that we can consider the Taylor polynomial $T_n(x)$. We expect to have

$$f(x) \approx T_n(x) \tag{1}$$

for x close to x_0. We hope that the *error* $E_n(x)$ given by

$$E_n(x) = f(x) - T_n(x) \tag{2}$$

is small if x is close to x_0. The following theorem gives some information on the size of $E_n(x)$, and for this reason the theorem is extremely important.

THEOREM 9.1   Taylor's theorem

Let $f(x)$ be defined for $x_0 - h < x < x_0 + k$, let its derivatives of orders less than or equal to $n + 1$ exist throughout that interval, and let $E_n(x) = f(x) - T_n(x)$. Then for each x in the interval, there exists a number c depending on x and strictly between x and x_0 (for $x \neq x_0$) such that

$$E_n(x) = \frac{f^{(n+1)}(c)}{(n+1)!}(x - x_0)^{n+1}. \tag{3} \quad \square$$

Exercise 1 asks you to show that for $n = 0$, Taylor's theorem actually reduces to the mean value theorem (Theorem 4.4). We will prove Taylor's theorem using Rolle's theorem (Theorem 4.3), which you recall is a special case of the mean value theorem, but we defer the proof to the end of this section.

Taylor's theorem is not difficult to remember. For f, x, and x_0 as described in the theorem, we have

$$f(x) = T_n(x) + E_n(x) = T_n(x) + \frac{f^{(n+1)}(c)}{(n+1)!}(x - x_0)^{n+1}.$$

Note that $E_n(x)$ is precisely what we would have for the "next" term of degree $n + 1$ in the Taylor polynomial $T_{n+1}(x)$, except that the derivative $f^{(n+1)}$ *must be computed at some c between x_0 and x instead of at x_0.*

We illustrate a few types of applications of Taylor's theorem. We are usually concerned with obtaining a *bound B* on the size of $E_n(x)$, that is, with finding a number $B > 0$ such that

$$|E_n(x)| \leq B.$$

EXAMPLE 1   Estimate $\sin 10° = \sin(\pi/18)$, using $T_3(x)$ at $x_0 = 0$. Then find a bound for the absolute error, $|E_3(\pi/18)|$.

Solution   (Remember that we developed calculus for trigonometric functions using *radian measure*.) Example 3 of the preceding section shows that for $\sin x$ at $x_0 = 0$, we have

$$T_7(x) = x - \frac{x^3}{3!} + \frac{x^5}{5!} - \frac{x^7}{7!}, \quad \text{so that} \quad T_3(x) = x - \frac{x^3}{3!}.$$

Then

$$T_3\left(\frac{\pi}{18}\right) = \frac{\pi}{18} - \frac{1}{6}\left(\frac{\pi}{18}\right)^3 \approx 0.1736468.$$

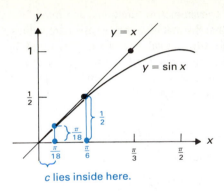

c lies inside here.

FIGURE 9.5 For

$0 < c < \pi/18, \quad \sin c < \sin(\pi/6) = \frac{1}{2};$

also,

$\sin c < \sin(\pi/18) < \pi/18.$

Now if $f(x) = \sin x$, then $f^{(4)}(x) = \sin x$ also. Then from formula (3),

$$\left| E_3\left(\frac{\pi}{18}\right) \right| = \left| \frac{\sin c}{4!} \left(\frac{\pi}{18}\right)^4 \right| \tag{4}$$

for some c where $0 < c < \pi/18$. Our task is to find a reasonable bound on $|\sin c|$ for $0 < c < \pi/18$. Of course $|\sin c| < 1$, but that is a rather crude bound for c in this interval. From Fig. 9.5, we see that $|\sin c| < \sin(\pi/6) = \frac{1}{2}$. We could also use the fact that $\sin x \le x$ for $x \ge 0$, as shown in the figure. Thus $|\sin c| < \sin(\pi/18) < \pi/18$ and we obtain, from Eq. (4),

$$\left| E_3\left(\frac{\pi}{18}\right) \right| < \frac{1}{4!}\left(\frac{\pi}{18}\right)^5 \approx 0.0000068.$$

Our calculator shows that $\sin(\pi/18) \approx 0.1736482$, so the actual error in our approximation is about 0.0000014, which is indeed less than the bound 0.0000068 that we obtained.

We could improve our bound on the error by noting that $T_4(x) = T_3(x)$ for $\sin x$ at $x_0 = 0$. Thus $T_4(\pi/18) = T_3(\pi/18)$, but formula (3) for $E_4(\pi/18)$ and the fact that $|\cos x| \le 1$ show that

$$\left| E_4\left(\frac{\pi}{18}\right) \right| = \left| \frac{\cos c}{5!}\left(\frac{\pi}{18}\right)^5 \right| < \frac{1}{5!}\left(\frac{\pi}{18}\right)^5 \approx 0.0000014.$$

Note the extra factor 5 in the denominator from the 5!, versus the 4! in the estimate in the preceding paragraph. This bound for the error is closer to the actual error than the one obtained using $E_3(\pi/18)$. ∎

EXAMPLE 2 Consider approximating $\sin 46°$ by using a Taylor polynomial $T_n(x)$ for $f(x) = \sin x$ at $x_0 = \pi/4$. Find a value of n such that the error in the approximation would be less than 0.00001.

Solution We need to have $|E_n(46\pi/180)| < 0.00001$. Since the derivatives of $\sin x$ are all either $\pm\sin x$ or $\pm\cos x$, we have $|f^{(n+1)}(c)| < 1$ for any n and c. Since $x - x_0 = \pi/180$, we have, from Eq. (2),

$$\left| E_n\left(\frac{46\pi}{180}\right) \right| < \frac{1}{(n+1)!}\left(\frac{\pi}{180}\right)^{n+1}.$$

Since $(\pi/180) < 1/50$, we have

$$\left| E_n\left(\frac{46\pi}{180}\right) \right| < \frac{1}{(n+1)!}\left(\frac{1}{50}\right)^{n+1} = \frac{1}{(n+1)!5^{n+1} \cdot 10^{n+1}}.$$

We try a few values of n and discover that for $n = 2$ we obtain

$$\left| E_2\left(\frac{46\pi}{180}\right) \right| < \frac{1}{3!5^3 \cdot 10^3} = \frac{1}{6 \cdot 125 \cdot 1000} < \frac{1}{100000} = 0.00001.$$

Thus we may use $n = 2$ with safety to achieve the desired accuracy. (Our calculator shows that $\sin 46° \approx 0.7193398$, while $T_2(46\pi/180) \approx 0.7193404$.) ∎

The preceding examples give some indication of how tables for the trigonometric functions were constructed. Also, rather than use space in the memory of an electronic computer to store tables of trigonometric functions, the manufacturer builds into the computer a program that the machine uses to estimate values of these functions as it needs them, as we estimated values in Examples 1 and 2.

Estimating $f(x)$ near x_0 using $T_1(x)$ at x_0 amounts to estimating $f(x)$ using the tangent line to its graph at x_0. This must then amount to the estimation that we discussed in Section 3.9. Indeed, if we set $\Delta x = x - x_0$ so $x = x_0 + \Delta x$, we obtain

$$T_1(x) = f(x_0) + f'(x_0)(x - x_0) = f(x_0) + f'(x_0)\Delta x.$$

We then have the linear approximation formula

$$f(x) = f(x_0 + \Delta x) \approx T_1(x_0 + \Delta x) = f(x_0) + f'(x_0)\Delta x, \qquad (5)$$

which is precisely the formula for linear approximation developed in Section 3.9, with a replaced by x_0. Theorem 9.1 now tells us that the error in this approximation is

$$E_1(x_0 + \Delta x) = \frac{f''(c)}{2!}(\Delta x)^2 \qquad (6)$$

for some c between x_0 and $x_0 + \Delta x$.

EXAMPLE 3   Use Eq. (5) to estimate $\sqrt{101}$, and find a bound for the error.

Solution   We let $f(x) = x^{1/2}$, $x_0 = 100$, and $\Delta x = 1$. Our estimate is

$$T_1(x_0 + \Delta x) = f(x_0) + f'(x_0)\,\Delta x.$$

Since $f'(x) = \frac{1}{2}x^{-1/2}$, we obtain

$$T_1(101) = \sqrt{100} + \frac{1}{2}\cdot\frac{1}{\sqrt{100}}\cdot 1 = 10 + \frac{1}{20} = 10.05$$

as our estimate for $\sqrt{101}$.

Now $f''(x) = (-\frac{1}{4})x^{-3/2}$, so for the error (6), we have

$$|E_1(101)| = \frac{1}{2!}\cdot\left|-\frac{1}{4}c^{-3/2}\right|\cdot 1 = \frac{1}{8}\cdot\frac{1}{c^{3/2}}$$

for some c where $100 < c < 101$. But for such c, the largest values of $1/c^{3/2}$ must occur where $c^{3/2}$ is smallest, that is, for c close to 100, so

$$\frac{1}{c^{3/2}} < \frac{1}{100^{3/2}} = \frac{1}{1000}.$$

Thus

$$|E_1(101)| < \frac{1}{8}\cdot\frac{1}{1000} = \frac{1}{8000} = 0.000125.$$

Therefore our estimate 10.05 for $\sqrt{101}$ is accurate to at least three decimal places.   ■

The preceding example illustrates again the *typical procedure* for finding a bound on $|E_n(x)|$. We don't know the *precise value of c*, so we attempt to find the *maximum possible value* that $|f^{(n+1)}(c)|$ can have for any c *over the entire inter-*

val from x_0 to x. If the exact maximum of $\left|f^{(n+1)}(c)\right|$ is hard to find, we find the best easily computed bound we can for $\left|f^{(n+1)}(c)\right|$.

In our next example, we consider $\lim_{n\to\infty} E_n(x)$.* In computing this limit, it is sometimes useful to know that for every real number a, we have $\lim_{n\to\infty} |a|^n/n! = 0$. We have mentioned that the exponential a^n increases very rapidly as $n\to\infty$ provided that $a > 1$. But $n!$ increases even faster! Simply stated, the reason for this is that both a^n and $n!$ are defined as the product of n numbers. For a^n, all these numbers are the same no matter how large n is, while for $n!$ most of them become much larger than $|a|$ as $n\to\infty$. If n is a great deal larger than $|a|$, then $n!$ is a lot larger than $|a|^n$. To proceed rigorously, let m be an integer such that $m > 2|a|$. Then

$$\frac{|a|^n}{n!} = \frac{|a|^m}{m!} \cdot \frac{|a|^{n-m}}{n!/m!} \leq \frac{|a|^m}{m!} \cdot \frac{|a|^{n-m}}{m^{n-m}} \leq \frac{|a|^m}{m!} \cdot \frac{1}{2^{n-m}},$$

and we see at once that we indeed have $\lim_{n\to\infty} |a|^n/n! = 0$. We say that $n!$ *dominates* $|a^n|$ in size.

EXAMPLE 4   Consider $T_n(x)$ for e^x at $x_0 = 0$. Show that for *every* real number a, we have

$$\lim_{n\to\infty} \left|e^a - T_n(a)\right| = 0.$$

This will show that for any a, we can approximate e^a accurately by $T_n(a)$ if n is large enough.

Solution   Since all derivatives of $f(x) = e^x$ are again e^x, we have

$$E_n(a) = \frac{e^c}{(n+1)!} a^{n+1}$$

for some c between 0 and a. For $a > 0$ we have $e^c < e^a$, and for $a \leq 0$ we have $e^c \leq e^0 = 1$. Thus

$$\left|e^a - T_n(a)\right| = \left|E_n(a)\right| \leq \begin{cases} \left|\dfrac{e^a \cdot a^{n+1}}{(n+1)!}\right| & \text{for } a > 0, \\[3ex] \left|\dfrac{a^{n+1}}{(n+1)!}\right| & \text{for } a \leq 0. \end{cases}$$

Now $n!$ dominates $|a^n|$ as $n\to\infty$, so we have

$$\lim_{n\to\infty} \frac{|a|^{n+1}}{(n+1)!} = 0,$$

which shows that $\lim_{n\to\infty} \left|E_n(a)\right| = 0$.   ∎

EXAMPLE 5   Consider $T_n(x)$ for $f(x) = \ln x$ at $x_0 = 1$. Show that $\lim_{n\to\infty} \left|(\ln 2) - T_n(2)\right| = 0$, so that $\ln 2$ can be approximated by $T_n(2)$ when n is large enough.

---

*We will discuss limits of this type in detail in Section 10.1. Here we rely on the intuition you developed in Section 2.3 where we discussed $\lim_{x\to\infty} f(x)$.

Solution Computing $f^{(n+1)}(x)$, we find that

$$f(x) = \ln x, \qquad f'(x) = \frac{1}{x},$$

$$f''(x) = -\frac{1}{x^2}, \qquad f'''(x) = \frac{2}{x^3},$$

$$f^{(4)}(x) = -\frac{3!}{x^4}, \qquad f^{(5)}(x) = \frac{4!}{x^5},$$

$$\vdots$$

$$f^{(n+1)}(x) = \frac{(-1)^n n!}{x^{n+1}}.$$

Then for some c where $1 < c < 2$, we have

$$|(\ln 2) - T_n(2)| = |E_n(2)|$$

$$= \left| \frac{(-1)^n n!(2-1)^{n+1}}{c^{n+1}(n+1)!} \right| < \frac{n!}{1^{n+1}(n+1)!} = \frac{1}{n+1}.$$

Since $\lim_{n \to \infty} [1/(n+1)] = 0$, we must have $\lim_{n \to \infty} |(\ln 2) - T_n(2)| = 0$. ∎

Proof of Taylor's Theorem

Proof: For $x = x_0$, the theorem is obvious. Choose $x \neq x_0$ where $x_0 - h < x < x_0 + k$. We think of x as remaining fixed throughout this proof; in particular, we consider x to be a constant in any differentiation. For such a fixed x, there exists a unique number A such that

$$f(x) = f(x_0) + f'(x_0)(x - x_0) + \frac{f''(x_0)}{2!}(x - x_0)^2 + \cdots$$

$$+ \frac{f^{(n)}(x_0)}{n!}(x - x_0)^n + \frac{A}{(n+1)!}(x - x_0)^{n+1}. \tag{7}$$

Namely, we could solve Eq. (7) for A. We now define a function $F(t)$ of the variable t by

$$F(t) = f(x) - f(t) - f'(t)(x - t) - \frac{f''(t)}{2!}(x - t)^2 - \cdots$$

$$- \frac{f^{(n)}(t)}{n!}(x - t)^n - \frac{A}{(n+1)!}(x - t)^{n+1}.$$

We see from Eq. (7) that A was chosen in such a way that $F(x_0) = 0$. Note that $F(x) = 0$ also. We now apply Rolle's theorem to $F(t)$ and deduce that $F'(c) = 0$ for some c between x_0 and x. Now

$$F'(t) = -f'(t) - [f'(t)(-1) + (x - t)f''(t)]$$

$$- \left[f''(t)(x - t)(-1) + \frac{f'''(t)}{2!}(x - t)^2 \right] - \cdots$$

$$- \left[\frac{f^{(n)}(t)}{(n-1)!}(x - t)^{n-1}(-1) + \frac{f^{(n+1)}(t)}{n!}(x - t)^n \right] - \frac{A}{n!}(x - t)^n(-1).$$

Many terms cancel in this expression for $F'(t)$; in fact, the first term in each square bracket cancels with the last term of the preceding square bracket. After this cancellation, we are left with only

$$F'(t) = -\frac{f^{(n+1)}(t)}{n!}(x - t)^n + \frac{A}{n!}(x - t)^n;$$

thus

$$F'(c) = -\frac{f^{(n+1)}(c)}{n!}(x - c)^n + \frac{A}{n!}(x - c)^n = 0,$$

and, dividing through by the common factor $(x - c)^n/n!$, we obtain

$$-f^{(n+1)}(c) + A = 0,$$

so

$$A = f^{(n+1)}(c).$$

Substituting this expression for A in Eq. (7), we obtain Taylor's theorem. □

SUMMARY

1. Taylor's theorem states that, under suitable conditions,

$$f(x) = T_n(x) + E_n(x),$$

where the error expression (remainder term) $E_n(x)$ is given by

$$E_n(x) = \frac{f^{(n+1)}(c)}{(n + 1)!}(x - x_0)^{n+1}$$

for some c (depending on x) between x_0 and x.

2. If $\left| f^{(n+1)}(c) \right| \leq M$ for all c between x and x_0, then a bound for the error $E_n(x)$ in approximating $f(x)$ by $T_n(x)$ at x_0 is given by $|E_n(x)| \leq M|x - x_0|^{n+1}/(n + 1)!$.

EXERCISES 9.2

1. State the special case of Taylor's theorem for $n = 0$. What familiar theorem do you obtain?

2. (a) Estimate 2.98^3 using the linear approximation formula (5).
   (b) Find a bound for the error in your estimate in part (a).

3. (a) Estimate $\sqrt[3]{28}$ using the linear approximation formula (5).
   (b) Find a bound for the error in your estimate in part (a).

4. (a) Using linear approximation formula (5), estimate the change in volume of a cylindrical silo 20 ft high if the radius is increased from 3 ft to 3 ft 1 in.
   (b) Find a bound for the error in your estimate in part (a).

5. (a) Using linear approximation formula (5), estimate the change in the volume of a spherical ball of radius 2 ft if the radius is increased by 1 in.
   (b) Find a bound for the error in your estimate in part (a).

6. Repeat Exercise 5(a) and (b), but estimate the increase in the volume of the ball using the Taylor polynomial $T_2(r)$ at $r_0 = 2$, rather than estimating using linear approximation formula (5).
   (c) What would be the error if you estimated using $T_3(r)$?

7. Find a bound for the error if $T_8(x)$ at $x_0 = 0$ is used to estimate $\cos 3°$ by
   (a) finding a bound for $E_8(\pi/60)$;
   (b) finding a bound for $E_9(\pi/60)$. Why can we use $E_9(\pi/60)$ as a bound for the error in $T_8(\pi/60)$?

8. (a) Estimate $\tan 2°$ using the Taylor polynomial $T_2(x)$ at $x_0 = 0$.
   (b) Find a bound for the error in your estimate in part (a).

9. (a) Estimate $\sqrt{1.05} + 0.96^3$ using two Taylor polynomials of degree 2.
   (b) Find a bound for the error in your estimate in part (a).

10. (a) Using a calculator, estimate e by using $T_8(x)$ for e^x at $x_0 = 0$.
    (b) Find a bound for the error in your estimate in part (a).

11. (a) Using a calculator, estimate $\sin 15°$ using $T_6(x)$ for $\sin x$ at $x_0 = 0$.
    (b) Find a bound for the error in your estimate in part (a).

12. (a) Using a calculator, estimate $\cosh 1$ using $T_7(x)$ for $\cosh x$ at $x_0 = 0$.
    (b) Find a bound for the error in your estimate in part (a).

13. (a) Using a calculator, estimate $\sinh 1$ using $T_8(x)$ for $\sinh x$ at $x_0 = 0$.
    (b) Find a bound for the error in your estimate in part (a).

14. Use a calculator to find the smallest value of n such that approximation of $\sin x$ by $T_n(x)$ at $x_0 = 0$ can be used to find $\sin 1$ with error less than 10^{-6}. How does this value of n compare with that given by the error term in Taylor's formula if all derivatives of $\sin x$ at $x = c$ are replaced by 1 in bounding the error?

15. Repeat Exercise 14, but use $T_n(x)$ at $x_0 = 0$ for e^x to estimate $\sqrt{e}$, and answer the second part if all derivatives at $x = c$ of e^x are replaced by 2 in bounding the error.

In Exercises 16–24, show that $\lim_{n \to \infty} |f(x) - T_n(x)| = 0$ for the indicated function $f(x)$, point x_0 where $T_n(x)$ is centered, and indicated values of x.

16. $f(x) = e^{-x}$, $x_0 = 0$, all x
17. $f(x) = \sin x$, $x_0 = 0$, all x
18. $f(x) = \cos x$, $x_0 = 0$, all x
19. $f(x) = \sinh x$, $x_0 = 0$, all x
20. $f(x) = \cosh x$, $x_0 = 0$, all x
21. $f(x) = e^x - e^{3x}$, $x_0 = 0$, all x
22. $f(x) = \ln x$, $x_0 = 1$, $1 \le x \le 2$
23. $f(x) = x^{3/2}$, $x_0 = 1$, $1 \le x \le 2$
24. $f(x) = \sqrt{x + 2}$, $x_0 = 0$, $0 \le x \le 2$

9.3 INDETERMINATE QUOTIENT FORMS (0/0 TYPE)

In Chapter 2, we found limits of quotients $f(x)/g(x)$ in the case where both the numerator and denominator approach zero. Trivial but illustrative examples of such limits are

$$\lim_{x \to 0} \frac{2x}{3x} = \frac{2}{3}, \qquad \lim_{x \to 0} \frac{x^2}{2x} = 0, \qquad \text{and} \qquad \lim_{x \to 0} \frac{2x}{x^3} = \infty.$$

These three examples illustrate that $\lim_{x \to a} f(x)/g(x)$ cannot be determined if all we know is that $\lim_{x \to a} f(x) = \lim_{x \to a} g(x) = 0$. Under these circumstances, it may be that $\lim_{x \to a} f(x)/g(x)$ does not exist, or it may be any real number. Consequently, we say that, where $\lim_{x \to a} f(x) = \lim_{x \to a} g(x) = 0$, the expression $\lim_{x \to a} f(x)/g(x)$ is an **indeterminate form of type 0/0**. The limit of this quotient is not *determined* by knowing the limits of the numerator and of the denominator at the point. In Chapter 2, we attempted to find limits of such indeterminate quotient forms by canceling from the numerator and denominator a common factor that was causing them to approach zero. Such factors may be hard to find. In this section, we present a technique known as l'Hôpital's rule that does not require factoring.

L'Hôpital's Rule

We state and illustrate l'Hôpital's rule. Exercises 34 and 35 lead you step by step through its proof. After we have illustrated the rule, we will use Taylor polynomials to gain more insight into limits of 0/0-type indeterminate forms and to prove the rule for many functions f and g.

THEOREM 9.2   L'Hôpital's rule

If $\lim_{x \to a} f(x) = \lim_{x \to a} g(x) = 0$ and $\lim_{x \to a} f'(x)/g'(x)$ is either a finite number L or is ∞ or $-\infty$, then

$$\lim_{x \to a} \frac{f(x)}{g(x)} = \lim_{x \to a} \frac{f'(x)}{g'(x)}. \tag{1} \quad \square$$

EXAMPLE 1   In Example 9 of Section 2.1, we found

$$\lim_{x \to -1} \frac{x^2 - 1}{x^2 + 3x + 2}$$

by factoring. Find this limit using l'Hôpital's rule.

Solution   Note that $\lim_{x \to -1} (x^2 - 1) = 0$ and $\lim_{x \to -1} (x^2 + 3x + 2) = 0$, so we may attempt l'Hôpital's rule. Using the rule, we obtain

$$\lim_{x \to -1} \frac{x^2 - 1}{x^2 + 3x + 2} = \lim_{x \to -1} \frac{2x}{2x + 3} = \frac{-2}{1} = -2.$$

No factoring was necessary!   ∎

Warning: You must not apply l'Hôpital's rule to find $\lim_{x \to a} f(x)/g(x)$ unless $\lim_{x \to a} f(x) = \lim_{x \to a} g(x) = 0$.

$$\textbf{Wrong:} \quad \lim_{x \to 1} \frac{x^2 - 1}{x^2 + 1} = \lim_{x \to 1} \frac{2x}{2x} = \frac{2}{2} = 1.$$

While $\lim_{x \to 1} (x^2 - 1) = 0$, we have $\lim_{x \to 1} (x^2 + 1) = 1 \neq 0$. Thus we cannot expect l'Hôpital's rule to be valid. Indeed, its application gave the incorrect value 1 instead of the correct value 0.

EXAMPLE 2   Use l'Hôpital's rule to evaluate the fundamental limit $\lim_{x \to 0} (\sin x)/x$.

Solution   Since $\lim_{x \to 0} (\sin x) = \lim_{x \to 0} x = 0$, we can apply l'Hôpital's rule, obtaining

$$\lim_{x \to 0} \frac{\sin x}{x} = \lim_{x \to 0} \frac{\cos x}{1} = \frac{1}{1} = 1.$$

We point out that this computation cannot be considered to be another derivation of this famous limit since the fact that $\lim_{x \to 0} (\sin x)/x = 1$ was used in our proof that the derivative of $\sin x$ is $\cos x$.   ∎

HISTORICAL NOTE

L'HÔPITAL'S RULE appeared in 1696 in the first calculus textbook, *L'analyse des infiniment petits* (*Analysis of Infinitely Small Quantities*) by the French marquis Guillaume François Antoine de l'Hôpital (1661–1704). L'Hôpital's brief proof of the rule, using differentials, does not meet modern standards, but his examples were not trivial. The first one, communicated to him several years earlier by Johann Bernoulli, was to find the limit as x approaches a of

$$\frac{\sqrt{2a^3x - x^4} - a\sqrt[3]{a^2x}}{a - \sqrt[4]{ax^3}}.$$

In fact, though the rule is named for the marquis, it had been discovered by Bernoulli in the early 1690s. In 1691 l'Hôpital had asked Bernoulli to send him, for a good fee, lectures on the new calculus, articles about which were just beginning to appear in scientific journals. For a large monthly salary, Bernoulli continued sending l'Hôpital material about the calculus, including any new discoveries he made, and also agreed to give no one else access to them. In effect, Bernoulli was employed by l'Hôpital. Thus when l'Hôpital felt he understood the material well enough to publish a text, he had no compunction about using Bernoulli's organization and discoveries with only a bare acknowledgment.

Sometimes $\lim_{x \to a} f'(x)/g'(x)$ is again a 0/0-type indeterminate form. Under these circumstances, if f and g have derivatives of order 2, we can apply l'Hôpital's rule again, obtaining

$$\lim_{x \to a} \frac{f(x)}{g(x)} = \lim_{x \to a} \frac{f'(x)}{g'(x)} = \lim_{x \to a} \frac{f''(x)}{g''(x)},$$

provided that this final limit is either a number L or ∞ or $-\infty$. This chain can be continued to third derivatives, fourth derivatives, and so on, *as long as we keep obtaining 0/0-type forms*.

EXAMPLE 3   Use l'Hôpital's rule to find $\lim_{t \to 0} (t - \sin t)/t^3$.

Solution   The limit of both the numerator and the denominator is 0 as $t \to 0$, so we are dealing with a 0/0-type indeterminate form. L'Hôpital's rule yields

$$\lim_{t \to 0} \frac{t - \sin t}{t^3} = \lim_{t \to 0} \frac{1 - \cos t}{3t^2} = \lim_{t \to 0} \frac{\sin t}{6t} = \lim_{t \to 0} \frac{\cos t}{6} = \frac{1}{6}.$$

We kept applying l'Hôpital's rule until we no longer obtained a 0/0-type form. ■

Variants of L'Hôpital's Rule

It can be shown that l'Hôpital's rule is also valid for 0/0-type forms involving one-sided limits or limits as we approach ∞ or $-\infty$. (See Exercise 36.) Remember, we must always check that the limits of both the numerator and the denominator are zero. Our next example illustrates the case for the limit as $x \to \infty$, and it also illustrates that we should always simplify as much as possible before applying l'Hôpital's rule.

EXAMPLE 4   Find $\lim_{x \to \infty} [(e^{1/x} - 1)/(1/x)]$.

Solution   We note that both $e^{1/x} - 1$ and $1/x$ approach 0 as $x \to \infty$, so we can apply l'Hôpital's rule. We obtain

$$\lim_{x \to \infty} \frac{e^{1/x} - 1}{1/x} = \lim_{x \to \infty} \frac{-(1/x^2)e^{1/x}}{-1/x^2}$$
$$= \lim_{x \to \infty} e^{1/x} = e^0 = 1. \qquad \textit{Simplifying}$$

Note that the second limit is still a 0/0-type form, but we simplified it as indicated to obtain a limit that we could evaluate. If we had simply kept applying l'Hôpital's rule to successive 0/0-type forms without simplifying, the process would never have ended. ■

EXAMPLE 5   Find $\lim_{x \to 0+} (e^x - 1)/\sqrt{x}$.

Solution   This is a one-sided limit. Since $\lim_{x \to 0+} (e^x - 1) = \lim_{x \to 0+} \sqrt{x} = 0$, we apply l'Hôpital's rule, obtaining

$$\lim_{x \to 0+} \frac{e^x - 1}{\sqrt{x}} = \lim_{x \to 0+} \frac{e^x}{1/(2\sqrt{x})} = \lim_{x \to 0+} e^x 2\sqrt{x} = 1 \cdot 2 \cdot 0 = 0.$$

Once again, we simplified our quotient after applying l'Hôpital's rule. ■

L'Hôpital's Rule and Taylor Polynomials

Consider a polynomial function

$$h(x) = c_0 + c_1(x - a) + c_2(x - a)^2 + \cdots + c_n(x - a)^n.$$

Suppose that $c_0 \neq 0$. Then c_0 is the *dominant term* of the polynomial for x very close to a, because if x is close enough to a, then the magnitudes $|c_i(x - a)^i|$ are all very small compared with $|c_0|$. On the other hand, if $c_0 = 0$ and $c_1 \neq 0$, then $c_1(x - a)$ is the dominant term of the polynomial for x very close to a, because $c_2(x - a)^2, \ldots, c_n(x - a)^n$ are all of small magnitude compared with $|c_1(x - a)|$ for x very close to a. Continuing this argument, we see that $h(x)$ *is dominated by its nonzero term of lowest degree for x very close to a.*

Suppose now that $f(x)$ and $g(x)$ satisfy the requirements of Taylor's theorem and have nonzero Taylor polynomials at $x = a$. Let n be the smallest positive integer such that $f^{(n)}(a) \neq 0$ and let m be the smallest positive integer such that $g^{(m)}(a) \neq 0$. By Taylor's theorem, we have

$$f(x) = \frac{f^{(n)}(a)}{n!}(x - a)^n + \frac{f^{(n+1)}(c_1)}{(n + 1)!}(x - a)^{n+1}$$

and

$$g(x) = \frac{g^{(m)}(a)}{m!}(x - a)^m + \frac{g^{(m+1)}(c_2)}{(m + 1)!}(x - a)^{m+1}$$

for x sufficiently close to a and some c_1 and c_2 between a and x. Then

$$f(x) = \frac{f^{(n)}(a)}{n!}(x - a)^n \left[1 + \frac{1}{n + 1} \cdot \frac{f^{(n+1)}(c_1)}{f^{(n)}(a)}(x - a) \right], \tag{2}$$

and a similar expression holds for $g(x)$. Now suppose that the derivatives $f^{(n+1)}(x)$ and $g^{(m+1)}(x)$ are bounded near a, so that $|f^{(n+1)}(x)|$ and $|g^{(m+1)}(x)|$ are both less than some number M for x sufficiently close to a. Then the term in square brackets in Eq. (2) approaches 1 as $x \to a$, and we see that

$$\lim_{x \to a} \frac{f(x)}{g(x)} = \lim_{x \to a} \frac{\dfrac{f^{(n)}(a)}{n!}(x - a)^n}{\dfrac{g^{(m)}(a)}{m!}(x - a)^m} = \lim_{x \to a} \frac{(m!)\, f^{(n)}(a)}{(n!)\, g^{(m)}(a)}(x - a)^{n-m}. \tag{3}$$

Thus when Taylor's theorem holds and these derivatives are bounded near a, we see that $\lim_{x \to a} f(x)/g(x)$ is equal to the limit as $x \to a$ of the quotients of the dominating terms of the nonzero Taylor polynomials at $x = a$. In particular, if $m = n$ in Eq. (3), then this limit is $f^{(n)}(a)/g^{(n)}(a)$. Of course, this is the result that l'Hôpital's rule would give, and it proves the rule for such functions.

We illustrate this discussion with some examples. For these examples and some of the exercises, we note that the nth Taylor polynomials for the functions $\sin x$, $\cos x$, e^x, and $1/(1 - x)$ at 0 are the terms of degree less than or equal to n from the sum shown after the function as follows:

$$\sin x: \qquad x - \frac{x^3}{3!} + \frac{x^5}{5!} - \frac{x^7}{7!} + \frac{x^9}{9!} - \cdots, \tag{4}$$

$$\cos x: \qquad 1 - \frac{x^2}{2!} + \frac{x^4}{4!} - \frac{x^6}{6!} + \frac{x^8}{8!} - \cdots, \tag{5}$$

$$e^x: \qquad 1 + x + \frac{x^2}{2!} + \frac{x^3}{3!} + \frac{x^4}{4!} + \frac{x^5}{5!} + \cdots, \tag{6}$$

$$\frac{1}{1-x}: \qquad 1 + x + x^2 + x^3 + x^4 + x^5 + \cdots. \tag{7}$$

This is easily verified using the definition of the Taylor polynomials in Section 9.1. Note that each derivative of every order n of $\sin x$, $\cos x$, e^x, and $1/(1-x)$ is bounded near $x = 0$.

EXAMPLE 6 Find $\lim_{x \to 0} x/[(\sin x) - x]$ without using l'Hôpital's rule.

Solution The Taylor polynomial $T_n(x)$ for $(\sin x) - x$ at $x = 0$, where $n \geq 1$, is obtained by taking the nth Taylor polynomial for $\sin x$ and subtracting the nth Taylor polynomial for x there. (See Exercise 30 of Section 9.1.) Of course, the nth Taylor polynomial for x at $x = 0$ is x. Thus the nonzero Taylor polynomials for $(\sin x) - x$ have dominant term $-x^3/3!$, and we have

$$\lim_{x \to 0} \frac{x}{(\sin x) - x} = \lim_{x \to 0} \frac{x}{-x^3/3!} = \lim_{x \to 0} \frac{-6}{x^2} = -\infty. \qquad \blacksquare$$

EXAMPLE 7 Find $\lim_{x \to 0} (e^x - 1 - x)/x^2$ without using l'Hôpital's rule.

Solution Reasoning as in Example 6 and using expression (6), we see that the dominant term of nonzero Taylor polynomials of $e^x - 1 - x$ at $x = 0$ is $x^2/2! = x^2/2$. Therefore

$$\lim_{x \to 0} \frac{e^x - 1 - x}{x^2} = \lim_{x \to 0} \frac{x^2/2}{x^2} = \frac{1}{2}.$$

Since the dominant terms of the nonzero Taylor polynomials are of degree 2, it would take two applications of l'Hôpital's rule to find this limit. $\qquad \blacksquare$

It is their understanding of this role of dominant terms of nonzero Taylor polynomials that enables instructors to make up limit problems that require many differentiations to solve using l'Hôpital's rule.

EXAMPLE 8 Find a 0/0-type indeterminate form problem that requires six repeated applications to solve if l'Hôpital's rule is used.

Solution Since the Taylor polynomials for $\sin x$ at $x = 0$ start

$$x - \frac{x^3}{3!} + \frac{x^5}{5!}, \tag{8}$$

it seems likely that the Taylor polynomials of $\sin x^2$ at $x = 0$ start

$$x^2 - \frac{x^6}{3!} + \frac{x^{10}}{5!},$$

which we obtain from the polynomial (8) by replacing x by x^2. (See Exercise 28 in Section 9.1.) This can indeed be shown. Thus the dominant term of the nonzero Taylor polynomials for $x^2 - \sin x^2$ is $x^6/3!$. Therefore

$$\lim \frac{x^2 - \sin x^2}{x^6} = \lim \frac{x^6/3!}{x^6} = \frac{1}{6},$$

but to find this limit using l'Hôpital's rule would require six applications of the rule since the first nonzero derivatives are the sixth derivatives. ■

SUMMARY

1. An indeterminate form of 0/0 type is $\lim_{x \to a} f(x)/g(x)$ where $\lim_{x \to a} f(x) = \lim_{x \to a} g(x) = 0$.

2. *L'Hôpital's Rule:* If $\lim_{x \to a} f(x)/g(x)$ is an indeterminate form of 0/0 type and $\lim_{x \to a} f'(x)/g'(x)$ is either a real number L or ∞ or $-\infty$, then

$$\lim_{x \to a} \frac{f(x)}{g(x)} = \lim_{x \to a} \frac{f'(x)}{g'(x)}.$$

   L'Hôpital's rule is also valid for one-sided limits and limits at ∞ and $-\infty$.

3. The nonzero term of lowest degree of a polynomial function centered at a dominates the other terms for x very close to a.

4. If $f(x)$ and $g(x)$ satisfy the hypotheses for Taylor's theorem at $x = a$ and if their derivatives of all orders are bounded for all x sufficiently close to a, then $\lim_{x \to a} f(x)/g(x)$ is equal to the limit of the corresponding quotient of the dominant terms of their nonzero Taylor polynomials at a.

EXERCISES 9.3

In Exercises 1–22, find the limit using l'Hôpital's rule, if the rule is appropriate.

1. $\lim_{x \to 1} \dfrac{3x^2 + 2x - 5}{x^4 + 3x^2 - 4}$

2. $\lim_{t \to -1} \dfrac{5t^4 + t - 4}{2t^3 + 5t + 7}$

3. $\lim_{u \to 0} \dfrac{e^u - 1}{u^2 + 5u}$

4. $\lim_{x \to 0} \dfrac{\sin 2x}{e^{3x} - 1}$

5. $\lim_{t \to 1} \dfrac{\ln t}{1 - t}$

6. $\lim_{x \to 2} \dfrac{\ln(x/2)}{x^3 - 2x - 4}$

7. $\lim_{x \to \pi/2} \dfrac{\cos x}{x - \pi/2}$

8. $\lim_{x \to 3} \dfrac{x - \sqrt{3x}}{27 - x^3}$

9. $\lim_{x \to 0} \dfrac{\cos x}{x^2}$

10. $\lim_{t \to 0} \dfrac{\ln(\cos t)}{\sin^2 t}$

11. $\lim_{x \to 0} \dfrac{\sin x}{e^x - e^{-x}}$

12. $\lim_{u \to 0} \dfrac{\sin u^2}{\cos^2 u - 1}$

13. $\lim_{t \to 0} \dfrac{\sin t - t}{e^{t^2} - 1}$

14. $\lim_{x \to 0} \dfrac{\sin x^2}{e^x - 1 - x}$

15. $\lim_{x \to 0^+} \dfrac{\ln(1 - x)}{x^2}$

16. $\lim_{x \to 0^-} \dfrac{\sin x - x}{x^4}$

17. $\lim_{u \to 0^+} \dfrac{\cos u - 1}{u^5}$

18. $\lim_{x \to 0^-} \dfrac{\cos x - 1}{\sin x - x}$

19. $\lim\limits_{x\to\infty} \dfrac{\tan(1/x)}{2/x}$

20. $\lim\limits_{t\to\infty} \dfrac{\cos(1/t)-1}{e^{1/t^2}-1}$

21. $\lim\limits_{u\to\infty} \dfrac{\sin(2/u)}{\cos(1/u)-1}$

22. $\lim\limits_{x\to-\infty} \dfrac{e^{1/x}-1}{1/x^2}$

In Exercises 23–32, find the limit by finding the quotient of the dominant terms of the nonzero Taylor polynomials at the point where the limit is taken. Use the sums (4)–(7) and see Examples 6, 7, and 8. You may assume that if $T_n(x)$ is the nth Taylor polynomial of $f(x)$, then the nth Taylor polynomial of $f(rx)$ is $T_n(rx)$ and the (mn)th Taylor polynomial of $f(x^m)$ is $T_n(x^m)$, where m is a positive integer and r is a constant.

23. $\lim\limits_{x\to0+} \dfrac{\cos x-1}{x^3}$

24. $\lim\limits_{t\to0+} \dfrac{t-\sin t}{t^4}$

25. $\lim\limits_{u\to0} \dfrac{(\cos u^3)-1}{(\sin u^2)-u^2}$

26. $\lim\limits_{x\to0-} \dfrac{\dfrac{1}{1-x}-1}{\cos x-1}$

27. $\lim\limits_{x\to0} \dfrac{\dfrac{1}{1-x^3}-1}{x-\sin x}$

28. $\lim\limits_{t\to0} \dfrac{(\sin 2t)-2t}{\dfrac{1}{1-t^3}-1}$

29. $\lim\limits_{u\to0} \dfrac{(\cos u^4)-1}{\dfrac{1}{1-u^6}-1}$

30. $\lim\limits_{s\to0} \dfrac{e^{2s^3}-1}{s-\sin s}$

31. $\lim\limits_{x\to0} \dfrac{e^{x^4}-1}{1-\cos x^2}$

32. $\lim\limits_{t\to0} \dfrac{\sin t^2-t^2}{e^{t^3}-1-t^3}$

33. Under the conditions for which Eq. (3) is true, describe $\lim_{x\to a+} f(x)/g(x)$ and $\lim_{x\to a-} f(x)/g(x)$ in the following cases.
(a) $n>m$ (b) $n=m$ (c) $n<m$ and $n-m$ odd
(d) $n<m$ and $n-m$ even

34. *Cauchy mean value theorem:* Let f and g be continuous on $[a,b]$ and differentiable for $a<x<b$. Let $g'(x)\neq0$ for $a<x<b$. Let
$$h(x)=(f(b)-f(a))g(x)-(g(b)-g(a))f(x).$$
(a) Show that $h(x)$ satisfies the hypotheses for Rolle's theorem (Theorem 4.3) on $[a,b]$. (Don't forget the continuity and differentiability hypotheses.)
(b) Deduce from Rolle's theorem that there exists c, where $a<c<b$, such that
$$(f(b)-f(a))g'(c)=(g(b)-g(a))f'(c).$$

(c) Using the hypotheses of this problem and Rolle's theorem, show that $g(b)-g(a)\neq0$.
(d) *Cauchy mean value theorem:* Show that part (b) can be rewritten
$$\frac{f(b)-f(a)}{g(b)-g(a)}=\frac{f'(c)}{g'(c)}$$
for some c, where $a<c<b$.

35. *L'Hôpital's rule:* Let f and g be continuous and differentiable in a neighborhood $a-h<x<a+h$ of a. Suppose $f(a)=g(a)=0$. Let $g'(x)\neq0$ in this neighborhood, except possibly at a. Finally, suppose $\lim_{x\to a} f'(x)/g'(x)$ exists or is ∞ or $-\infty$.
(a) By applying the Cauchy mean value theorem (Exercise 34) to the interval $[a,x]$, where $a<x<a+h$, show that
$$\lim_{x\to a+}\frac{f(x)}{g(x)}=\lim_{x\to a+}\frac{f'(x)}{g'(x)}.$$
(b) By applying the Cauchy mean value theorem to the interval $[x,a]$ for $a-h<x<a$, show that
$$\lim_{x\to a-}\frac{f(x)}{g(x)}=\lim_{x\to a-}\frac{f'(x)}{g'(x)}.$$
(c) *L'Hôpital's rule:* Deduce that
$$\lim_{x\to a}\frac{f(x)}{g(x)}=\lim_{x\to a}\frac{f'(x)}{g'(x)}.$$

36. *L'Hôpital's rule at ∞:* Suppose $\lim_{x\to\infty}(f(x))/(g(x))$ corresponds to a $0/0$-type indeterminate form. Let $x=1/u$, $F(u)=f(1/u)$, and $G(u)=g(1/u)$.
(a) Show that $\lim_{u\to0+}(F(u))/(G(u))$ corresponds to a $0/0$-type indeterminate form.
(b) *L'Hôpital's rule at ∞:* Assuming that the hypotheses in Exercise 35 for l'Hôpital's rule hold for F and G at 0, show that
$$\lim_{x\to\infty}\frac{f(x)}{g(x)}=\lim_{x\to\infty}\frac{f'(x)}{g'(x)}.$$

In Exercises 37–42, try to find the limit in the indicated exercise using a calculator or computer rather than the methods of this section.

37. Exercise 3   **38.** Exercise 10
39. Exercise 13   **40.** Exercise 20
41. Exercise 25   **42.** Exercise 28

9.4 OTHER INDETERMINATE FORMS

In the preceding section, we discussed the $0/0$-type indeterminate forms and saw how l'Hôpital's rule sometimes can be used to evaluate such limits. In this section, we continue our discussion with other indeterminate quotient forms, followed by indeterminate product, sum, difference, and exponential forms. We start with indeterminate quotient forms because many other types of indeterminate forms can be reduced to quotient forms for evaluation.

Indeterminate Quotient Forms

We have seen that $0/0$ is an indeterminate quotient form. The limits

$$\lim_{x \to 2+} \frac{1/(x - 2)}{2/(x - 2)} = \frac{1}{2} \quad \text{and} \quad \lim_{x \to 2+} \frac{1/(x - 2)}{3/(x - 2)^2} = 0,$$

where both the numerator and denominator in each limit approach ∞ as $x \to 2+$, show that we should consider ∞/∞ to be an indeterminate form.

Note that $0/\infty$ is not an indeterminate form, for if $\lim_{x \to a} f(x) = 0$ and $\lim_{x \to a} g(x) = \infty$, then $\lim_{x \to a} f(x)/g(x) = 0$. Also, $2/0$ is not an indeterminate form, for if $\lim_{x \to a} f(x) = 2$ and $\lim_{x \to a} g(x) = 0$, then $\lim_{x \to a} f(x)/g(x)$ is always undefined, and the quotient $f(x)/g(x)$ becomes large in absolute value as x approaches a.

Indeterminate Quotient Forms

$$\frac{0}{0}, \qquad \pm\frac{\infty}{\infty}$$

It can be shown that l'Hôpital's rule, stated in Theorem 9.2, also is valid for the ∞/∞-type indeterminate forms. Exercise 49 shows this for some of these forms. Just as for the $0/0$-type forms, the limit can be one-sided or at ∞ or $-\infty$ as well as at any point a.

EXAMPLE 1   We have often said that the exponential function e^x increases faster for large x than any monomial function x^n does. Demonstrate this by showing that $\lim_{x \to \infty} e^x/x^n = \infty$ for every positive integer n.

Solution   Since $\lim_{x \to \infty} e^x = \infty$ and $\lim_{x \to \infty} x^n = \infty$, we can apply l'Hôpital's rule. We obtain

$$\lim_{x \to \infty} \frac{e^x}{x^n} = \lim_{x \to \infty} \frac{e^x}{nx^{n-1}}$$

$$= \lim_{x \to \infty} \frac{e^x}{n(n-1)x^{n-2}} = \cdots = \lim_{x \to \infty} \frac{e^x}{n!} = \infty,$$

where we have applied l'Hôpital's rule a total of n times.   ∎

It is convenient that l'Hôpital's rule can be used to handle both indeterminate quotient forms $0/0$ and ∞/∞, for as we will see, other types of indeterminate forms often can be modified to appear as quotient types.

Indeterminate Product Forms

Turning to products, we find that the limits

$$\lim_{x \to 1+} (x - 1)\frac{3}{x - 1} = 3 \quad \text{and} \quad \lim_{x \to 1+} (x - 1)^2\left(\frac{2}{x - 1}\right) = 0$$

show that we should consider $0 \cdot \infty$ to be an indeterminate form.

Indeterminate Product Forms

$$0 \cdot \infty, \qquad 0(-\infty), \qquad \infty \cdot 0, \qquad (-\infty)0$$

The indeterminate product forms are converted to quotient forms as follows. Suppose that $\lim_{x \to a} f(x) = 0$ and $\lim_{x \to a} g(x) = \infty$. Then

$$\lim_{x \to a} [f(x) \cdot g(x)] = \lim_{x \to a} \frac{f(x)}{1/g(x)},$$

and the second limit is an indeterminate form of $0/0$ type. A mathematically unjustified but mnemonically helpful way to remember how to convert a $0 \cdot \infty$-type problem to a $0/0$-type problem is to write $0 \cdot \infty$ as $0/(1/\infty)$, which is of type $0/0$. Symbolically, we write

$$0 \cdot \infty \sim \frac{0}{1/\infty} \sim \frac{0}{0}.$$

EXAMPLE 2   Compute $\lim_{x \to 0+} (\cot x)[\ln(1 - x)]$.

Solution   Since $\lim_{x \to 0+} \cot x = \infty$ and $\lim_{x \to 0+} \ln(1 - x) = 0$, this is an $\infty \cdot 0$-type form. We convert it to a $0/0$-type problem and apply l'Hôpital's rule:

$$\lim_{x \to 0+} (\cot x)[\ln(1 - x)] = \lim_{x \to 0+} \frac{\ln(1 - x)}{1/(\cot x)} = \lim_{x \to 0+} \frac{\ln(1 - x)}{\tan x}$$

$$= \lim_{x \to 0+} \frac{-1/(1 - x)}{\sec^2 x} = \frac{-1}{1} = -1. \qquad \blacksquare$$

EXAMPLE 3   Find $\lim_{x \to 0+} x^2 e^{1/x}$.

Solution   This is a $0 \cdot \infty$-type form, and we write

$$\lim_{x \to 0+} x^2 e^{1/x} = \lim_{x \to 0+} \frac{x^2}{e^{-1/x}}$$

to convert it to a $0/0$-type form. Then

$$\lim_{x \to 0+} \frac{x^2}{e^{-1/x}} = \lim_{x \to 0+} \frac{2x}{e^{-1/x} \cdot 1/x^2} = \lim_{x \to 0+} \frac{2x^3}{e^{-1/x}}.$$

This limit is more complicated; obviously we are going "the wrong way."

Starting again and converting to an ∞/∞-type form, we have

$$\lim_{x \to 0+} x^2 e^{1/x} = \lim_{x \to 0+} \frac{e^{1/x}}{1/x^2}$$

$$= \lim_{x \to 0+} \frac{e^{1/x}(-1/x^2)}{-2/x^3} = \lim_{x \to 0+} \frac{e^{1/x}}{2/x}$$

$$= \lim_{x \to 0+} \frac{e^{1/x}(-1/x^2)}{-2/x^2} = \lim_{x \to 0+} \frac{e^{1/x}}{2} = \infty. \quad \blacksquare$$

EXAMPLE 4   We repeat the previous example but use the substitution $x = 1/t$. This device converts a limit as $x \to 0+$ to a limit as $t \to \infty$.

Solution   We have

$$\lim_{x \to 0+} x^2 e^{1/x} = \lim_{t \to \infty} \frac{1}{t^2} e^t = \lim_{t \to \infty} \frac{e^t}{t^2} = \lim_{t \to \infty} \frac{e^t}{2t} = \lim_{t \to \infty} \frac{e^t}{2} = \infty.$$

Clearly, this is easier than the computation in Example 3. $\blacksquare$

Indeterminate Sum and Difference Forms

From

$$\lim_{x \to a+} \left(\frac{1}{x - a} - \frac{1}{x - a} \right) = 0$$

and

$$\lim_{x \to a+} \left(\frac{1}{x - a} - \frac{1 + 2a - 2x}{x - a} \right) = \lim_{x \to a+} \frac{2(x - a)}{x - a} = 2,$$

we see that $\infty - \infty$ should be considered an indeterminate form.

Indeterminate Sum and Difference Forms

$$-\infty + \infty, \qquad \infty - \infty$$

We attempt to change such a form to an indeterminate quotient form by algebraic means. In the following example, this amounts to finding a common denominator and performing the subtraction.

EXAMPLE 5   Compute

$$\lim_{x \to 0+} \left(\frac{2}{x} - \frac{x + 1}{x - x^2} \right),$$

which is clearly an $(\infty - \infty)$-type problem.

Solution   Performing the subtraction and applying l'Hôpital's rule to the

0/0-type form that results, we have

$$\lim_{x \to 0+} \left(\frac{2}{x} - \frac{x+1}{x-x^2} \right) = \lim_{x \to 0+} \frac{2(x-x^2) - x(x+1)}{x^2 - x^3}$$

$$= \lim_{x \to 0+} \frac{x - 3x^2}{x^2 - x^3} = \lim_{x \to 0+} \frac{1 - 6x}{2x - 3x^2} = \infty. \qquad ■$$

Indeterminate Exponential Forms

Finally, indeterminate exponential forms arise from expressions of the form $\lim_{x \to a} f(x)^{g(x)}$. An example of such an indeterminate form is furnished by

$$\lim_{x \to \infty} \left(1 + \frac{1}{x} \right)^x.$$

Note that as $x \to \infty$, the function $1 + 1/x$ is greater than 1 but is approaching 1. We know that 1 raised to any power again yields 1 but that a number just greater than 1 raised to a sufficiently large power can yield a very large number. Thus it is impossible to know the value of this limit without some more work. The limit is of indeterminate form 1^∞. We will show in Example 6 that

$$\lim_{x \to \infty} \left(1 + \frac{1}{x} \right)^x = e. \qquad (1)$$

This is a very famous limit, which we treated intuitively in Section 3.6.

Let us try to find all types of indeterminate exponential forms arising from $\lim_{x \to a} f(x)^{g(x)}$. Recall that we have defined the exponential r^s for all s only if $r > 0$; hence we assume that $f(x) > 0$ for x near a but $x \neq a$. Since the logarithm function is continuous and is the inverse of the exponential function, we see that if $\lim_{x \to a} \ln(f(x)^{g(x)}) = b$, then $\lim_{x \to a} f(x)^{g(x)} = e^b$. Now

$$\ln[f(x)^{g(x)}] = g(x)\ln(f(x)),$$

so for $\lim_{x \to a} f(x)^{g(x)}$ to give rise to an indeterminate exponential form, the product $g(x)\ln(f(x))$ must give rise to one of the indeterminate product forms $0 \cdot \infty$, $\infty \cdot 0$, $0(-\infty)$, or $(-\infty)0$ at $x = a$. The product $g(x)\ln(f(x))$ gives rise to $0 \cdot \infty$ at $x = a$ if $\lim_{x \to a} g(x) = 0$ and $\lim_{x \to a} \ln(f(x)) = \infty$, in which case $\lim_{x \to a} f(x) = \infty$ also. Thus $0 \cdot \infty$ gives rise to the indeterminate exponential form ∞^0. Similarly, the product form $\infty \cdot 0$ gives rise to the exponential form 1^∞, while $0(-\infty)$ gives rise to the form 0^0 and $(-\infty)0$ gives rise to $1^{-\infty}$.

Indeterminate Exponential Forms

$$0^0, \qquad 1^\infty, \qquad 1^{-\infty}, \qquad \infty^0$$

This analysis of indeterminate exponential forms also indicates how to go about evaluating them. Namely, if we take the logarithm of the exponential expression, we obtain an indeterminate product form that we can in turn convert to an indeterminate quotient form. *It is essential to remember that if the limit of* *the logarithm of an expression is c, then the limit of the original expression is e^c.* We illustrate by demonstrating the famous Eq. (1).

EXAMPLE 6   Compute the important limit

$$\lim_{x \to \infty} \left(1 + \frac{1}{x}\right)^x,$$

which is an indeterminate form of 1^∞ type.

Solution   Taking the logarithm, we compute instead

$$\lim_{x \to \infty} \left[x \ln\left(1 + \frac{1}{x}\right)\right],$$

which is an indeterminate product form of $\infty \cdot 0$ type. We now convert this to an indeterminate quotient form. *When performing this step, always leave the logarithmic expression in the numerator so that the logarithm will disappear after one application of l'Hôpital's rule.* We obtain a 0/0-type form to which we then apply l'Hôpital's rule:

$$\lim_{x \to \infty} \frac{\ln[1 + (1/x)]}{1/x} = \lim_{x \to \infty} \frac{\dfrac{1}{1 + (1/x)} \cdot \dfrac{-1}{x^2}}{-1/x^2} = \lim_{x \to \infty} \frac{1}{1 + (1/x)} = \frac{1}{1} = 1.$$

Since the limit of the *logarithm* approaches 1, the limit of our original form is $e^1 = e$. (See the italicized comment just before this example.)   ∎

EXAMPLE 7   Compute $\lim_{x \to \infty} x^{1/x}$.

Solution   Taking logarithms, we try to compute

$$\lim_{x \to \infty} \frac{1}{x}(\ln x) = \lim_{x \to \infty} \frac{\ln x}{x}.$$

This is a limit at ∞ of type ∞/∞, and we find, by l'Hôpital's rule,

$$\lim_{x \to \infty} \frac{\ln x}{x} = \lim_{x \to \infty} \frac{1/x}{1} = \frac{0}{1} = 0.$$

Since $\ln x^{1/x} \to 0$ at ∞, we must have

$$\lim_{x \to \infty} x^{1/x} = e^0 = 1.$$ ∎

SUMMARY

1. Limits of sums, products, quotients, and exponentials involving two functions are called indeterminate forms if they lead formally to an expression of one of the following types:

$$\frac{0}{0}, \quad \frac{\infty}{\infty}, \quad 0 \cdot \infty, \quad 0(-\infty), \quad \infty \cdot 0, \quad (-\infty)0,$$

$$(-\infty) + \infty, \quad \infty - \infty, \quad 0^0, \quad 1^\infty, \quad 1^{-\infty}, \quad \infty^0.$$

2. Product indeterminate forms are reduced to the quotient type by the algebraic device symbolized by

$$0 \cdot \infty \sim \frac{0}{1/\infty} \sim \frac{0}{0} \quad \text{or} \quad 0 \cdot \infty \sim \frac{\infty}{1/0} \sim \frac{\infty}{\infty}.$$

Exponential types are reduced to product types by taking a logarithm; as a symbolic illustration, $\ln(0^0) \sim 0(\ln 0) \sim 0(-\infty)$.

3. The l'Hôpital's rule procedure can be used to find limits of ∞/∞ type as well as $0/0$ type.

EXERCISES 9.4

In Exercises 1–48, find the limit by l'Hôpital's rule.

1. $\lim\limits_{x \to \infty} \dfrac{x}{e^x}$

2. $\lim\limits_{x \to \infty} \dfrac{x^{20}}{e^x}$

3. $\lim\limits_{x \to \infty} \dfrac{\ln x}{x}$

4. $\lim\limits_{x \to \infty} \dfrac{(\ln x)^2}{x}$

5. $\lim\limits_{x \to \infty} \dfrac{(\ln x)^{100}}{x}$

6. $\lim\limits_{x \to \infty} \dfrac{x \ln x}{\ln(xe^x)}$

7. $\lim\limits_{x \to \infty} \dfrac{e^{2+\ln x}}{3x + 4}$

8. $\lim\limits_{x \to 0} \left(\dfrac{\sin x}{x \cos x - x} - \dfrac{1}{\cos x - 1} \right)$

9. $\lim\limits_{x \to 0} \left(\dfrac{1}{x} - \dfrac{1}{\sin x} \right)$

10. $\lim\limits_{x \to 0} \left(\dfrac{1}{x} - \dfrac{1}{e^x - 1} \right)$

11. $\lim\limits_{x \to 0} \left[\dfrac{2}{x^2} + \dfrac{1}{(\cos x) - 1} \right]$

12. $\lim\limits_{x \to 0} (e^{x+1} - x)$

13. $\lim\limits_{x \to 0+} [\ln(2x + 1) - \ln x]$

14. $\lim\limits_{x \to \infty} [\ln(2 + e^x) - x]$

15. $\lim\limits_{x \to \infty} (e^{1+\ln x} - 3x)$

16. $\lim\limits_{x \to \infty} (\sqrt{x^2 + x} - x)$

17. $\lim\limits_{x \to \infty} (\sqrt{x^2 + 1} - \sqrt{x^2 + 2x})$

18. $\lim\limits_{x \to 0} (\cot 2x^2)[\ln(1 - x^2)]$

19. $\lim\limits_{x \to \infty} \left(1 - \dfrac{1}{x^2} \right)^{\cot(1/x)}$

20. $\lim\limits_{x \to 0+} \left(\dfrac{1}{1 - x} \right)^{-1/x^2}$

21. $\lim\limits_{x \to 0+} x^x$

22. $\lim\limits_{x \to 0+} x(\ln x)$

23. $\lim\limits_{x \to 0+} \dfrac{\ln(\sin x)}{\csc x}$

24. $\lim\limits_{x \to 0} \dfrac{\ln(\cos x)}{\sin^2 x}$

25. $\lim\limits_{x \to \infty} \dfrac{e^{2x}}{e^x}$

26. $\lim\limits_{x \to 1+} \dfrac{\ln(x^2 - 1)}{\ln(3x^2 + 3x - 6)}$

27. $\lim\limits_{x \to 1+} (1 - x)[\ln(\ln x)]$

28. $\lim\limits_{x \to 0+} \dfrac{\ln(e^x - 1)}{\ln x}$

29. $\lim\limits_{x \to \pi/2} \dfrac{e^x - e^{\pi/2}(x - \pi/2)}{\cos^2 x}$

30. $\lim\limits_{x \to \pi/2+} \dfrac{\sec x}{x - \pi/2}$

31. $\lim\limits_{x \to \pi} \dfrac{\sin^2 x}{(x - \pi)^2}$

32. $\lim\limits_{x \to -\pi/2} \dfrac{1 + \sin x}{\cos^2 x}$

33. $\lim\limits_{x \to 1} \dfrac{(\ln x)^2}{x^2 - 2x + 1}$

34. $\lim\limits_{x \to \infty} \left(1 - \dfrac{1}{x} \right)^x$

35. $\lim\limits_{x \to \infty} \left(1 + \dfrac{2}{x} \right)^x$

36. $\lim\limits_{x \to \infty} \left(\dfrac{x}{x + 1} \right)^x$

37. $\lim\limits_{x \to \infty} \left(1 + \dfrac{1}{x^2} \right)^x$

38. $\lim\limits_{x \to \infty} \left(1 - \dfrac{1}{x} \right)^{x^2}$

39. $\lim\limits_{x \to (\pi/2)+} (\cos x) \ln\left(x - \dfrac{\pi}{2} \right)$

40. $\lim\limits_{x \to \infty} \left(1 - \dfrac{1}{x} \right)^{2x}$

41. $\lim\limits_{x \to \infty} \left(1 + \dfrac{1}{x} \right)^{x^2}$

42. $\lim\limits_{x \to \pi/2} (\sin x)^{1/(\pi - 2x)}$

43. $\lim\limits_{x \to \infty} \left(\dfrac{3x + 2}{2x} \right)^x$

44. $\lim\limits_{x \to \infty} (1 + x^2)^{1/x}$

45. $\lim\limits_{x \to \infty} (x + e^x)^{3/x}$

46. $\lim\limits_{x \to 0+} (\sec x)^{\cot x}$

47. $\lim\limits_{x \to 0+} x^3 e^{1/x}$

48. $\lim\limits_{x \to 1+} (x - 1)^{\ln x}$

49. Suppose that f and g are differentiable functions for $x \neq a$ and that $\lim_{x \to a} f(x) = \lim_{x \to a} g(x) = \infty$ and $\lim_{x \to a} f(x)/g(x) = L$, where L is a nonzero real number. Show that $\lim_{x \to a} f(x)/g(x) = \lim_{x \to a} f'(x)/g'(x)$. [*Hint:* Apply l'Hôpital's rule to the equivalent quotient form $\lim_{x \to a} [(1/g(x))/(1/f(x))]$, which is of $0/0$ type.]

In Exercises 50–54, try to find the limit in the indicated exercise using a calculator or computer rather than the methods of this section.

50. Exercise 7

51. Exercise 11

52. Exercise 9

53. Exercise 5 (This should be illuminating!)

54. Exercise 26

9.5 IMPROPER INTEGRALS

We now examine another aspect of the behavior of a function at a point b or at ∞. Even if $\lim_{x \to b} f(x) = \infty$, does the function approach ∞ slowly enough so that $\int_a^b f(x)\, dx$ makes some sort of sense, and does $\int_a^\infty f(x)\, dx$ make sense? Such integrals are known as *improper integrals*. Geometrically, these improper integrals are concerned with areas of regions that are infinitely wide or infinitely high in altitude.

Some of the outcomes of our study may seem very surprising. Exercise 31 will give an example of a solid of finite volume but infinite surface area! Just imagine, such a solid could be formed using a finite amount of plaster of paris, but then there is not enough red paint in the world to paint it red. To make it red, we would have to mix red coloring in the plaster before we formed the solid. We find such ideas intriguing.

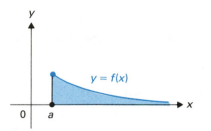

FIGURE 9.6 $\int_a^\infty f(x)\, dx$ is the area of the shaded region, if it is finite.

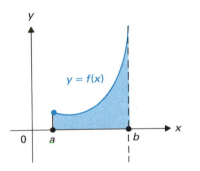

FIGURE 9.7 $\int_a^b f(x)\, dx$ is the area of the shaded region, if it is finite.

Basic Types of Improper Integrals

Thus far, our discussion of a definite integral $\int_a^b f(x)\, dx$ has been confined to integrals of continuous functions over a closed interval. We now turn to such integrals as

$$\int_a^\infty f(x)\, dx \qquad \text{and} \qquad \int_a^b f(x)\, dx,$$

where $\lim_{x \to b^-} |f(x)| = \infty$. These are *improper integrals*. If $f(x) \geq 0$, the first integral is an attempt to find the area under the graph "from a to ∞," as shown in Fig. 9.6. The second integral is an attempt to find the area of the shaded region in Fig. 9.7.

If f is continuous for all $x \geq a$, as shown in Fig. 9.8, then $F(h) = \int_a^h f(x)\, dx$ is defined for $h \geq a$. The definition that follows is the natural attempt to make sense out of $\int_a^\infty f(x)\, dx$.

DEFINITION 9.2

Improper integrals
$\int_a^\infty f(x)\, dx$ and $\int_{-\infty}^a f(x)\, dx$

Let f be continuous for $x \geq a$. The **improper integral** $\int_a^\infty f(x)\, dx$ is defined by

$$\int_a^\infty f(x)\, dx = \lim_{h \to \infty} \int_a^h f(x)\, dx. \qquad (1)$$

If this limit exists, the integral is said to **converge,** whereas the integral **diverges** if the limit is not a finite number.

The **improper integral** $\int_{-\infty}^a f(x)\, dx$ is defined in a similar fashion by

$$\int_{-\infty}^a f(x)\, dx = \lim_{h \to -\infty} \int_h^a f(x)\, dx \qquad (2)$$

and **converges** or **diverges** according to whether the limit exists.

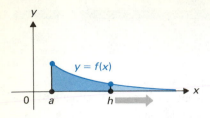

FIGURE 9.8

$$\int_a^\infty f(x)\,dx = \lim_{h \to \infty} \int_a^h f(x)\,dx$$

EXAMPLE 1 Find the improper integral $\int_1^\infty (1/x^2)\,dx$, if it converges.

Solution We have

$$\int_1^\infty \frac{1}{x^2}\,dx = \lim_{h \to \infty} \int_1^h \frac{1}{x^2}\,dx = \lim_{h \to \infty} \left. -\frac{1}{x} \right]_1^h$$

$$= \lim_{h \to \infty} \left[-\frac{1}{h} - (-1) \right] = 0 + 1 = 1.$$

Thus the integral converges to 1. ∎

EXAMPLE 2 Find the improper integral $\int_1^\infty (1/x)\,dx$, if it converges.

Solution We have

$$\int_1^\infty \frac{1}{x}\,dx = \lim_{h \to \infty} \int_1^h \frac{1}{x}\,dx = \lim_{h \to \infty} \ln|x| \Big]_1^h = \lim_{h \to \infty} \ln|h| - \ln 1 = \infty.$$

Thus the integral diverges. ∎

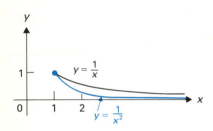

FIGURE 9.9 $1/x^2 < 1/x$ for $x > 1$

The graphs of $1/x^2$ and $1/x$ are shown in Fig. 9.9. Note that $1/x^2 < 1/x$ for $x > 1$. Intuitively, we see that for a *decreasing function f* with graph of the type shown in Fig. 9.6, the integral $\int_a^\infty f(x)\,dx$ converges only if the graph approaches the x-axis fast enough as $x \to \infty$. That is, $f(x)$ must approach zero fast enough as $x \to \infty$. Example 1 showed that $1/x^2$ does approach zero fast enough as $x \to \infty$ to make $\int_1^\infty (1/x^2)\,dx$ converge. Example 2 showed that $1/x$ does not approach zero fast enough as $x \to \infty$ to make $\int_1^\infty (1/x)\,dx$ converge.

EXAMPLE 3 The velocity of a particle at time t sec moving on the x-axis (units in feet) is given by

$$v(t) = \frac{dx}{dt} = \frac{1}{1 + t^2} \text{ ft/sec} \qquad \text{for } t \geq 0.$$

Find the limiting total distance the body travels throughout eternity, that is, if it is allowed to travel forever.

Solution The total distance the body has traveled at time $t = h$ for $h \geq 0$ is given by

$$\int_0^h |v(t)|\,dt = \int_0^h \frac{1}{1 + t^2}\,dt \text{ ft.}$$

Thus the total distance it travels throughout eternity approaches

$$\int_0^\infty \frac{dt}{1 + t^2} = \lim_{h \to \infty} \int_0^h \frac{dt}{1 + t^2} = \lim_{h \to \infty} \tan^{-1}t \Big]_0^h$$

$$= \lim_{h \to \infty} (\tan^{-1}h - \tan^{-1}0) = \frac{\pi}{2} - 0 = \frac{\pi}{2} \text{ ft.} \quad ∎$$

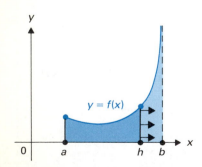

FIGURE 9.10

$$\int_a^b f(x)\,dx = \lim_{h \to b^-} \int_a^h f(x)\,dx$$

If f is continuous for $a \leq x < b$, as shown in Fig. 9.10, then $\int_a^h f(x)\,dx$ for $a \leq h < b$ makes perfectly good sense. We give the following definition analogous to Definition 9.2.

DEFINITION 9.3

Improper integral $\int_a^b f(x)\,dx$

Let f be continuous for $a \le x < b$, while f is either undefined or not continuous at $x = b$. The **improper integral** $\int_a^b f(x)\,dx$ is defined by

$$\int_a^b f(x)\,dx = \lim_{h \to b-} \int_a^h f(x)\,dx. \tag{3}$$

If the limit exists, the integral is said to **converge,** while the integral **diverges** if the limit is not a finite number.

EXAMPLE 4   Find the improper integral $\int_0^1 1/(1-x)\,dx$, if it converges.

Solution   We have

$$\int_0^1 \frac{1}{1-x}\,dx = \lim_{h \to 1-} \int_0^h \frac{1}{1-x}\,dx = \lim_{h \to 1-} -\ln|1-x|\Big]_0^h$$

$$= \lim_{h \to 1-} [-\ln|1-h| + \ln 1] = \infty.$$

Thus the integral diverges.   ∎

EXAMPLE 5   Find the improper integral $\int_0^1 (1/\sqrt{1-x})\,dx$, if it converges.

Solution   We have

$$\int_0^1 \frac{1}{\sqrt{1-x}}\,dx = \lim_{h \to 1-} \int_0^h \frac{1}{\sqrt{1-x}}\,dx = \lim_{h \to 1-} -2\sqrt{1-x}\Big]_0^h$$

$$= \lim_{h \to 1-} [-2\sqrt{1-h} + 2] = 2.$$

Thus the integral converges to 2.   ∎

Suppose that f is continuous for $a < x \le b$, while f is either undefined or not continuous at $x = a$, as shown in Fig. 9.11. By analogy with Definition 9.3, we define

$$\int_a^b f(x)\,dx = \lim_{h \to a+} \int_h^b f(x)\,dx, \tag{4}$$

if the limit exists.

Figures 9.7, 9.10, and 9.11 all illustrate improper integrals where $|f(x)| \to \infty$ as we approach a limit of integration. There are other possibilities.

EXAMPLE 6   Discuss $\int_0^1 \sin(1/x)\,dx$.

Solution   We consider this integral to be *improper* since $\sin(1/x)$ is not defined where $x = 0$. An attempt to sketch the graph of $\sin(1/x)$ for $0 < x \le 1$ is shown in Fig. 9.12. All the waves of the $\sin x$ graph for $x > 1$ are thrown between 0 and 1 by the reciprocal transformation $1/x$. We do not yet have the technique to decide whether $\int_0^1 \sin(1/x)\,dx$ converges. Techniques introduced in Section 10.5 can be used to show that the integral does converge.   ∎

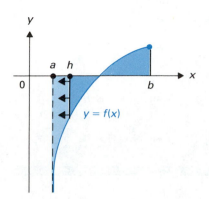

FIGURE 9.11

$$\int_a^b f(x)\,dx = \lim_{h \to a+} \int_h^b f(x)\,dx$$

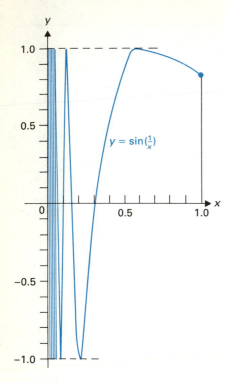

FIGURE 9.12   Attempt to graph
$$y = \sin(1/x) \quad \text{for } 0 < x \le 1.$$
The graph oscillates an infinite number
of times as $x \to 0$.

EXAMPLE 7   Discuss $\int_0^1 (\sin x)/x \, dx$.

Solution   The integral is improper since $(\sin x)/x$ is not defined where $x = 0$.
We know that

$$\lim_{x \to 0} \frac{\sin x}{x} = 1.$$

Thus if we let

$$g(x) = \begin{cases} \dfrac{\sin x}{x} & \text{for } 0 < x \le 1, \\ 1 & \text{for } x = 0, \end{cases}$$

then $g(x) = (\sin x)/x$ for $0 < x \le 1$, and g is continuous for $0 \le x \le 1$. Since

$$\int_h^1 \frac{\sin x}{x} \, dx = \int_h^1 g(x) \, dx \qquad \text{for } 0 < h \le 1,$$

we see that

$$\lim_{h \to 0+} \int_h^1 \frac{\sin x}{x} \, dx = \lim_{h \to 0+} \int_h^1 g(x) \, dx = \int_0^1 g(x) \, dx,$$

so $\int_0^1 (\sin x)/x \, dx$ does converge. Using Simpson's rule with $n = 40$ and a calculator to estimate $\int_{0.0001}^1 [(\sin x)/x] \, dx$, we find

$$\int_0^1 \frac{\sin x}{x} \, dx \approx 0.946.$$ ∎

Other Types of Improper Integrals

Letting f be continuous except where indicated, we list the basic types of improper integrals, presented in the previous subsection.

Basic Types of Improper Integrals

$$\int_a^\infty f(x) \, dx = \lim_{h \to \infty} \int_a^h f(x) \, dx$$

$$\int_{-\infty}^b f(x) \, dx = \lim_{h \to -\infty} \int_h^b f(x) \, dx$$

$$\int_a^b f(x) \, dx = \lim_{h \to b-} \int_a^h f(x) \, dx, \qquad f \text{ undefined at } b$$

$$\int_a^b f(x) \, dx = \lim_{h \to a+} \int_h^b f(x) \, dx, \qquad f \text{ undefined at } a$$

We consider ∞ and $-\infty$ as well as b and a in the last two integrals listed to be "bad points" causing the integral to be improper. Some improper integrals involve more than one "bad point." For an improper integral involving only a

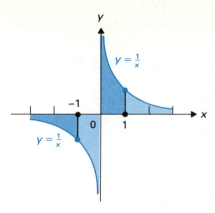

FIGURE 9.13 Breaking $\int_{-\infty}^{\infty} (1/x)\,dx$ into a sum of four basic types of improper integrals.

finite number of "bad points," we can express the integral as a sum of integrals of the basic types listed. Such an integral **converges** only if *each* of the basic types of which it is composed converges. If even one of the basic types involved diverges, then the whole integral is said to **diverge.** This is easier to illustrate by examples than to express in words.

EXAMPLE 8   Express $\int_{-\infty}^{\infty} (1/x)\,dx$ as a sum of basic types and determine whether the integral converges or diverges.

Solution   The graph of $1/x$ is shown in Fig. 9.13. "Bad points" for $\int_{-\infty}^{\infty} (1/x)\,dx$ are $-\infty$, 0, and ∞. The integral can be expressed as a sum of basic types by

$$\int_{-\infty}^{\infty} \frac{1}{x}\,dx = \int_{-\infty}^{-1} \frac{1}{x}\,dx + \int_{-1}^{0} \frac{1}{x}\,dx + \int_{0}^{1} \frac{1}{x}\,dx + \int_{1}^{\infty} \frac{1}{x}\,dx.$$

The points -1 and 1 simply serve as convenient "good points" to separate the "bad points." We could replace -1 by -10 and 1 by 57, for example, if we wished. Example 2 showed that $\int_{1}^{\infty} (1/x)\,dx$ diverges. Consequently, $\int_{-\infty}^{\infty} (1/x)\,dx$ diverges, since at least one of the basic types of which it is composed diverges. ∎

EXAMPLE 9   Express $\int_{-\infty}^{\infty} 1/(1 + x^2)\,dx$ as a sum of basic types and determine whether the integral converges or diverges.

Solution   The graph of $1/(1 + x^2)$ is shown in Fig. 9.14. We have

$$\int_{-\infty}^{\infty} \frac{dx}{1 + x^2} = \int_{-\infty}^{0} \frac{dx}{1 + x^2} + \int_{0}^{\infty} \frac{dx}{1 + x^2}$$

$$= \lim_{h \to -\infty} \int_{h}^{0} \frac{dx}{1 + x^2} + \lim_{h \to \infty} \int_{0}^{h} \frac{dx}{1 + x^2}$$

$$= \lim_{h \to -\infty} \tan^{-1}x \Big]_{h}^{0} + \lim_{h \to \infty} \tan^{-1}x \Big]_{0}^{h}$$

$$= \lim_{h \to -\infty} (0 - \tan^{-1}h) + \lim_{h \to \infty} (\tan^{-1}h - 0)$$

$$= 0 - \left(-\frac{\pi}{2}\right) + \left(\frac{\pi}{2} - 0\right) = \pi.$$

Thus the integral converges to π. ∎

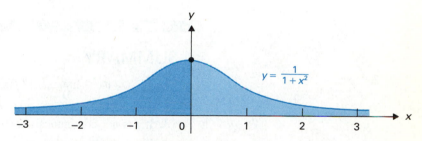

FIGURE 9.14 $\displaystyle\int_{-\infty}^{\infty} [1/(1 + x^2)]\,dx = \int_{-\infty}^{0} [1/(1 + x^2)]\,dx + \int_{0}^{\infty} [1/(1 + x^2)]\,dx$

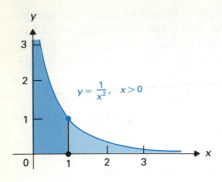

FIGURE 9.15

$$\int_0^\infty (1/x^2)\,dx$$

$$= \int_0^1 (1/x^2)\,dx + \int_1^\infty (1/x^2)\,dx$$

EXAMPLE 10 Express $\int_0^\infty (1/x^2)\,dx$ as a sum of basic types and determine whether the integral converges.

Solution The graph of $1/x^2$ for $x > 0$ is shown in Fig. 9.15. We have

$$\int_0^\infty \frac{1}{x^2}\,dx = \int_0^1 \frac{1}{x^2}\,dx + \int_1^\infty \frac{1}{x^2}\,dx.$$

Example 1 showed that $\int_1^\infty (1/x^2)\,dx$ converges to 1. Now

$$\int_0^1 \frac{1}{x^2}\,dx = \lim_{h \to 0+} \int_h^1 x^{-2}\,dx = \lim_{h \to 0+} \left.\frac{-1}{x}\right]_h^1 = \lim_{h \to 0+} \left(-1 + \frac{1}{h}\right) = \infty.$$

Thus $\int_0^1 (1/x^2)\,dx$ diverges, so $\int_0^\infty (1/x^2)\,dx$ diverges. ∎

Our mathematical life has abruptly become more complicated. Now we have to examine every definite integral for a limit of ∞ or $-\infty$, or for a point where the integrand is not defined, to decide whether it is improper before we try to evaluate it. The following example shows the danger if we do not do this but just blindly integrate and evaluate.

EXAMPLE 11 Find $\int_{-1}^1 (1/x)\,dx$.

Wrong Solution We have

$$\int_{-1}^1 \frac{1}{x}\,dx = \left.\ln|x|\right]_{-1}^1 = (\ln 1) - (\ln 1) = 0 - 0 = 0.$$

Correct Solution The integral is improper because $1/x$ is not defined where $x = 0$. Expressing the integral as a sum of basic types, we have

$$\int_{-1}^1 \frac{1}{x}\,dx = \int_{-1}^0 \frac{1}{x}\,dx + \int_0^1 \frac{1}{x}\,dx.$$

Now

$$\int_{-1}^0 \frac{1}{x}\,dx = \lim_{h \to 0-} \left.\ln|x|\right]_{-1}^h = \lim_{h \to 0-} (\ln|h| - \ln 1)$$

$$= \lim_{h \to 0-} \ln|h| = -\infty.$$

Thus $\int_{-1}^1 (1/x)\,dx$ diverges. ∎

SUMMARY

1. $\int_a^\infty f(x)\,dx = \lim_{h \to \infty} \int_a^h f(x)\,dx$
2. $\int_{-\infty}^b f(x)\,dx = \lim_{h \to -\infty} \int_h^b f(x)\,dx$
3. If f is not continuous at b, then $\int_a^b f(x)\,dx = \lim_{h \to b-} \int_a^h f(x)\,dx$.
4. If f is not continuous at a, then $\int_a^b f(x)\,dx = \lim_{h \to a+} \int_h^b f(x)\,dx$.
5. The improper integrals listed in 1–4 are said to converge if the limits are finite. Otherwise, they diverge.

6. An improper integral involving a finite number of ''bad points'' can be expressed as a sum of the basic types (1–4 above). The entire integral converges only if each individual basic type involved converges.

7. You should now examine every definite integral for a limit of ∞ or $-\infty$, or a point where the integrand is undefined, to see whether the integral is improper before you attempt to evaluate it.

EXERCISES 9.5

In Exercises 1–12, express the improper integrals as a sum of basic types described in the text, using limits. You need not establish the convergence or divergence of the integrals.

1. $\int_0^1 \dfrac{dx}{x^2 - 1}$

2. $\int_0^\pi \tan x \, dx$

3. $\int_1^\infty \dfrac{dx}{x - 1}$

4. $\int_0^1 \dfrac{dx}{x^2 + x}$

5. $\int_{-\infty}^\infty e^{-x^2} \, dx$

6. $\int_{-1}^1 \dfrac{dx}{x^2 - 4}$

7. $\int_0^2 \dfrac{dx}{x^3 + 1}$

8. $\int_0^4 \dfrac{dx}{x^2 - 2x - 3}$

9. $\int_0^\infty \dfrac{dx}{\sqrt{x + x^2}}$

10. $\int_{-\infty}^\infty \dfrac{dx}{\sqrt[3]{x + x^3}}$

11. $\int_0^{2\pi} \sec x \, dx$

12. $\int_0^\pi \dfrac{dx}{\sin x - \cos x}$

In Exercises 13–28, determine whether or not the improper integral converges, and find its value if it converges.

13. $\int_1^\infty \dfrac{1}{x^3} \, dx$

14. $\int_1^\infty \dfrac{1}{\sqrt{x}} \, dx$

15. $\int_0^\infty \dfrac{1}{x^2 + 1} \, dx$

16. $\int_{-\infty}^0 \dfrac{x^2}{x^3 + 1} \, dx$

17. $\int_{-\infty}^0 e^x \, dx$

18. $\int_0^1 \dfrac{1}{x^{2/3}} \, dx$

19. $\int_0^2 \dfrac{1}{\sqrt{2 - x}} \, dx$

20. $\int_1^2 \dfrac{1}{(x - 1)^2} \, dx$

21. $\int_{-1}^1 \dfrac{1}{x^{2/5}} \, dx$

22. $\int_0^\infty \dfrac{1}{\sqrt{x}} \, dx$

23. $\int_{-\infty}^\infty |x| e^{-x^2} \, dx$

24. $\int_0^1 \dfrac{1}{\sqrt{2x - x^2}} \, dx$

25. $\int_{-\infty}^\infty e^{-x} \cos x \, dx$

26. $\int_0^\infty e^{-x} \sin x \, dx$

27. $\int_0^{\pi/2} \tan x \, dx$

28. $\int_0^{\pi/2} \sqrt{\cos x} \cot x \, dx$

29. Show that for $a > 0$, the integral $\int_a^\infty (1/x^p) \, dx$ converges if $p > 1$ and diverges if $p \le 1$.

30. Show that

$$\int_a^b \left[\frac{1}{(x - a)^p} \right] dx \quad \text{and} \quad \int_a^b \left[\frac{1}{(b - x)^p} \right] dx$$

converge if $p < 1$ and diverge if $p \ge 1$.

31. Consider the region under the graph of $1/x$ over the half-line $x \ge 1$.
 (a) Does the region have finite area?
 (b) Show that the unbounded solid obtained by revolving the region about the x-axis has finite volume, and compute this volume.
 (c) Does the solid of finite volume described in part (b) have finite surface area?

32. Consider the region under the graph of $1/\sqrt{x}$ over the half-open interval $0 < x \le 1$.
 (a) Does the region have finite area?
 (b) Show that the unbounded solid obtained by revolving the region about the y-axis has finite volume, and compute this volume.
 (c) Does the solid of finite volume described in part (b) have finite surface area?

33. The position (x, y) of a body in the plane at time t is given by $x = e^{-t}$, $y = e^{-t}$. Find the total distance the body travels throughout time $t \ge 0$.

34. Two electrons a distance s apart repel each other with a force k/s^2, where k is some constant. If an electron is fixed at the point 2 on the x-axis, find the work done by the force in moving another electron from the point 4 ''all the way out to infinity.''

35. With reference to Exercise 34, suppose one electron is fixed at the point 0 and another is fixed at the point 1. Find the work done by the forces in moving a third electron from the point 2 ''all the way out to infinity.''

36. A body outside the earth is attracted to the earth by a gravitational force inversely proportional to the square of the distance from the body to the center of the earth. Assume the earth is a ball of radius 4000 mi. Find the work done in moving a body weighing 10 lb at the surface of the earth from the surface of the earth to beyond the earth's gravitational field (that is, "to infinity"). Neglect the influence of other celestial bodies.

37. Suppose a spaceship could achieve a velocity v that increases exponentially as a function of the distance s traveled, so that $v = Ae^{ks}$ for some positive constants k and A.
    (a) Show that such a spaceship could travel an infinite distance in a finite time. [*Hint:* Since the time required for

a journey is Distance/Velocity if the velocity is constant, we see that the time required to travel a distance c is $\int_0^c (1/v(s)) \, ds$ if the velocity is not constant.]
    (b) Show that the acceleration dv/dt of such a spaceship is directly proportional to the square of its velocity.
    (c) Show that the spaceship could still achieve an infinite distance in a finite time with a velocity increasing more slowly than an exponential function of the total distance traveled.

9.6 TESTS FOR CONVERGENCE OF IMPROPER INTEGRALS

This section closely parallels work that we will be doing with infinite series in Chapter 10. The ideas are very important and different from anything we have done so far. Developing convergence tests here for improper integrals as well as in Chapter 10 for series gives us more opportunity to understand them and work with them.

The p-Integrals

We will develop methods for trying to establish the convergence or divergence of an improper integral without actual integration. Basically, we will compare the integral in question with another improper integral whose behavior (convergence or divergence) we already know. Of course, we can do this only if we have at least a small stockpile of improper integrals whose behavior we know. Theorem 9.3 provides such a stockpile. The improper integrals appearing in the theorem are known as the p-integrals. You were asked to prove the theorem in Exercises 29 and 30 of the preceding section.

THEOREM 9.3 p-integrals

1. For $a > 0$, the improper integral $\int_a^\infty (1/x^p) \, dx$ converges if $p > 1$ and diverges if $p \leq 1$.
2. The improper integrals $\int_a^b [1/(b - x)^p] \, dx$ and $\int_a^b [1/(x - a)^p] \, dx$ converge if $p < 1$ and diverge if $p \geq 1$. □

Intuitively, we think of $p = 1$ in Theorem 9.3 as being the transition exponent between convergent and divergent integrals. Figure 9.16 shows the graph of $1/x$ and graphs of $1/x^p$ for $p > 1$ and for $p < 1$. The theorem asserts that the area of the dark-shaded region, corresponding to $p > 1$, remains finite over the interval $1 \leq x < \infty$, while the area of the total shaded region becomes infinite. Figure 9.17 similarly illustrates Theorem 9.3 for $\int_a^b [1/(x - a)^p] \, dx$, but this time the region has finite area if $p < 1$.

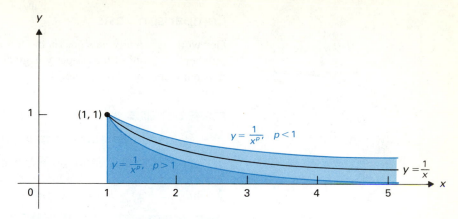

FIGURE 9.16 The dark-shaded region has finite area over $1 \le x < \infty$; the total shaded region has infinite area over $1 \le x < \infty$.

EXAMPLE 1 Establish the convergence or divergence of the integral $\int_1^\infty [1/(x-1)^2]\, dx$.

Solution Splitting the integral into basic types, we have

$$\int_1^\infty \frac{dx}{(x-1)^2} = \int_1^2 \frac{dx}{(x-1)^2} + \int_2^\infty \frac{dx}{(x-1)^2}.$$

By part 2 of Theorem 9.3, the integral $\int_1^2 [1/(x-1)^2]\, dx$ diverges as a p-integral at $a = 1$ with $p = 2$. Thus $\int_1^\infty [1/(x-1)^2]\, dx$ diverges. ■

EXAMPLE 2 Establish the convergence or divergence of

$$\int_{-3}^\infty \frac{dx}{\sqrt{x+3}}.$$

Solution Splitting the integral into basic types, we have

$$\int_{-3}^\infty \frac{dx}{\sqrt{x+3}} = \int_{-3}^0 \frac{dx}{\sqrt{x+3}} + \int_0^\infty \frac{dx}{\sqrt{x+3}}.$$

By part 2 of Theorem 9.3, we see that

$$\int_{-3}^0 \frac{dx}{\sqrt{x+3}} = \int_{-3}^0 \frac{dx}{(x+3)^{1/2}}$$

converges as a p-integral at $a = -3$ with $p = \frac{1}{2}$.

Now $\int_0^\infty (1/\sqrt{x+3})\, dx$ is not precisely in the form of part 1 of Theorem 9.3. However, we can put it in that form by the substitution $u = x + 3$, $du = dx$. Then

$$\int_0^\infty \frac{dx}{\sqrt{x+3}} = \int_3^\infty \frac{du}{\sqrt{u}},$$

which diverges as a p-integral at ∞ with $p = \frac{1}{2}$. Thus $\int_{-3}^\infty (1/\sqrt{x+3})\, dx$ also diverges. ■

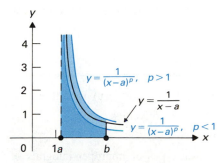

FIGURE 9.17 The dark-shaded region has finite area over $a < x \le b$; the total shaded region has infinite area over $a < x \le b$.

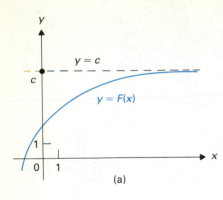

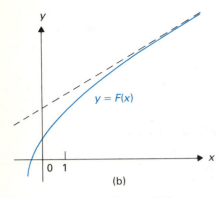

FIGURE 9.18   (a) $\lim_{x \to \infty} F(x) = c$; (b) $\lim_{x \to \infty} F(x) = \infty$.

Comparison Tests

Our work here really depends on the following fundamental property of the real numbers. A proof of the property is given in more advanced courses, where the real numbers are "constructed."

Fundamental Property of the Real Numbers

Let $F(x)$ be defined for all sufficiently large x, and suppose $F(x_1) \leq F(x_2)$ if $x_1 < x_2$. (Such a function is monotone increasing.) Then either $\lim_{x \to \infty} F(x) = c$ for some finite number c or $\lim_{x \to \infty} F(x) = \infty$.

Figure 9.18 illustrates these two possible types of behavior for $F(x)$. The fundamental property certainly seems reasonable. It says that something that is increasing either remains bounded, in which case it approaches a limiting magnitude c from below, or else becomes arbitrarily large, that is, approaches ∞.

Suppose now that f is a continuous function and $f(x) \geq 0$ for all $x \geq a$. Then

$$F(h) = \int_a^h f(x) \, dx$$

is a monotone-increasing function of h. Consequently, either $\lim_{h \to \infty} F(h) = c$ for a finite number c or $\lim_{h \to \infty} F(h) = \infty$. That is, either $\int_a^\infty f(x) \, dx$ converges to a finite number c or diverges to ∞. The same result holds for the other basic types of improper integrals, described in Section 9.5.

THEOREM 9.4   Improper integrals of nonnegative functions

Let f be a continuous function such that $f(x) \geq 0$ for all x in the domain of f. Then each of the basic types of improper integrals of $f(x)$, as described in Section 9.5, either converges to a finite number c or diverges to ∞.   □

We are now ready for the main goal of this section, comparison tests. The following comparison test is illustrated in Fig. 9.19.

THEOREM 9.5   Standard comparison test (SCT)

1. Let f and g be continuous, and suppose $0 \leq f(x) \leq g(x)$ for all $x \geq a$.
   (a) If $\int_a^\infty g(x) \, dx$ converges, then $\int_a^\infty f(x) \, dx$ converges.
   (b) If $\int_a^\infty f(x) \, dx$ diverges, then $\int_a^\infty g(x) \, dx$ diverges.
2. Analogous results hold for the other basic types of improper integrals of *nonnegative* functions:
   (a) If the improper integral of the larger function converges, then the corresponding improper integral of the smaller function converges.
   (b) If the improper integral of the smaller function diverges, then the corresponding improper integral of the larger function diverges.

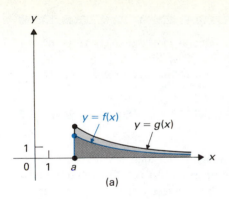

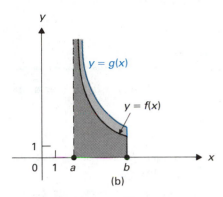

FIGURE 9.19 (a) If $\int_a^\infty g(x)\,dx$ converges, then $\int_a^\infty f(x)\,dx$ converges; (b) if $\int_a^b f(x)\,dx$ diverges, then $\int_a^b g(x)\,dx$ diverges.

Proof: We demonstrate part 1 of the theorem. The proof of part 2 is analogous. Suppose $0 \le f(x) \le g(x)$ for $x \ge a$. We have

$$\int_a^h f(x)\,dx \le \int_a^h g(x)\,dx. \qquad (1)$$

Now suppose $\int_a^\infty g(x)\,dx$ converges, so that

$$\lim_{h \to \infty} \int_a^h g(x)\,dx = c$$

for some finite number c. Then by Eq. (1), $\lim_{h \to \infty} \int_a^h f(x)\,dx = \infty$ is impossible, so by Theorem 9.4, the integral $\int_a^\infty f(x)\,dx$ converges also. If, on the other hand, $\int_a^\infty f(x)\,dx$ diverges to ∞, then Eq. (1) shows immediately that $\int_a^\infty g(x)\,dx$ must diverge also. $\qquad \square$

EXAMPLE 3 Show that

$$\int_4^\infty \frac{dx}{x^2 + 10x + 23}$$

converges.

Solution We have

$$\frac{1}{x^2 + 10x + 23} < \frac{1}{x^2} \qquad \text{for } x \ge 4.$$

Now $\int_4^\infty (1/x^2)\,dx$ converges as a p-integral at ∞ with $p = 2$. Consequently, $\int_4^\infty [1/(x^2 + 10x + 23)]\,dx$ converges by the SCT. ∎

EXAMPLE 4 Show that

$$\int_0^2 \frac{(x + 7)}{(x - 2)}\,dx$$

diverges.

Solution We have

$$\frac{1}{x - 2} < \frac{x + 7}{x - 2} \qquad \text{for } 0 \le x < 2.$$

Now $\int_0^2 [1/(x - 2)]\,dx$ diverges as a p-integral at $b = 2$ with $p = 1$. Thus $\int_0^2 [(x + 7)/(x - 2)]\,dx$ diverges by the SCT. ∎

Suppose we wish to establish the convergence or divergence of

$$\int_5^\infty \frac{dx}{x^2 - 3}.$$

We have

$$\frac{1}{x^2} < \frac{1}{x^2 - 3} \qquad \text{for } x \ge 5. \qquad (2)$$

Now $\int_5^\infty (1/x^2)\,dx$ converges as a p-integral at ∞ with $p = 2$. However, the inequality (2) goes the wrong way for the SCT.

> An integral larger than a convergent integral may either converge or diverge, depending on how much larger it is. Similarly, an integral smaller than a divergent integral may either converge or diverge, depending on how much smaller it is.

We are suspicious that $1/(x^2 - 3)$ is not enough larger than $1/x^2$ to cause divergence of $\int_5^\infty [1/(x^2 - 3)]\, dx$. The limit comparison test (LCT), which we now present, shows that this is the case. While the new test seems to be much more powerful and useful than the SCT of Theorem 9.5, we will see that it is really a straightforward corollary of the theorem.

THEOREM 9.6   Limit comparison test (LCT)

1. Let f and g be continuous and let $f(x) > 0$ and $g(x) > 0$ for $x \geq a$. If

$$\lim_{x \to \infty} \frac{f(x)}{g(x)} = c > 0$$

   for some positive constant c, then $\int_a^\infty f(x)\, dx$ and $\int_a^\infty g(x)\, dx$ behave the same way; that is, they both converge or both diverge.

2. Analogous results hold for the other basic types of improper integrals, corresponding to other possible "bad points" $-\infty$, a, and b. Let $f(x)$ and $g(x)$ be positive and continuous. If

$$\lim_{x \to \text{``bad point''}} \frac{f(x)}{g(x)} = c > 0,$$

   then the improper integrals of $f(x)$ and of $g(x)$ with the same limits of integration either both converge or both diverge.

Proof: Again, we prove just part 1. We have

$$\lim_{x \to \infty} \frac{f(x)}{g(x)} = c > 0.$$

Thus $f(x)/g(x)$ is very close to c for x large enough. That is, there is some number η such that

$$\frac{c}{2} < \frac{f(x)}{g(x)} < 2c \qquad \text{for } x \geq \eta,$$

so

$$\frac{c}{2} \cdot g(x) < f(x) < 2c \cdot g(x) \qquad \text{for } x \geq \eta. \tag{3}$$

Now

$$\int_a^\infty f(x)\, dx = \int_a^\eta f(x)\, dx + \int_\eta^\infty f(x)\, dx,$$

and we see that $\int_a^\infty f(x)\,dx$ converges or diverges precisely when $\int_\eta^\infty f(x)\,dx$ converges or diverges, since $\int_a^\eta f(x)\,dx$ is finite. The same holds true for $\int_a^\infty g(x)\,dx$. Now suppose $\int_\eta^\infty f(x)\,dx$ converges. Then by Eq. (3) and the SCT, we see that $\int_\eta^\infty (c/2)g(x)\,dx$ converges, and consequently $\int_\eta^\infty g(x)\,dx$ converges, since $c \ne 0$. On the other hand, if $\int_\eta^\infty f(x)\,dx$ diverges, then by Eq. (3) and the SCT, the integral $\int_\eta^\infty 2c \cdot g(x)\,dx$ diverges, and consequently $\int_\eta^\infty g(x)\,dx$ diverges, since $c \ne 0$. This concludes the proof of part 1. □

We now give examples illustrating the great power of the SCT and the LCT.

EXAMPLE 5 Establish the convergence or divergence of the integral $\int_5^\infty [1/(x^2 - 3)]\,dx$, discussed prior to Theorem 9.6.

Solution We let $f(x) = 1/(x^2 - 3)$ and $g(x) = 1/x^2$. We have

$$\lim_{x\to\infty} \frac{f(x)}{g(x)} = \lim_{x\to\infty} \frac{\dfrac{1}{x^2-3}}{\dfrac{1}{x^2}} = \lim_{x\to\infty} \frac{x^2}{x^2-3} = \lim_{x\to\infty} \frac{x^2}{x^2} = 1 > 0.$$

(Recall from Section 2.3 that the limit as $x \to \infty$ of a quotient of polynomials in x equals the limit of the quotient of the monomial terms of highest degree.) Since $\int_5^\infty (1/x^2)\,dx$ converges as a p-integral at ∞ with $p = 2$, we see that the integral $\int_5^\infty [1/(x^2 - 3)]\,dx$ converges by the LCT. ■

Example 5 indicates that for a basic type of improper integral at ∞ or $-\infty$, we can replace any polynomial expression by the monomial term of greatest degree, according to the LCT. The LCT is very powerful.

EXAMPLE 6 Establish the convergence or divergence of

$$\int_1^\infty \frac{x^3 - 2x^2 + x - 1}{2x^4 + 3x + 2}\,dx.$$

Solution As $x \to \infty$, the LCT shows that this integral behaves like

$$\int_1^\infty \frac{x^3}{2x^4}\,dx = \frac{1}{2}\int_1^\infty \frac{dx}{x}.$$

(See the remarks preceding this example.) Now $\int_1^\infty (1/x)\,dx$ diverges as a p-integral at ∞ with $p = 1$. Thus the given integral diverges, by the LCT. ■

EXAMPLE 7 Establish the convergence or divergence of

$$\int_0^\infty \frac{dx}{\sqrt{x} + x^2}.$$

Solution Splitting the integral into its basic types, we have

$$\int_0^\infty \frac{dx}{\sqrt{x} + x^2} = \int_0^1 \frac{dx}{\sqrt{x} + x^2} + \int_1^\infty \frac{dx}{\sqrt{x} + x^2}.$$

By the remarks preceding Example 6 and by the LCT, we see that

$$\int_1^\infty \frac{dx}{\sqrt{x} + x^2}$$

behaves like $\int_1^\infty (1/x^2)\,dx$, which converges. Near zero, a sum of powers of x is dominated by its term of *lowest* degree. Thus we expect $\sqrt{x} + x^2$ to behave like $\sqrt{x}$ near zero. We check this out carefully, using the LCT, this one time:

$$\lim_{x\to 0} \frac{1/\sqrt{x}}{\dfrac{1}{\sqrt{x} + x^2}} = \lim_{x\to 0} \frac{\sqrt{x} + x^2}{\sqrt{x}} = \lim_{x\to 0} \frac{\sqrt{x}(1 + x^{3/2})}{\sqrt{x}}$$

$$= \lim_{x\to 0} (1 + x^{3/2}) = 1 > 0.$$

Thus

$$\int_0^1 \frac{dx}{\sqrt{x} + x^2}$$

behaves like $\int_0^1 (1/\sqrt{x})\,dx$, which converges as a p-integral at $a = 0$ with $p = \frac{1}{2}$.

Since both of the basic types of integrals converge, we see that our original integral $\int_0^\infty [1/(\sqrt{x} + x^2)]\,dx$ converges. ∎

EXAMPLE 8 Establish the convergence or divergence of

$$\int_0^\infty \frac{|\sin x|}{x^{4/3}}\,dx.$$

Solution Splitting the integral into its basic types, we have

$$\int_0^\infty \frac{|\sin x|}{x^{4/3}}\,dx = \int_0^1 \frac{|\sin x|}{x^{4/3}}\,dx + \int_1^\infty \frac{|\sin x|}{x^{4/3}}\,dx.$$

Now $|\sin x| \le 1$, so

$$\frac{|\sin x|}{x^{4/3}} \le \frac{1}{x^{4/3}} \qquad \text{for } x \ge 1. \tag{4}$$

Now $\int_1^\infty (1/x^{4/3})\,dx$ converges as a p-integral at ∞ with $p = \frac{4}{3}$. By Eq. (4) and the SCT, we see that $\int_1^\infty [|\sin x|/x^{4/3}]\,dx$ converges.

If $0 \le x \le 1$, we have $\sin x \ge 0$, so the absolute value is not needed as we work near zero. We have

$$\frac{\sin x}{x^{4/3}} = \frac{\sin x}{x} \cdot \frac{1}{x^{1/3}}.$$

Since $\lim_{x\to 0} [(\sin x)/x] = 1$, we see by the LCT that $\int_0^1 [(\sin x)/x^{4/3}]\,dx$ behaves like $\int_0^1 (1/x^{1/3})\,dx$, which converges as a p-integral at $a = 0$ with $p = \frac{1}{3}$. Thus $\int_0^1 [(\sin x)/x^{4/3}]\,dx$ converges.

Since both of the basic types of integrals of which it is composed converge, we see that our original integral $\int_0^\infty [|\sin x|/x^{4/3}]\,dx$ converges. ∎

EXAMPLE 9 Establish the convergence or divergence of

$$\int_3^\infty \frac{dx}{\sqrt{x^3 - 2x^2 - 3x}}.$$

Solution Since $x^3 - 2x^2 - 3x = x(x + 1)(x - 3)$, we see that the improper integral has the splitting

$$\int_3^\infty \frac{dx}{\sqrt{x^3 - 2x^2 - 3x}} = \int_3^4 \frac{dx}{\sqrt{x^3 - 2x^2 - 3x}} + \int_4^\infty \frac{dx}{\sqrt{x^3 - 2x^2 - 3x}}$$

into basic types. Now

$$\frac{1}{\sqrt{x^3 - 2x^2 - 3x}} = \frac{1}{\sqrt{x^2 + x}\sqrt{x - 3}}.$$

Since

$$\lim_{x \to 3} \frac{1}{\sqrt{x^2 + x}} = \frac{1}{\sqrt{12}} > 0,$$

the LCT shows that

$$\int_3^4 \frac{dx}{\sqrt{x^3 - 2x^2 - 3x}}$$

behaves like

$$\int_3^4 \frac{dx}{\sqrt{x - 3}},$$

which converges as a p-integral at $a = 3$ with $p = \frac{1}{2}$.

The LCT shows that the integral $\int_4^\infty (1/\sqrt{x^3 - 2x^2 - 3x})\, dx$ behaves like $\int_4^\infty (1/x^{3/2})\, dx$, which converges as a p-integral at ∞ with $p = \frac{3}{2}$. Thus $\int_3^\infty (1/\sqrt{x^3 - 2x^2 - 3x})\, dx$ converges. ∎

SUMMARY

1. For $a > 0$, the improper integral $\int_a^\infty (1/x^p)\, dx$ converges if $p > 1$ and diverges if $p \leq 1$.

2. The improper integrals $\int_a^b [1/(x - a)^p]\, dx$ and $\int_a^b [1/(b - x)^p]\, dx$ converge if $p < 1$ and diverge if $p \geq 1$.

3. *Standard comparison test (SCT):* Let $0 \leq f(x) \leq g(x)$. If an improper integral of $g(x)$ converges, then the improper integral of $f(x)$, having the same limits, converges. If an improper integral of $f(x)$ diverges, then the same improper integral of $g(x)$ diverges.

4. *Limit comparison test (LCT):* Let $f(x) > 0$ and $g(x) > 0$. Suppose that basic types of improper integrals of $f(x)$ and $g(x)$ have the same limits of integration, and the two integrals have a common "bad point" of $-\infty$, a, b, or ∞. If

$$\lim_{x \to \text{"bad point"}} \frac{f(x)}{g(x)} = c > 0,$$

then the improper integrals of $f(x)$ and $g(x)$ either both converge or both diverge.

EXERCISES 9.6

In Exercises 1–40, decide whether the integral converges or diverges. Be prepared to give reasons for your answers. You need not compute the value of the integral if it converges.

1. $\displaystyle\int_{-\infty}^{2} \frac{dx}{\sqrt{2-x}}$

2. $\displaystyle\int_{-3}^{10} \frac{dx}{\sqrt{x+3}}$

3. $\displaystyle\int_{-1}^{1} \frac{dx}{x^{2/3}}$

4. $\displaystyle\int_{-\infty}^{\infty} \frac{dx}{x^{2/3}}$

5. $\displaystyle\int_{-\infty}^{\infty} \frac{dx}{x^{4/3}}$

6. $\displaystyle\int_{2}^{\infty} \frac{x^3 + 3x + 2}{x^5 - 8x}\, dx$

7. $\displaystyle\int_{1}^{\infty} \frac{x^4 - 3x^2 + 8}{x^5 + 7x}\, dx$

8. $\displaystyle\int_{1}^{\infty} \frac{x^2 + 2}{\sqrt{x^5 + 3x^2}}\, dx$

9. $\displaystyle\int_{1}^{\infty} \frac{x^2 + 3x + 1}{\sqrt{x^7 + 2x}}\, dx$

10. $\displaystyle\int_{3}^{\infty} \frac{dx}{x^2 - 3x - 4}$

11. $\displaystyle\int_{-\infty}^{\infty} \frac{dx}{x^2 + 4x + 4}$

12. $\displaystyle\int_{-\infty}^{\infty} \frac{dx}{x^2 + 4x + 5}$

13. $\displaystyle\int_{2}^{\infty} \frac{dx}{x^2 - x}$

14. $\displaystyle\int_{2}^{\infty} \frac{dx}{x^2 - 4}$

15. $\displaystyle\int_{2}^{\infty} \frac{dx}{\sqrt{x^2 - 4}}$

16. $\displaystyle\int_{2}^{\infty} \frac{dx}{(x^2 - 4)^{2/3}}$

17. $\displaystyle\int_{0}^{\infty} \frac{dx}{x + x^2}$

18. $\displaystyle\int_{0}^{\infty} \frac{dx}{x^{2/3} + x^{3/2}}$

19. $\displaystyle\int_{-\infty}^{\infty} \frac{dx}{|x^2 - 4|}$

20. $\displaystyle\int_{-\infty}^{\infty} \frac{dx}{(x^2 - 4)^{2/3}}$

21. $\displaystyle\int_{0}^{2} \frac{dx}{\sqrt{2x - x^2}}$

22. $\displaystyle\int_{0}^{2} \frac{dx}{\sqrt{2x^2 - x^3}}$

23. $\displaystyle\int_{0}^{1} \frac{dx}{\sqrt{x} - 1}$

24. $\displaystyle\int_{0}^{\infty} \frac{dx}{x^2 - \sqrt{x}}$

25. $\displaystyle\int_{2}^{\infty} \frac{dx}{x(\ln x)}$

26. $\displaystyle\int_{1}^{2} \frac{dx}{x(\ln x)}$

27. $\displaystyle\int_{1}^{2} \frac{dx}{x\sqrt{\ln x}}$

28. $\displaystyle\int_{2}^{\infty} \frac{dx}{x(\ln x)^2}$

29. $\displaystyle\int_{1}^{\infty} \frac{dx}{x[\sqrt{\ln x} + (\ln x)^2]}$

30. $\displaystyle\int_{1}^{\infty} \frac{|\cos x|}{x^2}\, dx$

31. $\displaystyle\int_{0}^{\infty} \frac{|\cos x|}{x^2}\, dx$

32. $\displaystyle\int_{0}^{\pi} \frac{|\cos x|}{\sqrt{x}}\, dx$

33. $\displaystyle\int_{0}^{\pi} \frac{\sin x}{x^{3/2}}\, dx$

34. $\displaystyle\int_{0}^{\infty} \frac{|\sin x|}{x^{3/2}}\, dx$

35. $\displaystyle\int_{0}^{\pi} \frac{\sin x}{x^{5/2}}\, dx$

36. $\displaystyle\int_{0}^{\pi} \frac{\sqrt{\sin x}}{x^{7/4}}\, dx$

37. $\displaystyle\int_{0}^{\pi} \frac{\sin x}{x^{7/4}}\, dx$

38. $\displaystyle\int_{0}^{\pi/4} \frac{\tan x}{x^{3/2}}\, dx$

39. $\displaystyle\int_{0}^{\pi/4} \frac{\tan x}{x^2}\, dx$

40. $\displaystyle\int_{0}^{\pi/2} \frac{\tan x}{x^{3/2}}\, dx$

The following theorem is sometimes useful for establishing convergence of an improper integral of a function that assumes both positive and negative values.

THEOREM 9.7   Absolute convergence

Let f be continuous. If an improper integral of $|f(x)|$ converges, then the improper integral with the same limits of $f(x)$ also converges. In this case, we say the integral of $f(x)$ **converges absolutely.**   □

In Exercises 41–50, decide whether the given integral converges absolutely.

41. $\displaystyle\int_{1}^{\infty} \frac{\sin x}{x^2}\, dx$

42. $\displaystyle\int_{0}^{\infty} \frac{\cos x}{x^2}\, dx$

43. $\displaystyle\int_{0}^{\infty} \frac{\sin x}{x^2}\, dx$

44. $\displaystyle\int_{0}^{\infty} \frac{\sin x}{x^{3/2}}\, dx$

45. $\displaystyle\int_{0}^{\infty} \frac{\cos x^2}{x^2 + 1}\, dx$

46. $\displaystyle\int_{0}^{\infty} \frac{\sin x^2}{x^2}\, dx$

47. $\displaystyle\int_{0}^{\infty} \frac{\sin x^2}{x^3}\, dx$

48. $\displaystyle\int_{0}^{\infty} \frac{\sin x^2}{x^{5/2}}\, dx$

49. $\displaystyle\int_{0}^{1} \frac{\sin(1/x)}{\sqrt{x}}\, dx$

50. $\displaystyle\int_{0}^{\infty} \frac{\sin(1/x)}{\sqrt{x} + x}\, dx$

SUPPLEMENTARY EXERCISES FOR CHAPTER 9

1. Express $x^2 - 4x + 5$ as a polynomial centered at $x_0 = -1$.

2. Express $4(x - 1)^3 + 3(x - 1)^2 + 7(x - 1) - 5$ as a polynomial centered at $x_0 = -1$.

3. Find the coefficient of $(x + 1)^3$ if $x^6 - 5x^4 + 3x^3 - x$ is expressed as a polynomial centered at $x_0 = -1$.

4. Express $x^4 - 2(x - 3)^3 + (2x - 4)^2 - 4x + 3$ as a polynomial centered at $x_0 = 2$.

5. Find the third Taylor polynomial for $1/(x + 1)$ at $x_0 = 1$.

6. Find the fourth Taylor polynomial $T_4(x)$ for $f(x) = \sin x$ at $x_0 = \pi/3$.

7. Find the fourth Taylor polynomial $T_4(x)$ for $f(x) = \cos 2x$ at $x_0 = \pi/4$.

8. Find the fourth Taylor polynomial $T_4(x)$ for $f(x) = e^{2x}$ centered at $x_0 = 1$.

9. Approximate $\sqrt{e}$ using the Taylor polynomial $T_4(x)$ for e^x centered at $x_0 = 0$, and find a bound for the absolute error.

10. Use linear approximation to estimate $9.05^{3/2}$, and find a bound for the absolute error.

11. Approximate $\ln 1.04$ using the Taylor polynomial $T_3(x)$ for $\ln x$ at $x_0 = 1$, and find a bound for the error.

12. Approximate $\sinh 0.1$ using the Taylor polynomial $T_4(x)$ at $x_0 = 0$, and find a bound for the error.

13. Give the formula for the error term $E_n(x)$ in Taylor's theorem for $f(x)$ at $x = x_0$, and explain the meaning of any letter used that does not appear in the statement of this exercise.

14. Use Taylor's theorem to find b such that for $|x| < b$, we have $|1 - \cos x| < 0.0001$. [*Hint:* Note that $T_0(x) = T_1(x)$ for $\cos x$ at $x_0 = 0$.]

15. Explain why for any given value of x, the Taylor polynomial $T_n(x)$ for $\sin x$ at $x_0 = 0$ will give an accurate estimate for $\sin x$ if n is sufficiently large.

16. Use Taylor's theorem to find b such that for $|x| < b$, we have $|(x - x^3/6) - \sin x| < 0.01$. [*Hint:* Note that $T_3(x) = T_4(x)$ for $\sin x$ at $x_0 = 0$.]

17. Find $\lim\limits_{x \to 1} \dfrac{x^6 - 2x^2 + 1}{8x^5 - 5x^3 - 3}$.

18. Find $\lim\limits_{x \to 0} \dfrac{e^{3x} - 1 - 3x}{1 - \cos x}$.

19. Find $\lim\limits_{x \to 1} \dfrac{x^3 - 7x^2 + 6}{x^4 - 2x^3 + 2}$.

20. Find $\lim\limits_{x \to \infty} \dfrac{e^{3x+10}}{e^{3x}}$.

21. Use Taylor polynomials to find $\lim\limits_{x \to 0} \dfrac{\cos x^2 - 1}{x \sin x^3}$.

22. Find $\lim\limits_{x \to 0} \dfrac{(\sin x^4) - x^4(\cos x^4)}{x^{12}}$.

23. Use Taylor polynomials to find $\lim\limits_{x \to 0} \dfrac{e^{x^3} - 1}{x - \sin x}$.

24. Find $\lim\limits_{x \to 0} \dfrac{x(e^{x^5} - 1)}{x^2 - \sin x^2}$.

25. Find $\lim\limits_{x \to \infty} \left(\dfrac{2x^3 - 3x + 1}{3x^2} - \dfrac{4x^2 + 2x - 3}{6x} \right)$.

26. Find $\lim\limits_{x \to \pi/2} \left(x - \dfrac{\pi}{2} \right)^2 \tan x$.

27. Find $\lim\limits_{x \to \infty} (\ln x)\left(\ln \dfrac{x + 1}{x} \right)$.

28. Find $\lim\limits_{x \to 0+} \left(1 + \dfrac{2}{\sin x} \right)^x$.

29. Find $\lim\limits_{x \to \infty} \left(\dfrac{2x + 3}{2x} \right)^x$.

30. Find $\lim\limits_{x \to \infty} \left(\dfrac{2x}{3x + 1} \right)^{\sqrt{x}}$.

31. Find $\lim\limits_{x \to 0+} (1 + \sin 2x)^{\csc x}$.

32. Find $\displaystyle\int_1^\infty \dfrac{1}{1 + x^2}\, dx$, if the integral converges.

33. Find $\displaystyle\int_0^2 \dfrac{x}{\sqrt{2 - x}}\, dx$, if the integral converges.

34. Find $\displaystyle\int_2^\infty \dfrac{dx}{x(\ln x)^2}$, if the integral converges.

35. Find $\displaystyle\int_{-\infty}^\infty \dfrac{e^x}{1 + e^{2x}}\, dx$, if the integral converges.

36. Find $\displaystyle\int_0^\infty \dfrac{dx}{\sqrt{x} + x^{3/2}}$, if the integral converges.

In Exercises 37–48, determine the convergence or divergence of the following improper integrals.

37. $\displaystyle\int_{-\infty}^\infty \dfrac{1}{x^2}\, dx$

38. $\displaystyle\int_2^\infty \dfrac{x + 7}{x^2 - 1}\, dx$

39. $\displaystyle\int_1^2 \dfrac{\sqrt{2 - x}}{x^2 - 4}\, dx$

40. $\displaystyle\int_0^\infty \dfrac{1}{x^3}\, dx$

41. $\displaystyle\int_{-1}^4 \dfrac{x}{(x - 1)^{2/3}}\, dx$

42. $\displaystyle\int_{-\infty}^\infty \dfrac{1}{1 + x^3}\, dx$

43. $\displaystyle\int_0^\infty \dfrac{dx}{\sqrt{x}(x + 1)}$

44. $\displaystyle\int_0^1 \dfrac{x}{\sin^2 x}\, dx$

45. $\displaystyle\int_{-\infty}^\infty \dfrac{\cos^2 x}{1 + x^2}\, dx$

46. $\displaystyle\int_1^2 \dfrac{\sin x}{\sqrt{x - 1}}\, dx$

47. $\displaystyle\int_0^\infty \dfrac{\sin^2(1/x)}{\sqrt{x}}\, dx$

48. $\displaystyle\int_{-\infty}^\infty \dfrac{dx}{1 + x^5}$

INFINITE SERIES

10

In the preceding chapter, we discussed the Taylor polynomial $T_n(x)$ associated with a function f having derivatives of order less than or equal to n. If f has derivatives of all orders, we can let n approach ∞ and obtain a "polynomial function of infinite degree" associated with f. The first five sections of this chapter deal with the problem of evaluating such a polynomial function of infinite degree at a point, which amounts to adding up infinitely many numbers. The final three sections deal with properties of such functions and with applications.

10.1 SEQUENCES

Motivation for Series and Sequences

We mention again that it is straightforward to work with polynomial functions. In Example 3 of Section 9.1, we found the Taylor polynomial $T_7(x)$ for $\sin x$ at $x_0 = 0$, namely

$$T_7(x) = x - \frac{x^3}{3!} + \frac{x^5}{5!} - \frac{x^7}{7!}.$$

We will see in this chapter that $\sin x$, while not a polynomial function, can be considered to be actually equal to the following infinite sum:

$$\sin x = x - \frac{x^3}{3!} + \frac{x^5}{5!} - \frac{x^7}{7!} + \frac{x^9}{9!} - \frac{x^{11}}{11!} + \cdots. \tag{1}$$

Replacing x by x^2, we obtain

$$\sin x^2 = x^2 - \frac{x^6}{3!} + \frac{x^{10}}{5!} - \frac{x^{14}}{7!} + \frac{x^{18}}{9!} - \frac{x^{22}}{11!} + \cdots. \tag{2}$$

It can be shown that we can't find $\int \sin x^2 \, dx$ as a combination of elementary functions. However, we will see that we can use Eq. (2) to integrate $\sin x^2$ as an "infinite polynomial."

If we use Eq. (1) to compute sin 1 (the sine of one radian), replacing x by 1, then we have to compute the infinite sum

$$1 - \frac{1}{3!} + \frac{1}{5!} - \frac{1}{7!} + \frac{1}{9!} - \frac{1}{11!} + \cdots \qquad (3)$$

to find the answer. This leads us to study infinite sums such as (3), or *infinite series,* as they are called. Of course, we really have this idea embodied in our notation for numbers; for example,

$$\frac{1}{3} = 0.33333 \ldots = \frac{3}{10} + \frac{3}{100} + \frac{3}{1000} + \frac{3}{10000} + \frac{3}{100000} + \cdots$$

and

$$e = 2.71828 \ldots$$

$$= 2 + \frac{7}{10} + \frac{1}{100} + \frac{8}{1000} + \frac{2}{10000} + \frac{8}{100000} + \cdots.$$

EXAMPLE 1 Use Eq. (1) and a calculator to estimate sin 1.

Solution Our calculator shows that

$$1 = 1.0000000000$$

$$1 - \frac{1}{3!} = 0.8333333333$$

$$1 - \frac{1}{3!} + \frac{1}{5!} = 0.8416666667$$

$$1 - \frac{1}{3!} + \frac{1}{5!} - \frac{1}{7!} = 0.841468254$$

$$1 - \frac{1}{3!} + \frac{1}{5!} - \frac{1}{7!} + \frac{1}{9!} = 0.8414710097$$

$$1 - \frac{1}{3!} + \frac{1}{5!} - \frac{1}{7!} + \frac{1}{9!} - \frac{1}{11!} = 0.8414709846$$

$$1 - \frac{1}{3!} + \frac{1}{5!} - \frac{1}{7!} + \frac{1}{9!} - \frac{1}{11!} + \frac{1}{13!} = 0.8414709848$$

$$\vdots$$

From Eq. (1), we expect the *sequence of numbers* in the column to the right of the equal signs to get closer and closer to sin 1 as we go down the column. Our calculator gives sin 1 = 0.8414709848, so our last computation from Eq. (1) gives sin 1 as accurately as our calculator can display it. ∎

Calculators and computers do not have huge tables for $\sin x$ stored in them. When we use them to evaluate $\sin x$, they simply evaluate some suitable algebraic formula that approximates $\sin x$ correct to the number of digits the calculator works with. Example 1 indicates how this might be done.

EXAMPLE 2 Try to find the "sum" of the infinite series of constants

$$1 + \frac{1}{2} + \frac{1}{4} + \frac{1}{8} + \frac{1}{16} + \cdots. \tag{4}$$

Solution We try to add up all the numbers in Eq. (4), and we obtain successively

$$1 = 1$$

$$1 + \frac{1}{2} = \frac{3}{2}$$

$$1 + \frac{1}{2} + \frac{1}{4} = \frac{7}{4}$$

$$1 + \frac{1}{2} + \frac{1}{4} + \frac{1}{8} = \frac{15}{8}$$

$$\frac{15}{8} + \frac{1}{16} = \frac{31}{16}$$

$$\vdots$$

This leads us to consider the *sequence of numbers*

$$1, \quad \frac{3}{2}, \quad \frac{7}{4}, \quad \frac{15}{8}, \quad \frac{31}{16}, \quad \cdots \tag{5}$$

Figure 10.1 indicates that each number in the sequence is halfway between the preceding number and 2. Thus it is clear that the numbers in the sequence (5) get very close to 2 as we continue through the sequence. This suggests that the "sum" of the infinite series (4) should be 2. ∎

From Examples 1 and 2, it is natural to try to define the sum of an infinite *series* of numbers by examining the *sequence* of numbers obtained by adding more and more terms of the series. In the following subsection we discuss sequences and their limits. We then will use these ideas to discuss series in Section 10.2.

Sequences

An example of a sequence was given in (5) above. We naively think of a sequence as an endless row of numbers separated by commas, symbolically,

$$a_1, \quad a_2, \quad a_3, \quad \ldots, \quad a_n, \quad \ldots, \tag{6}$$

where each a_i is a real number. Sequence (6) is also written $\{a_n\}$ for brevity. Since a sequence has one term for each positive integer, we can also consider a sequence to be a real-valued function ϕ with domain the set of positive integers.

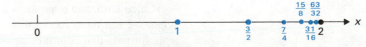

FIGURE 10.1 The sequence $1, \frac{3}{2}, \frac{7}{4}, \frac{15}{8}, \frac{31}{16}, \frac{63}{32}, \ldots$ converges to 2.

For sequence (6), we then have

$$\phi(1) = a_1, \qquad \phi(2) = a_2, \qquad \phi(3) = a_3, \qquad \ldots \ .$$

We may regard sequence (5) as a function ϕ where

$$\phi(1) = 1, \qquad \phi(2) = \frac{3}{2}, \qquad \phi(3) = \frac{7}{4}, \qquad \ldots \ .$$

DEFINITION 10.1

Sequence

A **sequence** is a real-valued function ϕ with domain the set of positive integers. If $\phi(n) = a_n$, the sequence is denoted by $\{a_n\}$ or by $a_1, a_2, \ldots ,$ $a_n, \ldots \ .$

We want to find the limit of a sequence if it exists. Naively defined, the limit of $\{a_n\}$ is c if the terms a_n get just as close to c as we wish, provided that n is large enough. This is a vague statement similar to the one we used to introduce $\lim_{x \to a} f(x) = c$. Here is a precise definition.

DEFINITION 10.2

Limit of a sequence

The **limit of a sequence** $\{a_n\}$ is c if for each $\varepsilon > 0$ there exists an integer N such that $|a_n - c| < \varepsilon$ if $n > N$. We write "$\lim_{n \to \infty} \{a_n\} = c$" or "$\lim_{n \to \infty} a_n = c$," and we say that the sequence $\{a_n\}$ **converges to** c. A sequence that has no limit **diverges.**

We leave to the exercises (Exercise 38) the straightforward proof that a sequence can't converge to two different values.

Figure 10.2 illustrates Definition 10.2. We mark c on the x-axis and then mark $c - \varepsilon$ and $c + \varepsilon$. The definition asserts that we must have all terms beyond some point in the sequence falling within the bracketed interval between $c - \varepsilon$ and $c + \varepsilon$. The N in the definition is simply a position indicator, giving the position in the sequence beyond which all terms, that is, $a_{N+1}, a_{N+2}, a_{N+3}, \ldots ,$ must be in the bracketed interval. As illustrated in the next example, the smaller ε is, the larger we expect N to have to be.

EXAMPLE 3   Figure 10.3 shows some terms of a sequence $\{a_n\}$ that has limit 4. Assume that each term of the sequence is closer to 4 than the preceding term is. Find the smallest possible N in Definition 10.2 corresponding to $\varepsilon = 3$, $\varepsilon = 1$, and $\varepsilon = \frac{1}{2}$.

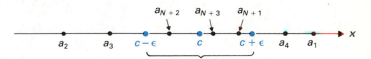

FIGURE 10.2   Terms $a_{N+1}, a_{N+2}, a_{N+3}, \ldots ,$ falling between $c - \varepsilon$ and $c + \varepsilon$ for Definition 10.2.

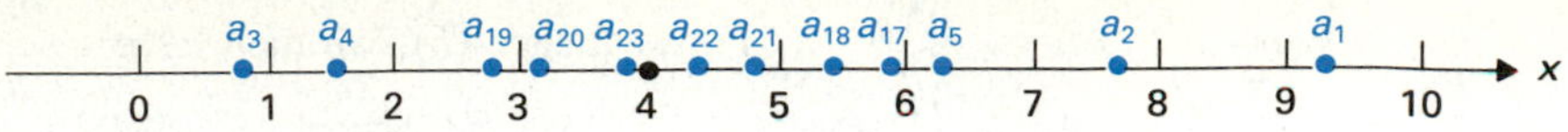

FIGURE 10.3 Some terms of a sequence $\{a_n\}$ that converges to 4.

Solution

$\varepsilon = 3$ We need to have terms fall between $4 - 3 = 1$ and $4 + 3 = 7$. Now a_3 is not between 1 and 7, but a_4 is. Since we are told that terms of the sequence become consecutively closer to 4, we see that we can take $N = 3$, since for $n > 3$ we have a_n between 1 and 7.

$\varepsilon = 1$ We need to have terms between $4 - 1 = 3$ and $4 + 1 = 5$. Figure 10.3 shows that a_{19} is not between 3 and 5, but a_{20} is. Thus we have $N = 19$.

$\varepsilon = \frac{1}{2}$ We need to have terms between $4 - \frac{1}{2} = \frac{7}{2}$ and $4 + \frac{1}{2} = \frac{9}{2}$. Figure 10.3 shows that a_{21} is not between $\frac{7}{2}$ and $\frac{9}{2}$, but a_{22} is. Thus we have $N = 21$.   ∎

EXAMPLE 4   For the sequence

$$1, \quad \frac{3}{2}, \quad \frac{7}{4}, \quad \frac{15}{8}, \quad \frac{31}{16}, \quad \cdots$$

in sequence (5), describe how to find an N in Definition 10.2 for a given $\varepsilon > 0$.

Solution   For this sequence, which has limit 2, $a_1 = 1$ is 1 less than 2, $a_2 = \frac{3}{2}$ is $\frac{1}{2}$ less than 2, $a_3 = \frac{7}{4}$ is $\frac{1}{4}$ less than 2, and so on, as illustrated in Fig. 10.1. We then see that

$$a_n = 2 - \frac{1}{2^{n-1}}.$$

Thus

$$|a_n - 2| = \frac{1}{2^{n-1}},$$

so, given $\varepsilon > 0$, if we choose N so that $(\frac{1}{2})^{N-1} < \varepsilon$, then

$$|a_n - 2| = \frac{1}{2^{n-1}} < \varepsilon$$

for $n > N$. Solving $(\frac{1}{2})^{N-1} < \varepsilon$ for N, we obtain

$$\left(\frac{1}{2}\right)^{N-1} < \varepsilon,$$

$$(N - 1)\left(\ln \frac{1}{2}\right) < \ln \varepsilon, \qquad \textit{Taking } \ln \textit{ of both sides}$$

$$(N - 1)(-\ln 2) < \ln \varepsilon, \qquad \ln(\tfrac{1}{2}) = \ln 1 - \ln 2$$

$$N - 1 > -\frac{\ln \varepsilon}{\ln 2},$$

$$N > 1 - \frac{\ln \varepsilon}{\ln 2}.$$

For example, with $\varepsilon = 0.001$, our calculator gives $N > 1 - (-9.966) = 10.966$. Thus we can take $N = 11$ for $\varepsilon = 0.001$. ∎

Frequently, we will describe a sequence $\{a_n\}$ by giving a formula in terms of n for the "nth term" a_n of the sequence. For example, the sequence discussed in Example 4 is the sequence where $a_n = 2 - (\frac{1}{2})^{n-1}$, or, more briefly, the sequence

$$\left\{2 - \frac{1}{2^{n-1}}\right\}.$$

Example 4 shows that

$$\lim_{n \to \infty} \left(2 - \frac{1}{2^{n-1}}\right) = 2 - 0 = 2.$$

Finding the limit of a sequence $\{a_n\}$, where a_n is given by a formula involving n, is much like finding $\lim_{x \to \infty} f(x)$ where $f(x)$ is given by a formula. We do not expect you to exhibit N corresponding to ε to prove convergence to the limit in your routine work.

EXAMPLE 5 Find the limit of the sequence

$$\left\{\frac{1}{n}\right\} = 1, \quad \frac{1}{2}, \quad \frac{1}{3}, \quad \frac{1}{4}, \quad \cdots , \quad \frac{1}{n}, \quad \cdots .$$

Solution We must find $\lim_{n \to \infty} 1/n$. Clearly

$$\lim_{n \to \infty} \frac{1}{n} = 0.$$

Thus the sequence converges to 0. ∎

EXAMPLE 6 Establish the convergence or divergence of the sequence

$$\{(-1)^{n+1}\} = 1, \quad -1, \quad 1, \quad -1, \quad 1, \quad -1, \quad \cdots .$$

Solution The sequence diverges, since the terms of the sequence do not approach and stay close to any number c as we continue along the sequence. We leave an ε,N-proof of this fact to the exercises (Exercise 42). ∎

EXAMPLE 7 Find the limit of the sequence $\{(3n^2 - 3)/(5n - 2n^2)\}$, if it converges.

Solution We must find

$$\lim_{n \to \infty} \frac{3n^2 - 3}{5n - 2n^2}.$$

We use the technique we would use to find $\lim_{x \to \infty} f(x)$ for $f(x) = (3x^2 - 3)/(5x - 2x^2)$. Our work in Section 2.3 showed that for a quotient of polynomials, we can compute the limit at ∞ and $-\infty$ by keeping only the monomial terms of highest degree. Thus

$$\lim_{n \to \infty} \frac{3n^2 - 3}{5n - 2n^2} = \lim_{n \to \infty} \frac{3n^2}{-2n^2} = -\frac{3}{2},$$

so the sequence converges to $-\frac{3}{2}$. ∎

EXAMPLE 8 Find the limit of the sequence $\{n \sin(2/n)\}$, if it converges.

Solution We wish to find $\lim_{n \to \infty}[n \sin(2/n)]$. Since $\lim_{n \to \infty} \sin(2/n) = \sin 0 = 0$, we see that we run into an indeterminate expression $\infty \cdot 0$. We replace the variable n, which we think of as assuming only integer values, by the continuous variable x, which can assume any real value, and use l'Hôpital's rule. We obtain

$$\lim_{x \to \infty} \left[x \sin\left(\frac{2}{x}\right) \right] = \lim_{x \to \infty} \frac{\sin(2/x)}{1/x}$$

$$= \lim_{x \to \infty} \frac{\cos(2/x)(-2/x^2)}{-1/x^2}$$

$$= \lim_{x \to \infty} \left[2 \cos\left(\frac{2}{x}\right) \right] = 2(\cos 0) = 2 \cdot 1 = 2.$$

Thus we must also have $\lim_{n \to \infty}[n \sin(2/n)] = 2$, so the sequence converges to 2. ∎

Just as for a function, we have the notions of $\lim_{n \to \infty} a_n = \infty$ and $\lim_{n \to \infty} a_n = -\infty$. We ask you to state these definitions for sequences in Exercises 1 and 2.

Let us emphasize that a sequence is said to *converge* if and only if it has a *finite* limit. A sequence that does not converge to a finite limit *diverges*.

EXAMPLE 9 Determine the convergence or divergence of the sequences $\{n^2\}$ and $\{(-n^2 + 2)/(n + 1)\}$.

Solution We have

$$\lim_{n \to \infty} n^2 = \infty$$

while

$$\lim_{n \to \infty} \frac{-n^2 + 2}{n + 1} = \lim_{n \to \infty} \frac{-n^2}{n} = \lim_{n \to \infty} (-n) = -\infty.$$

Thus the sequence $\{n^2\}$ diverges to ∞ and the sequence $\{(-n^2 + 2)/(n + 1)\}$ diverges to $-\infty$. ∎

EXAMPLE 10 Discuss the convergence or divergence of the sequence $\{(-1)^n \cdot n\}$.

Solution The sequence

$$\{(-1)^n \cdot n\} = -1, \quad 2, \quad -3, \quad 4, \quad -5, \quad 6, \quad -7, \quad \ldots$$

diverges. However, it does not diverge to either ∞ or $-\infty$, since for n even the terms approach ∞, and for n odd the terms approach $-\infty$. ∎

TABLE 10.1	
n	$\sqrt{n^2 + n} - n$
100	0.499376558
200	0.499584026
400	0.499750250
800	0.499861188
1600	0.499926492
3200	0.499962127
6400	0.499980771
12800	0.499990310
25600	0.499995136
51200	0.499997563
102400	0.499998780
204800	0.499999390

EXAMPLE 11 Find the limit of the sequence

$$\left\{\frac{\cos n}{\ln n}\right\}, \qquad n > 1,$$

if the sequence converges.

Solution The restriction $n > 1$ is made since $\ln 1 = 0$ and we cannot have zero in a denominator. Now $|\cos n| \le 1$ and we know that $\ln n$ becomes arbitrarily large as $n \to \infty$, although it increases slowly. Thus the terms of the sequence eventually become arbitrarily close to zero, with some terms negative and others positive. The sequence thus converges to zero. ∎

A calculator or computer can be used to try to estimate the limit of a sequence.

EXAMPLE 12 Use a calculator or computer to estimate the limit of the sequence $\{\sqrt{n^2 + n} - n\}$.

Solution A computer program produced the data in Table 10.1. Based on these data, we estimate the limit of the sequence to be 0.5. ∎

SUMMARY

1. A sequence is an endless row of numbers

$$a_1, \quad a_2, \quad a_3, \quad \ldots, \quad a_n, \quad \ldots$$

separated by commas. (Repetition of numbers is allowed.) Alternatively, it is a real-valued function with domain the set of positive integers.

2. $\lim_{n \to \infty} a_n = c$ if for each $\varepsilon > 0$ there exists a positive integer N depending on ε such that $|a_n - c| < \varepsilon$ if $n > N$.

3. If $\lim_{n \to \infty} a_n$ exists (is a *finite* number c), then the sequence converges; otherwise it diverges.

EXERCISES 10.1

1. Define what is meant by $\lim_{n \to \infty} a_n = \infty$.

2. Define what is meant by $\lim_{n \to \infty} b_n = -\infty$.

In Exercises 3–36, find the limit of the sequence if the sequence converges or has limit ∞ or $-\infty$.

3. $\left\{\dfrac{1}{2n}\right\}$

4. $1, 0, \frac{1}{2}, 0, \frac{1}{3}, 0, \frac{1}{4}, 0, \ldots$

5. $\frac{1}{2}, \frac{2}{3}, \frac{3}{4}, \frac{4}{5}, \ldots$

6. $\left\{\dfrac{n^2 + 1}{3n}\right\}$

7. $\left\{\dfrac{2n - \sqrt{n}}{n}\right\}$

8. $\left\{\dfrac{3n^2 - 2n}{2n^3}\right\}$

9. $\left\{\dfrac{2^n}{n!}\right\}$

10. $\left\{\dfrac{(n + 3)!}{n^3 n!}\right\}$

11. $\left\{\left(-\dfrac{1}{2}\right)^n\right\}$

12. $\{\sin^n 1\}$

13. $1, 2, 1, -3, 1, 4, 1, -5, 1, 6, \ldots$

14. $\{e^n\}$

15. $\{e^{-n/(\sqrt{n}+1)}\}$

16. $\left\{\dfrac{e^{2n}}{n!}\right\}$

17. $\left\{\dfrac{en^2}{n!}\right\}$

18. $\{e^{\sin n}\}$

19. $\{\ln n\}$

20. $\left\{\dfrac{1}{\ln(n+1)}\right\}$

21. $\left\{\ln\left(\dfrac{2n^2+1}{3n+2n^2}\right)\right\}$

22. $\left\{\left[\ln\left(\dfrac{2n^2+3}{n^2+n}\right)\right]^n\right\}$

23. $\left\{\left[\ln\left(\dfrac{3n^2+2}{n^2+n}\right)\right]^n\right\}$

24. $\left\{\dfrac{\sin n}{\ln n}\right\}, \quad n > 1$

25. $\{\sin n\}$

26. $\left\{\sin \dfrac{1}{n}\right\}$

27. $\left\{\dfrac{n + \cos n}{n}\right\}$

28. $\{n^{\csc n}\}$

29. $\{n^{\sin(1/n)}\}$

30. $\left\{n \sin\left(\dfrac{1}{n}\right)\right\}$

31. $\{n^{1/(2n)}\}$

32. $\left\{n \ln\left(\dfrac{n+1}{n}\right)\right\}$

33. $\left\{\dfrac{n^n}{n!}\right\}$

34. $\left\{\dfrac{2^n + 5^n}{3^n + 5^n}\right\}$

35. $\left\{\dfrac{n3^n}{5^n}\right\}$

36. $\left\{\dfrac{2^{2n} + 2^{3n}}{3^{2n}}\right\}$

37. A ball dropped from a height of 40 ft bounces repeatedly, each time to half the height of the preceding bounce. In the next section, we will be able to show that the total distance the ball has traveled at the instant it bounces for the nth time is $40(3 - 1/2^{n-2})$ ft. Find the total distance the ball bounces if it is allowed to bounce forever.

38. Show that a sequence $\{a_n\}$ can't converge to two different limits.

39. Give an ε,N-proof that the sequence $1, 1, 1, \ldots, 1, \ldots$ converges.

40. Give an ε,N-proof that the sequence $1, -\frac{1}{2}, \frac{1}{3}, -\frac{1}{4}, \frac{1}{5}, -\frac{1}{6},$ $\ldots$ converges.

41. Give an ε,N-proof that the sequence

$$\frac{1}{\sqrt{2}}, \quad \frac{1}{\sqrt{3}}, \quad \frac{1}{\sqrt{4}}, \quad \ldots, \quad \frac{1}{\sqrt{n+1}}, \quad \ldots$$

converges.

42. Give an ε,N-proof that the sequence

$$1, \quad -1, \quad 1, \quad -1, \quad 1, \quad \ldots, \quad (-1)^{n+1}, \quad \ldots$$

diverges.

43. Give an ε,N-proof that the sequence $1, 2, 3, 4, \ldots, n, \ldots$ diverges.

44. We will often use the fact that if $|a| < 1$, then $\{a^n\}$ converges to zero. Convince yourself that $\lim_{n\to\infty} a^n = 0$ without giving a formal ε,N-proof. [*Suggestion:* Find $\lim_{n\to\infty} \ln(|a|^n)$.]

45. Give an ε,N-proof of the limit in Exercise 44. Give a formula for N in terms of ε.

46. It can be shown that $\lim_{n\to\infty} (1 + 1/n)^n = e$. Use a calculator to find the smallest integer N such that

$$|(1 + 1/N)^N - e| < \varepsilon$$

for the following values of ε.
  (a) 0.1   (b) 0.01   (c) 0.001   (d) 0.0001

In Exercises 47–51, attempt to find the limit of the sequence using a calculator or computer. Exercises 50 and 51 indicate one type of problem that may arise.

47. $\left\{\left(1 - \dfrac{1}{n}\right)^n\right\}$

48. $\left\{\left(1 + \dfrac{1}{2n}\right)^n\right\}$

49. $\{(1 + n)^{1/n}\}$

50. $\{n(\sqrt[n]{2} - 1)\}$

51. $\left\{n^2\left[1 - \cos\left(\dfrac{1}{n}\right)\right]\right\}$

EXPLORING CALCULUS

G: limit demonstration

10.2 SERIES

The Sum of a Series

It is mathematically imprecise to define a *sequence* of constants to be an endless row of numbers

$$a_1, \quad a_2, \quad \ldots, \quad a_n, \quad \ldots \tag{1}$$

separated by commas, although this is the way we sometimes think of a sequence in our work. In Section 10.1, we defined a *sequence* to be a *function*. It would be similarly imprecise to define an infinite *series* of constants to be an endless row of numbers

$$a_1 + a_2 + \cdots + a_n + \cdots \tag{2}$$

with plus signs between them, although this is often the way we think of a series in our work. We saw in the previous section that in attempting to find the "sum" of the series (2), we are led to consider the limit of the sequence

$$s_1, \quad s_2, \quad \ldots, \quad s_n, \quad \ldots, \tag{3}$$

where $s_1 = a_1$, $s_2 = a_1 + a_2$, and, in general, $s_n = a_1 + \cdots + a_n$. From a mathematical standpoint, the precise thing to do is to define the *series* (2) to *be* the sequence in (3); we have already based the notion of a sequence on the notion of a function.

DEFINITION 10.3

Infinite series

Let $a_1, a_2, \ldots, a_n, \ldots$ be a sequence and let $s_n = a_1 + \cdots + a_n$ for all positive integers n. The **infinite series**

$$\sum_{n=1}^{\infty} a_n = a_1 + a_2 + \cdots + a_n + \cdots$$

is the sequence $\{s_n\}$. The number a_n is the ***n*th term of the series** $\sum_{n=1}^{\infty} a_n$, and s_n is the ***n*th partial sum of the series.**

DEFINITION 10.4

Sum of a series

Let $\sum_{n=1}^{\infty} a_n$ be an infinite series. The **sum of the series** is the limit of the sequence $\{s_n\}$ of partial sums, if this limit exists. If $\lim_{n \to \infty} s_n$ is a finite number c, then the series $\sum_{n=1}^{\infty} a_n$ **converges to** c. If $\lim_{n \to \infty} s_n$ is ∞, $-\infty$, or is undefined, then the series **diverges.**

EXAMPLE 1   Determine the convergence or divergence of the series

$$0 + 0 + 0 + 0 + \cdots.$$

Solution Computing partial sums, we have $0 = 0$, $0 + 0 = 0$, $0 + 0 + 0 = 0$, ... Thus the sequence of partial sums is

$$0, \quad 0, \quad 0, \quad 0, \quad 0, \quad \dots \,,$$

which converges to 0. Thus the series converges to 0.   ■

EXAMPLE 2 Determine the convergence or divergence of the series

$$1 - 1 + 1 - 1 + 1 - 1 + \cdots.$$

Solution We find that the sequence of partial sums is

$$1, \quad 0, \quad 1, \quad 0, \quad 1, \quad 0, \quad \dots \, .$$

Since this sequence diverges, the series diverges.   ■

Harmonic and Geometric Series

In the next section, we will develop comparison tests for convergence or divergence of series, similar to the comparison tests for improper integrals in Section 9.6. Such comparison tests are useful only if we have a stockpile of series whose behavior (convergence or divergence) is known. The series discussed now contribute to such a stockpile.

The series

$$\sum_{n=1}^{\infty} \frac{1}{n} = 1 + \frac{1}{2} + \frac{1}{3} + \frac{1}{4} + \cdots + \frac{1}{n} + \cdots \tag{4}$$

is the **harmonic series.** This series diverges, as the following argument shows. We group together certain terms of the series, using parentheses, so that the series appears as

$$(1) + \left(\frac{1}{2}\right) + \left(\frac{1}{3} + \frac{1}{4}\right) + \left(\frac{1}{5} + \cdots + \frac{1}{8}\right) + \left(\frac{1}{9} + \cdots + \frac{1}{16}\right) + \cdots. \tag{5}$$

The sum of the terms in each parenthesis in series (5) is greater than or equal to $\frac{1}{2}$; for example,

$$\frac{1}{5} + \frac{1}{6} + \frac{1}{7} + \frac{1}{8} > \frac{1}{8} + \frac{1}{8} + \frac{1}{8} + \frac{1}{8} = \frac{1}{2}.$$

This shows that for the harmonic series, we have

$$s_{2^n} \geq 1 + \frac{1}{2}n,$$

and it is then clear that the series diverges to ∞.

We turn now to the series

$$\sum_{n=0}^{\infty} ar^n = a + ar + ar^2 + \cdots + ar^n + \cdots. \tag{6}$$

The series (6) is the **geometric series** with **initial term** a and **ratio** r; it is a very important series. We have

$$s_n = a + ar + \cdots + ar^{n-1}$$

and, consequently,

$$rs_n = ar + \cdots + ar^{n-1} + ar^n.$$

Subtracting, we obtain

$$s_n - rs_n = a - ar^n,$$

so if $r \neq 1$,

$$s_n = \frac{a - ar^n}{1 - r} = \frac{a}{1 - r} - \frac{ar^n}{1 - r}.$$

Thus

$$\lim_{n \to \infty} s_n = \lim_{n \to \infty} \left(\frac{a}{1 - r} - \frac{ar^n}{1 - r} \right) = \frac{a}{1 - r} - a \left(\lim_{n \to \infty} \frac{r^n}{1 - r} \right). \qquad (7)$$

The value of the limit in Eq. (7) depends on the size of r. We have

$$\lim_{n \to \infty} \frac{r^n}{1 - r} = \begin{cases} 0, & \text{if } -1 < r < 1, \\ -\infty, & \text{if } r > 1, \\ \text{undefined}, & \text{if } r \leq -1. \end{cases}$$

If $r = 1$, then the series (6) reduces to

$$a + a + \cdots + a + \cdots, \qquad (8)$$

which obviously does not converge if $a \neq 0$. Thus the geometric series (6), where $a \neq 0$, converges if $|r| < 1$ and diverges if $|r| \geq 1$. For $|r| < 1$, the sum of the series is found from Eq. (7) to be

$$\lim_{n \to \infty} s_n = \frac{a}{1 - r} - a \left(\lim_{n \to \infty} \frac{r^n}{1 - r} \right)$$
$$= \frac{a}{1 - r} - 0 = \frac{a}{1 - r}.$$

We summarize the results of this section in the following theorem for easy reference.

THEOREM 10.1   Harmonic and geometric series

1. The *harmonic series* $\sum_{n=1}^{\infty} 1/n$ diverges to ∞.

2. The *geometric series*

$$\sum_{n=0}^{\infty} ar^n \begin{cases} \text{converges to } \dfrac{a}{1 - r} & \text{if } |r| < 1, \\ \text{diverges} & \text{if } |r| \geq 1 \text{ and } a \neq 0. \end{cases} \qquad \square$$

EXAMPLE 3   Find the sum of the series

$$\sum_{n=0}^{\infty} \frac{1}{2^n} = 1 + \frac{1}{2} + \frac{1}{4} + \cdots + \frac{1}{2^n} + \cdots$$

discussed in Example 2, Section 10.1.

Solution   This is a geometric series with $a = 1$ and $r = \frac{1}{2}$. By Theorem 10.1, the series converges to

$$\frac{a}{1-r} = \frac{1}{1-\frac{1}{2}} = 2,$$

which we guessed to be the case in Section 10.1.   ∎

EXAMPLE 4   Find the sum of the series

$$\sum_{n=2}^{\infty} \frac{(-1)^n}{3^n},$$

if it converges.

Solution   This series

$$\frac{1}{9} - \frac{1}{27} + \frac{1}{81} - \frac{1}{243} + \cdots$$

is a geometric series with initial term $a = \frac{1}{9}$ and ratio $r = -\frac{1}{3}$. Since $|r| = \frac{1}{3} < 1$, Theorem 10.1 shows that the series converges to

$$\frac{a}{1-r} = \frac{\frac{1}{9}}{1-(-\frac{1}{3})} = \frac{\frac{1}{9}}{\frac{4}{3}} = \frac{1}{9} \cdot \frac{3}{4} = \frac{1}{12}.$$ ∎

EXAMPLE 5   Find the sum of the series

$$\sum_{n=0}^{\infty} \frac{2^{3n}}{7^n},$$

if the series converges.

Solution   Since $2^{3n} = (2^3)^n = 8^n$, we have

$$\sum_{n=0}^{\infty} \frac{2^{3n}}{7^n} = \sum_{n=0}^{\infty} \frac{8^n}{7^n} = \sum_{n=0}^{\infty} \left(\frac{8}{7}\right)^n.$$

This is a geometric series with ratio $r = \frac{8}{7} > 1$, so by Theorem 10.1 the series diverges.   ∎

EXAMPLE 6   A ball bounces in such a way that the length of time to complete the next bounce (the time between consecutive strikes on the floor) is always $\frac{2}{5}$ of the time consumed by the present bounce. If the first bounce takes 3 sec, find the elapsed time from the instant the ball first touches the floor until it comes to rest.

Solution   The first bounce takes 3 sec, the next bounce takes $(\frac{2}{5})3$ sec, the following one takes $(\frac{2}{5})(\frac{2}{5})3$ sec, and so on. Thus we must sum the series

$$3 + 3\left(\frac{2}{5}\right) + 3\left(\frac{2}{5}\right)^2 + 3\left(\frac{2}{5}\right)^3 + \cdots.$$

This is a convergent geometric series with initial term 3 and ratio $\frac{2}{5}$, so the sum of the series is

$$\frac{3}{1 - \frac{2}{5}} = \frac{3}{\frac{3}{5}} = \frac{3}{1} \cdot \frac{5}{3} = 5.$$

Thus the ball comes to rest 5 sec after the time it first touches the floor. ∎

EXAMPLE 7 Express the repeating decimal 4.277777 . . . as a rational number (quotient n/m of integers where $m \neq 0$).

Solution We have

$$4.277777 \ldots = \frac{42}{10} + \left(\frac{7}{100} + \frac{7}{1000} + \frac{7}{10000} + \cdots \right).$$

The sum in brackets is a geometric series with initial term $a = \frac{7}{100}$ and ratio $r = \frac{1}{10}$. Thus

$$4.277777 \ldots = \frac{42}{10} + \frac{\frac{7}{100}}{1 - \frac{1}{10}}$$

$$= \frac{42}{10} + \frac{7}{100} \cdot \frac{10}{9} = \frac{42}{10} + \frac{7}{90}$$

$$= \frac{378 + 7}{90} = \frac{385}{90} = \frac{77}{18}.$$ ∎

Exercises 42–48 continue the idea of Example 7 and indicate that the rational numbers are precisely the real numbers whose unending decimal expansion has a repeating pattern.

Algebra of Sequences and Series

There are natural ways to "add" two sequences or two series and to "multiply by a constant" a sequence or a series.

DEFINITION 10.5

Sequence algebra

Let $\{s_n\}$ and $\{t_n\}$ be sequences and let c be any number. The sequence $\{s_n + t_n\}$ with nth term $s_n + t_n$ is the **sum of $\{s_n\}$ and $\{t_n\}$**, while the sequence $\{cs_n\}$ with nth term cs_n is the **product of the constant c and $\{s_n\}$**.

DEFINITION 10.6

Series algebra

Let $\sum_{n=1}^{\infty} a_n$ and $\sum_{n=1}^{\infty} b_n$ be series of constants, and let c be any number. The series $\sum_{n=1}^{\infty} (a_n + b_n)$ with nth term $a_n + b_n$ results from **adding the series $\sum_{n=1}^{\infty} a_n$ and $\sum_{n=1}^{\infty} b_n$**, while the series $\sum_{n=1}^{\infty} ca_n$ with nth term ca_n results from **multiplying the series $\sum_{n=1}^{\infty} a_n$ by the constant c**.

It is important to note that if s_n is the nth partial sum of $\sum_{n=1}^{\infty} a_n$, and t_n is the nth partial sum of $\sum_{n=1}^{\infty} b_n$, then the nth partial sum of $\sum_{n=1}^{\infty} (a_n + b_n)$ is $s_n + t_n$, while the nth partial sum of $\sum_{n=1}^{\infty} ca_n$ is cs_n.

Throughout this chapter, we are interested primarily in whether sequences and series converge or diverge. After the preceding definitions, we at once ask ourselves whether the sequence obtained by adding two convergent sequences still converges and whether the sequence obtained by multiplying a convergent sequence by a constant still converges. We then ask the analogous questions for series. The answers to these questions are contained in the following theorem and its corollaries; the answers are intuitively obvious.

THEOREM 10.2   Convergence of sums and multiples

If the sequence $\{s_n\}$ converges to s and the sequence $\{t_n\}$ converges to t, then the sequence $\{s_n + t_n\}$ converges to $s + t$ and the sequence $\{cs_n\}$ converges to cs for all c.

Proof: Let $\varepsilon > 0$ be given. Find an integer N_1 such that $|s_n - s| < \varepsilon/2$ for $n > N_1$, and find an integer N_2 such that $|t_n - t| < \varepsilon/2$ for $n > N_2$. Let N be the maximum of N_1 and N_2. Then for $n > N$, we have simultaneously

$$-\frac{\varepsilon}{2} < s_n - s < \frac{\varepsilon}{2}, \qquad -\frac{\varepsilon}{2} < t_n - t < \frac{\varepsilon}{2}.$$

Adding, we obtain

$$-\varepsilon < (s_n + t_n) - (s + t) < \varepsilon, \qquad \text{or} \qquad |(s_n + t_n) - (s + t)| < \varepsilon$$

for $n > N$. Thus the sequence $\{s_n + t_n\}$ converges to $s + t$.

If $c = 0$, then $\{cs_n\}$ is the sequence $0, 0, 0, \ldots$, which clearly converges to $0 = 0 \cdot s$. If $c \neq 0$, we can find N_3 such that

$$|s_n - s| < \frac{\varepsilon}{|c|}$$

for $n > N_3$. Then for $n > N_3$, we have

$$-\frac{\varepsilon}{|c|} < s_n - s < \frac{\varepsilon}{|c|}.$$

Multiplying by c, we obtain for $c < 0$ as well as $c > 0$,

$$-\varepsilon < cs_n - cs < \varepsilon,$$

so $|cs_n - cs| < \varepsilon$ for $n > N_3$. Thus the sequence $\{cs_n\}$ converges to cs. □

COROLLARY 1   Convergence and series algebra

If $\sum_{n=1}^{\infty} a_n$ converges to a and $\sum_{n=1}^{\infty} b_n$ converges to b, then $\sum_{n=1}^{\infty} (a_n + b_n)$ converges to $a + b$ and $\sum_{n=1}^{\infty} ca_n$ converges to ca for all c.

Proof: The proof is immediate from the preceding theorem and the observation that if $\{s_n\}$ is the sequence of partial sums of $\sum_{n=1}^{\infty} a_n$ and $\{t_n\}$ is the sequence of partial sums of $\sum_{n=1}^{\infty} b_n$, then $\{s_n + t_n\}$ is the sequence of partial sums of $\sum_{n=1}^{\infty} (a_n + b_n)$ and $\{cs_n\}$ is the sequence of partial sums of $\sum_{n=1}^{\infty} ca_n$. □

COROLLARY 2   Divergence and series algebra

If $\sum_{n=1}^{\infty} a_n$ diverges, then for any $c \neq 0$, the series $\sum_{n=1}^{\infty} ca_n$ diverges also.

Proof: If $\sum_{n=1}^{\infty} (ca_n)$ converges, then by Corollary 1, the series

$$\sum_{n=1}^{\infty} \frac{1}{c} (ca_n) = \sum_{n=1}^{\infty} a_n$$

would also converge, contrary to hypothesis. $\square$

EXAMPLE 8   Establish the convergence or divergence of the series $\sum_{n=1}^{\infty} 1/(2n)$.

Solution   The series diverges by Corollary 2, for it is the divergent harmonic series multiplied by $\frac{1}{2}$.   ■

EXAMPLE 9   Find the sum of the series

$$\sum_{n=0}^{\infty} \left(\frac{1}{2^n} - \frac{2}{3^n} \right),$$

if it converges.

Solution   The two geometric series $\sum_{n=0}^{\infty} (\frac{1}{2})^n$ and $\sum_{n=0}^{\infty} (\frac{1}{3})^n$ with ratios $\frac{1}{2}$ and $\frac{1}{3}$ both converge. Corollary 1 then shows that

$$\sum_{n=0}^{\infty} \left(\frac{1}{2^n} - \frac{2}{3^n} \right) = \sum_{n=0}^{\infty} \left(\frac{1}{2} \right)^n - 2 \sum_{n=0}^{\infty} \left(\frac{1}{3} \right)^n$$

converges. Now

$$\sum_{n=0}^{\infty} \frac{1}{2^n} \text{ converges to } \frac{1}{1 - \frac{1}{2}} = 2$$

and

$$\sum_{n=0}^{\infty} \frac{1}{3^n} \text{ converges to } \frac{1}{1 - \frac{1}{3}} = \frac{3}{2},$$

so

$$\sum_{n=0}^{\infty} \left(\frac{1}{2^n} - \frac{2}{3^n} \right)$$

converges to $2 + (-2)\frac{3}{2} = 2 - 3 = -1$.   ■

EXAMPLE 10   Find the sum of the series

$$\sum_{n=1}^{\infty} \frac{5^n + 3 \cdot 2^{3n}}{9^n}.$$

if it converges.

Solution Now

$$\frac{5^n + 3 \cdot 2^{3n}}{9^n} = \left(\frac{5}{9}\right)^n + 3\left(\frac{8}{9}\right)^n.$$

Since $\sum_{n=1}^{\infty} \left(\frac{5}{9}\right)^n$ and $\sum_{n=1}^{\infty} \left(\frac{8}{9}\right)^n$ both converge, we see by Corollary 1 that

$$\sum_{n=1}^{\infty} \frac{5^n + 3 \cdot 2^{3n}}{9^n} = \sum_{n=1}^{\infty} \left(\frac{5}{9}\right)^n + 3 \sum_{n=1}^{\infty} \left(\frac{8}{9}\right)^n$$

$$= \frac{\frac{5}{9}}{1 - \frac{5}{9}} + 3 \cdot \frac{\frac{8}{9}}{1 - \frac{8}{9}}$$

$$= \frac{5}{9} \cdot \frac{9}{4} + 3 \cdot \frac{8}{9} \cdot \frac{9}{1} = \frac{5}{4} + 24 = \frac{101}{4}. \quad\blacksquare$$

You may be expecting an example on determining the convergence or divergence of a series using a calculator or computer. In general, this is not possible to do. If we *know* a series converges, a calculator might be useful to estimate the sum of the series, assuming we know that it converges "fast enough." To illustrate the difficulty of establishing the convergence or divergence of a series using a calculator, consider the harmonic series $\sum_{n=1}^{\infty} 1/n$. In Section 10.4, we will see how to show that

$$9.21 \le \sum_{n=1}^{10000} \frac{1}{n} \le 10.21.$$

This indicates that it would be impossible to establish the divergence to ∞ of $\sum_{n=1}^{\infty} 1/n$ using a calculator. The convergence or divergence of $\sum_{n=1}^{\infty} a_n$ can't be determined by the values of any *finite* number of partial sums, and this is all a computer can find.

SUMMARY

1. Associated with a series

$$\sum_{n=1}^{\infty} a_n = a_1 + a_2 + a_3 + \cdots + a_n + \cdots$$

is the sequence of partial sums $\{s_n\}$, where

$$s_1 = a_1, \quad s_2 = a_1 + a_2, \quad s_3 = a_1 + a_2 + a_3, \quad \ldots.$$

The series converges to the sum c if $\lim_{n\to\infty} s_n = c$, and the series diverges if $\{s_n\}$ is a divergent sequence.

2. Two sequences (series) may be added term by term, and each may be multiplied by a constant. If each of two sequences (series) converges, then their sum converges to the sum of the limits. A constant multiple of a convergent sequence (series) converges to that multiple of the limit of the original sequence (series).

3. The harmonic series $\sum_{n=1}^{\infty} 1/n$ diverges.

4. The geometric series $\sum_{n=0}^{\infty} ar^n$ converges to $a/(1-r)$ if $|r| < 1$ and diverges if $|r| \geq 1$ and $a \neq 0$.

EXERCISES 10.2

1. Consider the series $1 + 0 - 1 + 0 + 1 + 0 - 1 + 0 + \cdots$. Find the indicated partial sums.
   (a) s_1 (b) s_2 (c) s_3 (d) s_4
   (e) s_8 (f) s_{15} (g) s_{122}

2. Find the first four partial sums of the harmonic series $\sum_{n=1}^{\infty} 1/n$.

3. Find the first five terms $a_1, \ldots, a_5$ of the series having as sequence of partial sums $\{s_n\} = \frac{1}{2}, \frac{1}{3}, \frac{1}{4}, \frac{1}{5}, \frac{1}{6}, \ldots$.

4. Find the first six terms of the series having as sequence of partial sums $\{s_n\} = \frac{1}{2}, \frac{2}{3}, \frac{3}{4}, \frac{4}{5}, \frac{5}{6}, \frac{6}{7}, \ldots$.

In Exercises 5–27, determine whether the series converges or diverges and find the sum of the series if the series converges.

5. $\sum_{n=1}^{\infty} \dfrac{1}{n}$

6. $\sum_{n=0}^{\infty} \dfrac{1}{3^n}$

7. $\sum_{n=1}^{\infty} \dfrac{1}{2^n}$

8. $\sum_{n=2}^{\infty} \dfrac{1}{2^n}$

9. $\sum_{n=1}^{\infty} \dfrac{-1}{2n}$

10. $\sum_{n=1}^{\infty} (-1)^n$

11. $\sum_{n=1}^{\infty} \dfrac{n+1}{n}$

12. $\sum_{n=1}^{\infty} \dfrac{3}{10^n}$

13. $\sum_{n=1}^{\infty} \dfrac{4}{(-2)^n}$

14. $\sum_{n=1}^{\infty} \dfrac{1}{n+1}$

15. $\sum_{n=0}^{\infty} e^{-2n}$

16. $\sum_{n=0}^{\infty} \dfrac{1}{(-5)^{n+3}}$

17. $\sum_{n=0}^{\infty} \dfrac{3^{2n+1}}{8^n}$

18. $\sum_{n=0}^{\infty} \dfrac{3^{2n+1}}{10^n}$

19. $\sum_{n=1}^{\infty} \dfrac{2^n}{3^{2n+1}}$

20. $\sum_{n=0}^{\infty} 5(\ln 2)^n$

21. $\sum_{n=0}^{\infty} 4\left(\ln \dfrac{1}{2}\right)^n$

22. $\sum_{n=0}^{\infty} 2\left(\ln \dfrac{1}{3}\right)^n$

23. $\sum_{n=0}^{\infty} \left(\dfrac{1}{2^n} + \dfrac{1}{3^n}\right)$

24. $\sum_{n=0}^{\infty} \left(7 \cdot \dfrac{1}{3^n} - 4 \cdot \dfrac{1}{2^n}\right)$

25. $\sum_{n=0}^{\infty} \dfrac{2^n + 3^n}{4^n}$

26. $\sum_{n=0}^{\infty} \dfrac{3^{n+1} - 7 \cdot 5^n}{10^n}$

27. $\sum_{n=1}^{\infty} \dfrac{8^n + 9^n}{10^n}$

28. *Variant of Achilles and the tortoise.* A driver is speeding down a straight road at 70 mph. A state trooper 1 mi behind is chasing the speeder at 80 mph.
   (a) Assuming the speeds remain the same, how long will it take the trooper to catch the speeder? (You should not need to use series to solve this part.)
   (b) The speeder, being an optimistic philosopher, reasons as follows:

   When the trooper reaches my present position, I will no longer be here but at a new position, so he will not have caught me. When he reaches my new position, I will again have moved on to another new position, so he still will not have caught me. Since I can continue that argument an infinite number of times, the trooper will never catch me.

   Find the flaw in the speeder's argument.

29. A ball is dropped from a height of 30 ft. Each time it hits the ground, it rebounds to $\frac{1}{3}$ of the height it attained on the preceding bounce. Find the total distance the ball travels before coming to rest.

30. An unscrupulous mathematician tells a farmer, "I will work for you in your fields today all day for just one penny, as long as you agree to hire me for a total of 25 days this month and to double my wages each day." The farmer agrees.
   (a) Express the amount the farmer must pay the mathematician for the 25 days of work as the first 25 terms of an infinite series.
   (b) Find the sum of the 25 terms of part (a). [*Hint:* Use the formula for the nth partial sum of a geometric series just before Eq. (7) of the text.]

31. Show that if one could build a spaceship with the property that the time required to travel the next mile is k times the time required to cover the last mile for some fixed positive constant $k < 1$, then one could travel an infinite distance in a finite time.

32. Consider the series

$$\sum_{n=1}^{\infty} \left(\frac{1}{n} - \frac{1}{n+1} \right).$$

(a) Compute the first four partial sums of the series.
(b) Find a formula for the nth partial sum s_n of the series. (The series is known as a "telescoping series." Can you guess why?)
(c) Show that the series converges, and find the sum of the series.

33. Consider the series

$$\sum_{n=1}^{\infty} \left(\ln \frac{n+1}{n} \right).$$

(a) Show that the series can be viewed as a telescoping series (see Exercise 32, and compute the nth partial sum s_n). [*Hint:* Use a property of the function ln.]
(b) Show that the series diverges.

34. Following the idea of Exercise 32, find the sum of the series

$$\sum_{n=1}^{\infty} \left(\frac{1}{n} - \frac{1}{n+2} \right).$$

35. Following the idea of Exercise 32, find the sum of the series

$$\sum_{n=1}^{\infty} \left(\frac{1}{n} - \frac{1}{n+3} \right).$$

36. (a) Decompose $1/(n^2 + 3n + 2)$ into partial fractions.
(b) Following the idea of Exercise 32, find the sum of the series

$$\sum_{n=1}^{\infty} \frac{1}{n^2 + 3n + 2}.$$

37. (a) Find the partial fraction decomposition of $1/(n^2 - 1)$.
(b) Following the idea of Exercise 32, find the sum of the series

$$\sum_{n=4}^{\infty} \frac{1}{n^2 - 1}.$$

38. Find the fallacy in the following argument. If $\{s_n\}$ converges to c and $\{t_n\}$ converges to d, then $\{s_n t_n\}$ converges to cd. If s_n is the nth partial sum of $\sum_{n=1}^{\infty} a_n$, and t_n is the nth partial sum of $\sum_{n=1}^{\infty} b_n$, then $s_n t_n$ is the nth partial sum of $\sum_{n=1}^{\infty} a_n b_n$. Therefore if $\sum_{n=1}^{\infty} a_n$ converges to c, and $\sum_{n=1}^{\infty} b_n$ converges to d, then $\sum_{n=1}^{\infty} a_n b_n$ converges to cd.

39. Give an example of two divergent sequences whose sum converges.

40. Give an example of two divergent series whose sum converges.

41. If $\sum_{n=1}^{\infty} a_n$ converges and $\sum_{n=1}^{\infty} b_n$ diverges, what can be said concerning the convergence or divergence of $\sum_{n=1}^{\infty} (a_n + b_n)$?

In Exercises 42–47, find a rational number equal to the given repeating decimal. (See Example 7.)

42. 0.222222 . . .   **43.** 8.2333333 . . .

44. 11.31888888 . . .   **45.** 0.1212121212 . . .

46. 7.1465465465 . . .   **47.** 273.14653653653 . . .

48. Exercises 42–47 indicate that every repeating decimal represents a rational number. Show, conversely, that the decimal expansion of any rational number tails off in a repeating pattern. [*Hint:* Think of using long division to find the decimal form of a rational number. Argue that some "remainder" obtained in the division must eventually repeat and that the "quotients" obtained must therefore eventually occur in a repeating pattern.]

49. Consider 4.13000000 . . . and 4.12999999 . . . .
(a) Which number do you think is larger?
(b) Show that the numbers are equal by arguing that their difference must be zero. [*Hint:* If (4.13000000 . . .) − (4.12999999 . . .) > 0, then its decimal expansion must have a nonzero digit at some finite number of places beyond the decimal point. Show that this is impossible.]
(c) Show that the numbers are equal by summing a geometric series to show that 4.12999999 . . . = $\frac{413}{100}$.

> **EXPLORING CALCULUS**
>
> **4:** series partial sums

10.3 COMPARISON TESTS AND THE ROOT TEST

The comparison tests in this section parallel very closely the comparison tests for convergence or divergence of an improper integral of a positive function, discussed in Section 9.6. In Section 10.4, we show that the convergence or divergence of many series depends on the convergence or divergence of an improper integral, so it is not surprising that the tests for improper integrals can be carried over to series. The comparison tests in this section will provide our most general tests for convergence or divergence of a series. All the tests developed in the rest of this chapter will be based on these comparison tests.

We open the section by showing that the first few hundred, or million, terms of a series do not affect its convergence or divergence. We indicated this in Section 10.2, when we explained that computation of a finite number of partial sums can never be used as a criterion for convergence or divergence of a series.

Insertion or Deletion of Terms in a Series

The insertion or deletion of a finite number of terms in a series cannot affect whether the series converges or diverges, although if the series converges, the sum of the series may be affected. In particular, suppose the first few terms of a series do not conform to a pattern present in the rest of the series. Then we can neglect those first few terms when studying the convergence or divergence of the series. After an example, we state this property of series as a theorem and leave the straightforward proof as an exercise.

EXAMPLE 1 Find the sum of the series

$$\pi - 3 + 17 + 1 + \tfrac{1}{2} + \tfrac{1}{4} + \tfrac{1}{8} + \tfrac{1}{16} + \cdots.$$

Solution The geometric series

$$1 + \tfrac{1}{2} + \tfrac{1}{4} + \tfrac{1}{8} + \tfrac{1}{16} + \cdots \tag{1}$$

converges to 2. The series

$$\pi - 3 + 17 + 1 + \tfrac{1}{2} + \tfrac{1}{4} + \tfrac{1}{8} + \tfrac{1}{16} + \cdots,$$

obtained from series (1) by inserting three additional terms at the beginning, also converges and clearly must converge to

$$(\pi - 3 + 17) + 2 = \pi + 16. \qquad \blacksquare$$

THEOREM 10.3   Changing a finite number of terms

Suppose that each of two series $\sum_{n=1}^{\infty} a_n$ and $\sum_{n=1}^{\infty} b_n$ contains all but a finite number of terms of the other in the same order. That is, suppose that there exist N and k such that $b_n = a_{n+k}$ for all $n > N$. Then either both series converge or both series diverge. $\qquad \square$

The condition in Theorem 10.3 that the common terms in the two series be "in the same order" is very important. Later work will show that the series

$$1 - \tfrac{1}{2} + \tfrac{1}{3} - \tfrac{1}{4} + \tfrac{1}{5} - \tfrac{1}{6} + \tfrac{1}{7} - \tfrac{1}{8} + \tfrac{1}{9} - \tfrac{1}{10} + \tfrac{1}{11} - \cdots$$

converges, and it can be proved that a *divergent* series can be found that contains *exactly the same terms* but in a different order.

If $\lim_{n \to \infty} a_n \neq 0$, then $\sum_{n=1}^{\infty} a_n$ Diverges

The following theorem is very important and is frequently misused. It shows that some series diverge, but it can *never* be used to show convergence of a series.

THEOREM 10.4   Divergence of $\Sigma_{n=1}^{\infty} a_n$ if $\lim_{n \to \infty} a_n \neq 0$

If the series $\Sigma_{n=1}^{\infty} a_n$ converges, then $\lim_{n \to \infty} a_n = 0$. Equivalently phrased, if $\lim_{n \to \infty} a_n \neq 0$, then $\Sigma_{n=1}^{\infty} a_n$ diverges.

Proof: Let $\Sigma_{n=1}^{\infty} a_n$ converge to c, and let $\varepsilon > 0$ be given. Then there exists a positive integer N such that for all $n > N$, the nth partial sum s_n satisfies

$$|s_n - c| < \varepsilon/2.$$

In particular, if $n > N + 1$, we have

$$|s_{n-1} - c| < \frac{\varepsilon}{2} \quad \text{and} \quad |s_n - c| < \frac{\varepsilon}{2}.$$

Since s_{n-1} and s_n are both within the distance $\varepsilon/2$ of c, they must be a distance at most ε from each other. That is,

$$|s_n - s_{n-1}| < \varepsilon$$

for $n > N + 1$. But

$$s_n - s_{n-1} = (a_1 + \cdots + a_{n-1} + a_n) - (a_1 + \cdots + a_{n-1}) = a_n,$$

so $|a_n| < \varepsilon$ for $n > N + 1$. Hence $\lim_{n \to \infty} a_n = 0$. $\square$

EXAMPLE 2   Establish the convergence or divergence of the series

$$\sum_{n=1}^{\infty} \frac{n^2}{5n^2 + 100n}.$$

Solution   Theorem 10.4 tells us that the series

$$\sum_{n=1}^{\infty} \frac{n^2}{5n^2 + 100n}$$

diverges, for

$$\lim_{n \to \infty} \frac{n^2}{5n^2 + 100n} = \frac{1}{5} \neq 0. \qquad \blacksquare$$

EXAMPLE 3   Establish the convergence or divergence of the series

$$\sum_{n=1}^{\infty} \frac{4 - n^{3/2}}{n + 5}.$$

Solution   We have

$$\lim_{n \to \infty} \frac{4 - n^{3/2}}{n + 5} = \lim_{n \to \infty} \frac{-n^{3/2}}{n} = \lim_{n \to \infty} (-\sqrt{n}) = -\infty.$$

Thus the series diverges, by Theorem 10.4. $\blacksquare$

EXAMPLE 4   Establish the convergence or divergence of the series $\Sigma_{n=0}^{\infty} \sin n$.

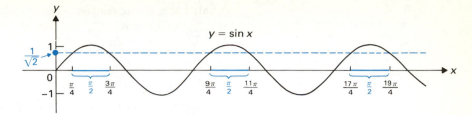

FIGURE 10.4 Since $\pi/2 > 1$, each colored interval contains an integer n such that $\sin n > 1/\sqrt{2}$.

Solution Since $3\pi/4 - \pi/4 = \pi/2 > 1$, we see that the interval $[\pi/4, 3\pi/4]$ must contain at least one integer n. (Of course, it contains both 1 and 2.) A similar argument shows that, for each positive integer m, the interval $[2m\pi + (\pi/4), 2m\pi + (3\pi/4)]$ contains at least one integer n. As shown in Fig. 10.4, for any integer n in one of these intervals, we have $\sin n \geq 1/\sqrt{2}$. Thus $\lim_{n \to \infty} \sin n \neq 0$, so $\sum_{n=0}^{\infty} \sin n$ diverges, by Theorem 10.4. ∎

The Standard Comparison Test (SCT)

During the course of this chapter, we will be introducing certain "tests" for convergence or divergence of a series of constants. Most of these tests depend on the *comparison* of the series with another series that is either known to converge or known to diverge. (This is one reason why it is important to build a "stockpile" of series whose convergence or divergence is known.) Our work in the rest of this section parallels our treatment of comparison tests for improper integrals in Section 9.6.

Comparison tests for series with nonnegative terms follow from the fundamental property of the real numbers presented here. A proof of this property is given in more advanced courses where the real numbers are "constructed."

Fundamental Property of the Real Numbers

Let $\{s_n\}$ be a sequence of numbers such that $s_{n+1} \geq s_n$ for $n = 1, 2, 3,$ . . . . (Such a sequence is **monotone increasing.**) Then either $\{s_n\}$ converges to some c or $\lim_{n \to \infty} s_n = \infty$.

Our fundamental property asserts that the only way a monotone-increasing sequence $\{s_n\}$ can *fail* to converge is to diverge to ∞. Note that if $\sum_{n=1}^{\infty} a_n$ is a series of *nonnegative* terms, then the sequence $\{s_n\}$ of partial sums is monotone increasing.

THEOREM 10.5   Standard comparison test (SCT)

Let $\sum_{n=1}^{\infty} a_n$ and $\sum_{n=1}^{\infty} b_n$ be series of nonnegative terms such that $a_n \leq b_n$ for $n = 1, 2, 3,$ . . . . If $\sum_{n=1}^{\infty} b_n$ converges, then $\sum_{n=1}^{\infty} a_n$ converges also, while if $\sum_{n=1}^{\infty} a_n$ diverges, then $\sum_{n=1}^{\infty} b_n$ diverges also.

Proof: Let s_n be the nth partial sum of $\sum_{n=1}^{\infty} a_n$ and let t_n be the nth partial sum of $\sum_{n=1}^{\infty} b_n$. Suppose $\sum_{n=1}^{\infty} b_n$ converges to c, so that $\lim_{n\to\infty} t_n = c$. From $a_n \leq b_n$, we see at once that

$$s_n \leq t_n$$

for $n = 1, 2, 3, \ldots$. Since $\lim_{n\to\infty} t_n = c$ and $s_n \leq t_n$, we see that $\lim_{n\to\infty} s_n = \infty$ is impossible, so by the fundamental property, the sequence $\{s_n\}$ must converge also. This means that $\sum_{n=1}^{\infty} a_n$ converges.

Suppose that $\sum_{n=1}^{\infty} a_n$ diverges. Since $\{s_n\}$ is monotone increasing and diverges, we must have $\lim_{n\to\infty} s_n = \infty$ by the fundamental property. Since $t_n \geq s_n$, then $\lim_{n\to\infty} t_n = \infty$, so $\sum_{n=1}^{\infty} b_n$ diverges also. $\square$

Theorem 10.5 is sometimes summarized by saying that, for series of nonnegative terms, a series "smaller" than a known convergent series also converges, while a series "larger" than a known divergent series must diverge also. If a series is "smaller" than a known divergent series, it may either converge or diverge, depending on how much "smaller" it is. Similarly, a series that is "larger" than a known convergent series may converge or diverge, depending on how much "larger" it is. It is important for us to remember that the comparison test works only in the stated direction.

EXAMPLE 5   Establish the convergence or divergence of the series

$$\sum_{n=1}^{\infty} \frac{1}{2^{n-1} + 1}.$$

Solution   We know that the series

$$1 + \frac{1}{2} + \frac{1}{4} + \frac{1}{8} + \cdots + \frac{1}{2^{n-1}} + \cdots$$

converges. Therefore the "smaller" series

$$\frac{1}{2} + \frac{1}{3} + \frac{1}{5} + \frac{1}{9} + \cdots + \frac{1}{2^{n-1} + 1} + \cdots$$

converges, by the SCT.   ∎

EXAMPLE 6   Establish the convergence or divergence of the series $\sum_{n=1}^{\infty} 1/\sqrt{n}$.

Solution   We know that the harmonic series

$$1 + \frac{1}{2} + \frac{1}{3} + \frac{1}{4} + \cdots + \frac{1}{n} + \cdots$$

diverges. Since $\sqrt{n} \leq n$, we have $1/n \leq 1/\sqrt{n}$. Therefore the "larger" series

$$1 + \frac{1}{\sqrt{2}} + \frac{1}{\sqrt{3}} + \frac{1}{\sqrt{4}} + \cdots + \frac{1}{\sqrt{n}} + \cdots$$

diverges, by the SCT.   ∎

Since a finite number of terms can be inserted in a series or deleted from it without affecting its convergence or divergence (Theorem 10.3), we can weaken the hypotheses of the SCT and require only that $a_n \leq b_n$ for all but a finite number of positive integers n.

From Theorem 10.5 and our work with geometric series, we can quickly develop the *root test* for convergence of a series $\sum_{n=1}^{\infty} a_n$. Suppose $a_n > 0$ for n sufficiently large, and suppose that $\lim_{n \to \infty} \sqrt[n]{a_n} = L$. If $L < 1$, we choose a number r between L and 1. For example, we might take $r = (L + 1)/2$. Then $\sqrt[n]{a_n} < r$ for all n sufficiently large, and consequently $a_n < r^n$ for all n sufficiently large. Since $\sum_{n=1}^{\infty} r^n$ converges as a geometric series with ratio $r < 1$, Theorem 10.5 shows that $\sum_{n=1}^{\infty} a_n$ converges also. However, if $L > 1$ and we choose r between 1 and L, we see that $a_n > r^n > 1$ for all n sufficiently large, so the series $\sum_{n=1}^{\infty} a_n$ diverges by Theorem 10.4. Finally, we show that if $L = 1$, we can't conclude just on that basis whether or not the series converges. Example 7 of Section 9.4 shows that $\lim_{x \to \infty} x^{1/x} = 1$, so $\lim_{n \to \infty} n^{1/n} = 1$. Hence

$$\lim_{n \to \infty} \left(\frac{1}{n}\right)^{1/n} = \frac{1}{1} = 1 \quad \text{and} \quad \lim_{n \to \infty} \left(\frac{1}{n^2}\right)^{1/n} = \lim_{n \to \infty} \left[\left(\frac{1}{n}\right)^{1/n}\right]^2 = 1.$$

This shows that for both $a_n = 1/n$ and $b_n = 1/n^2$, we have $\lim_{n \to \infty} \sqrt[n]{a_n} = \lim_{n \to \infty} \sqrt[n]{b_n} = 1$. However, we know that the harmonic series $\sum_{n=1}^{\infty} 1/n$ diverges, and we will show in the next section that $\sum_{n=1}^{\infty} 1/n^2$ converges. Thus we cannot determine convergence if we know only that $L = 1$.

We state the test that we just described as a corollary of Theorem 10.5, the SCT.

COROLLARY   Root test

Let $a_n > 0$ for all n sufficiently large, and let $\lim_{n \to \infty} \sqrt[n]{a_n} = L$.

1. If $L < 1$, then $\sum_{n=1}^{\infty} a_n$ converges.
2. If $L > 1$, then $\sum_{n=1}^{\infty} a_n$ diverges.
3. If $L = 1$, the series $\sum_{n=1}^{\infty} a_n$ may converge or may diverge.

EXAMPLE 7   Apply the root test to the series

$$\sum_{n=1}^{\infty} \left(\frac{5000}{\sqrt{n}}\right)^n.$$

Solution   Let $a_n = (5000/\sqrt{n})^n$. Then $\sqrt[n]{a_n} = 5000/\sqrt{n}$ and

$$\lim_{n \to \infty} \sqrt[n]{a_n} = \lim_{n \to \infty} \frac{5000}{\sqrt{n}} = 0 < 1,$$

so the series converges.   ∎

EXAMPLE 8   Apply the root test to the series

$$\sum_{n=1}^{\infty} \left(\frac{4n + 1}{3n + 200}\right)^n.$$

Solution For a_n in this series, we have $\sqrt[n]{a_n} = (4n + 1)/(3n + 200)$. Our work in Section 2.3 then shows that $\lim_{n\to\infty} [(4n + 1)/(3n + 200)] = \lim_{n\to\infty} [(4n)/(3n)] = \frac{4}{3} > 1$, so this series diverges by the root test. ■

The Limit Comparison Test (LCT)

The series

$$\sum_{n=1}^{\infty} \frac{n^2 + 3n}{2n^3 - n^2}$$

diverges because

$$\frac{n^2 + 3n}{2n^3 - n^2} > \frac{n^2}{2n^3} = \frac{1}{2n},$$

and $\sum_{n=1}^{\infty} 1/(2n)$ diverges since it is "half" the harmonic series. Rather than fuss about such a precise inequality for the SCT, the mathematician usually says that the given series diverges since $(n^2 + 3n)/(2n^3 - n^2)$ is of order of magnitude $n^2/(2n^3) = 1/(2n)$ for large n, and $\sum_{n=1}^{\infty} 1/(2n)$ diverges. The justification for this argument is given in the following theorem.

THEOREM 10.6   **Limit comparison test (LCT)**

Let $\sum_{n=1}^{\infty} a_n$ and $\sum_{n=1}^{\infty} b_n$ be series of nonnegative terms, with $a_n \neq 0$ for all sufficiently large n, and suppose that $\lim_{n\to\infty} b_n/a_n = c > 0$. Then the two series either both converge or both diverge.

Proof: Since $\lim_{n\to\infty} b_n/a_n = c > 0$, there exists N such that for $n > N$, we have $c/2 < b_n/a_n < 3c/2$, or

$$\frac{c}{2} a_n < b_n < \frac{3c}{2} a_n. \tag{2}$$

If $\sum_{n=1}^{\infty} a_n$ converges, then $\sum_{n=1}^{\infty} (3c/2)a_n$ converges by Corollary 1 of Theorem 10.2. Since inequality (2) provides a "comparison test" for all but a finite number of terms b_n, we see from the SCT and the remarks following Example 8 that $\sum_{n=1}^{\infty} b_n$ converges also. Similarly, if $\sum_{n=1}^{\infty} a_n$ diverges, then $\sum_{n=1}^{\infty} (c/2)a_n$ diverges, so by inequality (2) and the SCT, we see that $\sum_{n=1}^{\infty} b_n$ diverges. □

We saw in Section 2.3 that the limit at ∞ of a quotient of polynomial functions in x is equal to the limit of the quotient of the dominating monomial terms of highest degree in the numerator and denominator. Of course, the same is true for the limit at ∞ of a quotient of polynomials in n.

EXAMPLE 9   Determine the convergence or divergence of the series

$$\sum_{n=1}^{\infty} \frac{2n^3 - 3n^2}{7n^4 + 100n^3 + 7}.$$

Solution For large n, the nth term of the series is of order of magnitude $2n^3/(7n^4)$, or $2/(7n)$. More precisely,

$$\lim_{n\to\infty} \frac{(2n^3 - 3n^2)/(7n^4 + 100n^3 + 7)}{2/(7n)} = \lim_{n\to\infty} \frac{14n^4 - 21n^3}{14n^4 + 200n^3 + 14} = 1.$$

Thus by the LCT, the given series diverges, since $\sum_{n=1}^{\infty} 2/(7n)$ is $\frac{2}{7}$ times the harmonic series and therefore diverges. ∎

We give one more example of the way in which the LCT is used to determine convergence or divergence.

EXAMPLE 10 Establish the convergence or divergence of

$$\sum_{n=1}^{\infty} \frac{(4n^3 + 5)\sin(1/n)}{n^2 3^n}.$$

Solution We know that

$$\lim_{n\to\infty}\left[n\sin\left(\frac{1}{n}\right)\right] = \lim_{n\to\infty} \frac{\sin(1/n)}{1/n} = 1.$$

Thus

$$\lim_{n\to\infty} \frac{(4n^3 + 5)\sin(1/n)}{n^2} = \lim_{n\to\infty}\left[\frac{4n^3 + 5}{n^3} n\sin\left(\frac{1}{n}\right)\right]$$

$$= 4 \cdot 1 = 4 > 0.$$

The LCT with

$$a_n = \frac{(4n^3 + 5)\sin(1/n)}{n^2 3^n} \qquad \text{and} \qquad b_n = \frac{1}{3^n},$$

shows that the given series behaves like $\sum_{n=1}^{\infty} 1/3^n$, which converges as a geometric series with ratio $\frac{1}{3}$. ∎

SUMMARY

1. The convergence or divergence of a series (or sequence) is not changed by inserting, deleting, or altering any *finite* number of terms.

2. If $\lim_{n\to\infty} a_n \neq 0$, then $\sum_{n=1}^{\infty} a_n$ diverges.

3. *Standard comparison test (SCT):* If $0 \leq a_n \leq b_n$ for all sufficiently large n, and $\sum_{n=1}^{\infty} b_n$ converges, then $\sum_{n=1}^{\infty} a_n$ converges. On the other hand, if $\sum_{n=1}^{\infty} a_n$ diverges, then $\sum_{n=1}^{\infty} b_n$ diverges.

4. *Limit comparison test (LCT):* If $a_n > 0$ for sufficiently large n and $b_n \geq 0$ for all n, and $\lim_{n\to\infty} b_n/a_n = c > 0$, then the series $\sum_{n=1}^{\infty} a_n$ and $\sum_{n=1}^{\infty} b_n$ either both converge or both diverge.

5. *Root test:* If $a_n > 0$ for sufficiently large n and $\lim_{n \to \infty} \sqrt[n]{a_n} = L$, then $\sum_{n=1}^{\infty} a_n$ converges if $L < 1$ and diverges if $L > 1$. If $L = 1$, convergence is not determined.

EXERCISES 10.3

In Exercises 1–4, determine whether the given series converges or diverges; if it is convergent, find its sum.

1. $1 - 2 + \dfrac{1}{2} + \sqrt{5} + \dfrac{1}{4} + \dfrac{1}{8} + \dfrac{1}{16} + \cdots$
$$+ \frac{1}{2^{n-3}} + \cdots, \qquad \text{where } n \geq 5$$

2. $1 + \dfrac{1}{2} + \dfrac{1}{4} + \dfrac{1}{8} + \dfrac{1}{16} + \dfrac{1}{32} + \dfrac{2}{32} + \dfrac{3}{32} + \cdots$
$$+ \frac{n-5}{32} + \cdots, \qquad \text{where } n \geq 6$$

3. $\sqrt{2} + 1 - \dfrac{1}{3} - \sqrt{3} + \dfrac{1}{9} - \dfrac{1}{27} + \dfrac{1}{81} - \cdots$
$$+ (-1)^{n+1} \frac{1}{3^{n-3}} + \cdots, \qquad \text{where } n \geq 5$$

4. $2 - 5 + 7 - \dfrac{1}{2} + 1 - \dfrac{1}{4} + \dfrac{1}{3} - \dfrac{1}{8} + \dfrac{1}{9} - \dfrac{1}{16}$
$$+ \frac{1}{27} - \frac{1}{32} + \frac{1}{81} - \cdots$$
$(1/3^{(n-5)/2}$ for odd $n \geq 5$; $-1/2^{(n/2-1)}$ for even $n \geq 4)$

5. Mark each of the following true or false.
_____ (a) The SCT and LCT as stated in the text hold only for series of nonnegative terms.
_____ (b) The test "if $\lim_{n \to \infty} a_n \neq 0$, then $\sum_{n=1}^{\infty} a_n$ diverges" holds only for series of nonnegative terms.
_____ (c) If $\lim_{n \to \infty} a_n = 0$, then $\sum_{n=1}^{\infty} a_n$ converges.
_____ (d) If $\sum_{n=1}^{\infty} a_n$ converges, then $\lim_{n \to \infty} a_n = 0$.
_____ (e) If $a_n \leq b_n$, and $\sum_{n=1}^{\infty} b_n$ converges, then $\sum_{n=1}^{\infty} a_n$ converges.
_____ (f) If $0 \leq a_n \leq b_n$, and $\sum_{n=1}^{\infty} b_n$ converges, then $\sum_{n=1}^{\infty} a_n$ converges.
_____ (g) If $0 \leq a_n \leq b_n$, and $\sum_{n=1}^{\infty} a_n$ diverges, then $\sum_{n=1}^{\infty} b_n$ diverges.
_____ (h) If $a_n \leq b_n \leq 0$, and $\sum_{n=1}^{\infty} b_n$ converges, then $\sum_{n=1}^{\infty} a_n$ converges.
_____ (i) If $a_n \leq b_n \leq 0$, and $\sum_{n=1}^{\infty} b_n$ diverges, then $\sum_{n=1}^{\infty} a_n$ diverges.
_____ (j) If $a_n \leq b_n \leq 0$, and $\sum_{n=1}^{\infty} a_n$ converges, then $\sum_{n=1}^{\infty} b_n$ converges.

In Exercises 6–42, classify the series as convergent or divergent, and indicate a reason for your answer. (You should do all these exercises and try to develop facility in ascertaining the convergence or divergence of the series at a glance.)

6. $\displaystyle\sum_{n=1}^{\infty} \frac{1}{10n}$

7. $\displaystyle\sum_{n=1}^{\infty} \frac{\sqrt{n}}{n \cdot 2^n}$

8. $\displaystyle\sum_{n=1}^{\infty} (\ln n)$

9. $\displaystyle\sum_{n=2}^{\infty} \frac{\cos^2 n}{2^n}$

10. $\displaystyle\sum_{n=3}^{\infty} \frac{1}{n-1}$

11. $\displaystyle\sum_{n=1}^{\infty} \frac{1}{n + 2^n}$

12. $\displaystyle\sum_{n=1}^{\infty} \frac{1}{\sqrt[3]{n}}$

13. $\displaystyle\sum_{n=1}^{\infty} \frac{n^3}{3n^3 + n^2}$

14. $\displaystyle\sum_{n=1}^{\infty} \left(\frac{1}{n} - \frac{1}{2^n} \right)$

15. $\displaystyle\sum_{n=1}^{\infty} \frac{|\sin n|}{5^n}$

16. $\displaystyle\sum_{n=1}^{\infty} \frac{\sqrt{n}}{n + 17}$

17. $\displaystyle\sum_{n=1}^{\infty} \frac{8^n + 9^n}{10^n}$

18. $\displaystyle\sum_{n=1}^{\infty} \frac{n^2 - 3n}{4n^3 + n^2}$

19. $\displaystyle\sum_{n=1}^{\infty} \frac{n + 2}{n^2 + 3}$

20. $\displaystyle\sum_{n=1}^{\infty} \frac{1}{\sqrt[n]{100}}$

21. $\displaystyle\sum_{n=1}^{\infty} \frac{n^2 e^{-2n}}{3n^2 - 2n}$

22. $\displaystyle\sum_{n=1}^{\infty} \frac{e^n}{n^3 + 4n}$

23. $\displaystyle\sum_{n=1}^{\infty} \frac{1}{\sin^2 n}$

24. $\displaystyle\sum_{n=1}^{\infty} \frac{2^n}{3^{2n+1}}$

25. $\displaystyle\sum_{n=1}^{\infty} \sqrt[n]{n}$

26. $\displaystyle\sum_{n=1}^{\infty} \frac{1}{\sqrt[n]{n}}$

27. $\displaystyle\sum_{n=1}^{\infty} \frac{n^2 - 1}{10n^2(\ln 2)^n}$

28. $\displaystyle\sum_{n=1}^{\infty} \frac{\sqrt{n + 1000}}{\sqrt{n}(\ln 3)^n}$

29. $\displaystyle\sum_{n=1}^{\infty} \frac{(n^2 + 1)\cos^2 n}{4n^2 + 3n + 2}$

30. $\displaystyle\sum_{n=2}^{\infty} \left(\frac{\ln n}{\ln n^2} \right)^n$

31. $\displaystyle\sum_{n=1}^{\infty} n \sin\left(\frac{1}{n^2} \right)$

32. $\displaystyle\sum_{n=1}^{\infty} \frac{n^{3/2} - 1}{1000n^2 + 300n + 50}$ **33.** $\displaystyle\sum_{n=1}^{\infty} \left(\frac{n-1}{2n+3}\right)^n$ **38.** $\displaystyle\sum_{n=1}^{\infty} \frac{\sqrt{n^5 + 5n - 1}}{4n^3 + 7}$ **39.** $\displaystyle\sum_{n=1}^{\infty} (n^2 + 3)\sin\left(\frac{1}{n^2}\right)$

34. $\displaystyle\sum_{n=1}^{\infty} \frac{100^n}{n^n}$ **35.** $\displaystyle\sum_{n=1}^{\infty} \left(\frac{n+30}{2n-1}\right)^n$ **40.** $\displaystyle\sum_{n=1}^{\infty} \left(\frac{2n^3 + n^2 + 1}{2n^3 + 100n}\right)^n$ **41.** $\displaystyle\sum_{n=1}^{\infty} \frac{100^{2n+3}}{n^{n/2}}$

36. $\displaystyle\sum_{n=1}^{\infty} \left[n^2\sin\left(\frac{2n}{n^3 + n^2}\right)\right]^n$ **37.** $\displaystyle\sum_{n=1}^{\infty} \frac{n(\sin^2 n)}{50n + 100}$ **42.** $\displaystyle\sum_{n=1}^{\infty} \frac{n^2 \cdot 2^{2n-10}}{(n^2 + 100n)3^n}$

10.4   THE INTEGRAL AND RATIO TESTS

The tests given in this section are all derived from the SCT, which can be viewed as the most general test we consider.

The Integral Test

We discussed convergence of improper integrals of the form $\int_1^\infty f(x)\, dx$ in Sections 9.5 and 9.6. If the nth term a_n of a series $\sum_{n=1}^{\infty} a_n$ is given by a formula in terms of n, then we may replace n by x in that formula and obtain a function $f(x)$. For example, the series $\sum_{n=1}^{\infty} 1/n^2$ gives rise to the function $f(x) = 1/x^2$ in this way. For many important series, there is a close relationship between the behavior of $\sum_{n=1}^{\infty} a_n$ and the behavior of $\int_1^\infty f(x)\, dx$. This relationship is explained in the following theorem.

THEOREM 10.7   The integral test

Let $\sum_{n=1}^{\infty} a_n$ be a series of nonnegative terms. Suppose also that f is a continuous function for $x \geq 1$ such that

1. $f(n) = a_n$ for $n = 1, 2, 3, \ldots$, and

2. f is monotone decreasing for $x \geq 1$; that is, $f(x_1) \geq f(x_2)$ if $1 \leq x_1 \leq x_2$.

Then $\sum_{n=1}^{\infty} a_n$ converges if and only if $\int_1^\infty f(x)\, dx$ converges.

Proof: The proof follows from our previous work and from Figs. 10.5 and 10.6. Let s_n be the nth partial sum of the series $\sum_{n=1}^{\infty} a_n$. Now $\int_1^n f(x)\, dx$ is the area of the region under the graph of f over $[1, n]$. If we approximate $\int_1^n f(x)\, dx$ by the upper and lower sums given by the rectangles with bases of length 1 in Figs. 10.5 and 10.6, we obtain

$$a_2 + \cdots + a_n \leq \int_1^n f(x)\, dx \leq a_1 + a_2 + \cdots + a_{n-1}. \tag{1}$$

Note that the relation (1) depends on the fact that f is monotone decreasing for $x \geq 1$. As illustrated in Fig. 10.7, the relation (1) need not hold if $f(x)$ is not

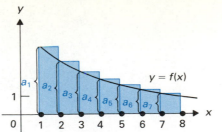

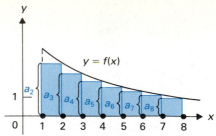

FIGURE 10.5 Upper sum approximation: $\int_1^8 f(x)\,dx \leq a_1 + a_2 + \cdots + a_7$.

FIGURE 10.6 Lower sum approximation: $a_2 + a_3 + \cdots + a_8 \leq \int_1^8 f(x)\,dx$.

monotone decreasing. From the relation (1), we obtain

$$s_n - a_1 \leq \int_1^n f(x)\,dx \leq s_{n-1}. \tag{2}$$

Since $f(x) \geq 0$ for $x \geq 1$, the sequence $\{\int_1^n f(x)\,dx\}$ is clearly monotone increasing. Therefore, if $\int_1^\infty f(x)\,dx$ converges, then by the fundamental property on page 493, the sequence $\{\int_1^n f(x)\,dx\}$ must converge. From the relation (2), using a comparison test, we find that the sequence $\{s_n - a_1\}$ converges. But then the sequence $\{s_n\}$ converges, by Theorem 10.2, so the series $\sum_{n=1}^\infty a_n$ converges.

On the other hand, if $\{s_n\}$ converges, then the sequence $\{\int_1^n f(x)\,dx\}$ must converge by the relation (2), using the SCT. Since $F(t) = \int_1^t f(x)\,dx$ is a monotone-increasing function of t, it is clear that $\int_1^\infty f(x)\,dx$ must converge also. $\square$

We regard the integral test as asserting that for $\sum_{n=1}^\infty a_n$ and f as described, the series "behaves like" $\int_1^\infty f(x)\,dx$, as far as convergence or divergence is concerned.

EXAMPLE 1 Show the divergence of the harmonic series $\sum_{n=1}^\infty 1/n$ using the integral test.

Solution If $f(x) = 1/x$, then the conditions of the integral test are satisfied. Now

$$\lim_{h \to \infty} \int_1^h \frac{1}{x}\,dx = \lim_{h \to \infty} (\ln x)\Big]_1^h = \lim_{h \to \infty} (\ln h - \ln 1) = \lim_{h \to \infty} (\ln h) = \infty.$$

Thus the integral diverges, so the series diverges also. ∎

One type of series handled by the integral test is so important that we treat it formally in a corollary.

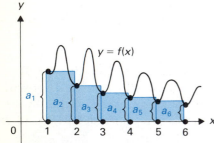

FIGURE 10.7

$$\int_1^n f(x)\,dx \leq a_1 + a_2 + \cdots + a_{n-1}$$

need not hold if $f(x)$ is not monotone decreasing.

COROLLARY   *p*-series test

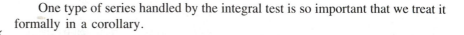

Let p be any real number. The series $\sum_{n=1}^\infty 1/n^p$ converges if $p > 1$ and diverges for $p \leq 1$.

Proof: If $p > 1$, then we define f by $f(x) = 1/x^p$. Then f is a continuous function that satisfies the conditions of the integral test, and the behavior of the series $\sum_{n=1}^{\infty} 1/n^p$ is the same as the behavior of $\int_1^{\infty} (1/x^p)\, dx$. By Theorem 9.3, we know that $\int_1^{\infty} 1/x^p\, dx$ converges for $p > 1$.

If $p \leq 1$, then $n^p \leq n$, so $1/n^p \geq 1/n$, and the series diverges by comparison with the harmonic series. $\qquad\square$

The preceding result is especially useful when combined with the LCT (Section 10.3), as illustrated now.

EXAMPLE 2 Establish the convergence or divergence of

$$\sum_{n=1}^{\infty} \frac{n + 20}{n^3 + 20n + 1}.$$

Solution By the LCT, the series behaves like $\sum_{n=1}^{\infty} 1/n^2$, which is a "$p$-series" with $p = 2$, and hence converges. $\qquad\blacksquare$

As Example 2 indicates, the LCT and the p-series test enable us to determine at a glance whether or not $\sum_{n=1}^{\infty} a_n$ converges if a_n is a quotient of polynomial functions in n. The same technique is valid when radical expressions are involved.

EXAMPLE 3 Determine the convergence or divergence of

$$\sum_{n=1}^{\infty} \frac{(2n^5 + 4n^2 + 2)^{1/3}}{5n^3 + 2n}.$$

Solution Keeping only the monomial terms of highest degree and neglecting their coefficients, which is permissible by the LCT, the series behaves like

$$\sum_{n=1}^{\infty} \frac{(n^5)^{1/3}}{n^3} = \sum_{n=1}^{\infty} \frac{n^{5/3}}{n^3} + \sum_{n=1}^{\infty} \frac{1}{n^{4/3}}.$$

This series converges by the p-series test, with $p = \frac{4}{3} > 1$. $\qquad\blacksquare$

EXAMPLE 4 Establish the convergence or divergence of

$$\sum_{n=1}^{\infty} \sin\!\left(\frac{2}{n^2 + 4n}\right).$$

Solution Motivated by the fact that $\lim_{x \to 0} [(\sin x)/x] = 1$, we write

$$\sum_{n=1}^{\infty} \sin\!\left(\frac{2}{n^2 + 4n}\right) = \sum_{n=1}^{\infty} \frac{2 \sin\!\left(\dfrac{2}{n^2 + 4n}\right)}{(n^2 + 4n)\dfrac{2}{n^2 + 4n}}.$$

The LCT then shows that the series behaves the same as

$$\sum_{n=1}^{\infty} \frac{2}{n^2 + 4n},$$

or, using the LCT again, as

$$\sum_{n=1}^{\infty} \frac{1}{n^2}.$$

This series converges as a *p*-series with $p = 2 > 1$.   ∎

Estimating the Sum of a Series of Monotone-Decreasing Terms by an Integral

The proof of the integral test really gives more information than indicated by the statement of the test because the relation (2) gives a very useful estimate for the partial sum s_n of a series, provided that a function f can be found satisfying the hypotheses of the test. Namely, using the relation (2) and the fact that $s_{n-1} \leq s_n$, we obtain

$$\int_1^n f(x)\, dx \leq s_n \leq \left(\int_1^n f(x)\, dx \right) + a_1. \tag{3}$$

If the series and integral converge, then we obtain from the relation (3) upon letting $n \to \infty$

$$\int_1^{\infty} f(x)\, dx \leq \sum_{n=1}^{\infty} a_n \leq a_1 + \int_1^{\infty} f(x)\, dx. \tag{4}$$

We illustrate the use of the relations (3) and (4) by examples.

EXAMPLE 5   Use the relation (3) to estimate the partial sums s_n of the harmonic series $\sum_{n=1}^{\infty} 1/n$.

Solution   From the relation (3), with $f(x) = 1/x$, we see that for the partial sum s_n of the harmonic series $1 + \frac{1}{2} + \frac{1}{3} + \cdots$, we have

$$\int_1^n \frac{1}{x}\, dx \leq s_n \leq \left(\int_1^n \frac{1}{x}(x)\, dx \right) + 1.$$

Since $\int_1^n (1/x)\, dx = \ln n - \ln 1 = \ln n$, we have

$$\ln n \leq s_n \leq (\ln n) + 1.$$

In particular, since $\ln 10{,}000 \approx 9.21$, we obtain the estimate

$$9.21 \leq \sum_{n=1}^{10{,}000} \frac{1}{n} \leq 10.21$$

stated on page 488.   ∎

EXAMPLE 6   Use the relation (4) to estimate $\sum_{n=1}^{\infty} 1/n^2$.

Solution   We have

$$\int_1^{\infty} \frac{1}{x^2}\, dx = \lim_{h \to \infty} -\frac{1}{x}\Big]_1^h = \lim_{h \to \infty} \left(-\frac{1}{h} + 1 \right) = 1.$$

We then obtain from the relation (4),

$$1 \le \sum_{n=1}^{\infty} \frac{1}{n^2} \le 1 + 1 = 2.$$ ■

We may replace our lower summation index and integral limit 1 in relation (4) by any positive integer r and obtain

$$\int_r^{\infty} f(x) \, dx \le \sum_{n=r}^{\infty} a_n \le a_r + \int_r^{\infty} f(x) \, dx. \tag{5}$$

Suppose $a_r \le \varepsilon$. Since $\sum_{n=1}^{\infty} a_n = \sum_{n=1}^{r-1} a_n + \sum_{n=r}^{\infty} a_n$, the relation (5) shows that

$$\sum_{n=1}^{\infty} a_n \approx \sum_{n=1}^{r-1} a_n + \int_r^{\infty} f(x) \, dx, \tag{6}$$

with error of magnitude at most ε. We could use a calculator or computer to find $\sum_{n=1}^{r-1} a_n$ if r is not too large.

EXAMPLE 7 Continuing Example 6, find an estimate for $\sum_{n=1}^{\infty} 1/n^2$ with error ≤ 0.001.

Solution To have $a_r = 1/r^2 \le 0.001$, we must have $r^2 \ge 1000$, or $r \ge \sqrt{1000} \approx 31.6$. Thus we may take $r = 32$. Our calculator shows that $\sum_{n=1}^{31} 1/n^2 \approx 1.61319$. Now

$$\int_{32}^{\infty} \frac{1}{x^2} \, dx = \lim_{h \to \infty} -\frac{1}{x} \Big]_{32}^{h} = \lim_{h \to \infty} \left(-\frac{1}{h} + \frac{1}{32} \right) = \frac{1}{32}.$$

Then the relation (6) shows that

$$\sum_{n=1}^{\infty} \frac{1}{n^2} \approx 1.61319 + \frac{1}{32} \approx 1.64444,$$

and the error is at most 0.001. ■

The Ratio Test

Recall that if there is any hope that $\sum_{n=1}^{\infty} a_n$ converges, we must have $\lim_{n \to \infty} a_n = 0$. The ratio test studies the rate at which terms a_n decrease as n increases and is based on a comparison of the series $\sum_{n=1}^{\infty} a_n$ with a suitable geometric series. The (geometric) decrease from the term a_n to the next term a_{n+1} in $\sum_{n=1}^{\infty} a_n$ can be measured by the *ratio* a_{n+1}/a_n.

THEOREM 10.8   Ratio test

Let $\sum_{n=1}^{\infty} a_n$ be a series of positive terms and suppose that $\lim_{n \to \infty} a_{n+1}/a_n$ exists and is r. Then $\sum_{n=1}^{\infty} a_n$ converges if $r < 1$ and diverges if $r > 1$. (If $r = 1$, further information is necessary to determine the convergence or divergence of the series.)

Proof: Suppose $\lim_{n\to\infty} a_{n+1}/a_n = r < 1$. Find $\delta > 0$ such that

$$r < r + \delta < 1.$$

For example, we could let $\delta = (1 - r)/2$. Since $\lim_{n\to\infty} a_{n+1}/a_n = r$, there exists an integer N such that

$$r - \delta < \frac{a_{n+1}}{a_n} < r + \delta$$

for $n > N$. Multiplying by a_n, we find that

$$a_{n+1} < (r + \delta)a_n \tag{7}$$

for $n > N$. Using relation (7) repeatedly, starting with $n = N + 1$, we obtain

$$a_{N+2} < (r + \delta)a_{N+1},$$
$$a_{N+3} < (r + \delta)a_{N+2} < (r + \delta)^2 a_{N+1}$$
$$a_{N+4} < (r + \delta)a_{N+3} < (r + \delta)^3 a_{N+1}, \ldots.$$

Thus each term of the series

$$a_{N+1} + a_{N+2} + a_{N+3} + \cdots + a_{N+k} + \cdots \tag{8}$$

is less than or equal to the corresponding term of the series

$$a_{N+1} + a_{N+1}(r + \delta) + a_{N+1}(r + \delta)^2 + \cdots + a_{N+1}(r + \delta)^{k-1} + \cdots. \tag{9}$$

But series (9) is a geometric series with ratio $r + \delta < 1$, and it converges. By the SCT, we see that the series (8) converges. Since the series (8) is $\sum_{n=1}^{\infty} a_n$ with a finite number of terms deleted, Theorem 10.3 shows that $\sum_{n=1}^{\infty} a_n$ converges also.

Now suppose that $r > 1$. Then for all sufficiently large n,

$$\frac{a_{n+1}}{a_n} > 1;$$

that is,

$$a_{n+1} > a_n$$

for n sufficiently large. This means that after a while, the terms of $\sum_{n=1}^{\infty} a_n$ *increase*. Thus $\lim_{n\to\infty} a_n = 0$ is impossible, so by Theorem 10.4, the series $\sum_{n=1}^{\infty} a_n$ must diverge. $\square$

We should comment on the observation in parentheses in the statement of the ratio test. For the series $\sum_{n=1}^{\infty} 1/n^2$, we have

$$r = \lim_{n\to\infty} \frac{a_{n+1}}{a_n} = \lim_{n\to\infty} \frac{1/(n + 1)^2}{1/n^2} = \lim_{n\to\infty} \left(\frac{n}{n + 1}\right)^2 = 1,$$

and for the series $\sum_{n=1}^{\infty} 1/n$, we also have

$$r = \lim_{n\to\infty} \frac{a_{n+1}}{a_n} = \lim_{n\to\infty} \frac{1/(n + 1)}{1/n} = \lim_{n\to\infty} \frac{n}{n + 1} = 1.$$

In both cases, the *limiting ratio r exists and is* 1, but by the *p*-series test, the series $\sum_{n=1}^{\infty} 1/n^2$ converges, while the harmonic series $\sum_{n=1}^{\infty} 1/n$ diverges. This illustrates that for a series having a limiting ratio $r = 1$, more information must be obtained before the convergence or divergence of the series can be determined.

The ratio test is especially useful in handling series whose nth term a_n is given by a formula involving a constant to the nth power (such as 3^n) or involving a factor $n!$ where

$$n! = (n)(n - 1) \cdots (3)(2)(1).$$

It is worth mentioning that for any constant $c > 1$, the exponential c^n increases much faster as $n \to \infty$ than a polynomial in n of any degree s. This is true in the strong sense that

$$\lim_{n \to \infty} \frac{c^n}{n^s} = \infty \qquad \text{and} \qquad \lim_{n \to \infty} \frac{n^s}{c^n} = 0$$

for any s. We can use l'Hôpital's rule to show that

$$\lim_{x \to \infty} \frac{c^x}{x^s} = \infty.$$

It is also worth recalling that $n!$ increases much faster than c^n, once again in the strong sense that

$$\lim_{n \to \infty} \frac{n!}{c^n} = \infty \qquad \text{and} \qquad \lim_{n \to \infty} \frac{c^n}{n!} = 0.$$

Write

$$\frac{n!}{c^n} = \frac{n}{c} \cdot \frac{n - 1}{c} \cdot \frac{n - 2}{c} \cdots \frac{r}{c} \cdot \frac{3}{c} \cdot \frac{2}{c} \cdot \frac{1}{c},$$

where r is the largest integer such that $r < 2c$. Then

$$\frac{n!}{c^n} \geq 2 \cdot 2 \cdot 2 \cdots 2 \cdot \frac{r}{c} \cdot \frac{r - 1}{c} \cdots \frac{3}{c} \cdot \frac{2}{c} \cdot \frac{1}{c} = 2^{n-r} \left(\frac{r}{c} \cdots \frac{3}{c} \cdot \frac{2}{c} \cdot \frac{1}{c} \right).$$

Consequently, $\lim_{n \to \infty} n!/c^n = \infty$.

Summarizing, for large n, we expect $n!$ to "dominate" c^n for any c, and c^n for $c > 1$ in turn "dominates" any polynomial in n for large n.

We now give some applications of the ratio test that illustrate these ideas.

EXAMPLE 8 Find the limiting ratio r of the series

$$\sum_{n=1}^{\infty} \frac{2^n}{n!}.$$

Solution   We have

$$r = \lim_{n \to \infty} \frac{a_{n+1}}{a_n} = \lim_{n \to \infty} \frac{[2^{n+1}/(n+1)!]}{2^n/n!} = \lim_{n \to \infty} \left(\frac{2^{n+1}}{(n+1)!} \cdot \frac{n!}{2^n} \right)$$

$$= \lim_{n \to \infty} \frac{2 \cdot n!}{(n+1)n!} = \lim_{n \to \infty} \frac{2}{n+1} = 0.$$

Since r exists and is $0 < 1$, we see that the series converges.

Note our use of the relation $(n+1)! = (n+1)n!$. This relation is used often.

Example 8 illustrates that $n!$ increases with n so much faster than 2^n that $\sum_{n=1}^{\infty} 2^n/n!$ converges. You should try to develop an intuitive feeling for such behavior of a series. Note that the numerator 2^n of the nth term of the series contributed the 2 in the numerator of the ratio. The $n!$ in the denominator contributed the $n+1$ in the denominator of the ratio, before the limit was taken.   ■

EXAMPLE 9   Determine the convergence or divergence of the series

$$\sum_{n=1}^{\infty} \frac{n^{25}}{3^n}.$$

Solution   We know that 3^n increases more rapidly than n^{25} for large n, and we wonder whether 3^n increases rapidly enough to completely dominate n^{25} and make our series behave like a geometric series with ratio $\frac{1}{3}$.

We find that

$$r = \lim_{n \to \infty} \frac{a_{n+1}}{a_n} = \lim_{n \to \infty} \frac{(n+1)^{25}/3^{n+1}}{n^{25}/3^n} = \lim_{n \to \infty} \left[\frac{(n+1)^{25}}{n^{25}} \cdot \frac{3^n}{3^{n+1}} \right]$$

$$= \lim_{n \to \infty} \left(\frac{n+1}{n} \right)^{25} \cdot \frac{1}{3} = 1^{25} \cdot \frac{1}{3} = \frac{1}{3}.$$

Thus r exists and is $\frac{1}{3} < 1$, so our series does indeed converge.   ■

Example 9 shows that if n is large enough, 3^n dominates n^{25} to such an extent that $\sum_{n=1}^{\infty} n^{25}/3^n$ converges and, indeed, "behaves" like a geometric series with ratio $\frac{1}{3}$. The 3^n in the denominator of the series contributed the 3 in the denominator of the ratio.

Example 9 also illustrates that a "polynomial part" of a formula giving the nth term a_n of a series contributes just a factor 1 to the limiting ratio and hence is never significant in a ratio test. (Note the contribution $1^{25} = 1$ from n^{25} in our computation.) To see this in more detail, let $p(n)$ be a polynomial function of n. Then $p(n+1)$ and $p(n)$ are polynomials of the same degree and even have the same coefficients of the terms of highest degree. As shown in Section 2.3, we then must have

$$\lim_{n \to \infty} \frac{p(n+1)}{p(n)} = 1.$$

EXAMPLE 10   Establish the convergence or divergence of the series

$$\sum_{n=1}^{\infty} \frac{n!}{n^{100} \cdot 5^n}.$$

Solution   Example 8 indicates that the term $n!$ will contribute to the ratio a factor $n + 1$ after cancellation, in the same (numerator or denominator) position. The same example indicates that 5^n will contribute a factor 5 in the same (numerator or denominator) position in the ratio. We just showed that the monomial term n^{100} will contribute a factor 1 after taking the limit in the ratio test. Thus the limiting ratio for this series will be

$$\lim_{n\to\infty} \frac{n + 1}{5} = \infty.$$

Therefore the series diverges. The $n!$ dominates both n^{100} and 5^n.   ∎

EXAMPLE 11   Determine the convergence or divergence of the series

$$\sum_{n=2}^{\infty} \frac{100^n(n^2 + 3)}{n!}.$$

Solution   An analysis like that in Example 10 shows that the limiting ratio will be

$$\lim_{n\to\infty} \frac{100}{n + 1} = 0 < 1.$$

Thus the series converges. The $n!$ dominates both 100^n and $n^2 + 3$.   ∎

EXAMPLE 12   Establish the convergence or divergence of the series

$$\sum_{n=1}^{\infty} \frac{n^5 \cdot 2^{n+1}}{3^n}.$$

Solution   We may rewrite our series as

$$2 \sum_{n=1}^{\infty} \frac{n^5 \cdot 2^n}{3^n}.$$

Our arguments above show that the limiting ratio will be $\frac{2}{3} < 1$, because the n^5 contributes a factor 1, the 2^n a factor 2 in the numerator, and the 3^n a factor 3 in the denominator. Thus the series converges. The geometric terms 2^n and 3^n dominate n^5.   ∎

EXAMPLE 13   Establish the convergence or divergence of the series

$$\sum_{n=1}^{\infty} \frac{2^{3n-1}}{3^{2n+4}}.$$

Solution   We may write the series as

$$\left(\frac{1}{2} \cdot \frac{1}{3^4}\right) \sum_{n=1}^{\infty} \frac{2^{3n}}{3^{2n}} = \frac{1}{2 \cdot 3^4} \sum_{n=1}^{\infty} \frac{8^n}{9^n}.$$

We see at once that the series converges as a geometric series with ratio $\frac{8}{9}$. ∎

EXAMPLE 14   Establish the convergence or divergence of the series

$$\sum_{n=1}^{\infty} \frac{(2n)!}{100^n \cdot n!}.$$

Solution   Since we have not worked out a ratio test involving a factor $(2n)!$, we will have to write out that portion of the ratio to see what this factor contributes. Since 100^n contributes 100 in the denominator of the ratio and $n!$ contributes $n + 1$ in the denominator, we have

$$\lim_{n \to \infty} \frac{a_{n+1}}{a_n} = \lim_{n \to \infty} \left[\frac{1}{(100)(n+1)} \cdot \frac{(2n+2)!}{(2n)!} \right].$$

Now $(2n + 2)! = (2n + 2)(2n + 1)(2n)!$, so we obtain upon cancellation

$$\lim_{n \to \infty} \frac{a_{n+1}}{a_n} = \lim_{n \to \infty} \frac{(2n+2)(2n+1)}{100(n+1)} = \infty.$$

Thus the series diverges. ∎

SUMMARY

1. *Integral test:* Let $\sum_{n=1}^{\infty} a_n$ be a series of nonnegative terms. Suppose also that $f(x)$ is a continuous function for $x \geq 1$ such that

   (a) $f(n) = a_n$ for $n = 1, 2, 3, \ldots$ and

   (b) f is monotone decreasing for $x \geq 1$; that is, $f(x_1) \geq f(x_2)$ if $1 \leq x_1 \leq x_2$.

   Then $\sum_{n=1}^{\infty} a_n$ converges if and only if $\int_1^{\infty} f(x)\, dx$ converges.

2. *p-series test:* Let p be a real number. The series $\sum_{n=1}^{\infty} 1/n^p$ converges if $p > 1$ and diverges if $p \leq 1$.

3. If $\sum_{n=1}^{\infty} a_n$ and $f(x)$ satisfy the conditions of the integral test and the series converges, then its sum may be estimated by

$$\sum_{n=1}^{\infty} a_n \approx \sum_{n=1}^{r-1} a_n + \int_r^{\infty} f(x)\, dx.$$

   The actual value of $\sum_{n=1}^{\infty} a_n$ exceeds this estimate by at most a_r.

4. *Ratio test:* Let $\sum_{n=1}^{\infty} a_n$ be a series of positive terms and suppose that $\lim_{n \to \infty} a_{n+1}/a_n$ exists and is r. Then $\sum_{n=1}^{\infty} a_n$ converges if

$r < 1$ and diverges if $r > 1$. (If $r = 1$, further information is necessary to determine the convergence or divergence of the series.)

5. Suppose a_n in $\Sigma_{n=1}^{\infty} a_n$ is given by a formula involving n, containing factors in either the numerator or denominator of the form b^n, $n!$, or $p(n)$, where $p(n)$ is a polynomial function of n. The individual contributions of these factors in the ratio test are as follows.

b^n: Contributes a factor b in the same (numerator or denominator) position.

$n!$: Contributes a factor $n + 1$ in the same position, before $\lim_{n \to \infty}$ is computed.

$p(n)$: Contributes a factor 1 after $\lim_{n \to \infty}$ is computed.

Using these ideas, we see at a glance that

$$\sum_{n=1}^{\infty} \frac{2^n \cdot n^4}{n!} \quad \text{leads to} \quad r = \lim_{n \to \infty} \frac{2}{n + 1} = 0$$

in the ratio test, while

$$\sum_{n=1}^{\infty} \frac{n!(n^7 + n^3)}{(n + 2)! \cdot 5^n} \quad \text{leads to} \quad r = \lim_{n \to \infty} \frac{n + 1}{(n + 3) \cdot 5} = \frac{1}{5}.$$

EXERCISES 10.4

While you should be able to ascertain the convergence or divergence of each series in Exercises 1–6 at a glance, use the integral test, for practice, to discover whether the series converges.

1. $\displaystyle\sum_{n=1}^{\infty} \frac{1}{3n}$

2. $\displaystyle\sum_{n=1}^{\infty} \frac{1}{n^2 + 1}$

3. $\displaystyle\sum_{n=1}^{\infty} \frac{1}{4n - 1}$

4. $\displaystyle\sum_{n=1}^{\infty} \frac{n + 1}{n^2 + 2n - 2}$

5. $\displaystyle\sum_{n=1}^{\infty} \frac{n}{(n + 1)^3}$

6. $\displaystyle\sum_{n=1}^{\infty} \frac{1}{e^n}$

7. Note that $1/n^2 \le 1/[n(\ln n)] \le 1/n$ for $n \ge 3$, and $\Sigma_{n=2}^{\infty} 1/n^2$ converges while $\Sigma_{n=2}^{\infty} 1/n$ diverges. Use the integral test to show that $\Sigma_{n=2}^{\infty} 1/[n(\ln n)]$ diverges, and try to file this result in your memory with other series that you know diverge.

8. Use the integral test to show that the series $\Sigma_{n=2}^{\infty} 1/[n(\ln n)^2]$ converges.

While you should be able to ascertain the convergence or divergence of each series in Exercises 9–15 at a glance, write out the ratio test, as in Examples 8 and 9, to determine whether the series converges.

9. $\displaystyle\sum_{n=1}^{\infty} \frac{n^2 \cdot 2^n}{n!}$

10. $\displaystyle\sum_{n=1}^{\infty} \frac{n^3 + 3n}{2^n}$

11. $\displaystyle\sum_{n=1}^{\infty} \frac{5^{n+1}}{n^3 \cdot 4^{n+2}}$

12. $\displaystyle\sum_{n=1}^{\infty} \frac{(n + 1)!}{100^{n+10}}$

13. $\displaystyle\sum_{n=1}^{\infty} \frac{n^2 \cdot 5^{n+1}}{3^{2n-1}}$

14. $\displaystyle\sum_{n=1}^{\infty} \frac{(n + 3)(n + 7)}{n!}$

15. $\displaystyle\sum_{n=1}^{\infty} \frac{(n + 5)!}{n^2 \cdot n! \cdot 2^n}$

16. Obviously $n^n > n!$ for large n. Let's discover whether n^n dominates $n!$ enough to make $\Sigma_{n=1}^{\infty} n!/n^n$ converge.
   (a) Show that if n is even, then

$$\frac{n!}{n^n} < \left(\frac{1}{2}\right)^{n/2} = \left(\frac{1}{\sqrt{2}}\right)^n.$$

[*Hint:* After you are "halfway through the $n!$," the factors in $n!$ are less than $n/2$.]

(b) Show that if n is odd, then $n!/n^n \le (\frac{1}{2})^{(n-1)/2} = (1/\sqrt{2})^{n-1}$. [*Hint:* After you are "down to $(n-1)/2$" in the $n!$, the ratio of each remaining factor of $n!$ to n is less than $(n-1)/(2n)$.]

(c) From parts (a) and (b), conclude that n^n does indeed dominate $n!$ to such an extent that $\sum_{n=1}^{\infty} n!/n^n$ converges.

In Exercises 17–48, try to determine by "inspection" (without writing out any computations, as illustrated in Examples 10–13) whether the series converges. Make use of the results in Exercises 7, 8, and 16 where appropriate. Write out a formal test only if you get stuck.

17. $\sum_{n=1}^{\infty} \dfrac{3n^2 + 3n}{n^4 + 2}$

18. $\sum_{n=1}^{\infty} \dfrac{n}{\sqrt{n^3 + 3n}}$

19. $\sum_{n=1}^{\infty} \dfrac{1}{\sqrt{n + 3}}$

20. $\sum_{n=1}^{\infty} \dfrac{n^{10}}{2^n}$

21. $\sum_{n=1}^{\infty} \dfrac{\ln n}{n}$

22. $\sum_{n=2}^{\infty} \dfrac{n-1}{n^2(\ln n)}$

23. $\sum_{n=1}^{\infty} \dfrac{1}{\ln(n^2 + 4n)}$

24. $\sum_{n=1}^{\infty} \dfrac{n^n}{n^3 \cdot n!}$

25. $\sum_{n=1}^{\infty} \dfrac{\sqrt{3n^2 + 6n - 1}}{4n^3 - 3n}$

26. $\sum_{n=1}^{\infty} \dfrac{n^3 \cdot 4^n}{n!}$

27. $\sum_{n=1}^{\infty} \dfrac{2^{3n+1}}{n!}$

28. $\sum_{n=1}^{\infty} \dfrac{3^{2n+4}}{10^n}$

29. $\sum_{n=1}^{\infty} \dfrac{2^{3n-1}}{8^{n+6}}$

30. $\sum_{n=1}^{\infty} \dfrac{n!}{(2n)!}$

31. $\sum_{n=1}^{\infty} \dfrac{n^2 \cdot n!}{(n + 3)!}$

32. $\sum_{n=1}^{\infty} \dfrac{n! + 3^n}{(n + 1)!}$

33. $\sum_{n=1}^{\infty} \dfrac{n! + 2^n}{(n + 2)!}$

34. $\sum_{n=1}^{\infty} \dfrac{(n + 3)! - n!}{2^n}$

35. $\sum_{n=1}^{\infty} \dfrac{(n + 3)! - (n + 1)!}{2^n \cdot (n + 2)!}$

36. $\sum_{n=1}^{\infty} \dfrac{(n + 3)! - (n - 1)!}{(n + 4)!}$

37. $\sum_{n=1}^{\infty} \dfrac{(n + 3)! - (n + 1)!}{(n + 5)!}$

38. $\sum_{n=1}^{\infty} \dfrac{n!}{3^{2n}}$

39. $\sum_{n=1}^{\infty} \dfrac{n!}{3^{n^2}}$

40. $\sum_{n=1}^{\infty} \dfrac{1}{2^{\ln n}}$

41. $\sum_{n=1}^{\infty} \dfrac{1}{3^{\ln n}}$

42. $\sum_{n=1}^{\infty} \dfrac{1}{2^{\ln n^2}}$

43. $\sum_{n=1}^{\infty} \dfrac{1}{(\sqrt{2})^{\ln n^2}}$

44. $\sum_{n=2}^{\infty} \dfrac{1}{n(\ln n)^{3/2}}$

45. $\sum_{n=1}^{\infty} \sin\left(\dfrac{n}{n^2 + 1}\right)$

46. $\sum_{n=1}^{\infty} \sin\left(\dfrac{\sqrt{n}}{n^2 + 1}\right)$

47. $\sum_{n=1}^{\infty} \dfrac{(2n)!}{3^{n^2}}$

48. $\sum_{n=1}^{\infty} \dfrac{n^n}{3^{n^2}}$

49. Find n such that the partial sum s_n of the harmonic series ≥ 1000. (See Example 5.)

In Exercises 50–55, use a calculator or computer to estimate the sum of the series to the accuracy indicated.

50. $\sum_{n=1}^{\infty} \dfrac{1}{n^3}$, error ≤ 0.0001

51. $\sum_{n=0}^{\infty} \dfrac{1}{n^2 + 1}$, error ≤ 0.001

52. $\sum_{n=0}^{\infty} \dfrac{8n}{n^4 + 1}$, error ≤ 0.001

53. $\sum_{n=1}^{\infty} \dfrac{1}{n^{3/2}}$, error ≤ 0.01

54. $\sum_{n=1}^{\infty} \dfrac{10}{n^4}$, error ≤ 0.00001

55. $\sum_{n=2}^{\infty} \dfrac{5}{n(\ln n)^2}$, error ≤ 0.01

10.5 ALTERNATING SERIES TEST; ABSOLUTE CONVERGENCE

Alternating Series

We have established several tests for the convergence of a series of nonnegative terms. Analogous tests may be used for series all of whose terms are less than or equal to zero, since $\sum_{n=1}^{\infty} a_n$ converges to s if and only if $\sum_{n=1}^{\infty} (-a_n)$ converges to $-s$. A finite number of negative (or positive) terms can be neglected in establishing the convergence or divergence of a series. However, if a series contains infinitely many positive and infinitely many negative terms, the situation becomes more complicated. One type of series often encountered is an *alternating series,* in which the terms are alternately positive and negative. For example, the series

$$1 - 2 + 3 - 4 + 5 - 6 + \cdots$$

is an alternating series. This series diverges since the nth term does not approach 0 as $n \to \infty$.

The following test establishes the convergence of certain alternating series. While the class of series covered by the test may seem very restrictive, such series occur quite often in practice.

THEOREM 10.9   Alternating series test

Let $\sum_{n=1}^{\infty} a_n$ be a series such that

1. the series is alternating,
2. $|a_{n+1}| \le |a_n|$ for all n, and
3. $\lim_{n \to \infty} a_n = 0$.

Then the series converges.   □

An ε,N-proof of the alternating series test will be given in a moment. First, we give an intuitive explanation of why a series that satisfies this test converges.

Suppose $\sum_{n=1}^{\infty} a_n$ satisfies the alternating series test, and suppose $a_1 > 0$. To make our argument simpler, we suppose that condition 2 of the test is satisfied in the strong form $|a_{n+1}| < |a_n|$. Now $s_1 = a_1$, and

$$s_2 = s_1 + a_2 < s_1,$$

since $a_2 < 0$. However, since $|a_2| < |a_1|$, we see that s_2 is still positive, as shown in Fig. 10.8. Next

$$s_3 = s_2 + a_3 > s_2,$$

since $a_3 > 0$. Since $|a_3| < |a_2|$, we see that $s_3 < s_1$. In a similar fashion we find that the partial sums satisfy

$$s_2 < s_4 < s_6 < s_8 < \cdots < \cdots < s_7 < s_5 < s_3 < s_1,$$

as shown in Fig. 10.8. The sequence of partial sums thus oscillates back and forth. The size of the oscillation from s_{n-1} to s_n is $|a_n|$, which decreases by our assumption that $|a_{n+1}| < |a_n|$. If the magnitude of the oscillations does not approach zero, then we could have the situation shown in Fig. 10.9, where the oscillations always bridge a gap between two numbers L_1 and L_2, and the

FIGURE 10.8   Partial sum sequence terms of an alternating series $\sum_{n=1}^{\infty} a_n$, where $|a_{n+1}| < |a_n|$.

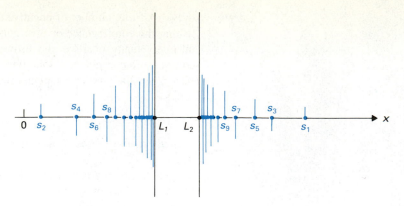

FIGURE 10.9   Possible behavior for alternating series $\sum_{n=1}^{\infty} a_n$ if $|a_{n+1}| < |a_n|$, but oscillations remain large.

sequence $\{s_n\}$ would not converge. However, condition 3 of the test tells us that this cannot happen because $\lim_{n\to\infty} a_n = 0$. Thus the situation must be as shown in Fig. 10.10, where $\{s_n\}$ and therefore $\sum_{n=1}^{\infty} a_n$ approach a limit L.

As shown in Fig. 10.10, the partial sums s_n jump back and forth across L as n increases. We obtain at once the following error estimate.

Error Estimate

If $\sum_{n=1}^{\infty} a_n$ satisfies the alternating series test and converges to L, then

$$|s_n - L| \le |a_{n+1}|.$$

That is, the error in approximating L by the sum of the first n terms of the series is at most the magnitude of the next term.

EXAMPLE 1   Show that the alternating harmonic series

$$\sum_{n=1}^{\infty} (-1)^n \cdot \frac{1}{n}$$

converges.

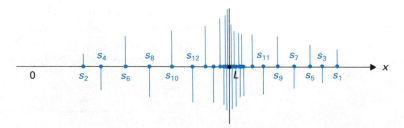

FIGURE 10.10   Behavior of an alternating series $\sum_{n=1}^{\infty} a_n$ if $|a_{n+1}| < |a_n|$ and $\lim_{n\to\infty} a_n = 0$.

Solution The series is alternating, $1/(n + 1) < 1/n$, and $\lim_{n\to\infty} 1/n = 0$. Thus the conditions of the alternating series test are satisfied, and the series converges. Exercise 37 in Section 10.7 will show that the sum of the series is ln 2. ■

EXAMPLE 2 Show that the series

$$\sum_{n=1}^{\infty} (-1)^n \cdot \frac{1}{n - \frac{7}{2}}$$

converges.

Solution The first few terms of the series are

$$\frac{2}{5} - \frac{2}{3} + 2 + 2 - \frac{2}{3} + \frac{2}{5} - \frac{2}{7} + \frac{2}{9} - \frac{2}{11} + \cdots.$$

The series satisfies the alternating series test starting with the fourth term. Since the convergence or divergence of a series is not affected by a finite number of terms, the series converges. ■

EXAMPLE 3 Show that the series

$$\sum_{n=1}^{\infty} (-1)^n \frac{\ln n}{n}$$

converges.

Solution The series is alternating. Since

$$\frac{d}{dx}\left(\frac{\ln x}{x}\right) = \frac{x \cdot (1/x) - \ln x}{x^2} = \frac{1 - \ln x}{x^2} < 0 \qquad \text{for } x > e,$$

we have $|a_{n+1}| < |a_n|$ if $n \geq 3$. L'Hôpital's rule shows that $\lim_{n\to\infty} [(\ln n)/n] = 0$. (The graph of ln x, with slope approaching 0 as $x \to \infty$, increases extremely slowly compared with the graph of x, as shown in Fig. 10.11.) The series satisfies the alternating series test for $n \geq 3$ and consequently converges. ■

EXAMPLE 4 Show that $\sum_{n=1}^{\infty} (-1)^n/n^3$ converges, and find the sum of the series with error ≤ 0.001.

Solution The series satisfies the alternating series test, so it converges. Using the error estimate given before Example 1 and our calculator, we find that

$$\sum_{n=1}^{\infty} (-1)^n \frac{1}{n^3} \approx -1 + \frac{1}{2^3} - \frac{1}{3^3} + \frac{1}{4^3} - \frac{1}{5^3} + \frac{1}{6^3} - \frac{1}{7^3} + \frac{1}{8^3} - \frac{1}{9^3}$$

$$\approx -0.9021,$$

with error at most $1/10^3 = 0.001$. ■

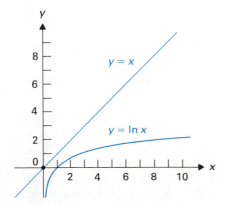

FIGURE 10.11 x increases so much faster than ln x that

$$\lim_{x\to\infty} (\ln x)/x = 0.$$

We emphasize that *all three* conditions of the alternating series test must hold before we can conclude that the series converges. We leave as exercises the construction of examples to show that if any one of these three conditions is dropped, a series can be found that satisfies the remaining two conditions but diverges. (See Exercises 1, 2, and 3.)

Proof of the Alternating Series Test

Proof: Let s_n be the nth partial sum of the series $\sum_{n=1}^{\infty} a_n$. We may suppose that $a_1 > 0$, so that

$$a_1 > 0, \qquad a_2 < 0, \qquad a_3 > 0, \quad \ldots .$$

(A similar proof will hold if $a_1 < 0$.) We have

$$s_2 = (a_1 + a_2),$$
$$s_4 = (a_1 + a_2) + (a_3 + a_4),$$
$$s_6 = s_4 + (a_5 + a_6),$$
$$s_8 = s_6 + (a_7 + a_8),$$
$$\vdots \qquad \vdots$$

and the sequence $s_2, s_4, s_6, s_8, \ldots$ is monotone increasing, for each term in parentheses is nonnegative by condition 2 of the test. (See Fig. 10.8.) Since

$$s_{2n} = a_1 + (a_2 + a_3) + \cdots + (a_{2n-2} + a_{n-1}) + a_{2n} \tag{1}$$

and each term in parentheses in Eq. (1) is nonpositive by condition 2, and since $a_{2n} < 0$, we see that

$$s_{2n} \leq a_1.$$

Thus $s_2, s_4, s_6, s_8, \ldots$ is a monotone-increasing sequence that is bounded above and therefore converges to a number L by the fundamental property (page 493).

We claim that $\sum_{n=1}^{\infty} a_n$ converges to L. Let $\varepsilon > 0$ be given and find N such that $|a_n| < \varepsilon/2$ for $n > N$ and also $|s_{2m} - L| < \varepsilon/2$ for $2m > N$. This is possible by condition 3 and the preceding paragraph. If n is even and $n > N$, then $|s_n - L| < \varepsilon/2 < \varepsilon$ by choice of N. If n is odd and $n > N$, then

$$s_n = s_{n+1} - a_{n+1},$$

so

$$|s_n - L| = |(s_{n+1} - L) - a_{n+1}| \leq |s_{n+1} - L| + |a_{n+1}| \leq \frac{\varepsilon}{2} + \frac{\varepsilon}{2} = \varepsilon$$

by choice of N. Thus $\{s_n\}$ converges to L, so $\sum_{n=1}^{\infty} a_n$ converges and has sum L. $\square$

Absolute Convergence

Let $\sum_{n=1}^{\infty} a_n$ be a series containing both positive and negative terms. The series $\sum_{n=1}^{\infty} |a_n|$ contains only nonnegative terms; we could apply some of the tests developed in the preceding sections to $\sum_{n=1}^{\infty} |a_n|$, and we might be able to establish its convergence or divergence. The next theorem shows that if $\sum_{n=1}^{\infty} |a_n|$ converges, then $\sum_{n=1}^{\infty} a_n$ converges also. It is important to note that if $\sum_{n=1}^{\infty} |a_n|$ diverges, then $\sum_{n=1}^{\infty} a_n$ may diverge or may converge. We illustrate this following the theorem.

THEOREM 10.10   **Absolute convergence test**

Let $\sum_{n=1}^{\infty} a_n$ be any infinite series. If $\sum_{n=1}^{\infty} |a_n|$ converges, then $\sum_{n=1}^{\infty} a_n$ converges.

Proof: We define a new series $\sum_{n=1}^{\infty} u_n$ by replacing the negative terms of $\sum_{n=1}^{\infty} a_n$ by zeros, and a new series $\sum_{n=1}^{\infty} v_n$ by replacing the positive terms of $\sum_{n=1}^{\infty} a_n$ by zeros. That is, we let

$$u_n = \begin{cases} a_n, & \text{if } a_n \ge 0, \\ 0, & \text{if } a_n < 0, \end{cases} \quad \text{and} \quad v_n = \begin{cases} a_n, & \text{if } a_n \le 0, \\ 0, & \text{if } a_n > 0. \end{cases}$$

Note that

$$\sum_{n=1}^{\infty} a_n = \sum_{n=1}^{\infty} (u_n + v_n).$$

Suppose that $\sum_{n=1}^{\infty} |a_n|$ converges. Now $\sum_{n=1}^{\infty} u_n$ is a series of nonnegative terms and $u_n \le |a_n|$, so $\sum_{n=1}^{\infty} u_n$ converges by a comparison test. Similarly, $\sum_{n=1}^{\infty} (-v_n)$ is a series of nonnegative terms and $-v_n \le |a_n|$, so $\sum_{n=1}^{\infty} (-v_n)$ converges by a comparison test also. We then find that the series $(-1) \sum_{n=1}^{\infty} (-v_n) = \sum_{n=1}^{\infty} v_n$ converges, and therefore the series

$$\sum_{n=1}^{\infty} (u_n + v_n) = \sum_{n=1}^{\infty} a_n$$

converges. This is what we wished to prove.   □

EXAMPLE 5   Determine the convergence or divergence of the series

$$1 + \frac{1}{2^2} - \frac{1}{3^2} + \frac{1}{4^2} + \frac{1}{5^2} - \frac{1}{6^2} + \frac{1}{7^2} + \frac{1}{8^2} - \frac{1}{9^2} + \cdots,$$

with two positive terms followed by a negative term.

Solution   The series does not satisfy the alternating series test. However, the series does converge, for the corresponding series of absolute values is the series $\sum_{n=1}^{\infty} 1/n^2$, which is a convergent p-series with $p = 2 > 1$.   ∎

We emphasize again that the absolute convergence test does not say anything about the behavior of $\sum_{n=1}^{\infty} a_n$ if $\sum_{n=1}^{\infty} |a_n|$ diverges. To illustrate, both the series $1 - 2 + 3 - 4 + 5 - 6 + \cdots$ and the corresponding series $1 + 2 + 3 + 4 + 5 + 6 + \cdots$ of absolute values diverge since the nth terms do not approach zero. *However, the alternating harmonic series*

$$1 - \frac{1}{2} + \frac{1}{3} - \frac{1}{4} + \frac{1}{5} - \frac{1}{6} + \cdots$$

converges (by the alternating series test), while the corresponding series of absolute values is the harmonic series

$$1 + \frac{1}{2} + \frac{1}{3} + \frac{1}{4} + \frac{1}{5} + \frac{1}{6} + \cdots,$$

which diverges. We now describe terminology used in this connection.

DEFINITION 10.7

Absolute and conditional convergence

A series $\sum_{n=1}^{\infty} a_n$ **converges absolutely** (or is **absolutely convergent**) if the series $\sum_{n=1}^{\infty} |a_n|$ converges. If $\sum_{n=1}^{\infty} a_n$ converges and $\sum_{n=1}^{\infty} |a_n|$ diverges, then $\sum_{n=1}^{\infty} a_n$ **converges conditionally** (or is **conditionally convergent**).

Note that every convergent series of nonnegative terms is absolutely convergent, since it is identical with the corresponding series of absolute values.

EXAMPLE 6   Classify the series $\sum_{n=1}^{\infty} (-1)^n / n!$ as absolutely convergent, conditionally convergent, or divergent.

Solution   The series $\sum_{n=1}^{\infty} 1/n!$ converges by the ratio test. Thus $\sum_{n=1}^{\infty} (-1)^n / n!$ is absolutely convergent.   ∎

We have often heard students say that a series converges conditionally "because it satisfies the alternating series test."

> Satisfying the alternating series test is not enough to guarantee that a series converges conditionally. The series of absolute values must also diverge.

For example, the series $\sum_{n=1}^{\infty} (-1)^n / n!$ of Example 6 satisfies the alternating series test. However, the series $\sum_{n=1}^{\infty} 1/n!$ of absolute values converges, so the series is absolutely convergent.

EXAMPLE 7   Classify the alternating harmonic series $\sum_{n=1}^{\infty} (-1)^n / n$ as absolutely convergent, conditionally convergent, or divergent.

Solution   The harmonic series $\sum_{n=1}^{\infty} 1/n$ of absolute values diverges. The alternating harmonic series $\sum_{n=1}^{\infty} (-1)^n / n$ converges by the alternating series test. Thus the series is conditionally convergent.   ∎

When classifying a series as absolutely convergent, conditionally convergent, or divergent, we should *first* check whether the series converges absolutely. If it does converge absolutely, we are done. If it does not, we try the alternating series test. It is a waste of time to check the alternating series test first.

EXAMPLE 8   Classify the series

$$\sum_{n=1}^{\infty} (-1)^n \frac{n^2 + n}{n^4 + 3n^2 + 1}$$

as absolutely convergent, conditionally convergent, or divergent.

Solution The series of absolute values behaves like $\Sigma_{n=1}^{\infty} 1/n^2$ by the LCT, and this series converges as a *p*-series with $p = 2 > 1$. Thus the original series converges absolutely. ∎

EXAMPLE 9 Classify the series

$$\sum_{n=1}^{\infty} (-1)^n \frac{2n^2 + 1}{n^2 + 100n}$$

as absolutely convergent, conditionally convergent, or divergent.

Solution The series of absolute values diverges since

$$\lim_{n \to \infty} \frac{2n^2 + 1}{n^2 + 100n} = 2 \neq 0.$$

Thus $\lim_{n \to \infty} a_n \neq 0$, so the original series $\Sigma_{n=1}^{\infty} a_n$ also diverges. That the series is alternating is irrelevant. ∎

EXAMPLE 10 Classify the series $\Sigma_{n=1}^{\infty} (-1)^n \sin(1/n)$ as absolutely convergent, conditionally convergent, or divergent.

Solution We have

$$\sum_{n=1}^{\infty} \sin\left(\frac{1}{n}\right) = \sum_{n=1}^{\infty} \frac{1}{n} \cdot \frac{\sin(1/n)}{1/n}.$$

By the LCT, this series behaves like $\Sigma_{n=1}^{\infty} 1/n$, which diverges. Thus the series does not converge absolutely. A glance at the graph of $\sin x$ in Fig. 10.12 shows that the original series is alternating, $\sin[1/(n + 1)] < \sin(1/n)$, and $\lim_{n \to \infty} \sin(1/n) = 0$. Thus the series satisfies the alternating series test and converges. Since it did not converge absolutely, the series is conditionally convergent. ∎

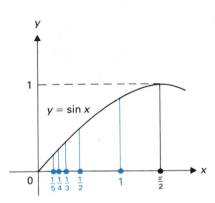

FIGURE 10.12

$$\sin[1/(n + 1)] < \sin(1/n),$$
$$\lim_{n \to \infty} \sin(1/n) = 0$$

SUMMARY

1. An alternating series is one containing alternately positive and negative terms.

2. *Alternating series test:* Let $\Sigma_{n=1}^{\infty} a_n$ be a series such that
   (a) the series is alternating,
   (b) $|a_{n+1}| \leq |a_n|$ for all n, and
   (c) $\lim_{n \to \infty} a_n = 0$.
   Then the series converges.

3. For a series $\Sigma_{n=1}^{\infty} a_n$ that satisfies the alternating series test, the estimate of $\Sigma_{n=1}^{\infty} a_n$ by the *n*th partial sum s_n has error of magnitude at most $|a_{n+1}|$.

4. *Absolute convergence test:* Let $\Sigma_{n=1}^{\infty} a_n$ be any infinite series. If $\Sigma_{n=1}^{\infty} |a_n|$ converges, then $\Sigma_{n=1}^{\infty} a_n$ converges; the series is said to be absolutely convergent.

5. A conditionally convergent series is one that converges but is not absolutely convergent.

EXERCISES 10.5

1. Give an example of a divergent series that satisfies conditions 1 and 2 of the alternating series test.

2. Give an example of a divergent series that satisfies conditions 1 and 3 of the alternating series test.

3. Give an example of a divergent series that satisfies conditions 2 and 3 of the alternating series test.

4. Show that every convergent series of nonpositive terms converges absolutely.

5. Mark each of the following true or false.
  ____ (a) Every convergent series is absolutely convergent.
  ____ (b) Every absolutely convergent series is convergent.
  ____ (c) If $\sum_{n=1}^{\infty} a_n$ is conditionally convergent, then $\sum_{n=1}^{\infty} |a_n|$ diverges.
  ____ (d) If $\sum_{n=1}^{\infty} |a_n|$ diverges, then $\sum_{n=1}^{\infty} a_n$ is conditionally convergent.
  ____ (e) Every alternating series converges.
  ____ (f) Every convergent alternating series is conditionally convergent.

In Exercises 6–39, classify the series as either absolutely convergent, conditionally convergent, or divergent.

6. $\displaystyle\sum_{n=1}^{\infty} (-1)^n \frac{1}{4n}$

7. $\displaystyle\sum_{n=1}^{\infty} (-1)^n \frac{1}{4n^2}$

8. $\displaystyle\sum_{n=1}^{\infty} (-1)^{n+1} \frac{\cos^2 n}{2^n}$

9. $\displaystyle\sum_{n=1}^{\infty} \frac{-1}{\sqrt[3]{n}}$

10. $\displaystyle\sum_{n=4}^{\infty} \frac{2 - n^2}{(n-3)^2}$

11. $\displaystyle\sum_{n=1}^{\infty} (-1)^n \frac{n}{n^2 - 10{,}001}$

12. $\displaystyle\sum_{n=1}^{\infty} (-1)^{n-1} \frac{n^2}{n^4 - 10{,}001}$

13. $\displaystyle\sum_{n=1}^{\infty} \frac{(-1)^{2n+1}}{\sqrt{n}}$

14. $\displaystyle\sum_{n=1}^{\infty} (-1)^{2n} \frac{n^2 + 3n}{n^3 + 4}$

15. $\displaystyle\sum_{n=1}^{\infty} (-1)^n \frac{\sin^2 n}{n^2}$

16. $\displaystyle\sum_{n=1}^{\infty} (-1)^{n+1} \frac{\sin^2 n}{n^{3/2}}$

17. $\displaystyle\sum_{n=1}^{\infty} \frac{\cos n}{n^{3/2}}$

18. $\displaystyle\sum_{n=1}^{\infty} \frac{\sin n}{n^2}$

19. $\displaystyle\sum_{n=1}^{\infty} (-1)^n \frac{1}{\sin^2 n}$

20. $\displaystyle\sum_{n=2}^{\infty} (-1)^n \frac{1}{\ln n}$

21. $\displaystyle\sum_{n=2}^{\infty} (-1)^n \frac{1}{n(\ln n)}$

22. $\displaystyle\sum_{n=2}^{\infty} (-1)^n \frac{1}{n(\ln n)^2}$

23. $\displaystyle\sum_{n=1}^{\infty} (-1)^n \frac{\ln n}{n}$

24. $\displaystyle\sum_{n=1}^{\infty} \frac{\ln(1/n)}{n}$

25. $\displaystyle\sum_{n=2}^{\infty} (-1)^n \frac{\ln(n^n)}{n^2}$

26. $\displaystyle\sum_{n=1}^{\infty} (-1)^n \frac{n^3 \cdot 2^n}{n!}$

27. $\displaystyle\sum_{n=1}^{\infty} (-1)^n \frac{n!}{100^n}$

28. $\displaystyle\sum_{n=1}^{\infty} (-1)^n \frac{n!}{n^n}$

29. $\displaystyle\sum_{n=1}^{\infty} (-1)^{n+1} \frac{n^4 \cdot 3^{2n}}{10^n}$

30. $\displaystyle\sum_{n=1}^{\infty} (-1)^{n-1} \frac{(n+1)! - n!}{(n+3)!}$

31. $\displaystyle\sum_{n=1}^{\infty} (-1)^{n+2} \frac{(n+2)! - n!}{(n+3)!}$

32. $\displaystyle\sum_{n=1}^{\infty} (-1)^{n+1} n \sin\left(\frac{1}{n}\right)$

33. $\displaystyle\sum_{n=1}^{\infty} (-1)^{n+3} \frac{\sin(1/n)}{n}$

34. $\displaystyle\sum_{n=1}^{\infty} (-1)^n \csc\left(\frac{1}{n}\right)$

35. $\displaystyle\sum_{n=1}^{\infty} (-1)^n \frac{\csc(1/n)}{n}$

36. $\displaystyle\sum_{n=1}^{\infty} (-1)^n \frac{\csc(1/n)}{n^2}$

37. $\displaystyle\sum_{n=1}^{\infty} \frac{\sin(n\pi/2)}{n}$

38. $\displaystyle\sum_{n=1}^{\infty} \frac{1}{\sqrt{n}}\tan\frac{(2n+1)\pi}{4}\Big]$

39. $\displaystyle\sum_{n=1}^{\infty} \frac{1}{n}\sin\left[\frac{(2n+1)\pi}{4}\right]$

40. Show that $\sum_{n=1}^{\infty}(1/n)\sin(n\pi/4)$ converges. [*Hint:* Express the series as a sum of convergent series.]

41. Let $\sum_{n=1}^{\infty} a_n$ be a series and let $\sum_{n=1}^{\infty} u_n$ be the series of positive terms (and zeros) and $\sum_{n=1}^{\infty} v_n$ the series of negative terms (and zeros) defined in the proof of Theorem 10.10.
(a) Show that if $\sum_{n=1}^{\infty} u_n$ and $\sum_{n=1}^{\infty} v_n$ both converge, then $\sum_{n=1}^{\infty} a_n$ converges.
(b) Show that if one of $\sum_{n=1}^{\infty} u_n$ and $\sum_{n=1}^{\infty} v_n$ converges while the other diverges, then $\sum_{n=1}^{\infty} a_n$ diverges.
(c) Show by an example that it is possible that $\sum_{n=1}^{\infty} a_n$ converges while both $\sum_{n=1}^{\infty} u_n$ and $\sum_{n=1}^{\infty} v_n$ diverge.

42. Give an example of a conditionally convergent series that is not an alternating series and does not become an alternating series upon deletion of a finite number of terms.

In Exercises 43–48, find the smallest value of n for which you know that the series can be estimated by s_n with error indicated, using number 3 of the Summary.

43. $\displaystyle\sum_{n=1}^{\infty} (-1)^n\frac{1}{n}$, error ≤ 0.01

44. $\displaystyle\sum_{n=1}^{\infty} (-1)^n\frac{1}{n^2}$, error ≤ 0.01

45. $\displaystyle\sum_{n=1}^{\infty} (-1)^n\frac{1}{n^3}$, error ≤ 0.008

46. $\displaystyle\sum_{n=1}^{\infty} (-1)^n\frac{1}{n^2+200}$, error ≤ 0.0001

47. $\displaystyle\sum_{n=1}^{\infty} (-1)^n\frac{1}{n!}$, error ≤ 0.001

48. $\displaystyle\sum_{n=1}^{\infty} (-1)^n\frac{1}{(2n+1)!}$, error ≤ 0.0001

In Exercises 49–54, use a calculator or computer to find the sum of the series with error of magnitude indicated.

49. $\displaystyle\sum_{n=1}^{\infty} (-1)^n\frac{1}{n^2}$, error ≤ 0.001

50. $\displaystyle\sum_{n=1}^{\infty} (-1)^n\frac{1}{n^4}$, error ≤ 0.0001

51. $\displaystyle\sum_{n=1}^{\infty} (-1)^n\frac{10}{n^2+1}$, error ≤ 0.01

52. $\displaystyle\sum_{n=2}^{\infty} (-1)^n\frac{1}{(\ln n)^3}$, error ≤ 0.05

53. $\displaystyle\sum_{n=1}^{\infty} (-1)^n\frac{1}{n!}$, error ≤ 0.0001

54. $\displaystyle\sum_{n=1}^{\infty} (-1)^n\frac{1}{(2n+1)!}$, error ≤ 0.0001

10.6 POWER SERIES

The Function Represented by a Power Series

Among the most important series are *power series,* and most of our work will be with these series. Power series are precisely the "polynomial functions of infinite degree" that we mentioned at the start of this chapter. We will see that

$$\sin x = x - \frac{x^3}{3!} + \frac{x^5}{5!} - \frac{x^7}{7!} + \frac{x^9}{9!} - \frac{x^{11}}{11!} + \cdots$$

for any value of x. We will also see that

$$\frac{1}{x} = 1 - (x-1) + (x-1)^2 - (x-1)^3 + (x-1)^4 - (x-1)^5 + \cdots$$

for any x such that $0 < x < 2$. The series for $\sin x$ is in powers of $x = x - 0$, and the series for $1/x$ is in powers of $x - 1$.

DEFINITION 10.8

Power series centered at x_0

A **power series centered at** x_0 is a series of the form

$$\sum_{n=0}^{\infty} a_n(x - x_0)^n = a_0 + a_1(x - x_0) + \cdots + a_n(x - x_0)^n + \cdots. \quad (1)$$

The constants a_i are the **coefficients of the series;** in particular, a_0 is the **constant term of the series.**

It will be convenient to change terminology slightly and to consider the constant term a_0 of Eq. (1) to be the *zeroth term of the series* and the term $a_n(x - x_0)^n$ to be the *n*th *term*. With this convention, the *n*th term of a power series becomes the term with exponent n.

At each value of x, the power series (1) becomes a series of constants that may or may not converge. The sum of such a convergent series of constants generally varies with the value of x and is a function of x. This function is the **sum function** of the series. The set of all values of x for which the power series converges is the **domain of the sum function** defined by the series; the value of the function at each such point is the sum of the series for that value of x.

The power series (1) should be regarded as an attempt to describe a function *locally,* near x_0. In translated coordinates with $\Delta x = x - x_0$, the series becomes $\sum_{n=0}^{\infty} a_n(\Delta x)^n$. The series (1) converges for $x = x_0$ because at x_0 the series becomes

$$a_0 + a_1 \cdot 0 + a_2 \cdot 0 + \cdots + a_n \cdot 0 + \cdots,$$

which converges to a_0. It is possible that this is the only point at which the series converges, as the following example shows, but such series are of little importance for us.

EXAMPLE 1   Show that $\sum_{n=1}^{\infty} (n!)x^n$ converges only for $x = 0$.

Solution   If $x = a$, we obtain the series $\sum_{n=1}^{\infty} (n!)a^n$. As shown in Section 10.5, the ratio test for absolute convergence yields the limiting ratio

$$\lim_{n \to \infty} |(n + 1)a| = \infty, \qquad \text{if } a \neq 0.$$

Thus if $a \neq 0$, the terms of the series eventually increase in magnitude, so the *n*th term does not approach zero. Therefore the series converges only if $x = a = 0$. ∎

The Radius of Convergence of a Power Series

Theorem 10.11 shows that the domain of convergence of a power series centered at x_0 has x_0 as center point. The theorem is stated and proved in the case where $x_0 = 0$, that is, for a series $\sum_{n=0}^{\infty} a_n x^n$. See the remark after the proof of the theorem.

THEOREM 10.11   Absolute convergence

If a power series $\sum_{n=0}^{\infty} a_n x^n$ converges for $x = c \neq 0$, then the series converges absolutely for all x such that $|x| < |c|$. If the series diverges at $x = d$, then the series diverges for all x such that $|x| > |d|$.

Proof: Suppose that $\sum_{n=0}^{\infty} a_n c^n$ converges, and let $|b| < |c|$. Since $\sum_{n=0}^{\infty} a_n c^n$ converges, we must have $\lim_{n \to \infty} a_n c^n = 0$; in particular, $|a_n c^n| < 1$, or

$$|a_n| < \frac{1}{|c^n|}$$

for n sufficiently large. We then obtain

$$|a_n b^n| = |a_n| \cdot |b^n| < \left| \frac{b^n}{c^n} \right| = \left| \frac{b}{c} \right|^n$$

for n sufficiently large. Recall that we are assuming $|b/c| < 1$. Thus the series $\sum_{n=0}^{\infty} |a_n b^n|$ is, term for term, less than the convergent geometric series $\sum_{n=0}^{\infty} |b/c|^n$ for n sufficiently large, so $\sum_{n=0}^{\infty} |a_n b^n|$ converges by the comparison test. This shows that $\sum_{n=0}^{\infty} a_n x^n$ converges absolutely for all x such that $|x| < |c|$.

The assertion regarding divergence is really the contrapositive of the assertion we just proved for convergence. Suppose that $\sum_{n=0}^{\infty} a_n d^n$ diverges, and let $|h| > |d|$. Convergence of $\sum_{n=0}^{\infty} a_n h^n$ would imply convergence of $\sum_{n=0}^{\infty} a_n d^n$ by the first part of our proof. But this would contradict our hypothesis, so $\sum_{n=0}^{\infty} a_n h^n$ diverges also. ☐

We stated and proved the theorem for a power series centered at $x_0 = 0$. It is immediate from the theorem that if $\sum_{n=0}^{\infty} a_n (\Delta x)^n$ converges for $\Delta x = c$, then the series converges for $|\Delta x| < |c|$. Putting $\Delta x = x - x_0$ as usual, we see that if a series $\sum_{n=0}^{\infty} a_n (x - x_0)^n$ converges at $x = x_0 + c$, that is, for $\Delta x = c$, then it converges for all x such that $|x - x_0| < |c|$.

The important corollary that follows seems very plausible from our theorem. A careful proof depends on the fundamental property of the real numbers (page 493) and is not given here.

COROLLARY   Radius of convergence

For a power series $\sum_{n=0}^{\infty} a_n (x - x_0)^n$, exactly one of the three following alternatives holds.

(a) The series converges at x_0 only.

(b) The series converges for all x.

(c) There exists R such that the series converges for $|x - x_0| < R$, that is, for $x_0 - R < x < x_0 + R$, and diverges for $|x - x_0| > R$.

The number R that appears in case (c) of the corollary is the **radius of convergence of the series.** In case (a), it is natural to say that the radius of convergence is zero and in case (b) that the radius of convergence is ∞. In case (c), the behavior of the series at the endpoints of the interval $x_0 - R < x < x_0 + R$ depends on the individual series; certain series converge at both endpoints, others diverge at both endpoints, and some converge at one endpoint and diverge at the other.

EXAMPLE 2   Find the radius of convergence of the geometric series $\sum_{n=0}^{\infty} x^n$.

Solution   Our work in Section 10.2 shows that this series converges to $1/(1-x)$ for $|x| < 1$. The radius of convergence of the series is 1. The series diverges for $x = 1$ and $x = -1$, since the nth term does not approach 0 as $n \to \infty$ at these points.   ∎

For many power series, the radius of convergence can be computed by using the ratio test. The technique is best illustrated by examples. By Theorem 10.11, if a power series converges at $x - x_0 = c$, it converges absolutely for $|x - x_0| < |c|$, so we try to compute the limit of the *absolute value* of the ratio.

EXAMPLE 3   Find the radius of convergence of the series

$$\sum_{n=1}^{\infty} \frac{x^n}{n \cdot 2^n}.$$

Solution   The absolute value of the ratio of the $(n+1)$st term to the nth is

$$\left| \frac{x^{n+1}/[(n+1)2^{n+1}]}{x^n/(n \cdot 2^n)} \right| = \left| \frac{x^{n+1}}{(n+1)2^{n+1}} \cdot \frac{n \cdot 2^n}{x^n} \right| = \left| \frac{nx}{(n+1)2} \right|.$$

The limit as $n \to \infty$ is

$$\lim_{n \to \infty} \left| \frac{nx}{(n+1)2} \right| = \left| \frac{x}{2} \right|.$$

Thus the series converges for $|x/2| < 1$, or $|x| < 2$. The radius of convergence is therefore 2, and the series converges at least for $-2 < x < 2$.

To see what happens at the endpoint 2 of the interval $-2 < x < 2$, examine the series

$$\sum_{n=1}^{\infty} \frac{2^n}{n \cdot 2^n} = \sum_{n=1}^{\infty} \frac{1}{n}.$$

This is the harmonic series, and it diverges. At the endpoint -2, we obtain as series

$$\sum_{n=1}^{\infty} \frac{(-2)^n}{n \cdot 2^n} = \sum_{n=1}^{\infty} (-1)^n \cdot \frac{1}{n},$$

which is the convergent alternating harmonic series. Thus the series converges for $-2 \leq x < 2$. This interval is the **interval of convergence of the series.**   ∎

Example 3 illustrates the usual procedure for finding the *radius* and *interval* of convergence for a power series where the limit of the ratio exists. The two series of constants corresponding to the endpoints of the interval must be examined separately.

> The ratio test never determines the behavior at the endpoints of the interval of convergence, for the limit of the ratio at these endpoints will be 1.

We illustrate with an example for a power series centered at a point $x_0 \neq 0$.

EXAMPLE 4 Determine the interval of convergence of the series

$$\sum_{n=1}^{\infty} \frac{(x-3)^{2n}}{n^2 \cdot 5^n},$$

which is a power series centered at $x_0 = 3$.

Solution For the ratio, we obtain

$$\left| \frac{(x-3)^{2(n+1)}}{(n+1)^2 \cdot 5^{n+1}} \cdot \frac{n^2 \cdot 5^n}{(x-3)^{2n}} \right| = \left| \frac{n^2(x-3)^2}{(n+1)^2 \cdot 5} \right|.$$

We have

$$\lim_{n \to \infty} \left| \frac{n^2(x-3)^2}{(n+1)^2 \cdot 5} \right| = \left| \frac{(x-3)^2}{5} \right|.$$

Thus the series converges if $|(x-3)^2/5| < 1$, or $|x-3|^2 < 5$. This is equivalent to $|x-3| < \sqrt{5}$. The radius of convergence about $x_0 = 3$ is thus $\sqrt{5}$, and the series converges at least for $3 - \sqrt{5} < x < 3 + \sqrt{5}$.

Turning to the endpoints, at $3 - \sqrt{5}$ the series becomes

$$\sum_{n=1}^{\infty} \frac{(-\sqrt{5})^{2n}}{5^n \cdot n^2} = \sum_{n=1}^{\infty} \frac{5^n}{5^n \cdot n^2}$$

$$= \sum_{n=1}^{\infty} \frac{1}{n^2}.$$

This series converges as a p-series with $p = 2 > 1$. The same series is obtained at $3 + \sqrt{5}$, so the series converges at both endpoints, and the interval of convergence is

$$[3 - \sqrt{5}, \; 3 + \sqrt{5}]. \qquad \blacksquare$$

The preceding examples indicate that the term $(x - x_0)^n$ in a power series $\sum_{n=0}^{\infty} a_n(x - x_0)^n$ contributes $|x - x_0|$ to the limiting absolute value ratio, the term $(x - x_0)^{2n}$ contributes $|x - x_0|^2$, and so on. Of course, we expect this from our work in Section 10.4. We should be able to use the technique described in number 5 of the Summary of Section 10.4 to find the radius of convergence for many power series without actually writing out the ratio test in detail.

EXAMPLE 5 Find the interval of convergence of the series $\sum_{n=1}^{\infty} (x^n/n!)$.

Solution Since x^n contributes x to the ratio and $n!$ contributes $n + 1$, the limit of the absolute value ratio is

$$\lim_{n \to \infty} \left| \frac{x}{n + 1} \right| = 0 < 1$$

for all x. Thus the series converges for all x; its interval of convergence is $-\infty < x < \infty$. ∎

EXAMPLE 6 Find the interval of convergence of the series

$$\sum_{n=1}^{\infty} \frac{(-2)^n(x + 3)^n}{n}.$$

Solution Now $(-2)^n$ contributes -2 and $(x + 3)^n$ contributes $x + 3$ to the ratio. Also, n contributes 1 after taking the limit as $n \to \infty$. (See the Summary of Section 10.4). We see that the absolute value limiting ratio is

$$|(-2)(x + 3)| = 2|x + 3|.$$

Thus the series converges for

$$2|x + 3| < 1, \quad \text{or} \quad |x + 3| < \frac{1}{2}.$$

The radius of convergence is $\frac{1}{2}$, and the series converges for

$$-3 - \frac{1}{2} < x < -3 + \frac{1}{2}, \quad \text{or} \quad -\frac{7}{2} < x < -\frac{5}{2}.$$

For $x = -\frac{7}{2}$, we obtain the series

$$\sum_{n=1}^{\infty} \frac{(-2)^n(-\frac{1}{2})^n}{n} = \sum_{n=1}^{\infty} \frac{1}{n},$$

which diverges. For $x = -\frac{5}{2}$, we obtain the series

$$\sum_{n=1}^{\infty} \frac{(-2)^n(\frac{1}{2})^n}{n} = \sum_{n=1}^{\infty} \frac{(-1)^n}{n},$$

which converges. Thus the interval of convergence is $-\frac{7}{2} < x \le -\frac{5}{2}$. ∎

Geometric Power Series

The series

$$a + ar + ar^2 + ar^3 + \cdots + ar^n + \cdots$$

converges to $a/(1 - r)$ if $|r| < 1$. We can sometimes "work backwards" using this sum formula, starting with a function and finding a geometric power series that equals the function in its interval of convergence.

EXAMPLE 7   Find a power series centered at $x_0 = 0$ that equals $2/(1 + x)$ in its interval of convergence.

Solution   We put $2/(1 + x)$ in the form $a/(1 - r)$ by writing

$$\frac{2}{1 + x} = \frac{2}{1 - (-x)}.$$

We recognize this as the sum of a geometric series with initial term 2 and ratio $-x$. The series is

$$2 - 2x + 2x^2 - 2x^3 + 2x^4 - \cdots + (-1)^n 2 x^n + \cdots.$$

The series converges to $2/(1 + x)$ for $|x| < 1$.   ∎

EXAMPLE 8   Find a power series centered at $x_0 = 0$ that equals $5/(3 - x)$ in its interval of convergence.

Solution   We put $5/(3 - x)$ into the form $a/(1 - r)$:

$$\frac{5}{3 - x} = \frac{1}{3} \cdot \frac{5}{1 - \dfrac{x}{3}} = \frac{\frac{5}{3}}{1 - \dfrac{x}{3}}.$$

This is the sum of a geometric series with initial term $\frac{5}{3}$ and ratio $x/3$. The series is

$$\frac{5}{3} + \frac{5}{3^2}x + \frac{5}{3^3}x^2 + \frac{5}{3^4}x^3 + \cdots + \frac{5}{3^{n+1}}x^n + \cdots = \sum_{n=0}^{\infty} \frac{5}{3}\left(\frac{x}{3}\right)^n.$$

The series converges for $|x/3| < 1$, or $|x| < 3$.   ∎

EXAMPLE 9   Find a power series centered at $x_0 = 1$ that equals $5/(3 - x)$ in its interval of convergence.

Solution   As in Example 8, we put $5/(3 - x)$ in the form $a/(1 - r)$, but this time we want the ratio r in terms of $x - 1$ rather than x. We have

$$\frac{5}{3 - x} = \frac{5}{3 - [(x - 1) + 1]} = \frac{5}{2 - (x - 1)}$$

$$= \frac{1}{2} \cdot \frac{5}{1 - \dfrac{x - 1}{2}} = \frac{\frac{5}{2}}{1 - \dfrac{x - 1}{2}}.$$

This is the sum of a geometric series with initial term $\frac{5}{2}$ and ratio $(x - 1)/2$. The series is

$$\frac{5}{2} + \frac{5}{2^2}(x - 1) + \frac{5}{2^3}(x - 1)^2 + \cdots + \frac{5}{2^{n+1}}(x - 1)^n + \cdots$$

$$= \sum_{n=0}^{\infty} \frac{5}{}$$

which converges for $|(x - 1)/2| < 1$, or $|x - 1| < 2$.   ∎

SUMMARY

1. A power series centered at x_0 is a series of the form

$$a_0 + a_1(x - x_0) + a_2(x - x_0)^2 + \cdots + a_n(x - x_0)^n + \cdots.$$

The function having as domain all values of x for which the series converges, and having as value at each such x the sum of the series, is called the sum function defined by the series.

2. The series in number (1) may converge for $x = x_0$ only, or it may converge for all x, or it may converge if $|x - x_0| < R$ and diverge if $|x - x_0| > R$ for some $R > 0$. Such a number R is the radius of convergence of the series.

3. The radius R of convergence of a power series centered at x_0 can often be found by

   (a) forming the absolute value of the ratio of the $(n + 1)$st term divided by the nth term,

   (b) computing the limit of this ratio as $n \to \infty$,

   (c) setting the resulting limit less than 1,

   (d) solving the resulting inequality for $|x - x_0|$, obtaining an expression of the form $|x - x_0| < R$. The radius of convergence is then R.

4. To determine the interval of convergence of a power series centered at x_0 after the radius R of convergence has been found, substitute $x = x_0 - R$ and $x = x_0 + R$ to obtain two series of constants. Test these series for convergence to determine which of the endpoints, $x_0 - R$ or $x_0 + R$, should be included with $|x - x_0| < R$ to obtain the interval of convergence. The ratio test should never be tried at these endpoint series, for the limiting ratio will always be 1.

5. If a function $f(x)$ can be written in the form $a/(1 - r)$ where both a and r are monomial functions of $x - x_0$, then we can find a geometric power series centered at x_0 whose sum equals $f(x)$ in its interval of convergence.

EXERCISES 10.6

In Exercises 1–8, use the ratio test to find the radius of convergence of the series. Then find the interval of convergence (including endpoints).

1. $\displaystyle\sum_{n=1}^{\infty} \frac{x^n}{n}$

2. $\displaystyle\sum_{n=1}^{\infty} \frac{x^{2n+1}}{n}$

3. $\displaystyle\sum_{n=1}^{\infty} (-1)^n \frac{x^n}{n}$

4. $\displaystyle\sum_{n=0}^{\infty} (-1)^n \frac{x^{2n}}{n + 3}$

5. $\displaystyle\sum_{n=0}^{\infty} \frac{(x - 1)^n}{n^2 + 1}$

6. $\displaystyle\sum_{n=0}^{\infty} \frac{(x + 2)^n}{3^n}$

7. $\displaystyle\sum_{n=1}^{\infty} \frac{(x+1)^n}{n \cdot 5^{n+2}}$

8. $\displaystyle\sum_{n=2}^{\infty} \frac{(-1)^n(x-3)^{n+1}}{(\ln n) \cdot 2^{n-1}}$

In Exercises 9–27, proceed as above, but this time try to find the radius of convergence without explicitly writing out the ratio, as illustrated after Example 4 in the text; write out the ratio test only if you have to.

9. $\displaystyle\sum_{n=0}^{\infty} \frac{x^n}{3^n}$

10. $\displaystyle\sum_{n=1}^{\infty} \frac{2^{n+1}x^n}{n \cdot 3^n}$

11. $\displaystyle\sum_{n=0}^{\infty} \frac{(n+1)x^n}{(n^3+4)2^{2n+1}}$

12. $\displaystyle\sum_{n=2}^{\infty} \frac{(-2)^n x^n}{n(\ln n)}$

13. $\displaystyle\sum_{n=0}^{\infty} \frac{\sqrt{n}(x-1)^n}{n+1}$

14. $\displaystyle\sum_{n=1}^{\infty} \frac{(x-2)^{2n}}{n \cdot 9^n}$

15. $\displaystyle\sum_{n=1}^{\infty} \frac{n^2(x+4)^n}{n!}$

16. $\displaystyle\sum_{n=0}^{\infty} n^2(x-3)^n$

17. $\displaystyle\sum_{n=1}^{\infty} n!(x+5)^{2n+1}$

18. $\displaystyle\sum_{n=1}^{\infty} \frac{(x+4)^{n+1}}{n^2 \cdot 3^n}$

19. $\displaystyle\sum_{n=1}^{\infty} \frac{(-2)^n(x+3)^{n+1}}{\sqrt{n}}$

20. $\displaystyle\sum_{n=1}^{\infty} \frac{3^n(x-2)^{2n+1}}{n!}$

21. $\displaystyle\sum_{n=1}^{\infty} \frac{(2x-4)^n}{n^{3/2} \cdot 3^n}$

22. $\displaystyle\sum_{n=1}^{\infty} \frac{(3x-12)^{2n}}{\sqrt{n}}$

23. $\displaystyle\sum_{n=0}^{\infty} \frac{(2x+6)^n}{(n+1)3^n}$

24. $\displaystyle\sum_{n=1}^{\infty} \frac{(2x-1)^n}{n^2 \cdot 2^{2n+1}}$

25. $\displaystyle\sum_{n=1}^{\infty} \frac{n!(3x+6)^n}{(n+2)!}$

26. $\displaystyle\sum_{n=0}^{\infty} \frac{(5x-10)^{2n}}{(n+2)4^n}$

27. $\displaystyle\sum_{n=0}^{\infty} (-1)^n \frac{\sqrt{n}(2x+4)^{2n+1}}{n+5}$

In Exercises 28–31, give a power series that has the given interval as interval of convergence, including or excluding endpoints as indicated. (Many answers are possible.)

28. $[0, 6]$

29. $1 < x < 4$

30. $-2 < x \le 4$

31. $-5 \le x < -1$

In Exercises 32–46, find a geometric power series centered at the indicated point x_0 that equals the given function in its interval of convergence.

32. $\dfrac{1}{1+x}$ at $x_0 = 0$

33. $\dfrac{1}{1-x^2}$ at $x_0 = 0$

34. $\dfrac{3}{1-2x}$ at $x_0 = 0$

35. $\dfrac{5}{3x-1}$ at $x_0 = 0$

36. $\dfrac{2x}{x^3-1}$ at $x_0 = 0$

37. $\dfrac{2}{4-x}$ at $x_0 = 0$

38. $\dfrac{3x}{9+x}$ at $x_0 = 0$

39. $\dfrac{4x^2}{2+3x}$ at $x_0 = 0$

40. $\dfrac{2x^3}{3-2x}$ at $x_0 = 0$

41. $\dfrac{1}{1+x}$ at $x_0 = 1$

42. $\dfrac{3}{2+x}$ at $x_0 = -1$

43. $\dfrac{2}{6+x}$ at $x_0 = 2$

44. $\dfrac{2}{3+2x}$ at $x_0 = -1$

45. $\dfrac{(x-3)^2}{4-2x}$ at $x_0 = 3$

46. $\dfrac{2(x+2)^3}{3x-5}$ at $x_0 = -2$

Exercises 47–56 deal with series of the form

$$\sum_{n=0}^{\infty} \frac{a_n}{(x-x_0)^n}.$$

47. Show that if $\sum_{n=0}^{\infty} a_n/x^n$ converges at $x = c$, then the series converges absolutely for all x such that $|x| > |c|$. [*Hint:* Let $u = 1/x$ and apply Theorem 10.11.]

48. Show that if $\sum_{n=0}^{\infty} a_n/x^n$ diverges at $x = d$, then it diverges for all x such that $|x| < |d|$. [*Hint:* Use Exercise 47.]

In Exercises 49–56, apply the absolute value ratio test and find all values of x for which the series converges, including "endpoints."

49. $\displaystyle\sum_{n=0}^{\infty} \frac{2^n}{x^n}$

50. $\displaystyle\sum_{n=0}^{\infty} \frac{3^n}{(n+1)x^n}$

51. $\displaystyle\sum_{n=1}^{\infty} \frac{n!}{x^n}$

52. $\displaystyle\sum_{n=1}^{\infty} \frac{5^n}{n!x^n}$

53. $\displaystyle\sum_{n=1}^{\infty} \frac{4^{n+3}}{n^2(x-2)^n}$

54. $\displaystyle\sum_{n=1}^{\infty} \frac{(-1)^n 5^{4n-1}}{n(x+3)^{2n}}$

55. $\displaystyle\sum_{n=1}^{\infty} \frac{16^n}{(3x-12)^{2n}}$

56. $\displaystyle\sum_{n=1}^{\infty} \frac{n!+4n}{(n+2)!(2x+6)^n}$

10.7   TAYLOR SERIES

Taylor Series

Taylor's theorem suggests that, for a function having derivatives of *all* orders in a neighborhood $x_0 - h < x < x_0 + h$ of x_0, we consider the power series

$$\sum_{n=0}^{\infty} \frac{f^{(n)}(x_0)}{n!}(x - x_0)^n$$

centered at x_0. This series is the **Taylor series of $f(x)$ at x_0.** (If $x_0 = 0$, the series is often called the **Maclaurin series of $f(x)$.**)

When we speak of "the Taylor series of f at x_0," we will understand that f has derivatives of all orders in some neighborhood of x_0. The Taylor series **represents $f(x)$ at x_1** if the series converges to $f(x_1)$ when $x = x_1$. The Taylor series of a function f at x_0 certainly represents f at x_0, because it converges at x_0 to its constant term, $f(x_0)$.

Unfortunately, it is not always true that the Taylor series represents f throughout the neighborhood of x_0 where the derivatives exist. In extreme cases, the series may represent f only at the point x_0 itself. One can show that the function f defined by

$$f(x) = \begin{cases} e^{-1/x^2} & \text{for } x \neq 0, \\ 0 & \text{for } x = 0 \end{cases}$$

has derivatives of all orders everywhere; in particular, one can show that $f^{(n)}(0) = 0$ for all n. The Taylor series of f at $x_0 = 0$ is therefore

$$0 + 0x + 0x^2 + 0x^3 + \cdots + 0x^n + \cdots,$$

which represents f only at 0.

A necessary and sufficient condition for the Taylor series of f at x_0 to represent f at a point $x_1 \neq x_0$ is readily obtained from Taylor's theorem.

THEOREM 10.12   Representation

Let f have derivatives of all orders in a neighborhood $x_0 - h < x < x_0 + k$ of x_0. The Taylor series of f at x_0 represents f at x_1 in this neighborhood if and only if $\lim_{n \to \infty} E_n(x_1) = 0$, where

$$E_n(x_1) = \frac{f^{(n+1)}(c)}{(n+1)!}(x_1 - x_0)^{n+1}$$

is the error term in Taylor's theorem (page 436).

Proof: For the Taylor series, the nth partial sum s_n at $x = x_1$ is given by

$$s_n(x_1) = T_n(x_1) = \sum_{i=0}^{n} \frac{f^{(i)}(x_0)}{i!}(x_1 - x_0)^i.$$

By Taylor's theorem, we have

$$f(x_1) - s_n(x_1) = f(x_1) - T_n(x_1) = E_n(x_1).$$

BROOK TAYLOR (1685–1731) published what is now called the Taylor series as proposition 7, corollary 2 of his 1715 text *Method of Increments, Direct and Inverse,* in which he gave new explanations of Newton's calculus and showed its usefulness in, among other problems, summing series, determining tangents and areas, finding the motion of a plucked string, and determining the refraction of light. Though the Taylor series had earlier appeared in some form in works of James Gregory (1638–1675), Newton, Leibniz, Johann Bernoulli, and Abraham de Moivre (1667–1754), Taylor was the first to call attention to its significance and its use in solving problems in analysis.

Taylor studied at St. John's College, Cambridge, and although he received law degrees in 1709 and 1714, it is doubtful that he ever practiced law. Instead, his mathematical talents gained him election as a fellow of the Royal Society in 1712. In the next six years he wrote several mathematical articles as well as two books. The last decade of his life, during which he concentrated largely on art, philosophy, and music, was marked by poor health and the emotional strain of the deaths of his first and second wives in childbirth.

Thus for any $\varepsilon > 0$, we have $|f(x_1) - s_n(x_1)| < \varepsilon$ if and only if $|E_n(x_1)| < \varepsilon$, so the conditions $\lim_{n \to \infty} s_n(x_1) = f(x_1)$ and $\lim_{n \to \infty} E_n(x_1) = 0$ are equivalent. $\qquad \square$

The following corollary of Theorem 10.12 will usually suffice for our purposes.

COROLLARY Simultaneously bounded derivatives

Let f have derivatives of all orders in $x_0 - h < x < x_0 + h$. If there is a number $B > 0$ such that $|f^{(n)}(x)| \le B$ for all positive integers n and for all x such that $x_0 - h < x < x_0 + h$, then the Taylor series of f at x_0 represents f throughout the neighborhood $x_0 - h < x < x_0 + h$.

Proof: By hypothesis, we have, for $x_0 - h < x_1 < x_0 + h$,

$$\lim_{n \to \infty} |E_n(x_1)| = \lim_{n \to \infty} \left| f^{(n+1)}(c) \frac{(x_1 - x_0)^{n+1}}{(n+1)!} \right|$$

$$\le B \left(\lim_{n \to \infty} \frac{|x_1 - x_0|^{n+1}}{(n+1)!} \right)$$

$$= B \cdot 0 = 0,$$

so, by Theorem 10.12, the Taylor series of f represents f at x_1. $\qquad \square$

EXAMPLE 1 Find a power series centered at $x_0 = 0$ representing e^x.

Solution All derivatives of the function e^x are again e^x, and the Taylor series of e^x at $x_0 = 0$ is readily found to be $\sum_{n=0}^{\infty} x^n/n!$. Since e^x is bounded by e^b in every interval $-b < x < b$, the corollary shows that the series converges to e^x for all x; that is,

$$e^x = 1 + x + \frac{x^2}{2!} + \frac{x^3}{3!} + \cdots + \frac{x^n}{n!} + \cdots \qquad (1)$$

for all x. Remember this series for e^x; it occurs frequently. $\qquad \blacksquare$

EXAMPLE 2 Find power series centered at $x_0 = 0$ representing $\sin x$ and $\cos x$.

Solution Derivatives of sine and cosine are again just sine and cosine functions, and these are bounded by 1 everywhere. The Taylor series for these functions at any x_0 therefore represent them everywhere. Computing the Taylor series at $x_0 = 0$, we find that

$$\sin x = x - \frac{x^3}{3!} + \frac{x^5}{5!} - \frac{x^7}{7!} + \cdots + (-1)^n \frac{x^{2n+1}}{(2n+1)!} + \cdots \qquad (2)$$

and

$$\cos x = 1 - \frac{x^2}{2!} + \frac{x^4}{4!} - \frac{x^6}{6!} + \cdots + (-1)^n \frac{x^{2n}}{(2n)!} + \cdots \qquad (3)$$

for all x. Remember these series also. $\qquad \blacksquare$

Differentiation and Integration of Power Series

We now turn to the study of functions that are represented throughout some neighborhood of x_0 by a power series at x_0. Such functions are very important and are called **analytic at x_0**. A function is **analytic** if it is analytic at each point in its domain.

EXAMPLE 3   Show that e^x, $\sin x$, and $\cos x$ are analytic.

Solution   Examples 1 and 2 show that these functions are analytic at $x_0 = 0$. The arguments in those examples concerning the bounds on their derivatives can be made in a neighborhood of any point. Thus e^x, $\sin x$, and $\cos x$ are analytic everywhere.   ∎

It can be shown that if f is analytic at x_0 and is represented by a power series throughout $x_0 - h < x < x_0 + h$, then f is analytic at every point in $x_0 - h < x < x_0 + h$. Thus, since the Taylor series for e^x, $\sin x$, and $\cos x$ at $x_0 = 0$ represent these functions for all x, we see again that these three functions are analytic at each point. Of course, the coefficients are different in two Taylor series representing one function like e^x if the series are centered at different points.

The calculus of analytic functions ("infinite polynomial functions") is much like the calculus of the polynomial functions. If f is analytic at x_0, then f has derivatives of all orders at x_0, and these derivatives can be computed by differentiating the series at x_0 representing f, just as we would differentiate a polynomial. Antiderivatives can be found similarly. The next theorem states this formally. A proof is left to a more advanced course.

THEOREM 10.13   Term-by-term differentiation and integration

Let f be analytic at x_0 and let $\sum_{n=0}^{\infty} a_n(x - x_0)^n$ represent f in a neighborhood $x_0 - h < x < x_0 + h$, so that

$$f(x) = \sum_{n=0}^{\infty} a_n(x - x_0)^n$$

in that neighborhood. Then f has derivatives of all orders throughout this neighborhood, and the derivatives at any x in the neighborhood may be computed by differentiating the series term by term. Thus

$$f'(x) = \sum_{n=1}^{\infty} n \cdot a_n(x - x_0)^{n-1}$$

for $x_0 - h < x < x_0 + h$. The indefinite integral of f in this same neighborhood can be obtained by integrating the series term by term:

$$\int f(x)\, dx = C + \sum_{n=0}^{\infty} \frac{a_n}{n + 1}(x - x_0)^{n+1},$$

where C is an arbitrary constant.   □

EXAMPLE 4 Illustrate term-by-term differentiation for the series for e^x at $x_0 = 0$.

Solution By Eq. (1), we have

$$e^x = 1 + x + \frac{x^2}{2!} + \frac{x^3}{3!} + \cdots + \frac{x^n}{n!} + \cdots .$$

The derivative of this series is

$$0 + 1 + \frac{2x}{2!} + \frac{3x^2}{3!} + \cdots + \frac{nx^{n-1}}{n!} + \cdots$$

$$= 1 + x + \frac{x^2}{2!} + \cdots + \frac{x^{n-1}}{(n-1)!} + \cdots ,$$

which is the same series and again represents e^x. We expect this since the derivative of e^x is again e^x. ∎

EXAMPLE 5 Illustrate Theorem 10.13 by differentiating and integrating the series representing $1/(1-x)$ near $x_0 = 0$.

Solution The function $1/(1-x)$ is analytic at zero, for we know that the geometric series

$$\sum_{n=0}^{\infty} x^n = 1 + x + x^2 + x^3 + \cdots + x^n + \cdots$$

converges to $1/(1-x)$ in $-1 < x < 1$. Thus

$$\frac{d}{dx}\left(\frac{1}{1-x}\right) = \frac{1}{(1-x)^2}$$

is analytic at zero, and we obtain by differentiation

$$\frac{1}{(1-x)^2} = \sum_{n=1}^{\infty} nx^{n-1}$$

$$= 1 + 2x + 3x^2 + 4x^3 + \cdots + nx^{n-1} + \cdots$$

for $-1 < x < 1$.

Since $\int 1/(1-x)\,dx = -\ln|1-x| + C$, we find by integrating the geometric series that

$$-\ln(1-x) = k + \sum_{n=0}^{\infty} \frac{x^{n+1}}{n+1}$$

$$= k + x + \frac{x^2}{2} + \frac{x^3}{3} + \cdots + \frac{x^n}{n} + \cdots$$

for a constant k and $-1 < x < 1$. Putting $x = 0$, we have $k = -\ln 1 = 0$, so

$$\ln(1-x) = -x - \frac{x^2}{2} - \frac{x^3}{3} - \cdots - \frac{x^n}{n} - \cdots \qquad (4)$$

for $-1 < x < 1$. ∎

Uniqueness of Power Series Representation

It is an important fact that if f is analytic at x_0, then there is *only one* power series at x_0 that represents f throughout a neighborhood of x_0. This must be the Taylor series, for the coefficients of a power series representing the function are determined by the derivatives of the function to be precisely the coefficients in the Taylor series, as indicated on page 429.

Using this uniqueness, we can find the Taylor series for many functions f without differentiating f to compute coefficients. We illustrate with examples.

EXAMPLE 6   Find the power series at $x_0 = 0$ representing $\sin x^2$.

Solution   We know that

$$\sin x = x - \frac{x^3}{3!} + \cdots + (-1)^n \frac{x^{2n+1}}{(2n+1)!} + \cdots$$

for all x. Replacing x by x^2, we have

$$\sin x^2 = x^2 - \frac{x^6}{3!} + \cdots + (-1)^n \frac{x^{4n+2}}{(2n+1)!} + \cdots \tag{5}$$

for all x. Therefore the series (5) must be the Taylor series for $\sin x^2$ at $x_0 = 0$. If we try to find the Taylor series by differentiating $\sin x^2$ repeatedly, we appreciate the quick way we obtained series (5).   ■

EXAMPLE 7   If $f(x) = \sin x^2$, find $f^{(6)}(0)$.

Solution   The power series for $\sin x^2$ at $x_0 = 0$ is given in series (5). We know this series must be the Taylor series, so the coefficient of x^6 must be the Taylor coefficient $f^{(6)}(0)/6!$. From series (5), we then obtain

$$\frac{f^{(6)}(0)}{6!} = -\frac{1}{3!},$$

so

$$f^{(6)}(0) = -\frac{1}{3!} \cdot 6! = -6 \cdot 5 \cdot 4 = -120.$$ ■

EXAMPLE 8   Find the Taylor series for $\ln(1 - x)$ at $x_0 = 0$.

Solution   In Eq. (4), we obtained a power series centered at $x_0 = 0$ that represented $\ln(1 - x)$ for $-1 < x < 1$. By uniqueness, this must be the Taylor series.   ■

EXAMPLE 9   Find the Taylor series for $\tan^{-1} x$ at $x_0 = 0$.

Solution   Now $d(\tan^{-1} x)/dx = 1/(1 + x^2)$. We can represent $1/(1 + x^2)$ as the sum of a geometric series with initial term 1 and ratio $-x^2$:

$$\frac{1}{1 + x^2} = 1 - x^2 + x^4 - x^6 + \cdots + (-1)^n x^{2n} + \cdots \tag{6}$$

for $-1 < x < 1$. Integrating Eq. (6), we find that

$$\tan^{-1}x = k + x - \frac{x^3}{3} + \frac{x^5}{5} - \frac{x^7}{7} + \cdots + (-1)^n \frac{x^{2n+1}}{2n+1} + \cdots$$

for some constant k. Putting $x = 0$, we see that $k = \tan^{-1}0 = 0$, so

$$\tan^{-1}x = x - \frac{x^3}{3} + \frac{x^5}{5} - \frac{x^7}{7} + \cdots + (-1)^n \frac{x^{2n+1}}{2n+1} + \cdots \qquad (7)$$

for $-1 < x < 1$. The series (7) must be the Taylor series of $\tan^{-1}x$ at $x_0 = 0$. If we try to compute the series (7) by differentiating $\tan^{-1}x$ repeatedly, we appreciate the quick way we obtained the series (7). ∎

EXAMPLE 10 Let $f(x) = \tan^{-1}(x^2)$. Find $f^{(10)}(0)$.

Solution From the series (7), we see that the Taylor series for $\tan^{-1}(x^2)$ must start

$$\tan^{-1}(x^2) = x^2 - \frac{x^6}{3} + \frac{x^{10}}{5} - \frac{x^{14}}{7} + \cdots.$$

Using the formula for the Taylor coefficient of x^{10}, we obtain

$$\frac{f^{(10)}(0)}{10!} = \frac{1}{5},$$

so

$$f^{(10)}(0) = \frac{10!}{5}. \qquad ∎$$

We take a moment to give a preview of coming attractions, if you ever study *functions of complex variables*. The function $1/(1-x)$ can be expressed as the sum of a geometric power series near $x_0 = 0$ by

$$\frac{1}{1-x} = 1 + x + x^2 + x^3 + \cdots + x^n + \cdots. \qquad (8)$$

This series converges only for $-1 < x < 1$, but we are not surprised because it represents the function $1/(1-x)$, which "blows up" at $x = 1$. The function $1/(1 + x^2)$ has derivatives of all orders everywhere and can be shown to be analytic at every point x_0. However, its Taylor series (6) at $x_0 = 0$ still only converges to $1/(1 + x^2)$ for $-1 < x < 1$. To fully appreciate why this happens, one must study complex analysis. The function $1/(1 + x^2)$ "blows up" at $x = i$ and $x = -i$, and in complex analysis one sees that the numbers i and $-i$ have distance 1 from $x_0 = 0$. It is for this reason that the radius of convergence of the Taylor series for $1/(1 + x^2)$ at $x_0 = 0$ is only 1.

Multiplication and Division of Power Series

Two power series at x_0 representing functions f and g in $x_0 - h < x < x_0 + h$ can be multiplied and divided as "infinite" polynomials, to yield power series representing the functions fg and f/g in neighborhoods of x_0, with the obvious restriction that $g(x_0) \neq 0$. We state this as a theorem without proof. The functions f and g can be approximated at each point of $x_0 - h < x < x_0 + h$ as closely as we like by polynomial partial sums of the series, so the theorem seems reasonable.

THEOREM 10.14   Series multiplication and division

Let series $\sum_{n=0}^{\infty} a_n(x - x_0)^n$ and $\sum_{n=0}^{\infty} b_n(x - x_0)^n$ converge to functions f and g, respectively, in $x_0 - h < x < x_0 + h$. Then the product series (called the **Cauchy product**)

$$a_0 b_0 + (a_0 b_1 + a_1 b_0)(x - x_0) + (a_0 b_2 + a_1 b_1 + a_2 b_0)(x - x_0)^2 + \cdots,$$

whose nth coefficient is $\sum_{i=0}^{n} (a_i b_{n-i})$, represents fg throughout $x_0 - h < x < x_0 + h$. Also, if $b_0 = g(x_0) \neq 0$, the series

$$\frac{a_0}{b_0} + \frac{a_1 b_0 - a_0 b_1}{b_0{}^2}(x - x_0) + \cdots,$$

obtained by long division, represents f/g in some neighborhood $x_0 - \delta < x < x_0 + \delta$. □

We are accustomed to polynomial long division, where we write the polynomials with the terms of highest degree first and divide only until we obtain a remainder of lower degree than the divisor. *In series long division, we write the terms of lowest degree first and the division may never terminate;* symbolically,

$$
\require{enclose}
\begin{array}{r}
\frac{a_0}{b_0} + \frac{a_1 b_0 - a_0 b_1}{b_0{}^2}(x - x_0) + \cdots \\[2ex]
b_0 + b_1(x - x_0) + \cdots \enclose{longdiv}{\; a_0 \qquad\qquad a_1(x - x_0) + \cdots} \\[2ex]
a_0 \qquad \frac{a_0 b_1}{b_0}(x - x_0) + \cdots \\[2ex]
\hline
0 + \frac{a_1 b_0 - a_0 b_1}{b_0}(x - x_0) + \cdots
\end{array}
$$

etc.

An illustration of such division appears in Example 12.

EXAMPLE 11 Use series multiplication to find the first few terms of the Taylor series for $e^x \sin x$ at $x_0 = 0$.

Solution Since

$$e^x = 1 + x + \frac{x^2}{2} + \frac{x^3}{6} + \frac{x^4}{24} + \frac{x^5}{120} + \cdots$$

and

$$\sin x = x - \frac{x^3}{6} + \frac{x^5}{120} - \cdots,$$

we obtain

$$e^x \sin x = x + x^2 + \left(\frac{x^3}{2} - \frac{x^3}{6}\right) + \left(\frac{x^4}{6} - \frac{x^4}{6}\right)$$

$$+ \left(\frac{x^5}{24} - \frac{x^5}{12} + \frac{x^5}{120}\right) + \cdots$$

$$= x + x^2 + \frac{x^3}{3} - \frac{x^5}{30} + \cdots$$

for all x.

EXAMPLE 12 Find the Taylor series for $(1 + x^2)/(1 - x)$ at $x_0 = 0$ in two ways.

Solution The computation using series division is

$$
\begin{array}{r}
1 + x + 2x^2 + 2x^3 + \cdots \\
1 - x \overline{\smash{\big)}\ 1 \quad\ \ + x^2} \\
\underline{1 - x} \\
x + x^2 \\
\underline{x - x^2} \\
2x^2 \\
2x^2 - 2x^3 \\
\underline{\ 2x^3} \\
2x^3 - 2x^4 \\
\underline{\ 2x^4}
\end{array}
$$

etc.

We thus obtain

$$\frac{1 + x^2}{1 - x} = 1 + x + 2x^2 + 2x^3 + 2x^4 + \cdots + 2x^n + \cdots$$

for $-1 < x < 1$. Or we could multiply by $1 + x^2$ the geometric series

$$1 + x + x^2 + \cdots + x^n + \cdots$$

for $1/(1 - x)$. The computation is

$$
\begin{array}{l}
1 + x + x^2 + x^3 + \cdots + x^n + x^{n+1} + \cdots \\
\qquad\qquad\qquad\qquad\qquad\qquad\quad \times\ 1 + x^2 \\
\hline
1 + x + x^2 + x^3 + \cdots + x^n + x^{n+1} + \cdots \\
\qquad\quad\ x^2 + x^3 + \cdots + x^n + x^{n+1} + \cdots \\
\hline
1 + x + 2x^2 + 2x^3 + \cdots + 2x^n + 2x^{n+1} + \cdots
\end{array}
$$

and we obtain the same series, as we must by uniqueness. If we try to compute the Taylor series for $(1 + x^2)/(1 - x)$ at $x_0 = 0$ by repeated differentiation, we appreciate the quick ways we found it here.

Sometimes we can recognize an elementary function that a given series represents.

EXAMPLE 13   Find the elementary function represented by the series

$$10x + 2x^2 - \frac{x^3}{3!} + \frac{x^5}{5!} - \frac{x^7}{7!} + \cdots + (-1)^n \frac{x^{2n+1}}{(2n+1)!} + \cdots \text{ for } n \geq 1.$$

Solution   If the first two terms, $10x + 2x^2$, of the series were replaced by x, we would have the series for $\sin x$. Since

$$10x + 2x^2 = x + (9x + 2x^2),$$

we see that the given series represents $9x + 2x^2 + \sin x$.   ■

EXAMPLE 14   Find the elementary function represented by the series

$$3 + x - \frac{x^3}{3} + \frac{x^5}{5} - \frac{x^7}{7} + \cdots + (-1)^n \frac{x^{2n+1}}{2n+1} + \cdots$$

for $n \geq 1$ and $|x| < 1$.

Solution   We can get rid of the troublesome denominators by differentiating. Let the given series represent $f(x)$. Then

$$f'(x) = 1 - x^2 + x^4 - x^6 + \cdots + (-1)^n x^{2n} + \cdots.$$

We recognize this series as geometric with initial term 1 and ratio $-x^2$. Thus

$$f'(x) = \frac{1}{1 - (-x^2)} = \frac{1}{1 + x^2}.$$

Integrating, we find that $f(x) = k + \tan^{-1}x$ for some k and $|x| < 1$. From our original series for $f(x)$, we see that $f(0) = 3$, so $3 = k + \tan^{-1}(0) = k$. Thus the series represents $3 + \tan^{-1}x$ for $|x| < 1$.   ■

SUMMARY

1. If $f(x)$ has derivatives of all orders at x_0, then

$$\sum_{n=0}^{\infty} \frac{f^{(n)}(x_0)}{n!}(x - x_0)^n$$

is the Taylor series of $f(x)$ at x_0. (Here $0! = 1$ and $f^{(0)}(x) = f(x)$.)

2. The Taylor series of $f(x)$ at x_0 represents $f(x)$ at x_1 if and only if $\lim_{n \to \infty} E_n(x_1) = 0$. This condition will always hold if all derivatives between x_0 and x_1 are bounded by the same constant B.

3. A function is analytic at x_0 if it is represented by some power series in some neighborhood of x_0. It is analytic if it is analytic at each point in its domain.

4. If $f(x)$ is represented by a power series in an open interval, then $f'(x)$ is represented in that interval by the term-by-term derivative of that power series, and $\int f(x)\,dx$ by the term-by-term antiderivative of that power series, plus an arbitrary constant.

5. The only power series at x_0 that can represent $f(x)$ in a neighborhood of x_0 is the Taylor series of $f(x)$.

6. Series at x_0 representing $f(x)$ and $g(x)$ in a common neighborhood of x_0 can be multiplied (as infinite polynomials) to obtain the series representing $f(x)g(x)$ in that neighborhood and divided if $g(x_0) \neq 0$ to represent $f(x)/g(x)$ in *some* neighborhood of x_0.

7. It is convenient to remember the following series.

$$e^x = 1 + x + \frac{x^2}{2!} + \frac{x^3}{3!} + \cdots + \frac{x^n}{n!} + \cdots \qquad \text{for all } x$$

$$\sin x = x - \frac{x^3}{3!} + \frac{x^5}{5!} - \cdots + (-1)^n \frac{x^{2n+1}}{(2n+1)!} + \cdots \qquad \text{for all } x$$

$$\cos x = 1 - \frac{x^2}{2!} + \frac{x^4}{4!} - \cdots + (-1)^n \frac{x^{2n}}{(2n)!} + \cdots \qquad \text{for all } x$$

$$\frac{1}{1-x} = 1 + x + x^2 + x^3 + \cdots + x^n + \cdots \qquad \text{for } -1 < x < 1$$

$$\ln(1-x) = -x - \frac{x^2}{2} - \frac{x^3}{3} - \cdots - \frac{x^n}{n} - \cdots \qquad \text{for } -1 < x < 1$$

EXERCISES 10.7

1. Mark each of the following true or false.
   _____ (a) If f has derivatives of all orders throughout some neighborhood of x_0, then f is analytic at x_0.
   _____ (b) If f is analytic at x_0, then f has derivatives of all orders throughout some neighborhood of x_0.
   _____ (c) If f and g are analytic at x_0, then $f + g$ is analytic at x_0.
   _____ (d) Every power series represents a function that is analytic at every point (except possibly endpoints) of the interval of convergence of the series.
   _____ (e) There is at most one power series at x_0 that represents a given function f at x_0.
   _____ (f) There is at most one power series at x_0 that represents a given function f throughout some neighborhood of x_0.

2. Is the function $\sqrt{x}$ analytic at $x_0 = 0$? Why?

In Exercises 3–24, find as many terms of the Taylor series of the function at the given point as you conveniently can in the easiest way you can.

3. $x^2 + e^x$ at $x_0 = 0$

4. $1 + x^3 - \sin x$ at $x_0 = 0$

5. $x \sin x$ at $x_0 = 0$

6. $\cos x$ at $x_0 = \pi$

7. $\dfrac{x}{1-x}$ at $x_0 = 0$

8. e^{-x^2} at $x_0 = 0$

9. $\cos x^3$ at $x_0 = 0$

10. $\dfrac{2x + 3x^2}{1 + 4x}$ at $x_0 = 0$

11. $e^x \cos x$ at $x_0 = 0$

12. e^{3x} at $x_0 = 0$

13. $\dfrac{1}{(1+x)^2}$ at $x_0 = 0$

14. $\ln(\cos x)$ at $x_0 = 0$

15. $\sec x$ at $x_0 = 0$

16. $\ln x$ at $x_0 = 2$

17. $\dfrac{e^x}{1-x}$ at $x_0 = 0$

18. $\dfrac{e^x + e^{-x}}{2}$ at $x_0 = 0$

19. $\dfrac{e^x - e^{-x}}{2}$ at $x_0 = 0$

20. $\sqrt{x}$ at $x_0 = 1$

21. $\dfrac{1}{e^x}$ at $x_0 = 0$

22. $\dfrac{x-1}{x^3}$ at $x_0 = 1$

23. $\sec x \tan x$ at $x_0 = 0$

24. $\dfrac{1}{2-x}$ at $x_0 = 0$

In Exercises 25–32, use series methods to find the indicated derivative of the given function.

25. $f(x) = \dfrac{x^2}{1 + x^2}$, find $f^{(4)}(0)$

26. $f(x) = \sin(3x^2)$, find $f^{(6)}(0)$

27. $f(x) = e^{-x^2/4}$, find $f^{(4)}(0)$

28. $f(x) = \tan^{-1}\left(\dfrac{x^3}{2}\right)$, find $f^{(3)}(0)$

29. $f(x) = \tan^{-1}\left(\dfrac{x^2}{4}\right)$, find $f^{(10)}(0)$

30. $f(x) = \ln(1 - 2x^2)$, find $f^{(6)}(0)$

31. $f(x) = x^4\sin 2x$, find $f^{(5)}(0)$

32. $f(x) = \dfrac{\tan^{-1}x^2}{1 - x^2}$, find $f^{(4)}(0)$

33. (a) Find the terms for $n \leq 5$ of the Taylor series of $\sin x \cos x$ at $x_0 = 0$ by series multiplication.
(b) Find the Taylor series of $\sin x \cos x$ at $x_0 = 0$ by the use of the identity $\sin x \cos x = (\sin 2x)/2$.

34. Find the Taylor series of $1/x$ at $x_0 = 2$ by
(a) differentiating $1/x$ repeatedly to compute the coefficients,
(b) using the identity

$$\frac{1}{x} = \frac{1}{2} \cdot \frac{1}{1 - [-(x - 2)/2]}$$

and expanding in a geometric series.

35. Use the technique suggested by Exercise 34(b) to find the Taylor series of $1/x$ at $x = -1$; you have to find the appropriate identity.

36. Find the terms for $n \leq 3$ of the Taylor series of $\tan x$ at $x_0 = 0$ by dividing the series for $\sin x$ by the series for $\cos x$.

37. (a) Obtain the series expansion

$$\ln(1 + x) = x - \frac{x^2}{2} + \cdots + (-1)^{n+1}\frac{x^n}{n} + \cdots$$

for $-1 < x < 1$. [*Hint:* Use Eq. (4) or integrate the geometric series for $1/(1 + x) = 1/(1 - (-x))$.]
(b) Show that the series in part (a) converges for $x = 1$.
(c) Show that the alternating harmonic series

$$1 - \frac{1}{2} + \frac{1}{3} - \frac{1}{4} + \cdots$$

converges to $\ln 2$. [*Hint:* Since 1 is an endpoint of the interval of convergence of the series in part (a), Theorem 10.13 cannot be used. You must check $\lim_{n\to\infty} E_n(1)$ for the function $\ln(1 + x)$.]

38. Find the series at $x_0 = 0$ representing the function f defined by

$$f(x) = \int_0^x \ln(1 - t)\, dt$$

for $-1 < x < 1$.

39. Find the series expansion at $x_0 = 0$ for the indefinite integral of e^{x^2}.

40. Find the series at $x_0 = 0$ representing the function f defined by

$$f(x) = \int_0^x [1/(1 - t^3)]\, dt$$

for $-1 < x < 1$.

41. Find the series at $x_0 = 0$ representing the function f defined by $f(x) = \pi + \int_0^x \cos t^2\, dt$ for all x.

42. Find the series at $x_0 = 0$ representing the function f defined by

$$f(x) = \int_0^x \frac{(3 + t^2)}{(1 + 2t)}\, dt$$

for $-\frac{1}{2} < x < \frac{1}{2}$.

43. (a) Proceeding purely formally, find the Taylor series at $x_0 = 0$ of e^{ix} and of e^{-ix}, where $i^2 = -1$.
(b) From part (a), "derive" Euler's formula $e^{ix} = \cos x + i(\sin x)$.
(c) From part (a), "derive" the formula $e^{-ix} = \cos x - i(\sin x)$.
(d) From parts (b) and (c), find formulas for $\cos x$ and $\sin x$ in terms of the complex exponential function.
(e) Compare the formulas for $\sin x$ and $\cos x$ found in part (d) with the formulas for $\sinh x$ and $\cosh x$ in terms of the exponential function.

Exercises 44–55 give you practice in series recognition. The given series represents a familiar elementary function in its interval of convergence. Find the function.

44. $1 - x + x^2 - x^3 + \cdots + (-1)^n x^n + \cdots$

45. $1 + x + x^2 - x^3 + x^4 - x^5 + \cdots + (-1)^n x^n + \cdots$
for $n \geq 2$

46. $1 - x + \dfrac{x^2}{2!} - \dfrac{x^3}{3!} + \dfrac{x^4}{4!} - \cdots + (-1)^n\dfrac{x^n}{n!} + \cdots$

47. $1 + 3x - \dfrac{x^2}{2!} + \dfrac{x^4}{4!} - \cdots + (-1)^n\dfrac{x^{2n}}{(2n)!} + \cdots$ for $n \geq 1$

48. $1 - \dfrac{x}{2} + \dfrac{x^2}{4} - \dfrac{x^3}{8} + \cdots + \dfrac{(-1)^n}{2^n}x^n + \cdots$

49. $1 + x - \dfrac{x^2}{2!} - \dfrac{x^3}{3!} + \dfrac{x^4}{4!} + \dfrac{x^5}{5!} - \dfrac{x^6}{6!} - \dfrac{x^7}{7!} + \cdots$

50. $1 - x^3 + x^6 - x^9 + x^{12} - \cdots + (-1)^n x^{3n} + \cdots$

51. $1 + 2x + \dfrac{2! + 1}{2!}x^2 + \dfrac{3! + 1}{3!}x^3 + \cdots + \dfrac{n! + 1}{n!}x^n + \cdots$
for $n \geq 2$

52. $-1 + 2x - 3x^2 + 4x^3 - \cdots + (-1)^{n+1}(n + 1)x^n + \cdots$
[*Hint:* Integrate the series.]

53. $2 + 3 \cdot 2x + 4 \cdot 3x^2 + 5 \cdot 4x^3 + \cdots$
$+ (n + 2)(n + 1)x^n + \cdots$

54. $x^3 - \dfrac{x^7}{5!} + \dfrac{x^9}{7!} - \cdots + (-1)^n\dfrac{x^{2n+1}}{(2n - 1)!} + \cdots$ for $n \geq 3$

55. $4x^4 - 8x^6 + 16x^8 - 32x^{10} + \cdots$
$+ (-1)^{n+1}2^{n+1}x^{2n+2} + \cdots$ for $n \geq 2$

10.8 BINOMIAL SERIES; COMPUTATIONS

In this final section on series, we present another important class of series, binomial series. We then show how series can be used to compute definite integrals like

$$\int_0^1 \sin x^2 \, dx,$$

where the corresponding indefinite integral cannot be expressed in terms of elementary functions.

Binomial Series

The binomial theorem states that for any numbers a and b and any positive integer n, we have

$$(a + b)^n = a^n + na^{n-1}b + \frac{n(n-1)}{2!}a^{n-2}b^2 + \cdots$$

$$+ \frac{n(n-1)\cdots(n-k+1)}{k!}a^{n-k}b^k + \cdots + b^n. \tag{1}$$

EXAMPLE 1 Expand $(a + b)^4$ by the binomial theorem.

Solution We have

$$(a + b)^4 = a^4 + 4a^3b + \frac{4 \cdot 3}{2 \cdot 1}a^2b^2 + \frac{4 \cdot 3 \cdot 2}{3 \cdot 2 \cdot 1}ab^3 + \frac{4 \cdot 3 \cdot 2 \cdot 1}{4 \cdot 3 \cdot 2 \cdot 1}b^4$$

$$= a^4 + 4a^3b + 6a^2b^2 + 4ab^3 + b^4. \qquad \blacksquare$$

We can write Eq. (1) in the form

$$(a + b)^n = \sum_{k=0}^{n} \binom{n}{k} a^{n-k}b^k, \tag{2}$$

where we define the *binomial coefficients* $\binom{n}{k}$ by

$$\binom{n}{k} = \begin{cases} \dfrac{n(n-1)\cdots(n-k+1)}{k!} & \text{for } k > 0, \\ 1 & \text{for } k = 0. \end{cases} \tag{3}$$

If we let $a = 1$ and $b = x$ in Eq. (2), we obtain

$$(1 + x)^n = \sum_{k=0}^{n} \binom{n}{k} x^k. \tag{4}$$

EXAMPLE 2 Expand $(1 + x)^5$.

Solution We have

$$(1 + x)^5 = 1 + 5x + \frac{5 \cdot 4}{2 \cdot 1}x^2 + \frac{5 \cdot 4 \cdot 3}{3 \cdot 2 \cdot 1}x^3 + \frac{5 \cdot 4 \cdot 3 \cdot 2}{4 \cdot 3 \cdot 2 \cdot 1}x^4$$

$$+ \frac{5 \cdot 4 \cdot 3 \cdot 2 \cdot 1}{5 \cdot 4 \cdot 3 \cdot 2 \cdot 1}x^5$$

$$= 1 + 5x + 10x^2 + 10x^3 + 5x^4 + x^5. \qquad \blacksquare$$

The sum in Eq. (4) must (by uniqueness) be the Taylor series for $(1 + x)^n$ at $x_0 = 0$. We can also verify this directly by differentiation, readily finding that

$$D^k(1 + x)^n = n(n - 1) \cdots (n - k + 1)(1 + x)^{n-k},$$

so

$$D^k(1 + x)^n \big|_{x=0} = n(n - 1) \cdots (n - k + 1).$$

Since n is a positive *integer*, $(1 + x)^n$ is a polynomial of degree n, so $D^k(1 + x)^n = 0$ for $k > n$. For n a positive integer, we thus expect $\binom{n}{k} = 0$ for $k > n$, and this is readily seen from Eq. (3); namely,

$$\binom{n}{k} = \frac{n(n - 1) \cdots (n - n) \cdots (n - k + 1)}{k!}$$

$$= \frac{0}{k!} = 0$$

for $k > n$.

We generalize Eq. (3), and define the **binomial coefficient** $\binom{p}{k}$ for any real number p and integer $k \geq 0$ to be given by

$$\binom{p}{k} = \begin{cases} \dfrac{p(p - 1) \cdots (p - k + 1)}{k!} & \text{for } k > 0, \\ 1 & \text{for } k = 0. \end{cases}$$

EXAMPLE 3 Find the binomial coefficient

$$\binom{3/2}{5}.$$

Solution We have

$$\binom{3/2}{5} = \frac{\dfrac{3}{2} \cdot \dfrac{1}{2} \cdot -\dfrac{1}{2} \cdot -\dfrac{3}{2} \cdot -\dfrac{5}{2}}{5 \cdot 4 \cdot 3 \cdot 2 \cdot 1}$$

$$= \frac{\dfrac{-3}{2^5}}{4 \cdot 2 \cdot 1} = \frac{-3}{8 \cdot 32} = \frac{-3}{256}. \qquad \blacksquare$$

Note that if p is not a nonnegative integer, then no factor in the numerator of $\binom{p}{k}$ is zero, even if $k > p$. We are led to consider the series analogue of the sum (4) for any real number p.

DEFINITION 10.9

Binomial series

For any p, the **binomial series for** $(1 + x)^p$ is

$$\sum_{k=0}^{\infty} \binom{p}{k} x^k = 1 + px + \frac{p(p - 1)}{2!} x^2 + \cdots$$

$$+ \frac{p(p - 1) \cdots (p - k + 1)}{k!} x^k + \cdots . \qquad (5)$$

(We use k rather than n for summation index to avoid confusion with the use of n in Eq. 1.)

It is easily checked that the series (5) is the Taylor series for $(1 + x)^p$ at $x_0 = 0$; we find that

$$D^k(1 + x)^p \big|_{x=0} = p(p - 1) \cdots (p - k + 1)(1 + x)^{p-k} \big|_{x=0}$$
$$= p(p - 1) \cdots (p - k + 1).$$

We would like to find the radius of convergence of the binomial series (5) and to determine whether the series does represent $(1 + x)^p$ in a neighborhood of zero. Of course, if p is a nonnegative integer, then the series contains only a finite number of terms with nonzero coefficients, has radius of convergence ∞, and represents $(1 + x)^p$ by sum (4). We now suppose that p is not a nonnegative integer and compute the ratio

$$\frac{\binom{p}{k + 1} x^{k+1}}{\binom{p}{k} x^k} = \frac{[(p(p - 1) \cdots (p - k))/(k + 1)!] x^{k+1}}{[(p(p - 1) \cdots (p - k + 1))/k!] x^k} = \frac{p - k}{k + 1} x.$$

We obtain

$$\lim_{k \to \infty} \left| \frac{\binom{p}{k + 1} x^{k+1}}{\binom{p}{k} x^k} \right| = \lim_{k \to \infty} \left| \frac{p - k}{k + 1} x \right| = |-x| = |x|.$$

The radius of convergence of the binomial series for $(1 + x)^p$, where p is not a nonnegative integer, is therefore 1; the series converges if $|x| < 1$ or if $-1 < x < 1$.

To show that the series (5) represents $(1 + x)^p$ for $-1 < x < 1$, we could try to show that $\lim_{k \to \infty} E_k(x_1) = 0$ for $-1 < x_1 < 1$, but the following argument is less tedious. We set

$$f(x) = \sum_{k=0}^{\infty} \binom{p}{k} x^k \qquad (6)$$

for $-1 < x < 1$. Then we may differentiate term by term to obtain

$$f'(x) = \sum_{k=0}^{\infty} \binom{p}{k} k x^{k-1} = \sum_{k=1}^{\infty} \frac{p(p - 1) \cdots (p - k + 1)}{(k - 1)!} x^{k-1} \qquad (7)$$

From the series (6) and (7), it is straightforward to verify that

$$pf(x) = (1 + x)f'(x) \tag{8}$$

(see Exercise 49). From Eq. (8), we obtain

$$\int \frac{f'(x)}{f(x)}\,dx = \int \frac{p}{1+x}\,dx, \tag{9}$$

or, taking the antiderivatives,

$$\ln|f(x)| = p\,\ln|1 + x| + C = \ln|1 + x|^p + C \tag{10}$$

for $-1 < x < 1$. From the series (6), we see that $f(0) = 1$, so putting $x = 0$ in Eq. (10), we obtain

$$0 = \ln 1 = \ln(1 + 0)^p + C = \ln 1 + C = C.$$

Thus $C = 0$ and $\ln|f(x)| = \ln|1 + x|^p$. Therefore we have $|f(x)| = |1 + x|^p$, so $f(x) = \pm(1 + x)^p$. However, $f(0) = 1$, so the positive sign is appropriate, and

$$f(x) = (1 + x)^p$$

for $-1 < x < 1$, which is what we wished to show. We summarize these results in a theorem.

THEOREM 10.15   Binomial representation

If p is not a nonnegative integer, then the binomial series $\sum_{k=0}^{\infty} \binom{p}{k} x^k$ has radius of convergence 1 and represents $(1 + x)^p$ for $-1 < x < 1$. For a nonnegative integer n, the (finite) binomial series $\sum_{k=0}^{n} \binom{n}{k} x^k$ converges for all x to $(1 + x)^n$.

$\square$

EXAMPLE 4   Find the binomial series for $(1 + x)^{1/2}$ and use it to estimate $\sqrt{\frac{3}{2}}$.

Solution   We have

$$(1 + x)^{1/2} = 1 + \frac{1}{2}x + \frac{\frac{1}{2}\cdot -\frac{1}{2}}{2!}x^2 + \frac{\frac{1}{2}\cdot -\frac{1}{2}\cdot -\frac{3}{2}}{3!}x^3 + \cdots$$

$$= 1 + \frac{1}{2}x - \frac{1}{2^2 \cdot 2!}x^2 + \frac{3}{2^3 \cdot 3!}x^3 - \cdots$$

$$+ (-1)^{k-1}\frac{3 \cdot 5 \cdot 7 \cdots (2k - 3)}{2^k \cdot k!}x^k + \cdots,$$

for $k \geq 1$ and for $-1 < x < 1$. Putting $x = \frac{1}{2}$ and using the terms of exponent less than or equal to 3, we obtain the estimate

$$\sqrt{\frac{3}{2}} \approx 1 + \frac{1}{4} - \frac{1}{32} + \frac{1}{128} \approx 1.2266.$$

When $x = \frac{1}{2}$, our series is alternating with terms of decreasing size, so the error in our estimate is less than the next term $5/2^{11} \approx 0.0024$. The actual value of $\sqrt{\frac{3}{2}}$ to six decimal places is 1.224745. ∎

EXAMPLE 5 Find the terms of degree less than or equal to 9 of the series for $\sin^{-1}x$ at $x_0 = 0$.

Solution Differentiating and using the binomial series, we have

$$\frac{d(\sin^{-1}x)}{dx} = \frac{1}{\sqrt{1 - x^2}} = (1 - x^2)^{-1/2}$$

$$= 1 - \frac{1}{2}(-x^2) + \frac{-\frac{1}{2} \cdot -\frac{3}{2}}{2 \cdot 1}(-x^2)^2$$

$$+ \frac{-\frac{1}{2} \cdot -\frac{3}{2} \cdot -\frac{5}{2}}{3 \cdot 2 \cdot 1}(-x^2)^3$$

$$+ \frac{-\frac{1}{2} \cdot -\frac{3}{2} \cdot -\frac{5}{2} \cdot -\frac{7}{2}}{4 \cdot 3 \cdot 2 \cdot 1}(-x^2)^4 + \cdots$$

$$= 1 + \frac{1}{2}x^2 + \frac{3}{8}x^4 + \frac{5}{16}x^6 + \frac{35}{128}x^8 + \cdots$$

for $-1 < x < 1$. Integrating, we find that

$$\sin^{-1}x = k + x + \frac{1}{6}x^3 + \frac{3}{40}x^5 + \frac{5}{112}x^7 + \frac{35}{1152}x^9 + \cdots$$

for some constant k. Since $\sin^{-1}0 = 0$, we see that $k = 0$, so

$$\sin^{-1}x = x + \frac{1}{6}x^3 + \frac{3}{40}x^5 + \frac{5}{112}x^7 + \frac{35}{1152}x^9 + \cdots$$

for $-1 < x < 1$. ∎

Estimating Integrals

Suppose we wish to compute $\int_a^b f(x)\, dx$. Perhaps it is hard to find an elementary function that is an antiderivative of f, even if f is a known elementary function. For example, it can be shown that no antiderivative of e^{-x^2} is an elementary function. To estimate $\int_a^b f(x)\, dx$, we might take a power series $\sum_{n=0}^{\infty} a_n(x - x_0)^n$ that converges to f in $[a, b]$ and compute

$$\int_a^b \left[\sum_{n=0}^{\infty} a_n(x - x_0)^n \right] dx$$

instead. The theory of power series shows that this integral can be computed by term-by-term integration. As another alternative, we might just use a partial sum s_n of the series to estimate f in $[a, b]$, and estimate $\int_a^b f(x)\, dx$ by $\int_a^b s_n(x)\, dx$. In this case, we would like a bound for our error. If f and g are any continuous functions defined on $[a, b]$, it follows easily from the definition of the definite integral that

$$\text{if } \quad |f(x) - g(x)| < \varepsilon \quad \text{ for all } x \text{ in } [a, b],$$

$$\text{then } \quad \left| \int_a^b f(x)\, dx - \int_a^b g(x)\, dx \right| < \varepsilon(b - a)$$

(see Exercise 50). In the examples that follow, we illustrate both estimation by integrating a series and estimation by integrating a partial sum.

EXAMPLE 6 Estimate $\int_0^1 e^{-x^2}\,dx$ by integrating a series.

Solution If we replace x by $-x^2$ in the well-known series for e^x, we find that

$$e^{-x^2} = 1 - x^2 + \frac{x^4}{2!} - \frac{x^6}{3!} + \cdots + (-1)^n \frac{x^{2n}}{n!} + \cdots$$

for all x. Thus

$$\int_0^1 e^{-x^2}\,dx = \int_0^1 \left(1 - x^2 + \frac{x^4}{2!} - \frac{x^6}{3!} + \frac{x^8}{4!} - \cdots + (-1)^n \frac{x^{2n}}{n!} + \cdots \right) dx$$

$$= \left(x - \frac{x^3}{3} + \frac{x^5}{5 \cdot 2!} - \frac{x^7}{7 \cdot 3!} + \frac{x^9}{9 \cdot 4!} - \cdots \right.$$

$$\left. + \frac{(-1)^n x^{2n+1}}{(2n+1)n!} + \cdots \right) \Bigg]_0^1$$

$$= 1 - \frac{1}{3} + \frac{1}{5 \cdot 2!} - \frac{1}{7 \cdot 3!} + \frac{1}{9 \cdot 4!} - \cdots + \frac{(-1)^n}{(2n+1)n!} + \cdots .$$

Our answer is the sum of an infinite series of constants and may in turn be approximated by a partial sum of the series. Since the series is alternating with terms of decreasing size, the error in approximating this series by a partial sum is less than the size of the next term. Thus

$$\int_0^1 e^{-x^2}\,dx \approx 1 - \frac{1}{3} + \frac{1}{5 \cdot 2!} - \frac{1}{7 \cdot 3!} + \frac{1}{9 \cdot 4!} \approx 0.7475,$$

with error less than

$$\frac{1}{11 \cdot 5!} \approx 0.0008. \qquad \blacksquare$$

EXAMPLE 7 We know that

$$\int_0^{1/2} (1 - x^2)^{-1/2}\,dx = \sin^{-1}x \Bigg]_0^{1/2}$$

$$= \sin^{-1}\frac{1}{2} - \sin^{-1}0 = \frac{\pi}{6} - 0 = \frac{\pi}{6}.$$

Integrate a partial sum of the binomial series representing $(1 - x^2)^{-1/2}$ to estimate $\pi/6$ and then $\pi = 6(\pi/6)$.

Solution We have

$$(1 - x^2)^{-1/2} = 1 + \left(-\frac{1}{2} \right)(-x^2) + \frac{-\frac{1}{2} \cdot -\frac{3}{2}}{2!}(-x^2)^2$$

$$+ \frac{-\frac{1}{2} \cdot -\frac{3}{2} \cdot -\frac{5}{2}}{3!}(-x^2)^3 + \cdots$$

$$= 1 + \frac{1}{2}x^2 + \frac{3}{2^2 \cdot 2!}x^4 + \frac{3 \cdot 5}{2^3 \cdot 3!}x^6$$

$$+ \left(\frac{3 \cdot 5 \cdot 7}{2^4 \cdot 4!}x^8 + \frac{3 \cdot 5 \cdot 7 \cdot 9}{2^5 \cdot 5!}x^{10} + \cdots \right)$$

for $-1 < x < 1$. The last portion of the series that we have placed in parentheses has all coefficients less than or equal to 1 and is therefore, term for term, less than the series

$$x^8 + x^{10} + \cdots = x^8(1 + x^2 + x^4 + \cdots).$$

Since we will be integrating for $0 \le x \le \frac{1}{2}$, we have

$$x^8(1 + x^2 + x^4 + \cdots) \le \frac{1}{2^8}\left(1 + \frac{1}{4} + \frac{1}{4^2} + \cdots\right)$$

$$= \frac{1}{2^8} \cdot \frac{1}{1 - \frac{1}{4}}$$

$$= \frac{1}{2^8} \cdot \frac{4}{3} = \frac{1}{192}.$$

Thus we have

$$\frac{\pi}{6} = \int_0^{1/2} (1 - x^2)^{-1/2}\, dx \approx \int_0^{1/2} \left(1 + \frac{1}{2}x^2 + \frac{3}{8}x^4 + \frac{15}{48}x^6\right) dx$$

$$= \left(x + \frac{1}{6}x^3 + \frac{3}{40}x^5 + \frac{15}{336}x^7\right)\Big]_0^{1/2}$$

$$= \frac{1}{2} + \frac{1}{48} + \frac{3}{1280} + \frac{15}{43008} \approx 0.52353,$$

with error at most

$$\frac{1}{192}\left(\frac{1}{2} - 0\right) = \frac{1}{384} \approx 0.00260.$$

Therefore,

$$\pi = 6\left(\frac{\pi}{6}\right) \approx 6(0.52353) = 3.14118$$

with error at most $6(0.00260) = 0.01560$. Since the value of π to five decimal places is 3.14159, our actual error in estimating π was only about 0.00041. The reason we obtained such a crude bound for the error is that we estimated the series $x^8 + x^{10} + \cdots$ throughout $[0, \frac{1}{2}]$ by its greatest value at $x = \frac{1}{2}$. ∎

SUMMARY

1. If p is not a nonnegative integer, the binomial series $\sum_{k=0}^{\infty} \binom{p}{k}x^k$ has radius of convergence 1 and represents $(1 + x)^p$ for $-1 < x < 1$. For a nonnegative integer n, the (finite-length) binomial series $\sum_{k=0}^{n} \binom{n}{k}x^k$ converges for all x to $(1 + x)^n$.

2. Integrals $\int_a^b f(x)\, dx$ such as $\int_a^b e^{-x^2}\, dx$, which cannot be found in the usual way, can sometimes be estimated by taking a series expression for $f(x)$ and integrating the series.

EXERCISES 10.8

In Exercises 1–12, compute the binomial coefficient.

1. $\binom{4}{2}$ 2. $\binom{7}{3}$ 3. $\binom{8}{7}$ 4. $\binom{10}{0}$

5. $\binom{5}{5}$ 6. $\binom{5}{7}$ 7. $\binom{3.5}{3}$ 8. $\binom{-1.5}{2}$

9. $\binom{7.5}{0}$ 10. $\binom{0.5}{4}$ 11. $\binom{1/3}{3}$ 12. $\binom{-4}{4}$

13. What is the Taylor series for $(1 + x)^0$ at $x_0 = 0$? Check that the coefficient of x^k in this series is indeed the binomial coefficient $\binom{0}{k}$ as defined on page 540.

In Exercises 14–23, find the first five terms of the binomial series that represents the given function in a neighborhood of zero. Give the radius of convergence in each case.

14. $(1 + x)^{3/2}$ 15. $(1 + x)^{-2}$
16. $(1 - x)^{1/2}$ 17. $(1 - x^2)^{1/3}$
18. $\left(1 + \dfrac{x}{2}\right)^{2/3}$ 19. $(1 + 3x)^{5/3}$
20. $(4 + x)^{1/2}$ [Hint: $(4 + x)^{1/2} = 2[1 + x/4]^{1/2}$.]
21. $(25 - x^2)^{-1/2}$
22. $(8 - x)^{-1/3}$
23. $(8 + x^3)^{4/3}$
24. Estimate $\sqrt{2}$ by finding a partial sum with $x = \frac{1}{2}$ of a binomial series that represents $(1 - x)^{-1/2}$ in a neighborhood of the origin.

In Exercises 25–28, indicate how you could estimate the given quantity by using a suitable binomial series. (You need not compute an estimate.) To illustrate, Exercise 24 shows how $\sqrt{2}$ could be estimated (rather inefficiently) using a binomial series.

25. $\sqrt{2/3}$ 26. $\sqrt{5}$
27. $\sqrt[3]{2}$ 28. $\sqrt[4]{17}$

In Exercises 29–38, find as many terms of the series for the given function at $x_0 = 0$ as you conveniently can.

29. $\sinh^{-1}x$ [Hint: $D(\sinh^{-1}x) = 1/\sqrt{1 + x^2}$.]
30. $\sin^{-1}x^2$ 31. $x/\sqrt{1 - x^2}$
32. $(1 + x^2)\sqrt{1 - x}$ 33. $\sqrt{1 + x}/(1 - x)$

34. $\dfrac{\sqrt{1 - x}}{1 + x}$ 35. $\sqrt{1 + x}\sqrt{1 + 2x}$

36. $x^2\sqrt{16 - x^2}$ 37. $\dfrac{x^3}{\sqrt{4 - x^2}}$

38. $\dfrac{\sqrt{1 - x^2}}{\sqrt{4 + x^2}}$

In Exercises 39–48, estimate the integral by series methods and find a bound for your error. (You need not simplify your answers or put them in decimal form.)

39. $\displaystyle\int_0^1 \sin x^2 \, dx$ 40. $\displaystyle\int_0^{1/2} e^{x^2} \, dx$

41. $\displaystyle\int_0^1 x^3\cos x^2 \, dx$ 42. $\displaystyle\int_0^{1/2} (1 + x^2)^{1/3} \, dx$

43. $\displaystyle\int_0^{0.1} e^{-x^2} \, dx$

44. $\displaystyle\int_0^1 \sqrt{16 - x^4} \, dx$ $\left[\text{Hint: } \sqrt{16 - x^4} = 4 \cdot \sqrt{1 - \dfrac{x^4}{16}}.\right]$

45. $\displaystyle\int_0^{1/2} \cos x^3 \, dx$ 46. $\displaystyle\int_0^1 e^{-x^3} \, dx$

47. $\displaystyle\int_{1/2}^{3/2} \cos(x - 1)^2 \, dx$ 48. $\displaystyle\int_0^1 x^2\sqrt{25 - x^2} \, dx$

49. Verify Eq. (8) on page 542.

50. Use the definition of $\int_a^b f(x) \, dx$ to show that if f and g are continuous in $[a, b]$ and $|f(x) - g(x)| < \varepsilon$ for all x in $[a, b]$, then

$$\left|\int_a^b f(x) \, dx - \int_a^b g(x) \, dx\right| < \varepsilon(b - a).$$

▦ Use a calculator or computer for Exercises 51 and 52.

51. Find the absolute numerical difference between the estimate for $\int_0^1 e^{x^2} \, dx$ using Simpson's rule with $n = 20$ and integrating the first eight nonzero terms of the series for e^{x^2}.

52. Repeat Exercise 51 for $\int_0^1 \cos x^2 \, dx$, but use the first five nonzero terms of the series for $\cos x^2$.

SUPPLEMENTARY EXERCISES FOR CHAPTER 10

1. Define what is meant by $\lim_{n\to\infty} a_n = c$.

2. Find the limit of the given sequence if it converges or has limit ∞ or $-\infty$.

   (a) $\left\{\dfrac{4n^2 - 2n}{3 - 5n^2}\right\}$ (b) $\left\{\dfrac{n^3}{e^n}\right\}$

3. Give an ε,N-proof that the sequence $\{(n - 1)/n\}$ converges.

4. Find the limit of the given sequence if it converges or has limit ∞ or $-\infty$.

   (a) $\left\{\dfrac{\sqrt{n} - n^2}{3n - 7}\right\}$ (b) $\left\{\dfrac{1}{n}\sin\left(\dfrac{1}{n}\right)\right\}$

5. Give a careful definition of $\lim_{n\to\infty} a_n = \infty$.

6. Find the limit of the given sequence if it converges or has limit ∞ or $-\infty$.

(a) $\left\{ \dfrac{n^2 + 1}{n} - \dfrac{n^3 + n^2}{n^2 + 1} \right\}$ (b) $\left\{ \dfrac{3^n - 5^n}{4^n} \right\}$

7. Give an example of sequences $\{s_n\}$ and $\{t_n\}$, both of which diverge, such that $\{s_n + t_n\}$ converges to 3.

8. Find the limit of the given sequence if it converges or has limit ∞ or $-\infty$.

(a) $\left\{ \dfrac{\sin n}{\sqrt{n}} \right\}$ (b) $\left\{ \dfrac{\csc^2(1/n)}{n^{3/2}} \right\}$

9. Find the sum of the series

$$\sum_{n=0}^{\infty} \frac{3^{n+1}}{4^{n-1}}$$

if the series converges.

10. Express the repeating decimal $4.731313131\ldots$ as a fraction.

11. A ball has the property that when it is dropped, it rebounds to $\frac{1}{3}$ of its height on the previous bounce. Find the height from which the ball must be dropped if the total distance it is to travel is 60 ft.

12. If possible, find r such that the sum of the series

$$3 - 3r + 3r^2 - 3r^3 + \cdots + (-1)^n 3r^n + \cdots$$

is 7.

13. Find the sum of the series

$$\sum_{n=0}^{\infty} \frac{3^n - 2^{n+2}}{5^n},$$

if it converges.

14. Express the repeating decimal $5.2107107107107\ldots$ as a fraction.

15. A ball has the property that when it is dropped, it rebounds to k times its height on the preceding bounce. If the ball travels a total distance of 400 ft when dropped from a height of 100 ft, find the value of k.

16. Is $3.01001000100001000001\ldots$ a rational number? Why or why not?

In Exercises 17–32, classify the series as convergent or divergent, and give a reason for your answer.

17. $\displaystyle\sum_{n=1}^{\infty} \frac{n^2 + 3n}{400n + n^2}$ **18.** $\displaystyle\sum_{n=2}^{\infty} \frac{n^2 - 1}{1 + n^3}$

19. $\displaystyle\sum_{n=1}^{\infty} \frac{1}{\sqrt{n} + 17}$ **20.** $\displaystyle\sum_{n=1}^{\infty} \frac{\sqrt{n} + 3}{n^2 + 4n}$

21. $\displaystyle\sum_{n=1}^{\infty} \frac{n^2 + 3}{n^2 \cdot 3^{n-1}}$ **22.** $\displaystyle\sum_{n=1}^{\infty} \frac{3^n + 4^n}{5^n}$

23. $\displaystyle\sum_{n=1}^{\infty} \frac{n^2 + \sin n}{n^4 + 3n}$ **24.** $\displaystyle\sum_{n=1}^{\infty} \frac{1}{\cos^2 n}$

25. $\displaystyle\sum_{n=1}^{\infty} \frac{\sqrt{n^5 + 4n}}{(n^2 + n + 3)^2}$ **26.** $\displaystyle\sum_{n=1}^{\infty} \sin\!\left(\frac{1}{n^2 + 1} \right)$

27. $\displaystyle\sum_{n=1}^{\infty} \frac{3^{2n+1}}{(n + 4)9^n}$ **28.** $\displaystyle\sum_{n=1}^{\infty} \frac{(n + 3)! - (n + 1)!}{(n + 4)!}$

29. $\displaystyle\sum_{n=1}^{\infty} \frac{(3^n - 2^n)^2}{7^n}$ **30.** $\displaystyle\sum_{n=1}^{\infty} \frac{1}{(2.7)^{\ln n}}$

31. $\displaystyle\sum_{n=1}^{\infty} \frac{(n + 4)! + (n + 3)!}{(n + 1)! \cdot 2^{n/2}}$ **32.** $\displaystyle\sum_{n=2}^{\infty} \frac{\sqrt{\ln(n - 1)}}{n(\ln n)^2}$

33. State the integral test, giving all hypotheses.

34. Use the integral test to establish the convergence or divergence of

$$\sum_{n=2}^{\infty} \frac{1}{n^2 - n}.$$

35. Use the integral test to establish the convergence or divergence of

$$\sum_{n=2}^{\infty} \frac{n + 1}{n^2 + 1}.$$

36. Use the integral test to establish the convergence or divergence of

$$\sum_{n=1}^{\infty} \frac{e^n}{1 + e^{2n}}.$$

37. State the ratio test, giving all hypotheses.

38. Write out the ratio test to establish the convergence or divergence of

$$\sum_{n=1}^{\infty} \frac{n^2 \cdot 2^n}{e^{n+1}}.$$

39. Write out the ratio test to establish the convergence or divergence of

$$\sum_{n=1}^{\infty} \frac{n!}{100^n}.$$

40. Write out the ratio test to establish the convergence or divergence of

$$\sum_{n=1}^{\infty} \frac{4^{3n}}{n^2 7^{2n+4}}.$$

In Exercises 41–56, classify the series as absolutely convergent, conditionally convergent, or divergent, and indicate the reason for your answer.

41. $\displaystyle\sum_{n=1}^{\infty} \frac{(-1)^n}{\sqrt{n} + 3}$ **42.** $\displaystyle\sum_{n=2}^{\infty} \frac{(-1)^n}{n(\ln n)^2}$

43. $\displaystyle\sum_{n=1}^{\infty} (-1)^n \frac{\cos(1/n)}{n^{3/2}}$ **44.** $\displaystyle\sum_{n=1}^{\infty} (-1)^n \frac{3^n}{n^{100}}$

45. $\displaystyle\sum_{n=2}^{\infty} \frac{(-1)^n}{n(\ln n)}$

46. $\displaystyle\sum_{n=1}^{\infty} (-1)^n \frac{\cos^2 n}{n^{3/2}}$

47. $\displaystyle\sum_{n=1}^{\infty} (-1)^n \frac{n^n}{n!}$

48. $\displaystyle\sum_{n=1}^{\infty} (-1)^n \frac{\ln n}{n}$

49. $\displaystyle\sum_{n=1}^{\infty} (-1)^n \frac{\sqrt{n}}{n^2 + 1}$

50. $\displaystyle\sum_{n=1}^{\infty} \frac{\sin(3n\pi/2)}{n}$

51. $\displaystyle\sum_{n=1}^{\infty} \frac{(-3)^{n+1}}{(\sqrt{3})^{3n}}$

52. $\displaystyle\sum_{n=1}^{\infty} (-1)^n \frac{\ln(n^2)}{n}$

53. $\displaystyle\sum_{n=1}^{\infty} (-1)^n \frac{1}{n^4 \cdot 3^{-n}}$

54. $\displaystyle\sum_{n=1}^{\infty} \frac{\cos n\pi}{\sqrt{n}}$

55. $\displaystyle\sum_{n=1}^{\infty} (-1)^n \ln\left(\frac{1}{n^2}\right)$

56. $\displaystyle\sum_{n=1}^{\infty} (-1)^n \frac{(n!)^2 \cdot 5^n}{(3n)!}$

57. Find the interval of convergence, including endpoints, of the series
$$\sum_{n=1}^{\infty} \frac{n}{(n^2 + 1)3^n}(x + 5)^n.$$

58. Find an expression for the nth term $a_n(x - x_0)^n$ of a power series having as interval of convergence $-1 \le x < 3$. (Many answers are possible.)

59. Find the interval of convergence, including endpoints, of the series
$$\sum_{n=1}^{\infty} \frac{(2x + 1)^{2n}}{n^2 \cdot 3^n}.$$

60. Find the radius of convergence of the series
$$\sum_{n=1}^{\infty} \frac{(x - 4)^{2n+1}}{3^{4n-3}}.$$

61. Find the interval of convergence, including endpoints, of the series
$$\sum_{n=0}^{\infty} \frac{100^n(x - 1)^n}{n!}.$$

62. Find the radius of convergence of the series
$$\sum_{n=0}^{\infty} \frac{(4x - 2)^{2n}}{3^n}.$$

63. Find the interval of convergence, including endpoints, of the series
$$\sum_{n=0}^{\infty} \frac{n!(x - 3)^{5n}}{1000^n}.$$

64. Find a power series with interval of convergence $0 \le x \le 100$. (Many answers are possible.)

65. (a) Define the notion of an analytic function of a real variable.
   (b) Find the Taylor series for $x/(1 - x^2)$ at $x_0 = 0$ in the easiest way you can.

66. Use series long division to find the first three terms of the series for $(\sin x)/e^x$ at $x_0 = 0$.

67. Find the Taylor series for $x^3 \tan^{-1} x^2$ at $x_0 = 0$ in the easiest way you can.

68. Use series multiplication to find the first four nonzero terms of the Taylor series for $e^x \sin 2x$ at $x_0 = 0$.

69. If $f(x) = x^4/(1 - 2x^2)$, find $f^{(10)}(0)$.

70. Find $\displaystyle\sum_{n=1}^{\infty} \frac{3^n}{n!}$.

71. Find as many terms as you conveniently can of the Taylor series for $(1 - x^2)/(1 + x^2)$ at $x_0 = 1$.

72. Find the function represented near 0 by the series
$$x + \frac{x^3}{3} + \frac{x^5}{5} + \frac{x^7}{7} + \frac{x^9}{9} + \cdots.$$

73. Use the binomial series expansion to give the first five nonzero terms of the series for $\sqrt{1 + x^2}$ at $x_0 = 0$.

74. Use series techniques to find $\int_0^1 \sin x^2 \, dx$, with error at most 0.001.

75. Give the first four terms of the binomial series representing $(1 - x/2)^{-1/2}$ in a neighborhood of $x_0 = 0$.

76. Use series methods to estimate
$$\int_0^1 \frac{1 - \cos x^2}{x} \, dx,$$
with error at most 0.001.

77. Give the first four terms of the Taylor series representing $(8 + x)^{2/3}$ near $x_0 = 0$.

78. Use series techniques to estimate $\int_0^{1/2} \cos x^3 \, dx$, with error at most 0.00001.

79. Compute $\dbinom{5/3}{4}$.

80. Find the first four terms of the Taylor series for $x^4/\sqrt{1 - x}$ at $x_0 = 0$.

PLANE CURVES AND POLAR COORDINATES

11

Thus far we have worked primarily with plane curves that are graphs of functions. Recall that the graph of a function $y = f(x)$ has the property that a vertical line meets the graph in at most one point. Many important plane curves do not satisfy this property; a circle is one of them. This chapter is concerned with describing such curves and studying them using calculus. A detailed study of motion on such curves appears in Chapter 13.

In Section 11.1, we see how to sketch the graph of a second-degree equation in x and y in *standard form*. The resulting curves are either ellipses (or circles), parabolas, hyperbolas, or degenerate cases of these *conic sections*. It can be shown that every second-degree equation in x and y describes such a curve.

Sections 11.2 and 11.3 introduce the notion of *parametrization* of a curve, which can be regarded as describing a curve by giving the position at every time t of a body that travels the entire curve. We will see how to find the slope and arc length of a curve described in this way. Section 11.4 deals with measuring how sharply a curve bends at each point on the curve. We will develop formulas for measuring this *curvature* at a point for many curves.

Sections 11.5–11.8 introduce new coordinates, *polar coordinates*, for points in the plane. This coordinate system simplifies the equations of some curves, such as circles, that arise in applications. After practicing sketching such curves, we will see how to use calculus to measure areas of regions they bound, their slope at each point, and their length.

11.1 SKETCHING CONIC SECTIONS

Let two congruent right circular cones in space be placed vertex to vertex to form a double cone, as shown in Fig. 11.1. A plane can intersect this cone in three types of curves. Figure 11.1(a) shows the closed-curve type of intersection,

HISTORICAL NOTE

APOLLONIUS is the author of the only surviving Greek text on the conic sections. He was born in Perga, a town in Asia Minor, in the third century B.C. and in his youth studied in Alexandria with the successors of Euclid. He probably spent most of his life in Alexandria studying, teaching, and writing. Apollonius's massive *Conics* can be considered the culmination of Greek mathematics. It is difficult for mathematicians today to comprehend how Apollonius could discover and prove hundreds of beautiful and difficult theorems without modern algebraic symbolism. Nevertheless, he succeeded.

Apollonius defined the conics as sections of a double cone (see Fig. 11.1). But he immediately derived from his definitions relations between the coordinates of points on the curves equivalent to modern equations. From these relations he derived the various properties of the conics. Among these properties are many generally derived today using calculus techniques. For example, Apollonius showed how to construct tangent and normal lines to the conic sections and also derived the important result that the lines from the foci of an ellipse drawn to a point on the curve make equal angles with the tangent at that point.

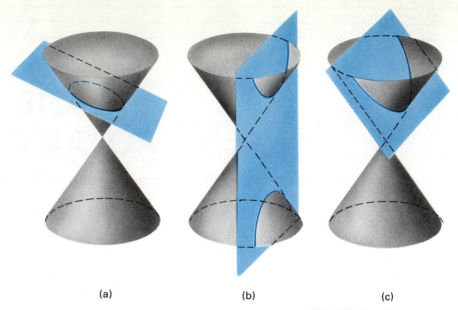

(a)   (b)   (c)

FIGURE 11.1   Conic sections: (a) elliptic section; (b) hyperbolic section; (c) parabolic section.

which is an *ellipse*. Figure 11.1(b) shows the intersection giving a two-piece, open-ended curve, which is a *hyperbola*. The one-piece, open-ended curve in Fig. 11.1(c) is a *parabola*.

In this section, we sketch these curves, called *conic sections,* starting with their equations with respect to carefully chosen x- and y-axes in the plane of intersection.

The Ellipse

The equation

$$\frac{x^2}{a^2} + \frac{y^2}{b^2} = 1 \tag{1}$$

describes an ellipse, shown in Fig. 11.2. Setting $x = 0$, we see that the curve meets the y-axis at b and $-b$. Setting $y = 0$, we find that the x-intercepts are a and $-a$. If $a = b$, the ellipse becomes a circle.

Recall from Section 1.1 that if we choose translated $\bar{x}, \bar{y}$-axes at a new origin (h, k), then

$$\bar{x} = x - h \quad \text{and} \quad \bar{y} = y - k.$$

It follows that an equation of the form

$$\frac{(x - h)^2}{a^2} + \frac{(y - k)^2}{b^2} = 1 \tag{2}$$

can be rewritten as

$$\frac{\bar{x}^2}{a^2} + \frac{\bar{y}^2}{b^2} = 1$$

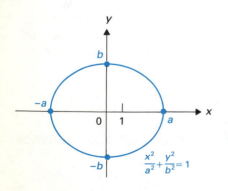

FIGURE 11.2   An ellipse $x^2/a^2 + y^2/b^2 = 1$.

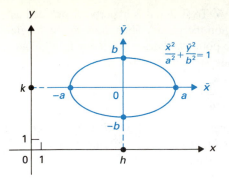

FIGURE 11.3 The ellipse

$$(x - h)^2/a^2 + (y - k)^2/b^2 = 1$$

becomes

$$\bar{x}^2/a^2 + \bar{y}^2/b^2 = 1$$

using translated axes at (h, k).

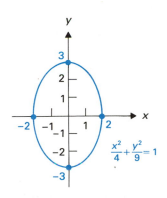

FIGURE 11.4 The ellipse

$$9x^2 + 4y^2 = 36.$$

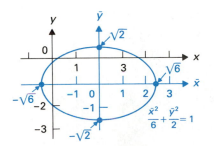

FIGURE 11.5 The ellipse

$$x^2 + 3y^2 - 4x + 6y = -1.$$

with respect to the translated $\bar{x},\bar{y}$-axes. This shows that Eq. (2) describes an ellipse with center at (h, k), as sketched in Fig. 11.3.

EXAMPLE 1 Sketch the curve $9x^2 + 4y^2 = 36$.

Solution Dividing through by 36, we obtain

$$\frac{x^2}{4} + \frac{y^2}{9} = 1,$$

which is the ellipse shown in Fig. 11.4. The x-intercepts are $\pm\sqrt{4} = \pm 2$, and the y-intercepts are $\pm\sqrt{9} = \pm 3$. ∎

EXAMPLE 2 Sketch the curve with equation

$$x^2 + 3y^2 - 4x + 6y = -1.$$

Solution Completing the square, we obtain

$$(x^2 - 4x) + 3(y^2 + 2y) = -1,$$
$$(x - 2)^2 + 3(y + 1)^2 = 4 + 3 - 1 = 6,$$
$$\frac{(x - 2)^2}{6} + \frac{(y + 1)^2}{2} = 1,$$

which is in the form of Eq. (2). Setting $\bar{x} = x - 2$ and $\bar{y} = y + 1$, we obtain the equation

$$\frac{\bar{x}^2}{6} + \frac{\bar{y}^2}{2} = 1.$$

This is the equation of an ellipse with center $(h, k) = (2, -1)$, as shown in Fig. 11.5. ∎

EXAMPLE 3 Describe the curve with equation

$$2x^2 + 5y^2 - 4x + 10y = -8.$$

Solution Completing the square, we obtain

$$2(x^2 - 2x) + 5(y^2 + 2y) = -8,$$
$$2(x - 1)^2 + 5(y + 1)^2 = 2 + 5 - 8 = -1.$$

Setting $\bar{x} = x - 1$ and $\bar{y} = y + 1$, we have

$$2\bar{x}^2 + 5\bar{y}^2 = -1.$$

Now for any point $(\bar{x}, \bar{y})$, we must have $2\bar{x}^2 + 5\bar{y}^2 \geq 0$, so no point in the plane satisfies the equation. Our "curve" is empty. ∎

The Hyperbola

The equation

$$\frac{x^2}{a^2} - \frac{y^2}{b^2} = 1 \tag{3}$$

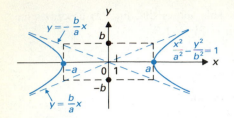

FIGURE 11.6   A hyperbola
$x^2/a^2 - y^2/b^2 = 1$.

describes a hyperbola, shown in Fig. 11.6. To sketch this hyperbola, proceed as follows. Mark $\pm a$ on the x-axis and $\pm b$ on the y-axis. Then draw the rectangle crossing at right angles the axes at those points and draw and extend the diagonals of that rectangle. The hyperbola has these diagonals

$$y = \pm \frac{b}{a}x$$

as *asymptotes*. The hyperbola crosses the x-axis at $\pm a$ but does not meet the y-axis, because if $x = 0$, then Eq. (3) reduces to $y^2 = -b^2$, which has no real solution. If we solve Eq. (3) for y, we obtain

$$y = \pm b \sqrt{(x^2/a^2) - 1}$$

and

$$\lim_{x \to \infty} \left[b\sqrt{(x^2/a^2) - 1} - \frac{b}{a}x \right] = 0,$$

which shows that the lines $y = \pm(b/a)x$ are indeed asymptotes of the curve.

If the negative sign in Eq. (3) appears with the other term x^2/a^2 instead, as in

$$-\frac{x^2}{a^2} + \frac{y^2}{b^2} = 1, \qquad \text{(4)}$$

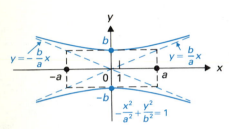

FIGURE 11.7   A hyperbola
$-(x^2/a^2) + y^2/b^2 = 1$.

then the asymptotes are still $y = \pm(b/a)x$, but now the hyperbola crosses the y-axis at $\pm b$ and does not meet the x-axis, as shown in Fig. 11.7.

EXAMPLE 4   Sketch the hyperbola

$$\frac{y^2}{4} - \frac{x^2}{9} = 1.$$

Solution   The sketch is shown in Fig. 11.8.   ∎

The device of completing the square can be used again to sketch a hyperbola whose center is not $(0, 0)$.

EXAMPLE 5   Sketch $2x^2 - 3y^2 - 4x - 6y = 13$.

Solution   Completing the square, we obtain

$$2(x^2 - 2x) - 3(y^2 + 2y) = 13,$$
$$2(x - 1)^2 - 3(y + 1)^2 = 2 - 3 + 13 = 12,$$
$$\frac{(x - 1)^2}{6} - \frac{(y + 1)^2}{4} = 1.$$

Setting $\bar{x} = x - 1$ and $\bar{y} = y + 1$, we have

$$\frac{\bar{x}^2}{6} - \frac{\bar{y}^2}{4} = 1.$$

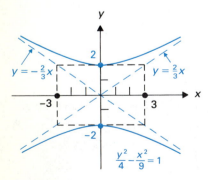

FIGURE 11.8   The hyperbola
$y^2/4 - x^2/9 = 1$
with asymptotes $y = \pm\frac{2}{3}x$.

We see that we have a hyperbola with center $(h, k) = (1, -1)$ and crossing the $\bar{x}$-axis at $\pm\sqrt{6}$, as shown in Fig. 11.9.   ∎

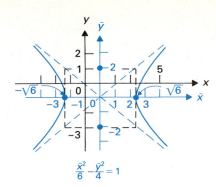

FIGURE 11.9 The hyperbola

$$2x^2 - 3y^2 - 4x - 6y = 13$$

centered at $(h, k) = (1, -1)$.

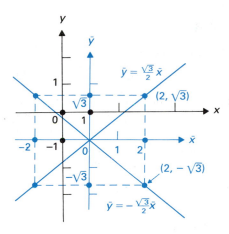

FIGURE 11.10 The degenerate hyperbola

$$3x^2 - 4y^2 - 6x - 8y = 1,$$
$$\bar{x}^2/4 - \bar{y}^2/3 = 0, \quad \bar{y} = \pm(\sqrt{3}/2)\bar{x}.$$

EXAMPLE 6 Sketch the curve $3x^2 - 4y^2 - 6x - 8y = 1$.

Solution Completing the square, we have

$$3(x^2 - 2x) - 4(y^2 + 2y) = 1,$$
$$3(x - 1)^2 - 4(y + 1)^2 = 3 - 4 + 1 = 0.$$

Setting $\bar{x} = x - 1$ and $\bar{y} = y + 1$, we obtain the equation

$$3\bar{x}^2 - 4\bar{y}^2 = 0,$$

$$\bar{y}^2 = \frac{3}{4}\bar{x}^2,$$

$$\bar{y} = \pm\frac{\sqrt{3}}{2}\bar{x}.$$

We see that our "curve" consists of the two intersecting lines shown in Fig. 11.10. We view these intersecting lines as a *degenerate hyperbola*. Such a pair of intersecting lines is obtained if we intersect the cone in Fig. 11.1 with a vertical plane through the origin. ∎

The Parabola

The curves

$$y = 4x^2, \qquad y = -x^2,$$
$$x = \frac{1}{4}y^2, \qquad x = -4y^2$$

are parabolas with vertices at the origin, as shown in Fig. 11.11. The size of the coefficient of the quadratic term controls how fast the parabola "opens." If the vertex of $y = 4x^2$ were moved to (h, k), the equation of the curve would become

$$y - k = 4(x - h)^2.$$

As indicated in Section 1.4, any polynomial equation in x and y that is quadratic in one of the variables and linear in the other describes a parabola.

EXAMPLE 7 Sketch $3y + 4x^2 + 8x = 5$.

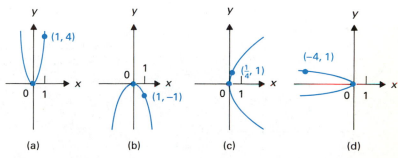

FIGURE 11.11 Four types of parabolas with vertices at the origin: (a) $y = 4x^2$; (b) $y = -x^2$; (c) $x = \frac{1}{4}y^2$; (d) $x = -4y^2$.

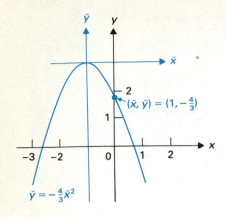

FIGURE 11.12 The parabola
$3y + 4x^2 + 8x = 5$.

Solution Completing the square yields

$$3y + 4(x^2 + 2x) = 5,$$
$$3y + 4(x + 1)^2 = 4 + 5 = 9,$$
$$3y - 9 = -4(x + 1)^2,$$
$$3(y - 3) = -4(x + 1)^2,$$
$$y - 3 = -\frac{4}{3}(x + 1)^2.$$

Setting $\bar{x} = x + 1$ and $\bar{y} = y - 3$, we obtain

$$\bar{y} = -\frac{4}{3}\bar{x}^2.$$

This is a parabola with vertex at $(h, k) = (-1, 3)$, opening downward, and symmetric about the line $x = -1$, as shown in Fig. 11.12.   ∎

The General Second-Degree Equation

We state without proof that every second-degree plane curve

$$Ax^2 + Bxy + Cy^2 + Dx + Ey + F = 0 \tag{5}$$

has as graph a conic section (possibly empty or degenerate). To obtain an equation of the type we have discussed in this section, it may be necessary to choose new axes in the plane. It can be shown that by rotating the x,y-axes through an appropriate angle θ to become new x',y'-axes (see Fig. 11.13), we can obtain an x',y'-version of Eq. (5) in which the coefficient B' of $x'y'$ is zero. That is, the x',y'-version of Eq. (5) has the form

$$A'(x')^2 + C'(y')^2 + D'x' + E'y' + F' = 0. \tag{6}$$

As illustrated in this section, the curve can then be sketched by completing the square and translating axes, and is a (possibly empty or degenerate) conic section.

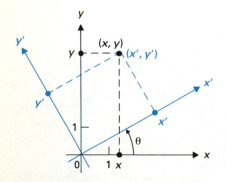

FIGURE 11.13 Rotation of axes through θ to new x',y'-axes.

SUMMARY

1. An equation of the form

$$\frac{(x - h)^2}{a^2} + \frac{(y - k)^2}{b^2} = 1$$

describes an ellipse with center at (h, k), meeting the $\bar{x}$-axis at $\pm a$ and the $\bar{y}$-axis at $\pm b$, where $\bar{x} = x - h$ and $\bar{y} = y - k$.

2. Equations of the form

$$\frac{(x - h)^2}{a^2} - \frac{(y - k)^2}{b^2} = 1$$

or

$$-\frac{(x - h)^2}{a^2} + \frac{(y - k)^2}{b^2} = 1$$

describe hyperbolas with center at (h, k). To sketch them, draw the rectangle with center (h, k) crossing at right angles the $\bar{x}$-axis at $\pm a$ and the $\bar{y}$-axis at $\pm b$. Draw and extend the diagonals of the rectangle. The first hyperbola crosses the $\bar{x}$-axis at $\pm a$, while the second hyperbola crosses the $\bar{y}$-axis at $\pm b$. Both hyperbolas have the diagonals of the rectangle as asymptotes.

3. Equations of the form

$$y - k = c(x - h)^2 \qquad \text{and} \qquad x - h = c(y - k)^2$$

describe parabolas with vertices at (h, k). The magnitude $|c|$ controls how fast the parabola "opens," and the sign of c determines the direction in which it opens.

4. Any equation of the form $Ax^2 + Cy^2 + Dx + Ey + F = 0$ represents a (possibly degenerate) conic section. The curve may be sketched by completing the square to obtain one of the forms described above.

EXERCISES 11.1

In Exercises 1–30, sketch the given curve.

1. $x^2 + 4y^2 = 16$
2. $x^2 - 4y^2 = 16$
3. $4y^2 - x^2 = 16$
4. $x^2 = -4y$
5. $x = 4y^2$
6. $9x^2 = -y$
7. $9x = y^2$
8. $4x^2 - 9y^2 = -36$
9. $4x^2 - 9y^2 = 36$
10. $x^2 + 4y^2 = -4$
11. $5x^2 + 2y^2 = 50$
12. $2x^2 - 3y^2 + 4x + 12y = 0$
13. $x^2 - y^2 + 6x + 2y = -3$
14. $4x^2 - 8x + 2y = 5$
15. $x^2 - 4x - 4y = 0$
16. $x^2 + 9y^2 - 4x + 18y = -4$
17. $x^2 + 4y^2 + 2x - 8y = -1$
18. $x^2 + y^2 - 4y = 9$
19. $4x^2 + 4y^2 - 8x + 8y = 1$
20. $x^2 + 2y^2 - 4x + 12y = -24$
21. $3x^2 + y^2 + 6x = -5$
22. $4y^2 + x + 8y = 0$
23. $9x^2 - 18x + y = -4$
24. $3x^2 - y^2 + 18x + 4y = -23$
25. $x^2 - 4y^2 + 2x + 16y = 15$
26. $x^2 - y^2 + 4x - 2y = 1$
27. $4x^2 + 2y^2 + 8x - 4y = -6$
28. $4x^2 + y^2 + 6y = -5$

29. $3x^2 - 4y^2 - 6x - 8y = 0$
30. $-x^2 + 2y^2 + 2x + 8y = -1$
31. For every real number t, the point $(\cos t, \sin t)$ lies on the circle $x^2 + y^2 = 1$. The functions sine and cosine are *circular functions*. Can you think of a reason why the hyperbolic sine and the hyperbolic cosine are called "hyperbolic functions"?

32. Each nondegenerate conic section has an associated nonnegative number, called its **eccentricity** and usually denoted by e, defined as follows for $a > 0$ and $b > 0$:

Parabola: $e = 1$

Ellipse: $\dfrac{x^2}{a^2} + \dfrac{y^2}{b^2} = 1$ or $\dfrac{x^2}{b^2} + \dfrac{y^2}{a^2} = 1$, $a \geq b$:

$$e = \frac{\sqrt{a^2 - b^2}}{a}$$

Hyperbola: $\dfrac{x^2}{a^2} - \dfrac{y^2}{b^2} = 1$ or $\dfrac{y^2}{a^2} - \dfrac{x^2}{b^2} = 1$:

$$e = \frac{\sqrt{a^2 + b^2}}{a}$$

(a) Show that for an ellipse, we have $e < 1$. Draw an ellipse having eccentricity close to 1 and an ellipse having eccentricity close to 0. What is the eccentricity of a circle?
(b) Show that for a hyperbola, we have $e > 1$. Draw a hyperbola having eccentricity close to 1 and a hyperbola having large eccentricity.

11.2 PARAMETRIC CURVES

Let a body be moving in the x,y-plane, perhaps in the direction of the arrows on the curve shown in Fig. 11.14. The curve need not be the graph of a function. The x-coordinate of the body's position at time t is some function $x = h(t)$, while the y-coordinate of the position is $y = k(t)$. The equations

$$x = h(t), \qquad y = k(t), \tag{1}$$

are **parametric equations** of the curve, and t is the time **parameter.** We can view the equations (1) as a rule for picking up all or parts of the t-axis and setting it down in the x,y-plane. That is, the equations describe a function with domain part of the real line and range in the plane.

Describing a plane curve using the parametric form (1) provides more information than describing it using an x,y-equation. Equations (1) tell us not only what part of the plane the curve occupies but also where we are on the curve at every time t in the domain of the functions h and k. If you have to intercept some body traveling on a curve, it is very handy to know where it is at all times. We will give a detailed discussion of motion on curves described parametrically in Chapter 13. Even if we are just discussing a curve defined by parametric equations of the form (1) with no physical context, we may still refer to the parameter t as *time*. Of course, letters other than t can be used as parameter, as we illustrate in some of the examples that follow.

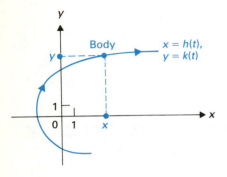

FIGURE 11.14 x- and y-projections of a body moving in the plane on a curve $x = h(t)$, $y = k(t)$.

Sketching and Parametrizing

Curves given parametrically can sometimes be sketched by eliminating the parameter t to obtain a (we hope familiar) x,y-equation.

EXAMPLE 1   Sketch the plane curve given parametrically by

$$x = a \cos t, \qquad y = a \sin t.$$

Solution   Squaring and adding the equations, we find that

$$x^2 + y^2 = a^2\cos^2 t + a^2\sin^2 t = a^2(\cos^2 t + \sin^2 t) = a^2 \cdot 1 = a^2.$$

Thus each point (x, y) of the curve lies on the circle $x^2 + y^2 = a^2$, shown in Fig. 11.15. We may view these parametric equations as picking up the t-axis and wrapping it around and around this circle, with $t = 0$ placed at $(a, 0)$. ■

One interpretation of the parameter t in Example 1 is the angle shown in Fig. 11.16. We see from the figure that

$$x = a \cos t, \qquad y = a \sin t. \tag{2}$$

We frequently want to parametrize a circle, so remember the parametrization (2). Also remember that $t = 0$ corresponds to the point $(a, 0)$, and increasing t corresponds to going *counterclockwise* around the circle.

When sketching a parametric curve $x = h(t)$, $y = k(t)$ using an x,y-equation obtained by eliminating the parameter, we must be careful. The nature of the

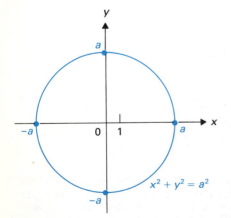

FIGURE 11.15   The circle $x^2 + y^2 = a^2$.

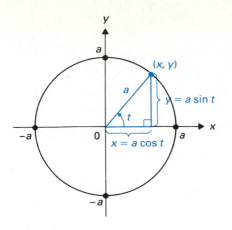

FIGURE 11.16 Parametrizing a circle by the central angle t.

functions $x = h(t)$ and $y = k(t)$ may have placed restrictions on x and y that are not apparent from the x,y-equation.

EXAMPLE 2 Sketch the parametric curve

$$x = \sin t, \qquad y = \sin t.$$

Solution Elimination of t from the parametric equations yields $y = x$. However, since $-1 \le \sin t \le 1$ for all values of t, we see that the points of the curve given by these parametric equations all lie on the segment of the line $y = x$ joining $(-1, -1)$ and $(1, 1)$, as shown in Fig. 11.17. An object whose position in the plane at time t is given by these equations travels back and forth along this line segment. ∎

EXAMPLE 3 Sketch the parametric curve

$$x = t^2, \qquad y = t^4 + 1.$$

Solution Since $t^2 = x$, we see that

$$y = t^4 + 1 = (t^2)^2 + 1 = x^2 + 1.$$

However, since $t^2 \ge 0$, we see that $x \ge 0$, so we obtain only the right-hand half of the parabola $y = x^2 + 1$, as shown in Fig. 11.18. ∎

Sometimes a curve given by an equation in x and y can be parametrized in terms of some quantity (parameter) arising naturally from the curve or from some physical consideration. For such a parametrization to be accomplished, there must be a *unique* point on the curve corresponding to each value of the parameter. The next three examples illustrate this idea.

EXAMPLE 4 Parametrize the parabola $y = x^2$, taking as parameter the slope m of the parabola at each point.

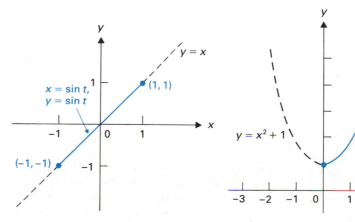

FIGURE 11.17 The parametric curve $x = \sin t$, $y = \sin t$.

FIGURE 11.18 The parametric curve $x = t^2$, $y = t^4 + 1$.

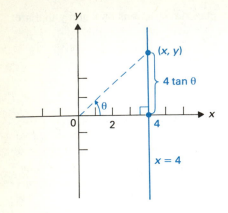

FIGURE 11.19 Parametrizing the line $x = 4$ using the angle of inclination θ.

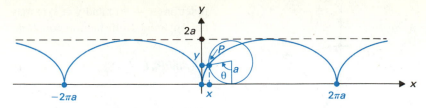

FIGURE 11.20 A cycloid to be parametrized using the angle θ through which the circle has rolled from the origin.

Solution Our problem is to express both x and y in terms of the slope m. At a point (x, y) on the parabola, the slope is given by $m = y' = 2x$, so $x = m/2$. Since $y = x^2$, we obtain the parametrization

$$x = \tfrac{1}{2}m, \qquad y = \tfrac{1}{4}m^2$$

of the parabola. ∎

EXAMPLE 5 Parametrize the line $x = 4$, taking as parameter the angle of inclination θ at the origin from the x-axis to the point (x, y) on the line, as shown in Fig. 11.19.

Solution Our problem is to express both x and y in terms of θ. From Fig. 11.19, we see that $x = 4$, while $y/4 = \tan \theta$, so $y = 4 \tan \theta$. Since we want the parameter θ to be the angle of inclination, we should restrict θ to the interval $-\pi/2 < \theta < \pi/2$. Our parametrization is therefore

$$x = 4, \qquad y = 4 \tan \theta \qquad \text{for } -\frac{\pi}{2} < \theta < \frac{\pi}{2}. \quad ∎$$

EXAMPLE 6 Consider the plane curve traced by a point P on a circle of radius a as the circle rolls along the x-axis. This curve is a *cycloid*. We assume that the point P on the circle touches the x-axis at the points 0 and $\pm 2n\pi a$ as the circle rolls along, as shown in Fig. 11.20.

Parametrize this cycloid in terms of the angle θ through which the circle has rolled, starting with $\theta = 0$ when P is at the origin, as shown in Fig. 11.20.

Solution We assume that θ is taken to be positive as the circle rolls to the right. From the detail of the rolling circle shown in Fig. 11.21, we obtain

$$x = a\theta - a \sin \theta, \qquad y = a - a \cos \theta$$

as the desired parametric equations. ∎

FIGURE 11.21

$$x = a\theta - a \sin \theta,$$
$$y = a - a \cos \theta.$$

The cycloid discussed in Example 6 has some very fascinating physical properties. Let a point Q be given in the fourth quadrant, and imagine a drop of water placed at $(0, 0)$ that slides (without friction) along a curve from the origin $(0, 0)$ to Q, subject only to a force mg downward due to gravity, as shown in Fig. 11.22. What shape should the curve have so that the drop slides from $(0, 0)$ to Q in the least possible time? (This is the *brachistochrone problem*; the name comes from two Greek words meaning ''shortest time.'') Initially, we might

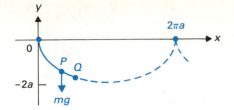

FIGURE 11.22 A drop of water at P sliding from $(0, 0)$ to Q subject only to the force mg of gravity.

think that the drop should slide along the straight-line segment joining $(0, 0)$ and Q, but after a little thought it might seem reasonable that the curve should drop more steeply at first and allow the drop to gain speed more quickly. It can be shown that the "smooth" curve (with no sharp corners) corresponding to the shortest time is a portion of a single arc of an inverted cycloid

$$x = a\theta - a \sin \theta, \qquad y = a \cos \theta - a$$

generated by a point P on a circle of radius a rolling on the *under* side of the x-axis, with P starting at the origin. The cycloid is thus the solution to the brachistochrone problem.

It can be shown that the inverted cycloid is also the solution of the *tautochrone problem* (meaning "same time"), for if a drop of water is placed at a point other than the low point on an arc of an inverted cycloid, the time required for it to slide to the low point of the arc is independent of the initial point where the drop is placed.

SUMMARY

1. A parametric curve can sometimes be sketched by eliminating the parameter to obtain an x,y-equation, but care must be taken to note any restrictions on x and y imposed by the parametric equations.

EXERCISES 11.2

In Exercises 1–12, sketch the curve having the given parametric representation.

1. $x = \sin t, y = \cos t$ for $0 \leq t \leq \pi$

2. $x = 2 \cos t, y = 3 \sin t$ for all t

3. $x = \tan t, y = \sec t$ for $-\dfrac{\pi}{2} < t < \dfrac{\pi}{2}$

4. $x = t^2, y = t + 1$ for all t

5. $x = t^2, y = t^3$ for all t

6. $x = \sin t, y = \cos 2t$ for $0 \leq t \leq 2\pi$

7. $x = t - 3, y = t^2 + 1$ for all t

8. $x = 3 \cosh t, y = 2 \sinh t$ for all t

9. $x = t - 1, y = \ln t$ for $0 < t < \infty$

10. $x = \sin t, y = 1 + \sin^2 t$ for all t

11. $x = t^2, y = t^2 + 1$ for all t

12. $x = t^2, y = t^4$ for all t

13. Parametrize the curve $y = e^x$, taking as parameter the slope m of the curve at (x, y).

14. Parametrize the curve $y = \sqrt{x}$, taking as parameter the slope s of the normal to the curve at (x, y).

15. Parametrize the cubic $y = x^3$ in terms of the second derivative s at the point (x, y) on the curve.

16. Parametrize the curve $y = \sin x$ in terms of degree measure ϕ of the angle with radian measure x.

17. Parametrize the half-line $y = 3, x \geq 0$, taking as parameter the angle of inclination θ at the origin from the x-axis to the point (x, y) on the half-line.

18. Parametrize the half-line in Exercise 17, taking as parameter the distance s from (x, y) to the origin.

19. Can you parametrize the entire graph of $y = x^4$ taking as parameter
(a) the first derivative m at a point (x, y)?
(b) the second derivative s at a point (x, y)?
(c) the third derivative t at a point (x, y)?
(d) the fourth derivative f at a point (x, y)?
Give reasons for your answers.

20. Can you parametrize the curve $y = x^2$ by taking as parameter the distance from (x, y) to $(0, 0)$? Why?

21. Reparametrize the arc $x = a \cos \theta$, $y = a \sin \theta$ for $\pi/4 \le \theta \le 3\pi/4$, taking as parameter the slope m of the arc.

22. Parametrize the curve $y = \sqrt{x}$, taking as parameter the distance d from (x, y) to $(0, 0)$.

23. A circular disk of radius a rolls along the x-axis. Find parametric equations of the curve traced by a point P on the disk a distance b from the center of the disk, where $0 \le b \le a$. Assume that the point P falls on the interval $[0, a]$ on the y-axis when the disk touches the origin, and take as parameter the angle θ through which the disk rolls from its position at the origin. (This curve is a *trochoid*. See Fig. 11.23. For $b = a$, we obtain the cycloid, while $b = 0$ yields the line $y = a$.)

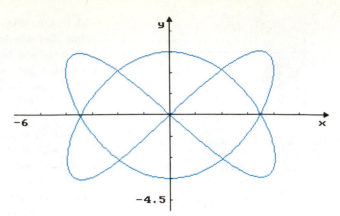

FIGURE 11.24
Lissajous figure $x = 4 \sin 2t$, $y = 3 \cos 3t$.

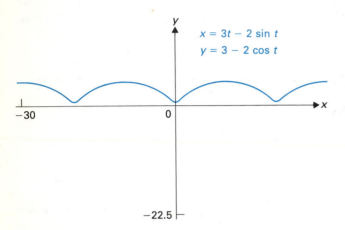

FIGURE 11.23
The trochoid $x = 3t - 2 \sin t$, $y = 3 - 2 \cos t$.

24. A curve given by parametric equations

$$x = a \sin \omega_1 t, \qquad y = b \cos \omega_2 t$$

is called a **Lissajous figure.** These curves occur in the study of electricity. Two Lissajous figures are shown in Figs. 11.24 and 11.25 as they would appear on a PC monitor. Sketch the Lissajous figure given by the equations $x = \sin t$, $y = \cos 2t$.

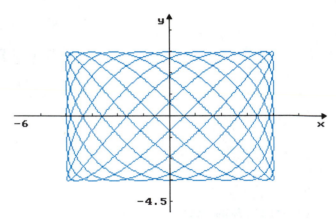

FIGURE 11.25
Lissajous figure $x = 4 \sin 7t$, $y = 3 \cos 11t$.

 EXPLORING CALCULUS

1: parametric graphs

11.3 SLOPE AND ARC LENGTH FOR PARAMETRIC CURVES

Slope of a Parametric Curve

We now use differential calculus and find the slope of a tangent line to a curve defined parametrically by $x = h(t)$, $y = k(t)$. Let us think for the moment of these equations as giving the position at time t of a body moving in the plane. If

h and k are differentiable, then the derivative $dx/dt = h'(t)$ gives the velocity of the body in the x-direction (the x-component of the velocity), and $dy/dt = k'(t)$ gives the y-component of the velocity. If, furthermore, these derivatives are continuous, then the body moves in a smooth fashion, with no abrupt jerks. Finally, suppose that there is no value of t where $h'(t) = k'(t) = 0$, so that it is impossible for the body to come to a stop and then to start again in a different direction. That is, our curve can have no sharp points. Such a curve $x = h(t)$, $y = k(t)$ where $h'(t)$ and $k'(t)$ are continuous and never simultaneously zero is a **smooth curve.**

Let $x = h(t)$, $y = k(t)$ describe a smooth curve and suppose that a piece of the curve for $a < t < b$ is the graph of a differentiable function of x. Then by the chain rule, we have

$$\frac{dy}{dt} = \frac{dy}{dx} \cdot \frac{dx}{dt}.$$

Therefore at a point (x, y) on the curve given by a value t where $a < t < b$ and $dx/dt \neq 0$, we have

$$\frac{dy}{dx} = \frac{dy/dt}{dx/dt}. \tag{1}$$

If the curve is smooth and $dx/dt \neq 0$ for $a < t < b$, then dx/dt must have the same sign throughout that interval. Consequently, x is either an increasing or a decreasing function of t for $a < t < b$, so the portion of the curve corresponding to those values of t is the graph of some function $y = f(x)$.

EXAMPLE 1 Find the tangent and normal lines to the curve $x = \cos t$, $y = \sin t$ at the point $(\sqrt{3}/2, \frac{1}{2})$ corresponding to $t_1 = \pi/6$.

Solution We have

$$\left.\frac{dy}{dx}\right|_{t=\pi/6} = \left.\frac{dy/dt}{dx/dt}\right|_{t=\pi/6} = \frac{\cos(\pi/6)}{-\sin(\pi/6)} = \frac{\sqrt{3}/2}{-\frac{1}{2}} = -\sqrt{3}.$$

The tangent line thus has x,y-equation $y = -\sqrt{3}x + 2$, and the normal line has equation $y = (1/\sqrt{3})x$. ∎

If dx/dt is zero and dy/dt nonzero at $t = t_1$, then the curve has a vertical tangent at the corresponding point (x_1, y_1).

EXAMPLE 2 Find all vertical tangent lines to the parametric curve

$$x = t^2 - 4t + 3, \qquad y = t - 1.$$

Solution We have

$$\frac{dx}{dt} = 2t - 4 \qquad \text{and} \qquad \frac{dy}{dt} = 1.$$

Thus $dx/dt = 0$ when $t = 2$, and $dy/dt \neq 0$ there. Consequently, we have a vertical tangent line when $t = 2$, or at the point $(x, y) = (-1, 1)$. The equation of this vertical line is of course $x = -1$. ∎

Suppose now that the curve is not smooth and that dx/dt and dy/dt exist but are both zero at $t = t_1$. In that case, the slope is given by the limit of the quotient (1) as $t \to t_1$ if the limit exists.

EXAMPLE 3   Find the slope when $t = 0$ of the parametric curve

$$x = t^2, \qquad y = \cos t.$$

Solution   If $x = t^2$ and $y = \cos t$, then

$$\frac{dy}{dx} = \frac{dy/dt}{dx/dt} = \frac{-\sin t}{2t}.$$

The slope of the curve when $t = 0$ is given by

$$\lim_{t \to 0} \frac{-\sin t}{2t} = \lim_{t \to 0} \left(-\frac{1}{2} \cdot \frac{\sin t}{t} \right) = -\frac{1}{2} \cdot 1 = -\frac{1}{2}. \qquad \blacksquare$$

If $h(t)$ and $k(t)$ have derivatives of sufficiently high order, then it is a straightforward (but often tedious) matter to compute the higher-order derivatives d^2y/dx^2, d^3y/dx^3, and so on. They can be computed successively using the following formulas:

$$\frac{d^2y}{dx^2} = \frac{\dfrac{d(dy/dx)}{dt}}{dx/dt}, \qquad \frac{d^3y}{dx^3} = \frac{\dfrac{d(d^2y/dx^2)}{dt}}{dx/dt}, \qquad \frac{d^4y}{dx^4} = \frac{\dfrac{d(d^3y/dx^3)}{dt}}{dx/dt},$$

and so on.

EXAMPLE 4   Find d^2y/dx^2 for the curve $x = \cos t$, $y = \sin t$.

Solution   We have

$$\frac{dy}{dx} = \frac{dy/dt}{dx/dt} = \frac{\cos t}{-\sin t} = -\cot t.$$

Then

$$\frac{d^2y}{dx^2} = \frac{\dfrac{d(dy/dx)}{dt}}{dx/dt} = \frac{\csc^2 t}{-\sin t} = -\csc^3 t. \qquad \blacksquare$$

EXAMPLE 5   Find d^3y/dx^3 if $x = t^2 - 3t + 4$ and $y = t - 1$.

Solution   We have

$$\frac{dy}{dx} = \frac{dy/dt}{dx/dt} = \frac{1}{2t - 3} = (2t - 3)^{-1}.$$

Then

$$\frac{d^2y}{dx^2} + \frac{\dfrac{d(dy/dx)}{dt}}{dx/dt} = \frac{-1(2t - 3)^{-2} \cdot 2}{2t - 3} = -2(2t - 3)^{-3}.$$

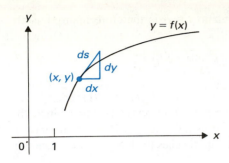

FIGURE 11.26 The differential triangle of a graph $y = f(x)$ at (x, y).

Finally,

$$\frac{d^3y}{dx^3} = \frac{\dfrac{d(d^2y/dx^2)}{dt}}{dx/dt}$$

$$= \frac{6(2t - 3)^{-4} \cdot 2}{2t - 3} = \frac{12}{(2t - 3)^5}. \quad \blacksquare$$

Length of a Parametric Curve

We start by using differential notation of Section 6.5 to suggest the appropriate integral giving the length of a smooth parametric curve. Then we justify the integral as a limit of Riemann sums.

Figure 11.26 shows again the differential triangle for the graph of a function that was presented in Section 6.5. From this triangle, we recall that the differential of arc length is given by

$$ds = \sqrt{(dx)^2 + (dy)^2}. \quad (2)$$

We can consider this same differential triangle at a point of a smooth parametric curve given by equations

$$x = h(t), \qquad y = k(t)$$

where $h'(t)$ and $k'(t)$ are continuous and never simultaneously zero. A formal manipulation of Leibniz notation from Eq. (2) leads to

$$ds = \sqrt{(dx)^2 + (dy)^2} = \sqrt{\left(\frac{dx}{dt}\right)^2 + \left(\frac{dy}{dt}\right)^2} \, dt. \quad (3)$$

Thus we expect the total distance s traveled on the curve from time $t = a$ to time $t = b$ to be

$$s = \int_{t=a}^{t=b} \sqrt{\left(\frac{dx}{dt}\right)^2 + \left(\frac{dy}{dt}\right)^2} \, dt. \quad (4)$$

This is the way we *remember* Eqs. (3) and (4), but it cannot be regarded as a proof.

To prove Eq. (4), form a partition P of the interval $[a, b]$ on the t-axis into n subintervals, so that $a = t_0 < t_1 < \cdots < t_n = b$. We approximate the arc length by adding the lengths of chords of the curve. The ith chord, shown in Fig. 11.27, joins the points

$$(h(t_{i-1}), k(t_{i-1})) \qquad \text{and} \qquad (h(t_i), k(t_i)).$$

Let $\Delta t_i = t_i - t_{i-1}$. The length of this chord is

$$\sqrt{(h(t_i) - h(t_{i-1}))^2 + (k(t_i) - k(t_{i-1}))^2}$$

$$= \sqrt{\left(\frac{h(t_i) - h(t_{i-1})}{t_i - t_{i-1}}\right)^2 + \left(\frac{k(t_i) - k(t_{i-1})}{t_i - t_{i-1}}\right)^2} \cdot \Delta t_i.$$

By the mean value theorem, there exist points c_i and c_i' between t_{i-1} and t_i such that this last expression becomes

$$\sqrt{h'(c_i)^2 + k'(c_i')^2} \cdot \Delta t_i.$$

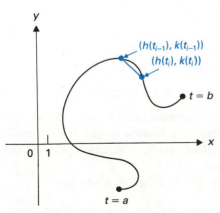

FIGURE 11.27 Chord of length $\sqrt{(h(t_i) - h(t_{i-1}))^2 + (k(t_i) - k(t_{i-1}))^2}$.

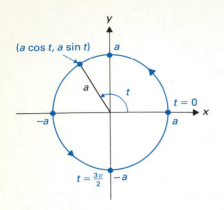

FIGURE 11.28   The circle $x^2 + y^2 = a^2$ parametrized as $x = a \cos t$, $y = a \sin t$ for $0 \le t \le 2\pi$.

The total length of the curve from $t = a$ to $t = b$ is therefore approximately

$$\sum_{i=1}^{n} \left[\sqrt{h'(c_i)^2 + k'(c_i')^2} \cdot \Delta t_i \right].$$

This sum is *almost* a Riemann sum for the function $\sqrt{h'(t)^2 + k'(t)^2}$ over $[a, b]$. The only problem is the *two* points c_i and c_i' in the ith interval used to evaluate the two summands under the radical. But there is a theorem of Bliss, mentioned in Section 6.3, that also says that such a sum approaches the integral of the function as $\|P\| \to 0$. This establishes Eq. (4).

EXAMPLE 6   Verify the formula $C = 2\pi a$ for the circumference of a circle having radius a.

Solution   We take as parametric equations for the circle

$$x = a \cos t, \qquad y = a \sin t \qquad \text{for } 0 \le t \le 2\pi.$$

See Fig. 11.28. Since $dx/dt = -a \sin t$ and $dy/dt = a \cos t$, we see from Eq. (4) that the length of the circle is

$$s = \int_0^{2\pi} \sqrt{(-a \sin t)^2 + (a \cos t)^2}\, dt$$

$$= \int_0^{2\pi} \sqrt{a^2}\, dt = a \int_0^{2\pi} (1)\, dt$$

$$= 2\pi a. \qquad \blacksquare$$

EXAMPLE 7   Find the length of the parametric curve $x = \sin^3 t$, $y = \cos^3 t$ for $0 \le t \le \pi$, shown in Fig. 11.29.

Solution   This example illustrates a tricky point. We have

$$\frac{dx}{dt} = 3 \sin^2 t \cos t, \qquad \frac{dy}{dt} = -3 \cos^2 t \sin t.$$

Thus

$$ds = \sqrt{\left(\frac{dx}{dt}\right)^2 + \left(\frac{dy}{dt}\right)^2}\, dt$$

$$= \sqrt{9 \sin^4 t \cos^2 t + 9 \cos^4 t \sin^2 t}\, dt$$

$$= \sqrt{(9 \sin^2 t \cos^2 t)(\sin^2 t + \cos^2 t)}\, dt$$

$$= \sqrt{9 \sin^2 t \cos^2 t}\, dt = |3 \sin t \cos t|\, dt.$$

The absolute value is the "tricky point." The formula for ds involves the *positive* square root; the short tangent line segments have *positive* length. If the expression under the radical turns out to be a perfect square, be sure always to use the *positive* square root.

Continuing our solution, we see that the arc length is given by

$$s = \int_0^{\pi} |3 \sin t \cos t|\, dt.$$

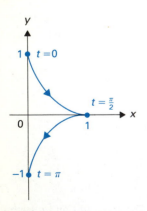

FIGURE 11.29   The curve $x = \sin^3 t$, $y = \cos^3 t$ for $0 \le t \le \pi$.

Now $3 \sin t \cos t$ is positive for $0 < t < \pi/2$ and negative for $\pi/2 < t < \pi$. Thus

$$s = \int_0^{\pi/2} 3 \sin t \cos t \, dt + \int_{\pi/2}^{\pi} -3 \sin t \cos t \, dt$$

$$= \frac{3 \sin^2 t}{2} \Bigg]_0^{\pi/2} - \frac{3 \sin^2 t}{2} \Bigg]_{\pi/2}^{\pi}$$

$$= \left(\frac{3}{2} - 0 \right) - \left(0 - \frac{3}{2} \right) = \frac{6}{2} = 3.$$

Note that if we had computed $\int_0^\pi 3 \sin t \cos t \, dt$, we would have obtained 0, which is obviously incorrect. ∎

EXAMPLE 8 Use Simpson's rule with $n = 20$ and a calculator or computer to estimate the length of the parametric curve $x = t^2$, $y = \sin t$ for $0 \le t \le 2\pi$.

Solution We have

$$ds = \sqrt{\left(\frac{dx}{dt} \right)^2 + \left(\frac{dy}{dt} \right)^2} \, dt = \sqrt{4t^2 + \cos^2 t} \, dt.$$

Our calculator shows that

$$\text{Arc length} = \int_0^{2\pi} \sqrt{4t^2 + \cos^2 t} \, dt \approx 40.051. \qquad ∎$$

In Section 6.6, we showed that the area of the surface generated by revolving a curve about an axis is given by the integral $\int_a^b 2\pi r \, ds$, where r is the *radius of revolution* for a small piece of arc of length approximately ds as it swings around the axis. (See Fig. 6.50 in Section 6.6.) If the curve is given in terms of parametric equations so that ds is expressed in terms of a parameter t, then we must also express r as a function of t.

EXAMPLE 9 Derive the formula $A = 4\pi a^2$ for the surface of a sphere of radius a.

Solution We rotate the top half of the circle $x = a \cos t$, $y = a \sin t$ shown in Fig. 11.28 about the x-axis. This portion of the circle corresponds to values of t where $0 \le t \le \pi$. The radius of revolution r about the x-axis is $y = a \sin t$. As in Example 6, we find that

$$ds = \sqrt{\left(\frac{dx}{dt} \right)^2 + \left(\frac{dy}{dt} \right)^2} \, dt = \sqrt{(-a \sin t)^2 + (a \cos t)^2} \, dt = a \, dt.$$

Thus the area of the sphere of radius a is given by

$$\int_0^\pi 2\pi(a \sin t)(a \, dt) = 2\pi a^2 (-\cos t) \Bigg]_0^\pi$$

$$= 2\pi a^2 [-(-1) - (-1)]$$

$$= 2\pi a^2 (2) = 4\pi a^2. \qquad ∎$$

SUMMARY

1. If $x = h(t)$ and $y = k(t)$ and if h and k have derivatives of sufficiently high order, then

$$\frac{dy}{dx} = \frac{dy/dt}{dx/dt}, \qquad \frac{d^2y}{dx^2} = \frac{\dfrac{d(dy/dx)}{dt}}{dx/dt}, \qquad \frac{d^3y}{dx^3} = \frac{\dfrac{d(d^2y/dx^2)}{dt}}{dx/dt},$$

and so on, where $dx/dt \neq 0$.

2. The differential of arc length is $ds = \sqrt{(dx/dt)^2 + (dy/dt)^2}\, dt$ and

$$\text{Arc length} = \int_a^b ds = \int_a^b \sqrt{\left(\frac{dx}{dt}\right)^2 + \left(\frac{dy}{dt}\right)^2}\, dt.$$

EXERCISES 11.3

In Exercises 1–11, find the slope of the curve, with the given parametric representation, at the point on the curve corresponding to the indicated value of the parameter.

1. $x = t^3$, $y = t^2$ at any time t

2. $x = 1/t^2$, $y = (t + 1)/(t - 1)$ at $t = 2$

3. $x = \sqrt{t^2 + 1}$, $y = (t^2 - 1)^3$ at $t = 1$

4. $x = (2t - 4)^2$, $y = 1 + (1/t)$ at $t = 2$

5. $x = (3t + 7)^{3/2}$, $y = t - 100$ at $t = 3$

6. $x = \sqrt{4t + 5}$, $y = 1/(t^2 - 4t)$ at $t = 5$

7. The cycloid $x = a\theta - a \sin\theta$, $y = a - a\cos\theta$, $\theta = \pi/4$

8. $x = t^2 - 3t$, $y = \sin 2t$, where $t = 0$

9. $x = \sinh t$, $y = \cosh 2t$, where $t = 0$

10. $x = e^t$, $y = \ln(t + 1)$, where $t = 0$

11. $x = t^3 - 3t^2 + 3t - 5$, $y = 4(t - 1)^3$, where $t = 1$

12. Find the equation of the tangent line to the curve $x = t^2 - 3t$, $y = \sqrt{t}$, at the point where $t = 4$.

13. Find the equation of the tangent line to the curve $x = 3t + 4$, $y = t^2 - 3t$ at the point where $x = -5$.

14. Find the equation of the tangent line to the curve with parametric equations $x = 3t^2 - 10t$, $y = \sqrt{t + 1}$, when $t = 3$.

15. Find the equation of the normal line to the curve with parametric equations

$$x = \frac{2t}{t^2 + 1}, \qquad y = 2t - 3$$

16. Find the equation of the normal line to the curve $x = (t^2 - 3)^2$, $y = (t^2 + 3)/(t - 1)$, where $t = 2$.

17. Find the equation of the normal line to the curve $x = \sqrt{2t^2 + 1}$, $y = 4t - 5$, where $y = 3$.

18. Find all points where the parametric curve $x = 3t - 1$, $y = t^3 - 3t^2 - 9t + 1$ has a horizontal tangent.

19. Find all points where the parametric curve $x = t^2 + 2t + 3$, $y = e^t - t$ has a horizontal tangent.

20. Find all points where the parametric curve $x = t^2 - 2t$, $y = e^t - t$ has a vertical tangent.

21. Find all points where the curve $x = \cos 2t$, $y = \sin t$ has a vertical tangent.

22. Find the minimum distance from a point on the parametric curve $x = t - 3$, $y = t + 1$ to the origin.

23. Find the maximum distance from a point on the parametric curve $x = 1/(1 + t^2)$, $y = 2t/(1 + t^2)$ to the origin.

24. Let $x = \sqrt{t}$ and $y = t^3$. Find dy/dx and d^2y/dx^2 in terms of t.

In Exercises 25–28, the parametric equations represent a curve that is the graph of a function in a neighborhood of the point corresponding to the indicated value of the parameter. Find d^2y/dx^2 at this point.

25. $x = t^2$, $y = t^3 - 2t^2 + 5$, where $t = 1$

26. $x = t^2$, $y = 1/(t + 1)$, where $t = 1$

27. $x = \sin 3t$, $y = e^t$, where $t = 0$

28. $x = \ln(t + 3)$, $y = \cos 2t$, where $t = 0$

29. If $x = 3t^2 - 1$ and $y = 4t + 1$, find d^3y/dx^3 in terms of t.

30. If $x = 3t + 2$ and $y = \sin 2t$, find d^4y/dx^4 in terms of t.

31. If $x = 2t - 5$ and $y = e^{3t}$, find d^5y/dx^5 in terms of t.

In Exercises 32–41, find the length of the curve with the given parametric representation.

32. $x = 4\sin(t/2)$, $y = 4\cos(t/2)$ from $t = 0$ to $t = \pi$

33. $x = t^2$, $y = \frac{2}{3}(2t + 1)^{3/2}$ from $t = 0$ to $t = 4$

34. $x = a\cos^3 t$, $y = a\sin^3 t$ from $t = 0$ to $t = \pi/2$

35. $x = a\sin^3 2t$, $y = a\cos^3 2t$ from $t = 0$ to $t = \pi$

36. $x = (t - 1)^2$, $y = \frac{8}{3}t^{3/2}$ from $t = 1$ to $t = 2$

37. $x = \frac{2}{5}t^{5/2} + 4$, $y = \frac{1}{2}t^2 - 1$ from $t = 0$ to $t = 3$ (Use the integral table on the endpapers of this book.)

38. $x = \frac{2}{7}t^{7/2} - 2$, $y = (1/\sqrt{3})t^3$ from $t = 0$ to $t = 1$ (Use the integral table.)

39. $x = t^2$, $y = \frac{2}{3}(2t + 1)^{3/2}$ from $t = 0$ to $t = 4$

40. $x = t$, $y = \ln(\cos t)$ from $t = 0$ to $t = \pi/4$

41. $x = a\cos^3 t$, $y = a\sin^3 t$ from $t = 0$ to $t = \pi/2$

42. Find the approximate length of the arc of the curve $x = 4t$, $y = \sin t$ from $t = 0$ to $t = 0.05$.

43. Find the approximate length of the arc of the curve $x = 4\cos t$, $y = 5t + 1$ from $t = 0$ to $t = 0.05$.

44. Find the approximate value of $t_1 > 0$ such that the length of the curve $x = \sin t$, $y = \tan t$ from $t = 0$ to $t = t_1$ is 0.1.

45. Find the length of one arch of the cycloid $x = a\theta - a\sin\theta$, $y = a - a\cos\theta$. [*Hint:* Integrate $\int \sqrt{1 - \cos\theta}\, d\theta$ by multiplying numerator and denominator by $\sqrt{1 + \cos\theta}$.]

In Exercises 46–52, find the area of the surface generated by revolving the given parametric curve about the indicated axis.

46. $x = a\cos t$, $y = a\sin t$ for $0 \le t \le 2\pi$ about the line $x = b$ for $b > a$. (This gives the surface area of a torus.)

47. $x = (t - 1)^2$, $y = \frac{8}{3}t^{3/2}$ from $t = 1$ to $t = 2$ about the line $x = -1$

48. $x = a\cos^3 t$, $y = a\sin^3 t$ from $t = 0$ to $t = \pi/4$ about the x-axis

49. $x = t$, $y = t^2 - 2$ from $t = 0$ to $t = 3$ about the y-axis

50. $x = t^2/2 + 2t$, $y = t + 1$ from $t = 0$ to $t = 2$ about the line $y = -1$

51. $x = 2t + 1$, $y = t^2 - 3t$ from $t = 3$ to $t = 4$ about the line $x = 4$

52. $x = 4\cos(t/2)$, $y = 4\sin(t/2)$ from $t = 0$ to $t = 2\pi$ about the line $y = -2$

In Exercises 53–56, use Simpson's rule to estimate the length of the indicated parametric arc.

53. $x = t^3 - 3t$, $y = 1/(t + 2)$ for $-1 \le t \le 1$

54. $x = \sinh t$, $y = \cosh t$ for $0 \le t \le 2$

55. $x = \sin t$, $y = \cos 3t$ for $0 \le t \le 2\pi$

56. $x = \ln t$, $y = e^t$ for $1 \le t \le 4$

11.4 CURVATURE

What Is Curvature?

We are interested in the rate at which a plane curve bends (or "curves") as we travel along the curve. We will attempt to give a numerical measure of such a rate of turning at a point on the curve; this number will be the *curvature of the curve at the point*. Thus the curvature of a curve at a point is to be a number; the more the curve bends at the point, the larger the number will be. It would be natural for us to expect curvature to satisfy the following three conditions.

Condition 1 Since a straight line does not bend at all, we would like its curvature at each point to be zero.

Condition 2 A circle "curves" at a uniform rate; we would like the curvature of a circle at one point to be the same as the curvature of the circle at every other point on the circle and the same as the curvature at each point of any other circle of the same radius. This would allow us to speak of the *curvature of a circle of radius a*.

Condition 3 The smaller the radius of a circle, the more the circle "curves" at each point. We would like the curvature of a small circle to be greater than the curvature of a circle of larger radius.

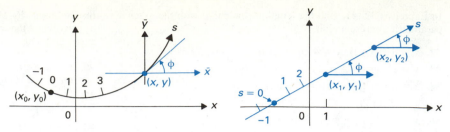

FIGURE 11.30 $\kappa = |d\phi/ds|$ measures the rate the curve bends at (x, y).

FIGURE 11.31 ϕ remains constant for a straight line, so $\kappa = |d\phi/ds| = 0$.

Consider a curve defined by one or more differentiable functions and having a tangent line at each point. We choose a point (x_0, y_0) on our curve from which we measure arc length s along the curve, so that (x_0, y_0) corresponds to $s = 0$, as shown in Fig. 11.30. Let ϕ be the angle shown in the figure, measured from the positive $\bar{x}$-axis at (x, y) to the direction corresponding to increasing s along the tangent line. The rate at which the curve bends at (x, y) can be described in terms of the rate at which the tangent line is turning as we travel along the curve at (x, y), and this in turn can be measured by the rate at which the angle ϕ is changing at (x, y). We want curvature to be intrinsic to the point set or *trace* of the curve and not dependent on the rate at which we may be traveling along the curve. Thus it would not be appropriate to let the curvature be the rate of change of ϕ per unit change in *time* as we travel along the curve. It is intuitively apparent that the notion of arc length is intrinsic to the trace, and we let the curvature of the curve at a point be the rate of change of ϕ per unit change in arc length along the curve at that point.

DEFINITION 11.1

Curvature

The **curvature** κ of a curve at (x, y) is $|d\phi/ds|$, where ϕ is the angle in Fig. 11.30 and s is arc length measured along the curve.*

EXAMPLE 1 Find the curvature of a straight line.

Solution Since the angle ϕ along a straight line is a constant function of the arc length s (see Fig. 11.31), we have $d\phi/ds = 0$, so the curvature of a straight line at each point is zero. Thus condition 1 is satisfied by our definition of curvature. ∎

EXAMPLE 2 Find the curvature of a circle of radius a at a point on the circle.

Solution Measure arc length on the circle $x^2 + y^2 = a^2$ in the counterclockwise direction, starting with $s = 0$ at $(a, 0)$, as shown in Fig. 11.32. From this

---

*Some texts define the curvature to be the signed quantity $d\phi/ds$, but we take a definition that specializes the notion of curvature for a space curve, which will be introduced in Chapter 13. The interpretation of the sign of $d\phi/ds$ is explained in Exercise 17.

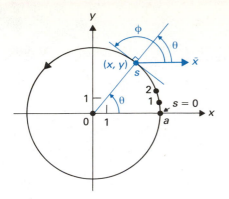

FIGURE 11.32 s at (x, y) is $a\theta$; $\phi = \theta + \pi/2 = s/a + \pi/2$; $\kappa = |d\phi/ds| = 1/a$.

figure, we see that at a point (x, y) on the circle corresponding to a central angle θ and hence to an arc length $s = a\theta$, we have $\phi = \theta + (\pi/2) + 2n\pi$. (The value of n depends on the quadrant of θ; for θ in the first quadrant as in Fig. 11.31, we have $n = 0$.) Therefore

$$\phi = \theta + \frac{\pi}{2} + 2n\pi = \frac{1}{a}s + \frac{\pi}{2} + 2n\pi,$$

so

$$\kappa = \frac{d\phi}{ds} = \frac{1}{a}.$$

Thus the curvature at any point of a circle of radius a is the reciprocal $1/a$ of the radius. We see that conditions 2 and 3 are satisfied by our definition of curvature. ∎

The preceding example suggests the following definition.

DEFINITION 11.2

Radius of curvature

If a curve has curvature $\kappa \neq 0$ at a point, then the **radius of curvature** of the curve at this point is $\rho = 1/\kappa$.

The Formula for the Curvature of $y = f(x)$

Let a curve be the graph $y = f(x)$ for a *twice*-differentiable function f, and let the direction of increasing arc length measured from (x_0, y_0) be the direction of increasing x, so that

$$s = \int_{x_0}^{x} \sqrt{1 + f'(t)^2} \, dt. \tag{1}$$

For the angle ϕ shown in Fig. 11.30, we have

$$\tan \phi = f'(x)$$

at any point (x, y) on the graph. Thus in a neighborhood of (x, y),

$$\phi = b + \tan^{-1} f'(x), \tag{2}$$

where the constant b occurs since ϕ may not fall in the principal value range of the inverse tangent function. (In Fig. 11.30, we see that $b = 0$.) Since f is a twice-differentiable function of x, we see from Eq. (2) that ϕ is a differentiable function of x. From Eq. (1) we see that s is a differentiable function of x; since $ds/dx \neq 0$, it can then be shown that x appears as a differentiable function of s. Under these conditions, ϕ then appears as a differentiable function of s, and, by the chain rule,

$$\frac{d\phi}{ds} = \frac{d\phi}{dx} \cdot \frac{dx}{ds}. \tag{3}$$

From Eq. (2),

$$\frac{d\phi}{dx} = \frac{1}{1 + (f'(x))^2} \cdot f''(x), \tag{4}$$

and from Eq. (1),

$$\frac{ds}{dx} = \sqrt{1 + (f'(x))^2}.$$

Thus

$$\frac{dx}{ds} = \frac{1}{\sqrt{1 + (f'(x))^2}}, \tag{5}$$

and, from Eqs. (3), (4), and (5),

$$\kappa = \left| \frac{d\phi}{ds} \right| = \left| \frac{f''(x)}{[1 + (f'(x))^2]^{3/2}} \right|. \tag{6}$$

It is straightforward to check that $d\phi/ds > 0$ where, when traveling along the curve in the direction of increasing s, the curve bends to the left (ϕ is increasing), and $d\phi/ds < 0$ where the curve bends to the right (ϕ is decreasing). See Exercise 17.

THEOREM 11.1   Curvature of a graph

The curvature of the graph of a twice-differentiable function f at a point (x, y) on the graph is given by

$$\kappa = \left| \frac{d\phi}{ds} \right| = \left| \frac{f''(x)}{[1 + (f'(x))^2]^{3/2}} \right|. \tag{7}$$

(Geometrically, $d\phi/ds$ is positive at a point if, when traveling along the curve in the direction of increasing s, the curve bends to the left, and $d\phi/ds$ is negative if the curve bends to the right.)   □

EXAMPLE 3   Find the curvature and radius of curvature of the parabola $y = x^2$ at the origin.

Solution   For $f(x) = x^2$, we have

$$\kappa = \left| \frac{f''(x)}{[1 + (f'(x))^2]^{3/2}} \right| = \frac{2}{(1 + 4x^2)^{3/2}},$$

so

$$\kappa \big|_{(0, 0)} = 2.$$

The radius of curvature is $\rho = 1/\kappa = \frac{1}{2}$.   ∎

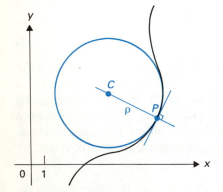

FIGURE 11.33   The osculating circle is the one best fitting the curve at P.

At a point P on a curve where the radius of curvature is ρ, go along the normal line on the concave side of the curve the distance ρ, arriving at a point C on the normal. The circle with center C and radius ρ is clearly the circle that best fits the curve at P, as shown in Fig. 11.33.

DEFINITION 11.3

Osculating circle and center of curvature of a curve

Let a curve have radius of curvature ρ at a point (x, y). The **osculating circle of the curve at** (x, y) is the circle of radius ρ through (x, y) having the same tangent as the curve at (x, y) and having center on the concave side of the curve. The center of this circle is the **center of curvature of the curve at** (x, y).

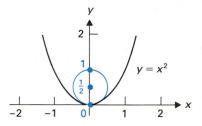

FIGURE 11.34 The osculating circle to $y = x^2$ at $(0, 0)$.

EXAMPLE 4 Find the center of curvature and the equation of the osculating circle to the graph of $y = x^2$ at the origin.

Solution From Example 3 and Fig. 11.34, we can see that the center of curvature of the parabola $y = x^2$ at $(0, 0)$ is $(0, \frac{1}{2})$, and the osculating circle has equation

$$x^2 + \left(y - \frac{1}{2}\right)^2 = \frac{1}{4}. \qquad \blacksquare$$

EXAMPLE 5 Find the curvature and radius of curvature of the ellipse $x^2/25 + y^2/9 = 1$ at the point $(0, 3)$.

Solution Near $(0, 3)$, the ellipse defines y as a function f of x. We find $f'(x) = dy/dx$ and $f''(x) = d^2y/dx^2$ for Eq. (7) by implicit differentiation:

$$\frac{x^2}{25} + \frac{y^2}{9} = 1, \qquad \frac{2x}{25} + \frac{2y}{9} \cdot \frac{dy}{dx} = 0,$$

$$\frac{dy}{dx} = -\frac{9x}{25y}, \qquad \frac{d^2y}{dx^2} = -\frac{25y \cdot 9 - 9x[25(dy/dx)]}{(25y)^2}.$$

We obtain

$$\left.\frac{dy}{dx}\right|_{(0, 3)} = 0 \qquad \text{and} \qquad \left.\frac{d^2y}{dx^2}\right|_{(0, 3)} = -\frac{25 \cdot 3 \cdot 9 - 0}{(25 \cdot 3)^2} = -\frac{3}{25}.$$

Thus

$$\kappa = \frac{|d^2y/dx^2|}{[1 + (dy/dx)^2]^{3/2}} = \frac{\frac{3}{25}}{(1 + 0^2)^{3/2}} = \frac{3}{25}.$$

The radius of curvature is $\rho = 1/\kappa = \frac{25}{3}$. $\qquad \blacksquare$

EXAMPLE 6 Find the center of curvature and the equation of the osculating circle at the point $(0, 3)$ of the ellipse in Example 5.

Solution From Fig. 11.35, we see that the center of curvature is the point $(0, 3 - \frac{25}{3}) = (0, -\frac{16}{3})$. The osculating circle then has equation

$$(x - 0)^2 + \left(y + \frac{16}{3}\right)^2 = \left(\frac{25}{3}\right)^2, \quad \text{or} \quad 3x^2 + 3y^2 + 32y^2 = 123. \qquad \blacksquare$$

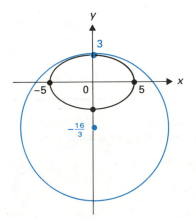

FIGURE 11.35 The osculating circle to the ellipse $x^2/25 + y^2/9 = 1$ at $(0, 3)$.

If the normal line to a curve at a point is not horizontal or vertical, we may not be able to find the center of curvature from ρ by inspection, as we did in

Examples 4 and 6. Formulas given just before Exercise 30 are convenient for finding the center of curvature in such a case.

The Curvature Formula in Parametric Form

Let a curve be given parametrically by $x = h(t)$, $y = k(t)$, where h and k are twice-differentiable functions of t. We have previously defined the arc length function

$$s(t) = \int_{t_0}^{t} \sqrt{h'(u)^2 + k'(u)^2}\, du. \tag{8}$$

If $ds/dt \neq 0$, then it can be shown that t appears as a differentiable function of s, with

$$\frac{dt}{ds} = \frac{1}{\sqrt{(h'(t))^2 + (k'(t))^2}}. \tag{9}$$

For the angle ϕ shown in Fig. 11.30, we obtain

$$\phi = b + \tan^{-1} \frac{k'(t)}{h'(t)} \tag{10}$$

as analogue of Eq. (2), so ϕ is a differentiable function of t. Under these conditions,

$$\frac{d\phi}{ds} = \frac{d\phi}{dt} \cdot \frac{dt}{ds}$$

$$= \frac{1}{1 + (k'(t)/h'(t))^2} \cdot \frac{h'(t)k''(t) - k'(t)h''(t)}{(h'(t))^2} \cdot \frac{1}{\sqrt{(h'(t))^2 + (k'(t))^2}}$$

$$= \frac{h'(t)k''(t) - k'(t)h''(t)}{[(h'(t))^2 + (k'(t))^2]^{3/2}}. \tag{11}$$

Formula (11) looks less cumbersome if we let

$$\dot{x} = \frac{dx}{dt} = h'(t) \qquad \text{and} \qquad \dot{y} = \frac{dy}{dt} = k'(t)$$

and denote second derivatives with respect to t by $\ddot{x}$ and $\ddot{y}$. This *Newtonian notation* appears often in physics texts. We then obtain formula (12), given in the following theorem.

THEOREM 11.2   Curvature of a parametric curve

Let h and k be twice-differentiable functions. The curvature of the parametric curve $x = h(t)$, $y = k(t)$ is given at a point corresponding to a value t by

$$\kappa = \left| \frac{d\phi}{ds} \right| \tag{12}$$

$$= \left| \frac{\dot{x}\ddot{y} - \dot{y}\ddot{x}}{(\dot{x}^2 + \dot{y}^2)^{3/2}} \right|. \qquad \square$$

EXAMPLE 7 Find the curvature of the ellipse $x = 3 \cos t$, $y = 4 \sin t$ at the point $(3, 0)$ corresponding to $t = 0$, taking the direction of increasing s counterclockwise to coincide with the direction of increasing t.

Solution From formula (12), we obtain

$$\kappa = \left| \frac{\dot{x}\ddot{y} - \dot{y}\ddot{x}}{(\dot{x}^2 + \dot{y}^2)^{3/2}} \right|$$

$$= \left| \frac{12 \sin^2 t - (-12 \cos^2 t)}{(9 \sin^2 t + 16 \cos^2 t)^{3/2}} \right|.$$

For $t = 0$ corresponding to the point $(3, 0)$, we obtain

$$\kappa|_{t=0} = \frac{12}{16^{3/2}} = \frac{12}{64} = \frac{3}{16}.$$ ∎

EXAMPLE 8 Find the curvature and radius of curvature of the curve $x = t^2 - 3t$, $y = t^3$ at the point where $t = 1$.

Solution We have

$$\dot{x}|_1 = (2t - 3)|_1 = -1, \qquad \dot{y}|_1 = 3t^2|_1 = 3,$$
$$\ddot{x}|_1 = 2, \qquad \ddot{y}|_1 = 6t|_1 = 6.$$

Thus from formula (12),

$$\kappa = \left| \frac{-1 \cdot 6 - 3 \cdot 2}{(1 + 9)^{3/2}} \right| = \left| \frac{-12}{10\sqrt{10}} \right| = \frac{6}{5\sqrt{10}}.$$

The radius of curvature is $\rho = 1/\kappa = 5\sqrt{10}/6$. ∎

Sometimes we may want to find the curvature of a graph given in the form $x = g(y)$ rather than $y = f(x)$. In such a case,

$$\kappa = \left| \frac{g''(y)}{[1 + (g'(y))^2]^{3/2}} \right|, \tag{13}$$

which is a formula parallel to Eq. (6). Of course, we would expect this to be the case; κ is intrinsic to the trace of the curve and independent of the choice of axes. For another way of justifying Eq. (13), we put $x = g(y)$ in parametric form as

$$x = g(t), \qquad y = t.$$

Then

$$\dot{x} = g'(t), \qquad \ddot{x} = g''(t), \qquad \dot{y} = 1, \qquad \ddot{y} = 0.$$

Formula (12) yields

$$\kappa = \left| \frac{-1 \cdot g''(t)}{[(g'(t))^2 + 1]^{3/2}} \right| = \left| \frac{g''(t)}{[1 + (g'(t))^2]^{3/2}} \right|.$$

Since $t = y$, we at once obtain Eq. (13).

EXAMPLE 9   Find the curvature and radius of curvature of the curve $x = y^3 - 3y^2 + 2$ at the point $(-2, 2)$.

Solution   We have

$$\frac{dx}{dy}\bigg|_2 = (3y^2 - 6y)\big|_2 = 12 - 12 = 0,$$

$$\frac{d^2x}{dy^2}\bigg|_2 = (6y - 6)\big|_2 = 12 - 6 = 6.$$

Thus from Eq. (13),

$$\kappa = \left|\frac{6}{(1 + 0^2)^{3/2}}\right| = 6 \quad \text{and} \quad \rho = \frac{1}{\kappa} = \frac{1}{6}. \qquad \blacksquare$$

SUMMARY

1. Curvature $\kappa = |d\phi/ds|$, where ϕ is the angle (measured counter-clockwise) from the horizontal to the tangent to the curve, and s is arc length.

2. If $y = f(x)$, then

$$\kappa = \left|\frac{d^2y/dx^2}{[1 + (dy/dx)^2]^{3/2}}\right|.$$

3. If $x = g(y)$, then

$$\kappa = \left|\frac{d^2x/dy^2}{[1 + (dx/dy)^2]^{3/2}}\right|.$$

4. If $x = h(t)$ and $y = k(t)$, then

$$\kappa = \left|\frac{\dot{x}\ddot{y} - \dot{y}\ddot{x}}{(\dot{x}^2 + \dot{y}^2)^{3/2}}\right|,$$

where $\dot{x} = dx/dt$, $\ddot{x} = d^2x/dt^2$, and $\dot{y}$ and $\ddot{y}$ are similarly defined.

5. The radius of curvature ρ at a point on a curve is equal to $1/\kappa$.

6. The osculating circle to a curve at a point is the circle with center on the concave side of the curve, radius $\rho = 1/\kappa$, and having the same tangent line as the curve at that point.

EXERCISES 11.4

In Exercises 1–16, find the curvature κ at the indicated point of the curve having the given x,y-equation or parametric equations.

1. $y = \sin x$ at $(\pi/2, 1)$

2. $xy = 1$ at $(1, 1)$

3. $y = \ln x$ at $(1, 0)$

4. $y = \dfrac{1}{1 + x^2}$ at $(0, 1)$

5. $y = x^4 - 3x^3 + 4x^2 - 2x + 2$ at $(1, 2)$

6. $x = e^y - 2y$ at $(1, 0)$

7. $x = \sin 2y - 3 \cos 2y$ at $(3, \pi/2)$

8. $x^2 - y^2 = 16$ at $(5, 3)$

9. $x = \sqrt{25 - y^2}$ at $(3, -4)$

10. $x^2y + 2xy^2 = 3$ at $(1, 1)$

11. $(x^2/16) - (y^2/25) = 1$ at $(4, 0)$ [*Hint:* The hyperbola defines x as a function of y near $(4, 0)$.]

12. $x = t^2 + 3t$, $y = t^4 - 3t^2$, where $t = -1$

13. $x = 4 \sin t$, $y = 5 \cos t$, where $t = 0$

14. The cycloid $x = a(\theta - \sin \theta)$, $y = a(1 - \cos \theta)$, where $\theta = \pi$

15. $x = e^t$, $y = t^2$, where $t = 0$

16. $x = \cosh t$, $y = \sinh t$, where $t = 0$

17. Give rate-of-change arguments that $d\phi/ds$ at a point of a curve is *positive* if the curve bends to the left as you travel in the direction of increasing s at the point and is *negative* if the curve bends to the right.

18. Discuss the curvature at the origin of the graphs of the monomial functions ax^n for integers $n \geq 1$.

19. What is the curvature of the graph of a twice-differentiable function at an inflection point on the graph?

20. The osculating circle to a given curve at a point (x, y) has the same tangent line and the same radius of curvature as the given curve at (x, y).
    (a) Show that the given curve and the circle have the same curvature κ at (x, y).
    (b) Argue from part (a) that if both the given curve and the circle are graphs of functions in a neighborhood of (x, y), then these two functions have the same first derivatives and the same second derivatives for this value x.

21. Prove that the only twice-differentiable functions whose graphs have curvature zero at each point are those functions having straight lines as graphs.

22. Find the equation of the osculating circle to the curve $y = \ln x$ at $(1, 0)$.

23. Find the equation of the osculating circle to the curve $x^2 + y^2 - 4x + 2y = 0$ at the point $(3, 1)$.

24. Find the equation of the osculating circle to the hyperbola $(x^2/4) - (y^2/9) = 1$ at the point $(-2, 0)$.

25. Let κ_n be the curvature of $y = x^n$ at $(1, 1)$. Find $\lim_{n \to \infty} \kappa_n$.

26. Find the point(s) where the curve $y = x^2$ has maximum curvature.

27. Find the point(s) where the curve $y = x^3$ has (a) maximum curvature, (b) minimum curvature.

28. For $n > 2$, the curve $y = x^n$ has two points of maximum curvature, one point A_n in the first quadrant and one point B_n in either the second or third quadrant.
    (a) Find the point in the plane that A_n approaches as $n \to \infty$; that is, find $\lim_{n \to \infty} A_n$.
    (b) Find $\lim_{n \to \infty} B_n$.
    [*Hint:* Don't try to solve this problem analytically. Just think in terms of the graphs.]

29. Find the point of maximum curvature of $x = \cosh t$, $y = \sinh t$.

Let the graph $y = f(x)$ of a twice-differentiable function f have nonzero curvature at a point (x, y). It can be shown that the coordinates (α, β) of the center of curvature of the graph at (x, y) are given by

$$\alpha = x - y' \frac{1 + (y')^2}{y''}, \qquad \beta = y + \frac{1 + (y')^2}{y''}.$$

In Exercises 30–33, use these formulas to find the equation of the osculating circle to the given curve at the indicated point.

30. $y = x^2$ at $(1, 1)$

31. $xy = 1$ at $(1, 1)$

32. $y = \tan x$ at $(\pi/4, 1)$

33. $x = t^2$, $y = t^3$, where $t = -1$

The set of all centers of curvature of a given curve is the **evolute of the curve**. The given curve is an **involute** of this set of centers of curvature. (While a curve has only one evolute, a single set may have many involutes, that is, it may be the evolute of many curves, as illustrated by the next exercise.)

34. (a) What is the evolute of a circle?
    (b) What are the involutes of a point?

35. Find parametric equations for the evolute of the parabola $y = x^2$. See Fig. 11.36. Use the formulas given before Exercise 30.

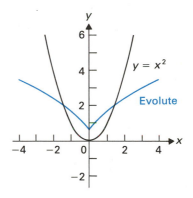

FIGURE 11.36
The evolute of the parabola $y = x^2$.

36. Find parametric equations for the evolute of the ellipse $x = 2 \sin t$, $y = \cos t$, using the formulas given before Exercise 30.

EXPLORING CALCULUS

N: calculus topics quiz (topic 6)
V: osculating graphic
W: evolute graphic
X: osculate + evolute graphic
Y: evolute-envelope graphic

11.5 POLAR COORDINATES

FIGURE 11.37 The polar axis.

Lines through the origin and circles with center at the origin occur often in applications. For example, temperature in the plane due to a source of heat at the origin is constant on each circle with center at the origin. A nonvertical line through the origin has quite a simple equation, $y = mx$. A circle with center $(0, 0)$ has the more complicated equation $x^2 + y^2 = a^2$. In polar coordinates, a point has coordinates (r, θ). We will see that *every* line through the origin has an equation of the form $\theta = c$ and the circle with center $(0, 0)$ and radius a has equation $r = a$. These equations are simpler than their Cartesian counterparts, and consequently polar coordinates are especially useful when dealing with lines through the origin and circles with center $(0, 0)$.

We now describe the polar coordinate system and its relation to x,y-coordinates. Choose any point O in the Euclidean plane as the *pole* for our coordinate system and any half-line emanating from O as the *polar axis*. It is conventional in sketching to let the polar axis be a horizontal half-line, extending to the right, as shown in Fig. 11.37. We choose a scale on the polar axis, as indicated in the figure.

Let P be any point in the plane, and rotate the polar axis through an angle θ so that the rotated axis passes through the point P, as indicated in Fig. 11.38. We will always let positive values of θ correspond to *counterclockwise* rotation. The point P falls at a number r on the scale of the rotated axis, and r and θ are **polar coordinates** for the point P. We denote P by (r, θ).

In a Cartesian coordinate system, each point in the plane corresponds to a *unique* ordered pair (x, y) of numbers. A feature of a polar coordinate system is that each point has an *infinite* number of polar coordinate pairs. As indicated in Fig. 11.39, if a point P has polar coordinates (r, θ) it also has polar coordinates $(r, \theta + 2n\pi)$ for each integer n. It is conventional to allow negative values for r on the extension of the polar axis through the pole. Figure 11.40 shows that the same point P also has polar coordinates $(-r, \theta + \pi + 2n\pi)$ for all integers n. All polar coordinates of a point (r, θ) different from O are of the form

$$(r, \theta + 2n\pi) \qquad \text{or} \qquad (-r, \theta + \pi + 2n\pi).$$

Note that the pole O has coordinates $(0, \theta)$ for every real number θ.

We will simultaneously consider a Cartesian and a polar coordinate system for the plane. It is conventional to let the pole be at the Euclidean origin $(0, 0)$ with the polar axis falling on the positive x-axis. We will always follow this convention. From Fig. 11.41, we see that the Cartesian x,y-coordinates can be

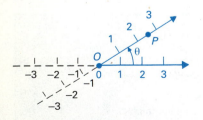

FIGURE 11.38 The polar axis rotated to go through P.

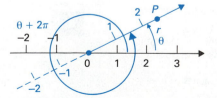

FIGURE 11.39 P has polar coordinates $(r, \theta + 2n\pi)$ for all integers n.

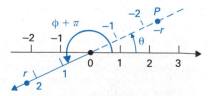

FIGURE 11.40 P also has coordinates $(-r, \theta + \pi + 2n\pi)$ for all integers n.

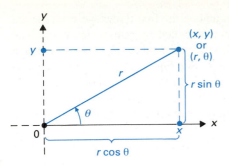

FIGURE 11.41

$x = r \cos \theta$; $y = r \sin \theta$;
$r^2 = x^2 + y^2$; $\theta = \tan^{-1}(y/x)$

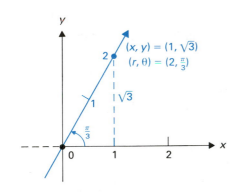

FIGURE 11.42

$x = 2 \cos(\pi/3) = 1$,
$y = 2 \sin(\pi/3) = \sqrt{3}$

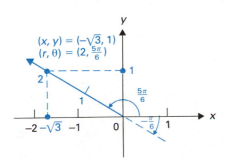

FIGURE 11.43   Cartesian $(-\sqrt{3}, 1)$ is
polar $(2, 5\pi/6)$ or $(-2, -\pi/6)$.

expressed in terms of polar r, θ-coordinates by the equations

$$x = r \cos \theta, \quad y = r \sin \theta. \tag{1}$$

It also follows at once from the figure that

$$r^2 = x^2 + y^2, \qquad \theta = \tan^{-1}\frac{y}{x}, \tag{2}$$

provided that we select θ such that $-\pi/2 < \theta < \pi/2$. Equations (1) and (2) are useful in changing from one coordinate system to the other.

EXAMPLE 1   Find Cartesian x, y-coordinates for the point with polar coordinates $(r, \theta) = (2, \pi/3)$, shown in Fig. 11.42.

Solution   From Eqs. (1), we find that

$$(x, y) = \left(2 \cos \frac{\pi}{3}, 2 \sin \frac{\pi}{3}\right) = (1, \sqrt{3}). \quad \blacksquare$$

EXAMPLE 2   Find all polar r, θ-coordinates for the point with Cartesian coordinates $(x, y) = (-\sqrt{3}, 1)$.

Solution   From Eqs. (2), the point with Cartesian coordinates $(x, y) = (-\sqrt{3}, 1)$ has polar coordinates (r, θ) such that

$$r^2 = x^2 + y^2 = 4$$

and

$$\theta = \tan^{-1}\frac{y}{x} = \tan^{-1}\left(-\frac{1}{\sqrt{3}}\right) = -\frac{\pi}{6}.$$

Since $(-\sqrt{3}, 1)$ lies in the second Cartesian quadrant, we see that $r = -2$, and the point has polar coordinates $(r, \theta) = (-2, -\pi/6)$. Of course, the point also has polar coordinates

$$\left(-2, -\frac{\pi}{6} + 2n\pi\right) \quad \text{and} \quad \left(2, \frac{5\pi}{6} + 2n\pi\right)$$

for all integers n. This also may be seen graphically, in Fig. 11.43. $\quad \blacksquare$

EXAMPLE 3   The force of attraction between two bodies has magnitude directly proportional to the product of their masses and inversely proportional to the square of the distance between them. A satellite of mass m is in planar orbit about the earth. Imagine the earth to be at the pole for polar coordinates in this plane. If the earth has mass M, express the magnitude of the force the earth exerts on the satellite in terms of the polar coordinate position (r, θ) of the satellite.

Solution   With the earth at the pole, the distance from the earth to the satellite is given by the polar coordinate r. Thus the force of attraction has magnitude

$$\frac{GmM}{r^2}$$

for some constant G (the constant of gravitational attraction). (In Section 13.3, we make a more detailed study of this force, using polar coordinates.) $\quad \blacksquare$

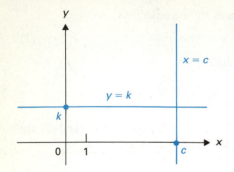

FIGURE 11.44   Level coordinate "curves" in Cartesian coordinates.

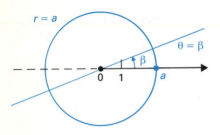

FIGURE 11.45   Level coordinate "curves" in polar coordinates.

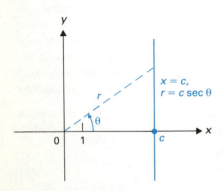

FIGURE 11.46   Cartesian: $x = c$; polar: $r = c \sec \theta$.

In a coordinate system for the plane, among the first things we examine are the *level coordinate curves* of the system, obtained by putting the coordinate variables equal to constants. In a Cartesian system, a level curve is either a vertical line $x = c$ or a horizontal line $y = k$. See Fig. 11.44. In a polar coordinate system, a level curve is either $r = a$, which is a circle about the pole of radius a, or $\theta = \beta$, which is a line through the pole, as indicated in Fig. 11.45. This suggests that polar coordinates might be useful in handling circles and lines through the origin.

Equations (1) and (2) can be used to express a Cartesian characterization of a curve in polar form, and vice versa.

EXAMPLE 4   Find a polar equation for the vertical line $x = c$, shown in Fig. 11.46.

Solution   Since $x = r \cos \theta$, we obtain from $x = c$

$$r \cos \theta = c \qquad \text{or} \qquad r = c \sec \theta.$$

The line is traced once for $-\pi/2 < \theta < \pi/2$.   ∎

EXAMPLE 5   Sketch the polar curve $r = a \sin \theta$ by changing it to Cartesian form.

Solution   If $r \neq 0$, our equation is equivalent to $r^2 = ar \sin \theta$. Since $r = 0$ if $\theta = 0$ in our original equation, we see that our new equation actually gives exactly the original curve. By Eqs. (1) and (2), the equation $r^2 = ar \sin \theta$ has Cartesian form

$$x^2 + y^2 = ay, \qquad \text{or} \qquad x^2 + \left(y - \frac{a}{2}\right)^2 = \frac{a^2}{4}.$$

We see that the curve is the circle with center at the point $(x, y) = (0, a/2)$ and radius $a/2$, as shown in Fig. 11.47.

We could have found the x,y-equation without multiplying $r = a \sin \theta$ through by r, although it is a bit more work. Starting with $r = a \sin \theta$, Eqs. (2) yield

$$\sqrt{x^2 + y^2} = a \sin\left(\tan^{-1} \frac{y}{x}\right).$$

From the triangle in Fig. 11.48, we see that $\sin(\tan^{-1}(y/x)) = y/\sqrt{x^2 + y^2}$. Thus our equation becomes

$$\sqrt{x^2 + y^2} = a \frac{y}{\sqrt{x^2 + y^2}},$$

or

$$x^2 + y^2 = ay,$$

which is the form we obtained before.   ∎

EXAMPLE 6   Find the polar coordinate equation of the ellipse with x,y-equation $x^2 + 3y^2 = 10$.

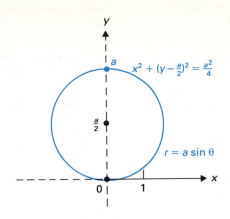

FIGURE 11.47 Polar: $r = a \sin \theta$; Cartesian: $x^2 + y^2 = ay$.

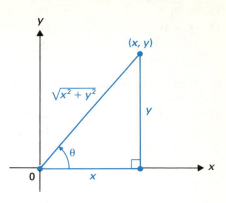

FIGURE 11.48 $\theta = \tan^{-1}(y/x)$; $\sin \theta = y/\sqrt{x^2 + y^2}$.

Solution We use Eqs. (1) and obtain

$$r^2 \cos^2\theta + 3(r^2 \sin^2\theta) = 10, \quad \text{or} \quad r^2(\cos^2\theta + 3 \sin^2\theta) = 10.$$

Since $\cos^2\theta = 1 - \sin^2\theta$, we may write our equation in the form

$$r^2(1 + 2 \sin^2\theta) = 10, \quad \text{or} \quad r^2 = \frac{10}{1 + 2 \sin^2\theta}. \quad \blacksquare$$

SUMMARY

1. A point with the polar coordinates (r, θ) also has as polar coordinates $(r, \theta + 2n\pi)$ and $(-r, \theta + \pi + 2n\pi)$ for every integer n.

2. The equations for changing from x,y-coordinates to polar r,θ-coordinates and vice versa are

$$x = r \cos \theta, \quad r^2 = x^2 + y^2,$$

$$y = r \sin \theta, \quad \theta = \tan^{-1}\frac{y}{x}, \quad -\frac{\pi}{2} < \theta < \frac{\pi}{2}.$$

EXERCISES 11.5

In Exercises 1–10, find the Cartesian coordinates of the points with the given polar coordinates.

1. $(4, \pi/4)$

2. $(0, 5\pi/7)$

3. $(6, -\pi/2)$

4. $(3, 5\pi/6)$

5. $(-2, \pi/4)$

6. $(-1, 2\pi/3)$

7. $(-4, 3\pi)$

8. $(-2, 11\pi/4)$

9. $(0, \pi/12)$

10. $(3, -\pi/4)$

In Exercises 11–20, find all polar coordinates of the points with the given Cartesian coordinates.

11. $(2, 2)$

12. $(-1, \sqrt{3})$

13. $(0, 0)$

14. $(-3, -3)$

15. $(3, -4)$

16. $(-3, 0)$

17. $(-2, -3)$

18. $(0, -5)$

19. $(2\sqrt{3}, -2)$

20. $(-1, 2)$

In Exercises 21–30, find the polar coordinate equation of the curve with the given x,y-equation.

21. $2x + 3y = 5$

22. $y^2 = 8x$

23. $x^2 - y^2 = 1$

24. $x^2 + y^2 = 3$

25. $2x^2 + y^2 = 4$

26. $x^3 + xy^2 = 3$

27. $x^2y + y^3 = -4$

28. $x^2 + y^2 - 8x = 0$

29. $x^2 + y^2 + 4y = 0$

30. $y = -x$

In Exercises 31–40, find the x,y-equation of the curve with the given polar coordinate equation.

31. $r = 8$

32. $r = 2a \cos \theta$

33. $r = -3a \sin \theta$

34. $r = 4 \csc \theta$

35. $r = -3 \sec \theta$

36. $r^2 = 2a \sin \theta$

37. $r^2 = -2a \cos \theta$

38. $r^3 = \sin \theta$

39. $r^3 = 4 \csc \theta$

40. $r^4 = 3 \sec^2 \theta$

41. Show that the distance d in the plane between points having polar coordinates (r_1, θ_1) and (r_2, θ_2) is

$$\sqrt{r_1{}^2 - 2r_1r_2 \cos(\theta_1 - \theta_2) + r_2{}^2}.$$

[*Hint:* Use the distance formula in Cartesian form and Eqs. (1).]

In Exercises 42–50, sketch the curve or shade the region in the plane satisfying the polar coordinate relations.

42. $r = 0$

43. $r = 3$

44. $0 \leq r \leq 4$

45. $2 \leq r \leq 5$

46. $\theta = \dfrac{\pi}{3}$

47. $\theta = -\dfrac{\pi}{4}, r \geq 0$

48. $\dfrac{\pi}{4} \leq \theta \leq \dfrac{3\pi}{4}$

49. $\dfrac{\pi}{4} \leq \theta \leq \dfrac{3\pi}{4}, r \geq 0$

50. $2 \leq r \leq 3, \dfrac{\pi}{6} \leq \theta \leq \dfrac{\pi}{3}$

11.6 SKETCHING CURVES IN POLAR COORDINATES

Sketching curves given by a polar coordinate equation is not always an easy task. However, some attractive and intriguing curves are obtained. A computer program for sketching is very helpful. We mention a few general principles, sketch some curves, and discuss finding points of intersection of curves.

General Principles

If a polar coordinate curve is bounded, so that $|r| < b$ for some number b at all points (r, θ) on the curve, then we would like to be sure to plot the points on the curve where $|r|$ is a local maximum or minimum, that is, where we are locally as far from or close to the origin as possible. Consider, for example, the equations $r = a + b \cos \theta$ and $r = a + b \sin \theta$. Since maximum and minimum values of $\cos \theta$ and $\sin \theta$ occur among values of θ that are integer multiples of $\pi/2$, we would like to be sure to plot points on these curves corresponding to such values of θ, that is, we want to plot points at least "every 90°." Since maximum and minimum values of $\cos 2\theta$ and $\sin 2\theta$ occur among values of θ that are integer multiples of $\pi/4$, we want to plot points on the curves $r = a + b \cos 2\theta$ and $r = a + b \sin 2\theta$ at least "every 45°." We make a similar argument whenever the polar equation contains the term $\cos n\theta$ or $\sin n\theta$.

Since each point of the plane has an infinite number of polar coordinates, symmetry about the x-axis, y-axis, or the origin may be hard to establish from the equation. We mention some relations that may be useful.

Symmetry about the x-axis If the same polar equation is obtained by replacing θ by $-\theta$ or by replacing θ by $\pi - \theta$ and r by $-r$, then the curve is symmetric

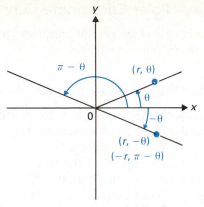

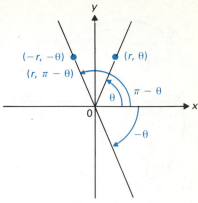

(a) Symmetry about the x-axis

(b) Symmetry about the y-axis

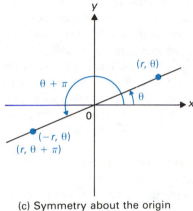

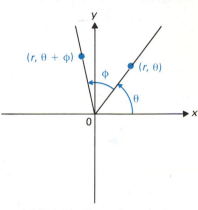

(c) Symmetry about the origin

(d) Rotation through angle φ

FIGURE 11.49

about the x-axis. See Fig. 11.49(a). *Example:* $r = a + b \cos \theta$ describes a curve symmetric about the x-axis since $\cos(-\theta) = \cos \theta$.

Symmetry about the y-axis   If the same polar equation is obtained by replacing θ by $\pi - \theta$ or by replacing θ by $-\theta$ and r by $-r$, then the curve is symmetric about the y-axis. See Fig. 11.49(b). *Examples:* $r = a + b \sin \theta$ describes a curve symmetric about the y-axis since $\sin(\pi - \theta) = \sin \theta$, and so does $r = \theta$ since $-r = -\theta$ is an equivalent equation.

Symmetry about the origin   If the same polar equation is obtained by replacing r by $-r$ or by replacing θ by $\theta + \pi$, then the curve is symmetric about the origin. See Fig. 11.49(c). *Examples:* $r^2 = a^2 \sin 2\theta$ describes a curve symmetric about the origin, as does $r = a \cos 2\theta$ since $\cos 2(\theta + \pi) = \cos(2\theta + 2\pi) = \cos 2\theta$.

Rotation through φ   If θ is replaced by $\theta - \phi$ in a polar coordinate equation, the curve described by the new equation can be obtained from the curve described by the original equation by rotating the original curve about the origin counterclockwise through the angle ϕ. See Fig. 11.49(d). *Example:* The curve described by $r = a + b \sin \theta$ can be obtained from the curve described by $r = a + b \cos \theta$ by rotating the latter curve counterclockwise 90°, since $\cos(\theta - \pi/2) = \sin \theta$.

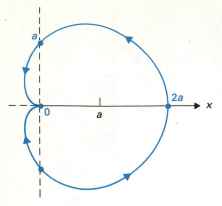

FIGURE 11.50 The cardioid
$r = a(1 + \cos \theta)$.

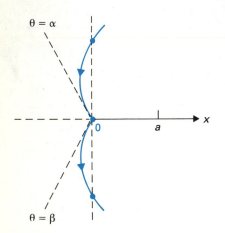

FIGURE 11.51 An *incorrect* sketch of $r = a(1 + \cos \theta)$ at the origin: The curve should come in tangent to the negative x-axis there.

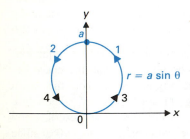

FIGURE 11.52

Some Polar Coordinate Curves

Making use of the general principles just presented, we proceed to sketch a few curves in polar coordinates.

EXAMPLE 1   Sketch the *cardioid* $r = a(1 + \cos \theta)$.

Solution   Since $\cos(-\theta) = \cos \theta$, the curve is symmetric about the x-axis. Note that $r \geq 0$ for all θ since $\cos \theta \geq -1$. Thus we may always measure r along the *positive* rotated polar axis.

Since θ appears in the equation only in "$\cos \theta$," we plot points (r, θ) "every 90°," starting with $\theta = 0$. The "90° lines" are marked with dashes in Fig. 11.50. Of course, our curve $r = a(1 + \cos \theta)$ repeats itself after θ runs through 2π radians.

We find that for $\theta = 0$, we have $r = a(1 + \cos 0) = 2a$; we mark this point on the polar axis. As θ increases to $\pi/2$, $\cos \theta$ decreases from 1 to 0 and r decreases from $2a$ to a. This enables us to sketch the first-quadrant portion of our curve, as shown in Fig. 11.50. To establish the actual shape of the curve in the first quadrant, we could plot (r, θ) for a few more values of θ, say $\theta = \pi/6$, $\pi/4$, and $\pi/3$. We will not bother to plot these points but will content ourselves with rough sketches. We will see in Section 11.8 that the curve makes an angle of 45° with the y-axis where it crosses it.

As θ increases from $\pi/2$ to π, we see that $\cos \theta$ decreases from 0 to -1, so r decreases from a to 0; this enables us to sketch the second-quadrant portion of the curve. The curve actually comes into the origin tangent to the negative x-axis. A common error is to draw the curve near the origin as shown in Fig. 11.51. If this were the curve, r would have value zero for $\alpha \leq \theta \leq \beta$, and, of course, $r = 0$ only if $\theta = \pi$.

In a similar fashion, we see that r increases from 0 to a through the third quadrant and from a to $2a$ through the fourth quadrant. The origin of the name *cardioid* for this curve is clear from the shape of the curve. The arrows in Fig. 11.50 indicate the direction of increasing θ along the curve.   ■

EXAMPLE 2   Sketch the curve $r = a \sin \theta$.

Solution   Since $\sin(-\theta) = -\sin \theta$, we see that the equation $-r = a \sin(-\theta)$ is equivalent to $r = a \sin \theta$, so the curve is symmetric about the y-axis. We sketch this curve "every 90°."

When $\theta = 0$, we have $r = \sin 0 = 0$, so we start at the origin. When $\theta = \pi/2$, we obtain $r = a$, so r increases from 0 to a as θ goes from 0 to $\pi/2$, giving rise to the curve (colored arrow 1) in the first quadrant shown in Fig. 11.52. As θ increases from $\pi/2$ to π, the values for r decrease from a back to 0, giving the portion of the curve (colored arrow 2) in the second quadrant. Now as θ goes from π to $3\pi/2$, $r = a \sin \theta$ is *negative,* so the curve lies in the quadrant *diagonally opposite* the third quadrant, that is, back in the first quadrant. Thus the first-quadrant portion of the curve is simply retraced (black arrow 3) as θ goes from π to $3\pi/2$. Similarly, the second-quadrant portion is retraced (black arrow 4) as θ goes from $3\pi/2$ to 2π, since $r = a \sin \theta$ is negative there also. Consequently, Fig. 11.52 gives the complete curve. The curve is actually a circle with radius $a/2$, as you can verify by changing the equation $r = a \sin \theta$ to $r^2 = a(r \sin \theta)$ to $x^2 + y^2 = ay$.   ■

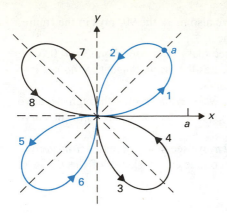

FIGURE 11.53 The four-leaved rose
$r = a \sin 2\theta$.

EXAMPLE 3   Sketch the *four-leaved rose* $r = a \sin 2\theta$.

Solution   This curve is actually symmetric

about the *x*-axis (replace θ by $\pi - \theta$ and r by $-r$),
about the *y*-axis (replace θ by $-\theta$ and r by $-r$), and
about the origin (replace θ by $\theta + \pi$).

It is hard to spot all these symmetries from the equation; we prefer simply to sketch the curve and then use the symmetries in our sketch as a partial check for the accuracy of the sketch.

Since $a \sin 2\theta$ assumes both positive and negative values, we see that r can be both positive and negative.

To help you follow the figures, we draw portions of the curve corresponding to $r \geq 0$ in color and portions corresponding to $r < 0$ in black in this section.

We are interested in plotting points where r assumes maximum and minimum values or becomes zero; this occurs every 90° for 2θ, or every 45° for θ. Thus we plot the curve "every 45°" and start by marking the dashed "45° lines" in Fig. 11.53.

As θ increases from 0 to $\pi/4$, we see that r increases from 0 to a; this gives the arc of the curve we have numbered 1 in Fig. 11.53. As θ increases from $\pi/4$ to $\pi/2$, r decreases from a to 0, giving the arc numbered 2.

Now as θ increases from $\pi/2$ to $3\pi/4$, we see that r runs through the *negative* values from 0 to $-a$, and our curve therefore lies in the "diagonally opposite wedge" and forms the black arc numbered 3 in the figure. We can check that as θ continues to increase up to 2π, we obtain in succession the arcs numbered 4 through 8 for each 45° increment in θ. The arrows on the arcs indicate the direction of increasing θ.   ■

EXAMPLE 4   Sketch the *spiral of Archimedes*, $r = \theta$.

Solution   This time, the curve does not repeat after θ runs through 2π radians, for r increases without bound as θ increases. For $\theta \geq 0$, we have $r \geq 0$, and we obtain the colored spiral in Fig. 11.54. For $\theta < 0$, we have $r < 0$, and we obtain the black spiral in the figure. We note the symmetry in the *y*-axis in our figure, which occurs since the equation $r = \theta$ is obtained again if r is replaced by $-r$ and θ is replaced by $-\theta$.   ■

Curves having equations $r = a + b \cos \theta$ (or $r = a + b \sin \theta$ obtained by rotating 90°) are known as *limaçons*. There are essentially four types of limaçons, depending on the relative magnitudes of a and b. The *cardioid* in Example 1 is one type, obtained when $|a| = |b|$. The next example shows the *limaçon with inner loop*. The *dimpled limaçon* and the *convex limaçon* are the subjects of Exercise 8 in Section 11.8.

EXAMPLE 5   Sketch the *limaçon with inner loop* $r = a(1 + 2 \cos \theta)$.

Solution   We first note that r will be negative when $\cos \theta < -\frac{1}{2}$. A moment's thought shows that $\cos \theta < -\frac{1}{2}$ when $2\pi/3 < \theta < 4\pi/3$, so we sketch the rays

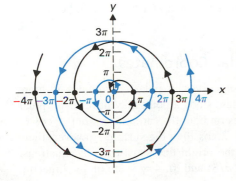

FIGURE 11.54 The spiral of Archimedes $r = \theta$.

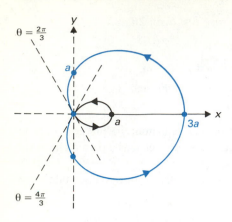

FIGURE 11.55 The limaçon

$$r = a(1 + 2 \cos \theta).$$

$\theta = 2\pi/3$ and $\theta = 4\pi/3$ in Fig. 11.55. We also mark the 90° rays in the figure, since we want to let θ jump 90° at a time.

As θ increases from 0 to $2\pi/3$, we see that r decreases from $3a$ to 0, with value a when $\theta = \pi/2$. The curve approaches the origin tangent to the ray $\theta = 2\pi/3$.

As θ increases from $2\pi/3$ to $4\pi/3$, we see that r runs through negative values, with value $-a$ when $\theta = \pi$. This gives us the black portion of the curve in Fig. 11.55.

Finally, as θ increases from $4\pi/3$ to 2π, we see that r increases from zero to $3a$, with value a when $\theta = 3\pi/2$. This completes the sketch of the curve in Fig. 11.55. ∎

EXAMPLE 6 Sketch the *lemniscate* $r^2 = a^2 \sin 2\theta$.

Solution Since r^2 can't be negative, we note that there will be no curve when $\sin 2\theta < 0$, which occurs for $\pi/2 < \theta < \pi$ and $3\pi/2 < \theta < 2\pi$. Note also that when $\sin 2\theta > 0$, we may have r either negative or positive, because $r = \pm a \sqrt{\sin 2\theta}$. We should sketch this curve "every 45°," since as θ jumps 45°, 2θ jumps 90°. We dash the 45° rays in Fig. 11.56.

We first sketch the part of the curve corresponding to $r \geq 0$, where $r = a\sqrt{\sin 2\theta}$. As θ increases from 0 to $\pi/4$, r increases from 0 to a. Then r decreases to zero again at $\theta = \pi/2$. There is no curve for $\pi/2 < \theta < \pi$. As θ increases from π to $5\pi/4$, r increases from 0 to a. Then r decreases from a to 0 as θ increases from $5\pi/4$ to $3\pi/2$. Finally, there is no curve for $3\pi/2 < \theta < 2\pi$.

For $r = a\sqrt{\sin 2\theta}$, we obtained in succession as θ increased the color-numbered arcs 1, 2, 3, and 4 in Fig. 11.56. We can easily see that for $r = -a\sqrt{\sin 2\theta}$, we obtain in succession the black-numbered arcs 1, 2, 3, and 4 as θ increases. Thus negative values of r do not give any additional curve. ∎

EXAMPLE 7 Sketch the curve $r = a \sin(\theta/2)$.

Solution We must sketch this curve for $0 \leq \theta \leq 4\pi$, since $\sin(\theta/2)$ has period 4π. We plot the curve every 180°, because $\theta/2$ jumps 90° when θ jumps 180°.

As θ increases from 0 to π, we see that r increases from 0 to a, giving the arc numbered 1 in Fig. 11.57. Then r decreases from a to 0 as θ increases from π to 2π, giving the arc numbered 2.

For $2\pi < \theta < 4\pi$, we see that r is negative. The arc numbered 3 in Fig. 11.57 is traced as θ increases from 2π to 3π, and the arc numbered 4 is traced as θ increases from 3π to 4π. ∎

Intersections of Curves in Polar Coordinates

The problem of finding intersections of curves in polar coordinates is more complicated than in Cartesian coordinates, since a point may have many polar coordinates. For example, the origin lies on both the cardioid $r = a(1 + \cos \theta)$ in Fig. 11.50 and the limaçon $r = a(1 + 2 \cos \theta)$ in Fig. 11.55. However, there is no *one* set of coordinates (r, θ) for the origin that satisfies *both* equations, since $(0, \pi)$ is on the cardioid and $(0, 2\pi/3)$ and $(0, 4\pi/3)$ are on the limaçon.

The origin is a point of intersection of two polar curves if, for each curve separately, r can be 0 for some value of θ.

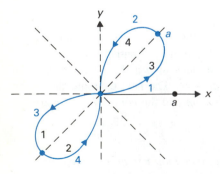

FIGURE 11.56 The leminscate

$$r^2 = a^2 \sin 2\theta.$$

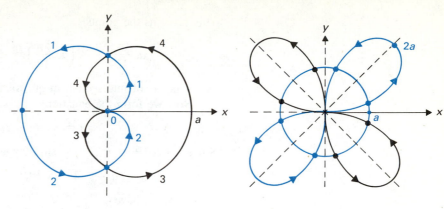

FIGURE 11.57 The curve
$$r = a \sin(\theta/2).$$

FIGURE 11.58 The eight points of intersection of the circle $r = a$ and the rose $r = 2a \sin 2\theta$.

Figure 11.58 shows the points of intersection of the circle $r = a$ and the rose $r = 2a \sin 2\theta$. There are eight points of intersection in all. The points where a colored arc of the rose meets the circle have coordinates (r, θ) for $r > 0$ satisfying both $r = a$ and $r = 2a \sin 2\theta$ simultaneously. However, the points where a black arc of the rose meets the circle have r-coordinate a on the circle and r-coordinate $-a$ on the rose. Thus simultaneous algebraic solution of $r = a$ and $r = 2a \sin 2\theta$ would miss the four points of intersection in quadrants II and IV in Fig. 11.58. Taking these problems into account, we come up with the following rule.

To Find Intersections of Polar Curves

Find (r, θ) satisfying the first equation for which some points

$$(r, \theta + 2n\pi) \qquad \text{or} \qquad (-r, \theta + \pi + 2n\pi)$$

satisfy the second equation. Check separately to see if the origin lies on both curves, that is, if r can be 0 for both curves. Use a sketch as an aid when possible.

EXAMPLE 8 Find the points of intersection of $r = a$ and $r = 2a \sin 2\theta$, shown in Fig. 11.58.

Solution First we solve $a = 2a \sin 2(\theta + 2n\pi)$. Since $\sin(2\theta + 4n\pi) = \sin 2\theta$, we may simplify to

$$a = 2a \sin 2\theta, \qquad \text{or} \qquad \sin 2\theta = \tfrac{1}{2}.$$

We want all values of θ from 0 to 2π, so we should find all 2θ from 0 to 4π such that $\sin 2\theta = \tfrac{1}{2}$. We have

$$2\theta = \frac{\pi}{6}, \frac{5\pi}{6}, \frac{13\pi}{6}, \frac{17\pi}{6}, \qquad \text{so} \qquad \theta = \frac{\pi}{12}, \frac{5\pi}{12}, \frac{13\pi}{12}, \frac{17\pi}{12}.$$

This gives us the points

$$\left(a, \frac{\pi}{12}\right), \qquad \left(a, \frac{5\pi}{12}\right), \qquad \left(a, \frac{13\pi}{12}\right), \qquad \left(a, \frac{17\pi}{12}\right)$$

as the points of intersection in quadrants I and III of Fig. 11.58.

Now we form the equation

$$a = -2a \sin 2(\theta + \pi + 2n\pi).$$

Since $\sin 2(\theta + \pi + 2n\pi) = \sin 2\theta$ again, we obtain

$$a = -2a \sin 2\theta, \qquad \text{or} \qquad \sin 2\theta = -\tfrac{1}{2}.$$

This time we have

$$2\theta = \frac{7\pi}{6}, \frac{11\pi}{6}, \frac{19\pi}{6}, \frac{23\pi}{6}, \qquad \text{so} \qquad \theta = \frac{7\pi}{12}, \frac{11\pi}{12}, \frac{19\pi}{12}, \frac{23\pi}{12}.$$

These yield the points of intersection

$$\left(a, \frac{7\pi}{12}\right), \qquad \left(a, \frac{11\pi}{12}\right), \qquad \left(a, \frac{19\pi}{12}\right), \qquad \left(a, \frac{23\pi}{12}\right)$$

in quadrants II and IV in Fig. 11.58.

Finally, we check whether the origin is a point of intersection. Since $r \neq 0$ on the circle $r = a$, we see that the origin is not a point of intersection. ∎

EXAMPLE 9   Find all points of intersection of

$$r = 3 + 2 \sin \theta \qquad \text{and} \qquad r = \cos 2\theta.$$

Solution   We first try to find coordinates (r, θ) that satisfy the first equation, while coordinates $(r, \theta + 2n\pi)$ satisfy the second equation. That is, we find solutions of

$$3 + 2 \sin \theta = \cos 2(\theta + 2n\pi).$$

We obtain

$$3 + 2 \sin \theta = \cos(2\theta + 4n\pi)$$
$$= \cos 2\theta = 1 - 2 \sin^2\theta.$$

This yields

$$\sin^2\theta + \sin \theta + 1 = 0,$$

or, solving by the quadratic formula,

$$\sin \theta = \frac{-1 \pm \sqrt{1 - 4}}{2},$$

which gives no real solutions.

Now we try to find coordinates (r, θ) satisfying the first equation, while coordinates $(-r, \theta + \pi + 2n\pi)$ satisfy the second. We obtain the equation

$$3 + 2 \sin \theta = -\cos(2\theta + 2\pi + 4n\pi) = -\cos 2\theta$$
$$= 2 \sin^2\theta - 1,$$

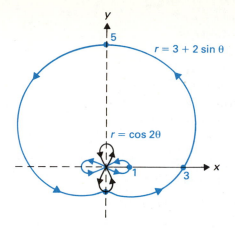

FIGURE 11.59 The point of intersection of $r = \cos 2\theta$ and $r = 3 + 2 \sin \theta$.

which yields

$$\sin^2\theta - \sin\theta - 2 = 0$$
$$(\sin\theta - 2)(\sin\theta + 1) = 0.$$

Since $\sin\theta = 2$ is impossible, we are left with $\sin\theta = -1$, or $\theta = 3\pi/2$. Substituting in $r = 3 + 2\sin\theta$, we obtain the point $(1, 3\pi/2)$ of intersection.

Finally we check whether the origin lies on both curves. Now $3 + 2\sin\theta$ is never 0, since $-1 \leq \sin\theta \leq 1$, so the origin does not lie on even the first curve. Thus the point $(1, 3\pi/2)$ is the only point of intersection of the curves, which are sketched in Fig. 11.59. ∎

SUMMARY

1. When plotting a polar curve $r = f(\theta)$, plot those points where r assumes relative maximum or minimum values or becomes 0. For example, $r = 4 \cos 3\theta$ should be plotted in increments of 30° in θ, starting with $\theta = 0$.

2. To find points of intersection of two polar curves, find (r, θ) satisfying the first equation for which some points $(r, \theta + 2n\pi)$ or $(-r, \theta + \pi + 2n\pi)$ satisfy the second equation. Check separately to see if the origin lies on both curves, that is, if r can be 0. Sketch the curves.

EXERCISES 11.6

In Exercises 1–30, sketch the curve having the given polar equation. Verify algebraically any symmetry about the x-axis, y-axis, or origin that you find in your sketch.

1. $r = a\theta$ (spiral of Archimedes)
2. $r\theta = a$ (hyperbolic spiral)
3. $r = ae^{\theta}$ (logarithmic spiral)
4. $r = a(1 + \sin\theta)$
5. $r = a(1 - \cos\theta)$
6. $r = 1 + 2\sin\theta$
7. $r = 3 + 2\cos\theta$
8. $r = 2 + 3\cos\theta$
9. $r = a(4 - 2\sin\theta)$
10. $r = 2a\cos\theta$
11. $r = 2a\sin\theta$
12. $r = 3\csc\theta$
13. $r = -2\sec\theta$
14. $r^2 = 9\sec^2\theta$
15. $r = a\cos 2\theta$
16. $r = a\cos 3\theta$
17. $r = a\sin 3\theta$
18. $r = a\sin 4\theta$
19. $r = a\sin(\theta/2)$
20. $r = a\cos(\theta/2)$
21. $r = a\cos(3\theta/2)$

22. $r = 4$
23. $r^2 = a^2\cos 2\theta$
24. $r^2 = a^2\sin 3\theta$
25. $r^2 = a^2\cos 3\theta$
26. $r = a(1 + \cos(\theta/2))$
27. $r = a(1 + \cos 2\theta)$
28. $r = a(1 - \sin 3\theta)$
29. $r = a(2 + \cos 3\theta)$
30. $r = a(2 + |\sin 3\theta|)$

31. Let $r = 2 - \csc\theta$.
    (a) Show that $y = 2/(2 - r) - 1$. [*Hint:* Write $\csc\theta = 1/\sin\theta$, and remember that $r\sin\theta = y$.]
    (b) Show that the line $y = -1$ is a horizontal asymptote of the curve by computing $\lim_{r\to\infty} y$ from part (a).
    (c) Sketch the curve.

32. Following the idea of Exercise 31, sketch the polar curve $r = \sqrt{2} + \sec\theta$.

33. Let $r = a(1 + \cos\theta)$.
    (a) Recall that $x = r\cos\theta$ and $y = r\sin\theta$. Express x and y in terms of θ only.

(b) Regarding your answer to part (a) as parametric equations of a curve, find the slope of the tangent line to the cardioid where $\theta = \pi/6$.

34. As in Exercise 33, find the slope of the polar curve $r = a \cos \theta$ where $\theta = \pi/3$.

35. Following the idea of Exercise 33, find all points where the cardioid $r = a(1 + \cos \theta)$ has a vertical tangent.

36. Following the idea of Exercise 33, find all points where the cardioid $r = a(1 + \cos \theta)$ has a horizontal tangent.

In Exercises 37–46, find all points of intersection of the given curves.

37. $r = \theta/\pi$ and $r = \frac{1}{4}$

38. $\theta = \pi/6$ and $r = 2$

39. $r = a \sin \theta$ and $r = a \cos \theta$

40. $r = 2a \sin \theta$ and $r = a$

41. $r = a$ and $r^2 = 2a^2 \sin 2\theta$

42. $r = a \sin 2\theta$ and $r = a \cos \theta$

43. $r = a$ and $r = 2a \cos 2\theta$

44. $r = 1 - \cos \theta$ and $r = 1 + 2 \cos \theta$

45. $r = \cos(\theta/2)$ and $r = \sqrt{3}/2$

46. $r = a \cos 2\theta$ and $r = a(1 + \cos \theta)$

11.7 AREA IN POLAR COORDINATES

Let $r = f(\theta)$ be a continuous function of θ. We want to find the area A of a region bounded by the polar curve $r = f(\theta)$ and by two rays $\theta = \alpha$ and $\theta = \beta$, as shown in Fig. 11.60. Our treatment parallels the development of the integral in Chapter 5.

Assume $r \geq 0$ for θ in $[\alpha, \beta]$. We form a partition P of the interval $[\alpha, \beta]$ into n subintervals of size $\Delta\theta_i$, as shown in Fig. 11.61 for $n = 8$. Let M_i be the maximum value of r for θ in the ith subinterval, as indicated in Fig. 11.62. In that figure, we have shaded in color a wedge of constant radius M_i and central angle $\Delta\theta_i$. This wedge has area equal to the sector of the circle shown in Fig. 11.63. The area of this sector is the fraction $\Delta\theta_i/(2\pi)$ of the area of the entire circle so it has area

$$\frac{\Delta\theta_i}{2\pi} \cdot \pi M_i^2 = \frac{1}{2} M_i^2 \Delta\theta_i.$$

For the area in Fig. 11.60, we see immediately that

$$\text{Area} \leq \sum_{i=1}^{n} \frac{1}{2} M_i^2 \Delta\theta_i. \tag{1}$$

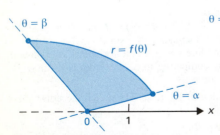

FIGURE 11.60 The region bounded by $r = f(\theta)$ and the rays $\theta = \alpha$, $\theta = \beta$.

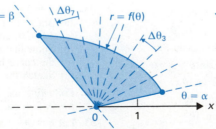

FIGURE 11.61 Subdivision of $[\alpha, \beta]$ into $n = 8$ subintervals of length $\Delta\theta_i$.

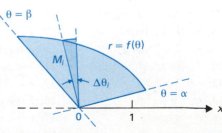

FIGURE 11.62 Wedge of constant radius M_i and central angle $\Delta\theta i$.

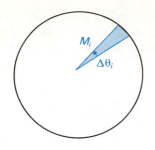

FIGURE 11.63 The area of the sector is $(\Delta\theta_i/2\pi) \cdot \pi M_i^2 = \frac{1}{2}M_i^2 \, \Delta\theta_i$.

In exactly the same fashion, we let m_i be the minimum value of r for θ in the ith subinterval, as indicated in Fig. 11.64. We then find that

$$\sum_{i=1}^{n} \frac{1}{2}m_i^2 \Delta\theta_i \leq \text{Area}. \qquad (2)$$

We recognize the sums in relations (1) and (2) as the upper and lower sums for the function $\frac{1}{2}(f(\theta))^2$. As $\|P\| \to 0$, both of these sums approach $\int_\alpha^\beta \frac{1}{2}(f(\theta))^2 \, d\theta$. Thus relations (1) and (2) show that

$$\text{Area} = \int_\alpha^\beta \frac{1}{2}r^2 \, d\theta = \int_\alpha^\beta \frac{1}{2}(f(\theta))^2 \, d\theta, \qquad (3)$$

at least for the case where $r \geq 0$. In case the curve in Fig. 11.60 actually corresponds to negative values of r, we would let M_i and m_i be the maximum and minimum values, respectively, of $|r| = |f(\theta)|$. Since $|r|^2 = r^2$, we again obtain formula (3). In setting up an integral using differential notation, we regard the *differential wedge* in Fig. 11.65 as having area $\frac{1}{2}r^2/d\theta$.

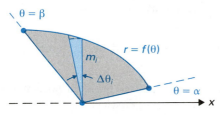

FIGURE 11.64 Wedge of constant radius m_i and central angle $\Delta\theta_i$.

EXAMPLE 1 Find the area of the region that is bounded by the cardioid $r = a(1 + \cos\theta)$, shown in Fig. 11.65.

Solution We think of the differential wedge shaded in color in Fig. 11.65 as having area $\frac{1}{2}r^2 \, d\theta$. This is the way we remember formula (3).

Since the curve is traced completely as θ varies from 0 to 2π, we want to add the areas of these wedges with an integral as θ goes from 0 to 2π. Expressing r in terms of θ, we obtain from formula (3)

$$
\begin{aligned}
\text{Area} &= \int_0^{2\pi} \frac{1}{2}r^2 \, d\theta = \int_0^{2\pi} \frac{1}{2}[a(1 + \cos\theta)]^2 \, d\theta \\
&= \frac{1}{2}a^2 \int_0^{2\pi} (1 + 2\cos\theta + \cos^2\theta) \, d\theta \\
&= \frac{1}{2}a^2 \int_0^{2\pi} \left[1 + 2\cos\theta + \frac{1}{2}(1 + \cos 2\theta)\right] d\theta \\
&= \frac{1}{2}a^2 \int_0^{2\pi} \left(\frac{3}{2} + 2\cos\theta + \frac{1}{2}\cos 2\theta\right) d\theta \\
&= \frac{1}{2}a^2 \left(\frac{3}{2}\theta + 2\sin\theta + \frac{1}{4}\sin 2\theta\right)\Bigg]_0^{2\pi} \\
&= \frac{1}{2}a^2 \left(\frac{3}{2} \cdot 2\pi\right) - \frac{1}{2}a^2 \cdot 0 = \frac{3}{2}\pi a^2. \qquad \blacksquare
\end{aligned}
$$

A common mistake is to always integrate from 0 to 2π when finding the area of a region using polar coordinates.

Think carefully just what interval for θ is needed for the wedges to sweep out the region just once.

The following example illustrates the problem.

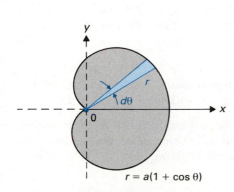

FIGURE 11.65 Differential wedge of area $\frac{1}{2}r^2 \, d\theta$.

EXAMPLE 2 Find the area of the region bounded by the circle $r = 2a\cos\theta$, shown in Fig. 11.66, by integration in polar coordinates.

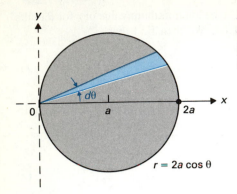

FIGURE 11.66 Differential wedge of area $\frac{1}{2}r^2\,d\theta$.

Solution We know the area is πa^2, since the circle has radius a. Now the entire circle is traced once as θ goes from 0 to π. (The circle would be traced *twice* as θ goes from 0 to 2π.) Thus to get the area of the circle, we may integrate as θ goes from 0 to π. We have

$$\text{Area} = \int_0^\pi \frac{1}{2}(2a\cos\theta)^2\,d\theta$$

$$= 2a^2 \int_0^\pi \cos^2\theta\,d\theta$$

$$= 2a^2 \int_0^\pi \frac{1}{2}(1+\cos 2\theta)\,d\theta$$

$$= a^2\left(\theta + \frac{1}{2}\sin 2\theta\right)\Big]_0^\pi$$

$$= a^2\pi - 0 = \pi a^2.$$

If we had integrated blindly from 0 to 2π, we would have obtained the incorrect result $2\pi a^2$. ∎

Symmetry is often useful, as illustrated in the next example.

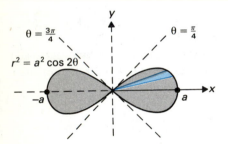

FIGURE 11.67 Differential wedge of area $\frac{1}{2}r^2\,d\theta$.

EXAMPLE 3 Find the total area of the regions bounded by the lemniscate $r^2 = a^2\cos 2\theta$ shown in Fig. 11.67.

Solution By symmetry, we may find the area of the first-quadrant portion of the region and multiply by 4. We thus obtain from relation (1)

$$4\int_0^{\pi/4} \frac{1}{2}r^2\,d\theta = 4\int_0^{\pi/4} \frac{1}{2}a^2\cos 2\theta\,d\theta$$

$$= 2a^2 \int_0^{\pi/4} \cos 2\theta\,d\theta$$

$$= 2a^2\,\frac{1}{2}\sin 2\theta\Big]_0^{\pi/4}$$

$$= a^2\left(\sin\frac{\pi}{2}\right) - a^2(\sin 0) = a^2$$

as our desired area. Note that r is not defined for $\pi/4 < \theta < 3\pi/4$ and $5\pi/4 < \theta < 7\pi/4$. Blindly integrating from 0 to 2π would give the incorrect answer 0. ∎

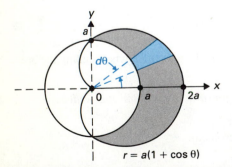

FIGURE 11.68 Differential region of area $\frac{1}{2}(r_{\text{cardioid}})^2\,d\theta - \frac{1}{2}(r_{\text{circle}})^2\,d\theta$.

EXAMPLE 4 Find the area of the region that is inside the cardioid $r = a(1 + \cos\theta)$ but outside the circle $r = a$, shown shaded in Fig. 11.68.

Solution We may double the area of the first-quadrant portion of the region, which is traced out as θ varies from 0 to $\pi/2$. Using differential notation, the area of the dark partial wedge of central angle $d\theta$ shown in Fig. 11.68 is approximately

$$\frac{1}{2}(r_{\text{cardioid}})^2\,d\theta - \frac{1}{2}(r_{\text{circle}})^2\,d\theta.$$

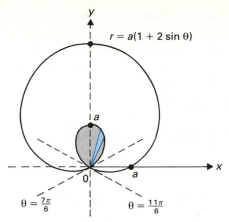

$r = a(1 + 2 \sin \theta)$

$\theta = \dfrac{7\pi}{6}$ $\theta = \dfrac{11\pi}{6}$

FIGURE 11.69 Differential wedge of area $\frac{1}{2}r^2\,d\theta$.

We obtain

$$2 \cdot \frac{1}{2} \int_0^{\pi/2} [a^2(1 + \cos\theta)^2 - a^2]\,d\theta$$

$$= \int_0^{\pi/2} (a^2 + 2a^2 \cos\theta + a^2 \cos^2\theta - a^2)\,d\theta$$

$$= a^2 \int_0^{\pi/2} (2\cos\theta + \cos^2\theta)\,d\theta$$

$$= a^2 \left(2\sin\theta + \frac{\theta}{2} + \frac{\sin 2\theta}{4} \right)\Big]_0^{\pi/2}$$

$$= a^2 \left(2 + \frac{\pi}{4} \right) - a^2(0)$$

$$= 2a^2 + \frac{\pi a^2}{4}$$

as the desired area.

EXAMPLE 5 Find the area of the region bounded by the inner loop of the limaçon $r = a(1 + 2 \sin\theta)$, shown in Fig. 11.69.

Solution We may double the area inside the first-quadrant part of the loop, swept out as θ goes from $7\pi/6$ to $3\pi/2$, or we can simply integrate from $7\pi/6$ to $11\pi/6$. Since we can evaluate trigonometric functions more easily at $3\pi/2$ than at $11\pi/6$, we select the first alternative. We have

$$\text{Area} = 2 \int_{7\pi/6}^{3\pi/2} \frac{1}{2}[a(1 + 2\sin\theta)]^2\,d\theta$$

$$= a^2 \int_{7\pi/6}^{3\pi/2} (1 + 4\sin\theta + 4\sin^2\theta)\,d\theta$$

$$= a^2 \int_{7\pi/6}^{3\pi/2} [1 + 4\sin\theta + 2(1 - \cos 2\theta)]\,d\theta$$

$$= a^2 \int_{7\pi/6}^{3\pi/2} (3 + 4\sin\theta - 2\cos 2\theta)\,d\theta$$

$$= a^2(3\theta - 4\cos\theta - \sin 2\theta)\Big]_{7\pi/6}^{3\pi/2}$$

$$= a^2 \left(\frac{9\pi}{2} - 0 - 0 \right) - a^2 \left[\frac{7\pi}{2} - 4 \left(-\frac{\sqrt{3}}{2} \right) - \frac{\sqrt{3}}{2} \right]$$

$$= a^2 \left(\pi - \frac{3\sqrt{3}}{2} \right).$$

SUMMARY

To find the area of a region bounded by polar curves:

STEP 1. Draw a figure.

STEP 2. Draw polar rays corresponding to a small increment $d\theta$ in θ.

STEP 3. Write down the area of the resulting wedge. In differential notation, a wedge with vertex at the origin, small central angle $d\theta$, and going out to $r = f(\theta)$ has approximate area $dA = \frac{1}{2}r^2\,d\theta = \frac{1}{2}f(\theta)^2\,d\theta$.

STEP 4. Integrate between the appropriate limits.

EXERCISES 11.7

In Exercises 1–21, find the area of the indicated region, using integration in polar coordinates.

1. The region bounded by $r = a$ (Does this problem constitute a "proof" that the area of a disk of radius a is πa^2? Why?)

2. The region bounded by the spiral $r = a\theta$ for $\pi/2 \le \theta \le 3\pi/2$ and the rays $\theta = \pi/2$ and $\theta = 3\pi/2$

3. The region bounded by the spiral $r = ae^\theta$ for $0 \le \theta \le \pi$ and the x-axis

4. The region bounded by $r\theta = 1$ for $\pi/2 \le \theta \le \pi$ and the rays $\theta = \pi/2$ and $\theta = \pi$

5. The region inside the cardioid $r = a(1 + \sin\theta)$

6. The region bounded by $r = 3 + 2\cos\theta$

7. A region bounded by one leaf of the rose $r = a\cos 2\theta$

8. A region bounded by one leaf of the rose $r = a\sin 4\theta$

9. A region inside one whole loop, from $(0, 0)$ back to $(0, 0)$, of $r = a(\cos 3\theta/2)$

10. The total region inside $r^2 = a^2\sin 2\theta$

11. The total region inside $r^2 = a^2\sin 3\theta$

12. The region inside both $r = 2a\cos\theta$ and $r = 2a\sin\theta$

13. The total region inside the rose $r = 2a\cos 2\theta$ but outside the circle $r = a$

14. The region inside both the cardioid $r = a(1 + \cos\theta)$ and $r = a$

15. The region between the loops of the limaçon
$$r = a(1 - 2\cos\theta)$$

16. The region in the upper half-plane having a portion of $r = \sin(\theta/2)$ as its total boundary

17. The region in the right half-plane having a portion of $r = a\sin(\theta/2)$ as its total boundary

18. The smaller region bounded by $r = 4$ and $r = 2\sec\theta$

19. The region bounded by $r = a(4 - \sin^2\theta)$

20. The total region bounded by $r = a|\cos\theta|$

21. The region inside $r = a(3 + |\sin 3\theta|)$

In Exercises 22–26, use a calculator and Simpson's rule to estimate the area of the region.

22. The region inside the ellipse $r = 3/(2 - \cos\theta)$

23. The region inside the ellipse $r = 6/(5 + \sin\theta)$

24. The region inside $r = 2 + \cos(\theta^2/2\pi)$, $0 \le \theta \le 2\pi$

25. The region bounded by $r = 10/(2 + \cos^2\theta)$

26. The region bounded by $r = (4 + \sin^2\theta)/(2 + \cos^4\theta)$

11.8 THE TANGENT LINE AND ARC LENGTH

The Slope of the Tangent Line

Let $r = f(\theta)$, where f is differentiable. We would like to find the slope dy/dx of the tangent line to the graph at a point where y appears as a differentiable function of x.

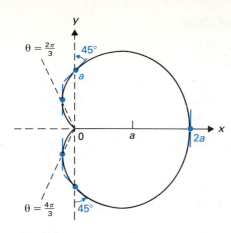

$\theta = \frac{2\pi}{3}$ 45°

a

0   a $2a$

$\theta = \frac{4\pi}{3}$ 45°

FIGURE 11.70   Some tangent lines to the cardioid $r = a(1 + \cos \theta)$.

Recall that

$$x = r \cos \theta, \qquad y = r \sin \theta.$$

Substituting $f(\theta)$ for r, we have

$$x = f(\theta) \cdot \cos \theta, \qquad y = f(\theta) \cdot \sin \theta. \qquad (1)$$

Equations (1) can be regarded as parametric equations for the curve $r = f(\theta)$, where θ is the parameter. From Section 11.3, we see that

$$\frac{dy}{dx} = \frac{dy/d\theta}{dx/d\theta}, \qquad (2)$$

where $dx/d\theta \neq 0$. We actually anticipated Eqs. (1) and (2) in Exercises 33–36 of Section 11.6.

EXAMPLE 1   Show that the cardioid $r = a(1 + \cos \theta)$ in Fig. 11.70 makes an angle of 45° with the y-axis where it crosses it.

Solution   We want to show that the tangent lines to the cardioid at the points $(r, \theta) = (a, \pi/2)$ and $(a, 3\pi/2)$ have slope ± 1. Equations (1) become

$$x = a(1 + \cos \theta)(\cos \theta) = a(\cos \theta + \cos^2 \theta),$$

$$y = a(1 + \cos \theta)(\sin \theta) = a(\sin \theta + \sin \theta \cos \theta).$$

From Eq. (2), we have

$$\frac{dy}{dx} = \frac{dy/d\theta}{dx/d\theta} = \frac{a(\cos \theta - \sin^2 \theta + \cos^2 \theta)}{a(-\sin \theta - 2 \sin \theta \cos \theta)}.$$

Thus

$$\left. \frac{dy}{dx} \right|_{\theta = \pi/2} = \frac{a(0 - 1 + 0)}{a(-1 - 0)} = 1$$

and

$$\left. \frac{dy}{dx} \right|_{\theta = 3\pi/2} = \frac{a(0 - 1 + 0)}{a(1 - 0)} = -1. \qquad ∎$$

EXAMPLE 2   Find all points where the tangent line to the cardioid

$$r = a(1 + \cos \theta)$$

is vertical.

Solution   From Eq. (2), we see that the tangent line is vertical at any point where $dx/d\theta = 0$ and $dy/d\theta \neq 0$. From Example 1, we see that

$$\frac{dx}{d\theta} = a(-\sin \theta - 2 \sin \theta \cos \theta)$$

$$= a(\sin \theta)(-1 - 2 \cos \theta) = 0$$

when $\theta = 0$, π, $2\pi/3$, and $4\pi/3$. We see that $dy/d\theta \neq 0$ at $\theta = 0$, $2\pi/3$, and $4\pi/3$. However, $dy/d\theta = 0$ when $\theta = \pi$, so we must examine this point further.

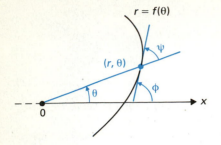

FIGURE 11.71 $\phi = \theta + \psi$

From Example 1 and l'Hôpital's rule, we obtain

$$\frac{dy}{dx}\bigg|_{\theta=\pi} = \lim_{\theta \to \pi} \frac{\cos\theta - \sin^2\theta + \cos^2\theta}{-\sin\theta - 2\sin\theta\cos\theta}$$

$$= \lim_{\theta \to \pi} \frac{-\sin\theta - 2\sin\theta\cos\theta - 2\cos\theta\sin\theta}{-\cos\theta + 2\sin^2\theta - 2\cos^2\theta} = \frac{0}{-1} = 0.$$

Thus the tangent line at $(r, \theta) = (0, \pi)$ is horizontal and not vertical. The tangent lines at $(r, \theta) = (2a, 0)$, $(a/2, 2\pi/3)$, and $(a/2, 4\pi/3)$ are vertical, as shown in Fig. 11.70. ■

The Angle ψ from the Radial Line to the Tangent Line

Let $r = f(\theta)$ where f is differentiable. At a point (r, θ) on the curve, we now find the angle ψ measured counterclockwise from the radial line to the tangent line, as shown in Fig. 11.71. For the angle ϕ shown in the figure, we have $\tan\phi = dy/dx$. We just saw how to find dy/dx, so we know how to find $\tan\phi$ and therefore ϕ.

By plane geometry, we know that

$$\phi = \theta + \psi. \tag{3}$$

We will show that

$$\tan\psi = \frac{r}{dr/d\theta} = \frac{f(\theta)}{f'(\theta)}, \tag{4}$$

if $f'(\theta) \neq 0$. Figure 11.72 indicates how formula (4) can be remembered. The figure shows the *differential right triangle* for polar coordinates. This triangle has as hypotenuse the segment of the tangent line at (r, θ) between the radial lines corresponding to θ and $\theta + d\theta$. The side of the triangle opposite the angle ψ is really a short arc of a circle, but we pretend it is a straight line of approximate length $r\, d\theta$. For small $d\theta$, the legs of the "right triangle" are approximately of lengths dr and $r\, d\theta$, which at once suggests that

$$\tan\psi = \frac{r\, d\theta}{dr} = \frac{r}{dr/d\theta}.$$

For a careful derivation of formula (4), note from Eq. (3) that

$$\tan\psi = \tan(\phi - \theta) = \frac{\tan\phi - \tan\theta}{1 + \tan\phi\tan\theta}. \tag{5}$$

Now $\tan\phi = dy/dx$, and from the parametric equations

$$x = r\cos\theta = f(\theta)\cos\theta, \qquad y = r\sin\theta = f(\theta)\sin\theta,$$

we obtain

$$\tan\phi = \frac{dy}{dx} = \frac{dy/d\theta}{dx/d\theta}$$

$$= \frac{r\cos\theta + (dr/d\theta)\sin\theta}{-r\sin\theta + (dr/d\theta)\cos\theta}. \tag{6}$$

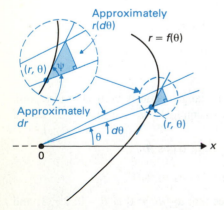

FIGURE 11.72   The differential triangle of polar coordinates.

Substituting in Eq. (5) the value for $\tan\phi$ found in Eq. (6), and putting $\tan\theta =$

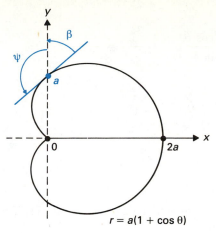

FIGURE 11.73 $\beta = \pi - \psi$

$(\sin \theta)/(\cos \theta)$, we obtain a compound quotient that we can show yields

$$\tan \psi = \frac{r \cos^2\theta + (dr/d\theta) \sin \theta \cos \theta + r \sin^2\theta - (dr/d\theta) \sin \theta \cos \theta}{-r \sin \theta \cos \theta + (dr/d\theta) \cos^2\theta + r \sin \theta \cos \theta + (dr/d\theta)\sin^2\theta}$$

$$= \frac{r}{dr/d\theta}.$$

EXAMPLE 3 Show again that the cardioid $r = a(1 + \cos \theta)$ meets the y-axis at 45° by finding the acute angle β shown in Fig. 11.73.

Solution From Fig. 11.73, we see that we can reduce our problem to finding the angle ψ, for the angle β shown in the figure is then given by $\beta = \pi - \psi$. We have

$$\tan \psi = \left.\frac{r}{dr/d\theta}\right|_{\theta = \pi/2} = \left.\frac{a(1 + \cos \theta)}{-a \sin \theta}\right|_{\pi/2} = \frac{a}{-a} = -1.$$

Thus $\psi = 3\pi/4$, and $\beta = \pi - 3\pi/4 = \pi/4$.   ∎

From Fig. 11.74, we see that the angle β between curves $r = f_1(\theta)$ and $r = f_2(\theta)$ can be computed by finding

$$\tan \beta = \tan(\psi_2 - \psi_1) = \frac{\tan \psi_2 - \tan \psi_1}{1 + \tan \psi_2 \tan \psi_1}. \tag{7}$$

EXAMPLE 4 Find the angle β between the cardioid $r = 2 - 2 \cos \theta$ and the curve $r = 2 + \cos \theta$ at the point $(r, \theta) = (2, \pi/2)$, shown in Fig. 11.75.

Solution From Fig. 11.75 and formula (4), we have

$$\tan \psi_1 = \left.\frac{r}{dr/d\theta}\right|_{\pi/2} = \left.\frac{2 - 2 \cos \theta}{2 \sin \theta}\right|_{\pi/2} = \frac{2}{2} = 1$$

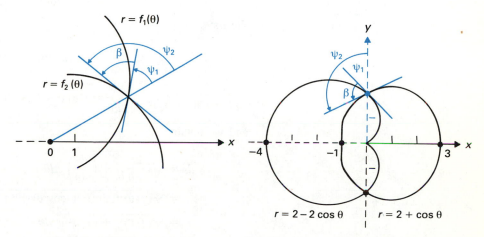

FIGURE 11.74 $\beta = \psi_2 - \psi_1$

FIGURE 11.75 The angle $\beta = \psi_2 - \psi_1$ of intersection of the curves.

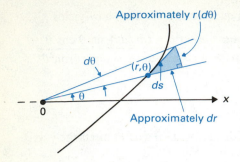

FIGURE 11.76 From the differential triangle, $ds = \sqrt{(dr)^2 + (r\,d\theta)^2}$.

and

$$\tan \psi_2 = \left.\frac{r}{dr/d\theta}\right|_{\pi/2} = \left.\frac{2 + \cos \theta}{-\sin \theta}\right|_{\pi/2} = \frac{2}{-1} = -2.$$

From Eq. (7), we have

$$\tan \beta = \frac{-2 - 1}{1 + (-2)(1)} = \frac{-3}{-1} = 3,$$

so $\beta = \tan^{-1} 3 \approx 71.565°$. ∎

Arc Length in Polar Coordinates

In Fig. 11.76, we shade again the differential right triangle shown in Fig. 11.72. From this triangle, we obtain the estimate

$$ds = \sqrt{(dr)^2 + (r\,d\theta)^2} = \sqrt{\left(\frac{dr}{d\theta}\right)^2 + r^2}\, d\theta \qquad (8)$$

for the length of the tangent line segment to the curve. To estimate arc length, we may add lengths of tangent line segments, and from Eq. (8), we expect the arc length of a continuously differentiable polar curve $r = f(\theta)$ from (r_1, θ_1) to (r_2, θ_2) to be given by

$$\text{Arc length} = \int_{\theta_1}^{\theta_2} \sqrt{\left(\frac{dr}{d\theta}\right)^2 + r^2}\, d\theta. \qquad (9)$$

For a careful derivation of Eq. (8), simply note that the polar curve $r = f(\theta)$ is defined parametrically by

$$x = r \cos \theta = f(\theta) \cos \theta,$$
$$y = r \sin \theta = f(\theta) \sin \theta,$$

and use the parametric formula

$$ds = \sqrt{\left(\frac{dx}{d\theta}\right)^2 + \left(\frac{dy}{d\theta}\right)^2}\, d\theta.$$

We have

$$\frac{dx}{d\theta} = -f(\theta) \sin \theta + f'(\theta) \cos \theta,$$

$$\frac{dy}{d\theta} = f(\theta) \cos \theta + f'(\theta) \sin \theta,$$

and we obtain

$$\begin{aligned}
ds &= \sqrt{\left(\frac{dx}{d\theta}\right)^2 + \left(\frac{dy}{d\theta}\right)^2}\, d\theta \\
&= \sqrt{(f(\theta))^2(\sin^2\theta + \cos^2\theta) + (f'(\theta))^2(\sin^2\theta + \cos^2\theta)}\, d\theta \\
&= \sqrt{(f(\theta))^2 + (f'(\theta))^2}\, d\theta \\
&= \sqrt{r^2 + \left(\frac{dr}{d\theta}\right)^2}\, d\theta.
\end{aligned}$$

EXAMPLE 5   Find the length of the spiral $r = \theta$ shown in Fig. 11.77 from $\theta = 0$ to $\theta = 2\pi$.

Solution   The arc length is given by the integral

$$\int_0^{2\pi} \sqrt{r^2 + \left(\frac{dr}{d\theta}\right)^2}\, d\theta = \int_0^{2\pi} \sqrt{\theta^2 + 1}\, d\theta$$

$$= \left(\frac{\theta}{2}\sqrt{1 + \theta^2} + \frac{1}{2}\ln(\theta + \sqrt{1 + \theta^2})\right)\Bigg]_0^{2\pi}$$

$$= \pi\sqrt{1 + 4\pi^2} + \frac{1}{2}\ln(2\pi + \sqrt{1 + 4\pi^2}).$$ ∎

EXAMPLE 6   Find the area of the surface generated when the cardioid $r = a(1 + \cos\theta)$ shown in Fig. 11.70 is revolved about the x-axis.

Solution   The surface area is given by

$$\int_0^{\pi} 2\pi y\, ds = \int_0^{\pi} 2\pi(r\sin\theta) \sqrt{r^2 + \left(\frac{dr}{d\theta}\right)^2}\, d\theta$$

$$= \int_0^{\pi} 2\pi a(1 + \cos\theta)(\sin\theta)\sqrt{a^2(1 + \cos\theta)^2 + a^2\sin^2\theta}\, , d\theta$$

$$= 2\pi a^2 \int_0^{\pi} (1 + \cos\theta)(\sin\theta)\sqrt{2 + 2\cos\theta}\, d\theta$$

$$= 2\sqrt{2}\pi a^2 \int_0^{\pi} (1 + \cos\theta)^{3/2}(\sin\theta)\, d\theta$$

$$= -2\sqrt{2}\pi a^2\frac{2}{5}(1 + \cos\theta)^{5/2}\Bigg]_0^{\pi} = \frac{32}{5}\pi a^2.$$ ∎

EXAMPLE 7   Use a calculator and Simpson's rule to estimate the length of the curve $r = a(2 - \cos\theta)$ in Fig. 11.78.

Solution   It is probably more accurate to estimate the length of the curve for $0 \le \theta \le \pi$ using a value for n in Simpson's rule and to double the answer than to

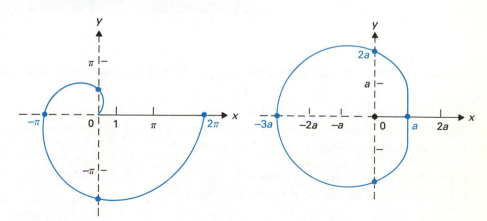

FIGURE 11.77   The spiral $r = \theta$ for $0 \le \theta \le 2\pi$.

FIGURE 11.78   The curve $r = a(2 - \cos\theta)$.

estimate the integral from 0 to 2π using the same n. Now $dr/d\theta = a\sin\theta$, so the length is given by

$$2\int_0^\pi \sqrt{a^2\sin^2\theta + a^2(2 - \cos\theta)^2}\, d\theta$$

$$= 2a\int_0^\pi \sqrt{\sin^2\theta + 4 - 4\cos\theta + \cos^2\theta}\, d\theta$$

$$= 2a\int_0^\pi \sqrt{5 - 4\cos\theta}\, d\theta \approx (13.36489)a. \qquad\blacksquare$$

SUMMARY

1. If $r = f(\theta)$, then $x = f(\theta)\cos\theta$ and $y = f(\theta)\sin\theta$. Also,

$$\frac{dy}{dx} = \frac{dy/d\theta}{dx/d\theta}.$$

2. If $r = f(\theta)$ is differentiable and ψ is the angle from the radius vector to the tangent of the polar curve $r = f(\theta)$, then

$$\tan\psi = \frac{r}{dr/d\theta} = \frac{f(\theta)}{f'(\theta)}.$$

3. The angle β between polar curves $r = f_1(\theta)$ and $r = f_2(\theta)$ at a point of intersection is given by

$$\tan\beta = \frac{\tan\psi_2 - \tan\psi_1}{1 + \tan\psi_1\tan\psi_2}.$$

4. If $r = f(\theta)$ is continuously differentiable, then the length of the polar curve from θ_1 to θ_2 is

$$s = \int_{\theta_1}^{\theta_2} \sqrt{\left(\frac{dr}{d\theta}\right)^2 + r^2}\, d\theta.$$

EXERCISES 11.8

1. Find the slope of the circle $r = a\sin\theta$, where $\theta = \pi/3$.

2. Find the slope of the limaçon $r = 1 + 2\cos\theta$, where $\theta = 5\pi/6$.

3. Find the slope of the spiral $r = \theta$, where $\theta = \pi/2$.

4. Find the slope of the curve $r = 1 + \cos(\theta/2)$, where $\theta = \pi/2$.

5. Find all points where the circle $r = a\cos\theta$ has slope 1.

6. Find all points where the limaçon $r = 1 + 2\sin\theta$ has a horizontal tangent.

7. Find all points where the limaçon $r = 1 + 2\sin\theta$ has a vertical tangent. [*Hint:* Express $dx/d\theta$ in terms of $\sin\theta$ and solve $dx/d\theta = 0$ by the quadratic formula.]

8. The cardioid $r = a + a\cos\theta$ has a sharp point when $\theta = \pi$.
   (a) Show that the curve $r = a + b\cos\theta$, with $a > b > 0$, has no sharp point where $\theta = \pi$.
   (b) Find the values of b, where $0 < b < a$, for which the curve $r = a + b\cos\theta$ still has a "dimple" where $\theta = \pi$.

9. Find the acute angle that the hyperbolic spiral $r\theta = a$ makes with the y-axis at the point $(r, \theta) = (2a/\pi, \pi/2)$.

10. Find the angle that the polar curve $r = 2 + 3 \sin \theta$ makes with the x-axis at the point $(r, \theta) = (2, 0)$.

11. Find the acute angle that the polar curve $r = a \cos(\theta/2)$ makes with the y-axis each time it crosses it at a point other than the origin.

12. Let ψ be the angle at (r, θ) on the spiral $r = a\theta$, as described in the text. Find $\lim_{\theta \to \infty} \psi$.

13. Show that the spiral $r = ae^\theta$ makes a constant angle with the radial line at each point on the spiral, and find the angle.

14. Find the angle between the circles $r = 2a \cos \theta$ and $r = 2a \sin \theta$ at the point $(r, \theta) = (a\sqrt{2}, \pi/4)$ of intersection.

15. Find the angle between the circle $r = a$ and the four-leaved rose $r = 2a \cos 2\theta$ at the point of intersection $(r, \theta) = (a, \pi/6)$.

16. Find the acute angle between the cardioid $r = a(1 + \cos \theta)$ and the limaçon $r = a(1 + 2 \cos \theta)$ at the point $(a, \pi/2)$.

17. Find the acute angle at which the polar curve $r = a \sin(\theta/2)$ meets itself corresponding to $\theta = \pi/2$ and $\theta = 7\pi/2$.

18. Find the length of the parabolic spiral $r = a\theta^2$ from $\theta = 0$ to $\theta = 2\pi$.

19. Find the total length of the cardioid $r = a(1 + \sin \theta)$. [*Hint:* Evaluate the integral by multiplying the integrand by

$$\frac{\sqrt{2 - 2 \sin \theta}}{\sqrt{2 - 2 \sin \theta}}.]$$

20. Express as an integral the length of the polar curve $r = a \cos(\theta/2)$ from $\theta = 0$ to $\theta = \pi$.

21. Express as an integral the total length of the three-leaved rose $r = a \sin 3\theta$.

22. Find the area of the surface generated when the circle $r = 2a \sin \theta$ is revolved about the x-axis.

23. Find the area of the surface generated when the cardioid $r = 2 + 2 \sin \theta$ is revolved about the y-axis.

24. Express as an integral the area of the surface generated when the arc of the spiral $r = \theta$ from $\theta = 0$ to $\theta = \pi$ is revolved about the y-axis.

25. Let f be a twice-differentiable function. Show that the curvature κ of the curve $r = f(\theta)$ at a point (r, θ) is given by the formula

$$\kappa = \left| \frac{(f(\theta))^2 + 2(f'(\theta))^2 - f(\theta) f''(\theta)}{[(f(\theta))^2 + (f'(\theta))^2]^{3/2}} \right|.$$

26. Use the formula in Exercise 25 to find the curvature of the cardioid $r = a(1 + \cos \theta)$, where $\theta = 0$.

27. Use the formula in Exercise 25 to find the radius of curvature of the limaçon $r = a(1 + 2 \sin \theta)$, where $\theta = 3\pi/2$.

⊞ Use a calculator and Simpson's rule in Exercises 28–32.

28. Estimate the arc length in Exercise 20.

29. Estimate the arc length in Exercise 21.

30. Estimate the surface area in Exercise 24.

31. Estimate the area of the surface generated when the polar curve $r = 3 + 2 \sin \theta$ is revolved about the line $y = -2$.

32. Estimate the area of the surface generated when the limaçon $r = 1 + 2 \cos \theta$ is revolved about the line $x = -1$.

33. Fly A is located at $(x, y) = (1, 1)$, while fly B is at $(-1, 1)$, fly C is at $(-1, -1)$, and fly D is at $(1, -1)$. The flies all crawl at the same rate of 1 unit distance per unit time. The flies all start crawling at the same instant, with fly A always crawling directly toward B, B directly toward C, C directly toward D, and D directly toward A. See Fig. 11.79.
(a) Find the point at which the flies meet.
(b) How long do the flies crawl before they meet?
(c) Find the polar coordinate equation of the path traveled by fly A. Sketch the path. [*Hint:* Find the angle ψ at a point on this path, and solve the differential equation $r = (\tan \psi) \, dr/d\theta.$]
(d) Find the length of the path traveled by fly A.
(e) What physiological problem will fly A encounter as it travels the path found in part (c) in the time found in part (b)?

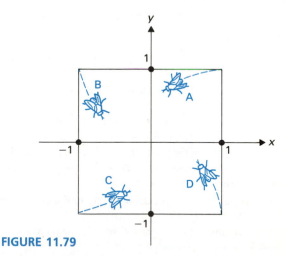

FIGURE 11.79

SUPPLEMENTARY EXERCISES FOR CHAPTER 11

1. Sketch the curve $x^2 - 4y^2 = 16$.

2. Sketch the curve $x + 4y^2 - 8y = -12$.

3. Sketch the curve $3x^2 + y^2 - 12x + 4y = -4$.

4. Sketch the curve $x^2 + 10x + 4y = 3$.

5. Sketch the curve $2x^2 - y^2 - 4x + 6y = 1$.

6. Sketch the curve $4x^2 + y^2 - x + 3y = \frac{111}{64}$.

7. Sketch the curve $4x^2 + 9y^2 = 36$.

8. Sketch the curve $x^2 + 3y^2 - 2x + 12y + 13 = 0$.

9. Sketch the curve $y^2 - 4x^2 + 8x - 6y = 11$.

10. Sketch the curve $4x^2 - 8x + 2y = 0$.

11. Sketch the curve $x^2 - 4y^2 + 2x - 8y - 3 = 0$.

12. Sketch the curve $2x^2 + 3y^2 - 4x - 27y + 30 = 0$.

13. Sketch the curve described parametrically by $x = t^2$, $y = t^2 - 1$.

14. Sketch the curve described parametrically by $x = 1 + 3 \sin t$, $y = 2 - \cos t$.

15. Sketch the parametric curve $x = 4 \sin^2 t$, $y = 4 \cos^2 t$.

16. Sketch the curve defined parametrically by $x = e^t + 1$, $y = e^{3t}$.

17. Find dy/dx and d^2y/dx^2 at $t = 1$ if $x = t^2$, $y = t^3 - 2t$.

18. Find dy/dx and d^2y/dx^2 at $t = \pi/3$ if $x = \sin t$, $y = \cos 2t$.

19. Find d^4y/dx^4 in terms of t if $x = t^2$ and $y = t^4 + 3t^3$.

20. Find the point(s) on the parametric curve $x = t^2$, $y = t^3 + 4t^2$ where the slope is 1.

21. Parametrize the curve $y = x^3$, taking as parameter the y-coordinate h at each point on the curve.

22. Parametrize the curve $y = e^{-x/2}$, taking as parameter the slope m of the curve at each point.

23. Reparametrize the curve $x = t^2 - 1$, $y = t^4 + 2t^2 - 3$, taking as parameter the slope m of the curve at each point.

24. Parametrize the curve $y = e^x$, taking as parameter t the square of the y-coordinate.

25. Express as an integral the length of the curve $x = \sin t$, $y = t^2 - 2$ from $t = 0$ to $t = 3$.

26. Find the point (x_0, y_0) on the curve $x = t$, $y = \frac{2}{3}t^{3/2}$ such that the distance from the origin to (x_0, y_0) measured along the curve is 42.

27. Find the length of the parametric curve $x = t^2 + 5$, $y = t^3 - 7$ from $t = 0$ to $t = 3$.

28. Use Simpson's rule and a calculator to estimate the length of the curve $x = t^2$, $y = t + \ln t$ for $1 \le t \le 3$.

29. Find the curvature of the parabola $y = x^2$ at the point $(2, 4)$.

30. Find the curvature of the curve $x = \sin 2t$, $y = \cos 3t$ at the point $t = \pi/6$.

31. Find the equation of the osculating circle to the curve $y = \cos x$ at the point $(0, 1)$.

32. Find the curvature of the curve $y = \sin x$ at the point where $x = \pi/4$.

33. Find the curvature of the curve $x = t^2$, $y = t^3 + 2t$ at the point where $t = 1$.

34. Find the equation of the osculating circle to the curve $x = 2 \cos t$, $y = 3 \sin t$ at the point $t = \pi/2$.

35. Find the curvature of the curve $x = (\ln y)^2$ at the point $(0, 1)$.

36. Find the point where the graph of $y = e^x$ has maximum curvature.

37. Find the radius of curvature of the parametric curve $x = 1/t$, $y = t^2 + 3t$, where $t = 2$.

38. Find all points on $y = x^2$ where the curvature is $1/\sqrt{2}$.

39. Find the equation of the osculating circle to the parabola $x^2 + 4x + 2y - 6 = 0$ at its vertex.

40. Find the curvature of the parametric curve $x = \sinh 2t$, $y = \cosh 3t$ at the point where $t = 0$.

41. Find *all* polar coordinates of the point $(-\sqrt{3}, 1)$.

42. Find *all* polar coordinates of the point $(1, -1)$.

43. Find *all* polar coordinates of the point $(x, y) = (-4, -4\sqrt{3})$.

44. Find *all* polar coordinates of the point $(x, y) = (0, 0)$.

45. Find x,y-coordinates of the point $(r, \theta) = (-5, \pi/3)$.

46. Find x,y-coordinates of the point $(r, \theta) = (3, 3\pi/4)$.

47. Find x,y-coordinates of the point $(r, \theta) = (0, 3\pi/17)$.

48. Find the polar coordinate equation of the ellipse $4x^2 + 9y^2 = 1$.

49. Find the polar coordinate equation of the hyperbola $x^2 - y^2 + 4x = 9$.

50. Find the polar coordinate equation of the line $x - 3y = 5$.

51. Find the x,y-equation of the polar curve $r = \sin \theta + \cos \theta$.

52. Find the x,y-equation of the polar curve $r^2 = 2 + \sin 2\theta$.

53. Find the x,y-equation of the polar curve $r^2 = \cos 2\theta$.

54. Sketch the curve with polar coordinate equation $r = a(1 + 2 \sin \theta)$.

55. Sketch the curve with polar coordinate equation $r = a \sin 2\theta$.

56. Sketch the curve with polar coordinate equation $r = -3 \sec \theta$.

57. Sketch the curve with polar coordinate equation $r = 1 + \cos(\theta/2)$.

58. Sketch the polar curve $r = 4 - 3|\cos \theta|$.

59. Sketch the polar curve $r = 1 + \cos 3\theta$.

60. Find all points of intersection of $r^2 = a^2 \sin \theta$ and $r = a/\sqrt{2}$.

61. Find all points of intersection of $r^2 = a^2 \cos 2\theta$ and $r = a/\sqrt{2}$.

62. Find all points of intersection of $r = 2a \sin \theta$ and $r = a$.

63. Find the area inside $r = a(1 + \frac{1}{2} \sin \theta)$ and outside the circle $r = a$.

64. Find the total area enclosed by the polar curve $r^2 = 1 + \sin^2 \theta$.

65. Find the total area of the region inside the curve $r^2 = 1 + \cos 2\theta$.

66. Find the total area of the regions enclosed by the curve $r = a|\sin \theta|$.

67. Find the slope of the circle $r = 2a \sin \theta$, where $\theta = 2\pi/3$.

68. Find the points where the curve $r = 1 - \cos \theta$ has a horizontal tangent line.

69. Find the slope of the curve $r = a \cos(\pi/2)$, where $\theta = \pi/3$.

70. Find the slope of the tangent line to the tip of the leaf of the rose $r = a \sin 3\theta$ lying in the first quadrant.

71. Find the angle between $r^2 = a^2 \sin \theta$ and $r = a/\sqrt{2}$ at their first-quadrant point of intersection.

72. Find the angle between the cardioids $r = a(1 + \cos \theta)$ and $r = -a(1 + \cos \theta)$ at their point of intersection in the upper half-plane.

73. Find the acute angle that the lemniscate $r^2 = 4 \sin 2\theta$ makes with the ray $\theta = \pi/6$.

74. Find the length of the arc of the curve $r = a \cos^2(\theta/2)$ from $\theta = 0$ to $\theta = \pi/2$.

75. Find the length of the arc of the spiral $r = e^{2\theta}$ from $\theta = 0$ to $\theta = 2\pi$.

76. Use a calculator to estimate the total length of the curve $r = 5 \sin 4\theta$.

77. Use a calculator to estimate the total length of the curve $r = 2 + \sin 2\theta$.

78. Use a calculator to estimate the area of the surface generated when the cardioid $r = 1 + \cos \theta$ is revolved about the line $y = 2$.

79. Let f be a twice-differentiable function. Show that $|\kappa(x)| \leq |f''(x)|$, where $\kappa(x)$ is the curvature of the graph at $(x, f(x))$.

80. Let $f(x) = 1 + x + (x/2)^2 + \cdots + (x/n)^n + \cdots$. Find the curvature of the graph where $x = 0$.

81. Find the length of the arc of the curve $x = a \cos^3 t$, $y = a \sin^3 t$ for $0 \leq t \leq 2\pi/3$.

COORDINATES IN SPACE AND VECTORS

12

This chapter begins our introduction to analytic geometry in space. Sections 12.3–12.5 present the notion of a *vector* and develop some *vector algebra*. Chapter 12 plays a role in the study of multivariable calculus similar to the role Chapter 1 played in studying single-variable calculus. The main difference is our introduction here of vector methods. It is not essential to use vector methods for multivariable calculus, but they are very neat and convenient. Vector methods started appearing in the undergraduate calculus sequence in the 1940s, but they were known many years before that. They could have been used perfectly well in Chapter 1, in which case Chapter 14 would seem a much more natural extension of the single-variable calculus. However, it is traditional to introduce them at this point rather than ask the beginning calculus student to be concerned with them. They have a long history of use in physics and engineering.

12.1 RECTANGULAR COORDINATES AND CYLINDRICAL SURFACES

Rectangular Coordinates

We know how to describe the location of a point in the plane using an ordered pair (x, y) of real numbers. The location of a point in space can be described using an ordered triple (x, y, z) of numbers. We set up a *rectangular* (or *Cartesian*) system of coordinates as follows. Select some point of space as *origin* and imagine three coordinate axes, any two of which are perpendicular, through this point. Figure 12.1 shows only half of each of these x-, y-, and z-axes for clarity. The plane containing the x- and y-coordinate axes is the *x,y-coordinate plane*, and the *x,z-coordinate plane* and *y,z-coordinate plane* are similarly defined.

The three coordinate planes naturally divide space into eight parts or *octants* according to whether the coordinates are positive or negative. Symbolically, the

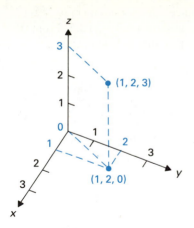

FIGURE 12.1 The Cartesian point $(1, 2, 3)$ in space.

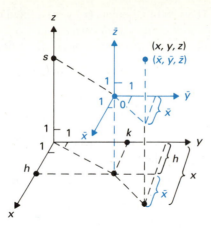

FIGURE 12.2 Axes translated to (h, k, s); $\bar{x} = x - h$, $\bar{y} = y - k$, $\bar{z} = z - s$.

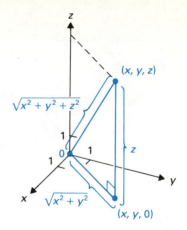

FIGURE 12.3 The distance from (x, y, z) to $(0, 0, 0)$ is $\sqrt{x^2 + y^2 + z^2}$.

possible combinations of coordinate signs are

$$(+, +, +), \quad (+, +, -), \quad (+, -, +), \quad (+, -, -),$$
$$(-, +, +), \quad (-, +, -), \quad (-, -, +), \quad (-, -, -).$$

The portion where all coordinates are positive, that is, the $(+, +, +)$ part, is called the *first octant*. We have never seen any attempt to number the other octants.

We can use translated $\bar{x}, \bar{y}, \bar{z}$-coordinates in space, just as we used $\bar{x}, \bar{y}$-coordinates in the plane. Figure 12.2 shows translated axes at a new origin (h, k, s). The equations

$$\bar{x} = x - h, \qquad \bar{y} = y - k, \qquad \bar{z} = z - s \qquad (1)$$

for transforming coordinates are natural extensions of the formulas we are familiar with for translation in the plane. We simply have one more equation to handle the third coordinate.

From Fig. 12.3, it is clear that the distance from the origin to the point (x, y, z) is $\sqrt{x^2 + y^2 + z^2}$. We can compute the distance from (x_1, y_1, z_1) to a point (x_2, y_2, z_2). At (x_1, y_1, z_1), we take translated $\bar{x}, \bar{y}, \bar{z}$-coordinates as shown in Fig. 12.4. From Eqs. (1), the point $(x, y, z) = (x_2, y_2, z_2)$ has translated coordinates $(\bar{x}, \bar{y}, \bar{z}) = (x_2 - x_1, y_2 - y_1, z_2 - z_1)$. The distance from this point to the new origin is then $\sqrt{\bar{x}^2 + \bar{y}^2 + \bar{z}^2}$, so we see that the distance d between (x_1, y_1, z_1) and (x_2, y_2, z_2) is given by

$$d = \sqrt{(x_2 - x_1)^2 + (y_2 - y_1)^2 + (z_2 - z_1)^2}. \qquad (2)$$

This is an easily remembered extension of the formula for the distance between points (x_1, y_1) and (x_2, y_2) in the plane.

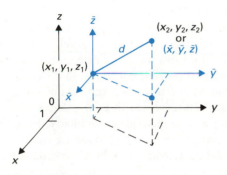

FIGURE 12.4
$$d = \sqrt{\bar{x}^2 + \bar{y}^2 + \bar{z}^2}$$
$$= \sqrt{(x_2 - x_1)^2 + (y_2 - y_1)^2 + (z_2 - z_1)^2}$$

EXAMPLE 1 Find the distance between $(-1, 3, 2)$ and $(1, 1, 3)$.

Solution By Eq. (2), the distance is

$$\sqrt{[1 - (-1)]^2 + (1 - 3)^2 + (3 - 2)^2} = \sqrt{2^2 + (-2)^2 + 1^2}$$
$$= \sqrt{9} = 3. \quad \blacksquare$$

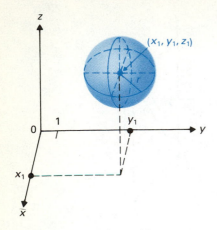

FIGURE 12.5 Sphere with center (x_1, y_1, z_1) and radius r.

The sets of all points (x, y, z) a fixed distance r from a point (x_1, y_1, z_1) is a *sphere of radius r with center at* (x_1, y_1, z_1), as shown in Fig. 12.5. From the distance formula, we see that (x, y, z) lies on this sphere if and only if

$$\sqrt{(x - x_1)^2 + (y - y_1)^2 + (z - z_1)^2} = r,$$

or

$$(x - x_1)^2 + (y - y_1)^2 + (z - z_1)^2 = r^2. \tag{3}$$

Equation (2) is thus the equation of a sphere. By completing the square, we see that any equation

$$x^2 + y^2 + z^2 + ax + by + cz = d$$

describes a sphere (possibly degenerating to a point) if the equation has a non-empty solution set in space.

EXAMPLE 2 Find the center and radius of the sphere

$$x^2 + y^2 + z^2 - 6x + 4y = -9$$

and then sketch the sphere.

Solution The steps for completing the square are

$$(x^2 - 6x) + (y^2 + 4y) + z^2 = -9,$$
$$(x - 3)^2 + (y + 2)^2 + (z - 0)^2 = -9 + 9 + 4 = 4.$$

Thus the sphere has center $(3, -2, 0)$ and radius 2. Its sketch is shown in Fig. 12.6. ∎

EXAMPLE 3 Find the solution set of the equation

$$x^2 + y^2 + z^2 + 2x - 4z = -6.$$

Solution Completing the square, we obtain

$$(x^2 + 2x) + y^2 + (z^2 - 4z) = -6,$$
$$(x + 1)^2 + y^2 + (z - 2)^2 = -6 + 1 + 4 = -1.$$

Since a sum of squares cannot be negative, the solution set is empty. ∎

In any coordinate system, we want to know the *level coordinate sets*. Such a set is described by setting one of the coordinate variables equal to a constant; it consists of the points where the coordinate achieves that constant ''level.'' Level coordinate sets are the geometric configurations that the coordinate system can handle most easily, since they have the simplest possible equations. In x,y-coordinates for the plane, the equations $x = a$ and $y = b$ describe a vertical line and a horizontal line, respectively. From Fig. 12.7(a), we see that in space, $x = a$ describes a vertical plane, parallel to the y,z-coordinate plane. Similarly, Figs. 12.7(b) and (c) show that $y = b$ gives a vertical plane parallel to the x,z-coordinate plane, and $z = c$ describes a horizontal plane parallel to the x,y-coordinate plane.

In the Cartesian plane, a single linear equation gives a *line*. We will show later that in space, a single linear equation describes a *plane*. The level coordi-

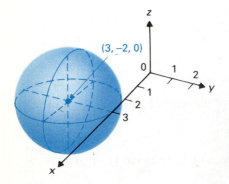

FIGURE 12.6 The sphere $(x - 3)^2 + (y + 2)^2 + z^2 = 4$.

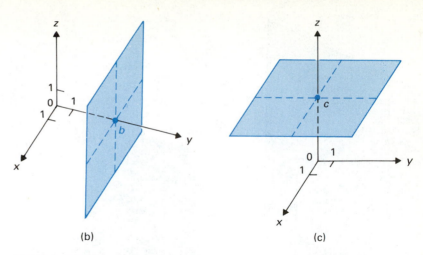

(a)   (b)   (c)

FIGURE 12.7 Level coordinate sets for rectangular coordinates: (a) the plane $x = a$; (b) the plane $y = b$; (c) the plane $z = c$.

nate planes $x = a$, $y = b$, $z = c$ are our first illustrations of this. We given another illustration.

EXAMPLE 4 Sketch the solution set of $x + y = 1$ in space.

Solution First we draw the line $x + y = 1$ in the x,y-coordinate plane in Fig. 12.8. The equation places no restriction on z, so any point directly above or below this line also satisfies $x + y = 1$. The solution set is thus the vertical plane containing the line $x + y = 1$ in the x,y-coordinate plane, as shown in Fig. 12.8.

■

Cylinders

The points in space satisfying a single equation in x, y, and z usually form a surface. If one of the variables x, y, and z is missing from the equation, then the surface is a cylinder. A **cylinder** in space is a surface that can be generated by a line that moves along a plane curve, keeping a fixed direction. The parallel lines (rulings) on the cylinder corresponding to positions of the generating line are the *elements of the cylinder*. To illustrate, suppose the variable x is missing in an equation $F(x, y, z) = 0$. If (a, b, c) lies on the surface, then so will (x, b, c) for all x, that is, the line through (a, b, c) parallel to the x-axis. Consequently, the surface is a cylinder with elements parallel to the x-axis and meeting the y,z-plane in the curve $F(0, y, z) = 0$. Similar results hold if y or z is missing. Thus, the surface in space described by an equation in two of the variables x, y, and z is a cylinder. The cylinder is generated by a line parallel to the axis of the missing variable as the line moves along the curve the equation describes in the coordinate plane of present variables.

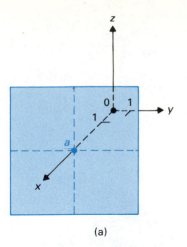

The line $x + y = 1$ in the x,y-coordinate plane

FIGURE 12.8 The plane $x + y = 1$ in space.

EXAMPLE 5 Sketch the *cylinder* $x + y = 1$ in space.

Solution This cylinder is the plane in the preceding example. We show this vertical plane with rulings in Fig. 12.9. The plane may be regarded as a cylinder

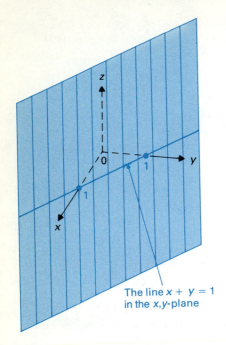

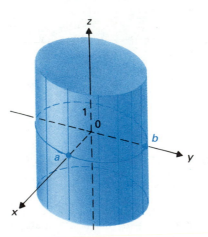

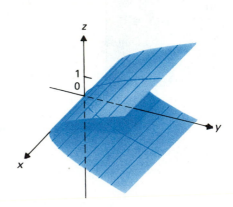

FIGURE 12.9 The plane $x + y = 1$ viewed as a cylinder with rulings parallel to the z-axis.

FIGURE 12.10 The elliptic cylinder $x^2/a^2 + y^2/b^2 = 1$.

FIGURE 12.11 The parabolic cylinder $z^2 = 4py$.

generated by a line parallel to the z-axis moving along the line $x + y = 1$ in the x,y-plane. The rulings show some of the positions of the generating line. ∎

EXAMPLE 6   Sketch the *elliptic cylinder* $x^2/a^2 + y^2/b^2 = 1$ in space.

Solution   This cylinder is sketched in Fig. 12.10. It is generated by a line parallel to the z-axis moving along the ellipse $x^2/a^2 + y^2/b^2 = 1$ in the x,y-plane. ∎

EXAMPLE 7   Sketch the *parabolic cylinder* $z^2 = 4py$ in space.

Solution   The cylinder, shown in Fig. 12.11, is generated by a line parallel to the x-axis moving along the parabola $z^2 = 4py$ in the y,z-plane. ∎

SUMMARY

1. In rectangular coordinates in space, a point has coordinates (x, y, z).

2. The distance from (x_1, y_1, z_1) to (x_2, y_2, z_2) in rectangular coordinates is

$$d = \sqrt{(x_2 - x_1)^2 + (y_2 - y_1)^2 + (z_2 - z_1)^2}.$$

3. The rectangular equation of a sphere in space with center (x_1, y_1, z_1) and radius r is

$$(x - x_1)^2 + (y - y_1)^2 + (z - z_1)^2 = r^2.$$

4. An equation containing only two of the three space variables x, y, z has as graph a cylinder with the axis of the cylinder parallel to the axis of the missing variable. To illustrate, suppose x and y are present in the equation. The cylinder intersects the x,y-coordinate plane in the curve given by the equation and is parallel to the z-axis.

EXERCISES 12.1

In Exercises 1–8, plot the given points in space.

1. $(0, 0, -3)$ **2.** $(1, -1, 0)$

3. $(-1, 0, 4)$ **4.** $(2, -1, 1)$

5. $(2, 1, -1)$ **6.** $(-2, 2, -1)$

7. $(-2, -2, 3)$ **8.** $(-1, -1, -1)$

In Exercises 9–17, sketch all points in space satisfying the given equation.

9. $x = 2$ **10.** $z = 3$

11. $x = y$ **12.** $y^2 = 1$

13. $x = z$ **14.** $y + z = 1$

15. $x - y = 1$ **16.** $x = y = z$

17. $x = -y = z$

While we have not defined a line or a plane in space, use your geometric intuition in Exercises 18–23 to find the desired point.

18. The point such that the line segment joining it to $(-2, 1, -4)$ is bisected by and perpendicular to the plane $x = 0$

19. The point such that the line segment joining it to $(-1, \pi, \sqrt{2})$ has the origin as midpoint

20. The point such that the line segment joining it to $(-1, 4, -3)$ has $(-1, 2, -3)$ as midpoint

21. The point in the plane $y = 2$ that is closest to the point $(-1, -5, 2)$

22. The point on the z-axis closest to the point $(3, -1, 2)$

23. The midpoint of the line segment joining $(-1, 4, 3)$ and $(3, -2, 5)$

24. Let $(-1, 2, 1)$ be chosen as origin for $\bar{x},\bar{y},\bar{z}$-coordinates. Express each point in terms of these new translated coordinates.
(a) $(1, -2, 1)$ (b) $(-3, 4, 0)$ (c) $(5, -1, 2)$

In Exercises 25–29, find the distance between the given points.

25. $(-1, 0, 4)$ and $(1, 1, 6)$ **26.** $(2, -1, 3)$ and $(0, 1, 7)$

27. $(3, -1, 0)$ and $(-1, 2, 6)$

28. $(-2, 1, 4)$ and $(-3, 2, 8)$

29. $(0, 1, 8)$ and $(-3, 1, 12)$

30. Find the equation of the sphere with center $(-2, 1, 0)$ and radius 4.

31. Find the equation of the sphere with center $(-1, 2, 4)$ and passing through the point $(2, -1, 5)$.

32. Find the equation of the sphere having $(-1, 2, 6)$ and $(1, 6, 0)$ as endpoints of a diameter.

In Exercises 33–36, find the center and radius of the given sphere, if it is not empty.

33. $x^2 + y^2 + z^2 - 2x + 2y = 0$

34. $x^2 + y^2 + z^2 - 6x - 4y + 8z + 4 = 0$

35. $x^2 + y^2 + z^2 - x + 2y - z + 4 = 0$

36. $x^2 + y^2 + z^2 + 4x - 2y + 4z + 9 = 0$

In Exercises 37–46, sketch the given cylinder. Draw some rulings on the cylinder.

37. $y^2 + z^2 - 4 = 0$ **38.** $x^2 + 2x + y^2 = 0$

39. $y^2 - z = 0$ **40.** $xz - 1 = 0$

41. $4x - y^2 + 2y + 3 = 0$ **42.** $x + 2z = 4$

43. $x + z = -2$ **44.** $y = e^{-x}$

45. $x = \sin y$ **46.** $x = -\ln z$

47. A boat traveling down the middle of a straight river approaches a bridge for a highway going straight across the river 20 ft above the river. When the boat is 60 ft from the point directly under the middle of the bridge, a car, driven in the middle of the road, is approaching the bridge at a point 30 ft from the center of the bridge. See Fig. 12.12. Find the distance from the boat to the car at that instant.

48. Suppose the boat in Exercise 47 is traveling 10 ft/sec and the car 40 ft/sec. Find the rate of change of the distance between them at the instant given in Exercise 47.

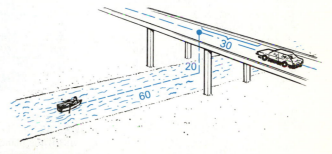

FIGURE 12.12

12.2   CYLINDRICAL AND SPHERICAL COORDINATES

Cylindrical Coordinates

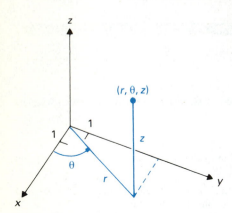

FIGURE 12.13 r, θ, and z in cylindrical coordinates.

We can locate a point in space by specifying the x,y-coordinates of its position using polar r,θ-coordinates and specifying its height by the z-coordinate. The point then has coordinates (r, θ, z) as well as coordinates (x, y, z). Of course, the coordinates r and θ are not unique, being the usual polar coordinates. Such coordinates are shown in Fig. 12.13.

Figure 12.14(a) shows that the level coordinate set $r = a$ is a cylinder of radius a about the z-axis, for $r = a$ has a circle in the x,y-plane as solution set and there is no restriction placed on z. Consequently, (a, θ, z) is on the solution set for all θ and all z. For this reason, these r,θ,z-coordinates are called *cylindrical coordinates*. Figure 12.14(b) shows that the level coordinate set $\theta = \alpha$ is a vertical plane. The level coordinate set $z = c$ is the plane shown in Fig. 12.7(c).

Since we know how to change from polar r,θ-coordinates to x,y-coordinates in the plane, we know how to change back and forth from cylindrical to rectangular coordinates in space. Namely,

$$x = r \cos \theta, \qquad y = r \sin \theta, \qquad z = z \tag{1}$$

and

$$r^2 = x^2 + y^2, \qquad \theta = \tan^{-1}\left(\frac{y}{x}\right)[+ \, n\pi], \qquad z = z. \tag{2}$$

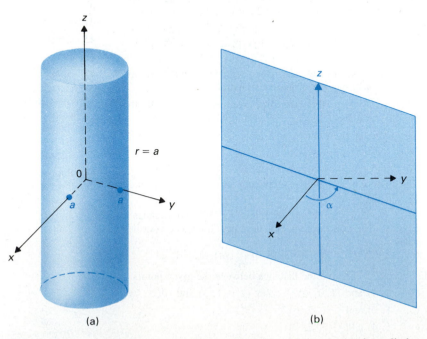

(a)   (b)

FIGURE 12.14   Level coordinate sets for cylindrical coordinates: (a) the cylinder $r = a$; (b) the plane $\theta = \alpha$.

The $[+ n\pi]$ in Eqs. (2) appears in case the value for the inverse tangent functions does not lie in the appropriate quadrant. Suppose, for example, we always wish to have θ in the range $0 \leq \theta < 2\pi$. Then for a point (x,y) in the first quadrant, we may take $n = 0$. For points in the second or third quadrant, we need $n = 1$, while for a point in the fourth quadrant, we need $n = 2$.

EXAMPLE 1 Transform the equation of the sphere $x^2 + y^2 + z^2 = 4$ to a cylindrical coordinate form.

Solution Since $x^2 + y^2 = r^2$, the sphere is described by $r^2 + z^2 = 4$ in cylindrical coordinates. ■

EXAMPLE 2 Find cylindrical coordinates of the point $(x, y, z) = (1, -1, 3)$.

Solution From Eqs. (2) and Fig. 12.15, we see that $r^2 = 1^2 + (-1)^2 = 2$, so $r = \sqrt{2}$. Then $\theta = \tan^{-1}(y/x) = \tan^{-1}(-1) = -\pi/4$. Of course, $z = 3$. The point is thus $(\sqrt{2}, -\pi/4, 3)$ or $(\sqrt{2}, 7\pi/4, 3)$ in cylindrical coordinates. ■

Spherical Coordinates

Another useful coordinate system in space is the *spherical coordinate* system, in which a point has coordinates (ρ, ϕ, θ), as indicated in Fig. 12.16. The coordinate ρ is the length of the line segment joining the point and the origin. We will restrict ourselves to values of $\rho \geq 0$, considering ρ to be a nondirected distance. This is in contrast to our use of r in polar coordinates. We will be using spherical coordinates chiefly with integrals, and $\rho \geq 0$ will suffice for that purpose.

The angle ϕ is the angle from the z-axis to the line segment from the origin to the point, shown in Fig. 12.16. We will also consider ϕ to be a nondirected measurement, and we see from Fig. 12.16 that we may always take ϕ in the range $0 \leq \phi \leq \pi$. Finally, θ is the same angle as in cylindrical coordinates. Later, when integrating using spherical coordinates, we generally let θ have values from 0 to 2π. However, we may on occasion use negative values of θ when using cylindrical coordinates.

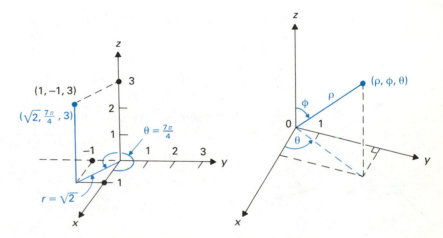

FIGURE 12.15 $(x, y, z) = (1, -1, 3)$, $(r, \theta, z) = (\sqrt{2}, 7\pi/4, 3)$

FIGURE 12.16 ρ, ϕ, and θ, in spherical coordinates.

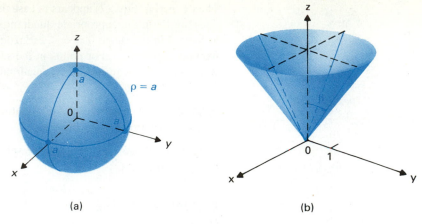

FIGURE 12.17 Level coordinate sets for spherical coordinates: (a) the sphere $\rho = a$; (b) the cone $\phi = \beta$.

The level coordinate surface $\rho = a$ is the sphere shown in Fig. 12.17(a); hence the name "spherical coordinates." The level coordinate surface $\phi = \beta$ is the cone shown in Fig. 12.17(b). The level coordinate set $\theta = \alpha$ is the plane shown in Fig. 12.14(b). Spherical coordinates are especially useful when dealing with spheres or cones.

We need to express x, y, and z in terms of ρ, ϕ, and θ, and conversely. From Fig. 12.18, we see that

$$x = \rho \sin \phi \cos \theta, \qquad y = \rho \sin \phi \sin \theta,$$
$$z = \rho \cos \phi. \tag{3}$$

Since ρ is the distance from the point (x, y, z) to the origin, it is obvious that $\rho^2 = x^2 + y^2 + z^2$. From Fig. 12.18, we can see that

$$\rho = \sqrt{x^2 + y^2 + z^2}, \qquad \phi = \cos^{-1}\left(\frac{z}{\sqrt{x^2 + y^2 + z^2}}\right),$$
$$\theta = \tan^{-1}\left(\frac{y}{x}\right)[+ n\pi]. \tag{4}$$

The $[+ n\pi]$ in Eqs. (4) is included in case the value of the inverse function does not lie in the appropriate quadrant.

EXAMPLE 3 Find spherical coordinates of the point $(x, y, z) = (-1, -1, -1)$.

Solution We obtain from Eqs. (4)

$$\rho = \sqrt{(-1)^2 + (-1)^2 + (-1)^2} = \sqrt{3},$$
$$\phi = \cos^{-1}\left(\frac{-1}{\sqrt{1 + 1 + 1}}\right) = \cos^{-1}\left(\frac{-1}{\sqrt{3}}\right),$$
$$\theta = \tan^{-1}\left(\frac{-1}{-1}\right) + \pi = \tan^{-1}1 + \pi = \frac{5\pi}{4}.$$

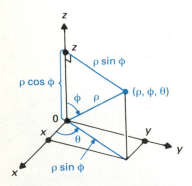

FIGURE 12.18 Relations between rectangular and spherical coordinates.

Our choice of $\tan^{-1}1 + \pi$ rather than $\tan^{-1}1$ for θ comes from the fact that we require $\rho > 0$ for spherical coordinates. As shown in Fig. 12.19, the point

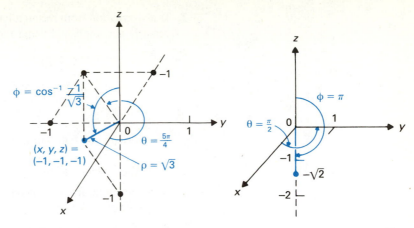

FIGURE 12.19 Spherical coordinates. $(\sqrt{3}, \cos^{-1}(-1/\sqrt{3}), 5\pi/4)$ for $(x, y, z) = (-1, -1, -1)$.

FIGURE 12.20 Spherical $(\sqrt{2}, \pi, \pi/2)$ is rectangular $(0, 0, -\sqrt{2})$.

$(-1, -1, -1)$ lies under the *third* quadrant of the x,y-coordinate plane. So we *must* have a third-quadrant angle for θ. Since $\tan^{-1}1$ is a *first*-quadrant angle, we must add π. Thus the spherical coordinates are $(\sqrt{3}, \cos^{-1}(-1/\sqrt{3}), 5\pi/4)$, as shown in Fig. 12.19. ∎

EXAMPLE 4 Find x,y,z-coordinates of the point $(\rho, \phi, \theta) = (\sqrt{2}, \pi, \pi/2)$.

Solution This time we use Eqs. (3) and obtain

$$x = \rho \sin \phi \cos \theta = \sqrt{2} \cdot 0 \cdot 0 = 0,$$
$$y = \rho \sin \phi \sin \theta = \sqrt{2} \cdot 0 \cdot 1 = 0,$$
$$z = \rho \cos \phi = \sqrt{2} \cdot (-1) = -\sqrt{2}.$$

Thus the point is $(x, y, z) = (0, 0, -\sqrt{2})$, as shown in Fig. 12.20. ∎

Both Examples 3 and 4 could have been solved just by referring to the figures, without using Eqs. (3) and (4). It is a good idea to do some simple problems from the figures occasionally to be sure you understand the geometric meaning of ρ, ϕ, and θ. Some exercises ask you to do this.

EXAMPLE 5 Find the spherical coordinate equation of the plane $z = 2$.

Solution From Eqs. (3), we see that the equation becomes $\rho \cos \phi = 2$, or $\rho = 2 \sec \phi$. ∎

SUMMARY

1. In cylindrical coordinates in space, a point has coordinates (r, θ, z), where r and θ are the usual polar coordinates in the x,y-plane.

2. Transformation from rectangular to cylindrical coordinates and vice versa is accomplished using the formulas

$$x = r \cos \theta, \qquad\qquad r = \sqrt{x^2 + y^2},$$

$$y = r \sin \theta, \quad \text{and} \quad \theta = \tan^{-1}\left(\frac{y}{x}\right)[+ n\pi],$$

$$z = z, \qquad\qquad z = z.$$

3. In spherical ρ, ϕ, θ-coordinates for a point in space,

- ρ is the distance to the origin,
- ϕ is the angle from the positive z-axis to the ray from the origin, where $0 \leq \phi \leq \pi$,
- θ is the same angle as for cylindrical coordinates.

4. Transformation from rectangular to spherical coordinates and vice versa is accomplished using the formulas

$$x = \rho \sin \phi \cos \theta, \qquad\qquad \rho = \sqrt{x^2 + y^2 + z^2},$$

$$y = \rho \sin \phi \sin \theta, \quad \text{and} \quad \phi = \cos^{-1}\left(\frac{z}{\sqrt{x^2 + y^2 + z^2}}\right),$$

$$z = \rho \cos \phi, \qquad\qquad \theta = \tan^{-1}\left(\frac{y}{x}\right)[+ n\pi].$$

EXERCISES 12.2

In Exercises 1–8, plot the point with the given cylindrical coordinates (r, θ, z).

1. $(2, \pi/2, 2)$ **2.** $(1, 3\pi/4, -1)$

3. $(0, 7\pi/6, -3)$ **4.** $(0, -\pi/4, 2)$

5. $(-3, \pi/4, 3)$ **6.** $(2, 0, -1)$

7. $(3, 0, 1)$ **8.** $(4, \pi, 0)$

9. Sketch in space the set of points satisfying the equation $\theta = \pi/4$ in cylindrical coordinates.

10. Sketch in space the set of points satisfying the equation $r = 2$ in cylindrical coordinates.

11. Sketch in space the set of points satisfying the equation $r = z$ in cylindrical coordinates.

12. (a) Transform the x,y,z-equation $x^2 + y^2 = 9$ into cylindrical coordinates.
    (b) Using your answer to part (a), sketch all points in space satisfying the equation.

In Exercises 13–20, plot the points in space with the given spherical coordinates (ρ, ϕ, θ).

13. $(2, 0, 0)$ **14.** $(2, 0, \pi/2)$

15. $(0, \pi/4, \pi/2)$ **16.** $(3, \pi/4, \pi/2)$

17. $(2, 3\pi/4, 7\pi/4)$ **18.** $(1, \pi, \pi/4)$

19. $(3, \pi, \pi)$ **20.** $(3, \pi/2, 3\pi/2)$

21. Sketch in space the set of points satisfying the equation $\phi = \pi/4$ in spherical coordinates.

22. Sketch in space the set of points satisfying $\rho = 3$ in spherical coordinates.

23. Sketch in space the set of points satisfying $\rho \sin \phi = 2$ in spherical coordinates.

24. Find *all* cylindrical coordinates of the point $(1, 1, 1)$.

25. Find all cylindrical coordinates of the Cartesian point $(1, -\sqrt{3}, 5)$.

In Exercises 26–39, use a figure rather than transformation equations to find the indicated coordinates. (For cylindrical coordinates, give r-values ≥ 0 and θ-values from 0 to 2π.)

26. Rectangular coordinates for cylindrical $(\sqrt{2}, \pi/4, 3)$

27. Rectangular coordinates for cylindrical $(4, 2\pi/3, -2)$

28. Rectangular coordinates for cylindrical $(0, \pi, -1)$

29. Cylindrical coordinates for rectangular $(0, 1, 1)$

30. Cylindrical coordinates for rectangular $(-1, 1, -5)$

31. Cylindrical coordinates for rectangular $(\sqrt{3}, -1, 2)$

32. Spherical coordinates for rectangular $(1, 0, 0)$

33. Spherical coordinates for rectangular $(0, 0, -4)$

34. Spherical coordinates for rectangular $(1, 1, 1)$

35. Spherical coordinates for rectangular $(-3, -4, 5)$

36. Rectangular coordinates for spherical $(2, \pi/4, -\pi)$

37. Rectangular coordinates for spherical $(0, 3\pi/4, \pi/6)$

38. Rectangular coordinates for spherical $(4, \pi/2, \pi/3)$

39. Rectangular coordinates for spherical $(6, 2\pi/3, \pi)$

40. Express the cylindrical coordinates r, θ, and z in terms of the spherical coordinates ρ, ϕ, and θ.

41. Express the spherical coordinates ρ, ϕ, and θ in terms of the cylindrical coordinates r, θ, and z.

In Exercises 42–49, use the transformation equations to find the indicated coordinates. (For cylindrical coordinates, give r-values ≥ 0 and θ-values from 0 to 2π.)

42. Spherical coordinates for rectangular $(1, 2, -2)$

43. Spherical coordinates for rectangular $(3, 4, 0)$

44. Spherical coordinates for rectangular $(0, 1, -\sqrt{3})$

45. Cylindrical coordinates for rectangular $(1, -1, 3)$

46. Cylindrical coordinates for rectangular $(-\sqrt{3}, 1, 4)$

47. Rectangular coordinates for spherical $(2, \pi/4, 2\pi/3)$

48. Rectangular coordinates for spherical $(1, 5\pi/6, 4\pi/3)$

49. Rectangular coordinates for cylindrical $(3, 7\pi/6, -5)$

50. Find the spherical coordinate equation for $x^2 + y^2 + z^2 = 25$ by using the transformation equations.

51. Find the volume of the region described in spherical coordinates by $2 \leq \rho \leq 5$ and $0 \leq \phi \leq \pi/2$.

52. Describe in terms of x,y,z-coordinates all points in space where the cylindrical r-coordinate is equal to the spherical ρ-coordinate.

53. Describe in terms of spherical coordinates all points in space where $x + y = 1$.

54. Describe in terms of spherical coordinates all points in space where $z^2 = 4$.

12.3 VECTORS IN THE PLANE AND IN SPACE

Represented geometrically, a vector in the plane is nothing more than an arrow that has a *length* and points in a *direction*. The notion of direction will be of great concern to us when we study calculus of functions of more than one variable in Chapter 14. If $y = f(x)$, the derivative $f'(c)$ gives the rate the function changes at c *in the direction of increasing x*. Of course, at a point c on the x-axis, there are only two directions one can go in the domain of the function, the direction of *increasing x* (to the right) and the direction of *decreasing x* (to the left). If $f'(c) = 3$, the function is *increasing* at a rate of 3 as we go to the right and is *decreasing* at a rate of 3 (or increasing at a rate of -3) as we go to the left. With only these two directions to worry about, it was not necessary to introduce vectors to talk about direction in single-variable calculus.

On the other hand, if $z = f(x, y)$ is a function of two variables having as domain the x,y-coordinate plane, than we can travel in infinitely many directions from a point (a, b) in the domain. In Chapter 14, we will see how to find the rate at which $f(x, y)$ changes as we travel in *any* of these directions. Vectors will be very handy when we are finding all those rates of change (derivatives).

We have also described the *direction* of a line in the plane by its *slope m*. There is only one catch to this description: A vertical line has no slope, so the slope does not really do the entire job for us. But the direction of a vertical line poses no problem when we use vectors. We can consider an arrow pointing straight up (or down) as easily as in any other direction.

Vectors have long been used in physics and engineering to describe *direction* and *magnitude* (represented by the *length* of the vector). The magnitude and direction of a force acting on a body is often represented by an arrow, as we will explain later in this section. Also, the magnitude and direction of the electric

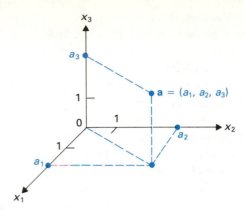

FIGURE 12.21 The point **a** in space.

current induced in a wire as it moves through a magnetic field may be represented by a vector.

Subscript Notation for Coordinates

When we need more and more variables, it is convenient to use a single letter with different subscripts rather than different letters. For example, we might use

$$x_1, x_2, x_3, x_4 \quad \text{in place of} \quad x, y, z, w$$

as variables. If we had more than 26 variables, we would run out of lowercase letters, but we will never run out of subscripts! Thus we will often write

$$(a_1, a_2) \quad \text{in place of} \quad (a, b),$$
$$(x_1, x_2) \quad \text{in place of} \quad (x, y),$$
$$(a_1, a_2, a_3) \quad \text{in place of} \quad (a, b, c),$$
$$(x_1, x_2, x_3) \quad \text{in place of} \quad (x, y, z).$$

As we work with more coordinates, lengthy notations such as (a_1, a_2, a_3) are time-consuming to write and may cause printing problems if a number of such notations appear in a single formula. We will often use a single boldface letter **a** to denote such a point. The number of coordinates will always be either explicitly stated or clear from the context. For example, the point **a** in space is (a_1, a_2, a_3) as shown in Fig. 12.21, and the point **b** in the plane is (b_1, b_2) as shown in Fig. 12.22. Such compact notation is one advantage of using a single letter of the alphabet with subscripts. We use a boldface zero for the origin; in the plane, **0** $= (0, 0)$, while in space, **0** $= (0, 0, 0)$.

Vectors in the Plane

Each point **a** in the plane naturally describes a numerical *magnitude*, the distance $\sqrt{a_1^2 + a_2^2}$ from **0** to **a**, and a *direction*, the direction from **0** to **a**. In the terminology of classical mechanics, any quantity that has associated with it a magnitude and a direction is called a *vector*. We will use this classical terminology and consider **a** to be a **vector** in space as well as a point in space. Thus when taking the vector viewpoint, we focus our attention on the *magnitude* and *direction* described by **a**.

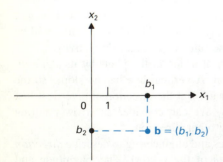

FIGURE 12.22 The point **b** in the plane.

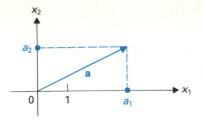

FIGURE 12.23 The vector **a** in the plane.

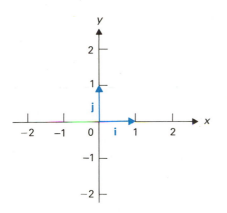

FIGURE 12.24 Unit coordinate vectors **i** and **j** in the plane.

It is natural to associate with a vector **a** an arrow from the origin to the point **a**, as in Fig. 12.23. The length of the arrow is the **magnitude** or **length** of the vector and is denoted by $|\mathbf{a}|$, so that

$$|\mathbf{a}| = \sqrt{a_1{}^2 + a_2{}^2}$$

The arrow points in the *direction* of **a**. When writing coordinates, we will use $\langle a_1, a_2 \rangle$ for the vector **a**, with "pointed brackets" suggesting arrows:

the point **a** is (a_1, a_2),

the vector **a** is $\langle a_1, a_2 \rangle$.

For the vector $\mathbf{a} = \langle a_1, a_2 \rangle$, we call a_i the **ith component** of the vector. We call $\mathbf{0} = \langle 0, 0 \rangle$ the **zero vector;** it has magnitude zero, but it is the only vector that has no unique direction associated with it. A vector of length 1 is a **unit vector.** The vectors

$$\mathbf{i} = \langle 1, 0 \rangle \qquad \text{and} \qquad \mathbf{j} = \langle 0, 1 \rangle$$

are the **unit coordinate vectors** in the plane and are shown in Fig. 12.24.

We emphasize that the mathematical definition of a *vector* is identical to the definition of a *point;* each is an ordered pair of real numbers. The names "vector" and "point" indicate different geometric interpretations for such a pair. If you ask mathematicians to show pictorially the *vector* $\langle 1, 2 \rangle$ in the plane, they will draw the arrow indicating length and direction shown in Fig. 12.25. On the other hand, if you ask them to show pictorially the *point* $(3, -2)$, they will make the large dot shown in the figure, just indicating a position.

The vector $\mathbf{b} = \langle 1, 3 \rangle$ in the plane is shown in Fig. 12.26. If we choose translated $\bar{x}, \bar{y}$-coordinates at the point $(x, y) = (-4, 2)$, then the point $(x, y) = (-3, 5)$ has translated coordinates $(\bar{x}, \bar{y}) = (1, 3)$. The vector from the translated origin to $(\bar{x}, \bar{y}) = (1, 3)$ has the same magnitude and direction as **b**; the arrows have the same length and are parallel. Since we consider magnitude and direction to be the distinguishing characteristics of a vector, it is natural to call this vector, having $(x, y) = (-4, 2)$ as point of origin, **b** also. That is, we consider two vectors to be **equal** if they have the same magnitude and direction, even though they emanate from different points. Figure 12.27 gives an illustration of this. Sometimes the term "free vectors" is used to indicate this independence from the point of origin.

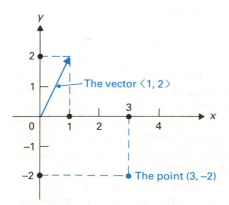

FIGURE 12.25 Draw an arrow for a vector and a dot for a point.

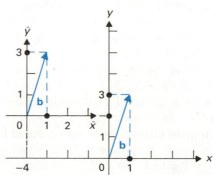

FIGURE 12.26 The two vectors **b** = $\langle 1, 3 \rangle$ have the same magnitude and direction and are considered equal.

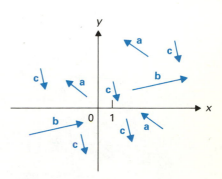

FIGURE 12.27 Vectors labeled **a** are equal, as are vectors labeled **b** and vectors labeled **c**.

EXAMPLE 1 Show that the vectors $\mathbf{a} = \langle 0, 1 \rangle$ and $\mathbf{b} = \langle \frac{1}{2}, \sqrt{3}/2 \rangle$ are both unit vectors.

Solution A unit vector is one of length 1. We indeed have

$$|\mathbf{a}| = \sqrt{0^2 + 1^2} = 1$$

and

$$|\mathbf{b}| = \sqrt{\left(\frac{1}{2}\right)^2 + \left(\frac{\sqrt{3}}{2}\right)^2} = \sqrt{\frac{4}{4}} = 1. \qquad \blacksquare$$

The Algebra of Vectors

Addition of numbers is a very important operation. It was the first topic in mathematics we ever studied. We think of the real numbers as filling up a line. Addition of numbers on the line has a very important generalization to addition in the plane. Addition in the plane is usually phrased in the language of vectors. We will follow this convention.

DEFINITION 12.1

Vector addition

Let $\mathbf{a} = \langle a_1, a_2 \rangle$ and $\mathbf{b} = \langle b_1, b_2 \rangle$ be vectors in the plane. The **sum** of $\mathbf{a}$ and $\mathbf{b}$ is

$$\mathbf{a} + \mathbf{b} = \langle a_1 + b_1, a_2 + b_2 \rangle.$$

EXAMPLE 2 Let $\mathbf{a} = \langle 0, -1 \rangle$ and $\mathbf{b} = \langle 4, 2 \rangle$. Find $\mathbf{a} + \mathbf{b}$.

Solution We have $\mathbf{a} + \mathbf{b} = \langle 0, -1 \rangle + \langle 4, 2 \rangle = \langle 4, 1 \rangle$. $\qquad \blacksquare$

We can interpret vector addition geometrically, in terms of arrows. There are two ways that we can arrive at the arrow representing $\mathbf{a} + \mathbf{b}$ from arrows representing $\mathbf{a}$ and representing $\mathbf{b}$. One way is to find the diagonal of the parallelogram that has $(0, 0)$ as a vertex and the arrows represented by $\mathbf{a}$ and $\mathbf{b}$ as sides emanating from $(0, 0)$. As indicated in Fig. 12.28, the diagonal arrow of the parallelogram that starts at $(0, 0)$ represents the vector $\mathbf{a} + \mathbf{b}$.

For an alternative method to represent $\mathbf{a} + \mathbf{b}$, choose the point (a_1, a_2) as new origin and draw the arrow representing the *translated vector* $\mathbf{b}$. This arrow starts at the tip of the original vector $\mathbf{a}$ and goes to the point with translated coordinates (b_1, b_2). The tip of this translated vector $\mathbf{b}$ then falls at the same point as the tip of the desired vector $\mathbf{a} + \mathbf{b}$ emanating from the original origin, $(0, 0)$, as shown in Fig. 12.29. As we remarked before, we consider the vector $\mathbf{b}$ from the original origin and the vector $\mathbf{b}$ emanating from (a_1, a_2) in Fig. 12.29 to be equal. They have the same magnitude and direction.

We turn now to another operation in vector algebra. When dealing with vectors, we often refer to a real number as a **scalar** to distinguish it from the vectors. As our second algebraic operation involving vectors, we define the product of a scalar r and a vector $\mathbf{a} = \langle a_1, a_2 \rangle$.

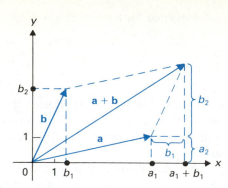

FIGURE 12.28 **a** + **b** as the diagonal of the parallelogram.

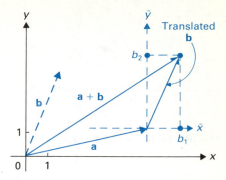

FIGURE 12.29 Finding **a** + **b** by starting **b** at the tip of **a**.

DEFINITION 12.2

Scalar multiplication

The **product** $r\mathbf{a}$ of the scalar r and the vector $\mathbf{a} = \langle a_1, a_2 \rangle$ is the vector

$$r\mathbf{a} = \langle ra_1, ra_2 \rangle.$$

EXAMPLE 3 Let $\mathbf{a} = \langle 3, -1 \rangle$. Find $2\mathbf{a}$.

Solution We have $2\mathbf{a} = 2\langle 3, -1 \rangle = \langle 6, -2 \rangle$. ■

From Definition 12.2, we see at once that for any real number r and vector $\mathbf{a} = \langle a_1, a_2 \rangle$, we have

$$|r\mathbf{a}| = |\langle ra_1, ra_2 \rangle| = \sqrt{(ra_1)^2 + (ra_2)^2}$$
$$= \sqrt{r^2} \cdot \sqrt{a_1^2 + a_2^2}.$$

But $\sqrt{r^2} = |r|$, and $\sqrt{a_1^2 + a_2^2} = |\mathbf{a}|$. Consequently,

$$|r\mathbf{a}| = |r| \cdot |\mathbf{a}|. \tag{1}$$

Thus if we wish to describe $r\mathbf{a}$ in terms of *length* and *direction*, Eq. (1) shows that the length of the product $r\mathbf{a}$ is $|r|$ times the length of **a**. Figure 12.30 indicates that we should consider $r\mathbf{a}$ to have the same direction as **a** if $r > 0$ and the opposite direction if $r < 0$.

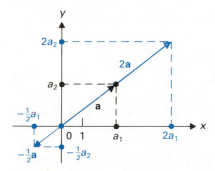

FIGURE 12.30 Multiplying the vector **a** by 2 and by $-\frac{1}{2}$.

EXAMPLE 4 Let $|\mathbf{a}| = 5$. Describe the magnitude of $3\mathbf{a}$ and $-7\mathbf{a}$ and compare their direction with that of **a**.

Solution From Eq. (1), we have

$$|3\mathbf{a}| = |3| \cdot |\mathbf{a}| = 3 \cdot 5 = 15$$

and

$$|-7\mathbf{a}| = |-7| \cdot |\mathbf{a}| = 7 \cdot 5 = 35.$$

Now $3 > 0$ and $-7 < 0$, so $3\mathbf{a}$ has the same direction as **a**, and $-7\mathbf{a}$ has the opposite direction. ■

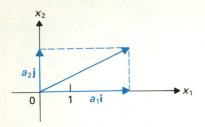

FIGURE 12.31 $\mathbf{a} = a_1\mathbf{i} + a_2\mathbf{j}$

Note that for every vector $\mathbf{a} = \langle a_1, a_2 \rangle$, we have

$$\mathbf{a} = \langle a_1, a_2 \rangle$$
$$= a_1\langle 1, 0 \rangle + a_2\langle 0, 1 \rangle$$
$$= a_1\mathbf{i} + a_2\mathbf{j},$$

as illustrated in Fig. 12.31. Such $\mathbf{i}$, $\mathbf{j}$ expressions for vectors are used extensively. We will use this notation wherever it is not too cumbersome, so that you will become familiar with it. This notation is widely used in the physical sciences.

Note that nonzero vectors $\mathbf{a}$ and $\mathbf{b}$ are *parallel* if and only if $\mathbf{b} = r\mathbf{a}$ for some real number r. Thus $r\mathbf{a}$ is a vector parallel to $\mathbf{a}$ of length $|r| \cdot |\mathbf{a}|$ and having the *same* direction as $\mathbf{a}$ if $r > 0$ and the *opposite* direction if $r < 0$.

EXAMPLE 5 Show that the vectors $\mathbf{a} = \mathbf{i} - 3\mathbf{j}$ and $\mathbf{b} = 2\mathbf{i} - 6\mathbf{j}$ are parallel and have the same direction.

Solution If they are parallel, then $\mathbf{b} = r\mathbf{a}$ for some r. Looking at the coefficients of $\mathbf{i}$ in the two vectors, we see that we would have to have $r = 2$. We find that

$$2\mathbf{a} = 2(\mathbf{i} - 3\mathbf{j}) = 2\mathbf{i} - 6\mathbf{j} = \mathbf{b},$$

so the vectors are parallel. Since $2 > 0$, they have the same direction. ∎

We now define the *difference* $\mathbf{b} - \mathbf{a}$ of vectors $\mathbf{a}$ and $\mathbf{b}$ by

$$\mathbf{b} - \mathbf{a} = \mathbf{b} + (-1)\mathbf{a}.$$

We have already defined vector addition and the product $(-1)\mathbf{a}$, and, since $\mathbf{a} + (\mathbf{b} - \mathbf{a}) = \mathbf{b}$, we see that

$$\mathbf{b} - \mathbf{a} \text{ is the vector that when added to } \mathbf{a} \text{ yields } \mathbf{b}.$$

A translated coordinate representation of $\mathbf{b} - \mathbf{a}$, starting from the tip of $\mathbf{a}$ as translated origin, therefore ends at (b_1, b_2), as shown in Fig. 12.32.

EXAMPLE 6 Find the vector that, as an arrow, starts from the point $(1, -4)$ and ends at the point $(-3, 2)$ in the plane.

Solution We let $\mathbf{a} = \mathbf{i} - 4\mathbf{j}$ and $\mathbf{b} = -3\mathbf{i} + 2\mathbf{j}$. We have just seen that $\mathbf{b} - \mathbf{a}$ is the vector starting from the point $\mathbf{a} = (1, -4)$ that reaches to the point $\mathbf{b} = (-3, 2)$. Thus the desired vector is

$$\mathbf{b} - \mathbf{a} = (-3\mathbf{i} + 2\mathbf{j}) - (\mathbf{i} - 4\mathbf{j})$$
$$= -4\mathbf{i} + 6\mathbf{j}. \qquad ∎$$

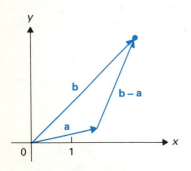

FIGURE 12.32 $\mathbf{b} - \mathbf{a}$ goes from the tip of $\mathbf{a}$ to the tip of $\mathbf{b}$, for

$$\mathbf{a} + (\mathbf{b} - \mathbf{a}) = \mathbf{b}.$$

In summary, vector addition, subtraction, and multiplication by a scalar are straightforward to perform; we simply carry out the corresponding numerical operations in each component.

We list some algebraic laws that hold for vector algebra. Their proofs are left to the exercises (see Exercises 48–52).

THEOREM 12.1   Properties of vector operations

For all vectors **a**, **b**, and **c** and all scalars r and s, the following laws hold.

$(\mathbf{a} + \mathbf{b}) + \mathbf{c} = \mathbf{a} + (\mathbf{b} + \mathbf{c})$	*Associativity of addition*
$\mathbf{a} + \mathbf{b} = \mathbf{b} + \mathbf{a}$	*Commutativity of addition*
$r(s\mathbf{a}) = (rs)\mathbf{a}$	*Associativity of multiplication by scalars*
$(r + s)\mathbf{a} = r\mathbf{a} + s\mathbf{a}$	*A right distributive law*
$r(\mathbf{a} + \mathbf{b}) = r\mathbf{a} + r\mathbf{b}$	*A left distributive law* □

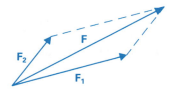

FIGURE 12.33   Forces $\mathbf{F}_1$ and $\mathbf{F}_2$ applied to an object are equivalent to the single force $\mathbf{F} = \mathbf{F}_1 + \mathbf{F}_2$.

A Physical Model for Vectors

When studying motion, a physicist may use a vector to represent a *force*. Suppose, for example, that we are moving some object by pushing it. The motion of the object is influenced by the *direction* in which we are pushing and how *hard* we are pushing. Thus the force with which we are pushing can be conveniently represented by a vector, which we regard as having the *direction* in which we are pushing and a *length* representing how hard we are pushing. If we double the force with which we push, the force vector doubles in length; this corresponds to the multiplication of the force vector by the scalar 2.

Suppose that two people are pushing on an object with forces that correspond to vectors $\mathbf{F}_1$ and $\mathbf{F}_2$, as shown in Fig. 12.33. It can be shown that the motion of the object due to these combined forces is the same as the motion that would result if only one person were pushing with a force expressed by a vector that is the diagonal of the parallelogram with arrow vectors $\mathbf{F}_1$ and $\mathbf{F}_2$ as adjacent sides (see Fig. 12.33). Thus this *resultant force vector* is precisely the vector $\mathbf{F}_1 + \mathbf{F}_2$.

EXAMPLE 7   A body in the plane is traveling subject only to the two constant forces $\mathbf{F}_1 = 2\mathbf{i} + 3\mathbf{j}$ and $\mathbf{F}_2 = -\mathbf{i} + \mathbf{j}$. The body passes through the point $(1, 3)$. Find the equation of the line on which the body travels.

Solution   The resultant force vector is

$$\mathbf{F} = \mathbf{F}_1 + \mathbf{F}_2 = (2\mathbf{i} + 3\mathbf{j}) + (-\mathbf{i} + \mathbf{j}) = \mathbf{i} + 4\mathbf{j},$$

as shown in Fig. 12.34. The body moves in the direction given by this resultant vector, along the line containing the arrow of this vector viewed as starting at the point $(1, 3)$. This line has slope

$$\frac{\mathbf{j}\text{-coefficient}}{\mathbf{i}\text{-coefficient}} = \frac{4}{1} = 4.$$

The equation of the line is then

$$y - 3 = 4(x - 1), \qquad \text{or} \qquad 4x - y = 1. \qquad \blacksquare$$

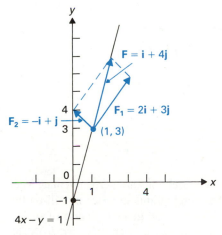

FIGURE 12.34 $\mathbf{F} = \mathbf{F}_1 + \mathbf{F}_2$ gives the direction of motion.

Vectors in Space

Everything we have done for vectors in the plane has a natural analogue for vectors in space; we simply include a third component. Thus a vector $\mathbf{a} = \langle a_1, a_2, a_3 \rangle$ in space is viewed geometrically as an arrow emanating from the

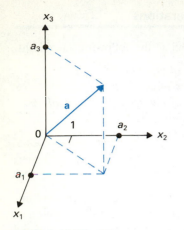

FIGURE 12.35 The vector **a** in space.

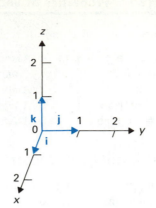

FIGURE 12.36 Unit coordinate vectors **i**, **j**, and **k** in space.

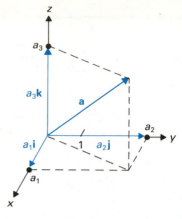

FIGURE 12.37 $\mathbf{a} = a_1\mathbf{i} + a_2\mathbf{j} + a_3\mathbf{k}$.

origin, as shown in Fig. 12.35. The length of **a** is given by

$$|\mathbf{a}| = \sqrt{a_1{}^2 + a_2{}^2 + a_3{}^2}.$$

Unit vectors are those of length 1. The **unit coordinate vectors** in space are

$$\mathbf{i} = \langle 1, 0, 0 \rangle, \qquad \mathbf{j} = \langle 0, 1, 0 \rangle, \qquad \text{and} \qquad \mathbf{k} = \langle 0, 0, 1 \rangle,$$

as shown in Fig. 12.36, and the zero vector is $\mathbf{0} = \langle 0, 0, 0 \rangle$. Addition and scalar multiplication are defined in the natural way; we just add third components also and multiply the third component by a scalar. The vector $\mathbf{a} = \langle a_1, a_2, a_3 \rangle$ can be expressed as $a_1\mathbf{i} + a_2\mathbf{j} + a_3\mathbf{k}$, as shown in Fig. 12.37. The properties of vector operations given in Theorem 12.1 are valid in space as well as in the plane. In advanced work, vectors with even more than three components are commonly used.

EXAMPLE 8 Find all values of c such that $\mathbf{a} = \langle \frac{1}{3}, -\frac{2}{3}, c \rangle$ is a unit vector.

Solution We must have

$$1 = |\mathbf{a}| = \sqrt{\left(\frac{1}{3}\right)^2 + \left(-\frac{2}{3}\right)^2 + c^2}$$

$$= \sqrt{\frac{5}{9} + c^2}.$$

Thus $\frac{5}{9} + c^2 = 1$, so $c^2 = 1 - \frac{5}{9} = \frac{4}{9}$, and $c = \pm\frac{2}{3}$. ∎

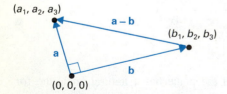

FIGURE 12.38 For a 90° angle at $(0, 0, 0)$, we must have

$$|\mathbf{a} - \mathbf{b}|^2 = |\mathbf{a}|^2 + |\mathbf{b}|^2.$$

Perpendicular Vectors

It is very important for us to know when two directions are perpendicular. Vectors provide a simple criterion for this. Any three points in space not all on the same line determine a plane. Imagine that Fig. 12.38 gives a picture of such a plane determined by three points, $(0, 0, 0)$, (a_1, a_2, a_3), and (b_1, b_2, b_3). The vectors **a** and **b** of Fig. 12.38 are perpendicular if and only if the Pythagorean relation holds, that is, if and only if

$$|\mathbf{a}|^2 + |\mathbf{b}|^2 = |\mathbf{a} - \mathbf{b}|^2. \qquad (2)$$

By definition of the length of a vector, we see that this is true when

$$a_1{}^2 + a_2{}^2 + a_3{}^2 + b_1{}^2 + b_2{}^2 + b_3{}^2$$
$$= (a_1 - b_1)^2 + (a_2 - b_2)^2 + (a_3 - b_3)^2. \tag{3}$$

Squaring the terms on the right-hand side of Eq. (3) and canceling, we obtain the condition

$$0 = -2a_1b_1 - 2a_2b_2 - 2a_3b_3, \quad \text{or} \quad a_1b_1 + a_2b_2 + a_3b_3 = 0.$$

> Vectors **a** and **b** in space are *perpendicular* if and only if
>
> $$a_1b_1 + a_2b_2 + a_3b_3 = 0. \tag{4}$$

Of course, the corresponding result with just two components holds for vectors in the plane.

EXAMPLE 9 Determine whether or not the vectors $\mathbf{a} = -\mathbf{i} + 3\mathbf{j} + 2\mathbf{k}$ and $\mathbf{b} = 5\mathbf{i} - \mathbf{j} + 4\mathbf{k}$ in space are perpendicular.

Solution We have

$$-1 \cdot 5 + 3 \cdot -1 + 2 \cdot 4 = -5 - 3 + 8 = 0,$$

so **a** and **b** are indeed perpendicular. ∎

According to condition (4), the zero vector **0** is perpendicular to *every* vector in space. For this reason, it is convenient to think of **0** as having *all directions* rather than *no direction*.

SUMMARY

Formulas are given for vectors in space, but are equally valid for vectors in the plane when third components are deleted.

1. Each point $\mathbf{a} = (a_1, a_2, a_3)$ determines a vector $\mathbf{a} = \langle a_1, a_2, a_3 \rangle$ having direction from the origin to the point (a_1, a_2, a_3) and length $|\mathbf{a}| = \sqrt{a_1{}^2 + a_2{}^2 + a_3{}^2}$. A unit vector has length 1. The zero vector is $\mathbf{0} = \langle 0, 0, 0 \rangle$.

2. Addition of vectors is given by adding corresponding coordinates, that is, $\mathbf{a} + \mathbf{b} = \langle a_1, a_2, a_3 \rangle + \langle b_1, b_2, b_3 \rangle = \langle a_1 + b_1, a_2 + b_2, a_3 + b_3 \rangle$.

3. In space, the unit coordinate vectors are often written

$$\mathbf{i} = \langle 1, 0, 0 \rangle, \quad \mathbf{j} = \langle 0, 1, 0 \rangle, \quad \mathbf{k} = \langle 0, 0, 1 \rangle,$$

so $\langle a_1, a_2, a_3 \rangle = a_1\mathbf{i} + a_2\mathbf{j} + a_3\mathbf{k}$. In the plane, we use just $\mathbf{i} = \langle 1, 0 \rangle$ and $\mathbf{j} = \langle 0, 1 \rangle$.

4. Multiplication of a vector **a** by a scalar (real number) r is given by

$$r\mathbf{a} = r(a_1\mathbf{i} + a_2\mathbf{j} + a_3\mathbf{k}) = (ra_1)\mathbf{i} + (ra_2)\mathbf{j} + (ra_3)\mathbf{k}.$$

5. $\mathbf{b} - \mathbf{a} = (b_1 - a_1)\mathbf{i} + (b_2 - a_2)\mathbf{j} + (b_3 - a_3)\mathbf{k}$

6. $\mathbf{a} + \mathbf{b}$ is the vector along the diagonal of the parallelogram having **a** and **b** as coterminous edges.

7. $\mathbf{b} - \mathbf{a}$ is the vector starting at the tip of **a** and reaching to the tip of **b**.

8. Two nonzero vectors **a** and **b** are parallel if $\mathbf{b} = r\mathbf{a}$ for some scalar r.

9. Two vectors **a** and **b** are perpendicular if $a_1b_1 + a_2b_2 + a_3b_3 = 0$.

EXERCISES 12.3

1. Let $\mathbf{a} = 2\mathbf{i} - \mathbf{j}$ and $\mathbf{b} = -3\mathbf{i} - 2\mathbf{j}$ be vectors in the plane. Sketch, using arrows, the vectors **a**, **b**, $\mathbf{a} + \mathbf{b}$, $\mathbf{a} - \mathbf{b}$, and $-\frac{4}{3}\mathbf{a}$.

2. Repeat Exercise 1 for the vectors $\mathbf{a} = 3\mathbf{i} + \mathbf{j}$ and $\mathbf{b} = -\mathbf{i} + 2\mathbf{j}$.

In Exercises 3–10, let $\mathbf{a} = -\mathbf{i} + 3\mathbf{j} - 2\mathbf{k}$, $\mathbf{b} = 4\mathbf{i} - \mathbf{k}$, and $\mathbf{c} = -3\mathbf{i} - \mathbf{j} + 2\mathbf{k}$. Compute the indicated vector.

3. $3\mathbf{a}$

4. $-2\mathbf{c}$

5. $\mathbf{a} + \mathbf{b}$

6. $3\mathbf{b} - 2\mathbf{c}$

7. $\mathbf{a} + 2(\mathbf{b} - 3\mathbf{c})$

8. $3(\mathbf{a} - 2\mathbf{b})$

9. $4(3\mathbf{a} + 5\mathbf{b})$

10. $2\mathbf{a} - (3\mathbf{b} - \mathbf{c})$

In Exercises 11–16, let $\mathbf{a} = 3\mathbf{i} - 2\mathbf{j} + 2\mathbf{k}$ and $\mathbf{b} = -\mathbf{i} + 4\mathbf{j} + \mathbf{k}$. Compute the indicated quantity.

11. $|\mathbf{a}|$

12. $|\mathbf{a} + \mathbf{b}|$

13. $|\mathbf{a}| + |\mathbf{b}|$

14. $|-2\mathbf{a}|$

15. $|\mathbf{b}| - |\mathbf{a}|$

16. $|\mathbf{b} - 3\mathbf{a}|$

In Exercises 17–20, find the vector represented by an arrow reaching from the first point to the second.

17. From $(-1, 3)$ to $(4, 2)$

18. From $(-5, -8)$ to $(3, -2)$

19. From $(-3, 2, 5)$ to $(4, -2, -6)$

20. From $(3, -6, -2)$ to $(1, 4, -8)$

In Exercises 21–26, determine whether the pair of vectors is parallel, perpendicular, or neither. If two vectors are parallel, state whether they have the same or opposite directions.

21. $3\mathbf{i} - \mathbf{j}$ and $4\mathbf{i} + 12\mathbf{j}$

22. $-2\mathbf{i} + 6\mathbf{j}$ and $4\mathbf{i} - 12\mathbf{j}$

23. $3\mathbf{i} - \mathbf{j}$ and $4\mathbf{i} + 3\mathbf{j} + 2\mathbf{k}$

24. $2\mathbf{i} - 3\mathbf{j} + \mathbf{k}$ and $8\mathbf{i} + 2\mathbf{j} - 10\mathbf{k}$

25. $\sqrt{2}\mathbf{i} + \sqrt{18}\mathbf{j} - \sqrt{8}\mathbf{k}$ and $2\mathbf{i} + 6\mathbf{j} - 4\mathbf{k}$

26. $6\mathbf{i} - 9\mathbf{j} + 3\mathbf{k}$ and $-4\mathbf{i} + 6\mathbf{j} - 2\mathbf{k}$

In Exercises 27–36, find all values of c, if there are any, such that the given statement is true.

27. $2\mathbf{i} + c\mathbf{j}$ is parallel to $4\mathbf{i} + 6\mathbf{j}$

28. $2\mathbf{j} + c\mathbf{j}$ is parallel to $-5\mathbf{i} + 3\mathbf{j}$

29. $2\mathbf{i} + c\mathbf{j}$ is parallel to $3\mathbf{i}$

30. $2\mathbf{i} + c\mathbf{j}$ is parallel to $3\mathbf{j}$

31. $2\mathbf{i} + c\mathbf{j}$ is perpendicular to $3\mathbf{i} + 4\mathbf{j}$

32. $2\mathbf{i} + c\mathbf{j}$ is perpendicular to $8\mathbf{i} - c\mathbf{j}$

33. $2\mathbf{i} + c\mathbf{j}$ is perpendicular to $8\mathbf{i} + c\mathbf{j}$

34. $c\mathbf{i} + 2\mathbf{j} - \mathbf{k}$ is perpendicular to $\mathbf{j} - 4\mathbf{k}$

35. $c\mathbf{i} + 2\mathbf{j} - \mathbf{k}$ is perpendicular to $\mathbf{i} - 3\mathbf{k}$

36. $c\mathbf{i} + 2\mathbf{j} - \mathbf{k}$ is perpendicular to $-5\mathbf{i} + \mathbf{j} + 2\mathbf{k}$

37. Find a unit vector in space parallel to $\mathbf{i} - \mathbf{j} + 3\mathbf{k}$ and having the same direction.

38. Find two unit vectors in the plane perpendicular to $3\mathbf{i} - 4\mathbf{j}$.

39. Find two unit vectors in space that are not parallel and each of which is perpendicular to $-2\mathbf{i} + \mathbf{j} + 2\mathbf{k}$.

40. Find a unit vector parallel to the line tangent to $y = x^2$ at $(1, 1)$ and pointing in the direction of increasing x.

41. Find a unit vector normal to the tangent line to $y = x^4$ at $(-1, 1)$ and pointing in the direction of increasing y.

42. A body in the plane is moving subject only to the constant force vectors $\mathbf{F}_1 = \mathbf{i} + 3\mathbf{j}$ and $\mathbf{F}_2 = 2\mathbf{i} - \mathbf{j}$. If the body travels through $(1, 2)$, find the equation of the line on which it travels.

43. Repeat Exercise 42 for the force vectors $\mathbf{F}_1 = -3\mathbf{i} + 2\mathbf{j}$ and $\mathbf{F}_2 = 5\mathbf{i} - 4\mathbf{j}$ and the point $(-1, -3)$.

44. Repeat Exercise 42 for the force vectors $\mathbf{F}_1 = 3\mathbf{i} + 4\mathbf{j}$, $\mathbf{F}_2 = -\mathbf{i} + 2\mathbf{j}$, and $\mathbf{F}_3 = -3\mathbf{i} - 2\mathbf{j}$ and the point $(2, -1)$.

45. Repeat Exercise 42 for the force vectors $\mathbf{F}_1 = \mathbf{i} - 3\mathbf{j}$, $\mathbf{F}_2 = 4\mathbf{i} - 2\mathbf{j}$, and $\mathbf{F}_3 = -5\mathbf{i} + 5\mathbf{j}$ and the point $(-1, 4)$.

46. Show that $(1, -5)$, $(9, -11)$, and $(4, -1)$ are vertices of a right triangle in the plane.

47. Show that $(1, -1, 4)$, $(3, -2, 4)$, $(-4, 2, 6)$, and $(-2, 1, 6)$ are vertices of a parallelogram in space.

In Exercises 48–52, let $\mathbf{a} = a_1\mathbf{i} + a_2\mathbf{j} + a_3\mathbf{k}$, $\mathbf{b} = b_1\mathbf{i} + b_2\mathbf{j} + b_3\mathbf{k}$, and $\mathbf{c} = c_1\mathbf{i} + c_2\mathbf{j} + c_3\mathbf{k}$ be any vectors, and let r and s be any scalars. Prove the indicated property in each exercise.

48. $(\mathbf{a} + \mathbf{b}) + \mathbf{c} = \mathbf{a} + (\mathbf{b} + \mathbf{c})$

49. $\mathbf{a} + \mathbf{b} = \mathbf{b} + \mathbf{a}$

50. $r(s\mathbf{a}) = (rs)\mathbf{a}$

51. $(r + s)\mathbf{a} = r\mathbf{a} + s\mathbf{a}$

52. $r(\mathbf{a} + \mathbf{b}) = r\mathbf{a} + r\mathbf{b}$

12.4 THE DOT PRODUCT OF VECTORS

It is often important to know the *angle* between two vectors, viewing the vectors as arrows starting from the same point. The direction a sailboat is sailing and the direction from which the wind is blowing can both be represented by vectors. The angle between these vectors measures how close the boat is sailing into the wind.

We know how to determine whether two vectors are parallel (at angles of 0 or π) and whether two vectors are perpendicular (at an angle of $\pi/2$). In this section, we will derive a formula for the angle between any two nonzero vectors. This formula will involve the *dot product* of the vectors, which is a scalar quantity readily computed from the components of the vectors. The dot product will also enable us to find what part of a force vector can be considered to be acting in a particular direction. For example, if a number of people are pushing a stalled car, some may not be able to get directly behind the car to push but may be pushing at an angle. How much of their effort actually goes toward moving the car forward? The dot product will enable us to answer this question.

In Section 12.5, we take up another way of multiplying two vectors in space, their cross product. The cross product is a *vector* quantity, while the dot product is a *scalar*.

The Dot Product

Let $\mathbf{a}$ and $\mathbf{b}$ be nonparallel and nonzero vectors. We can view $\mathbf{a}$ and $\mathbf{b}$ geometrically as arrows starting from the origin, as shown in Fig. 12.39. We have not shown any coordinate axes in this figure, which is valid for vectors in the plane and for vectors in space. The vector $\mathbf{a}$ gives the direction for a line through the origin, labeled the "line along $\mathbf{a}$" in Fig. 12.39. Similarly, the vector $\mathbf{b}$ gives the direction for the line along $\mathbf{b}$. There is a unique plane containing these two lines, which we take as the plane of our page in Fig. 12.39.

We are interested in finding the angle θ between $\mathbf{a}$ and $\mathbf{b}$, shown in Fig. 12.39. We develop a formula for $\cos \theta$ in terms of the components of $\mathbf{a}$ and of $\mathbf{b}$. We make our computations for vectors

$$\mathbf{a} = a_1\mathbf{i} + a_2\mathbf{j} + a_3\mathbf{k} \qquad \text{and} \qquad \mathbf{b} = b_1\mathbf{i} + b_2\mathbf{j} + b_3\mathbf{k}$$

in space. Dropping the third components, we obtain a formula for the $\cos \theta$ if the vectors lie in the x,y-plane.

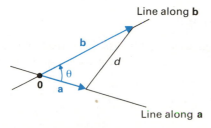

FIGURE 12.39 The angle θ between $\mathbf{a}$ and $\mathbf{b}$.

To find $\cos \theta$, we apply the law of cosines to the triangle shown in Fig. 12.39 and obtain

$$d^2 = |\mathbf{a}|^2 + |\mathbf{b}|^2 - 2|\mathbf{a}|\,|\mathbf{b}|\cos \theta. \tag{1}$$

From the preceding section, we know that $d^2 = |\mathbf{b} - \mathbf{a}|^2$. We can compute $|\mathbf{b} - \mathbf{a}|^2$, $|\mathbf{a}|$, and $|\mathbf{b}|$ in terms of the components a_i and b_i of the vectors $\mathbf{a}$ and $\mathbf{b}$, so we can use Eq. (1) to find $\cos \theta$ in terms of these components. We have

$$d^2 = |\mathbf{b} - \mathbf{a}|^2 = (b_1 - a_1)^2 + (b_2 - a_2)^2 + (b_3 - a_3)^2,$$

while

$$|\mathbf{a}|^2 = a_1{}^2 + a_2{}^2 + a_3{}^2 \qquad \text{and} \qquad |\mathbf{b}|^2 = b_1{}^2 + b_2{}^2 + b_3{}^2.$$

Substituting these quantities in Eq. (1), squaring the terms $(b_i - a_i)^2$ in d^2, and canceling the terms a_i^2 and b_i^2 from both sides of the resulting equation, we obtain

$$-2a_1 b_1 - 2a_2 b_2 - 2a_3 b_3 = -2|\mathbf{a}|\,|\mathbf{b}|\cos \theta. \tag{2}$$

Thus

$$\cos \theta = \frac{a_1 b_1 + a_2 b_2 + a_3 b_3}{|\mathbf{a}|\,|\mathbf{b}|}. \tag{3}$$

The numerator in Eq. (3) is familiar; we saw on page 621 that vectors $\mathbf{a}$ and $\mathbf{b}$ are perpendicular if and only if $a_1 b_1 + a_2 b_2 + a_3 b_3 = 0$. Note that this is consistent with Eq. (3); the vectors should be perpendicular if and only if $\theta = \pi/2$ or $\theta = 3\pi/2$, so that $\cos \theta = 0$. The number $a_1 b_1 + a_2 b_2 + a_3 b_3$ appearing in Eq. (3) is so important that it is given a special name, the *dot* (or *scalar* or *inner*) *product of $\mathbf{a}$ and $\mathbf{b}$*. (The result of this product of $\mathbf{a}$ and $\mathbf{b}$ is a *scalar*.) From Eq. (3), we obtain

$$a_1 b_1 + a_2 b_2 + a_3 b_3 = |\mathbf{a}|\,|\mathbf{b}|\cos \theta. \tag{4}$$

Note that if either $\mathbf{a}$ or $\mathbf{b}$ is $\mathbf{0}$, so that θ is undefined, then $|\mathbf{a}|$ or $|\mathbf{b}|$ is zero, and Eq. (4) still holds formally.

DEFINITION 12.3

Dot product

The **dot product** $\mathbf{a} \cdot \mathbf{b}$ of vectors $\mathbf{a}$ and $\mathbf{b}$ in space is

$$\mathbf{a} \cdot \mathbf{b} = a_1 b_1 + a_2 b_2 + a_3 b_3. \tag{5}$$

Equation (4) shows that

$$\mathbf{a} \cdot \mathbf{b} = |\mathbf{a}|\,|\mathbf{b}|\cos \theta, \tag{6}$$

where θ is the angle between $\mathbf{a}$ and $\mathbf{b}$. This gives us a geometric interpretation of the dot product. If we are interested in finding the ang¹ we rewrite Eq. (6) as

$$\cos \theta = \frac{\mathbf{a} \cdot \mathbf{b}}{|\mathbf{a}|\,|\mathbf{b}|}, \qquad \text{or} \qquad \theta = \cos^{-1}\frac{\mathbf{a} \cdot \mathbf{b}}{|\mathbf{a}|\,|\mathbf{b}|}. \tag{7}$$

EXAMPLE 1 Compute $\mathbf{a} \cdot \mathbf{b}$ for the vectors

$$\mathbf{a} = \mathbf{i} - 4\mathbf{j} + 3\mathbf{k} \qquad \text{and} \qquad \mathbf{b} = 6\mathbf{i} - 2\mathbf{j} - \mathbf{k}.$$

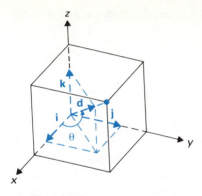

FIGURE 12.40 $\mathbf{d} = \mathbf{i} + \mathbf{j} + \mathbf{k}$ is directed along the diagonal of the cube.

Solution   We have

$$\mathbf{a} \cdot \mathbf{b} = (1)(6) + (-4)(-2) + (3)(-1) = 6 + 8 - 3 = 11. \qquad \blacksquare$$

EXAMPLE 2   Find the angle between the vectors $\mathbf{a} = \mathbf{i} - 4\mathbf{j}$ and $\mathbf{b} = 3\mathbf{i} + 2\mathbf{j}$ in the plane.

Solution   We have

$$\theta = \cos^{-1} \frac{\mathbf{a} \cdot \mathbf{b}}{|\mathbf{a}| \, |\mathbf{b}|} = \cos^{-1} \frac{3 - 8}{\sqrt{17}\sqrt{13}}$$

$$= \cos^{-1} \frac{-5}{\sqrt{17}\sqrt{13}} \approx 109.65°. \qquad \blacksquare$$

EXAMPLE 3   Find the angle θ that the diagonal of a cube in space makes with an edge of the cube.

Solution   We may take a cube with a vertex at the origin and with edges falling on the positive coordinate axes, as shown in Fig. 12.40. Then $\mathbf{i}$, $\mathbf{j}$, and $\mathbf{k}$ are vectors along edges of the cube, while a vector along a diagonal is

$$\mathbf{d} = \mathbf{i} + \mathbf{j} + \mathbf{k}.$$

We have

$$\theta = \cos^{-1} \frac{\mathbf{i} \cdot \mathbf{d}}{|\mathbf{i}| \, |\mathbf{d}|} = \cos^{-1} \frac{1}{1 \cdot \sqrt{3}}$$

$$= \cos^{-1} \frac{1}{\sqrt{3}} \approx 54.74°. \qquad \blacksquare$$

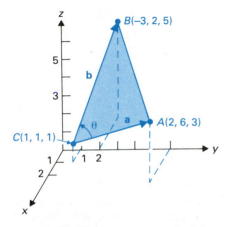

FIGURE 12.41   Angle $ACB = \theta$ of the triangle with the given vertices.

EXAMPLE 4   Find the angle BCA of the triangle in space with vertices $A(2, 6, 3)$, $B(-3, 2, 5)$, and $C(1, 1, 1)$.

Solution   We find the vector $\mathbf{a}$ reaching from C to A and the vector $\mathbf{b}$ reaching from C to B, as shown in Fig. 12.41. From the preceding section, we know that

$$\mathbf{a} = \text{Vector from } C \text{ to } A = \mathbf{i} + 5\mathbf{j} + 2\mathbf{k},$$

$$\mathbf{b} = \text{Vector from } C \text{ to } B = -4\mathbf{i} + \mathbf{j} + 4\mathbf{k}.$$

From Eq. (7), we have

$$\cos \theta = \frac{\mathbf{a} \cdot \mathbf{b}}{|\mathbf{a}| \, |\mathbf{b}|} = \frac{-4 + 5 + 8}{\sqrt{30}\sqrt{33}} = \frac{9}{3\sqrt{110}} = \frac{3}{\sqrt{110}}.$$

Then

$$\theta = \cos^{-1} \frac{3}{\sqrt{110}} \approx 73.38°. \qquad \blacksquare$$

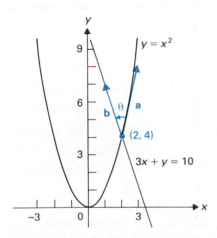

FIGURE 12.42   The angle θ between $y = x^2$ and $3x + y = 10$ at $(2, 4)$.

EXAMPLE 5   Find the acute angle at which the line $3x + y = 10$ meets the parabola $y = x^2$ at the point $(2, 4)$.

Solution   We try to find a vector $\mathbf{a}$ with direction tangent to the parabola at $(2, 4)$ and a vector $\mathbf{b}$ along the line at $(2, 4)$, as shown in Fig. 12.42. The angle between the line and the parabola is then the angle θ between **a** and **b** in the figure.

The slope of the parabola at $(2, 4)$ is given by

$$\frac{dy}{dx}\bigg|_2 = 2x\big|_2 = 4,$$

so we see that the vector

$$\mathbf{a} = \mathbf{i} + 4\mathbf{j}$$

has direction tangent to the parabola. The line $y = -3x + 10$ has slope -3, so the vector $\mathbf{i} - 3\mathbf{j}$ is directed along the line. It appears from Fig. 12.42 that we want a vector with negative **i**-component for **b**, for we want the acute angle θ. Thus we take

$$\mathbf{b} = -\mathbf{i} + 3\mathbf{j}.$$

Equation (7) then yields

$$\cos \theta = \frac{\mathbf{a} \cdot \mathbf{b}}{|\mathbf{a}|\,|\mathbf{b}|}$$

$$= \frac{-1 + 12}{\sqrt{17}\sqrt{10}} = \frac{11}{\sqrt{170}}.$$

Thus

$$\theta = \cos^{-1}\frac{11}{\sqrt{170}} \approx 32.47°. \qquad \blacksquare$$

Algebraic Properties of the Dot Product

Theorem 12.2 lists some of the algebraic properties of the dot product. We observe the usual convention that an algebraic operation written in multiplicative notation is performed before one written in additive notation, in the absence of parentheses. For example,

$$\mathbf{a} \cdot \mathbf{b} + \mathbf{a} \cdot \mathbf{c} = (\mathbf{a} \cdot \mathbf{b}) + (\mathbf{a} \cdot \mathbf{c}).$$

THEOREM 12.2   Properties of the dot product

Let **a**, **b**, and **c** be vectors with the same number of components and let r be a scalar. Then

(a) $\mathbf{a} \cdot \mathbf{a} \geq 0$ and $\mathbf{a} \cdot \mathbf{a} = 0$ if and only if $\mathbf{a} = \mathbf{0}$ *Nonnegative property*

(b) $\mathbf{a} \cdot \mathbf{b} = \mathbf{b} \cdot \mathbf{a}$ *Commutative property*

(c) $\mathbf{a} \cdot (\mathbf{b} + \mathbf{c}) = \mathbf{a} \cdot \mathbf{b} + \mathbf{a} \cdot \mathbf{c}$ *Distributive property*

(d) $(r\mathbf{a}) \cdot \mathbf{b} = \mathbf{a} \cdot (r\mathbf{b}) = r(\mathbf{a} \cdot \mathbf{b})$ *Homogeneous property*

(e) $\mathbf{a} \cdot \mathbf{a} = |\mathbf{a}|^2$ *Length property*

(f) $\mathbf{a} \cdot \mathbf{b} = 0$ if and only if **a** and **b** are perpendicular   *Perpendicular property*   □

Proofs of properties (a), (b), (c), and (d) are straightforward from formula (5) for $\mathbf{a} \cdot \mathbf{b}$ in terms of the components of **a** and of **b**. Illustrating with

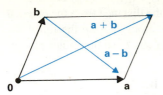

FIGURE 12.43 The parallelogram has $\mathbf{a} + \mathbf{b}$ and $\mathbf{a} - \mathbf{b}$ as vector diagonals.

the proof of property (a), we have, for vectors in space,

$$\mathbf{a} \cdot \mathbf{a} = a_1 a_1 + a_2 a_2 + a_3 a_3 = a_1^2 + a_2^2 + a_3^2 \geq 0,$$

and this sum of squares is zero if and only if each $a_i = 0$, that is, if and only if $\mathbf{a} = \mathbf{0}$. We leave the proofs of properties (b), (c), and (d) to the exercises (see Exercises 37, 38, and 39).

Properties (e) and (f) are really restatements of previous definitions in the notation of the dot product. We defined the length of a vector $\mathbf{a}$ in space to be $\sqrt{a_1^2 + a_2^2 + a_3^2} = \sqrt{\mathbf{a} \cdot \mathbf{a}}$, and we also defined $\mathbf{a}$ and $\mathbf{b}$ to be perpendicular vectors if and only if $\mathbf{a} \cdot \mathbf{b} = a_1 b_1 + a_2 b_2 + a_3 b_3 = 0$. Recall that we defined the vector $\mathbf{0}$ to be perpendicular to every vector.

The properties of the dot product are very important, and many consequences can be derived from them. We will give a geometric illustration.

EXAMPLE 6 Show that the sum of the squares of the lengths of the diagonals of a parallelogram is equal to the sum of the squares of the lengths of the sides. (This is the *parallelogram relation*.)

Solution We take our parallelogram with a vertex at the origin and vectors $\mathbf{a}$ and $\mathbf{b}$ as coterminous sides, as shown in Fig. 12.43. The lengths of the diagonals are then $|\mathbf{a} + \mathbf{b}|$ and $|\mathbf{a} - \mathbf{b}|$. Using Theorem 12.2, we have

$$\begin{aligned}|\mathbf{a} + \mathbf{b}|^2 + |\mathbf{a} - \mathbf{b}|^2 &= (\mathbf{a} + \mathbf{b}) \cdot (\mathbf{a} + \mathbf{b}) + (\mathbf{a} - \mathbf{b}) \cdot (\mathbf{a} - \mathbf{b}) \\ &= \mathbf{a} \cdot \mathbf{a} + 2\mathbf{a} \cdot \mathbf{b} + \mathbf{b} \cdot \mathbf{b} + \mathbf{a} \cdot \mathbf{a} - 2\mathbf{a} \cdot \mathbf{b} + \mathbf{b} \cdot \mathbf{b} \\ &= 2(\mathbf{a} \cdot \mathbf{a}) + 2(\mathbf{b} \cdot \mathbf{b}) \\ &= 2|\mathbf{a}|^2 + 2|\mathbf{b}|^2, \end{aligned}$$

which is what we wished to prove. You may think that we have used only property (e) of Theorem 12.2, but we also used properties (b), (c), and (d), as we ask you to show in Exercise 40. ∎

Vector Projection

Let $\mathbf{a}$ and $\mathbf{b}$ be nonzero vectors in space. Then

$$\frac{1}{|\mathbf{a}|}\mathbf{a} \qquad \text{and} \qquad \frac{1}{|\mathbf{b}|}\mathbf{b}$$

are unit vectors in the directions of $\mathbf{a}$ and $\mathbf{b}$, respectively. We write such vectors as

$$\frac{\mathbf{a}}{|\mathbf{a}|} \qquad \text{and} \qquad \frac{\mathbf{b}}{|\mathbf{b}|}$$

in what follows; for a nonzero scalar r and vector $\mathbf{a}$, we define

$$\frac{\mathbf{a}}{r} = \frac{1}{r}\mathbf{a}.$$

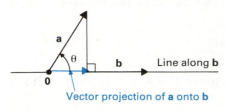

FIGURE 12.44 Vector projection of $\mathbf{a}$ onto $\mathbf{b}$ is $|\mathbf{a}| (\cos \theta)(\mathbf{b}/|\mathbf{b}|)$; scalar component of $\mathbf{a}$ along $\mathbf{b}$ is $|\mathbf{a}| (\cos \theta)$.

In Fig. 12.44, we again imagine that the plane containing both $\mathbf{a}$ and $\mathbf{b}$ is the plane of the page and show the angle θ between $\mathbf{a}$ and $\mathbf{b}$. In the figure, we have also labeled the *vector projection of $\mathbf{a}$ onto $\mathbf{b}$*, or *on the line along $\mathbf{b}$*. This vector has the direction of $\mathbf{b}$ if θ is an acute angle and the direction of $-\mathbf{b}$ if θ is an obtuse angle. The length of this vector projection is the distance from the origin **0** to the foot of the perpendicular dropped to the line along **b** from the tip of the vector **a**.* From Fig. 12.44, this (signed) length is $|\mathbf{a}|\cos\theta$, so the vector projection is

$$(|\mathbf{a}|\cos\theta)\frac{\mathbf{b}}{|\mathbf{b}|} = |\mathbf{a}|\,|\mathbf{b}|(\cos\theta)\frac{\mathbf{b}}{|\mathbf{b}|^2}$$

$$= \frac{\mathbf{a}\cdot\mathbf{b}}{|\mathbf{b}|^2}\mathbf{b} = \left(\frac{\mathbf{a}\cdot\mathbf{b}}{\mathbf{b}\cdot\mathbf{b}}\right)\mathbf{b},$$

where we have used the fact that $\mathbf{b}\cdot\mathbf{b} = |\mathbf{b}|^2$ from property (e) of Theorem 12.2. The number

$$|\mathbf{a}|(\cos\theta) = \frac{|\mathbf{a}|\,|\mathbf{b}|\cos\theta}{|\mathbf{b}|} = \frac{\mathbf{a}\cdot\mathbf{b}}{|\mathbf{b}|}$$

is the (signed) length of the vector projection. In summary, if $\mathbf{b}\neq\mathbf{0}$, then we have

$$\frac{\mathbf{a}\cdot\mathbf{b}}{\mathbf{b}\cdot\mathbf{b}}\mathbf{b} = \textbf{Vector projection} \text{ of } \mathbf{a} \text{ onto } \mathbf{b}, \tag{8}$$

$$\frac{\mathbf{a}\cdot\mathbf{b}}{|\mathbf{b}|} = \textbf{Scalar component} \text{ of } \mathbf{a} \text{ along } \mathbf{b}. \tag{9}$$

EXAMPLE 7 Find the vector projection of $\mathbf{a} = 2\mathbf{i} - \mathbf{j} + 3\mathbf{k}$ onto $\mathbf{b} = \mathbf{i} - 3\mathbf{j} - \mathbf{k}$ and the scalar component of **a** along **b**.

Solution The vector projection of **a** onto **b** is

$$\frac{\mathbf{a}\cdot\mathbf{b}}{|\mathbf{b}|^2}\mathbf{b} = \frac{2+3-3}{1+9+1}(\mathbf{i} - 3\mathbf{j} - \mathbf{k})$$

$$= \frac{2}{11}(\mathbf{i} - 3\mathbf{j} - \mathbf{k})$$

$$= \frac{2}{11}\mathbf{i} - \frac{6}{11}\mathbf{j} - \frac{2}{11}\mathbf{k}.$$

The scalar component of **a** along **b** is

$$\frac{\mathbf{a}\cdot\mathbf{b}}{|\mathbf{b}|} = \frac{2}{\sqrt{11}}. \qquad\blacksquare$$

If **b** is a unit vector so that $|\mathbf{b}| = 1$, then the vector projection of **a** onto **b** takes the simpler form

$$(\mathbf{a}\cdot\mathbf{b})\mathbf{b}.$$

EXAMPLE 8 Show that the vector projection of $\mathbf{a} = 3\mathbf{i} - 4\mathbf{j}$ along **i** is $3\mathbf{i}$ and along **j** is $-4\mathbf{j}$.

---

*For this reason, this vector projection is also called the orthogonal (perpendicular) projection.

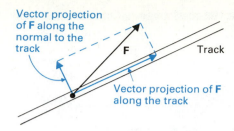

Vector projection of **F** along the normal to the track

F

Track

Vector projection of **F** along the track

FIGURE 12.45 Force vector **F** broken into perpendicular projections.

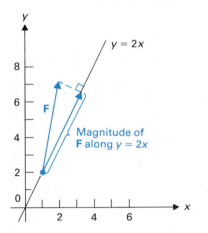

FIGURE 12.46 The scalar component (or magnitude) of **F** along $y = 2x$.

Solution By Eq. (8), the vector projection of **a** along **i** is

$$(\mathbf{a} \cdot \mathbf{i})\mathbf{i} = (3 + 0)\mathbf{i} = 3\mathbf{i}$$

and along **j** is

$$(\mathbf{a} \cdot \mathbf{j})\mathbf{j} = (0 - 4)\mathbf{j} = -4\mathbf{j}.$$

Of course the scalar components of **a** along **i** and **j** are just 3 and -4, respectively. ∎

Suppose a body is constrained to move along a track, as shown in Fig. 12.45. It can be shown that if a force **F** is applied to the body, as shown in the figure, then the body moves along the track as though **F** were replaced by its vector projection along the track. This projection along the track is the portion of the force vector that actually moves the body along the track. The vector projection of **F** along the perpendicular to the track acts to try to tip the body off the track. Note that **F** in Fig. 12.45 is the sum of these two vector projections.

EXAMPLE 9 Let a body in the plane be constrained to move on a track along the line $y = 2x$. If a force vector $\mathbf{F} = \mathbf{i} + 5\mathbf{j}$ acts on the body, find the magnitude of the force that actually moves the body along the track.

Solution Since we are asked for the magnitude of the force along the track, we want the scalar component of **F** along the line $y = 2x$. A vector along the line is $\mathbf{a} = \mathbf{i} + 2\mathbf{j}$, as shown in Fig. 12.46. By Eq. (9), the scalar component of **F** along the line is

$$\frac{\mathbf{F} \cdot \mathbf{a}}{|\mathbf{a}|} = \frac{1 + 10}{\sqrt{5}} = \frac{11}{\sqrt{5}}.$$

∎

SUMMARY

1. Let **a** and **b** be vectors in space or in the plane. The dot (scalar) product of **a** and **b** is the number

$$\mathbf{a} \cdot \mathbf{b} = a_1 b_1 + a_2 b_2 + a_3 b_3.$$

2. The dot product of **a** and **b** is described geometrically by

$$\mathbf{a} \cdot \mathbf{b} = |\mathbf{a}|\,|\mathbf{b}|\cos \theta,$$

where θ is the angle between **a** and **b**.

3. Algebraic properties of the dot product are listed in Theorem 12.2.

4. The vector projection of **a** onto **b** if $\mathbf{b} \neq \mathbf{0}$ is the vector

$$\frac{\mathbf{a} \cdot \mathbf{b}}{\mathbf{b} \cdot \mathbf{b}}\mathbf{b},$$

and the number $(\mathbf{a} \cdot \mathbf{b})/|\mathbf{b}|$ is the scalar component of **a** along **b**.

EXERCISES 12.4

In Exercises 1–10, find the angle between the vectors.

1. $\mathbf{i} + 4\mathbf{j}$ and $-8\mathbf{i} + 2\mathbf{j}$
2. $3\mathbf{i} + 2\mathbf{j} - 2\mathbf{k}$ and $4\mathbf{j} + \mathbf{k}$
3. $\mathbf{k}$ and $\mathbf{i} - \mathbf{k}$
4. $3\mathbf{i} + 4\mathbf{j}$ and $-\mathbf{i}$
5. $\left(\cos\dfrac{\pi}{3}\right)\mathbf{i} - \left(\sin\dfrac{\pi}{3}\right)\mathbf{j}$ and $\left(\cos\dfrac{\pi}{4}\right)\mathbf{i} + \left(\sin\dfrac{\pi}{4}\right)\mathbf{j}$.
   [*Hint:* Make use of a trigonometric identity.]
6. $\left(\sin\dfrac{\pi}{6}\right)\mathbf{i} + \left(\cos\dfrac{\pi}{6}\right)\mathbf{j}$ and $\left(\cos\dfrac{\pi}{4}\right)\mathbf{i} + \left(\sin\dfrac{\pi}{4}\right)\mathbf{j}$.
   [*Hint:* As in Exercise 5.]
7. $\left(\sin\dfrac{\pi}{3}\right)\mathbf{i} + \left(\cos\dfrac{\pi}{3}\right)\mathbf{j}$ and $\left(\cos\dfrac{\pi}{5}\right)\mathbf{i} - \left(\sin\dfrac{\pi}{5}\right)\mathbf{j}$.
   [*Hint:* As in Exercise 5.]
8. $\mathbf{i} - 3\mathbf{j} + 4\mathbf{k}$ and $10\mathbf{i} + 6\mathbf{j} + 2\mathbf{k}$
9. $3\mathbf{i} - \mathbf{j} + 2\mathbf{k}$ and $-9\mathbf{i} + 3\mathbf{j} - 6\mathbf{k}$
10. $\mathbf{i} - 2\mathbf{j} + 2\mathbf{k}$ and $-3\mathbf{i} - 6\mathbf{j} - 2\mathbf{k}$
11. Find the angle ACB of the triangle with vertices $A(0, 1, 6)$, $B(2, 3, 0)$, and $C(-1, 3, 4)$.
12. Find the angle between the line through $(-1, 2, 4)$ and $(3, 4, 0)$ and the line also through $(-1, 2, 4)$ that passes through $(5, 7, 2)$.
13. Find the acute angle between the lines $x - 2y = -4$ and $2x + 3y = 13$ at their point of intersection.
14. Find two lines in the plane through $(1, -4)$ that intersect the line $y = -2x + 7$ at an angle of $45°$.
15. Find the acute angle between the curves $y = x^2$ and $y = 2 - x^2$ at a point of intersection.
16. Find the acute angle between the curves $y = x^2$ and $y = x^3$ at the point $(1, 1)$ of intersection.

In Exercises 17–24, find the vector projection of the first vector onto the second vector and the scalar component of the first vector along the second.

17. $6\mathbf{i} - 3\mathbf{j} + 12\mathbf{k}$ on $\frac{1}{3}\mathbf{i} - \frac{2}{3}\mathbf{j} + \frac{2}{3}\mathbf{k}$
18. $\mathbf{i} + 3\mathbf{j} + 4\mathbf{k}$ on $\mathbf{j}$
19. $\mathbf{j}$ on $\mathbf{i} + 3\mathbf{j} + 4\mathbf{k}$
20. $2\mathbf{i} - \mathbf{j}$ on $-2\mathbf{i} + 3\mathbf{j}$
21. $3\mathbf{i} + \mathbf{j} - 2\mathbf{k}$ on $4\mathbf{i} + 2\mathbf{j} + 7\mathbf{k}$
22. $\mathbf{a} = -\mathbf{i} + \mathbf{j} + 3\mathbf{k}$ on $\mathbf{b} = 3\mathbf{i} - 2\mathbf{j} + \mathbf{k}$
23. $\mathbf{i} + \mathbf{j} + \mathbf{k}$ on $\mathbf{i} + \mathbf{j}$
24. $\mathbf{i} + \mathbf{j}$ on $\mathbf{i} + \mathbf{j} + \mathbf{k}$
25. Let $\mathbf{a}$ and $\mathbf{b}$ be vectors with $\mathbf{b} \neq \mathbf{0}$ and let $\mathbf{c}$ be the vector projection of $\mathbf{a}$ onto $\mathbf{b}$. Show that $\mathbf{a} - \mathbf{c}$ is perpendicular to $\mathbf{b}$.
26. A person is pushing a box across a floor with a force vector of magnitude 20 lb directed at an angle of $30°$ below the horizontal. Find the magnitude of the component of the

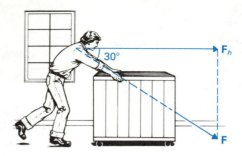

FIGURE 12.47

force vector actually moving the box along the floor. See Fig. 12.47.

27. A body in the plane is constrained to move on the curve $y = 3x^2$. The body is moved in the direction of increasing x by a constant force vector $\mathbf{F} = 50\mathbf{j}$. Find the magnitude of the component of $\mathbf{F}$ actually moving the body along the curve at the instant when the body is at $(1, 3)$.
28. (a) Answer Exercise 27 if the body is at the point (x, y) on $y = 3x^2$.
   (b) What is the magnitude of the component of $\mathbf{F}$ actually moving the body when it is at the origin?
   (c) Find the limit of the magnitude of the component of $\mathbf{F}$ moving the body as $x \to \infty$.
29. A 100-lb weight is suspended by a rope passed through an eyelet on top of the weight and making angles of $30°$ with the vertical, as shown in Fig. 12.48. Find the tension (magnitude of the force vector) along the rope. [*Hint:* The sum of the force vectors along the two halves of the rope at the eyelet must be an upward vertical vector of magnitude 100.]

FIGURE 12.48
Both halves of the rope make an angle of $30°$ with the vertical.

30. (a) Answer Exercise 29 if each half of the rope makes an angle of α with the vertical at the eyelet.
    (b) Find the tension in the rope if both sides are vertical ($\alpha = 0$).
    (c) What happens if an attempt is made to stretch the rope out straight (horizontal) while the 100-lb weight hangs on it?

31. Suppose a weight of 100 lb is suspended by *two* ropes *tied* at an eyelet on top of the weight, as shown in Fig. 12.49. Let the angles the ropes make with the vertical be α and β, as shown in the figure. Let the tension in the ropes be T_1 for the right-hand rope and T_2 for the left-hand rope.
    (a) Argue that the force vector $\mathbf{F}_1$ shown in Fig. 12.49 is $T_1(\sin \alpha)\mathbf{i} + T_1(\cos \alpha)\mathbf{j}$.
    (b) Find the corresponding expression for $\mathbf{F}_2$ in terms of T_2 and β.
    (c) If the system is in equilibrium, $\mathbf{F}_1 + \mathbf{F}_2 = 100\mathbf{j}$, so $\mathbf{F}_1 + \mathbf{F}_2$ must have $\mathbf{i}$-component 0 and $\mathbf{j}$-component 100. Write two equations reflecting this fact, using your answers to parts (a) and (b).
    (d) Find T_1 and T_2 if $\alpha = 45°$ and $\beta = 30°$.

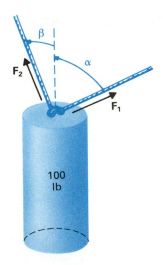

FIGURE 12.49
Two ropes tied at the eyelet and making angles of α and β with the vertical.

32. Use vector methods to show that the diagonals of a rhombus (parallelogram with equal sides) are perpendicular. [*Hint:*

Use a figure like Fig. 12.43 and show that
$$(\mathbf{a} + \mathbf{b}) \cdot (\mathbf{a} - \mathbf{b}) = 0.]$$

33. Use vector methods to show that the midpoint of the hypotenuse of a right triangle is equidistant from the three vertices. [*Hint:* See Fig. 12.50. Show that
$$\left| \frac{\mathbf{a} + \mathbf{b}}{2} \right| = \left| \frac{\mathbf{a} - \mathbf{b}}{2} \right|.]$$

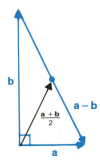

FIGURE 12.50 Vector $(\mathbf{a} + \mathbf{b})/2$ to the midpoint of the hypotenuse.

34. Show that the vectors $|\mathbf{a}|\mathbf{b} + |\mathbf{b}|\mathbf{a}$ and $|\mathbf{a}|\mathbf{b} - |\mathbf{b}|\mathbf{a}$ are perpendicular.

35. Show that the vector $|\mathbf{a}|\mathbf{b} + |\mathbf{b}|\mathbf{a}$ bisects the angle between $\mathbf{a}$ and $\mathbf{b}$.

36. Show that for $\mathbf{a}$, $\mathbf{b}$, and $\mathbf{c}$, the equation
$$\mathbf{a} \cdot \mathbf{b} = \mathbf{a} \cdot \mathbf{c},$$
where $\mathbf{a} \neq \mathbf{0}$, need not imply $\mathbf{b} = \mathbf{c}$.

37. Using the formula for the dot product, show that
$$\mathbf{a} \cdot \mathbf{b} = \mathbf{b} \cdot \mathbf{a}$$
for all $\mathbf{a}$ and $\mathbf{b}$ in space.

38. Using the formula for the dot product, show that
$$\mathbf{a} \cdot (\mathbf{b} + \mathbf{c}) = \mathbf{a} \cdot \mathbf{b} + \mathbf{a} \cdot \mathbf{c}$$
for all $\mathbf{a}$, $\mathbf{b}$, and $\mathbf{c}$ in space.

39. Using the formula for the dot product, show that $(r\mathbf{a}) \cdot \mathbf{b} = \mathbf{a} \cdot (r\mathbf{b}) = r(\mathbf{a} \cdot \mathbf{b})$ for all $\mathbf{a}$ and $\mathbf{b}$ in space and all scalars r.

40. Find where properties (b), (c), and (d) of Theorem 12.2 were used in the proof of the parallelogram relation in Example 6.

12.5 THE CROSS PRODUCT AND TRIPLE PRODUCTS

In the preceding section, we introduced the *dot product* of two vectors. The dot product is a *scalar* quantity. Using the dot product, we can quickly find the angle between vectors and vector and scalar projections of one vector along another.

In this section, we study the *cross product* of two vectors in space. The cross product is a *vector* quantity. In this chapter, we use the cross product to find a vector perpendicular to each of two given vectors. The cross product of two vectors is computed using a symbolic *determinant*. We open this section with the information about matrices and determinants that we need to understand the cross product. The section closes with an application of the dot and cross products to finding the area of a triangle in space and the volume of a skew box in space.

Review of 2 × 2 and 3 × 3 Determinants

A *square matrix* is a square array of numbers. For example,

$$\begin{pmatrix} -3 & 4 \\ 2 & 6 \end{pmatrix}$$

is a 2 × 2 (read "two-by-two") matrix, and

$$\begin{pmatrix} -1 & 0 & 4 \\ 2 & 1 & 0 \\ 3 & -4 & 5 \end{pmatrix}$$

is a 3 × 3 matrix. Each square matrix has associated with it a number, called the *determinant* of the matrix. The determinant is denoted by vertical lines rather than large parentheses on the sides of the array. The determinant of a 2 × 2 matrix is defined to be

$$\begin{vmatrix} a_1 & a_2 \\ b_1 & b_2 \end{vmatrix} = a_1 b_2 - a_2 b_1. \tag{1}$$

The determinant of a 3 × 3 matrix is defined in terms of the determinants of 2 × 2 matrices, as follows:

$$\begin{vmatrix} a_1 & a_2 & a_3 \\ b_1 & b_2 & b_3 \\ c_1 & c_2 & c_3 \end{vmatrix} = a_1 \begin{vmatrix} b_2 & b_3 \\ c_2 & c_3 \end{vmatrix} - a_2 \begin{vmatrix} b_1 & b_3 \\ c_1 & c_3 \end{vmatrix} + a_3 \begin{vmatrix} b_1 & b_2 \\ c_1 & c_2 \end{vmatrix}. \tag{2}$$

The coefficients of the three determinants on the right-hand side of formula (2) are the entries of the first row of the original 3 × 3 matrix, with alternate plus and minus signs. The first determinant on the right-hand side in formula (2) is the determinant of the 2 × 2 matrix obtained by crossing out the row and column in which the coefficient a_1 appears in the 3 × 3 matrix. The second determinant is obtained by crossing out the row and column in which a_2 appears, and so on.

EXAMPLE 1   Find

$$\begin{vmatrix} 7 & -2 \\ 4 & 3 \end{vmatrix}.$$

Solution   We have

$$\begin{vmatrix} 7 & -2 \\ 4 & 3 \end{vmatrix} = 7 \cdot 3 - (-2)4 = 21 + 8 = 29. \qquad \blacksquare$$

EXAMPLE 2   Find

$$\begin{vmatrix} 2 & 3 & 5 \\ -4 & 2 & 6 \\ 1 & 0 & 3 \end{vmatrix}.$$

Solution   We have

$$\begin{vmatrix} 2 & 3 & 5 \\ -4 & 2 & 6 \\ 1 & 0 & 3 \end{vmatrix} = 2\begin{vmatrix} 2 & 6 \\ 0 & 3 \end{vmatrix} - 3\begin{vmatrix} -4 & 6 \\ 1 & 3 \end{vmatrix} + 5\begin{vmatrix} -4 & 2 \\ 1 & 0 \end{vmatrix}$$

$$= 2(6 - 0) - 3(-12 - 6) + 5(0 - 2)$$

$$= 12 + 54 - 10 = 56. \qquad\blacksquare$$

The only facts we need to know about determinants for our work with this text are given in the following theorem.

THEOREM 12.3   Properties of determinants

(a) If any two rows of a square matrix are the same, then the determinant of the matrix is zero.

(b) If any two rows of a square matrix are interchanged, the determinant of the new matrix differs from the determinant of the original matrix only in sign.

$\square$

We will actually need special cases of Theorem 12.3 only for 3×3 matrices, where for part (a) one of the two identical rows is the first row and for part (b) the second and third rows are interchanged. We ask you to prove these special cases in Exercises 9 and 10 by computing that

$$\begin{vmatrix} a_1 & a_2 & a_3 \\ a_1 & a_2 & a_3 \\ c_1 & c_2 & c_3 \end{vmatrix} = 0,$$

$$\begin{vmatrix} a_1 & a_2 & a_3 \\ b_1 & b_2 & b_3 \\ a_1 & a_2 & a_3 \end{vmatrix} = 0,$$

and

$$\begin{vmatrix} a_1 & a_2 & a_3 \\ b_1 & b_2 & b_3 \\ c_1 & c_2 & c_3 \end{vmatrix} = -\begin{vmatrix} a_1 & a_2 & a_3 \\ c_1 & c_2 & c_3 \\ b_1 & b_2 & b_3 \end{vmatrix}.$$

The Cross Product of Vectors

Let $\mathbf{a} = a_1\mathbf{i} + a_2\mathbf{j} + a_3\mathbf{k}$ and $\mathbf{b} = b_1\mathbf{i} + b_2\mathbf{j} + b_3\mathbf{k}$ be vectors in space.

DEFINITION 12.4

Cross product of vectors

The **cross product a × b** of **a** and **b** is the vector found by computing a symbolic determinant as follows:

$$\mathbf{a} \times \mathbf{b} = \begin{vmatrix} \mathbf{i} & \mathbf{j} & \mathbf{k} \\ a_1 & a_2 & a_3 \\ b_1 & b_2 & b_3 \end{vmatrix}$$

(3)

$$= \begin{vmatrix} a_2 & a_3 \\ b_2 & b_3 \end{vmatrix} \mathbf{i} - \begin{vmatrix} a_1 & a_3 \\ b_1 & b_3 \end{vmatrix} \mathbf{j} + \begin{vmatrix} a_1 & a_2 \\ b_1 & b_2 \end{vmatrix} \mathbf{k}.$$

This cross product is also known as the **vector product,** since **a × b** is a vector quantity.

EXAMPLE 3 Find **a × b** if

$$\mathbf{a} = 3\mathbf{i} - 2\mathbf{j} + \mathbf{k} \qquad \text{and} \qquad \mathbf{b} = -2\mathbf{i} + 3\mathbf{j} + 4\mathbf{k}.$$

Solution We have

$$\mathbf{a} \times \mathbf{b} = \begin{vmatrix} \mathbf{i} & \mathbf{j} & \mathbf{k} \\ 3 & -2 & 1 \\ -2 & 3 & 4 \end{vmatrix}$$

$$= \begin{vmatrix} -2 & 1 \\ 3 & 4 \end{vmatrix} \mathbf{i} - \begin{vmatrix} 3 & 1 \\ -2 & 4 \end{vmatrix} \mathbf{j} + \begin{vmatrix} 3 & -2 \\ -2 & 3 \end{vmatrix} \mathbf{k}$$

$$= -11\mathbf{i} - 14\mathbf{j} + 5\mathbf{k}. \qquad \blacksquare$$

EXAMPLE 4 Show that

$$\mathbf{i} \times \mathbf{j} = \mathbf{k}, \qquad \mathbf{j} \times \mathbf{k} = \mathbf{i}, \qquad \text{and} \qquad \mathbf{k} \times \mathbf{i} = \mathbf{j}.$$

Solution We show that **i × j = k**. The other two computations are similar. We have

$$\mathbf{i} \times \mathbf{j} = \begin{vmatrix} \mathbf{i} & \mathbf{j} & \mathbf{k} \\ 1 & 0 & 0 \\ 0 & 1 & 0 \end{vmatrix}$$

$$= \begin{vmatrix} 0 & 0 \\ 1 & 0 \end{vmatrix} \mathbf{i} - \begin{vmatrix} 1 & 0 \\ 0 & 0 \end{vmatrix} \mathbf{j} + \begin{vmatrix} 1 & 0 \\ 0 & 1 \end{vmatrix} \mathbf{k}$$

$$= 0\mathbf{i} + 0\mathbf{j} + \mathbf{k} = \mathbf{k}. \qquad \blacksquare$$

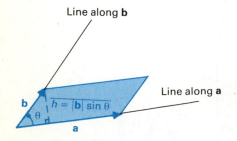

FIGURE 12.51

Area $= |\mathbf{a}|h = |\mathbf{a}|\,|\mathbf{b}|(\sin \theta).$

The important geometric properties of any vector are its *length* and its *direction*. In Fig. 12.51, we take the plane determined by the lines along **a** and **b** as the plane of the page of the text and shade the parallelogram having **a** and **b** as two edges. We claim that the length $|\mathbf{a} \times \mathbf{b}|$ is equal to the area of this shaded

THE DOT AND CROSS PRODUCTS of vectors grew out of the invention of quaternions by the Irish mathematician William Rowan Hamilton (1805–1865). Hamilton was a child prodigy: By the age of fourteen he was fluent in eight foreign languages, including Arabic and Sanskrit. In 1827, while still an undergraduate at Trinity College in Dublin, he was appointed professor of astronomy and also Royal Astronomer of Ireland, positions he held until the end of his life.

Hamilton for many years sought an algebra for vectors in three-dimensional space, that is, a way of adding and multiplying triples of numbers that would satisfy reasonable algebraic laws. In 1843, he finally succeeded in discovering such a result, but for quadruples rather than triples. These quaternions, which can be written as $w + x\mathbf{i} + y\mathbf{j} + z\mathbf{k}$, with w, x, y, z real numbers, have a multiplication generated by the formulas $\mathbf{i}^2 = \mathbf{j}^2 = \mathbf{k}^2 = \mathbf{ijk} = -1$. In particular, the product of the two quaternions $\alpha = x\mathbf{i} + y\mathbf{j} + z\mathbf{k}$ and $\beta = x'\mathbf{i} + y'\mathbf{j} + z'\mathbf{k}$ is

$$-xx' - yy' - zz' +$$
$$(yz' - zy')\mathbf{i} + (zx' - xz')\mathbf{j} +$$
$$(xy' - yx')\mathbf{k}.$$

The scalar part of this result is the negative of the modern dot product of α and β, and the "vector" part is the modern cross product.

In terms of three-dimensional vectors themselves, these products first appeared in the pamphlet *Elements of Vector Analysis* by the American physicist and Yale professor Josiah Willard Gibbs (1839–1903), which he prepared for use by his students in electricity and magnetism.

parallelogram, and we compute the area to verify this. Referring to Fig. 12.51, we have

$$\text{Area} = (\text{Length of base})(\text{Altitude})$$
$$= |\mathbf{a}|h = |\mathbf{a}| \cdot |\mathbf{b}| \sin \theta.$$

Therefore

$$\begin{aligned}
\text{Area}^2 &= |\mathbf{a}|^2|\mathbf{b}|^2\sin^2\theta \\
&= |\mathbf{a}|^2|\mathbf{b}|^2(1 - \cos^2\theta) \\
&= |\mathbf{a}|^2|\mathbf{b}|^2 - (|\mathbf{a}| \cdot |\mathbf{b}| \cos \theta)^2 \\
&= |\mathbf{a}|^2|\mathbf{b}|^2 - (\mathbf{a} \cdot \mathbf{b})^2 \\
&= (a_1{}^2 + a_2{}^2 + a_3{}^2)(b_1{}^2 + b_2{}^2 + b_3{}^2) - (a_1b_1 + a_2b_2 + a_3b_3)^2.
\end{aligned}$$

A bit of straightforward pencil-pushing shows that this boils down to

$$\begin{aligned}
\text{Area}^2 &= (a_2b_3 - a_3b_2)^2 + (a_1b_3 - a_3b_1)^2 + (a_1b_2 - a_2b_1)^2 \\
&= \begin{vmatrix} a_2 & a_3 \\ b_2 & b_3 \end{vmatrix}^2 + \begin{vmatrix} a_1 & a_3 \\ b_1 & b_3 \end{vmatrix}^2 + \begin{vmatrix} a_1 & a_2 \\ b_1 & b_2 \end{vmatrix}^2.
\end{aligned}$$

But this is the square of the length of the cross product $\mathbf{a} \times \mathbf{b}$, so $\text{Area}^2 = |\mathbf{a} \times \mathbf{b}|^2$ and $\text{Area} = |\mathbf{a} \times \mathbf{b}|$. This shows that

$$|\mathbf{a} \times \mathbf{b}| = \text{Area of parallelogram} = |\mathbf{a}| \cdot |\mathbf{b}|\sin \theta. \qquad (4)$$

Finally, we want to know the direction of $\mathbf{a} \times \mathbf{b}$. First, we see that $\mathbf{a} \times \mathbf{b}$ is perpendicular to both $\mathbf{a}$ and $\mathbf{b}$ and therefore perpendicular to the plane containing the parallelogram shaded in Fig. 12.51. We need only show that $\mathbf{a} \cdot (\mathbf{a} \times \mathbf{b}) = 0$ and $\mathbf{b} \cdot (\mathbf{a} \times \mathbf{b}) = 0$. Referring to Eq. (3), where $\mathbf{a} \times \mathbf{b}$ is defined, we see that

$$\mathbf{a} \cdot (\mathbf{a} \times \mathbf{b}) = a_1\begin{vmatrix} a_2 & a_3 \\ b_2 & b_3 \end{vmatrix} - a_2\begin{vmatrix} a_1 & a_3 \\ b_1 & b_3 \end{vmatrix} + a_3\begin{vmatrix} a_1 & a_2 \\ b_1 & b_2 \end{vmatrix}.$$

But this is equal to the determinant

$$\begin{vmatrix} a_1 & a_2 & a_3 \\ a_1 & a_2 & a_3 \\ b_1 & b_2 & b_3 \end{vmatrix} = 0,$$

which is 0 since the first and second rows are the same (Theorem 12.3). Thus $\mathbf{a} \cdot (\mathbf{a} \times \mathbf{b}) = 0$ and $\mathbf{a}$ is perpendicular to $\mathbf{a} \times \mathbf{b}$. A similar argument shows that $\mathbf{b} \cdot (\mathbf{a} \times \mathbf{b}) = 0$; this time the determinant has its first and third rows the same.

We know know that $\mathbf{a} \times \mathbf{b}$ has length $|\mathbf{a}| \cdot |\mathbf{b}|\sin \theta$ and direction perpendicular to the plane determined by $\mathbf{a}$ and $\mathbf{b}$. There are two vectors of this length perpendicular to the plane; one is the negative of the other. One of them is $\mathbf{a} \times \mathbf{b}$, and the other is $-(\mathbf{a} \times \mathbf{b})$, which we claim equals $\mathbf{b} \times \mathbf{a}$. Referring to Eq. (3), we see that the determinants of the 2×2 matrices used to find $\mathbf{b} \times \mathbf{a}$ are the ones used to find $\mathbf{a} \times \mathbf{b}$ with the rows interchanged. Theorem 12.3 thus shows at once that

$$\mathbf{b} \times \mathbf{a} = -(\mathbf{a} \times \mathbf{b}). \qquad (5)$$

We summarize what we have done and a bit more in a theorem.

THEOREM 12.4 Magnitude and direction of a × b

The vector **a × b** has length given by

$$|\mathbf{a} \times \mathbf{b}| = |\mathbf{a}| \cdot |\mathbf{b}| \sin \theta,$$

where θ satisfying $0 \leq \theta \leq \pi$ is the angle between **a** and **b**. The direction of **a × b** is perpendicular to both **a** and **b** in the direction in which the thumb of the right hand points as the fingers curl through θ from **a** to **b**. This manner of describing the direction of **a × b** is known as the "right-hand rule." It is illustrated in Fig. 12.52. □

The only part of the theorem that we have not proved is the "right-hand rule" part. We will not prove this; we can illustrate it using

$$\mathbf{i} \times \mathbf{j} = \mathbf{k}, \qquad \mathbf{j} \times \mathbf{k} = \mathbf{i}, \qquad \text{and} \qquad \mathbf{k} \times \mathbf{i} = \mathbf{j},$$

which were shown in Example 4.

EXAMPLE 5 Find all possible cross products of two of the vectors **i**, **j**, and **k**.

Solution From Example 4, we have

$$\mathbf{i} \times \mathbf{j} = \mathbf{k}, \qquad \mathbf{j} \times \mathbf{k} = \mathbf{i}, \qquad \text{and} \qquad \mathbf{k} \times \mathbf{i} = \mathbf{j}. \tag{6}$$

Therefore, from Eq. (5),

$$\mathbf{j} \times \mathbf{i} = -\mathbf{k}, \qquad \mathbf{k} \times \mathbf{j} = -\mathbf{i}, \qquad \text{and} \qquad \mathbf{i} \times \mathbf{k} = -\mathbf{j}. \tag{7}$$

■

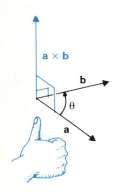

Right-hand rule for
a × b, $0 < \theta < \pi$.

We can remember Eqs. (6) and (7) by writing the sequence

$$\mathbf{i}, \quad \mathbf{j}, \quad \mathbf{k}, \quad \mathbf{i}, \quad \mathbf{j}, \quad \mathbf{k}.$$

The cross product of two consecutive vectors in left-to-right order is the next one to the right, while the cross product in right-to-left order is the negative of the next vector to the left.

Equation (4) gives us a straightforward way to find the area of a parallelogram or a triangle.

EXAMPLE 6 Find the area of the parallelogram in the plane having **i + 3j** and **4i + 2j** as coterminous edges.

Solution In Fig. 12.53, we have shown a parallelogram with these vectors emanating from the origin as edges. Equation (4) involves the cross product, which is defined for vectors in space. However, we can view the x,y-plane as part of space and the given vectors as

$$\mathbf{a} = \mathbf{i} + 3\mathbf{j} + 0\mathbf{k} \qquad \text{and} \qquad \mathbf{b} = 4\mathbf{i} + 2\mathbf{j} + 0\mathbf{k}.$$

We then have

$$\mathbf{a} \times \mathbf{b} = \begin{vmatrix} \mathbf{i} & \mathbf{j} & \mathbf{k} \\ 1 & 3 & 0 \\ 4 & 2 & 0 \end{vmatrix} = 0\mathbf{i} + 0\mathbf{j} - 10\mathbf{k}.$$

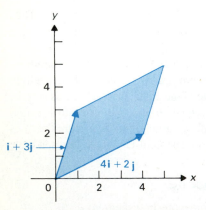

FIGURE 12.53 Parallelogram with adjacent edges **i + 3j** and **4i + 2j**.

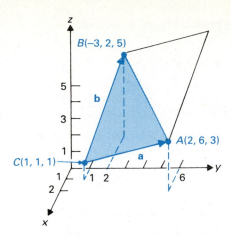

FIGURE 12.54 Triangle with vertices (2, 6, 3), (−3, 2, 5), and (1, 1, 1).

From Eq. (4), we have

$$\text{Area of parallelogram} = |\mathbf{a} \times \mathbf{b}| = \sqrt{0^2 + 0^2 + (-10)^2} = 10. \quad \blacksquare$$

EXAMPLE 7 Find the area of the triangle in space with vertices $A(2, 6, 3)$, $B(-3, 2, 5)$, and $C(1, 1, 1)$.

Solution The triangle is shown in Fig. 12.54. We let $\mathbf{a}$ be the vector from C to A and $\mathbf{b}$ the vector from C to B, as shown in the figure. Then

$$\mathbf{a} = \mathbf{i} + 5\mathbf{j} + 2\mathbf{k} \quad \text{and} \quad \mathbf{b} = -4\mathbf{i} + \mathbf{j} + 4\mathbf{k}.$$

The area of the triangle is half the area of the parallelogram having $\mathbf{a}$ and $\mathbf{b}$ as coterminous edges. We find that

$$\mathbf{a} \times \mathbf{b} = \begin{vmatrix} \mathbf{i} & \mathbf{j} & \mathbf{k} \\ 1 & 5 & 2 \\ -4 & 1 & 4 \end{vmatrix} = 18\mathbf{i} - 12\mathbf{j} + 21\mathbf{k}.$$

Then

$$\text{Area of triangle} = \frac{1}{2}|\mathbf{a} \times \mathbf{b}| = \frac{1}{2}\sqrt{18^2 + 12^2 + 21^2}$$

$$= \frac{1}{2}\sqrt{909} \approx 15.075. \quad \blacksquare$$

EXAMPLE 8 Find a unit vector perpendicular to both $\mathbf{a} = \mathbf{i} + \mathbf{j} + 2\mathbf{k}$ and $\mathbf{b} = -\mathbf{i} + \mathbf{j} - 3\mathbf{k}$.

Solution By Theorem 12.4, the vector $\mathbf{a} \times \mathbf{b}$ is perpendicular to both $\mathbf{a}$ and $\mathbf{b}$. Now

$$\mathbf{a} \times \mathbf{b} = \begin{vmatrix} \mathbf{i} & \mathbf{j} & \mathbf{k} \\ 1 & 1 & 2 \\ -1 & 1 & -3 \end{vmatrix} = -5\mathbf{i} + \mathbf{j} + 2\mathbf{k}.$$

To form a unit vector, we multiply $\mathbf{a} \times \mathbf{b}$ by the reciprocal of its length

$$|\mathbf{a} \times \mathbf{b}| = |-5\mathbf{i} + \mathbf{j} + 2\mathbf{k}| = \sqrt{25 + 1 + 4} = \sqrt{30}.$$

We obtain as answer

$$-\frac{5}{\sqrt{30}}\mathbf{i} + \frac{1}{\sqrt{30}}\mathbf{j} + \frac{2}{\sqrt{30}}\mathbf{k}.$$

Another possible answer is $(5/\sqrt{30})\mathbf{i} - (1/\sqrt{30})\mathbf{j} - (2/\sqrt{30})\mathbf{k}$, which corresponds to computing $\mathbf{b} \times \mathbf{a}$ rather than $\mathbf{a} \times \mathbf{b}$. $\quad \blacksquare$

EXAMPLE 9 Show that if $\mathbf{a}$ and $\mathbf{b}$ are parallel, then $\mathbf{a} \times \mathbf{b} = \mathbf{0}$.

Solution If $\mathbf{a}$ and $\mathbf{b}$ are parallel, then the angle θ between them is either 0 or π. By Theorem 12.4, the magnitude of $\mathbf{a} \times \mathbf{b}$ is then

$$|\mathbf{a} \times \mathbf{b}| = |\mathbf{a}| \, |\mathbf{b}|\sin \theta = |\mathbf{a}| \, |\mathbf{b}| \cdot 0 = 0.$$

Thus $\mathbf{a} \times \mathbf{b}$ must be the zero vector. $\quad \blacksquare$

Equation (5) shows that taking cross products is not a commutative operation. However, it is true that

$$\mathbf{a} \times (\mathbf{b} + \mathbf{c}) = \mathbf{a} \times \mathbf{b} + \mathbf{a} \times \mathbf{c} \tag{8}$$

and

$$(k\mathbf{a}) \times \mathbf{b} = \mathbf{a} \times (k\mathbf{b}) = k(\mathbf{a} \times \mathbf{b}) \tag{9}$$

for any vectors $\mathbf{a}$, $\mathbf{b}$, $\mathbf{c}$ in space and any scalar k. The proofs of Eqs. (8) and (9) are straightforward computations with components.

Triple Products

We now know two ways to take a product of vectors $\mathbf{a}$ and $\mathbf{b}$ in space. We can find $\mathbf{a} \cdot \mathbf{b}$, which is a scalar, or $\mathbf{a} \times \mathbf{b}$, which is a vector. It is natural to try to multiply three vectors, $\mathbf{a}$, $\mathbf{b}$, and $\mathbf{c}$, in space. The product $\mathbf{a} \cdot (\mathbf{b} \cdot \mathbf{c})$ makes no sense, for $\mathbf{a}$ is a vector and $\mathbf{b} \cdot \mathbf{c}$ is a scalar. However, $\mathbf{a} \times (\mathbf{b} \times \mathbf{c})$ makes sense, for both $\mathbf{a}$ and $\mathbf{b} \times \mathbf{c}$ are vectors. This product $\mathbf{a} \times (\mathbf{b} \times \mathbf{c})$ is the *triple vector product*. It is readily computed using the formula

$$\mathbf{a} \times (\mathbf{b} \times \mathbf{c}) = (\mathbf{a} \cdot \mathbf{c})\mathbf{b} - (\mathbf{a} \cdot \mathbf{b})\mathbf{c}, \tag{10}$$

which we ask you to establish in Exercise 42. In Exercise 47, you are asked to convince yourself that

$$\mathbf{a} \times (\mathbf{b} \times \mathbf{c}) \neq (\mathbf{a} \times \mathbf{b}) \times \mathbf{c}$$

unless $\mathbf{a}$, $\mathbf{b}$, and $\mathbf{c}$ are carefully chosen. That is, the triple cross product is not associative.

EXAMPLE 10   If $\mathbf{a} = 2\mathbf{i} - 3\mathbf{j} + 4\mathbf{k}$ while $\mathbf{b} = 3\mathbf{i} - \mathbf{j} - 2\mathbf{k}$ and $\mathbf{c} = -3\mathbf{i} - 5\mathbf{j} - \mathbf{k}$, find $\mathbf{a} \times (\mathbf{b} \times \mathbf{c})$.

Solution   From Eq. (10), we have

$$\mathbf{a} \times (\mathbf{b} \times \mathbf{c}) = (\mathbf{a} \cdot \mathbf{c})\mathbf{b} - (\mathbf{a} \cdot \mathbf{b})\mathbf{c}$$
$$= 5\mathbf{b} - \mathbf{c} = 18\mathbf{i} - 9\mathbf{k}. \qquad \blacksquare$$

The product $\mathbf{a} \cdot (\mathbf{b} \times \mathbf{c})$ also makes sense, but the answer this time is a scalar. Consequently $\mathbf{a} \cdot (\mathbf{b} \times \mathbf{c})$ is called the *triple scalar product*. Let $\mathbf{a}$, $\mathbf{b}$, and $\mathbf{c}$ be as shown in Fig. 12.55. The vectors are coterminous edges of a box, shown shaded in the figure. For such a box, we have

$$\text{Volume} = (\text{Area of base})(\text{Altitude}).$$

Using the vectors and angles shown in Fig. 12.55, we see that

$$\text{Volume} = (\text{Area of base})(|\mathbf{a}| \cos \phi).$$

Now the area of the base is $|\mathbf{b} \times \mathbf{c}|$, by Eq. (4). Therefore

$$\text{Volume} = |\mathbf{b} \times \mathbf{c}| \cdot |\mathbf{a}| \cos \phi.$$

But $\mathbf{b} \times \mathbf{c}$ is perpendicular to the base of the box, and ϕ is the angle between $\mathbf{b} \times \mathbf{c}$ and $\mathbf{c}$. Therefore

$$|\mathbf{b} \times \mathbf{c}| \cdot |\mathbf{a}| \cos \phi = \mathbf{a} \cdot (\mathbf{b} \times \mathbf{c}).$$

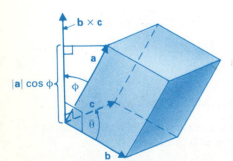

FIGURE 12.55   Box with $\mathbf{a}$, $\mathbf{b}$, $\mathbf{c}$ as edges and volume $\mathbf{a} \cdot (\mathbf{b} \times \mathbf{c})$.

If the order of **b** and **c** is reversed, we obtain $\mathbf{a} \cdot (\mathbf{c} \times \mathbf{b}) = -\mathbf{a} \cdot (\mathbf{b} \times \mathbf{c})$, but of course the box given by **a**, **c**, **b** is the same one as that given by **a**, **b**, **c**. Therefore we have the formula

$$\text{Volume} = |\mathbf{a} \cdot (\mathbf{b} \times \mathbf{c})|. \tag{11}$$

There is a very quick way to compute $\mathbf{a} \cdot (\mathbf{b} \times \mathbf{c})$. Form the matrix having the vectors **a**, **b**, and **c** as the first, second, and third rows, respectively. Then $\mathbf{a} \cdot (\mathbf{b} \times \mathbf{c})$ is the determinant of this matrix. To see why this is true, note that if $\mathbf{a} = a_1\mathbf{i} + a_2\mathbf{j} + a_3\mathbf{k}$ and $\mathbf{d} = d_1\mathbf{i} + d_2\mathbf{j} + d_3\mathbf{k}$, then $\mathbf{a} \cdot \mathbf{d} = d_1a_1 + d_2a_2 + d_3a_3$ can be found by formally replacing the **i**, **j**, and **k** in the expression for **d** by a_1, a_2, and a_3, respectively. If we do this for

$$\mathbf{d} = \mathbf{b} \times \mathbf{c} = \begin{vmatrix} \mathbf{i} & \mathbf{j} & \mathbf{k} \\ b_1 & b_2 & b_3 \\ c_1 & c_2 & c_3 \end{vmatrix}$$

to form $\mathbf{a} \cdot \mathbf{d} = \mathbf{a} \cdot (\mathbf{b} \times \mathbf{c})$, we obtain

$$\mathbf{a} \cdot (\mathbf{b} \times \mathbf{c}) = \begin{vmatrix} a_1 & a_2 & a_3 \\ b_1 & b_2 & b_3 \\ c_1 & c_2 & c_3 \end{vmatrix}. \tag{12}$$

EXAMPLE 11   Find the volume of the box in space having as coterminous edges the vectors $\mathbf{a} = \mathbf{i} - 2\mathbf{j} + \mathbf{k}$, $\mathbf{b} = 2\mathbf{i} + 3\mathbf{j} - 2\mathbf{k}$, and $\mathbf{c} = -\mathbf{i} + 3\mathbf{j} - 2\mathbf{k}$.

Solution   We have

$$\mathbf{a} \cdot (\mathbf{b} \times \mathbf{c}) = \begin{vmatrix} 1 & -2 & 1 \\ 2 & 3 & -2 \\ -1 & 3 & -2 \end{vmatrix} = 1(0) - (-2)(-6) + 1(9) = -3.$$

Therefore,

$$\text{Volume} = |\mathbf{a} \cdot (\mathbf{b} \times \mathbf{c})| = |-3| = 3. \quad \blacksquare$$

SUMMARY

1. $$\begin{vmatrix} a_1 & a_2 \\ b_1 & b_2 \end{vmatrix} = a_1b_2 - a_2b_1,$$

$$\begin{vmatrix} a_1 & a_2 & a_3 \\ b_1 & b_2 & b_3 \\ c_1 & c_2 & c_3 \end{vmatrix} = a_1 \begin{vmatrix} b_2 & b_3 \\ c_2 & c_3 \end{vmatrix} - a_2 \begin{vmatrix} b_1 & b_3 \\ c_1 & c_3 \end{vmatrix} + a_3 \begin{vmatrix} b_1 & b_2 \\ c_1 & c_2 \end{vmatrix}$$

2. $$\mathbf{a} \times \mathbf{b} = \begin{vmatrix} \mathbf{i} & \mathbf{j} & \mathbf{k} \\ a_1 & a_2 & a_3 \\ b_1 & b_2 & b_3 \end{vmatrix}$$

3. The cross product $\mathbf{a} \times \mathbf{b}$ has length $|\mathbf{a}| \cdot |\mathbf{b}| \sin \theta$, where θ is the angle between **a** and **b**. This length equals the area of the parallelogram having **a** and **b** as adjacent sides.

4. The direction of $\mathbf{a} \times \mathbf{b}$ is perpendicular to the plane of $\mathbf{a}$ and $\mathbf{b}$ and in the direction given by the right-hand rule, illustrated in Fig. 12.52.

5. For any vectors $\mathbf{a}$, $\mathbf{b}$, $\mathbf{c}$ in space and any scalar k,

$$\mathbf{a} \times \mathbf{b} = -\mathbf{b} \times \mathbf{a},$$
$$\mathbf{a} \times (\mathbf{b} + \mathbf{c}) = \mathbf{a} \times \mathbf{b} + \mathbf{a} \times \mathbf{c},$$
$$(k\mathbf{a}) \times \mathbf{b} = \mathbf{a} \times (k\mathbf{b}) = k(\mathbf{a} \times \mathbf{b}).$$

6. $$\mathbf{i} \times \mathbf{j} = \mathbf{k}, \quad \mathbf{j} \times \mathbf{k} = \mathbf{i}, \quad \mathbf{k} \times \mathbf{i} = \mathbf{j},$$

while

$$\mathbf{j} \times \mathbf{i} = -\mathbf{k}, \quad \mathbf{k} \times \mathbf{j} = -\mathbf{i}, \quad \mathbf{i} \times \mathbf{k} = -\mathbf{j}.$$

7. The triple scalar product $\mathbf{a} \cdot (\mathbf{b} \times \mathbf{c})$ and the triple vector product $\mathbf{a} \times (\mathbf{b} \times \mathbf{c})$ are most easily computed using

$$\mathbf{a} \cdot (\mathbf{b} \times \mathbf{c}) = \begin{vmatrix} a_1 & a_2 & a_3 \\ b_1 & b_2 & b_3 \\ c_1 & c_2 & c_3 \end{vmatrix}$$

and

$$\mathbf{a} \times (\mathbf{b} \times \mathbf{c}) = (\mathbf{a} \cdot \mathbf{c})\mathbf{b} - (\mathbf{a} \cdot \mathbf{b})\mathbf{c}.$$

8. $|\mathbf{a} \cdot (\mathbf{b} \times \mathbf{c})|$ is the volume of the box having $\mathbf{a}$, $\mathbf{b}$, and $\mathbf{c}$ as adjacent edges.

EXERCISES 12.5

In Exercises 1–8, find the indicated determinant.

1. $\begin{vmatrix} -1 & 3 \\ 5 & 0 \end{vmatrix}$

2. $\begin{vmatrix} -1 & 0 \\ 0 & 7 \end{vmatrix}$

3. $\begin{vmatrix} 0 & -3 \\ 5 & 0 \end{vmatrix}$

4. $\begin{vmatrix} 21 & -4 \\ 10 & 7 \end{vmatrix}$

5. $\begin{vmatrix} 1 & 4 & -2 \\ 3 & 13 & 0 \\ 2 & -1 & 3 \end{vmatrix}$

6. $\begin{vmatrix} 2 & -5 & 3 \\ 1 & 3 & 4 \\ -2 & 3 & 7 \end{vmatrix}$

7. $\begin{vmatrix} 1 & -2 & 7 \\ 0 & 1 & 4 \\ 1 & 0 & 3 \end{vmatrix}$

8. $\begin{vmatrix} 2 & -1 & 1 \\ -1 & 0 & 3 \\ 2 & 1 & -4 \end{vmatrix}$

9. Show by direct computation that

(a) $\begin{vmatrix} a_1 & a_2 & a_3 \\ a_1 & a_2 & a_3 \\ c_1 & c_2 & c_3 \end{vmatrix} = 0,$ (b) $\begin{vmatrix} a_1 & a_2 & a_3 \\ b_1 & b_2 & b_3 \\ a_1 & a_2 & a_3 \end{vmatrix} = 0.$

10. Show by direct computation that

$$\begin{vmatrix} a_1 & a_2 & a_3 \\ b_1 & b_2 & b_3 \\ c_1 & c_2 & c_3 \end{vmatrix} = -\begin{vmatrix} a_1 & a_2 & a_3 \\ c_1 & c_2 & c_3 \\ b_1 & b_2 & b_3 \end{vmatrix}.$$

In Exercises 11–16, find $\mathbf{a} \times \mathbf{b}$.

11. $\mathbf{a} = 2\mathbf{i} - \mathbf{j} + 3\mathbf{k}, \mathbf{b} = \mathbf{i} + 2\mathbf{j}$

12. $\mathbf{a} = -5\mathbf{i} + \mathbf{j} + 4\mathbf{k}, \mathbf{b} = 2\mathbf{i} + \mathbf{j} - 3\mathbf{k}$

13. $\mathbf{a} = -\mathbf{i} + 2\mathbf{j} + 4\mathbf{k}, \mathbf{b} = 2\mathbf{i} - 4\mathbf{j} - 8\mathbf{k}$

14. $\mathbf{a} = \mathbf{i} - \mathbf{j} + \mathbf{k}, \mathbf{b} = 3\mathbf{i} - 2\mathbf{j} + 7\mathbf{k}$

15. $\mathbf{a} = 2\mathbf{i} - 3\mathbf{j} + 5\mathbf{k}, \mathbf{b} = 4\mathbf{i} - 5\mathbf{j} + \mathbf{k}$

16. $\mathbf{a} = -2\mathbf{i} + 3\mathbf{j} - \mathbf{k}, \mathbf{b} = 4\mathbf{i} - 6\mathbf{j} + \mathbf{k}$

In Exercises 17–21, find the area of the parallelogram having the given vectors as edges. If the vectors are in the plane, regard them as being in space with the coefficient of $\mathbf{k}$ being zero.

17. $-\mathbf{i} + 4\mathbf{j}$ and $2\mathbf{i} + 3\mathbf{j}$

18. $-5\mathbf{i} + 3\mathbf{j}$ and $\mathbf{i} + 7\mathbf{j}$

19. $\mathbf{i} + 3\mathbf{j} - 5\mathbf{k}$ and $2\mathbf{i} + 4\mathbf{j} - \mathbf{k}$

20. $2\mathbf{i} - \mathbf{j} + \mathbf{k}$ and $\mathbf{i} + 3\mathbf{j} - \mathbf{k}$

21. $\mathbf{i} - 4\mathbf{j} + \mathbf{k}$ and $2\mathbf{i} + 3\mathbf{j} - 2\mathbf{k}$

In Exercises 22–29, find the area of the given geometric configuration.

22. The triangle with vertices $(-1, 2)$, $(3, -1)$, and $(4, 3)$

23. The triangle with vertices $(3, -4)$, $(1, 1)$, and $(5, 7)$

24. The triangle with vertices $(2, 1, -3)$, $(3, 0, 4)$, and $(1, 0, 5)$

25. The triangle with vertices $(3, 1, -2)$, $(1, 4, 5)$, and $(2, 1, -4)$

26. The triangle in the plane bounded by the lines $y = x$, $y = -3x + 8$, and $3y + 5x = 0$

27. The parallelogram with vertices $(1, 3)$, $(-2, 6)$, $(1, 11)$, and $(4, 8)$

28. The parallelogram with vertices $(1, 0, 1)$, $(3, 1, 4)$, $(0, 2, 9)$, and $(-2, 1, 6)$

29. The parallelogram in the plane bounded by the lines $x - 2y = 3$, $x - 2y = 8$, $2x + 3y = -1$, and $2x + 3y = -5$

In Exercises 30–33, find $\mathbf{a} \cdot (\mathbf{b} \times \mathbf{c})$ and $\mathbf{a} \times (\mathbf{b} \times \mathbf{c})$.

30. $\mathbf{a} = \mathbf{i} + 2\mathbf{j} - 3\mathbf{k}$, $\mathbf{b} = 4\mathbf{i} - \mathbf{j} + 2\mathbf{k}$, and $\mathbf{c} = 3\mathbf{i} + \mathbf{k}$

31. $\mathbf{a} = -\mathbf{i} + \mathbf{j} + 2\mathbf{k}$, $\mathbf{b} = \mathbf{i} + \mathbf{k}$, and $\mathbf{c} = 3\mathbf{i} - 2\mathbf{j} + 5\mathbf{k}$

32. $\mathbf{a} = \mathbf{i} - 3\mathbf{k}$, $\mathbf{b} = -\mathbf{i} + 4\mathbf{j}$, and $\mathbf{c} = \mathbf{i} + \mathbf{j} + \mathbf{k}$

33. $\mathbf{a} = 4\mathbf{i} - \mathbf{j} + 2\mathbf{k}$, $\mathbf{b} = 3\mathbf{i} + 5\mathbf{j} - 2\mathbf{k}$, and $\mathbf{c} = \mathbf{i} - 3\mathbf{j} + \mathbf{k}$

In Exercises 34–37, find the volume of the box having the given vectors as adjacent edges.

34. $-\mathbf{i} + 4\mathbf{j} + 7\mathbf{k}$, $3\mathbf{i} - 2\mathbf{j} - \mathbf{k}$, and $4\mathbf{i} + 2\mathbf{k}$

35. $2\mathbf{i} + \mathbf{j} - 4\mathbf{k}$, $3\mathbf{i} - \mathbf{j} + 2\mathbf{k}$, and $\mathbf{i} + 3\mathbf{j} - 10\mathbf{k}$

36. $-2\mathbf{i} + \mathbf{j}$, $3\mathbf{i} - 4\mathbf{j} + \mathbf{k}$, and $\mathbf{i} - 2\mathbf{k}$

37. $3\mathbf{i} - \mathbf{j} + 4\mathbf{k}$, $\mathbf{i} - 2\mathbf{j} + 7\mathbf{k}$, and $5\mathbf{i} - 3\mathbf{j} + 10\mathbf{k}$

In Exercises 38–41, find the volume of the tetrahedron having the given vertices. [*Hint:* How is the volume of a tetrahedron having three vectors from one point as edges related to the volume of the box having the same three vectors as adjacent edges?]

38. $(-3, 0, 1)$, $(4, 2, 1)$, $(0, 1, 7)$, and $(1, 1, 1)$

39. $(0, 1, 1)$, $(8, 2, -7)$, $(3, 1, 6)$, and $(-4, -2, 0)$

40. $(-1, 1, 2)$, $(3, 1, 4)$, $(-1, 6, 0)$, and $(2, -1, 5)$

41. $(-1, 2, 4)$, $(2, -3, 0)$, $(-4, 2, -1)$, and $(0, 3, -2)$

42. Let $\mathbf{a} = a_1\mathbf{i} + a_2\mathbf{j} + a_3\mathbf{k}$, $\mathbf{b} = b_1\mathbf{i} + b_2\mathbf{j} + b_3\mathbf{k}$, and $\mathbf{c} = c_1\mathbf{i} + c_2\mathbf{j} + c_3\mathbf{k}$. Verify that $\mathbf{a} \times (\mathbf{b} \times \mathbf{c}) = (\mathbf{a} \cdot \mathbf{c})\mathbf{b} - (\mathbf{a} \cdot \mathbf{b})\mathbf{c}$. (This problem involves merely a lot of tedious algebra.)

43. Use the properties of the cross product to find a formula, similar to that in the preceding exercise, for the computation of $(\mathbf{a} \times \mathbf{b}) \times \mathbf{c}$.

44. Use the results of Exercises 42 and 43 to express $(\mathbf{a} \times \mathbf{b}) \times (\mathbf{c} \times \mathbf{d})$ in each of the forms $h\mathbf{a} + k\mathbf{b}$ and $r\mathbf{c} + s\mathbf{d}$ for scalars h, k, r, and s.

45. Show that, for any vectors $\mathbf{a}$, $\mathbf{b}$ in space, $\mathbf{a} \cdot (\mathbf{a} \times \mathbf{b}) = 0$.

46. Show that $\mathbf{i} \times (\mathbf{i} \times \mathbf{k}) = -\mathbf{k}$, while $(\mathbf{i} \times \mathbf{i}) \times \mathbf{k} = \mathbf{0}$, illustrating that the cross product is not associative.

47. Consider the triple vector products $\mathbf{a} \times (\mathbf{b} \times \mathbf{c})$ and $(\mathbf{a} \times \mathbf{b}) \times \mathbf{c}$ for vectors $\mathbf{a}$, $\mathbf{b}$, and $\mathbf{c}$ in space.
    (a) Argue geometrically that $\mathbf{a} \times (\mathbf{b} \times \mathbf{c})$ is a vector in the plane containing $\mathbf{b}$ and $\mathbf{c}$.
    (b) Argue geometrically that $(\mathbf{a} \times \mathbf{b}) \times \mathbf{c}$ is a vector in the plane containing $\mathbf{a}$ and $\mathbf{b}$.
    (c) Use parts (a) and (b) to argue that the cross product is not associative.

12.6 LINES

This section develops parametric equations for lines in space. If we had covered vectors in Chapter 1, we could have presented lines in the plane parametrically also. The present section then would seem really trivial; we would simply put in one new equation for the z-component. Using parametric equations, it is possible to describe a line (a one-dimensional "flat" piece) in space of any finite dimension.

Parametric Equations for a Line

We are accustomed to thinking of a line as being determined by two points. While this is perfectly correct, it will be useful for us to think of a line as being determined by one *point* on the line and the *direction* of the line. Of course, the

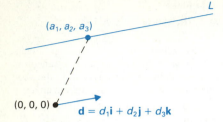

FIGURE 12.56 The line L through (a_1, a_2, a_3) with parallel vector **d**.

direction of a line can be specified in terms of a nonzero vector parallel to the line.

Let (a_1, a_2, a_3) be a point on a line L in space. If the origin $(0, 0, 0)$ is not on L, there is a unique plane in space containing the origin and L. If the origin lies on L, there are many such planes.

In Fig. 12.56, we take the plane of the page to be a plane containing L and the origin. The figure shows the point (a_1, a_2, a_3) on the line L and a vector $\mathbf{d} = d_1\mathbf{i} + d_2\mathbf{j} + d_3\mathbf{k}$ parallel to the line. We consider any such vector **d** to be a *direction vector* for the line. The line L is the unique line through (a_1, a_2, a_3) parallel to **d**.

We want to find all points (x_1, x_2, x_3) on the line L. (Subscripted coordinates make things clearer.) In Fig. 12.57, we show the vectors **a** from $(0, 0, 0)$ to (a_1, a_2, a_3) and **x** from $(0, 0, 0)$ to a point (x_1, x_2, x_3) on the line. The vector along the line from (a_1, a_2, a_3) to (x_1, x_2, x_3) is parallel to **d**. Consequently, it must be of the form $t\mathbf{d}$ for some scalar t, as shown in Fig. 12.57. From the figure, we have

$$\mathbf{x} = \mathbf{a} + t\mathbf{d}. \tag{1}$$

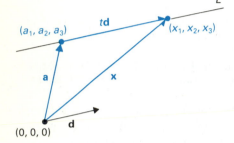

FIGURE 12.57 $\mathbf{x} = \mathbf{a} + t\mathbf{d}$

Equation (1) is a **vector equation of the line.** As t runs through all values, we obtain all vectors from the origin to points on the line. Equation (1) makes the line L into a t-axis, as shown in Fig. 12.58. When $t = 0$, Eq. (1) gives the vector **a** to the point (a_1, a_2, a_3) on the line, so this point corresponds to the t-origin, as shown in Fig. 12.58. The point corresponding to $t = 1$ is the tip of the vector $\mathbf{a} + 1\mathbf{d}$, and so on. Of course, the t-scale on L will be different from the scale on the coordinate axes unless **d** is a unit vector.

If we write Eq. (1) in coordinate form, we obtain

$$\langle x_1, x_2, x_3 \rangle = \langle a_1, a_2, a_3 \rangle + t \langle d_1, d_2, d_3 \rangle. \tag{2}$$

Equating first, second, and third coordinates, we obtain

$$\begin{aligned} x_1 &= a_1 + d_1 t, \\ x_2 &= a_2 + d_2 t, \\ x_3 &= a_3 + d_3 t \end{aligned} \tag{3}$$

as scalar parametric equations of the line.

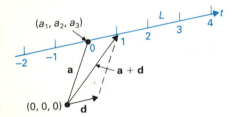

FIGURE 12.58 The line L becomes a t-axis.

We summarize our work in a definition. We go back to x, y, z in place of x_1, x_2, x_3 for ease in writing as we work problems.

DEFINITION 12.5

Line in space

Let (a_1, a_2, a_3) be a point and let $\mathbf{d} = d_1\mathbf{i} + d_2\mathbf{j} + d_3\mathbf{k}$ be a vector. The **line** through (a_1, a_2, a_3) with **direction vector d** is the set of all points (x, y, z) such that, for some scalar t,

$$\begin{aligned} x &= a_1 + d_1 t, \\ y &= a_2 + d_2 t, \\ z &= a_3 + d_3 t. \end{aligned} \tag{4}$$

Equations (4) are **parametric equations** for the line, and t is the **parameter.**

It is also possible to describe a line as an intersection of planes each of which is parallel to a coordinate axis. This leads to the **symmetric equations** of a line, considered in Exercise 55 of Section 12.7. We consider the parametric equations to be preferable.

In Chapter 1, we found the single linear equation for a nonvertical line in the plane by following the step sequence

$$\begin{array}{l} \textbf{Point} \\ \textbf{Slope} \\ \textbf{Equation} \end{array} \qquad (5)$$

Students are sometimes drilled that a single linear equation gives a line. This is true in the plane, but not in space. We will see in the next section that a single linear equation in space gives a plane. Lines in space of any dimension are readily described parametrically. We use $\parallel$ as a symbol for "parallel."

Our step sequence for finding parametric equations of a line is

$$\begin{array}{l} \textbf{Point} \\ \parallel \textbf{ vector} \\ \textbf{Equations} \end{array} \qquad (6)$$

EXAMPLE 1   Find parametric equations of the line through $(-1, 0, 2)$ having $\mathbf{d} = 2\mathbf{i} - 3\mathbf{j} + \mathbf{k}$ as direction vector.

Solution   Following the step sequence (6), we have from Eqs. (4)

> **Point** $(-1, 0, 2)$
> $\parallel$ **vector** $2\mathbf{i} - 3\mathbf{j} + \mathbf{k}$
> **Equations** $x = -1 + 2t,$
> $\qquad\qquad\quad y = 0 - 3t,$
> $\qquad\qquad\quad z = 2 + t$ ∎

EXAMPLE 2   Find the point on the line in Example 1 with z-coordinate 6.

Solution   This example emphasizes that as t runs through all values, we obtain all points on the line. To have z-coordinate 6, we must have

$$6 = 2 + t, \qquad \text{or} \qquad t = 4.$$

When $t = 4$, we obtain the point $(7, -12, 6)$ on the line.   ∎

The next example shows how to find the line through two points.

EXAMPLE 3   Find parametric equations of the line through $(1, -2, 3)$ and $(0, 1, 4)$.

Solution   We have two points from which to choose. For direction vector, we may take the vector $\langle 0, 1, 4\rangle - \langle 1, -2, 3\rangle$ from $(1, -2, 3)$ to $(0, 1, 4)$.

> **Point** $(1, -2, 3)$
>
> ‖ **vector** $-\mathbf{i} + 3\mathbf{j} + \mathbf{k}$
>
> **Equations** $x = 1 - t,$
> $\qquad\qquad y = -2 + 3t,$
> $\qquad\qquad z = 3 + t$ ∎

Parametric equations for a line are not unique. We can use *any* point on the line and *any* vector $\mathbf{d}$ parallel to the line. To illustrate, with $t = 2$, we see that $(-1, 4, 5)$ is on the line in Example 3. Multiplying the direction vector $-\mathbf{i} + 3\mathbf{j} + \mathbf{k}$ by -3, we obtain

$$x = -1 + 3t, \qquad y = 4 - 9t, \qquad z = 5 - 3t$$

as another system of parametric equations for the same line.

Different systems of parametric equations for a line correspond geometrically to different choices for origin and scale on the line as a t-axis in Fig. 12.58.

EXAMPLE 4   Find parametric equations for the line $x - 3y = 7$ in the plane.

Solution   We find any point on the line; $(4, -1)$ will do. The line has slope $\frac{1}{3}$, so $\mathbf{i} + (\frac{1}{3})\mathbf{j}$ is parallel to the line. To eliminate fractions, we multiply by 3 and take the parallel vector $3\mathbf{i} + \mathbf{j}$.

> **Point** $(4, -1)$
>
> ‖ **vector** $3\mathbf{i} + \mathbf{j}$
>
> **Equations** $x = 4 + 3t,$
> $\qquad\qquad y = -1 + t$ ∎

Two lines in the plane intersect unless they are parallel. Two lines in space usually don't intersect. The next two examples show how to find points of intersection.

EXAMPLE 5   Find the point of intersection of the lines

$$\begin{array}{ccc} x = 1 + 2t, & & x = 5 + 3s, \\ & \text{and} & \\ y = 2 - 3t, & & y = 3 - 4s \end{array}$$

in the plane.

Solution   We used s as parameter for the second line, because we can't expect the point of intersection to correspond to the same values of the parameters on the two lines. (We will find in a moment that the lines intersect at the point $(-37, 59)$, which corresponds to $t = -19$ in the first line and $s = -14$ in the second. Trying to use the same parameter t for both lines would lead to hopeless algebraic confusion.)

Equating x- and y-coordinates, we have

$$\begin{cases} 1 + 2t = 5 + 3s, \\ 2 - 3t = 3 - 4s, \end{cases} \quad \text{or} \quad \begin{cases} 2t - 3s = 4, \\ -3t + 4s = 1. \end{cases}$$

Multiplying the first equation by 3 and the second by 2 and adding, we have

$$-s = 14 \quad \text{or} \quad s = -14.$$

If we trust our work, we can simply put $s = -14$ in the second given system of equations, obtaining

$$x = 5 + 3(-14) = -37,$$
$$y = 3 - 4(-14) = 59,$$

giving the point $(-37, 59)$ of intersection. As a check, we find that when $s = -14$, we have $t = -19$, which also yields $(-37, 59)$ when substituted in the equations for the first line.   ∎

EXAMPLE 6   Determine whether the line through $(-1, 2, 0)$ and $(2, 1, 4)$ intersects the line through $(1, 2, 1)$ and $(-14, 5, -14)$.

Solution   We obtain parametric equations for the two lines as in Example 3.

First line	**Second line**
Point $(-1, 2, 0)$	**Point** $(1, 2, 1)$
∥ vector $3\mathbf{i} - \mathbf{j} + 4\mathbf{k}$	**∥ vector** $-15\mathbf{i} + 3\mathbf{j} - 15\mathbf{k}$
	or $-5\mathbf{i} + \mathbf{j} - 5\mathbf{k}$
Equations $x = -1 + 3t,$	**Equations** $x = 1 - 5s,$
$y = 2 - t,$	$y = 2 + s,$
$z = 4t$	$z = 1 - 5s$

Equating coordinates, we have

$$\begin{cases} -1 + 3t = 1 - 5s, \\ 2 - t = 2 + s, \\ 4t = 1 - 5s, \end{cases} \quad \text{or} \quad \begin{cases} 3t + 5s = 2, \\ -t - s = 0, \\ 4t + 5s = 1. \end{cases}$$

We should be able to solve for t and s using just two of these three equations. Subtracting the third equation from the first, we have

$$-t = 1, \quad \text{or} \quad t = -1.$$

Then from the first equation, we obtain $s = 1$. We would not expect these values for t and s to satisfy the second equation, $-t - s = 0$, but of course this example was fixed, so they do. Thus these two lines do intersect, and, setting $t = -1$ in the equations for the first line, we obtain $(-4, 3, -4)$ as the point of intersection.   ∎

EXAMPLE 7   Find the point on the line through $(2, 3, 1)$ and $(-1, 4, 2)$ that is closest to the origin.

Solution We find parametric equations for the line:

Point $(2, 3, 1)$

‖ vector $-3\mathbf{i} + \mathbf{j} + \mathbf{k}$

Equations $x = 2 - 3t,$
$$y = 3 + t,$$
$$z = 1 + t$$

We want to minimize $s = \sqrt{x^2 + y^2 + z^2}$. Equivalently, we minimize $s^2 = x^2 + y^2 + z^2$ to avoid working with the radical. We have

$$s^2 = x^2 + y^2 + z^2 = (2 - 3t)^2 + (3 + t)^2 + (1 + t)^2.$$

Taking a derivative, we have

$$\frac{d(s^2)}{dt} = 2(2 - 3t)(-3) + 2(3 + t) + 2(1 + t) = 22t - 4 = 0,$$

when $t = \frac{2}{11}$. Since the second derivative, 22, is positive, this corresponds to the minimum distance. The point corresponding to $t = \frac{2}{11}$ is $(\frac{16}{11}, \frac{35}{11}, \frac{13}{11})$. ∎

The Angle Between Intersecting Lines

One nice thing about *parametric* equations for a line is that the direction of the line is readily found from the equations. Naturally two lines are *parallel* if they have parallel direction vectors and are *perpendicular* if they meet and have perpendicular direction vectors. Of course, two lines in the plane either are parallel or intersect, but this need not be true in space.

DEFINITION 12.6

Angle

The **angle between intersecting lines** is the angle between their direction vectors.

EXAMPLE 8 Find the angle between the following lines in the plane:

$$\begin{cases} x = 3 + 4t, \\ y = -2 + t, \end{cases} \quad \text{and} \quad \begin{cases} x = -1 - 3t, \\ y = 5 + 12t. \end{cases}$$

Solution Direction vectors for these lines are $\mathbf{d} = 4\mathbf{i} + \mathbf{j}$ and $\mathbf{d}' = -3\mathbf{i} + 12\mathbf{j}$. Now $\mathbf{d} \cdot \mathbf{d}' = -12 + 12 = 0$, so the vectors, and therefore the lines, are perpendicular. ∎

EXAMPLE 9 Find the acute angle between the lines

$$\begin{cases} x = -2 + 2t, \\ y = 3 - 4t, \\ z = -4 + t, \end{cases} \quad \text{and} \quad \begin{cases} x = -2 - t, \\ y = 3 + 2t, \\ z = -4 + 3t, \end{cases}$$

which meet at the point $(-2, 3, -4)$ in space.

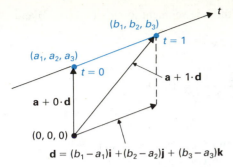

FIGURE 12.59 The line segment from (a_1, a_2, a_3) to (b_1, b_2, b_3).

Solution Direction vectors for these lines are $\mathbf{d} = 2\mathbf{i} - 4\mathbf{j} + \mathbf{k}$ and $\mathbf{d}' = -\mathbf{i} + 2\mathbf{j} + 3\mathbf{k}$. Now $\mathbf{d} \cdot \mathbf{d}' = -2 - 8 + 3 = -7$. We change the direction vector of the first line to $-\mathbf{d} = -2\mathbf{i} + 4\mathbf{j} - \mathbf{k}$ to obtain a positive dot product, corresponding to an acute angle θ. Then

$$\cos \theta = \frac{(-\mathbf{d}) \cdot \mathbf{d}'}{|-\mathbf{d}| \cdot |\mathbf{d}'|}$$

$$= \frac{7}{\sqrt{21}\sqrt{14}} = \frac{1}{\sqrt{6}}.$$

Therefore,

$$\theta = \cos^{-1}\left(\frac{1}{\sqrt{6}}\right) \approx 65.91°.$$

Line Segments

We describe how to find the points on the line segment from (a_1, a_2, a_3) to (b_1, b_2, b_3) in space. (We could also drop one coordinate and work in the plane.) In Fig. 12.59, we consider the plane of the page to be the plane containing the origin and the line through the two points. We take as direction vector for this line

$$\mathbf{d} = (b_1 - a_1)\mathbf{i} + (b_2 - a_2)\mathbf{j} + (b_3 - a_3)\mathbf{k}.$$

Parametric equations for the line are

$$x = a_1 + (b_1 - a_1)t,$$
$$y = a_2 + (b_2 - a_2)t,$$
$$z = a_3 + (b_3 - a_3)t.$$

Our line becomes a t-axis, as shown. From the figure (or equations), we see that $t = 0$ corresponds to (a_1, a_2, a_3) and $t = 1$ to (b_1, b_2, b_3) on the line. Thus we obtain all points on the line segment between these points by taking all values of t where $0 \leq t \leq 1$. Thinking of the line as a t-axis, we see that $t = \frac{1}{2}$ gives the midpoint of the line segment, $t = \frac{1}{3}$ gives the point one-third of the way from (a_1, a_2, a_3) to (b_1, b_2, b_3), and so on. We summarize in a definition.

DEFINITION 12.7

Segments and midpoints

Let (a_1, a_2, a_3) and (b_1, b_2, b_3) be two points in space (or the plane). The **line segment** joining them consists of all points (x, y, z) such that, for some value of t, where $0 \leq t \leq 1$,

$$x = a_1 + (b_1 - a_1)t,$$
$$y = a_2 + (b_2 - a_2)t, \qquad (7)$$
$$z = a_3 + (b_3 - a_3)t.$$

The **midpoint** of the line segment is the point

$$\left(\frac{a_1 + b_1}{2}, \frac{a_2 + b_2}{2}, \frac{a_3 + b_3}{2}\right).$$

EXAMPLE 10 Find the midpoint of the line segment joining $(-1, 3, 2)$ and $(3, 1, -1)$.

Solution The midpoint is

$$\left(\frac{-1 + 3}{2}, \frac{3 + 1}{2}, \frac{2 - 1}{2}\right) = \left(1, 2, \frac{1}{2}\right).$$

EXAMPLE 11 Find the point one-third of the way from $(-1, 3, 2)$ to $(3, 1, -1)$ on the line segment joining the two points.

Solution Equations (7) become

$$x = -1 + 4t,$$
$$y = 3 - 2t,$$
$$z = 2 - 3t.$$

Setting $t = \frac{1}{3}$, we obtain the point $(\frac{1}{3}, \frac{7}{3}, 1)$.

SUMMARY

1. A line in the plane or space is determined by a point on the line and a vector **d** parallel to the line.

2. If a line goes through a point (a_1, a_2, a_3) and has direction given by the nonzero vector

$$\mathbf{d} = d_1\mathbf{i} + d_2\mathbf{j} + d_3\mathbf{k},$$

then parametric equations for the line are

$$x = a_1 + d_1t,$$
$$y = a_2 + d_2t,$$
$$z = a_3 + d_3t.$$

All values of the parameter t give all points on the line.

3. Parametric equations of the line through (a_1, a_2, a_3) and (b_1, b_2, b_3) are

$$x = a_1 + (b_1 - a_1)t,$$
$$y = a_2 + (b_2 - a_2)t,$$
$$z = a_3 + (b_3 - a_3)t.$$

The first two equations are used for a line in the plane. The points obtained by restricting t to the interval $0 \le t \le 1$ make up the line segment joining (a_1, a_2, a_3) and (b_1, b_2, b_3). The midpoint is obtained when $t = \frac{1}{2}$.

EXERCISES 12.6

In Exercises 1–24, find parametric equations for the indicated line.

1. Through $(3, -2)$ with direction vector $\mathbf{d} = -8\mathbf{i} + 4\mathbf{j}$
2. Through $(-4, -1)$ with direction vector $\mathbf{d} = 2\mathbf{i} + \mathbf{j}$
3. Through $(4, -1, 0)$ with direction vector $\mathbf{d} = \mathbf{i} - 3\mathbf{j} + \mathbf{k}$
4. Through $(0, 0, 0)$ with direction vector $\mathbf{d} = 2\mathbf{i} - 3\mathbf{j} + \mathbf{k}$
5. Through $(1, -1, 1)$ with direction vector $\mathbf{d} = \mathbf{i} + 4\mathbf{k}$
6. Through $(-1, 4, 0)$ with direction vector $\mathbf{d} = \mathbf{j}$
7. Through $(1, -4)$ parallel to the line $x = 3 - 2t$, $y = 4 + 5t$
8. Through $(2, -3)$ perpendicular to the line $x = 4 - t$, $y = 3 + 2t$
9. Through $(3, -1, 4)$ parallel to the line $x = 3 - 2t$, $y = 4 + 3t$, $z = 6 - 8t$
10. Through $(4, -1, 2)$ parallel to the line $x = -2t$, $y = 5$, $z = 3 + 4t$
11. The line $x - 2y = 5$ in the plane
12. The line $7x - 3y = 2$ in the plane
13. Through $(-1, 5)$ and perpendicular to $3x + y = 7$ in the plane
14. Through $(-2, -3)$ and parallel to $4x - 5y = 8$ in the plane
15. Through $(-1, 4)$ and $(2, 6)$
16. Through $(3, -5)$ and $(6, -2)$
17. Through $(-1, 4, 0)$ and $(8, 1, -4)$
18. Through $(8, -1, 2)$ and $(0, 0, 0)$
19. Through $(4, -1, 3)$ and $(-1, 4, 6)$
20. Through $(2, -1, 4)$ and $(-1, 1, -1)$
21. Through $(2, 1, 4)$ and perpendicular to the x,z-coordinate plane
22. Through $(-1, 4, 6)$ and perpendicular to the y,z-coordinate plane
23. Through $(-1, 2, 3)$ perpendicular to both the lines $x = -1 + 3t$, $y = 2$, $z = 3 - t$ and $x = -1 - t$, $y = 2 + 3t$, $z = 3 + t$. [*Hint:* Use a cross product.]
24. Through $(2, -1, 4)$ perpendicular to the lines $x = 2 + 3t$, $y = -1 - 2t$, $z = 4 + t$ and $x = 2 - t$, $y = -1 + t$, $z = 4 - 5t$. [*Hint:* As in Exercise 23.]
25. Find the point on the line through $(-1, 5)$ and $(2, 6)$ having x-coordinate -7.
26. Find the point on the line through $(1, -2)$ and $(7, -5)$ having y-coordinate 7.
27. Find the point on the line through $(-1, 4, 6)$ and $(2, 1, 3)$ having y-coordinate -2.
28. Find the point on the line through $(2, -1, 2)$ and $(0, 1, 5)$ having z-coordinate 7.

In Exercises 29–34, find all points of intersection of the given pair of lines, if there are any.

29. $x = 5 + 3t$, $y = -6 - 5t$ and $x = -5 + 4s$, $y = 5 - s$ in the plane
30. $x = 6 - 2t$, $y = -4 + 4t$ and $x = -3 + 3s$, $y = 7 - 6s$ in the plane
31. $x = -3 + 3t$, $y = -11 + 9t$ and $x = 12 - 6s$, $y = -6 + 2s$ in the plane
32. $x = 4 + t$, $y = 2 - 3t$, $z = -3 + 5t$ and $x = 11 + 3s$, $y = -9 - 4s$, $z = -4 - 3s$ in space
33. $x = 11 + 3t$, $y = -3 - t$, $z = 4 + 3t$ and $x = 6 - 2s$, $y = -2 + s$, $z = -15 + 7s$ in space
34. $x = -2 + t$, $y = 3 - 2t$, $z = 1 + 5t$ and $x = 7 - 3s$, $y = 1 + 2s$, $z = 4 - s$ in space

In Exercises 35–38, find the acute angle between the pairs of intersecting lines.

35. $x = 3 - 4t$, $y = 2 + 3t$ and $x = 5 - t$, $y = 7 + 2t$ in the plane
36. $x = 2 - 3t$, $y = 4 + t$ and $x = 3 - t$, $y = 2 + 6t$ in the plane
37. $x = 2 - 4t$, $y = -1 - t$, $z = 3 + 5t$ and $x = 2 - t$, $y = -1 + 2t$, $z = 3$ in space
38. $x = -2t$, $y = 3 - 2t$, $z = 1$ and $x = -3t$, $y = 3 + 4t$, $z = 1 - 3t$ in space
39. Find the acute angle that the line $x = t$, $y = t$, $z = t$ in space makes with the coordinate axes.
40. Let a line in space pass through the origin and make angles of α, β, and γ with the three positive coordinate axes. Show that $\cos^2\alpha + \cos^2\beta + \cos^2\gamma = 1$.
41. Find the cosines of the angles that the direction vector of the line

$$x = a_1 + d_1 t,$$
$$y = a_2 + d_2 t,$$
$$z = a_3 + d_3 t$$

makes with the vectors $\mathbf{i}$, $\mathbf{j}$, and $\mathbf{k}$ along the positive coordinate axes. These cosines are called the *direction cosines* of the line. They are usually denoted by $\cos \alpha$, $\cos \beta$, and $\cos \gamma$, respectively. (See Exercise 40.)

42. Find the point on the line $x = 3 + t$, $y = 4 - 2t$ in the plane closest to the origin.
43. Find the point on the line $x = 5 - t$, $y = 3 - 4t$ in the plane closest to the point $(-1, 4)$.
44. Find the point on the line $x = t$, $y = 4 - 3t$, $z = 5 + 2t$ in space closest to the origin.
45. Find the point on the line $x = 5 + t$, $y = 2 - 3t$, $z = 1 + t$ in space closest to the point $(-1, 4, 1)$.
46. Find the midpoint of the line segment joining $(3, -1, 6)$ and $(0, -3, -1)$.
47. Find the midpoint of the line segment joining $(2, -1, 4)$ and $(3, 7, -2)$.

48. Find the point in space one-fourth of the way from $(-2, 1, 3)$ to $(0, -5, 2)$ on the line segment joining them.

49. Find the point in space two-thirds of the distance from $(3, -1, 2)$ to $(2, 5, 8)$ on the line segment joining them.

50. Find the point in the plane on the line segment joining $(-1, 3)$ and $(2, 5)$ that is twice as close to $(-1, 3)$ as to $(2, 5)$.

51. Let (a_1, a_2, a_3) and (b_1, b_2, b_3) be any two points in space. Show that the line segment joining the points consist of all points (x_1, x_2, x_3), where $x_i = (1 - t)a_i + tb_i$ for $0 \le t \le 1$, $i = 1, 2, 3$.

52. Find the point in space on the line through $(0, 2, 1)$ and $(2, 1, 3)$ that is 5 units beyond $(2, 1, 3)$ in the direction of increasing x. [*Hint:* Think of making the line into a *t*-axis with $(2, 1, 3)$ the origin, as in Fig. 12.58, but use a *unit* direction vector **d**.]

53. Find the point on the line through $(-1, 3, 2)$ and $(4, -1, 0)$ that is $\sqrt{5}$ units from $(-1, 3, 2)$ in the direction of decreasing y. [*Hint:* As in Exercise 52.]

54. Find the point on the line $x = -1 + 2t$, $y = 4 + 5t$, $z = -3 - t$ that is 5 units from the point $(-5, -6, -1)$ in the direction of increasing z. [*Hint:* As in Exercise 52.]

55. Find the point on the line $x = 2 + 3t$, $y = 4 - t$, $z = 6$ that is 4 units from $(-1, 5, 6)$ in the direction of decreasing x. [*Hint:* As in Exercise 52.]

12.7 PLANES

The Equation of a Plane

Let (a_1, a_2, a_3) be a point in space and, for the moment, use subscripted variable notation. The line with parametric equations

$$x_1 = a_1 + d_1 t, \qquad x_2 = a_2 + d_2 t, \qquad x_3 = a_3 + d_3 t$$

can be characterized as the set of all points (x_1, x_2, x_3) such that the vector

$$\mathbf{x} - \mathbf{a} = (x_1 - a_1)\mathbf{i} + (x_2 - a_2)\mathbf{j} + (x_3 - a_3)\mathbf{k}$$

is *parallel* to the direction vector $\mathbf{d} = d_1\mathbf{i} + d_2\mathbf{j} + d_3\mathbf{k}$ of the line. See Fig. 12.60. Consider now the set of all (x_1, x_2, x_3) such that the vector

$$\mathbf{x} - \mathbf{a} = (x_1 - a_1)\mathbf{i} + (x_2 - a_2)\mathbf{j} + (x_3 - a_3)\mathbf{k}$$

is *perpendicular* to a vector $\mathbf{d} = d_1\mathbf{i} + d_2\mathbf{j} + d_3\mathbf{k}$, where not all d_i are zero. These two vectors are perpendicular if and only if

$$d_1(x_1 - a_1) + d_2(x_2 - a_2) + d_3(x_3 - a_3) = 0. \tag{1}$$

We try to see whether this concept gives some familiar geometric configuration in space and start by looking at the similarly defined set with equation $d_1(x_1 - a_1) + d_2(x_2 - a_2) = 0$ in the plane. From Fig. 12.61, we see that all such points (x_1, x_2) in the plane lie on a line through (a_1, a_2) that is perpendicular to **d**. From Fig. 12.62, we see that all such points (x_1, x_2, x_3) in space lie in a plane through (a_1, a_2, a_3). This suggests the following definition.

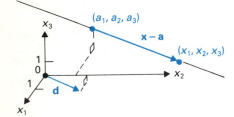

FIGURE 12.60 Points (x_1, x_2, x_3), where $(\mathbf{x} - \mathbf{a}) \parallel \mathbf{d}$ lie on a line.

DEFINITION 12.8

Plane in space

Let (a_1, a_2, a_3) be a point in space and $\mathbf{d} = d_1\mathbf{i} + d_2\mathbf{j} + d_3\mathbf{k}$ be a nonzero vector. The **plane** through (a_1, a_2, a_3) with normal vector **d** is the set of all points (x, y, z) satisfying

$$d_1(x - a_1) + d_2(y - a_2) + d_3(z - a_3) = 0. \tag{2}$$

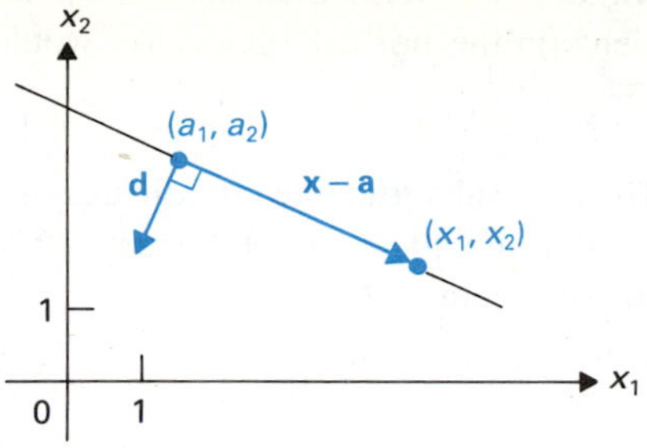

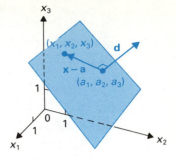

FIGURE 12.61 Points (x_1, x_2) in the plane where $(\mathbf{x} - \mathbf{a}) \perp \mathbf{d}$ lie on a line.

FIGURE 12.62 Points (x_1, x_2, x_3) in space where $(\mathbf{x} - \mathbf{a}) \perp \mathbf{d}$ lie on a plane.

We emphasize that the vector giving "direction" for the plane in Fig. 12.62 is *perpendicular* to the plane. We used the symbol $\parallel$ for "parallel," and we will use the symbol $\perp$ for "perpendicular." The terms "normal" and "orthogonal" are also used for perpendicular. You should get accustomed to all three terms.

A step sequence for finding an equation of a plane is

Point

$\perp$ vector　　　　　　　　　　　　　　(3)

Equation

EXAMPLE 1 Find an equation of the plane in space that passes through $(-1, 2, 1)$ and has the normal vector $\mathbf{d} = \mathbf{i} - 3\mathbf{j} + 2\mathbf{k}$.

Solution Following the step sequence (3), we have

Point $(-1, 2, 1)$

$\perp$ vector $\mathbf{i} - 3\mathbf{j} + 2\mathbf{k}$

Equation $1[x - (-1)] + (-3)(y - 2) + 2(z - 1) = 0$

This equation may be written in the form

$$x - 3y + 2z = -5.$$ ■

Of course, if $\mathbf{d}$ is a normal vector to a plane, then $r\mathbf{d}$ is also a normal vector to the same plane for all nonzero r.

Just as in Example 1, Eq. (2) can always be rewritten in the form

$$d_1 x + d_2 y + d_3 z = c,$$

where $c = d_1 a_1 + d_2 a_2 + d_3 a_3$ is a real number. Referring back to Example 1, we would normally write our final answer

$$x - 3y + 2z = -5$$

without the intermediate step. Since the coefficients of x, y, and z are the coefficients in the normal vector, we would at once write

$$x - 3y + 2z = \underline{\hspace{1cm}}. \tag{4}$$

Then we substitute the coordinates of the point $(-1, 2, 1)$ in Eq. (4) to find the constant for the right-hand side. Substitution yields $(-1) - 3(2) + 2(1) = -5$, so we obtain

$$x - 3y + 2z = -5. \tag{5}$$

We have no trouble writing this answer (5) down at once, doing all the steps in our head.

Now we will show that every linear equation of the form $d_1 x + d_2 y + d_3 z = c$, where not all d_i are zero, indeed has as solution set some plane in space. We may suppose that $d_1 \neq 0$. We choose any numbers a_2 and a_3 and let

$$a_1 = \frac{c - d_2 a_2 - d_3 a_3}{d_1}.$$

Then $d_1 a_1 + d_2 a_2 + d_3 a_3 = c$. Now for any point (x, y, z) such that

$$d_1 x + d_2 y + d_3 z = c,$$

we have the two equations

$$d_1 x + d_2 y + d_3 z = c,$$
$$d_1 a_1 + d_2 a_2 + d_3 a_3 = c.$$

Subtracting, we obtain

$$d_1(x - a_1) + d_2(y - a_2) + d_3(z - a_3) = 0.$$

Thus (x, y, z) lies on the plane through (a_1, a_2, a_3) with normal vector $\mathbf{d} = d_1 \mathbf{i} + d_2 \mathbf{j} + d_3 \mathbf{k}$. Dropping one coordinate, we see that we can view the solution set in the x,y-plane of a linear equation in x and y similarly.

EXAMPLE 2   Find vectors normal and parallel to the line $x - 3y = 4$ in the plane.

Solution   A normal vector is $\mathbf{d} = \mathbf{i} - 3\mathbf{j}$, having as coefficients of $\mathbf{i}$ and $\mathbf{j}$ the coefficients of x and y, respectively. A vector parallel to the line is a vector orthogonal to $\mathbf{d}$. We can quickly find a vector orthogonal to one in the plane by interchanging the coefficients of $\mathbf{i}$ and $\mathbf{j}$ and then changing the sign of one of them. (Check that the dot product of the vectors is then zero.) Thus $3\mathbf{i} + \mathbf{j}$ is orthogonal to $\mathbf{d} = \mathbf{i} - 3\mathbf{j}$, so $3\mathbf{i} + \mathbf{j}$ is parallel to the line. ■

EXAMPLE 3   Describe the plane $z = 0$ in light of the preceding discussion.

Solution   The plane $z = 0$ contains the origin. It has $\mathbf{k}$ as a normal vector. Thus it must be the x,y-coordinate plane. ■

We can check that the following definitions coincide with our intuitive ideas of parallel and perpendicular.

Parallel and perpendicular planes

Two planes in space are **parallel** if normal vectors for the two planes are parallel to each other. The planes are **perpendicular** if their normal vectors are orthogonal to each other. The **angle between** two planes is the angle between their normal vectors. A line and a plane are **parallel** if a parallel vector for the line is orthogonal to a normal vector for the plane. A line and a plane are **perpendicular** if a parallel vector for the line is also a normal vector for the plane.

Computations

The following examples illustrate a few of the many types of problems we can now solve. Keep these requirements in mind.

Data for Lines and Planes

(a) To find parametric equations for a line, we need to know a point on the line and a parallel vector to the line.

(b) To find an equation of a plane, we need to know a point in the plane and a normal vector to the plane.

EXAMPLE 4   Find parametric equations for the line in space passing through the point $(1, 2, -1)$ and perpendicular to the plane with equation

$$3x + 5y - z = 6.$$

Solution   We know a point $(1, 2, -1)$ on the line, and we need a direction vector for the line. Since the line is to be perpendicular to the plane given by $3x + 5y - z = 6$, we see that $\mathbf{d} = 3\mathbf{i} + 5\mathbf{j} - \mathbf{k}$ is a direction vector for the line. Following the step sequence for a line, we have

> **Point** $(1, 2, -1)$
> $\parallel$ **vector** $3\mathbf{i} + 5\mathbf{j} - \mathbf{k}$
> **Equations** $x = 1 + 3t,$
> $\qquad\qquad\quad y = 2 + 5t,$
> $\qquad\qquad\quad z = -1 - t$ ∎

EXAMPLE 5   Find the equation of the plane through $(-1, 5, 2)$ and parallel to the plane $x - 3y + 5z = 8$.

Solution   By Definition 12.9, parallel planes have their normal vectors parallel to each other. Thus we can take the coefficients of x, y, and z in the given equation as coefficients of **i**, **j**, and **k** for the desired normal vector. We then have

> **Point** $(-1, 5, 2)$
> ∥ **vector** $\mathbf{i} - 3\mathbf{j} + 5\mathbf{k}$
> **Equation** $x - 3y + 5z = -6$ ∎

Example 5 makes it clear that as c varies, the equation $x - 3y + 5z = c$ gives parallel planes. We can think of the planes as "sliding along" perpendicular to the line along the vector $\mathbf{d} = \mathbf{i} - 3\mathbf{j} + 5\mathbf{k}$ as c varies.

EXAMPLE 6 Find all points of intersection in space of the plane with equation

$$3x + 5y - z = -2$$

and the line with parametric equations

$$x = -3 + 2t, \qquad y = 4 + t, \qquad z = -1 - 3t.$$

Solution If (x, y, z) lies on both the line and the plane, then, substituting, we must have

$$3(-3 + 2t) + 5(4 + t) - (-1 - 3t) = -2,$$

so

$$14t = -14 \qquad \text{and} \qquad t = -1.$$

Thus the only point of intersection is $(-5, 3, 2)$. ∎

EXAMPLE 7 Find the intersection of the plane

$$x - y + 2z = 10$$

with the line

$$x = 3 + 2t, \qquad y = 4 + 8t, \qquad \text{and} \qquad z = -1 + 3t.$$

Solution Proceeding as in Example 6, we obtain

$$(3 + 2t) - (4 + 8t) + 2(-1 + 3t) = 10,$$
$$0 \cdot t = 13.$$

This equation has no solutions, so the line is parallel to the plane but not in the plane. ∎

If the plane in Example 7 instead has equation $x - y + 2z = -3$, we would obtain the equation

$$0 \cdot t = 0,$$

which is satisfied for all values of t. In this case, the line lies in the plane $x - y + 2z = -3$.

EXAMPLE 8 Find the acute angle θ between the planes $x - 3y + z = 4$ and $3x + 2y + 4z = 6$.

Solution Normal vectors for the planes are

$$\mathbf{d} = \mathbf{i} - 3\mathbf{j} + \mathbf{k} \qquad \text{and} \qquad \mathbf{d}' = 3\mathbf{i} + 2\mathbf{j} + 4\mathbf{k}.$$

Now $\mathbf{d} \cdot \mathbf{d}' = 3 - 6 + 4 = 1$, so

$$\cos\theta = \frac{\mathbf{d} \cdot \mathbf{d}'}{|\mathbf{d}| \cdot |\mathbf{d}'|}$$
$$= \frac{1}{\sqrt{11}\sqrt{29}}.$$

Therefore

$$\theta = \cos^{-1}\left(\frac{1}{\sqrt{11}\sqrt{29}}\right)$$
$$\approx 86.79°. \qquad \blacksquare$$

EXAMPLE 9 Find the equation of the plane in space that contains the points $(1, -1, 1)$, $(2, 3, -4)$, and $(0, 1, -2)$.

Solution To find the equation of a plane, we want to know a vector normal to the plane. Taking $(1, -1, 1)$ as new origin, we see that the vectors $\langle 2, 3, -4 \rangle - \langle 1, -1, 1 \rangle$ from $(1, -1, 1)$ to $(2, 3, -4)$ and $\langle 0, 1, -2 \rangle - \langle 1, -1, 1 \rangle$ from $(1, -1, 1)$ to $(0, 1, -2)$ lie in the plane. Thus, starting at $(1, -1, 1)$,

$$\mathbf{i} + 4\mathbf{j} - 5\mathbf{k} \qquad \text{and} \qquad -\mathbf{i} + 2\mathbf{j} - 3\mathbf{k}$$

are vectors in the desired plane. A normal vector to the plane must be perpendicular to both of these vectors; their cross product will be such a vector. The symbolic determinant

$$\begin{vmatrix} \mathbf{i} & \mathbf{j} & \mathbf{k} \\ 1 & 4 & -5 \\ -1 & 2 & -3 \end{vmatrix} = -2\mathbf{i} + 8\mathbf{j} + 6\mathbf{k}$$

shows that $\mathbf{i} - 4\mathbf{j} - 3\mathbf{k}$ is a normal vector to the plane. Following the step sequence (3), we have

Point $(1, -1, 1)$
⊥ vector $\mathbf{i} - 4\mathbf{j} - 3\mathbf{k}$
Equation $x - 4y - 3z = 2$ $\qquad \blacksquare$

EXAMPLE 10 Find parametric equations for the line of intersection of the planes $x - 2y + z = 4$ and $2x + 3y - z = 7$.

Solution We show two intersection planes (not the given ones) in Fig. 12.63. The figure indicates that a parallel vector to the line of intersection is orthogonal to both the normal vectors to the two planes. Such a vector can be found by

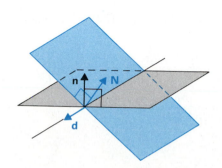

FIGURE 12.63 The vector $\mathbf{d}$ along the line of intersection is perpendicular to the normals $\mathbf{n}$ and $\mathbf{N}$ to the planes.

taking the cross product of the normal vectors to the planes. We obtain

$$\begin{vmatrix} \mathbf{i} & \mathbf{j} & \mathbf{k} \\ 1 & -2 & 1 \\ 2 & 3 & -1 \end{vmatrix} = -\mathbf{i} + 3\mathbf{j} + 7\mathbf{k}.$$

We still have to find a point on the line. We need a point in both planes. Adding the equations for the planes, we have

$$\begin{aligned} x - 2y + z &= 4 \\ \underline{2x + 3y - z} &= \underline{7} \\ 3x + y &= 11. \end{aligned}$$

We can take $x = 3$ and $y = 2$ to satisfy this equation. Substituting back into either of the original equations, we find that $z = 5$. Therefore we obtain as answer

Point $(3, 2, 5)$

‖ **vector** $-\mathbf{i} + 3\mathbf{j} + 7\mathbf{k}$

Equations $x = 3 - t$,

$y = 2 + 3t$,

$z = 5 + 7t$ ∎

The Distance from a Point to a Plane

Let $d_1 x + d_2 y + d_3 z = c$ be the equation of a plane in space, and let (a_1, a_2, a_3) be a point in space. We would like to find the distance from (a_1, a_2, a_3) to the plane. This distance is measured along the line through (a_1, a_2, a_3) and perpendicular to the plane, since this gives the shortest distance.

Let (b_1, b_2, b_3) be any point on the plane, so that

$$d_1 b_1 + d_2 b_2 + d_3 b_3 = c. \tag{6}$$

From Fig. 12.64, we see that the distance s we are looking for is the magnitude of the scalar component of the vector

$$\mathbf{v} = (a_1 - b_1)\mathbf{i} + (a_2 - b_2)\mathbf{j} + (a_3 - b_3)\mathbf{k}$$

from (b_1, b_2, b_3) to (a_1, a_2, a_3) along $\mathbf{d} = d_1 \mathbf{i} + d_2 \mathbf{j} + d_3 \mathbf{k}$. The scalar component of $\mathbf{v}$ along $\mathbf{d}$ is

$$\begin{aligned} \frac{\mathbf{d} \cdot \mathbf{v}}{|\mathbf{d}|} &= \frac{d_1(a_1 - b_1) + d_2(a_2 - b_2) + d_3(a_3 - b_3)}{\sqrt{d_1^2 + d_2^2 + d_3^2}} \\ &= \frac{d_1 a_1 + d_2 a_2 + d_3 a_3 - (d_1 b_1 + d_2 b_2 + d_3 b_3)}{\sqrt{d_1^2 + d_2^2 + d_3^2}} \\ &= \frac{d_1 a_1 + d_2 a_2 + d_3 a_3 - c}{\sqrt{d_1^2 + d_2^2 + d_3^2}}, \end{aligned}$$

where we used Eq. (6) to obtain the last expression. Thus the distance s from (a_1, a_2, a_3) to $d_1 x + d_2 y + d_3 z = c$ is given by

$$s = \frac{|d_1 a_1 + d_2 a_2 + d_3 a_3 - c|}{\sqrt{d_1^2 + d_2^2 + d_3^2}} \tag{7}$$

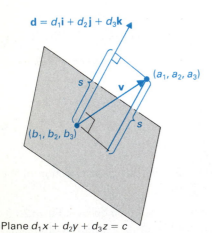

$\mathbf{d} = d_1\mathbf{i} + d_2\mathbf{j} + d_3\mathbf{k}$

(a_1, a_2, a_3)

(b_1, b_2, b_3)

Plane $d_1 x + d_2 y + d_3 z = c$

FIGURE 12.64 The distance s from (a_1, a_2, a_3) to the plane is the scalar projection of $\mathbf{v}$ along $\mathbf{d}$.

Of course, the same derivation using two components gives the formula

$$\frac{|d_1 a_1 + d_2 a_2 - c|}{\sqrt{d_1^2 + d_2^2}}$$

as the distance in the x,y-plane from (a_1, a_2) to the line $d_1 x + d_2 y = c$.

EXAMPLE 11   Find the distance from the point $(-9, 6, 2)$ to the plane with equation $5x - 2y + z = 2$.

Solution   The preceding discussion shows that this distance is

$$\frac{|-45 - 12 + 2 - 2|}{\sqrt{25 + 4 + 1}} = \frac{|-57|}{\sqrt{30}} = \frac{57}{\sqrt{30}}. \qquad \blacksquare$$

EXAMPLE 12   Find the distance from $(-1, 4)$ to $x - 3y = 8$ in the plane.

Solution   We obtain from Eq. (7) the distance

$$\frac{|-1 - 12 - 8|}{\sqrt{1 + 9}} = \frac{|-21|}{\sqrt{10}} = \frac{21}{\sqrt{10}}. \qquad \blacksquare$$

SUMMARY

1. A plane is determined by a point in the plane and a normal (perpendicular) vector.

2. The plane through (a_1, a_2, a_3) and having normal vector $d_1 \mathbf{i} + d_2 \mathbf{j} + d_3 \mathbf{k}$ has equation

$$d_1(x - a_1) + d_2(y - a_2) + d_3(z - a_3) = 0.$$

3. The equation $d_1 x + d_2 y + d_3 z = c$, where not all d_i are zero, has as solution set in space a plane with $d_1 \mathbf{i} + d_2 \mathbf{j} + d_3 \mathbf{k}$ as normal vector.

4. The line $d_1 x + d_2 y = c$ in the plane has $d_1 \mathbf{i} + d_2 \mathbf{j}$ as normal vector.

5. Planes are parallel if their normal vectors are parallel and perpendicular if their normal vectors are perpendicular. The angle between two planes is the angle between their normal vectors.

6. The distance from a point (a_1, a_2, a_3) to the plane $d_1 x + d_2 y + d_3 z = c$ is

$$\frac{|d_1 a_1 + d_2 a_2 + d_3 a_3 - c|}{\sqrt{d_1^2 + d_2^2 + d_3^2}}$$

Similarly, the distance from (a_1, a_2) to the line $d_1 x + d_2 y = c$ in the plane is

$$\frac{|d_1 a_1 + d_2 a_2 - c|}{\sqrt{d_1^2 + d_2^2}}.$$

EXERCISES 12.7

In Exercises 1–29, find the equation of the indicated plane in space.

1. Through $(-1, 4, 2)$ with normal vector $\mathbf{i} - 2\mathbf{j} + \mathbf{k}$
2. Through $(1, -3, 0)$ with normal vector $\mathbf{j} + 4\mathbf{k}$
3. Through $(2, -1, 3)$ with normal vector $\mathbf{i}$
4. Through $(1, 2, -5)$ with normal vector $\mathbf{i} + 2\mathbf{j} - 4\mathbf{k}$
5. Through $(3, -4, 2)$ with normal vector $4\mathbf{i} - 2\mathbf{j} + \mathbf{k}$
6. Through $(3, -1, 4)$ and parallel to $2x - 4y + 5z = 7$
7. Through $(1, 1, -1)$ and parallel to $4x - 3y + 4z = 10$
8. Through $(5, -2, 1)$ and parallel to $-3x + 4y - z = 13$
9. Through $(-1, 4, -3)$ and orthogonal to the line $x = 3 - 7t$, $y = 4 + t$, $z = 2t$
10. Through $(2, -2, 4)$ and orthogonal to the line $x = t$, $y = 4 - t$, $z = 5$
11. Through $(-2, 1, 5)$ and orthogonal to the line $x = 3 - 2t$, $y = 3t$, $z = 4 - t$
12. Through the origin and orthogonal to the line through $(-1, 3, 0)$ and $(2, -4, 3)$
13. Through $(1, -1, 1)$ and orthogonal to the line through $(4, -1, 2)$ and $(3, 1, -5)$
14. Through $(0, 3, -1)$ and orthogonal to the line through $(2, -1, 7)$ and $(4, 2, -6)$
15. Through the points $(1, 0, 0)$, $(0, 1, 0)$, and $(0, 0, 1)$
16. Through the points $(1, 0, 1)$, $(-1, 2, 0)$, and $(0, 1, 3)$
17. Through the points $(2, 1, 5)$, $(-1, -2, 1)$, and $(3, -1, 4)$
18. Through the points $(-1, 4, 2)$, $(1, 2, -1)$, and $(3, -2, 0)$
19. Through the points $(2, -3, 1)$, $(5, -2, 7)$, and $(-1, 1, 0)$
20. Through the points $(3, -2, 3)$, $(1, 1, 2)$, and $(-1, -5, -2)$
21. Containing the lines $x = -1 + 3t$, $y = 4 + t$, $z = 2 - 2t$ and $x = -1 + s$, $y = 4$, $z = 2 + 7s$
22. Containing the lines $x = 4 - t$, $y = 3 + 7t$, $z = 1 + 2t$ and $x = 4 + 2s$, $y = 3 + s$, $z = 1 - 3s$
23. Containing the lines $x = -1 + t$, $y = 3 - 2t$, $z = 2$ and $x = -1 - 3s$, $y = 3 + 5s$, $z = 2 + 4s$
24. Containing the line $x = 3 + 2t$, $y = 1 - t$, $z = 4 - 2t$ and the point $(3, -1, 2)$
25. Containing the line $x = 2 - t$, $y = 4 + 5t$, $z = 3 - 7t$ and the point $(4, -1, 3)$
26. Containing the line $x = 3t$, $y = 5$, $z = 4 + t$ and the point $(-2, 1, 4)$
27. Containing $(-1, 4, 3)$ and the line of intersection of the planes $2x - 3y + 4z = 7$ and $x - 2z = 5$
28. Containing $(2, -1, 5)$ and the line of intersection of the planes $3x - 4y + z = 5$ and $2x + y - 2z = 4$
29. Containing $(-5, 3, 8)$ and the line of intersection of the planes $x + z = 4$ and $y = 2$

In Exercises 30–36, find the acute angle between the given lines or planes.

30. $x - 3y = 7$ and $2x + 4y = 1$ in the plane
31. $3x + 2y = -7$ and $6x + 4y = 2$ in the plane
32. $3x + 4y - z = 1$ and $x - 2y = 3$ in space
33. $4x - 7y + z = 3$ and $3x + 2y + 2z = 17$ in space
34. $x = 4$ and $x + z = 2$ in space
35. $x - 2y + 2z = 3$ and $x = 2 - 6t$, $y = 4 + 3t$, $z = 1 - 2t$ in space
36. $3x - z = 4$ and $x = -t$, $y = 2 + 2t$, $z = 3 - t$ in space

In Exercises 37–40, find all points of intersection of the given line and plane in space.

37. $x = 5 + t$, $y = -3t$, $z = -2 + 4t$ and $x - 3y + 2z = -35$
38. $x = 2$, $y = 4 - 3t$, $z = 3t$ and $4x - 2y + 3z = 30$
39. $x = 3 - t$, $y = 4 - 5t$, $z = 6 - 3t$ and $2x - 4y + 6z = 5$
40. $x = 3t$, $y = 4 - t$, $z = 5 - 5t$ and $6x - 2y + 4z = 12$

In Exercises 41–46, find parametric equations of the indicated line.

41. Through $(-2, 1, 0)$ and orthogonal to the plane $x - 2y + 4z = 3$
42. Through $(1, 4, -3)$ and orthogonal to the plane $2x + y - 7z = 1$
43. Through $(1, -1, 3)$ and orthogonal to the plane $4x - 3y + z = 11$
44. The line of intersection of $x - 3y + z = 7$ and $3x + 2y + z = -1$
45. The line of intersection of $4x - 2y + z = 3$ and $-3x + 2y + z = 4$
46. The line of intersection of $x + y - 2z = 7$ and $2x + y - z = -2$

In Exercises 47–52, find the distance from the given point to the given line or plane.

47. $(-1, 3)$ to $3x - 4y = 4$ 48. $(2, 4)$ to $x - 3y = 4$
49. $(2, -1)$ to the line with parametric equations $x = 3 - t$, $y = 2 + 4t$ [*Hint:* Obtain the x,y-equation for the line by eliminating t from the parametric equations.]
50. $(1, 3, -1)$ to $2x + y + z = 4$
51. $(-1, 4, 3)$ to $x - 2y + 2z = 3$
52. $(4, 1, -2)$ to $3x + 6y - 2z = 4$
53. Show that the equation of the line in the plane through two distinct points (a_1, a_2) and (b_1, b_2) is given by

$$\begin{vmatrix} x & y & 1 \\ a_1 & a_2 & 1 \\ b_1 & b_2 & 1 \end{vmatrix} = 0.$$

[*Hint:* Do not expand the determinant, but argue that a linear equation in x and y is obtained and that the equation is satisfied if $(x, y) = (a_1, a_2)$ or $(x, y) = (b_1, b_2)$.]

54. Use the method suggested by Exercise 53 to find the equation of the line through $(1, -4)$ and $(2, 3)$ in the plane.

55. Let $\mathbf{d} = d_1\mathbf{i} + d_2\mathbf{j} + d_3\mathbf{k}$ be a vector with all coefficients not equal to zero.
(a) Show that the line through (a_1, a_2, a_3) with parallel direction vector $\mathbf{d}$ consists of all (x, y, z) such that
$$\frac{x - a_1}{d_1} = \frac{y - a_2}{d_2} = \frac{z - a_3}{d_3}.$$
[*Hint:* Write parametric equations for the line and eliminate the parameter.]
(b) From part (a), find three planes containing the line. [The equations in part (a) are called *symmetric equations* of the line. The line appears as an intersection of planes. Some texts emphasize these symmetric equations. We consider the parametric equations to be much better, since one or two zero coefficients in $\mathbf{d}$ cause no trouble. Whenever we have tried using symmetric equations, some of our students write the line through $(1, -1, 4)$ with direction vector $4\mathbf{i} - \mathbf{k}$ as
$$\frac{x - 1}{4} = \frac{y + 1}{0} = \frac{z - 4}{-1},$$
which is of course nonsense. A correct symmetric description is
$$\frac{x - 1}{4} = \frac{z - 4}{-1}, \qquad y = -1.]$$

12.8 QUADRIC SURFACES

A **quadric surface** in space is the solution set of a polynomial equation in x, y, and z of degree 2. We have seen that if one of the three variables is missing, the surface is a cylinder, which is fairly easy to sketch. The cylinders $x^2/a^2 + y^2/b^2 = 1$ and $z^2 = 4py$ in Examples 6 and 7 of Section 12.1 are quadric surfaces. We now restrict ourselves to the case where all three variables are present.

As an aid in sketching such surfaces, examine the curves in which they intersect planes parallel to the coordinate planes. Note that $x = x_0$ describes in space a plane parallel to the y,z-coordinate plane. Similarly $y = y_0$ is a plane parallel to the x,z-coordinate plane, and $z = z_0$ is a plane parallel to the x,y-coordinate plane. Each such plane meeting a quadric surface intersects it in an ellipse, hyperbola, or parabola. The coordinate planes themselves, $x = 0$, $y = 0$, and $z = 0$, are especially useful.

As for second-degree plane curves, the device of completing squares and choosing a new origin can often be used to simplify the sketching of a quadric surface. We assume that this now causes no difficulty, and the examples that follow start with eqations where the completion of squares is unnecessary. These examples exhibit some types of quadric surfaces.

EXAMPLE 1 Sketch the *ellipsoid*
$$\frac{x^2}{a^2} + \frac{y^2}{b^2} + \frac{z^2}{c^2} = 1.$$

Solution The surface is sketched in Fig. 12.65. We set $x = 0$, which corresponds to intersecting the surface with the y,z-plane, and obtain the equation
$$\frac{y^2}{b^2} + \frac{z^2}{c^2} = 1,$$
which is an ellipse. This ellipse is the *trace* of the surface in the y,z-plane.

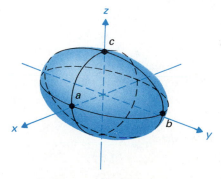

FIGURE 12.65 The ellipsoid
$x^2/a^2 + y^2/b^2 + z^2/c^2 = 1$.

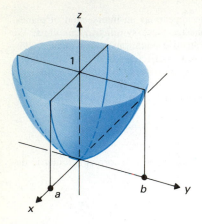

FIGURE 12.66 The elliptic paraboloid $z = x^2/a^2 + y^2/b^2$.

We now set $x = x_0$, which corresponds to intersecting the surface with the plane $x = x_0$, parallel to the y,z-plane. The equation becomes

$$\frac{y^2}{b^2} + \frac{z^2}{c^2} = 1 - \frac{x_0^2}{a^2}.$$

If $|x_0| > a$, the right-hand side of the equation is negative, and the intersection of the surface with the plane $x = x_0$ is empty. If $-a < x_0 < a$, we obtain the equation of an ellipse in the plane $x = x_0$. The largest ellipse occurs when $x_0 = 0$, and the ellipses become smaller as $|x_0| \to a$. When $x_0 = \pm a$, we have the points $(\pm a, 0, 0)$ as the only solution.

Similar results hold for planes $y = y_0$ if $-b < y_0 < b$ and for planes $z = z_0$ if $-c < z_0 < c$. ∎

EXAMPLE 2 Sketch the *elliptic paraboloid* $z = x^2/a^2 + y^2/b^2$.

Solution The sketch is shown in Fig. 12.66. The plane $z = z_0$ does not meet the surface if $z_0 < 0$; it meets the surface in a point if $z_0 = 0$; and it meets the surface in an ellipse if $z_0 > 0$. The surface thus has, as horizontal cross sections, ellipses that expand as we go higher.

If we set $x = x_0$, we obtain the equation

$$z = \frac{x_0^2}{a^2} + \frac{y^2}{b^2},$$

which is linear in z and quadratic in y. This must be the equation of a parabola. In particular, $x_0 = 0$ yields the trace parabola $b^2 z = y^2$ in the y,z-plane. Setting $y = 0$, we obtain the trace parabola $a^2 z = x^2$ in the x,z-plane. ∎

EXAMPLE 3 Sketch the *elliptic cone* $z^2 = x^2/a^2 + y^2/b^2$.

Solution The surface is shown in Fig. 12.67. Setting $z = z_0$ yields elliptical sections; the larger $|z_0|$, the larger the ellipse. Setting $y = 0$, we obtain the lines $z = \pm x/a$ in the x,z-plane, and setting $x = 0$, we have the lines $z = \pm y/b$ in the y,z-plane. Setting x or y equal to nonzero constants gives hyperbolic sections. (Recall that a hyperbola is a *conic section*.) ∎

EXAMPLE 4 Sketch the *hyperboloid of two sheets*

$$\frac{z^2}{c^2} - 1 = \frac{x^2}{a^2} + \frac{y^2}{b^2}.$$

Solution The sketch is shown in Fig. 12.68. Clearly, the plane $z = z_0$ has empty intersection with the surface if $|z_0| < c$. (Exercise 1 deals with sections by planes parallel to the coordinate planes.) ∎

EXAMPLE 5 Sketch the *hyperboloid of one sheet*

$$\frac{z^2}{c^2} + 1 = \frac{x^2}{a^2} + \frac{y^2}{b^2}.$$

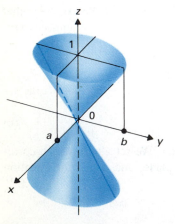

FIGURE 12.67 The elliptic cone $z^2 = x^2/a^2 + y^2/b^2$.

Solution The surface is sketched in Fig. 12.69. Again, we leave the discussion of planar sections as an exercise (see Exercise 2.) ∎

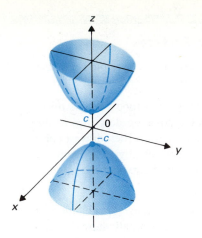

FIGURE 12.68 The hyperboloid of two sheets $z^2/c^2 - 1 = x^2/a^2 + y^2/b^2$.

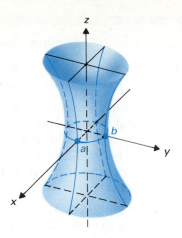

FIGURE 12.69 The hyperboloid of one sheet $z^2/c^2 + 1 = x^2/a^2 + y^2/b^2$.

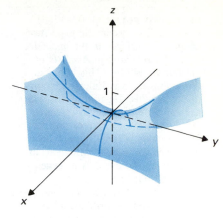

FIGURE 12.70 The hyperbolic paraboloid $z = y^2/b^2 - x^2/a^2$.

EXAMPLE 6 Sketch the *hyperbolic paraboloid* $z = y^2/b^2 - x^2/a^2$.

Solution The sketch is given in Fig. 12.70. This is not an easy surface for a person without artistic ability to sketch. We find that a quadric surface is fairly easy to sketch if the equation can be put in a form where one side consists of a *sum* of two quadratic terms. As the other side assumes constant values, we obtain elliptical sections, and elliptical sections are not hard to sketch and visualize. But this cannot be done with the equation in this example.

The plane $z = z_0$ meets the surface in a hyperbola that "opens" in the y-direction if $z_0 > 0$ and in the x-direction if $z_0 < 0$, while the plane $z = 0$ meets the surface in the degenerate hyperbola consisting of two intersecting lines. A plane $x = x_0$ meets the surface in a parabola "opening upward" while a plane $y = y_0$ meets the surface in a parabola "opening downward." ∎

Sketching Quadric Surface

The steps listed on the next page may be helpful in sketching quadric surfaces.

EXAMPLE 7 Sketch the surface $4x = y^2$ in space.

Solution Since z is missing, this is a cylinder, indeed, a parabolic cylinder since the equation describes a parabola in the x,y-plane. We draw the parabola in the x,y-plane and copies of it above and below in parallel planes (see Fig. 12.71). Then we draw some of the rulings parallel to the axis of the missing variable, that is, the z-axis. ∎

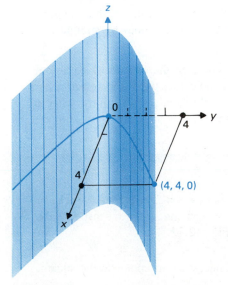

FIGURE 12.71 The parabolic cylinder $4x = y^2$.

EXAMPLE 8 Sketch the surface $x^2 + y + z^2 = 0$.

Solution This is not a cylinder. We rearrange the equation in the form

$$-y = x^2 + z^2.$$

We see at once that for $y > 0$, we have no surface. For $y < 0$, we have circular

HISTORICAL NOTE

LEONHARD EULER (1707–1783) gave the first systematic exposition of quadric surfaces in the appendix to his *Introduction to Analysis of the Infinite* (1748). This book, in two volumes, was Euler's version of precalculus. He followed it with a volume on differential calculus (1755) and three volumes on integral calculus (1768–1770), works that were immensely influential in introducing the calculus to generations of students.

Born in Basel, Switzerland, Euler showed his brilliance early, graduating from the University of Basel when he was fifteen. Though his father preferred that he prepare for the ministry, Euler convinced Johann Bernoulli to tutor him privately in mathematics, whereupon Bernoulli soon persuaded Euler's father to allow his son to become a mathematician. Though Euler was rejected for a position at the University of Basel in 1726, partly because of his youth, he was immediately thereafter, on the recommendation of Bernoulli's son Daniel, invited to Russia to accept the only vacant position, in medicine and physiology, at the newly founded St. Petersburg Academy of Sciences. Within a few years he was appointed to a position in mathematics, which he held until 1741 when he accepted the invitation of Frederick II of Prussia to join the Berlin Academy. Euler returned to Russia at the request of Catherine the Great in 1766 and continued his work there until his death.

Sketching Quadric Surfaces

STEP 1. Complete squares and translate axes if any variable is present in both linear and quadratic terms.

STEP 2. Decide whether the surface is a cylinder (one variable missing). If it is, sketch the curve the equation describes in the coordinate plane of the present variables. Then draw a copy or two of the curve in parallel planes. Finally, draw some of the element lines (rulings) parallel to the axis of the missing variable. See Example 7.

STEP 3. If the surface is not a cylinder, try to arrange the equation so that one side consists of a *sum* of two quadratic terms. If this can be done, then setting the variable on the other side equal to constants yields elliptical or circular (or empty) sections. Visualize how these elliptical sections expand or shrink for different values of the constants. Then set the variables in the sum of the quadratic terms equal to zero individually, to see what the curves look like through the "ends" of the ellipses. See Example 8.

STEP 4. If neither step 2 nor step 3 is the case and there are two quadratic terms, then the surface is a hyperbolic paraboloid. With Fig. 12.70 as a guide, sketch the parabola opening "upward," in the positive direction of the axis of the linear variable. (Of course, the saddle surface is oriented to stand on end if this linear variable is x or y, so "upward" might be "sideways.") Similarly, sketch the "downward"-opening parabola, which meets the first one at right angles at their common vertex. Then draw the two hyperbolas that meet the ends of the two parabolas you have drawn; a total of four curves is involved. Finally, join the ends of the "upper" hyperbola to the ends of the "lower" ones by straight-line segments.

STEP 5. As a general rule, try to visualize the entire surface before sketching it. Dash all axes and curves until you are sure what you can "see" and make unbroken, and determine what is behind another part of the surface and thus must be left dashed.

sections. These circular sections expand as $y \to -\infty$. The planes $x = 0$ and $z = 0$ intersect the surface in the parabolas $y = -z^2$ and $y = -x^2$, respectively. We have the circular paraboloid shown in Fig. 12.72. ∎

We give an example in which it is necessary to complete the square.

EXAMPLE 9 Sketch the surface $-16x^2 + 4y^2 - z^2 - 8y + 4z = 0$.

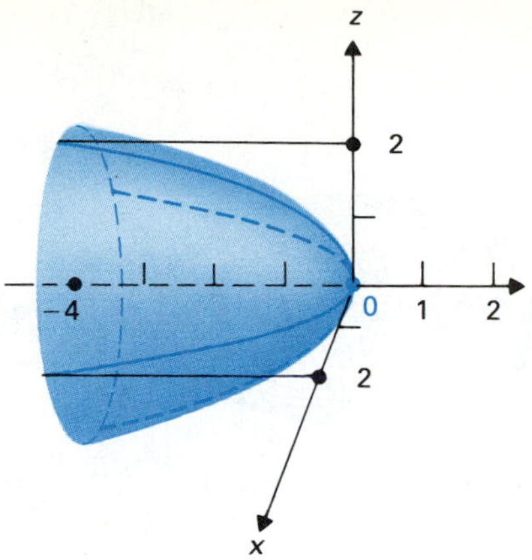

FIGURE 12.72 The circular paraboloid $-y = x^2 + z^2$.

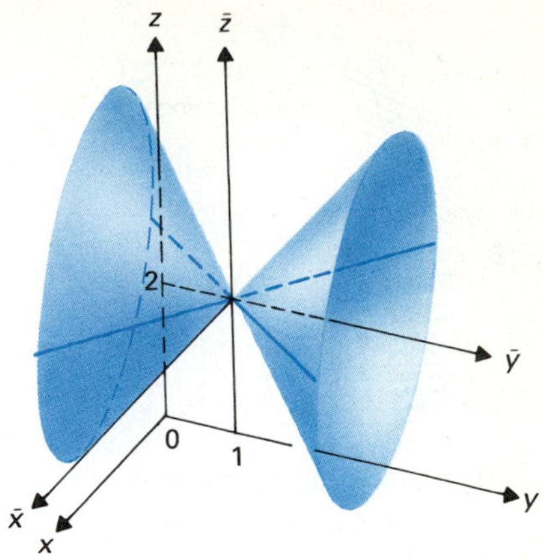

FIGURE 12.73 The elliptic cone $4(y - 1)^2 = 16x^2 + (z - 2)^2$.

Solution We complete squares and arrange our equation as follows:

$$-16x^2 + 4(y^2 - 2y) - (z^2 - 4z) = 0,$$

$$-16x^2 + 4(y - 1)^2 - (z - 2)^2 = 4 - 4 = 0,$$

$$4(y - 1)^2 = 16x^2 + (z - 2)^2,$$

$$4\bar{y}^2 = 16\bar{x}^2 + \bar{z}^2,$$

where $\bar{x} = x$, $\bar{y} = y - 1$, $\bar{z} = z - 2$. In Fig. 12.73, we take $\bar{x},\bar{y},\bar{z}$-axes at $(0, 1, 2)$. Planes $\bar{y} = c$ meet the surface in elliptical sections, which increase in size as $|c|$ increases. Setting $\bar{x} = 0$, we see that the $\bar{y},\bar{z}$-plane meets the surface in the two lines $\bar{z} = \pm 2\bar{y}$. Setting $\bar{y} = 0$, we obtain only the point $(\bar{x},\bar{y},\bar{z}) = (0, 0, 0)$. Setting $\bar{z} = 0$, we obtain the two lines $\bar{y} = \pm 2\bar{x}$ in the $\bar{x},\bar{y}$-plane. The surface is an elliptic cone. ■

It is not easy to draw a picture representing a three-dimensional object on a paper plane. Even though we perceive our environment to be three-dimensional, many people have difficulty visualizing some of the quadric surfaces. The present custom is to use computer graphics to draw three-dimensional objects, again just on a plane surface (see Figs. 12.74–12.78). Fifty years ago, however, every mathematics department had a collection of string and plaster models of quadric surfaces, showing intersections with the coordinate planes. These models, which

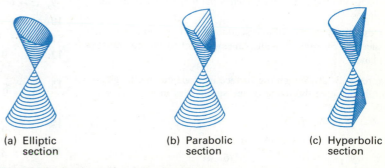

(a) Elliptic section

(b) Parabolic section

(c) Hyperbolic section

FIGURE 12.74 Sections of a right circular cone.

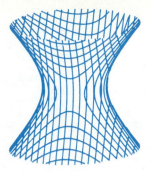

FIGURE 12.75 Parabolic cylinder.

FIGURE 12.76 Ellipsoid.

FIGURE 12.77 Hyperboloid of one sheet.

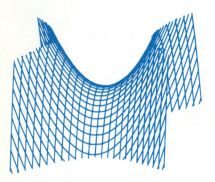

FIGURE 12.78 Hyperbolic paraboloid.

are now practically collectors' items, were taken into the classroom as teaching aids. It is doubtful whether the computer graphics convey more than a good artist's sketch; certainly they are not as informational as the string or plaster models. Fashions change.

SUMMARY

1. To sketch quadric surfaces, first complete squares where possible. If possible, put the equation in a form where one side is a sum of two quadratic terms, yielding elliptical cross sections as the other side assumes constant values. Sketch the trace curves in the coordinate planes $x = 0$, $y = 0$, and $z = 0$. See Figs. 12.65–12.70 for the possible types of surfaces that are not cylinders.

EXERCISES 12.8

1. Describe the curves of intersection of the hyperboloid of two sheets in Example 4 with planes parallel to the coordinate planes.

2. Describe the curves of intersection of the hyperboloid of one sheet in Example 5 with planes parallel to the coordinate planes.

In Exercises 3–20, sketch the surface in space having the given equation and give the name of the surface, as in Figs. 12.65–12.70.

3. $y^2 - x^2 - z^2 = 0$
4. $x^2 + 4y^2 - z^2 - 8 + 16 = 0$
5. $x^2 + y^2 - z = 0$
6. $4x^2 + 9y^2 - 6z = 0$
7. $9x^2 + 36y^2 + 4z^2 - 36 = 0$

8. $x^2 + y^2 + z^2 - 2x + 4y - 6z - 2 = 0$
9. $x^2 - y^2 - 2y - z = 0$
10. $x^2 + y^2 - z^2 + 2x - 8 = 0$
11. $16x^2 + 16y^2 - z^2 - 64y + 73 = 0$
12. $36x - 9y^2 - 16z^2 = 0$
13. $\dfrac{x^2}{4} - \dfrac{y^2}{9} + z^2 + 1 = 0$
14. $\dfrac{x^2}{4} - \dfrac{y^2}{25} - \dfrac{z^2}{9} + 4 = 0$
15. $\dfrac{x^2}{4} + \dfrac{y^2}{9} + z - 3 = 0$
16. $\dfrac{x^2}{4} - \dfrac{y^2}{9} + z - 1 = 0$
17. $2x^2 + 3y^2 + 4z^2 - 24 = 0$
18. $x^2 - 4y^2 + 16z^2 = 0$

19. $\dfrac{x^2}{4} - \dfrac{y^2}{25} - \dfrac{z^2}{9} - 4 = 0$ **20.** $x^2 - 4y^2 + z^2 - 8 = 0$

In Exercises 21–35, give the descriptive name of the quadric surface without sketching it.

21. $x - 4y^2 - 2z^2 + 3 = 0$

22. $x^2 - 3y^2 + 4z^2 - 2x + 6y - 8z + 2 = 0$

23. $x^2 - 3y^2 + 4z^2 - 2x + 6y - 8z + 3 = 0$

24. $x^2 - 3y^2 + 4z^2 - 2x + 6y - 8z + 1 = 0$

25. $x + 4y^2 - 3z^2 - 2y + z - 5 = 0$

26. $x^2 + 2y^2 + 4z^2 - 2x + 2 = 0$

27. $x^2 + 2y^2 + 4z^2 - 2x + 1 = 0$

28. $x^2 + 2y^2 + 4z^2 - 2x = 0$

29. $x^2 - z^2 + 2x - 4z + 3 = 0$

30. $2x^2 - y^2 - z^2 = 0$ **31.** $x^2 + 2z^2 - 2x = 0$

32. $2x^2 - 3y - 5z^2 + 6z - 4 = 0$

33. $3x - 2y^2 + 5y - 3 = 0$

34. $3x^2 - 4y + 3z^2 - 6x + 8 = 0$

35. $2x^2 + 3y^2 - z^2 + 2z = 0$

SUPPLEMENTARY EXERCISES FOR CHAPTER 12

1. Find the distance between $(-1, 2, 4)$ and $(3, 8, -2)$.

2. Find c so that the distance from $(c, -2, 3)$ to $(1, 4, -5)$ is 15.

3. Find all values c such that $(-1, 3, 0)$, $(2, 1, -2)$, and $(1, 1, c)$ are vertices of a right triangle with right angle at the third point.

4. Find the equation of the sphere having the line segment from $(-1, 4, 2)$ to $(-3, -2, 6)$ as a diameter.

5. Find the equation of the sphere having $(1, -3, 5)$ and $(-3, 5, 7)$ as endpoints of a diameter.

6. Find the center and radius of the sphere $x^2 + y^2 + z^2 - 2x + 4z = 4$.

7. Sketch the surface $z = y^2 + 1$.

8. Sketch the surface $x = -\sqrt{y}$ in space.

9. Find cylindrical coordinates of $(x, y, z) = (1, -\sqrt{3}, -2)$.

10. Sketch the solution set of the cylindrical coordinate equation $r = 4$.

11. Find rectangular coordinates of the point $(\rho, \phi, \theta) = (3, 5\pi/6, 3\pi/4)$.

12. Find spherical coordinates of the point $(x, y, z) = (0, 1, -1)$.

13. Find spherical coordinates of the point with cylindrical coordinates $(2, 2\pi/3, -2)$.

14. Find cylindrical coordinates of the point with spherical coordinates $(4, 5\pi/6, 3\pi/4)$.

15. Find the length of the vector $2\mathbf{i} - 3\mathbf{j} + \mathbf{k}$.

16. Find all values of c such that the length of $\mathbf{i} - 2\mathbf{j} + c\mathbf{k}$ is 5.

17. Find a unit vector having direction from $(-1, 3, 2)$ toward $(1, 4, 0)$.

18. Find c_2 so that the vectors $\mathbf{i} - c_2\mathbf{j} + 3\mathbf{k}$ and $4\mathbf{i} + 2\mathbf{j} - 7\mathbf{k}$ are perpendicular.

19. Let $\mathbf{a} = \mathbf{i} - 2\mathbf{j} + 3\mathbf{k}$, $\mathbf{b} = 4\mathbf{i} - \mathbf{j} + 2\mathbf{k}$, and $\mathbf{c} = 2\mathbf{i} - 3\mathbf{k}$. Find s such that $\mathbf{a} + s\mathbf{b}$ is perpendicular to $\mathbf{c}$.

20. Find all unit vectors in the plane orthogonal to $3\mathbf{i} - 4\mathbf{j}$.

21. Find the angle between the vectors $2\mathbf{i} - 3\mathbf{j} + \mathbf{k}$ and $-\mathbf{i} + 2\mathbf{j} + 3\mathbf{k}$.

22. Find the angle between the vectors $\mathbf{i} - 3\mathbf{j} + \mathbf{k}$ and $2\mathbf{i} - \mathbf{j} + 2\mathbf{k}$.

23. Find all values of c such that the angle between $\mathbf{i} - \mathbf{j} + \mathbf{k}$ and $\mathbf{i} + \mathbf{j} + c\mathbf{k}$ is $\pi/3$.

24. Find the vector projection of $2\mathbf{i} - 3\mathbf{j} + \mathbf{k}$ on $3\mathbf{i} - 4\mathbf{k}$.

25. Find the vector projection of $3\mathbf{i} - 4\mathbf{j} + 2\mathbf{k}$ onto $\mathbf{i} - 2\mathbf{j} + \mathbf{k}$.

26. Find the scalar component of $\mathbf{i} - 3\mathbf{j} + 2\mathbf{k}$ along $2\mathbf{i} + \mathbf{j} - 3\mathbf{k}$.

27. A body in the plane is constrained to move on the curve $y = x^3 - x^2 + 2x$, subject to the force vector $\mathbf{F} = \mathbf{i} + \mathbf{j}$. Find the scalar component of $\mathbf{F}$ actually moving the body on the curve at the instant when $x = 2$.

28. Let $\mathbf{a}$ and $\mathbf{b}$ be parallel unit vectors with opposite direction in space. Simplify each of the following as much as possible from this information.
(a) $\mathbf{a} \cdot \mathbf{b}$ (b) $\mathbf{a} \times \mathbf{b}$
(c) $(\mathbf{a} \cdot \mathbf{a})\mathbf{b}$ (d) $|\mathbf{b} - \mathbf{a}|$

29. Find $\mathbf{a} \times \mathbf{b}$ if $\mathbf{a} = \mathbf{i} + 2\mathbf{j} - 3\mathbf{k}$ and $\mathbf{b} = 4\mathbf{i} - \mathbf{j} + \mathbf{k}$.

30. Find the area of the parallelogram in space having as adjacent edges $\mathbf{i} + 2\mathbf{j} - \mathbf{k}$ and $2\mathbf{i} - 3\mathbf{j} + 4\mathbf{k}$.

31. Find the area of the triangle in the plane with vertices $(-1, 4)$, $(2, -3)$, and $(4, -2)$.

32. Find the area of the triangle in space with vertices $(-1, 3, 0)$, $(1, 0, 5)$, and $(-5, 3, -2)$.

33. If $\mathbf{a} = \mathbf{i} - 3\mathbf{j} + \mathbf{k}$, $\mathbf{b} = 2\mathbf{i} - \mathbf{k}$, and $\mathbf{c} = 4\mathbf{i} + \mathbf{j} - 3\mathbf{k}$, find $\mathbf{b} \times (\mathbf{a} \times \mathbf{c})$.

34. If $\mathbf{a} = \mathbf{i} - 2\mathbf{j} + \mathbf{k}$, $\mathbf{b} = -2\mathbf{i} + 3\mathbf{j} - \mathbf{k}$, and $\mathbf{c} = 4\mathbf{i} - 2\mathbf{j} + 2\mathbf{k}$, compute $(\mathbf{a} \times \mathbf{b}) \times \mathbf{c}$.

35. Let $\mathbf{a}$ and $\mathbf{b}$ be perpendicular unit vectors in space. Simplify each of the following as much as possible from this information.
(a) $(\mathbf{a} \times \mathbf{b}) \cdot \mathbf{a}$ (b) $(\mathbf{a} \times \mathbf{b}) \times \mathbf{a}$ (c) $\mathbf{a} \times (\mathbf{b} \times \mathbf{b})$

36. Find the volume of the box in space having the vectors

$\mathbf{i} - 3\mathbf{j} + \mathbf{k}$, $2\mathbf{i} - 3\mathbf{j} + 2\mathbf{k}$, $-3\mathbf{i} + 5\mathbf{j} - 2\mathbf{k}$ as coterminous edges.

37. Find the volume of the box in space with one corner at $(-1, 3, 4)$ and adjacent corners at $(0, -1, 2)$, $(3, -1, 4)$, and $(-1, 2, -1)$.

38. Find all values of c such that the box having the vectors $\mathbf{i} - 2\mathbf{j} + \mathbf{k}$, $3\mathbf{i} + 4\mathbf{j} + 2\mathbf{k}$, and $c\mathbf{i} - 3\mathbf{j} + 5\mathbf{k}$ as coterminous edges has area 20.

39. Find the area of the tetrahedron in space with vertices $(-1, 4, 1)$, $(0, 0, 3)$, $(5, -1, 2)$, and $(1, 4, -3)$.

40. Find the point on the line $x = 2 - 3t$, $y = 4 + t$, $z = 2 - 5t$ having z-coordinate 17.

41. Find all points on the line $x = 1 - 2t$, $y = 3 + 4t$, $z = 2 - t$, whose distance from $(1, 6, 3)$ is $\sqrt{18}$.

42. Find parametric equations of the line through $(-5, 1, 2)$ and $(2, -1, 7)$.

43. Find parametric equations of the line through $(-1, 5, 2)$ and $(3, -1, 4)$.

44. Find parametric equations of the line in the plane through $(-1, 3)$ and parallel to the line $x - 2y = 4$.

45. Find the point of intersection (if any) of the lines $x = -4 + 2t$, $y = 3 - 7t$, $z = 1 - 3t$ and $x = 6 + 2s$, $y = -2 + 3s$, $z = 7 + 4s$.

46. Find parametric equations of the line through $(-1, 2, 7)$ and orthogonal to the plane through $(1, -4, 2)$, $(2, -3, 4)$, and $(0, 1, 0)$.

47. Find parametric equations of the line perpendicular to the plane $x - 2y + 3z = 8$ and passing through the origin.

48. Find the equation of the plane through $(-2, 5, 4)$ and parallel to the plane $3x - 4y + 7z = 0$.

49. Find the equation of the plane through $(-1, 4, 3)$ and parallel to the plane $2x - 3y + z = 11$.

50. Find the equation of the plane through the origin and orthogonal to the line through $(-1, 4, 3)$ and $(2, 0, -1)$.

51. Find the equation of the plane passing through $(-1, 1, 4)$ and perpendicular to the line $x = 2 + t$, $y = 3 - 4t$, $z = -7 + 2t$.

52. Find the equation of the plane through the points $(-1, 3, 2)$, $(4, 0, 1)$, and $(0, -3, 5)$.

53. Find the equation of the plane containing both $(-1, 2, 0)$ and the line $x = 3 + 4t$, $y = 2 - t$, $z = t$.

54. Find the equation of the plane containing the point $(1, -2, 4)$ and the line of intersection of the planes $2x - 3y + z = 4$ and $x - 3y + 2z + 5$.

55. Find the equation of the plane containing the two lines $x = 2 + t$, $y = -1 + 2t$, $z = 3 - 4t$ and $x = 2 - 3t$, $y = -1 + 4t$, $z = 3 - t$ that intersect at $(2, -1, 3)$.

56. Classify the given planes as parallel, perpendicular, or neither.
    (a) $x - 3y + 4z = 7$, $-3x + y + 2z = 11$
    (b) $2x + y + 3z = 8$, $4x + 7y - 5z = -3$
    (c) $-4x + 6y - 12z = 7$, $2x - 3y + 6z = 5$

57. Find the point of intersection of the line $x = 2 - t$, $y = 4 + t$, $z = 1 - 3t$, with the plane $x - 3y + 4z = -1$.

58. Find the point of intersection of the line $x = 4 + t$, $y = 3 - t$, $z = 2 + 3t$ with the plane $2x + y + 3z = -3$.

59. Find the point where the line of intersection of the planes $x - y + 2z = 4$ and $2x + y - z = -4$ meets the plane $3x - 4y + 5z = 8$.

60. Find the distance from $(-2, 1)$ to the line $x = 4 + t$, $y = 2 - 3t$ in the plane.

61. Find the distance from $(-1, 1, 3)$ to the plane $2x + y - 2z = 4$.

62. Find all values of c such that the point $(-3, 2, 1)$ is 4 units from the plane $2x - y + 2z = c$.

63. Sketch the surface $x^2/4 + y^2/9 = 1 + z^2$.

64. Sketch the surface $y/4 = x^2/4 + z^2/9$.

65. Sketch the surface $x^2 + 4y^2 - z^2 - 2x + 16y + 17 = 0$.

66. Find the distance between the planes $x - 2y + 3z = 10$ and $4x - 8y + 12z = -7$.

67. Find the (shortest) distance between the lines $x = 1 - t$, $y = 2 + 3t$, $z = -1 + 2t$, and $x = 4 + 2t$, $y = -3 + t$, $z = -2 + 5t$.

68. Find the equation of the plane containing the intersection of the sphere with center $(-1, 2, 4)$ and radius 5 with the sphere with center $(1, -1, 3)$ and radius 3.

69. Give a geometric discussion of the accuracy of the statement "Three simultaneous linear equations in three unknowns have a unique solution."

70. Find the shortest distance from the line $x = 7 - 3t$, $y = -5 + 4t$, $z = 6 + 2t$ to the sphere $x^2 + y^2 + z^2 = 9$.

71. Consider the parallelogram in the plane having $\mathbf{a} = a_1\mathbf{i} + a_2\mathbf{j}$ and $\mathbf{b} = b_1\mathbf{i} + b_2\mathbf{j}$ as continuous edges emanating from the origin, as in Fig. 12.51. Show that the area of this parallelogram is the absolute value of the determinant.

$$\begin{vmatrix} a_1 & a_2 \\ b_1 & b_2 \end{vmatrix}.$$

VECTOR STUDY OF MOTION AND CURVES

13

This chapter builds on the notion of a *vector* introduced in Section 12.3 and provides a few introductory lessons on the differential geometry of curves in the plane or in space. In Sections 13.1 and 13.2, we develop calculus for vectors and use it to study motion in the plane and in space. Section 13.3 gives an important application to physics and astronomy. In Section 13.4, we make a more detailed study of motion in space and of space curves.

13.1 VELOCITY AND ACCELERATION VECTORS

The Position and Velocity Vectors

Let $x = h(t)$ and $y = k(t)$ be a parametric representation of a smooth curve, so that h' and k' exist, are continuous, and are never simultaneously zero. We may view these parametric equations as giving the position of a body in the plane at time t. We let $\mathbf{r}$ be the vector from the origin to a point (x, y) on the curve, shown in Fig. 13.1. This vector

$$\mathbf{r} = x\mathbf{i} + y\mathbf{j} = h(t)\mathbf{i} + k(t)\mathbf{j} \tag{1}$$

is the **position vector** of the body at time t.

We can view Eq. (1) as describing a function f that carries a portion of a t-axis into the x,y-plane, so that

$$(x, y) = f(t) = (h(t), k(t)).$$

Such a function, having as values points (which may also be regarded as vectors) in a space of dimension greater than one, is often called a **vector-valued function.**

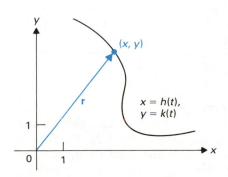

FIGURE 13.1 The position vector $\mathbf{r} = x\mathbf{i} + y\mathbf{j}$.

667

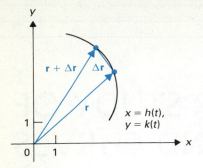

FIGURE 13.2 The increment vector $\Delta \mathbf{r}$.

A change Δt in t produces a change $\Delta \mathbf{r}$ in $\mathbf{r}$, where $\Delta \mathbf{r}$ is as shown in Fig. 13.2. The vector $\Delta \mathbf{r}$ is directed along a chord of the curve. Let the notation $\Delta \mathbf{r}/\Delta t$ be understood to mean the scalar $1/\Delta t$ times $\Delta \mathbf{r}$. We may write $\Delta \mathbf{r} = (\Delta x)\mathbf{i} + (\Delta y)\mathbf{j}$. Then

$$\frac{\Delta \mathbf{r}}{\Delta t} = \frac{\Delta x}{\Delta t}\mathbf{i} + \frac{\Delta y}{\Delta t}\mathbf{j}.$$

Taking the limit as $\Delta t \to 0$, we define the **vector derivative** by

$$\frac{d\mathbf{r}}{dt} = \lim_{t \to 0} \frac{\Delta \mathbf{r}}{\Delta t}$$

$$= \frac{dx}{dt}\mathbf{i} + \frac{dy}{dt}\mathbf{j}. \tag{2}$$

Note that dx/dt and dy/dt exist, since we are assuming that $x = h(t)$ and $y = k(t)$ are differentiable functions.

We now study the vector $d\mathbf{r}/dt$. The first things we want to know about a vector are its length and direction. From Eq. (2), we have

$$\left|\frac{d\mathbf{r}}{dt}\right| = \sqrt{\left(\frac{dx}{dt}\right)^2 + \left(\frac{dy}{dt}\right)^2}. \tag{3}$$

From the formula

$$s(t_1) = \int_{t_0}^{t_1} \sqrt{\left(\frac{dx}{dt}\right)^2 + \left(\frac{dy}{dt}\right)^2} \, dt$$

for the arc length traveled from starting time t_0 to time t_1, we obtain

$$\frac{ds}{dt} = \sqrt{\left(\frac{dx}{dt}\right)^2 + \left(\frac{dy}{dt}\right)^2}, \tag{4}$$

so Eq. (3) yields

$$\left|\frac{d\mathbf{r}}{dt}\right| = \frac{ds}{dt}. \tag{5}$$

Now ds/dt has the physical interpretation of the instantaneous rate of change of arc length per unit change in time as the body travels along the curve, so ds/dt is the **speed** of the body along the curve.

Turning to the direction of $d\mathbf{r}/dt$, the limiting direction of $\Delta \mathbf{r}$ as $\Delta t \to 0$ is tangent to the curve. Consequently $d\mathbf{r}/dt$ has direction tangent to the curve in the direction given by increasing t. See Fig. 13.3. We can also see this from Eq. (2), since, where $h'(t) \neq 0$, the "slope" of the vector arrow $d\mathbf{r}/dt$ is

$$\frac{dy/dt}{dx/dt} = \frac{dy}{dx},$$

which is the slope of the tangent to the curve.

Summarizing, at any instant the vector $d\mathbf{r}/dt$ points in the *direction* in which the body is moving and has *length* equal to the speed of the body. We think of *velocity* as a directed quantity, reflecting not only *speed* but the *direction* of motion. We see that $d\mathbf{r}/dt$ does precisely that.

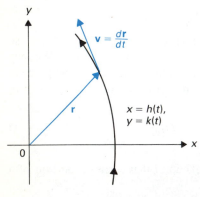

FIGURE 13.3 $\mathbf{v} = d\mathbf{r}/dt$ is the tangent to the curve in the direction of increasing t.

DEFINITION 13.1

Velocity vector

If $\mathbf{r} = h(t)\mathbf{i} + k(t)\mathbf{j}$ is the position vector of a body traveling on a smooth curve, then

$$\mathbf{v} = \frac{d\mathbf{r}}{dt} = h'(t)\mathbf{i} + k'(t)\mathbf{j} \qquad (6)$$

is the **velocity vector** (or simply the **velocity**) of the body at time t.

EXAMPLE 1 Let the position of a body in the plane at time t be given by $x = \cos t$, $y = \sin t$. Find the position vector, velocity vector, and speed at time t.

Solution Since $x^2 + y^2 = \cos^2 t + \sin^2 t = 1$, we know that the body travels on the circle of radius 1 with center at the origin, shown in Fig. 13.4. The position vector is

$$\mathbf{r} = (\cos t)\mathbf{i} + (\sin t)\mathbf{j}.$$

The velocity vector is

$$\mathbf{v} = \frac{d\mathbf{r}}{dt} = (-\sin t)\mathbf{i} + (\cos t)\mathbf{j}.$$

Since $\mathbf{r}$ is along a radial line of the circle and $\mathbf{v}$ is tangent to the circle, $\mathbf{r}$ and $\mathbf{v}$ should be orthogonal. We check this by computing

$$\mathbf{r} \cdot \mathbf{v} = (\cos t)(-\sin t) + (\sin t)(\cos t) = 0.$$

Finally, the speed is given by

$$\text{Speed} = |\mathbf{v}| = \sqrt{\sin^2 t + \cos^2 t} = 1.$$

Thus the speed is constant for this motion on the circle. In Fig. 13.4, we were careful to make $|\mathbf{v}| = |\mathbf{r}| = 1$. ∎

Since $\mathbf{v}$ is tangent to the curve $x = h(t)$, $y = k(t)$ in the direction corresponding to increasing t, we see that, for $|\mathbf{v}| \neq 0$,

$$\mathbf{t} = \frac{1}{|\mathbf{v}|}\mathbf{v} \qquad (7)$$

is a *unit* vector tangent to the curve in the direction of increasing t. Since $|\mathbf{v}| = ds/dt$, we obtain, from Eq. (7),

$$\mathbf{v} = \frac{ds}{dt}\mathbf{t}. \qquad (8)$$

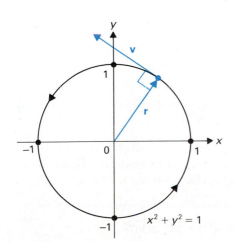

FIGURE 13.4
$\mathbf{r} = (\cos t)\mathbf{i} + (\sin t)\mathbf{j}$;
$\mathbf{v} = (-\sin t)\mathbf{i} + (\cos t)\mathbf{j}$;
$\mathbf{v} \perp \mathbf{r}$ and $|\mathbf{v}| = 1$

Equation (8) is a compact description of the velocity vector $\mathbf{v}$ as tangent to the curve and of magnitude ds/dt.

Exactly the same analysis can be made for a space curve

$$x = h(t), \qquad y = k(t), \qquad z = g(t),$$

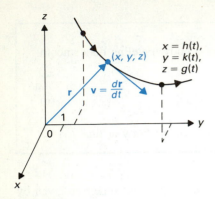

FIGURE 13.5 Position vector **r** and velocity vector **v** to a curve in space.

where $h(t)$, $k(t)$, and $g(t)$ all have continuous derivatives. See Fig. 13.5. The position vector is

$$\mathbf{r} = h(t)\mathbf{i} + k(t)\mathbf{j} + g(t)\mathbf{k}, \tag{9}$$

and the velocity vector is

$$\mathbf{v} = \frac{d\mathbf{r}}{dt} = \frac{dx}{dt}\mathbf{i} + \frac{dy}{dt}\mathbf{j} + \frac{dz}{dt}\mathbf{k}. \tag{10}$$

Since the direction of **v** is determined as the limiting direction of a chord of the curve corresponding to a vector $\Delta \mathbf{r}$ as $\Delta t \rightarrow 0$, we again see that **v** is tangent to the curve and points in the direction given by increasing t.

EXAMPLE 2   Find the equation of the line tangent to the space curve with parametric equations

$$x = t^2, \qquad y = \frac{1}{t^2 + 1}, \qquad \text{and} \qquad z = \ln t$$

at the point where $t = 1$.

Solution   To find parametric equations of a line, we need to know a point on the line and a direction vector parallel to the line. Setting $t = 1$, we find the point $(1, \frac{1}{2}, 0)$. Taking the position vector

$$\mathbf{r} = t^2\mathbf{i} + \frac{1}{t^2 + 1}\mathbf{j} + (\ln t)\mathbf{k},$$

we find that

$$\mathbf{v} = \frac{d\mathbf{r}}{dt} = (2t)\mathbf{i} + \frac{-2t}{(t^2 + 1)^2}\mathbf{j} + \frac{1}{t}\mathbf{k}$$

is tangent to the curve. We can find the desired direction vector by finding $\mathbf{v}|_{t=1}$. Thus we have

> **Point** $(1, \frac{1}{2}, 0)$
>
> **‖ vector** $2\mathbf{i} - \frac{2}{4}\mathbf{j} + \mathbf{k}$
>
> **Equations** $x = 1 + 2t,\ y = \frac{1}{2} - \frac{1}{2}t,\ z = t$ ■

We now turn to the length of a space curve with position vector (9). Using differential notation, a small change dt in t produces approximate changes dx in x, dy in y, and dz in z, so the approximate change in arc length is:

$$ds = \sqrt{(dx)^2 + (dy)^2 + (dz)^2}$$
$$= \sqrt{(dx/dt)^2 + (dy/dt)^2 + (dz/dt)^2}\ dt. \tag{11}$$

(A rigorous demonstration of Eq. 11 can be given just as for length of parametrically described plane curves, which was treated in Section 11.3.) Thus Eq. (11) gives the *differential of arc length* for a space curve; the length of arc at time t_1, measured from a point where $t = t_0$, is

$$s(t_1) = \int_{t_0}^{t_1} \sqrt{(dx/dt)^2 + (dy/dt)^2 + (dz/dt)^2}\ dt. \tag{12}$$

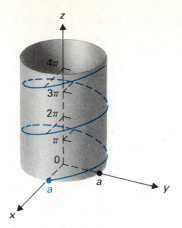

FIGURE 13.6 The helix $x = a \cos t$, $y = a \sin t$, $z = t$.

Once again, the length of the velocity vector is the speed, for

$$|\mathbf{v}| = \sqrt{(dx/dt)^2 + (dy/dt)^2 + (dz/dt)^2} = \frac{ds}{dt}. \tag{13}$$

The unit tangent vector $\mathbf{t}$ is again given by Eq. (7), and Eq. (8) is still valid.

EXAMPLE 3   The space curve

$$x = a \cos t, \qquad y = a \sin t, \qquad z = t$$

lies on the cylinder $x^2 + y^2 = a^2$. This curve, which is a *helix*, is sketched in Fig. 13.6. Find the vectors $\mathbf{r}$ and $\mathbf{v}$, the speed, and the length of one "turn" of the helix about the cylinder.

Solution   We have

$$\mathbf{r} = (a \cos t)\mathbf{i} + (a \sin t)\mathbf{j} + t\mathbf{k},$$
$$\mathbf{v} = -(a \sin t)\mathbf{i} + (a \cos t)\mathbf{j} + \mathbf{k},$$
$$\frac{ds}{dt} = |\mathbf{v}| = \sqrt{a^2\sin^2 t + a^2\cos^2 t + 1} = \sqrt{a^2 + 1}.$$

The length of one "turn" of the helix is therefore

$$\int_0^{2\pi} \sqrt{a^2 + 1}\, dt = 2\pi\sqrt{a^2 + 1}. \qquad \blacksquare$$

The Acceleration Vector

When we studied motion on a line in Section 3.7, we let $x = h(t)$ describe the position of a body on a line at time t. We defined the velocity to be the signed quantity $h'(t)$. We would now use the vector notation $\mathbf{v} = h'(t)\mathbf{i}$. We then defined the acceleration to be $h''(t)$. Again, we would now use $\mathbf{a} = h''(t)\mathbf{i}$. The following definition is an extension to the plane and to space. The definition is given for the plane case but is equally valid, with one more component, for space.

DEFINITION 13.2

Acceleration vector

If the position vector of a body at time t is $\mathbf{r} = h(t)\mathbf{i} + k(t)\mathbf{j}$, where h and k are twice-differentiable functions of t, then

$$\mathbf{a} = \frac{d\mathbf{v}}{dt} = \frac{d^2\mathbf{r}}{dt^2} = \frac{d^2x}{dt^2}\mathbf{i} + \frac{d^2y}{dt^2}\mathbf{j} \tag{14}$$

is the **acceleration vector** (or simply the **acceleration**) of the body at time t.

As in the case of the velocity vector, we would like to know the *magnitude* and *direction* of the acceleration vector. From Eq. (14), we obtain

$$|\mathbf{a}| = \sqrt{(h''(t))^2 + (k''(t))^2}. \tag{15}$$

We leave a detailed discussion of the direction of the acceleration vector $\mathbf{a}$ to the next section. We mention now only the information given by Newton's law,

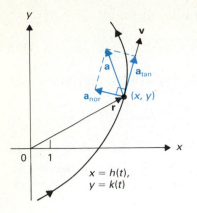

FIGURE 13.7 $\mathbf{a} = \mathbf{a}_{\text{tan}} + \mathbf{a}_{\text{nor}}$; speed is increasing.

$\mathbf{F} = m\mathbf{a}$. Here $\mathbf{F}$ is the *force vector*, $\mathbf{a}$ is the *acceleration* of the body on which $\mathbf{F}$ acts, and m is the *mass* of the body.

Look at the plane motion in Fig. 13.7. At a point where $\mathbf{v} \neq \mathbf{0}$, we can break the acceleration vector $\mathbf{a}$ into vector components $\mathbf{a}_{\text{tan}}$ along $\mathbf{v}$ and $\mathbf{a}_{\text{nor}}$ orthogonal to $\mathbf{v}$, such that

$$\mathbf{a} = \mathbf{a}_{\text{tan}} + \mathbf{a}_{\text{nor}}. \tag{16}$$

To cause the body to travel a *curved* path, the vector component $m\mathbf{a}_{\text{nor}}$ of the force vector $\mathbf{F}$ must be nonzero and points toward the concave side of the curve. A nonzero vector component $m\mathbf{a}_{\text{tan}}$ of $\mathbf{F}$ having the direction of $\mathbf{v}$, as in Fig. 13.7, causes the speed to *increase*. A nonzero component $m\mathbf{a}_{\text{tan}}$ of $\mathbf{F}$ having direction opposite to $\mathbf{v}$, as shown in Fig. 13.8, causes a *decrease* in speed. If the speed remains constant, then $\mathbf{a}_{\text{tan}}$ should be $\mathbf{0}$, and $\mathbf{a} = \mathbf{a}_{\text{nor}}$ is normal to the curve at each point, and it points toward the concave side. All this seems clear from the physical meaning of $\mathbf{F} = m\mathbf{a}$.

EXAMPLE 4   Discuss the direction and magnitude of the acceleration vector $\mathbf{a}$ for the motion on the circle in Example 1 with position vector

$$\mathbf{r} = (\cos t)\mathbf{i} + (\sin t)\mathbf{j}.$$

Solution   We saw in Example 1 that

$$\mathbf{v} = \frac{d\mathbf{r}}{dt} = (-\sin t)\mathbf{i} + (\cos t)\mathbf{j}$$

and $|\mathbf{v}| = 1$. We have

$$\mathbf{a} = (-\cos t)\mathbf{i} - (\sin t)\mathbf{j},$$

so $|\mathbf{a}| = \sqrt{\cos^2 t + \sin^2 t} = 1$. We easily see that $\mathbf{a} \cdot \mathbf{v} = 0$, so $\mathbf{a} = \mathbf{a}_{\text{nor}}$ is orthogonal to $\mathbf{v}$ in this case where the speed is constant. Note that $\mathbf{a} = -\mathbf{r}$, so $\mathbf{a}$ is directed toward the center of the circle in Fig. 13.4.   ∎

EXAMPLE 5   Let the position vector of a body in the plane at time t be given by

$$\mathbf{r} = (t^2 + t)\mathbf{i} - t^3\mathbf{j}.$$

Find the vector components $\mathbf{a}_{\text{tan}}$ and $\mathbf{a}_{\text{nor}}$ of the acceleration vector that are tangent and normal to the curve at time $t = 1$. Determine whether the speed is increasing or decreasing there.

Solution   We obtain

$$\mathbf{r}|_{t=1} = [(t^2 + t)\mathbf{i} - t^3\mathbf{j}]|_1 = 2\mathbf{i} - \mathbf{j},$$
$$\mathbf{v}|_{t=1} = [(2t + 1)\mathbf{i} - 3t^2\mathbf{j}]|_1 = 3\mathbf{i} - 3\mathbf{j},$$
$$\mathbf{a}|_{t=1} = [2\mathbf{i} - (6t)\mathbf{j}]|_1 = 2\mathbf{i} - 6\mathbf{j}.$$

Now $\mathbf{a}_{\text{tan}}$ is the vector projection of $\mathbf{a}$ along $\mathbf{v}$, since $\mathbf{v}$ is tangent to the curve. Thus

$$\mathbf{a}_{\text{tan}} = \left(\frac{\mathbf{a} \cdot \mathbf{v}}{\mathbf{v} \cdot \mathbf{v}}\right)\mathbf{v}.$$

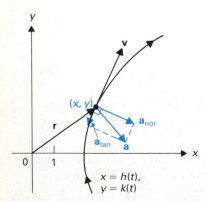

FIGURE 13.8 $\mathbf{a} = \mathbf{a}_{\text{tan}} + \mathbf{a}_{\text{nor}}$; speed is decreasing.

When $t = 1$,

$$\mathbf{a}_{\text{tan}}\big|_{t=1} = \frac{6 + 18}{9 + 9}(3\mathbf{i} - 3\mathbf{j}) = \frac{4}{3}(3\mathbf{i} - 3\mathbf{j}) = 4\mathbf{i} - 4\mathbf{j}.$$

Since $\mathbf{a}_{\text{nor}} = \mathbf{a} - \mathbf{a}_{\text{tan}}$, we see that

$$\mathbf{a}_{\text{nor}}\big|_{t=1} = (2\mathbf{i} - 6\mathbf{j}) - (4\mathbf{i} - 4\mathbf{j}) = -2\mathbf{i} - 2\mathbf{j}.$$

Since $\mathbf{a} \cdot \mathbf{v} = 24 > 0$, we see that $\mathbf{a}_{\text{tan}}$ and $\mathbf{v}$ have the same direction, so the speed is increasing. ∎

EXAMPLE 6 Find the acceleration vector $\mathbf{a}$ and its magnitude $|\mathbf{a}|$ for the motion in Example 3 on a helix, where the position vector is

$$\mathbf{r} = (a \cos t)\mathbf{i} + (a \sin t)\mathbf{j} + t\mathbf{k}.$$

Solution Taking first and second derivatives of $\mathbf{r}$, we have

$$\mathbf{v} = (-a \sin t)\mathbf{i} + (a \cos t)\mathbf{j} + \mathbf{k},$$
$$\mathbf{a} = (-a \cos t)\mathbf{i} - (a \sin t)\mathbf{j}.$$

Then $|\mathbf{a}| = \sqrt{a^2\cos^2 t + a^2\sin^2 t} = a$. Note that for this motion, $\mathbf{a} \cdot \mathbf{v} = 0$, so $\mathbf{a}$ is orthogonal to $\mathbf{v}$. We therefore expect the speed to be constant, and Example 3 showed this to be the case. ∎

We have seen how to find the velocity and acceleration vectors if the position vector is known. Equally important is finding the velocity and position vectors if the acceleration vector is known. Many times, we know what force vector is being applied, and we want to find the position vector of the body. Of course, this can be done by integrating.

EXAMPLE 7 Let the force vector

$$\mathbf{F} = (12t)\mathbf{i} + (4 \sin t)\mathbf{j} + (2 \cos 2t)\mathbf{k}$$

act on a body of mass 2. If the body has initial position and velocity vectors

$$\mathbf{r}\big|_{t=0} \, \mathbf{i} - 2\mathbf{j} + \mathbf{k} \qquad \text{and} \qquad \mathbf{v}\big|_{t=0} = -3\mathbf{i} - 2\mathbf{k}$$

when $t = 0$, find the velocity and position vectors at time t.

Solution Since $\mathbf{F} = m\mathbf{a}$ and $m = 2$, we see that

$$\mathbf{a} = (6t)\mathbf{i} + (2 \sin t)\mathbf{j} + (\cos 2t)\mathbf{k}.$$

Integrating the coefficients of $\mathbf{i}$, $\mathbf{j}$, and $\mathbf{k}$, we obtain

$$\mathbf{v} = (3t^2 + C_1)\mathbf{i} - (2 \cos t + C_2)\mathbf{j} + (\tfrac{1}{2} \sin 2t + C_3)\mathbf{k}.$$

When $t = 0$, we must have

$$\mathbf{v}\big|_{t=0} = C_1\mathbf{i} - (2 + C_2)\mathbf{j} + C_3\mathbf{k} = -3\mathbf{i} - 2\mathbf{k}.$$

Thus $C_1 = -3$, $C_2 = -2$, and $C_3 = -2$. We now have

$$\mathbf{v} = (3t^2 - 3)\mathbf{i} - (2 \cos t - 2)\mathbf{j} + (\tfrac{1}{2} \sin 2t - 2)\mathbf{k}.$$

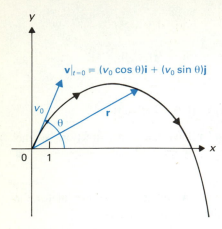

FIGURE 13.9 Path of a projectile fired from the origin with muzzle speed v_0 at the angle of inclination θ.

Thus

$$\mathbf{r} = (t^3 - 3t + C_4)\mathbf{i} - (2 \sin t - 2t + C_5)\mathbf{j} - (\tfrac{1}{4} \cos 2t + 2t + C_6)\mathbf{k}.$$

When $t = 0$, we must have

$$\mathbf{r}|_{t=0} = C_4\mathbf{i} - C_5\mathbf{j} - (\tfrac{1}{4} + C_6)\mathbf{k} = \mathbf{i} - 2\mathbf{j} + \mathbf{k},$$

so $C_4 = 1$, $C_5 = 2$, and $C_6 = -\tfrac{5}{4}$. Therefore

$$\mathbf{r} = (t^3 - 3t + 1)\mathbf{i} - (2 \sin t - 2t + 2)\mathbf{j} - (\tfrac{1}{4} \cos 2t + 2t - \tfrac{5}{4})\mathbf{k}. \qquad ■$$

EXAMPLE 8 A projectile is fired from a cannon with muzzle speed v_0 ft/sec at an angle of elevation θ with the horizontal. Neglecting air resistance, the acceleration after firing remains constant at $-32\mathbf{j}$, where units are in ft/sec^2. See Fig. 13.9. Find the path of the projectile.

Solution We choose our origin at the muzzle of the cannon, as shown in Fig. 13.9. Finding the position vector at time t is a good way to describe the path of the projectile. When $t = 0$, we have

$$\mathbf{r}|_{t=0} = 0\mathbf{i} + 0\mathbf{j}$$

and

$$\mathbf{v}|_{t=0} = (v_0 \cos \theta)\mathbf{i} + (v_0 \sin \theta)\mathbf{j}.$$

From $\mathbf{a} = -32\mathbf{j}$, we obtain

$$\mathbf{v} = C_1\mathbf{i} + (C_2 - 32t)\mathbf{j}.$$

When $t = 0$, we must have

$$\mathbf{v}|_{t=0} = C_1\mathbf{i} + C_2\mathbf{j} = (v_0 \cos \theta)\mathbf{i} + (v_0 \sin \theta)\mathbf{j},$$

so $C_1 = v_0 \cos \theta$ and $C_2 = v_0 \sin \theta$. Thus

$$\mathbf{v} = (v_0 \cos \theta)\mathbf{i} + (v_0 \sin \theta - 32t)\mathbf{j}.$$

Consequently,

$$\mathbf{r} = [(v_0 \cos \theta)t + C_3]\mathbf{i} + [(v_0 \sin \theta)t - 16t^2 + C_4]\mathbf{j}.$$

When $t = 0$, we must have

$$\mathbf{r}|_{t=0} = C_3\mathbf{i} + C_4\mathbf{j} = 0\mathbf{i} + 0\mathbf{j},$$

so $C_3 = C_4 = 0$. Thus

$$\mathbf{r} = [(v_0 \cos \theta)t]\mathbf{i} + [(v_0 \sin \theta)t - 16t^2]\mathbf{j}. \qquad (17)$$

Of course this position vector is valid only until the projectile hits the earth. See the next example. ■

EXAMPLE 9 Suppose the projectile in Example 8 is fired from a cannon at the edge of a cliff, with the muzzle of the cannon 96 ft above a flat plain below. If the muzzle speed is 160 ft/sec and the angle of fire is 30° above the horizontal, find the distance from the base of the cliff where the projectile hits. (This distance is called the *range*.)

Solution The projectile hits the ground when the $\mathbf{j}$-coefficient of the position vector is -96. Using Eq. (17) in Example 8 and taking $v_0 = 160$ ft/sec and

$\sin \theta = \sin 30° = \frac{1}{2}$, we see that

$$160 \cdot \tfrac{1}{2}t - 16t^2 = -96,$$
$$16t^2 - 80t - 96 = 0,$$
$$t^2 - 5t - 6 = 0,$$
$$(t - 6)(t + 1) = 0.$$

Thus the projectile hits the ground when $t = 6$ sec. When $t = 6$, the **i**-coefficient of **r** is

$$160 \cdot \frac{\sqrt{3}}{2} \cdot 6 = 480\sqrt{3} \text{ ft.}$$

This is the range of the projectile. ∎

Differentiation of Products of Vectors

The previous discussion has shown that differentiation of a vector can be accomplished by just differentiating each component of the vector. Let **a** and **b** be differentiable vector functions of t in space, and let $f(t)$ be a differentiable scalar function. We can form the products $f(t)\mathbf{a}$, $\mathbf{a} \cdot \mathbf{b}$, and $\mathbf{a} \times \mathbf{b}$. As we might guess, the usual product formula holds for differentiation of all these products. Namely,

$$\frac{d(f(t)\mathbf{a})}{dt} = f(t)\frac{d\mathbf{a}}{dt} + f'(t)\mathbf{a}, \tag{18}$$

$$\frac{d(\mathbf{a} \cdot \mathbf{b})}{dt} = \mathbf{a} \cdot \frac{d\mathbf{b}}{dt} + \frac{d\mathbf{a}}{dt} \cdot \mathbf{b}, \tag{19}$$

$$\frac{d(\mathbf{a} \times \mathbf{b})}{dt} = \mathbf{a} \times \frac{d\mathbf{b}}{dt} + \frac{d\mathbf{a}}{dt}\mathbf{a} \times \mathbf{b}. \tag{20}$$

*Since the cross product is not commutative, we must be careful to always have the "**a**-factor" first, before the "**b**-factor" in Eq. (20).*

We give the proof of Eq. (19). Proofs of the others are similar. Using properties of the dot product $\mathbf{a} \cdot \mathbf{b}$, we see that a change Δt in t produces a change

$$(\mathbf{a} + \Delta\mathbf{a}) \cdot (\mathbf{b} + \Delta\mathbf{b}) - \mathbf{a} \cdot \mathbf{b} = \mathbf{a} \cdot \mathbf{b} + \mathbf{a} \cdot \Delta\mathbf{b} + \Delta\mathbf{a} \cdot \mathbf{b} + \Delta\mathbf{a} \cdot \Delta\mathbf{b} - \mathbf{a} \cdot \mathbf{b}$$
$$= \mathbf{a} \cdot \Delta\mathbf{b} + \Delta\mathbf{a} \cdot \mathbf{b} + \Delta\mathbf{a} \cdot \Delta\mathbf{b}$$

in $\mathbf{a} \cdot \mathbf{b}$. Thus the derivative $d(\mathbf{a} \cdot \mathbf{b})/dt$ is

$$\frac{d(\mathbf{a} \cdot \mathbf{b})}{dt} = \lim_{\Delta t \to 0}\left(\mathbf{a} \cdot \frac{\Delta\mathbf{b}}{\Delta t} + \frac{\Delta\mathbf{a}}{\Delta t} \cdot \mathbf{b} + \frac{\Delta\mathbf{a}}{\Delta t} \cdot \Delta\mathbf{b}\right)$$

$$= \mathbf{a} \cdot \frac{d\mathbf{b}}{dt} + \frac{d\mathbf{a}}{dt} \cdot \mathbf{b} + \frac{d\mathbf{a}}{dt} \cdot \mathbf{0} = \mathbf{a} \cdot \frac{d\mathbf{b}}{dt} + \frac{d\mathbf{a}}{dt} \cdot \mathbf{b},$$

which is Eq. (19).

We ask you to illustrate Eqs. (18), (19), and (20) in the exercises.

SUMMARY

Assume that you have parametric equations $x = h(t)$, $y = k(t)$ of a plane curve, or $x = h(t)$, $y = k(t)$, $z = g(t)$ for a space curve, where all the functions of t are twice differentiable.

1. The position vector to the curve is

$$\mathbf{r} = h(t)\mathbf{i} + k(t)\mathbf{j} \quad \text{for a plane curve,}$$

$$\mathbf{r} = h(t)\mathbf{i} + k(t)\mathbf{j} + g(t)\mathbf{k} \quad \text{for a space curve.}$$

2. The velocity vector is

$$\mathbf{v} = \frac{d\mathbf{r}}{dt} = \frac{dx}{dt}\mathbf{i} + \frac{dy}{dt}\mathbf{j} \quad \text{for a plane curve,}$$

$$\mathbf{v} = \frac{d\mathbf{r}}{dt} = \frac{dx}{dt}\mathbf{i} + \frac{dy}{dt}\mathbf{j} + \frac{dz}{dt}\mathbf{k} \quad \text{for a space curve.}$$

3. The length of the velocity vector is

$$|\mathbf{v}| = \frac{ds}{dt} = \text{Speed along the curve.}$$

4. The distance traveled along a space curve from time t_0 to t_1 is

$$s(t_1) = \int_{t_0}^{t_1} \sqrt{\left(\frac{dx}{dt}\right)^2 + \left(\frac{dy}{dt}\right)^2 + \left(\frac{dz}{dt}\right)^2} \, dt.$$

5. The direction of $\mathbf{v}$ is tangent to the curve and in the direction corresponding to increasing t.

6. The acceleration vector is

$$\mathbf{a} = \frac{d^2\mathbf{r}}{dt^2} = \frac{d\mathbf{v}}{dt} = \frac{d^2x}{dt^2}\mathbf{i} + \frac{d^2y}{dt^2}\mathbf{j} \quad \text{for a plane curve,}$$

$$\mathbf{a} = \frac{d^2\mathbf{r}}{dt^2} = \frac{d\mathbf{v}}{dt} = \frac{d^2x}{dt^2}\mathbf{i} + \frac{d^2y}{dt^2}\mathbf{j} + \frac{d^2z}{dt^2}\mathbf{k} \quad \text{for a space curve.}$$

7. We may write $\mathbf{a} = \mathbf{a}_{\text{tan}} + \mathbf{a}_{\text{nor}}$, where $\mathbf{a}_{\text{tan}}$ and $\mathbf{a}_{\text{nor}}$ are the vector components of $\mathbf{a}$ tangent and normal to the curve. We have

$$\mathbf{a}_{\text{tan}} = [(\mathbf{a} \cdot \mathbf{v})/(\mathbf{v} \cdot \mathbf{v})]\mathbf{v} \quad \text{and} \quad \mathbf{a}_{\text{nor}} = \mathbf{a} - \mathbf{a}_{\text{tan}}.$$

8. For differentiable vector functions $\mathbf{a}$ and $\mathbf{b}$ of t and a differentiable scalar function $f(t)$, we have

$$\frac{d(f(t)\mathbf{a})}{dt} = f(t)\frac{d\mathbf{a}}{dt} + f'(t)\mathbf{a}, \qquad \frac{d(\mathbf{a} \cdot \mathbf{b})}{dt} = \mathbf{a} \cdot \frac{d\mathbf{b}}{dt} + \frac{d\mathbf{a}}{dt} \cdot \mathbf{b},$$

$$\frac{d(\mathbf{a} \times \mathbf{b})}{dt} = \mathbf{a} \times \frac{d\mathbf{b}}{dt} + \frac{d\mathbf{a}}{dt} \times \mathbf{b}.$$

EXERCISES 13.1

1. Let $\mathbf{r}(t)$ be the position vector from the origin to a point

$$(x, y) = (h(t), k(t))$$

on a smooth curve. Show that if s is the arc length along the curve defined as usual, then $d\mathbf{r}/ds = \mathbf{t}$, where $\mathbf{t}$ is the *unit* tangent vector to the curve in the direction of increasing t.

2. Let $\mathbf{b}$ be a differentiable vector function of t.
   (a) Show that if $|\mathbf{b}|$ is a constant function of t, then $\mathbf{b}$ is orthogonal to $d\mathbf{b}/dt$. [*Hint:* Differentiate $\mathbf{b} \cdot \mathbf{b}$.]
   (b) State the interpretation of part (a) if $\mathbf{b} = \mathbf{r}$, the position vector of a body in motion.
   (c) State the interpretation of part (a) if $\mathbf{b} = \mathbf{v}$, the velocity vector of a body in motion.

In Exercises 3–13, the position vector $\mathbf{r}$ at time t of a body moving on a curve is given. Find the following at the indicated time t_0: (a) the velocity vector of the body, (b) the speed of the body, (c) the acceleration vector of the body.

3. $\mathbf{r} = 2t\mathbf{i} + (3t - 1)\mathbf{j}$ at $t_0 = 0$
4. $\mathbf{r} = (3t + 1)\mathbf{i} + t^2\mathbf{j}$ at $t_0 = 1$
5. $\mathbf{r} = (\sin t)\mathbf{i} + (\cos 2t)\mathbf{j}$ at $t_0 = \pi$
6. $\mathbf{r} = e^t\mathbf{i} + t^2\mathbf{j}$ at $t_0 = 0$
7. $\mathbf{r} = (\ln t)\mathbf{i} + (\cosh(t - 1))\mathbf{j}$ at $t_0 = 1$
8. $\mathbf{r} = (e^t \sin t)\mathbf{i} + (e^t \cos t)\mathbf{j}$ at $t_0 = 0$
9. $\mathbf{r} = \dfrac{1}{t}\mathbf{i} + \dfrac{1}{t^2}\mathbf{j}$ at $t_0 = 1$
10. $\mathbf{r} = (\ln(\sin t))\mathbf{i} + (\ln(\cos t))\mathbf{j}$ at $t_0 = \pi/4$
11. $\mathbf{r} = t^2\mathbf{i} + t^3\mathbf{j} - (t + 1)\mathbf{k}$ at $t_0 = 2$
12. $\mathbf{r} = e^t\mathbf{i} - (\sin t)\mathbf{j} + (\cos t)\mathbf{k}$ at $t_0 = 0$
13. $\mathbf{r} = \left(1 + \dfrac{1}{t}\right)\mathbf{i} + \left(1 - \dfrac{1}{t}\right)\mathbf{j} + t^2\mathbf{k}$ at $t_0 = 1$

14. Prove that if we let

$$\mathbf{a}_{\text{tan}} = [(\mathbf{a} \cdot \mathbf{v})/(\mathbf{v} \cdot \mathbf{v})]\mathbf{v}$$

and define $\mathbf{a}_{\text{nor}} = \mathbf{a} - \mathbf{a}_{\text{tan}}$, then $\mathbf{a}_{\text{nor}}$ is indeed orthogonal to $\mathbf{v}$. [*Hint:* Compute a dot product.]

In Exercises 15–20, find tangential and normal vector components $\mathbf{a}_{\text{tan}}$ and $\mathbf{a}_{\text{nor}}$ of the acceleration vector $\mathbf{a}$ at the given point, so that $\mathbf{a} = \mathbf{a}_{\text{tan}} + \mathbf{a}_{\text{nor}}$. State whether the speed is increasing or decreasing at that point. See Example 5.

15. $\mathbf{r} = (t^2 - 3t)\mathbf{i} + t^3\mathbf{j}$, where $t = -1$
16. $\mathbf{r} = te^t\mathbf{i} - (\cos t)\mathbf{j}$, where $t = 0$
17. $\mathbf{r} = \dfrac{1}{t}\mathbf{i} + (\ln t)\mathbf{j}$, where $t = 1$
18. $\mathbf{r} = t^2\mathbf{i} + t^3\mathbf{j} - t^4\mathbf{k}$, where $t = -1$
19. $\mathbf{r} = (\sin t)\mathbf{i} + (\cos 2t)\mathbf{j} - t^2\mathbf{k}$, where $t = \pi$
20. $\mathbf{r} = (t^3 - 4t^2)\mathbf{i} + \dfrac{8}{t}\mathbf{j} - (t^2 - 4t)\mathbf{k}$, where $t = 2$

In Exercises 21–24, find parametric equations of the line tangent to the given space curve at the given point.

21. $x = t + \cosh t$, $y = \sinh t$, $z = t^3$, where $t = 0$
22. $x = \sin t$, $y = \cos t$, $z = t$, where $t = \pi$
23. $x = \dfrac{1}{t^2 + 1}$, $y = \dfrac{8}{t}$, $z = t^3 - 2$, where $t = 1$
24. $x = \sinh 2t$, $y = \ln(t^2 + t + 1)$, $z = \dfrac{1}{t - 1}$, where $t = 0$

In Exercises 25–28, find the equation of the plane normal to the given space curve at the given point.

25. The curve and point in Exercise 21
26. The curve and point in Exercise 22
27. The curve and point in Exercise 23
28. The curve and point in Exercise 24

In Exercises 29–33, find the length of the indicated portion of the space curve. (Estimate using a calculator and Simpson's rule in Exercises 32 and 33.)

29. $x = a \cos 2t$, $y = t$, $z = a \sin 2t$ for $0 \le t \le 2\pi$
30. $x = 2t$, $y = t^2$, $z = -t^2$ for $0 \le t \le 2$
31. $x = 3t^2$, $y = 2t^3$, $z = 3t$ for $0 \le t \le 4$
32. $x = t^2$, $y = 2t + 1$, $z = \sin \pi t$ for $0 \le t \le 1$
33. $x = \sqrt{t^2 + 4}$, $y = t^3$, $z = \dfrac{1}{1 + t^2}$ for $0 \le t \le 2$

34. Let $\mathbf{a} = (t \sin t)\mathbf{i} - t^2\mathbf{j} + te^t\mathbf{k}$ and let $f(t) = 1/t$. Illustrate Eq. (18) of the text by computing $d(f(t)\mathbf{a})/dt$ in two ways.

35. Let

$$\mathbf{a} = t^2\mathbf{i} - (3t + 1)\mathbf{j} \quad \text{and} \quad \mathbf{b} = (2t)\mathbf{i} - t^3\mathbf{j}$$

in the plane. Illustrate Eq. (19) of the text by computing $d(\mathbf{a} \cdot \mathbf{b})/dt$ in two ways.

36. Let

$$\mathbf{a} = 3t\mathbf{i} - (4t + 1)\mathbf{j} + t^2\mathbf{k}$$

and

$$\mathbf{b} = (t^2 - 2)\mathbf{i} + 5t\mathbf{j} - 6t\mathbf{k}$$

in space. Illustrate Eq. (20) of the text by computing $d(\mathbf{a} \times \mathbf{b})/dt$ in two ways.

Exercises 37–43 refer to the path of a projectile having muzzle speed v_0 ft/sec and fired with angle of inclination θ to the horizontal. See Examples 8 and 9. Air resistance is neglected in all these problems.

37. Show that the path of the projectile is a parabola.
38. Show that the projectile returns to the same altitude at which it was fired at time $(v_0 \sin \theta)/16$ sec.

39. If a projectile is fired at ground level over a level plain, show that its range is $(v_0^2 \sin 2\theta)/32$ ft.

40. Use the result of Exercise 39 to find the angle θ of inclination for maximum ground-level range with a fixed muzzle speed v_0.

41. A baseball player bats a ball at an angle of inclination 15° (see Fig. 13.10), and the ball is caught at the same height 256 ft away. Find the speed with which the ball left the bat.

15°

FIGURE 13.10

42. Show that the maximum altitude attained by a projectile fired at ground level is $(v_0^2 \sin^2\theta)/64$ ft.

43. A projectile fired at ground level 800 ft from the bottom of a 200-ft cliff is to reach the top of the cliff. Find the minimum muzzle speed v_0 required and the corresponding angle θ of inclination at which the projectile should be fired.

44. The vector $\mathbf{r} = a(\cos t)\mathbf{i} + a(\sin t)\mathbf{j}$ is the position vector for motion on the circle $x^2 + y^2 = a^2$, where the body makes one revolution around the circle in 2π units of time. Modify $\mathbf{r}$ so that the body makes β revolutions per unit time.

45. A body travels at constant speed making β revolutions per second around a circle of radius a. Show that the magnitude of the force required to hold the body on the circle is directly proportional to the mass m, radius a, and β^2.

46. A child whirls a $\frac{1}{2}$-lb stone tied to a string around and around his head in a circle of radius 4 ft. If the string breaks at any pull exceeding 25 lb, find the maximum number of revolutions per second that the stone can make without breaking the string. (A mass of 1 slug weighs 32 lb.)

 EXPLORING CALCULUS

2: velocity and acceleration vectors

13.2 TANGENTIAL AND NORMAL MAGNITUDES OF ACCELERATION

Suppose we are passengers in a car and we are instructing the driver which way to go at an intersection. We rarely say "turn north," or "turn south," or "continue west," using a global coordinate system for our earth. We are much more likely to say "turn right," or "turn left," or "continue straight ahead." We have in mind two perpendicular axes that travel right along with us: the "ahead/behind" axis and the "left/right" axis.

In the context just discussed, we think of an "ahead" unit vector and a "left" unit vector as forming an *orthogonal frame of unit vectors* at each instant. (We could equally well have chosen "ahead" and "right," or "behind" and "left," and so on.) In this section, we discuss such an orthogonal frame of two unit vectors $\mathbf{t}$ and $\mathbf{n}$, at each point of a smooth plane curve described by a position vector function $\mathbf{r}(t)$. As you might guess, $\mathbf{t}$ will point in a direction *tangent* to the curve, and $\mathbf{n}$ will point in a *normal* direction to the curve. To a bug crawling on the curve, $\mathbf{t}$ seems to point *ahead* at any instant, while $\mathbf{n}$ points either *right* or *left*. We will also describe $\mathbf{t}$ and $\mathbf{n}$ for space curves. Of course, a bug traveling on a curve in space needs another vector orthogonal to both $\mathbf{t}$ and $\mathbf{n}$ to complete a space frame. In Section 13.4, we will define this other vector to be $\mathbf{b} = \mathbf{t} \times \mathbf{n}$ and make a more detailed study of the space situation.

After discussing the orthogonal frame $\mathbf{t}$ and $\mathbf{n}$ for curves in the plane or in space, we will express the velocity vector $\mathbf{v}$ and the acceleration vector $\mathbf{a}$ associated with a moving body in terms of this moving orthogonal frame.

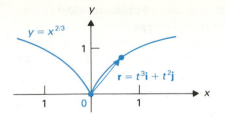

FIGURE 13.11 $\mathbf{v} = 3t^2\mathbf{i} + 2t\mathbf{j}$ is $\mathbf{0}$ where $t = 0$; there is a sharp point at the origin.

The Orthogonal Vector Frame **t** and **n**

Let $h''(t)$ and $k''(t)$ be continuous, and let $\mathbf{r} = h(t)\mathbf{i} + k(t)\mathbf{j}$ be the position vector for motion in the plane. We require that $\mathbf{v} \neq \mathbf{0}$ at any point of the arc. This means that $h'(t)$ and $k'(t)$ must never be simultaneously zero, so that the curve is *smooth*. If $\mathbf{v} = \mathbf{0}$ at some point, then the body in motion comes to a stop at that point and could proceed in a new direction, making a sharp point on the curve. This explains the use of the term *smooth*. For example, $\mathbf{r} = t^3\mathbf{i} + t^2\mathbf{j}$ is not smooth where $t = 0$. The curve has x,y-equation $y^3 = x^2$ or $y = x^{2/3}$ and has a sharp point at the origin, as shown in Fig. 13.11. In space, where $\mathbf{r} = h(t)\mathbf{i} + k(t)\mathbf{j} + g(t)\mathbf{k}$, we require that $h'(t)$, $k'(t)$, and $g'(t)$ never be simultaneously zero for a curve to be smooth.

On a smooth curve in the plane or in space, we let

$$\mathbf{t} = \frac{1}{|\mathbf{v}|}\mathbf{v}. \qquad (1)$$

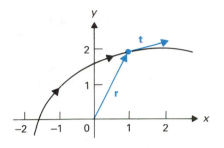

FIGURE 13.12 Unit tangent vector $\mathbf{t} = \mathbf{v}/|\mathbf{v}|$.

Then $\mathbf{t}$ is a unit vector tangent to the curve in the direction corresponding to increasing t, shown in Fig. 13.12. To a bug crawling along the curve, $\mathbf{t}$ looks to be "straight ahead" at any instant. Since $\mathbf{t}$ is a unit vector, we have

$$|\mathbf{t}| = 1.$$

Now $\mathbf{t}$ can be viewed as a vector-valued function of the parameter t, the *arc length s* along the curve, or other possible parameters. Exercise 2 of the preceding section showed that the derivative of a vector-valued function of constant magnitude is always orthogonal to the vector. This is so important that we state it again as a theorem and give the proof. While t is used as parameter in the theorem, the argument is the same for any parameter.

THEOREM 13.1   Derivatives of vectors of constant magnitude

Let $\mathbf{b}(t)$ be a differentiable vector-valued function of constant magnitude. Then $d\mathbf{b}/dt$ is orthogonal to $\mathbf{b}$.

Proof: From the preceding section, we know that

$$\frac{d(\mathbf{b} \cdot \mathbf{b})}{dt} = \mathbf{b} \cdot \frac{d\mathbf{b}}{dt} + \frac{d\mathbf{b}}{dt} \cdot \mathbf{b} = 2\left(\mathbf{b} \cdot \frac{d\mathbf{b}}{dt}\right).$$

But $\mathbf{b} \cdot \mathbf{b} = |\mathbf{b}|^2$ is constant by hypothesis, so its derivative is zero. Consequently, $\mathbf{b} \cdot (d\mathbf{b}/dt) = 0$, so $d\mathbf{b}/dt$ is orthogonal to $\mathbf{b}$. □

We are now ready to define the second unit vector $\mathbf{n}$ of the orthogonal vector frame $\mathbf{t}, \mathbf{n}$ that we carry along with us as we travel a curve in the plane or in space. Since $\mathbf{t}$ has constant length 1, Theorem 13.1 shows that we could define $\mathbf{n}$ using the derivative of $\mathbf{t}$ with respect to any parameter that determines $\mathbf{t}$. It will be convenient for us to use the arc length parameter s, for we are going to run into the *curvature* of the curve. Recall from Section 11.3 that the curvature κ of a plane curve is defined to be

$$\kappa = \left|\frac{d\phi}{ds}\right|,$$

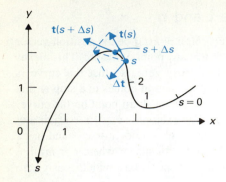

FIGURE 13.13 $\Delta \mathbf{t} = \mathbf{t}(s + \Delta s) - \mathbf{t}(s)$ points toward the concave side of the curve.

where ϕ at any point on the curve is the angle measured counterclockwise from the horizontal to the tangent line to the curve at that point.

Let $\mathbf{t}$ be the unit vector tangent to a curve in the plane or in space as defined by Eq. (1). By Theorem 13.1, $d\mathbf{t}/ds$ is orthogonal to $\mathbf{t}$. If $d\mathbf{t}/ds \neq \mathbf{0}$, we select a unit vector $\mathbf{n}$ orthogonal to $\mathbf{t}$ by choosing

$$\mathbf{n} = \frac{d\mathbf{t}/ds}{|d\mathbf{t}/ds|}. \tag{2}$$

Figure 13.13 shows $\Delta \mathbf{t}$ for a plane curve. As indicated in the figure,

$$\Delta \mathbf{t} = \mathbf{t}(s + \Delta s) - \mathbf{t}(s)$$

points toward the *concave* side of the curve. Thus $d\mathbf{t}/ds = \lim_{\Delta s \to 0} \Delta \mathbf{t}/\Delta s$ is a vector orthogonal to $\mathbf{t}$ on the *concave* side of the curve, and the unit vector $\mathbf{n}$ in that direction therefore also points to the concave side, as shown in Fig. 13.14.

We could give an example, computing $\mathbf{n}$ at a point on a curve, using definition (2) for $\mathbf{n}$. However, we will find an easier method for computing $\mathbf{n}$ in a moment, so we delay this illustration.

It is a bit more complicated to describe just which way $\mathbf{n}$ points for a space curve, because any unit vector in the entire plane normal to $\mathbf{t}$, shown in Fig. 13.15, is a vector orthogonal to $\mathbf{t}$. The vector $\mathbf{n}$ is called the **principal normal vector**. The plane containing $\mathbf{t}$ and $\mathbf{n}$ at a point is called the **osculating plane** at that point. In Fig. 13.16, we show a triangle having as two sides $\mathbf{t}(s)$ and $\mathbf{t}(s + \Delta s)$ starting at the point s and, as third side, $\Delta \mathbf{t} = \mathbf{t}(s + \Delta s) - \mathbf{t}(s)$. The vector $\mathbf{t}(s + \Delta s)$ has been translated from its natural position at $s + \Delta s$ on the curve back to the position s. This triangle determines a plane. As $\Delta s \to 0$, it seems reasonable that the limiting position of these planes is the plane most nearly containing the curve near the point s. Since $d\mathbf{t}/ds = \lim_{\Delta s \to 0} \Delta \mathbf{t}/\Delta s$ and $\mathbf{n}$ and $d\mathbf{t}/ds$ are parallel, we see that this limiting plane is our osculating plane, containing $\mathbf{t}$ and $\mathbf{n}$.

We would like to compute the equation of an osculating plane. To find the equation of any plane, we want to know a point in the plane and a normal vector to the plane. How can we find such a normal vector? One way is to take a cross

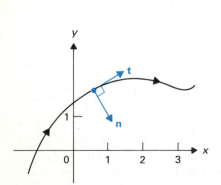

FIGURE 13.14 The orthogonal frame $\mathbf{t}$ and $\mathbf{n}$ at a point on the curve where $d\mathbf{t}/ds \neq 0$.

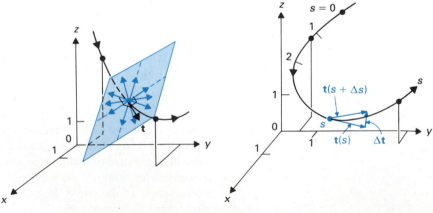

FIGURE 13.15 In the plane normal to the curve, every vector is orthogonal to $\mathbf{t}$.

FIGURE 13.16 The triangle with sides $\mathbf{t}(s)$, $\mathbf{t}(s + \Delta s)$, and $\Delta \mathbf{t}$.

product of two nonparallel vectors in the plane, say **t** and **n**. Since **v** and **t** are parallel, we can use **v** and **n** just as well. Recall from Eq. (18), Section 13.1, that

$$\frac{d(f(s)\mathbf{t})}{ds} = f(s)\frac{d\mathbf{t}}{ds} + f'(s)\mathbf{t}.$$

This equation shows that $d(f(s)\mathbf{t})/ds$ lies in the same plane as $d\mathbf{t}/ds$ and **t**. Since **v** and **t** differ by the scalar function factor $f(s) = |\mathbf{v}|$, this argument shows that $d\mathbf{v}/ds$ lies in the same plane as **t** and $d\mathbf{t}/ds$, that is, $d\mathbf{v}/ds$ is also in the osculating plane. Finally, since

$$\frac{d\mathbf{v}}{ds} = \frac{dt}{ds}\cdot\frac{d\mathbf{v}}{dt} = \frac{dt}{ds}\mathbf{a},$$

we see that **a** lies in the osculating plane. Since **v** and **a** are readily computed, we take the cross product of them, rather than of **t** and **n**, to find a normal vector to the osculating plane.

EXAMPLE 1 Find the equation of the osculating plane to the twisted cubic $x = 3t^2$, $y = 2t^3$, $z = 3t$, where $t = 1$.

Solution We have

$$\mathbf{r}|_1 = (3t^2\mathbf{i} + 2t^3\mathbf{j} + 3t\mathbf{k})|_1 = 3\mathbf{i} + 2\mathbf{j} + 3\mathbf{k},$$
$$\mathbf{v}|_1 = (6t\mathbf{i} + 6t^2\mathbf{j} + 3\mathbf{k})|_1 = 6\mathbf{i} + 6\mathbf{j} + 3\mathbf{k},$$
$$\mathbf{a}|_1 = (6\mathbf{i} + 12t\mathbf{j})|_1 = 6\mathbf{i} + 12\mathbf{j}.$$

Then when $t = 1$, we see that $\frac{1}{3}\mathbf{v} = 2\mathbf{i} + 2\mathbf{j} + \mathbf{k}$ and $\frac{1}{6}\mathbf{a} = \mathbf{i} + 2\mathbf{j}$ lie in the osculating plane. Consequently,

$$\begin{vmatrix} \mathbf{i} & \mathbf{j} & \mathbf{k} \\ 2 & 2 & 1 \\ 1 & 2 & 0 \end{vmatrix} = -2\mathbf{i} + \mathbf{j} + 2\mathbf{k}$$

is normal to the plane. We have

Point $(3, 2, 3)$
$\perp$ vector $-2\mathbf{i} + \mathbf{j} + 2\mathbf{k}$
Equation $-2x + y + 2z = 2$ ∎

$|\mathbf{a}_{\text{tan}}|$ and $|\mathbf{a}_{\text{nor}}|$

In Section 13.1, we learned to break the acceleration vector **a** into tangential and normal vector components $\mathbf{a}_{\text{tan}}$ and $\mathbf{a}_{\text{nor}}$, where

$$\mathbf{a}_{\text{tan}} = \frac{\mathbf{a}\cdot\mathbf{v}}{\mathbf{v}\cdot\mathbf{v}}\mathbf{v} \quad \text{and} \quad \mathbf{a}_{\text{nor}} = \mathbf{a} - \mathbf{a}_{\text{tan}}. \tag{3}$$

We now study $|\mathbf{a}_{\text{tan}}|$ and $|\mathbf{a}_{\text{nor}}|$ a bit more.

For a smooth curve in the plane, let ϕ be the angle that **t** makes with the horizontal, shown in Fig. 13.17. Recall that the curvature κ is given by

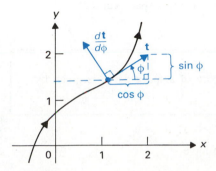

FIGURE 13.17 $\mathbf{t} = (\cos\phi)\mathbf{i} + (\sin\phi)\mathbf{j}$; $d\mathbf{t}/d\phi = (-\sin\phi)\mathbf{i} + (\cos\phi)\mathbf{j}$.

$$\kappa = \left|\frac{d\phi}{ds}\right|. \tag{4}$$

We may write **t** as

$$\mathbf{t} = (\cos \phi)\mathbf{i} + (\sin \phi)\mathbf{j}. \tag{5}$$

Then

$$\frac{d\mathbf{t}}{d\phi} = (-\sin \phi)\mathbf{i} + (\cos \phi)\mathbf{j}.$$

By Theorem 13.1, $d\mathbf{t}/d\phi$ is orthogonal to **t**, and it is a unit vector since $\sin^2\phi + \cos^2\phi = 1$. Recall the trigonometric identities

$$\sin\left(\phi + \frac{\pi}{2}\right) = \cos \phi,$$

$$\cos\left(\phi + \frac{\pi}{2}\right) = -\sin \phi.$$

Thus $d\mathbf{t}/d\phi$ may be obtained by replacing ϕ by $\phi + \pi/2$ in Eq. (5). This shows that $d\mathbf{t}/d\phi$ can be obtained by rotating **t** *counterclockwise* through 90°. Consequently, $d\mathbf{t}/d\phi$ is either **n** or $-\mathbf{n}$.

By the chain rule, we see that

$$\frac{d\mathbf{t}}{ds} = \frac{d\phi}{ds} \cdot \frac{d\mathbf{t}}{d\phi}. \tag{6}$$

If the arc bends to the left as s increases, then ϕ is increasing, so $d\phi/ds = \kappa$, and $d\mathbf{t}/d\phi = \mathbf{n}$, since the concave side of the arc is on the left. If the arc bends to the right, then $d\phi/ds = -\kappa$ and $d\mathbf{t}/d\phi = -\mathbf{n}$. Thus, in either case, Eq. (6) shows that

$$\frac{d\mathbf{t}}{ds} = \frac{d\phi}{ds} \cdot \frac{d\mathbf{t}}{d\phi} = \kappa\mathbf{n}. \tag{7}$$

We have not defined curvature for a space curve. Recall that we defined the principal normal vector **n** for a space curve to be

$$\mathbf{n} = \frac{d\mathbf{t}/ds}{|d\mathbf{t}/ds|}, \tag{8}$$

if $d\mathbf{t}/ds \neq \mathbf{0}$. Equation (7) shows that in the plane, we have

$$\kappa = \left|\frac{d\mathbf{t}}{ds}\right|. \tag{9}$$

We take Eq. (9) as the *definition* of curvature for a space curve.

DEFINITION 13.3

Curvature of a space curve

The **curvature** κ at a point on a smooth arc of a space curve is given by $\kappa = |d\mathbf{t}/ds|$.

Thus for both plane and space curves we have

$$\frac{d\mathbf{t}}{ds} = \kappa\mathbf{n}. \tag{10}$$

Recall that for both a plane curve and a space curve,

$$\mathbf{v} = \frac{ds}{dt}\mathbf{t}. \tag{11}$$

Differentiating Eq. (11), using the product and chain rules, we have

$$\mathbf{a} = \frac{d\mathbf{v}}{dt} = \frac{ds}{dt}\cdot\frac{d\mathbf{t}}{dt} + \frac{d^2s}{dt^2}\mathbf{t}$$

$$= \frac{ds}{dt}\left(\frac{ds}{dt}\cdot\frac{d\mathbf{t}}{ds}\right) + \frac{d^2s}{dt^2}\mathbf{t} = \frac{d^2s}{dt^2}\mathbf{t} + \left(\frac{ds}{dt}\right)^2\frac{d\mathbf{t}}{ds}.$$

Using Eq. (10), we then obtain

$$\mathbf{a} = \frac{d^2s}{dt^2}\mathbf{t} + \kappa\left(\frac{ds}{dt}\right)^2\mathbf{n}. \tag{12}$$

Comparing Eq. (12) and the decomposition $\mathbf{a} = \mathbf{a}_{\text{tan}} + \mathbf{a}_{\text{nor}}$, we see that the magnitudes of the tangential and normal components of acceleration are given by

$$|\mathbf{a}_{\text{tan}}| = \left|\frac{d^2s}{dt^2}\right| \quad \text{and} \quad |\mathbf{a}_{\text{nor}}| = \kappa\left(\frac{ds}{dt}\right)^2 = \kappa|v|^2. \tag{13}$$

The relations in Eq. (13) have a nice physical interpretation. By Newton's second law of motion, the force vector $\mathbf{F}$ governing a body of mass m moving on a plane curve with acceleration $\mathbf{a}$ is given by $\mathbf{F} = m\mathbf{a}$. From Eq. (13), we see that the component of the force tangential to the path of the body controls the rate of change of the speed of the body, for

$$m|\mathbf{a}_{\text{tan}}| = m\left|\frac{d^2s}{dt^2}\right| = m\left|\frac{d(ds/dt)}{dt}\right|.$$

On the other hand, the component of the force normal to the direction of a body controls the curvature of the path on which the body travels. This seems intuitively reasonable. The formula for $|\mathbf{a}_{\text{nor}}|$ in Eq. (13) shows that, when we are driving a car around an unbanked curve, the magnitude of the force normal to the curve exerted by the road on the wheels is

$$|\mathbf{F}_{\text{nor}}| = m|\mathbf{a}_{\text{nor}}| = m\kappa|\mathbf{v}|^2 = \frac{m}{\rho}|\mathbf{v}|^2,$$

where $\rho = 1/\kappa$ is the radius of curvature of the curve. This force is directly proportional to the curvature of the curve (inversely proportional to the radius of curvature) and directly proportional to the *square* of the speed of the car. The larger the curvature (smaller the radius of curvature), the more force is required. If the speed of the car is doubled, the force required normal to the curve to "hold the car on the road" is *quadrupled*. If we take a sharp unbanked curve too fast, then the available force due to friction normal to the wheels is not sufficient for the car to make the curve; we skid and go off the road.

Computations

We summarize in Table 13.1 straightforward ways to compute all the things defined in this section and the preceding one. Everything is readily found from $\mathbf{v}$ and $\mathbf{a}$, which are themselves usually readily computed. The computations should be made in the order in the table.

TABLE 13.1   Formulas for computations

Position vector	$\mathbf{r}$ (given)	(14)
Velocity vector	$\mathbf{v} = \dfrac{d\mathbf{r}}{dt}$	(15)
Speed	$\|\mathbf{v}\| = \dfrac{ds}{dt}$	(16)
Acceleration vector	$\mathbf{a} = \dfrac{d\mathbf{v}}{dt} = \dfrac{d^2\mathbf{r}}{dt^2}$	(17)
Tangential vector component of $\mathbf{a}$	$\mathbf{a}_{\text{tan}} = \dfrac{(\mathbf{a} \cdot \mathbf{v})}{(\mathbf{v} \cdot \mathbf{v})}\mathbf{v}$	(18)
Normal vector component of $\mathbf{a}$	$\mathbf{a}_{\text{nor}} = \mathbf{a} - \mathbf{a}_{\text{tan}}$	(19)
Unit tangent vector if $\mathbf{v} \neq \mathbf{0}$	$\mathbf{t} = \dfrac{1}{\|\mathbf{v}\|}\mathbf{v}$	(20)
Principal normal vector if $\mathbf{a}_{\text{nor}} \neq \mathbf{0}$	$\mathbf{n} = \dfrac{1}{\|\mathbf{a}_{\text{nor}}\|}\mathbf{a}_{\text{nor}}$	(21)
Rate of change of speed	$\dfrac{d^2s}{dt^2} = \pm\|\mathbf{a}_{\text{tan}}\|$, negative if $\mathbf{a} \cdot \mathbf{v} < 0$	(22)
Curvature if $\mathbf{v} \neq \mathbf{0}$	$\kappa = \dfrac{\|\mathbf{a}_{\text{nor}}\|}{(\mathbf{v} \cdot \mathbf{v})} = \dfrac{\|\mathbf{a}_{\text{nor}}\|}{\|\mathbf{v}\|^2}$	(23)
Osculating plane if $\mathbf{v}$ and $\mathbf{a}$ are not parallel	Point (given), normal vector $\mathbf{v} \times \mathbf{a}$	(24)

The formulas in Table 13.1 can be verified from the work we have done. To illustrate, Eq. (13) tells us that $\|\mathbf{a}_{\text{nor}}\| = \kappa(ds/dt)^2$. Since $(ds/dt)^2 = \|\mathbf{v}\|^2 = \mathbf{v} \cdot \mathbf{v}$, we obtain the formula for κ in Eq. (23). For another illustration, note that if $\mathbf{a}_{\text{tan}}$ has opposite direction to $\mathbf{v}$, then the speed is decreasing, so $d^2s/dt^2 < 0$, while $d^2s/dt^2 > 0$ when $\mathbf{a}_{\text{tan}}$ has the same direction as $\mathbf{v}$. Since $\|\mathbf{a}_{\text{tan}}\| = \|d^2s/dt^2\|$ and $\mathbf{a}_{\text{tan}}$ has opposite direction to $\mathbf{v}$ precisely when $\mathbf{a} \cdot \mathbf{v} < 0$, we obtain Eq. (22).

EXAMPLE 2   Let the position vector of a body in the plane at time t be

$$\mathbf{r} = (1 + \cos t - \sin t)\mathbf{i} + (\sin t + \cos t)\mathbf{j}.$$

Find all the things listed in Table 13.1, except the equation of the osculating plane, at any time t.

Solution   From Eqs. (15) and (16),

$$\mathbf{v} = (-\sin t - \cos t)\mathbf{i} + (\cos t - \sin t)\mathbf{j},$$

$$\text{Speed} = \|\mathbf{v}\| = \sqrt{2\,\sin^2 t + 2\,\cos^2 t} = \sqrt{2}.$$

By Eq. (17),

$$\mathbf{a} = (-\cos t + \sin t)\mathbf{i} - (\sin t + \cos t)\mathbf{j}.$$

We easily see that $\mathbf{a} \cdot \mathbf{v} = 0$. Thus Eq. (18) yields

$$\mathbf{a}_{\text{tan}} = \mathbf{0}.$$

From Eq. (19), we have

$$\mathbf{a}_{nor} = \mathbf{a} - \mathbf{0} = (-\cos t + \sin t)\mathbf{i} - (\sin t + \cos t)\mathbf{j}.$$

Equation (20) shows that

$$\mathbf{t} = \frac{1}{\sqrt{2}}\mathbf{v}$$

$$= \frac{1}{\sqrt{2}}[(-\sin t - \cos t)\mathbf{i} + (\cos t - \sin t)\mathbf{j}].$$

We easily see that

$$|\mathbf{a}_{nor}| = \sqrt{2 \sin^2 t + 2 \cos^2 t} = \sqrt{2}.$$

Equation (21) becomes

$$\mathbf{n} = \frac{1}{\sqrt{2}}[(-\cos t + \sin t)\mathbf{i} - (\sin t + \cos t)\mathbf{j}].$$

Since $ds/dt = \sqrt{2}$, we see that

$$\frac{d^2 s}{dt^2} = 0.$$

Finally, Eq. (23) tells us that

$$\kappa = \frac{\sqrt{2}}{(\sqrt{2})^2} = \frac{1}{\sqrt{2}}.$$

Our curve has constant curvature $1/\sqrt{2}$, so it must be a circle of radius $\sqrt{2}$.

■

Generally, the computations at a single point are less tedious than the computations for any time t that we did in the preceding example.

EXAMPLE 3 Find all the things listed in Table 13.1 for the twisted cubic $x = 3t^2$, $y = 2t^3$, $z = 3t$ of Example 1, where $t = 1$.

Solution From Example 1, we know that when $t = 1$,

$$\mathbf{r} = 3\mathbf{i} + 2\mathbf{j} + 3\mathbf{k},$$
$$\mathbf{v} = 6\mathbf{i} + 6\mathbf{j} + 3\mathbf{k},$$
$$\mathbf{a} = 6\mathbf{i} + 12\mathbf{j}.$$

Then

$$\text{Speed} = |\mathbf{v}| = \sqrt{36 + 36 + 9} = \sqrt{81} = 9.$$

We have

$$\mathbf{a}_{tan} = \frac{\mathbf{a} \cdot \mathbf{v}}{\mathbf{v} \cdot \mathbf{v}}\mathbf{v} = \frac{108}{81}\mathbf{v} = \frac{4}{3}\mathbf{v} = 8\mathbf{i} + 8\mathbf{j} + 4\mathbf{k},$$

and

$$\mathbf{a}_{nor} = \mathbf{a} - \mathbf{a}_{tan} = -2\mathbf{i} + 4\mathbf{j} - 4\mathbf{k}.$$

Now

$$t = \frac{1}{|\mathbf{v}|}\mathbf{v} = \frac{1}{9}\mathbf{v} = \frac{2}{3}\mathbf{i} + \frac{2}{3}\mathbf{j} + \frac{1}{3}\mathbf{k},$$

and

$$\mathbf{n} = \frac{1}{|\mathbf{a}_{\text{nor}}|}\mathbf{a}_{\text{nor}}$$

$$= \frac{1}{6}\mathbf{a}_{\text{nor}} = -\frac{1}{3}\mathbf{i} + \frac{2}{3}\mathbf{j} - \frac{2}{3}\mathbf{k}.$$

We have

$$\frac{d^2s}{dt^2} = |a_{\text{tan}}| = 12$$

since $\mathbf{a} \cdot \mathbf{v} > 0$, and

$$\kappa = \frac{|\mathbf{a}_{\text{nor}}|}{|\mathbf{v}|^2} = \frac{6}{81} = \frac{2}{27}.$$

We found the equation of the osculating plane in Example 1 to be

$$-2x + y + 2z = 2.$$

SUMMARY

Let $\mathbf{t}$ and $\mathbf{n}$ be tangent and normal unit vectors to a twice-differentiable smooth plane curve at a point, with t pointing in the direction corresponding to increasing t and $\mathbf{n}$ pointing toward the concave side of the curve.

1. $d\mathbf{t}/ds = \kappa\mathbf{n}$, where κ is the curvature of the curve

2. $\mathbf{v} = (ds/dt)\mathbf{t}$

3. $\mathbf{a} = d^2\mathbf{r}/dt^2 = (d^2s/dt^2)\mathbf{t} + \kappa(ds/dt)^2\mathbf{n}$,

   so the tangential and normal scalar components of acceleration are, respectively,

   $$\frac{d^2s}{dt^2} \quad \text{and} \quad \kappa\left(\frac{ds}{dt}\right)^2.$$

Now consider a twice-differentiable smooth space curve.

4. $\mathbf{t} = (1/|\mathbf{v}|)\mathbf{v}$

5. The principal normal vector $\mathbf{n}$ is the unit vector in the direction of $d\mathbf{t}/ds$ if $d\mathbf{t}/ds \neq \mathbf{0}$.

6. The osculating plane at a point is the plane containing $\mathbf{t}$ and $\mathbf{n}$ if $d\mathbf{t}/ds \neq \mathbf{0}$. It is the plane that most nearly contains the curve close to the point.

7. The curvature $\kappa = |d\mathbf{t}/ds|$

8. Number 3 also holds for space curves.

9. See Table 13.1 for straightforward ways to compute these things.

EXERCISES 13.2

In Exercises 1–5, find the indicated quantity when $t = 0$ for the motion in the plane having position vector

$$\mathbf{r} = a(\sin 2t)\mathbf{i} + b(\cos 3t)\mathbf{j}.$$

1. The vector $\mathbf{t}$

2. The vector $\mathbf{n}$

3. The rate of change of speed

4. $|\mathbf{a}_{\text{nor}}|$ **5.** The curvature

In Exercises 6–10, find the indicated quantity when $t = 1$ for the motion in the plane having position vector

$$\mathbf{r} = (t^2 + 2t)\mathbf{i} + \left(4t + \frac{1}{t}\right)\mathbf{j}.$$

6. The vector $\mathbf{t}$

7. The vector $\mathbf{n}$

8. The rate of change of speed

9. $|\mathbf{a}_{\text{nor}}|$

10. The curvature

In Exercises 11–15, find the indicated quantity when $t = 0$ for the motion in the plane having position vector

$$\mathbf{r} = e^{2t}\mathbf{i} + \ln(1 + t)\mathbf{j}.$$

11. The vector $\mathbf{t}$

12. The vector $\mathbf{n}$

13. The rate of change of speed

14. $|\mathbf{a}_{\text{nor}}|$ **15.** The curvature

In Exercises 16–22, find the indicated quantity when $t = 1$ for the motion in space with position vector

$$\mathbf{r} = \left(\frac{1}{t}\right)\mathbf{i} + \left(\frac{1}{t^2}\right)\mathbf{j} + \left(\frac{2}{t}\right)\mathbf{k}.$$

16. The vector $\mathbf{t}$

17. The speed

18. The vector $\mathbf{n}$

19. The equation of the osculating plane

20. The rate of change of speed

21. $|\mathbf{a}_{\text{nor}}|$

22. The curvature

In Exercises 23–29, find the indicated quantity when $t = 0$ for the motion in space with position vector

$$\mathbf{r} = (e^t \sin t)\mathbf{i} + (e^t \cos t)\mathbf{j} + e^{2t}\mathbf{k}.$$

23. The vector $\mathbf{t}$

24. The speed   **25.** The vector $\mathbf{n}$

26. The equation of the osculating plane

27. The rate of change of speed

28. $|\mathbf{a}_{\text{nor}}|$ **29.** The curvature

In Exercises 30–36, find the indicated quantity when $t = \pi/4$ for the motion in space with position vector

$$\mathbf{r} = [\ln(\sin t)]\mathbf{i} + [\ln(\cos t)]\mathbf{j} + (\sin 2t)\mathbf{k}.$$

30. The vector $\mathbf{t}$ **31.** The speed

32. The vector $\mathbf{n}$

33. The equation of the osculating plane

34. The rate of change of speed

35. $|\mathbf{a}_{\text{nor}}|$

36. The curvature

37. Show that if the curvature of a space curve is zero at a point, then the osculating plane to the curve at the point is not defined.

38. Consider the motion in space with the position vector $\mathbf{r} = (3 + 7t)\mathbf{i} + (4 - t)\mathbf{j} + (1 + t)\mathbf{k}$. Without using any calculus, explain why there is no unique osculating plane at any point of the path.

39. A body in space of mass 3 is moving subject to the force vector

$$\mathbf{F} = 6(\cos t)\mathbf{i} + 9(\sin t)\mathbf{j}.$$

If $\mathbf{v}|_{t=0} = \mathbf{i} - 3\mathbf{j} + 4\mathbf{k}$, find
(a) the rate of change of speed when $t = \pi$,
(b) the curvature of the path when $t = \pi$.

40. A body of mass 2 in space moves subject to the force vector

$$\mathbf{F} = \frac{24}{t^4}\mathbf{j} - \frac{12}{t^3}\mathbf{k}$$

for $t \geq 1$. If $\mathbf{v}|_{t=2} = 2\mathbf{i} + \mathbf{j}$, find
(a) the speed when $t = 2$,
(b) the rate of change of speed when $t = 2$,
(c) the curvature of the path when $t = 2$.

13.3 POLAR VECTOR ANALYSIS AND KEPLER'S LAWS (OPTIONAL)

Velocity and Acceleration in Polar Coordinates

At each point in the polar coordinate plane except the pole (origin), we let $\mathbf{u}_r$ be the unit vector pointing directly away from the origin. We then let $\mathbf{u}_\theta$ be the unit vector normal to $\mathbf{u}_r$ in the direction of increasing θ, as illustrated in Fig. 13.18. As shown in Fig. 13.19, we have

$$\mathbf{u}_r = (\cos \theta)\mathbf{i} + (\sin \theta)\mathbf{j}, \tag{1}$$

while

$$\mathbf{u}_\theta = (-\sin \theta)\mathbf{i} + (\cos \theta)\mathbf{j}. \tag{2}$$

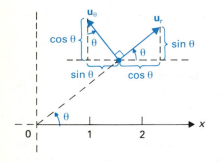

FIGURE 13.18 Orthogonal unit vectors $\mathbf{u}_r$ and $\mathbf{u}_\theta$ at points other than the origin.

EXAMPLE 1 Express $\mathbf{u}_r$ and $\mathbf{u}_\theta$ at the point $(r, \theta) = (5, \pi/3)$ in terms of $\mathbf{i}$ and $\mathbf{j}$.

Solution From Eqs. (1) and (2), we have

$$\mathbf{u}_r = \left(\cos \frac{\pi}{3}\right)\mathbf{i} + \left(\sin \frac{\pi}{3}\right)\mathbf{j} = \frac{1}{2}\mathbf{i} + \frac{\sqrt{3}}{2}\mathbf{j}$$

and

$$\mathbf{u}_\theta = \left(-\sin \frac{\pi}{3}\right)\mathbf{i} + \left(\cos \frac{\pi}{3}\right)\mathbf{j} = -\frac{\sqrt{3}}{2}\mathbf{i} + \frac{1}{2}\mathbf{j}. \qquad \blacksquare$$

Let $r = f(\theta)$, where f is a twice-differentiable function, and consider a body moving along the curve $r = f(\theta)$. We wish to express the velocity and acceleration vectors of the body at time t in terms of the unit vector $\mathbf{u}_r$ directed away from the origin and the perpendicular unit vector $\mathbf{u}_\theta$ in the direction of increasing θ, as shown in Fig. 13.20. From Eqs. (1) and (2), we at once obtain

$$\frac{d\mathbf{u}_r}{d\theta} = (-\sin \theta)\mathbf{i} + (\cos \theta)\mathbf{j} = \mathbf{u}_\theta \tag{3}$$

and

$$\frac{d\mathbf{u}_\theta}{d\theta} = (-\cos \theta)\mathbf{i} + (-\sin \theta)\mathbf{j} = -\mathbf{u}_r. \tag{4}$$

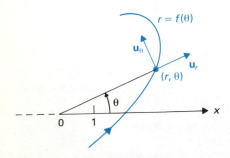

FIGURE 13.19 $\mathbf{u}_r = (\cos \theta)\mathbf{i} + (\sin \theta)\mathbf{j}$; $\mathbf{u}_\theta = (-\sin \theta)\mathbf{i} + (\cos \theta)\mathbf{j}$

We can now obtain the desired formulas for the velocity vector $\mathbf{v}$ and acceleration vector $\mathbf{a}$ by differentiating the position vector

$$\mathbf{r} = r\mathbf{u}_r \tag{5}$$

with respect to t. Using a product rule, the chain rule, and Eq. (3), we obtain from Eq. (5)

$$\mathbf{v} = \frac{d\mathbf{r}}{dt} = r\frac{d\mathbf{u}_r}{dt} + \frac{dr}{dt}\mathbf{u}_r$$

$$= r\frac{d\theta}{dt} \cdot \frac{d\mathbf{u}_r}{d\theta} + \frac{dr}{dt}\mathbf{u}_r = r\frac{d\theta}{dt}\mathbf{u}_\theta + \frac{dr}{dt}\mathbf{u}_r,$$

FIGURE 13.20 The vectors $\mathbf{u}_r$ and $\mathbf{u}_\theta$ on the path of a body in motion.

so

$$\mathbf{v} = \dot{r}\mathbf{u}_r + r\dot{\theta}\mathbf{u}_\theta, \tag{6}$$

where a dot over a variable denotes a derivative with respect to time. Differentiating Eq. (6), using a product rule, chain rules, and Eqs. (3) and (4), we have

$$\mathbf{a} = \frac{d\mathbf{v}}{dt} = \dot{r}\frac{d\mathbf{u}_r}{dt} + \ddot{r}\mathbf{u}_r + r\dot{\theta}\frac{d\mathbf{u}_\theta}{dt} + (r\ddot{\theta} + \dot{r}\dot{\theta})\mathbf{u}_\theta$$

$$= \dot{r}\frac{d\theta}{dt}\cdot\frac{d\mathbf{u}_r}{d\theta} + \ddot{r}\mathbf{u}_r + r\dot{\theta}\frac{d\theta}{dt}\cdot\frac{d\mathbf{u}_\theta}{d\theta} + (r\ddot{\theta} + \dot{r}\dot{\theta})\mathbf{u}_\theta$$

$$= \dot{r}\dot{\theta}\mathbf{u}_\theta + \ddot{r}\mathbf{u}_r + r\dot{\theta}^2(-\mathbf{u}_r) + (r\ddot{\theta} + \dot{r}\dot{\theta})\mathbf{u}_\theta,$$

which yields

$$\mathbf{a} = (\ddot{r} - r\dot{\theta}^2)\mathbf{u}_r + (r\ddot{\theta} + 2\dot{r}\dot{\theta})\mathbf{u}_\theta. \tag{7}$$

EXAMPLE 2 The polar coordinate position of a body in the plane at time $t \geq 0$ is given by $(r, \theta) = (t^2, \sqrt{t})$. Find the velocity and acceleration vectors at time $t > 0$.

Solution We have

$$r = t^2, \qquad \theta = t^{1/2},$$
$$\dot{r} = 2t, \qquad \dot{\theta} = \tfrac{1}{2}t^{-1/2},$$
$$\ddot{r} = 2, \qquad \ddot{\theta} = -\tfrac{1}{4}t^{-3/2}.$$

Using Eqs. (6) and (7), we obtain

$$\mathbf{v} = 2t\mathbf{u}_r + \tfrac{1}{2}t^{3/2}\mathbf{u}_\theta$$

and

$$\mathbf{a} = (2 - \tfrac{1}{4}t)\mathbf{u}_r + (-\tfrac{1}{4}\sqrt{t} + 2\sqrt{t})\mathbf{u}_\theta$$
$$= (2 - \tfrac{1}{4}t)\mathbf{u}_r + \tfrac{7}{4}\sqrt{t}\mathbf{u}_\theta. \qquad \blacksquare$$

EXAMPLE 3 For the motion in Example 2, find the speed when $t = 4$.

Solution We know that

$$\text{Speed} = |\mathbf{v}|.$$

From Example 2,

$$\mathbf{v}\big|_{t=4} = 8\mathbf{u}_r + 4\mathbf{u}_\theta.$$

Since $\mathbf{u}_r$ and $\mathbf{u}_\theta$ are *orthogonal unit vectors*, the speed when $t = 4$ is

$$\text{Speed} = |8\mathbf{u}_r + 4\mathbf{u}_\theta| = \sqrt{64 + 16} = \sqrt{80} = 4\sqrt{5}. \qquad \blacksquare$$

EXAMPLE 4 Find the rate of change of speed for the body in Example 2 when $t = 4$.

Solution We have, from Example 2,

$$\text{Speed} = \frac{ds}{dt} = |\mathbf{v}| = \sqrt{4t^2 + (t^3/4)}.$$

The rate of change of speed is then

$$\frac{d^2s}{dt^2} = \frac{1}{2} \cdot \frac{8t + (3t^2/4)}{\sqrt{4t^2 + (t^3/4)}}.$$

When $t = 4$, we obtain

$$\frac{d^2s}{dt^2}\bigg|_{t=4} = \frac{1}{2} \cdot \frac{32 + 12}{\sqrt{64 + 16}} = \frac{22}{4\sqrt{5}} = \frac{11}{2\sqrt{5}}. \qquad ■$$

EXAMPLE 5   For the body in Example 2, find the curvature of the path when $t = 4$.

Solution   From Example 2, we have

$$\mathbf{a}\big|_{t=4} = \mathbf{u}_r + \frac{7}{2}\mathbf{u}_\theta.$$

Then

$$|\mathbf{a}|^2\big|_{t=4} = 1 + \frac{49}{4} = \frac{53}{4}.$$

From our work in the previous section, we know that

$$|\mathbf{a}|^2 = |\mathbf{a}_{\tan}|^2 + |\mathbf{a}_{\text{nor}}|^2 = \left(\frac{d^2s}{dt^2}\right)^2 + \left(\kappa\left(\frac{ds}{dt}\right)^2\right)^2.$$

From Examples 3 and 4, when $t = 4$, this equation becomes

$$\frac{53}{4} = \frac{121}{20} + (\kappa \cdot 80)^2,$$

so

$$(\kappa \cdot 80)^2 = \frac{53}{4} - \frac{121}{20} = \frac{265 - 121}{20} = \frac{144}{20}.$$

Thus

$$\kappa = \frac{1}{80} \cdot \frac{12}{\sqrt{20}} = \frac{3}{40\sqrt{5}}. \qquad ■$$

Central Force Fields and Kepler's Laws

In this section, we work in polar coordinates and show how we can derive the planetary laws of Johannes Kepler (1571–1630) from Newton's law of gravitation. Before stating Kepler's laws, we explain some terminology connected with the ellipse with equation

$$\frac{(x - c)^2}{a^2} + \frac{y^2}{b^2} = 1, \qquad \text{where } a > b \quad \text{and} \quad c = \sqrt{a^2 - b^2},$$

shown in Fig. 13.21. The horizontal line segment of length $2a$ dividing the ellipse into two equal portions is the **major axis** of the ellipse, while the shorter, vertical line segment of length $2b$ dividing the ellipse equally is the **minor axis.**

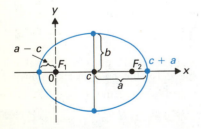

FIGURE 13.21   Ellipse with foci F_1 (at the origin) and F_2.

The two points on the major axis of the ellipse at a distance c from the center of the ellipse are the **foci** of the ellipse. The ellipse shown in Fig. 13.21 has one focus at the origin. (It can be shown that if an ellipse is constructed of a wire that reflects light, then light in the plane emanating from a source at one focus is all reflected to the other focus. This explains the reason for the term *focus*.)

We now have the background to understand Newton's law of gravitation and Kepler's laws. These laws are as follows:

Newton's Law of Universal Gravitation

The force F of attraction between two bodies of masses m and M is given by

$$F = \frac{GmM}{d^2},$$

where G is a "universal" constant of gravitational attraction and d is the distance between the bodies.

Kepler's Laws

1. The orbit of each planet is an ellipse with the sun at a focus.

2. A line segment joining the sun and a particular planet sweeps out regions of equal areas during equal time intervals. (See Fig. 13.22.)

3. Let the elliptical orbit of a planet have a major axis of length $2a$ and let T be the **period** of the planet (that is, the time required for one complete revolution about the sun). Then the ratio a^3/T^2 is the same for all planets of our sun.

We derive Kepler's laws from Newton's law. This reverses the historical order. Newton actually derived his inverse square law from Kepler's laws. Kepler, in turn, formulated his laws on the basis of his analysis of astronomical observations by Tycho Brahe (1546–1601). After some exercises illustrating the material in the text, we give a sequence of exercises that constitute a detailed outline for the derivation of Newton's inverse square law from Kepler's three laws.

We will consider a body of mass M to be fixed at the pole O of a polar coordinate system and will examine the motion of a body of mass m, which is subject only to the force of gravitational attraction toward the body at O, as specified by Newton's law. Newton's second law of motion in vector form is

$$\mathbf{F} = m\mathbf{a}.$$

Since the force is directed entirely in the direction $-\mathbf{u}_r$ toward O, we see from

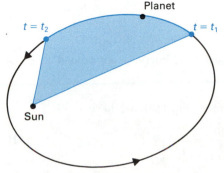

FIGURE 13.22 Region swept out from time t_1 to t_2, referred to in Kepler's second law.

JOHANNES KEPLER (1571–1630) became acquainted with Copernicus's heliocentric theory while at the University of Tübingen and convinced himself that in essence it represented the correct system of the world. Copernicus's details could not be correct, however, since predictions based on them were not accurate. To work out better details, Kepler studied the planetary observations of Tycho Brahe (1546–1601), in particular his records of the planet Mars. After eight years, Kepler finally discovered the laws of planetary motion.

In his *New Astronomy* of 1609 Kepler relates the story of his false starts, his stupid mistakes, and his perseverance until he reached his goal. He had begun by assuming that the orbit of Mars was a circle centered somewhere near the sun. But his faith in Tycho's observations was so great that when his calculations resulted in observations differing from theory by as little as 8' of arc, he decided to drop his preconceived notion and begin a new calculation of the orbit.

He discovered the second law of planetary motion first, from an incorrect argument, and then, virtually by accident, the first law. The third law, which Kepler did not derive but merely stated, first appears in his *Harmonies of the World* (1619) along with various other supposed laws of planetary motion by which Kepler attempted to show that "the movements of the heavens [form] a certain everlasting polyphony," the "music of the spheres."

Eq. (7) that

$$r\ddot{\theta} + 2\dot{r}\dot{\theta} = 0. \tag{8}$$

We obtain, from Eq. (7) and Newton's law of universal gravitation,

$$m(\ddot{r} - r\dot{\theta}^2) = -\frac{mMG}{r^2},$$

or

$$\ddot{r} - r\dot{\theta}^2 = -\frac{MG}{r^2} = -\frac{k}{r^2}, \tag{9}$$

where $k = MG > 0$ is a constant independent of the mass m of the moving body.

We derive Kepler's laws from Eqs. (8) and (9). Kepler's second law is the easiest one to derive.

Derivation of Kepler's Second Law: From Eq. (8), we see that

$$\frac{d}{dt}\left(\frac{1}{2}r^2\dot{\theta}\right) = \frac{1}{2}r^2\ddot{\theta} + r\dot{r}\dot{\theta} = \frac{1}{2}r(r\ddot{\theta} + 2\dot{r}\dot{\theta}) = 0.$$

Thus

$$r^2\dot{\theta} = K \tag{10}$$

for some constant K. Since we are assuming that $r \neq 0$, if $K = 0$ we must have $\dot{\theta} = 0$ and the body is moving in a straight line, either directly toward or directly away from the origin at the center of the force field. We assume from now on that $K \neq 0$.

Now the area A of the region swept out by the line segment joining O and the moving body from time $t = t_0$ where $\theta = \theta_0$ to time t is given by

$$A = \int_{\theta_0}^{\theta(t)} \frac{1}{2}r^2 d\theta = \int_{\theta_0}^{\theta(t)} \frac{1}{2}(f(\theta))^2 \, d\theta.$$

Thus

$$\frac{dA}{dt} = \frac{dA}{d\theta} \cdot \frac{d\theta}{dt} = \frac{1}{2}r^2\frac{d\theta}{dt} = \frac{1}{2}r^2\dot{\theta} = \frac{1}{2}K. \tag{11}$$

From Eq. (11), we have

$$A = \frac{1}{2}Kt + C,$$

from which it follows at once that equal time intervals correspond to equal increments in A.

Derivation of Kepler's First Law: From Eq. (10), we have

$$r\dot{\theta}^2 = \frac{(r^2\dot{\theta})^2}{r^3} = \frac{K^2}{r^3}. \tag{12}$$

Using Eq. (9), we then have

$$\ddot{r} - \frac{K^2}{r^3} = -\frac{k}{r^2}. \tag{13}$$

Multiplication of Eq. (13) by $\dot{r}$ yields

$$\ddot{r}\dot{r} - K^2\frac{\dot{r}}{r^3} = -k\frac{\dot{r}}{r^2}. \tag{14}$$

Integrating Eq. (13) with respect to t and multiplying by 2, we obtain

$$\dot{r}^2 + K^2\frac{1}{r^2} = 2k\frac{1}{r} + C. \tag{15}$$

If we now let $p = 1/r$, then $r = 1/p$ and $\dot{r} = -\dot{p}/p^2$. We may then write Eq. (15) as

$$\frac{\dot{p}^2}{p^4} + K^2p^2 = 2kp + C. \tag{16}$$

From $r^2\dot{\theta} = K$ in Eq. (10), we obtain $\dot{\theta}/p^2 = K$, so

$$K^2\left(\frac{dp}{d\theta}\right)^2 = \frac{1}{p^4}\left(\frac{d\theta}{dt}\right)^2\left(\frac{dp}{d\theta}\right)^2 = \frac{1}{p^4}\left(\frac{dp}{dt}\right)^2 = \frac{\dot{p}^2}{p^4}. \tag{17}$$

From Eqs. (16) and (17),

$$K^2\left[\left(\frac{dp}{d\theta}\right)^2 + p^2\right] = 2kp + C. \tag{18}$$

Differentiation of Eq. (18) with respect to θ yields

$$K^2\left[2\left(\frac{dp}{d\theta}\right)\frac{d^2p}{d\theta^2} + 2p\frac{dp}{d\theta}\right] = 2k\frac{dp}{d\theta},$$

or

$$2K^2\frac{dp}{d\theta}\left(\frac{d^2p}{d\theta^2} + p - \frac{k}{K^2}\right) = 0. \tag{19}$$

If $dp/d\theta = 0$, then $p = 1/r = a$ for some constant a, so $r = 1/a$ and the orbit of the body is a circle with center O, which is in accord with Kepler's first law. Suppose

$$\frac{d^2p}{d\theta^2} + p - \frac{k}{K^2} = 0. \tag{20}$$

In Chapter 17 we will see that the solution of the differential equation

$$\frac{d^2p}{d\theta^2} + p = \frac{k}{K^2} \tag{21}$$

describing the motion of the body must be given by

$$p = A\cos\theta + B\sin\theta + \frac{k}{K^2}.$$

If both A and B are 0, then $p = k/K^2$, so $r = K^2/k$ is constant and the body is in a circular orbit. We assume from now on that this is not the case. We can then express p as

$$p = \sqrt{A^2 + B^2}\left(\frac{A}{\sqrt{A^2 + B^2}}\cos\theta + \frac{B}{\sqrt{A^2 + B^2}}\sin\theta\right) + \frac{k}{K^2} \tag{22}$$

for some constants A and B. We can find an angle γ such that

$$\frac{A}{\sqrt{A^2 + B^2}} = -\cos \gamma \quad \text{and} \quad \frac{B}{\sqrt{A^2 + B^2}} = -\sin \gamma. \tag{23}$$

Let $E = \sqrt{A^2 + B^2}$. From Eqs. (22) and (23), we obtain

$$p = -E \cos(\theta - \gamma) + \frac{k}{K^2}. \tag{24}$$

By choosing a new polar axis, if necessary, we can assume that $\gamma = 0$, so that Eq. (24) becomes

$$\frac{1}{r} = -E \cos \theta + \frac{k}{K^2}, \tag{25}$$

which yields

$$r = \frac{1}{(k/K^2) - E \cos \theta} = \frac{K^2/k}{1 - (K^2 E/k) \cos \theta}. \tag{26}$$

Note that $K^2 E/k > 0$. Setting $e = K^2 E/k$, we obtain

$$r = \frac{e(1/E)}{1 - e \cos \theta}. \tag{27}$$

We proceed to express this polar equation in rectangular coordinates:

$$r - er \cos \theta = \frac{e}{E},$$

$$\sqrt{x^2 + y^2} - ex = \frac{e}{E},$$

$$x^2 + y^2 = e^2\left(x + \frac{1}{E}\right)^2,$$

$$(1 - e^2)x^2 + y^2 - 2\left(\frac{e^2}{E}\right)x = \frac{e^2}{E^2}. \tag{28}$$

We recognize this as the equation of an ellipse if $0 < e < 1$, of a parabola if $e = 1$, and of a hyperbola if $e > 1$. (The number e is the **eccentricity** of the conic section.) This shows that a body traveling in space subject only to a central force field, and not traveling in a straight line directly toward or away from the center of the field, is traveling on a conic section. In particular, if the orbit is a closed path (as for the planets orbiting our sun) the orbit must be an ellipse, so we have $0 < e < 1$.

We now suppose that Eq. (28) describes an ellipse, so that $0 < e < 1$. We put Eq. (28) in the form $(x - h)^2/a^2 + y^2/b^2 = 1$. Completing the square on the x-terms in Eq. (28), we have

$$(1 - e^2)\left[x^2 - \frac{2e^2}{E(1 - e^2)}x\right] + y^2 = \frac{e^2}{E^2},$$

$$(1 - e^2)\left[x - \frac{e^2}{E(1 - e^2)}\right]^2 + y^2 = \frac{e^2}{E^2} + \frac{e^4}{E^2(1 - e^2)} = \frac{e^2}{E^2(1 - e^2)}.$$

From this last equation, we find that

$$a^2 = \frac{e^2}{E^2(1 - e^2)^2} \quad \text{and} \quad b^2 = \frac{e^2}{E^2(1 - e^2)}. \tag{29}$$

Since $0 < e < 1$, we see at once that $a^2 > b^2$, so the major axis of the ellipse is horizontal and of length $2a = 2e/[E(1 - e^2)]$. Note that

$$c^2 = a^2 - b^2 = \frac{e^2}{E^2(1 - e^2)}\left[\frac{1}{1 - e^2} - 1\right] = \frac{e^4}{E^2(1 - e^2)^2}.$$

Thus

$$a - c = \frac{e}{E(1 - e^2)} - \frac{e^2}{E(1 - e^2)} = \frac{e}{E} \cdot \frac{1 - e}{1 - e^2} = \frac{e}{E(1 + e)}.$$

Since this is the value for r obtained by setting $\theta = \pi$ in Eq. (27), we see from Fig. 13.21 that a focus of this ellipse is indeed at the pole O. This completes the demonstration of Kepler's first law.

Derivation of Kepler's Third Law: By integration, we find that the area of the region enclosed by the ellipse $(x^2/a^2) + (y^2/b^2) = 1$ of major axis $2a$ and minor axis $2b$ is πab. If T is the period of the orbit of our planet, then Eq. (11) shows that the area swept out by the line segment joining O and the planet during one revolution is given by

$$\int_0^T \frac{dA}{dt}\, dt = \int_0^T \frac{1}{2}K\, dt = \frac{1}{2}Kt\bigg]_0^T = \frac{1}{2}KT.$$

Thus we have

$$\frac{1}{2}KT = \pi ab. \tag{30}$$

From Eq. (30) and Eq. (29), we find that

$$\frac{a^3}{T^2} = \frac{a^3}{\dfrac{4\pi^2 a^2 b^2}{K^2}} = \frac{aK^2}{4\pi^2 b^2} = \frac{K^2}{4\pi^2} \cdot \frac{\dfrac{e}{E(1 - e^2)}}{\dfrac{e^2}{E^2(1 - e^2)}} \tag{31}$$

$$= \frac{K^2}{4\pi^2} \cdot \frac{E}{e} = \frac{k}{4\pi^2} = \frac{GM}{4\pi^2}.$$

This gives Kepler's third law, because $GM/(4\pi^2)$ is independent of the mass m of the planet orbiting the sun of mass M.

SUMMARY

1. For unit vectors $\mathbf{u}_r$ and $\mathbf{u}_\theta$ in polar coordinates,

Position vector $= \mathbf{r} = r\,\mathbf{u}_r$,

Velocity vector $= \mathbf{v} = \dot{r}\,\mathbf{u}_r + r\dot{\theta}\,\mathbf{u}_\theta$,

Acceleration vector $= \mathbf{a} = (\ddot{r} - r\dot{\theta}^2)\mathbf{u}_r + (r\ddot{\theta} + 2\dot{r}\dot{\theta})\mathbf{u}_\theta$.

2. Newton's and Kepler's laws are given on page 691.

EXERCISES 13.3

In sequential Exercises 1–5, find the indicated quantity when $t = 1$ for the motion in the plane where the polar coordinate position at time t is $(r, \theta) = (t^3 - 2t, t - 1)$.

1. $\mathbf{u}_r$ and $\mathbf{u}_\theta$ in terms of $\mathbf{i}$ and $\mathbf{j}$

2. The velocity and acceleration vectors
   (a) in terms of $\mathbf{u}_r$ and $\mathbf{u}_\theta$ (b) in terms of $\mathbf{i}$ and $\mathbf{j}$

3. The speed

4. The rate of change of speed

5. The curvature of the path

In sequential Exercises 6–10, find the indicated quantity when $t = \pi/3$ for the motion in the plane where the polar coordinate position at time t is $(r, \theta) = (2a \cos t, t)$.

6. $\mathbf{u}_r$ and $\mathbf{u}_\theta$ in terms of $\mathbf{i}$ and $\mathbf{j}$

7. The velocity and acceleration vectors
   (a) in terms of $\mathbf{u}_r$ and $\mathbf{u}_\theta$ (b) in terms of $\mathbf{i}$ and $\mathbf{j}$

8. The speed

9. The rate of change of speed

10. The curvature of the path

In sequential Exercises 11–15, find the indicated quantity when $t = 5\pi/6$ for the motion in the plane where the polar coordinate position at time t is $(r, \theta) = (a + a \sin t, t)$.

11. $\mathbf{u}_r$ and $\mathbf{u}_\theta$ in terms of $\mathbf{i}$ and $\mathbf{j}$

12. The velocity and acceleration vectors
   (a) in terms of $\mathbf{u}_r$ and $\mathbf{u}_\theta$
   (b) in terms of $\mathbf{i}$ and $\mathbf{j}$

13. The speed

14. The rate of change of speed

15. The curvature of the path

16. The value of the universal gravitational constant G was calculated by Cavendish in 1798. Explain how one might use Eq. (31) to calculate the mass of the sun.

17. Argue from Eq. (31) that the period of a body in an elliptical orbit in a given central force field can be found if the length of the major axis of the ellipse is known.

18. The *apogee* of an earth satellite is its maximum distance from the surface of the earth, and the *perigee* is its minimum distance from the surface. Use Eq. (31) to argue that the period of an earth satellite is completely determined by the apogee and perigee of the satellite. [*Hint:* Let the radius of the earth be R and the length of the major axis of the elliptical orbit be $2a$. Find a relation between R, a, the apogee, and the perigee.]

19. Would you expect the ratio a^3/T^2 of Eq. (31) to be the same for the earth in revolution about the sun as for a moon in revolution about the planet Jupiter? Explain.

Exercises 20–33 deal with the derivation of the inverse square property and the formulation of Newton's law of universal gravitation from Kepler's laws. Suppose that planets describe planar orbits about the sun in accordance with Kepler's laws, and let the position of a certain planet with respect to a polar coordinate system with pole at the sun be given by $r = r(t)$, $\theta = \theta(t)$.

20. Deduce from Kepler's second law that if $A = A(t)$ is the area of the region swept out by the line segment joining the sun and the planet from time t_0 to time t, then $dA/dt = \beta$ for some constant β.

21. Deduce from Exercise 20 and the formula for area in polar coordinates that $r^2\dot{\theta} = 2\beta$.

22. Deduce from Exercise 21 that $r\ddot{\theta} + 2\dot{r}\dot{\theta} = 0$.

23. Deduce from Exercise 22 that the acceleration of the planet is toward the sun at all times. (This shows that we are indeed in a "central force field" situation.)

24. Let the orbit of the planet be an ellipse

$$r = \frac{B}{1 - e \cos \theta}$$

for $0 < e < 1$ and $B > 0$. (Recall the equivalence of Eqs. 27 and 28.) This is Kepler's first law. Show that

$$\dot{r}(1 - e \cos \theta) + re\dot{\theta} \sin \theta = 0.$$

25. Multiply the result of Exercise 24 by r and use Exercise 21 to show that

$$B\dot{r} + 2\beta e \sin \theta = 0.$$

26. Deduce by differentiating the result in Exercise 25 and using Exercise 21 that

$$\ddot{r} = -\frac{4\beta^2}{r^2} \cdot \frac{e \cos \theta}{B} = \frac{4\beta^2}{r^2}\left(\frac{1}{r} - \frac{1}{B}\right).$$

27. Deduce from Exercise 21 that $r\dot{\theta}^2 = 4\beta^2/r^3$, and conclude from Exercise 26 that

$$\ddot{r} - r\dot{\theta}^2 = -\frac{4\beta^2}{B} \cdot \frac{1}{r^2}.$$

28. Use the result of Exercise 23 to argue that the force on a planet of mass m is of magnitude

$$\frac{4m\beta^2/B}{r^2}$$

and is directed toward the sun. (This gives the inverse square property.)

29. Let $2a$ be the length of the major axis and $2b$ the length of the minor axis of the ellipse in Exercise 24 and show that

$$B = a(1 - e^2) \qquad \text{and} \qquad b = a\sqrt{1 - e^2}.$$

30. Show from Exercise 20 that if the planet has elliptical orbit of period T, then the area of the ellipse is βT.

31. Deduce from Exercise 30 that $\beta = \pi ab/T$.

32. Deduce from Exercises 29 and 31 that the force in Exercise 28 may be written as

$$\frac{4\pi^2 a^3 m}{T^2 r^2}.$$

33. Conclude from Exercise 32 and Kepler's third law that the force of attraction of the sun per unit mass of planet is independent of the planet ("universal"), depending only on the distance from the planet to the sun.

34. The ratio a^3/T^2 can be measured astronomically for various

situations (a planet orbiting the sun or a moon orbiting a planet) in our solar system. Show that astronomical indications that a^3/T^2 is proportional to the mass at the center of the central force field, together with Exercises 32 and 33, would lead to prediction of Newton's universal law of gravitation.

13.4 THE BINORMAL VECTOR AND TORSION

Section 13.2 introduced the orthogonal unit vector frame, **t** and **n**, for both curves in the plane and space curves. In space, we need another unit vector, orthogonal to both **t** and **n**, to complete a three-dimensional space frame. In this section, we introduce such a vector, the *binormal vector* **b**.

Recall that at a point on a space curve, the vectors **t** and **n** lie in the osculating plane to the curve at that point, which is the plane that most nearly contains the curve near the point. The vector **b** is then normal to the osculating plane. The rate that **b** "turns" as we travel along the curve is a measure of how quickly the curve twists away from the osculating plane. We are led to introduce the notion of the *torsion* (twist) of a space curve. The section concludes with the *Frenet formulas*, which summarize the relationships of the vectors **t**, **n**, and **b** to each other.

In this section, we always assume that we are working on a smooth arc of a space curve with parametrizing functions having continuous derivatives of all orders that we desire to take.

The Binormal Vector

On page 698 we list for quick reference the vector geometry concepts we developed for space curves in Section 13.2.

We illustrate the computation of **t** and **n** again for a helix.

EXAMPLE 1 Find **t** and **n** for the helix with position vector

$$\mathbf{r} = (a \cos t)\mathbf{i} + (a \sin t)\mathbf{j} + t\mathbf{k}.$$

Solution We have

$$\mathbf{v} = (-a \sin t)\mathbf{i} + (a \cos t)\mathbf{j} + \mathbf{k}, \qquad \mathbf{a} = (-a \cos t)\mathbf{i} - (a \sin t)\mathbf{j}.$$

Since $|\mathbf{v}| = \sqrt{a^2\sin^2 t + a^2\cos^2 t + 1} = \sqrt{a^2 + 1}$, we have

$$\mathbf{t} = \frac{1}{|\mathbf{v}|}\mathbf{v} = \frac{1}{\sqrt{a^2 + 1}}[(-a \sin t)\mathbf{i} + (a \cos t)\mathbf{j} + \mathbf{k}].$$

Now $\mathbf{a} \cdot \mathbf{v} = a^2\sin t \cos t - a^2\sin t \cos t = 0$, so **a** is normal to **v**. Since $|\mathbf{a}| = a$, we see in this case that

$$\mathbf{n} = \frac{1}{|\mathbf{a}|}\mathbf{a} = \frac{1}{a}[-(a \cos t)\mathbf{i} - (a \sin t)\mathbf{j}] = -(\cos t)\mathbf{i} - (\sin t)\mathbf{j}.$$

Vector Formulas for Space Curves

$\mathbf{r} = x\mathbf{i} + y\mathbf{j} + z\mathbf{k}$	*Position vector*
$\mathbf{v} = \dfrac{d\mathbf{r}}{dt}$	*Velocity*
$\dfrac{ds}{dt} = |\mathbf{v}|$	*Speed*
$\mathbf{a} = \dfrac{d\mathbf{v}}{dt} = \dfrac{d^2\mathbf{r}}{dt^2}$	*Acceleration*
$\mathbf{t} = \dfrac{1}{|\mathbf{v}|}\mathbf{v}$ if $\mathbf{v} \neq \mathbf{0}$	*Unit tangent vector*
$\mathbf{n} = \dfrac{1}{|d\mathbf{t}/ds|}\cdot\dfrac{d\mathbf{t}}{ds}$ if $\dfrac{d\mathbf{t}}{ds} \neq \mathbf{0}$	*Principal normal vector*
$\kappa = \left|\dfrac{d\mathbf{t}}{ds}\right|$	*Curvature*
$\mathbf{a}_{\tan} = \left(\dfrac{d^2s}{dt^2}\right)\mathbf{t} = \dfrac{\mathbf{v}\cdot\mathbf{a}}{\mathbf{v}\cdot\mathbf{v}}\mathbf{v}$	*Tangential vector component of* $\mathbf{a}$
$\mathbf{a}_{\text{nor}} = \mathbf{a} - \mathbf{a}_{\tan} = \kappa\left(\dfrac{ds}{dt}\right)^2\mathbf{n}$	*Normal vector component of* $\mathbf{a}$
Plane at the point with normal vector $\mathbf{t} \times \mathbf{n}$ or $\mathbf{v} \times \mathbf{a}$	*Osculating plane*

Note that $\mathbf{n}$ is directed toward the z-axis, as shown in Fig. 13.23. ∎

Now consider a point on the space curve where $d\mathbf{t}/ds \neq 0$, which is equivalent to saying that $\kappa \neq 0$. At such a point, $\mathbf{t}$ and $\mathbf{n}$ are both defined. We then define the **binormal vector b** at the point to be

$$\mathbf{b} = \mathbf{t} \times \mathbf{n}. \tag{1}$$

Therefore

$$|\mathbf{b}| = |\mathbf{t}|\,|\mathbf{n}|\sin 90° = 1,$$

so $\mathbf{b}$ is a unit vector perpendicular to both $\mathbf{t}$ and $\mathbf{n}$. The sequence

$$\mathbf{t}, \quad \mathbf{n}, \quad \mathbf{b}$$

of vectors is a right-hand triple of orthogonal unit vectors at each point on the space curve and may be regarded as a local coordinate 3-frame at each point on the curve (see Fig. 13.24). To a bug crawling on the curve in the direction of increasing t so that $\mathbf{t}$ points "straight ahead" and $\mathbf{n}$ points "left," it appears that $\mathbf{b}$ points "up." Of course,

$$\mathbf{b} = \mathbf{t} \times \mathbf{n}, \quad \mathbf{t} = \mathbf{n} \times \mathbf{b}, \quad \text{and} \quad \mathbf{n} = \mathbf{b} \times \mathbf{t}. \tag{2}$$

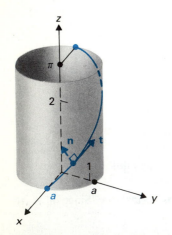

FIGURE 13.23 $\mathbf{t}$ and $\mathbf{n}$ on the helix $x = a\cos t$, $y = a\sin t$, $z = t$.

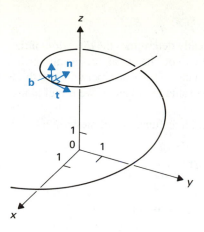

FIGURE 13.24 Orthogonal local unit vector frame **t, n, b** on a curve.

EXAMPLE 2 Find **b** for the helix in Example 1.

Solution We see from the example that

$$\mathbf{b} = \mathbf{t} \times \mathbf{n} = \frac{1}{\sqrt{a^2 + 1}} \begin{vmatrix} \mathbf{i} & \mathbf{j} & \mathbf{k} \\ -a \sin t & a \cos t & 1 \\ -\cos t & -\sin t & 0 \end{vmatrix}$$

$$= \frac{1}{\sqrt{a^2 + 1}}[(\sin t)\mathbf{i} - (\cos t)\mathbf{j} + a\mathbf{k}]. \qquad ▪$$

Note that at a point on a space curve, **b** is a normal vector to the osculating plane to the curve. Recall that **v** × **a** is also such a normal vector. We argue that **v** × **a** has the same direction as **b**. Now the osculating plane doesn't really contain the curve near the point, but the curve clings very close to it for a short distance. In Fig. 13.25, we suppose the osculating plane is the plane of the page, and we draw in this plane a short piece of the curve that is closest to the space curve. Since **a** and **n** both point toward the concave side of this curve, we see that **b** = **t** × **n** and **v** × **a** have the same direction. Thus we can compute **b** without going after **t** and **n**, using

$$\mathbf{b} = \frac{1}{|\mathbf{v} \times \mathbf{a}|}(\mathbf{v} \times \mathbf{a}). \qquad (3)$$

EXAMPLE 3 Find **b** when $t = 1$ for the space curve with position vector

$$\mathbf{r} = (t^2 - 3t)\mathbf{i} + t^3\mathbf{j} - (t^2 + 2t)\mathbf{k}.$$

Solution We have

$$\mathbf{v} = (2t - 3)\mathbf{i} + 3t^2\mathbf{j} - (2t + 2)\mathbf{k}$$

and

$$\mathbf{a} = 2\mathbf{i} + 6t\mathbf{j} - 2\mathbf{k}.$$

When $t = 1$, we obtain

$$\mathbf{v} = -\mathbf{i} + 3\mathbf{j} - 4\mathbf{k} \qquad \text{and} \qquad \mathbf{a} = 2\mathbf{i} + 6\mathbf{j} - 2\mathbf{k}.$$

Consequently, when $t = 1$,

$$\mathbf{v} \times \mathbf{a} = \begin{vmatrix} \mathbf{i} & \mathbf{j} & \mathbf{k} \\ -1 & 3 & -4 \\ 2 & 6 & -2 \end{vmatrix}$$

$$= 18\mathbf{i} - 10\mathbf{j} - 12\mathbf{k} = 2(9\mathbf{i} - 5\mathbf{j} - 6\mathbf{k}).$$

Finally, when $t = 1$,

$$\mathbf{b} = \frac{1}{|\mathbf{v} \times \mathbf{a}|}(\mathbf{v} \times \mathbf{a})$$

$$= \frac{1}{\sqrt{142}}(9\mathbf{i} - 5\mathbf{j} - 6\mathbf{k}). \qquad ▪$$

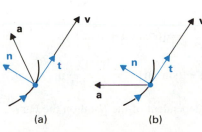

(a)   (b)

FIGURE 13.25 **t** × **n** and **v** × **a** have the same direction: (a) increasing speed case; (b) decreasing speed case.

Torsion

Now **b** is normal to the osculating plane and really determines the *direction* of the plane. Since **b** is a unit vector, its length does not change as we go along the curve, but its direction changes. The rate of change $d\mathbf{b}/ds$ of **b** per unit length of curve should be a measure of the twist of the osculating plane as we travel along the curve.

Recall that $d\mathbf{t}/ds = \kappa\mathbf{n}$. Viewing **t**, **n**, and **b** as functions of arc length s and differentiating, we obtain

$$\frac{d\mathbf{b}}{ds} = \frac{d(\mathbf{t} \times \mathbf{n})}{ds} = \mathbf{t} \times \frac{d\mathbf{n}}{ds} + \frac{d\mathbf{t}}{ds} \times \mathbf{n}$$

$$= \mathbf{t} \times \frac{d\mathbf{n}}{ds} + (\kappa\mathbf{n}) \times \mathbf{n} = \mathbf{t} \times \frac{d\mathbf{n}}{ds}. \tag{4}$$

Since **n** is a vector of constant magnitude 1, Theorem 13.1 shows that **n** is orthogonal to $d\mathbf{n}/ds$. Equation (4) shows that $d\mathbf{b}/ds$ has direction orthogonal to both **t** and $d\mathbf{n}/ds$. Since **n** is orthogonal to both **t** and $d\mathbf{n}/ds$, we see that $d\mathbf{b}/ds$ is parallel to **n**. Hence

$$\frac{d\mathbf{b}}{ds} = -\tau\mathbf{n} \tag{5}$$

for some (positive or negative) constant τ. Since $d\mathbf{b}/ds$ measures the rate at which the osculating plane *twists* per unit change in arc length, the number τ is called the **torsion** of the curve at the point. The negative sign is introduced in Eq. (5) so that *positive* torsion corresponds to the vector **b** turning in the direction of $-\mathbf{n}$ (like a right-threaded screw) as we travel the curve in the direction given by the tangent vector **t**.

EXAMPLE 4 Find the torsion τ for the helix in Examples 1 and 2.

Solution From Example 1, we see that

$$\frac{ds}{dt} = |\mathbf{v}| = \sqrt{a^2 + 1}, \qquad \text{so} \qquad \frac{dt}{ds} = \frac{1}{\sqrt{a^2 + 1}}.$$

From Example 2, we know that

$$\mathbf{b} = \frac{1}{\sqrt{a^2 + 1}}[(\sin t)\mathbf{i} - (\cos t)\mathbf{j} + a\mathbf{k}].$$

By the chain rule and Example 1,

$$\frac{d\mathbf{b}}{ds} = \frac{dt}{ds} \cdot \frac{d\mathbf{b}}{dt} = \frac{1}{\sqrt{a^2 + 1}} \cdot \frac{1}{\sqrt{a^2 + 1}}[(\cos t)\mathbf{i} + (\sin t)\mathbf{j}]$$

$$= \frac{1}{a^2 + 1}[(\cos t)\mathbf{i} + (\sin t)\mathbf{j}] = -\frac{1}{a^2 + 1}\mathbf{n}.$$

Thus the torsion of the helix is constant, independent of the point on the curve, and is given by

$$\tau = \frac{1}{a^2 + 1}. \qquad\blacksquare$$

Formulas for Computing κ and τ

We have had some practice computing κ using the formula $\kappa = |\mathbf{a}_{nor}|/|\mathbf{v}|^2$ in Section 13.2. For the helix in Example 4, we found τ readily from its definition in Eq. (5). We give alternative formulas for κ and τ that are often easier to use. The sequence of Exercises 27–30 leads to the formulas

$$\kappa = \frac{|\dot{\mathbf{r}} \times \ddot{\mathbf{r}}|}{|\dot{\mathbf{r}}|^3} \qquad \text{and} \qquad \tau = \frac{(\dot{\mathbf{r}} \times \ddot{\mathbf{r}}) \cdot \dddot{\mathbf{r}}}{|\dot{\mathbf{r}} \times \ddot{\mathbf{r}}|^2}. \tag{6}$$

EXAMPLE 5 Find the curvature κ and torsion τ for the curve with position vector

$$\mathbf{r} = t^2\mathbf{i} - (3t + 1)\mathbf{j} + t^3\mathbf{k}$$

at the point where $t = 1$.

Solution Now

$$\dot{\mathbf{r}} = 2t\mathbf{i} - 3\mathbf{j} + 3t^2\mathbf{k},$$
$$\ddot{\mathbf{r}} = 2\mathbf{i} + 6t\mathbf{k},$$
$$\dddot{\mathbf{r}} = 6\mathbf{k}.$$

Thus at $t = 1$,

$$\dot{\mathbf{r}} = 2\mathbf{i} - 3\mathbf{j} + 3\mathbf{k},$$
$$\ddot{\mathbf{r}} = 2\mathbf{i} + 6\mathbf{k},$$
$$\dddot{\mathbf{r}} = 6\mathbf{k},$$

$$\dot{\mathbf{r}} \times \ddot{\mathbf{r}} = \begin{vmatrix} \mathbf{i} & \mathbf{j} & \mathbf{k} \\ 2 & -3 & 3 \\ 2 & 0 & 6 \end{vmatrix} = -18\mathbf{i} - 6\mathbf{j} + 6\mathbf{k},$$

$$|\dot{\mathbf{r}} \times \ddot{\mathbf{r}}| = 6\sqrt{9 + 1 + 1} = 6\sqrt{11},$$
$$|\dot{\mathbf{r}}| = \sqrt{4 + 9 + 9} = \sqrt{22}.$$

Hence

$$\kappa = \frac{|\dot{\mathbf{r}} \times \ddot{\mathbf{r}}|}{|\dot{\mathbf{r}}|^3} = \frac{6\sqrt{11}}{22\sqrt{22}} = \frac{3}{11\sqrt{2}}$$

and

$$\tau = \frac{(\dot{\mathbf{r}} \times \ddot{\mathbf{r}}) \cdot \dddot{\mathbf{r}}}{|\dot{\mathbf{r}} \times \ddot{\mathbf{r}}|^2} = \frac{(-18)(0) + (-6)(0) + 6 \cdot 6}{36 \cdot 11}$$

$$= \frac{36}{36 \cdot 11} = \frac{1}{11}. \qquad \blacksquare$$

The Frenet Formulas

Recall that

$$\frac{d\mathbf{t}}{ds} = \kappa\mathbf{n} \qquad \text{and} \qquad \frac{d\mathbf{b}}{ds} = -\tau\mathbf{n}. \tag{7}$$

We would like to find $d\mathbf{n}/ds$ so that we know the rate of change of all three of our local coordinate unit vectors with respect to arc length as we move along the curve. We differentiate

$$\mathbf{n} = \mathbf{b} \times \mathbf{t}$$

with respect to s and obtain

$$\frac{d\mathbf{n}}{ds} = \mathbf{b} \times \frac{d\mathbf{t}}{ds} + \frac{d\mathbf{b}}{ds} \times \mathbf{t}. \tag{8}$$

Making use of Eqs. (2) and (7), we obtain

$$\frac{d\mathbf{n}}{ds} = \mathbf{b} \times (\kappa\mathbf{n}) - (\tau\mathbf{n}) \times \mathbf{t} = \kappa(\mathbf{b} \times \mathbf{n}) - \tau(\mathbf{n} \times \mathbf{t}) = -\kappa\mathbf{t} + \tau\mathbf{b}. \tag{9}$$

Equations (7) and (9) are known as the *Frenet formulas*. We collect them in one place.

Frenet Formulas

$$\frac{d\mathbf{t}}{ds} = \kappa\mathbf{n}, \qquad \frac{d\mathbf{n}}{ds} = -\kappa\mathbf{t} + \tau\mathbf{b}, \qquad \frac{d\mathbf{b}}{ds} = -\tau\mathbf{n}$$

In Exercise 35, you are asked to use the Frenet formulas to show that the shape of a space curve is completely determined by the curvature and the torsion as functions of the arc length traveled.

SUMMARY

1. For a space curve with unit tangent vector $\mathbf{t}$ and principal normal vector $\mathbf{n}$, the unit binormal vector is defined by $\mathbf{b} = \mathbf{t} \times \mathbf{n}$.

2. $\mathbf{b}$ may be computed using

$$\mathbf{b} = \frac{1}{|\mathbf{v} \times \mathbf{a}|}(\mathbf{v} \times \mathbf{a}).$$

3. The torsion τ of the curve is defined by

$$\frac{d\mathbf{b}}{ds} = -\tau\mathbf{n}.$$

4. Curvature κ and torsion τ may also be found from the position vector $\mathbf{r}$ using

$$\kappa = \frac{|\dot{\mathbf{r}} \times \ddot{\mathbf{r}}|}{|\dot{\mathbf{r}}|^3} \qquad \text{and} \qquad \tau = \frac{(\dot{\mathbf{r}} \times \ddot{\mathbf{r}}) \cdot \dddot{\mathbf{r}}}{|\dot{\mathbf{r}} \times \ddot{\mathbf{r}}|^2}.$$

5. The Frenet formulas are

$$\frac{d\mathbf{t}}{ds} = \kappa\mathbf{n}, \qquad \frac{d\mathbf{n}}{ds} = -\kappa\mathbf{t} + \tau\mathbf{b}, \qquad \frac{d\mathbf{b}}{ds} = -\tau\mathbf{n}.$$

EXERCISES 13.4

1. The torsion of a straight line is not defined by $d\mathbf{b}/ds = -\tau\mathbf{n}$ since $\mathbf{n}$ and $\mathbf{b}$ are not defined for a line in space. What do you think the torsion of a straight line should be defined to be?

Exercises 2–6 are concerned with the space curve

$$x = 2t, \quad y = t^2, \quad z = -t^2.$$

2. Find the vectors $\mathbf{t}$, $\mathbf{n}$, and $\mathbf{b}$ at a point on the curve in terms of the parameter t.

3. Find the curvature κ of the curve in terms of the parameter t.

4. Find the equation of the osculating plane to the curve for a value t_0 of the parameter.

5. Find the torsion τ of the curve in terms of the parameter t.

6. From the answers to the two preceding exercises, it appears that the curve lies in a plane. Deduce this directly from the equations of the curve.

Exercises 7–10 are concerned with the twisted cubic

$$x = 3t, \quad y = 3t^2, \quad z = 2t^3.$$

Find the indicated quantity when $t = 1$.

7. The binormal vector $\mathbf{b}$

8. The equation of the osculating plane

9. The curvature κ

10. The torsion τ

Exercises 11–14 are concerned with the space curve

$$x = \frac{1}{t}, \quad y = \frac{1}{t^2}, \quad z = \frac{2}{t},$$

where $t > 0$. Find the indicated quantity when $t = 1$.

11. The binormal vector $\mathbf{b}$

12. The equation of the osculating plane

13. The curvature κ

14. The torsion τ

Exercises 15–18 are concerned with the space curve

$$x = e^t \sin t, \quad y = e^t \cos t, \quad z = e^{2t}.$$

Find the indicated quantity when $t = 0$.

15. The binormal vector $\mathbf{b}$

16. The equation of the osculating plane

17. The curvature κ

18. The torsion τ

Exercises 19–22 are concerned with the space curve

$$x = \ln(\sin t), \quad y = \ln(\cos t), \quad z = \sin 2t$$

for $0 < t < \pi/2$. Find the indicated quantity when $t = \pi/4$.

19. The binormal vector $\mathbf{b}$

20. The equation of the osculating plane

21. The curvature κ

22. The torsion τ

Exercises 23–26 are concerned with the space curve

$$x = \cos t, \quad y = e^{2t}, \quad z = (t + 1)^3.$$

Find the indicated quantity when $t = 0$.

23. The binormal vector $\mathbf{b}$

24. The equation of the osculating plane

25. The curvature κ

26. The torsion τ

27. Show that $\ddot{\mathbf{r}} = \ddot{s}\mathbf{t} + \dot{s}^2\kappa\mathbf{n}$. [*Hint:* Differentiate $\dot{\mathbf{r}} = \dot{s}\mathbf{t}$.]

28. Show that $\dddot{\mathbf{r}} = (\dddot{s} - \dot{s}^3\kappa^2)\mathbf{t} + (3\dot{s}\ddot{s}\kappa + \dot{s}^2\dot{\kappa})\mathbf{n} + \dot{s}^3\kappa\tau\mathbf{b}$. [*Hint:* Differentiate the result in Exercise 27.]

29. Deduce from the preceding two exercises that
(a) $\dot{\mathbf{r}} \times \ddot{\mathbf{r}} = \dot{s}^3\kappa\mathbf{b}$,   (b) $(\dot{\mathbf{r}} \times \ddot{\mathbf{r}}) \cdot \dddot{\mathbf{r}} = \dot{s}^6\kappa^2\tau$.

[*Hint:* Since $\mathbf{t}$, $\mathbf{n}$, $\mathbf{b}$ form a right-hand perpendicular unit 3-frame at each point, they may be used in the role of $\mathbf{i}$, $\mathbf{j}$, and $\mathbf{k}$ in computations of the cross product.]

30. Deduce from the preceding exercise that

(a) $\kappa = \dfrac{|\dot{\mathbf{r}} \times \ddot{\mathbf{r}}|}{|\dot{\mathbf{r}}|^3}$,   (b) $\tau = \dfrac{(\dot{\mathbf{r}} \times \ddot{\mathbf{r}}) \cdot \dddot{\mathbf{r}}}{|\dot{\mathbf{r}} \times \ddot{\mathbf{r}}|^2}$.

31. Show that if we let $\boldsymbol{\delta} = \tau\mathbf{t} + \kappa\mathbf{b}$ (the *Darboux vector*), then the Frenet formulas take the symmetric form

$$\frac{d\mathbf{t}}{ds} = \boldsymbol{\delta} \times \mathbf{t}, \qquad \frac{d\mathbf{n}}{ds} = \boldsymbol{\delta} \times \mathbf{n},$$

$$\frac{d\mathbf{b}}{ds} = \boldsymbol{\delta} \times \mathbf{b}.$$

32. Show that

$$\left|\frac{d\mathbf{n}}{ds}\right|^2 = \left|\frac{d\mathbf{t}}{ds}\right|^2 + \left|\frac{d\mathbf{b}}{ds}\right|^2$$

33. Show that if the curvature of a smooth space curve is 0 at each point, then the curve is a straight line. [*Hint:* Use the chain rule to show that $\mathbf{t} = d\mathbf{r}/ds$. Argue that $\kappa = 0$ then implies that x, y, and z are linear functions of the parameter s of arc length.]

34. Show that a space curve with torsion zero at each point lies in a plane. [*Hint:* Deduce that $\mathbf{b}$ is a constant vector. Show that $d(\mathbf{b} \cdot \mathbf{r})/ds = 0$, and conclude that $\mathbf{b} \cdot (\mathbf{r}(t) - \mathbf{r}(t_0)) = 0$, so that the curve lies in the plane through the point where $t = t_0$ having orthogonal vector $\mathbf{b}$.]

35. *Fundamental theorem:* Use the Frenet formulas to show that two space curves having the same curvature and torsion for each value of the arc length parameter s are congruent. [*Hint:* You may suppose that the position vectors of the curves are given by $\mathbf{r}(s)$ and $\bar{\mathbf{r}}(s)$ and that

$$\mathbf{r}(0) = \bar{\mathbf{r}}(0), \quad \mathbf{t}(0) = \bar{\mathbf{t}}(0), \quad \mathbf{n}(0) = \bar{\mathbf{n}}(0),$$

and

$$\mathbf{b}(0) = \bar{\mathbf{b}}(0).$$

You must then show that $\mathbf{r}(s) = \bar{\mathbf{r}}(s)$. Set

$$w = \mathbf{t} \cdot \bar{\mathbf{t}} + \mathbf{n} \cdot \bar{\mathbf{n}} + \mathbf{b} \cdot \bar{\mathbf{b}},$$

and show that $dw/ds = 0$. Show that $\mathbf{w}(0) = 3$, and conclude that

$$\mathbf{t} = \bar{\mathbf{t}}, \quad \mathbf{n} = \bar{\mathbf{n}}, \quad \text{and} \quad \mathbf{b} = \bar{\mathbf{b}}$$

for all values of s. From $\mathbf{t} = \bar{\mathbf{t}}$, deduce that $\mathbf{r} = \bar{\mathbf{r}} + \mathbf{c}$, and, taking $s = 0$, that $\mathbf{c} = \mathbf{0}$, so that $\mathbf{r}(s) = \bar{\mathbf{r}}(s)$.]

SUPPLEMENTARY EXERCISES FOR CHAPTER 13

Exercises 1–8 concern a body moving in the plane with position at time t described parametrically by $x = \sin 2t$, $y = \cos t$. Find the indicated quantity when $t = \pi/4$.

1. The position vector $\mathbf{r}$
2. The velocity vector $\mathbf{v}$
3. The speed
4. The acceleration vector $\mathbf{a}$
5. The tangential vector component of $\mathbf{a}$
6. The rate of change of speed
7. The normal vector component of $\mathbf{a}$
8. The curvature κ

Exercises 9–14 concern the motion of a body in the plane with position vector

$$\mathbf{r} = [\ln(\sin 2t)]\mathbf{i} + [\ln(\cos 2t)]\mathbf{j}.$$

Find the indicated quantity when $t = \pi/8$.

9. The speed
10. The unit tangent vector $\mathbf{t}$
11. The unit normal vector $\mathbf{n}$
12. The rate of change of speed
13. The magnitude $|\mathbf{a}_{nor}|$ of the normal vector component of $\mathbf{a}$
14. The curvature κ
15. Let a body be moving in the plane subject to some force. Show that if

$$\frac{d^2s/dt^2}{\kappa(ds/dt)^2} = 1$$

at all times t, then the force must be directed at an angle of 45° to the direction of motion of the body at every time t.

16. For a body traveling a circular track at 20 ft/sec, a force of 500 lb perpendicular to the track is required to keep the body from leaving the track. If the maximum force the track can exert against the body in this perpendicular direction is 4500 lb, how fast can the body travel without leaving the track?

Exercises 17–21 concern a body moving in space with position at time t described parametrically by $x = t$, $y = 3 \sin t$, $z = -3 \cos t$. Find the indicated quantity when $t = \pi/4$.

17. The velocity vector $\mathbf{v}$
18. The speed
19. The equation of the osculating plane
20. The rate of change of speed
21. The unit normal vector $\mathbf{n}$

Exercises 22–26 concern the motion of a body in space with position vector

$$\mathbf{r} = t^3\mathbf{i} + 2t\mathbf{j} + t^2\mathbf{k}.$$

Find the indicated quantity when $t = -1$.

22. The speed
23. The unit tangent vector $\mathbf{t}$
24. The tangential vector component $\mathbf{a}_{tan}$ of $\mathbf{a}$
25. The rate of change of speed
26. The equation of the osculating plane

In Exercises 27–29, compute the indicated quantity for the curve in Exercises 17–21.

27. The curvature κ

28. The unit binormal vector **b**

29. The torsion τ

In Exercises 30–32, compute the indicated quantity for the curve traveled in Exercises 22–26.

30. The unit binormal vector **b**

31. The curvature κ

32. The torsion τ

33. What type of curve is traveled by a body moving subject to a central force field?

34. Give the polar formula for $\mathbf{u}_r$ and $\mathbf{u}_\theta$ components of acceleration for a body traveling a curve in the plane.

35. State Newton's law of universal gravitation.

36. If the polar coordinate position of a body in the plane at time t is given by $(r, \theta) = (\sqrt{t}, \sqrt{3t + 4})$, find the velocity and acceleration vectors in terms of the unit vectors $\mathbf{u}_r$ and $\mathbf{u}_\theta$ at time $t = 4$.

37. State Kepler's laws.

38. Two satellites A and B are traveling elliptic orbits about the same body. If the major axis of A's orbit is four times the length of the major axis of B's orbit, and if A's period is 48 hours, find B's period.

39. Explain how the torsion of a space curve is defined.

40. Explain how the curvature of a space curve is defined.

MULTIVARIABLE DIFFERENTIAL CALCULUS

14

Until now, we have managed to work primarily with functions of a single variable, usually symbolized by $y = f(x)$. It is amazing how much we were able to accomplish with this restriction, for so many things depend on more than one parameter. To take a practical illustration, a tailor making a new suit for a person needs to know many parameters, including measurements of waist, hips, inseam, sleeve, chest, shoulders, and so on. For a geometric illustration, note that the volume $V = \pi r^2 h$ of a cylinder depends on two parameters, the radius r and the altitude h of the cylinder.

In this chapter, we study instantaneous rates of change (derivatives) for functions of more than one variable. The notion of a limit is involved just as for a function of one variable. As we point out in Section 14.1, limits can be difficult to compute for functions of more than one variable. It is indeed fortunate that for most functions we will encounter, we can reduce the computation of the limits needed for rates of change to the one-variable case by repeatedly setting all but one of the variables equal to a constant. As we will see, this allows us still to use all of the differentiation formulas developed for functions of one variable as we do differential calculus for functions of several variables.

14.1 MULTIVARIABLE FUNCTIONS AND PARTIAL DERIVATIVES

Functions of More than One Variable

Recall that a **function** f is a rule that assigns to each element of a set, the **domain** of the function, an element of some set. If the domain of f consists of points (x, y) in the plane, we say that f is a **function of two real variables.** If the domain of g consists of points (x, y, z) in space, we have a **function g of three real variables.** In this chapter, we are interested primarily in *real-valued functions,* which assign to each point in their domains a real number. We can write such a function of two variables as $z = f(x, y)$ and such a function of three variables as $w = g(x, y, z)$. When discussing functions of many variables, we often turn to subscripted notation and write $y = f(x_1, x_2, \ldots, x_n)$ for a function of n variables. We will see that it is sometimes convenient to view y as a function of a single *vector variable* $\mathbf{x} = \langle x_1, x_2, \ldots, x_n \rangle$ and to use the notation $y = f(\mathbf{x})$.

Just as in the single-variable case, the domain of a real-valued function of real variables that is defined by a formula is understood to be all values of the variables at which the formula can be evaluated, unless the domain is explicitly given. Thus the domain of f where

$$f(x, y) = \frac{xy}{x^2 + y^2}$$

consists of all points in the x,y-plane except the point $(0, 0)$, and the domain of g where

$$g(x, y, z) = \frac{\sin yz}{\sqrt{4 - x^2 - y^2 - z^2}}$$

consists of all points in x,y,z-space such that $x^2 + y^2 + z^2 < 4$.

DEFINITION 14.1

Graph

The **graph** of $z = f(x, y)$ consists of all points (x, y, z) in space such that $z = f(x, y)$.

The graph of a function $w = g(x, y, z)$ or of $y = h(\mathbf{x}) = h(x_1, x_2, \ldots, x_n)$ is similarly defined.

Pictures helped us understand calculus of functions of one variable. We often sketched a graph of a function f, consisting of points (x, y) in the plane, where $y = f(x)$. The helpful graph for a function of two variables, which we generally write as $z = f(x, y)$, consists of all points (x, y, z) in space, where $z = f(x, y)$. We will also deal with functions of three or more variables. A function of three variables is often written as $w = f(x, y, z)$. This time, the graph consists of points (x, y, z, w) in four-dimensional space such that $w = f(x, y, z)$. This is almost hopeless to sketch on a page, and we will make no attempt to do so. However, the concept is no more mysterious than for functions of one or two variables.

EXAMPLE 1   Sketch the graph of

$$z = f(x, y) = \sqrt{a^2 - x^2 - y^2}.$$

Solution   Since $x^2 + y^2 + z^2 = a^2$ is the equation of a sphere with center at the origin and radius a we see that $z = f(x, y) = \sqrt{a^2 - x^2 - y^2}$ has the upper hemisphere as its graph. The graph is shown in Fig. 14.1. ∎

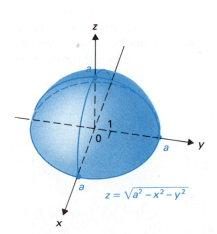

$z = \sqrt{a^2 - x^2 - y^2}$

FIGURE 14.1   The graph of $f(x, y) = \sqrt{a^2 - x^2 - y^2}$.

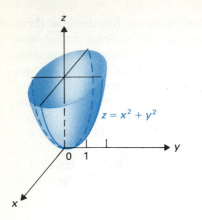

FIGURE 14.2 The graph of $f(x, y) = x^2 + y^2$.

EXAMPLE 2 Sketch the graph of $z = f(x, y) = x^2 + y^2$.

Solution The graph is the circular paraboloid shown in Fig. 14.2.   ■

One of the reasons we spent time sketching quadric surfaces in Section 12.8 was to be able to sketch the graphs of a few nonlinear functions of two variables. Of course, the graph of every linear function $z = g(x, y) = ax + by + c$ is a plane in space.

Limits and Continuity

We now restrict our discussion to real-valued functions of real variables. The limit at a point of a function of several variables and the concept of a continuous function are defined much as for a function of one variable. We give an intuitive definition of a limit for a function $z = f(x, y)$. This intuitive definition is similar to Definition 2.1. A rigorous definition would be analogous to Definition 2.7.

DEFINITION 14.2

Limit of f at (a, b) (intuitive)

Let $z = f(x, y)$ be a function whose domain includes points that are arbitrarily close to (a, b) but different from (a, b). The **limit of f at** (a, b) **is** L, written $\lim_{(x,y)\to(a,b)} f(x, y) = L$, if $f(x, y)$ can be made as *close* to L as we wish by taking any (x, y) *close enough* to (a, b), but different from (a, b), in the domain of f.

Computation of limits can be much more complicated for functions of two or more variables than for functions of a single variable. When finding $\lim_{x\to a} f(x)$, we can approach the point a only along one line, the x-axis. When finding $\lim_{(x,y)\to(a,b)} f(x, y)$, we can approach (a, b) along any one of an infinite number of lines through (a, b) in the plane, along any curve through (a, b), or, indeed, along any sequence of points in the plane that gets close to (a, b). The values $f(x, y)$ as (a, b) is approached in all these ways must become as close as we please to L in order to have $\lim_{(x,y)\to(a,b)} f(x, y) = L$. Fortunately, the limits we will need to compute to find rates of change can be reduced to the one-variable case. We give two examples involving more general limits but leave their further study to a course in advanced calculus.

EXAMPLE 3 Show that

$$\lim_{(x,y)\to(0,0)} \frac{xy}{x^2 + y^2}$$

does not exist.

Solution Let us consider the values of $f(x, y) = xy/(x^2 + y^2)$ on the lines $y = mx$ as m varies. Substituting $y = mx$, we find that

$$\frac{xy}{x^2 + y^2} = \frac{mx^2}{x^2 + m^2x^2} = \frac{m}{1 + m^2} \qquad \text{if} \quad x \neq 0.$$

Thus at every point but $(0, 0)$ on the line through the origin with slope m, we have $f(x, y) = m/(1 + m^2)$. For example, $f(x, y)$ will be 0 arbitrarily close to the origin on the x-axis where $m = 0$, but it will be $\frac{2}{5}$ arbitrarily close to the origin on the line $y = 2x$. Thus there is no *single* value that $f(x, y)$ approaches as $(x, y) \to (0, 0)$, so the limit does not exist as we approach the origin. ∎

Sums, products, and quotients of functions of the same variables are defined just as for the case of one variable. If the limits of f and g at a point both exist, then so do the limits of $f + g$, fg, and f/g at the point, provided the limit of g there is nonzero. Of course, the limit of $f + g$ is the limit of f plus the limit of g, and so on. Thus the sum, product, and quotient properties for limits continue to hold for functions of the same variables.

EXAMPLE 4 Find the limit of the function $f(x, y) = xy/(x^2 + y^2)$ of Example 3 at the point $(2, -3)$.

Solution Let $h(x, y) = x$ and $k(x, y) = y$. Surely $h(x, y)$ approaches 2 and $k(x, y)$ approaches -3 as (x, y) approaches $(2, -3)$. By the product property of limits, the numerator xy in the formula for $f(x, y)$ approaches $(2)(-3) = -6$. In a similar fashion, the product and sum properties of limits show that the denominator $x^2 + y^2$ approaches $2^2 + (-3)^2 = 13$. The quotient property then shows that f has limit $-\frac{6}{13}$ at the point $(2, -3)$. ∎

DEFINITION 14.3

Continuity

A function $z = f(x, y)$ is **continuous at a point** (a, b) if

1. (a, b) is in the domain of f,
2. $\lim_{(x,y)\to(a,b)} f(x, y)$ exists,
3. $\lim_{(x,y)\to(a,b)} = f(a, b)$.

The function f is **continuous** if it is continuous at every point in its domain.

Intuitively, we think of the graph of a continuous function as an unbroken surface, like the one shown in Fig. 14.3. As a consequence of the limit properties, we see that the sums, products, and quotients of continuous functions of the same variables are again continuous. Of course, quotients are not defined where the function in the denominator has the value 0.

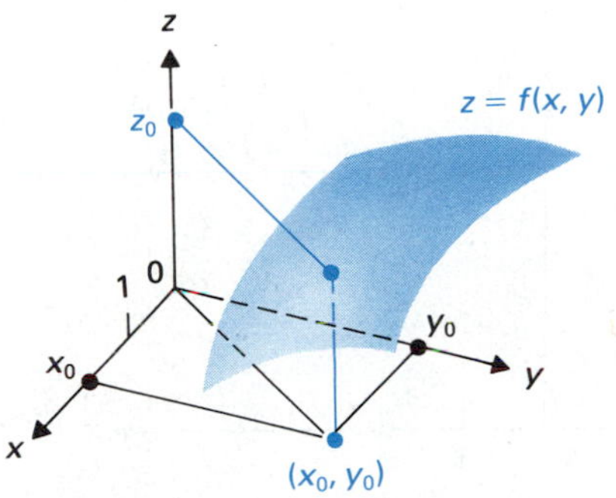

Partial Derivatives

If $y = f(x)$, the derivative $f'(x)$ gives the rate of change of y with respect to x. For a function of one variable, there are only two directions we can travel from a point in the domain, in the direction of increasing x or in the direction of decreasing x. We can describe these two directions by the vectors $\mathbf{i}$ and $-\mathbf{i}$, respectively, as shown in Fig. 14.4.

Suppose the rate of change of $y = f(x)$ at a is 3 in the direction of $\mathbf{i}$, so that $f'(a) = 3$ as in Fig. 14.5. Then the rate of change in the direction of $-\mathbf{i}$ is -3;

FIGURE 14.3 The graph of a continuous function $f(x, y)$.

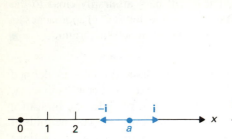

FIGURE 14.4 **i** and $-$**i** give the only directions on the x-axis from a.

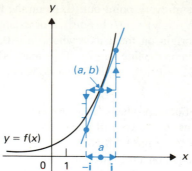

FIGURE 14.5 Rate of change of $f(x)$ at a is 3 in the direction **i**, -3 in the direction $-$**i**.

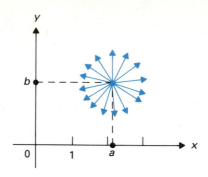

FIGURE 14.6 There are many directions from (a, b) in the plane.

that is, the function is *decreasing* at an instantaneous rate of 3 units per unit *decrease* in x there.

If $z = f(x, y)$, we can also attempt to find some rates of change. This time there are many directions we can go from a point (a, b) in the domain. See Fig. 14.6. We will eventually show how to find the rate at which z changes in each of these directions.

First we will find the rate of change in the direction parallel to the x-axis given by increasing x as y is held constant. That is, we will find the rate of change at (a, b) in the direction given by the vector **i** at that point. Suppose $z = f(x, y)$ has as graph the surface shown in Fig. 14.7. The portion of the surface over the line $y = b$ in the x,y-plane is a curve on the surface, as shown in the figure. On the curve, the height z appears as a function $g(x)$ of x only, as indicated in Fig. 14.8. This function $g(x)$ is, of course, found by setting $y = b$ in $z = f(x, y)$, that is,

$$z = g(x) = f(x, b).$$

Since we have a function of one variable $z = g(x)$, we can find the rate of change $g'(x)$ of z with respect to x. This time we denote the rate of change by $\partial z / \partial x$ rather than dz/dx. The "round d's" remind us that z was originally a function of more than one variable. We call $\partial z / \partial x$ the **partial derivative** of z with respect

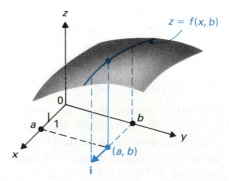

FIGURE 14.7 The portion of the graph where $y = b$.

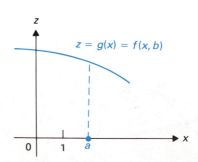

FIGURE 14.8 The graph of $z = g(x) = f(x, b)$.

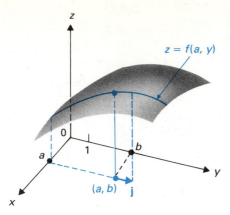

FIGURE 14.9   The portion of the graph where $x = a$.

to x. The value of this derivative at $x = a$ (recall that $y = b$) is written

$$\left.\frac{\partial z}{\partial x}\right|_{(a,b)} \quad \text{or} \quad \left.\frac{\partial f}{\partial x}\right|_{(a,b)} \quad \text{or} \quad f_x(a, b).$$

We use the function $z = h(y) = f(a, y)$ obtained by holding x constant at a to find the rate of change $h'(y)$ of $z = f(x, y)$ at (a, b) in the direction given by the vector $\mathbf{j}$ at that point. See Fig. 14.9. The notations are

$$\left.\frac{\partial z}{\partial y}\right|_{(a,b)} \quad \text{or} \quad \left.\frac{\partial f}{\partial y}\right|_{(a,b)} \quad \text{or} \quad f_y(a, b).$$

The derivative of $g(x) = f(x, b)$ at a is defined as the *limit*

$$\lim_{\Delta x \to 0} \frac{g(a + \Delta x) - g(a)}{\Delta x}.$$

Note that $g(a + \Delta x) - g(a) = f(a + \Delta x, b) - f(a, b)$. We are led to the following definition of the partial derivatives as limits.

DEFINITION 14.4

Partial derivative

Let $f(x, y)$ be defined everywhere inside some sufficiently small circle with (a, b) as center. Then the **partial derivative of f with respect to x** is

$$\left.\frac{\partial f}{\partial x}\right|_{(a,b)} = f_x(a, b) = \lim_{\Delta x \to 0} \frac{f(a + \Delta x, b) - f(a, b)}{\Delta x}$$

and the **partial derivative of f with respect to y** is

$$\left.\frac{\partial f}{\partial y}\right|_{(a,b)} = f_y(a, b) = \lim_{\Delta y \to 0} \frac{f(a, b + \Delta y) - f(a, b)}{\Delta y},$$

if these limits exist.

Computation of Partial Derivatives

We can regard $f_x(a, b)$ as the derivative at a of the function g given by $g(x) = f(x, b)$, that is, the function of one real variable x obtained from f by keeping $y = b$ and allowing only x to vary. This means that partial derivatives can be computed using the techniques for finding derivatives of functions of one variable. Such computations are illustrated in the following examples.

EXAMPLE 5   Let f be the function given by $f(x, y) = x^2 + 3xy + 2y^3$. Find

$$f_x(1, 2) = \left.\frac{\partial f}{\partial x}\right|_{(1,2)} \quad \text{and} \quad f_y(1, 2) = \left.\frac{\partial f}{\partial y}\right|_{(1,2)}.$$

Solution   Now $f_x(1, 2)$ can be viewed as the derivative at $x = 1$ of the function g obtained by setting $y = 2$ in the polynomial $x^2 + 3xy + 2y^3$. That is,

$$g(x) = x^2 + 3x(2) + 2(2^3) = x^2 + 6x + 16.$$

The derivative of $x^2 + 6x + 16$ is $2x + 6$, and the value of $2x + 6$ when $x = 1$

is 8. Hence

$$f_x(1, 2) = \frac{\partial f}{\partial x}\bigg|_{(1,2)} = 8.$$

Similarly, putting $x = 1$, we find that $f_y(1, 2)$ is the derivative, when $y = 2$, of $1 + 3y + 2y^3$. Since

$$\frac{d(1 + 3y + 2y^3)}{dy} = 3 + 6y^2,$$

we obtain

$$f_y(1, 2) = \frac{\partial f}{\partial y}(1, 2) = 3 + 6 \cdot 2^2 = 27. \qquad \blacksquare$$

We could simplify the computation of the partial derivatives in Example 5 by noting that we can compute $f_x(x, y)$ for any point (x, y) by differentiating $x^2 + 3xy + 2y^3$ "with respect to x only," treating y as a constant. The notation "$\partial f/\partial x$" is a practical way to indicate this derivative "with respect to x."

EXAMPLE 6 Repeat Example 5, using the technique just described.

Solution Differentiating $f(x, y) = x^2 + 3xy + 2y^3$ with respect to x, regarding y as a constant, we obtain

$$\frac{\partial f}{\partial x} = \frac{\partial(x^2 + 3xy + 2y^3)}{\partial x} = 2x + 3y,$$

for $\partial(2y^3)/\partial x = 0$, since we think of y as a constant. Similarly,

$$\frac{\partial f}{\partial y} = \frac{\partial(x^2 + 3xy + 2y^3)}{\partial y} = 3x + 6y^2.$$

Therefore

$$\frac{\partial f}{\partial x}(1, 2) = 2 \cdot 1 + 3 \cdot 2 = 8 \quad \text{and} \quad \frac{\partial f}{\partial y}(1, 2) = 3 \cdot 1 + 6 \cdot 2^2 = 27. \qquad \blacksquare$$

In practice, we always find partial derivatives by the technique illustrated in Example 6, and the ∂-notation is obviously suggestive here; we differentiate "with respect to x" or "with respect to y." For instance, to find $\partial f/\partial x$, just differentiate f with respect to x only, pretending that y is constant.

EXAMPLE 7 Find $\partial z/\partial x$ and $\partial z/\partial y$ if $z = xe^{xy^2}$.

Solution Regarding y as a constant, we obtain

$$\frac{\partial z}{\partial x} = x \cdot \frac{\partial(e^{xy^2})}{\partial x} + e^{xy^2} \cdot \frac{\partial(x)}{\partial x} = x \cdot e^{xy^2} \cdot y^2 + e^{xy^2} \cdot 1$$
$$= e^{xy^2}(xy^2 + 1).$$

Similarly, regarding x as a constant, we have

$$\frac{\partial z}{\partial y} = x \cdot e^{xy^2} \cdot 2xy = 2x^2 ye^{xy^2}. \qquad \blacksquare$$

As derivatives, $\partial f/\partial x$ and $\partial f/\partial y$ give the rates of change of $f(x, y)$ as we go in the directions of the vectors $\mathbf{i}$ and $\mathbf{j}$, respectively. We give two-variable approaches to some problems that we could solve also by one-variable techniques.

EXAMPLE 8 Find the rate of change of the volume with respect to the radius r of a right circular cylinder when the radius is 4 ft if the height remains constant at 20 ft.

Solution The volume of the cylinder is given by the formula

$$V = f(r, h) = \pi r^2 h.$$

The rate of change of V per unit increase in r as h is held constant is precisely $\partial V/\partial r$. When $r = 4$ and $h = 20$, we have

$$\left.\frac{\partial V}{\partial r}\right|_{(4,20)} = 2\pi r h|_{(4,20)} = 2\pi \cdot 4 \cdot 20$$

$$= 160\pi \ \text{ft}^3/(\text{ft increase in } r). \quad \blacksquare$$

Functions of More Variables

If (a, b, c) is a point in the domain of a function $w = f(x, y, z)$, then we can try to find the partial derivatives of w in the directions of $\mathbf{i}$, $\mathbf{j}$, and $\mathbf{k}$ at (a, b, c). To find the derivative in the direction given by $\mathbf{i}$, we should set $y = b$ and $z = c$, allowing only x to vary. That is, we hold y and z constant and differentiate just with respect to x to find

$$\left.\frac{\partial w}{\partial x}\right|_{(a,b,c)} \quad \text{or} \quad f_x(a, b, c).$$

Similarly, we can find $\partial w/\partial y$ and $\partial w/\partial z$.

EXAMPLE 9 Find the three partial derivatives for $w = x \sin yz$.

Solution We see that

$$\frac{\partial w}{\partial x} = \sin yz, \qquad \frac{\partial w}{\partial y} = xz \cos yz,$$

$$\frac{\partial w}{\partial z} = xy \cos yz. \quad \blacksquare$$

For functions of two or more variables, it is sometimes handy to use the subscripted notation presented in Chapter 12, where a point in the plane is (x_1, x_2) and a point in space is (x_1, x_2, x_3). The notation $\mathbf{x}$ may be used to represent either of these points; the dimension should be clear from the context. Thus a function of three variables may be written

$$y = f(\mathbf{x}) = f(x_1, x_2, x_3).$$

Of course, then we can compute all the partial derivatives

$$\frac{\partial y}{\partial x_1}, \qquad \frac{\partial y}{\partial x_2}, \qquad \frac{\partial y}{\partial x_3},$$

and $\partial y/\partial x_i$ is also written $\partial f/\partial x_i$ or $f_{x_i}(x_1, x_2, x_3)$.

EXAMPLE 10   Find f_{x_1} and f_{x_3} if $f(x) = x_1 x_2^2 / x_3$.

Solution   Regarding x_2 and x_3 as constants, we have

$$f_{x_1} = \frac{\partial f}{\partial x_1} = \frac{\partial [(x_2^2 / x_3) x_1]}{\partial x_1} = \frac{x_2^2}{x_3}.$$

Regarding x_1 and x_2 as constants, we have

$$f_{x_3} = \frac{\partial f}{\partial x_3} = \frac{\partial [x_1 x_2^2 (1/x_3)]}{\partial x_3}$$

$$= x_1 x_2^2 \cdot \frac{-1}{x_3^2} = -\frac{x_1 x_2^2}{x_3^2}. \qquad \blacksquare$$

Higher-Order Derivatives

Let $z = f(x, y)$. Then $\partial z / \partial x = f_x(x, y)$ is again a function of two variables and we can attempt to compute its partial derivatives with respect to either x or y. These are *second-order derivatives* of our original function $f(x, y)$. The notations are

$$\frac{\partial}{\partial x}\left(\frac{\partial z}{\partial x}\right) = \frac{\partial^2 z}{\partial x^2} = f_{xx}(x, y) \qquad \text{and} \qquad \frac{\partial}{\partial y}\left(\frac{\partial z}{\partial x}\right) = \frac{\partial^2 z}{\partial y \, \partial x} = f_{xy}(x, y).$$

The derivative f_{xy} is sometimes called a *mixed* or *cross* second partial derivative, since derivatives with respect to more than one variable are involved.

Of course, we could equally well find first $\partial z / \partial y$ and then the second-order derivatives

$$\frac{\partial}{\partial y}\left(\frac{\partial z}{\partial y}\right) = \frac{\partial^2 z}{\partial y^2} = f_{yy}(x, y) \qquad \text{and} \qquad \frac{\partial}{\partial x}\left(\frac{\partial z}{\partial y}\right) = \frac{\partial^2 z}{\partial x \, \partial y} = f_{yx}(x, y).$$

Note that f_{xy} means $(f_x)_y$ while f_{yx} means $(f_y)_x$.

It is a theorem that, if $z = f(x, y)$ has continuous second partial derivatives, then the "mixed" partial derivatives are equal, that is,

$$\frac{\partial^2 z}{\partial y \, \partial x} = \frac{\partial^2 z}{\partial x \, \partial y}.$$

This hypothesis of continuity is true for all elementary functions with which we will work.

EXAMPLE 11   Illustrate that $f_{xy} = f_{yx}$ for $f(x, y) = \sin(x^2 y)$.

Solution   Now

$$f_x(x, y) = 2xy \cos(x^2 y),$$

so

$$f_{xy}(x, y) = -2x^3 y \sin(x^2 y) + 2x \cos(x^2 y).$$

Differentiating in the other order, we have

$$f_y(x, y) = x^2 \cos(x^2 y),$$

so

$$f_{yx}(x, y) = -2x^3y \sin(x^2y) + 2x \cos(x^2y).$$

Thus $f_{xy} = f_{yx}$. ∎

Of course, we can continue taking derivatives. Thus $\partial^3 f/\partial y^2 \partial x$ means the third partial derivative of $f(x, y)$, first with respect to x and then twice more with respect to y. We have

$$\frac{\partial^3 f}{\partial y^2 \, \partial x} = \frac{\partial^3 f}{\partial y \, \partial x \, \partial y} = \frac{\partial^3 f}{\partial x \, \partial y^2}$$

for functions with continuous partial derivatives of order 3. Similar notations are used for functions of more than two variables. We state conditions for equality of mixed partial derivatives in the general case as a formal theorem. The proof is not given here.

THEOREM 14.1   Equality of mixed partials

Two mixed partial derivatives of the same order are equal if all partial derivatives of that order are continuous and the total number of differentiations with respect to each variable is the same in one mixed partial as in the other. □

Occasionally, Theorem 14.1 can be used to simplify the computation of a higher-order derivative.

EXAMPLE 12 Let $w = f(x, y, z) = xy/(y^2 + z^4)$. Find f_{yzxx}.

Solution The notation f_{yzxx} means to differentiate first with respect to y, then with respect to z, and finally twice with respect to x. By Theorem 14.1, we can change the order and compute instead f_{xxyz}. Since

$$f_x = \frac{y}{y^2 + z^4} \qquad \text{and} \qquad f_{xx} = 0,$$

we see that $f_{xxyz} = 0$, so $f_{yzxx} = 0$ also. ∎

SUMMARY

1. If $z = f(x, y)$, then $\partial z/\partial x = f_x(x, y)$ is the partial derivative of $f(x, y)$ in the direction $\mathbf{i}$ corresponding to increasing x as y is held constant. It is computed by the usual differentiation methods, regarding y as a constant. The partial derivative $\partial z/\partial y = f_y(x, y)$ is similarly defined and computed.

2. If $y = f(\mathbf{x}) = f(x_1, x_2, x_3)$, then $\partial y/\partial x_i$ is computed by differentiating with respect to x_i only, treating all other variables as though they were constants in the differentiation.

3. If $z = f(x, y)$, then second-order partial derivatives are

$$\frac{\partial}{\partial x}\left(\frac{\partial z}{\partial x}\right) = \frac{\partial^2 z}{\partial x^2} = f_{xx}(x, y),$$

$$\frac{\partial}{\partial y}\left(\frac{\partial z}{\partial y}\right) = \frac{\partial^2 z}{\partial y^2} = f_{yy}(x, y),$$

together with the mixed partial derivatives

$$\frac{\partial}{\partial y}\left(\frac{\partial z}{\partial x}\right) = \frac{\partial^2 z}{\partial y\, \partial x} = f_{xy}(x, y)$$

and

$$\frac{\partial}{\partial x}\left(\frac{\partial z}{\partial y}\right) = \frac{\partial^2 z}{\partial x\, \partial y} = f_{yx}(x, y).$$

For all such functions with continuous second partial derivatives, we have $\partial^2 z/\partial x\, \partial y = \partial^2 z/\partial y\, \partial x$. Similar notations are used for more variables and higher-order derivatives. If all partial derivatives of a certain order are continuous, then mixed partials of that order are equal if differentiation with respect to each variable occurs the same total number of times. For example, if $w = f(x, y, z)$, we have $\partial^4 w/\partial x\, \partial y\, \partial x\, \partial z = \partial^4 w/\partial z\, \partial y\, \partial x^2$.

EXERCISES 14.1

In Exercises 1–6, find the limit of the indicated function at the given point, if the limit exists. Use the symbol ∞ or $-\infty$ to indicate behavior where appropriate.

1. $f(x, y) = x/y$ at $(0, 0)$

2. $g(r, s) = r/(s^2 + 1)$ at $(0, 0)$

3. $f(x, y) = \dfrac{x^2 y}{x^2 + y^2}$ at $(0, 0)$

4. $g(s, t) = \dfrac{s^2 t}{s^2 + t^2}$ at $(1, 3)$

5. $h(x, y, z) = \dfrac{x^2 + y^2}{2z}$ at $(0, 1, 1)$

6. $k(r, s, t) = \dfrac{2r + 3s - t}{\ln(r^2 + t^2)}$ at $(0, 1, 0)$

In Exercises 7–27, find $f_x(x, y)$ and $f_y(x, y)$ for the given function $f(x, y)$. You need not simplify the answers.

7. $3x + 4y$

8. xy

9. $x^2 + y^2$

10. $xy^3 + x^2 y^2$

11. $e^{x/y}$

12. $\dfrac{x^2 + 3x + 1}{y}$

13. $xy^2 + \dfrac{3x^2}{y^3}$

14. $(3xy^2 - 2x^2 y)^3$

15. $(x^2 + 2xy)(y^3 + x^2)$

16. $(xy)^3(x^2 - y^3)^2$

17. $\dfrac{x^2 + y^2}{x^2 + y}$

18. $\sin xy$

19. $\tan(x^2 + y^2)$

20. $e^{xy}\cos x^2$

21. $e^{xy^2}\sec(x^2 y)$

22. $\ln(2x + 3y)$

23. $\ln(2x + y) \cdot \cot y^2$

24. $(\sin x^2 + \cos y^2)^5$

25. $y \sec^3 x + xy^2$

26. $\ln[\sin(xy)]$

27. $\tan^{-1}(xy^2)$

In Exercises 28–36, find the rate of change of the function in the indicated direction at the given point.

28. $f(x, y) = \sqrt{x^2 - y^2}$ at $(5, -3)$ in the direction of $\mathbf{j}$

29. $f(x, y) = \dfrac{x + y}{y^2}$ at $(1, -1)$ in the direction of $\mathbf{i}$

30. $f(x, y, z) = \dfrac{x}{\sqrt{x^2 + y^2 + z^2}}$ at $(-2, 1, -2)$ in the direction of $\mathbf{j}$

31. $f(x, y, z) = x \sin y^2 z$ at $(2, 3, 0)$ in the direction of $\mathbf{k}$

32. $f(x, y, z) = x^2 e^{xz}$ at $(3, -2, 4)$ in the direction of $\mathbf{j}$

33. $f(x, y) = (x^3 - 2xy)^2$ at $(-2, 3)$ in the direction of $-\mathbf{i}$

34. $f(x, y) = \dfrac{\cos xy}{x + y^2}$ at $(0, 4)$ in the direction of $-\mathbf{j}$

35. $f(x, y, z) = \dfrac{y + z}{x - z}$ at $(-1, 3, 1)$ in the direction of $-\mathbf{k}$

36. $f(x, y, z) = \dfrac{e^{xz}}{y + z}$ at $(2, -1, 0)$ in the direction of $-\mathbf{j}$

In Exercises 37–44, find the indicated partial derivative of the given function f. You need not simplify answers.

37. $xyz;\ \dfrac{\partial^2 f}{\partial z^2}$

38. $x^2 yz;\ \dfrac{\partial^2 f}{\partial z\, \partial x}$

39. $\dfrac{xy}{z};\ f_{xyz}$

40. $xe^{yz};\ f_{zx}$

41. $\dfrac{x^3 + y}{y^3 + yz};\ \dfrac{\partial^4 f}{\partial x^4}$

42. $\ln[\cos(2x + y - 3z)];\ \dfrac{\partial^3 f}{\partial z\, \partial x^2}$

43. $\dfrac{xyz}{x^4 + y^4};\ f_{xyzz}$

44. $xz \cos(y^2 z^2);\ f_{yzxx}$

In Exercises 45–48, verify that $f_{xy} = f_{yx}$ by computing the two second partial derivatives for the given function f.

45. $x^2 y$

46. $x^3 y^2 + (x^2/y)$

45. $\ln(xy^2)$

48. $\sin^{-1}(xy^2)$

49. Let $f(x, y) = \sin xy$. Show that $x^2 \cdot f_{xx} = y^2 \cdot f_{yy}$.

50. Verify that $f_{xxy} = f_{xyx}$ if $f(x, y) = x^3 y^2 + (x^2/y^3)$.

51. Find the rate of change of the volume of a cone with respect only to the altitude h when the altitude is 8 ft if the radius remains constant at 3 ft.

52. A positive charge at the origin exerts a force of attraction on a negative charge that is inversely proportional to the square of the distance between them. The *Newtonian potential function* for the force is $-k\sqrt{x^2 + y^2 + z^2}$ for some constant k. Find the rate of change of this potential function at the point $(-1, 2, -2)$ with respect to y only.

53. A silo is in the shape of a right circular cylinder surmounted by a hemisphere. Find the rate of change of the volume of the silo with respect only to the radius when $r = 6$ ft if the height of the cylinder remains constant at 25 ft.

54. A family lives at the intersection of a north–south road with an east–west road. On a calm day, a train is derailed at a point 500 yd south and 700 yd west of the intersection and is releasing toxic gas. Which way should the family drive to increase their distance from the gas leak as rapidly as possible?

55. Consider the vector $\mathbf{i} + 2\mathbf{j} - 2\mathbf{k}$ from the origin to the point $(x, y, z) = (1, 2, -2)$. Determine whether the length of this vector is increased most rapidly by increasing the x-, y-, or z-coordinate. Why?

14.2 TANGENT PLANES AND APPROXIMATIONS

If $y = g(x)$ is a differentiable function of one variable, then we can approximate g at each point of its domain by a linear function. The linear approximation at each point has as graph the line tangent to the graph of g at that point, as shown in Fig. 14.10.

Linear functions are readily handled. The replacement of a nonlinear problem by a good linear approximation to it is central to calculus. To illustrate, in Section 4.7, we used Newton's method to replace a nonlinear equation $g(x) = 0$ by a succession of readily solved linear equations, and we obtained very accurate estimates for solutions of our original equation. We now turn to a parallel discussion for $z = f(x, y)$.

Tangent Planes

A linear equation in three variables gives a *plane* in space. Clearly, the analogue of a tangent line to the curve that is the graph of $y = g(x)$ in Fig. 14.10 is a tangent plane to the surface that is the graph of $z = f(x, y)$ in Fig. 14.11. Recall that $g(x) = |x|$ does not have a tangent line at the origin. Similarly, there are functions of two variables whose graphs do not have tangent planes at some

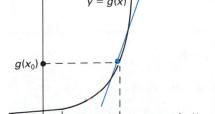

FIGURE 14.10 Graph of the linear approximation to $y = g(x)$ at x_0.

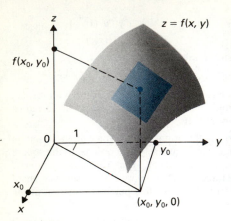

FIGURE 14.11   Plane tangent to the graph of $f(x, y)$ over $(x_0, y_0, 0)$.

points. That is, there are no planes clinging really closely to the graph of the function near those points. In the next section, we will show that tangent planes do exist if the function has continuous first-order partial derivatives. In this section, we will assume that this is so and proceed to discover what the equation of the tangent plane must be.

To find the equation of a plane, we need to know a point on the plane, which we suppose to be a given point on the surface. We also need to know a vector normal to the plane. To do this, we first find two vectors that we can regard as lying in the plane if they start from the point of tangency. Then we can find a normal vector by taking their cross product.

The derivative of a function $y = g(x)$ of one variable is the slope of the tangent line to the graph. The vector

$$\mathbf{i} + \frac{dy}{dx}\mathbf{j}$$

in Fig. 14.12, starting from a point (x, y) on the curve, is tangent to the curve.

For a function of two variables $z = f(x, y)$, the partial derivatives are related to the plane tangent to the graph. Since $\partial z / \partial x$ gives the rate of increase in z with respect to x as y is held constant, the vector

$$\mathbf{i} + \frac{\partial z}{\partial x}\mathbf{k}$$

emanating from a point (x, y, z) on the surface is tangent to the surface. That is, the vector is tangent to the curve on the surface obtained by holding y constant, as shown in Fig. 14.13. Similarly,

$$\mathbf{j} + \frac{\partial z}{\partial y}\mathbf{k}$$

is tangent to the surface there also. Therefore, their cross product

$$\begin{vmatrix} \mathbf{i} & \mathbf{j} & \mathbf{k} \\ 1 & 0 & \dfrac{\partial z}{\partial x} \\ 0 & 1 & \dfrac{\partial z}{\partial y} \end{vmatrix} = -\frac{\partial z}{\partial x}\mathbf{i} - \frac{\partial z}{\partial y}\mathbf{j} + \mathbf{k}$$

is normal to the plane there. We choose to multiply by -1.

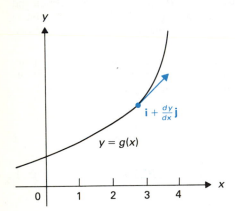

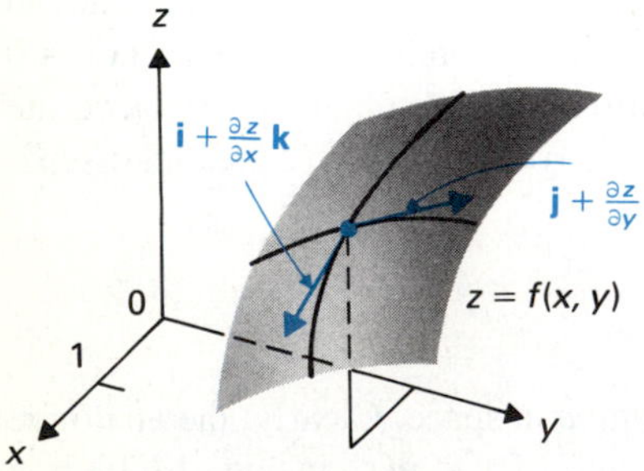

FIGURE 14.12   The vector $\mathbf{i} + (dy/dx)\mathbf{j}$ is tangent to the graph of $g(x)$ at each point.

FIGURE 14.13   The vectors $\mathbf{i} + (\partial z/\partial x)\mathbf{k}$ and $\mathbf{j} + (\partial z/\partial y)\mathbf{k}$ are tangent to the surface at each point.

The vector

$$\frac{\partial z}{\partial x}\mathbf{i} + \frac{\partial z}{\partial y}\mathbf{j} - \mathbf{k}$$

is normal to the graph of $z = f(x, y)$ at each point.

EXAMPLE 1   Let $z = f(x, y) = x^2 + 3xy + 2y^3$. Find the equation of the tangent plane at $(1, 2, 23)$.

Solution   We have

$$\frac{\partial z}{\partial x} = 2x + 3y \qquad \text{and} \qquad \frac{\partial z}{\partial y} = 3x + 6y^2,$$

so at $(1, 2, 23)$ we have

$$\left.\frac{\partial z}{\partial x}\right|_{(1,2)} = 2 + 6 = 8$$

and

$$\left.\frac{\partial z}{\partial y}\right|_{(1,2)} = 3 + 24 = 27.$$

Therefore a vector normal to the plane is $8\mathbf{i} + 27\mathbf{j} - \mathbf{k}$. The equation of the plane is found by

Point $(1, 2, 23)$

⊥ vector $8\mathbf{i} + 27\mathbf{j} - \mathbf{k}$

Equation $8x + 27y - z = 39$ ∎

Of course, we can also find parametric equations of the normal line to a surface if we know a point and a vector parallel to the line, that is, normal to the surface.

EXAMPLE 2   Find parametric equations of the normal line to the surface $z = x^2 + 3xy + 2y^3$ of Example 1 at the point $(1, 2, 23)$.

Solution   By Example 1, we have

Point $(1, 2, 23)$

∥ vector $8\mathbf{i} + 27\mathbf{j} - \mathbf{k}$

Equations $x = 1 + 8t,$

$\qquad\qquad\quad y = 2 + 27t,$

$\qquad\qquad\quad z = 23 - t$ ∎

Approximations

In Section 3.9, we saw how to approximate $g(x_0 + \Delta x)$, where $g(x_0)$ is easily computed and Δx is small. Geometrically, we replaced the curve by its tangent line at $(x_0, g(x_0))$. We then found the value y on this tangent line corresponding to $x = x_0$, as shown in Fig. 14.14.

We are now working with a plane tangent to the surface graph of $z = f(x, y)$ at $(x_0, y_0, f(x_0, y_0))$. To strengthen our geometric intuition in preparation for the differential in the next section, we treat geometrically the analogous approximation of $f(x_0 + \Delta x, y_0 + \Delta y)$, where $f(x_0, y_0)$ is readily computed and where Δx and Δy are small. This approximation will be treated again in the language of differentials in the next section, where we will describe conditions on f that guarantee that the approximation is a good one for sufficiently small values Δx and Δy.

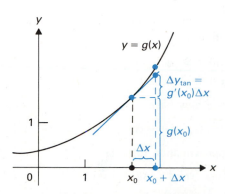

FIGURE 14.14   Recall the approximation $g(x_0 + \Delta x) \approx g(x_0) + g'(x_0)\, \Delta x$.

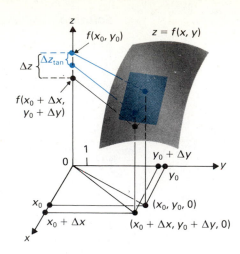

FIGURE 14.15 $\Delta z = f(x_0 + \Delta x, y_0 + \Delta y) - f(x_0, y_0)$; $\Delta z_{\tan}$ = Change in height of the tangent plane.

Let $z = f(x, y)$. The equation of the tangent plane at a point (x_0, y_0, z_0) on the graph is found as follows:

> **Point** (x_0, y_0, z_0)
>
> **⊥ vector** $f_x(x_0, y_0)\mathbf{i} + f_y(x_0, y_0)\mathbf{j} - \mathbf{k}$
>
> **Equation** $f_x(x_0, y_0)(x - x_0) + f_y(x_0, y_0)(y - y_0) - (z - z_0) = 0$

We write this equation in the form

$$z - z_0 = f_x(x_0, y_0)(x - x_0) + f_y(x_0, y_0)(y - y_0). \tag{1}$$

We set $\Delta x = x - x_0$, $\Delta y = y - y_0$, and $\Delta z_{\tan} = z - z_0$ in Eq. (1). We use $\Delta z_{\tan}$ rather than Δz since it represents the change in z for the tangent plane, and we regard Δz as representing the change in z for $z = f(x, y)$. See Fig. 14.15. Then Eq. (1) becomes

$$\Delta z_{\tan} = f_x(x_0, y_0)\Delta x + f_y(x_0, y_0)\Delta y. \tag{2}$$

We can regard Eq. (2) as the equation of the tangent plane with respect to local $\Delta x, \Delta y, \Delta z$-axes, as shown in Fig. 14.16.

Recall that for a function $y = g(x)$ of one variable, we have $\Delta y_{\tan} = g'(x_0)\,\Delta x$, which leads to the approximation

$$g(x_0 + \Delta x) \approx g(x_0) + \Delta y_{\tan}$$
$$= g(x_0) + g'(x_0)\,\Delta x$$

shown in Fig. 14.14. Correspondingly, we now have

$$f(x_0 + \Delta x, y_0 + \Delta y) \approx f(x_0, y_0) + \Delta z_{\tan}. \tag{3}$$

From Eqs. (2) and (3), we obtain the approximation formula

$$f(x_0 + \Delta x, y_0 + \Delta y) \approx f(x_0, y_0) + f_x(x_0, y_0)\,\Delta x \\ + f_y(x_0, y_0)\,\Delta y, \tag{4}$$

where the approximation is best for small values of Δx and Δy.

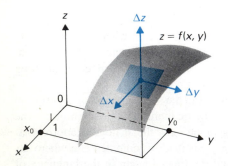

FIGURE 14.16 The tangent plane has equation

$$\Delta z = f_x(x_0, y_0)\,\Delta x + f_y(x_0, y_0)\,\Delta y.$$

EXAMPLE 3 Estimate $1.05^2 \cdot 2.99^3$.

Solution Let $f(x, y) = x^2 y^3$. We know $f(1, 3) = 1^2 \cdot 3^3 = 27$. Let $\Delta x = 0.05$ and $\Delta y = -0.01$. Now

$$f_x = 2xy^3 \qquad \text{and} \qquad f_y = 3x^2 y^2.$$

Then by the approximation (4), we have

$$\begin{aligned}
f(1.05, 2.99) &= f(1 + \Delta x, 3 + \Delta y) \\
&\approx 27 + f_x(1, 3)(0.05) + f_y(1, 3)(-0.01) \\
&= 27 + 54(0.05) + 27(-0.01) = 27 + 2.70 - 0.27 \\
&= 29.43.
\end{aligned}$$

Actually, $1.05^2 \cdot 2.99^3 = 29.4708161475$. ∎

EXAMPLE 4 Find the approximate change in volume of a right circular cone of base radius 4 and altitude 10 if the radius of the base is decreased by 0.05 and the altitude is increased by 0.2.

Solution The volume V of a cone is given by

$$V = f(r, h) = \frac{1}{3}\pi r^2 h.$$

The approximate change in V is given by

$$\begin{aligned}
\Delta V_{\tan} &= \frac{\partial V}{\partial r}\Delta r + \frac{\partial V}{\partial h}\Delta h \\
&= \frac{2}{3}\pi r h\, \Delta r + \frac{1}{3}\pi r^2\, \Delta h.
\end{aligned}$$

Taking

$$r = 4, \qquad \Delta r = -0.05, \qquad h = 10, \qquad \text{and} \qquad \Delta h = 0.2,$$

we obtain as answer

$$\begin{aligned}
\Delta V_{\tan} &= \left(\frac{2}{3}\pi \cdot 40\right)(-0.05) + \left(\frac{1}{3}\pi \cdot 16\right)(0.2) \\
&= \frac{2}{3}\pi(-2 + 1.6) \\
&= \frac{2}{3}\pi(-0.4) = -\frac{8\pi}{30}.
\end{aligned}$$
∎

Similar results hold for functions of more variables. For example, if $w = f(x, y, z)$, then the analogue of the approximation (4) is

$$\begin{aligned}
f(x_0 + \Delta x, y_0 + \Delta y, z_0 + \Delta z) &\approx f(x_0, y_0, z_0) + f_x(x_0, y_0, z_0)\, \Delta x \\
&\quad + f_y(x_0, y_0, z_0)\, \Delta y + f_z(x_0, y_0, z_0)\, \Delta z.
\end{aligned} \tag{5}$$

Such formulas can be written more compactly using boldface notation with subscripted variables. For example, the approximation near $\mathbf{a} = (a_1, a_2, a_3)$ of

$$y = f(\mathbf{x}) = f(x_1, x_2, x_3)$$

corresponding to a change $\Delta\mathbf{x} = (\Delta x_1, \Delta x_2, \Delta x_3)$ is given by

$$f(\mathbf{a} + \Delta\mathbf{x}) \approx f(\mathbf{a}) + f_{x_1}(\mathbf{a})\,\Delta x_1 + f_{x_2}(\mathbf{a})\,\Delta x_2 + f_{x_3}(\mathbf{a})\,\Delta x_3, \qquad \textbf{(6)}$$

for a vector $\Delta\mathbf{x}$ of sufficiently small magnitude. The use of boldface notation and subscripted variables often makes the structure of formulas easier to follow. Formulas just like (6) hold for approximation of functions of one or two variables also; it just depends on the subscript of the "last x."

EXAMPLE 5 Estimate $\sqrt{2.01^2 + 1.98^2 + 1.05^2}$.

Solution We let $f(x, y, z) = \sqrt{x^2 + y^2 + z^2}$ and use approximation (5) with

$$x_0 = 2, \qquad y_0 = 2, \qquad z_0 = 1$$
$$\Delta x = 0.01, \qquad \Delta y = -0.02, \qquad \Delta z = 0.05.$$

We readily find that

$$f_x(2, 2, 1) = \frac{x}{\sqrt{x^2 + y^2 + z^2}}\bigg|_{(2,2,1)} = \frac{2}{3},$$

and, similarly,

$$f_y(2, 2, 1) = \frac{2}{3}$$

and

$$f_z(2, 2, 1) = \frac{1}{3}.$$

Then approximation (5) becomes

$$\sqrt{2.01^2 + 1.98^2 + 1.05^2} \approx 3 + \frac{2}{3}(0.01) + \frac{2}{3}(-0.02) + \frac{1}{3}(0.05)$$

$$= 3 + \frac{0.03}{3} = 3.01. \qquad \blacksquare$$

Problems like those in Examples 3, 4, and 5 are common in calculus texts but are more readily solved today using calculators and computers. It is a triviality to compute that

$$\sqrt{2.01^2 + 1.98^2 + 1.05^2} \approx 3.010481689$$

using a calculator. It takes much less time than working through Example 5. Such examples serve to illustrate the approximations (4) and (5). It is more practical to use the approximation (3) to estimate the changes Δx and Δy needed to produce a desired change Δz.

EXAMPLE 6 A conical pile of sand has base radius 6 ft and is 15 ft high. An additional 10 ft^3 of sand is dropped onto the top of the pile to form a new conical pile. If the altitude of the pile increases by 1 in., estimate the change in the radius of the pile.

Solution The volume of a cone is given by $V = \frac{1}{3}\pi r^2 h$. We want to estimate Δr to provide a change $\Delta V = 10$ and $\Delta h = \frac{1}{12}$ when $r = 6$ and $h = 15$. As in

Example 4, we have

$$\Delta V_{\text{tan}} = \frac{2}{3} \pi r h \, \Delta r + \frac{1}{3} \pi r^2 \, \Delta h.$$

Putting in the appropriate values, we find that

$$10 = \frac{2}{3} \pi \cdot 90 \, \Delta r + \frac{1}{3} \pi \cdot 36 \frac{1}{12} = 60\pi \, \Delta r + \pi.$$

Thus $60\pi \, \Delta r = 10 - \pi$, so $\Delta r = (10 - \pi)/(60\pi) \approx 0.036$ ft, or about 0.437 in. ■

Newton's Method for Functions of Two Variables*

Let $z = f(x, y)$. We will write Δf_{tan} for Δz_{tan} here. Then Eq. (2) becomes

$$\Delta f_{\text{tan}} = f_x(x_0, y_0) \, \Delta x + f_y(x_0, y_0) \, \Delta y. \tag{7}$$

If we use Eq. (7) to produce a desired change Δf_{tan} at a point (x_0, y_0), there are generally many choices for Δx and Δy. If the partial derivatives are nonzero, we can set Δx equal to any number and solve for Δy. Indeed, we expect to be able to solve *two* equations like Eq. (7) *simultaneously* for Δx and Δy.

Let $z = g(x, y)$ be another function. The surfaces that are the graphs of $f(x, y)$ and $g(x, y)$ generally intersect in a curve, as shown in Fig. 14.17. This curve may go through the x,y-plane at one point, many points, or perhaps no point at all. To find such points amounts to solving the system of simultaneous equations

$$\begin{cases} f(x, y) = 0, \\ g(x, y) = 0. \end{cases} \tag{8}$$

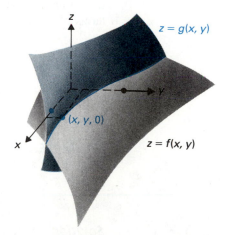

FIGURE 14.17   The curve of intersection of the surfaces $z = f(x, y)$ and $z = g(x, y)$.

---

*This topic formerly was omitted from calculus texts because it involves quite a bit of computation. With calculators and computers, it is now entirely feasible. We urge students at least to read it to appreciate the real importance of the approximations (4) and (5).

Simultaneous equations like Eq. (8) are often very tough to solve if they are not linear. But

$$\begin{cases} \Delta f_{\tan} = f_x(x_0, y_0)\,\Delta x + f_y(x_0, y_0)\,\Delta y, \\ \Delta g_{\tan} = g_x(x_0, y_0)\,\Delta x + g_y(x_0, y_0)\,\Delta y \end{cases} \tag{9}$$

is a *linear* system in unknowns Δx and Δy and is readily solved!

Newton's method for two variables makes use of the system (9). The steps of the method are as follows.

Newton's Method for Two Variables

STEP 1. Try to find a reasonably good first approximate solution point (a_1, b_1) for the system (8).

STEP 2. Set $\Delta f_{\tan} = -f(a_1, b_1)$ and $\Delta g_{\tan} = -g(a_1, b_1)$ in system (9), obtaining the system

$$\begin{cases} f_x(a_1, b_1)\,\Delta x + f_y(a_1, b_1)\,\Delta y = -f(a_1, b_1), \\ g_x(a_1, b_1)\,\Delta x + g_y(a_1, b_1)\,\Delta y = -g(a_1, b_1). \end{cases} \tag{10}$$

STEP 3. Solve the linear system (10) for Δx and Δy.

STEP 4. Let $a_2 = a_1 + \Delta x$ and $b_2 = b_1 + \Delta y$, to obtain the next approximate solution point (a_2, b_2) for the system (8).

STEP 5. Start step 2 again with (a_2, b_2) in place of (a_1, b_1), and so on.

In Exercise 39, we ask you to show that recursion formulas for a_{i+1} and b_{i+1} in terms of a_i and b_i are

$$a_{i+1} = a_i - \left. \left(\frac{f \cdot g_y - g \cdot f_y}{f_x g_y - g_x f_y} \right) \right|_{(a_i, b_i)},$$

$$\tag{11}$$

$$b_{i+1} = b_i - \left. \left(\frac{g \cdot f_x - f \cdot g_x}{f_x g_y - g_x f_y} \right) \right|_{(a_i, b_i)}.$$

EXAMPLE 7 Use one iteration of Newton's method to estimate a solution point of the system

$$\begin{cases} f(x, y) = x^2 + 3xy + y^2 - 9 = 0, \\ g(x, y) = x^3 - y^3 + 10 = 0. \end{cases}$$

Solution A few moments of experimentation show that $f(1, 2) = 2$ and $g(1, 2) = 3$. Since these function values are not too big, we take $(a_1, b_1) = (1, 2)$ as our first-approximation solution point.

We must also compute f_x, f_y, g_x, and g_y at $(1, 2)$. We have

$$f_x(1, 2) = (2x + 3y)\big|_{(1,2)} = 8,$$

$$f_y(1, 2) = (3x + 2y)\big|_{(1,2)} = 7,$$
$$g_x(1, 2) = 3x^2\big|_{(1,2)} = 3,$$
$$g_y(1, 2) = -3y^2\big|_{(1,2)} = -12.$$

We could form the system (10) and solve it, but we prefer to simply substitute in the recursion formulas (11), obtaining

$$a_2 = 1 - \frac{(2)(-12) - (3)(7)}{(8)(-12) - (3)(7)}$$

$$= 1 - \frac{-45}{-117} = \frac{72}{117} \approx 0.6154$$

and

$$b_2 = 2 - \frac{(3)(8) - (2)(3)}{-117}$$

$$= 2 - \frac{18}{-117} = \frac{252}{117} \approx 2.1538.$$

Thus our answer is $(a_2, b_2) = (\frac{72}{117}, \frac{252}{117})$. Our calculator now shows that $f(a_2, b_2) \approx -0.006$ and $g(a_2, b_2) \approx 0.241$, so the function values are closer to zero than at (a_1, b_1). ∎

EXAMPLE 8 Use a calculator and Newton's method to estimate a solution point of the system of equations in Example 7, starting with $(a_1, b_2) = (1, 2)$.

Solution We have a programmable calculator, which is very convenient for this problem. We use some of its memory registers as follows.

Register no.:	1	2	3	4	5	6	7	8
Quantity:	a_i	b_i	f	g	f_x	f_y	g_x	g_y

That is, we program the calculator to compute $f(a_i, b_i)$ using the values in registers 1 and 2 and then store the result in register 3. Then the program computes $g(a_i, b_i)$ from the values in registers 1 and 2 and stores the result in register 4, and so on. After register 8 is computed, the program computes a_{i+1} and b_{i+1} using formulas (11), storing them in registers 1 and 2, respectively. We then arrange for the run to stop momentarily so we can copy down a_{i+1}, b_{i+1}, $f(a_i, b_i)$, and $g(a_i, b_i)$, which are available from registers 1, 2, 3, and 4, respectively. The program then resets the calculator to the start of the program to find the next iterate. Having programmed our calculator to perform this sequence of steps, we enter 1 in register 1 and 2 in register 2 and then run the program. We obtain the data in Table 14.1. Thus we see that

$$(0.6031892778, 2.170081386)$$

is very close to a solution point of the system. ∎

The iterates in formulas (11) may not converge to a solution, so it is best to always check $f(a_i, b_i)$ and $g(a_i, b_i)$ as in Table 14.1 to be sure they are approaching zero.

TABLE 14.1

i	a_i	b_i	$f(a_i, b_i)$	$g(a_i, b_i)$
1	1	2	2	3
2	0.6153846154	2.153846154	-0.0059171598	0.241238052
3	0.6030896945	2.170176343	-0.000184495	-0.0014502708
4	0.6031892745	2.170081389	-0.0000000094	-0.0000000408
5	0.6031892778	2.170081386	-7×10^{-12}	1×10^{-11}
6	0.6031892778	2.170081386		

Of course, Newton's method can be used to solve a system of three equations

$$\begin{cases} f(x, y, z) = 0, \\ g(x, y, z) = 0, \\ h(x, y, z) = 0 \end{cases}$$

in three unknowns. The problem is reduced at each iteration to solving a system of three *linear* equations in Δx, Δy, and Δz. In general, we can attempt the method with any system that has the same number of equations as unknowns. It is less work to use a computer than a calculator with Newton's method.

SUMMARY

1. A vector normal to the surface given by $z = f(x, y)$ at any point (x, y, z) on the surface is

$$\frac{\partial z}{\partial x}\mathbf{i} + \frac{\partial z}{\partial y}\mathbf{j} - \mathbf{k}.$$

2. In view of the vector just described, we can find the tangent plane and normal line to the graph of the function, for they are determined by the point on the graph and a vector perpendicular to the graph.

3. The approximate value of $f(x_0 + \Delta x, y_0 + \Delta y)$, given by the height of the tangent plane over $(x_0 + \Delta x, y_0 + \Delta y)$, is

$$f(x_0 + \Delta x, y_0 + \Delta y) \approx f(x_0, y_0) + f_x(x_0, y_0)\,\Delta x + f_y(x_0, y_0)\,\Delta y$$

for sufficiently small Δx and Δy.

4. If $y = f(\mathbf{x}) = f(x_1, x_2, x_3)$ while $\mathbf{a} = (a_1, a_2, a_3)$ and

$$\mathbf{\Delta x} = (\Delta x_1, \Delta x_2, \Delta x_3),$$

then

$$f(\mathbf{a} + \mathbf{\Delta x}) \approx f(\mathbf{a}) + f_{x_1}(\mathbf{a})\,\Delta x_1 + f_{x_2}(\mathbf{a})\,\Delta x_2 + f_{x_3}(\mathbf{a})\,\Delta x_3$$

for $\mathbf{\Delta x}$ of sufficiently small magnitude.

5. In Newton's method for finding approximate solutions (a_i, b_i) of simultaneous equations $f(x, y) = 0$ and $g(x, y) = 0$, the recursion relations are

$$a_{i+1} = a_i - \frac{f \cdot g_y - g \cdot f_y}{f_x g_y - g_x f_y} \quad \text{and} \quad b_{i+1} = b_i - \frac{g \cdot f_x - f \cdot g_x}{f_x g_y - g_x f_y},$$

where the functions and their derivatives are evaluated at (a_i, b_i).

EXERCISES 14.2

In Exercises 1–6, find the equation of the plane tangent to the graph of the given function at the indicated point.

1. $f(x, y) = xy + 3y^2$ at $(-2, 3, 21)$

2. $f(x, y) = \sin xy$ at $(1, \pi/2, 1)$

3. $f(x, y) = x^2/(x + y)$ at $(2, 2, 1)$

4. $f(x, y) = \ln(x^2 + y)$ at $(1, 0, 0)$

5. $f(x_1, x_2) = x_1^2 x_2 - 4x_1 x_2^3$ at $(-2, 1, 12)$

6. $f(x_1, x_2) = x_1^2 e^{x_1 x_2}$ at $(3, 0, 9)$

In Exercises 7–12, find parametric equations of the line normal to the surface at the indicated point.

7. $z = x^3 y + xy^2$ at $(1, -1, 0)$

8. $z = \ln(x^2 + y^2)$ at $(3, 4, \ln 25)$

9. $z = xy/(x + y)$ at $(2, -3, 6)$

10. $z = y \tan^{-1}(xy)$ at $(1, -1, \pi/4)$

11. $x_3 = \sqrt{x_1^2 + 4x_2^2}$ at $(-3, 2, 5)$

12. $x_3 = x_2 \sinh(x_1 + x_2)$ at $(-2, 2, 0)$

13. Find the point on the surface $z = x^2 + y^2$ where the tangent plane is parallel to the plane $6x - 4y + 2z = 5$.

14. Find the point on the graph of $f(x, y) = x^2 - y^2$ where the tangent plane is parallel to the plane $3x + 4y + z = 6$.

15. Find the point on the surface $z = xy$ where the normal line is parallel to the line $x = 3 - 2t$, $y = 4 + 5t$, $z = 3 + 3t$.

16. Find the point on the graph of $f(x, y) = xy + x^2$ where the normal line is parallel to the line $x = 4 - 3t$, $y = 5 + 8t$, $z = 1 - 2t$.

17. Find all points on the surface $z = x^2 y$ where the tangent plane is orthogonal to the line $x = 2 - 6t$, $y = 3 - 12t$, $z = 2 + 3t$.

18. Find all points on the surface $z = x/y^2$ where the normal line is orthogonal to the plane $12x + 6y - 3z = 1$.

In Exercises 19–22, use calculus to estimate the indicated quantity.

19. $\sqrt{(4.04)(0.95)}$

20. $(2.01)(1.98)^3 + (2.01)^2(1.98)$

21. $(\cos 1°)(\tan 44°)$

22. $\sqrt{1.97^2 + 2.02^2 + 1.05^2}$

23. A rectangular box has inside measurements of 14-in. width, 20-in. length, and 8-in. height. Estimate the volume of material used in construction of the box if the sides and bottom are $\frac{1}{8}$ in. thick and the box has no top.

24. A cylindrical silo with a hemispherical cap has volume

$$V = \pi r^2 h + \tfrac{2}{3}\pi r^3,$$

where h is the height and r the radius of the cylinder. Estimate the change in volume of a silo of 6-ft radius and 30-ft height if the radius is increased by 4 in. and the height is decreased by 6 in.

25. The magnitude of the centripetal acceleration of a body moving on a circle of radius r with constant speed v is v^2/r. A body is moving at a speed of 10 ft/sec on a circle of radius 2 ft. Find the approximate change in the magnitude of the acceleration if the speed is increased 0.5 ft/sec and the radius is decreased 1 in.

26. For the body in Exercise 25, suppose the magnitude of the centripetal force is kept constant. Find the approximate change in the velocity if the radius r is increased by 1 in.

27. The pressure P in lb/ft^2 of a certain gas at temperature T degrees in a container of variable volume V ft^3 is given by $P = 8T/V$. Suppose the temperature is 20° and the volume is 10 ft^3. If the temperature is increased by 0.5°, estimate the change in volume that produces a decrease in pressure of 0.25 lb/ft^2.

28. The voltage drop V, measured in volts, across a certain conductor of variable resistance R ohms is IR, where I, measured in amperes, is the current flowing through the conductor. At a certain instant, a current of 1 amp is flowing through a resistance of 200 ohms. If the resistance is decreased by 2 ohms and the voltage is decreased by 4 volts, estimate the change in the current.

If $z = f(x, y)$, then small changes Δx in x and Δy in y produce approximately a $100(\Delta z_{\text{tan}})/z$ percent change in z. Exercises 29–32 deal with this phrasing in terms of percent change.

29. Find the approximate percent change in $z = x^2 + xy$ when $x = 2$ and $y = 4$ if x is increased by 5% and y is decreased by 6%.

30. If each dimension of a rectangular box is increased by $a\%$, where a is small, find the approximate percent change in the volume of the box.

31. If both the radius of the base and the altitude of a right circular cylinder are increased by $a\%$, where a is small, find the approximate percent change in the volume of the cylinder.

32. Find the approximate percent change in the volume of a right circular cone if the altitude is increased by 2% and the radius decreased by 0.5%.

In Exercises 33–38, use Newton's method to find the next approximate solution (a_2, b_2) to the given system of two equations, starting with the given approximate solution (a_1, b_1).

33. $(a_1, b_1) = (6, -1)$, $\begin{cases} x^2 + xy - 30 = 0 \\ xy^2 - 2y^3 - 20 = 0 \end{cases}$

34. $(a_1, b_1) = (10, 0)$, $\begin{cases} x + y - 10 = 0 \\ x^2y + 5 = 0 \end{cases}$

35. $(a_1, b_1) = (1, 1)$, $\begin{cases} x^3 - 3x^2y - 3y^3 + 4 = 0 \\ x^2y - y^3 - 1 = 0 \end{cases}$

36. $(a_1, b_1) = (0, -2)$, $\begin{cases} e^{xy} - y - 4 = 0 \\ x^2 - 2y^2 + 5 = 0 \end{cases}$

37. $(a_1, b_1) = (0, 0)$, $\begin{cases} e^x - e^y + 1 = 0 \\ x + 3y - 2 = 0 \end{cases}$

38. $(a_1, b_1) = (0, 2)$, $\begin{cases} 8 \sin(xy) - e^{xy} + 2 = 0 \\ x^3 - y^3 + 6 = 0 \end{cases}$

39. Derive the recursion formulas (11) for the two-variable case of Newton's method.

In Exercises 40–45, use a programmable calculator or a computer and Newton's method. Determine accuracy to as many places as you can for a solution point for the indicated system, starting from the given first approximation.

40. The system and point in Exercise 33

41. The system and point in Exercise 34

42. The system and point in Exercise 35

43. The system and point in Exercise 36

44. The system and point in Exercise 37

45. The system and point in Exercise 38

EXPLORING CALCULUS

5: multivariable Newton's method

14.3 THE DIFFERENTIAL AND THE GRADIENT VECTOR

In the preceding section, we learned how to find the equation of a tangent plane to the graph of a function $f(x, y)$, and we regarded the plane as giving a good linear approximation to the function f near the point of tangency. We then practiced using this approximation. In the present section, we will justify our use of the tangent-plane approximation and give some terminology and notation.

The Linear Approximation Theorem

It is instructive to recall our work in Section 3.9 for a function $y = f(x)$ of one variable. The hypotheses that f is defined in an interval centered at x_0 and that $f'(x_0)$ exists were all we needed to show that $f(x)$ has a good linear approximation near x_0. Recall that

$$\Delta y = f(x_0 + \Delta x) - f(x_0) \tag{1}$$

and

$$\Delta y_{\tan} = f'(x_0)\,\Delta x. \tag{2}$$

We let

$$E(\Delta x) = \Delta y - \Delta y_{\tan}, \tag{3}$$

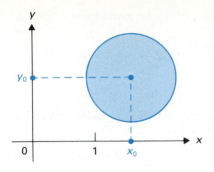

FIGURE 14.18 Disk centered at (x_0, y_0).

so that $E(\Delta x)$ is the error in the approximation of Δy by Δy_{tan}. We then showed that

$$\lim_{\Delta x \to 0} \frac{E(\Delta x)}{\Delta x} = 0. \tag{4}$$

Equation (4) means that the error in approximating Δy by Δy_{tan} is not merely small as $\Delta x \to 0$ but is even small compared with the size of Δx. We defined the function $\varepsilon_1(\Delta x)$ by

$$\varepsilon_1(\Delta x) = \begin{cases} \dfrac{E(\Delta x)}{\Delta x} & \text{for } \Delta x \neq 0, \\ 0 & \text{for } \Delta x = 0. \end{cases} \tag{5}$$

From Eq. (4), we have

$$\lim_{\Delta x \to 0} \varepsilon_1(\Delta x) = 0. \tag{6}$$

Then

$$\begin{aligned} f(x_0 + \Delta x) = f(x_0) + \Delta y &= f(x_0) + \Delta y_{tan} + E(\Delta x) \\ &= f(x_0) + f'(x_0)\,\Delta x + E(\Delta x), \end{aligned}$$

so

$$f(x_0 + \Delta x) = f(x_0) + f'(x_0)\,\Delta x + \varepsilon_1(\Delta x) \cdot \Delta x, \tag{7}$$
$$\text{where } \lim_{\Delta x \to 0} \varepsilon_1(\Delta x) = 0.$$

The relation (7) was crucial to our proof of the chain rule in Section 3.9.

We wish to parallel this development for a function $z = f(x, y)$ of two variables. Let $f(x, y)$ be defined in a disk centered at (x_0, y_0), as shown in Fig. 14.18. With a function of one variable, we needed *only the existence* of the derivative at the center point x_0 to derive the relation (7). The big difference for two or more variables is the need to strengthen this hypothesis. We now assume that f_x and f_y exist and are *continuous* throughout a disk centered at (x_0, y_0). We proceed to develop the analogues of Eqs. (1)–(7). We let

$$\Delta z = f(x_0 + \Delta x, y_0 + \Delta y) - f(x_0, y_0) \tag{8}$$

and

$$\Delta z_{tan} = f_x(x_0, y_0) \cdot \Delta x + f_y(x_0, y_0) \cdot \Delta y. \tag{9}$$

We also let

$$E(\Delta x, \Delta y) = \Delta z - \Delta z_{tan}, \tag{10}$$

so that $E(\Delta x, \Delta y)$ is the error in the approximation of Δz by Δz_{tan}. Our goal is to show that $E(\Delta x, \Delta y)$ becomes small compared with the size of Δx and Δy as both $\Delta x \to 0$ and $\Delta y \to 0$, analogous to Eq. (4).

Now

$$\begin{aligned} \Delta z &= f(x_0 + \Delta x, y_0 + \Delta y) - f(x_0, y_0) \\ &= [f(x_0 + \Delta x, y_0 + \Delta y) - f(x_0, y_0 + \Delta y)] \\ &\quad + [f(x_0, y_0 + \Delta y) - f(x_0, y_0)]. \end{aligned} \tag{11}$$

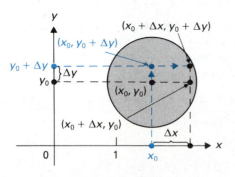

FIGURE 14.19 The change in $f(x, y)$ going from (x_0, y_0) to

$$(x_0 + \Delta x, y_0 + \Delta y)$$

is the sum of the changes involving $(x_0, y_0 + \Delta y)$ as an intermediate point.

Relation (11) is a "subtract-and-add trick" using the point $(x_0, y_0 + \Delta y)$ shown in Fig. 14.19. The point $(x_0 + \Delta x, y_0)$ could have been used equally well. Let

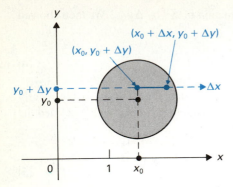

FIGURE 14.20 $f(x, y)$ is a function of the variable Δx only on the line segment from $(x_0, y_0 + \Delta y)$ to $(x_0 + \Delta x, y_0 + \Delta y)$.

the function $\varepsilon_1(\Delta x, \Delta y)$ of the variables Δx and Δy be given by

$$\varepsilon_1 = \begin{cases} \dfrac{f(x_0 + \Delta x, y_0 + \Delta y) - f(x_0, y_0 + \Delta y)}{\Delta x} - f_x(x_0, y_0) \text{ for } \Delta x \neq 0, \\ 0 \qquad \text{for } \Delta x = 0. \end{cases} \quad (12)$$

(We suppress the Δx and Δy in $\varepsilon_1(\Delta x, \Delta y)$ here and in some of the sequel for reasons of space.) From Eq. (12), we see that

$$f(x_0 + \Delta x, y_0 + \Delta y) - f(x_0, y_0 + \Delta y) = f_x(x_0, y_0) \cdot \Delta x + \varepsilon_1 \cdot \Delta x. \quad (13)$$

The numerator of the quotient appearing in Eq. (12) is a function of the single variable Δx on the line segment shown in Fig. 14.20, where Δy is constant. We can apply the mean value theorem to this function over this line segment, obtaining

$$\frac{f(x_0 + \Delta x, y_0 + \Delta y) - f(x_0, y_0 + \Delta y)}{\Delta x} = f_x(c, y_0 + \Delta y) \quad (14)$$

for some point $(c, y_0 + \Delta y)$ on the line segment, shown in Fig. 14.20. Using Eq. (12), we therefore have

$$\varepsilon_1 = f_x(c, y_0 + \Delta y) - f_x(x_0, y_0),$$

where c is between x_0 and $x_0 + \Delta x$. *Since f_x is assumed to be continuous*, we see that

$$\lim_{\Delta x, \Delta y \to 0} \varepsilon_1 = f_x(x_0, y_0) - f_x(x_0, y_0) = 0.$$

Now let

$$\varepsilon_2 = \begin{cases} \dfrac{f(x_0, y_0 + \Delta y) - f(x_0, y_0)}{\Delta y} - f_y(x_0, y_0) \qquad \text{for } \Delta y \neq 0, \\ 0 \qquad \text{for } \Delta y = 0. \end{cases} \quad (15)$$

Then

$$f(x_0, y_0 + \Delta y) - f(x_0, y_0) = f_y(x_0, y_0) \cdot \Delta y + \varepsilon_2 \cdot \Delta y. \quad (16)$$

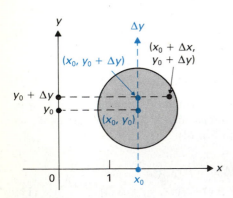

FIGURE 14.21 $f(x, y)$ is a function of the variable Δy only on the line segment from (x_0, y_0) to $(x_0, y_0 + \Delta y)$.

An argument using the mean value theorem as we did above, but applied to the function of Δy on the line segment joining (x_0, y_0) to $(x_0, y_0 + \Delta y)$ in Fig. 14.21, shows that $\lim_{\Delta x, \Delta y \to 0} \varepsilon_2 = 0$. Substituting the expressions in Eqs. (13) and (16) for the bracketed expressions in relation (11), we see that

$$\Delta z = f_x(x_0, y_0) \cdot \Delta x + f_y(x_0, y_0) \cdot \Delta y + \varepsilon_1 \cdot \Delta x + \varepsilon_2 \cdot \Delta y, \quad (17)$$

$$\text{where } \lim_{\Delta x, \Delta y \to 0} \varepsilon_1 = \lim_{\Delta x, \Delta y \to 0} \varepsilon_2 = 0.$$

Our error for the approximation of Δz by Δz_{tan} is thus

$$E(\Delta x, \Delta y) = \varepsilon_1(\Delta x, \Delta y) \cdot \Delta x + \varepsilon_2(\Delta x, \Delta y) \cdot \Delta y,$$

and since ε_1 and ε_2 both approach zero as $\Delta x \to 0$ and $\Delta y \to 0$, we see that this error is small compared with the size of Δx and Δy. Thus the relation (17) achieves our goal and justifies our approximation of Δz by Δz_{tan} for small Δx and Δy if f has continuous first partial derivatives. The relation (17) will enable us to prove the chain rule in Section 14.4 and is so important that we state it as a theorem.

THEOREM 14.2   Linear approximation

Let $f(x, y)$ have continuous partial derivatives f_x and f_y inside a circle of radius $r > 0$ with center (x_0, y_0). Then there exist functions $\varepsilon_1(\Delta x, \Delta y)$ and $\varepsilon_2(\Delta x, \Delta y)$, defined for $(\Delta x)^2 + (\Delta y)^2 < r^2$, such that

$$\lim_{\Delta x, \Delta y \to 0} \varepsilon_1 = \lim_{\Delta x, \Delta y \to 0} \varepsilon_2 = 0$$

and such that

$$f(x_0 + \Delta x, y_0 + \Delta y) = f(x_0, y_0) + f_x(x_0, y_0)\, \Delta x + f_y(x_0, y_0)\, \Delta y$$
$$+ \varepsilon_1(\Delta x, \Delta y) \cdot \Delta x + \varepsilon_2(\Delta x, \Delta y) \cdot \Delta y. \qquad \Box$$

EXAMPLE 1   Find $\varepsilon_1(\Delta x, \Delta y)$ and $\varepsilon_2(\Delta x, \Delta y)$ in Theorem 14.2 for $f(x, y) = x^2 + xy$ at the point $(1, 2)$. Show directly that

$$\lim_{\Delta x, \Delta y \to 0} \varepsilon_1 = \lim_{\Delta x, \Delta y \to 0} \varepsilon_2 = 0.$$

Solution   Now $f_x = 2x + y$, so $f_x(1, 2) = 4$. By Eq. (12), we have

$$\varepsilon_1(\Delta x, \Delta y) = \frac{f(1 + \Delta x, 2 + \Delta y) - f(1, 2 + \Delta y)}{\Delta x} - 4$$

$$= \frac{[(1 + \Delta x)^2 + (1 + \Delta x)(2 + \Delta y)] - [1^2 + 2 + \Delta y]}{\Delta x} - 4$$

$$= \frac{1 + 2(\Delta x) + (\Delta x)^2 + 2 + 2(\Delta x) + \Delta y + (\Delta x)(\Delta y) - 3 - \Delta y - 4(\Delta x)}{\Delta x}$$

$$= \frac{(\Delta x)^2 + (\Delta x)(\Delta y)}{\Delta x} = \Delta x + \Delta y.$$

Since $f_y(1, 2) = 1$, we see from Eq. (15) that

$$\varepsilon_2(\Delta x, \Delta y) = \frac{f(1, 2 + \Delta y) - f(1, 2)}{\Delta y} - 1$$

$$= \frac{1 + 2 + \Delta y - 3}{\Delta y} - 1 = \frac{\Delta y - \Delta y}{\Delta y} = 0.$$

Then $\lim_{\Delta x, \Delta y \to 0} \varepsilon_1 = \lim_{\Delta x, \Delta y \to 0} (\Delta x + \Delta y) = 0$, and of course

$$\lim_{\Delta x, \Delta y \to 0} \varepsilon_2 = 0. \qquad \blacksquare$$

If $w = f(x, y, z)$ has continuous first partial derivatives in a ball having (x_0, y_0, z_0) as center, the analogue of Theorem 14.2 holds with functions $\varepsilon_1, \varepsilon_2$, and ε_3 of $\Delta x, \Delta y$, and Δz. In fact, the hypothesis of continuous first partial derivatives is sufficient to guarantee that a function $f(x_1, x_2, \ldots, x_n)$ of n variables has good local linear approximations.

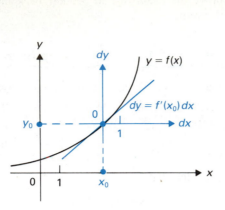

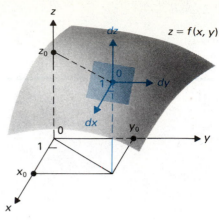

FIGURE 14.22 $dy = f'(x_0)\,dx$ is the equation of the tangent line at (x_0, y_0).

FIGURE 14.23

$$dz = f_x(x_0, y_0)\,dx + f_y(x_0, y_0)\,dy$$

is the equation of the tangent plane at (x_0, y_0, z_0).

The Differential and the Gradient Vector

If $y = f(x)$, the *differential dy* of f at x_0 is the linear function

$$dy = f'(x_0)\,dx \tag{18}$$

of the independent variable dx. Note that Eq. (18) is the equation of the tangent line to the graph of $y = f(x)$, where $x = x_0$, with respect to local dx,dy-axes at the point (x_0, y_0), as shown in Fig. 14.22. The analogue of Fig. 14.22 for a function of two variables is Fig. 14.23, and we see that

$$dz = f_x(x_0, y_0)\,dx + f_y(x_0, y_0)\,dy \tag{19}$$

is the equation with respect to local dx,dy,dz-axes of the plane tangent to $z = f(x, y)$ at (x_0, y_0, z_0), if the tangent plane exists. This suggests the following definition.

DEFINITION 14.5

Differential

Let $z = f(x, y)$ have partial derivatives f_x and f_y at (x_0, y_0). Let dx and dy be independent variables. The **differential** dz (or df) **of** f **at** (x_0, y_0) is

$$dz = f_x(x_0, y_0)\,dx + f_y(x_0, y_0)\,dy.$$

If f_x and f_y exist at all points in the domain of f, then the **differential** dz (or df) is

$$dz = f_x(x, y)\,dx + f_y(x, y)\,dy.$$

The differential defined in Definition 14.5 is sometimes called the *total differential*.

EXAMPLE 2 Find the differential dz if $z = x^3y + xy^2$.

Solution   We have

$$\frac{\partial z}{\partial x} = 3x^2y + y^2 \qquad \text{and} \qquad \frac{\partial z}{\partial y} = x^3 + 2xy,$$

so

$$dz = (3x^2y + y^2)\,dx + (x^3 + 2xy)\,dy. \qquad\blacksquare$$

The natural extension of Definition 14.5 is made for functions of more than two variables. If $y = f(\mathbf{x}) = f(x_1, x_2, \ldots, x_n)$, then

$$dy = f_{x_1}\,dx_1 + f_{x_2}\,dx_2 + \cdots + f_{x_n}\,dx_n. \tag{20}$$

EXAMPLE 3   Find dy if $y = f(x_1, x_2, x_3) = x_1{}^2 + x_2 e^{x_3}$.

Solution   We have $f_{x_1} = 2x_1$, $f_{x_2} = e^{x_3}$, and $f_{x_3} = x_2 e^{x_3}$, so

$$dy = 2x_1\,dx_1 + e^{x_3}\,dx_2 + x_2 e^{x_3}\,dx_3. \qquad\blacksquare$$

EXAMPLE 4   Find dw if $w = z\tan^{-1}(xy)$.

Solution   We have

$$\frac{\partial w}{\partial x} = \frac{yz}{1 + x^2y^2}, \qquad \frac{\partial w}{\partial y} = \frac{xz}{1 + x^2y^2}, \qquad \text{and} \qquad \frac{\partial w}{\partial z} = \tan^{-1}(xy).$$

Consequently,

$$dw = \frac{yz}{1 + x^2y^2}\,dx + \frac{xz}{1 + x^2y^2}\,dy + \tan^{-1}(xy)\,dz. \qquad\blacksquare$$

Recall that a function of one variable is *differentiable at a point* if the derivative exists there. If functions f and g of one variable are differentiable, then the chain rule asserts that the composite function $f(g(x))$ is also differentiable. We would like this to be true for functions of two or more variables. We will need the relation (17) to prove the chain rule, and since it does not follow from the mere existence of first partial derivatives, we take the relation (17) as the defining criterion for a function of two variables to be differentiable at a point.

DEFINITION 14.6

Differentiable function

A function $f(x, y)$ is **differentiable at** (x_0, y_0) if it is defined and its first partial derivatives exist in a disk having (x_0, y_0) as center, and if functions $\varepsilon_1(\Delta x, \Delta y)$ and $\varepsilon_2(\Delta x, \Delta y)$ exist so that relation (17) holds. The function is **differentiable** if it is differentiable at every point in its domain.

The definition asserts that a function is differentiable at a point if it has a good local linear approximation there. In view of Theorem 14.2, functions having continuous first partial derivatives are differentiable.

The approximation problems of the previous section can all be rephrased in terms of "approximation by differentials" just as explained in Section 3.9, and the approximation formula becomes

$$f(x_0 + dx, y_0 + dy) \approx f(x_0, y_0) + f_x(x_0, y_0)\, dx + f_y(x_0, y_0)\, dy. \qquad \textbf{(21)}$$

EXAMPLE 5 Approximate $2.98^4/1.03^2$ using a differential.

Solution We use Eq. (21) with $f(x, y) = x^4/y^2$, $(x_0, y_0) = (3, 1)$, $dx = -0.02$, and $dy = 0.03$. Then

$$f_x(3, 1) = \frac{4x^3}{y^2}\bigg|_{(3,1)} = 108 \qquad \text{and} \qquad f_y(3, 1) = \frac{-2x^4}{y^3}\bigg|_{(3,1)} = -162.$$

We obtain from Eq. (21)

$$\frac{2.98^4}{1.03^2} \approx 81 + 108(-0.02) - 162(0.03)$$

$$= 81 - 2.16 - 4.86 = 73.98. \qquad \blacksquare$$

For $z = f(x, y)$, we have discussed *partial* derivatives at a point. We proceed to define a *vector* that we will see can appropriately be considered to be the "*total* derivative" of a differentiable function at a point.

DEFINITION 14.7

Gradient vector

Let $z = f(x, y)$ and let f_x and f_y exist at (x_0, y_0). The **gradient vector** $\nabla f(x_0, y_0)$ is the vector

$$\nabla f(x_0, y_0) = f_x(x_0, y_0)\mathbf{i} + f_y(x_0, y_0)\mathbf{j}.$$

The symbol ∇ in the preceding definition is read "del." It is often convenient to think of ∇ as a symbolic operator:

$$\nabla = \frac{\partial}{\partial x}\mathbf{i} + \frac{\partial}{\partial y}\mathbf{j}.$$

To find ∇ of a function, we take its partial with respect to x times $\mathbf{i}$ plus its partial with respect to y times $\mathbf{j}$.

EXAMPLE 6 If $z = f(x, y) = x^2y + xy^3$, find ∇f at $(1, -2)$.

Solution We have

$$f_x(1, -2) = (2xy + y^3)|_{(1,-2)} = -12$$

and

$$f_y(1, -2) = (x^2 + 3xy^2)|_{(1,-2)} = 13.$$

Thus $\nabla f(1, -2) = -12\mathbf{i} + 13\mathbf{j}$. $\qquad \blacksquare$

The notion of the gradient vector extends naturally to a function of more than two variables; we simply have more components in the vector.

EXAMPLE 7 Find ∇g for $w = g(x, y, z) = (x + y)/z$.

Solution   We have

$$g_x = \frac{1}{z}, \qquad g_y = \frac{1}{z}, \qquad \text{and} \qquad g_z = -\frac{x+y}{z^2}.$$

Thus

$$\nabla g = \frac{1}{z}\mathbf{i} + \frac{1}{z}\mathbf{j} - \frac{x+y}{z^2}\mathbf{k}. \qquad \blacksquare$$

We now indicate why ∇f plays the role of a total derivative of f. We know that f_x and f_y give the rates of change of $f(x, y)$ in only two of the many possible directions at a point. Namely, f_x gives the rate of change in the direction of $\mathbf{i}$, and f_y in the direction of $\mathbf{j}$. We will see in Section 14.5 that if we know ∇f at a point for a differentiable function f, then we can easily find the rate of change of f in *every* possible direction at that point. Thus we think of ∇f as the *derivative* (or *total derivative*) at the point.

For another insight into the role of ∇f, recall that if $y = f(x)$, then

$$dy = f'(x)\,dx. \qquad (22)$$

We will use subscript notation for a function $y = f(\mathbf{x}) = f(x_1, x_2)$ of two variables. At a point $\mathbf{x} = (x_1, x_2)$, we have

$$\nabla f(\mathbf{x}) = f_{x_1}(\mathbf{x})\mathbf{i} + f_{x_1}(\mathbf{x})\mathbf{j}.$$

If we let the differential vector $\mathbf{dx}$ be

$$\mathbf{dx} = dx_1\mathbf{i} + dx_2\mathbf{j},$$

then the differential $dy = f_{x_1}(\mathbf{x})\,dx_1 + f_{x_1}(\mathbf{x})\,dx_1$ appears as the dot product

$$dy = \nabla f(\mathbf{x}) \cdot \mathbf{dx}. \qquad (23)$$

Comparing Eqs. (22) and (23), we see that ∇f seems to play the role of $f'(x)$. It would be wonderful if mathematicians used either the notation ∇f in place of f' for a function of one variable or the notation $\mathbf{f}'$ in place of ∇f for the gradient vector. However, that is not the way the notation developed.

The gradient vector ∇f is the first vector we have introduced without promptly discussing its magnitude and direction. We will show in Section 14.5 that $\nabla f(x_0, y_0)$ points in the direction of maximum rate of change of the function f at (x_0, y_0) and has magnitude equal to that maximum rate of change.

SUMMARY

1. If f_x and f_y are continuous, then

$$f(x_0 + \Delta x, y_0 + \Delta y) = f(x_0, y_0) + f_x(x_0, y_0)\,\Delta x$$
$$+ f_y(x_0, y_0)\,\Delta y + \varepsilon_1 \cdot \Delta x + \varepsilon_2 \cdot \Delta y,$$

where ε_1 and ε_2 are functions of Δx and Δy that both approach zero as $\Delta x \to 0$ and $\Delta y \to 0$.

2. The differential dz or df of $z = f(x, y)$ at a point (x_0, y_0) is

$$dz = f_x(x_0, y_0)\,dx + f_y(x_0, y_0)\,dy.$$

3. The differential of $y = f(\mathbf{x}) = f(x_1, x_2, \ldots, x_n)$ is

$$dy = f_{x_1}(\mathbf{x})\, dx_1 + f_{x_2}(\mathbf{x})\, dx_2 + \cdots + f_{x_n}(\mathbf{x})\, dx_n.$$

4. If a function has continuous first partial derivatives, its differential at a point is a good local linear approximation to the function near that point.

5. The gradient vector ∇f of $f(x, y)$ at a point (x_0, y_0) is

$$\nabla f(x_0, y_0) = f_x(x_0, y_0)\mathbf{i} + f_y(x_0, y_0)\mathbf{j}.$$

More components are added for functions of more variables.

EXERCISES 14.3

Sequential Exercises 1–5 deal with the function $z = f(x, y) = \sqrt{2|xy|}$.

1. What is the value of the function on the x-axis and on the y-axis?

2. Find $f_x(0, 0)$ and $f_y(0, 0)$.

3. What plane contains the point $(0, 0, f(0, 0))$ and has as normal vector $f_x(0, 0)\mathbf{i} + f_y(0, 0)\mathbf{j} - \mathbf{k}$?

4. Show that the portions of the graph lying over the lines $y = x$ and $y = -x$ in the x,y-plane consist of four straight rays leaving the origin at an angle of inclination of $45°$ with the x,y-plane.

5. Write relation (17) at $(x_0, y_0) = (0, 0)$ for $\Delta y = \Delta x$, and deduce that $f(x, y)$ is not differentiable at $(0, 0)$.

In Exercises 6–10, find the functions $\varepsilon_1(\Delta x, \Delta y)$ and $\varepsilon_2(\Delta x, \Delta y)$ defined in Eqs. (12) and (15), and show directly that they approach zero as $\Delta x \to 0$ and $\Delta y \to 0$.

6. $f(x, y) = x^2 + y^2$ at $(0, 0)$

7. $f(x, y) = x^2 + y^2$ at $(1, -1)$

8. $f(x, y) = x/y$ at $(1, 1)$

9. $f(x, y) = 1/(xy)$ at $(-1, 1)$

10. $f(x, y) = x^2 - 2xy + 3x - y^2$ at a general point (x, y)

In Exercises 11–16, find the differential of the indicated function at the given point.

11. $f(x, y) = x^3 y^2 + 2x^2 y$ at $(2, 3)$

12. $f(x, y) = x^2 - 2y^2 + \tan(xy) + (1/x)$ at $(-2, 0)$

13. $f(x, y) = 2x + x \cos y + x \sin y$ at $(3, \pi)$

14. $f(x, y, z) = x^2 + 2yz$ at $(4, -1, 2)$

15. $f(x, y, z) = \ln(xy) + e^{yz} + \sin(xz)$ at $(2, 4, \pi)$

16. $f(x) = 3x^2 + \sec x + \ln x$ at π

17. Let $f(x, y, z) = xy + \sin z$. Use a differential to estimate $f(0.98, 2.03, 0.05)$.

18. Let $f(x, y) = e^{xy} + \sin(xy) + 4y$. Use a differential to estimate $f(0.02, 4.97)$.

19. Let $z = f(x, y) = x/y^2$. When $x = 1$ and $y = 2$, use a differential to estimate the change in x required to keep z constant if y is decreased by 0.05.

20. Let $z = f(x, y) = x/(x + y)$. When $x = -1$ and $y = 3$, use a differential to estimate the change in x that produces an increase of 0.2 in z if y is increased by 0.05.

In Exercises 21–26, find the gradient vector of the given function at the indicated point.

21. $f(x, y) = x^3 - 3xy^2$ at $(1, 1)$

22. $f(x, y) = xe^y$ at $(2, 0)$

23. $f(x, y, z) = \dfrac{x - y}{z}$ at $(1, -1, 3)$

24. $f(x, y, z) = xy^3 + ye^{xz}$ at $(-1, 2, 0)$

25. $f(x) = \cos\left(\dfrac{x}{2}\right) + 2 \sin x$ at $x = \pi$

26. $f(x, y, z) = xy^2 + \tan^{-1}(xz)$ at $(-1, 4, 1)$

In Exercises 27–30, describe all points in the plane where the gradient vector of $f(x, y) = x^2 + y^2 + 2x$ has the given property.

27. A unit vector

28. Parallel to the vector $\mathbf{i} - 2\mathbf{j}$

29. Orthogonal to the vector $3\mathbf{i} - 4\mathbf{j}$

30. Orthogonal to the position vector from the origin to the point

In Exercises 31–34, describe all points in space where the gradient vector of $f(x, y, z) = xy + z^2$ has the given property.

31. Magnitude 2

32. Parallel to the vector $2\mathbf{i} - 3\mathbf{j} + \mathbf{k}$

33. Orthogonal to the vector $\mathbf{i} - 3\mathbf{j} + 2\mathbf{k}$

34. Directed toward the origin

14.4   CHAIN RULES

Let $y = f(x)$ and $x = g(t)$ both be differentiable functions. In Section 3.8, we showed that the derivative of the *composite* function $y = (f \circ g)(t) = f(g(t))$ can be found by the *chain rule*

$$(f \circ g)'(t) = f'(g(t))g'(t), \qquad \text{or} \qquad \frac{dy}{dt} = \frac{dy}{dx} \cdot \frac{dx}{dt}.$$

In this section, we study chain rules for derivatives of composite functions composed of functions of more than one variable. The rules are a bit more complicated than for the single-variable case. At the end of this section, we indicate how vector notation can be used to make the chain rules all look just like the one above for the single-variable case.

Suppose $z = f(x, y)$ and $x = g_1(t)$, $y = g_2(t)$, so t determines z. Suppose also that the derivatives $\partial z/\partial x$, $\partial z/\partial y$, dx/dt, and dy/dt all exist and are *continuous*. We want to find dz/dt, since z appears as a composite function of the one variable t. The rate of change of z with respect to t can be expressed in terms of the rates at which z changes with respect to x and to y and the rates at which x and y change with respect to t. This is surely not too surprising. The total rate of change of z is the sum of the rates of change due to the changing quantities x and y. These rates of change due to x and y individually are, by the chain rule in Chapter 3,

$$\frac{\partial z}{\partial x} \cdot \frac{dx}{dt} \qquad \text{and} \qquad \frac{\partial z}{\partial y} \cdot \frac{dy}{dt}.$$

Thus, although we have not proved it yet, the valid formula

$$\frac{dz}{dt} = \frac{\partial z}{\partial x} \cdot \frac{dx}{dt} + \frac{\partial z}{\partial y} \cdot \frac{dy}{dt} \tag{1}$$

seems reasonable.

We can prove the formula (1) from our work in the last section. Theorem 14.2 shows that

$$\Delta z = f_x(x, y)\, \Delta x + f_y(x, y)\, \Delta y + \varepsilon_1 \cdot \Delta x + \varepsilon_2 \cdot \Delta y, \tag{2}$$

where $\varepsilon_1 \to 0$ and $\varepsilon_2 \to 0$ as $\Delta x \to 0$ and $\Delta y \to 0$. Thus

$$\frac{dz}{dt} = \lim_{\Delta t \to 0} \frac{\Delta z}{\Delta t} = \lim_{\Delta t \to 0} \left(f_x(x, y)\frac{\Delta x}{\Delta t} + f_y(x, y)\frac{\Delta y}{\Delta t} + \varepsilon_1 \frac{\Delta x}{\Delta t} + \varepsilon_2 \frac{\Delta y}{\Delta t} \right)$$

$$= f_x(x, y)\frac{dx}{dt} + f_y(x, y)\frac{dy}{dt} + 0 \cdot \frac{dx}{dt} + 0 \cdot \frac{dy}{dt}$$

$$= \frac{\partial z}{\partial x} \cdot \frac{dx}{dt} + \frac{\partial z}{\partial y} \cdot \frac{dy}{dt},$$

which substantiates formula (1).

A similar argument shows that if $w = f(x, y, z)$ and $x = g_1(t)$, $y = g_2(t)$, and $z = g_3(t)$, then

$$\frac{dw}{dt} = \frac{\partial w}{\partial x} \cdot \frac{dx}{dt} + \frac{\partial w}{\partial y} \cdot \frac{dy}{dt} + \frac{\partial w}{\partial z} \cdot \frac{dz}{dt}.$$

In subscripted notation, if $y = f(\mathbf{x}) = f(x_1, x_2, x_3)$ and $x_1 = g_1(t)$, $x_2 = g_2(t)$, $x_3 = g_3(t)$, then

$$\frac{dy}{dt} = \frac{\partial y}{\partial x_1} \cdot \frac{dx_1}{dt} + \frac{\partial y}{\partial x_2} \cdot \frac{dx_2}{dt} + \frac{\partial y}{\partial x_3} \cdot \frac{dx_3}{dt}.$$

EXAMPLE 1   Let $z = x^2 + (x/y)$, where $x = t^2 - 3t$ and $y = 3t - 5$. Find dz/dt, when $t = 2$, in two ways:

(a) by expressing z directly as a function of t and
(b) using the chain rule (1).

Solution

(a) We have

$$z = x^2 + \frac{x}{y} = (t^2 - 3t)^2 + \frac{t^2 - 3t}{3t - 5}.$$

Thus

$$\frac{dz}{dt} = 2(t^2 - 3t)(2t - 3) + \frac{(3t - 5)(2t - 3) - (t^2 - 3t)3}{(3t - 5)^2}.$$

Setting $t = 2$, we obtain

$$\left.\frac{dz}{dt}\right|_{t=2} = 2(-2)(1) + \frac{(1)(1) - (-2)(3)}{1^2} = -4 + 7 = 3.$$

(b) Using the chain rule (1), we have

$$\frac{dz}{dt} = \frac{\partial z}{\partial x} \cdot \frac{dx}{dt} + \frac{\partial z}{\partial y} \cdot \frac{dy}{dt} = \left(2x + \frac{1}{y}\right)(2t - 3) + \left(\frac{-x}{y^2}\right)3.$$

When $t = 2$, we see that $x = -2$ and $y = 1$. Substituting in these values, we obtain

$$\left.\frac{dz}{dt}\right|_{t=2} = (-4 + 1)(1) + \left(\frac{2}{1}\right)3 = -3 + 6 = 3. \quad \blacksquare$$

EXAMPLE 2   The volume V of a right circular cylinder of radius r and height h is given by $V = \pi r^2 h$. The volume is increasing at a rate of 72π in^3/min while the height is decreasing at a rate of 4 in./min. Find the rate of increase of the radius when the height is 3 in. and the radius is 6 in.

Solution   We have

$$\frac{dV}{dt} = \frac{\partial V}{\partial r} \cdot \frac{dr}{dt} + \frac{\partial V}{\partial h} \cdot \frac{dh}{dt},$$

so

$$\frac{dV}{dt} = 2\pi r h \frac{dr}{dt} + \pi r^2 \frac{dh}{dt}.$$

We know that $dV/dt = 72\pi$ and $dh/dt = -4$. (The negative sign occurs because h is decreasing.) We want to find dr/dt when $r = 6$ and $h = 3$. Substituting, we obtain

$$72\pi = 2\pi(6)(3)\frac{dr}{dt} + \pi(6^2)(-4),$$

so

$$36\pi\frac{dr}{dt} = 216\pi.$$

Hence $dr/dt = 216\pi/36\pi = 6$ in./min. ■

Now let $z = f(x, y)$, $x = g_1(s, t)$, and $y = g_2(s, t)$. This time z appears as the composite function of *two* variables, s and t, and we are interested in the *partial* derivatives $\partial z/\partial s$ and $\partial z/\partial t$. But Eq. (2) is still valid, and we divide by the increment Δt and take the limit as s is held constant, to find $\partial z/\partial t$. That is, we obtain from Eq. (2)

$$\frac{\partial z}{\partial t} = \lim_{\Delta t \to 0} \frac{\Delta z}{\Delta t} = f_x(x, y)\frac{\partial x}{\partial t} + f_y(x, y)\frac{\partial y}{\partial t} + 0 \cdot \frac{\partial x}{\partial t} + 0 \cdot \frac{\partial y}{\partial t}$$

$$= \frac{\partial z}{\partial x} \cdot \frac{\partial x}{\partial t} + \frac{\partial z}{\partial y} \cdot \frac{\partial y}{\partial t}.$$

Thus the derivatives dx/dt and dy/dt in formula (1) simply become partial derivatives in this case. There are so many different types of situations where a chain rule applies that it is awkward to state one all-inclusive theorem. We state one special case as a theorem.

THEOREM 14.3   Chain rule

Let $w = f(x, y, z)$ be a differentiable function. Let x, y, and z all be differentiable functions of s and t. Then w is a differentiable composite function of s and t with first partial derivatives

$$\frac{\partial w}{\partial s} = \frac{\partial w}{\partial x} \cdot \frac{\partial x}{\partial s} + \frac{\partial w}{\partial y} \cdot \frac{\partial y}{\partial s} + \frac{\partial w}{\partial z} \cdot \frac{\partial z}{\partial s}$$

and

$$\frac{\partial w}{\partial t} = \frac{\partial w}{\partial x} \cdot \frac{\partial x}{\partial t} + \frac{\partial w}{\partial y} \cdot \frac{\partial y}{\partial t} + \frac{\partial w}{\partial z} \cdot \frac{\partial z}{\partial t}.$$ □

EXAMPLE 3   Consider the situation where

$$w = f(x, y, z) = xy^2 + ze^{x^2},$$

while $x = u$, $y = v - 1$, and $z = uv$. Compute $\partial w/\partial u$ at $(0, 2)$ in two ways:

(a) by expressing w directly as a function of u and v and
(b) by using the chain rule.

Solution

(a) Expressing w directly as a function of u and v, we have

$$w = f(u, v - 1, uv)$$
$$= u(v - 1)^2 + uve^{u^2}.$$

Thus

$$\frac{\partial w}{\partial u} = (v - 1)^2 + 2u^2 ve^{u^2} + ve^{u^2}.$$

Therefore

$$\left.\frac{\partial w}{\partial u}\right|_{(0,2)} = 1 + 0 + 2 \cdot e^0$$
$$= 1 + 2 = 3.$$

(b) To use the chain rule, we note that when $(u, v) = (0, 2)$,

$$(x, y, z) = (0, 1, 0).$$

If $w = xy^2 + ze^{x^2}$, then

$$\frac{\partial w}{\partial u} = \frac{\partial w}{\partial x} \cdot \frac{\partial x}{\partial u} + \frac{\partial w}{\partial y} \cdot \frac{\partial y}{\partial u} + \frac{\partial w}{\partial z} \cdot \frac{\partial z}{\partial u}$$
$$= (y^2 + 2xze^{x^2})(1) + (2xy)(0) + (e^{x^2})(v).$$

Thus

$$\left.\frac{\partial w}{\partial u}\right|_{(u,v)=(0,2)} = (1 + 0)(1) + 0 + (1)(2) = 1 + 2 = 3. \quad \blacksquare$$

EXAMPLE 4 Suppose that $w = f(u, v)$ is a differentiable function. Let $u = ax + by$ and $v = ax - by$. Show that

$$\frac{\partial w}{\partial x} \cdot \frac{\partial w}{\partial y} = ab\left[\left(\frac{\partial w}{\partial u}\right)^2 - \left(\frac{\partial w}{\partial v}\right)^2\right].$$

Solution   We have

$$\frac{\partial w}{\partial x} = \frac{\partial w}{\partial u} \cdot \frac{\partial u}{\partial x} + \frac{\partial w}{\partial v} \cdot \frac{\partial v}{\partial x} = a\frac{\partial w}{\partial u} + a\frac{\partial w}{\partial v}.$$

Also

$$\frac{\partial w}{\partial y} = \frac{\partial w}{\partial u} \cdot \frac{\partial u}{\partial y} + \frac{\partial w}{\partial v} \cdot \frac{\partial v}{\partial y} = b\frac{\partial w}{\partial u} - b\frac{\partial w}{\partial v}.$$

Then

$$\frac{\partial w}{\partial x} \cdot \frac{\partial w}{\partial y} = \left(a\frac{\partial w}{\partial u} + a\frac{\partial w}{\partial v}\right)\left(b\frac{\partial w}{\partial u} - b\frac{\partial w}{\partial v}\right) = ab\left[\left(\frac{\partial w}{\partial u}\right)^2 - \left(\frac{\partial w}{\partial v}\right)^2\right]. \quad \blacksquare$$

We will consider a case involving subscripted variables. Suppose that $y = f(\mathbf{x}) = f(x_1, x_2, x_3)$ and $x_1 = g_1(t_1, t_2)$, $x_2 = g_2(t_1, t_2)$, and $x_3 = g_3(t_1, t_2)$.

Then

$$\frac{\partial y}{\partial t_1} = \frac{\partial y}{\partial x_1} \cdot \frac{\partial x_1}{\partial t_1} + \frac{\partial y}{\partial x_2} \cdot \frac{\partial x_2}{\partial t_1} + \frac{\partial y}{\partial x_3} \cdot \frac{\partial x_3}{\partial t_1}. \tag{3}$$

Suppose we introduce the Leibniz-type notation

$$\frac{\partial y}{\partial \mathbf{x}} = \frac{\partial y}{\partial x_1}\mathbf{i} + \frac{\partial y}{\partial x_2}\mathbf{j} + \frac{\partial y}{\partial x_3}\mathbf{k}$$

and

$$\frac{\partial \mathbf{x}}{\partial t_1} = \frac{\partial x_1}{\partial t_1}\mathbf{i} + \frac{\partial x_2}{\partial t_1}\mathbf{j} + \frac{\partial x_3}{\partial t_1}\mathbf{k}.$$

In both of these Leibniz-type notations, the boldface part of the symbol indicates the variable whose subscripts change to give the components of the vectors. Now Eq. (3) appears as a dot product

$$\frac{\partial y}{\partial t_1} = \frac{\partial y}{\partial \mathbf{x}} \cdot \frac{\partial \mathbf{x}}{\partial t_1}. \tag{4}$$

We have recovered our old chain rule formula by using vector notations, subscripted variables, and a dot product. Using subscripted variables illuminates the structure of these formulas for functions of more than one variable.

EXAMPLE 5 If $y = f(\mathbf{x}) = x_1^2 x_2^3$, where $x_1 = t_1^2 - 2t_2 + t_3$ and $x_2 = t_1 t_2 - t_3^2$, use Eq. (4) to find $\partial y/\partial t_1$ where $(t_1, t_2, t_3) = (-1, 1, 2)$.

Solution When $(t_1, t_2, t_3) = (-1, 1, 2)$, we see that $(x_1, x_2) = (1, -5)$. Now

$$\frac{\partial y}{\partial \mathbf{x}} = \frac{\partial y}{\partial x_1}\mathbf{i} + \frac{\partial y}{\partial x_2}\mathbf{j}$$
$$= 2x_1 x_2^3\mathbf{i} + 3x_1^2 x_2^2\mathbf{j},$$

so

$$\frac{\partial y}{\partial \mathbf{x}}\bigg|_{(1,-5)} = -250\mathbf{i} + 75\mathbf{j}.$$

We also see that

$$\frac{\partial \mathbf{x}}{\partial t_1} = \frac{\partial x_1}{\partial t_1}\mathbf{i} + \frac{\partial x_2}{\partial t_1}\mathbf{j} = 2t_1\mathbf{i} + t_2\mathbf{j}$$

so

$$\frac{\partial \mathbf{x}}{\partial t_1}\bigg|_{(-1,1,2)} = -2\mathbf{i} + \mathbf{j}.$$

Thus when $(t_1, t_2, t_3) = (-1, 1, 2)$, we have

$$\frac{\partial y}{\partial t_1} = \frac{\partial y}{\partial \mathbf{x}} \cdot \frac{\partial \mathbf{x}}{\partial t_1} = (-250\mathbf{i} + 75\mathbf{j}) \cdot (-2\mathbf{i} + \mathbf{j})$$
$$= 500 + 75 = 575. \qquad \blacksquare$$

SUMMARY

1. If $z = f(x, y)$ and $x = g_1(t)$, $y = g_2(t)$, then

$$\frac{dz}{dt} = \frac{\partial z}{\partial x} \cdot \frac{dx}{dt} + \frac{\partial z}{\partial y} \cdot \frac{dy}{dt}.$$

With subscripted variables, if $y = f(\mathbf{x}) = f(x_1, x_2, x_3)$ and $x_1 = g_1(t)$, $x_2 = g_2(t)$, $x_3 = g_3(t)$, then

$$\frac{dy}{dt} = \frac{\partial y}{\partial x_1} \cdot \frac{dx_1}{dt} + \frac{\partial y}{\partial x_2} \cdot \frac{dx_2}{dt} + \frac{\partial y}{\partial x_3} \cdot \frac{dx_3}{dt}.$$

2. If $z = f(x, y)$ and $x = g_1(s, t)$, $y = g_2(s, t)$, then

$$\frac{\partial z}{\partial t} = \frac{\partial z}{\partial x} \cdot \frac{\partial x}{\partial t} + \frac{\partial z}{\partial y} \cdot \frac{\partial y}{\partial t}.$$

3. In vector notation, if $y = f(\mathbf{x}) = f(x_1, x_2, x_3)$ while $x_1 = g_1(t_1, t_2)$, $x_2 = g_2(t_1, t_2)$, and $x_3 = g_3(t_1, t_2)$, then

$$\frac{\partial y}{\partial t_k} = \frac{\partial y}{\partial x_1} \cdot \frac{\partial x_1}{\partial t_k} + \frac{\partial y}{\partial x_2} \cdot \frac{\partial x_2}{\partial t_k} + \frac{\partial y}{\partial x_3} \cdot \frac{\partial x_3}{\partial t_k}.$$

The partial derivative $\partial y/\partial t_k$ is given by the dot product

$$\frac{\partial y}{\partial t_k} = \frac{\partial y}{\partial \mathbf{x}} \cdot \frac{\partial \mathbf{x}}{\partial t_k},$$

where

$$\frac{\partial y}{\partial \mathbf{x}} = \frac{\partial y}{\partial x_1}\mathbf{i} + \frac{\partial y}{\partial x_2}\mathbf{j} + \frac{\partial y}{\partial x_3}\mathbf{k}$$

and

$$\frac{\partial \mathbf{x}}{\partial t_k} = \frac{\partial x_1}{\partial t_k}\mathbf{i} + \frac{\partial x_2}{\partial t_k}\mathbf{j} + \frac{\partial x_3}{\partial t_k}\mathbf{k}.$$

EXERCISES 14.4

1. Let $z = x^2 + (1/y^2)$, $x = t^2$, and $y = t + 1$.
   (a) Find x, y, and z when $t = 1$.
   (b) Find $dz/dt|_{t=1}$ using a chain rule.
   (c) Express z as a function of t by substitution.
   (d) Find $dz/dt|_{t=1}$ by differentiating your answer to part (c).

2. Let $w = f(u, v) = u^2/v$, $u = s^2 - 3s + 1$, and $v = s^3 - s^2$.
   (a) Find u, v, and w when $s = 2$.
   (b) Find $dw/ds|_{s=2}$ using a chain rule.
   (c) Express w as a function of s by substitution.
   (d) Find $dw/ds|_{s=2}$ by differentiating your answer to part (c).

3. Let $z = f(x, y) = xy^2$, $x = r \cos \theta$, and $y = r \sin \theta$.
   (a) Find x, y, and z when $(r, \theta) = (2, 3\pi/4)$.
   (b) Find $\partial z/\partial r$ and $\partial z/\partial \theta$ when $(r, \theta) = (2, 3\pi/4)$.
   (c) Express z as a function of r and θ by substitution.
   (d) Find $\partial z/\partial r$ and $\partial z/\partial \theta$ when $(r, \theta) = (2, 3\pi/4)$ by differentiating your answer to part (c).

4. Let $w = f(x, y, z) = xz^2 + y/z$, $x = 2s + 3t$, $y = s^2 - t^2$, and $z = s^2t$.
   (a) Find x, y, z, and w when $(s, t) = (-1, 1)$.
   (b) Find $\partial w/\partial s$ and $\partial w/\partial t$ when $(s, t) = (-1, 1)$.
   (c) Express w as a function of s and t by substitution.
   (d) Find $\partial w/\partial s$ and $\partial w/\partial t$ when $(s, t) = (-1, 1)$ by differentiating your answer to part (c).

In Exercises 5–14, use a chain rule to find the indicated derivative at the given point.

5. $z = x^2 - 2xy + xy^3$, $x = t^3 + 1$, $y = 1/t$; find dz/dt when $t = 1$.

6. $w = xy^2 + z^3$, $x = 2s - 1$, $y = s^3$, $z = s - 4$; find dw/ds when $s = -1$.

7. $w = \sin(uv)$, $u = x^2 - 2x$, $v = x^3 - 5x$; find dw/dx when $x = 2$.

8. $w = x \sin(yz)$, $x = 2t + 1$, $y = 3t^2$, $z = \pi t/2$; find dw/dt when $t = -1$.

9. $z = x^2/y^3$, $x = 2t + 3s$, $y = 3t^3 + 2s$; find $\partial z/\partial t$ when $(t, s) = (-1, 1)$.

10. $w = \tan^{-1}(uv)$, $u = 2x + 3y$, $v = x^2 + y^2$; find $\partial w/\partial y$ when $(x, y) = (-1, 1)$.

11. $w = x^2 + yz$, $x = uv$, $y = u - v$, $z = 2u^2v$; find $\partial w/\partial u$ when $(u, v) = (-1, 2)$.

12. $y = x_1{}^2 - 3x_1x_2$, $x_1 = (t_1 + 1)/t_2$, $x_2 = t_2e^{t_1}$; find $\partial y/\partial t_1$ when $(t_1, t_2) = (0, -1)$.

13. $z = y_1e^{y_2}$, $y_1 = x_1 - x_2 - x_3$, $y_2 = 2x_1 + x_3$; find $\partial z/\partial x_2$ when $(x_1, x_2, x_3) = (1, 3, -2)$.

14. $w = y_1{}^2 + \sin(y_2 y_3) - e^{y_1}$, $y_1 = x_1x_3$, $y_2 = \ln(x_3{}^2 + 1)$, $y_3 = x_2 \cos x_3$; find $\partial w/\partial x_2$, where $(x_1, x_2, x_3) = (-1, 2, 0)$.

In Exercises 15–18, let $z = f(x, y)$ be a differentiable function, and let $x = t^2 - 2s$ and $y = 3t + s^2$.

15. find $f_x(-1, 4)$ if $f_y(-1, 4) = 5$ and $\partial z/\partial t = 3$ when $(t, s) = (1, 1)$.

16. Find $f_y(1, 3)$ if $f_x(1, 3) = 2$ and $\partial z/\partial t = -4$ when $(t, s) = (1, 0)$.

17. Find $f_x(-1, 4)$ and $f_y(-1, 4)$ if $\partial z/\partial t = 2$ and $\partial z/\partial s = -1$ when $(t, s) = (1, 1)$.

18. Find $f_x(1, 3)$ and $f_y(1, 3)$ if $\partial z/\partial t = 0$ and $\partial z/\partial s = -1$ when $(t, s) = (1, 0)$.

19. If the radius of a circular cylinder is increasing at a rate of 4 in./min while the length is increasing at a rate of 8 in./min, find the rate of change of the volume of the cylinder when the radius is 10 in. and the length is 50 in.

20. The voltage drop V, measured in volts, across a certain conductor of variable resistance R ohms is IR, where I, mea-

sured in amperes, is the current flowing through the conductor. The current increases at a constant rate of 2 amp/sec while the voltage drop is kept constant by decreasing the resistance as the current increases. Find the rate of change of the resistance when $I = 5$ amp and $R = 1000$ ohms.

21. The moment of inertia I about an axis of a body of mass m and distance s from the axis is given by $I = ms^2$. Find the rate of change of the moment of inertia about the axis when $s = 50$ if the mass remains constant while the distance s is decreasing at a rate of 3 units length per unit time.

22. Answer Exercise 21 if the body is gaining mass at a rate of 2 units mass per unit time and has mass 20 when $s = 50$, while the other data remain the same.

23. The pressure P in lb/ft^2 of a certain gas at temperature T degrees in a container of variable volume V ft^3 is given by $P = 8T/V$. The temperature is increased at a rate of $5°$/min while the volume of the container is increased at a rate of 2 ft^3/min. If, at time t_0, the temperature was $20°$ and the volume was 10 ft^3, find the rate of change of the pressure 5 min later.

24. By Newton's law of gravitation, the force of attraction between two bodies of masses m_1 and m_2 is Gm_1m_2/s^2, where s is the distance between the bodies and G is the universal gravitational constant. Find the rate of change of the force of attraction for two bodies of constant masses of 10^4 and 10^7 units that are 10^4 units distance apart and are separating at a rate of 10^2 units distance per unit time. (Assume the given units are compatible and don't worry about the value of G or the name of the units in the answer.)

25. Repeat Exercise 24 if the first body is gaining mass at a rate of 30 units mass per unit time and the second body is losing mass at the rate of 80 units mass per unit time, while the other data remain the same.

In Exercises 26–34, assume that all functions encountered satisfy enough differentiability conditions to enable you to use any chain rule you wish.

26. If $w = f(u)$ and $u = ax + by$, show that $b(\partial w/\partial x) = a(\partial w/\partial y)$.

27. Obtain a result similar to that in Exercise 26 for $w = f(u)$ and $u = ax + by + cz$.

28. If $w = f(u, v)$, $u = x + y$, and $v = 2x - 2y$, show that
$$\frac{\partial w}{\partial x} \cdot \frac{\partial w}{\partial y} = \left(\frac{\partial f}{\partial u}\right)^2 - 4\left(\frac{\partial f}{\partial v}\right)^2.$$

29. If $w = f(u)$ and $u = xy^2$, show that $2x(\partial w/\partial x) - y(\partial w/\partial y) = 0$.

30. If $w = f(u) + g(v)$, $u = ax + by$, and $v = ax - by$, show that $b^2(\partial^2 w/\partial x^2) = a^2(\partial^2 w/\partial y^2)$.

31. If $w = f(u)$ and $u = x/y$, show that $x(\partial w/\partial x) + y(\partial w/\partial y) = 0$.

32. Let $z = f(x, y)$ and let x and y be functions of t. Find a formula for d^2z/dt^2.

33. Let $z = f(x, y)$ and let x and y be functions of t and s. Find a formula for $\partial^2 z/\partial s^2$.

34. Let $z = f(x, y)$ and let x and y be functions of t and s. Find a formula for $\partial^2 z / \partial s\, \partial t$.

35. If f is a differentiable function of two variables and $f(tx, ty) = t^2 f(x, y)$ for all t, show that $x \cdot f_x(x, y) + y \cdot f_y(x, y) = 2 \cdot f(x, y)$. (Such a function is *homogeneous of degree* 2.) [*Hint:* Differentiate the equation $f(tx_0, ty_0) = t^2 f(x_0, y_0)$ with respect to t, and put $t = 1$.]

36. Generalize the conclusion in Exercise 35 in the case that $f(tx, ty) = t^k f(x, y)$ for all t. (Such a function is *homogeneous of degree k.*)

37. Show that the result in Exercise 31 is a special case of your generalization in Exercise 36.

38. Generalize the conclusion in Exercise 36 for a differentiable function of n variables.

14.5 THE DIRECTIONAL DERIVATIVE

Derivatives in All Directions

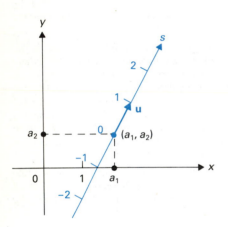

FIGURE 14.24 **u** gives a direction at (a_1, a_2).

Let (a_1, a_2) be a point in the x,y-plane in the domain of $f(x, y)$. We want to find the instantaneous rate at which $f(x, y)$ increases per unit change along a direction given by a unit vector $\mathbf{u} = u_1 \mathbf{i} + u_2 \mathbf{j}$ at (a_1, a_2), as indicated in Fig. 14.24. We think of taking an s-axis, with origin at (a_1, a_2) in the plane, in the direction given by $\mathbf{u}$, as shown in the figure. Figure 14.25 shows that the curve on the surface $z = f(x, y)$ lying directly over an s-axis is the graph of a function $z = h(s)$. The rate of change of z with respect to s at (a_1, a_2) is the derivative $h'(s) = dz/ds$ evaluated at $s = 0$. If we choose the scale on the s-axis to be the same as on the coordinate axes, then $h'(s)$ gives the rate of change of z in the direction given by the vector $\mathbf{u}$ at (a_1, a_2). This derivative $h'(s)$ is called the **directional derivative** of f at (a_1, a_2) in the direction given by $\mathbf{u}$. We use the notation $f_\mathbf{u}(a_1, a_2)$.

We can compute dz/ds using a chain rule, because z is a function of x and y and it is straightforward to express x and y in terms of s. It is crucial to remember, however, that we want the *scale* on the s-axis to be the same as on our coordinate axes. We want dz/ds to be the rate of change of z with respect to *distance*. Let the origin on the s-axis be at the point (a_1, a_2), as shown in Fig. 14.26. *Since $\mathbf{u}$ is a unit vector,* we can measure off our desired scale on the s-axis in terms of multiples of $\mathbf{u}$, where we think of $\mathbf{u}$ as starting from (a_1, a_2). As indicated in Fig. 14.26, for a point (x, y) on the s-axis corresponding to a value s, we have

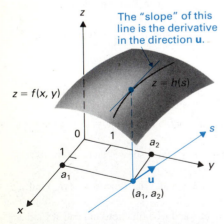

FIGURE 14.25 The directional derivative dz/ds is the "slope" of the surface in the direction **u**.

$$x\mathbf{i} + y\mathbf{j} = (a_1\mathbf{i} + a_2\mathbf{j}) + s\mathbf{u} = a_1\mathbf{i} + a_2\mathbf{j} + s(u_1\mathbf{i} + u_2\mathbf{j})$$
$$= (a_1 + u_1 s)\mathbf{i} + (a_2 + u_2 s)\mathbf{j}.$$

Therefore,

$$x = a_1 + u_1 s, \qquad y = a_2 + u_2 s.$$

Then the chain rule shows that

$$\frac{dz}{ds} = \frac{\partial z}{\partial x} \cdot \frac{dx}{ds} + \frac{\partial z}{\partial y} \cdot \frac{dy}{ds} = \frac{\partial z}{\partial x} u_1 + \frac{\partial z}{\partial y} u_2. \tag{1}$$

When $s = 0$, we have

$$\left. \frac{dz}{ds} \right|_{s=0} = u_1 f_x(a_1, a_2) + u_2 f_y(a_1, a_2). \tag{2}$$

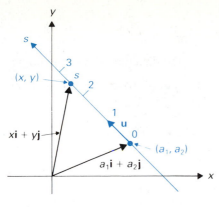

FIGURE 14.26

$$x\mathbf{i} + y\mathbf{j} = (a_1\mathbf{i} + a_2\mathbf{j}) + s\mathbf{u}$$

If $w = f(x, y, z)$, then the directional derivative of w at $\mathbf{a} = (a_1, a_2, a_3)$ in the direction given by the *unit* vector $\mathbf{u} = u_1\mathbf{i} + u_2\mathbf{j} + u_3\mathbf{k}$ is, of course,

$$\left.\frac{dw}{ds}\right|_{s=0} = u_1 f_x(\mathbf{a}) + u_2 f_y(\mathbf{a}) + u_3 f_z(\mathbf{a}). \tag{3}$$

We also denote the directional derivative given in Eq. (2) by $f_{\mathbf{u}}(\mathbf{a})$, the derivative of $f(x, y)$ at $\mathbf{a} = (a_1, a_2)$ in the direction given by the unit vector $\mathbf{u} = u_1\mathbf{i} + u_2\mathbf{j}$. Thus

$$f_{\mathbf{u}}(\mathbf{a}) = u_1 f_x(\mathbf{a}) + u_2 f_y(\mathbf{a}). \tag{4}$$

Using subscripted variables, if $y = f(\mathbf{x}) = f(x_1, x_2, x_3)$, while $\mathbf{a} = (a_1, a_2, a_3)$ and $\mathbf{u} = u_1\mathbf{i} + u_2\mathbf{j} + u_3\mathbf{k}$, then

$$f_{\mathbf{u}}(\mathbf{a}) = u_1 f_{x_1}(\mathbf{a}) + u_2 f_{x_2}(\mathbf{a}) + u_3 f_{x_3}(\mathbf{a}). \tag{5}$$

Recall that the gradient vector $\nabla f(\mathbf{a})$ for a function $f(x, y)$ at a point $\mathbf{a} = (a_1, a_2)$ is given by

$$\nabla f(\mathbf{a}) = f_x(\mathbf{a})\mathbf{i} + f_y(\mathbf{a})\mathbf{j}. \tag{6}$$

Since $\mathbf{u} = u_1\mathbf{i} + u_2\mathbf{j}$, we see that Eq. (4) can be written as a dot product,

$$f_{\mathbf{u}}(\mathbf{a}) = \nabla f(\mathbf{a}) \cdot \mathbf{u}. \tag{7}$$

We will always think in terms of Eq. (7) when actually computing a directional derivative. Since $\mathbf{u}$ is a unit vector, we see from Eq. (7) that $f_{\mathbf{u}}(\mathbf{a})$ is the scalar component of $\nabla f(\mathbf{a})$ along $\mathbf{u}$.

To Find a Directional Derivative of f at Point a

STEP 1. Find a *unit* vector $\mathbf{u}$ in the direction of the derivative.

STEP 2. Find the gradient vector $\nabla f(\mathbf{a})$.

STEP 3. The directional derivative is $f_{\mathbf{u}}(\mathbf{a}) = \nabla f(\mathbf{a}) \cdot \mathbf{u}$.

EXAMPLE 1 Use the steps just highlighted to find the directional derivatives of a function $f(x, y)$ at $\mathbf{a} = (a_1, a_2)$ in the positive coordinate directions.

Solution A unit vector in the direction of increasing x is $\mathbf{i}$ and a unit vector in the direction of increasing y is $\mathbf{j}$. Since

$$\nabla f(\mathbf{a}) = f_x(\mathbf{a})\mathbf{i} + f_y(\mathbf{a})\mathbf{j},$$

we have

$$f_{\mathbf{i}}(\mathbf{a}) = \nabla f(\mathbf{a}) \cdot \mathbf{i} = f_x(\mathbf{a})$$

and

$$f_{\mathbf{j}}(\mathbf{a}) = \nabla f(\mathbf{a}) \cdot \mathbf{j} = f_y(\mathbf{a}).$$

Of course, this just verifies our intuitive understanding of $f_x(\mathbf{a})$ and $f_y(\mathbf{a})$ as derivatives in the positive coordinate directions. ∎

EXAMPLE 2   Let f be defined by

$$f(x, y) = x^2 + 3xy^2.$$

Find the directional derivative of f at $(1, 2)$ in the direction toward the origin.

Solution   A vector in the direction from $(1, 2)$ to $(0, 0)$ is $-\mathbf{i} - 2\mathbf{j}$, so a *unit* vector in this direction is therefore

$$\mathbf{u} = -\frac{1}{\sqrt{5}}\mathbf{i} - \frac{2}{\sqrt{5}}\mathbf{j}.$$

We find that

$$f_x(1, 2) = (2x + 3y^2)\big|_{(1,2)} = 14$$

and

$$f_y(1, 2) = 6xy\big|_{(1,2)} = 12.$$

Thus

$$\nabla f(1, 2) = 14\mathbf{i} + 12\mathbf{j}.$$

Therefore the directional derivative is given by

$$f_{\mathbf{u}}(1, 2) = \nabla f(1, 2) \cdot \mathbf{u} = -\frac{1}{\sqrt{5}}(14) + \left(-\frac{2}{\sqrt{5}}\right)(12) = -\frac{38}{\sqrt{5}}. \quad\blacksquare$$

EXAMPLE 3   Suppose the temperature for $x^2 + y^2 + z^2 > 1$ in space is given by

$$T(x, y, z) = \frac{180}{x^2 + y^2 + z^2} \text{ degrees.}$$

Find the rate of change of temperature with respect to distance at $(1, -1, 4)$ in the direction toward the point $(2, -3, 2)$.

Solution   The vector from $(1, -1, 4)$ to $(2, -3, 2)$ is $\mathbf{i} - 2\mathbf{j} - 2\mathbf{k}$. Since this vector has length $\sqrt{1^2 + (-2)^2 + (-2)^2} = 3$, a unit vector with the direction indicated is

$$\mathbf{u} = \frac{1}{3}\mathbf{i} - \frac{2}{3}\mathbf{j} - \frac{2}{3}\mathbf{k}.$$

We find that

$$\frac{\partial T}{\partial x}\bigg|_{(1,-1,4)} = \frac{180(-2x)}{(x^2 + y^2 + z^2)^2}\bigg|_{(1,-1,4)} = \frac{-360}{18^2} = -\frac{10}{9}.$$

Similarly, we find that

$$\frac{\partial T}{\partial y}\bigg|_{(1,-1,4)} = \frac{10}{9} \quad\text{and}\quad \frac{\partial T}{\partial z}\bigg|_{(1,-1,4)} = -\frac{40}{9}.$$

Thus

$$\nabla T(1, -1, 4) = -\frac{10}{9}\mathbf{i} + \frac{10}{9}\mathbf{j} - \frac{40}{9}\mathbf{k}.$$

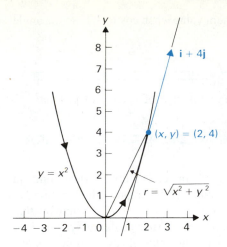

FIGURE 14.27 As a body moves on $y = x^2$ in the direction of increasing x, its direction at $(2, 4)$ is given by $\mathbf{i} + 4\mathbf{j}$.

Consequently, the rate of change of temperature is

$$T_{\mathbf{u}}(1, -1, 4) = \nabla T(1, -1, 4) \cdot \mathbf{u} = -\frac{10}{27} - \frac{20}{27} + \frac{80}{27} = \frac{50}{27}. \quad \blacksquare$$

EXAMPLE 4 A body in the plane moves on the parabola $y = x^2$ in the direction of increasing x, as shown in Fig. 14.27. Find the rate of change with respect to arc length s of the distance r of the body from the origin when the body is at the point $(2, 4)$.

Solution We need the directional derivative of $r = f(x, y) = \sqrt{x^2 + y^2}$ at $(2, 4)$ in the direction tangent to the parabola and in which x is increasing. Since $dy/dx = 2x = 4$ when $x = 2$, we see that a vector tangent to the curve is $\mathbf{i} + 4\mathbf{j}$. Since this vector has positive coefficient of $\mathbf{i}$, it points in the direction tangent to the curve and in which x increases. A unit vector in this direction is

$$\mathbf{u} = \frac{1}{\sqrt{17}}\mathbf{i} + \frac{4}{\sqrt{17}}\mathbf{j}.$$

Now

$$\frac{\partial f}{\partial x} = \frac{x}{\sqrt{x^2 + y^2}} \quad \text{and} \quad \frac{\partial f}{\partial y} = \frac{y}{\sqrt{x^2 + y^2}}.$$

At $(x, y) = (2, 4)$, we see that

$$\nabla f(2, 4) = \frac{2}{\sqrt{20}}\mathbf{i} + \frac{4}{\sqrt{20}}\mathbf{j} = \frac{1}{\sqrt{5}}\mathbf{i} + \frac{2}{\sqrt{5}}\mathbf{j}.$$

Thus

$$f_{\mathbf{u}}(2, 4) = \nabla f(2, 4) \cdot \mathbf{u} = \frac{1}{\sqrt{85}} + \frac{8}{\sqrt{85}} = \frac{9}{\sqrt{85}}. \quad \blacksquare$$

The Magnitude and Direction of ∇f

We saw that

$$f_{\mathbf{u}}(\mathbf{a}) = \nabla f(\mathbf{a}) \cdot \mathbf{u}. \tag{8}$$

Recall that a dot product of two vectors is equal to the product of their lengths and the cosine of the angle θ between them. See Fig. 14.28. Thus

$$f_{\mathbf{u}}(\mathbf{a}) = |\nabla f(\mathbf{a})| \, |\mathbf{u}|(\cos \theta) = |\nabla f(\mathbf{a})|(\cos \theta), \tag{9}$$

since $|\mathbf{u}| = 1$. Now $|\nabla f(\mathbf{a})|(\cos \theta)$ is a function of θ at the point $\mathbf{a}$. It has its maximum value when $\cos \theta = 1$, and that maximum is then $|\nabla f(\mathbf{a})|$. This shows that

the magnitude of $\nabla f(\mathbf{a})$ is the maximum of the directional derivatives of f at $\mathbf{a}$ in all possible directions.

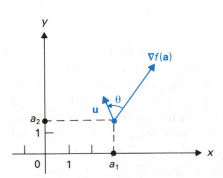

FIGURE 14.28

$$\nabla f(\mathbf{a}) \cdot \mathbf{u} = |\nabla f(\mathbf{a})| \, |\mathbf{u}|(\cos \theta)$$
$$= |\nabla f(\mathbf{a})|(\cos \theta)$$

Since $f_{\mathbf{u}}(\mathbf{a})$ in Eq. (9) assumes this maximum value when $\cos \theta = 1$, so the angle θ between $f(\mathbf{a})$ and $\mathbf{u}$ is 0, we see that

> $\nabla f(\mathbf{a})$ points in the direction of maximum rate of change of f at the point $\mathbf{a}$.

We now understand the use of the term "gradient vector" for $\nabla f(\mathbf{a})$. It points in the direction of maximum steepness or "grade" of the graph of the function f at the point and has length equal to the slope or "grade" of the climb in that direction. Along any curve in the domain of f and perpendicular to ∇f at each of its points, the directional derivative is 0. Thus the value of $f(\mathbf{x})$ does not change along such a curve.

EXAMPLE 5   Find the maximum rate of change with respect to distance, taken over all possible directions, of $f(x, y) = x^2 e^y + y \tan^{-1}z$ at the point $(1, 0, 0)$.

Solution   The gradient vector of f at $(1, 0, 0)$ is given by

$$\nabla f(1, 0, 0) = f_x(1, 0, 0)\mathbf{i} + f_y(1, 0, 0)\mathbf{j} + f_z(1, 0, 0)\mathbf{k} = 2\mathbf{i} + \mathbf{j}.$$

Thus the direction of maximum rate of increase of $f(x, y, z)$ at $(1, 0, 0)$ is given by the vector $2\mathbf{i} + \mathbf{j}$, and this maximum rate of increase is

$$|\nabla f(1, 0, 0)| = |2\mathbf{i} + \mathbf{j}| = \sqrt{5}. \qquad \blacksquare$$

We observe from Eq. (9) that $f_{\mathbf{u}}(\mathbf{a})$ is minimum at $\mathbf{a}$ when $\cos \theta = -1$, when $\theta = \pi$. Of course the direction is then opposite to $\nabla f(\mathbf{a})$. Thus $-\nabla f(\mathbf{a})$ points in the direction of *minimum* rate of *increase* (or *maximum* rate of *decrease*) of f at $\mathbf{a}$, and the rate of change in that direction is $-|\nabla f(\mathbf{a})|$. This is intuitively clear. If a function has a maximum rate of change of 5 in some direction, then the rate of change in the opposite direction is -5, and that is the minimum rate of change.

EXAMPLE 6   Consider the points (x, y, z) in the first octant in space, where x, y, and z are all positive. The volume of a rectangular box having length x, width y, and height z is $V = xyz$. We wish to travel in space from $(6, 3, 2)$ in the direction corresponding to the greatest rate of *decrease* of volume of the box with respect to distance traveled in space. Describe the relative rates of decrease of the length, width, and height of the box as we leave the point $(6, 3, 2)$ in that direction.

Solution   The volume of a box of length x, width y, and height z is

$$V = f(x, y, z) = xyz.$$

We see that

$$\nabla f(6, 3, 2) = (yz\mathbf{i} + xz\mathbf{j} + xy\mathbf{k})\big|_{(6,3,2)} = 6\mathbf{i} + 12\mathbf{j} + 18\mathbf{k}.$$

Then

$$-\nabla f(6, 3, 2) = -6\mathbf{i} - 12\mathbf{j} - 18\mathbf{k}.$$

This vector points in the direction for maximum rate of decrease of volume. To achieve this maximum rate of decrease, we should start at (6, 3, 2) to decrease the width twice as rapidly as the length and decrease the height three times as rapidly as the length. ∎

SUMMARY

1. The directional derivative of a function f at a point $\mathbf{a}$ in the direction of a unit vector $\mathbf{u}$ is the instantaneous rate of change at $\mathbf{a}$ of function values z with respect to distance s as we travel in the direction given by $\mathbf{u}$. Notations are $dz/ds|_{\mathbf{a}}$, $f_{\mathbf{u}}(\mathbf{a})$, and $D_{\mathbf{u}}f(\mathbf{a})$.

2. If f is differentiable at $\mathbf{a}$, the directional derivative just described is found by the formula

$$f_{\mathbf{u}}(\mathbf{a}) = \nabla f(\mathbf{a}) \cdot \mathbf{u}.$$

It is very important to remember that $\mathbf{u}$ must be a *unit* vector.

3. The gradient vector $\nabla f(\mathbf{a})$ points in the direction of maximum rate of increase of the function at $\mathbf{a}$ and has magnitude equal to that rate of increase.

EXERCISES 14.5

In Exercises 1–20, find the directional derivative of the indicated function f at the given point in the given direction.

1. $f(x, y) = x^2 - 3xy$ at $(1, -2)$ in the direction of $3\mathbf{i} - 4\mathbf{j}$

2. $f(x, y) = \sin(xy)$ at $(\pi, -1)$ in the direction of $\mathbf{i} - \mathbf{j}$

3. $f(x, y, z) = xy/z$ at $(2, -3, -1)$ in the direction of $-\mathbf{i} + 2\mathbf{j} + 2\mathbf{k}$

4. $f(x, y, z) = x^2 + 2xz^2$ at $(1, 5, -1)$ in the direction of $\mathbf{i} - \mathbf{j} + \mathbf{k}$

5. $f(x, y) = x^2 - 3xy^3$ at $(-2, 1)$ toward the origin

6. $f(x, y) = \sin^{-1}(x/y)$ at $(3, 5)$ toward $(4, 4)$

7. $f(x_1, x_2, x_3) = x_1 x_2^2 / e^{x_3}$ at $(1, 3, 0)$ toward $(1, 3, -1)$

8. $f(x_1, x_2, x_3) = x_2 \tan^{-1}(x_1 x_3)$ at $(-1, 2, 1)$ in the direction toward $(3, 1, -1)$

9. $f(x, y) = \tan^{-1}(x^2 + y^2)$ at $(1, -2)$ in the direction of increasing y along the line $y = 3x - 5$

10. $f(x, y) = x^2 e^{xy}$ at $(3, 0)$ in the direction of decreasing x along the normal to the line $3x - 2y = 9$

11. $f(x, y) = xe^{x+y}$ at $(-3, 3)$ in the direction of decreasing y normal to the curve $y = x^2 + 3x + 3$

12. $f(x, y) = x^2 + y \sinh(xy)$ at $(2, 0)$ in the direction of increasing x tangent to the curve $y = \sqrt{x + 7} - 3$

13. $f(x, y, z) = x^2 z/y$ at $(2, -1, 3)$ in the direction of increasing z normal to the plane $x + 2y - 2z = -6$

14. $f(x, y, z) = x^2 + 3yz$ at $(1, -1, 2)$ in the direction of increasing y normal to the surface $z = x^2 + y^2$

15. $f(x, y, z) = x^2 y - 3xz^2$ at $(1, -1, 1)$ in the direction of decreasing x normal to the surface $z = xy^2$

16. $f(x, y, z) = x^2 + \ln(y^2 + z)$ at $(3, 2, -2)$ in the direction of decreasing z normal to the surface $z = xy - 3x + 4y - 7$

17. $f(x, y, z) = z/(x^2 + y^2)$ at $(-1, 1, 3)$ along the line $x = -1 - 2t$, $y = 1 + t$, $z = 3 + 2t$ in the direction of decreasing t

18. $f(x, y, z) = y^2 \sin(xz^2)$ at $(0, -2, 2)$ along the line $x = 6t$, $y = -2 + 2t$, $z = 2 - 3t$ in the direction of increasing t

19. $f(x, y, z) = xy^2 e^z$ at $(2, -1, 0)$ in the direction of increasing y along the line of intersection of the planes $x - y + z = 3$ and $2x + y - z = 3$

20. $f(x, y, z) = (xy + y^2)/z$ at $(2, -1, -1)$ in the direction of decreasing x along the line of intersection of the planes $3x - y + z = 6$ and $2x - 3y - z = 8$

21. One side of the roof of a cathedral has the shape of the plane $z = 30 + 0.8y$ for $0 \le x \le 100$ and $0 \le y \le 40$. Find the rate at which the height to the roof is increasing over the head of a person at $(35, 25, 0)$ walking toward $(0, 40, 0)$.

22. If the temperature in space is $(y - z)/(1 + x^2 + y^2 + z^2)$ degrees at the point (x, y, z), find the rate of change of temperature at $(1, 0, -1)$ in the direction toward $(1, 1, 1)$.

In Exercises 23–26, find (a) a vector in the direction in which the function increases most rapidly at the given point, (b) that maximum rate of increase, (c) a vector in the direction for minimum rate of change at the point, and (d) that minimum rate of change.

23. $f(x, y) = x^2 + x \ln(y)$ at $(2, 1)$

24. $f(x, y) = xy^2 + e^{xy}$ at $(0, 2)$

25. $f(x, y, z) = x^2 y + \tan^{-1}(xz)$ at $(1, -2, 1)$

26. $f(x, y, z) = (x + y)/z^2$ at $(2, -1, 2)$

27. Let f be differentiable at **a** and have a local maximum at **a**. What can be said concerning the directional derivatives of f in all directions at **a**?

28. Give an example to show that it is possible to have $f(x, y)$ such that $f_\mathbf{u}(0, 0) = 0$ for all unit vectors **u** and still have neither a local maximum nor a local minimum for f at $(0, 0)$.

29. Consider a surface in space with equation $f(x, y, z) = c$, where c is some constant.
    (a) Argue that since the derivative is a rate of change, the directional derivative of f at any point on the surface in any direction tangent to the surface should be zero.
    (b) Argue from part (a) that ∇f should be normal to the surface at each point on the surface.

You know that for a suitable function f, the gradient vector $\nabla f(\mathbf{a})$ points in the direction of maximum rate of change of $f(\mathbf{x})$ at **a** if $\nabla f(\mathbf{a}) \ne \mathbf{0}$. If $\nabla f(\mathbf{a}) = \mathbf{0}$, then the rate of increase of $f(\mathbf{x})$ in all directions at **a** is zero. In Exercises 30–36, find the direction or directions in which you should travel from the given point to attain maximum increase $f(\mathbf{x})$ over a short distance. The text gives no instruction in this; just think about it in each case.

30. $f(x) = x^2$ at 0

31. $f(x) = x^3$ at 0

32. $f(x, y) = x^2 + y^2$ at $(0, 0)$

33. $f(x, y) = x^2 - y^2$ at $(0, 0)$

34. $f(x, y) = x^2 + 2xy + y^2$ at $(0, 0)$

35. $f(x, y) = x^3 - x^2 + y^2$ at $(0, 0)$

36. $f(x, y, z) = 5$ at $(1, -2, 7)$

14.6 DIFFERENTIATION OF IMPLICIT FUNCTIONS

Section 3.9 treated differentiation of implicitly defined functions of one variable, finding dy/dx given an equation of the form $G(x, y) = c$. We refresh our memory with an example.

EXAMPLE 1 Let $x^2 y^3 + 2xy^2 - x^3 = 3$ and find dy/dx by differentiating implicitly.

Solution We have

$$x^2 \cdot 3y^2 \frac{dy}{dx} + 2xy^3 + 2x \cdot 2y \frac{dy}{dx} + 2y^2 - 3x^2 = 0,$$

so

$$\frac{dy}{dx} = \frac{3x^2 - 2xy^3 - 2y^2}{3x^2 y^2 + 4xy}. \qquad \blacksquare$$

An equation of the form $G(x, y, z) = c$ may define z implicitly as one or more functions of both x and y. For example, $x^2 + y^2 + z^2 = 16$ gives rise to the

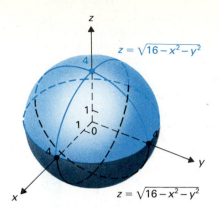

FIGURE 14.29 Graphs of two functions defined implicitly by

$$x^2 + y^2 + z^2 = 16.$$

functions

$$z = \sqrt{16 - x^2 - y^2} \qquad \text{and} \qquad z = -\sqrt{16 - x^2 - y^2},$$

whose graphs are shaded in different colors in Fig. 14.29. We will not describe exactly when $G(x, y, z) = c$ gives such implicitly defined functions. For functions we will use here, the solution set of $G(x, y, z) = c$ is a surface in space, and a piece of such a surface containing a point (x_0, y_0, z_0) can often be regarded as the graph of a function $z = f(x, y)$ such that $z_0 = f(x_0, y_0)$. In this case, we would like to find $\partial z/\partial x$ and $\partial z/\partial y$. This can again be accomplished by implicit differentiation.

EXAMPLE 2   Let $x^2 z + yz^3 - 2xy^2 = -9$. Find $\partial z/\partial x$ at the point $(1, -2, 1)$ on the surface.

Solution   We differentiate implicitly with respect to x, thinking of y as a constant, obtaining

$$x^2 \frac{\partial z}{\partial x} + z \cdot 2x + y \cdot 3z^2 \frac{\partial z}{\partial x} - 2y^2 = 0.$$

Then

$$\frac{\partial z}{\partial x} = \frac{2y^2 - 2xz}{x^2 + 3yz^2},$$

so

$$\left. \frac{\partial z}{\partial x} \right|_{(1,-2,1)} = \frac{8 - 2}{1 - 6} = -\frac{6}{5}. \qquad \blacksquare$$

There is a nice formula that avoids the technique of implicit differentiation. Suppose $G(x, y, z) = c$ defines $z = f(x, y)$. Then the equations

$$w = G(x, y, z), \tag{1}$$

$$x = x, \qquad y = y, \qquad z = f(x, y) \tag{2}$$

define w as a composite function $g(x, y)$ of the *two* variables x and y. Furthermore, as a function of the two independent variables x and y in Eqs. (2), w remains constant with value c, because

$$w = g(x, y) = G(x, y, f(x, y)) = G(x, y, z) = c$$

since $z = f(x, y)$ was chosen so that $G(x, y, z) = c$. Thus $\partial g/\partial x = 0$ and $\partial g/\partial y = 0$. By the chain rule for the composition of functions given by Eqs. (1) and (2),

$$\frac{\partial g}{\partial x} = \frac{\partial G}{\partial x} \cdot \frac{\partial x}{\partial x} + \frac{\partial G}{\partial y} \cdot \frac{\partial y}{\partial x} + \frac{\partial G}{\partial z} \cdot \frac{\partial z}{\partial x}$$

$$= \frac{\partial G}{\partial x} \cdot 1 + \frac{\partial G}{\partial y} \cdot 0 + \frac{\partial G}{\partial z} \cdot \frac{\partial z}{\partial x}. \tag{3}$$

(Note that $\partial y/\partial x = 0$ from Eq. 2, where y is regarded as a function of x and y.)

Since $\partial g/\partial x = 0$, we obtain from Eq. (3)

$$\frac{\partial G}{\partial x} + \frac{\partial G}{\partial z} \cdot \frac{\partial z}{\partial x} = 0,$$

so

$$\frac{\partial z}{\partial x} = -\frac{\partial G/\partial x}{\partial G/\partial z}. \tag{4}$$

Formula (4) for $\partial z/\partial x$ is valid wherever $\partial G/\partial z \neq 0$. It can be shown that, for a nice function G, the condition $\partial G/\partial z \neq 0$ guarantees that $G(x, y, z) = c$ does define z implicitly as a function of x and y.

EXAMPLE 3 Do Example 2 again using formula (4).

Solution From

$$G(x, y, z) = x^2 z + yz^3 - 2xy^2 = -9,$$

we obtain

$$\frac{\partial G}{\partial x} = 2xz - 2y^2 \qquad \text{and} \qquad \frac{\partial G}{\partial z} = x^2 + 3yz^2,$$

so

$$\frac{\partial z}{\partial x} = -\frac{2xz - 2y^2}{x^2 + 3yz^2} = \frac{2y^2 - 2xz}{x^2 + 3yz^2}.$$

The derivative is then computed at $(1, -2, 1)$ as in Example 2, but the messy implicit differentiation is gone. ■

The formula

$$\frac{\partial z}{\partial x} = -\frac{\partial G/\partial x}{\partial G/\partial z}$$

is not hard to remember in this Leibniz notation; the ∂G's "cancel" to give what we want, *but we must remember the negative sign also.* This is one place where the Leibniz notation lets us down just a bit.

Of course, a similar argument shows that

$$\frac{\partial z}{\partial y} = -\frac{\partial G/\partial y}{\partial G/\partial z}.$$

Indeed, if $G(x_1, x_2, x_3) = c$ and we solve for x_i in terms of the other x's, similar arguments show that

$$\frac{\partial x_i}{\partial x_j} = -\frac{\partial G/\partial x_j}{\partial G/\partial x_i}, \qquad i \neq j. \tag{5}$$

We can even use the formula to solve the implicit differentiation problems in Section 3.8, finding dy/dx if $G(x, y) = c$, for

$$\frac{dy}{dx} = -\frac{\partial G/\partial x}{\partial G/\partial y}. \tag{6}$$

EXAMPLE 4 Solve Example 1 again using Eq. (6), finding dy/dx if $G(x, y) = x^2y^3 + 2xy^2 - x^3 = 3$.

Solution We have

$$\frac{\partial G}{\partial x} = 2xy^3 + 2y^2 - 3x^2 \quad \text{and} \quad \frac{\partial G}{\partial y} = 3x^2y^2 + 4xy,$$

so

$$\frac{dy}{dx} = -\frac{2xy^3 + 2y^2 - 3x^2}{3x^2y^2 + 4xy} = \frac{3x^2 - 2xy^3 - 2y^2}{3x^2y^2 + 4xy},$$

as obtained in Example 1. ∎

Recall that a vector normal to a surface $z = f(x, y)$ given implicitly by $G(x, y, z) = c$ is

$$\frac{\partial z}{\partial x}\mathbf{i} + \frac{\partial z}{\partial y}\mathbf{j} - \mathbf{k} = -\frac{\partial G/\partial x}{\partial G/\partial z}\mathbf{i} - \frac{\partial G/\partial y}{\partial G/\partial z}\mathbf{j} - \mathbf{k}. \tag{7}$$

Multiplying the vector in Eq. (7) by $-\partial G/\partial z$, we obtain, as a vector normal to the surface $G(x, y, z) = c$,

$$\frac{\partial G}{\partial x}\mathbf{i} + \frac{\partial G}{\partial y}\mathbf{j} + \frac{\partial G}{\partial z}\mathbf{k}. \tag{8}$$

But Eq. (8) is the gradient vector ∇G. The surface $G(x, y, z) = c$ is called a **level surface** of the function $G(x, y, z)$, for it consists of all points (x, y, z) where the function attains the "level" c.

> The gradient vector of a function at a point is normal to the level surface of the function through that point. (9)

Of course, for a function $G(x, y)$ of just two variables, $G(x, y) = c$ is a curve in the plane, called a **level curve** of G. Again, the gradient vector ∇G is perpendicular to the level curve at each point. This follows at once from the statement (9) if we consider the solution set of $G(x, y) = c$ in space, which is a vertical cylinder intersecting the x,y-plane in the curve $G(x, y) = c$.

EXAMPLE 5 Let $G(x, y) = y - x^2$. Use the statement (9) to find a vector normal to the level curve $G(x, y) = -1$ at the point $(2, 3)$. Sketch the curve and vector.

Solution The level curve $G(x, y) = -1$ is the plane curve $y - x^2 = -1$ or $y = x^2 - 1$. A perpendicular vector to the curve at $(2, 3)$ is given by $\nabla G(2, 3)$. We obtain

$$\frac{\partial G}{\partial x} = -2x \quad \text{and} \quad \frac{\partial G}{\partial y} = 1,$$

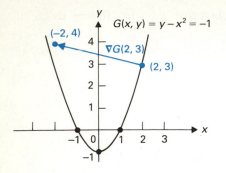

FIGURE 14.30 $\nabla G(2, 3) = -4\mathbf{i} + \mathbf{j}$ is normal to the level curve of

$$G(x, y) = y - x^2$$

through $(2, 3)$.

so $\nabla G = -2x\mathbf{i} + \mathbf{j}$, and $\nabla G(2, 3) = -4\mathbf{i} + \mathbf{j}$. The curve and this vector are shown in Fig. 14.30.   ∎

EXAMPLE 6   Find the tangent plane and normal line to the surface $x^2yz + x^2y^3 + \sin(x^2z) = 8$ at the point $(-1, 2, 0)$.

Solution   Setting $G(x, y, z) = x^2yz + x^2y^3 + \sin(x^2z)$, we have

$$G_x(-1, 2, 0) = [2xyz + 2xy^3 + 2xz \cos(x^2z)]|_{(-1,2,0)} = -16,$$
$$G_y(-1, 2, 0) = (x^2z + 3x^2y^2)|_{(-1,2,0)} = 12,$$
$$G_z(-1, 2, 0) = [x^2y + x^2\cos(x^2z)]|_{(-1,2,0)} = 3.$$

Therefore the normal vector (8) is $-16\mathbf{i} + 12\mathbf{j} + 3\mathbf{k}$. The tangent plane at $(-1, 2, 0)$ has equation

$$16x - 12y - 3z = -40,$$

and the normal line has parametric equations

$$x = -1 - 16t, \qquad y = 2 + 12t, \qquad z = 3t.$$ ∎

Let $T(x, y, z)$ give the temperature at points (x, y, z) in space. A surface $T(x, y, z) = c$, where the temperature remains constant at c, is called an **isothermal surface.**

EXAMPLE 7   Let the temperature at (x, y, z) be given by

$$T(x, y, z) = \frac{100}{2 + x^2 + y^2 + z^2 + 2y} \text{ degrees.}$$

Find

(a) the temperature at $(2, 1, -1)$,
(b) the equation of the isothermal surface through $(2, 1, -1)$, and sketch the surface,
(c) the direction for maximum rate of change of temperature at $(2, 1, -1)$ and the rate of change in that direction.

Solution

(a) The temperature at $(2, 1, -1)$ is

$$T(2, 1, -1) = \frac{100}{2 + 4 + 1 + 1 + 2} = \frac{100}{10} = 10 \text{ degrees.}$$

(b) The equation of the isothermal surface through $(2, 1, -1)$ is $T(x, y, z) = 10$, or

$$\frac{100}{2 + x^2 + y^2 + z^2 + 2y} = 10,$$
$$2 + x^2 + y^2 + z^2 + 2y = 10,$$
$$x^2 + (y^2 + 2y) + z^2 = 8,$$
$$x^2 + (y + 1)^2 + z^2 = 1 + 8 = 9.$$

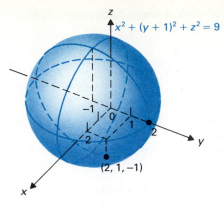

FIGURE 14.31 The isothermal surface of $T(x, y, z) = 100/(2 + x^2 + y^2 + z^2 + 2y)$ through $(2, 1, -1)$.

The surface is therefore a sphere with center $(0, -1, 0)$ and radius 3, sketched in Fig. 14.31.

(c) The gradient $\nabla T(2, 1, -1)$ gives the direction for maximum rate of change of temperature at the point $(2, 1, -1)$. We have

$$\frac{\partial T}{\partial x} = \frac{-200x}{(2 + x^2 + y^2 + z^2 + 2y)^2},$$

$$\frac{\partial T}{\partial y} = \frac{-200y - 200}{(2 + x^2 + y^2 + z^2 + 2y)^2},$$

and

$$\frac{\partial T}{\partial z} = \frac{-200z}{(2 + x^2 + y^2 + z^2 + 2y)^2}.$$

We readily find that

$$\nabla T(2, 1, -1) = -4\mathbf{i} - 4\mathbf{j} + 2\mathbf{k},$$

so $-2\mathbf{i} - 2\mathbf{j} + \mathbf{k}$ is a vector at $(2, 1, -1)$ pointing in the direction of maximum rate of change. This maximum rate of change is $|\nabla T| = \sqrt{4^2 + 4^2 + 2^2} = \sqrt{36} = 6$. ∎

SUMMARY

1. If $G(x, y, z) = c$ defines $z = f(x, y)$, then $\partial z/\partial x$ and $\partial z/\partial y$ can be found using either implicit differentiation or the formulas

$$\frac{\partial z}{\partial x} = -\frac{\partial G/\partial x}{\partial G/\partial z} \quad \text{and} \quad \frac{\partial z}{\partial y} = -\frac{\partial G/\partial y}{\partial G/\partial z}.$$

2. If $G(x, y) = c$ defines $y = f(x)$, then

$$\frac{dy}{dx} = -\frac{\partial G/\partial x}{\partial G/\partial y}.$$

3. If $G(\mathbf{x}) = G(x_1, x_2, x_3) = c$ defines x_i as a function of the other x's, then $\partial x_i/\partial x_j$ for $i \neq j$ can be found by implicit differentiation or by using the formula

$$\frac{\partial x_i}{\partial x_j} = -\frac{\partial G/\partial x_j}{\partial G/\partial x_i}.$$

4. Given a level surface $G(x, y, z) = c$, the gradient vector

$$\nabla G = \frac{\partial G}{\partial x}\mathbf{i} + \frac{\partial G}{\partial y}\mathbf{j} + \frac{\partial G}{\partial z}\mathbf{k}$$

is normal to the surface at each point on the surface.

EXERCISES 14.6

In Exercises 1–16, find the desired derivative using formulas like those in Eqs. (4) and (5).

1. $dy/dx|_{(1,-1)}$ if $x^2y - x\sin(\pi y) - y^3 = 0$

2. $dx/dy|_{(0,2)}$ if $e^{xy} - (3xy + 2y)^3 = -63$

3. $dx/dy|_{(1,1)}$ if $3x^2y + \ln(xy) = 3$

4. $dy/dx|_{(2,-1)}$ if $(2x + y)^3 + (x^2/y) = 23$

5. $\partial z/\partial x|_{(1,-1,2)}$ if $(2xy/z)+z/(x^2+y^2) = 0$

6. $\partial z/\partial x|_{(-1,2,0)}$ if $x\sin(yz) - 3x^2z + ye^z = 2$

7. $\partial u/\partial v$ if $u^2w^3 - 2u^3v^2 = 3v - 4$

8. $\partial v/\partial w$ if $uv^2 - \tan^{-1}(vw^2) = 8$

9. $\partial x_3/\partial x_2|_{(1,-1,1)}$ if $x_1{}^2 - 2x_2x_3{}^4 = 3x_1x_3{}^2$

10. $\partial x_1/\partial x_3|_{(1,2,\pi)}$ if $\sin(x_1x_3) - x_1{}^3x_2{}^2 = -4$

11. $\partial y/\partial z|_{(0,1,3)}$ if $\tan^{-1}(x + y) + \ln(xz + y) = \pi/4$

12. $\partial x/\partial z|_{(3,0,0)}$ if $\sin^{-1}z + x^2e^{y^2} = 12 - xe^y$

13. $\partial y/\partial x|_{(2,0,2)}$ if $\ln(xy + 1) + 3xz^2 = 24 + \tan(x - z)$

14. $\partial u/\partial y$ if $x^2y^3 - xu^2 + 2vu^3 - v^2y^4 = 1$

15. $\partial x/\partial w$ if $x^2\sin(vw) - e^{xu} = 0$

16. $\partial y/\partial v$ if $\ln(xy) + 3xv^2 - e^{yv^2} = 0$

In Exercises 17–22, use implicit differentiation to find the desired derivative.

17. $\partial x/\partial z|_{(-1,0,1)}$ if $\cos(x^2y) - yz^3 + xz^2 = 0$

18. $\partial x_2/\partial x_3|_{(0,1,-1)}$ if $x_1x_2{}^3 - 3x_2x_3 + x_1x_3{}^4 = 3$

19. $\partial z/\partial y|_{(-1,2,0)}$ if $e^{xyz} - \ln(xz + 1) = 1$

20. $\partial x_3/\partial x_1|_{(1,2,0)}$ if $x_2{}^2 - 4x_1x_2{}^2 = 7x_1x_3 - 12x_1{}^3$

21. $\partial u/\partial v$ if $x^2v + yu^2 - (xv/u) = 1$

22. $\partial v/\partial x$ if $u^2vx^3 - \tan^{-1}(2x + u) = 6$

In Exercises 23–28, find the equation of the tangent line or plane and parametric equations of the normal line to the given curve or surface at the given point.

23. $x^3y - 3y^2 = -2$ at $(1, 1)$

24. $xy^3 - (x^2/y) = -2$ at $(2, 1)$

25. $3xe^y + xy^3 = 2 + x$ at $(1, 0)$

26. $x\sin y + x^2e^z = 4$ at $(2, \pi, 0)$

27. $xz^2 + (2x - z)^2/y^3 = 19$ at $(2, 1, 3)$

28. $\sin(x^2y) + (3x + 2z)^5 = x$ at $(-1, 0, 1)$

29. Find the directional derivative of $f(x, y) = x^2 - 3xy^3$ at $(-1, 1)$ in the direction of increasing x normal to the curve $xy^3 - 3xy = 2$.

30. Find the directional derivative of the function $f(x, y) = (x/y^2) + (x + 2y)^3$ at $(3, -1)$ in the direction of increasing y normal to the curve $x^2 - 3xy + y^3 = 17$.

31. Find the directional derivative of $f(x, y, z) = (xy + y^2z)/x^3$ at $(1, 2, -1)$ in the direction of decreasing z normal to the surface $x^2z - yz^3 = 1$.

32. Find the directional derivative of $f(x, y, z) = x^2e^{yz}$ at $(1, 0, 2)$ in the direction of decreasing y normal to the surface $z + xyz^3 = 2$.

In Exercises 33–36, the function T gives the temperature in the plane or in space. Find
(a) the temperature at the given point,
(b) the equation of the isothermal curve or surface through the given point,
(c) the direction for maximum rate of change of temperature at the given point,
(d) the maximum change of temperature at the given point.

33. $T(x, y) = 80/(2 + x^2 + y^2)$ at $(2, -2)$

34. $T(x, y) = (100 + x^2 + y^2)/(15 + x^4 + y^4)$ at $(1, 1)$

35. $T(x, y, z) = 27/(1 + x^2 + y^2) + 36/(1 + y^2 + z^2)$ at $(1, -1, 1)$

36. $T(x, y, z) = \ln(1 + x^2 + y^2 + z^2)$ at $(1, -1, 2)$

37. Show that if $G(x, y)$ has continuous second partial derivatives and if the curve $G(x, y) = c$ defines y as a twice-differentiable function of x, then at a point (x, y) where $G_y(x, y) \neq 0$, we have

$$\frac{d^2y}{dx^2} = -\frac{G_y{}^2G_{xx} - 2G_xG_yG_{xy} + G_x{}^2G_{yy}}{G_y{}^3}.$$

38. Use the result in Exercise 37 to compute $d^2y/dx^2|_{(0,1)}$ for y defined implicitly as a function of x by $x^3 - 3xy^2 + 4y^3 = 4$.

39. Let $G(x, y, z)$ have continuous second partial derivatives, and let $G(x, y, z) = c$ define z as a function of x and y. Find a formula like that in Exercise 37 for $\partial^2z/\partial y^2$.

40. Repeat Exercise 39, but find a formula for $\partial^2z/\partial x\,\partial y$.

41. Show that the curves

$$5x^4y - 10x^2y^3 + y^5 = 4$$

and

$$x^5 - 10x^3y^2 + 5xy^4 = -4$$

are orthogonal at all points of intersection.

42. Show that the surfaces

$$x^2 - 2y^2 + z^2 = 0 \qquad \text{and} \qquad xyz = 1$$

are orthogonal at all points of intersection. (Surfaces are *orthogonal* at a point of intersection if their normal lines at that point are orthogonal.)

43. Show that the surfaces

$$x + y^2 + 2z^3 = 4$$

and

$$12x - 3(\ln y) + z^{-1} = 13$$

are orthogonal at all points of intersection.

14.7 MAXIMA AND MINIMA

In Section 2.4, we stated that a *continuous* function $f(x)$ attains a maximum value and also a minimum value on each *closed interval* in its domain. The situation for functions of more than one variable is similar, although more complicated to explain carefully; we make a few intuitive introductory remarks. We consider a region in the plane to be **bounded** if it is contained inside some sufficiently large circle so that it doesn't extend infinitely far in any direction. Such a region is **closed** if it contains all points that we would consider to be part of the *boundary* of the region. (We will not give a formal definition of the boundary.) It can be shown that a *continuous* function $f(x, y)$ attains a maximum value and also a minimum value on each *closed, bounded* region in its domain. This is the analogue for $f(x, y)$ of the result stated above for $f(x)$. The boundary of $[a, b]$ is considered to consist of the two points a and b, so $[a, b]$ is *closed* since it includes those points. We restricted the theorem for $f(x)$ to an *interval*, which does not extend to infinity and is hence *bounded*. The theorem for functions of more than two variables is similar in every way.

In Section 4.3, we learned to find such extrema for $f(x)$ on $[a, b]$: We found *local extrema* inside the interval and compared function values there with the values $f(a)$ and $f(b)$ at points on the boundary. In the present section, we consider only the problem of finding the *local extrema* for a function of two or more variables, and we restrict ourselves to quite simple cases. Sometimes these local extrema are obviously maximum or minimum values for a function in its entire domain.

Finding Local Extrema

Let $z = f(x, y)$ and suppose $f(x, y) \leq f(x_0, y_0)$ for all (x, y) inside some sufficiently small circle with center at (x_0, y_0). Then $f(x, y)$ has a **local maximum** or **relative maximum** of $f(x_0, y_0)$ at the point (x_0, y_0). See Fig. 14.32. Of course, if the inequality were reversed so that $f(x, y) \geq f(x_0, y_0)$ for all such (x_0, y_0), then $f(x_0, y_0)$ would be a **local minimum** or **relative minimum.** This is an obvious generalization of the same notions for a function of one variable; and still further generalizations to functions of three variables are clear.

We want to find such local maxima and minima of $z = f(x, y)$. Suppose that $f_x(x, y)$ and $f_y(x, y)$ exist. If $f(x_0, y_0)$ is a local maximum, then the function $g(x) = f(x, y_0)$ shown in Fig. 14.33 has a local maximum $g(x_0)$ at x_0. Consequently,

$$g'(x_0) = f_x(x_0, y_0) = 0.$$

Also, we must have a local maximum at y_0 of $h(y) = f(x_0, y)$, so

$$h'(y_0) = f_y(x_0, y_0) = 0.$$

A similar argument shows that first partial derivatives must be zero if $f(x_0, y_0)$ is a local minimum, and, indeed, the same results hold for functions of more than two variables.

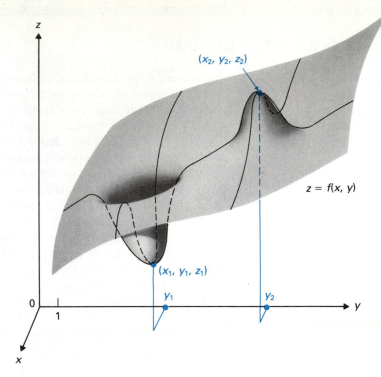

FIGURE 14.32 $f(x, y)$ has a local minimum of z_1 at (x_1, y_1) and a local maximum of z_2 at (x_2, y_2).

THEOREM 14.4   Partial derivatives at local extrema

If a function has first-order partial derivatives, then these derivatives will be zero at any point where the function has a local maximum or a local minimum.

□

Consequently, we can find all *candidates* for local extrema of differentiable functions by finding those points where all first-order partial derivatives are zero simultaneously. We regard such points as **critical points.**

So far the situation is just the same as for a function of one variable. From here on, more complicated things can happen for functions of two or more variables, as the following two examples show.

EXAMPLE 1   Test $f(x, y) = 1 - x^2 + y^2$ for local maxima or minima.

Solution   Using Theorem 14.4, we find that

$$f_x = -2x = 0 \quad \text{only if } x = 0 \qquad \text{and} \qquad f_y = 2y = 0 \quad \text{only if } y = 0.$$

Thus $(0, 0)$ is our only candidate for a local extremum. Clearly f has neither a local maximum nor a local minimum at $(0, 0)$, for $f(0, 0) = 1$, but $f(x_1, 0) < 1$ and $f(0, y_1) > 1$ for nonzero x_1 and y_1 close to zero. The graph of f is shown in Fig. 14.34. ∎

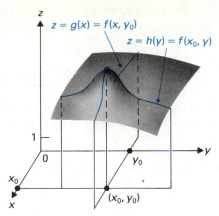

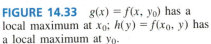

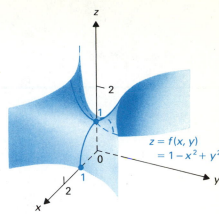

FIGURE 14.33 $g(x) = f(x, y_0)$ has a local maximum at x_0; $h(y) = f(x_0, y)$ has a local maximum at y_0.

FIGURE 14.34 $f_x(0, 0) = f_y(0, 0)$, but f has neither a local maximum nor a local minimum at $(0, 0)$.

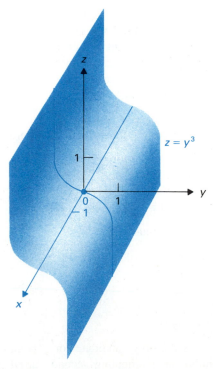

FIGURE 14.35 The graph of $f(x, y) = y^3$ has a "saddle point" at every point on the x-axis; $f_x = f_y = 0$ there, but there is neither a local maximum nor a local minimum.

The surface in Fig. 14.34 is shaped like a saddle near the origin. It is customary to refer to a point of the graph of $z = f(x, y)$ where $f_x = 0$ and $f_y = 0$ but where there is no local extremum as a **saddle point.** Figure 14.35 shows the graph of $z = f(x, y) = -y^3$. We see that $f_x = f_y = 0$ where $y = 0$, which corresponds to all points of the graph on the entire x-axis. The surface does not actually look like a saddle near any of these points, but it is nevertheless customary to call them all *saddle points*.

For a function of one variable with first derivative zero at x_0, the sign of a nonzero second derivative at x_0 determines whether the function has a local minimum or a local maximum at x_0. The situation is not so simple for functions of more than one variable, as we show in the next example.

EXAMPLE 2 Show that $f(x, y) = x^2 + 4xy + y^2$ has the property that

$$f_x(0, 0) = 0, \qquad f_y(0, 0) = 0,$$
$$f_{xx}(0, 0) > 0, \qquad f_{xy}(0, 0) > 0, \qquad f_{yy}(0, 0) > 0,$$

but that the function has no local maximum or minimum at $(0, 0)$.

Solution We compute that

$$f_x(0, 0) = (2x + 4y)|_{(0,0)} = 0, \qquad f_y(0, 0) = (4x + 2y)|_{(0,0)} = 0,$$
$$f_{xx} = 2 > 0, \qquad f_{xy} = 4 > 0, \qquad f_{yy} = 2 > 0.$$

However, the function f still does not have a local extremum at $(0, 0)$. To see this, note that on the line $y = x$, we have

$$f(x, y) = f(x, x) = x^2 + 4x^2 + x^2 = 6x^2 \geq 0,$$

so $f(x, y) > 0$ on the line $y = x$ except at $(0, 0)$. However, on the line $y = -x$,

$$f(x, y) = f(x, -x) = x^2 - 4x^2 + x^2 = -2x^2 \leq 0,$$

so $f(x, y) < 0$ on the line $y = -x$ except at $(0, 0)$. Since $f(0, 0) = 0$, we see that f has neither a local maximum nor a local minimum at $(0, 0)$. ∎

We state without proof a second-order derivative test for a local maximum or minimum of a function of two variables. A proof can be found in any advanced calculus text.

THEOREM 14.5   Test for local extrema

Let $f(x, y)$ be a function of two variables with continuous partial derivatives of orders less than or equal to 2 in some disk with center (x_0, y_0), and suppose that

$$f_x(x_0, y_0) = f_y(x_0, y_0) = 0,$$

while not all second-order partial derivatives are zero at (x_0, y_0). Let

$$A = f_{xx}(x_0, y_0), \qquad B = f_{xy}(x_0, y_0), \qquad \text{and} \qquad C = f_{yy}(x_0, y_0).$$

(a) If $AC - B^2 > 0$, the function has a local maximum at (x_0, y_0) if $A < 0$ and a local minimum there if $A > 0$.

(b) If $AC - B^2 < 0$, the function has neither a local maximum nor a local minimum at (x_0, y_0); it has a saddle point there.

(c) If $AC - B^2 = 0$, more work is needed to determine whether the function has a local maximum, a local minimum, or neither at (x_0, y_0). □

Referring to Example 2, we see that there $A = 2$, $B = 4$, and $C = 2$, so $AC - B^2 = 4 - 16 = -12 < 0$. Consequently, $f(x, y) = x^2 + 4xy + y^2$ has neither a local maximum nor a local minimum at $(0, 0)$; the graph has a saddle point there. In Exercise 27, we ask you to demonstrate part (c) of Theorem 14.5 by giving examples of three functions, each having the origin as a critical point with $AC - B^2 = 0$ there. One function is to have a local maximum at $(0, 0)$, another is to have a local minimum there, and the third is to have neither a local maximum nor a local minimum there.

EXAMPLE 3   Find all local minima and maxima of the function f where

$$f(x, y) = x^2 - 2xy + 2y^2 - 2x + 2y + 4.$$

Solution   We have

$$f_x(x, y) = 2x - 2y - 2 \qquad \text{and} \qquad f_y(x, y) = -2x + 4y + 2.$$

In order for both partial derivatives to be zero, we must have

$$2x - 2y - 2 = 0,$$
$$-2x + 4y + 2 = 0.$$

Adding these equations, we find that

$$2y = 0, \qquad \text{or} \qquad y = 0.$$

Then we must have $x = 1$. Therefore $(1, 0)$ is the only candidate for a point where f can have a local maximum or minimum. Computing second partial derivatives, we have

$$A = f_{xx}(1, 0) = 2, \qquad B = f_{xy}(1, 0) = -2, \qquad C = f_{yy}(1, 0) = 4.$$

Therefore $AC - B^2 = 8 - 4 = 4 > 0$. Since $A > 0$, we know by Theorem 14.5 that f has a local minimum of $f(1, 0) = 3$ at $(1, 0)$. ∎

EXAMPLE 4 Find all local maxima and minima of the function f where

$$f(x, y) = x^3 - y^3 - 3xy + 4.$$

Solution We have

$$f_x(x, y) = 3x^2 - 3y \qquad \text{and} \qquad f_y(x, y) = -3y^2 - 3x.$$

Setting these derivatives equal to 0, we obtain from the first $y = x^2$, and substituting into the second we have $x^4 + x = 0$. Now

$$x^4 + x = x(x^3 + 1)$$
$$= x(x + 1)(x^2 - x + 1) = 0$$

has only $x = 0$ and $x = -1$ as real solutions.

At $x = 0$, we obtain $y = 0$, and

$$A = f_{xx}(0, 0) = 0, \qquad B = f_{xy}(0, 0) = -3, \qquad C = f_{yy}(0, 0) = 0.$$

In this case, $AC - B^2 = -9 < 0$, so f has neither a local maximum nor a local minimum at $(0, 0)$.

At $x = -1$, we have $y = 1$, and

$$A = f_{xx}(-1, 1) = -6, \qquad B = f_{xy}(-1, 1) = -3,$$
$$C = f_{yy}(-1, 1) = -6,$$

so $AC - B^2 = 36 - 9 = 27 > 0$. Since $A < 0$, we see in this case that f has a local maximum of $f(-1, 1) = 5$ at $(-1, 1)$. ∎

The technique described in Theorems 14.4 and 14.5 and used in Examples 3 and 4 can be a lot of work in some cases. Sometimes the answer is obvious without the work.

EXAMPLE 5 Find all local maxima and minima of the function

$$f(x, y) = \sin(xy).$$

Solution We could find where $f_x = f_y = 0$ and then check $AC - B^2$ at those candidate points. We would discover it involves more work than in Examples 3 and 4, and we would also run into some cases where $AC - B^2 = 0$.

We can readily find the answer by inspection. We know that $\sin(xy)$ has maxima of 1 where $xy = (\pi/2) + 2n\pi$, which occurs on the hyperbolas drawn in color in Fig. 14.36. Minima of -1 occur where $xy = (3\pi/2) + 2n\pi$, which occurs on the hyperbolas shown in black in Fig. 14.36. ∎

We will do little toward finding local extrema for functions of more than two variables. Theorem 14.4 is true for any number of variables; all first-order partial derivatives must be zero at a local extremum. However, Theorem 14.5 is a specialized test for functions of just two variables.

There is one situation we can handle for a function f of more than two variables. That is the case where the function can be expressed as a sum of functions of at most two variables, with no variable appearing in more than one of the summand functions. For example, perhaps

$$f(x, y, z) = g(x) + h(y) + k(z), \tag{1}$$

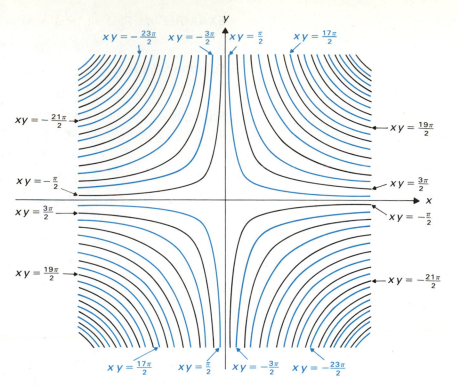

FIGURE 14.36 $\sin(xy)$ attains a maximum of 1 on the colored curves $xy = \pi/2 + 2n\pi$ and a minimum of -1 on the black curves $xy = 3\pi/2 + 2n\pi$.

or

$$f(x, y, z) = g(x, y) + h(z), \qquad (2)$$

or

$$f(w, x, y, z) = g(w, y) + h(x, z). \qquad (3)$$

In cases like Eqs. (1), (2), and (3), we can see that f can have a local maximum only at a point where each individual summand function also has a local maximum for corresponding coordinate values. Take, for illustration, case (2), where

$$f(x, y, z) = g(x, y) + h(z).$$

It is obvious that if $f(x_0, y_0, z_0)$ is a local maximum, then $g(x_0, y_0)$ and $h(z_0)$ must also be local maxima of those functions, and conversely. If, however, $g(x_0, y_0)$ gives a local maximum and $h(z_0)$ a local minimum, then $f(x_0, y_0, z) \geq f(x_0, y_0, z_0)$ for z close to z_0, while $f(x, y, z_0) \leq f(x_0, y_0, z_0)$ for x close to x_0 and y close to y_0. By such arguments, we conclude the following.

A function of the type in Eq. (1), (2), or (3) has a local extremum at a point if and only if each of the summand functions has the same type of local extremum at the points with corresponding coordinate values.

EXAMPLE 6 Find all local maxima and minima of the function

$$f(x, y, z) = x^3 + y^2 + z^2 - 12x - 2y + 6z.$$

Solution This function is of the form of Eq. (1), and we may write

$$f(x, y, z) = (x^3 - 12x) + (y^2 - 2y) + (z^2 + 6z).$$

We find that

$$f_x = 3x^2 - 12 = 0, \qquad \text{if } x = \pm 2,$$
$$f_y = 2y - 2 = 0, \qquad \text{if } y = 1,$$
$$f_z = 2z + 6 = 0, \qquad \text{if } z = -3.$$

Thus the only candidates for local extrema are

$$(2, 1, -3) \qquad \text{and} \qquad (-2, 1, -3).$$

Since $f_{xx} = 6x$, we see that $x^3 - 12x$ has a local maximum where $x = -2$ and a local minimum where $x = 2$. Since $f_{yy} = 2 > 0$ and $f_{zz} = 2 > 0$, we see that $y^2 - 2y$ and $z^2 + 6z$ have local minima where $y = 1$ and $z = -3$, respectively. The point where all three summand functions have the same type of extremum is therefore $(2, 1, -3)$, where they all have a local minimum. Thus $f(x, y, z)$ has a local minimum of -26 at $(2, 1, -3)$. There is neither a local minimum nor a local maximum at $(-2, 1, -3)$. ∎

EXAMPLE 7 Find all local maxima and minima of the function

$$f(x, y, z) = x^3 - y^3 - z^3 - 3xy + 3z + 4.$$

Solution We can write the function in the form of Eq. (2) as

$$f(x, y, z) = (x^3 - y^3 - 3xy + 4) + (3z - z^3).$$

We saw in Example 4 that $x^3 - y^3 - 3xy + 4$ has a local maximum of 5 at $(x, y) = (-1, 1)$ as its only local extremum. We easily see that $3z - z^3$ has a local minimum where $z = -1$ and a local maximum where $z = 1$. Thus the only local extremum of $f(x, y, z)$ occurs at $(-1, 1, 1)$, where the function has a local maximum of $5 + 2 = 7$. ∎

Applications

As an aid to solving maximum/minimum word problems involving several variables, we give a step-by-step procedure on page 764, modifying slightly the outline we gave in Section 4.6.

EXAMPLE 8 Show that the rectangular box of given volume V_0 with minimum surface area is a cube.

Solution

STEP 1. We take the box with one corner at the origin and opposite corner at (x, y, z), as shown in Fig. 14.37. The box then has dimensions x, y, and z and volume $V_0 = xyz$. We wish to minimize the surface area S of the box.

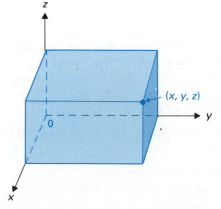

FIGURE 14.37 Rectangular box with opposite vertices $(0, 0, 0)$ and (x, y, z).

Solving a Maximum/Minimum Word Problem

STEP 1.   Draw a figure where appropriate and assign letter variables. Decide what to maximize or minimize.

STEP 2.   Express this quantity to be maximized or minimized as a function f of *independent* variables. (You may need algebra to do this.)

STEP 3.   Find all points where the first partial derivatives of f are simultaneously zero.

STEP 4.   Decide whether the desired maximum or minimum occurs at one of the points you found in step 3. Frequently, it is clear that a maximum or minimum exists from the nature of the problem, and if step 3 gives only one candidate, that candidate provides the answer. If f is a function of two variables, try the $AC - B^2$ test in Theorem 14.5.

STEP 5.   Put the answer in the requested form.

STEP 2.   We have $S = 2xy + 2xz + 2yz$. Since $xyz = V_0$, we have $z = V_0/(xy)$, and we can express S as a function of independent variables x and y by

$$S = f(x, y) = 2xy + \frac{2V_0}{y} + \frac{2V_0}{x} \qquad \text{for } x, y \text{ both} > 0.$$

STEP 3.   We have

$$\frac{\partial S}{\partial x} = 2y - \frac{2V_0}{x^2} \qquad \text{and} \qquad \frac{\partial S}{\partial y} = 2x - \frac{2V_0}{y^2}.$$

Setting these partial derivatives equal to zero, we obtain the system

$$\begin{cases} 2x^2 y - 2V_0 = 0, \\ 2xy^2 - 2V_0 = 0, \end{cases} \quad \text{or} \quad \begin{cases} x^2 y = V_0, \\ xy^2 = V_0. \end{cases}$$

Thus $x^2 y = xy^2$, so $x = y$. Then $x^3 = y^3 = V_0$, so $x = y = \sqrt[3]{V_0}$. Finally, $z = V_0/(xy) = V_0/(V_0^{2/3}) = \sqrt[3]{V_0}$, so $x = y = z = \sqrt[3]{V_0}$.

STEP 4.   This is a problem where it is obvious that a minimum does exist, and we have only one candidate for it. That is, it is obvious that a box with huge base and very small altitude has a large surface area, and so on. Alternatively, with $x = y = \sqrt[3]{V_0}$, the test of Theorem 14.5 gives $AC - B^2 = 4 \cdot 4 - 2^2 = 12 > 0$, so we have a minimum.

STEP 5.   Since $x = y = z = \sqrt[3]{V_0}$, our box of minimum surface area is indeed a cube.   ∎

EXAMPLE 9   Find the point on the plane $x - 2y + z = 5$ that is closest to $(5, -3, 6)$.

Solution

STEP 1. We want to minimize the distance s from $(5, -3, 6)$ to a point (x, y, z) on the plane $x - 2y + z = 5$.

STEP 2. We have $s = \sqrt{(x - 5)^2 + (y + 3)^2 + (z - 6)^2}$. Minimizing s is equivalent to minimizing

$$s^2 = (x - 5)^2 + (y + 3)^3 + (z - 6)^2,$$

which eliminates the radical from the problem. Since $z = 5 - x + 2y$, we obtain

$$s^2 = f(x, y) = (x - 5)^2 + (y + 3)^2 + (5 - x + 2y - 6)^2$$
$$= (x - 5)^2 + (y + 3)^2 + (-x + 2y - 1)^2$$

as a function of two independent variables x and y.

STEP 3. We have

$$\frac{\partial(s^2)}{\partial x} = 2(x - 5) - 2(-x + 2y - 1)$$

and

$$\frac{\partial(s^2)}{\partial y} = 2(y + 3) + 4(-x + 2y - 1).$$

Setting both of these partial derivatives equal to zero, we obtain the system

$$\begin{cases} 4x - 4y - 8 = 0, \\ -4x + 10y + 2 = 0, \end{cases} \quad \text{or} \quad \begin{cases} 2x - 2y = 4, \\ -2x + 5y = -1. \end{cases}$$

Adding these equations, we obtain $3y = 3$, so $y = 1$. We then find that $x = 3$ and $z = 5 - x + 2y = 4$.

STEP 4. It is geometrically obvious that there is a point on the plane $x - 2y + z = 5$ closest to $(5, -3, 6)$, and we have only the one candidate $(3, 1, 4)$. Alternatively, we can use Theorem 14.5, and we find that

$$AC - B^2 = 4 \cdot 10 - (-4)^2 = 24 > 0.$$

Since $A = 4 > 0$, we have a minimum.

STEP 5. We are asked to find the point on the plane that is closest to $(5, -3, 6)$, rather than the minimum distance, so our answer is $(3, 1, 4)$. ∎

SUMMARY

1. If a function has a local maximum or minimum at a point where the first-order partial derivatives exist, then these first-order partial derivatives must be zero there.

2. Let $f(x, y)$ be a function of two variables with continuous partial derivatives of orders less than or equal to 2 in a neighborhood of (x_0, y_0), and suppose that

$$f_x(x_0, y_0) = f_y(x_0, y_0) = 0,$$

while not all second-order partial derivatives are zero at (x_0, y_0). Let

$$A = f_{xx}(x_0, y_0), \qquad B = f_{xy}(x_0, y_0), \qquad \text{and} \qquad C = f_{yy}(x_0, y_0).$$

Then the function $f(x, y)$ has either a local maximum or a local minimum at (x_0, y_0) if $AC - B^2 > 0$. In this case, the function has a local minimum if $A > 0$ and a local maximum if $A < 0$. If $AC - B^2 < 0$, then $f(x, y)$ has neither a local maximum nor a local minimum at (x_0, y_0), but rather it has a saddle point there. If $AC - B^2 = 0$, the function may or may not have a local extremum at (x_0, y_0).

3. A function that can be expressed as a sum of functions of distinct variables has a local extremum at a point if and only if each summand function has the same type of local extremum for those values of the variables.

4. An abbreviated outline to aid in solving extremum word problems is as follows:

STEP 1. Decide what to maximize or minimize.

STEP 2. Express it as a function f of *independent* variables.

STEP 3. Find where all first partial derivatives of f are simultaneously zero.

STEP 4. Test for extrema at the points found in step 3.

STEP 5. Put the answer in the requested form.

EXERCISES 14.7

In Exercises 1–26, find all local maxima and minima of the function.

1. $x^2 + y^2 - 4$
2. $xy + 3$
3. $x^2 + y^2 + 4x - 2y + 3$
4. $x^2 - y^2 + 2x + 8y - 7$
5. $x^2 + y^2 + 4xy - 2x + 6y$
6. $3x^2 + y^2 - 3xy + 6x - 4y$
7. $x^3 + 2y^3 - 3x^2 - 24y + 16$
8. $x^3 + y^3 + 3xy - 6$
9. $x^4 - 2x^2y + 2y^2 - 8y$
10. $x^3 - 3xy + y^2 - 2y + 2$
11. $4y^3 + 6xy - x^2 - 4x + 10$
12. $x^3 - 3x^2y - y^2 + 10y - 3$
13. $\ln(x^2 + 2xy + 2y^2 - 2x - 8y + 20)$ [*Hint:* Use the fact that $\ln u$ is an increasing function of u.]
14. $e^{4xy - x^2 - 5y^2 + 2y + 6}$ [*Hint:* Similar to that in Exercise 13.]
15. $\cos(xy - 1)$
16. $\sin(y + x^2)$
17. $6/(2x^2 + 2xy + y^2 - 2x + 3)$

18. $8/(4xy - 4x^2 - 2y^2 - 4y)$

19. $x^4 + 2y^2 + 3z^2 - 2x^2 + 4y - 12z + 3$

20. $x^4 + y^2 + 2z^2 - 2x^2 - 4y - 12z + 5$

21. $x^4 + y^4 - z^4 - 2x^2 - 8y^2 + 2z^2 + 10$

22. $2x^2 - 2y^2 + 4yz - 3z^2 - x^4 + 5$

23. $z^4 - 3x^2 - 4y^2 + 4xy - 2z^2 - 11$

24. $3x^2 + 5y^2 + z^2 - 8xy + 6x - 8y - 4z + 9$

25. $x^4 - 2x^2y - 6x^2 + 2y^2 + 2z^2 + w^2 + 2zw + 4y + 4z + 18$

26. $z^4 - 2xz^2 + x^2 + y^2 + z^2 + 3w^2 - 4yw - 2z + 2w - 5$

27. Consider a function f of two variables having continuous partial derivatives of orders less than or equal to 2 with $f_x(0, 0) = f_y(0, 0) = 0$, and let

$$A = f_{xx}(0, 0), \qquad B = f_{xy}(0, 0), \text{ and } \qquad C = f_{yy}(0, 0).$$

Suppose that $AC - B^2 = 0$ with A, B, and C not all zero.
(a) Give an example of such a function f with a local maximum (0, 0).
(b) Give an example of such a function f with a local minimum at (0, 0).
(c) Give an example of such a function f with neither a local maximum nor a local minimum at (0, 0).
(d) What is the significance of this exercise?

In Exercises 28–32, use common sense to find a point at which the function assumes its maximum value on the square where $-1 \le x \le 1$ and $-1 \le y \le 1$. Then find a point where it assumes its minimum value on this square.

28. $x^2 + y^2$

29. xy

30. $y - 2x$

31. $x^2 + y^2 - xy$

32. $x^2 - y^2 + y$

Work Exercises 33–43 by the methods of this section.

33. Find the point on the plane $4x - 3y + z = 10$ closest to the points $(17, -7, -1)$.

34. Find the point on the plane $3x - 2y + 4z = -7$ closest to the point $(-5, 7, -9)$.

35. A rectangular box with open top is to have volume 108 ft³. Find the dimensions of the box for minimum surface area.

36. A rectangular box with open top is to have volume 192 ft³. Find the dimensions for minimum cost if material for the ends costs \$6/ft², for the sides \$3/ft², and for the bottom \$9/ft².

37. Find three positive real numbers whose sum is 9 and whose product is maximum.

38. Find the minimum distance from the origin to a point on the surface $xyz^2 = 32$.

39. Find the minimum distance from the origin to a point on the surface $x^2yz^3 = 162\sqrt{3}$.

40. Find the maximum possible volume of a rectangular box having one vertex at the origin and the diagonally opposite vertex at a point (x, y, z) in the first octant, as shown in Fig. 14.37, if (x, y, z) lies on the plane $x + 4y + 3z = 18$.

41. Find the volume of the rectangular box of maximum volume that can be inscribed in the ellipsoid $(x^2/a^2) + (y^2/b^2) + (z^2/c^2) = 1$ with sides parallel to the coordinate planes.

42. A long rectangular sheet of tin 12 in. wide is to be bent to form a gutter with cross section in the shape of an isosceles trapezoid, as shown in Fig. 14.38. Find the length x and the angle θ shown in the figure if the gutter has maximum carrying capacity.

FIGURE 14.38
Cross section of gutter.

43. *Method of least squares:* Find the line $y = mx + b$ that most nearly contains the points $(2, 3)$, $(4, 2)$, and $(6, 4)$ in the sense that the sum of the squares of the *deviations* d_1, d_2, and d_3 shown in Fig. 14.39 is minimum. [*Hint:* The unknowns are m and b.] (This method is often used to find the best linear fit to measure data that we believe should be linear but that are actually not linear due to measurement errors.)

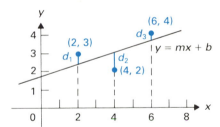

FIGURE 14.39
Deviations d_1, d_2, d_3 from the line $y = mx + b$ to three data points.

44. Consider the line $y = mx + b$ that best fits data points $(x_1, y_1), (x_2, y_2), \ldots, (x_n, y_n)$ in the least-squares sense of Exercise 43. Show that m and b satisfy the system of equations

$$\begin{cases} Am + nb = B, \\ Cm + Ab = D, \end{cases}$$

where

$$A = \Sigma_{i=1}^n x_i, \quad B = \Sigma_{i=1}^n y_i, \quad C = \Sigma_{i=1}^n x_i^2,$$

and

$$D = \Sigma_{i=1}^n x_iy_i.$$

45. Use the result of Exercise 44 to find the best linear fit to the data points $(-2, 4)$, $(-1, 3)$, $(1, 0)$, $(3, -1)$, $(4, -2)$ in the least-squares sense.

14.8 LAGRANGE MULTIPLIERS

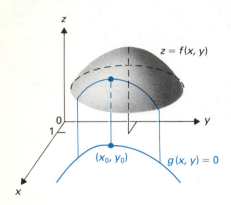

FIGURE 14.40 $f(x, y)$ restricted to the curve $g(x, y) = 0$ has a local maximum at (x_0, y_0).

The function $x^2 + y^2$ does not assume a maximum value in the entire plane, because x and y may become arbitrarily large there, so $x^2 + y^2$ may also become arbitrarily large. However, $x^2 + y^2$ does attain a maximum value for points (x, y) restricted to the ellipse $(x^2/4) + (y^2/9) = 1$; the ellipse is a closed and bounded subset of the plane. In this section, we present a useful method for finding the maximum or minimum value attained by a function, subject to some restriction on the domain of the function. This problem is of practical importance. For example, one may want to find the maximum current that may flow in a circuit containing several resistors under the restriction that the sum of the resistances have some constant value.

Suppose we want to maximize or minimize a function $f(x, y)$ subject to some relation $g(x, y) = 0$. The relation $g(x, y) = 0$ is called a **side condition** or a **constraint.** Geometrically, $g(x, y) = 0$ describes a curve in the x,y-plane, shown in Fig. 14.40. We wish to find local extrema of $f(x, y)$ where we restrict (x, y) to points on this curve. If $f(x_0, y_0)$ is a local maximum subject to $g(x, y) = 0$ as in Fig. 14.40, then $f(x_0, y_0) \geq f(x, y)$ for points (x, y) sufficiently close to (x_0, y_0) *but on the curve $g(x, y) = 0$ through (x_0, y_0).* It need *not* be true that $f(x_0, y_0) \geq f(x, y)$ for (x, y) close to (x_0, y_0) but off the curve.

This is really not a new concept for us. We first met problems of this type in Section 4.6. There we solved $g(x, y) = 0$ for one of the variables in terms of the other and then substituted to express f as a function of just one of the variables, x or y, along the curve. We give an example as a reminder.

EXAMPLE 1 Describe a method to find the dimensions of the rectangle of maximum area that can be inscribed in a semicircle of radius a.

Solution From Fig. 14.41, we see that the problem is to maximize $f(x, y) = 2xy$ subject to the constraint $x^2 + y^2 = a^2$. In Section 4.6, we would have solved this problem by using the constraint to express the area as a function of one *independent* variable as follows:

$$\text{Area} = 2xy,$$
$$x^2 + y^2 - a^2 = 0, \quad \text{so} \quad y = \sqrt{a^2 - x^2},$$
$$\text{Area} = 2x\sqrt{a^2 - x^2}.$$

We then took the derivative of this area function, set it equal to zero, solved that equation, and so on. We continue this example in a moment. ◼

When discussing word problems in Section 14.7, we generalized the technique of Example 1 to cases involving more variables. We expressed the quantity we wish to maximize or minimize as a function of *independent* variables. As in Example 1, this involves solving constraint equations for one variable in terms of others. In itself, that algebraic problem can be tough to execute. The method of Lagrange multipliers that we now present avoids this algebraic problem with the constraint equations.

We wish to find local extrema of $f(x, y)$ subject to $g(x, y) = 0$. The constraint $g(x, y) = 0$ is a curve in the plane. Since this curve is a level curve of

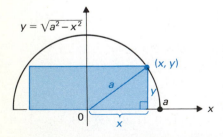

FIGURE 14.41 Rectangle inscribed in a semicircle of radius a.

$g(x, y)$, we know that

$$\nabla g = \frac{\partial g}{\partial x}\mathbf{i} + \frac{\partial g}{\partial y}\mathbf{j} \tag{1}$$

is perpendicular to the curve at each point on the curve. See Fig. 14.42. Turning to the function $f(x, y)$ to be maximized or minimized, we know that

$$\nabla f = \frac{\partial f}{\partial x}\mathbf{i} + \frac{\partial f}{\partial y}\mathbf{j} \tag{2}$$

points in the direction of maximum increase of the function $f(x, y)$ at each point. Furthermore, if $\mathbf{u}$ is a unit vector,

$$\nabla f \cdot \mathbf{u} = \text{Directional derivative of } f \text{ in the direction } \mathbf{u}. \tag{3}$$

Now at a point (x_0, y_0) on the curve $g(x, y) = 0$, where $f(x, y)$ has a local extremum *when considered only on the curve,* the directional derivative of f along the curve must be zero. (This is clear if we regard the curve as a bent s-axis and regard f as a function of the one variable s on the curve.) That is, ∇f must be normal to the curve at that point. Consequently, ∇f and ∇g must be *parallel* at (x_0, y_0), as illustrated in Fig. 14.42. Therefore there must exist λ such that

$$\nabla f = \lambda(\nabla g). \tag{4}$$

Equation (4), together with $g(x\, y) = 0$, leads to the three conditions

$$\frac{\partial f}{\partial x} = \lambda\frac{\partial g}{\partial x}, \qquad \frac{\partial f}{\partial y} = \lambda\frac{\partial g}{\partial y}, \qquad g(x, y) = 0. \tag{5}$$

We then solve the three Eqs. (5) for the three unknowns x, y, and λ to find candidates (x, y) for points where local extrema occur.

You may quite properly observe that solving Eqs. (5) simultaneously looks as hard as using the substitution technique in Example 1. For the problems we can handle with pencil and paper, that is usually true. However, using a computer, techniques such as Newton's method are available to solve a system of n equations in n unknowns, even if the equations are not linear. See the subsection on Newton's method in Section 14.2. The substitution technique, requiring us to express the quantity we want to maximize or minimize as a function of independent variables, is not as feasible on a computer at this time.

Equations (5) are the conditions of the *method of Lagrange multipliers*. The method itself is nothing more than a handy device for obtaining the conditions (5). Let

$$L(x, y, \lambda) = f(x, y) - \lambda g(x, y). \tag{6}$$

The conditions (5) are then equivalent to the following conditions, in the same order:

$$\frac{\partial L}{\partial x} = 0, \qquad \frac{\partial L}{\partial y} = 0, \qquad \frac{\partial L}{\partial \lambda} = 0. \tag{7}$$

The variable λ is called the *Lagrange multiplier*.

A point satisfying Eqs. (7) is a *candidate* for a point where $f(x, y)$ has a local maximum or minimum value, subject to $g(x, y) = 0$. If we know that such a local maximum or minimum exists, and if we can find all solutions of Eqs. (7), then computation of $f(x, y)$ at those points may indicate which is the desired extremum.

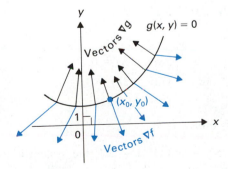

FIGURE 14.42 ∇f is parallel to ∇g at (x_0, y_0).

HISTORICAL NOTE

JOSEPH LOUIS LAGRANGE (1736–1813) developed the notion of Lagrange multipliers in his most famous work, *Analytical Mechanics* (1788), in which he extended and formalized Newton's work on mechanics. Lagrange was proud that he had been able to reduce mechanics to pure analysis; thus the work contained not a single diagram.

In calculus, Lagrange is best known for his attempt to base the subject on the notion of a power series. In his *Theory of Analytic Functions* (1797), he assumed that any function can be expressed as a power series around any value x in its domain:

$$f(x + h) = f(x) + p(x)h + q(x)h^2 + r(x)h^3 + \cdots.$$

It followed that the derivative $f'(x)$, the notation that Lagrange introduced, was precisely $p(x)$. Thus there was no necessity for limits or infinitesimals, ideas that no one at the time had rigorously defined. Unfortunately for Lagrange, mathematicians soon discovered that some functions cannot be expanded as power series.

EXAMPLE 2   Continue Example 1, and solve the problem of maximizing the function $f(x, y) = 2xy$, subject to $x^2 + y^2 - a^2 = 0$, using a Lagrange multiplier.

Solution   First we form

$$L(x, y, \lambda) = 2xy - \lambda(x^2 + y^2 - a^2).$$

Then we set

$$\frac{\partial L}{\partial x} = 2y - 2x\lambda = 0, \tag{8}$$

$$\frac{\partial L}{\partial y} = 2x - 2y\lambda = 0, \tag{9}$$

$$\frac{\partial L}{\partial \lambda} = -x^2 - y^2 + a^2 = 0. \tag{10}$$

Substituting $y = x\lambda$ and $x = y\lambda$, obtained from the first two conditions, in Eq. (10), we have

$$-y^2\lambda^2 - x^2\lambda^2 + a^2 = 0, \quad \text{or} \quad -(x^2 + y^2)\lambda^2 + a^2 = 0.$$

From Eq. (10), we then know that

$$-a^2\lambda^2 + a^2 = 0 \quad \text{or} \quad a^2(-\lambda^2 + 1) = 0,$$

so

$$\lambda^2 = 1 \quad \text{and} \quad \lambda = \pm 1.$$

The value $\lambda = -1$ would give $y = -x$, which is impossible for our geometric problem. Thus $\lambda = 1$, and $y = x$. From Eq. (10), we obtain

$$2x^2 = a^2, \quad \text{so} \quad x = \frac{a}{\sqrt{2}},$$

so $(x, y) = (a/\sqrt{2}, a/\sqrt{2})$ gives the maximum. That is, we know from geometry that a maximum exists, and we found only one candidate for it.   ∎

EXAMPLE 3   Find local extrema of $f(x, y) = 3x^2 + y^3$ on the circle $x^2 + y^2 = 9$.

Solution   We form

$$L(x, y, \lambda) = (3x^2 + y^3) - \lambda(x^2 + y^2 - 9).$$

The conditions (7) become

$$L_x = 6x - 2\lambda x = 0, \qquad\qquad 2x(3 - \lambda) = 0,$$
$$L_y = 3y^2 - 2\lambda y = 0, \quad \text{or} \quad y(3y - 2\lambda) = 0,$$
$$L_\lambda = -(x^2 + y^2 - 9) = 0, \qquad\qquad x^2 + y^2 = 9.$$

The first equation tells us that either $x = 0$ or $\lambda = 3$.

Case $x = 0$ From $x^2 + y^2 = 9$, we have $y = \pm 3$. If $\lambda = \pm\frac{9}{2}$, the system is satisfied. This gives the candidate points

$$(0, 3) \quad \text{and} \quad (0, -3).$$

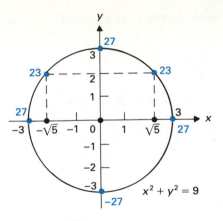

FIGURE 14.43 Values of

$$f(x, y) = 3x^2 + y^3$$

at candidate points for local extrema on $x^2 + y^2 = 9$.

Case $\lambda = 3$ From the second equation, we have $y = 0$ or $y = 2$, giving the candidate points

$$(3, 0), \quad (-3, 0), \quad (\sqrt{5}, 2), \quad \text{and} \quad (-\sqrt{5}, 2).$$

It is convenient to sketch the circle $x^2 + y^2 = 9$ and label the candidate points with the function values $f(x, y)$ there, as shown in color in Fig. 14.43. We see that $f(x, y)$ restricted to the circle has

local maxima of 27 at $(-3, 0), (0, 3),$ and $(3, 0)$

and

local minima of 23 at $(\pm\sqrt{5}, 2)$ and of -27 at $(0, -3)$.   ∎

We can make analogous use of Lagrange multipliers to handle situations with more variables or more constraints. For example, to maximize $f(x, y, z)$ subject to $g(x, y, z) = 0$, the gradient

$$\nabla f = \frac{\partial f}{\partial x}\mathbf{i} + \frac{\partial f}{\partial y}\mathbf{j} + \frac{\partial f}{\partial z}\mathbf{k}$$

must be perpendicular to the level surface $g(x, y, z) = 0$, and, consequently, ∇f must be parallel to ∇g, so again $\nabla f = \lambda(\nabla g)$. See Fig. 14.44. The four conditions

$$\frac{\partial f}{\partial x} = \lambda\frac{\partial g}{\partial x}, \qquad \frac{\partial f}{\partial y} = \lambda\frac{\partial g}{\partial y}, \qquad \frac{\partial f}{\partial z} = \lambda\frac{\partial g}{\partial z}, \qquad g(x, y, z) = 0$$

in $x, y, z,$ and λ can again be concisely expressed as

$$\frac{\partial L}{\partial x} = 0, \qquad \frac{\partial L}{\partial y} = 0, \qquad \frac{\partial L}{\partial z} = 0, \qquad \frac{\partial L}{\partial \lambda} = 0, \qquad \textbf{(11)}$$

where

$$L(x, y, z, \lambda) = f(x, y, z) - \lambda g(x, y, z).$$

EXAMPLE 4   Use Lagrange multipliers to find the point on the plane $2x - 2y + z = 4$ that is closest to the origin.

Solution   We want to minimize $\sqrt{x^2 + y^2 + z^2}$ subject to $2x - 2y + z - 4 = 0$. To make things easier, we minimize the square of the distance $x^2 + y^2 + z^2$ subject to $2x - 2y + z - 4 = 0$.

Let $L(x, y, z, \lambda) = x^2 + y^2 + z^2 - \lambda(2x - 2y + z - 4)$. Then the conditions (11) are

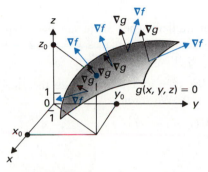

FIGURE 14.44 ∇f is parallel to ∇g at (x_0, y_0, z_0).

$$\frac{\partial L}{\partial x} = 2x - 2\lambda = 0,$$

$$\frac{\partial L}{\partial y} = 2y + 2\lambda = 0,$$

$$\frac{\partial L}{\partial z} = 2z - \lambda = 0,$$

$$\frac{\partial L}{\partial \lambda} = -2x + 2y - z + 4 = 0.$$

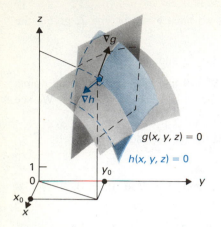

FIGURE 14.45   The plane normal to the curve of intersection contains ∇g and ∇h.

Substituting the values for $2x$, $2y$, and z from the first three conditions into the fourth condition yields

$$-2\lambda - 2\lambda - \frac{\lambda}{2} + 4 = 0, \quad \text{or} \quad -\frac{9}{2}\lambda + 4 = 0, \quad \text{or} \quad \lambda = \frac{8}{9}.$$

Therefore $x = \frac{8}{9}$, $y = -\frac{8}{9}$, and $z = \frac{4}{9}$, so $(\frac{8}{9}, -\frac{8}{9}, \frac{4}{9})$ is the desired point, and the distance from this point to the origin is

$$\sqrt{\frac{64 + 64 + 16}{81}} = 4 \cdot \sqrt{\frac{4 + 4 + 1}{81}} = 4 \cdot \sqrt{\frac{1}{9}} = \frac{4}{3}. \quad \blacksquare$$

For another case, suppose we wish to maximize or minimize $f(x, y, z)$ subject to *two* constraints $g(x, y, z) = 0$ and $h(x, y, z) = 0$. The two constraints $g(x, y, z) = 0$ and $h(x, y, z) = 0$ define the curve of intersection of the two level surfaces. See Fig. 14.45. At a point where $f(x, y, z)$ is maximum or minimum, subject to these constraints, the directional derivative along this curve must be zero. Consequently, ∇f must be perpendicular to the curve. Since both ∇g and ∇h are perpendicular to the curve, it must be that ∇f lies in the plane determined by the vectors* ∇g and ∇h. This time there are *two* Lagrange multipliers. We must have

$$\nabla f = \lambda_1(\nabla g) + \lambda_2(\nabla h)$$

for some λ_1 and λ_2. This leads to

$$\frac{\partial f}{\partial x} = \lambda_1 \frac{\partial g}{\partial x} + \lambda_2 \frac{\partial h}{\partial x}, \qquad \frac{\partial f}{\partial y} = \lambda_1 \frac{\partial g}{\partial y} + \lambda_2 \frac{\partial h}{\partial y}, \qquad \frac{\partial f}{\partial z} = \lambda_1 \frac{\partial g}{\partial z} + \lambda_2 \frac{\partial h}{\partial z},$$

and

$$g(x, y, z) = 0, \qquad h(x, y, z) = 0.$$

These are five equations in five unknowns and can be written as

$$\frac{\partial L}{\partial x} = 0, \qquad \frac{\partial L}{\partial y} = 0, \qquad \frac{\partial L}{\partial z} = 0, \qquad \frac{\partial L}{\partial \lambda_1} = 0, \qquad \frac{\partial L}{\partial \lambda_2} = 0, \quad (12)$$

where

$$L(x, y, z, \lambda_1, \lambda_2) = f(x, y, z) - \lambda_1 g(x, y, z) - \lambda_2 h(x, y, z).$$

Of course, solving these equations (12) can be very messy.

EXAMPLE 5   To illustrate Lagrange multipliers when there are two constraints, find the point on the line of intersection of the planes $x - y = 2$ and $x - 2z = 4$ that is closest to the origin.

Solution   This time we want to minimize $x^2 + y^2 + z^2$ subject to the side conditions $x - y - 2 = 0$ and $x - 2z - 4 = 0$. We form

$$L(x, y, z, \lambda_1, \lambda_2) = x^2 + y^2 + z^2 - \lambda_1(x - y - 2) - \lambda_2(x - 2z - 4).$$

_____

*One hopes these will not be parallel.

The conditions (12) are

$$\frac{\partial L}{\partial x} = 2x - \lambda_1 - \lambda_2 = 0,$$

$$\frac{\partial L}{\partial y} = 2y + \lambda_1 = 0,$$

$$\frac{\partial L}{\partial z} = 2z + 2\lambda_2 = 0,$$

$$\frac{\partial L}{\partial \lambda_1} = -x + y + 2 = 0,$$

$$\frac{\partial L}{\partial \lambda_2} = -x + 2z + 4 = 0.$$

The second and third conditions give

$$\lambda_1 = -2y \qquad \text{and} \qquad \lambda_2 = -z,$$

so the first condition becomes $2x + 2y + z = 0$. We then have

$$2x + 2y + z = 0,$$
$$-x + y = -2,$$
$$-x + 2z = -4.$$

The last two equations may be written as $y = x - 2$ and $z = (x - 4)/2$. Substitution of these values into the first equation gives

$$2x + 2(x - 2) + \frac{x - 4}{2} = 0, \qquad \text{or} \qquad \frac{9}{2}x - 6 = 0, \qquad \text{or} \qquad x = \frac{4}{3}.$$

Consequently, $y = -\frac{2}{3}$ and $z = -\frac{4}{3}$. The desired point is therefore $(\frac{4}{3}, -\frac{2}{3}, -\frac{4}{3})$. ∎

SUMMARY

1. To maximize or minimize a function $f(x, y)$ subject to a constraint $g(x, y) = 0$, form $L(x, y, \lambda) = f(x, y) - \lambda g(x, y)$. Candidates for points (x, y) where extrema occur are such that x, y, and λ satisfy the three simultaneous conditions

$$\frac{\partial L}{\partial x} = 0, \qquad \frac{\partial L}{\partial y} = 0, \qquad \frac{\partial L}{\partial \lambda} = 0.$$

For a function $f(x, y, z)$ subject to a constraint $g(x, y, z) = 0$, form $L(x, y, z, \lambda) = f(x, y, z) - \lambda g(x, y, z)$ and solve simultaneously the four conditions

$$\frac{\partial L}{\partial x} = 0, \qquad \frac{\partial L}{\partial y} = 0, \qquad \frac{\partial L}{\partial z} = 0, \qquad \frac{\partial L}{\partial \lambda} = 0.$$

2. If more than one constraint is present, use Lagrange multipliers, $\lambda_1, \lambda_2, \ldots$, equal in number to the number of constraints. To illustrate with $f(x, y, z)$ subject to constraints $g(x, y, z) = 0$ and $h(x, y, z) = 0$, form

$$L(x, y, z, \lambda_1, \lambda_2) = f(x, y, z) - \lambda_1 g(x, y, z) - \lambda_2 h(x, y, z).$$

The conditions are then

$$\frac{\partial L}{\partial x} = 0, \qquad \frac{\partial L}{\partial y} = 0, \qquad \frac{\partial L}{\partial z} = 0, \qquad \frac{\partial L}{\partial \lambda_1} = 0, \qquad \frac{\partial L}{\partial \lambda_2} = 0.$$

EXERCISES 14.8

In Exercises 1–16, find local extrema of the function f subject to the given constraints, using Lagrange multipliers.

1. $f(x, y) = x + y$; constraint $x^2 + y^2 = 4$

2. $f(x, y) = xy$; constraint $x^2 + y^2 = 4$

3. $f(x, y) = x - y$; constraint $x^2 + 2y^2 = 24$

4. $f(x, y) = x + 2y$; constraint $2x^2 + y^2 = 36$

5. $f(x, y) = x^2 - 3y^2$; constraint $x^2 + 4y^2 = 36$

6. $f(x, y) = x^3 + 2y^2$; constraint $x^2 + y^2 = 4$

7. $f(x, y) = x^2 + 4y^3$; constraint $2x + 4y = 2$

8. $f(x, y, z) = x^2 + y^2 - z^2$; constraint $2x - y + 3z = 4$

9. $f(x, y, z) = x^2 - 2y^2 - 2z^2$; constraint $3x - y - z = 32$

10. $f(x, y, z) = x + y - z$; constraint $x^2 + y^2 + z^2 = 12$

11. $f(x, y, z) = 2x - 3y + z$; constraint $x^2 + y^2 + 3z^2 = 120$

12. $f(x, y, z) = x^2 + y^2 + z^2$; constraints $x^2 + y^2 = 1$, $y - z = 0$

13. $f(x, y, z) = x - y + z^2$; constraints $y^2 + z^2 = 1$, $x + y = 2$

14. $f(x, y, z) = xyz$; constraints $x - y = 0$, $z = y - 2$

15. $f(x, y, z) = x^2 + y^2 + z^2$; constraints $z^2 = x^2 + y^2$, $x - y + z = 1$

16. $f(x, y, z) = xyz$; constraints $x^2 + y^2 + z^2 = 8$, $x + y = 0$

Solve Exercises 17–25 using Lagrange multipliers.

17. Find the point on the plane $2x - 3y + 6z = 5$ that is closest to the origin.

18. Find the point on the plane $3x - 2y + z = 5$ that is closest to $(4, -8, 5)$.

19. A rectangular box has a square base and open top. Find the dimensions for minimum surface area if the volume is to be 108 in^3.

20. An open rectangular box has ends costing $6/ft^2$, sides costing $4/ft^2$, and a bottom costing $10/ft^2$. Find the dimensions for minimum cost if the volume of the box is 120 ft^3.

21. Find the maximum possible volume of a right circular cone inscribed in a sphere of radius a.

22. The sides of a closed cylindrical container cost twice as much per square foot as the ends. Find the ratio of the radius to the altitude of the cylinder for the cheapest such container having a fixed volume.

23. A pentagon consists of a rectangle surmounted by an isosceles triangle. Find the dimensions of the pentagon having the maximum area if the perimeter is to be P.

24. Find the point on the curve of intersection of $x^2 + z^2 = 4$ and $x - y = 8$ that is farthest from the origin.

25. Find the point on the line of intersection of the planes $x - y = 4$ and $y + 3z = 6$ that is closest to $(-1, 3, 2)$.

SUPPLEMENTARY EXERCISES FOR CHAPTER 14

1. If $w = f(x, y, z) = xz^2 + y^2 e^{xz}$, find the following.
(a) $\partial w/\partial x$ (b) $\partial^2 w/\partial x^2$ (c) $\partial^2 w/\partial x\, \partial y$

2. If $y = x_1 \sin x_2 x_3 - x_2{}^2 x_1{}^3$, find the following.
(a) $\partial y/\partial x_2$ (b) $\partial^2 y/\partial x_1\, \partial x_3$

3. Find $\partial z/\partial x$ if $z = x^2 e^{\sin(xy)}$.

4. If the temperature at a point in space is $T(x, y, z) = 500/(1 + x^2 + y^2 + z^2)$, find the rate of change of temperature at $(-1, 2, -2)$ in the positive coordinate directions.

5. Find the equation of the plane tangent to the surface $z = (x + y)/(x - y)$ at the point $(2, 1, 3)$.

6. Find all points on the surface $z = x^2y + y$ where the tangent plane is parallel to the plane $3x - 10y + 2z = 7$.

7. Find the equation of the line normal to the surface

$$z = \frac{x^2 - y}{x + y}$$

at the point $(2, 1, 1)$.

8. Find all points on the surface $z = x^2 - 2y^3$ where the normal line is parallel to the line through $(-1, 4, 2)$ and $(1, -2, 0)$.

9. Find the differential df at $(0, -1, 2)$ of $f(x, y, z) = z \sin(xz) + z^2e^{xy}$.

10. Let $w = f(x, y, z) = x^2y - xy^2z^3$. Find the gradient vector $\nabla f(1, -1, 2)$.

11. Let $y = f(x_1, x_2) = x_2{}^3 + x_1 \cos x_2$. Find the gradient vector $\nabla f(1, 0)$.

12. Use a differential to estimate $\sqrt{2.95^3 + 3.01^2}$.

13. Use a differential to estimate $2.03^2 \cos(-0.05)$.

14. Use a differential to estimate

$$\frac{2.05^4 - 3.97^2}{1.08^5}.$$

15. A right circular cone has base radius 3 ft and altitude 6 ft. Use a differential to estimate the change in r to produce an increase of 0.5 ft^3 in volume if the altitude is decreased by 4 in.

16. Let z be a differentiable function of u and v, let u and v be differentiable functions of x and y, and let x and y be differentiable functions of t. Give the chain rule formula for dz/dt.

17. Let $z = f(x, y)$ be differentiable, and let $x = t^2 + 2t$ and $y = (t + 1)/t$. If $dz/dt = 5$ when $t = -1$, find $f_y(-1, 0)$.

18. If $z = u^2 - 3uv$ while $u = (x + y)/(y + 1)$ and $v = xy$, use a chain rule to compute $\partial z/\partial y$ at the point $(x, y) = (3, 1)$.

19. Let z be defined implicitly as a function of x and y near $(2, 1, -1)$ by $x^2z + xyz^3 = -6$. If $x = t^3 - 3t$ and $y = 3t^2 - 6t + 1$, use a chain rule to find dz/dt when $t = 2$.

20. Let $z = f(u, v) = u^2/v^3$, and let $u = t^2 - 3t - 7$ and $v = t^3 + 7$. Use a chain rule to find dz/dt when $t = -2$.

21. Let $z = f(x, y) = e^{x^2y}$, while $x = 2st$ and $y = s^2 + t^3$. Use a chain rule to find $\partial z/\partial t$ at the points $(s, t) = (1, -1)$.

22. Let $z = f(x, y)$ be a differentiable function, and let $x = ts - 2t$ and $y = t^2 - 3ts$. If $\partial z/\partial t = 3$ and $\partial z/\partial s = -4$ at $(t, s) = (2, 1)$ find $f_x(-2, -2)$ and $f_y(-2, -2)$.

23. Let $z = f(x, y)$, while $x = g_1(t)$ and $y = g_2(t)$. If $g_1(1) = -3$ and $g_2(1) = 4$, while

$$\left.\frac{dz}{dt}\right|_{t=1} = 2, \qquad \left.\frac{dx}{dt}\right|_{t=1} = 4,$$

$$\left.\frac{dy}{dt}\right|_{t=1} = -1, \qquad \left.\frac{\partial z}{\partial x}\right|_{(-3,4)} = 3,$$

find $f_y(-3, 4)$.

24. A circular pond covers the disk $x^2 + y^2 \leq 100{,}000$. The depth of water in the pond at a point (x, y) in the disk is $100 - (x^2 + y^2)^{2/5}$ in the same units as the coordinates. Find the rate of change of depth with respect to distance for a swimmer at the point $(4, -4)$ swimming toward the point $(-26, -44)$.

25. Let $z = f(x, y)$ be differentiable, and let x and y be differentiable functions of t. Find a formula for d^2z/dt^2.

26. Find the directional derivative of $x^3y + (x/y)$ at $(-1, 1)$ in the direction toward $(2, -3)$.

27. Find the directional derivative of $z = f(x, y) = x^2y^3 - 3y^2$ at $(-3, 4)$ in the direction toward the origin.

28. Find the directional derivative of $w = f(x, y, z) = xy^2z^3$ at $(2, -1, 1)$ in the direction toward $(0, 1, 0)$.

29. Find the direction for maximum rate of increase of $f(x, y, z) = y^2\sin xz^2$ at the point $(0, 3, -1)$ and the magnitude of this maximum rate of increase.

30. Find all points (x, y) where $z = f(x, y) = x^2 - y^2 = 8$ and where the maximum value of the directional derivative of f over all possible directions is also 8.

31. Find the direction of the maximum rate of increase of $f(x, y) = x^2e^{xy}$ at the point $(3, 0)$, and then find the magnitude of this rate of change.

32. For the temperature function

$$T(x, y, z) = \frac{500}{1 + x^2 + y^2 + z^2},$$

(a) find a vector in the direction for maximum rate of change at the point $(-1, 2, -2)$,

(b) find the maximum rate of change over all possible directions at $(-1, 2, -2)$.

33. Find $\partial z/\partial y$ if $x^2y + xz^3 + y^2z^2 = 3$.

34. Let y be defined implicitly as a function of x and z by $x^2y - 3xz^2 + y^3 = -13$. Find $\partial y/\partial x$ at $(1, -2, 1)$.

35. If x is defined implicitly as a function of y and z by $y^2\cos x + xz^3 - 3y^2z = -5$, find $\partial x/\partial z$ at $(0, 1, 2)$.

36. Let z be defined implicitly as a differentiable function of x and y by $G(x, y, z) = 0$, and let x and y be defined implicitly as differentiable functions of u and v by $H(x, u, v) = 0$ and $K(y, u, v) = 0$. Find a formula for $\partial z/\partial u$.

37. Find the equation of the plane tangent to the surface $x^3y + y^3z + z^3x = -1$ at the point $(1, 1, -1)$.

38. Find parametric equations of the line normal to the surface $x^2y - 3xz^2 + y^3 = -13$ at the point $(1, -2, 1)$.

39. (a) Find the equation of the level surface of $G(x, y, z) = x^3y - xyz^2 + z^3$ through the point $(-1, 1, 2)$.

(b) Find the equation of the tangent plane to the level surface in part (a) at $(-1, 1, 2)$.

40. (a) Find the equation of the level curve of $G(x, y) = x^3 + 2xy + e^{xy}$ that passes through the point $(2, 0)$.

(b) Find parametric equations of the normal line to the level curve in part (a) at $(2, 0)$.

41. Find all relative maxima and minima of the function

$$f(x, y) = xy - x^2 - 2y^2 + 3x - 5y - 6.$$

42. Find all local maxima and minima of the function

$$f(x, y) = 2x^2 - xy + y - y^2 - 7x + 3.$$

43. Find all local maxima and local minima of

$$x^2 - y^2 + 2z^2 + 2xz + 6x + 4y - 2z + 5.$$

44. Find all local maxima and local minima of the function

$$f(x, y, z) = z^3 + x^2 + 3y^2 - 3z^2 - 2xy + 3.$$

45. Find the maximum possible volume of a rectangular box having one vertex at $(0, 0, 0)$ and the diagonally opposite vertex at a point in the first octant on the plane $x + y + 2z = 6$.

46. Find the minimum distance from the origin to a point on the surface $xy^2z^2 = 4$.

47. Use the method of Lagrange multipliers to maximize $x^2 + y + z^2$ over the ellipsoid

$$x^2 + y^2 + 4z^2 = 4.$$

48. Use the method of Lagrange multipliers to find the point on the line of intersection of the planes

$$x - 2y + z = 4 \quad \text{and} \quad 2x + y - z = 8$$

that is closest to the origin.

49. Use the method of Lagrange multipliers to maximize $2x + 3y - 4$ on the circle $x^2 + y^2 = 13$.

50. Use the method of Lagrange multipliers to find the area of the largest isosceles triangle that can be inscribed in a circle of radius a.

51. Use Lagrange multipliers to maximize $x + y^2$ over the ellipse $x^2 + 4y^2 = 4$.

52. Use the method of Lagrange multipliers to find the point on the intersection of the spheres $x^2 + y^2 + z^2 = 4$ and $(x - 2)^2 + y^2 + z^2 = 12$ that is closest to $(1, 1, 0)$. Find also the point farthest from $(1, 1, 0)$.

53. Use the method of Lagrange multipliers to find all local extrema of $x^2 + y^2 + z^2$ subject to the constraints $x^2 + y^2 = 4$ and $x + z = 1$.

54. The strength of a beam of rectangular cross section is directly proportional to the width and to the square of the depth. Use the method of Lagrange multipliers to find the dimensions of the strongest beam that can be cut from a circular log of radius a inches.

55. Find the appropriate "subtract-and-add trick" that could be used to prove the three-variable analogue of Theorem 14.2 for a function $w = f(x, y, z)$. That is, give the analogue of Eq. (11) of Section 14.3 for $w = f(x, y, z)$.

Exercises 56–63 introduce power series representation for a function of two variables.

56. Consider the polynomial function

$$\begin{aligned} P(x, y) = a_{00} &+ a_{10}(x - x_0) + a_{01}(y - y_0) + \cdots \\ &+ a_{ij}(x - x_0)^i(y - y_0)^j + \cdots \\ &+ a_{0n}(y - y_0)^n \end{aligned}$$

for all nonnegative integers i and j, where $i + j \le n$. Let $f(x, y)$ have continuous partial derivatives of all orders less than or equal to n at (x_0, y_0). Find a formula for a_{ij} if $P(x, y)$ has the same partial derivatives as $f(x, y)$ of all orders less

than or equal to n at (x_0, y_0); that is, if

$$\left.\frac{\partial^{i+j}P}{\partial x^i\partial y^j}\right|_{(x_0, y_0)} = \left.\frac{\partial^{i+j}f}{\partial x^i\partial y^j}\right|_{(x_0, y_0)}.$$

Taylor polynomials

We define the **nth Taylor polynomial** $T_n(x, y)$ for $f(x, y)$ at (x_0, y_0) to be

$$T_n(x, y)$$

$$= \sum_{i+j \le n} \left[\frac{1}{i!j!} \cdot \left.\frac{\partial^{i+j}f}{\partial x^i\partial y^j}\right|_{(x_0, y_0)} \cdot (x - x_0)^i(y - y_0)^j\right]. \quad \textbf{(1)}$$

57. Find the Taylor polynomial $T_3(x, y)$ for $\sin(x + y)$ at $(0, 0)$.

58. Explain how the answer to Exercise 57 is predictable in terms of Taylor series for a function of one variable.

59. Following the idea of Exercise 58, predict the polynomial $T_4(x, y)$ for e^{xy} at $(0, 0)$. Then verify your answer by computing Eq. (1) with $n = 4$.

60. Find the Taylor polynomial $T_4(x, y)$ for e^{xy} at $(0, 1)$.

Taylor's theorem

Let $f(x, y)$ have continuous partial derivatives of all orders less than or equal to $n + 1$ in some disk with center at (x_0, y_0). Then for each $(x, y) \ne (x_0, y_0)$ in the disk, there exists (c_1, c_2), depending on x and y and strictly between (x_0, y_0) and (x, y) on the line segment joining them, such that

$$E_n(x, y) = \sum_{\substack{i,j \\ i+j=n+1}} \left[\frac{1}{i!j!} \cdot \left.\frac{\partial^{i+j}f}{\partial x^i\partial y^j}\right|_{(c_1, c_2)} \cdot (x - x_0)^i(y - y_0)^j\right],$$

where $E_n(x, y) = f(x, y) - T_n(x, y)$. $\qquad\square$

61. Estimate $1.02^2 \ln 0.97$ using a differential, and then use Taylor's theorem to find a bound for the error.

62. Use the Taylor polynomial $T_2(x, y)$ at $(0, 1)$ for $y^3 \cos x$ to estimate $1.03^3 \cos(-0.02)$, and then find a bound for the error.

63. Let f be a function of two variables with continuous coordinate derivatives of all orders less than or equal to n. Let us introduce the *operation notations*

$$\left(\frac{\partial}{\partial x}h + \frac{\partial}{\partial y}k\right)f = \frac{\partial f}{\partial x}h + \frac{\partial f}{\partial y}k,$$

$$\left(\frac{\partial}{\partial x}h + \frac{\partial}{\partial y}k\right)^2 f = \frac{\partial^2 f}{\partial x^2}h^2 + 2\frac{\partial^2 f}{\partial x\,\partial y}hk + \frac{\partial^2 f}{\partial y^2}k^2,$$

$$\vdots \qquad\qquad \vdots$$

$$\left(\frac{\partial}{\partial x}h + \frac{\partial}{\partial y}k\right)^n f = \sum_{i=0}^{n}\binom{n}{i}\frac{\partial^n f}{\partial x^i\,\partial y^{n-i}}h^ik^{n-i}.$$

Show that the nth Taylor polynomial for f at (x_0, y_0) is given by

$$T_n(x, y) = f(x_0, y_0)$$

$$+ \sum_{i=1}^{n}\frac{1}{i!}\left(\left.\frac{\partial}{\partial x}\right|_{(x_0, y_0)}(x - x_0) + \left.\frac{\partial}{\partial y}\right|_{(x_0, y_0)}(y - y_0)\right)^i f.$$

Exercises 64–66 are designed to shed a little light on the criterion $AC - B^2 > 0$ given in Theorem 14.5 for a local maximum or local minimum of $f(x, y)$.

64. Taylor's theorem for functions $f(x, y)$ of two variables was stated before Exercise 61.
   (a) Write the expression for $E_1(x, y)$ given in the theorem.
   (b) Suppose $f_x(x_0, y_0) = f_y(x_0, y_0) = 0$. Show that
   (i) $f(x_0, y_0)$ is a local minimum if $E_1(x, y) \geq 0$ for all (x, y) in some small disk with center at (x_0, y_0);
   (ii) $f(x_0, y_0)$ is a local maximum if $E_1(x, y) \leq 0$ for all (x, y) in some small disk with center at (x_0, y_0);
   (iii) $f(x_0, y_0)$ is neither a local minimum nor a local maximum if $E_1(x, y)$ assumes both positive and negative values in every small disk with center at (x_0, y_0).

65. Let $f(x, y)$ have continuous second partial derivatives, and let $A = f_{xx}(x_0, y_0)$, $B = f_{xy}(x_0, y_0)$, and $C = f_{yy}(x_0, y_0)$. Convince yourself that if

$$A(\Delta x)^2 + 2B(\Delta x)(\Delta y) + C(\Delta y)^2 > 0$$

for all choices of $(\Delta x, \Delta y) \neq (0, 0)$, then $E_1(x, y) \geq 0$ for all (x, y) in some small disk with center at (x_0, y_0). [*Hint:*

Think of $\Delta x = x - x_0$, $\Delta y = y - y_0$, and set $\Delta y = t(\Delta x)$.] What would be true if

$$A(\Delta x)^2 + 2B(\Delta x)(\Delta y) + C(\Delta y)^2 < 0$$

for $(\Delta x, \Delta y) \neq (0, 0)$? If it were sometimes positive and sometimes negative?

66. Show that $A(\Delta x)^2 + 2B(\Delta x)(\Delta y) + C(\Delta y)^2 > 0$ for all $(\Delta x, \Delta y) \neq (0, 0)$ if $AC - B^2 > 0$ and $A > 0$ or $C > 0$. [*Hint:* Note that

$$A(A(\Delta x)^2 + 2B(\Delta x)(\Delta y) + C(\Delta y)^2)$$
$$= (A(\Delta x) + B(\Delta y))^2 + (AC - B^2)(\Delta y)^2.]$$

67. Suppose $f(x, y)$ has continuous partial derivatives of all orders less than or equal to n. Suppose, further, that all partial derivatives of order less than n are zero at (x_0, y_0), but some partial derivative of order n is nonzero at (x_0, y_0).

(a) Show that if n is odd, then $f(x, y)$ has neither a local minimum nor a local maximum at (x_0, y_0).

(b) If n is even, show that Exercise 64(b) holds with $E_{n-1}(x, y)$ in place of $E_1(x, y)$.

MULTIPLE INTEGRALS

15

15.1 DOUBLE INTEGRALS OVER RECTANGULAR REGIONS

Integrals as Limits of Sums

We have tried to emphasize how important it is to regard an integral as a limit of sums. In an application involving an integral, we always consider what quantities we wish to add in order to arrive at the appropriate integral.

Let f be a continuous real-valued function defined on a region G. In Table 15.1 we give an outline showing on the left how we should regard the integral of f over G as a limit of sums. At the right, the table illustrates each step using the one-dimensional case $\int_a^b f(x)\,dx$ and *regular* Riemann sums where $\Delta x_i = (b - a)/n$.

FIGURE 15.1 Partition of $[a, b]$ into $n = 12$ equal-length subintervals.

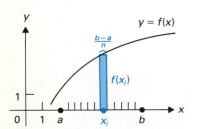

FIGURE 15.2 Strip of area $f(x) \cdot (b - a)/n$.

TABLE 15.1 The integral as a limit of Riemann sums		
	Integral of f over G	$\int_a^b f(x)\,dx$
STEP 1.	Form a partition P of G into subregions of small size such that values of f do not vary much over any single subregion.	Form a regular partition P of $[a, b]$ into n subintervals of equal length $(b - a)/n$. See Fig. 15.1.
STEP 2.	For each subregion, form the product of the value of f at some point in the subregion and the size of the subregion.	Form the products $$f(x_i) \cdot \left(\frac{b - a}{n}\right)$$ for x_i in the ith subinterval. See Fig. 15.2.

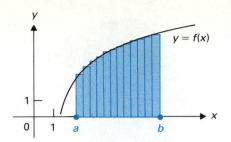

FIGURE 15.3 A Riemann sum R_P.

STEP 3. Add the products obtained in step 2. The result is an approximation to the integral.

Find the Riemann sum

$$R_P = \sum_{i=1}^{n} \left(f(x_i) \cdot \frac{b-a}{n} \right)$$

$$= \frac{b-a}{n} \sum_{i=1}^{n} f(x_i).$$

See Fig. 15.3.

STEP 4. The integral is the limit of sums found in step 3 as partitions become finer and finer in such a way that the variation of values of f on partition elements approaches zero.

$$\int_{a}^{b} f(x)\,dx = \lim_{\substack{\|P\| \to 0 \\ n \to \infty}} R_P$$

We apply the steps in Table 15.1 to describe the integral of a continuous function $f(x, y)$ over a rectangular region G, where $a \le x \le b$ and $c \le y \le d$, as shown in Fig. 15.4.

STEP 1. We partition $[a, b]$ into n subintervals of equal length and then do the same for $[c, d]$. This gives rise to a regular partition P of the region G into n^2 subrectangles, as illustrated in Fig. 15.4 for $n = 2$. Each subrectangle has area

$$\frac{(b-a)(d-c)}{n^2}.$$

STEP 2. Numbering the subrectangles from 1 to n^2 in some convenient fashion, we let (x_i, y_i) be a point in the ith subrectangle, as shown in Fig. 15.4. We then form the n^2 products

$$f(x_i, y_i) \cdot \frac{(b-a)(d-c)}{n^2}.$$

Each such product gives the volume of a column, as illustrated in Fig. 15.5.

STEP 3. We form the regular **Riemann sum**

$$R_P = \frac{(b-a)(d-c)}{n^2} \cdot \sum_{i=1}^{n^2} f(x_i, y_i). \tag{1}$$

See Fig. 15.6.

STEP 4. We regard $\|P\|$ as the maximum of all distances between two points that are in the same subrectangle in the partition P. The integral is the limit of the Riemann sums R_P as $n \to \infty$ so that $\|P\| \to 0$. This limit can be shown to exist.

We have been led to the following definition.

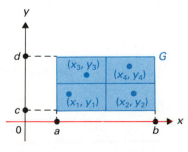

FIGURE 15.4 Partition of a rectangular region G for a regular Riemann sum with $n = 2$.

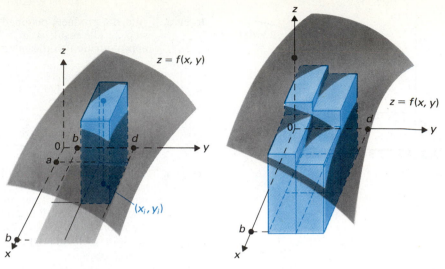

FIGURE 15.5 Column of volume $f(x_i, y_i) \cdot [(b - a)(d - c)]/n$ for $n = 2$.

FIGURE 15.6 A Riemann sum R_P for $n = 2$.

DEFINITION 15.1

Definite Integral

Let $f(x, y)$ be continuous in the rectangular region G, where $a \leq x \leq b$ and $c \leq y \leq d$. The **definite integral of f over G** is

$$\int\int f(x, y) \, dx \, dy = \lim_{\substack{\|P\| \to 0 \\ n \to \infty}} R_P$$

for the Riemann sums R_P in Eq. (1).

The double integral sign $\int\int$ in Definition 15.1 is used to indicate that the domain of f is two-dimensional. The heuristic geometric meaning of the integrand $f(x, y) \, dx \, dy$ is shown in Fig. 15.7.

It is geometrically clear that if $f(x, y) \geq 0$, then $\int\int_G f(x, y) \, dx \, dy$ is equal to the volume of the solid having G as base, vertical plane sides, and a portion of the surface $z = f(x, y)$ as top, as illustrated in Fig. 15.8. If $f(x, y)$ is sometimes positive and sometimes negative, then $\int\int_G f(x, y) \, dx \, dy$ gives the volume of the portion of the solid above the x,y-plane minus the volume of the portion below the x,y-plane, as shown in Fig. 15.9.

EXAMPLE 1 Let G be the rectangular region $1 \leq x \leq 3, 0 \leq y \leq 1$ shown in Fig. 15.10. Estimate

$$\int\int_G (x^2 + 3xy) \, dx \, dy$$

using the regular partition R_P with $n = 2$, taking as points (x_i, y_i) the midpoints of the subrectangles.

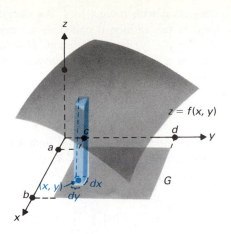

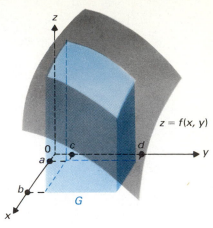

FIGURE 15.7 $f(x, y) \, dx \, dy$ is the volume of the column.

FIGURE 15.8 $\iint_G f(x, y) \, dx \, dy$ is the volume of the solid.

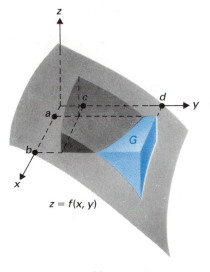

FIGURE 15.9 $\iint_G f(x, y) \, dx \, dy =$ (Volume of black solid) − (Volume of blue solid)

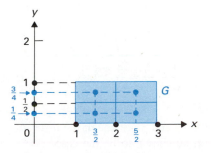

FIGURE 15.10 Partition for the midpoint sum R_P.

Solution The partition and points are shown in Fig. 15.10. The Riemann sum R_P in Eq. (1) becomes

$$R_P = \frac{(3-1)(1-0)}{2^2}\left[f\left(\frac{3}{2}, \frac{1}{4}\right) + f\left(\frac{5}{2}, \frac{1}{4}\right) + f\left(\frac{3}{2}, \frac{3}{4}\right) + f\left(\frac{5}{2}, \frac{3}{4}\right)\right]$$

$$= \frac{2}{4}\left[\left(\frac{9}{4} + \frac{9}{8}\right) + \left(\frac{25}{4} + \frac{15}{8}\right) + \left(\frac{9}{4} + \frac{27}{8}\right) + \left(\frac{25}{4} + \frac{45}{8}\right)\right]$$

$$= \frac{1}{2}\left(\frac{68}{4} + \frac{96}{8}\right) = \frac{1}{2}(17 + 12) = \frac{29}{2}. \quad \blacksquare$$

While regular partitions of a region G are often the most convenient, it is not essential that partitions be regular. If we let P be any partition of G into subregions and form a Riemann sum R_P of a continuous function $f(x, y)$ using P, then we still have

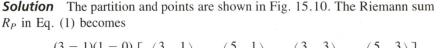

$$\iint_G f(x, y) \, dx \, dy = \lim_{\|P\| \to 0} R_P,$$

where $\|P\|$ is the maximum distance between any two points that are in the same subregion of the partition P. For convenience, we will continue to develop the theory of the integral for functions of more than one variable using regular partitions.

Iterated Integrals

How can we compute accurately an integral $\iint_G f(x, y) \, dx \, dy$ just defined? We need something like our fundamental theorem of calculus in Section 5.3, but for multiple integrals. Theorem 15.1, which we present in a moment, is our *fundamental theorem for computation*. The theorem states that $\iint_G f(x, y) \, dx \, dy$ can be computed by performing a *partial integration with respect to x* followed by an integration with respect to y. An integral computed by successive integration steps in this fashion is called an *iterated integral*.

Let $f(x, y)$ be a continuous function on a rectangular region $a \leq x \leq b$, $c \leq y \leq d$ in the plane. The *iterated integral* $\int_c^d \int_a^b f(x, y)\, dx\, dy$ is evaluated from "inside out" as

$$\int_c^d \int_a^b f(x, y)\, dx\, dy = \int_{y=c}^d \left(\int_{x=a}^b f(x, y)\, dx \right) dy. \tag{2}$$

That is, first integrate $f(x, y)$ with respect to x only, holding y constant, and evaluate from $x = a$ to $x = b$. This gives a function of y only, which is then integrated from $y = c$ to d.

EXAMPLE 2   Find $\displaystyle\int_0^1 \int_1^3 xy^2\, dx\, dy$.

Solution   We have

$$\int_0^1 \int_1^3 xy^2\, dx\, dy = \int_0^1 \left(\frac{x^2}{2} y^2 \Big]_{x=1}^{x=3} \right) dy = \int_0^1 \left(\frac{9}{2} y^2 - \frac{1}{2} y^2 \right) dy$$

$$= \int_0^1 4y^2\, dy = \frac{4}{3} y^3 \Big]_0^1 = \frac{4}{3}. \qquad \blacksquare$$

An iterated integral $\int_a^b \int_c^d f(x, y)\, dy\, dx$ is again computed from "inside out" as

$$\int_a^b \int_c^d f(x, y)\, dy\, dx = \int_{x=a}^b \left(\int_{y=c}^d f(x, y)\, dy \right) dx. \tag{3}$$

The first integration is with respect to y only, holding x constant, and when it is evaluated from $y = c$ to d, a function of x only is obtained. This function is then integrated from $x = a$ to b.

EXAMPLE 3   Compute the iterated integral

$$\int_1^3 \int_0^1 xy^2\, dy\, dx,$$

which is the integral in Example 2 in the "reverse order."

Solution   We have

$$\int_1^3 \int_0^1 xy^2\, dy\, dx = \int_1^3 \left(\frac{xy^3}{3} \Big]_{y=0}^{y=1} \right) dx = \int_1^3 \left(\frac{x}{3} - 0 \right) dx = \frac{x^2}{6} \Big]_1^3$$

$$= \frac{9}{6} - \frac{1}{6} = \frac{4}{3}. \qquad \blacksquare$$

Note that the iterated integrals in Examples 2 and 3 are equal; the *order* of integration does not seem to matter. This can be shown to be the case.

We will state the fundamental theorem for computation and then attempt to make it seem reasonable by geometric arguments.

THEOREM 15.1   Fundamental theorem for computation

Let $f(x, y)$ be continuous for (x, y) in the rectangular region G, where $a \le x \le b$ and $c \le y \le d$. Then

$$\int_c^d \int_a^b f(x, y) \, dx \, dy = \iint_G f(x, y) \, dx \, dy. \qquad \square$$

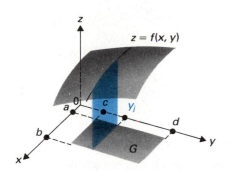

FIGURE 15.11 $\int_a^b f(x, y_j) \, dx$ is the area of the vertical surface.

Geometric explanation: The Leibniz differential notation suggests the following interpretation of $\int_c^d \int_a^b f(x, y) \, dx \, dy$. If we form $\int_a^b f(x, y_j) \, dx$, thinking of y_j as a constant, we obtain the area of the vertical shaded plane region shown in Fig. 15.11, which lies in the plane $y = y_j$ and under the surface $z = f(x, y)$ between $x = a$ and $x = b$. Multiplying by dy, we obtain the volume of the slab shown in Fig. 15.12, and the iterated integral

$$\int_c^d \left(\int_a^b f(x, y) \, dx \right) dy$$

yields the limit of the sum of the volumes of the slabs from $y = c$ to $y = d$ as their thickness approaches 0. Clearly the result for $f(x, y) \ge 0$ is the volume of the three-dimensional region under the surface $z = f(x, y)$ and over the rectangle G, which is precisely the geometric interpretation of $\iint_G f(x, y) \, dx \, dy$. Similar geometric considerations indicate that $\int_a^b \int_c^d f(x, y) \, dy \, dx$ must also be equal to $\iint_G f(x, y) \, dx \, dy$. We leave the sketching of figures like Figs. 15.11 and 15.12 to the exercises (see Exercise 2).

Theorem 15.1 is a powerful tool for computing $\iint_G f(x, y) \, dx \, dy$ when the integration can be carried out.

EXAMPLE 4   Let G be the rectangular region $1 \le x \le 3$, $0 \le y \le 1$. In Example 1, we estimated $\iint_G (x^2 + 3xy) \, dx \, dy$ using the midpoint Riemann sum R_P for the regular partition P with $n = 2$. Find the exact value of this integral.

Solution   By Theorem 15.1, the integral is

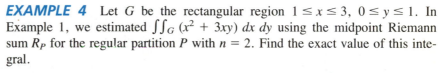

$$\int_0^1 \int_1^3 (x^2 + 3xy) \, dx \, dy = \int_0^1 \left(\int_{x=1}^3 (x^2 + 3xy) \, dx \right) dy$$

$$= \int_0^1 \left(\frac{x^3}{3} + \frac{3}{2} x^2 y \right) \Bigg]_{x=1}^3 dy$$

$$= \int_0^1 \left[\left(9 + \frac{27}{2} y \right) - \left(\frac{1}{3} + \frac{3}{2} y \right) \right] dy$$

$$= \int_0^1 \left(\frac{26}{3} + 12y \right) dy$$

$$= \left(\frac{26}{3} y + 6y^2 \right) \Bigg]_0^1 = \frac{26}{3} + 6$$

$$= \frac{26 + 18}{3} = \frac{44}{3} \approx 14.67.$$

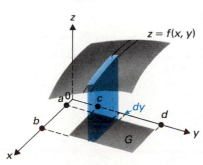

FIGURE 15.12 $[\int_a^b f(x, y) \, dx] \, dy$ is the volume of the vertical slab.

In Example 1, we obtained 14.5 as estimate.   ■

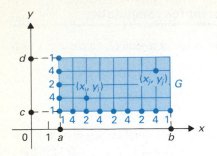

FIGURE 15.13 Partition of G for Simpson's rule with $m = 8$, $n = 4$; the coefficient of $f(x_i, y_i)$ is $2 \cdot 4 = 8$, and of $f(x_j, y_j)$ is $4 \cdot 4 = 16$.

We would generally omit the large parentheses in the first computation line of Example 4 in computing an iterated integral like $\int_a^b \int_c^d f(x, y)\, dy\, dx$. Just remember to compute from "inside out" and be sure to evaluate using the correct variables. We give another illustration.

EXAMPLE 5 Find $\int_0^4 \int_0^\pi x^2 \sin y\, dy\, dx$.

Solution We have

$$\int_0^4 \int_0^\pi x^2 \sin y\, dy\, dx = \int_0^4 (-x^2 \cos y)\Big]_0^\pi dx = \int_0^4 [-x^2(-1) + x^2(1)]\, dx$$

$$= \int_0^4 2x^2\, dx = \frac{2x^3}{3}\Big]_0^4 = \frac{128}{3}. \qquad \blacksquare$$

Simpson's Rule for Double Integrals (Optional)

Numerical estimation of multiple integrals is a much tougher job than for an integral of a function of one variable. As the dimension of the domain of f increases, the number of points required for accurate estimation increases exponentially.

We describe Simpson's rule for estimating $\int_a^b \int_c^d f(x, y)\, dx\, dy$. In Fig. 15.13, we show a rectangular region $a \leq x \leq b$, $c \leq y \leq d$. We divide $[a, b]$ into m equal subintervals and $[c, d]$ into n equal subintervals, where m and n are both even. In Fig. 15.13, $m = 8$ and $n = 4$. This gives us a natural grid on the rectangle. At the points along the bottom and left edges of the rectangle, we write the coefficients $1, 4, 2, 4, 2, 4, 2, \ldots, 4, 1$ of function values that appear in Simpson's rule for a function of one variable. For two variables, we follow the steps given at the bottom of the page.

It is not difficult to see why these steps give an estimate for the integral. If we add the products

$$\text{Bottom coefficient} \cdot f(x_i, y_i)$$

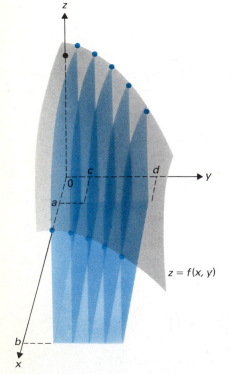

FIGURE 15.14 Areas of vertical surfaces estimated using Simpson's rule on grid lines parallel to the x-axis.

Simpson's Rule for $\int_c^d \int_a^b f(x, y)\, dx\, dy$

STEP 1. Find $f(x_i, y_i)$ at each of the $(m + 1)(n + 1)$ grid points in the rectangle.

STEP 2. Multiply each function value $f(x_i, y_i)$ in step 1 by the *product* of the coefficient at the left of (x_i, y_i) and the one underneath (x_i, y_i), as illustrated in Fig. 15.14.

STEP 3. Add the results in step 2.

STEP 4. Multiply the sum in step 3 by

$$\frac{b - a}{3m} \cdot \frac{d - c}{3n}$$

to obtain the estimate for the integral.

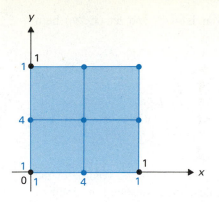

FIGURE 15.15 Grid points and co-efficients for Simpson's rule with $m = n = 2$.

separately for each row of the grid and multiply each such sum by $(b - a)/(3m)$, we obtain our old Simpson's rule estimate for the areas of the vertical surfaces up to $z = f(x, y)$, as shown in Fig. 15.14. To estimate the volume, we then use our old Simpson's rule in the y-direction. That is, we add these surface areas, multiplied by the coefficients at the left of the rectangle, and then multiply the sum by $(d - c)/(3n)$. We can readily convince ourselves that this is equivalent to the steps given on the previous page.

EXAMPLE 6   Use a calculator and Simpson's rule with $m = n = 2$ to estimate

$$\int_0^1 \int_0^1 (x + 1)^y \, dx \, dy.$$

Solution   Let $f(x, y) = (x + 1)^y$. Using Fig. 15.15 and the steps outlined on page 784, we obtain the estimate

$$\int_0^1 \int_0^1 (x + 1)^y \, dx \, dy \approx \frac{1}{6} \cdot \frac{1}{6}\left[f(0, 0) + 4f\left(\frac{1}{2}, 0\right) + f(1, 0) + 4f\left(0, \frac{1}{2}\right)\right.$$

$$+ 16f\left(\frac{1}{2}, \frac{1}{2}\right) + 4f\left(1, \frac{1}{2}\right) + f(0, 1)$$

$$\left. + 4f\left(\frac{1}{2}, 1\right) + f(1, 1)\right] \approx 1.229243672.$$

We also estimated using Simpson's rule with $m = n = 4$ and obtained as estimate 1.229271648. Once again, it appears that Simpson's rule gives quite accurate estimates for low values of m and n. ∎

A better way to estimate the integral in Example 6 is to carry out the first integration, which is feasible, and then use Simpson's rule for a function of a single variable the rest of the way. We do this in the next example.

EXAMPLE 7   Estimate $\int_0^1 \int_0^1 (x + 1)^y \, dx \, dy$ by the method just described.

Solution   We have

$$\int_0^1 \int_0^1 (x + 1)^y \, dx \, dy = \int_0^1 \frac{(x + 1)^{y+1}}{y + 1}\Bigg]_0^1 \, dy = \int_0^1 \frac{2^{y+1} - 1}{y + 1} \, dy.$$

Taking up our calculator, we use Simpson's rule with $n = 20$ for this integral and obtain as estimate 1.229274137. With $n = 40$, we obtain as estimate 1.229274135. This indicates that our estimate in Example 6 with $m = n = 2$ was accurate to five significant figures, and the estimate with $n = m = 4$ was accurate to six significant figures. ∎

SUMMARY

Let $f(x, y)$ be continuous in the rectangular region G, where $a \leq x \leq b$ and $c \leq y \leq d$. Form a regular partition P of G into n^2 subrectangles by subdividing both intervals $[a, b]$ and $[c, d]$ into n subintervals of equal length. Number the subrectangles from 1 to n^2, and let (x_i, y_i) be a point in the ith subrectangle.

1. A regular Riemann sum is

$$R_P = \frac{(b - a)(d - c)}{n^2} \cdot \sum_{i=1}^{n^2} f(x_i, y_i)$$

2. $\iint_G f(x, y) \, dx \, dy = \lim_{\substack{\|P\| \to 0 \\ n \to \infty}} R_P$

3. If $f(x, y) \geq 0$ in G, then $\iint_G f(x, y) \, dx \, dy$ is the volume of the solid with base G, vertical plane sides, and a portion of the graph of $z = f(x, y)$ as top.

4. *Fundamental theorem of computation*

$$\iint_G f(x, y) \, dx \, dy = \int_c^d \int_a^b f(x, y) \, dx \, dy,$$

5. If an iterated integral cannot be evaluated by integration, it may be estimated using a midpoint Riemann sum R_P. If only the last step of integration cannot be done, Simpson's rule can be used at that stage to obtain an estimate.

6. If none of the integration steps can be done for an iterated integral, Simpson's rule for multiple integrals, explained on pages 784–785, can be used to estimate the integral.

EXERCISES 15.1

1. We have treated only a very restrictive case of double integrals of continuous functions in this section. Can you guess the restriction to which we are referring?

2. Sketch figures similar to Figs. 15.11 and 15.12 to illustrate geometrically that

$$\int_a^b \int_c^d f(x, y) \, dy \, dx = \iint_G f(x, y) \, dx \, dy,$$

where G is the rectangle $a \leq x \leq b, c \leq y \leq d$ and f is continuous over G.

3. We have defined iterated integrals $\int_c^d \int_a^b f(x, y) \, dx \, dy$ and $\int_a^b \int_c^d f(x, y) \, dy \, dx$ over G where $a \leq x \leq b, c \leq y \leq d$ for f continuous on G. We could also consider

$$\int_d^c \int_a^b f(x, y) \, dx \, dy,$$

where, again, the integral is to be computed from the "inside outward." List all such iterated integrals of f arising from G, and compare the values of these integrals. [*Hint:* There are eight of them.]

In Exercises 4–9, estimate the given integral by computing the midpoint Riemann sum R_P for the regular partition into n^2 subrectangles for the given value of n.

4. $\int_1^5 \int_0^2 (x + y) \, dx \, dy$ with $n = 2$

5. $\int_0^4 \int_{-1}^1 xy \, dx \, dy$ with $n = 3$

6. $\int_0^4 \int_1^3 \frac{x}{y} \, dy \, dx$ with $n = 2$

7. $\int_0^2 \int_1^5 \frac{y}{x + y} \, dy \, dx$ with $n = 2$

8. $\int_0^4 \int_1^5 \frac{xy}{x + y} \, dx \, dy$ with $n = 4$

9. $\int_1^9 \int_0^8 (x + xy) \, dy \, dx$ with $n = 4$

In Exercises 10–23, compute the given iterated integral.

10. $\displaystyle\int_1^5 \int_0^2 (x + y)\, dx\, dy$

11. $\displaystyle\int_0^4 \int_{-1}^1 xy\, dx\, dy$

12. $\displaystyle\int_0^4 \int_1^3 \frac{x}{y}\, dy\, dx$

13. $\displaystyle\int_1^9 \int_0^8 (x + xy)\, dy\, dx$

14. $\displaystyle\int_1^4 \int_0^2 (x + y^2)\, dx\, dy$

15. $\displaystyle\int_1^3 \int_{-1}^2 x^2 y\, dy\, dx$

16. $\displaystyle\int_0^\pi \int_0^{\pi/2} x \sin y\, dx\, dy$

17. $\displaystyle\int_0^2 \int_0^{\pi/2} x \sin y\, dy\, dx$

18. $\displaystyle\int_0^2 \int_0^\pi x \sin^2 y\, dy\, dx$

19. $\displaystyle\int_1^{e^2} \int_1^e \ln(xy)\, dx\, dy$

20. $\displaystyle\int_1^3 \int_0^1 xye^x\, dx\, dy$

21. $\displaystyle\int_0^1 \int_1^5 \frac{y + 1}{e^x}\, dy\, dx$

22. $\displaystyle\int_0^1 \int_0^1 \frac{y}{1 + x^2 y^2}\, dx\, dy$

23. $\displaystyle\int_0^{\sqrt{2}} \int_0^1 \frac{x}{\sqrt{4 - x^2 y^2}}\, dy\, dx$

🖩 In Exercises 24–28, estimate the iterated integral by integrating as far as you can and then using Simpson's rule with a calculator or computer to estimate the final integral.

24. $\displaystyle\int_0^2 \int_1^3 x^y\, dx\, dy$

25. $\displaystyle\int_0^{\sqrt{\pi}} \int_0^1 x^2 \cos(x^2 y)\, dy\, dx$

26. $\displaystyle\int_0^\pi \int_0^1 \frac{x \sin x}{1 + x^2 y^2}\, dy\, dx$

27. $\displaystyle\int_0^1 \int_0^2 y^2 e^{x + y^2}\, dx\, dy$

28. $\displaystyle\int_1^3 \int_1^2 (xy)^x\, dy\, dx$

🖩 In Exercises 29–32, use a calculator or computer and Simpson's rule for double integrals with the given values of m and n to estimate the given integral.

29. $\displaystyle\int_0^1 \int_0^1 e^{x^2 + y^2}\, dx\, dy$, $m = n = 2$

30. Repeat Exercise 29 for $m = n = 4$.

31. $\displaystyle\int_2^4 \int_1^2 \frac{x^4 + y^4}{1 + 2\sqrt{x} + 3\sqrt{y}}\, dx\, dy$, $m = 2$ intervals for $[1, 2]$, $n = 4$ intervals for $[2, 4]$

32. Repeat Exercise 31 using $m = 4$ intervals for $[1, 2]$ and $n = 8$ intervals for $[2, 4]$.

> **⊙ EXPLORING CALCULUS**
>
> **U:** Simpson's rule compute

15.2 DOUBLE INTEGRALS OVER MORE GENERAL REGIONS

Integrals as Limits of Sums

A region G in the plane is **bounded** if it is contained in some sufficiently large rectangle. Naively stated, bounded regions are those that do not extend to infinity in any direction. We will consider "nice" bounded regions, like the region G in Fig. 15.16, where we have a natural idea of the *boundary* (bounding curves) of the region and the *interior* (inside) of the region. A region is **closed** if the boundary is considered to be part of the region.

Let G be a closed, bounded region in the plane of the nice type just mentioned. We indicate how to regard

$$\iint_G f(x, y)\, dx\, dy$$

as a limit of sums. We need only show how to perform step 1 of the four-step outline in Section 15.1.

For step 1, we form a partition P of G into subregions of small size. Since G is bounded, we can enclose it in some rectangle with sides parallel to the coordinate axes, as shown in Fig. 15.17. We subdivide this rectangle into n^2 subrectangles precisely as in Section 15.1. The grid on the rectangle then chops the region G into small pieces, shaded red in Fig. 15.17. This completes step 1.

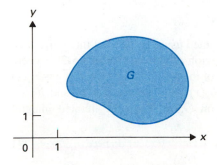

FIGURE 15.16 A nonrectangular region G.

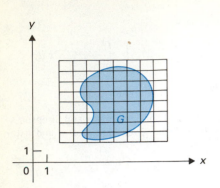

FIGURE 15.17 The grid on the rectangle ($n = 8$) gives a partition of G.

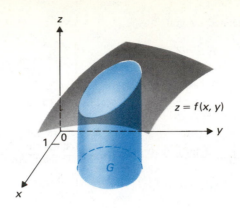

FIGURE 15.18 $\iint_G f(x, y)\, dx\, dy$ is the volume of the blue-shaded solid.

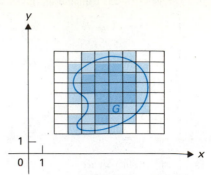

FIGURE 15.19 $\iint_G f(x, y)\, dx\, dy$ equals the limit as $n \to \infty$ of sums taken over just the blue-shaded rectangles.

For step 2, multiply the area of each subregion of G by $f(x, y)$ for some point (x, y) in the subregion. In step 3, we add these products, obtaining an approximation to $\iint_G f(x, y)\, dx\, dy$. The **definite integral** $\iint_G f(x, y)\, dx\, dy$ is defined as the limit of these approximations as $n \to \infty$, so that $\|P\| \to 0$. It is clear that for $f(x, y) \geq 0$, the integral $\iint_G f(x, y)\, dx\, dy$ is equal to the volume of the solid having base G and part of the surface $z = f(x, y)$ as top, as in Fig. 15.18.

Accurate estimation of an integral using such sums is generally not feasible, since some subregions in these partitions of G are not rectangular, and their areas may not be known. It can be shown that if the total length of the curves bounding G is finite, then the integral is actually equal to the limit of sums taken over only those subregions of G that are entire rectangles. Such subregions are shaded darkest in Fig. 15.19. Specifically, it can be shown that if the boundary has finite length, the sum of the areas of the rectangles containing portions of the boundary, shaded lightest in Fig. 15.19, approaches zero as $n \to \infty$. Thus omitting them does not affect the limit of the approximating sums. However, for small values of n that we like to use for approximations, as in Fig. 15.19, omission of these subregions can ruin the accuracy of our estimate. We will not bother to compute such sums as estimates. At the end of this section, we show how to use Simpson's rule if it is necessary to estimate integrals over regions that are not rectangular.

Iterated Integrals over Regions

The integrals just defined are usually computed using an iterated integral, much as in Section 15.1. Consider the plane region G shown in Fig. 15.20, where $a \leq x \leq b$ and where the "lower" portion of the boundary is the curve $y = h_1(x)$ and the "top" portion is the curve $y = h_2(x)$. We assume that h_1 and h_2 are continuous functions. If f is continuous on G, we form the iterated integral

$$\int_a^b \int_{h_1(x)}^{h_2(x)} f(x, y)\, dy\, dx = \int_a^b \left(\int_{h_1(x)}^{h_2(x)} f(x, y)\, dy \right) dx. \tag{1}$$

The computation of Eq. (1) is again to be made from the "inside outward," as

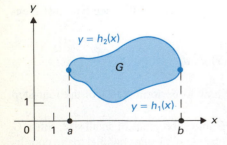

FIGURE 15.20 Upper and lower boundary curves of G.

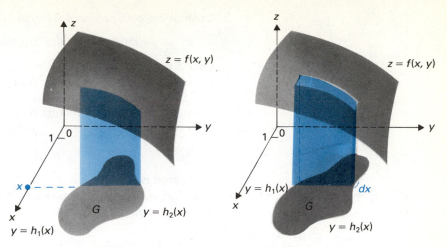

FIGURE 15.21 $\int_{h_1(x)}^{h_2(x)} f(x,\ y)\ dy$ is the area of the color-shaded surface.

FIGURE 15.22 $\int_{h_1(x)}^{h_2(x)} f(x,\ y)\ dy$ is the volume of the slab.

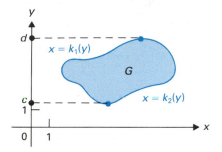

FIGURE 15.23 Left and right boundary curves of G.

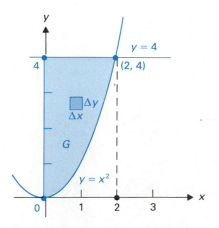

FIGURE 15.24 The region G in the first quadrant bounded by $y = x^2$, $y = 4$, and the y-axis.

indicated by the parentheses. The first integration is performed with respect to y only, and the first limits depend on x. For a particular value x, the integral

$$\int_{h_1(x)}^{h_2(x)} f(x_i,\ y)\ dy$$

gives us the area of the vertical color-shaded surface in Fig. 15.21. Heuristically, this area is then multiplied by dy, giving the volume of the slab shaded in color in Fig. 15.22. Integrating from a to b gives us the volume of the solid with base G and top a portion of the graph of $z = f(x,\ y)$, like the solid shown in Fig. 15.18. This same volume is given by $\iint_G f(x,\ y)\ dx\ dy$. This suggests the "fundamental theorem for computation"

$$\iint_G f(x,\ y)\ dx\ dy = \int_a^b \int_{h_1(x)}^{h_2(x)} f(x,\ y)\ dy\ dx, \qquad (2)$$

which is the analogue of Theorem 15.1 in the last section, but for more general regions that a rectangle.

We could also integrate in the reverse order, in which case we consider the "left" and "right" boundary curves of G to be given by $x = k_1(y)$ and $x = k_2(y)$ for $c \le y \le d$, as shown in Fig. 15.23. Our integral then takes the form

$$\int_c^d \int_{k_1(y)}^{k_2(y)} f(x,\ y)\ dx\ dy = \int_c^d \left(\int_{k_1(y)}^{k_2(y)} f(x,\ y)\ dx \right) dy.$$

In forming an iterated integral to compute $\iint_G f(x,\ y)\ dx\ dy$, *draw a sketch of the region G*. The appropriate limits for the iterated integral are found by studying the sketch.

EXAMPLE 1 Let G be the plane region in the first quadrant bounded by $y = x^2$, $y = 4$, and the y-axis. Find $\iint_G x^2 y\ dx\ dy$.

Solution The region G is shown in Fig. 15.24, where we also draw a rectangle of area $\Delta A = \Delta x \Delta y$. We show how to find and compute appropriate iterated

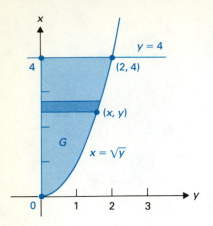

FIGURE 15.25 $\left(\int_0^{\sqrt{y}} x^2y \, dx\right) \Delta y$ sums the contributions $x^2y \, \Delta x \, \Delta y$ on this strip.

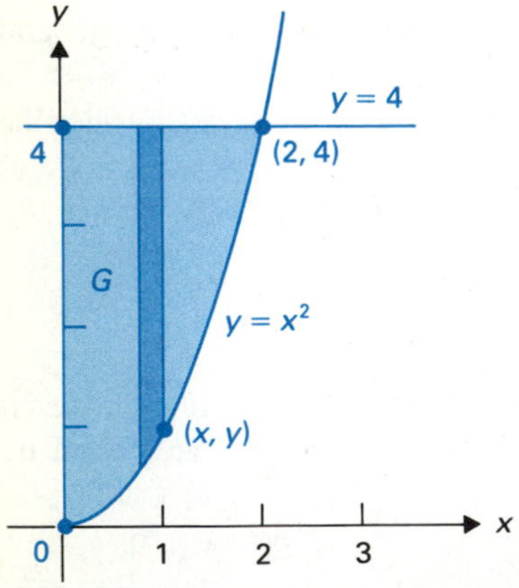

integrals for both orders of integration. In both cases,

we think of summing contributions $f(x, y) \, \Delta x \Delta y$ over G.

Case 1 x,y-order If we integrate first with respect to x, holding y constant, we are adding contributions $x^2y \, \Delta x \, \Delta y$ in the x-direction, along the strip shown in Fig. 15.25. In terms of its y-height, the left end of the strip has x-coordinate zero and the right end has x-coordinate $x = \sqrt{y}$. Thus we arrive at

$$\left(\int_0^{\sqrt{y}} x^2y \, dx\right) \Delta y$$

as the total contribution along the strip in Fig. 15.25. We then sum such strip contributions in the y-direction to sweep out the entire region G. The smallest y-value is 0 and the largest is 4. Thus we obtain the iterated integral

$$\int_0^4 \int_0^{\sqrt{y}} x^2y \, dx \, dy = \int_0^4 \frac{x^3}{3}y \Big|_{x=0}^{\sqrt{y}} dy$$

$$= \int_0^4 \frac{y^{5/2}}{3} \, dy = \frac{1}{3} \cdot \frac{2}{7}y^{7/2}\Big]_0^4 = \frac{2}{21}4^{7/2} = \frac{256}{21}.$$

Case 2 y,x-order Integrating first with respect to y, we sum contributions $x^2y \, \Delta x \Delta y$ along the vertical strip shown in Fig. 15.26. In terms of the left–right position x of the strip, the y-value at the bottom is $y = x^2$ and the y-value at the top is $y = 4$. We are led to

$$\left(\int_{x^2}^4 x^2y \, dy\right) \Delta x$$

We then add the contributions of these strips in the x-direction. Since the smallest x-value is 0 and the largest is 2, we obtain the integral

$$\int_0^2 \int_{x^2}^4 x^2y \, dy \, dx = \int_0^2 \left(\int_{x^2}^4 x^2y \, dy\right) dx = \int_0^2 x^2\frac{y^2}{2}\Big]_{y=x^2}^{y=4} dx$$

$$= \int_0^2 \left(8x^2 - \frac{1}{2}x^6\right) dx$$

$$= \left(\frac{8}{3}x^3 - \frac{1}{14}x^7\right)\Big]_0^2$$

$$= \frac{64}{3} - \frac{64}{7} = \frac{256}{21}. \qquad \blacksquare$$

When we are evaluating $\iint_G f(x, y) \, dx \, dy$ using an iterated integral, one order of integration might be impossible, while the other might be more straightforward. We give an example.

EXAMPLE 2 Let G be the region bounded by $y = x$, $x = 1$, and the x-axis. Find $\iint_G e^{x^2} \, dx \, dy$.

Solution The region G is shown in Fig. 15.27. If we integrate first in the x-direction, over the horizontal strip in Fig. 15.28, then an argument analogous

FIGURE 15.26 $\left(\int_{x^2}^4 x^2y \, dy\right) \Delta x$ sums the contributions $x^2y \, \Delta y \, \Delta x$ over this strip.

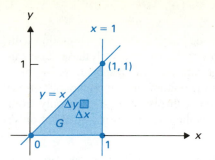

FIGURE 15.27 Area $\Delta A = \Delta x\,\Delta y$ in G.

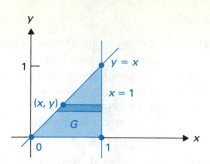

FIGURE 15.28 $\left(\int_y^1 e^{x^2}\,dx\right)\Delta y$ sums $e^{x^2}\,\Delta x\,\Delta y$ over this strip.

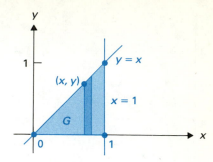

FIGURE 15.29 $\left(\int_0^x e^{x^2}\,dy\right)\Delta x$ sums $e^{x^2}\,\Delta x\,\Delta y$ over this strip.

to that in Example 1 shows that the appropriate iterated integral is

$$\int_0^1 \int_y^1 e^{x^2}\,dx\,dy.$$

However, we can't find $\int e^{x^2}\,dx$.

The other order of integration, with respect to y first, sums along the vertical strip in Fig. 15.29. This leads to the integral

$$\int_0^1 \int_0^x e^{x^2}\,dy\,dx = \int_0^1 ye^{x^2}\Big]_{y=0}^x\,dx = \int_0^1 xe^{x^2}\,dx = \frac{1}{2}e^{x^2}\Big]_0^1 = \frac{1}{2}(e-1). \quad \blacksquare$$

Changing the Order of Integration

It is important to be able to change the order of integration in a given iterated integral. An outline of a procedure by which this may be done for double integrals is as follows.

STEP 1. *Draw a sketch* and shade the region of the plane over which integration takes place. Draw a little rectangle of dimensions Δx by Δy in the shaded region.

STEP 2. Convert your given iterated integral into one with the order of integration reversed *by looking at your sketch* and writing down the appropriate limits. The limits on the inside integral sign are functions of the remaining variable of integration, while the limits on the outside integral sign are always constants.

EXAMPLE 3 Reverse the order of integration for the integral

$$\int_0^2 \int_{x^3}^{4x} x^2 y\,dy\,dx.$$

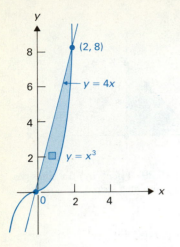

FIGURE 15.30   Area $\Delta A = \Delta x \, \Delta y$.

Solution

STEP 1.   Starting with the limits on the inside integral, the first integration with respect to y goes from $y = x^3$ to $y = 4x$, so we sketch these curves in Fig. 15.30. Since this first integration was with respect to y, we think of $y = x^3$ and $y = 4x$ as forming the bottom and top boundaries of our region. Now the final integration with respect to x goes only from $x = 0$ to $x = 2$. Thus the region is the one in the first quadrant only, shaded in Fig. 15.30.

STEP 2.   To reverse the order of integration, we wish to integrate first with respect to x. Pushing our little rectangle as far to the left (negative x-direction) as it will go, it is always stopped by the line $y = 4x$. Since these inside limits must be expressed as functions of y, we write $y = x^3$ as $x = \sqrt[3]{y}$ and $y = 4x$ as $x = y/4$. This leads to

$$\left(\int_{y/4}^{\sqrt[3]{y}} x^2 y \, dx \right) \Delta y.$$

This integral corresponds to adding the contributions $x^2 y \, \Delta x \, \Delta y$ over the horizontal strip shown in Fig. 15.31. Now we must add the contributions of these strips from a strip at the bottom of the region, where $y = 0$, to a strip at the top, where $y = 8$. Thus we arrive at

$$\int_0^8 \int_{y/4}^{\sqrt[3]{y}} x^2 y \, dx \, dy$$

as the desired integral.   ∎

EXAMPLE 4   Reverse the order of integration for the integral

$$\int_{-3}^3 \int_0^{2\sqrt{1 - y^2/9}} \cos xy^2 \, dx \, dy.$$

Solution

STEP 1.   Taking the limits of the inside integral, we see that x goes from $x = 0$ (the y-axis) to $x = 2\sqrt{1 - y^2/9}$. Now $x = 2\sqrt{1 - y^2/9}$ is the right-hand portion of the curve $x^2 = 4(1 - y^2/9)$, or $x^2/4 = 1 - y^2/9$, which is the ellipse

$$\frac{x^2}{4} + \frac{y^2}{9} = 1.$$

Thus the region of integration is the right-hand half of the region bounded by this ellipse, shown in Fig. 15.32.

STEP 2.   We want to integrate first with respect to y, letting our little rectangle sweep out the vertical strip shown in Fig. 15.32. Solving for y in the equation of the ellipse, we obtain $y = \pm 3\sqrt{1 - x^2/4}$, so we are led to the inside integral

$$\int_{-3\sqrt{1 - x^2/4}}^{3\sqrt{1 - x^2/4}} \cos xy^2 \, dy.$$

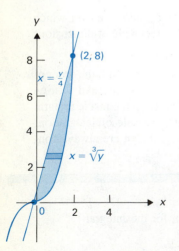

FIGURE 15.31 $(\int_{y/4}^{\sqrt[3]{y}} x^2 y \, dx) \, \Delta y$ sums $x^2 y \, \Delta x \, \Delta y$ over this strip.

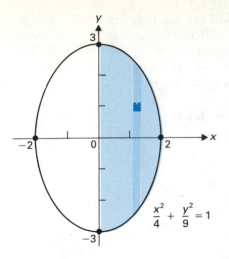

FIGURE 15.32

$$\left[\int_{-3\sqrt{1-x^2/4}}^{3\sqrt{1-x^2/4}} \cos xy^2 \, dy \right] \Delta x$$

sums $\cos x^2 y \, \Delta y \, \Delta x$ over this strip.

Then we add these contributions of the vertical strips from $x = 0$ to $x = 2$, arriving at

$$\int_0^2 \int_{-3\sqrt{1-x^2/4}}^{3\sqrt{1-x^2/4}} \cos xy^2 \, dy \, dx. \qquad \blacksquare$$

Areas and Volumes by Double Integration

Let G be a closed, bounded region in the plane. It is geometrically clear that for the constant function 1, the integral

$$\iint_G 1 \, dx \, dy$$

gives the area of the region. The integral is usually computed using an iterated integral. We give two examples.

EXAMPLE 5   We solve, using "double integrals," an area problem we could easily have solved in Chapter 6. Find the area of the plane region in the first quadrant of the plane bounded by the curves $y = x^3$ and $y = \sqrt{x}$.

Solution   The region is sketched in Fig. 15.33. Using Leibniz differential notation, we think of $dx \, dy = dy \, dx$ as the area of a small rectangle in the region and use an integral to add up the areas of such rectangles over the region as $dx \to 0$ and $dy \to 0$. The "lower" and "upper" bounding curves of the region are $y = x^3$ and $y = \sqrt{x}$, respectively, so our iterated integral is

$$\int_0^1 \int_{x^3}^{\sqrt{x}} dy \, dx = \int_0^1 \left(\int_{x^3}^{\sqrt{x}} dy \right) dx = \int_0^1 y \Big]_{x^3}^{\sqrt{x}} dx = \int_0^1 (\sqrt{x} - x^3) \, dx$$

$$= \left(\frac{2}{3} x^{3/2} - \frac{x^4}{4} \right) \Big]_0^1 = \frac{2}{3} - \frac{1}{4} = \frac{5}{12}.$$

Note that the integral $\int_0^1 (\sqrt{x} - x^3) \, dx$ that occurs in the middle of the computation above is the one we would have started with in Chapter 6 to compute the area.

If we wish to compute the iterated integral in the other order, we must find the "left" and "right" bounding curves of our region, and we obtain $x = y^2$ and $x = y^{1/3}$, respectively. The iterated integral in this order is therefore

$$\int_0^1 \int_{y^2}^{y^{1/3}} dx \, dy = \int_0^1 x \Big]_{y^2}^{y^{1/3}} dy = \int_0^1 (y^{1/3} - y^2) \, dy$$

$$= \left(\frac{3}{4} y^{4/3} - \frac{y^3}{3} \right) \Big]_0^1 = \frac{3}{4} - \frac{1}{3} = \frac{5}{12}.$$

Of course, we obtained the same answer. $\blacksquare$

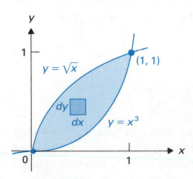

FIGURE 15.33   Differential area $dA = dx \, dy$.

EXAMPLE 6   Use double integration to find the volume of the region in space bounded above by the surface $z = 1 - x^2 - y^2$, on the sides by the planes $x = 0$, $y = 0$, $x + y = 1$, and below by the plane $z = 0$. The region is sketched in Fig. 15.34.

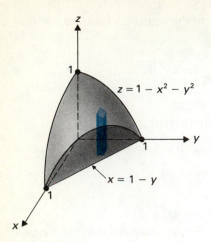

FIGURE 15.34 Column of "differential" volume

$$dV = (1 - x^2 - y^2)\, dx\, dy.$$

Solution We think of adding volumes of columns of height $z = 1 - x^2 - y^2$ and with rectangular cross sections of "differential" area $dx\, dy$ as shown in Fig. 15.34. This "differential" volume is thus $dV = (1 - x^2 - y^2)\, dx\, dy$, and we sum these contributions to the volume over the triangular region in the x,y-plane that forms the base of the solid in space. We obtain

$$= \int_0^1 \int_0^{1-y} (1 - x^2 - y^2)\, dx\, dy$$

$$= \int_0^1 \left(x - \frac{x^3}{3} - y^2 x \right) \Bigg]_0^{1-y} dy$$

$$= \int_0^1 \left((1 - y) - \frac{(1-y)^3}{3} - y^2(1 - y) \right) dy$$

$$= \int_0^1 \left(1 - y - y^2 + y^3 - \frac{(1-y)^3}{3} \right) dy$$

$$= \left(y - \frac{y^2}{2} - \frac{y^3}{3} + \frac{y^4}{4} + \frac{(1-y)^4}{12} \right) \Bigg]_0^1$$

$$= 1 - \frac{1}{2} - \frac{1}{3} + \frac{1}{4} - \frac{1}{12}$$

$$= 1 - \frac{6 + 4 - 3 + 1}{12} = 1 - \frac{8}{12} = \frac{1}{3}. \quad \blacksquare$$

Simpson's Rule (Optional)

We give an example that indicates how to estimate a double iterated integral if we can't perform even the first integration. An analogous method can be used for triple integrals.

EXAMPLE 7 Use a calculator or computer to estimate

$$\int_0^1 \int_0^{\sqrt{y}} \sin(\sqrt{xy})\, dx\, dy.$$

Solution The region of integration is sketched in Fig. 15.35. We decide to use Simpson's rule first in the x-direction with $m = 4$ and then in the y-direction with $n = 4$. We have labeled the points on the y-axis corresponding to partitioning the interval $[0, 1]$ into $n = 4$ equal subintervals. For each of these five values

$$y_i = 0,\ 0.25,\ 0.5,\ 0.75,\ 1,$$

we estimate the integral

$$\int_0^{\sqrt{y_i}} \sin(\sqrt{xy_i})\, dx$$

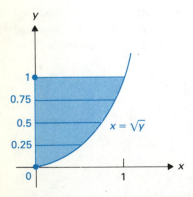

FIGURE 15.35 Table 15.2 contains

$$\int_0^{\sqrt{y_i}} \sin\sqrt{xy_i}\, dx$$

across horizontal segments.

using Simpson's rule with $m = 4$. The results are shown in Table 15.2. We then use Simpson's rule with $n = 4$ to "integrate" the entries in the second column of the table. That is, we multiply each entry by the corresponding Simpson's coefficient in the third column, add these products, and multiply the sum by

$$\frac{1 - 0}{3n} = \frac{1}{12}.$$

TABLE 15.2

y_i	$\displaystyle\int_0^{\sqrt{y_i}} \sin(\sqrt{xy_i})\, dx$	Simpson's coefficients
0	0	1
0.25	0.1145903722	4
0.5	0.2662387945	2
0.75	0.4286623777	4
1.0	0.5921239119	1

The result is 0.2748010417, which is our estimate for the integral. We also worked it out on our calculator for $m = n = 8$ and obtained 0.2774440753. The reason this technique works was indicated in the final part of Section 15.1. ∎

For iterated integrals where all but the final integration can be performed, we use Simpson's rule at that last stage to estimate the integral.

EXAMPLE 8   Estimate $\int_0^1 \int_{x^2}^x e^{x^2}\, dy\, dx$.

Solution   We have

$$\int_0^1 \int_{x^2}^x e^{x^2}\, dy\, dx = \int_0^1 \left. y e^{x^2} \right]_{x^2}^x dx = \int_0^1 (x e^{x^2} - x^2 e^{x^2})\, dx$$

$$= \frac{1}{2} e^{x^2} \Big]_0^1 - \int_0^1 x^2 e^{x^2}\, dx = \frac{1}{2}(e - 1) - \int_0^1 x^2 e^{x^2}\, dx.$$

Simpson's rule with $n = 40$ shows that

$$\int_0^1 x^2 e^{x^2}\, dx \approx 0.6278154418.$$

Thus we obtain the estimate

$$\frac{1}{2}(e - 1) - 0.6278154418 \approx 0.2313254725$$

for the integral. ∎

SUMMARY

1. A bounded plane region G is one that can be enclosed in a rectangle. The region is closed if the region includes its boundary.

2. Let a rectangle containing G be partitioned into n^2 subrectangles of equal size, as in Section 15.1. This rectangular grid partitions G also. Sums approximating $\iint_G f(x, y)\, dx\, dy$ are defined as in

Section 15.1. The integral is defined as the limit as $n \to \infty$ of these approximating sums. For a nonrectangular region G, these sums may be hard to actually compute.

3. *Fundamental theorem:* Iterated integrals are used for computation of an integral over a region.

4. Draw a sketch and work from that sketch when changing the order of integration.

5. The area of a region G is found by integrating the constant function 1 over G.

6. Simpson's rule for estimating double integrals over nonrectangular regions is illustrated in Example 7.

EXERCISES 15.2

In Exercises 1–6, compute the given iterated integral.

1. $\displaystyle\int_0^1 \int_0^{\sqrt{1-y^2}} 4xy \, dx \, dy$

2. $\displaystyle\int_0^\pi \int_0^{\sin x} x \, dy \, dx$

3. $\displaystyle\int_1^e \int_0^{\ln x} \frac{y}{x} \, dy \, dx$

4. $\displaystyle\int_9^1 \int_{-\sqrt{y}}^{y} (x + y^2) \, dx \, dy$

5. $\displaystyle\int_0^2 \int_{1-x}^{1+x} x^2 \, dy \, dx$

6. $\displaystyle\int_0^{\sqrt{2}} \int_y^{\sqrt{4-y^2}} y \, dx \, dy$

In Exercises 7–13, find the integral of the given function f over the indicated region G.

7. $f(x, y) = y$; G the plane region bounded by $y = 1 - x^2$ and the x-axis

8. $f(x, y) = xy$; G the quarter-disk $x^2 + y^2 \le a^2$ in the first quadrant

9. $f(x, y) = x$; G the plane region bounded by $y = x^2$ and $y = x + 2$

10. $f(x, y) = e^{x^2}$; G the plane region bounded by $y = x$, $y = -x$, and $x = 4$

11. $f(x, y) = \sin y^2$; G the plane region bounded by $y = x$, $y = \sqrt{\pi}$, and $x = 0$

12. $f(x, y) = y \sin^2 x^2$; G the plane region in the first quadrant bounded by $x = y^2$, $x = \sqrt{\pi}$, and $y = 0$

13. $f(x, y) = ye^{x^2}$; G the plane region bounded by $y = x^{3/2}$, $y = 0$, and $x = 1$

In Exercises 14–23, sketch the region of integration for the given iterated integral and then write an equal iterated integral (or an equal sum of iterated integrals) with the integration in the "reverse order." (You are not asked to compute the integral.)

14. $\displaystyle\int_0^1 \int_{-\sqrt{1-y^2}}^{\sqrt{1-y^2}} 4xy \, dx \, dy$

15. $\displaystyle\int_0^2 \int_0^{x^2} (\sin xy) \, dy \, dx$

16. $\displaystyle\int_0^1 \int_1^{e^y} x^2 \, dx \, dy$

17. $\displaystyle\int_0^1 \int_{x^2}^{x} y \cos x \, dy \, dx$

18. $\displaystyle\int_{-3}^1 \int_{-\sqrt{1-y}}^{\sqrt{1-y}} e^{xy} \, dx \, dy$

19. $\displaystyle\int_0^1 \int_{2y-2}^{1-y} x \cos^2 y \, dx \, dy$

20. $\displaystyle\int_{-1}^1 \int_{1-y^2}^{y^2-1} x^2 y^3 \, dx \, dy$

21. $\displaystyle\int_0^4 \int_{-x}^{x} x^2 y^2 \, dy \, dx$

22. $\displaystyle\int_0^{\pi/2} \int_0^{(\pi/4)(\sin x)} (x + y) \, dy \, dx$

23. $\displaystyle\int_0^\pi \int_0^{\sin x} x^2 y \, dy \, dx$

In Exercises 24–31, use a double integral to find the area or volume of the indicated region.

24. The region of the plane bounded by the curves $y = 0$, $y = 1 + x$, and $y = \sqrt{1 - x}$

25. The region of the plane bounded by the curves $y = \sin x$, $x = \pi/2$, and $y = x$

26. The region of the plane bounded by $y = \ln x$, $y = 1 - x$, and $y = 1$

27. The region of the plane bounded by $y = e^x$, $x + y = e + 1$, and $x = 0$

28. The region of space bounded by $z = 2x^2$, $z = 8$, $y = 0$, and $y = 4$

29. The region of space bounded by $z = \cos x$ for $-\pi/2 \le x \le \pi/2$, $z = 0$, $y = -1$, and $y = 2$

30. The region of space bounded by $z = 4 - x^2 - y^2$ and $z = x^2 + y^2 - 4$

31. The region of space bounded by $z = 1 + x^2 + y^2$, $y = 1 - x^2$, $y = 0$, and $z = 0$

In Exercises 32–35, carry the integration as far as you can and then use a calculator or computer and Simpson's rule to estimate the integral of the given function f over the indicated region.

32. $f(x, y) = (x + 1)^y$ over the plane region bounded by $y = 0$, $y = x^2$, and $x = 1$

33. $f(x, y) = e^{-y-x^2}$ over the plane region bounded by $x = 0$, $y = 0$, and $x + y = 1$

34. $f(x, y) = \sin(xy^2)$ over the plane region bounded by $x = 1 - y^2$ and $x = 0$

35. $f(x, y) = (100x^3y^2)/(1 + x^4y^2)$ over the quarter-disk $x^2 + y^2 = 1$ in the first quadrant.

In Exercises 36–39, use Simpson's rule, illustrated in Example 7, to estimate the integral, taking $m = n = 4$.

36. $\displaystyle\int_0^1 \int_0^{\sqrt{1-y^2}} e^{-x^2-y^2} \, dx \, dy$

37. $\displaystyle\int_0^{0.4} \int_0^{0.4-y} [\ln(3 + x)]^y \, dx \, dy$

38. $\displaystyle\int_0^1 \int_0^x \sin\left(\frac{1 + x^2}{1 + y^2}\right) dy \, dx$

39. $\displaystyle\int_0^1 \int_{x^3}^x \left(\frac{1 + x}{1 + y}\right)^{xy} dy \, dx$

15.3 DOUBLE INTEGRATION IN POLAR COORDINATES

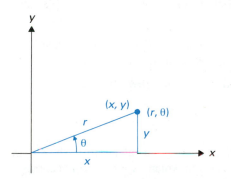

FIGURE 15.36 $x = r \cos \theta$, $y = r \sin \theta$.

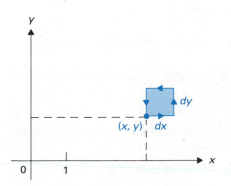

FIGURE 15.37   In Cartesian coordinates, $dA = (dx)(dy)$.

Consider polar r,θ-coordinates in the plane. Recall that the transformation from polar r,θ-coordinates to rectangular x,y-coordinates is given by

$$x = r \cos \theta,$$
$$y = r \sin \theta, \tag{1}$$

as indicated in Fig. 15.36.

Suppose we wish to integrate a continuous function over a region G in the plane. We may express our function either in terms of the independent variables x and y or in terms of r and θ. If the function is expressed by $f(x, y)$ in terms of x and y, then in terms of r,θ-coordinates it is

$$h(r, \theta) = f(r \cos \theta, r \sin \theta).$$

We develop double integration in polar coordinates using the heuristic Leibniz differential notation. The differential area element, $dA = dx \, dy$, should be replaced by one in polar coordinates. Now the rectangular area element, $dx \, dy$, is obtained by starting at a point in the region, increasing x by dx, increasing y by dy, then decreasing x by dx, and finally decreasing y by dy. We show this in Fig. 15.37, where the arrows indicate the way we generate this Cartesian differential area element, starting at (x, y).

Figure 15.38 shows the generation of the polar differential area element in the same fashion. We start at a point (r, θ), increase r by dr, increase θ by $d\theta$, decrease r by dr, and finally decrease θ by $d\theta$.

If we pretend that the shaded differential area element in Fig. 15.38 is a rectangle, then its area becomes $(r \, d\theta) \, dr$, and we have

$$dA = r \, dr \, d\theta \tag{2}$$

for polar coordinates. *It is important to note that dA is not just dr dθ.*

We have attempted to make it seem reasonable that the integral of $f(x, y)$

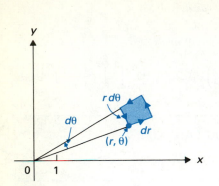

FIGURE 15.38   In polar coordinates, $dA = (r \, d\theta)(dr)$.

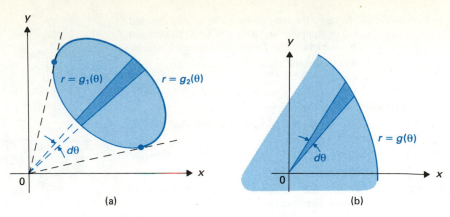

FIGURE 15.39   (a) $(\int_{g_1(\theta)}^{g_2(\theta)} h(r, \theta)r \, dr) \, d\theta$ sums contributions over this tapered strip; (b) $(\int_0^{g(\theta)} h(r, \theta)r \, dr) \, d\theta$ sums contributions over this wedge.

over a region G is given by

$$\iint_G f(x, y) \, dx \, dy = \iint_G f(r \cos \theta, r \sin \theta) \, r \, dr \, d\theta \qquad (3)$$

in polar coordinates. A careful proof is left to a more advanced course. (See also Example 2 in Section 15.9.)

When finding the appropriate polar coordinate limits for an iterated integral over a region G, it is important to draw a sketch of G.

Suppose we integrate in the r,θ-order. Then the first integral

$$\left(\int_{g_1(\theta)}^{g_2(\theta)} h(r, \theta)r \, dr \right) d\theta$$

adds up contributions $h(r, \theta)r \, dr \, d\theta$ over the tapered strip shown in Fig. 15.39(a). Often, the lower limit is zero, in which case the strip becomes a wedge from the origin, shown in Fig. 15.39(b).

Suppose, on the other hand, we integrate in the θ,r-order. Then the first integral

$$\left(\int_{k_1(r)}^{k_2(r)} h(r, \theta)r \, d\theta \right) dr$$

gives the curved strip in Fig. 15.40. In practice, we almost always integrate in the r,θ-order.

Recall that integration of the constant function 1 over a region gives the area of the region.

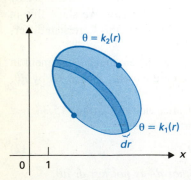

FIGURE 15.40 $(\int_{k_1(r)}^{k_2(r)} h(r, \theta)r \, d\theta) \, dr$ sums contributions over this curved strip.

EXAMPLE 1   Find the area of the region bounded by the first loop of the spiral $r = \theta$, for $0 \le \theta \le 2\pi$, and the positive x-axis.

Solution   The region is sketched in Fig. 15.41. From the figure, we obtain the integral

$$\int_0^{2\pi} \int_0^{\theta} 1 \cdot r \, dr \, d\theta = \int_0^{2\pi} \frac{r^2}{2} \Big|_0^{\theta} \, d\theta = \int_0^{2\pi} \frac{\theta^2}{2} \, d\theta = \frac{\theta^3}{6} \Big]_0^{2\pi} = \frac{8\pi^3}{6} = \frac{4}{3}\pi^3.$$

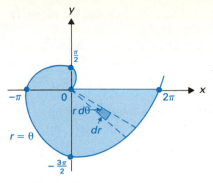

FIGURE 15.41

$$dA = r \, dr \, d\theta; \quad A = \int_0^{2\pi} \int_0^\theta r \, dr \, d\theta$$

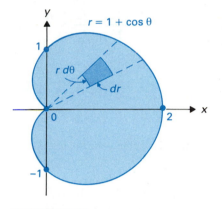

FIGURE 15.42

$$dA = r \, dr \, d\theta; \quad A = \int_0^{2\pi} \int_0^{1+\cos\theta} r \, dr \, d\theta$$

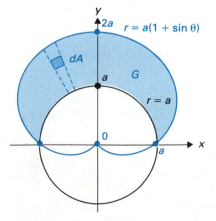

FIGURE 15.43

$$\iint_G h(r, \theta) r \, dr \, d\theta$$

$$= \int_0^\pi \int_a^{a(1+\sin\theta)} h(r, \theta) r \, dr \, d\theta$$

Note that the integral $\int_0^{2\pi} \frac{1}{2}\theta^2 \, d\theta$ in the computation is the one we would have started with in Chapter 11 to find this area. ∎

EXAMPLE 2 Find the area of the region G bounded by the cardioid $r = 1 + \cos\theta$, shown in Fig. 15.42, by integrating the constant function 1 over this region.

Solution By symmetry, we can find the area of the upper half of the region and double it for our answer. From Fig. 15.42, we see that the area is given by

$$2\int_0^\pi \int_0^{1+\cos\theta} 1 \cdot r \, dr \, d\theta = 2\int_0^\pi \frac{1}{2}r^2\bigg]_0^{1+\cos\theta} d\theta = 2\int_0^\pi \frac{1}{2}(1 + \cos\theta)^2 \, d\theta$$

$$= \int_0^\pi (1 + 2\cos\theta + \cos^2\theta) \, d\theta$$

$$= \left(\theta + 2\sin\theta + \frac{\theta}{2} + \frac{\sin 2\theta}{4}\right)\bigg]_0^\pi$$

$$= \pi + \frac{\pi}{2} = \frac{3\pi}{2}.$$

Note that the integral $2\int_0^\pi \frac{1}{2}(1 + \cos\theta)^2 \, d\theta$ in our computation is the one we would have started with in Chapter 11 to find the area. ∎

EXAMPLE 3 Integrate the polar function $h(r, \theta) = 1/r$ over the region inside the cardioid $r = a(1 + \sin\theta)$ and outside the circle $r = a$.

Solution The region of integration is shown in Fig. 15.43. From the figure, we obtain the integral

$$\int_0^\pi \int_a^{a(1+\sin\theta)} \frac{1}{r} \cdot r \, dr \, d\theta = \int_0^\pi \int_a^{a(1+\sin\theta)} 1 \, dr \, d\theta$$

$$= \int_0^\pi r\bigg]_a^{a(1+\sin\theta)} d\theta = \int_0^\pi [a(1 + \sin\theta) - a] \, d\theta$$

$$= \int_0^\pi a\sin\theta \, d\theta = a(-\cos\theta)\bigg]_0^\pi = a(1) - a(-1) = 2a.$$ ∎

We should also learn to change a double integral from rectangular to polar coordinates. Remember that the differential area $dA = dx \, dy$ must be changed to $r \, dr \, d\theta$. If we are integrating a function $f(x, y)$ over a region, we replace x by $r\cos\theta$ and replace y by $r\sin\theta$. Step 1 for changing variables given in Section 15.2 is still very important: *draw a sketch*.

EXAMPLE 4 Change

$$\int_{-2}^2 \int_0^{\sqrt{4-x^2}} x^2 y \, dy \, dx$$

to polar coordinates, and evaluate the resulting integral.

Solution We start by drawing a sketch of the region of integration. The curve $y = \sqrt{4 - x^2}$ is of course the top half of the circle $x^2 + y^2 = 4$, and since the

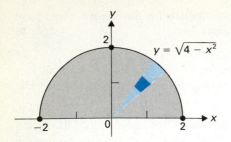

FIGURE 15.44 $\int_0^\pi \int_0^2$ ———— $r\, dr\, d\theta$ sweeps out this half-disk.

y-limit starts at 0 and the *x*-limits run from -2 to 2, we see that the region of integration is the upper half-disk shown in Fig. 15.44.

The limits of integration over a half-disk are constant when using polar coordinates. Replacing $dy\, dx$ by $r\, dr\, d\theta$ and x by $r \cos \theta$ and y by $r \sin \theta$, we see that

$$\int_{-2}^{2} \int_{0}^{\sqrt{4-x^2}} x^2 y \, dy \, dx = \int_{0}^{\pi} \int_{0}^{2} (r \cos \theta)^2 (r \sin \theta) \, r \, dr \, d\theta$$

$$= \int_{0}^{\pi} \int_{0}^{2} r^4 \cos^2 \theta \sin \theta \, dr \, d\theta$$

$$= \int_{0}^{\pi} \frac{r^5}{5} \bigg]_{0}^{2} \cos^2 \theta \sin \theta \, d\theta$$

$$= \int_{0}^{\pi} \frac{32}{5} \cos^2 \theta \sin \theta \, d\theta = -\frac{32}{5} \frac{\cos^3 \theta}{3} \bigg]_{0}^{\pi}$$

$$= -\frac{32}{5}\left(-\frac{1}{3}\right) - \left(-\frac{32}{5}\right)\frac{1}{3} = \frac{32}{15} + \frac{32}{15} = \frac{64}{15}. \quad \blacksquare$$

SUMMARY

1. The area of the differential area element in polar coordinates is

$$dA = r \cdot dr \, d\theta.$$

2. An integral $\iint f(x, y) \, dx \, dy$ with appropriate x,y-limits becomes in polar coordinates $\iint f(r \cos \theta, r \sin \theta) \cdot r \, dr \, d\theta$ with appropriate polar limits for the region.

EXERCISES 15.3

1. An integral in polar coordinates is of the form

$$\int_{\alpha}^{\beta} \int_{a}^{b} h(r, \theta) r \, dr \, d\theta$$

for $0 \le \alpha < \theta \le 2\pi$ and $0 \le a \le b$. Sketch the region of integration.

In Exercises 2–10, find the area of the plane region, using double integration in polar coordinates.

2. The disk $x^2 + y^2 \le a^2$

3. The annular ring $a^2 \le x^2 + y^2 \le b^2$ for $0 \le a \le b$

4. The region bounded by the positive x-axis for $2\pi a \le x \le 4\pi a$ and the portion of the spiral $r = a\theta$ for $2\pi \le \theta \le 4\pi$

5. The region inside the cardioid $r = a(1 + \cos \theta)$ and outside the circle $r = a$

6. The region inside one loop of the four-leaved rose $r = a \sin 2\theta$

7. The region in the first quadrant bounded by $x^2 + y^2 = a^2$, $y = 0$, and $x = a/2$

8. The region inside the larger loop and outside the smaller loop of the limaçon $r = a(1 + 2 \cos \theta)$

9. The region inside the circle $r = a \cos \theta$ and outside the cardioid $r = a(1 - \cos \theta)$

10. The total region inside the lemniscate $r^2 = 2a^2 \cos 2\theta$ and outside the circle $r = a$

11. Find the integral of the polar function $h(r, \theta) = r \sin^2\theta$, $r \geq 0$, over the closed disk bounded by $r = a$.

12. Find the integral of the polar function $h(r, \theta) = \cos \theta$, $r \geq 0$, over the region bounded by the cardioid $r = a(1 + \sin \theta)$.

In Exercises 13–20, change the integral to polar coordinates and then evaluate the transformed version.

13. $\displaystyle\int_0^a \int_0^{\sqrt{a^2-y^2}} xy \; dx \; dy$

14. $\displaystyle\int_{-2}^2 \int_{-\sqrt{4-y^2}}^{\sqrt{4-y^2}} (x^2 + y^2) \; dx \; dy$

15. $\displaystyle\int_{-1}^1 \int_{-\sqrt{1-y^2}}^{\sqrt{1-y^2}} e^{-x^2-y^2} \; dx \; dy$

16. $\displaystyle\int_0^2 \int_{-\sqrt{4-x^2}}^{\sqrt{4-x^2}} \frac{1}{1 + x^2 + y^2} \; dy \; dx$

17. $\displaystyle\int_0^{3/\sqrt{2}} \int_y^{\sqrt{9-y^2}} y \; dx \; dy$

18. $\displaystyle\int_{-2}^2 \int_2^{\sqrt{8-x^2}} \frac{1}{y} \; dy \; dx$

19. $\displaystyle\int_0^4 \int_0^x (x^2 + y^2) \; dy \; dx$

20. $\displaystyle\int_0^2 \int_{-y}^y (x^2 + y^2) \; dx \; dy$

15.4 SURFACE AREA

Let $z = f(x, y)$ be a function of two variables with continuous partial derivatives. Let G be a closed bounded region in the domain of f. The graph of f over G is then a surface in space, as indicated in Fig. 15.45. We attempt to find the area of this surface, continuing to use the heuristic differential notation.

The situation is analogous to finding the length of a curve lying over an interval $[a, b]$ on the x-axis, as shown in Fig. 15.46. In that case, we use a short tangent line segment of length ds to approximate a short length of curve. The tangent segment of length ds lies over a segment of length dx on the x-axis, as shown in Fig. 15.46. We have a formula for ds in terms of dx, namely

$$ds = \sqrt{1 + (f'(x))^2} \; dx.$$

For surfaces, we approximate the area of a small piece of surface by the area of a piece of a tangent *plane* to the surface. We would like to find a formula for the area dS of a piece of tangent plane lying over a differential rectangle of area $dx \; dy$ in the x,y-plane.

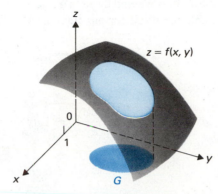

FIGURE 15.45 Portion of the surface $z = f(x, y)$ lying over G.

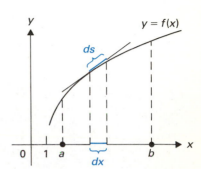

FIGURE 15.46 Approximation ds to arc length lying over the segment of length dx.

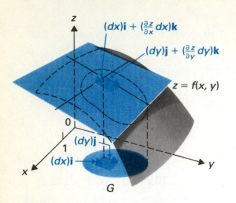

FIGURE 15.47 The area dS of the parallelogram in the tangent plane is the magnitude of the cross product of the vectors shown.

Figure 15.47 shows the tangent plane to $z = f(x, y)$ at a point (x, y, z) and also shows a parallelogram in the tangent plane that lies over a differential rectangle of area $dx\,dy$ in the x,y-plane. We can find the vectors along the two edges of the parallelogram shown in Fig. 15.47. The vector lying over the line segment of length dx in the x,y-plane must have dx as coefficient of $\mathbf{i}$. Since $\partial z/\partial x$ gives the "slope" in the x-direction of the tangent plane, we see that this vector is

$$(dx)\mathbf{i} + \left(\frac{\partial z}{\partial x}\,dx\right)\mathbf{k}.$$

A similar argument shows that the vector lying over the line segment of length dy is

$$(dy)\mathbf{j} + \left(\frac{\partial z}{\partial y}\,dy\right)\mathbf{k}.$$

The area of this parallelogram is the magnitude of the cross product of the vectors. Computing, we obtain the cross product

$$\begin{vmatrix} \mathbf{i} & \mathbf{j} & \mathbf{k} \\ dx & 0 & \dfrac{\partial z}{dx}\,dx \\ 0 & dy & \dfrac{\partial z}{\partial y}\,dy \end{vmatrix} = -\left(\frac{\partial z}{dx}\,dx\,dy\right)\mathbf{i} - \left(\frac{\partial z}{\partial y}\,dx\,dy\right)\mathbf{j} + (dx\,dy)\mathbf{k}.$$

The length of this cross product is the *differential element dS of surface area*, so

$$dS = \sqrt{\left(\frac{\partial z}{\partial x}\right)^2 + \left(\frac{\partial z}{\partial y}\right)^2 + 1}\ dx\,dy. \tag{1}$$

Consequently, we have

$$\text{Surface area} = \iint_G \sqrt{\left(\frac{\partial z}{\partial x}\right)^2 + \left(\frac{\partial z}{\partial y}\right)^2 + 1}\ dx\,dy. \tag{2}$$

EXAMPLE 1 Find the area of the sphere $x^2 + y^2 + z^2 = a^2$.

Solution We find the area of the top hemisphere $z = \sqrt{a^2 - x^2 - y^2}$ and then double the result for our final answer. Computing, we find that

$$\frac{\partial z}{\partial x} = \frac{-x}{\sqrt{a^2 - x^2 - y^2}} \quad \text{and} \quad \frac{\partial z}{\partial y} = \frac{-y}{\sqrt{a^2 - x^2 - y^2}}.$$

Hence

$$dS = \sqrt{\left(\frac{\partial z}{\partial x}\right)^2 + \left(\frac{\partial z}{\partial y}\right)^2 + 1}\ dx\,dy = \sqrt{\frac{x^2 + y^2}{a^2 - x^2 - y^2} + 1}\ dx\,dy$$

$$= \sqrt{\frac{x^2 + y^2 + a^2 - x^2 - y^2}{a^2 - x^2 - y^2}}\ dx\,dy$$

$$= \sqrt{\frac{a^2}{a^2 - x^2 - y^2}}\ dx\,dy$$

$$= \frac{a}{\sqrt{a^2 - x^2 - y^2}}\ dx\,dy.$$

Since we will be integrating over the disk $x^2 + y^2 \leq a^2$ (or $r \leq a$), and since $x^2 + y^2 = r^2$ appears in our computation of dS, we change to polar coordinates, where

$$\frac{a}{\sqrt{a^2 - x^2 - y^2}} = \frac{a}{\sqrt{a^2 - r^2}}.$$

We thus form the integral

$$\int_0^{2\pi} \int_0^a \frac{a}{\sqrt{a^2 - r^2}} \, r \, dr \, d\theta.$$

We should point out that our integrand is undefined for $r = a$, so this is an improper integral of two variables, which we really have not discussed. (Geometrically, this happens because our surface is perpendicular to the x,y-plane at $r = a$, so $\partial z/\partial x$ and $\partial z/\partial y$ are undefined there.) In straightforward analogy with our improper integrals of a function of one variable, we compute

$$\int_0^{2\pi} \left(\lim_{h \to a-} \int_0^h \frac{a}{\sqrt{a^2 - r^2}} \, r \, dr \right) d\theta = \int_0^{2\pi} \lim_{h \to a-} \left. (-a\sqrt{a^2 - r^2}) \right]_0^h d\theta$$

$$= \int_0^{2\pi} \left[\lim_{h \to a-} (-a\sqrt{a^2 - h^2}) + a^2 \right] d\theta$$

$$= \int_0^{2\pi} a^2 \, d\theta = a^2\theta \Big]_0^{2\pi} = 2\pi a^2.$$

Doubling, we obtain $4\pi a^2$ as the area of the sphere. ∎

The radical appearing in formula (1) for dS often makes integration to find surface area difficult. Sometimes it is necessary to use a table or to find a numerical estimate for the surface area.

EXAMPLE 2   Find the area of the portion of the surface $z = x^2 + y$ that lies over the region $0 \leq x \leq 1$, $0 \leq y \leq 1$ in the x,y-plane.

Solution   We see from formula (1) that $dS = \sqrt{4x^2 + 1 + 1} \, dx \, dy = \sqrt{2 + 4x^2} \, dx \, dy$. Integrating first with respect to y and then using a table, we find that the surface area is

$$\int_0^1 \int_0^1 \sqrt{2 + 4x^2} \, dy \, dx = \int_0^1 \left. (\sqrt{2 + 4x^2})y \right]_0^1 dx = \int_0^1 \sqrt{2 + 4x^2} \, dx$$

$$= \int_0^1 \sqrt{1 + (\sqrt{2}x)^2} \sqrt{2} \, dx$$

$$= \left(\frac{\sqrt{2}x}{2} \sqrt{1 + (\sqrt{2}x)^2} \right.$$

$$\left. + \frac{1}{2} \ln\left(\sqrt{2}x + \sqrt{1 + (\sqrt{2}x)^2} \right) \right) \Big]_0^1$$

$$= \frac{\sqrt{2}}{2}\sqrt{3} + \frac{1}{2} \ln(\sqrt{2} + \sqrt{3}) - 0$$

$$= \frac{1}{2}\left(\sqrt{6} + \ln(\sqrt{2} + \sqrt{3}) \right). \quad ∎$$

Sometimes it is more convenient to integrate over the projection of a surface on the x,z-plane or y,z-plane rather than the x,y-plane. We then make the obvious modification in formula (1) for dS.

EXAMPLE 3 Use a calculator or computer to estimate the area of the portion of the surface $x + y + z^3 = 5$, where $0 \le z \le 1$ and $-1 \le y \le 0$.

Solution We write the surface equation in the form $x = f(y, z) = 5 - y - z^3$. Interchanging the roles of x and z in formula (1), we have

$$dS = \sqrt{\left(\frac{\partial x}{\partial z}\right)^2 + \left(\frac{\partial x}{\partial y}\right)^2 + 1} \; dy \; dz = \sqrt{9z^4 + 1 + 1} \; dy \; dz.$$

Thus the surface area is given by

$$\int_0^1 \int_{-1}^0 \sqrt{9z^4 + 2} \; dy \; dz = \int_0^1 (\sqrt{9z^4 + 2})y \Big]_{-1}^0 \; dz$$

$$= \int_0^1 \sqrt{9z^4 + 2} \; dz.$$

Using Simpson's rule with $n = 40$, we obtain the estimate 1.870345415 for this integral and the surface area. ∎

Sometimes a surface is given in the form $F(x, y, z) = 0$ rather than in the form $z = f(x, y)$. Recall that if F has continuous partial derivatives and $\partial F/\partial z$ does not assume the value zero in a neighborhood of a point, then the surface does define an implicit function $z = f(x, y)$ in a neighborhood of the point, and, furthermore,

$$\frac{\partial z}{\partial x} = -\frac{\partial F/\partial x}{\partial F/\partial z} \quad \text{and} \quad \frac{\partial z}{\partial y} = -\frac{\partial F/\partial y}{\partial F/\partial z}.$$

We then obtain

$$\sqrt{\left(\frac{\partial z}{\partial x}\right)^2 + \left(\frac{\partial z}{\partial y}\right)^2 + 1} = \sqrt{\left(\frac{\partial F/\partial x}{\partial F/\partial z}\right)^2 + \left(\frac{\partial F/\partial y}{\partial F/\partial z}\right)^2 + 1}$$

$$= \frac{\sqrt{(\partial F/\partial x)^2 + (\partial F/\partial y)^2 + (\partial F/\partial z)^2}}{|\partial F/\partial z|}.$$

This gives us the formula

$$dS = \frac{\sqrt{F_x^2 + F_y^2 + F_z^2}}{|F_z|} \; dx \; dy = \frac{|\nabla F|}{|F_z|} \tag{3}$$

in place of formula (1). Of course, in using formula (3) to find surface area, we have to express all the partial derivatives as functions of x and y only, which may be difficult. We repeat Example 1, finding the area of a sphere, but using formula (3) to find dS.

EXAMPLE 4 Find the area of the sphere $x^2 + y^2 + z^2 = a^2$, using formula (3) to compute dS.

Solution Let $F(x, y, z) = x^2 + y^2 + z^2 - a^2$. From formula (3), we have

$$dS = \frac{\sqrt{(2x)^2 + (2y)^2 + (2z)^2}}{|2z|} \, dx \, dy = \frac{\sqrt{4(x^2 + y^2 + z^2)}}{|2z|} \, dx \, dy$$

$$= \frac{\sqrt{4a^2}}{2|z|} \, dx \, dy = \frac{a}{|z|} \, dx \, dy = \frac{a}{\sqrt{a^2 - x^2 - y^2}} \, dx \, dy.$$

We have obtained the same formula for dS as in Example 1, and from this point the computations are identical. ∎

SUMMARY

1. The area of a surface consisting of part of a graph $z = f(x, y)$ is equal to the integral

$$\iint_G \sqrt{\left(\frac{\partial z}{\partial x}\right)^2 + \left(\frac{\partial z}{\partial y}\right)^2 + 1} \, dx \, dy$$

evaluated over the region G in the x,y-plane under the surface.

2. The area of a surface consisting of part of a surface $F(x, y, z) = c$ is equal to the integral

$$\iint_G \frac{\sqrt{(\partial F/\partial x)^2 + (\partial F/\partial y)^2 + (\partial F/\partial z)^2}}{|\partial F/\partial z|} \, dx \, dy$$

evaluated over the region G in the x,y-plane under the surface.

3. The differential element of surface area is

$$dS = \sqrt{\left(\frac{dz}{\partial x}\right)^2 + \left(\frac{dz}{dy}\right)^2 + 1} \, dx \, dy \qquad \text{for } z = f(x, y)$$

and

$$dS = \frac{\sqrt{(\partial F/\partial x)^2 + (\partial F/\partial y)^2 + (\partial F/\partial z)^2}}{|\partial F/\partial z|} \, dx \, dy$$

$$= \frac{|\nabla F|}{|\partial F/\partial z|} \, dx \, dy \qquad \text{for } F(x, y, z) = c.$$

EXERCISES 15.4

In Exercises 1–16, find the area of the indicated surface. Use the integral table where necessary.

1. The portion of the plane $x + 2y + 2z = 10$ inside the cylinder $x^2 + y^2 = 4$

2. The portion of the plane $6x + 3y + 2z = 100$ inside the cylinder $y^2 + z^2 = 9$

3. The portion of the plane $4x + 3y + 5z = 50$ lying over the region in the x,y-plane bounded by $y = x$ and $y = x^2$

4. The portion of the surface $x + 2y + 2z^{3/2} = 100$ where $0 \le x \le 4$ and $0 \le z \le 3$

5. The portion of the surface $4x^{3/2} + 3y + 4z = 8$, where $0 \le x \le 4$ and $0 \le y \le 2$

6. The portion of the surface $z = 4x^2 + 3y$, where $0 \le x \le 1$ and $0 \le y \le 1$

7. The portion of the surface $4x + 3y^2 + 2z = 10$, where $0 \le y \le 1$ and $0 \le z \le 2$

8. The portion of the surface $z = \frac{2}{3}(x^{3/2} + y^{3/2})$, where $0 \le x \le 1$ and $0 \le y \le 2$

9. The portion of the surface $z = x^2 + y^2$ inside the cylinder $x^2 + y^2 = a^2$

10. The portion of the surface $4x - y^2 + z^2 = 8$ inside the cylinder $y^2 + z^2 = 25$

11. The portion of the hemisphere $z = -\sqrt{25 - x^2 - y^2}$ inside the cylinder $x^2 + y^2 = 9$

12. The portion of the sphere $x^2 + y^2 + z^2 = a^2$ inside the cone $z = \sqrt{x^2 + y^2}$

13. The portion of the surface $ax = z^2 - y^2$ inside the cylinder $y^2 + z^2 = a^2$

14. The portion of the sphere $x^2 + y^2 + z^2 = a^2$ inside the cylinder $x^2 + z^2 = az$

15. The surface of the solid bounded by $z = 4 - x^2 - y^2$ and by $z = -4 + x^2 + y^2$

16. The surface of the solid bounded above by $z = 4 - x^2 - y^2$, below by $z = -\sqrt{8 - x^2 - y^2}$, and on the sides by $x^2 + y^2 = 4$

In Exercises 17–20, use a calculator or computer and Simpson's rule to estimate the area of the indicated surface.

17. The portion of the surface $4x + 3y^3 + 5z = 100$, where $0 \le x \le 5$ and $0 \le y \le 1$

18. The portion of the surface $3x + 4y^{3/2} + z^3 = 16$, where $0 \le y \le 4$ and $0 \le z \le 1$

19. The portion of the hemisphere $z = \sqrt{25 - x^2 - y^2}$, where $-1 \le x \le 1$ and $-1 \le y \le 1$

20. The portion of the paraboloid $z = x^2 + y^2$, for $-2 \le x \le 2$ and $-2 \le y \le 2$

15.5 TRIPLE INTEGRALS IN RECTANGULAR AND CYLINDRICAL COORDINATES

Triple integrals arise when we integrate a function of three variables over a region of space.

Triple Integrals over Rectangular Boxes

If $f(x, y, z)$ is continuous on a rectangular box G for $a \le x \le b$, $c \le y \le d$, $r \le z \le s$ in space, then $\iiint_G f(x, y, z)\, dx\, dy\, dz$ is defined as the limit of regular Riemann sums

$$R_P = \frac{(b - a)(d - c)(s - r)}{n^3} \cdot \sum_{i=1}^{n} f(x_i, y_i, z_i) \tag{1}$$

obtained by following the steps in our outline at the start of Section 15.1 just as we obtained Eq. (1) there.

EXAMPLE 1 Use a calculator and the midpoint Riemann sum with $n = 2$ to estimate $\iiint_G x^{y+2z}\, dx\, dy\, dz$, where G is the rectangular region $1 \le x \le 2$, $0 \le y \le 2$, $0 \le z \le 1$ shown in Fig. 15.48(a).

Solution Figure 15.48(b) indicates how to find the midpoint coordinates. We have

$$\frac{(b - a)(d - c)(s - r)}{n^3} = \frac{(2 - 1)(2 - 0)(1 - 0)}{2^3} = \frac{1}{4}.$$

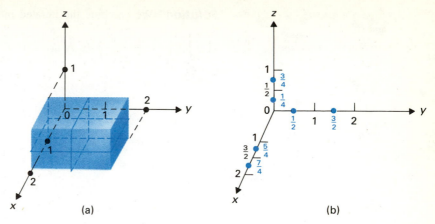

FIGURE 15.48 Partition for $n = 2$ using midpoints: (a) the box $1 \leq x \leq 2$, $0 \leq y \leq 2$, $0 \leq z \leq 1$ partitioned for $n = 2$; (b) finding midpoint coordinates.

We should sum the values of $f(x, y, z) = x^{y+2z}$ at the eight points

$$\left(\frac{5}{4}, \frac{1}{2}, \frac{1}{4}\right), \qquad \left(\frac{5}{4}, \frac{1}{2}, \frac{3}{4}\right), \qquad \left(\frac{5}{4}, \frac{3}{2}, \frac{1}{4}\right), \qquad \left(\frac{5}{4}, \frac{3}{2}, \frac{3}{4}\right),$$

$$\left(\frac{7}{4}, \frac{1}{2}, \frac{1}{4}\right), \qquad \left(\frac{7}{4}, \frac{1}{2}, \frac{3}{4}\right), \qquad \left(\frac{7}{4}, \frac{3}{2}, \frac{1}{4}\right), \qquad \left(\frac{7}{4}, \frac{3}{2}, \frac{3}{4}\right)$$

and multiply this sum by $\frac{1}{4}$. Our calculator gives

$$\frac{1}{4}(19.5625) = 4.890625$$

for this Riemann sum.   ■

Of course, better accuracy in Example 1 can be obtained by using a value of n greater than 2, but the number n^3 of points quickly becomes large. To handle the sum for $n = 10$ with 1000 points, we would need to write a short computer program rather than rely on our calculator.

To evaluate $\iiint_G f(x, y, z) \, dx \, dy \, dz$ over a rectangular box G, where $a \leq x \leq b$, $c \leq y \leq d$, and $r \leq z \leq s$, we may use a triple iterated integral

$$\int_r^s \int_c^d \int_a^b f(x, y, z) \, dx \, dy \, dz = \int_r^s \left(\int_{y=c}^d \left(\int_{x=a}^b f(x, y, z) \, dx \right) dy \right) dz. \quad \textbf{(2)}$$

Again, the integral is computed from "inside out." Also, it can be shown that the same value is obtained for all orders of integration; that is,

$$\int_r^s \int_c^d \int_a^b f(x, y, z) \, dx \, dy \, dz = \int_a^b \int_r^s \int_c^d f(x, y, z) \, dy \, dz \, dx,$$

and so on. We ask you to discuss possible orders of integration in Exercise 1.

EXAMPLE 2 Let G be the rectangular box $1 \leq x \leq 3$, $0 \leq y \leq 2$, $1 \leq z \leq 2$ in space. Find $\iiint_G [(xy - y^2)/z] \, dx \, dy \, dz$.

Solution We compute the iterated integral

$$\int_1^2 \int_0^2 \int_1^3 \left(\frac{xy - y^2}{z}\right) dx\, dy\, dz = \int_1^2 \int_0^2 \frac{1}{z}\left(\frac{x^2 y}{2} - xy^2\right)\bigg]_{x=1}^3 dy\, dz$$

$$= \int_1^2 \int_0^2 \frac{1}{z}\left[\left(\frac{9}{2}y - 3y^2\right) - \left(\frac{y}{2} - y^2\right)\right] dy\, dz$$

$$= \int_1^2 \int_0^2 \frac{1}{z}(4y - 2y^2)\, dy\, dz$$

$$= \int_1^2 \frac{1}{z}\left(2y^2 - \frac{2y^3}{3}\right)\bigg]_{y=0}^2 dz$$

$$= \int_1^2 \frac{1}{z}\left(8 - \frac{16}{3}\right) dz = \frac{8}{3}\ln|z|\bigg]_1^2$$

$$= \frac{8}{3}(\ln 2). \qquad \blacksquare$$

We should not think that numerical methods of integration such as midpoint Riemann sums are a waste of time. Let G be the rectangular box $1 \le x \le 2$, $0 \le y \le 2$, $0 \le z \le 1$. In Example 1, we estimated

$$\iiint_G x^{y+2z}\, dx\, dy\, dz$$

using $n = 2$. An iterated integral would be $\int_0^1 \int_0^2 \int_1^2 x^{y+2z}\, dx\, dy\, dz$. We cannot perform the integration to evaluate this integral. The first integration is not bad, but it leads to $\int [(2^{y+2z+1} - 1)/(y + 2z + 1)]\, dy$ for the second integral. We then turn to a numerical method.

Sometimes, it is best to carry integration as far as possible and to use a numerical estimate the rest of the way. An example follows.

EXAMPLE 3 Estimate

$$\int_1^2 \int_0^1 \int_1^3 xe^{yz}\, dx\, dy\, dz.$$

Solution We have

$$\int_1^2 \int_0^1 \int_1^3 xe^{yz}\, dx\, dy\, dz = \int_1^2 \int_0^1 \left(\frac{x^2}{2}e^{yz}\right)\bigg]_{x=1}^3 dy\, dz = \int_1^2 \int_0^1 4e^{yz}\, dy\, dz$$

$$= \int_1^2 \frac{4}{z}e^{yz}\bigg]_{y=0}^1 dz = \int_1^2 \frac{4}{z}(e^z - 1)\, dz.$$

At this stage, we are stuck. Using Simpson's rule for $n = 20$, we obtain from our calculator

$$\int_1^2 \frac{4}{z}(e^z - 1)\, dz \approx 9.463877538.$$

Simpson's rule with $n = 40$ yields 9.463877443, which we are confident is close to the correct answer. $\qquad \blacksquare$

Triple Integrals over More General Regions

Triple integrals over nonrectangular bounded regions of space can be defined just as we defined double integrals over nonrectangular regions in the plane in Section 15.2. Such triple integrals can be evaluated as iterated integrals. To find the appropriate limits on the integrals, it is advisable to draw a figure. People often have trouble finding the appropriate limits of integration when setting up an iterated triple integral. The next three examples illustrate the following general principle that applies to all iterated integrals.

> The *final* limits (on the left-hand integral sign) are always constant, and other limits may be functions of only those variables with respect to which integration will be performed *later*.
>
> *Wrong:* $\displaystyle\int_y^{y^2}\int_1^3 dy\, dx, \qquad \int_1^3\int_x^{x+5}\int_z^{z-y} yz^2\, dx\, dy\, dz.$

EXAMPLE 4   Let G be the region in the first octant bounded by the coordinate planes and the plane $x + y + z = 1$. Find $\iiint_G x\, dx\, dy\, dz$.

Solution   As always, we draw a sketch of G, shown in Fig. 15.49. The figure also shows a differential box of volume $dV = dx\, dy\, dz$ in the region. We will form a z,x,y-order iterated integral.

Integration with respect to z adds the contributions $x\, dz\, dx\, dy$ up the column in Fig. 15.50. At the bottom of the column, $z = 0$, while $z = 1 - x - y$ at the top. This leads to

$$\left(\int_0^{1-x-y} x\, dz\right) dx\, dy.$$

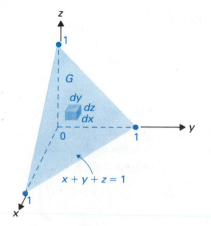

FIGURE 15.49   Differential volume $dV = dx\, dy\, dz$ in the region G.

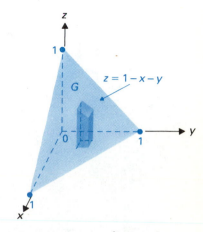

FIGURE 15.50 $[\int_0^{1-x-y} x\, dz]\, dx\, dy$ sums $x\, dx\, dy\, dz$ over this column.

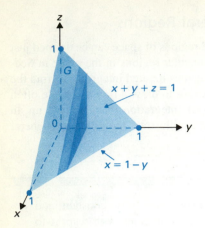

FIGURE 15.51

$[\int_0^{1-y} \int_0^{1-x-y} x \, dz \, dx] \, dy$

sums $x \, dx \, dy \, dz$ over this slab.

We now add these column contributions in the x-direction, along the slab shown in Fig. 15.51. The back of the slab is at $x = 0$. The front comes out to the line $x + y = 1$ in the x,y-plane. The equation of this line was obtained by setting $z = 0$ in the equation $x + y + z = 1$ of the front plane bounding the region. Thus $x = 1 - y$ at the front of the slab, and we are led to

$$\left(\int_0^{1-y} \int_0^{1-x-y} x \, dz \, dx \right) dy.$$

Finally, we add the contributions of these slabs in the y-direction, from the smallest y-value of zero to the largest y-value of 1. We obtain the iterated integral

$$\int_0^1 \int_0^{1-y} \int_0^{1-x-y} x \, dz \, dx \, dy = \int_0^1 \int_0^{1-y} xz \Big]_{z=0}^{1-x-y} dx \, dy$$

$$= \int_0^1 \int_0^{1-y} (x - x^2 - xy) \, dx \, dy$$

$$= \int_0^1 \int_0^{1-y} [x(1 - y) - x^2] \, dx \, dy$$

$$= \int_0^1 \left(\frac{x^2}{2}(1 - y) - \frac{x^3}{3} \right) \Big]_{x=0}^{1-y} dy$$

$$= \int_0^1 \left[\frac{(1 - y)^3}{2} - \frac{(1 - y)^3}{3} \right] dy$$

$$= \int_0^1 \frac{1}{6}(1 - y)^3 \, dy$$

$$= -\frac{1}{24}(1 - y)^4 \Big]_0^1 = 0 + \frac{1}{24} = \frac{1}{24}. \quad \blacksquare$$

EXAMPLE 5 Convert

$$\int_0^2 \int_{(3x/2)-3}^0 \int_0^{2-x+(2y/3)} xyz^2 \, dz \, dy \, dx$$

to the order $\iiint xyz^2 \, dx \, dz \, dy$.

Solution

STEP 1. The inside limits with respect to z show that the bottom of the region is the plane $z = 0$ and the top is the plane $z = 2 - x + (2y/3)$, or $3x - 2y + 3z = 6$. We sketch these planes in Fig. 15.52. The remaining limits of integration then show that the region of integration is the tetrahedron shaded in the figure.

STEP 2. In the new order, we wish to integrate first in the x-direction, from $x = 0$ out to the plane $3x - 2y + 3z = 6$, where $x = 2 + (2y/3) - z$. Thus we start with

$$\left(\int_0^{2+(2y/3)-z} xyz^2 \, dx \right) dz \, dy$$

corresponding to the horizontal box shown in Fig. 15.53. The next z-limits may be found from the back triangle of the region, in the y,z-plane, which has as base the line where $z = 0$ and as top the

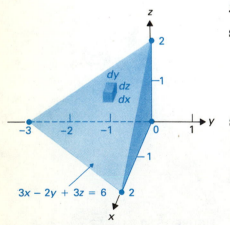

FIGURE 15.52 Differential volume $dV = dx \, dy \, dz$.

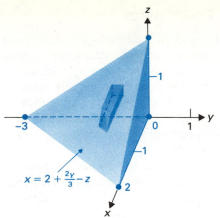

FIGURE 15.53

$$\left[\int_0^{2+(2y/3)-z} xyz^2\,dx\right]dz\,dy$$

sums $xyz^2\,dx\,dy\,dz$ over this column.

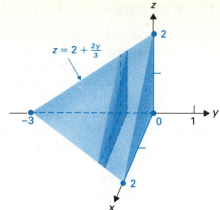

FIGURE 15.54

$$\left[\int_0^{2+(2y/3)}\int_0^{2+(2y/3)-z} xyz^2\,dx\,dz\right]dy$$

sums $xyz^2\,dx\,dy\,dz$ over this slab.

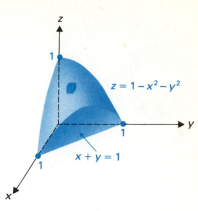

FIGURE 15.55 Differential volume $dV = dx\,dy\,dz$.

line where $-2y + 3z = 6$, or $z = 2 + (2y/3)$, obtained by setting $x = 0$ in $3x - 2y + 3z = 6$. We have arrived so far at

$$\left(\int_0^{2+(2y/3)}\int_0^{2+(2y/3)-z} xyz^2\,dx\,dz\right)dy,$$

corresponding to Fig. 15.54. Finally, the constant y-limits go from the minimum y-value of -3 to the maximum y-value of zero, so the desired integral is

$$\int_{-3}^{0}\int_0^{2+(2y/3)}\int_0^{2+(2y/3)-z} xyz^2\,dx\,dz\,dy. \qquad \blacksquare$$

EXAMPLE 6 Find the volume of the region in space bounded above by the surface $z = 1 - x^2 - y^2$, on the sides by the planes $x = 0$, $y = 0$, $x + y = 1$, and below by the plane $z = 0$. The region is sketched in Fig. 15.55.

Solution In differential notation, we think of adding the volumes of small rectangular boxes with edges having lengths dx, dy, and dz. An attempt to find our iterated integral by integrating in the x-direction first leads to problems in the x-limits, for the "lower" boxes would have to be summed from the plane $x = 0$ forward to the plane $x = 1 - y$, while the "higher" boxes would have to be summed from $x = 0$ forward to the surface $x = \sqrt{1 - y^2 - z}$, as indicated in Fig. 15.56. The same problem occurs if we integrate first in the y-direction. Thus we integrate in the order z, x, y and use Fig. 15.57 to obtain the iterated integral

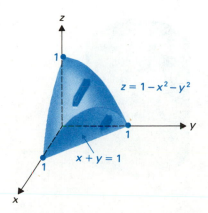

FIGURE 15.56 These columns in the x-direction have different front boundary surfaces.

$$\int_0^1\int_0^{1-y}\int_0^{1-x^2-y^2} dz\,dx\,dy = \int_0^1\int_0^{1-y}\left.z\right]_0^{1-x^2-y^2} dx\,dy$$

$$= \int_0^1\int_0^{1-y}(1 - x^2 - y^2)\,dx\,dy.$$

The computation now proceeds as in Example 6 of Section 15.2. $\blacksquare$

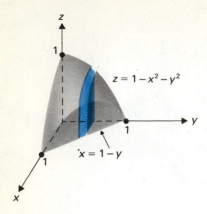

FIGURE 15.57

Column: $[\int_0^{1-x^2-y^2} dz] \, dx \, dy$
Slab: $[\int_0^{1-y} \int_0^{1-x^2-y^2} dz \, dx] \, dy$

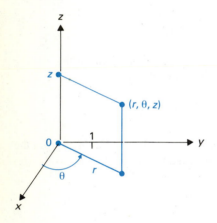

FIGURE 15.58 (r, θ, z) in cylindrical coordinates.

Triple Integrals in Cylindrical Coordinates

Recall that cylindrical r,θ,z-coordinates for space are formed by taking polar r,θ-coordinates in the x,y-plane and the usual rectangular z-coordinate, as indicated in Fig. 15.58. The solution set of $r = a$ is the cylinder $x^2 + y^2 = a^2$, shown in Fig. 15.59. Transformation to cylindrical coordinates is especially useful in integrating a Cartesian expression involving $x^2 + y^2$, or in integrating over regions bounded by surfaces with simple cylindrical coordinate equations. If an integral involves $x^2 + z^2$, we may take "cylindrical r,θ,y-coordinates," corresponding to polar r,θ-coordinates in the x,z-plane.

The differential volume element in cylindrical coordinates is shown in Fig. 15.60. The base of this differential element has the same size as the polar differential area element, as shown in the figure. Thus the cylindrical coordinate differential volume dV is

$$dV = r \, dr \, d\theta \, dz. \tag{3}$$

If $h(r, \theta, z)$ is a continuous cylindrical coordinate function defined on a region G in space, then the integral of $h(r, \theta, z)$ over G is

$$\iiint_G h(r, \theta, z) \cdot r \, dz \, dr \, d\theta. \tag{4}$$

The z,r,θ-order in integral (4) is often the most convenient order of integration.

EXAMPLE 7 Integrate $f(x, y, z) = x^2 z$ over the region in space bounded by $x^2 + y^2 = 4$, $z = 0$, and $z = 4$.

Solution The region is a solid cylinder, shown in Fig. 15.61. We change to cylindrical coordinates. Now $x^2 z = (r \cos \theta)^2 z$. Thus we integrate

$$h(r, \theta, z) = r^2 (\cos^2 \theta) z$$

over the cylinder. In Fig. 15.61, we have shown a differential volume element in the cylinder. We will integrate in the z,r,θ-order.

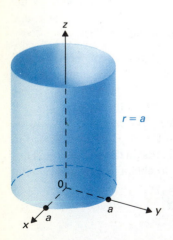

FIGURE 15.59 The cylinder $x^2 + y^2 = a^2$, or $r = a$.

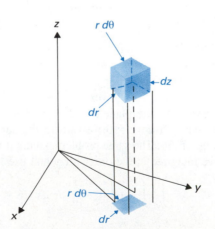

FIGURE 15.60 $dV = (r \, d\theta)(dr)(dz)$ in cylindrical coordinates.

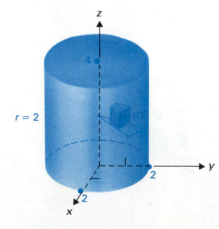

FIGURE 15.61 The solid cylinder $x^2 + y^2 \leq 4$, $0 \leq z \leq 4$.

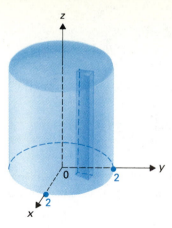

FIGURE 15.62

$$\int_0^4 h(r,\,\theta,\,z) r\,dz\,dr\,d\theta$$

goes over this column.

Integrating first with respect to z, we see that the smallest z-value is always 0, and the largest is always 4. We are led to

$$\left(\int_0^4 [r^2(\cos^2\theta)z]\cdot r\,dz\right)dr\,d\theta,$$

which sums contributions up the column shown in Fig. 15.62. The cylinder has cylindrical coordinate equation $r = 2$, and we next arrive at

$$\left(\int_0^2\int_0^4 r^2(\cos^2\theta)\cdot r\,dz\,dr\right)d\theta,$$

which sums contributions over the wedge shown in Fig. 15.63. Finally, we spin this wedge "all the way around" by letting θ increase from 0 to 2π, obtaining

$$\int_0^{2\pi}\int_0^2\int_0^4 [r^2(\cos^2\theta)z]\cdot r\,dz\,dr\,d\theta = \int_0^{2\pi}\int_0^2 r^3(\cos^2\theta)\frac{z^2}{2}\bigg]_{z=0}^4 dr\,d\theta$$

$$= \int_0^{2\pi}\int_0^2 8r^3\cos^2\theta\,dr\,d\theta = \int_0^{2\pi} 2r^4\bigg]_0^2\cos^2\theta\,d\theta = \int_0^{2\pi} 32\cos^2\theta\,d\theta$$

$$= \int_0^{2\pi} 16(1+\cos 2\theta)\,d\theta = (16\theta + 8\sin 2\theta)\bigg]_0^{2\pi} = 32\pi. \qquad\blacksquare$$

EXAMPLE 8 Find the volume of the solid in space bounded by $x^2 + z^2 = 9$, $y = 0$, and $y + z = 8$.

Solution We sketch the solid in Fig. 15.64. Since $x^2 + z^2 = 9$ is a circular cylinder perpendicular to the x,z-plane, we use cylindrical coordinates $(r,\,\theta,\,y)$, thinking of $(r,\,\theta)$ as polar coordinates in the x,z-plane, with $z = r\cos\theta$ and $x = r\sin\theta$. (We let z be the $r\cos\theta$ rather than x since the z,x,y-order gives a *right-hand* coordinate system, just like the familiar x,y,z-order.) Thus $y + z = 8$ becomes $y + r\cos\theta = 8$. From Fig. 15.64, we see that the volume is given by

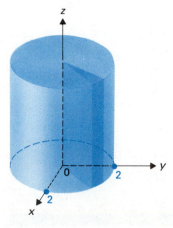

FIGURE 15.63

$$\int_0^2\int_0^4 h(r,\,\theta,\,z) r\,dz\,dr\,d\theta$$

goes over this wedge.

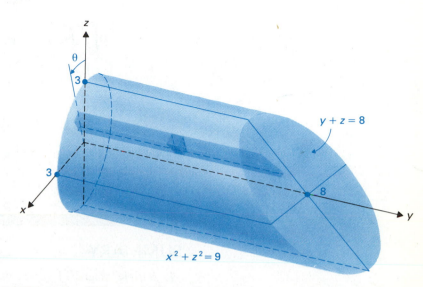

FIGURE 15.64 With cylindrical coordinates $(r,\,\theta,\,y)$, we have $dV = dy(r\,dr\,d\theta)$.

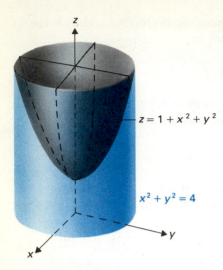

$z = 1 + x^2 + y^2$

$x^2 + y^2 = 4$

FIGURE 15.65 Solid bounded above by $z = 1 + x^2 + y^2$, on the sides by $x^2 + y^2 = 4$, and below by $z = 0$.

the integral

$$\int_0^{2\pi} \int_0^3 \int_0^{8 - r\cos\theta} 1 \cdot r \, dy \, dr \, d\theta = \int_0^{2\pi} \int_0^3 ry \Big]_{y=0}^{8 - r\cos\theta} dr \, d\theta$$

$$= \int_0^{2\pi} \int_0^3 (8r - r^2\cos\theta) \, dr \, d\theta = \int_0^{2\pi} \left(4r^2 - \frac{r^3}{3}\cos\theta\right)\Big]_{r=0}^3 d\theta$$

$$= \int_0^{2\pi} (36 - 9\cos\theta) \, d\theta = (36\theta - 9\sin\theta)\Big]_0^{2\pi} = 72\pi. \quad \blacksquare$$

EXAMPLE 9 Let G be the region bounded above by $z = 1 + x^2 + y^2$, below by $z = 0$, and on the sides by $x^2 + y^2 = 4$, as indicated in Fig. 15.65. Integrate $f(x, y, z) = x - y + z^2$ over G.

Solution Our z,x,y-integral would be

$$\int_{-2}^2 \int_{-\sqrt{4-y^2}}^{\sqrt{4-y^2}} \int_0^{1+x^2+y^2} (x - y + z^2) \, dz \, dx \, dy,$$

which is not too pleasant to evaluate. However, changing to cylindrical coordinates, we obtain the integral

$$\int_0^{2\pi} \int_0^2 \int_0^{1+r^2} (r\cos\theta - r\sin\theta + z^2)r \, dz \, dr \, d\theta$$

$$= \int_0^{2\pi} \int_0^2 \left((\cos\theta - \sin\theta)r^2 z + r\frac{z^3}{3}\right)\Big]_{z=0}^{z=1+r^2} dr \, d\theta$$

$$= \int_0^{2\pi} \int_0^2 \left((\cos\theta - \sin\theta)(r^2)(1 + r^2) + \frac{1}{3}r(1 + r^2)^3\right) dr \, d\theta$$

$$= \int_0^{2\pi} \left((\cos\theta - \sin\theta)\left(\frac{r^3}{3} + \frac{r^5}{5}\right) + \frac{1}{6} \cdot \frac{(1 + r^2)^4}{4}\right)\Big]_{r=0}^{r=2} d\theta$$

$$= \int_0^{2\pi} \left((\cos\theta - \sin\theta)\left(\frac{8}{3} + \frac{32}{5}\right) + \frac{5^4 - 1}{24}\right) d\theta$$

$$= \int_0^{2\pi} \left((\cos\theta - \sin\theta)\frac{136}{15} + \frac{624}{24}\right) d\theta$$

$$= \left(\frac{136}{15}(\sin\theta + \cos\theta) + \frac{624}{24}\theta\right)\Big]_0^{2\pi}$$

$$= \frac{136}{15}(1 - 1) + \frac{624}{24}2\pi = \frac{624}{12}\pi = 52\pi.$$

Perhaps you would like to perform the integration in rectangular coordinates and compare it with our cylindrical coordinate computation! $\quad \blacksquare$

SUMMARY

1. A triple integral $\iiint_G f(x, y, z) \, dx \, dy \, dz$ over a bounded region G in space can be evaluated as an iterated integral, analogous to the two-variable case.

2. The volume of the differential solid element in cylindrical coordinates is $dV = r \cdot dz\, dr\, d\theta$.

3. An integral $\iiint f(x, y, z)\, dx\, dy\, dz$ with appropriate x,y,z-limits becomes, in cylindrical coordinates,

$$\iiint f(r \cos \theta, r \sin \theta, z) \cdot r\, dz\, dr\, d\theta$$

with appropriate cylindrical coordinate limits for the region.

EXERCISES 15.5

1. Let f be a continuous function of three variables with domain containing a rectangular box $a_1 \le x \le b_1$, $a_2 \le y \le b_2$, $a_3 \le z \le b_3$.
   (a) Show that there are six possible "orders of integration" for an iterated integral of f over the box, where each integral has some a_i for lower limit and b_i for upper limit.
   (b) How many iterated integrals can you form for f over the box if the a_i's are not restricted to lower limits and the b_i's are not restricted to upper limits?

2. Estimate $\int_1^3 \int_0^2 \int_0^2 xyz^2\, dx\, dy\, dz$ using the regular midpoint Riemann sum with $n = 2$.

3. Estimate $\int_1^3 \int_1^3 \int_1^3 (1/xyz)\, dx\, dy\, dz$ using the regular midpoint Riemann sum with $n = 2$.

In Exercises 4–14, complete the integral.

4. $\displaystyle\int_4^6 \int_1^e \int_2^4 \frac{xy}{z}\, dy\, dz\, dx$

5. $\displaystyle\int_0^1 \int_2^3 \int_{-1}^4 xy^2z\, dx\, dy\, dz$

6. $\displaystyle\int_0^1 \int_{-1}^1 \int_1^2 (x^2 + yz)\, dz\, dx\, dy$

7. $\displaystyle\int_{-1}^3 \int_0^{\ln 2} \int_0^4 xze^y\, dx\, dy\, dz$

8. $\displaystyle\int_0^2 \int_0^1 \int_0^1 xyz\sqrt{2 - x^2 - y^2}\, dx\, dy\, dz$

9. $\displaystyle\int_0^\pi \int_0^1 \int_0^1 yz^2\sin(xyz)\, dx\, dy\, dz$

10. $\displaystyle\int_0^1 \int_0^1 \int_0^1 xy^3e^{xyz}\, dz\, dx\, dy$

11. $\displaystyle\int_0^1 \int_0^{1-y} \int_0^{x^2+y^2} y\, dz\, dx\, dy$

12. $\displaystyle\int_0^a \int_0^{\sqrt{a^2-y^2}} \int_0^{\sqrt{a^2-x^2-y^2}} x\, dz\, dx\, dy$

13. $\displaystyle\int_0^2 \int_0^{\sqrt{4-z^2}} \int_{y^2+z^2-4}^{4-y^2-z^2} 1\, dx\, dy\, dz$

14. $\displaystyle\int_0^1 \int_0^z \int_0^{y+z} yz\, dx\, dy\, dz$

In Exercises 15–19, find the integral of the function f over the region G.

15. $f(x, y, z) = y$; G the region in space bounded by $z = 1 - y^2$, $x = 0$, $x = 4$, and $z = 0$ [*Hint:* This integral can be found immediately by symmetry.]

16. $f(x, y, z) = y^2z$; G the region in space bounded by $x = 4 - y^2 - z^2$ and $x = 0$ [*Hint:* This integral can be found immediately by symmetry.]

17. $f(x, y, z) = x$; G the region in space bounded by the coordinate planes and the plane $x + y + 2z = 2$

18. $f(x, y, z) = yz$; G the region in space bounded by $z = x^2$, $z = 1$, $y = 0$, and $y = 2$

19. $f(x, y, z) = x$; G the region in space bounded by the coordinate planes and the planes $x + y = 1$ and $z = 4$

In Exercises 20–25, express the given triple integral in the indicated order. Do not integrate.

20. $\displaystyle\int_{-1}^1 \int_{-\sqrt{1-x^2}}^{\sqrt{1-x^2}} \int_0^{1-x^2-y^2} xyz^2\, dz\, dy\, dx$ in x,y,z-order

21. $\displaystyle\int_{-2}^1 \int_0^{\pi/2} \int_0^{\cos z} x \sin yz\, dx\, dz\, dy$ in y,z,x-order

22. $\displaystyle\int_{-1}^0 \int_0^{1+x} \int_0^{1+x-y} \sin(xy)\, dz\, dy\, dx$ in x,y,z-order

23. $\displaystyle\int_{-1}^0 \int_{-1-y}^0 \int_0^{1+x+y} 2xy\, dz\, dx\, dy$ in y,z,x-order

24. $\displaystyle\int_0^2 \int_1^3 \int_0^{4-y^2} z^2\, dz\, dx\, dy$ in x,y,z-order

25. $\displaystyle\int_2^5 \int_{-4}^0 \int_0^{16-x^2} (x + 1)\, dz\, dx\, dy$ in x,y,z-order

26. Express the volume of the ball $x^2 + y^2 + z^2 \leq a^2$ as an integral (do not integrate) in the integration order
(a) z,r,θ (b) z,θ,r (c) r,θ,z
(d) r,z,θ (e) θ,r,z (f) θ,z,r

In Exercises 27–33, find the volume of the given region in space, using triple integration in cylindrical coordinates.

27. The region bounded by the paraboloid $z = 4 - x^2 - y^2$ and the plane $z = 0$

28. The region bounded by the paraboloids $z = 9 - x^2 - y^2$ and $z = x^2 + y^2 - 9$

29. The region bounded by $x^2 + y^2 = 1$, $z = 0$ and $x + z = 4$

30. The region bounded by the paraboloid $x = 8 - y^2 - z^2$, the cylinder $y^2 + z^2 = 4$, and the plane $x = -1$ [*Hint:* Use x,r,θ-coordinates.]

31. The region bounded by the hemisphere $z = \sqrt{a^2 - x^2 - y^2}$ and the plane $z = b$ for $0 \leq b < a$

32. The region inside the semicircular cylinder bounded by $x = \sqrt{4 - z^2}$ and $x = 0$ and bounded on the ends by $y = 0$ and the hemisphere $y = \sqrt{16 - x^2 - y^2}$. [*Hint:* Use "cylindrical coordinates" (r, y, θ).]

33. The region bounded by the paraboloid $z = x^2 + y^2$, the plane $z = 0$, and the cylinder $x^2 + y^2 = 2x$

34. Find the integral of the cylindrical coordinate function $h(r, \theta, z) = rz \cos^2\theta$ for $r \geq 0$ over the region in space bounded by $r = a$, $z = 0$, and $z = 4$.

35. Find the integral of the cylindrical coordinate function $h(r, \theta, z) = rz^2$ for $r \geq 0$ over the region in space bounded by the cone $z^2 = x^2 + y^2$ and the plane $z = 4$.

In Exercises 36–41, transform the integral to cylindrical coordinates and then evaluate the transformed version.

36. $\displaystyle\int_0^1 \int_{-\sqrt{1-x^2}}^{\sqrt{1-x^2}} \int_0^3 x \, dz \, dy \, dx$

37. $\displaystyle\int_0^2 \int_{-\sqrt{4-y^2}}^{\sqrt{4-y^2}} \int_{1-y}^4 y^2 \, dz \, dx \, dy$

38. $\displaystyle\int_{-2}^2 \int_{-\sqrt{4-x^2}}^{\sqrt{4-x^2}} \int_0^{\sqrt{x^2+y^2}} z \, dz \, dy \, dx$

39. $\displaystyle\int_{-3}^3 \int_{-\sqrt{9-y^2}}^{\sqrt{9-y^2}} \int_{\sqrt{x^2+y^2}}^{12-x^2-y^2} z \, dz \, dx \, dy$

40. $\displaystyle\int_{-\sqrt{2}}^{\sqrt{2}} \int_{-\sqrt{2-z^2}}^{\sqrt{2-z^2}} \int_0^{1+y^2+z^2} z^2 \, dx \, dy \, dz$

41. $\displaystyle\int_0^2 \int_0^{\sqrt{4-x^2}} \int_0^{\sqrt{x^2+z^2}} y \, dy \, dz \, dx$

15.6   INTEGRATION IN SPHERICAL COORDINATES

Recall the spherical coordinate system, where a point has coordinates (ρ, ϕ, θ) as indicated in Fig. 15.66. The coordinate ρ is the length of the line segment joining the point and the origin, ϕ is the angle from the z-axis to this line segment, and θ is the same angle as in cylindrical coordinates. Note that $\rho = a$ describes a sphere with center at the origin and radius a, as indicated in Fig. 15.67.

Since ρ is the distance from the point to the origin, it is obvious that

$$\rho^2 = x^2 + y^2 + z^2, \tag{1}$$

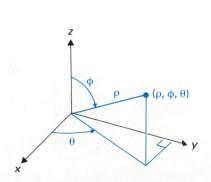

FIGURE 15.66   Spherical coordinates.

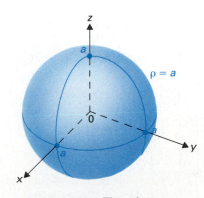

FIGURE 15.67   The sphere $\rho = a$.

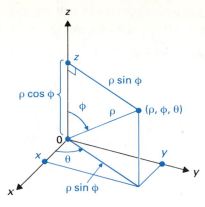

FIGURE 15.68

$$x = \rho \sin \phi \cos \theta,$$
$$y = \rho \sin \phi \sin \theta, \quad z = \rho \cos \phi$$

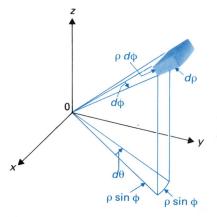

FIGURE 15.69 Solid differential element of volume

$$dV = (\rho \, d\phi)(d\rho)(\rho \sin \phi \, d\theta)$$
$$= \rho^2 \sin \phi \, d\rho \, d\phi \, d\theta.$$

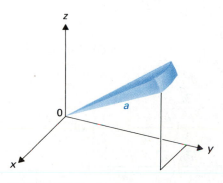

FIGURE 15.70 Integration with respect to ρ.

and consequently, transformation to spherical coordinates is useful in the triple integral of Cartesian expressions involving $x^2 + y^2 + z^2$ or integrals over regions bounded in part by spherical surfaces. Since $\phi = \alpha$ is a cone, we expect ρ, ϕ, θ-coordinates to be useful when cones are involved.

We need to express x, y, and z in terms of spherical ρ, ϕ, θ-coordinates so that we can express an integral $\iiint_G f(x, y, z) \, dx \, dy \, dz$ in terms of spherical coordinates. From Fig. 15.68, we recall that

$$x = \rho \sin \phi \cos \theta,$$
$$y = \rho \sin \phi \sin \theta, \tag{2}$$
$$z = \rho \cos \phi.$$

Again, we will use the heuristic differential notation. Increasing and decreasing ρ, ϕ, and θ by amounts $d\rho$, $d\phi$, and $d\theta$, we generate the differential element shown in Fig. 15.69. The volume of this element is approximately

$$(\rho \sin \phi \, d\theta)(d\rho)(\rho \, d\phi),$$

as shown in the figure. Therefore, the differential solid element in spherical coordinates has volume

$$dV = \rho^2 \sin \phi \, d\rho \, d\phi \, d\theta. \tag{3}$$

A real proof that Eq. (3) is appropriate is left to an advanced calculus course. (See also Exercise 8 in Section 15.9.)

Remember that we restrict ϕ to the range $0 \le \phi \le \pi$; thus $\sin \phi \ge 0$.

EXAMPLE 1 Find the volume of the ball bounded by the sphere

$$x^2 + y^2 + z^2 = a^2,$$

which has spherical coordinate equation $\rho = a$.

Solution We integrate the constant function 1 over this region, using the volume element (3) and limits in spherical coordinates. We integrate in the ρ, ϕ, θ-order, which frequently is the most convenient order. We think of the first integral, with respect to ρ, as adding up our volume elements to give the spike shown in Fig. 15.70. The next integration with respect to ϕ adds up the volumes of these spikes to give the volume of the wedge in Fig. 15.71, and the final integration with respect to θ from 0 to 2π adds up the volumes of these wedges to give the volume of the entire ball. Computing, we have

$$\int_0^{2\pi} \int_0^{\pi} \int_0^a (1)\rho^2 \sin \phi \, d\rho \, d\phi \, d\theta = \int_0^{2\pi} \int_0^{\pi} \frac{\rho^3}{3} \sin \phi \Big]_{\rho=0}^{\rho=a} d\phi \, d\theta$$

$$= \int_0^{2\pi} \int_0^{\pi} \frac{a^3}{3} \sin \phi \, d\phi \, d\theta$$

$$= \frac{a^3}{3} \int_0^{2\pi} - \cos \phi \Big]_0^{\pi} d\theta$$

$$= \frac{a^3}{3} \int_0^{2\pi} [-(-1) + 1] \, d\theta$$

$$= \frac{a^3}{3} 2\theta \Big]_0^{2\pi} = \frac{a^3}{3} 4\pi = \frac{4}{3} \pi a^3. \quad \blacksquare$$

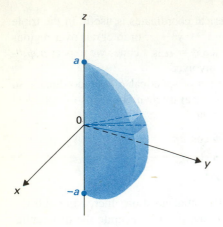

FIGURE 15.71 Subsequent integration with respect to ϕ.

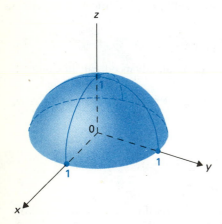

FIGURE 15.72 Half-ball bounded by $z = \sqrt{1 - x^2 - y^2}$ and $z = 0$.

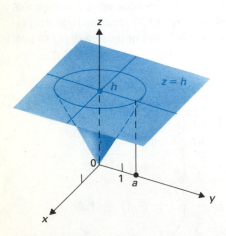

FIGURE 15.73 Cone with base radius a and altitude h; the cone has equation $\phi = \tan^{-1}(a/h)$.

EXAMPLE 2 Integrate the function $f(x, y, z) = z$ over the half-ball bounded above by $z = \sqrt{1 - x^2 - y^2}$ and below by $z = 0$, as indicated in Fig. 15.72.

Solution In spherical coordinates, $z = \rho \cos \phi$ by Eqs. (2), so our integral in spherical coordinate form is

$$\int_0^{2\pi} \int_0^{\pi/2} \int_0^1 (\rho \cos \phi)(\rho^2 \sin \phi)\, d\rho\, d\phi\, d\theta$$

$$= \int_0^{2\pi} \int_0^{\pi/2} \frac{\rho^4}{4} \cos \phi \sin \phi \bigg]_{\rho=0}^{\rho=1} d\phi\, d\theta$$

$$= \int_0^{2\pi} \int_0^{\pi/2} \frac{1}{4} \cos \phi \sin \phi\, d\phi\, d\theta$$

$$= \frac{1}{4} \int_0^{2\pi} \frac{\sin^2\phi}{2} \bigg]_0^{\pi/2} d\theta$$

$$= \frac{1}{4} \int_0^{2\pi} \frac{1}{2}\, d\theta = \frac{1}{8} \int_0^{2\pi} d\theta$$

$$= \frac{1}{8} \theta \bigg]_0^{2\pi} = \frac{1}{8} 2\pi = \frac{\pi}{4}.$$

Note that the rectangular coordinate integral of z over this half-ball is

$$\int_{-1}^1 \int_{-\sqrt{1-y^2}}^{\sqrt{1-y^2}} \int_0^{\sqrt{1-x^2-y^2}} z\, dz\, dx\, dy.$$

This integral is less pleasant to compute! ∎

EXAMPLE 3 Integrate in spherical coordinates to derive the formula $V = \frac{1}{3}\pi a^2 h$ for the volume of a solid right circular cone of altitude h and radius of base a.

Solution The solid bounded by $a^2 z^2 = h^2(x^2 + y^2)$ and $z = h$ is such a cone, as shown in Fig. 15.73. The plane $z = h$ becomes $\rho \cos \phi = h$, and the integral is

$$\int_0^{2\pi} \int_0^{\tan^{-1}(a/h)} \int_0^{h/\cos \phi} \rho^2 \sin \phi\, d\rho\, d\phi\, d\theta$$

$$= \int_0^{2\pi} \int_0^{\tan^{-1}(a/h)} \frac{\rho^3}{3} \sin \phi \bigg]_{\rho=0}^{\rho=h/\cos \phi} d\phi\, d\theta$$

$$= \int_0^{2\pi} \int_0^{\tan^{-1}(a/h)} \frac{h^3}{3} (\cos \phi)^{-3} \sin \phi\, d\phi\, d\theta$$

$$= \int_0^{2\pi} -\frac{h^3}{3} \cdot \frac{\sec^2\phi}{-2} \bigg]_0^{\tan^{-1}(a/h)} d\theta$$

$$= \int_0^{2\pi} -\frac{h^3}{3} \left(\frac{a^2 + h^2}{-2h^2} + \frac{1}{2} \right) d\theta$$

$$= 2\pi \left(-\frac{h^3}{3} \right) \left(\frac{a^2 + h^2 - h^2}{-2h^2} \right)$$

$$= 2\pi \left(-\frac{h^3}{3} \right) \left(\frac{a^2}{-2h^2} \right) = \frac{1}{3} \pi a^2 h. ∎$$

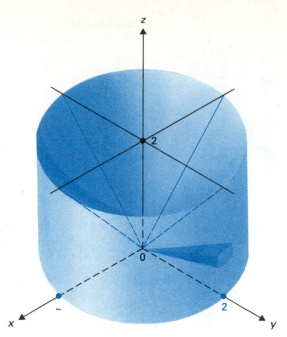

FIGURE 15.74 Region bounded by the cone $z = x^2 + y^2$ (or $\phi = \pi/4$), the cylinder $x^2 + y^2 = 4$ (or $\rho = 2 \csc \theta$), and the plane $z = 0$ (or $\phi = \pi/2$).

EXAMPLE 4 Transform the integral

$$\int_{-2}^{2} \int_{-\sqrt{4-y^2}}^{\sqrt{4-y^2}} \int_{0}^{\sqrt{x^2+y^2}} x^2 \, dz \, dx \, dy$$

to spherical coordinates.

Solution The lower z-limit, $z = 0$, gives the x,y-plane, while $z = \sqrt{x^2 + y^2}$ is the upper half of the cone $z^2 = x^2 + y^2$. We recognize that the x-limits and y-limits correspond to integrating over a disk with center at the origin in the x,y-plane and with radius 2. Thus the region of integration is the one in Fig. 15.74.

In Fig. 15.74, we have drawn the spherical coordinate differential volume element. If we integrate first with respect to ρ, we see that ρ goes from 0 to the cylinder $x^2 + y^2 = 4$. In spherical coordinates, this cylinder has equation

$$\rho^2 \sin^2\phi \, \cos^2\theta + \rho^2 \sin^2\phi \, \sin^2\theta = 4,$$

or $\rho^2 \sin^2\phi = 4$. Thus $\rho = 2/(\sin\theta) = 2 \csc\theta$. Consequently we start our integral with

$$\left(\int_{0}^{2\csc\phi} (\rho^2 \sin^2\phi \, \cos^2\theta) \rho^2 \sin\phi \, d\rho \right) d\phi \, d\theta.$$

Since the cone makes a 45° angle with the z-axis, ϕ goes from $\pi/4$ to $\pi/2$. Finally, we let θ go from 0 to 2π and obtain

$$\int_{0}^{2\pi} \int_{\pi/4}^{\pi/2} \int_{0}^{2\csc\phi} \rho^4 \sin^3\phi \, \cos^2\theta \, d\rho \, d\phi \, d\theta. \qquad \blacksquare$$

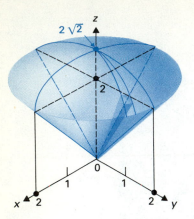

FIGURE 15.75 Region bounded by the hemisphere $z = \sqrt{8 - r^2}$ (or $\rho = 2\sqrt{2}$) and the cone $z = r$ (or $\phi = \pi/4$).

EXAMPLE 5 Transform the cylindrical coordinate integral

$$\int_0^{2\pi} \int_0^2 \int_r^{\sqrt{8-r^2}} r^2 z \, dz \, dr \, d\theta$$

to spherical coordinates.

Solution The lower z-limit, $z = r = \sqrt{x^2 + y^2}$, and the upper z-limit, $z = \sqrt{8 - r^2} = \sqrt{8 - x^2 - y^2}$, show that the region of integration is bounded below by a cone and above by the sphere shown in Fig. 15.75. Squaring the equations, we have $z^2 = r^2$ and $z^2 = 8 - r^2$. These yield $2z^2 = 8$, so $z = 2$. The intersection of these surfaces is thus in the plane $z = 2$ and has r,θ-equation $r = 2$, which is a circle of radius 2. The r-limit and θ-limit in the given integral then show that the region of integration is the entire solid bounded by the cone below and the hemisphere above, as shown in Fig. 15.75.

As indicated by Fig. 15.75, the ρ-limits for an integral in ρ,ϕ,θ-order are from 0 to $\sqrt{8} = 2\sqrt{2}$. The cone makes an angle of 45° with the z-axis. Now

$$r^2 z \, dz \, dr \, d\theta = (rz) \cdot r \, dz \, dr \, d\theta = (rz) \, dV.$$

Since

$$rz = (\rho \sin \phi)(\rho \cos \phi) = \rho^2 \sin \phi \cos \phi,$$

we see that the integral in spherical coordinates becomes

$$\int_0^{2\pi} \int_0^{\pi/4} \int_0^{2\sqrt{2}} \rho^4 \sin^2 \phi \cos \phi \, d\rho \, d\phi \, d\theta. \quad \blacksquare$$

SUMMARY

1. The meaning of spherical ρ,ϕ,θ-coordinates is given in Fig. 15.66.

2. Transformation from x,y,z-coordinates to ρ,ϕ,θ-coordinates is accomplished by

$$x = \rho \sin \phi \cos \theta,$$
$$y = \rho \sin \phi \sin \theta,$$
$$z = \rho \cos \phi.$$

3. The differential solid element in spherical coordinates has volume $dV = \rho^2 \sin \phi \, d\rho \, d\phi \, d\theta$.

4. An integral $\iiint f(x, y, z) \, dx \, dy \, dz$ with appropriate x,y,z-limits becomes, in spherical coordinates,

$$\iiint f(\rho \sin \phi \cos \theta, \rho \sin \phi \sin \theta, \rho \cos \phi) \cdot \rho^2 \sin \phi \, d\rho \, d\phi \, d\theta,$$

with appropriate spherical coordinate limits for the region chosen so that $0 \le \phi \le \pi$.

EXERCISES 15.6

In Exercises 1–8, find the volume of the given region by triple integration in spherical coordinates.

1. The region bounded by the cone $z^2 = x^2 + y^2$ and the hemisphere $z = \sqrt{16 - x^2 - y^2}$

2. The larger region bounded by the sphere $x^2 + y^2 + z^2 = 9$ and the cone $z = -\sqrt{x^2 + y^2}$

3. The region bounded above by the cone $z = \sqrt{x^2 + y^2}$, below by the cone $z = -2\sqrt{x^2 + y^2}$, and on the sides by the sphere $x^2 + y^2 + z^2 = 9$

4. The region between the cones $z^2 = x^2 + y^2$ and $3z^2 = x^2 + y^2$ and below the hemisphere $z = \sqrt{4 - x^2 - y^2}$

5. The region bounded by the hemisphere $y = \sqrt{4 - x^2 - z^2}$ and the planes $y = x$ and $y = \sqrt{3}x$

6. The region bounded on the sides by the cylinder $x^2 + y^2 = 4$, above by the cone $z = \sqrt{x^2 + y^2}$, and below by the plane $z = 0$

7. The region bounded by the hemisphere $z = \sqrt{a^2 - x^2 - y^2}$ and the plane $z = b$ for $0 \le b < a$

8. The region bounded below by the plane $z = b$, above by the plane $z = c$, and on the sides by the sphere $x^2 + y^2 + z^2 = a^2$ for $-a < b < c < a$ [*Hint:* Express as a difference of integrals.]

9. Find the integral of the spherical coordinate function $h(\rho, \phi, \theta) = \rho^2$ over the ball bounded by the sphere $x^2 + y^2 + z^2 = a^2$.

10. Find the integral of the spherical coordinate function $h(\rho, \phi, \theta) = \rho^2\cos\phi$ over the region bounded by the cone $z^2 = x^2 + y^2$ and the hemisphere $z = \sqrt{4 - x^2 - y^2}$.

In Exercises 11–20, transform the integrals into ones that use spherical coordinates. Do not evaluate the integrals.

11. $\int_{-2}^{2} \int_{-\sqrt{4-y^2}}^{\sqrt{4-y^2}} \int_{\sqrt{x^2+y^2}}^{\sqrt{8-x^2-y^2}} z\, dz\, dx\, dy$

12. $\int_{0}^{1} \int_{-\sqrt{1-x^2}}^{\sqrt{1-x^2}} \int_{\sqrt{3}\sqrt{x^2+y^2}}^{\sqrt{10-x^2-y^2}} x\, dz\, dy\, dx$

13. $\int_{0}^{3} \int_{0}^{\sqrt{9-x^2}} \int_{\sqrt{(x^2+y^2)/3}}^{\sqrt{3(x^2+y^2)}} xy\, dz\, dx\, dy$

14. $\int_{-2}^{0} \int_{-\sqrt{4-y^2}}^{0} \int_{-\sqrt{3(x^2+y^2)}}^{\sqrt{x^2+y^2}} x^2\, dz\, dx\, dy$

15. $\int_{-1}^{0} \int_{0}^{\sqrt{1-x^2}} \int_{3}^{\sqrt{10-x^2-y^2}} (y-x)\, dz\, dy\, dx$

16. $\int_{0}^{2\pi} \int_{0}^{4} \int_{r}^{4} r^2\, dz\, dr\, d\theta$

17. $\int_{0}^{\pi} \int_{0}^{3} \int_{0}^{\sqrt{3}r} r^2 z\, dz\, dr\, d\theta$

18. $\int_{0}^{\pi/2} \int_{0}^{4} \int_{-r}^{r/\sqrt{3}} r^3\, dz\, dr\, d\theta$

19. $\int_{0}^{2\pi} \int_{2}^{4} \int_{0}^{r} dz\, dr\, d\theta$

20. $\int_{0}^{2\pi} \int_{0}^{2} \int_{0}^{1} rz\, dz\, dr\, d\theta$

15.7 MASS AND MOMENTS

Mass

Imagine a physical body to occupy some region G in space. The *mass m* of the body is a numerical measure of the "amount of material" it contains. Near the surface of the earth, the *weight* of a body is *mg*, where *g* is the gravitational acceleration; one slug of mass weighs about 32 pounds.

The *mass density* of the body is the *mass per unit volume*. If the body is not homogeneous, the mass density may vary and be a function of the position within the body. To say that the mass density at a point (x_0, y_0, z_0) is $\sigma(x_0, y_0, z_0)$ is to say that if the body had the same composition everywhere that it has at (x_0, y_0, z_0), then its mass would be

$$\sigma(x_0, y_0, z_0) \cdot \text{Volume of } G.$$

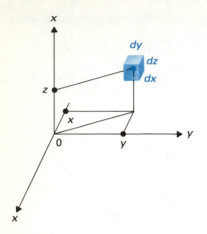

FIGURE 15.76 For mass density $\sigma(x, y, z)$, the differential element has approximate mass $dm = \sigma(x, y, z)\, dx\, dy\, dz$.

Using differential notation, suppose the mass density $\sigma(x, y, z)$ is a *continuous* function for (x, y, z) in G. If (x, y, z) is a point in the small box with edges of lengths dx, dy, and dz, shown in Fig. 15.76, then the approximate mass of the material in this box is

$$dm = \sigma(x, y, z)\, dx\, dy\, dz.$$

If we add all these small amounts of mass with an integral as dx, dy, and dz approach zero, we obtain

---

Mass of the Body

$$m = \iiint_G \sigma(x, y, z)\, dx\, dy\, dz. \tag{1}$$

Of course, in cylindrical and spherical coordinates, our differential volume elements are $r\, dz\, dr\, d\theta$ and $\rho^2 \sin\phi\, d\rho\, d\phi\, d\theta$, respectively.

If the physical body that we are considering is a flat plate lying in the x,y-plane, then the mass density can be considered to be a function $\sigma(x, y)$ of x and y, and the integral (1) becomes a double integral. The mass density in this case can be considered as *mass per unit area*. If we are considering a thin rod along the x-axis, the mass density becomes a function $\sigma(x)$ of just one variable, and the integral (1) is replaced by a single integral. If we are considering a thin wire covering some curve in the plane or in space, the density function is either $\sigma(x, y)$ or $\sigma(x, y, z)$, respectively, and the analogue of the integral (1) is a single integral $\int_C$ (Mass density) ds over the curve. The mass density in these one-dimensional cases can be interpreted as *mass per unit length*.

EXAMPLE 1 Let a rectangular plate covering the rectangle $0 \le x \le 2$, $0 \le y \le 1$ in the x,y-plane have mass density $\sigma(x, y) = x + y^2$ at (x, y). Find the mass of the plate.

Solution Consider a differential element of the plate containing a point (x, y) and having area $dA = dx\, dy$. The mass of this differential element is

$$dm = \sigma(x, y)\, dA = (x + y^2)\, dx\, dy.$$

Thus the mass of the plate is given by the integral

$$m = \int_0^1 \int_0^2 (x + y^2)\, dx\, dy = \int_0^1 \left(\frac{x^2}{2} + xy^2\right)\Bigg]_{x=0}^{2} dy$$

$$= \int_0^1 (2 + 2y^2)\, dy = \left(2y + \frac{2}{3}y^3\right)\Bigg]_0^1 = \left(2 + \frac{2}{3}\right) - 0 = \frac{8}{3}. \qquad \blacksquare$$

EXAMPLE 2 Let a wire lie on the curve $x = y^2$ from $y = 0$ to $y = 2$ and have mass density $\sigma(x, y) = ky$ for a constant k. Find the mass of the wire.

Solution We have

$$ds = \sqrt{\left(\frac{dx}{dy}\right)^2 + 1}\; dy = \sqrt{4y^2 + 1}\; dy,$$

so

$$m = \int_0^2 ky\sqrt{4y^2 + 1}\; dy = \frac{k}{8} \int_0^2 8y(4y^2 + 1)^{1/2}\; dy$$

$$= \frac{k}{8} \cdot \frac{2}{3}(4y^2 + 1)^{3/2}\Big]_0^2$$

$$= \frac{k}{12}(17\sqrt{17} - 1). \qquad \blacksquare$$

EXAMPLE 3 Let the mass density of a ball of radius a be proportional to the distance from the center of the ball. Find the mass of the ball if the mass density at a distance 1 unit from the center is k.

Solution If we take the center of the ball as origin, then the mass density is given by $\sigma(x, y, z) = k\sqrt{x^2 + y^2 + z^2}$. It is natural to use spherical coordinates in integrating over a ball; in terms of spherical coordinates, the mass density is given by

$$\sigma(\rho, \phi, \theta) = k\sqrt{x^2 + y^2 + z^2} = k\rho.$$

The mass is then

$$m = \int_0^{2\pi} \int_0^{\pi} \int_0^a k\rho \cdot \rho^2 \sin\phi\; d\rho\; d\phi\; d\theta$$

$$= \int_0^{2\pi} \int_0^{\pi} k\frac{\rho^4}{4}\Big]_0^a \sin\phi\; d\phi\; d\theta$$

$$= \int_0^{2\pi} \int_0^{\pi} k\frac{a^4}{4} \sin\phi\; d\phi\; d\theta = \frac{ka^4}{4} \int_0^{2\pi} -\cos\phi\Big]_0^{\pi} d\theta$$

$$= \frac{ka^4}{4} \int_0^{2\pi} [-(-1) + 1]\; d\theta$$

$$= \frac{2ka^4}{4} \int_0^{2\pi} d\theta = \frac{ka^4}{2}\theta\Big]_0^{2\pi} = k\pi a^4. \qquad \blacksquare$$

EXAMPLE 4 Let a flat sheet of material of constant thickness cover the region bounded by the cardioid $r = a(1 + \sin\theta)$ shown in Fig. 15.77, and let the area mass density of the sheet be proportional to the distance from the y-axis. Find the mass of the body.

Solution The mass density is

$$\sigma(x, y) = k|x|$$

$$= k|r\cos\theta|,$$

where k is a constant of proportionality. We make use of the symmetry of both

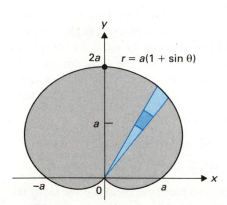

FIGURE 15.77 For $\sigma(x, y) = k|x|$, the differential element has approximate mass $k|r\cos\theta|r\, dr\, d\theta$.

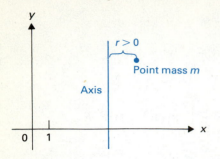

FIGURE 15.78 A point mass of magnitude m a signed distance r from an axis.

$\sigma(x, y)$ and the cardioid about the y-axis. The mass is given by the integral

$$
\begin{aligned}
m &= 2 \int_{-\pi/2}^{\pi/2} \int_0^{a(1+\sin\,\theta)} k(r\cos\theta) r\, dr\, d\theta \\
&= 2k \int_{-\pi/2}^{\pi/2} \frac{r^3}{3} \Big]_0^{a(1+\sin\,\theta)} \cos\theta\, d\theta \\
&= 2k \int_{-\pi/2}^{\pi/2} \frac{a^3}{3}(1+\sin\theta)^3 \cos\theta\, d\theta \\
&= \frac{2}{3}ka^3 \frac{(1+\sin\theta)^4}{4} \Big]_{-\pi/2}^{\pi/2} \\
&= \frac{2}{3}ka^3 \frac{2^4}{4} - \frac{2}{3}ka^3(0) \\
&= \frac{8ka^3}{3}.
\end{aligned}
$$

■

First Moments

A line (axis) in the plane divides the plane into two pieces. We will speak of these two pieces as being on one side or the other side of the axis, much as we speak of one side or the other side of a road. Let one side of the axis be designated as the *positive side* and the other as the *negative side*. We use the convention that for a vertical axis, the positive side is on the right, while for a horizontal axis, the positive side is above the axis.

Consider a point mass of magnitude m in the plane, and let r be the *signed* distance from the point mass to some particular axis, as shown in Fig. 15.78. That is, r is positive if the point mass is on the positive side of the axis and negative if it is on the negative side. The **first moment** M, or simply the **moment,** of the point mass about the axis is

$$M = mr.$$

The moment of a body consisting of several points of mass is the sum of the moments of the individual point masses.

Children encounter a problem in moments when they play on a seesaw. Figure 15.79 shows a seesaw consisting of a fiberglass plank of negligible mass supported on a fulcrum. A small child of mass m sits on the left end of the plank, and a larger child of mass M sits on the right end. For the choice of axes and coordinates shown in the figure, the sum of the moments of the children about the y-axis must be zero if the seesaw is to be balanced. That is, we must have

$$mx_1 + Mx_2 = 0.$$

Recall that x_1 is negative and x_2 is positive, so we must have

$$m|x_1| = M|x_2|.$$

Thus if $M = 2m$, we must have $|x_1| = 2|x_2|$, so the plank should be slid along the fulcrum so that the smaller child is twice as far from the fulcrum as the larger child.

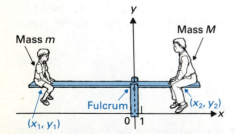

FIGURE 15.79 For a balanced seesaw, $mx_1 + Mx_2 = 0$, or $m|x_1| = M|x_2|$.

EXAMPLE 5 Let a body consist of a point mass of 3 at $(1, 4)$, a point mass of 2 at $(-2, 1)$, and a point mass of 5 at $(3, -4)$. Find (a) the moment M_x of the

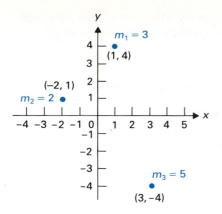

FIGURE 15.80 Three point masses in the plane.

body about the x-axis and (b) the moment M_y of the body about the y-axis.

Solution The body is shown in Fig. 15.80. For part (a), the moment M_x about the x-axis is

$$M_x = 3(4) + 2(1) + 5(-4) = -6.$$

For part (b), the moment M_y about the y-axis is

$$M_y = 3(1) + 2(-2) + 5(3) = 14. \qquad \blacksquare$$

Just as we can designate positive and negative sides of a line in the plane, we can designate positive and negative sides of a plane in space. The **first moment** of a point mass about such a plane in space is again the product of the mass m and the signed distance r from the point to the plane. The moment of a body consisting of several points of mass is again the sum of the moments of the individual point masses.

Now consider a body whose mass is not concentrated at a finite number of points (which is usually the case). We compute the first moment by adding up products of the masses of small pieces and the signed distances of the pieces from the axis (or plane) and taking the limit as the pieces become smaller and smaller. Of course, this leads to an integral. For a flat sheet of material in the plane, we let M_x and M_y be the first moments about the x-axis and y-axis, respectively. We then have

---

First Moments

$$M_x = \iint_G y \cdot \sigma(x, y) \, dx \, dy,$$

$$M_y = \iint_G x \cdot \sigma(x, y) \, dx \, dy. \tag{2}$$

For a body occupying a region G in space, the first moments M_{xy}, M_{yz}, and M_{xz} about the x,y-plane, y,z-plane, and x,z-plane, respectively, are given by

---

First Moments

$$M_{xy} = \iiint_G z \cdot \sigma(x, y, z) \, dx \, dy \, dz,$$

$$M_{yz} = \iiint_G x \cdot \sigma(x, y, z) \, dx \, dy \, dz, \tag{3}$$

$$M_{xz} = \iiint_G y \cdot \sigma(x, y, z) \, dx \, dy \, dz.$$

EXAMPLE 6 A wire of constant mass density $\sigma(x, y) = k$ covers the semicircle $x = 2 \cos t$, $y = 2 \sin t$ for $0 \le t \le \pi$. Find the first moment of the wire about the x-axis.

Solution The appropriate integral is of the form $\int_C ky \, ds$ over the curve. From the parametric equations for the semicircle, we find that

$$ds = \sqrt{\left(\frac{dx}{dt}\right)^2 + \left(\frac{dy}{dt}\right)^2} \, dt$$
$$= \sqrt{4 \sin^2 t + 4 \cos^2 t} \, dt = 2 \, dt.$$

Thus we obtain

$$M_x = \int_2^\pi k(2 \sin t)2 \, dt = -4k \cos t \Big]_0^\pi$$
$$= -4k(-1) - (-4k)(1) = 8k. \quad \blacksquare$$

EXAMPLE 7 Find the first moments about the x-axis and y-axis of a flat sheet of material covering the region bounded by the cardioid $r = a(1 + \sin \theta)$ shown in Fig. 15.81, if the area mass density is the constant k.

Solution By symmetry, the moment about the y-axis is zero, since the mass of a small piece is multiplied by the *signed* distance from the axis; a positive contribution of a piece on the right-hand side of the y-axis in Fig. 15.81 is counterbalanced by the negative contribution of the symmetric piece on the left-hand side. Since the signed distance from a point (x, y) to the x-axis is $y = r \sin \theta$, we obtain

$$M_x = 2 \int_{-\pi/2}^{\pi/2} \int_0^{a(1+\sin\,\theta)} (r \sin \theta)kr \, dr \, d\theta$$
$$= 2k \int_{-\pi/2}^{\pi/2} \frac{r^3}{3} \Big]_0^{a(1+\sin\,\theta)} \sin \theta \, d\theta$$
$$= 2k \int_{-\pi/2}^{\pi/2} \frac{a^3(1 + \sin \theta)^3}{3} \sin \theta \, d\theta$$
$$= \frac{2ka^3}{3} \int_{-\pi/2}^{\pi/2} (\sin \theta + 3 \sin^2\theta + 3 \sin^3\theta + \sin^4\theta) \, d\theta.$$

Since $\sin \theta = -\sin(-\theta)$ and $\sin^3\theta = -\sin^3(-\theta)$, their integrals over the interval $[-\pi/2, \pi/2]$ are zero. Our integral reduces to

$$\frac{2ka^3}{3} \int_{-\pi/2}^{\pi/2} (3 \sin^2\theta + \sin^4\theta) \, d\theta$$
$$= \frac{2ka^3}{3} \left(\frac{3\theta}{2} - \frac{3 \sin 2\theta}{4} + \frac{3\theta}{8} - \frac{\sin 2\theta}{4} + \frac{\sin 4\theta}{32} \right) \Big]_{-\pi/2}^{\pi/2}$$
$$= \frac{2ka^3}{3} \left[\frac{3\pi}{4} + \frac{3\pi}{16} - \left(-\frac{3\pi}{4} - \frac{3\pi}{16} \right) \right]$$
$$= \frac{2ka^3}{3} \left(\frac{3\pi}{2} + \frac{3\pi}{8} \right) = \frac{2ka^3}{3} \cdot \frac{15\pi}{8} = \frac{5k\pi a^3}{4}. \quad \blacksquare$$

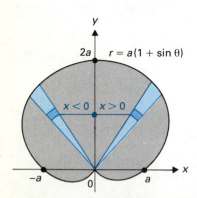

FIGURE 15.81 For $\sigma(x, y)$ symmetric about the y-axis, the sum of contributions of these two differential elements to M_y is zero.

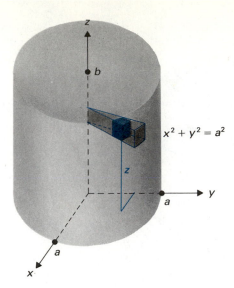

FIGURE 15.82

$\sigma(x, y, z) = kz;$

$$M_{xy} = \int_0^{2\pi} \int_0^a \int_0^b z(kz)r \, dz \, dr \, d\theta$$

EXAMPLE 8 Let a solid in space be bounded by the cylinder $x^2 + y^2 = a^2$ and the planes $z = 0$ and $z = b$. If the mass density at a height z above the x,y-plane is kz, find the first moments of the solid about the coordinate planes.

Solution The solid is illustrated in Fig. 15.82. Symmetry shows at once that

$$M_{xz} = M_{yz} = 0.$$

We use cylindrical coordinates and obtain

$$M_{xy} = \int_0^{2\pi} \int_0^a \int_0^b (z)(kz)r \, dz \, dr \, d\theta = k \int_0^{2\pi} \int_0^a \frac{z^3}{3}\Big]_0^b r \, dr \, d\theta$$

$$= k \int_0^{2\pi} \int_0^a \frac{b^3}{3} r \, dr \, d\theta = \frac{kb^3}{3} \int_0^{2\pi} \frac{r^2}{2}\Big]_0^a d\theta$$

$$= \frac{kb^3}{3} \int_0^{2\pi} \frac{a^2}{2} \, d\theta = \frac{ka^2 b^3}{6} \theta\Big]_0^{2\pi} = \frac{k\pi a^2 b^3}{3}. \qquad \blacksquare$$

Second Moments

The **second moment** I (or **moment of inertia**) about an axis of a "point mass" is the product of the mass and the *square* of the distance from the axis. A moment of inertia plays the role of mass in problems of rotational dynamics, such as computing kinetic energy of rotation, which is given by the formula

$$\text{K.E.} = \frac{1}{2}I\omega^2,$$

where ω is the angular speed of rotation. Computation of a moment of inertia often is accomplished by integration. We illustrate with an example.

EXAMPLE 9 Let a plate cover the first-quadrant region bounded by $y = x^3$ and $y = x$, and let the mass density at a point (x, y) be $\sigma(x, y) = \sqrt{y}$. Find the moment of inertia of the plate about the y-axis.

Solution The desired integral has the form $\iint_G x^2 \sigma(x, y) \, dx \, dy$ over the region G occupied by the plate, shown in Fig. 15.83. We obtain the integral

$$\int_0^1 \int_{x^3}^x x^2 \sqrt{y} \, dy \, dx = \int_0^1 x^2 \left(\frac{2}{3}y^{3/2}\right)\Big]_{x^3}^x dx = \int_0^1 \frac{2}{3}[x^2 x^{3/2} - x^2 x^{9/2}] \, dx$$

$$= \frac{2}{3} \int_0^1 (x^{7/2} - x^{13/2}) \, dx = \frac{2}{3}\left(\frac{2}{9}x^{9/2} - \frac{2}{15}x^{15/2}\right)\Big]_0^1$$

$$= \frac{2}{3}\left(\frac{2}{9} - \frac{2}{15}\right) = \frac{2}{3}\left(\frac{4}{45}\right) = \frac{8}{135}. \qquad \blacksquare$$

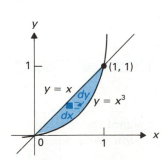

FIGURE 15.83 Plate bounded by $y = x$ and $y = x^3$.

EXAMPLE 10 Find the moment of inertia of a homogeneous ball of radius a and constant mass density k about a diameter.

Solution We take the center of the ball at the origin and let the z-axis be the diameter about which the moment of inertia is to be computed. The distance from

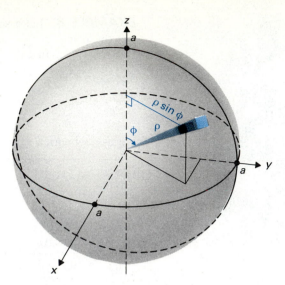

FIGURE 15.84 $\sigma(x, y, z) = k$, $I_z = \int_0^{2\pi} \int_0^{\pi} \int_0^a (\rho \sin \phi)^2 \cdot k \cdot \rho^2 \sin \phi \, d\rho \, d\phi \, d\theta$

a point (ρ, ϕ, θ) in spherical coordinates to the z-axis is easily seen to be $\rho \sin \phi$ (see Fig. 15.84). Thus

$$I = \int_0^{2\pi} \int_0^{\pi} \int_0^a (\rho \sin \phi)^2 k \rho^2 \sin \phi \, d\rho \, d\phi \, d\theta$$

$$= k \int_0^{2\pi} \int_0^{\pi} \frac{\rho^5}{5} \Big]_0^a \sin^3 \phi \, d\phi \, d\theta = k \int_0^{2\pi} \int_0^{\pi} \frac{a^5}{5} \sin^3 \phi \, d\phi \, d\theta$$

$$= \frac{ka^5}{5} \int_0^{2\pi} \int_0^{\pi} (1 - \cos^2 \phi) \sin \phi \, d\phi \, d\theta$$

$$= \frac{ka^5}{5} \int_0^{2\pi} \left(-\cos \phi + \frac{\cos^3 \phi}{3} \right) \Big]_0^{\pi} d\theta$$

$$= \frac{ka^5}{5} \int_0^{2\pi} \left[-(-1) - \frac{1}{3} - \left(-1 + \frac{1}{3} \right) \right] d\theta$$

$$= \frac{ka^5}{5} \int_0^{2\pi} \frac{4}{3} \, d\theta = \frac{4ka^5}{15} \, \theta \Big]_0^{2\pi} = \frac{8k\pi a^5}{15}. \qquad \blacksquare$$

SUMMARY

For mass density $\sigma(x, y)$ and area $dA = dx \, dy$ a signed (positive or negative) distance r from an axis, the following properties hold, where G is the plane region occupied by a thin plate.

1. Total mass $= m = \iint_G \sigma(x, y) \, dx \, dy$

2. First moment $= M = \iint_G r \, \sigma(x, y) \, dx \, dy$

3. Second moment (of inertia) $= I = \iint_G r^2 \sigma(x, y) \, dx \, dy$

Let a body in space occupy a region G.

4. The mass of the body is $m = \iiint_G \sigma(x, y, z)\, dx\, dy\, dz$, where $\sigma(x, y, z)$ is the mass density of the body at (x, y, z).

5. The first moment of the body about a plane is

$$\iiint_G (\text{Signed distance to plane}) \cdot \sigma(x, y, z)\, dx\, dy\, dz,$$

where the distance is from the differential element of volume $dx\, dy\, dz$ to the plane.

6. The second moment or moment of inertia of the body about an axis is

$$\iiint_G (\text{Distance from axis})^2 \cdot \sigma(x, y, z)\, dx\, dy\, dz,$$

where the distance is from the differential element of volume $dx\, dy\, dz$ to the axis.

EXERCISES 15.7

In Exercises 1–6, find the moment of the body consisting of the given point masses about the indicated axis or plane.

1. Mass 1 at $(-1, 2)$, 4 at $(1, 3)$, 2 at $(-3, -1)$, and 5 at $(4, -2)$ about the y-axis

2. Masses as in Exercise 1 about the x-axis

3. Mass 10 at $(0, 0)$, 4 at $(1, 3)$, 2 at $(-1, 4)$, 5 at $(-2, -3)$, and 1 at $(3, -1)$ about the x-axis

4. Mass 8 at $(-1, 5)$, 4 at $(1, -3)$, 2 at $(3, 2)$, and 5 at $(-4, -1)$ about the line $x = -3$

5. Mass 8 at $(2, -1, 3)$, 5 at $(-1, -4, 2)$ and 10 at $(-3, 1, 2)$ about the y,z-plane.

6. Masses as in Exercise 5 about the plane $y = -3$

7. Let a rectangular plate of mass density $\rho(x) = x(y - 1)$ at (x, y) cover the region bounded by $x = 1$, $x = 3$, $y = 2$, and $y = 6$. Find the mass of the plate.

8. Find the moment of the plate in Exercise 7 about the y-axis.

9. Find the moment of the plate in Exercise 7 about the line $x = 2$.

10. Find the moment of inertia of the plate in Exercise 7 about the line $x = -1$.

11. Find the moment of the plate in Exercise 7 about the x-axis.

12. Find the moment of the plate in Exercise 7 about the line $y = -1$.

13. Find the moment of the plate in Exercise 7 about the line $y = 4$.

14. The triangular region with vertices $(0, 0)$, $(a, 0)$, and $(0, b)$ for $a > 0$ and $b > 0$ is covered by a sheet of material whose mass density at a point (x, y) is kxy^2.
    (a) Find the mass of the sheet.
    (b) Find the first moment of the sheet about the x-axis.

15. Let the mass density at a point (x, y) of a plate covering the semicircular disk bounded by the x-axis and $y = \sqrt{1 - x^2}$ be $2(y + 1)$. Find the mass of the plate.

16. A metal plate is cut from a thin sheet of metal of constant mass density 3. If the plate just covers the first-quadrant region bounded by the curves $y = x^2$, $y = 1$, and $x = 0$, find the first moment of the plate about the y-axis.

17. For the plate in Exercise 16, find its first moment about the axis with equation $x = 2$.

18. For the plate in Exercise 16, find its moment of inertia about the x-axis.

19. Consider a thin rod of length a with constant mass density k (mass per unit length).
    (a) Find the first moment of the rod about an axis through an endpoint and perpendicular to the rod.
    (b) Find the moment of inertia of the rod about the axis described in part (a).

20. Repeat Exercise 19 in case the mass density is kx^2, where x is the distance along the rod from the axis of rotation.

21. Let a wire lie on $y = x^3$ from $x = 0$ to $x = 2$ and have constant mass density $\rho = 5$. Find the moment M_x of the wire about the x-axis.

22. Let a wire lie on $x = y^{3/2}$ from $y = 1$ to $y = 4$ and have mass density $\rho(y) = 3\sqrt{y}$. Express as an integral the first moment M_y of the wire about the y-axis.

23. Find the moment of inertia of a flat disk of radius a and constant mass density k about an axis perpendicular to the disk through its center.

24. Express as an integral the moment of inertia of a ball of radius a and constant mass density k (mass per unit volume) about a diameter of the ball.

25. Show that the first moment of a body in the plane about the line $x = -a$ is $M_y + ma$, where M_y is the moment about the y-axis. (This is known as the *parallel axis theorem*.)

Exercises 26–28 concern a thin plate covering the plane disk $x^2 + y^2 \leq a^2$. Let the mass density be proportional to the distance from the center of the disk, with mass density k at a distance 1 unit from the center.

26. Find the mass of the plate.

27. Find the first moment of the plate about the line $x = -a$.

28. Find the moment of inertia of the plate about an axis perpendicular to the plane through the origin.

Exercises 29–32 concern a solid cone in space bounded by $z = a$ and $z = \sqrt{x^2 + y^2}$. Let the mass density be proportional to the distance from the z-axis, with mass density k 1 unit away from the z-axis.

29. Find the mass of the solid.

30. Find the first moment of the solid about the plane $x = a$.

31. Find the first moment of the solid about the plane $z = -a$.

32. Find the moment of inertia of the solid about the z-axis.

Exercises 33 and 34 concern a solid in space bounded by the cylinder $y = a^2 - z^2$ and the planes $y = 0$, $x = 0$, and $x = b$. Let the mass density of the solid be $\sigma(x, y, z) = ky$.

33. Find the mass of the solid.

34. Express as an integral the moment of inertia of the solid about the y-axis.

35. Find the moment of inertia of a solid ball in space of radius a and constant mass density k about a line tangent to the ball.

15.8 CENTERS OF MASS, CENTROIDS, PAPPUS' THEOREM

Centers of Mass and Centroids

It can be proved that for a given body of mass m, there exists a unique point (not necessarily in the body) at which we may consider the entire mass of the body to be concentrated for the computation of *first moments* about *every* axis (or plane). This point is the **center of mass** (or **center of gravity**) of the body. If the (signed) distance from the center of mass of the body to an axis is s, then the first moment of the body about the axis is ms.

Suppose the body is a thin plate. We can think of the center of mass as the point in the plate where the plate could be balanced in a horizontal position on the point of a pencil, as illustrated in Fig. 15.85. For children on a balanced seesaw, as in Fig. 15.79, the center of mass is over the fulcrum. Since the center of mass is the point $(\bar{x}, \bar{y})$ at which we can consider all the mass to be concentrated for computation of first moments about the coordinate axes, we must have

$$M_x = m\bar{y} \quad \text{and} \quad M_y = m\bar{x}.$$

Hence

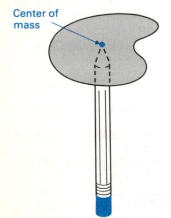

Center of mass

FIGURE 15.85 The center of mass is the balance point.

Center $(\bar{x}, \bar{y})$ of Mass

$$\bar{x} = \frac{M_y}{m} \quad \text{and} \quad \bar{y} = \frac{M_x}{m}. \tag{1}$$

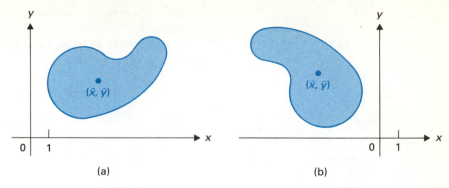

FIGURE 15.86 (a) Center of mass of a flat body; (b) center of mass of the same body in a different position.

It is a fact that

> the location of the center of mass in relation to the body is independent of the position of the body in the plane. **(2)**

See Fig. 15.86. We give some exercises that indicate the reason for this at the end of the section (Exercises 32 and 33). Also

> the first moment of the body about any *axis is the product of its mass and the (signed) distance from its center of mass to the axis.* **(3)**

We should warn you that, in general, there is *no* single point in a body at which the mass can be considered to be concentrated for computation of moments of inertia about *every* axis (see Exercise 34).

For a body in space, coordinates of the center of mass $(\bar{x}, \bar{y}, \bar{z})$ are given by

---

Center $(\bar{x}, \bar{y}, \bar{z})$ of Mass

$$\bar{x} = \frac{M_{yz}}{m}, \qquad \bar{y} = \frac{M_{xz}}{m}, \qquad \text{and} \qquad \bar{z} = \frac{M_{xy}}{m} \qquad (4)$$

---

in analogy with Eqs. (1). Analogues to statements (2) and (3) hold for a body in space.

If a body is homogeneous with constant mass density, the center of mass is also called the **centroid of the body,** or the **centroid of the region** that the body occupies.

To compute the center of mass, we simply form the quotients of the first moments by the mass, and we have illustrated how to compute mass and first moments in Section 15.7.

EXAMPLE 1 Find the centroid of the plane region bounded by the curves $y = x^2$ and $y = \sqrt{x}$.

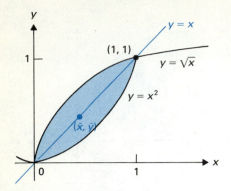

Solution The region is shown in Fig. 15.87. By symmetry, we see that $\bar{x} = \bar{y}$. We compute $\bar{x}$.

For a centroid of a region, we require constant mass density, which we may take to be 1. Then we have

$$m = \int_0^1 \int_{x^2}^{\sqrt{x}} 1 \cdot dy \, dx = \int_0^1 y \Big]_{x^2}^{\sqrt{x}} \, dx = \int_0^1 (\sqrt{x} - x^2) \, dx$$

$$= \left(\frac{2}{3}x^{3/2} - \frac{x^3}{3}\right)\Big]_0^1 = \frac{2}{3} - \frac{1}{3} = \frac{1}{3}.$$

Also

$$M_y = \int_0^1 \int_{x^2}^{\sqrt{x}} x \cdot 1 \, dy \, dx = \int_0^1 xy \Big]_{x^2}^{\sqrt{x}} \, dx = \int_0^1 (x^{3/2} - x^3) \, dx$$

$$= \left(\frac{2}{5}x^{5/2} - \frac{x^4}{4}\right)\Big]_0^1 = \frac{2}{5} - \frac{1}{4} = \frac{3}{20}.$$

Thus

$$\bar{x} = \frac{M_y}{m} = \frac{\frac{3}{20}}{\frac{1}{3}} = \frac{9}{20},$$

and $(\bar{x}, \bar{y}) = (\frac{9}{20}, \frac{9}{20})$. ∎

EXAMPLE 2 Consider a solid bounded by the cylinder $x^2 + y^2 = a^2$ and the planes $z = 0$ and $z = b$. Suppose the mass density of the solid at the point (x, y, z) is kz. Find the center of mass of the solid.

Solution In Example 8 of Section 15.7, we found that $M_{xz} = M_{yz} = 0$ and $M_{xy} = k\pi a^2 b^3/3$. It remains only to compute the mass, which is given by the integral

$$m = \int_0^{2\pi} \int_0^a \int_0^b kzr \, dz \, dr \, d\theta = k \int_0^{2\pi} \int_0^a \frac{z^2}{2}\Big]_0^b r \, dr \, d\theta = k \int_0^{2\pi} \int_0^a \frac{b^2}{2}r \, dr \, d\theta$$

$$= \frac{kb^2}{2} \int_0^{2\pi} \frac{r^2}{2}\Big]_0^a \, d\theta = \frac{kb^2}{2} \int_0^{2\pi} \frac{a^2}{2} \, d\theta = \frac{ka^2b^2}{4}\theta\Big]_0^{2\pi} = \frac{k\pi a^2 b^2}{2}.$$

Hence

$$\bar{z} = \frac{M_{xy}}{m} = \frac{(k\pi a^2 b^3/3)}{(k\pi a^2 b^2/2)} = \frac{2}{3}b,$$

so the center of mass is at the point $(0, 0, 2b/3)$. ∎

EXAMPLE 3 Find the first moment of the body in Example 2 about the plane $z = -b$.

Solution From Example 2, we know that the body has mass $m = k\pi a^2 b^2/2$ and center of mass $(0, 0, 2b/3)$. The distance from this center of mass to the plane $z = -b$ is $(2b/3) - (-b) = 5b/3$. Using property (3) for a body in space, we see that the desired first moment is given by

$$\frac{k\pi a^2 b^2}{2} \cdot \frac{5b}{3} = \frac{5k\pi a^2 b^3}{6}.$$ ∎

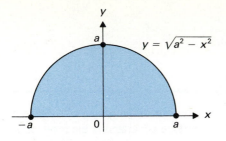

FIGURE 15.88 Half-disk of radius a.

EXAMPLE 4 Find the centroid of the half-disk bounded by the x-axis and the graph $y = \sqrt{a^2 - x^2}$ shown in Fig. 15.88.

Solution We can take mass density 1 and find the center of mass of the half-disk. We do not need to use an integral to find the mass, for we know that the area of a circle of radius a is πa^2, so the mass of this half-disk is $\pi a^2/2$. By symmetry, the x-coordinate of the centroid is $\bar{x} = 0$. We find the moment M_x about the x-axis using a double integral in polar coordinates:

$$M_x = \int_0^\pi \int_0^a (r \sin \theta)r \, dr \, d\theta = \int_0^\pi \frac{r^3}{3}\bigg]_0^a \sin \theta \, d\theta$$

$$= \frac{a^3}{3} \int_0^\pi \sin \theta \, d\theta = \frac{a^3}{3}(-\cos \theta)\bigg]_0^\pi = \frac{a^3}{3}[-(-1) - (-1)]$$

$$= \frac{a^3}{3}(2) = \frac{2}{3}a^3.$$

Therefore,

$$\bar{y} = \frac{M_x}{m} = \frac{2a^3/3}{\pi a^2/2} = \frac{4a}{3\pi}.$$

The centroid is $(\bar{x}, \bar{y}) = (0, 4a/3\pi)$. ∎

EXAMPLE 5 Find the center of mass of a homogeneous wire bent in the shape of a semicircle of radius a.

Solution We consider the wire to cover the curve

$$x = a \cos t, \qquad y = a \sin t, \qquad 0 \le t \le \pi,$$

shown in Fig. 15.89. Since the wire is homogeneous, we consider its mass density to be a constant k. By symmetry of the wire about the y-axis, we realize that $\bar{x} = 0$.

The mass m of the wire is obviously

$$m = k \cdot \text{Length of the wire} = k(\pi a).$$

We integrate to find M_x, which we need to compute $\bar{y}$. Now

$$ds = \sqrt{\left(\frac{dx}{dt}\right)^2 + \left(\frac{dy}{dt}\right)^2} \, dt = \sqrt{a^2\sin^2 t + a^2\cos^2 t} \, dt$$

$$= \sqrt{a^2(\sin^2 t + \cos^2 t)} \, dt = \sqrt{a^2} \, dt = a \, dt.$$

Consequently, $dm = k \, ds = ka \, dt$. Since the distance from the tangent line segment of length ds in Fig. 15.89 to the x-axis is $y = a \sin t$, we have

$$M_x = \int_0^\pi ka(\sin t)(a) \, dt = ka^2(-\cos t)\bigg]_0^\pi$$

$$= ka^2[-(-1) - (-1)] = 2ka^2.$$

Thus

$$\bar{y} = \frac{M_x}{m} = \frac{2ka^2}{\pi ak} = \frac{2a}{\pi}.$$

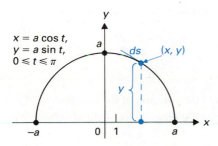

FIGURE 15.89 Tangent line segment of length ds to a semicircular wire.

The center of mass is $(\bar{x}, \bar{y}) = (0, 2a/\pi)$. This is *not* a point of the wire. ∎

Pappus' Theorem

The following is known as Pappus' theorem.

THEOREM 15.2   Pappus' theorem

1. If a plane arc of length L is revolved about an axis in the plane but not intersecting the arc, then the area of the surface generated is the product of the length L of the arc and the circumference of the circle described by the centroid of the arc.

2. If a plane region of area A is revolved about an axis in the plane but not intersecting the region, then the volume of the solid generated is the product of the area A and the circumference of the circle described by the centroid of the region. $\square$

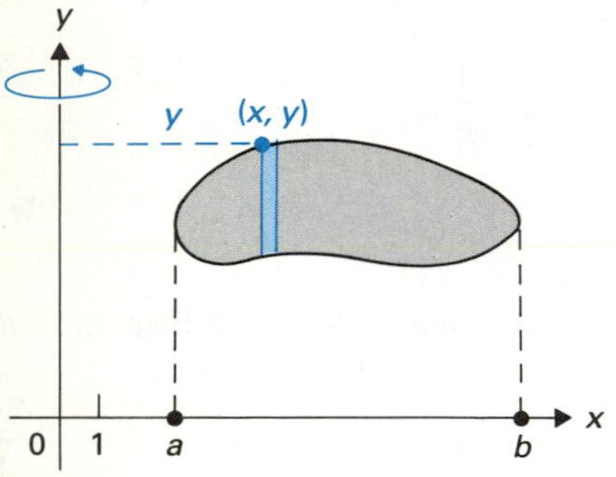

Obviously this theorem makes similar assertions for different dimensional objects. We now show why the second version is true. Consider the region shown in Fig. 15.90, and suppose it is revolved about the y-axis. Using the shell method of Section 6.3, we find the volume of the solid tube

$$V = \int_a^b 2\pi x \, dA = 2\pi \int_a^b x \, dA = 2\pi M_y, \tag{5}$$

where dA is the area of the color-shaded strip. But

$$\bar{x} = \frac{M_y}{A}, \tag{6}$$

so $M_y = \bar{x}A$. Thus from Eq. (5) we have

$$V = 2\pi\bar{x} \cdot A, \tag{7}$$

which is the assertion of Pappus' theorem.

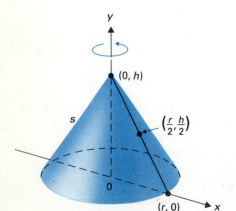

FIGURE 15.90   Differential strip of area dA.

EXAMPLE 6   Use Pappus' theorem to find the surface area of a cone.

Solution   If the line segment joining $(r, 0)$ and $(0, h)$ is revolved about the y-axis, it generates the surface of a cone (see Fig. 15.91). Let

$$s = \sqrt{r^2 + h^2}$$

be the slant height of the cone. The centroid of the line segment is at the point $(r/2, h/2)$, so, by Pappus' theorem, the surface area S of the cone is

$$S = \left(2\pi \cdot \frac{r}{2}\right)s = \pi r s. \qquad \blacksquare$$

FIGURE 15.91   The line segment with centroid $(r/2, h/2)$ generates the surface of the cone.

EXAMPLE 7   Use Pappus' theorem to find the centroid of the half-disk in Fig. 15.88, using the formula $V = \frac{4}{3}\pi a^3$ for the volume of the ball generated by revolving the disk about the x-axis.

Solution   We know the area of the half-disk is $\pi a^2/2$. If its centroid is $(0, \bar{y})$,

then

$$V = \tfrac{4}{3}\pi a^3 = 2\pi\bar{y} \cdot \tfrac{1}{2}\pi a^2 = \pi^2 a^2 \bar{y}.$$

Therefore $\bar{y} = 4a/(3\pi)$, as we found in Example 4. ∎

SUMMARY

1. The center of mass of a plane body is $(\bar{x}, \bar{y})$ where

$$\bar{x} = \frac{M_y}{m} \quad \text{and} \quad \bar{y} = \frac{M_x}{m}.$$

2. The center of mass of a body in space is $(\bar{x}, \bar{y}, \bar{z})$, where

$$\bar{x} = \frac{M_{yz}}{m}, \quad \bar{y} = \frac{M_{xz}}{m}, \quad \bar{z} = \frac{M_{xy}}{m}.$$

Here M_{yz} is the first moment about the y,z-plane, and so on.

The centroid of a region can be computed using the above formulas with constant mass density 1.

3. *Pappus' theorem:* If a plane region (arc) is revolved about an axis in its plane but not intersecting it, then the volume (area) generated is equal to the product of its area (length) and the circumference of the circle described by its centroid.

EXERCISES 15.8

1. Give an example of a plane region whose centroid is not a point in the region.

2. A rectangular sheet of material covers the part of the plane bounded by the coordinate axes and the lines $x = 2$ and $y = 4$. If the mass density of the material at (x, y) is $\sqrt{y}$, find the center of mass.

3. A square sheet of material covers the unit square bounded by the coordinate axes and the lines $x = 1$ and $y = 1$. If the mass density of the material at (x, y) is $x + 1$, find the center of mass.

4. Find the center of mass of a homogeneous sheet of material covering the region bounded by $y = x^2$ and $y = \sqrt{x}$.

5. Find the center of mass of a homogeneous sheet of material covering the region bounded by $y = x^2$ and $y = 1$.

6. Find the centroid of the plane region bounded by $y = x$ and $y = x^3$.

7. Find by integration the centroid of the triangular region whose vertices are $(0, 0)$, $(a, 0)$, $(0, b)$, where $a > 0$ and $b > 0$.

8. Find the centroid of the region bounded by the curves $y = \sqrt{a^2 - x^2}$, $x = a$, and $y = a$, where $a > 0$. [*Hint:* Use symmetry.]

9. Find the center of mass of a sheet of mass density $2y + 3$ that covers the plane region bounded by the curves $y = x^2$ and $y = 4$.

10. Find the centroid of the region bounded by the arc of $y = \sin x$, for $0 \le x \le \pi$, and the x-axis.

11. Find the center of mass of a thin rod on the x-axis covering the interval $[0, 4]$ if the mass density at $(x, 0)$ is $\sqrt{x}$.

12. A wire covers the semicircle $x = a \cos t$, $y = a \sin t$ for $0 \le t \le \pi$. The mass density of the wire at a point (x, y) is y. Find the center of mass of the wire.

Use a calculator for Exercises 13–16.

13. Find the center of mass of a sheet of material covering the disk $x^2 + (y - 4)^2 = 16$ if the mass density at (x, y) is $y + 1$.

14. Find the centroid of the region bounded by $y = \tan x$, $y = 0$, and $x = \pi/3$.

15. Find the center of mass of a homogeneous wire covering the curve $y = \sin x$ for $0 \leq x \leq \pi$.

16. Find the center of mass of a homogeneous wire covering the portion of the parabola $y = x^2$ for $-1 \leq x \leq 1$.

17. Let the half-disk $0 \leq y \leq \sqrt{1 - x^2}$ be covered by a sheet of material of constant mass density ρ. Find the first moment of the sheet about the line $x + y = 4$. [*Hint:* You know the center of mass of the sheet from Example 4.]

18. Let the portion of the face of a vertical dam that is covered by water be a plane region of area A ft^2 whose centroid is at a depth s ft below the surface of the water. Show that the force on the dam is $62.4sA$ lb. (See Section 6.7.)

19. (a) Find the x,y-coordinates of the centroid of the sector of the circle $0 \leq r \leq a$, where $0 \leq \theta \leq \alpha$ for $0 < \alpha \leq 2\pi$.
    (b) Find the limiting position of the centroid in part (a) as $\alpha \to 0$.

20. Find the centroid of the plane region inside the cardioid $r = a(1 + \cos \theta)$ and outside the circle $r = a$.

21. Let the area mass density at a point (x, y) of a flat body covering the square $0 \leq x \leq 1$, $0 \leq y \leq 1$ be xy^2.
    (a) Find the mass of the body.
    (b) Find the center of mass of the body.

22. Consider a flat body covering the plane region bounded by $y = x^2$ and $x = y^2$, and let the area mass density of the body at a point (x, y) be xy. Find the center of mass of the body. [*Hint:* Use symmetry.]

Exercises 23–26 concern a thin plate of mass density $\sigma(x, y)$ covering the upper half-disk $x^2 + y^2 \leq a^2$, $y \geq 0$.

23. Find the center of mass if $\sigma(x, y) = k$.

24. Find the center of mass if $\sigma(x, y) = kx^2y$.

25. Find the center of mass if $\sigma(x, y) = k\sqrt{x^2 + y^2}$.

26. (a) Find the center of mass if $\sigma(x, y) = k(x^2 + y^2)^{n/2}$ for $n \geq 0$.
    (b) Find the limiting position of the center of mass in part (a) as $n \to \infty$.

27. Find the centroid of the hemispherical region in space bounded by $z = \sqrt{a^2 - x^2 - y^2}$ and $z = 0$.

28. Find the centroid of the region in space bounded by $z = 0$, $x^2 + y^2 = 4$, and $z = 1 + x^2 + y^2$.

29. Find the centroid of the region in space bounded by $x = 0$ and $x = 4 - y^2 - z^2$.

Exercises 30 and 31 concern a solid of mass density $\sigma(\rho, \phi, \theta)$ that occupies the region in space bounded above by the sphere $\rho = a$ and below by the cone $\phi = \alpha$, where $0 < \alpha \leq \pi$.

30. (a) Find the center of mass if $\sigma(\rho, \phi, \theta) = k$.
    (b) Find the limiting position of the center of mass in part (a) as $\alpha \to 0$.
    (c) Use the answer to part (a) to find the centroid of the half-ball $0 \leq z \leq \sqrt{a^2 - x^2 - y^2}$.
    (d) Use the answer to part (a) to find the centroid of the solid bounded below by $z = \sqrt{x^2 + y^2}$ and above by $z = \sqrt{a^2 - x^2 - y^2}$.

31. (a) Find the center of mass if $\sigma(\rho, \phi, \theta) = k\rho^n$ for $n \geq 0$.
    (b) Find the limiting position of the center of mass in part (a) as $n \to \infty$.

32. (a) Show that the first moment of a body in the plane about the line $x = -a$ is $M_y + ma$. (This is known as the *parallel axis theorem*.)
    (b) Let a new origin (h, k) be chosen in the plane and let the x'-axis be the line $y = k$ and the y'-axis the line $x = h$. Argue from part (a) that the same location for the center of mass of a body in the plane, relative to the body, is obtained whether one computes coordinates of the center using the x-axis and y-axis or using the x'-axis and y'-axis.

33. State the analogue for space of Exercise 32(a).

34. Let a flat body of constant mass density k cover the unit square $0 \leq x \leq 1$, $0 \leq y \leq 1$ in the plane.
    (a) Find the moment of inertia of the body about the y-axis.
    (b) Find the moment of inertia of the body about the line $x = -a$.
    (c) Find a point (x_1, y_1) in the body such that the moment of inertia of the body about either the x-axis or the y-axis is the product of the mass and the square of the distance from (x_1, y_1) to the axis.
    (d) Find a point (x_2, y_2) in the body such that the moment of inertia of the body about either the line $x = -a$ or the line $y = -a$ is the product of the mass and the square of the distance from (x_2, y_2) to the line.
    (e) Compare the answers to parts (c) and (d), and comment on the result.

35. The **radius of gyration** R of a body about an axis is defined by $R = \sqrt{I/m}$, so that $I = mR^2$.
    (a) From the answer $k/3$ to Exercise 34(a), what is the radius of gyration about the y-axis of a homogeneous flat body covering the square $0 \leq x \leq 1$, $0 \leq y \leq 1$?
    (b) From the answer

$$\frac{k}{3}[(a + 1)^3 - a^3]$$

to Exercise 34(b), what is the radius of gyration about the line $x = -a$ of a homogeneous flat body covering this square?

15.9   CHANGE OF VARIABLES IN MULTIPLE INTEGRALS (OPTIONAL)

An Illustrative Example

Let G be the plane region bounded by the lines

$$x - y = 1, \quad x - y = -1, \quad 2x + y = 2, \quad \text{and} \quad 2x + y = -2$$

shown in Fig. 15.92. Suppose we wish to evaluate

$$\iint_G (x - y)^4 (2x + y)^2 \, dx \, dy. \tag{1}$$

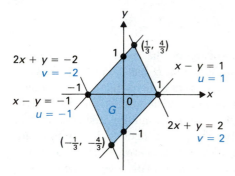

FIGURE 15.92 u and v limits for the region G in the x,y-plane.

This is a very tedious integral to evaluate using an iterated integral in x and y. Due to the shape of the region, integration in x,y-order would require two separate integrals, and integration in y,x-order would require three separate integrals. Moreover, we would have to multiply out $(x - y)^4 (2x + y)^2$ to perform the integration.

It seems reasonable to try the variable substitution

$$u = x - y, \qquad v = 2x + y \tag{2}$$

so that $(x - y)^4 (2x + y)^2$ becomes $u^4 v^2$ and our limits for u are from -1 to 1 and for v are from -2 to 2, as indicated by the colored equations in Fig. 15.92. But by what should we replace $dx \, dy$? Our experience with substitution in an integral $\int_a^b f(x) \, dx$ indicates that we probably should not just replace $dx \, dy$ by $du \, dv$. We answer this question by turning to Riemann sums.

From Eqs. (2), we see that $u + v = 3x$ and $2u - v = -3y$. Thus we have

$$x = \frac{1}{3}u + \frac{1}{3}v, \qquad y = -\frac{2}{3}u + \frac{1}{3}v. \tag{3}$$

We regard Eqs. (3) as giving a **transformation** T of the u,v-plane into the x,y-plane, as illustrated in Fig. 15.93. The rectangular region H bounded by

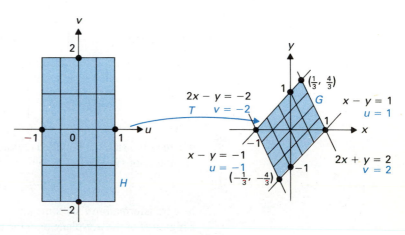

FIGURE 15.93 T carries partitioned H into partitioned G.

$u = -1, u = 1, v = -2, v = 2$ in the u,v-plane is carried into the region G in the x,y-plane, as indicated by the colored equations in Fig. 15.92. The partition of H into 16 rectangles of equal size in Fig. 15.93 is carried into a partition of G into 16 parallelograms of equal area.

We could estimate $\iint_G (x - y)^4(2x + y)^2 \, dx \, dy$ by a Riemann sum using the partition of G into 16 parallelograms, shown in Fig. 15.93. The area of G is the sum of the areas of two congruent triangles, one above and one below the x-axis. The area of each of these triangles is

$$\frac{1}{2}(\text{Length of base})(\text{Altitude}) = \frac{1}{2} \cdot 2 \cdot \frac{4}{3} = \frac{4}{3},$$

so G has area $\frac{8}{3}$, and each of the 16 parallelograms has area $\frac{1}{16} \cdot \frac{8}{3} = \frac{1}{6}$. On the other hand, each of the 16 rectangles in the region H has area $\frac{1}{16} \cdot 8 = \frac{1}{2}$. Thus each rectangle in H is carried into a parallelogram in G having only $\frac{1}{3}$ as much area. Consequently, if we express a Riemann sum arising from the partition of G in Fig. 15.93 in terms of the partition of H, we must multiply the area of each rectangle in H by $\frac{1}{3}$. For this illustration, where the equations describing T are linear, a partition of H into n^2 rectangles of equal area is always carried by T into a partition of G into n^2 parallelograms of equal area. (This may not be true if T is not linear.) Thus the *area change factor* $\frac{1}{3}$ we obtained did not depend on our use of $n = 4$ in our illustration. Taking the limit as $n \to \infty$, so that we form partitions of H into more and more, smaller and smaller rectangles, we see that we should replace $dx \, dy$ by $\frac{1}{3} du \, dv$ when making the variable substitution in Eqs. (2) in our integral. We thus obtain

$$\iint_G (x - y)^4(2x + y)^2 \, dx \, dy = \iint_H u^4 v^2 \frac{1}{3} \, du \, dv = \frac{1}{3} \int_{-2}^{2} \int_{-1}^{1} u^4 v^2 \, du \, dv$$

$$= \frac{1}{3} \int_{-2}^{2} \frac{u^5}{5} \Bigg]_{-1}^{1} v^2 \, dv = \frac{1}{3} \int_{-2}^{2} \frac{2}{5} v^2 \, dv$$

$$= \frac{2}{15} \cdot \frac{v^3}{3} \Bigg]_{-2}^{2} = \frac{2}{15} \cdot \frac{16}{3} = \frac{32}{45}.$$

Thus the integral (1) is readily evaluated using substitution.

Jacobians

With the preceding illustration to guide us, we turn to the general question of changing variables in a double integral

$$\iint_G F(x, y) \, dx \, dy, \tag{4}$$

where G is a suitable bounded region in the x,y-plane and $F(x, y)$ is continuous at all points of G. Let

$$x = f(u, v), \qquad y = g(u, v) \tag{5}$$

be a transformation T of some part of the u,v-plane into the x,y-plane that carries a region H of the u,v-plane onto the region G. We assume that T carries H onto G in a *one-to-one* fashion, so that no two distinct points of H are carried to the same point of G. Moreover, we assume that f and g have continuous partial derivatives throughout H. As in the preceding illustration, consideration of Riemann sums

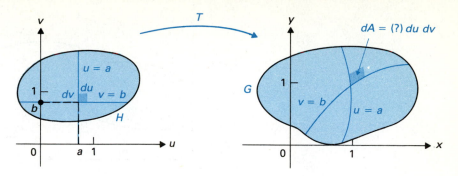

FIGURE 15.94 The image of a differential rectangle under T.

leads us to compare the area $du\ dv$ of a differential rectangle in H with the area of its image under T in G, shown in Fig. 15.94.

The horizontal line $v = b$ in the u,v-plane, shown at the left in Fig. 15.94, is carried by T into a curve in the x,y-plane, shown at the right in the figure. This curve has parametric equations

$$x = f(u,\ b), \qquad y = g(u,\ b),$$

where u is the parameter. We know that the velocity vector to this curve where $u = a$ is

$$\mathbf{c} = f_u(a,\ b)\mathbf{i} + g_u(a,\ b)\mathbf{j}. \tag{6}$$

Similarly,

$$\mathbf{d} = f_v(a,\ b)\mathbf{i} + g_v(a,\ b)\mathbf{j} \tag{7}$$

is the velocity vector at $v = b$ to the curve

$$x = f(a,\ v), \qquad y = g(a,\ v),$$

which is the image of the vertical line $u = a$ under T. If the analogous two velocity vectors were the same at all points of H as they are at $(a,\ b)$, then the partial derivatives of f and of g would be constant on H, so that f and g would be linear functions, like those in Eqs. (3). Under such circumstances, T would carry a *square* in H with sides of length 1 and parallel to the coordinate axes into a *parallelogram* in G having the vectors $\mathbf{c}$ and $\mathbf{d}$ along two adjacent edges (see Fig. 15.95). We could find the area of this parallelogram by computing the magnitude of the cross product

$$\mathbf{c} \times \mathbf{d} = \begin{vmatrix} \mathbf{i} & \mathbf{j} & \mathbf{k} \\ f_u(a,\ b) & g_u(a,\ b) & 0 \\ f_v(a,\ b) & g_v(a,\ b) & 0 \end{vmatrix} = \begin{vmatrix} f_u(a,\ b) & g_u(a,\ b) \\ f_v(a,\ b) & g_v(a,\ b) \end{vmatrix} \mathbf{k}.$$

Thus $|\mathbf{c} \times \mathbf{d}|$ is the absolute value at $(a,\ b)$ of the determinant

$$\begin{vmatrix} \dfrac{\partial x}{\partial u} & \dfrac{\partial y}{\partial u} \\[2mm] \dfrac{\partial x}{\partial v} & \dfrac{\partial y}{\partial v} \end{vmatrix}. \tag{8}$$

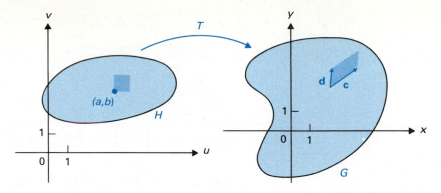

FIGURE 15.95 A square in H corresponds to a parallelogram in G.

We introduce the notation

$$\frac{\partial(x, y)}{\partial(u, v)} = \begin{vmatrix} \dfrac{\partial x}{\partial u} & \dfrac{\partial x}{\partial v} \\ \dfrac{\partial y}{\partial u} & \dfrac{\partial y}{\partial v} \end{vmatrix}, \tag{9}$$

which is equal to the determinant (8), for both are equal to $(\partial x/\partial u)(\partial y/\partial v) - (\partial x/\partial v)(\partial y/\partial u)$. The determinant (9) is the **Jacobian** of the transformation T given by $x = f(u, v)$, $y = g(u, v)$. Our arguments indicate that the absolute value of this Jacobian at (a, b) is the approximate factor by which the area of a very small rectangle in H containing (a, b) is changed when transformed by T, because the partial derivatives f_u, f_v, g_u, and g_v remain almost constant over such a small rectangle. Thus when approximating $\iint_G F(x, y) \, dx \, dy$ by a Riemann sum arising from a partition of H in the u,v-plane, we should multiply the area of each rectangle of the partition of H by the Jacobian of T, evaluated at some point of the rectangle, to get the approximate area of the corresponding part of the partition of G. Consequently, when making a variable substitution $x = f(u, v)$, $y = g(u, v)$ to evaluate the integral, we should replace $dx \, dy$ by

$$\left| \frac{\partial(x, y)}{\partial(u, v)} \right| du \, dv.$$

We summarize our work in a theorem.

THEOREM 15.3   Change of Variable in a Double Integral

Let $F(x, y)$ be continuous in a region G and let $x = f(u, v)$, $y = g(u, v)$ where f and g have continuous first partial derivatives. If the transformation of the u,v-plane into the x,y-plane carries a region H one to one onto G, then

$$\iint_G F(x, y) \, dx \, dy = \iint_H F(f(u, v), g(u, v)) \left| \frac{\partial(x, y)}{\partial(u, v)} \right| du \, dv. \tag{10}$$

$\square$

EXAMPLE 1 Compute the Jacobian of the transformation given by Eqs. (3), and compare the result with the area change factor obtained on page 838.

MULTIPLE INTEGRATION appeared in print beginning in the 1690s, but the first detailed discussion of the meaning of a double integral was in a paper of Leonhard Euler in 1769. Euler posed the question of how a double integral changes under a change of variables. Euler's first attempt at devising a change of variable formula was purely formal. He noted that if x, y and v, t are related by a clockwise rotation through angle θ, then

$$x = mt + \sqrt{1 - m^2}\, v,$$
$$y = \sqrt{1 - m^2}\, t - mv,$$

where $m = \cos \theta$. It follows that

$$dx = m\, dt + \sqrt{1 - m^2}\, dv,$$
$$dy = \sqrt{1 - m^2}\, dt - m\, dv.$$

Formal multiplication of these two equations gives

$$\begin{aligned} dx\, dy = {} & m\sqrt{1 - m^2}\,(dt)^2 \\ & + (1 - 2m^2)\, dt\, dv \\ & - m\sqrt{1 - m^2}\,(dv)^2, \end{aligned}$$

a result that Euler noted was obviously wrong and even meaningless. Realizing that $dx\, dy$ should equal $dt\, dv$ (since rotation does not change areas), he developed a new procedure, changing variables one at a time, which led him to the result of Theorem 15.3.

The corresponding result in three dimensions was developed by Joseph Louis Lagrange in 1773 in connection with a problem on determining the gravitational attraction that an elliptical spheroid exercises on any point on its surface or in its interior.

Solution The transformation is given by

$$x = \frac{1}{3}u + \frac{1}{3}v, \qquad y = -\frac{2}{3}u + \frac{1}{3}v.$$

We find that

$$\frac{\partial(x,\, y)}{\partial(u,\, v)} = \begin{vmatrix} \dfrac{\partial x}{\partial u} & \dfrac{\partial x}{\partial v} \\[2mm] \dfrac{\partial y}{\partial u} & \dfrac{\partial y}{\partial v} \end{vmatrix} = \begin{vmatrix} \dfrac{1}{3} & \dfrac{1}{3} \\[2mm] \dfrac{-2}{3} & \dfrac{1}{3} \end{vmatrix} = \frac{1}{9} + \frac{2}{9} = \frac{1}{3}.$$

This agrees with the area change factor found on page 838. ∎

EXAMPLE 2 Find the Jacobian of the transformation $x = r \cos \theta$, $y = r \sin \theta$, and justify the replacement of $dx\, dy$ by $r\, dr\, d\theta$ when transforming an integral from rectangular to polar coordinates.

Solution We have

$$\begin{aligned} \frac{\partial(x,\, y)}{\partial(r,\, \theta)} &= \begin{vmatrix} \dfrac{\partial x}{\partial r} & \dfrac{\partial x}{\partial \theta} \\[2mm] \dfrac{\partial y}{\partial r} & \dfrac{\partial y}{\partial \theta} \end{vmatrix} = \begin{vmatrix} \cos \theta & -r \sin \theta \\ \sin \theta & r \cos \theta \end{vmatrix} \\ &= r \cos^2 \theta + r \sin^2 \theta \\ &= r(\cos^2 \theta + \sin^2 \theta) = r(1) = r. \end{aligned}$$

Thus when changing from rectangular to polar coordinates in an integral, we should replace $dx\, dy$ by $|r|\, dr\, d\theta$. When integrating in polar coordinates, we always took r to be nonnegative, so the absolute value sign was not necessary. ∎

We emphasize the interpretation of the Jacobian of a transformation $x = f(u, v)$, $y = g(u, v)$. The absolute value of the Jacobian at each point is the **area change factor** of the transformation at that point. The Jacobian can be either positive or negative. It can be shown that it is positive at a point where the transformation *preserves rotation direction,* that is, where a very small circle traced counterclockwise about the point is carried into a curve again traced counterclockwise about the image point. In Exercise 9, we ask you to show that for a transformation $y = f(x)$ of the x-axis into the y-axis, the derivative $f'(x)$ has a similar interpretation as a change factor of length. If we were forced to come up with *one number* to serve as the "derivative" at (a, b) of a transformation of the u,v-plane into the x,y-plane, the Jacobian $\partial(x, y)/\partial(u, v)$ at (a, b) would be it! The *Jacobian matrix* (square array) whose determinant is $\partial(x, y)/\partial(u, v)$ should be regarded as the **matrix derivative** of the transformation. It contains all rate-of-change information for the transformation.

EXAMPLE 3 Find the area change factor of the transformation

$$x = u^2 - 2uv, \qquad y = uv^3 - 3v + 4$$

at the point $(u, v) = (2, -1)$.

Solution We simply compute the Jacobian

$$\frac{\partial(x, y)}{\partial(u, v)}$$

at the point $(u, v) = (2, -1)$. Computing partial derivatives at this point, we find that

$$\frac{\partial x}{\partial u}\bigg|_{(2,-1)} = (2u - 2v)\bigg|_{(2,-1)}$$
$$= 4 + 2 = 6,$$

$$\frac{\partial x}{\partial v}\bigg|_{(2,-1)} = -2u\bigg|_{(2,-1)}$$
$$= -4,$$

$$\frac{\partial y}{\partial u}\bigg|_{(2,-1)} = v^3\bigg|_{(2,-1)}$$
$$= -1,$$

$$\frac{\partial y}{\partial v}\bigg|_{(2,-1)} = (3uv^2 - 3)\bigg|_{(2,-1)}$$
$$= 6 - 3 = 3.$$

Thus

$$\frac{\partial(x, y)}{\partial(u, v)}\bigg|_{(2,-1)} = \begin{vmatrix} 6 & -4 \\ -1 & 3 \end{vmatrix} = 18 - 4 = 14,$$

so 14 is the area change factor at $(2, -1)$. Since $14 > 0$, rotation direction is preserved at $(u, v) = (2, -1)$. ∎

Changing Variables in Triple Integrals

Precisely the same analysis using Riemann sums can be made for changing variables in triple integrals. We suppose that we have a transformation T of u,v,w-space into x,y,z-space given by

$$x = f(u, v, w), \qquad y = g(u, v, w), \qquad z = h(u, v, w),$$

where f, g, and h have continuous first partial derivatives. If a region H in u,v,w-space is transformed one to one onto a region G in x,y,z-space, then the velocity vectors

$$f_u\mathbf{i} + g_u\mathbf{j} + h_u\mathbf{k}, \qquad f_v\mathbf{i} + g_v\mathbf{j} + h_v\mathbf{k}, \qquad \text{and} \qquad f_w\mathbf{i} + g_w\mathbf{j} + h_w\mathbf{k}$$

replace the velocity vectors in Eqs. (6) and (7), and can be regarded as emanating from a corner of a (skew) box at the point $T(u, v, w)$ in x,y,z-space. We know that the volume of this box is given by the absolute value of the triple scalar product

$$\begin{vmatrix} f_u & g_u & h_u \\ f_v & g_v & h_v \\ f_w & g_w & h_w \end{vmatrix}. \tag{11}$$

In Exercise 1, we ask you to show that this determinant is equal to the Jacobian

$$\frac{\partial(x, y, z)}{\partial(u, v, w)} = \begin{vmatrix} \dfrac{\partial x}{\partial u} & \dfrac{\partial x}{\partial v} & \dfrac{\partial x}{\partial w} \\[2mm] \dfrac{\partial y}{\partial u} & \dfrac{\partial y}{\partial v} & \dfrac{\partial y}{\partial w} \\[2mm] \dfrac{\partial z}{\partial u} & \dfrac{\partial z}{\partial v} & \dfrac{\partial z}{\partial w} \end{vmatrix}. \qquad (12)$$

The absolute value of this Jacobian is thus the **volume change factor** of T at (u, v, w). It can be shown that the Jacobian is positive if a very small right-threaded screw at (u, v, w) is carried by T into a (perhaps warped) right-threaded screw at the image point $T(u, v, w)$.

Applying this volume change factor to changing variables in an integral, suppose that $F(x, y, z)$ is a continuous function. Then we see that

$$\iiint_G F(x, y, z)$$
$$= \iiint_H F(f(u, v, w), \ g(u, v, w), \ h(u, v, w)) \left| \frac{\partial(x, y, z)}{\partial(u, v, w)} \right| du \, dv \, dw. \quad (13)$$

Exercise 8 asks you to compute the Jacobian of the transformation from rectangular coordinates to spherical coordinates, justifying the replacement of $dx \, dy \, dz$ by $\rho^2 \sin \phi \, d\rho \, d\phi \, d\theta$ when changing an integral from rectangular to spherical coordinates. We conclude with an example involving a *linear* transformation in space, quite similar to our opening illustrative example in the plane.

EXAMPLE 4 Make a variable transformation to compute

$$\iiint_G (x - y + z)^2(x + y + 2z)^3(2x - z)^2 \, dx \, dy \, dz,$$

where G is the skew box bounded by the planes

$$x - y + z = -1, \qquad x - y + z = 1,$$
$$x + y + 2z = 0, \qquad x + y + 2z = 2,$$
$$2x - z = -2, \qquad 2x - z = 1.$$

Solution Of course, we try the substitution

$$\begin{aligned} u &= x - y + z, \\ v &= x + y + 2z, \\ w &= 2x - z. \end{aligned} \qquad (14)$$

We need to compute

$$\frac{\partial(x, y, z)}{\partial(u, v, w)},$$

which at first suggests that we should solve for x, y, and z in terms of u, v, and w in Eqs. (14). We can avoid solving for x, y, and z by the following device.

We compute that

$$\frac{\partial(u, v, w)}{\partial(x, y, z)} = \begin{vmatrix} 1 & -1 & 1 \\ 1 & 1 & 2 \\ 2 & 0 & -1 \end{vmatrix}$$

$$= 1\begin{vmatrix} 1 & 2 \\ 0 & -1 \end{vmatrix} - (-1)\begin{vmatrix} 1 & 2 \\ 2 & -1 \end{vmatrix} + 1\begin{vmatrix} 1 & 1 \\ 2 & 0 \end{vmatrix}$$

$$= 1(-1) + 1(-5) + 1(-2) = -8,$$

so that as we go from x,y,z-space to u,v,w-space, the volume change factor is $|-8| = 8$. Consequently, as we go in the other direction, from u,v,w-space to x,y,z-space, the change factor is $\frac{1}{8}$. Therefore,

$$\left| \frac{\partial(x, y, z)}{\partial(u, v, w)} \right| = \frac{1}{8}.$$

Thus our integral becomes

$$\int_{-2}^{1} \int_{0}^{2} \int_{-1}^{1} (u^2 v^3 w^2) \frac{1}{8} \, du \, dv \, dw = \frac{1}{8} \int_{-2}^{1} \int_{0}^{2} \frac{u^3}{3} \Bigg]_{-1}^{1} v^3 w^2 \, dv \, dw$$

$$= \frac{1}{8} \cdot \frac{2}{3} \int_{-2}^{1} \frac{v^4}{4} \Bigg]_{0}^{2} w^2 \, dw$$

$$= \frac{1}{8} \cdot \frac{2}{3} \cdot 4 \frac{w^3}{3} \Bigg]_{-2}^{1}$$

$$= \frac{1}{3} \left(\frac{1}{3} - \frac{-8}{3} \right) = \frac{1}{3} \cdot \frac{9}{3} = 1. \quad \blacksquare$$

Let T be a continuously differentiable transformation of u,v,w-space into x,y,z-space. Arguing as in Example 4, we see that if (u, v, w) and (x, y, z) are corresponding points under T, then

$$\frac{\partial(x, y, z)}{\partial(u, v, w)} = \frac{1}{\dfrac{\partial(u, v, w)}{\partial(x, y, z)}},$$

provided that these Jacobians are nonzero. Of course, an analogous result holds for two-variable transformations. This property of Jacobians often saves us from solving for x, y, and z in terms of u, v, and w in cases where it is more natural to express u, v, and w in terms of x, y, and z as was the case in Example 4.

SUMMARY

1. Let a transformation T of the u,v-plane into the x,y-plane be given by $x = f(u, v)$, $y = g(u, v)$ where f and g have continuous partial derivatives. The magnitude of the Jacobian

$$\frac{\partial(x, y)}{\partial(u, v)} \qquad \text{at } (u, v) = (a, b)$$

is the area change factor of T at (a, b). If this Jacobian is positive at (a, b), then T preserves rotation direction there, while rotation direction is reversed if the Jacobian is negative.

2. Suppose that T defined in number 1 carries a region H of the u,v-plane one to one onto a region G of the x,y-plane. Then

$$\iint_G F(x, y) \, dx \, dy = \iint_H F(f(u, v), \, g(u, v)) \left| \frac{\partial(x, y)}{\partial(u, v)} \right| du \, dv.$$

3. At corresponding points (u, v) and (x, y) for T in number 1, we have

$$\frac{\partial(x, y)}{\partial(u, v)} = \frac{1}{\dfrac{\partial(u, v)}{\partial(x, y)}}$$

if these Jacobians are nonzero.

4. Analogous results hold for a transformation T of u,v,w-space to x,y,z-space.

EXERCISES 15.9

1. Show that

$$\begin{vmatrix} a & b & c \\ d & e & f \\ g & h & i \end{vmatrix} = \begin{vmatrix} a & d & g \\ b & e & h \\ c & f & i \end{vmatrix}$$

and deduce that the determinants (11) and (12) in the text are equal.

In Exercises 2–7, compute the Jacobian of the transformation defined by the given equations, and find the area or volume change factor at the indicated point.

2. $x = uv$, $y = \sin v$; point $(u, v) = (1, 0)$

3. $x = u^2 - v^2$, $y = 2uv$; point $(u, v) = (1, -2)$

4. $x = u/v$, $y = vu^2$; point $(u, v) = (-1, 2)$

5. $x = u - v + 2$, $y = v - w$, $z = u^2 - 2vw$; point $(u, v, w) = (-1, 1, 2)$

6. $x = 2u^2 - v^2 + 2w$, $y = \sin uvw$, $z = we^{uv}$; point $(u, v, w) = (0, -2, -2)$

7. $x = 3u^2vw$, $y = uvw^2$, $z = v/w$; point $(u, v, w) = (-1, 2, 1)$

8. Compute $\partial(x, y, z)/\partial(\rho, \phi, \theta)$ for the transformation equations from rectangular to spherical coordinates, and justify replacing $dx \, dy \, dz$ by $\rho^2 \sin \phi \, d\rho \, d\phi \, d\theta$ when changing an integral from rectangular to spherical coordinates.

9. Let $y = f(x)$ be a differentiable one-to-one function, and let $f'(x)$ be continuous.

(a) Argue from the graph of f that $|f'(a)|$ can be considered the length change factor at $x = a$ of the transformation of the x-axis into the y-axis given by $y = f(x)$.

(b) Give the interpretation of the sign of $f'(a)$ for the transformation of the x-axis into the y-axis.

In Exercises 10–16, transform the integral using an appropriate transformation, and evaluate the transformed integral.

10. $\int_{-1}^{0} \int_{y}^{-y} (x - y)(x + y)^4 \, dx \, dy$

11. $\iint_G (x - y)^4 \, dx \, dy$ where G is the parallelogram in the x,y-plane bounded by the lines $x - y = 1$, $x - y = 3$, $2x + 3y = 0$, and $2x + 3y = 5$

12. $\iint_G (x - y + 1)^2(2x + 3y - 1)^3 \, dx \, dy$ for the region G in Exercise 11, using the transformation $u = x - y$, $v = 2x + 3y$

13. $\iint_G (x + y)\cos^2(x - y) \, dx \, dy$ where G is the parallelogram having vertices $(0, -\pi)$, $(\pi, 0)$, $(2\pi, -\pi)$, and $(\pi, -2\pi)$

14. $\int_0^2 \int_{-x}^{x^2} 1/(x + y + 1)^2 \, dy \, dx$, letting $x = u + v$ and $y = u^2 - v$

15. $\iint_G (x - 2y + 3)^2 e^{2x+y+1} \, dx \, dy$ where G is the parallelogram bounded by the lines $x - 2y = 0$, $x - 2y = 3$, $2x + y = -1$, and $2x + y = 3$, letting $u = x - 2y + 3$ and $v = 2x + y + 1$

16. $\iiint_G (x - y - 2)^3(x + z)^2(2x - 3y + 2z)^2$ where G is the skew box bounded by the planes $x - y = 1$, $x - y = 2$, $x + z = -1$, $x + z = 1$, $2x - 3y + 2z = 0$, $2x - 3y + 2z = 1$, using $u = x - y$, $v = x + z$, and $w = 2x - 3y + 2z$

In Exercises 17–20, change variables to simplify the integral, and then evaluate the transformed integral.

17. $\int_0^a \int_0^{\sqrt{a^2-y^2}} (x^2 + y^2)^{3/2} \, dx \, dy$

18. $\int_{-1}^1 \int_{-2\sqrt{1-y^2}}^{2\sqrt{1-y^2}} (x^2 + 4y^2)^{3/2} \, dx \, dy$
[*Hint:* Use a transformation of the form $x = ar \cos \theta$, $y = br \sin \theta$ that simplifies $x^2 + 4y^2$.]

19. $\int_0^4 \int_0^{\sqrt{4-z^2/4}} \int_0^{\sqrt{4y^2+z^2}} x\sqrt{z^2 + 4y^2} \, dx \, dy \, dz$. [*Hint:* Use a transformation that simplifies $z^2 + 4y^2$, analogous to the type suggested in the hint of Exercise 18.]

20. $\int_0^{1/\sqrt{2}} \int_x^{\sqrt{1-x^2}} \int_0^{\sqrt{1-x^2-y^2}} z\sqrt{x^2 + y^2} \, dz \, dy \, dx$

21. Use *elliptic coordinates* $x = 3r \cos \theta$, $y = 4r \sin \theta$ to find the area bounded by the ellipse $x^2/9 + y^2/16 = 1$.

22. Use the elliptic coordinates in Exercise 21 to integrate $F(x, y) = x^2$ over the elliptical disk $x^2/9 + y^2/16 \le 1$.

23. Find the area of the region bounded by the curves
$$xy = 1, \quad xy = 4, \quad xy^2 = 1, \quad \text{and} \quad xy^2 = 2.$$
[*Hint:* Make use of number 3 of the Summary.]

24. Find the area of the region bounded by the curves $xy = 1$, $xy = 2$ and the lines $y = x$ and $2y = x$. [*Hint:* Write the equations of the lines as $x/y = 1$ and $x/y = 2$.]

25. Find the area of the first-quadrant region bounded by all four of the curves $y = x^2$, $y = 2x^2$, $x = y^2$, and $x = 2y^2$. [*Hint:* Write the equations as $y/x^2 = 1$, $y/x^2 = 2$, $x/y^2 = 1$, and $x/y^2 = 2$.]

26. Find the volume of the region bounded by the ellipsoid $x^2/a^2 + y^2/b^2 + z^2/c^2 = 1$. [*Hint:* Use the transformation $x = au$, $y = bv$, $z = cw$ and recall that a sphere of radius r has volume $\frac{4}{3}\pi r^3$.]

SUPPLEMENTARY EXERCISES FOR CHAPTER 15

1. Find the regular Riemann sum with $n = 2$, using midpoints of subrectangles, approximating the integral of the function $f(x, y) = xy^2$ over the rectangle $1 \le x \le 3$, $-1 \le y \le 3$.

2. Find the regular Riemann sum with $n = 2$, using midpoints of the subrectangles, approximating the integral of the function $f(x, y) = 3x - 2y$ over the rectangle $-1 \le x \le 3$, $1 \le y \le 3$.

3. Use Simpson's rule with a calculator or computer to estimate
$$\int_0^3 \int_0^1 \frac{x^2y}{y + \cos^2 y} \, dy \, dx.$$

4. Use a calculator or computer and Simpson's rule to estimate
$$\int_0^2 \int_0^1 \frac{y}{1 + \sin x^2} \, dx \, dy.$$

5. Compute $\int_0^1 \int_1^2 (x/y^2) \, dy \, dx$.

6. Compute $\int_4^{-2} \int_{-1}^3 (2xy - 3y^2) \, dy \, dx$.

7. Compute $\int_1^3 \int_0^2 (x + y)/(y + 1) \, dx \, dy$.

8. Compute $\int_0^1 \int_y^1 e^{x^2+1} \, dx \, dy$.

9. Compute $\int_{-1}^2 \int_1^4 (3xy^2 - 2y) \, dx \, dy$.

10. Express $\int_{-2}^2 \int_{x^2}^4 (x^2 - 3xy) \, dy \, dx$ in the form
$$\int\int (x^2 - 3xy) \, dx \, dy,$$
reversing the order of integration. Do not compute either integral.

11. Compute, using an iterated integral, the area of the region in the plane bounded by $y = x^2$ and $y = x$.

12. Compute, using an iterated integral, the area of the region in the plane bounded by $y = 1/x$ and $x + y = \frac{5}{2}$.

13. Find the integral of the polar function $h(r, \theta) = r \cos \theta$, $r \ge 0$, over the circle $r = 2a \cos \theta$.

14. Using double integration in polar coordinates, find the area of one loop of the rose $r = \cos 3\theta$.

15. Transform $\int_0^1 \int_x^1 (x^2 + y^2)^{3/2} \, dy \, dx$ into polar coordinates. Do not evaluate the integral.

16. Find the area of the surface $z = 16 - x^2 - y^2$ that lies above the plane $z = 12$.

17. Find the area of the portion of the surface of the sphere $x^2 + y^2 + z^2 = a^2$ that lies above the plane $z = b$ for $0 \le b \le a$.

18. Find the area of the portion of the plane $x - 3y + 2z = 4$ inside the elliptic cylinder $(x^2/4) + (y^2/9) = 1$.

19. Find the area of the portion of the sphere $x^2 + y^2 + z^2 = 16$ inside the cone $\sqrt{3}z = \sqrt{x^2 + y^2}$.

20. Express $\int_0^2 \int_0^{\sqrt{4-y^2}} \int_{x^2+y^2}^4 x^2z \, dz \, dx \, dy$ in the form
$$\int\int\int x^2z \, dx \, dy \, dz,$$
changing the order of integration. Do not evaluate either integral.

21. Express $\int_{-2}^2 \int_0^{\sqrt{4-y^2}} \int_0^{4-x^2-y^2} yz \, dz \, dx \, dy$ in the x,y,z-order.

22. Find the volume of the region in space bounded above by $z = 2x + y$, on the sides by $x = 0$, $y = 0$, and $x + y = 1$, and below by $z = -1$.

23. Compute, using an iterated integral, the volume of the region in space bounded below by $z = 0$, on the sides by $x = 0$, $y = 0$, and $x + y = 2$, and above by $z = x^2 + y^2$.

24. Express $\int_0^1 \int_0^3 \int_{\sqrt{1-z}}^{1-z} yz^2 \, dx \, dy \, dz$ in the z,x,y-order.

25. Find the volume of the region in space that is bounded by $z = x^2 + y^2$ and $z = \sqrt{x^2 + y^2}$.

26. Express as an iterated integral the volume of the region in space bounded below by $z = x^2 + y^2$ and above by the hemisphere $z - 4 = \sqrt{4 - x^2 - y^2}$. Do not compute the integral.

27. Transform the integral $\int_{\pi/2}^{\pi} \int_0^1 \int_0^{r^2} r^2 \, dz \, dr \, d\theta$ into rectangular coordinates. Do not evaluate the integral.

28. Find the integral of the cylindrical coordinate function $h(r, \theta, z) = rz \sin^2\theta, r \geq 0$, over the region bounded above by $z = 4 + x^2 + y^2$, below by $z = 0$, and on the sides by $x^2 + y^2 = 4$.

29. Use triple integration in cylindrical coordinates to find the volume of the solid bounded by the paraboloids $z = x^2 + y^2$ and $z = 8 - x^2 - y^2$.

30. Find the volume of the region bounded above by $z = \sqrt{16 - x^2 - y^2}$ and below by $z = \sqrt{3x^2 + 3y^2}$.

31. Transform $\int_0^{2\pi} \int_0^1 \int_{r^2}^{r} r \, dz \, dr \, d\theta$ into spherical coordinates. Do not evaluate the integral.

32. Integrate the spherical coordinate function

$$f(\rho, \phi, \theta) = \rho^2 \cos^2\theta$$

over the ball $\rho \leq a$.

33. Find the integral of the spherical coordinate function $h(\rho, \phi, \theta) = \rho \cos^2\theta$ over the ball $0 \leq \rho \leq a$.

34. Use triple integration in spherical coordinates to find the volume of the solid bounded by the cone $z^2 = 3x^2 + 3y^2$ and the hemisphere $z = \sqrt{16 - x^2 - y^2}$.

35. A thin plate covers the cardioid $r = a(1 + \cos\theta)$. If the mass density of the plate is $\sigma(r, \theta) = kr$, find the mass of the plate.

36. The mass density of a solid bounded above by $z = \sqrt{4 - x^2 - y^2}$ and below by $z = \sqrt{x^2 + y^2}$ is $\sigma(\rho, \phi, \theta) = k\rho$. Find the moment of inertia of the solid about the z-axis.

37. Find the moment of inertia of a solid ball $x^2 + y^2 + z^2 \leq a^2$ about the z-axis if the mass density of the ball at (x, y, z) is given by $|z|$.

38. Let a thin plate cover the disk $r \leq 4$ and have mass density $\sigma(r, \theta) = kr^2$. Find the absolute value of the first moment of the plate about the line $3x + 4y = 100$.

39. Find the centroid of the region bounded below by $z = 0$, above by $z = 1 - x^2$, and on the ends by $y = 0$ and $y = 4$.

40. Find the centroid of the region in space bounded by $z = -1$ and $z = 3 - x^2 - y^2$.

41. Use a change of variables to compute

$$\iint_G (x - y)^4 (3x + 2y)^5 \, dx \, dy$$

where G is the region bounded by the lines $x - y = 1$, $x - y = 3$, $3x + 2y = -1$, and $3x + 2y = 2$.

42. Use a change of variables to compute

$$\iiint_G (x + 2y - 1)^2 (x - z)^3 \, dx \, dy \, dz$$

where G is the region bounded by the planes $x + 2y = -1$, $x + 2y = 1$, $x - z = 0$, $x - z = 2$, $2x + y - z = 0$, and $2x + y - z = 3$.

43. Use a change of variables to compute

$$\int_{-1}^1 \int_0^{2\sqrt{1-y^2}} x^2 \, dx \, dy.$$

44. Use a change of variables to compute the area of the plane region bounded by $xy = 1$, $xy = 2$, $x\sqrt{y} = 1$, and $x\sqrt{y} = 3$.

45. Use integration to find the four-dimensional "volume" of the "4-ball of radius a" consisting of all (x, y, z, w) such that $x^2 + y^2 + z^2 + w^2 \leq a^2$. [*Hint:* Use (ρ, ϕ, θ, w)-coordinates where ρ, ϕ, and θ are the usual spherical coordinates replacing x, y, and z. This is analogous to cylindrical coordinates (r, θ, z) in space compared with the polar coordinates (r, θ) in the plane.]

46. Work Exercise 45 again, but this time use Pappus' theorem and the fact that the centroid of the half-ball

$$x^2 + y^2 + z^2 \leq a^2,$$

where $z \geq 0$, is $(0, 0, 3a/8)$. (See Exercise 27 of Section 15.8.)

47. Find the three-dimensional volume of the "3-sphere of radius a" consisting of all (x, y, z, w) such that $x^2 + y^2 + z^2 + w^2 = a^2$. [*Hint:* Recall that for a 3-ball $x^2 + y^2 + z^2 \leq a^2$ of volume $\frac{4}{3}\pi a^3$, the area of the surface of the 2-sphere $x^2 + y^2 + z^2 = a^2$ is $4\pi a^2$. Consider the relation between $V = \frac{4}{3}\pi r^3$ and $A = 4\pi r^2$ in terms of approximation of the volume V by a differential for a small change in radius. Then jump up one dimension and answer the new problem with essentially no work, using the answer to Exercise 45.]

If we partition the rectangle $0 \leq x \leq 4$, $-2 \leq y \leq 6$ into n^2 subrectangles of equal areas, then each subrectangle has area $32/n^2$. Using Riemann sums with the upper right corner of each subrectangle as the place to evaluate the function, we see that

$$\int_{-2}^6 \int_0^4 x^2 y \, dx \, dy = \lim_{n \to \infty} \sum_{i,j=1}^n \left[\left(\frac{4i}{n}\right)^2 \left(-2 + \frac{8j}{n}\right) \left(\frac{32}{n^2}\right) \right].$$

Exercises 48–51 give you some practice in writing double integrals as limits of double sums and in estimating some double sums using integrals.

48. Write a sum whose limit as $n \to \infty$ is equal to

$$\int_1^7 \int_{-3}^0 (x + 4y)^3 \, dx \, dy.$$

49. Repeat Exercise 48 for $\int_5^{10} \int_{-2}^2 (x^2 + 3xy) \, dy \, dx$.

50. Use an integral to estimate

$$\sum_{i,j=1}^{100} \left[\frac{3i}{100} \left(-1 + \left(\frac{2j}{100}\right)^2\right) \frac{6}{10{,}000} \right].$$

51. Use an integral to estimate $(8/10^7) \sum_{i,j=1}^{10} i^2 j^3$.

Exercises 52–54 deal with a simple algebraic device that is useful in multivariable calculus. We define a type of product of differential expressions, which is written using the symbol $\wedge$. All the properties of regular multiplication hold for this $\wedge$-product except that

$$dx \wedge dy = -dy \wedge dx$$

while

$$dx \wedge dx = dy \wedge dy = 0.$$

With three variables, we have the logical extension of such relations. For example,

$$dx \wedge dy \wedge dz = -dx \wedge dz \wedge dy$$

and

$$dy \wedge dz \wedge dy = 0.$$

Roughly speaking, a term containing the $\wedge$-product differentials like dx, dy, and so on is zero if two differentials are the same and changes sign if two differentials are interchanged. For a further illustration,

$$(x^2 \, dx + y \, dy) \wedge (2y \, dx - x^2 y \, dy)$$
$$= 2x^2 y \, dx \wedge dx - x^4 y \, dx \wedge dy + 2y^2 \, dy \wedge dx$$
$$- x^2 y^2 \, dy \wedge dy$$
$$= -x^4 y \, dx \wedge dy + 2y^2 \, dy \wedge dx$$
$$= -x^4 y \, dx \wedge dy - 2y^2 \, dx \wedge dy$$
$$= -(x^4 y + 2y^2) \, dx \wedge dy.$$

52. Compute

$$(x \, dx + x^2 y \, dy - xz \, dz) \wedge (yz \, dx + xz \, dy - z^2 \, dz)$$

as in the preceding illustration, simplifying as much as possible.

Exercises 53 and 54 indicate that this new multiplication permits changing variables in multiple integrals by *substitution*.

53. Recall that for polar coordinates, $x = r \cos \theta$ and $y = r \sin \theta$.
   (a) Compute dx and dy in terms of the polar coordinate variables.
   (b) Using part (a), compute $dx \wedge dy$ in terms of polar coordinates, simplifying as much as possible.

54. Recall that, for spherical coordinates,

$$x = \rho \sin \phi \cos \theta, \qquad y = \rho \sin \phi \sin \theta,$$
$$z = \rho \cos \phi.$$

   (a) Compute dx, dy, and dz in terms of spherical coordinates.
   (b) Using part (a), compute $dx \wedge dy \wedge dz$ in terms of spherical coordinates, simplifying as much as possible.

SOME TOPICS IN VECTOR CALCULUS

16

This chapter provides an introduction to some of the integral theorems of vector calculus. As we will indicate, these theorems are of great significance in physics, especially in the study of work and fluid dynamics.

The chapter opens with a section on differential forms that plays a role similar to the study we made of antiderivatives in Section 4.8, before we turned to integrals. The chapter concludes with three sections on divergence theorems, Green's theorem, and Stokes' theorem. All of these theorems are higher-dimensional analogues of the fundamental theorem of calculus that we presented in Section 5.3.

16.1   EXACT DIFFERENTIAL FORMS

Recall that if $H(x, y)$ has first partial derivatives, then the differential dH is given by

$$dH = H_x(x, y)\, dx + H_y(x, y)\, dy. \tag{1}$$

In this section, we consider expressions similar to Eq. (1), of the form

$$P(x, y)\, dx + Q(x, y)\, dy \tag{2}$$

for two variables, or

$$P(x, y, z)\, dx + Q(x, y, z)\, dy + R(x, y, z)\, dz \tag{3}$$

for three variables. We will see that the expressions (2) and (3) need not always be differentials of functions. For this reason, we consider them to be *differential forms* rather than *differentials*.

DEFINITION 16.1

Differential form

A **differential form** is an expression of type (2) or (3) for two or three variables and of type

$$f_1(\mathbf{x}) \, dx_1 + f_2(\mathbf{x}) \, dx_2 + \cdots + f_n(\mathbf{x}) \, dx_n$$

for n variables, where $\mathbf{x} = (x_1, x_2, \ldots, x_n)$.

In Section 5.5, we saw how to solve some differential equations by separating variables. For example, we can solve the equation $dy/dx = x^2/y^2$ as follows:

$$\frac{dy}{dx} = \frac{x^2}{y^2},$$

$$y^2 \, dy = x^2 \, dx,$$

$$\frac{y^3}{3} = \frac{x^3}{3} + C.$$

However, if we try that technique to solve the differential equation $dy/dx = -3x^2y/(x^3 + 6y)$, we run into a problem. We have

$$\frac{dy}{dx} = \frac{-3x^2y}{x^3 + 6y},$$

$$(x^3 + 6y) \, dy = -3x^2y \, dx. \tag{4}$$

We cannot "separate" the variables. However, let us write the equation in the form

$$3x^2y \, dx + (x^3 + 6y) \, dy = 0. \tag{5}$$

The left-hand side of Eq. (5) is a differential form. If it is the differential of a function $H(x, y)$, then Eq. (5) can be written as

$$dH = 0.$$

If we set $H(x, y) = C$, we then have the implicit general solution of the differential equation (4). That is, any differentiable function $y = f(x)$ defined implicitly by $H(x, y) = C$ is a solution of the differential equation (4).

EXAMPLE 1 Find $H(x, y)$ such that

$$dH = (3x^2y) \, dx + (x^3 + 6y) \, dy$$

and solve the differential equation (4).

Solution From Eq. (1), we see that we must have

$$\frac{\partial H}{\partial x} = 3x^2y \qquad \text{and} \qquad \frac{\partial H}{\partial y} = x^3 + 6y.$$

We can attempt to find $H(x, y)$ by *partial integration,* reversing the operation of partial differentiation.

Since $\partial H/\partial x = 3x^2y$, to find $H(x, y)$ we integrate $3x^2y$ with respect to x only, *treating y as a constant.* An indefinite integral is defined only up to a *constant.* For partial integration with respect to x, such a "constant" may be any function of y only. We obtain

$$H(x, y) = \int 3x^2 y \, dx = x^3 y + h(y), \tag{6}$$

where $h(y)$ is a function of y only.

Similarly, since we must have $\partial H/\partial y = x^3 + 6y$, we obtain

$$H(x, y) = \int (x^3 + 6y) \, dy = x^3 y + 3y^2 + k(x), \tag{7}$$

where $k(x)$ may be any function of x only.

Our task is now to find $h(y)$ and $k(x)$ so that the expressions (6) and (7) are identical. Clearly, we can take $h(y) = 3y^2$ and $k(x) = 0$. Thus $H(x, y) = x^3 y + 3y^2$ is a function such that dH is the given differential form. The differential equation (4) becomes $dH = 0$, and the equation has the implicit general solution $x^3 y + 3y^2 = C$. ∎

DEFINITION 16.2

Exact differential

A differential form

$$P(x, y) \, dx + Q(x, y) \, dy \tag{8}$$

is called an **exact differential** if there exists a function $H(x, y)$ such that the differential form is equal to dH.

In other words, an *exact differential* really is the *differential* of some function. Of course, the analogue of Definition 16.2 is made for functions of more than two variables.

Recall that the gradient ∇H of $H(x, y)$ is given by

$$\nabla H = H_x(x, y)\mathbf{i} + H_y(x, y)\mathbf{j}.$$

We see immediately that $P(x, y) \, dx + Q(x, y) \, dy$ is the differential dH of $H(x, y)$ if and only if $P(x, y)\mathbf{i} + Q(x, y)\mathbf{j}$ is the gradient ∇H of $H(x, y)$. That is, our work in this section with differential forms can be phrased in terms of gradient vectors.

Since

$$dH = \frac{\partial H}{\partial x} \, dx + \frac{\partial H}{\partial y} \, dy, \tag{9}$$

we see that, in order for Eq. (9) to be the same differential form as (8), we must have

$$\frac{\partial H}{\partial x} = P(x, y) \quad \text{and} \quad \frac{\partial H}{\partial y} = Q(x, y). \tag{10}$$

Now if $P(x, y)$ and $Q(x, y)$ have continuous partial derivatives, so that $H(x, y)$ has to have continuous second partial derivatives, we will have

$$\frac{\partial^2 H}{\partial y \, \partial x} = \frac{\partial^2 H}{\partial x \, \partial y}. \tag{11}$$

From Eqs. (10) and (11), we see that, in order for form (8) to be an exact differential, we must have

$$\frac{\partial P}{\partial y} = \frac{\partial^2 H}{\partial y\, \partial x} = \frac{\partial^2 H}{\partial x\, \partial y} = \frac{\partial Q}{\partial x}.$$

THEOREM 16.1   Necessary condition for exactness

If $P(x, y)$ and $Q(x, y)$ have continuous partial derivatives, then the differential form $P(x, y)\, dx + Q(x, y)\, dy$ can be exact only if

$$\frac{\partial P}{\partial y} = \frac{\partial Q}{\partial x}. \tag{12}$$

□

EXAMPLE 2 Test the differential form $x^2y\, dx + (x^2 - y^2)\, dy$ for exactness.

Solution Here $P(x, y) = x^2y$ and $Q(x, y) = x^2 - y^2$. Now

$$\frac{\partial P}{\partial y} = \frac{\partial(x^2y)}{\partial y} = x^2 \quad \text{and} \quad \frac{\partial Q}{\partial x} = \frac{\partial(x^2 - y^2)}{\partial x} = 2x.$$

Thus $\partial P/\partial y \neq \partial Q/\partial x$, so condition (12) is not satisfied. Consequently the differential form cannot be exact. ■

EXAMPLE 3 Test the differential form $(2xy + y^2)\, dx + (x^2 + 2xy)\, dy$ for exactness.

Solution This time

$$\frac{\partial P}{\partial y} = \frac{\partial(2xy + y^2)}{\partial y} = 2x + 2y \quad \text{and} \quad \frac{\partial Q}{\partial x} = \frac{\partial(x^2 + 2xy)}{\partial x} = 2x + 2y.$$

Thus $\partial P/\partial y = \partial Q/\partial x$, so condition (12) is satisfied. In this case, a little computation as in Example 1 shows that our differential form is indeed exact; it is dH for $H(x, y) = x^2y + xy^2$, because then

$$dH = \frac{\partial H}{\partial x}\, dx + \frac{\partial H}{\partial y}\, dy = (2xy + y^2)\, dx + (x^2 + 2xy)\, dy.$$ ■

The preceding example suggests that perhaps condition (12) not only is a *necessary* condition for $P(x, y)\, dx + Q(x, y)\, dy$ to be exact but also is a *sufficient* condition, at least if $P(x, y)$ and $Q(x, y)$ have continuous first partial derivatives. This is *not* always true, but it is true if the domain in the plane where $P(x, y)$ and $Q(x, y)$ are both defined has no "holes" in it. It can be demonstrated that

$$\frac{y}{x^2 + y^2}\, dx - \frac{x}{x^2 + y^2}\, dy,$$

which does satisfy condition (12), is not exact in its entire domain, which consists of the plane minus the origin. (See Exercise 57 of the supplementary exercises at the end of this chapter.) Here, the origin is a "hole" in the domain.

We will not concern ourselves with the niceties of finding regions as large as possible where a differential form that satisfies condition (12) is exact. We indicate informally that, if $P(x, y)\ dx + Q(x, y)\ dy$ is defined throughout some disk containing a point and $P(x, y)$ and $Q(x, y)$ are differentiable there with $\partial P/\partial y = \partial Q/\partial x$, then the differential form is exact in that disk. Our argument will indicate a four-step procedure to find all functions $H(x, y)$ such that

$$P(x, y)\ dx + Q(x, y)\ dy = dH.$$

Since we must have $\partial H/\partial x = P(x, y)$, we let $G(x, y)$ be some antiderivative of $P(x, y)$ with respect to x only, treating y as a constant. Such a function $G(x, y)$ is found by *partial integration* with respect to x, holding y constant, as in Example 1. Now an indefinite integral is defined only up to a constant. Since y is treated as a constant in this integration with respect to x, we see that the most general partial antiderivative of $P(x, y)$ with respect to x is of the form $G(x, y) + h(y)$ for an arbitrary function $h(y)$ of y only.

STEP 1. Let

$$H(x, y) = \int P(x, y)\ dx = G(x, y) + h(y), \tag{13}$$

where $G(x, y)$ is any computed antiderivative of $P(x, y)$ with respect to x only, treating y as a constant.

Our problem is to determine $h(y)$ such that $\partial H/\partial y = Q(x, y)$. Now

$$\frac{\partial H}{\partial y} = \frac{\partial G}{\partial y} + h'(y),$$

so $\partial G/\partial y + h'(y) = Q(x, y)$, or

$$h'(y) = Q(x, y) - \frac{\partial G}{\partial y}. \tag{14}$$

If $Q(x, y) - \partial G/\partial y$ is a continuous function of y *only*, we could set $h'(y)$ equal to it and integrate to find the desired $h(y) = \int h'(y)\ dy$. Now Eq. (14) is a function of y only, provided that

$$\frac{\partial}{\partial x}\left(Q(x, y) - \frac{\partial G}{\partial y}\right) = 0, \tag{15}$$

that is, if

$$\frac{\partial Q}{\partial x} - \frac{\partial^2 G}{\partial x\ \partial y} = 0. \tag{16}$$

But if $G(x, y)$ has continuous second-order partial derivatives throughout the neighborhood where we are working, then

$$\frac{\partial^2 G}{\partial x\ \partial y} = \frac{\partial^2 G}{\partial y\ \partial x} = \frac{\partial}{\partial y}\left(\frac{\partial G}{\partial x}\right) = \frac{\partial}{\partial y}(P(x, y)) = \frac{\partial P}{\partial y}.$$

Under these conditions, condition (16) is satisfied, for the left-hand side becomes

$$\frac{\partial Q}{\partial x} - \frac{\partial P}{\partial y},$$

which is zero by assumption. Thus we can complete our steps to find $H(x, y)$, as follows.

STEP 2. Compute $\partial H/\partial y = \partial G/\partial y + h'(y)$, set it equal to $Q(x, y)$, and solve for $h'(y)$.

STEP 3. Integrate to find $h(y) = \int h'(y)\, dy$.

STEP 4. The final answer is $H(x, y) = G(x, y) + h(y) + C$.

We state a theorem that seems reasonable in view of the work we just completed.

THEOREM 16.2   Test for exactness

If $P(x, y)$ and $Q(x, y)$ have continuous first partial derivatives in some disk, and if $\partial P/\partial x = \partial Q/\partial y$ in that disk, then there exists a function $H(x, y)$ defined in the disk such that $dH = P(x, y)\, dx + Q(x, y)\, dy$. □

EXAMPLE 4   Use the steps of the argument preceding Theorem 16.2 to show that

$$(2xy^3 + 6x)\, dx + (3x^2y^2 + 4y^3)\, dy$$

is an exact differential dH and to find the function $H(x, y)$.

Solution   Note that

$$\frac{\partial P}{\partial y} = \frac{\partial(2xy^3 + 6x)}{\partial y} = 6xy^2 \quad \text{and} \quad \frac{\partial Q}{\partial x} = \frac{\partial(3x^2y^2 + 4y^3)}{\partial x} = 6xy^2,$$

so the differential form is indeed exact. Then

STEP 1. $H(x, y) = \int P(x, y)\, dx = \int (2xy^3 + 6x)\, dx = x^2y^3 + 3x^2 + h(y)$;

STEP 2. $\partial H/\partial y = 3x^2y^2 + h'(y)$, so

$$3x^2y^2 + h'(y) = 3x^2y^2 + 4y^3, \quad \text{and} \quad h'(y) = 4y^3;$$

STEP 3. $h(y) = \int 4y^3\, dy = y^4 + C$;

STEP 4. $H(x, y) = x^2y^3 + 3x^2 + y^4 + C.$ ∎

We continue to suppose that we are dealing with **continuously differentiable** functions, that is, that first partial derivatives exist and are continuous. For a differential form in three variables

$$P(x, y, z)\, dx + Q(x, y, z)\, dy + R(x, y, z)\, dz$$

HISTORICAL NOTE

THE DIFFERENTIAL FORM $P\,dx +$ $Q\,dy$ was first considered in detail by Alexis Claude Clairaut (1713–1765). A child genius, he wrote a book on curves in space when he was sixteen and was elected to the Paris Academy of Sciences at the unprecedented age of seventeen. In 1743 he wrote a major work on the shape of the earth that showed mathematically why the earth's diameter is smaller at the poles than at the equator. Also in the 1740s he wrote two textbooks, one on geometry and one on algebra, that greatly improved the teaching of mathematics in France.

In 1739 Clairaut proved Theorems 16.1 and 16.2, the first by explicit calculation in terms of the function ax^my^n—since for him any function of two variables could be expressed as a series (possibly infinite) of such terms—and the second by reducing it to a problem of finding an antiderivative. Clairaut did not concern himself with the continuity condition of Theorem 16.2, however. Jean d'Alembert (1717–1783) noted in 1768 that the form

$$\frac{x}{x^2 + y^2}\,dy - \frac{y}{x^2 + y^2}\,dx$$

is not the differential of a function, even though the appropriate derivatives are equal, but it took nearly another century for this idea to be exploited.

to be exact, the corresponding conditions are

$$\frac{\partial P}{\partial y} = \frac{\partial Q}{\partial x}, \qquad \frac{\partial P}{\partial z} = \frac{\partial R}{\partial x}, \qquad \frac{\partial Q}{\partial z} = \frac{\partial R}{\partial y}. \tag{17}$$

Using the notation of subscripted variables, if $\mathbf{x} = (x_1, x_2, x_3)$, then

$$f_1(\mathbf{x})\,dx_1 + f_2(\mathbf{x})\,dx_2 + f_3(\mathbf{x})\,dx_3$$

is exact in some neighborhood of each point if and only if

$$\frac{\partial f_i}{\partial x_j} = \frac{\partial f_j}{\partial x_i} \qquad \text{for all } i \text{ and } j. \tag{18}$$

Again, the necessity of these relations follows from equality of mixed second-order partial derivatives. See Exercise 21.

EXAMPLE 5 Consider the differential form

$$(yz^2 - 6x \sin z)\,dx + (xz^2 - 3y^2\cos z)\,dy + (2xyz - 3x^2\cos z + y^3\sin z)\,dz,$$

which can be checked to satisfy the conditions (17). Find $H(x, y, z)$ so that the differential form is dH.

Solution We proceed in a fashion analogous to the steps preceding Theorem 16.2. From $\partial H/\partial x = yz^2 - 6x \sin z$, we have

$$H(x, y, z) = \int (yz^2 - 6x \sin z)\,dx$$

$$= xyz^2 - 3x^2\sin z + h(y, z).$$

Then $\partial H/\partial y = xz^2 + \partial h/\partial y$ yields

$$xz^2 + \frac{\partial h}{\partial y} = xz^2 - 3y^2\cos z,$$

so

$$\frac{\partial h}{\partial y} = -3y^2\cos z,$$

and

$$h(y, z) = \int -3y^2\cos z\,dy$$

$$= -y^3\cos z + k(z).$$

We are now down to

$$H(x, y, z) = xyz^2 - 3x^2\sin z - y^3\cos z + k(z).$$

Finally, $\partial H/\partial z = 2xyz - 3x^2\cos z + y^3\sin z + k'(z)$ yields

$$2xyz - 3x^2\cos z + y^3\sin z = 2xyz - 3x^2\cos z + y^3\sin z + k'(z),$$

so $k'(z) = 0$ and $k(z) = C$. Thus

$$H(x, y, z) = xyz^2 - 3x^2\sin z - y^3\cos z + C. \qquad \blacksquare$$

In Examples 4 and 5, we followed the steps of the argument preceding Theorem 16.2 to find the function H. Students often find the technique of Example 1 more natural. That technique is used in the remaining examples.

EXAMPLE 6 Find $H(x, y)$ such that

$$(2xy + ye^{xy} - 4x)\, dx + (x^2 + xe^{xy} - 3\sin y)\, dy$$

is dH, if the differential form is exact.

Solution If the form is dH, we must have

$$H_x(x, y) = 2xy + ye^{xy} - 4x,$$

so

$$H(x, y) = \int (2xy + ye^{xy} - 4x)\, dx = x^2y + e^{xy} - 2x^2 + h(y) \qquad \textbf{(19)}$$

for some function $h(y)$ of y only. Similarly,

$$H(x, y) = \int (x^2 + xe^{xy} - 3\sin y)\, dy = x^2y + e^{xy} + 3\cos y + k(x) \quad \textbf{(20)}$$

for some function $k(x)$ of x only. Comparison of forms (19) and (20) shows that we may take

$$H(x, y) = x^2y + e^{xy} - 2x^2 + 3\cos y + C$$

for any constant C, for this function is of both forms (19) and (20). Thus our differential form is exact, and we found $H(x, y)$. ∎

EXAMPLE 7 Find $H(x, y, z)$ such that

$$\left(2xyz + \frac{1}{z}\right) dx + (x^2z)\, dy + \left(x^2y - \frac{x}{z^2}\right) dz$$

is dH, if the differential form is exact.

Solution Arguing as in Example 6, we must have

$$H(x, y, z) = \int \left(2xyz + \frac{1}{z}\right) dx = x^2yz + \frac{x}{z} + h(y, z),$$

$$H(x, y, z) = \int (x^2z)\, dy = x^2yz + k(x, z),$$

and

$$H(x, y, z) = \int \left(x^2y - \frac{x}{z^2}\right) dz = x^2yz + \frac{x}{z} + g(x, y).$$

Comparison of these three expressions shows that

$$H(x, y, z) = x^2yz + \frac{x}{z} + C$$

satisfies all three conditions for any constant C. ∎

EXAMPLE 8 Find $H(u, v)$ such that

$$(v^2 - 2) \, du + (2uv + 3u) \, dv$$

is dH, if the differential form is exact.

Solution We must have

$$H(u, v) = \int (v^2 - 2) \, du = uv^2 - 2u + h(v)$$

and

$$H(u, v) = \int (2uv + 3u) \, dv = uv^2 + 3uv + k(u).$$

We see that this differential form is not exact, since there is no function $h(v)$ of v only that could supply the summand $3uv$ needed in the second expression for $H(u, v)$. ∎

EXAMPLE 9 Find the solution of the differential equation

$$\frac{dy}{dx} = -\frac{y}{x} - 2$$

that goes through the point $(1, 2)$.

Solution We rewrite the equation:

$$\frac{dy}{dx} = -\frac{y}{x} - 2 = -\frac{y + 2x}{x},$$

$$x \, dy = -(y + 2x) \, dx,$$

$$(y + 2x) \, dx + x \, dy = 0. \tag{21}$$

We can solve this equation if the differential form (21) is an exact differential dH. We must have

$$H(x, y) = \int (y + 2x) \, dx = xy + x^2 + h(y)$$

and

$$H(x, y) = \int x \, dy = xy + k(x).$$

We see that we may take

$$H(x, y) = xy + x^2,$$

and Eq. (21) then becomes $dH = 0$. Thus the general implicit solution is $H(x, y) = C$, or

$$xy + x^2 = C.$$

Evaluating C to find the solution containing the point $(1, 2)$, we have

$$1 \cdot 2 + 1^2 = C, \quad \text{so} \quad C = 3.$$

Thus

$$xy + x^2 = 3, \qquad \text{or} \qquad y = \frac{3 - x^2}{x}$$

is the desired solution.   ■

SUMMARY

1. A differential form $P(x, y)\, dx + Q(x, y)\, dy$ is exact if it is the differential of some function $H(x, y)$. A differential form $f_1(\mathbf{x})\, dx_1 + f_2(\mathbf{x})\, dx_2 + f_3(\mathbf{x})\, dx_3$ is exact if it is the differential of some function $H(\mathbf{x})$.

2. If P and Q are continuously differentiable throughout a disk, then $P(x, y)\, dx + Q(x, y)\, dy$ is exact there if and only if $\partial P/\partial y = \partial Q/\partial x$. Similarly, $P\, dx + Q\, dy + R\, dz$ is exact in a ball if and only if $\partial P/\partial y = \partial Q/\partial x$, $\partial P/\partial z = \partial R/\partial x$, and $\partial Q/\partial z = \partial R/\partial y$ there.

3. If $P(x, y)\, dx + Q(x, y)\, dy = dH$, then $H(x, y)$ may be found as follows:

   STEP 1.   Compute
   $$H(x, y) = \int P(x, y)\, dx = G(x, y) + h(y),$$
   where $G(x, y)$ is any partial antiderivative of $P(x, y)$ with respect to x, treating y as a constant.

   STEP 2.   Compute $\partial H/\partial y = \partial G/\partial y + h'(y)$, set it equal to $Q(x, y)$, and solve for $h'(y)$.

   STEP 3.   Integrate to find $h(y) = \int h'(y)\, dy$.

   STEP 4.   Then $H(x, y) = G(x, y) + h(y) + C$.

   See Examples 4 and 5. A similar technique is valid for differential forms involving more variables.

4. Another technique for finding $H(x, y)$ if $dH = P(x, y)\, dx + Q(x, y)\, dy$ is as follows:

   STEP 1.   Compute $H(x, y) = \int P(x, y)\, dx = G(x, y) + h(y)$ and $H(x, y) = \int Q(x, y)\, dy = F(x, y) + k(x)$.

   STEP 2.   Compare the results of step 1 and find a function $H(x, y)$ of both types.

   See Examples 6–8. A similar technique is valid for differential forms with more variables.

5. If a differential equation $P(x, y)\, dx + Q(x, y)\, dy = 0$ can be written in the form $dH = 0$, then the implicit general solution is $H(x, y) = C$.

EXERCISES 16.1

In Exercises 1–10, test whether the given differential form is exact inside disks using the criterion described in number 2 of the Summary. If it is exact, find the most general function H such that the differential form is dH, using the technique described in number 3 of the Summary.

1. $x^2\, dx - y\, dy$

2. $2xy\, dx + x^2\, dy$

3. $(3x - 2y)\, dx + (2x - 3y)\, dy$

4. $2xy\, dx + (x^2 - e^{-y})\, dy$

5. $(y\, \sec^2 xy)\, dx + (1 + x\, \sec^2 xy)\, dy$

6. $(2xy^3 - 3)\, dx + (3x^2y^2 + 4y)\, dy$

7. $\dfrac{-y}{x^2 + z^2}\, dy + \dfrac{x}{x^2 + z^2}\, dz$ for $x > 0$

8. $\dfrac{-z}{y^2 + z^2}\, dy + \dfrac{y}{y^2 + z^2}\, dz$ for $z > 0$

9. $(2xyz - 3y^2 + 2z^3)\, dx + (x^2 z - 6xy)\, dy$
   $+ (x^2 y + 8z + 6xz^2)\, dz$

10. $(yz\, \cos xyz - 3z^2)\, dx + (xz\, \cos xyz + 3y^2)\, dy$
    $+ (xy\, \cos xyz - 6xz)\, dz$

In Exercises 11–20, use the technique described in number 4 of the Summary to find the most general function H such that the given differential form is dH, if the form is exact.

11. $x^2 z\, dx - yz\, dy$

12. $2xz\, dx + x^2\, dz$

13. $\cos y\, dx + (1 - x\, \sin y)\, dy$

14. $y^2\, dx + \left(\dfrac{1}{y} + 2xy\right) dy$

15. $(e^y - y\, \cos xy)\, dx + (xe^y - x\, \cos xy)\, dy$

16. $\left[\dfrac{2u}{v} - \dfrac{v}{(u+v)^2} + \dfrac{1}{u^2}\right] du + \left[\dfrac{u}{(u+v)^2} - \dfrac{u^2}{v^2} - \dfrac{3}{v^2}\right] dv$

17. $\left[-2z\, \sin(2wz) + \dfrac{1}{(1+w)^2}\right] dw$
    $+ \left(\dfrac{1}{z} - 4w\, \sin wz\, \cos wz\right) dz$

18. $\left(x_2{}^3 - \dfrac{1}{x_3{}^2}\right) dx_1 + (3x_1 x_2{}^2 + 4x_2 x_3)\, dx_2$
    $+ \left(2x_2{}^2 + \dfrac{x_1}{x_3{}^3}\right) dx_3$

19. $(ze^{yz})\, dx + (xz^2 e^{yz})\, dy + (xyze^{yz})\, dz$

20. $(y\, \sin yz)\, dx + (xyz\, \cos yz + x\, \sin yz)\, dy + (xy^2 \cos yz)\, dz$

21. Let $\mathbf{x} = (x_1, x_2, \ldots, x_n)$ and let $f_1(\mathbf{x}), f_2(\mathbf{x}), \ldots, f_n(\mathbf{x})$ have continuous first partial derivatives. Show that if
    $$f_1(\mathbf{x})\, dx_1 + f_2(\mathbf{x})\, dx_2 + \cdots + f_n(\mathbf{x})\, dx_n$$
    is an exact differential form, then $\partial f_i/\partial x_j = \partial f_j/\partial x_i$ for each pair of subscripts i and j, where $1 \le i < j \le n$.

22. Consider a differential equation in x and y involving continuous functions that can be solved by separation of variables. Show that the equation can also be written in the form $dH = 0$ for some function $H(x, y)$.

In Exercises 23–30, find the solution of the given differential equation containing the given point.

23. $dy/dx = -y/(x + 2y)$; point $(-1, 3)$

24. $dy/dx = (3x^2 - y^2)/2xy$; point $(2, -3)$

25. $dy/dx = (2x - y\, \sin xy)/(x\, \sin xy)$; point $(1, \pi/2)$

26. $dy/dx = -(3x^2 + e^y)/(xe^y)$; point $(3, 0)$

27. $dy/dx = (2x + \sin y)/(3y^2 - x\, \cos y)$; point $(1, 0)$

28. $dy/dx = -(y^3 + y\, \sin xy)/(3xy^2 + x\, \sin xy + e^y)$; point $(0, 0)$

29. $dy/dx = (y + 2xy^2)/x$; point $(2, -1)$ [*Hint:* After writing the equation in differential form, divide by y^2.]

30. $dy/dx = -y/(x + x^2 y)$; point $(1, e)$ [*Hint:* After writing the equation in differential form, divide by $x^2 y^2$.]

16.2 LINE INTEGRALS

An integral $\int_a^b f(x)\, dx$ should be regarded as one-dimensional. The function f is integrated over a line segment, which is one-dimensional. In this section, we continue our study of one-dimensional integrals. We no longer restrict ourselves to a line segment but integrate over a portion of a curve γ in the plane or in space. Such integrals are called *line integrals:* the word "line" in this context should be viewed as indicating "dimension one" rather than meaning "straight."

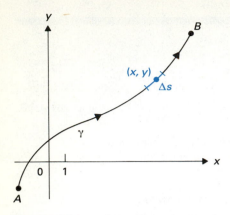

FIGURE 16.1 Partition element of γ of length Δs.

Recall that a *smooth curve* γ in the plane can be described parametrically by

$$x = h(t), \qquad y = k(t), \qquad \text{for } a \le t \le b, \tag{1}$$

where the functions h and k have continuous derivatives and those derivatives are never simultaneously zero for $a < t < b$. The point A corresponding to $t = a$ is the **initial point** of γ, and the point B, where $t = b$, is the **terminal point.** If A and B are the same point, the curve is a **loop.** We give a few illustrations of how integrals over such curves arise naturally.

Application 1: Consider a wire covering the smooth curve γ in Fig. 16.1. Suppose the mass density of the wire at a point (x, y) on the curve is $\sigma(x, y)$. Let γ be partitioned into *short* pieces over which $\sigma(x, y)$ remains almost constant. Suppose the arc of length Δs shown in Fig. 16.1 is such a piece. For the point (x, y) shown in Fig. 16.1, the approximate mass of this piece of wire of length Δs is $\sigma(x, y) \cdot \Delta s$. If we add up such products over a partition of the curve, we obtain an approximation to the mass of the wire. It is clear from our work in Chapter 15 that the actual mass of the wire should be given by an integral of $\sigma(x, y)$ over γ with respect to arc length. We denote this integral by $\int_\gamma \sigma(x, y)\, ds$, so we have

$$\text{Mass} = \int_\gamma \sigma(x, y)\, ds.$$

Application 2: Let $\kappa(x, y)$ be the curvature of γ in Fig. 16.1 at (x, y). If we sum terms of the form $\kappa(x, y) \cdot \Delta s$ over a partition of γ, we get an estimate for a *total measure* of the bend of the *entire curve*. Passing to an integral, we define

$$\textbf{(Total curvature of } \gamma) = \int_\gamma \kappa(x, y)\, ds.$$

Application 3: Let $\mathbf{F}(x, y) = P(x, y)\mathbf{i} + Q(x, y)\mathbf{j}$ be a force vector at each point (x, y) in a portion of the plane containing the curve γ in Fig. 16.2. Let $\mathbf{t}$ be the unit vector tangent to the curve at (x, y) in the direction corresponding to increasing time t. Then $\mathbf{F} \cdot \mathbf{t}$ is the scalar component of $\mathbf{F}$ along the curve. The product $(\mathbf{F} \cdot \mathbf{t})(\Delta s)$ for the curve in Fig. 16.2 is the approximate work done by $\mathbf{F}$ acting on a body moving on the arc of length Δs. Clearly the *total work* done on the body by $\mathbf{F}$ along the *entire curve* γ is

$$\text{Work} = \int_\gamma (\mathbf{F} \cdot \mathbf{t})\, ds. \tag{2}$$

Recall from Section 13.1 that

$$\mathbf{t} = \frac{1}{|\mathbf{v}|}\mathbf{v} = \frac{1}{ds/dt}\left(\frac{dx}{dt}\mathbf{i} + \frac{dy}{dt}\mathbf{j}\right)$$
$$= \frac{dx}{ds}\mathbf{i} + \frac{dy}{ds}\mathbf{j}.$$

Since $\mathbf{F}(x, y) = P(x, y)\mathbf{i} + Q(x, y)\mathbf{j}$, we have

$$\mathbf{F} \cdot \mathbf{t} = P(x, y)\frac{dx}{ds} + Q(x, y)\frac{dy}{ds}.$$

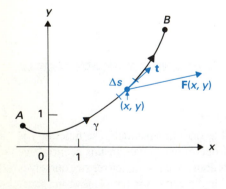

FIGURE 16.2 Force vector $\mathbf{F}(x, y)$ and unit tangent vector t at (x, y) in a partition element.

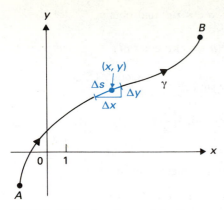

FIGURE 16.3 Δx, Δy, and Δs for a partition element.

Substituting this expression in formula (2), we can also express the work as

$$\text{Work} = \int_{\gamma} (P(x, y)\, dx + Q(x, y)\, dy). \tag{3}$$

This leads us to consider also integrals of functions with respect to x or y along a curve γ.

In the preceding applications, we have met integrals of the forms

$$\int_{\gamma} f(x, y)\, ds, \qquad \int_{\gamma} f(x, y)\, dx, \qquad \text{and} \qquad \int_{\gamma} f(x, y)\, dy. \tag{4}$$

It is important to remember that the integrals (4) come from limits of sums of terms of the form

$$f(x, y) \cdot \Delta s, \qquad f(x, y) \cdot \Delta x, \qquad \text{and} \qquad f(x, y) \cdot \Delta y, \tag{5}$$

for the point (x, y) and Δx, Δy, and Δs shown for γ in Fig. 16.3. As illustrated in the preceding applications, we think of products of the forms (5) in order to recognize what integral is appropriate. Also, we will see cases where line integrals can be evaluated quickly by recognizing that they correspond to limits of sums of terms of the forms (5).

To evaluate integrals of the forms (4), simply use the parametric equations

$$x = h(t), \qquad y = k(t), \qquad \text{for } a \le t \le b$$

for $\dot\gamma$, and substitute as follows:

$$f(x, y) = f(h(t), k(t)),$$
$$dx = h'(t)\, dt,$$
$$dy = k'(t)\, dt,$$
$$ds = \sqrt{\left(\frac{dx}{dt}\right)^2 + \left(\frac{dy}{dt}\right)^2}\, dt.$$

Then integrate the resulting expression from $t = a$ to $t = b$.

EXAMPLE 1 Let $f(x, y) = x^2 y$, and let γ be the semicircular curve

$$\gamma : x = a \cos t, \qquad y = a \sin t \qquad \text{for } 0 \le t \le \pi.$$

Find $\int_{\gamma} f(x, y)\, ds$ and $\int_{\gamma} f(x, y)\, dy$.

Solution Now

$$ds = \sqrt{\left(\frac{dx}{dt}\right)^2 + \left(\frac{dy}{dt}\right)^2}\, dt = \sqrt{a^2\sin^2 t + a^2\cos^2 t}\, dt = \sqrt{a^2}\, dt = a\, dt.$$

Thus

$$\int_{\gamma} f(x, y)\, ds = \int_{0}^{\pi} (a^2\cos^2 t)(a \sin t)(a\, dt)$$

$$= a^4 \int_{0}^{\pi} \cos^2 t \sin t\, dt = -a^4 \frac{\cos^3 t}{3}\Bigg]_{0}^{\pi}$$

$$= (-a^4)\left(-\frac{1}{3}\right) - (-a^4)\left(\frac{1}{3}\right) = \frac{2a^4}{3}.$$

For the other integral, we note that $dy = a \cos t \, dt$ and obtain

$$\int_\gamma f(x, y) \, dy = \int_\gamma (a^2\cos^2 t)(a \sin t)(a \cos t \, dt)$$

$$= a^4 \int_0^\pi \cos^3 t \sin t \, dt = -a^4 \frac{\cos^4 t}{4}\Big]_0^\pi$$

$$= (-a^4)\left(\frac{1}{4}\right) - (-a^4)\left(\frac{1}{4}\right) = 0. \qquad \blacksquare$$

Let γ be a smooth plane curve given parametrically by

$$x = h(t), \qquad y = k(t), \qquad \text{for } a \le t \le b.$$

We refer to the set of points $(h(t), k(t))$ in the plane for $a \le t \le b$ as the **trace** (or **path set) of** γ. The parametric equations $x = h(t)$, $y = k(t)$ are involved as part of γ. If we change the parametric equations, we consider that we have a curve *different* from γ, even if it has the same initial point A, terminal point B, and trace that γ has. However, the following theorem can be shown to be true.

THEOREM 16.3   Independence of parametrization

Let γ be a smooth curve covering each part of its trace only once, and let $f(x, y)$ be a continuous function with domain containing the trace of γ. Then the value of an integral $\int_\gamma f(x, y) \, ds$, $\int_\gamma f(x, y) \, dx$, or $\int_\gamma f(x, y) \, dy$ depends only on the initial point A, terminal point B, and the directed trace of γ. That is, two different parametrizations having the same trace covered in the same direction from A to B yield the same values for these integrals. $\square$

This independence of parametrization seems reasonable when we regard the integrals as limits of sums of terms of the forms (5). We illustrate Theorem 16.3 in the following example.

EXAMPLE 2   Consider the smooth curves

$$\gamma_1 : x = t, \qquad y = t^2 \qquad \text{for } 0 \le t \le 1$$

and

$$\gamma_2 : x = \sin t, \qquad y = \sin^2 t \qquad \text{for } 0 \le t \le \frac{\pi}{2}.$$

Both γ_1 and γ_2 are smooth curves from $(0, 0)$ to $(1, 1)$ and have as trace the portion of the parabola $y = x^2$ for $0 \le x \le 1$, shown in Fig. 16.4. Show that $\int_{\gamma_1} x \, ds = \int_{\gamma_2} x \, ds$.

Solution   For γ_1, we have

$$x = t \qquad \text{and} \qquad ds = \sqrt{\left(\frac{dx}{dt}\right)^2 + \left(\frac{dy}{dt}\right)^2} = \sqrt{1 + 4t^2} \, dt.$$

Therefore

$$\int_{\gamma_1} x \, ds = \int_0^1 t\sqrt{1 + 4t^2} \, dt = \frac{1}{8} \cdot \frac{2}{3}(1 + 4t^2)^{3/2}\Big]_0^1 = \frac{1}{12}(5^{3/2} - 1).$$

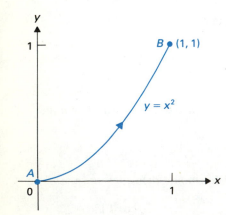

FIGURE 16.4   Trace of
$\gamma_1 : x = t, y = t^2, 0 \le t \le 1;$
$\gamma_2 : x = \sin t, y = \sin^2 t, 0 \le t \le \pi/2.$

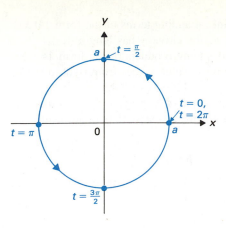

FIGURE 16.5 The circle $x = a \cos t$, $y = a \sin t$, $0 \leq t \leq 2\pi$.

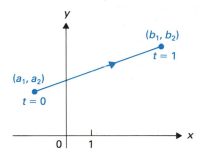

FIGURE 16.6 The line segment
$$x = a_1 + (b_1 - a_1)t,$$
$$y = a_2 + (b_2 - a_2)t, \; 0 \leq t \leq 1.$$

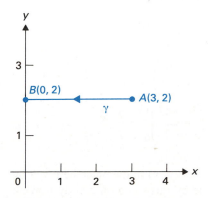

FIGURE 16.7 Trace of γ from (3, 2) to (0, 2).

For γ_2, we have

$$x = \sin t \qquad \text{and} \qquad ds = \sqrt{\cos^2 t + 4 \sin^2 t \cos^2 t} \; dt.$$

Thus

$$\int_{\gamma_2} x \, ds = \int_0^{\pi/2} \sin t \cos t \sqrt{1 + 4 \sin^2 t} \; dt$$

$$= \frac{1}{8} \cdot \frac{2}{3}(1 + 4 \sin t)^{3/2} \Big]_0^{\pi/2} = \frac{1}{12}(5^{3/2} - 1). \qquad \blacksquare$$

We have discussed line integrals of $f(x, y)$ over a smooth curve γ in the plane. Generalizations to integrals of a continuous function $f(x, y, z)$ over a smooth curve γ in space are apparent, and we will feel free to employ such integrals with no additional discussion.

The remainder of this section is devoted to examples involving line integrals over curves that are smooth or composed of a number of smooth pieces joined together. In view of Theorem 16.3, we may consider a line integral over a path set from a point A to a point B without specifying a parametrization. To evaluate such an integral, we have to be able to find a smooth parametrization with the path set as trace and covering no part of the trace more than once. We have found that students need plenty of practice doing this. In particular, remember that

$$x = a \cos t, \qquad y = a \sin t \qquad \text{for } 0 \leq t \leq 2\pi$$

is one *parametrization of the circle* with center (0, 0) and radius a. In this parametrization, the circle is traced once counterclockwise, with both initial and terminal point $(a, 0)$, as shown in Fig. 16.5. A *parametrization of the line segment* in Fig. 16.6 with initial point (a_1, a_2) and terminal point (b_1, b_2) is found by taking the parametric equations for the line, using our familiar steps but restricting t to [0, 1]:

> **Point** (a_1, a_2)
> $\parallel$ **vector** $(b_1 - a_1)\mathbf{i} + (b_2 - a_2)\mathbf{j}$
> **Segment** $\left. \begin{array}{l} x = a_1 + (b_1 - a_1)t \\ y = a_2 + (b_2 - a_2)t \end{array} \right\}$ for $0 \leq t \leq 1$

This section is devoted to computational technique. The next section continues with applications and relates the notions of exact differentials and line integrals.

Our next example illustrates how we may, on occasion, evaluate a line integral by regarding it as a limit of sums of products of one of the forms (5).

EXAMPLE 3 Let γ be a smooth curve with initial point (3, 2), terminal point (0, 2), and having as trace the line segment from (3, 2) to (0, 2), shown in Fig. 16.7. Assuming that no part of the trace is covered more than once, find

$$\int_\gamma (y^3 + 3y) \, ds, \qquad \int_\gamma (y^3 + 3y) \, dx, \qquad \text{and} \qquad \int_\gamma (y^3 + 3y) \, dy.$$

Solution Note that $y^3 + 3y$ has the constant value $2^3 + 3 \cdot 2 = 14$ on the trace of γ. When integrating with respect to s along γ, we always regard ds (and Δs) as *positive*. To find $\int_\gamma (y^3 + 3y) \, ds$, we think of adding terms of the form $14(\Delta s)$ along γ. Clearly we obtain $14 \cdot 3 = 42$, as the curve γ has length 3.

To find $\int_\gamma (y^3 + 3y) \, dx$, we think of adding products of the form $14 \cdot \Delta x$ along γ. Now Δx is *negative* as we go along γ, since γ goes from right to left in Fig. 16.7. Thus we obtain $\int_\gamma (y^3 + 3y) \, dx = 14(-3) = -42$.

Finally, $\Delta y = 0$ for an element of a partition of γ, so $\int_\gamma (y^3 + 3y) \, dy = 0$.

All these integrals could have been found using a smooth parametrization of γ, such as

$$x = 3 - 3t, \qquad y = 2, \qquad \text{for } 0 \le t \le 1,$$

but they were quickly evaluated from our understanding of the integrals (4) as limits of sums of terms in the forms (5). ∎

EXAMPLE 4   Find the mass of a wire covering the circle $x^2 + y^2 = 1$ with mass density $\sigma(x, y) = ky^2$.

Solution   The equations

$$x = h(t) = \cos t, \qquad y = k(t) = \sin t \qquad \text{for } 0 \le t \le 2\pi$$

parametrize a curve γ that covers the circle only once. Thus

$$m = \int_\gamma \sigma(x, y) \, ds = \int_\gamma ky^2 \, ds = \int_0^{2\pi} k \sin^2 t \sqrt{(-\sin t)^2 + (\cos t)^2} \, dt$$

$$= k \int_0^{2\pi} \sin^2 t \, dt = k \left(\frac{t}{2} - \frac{\sin 2t}{4} \right) \Bigg]_0^{2\pi} = k\pi. \quad ∎$$

EXAMPLE 5   Find the total curvature of a circle of radius a.

Solution   We know the curvature of the circle at any point is $\kappa = 1/a$. If γ is a curve covering the circle just once, we then know that the total curvature is given by $\int_\gamma (1/a) \, ds = (1/a) \int_\gamma ds$. Clearly, $\int_\gamma ds$ gives the arc length $2\pi a$ around the circle. Therefore the total curvature is $(1/a)(2\pi a) = 2\pi$. ∎

EXAMPLE 6   Find $\int_\gamma f(x, y) \, dx$ and $\int_\gamma f(x, y) \, dy$ if $f(x, y) = xy^2$ and γ is the curve joining $(0, 0)$ and $(1, 1)$ defined by

$$x = t, \qquad y = t^2 \qquad \text{for } 0 \le t \le 1.$$

Solution   Now $f(x, y) = xy^2$, so that $f(t, t^2) = t^5$. Therefore

$$\int_\gamma f(x, y) \, dx = \int_0^1 t^5 \cdot 1 \, dt = \frac{t^6}{6} \Bigg]_0^1 = \frac{1}{6},$$

while

$$\int_\gamma f(x, y) \, dy = \int_0^1 t^5 \cdot 2t \, dt = \frac{2t^7}{7} \Bigg]_0^1 = \frac{2}{7}. \quad ∎$$

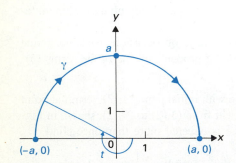

FIGURE 16.8

γ: $x = a \cos t$,

$\quad y = -a \sin t$, $\pi \le t \le 2\pi$

EXAMPLE 7   Let a force vector at each point (x, y) in the plane be given by $\mathbf{F}(x, y) = (x + y)\mathbf{i} + y^2\mathbf{j}$. Find the work done by this force on a body moving from $(-a, 0)$ to $(a, 0)$ along the top arc of the circle with center $(0, 0)$, as shown in Fig. 16.8.

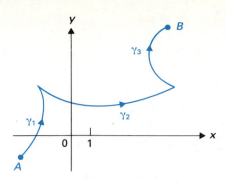

FIGURE 16.9 Piecewise-smooth curve
$\gamma = \gamma_1 + \gamma_2 + \gamma_3$.

Solution Let γ be a smooth parametrization of this semicircle. From formula (3), we have

$$\text{Work} = \int_\gamma [(x + y)\, dx + y^2\, dy].$$

We need a smooth parametrization of this semicircle. We know that $x = a \cos t$, $y = a \sin t$ parametrizes the circle in a *counterclockwise* direction. Since we want to go *clockwise,* we replace t by $-t$. Using some trigonometric identities, we see that we can consider γ to be given by

$$x = a \cos t, \qquad y = -a \sin t \qquad \text{for } \pi \le t \le 2\pi.$$

Then

$$\int_\gamma [(x + y)\, dx + y^2\, dy]$$

$$= \int_\pi^{2\pi} [(a \cos t - a \sin t)(-a \sin t) + (a^2 \sin^2 t)(-a \cos t)]\, dt$$

$$= a^2 \int_\pi^{2\pi} (-\sin t \cos t + \sin^2 t - a \sin^2 t \cos t)\, dt$$

$$= a^2 \int_\pi^{2\pi} \left[-\sin t \cos t + \frac{1}{2}(1 - \cos 2t) - a \sin^2 t \cos t \right] dt$$

$$= a^2 \left(-\frac{\sin^2 t}{2} + \frac{t}{2} - \frac{1}{4}\sin 2t - \frac{a}{3}\sin^3 t \right) \Bigg]_\pi^{2\pi}$$

$$= a^2 \left(\frac{2\pi}{2} \right) - a^2 \left(\frac{\pi}{2} \right)$$

$$= \frac{\pi a^2}{2}. \qquad \blacksquare$$

Finally, it is often useful to relax the condition that γ be smooth. (Remember that a curve is not smooth where it has a sharp point.) We will allow γ to be any curve that consists of a finite number of smooth arcs joined together, as illustrated in Fig. 16.9. Such a curve is called **piecewise smooth.** Let γ be a curve with smooth pieces γ_1, γ_2, and γ_3, as shown in the figure. It is natural to write $\gamma = \gamma_1 + \gamma_2 + \gamma_3$. We define

$$\int_\gamma f(x, y)\, ds = \int_{\gamma_1} f(x, y)\, ds + \int_{\gamma_2} f(x, y)\, ds + \int_{\gamma_3} f(x, y)\, ds,$$

and the integrals with respect to dx and dy are similarly defined.

EXAMPLE 8 Find $\int_\gamma xy^2\, dx$ and $\int_\gamma xy^2\, dy$ if $\gamma = \gamma_1 + \gamma_2$ is the piecewise-smooth curve joining $A(0, 0)$ and $B(1, 1)$ shown in Fig. 16.10.

Solution Here, γ_1 is the straight-line segment from $(0, 0)$ to $(1, 0)$, and γ_2 is the vertical-line segment from $(1, 0)$ to $(1, 1)$.

Since the value of xy^2 on γ_1 is 0 at each point, we see that

$$\int_{\gamma_1} xy^2\, dx = 0.$$

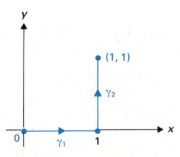

FIGURE 16.10 $\gamma = \gamma_1 + \gamma_2$

On γ_2, every short piece corresponds to $\Delta x = 0$, so also

$$\int_{\gamma_2} xy^2 \, dx = 0.$$

Consequently,

$$\int_{\gamma} xy^2 \, dx = \int_{\gamma_1} xy^2 \, dx + \int_{\gamma_2} xy^2 \, dx = 0 + 0 = 0.$$

Note that we discovered that this integral was 0 by thinking in terms of the contributions to the typical sums. We did not even bother to parametrize γ_1 or γ_2.

Turning to $\int_{\gamma} xy^2 \, dy$, we see that a little piece of γ_1 corresponds to a change $\Delta y = 0$, so

$$\int_{\gamma_1} xy^2 \, dy = 0.$$

But xy^2 is nonzero on most of γ_2, and $\Delta y \neq 0$ there, so there will be nonzero terms contributing to $\int_{\gamma_2} xy^2 \, dy$. We may take as parametrization of γ_2

$$x = 1, \qquad y = t \qquad \text{for } 0 \leq t \leq 1.$$

Then $dy = dt$, and

$$\int_{\gamma_2} xy^2 \, dy = \int_{\gamma_2} 1 \cdot t^2 \, dt = \frac{t^3}{3} \Big]_0^1 = \frac{1}{3},$$

so

$$\int_{\gamma} xy^2 \, dy = \int_{\gamma_1} xy^2 \, dy + \int_{\gamma_2} xy^2 \, dy = 0 + \frac{1}{3} = \frac{1}{3}. \qquad \blacksquare$$

We conclude with a space example.

EXAMPLE 9 Let γ_1 be the arc from $(2, 0, 0)$ to $(0, 0, 2)$ of the circle in the x,z-plane with center $(0, 0, 0)$, and let γ_2 be the straight-line segment from $(0, 0, 2)$ to $(1, 3, 1)$, as shown in Fig. 16.11. Find $\int_{\gamma_1 + \gamma_2} (x + y + z) \, dz$.

Solution A parametrization of γ_1 is given by

$$x = 2 \cos t, \qquad y = 0, \qquad z = 2 \sin t, \qquad \text{for } 0 \leq t \leq \frac{\pi}{2}.$$

Then

$$\int_{\gamma_1} (x + y + z) \, dz = \int_0^{\pi/2} (2 \cos t + 2 \sin t)(2 \cos t) \, dt$$

$$= 4 \int_0^{\pi/2} (\cos^2 t + \sin t \cos t) \, dt$$

$$= 4 \int_0^{\pi/2} \left[\frac{1}{2}(1 + \cos 2t) + \sin t \cos t \right] dt$$

$$= \left[2\left(t + \frac{1}{2} \sin 2t \right) + 2 \sin^2 t \right]_0^{\pi/2}$$

$$= \left[2\left(\frac{\pi}{2} \right) + 2 \right] - 0 = \pi + 2.$$

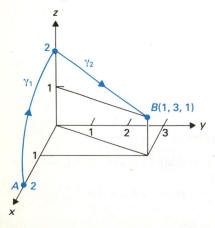

FIGURE 16.11 $\gamma = \gamma_1 + \gamma_2$

We now turn to γ_2. It is always easy to parametrize a line segment, where all we need is a *point* and a *parallel vector*. Take the initial point $(0, 0, 2)$ of γ_2 as the *point* and the vector $\mathbf{i} + 3\mathbf{j} - \mathbf{k}$ from $(0, 0, 2)$ to $(1, 3, 1)$ as the *direction vector*. The line segment is then covered for $0 \le t \le 1$. Thus γ_2 is given by $x = t$, $y = 3t$, $z = 2 - t$, for $0 \le t \le 1$. Then

$$\int_{\gamma_2} (x + y + z)\, dz = \int_0^1 (3t + 2)(-1)\, dt$$

$$= \left(-\frac{3t^2}{2} - 2t \right) \Bigg]_0^1 = -\frac{3}{2} - 2 = -\frac{7}{2}.$$

Therefore $\int_{\gamma_1 + \gamma_2} (x + y + z)\, dz = (\pi + 2) - \frac{7}{2} = \pi - \frac{3}{2}$. ∎

SUMMARY

Throughout the following, γ is a smooth curve in the plane given by $x = h(t)$, $y = k(t)$ for $a \le t \le b$, where $h(t)$ and $k(t)$ have continuous derivatives. Also, $f(x, y)$ is continuous.

1. $\int_\gamma f(x, y)\, ds$ is computed as

$$\int_a^b f(h(t), k(t)) \sqrt{\left(\frac{dx}{dt}\right)^2 + \left(\frac{dy}{dt}\right)^2}\, dt.$$

2. $\int_\gamma f(x, y)\, dx$ is computed as

$$\int_a^b f(h(t), k(t)) \left(\frac{dx}{dt}\right) dt.$$

3. $\int_\gamma f(x, y)\, dy$ is computed as

$$\int_a^b f(h(t), k(t)) \left(\frac{dy}{dt}\right) dt.$$

4. The value obtained in computing a line integral of a continuous function along the trace of a smooth curve from A to B is independent of the smooth parametrization used for the trace, provided that the parametrizations cover the trace in the same direction with each portion of the trace covered only once.

5. The mass of a wire covering γ with density function $\sigma(x, y)$ is $\int_\gamma \sigma(x, y)\, ds$.

6. The total curvature of a curve γ is $\int_\gamma \kappa\, ds$.

7. The work done by a force $\mathbf{F}(x, y) = P(x, y)\mathbf{i} + Q(x, y)\mathbf{j}$ acting on a body as it moves along γ is

$$\int_\gamma (\mathbf{F} \cdot \mathbf{t})\, ds = \int_\gamma (P(x, y)\, dx + Q(x, y)\, dy).$$

8. A line integral over a piecewise-smooth curve is the sum of the integrals over the smooth pieces.

EXERCISES 16.2

In Exercises 1–10, the curves are

γ_1 given by $x = t$, $y = t^2$ for $0 \le t \le 1$,

γ_2 given by $x = t + 1$, $y = 2t + 1$ for $0 \le t \le 1$,

γ_3 given by $x = \sin t$, $y = \cos t$, $z = -2t$ for $0 \le t \le \pi$.

Compute the indicated integrals.

1. $\int_{\gamma_1} x \, ds$

2. $\int_{\gamma_2} \sqrt{y} \, ds$

3. $\int_{\gamma_2} (x + y) \, dx$

4. $\int_{\gamma_1} (x - y) \, dy$

5. $\int_{\gamma_2} (x^2 \, dx - y \, dy)$

6. $\int_{\gamma_1} (xe^y \, dx + 3x^2 y \, dy)$

7. $\int_{\gamma_3} xy^2 \, ds$

8. $\int_{\gamma_3} (x \, dy + y \, dx + x \, dz)$

9. $\int_{\gamma_3} (xz \, dx + yz \, dy - 2xy \, dz)$

10. $\int_{\gamma_1 + \gamma_2} [(x^2 - 3y) \, dx - (y^2 + 3x) \, dy]$

11. Let γ be the space curve given by $x = 3t^2$, $y = 2t^3$, $z = 3t$ for $0 \le t \le 1$, and let $F(x, y, z) = 3x^2 yz$. Compute $\int_\gamma F(x, y, z) \, dx$.

12. Repeat Exercise 11, but compute $\int_\gamma F(x, y, z) \, dy$.

13. Let γ be the space curve given by $x = \sin t$, $y = \cos t$, $z = t$ for $0 \le t \le \pi/4$, and let $F(x, y, z) = 3x^2 yz$. Compute $\int_\gamma F(x, y, z) \, ds$.

14. Repeat Exercise 13, but compute $\int_\gamma F(x, y, z) \, dx$.

15. Let γ be a smooth curve having as trace the straight-line segment from $(1, 0)$ to $(3, 4)$. Find $\int_\gamma (x^2 + y^2) \, ds$.

16. Let $\gamma = \gamma_1 + \gamma_2$ be a piecewise-smooth curve where γ_1 has as trace the straight-line segment from $(0, 0)$ to $(2, 0)$ and γ_2 has as trace the straight-line segment from $(2, 0)$ to $(4, 2)$. Find $\int_\gamma (x^2 + xy) \, dx$.

17. Let $\gamma = \gamma_1 + \gamma_2$ be a piecewise-smooth curve where γ_1 has as trace the arc of the parabola $y = x^2$ from $(0, 0)$ to $(1, 1)$ and γ_2 has as trace the arc of the parabola $y = 2 - x^2$ from $(1, 1)$ to $(0, 2)$. Find $\int_\gamma (x + 2y) \, dy$.

18. Let $\gamma = \gamma_1 + \gamma_2$ be a piecewise-smooth curve where γ_1 has as trace the shorter arc of the circle $x^2 + y^2 = 4$ from $(2, 0)$ to $(0, 2)$ and γ_2 has as trace the straight-line segment from $(0, 2)$ to $(2, 0)$. Find $\int_\gamma (x \, dx - x^2 \, dy)$.

19. Let γ be a smooth curve having as trace the straight-line segment from $(2, -1, 3)$ to $(0, 1, 4)$. Find the integral $\int_\gamma (x + y + z^2) \, ds$.

20. Repeat Exercise 19, but find $\int_\gamma (x \, dx - y \, dy)$.

21. Let $\gamma = \gamma_1 + \gamma_2$ be a piecewise-smooth curve where γ_1 has as trace the straight-line segment from $(0, 0, 0)$ to $(1, 2, -1)$ and γ_2 has as trace the straight-line segment from $(1, 2, -1)$ to $(2, 1, 3)$. Find $\int_\gamma (x \, dy - y \, dx + z \, dz)$.

22. Let γ be a piecewise-smooth curve having as trace the unit square in the y,z-plane shown in Fig. 16.12, having the origin as both initial and terminal point. Find

$$\int_\gamma (yz \, dx - xz \, dy + xy \, dz).$$

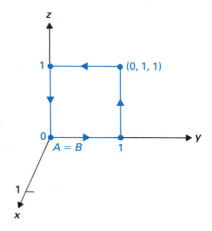

FIGURE 16.12
γ traces the square once counterclockwise.

23. Repeat Exercise 22, but find $\int_\gamma (xy \, dx - yz \, dy + y^2 \, dz)$.

24. Repeat Exercise 22, but find $\int_\gamma (xy - yz + y^2) \, ds$.

25. Let $\gamma = \gamma_1 + \gamma_2 + \gamma_3$, where γ_1 has as trace the shorter arc of the circle with center at the origin from $(0, -2, 0)$ to $(2, 0, 0)$, γ_2 has as trace the vertical line segment from $(2, 0, 0)$ to $(2, 0, 1)$, and γ_3 has as trace the shorter arc of the circle with center $(0, 0, 1)$ from $(2, 0, 1)$ to $(0, 2, 1)$. Find $\int_\gamma (x + 2y + z) \, ds$.

26. Let the mass density of a thin wire covering the portion of the parabola $y = x^2$ from $(0, 0)$ to $(2, 4)$ be $\sigma(x, y) = xy$. Find the total mass of the wire.

27. Let the mass density of a thin wire covering the semicircle $y = \sqrt{4 - x^2}$ be $\sigma(x, y) = y$. Find the mass of the wire.

28. Find the total curvature of the plane curve $y = x^2$ from $(0, 0)$ to $(2, 4)$.

29. Find the total curvature of the *entire* parabola $y = ax^2 + bx + c$.

30. Find the total curvature of the portion of the helix

$$x = a \cos t, \qquad y = a \sin t, \qquad z = t,$$

where $0 \le t \le 2\pi$.

In Exercises 31–34, find the work done by the given force $F(x, y)$ or $F(x, y, z)$ on a body as it moves along the given curve.

31. $F(x, y) = xy\mathbf{i} - y\mathbf{j}$; γ having as trace the straight-line segment from $(1, 2)$ to $(3, -1)$

32. $\mathbf{F}(x, y) = y\mathbf{i} - x\mathbf{j}$; γ having as trace the circle $x^2 + y^2 = a^2$ traced once clockwise.

33. $\mathbf{F}(x, y, z) = x\mathbf{i} + y\mathbf{j} - z\mathbf{k}$; γ having as trace the straight-line segment from $(1, -1, 2)$ to $(2, 1, 4)$

34. $\mathbf{F}(x, y, z) = x\mathbf{i} + z\mathbf{j} - y\mathbf{k}$; γ having as trace the rectangle shown in Fig. 16.13 with $(0, 0, 0)$ as both initial and terminal point

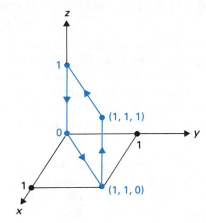

FIGURE 16.13
γ traces the rectangle once in the direction of the arrows.

35. In Theorem 16.3 on independence of parametrization, we required that the curve γ cover each part of its trace only once. Give an example of two smooth curves γ_1 and γ_2 having as trace the circle $x^2 + y^2 = 1$ and covering the circle in the same direction, but such that

$$\int_{\gamma_1} 1 \, ds \neq \int_{\gamma_2} 1 \, ds \qquad \text{and} \qquad \int_{\gamma_1} y \, dx \neq \int_{\gamma_2} y \, dx.$$

36. In Theorem 16.3 on independence of parametrization, we required that the curves cover their common trace in the same direction. Give an example of smooth curves γ_1 and γ_2 having as trace the circle $x^2 + y^2 = 1$ with each part of the circle between their common initial and terminal points covered only once, but such that

$$\int_{\gamma_1} 1 \, ds \neq \int_{\gamma_2} 1 \, ds \qquad \text{and} \qquad \int_{\gamma_1} y \, dx \neq \int_{\gamma_2} y \, dx.$$

 EXPLORING CALCULUS

Z: total curvature game

16.3 VECTOR FIELDS AND THEIR INTEGRALS OVER CURVES

This section introduces the notion of a *vector field* and indicates a few places where vector fields occur naturally in physics and fluid dynamics. Using the illustration of work done by a force field in moving a body along a path, we will see that we are led to consider *integrals of vector fields along curves*. We will immediately obtain line integrals of the type studied in the preceding section.

The Notion of a Vector Field

A **vector field** on a region G in the plane assigns to each point (x, y) in G a vector in the plane, which we visualize as emanating from (x, y). A vector field

$$\mathbf{F}(x, y) = P(x, y)\mathbf{i} + Q(x, y)\mathbf{j}$$

is **continuous** if P and Q are continuous functions. We use a boldface letter $\mathbf{F}$ to indicate that we are considering a vector field; we would use $\vec{F}$ in handwritten work.

Illustration 1 *Gradient vector field* We encountered the notion of a vector field in the plane in Chapter 14, where we discussed the gradient ∇f of a differentiable function f of two variables. To each point (x, y) in the domain of f, we

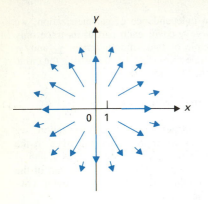

FIGURE 16.14 Force field of repulsion by a charge at the origin.

assign the gradient vector

$$\nabla f = \frac{\partial f}{\partial x}\mathbf{i} + \frac{\partial f}{\partial y}\mathbf{j}.$$

Recall that the direction at (x_0, y_0) of the gradient vector there is normal to the level curve of f through (x_0, y_0). Also, the gradient vector points in the direction of maximum rate of increase of $f(x, y)$ at (x_0, y_0) and has length equal to this maximum rate of increase.

Illustration 2 *Force field* An electrical charge at the origin in the plane exerts a force of repulsion on a like charge at any other point (x, y). This force at (x, y) may be represented by a vector of length equal to the magnitude of the force and having the direction of the force, namely, away from the origin. The vector field of these vectors for $(x, y) \neq (0, 0)$ is the *force field* of the charge. This vector field is continuous and grows weaker (with shorter vectors) as the distance from the origin increases (see Fig. 16.14).

Illustration 3 *Flux field* If a region G in the plane is covered with a flowing liquid or gas, then we can associate with each (x, y) in G and time t the velocity vector $\mathbf{V}(x, y)$ of the fluid flow at (x, y) at time t. This vector has the *direction* of the flow and *length* equal to the speed of the flow. The vector field $\mathbf{V}(x, y)$ is the *velocity field of the flow at time t*.

Let $\sigma(x, y)$ be the mass density of the fluid or gas at (x, y) at time t. The vector field $\mathbf{F}(x, y) = \sigma(x, y)\mathbf{V}(x, y)$ is the *flux field of the flow at time t*. The physical interpretation of $\mathbf{F}(x, y)$ is as follows. At (x, y), take a unit line segment orthogonal to $\mathbf{F}(x, y)$, as shown in Fig. 16.15. If the flow at all points on this line segment were just as it is at (x, y) at this time t, and if it did not change with time, then $|\mathbf{F}(x, y)|$ is the mass of fluid or gas flowing across the line segment per unit time. Of course, $\mathbf{F}(x, y)$ points in the direction of the flow at (x, y).

If the flux field does not vary with time, the flow is called *steady state*.

Integral of the Tangential Component of a Vector Field Along a Curve

Let $\mathbf{F}(x, y) = P(x, y)\mathbf{i} + Q(x, y)\mathbf{j}$ be a continuous vector field on a region G of the plane, and let

$$x = h(t), \qquad y = k(t), \qquad \text{for } a \leq t \leq b$$

be a smooth curve γ lying in G. Recall that the position vector to a point on γ at time t is

$$\mathbf{r} = x\mathbf{i} + y\mathbf{j} = h(t)\mathbf{i} + k(t)\mathbf{j}. \tag{1}$$

In the preceding section, we recalled that the *unit tangent vector* $\mathbf{t}$ (in the direction corresponding to increasing time) at each point on γ is given by

$$\mathbf{t} = \frac{dx}{ds}\mathbf{i} + \frac{dy}{ds}\mathbf{j}. \tag{2}$$

The scalar component of $\mathbf{F}$ along $\mathbf{t}$ is $\mathbf{F} \cdot \mathbf{t}$. The integral of this tangential scalar component of $\mathbf{F}$ along γ is

$$\int_{\gamma} (\mathbf{F} \cdot \mathbf{t})\, ds, \tag{3}$$

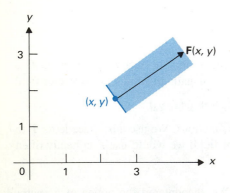

FIGURE 16.15 $|\mathbf{F}(x, y)|$ is the mass flowing across the unit line segment if the flow everywhere were the same as at (x, y).

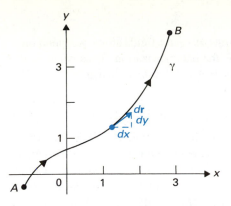

FIGURE 16.16   The tangent vector
$d\mathbf{r} = dx\,\mathbf{i} + dy\,\mathbf{j}$.

which may be written in the form

$$\int_\gamma (P(x, y)\,dx + Q(x, y)\,dy), \tag{4}$$

as we saw in the preceding section.

In our heuristic interpretation of differentials, the differential vector

$$d\mathbf{r} = dx\,\mathbf{i} + dy\,\mathbf{j} \tag{5}$$

shown in Fig. 16.16 has direction tangent to the curve and length $ds = \sqrt{(dx)^2 + (dy)^2}$. Note that

$$\begin{aligned}\mathbf{F} \cdot d\mathbf{r} &= (P(x, y)\mathbf{i} + Q(x, y)\mathbf{j}) \cdot (dx\,\mathbf{i} + dy\,\mathbf{j}) \\ &= P(x, y)\,dx + Q(x, y)\,dy.\end{aligned} \tag{6}$$

Comparing integral (4) and Eq. (6), we write the **integral of the tangential component of the vector field F along γ** as

$$\int_\gamma (\mathbf{F} \cdot \mathbf{t})\,ds = \int_\gamma (\mathbf{F} \cdot d\mathbf{r}). \tag{7}$$

In practice, Eq. (7) is calculated by expressing both $\mathbf{F}$ and $d\mathbf{r}$ in terms of the parameter t and dt.

If $\mathbf{F}$ is a force field, then the component of $\mathbf{F}$ tangent to a smooth curve γ is the portion of the force that acts to move a body along γ. The *work* done by this force field in moving a body along γ is then the integral with respect to arc length of this component, which we have seen is $\int_\gamma \mathbf{F} \cdot d\mathbf{r}$. Thus we may now write

$$\text{Work} = \int_\gamma \mathbf{F} \cdot d\mathbf{r}. \tag{8}$$

EXAMPLE 1   Find the work done by the force field

$$\mathbf{F}(x, y) = x^2\mathbf{i} + y^2\mathbf{j}$$

in moving a body from $(0, 0)$ to $(1, 1)$ if the position (x, y) of the body at time t on a curve γ is given by

$$x = t, \qquad y = t^2 \qquad \text{for } 0 \le t \le 1.$$

Solution   In terms of t, we have

$$\mathbf{F}(x, y) = \mathbf{F}(t, t^2) = t^2\mathbf{i} + t^4\mathbf{j},$$

and

$$d\mathbf{r} = dx\,\mathbf{i} + dy\,\mathbf{j} = dt\,(\mathbf{i} + 2t\mathbf{j}).$$

Thus

$$\begin{aligned}\int_\gamma \mathbf{F} \cdot d\mathbf{r} &= \int_0^1 (t^2\mathbf{i} + t^4\mathbf{j}) \cdot (\mathbf{i} + 2t\mathbf{j})\,dt \\ &= \int_0^1 (t^2 + 2t^5)\,dt \\ &= \left(\frac{t^3}{3} + \frac{2t^6}{6}\right)\Bigg]_0^1 = \frac{1}{3} + \frac{1}{3} = \frac{2}{3}.\end{aligned}$$

∎

The Curve $-\gamma$

Let γ be a smooth curve. Now line integrals of vector fields along γ depend only on the trace and direction of γ and not on the parametrization. Thus when working with γ, we can always assume that $0 \le t \le 1$ rather than $a \le t \le b$. We therefore suppose that γ is given by

$$\gamma : x = h(t), \qquad y = k(t) \qquad \text{for } 0 \le t \le 1, \tag{9}$$

with initial point A and terminal point B, as shown in Fig. 16.17(a). We denote by $-\gamma$ the curve in Fig. 16.17(b), which has the same trace as γ but is traveled in the opposite direction from B to A for $0 \le t \le 1$. In terms of time, a body moving on $-\gamma$ appears as though we took a moving picture of the body moving on γ and then ran the film backward. Wherever the body on γ appears at time t, the body on $-\gamma$ appears at time $1 - t$, as illustrated in Fig. 16.17. Thus parametric equations for $-\gamma$ are given by

$$-\gamma : x = h(1 - t), \qquad y = k(1 - t) \qquad \text{for } 0 \le t \le 1. \tag{10}$$

Note that

$$\frac{d}{dt}(h(1 - t)) = h'(1 - t)(-1) = -h'(1 - t).$$

If we let $u = 1 - t$ so that $du = -dt$, then

$$\int_{-\gamma} P(x, y) \, dx = \int_{t=0}^{1} P(h(1 - t), k(1 - t)) \cdot (-h'(1 - t)) \, dt$$

$$= \int_{u=1}^{0} P(h(u), k(u)) \cdot (-h'(u))(-du)$$

$$= \int_{1}^{0} P(h(u), k(u)) \cdot h'(u) \, du$$

$$= -\int_{0}^{1} P(h(u), k(u)) \cdot h'(u) \, du = -\int_{\gamma} P(x, y) \, dx.$$

A similar argument shows that

$$\int_{-\gamma} Q(x, y) \, dy = -\int_{\gamma} Q(x, y) \, dy.$$

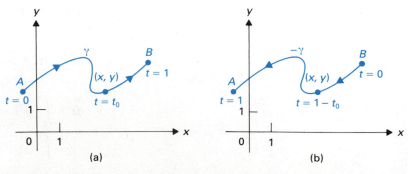

FIGURE 16.17   The curves γ and $-\gamma$: (a) γ: $x = h(t)$, $y = k(t)$, $0 \le t \le 1$; (b) $-\gamma$: $x = h(1 - t)$, $y = k(1 - t)$, $0 \le t \le 1$.

Thus we see that

$$\int_{-\gamma} (\mathbf{F} \cdot d\mathbf{r}) = -\int_{\gamma} (\mathbf{F} \cdot d\mathbf{r}). \qquad (11)$$

An intuitive way to see that Eq. (11) holds is to note that the unit tangent vectors at a point on γ and at the same point on $-\gamma$ have opposite direction. Since $\mathbf{F} \cdot d\mathbf{r} = (\mathbf{F} \cdot \mathbf{t}) \, ds$, and ds is always positive, we see that the sign of $\mathbf{F} \cdot d\mathbf{r}$ at a point on $-\gamma$ is opposite to its sign at the same point on γ.

EXAMPLE 2   Let γ be given by

$$\gamma : x = t - 1, \qquad y = 2t + 3 \qquad \text{for } 0 \le t \le 1.$$

Compute $\int_{\gamma} (xy \, dx - y \, dy)$ and $\int_{-\gamma} (xy \, dx - y \, dy)$, illustrating Eq. (11).

Solution   We have

$$\int_{\gamma} (xy \, dx - y \, dy) = \int_{0}^{1} [(2t^2 + t - 3) - (2t + 3)2] \, dt$$

$$= \int_{0}^{1} (2t^2 - 3t - 9) \, dt = \left(\frac{2t^3}{3} - \frac{3t^2}{2} - 9t \right) \Bigg]_{0}^{1}$$

$$= \frac{2}{3} - \frac{3}{2} - 9 = -\frac{5}{6} - 9 = -\frac{59}{6}.$$

From Eq. (10), we see that $-\gamma$ has the parametrization

$$-\gamma : x = -t, \qquad y = 5 - 2t \qquad \text{for } 0 \le t \le 1.$$

Then

$$\int_{-\gamma} (xy \, dx - y \, dy) = \int_{0}^{1} [(2t^2 - 5t)(-1) - (5 - 2t)(-2)] \, dt$$

$$= \int_{0}^{1} (-2t^2 + t + 10) \, dt = \left(-\frac{2t^3}{3} + \frac{t^2}{2} + 10t \right) \Bigg]_{0}^{1}$$

$$= -\frac{2}{3} + \frac{1}{2} + 10 = -\frac{1}{6} + 10 = \frac{59}{6},$$

which illustrates Eq. (11). ∎

Vector Fields in Space

We can extend the notion of a vector field in the plane to a vector field $\mathbf{F}(x, y, z) = P(x, y, z)\mathbf{i} + Q(x, y, z)\mathbf{j} + R(x, y, z)\mathbf{k}$ in space. The integral of the tangential component of $\mathbf{F}$ along a smooth space curve γ is given by

$$\int_{\gamma} \mathbf{F} \cdot d\mathbf{r} = \int_{\gamma} [P(x, y, z) \, dx + Q(x, y, z) \, dy + R(x, y, z) \, dz], \qquad (12)$$

where $d\mathbf{r} = dx\,\mathbf{i} + dy\,\mathbf{j} + dz\,\mathbf{k}$. Again, we intuitively regard $d\mathbf{r}$ as the differential tangent vector to the curve γ, and we see that $\int_{\gamma} \mathbf{F} \cdot d\mathbf{r} = \int_{\gamma} (\mathbf{F} \cdot \mathbf{t}) \, ds$ is again the integral of the tangential scalar component of $\mathbf{F}$ along γ. If $\mathbf{F}$ is a force field, then the integral (12) is the work done by $\mathbf{F}$ in moving a body along the space curve γ.

If γ is a space curve, then we regard $-\gamma$ as the curve with the same trace but traveled in the opposite direction, given by the natural extension of the parametrization (10) to space. Equation (11) of course continues to hold.

EXAMPLE 3 Find the work done by the force field $\mathbf{F} = z\mathbf{i} + x^2\mathbf{j} - 2y\mathbf{k}$ in moving a body along the curve γ with parametrization

$$x = t^2, \qquad y = 3t^3, \qquad z = 4t \qquad \text{for } 0 \leq t \leq 1.$$

Solution The work is given by

$$\int_\gamma \mathbf{F} \cdot d\mathbf{r} = \int_\gamma [z\,dx + x^2\,dy - 2y\,dz]$$

$$= \int_0^1 [(4t)(2t) + t^4(9t^2) - 6t^3(4)]\,dt$$

$$= \int_0^1 (8t^2 + 9t^6 - 24t^3)\,dt = \left(\frac{8}{3}t^3 + \frac{9}{7}t^7 + \frac{24}{4}t^4 \right) \Big]_0^1$$

$$= \frac{8}{3} + \frac{9}{7} + 6 = \frac{56 + 27 + 126}{21} = \frac{209}{21}. \qquad \blacksquare$$

EXAMPLE 4 Let γ_1 be the space curve from $(0, 0, 0)$ to $(1, 1, 1)$ with parametrization

$$x = t^2 e^{t-1}, \qquad y = \sin\left(\frac{\pi t}{2} \right), \qquad z = t^3 \qquad \text{for } 0 \leq t \leq 1.$$

Let $\mathbf{F}(x, y, z) = yz\mathbf{i} + xz\mathbf{j} + xy\mathbf{k}$. It can be shown that the integral of the tangential component of $\mathbf{F}$ is 0 along any piecewise-smooth curve (loop) that starts and ends at the same point (See Section 16.4.) Using this fact, find $\int_{\gamma_1} \mathbf{F} \cdot d\mathbf{r}$.

Solution Computing $\int_{\gamma_1} \mathbf{F} \cdot d\mathbf{r}$ from the given parametrization of γ_1 is not a pleasant task. Let γ_2 be the line segment path from $(0, 0, 0)$ to $(1, 1, 1)$ with parametrization $x = t, y = t, z = t$ for $0 \leq t \leq 1$. Then $\gamma_1 - \gamma_2$ is a piecewise-smooth loop starting and ending at $(0, 0, 0)$, so by the given property of $\mathbf{F}$, we know that $\int_{\gamma_1 - \gamma_2} \mathbf{F} \cdot d\mathbf{r} = 0$. But then $\int_{\gamma_1} \mathbf{F} \cdot d\mathbf{r} - \int_{\gamma_2} \mathbf{F} \cdot d\mathbf{r} = 0$, so that $\int_{\gamma_1} \mathbf{F} \cdot d\mathbf{r} = \int_{\gamma_2} \mathbf{F} \cdot d\mathbf{r}$, and we can compute the more easily evaluated integral $\int_{\gamma_2} \mathbf{F} \cdot d\mathbf{r}$ instead. We obtain

$$\int_{\gamma_2} \mathbf{F} \cdot d\mathbf{r} = \int_{\gamma_2} (yz\,dx + xz\,dy + xy\,dz) = \int_0^1 3t^2\,dt = t^3 \Big]_0^1 = 1. \qquad \blacksquare$$

Integral of the Normal Component of a Two-Dimensional Vector Field Along a Curve

Let $\mathbf{F}(x, y) = P(x, y)\mathbf{i} + Q(x, y)\mathbf{j}$ be the flux field of a flow in the plane, and let γ be a smooth curve in the plane. Recall that the flux field $\mathbf{F}(x, y)$ can be interpreted as measuring the mass of the flow per unit time through a line segment of unit length perpendicular to $\mathbf{F}(x, y)$ at (x, y). The integral $\int_\gamma \mathbf{F} \cdot d\mathbf{r} = \int_\gamma (\mathbf{F} \cdot \mathbf{t})\,ds$ that we discussed measures the mass flowing *tangent* to γ. It is equally important to know $\int_\gamma (\mathbf{F} \cdot \mathbf{n})\,ds$, where $\mathbf{n}$ is a unit vector normal to γ. This integral measures the mass of the flow *across* γ.

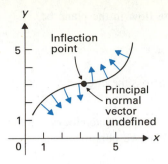

FIGURE 16.18 Principal normal vectors.

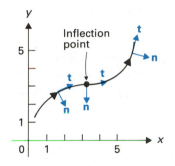

FIGURE 16.19 Vector **n** obtained by rotating **t** 90°.

In Section 13.2, we always considered **n** to be the *principal normal unit vector,* pointing out the concave side of a curve. For a curve with an inflection point, as shown in Fig. 16.18, principal normal vectors are undefined at the inflection point and change direction abruptly there. We want our normal vectors to keep pointing on *the same side* of our curve so that our integral will measure the flow across the curve from one side to the other. Thus we will not take **n** here as the principal unit normal vector but rather as the unit normal vector obtained by rotating the unit tangent vector **t** clockwise through 90°, as shown in Fig. 16.19. Note that if γ is a loop that goes counterclockwise around the boundary of a plane region G, as shown in Fig. 16.20, then **n** points *outward* from the region. Since $\mathbf{t} = (dx/ds)\mathbf{i} + (dy/ds)\mathbf{j}$ with $dx/ds < 0$ and $dy/ds > 0$ in Fig. 16.20, we see that we should take

$$\mathbf{n} = \frac{dy}{ds}\mathbf{i} - \frac{dx}{ds}\mathbf{j}. \tag{13}$$

You can check that Eq. (13) is valid for the other three possibilities for the signs of dx/ds and dy/ds. For this vector **n**, we then have

$$\int_{\gamma} (\mathbf{F} \cdot \mathbf{n}) \, ds = \int_{\gamma} [-Q(x, y) \, dx + P(x, y) \, dy]. \tag{14}$$

EXAMPLE 5 Let γ start at $(0, 0)$ and end at $(1, 2)$, with the line segment joining these points as its trace. Let $\mathbf{F}(x, y) = x^2 y \mathbf{i} - 2x\mathbf{j}$. Find $\int_{\gamma} (\mathbf{F} \cdot \mathbf{n}) \, ds$.

Solution The curve γ is shown in Fig. 16.21, where we show vectors **t** and **n**. A parametrization of γ is given by

$$x = t, \qquad y = 2t \qquad \text{for } 0 \le t \le 1.$$

Using Eq. (14), we find that

$$\int_{\gamma} (\mathbf{F} \cdot \mathbf{n}) \, ds = \int_{\gamma} [-Q \, dx + P \, dy]$$

$$= \int_{\gamma} [-(-2x) \, dx + x^2 y \, dy]$$

$$= \int_{0}^{1} [-(-2t) + t^2(2t)2] \, dt = \int_{0}^{1} (2t + 4t^3) \, dt$$

$$= (t^2 + t^4)]_{0}^{1} = 1 + 1 = 2. \qquad \blacksquare$$

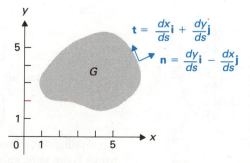

FIGURE 16.20 Vectors **t** and **n**.

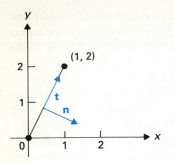

FIGURE 16.21   Vectors **t** and **n**.

EXAMPLE 6   Let the flux field of a steady-state flow in the plane be

$$\mathbf{F}(x, y) = \frac{x}{x^2 + y^2}\mathbf{i} + \frac{y}{x^2 + y^2}\mathbf{j} \quad \text{for } x^2 + y^2 \geq \frac{1}{2}.$$

Find the mass, per unit time, of the flow outward across the circle $x = 2 \cos t$, $y = 2 \sin t$ for $0 \leq t \leq 2\pi$, shown in Fig. 16.22.

Solution   Since the parametrization of the circle traces it counterclockwise, the mass of the flow outward will be given by Eq. (14). We obtain

$$\int_\gamma (\mathbf{F} \cdot \mathbf{n})\, ds = \int \left[-\frac{y}{x^2 + y^2}\, dx + \frac{x}{x^2 + y^2}\, dy \right]$$

$$= \int_0^{2\pi} \left[\frac{-2 \sin t}{4}(-2 \sin t) + \frac{2 \cos t}{4}(2 \cos t) \right] dt$$

$$= \int_0^{2\pi} \frac{4(\sin^2 t + \cos^2 t)}{4}\, dt = \int_0^{2\pi} 1\, dt = 2\pi. \quad\blacksquare$$

EXAMPLE 7   Find the integral of the outward normal component of the vector field **F** in Example 6 along the square η consisting of the line segments from $(1, 1)$ to $(-1, 1)$ to $(-1, -1)$ to $(1, -1)$ to $(1, 1)$, as shown in Fig. 16.22.

Solution   We break the integral

$$\int_\gamma \left[-\frac{y}{x^2 + y^2}\, dx + \frac{x}{x^2 + y^2}\, dy \right] \tag{15}$$

into four parts, corresponding to the line segments labeled L_1, L_2, L_3, and L_4 in Fig. 16.22. On L_1, we see that $dy = 0$ and $y = 1$, so the integral (15) over this part of η becomes

$$\int_{L_1} \frac{-1}{x^2 + 1}\, dx = \int_1^{-1} \frac{-dx}{x^2 + 1} = -\tan^{-1}x \Big]_1^{-1}$$

$$= -[\tan^{-1}(-1) - \tan^{-1}1] = -\left[-\frac{\pi}{4} - \frac{\pi}{4} \right] = \frac{\pi}{2}.$$

On L_2, we see that $dx = 0$ and $x = -1$, so our integral (15) over this part of η becomes

$$\int_{L_2} \frac{-1}{1 + y^2}\, dy = \int_1^{-1} \frac{-1}{1 + y^2}\, dy = \frac{\pi}{2}.$$

On L_3, we have $dy = 0$ and $y = -1$, and the contribution is

$$\int_{L_3} \frac{-(-1)}{x^2 + 1}\, dx = \int_{-1}^1 \frac{1}{x^2 + 1}\, dx = \frac{\pi}{2}.$$

On L_4, where $dx = 0$ and $x = 1$, the contribution is

$$\int_{L_4} \frac{1}{y^2 + 1}\, dy = \int_{-1}^1 \frac{1}{y^2 + 1}\, dy = \frac{\pi}{2}.$$

Adding these four contributions, we see that $\int_\gamma (\mathbf{F} \cdot \mathbf{n})\, ds = 4(\pi/2) = 2\pi$, which is the same value we obtained in Example 6. In Section 16.6, we will see that we obtained the same values because our flux field $\mathbf{F}(x, y) = P(x, y)\mathbf{i} + Q(x, y)\mathbf{j}$ satisfies $\partial P/\partial x + \partial Q/\partial y = 0$. $\quad\blacksquare$

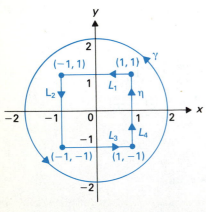

FIGURE 16.22   A circle and a square with center at $(0, 0)$.

SUMMARY

Throughout the following, γ is a piecewise-smooth curve, and $\mathbf{F}(x, y) = P(x, y)\mathbf{i} + Q(x, y)\mathbf{j}$ is a continuous vector field.

1. The integral of the tangential component of the vector field $\mathbf{F}$ along γ is

$$\int_\gamma (\mathbf{F} \cdot \mathbf{t}) \, ds = \int_\gamma \mathbf{F} \cdot d\mathbf{r} = \int_\gamma (P \, dx + Q \, dy).$$

2. If $\mathbf{F}$ is a force field, then $\int_\gamma \mathbf{F} \cdot d\mathbf{r}$ is the work done by the force on a body moving along γ.

3. For

$$\gamma : x = h(t), \qquad y = k(t) \qquad \text{for } 0 \le t \le 1,$$

the curve $-\gamma$ has the same trace traveled in the opposite direction and is given by

$$-\gamma : x = h(1 - t), \qquad y = k(1 - t) \qquad \text{for } 0 \le t \le 1.$$

4. $\int_{-\gamma} (\mathbf{F} \cdot d\mathbf{r}) = -\int_\gamma (\mathbf{F} \cdot d\mathbf{r})$

5. The analogues of numbers 1–4 are valid for continuous vector fields and piecewise-smooth curves in space.

6. The integral of the normal component of $\mathbf{F}$ over γ is

$$\int_\gamma (\mathbf{F} \cdot \mathbf{n}) \, ds = \int_\gamma [-Q(x, y) \, dx + P(x, y) \, dy].$$

Here $\mathbf{n}$ is the unit normal vector to γ obtained by rotating the unit tangent vector $\mathbf{t}$ clockwise 90°.

EXERCISES 16.3

In Exercises 1–6, the curves are

γ_1 given by $x = t$, $y = t^2$ for $0 \le t \le 1$,
γ_2 given by $x = t + 1$, $y = 2t + 1$ for $0 \le t \le 1$,

and the vector fields are

$$\mathbf{F}(x, y) = x^2\mathbf{i} + y^2\mathbf{j} \quad \text{and} \quad \mathbf{G}(x, y) = xy\mathbf{i} - x^2y\mathbf{j}.$$

Find the indicated integral.

1. $\displaystyle\int_{\gamma_1} \mathbf{F} \cdot d\mathbf{r}$

2. $\displaystyle\int_{\gamma_2} \mathbf{G} \cdot d\mathbf{r}$

3. $\displaystyle\int_{\gamma_1} (\mathbf{F} - \mathbf{G}) \cdot d\mathbf{r}$

4. $\displaystyle\int_{\gamma_2} (2\mathbf{F} + \mathbf{G}) \cdot d\mathbf{r}$

5. $\displaystyle\int_{\gamma_1 + \gamma_2} (\mathbf{F} + \mathbf{G}) \cdot d\mathbf{r}$

6. $\displaystyle\int_{-\gamma_1} \mathbf{G} \cdot d\mathbf{r}$

7. Let γ be the curve in the plane that consists of the line segment from $(-1, 2)$ to $(4, 3)$ traveled at constant speed for $0 \le t \le 1$. Find parametric equations for $-\gamma$.

8. Let γ be the curve in space that consists of the line segment from $(3, -1, 4)$ to $(2, 1, 0)$ traveled at constant speed for $0 \le t \le 1$. Find parametric equations for $-\gamma$.

Exercises 9–14 relate to the curves

$\gamma_1 : x = t, \qquad y = t^2 \qquad \text{for } 0 \le t \le 1,$
$\gamma_2 : x = 1 + t, \qquad y = 1 + 6t \qquad \text{for } 0 \le t \le 1,$
$\gamma_3 : x = t^2 - t, \qquad y = 2t^2 + t - 3 \qquad \text{for } 1 \le t \le 2.$

Decide whether or not the given expression is a single piecewise-smooth curve. If it is, find the initial point and the terminal point, and state whether or not it is a loop.

9. $\gamma_1 + \gamma_2$

10. $\gamma_1 + \gamma_2 + \gamma_3$

11. $\gamma_1 + \gamma_2 - \gamma_3$

12. $-\gamma_2 - \gamma_1 + \gamma_3$

13. $-\gamma_2 - \gamma_1$

14. $-\gamma_1 + \gamma_3 - \gamma_2$

15. Let $\mathbf{F}(x, y) = xy\mathbf{i} + x^2\mathbf{j}$ and let γ be the shorter arc from $(3, 0)$ to $(0, 3)$ of a circle with center at the origin. Find $\int_\gamma \mathbf{F} \cdot d\mathbf{r}$.

16. Let $\mathbf{F}(x, y, z) = (x^2 + z)\mathbf{i} + (x + y^2)\mathbf{j} + xz\mathbf{k}$ and let $\gamma = \gamma_1 + \gamma_2 + \gamma_3$ be the path consisting of three straight-line segments from $(0, 0, 0)$ to $(1, 0, 0)$ to $(1, 1, 1)$ to $(0, 1, 2)$. Find $\int_\gamma \mathbf{F} \cdot d\mathbf{r}$.

17. Let γ be the space curve given by $x = 3t^2$, $y = 2t^3$, $z = 3t$ for $0 \le t \le 1$ and let $F(x, y, z) = 3x^2yz$.
   (a) Compute $\int_\gamma dF$ by integration.
   (b) Compute the integral again, using Theorem 16.5 in the next section and the fact that dF is an exact differential.

18. Repeat Exercise 17(b) for the space curve γ given by $x = \sin t$, $y = \cos t$, $z = t$ for $0 \le t \le \pi/4$.

19. Solve the problem in Example 5 by finding $\mathbf{n}$ and ds explicitly, computing $\mathbf{F} \cdot \mathbf{n}$ and then integrating to find $\int_\gamma (\mathbf{F} \cdot \mathbf{n}) \, ds$.

20. Repeat Exercise 19, but for Example 6.

In Exercises 21–26, let γ_1, γ_2, $\mathbf{F}(x, y)$, and $\mathbf{G}(x, y)$ be as in Exercises 1–6. Find the indicated integral.

21. $\int_{\gamma_2} (\mathbf{F} \cdot \mathbf{n}) \, ds$

22. $\int_{\gamma_1} (\mathbf{G} \cdot \mathbf{n}) \, ds$

23. $\int_{\gamma_1 + \gamma_2} (\mathbf{G} \cdot \mathbf{n}) \, ds$

24. $\int_{\gamma_2 - \gamma_1} (\mathbf{F} \cdot \mathbf{n}) \, ds$

25. $\int_{-\gamma_2} (\mathbf{G} \cdot \mathbf{n}) \, ds$

26. $\int_{-\gamma_1} [(2\mathbf{F} - \mathbf{G}) \cdot \mathbf{n}] \, ds$

27. Proceed as in Example 7 to find the mass, per unit time, of the flow with flux field $\mathbf{F}(x, y) = 3xy^2\mathbf{i} - 2xy\mathbf{j}$ *outward* across the boundary curve η of the square shown in Fig. 16.22.

28. Find the mass, per unit time, of the flow with flux field $\mathbf{F}(x, y) = xy\mathbf{i} - x^2y\mathbf{j}$ *inward* across the boundary of the disk $x^2 + y^2 \le 1$.

29. Find the mass, per unit time, of the flow with flux field $\mathbf{F}(x, y) = (3x^2 + y)\mathbf{i} - (x - 2xy)\mathbf{j}$ *inward* across the boundary of the triangle with vertices $(0, 0)$, $(0, 1)$, and $(1, 0)$.

30. Find the mass, per unit time, of the flow with flux field $\mathbf{F}(x, y) = xy\mathbf{i} - 2y\mathbf{j}$ *outward* across the boundary of the region bounded by $y = x^2$ and $y = 1$.

16.4 INDEPENDENCE OF PATH

In the last two sections, we encountered integrals of differential forms over plane curves. For a continuous vector field $\mathbf{F}(x, y) = P(x, y)\mathbf{i} + Q(x, y)\mathbf{j}$, we have

$$\int_\gamma \mathbf{F} \cdot d\mathbf{r} = \int_\gamma (\mathbf{F} \cdot \mathbf{t}) \, ds = \int_\gamma [P(x, y) \, dx + Q(x, y) \, dy] \tag{1}$$

and

$$\int_\gamma (\mathbf{F} \cdot \mathbf{n}) \, ds = \int_\gamma [-Q(x, y) \, dx + P(x, y) \, dy]. \tag{2}$$

We now discuss further integrals of differential forms over plane curves γ, using the notation

$$\int_\gamma [P(x, y) \, dx + Q(x, y) \, dy].$$

We point out that our discussion has meaning for the integral in Eq. (2) also.

Suppose we pick up a box of books in the front hall, carry it up to the second floor, and set it on a desk. We have done some work against the gravitational force field of the earth. Now suppose we had picked up the box in the front hall, carried it up two flights to the third floor, changed our mind, and brought it back down to the second floor and set it on the desk. A physicist will tell us that we would have done the same amount of work on the box of books going via the third floor as when we stopped at the second floor the first time. The work done depends only on the weight of the box, its initial position, and its final position. It *does not depend on the path* used to transport the box. This is a very important property of the earth's gravitational field. *Not all vector fields have this property.* For what vector fields does this property hold?

Let $\mathbf{F}(x, y) = P(x, y)\mathbf{i} + Q(x, y)\mathbf{j}$ be a continuous vector field. For convenience, we shorten our notation to $\mathbf{F} = P\mathbf{i} + Q\mathbf{j}$ and $\mathbf{F} \cdot d\mathbf{r} = P \, dx + Q \, dy$. We are interested in the following question:

> Under what conditions does $\mathbf{F}$ have the property that for any piecewise-smooth curve γ in the domain of $\mathbf{F}$, the integral $\int_\gamma (\mathbf{F} \cdot d\mathbf{r})$ depends only on the initial point A and terminal point B of γ and not on the trace of γ from A to B?

DEFINITION 16.3

Independence of path

If $\mathbf{F}$ has the property just described, then we say that $\int_\gamma (\mathbf{F} \cdot d\mathbf{r})$ is **independent of the path.**

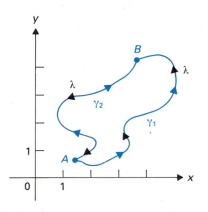

FIGURE 16.23 If γ_1 and γ_2 go from A to B, then $\lambda = \gamma_1 - \gamma_2$ is a loop from A to A, shown by the black arrows.

Let γ_1 and γ_2 be two piecewise-smooth curves from A to B as in Fig. 16.23. If $\int_\gamma (\mathbf{F} \cdot d\mathbf{r})$ is independent of the path, then we have

$$\int_{\gamma_1} (\mathbf{F} \cdot d\mathbf{r}) = \int_{\gamma_2} (\mathbf{F} \cdot d\mathbf{r}) \tag{3}$$

It follows that

$$\int_{\gamma_1 - \gamma_2} (\mathbf{F} \cdot d\mathbf{r}) = \int_{\gamma_1} (\mathbf{F} \cdot d\mathbf{r}) + \int_{-\gamma_2} (\mathbf{F} \cdot d\mathbf{r})$$
$$= \int_{\gamma_1} (\mathbf{F} \cdot d\mathbf{r}) - \int_{\gamma_2} (\mathbf{F} \cdot d\mathbf{r}) = 0. \tag{4}$$

Now $\lambda = \gamma_1 - \gamma_2$ is a piecewise-smooth curve with *both initial and terminal point A*. We call such a curve a **closed curve** or a **loop.** An integral around a closed curve λ is often denoted by $\oint_\gamma$ to call attention to the fact that λ is actually a loop.

Any piecewise-smooth loop from A to A can be regarded as $\gamma_1 - \gamma_2$ for piecewise-smooth curves γ_1 and γ_2. Simply choose a point B on the loop, as shown in Fig. 16.23, to define γ_1 and γ_2. From Eq. (4), we obtain at once the following theorem.

THEOREM 16.4   Independence of path and integrals over loops

Let $\mathbf{F}(x, y)$ be a continuous vector field. The integral $\int_\gamma (\mathbf{F} \cdot d\mathbf{r})$ is independent of the path for every piecewise-smooth curve γ in the domain of F if and only if for every loop λ in the domain of $\mathbf{F}$ we have $\oint_\lambda (\mathbf{F} \cdot d\mathbf{r}) = 0$.   □

EXAMPLE 1 Suppose that $\lambda = \gamma_1 - \gamma_2$ is a loop, and suppose that $\int_{\gamma_1} (\mathbf{F} \cdot d\mathbf{r}) = 18$ and $\int_{\gamma_2} (\mathbf{F} \cdot d\mathbf{r}) = 24$. Find $\oint_\lambda (\mathbf{F} \cdot d\mathbf{r})$ and $\oint_{-\lambda} (\mathbf{F} \cdot d\mathbf{r})$.

Solution   We have

$$\oint_\lambda (\mathbf{F} \cdot d\mathbf{r}) = \int_{\gamma_1 - \gamma_2} (\mathbf{F} \cdot d\mathbf{r}) = \int_{\gamma_1} (\mathbf{F} \cdot d\mathbf{r}) + \int_{-\gamma_2} (\mathbf{F} \cdot d\mathbf{r})$$

$$= \int_{\gamma_1} (\mathbf{F} \cdot d\mathbf{r}) - \int_{\gamma_2} (\mathbf{F} \cdot d\mathbf{r})$$

$$= 18 - 24 = -6.$$

We have $\oint_{-\lambda} (\mathbf{F} \cdot d\mathbf{r}) = -\oint_\lambda (\mathbf{F} \cdot d\mathbf{r}) = -(-6) = 6.$ ■

If γ is a smooth curve joining A and B in the plane and if $P(x, y)\, dx + Q(x, y)\, dy$ is an *exact* differential, then

$$\int_\gamma [P(x, y)\, dx + Q(x, y)\, dy]$$

depends only on A and B and is independent of which smooth curve γ is chosen joining A and B. To see that this is so, let the curve γ be given by $x = h(t)$, $y = k(t)$ for $a \le t \le b$. Also, let $H(x, y)$ be such that $dH = P(x, y)\, dx + Q(x, y)\, dy$. Then

$$\int_\gamma [P(x, y)\, dx + Q(x, y)\, dy] = \int_a^b \left(P(x, y)\frac{dx}{dt} + Q(x, y)\frac{dy}{dt} \right) dt$$

$$= \int_a^b \frac{d}{dt}(H(h(t), k(t)))\, dt$$

$$= H(h(b), k(b)) - H(h(a), k(a)).$$

Thus the integral depends only on the endpoints $A = (h(a), k(a))$ and $B = (h(b), k(b))$. We summarize in a theorem.

THEOREM 16.5   Integration of exact differentials

The line integral $\int_\gamma (P\, dx + Q\, dy)$ of an exact differential $dH = P\, dx + Q\, dy$ is independent of the path. Furthermore,

$$\int_\gamma (P\, dx + Q\, dy) = H(B) - H(A), \tag{5}$$

where A is the initial point and B the terminal point of γ.   □

EXAMPLE 2   Let γ be given by

$$\gamma: x = t^2 - 2t, \qquad y = \frac{t^4 + 3}{4} \qquad \text{for } 1 \le t \le 3.$$

Find $\int_\gamma [(2xy - 4x)\, dx + (x^2 + 2y)\, dy]$.

Solution   Using the methods of Section 16.1, we see that $(2xy - 4x)\, dx + (x^2 + 2y)\, dy$ is the exact differential $d(x^2y - 2x^2 + y^2)$. The initial point of γ when $t = 1$ is $A = (-1, 1)$, and the terminal point when $t = 3$ is $B = (3, 21)$.

We let $H(x, y) = x^2y - 2x^2 + y^2$, and by Theorem 16.5 we see that the desired integral is given by

$$H(3, 21) - H(-1, 1) = (9 \cdot 21 - 18 + 21^2) - (1 - 2 + 1)$$
$$= 189 - 18 + 441 = 612. \quad \blacksquare$$

The analogue for space of all our work in this section holds also.

EXAMPLE 3 Let γ be a piecewise-smooth curve with initial point $(3, -1, 2)$ and terminal point $(1, -1, 0)$ in space. Find

$$\int_\gamma (yz \, dx + xz \, dy + xy \, dz).$$

Solution Using the methods of Section 16.1, we see that the differential form being integrated is dH for $H(x, y, z) = xyz$. By the space analogue of Theorem 16.5, the desired integral is

$$H(1, -1, 0) - H(3, -1, 2) = 0 - (-6) = 6. \quad \blacksquare$$

Sometimes Theorem 16.5 can be used to help find $\int_\gamma (P \, dx + Q \, dy)$, even when $P \, dx + Q \, dy$ is not an exact differential form. The next example illustrates what we mean.

EXAMPLE 4 Let γ be given by

$$\gamma : x = t^2 - 2t, \qquad y = t^3 \qquad \text{for } 0 \le t \le 2.$$

Find $\int_\gamma [(3x^2y^2 - 3) \, dx + (2x^3y + x + 2y) \, dy]$.

Solution A check shows that the differential form to be integrated along γ is not exact. However, it may be written as

$$[(3x^2y^2 - 3) \, dx + (2x^3y + 2y) \, dy] + x \, dy,$$

where the differential form in the square bracket is exact and is dH for $H(x, y) = x^3y^2 - 3x + y^2$. The initial point of γ is $(0, 0)$, and the terminal point is $(0, 8)$. Thus we see that the desired integral can be evaluated as

$$H(0, 8) - H(0, 0) + \int_\gamma x \, dy = 64 - 0 + \int_0^2 (t^2 - 2t)3t^2 \, dt$$

$$= 64 + \int_0^2 (3t^4 - 6t^3) \, dt$$

$$= 64 + \left. \left(\frac{3t^5}{5} - \frac{6t^4}{4} \right) \right]_0^2$$

$$= 64 + \frac{96}{5} - 24 = \frac{296}{5}. \quad \blacksquare$$

We now show that independence of the path for line integrals of $\mathbf{F} = P\mathbf{i} + Q\mathbf{j}$ implies that $P \, dx + Q \, dy$ is exact. Let a region G be such that any two points in it can be joined by a piecewise-smooth curve and also such that for

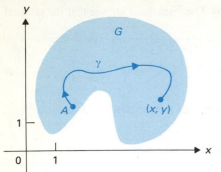

FIGURE 16.24

$$H(x, y) = \int_{\gamma} (P\,dx + Q\,dy)$$

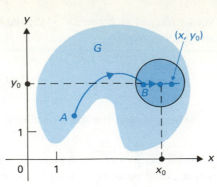

FIGURE 16.25

$$H(x, y_0) = H(B) + \int_{B}^{(x,\,y_0)} P(x, y_0)\,dx$$

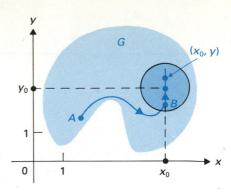

FIGURE 16.26

$$H(x_0, y) = H(B) + \int_{B}^{(x_0,\,y)} Q(x_0, y)\,dy$$

every point in G there is some small disk centered at that point contained entirely in G. Such a region G is a **connected open region.**

Let A be any point in the region G. At any point (x, y) in G, we define $H(x, y)$ as follows. Let γ be a piecewise-smooth curve in G from A to (x, y) as shown in Fig. 16.24. We define

$$H(x, y) = \int_{\gamma} (P\,dx + Q\,dy). \qquad (6)$$

By assumption, the value $H(x, y)$ is independent of the path γ chosen from A to (x, y).

We wish to demonstrate that at any point (x_0, y_0) in G, we have $H_x = P$ and $H_y = Q$, which will show that $P\,dx + Q\,dy$ is exact. Now some disk centered at (x_0, y_0) is contained in the open region G. Referring to Fig. 16.25, let B be a point at the left end of a horizontal line segment with midpoint (x_0, y_0) and contained in the disk. Let us denote by $\int_{B}^{(x_0, y_0)} (P\,dx + Q\,dy)$ the integral of the differential form along this line segment from B to a point (x, y_0) on it. See Fig. 16.25. *Since the integral of $P\,dx + Q\,dy$ is independent of the path*, we could use the path in Fig. 16.25 to find $H(x, y_0)$, obtaining

$$H(x, y_0) = H(B) + \int_{B}^{(x,y_0)} (P\,dx + Q\,dy) = H(B) + \int_{B}^{(x,y_0)} P(x, y_0)\,dx,$$

since $dy = 0$ on the line segment. By the fundamental theorem of calculus, we obtain

$$H_x(x_0, y_0) = P(x_0, y_0).$$

A similar argument using Fig. 16.26 shows that $H_y(x_0, y_0) = Q(x_0, y_0)$. Thus $dH = P\,dx + Q\,dy$, so $P\,dx + Q\,dy$ is an exact differential form.

THEOREM 16.6   Independence of path and exact differentials

If $\int_{\gamma} (P\,dx + Q\,dy)$ is independent of the path in a connected open region G, then $P\,dx + Q\,dy$ is an exact differential form.   □

Theorems 16.5 and 16.6 taken together show the following:

> In a connected open region G, the integral $\int_\gamma (P\,dx + Q\,dy)$ is independent of the path if and only if $P\,dx + Q\,dy$ is an exact differential form.

In view of Theorem 16.4, we then obtain

> $\oint_\lambda (P\,dx + Q\,dy) = 0$ for every loop λ in a connected open region G if and only if $P\,dx + Q\,dy$ is an exact differential form.

Recall that $dH = P\,dx + Q\,dy$ if and only if $\nabla H = P\mathbf{i} + Q\mathbf{j}$. Thus we also have the following:

> Line integrals of a vector field $\mathbf{F} = P\mathbf{i} + Q\mathbf{j}$ are independent of the path in a connected open region G if and only if $\mathbf{F} = \nabla H$ for some function H.

Of course, space analogues of all these statements are also true.

Applications

Let $\mathbf{F}$ be a continuous vector field on a connected open region in the plane or in space. If $\mathbf{F} = \nabla H$, then $\oint_\lambda (\mathbf{F} \cdot d\mathbf{r}) = 0$ around any loop λ in G. That is, the work done by the force field $\mathbf{F}$ moving a body around any loop in G is zero. Such a force field $\mathbf{F}$ is **conservative.** No work is done by such a force field acting on a body unless the position of the body is changed. The function $-H(x, y)$ in the plane, or $-H(x, y, z)$ in space, is called a **potential function of F.** A potential function gives the **potential energy** of the force field at each point. By Theorem 16.5, the work done by the conservative force field $\mathbf{F}$ in moving a body from A to B is

$$H(B) - H(A) = -H(A) - (-H(B)),$$

which is the *loss of potential energy from A to B*.

As mentioned in Section 16.1, $P\mathbf{i} + Q\mathbf{j}$ is a gradient vector field in a region having no holes in it if P and Q have continuous first partial derivatives and $\partial P/\partial y = \partial Q/\partial x$.

EXAMPLE 5   Consider the force field

$$\mathbf{F}(x, y) = (2x + \cos y)\mathbf{i} + (y - x \sin y)\mathbf{j}.$$

Show that $\mathbf{F}$ is conservative, find a potential function $-H(x, y)$, and find the loss in potential energy as a body is moved by the field from $(3, 0)$ to $(-2, \pi)$.

Solution   We have

$$\frac{\partial(2x + \cos y)}{\partial y} = -\sin y = \frac{\partial(y - x \sin y)}{\partial x},$$

so the field $\mathbf{F}$ is conservative. As in Section 16.1, we find that $\mathbf{F} = \nabla H$, where $H(x, y) = x^2 + x \cos y + (y^2/2)$. Therefore

$$-H(x, y) = -x^2 - x \cos y - \frac{y^2}{2}$$

is a potential function. The loss in potential energy from $(3, 0)$ to $(-2, \pi)$ is

$$-H(3, 0) - (-H(-2, \pi)) = H(-2, \pi) - H(3, 0)$$
$$= \left(4 + 2 + \frac{\pi^2}{2}\right) - (9 + 3) = -6 + \frac{\pi^2}{2}. \quad\blacksquare$$

The force of gravitational attraction of a body in space toward a mass at the origin is given by the force field

$$\mathbf{F}(x, y, z) = \frac{-K}{(x^2 + y^2 + z^2)^{3/2}}(x\mathbf{i} + y\mathbf{j} + z\mathbf{k})$$

for some constant K. Force vectors are directed toward the origin and are of magnitude inversely proportional to the square of the distance from the origin. It is easily checked that $\mathbf{F} = \nabla H$ where $H(x, y, z) = K/\sqrt{x^2 + y^2 + z^2}$. This shows that the gravitational attraction field $\mathbf{F}$ is a conservative force field in all of space except the origin, which is a connected open region. The work done by gravitational attraction in moving a body around any loop in space is zero. This is one aspect of the *law of conservation of energy* of physics. The potential function $-K/\sqrt{x^2 + y^2 + z^2}$ is the *Newtonian potential*.

Let F be the flux field of a steady-state flow in the plane or in space. If λ is a loop, then $\oint_\lambda (\mathbf{F} \cdot d\mathbf{r})$ is the **circulation of the flow around** λ. If $\mathbf{F} = \nabla H$, then the circulation of the flow around any loop is zero. Such a flux field $\mathbf{F}$ is called **irrotational.**

EXAMPLE 6   Find all functions $P(x, y)$ such that the flux field

$$\mathbf{F}(x, y) = P(x, y)\mathbf{i} + (xy^2 + 2x - 3y)\mathbf{j}$$

in the plane is irrotational.

Solution   Our work in Section 16.1 shows that $\mathbf{F}$ is irrotational if and only if

$$\frac{\partial P}{\partial y} = \frac{\partial(xy^2 + 2x - 3y)}{\partial x} = y^2 + 2.$$

Therefore we must have

$$P(x, y) = \int (y^2 + 2) \, dy$$

$$= \frac{y^3}{3} + 2y + k(x)$$

for some function $k(x)$ of x only. ∎

Suppose that the flux field $\mathbf{F}(x, y) = P(x, y)\mathbf{i} + Q(x, y)\mathbf{j}$ of a steady-state flow is such that $-Q(x, y) \, dx + P(x, y) \, dy$ is an exact differential. Then

$$\int_\lambda [-Q(x, y) \, dx + P(x, y) \, dy] = \int_\lambda (\mathbf{F} \cdot \mathbf{n}) \, ds = 0$$

for every loop λ, so the mass per unit time of the flow across any loop is zero. Such a flow is **incompressible.**

EXAMPLE 7 Find all differentiable functions $P(x, y)$ such that $\mathbf{F}(x, y) = P(x, y)\mathbf{i} + x^2y\mathbf{j}$ is an incompressible flow.

Solution For $-Q(x, y) \, dx + P(x, y) \, dy$ to be an exact differential form, we must have

$$\frac{\partial(-Q)}{\partial y} = \frac{\partial P}{\partial x}$$

as shown in Section 16.1. Thus

$$\frac{\partial P}{\partial x} = \frac{\partial(-Q)}{\partial y} = \frac{\partial(-x^2y)}{\partial y} = -x^2.$$

Integrating partially with respect to x, we find that

$$P(x, y) = -\frac{x^3}{3} + h(y),$$

where $h(y)$ is any continuously differentiable function of y only. ∎

SUMMARY

1. $\int_\gamma (\mathbf{F} \cdot d\mathbf{r})$ is independent of the path in a connected open region G, depending only on the initial point A and terminal point B for γ in G, if and only if $\oint_\lambda (\mathbf{F} \cdot d\mathbf{r}) = 0$ for every loop λ in G. (A loop or closed curve is one whose initial and terminal points coincide.)

2. $\int_\gamma (\mathbf{F} \cdot d\mathbf{r})$ is independent of the path in a connected open region G if and only if $P \, dx + Q \, dy$ is an exact differential form dH in G or, equivalently, if and only if $\mathbf{F}$ is a gradient vector field ∇H.

3. A force field $\mathbf{F}$ is called conservative if $\mathbf{F}$ is a gradient vector field ∇H. The function $-H(x, y)$ is then a potential function of $\mathbf{F}$.

4. If $\mathbf{F}$ is the flux field of a flow, then $\oint_\lambda \mathbf{F} \cdot d\mathbf{r}$ is the circulation of the flow around the loop λ. If $\mathbf{F}$ is a gradient field ∇H, then the field is irrotational.

5. A flux field $\mathbf{F}(x, y) = P(x, y)\mathbf{i} + Q(x, y)\mathbf{j}$ is irrotational if the form $-Q(x, y)\, dx + P(x, y)\, dy$ is an exact differential, so that $\oint_\lambda (\mathbf{F} \cdot \mathbf{n})\, ds = 0$ around every loop λ in the domain of $\mathbf{F}$.

EXERCISES 16.4

In Exercises 1–12, find $\int_\gamma \mathbf{F} \cdot d\mathbf{r}$ for the given vector field $\mathbf{F}$ along the given curve γ. Use the fact that $\mathbf{F}$ is either a gradient vector field or "almost" a gradient vector field. See Example 4.

1. $\mathbf{F}(x, y) = xy^2\mathbf{i} + x^2y\mathbf{j}$;
   $\gamma: x = 3t^2,\ y = 2t^3$ for $0 \le t \le 1$

2. $\mathbf{F}(x, y) = ((1/y) - x)\mathbf{i} + (2y - (x/y^2))\mathbf{j}$;
   $\gamma: x = 2t + 3,\ y = t^2 + 1$ for $0 \le t \le 1$

3. $\mathbf{F}(x, y) = (y^2 + 2x)\mathbf{i} + (2xy + x)\mathbf{j}$;
   $\gamma: x = \sin^2 t,\ y = \sin t$ for $0 \le t \le 3\pi/2$

4. $\mathbf{F}(x, y) = (ye^{xy} + 2xy^2 - y^2)\mathbf{i} + (xe^{xy} + 2x^2y)\mathbf{j}$;
   $\gamma: x = \cos 2t,\ y = \sin t$ for $0 \le t \le \pi/2$

5. $\mathbf{F}(x, y) = (2y \sin xy \cos xy + x)\mathbf{i} + (2x \sin xy \cos xy - x)\mathbf{j}$;
   $\gamma: x = 1 \cos t,\ y = 1 + \sin t$ for $0 \le t \le 2\pi$

6. $\mathbf{F}(x, y) = (e^y + 2xy^3 - 3 \sin x)\mathbf{i}$
   $+ (xe^y + 3x^2y^2 - \cos y^2)\mathbf{j}$;
   $\gamma: x = 2 + 3 \sin t,\ y = -4 + 2 \sin t$ for $0 \le t \le 2\pi$

7. $\mathbf{F}(x, y, z) = 2xyz\mathbf{i} + x^2z\mathbf{j} + x^2y\mathbf{k}$;
   $\gamma: x = 2t + 1,\ y = t^2,\ z = t - 3$ for $1 \le t \le 2$

8. $\mathbf{F}(x, y, z) = (y^2 - 6xz)\mathbf{i} + 2xy\mathbf{j} - 3x^2\mathbf{k}$;
   $\gamma: x = t^3 - 2t,\ y = (t + 2)/(t - 2),\ z = 3t - 1$
   for $0 \le t \le 1$

9. $\mathbf{F}(x, y, z) = (y/z)\mathbf{i} + ((x/z) + z^2)\mathbf{j} + (2yz - (xy/z^2))\mathbf{k}$;
   $\gamma: x = 4 + \sin \pi t,\ y = -3 - \cos \pi t,\ z = t^2 - 2t + 2$
   for $0 \le t \le 2$

10. $\mathbf{F}(x, y, z) = (yz + 2x)\mathbf{i} + (xz - y)\mathbf{j} + (xy + y - z)\mathbf{k}$;
    $\gamma: x = \cos t,\ y = \sin t,\ z = t^2 - \pi^2$ for $-\pi \le t \le \pi$

11. $\mathbf{F}(x, y, z) = (x + y)\mathbf{i} + ye^y\mathbf{j} - (z^3\sin z)\mathbf{k}$;
    $\gamma: x = \cos 2t,\ y = -\sin 2t,\ z = \cos t$ for $0 \le t \le 2\pi$

12. $\mathbf{F}(x, y, z) = (z + \sin x^2)\mathbf{i} + \cos^2 y^3\mathbf{j} + y\mathbf{k}$;
    $\gamma: x = t^2 + 1,\ y = \sin \pi t,\ z = \cos \pi t$ for $-1 \le t \le 1$

13. Suppose that the position of a body moving in the plane is $(\cos t, \sin 2t)$ at time t and that the body is subject to the force field $x\mathbf{i} + y\mathbf{j}$. Find the work done by the force field on the body from time $t = 0$ to time $t = \pi/2$.

14. Let the position of a moving body in space be $(3t^2, 2t^3, 3t)$ at time t, and let the body be subject to the force field $x\mathbf{i} + z\mathbf{j} + y\mathbf{k}$. Find the work done by the force field on the body from time $t = 1$ to time $t = 2$.

15. Let the flux field of a steady-state fluid flow in the plane be $x\mathbf{i} - y\mathbf{j}$. Find the circulation of the flow around the circle $x = \cos t,\ y = \sin t$ for $0 \le t \le 2\pi$.

16. Let the flux field of a steady-state fluid flow in the plane be $xy\mathbf{i} - x\mathbf{j}$. Find the circulation of the flow around the ellipse $x = 2 \cos t,\ y = 3 \sin t$ for $0 \le t \le 2\pi$.

17. Consider the conservative force field
    $$\mathbf{F}(x, y, z) = (2x + z^2)\mathbf{i} - 4y\mathbf{j} + 2xz\mathbf{k}.$$
    Find the loss of potential energy from $(2, 1, 0)$ to $(3, -1, 1)$.

18. Consider the conservative force field
    $$\mathbf{F}(x, y, z) = (y^2z - 2x)\mathbf{i} + (2xyz - z)\mathbf{j} + (xy^2 - y + 8z)\mathbf{k}.$$
    Find the loss of potential energy from $(1, -1, 1)$ to $(2, -3, 1)$.

In Exercises 19–22, find all differentiable functions $P(x, y)$ such that the given flux field of a steady-state flow is (a) irrotational, (b) incompressible.

19. $\mathbf{F}(x, y) = P(x, y)\mathbf{i} + (3x + y^2)\mathbf{j}$

20. $\mathbf{F}(x, y) = P(x, y)\mathbf{i} - (4y - 3xy^2)\mathbf{j}$

21. $\mathbf{F}(x, y) = P(x, y)\mathbf{i} - (\sin xy)\mathbf{j}$

22. $\mathbf{F}(x, y) = P(x, y)\mathbf{i} + (x \cos y)\mathbf{j}$

In Exercises 23–26, find all differentiable functions $Q(x, y)$ such that the given flux field of a steady-state flow is (a) irrotational, (b) incompressible.

23. $\mathbf{F}(x, y) = (x^2y^3 + 3xy)\mathbf{i} + Q(x, y)\mathbf{j}$

24. $\mathbf{F}(x, y) = (\sin x)\mathbf{i} + Q(x, y)\mathbf{j}$

25. $\mathbf{F}(x, y) = (3x - 2y)\mathbf{i} + Q(x, y)\mathbf{j}$

26. $\mathbf{F}(x, y) = e^{2x-y}\mathbf{i} + Q(x, y)\mathbf{j}$

16.5 SURFACE INTEGRALS

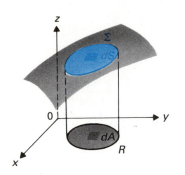

FIGURE 16.27 Differential surface element of area dS.

Consider a surface Σ in space lying over (or under) a region R in the x,y-plane, as shown in Fig. 16.27. We assume that the surface is **smooth,** that is, that it can be described by an equation like $z = f(x, y)$ or $G(x, y, z) = c$ where the functions have continuous partial derivatives. In Section 15.4, we found expressions for the differential element dS of surface area. We recall these expressions in Table 16.1. If the partial derivatives appearing in Table 16.1 are all continuous and if $h(x, y, z)$ is continuous for x, y, z on the surface over R, then the **integral of** $h(x, y, z)$ **over the surface** is

$$\iint_{\Sigma} h(x, y, z)\, dS = \iint_{R} h(x, y, f(x))\, dS. \tag{1}$$

An approximation of the integral (1) by a Riemann sum consists of summing products of the form

(Area of a piece of surface)($h(x, y, z)$ for (x, y, z) in the piece)

over a partition of the surface into small pieces.

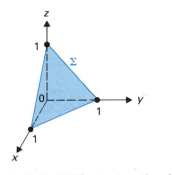

FIGURE 16.28 A triangular plate.

EXAMPLE 1 Let a thin plate cover the triangular region in space with vertices $(1, 0, 0)$, $(0, 1, 0)$, and $(0, 0, 1)$, and let the area mass density of the plate at (x, y, z) be $\sigma(x, y, z) = yz$. Find the mass of the plate.

Solution The triangular plate is sketched in Fig. 16.28. The plate covers the portion of the plane $x + y + z = 1$ lying in the first octant. Using the second formula for dS in Table 16.1, we find that

$$dS = \frac{\sqrt{1^2 + 1^2 + 1^2}}{|1|}\, dx\, dy = \sqrt{3}\, dx\, dy.$$

The plate lies over the triangular region R in the x,y-plane bounded by the x- and y-axes and the line $x + y = 1$, as shown in Fig. 16.29. We see that the mass of the plate is given by

$$\text{Mass} = \iint_{R} \sigma(x, y, z)\, dS = \int_{0}^{1} \int_{0}^{1-y} yz\sqrt{3}\, dx\, dy.$$

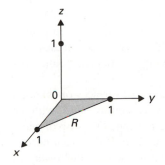

FIGURE 16.29 Triangular region R.

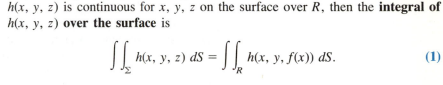

TABLE 16.1

Equation of the surface	Formula for dS
$z = f(x, y)$	$dS = \sqrt{f_x^2 + f_y^2 + 1}\, dx\, dy$
$G(x, y, z) = c$	$dS = \dfrac{\sqrt{G_x^2 + G_y^2 + G_z^2}}{|G_z|}\, dx\, dy$
	$= \dfrac{|\nabla G|}{|G_z|}\, dx\, dy$

FIGURE 16.30 Outward normal vectors on a sphere.

Since $z = 1 - x - y$ for points in the plate, we find that

$$\text{Mass} = \int_0^1 \int_0^{1-y} y(1 - x - y)\sqrt{3} \, dx \, dy = \sqrt{3} \int_0^1 \int_0^{1-y} (y - xy - y^2) \, dx \, dy$$

$$= \sqrt{3} \int_0^1 \left[y(1 - y) - \frac{(1 - y)^2}{2} y - y^2(1 - y) \right] dy$$

$$= \sqrt{3} \int_0^1 \left[y - y^2 - \frac{1}{2} y + y^2 - \frac{1}{2} y^3 - y^2 + y^3 \right] dy$$

$$= \sqrt{3} \int_0^1 \left(\frac{1}{2} y - y^2 + \frac{1}{2} y^3 \right) dy = \sqrt{3} \left(\frac{1}{4} - \frac{1}{3} + \frac{1}{8} \right) = \frac{\sqrt{3}}{24}. \quad \blacksquare$$

The integral in the preceding example was tedious to evaluate, even though dS did not contribute a square root of a variable quantity.

Integral of the Normal Component of a Vector Field over a Surface

Let $\mathbf{F}(x, y, z) = P(x, y, z)\mathbf{i} + Q(x, y, z)\mathbf{j} + R(x, y, z)\mathbf{k}$ be the flux field of a steady-state flow in space, and let Σ be a surface in space. Let $\mathbf{n}$ be a unit normal vector to Σ at (x, y, z) on Σ. Then $(\mathbf{F}(x, y, z) \cdot \mathbf{n}) \, dS$ gives the approximate mass that flows, per unit time, in the direction of $\mathbf{n}$ across a little piece of surface containing the point (x, y, z) and having area dS. Assuming that it makes sense to talk about *one side* of the surface and that the vectors $\mathbf{n}$ all point out the same side of Σ, we see that integrating $\mathbf{F} \cdot \mathbf{n}$ over Σ gives the mass that flows, per unit time, across Σ in the direction given by the vectors $\mathbf{n}$. This is called the **flux across** Σ in that direction.

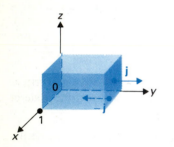

FIGURE 16.31 Surface of a box with an outward normal vector $\mathbf{j}$.

To compute $\iint_\Sigma (\mathbf{F} \cdot \mathbf{n}) \, dS$ over a surface Σ, we must know the vector $\mathbf{n}$ at each point of Σ. A smooth surface has two unit normal vectors, one the negative of the other, at each point of the surface. We will always assume that our surface Σ is composed of a finite number of smooth pieces on each of which $\mathbf{n}$ varies continuously, with no abrupt change in direction. For example, a sphere is one smooth surface, and we can choose either all normal vectors pointing outward, as in Fig. 16.30, or all normal vectors pointing inward. The surface of the box shown in Fig. 16.31 is composed of six smooth pieces, namely the top, bottom, left, right, front, and back sides. On the right side, we can choose either the outward unit normal vector $\mathbf{j}$ or the inward vector $-\mathbf{j}$ as $\mathbf{n}$, as shown in Fig. 16.31, and similarly for the other six sides. In practice, for a surface bounding a region in space, we usually choose either all inward or all outward normal vectors. For surfaces that do not enclose a region of space, we often describe the desired direction of the normal vectors by specifying the sign of their coefficient of $\mathbf{i}$, $\mathbf{j}$, or $\mathbf{k}$. For example, Fig. 16.32 shows the hemisphere $z = -\sqrt{4 - x^2 - y^2}$ with normal vectors $\mathbf{n}$ having nonnegative $\mathbf{k}$-coefficients.

We mention in passing that there are smooth *one-sided* surfaces on which it is impossible to choose unit normal vectors that vary continuously with no change in direction. A Möbius strip, made by joining the ends of a strip of paper with a half twist, illustrated in Fig. 16.33, is such a surface. If the black normal vectors vary continuously, starting at the point P with the one pointing toward us at the front and working around to the left, we wind up at the same point P with

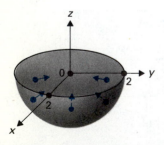

FIGURE 16.32 Normal vectors with positive $\mathbf{k}$-coefficients.

FIGURE 16.33 A Möbius strip.

TABLE 16.2

Equation of the surface	Formula for **n**
$z = f(x, y)$	$\mathbf{n} = \dfrac{\pm 1}{\sqrt{f_x^2 + f_y^2 + 1}}(f_x\mathbf{i} + f_y\mathbf{j} - \mathbf{k})$
$G(x, y, z) = c$	$\mathbf{n} = \dfrac{\pm 1}{\sqrt{G_x^2 + G_y^2 + G_z^2}}(G_x\mathbf{i} + G_y\mathbf{j} + G_z\mathbf{k}) = \dfrac{\pm\nabla G}{|\nabla G|}$

the vector shown in color pointing away from us. We will not concern ourselves further with such surfaces.

Assuming the direction of **n** has now been chosen, we turn to its computation. A normal vector to a surface given by $z = f(x, y)$ is $f_x\mathbf{i} + f_x\mathbf{j} - \mathbf{k}$, while a normal vector to a surface described by $G(x, y, z) = 0$ is $\nabla G = G_x\mathbf{i} + G_y\mathbf{j} + G_z\mathbf{k}$. Thus unit normal vectors for such surfaces are as shown in Table 16.2.

The radicals appearing in the denominator of the expressions for **n** in Table 16.2 can make $\mathbf{F} \cdot \mathbf{n}$ a fairly difficult expression, but remember that we want to integrate $(\mathbf{F} \cdot \mathbf{n})\, dS$. Table 16.1 shows that precisely these same radicals appear in the numerator in dS. Thus the radical expressions cancel when we compute $(\mathbf{F} \cdot \mathbf{n})\, dS$, and we wind up integrating either

$$(\mathbf{F} \cdot \mathbf{n})\, dS = \pm(Pf_x + Qf_y - R)\, dx\, dy \tag{2}$$

or

$$(\mathbf{F} \cdot \mathbf{n})\, dS = \pm\frac{PG_x + QG_y + RG_z}{|G_z|}\, dx\, dy \tag{3}$$

over a region R in the x,y-plane, assuming that we choose to project our surface Σ on that coordinate plane. The choice of sign depends on the direction of the normal unit vectors **n**. We illustrate with a few examples.

EXAMPLE 2 Let $\mathbf{F}(x, y, z) = x\mathbf{i} + y\mathbf{j} + z\mathbf{k}$ and let Σ be the hemisphere $x^2 + y^2 + z^2 = 4$, $z \geq 0$, with unit normal vectors **n** having positive **k**-coefficients. Find $\iint_\Sigma (\mathbf{F} \cdot \mathbf{n})\, dS$.

Solution The hemisphere and some normal vectors with positive **k**-coefficients are sketched in Fig. 16.34. The equation $G(x, y, z) = x^2 + y^2 + z^2 = 4$ for our hemisphere shows that formula (3) is appropriate for $(\mathbf{F} \cdot \mathbf{n})\, dS$. The sign of the **k**-coefficient of $G_x\mathbf{i} + G_y\mathbf{j} + G_z\mathbf{k}$ is determined by $G_z = 2z$. Since we want normal vectors with positive **k**-coefficients and since $z \geq 0$ on our surface, the plus sign in formula (3) is appropriate, and we obtain

$$(\mathbf{F} \cdot \mathbf{n})\, dS = \frac{x(2x) + y(2y) + z(2z)}{2z}\, dx\, dy = \frac{2x^2 + 2y^2 + 2z^2}{2z}\, dx\, dy.$$

Since we are on the sphere $x^2 + y^2 + z^2 = 4$, this simplifies to

$$(\mathbf{F} \cdot \mathbf{n})\, dS = \frac{8}{2z} = \frac{4}{\sqrt{4 - x^2 - y^2}}\, dx\, dy.$$

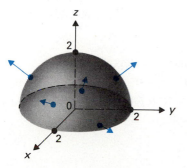

FIGURE 16.34 Hemisphere with normal vectors having positive **k**-coefficients.

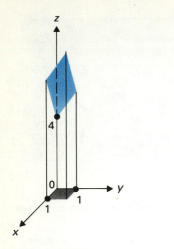

FIGURE 16.35

$z = 3x + 2y + 4$, $0 \le x \le 1$, $0 \le y \le 1$

We should integrate this expression over the disk in the x,y-plane with center at the origin and radius 1. Using polar coordinates, we obtain the integral

$$\int_0^{2\pi} \int_0^2 \frac{4}{\sqrt{4 - r^2}} r \, dr \, d\theta = -2 \int_0^{2\pi} \int_0^2 (4 - r^2)^{-1/2}(-2r) \, dr \, d\theta$$

$$= -2 \int_0^{2\pi} 2\sqrt{4 - r^2} \Big]_0^2 \, d\theta$$

$$= -2 \int_0^{2\pi} (0 - 4) \, d\theta$$

$$= 8 \int_0^{2\pi} d\theta = 8(2\pi) = 16\pi. \qquad \blacksquare$$

EXAMPLE 3 Let $\mathbf{F}(x, y, z) = y\mathbf{i} - xy\mathbf{j} + xz\mathbf{k}$. Let Σ be the portion of the plane $z = 3x + 2y + 4$ for $0 \le x \le 1$ and $0 \le y \le 1$. Let unit normal vectors in Σ be chosen with negative $\mathbf{j}$-coefficients. Find $\iint_\Sigma (\mathbf{F} \cdot \mathbf{n}) \, dS$.

Solution The portion of the plane is sketched in Fig. 16.35. This time, we will use formula 2 with $f(x, y) = 3x + 2y + 4$ to compute $(\mathbf{F} \cdot \mathbf{n}) \, dS$. Thus $(\mathbf{F} \cdot \mathbf{n}) \, dS = \pm[y(3) - xy(2) - xz]$. Since we want the normal vectors to have negative $\mathbf{j}$-coefficients and since $f_y = 2$ is positive, we see that we should use the negative sign. Now for z on the surface, we have $z = 3x + 2y + 4$ so $(\mathbf{F} \cdot \mathbf{n}) \, dS = -[3y - 2xy - x(3x + 2y + 4)] \, dx \, dy$, and we obtain the integral

$$\int_0^1 \int_0^1 (-3y + 4xy + 3x^2 + 4x) \, dx \, dy = \int_0^1 (-3y + 2y + 1 + 2) \, dy$$

$$= -\frac{1}{2} + 3 = \frac{5}{2}. \qquad \blacksquare$$

EXAMPLE 4 The flux field of a steady-state flow is $\mathbf{F}(x, y, z) = -xy\mathbf{i} + yz\mathbf{k}$. Find the flux of the flow *into* the cube shown in Fig. 16.36, having diagonally opposite vertices at $(0, 0, 0)$ and $(1, 1, 1)$.

Solution We must integrate $(\mathbf{F} \cdot \mathbf{n}) \, dS$ over the surface of this cube, taking unit vectors $\mathbf{n}$ pointing inward. We break the integral into six pieces, projecting the top and bottom of the box onto the x,y-plane, the left and right sides onto the x,z-plane, and the front and back onto the y,z-plane. The integrals are as follows.

Right side Here $\mathbf{n} = -\mathbf{j}$, so $\mathbf{F} \cdot \mathbf{n} = 0$ and the integral is 0.

Left side Since $\mathbf{n} = \mathbf{j}$, we again have $\mathbf{F} \cdot \mathbf{n} = 0$, so the integral is 0.

Front side Here $x = 1$ so $\mathbf{F}(x, y, z) = -y\mathbf{i} + yz\mathbf{k}$. Also, $\mathbf{n} = -\mathbf{i}$ and $dS = dy \, dz$, so $(\mathbf{F} \cdot \mathbf{n}) \, dS = y \, dy \, dz$. The integral is

$$\int_0^1 \int_0^1 y \, dy \, dz = \int_0^1 \frac{1}{2} \, dz = \frac{1}{2}.$$

Back side Here $x = 0$ so $\mathbf{F}(x, y, z) = yz\mathbf{k}$. Since $\mathbf{n} = \mathbf{i}$, we see that $\mathbf{F} \cdot \mathbf{n} = 0$, so the integral is 0.

Top side Here $z = 1$ so $\mathbf{F}(x, y, z) = -xy\mathbf{i} + y\mathbf{k}$. Also $\mathbf{n} = -\mathbf{k}$ and $dS =$

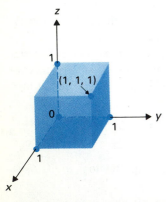

FIGURE 16.36 Cube with diagonally opposite vertices $(0, 0, 0)$ and $(1, 1, 1)$.

$dx\,dy$. Thus $(\mathbf{F} \cdot \mathbf{n})\,dS = -y\,dx\,dy$. The integral is

$$\int_0^1 \int_0^1 -y\,dy\,dz = \int_0^1 -\frac{1}{2}\,dz = -\frac{1}{2}.$$

Bottom side Here $z = 0$ so $\mathbf{F}(x, y, z) = -xy\mathbf{i}$. Since $\mathbf{n} = \mathbf{k}$, we see that $\mathbf{F} \cdot \mathbf{n} = 0$, so the integral is 0.

Summing these six contributions to the desired integral, we obtain

$$0 + 0 + \frac{1}{2} + 0 - \frac{1}{2} + 0 = 0.$$ ∎

SUMMARY

1. The integral of a continuous function $h(x, y, z)$ over a surface Σ is $\iint_\Sigma h(x, y, z)\,dS$. If the surface can be described by $z = f(x, y)$ or $G(x, y, z) = c$, then the formula for dS is as shown in Table 16.1.

2. Let Σ consist of smooth pieces of surface, on each of which unit normal vectors $\mathbf{n}$ vary continuously. Let $\mathbf{F} = P\mathbf{i} + Q\mathbf{j} + R\mathbf{k}$ be a continuous vector field. The integral of the normal component of $\mathbf{F}$ over Σ is $\iint_\Sigma (\mathbf{F} \cdot \mathbf{n})\,dS$. The expression $(\mathbf{F} \cdot \mathbf{n})\,dS$ can be formed as shown in Eq. (2) if the surface is given by $z = f(x, y)$ and as shown in Eq. (3) if Σ is given by $G(x, y, z) = c$.

EXERCISES 16.5

1. Find the mass of a triangular plate having vertices at $(1, 0, 0)$, $(0, 1, 0)$, and $(0, 0, 1)$ and area mass density $\sigma(x, y, z) = z$ at the point (x, y, z).

2. Find the center of mass of a homogeneous triangular plate having vertices at $(1, 0, 0)$, $(0, 1, 0)$, and $(0, 0, 1)$. [*Hint:* Find $\bar{z}$ and use symmetry. See Exercise 1.]

3. Find the mass of a thin plate covering the portion of the plane $z = x + y + 1$ for $0 \le x \le 1$, $0 \le y \le 1$ if the area mass density at (x, y, z) is $\sigma(x, y, z) = x$.

4. A sheet of material covers the hemisphere $z = \sqrt{4 - x^2 - y^2}$ and has area mass density $\sigma(x, y, z) = z$ at (x, y, z). Find the mass of the plate. [*Hint:* Divide the volume element $\rho^2 \sin\phi\,d\rho\,d\phi\,d\theta$ in spherical coordinates by its thickness $d\rho$ to find dS in spherical coordinates on this hemisphere.]

5. Find the moment of inertia about the z-axis of the hemispherical sheet of material in Exercise 4.

6. Find the centroid of the hemisphere $z = \sqrt{a^2 - x^2 - y^2}$.

7. Let $\mathbf{F}$ be a vector field tangent to a smooth surface Σ at each point of the surface, and let unit vectors $\mathbf{n}$ vary continuously over Σ. Find $\iint_\Sigma (\mathbf{F} \cdot \mathbf{n})\,dS$.

8. Let $\mathbf{F} = P\mathbf{i} + Q\mathbf{j} + R\mathbf{k}$ be a continuous vector field that is orthogonal to a smooth surface Σ at each point and let unit vectors $\mathbf{n}$ have opposite direction to $\mathbf{F}$ at each point on Σ. Express $\iint_\Sigma (\mathbf{F} \cdot \mathbf{n})\,dS$ in the form $\iint_\Sigma h(x, y, z)\,dS$, that is, find $h(x, y, z)$.

9. Let $\mathbf{F}$ be a continuous vector field such that $|\mathbf{F}(x, y, z)| \le 1$ for each (x, y, z) on a smooth surface Σ. Let unit vectors $\mathbf{n}$ on Σ vary continuously. Show that $|\iint_\Sigma (\mathbf{F} \cdot \mathbf{n})\,dS| \le$ Area of Σ.

10. Let Σ be the disk $x^2 + y^2 \le 9$, $z = 2$ in space, and let

$$\mathbf{F}(x, y, z) = (\sin x)\mathbf{i} - 2yz\mathbf{j} + 8\mathbf{k}.$$

Find $\iint_\Sigma (\mathbf{F} \cdot \mathbf{n})\,dS$ where the unit vectors $\mathbf{n}$ have positive $\mathbf{k}$-coefficients.

11. Let Σ be the portion of the plane $2x + y - 2z = 6$ for $0 \le x \le 1$, $0 \le y \le 1$. Let unit vectors $\mathbf{n}$ on Σ have negative $\mathbf{i}$-coefficients. Let $\mathbf{F} = z\mathbf{i} + 3xy\mathbf{j} + x\mathbf{k}$. Find $\iint_\Sigma (\mathbf{F} \cdot \mathbf{n})\, dS$.

12. Repeat Exercise 11 for the portion of the plane where $0 \le x \le 1$, $0 \le z \le 1$, keeping all other data the same.

13. Let $\mathbf{F}(x, y, z) = x\mathbf{i} + y\mathbf{j} + 2z\mathbf{k}$ be the flux field of a steady-state flow in space. Find the flux leaving the tetrahedron with vertices $(0, 0, 0)$, $(1, 0, 1)$, $(0, 2, 0)$, and $(0, 0, 1)$.

14. Let $\mathbf{F}(x, y, z) = x\mathbf{i} + y\mathbf{j} + z\mathbf{k}$ and let Σ be the sphere $x^2 + y^2 + z^2 = 4$. Find $\iint_\Sigma (\mathbf{F} \cdot \mathbf{n})\, dS$ where the unit normal vectors $\mathbf{n}$ point outward from the sphere.

15. Let $\mathbf{F}(x, y, z) = x^2\mathbf{i} + y^2\mathbf{j} + z^2\mathbf{k}$ be the flux field of a

steady-state flow. Find the flux leaving the region bounded by $z = x^2 + y^2$ and $z = 4$.

16. Let $\mathbf{F}(x, y, z) = 2x\mathbf{i} + y\mathbf{j} + 3z\mathbf{k}$. Find $\iint_\Sigma (\mathbf{F} \cdot \mathbf{n})\, dS$ where Σ is the surface of the unit cube in Fig. 16.36 with outward normal unit vectors.

17. Let R be the region in space bounded by $z = 4 - x^2 - y^2$ and $z = x^2 + y^2 - 4$, and let $\mathbf{F}(x, y, z) = y^2\mathbf{i} - x^2\mathbf{j} + x^2z\mathbf{k}$. Find the flux outward from the region. [*Hint:* Use symmetry.]

18. Repeat Exercise 17 for the region bounded by $z = 5 - x^2 - y^2$ and $x^2 + y^2 - 3$, which is one unit higher up the z-axis than the region in Exercise 17. [*Hint:* You can still use some symmetry.]

16.6 DIVERGENCE THEOREMS

The theorems to be studied in the rest of this chapter are all of the same nature. They all assert:

> *The integral of some quantity over the boundary of a region is equal to the integral of a related quantity over the region itself.* (1)

Weeks could be spent discussing what is meant by a region and by its boundary. Our presentation will be very intuitive. We will base the divergence theorems discussed in this section on physical considerations. A proof of the two-dimensional case appears in the next section.

Consider a steady-state flux field $\mathbf{F} = P\mathbf{i} + Q\mathbf{j} + R\mathbf{k}$ in space. We saw in the last section that the flux leaving a region G is equal to $\iint_\Sigma (\mathbf{F} \cdot \mathbf{n})\, dS$ where Σ is the surface bounding the region in space and the unit normal vectors $\mathbf{n}$ point outward from the region. Suppose now that the region G is partitioned into little boxes and that we can find the flux leaving each little box, where a positive value means that the net flow is out of the little box and a negative value means that the net flow is into the box. It seems reasonable that if we add the (signed) flux leaving all these little boxes, we obtain the total flux leaving the region G. We will obtain a triple integral representing the limit, as all boxes become smaller, of the sum just described. This triple integral should be equal to the integral of $(\mathbf{F} \cdot \mathbf{n})\, dS$ over the surface. That is, this triple integral over G is equal to a double integral over the boundary of G, which is a result of the form (1). This is the *three-dimensional divergence theorem*. First, we make essentially the same argument to derive the one-dimensional divergence theorem from one-dimensional flows, and the two-dimensional theorem from two-dimensional flows. In the one-dimensional case, we obtain again the fundamental theorem of calculus!

For the rest of this chapter, we will be dealing with **continuously differentiable functions,** which are those having continuous first partial derivatives. A **continuously differentiable vector field** is one whose component functions are continuously differentiable.

The One-Dimensional Divergence Theorem

We start by restating the fundamental theorem of calculus as it appears in the one-dimensional divergence theorem.

THEOREM 16.7   One-dimensional divergence theorem (fundamental theorem of calculus)

If $f(x)$ has a continuous derivative $f'(x)$ for all x in the one-dimensional region $[a, b]$, then

$$f(b) - f(a) = \int_a^b f'(x)\ dx. \qquad \square$$

When stated this way, the fundamental theorem is of the type described in statement (1). Surely $f(x)$ and $f'(x)$ are "related quantities," and the endpoints a and b can be viewed as the boundary of the one-dimensional region $[a, b]$. By suitable definition, $f(b) - f(a)$ can be considered to be the "integral" of $f(x)$ over this two-point boundary.

For a physical demonstration of Theorem 16.7, imagine a gas flowing through a long cylinder *with cross-sectional area* 1 and reaching from a to b on the x-axis, as shown in Fig. 16.37. In this idealistic situation, we will suppose that the velocity and mass density of the gas depend only on the location x along the cylinder and are independent of the up–down and front–back locations within a cross section of the cylinder. This is a model for *one-dimensional flow*. Both the velocity and mass density may vary with time, but we will concentrate on one particular instant and consider how the total mass of gas in the cylinder would change if it were always to flow just as it did at that instant.

The velocity could be represented by a vector function

$$\mathbf{V} = v(x)\mathbf{i}$$

and the mass density by a scalar function $\sigma(x)$. Recall from Section 16.3 that the vector function

$$\mathbf{F} = \sigma(x)v(x)\mathbf{i} = f(x)\mathbf{i}$$

is the **flux vector** of the flow. If the flow does not vary with time, then since the cylinder has 1 as cross-sectional area, $f(x)$ is a (signed) measure of the mass of gas that goes past x in 1 unit of time. The units for $f(x)$ might be in grams/cm²/sec. Since $\mathbf{i}$ is directed toward the right and since $\mathbf{F} = f(x)\mathbf{i}$, we see that $f(b)$ is the mass of gas *leaving* the right end of the cylinder in 1 unit time. That is, the gas is actually leaving at b if $f(b) > 0$ and is entering at b if $f(b) < 0$. Similarly, $f(a)$ is the mass of gas *entering* at the left end per unit time, actually entering if $f(a) > 0$ and leaving if $f(a) < 0$. No gas is entering or leaving through

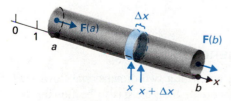

FIGURE 16.37 Flux vector $\mathbf{F}(x)$ for a one-dimensional flow through a cylinder of cross-sectional area 1.

the walls of the cylinder, only at the ends. Thus we see that

$$f(b) - f(a) = \begin{cases} \text{Decrease of mass of gas in the} \\ \text{cylinder per unit time.} \end{cases} \tag{2}$$

The same reasoning applied to the short element of cylinder from x to $x + \Delta x$ in Fig. 16.37 shows that

$$f(x + \Delta x) - f(x) = \begin{cases} \text{Decrease of mass of gas in this} \\ \text{cylindrical element per unit time.} \end{cases}$$

Therefore

$$\frac{f(x + \Delta x) - f(x)}{\Delta x} = \begin{cases} \text{Average decrease in mass of gas per} \\ \text{unit length of cylinder per unit time} \end{cases}$$

and

$$f'(x) = \lim_{\Delta x \to 0} \frac{f(x + \Delta x) - f(x)}{\Delta x}$$

$$= \begin{cases} \text{Decrease of mass of gas measured at} \\ x \text{ per unit length per unit time.} \end{cases}$$

Consequently $\int_a^b f'(x)\, dx$ has the following interpretation: The product $f'(x)\, dx$ may be viewed as the decrease in 1 unit time of mass of the gas over a short length dx of cylinder at x. Thus

$$\int_a^b f'(x)\, dx = \begin{cases} \text{Decrease of mass of gas over the} \\ \text{entire cylinder in 1 unit time.} \end{cases} \tag{3}$$

Comparison of Eqs. (2) and (3) shows that

$$f(b) - f(a) = \int_a^b f'(x)\, dx,$$

which is, of course, the fundamental theorem (Theorem 16.7).

EXAMPLE 1 Suppose that gas is flowing in the cylinder $y^2 + z^2 = 4$ for $-2 \le x \le 4$, with flux vector $\mathbf{F} = (3x - x^2)\mathbf{i}$ at a certain instant. Find the rate of decrease of mass of gas in the cylinder at that instant.

Solution By Eq. (2), the decrease of mass of gas per unit time in a cylinder *of cross-sectional area 1* would be $f(b) - f(a) = f(4) - f(-2)$ for $f(x) = 3x - x^2$. Now $f(4) - f(-2) = -4 - (-10) = 6$. The given cylinder has cross-sectional area $\pi r^2 = \pi 2^2 = 4\pi$. Thus the mass of gas leaving per unit time is $6 \cdot 4\pi = 24\pi$. ∎

The Two-Dimensional Divergence Theorem

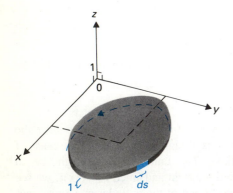

FIGURE 16.38 Parallel plates 1 unit apart.

The same ideas as those just presented, but with a two-dimensional flow, lead us to the two-dimensional divergence theorem. Imagine gas to be flowing between two identical parallel plates, placed 1 *unit distance apart* with the lower plate in the x,y-plane, as in Fig. 16.38. This time we suppose the *gas flow* is *two-dimensional,* so its velocity vector

$$\mathbf{V} = v_1(x, y)\mathbf{i} + v_2(x, y)\mathbf{j}$$

has no **k**-component and also does not depend on the position $0 \le z \le 1$ between the plates. Also, we assume that the mass density $\sigma(x, y)$ does not depend on z. The flux vector

$$\mathbf{F} = \sigma(x, y)\mathbf{V} = P(x, y)\mathbf{i} + Q(x, y)\mathbf{j},$$

where $P(x, y) = \sigma(x, y)v_1(x, y)$ and $Q(x, y) = \sigma(x, y)v_2(x, y)$, again measures the mass of flow of gas at each point per unit time. This flux vector has as direction the direction of the flow. Recall from Section 16.3 the meaning of $|\mathbf{F}|$, which we give again here. Imagine a small square placed perpendicular to $\mathbf{F}$ at a point, with the flux vector at the center of the square. A certain mass of gas flows through this square per unit time. The magnitude of the flux vector $\mathbf{F}$ is the limit of the quotients of these masses of gas divided by the areas of the squares, as the areas approach zero. That is, if the flow everywhere were the same as at (x_0, y_0), then $|\mathbf{F}(x_0, y_0)|$ would be the mass of gas flowing through a square of unit area, placed perpendicular to the flow, in 1 unit time.

Figure 16.39 shows the region G in the x,y-plane occupied by the lower of the two plates. The boundary of the region is denoted by ∂G, read "the boundary of G." Imagine that we travel along this curve, ∂G, so that the region G lies on our left-hand side, as indicated by the arrows on the curve. From previous work, we know that

$$\mathbf{t} = \frac{dx}{ds}\mathbf{i} + \frac{dy}{ds}\mathbf{j}$$

is a unit vector tangent to this curve at each point. As we showed in Section 16.3,

$$\mathbf{n} = \frac{dy}{ds}\mathbf{i} - \frac{dx}{ds}\mathbf{j}$$

is a unit vector normal to the curve and directed *outward* from the region.

Now look back at Fig. 16.38. How much gas is flowing out of the region between the plates through the little strip of height 1 and width ds along ∂G, shown in the figure? This outward flow is measured by the *normal scalar* component $\mathbf{F} \cdot \mathbf{n}$ of the flux vector. Thus the mass of gas per unit time coming through this strip of area ds (recall that the plates are 1 unit apart) is approximately $(\mathbf{F} \cdot \mathbf{n}) \, ds$. Consequently, the total mass of gas leaving the region between the plates per unit of time is

$$\oint_{\partial G} (\mathbf{F} \cdot \mathbf{n}) \, ds, \tag{4}$$

where $\oint_{\partial G}$ denotes the line integral once around the boundary of G in the direction given by the arrows in Fig. 16.39. But

$$\mathbf{F} \cdot \mathbf{n} = [P(x, y)\mathbf{i} + Q(x, y)\mathbf{j}] \cdot \left(\frac{dy}{ds}\mathbf{i} - \frac{dx}{ds}\mathbf{j} \right)$$

$$= P(x, y)\frac{dy}{ds} - Q(x, y)\frac{dx}{ds}.$$

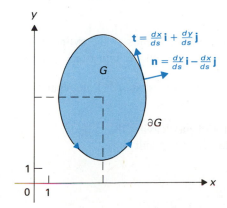

FIGURE 16.39 Unit tangent and normal vectors to ∂G.

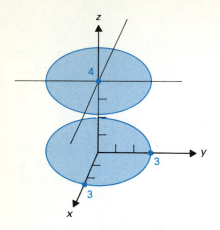

FIGURE 16.40 Parallel circular plates $x^2 + y^2 \leq 9$, where $z = 0$, $z = 4$.

Therefore integral (4) becomes

$$\oint_{\partial G} [-Q(x, y)\, dx + P(x, y)\, dy] = \begin{cases} \text{Mass of gas leaving the region} \\ \text{between the plates per unit time.} \end{cases} \quad (5)$$

EXAMPLE 2   Consider two parallel circular plates lying over $x^2 + y^2 \leq 9$ in the x,y-plane. One plate is in the x,y-plane and one is in the plane $z = 4$, as shown in Fig. 16.40. Suppose the flux vector of a two-dimensional flow is $\mathbf{F} = x\mathbf{i} + y\mathbf{j}$. Find the mass of gas leaving the region between the plates per unit time.

Solution   Let G be the disk $x^2 + y^2 \leq 9$ in the x,y-plane. Since our plates are 4 units apart, we see from formula (5) that the mass of gas leaving the region between the plates per unit time is

$$4 \oint_{\partial G} (-y\, dx + x\, dy).$$

Taking $x = 3 \cos t$, $y = 3 \sin t$ for $0 \leq t \leq 2\pi$ as parametrization for ∂G, we have

$$4 \oint_{\partial G} (-y\, dx + x\, dy) = 4 \int_0^{2\pi} [-(3 \sin t)(-3 \sin t) + (3 \cos t)(3 \cos t)]\, dt$$

$$= 4 \int_0^{2\pi} 9(\sin^2 t + \cos^2 t)\, dt$$

$$= 36 \int_0^{2\pi} dt = 72\pi. \qquad \blacksquare$$

We now compute in another way the mass of gas leaving, namely, by considering separately the contributions of $P(x, y)\mathbf{i}$ and $Q(x, y)\mathbf{j}$ to the flux vector $\mathbf{F}$. The vector $P(x, y)\mathbf{i}$ is the horizontal component of the flux. The region between the two plates over the strip of width dy shown in Fig. 16.41 is a cylinder having area of cross section $(dy)(1) = dy$. Referring back to formula (3), where we discussed flow along a cylinder, and considering just the vector component $P(x, y)\mathbf{i}$,

$$\int_{g_1(y)}^{g_2(y)} \frac{\partial P}{\partial x}(x, y)\, dx = \begin{cases} \text{Mass of gas leaving the ends of the cylinder} \\ \text{per unit area cross section per unit time.} \end{cases}$$

Since the cylinder has cross sectional area dy (the plates are still 1 unit apart), we have

$$\left(\int_{g_1(y)}^{g_2(y)} \frac{\partial P}{\partial x}(x, y)\, dx \right) dy = \begin{cases} \text{Mass of gas leaving the cylinder} \\ \text{at the ends per unit time.} \end{cases}$$

Therefore

$$\int_c^d \left(\int_{g_1(y)}^{g_2(y)} \frac{\partial P}{\partial x}(x, y)\, dx \right) dy = \iint_G \frac{\partial P}{\partial x}(x, y)\, dx\, dy$$

$$= \begin{cases} \text{Mass of gas leaving the region between} \\ \text{the plates per unit time due to } P(x, y)\mathbf{i}. \end{cases}$$

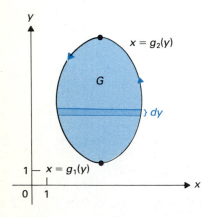

FIGURE 16.41   Strip of width dy.

A similar computation with a vertical cylindrical strip of width dx and the component $Q(x, y)\mathbf{j}$ of $\mathbf{F}$ shows that

$$\iint_G \frac{\partial Q}{\partial y}(x, y)\, dx\, dy = \begin{cases} \text{Mass of gas leaving the region between} \\ \text{the plates per unit time due to } Q(x, y)\mathbf{j}. \end{cases}$$

Thus

$$\iint_G \left[\frac{\partial P}{\partial x}(x, y) + \frac{\partial Q}{\partial y}(x, y) \right] dx\, dy$$

$$= \begin{cases} \text{Mass of gas leaving the region} \\ \text{between the plates per unit time.} \end{cases} \quad (6)$$

EXAMPLE 3   For the plates and flux vector in Example 2, use formula (6) to find the mass of gas leaving the region between the plates per unit time.

Solution   Since the plates are 4 units apart, we see from formula (6) that the answer is given by

$$4 \iint_G (1 + 1)\, dx\, dy = 8 \iint_G 1\, dx\, dy.$$

Since G is a disk of radius 3 with area 9π, our answer becomes $8 \cdot 9\pi = 72\pi$.
■

Of course, we obtained the same answer in Example 2 using formula (5) as we obtained in Example 3 using formula (6). Comparing formulas (5) and (6), we have

$$\iint_G \left[\frac{\partial P}{\partial x}(x, y) + \frac{\partial Q}{\partial y}(x, y) \right] dx\, dy = \oint_{\partial G} [-Q(x, y)\, dx + P(x, y)\, dy], \quad (7)$$

which is again a relation of the form (1). Equation (7) is the two-dimensional divergence theorem.

THEOREM 16.8   Two-dimensional divergence

For a suitable plane region G and for functions $P(x, y)$ and $Q(x, y)$ with continuous partial derivatives,

$$\oint_{\partial G} (-Q\, dx + P\, dy) = \iint_G \left(\frac{\partial P}{\partial x} + \frac{\partial Q}{\partial y} \right) dx\, dy. \quad (8)$$

In the line integral, ∂G is traced in the direction that keeps G on the left.
□

Equation (8) is often stated in vector form, reflecting the physical demonstration we gave for it. Let $\mathbf{F} = P(x, y)\mathbf{i} + Q(x, y)\mathbf{j}$ be a flux vector. We make use of the symbolic *gradient operator*

$$\nabla = \frac{\partial}{\partial x}\mathbf{i} + \frac{\partial}{\partial y}\mathbf{j},$$

where ∇ is read "del." Symbolically, we have

$$\nabla \cdot \mathbf{F} = \left(\frac{\partial}{\partial x}\mathbf{i} + \frac{\partial}{\partial y}\mathbf{j}\right) \cdot [P(x, y)\mathbf{i} + Q(x, y)\mathbf{j}]$$

$$= \frac{\partial}{\partial x}P(x, y) + \frac{\partial}{\partial y}Q(x, y)$$

$$= \frac{\partial P}{\partial x} + \frac{\partial Q}{\partial y}.$$

Also, we have seen that if $\mathbf{n}$ is the outward unit normal vector to the boundary, then

$$\oint_{\partial G} (-Q\, dx + P\, dy) = \oint_{\partial G} (\mathbf{F} \cdot \mathbf{n})\, ds.$$

Thus Eq. (8) becomes

$$\oint_{\partial G} (\mathbf{F} \cdot \mathbf{n})\, ds = \iint_G (\nabla \cdot \mathbf{F})\, dx\, dy. \tag{9}$$

The scalar $\nabla \cdot \mathbf{F}$ is the **divergence of F**. It measures the rate at which the gas diverges (leaves) at each point.

We labeled Theorem 16.8 a divergence theorem, and we just explained the reason for the adjective *divergence*. We do not claim that our physical arguments have proved the two-dimensional divergence theorem. In the next section, we prove Green's theorem, which has this divergence theorem as an immediate corollary.

EXAMPLE 4   Illustrate the two-dimensional divergence theorem for the functions

$$P(x, y) = x^3 \qquad \text{and} \qquad Q(x, y) = 2x + y^3$$

over the disk G given by $x^2 + y^2 \le a^2$.

Solution   Now

$$\frac{\partial P}{\partial x} = 3x^2 \qquad \text{and} \qquad \frac{\partial Q}{\partial y} = 3y^2,$$

so

$$\iint_G \left(\frac{\partial P}{\partial x} + \frac{\partial Q}{\partial y}\right) dx\, dy = \iint_G 3(x^2 + y^2)\, dx\, dy.$$

Changing to polar coordinates, we have

$$\iint_G 3(x^2 + y^2)\, dx\, dy = \int_0^{2\pi} \int_0^a 3r^2 \cdot r\, dr\, d\theta$$

$$= \int_0^{2\pi} 3\frac{r^4}{4}\Big]_0^a d\theta$$

$$= 3\frac{a^4}{4}\theta\Big]_0^{2\pi} = \frac{3a^4\pi}{2}.$$

The boundary ∂G may be parametrized by $x = a \cos\theta$, $y = a \sin\theta$ for

$0 \le \theta \le 2\pi$. The integral around the boundary becomes

$$\oint_{\partial G} (-Q \, dx + P \, dy) = \oint_{\partial G} [-(2x + y^3) \, dx + x^3 \, dy]$$

$$= \int_0^{2\pi} [-(2a \cos \theta + a^3 \sin^3 \theta)(-a \sin \theta \, d\theta) + a^3 \cos^3 \theta \cdot a \cos \theta \, d\theta]$$

$$= \int_0^{2\pi} [2a^2 (\sin \theta \cos \theta) + a^4 (\cos^4 \theta + \sin^4 \theta)] \, d\theta$$

$$= \int_0^{2\pi} \left[2a^2 \sin \theta \cos \theta + a^4 \left(\left(\frac{1 + \cos 2\theta}{2} \right)^2 + \left(\frac{1 - \cos 2\theta}{2} \right)^2 \right) \right] d\theta$$

$$= \int_0^{2\pi} \left[2a^2 \sin \theta \cos \theta + a^4 \left(\frac{\cos^2 2\theta}{2} + \frac{1}{2} \right) \right] d\theta$$

$$= \int_0^{2\pi} \left[2a^2 \sin \theta \cos \theta + a^4 \left(\frac{1 + \cos 4\theta}{4} + \frac{1}{2} \right) \right] d\theta$$

$$= \left[a^2 \sin^2 \theta + a^4 \left(\frac{1}{4} \theta + \frac{\sin 4\theta}{16} + \frac{1}{2} \theta \right) \right]_0^{2\pi}$$

$$= a^4 \left(\frac{2\pi}{4} + \frac{2\pi}{2} \right) - 0 = a^4 \frac{3\pi}{2} = \frac{3\pi a^4}{2}.$$

The same answer was obtained, illustrating the two-dimensional divergence theorem. Obviously, the area integral was much easier to evaluate than the line integral in this case. ∎

Often this two-dimensional divergence theorem is used to transform a line integral around a loop into a more easily computed area integral, or occasionally vice versa.

EXAMPLE 5 Let $P(x, y) = xy^2 - 3xy$ and $Q(x, y) = y/(1 + x^2)$. Compute $\oint_\lambda (-Q \, dx + P \, dy)$, where λ is the boundary of the unit square in Fig. 16.42, having opposite vertices at $(0, 0)$ and $(1, 1)$, traversed counterclockwise.

Solution Let G be the two-dimensional region enclosed by this square, so that $\lambda = \partial G$. By Theorem 16.8, we see that

$$\oint_\lambda (-Q \, dx + P \, dy) = \iint_G \left(\frac{\partial P}{\partial x} + \frac{\partial Q}{\partial y} \right) dx \, dy$$

$$= \int_0^1 \int_0^1 \left(y^2 - 3y + \frac{1}{x^2 + 1} \right) dx \, dy$$

$$= \int_0^1 (y^2 x - 3xy + \tan^{-1} x) \Big]_0^1 \, dy$$

$$= \int_0^1 \left(y^2 - 3y + \frac{\pi}{4} \right) dy$$

$$= \left(\frac{y^3}{3} - \frac{3y^2}{2} + \frac{\pi}{4} y \right) \Big]_0^1$$

$$= \frac{1}{3} - \frac{3}{2} + \frac{\pi}{4} = -\frac{7}{6} + \frac{\pi}{4}. \quad ∎$$

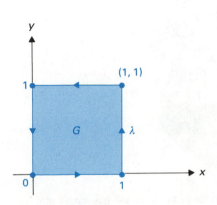

FIGURE 16.42 The loop $\gamma = \partial G$.

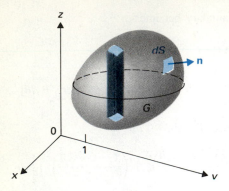

FIGURE 16.43 dS and $\mathbf{n}$ for $\iint_G (\mathbf{F} \cdot \mathbf{n}) \, dS$; column for $\iiint_G (\partial R / \partial z) \, dx \, dy \, dz$.

The Divergence Theorem

If we repeat the gas flow arguments we have made, but in space, we obtain the three-dimensional divergence theorem, which was classically referred to as the *divergence theorem*. Let G now be a region in space with boundary ∂G, as illustrated in Fig. 16.43. This time ∂G is a *surface*. Imagine gas flowing in space and able to flow in and out of G without impediment. (You can think of the boundary of G as not being physically present or, if you prefer, consider it to be a netting through which gas can flow unhindered.) Let

$$\mathbf{F} = P(x, y, z)\mathbf{i} + Q(x, y, z)\mathbf{j} + R(x, y, z)\mathbf{k}$$

the flux vector of the flow at (x, y, z) in the region. Let $\mathbf{n}$ be a unit normal vector outward from the boundary surface at a point and let a little differential piece of surface at that point have area dS. Then the mass of gas flowing out of the region G per unit time, that is, the flux, is given by

$$\iint_{\partial G} (\mathbf{F} \cdot \mathbf{n}) \, dS, \tag{10}$$

by reasoning like that in the previous subsection.

EXAMPLE 6 Let the flux vector for a given flow in space be

$$\mathbf{F}(x, y, z) = x\mathbf{i} + y\mathbf{j} + z\mathbf{k}.$$

Find the mass of gas leaving the ball $x^2 + y^2 + z^2 \le a^2$ per unit time.

Solution If G is the given ball, we need to find

$$\iint_{\partial G} (\mathbf{F} \cdot \mathbf{n}) \, dS.$$

Now ∂G is the sphere $x^2 + y^2 + z^2 = a^2$, and at a point on the sphere a unit normal vector outward is clearly $\mathbf{n} = (x/a)\mathbf{i} + (y/a)\mathbf{j} + (z/a)\mathbf{k}$, as shown in

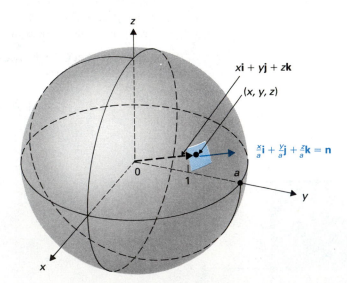

FIGURE 16.44 $x\mathbf{i} + y\mathbf{j} + z\mathbf{k}$ goes from $(0, 0, 0)$ to (x, y, z) and has length a; $\mathbf{n} = (1/a)(x\mathbf{i} + y\mathbf{j} + z\mathbf{k})$.

Fig. 16.44. (We could also find **n** using the fact that $\nabla(x^2 + y^2 + z^2)$ is orthogonal to $x^2 + y^2 + z^2 = a^2$.) Then

$$\mathbf{F} \cdot \mathbf{n} = (x\mathbf{i} + y\mathbf{j} + z\mathbf{k}) \cdot \left(\frac{x}{a}\mathbf{i} + \frac{y}{a}\mathbf{j} + \frac{z}{a}\mathbf{k} \right)$$

$$= \frac{1}{a}(x^2 + y^2 + z^2) = \frac{1}{a}(a^2) = a$$

at each point on the sphere. Since the area of the surface of the sphere is $4\pi a^2$, we have

$$\iint_{\partial G} (\mathbf{F} \cdot \mathbf{n}) \, dS = \iint_{\partial G} a \, dS$$

$$= a \cdot 4\pi a^2 = 4\pi a^3. \qquad \blacksquare$$

The contribution of the z-component $R(x, y, z)\mathbf{k}$ of **F** to the mass leaving G can be computed by finding the contribution in the cylinder of cross-sectional area $dx \, dy$, shown in Fig. 16.43, and then adding these contributions over the entire region G. We obtain

$$\iiint_G \frac{\partial R}{\partial z} \, dz \, dx \, dy$$

by reasoning just like that in the previous subsection. The total mass of gas leaving the region per unit time is thus

$$\iiint_G \left(\frac{\partial P}{\partial x} + \frac{\partial Q}{\partial y} + \frac{\partial R}{\partial z} \right) dx \, dy \, dz, \qquad \textbf{(11)}$$

which may also be written

$$\iiint_G \nabla \cdot \mathbf{F} \, dx \, dy \, dz, \qquad \textbf{(12)}$$

where this time

$$\nabla = \frac{\partial}{\partial x}\mathbf{i} + \frac{\partial}{\partial y}\mathbf{j} + \frac{\partial}{\partial z}\mathbf{k}.$$

EXAMPLE 7   Repeat Example 6, but use integral (11) to find the mass of gas leaving the ball per unit time.

Solution   Since $\mathbf{F} = x\mathbf{i} + y\mathbf{j} + z\mathbf{k}$, we have

$$\frac{\partial P}{\partial x} + \frac{\partial Q}{\partial y} + \frac{\partial R}{\partial z} = 1 + 1 + 1 = 3.$$

Thus integral (11) becomes

$$\iiint_G 3 \, dx \, dy \, dz.$$

This integral obviously equals three times the volume of the ball G with radius a. The answer is therefore $3 \cdot \frac{4}{3}\pi a^3 = 4\pi a^3.$ $\blacksquare$

The answer found in Example 6 using integral (10) is the same as that in Example 7 using integral (11). Comparison of integrals (10) and (12) gives us the classical *divergence theorem;* $\nabla \cdot \mathbf{F}$ is called the **divergence of F.**

THEOREM 16.9   Divergence theorem (three-dimensional)

If $\mathbf{F} = P(x, y, z)\mathbf{i} + Q(x, y, z)\mathbf{j} + R(x, y, z)\mathbf{k}$ is a vector field with continuously differentiable components over a suitable region G in space, then

$$\iint_{\partial G} (\mathbf{F} \cdot \mathbf{n}) \, dS = \iiint_G (\nabla \cdot \mathbf{F}) \, dx \, dy \, dz. \tag{13}$$

$\square$

EXAMPLE 8 Illustrate the divergence theorem for the vector field $\mathbf{F} = x^2\mathbf{i} + y^2\mathbf{j} + z^2\mathbf{k}$ over the region G bounded by $z = x^2 + y^2$ and $z = 4$, shown in Fig. 16.45.

Solution We first compute $\iint_{\partial G} (\mathbf{F} \cdot \mathbf{n}) \, dS$. Starting with the portion of the paraboloid $z = x^2 + y^2$, we find from Eq. (2) that

$$\mathbf{n} = \frac{1}{\sqrt{4x^2 + 4y^2 + 1}} (2x\mathbf{i} + 2y\mathbf{j} - \mathbf{k}).$$

Our sketch in Fig. 16.45 shows that the negative coefficient of $\mathbf{k}$ is appropriate for the outward normal. From Eq. (3), we have

$$dS = \sqrt{4x^2 + 4y^2 + 1} \, dx \, dy.$$

Therefore

$$(\mathbf{F} \cdot \mathbf{n}) \, dS = (2x^3 + 2y^3 - z^2) \, dx \, dy.$$

Since $z = x^2 + y^2$, we obtain

$$(\mathbf{F} \cdot \mathbf{n}) \, dS = [2x^3 + 2y^3 - (x^2 + y^2)^2] \, dx \, dy.$$

The projection of this portion of the paraboloid $z = x^2 + y^2$ for $0 \le z \le 4$ onto the x,y-plane is the disk $x^2 + y^2 \le 4$. The integral of $2x^3 + 2y^3$ over this disk is zero by symmetry, so we are left with the integral of $-(x^2 + y^2)^2$ over the disk. Changing to polar coordinates, we obtain the integral

$$\int_0^{2\pi} \int_0^2 (-r^4) r \, dr \, d\theta = \int_0^{2\pi} -\frac{r^6}{6} \Big]_0^2 \, d\theta = \int_0^{2\pi} -\frac{64}{6} \, d\theta$$

$$= 2\pi\left(-\frac{64}{6}\right) = -\frac{64}{3}\pi.$$

We now turn to the integral of $\mathbf{F} \cdot \mathbf{n}$ over the disk $x^2 + y^2 \le 4$ in the plane $z = 4$. Clearly, $\mathbf{n} = \mathbf{k}$ is the outward unit normal vector, as shown in Fig. 16.45. Then $\mathbf{F} \cdot \mathbf{n} = z^2 = 4^2 = 16$. The integral of $\mathbf{F} \cdot \mathbf{n}$ over the disk is then

$$16 \cdot \text{Area of the disk} = 16 \cdot 4\pi = 64\pi.$$

Adding the integrals of $\mathbf{F} \cdot \mathbf{n}$ over these two portions of ∂G, we have

$$\iint_{\partial G} (\mathbf{F} \cdot \mathbf{n}) \, dS = -\frac{64}{3}\pi + 64\pi = \frac{128}{3}\pi.$$

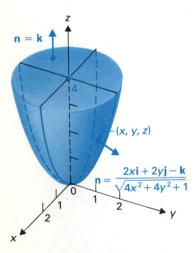

FIGURE 16.45 Outward unit normal vectors to the region bounded by $z = x^2 + y^2$ and $z = 4$.

THE DIVERGENCE THEOREM in three dimensions was first stated and proved in 1826 by Mikhail Ostrogradskii (1801–1861), a Russian mathematician who spent the late 1820s studying in Paris. Vector notation not being available, Ostrogradskii stated the theorem in the coordinate form

$$\int \left(a\frac{\partial p}{\partial x} + b\frac{\partial q}{\partial y} + c\frac{\partial r}{\partial z} \right)\omega$$

$$= \int (ap\cos\alpha + bq\cos\beta$$

$$+ cr\cos\gamma)\varepsilon,$$

where p, q, and r are differentiable functions of x, y, and z; a, b, and c are constants; ω is the element of volume of a region T; ε is the element of area of the boundary surface S; and α, β, and γ are the angles that the exterior normal to S makes with the positive x-axis, y-axis, and z-axis respectively.

We now compute

$$\iiint_G (\nabla \cdot \mathbf{F})\, dx\, dy\, dz = \iiint_G (2x + 2y + 2z)\, dx\, dy\, dz.$$

By symmetry, the integral over G of $2x + 2y$ is zero. We change to cylindrical coordinates and find that

$$\iiint_G 2z\, dz\, dx\, dy = \int_0^{2\pi}\int_0^2\int_{r^2}^4 2z\, r\, dz\, dr\, d\theta = \int_0^{2\pi}\int_0^2 z^2\Big]_{r^2}^4 r\, dr\, d\theta$$

$$= \int_0^{2\pi}\int_0^2 (16 - r^4)r\, dr\, d\theta = \int_0^{2\pi}\left(8r^2 - \frac{r^6}{6}\right)\Big]_0^2 d\theta$$

$$= 2\pi\left(32 - \frac{64}{6}\right) = \pi\left(64 - \frac{64}{3}\right) = \frac{128\pi}{3}. \qquad \blacksquare$$

EXAMPLE 9   Illustrate the divergence theorem for the vector field

$$\mathbf{F} = x\mathbf{i} + y\mathbf{j} + 2z\mathbf{k}$$

over the region G, which is the tetrahedron with vertices $(0, 0, 0)$, $(1, 0, 0)$, $(0, 2, 0)$, and $(0, 0, 1)$ shown in Fig. 16.46.

Solution   The volume integral is

$$\iiint_G \left(\frac{\partial P}{\partial x} + \frac{\partial Q}{\partial y} + \frac{\partial R}{\partial z}\right) dx\, dy\, dz = \iiint_G (1 + 1 + 2)\, dx\, dy\, dz$$

$$= 4(\text{Volume of tetrahedron})$$

$$= 4 \cdot \frac{1}{3} \cdot 1 \cdot 1 = \frac{4}{3}.$$

Now we turn to the integral $\iint_{\partial G} (\mathbf{F} \cdot \mathbf{n})\, dS$ over the four triangles that form the surface of the tetrahedron. For the triangle in the x,y-plane, $\mathbf{n} = -\mathbf{k}$, $z = 0$, and $dS = dx\, dy$, so the integral becomes

$$\int_0^1 \int_0^{2-2x} [(x\mathbf{i} + y\mathbf{j}) \cdot (-\mathbf{k})]\, dy\, dx = \int_0^1 \int_0^{2-2x} 0\, dy\, dx = 0.$$

The integrals over the triangles in the x,z-plane and in the y,z-plane are similarly zero.

We must now find the equation of the front plane of the region, determined by the points $(1, 0, 0)$, $(0, 2, 0)$, and $(0, 0, 1)$. A vector perpendicular to the plane is

$$(-\mathbf{i} + 2\mathbf{j}) \times (-\mathbf{i} + \mathbf{k}) = \begin{vmatrix} \mathbf{i} & \mathbf{j} & \mathbf{k} \\ -1 & 2 & 0 \\ -1 & 0 & 1 \end{vmatrix} = 2\mathbf{i} + \mathbf{j} + 2\mathbf{k},$$

so the equation of the plane is

$$2x + y + 2z = 2.$$

For this front triangle, we see from the equation $2x + y + 2z = 2$ of its plane that

$$\mathbf{n} = \frac{2}{3}\mathbf{i} + \frac{1}{3}\mathbf{j} + \frac{2}{3}\mathbf{k}$$

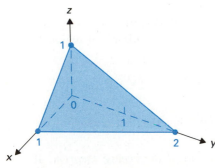

FIGURE 16.46   Tetrahedron with vertices $(0, 0, 0)$, $(1, 0, 0)$, $(0, 2, 0)$, and $(0, 0, 1)$.

and

$$dS = \left(\frac{\sqrt{4+1+4}}{2}\right) dx\ dy = \left(\frac{3}{2}\right) dx\ dy.$$

Therefore the surface integral over the front face of the tetrahedron becomes

$$\int_0^1 \int_0^{2-2x} \underbrace{\left[\frac{2}{3}x + \frac{1}{3}y + \frac{4}{3}\left(1 - x - \frac{y}{2}\right)\right]}_{\mathbf{F}\cdot\mathbf{n}} \underbrace{\frac{3}{2}}_{dS}\ dy\ dx$$

$$= \int_0^1 \int_0^{2-2x} \left(2 - x - \frac{y}{2}\right) dy\ dx$$

$$= \int_0^1 \left[2(2-2x) - x(2-2x) - \frac{1}{4}(2-2x)^2\right] dx$$

$$= \int_0^1 (3 - 4x + x^2)\ dx = \left(3x - 2x^2 + \frac{x^3}{3}\right)\Big]_0^1$$

$$= 3 - 2 + \frac{1}{3} = \frac{4}{3}.$$

Thus both sides of Eq. (1) are equal to $\frac{4}{3}$, illustrating the divergence theorem. ∎

We see that if $\mathbf{F} = P\mathbf{i} + Q\mathbf{j} + R\mathbf{k}$ is a flux vector such that the divergence $\nabla \cdot \mathbf{F} = \partial P/\partial x + \partial Q/\partial y + \partial R/\partial z = 0$ for all (x, y, z), then the total flux of mass outward (or inward) through the boundary of any suitable region is zero. Such a flow is said to be **incompressible.**

We see that Theorems 16.7, 16.8, and 16.9 are essentially the same theorem, but in different dimensions.

SUMMARY

1. The theorems to be studied in the rest of this chapter all relate the integral of a quantity over the boundary of a region to the integral of a related quantity over the region itself.

2. *Two-dimensional divergence theorem:* If $P(x, y)$ and $Q(x, y)$ are continuously differentiable throughout a suitable plane region G with boundary ∂G, then

$$\oint_{\partial G} (-Q\ dx + P\ dy) = \iint_G \left(\frac{\partial P}{\partial x} + \frac{\partial Q}{\partial y}\right) dx\ dy.$$

In the line integral, ∂G is traced in the direction that keeps G on the left.

3. *Vector statement of the two-dimensional divergence theorem:* If $\mathbf{F} = P(x, y)\mathbf{i} + Q(x, y)\mathbf{j}$ is a continuously differentiable vector

field throughout a suitable plane region G with boundary ∂G, then

$$\oint_{\partial G} (\mathbf{F} \cdot \mathbf{n}) \, ds = \iint_G (\nabla \cdot \mathbf{F}) \, dx \, dy,$$

where $\mathbf{n}$ is the outward unit normal vector at each point of ∂G and ∇ is the symbolic gradient operator

$$\nabla = \frac{\partial}{\partial x} \mathbf{i} + \frac{\partial}{\partial y} \mathbf{j}.$$

4. *Divergence theorem:* Let

$$\mathbf{F} = P(x, y, z)\mathbf{i} + Q(x, y, z)\mathbf{j} + R(x, y, z)\mathbf{k}$$

be a continuously differentiable vector field over a suitable region G in space with boundary surface ∂G. Then

$$\iint_{\partial G} (\mathbf{F} \cdot \mathbf{n}) \, dS = \iiint_G (\nabla \cdot \mathbf{F}) \, dx \, dy \, dz,$$

where $\mathbf{n}$ is the unit normal vector outward at each point of ∂G, dS is the differential element of surface area, and ∇ is the symbolic operator

$$\nabla = \frac{\partial}{\partial x} \mathbf{i} + \frac{\partial}{\partial y} \mathbf{j} + \frac{\partial}{\partial z} \mathbf{k}.$$

EXERCISES 16.6

1. Gas is flowing in the cylinder $y^2 + z^2 = 1$ for $0 \leq x \leq 8$, with flux vector at a certain instant equal to $\mathbf{F} = (4 + x^{2/3})\mathbf{i}$. Find the rate of decrease of mass of the gas in the cylinder at that instant.

2. Repeat Exercise 1 if the flux vector is

$$\mathbf{F} = \frac{16}{2 + x^{1/3}} \mathbf{i}.$$

3. Gas is flowing in the cylinder $-1 \leq x \leq 1$, $0 \leq z \leq 3$ for $-1 \leq y \leq 10$, with flux vector at a certain instant equal to $\mathbf{F} = (3y - 2)\mathbf{j}$. Find the rate of *increase* of mass of gas in the cylinder at that instant.

4. Gas is flowing in the cylinder bounded by $z = 1 - y^2$ and $z = 0$ for $0 \leq x \leq 4$. If the flux vector at a certain instant is

$$\mathbf{F} = (x^2 + 3x + 1)\mathbf{i},$$

find the rate of *increase* of gas in the cylinder at that instant.

In Exercises 5–10, illustrate the two-dimensional divergence theorem by computing

$$\oint_{\partial G} (-Q \, dx + P \, dy) \qquad \text{and} \qquad \iint_G \left(\frac{\partial P}{\partial x} + \frac{\partial Q}{\partial y} \right) dx \, dy$$

for the given $P(x, y)$, $Q(x, y)$ and the region G.

5. $P(x, y) = x$, $Q(x, y) = 2y$, where G is the square region with vertices $(1, 1)$, $(-1, 1)$, $(-1, -1)$, and $(1, -1)$

6. $P(x, y) = x^2y^2$, $Q(x, y) = 2x - 3y$, where G is the square region bounded by $x = 0$, $x = 1$, $y = 0$, and $y = 1$

7. $P(x, y) = y^2x$, $Q(x, y) = x^2y$, where G is the disk $x^2 + y^2 \leq 16$

8. $P(x, y) = x^2y$, $Q(x, y) = xy^2$, where G is the disk $x^2 + y^2 \leq 4$

9. $P(x, y) = y$, $Q(x, y) = xy$, where G is the region bounded by $y = x^2$ and $y = x$

10. $P(x, y) = x^2y + x$, $Q(x, y) = y^2 + 4x$, where G is the region bounded by $y = x^2$ and $y = 1$

As indicated by Examples 4 and 5, it may be that one of the integrals appearing in the two-dimensional divergence theorem is much easier to compute than the other. In Exercises 11–24, compute whichever integral is easier in Theorem 16.8.

11. $P(x, y) = xe^y$, $Q(x, y) = xy^2$, where G is the triangular region bounded by $x = 0$, $y = 0$, and $x + y = 2$

12. $P(x, y) = x/(1 + y^2)$, $Q(x, y) = y/(1 + x^2)$, where G is the triangular region bounded by $y = x$, $x = 1$, and $y = 0$

13. Two parallel identical plates are 3 units apart with the lower one having as boundary the circle $(x - 3)^2 + y^2 = 16$ in the x,y-plane. Gas flowing between the plates has as flux vector

$$(2x + y^3)\mathbf{i} + (e^x - 4y)\mathbf{j}$$

at a certain instant. Find the rate of decrease of mass of gas between the plates at that instant.

14. Repeat Exercise 13 if the plates are 2 units apart covering $x^2 + y^2 \leq 4$ and the flux vector is $\mathbf{F} = (xy^2 + 3x)\mathbf{i} + (x^2y - y)\mathbf{j}$.

15. Find the integral of the outward normal component of the vector field $\mathbf{F} = (3x - 2y)\mathbf{i} + (5x + 7y)\mathbf{j}$ *clockwise* around the rectangular region bounded by $x = 0$, $x = 3$, $y = 0$, and $y = 4$.

16. Repeat Exercise 15 if $\mathbf{F} = (2x + 4y^2)\mathbf{i} + (8x^2 - 5y)\mathbf{j}$.

17. Find the integral of the normal component of the vector field

$$\mathbf{F} = (x^2y + xy^2)\mathbf{i} + (xy)\mathbf{j}$$

counterclockwise around the triangle bounded by

$$x = 0, \quad y = 0, \quad \text{and} \quad y = 1 - x.$$

18. Find $\oint_\lambda (x^2y \, dy - y^2 \, dx)$, where λ is the boundary of the triangle bounded by $y = x$, $y = -x$, and $x = 1$, traced counterclockwise.

19. Find $\oint_\lambda (2xy \, dx + 3x^2 \, dy)$, where λ is the boundary of the region bounded by $y = \sqrt{x}$ and $y = x^2$, traced counterclockwise.

20. Find

$$\oint_\lambda \left(\frac{3^y}{y - 2} \, dx + \frac{2^x}{x + 1} \, dy \right),$$

where λ is the boundary of the square with vertices $(0, 0)$, $(0, 1)$, $(1, 0)$, and $(1, 1)$, traced counterclockwise.

21. Find

$$\oint_\lambda \left(\frac{y}{y^3 + 1} \, dx - \frac{x^2 + 4}{x^4 + 1} \, dy \right)$$

for λ as in Exercise 20.

22. Find

$$\oint_\lambda \left(\frac{y + 1}{y^2 + 3} \, dx - \frac{x + 1}{x^2 + 3} \, dy \right)$$

for the boundary λ of the triangle with vertices $(0, 0)$, $(1, 0)$, and $(1, 1)$ traced counterclockwise.

23. Find the integral of the divergence of the vector field $x\mathbf{i} + y\mathbf{j}$ over the ellipse $(x^2/a^2) + (y^2/b^2) = 1$. [*Hint*: A parametri-

zation of the ellipse is

$$x = a \cos t, \quad y = b \sin t$$

for $0 \leq t \leq 2\pi$.]

24. Find the integral of the normal component of the gradient vector field of the function $f(x, y) = x^2/y^2$ counterclockwise around the boundary of the rectangular region bounded by

$$x = 0, \quad x = 4, \quad y = 1, \quad \text{and} \quad y = 3.$$

In Exercises 25–29, let G be a region where the two-dimensional divergence theorem applies.

25. Show that the area of G is $\frac{1}{2} \oint_{\partial G} (x \, dy - y \, dx)$.

26. If $\int_{\partial G} [(3x + 2y) \, dy - (5y - 4x) \, dx] = 24$, find the area of G.

27. Repeat Exercise 26 if

$$\oint_{\partial G} [(y^3 - 4xy - 2x) \, dy - (x^4 + 2y^2 + 7y) \, dx] = 50.$$

28. If $\oint_{\partial G} (x \, dy - y \, dx) = 3$, $\oint_{\partial G} [y \, dx + (x^2 + x) \, dy] = 10$, and $\oint_{\partial G} [(4y^2 + 2y) \, dx + 2x \, dy] = 16$, find the centroid of G.

29. If

$$\oint_{\partial G} [(3x - y) \, dy - (x + 2y) \, dx] = 10,$$

$$\oint_{\partial G} [(4x - y) \, dx + (3x^2 - 2x) \, dy] = -18,$$

and

$$\oint_{\partial G} [(x^3 + 4y) \, dx + (4xy - y^3 + 2x) \, dy] = 25,$$

find the centroid of G.

30. Suppose that $\mathbf{F} = P(x, y)\mathbf{i} + Q(x, y)\mathbf{j}$ is a vector field such that $\partial P/\partial x = -\partial Q/\partial y$. Show that the integral of the normal component of $\mathbf{F}$ around ∂G is zero for any region G in the domain of $\mathbf{F}$ for which the two-dimensional divergence theorem applies.

31. Let G be a region where the two-dimensional divergence theorem applies. If $f(x, y)$ has G in its domain, find a line integral around ∂G equal to $\iint_G (\nabla \cdot \nabla f) \, dx \, dy$.

32. Follow the steps indicated to illustrate the divergence theorem for the vector field

$$\mathbf{F} = x^2z^2\mathbf{i} + xz\mathbf{j} + xy^2(z + 1)\mathbf{k}$$

and the unit cube G with one vertex at the origin and the diagonally opposite vertex at $(1, 1, 1)$.

(a) (i) Compute $\iint (\mathbf{F} \cdot \mathbf{n}) \, dS$ over the square face of G in the x,y-coordinate plane, where $z = 0$. Here

$$\mathbf{n} = -\mathbf{k} \quad \text{and} \quad dS = dx \, dy.$$

(ii) Compute $\iint (\mathbf{F} \cdot \mathbf{n}) \, dS$ over the square face of G in the plane $z = 1$. Here

$$\mathbf{n} = \mathbf{k} \quad \text{and} \quad dS = dx \, dy.$$

(iii) Proceed as in parts (i) and (ii) to find the integrals $\iint (\mathbf{F} \cdot \mathbf{n})\, dS$ over the square faces of G in the planes $y = 0$ and $y = 1$.

(iv) Proceed as in parts (i) and (ii) to find the integrals $\iint (\mathbf{F} \cdot \mathbf{n})\, dS$ over the square faces of G in the planes $x = 0$ and $x = 1$.

(v) Add up the six answers found in parts (i) through (iv) to give $\iint_{\partial G} (\mathbf{F} \cdot \mathbf{n})\, dS$.

(b) Compute $\iiint_G (\nabla \cdot \mathbf{F})\, dx\, dy\, dz$; the answer should be the same as in part (a)(v).

In Exercises 33–36, illustrate the divergence theorem for the given region G and vector field $\mathbf{F}$.

33. G the ball $x^2 + y^2 + z^2 \leq a^2$, $\mathbf{F} = x\mathbf{i} + y\mathbf{j} + z\mathbf{k}$

34. G the region bounded by $z = 1 - x^2 - y^2$ and $z = 0$, $\mathbf{F} = y\mathbf{j}$

35. G the region bounded by $z^2 = x^2 + y^2$ and $z = 4$, $\mathbf{F} = y\mathbf{i} - x\mathbf{j} - z\mathbf{k}$

36. G the half-ball $x^2 + y^2 + z^2 \leq a^2$ for $z \geq 0$, $\mathbf{F} = -y\mathbf{i} + x\mathbf{j} + \mathbf{k}$

In Exercises 37–40, use the divergence theorem to compute $\iint_{\partial G} (\mathbf{F} \cdot \mathbf{n})\, dS$.

37. $\mathbf{F} = (x + y)\mathbf{i} + (z^2 - 2y)\mathbf{j} + (x^2 - y^2 + 4z)\mathbf{k}$ and G the ball $x^2 + y^2 + z^2 \leq a$

38. $\mathbf{F} = xz\mathbf{i} + x^2\mathbf{j} + y^2\mathbf{k}$ and G the region bounded by $z = x^2 + y^2$ and $z = 1$

39. $\mathbf{F} = xz\mathbf{i} + xy\mathbf{j} + 2z\mathbf{k}$ and G the region bounded by $z = 4 - x^2 - y^2$ and $z = x^2 + y^2 - 4$

40. $\mathbf{F} = xy^2\mathbf{i} + yz^2\mathbf{j} + x^2z\mathbf{k}$ and G the region bounded by $z = \sqrt{a^2 - x^2 - y^2}$ and $z = \sqrt{x^2 + y^2}$

41. Let G be a three-dimensional region where the divergence theorem applies. Use the theorem to show that

$$\text{Volume of } G = \frac{1}{3} \iint_{\partial G} [(x\mathbf{i} + y\mathbf{j} + z\mathbf{k}) \cdot \mathbf{n}]\, dS.$$

42. (a) Generalizing Exercise 41, let G be a three-dimensional region where the divergence theorem applies and let $\mathbf{F}$ be a continuously differentiable vector field such that $\nabla \cdot \mathbf{F} = c$ throughout G, where c is a constant. Show that $\iint_{\partial G} (\mathbf{F} \cdot \mathbf{n})\, dS = c(\text{Volume of } G)$.

(b) Let G be a region in space where the divergence theorem applies and let

$$\mathbf{F} = (4x + 2x^2z + z^2y)\mathbf{i} + (3y - 2xyz + x^3z^2)\mathbf{j} + (2z - xz^2 + x^2y^3)\mathbf{k}.$$

If the volume of G is 6, find $\iint_{\partial G} (\mathbf{F} \cdot \mathbf{n})\, dS$.

43. Use the divergence theorem for the region $a^2 \leq x^2 + y^2 + z^2 \leq b^2$ to show that the flux of the gradient field $\mathbf{F}$ of the Newtonian potential function $(x^2 + y^2 + z^2)^{-1/2}$ across a sphere with center at the origin is independent of the radius of the sphere.

44. A vector field $\mathbf{F}$ on a region of space is *incompressible* if $\nabla \cdot \mathbf{F} = 0$ throughout the region. Show that the flux of an incompressible field $\mathbf{F}$ across the boundary of each ball inside the region is zero.

45. Often $\nabla \cdot (\nabla f)$ is written as $\nabla^2 f$. Show that for a suitable region G in space, we have

$$\iiint_G (\nabla^2 f)\, dx\, dy\, dz = \iint_{\partial G} (\nabla f \cdot \mathbf{n})\, dS$$

for a twice continuously differentiable function f of three variables.

16.7 GREEN'S THEOREM AND APPLICATIONS

In this section we prove Green's theorem, which again equates an integral of a function over a suitable plane region to an integral around the boundary of the region, just as the two-dimensional divergence theorem did. We will first prove Green's theorem for a *simple region* and then generalize the proof to integrals over more general regions. We will relate the theorem to work we did in Section 16.1 on exact differential forms. The two-dimensional divergence theorem, which we based only on physical arguments in the previous section, will be an immediate corollary of Green's theorem.

Once again, we emphasize that the fundamental theorem of calculus, Green's theorem, the divergence theorem, and Stokes' theorem (to be presented in Section 16.8) are all theorems of the same type. Indeed, the supplementary exercises at the end of this chapter indicate that they are all special cases of a very general theorem, the generalized Stokes' theorem.

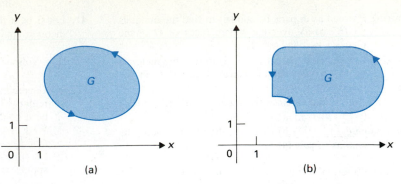

FIGURE 16.47   Two simple regions.

Green's Theorem

Again, we just assume we have the correct notion of a plane region G and its boundary ∂G, traced so that G is on the left. A region is **bounded** if it all lies inside some sufficiently large square and is **closed** if the boundary is considered to be part of the region.

It is usual to prove Green's theorem first for a special type of region G and then extend to more general regions by decomposing them into regions of the special type. We will call a region G **simple** if any line parallel to one of the coordinate axes *crosses* the boundary of G in at most two points. (We allow such a line to *coincide* with the boundary for a whole interval.) The regions shown in Fig. 16.47 are simple.

Let $\mathbf{F}(x, y) = P(x, y)\mathbf{i} + Q(x, y)\mathbf{j}$ be a flux vector field that is continuously differentiable on a region G in the plane. In Section 16.3, we saw that the integral of the *tangential component* of this flux vector field around ∂G is given by

$$\oint_{\partial G} (\mathbf{F} \cdot \mathbf{t}) \, ds = \oint_{\partial G} (\mathbf{F} \cdot d\mathbf{r})$$

$$= \oint_{\partial G} (P \, dx + Q \, dy). \tag{1}$$

This integral measures the *rotation* of the flow around ∂G.

THEOREM 16.10   Green's theorem for simple regions

Let G be a bounded, simple, closed region in the plane with boundary ∂G. If $P(x, y)$ and $Q(x, y)$ are continuously differentiable functions defined on G, then

$$\oint_{\partial G} [P(x, y) \, dx + Q(x, y) \, dy] = \iint_G \left(\frac{\partial Q}{\partial x} - \frac{\partial P}{\partial y} \right) dx \, dy. \tag{2}$$

In the line integral, ∂G is traced in the direction that keeps G on the left.

Proof: Since G is a simple region, ∂G can be split into a "top curve" and a "bottom curve," possibly separated by straight-line "sides," as shown in Fig. 16.48. The top and bottom curves may be parametrized by the parameter x;

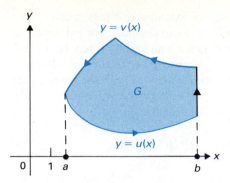

FIGURE 16.48 Top-bounding curve is $y = v(x)$; bottom-bounding curve is $y = u(x)$.

we let the equations of these curves be

$$y = u(x) \text{ for the bottom curve}$$

and

$$y = v(x) \text{ for the top curve,}$$

as shown in Fig. 16.48. The integral $\int P(x, y)\, dx$ over any portion of ∂G consisting of vertical line segments is zero, since x remains constant, so $dx = 0$ on any such segment. Consequently, from Fig. 16.48, we obtain

$$\oint_{\partial G} P(x, y)\, dx = \int_a^b P(x, u(x))\, dx + \int_b^a P(x, v(x))\, dx, \tag{3}$$

where the right-to-left integral sign $\int_b^a$ occurs since the top curve is traced from right to left. From Eq. (3), we have

$$\begin{aligned}
\oint_{\partial G} P(x, y)\, dx &= \int_a^b [P(x, u(x)) - P(x, v(x))]\, dx \\
&= \int_a^b -P(x, y)\Big]_{y=u(x)}^{y=v(x)}\, dx \\
&= \int_a^b \int_{u(x)}^{v(x)} -\frac{\partial P}{\partial y}\, dy\, dx = \iint_G -\frac{\partial P}{\partial y}\, dx\, dy.
\end{aligned} \tag{4}$$

A similar argument, which we ask you to give in Exercise 6, shows that

$$\oint_{\partial G} Q(x, y)\, dy = \iint_G \frac{\partial Q}{\partial x}\, dx\, dy. \tag{5}$$

From Eqs. (4) and (5), we have

$$\oint_{\partial G} [P(x, y)\, dx + Q(x, y)\, dy] = \iint_G \left(\frac{\partial Q}{\partial x} - \frac{\partial P}{\partial y} \right) dx\, dy,$$

which is the assertion of the theorem. $\square$

EXAMPLE 1 Let λ be the loop $x = a\cos t$, $y = a\sin t$ for $0 \le t \le 2\pi$. Use Green's theorem to find

$$\oint_\lambda [(2y + xy)\, dx + (3x - x^2)\, dy].$$

Solution The loop λ is precisely ∂G for the simple region G consisting of the disk $x^2 + y^2 \le a^2$. By Theorem 16.10, the desired integral is equal to

$$\iint_G \left[\frac{\partial(3x - x^2)}{\partial x} - \frac{\partial(2y + xy)}{\partial y} \right] dx\, dy = \iint_G (3 - 2x - 2 - x)\, dx\, dy$$

$$= \iint_G (1 - 3x)\, dx\, dy.$$

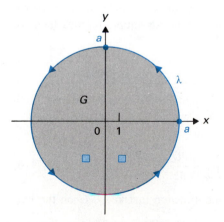

FIGURE 16.49 $\iint_G x\, dx\, dy = 0$ since contributions on these differential elements cancel each other.

Now $\iint_G 1 \cdot dx\, dy = \pi a^2$, the area of G. Also, $\iint_G x\, dx\, dy = 0$, since contributions from differential elements located symmetrically with respect to the y-axis cancel each other, as indicated in Fig. 16.49. Thus the answer is πa^2. ∎

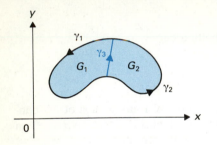

FIGURE 16.50 Breaking a region into two simple regions.

The result in Theorem 16.10 can easily be extended to a region that can be decomposed into a finite number of simple regions. Consider, for example, the region shown in Fig. 16.50. This region G is not simple, but it may be decomposed into the two simple regions G_1 and G_2 separated by the curve γ_3 shown in the figure. We recall some convenient algebraic notation. If γ_3 is the curve traced in the direction of the arrow in Fig. 16.50, let $-\gamma_3$ be the curve traced in the opposite direction. We know from Section 16.3 that

$$\int_{-\gamma_3} (P\,dx + Q\,dy) = -\int_{\gamma_3} (P\,dx + Q\,dy). \tag{6}$$

Symbolically, we write $\partial G_1 = \gamma_1 + \gamma_3$ and $\partial G_2 = \gamma_2 - \gamma_3$. In view of Eq. (6), we have symbolically

$$\oint_{\partial G_1} + \oint_{\partial G_2} = \oint_{\gamma_1+\gamma_3} + \oint_{\gamma_2-\gamma_3}$$

$$= \int_{\gamma_1} + \int_{\gamma_3} + \int_{\gamma_2} - \int_{\gamma_3} \tag{7}$$

$$= \oint_{\gamma_1+\gamma_2} = \oint_{\partial G},$$

where we have omitted the integrand $P\,dx + Q\,dy$ for brevity. Theorem 16.10 applied to the simple regions G_1 and G_2 tells us that

$$\oint_{\partial G_1} (P\,dx + Q\,dy) = \iint_{G_1} \left(\frac{\partial Q}{\partial x} - \frac{\partial P}{\partial y} \right) dx\,dy \tag{8}$$

and

$$\oint_{\partial G_2} (P\,dx + Q\,dy) = \iint_{G_2} \left(\frac{\partial Q}{\partial x} - \frac{\partial P}{\partial y} \right) dx\,dy. \tag{9}$$

Adding Eqs. (8) and (9), we see from Eq. (7) that

$$\oint_{\partial G} (P\,dx + Q\,dy) = \iint_{G} \left(\frac{\partial Q}{\partial x} - \frac{\partial P}{\partial y} \right) dx\,dy,$$

which is Green's theorem for the region G. Similar arguments can be used for any decomposition of a region into a finite number of simple regions, and we state the result as a corollary.

COROLLARY 1　Green's theorem for general regions

Let G be a bounded, closed region in the plane that can be decomposed into a finite number of simple regions. If $P(x, y)$ and $Q(x, y)$ are continuously differentiable on G, then

$$\oint_{\partial G} (P\,dx + Q\,dy) = \iint_{G} \left(\frac{\partial Q}{\partial x} - \frac{\partial P}{\partial y} \right) dx\,dy \tag{10}$$

In the line integral, ∂G is traced in the direction that keeps G on the left.

EXAMPLE 2　Let G be the annular region between the circles $x^2 + y^2 = 4$ and $x^2 + y^2 = 16$ shown in Fig. 16.51. The direction in which the boundary circles should be traced to keep G on the left is indicated in the figure. Now G is

FIGURE 16.51 Annular region broken into four simple regions by the axes.

HISTORICAL NOTE

GREEN'S THEOREM, though named after the English mathematician George Green (1793–1841), was neither stated nor proved by him. Though Green dealt with various three-dimensional results that could be reduced to "Green's theorem," he never stated the theorem himself. The first person to state it, in connection with the integral of a complex function around a closed curve, was Augustin Cauchy in 1846. The first proof was published in 1851 by Bernhard Riemann.

decomposed into four simple regions by the coordinate axes. Illustrate Green's theorem for G using $P(x, y) = xy$ and $Q(x, y) = -x$.

Solution   Let γ_1 be the clockwise-traced inner circle parametrized by

$$x = h_1(t) = 2 \cos t, \qquad y = k_1(t) = -2 \sin t \qquad \text{for } 0 \le t \le 2\pi,$$

and let γ_2 be the counterclockwise-traced outer circle parametrized by

$$x = h_2(t) = 4 \cos t, \qquad y = k_2(t) = 4 \sin t \qquad \text{for } 0 \le t \le 2\pi.$$

Then

$$\int_{\gamma_1 + \gamma_2} (xy \, dx - x \, dy) = \int_{\gamma_1} (xy \, dx - x \, dy) + \int_{\gamma_2} (xy \, dx - x \, dy)$$

$$= \int_0^{2\pi} (8 \cos t \sin^2 t + 4 \cos^2 t) \, dt$$

$$+ \int_0^{2\pi} (-64 \cos t \sin^2 t - 16 \cos^2 t) \, dt$$

$$= \int_0^{2\pi} (-56 \cos t \sin^2 t - 12 \cos^2 t) \, dt$$

$$= \left(-56 \frac{\sin^3 t}{3} - 12 \left(\frac{t}{2} + \frac{\sin 2t}{4} \right) \right) \Big]_0^{2\pi}$$

$$= -12 \frac{2\pi}{2} = -12\pi.$$

On the other hand, we obtain, using polar coordinates,

$$\iint_G \left[\frac{\partial(-x)}{\partial x} - \frac{\partial(xy)}{\partial y} \right] dx \, dy = \iint_G (-1 - x) \, dx \, dy$$

$$= \int_0^{2\pi} \int_2^4 (-1 - r \cos \theta) r \, dr \, d\theta$$

$$= \int_0^{2\pi} \left(-\frac{r^2}{2} - \frac{r^3}{3} \cos \theta \right) \Big]_{r=2}^{r=4} d\theta$$

$$= \int_0^{2\pi} \left(-6 - \frac{56}{3} \cos \theta \right) d\theta$$

$$= \left(-6\theta - \frac{56}{3} \sin \theta \right) \Big]_0^{2\pi} = -12\pi.$$

The same answer was obtained for each integral, illustrating Green's theorem. ∎

EXAMPLE 3   Let $P(x, y)$ and $Q(x, y)$ have continuous partial derivatives in the annular region G, where $4 \le x^2 + y^2 \le 16$, shown in Fig. 16.51. Suppose the counterclockwise integral of $P \, dx + Q \, dy$ around $x^2 + y^2 = 4$ is 5 and around $x^2 + y^2 = 16$ is -13. Find $\iint_G (\partial Q/\partial x - \partial P/\partial y)$.

Solution   For γ_1 and γ_2 shown in Fig. 16.51, we are given that

$$\oint_{-\gamma_1} (P \, dx + Q \, dy) = 5 \qquad \text{and} \qquad \oint_{\gamma_2} (P \, dx + Q \, dy) = -13.$$

By Green's theorem,

$$\iint_G \left(\frac{\partial Q}{\partial x} - \frac{\partial P}{\partial y} \right) dx\,dy = \oint_{\gamma_1} (P\,dx + Q\,dy) + \oint_{\gamma_2} (P\,dx + Q\,dy)$$
$$= -5 - 13 = -18. \qquad \blacksquare$$

As a second corollary, we prove the two-dimensional divergence theorem. The proof shows that the two theorems are really identical except for notation. The difference in notation reflects the two different physical interpretations: divergence across a boundary versus circulation around a boundary.

COROLLARY 2   Two-dimensional divergence theorem

Let G be a bounded, closed region in the plane for which Green's theorem is valid for any functions $P(x, y)$ and $Q(x, y)$ continuously differentiable on G. Then

$$\oint (-Q\,dx + P\,dy) = \iint_G \left(\frac{\partial P}{\partial x} + \frac{\partial Q}{\partial y} \right) dx\,dy. \qquad \textbf{(11)}$$

Proof: This corollary follows immediately from Corollary 1 by replacing $P(x, y)$ by the function $-Q(x, y)$ and replacing $Q(x, y)$ by the function $P(x, y)$ in Eq. (10). $\square$

Independence of Path

Recall that $\int_\gamma (P\,dx + Q\,dy)$ is *independent of path* in a connected open region if the integral depends only on the initial point A and the terminal point B of γ, not on its trace from A to B, for all piecewise-smooth curves γ. We recall as a theorem some things we proved or stated in Sections 16.1 and 16.4. We should keep them well in mind here, for they are closely related to Green's theorem. All curves mentioned in the theorem are to be piecewise smooth.

THEOREM 16.11   Equivalents to independence of path

If $P(x, y)$ and $Q(x, y)$ have continuous first partial derivatives in a connected open region, the following are equivalent.

(a) $\int_\gamma (P\,dx + Q\,dy)$ is independent of the path.

(b) $\oint_\lambda (P\,dx + Q\,dy) = 0$ around every loop λ.

(c) $P\,dx + Q\,dy$ is an exact differential form.

(d) The vector field $\mathbf{F} = P\mathbf{i} + Q\mathbf{j}$ is ∇H for some function $H(x, y)$.

If the region also has no holes in it, the above are also equivalent to

(e) $\partial P/\partial y = \partial Q/\partial x$. $\square$

The only part of Theorem 16.11 that we did not demonstrate earlier is that (e) implies (a)–(d) if the region has no holes in it. We can partially close this gap now. Note that Green's theorem shows that (e) implies (b) for any loop λ that is ∂G for a region G where Green's theorem applies. For if $\partial P/\partial y = \partial Q/\partial x$ in such

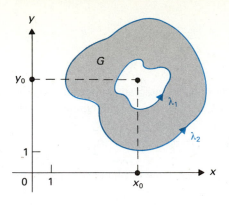

FIGURE 16.52 Two loops going once around (x_0, y_0) counterclockwise.

a region, then $\partial Q/\partial x - \partial P/\partial y = 0$, so

$$0 = \iint_G \left(\frac{\partial Q}{\partial x} - \frac{\partial P}{\partial y}\right) dx\, dy = \oint_{\partial G} (P\, dx + Q\, dy) = \oint_\lambda (P\, dx + Q\, dy).$$

Green's theorem can be an important tool for computing $\oint_\lambda (P\, dx + Q\, dy)$ in a case where $\partial P/\partial y = \partial Q/\partial x$ but where the loop λ goes around a "hole" in the domain of definition of P and Q. Suppose, for example, that $\partial P/\partial y = \partial Q/\partial x$ except at (x_0, y_0), and let λ_1 and λ_2 be two loops going once around (x_0, y_0) counterclockwise, as shown in Fig. 16.52. If the shaded region G between λ_1 and λ_2 in the figure is a region where Green's theorem applies, then

$$\oint_{\lambda_2} (P\, dx + Q\, dy) - \oint_{\lambda_1} (P\, dx + Q\, dy) = \oint_{\partial G} (P\, dx + Q\, dy)$$
$$= \iint_G \left(\frac{\partial Q}{\partial x} - \frac{\partial P}{\partial y}\right) dx\, dy$$
$$= \iint_G 0 \, dx\, dy = 0.$$

Thus

$$\oint_{\lambda_1} (P\, dx + Q\, dy) = \oint_{\lambda_2} (P\, dx + Q\, dy).$$

That is, the value of this line integral around a loop encircling such a "hole" once is independent of the loop. It can therefore be computed by choosing a special such loop to make the computation as easy as possible, perhaps a circle or a square.

EXAMPLE 4 Let λ be a loop going once around the origin counterclockwise. Find

$$\oint_\lambda \left(\frac{y}{x^2 + y^2}\, dx - \frac{x}{x^2 + y^2}\, dy\right).$$

Solution We take $P(x, y) = y/(x^2 + y^2)$ and $Q(x, y) = -x/(x^2 + y^2)$. We find that

$$\frac{\partial P}{\partial y} = \frac{x^2 - y^2}{(x^2 + y^2)^2} = \frac{\partial Q}{\partial x},$$

but P and Q are not defined at $(0, 0)$. We may compute the desired integral by choosing any loop λ going once around $(0, 0)$ counterclockwise. We illustrate with both a circle and a square.

Circle: We use the circle $\lambda_1: x = \cos t, y = \sin t, 0 \le t \le 2\pi$. Then

$$\oint_{\lambda_1} (P\, dx + Q\, dy) = \int_0^{2\pi} \left[\frac{\sin t}{1}(-\sin t) - \frac{\cos t}{1}(\cos t)\right] dt$$
$$= \int_0^{2\pi} (-\sin^2 t - \cos^2 t)\, dt = \int_0^{2\pi} (-1)\, dt = -2\pi.$$

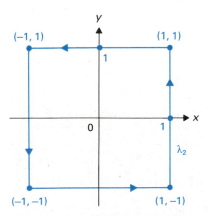

FIGURE 16.53 Square loop λ_2.

Square: We choose the square λ_2 with vertices $(1, 1), (-1, 1), (-1, -1)$, and $(1, -1)$, shown in Fig. 16.53. On the right-hand side of the square, $x = 1$ and

$dx = 0$, so our integral up this side is

$$\int_{-1}^{1} \frac{-1}{1 + y^2}\, dx = -\tan^{-1} y \Big]_{-1}^{1} = -\frac{\pi}{4} + \left(-\frac{\pi}{4}\right) = -\frac{\pi}{2}.$$

Along the top of the square, $y = 1$ and $dy = 0$. To integrate across the top from right to left (counterclockwise), we let x go from 1 to -1, obtaining

$$\int_{1}^{-1} \frac{1}{1 + x^2}\, dx = \tan^{-1} x \Big]_{1}^{-1} = -\frac{\pi}{4} - \frac{\pi}{4} = -\frac{\pi}{2}.$$

In a similar fashion, we find that the integrals down the left-hand side and across the bottom of the square are also each $-\pi/2$. Thus

$$\oint_{\lambda_2} (P\, dx + Q\, dy) = 4\left(-\frac{\pi}{2}\right) = -2\pi. \qquad \blacksquare$$

EXAMPLE 5   Find the integral of the differential form in Example 4 around the ellipse

$$\frac{(x - 10)^2}{4} + \frac{(y - 5)^2}{9} = 1$$

shown in Fig. 16.54.

Solution   Since the ellipse does not encircle the origin, the differential form is exact throughout the entire region bounded by this ellipse. Therefore its integral around the ellipse is zero. $\qquad \blacksquare$

Application to Circulation of a Flow

Let $\mathbf{F} = P(x, y)\mathbf{i} + Q(x, y)\mathbf{j}$ be the flux vector of a flow over a region G. We have seen a physical interpretation of the two-dimensional divergence theorem in terms of the divergence of the flow, which measures the rate at which mass is leaving G. This interpretation arises from integrating the outward *normal* component of the flux over the boundary of G.

Recall from Section 16.3 that the integral of the *tangential* component of the flux over the boundary measures the rotation or **circulation** of the flow around the boundary of G. A unit tangent vector is

$$\mathbf{t} = \frac{dx}{ds}\mathbf{i} + \frac{dy}{ds}\mathbf{j}.$$

The circulation around ∂G is

$$\oint_{\partial G} (\mathbf{F} \cdot \mathbf{t})\, ds = \oint_{\partial G} \left(P\frac{dx}{ds} + Q\frac{dy}{ds}\right) ds = \oint_{\partial G} (P\, dx + Q\, dy) = \oint_{\partial G} \mathbf{F} \cdot d\mathbf{r}.$$

By Theorem 16.10, we have

$$(\text{Circulation of flow around } \partial G) = \oint_{\partial G} (\mathbf{F} \cdot \mathbf{t})\, ds = \oint_{\partial G} (P\, dx + Q\, dy)$$

$$= \iint_{G} \left(\frac{\partial Q}{\partial x} - \frac{\partial P}{\partial y}\right) dx\, dy.$$

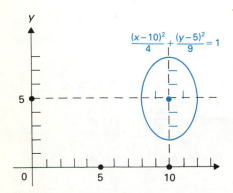

FIGURE 16.54 The ellipse does not contain the origin.

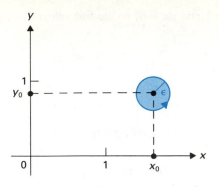

FIGURE 16.55 Disk D_ε of radius ε with center (x_0, y_0).

Let us take a very small disk D_ε of radius ε centered at a point (x_0, y_0) in G, as shown in Fig. 16.55. For ε very small, we have

$$\int_{\partial D_\varepsilon} (P \, dx + Q \, dy) = \iint_{D_\varepsilon} \left(\frac{\partial Q}{\partial x} - \frac{\partial P}{\partial y} \right) dx \, dy$$

$$\approx \pi \varepsilon^2 \left(\frac{\partial Q}{\partial x} - \frac{\partial P}{\partial y} \right) \Bigg]_{(x_0, y_0)}$$

Thus

$$\left(\frac{\partial Q}{\partial x} - \frac{\partial P}{\partial y} \right) \Bigg]_{(x_0, y_0)} \approx \frac{\text{Circulation around } \partial D_\varepsilon}{\text{Area of } D_\varepsilon},$$

and we expect the approximation to become very accurate as $\varepsilon \to 0$. Thus $\partial Q / \partial x - \partial P / \partial y$ can be viewed as measuring the *circulation per unit area* at each point (x, y). It thus measures the tendency of the flow to rotate or *curl* at each point. For this reason, we write

$$\text{Curl } \mathbf{F} = \frac{\partial Q}{\partial x} - \frac{\partial P}{\partial y}.^*$$

We then obtain

$$\oint_{\partial G} (\mathbf{F} \cdot \mathbf{t}) \, ds = \iint_G (\text{Curl } \mathbf{F}) \, dx \, dy. \tag{12}$$

If curl $\mathbf{F} = 0$ at all (x, y) in G, then the integral of the tangential component of $\mathbf{F}$ is zero about any closed path bounding a suitable region lying within G. That is, the total rotation or circulation around any such closed path is zero. Such a flow is called **irrotational.**

EXAMPLE 6 Let

$$\mathbf{F} = \frac{y}{1 + y^3} \mathbf{i} + \frac{3x}{2 + x^3} \mathbf{j}$$

and let G be the region bounded by the square with vertices $(0, 0)$, $(1, 0)$, $(1, 1)$, and $(0, 1)$, shown in Fig. 16.56. Find $\iint_G (\text{Curl } \mathbf{F}) \, dx \, dy$.

Solution Computation of

$$\text{Curl } \mathbf{F} = \partial \left(\frac{3x}{2 + x^3} \right) / \partial x - \partial \left(\frac{y}{1 + y^3} \right) / \partial y$$

is not particularly pleasant, so we try using Green's theorem. Now

$$\iint_G (\text{Curl } \mathbf{F}) \, dx \, dy = \oint_{\partial G} (\mathbf{F} \cdot \mathbf{t}) \, ds = \oint_{\partial G} (\mathbf{F} \cdot d\mathbf{r})$$

$$= \oint_{\partial G} \left(\frac{y}{1 + y^3} \, dx + \frac{3x}{2 + x^3} \, dy \right).$$

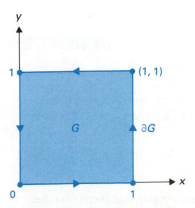

FIGURE 16.56 Unit square region G.

---

*In Section 16.8, we will define the *vector* **curl F** for a vector field in space. Our present *scalar* curl **F** is really the **k**-component of the vector **curl F**, if we regard **F** as being in space. Our scalar curl **F** is a temporary, expedient device.

On the right-hand side of the square, $x = 1$ and $dx = 0$, so the line integral up this side becomes

$$\int_0^1 \frac{3}{2+1}\, dy = \int_0^1 1\, dy = 1.$$

Similarly, we obtain

$$\int_1^0 \frac{1}{1+1}\, dx = \int_1^0 \frac{1}{2}\, dx = -\frac{1}{2}, \qquad \textit{Across the top}$$

$$\int_1^0 \frac{0}{2+0}\, dy = 0, \qquad \textit{Down the left}$$

$$\int_0^1 \frac{0}{1+0}\, dx = 0. \qquad \textit{Across the bottom}$$

Thus the desired integral has the value

$$1 - \frac{1}{2} + 0 + 0 = \frac{1}{2}. \qquad\blacksquare$$

Application to Work

We review the notions of work and potential energy. If

$$\mathbf{F} = P(x, y)\mathbf{i} + Q(x, y)\mathbf{j}$$

is regarded as a force field, then the integral of the tangential component of $\mathbf{F}$ along a curve is the work done by the force in moving a body along the curve. Once again, if $\partial Q/\partial x - \partial P/\partial y = 0$ for all (x, y), then the work done by the force in moving a body around any closed path bounding a suitable region with no holes is zero. In this context, such a force field $\mathbf{F}$ is said to be **conservative.** The work done by such a force field along a curve depends only on the endpoints of the curve; that is, we have independence of path. Note that $P\, dx + Q\, dy$ is then the exact differential of some function $H(x, y)$. Then $u(x, y) = -H(x, y)$ is called a **potential function** of the force field. The potential function gives the **potential energy** of the field at each point (x, y). By our work in Section 16.4, the work done by the force in moving a body from A to B is then

$$H(B) - H(A) = -H(A) - (-H(B)) = u(A) - u(B).$$

SUMMARY

1. *Green's theorem:* If $P(x, y)$ and $Q(x, y)$ are continuously differentiable and if G can be decomposed into a finite number of simple regions, then

$$\oint_{\partial G} (P\, dx + Q\, dy) = \iint_G \left(\frac{\partial Q}{\partial x} - \frac{\partial P}{\partial y} \right) dx\, dy.$$

In the line integral, ∂G is traced in the direction that keeps G on the left.

2. If $P(x, y)$ and $Q(x, y)$ have continuous first partial derivatives in a connected, open region, the following are equivalent.

(a) $\int_\gamma (P\ dx + Q\ dy)$ is independent of the path.

(b) $\oint_\lambda (P\ dx + Q\ dy) = 0$ around every loop λ.

(c) $P\ dx + Q\ dy$ is an exact differential form.

(d) The vector field $\mathbf{F} = P\mathbf{i} + Q\mathbf{j}$ is a gradient field ∇H for some function $H(x, y)$.

If the region also has no holes in it, the above are also equivalent to

(e) $\partial P/\partial y = \partial Q/\partial x$.

3. Suppose $\partial P/\partial y = \partial Q/\partial x$ except at a "hole" in a region. Then $\oint_\lambda (P\ dx + Q\ dy)$ is independent of the loop λ going once counterclockwise around the hole.

4. If $\mathbf{F} = P(x, y)\mathbf{i} + Q(x, y)\mathbf{j}$ is the flux vector of a flow, then

$$\oint_{\partial G} (\mathbf{F} \cdot \mathbf{t})\ ds = \iint_G (\text{Curl } \mathbf{F})\ dx\ dy,$$

where curl $\mathbf{F} = \partial Q/\partial x - \partial P/\partial y$. If curl $\mathbf{F} = 0$, then the flow is called irrotational.

5. The work done by a force field $\mathbf{F}$ along γ is $\int_\gamma (\mathbf{F} \cdot \mathbf{t})\ ds = \int_\gamma (\mathbf{F} \cdot d\mathbf{r})$.

6. A force field $\mathbf{F} = P(x, y)\mathbf{i} + Q(x, y)\mathbf{j}$ is conservative if $\partial Q/\partial x - \partial P/\partial y = 0$. For a conservative force field in a region with no holes in it, the work done is independent of the path, and $u(x, y) = -H(x, y)$ such that $d(H) = P\ dx + Q\ dy$ is a potential function of the force field. The work done by the force field in moving a body from A to B is then $H(B) - H(A) = u(A) - u(B)$.

EXERCISES 16.7

1. Let $P(x, y)$ and $Q(x, y)$ have continuous first partial derivatives. Let $\mathbf{F} = P\mathbf{i} + Q\mathbf{j}$ and let G be a region in the domain of $\mathbf{F}$ to which Green's theorem applies. Mark each of the following true or false.

____ (a) The theorem is concerned only with the integral of the tangential component of a vector field over ∂G.

____ (b) The theorem is concerned only with the integral of the normal component of a vector field over ∂G.

____ (c) The theorem is concerned with the integral of a differential form in two variables over ∂G.

____ (d) The theorem asserts that $\oint_{\partial G} (P\ dx + Q\ dy) = \iint_G [(\partial P/\partial x) + (\partial Q/\partial y)]\ dx\ dy$.

____ (e) The theorem asserts that $\oint_{\partial G} (P\ dx + Q\ dy) = \iint_G [(\partial Q/\partial x) - (\partial P/\partial y)]\ dx\ dy$.

____ (f) Green's theorem is mathematically equivalent to the two-dimensional divergence form.

____ (g) The integral of the tangential component of a force field $\mathbf{F}$ around any loop is always zero by the law of conservation of energy.

____ (h) The integral of the tangential component of a conservative force field $\mathbf{F}$ around any loop is zero.

____ (i) If $\partial P/\partial y = \partial Q/\partial x$, then $\oint_\lambda (\mathbf{F} \cdot d\mathbf{r}) = 0$ for any loop λ.

____ (j) If $\partial P/\partial y = \partial Q/\partial x$, then $\oint_\lambda (\mathbf{F} \cdot d\mathbf{r}) = 0$ for any loop λ that does not encircle any holes in the domain of $\mathbf{F}$.

2. Illustrate Green's theorem for $P(x, y) = x - 2y$, $Q(x, y) = y + x$, and G the annular region $1 \le x^2 + y^2 \le 4$.

3. Let G be the region between the two squares with center at the origin and sides of lengths 2 and 4 parallel to coordinate axes. Illustrate Green's theorem for $P(x, y) = y^3$ and $Q(x, y) = -x^3$; that is, for the vector field $\mathbf{F} = y^3\mathbf{i} - x^3\mathbf{j}$.

4. Let $\mathbf{F}$ be a vector field in the plane. Suppose the integrals $\oint (\mathbf{F} \cdot \mathbf{t}) \, ds$ taken counterclockwise about the circles $x^2 + y^2 = 100$ and $x^2 + y^2 = 25$ are 35 and -24, respectively. Find $\iint_G (\text{Curl } \mathbf{F}) \, dx \, dy$, where G is the region between the two circles.

5. Let $\mathbf{F}$ be a vector field in the plane, and let G be the annular region $1 \le x^2 + y^2 \le 4$. Let $\iint_G (\text{Curl } \mathbf{F}) \, dx \, dy = 10$ and $\oint_{\lambda_1} (\mathbf{F} \cdot d\mathbf{r}) = 7$, where λ_1 is the circle $x^2 + y^2 = 4$ traced once clockwise. Find $\oint_{\lambda_2} (\mathbf{F} \cdot d\mathbf{r})$, where λ_2 is the circle $x^2 + y^2 = 1$ traced once counterclockwise.

6. Prove Eq. (5) of the proof of Green's theorem in the text.

7. Let $\mathbf{E}$ and $\mathbf{F}$ be two vector fields on a simple, closed, bounded region G. Suppose that $\mathbf{E} = \mathbf{F}$ at each point on ∂G.
(a) Show that

$$\iint_G (\nabla \cdot \mathbf{E}) \, dx \, dy = \iint_G (\nabla \cdot \mathbf{F}) \, dx \, dy.$$

(b) Show that

$$\iint_G (\text{Curl } \mathbf{E}) \, dx \, dy = \iint_G (\text{Curl } \mathbf{F}) \, dx \, dy.$$

8. Let G be a simple, closed, bounded region and let $\mathbf{F}$ be a vector field on G.
(a) Suppose $\oint_{\lambda} (\mathbf{F} \cdot \mathbf{t}) \, ds = 0$ for every closed curve λ lying in G. Argue as best you can using just Green's theorem that curl $\mathbf{F} = 0$ at each (x, y) in G.
(b) State a result similar to that in part (a) under the hypothesis $\oint_{\lambda} (\mathbf{F} \cdot \mathbf{n}) \, ds = 0$.

9. Let a plane region G, which can be decomposed into a finite number of simple regions, be bounded by a piecewise-smooth curve λ. Using Green's theorem, show that the line integral $\oint_{\lambda} (-y \, dx + x \, dy)$ is equal to twice the area of G.

In Exercises 10–20, let λ be the ellipse $x = 2 \cos t$, $y = 3 \sin t$ for $0 \le t \le 2\pi$. For the given vector field $\mathbf{F}$, find $\oint_{\lambda} (\mathbf{F} \cdot \mathbf{t}) \, ds$ in the easiest way you can. You may use the fact that the ellipse $(x^2/a^2) + (y^2/b^2) = 1$ has area πab. Be sure to use the symmetry of the ellipse about the axes and origin, where such symmetry is helpful.

10. $\mathbf{F} = x^2\mathbf{i} + (y^3 + 2y)\mathbf{j}$

11. $\mathbf{F} = (x^2 + 2y)\mathbf{i} + (y^2 - 3x)\mathbf{j}$

12. $\mathbf{F} = (2xy - y)\mathbf{i} - (3x + x^2)\mathbf{j}$

13. $\mathbf{F} = (x^2y^2 - y)\mathbf{i} + (3x^2y^2 + 2x)\mathbf{j}$

14. $\mathbf{F} = (3xy^3 - 2x^3y - xy^2)\mathbf{i} + (x^4y^2 + x^3y^3 - 2x^2y^3)\mathbf{j}$

15. $\mathbf{F} = \dfrac{-2y}{x^2 + y^2}\mathbf{i} + \dfrac{2x}{x^2 + y^2}\mathbf{j}$

16. $\mathbf{F} = \dfrac{x}{x^2 + y^2}\mathbf{i} + \dfrac{y}{x^2 + y^2}\mathbf{j}$

17. $\mathbf{F} = \dfrac{x + y}{x^2 + y^2}\mathbf{i} + \dfrac{y - x}{x^2 + y^2}\mathbf{j}$

18. $\mathbf{F} = \dfrac{1 - y}{x^2 + (y - 1)^2}\mathbf{i} + \dfrac{x}{x^2 + (y - 1)^2}\mathbf{j}$

19. $\mathbf{F} = \dfrac{x^2y^2 - y}{x^2 + y^2}\mathbf{i} + \dfrac{x - xy}{x^2 + y^2}\mathbf{j}$

20. $\mathbf{F} = \dfrac{4 - y}{x^2 + (y - 4)^2}\mathbf{i} + \dfrac{x}{x^2 + (y - 4)^2}\mathbf{j}$

In Exercises 21–27, let $P(x, y)$ and $Q(x, y)$ have continuous first partial derivatives except at $(0, 0)$, $(0, 4)$, and $(6, 0)$ and let $\partial Q/\partial x = \partial P/\partial y$ except at those points. Let $\oint (P \, dx + Q \, dy)$ have these values taken counterclockwise once around these circles:

$$-5 \text{ taken around } x^2 + y^2 = 1,$$
$$3 \text{ taken around } x^2 + (y - 4)^2 = 1,$$
$$9 \text{ taken around } (x - 6)^2 + y^2 = 1.$$

Find $\oint_{\lambda} (P \, dx + Q \, dy)$ for the given loop with direction of increasing t.

21. λ given by $x = 5 \cos t$, $y = 5 \sin t$ for $0 \le t \le 2\pi$

22. λ given by $x = 3 + 4 \cos t$, $y = 4 \sin t$ for $0 \le t \le 2\pi$

23. λ given by $x = \cos t$, $y = \sin t$ for $0 \le t \le 4\pi$

24. λ given by $x = 8 \cos t$, $y = 8 \sin t$ for $0 \le t \le 2\pi$

25. λ given by $x = 3 \cos t$, $y = 2 - 3 \sin t$ for $0 \le t \le 2\pi$

26. λ given by $x = 2 + 3 \cos t$, $y = 3 - 3 \sin t$ for $0 \le t \le 4\pi$

27. λ given by $x = 2 + 2 \cos t$, $y = 2 + 2 \sin t$ for $0 \le t \le 2\pi$

28. Show that if $f(x, y)$ has continuous second partial derivatives, then curl $\nabla f = 0$.

29. Show that if ∇f is the flux vector of an incompressible flow, then f satisfies Laplace's equation $f_{xx} + f_{yy} = 0$.

30. Use Green's theorem to find the work done by the force field $\mathbf{F}(x, y) = (x^2 + y^2)\mathbf{i} - 2xy\mathbf{j}$ in moving a body counterclockwise around the square with vertices at $(0, 0)$, $(1, 0)$, $(1, 1)$, and $(0, 1)$.

31. Find all functions $g(x, y)$ such that the force field $\mathbf{F}(x, y) = 3xy^2\mathbf{i} + g(x, y)\mathbf{j}$ is conservative.

32. Let the flux vector of a plane flow be

$$\mathbf{F}(x, y) = (x^2 + y^2)\mathbf{i} + 2xy\mathbf{j}.$$

(a) Find the divergence of $\mathbf{F}$.
(b) Use the two-dimensional divergence theorem to find the flux of the flow across the border of the square with vertices $(0, 0)$, $(1, 0)$, $(1, 1)$, and $(0, 1)$.

33. For the flow in Exercise 32, use Green's theorem to find the circulation of the flow counterclockwise around the circle $x^2 + y^2 = 4$.

34. Show that every constant force field is conservative. What can be said concerning the work done by a constant force field in moving a body from a point A to a point B?

35. Show that every constant flux field in the plane gives a flow that is both irrotational and incompressible. Give physical interpretations of this result.

36. Let $\mathbf{E}$ and $\mathbf{F}$ be conservative force fields in a plane region G.
(a) Show that for any numbers a and b, the force field $a\mathbf{E} + b\mathbf{F}$ is conservative.

(b) If u is a potential function of **E** and v a potential function of **F** in G, describe all potential functions of a**E** + b**F**.

37. Consider the force field $\mathbf{F} = 2xy\mathbf{i} + x^2\mathbf{j}$ in the plane.
    (a) Show that the field is conservative.
    (b) Find a potential function for the field.
    (c) Find the potential energy $u(x, y)$ of the field such that $u(0, 0) = 5$.
    (d) Find the work done by the field in moving a body from the point $(1, -1)$ to the point $(2, 1)$.

38. Let **F** be a force field in the plane and let the position of a body in the plane at time t be given by $x = h_1(t)$, $y = h_2(t)$ for $a \le t \le b$. Let the position be A when $t = a$ and B when $t = b$. Let $W(A, B)$ be the work done by the field in moving the body from A to B for $a \le t \le b$. Show that

    $$W(A, B) = \tfrac{1}{2}m|\mathbf{v}(b)|^2 - \tfrac{1}{2}m|\mathbf{v}(a)|^2 = k(b) - k(a),$$

    where $k(t) = \tfrac{1}{2}m|\mathbf{v}(t)|^2$ is the *kinetic energy* of the body at time t. [*Hint:* Use $\mathbf{F}(t) = m\mathbf{a}(t) = m\mathbf{v}'(t)$ and $d\mathbf{r}/dt = \mathbf{v}(t)$ to express $W(A, B) = \int_a^b (\mathbf{F} \cdot d\mathbf{r})$ in terms of **v**.]

39. Continuing Exercise 38, show that if **F** is the force field for a potential function u, then $u(A) + k(A) = u(B) + k(B)$, so the sum of the potential and kinetic energies remains constant. (Note that it is permissible to write u and k as functions of positions in the plane since **F** is a conservative force field, so the work in Exercise 38 is independent of the path.) This is the reason why such a force field is called *conservative*.

40. A positive electric charge in the plane at the origin exerts a force of attraction on a negative unit charge in the plane that is inversely proportional to the square of the distance between the charges. Let the force of attraction on a negative unit charge at $(1, 0)$ be k units.
    (a) Find the force field created by the charge at the origin.
    (b) Show that the force field in part (a) is conservative by finding a potential function; the function is the *Newtonian potential*.
    (c) Using the answer to part (b), find the work done by the field in moving a unit negative charge from $(0, 2)$ to $(2, 3)$.

41. Let u be a potential function of a conservative force field **F** in the plane. Level curves of u are known as *equipotential curves*.
    (a) What geometric relation exists between the field **F** and its equipotential curves?
    (b) If γ_1 and γ_2 are equipotential curves, show that the work done by the field **F** in moving a body from a point P_1 on γ_1 to a point P_2 on γ_2 is independent of the choices of P_1 on γ_1 and P_2 on γ_2.

16.8 STOKES' THEOREM

Green's theorem is concerned with integrals over a plane region. Stokes' theorem is a generalization of Green's theorem to more general two-dimensional regions, which do not necessarily lie in a plane. That is, Stokes' theorem is concerned with integrals over a *surface* in space. We will not prove Stokes' theorem; the proof is best left to a more advanced course. Of course, we have proved a special case of it, namely Green's theorem, in the case where the "surface" lies in the x,y-plane.

The Curl of a Vector Field

Let $\mathbf{F} = P\mathbf{i} + Q\mathbf{j} + R\mathbf{k}$ be a differentiable vector field in space. The **curl of F** is the *vector* defined by the symbolic determinant

$$\mathbf{Curl\ F} = \nabla \times \mathbf{F} = \begin{vmatrix} \mathbf{i} & \mathbf{j} & \mathbf{k} \\ \dfrac{\partial}{\partial x} & \dfrac{\partial}{\partial y} & \dfrac{\partial}{\partial z} \\ P & Q & R \end{vmatrix}.$$

Note that **curl F** is a *vector*. Our definition of curl **F** as a scalar quantity in the preceding section was a temporary expedient in discussing rotation of a flow in the plane and is explained in Example 2 on the next page.

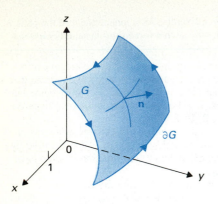

FIGURE 16.57 Surface G with directed boundary curve ∂G and unit normal vector **n**.

EXAMPLE 1 Find **curl F** if $\mathbf{F} = xy\mathbf{i} + y^2\mathbf{j} + yz^2\mathbf{k}$.

Solution We have

$$
\textbf{Curl F} = \nabla \times \mathbf{F} =
\begin{vmatrix}
\mathbf{i} & \mathbf{j} & \mathbf{k} \\
\dfrac{\partial}{\partial x} & \dfrac{\partial}{\partial y} & \dfrac{\partial}{\partial z} \\
xy & y^2 & yz^2
\end{vmatrix}
$$

$$
= z^2\mathbf{i} + 0\mathbf{j} - x\mathbf{k} = z^2\mathbf{i} - x\mathbf{k}. \qquad \blacksquare
$$

EXAMPLE 2 Find **curl F** if $R(x, y, z) = 0$, so $\mathbf{F} = P\mathbf{i} + Q\mathbf{j}$.

Solution We have

$$
\textbf{Curl F} = \nabla \times \mathbf{F} =
\begin{vmatrix}
\mathbf{i} & \mathbf{j} & \mathbf{k} \\
\dfrac{\partial}{\partial x} & \dfrac{\partial}{\partial y} & \dfrac{\partial}{\partial z} \\
P & Q & 0
\end{vmatrix}
$$

$$
= -\frac{\partial Q}{\partial z}\mathbf{i} + \frac{\partial P}{\partial z}\mathbf{j} + \left(\frac{\partial Q}{\partial x} - \frac{\partial P}{\partial y}\right)\mathbf{k}.
$$

We see that what we called curl **F** in Section 16.7 for $\mathbf{F} = P\mathbf{i} + Q\mathbf{j}$ was really the **k**-component of **curl F**, viewed as a vector in space. $\qquad \blacksquare$

Stokes' Theorem

We make no attempt to prove Stokes' theorem but will, roughly, state it. It is the generalization of Green's theorem to a two-dimensional surface G in space. For example, G might be the surface shown in Fig. 16.57. (The theorem is not true for some "one-sided" surfaces, like the Möbius strip in Fig. 16.33.) Figure 16.57 also shows a unit normal vector **n** at a point on G and indicates by arrows a direction around ∂G. (We use G rather than Σ for a surface to accentuate the parallel here with Sections 16.6 and 16.7.)

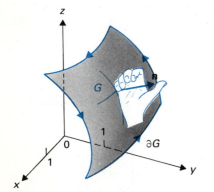

FIGURE 16.58 Right-hand rule relating the direction of **n** and the direction around ∂G.

Right-Hand Rule

For Stokes' theorem, the direction for **n** and the direction around ∂G are always related so that as the fingers of the right hand curve in the direction around ∂G, the thumb points in the direction of **n**.

This right-hand rule is illustrated in Fig. 16.58. To tie the rule in with G and ∂G for Green's theorem in the plane, note that it means that when we are walking along ∂G with our head in the direction of $\mathbf{n} = \mathbf{k}$, we have the region G on our left.

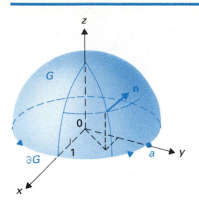

FIGURE 16.59 Unit normal vector

$$\mathbf{n} = (1/a)(x\mathbf{i} + y\mathbf{j} + z\mathbf{k})$$

to the hemisphere G; direction around ∂G.

THEOREM 16.12   Stokes' theorem

For a suitable surface G in space and a continuously differentiable vector field $\mathbf{F} = P\mathbf{i} + Q\mathbf{j} + R\mathbf{k}$, we have

$$\oint_{\partial G} (\mathbf{F} \cdot \mathbf{t}) \, ds = \iint_G [(\nabla \times \mathbf{F}) \cdot \mathbf{n}] \, dS. \qquad (1)$$

In the line integral, ∂G is traced in the direction with respect to $\mathbf{n}$ given by the right-hand rule.   □

In words, the integral of the tangential component of the vector field around ∂G is equal to the integral of the normal component of **curl F** over the surface G. If $\mathbf{F}$ is the flux field of a flow, this has the interpretation that the circulation around ∂G equals the integral of the normal component of the **curl** over G. The **curl** measures the rotation of the flow at each point, and the normal component measures the portion of this rotation that acts *tangent* to the surface. Remember that the direction of a *tangent* plane to a surface is specified by a *normal* direction to the surface.

EXAMPLE 3   Write Eq. (1) in Stokes' theorem in the case that the surface G lies in the x,y-plane and $\mathbf{F} = P\mathbf{i} + Q\mathbf{j}$, taking $\mathbf{n} = \mathbf{k}$.

Solution   If G lies in the x,y-plane and $\mathbf{n} = \mathbf{k}$, then $dS = dx \, dy$. If

$$\mathbf{F} = P\mathbf{i} + Q\mathbf{j},$$

then, as shown in Example 2,

$$\nabla \times \mathbf{F} = -\frac{\partial Q}{\partial z}\mathbf{i} + \frac{\partial P}{\partial z}\mathbf{j} + \left(\frac{\partial Q}{\partial x} - \frac{\partial P}{\partial y}\right)\mathbf{k}.$$

Equation (1) then becomes

$$\oint_{\partial G} (\mathbf{F} \cdot \mathbf{t}) \, ds = \iint_G \left(\frac{\partial Q}{\partial x} - \frac{\partial P}{\partial y}\right) dx \, dy.$$

This is the rotation form of Green's theorem, which is thus a special case of Stokes' theorem.   ∎

EXAMPLE 4   Illustrate Stokes' theorem if the surface G is the hemisphere $z = \sqrt{a^2 - x^2 - y^2}$, shown in Fig. 16.59 and

$$\mathbf{F} = yz\mathbf{i} - xz\mathbf{j} + 3\mathbf{k}.$$

Solution   Now

$$\mathbf{n} = \frac{1}{a}(x\mathbf{i} + y\mathbf{j} + z\mathbf{k}),$$

and Fig. 16.60 shows that we could use spherical coordinates and take

$$dS = (a \sin \phi \, d\theta)(a \, d\phi) = a^2 \sin \phi \, d\phi \, d\theta.$$

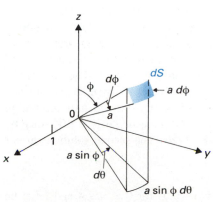

FIGURE 16.60 In spherical coordinates, we may take $dS = (a \sin \phi \, d\theta)(a \, d\phi)$ on the surface of the sphere of radius a with center $(0, 0, 0)$.

STOKES' THEOREM is closely connected with George Stokes (1819–1903). William Thomson (Lord Kelvin) (1824–1907) informed Stokes of the theorem in a letter of 1850. Four years later, Stokes posed its proof as a problem on the Smith's Prize Exam at Cambridge. It is not known whether any of the students solved the problem.

Also

$$\nabla \times \mathbf{F} = \begin{vmatrix} \mathbf{i} & \mathbf{j} & \mathbf{k} \\ \dfrac{\partial}{\partial x} & \dfrac{\partial}{\partial y} & \dfrac{\partial}{\partial z} \\ yz & -xz & 3 \end{vmatrix} = x\mathbf{i} + y\mathbf{j} - 2z\mathbf{k}.$$

Consequently,

$$(\nabla \times \mathbf{F}) \cdot \mathbf{n} = \frac{1}{a}(x^2 + y^2 - 2z^2) = \frac{1}{a}(x^2 + y^2 + z^2 - 3z^2).$$

$$= \frac{1}{a}(a^2 - 3z^2) = \frac{1}{a}(a^2 - 3a^2\cos^2\phi)$$

$$= a(1 - 3\cos^2\phi).$$

Then

$$\iint_G [(\nabla \times \mathbf{F}) \cdot \mathbf{n}] \, dS = \int_0^{2\pi} \int_0^{\pi/2} a(1 - 3\cos^2\phi) a^2 \sin\phi \, d\phi \, d\theta$$

$$= a^3 \int_0^{2\pi} (-\cos\phi + \cos^3\phi) \Big]_0^{\pi/2} d\theta$$

$$= a^3 \int_0^{2\pi} 0 \, d\theta = 0.$$

Turning to $\oint_{\partial G} (\mathbf{F} \cdot \mathbf{t}) \, ds$, we parametrize ∂G by $x = a \cos t$ and $y = a \sin t$ for $0 \le t \le 2\pi$. Then

$$\mathbf{t} = \frac{dx}{ds}\mathbf{i} + \frac{dy}{ds}\mathbf{j} + 0\mathbf{k}.$$

Since ∂G lies in the plane $z = 0$, we see that there $\mathbf{F} = 0\mathbf{i} + 0\mathbf{j} + 3\mathbf{k}$. Therefore $\mathbf{F} \cdot \mathbf{t} = 0$, so $\oint_{\partial G} (\mathbf{F} \cdot \mathbf{t}) \, ds = 0$ also, illustrating Stokes' theorem. ∎

We have seen that it is sometimes convenient to use Green's theorem or a divergence theorem to change from an integral over the boundary of a region to an integral over the region itself, or vice versa. Computing the integral is sometimes simplified in this way. Of course, Stokes' theorem can be used in the same way. In fact, Stokes' theorem presents even more possibilities. The problem of computing $\iint_G[(\nabla \times \mathbf{F}) \cdot \mathbf{n}] \, dS$ over a surface G can of course be transformed into computing $\oint_{\partial G} (\mathbf{F} \cdot \mathbf{t}) \, ds$. But it can also be transformed into computing $\iint_{G'}[(\nabla \times \mathbf{F}) \cdot \mathbf{n}] \, dS$ for *any suitable surface G' for which $\partial G' = \partial G$*, since by Stokes' theorem both integrals equal $\oint_{\partial G}(\mathbf{F} \cdot \mathbf{t}) \, ds$. The second solutions in our final examples illustrate this possibility.

EXAMPLE 5 Let G be the surface $z = x^2 + y^2$ for $0 \le z \le 4$ and let $\mathbf{n}$ be the unit normal vector pointing out the convex side of the surface. Let $\mathbf{F} = yz\mathbf{i} + xz^2\mathbf{j} - xyz\mathbf{k}$. Find $\iint_G[(\nabla \times \mathbf{F}) \cdot \mathbf{n}] \, dS$.

Solution 1 By Stokes' theorem, we may instead compute $\oint_{\partial G}(\mathbf{F} \cdot \mathbf{t}) \, ds = \oint_{\partial G}(\mathbf{F} \cdot d\mathbf{r})$. The surface is sketched in Fig. 16.61. We see that ∂G is the circle $x^2 + y^2 = 4$ in the plane $z = 4$, traced *clockwise* as viewed from above. A para-

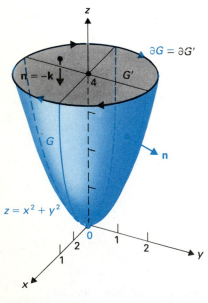

FIGURE 16.61

G: $z = x^2 + y^2$, $0 \le z \le 4$;
G': $x^2 + y^2 = 4$, $z = 4$, $\partial G = \partial G'$

metrization of ∂G is therefore

$$x = 2 \cos t, \qquad y = -2 \sin t, \qquad z = 4 \qquad \text{for } 0 \le t \le 2\pi.$$

(The minus sign in the y-equation is due to the *clockwise* tracing of the circle.) In the plane $z = 4$, we have $\mathbf{F} = 4y\mathbf{i} + 16x\mathbf{j} - 4xy\mathbf{k}$. Now $dz = 0$, and we see that

$$\oint_{\partial G} (\mathbf{F} \cdot d\mathbf{r}) = \oint_{\partial G} (4y\, dx + 16x\, dy)$$

$$= \int_0^{2\pi} [(-8 \sin t)(-2 \sin t) + (32 \cos t)(-2 \cos t)]\, dt$$

$$= \int_0^{2\pi} (16 \sin^2 t - 64 \cos^2 t)\, dt$$

$$= \int_0^{2\pi} [8(1 - \cos 2t) - 32(1 + \cos 2t)]\, dt$$

$$= 8 \int_0^{2\pi} (-3 - 5 \cos 2t)\, dt = 8\left(-3t - \frac{5}{2} \sin 2t\right)\Big]_0^{2\pi}$$

$$= 8(-6\pi) = -48\pi.$$

Solution 2 The disk G', where $0 \le x^2 + y^2 \le 4$, in the plane $z = 4$ with unit normal vector $\mathbf{n} = -\mathbf{k}$ has the same boundary as G. See Fig. 16.62. Thus we may also compute $\iint_{G'} [(\nabla \times \mathbf{F}) \cdot \mathbf{n}]\, dS$ to find the desired integral. We find that

$$\nabla \times \mathbf{F} = \begin{vmatrix} \mathbf{i} & \mathbf{j} & \mathbf{k} \\ \dfrac{\partial}{\partial x} & \dfrac{\partial}{\partial y} & \dfrac{\partial}{\partial z} \\ yz & xz^2 & -xyz \end{vmatrix}.$$

$$= (-xz - 2xz)\mathbf{i} + (y + yz)\mathbf{j} + (z^2 - z)\mathbf{k}.$$

In the plane $z = 4$, we have

$$\nabla \times \mathbf{F} = -12x\mathbf{i} + 5y\mathbf{j} + 12\mathbf{k}.$$

Since $\mathbf{n} = -\mathbf{k}$, we find that

$$(\nabla \times \mathbf{F}) \cdot \mathbf{n} = -12.$$

On this disk G' parallel to the x,y-plane, we have $dS = dx\, dy$. Thus

$$\iint_G [(\nabla \times \mathbf{F}) \cdot \mathbf{n}]\, dS = \iint_{G'} [(\nabla \times \mathbf{F}) \cdot \mathbf{n}]\, dx\, dy = \iint_{G'} (-12)\, dx\, dy$$

$$= -12 \cdot (\text{Area of } G') = -12 \cdot 4\pi = -48\pi. \qquad \blacksquare$$

EXAMPLE 6 Let G be the cylinder $x^2 + y^2 = 4$ for $1 \le x \le 5$, and let $\mathbf{F} = yz\mathbf{i} - xz\mathbf{j} + xyz\mathbf{k}$. If $\mathbf{n}$ points out the convex side of the cylinder, compute $\iint_G [(\nabla \times \mathbf{F}) \cdot \mathbf{n}]\, dS$.

Solution 1 The cylinder is shown in Fig. 16.62. Note that ∂G consists of *two* circles. The upper boundary circle $x^2 + y^2 = 4$ in the plane $z = 5$ is traced *clockwise* when viewed from above, according to the right-hand rule. The lower boundary circle $x^2 + y^2 = 4$ in the plane $z = 1$ is traced counterclockwise, viewed from above.

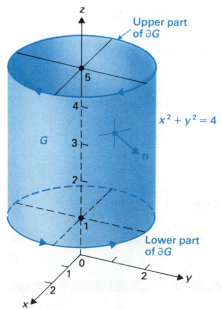

FIGURE 16.62 The boundary of this cylinder consists of two circles traced in opposite directions.

By Stokes' theorem,

$$\iint_G [(\nabla \times \mathbf{F}) \cdot \mathbf{n}] \, dS = \oint_{\partial G} (\mathbf{F} \cdot d\mathbf{r}).$$

Now in the plane $z = 5$, we have $\mathbf{F} = 5y\mathbf{i} - 5x\mathbf{j} + 5xy\mathbf{k}$. Since $dz = 0$ on both boundary circles, we have for the upper circle

$$\mathbf{F} \cdot d\mathbf{r} = 5(y \, dx - x \, dy).$$

We wish to integrate this expression *clockwise* around the circle $x^2 + y^2 = 4$. By the rotation form of Green's theorem, this integral is

$$-5 \oint_{x^2+y^2=4} (y \, dx - x \, dy) = -5 \iint_{x^2+y^2\leq 4} (-1-1) \, dx \, dy$$

$$= 10 \iint_{x^2+y^2\leq 4} 1 \cdot dx \, dy$$

$$= 10 \cdot 4\pi = 40\pi.$$

Similarly, in the plane $z = 1$, we have $\mathbf{F} = y\mathbf{i} - x\mathbf{j} + xy\mathbf{k}$. Since the part of ∂G in this plane is the circle $x^2 + y^2 = 4$ traced counterclockwise, the rotation form of Green's theorem gives us the integral value

$$\oint_{x^2+y^2=4} (y \, dx - x \, dy) = \iint_{x^2+y^2\leq 4} -2 \, dx \, dy = -2(4\pi) = -8\pi.$$

Thus our answer is $40\pi + (-8\pi) = 32\pi$.

Solution 2   Alternatively, we could let G_1 be the disk at the top of the cylinder with unit normal vector $-\mathbf{k}$ and G_2 the disk at the bottom with unit normal vector $\mathbf{k}$, as shown in Fig. 16.63. As explained before Example 5, we have

$$\iint_G [(\nabla \times \mathbf{F}) \cdot \mathbf{n}] \, dS = \iint_{G_1} [(\nabla \times \mathbf{F}) \cdot (-\mathbf{k})] \, dS$$

$$+ \iint_{G_2} [(\nabla \times \mathbf{F}) \cdot \mathbf{k}] \, dS.$$

Since we are going to compute a dot product with $\pm\mathbf{k}$, we need only compute the $\mathbf{k}$-component of $\nabla \times \mathbf{F}$, which we easily find to be $-2z\mathbf{k}$. Then

$$\iint_{G_1} [(\nabla \times \mathbf{F}) \cdot (-\mathbf{k})] \, dS = \iint_{G_1} 2z \, dS = \iint_{G_1} 10 \, dS = 10 \cdot 4\pi = 40\pi.$$

Similarly,

$$\iint_{G_2} [(\nabla \times \mathbf{F}) \cdot \mathbf{k}] \, dS = \iint_{G_2} -2z \, dS = \iint_{G_2} -2 \, dS = -8\pi.$$

This again gives the value $40\pi - 8\pi = 32\pi$ for the desired integral.   ∎

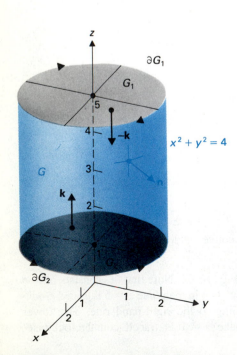

FIGURE 16.63 $\partial G = \partial G_1 + \partial G_2$

SUMMARY

Let $\mathbf{F} = P\mathbf{i} + Q\mathbf{j} + R\mathbf{k}$ be a continuously differentiable vector field.

1. **Curl F** $= \nabla \times \mathbf{F}$

2. *Stokes' theorem:* If G is a suitable surface in space and if the direction of the unit normal vector $\mathbf{n}$ and the direction around the boundary ∂G are related by the right-hand rule, then

$$\oint_{\partial G} (\mathbf{F} \cdot \mathbf{t}) \, ds = \iint_G [(\nabla \times \mathbf{F}) \cdot \mathbf{n}] \, dS.$$

3. When the value of the integrals in Stokes' theorem is desired, remember the integrals are also equal to $\iint_{G'} [(\nabla \times \mathbf{F}) \cdot \mathbf{n}] \, dS$ for any suitable surface G' for which $\partial G' = \partial G$.

EXERCISES 16.8

In Exercises 1–7, find **curl F**.

1. $\mathbf{F} = x\mathbf{i} - 2x\mathbf{j} + 5z\mathbf{k}$

2. $\mathbf{F} = x^2\mathbf{i} + y^2\mathbf{j} + z^2\mathbf{k}$

3. $\mathbf{F} = xyz\mathbf{i} - 2y^2z\mathbf{j} + 3z^4\mathbf{k}$

4. $\mathbf{F} = \dfrac{y}{x}\mathbf{i} - \dfrac{z}{y}\mathbf{j} + \dfrac{x}{z}\mathbf{k}$

5. $\mathbf{F} = \dfrac{y}{z}\mathbf{i} + \dfrac{z}{x}\mathbf{j} + \dfrac{x}{y}\mathbf{k}$

6. $\mathbf{F} = e^{xy}\mathbf{i} + e^{-yz}\mathbf{j} + e^{x^2}\mathbf{k}$

7. $\mathbf{F} = (\sin xy)\mathbf{i} + xe^y\mathbf{j} + (\ln yz)\mathbf{k}$

In Exercises 8–11, illustrate Stokes' theorem for the given surface G, direction of $\mathbf{n}$, and vector field $\mathbf{F}$.

8. G the paraboloid $z = x^2 + y^2$ for $0 \le z \le 1$, $\mathbf{n}$ pointing out the concave side, $\mathbf{F} = -y\mathbf{i} + x\mathbf{j} + z\mathbf{k}$

9. G the cone $z^2 = x^2 + y^2$ for $0 \le z \le 4$, $\mathbf{n}$ pointing out the convex side, $\mathbf{F} = -yz^2\mathbf{i} + xz^2\mathbf{j} + 2xyz\mathbf{k}$

10. G the portion of the sphere $x^2 + y^2 + z^2 = 5$ on or above the plane $z = 1$, $\mathbf{n}$ pointing out the convex side, $\mathbf{F} = 2y\mathbf{i} + 3x\mathbf{j} + xz\mathbf{k}$

11. G the triangular region consisting of the portion of the plane $2x + y + 2z = 2$ in the first octant, $\mathbf{n}$ with positive $\mathbf{k}$-coefficient, $\mathbf{F} = yz\mathbf{i} + y^2\mathbf{j} + xy\mathbf{k}$

In Exercises 12–23, compute the indicated integral in the easiest way you can. Make use of a divergence theorem or Stokes' theorem where appropriate, including the use of Stokes' theorem in number 3 of the Summary. You may use symmetry and the fact that the area of the region bounded by the ellipse $(x^2/a^2) + (y^2/b^2) = 1$ is πab.

12. $\iint_G [(\nabla \times \mathbf{F}) \cdot \mathbf{n}] \, dS$, where G is the surface $z = 4 - x^2 - y^2$ for $z \ge 0$, $\mathbf{n}$ has positive $\mathbf{k}$-coefficient, and $\mathbf{F} = x^2y\mathbf{i} + xyz\mathbf{j} + xz^2\mathbf{k}$

13. $\iint_G [(\nabla \times \mathbf{F}) \cdot \mathbf{n}] \, dS$, where G is the surface $36z = 4x^2 + 9y^2$ for $0 \le z \le 1$, $\mathbf{n}$ has positive $\mathbf{k}$-coefficient, and $\mathbf{F} = y\mathbf{i} + 3x\mathbf{j} - xyz\mathbf{k}$

14. $\iint_G [(\nabla \times \mathbf{F}) \cdot \mathbf{n}] \, dS$, where G is the cone $z^2 = x^2 + y^2$ for

$0 \le z \le 3$, $\mathbf{n}$ has negative $\mathbf{k}$-coefficient, and $\mathbf{F} = -yz\mathbf{i} + xz\mathbf{j} - [z/(1 + x^2 + y^2)]\mathbf{k}$

15. $\iint_G [(\nabla \times \mathbf{F}) \cdot \mathbf{n}] \, dS$, where G is the cylinder $x^2 + y^2 = 4$ for $0 \le z \le 3$, $\mathbf{n}$ points out the convex side, and $\mathbf{F} = -yz^2\mathbf{i} + xz^2\mathbf{j} + 2xyz\mathbf{k}$

16. $\iint_G [(\nabla \times \mathbf{F}) \cdot \mathbf{n}] \, dS$, where G is the cylinder $x^2 + y^2 = a^2$ for $0 \le z \le b$, $\mathbf{n}$ points out the convex side, and $\mathbf{F} = xz\mathbf{i} + (4 + z^2)\mathbf{j} + xe^{yz}\mathbf{k}$

17. $\iint_G (\mathbf{F} \cdot \mathbf{n}) \, dS$, where G is the sphere $x^2 + y^2 + z^2 = a^2$, $\mathbf{n}$ is the outward normal, and $\mathbf{F} = 4x\mathbf{i} + 5y\mathbf{j} - 3z\mathbf{k}$

18. $\iint_G (\mathbf{F} \cdot \mathbf{n}) \, dS$, where G is the surface of the region bounded by the cone $z = \sqrt{x^2 + y^2}$ and the plane $z = 2$, $\mathbf{n}$ points outward, and $\mathbf{F} = xy\mathbf{i} + yz\mathbf{j} + xyz^2\mathbf{k}$

19. $\iint_G (\mathbf{F} \cdot \mathbf{n}) \, dS$, where G is the surface of the region bounded by $z = 9 - x^2 - y^2$ and $z = x^2 + y^2 - 9$, $\mathbf{n}$ points inward, and $\mathbf{F} = (xy + 3x)\mathbf{i} + (y - 2xyz^2)\mathbf{j} + (z + 4xyz^3)\mathbf{k}$

20. $\iint_G (\mathbf{F} \cdot \mathbf{n}) \, dS$, where G is the surface of the region bounded by $y = x^2 + z^2 - 2$ and $y = -1$, $\mathbf{n}$ points inward, and $\mathbf{F} = (x^2y - 3xz)\mathbf{i} + (xyz + 3x^2y)\mathbf{j} + (x^3y^2z + z^3)\mathbf{k}$

21. $\iint_G [(\nabla \times \mathbf{F}) \cdot \mathbf{n}] \, dS$, where G is the cone $4x^2 = y^2 + 4z^2$ for $-1 \le x \le 2$, $\mathbf{n}$ points out the convex side, and $\mathbf{F} = x^3y^2\mathbf{i} + xyz^2\mathbf{j} + x^3y\mathbf{k}$

22. $\iint_G [(\nabla \times \mathbf{F}) \cdot \mathbf{n}] \, dS$, where G is the portion of the ellipsoid $x^2/4 + y^2 + z^2/9 = 5$ for $y \le 2$, $\mathbf{n}$ points out the convex side, and $\mathbf{F} = xy^2z^2\mathbf{i} + e^{xyz}\mathbf{j} + 4xy^2\mathbf{k}$

23. $\iint_G [(\nabla \times \mathbf{F}) \cdot \mathbf{n}] \, dS$, where G is the portion of the cylinder $x^2 + z^2 - 2x + 4z = 4$ for $-2 \le y \le 3$, $\mathbf{n}$ points out the concave side, and $\mathbf{F} = y^3z\mathbf{i} + 2xy\mathbf{j} - xy^2\mathbf{k}$

24. Let $\mathbf{F}$ be a continuously differentiable vector field on a surface G in space.
   (a) Show that if $\mathbf{F}$ is normal to the unit tangent vector $\mathbf{t}$ at each point on ∂G, then $\iint_G [(\mathbf{Curl\ F}) \cdot \mathbf{n}] \, dS = 0$.
   (b) Show that if the vector **curl F** is tangent to G at each point of G, then $\oint_{\partial G} (\mathbf{F} \cdot \mathbf{t}) \, ds = 0$.

25. A vector field $\mathbf{F}$ in a region of space is *irrotational* if **curl** $\mathbf{F} = \mathbf{0}$ throughout the region. Show that the circulation of an

irrotational vector field about every piecewise-smooth closed path bounding a suitable surface G lying within the region is zero.

26. (a) Show that $\nabla \cdot (\textbf{Curl F}) = 0$ for a twice continuously differentiable vector field in space.
   (b) Show that $\nabla \times (\nabla f) = \textbf{0}$ for a twice continuously differentiable function of three variables.

27. Argue as best you can from Stokes' theorem that if G is a surface that is the entire boundary of a suitable three-dimensional region, then

$$\iint_G [(\nabla \times \textbf{F}) \cdot \textbf{n}] \, dS = 0$$

for every continuously differentiable vector field $\textbf{F}$.

28. Let $\textbf{F}$ be a twice continuously differentiable vector field and let G be a three-dimensional region in space where the divergence theorem can be applied. Show that

$$\iint_{\partial G} [(\textbf{Curl F}) \cdot \textbf{n}] \, dS = \iiint_G \nabla \cdot (\textbf{Curl F}) \, dx \, dy \, dz.$$

29. Use the results of Exercises 27 and 28 to show that for $\textbf{F}$ and G as described in Exercise 28,

$$\iiint_G \nabla \cdot (\textbf{Curl F}) \, dx \, dy \, dz = 0.$$

30. Argue as best you can from Exercise 29, with no computation, that $\nabla \cdot (\nabla \times \textbf{F}) = 0$ for every twice continuously differentiable vector field $\textbf{F}$. You were asked to show this by computation in Exercise 26(a).

SUPPLEMENTARY EXERCISES FOR CHAPTER 16

1. Let $P(x, y)$ and $Q(x, y)$ have continuous partial derivatives in a region without any "holes" in it. State a criterion for $P(x, y) \, dx + Q(x, y) \, dy$ to be an exact differential.

2. Test whether

$$(y^2 e^x + 3x^2 y) \, dx + (2y e^x + x^3 + \sin y) \, dy$$

is an exact differential, and, if it is, find $H(x, y)$ such that the differential is dH.

3. Find $H(x, y)$ such that dH is

$$(2x \sin y - 3x^2 y) \, dx + (x^2 \cos y - x^3 + 4y^2) \, dy.$$

4. Find c and k such that the differential

$$(y^2 + kxy) \, dx + (3x^2 + cxy) \, dy$$

is exact.

5. If possible, find $H(x, y, z)$ such that

$$dH = (2xyz - 3x^2 y) \, dx + (x^2 z - x^3 + 2y) \, dy \\ + (x^2 y + 2y^2 z) \, dz.$$

6. Find the general solution of the differential equation

$$\frac{dy}{dx} = \frac{3 - 2xy}{x^2 + 1}.$$

7. Find the implicit solution of the differential equation $dy/dx = (2 - 3x^2 y)/(x^3 + 2y)$ that contains the point $(-1, 4)$.

8. Let $P(x, y)$ and $Q(x, y)$ have continuous first partial derivatives in the entire plane. If both $P \, dx + Q \, dy$ and $P \, dx - Q \, dy$ are exact differential forms, what can be said concerning P and Q?

9. A wire of variable density $\sigma(x, y) = 2x$ covers the curve

$$x = t^2 - 1, \qquad y = 3t + 1 \qquad \text{for } 1 \le t \le 2.$$

Express as an integral the moment of the wire about the y-axis.

10. Find the total curvature of the arc of the graph of $y = \sin x$ for $0 \le x \le \pi$.

11. Evaluate $\int_\gamma x \, ds$ if γ is the plane curve given by $x = 8t$, $y = 3t^2$ for $0 \le t \le 1$.

12. Let γ_1 have as path set the line segment from $(0, 0)$ to $(1, -2)$ and let γ_2 have as path set the line segment from $(1, -2)$ to $(1, 0)$. Find $\int_{\gamma_1 + \gamma_2} xy \, ds$.

13. Let γ_1 have as path set the line segment from $(0, 0)$ to $(1, -2)$ and let γ_2 have as path set the line segment from $(1, -2)$ to $(1, 0)$. Find $\int_{\gamma_1 + \gamma_2} xy \, ds$.

14. Let γ be given by $x = \cos t$, $y = \sin t$ for $0 \le t \le 2\pi$. Find $\int_{-\gamma} [(x^2 + y) \, dx - (y^3 + y) \, dy]$.

15. If γ is the curve $x = \sin t$, $y = 2 \cos t$ for $0 \le t \le \pi/4$, and $\textbf{F}(x, y)$ is the vector field $xy\textbf{i} - (x/y)\textbf{j}$, find $\int_\gamma \textbf{F} \cdot d\textbf{r}$.

16. Consider the force field in space $\textbf{F}(x, y, z) = xy\textbf{i} + (z/x)\textbf{j} + y^2 z\textbf{k}$. Find the work done by the field in moving a body along the curve $x = t$, $y = t$, $z = t - 1$, from $t = 1$ to $t = 2$.

17. Let $\textbf{F}(x, y) = (3x^2 - 2xy)\textbf{i} + (4y - x^2)\textbf{j}$ and let

$$\gamma : x = t^2 - 2t, \qquad y = 3t - 2 \qquad \text{for } 0 \le t \le 1.$$

Find $\int_\gamma \textbf{F} \cdot d\textbf{r}$.

18. Find the work done by the force field $\textbf{F}(x, y) = xy\textbf{i} - y\textbf{j}$ acting on a body moving from $(0, 0)$ to $(1, 1)$ along the parabola $y = x^2$.

19. Let the flux field of a flow in space be

$$\textbf{F}(x, y, z) = (2xyz + x - y)\textbf{i} + (x^2 z + y)\textbf{j} + (x^2 y - e^z)\textbf{k}.$$

Find the circulation of the flow around the loop

$$\lambda : x = a \cos t, \quad y = b \sin t, \quad z = c \sin t \quad 0 \le t \le 2\pi.$$

20. If the flux field of a fluid is $xy\textbf{i} + 2y\textbf{j}$, find the circulation of the fluid about the unit circle $x^2 + y^2 = 1$ in the counterclockwise direction. [*Hint:* Take the parametrization $x = \cos t$, $y = \sin t$ for $0 \le t \le 2\pi$.]

21. Let γ be given by $x = 2 \cos \pi t$, $y = 3 \sin \pi t$ for $0 \leq t \leq 1$. Find $-\gamma$.

22. Let γ be given by $x = \cos t$, $y = \sin t$ for $0 \leq t \leq 2\pi$. Find

$$\int_{-\gamma} [(x^2 + y) \, dx - (y^3 + y) \, dy].$$

23. The potential energy of the conservative force field

$$\mathbf{F}(x, y) = (2xy^3 - e^y)\mathbf{i} + (3x^2y^2 - xe^y)\mathbf{j}$$

at $(3, 0)$ is 15. Find the potential function associated with $\mathbf{F}$.

24. Consider the conservative force field

$$\mathbf{F}(x, y) = (x^2 + y^2)\mathbf{i} + 2xy\mathbf{j}.$$

Find the loss in potential energy of the field from $(3, 1)$ to $(1, -1)$.

25. Note that

$$\mathbf{F}(x, y) = (x^2 - 4xy)\mathbf{i} + (y^2 - 2x^2)\mathbf{j}$$

is a conservative force field. Find $\int_\gamma \mathbf{F} \cdot d\mathbf{r}$ where γ is any smooth curve from $(-1, 1)$ to $(1, -2)$.

26. Let γ be a smooth curve from $(-1, 2)$ to $(3, 4)$. When is $\int_\gamma (P(x, y) \, dx + Q(x, y) \, dy)$ independent of the choice of path joining these points?

27. Let the flux field for a flow in the x,y-plane be $\mathbf{F} = x^3\mathbf{i} + y^3\mathbf{j}$. Let λ be the circle $x^2 + y^2 = 9$ traced once counterclockwise. Find $\oint_\lambda (\mathbf{F} \cdot \mathbf{n}) \, ds$.

28. Find all functions $Q(x, y)$ such that $\mathbf{F} = x^2e^y\mathbf{i} + Q(x, y)\mathbf{j}$ is the flux field of an incompressible flow in the plane.

29. Let G be the ball $x^2 + y^2 + z^2 \leq a^2$. Let $\mathbf{F} = 2xyz^2\mathbf{i} - x^3y^2\mathbf{j} + z^3\mathbf{k}$. Find $\iint_{\partial G} (\mathbf{F} \cdot \mathbf{n}) \, dS$, where $\mathbf{n}$ is the outward unit normal vector.

30. State the vector form of the divergence theorem and explain its interpretation for the flux vector $\mathbf{F}$ of a flow.

31. Let $\mathbf{F}$ be the vector field

$$\mathbf{F} = x^3\mathbf{i} + y^3\mathbf{j} + z^2\mathbf{k}.$$

Use the divergence theorem to compute $\iint_{\partial G} (\mathbf{F} \cdot \mathbf{n}) \, dS$, where G is the region bounded by $z = x^2 + y^2$ and $z = 4$.

32. Let a ball of mass density 1 be bounded by the spherical surface with equation

$$x^2 + y^2 + z^2 = a^2.$$

Show that the moment of inertia of the ball about the z-axis is equal to

$$\iint_G \left[\left(\frac{x^3}{3}\mathbf{i} + \frac{y^3}{3}\mathbf{j} + 0\mathbf{k} \right) \cdot \mathbf{n} \right] dS,$$

where $\mathbf{n}$ is the outward normal to the sphere.

33. Let G be a region in space where the divergence theorem applies. Let $\mathbf{F} = 3x\mathbf{i} - 2y\mathbf{j} + 4z\mathbf{k}$. If $\iint_{\partial G} (\mathbf{F} \cdot \mathbf{n}) \, dS = 20$, find the volume of G.

34. State Green's theorem.

35. Use Green's theorem to compute

$$\oint_\lambda [xy \, dx + (x^3 + y^3) \, dy],$$

where λ is the boundary of the region G bounded by $y = x^2$ and $y = 4$.

36. Give a vector statement of Green's theorem.

37. Use Green's theorem to find the circulation of the flow with flux vector

$$\mathbf{F} = x\mathbf{i} + 3xy\mathbf{j}$$

about the boundary of the plane region bounded by

$$x = y^2 \quad \text{and} \quad y = x - 2.$$

38. Find all functions $P(x, y)$ such that

$$\mathbf{F} = P(x, y)\mathbf{i} + (x^2y^3 - 2x)\mathbf{j}$$

is the flux field of an irrotational flow in the plane.

39. Let $\mathbf{F} = P\mathbf{i} + Q\mathbf{j}$ be the continuously differentiable flux vector for a flow in the plane. Give the conditions for $\mathbf{F}$ to be (a) irrotational, (b) incompressible.

40. Let G be a region in the x,y-plane where Green's theorem applies. If the counterclockwise integral

$$\oint_{\partial G} (3y \, dx - 5x \, dy) = -28,$$

find the area of G.

41. Use Green's theorem to find the counterclockwise integral $\oint_{\partial G} (\mathbf{F} \cdot \mathbf{t}) \, ds$ if G is the region $0 \leq x \leq 2$, $0 \leq y \leq 1$, and $\mathbf{F} = -3xy\mathbf{i} + 4x^2y^3\mathbf{j}$.

42. Let $P(x, y) = (y^2 + 4x)/(y + 1)$ and let $Q(x, y) = (3^x + y^2)/(x + 2)$. Use Green's theorem to find

$$\int_0^1 \int_0^1 \left[\frac{\partial Q}{\partial x} - \frac{\partial P}{\partial y} \right] dx \, dy.$$

43. Let $\mathbf{F} = P(x)\mathbf{i} + Q(y)\mathbf{j}$ be a flux field for a flow in the plane. Under what conditions on P and Q is the flow (a) incompressible, (b) irrotational?

44. Let G be a region in the plane where the two-dimensional divergence theorem applies. Let $\mathbf{F} = 2xy\mathbf{i} + x^3\mathbf{j}$. If

$$\iint_G (\nabla \cdot F) \, dx \, dy = 12,$$

find the moment of G about the x-axis.

45. Let G be a region in the plane and let $\mathbf{F}(x, y) = \nabla H(x, y)$ be a continuously differentiable vector field. Find

$$\iint_G (\text{Curl } \mathbf{F}) \, dx \, dy.$$

46. (a) Give the condition for a force field $\mathbf{F} = P\mathbf{i} + Q\mathbf{j}$ in the plane to be conservative.
    (b) Show that the force field $\mathbf{F} = y^2\mathbf{i} + 2xy\mathbf{j}$ is conservative, and find a potential function u.
    (c) Find the work done by the force field in part (b) in moving a body from $(-1, 2)$ to $(3, -1)$.

47. Find **curl** $\mathbf{F}$ if $F = x\mathbf{i} + yz\mathbf{j} + xyz\mathbf{k}$.

48. Let $\mathbf{F} = (xy/z)\mathbf{i} - (x^2/y)\mathbf{j} + (yz/x^2)\mathbf{k}$. Find **curl** $\mathbf{F}$.

49. Let $\mathbf{F} = xyz^2\mathbf{i} + 2yz\mathbf{j} - x^3\mathbf{k}$.
    (a) Find **curl** $\mathbf{F}$.
    (b) Find the divergence of $\mathbf{F}$.

50. Let G be the portion of the cone $z = \sqrt{x^2 + y^2}$ for $1 \leq z \leq 4$, and let $\mathbf{n}$ point out the concave side of the cone. Let $\mathbf{F} = -y^3z\mathbf{i} + x^3z\mathbf{j} + e^{xyz}\mathbf{k}$. Find $\iint_G [(\text{Curl } \mathbf{F}) \cdot \mathbf{n}] \, dS$.

51. Let G be the cylindrical surface $x^2 + y^2 = 9$ for $0 \le z \le 2$, and let $\mathbf{F} = 4yz\mathbf{i} + 2xz\mathbf{j} - xy^2\mathbf{k}$. Use Stokes' theorem to find

$$\oint_{\partial G} [(\nabla \times \mathbf{F}) \cdot \mathbf{n}] \, dS,$$

where $\mathbf{n}$ points out the convex side of the cylinder.

52. Let $\mathbf{F} = 4\mathbf{i} + xz\mathbf{j} - xy\mathbf{k}$. Use Stokes' theorem to find $\iint_G [(\mathbf{Curl\ F}) \cdot \mathbf{n}] \, dS$, where G is the surface $x = y^2 + z^2$ for $0 \le x \le 9$, while vectors $\mathbf{n}$ are chosen so that their **i**-components are positive.

53. Let G be the surface $z = 4 + \sqrt{9 - x^2 - y^2}$ with normal vectors $\mathbf{n}$ having positive **k**-component. Let G' be the portion of the sphere $x^2 + y^2 + z^2 = 25$, where $z \le 4$, with outward normal vectors $\mathbf{n}$. Let $\mathbf{F}$ be a vector field. What relation holds between

$$\iint_G [(\mathbf{Curl\ F}) \cdot \mathbf{n}] \, dS$$

and

$$\iint_{G'} [(\mathbf{Curl\ F}) \cdot \mathbf{n}] \, dS?$$

Why?

54. Use Stokes' theorem to find $\iint_G [(\mathbf{Curl\ F}) \cdot \mathbf{n}] \, dS$ where $\mathbf{F} = xy\mathbf{i} + y\mathbf{j} + yz\mathbf{k}$ and G is the surface of the cube with no top and having diagonally opposite vertices at $(0, 0, 0)$ and $(1, 1, 1)$. Let $\mathbf{n}$ be the outward normal.

55. Let P, Q, R, and S be four functions of the four variables w, x, y, and z, all having continuous partial derivatives. State the criteria for $P \, dw + Q \, dx + R \, dy + S \, dz$ to be an exact differential in some region of four-space having no holes in it.

56. Using subscript notation, let $\mathbf{x} = (x_1, \ldots, x_n)$ and let F_1, $F_2, \ldots, F_n$ be n functions of $\mathbf{x}$ having continuous partial derivatives. Repeat Exercise 55, giving criteria for $F_1 dx_1 + F_2 dx_2 + \cdots + F_n dx_n$ to be exact in a suitable region. (It is much easier in this notation.)

57. Consider the differential form

$$\frac{y}{x^2 + y^2} \, dx - \frac{x}{x^2 + y^2} \, dy, \qquad (x, y) \ne (0, 0).$$

(a) Show that this differential form satisfies the criterion $\partial P/\partial y = \partial Q/\partial x$ for all $(x, y) \ne (0, 0)$.

(b) Show that, if the given differential form were dH for $H(x, y)$ and all $(x, y) \ne (0, 0)$, then we would have to have

$$H(x, y) = \tan^{-1}\left(\frac{x}{y}\right) + A \qquad \text{for } y > 0 \text{ and some } A$$

and

$$H(x, y) = \tan^{-1}\left(\frac{x}{y}\right) + B \qquad \text{for } y < 0 \text{ and some } B.$$

Show that there are no choices for A and B and no way of defining $H(x, y)$ on the positive and negative portions

of the x-axis that would give a differentiable function $H(x, y)$ for all $(x, y) \ne (0, 0)$. (This illustrates that the region should have no holes in it in order for the partial derivative criterion for exactness to guarantee exactness in the entire region.) [*Hint:* Show that we would have to have $B = A + \pi$, and $H(x, 0) = A + \pi/2$ for $x > 0$. Then show that it becomes impossible to define $H(x, 0)$ for $x < 0$ to make H even continuous on the negative x-axis, to say nothing of differentiable.]

58. Let G be a suitable surface in space and let $P(x, y, z)$, $Q(x, y, z)$, and $R(x, y, z)$ be continuously differentiable functions. Stokes' theorem is often stated in the form

$$\int_{\partial G} (P \, dx + Q \, dy + R \, dz)$$

$$= \iint_G \left[\left(\frac{\partial R}{\partial y} - \frac{\partial Q}{\partial z} \right) dy \, dz + \left(\frac{\partial P}{\partial z} - \frac{\partial R}{\partial x} \right) dz \, dx \right.$$

$$\left. + \left(\frac{\partial Q}{\partial x} - \frac{\partial P}{\partial y} \right) dx \, dy \right].$$

Convince yourself that this is equivalent to the statement

$$\int_{\partial G} (\mathbf{F} \cdot \mathbf{t}) \, ds = \iint_G [(\nabla \times \mathbf{F}) \cdot \mathbf{n}] \, dS$$

for $\mathbf{F} = P\mathbf{i} + Q\mathbf{j} + R\mathbf{k}$.

Exercises 59–65 use the notation of the ∧-product introduced in the last exercises in Chapter 15. Review that material before proceeding.

Table 16.3 indicates what is meant by a *differential form* ω, the *order* of the form, and its *exterior derivative* $d\omega$. Here P, Q, and R are differentiable functions of two or three variables (we give the table for three variables). For $P(x, y, z)$,

$$dP = \frac{\partial P}{\partial x} \, dx + \frac{\partial P}{\partial y} \, dy + \frac{\partial P}{\partial z} \, dz$$

as usual.

59. Show that if $\omega = P \, dx \wedge dy \wedge dz$, then $d\omega = 0$.

60. Note that if ω has order r, then $d\omega$ has order $r + 1$. Compute $d\omega$, simplifying as much as possible.
(a) $\omega = xy \, dx - x^2 \, dy$
(b) $\omega = xz \, dx \wedge dy - yz^3 \, dx \wedge dz$

All the main theorems in this chapter can be expressed in the following form, as Exercises 61–65 ask you to show.

GENERALIZED STOKES' THEOREM

If ω is a differential form of order r with continuously differentiable coefficients P, Q, and so on, and if G is a suitable $(r + 1)$-dimensional region, then

$$\int_{\partial G} \omega = \int_G d\omega. \tag{1}$$

In Eq. (1), a single integral sign is used for each integral, rather than double or triple integral signs as we have used in the past. The integrals of differential forms are defined so that

$$\int_G P(x, y) \, dx \wedge dy = \iint_G P(x, y) \, dx \, dy$$

$$= -\int_G P(x, y) \, dy \wedge dx,$$

$$\int_G P(x, y, z) \, dx \wedge dy \wedge dz = \iiint_G P(x, y, z) \, dx \, dy \, dz$$

$$= -\int_G P(x, y, z) \, dz \wedge dy \wedge dx,$$

and so on.

61. If G is a one-dimensional region consisting of a curve joining point A to point B, then ∂G is the symbolic expression $B - A$. The integral of a function f (a form of order 0) over ∂G is defined to be $f(B) - f(A)$. Show that if $\omega = f(x)$ and

G is the line segment $[a, b]$ from a to b, then Eq. (1) reduces to the fundamental theorem of calculus.

62. State Eq. (1) for the special case of $\omega = f(x, y, z)$ and G a curve γ joining $A = (a_1, a_2, a_3)$ to $B = (b_1, b_2, b_3)$ in space. (See Exercise 61.)

63. State Eq. (1) for $\omega = P(x, y) \, dx + Q(x, y) \, dy$ in the plane and G a plane region. Simplify the statement as much as possible. What familiar theorem do you obtain?

64. State Eq. (1) for $\omega = P \, dx + Q \, dy + R \, dz$ and G a surface in space. Simplify the statement as much as possible. What theorem do you obtain?

65. (a) State Eq. (1) for

$$\omega = P \, dy \wedge dz + Q \, dz \wedge dx + R \, dx \wedge dy,$$

simplifying as much as possible.

(b) Convince yourself that the statement obtained in part (a) is equivalent to the divergence theorem.

TABLE 16.3	Differential forms ω	
Order of ω	ω	$d\omega$
0	$P(x, y, z)$	$dP = \dfrac{\partial P}{\partial x} \, dx + \dfrac{\partial P}{\partial y} \, dy + \dfrac{\partial P}{\partial z} \, dz$
1	$P \, dx + Q \, dy + R \, dz$	$dP \wedge dx + dQ \wedge dy + dR \wedge dz$
2	$P \, dy \wedge dz + Q \, dz \wedge dx + R \, dx \wedge dy$	$dP \wedge dy \wedge dz + dQ \wedge dz \wedge dx + dR \wedge dx \wedge dy$
3	$P \, dx \wedge dy \wedge dz$	$dP \wedge dx \wedge dy \wedge dz$

LINEAR DIFFERENTIAL EQUATIONS

17

A linear differential equation is one of the form

$$p_0(x) \frac{d^n y}{dx^n} + p_1(x) \frac{d^{n-1} y}{dx^{n-1}} + \cdots + p_{n-1}(x) \frac{dy}{dx} + p_n(x)y = g(x).$$

They are called *linear* because no derivative appears to a power greater than 1 and no summand contains more than one derivative. Such equations occur in many applications, such as the study of damped vibrations. The **order** of such a differential equation is the order n of the derivative of highest order that appears. In Section 17.1, we study *first-order differential* equations, where $n = 1$. Sections 17.2 and 17.3 study equations of higher order, but only in the case that all the coefficient functions $p_i(x)$ are constant, for $i = 0, 1, 2, \ldots, n$.

17.1 FIRST-ORDER LINEAR EQUATIONS

In this section, we are concerned with finding all solutions in a neighborhood of some point x_0 of the first-order *linear* differential equation

$$p_0(x)y' + p_1(x)y = q(x), \tag{1}$$

where $p_0(x)$, $p_1(x)$, and $q(x)$ are known functions defined in some neighborhood of x_0. The equation is called *linear* because each summand contains at most one factor y or y', raised to just the first power.

We will restrict our discussion to the case where the coefficient $p_0(x)$ of y' in Eq. (1) takes on only nonzero values throughout some neighborhood of x_0. We may then divide Eq. (1) by $p_0(x)$ and set $p(x) = p_1(x)/p_0(x)$ and $g(x) = q(x)/p_0(x)$ to write Eq. (1) in the simpler form

$$y' + p(x)y = g(x). \tag{2}$$

We assume that the functions $p(x)$ and $g(x)$ in Eq. (2) are continuous near x_0 so that we can integrate them. This differential equation (2) is one whose solutions can be readily found; we can actually obtain a formula for the general solution.

The trick is to find a continuous function $\mu(x)$ that is nonzero in the neighborhood and that has the property

$$\mu(x)y' + \mu(x)p(x)y = v' \tag{3}$$

for some function v. (Such a function μ is an *integrating factor*.) Upon multiplication by $\mu(x)$, Eq. (2) would then reduce to the equivalent equation

$$v' = \mu(x)g(x), \tag{4}$$

which can be solved by a single integration. It has been found that

$$\mu(x) = e^{\int p(x)\, dx} \tag{5}$$

for any choice of antiderivative of $p(x)$ is such an integrating factor. (Observe that $\mu(x)$ is continuous and $\mu(x) \neq 0$.) To see that μ is indeed an integrating factor, note that if

$$v = \mu(x) \cdot y, \tag{6}$$

then

$$
\begin{aligned}
v' = \frac{d[\mu(x)]}{dx} \cdot y + \mu(x) \cdot y' &= \frac{d[e^{\int p(x)\, dx}]}{dx} \cdot y + \mu(x) \cdot y' \\
&= e^{\int p(x)\, dx} \cdot p(x) \cdot y + \mu(x) \cdot y' \\
&= \mu(x) \cdot p(x) \cdot y + \mu(x) \cdot y',
\end{aligned}
$$

which is Eq. (3). Thus multiplying Eq. (2) by $\mu(x)$, we obtain

$$v' = (\mu(x)y)' = \mu(x)g(x), \tag{7}$$

which yields

$$\mu(x)y = \int \mu(x)g(x)\, dx. \tag{8}$$

From Eq. (8), we obtain

$$y = \frac{1}{\mu(x)} \int \mu(x)g(x)\, dx. \tag{9}$$

If we take $\int_{x_0}^{x} \mu(t)g(t)\, dt$ as a particular antiderivative of $\mu(x)g(x)$, then Eq. (9) becomes

$$y = \frac{1}{\mu(x)} \left(\int_{x_0}^{x} \mu(t)g(t)\, dt + C \right). \tag{10}$$

We have almost proved the following theorem.

THEOREM 17.1   Existence theorem for $y' + p(x)y = g(x)$

Let $p(x)$ and $g(x)$ be continuous in a neighborhood of x_0 and let y_0 be any real number. Then there exists a *unique* solution $y = f(x)$ of the differential equation $y' + p(x)y = g(x)$ such that $y(x_0) = y_0$, and this solution satisfies the differential equation throughout the neighborhood. Furthermore, the general solution of the differential equation is

$$y = \frac{1}{\mu(x)} \int \mu(x)g(x)\, dx,$$

where

$$\mu(x) = e^{\int p(x)\, dx}.$$

Proof: We have already seen in Eq. (9) that the general solution is as stated in the theorem and is valid for all x in the neighborhood. It remains only to demonstrate the existence and uniqueness of a particular solution through the point (x_0, y_0). But putting $x = x_0$ and $y = y_0$ in the form of the general solution given in Eq. (10), we obtain the relation

$$y_0 = \frac{1}{\mu(x_0)}C, \qquad (11)$$

and hence $C = y_0\, \mu(x_0)$ yields the only solution through (x_0, y_0).   □

EXAMPLE 1   We showed in Example 4 of Section 5.5 that the general solution of the differential equation $y' = ky$ is $y = Ae^{kx}$, where A is an arbitrary constant that controls $y(0)$. Obtain it again by using the theorem.

Solution   Our equation is $y' - ky = 0$, so we have $p(x) = -k$ and $g(x) = 0$. Our integrating factor is thus

$$\mu(x) = e^{\int -k\, dx} = e^{-kx}.$$

The general solution is therefore

$$y = \frac{1}{e^{-kx}} \int e^{-kx} \cdot 0 \, dx = \frac{1}{e^{-kx}}C = Ce^{kx}.$$

This coincides with our previous result.   ■

EXAMPLE 2   Find the particular solution of the differential equation

$$y' + 3xy = x,$$

which passes through the point $(0, 4)$.

Solution   Here $p(x) = 3x$ and our integrating factor is

$$\mu(x) = e^{\int 3x\, dx} = e^{3x^2/2}.$$

By Theorem 17.1, the general solution is

$$y = \frac{1}{e^{3x^2/2}} \int xe^{3x^2/2}\, dx = \frac{1}{e^{3x^2/2}}\left(\frac{1}{3}e^{3x^2/2} + C\right) = \frac{1}{3} + Ce^{-3x^2/2}.$$

Putting $y = 4$ and $x = 0$, we find that

$$4 = \frac{1}{3} + C \cdot 1,$$

so $C = \frac{11}{3}$, and the desired particular solution is

$$y = \frac{1}{3} + \frac{11}{3}e^{-3x^2/2}. \qquad \blacksquare$$

The only problem in using Theorem 17.1 to solve $y' + p(x)y = g(x)$ is that the solution contains two integrals, namely

$$\mu(x) = e^{\int p(x)\, dx} \qquad \text{and} \qquad \int \mu(x)g(x)\, dx.$$

Sometimes it is impossible to evaluate one of these integrals in terms of elementary functions, even if $p(x)$ and $g(x)$ are themselves elementary functions. For example, if $p(x) = -x$ and $g(x) = 1$, then

$$\mu(x) = e^{-x^2/2} \qquad \text{and} \qquad \int \mu(x)g(x)\, dx = \int e^{-x^2/2}\, dx.$$

This last integral cannot be evaluated in terms of elementary functions. However, we have seen how this integral can be expressed as an infinite series. (This particular integral is so important in the theory of probability that $\int_0^x e^{-t^2/2}\, dt$ has actually been tabulated for many values of x.) There are many numerical methods for estimating an integral, so the presence of integrals in the general solution of the first-order linear differential equation is not really a serious problem in practical applications. Equations (10) and (11) show that the solution of $y' + p(x) \cdot y = g(x)$ through (x_0, y_0) is

$$y = \frac{1}{\mu(x)}\left(\int_{x_0}^x \mu(t)g(t)\, dt + \mu(x_0) \cdot y_0\right). \qquad (12)$$

EXAMPLE 3 Let $y = h(x)$ be the solution of the initial value problem

$$y' - xy = 1, \qquad y(1) = 4.$$

Estimate $h(2)$ using a programmable calculator or a computer and Simpson's rule with $n = 40$.

Solution For the equation $y' - xy = 1$, we have $p(x) = -x$, so

$$\mu(x) = e^{\int -x\, dx} = e^{-x^2/2}.$$

Using Eq. (12), the solution $y = h(x)$ through $(x_0, y_0) = (1, 4)$ is

$$y = h(x) = \frac{1}{e^{-x^2/2}}\left(\int_1^x e^{-t^2/2} \cdot 1\, dt + e^{-1/2} \cdot 4\right).$$

Therefore

$$h(2) = \frac{1}{e^{-2}}\left(\int_1^2 e^{-t^2/2}\, dt + 4e^{-1/2}\right) = e^2 \int_1^2 e^{-t^2/2}\, dt + 4e^{3/2}.$$

Our calculator and Simpson's rule with $n = 40$ yield

$$h(2) = e^2 \int_1^2 e^{-t^2/2} \, dt + 4e^{5/2} \approx e^2(0.3406636182) + 4e^{3/2}$$

$$\approx 20.44393887. \qquad \blacksquare$$

The exercises give some applications of these first-order linear differential equations to electric circuits, Newton's law of cooling, and mixing problems. We close with an example of a mixing problem to serve as a model for those exercises.

EXAMPLE 4   A tank initially contains 50 gal of brine in which 5 lb of salt is dissolved. Brine containing 1 lb of salt per gallon flows into the tank at a constant rate of 4 gal/min. The concentration of the brine in the tank is kept uniform throughout the tank by stirring. While the brine is entering the tank, brine from the tank is also being drawn off at a rate of 2 gal/min. Find the amount of salt in the tank after 25 min.

Solution   We let $y = f(t)$ be the number of pounds of salt in the tank at time t. Clearly,

$$\frac{dy}{dt} = (\text{Rate salt enters the tank}) - (\text{Rate salt leaves the tank}). \qquad \textbf{(13)}$$

Since 4 gal/min of brine containing 1 lb of salt per gallon is entering the tank, we see that

$$\text{Rate salt enters the tank} = 4 \text{ lb/min.}$$

The brine is drawn off at a rate of 2 gal/min and enters at a rate of 4 gal/min, so after t min the number of gallons of brine in the tank must be $50 + 2t$. The concentration of salt per gallon in the tank is therefore

$$\frac{y}{50 + 2t} \text{ lb/gal at time } t.$$

Since 2 gal/min are being drawn off, we see that

$$\text{Rate salt leaves the tank} = 2 \cdot \frac{y}{50 + 2t} = \frac{y}{25 + t} \text{ lb/min}$$

at time t. The differential equation (13) thus becomes

$$\frac{dy}{dt} = 4 - \frac{y}{25 + t}, \qquad \text{or} \qquad \frac{dy}{dt} + \frac{1}{25 + t}y = 4.$$

We take as integrating factor

$$\mu(t) = e^{\int [1/(25+t)] \, dt} = e^{\ln(25+t)} = 25 + t.$$

From Eq. (9), we obtain the general solution

$$y = \frac{1}{25 + t} \int (25 + t)4 \, dt$$

$$= \frac{1}{25 + t}(100t + 2t^2 + C).$$

Since $y = 5$ lb when $t = 0$ min, we have

$$5 = \frac{1}{25}C, \quad \text{so} \quad C = 125.$$

Consequently,

$$y = \frac{1}{25 + t}(2t^2 + 100t + 125),$$

so when $t = 25$ min we have

$$y = \frac{1}{50}[2(25)^2 + 100(25) + 125]$$

$$= 25 + 50 + 2.5 = 77.5 \text{ lb}.$$

This gives us the answer to our problem. ∎

SUMMARY

1. The general solution of the differential equation

$$y' + p(x)y = g(x),$$

where $p(x)$ and $g(x)$ satisfy suitable conditions described in the text, can be found as follows.

STEP 1. Compute the integrating factor

$$\mu(x) = e^{\int p(x)\, dx},$$

where $\int p(x)\, dx$ is any particular antiderivative of $p(x)$.

STEP 2. Upon multiplication by $\mu(x)\, dx$, the differential equation $y' + p(x)y = g(x)$ becomes

$$d(\mu(x) \cdot y) = \mu(x)g(x)\, dx.$$

STEP 3. The solution is then

$$\mu(x) \cdot y = \int \mu(x)g(x)\, dx,$$

or

$$y = \frac{1}{\mu(x)}\int \mu(x)g(x)\, dx.$$

2. The solution of $y' + p(x) \cdot y = g(x)$ through (x_0, y_0) is given by

$$y = h(x) = \frac{1}{\mu(x)}\left(\int_{x_0}^{x} \mu(t)g(t)\, dt + \mu(x_0) \cdot y_0\right),$$

where $\mu(x) = e^{\int p(x)\, dx}$. If $\mu(x)$ can be computed, then a calculator or computer and Simpson's rule can be used to estimate $h(x_1)$.

EXERCISES 17.1

In Exercises 1–10, find the general solution of the given differential equation.

1. $y' - xy = 0$

2. $y' - 3y = 2$

3. $y' + y = 3e^x$

4. $y' + 2y = x + e^{-3x}$

5. $y' + 2y = xe^{-2x} + 3$

6. $xy' + y = 2x \sin x; x > 0$ [*Hint:* Reduce to the form of Eq. (2) by dividing by x.]

7. $y' + (\cot x)y = 3x + 1, 0 < x < \pi$

8. $y' + (\sin x)y = 3 \sin x$

9. $y' + 2xy = x$

10. $y' - 2xy = x^3$

In Exercises 11–18, find the solution of the given initial value problem.

11. $y' - 3y = x + 2, y(0) = -1$

12. $xy' - y = x^3, y(1) = 5$

13. $(1 + x^2)y' + y = 3, y(0) = 2$

14. $y' - (\cos 2x)y = \cos 2x, y(\pi/2) = 3$

15. $y' - xy = x, y(0) = -3$

16. $y' + 2xy = e^{-x^2}, y(0) = 4$

17. $(1 + \sin x)y' + (\cos x)y = \sin^2 x, y(\pi/2) = 3$

18. $x^2y' + y = 1, y(1) = 0$

19. Find the general solution of the differential equation $y' - xy = x^2$ by using Theorem 17.1 and expressing the integral in series form.

20. Find the general solution of the differential equation
$$y' + 2x(\cot x^2)y = 3, \qquad 0 < x < \sqrt{\pi},$$
by using Theorem 17.1 and expressing the integral in series form.

21. Find the plane curve through $(2, 1)$ whose slope at (x, y) is $2 + (y/x)$.

22. Find the plane curve through $(1, -1)$ whose slope at (x, y) is $3 - (y/x)$.

23. A tank initially contains 60 gal of brine in which 15 lb of salt is dissolved. Brine containing 0.5 lb of salt per gallon flows into the tank at a constant rate of 6 gal/min. The concentration of brine in the tank is kept uniform by stirring, and the brine is drawn off at a rate of 3 gal/min. Find the amount of salt in the tank after 10 min.

24. A tank initially contains 40 gal of brine containing 0.5 lb of salt per gallon. Brine containing 1 lb of salt per gallon flows into the tank at a rate of 2 gal/min, while brine is drawn off at a rate of 6 gal/min. If the concentration of the solution in the tank is kept uniform by stirring, find
   (a) the amount of salt in the tank after 5 min,
   (b) the amount of salt in the tank after 10 min.

25. If $i(t)$ is the current at time t in an electrical circuit with constant resistance R, constant inductance L, and variable electromotive force $E(t)$, then it can be shown that
$$L\frac{di}{dt} + Ri = E(t).$$
   (a) Let the current at time $t = 0$ be i_0 and let $E(t)$ be a continuous function of the time t. Find an expression for i at time t.
   (b) Show that if E is constant, then for large values of t we have $i \approx E/R$ (so that Ohm's law is approximately true after a long time).
   (c) Describe the current after a long period of time if E diminishes exponentially, that is, if $E(t) = E_0 e^{-kt}$, where $E_0 = E(0)$ and $k > 0$.
   (d) Describe the behavior of the current as $t \to \infty$ if E is abruptly cut off at time t_0, so that $E(t) = 0$ for $t > t_0$.

26. According to Newton's law of cooling, the rate at which a body in a medium changes temperature is proportional to the difference between its temperature and the temperature of the medium.
   (a) Assuming that the temperature of the medium remains constant at a degrees and that the temperature T_0 of the body at time $t = 0$ is higher than that of the medium, express the temperature T of the body as a function of the time t for $t > 0$. (Let the constant of proportionality in Newton's law be $-k$.)
   (b) For $k > 0$, what is the approximate temperature of the body as $t \to \infty$?

27. If n is a constant different from 0 or 1, then the equation $y' + p(x)y = g(x)y^n$ is known as *Bernoulli's equation*. (For both $n = 0$ and $n = 1$, the equation is linear and can be solved as described in this section.)
   (a) Show that the substitution $v = y^{1-n}$ enables us to reduce the solution of Bernoulli's equation to a differential equation for v that is linear.
   (b) Use the result of part (a) and Theorem 17.1 to solve the differential equation
$$y' - 2xy = 5xy^3.$$

28. Use Exercise 27 to solve the initial value problem
$$y' + \frac{2}{x}y = \frac{1}{y}, \qquad y(1) = -2.$$

In Exercises 29–34, use a programmable calculator or computer and Simpson's rule to estimate $h(x_1)$ if $y = h(x)$ is the solution of the given initial value problem.

29. $y' + 2xy = \sin x, y(0) = 1, x_1 = 0.5$

30. $y' - x^2y = 1, y(1) = 4, x_1 = 2$

31. $y' - (\sin x)y = x^2, y(0) = 3, x_1 = -\dfrac{\pi}{4}$

32. $y' + \sqrt{x}\, y = x, y(4) = 2, x_1 = 3$

33. $y' - \dfrac{1}{2\sqrt{x}}y = x^3, y(4) = 0, x_1 = 4.5$

34. $y' + \dfrac{1}{1+x^2}y = x^2, y(1) = -1, x_1 = 2$

17.2 HOMOGENEOUS LINEAR EQUATIONS WITH CONSTANT COEFFICIENTS

We now turn to differential equations of order greater than 1. Recall that the **order** of a differential equation is the order of the highest-order derivative it contains. In this section and the next, we study linear differential equations with constant coefficients. Such equations occur frequently in applications.

The Existence Theorem

The *general linear differential equation of order n* is

$$p_0(x)y^{(n)} + p_1(x)y^{(n-1)} + \cdots + p_{n-1}(x)y' + p_n(x)y = g(x).$$

The equation is called *linear* because each term contains at most a single factor that is y or a derivative of y to only the first power. Theorem 17.2 describes the existence of solutions in the neighborhood of a point x_0 throughout which $p_0(x) \neq 0$. Where $p_0(x) \neq 0$, we may divide the equation by $p_0(x)$ and obtain 1 as coefficient of $y^{(n)}$. Thus it is no loss of generality at such a point x_0 to suppose that the equation has the form

$$y^{(n)} + p_1(x)y^{(n-1)} + \cdots + p_{n-1}(x)y' + p_n(x)y = g(x). \tag{1}$$

We state the main existence theorem without proof.

THEOREM 17.2   Existence theorem for the general linear equation

If $p_1(x), \ldots, p_n(x), g(x)$ are continuous in a neighborhood of x_0 and if $a_0, a_1, \ldots, a_{n-1}$ are any constants, then there is a *unique* solution $y = f(x)$ of Eq. (1) that is valid throughout the neighborhood and has the property

$$y(x_0) = a_0, \quad y'(x_0) = a_1, \quad \ldots, \quad y^{(n-1)}(x_0) = a_{n-1}. \qquad \square$$

The conditions

$$y(x_0) = a_0, \quad y'(x_0) = a_1, \quad \ldots, \quad y^{(n-1)}(x_0) = a_{n-1}$$

for the solution whose existence is asserted in Theorem 17.2 are called **initial conditions.** They all involve the values of y and its derivatives at the same *initial point* x_0. Equation (1) and these initial conditions constitute an **initial value problem.** Theorem 17.2 tells us that this initial value problem has a unique solution, provided the coefficients $p_i(x)$ and $g(x)$ in Eq. (1) are continuous. If $g(x) = 0$, then Eq. (1) is **homogeneous.**

In the case to be considered in the remainder of this section, the coefficient functions $p_1(x), \ldots, p_n(x)$ are constant functions and $g(x) = 0$. Then Eq. (1) takes the form

$$y^{(n)} + b_1 y^{(n-1)} + \cdots + b_{n-1}y' + b_n y = 0. \tag{2}$$

Equation (2) is a linear *homogeneous* differential equation with *constant coefficients.* In this section, we will see how the general solution of Eq. (2) can be obtained in terms of elementary functions by very simple algebraic means.

Polynomials in the Operator D

We may write Eq. (2) in the form

$$D^n y + b_1 D^{n-1} y + \cdots + b_{n-1} Dy + b_n y = 0, \tag{3}$$

where $Dy = y'$, $D^2 y = y''$, and so on. Proceeding purely formally, it is natural to factor out y in Eq. (3) and write the equation in the form

$$(D^n + b_1 D^{n-1} + \cdots + b_{n-1} D + b_n)y = 0, \tag{4}$$

or, more briefly,

$$P(D)y = 0, \tag{5}$$

where $P(D)$ is the polynomial $D^n + b_1 D^{n-1} + \cdots + b_{n-1} D + b_n$ in D.

Suppose that the polynomial $P(D)$ factors (in the sense of polynomial factorization) so that $P(D) = Q_1(D)Q_2(D)$ for polynomials $Q_1(D)$ and $Q_2(D)$ in D. It is not difficult to show that

$$P(D)y = Q_1(D)(Q_2(D)y), \tag{6}$$

where $y = f(x)$ has derivatives of all orders less than or equal to n. A careful proof of Eq. (6) can be given using mathematical induction; you will probably find it sufficiently convincing if we compute a special case to illustrate what we mean.

EXAMPLE 1   Illustrate that $D^2 - 3D + 2 = (D - 1)(D - 2)$, that is, that these polynomial operators have the same effect on any twice-differentiable function $y = f(x)$.

Solution   Computing, we find that for a twice-differentiable function $y = f(x)$, we have

$$
\begin{aligned}
(D - 1)((D - 2)y) &= (D - 1)(Dy - 2y) \\
&= D(Dy - 2y) - 1(Dy - 2y) \\
&= D^2 y - D(2y) - Dy + 2y \\
&= D^2 y - 2Dy - Dy + 2y \\
&= D^2 y - 3Dy + 2y \\
&= (D^2 - 3D + 2)y.
\end{aligned}
$$ ■

The result in Eq. (6) can easily be extended to more than two factors. Since polynomial multiplication is commutative (that is, it does not depend on the order of multiplication), we see at once from Eq. (6) that

$$Q_1(D)(Q_2(D)y) = Q_2(D)(Q_1(D)y) \tag{7}$$

for polynomials $Q_1(D)$ and $Q_2(D)$. Equation (7) can also be extended to any number of factors in any order.

Case 1: $P(D)$ Factors into Distinct Linear Factors

We first consider a differential equation of the form

$$P(D)y = 0, \tag{8}$$

where $P(D)$ is a product of *distinct* linear factors. In this case, we have

$$P(D) = (D - r_1)(D - r_2) \cdots (D - r_n), \tag{9}$$

where $r_i \neq r_j$ for $i \neq j$. Such an equation can be solved by repeated application of Theorem 17.1. The technique is best illustrated by an example.

EXAMPLE 2   Solve the homogeneous equation $y'' - 3y' + 2 = 0$.

Solution   Our equation can be written in the form

$$(D^2 - 3D + 2)y = (D - 1)(D - 2)y = 0. \tag{10}$$

If we let $u = (D - 2)y$, then we must have

$$(D - 1)u = 0. \tag{11}$$

Equation (11) is a linear first-order equation for u, and from Section 17.1 we have

$$u = \frac{1}{e^{-x}} \int e^{-x} \cdot 0 \, dx = C_1 e^x.$$

We then solve the equation

$$(D - 2)y = u = C_1 e^x$$

and obtain

$$y = \frac{1}{e^{-2x}} \int e^{-2x}(C_1 e^x) \, dx$$

$$= \frac{1}{e^{-2x}}(-C_1 e^{-x} + C_2)$$

$$= -C_1 e^x + C_2 e^{2x}.$$

If we write our linear factors in the reverse order so that Eq. (10) becomes

$$(D - 2)(D - 1)y = 0$$

and make the substitution $(D - 1)y = v$, we obtain first the e^{2x} part of the solution, namely $v = C_1 e^{2x}$. The equation $(D - 1)y = v$ then yields the same solution $y = C_1 e^{2x} + C_2 e^x$.   ■

It is clear that if $D - r$ is a factor of $P(D)$, then some solutions of $P(D)y = 0$ are given by $y = Ce^{rx}$, since

$$(D - r)(Ce^{rx}) = rCe^{rx} - rCe^{rx} = 0.$$

The argument in Example 2 can obviously be extended to give the following theorem.

THEOREM 17.3   Characteristic equation with distinct roots

Let $P(D) = (D - r_1)(D - r_2) \cdots (D - r_n)$, where $r_i \neq r_j$ for $i \neq j$. Then the general solution of the differential equation $P(D)y = 0$ is

$$y = C_1 e^{r_1 x} + C_2 e^{r_2 x} + \cdots + C_n e^{r_n x}. \qquad \square$$

EXAMPLE 3   Solve the differential equation

$$3y''' - 2y'' - y' = 0.$$

Solution   We write our equation in the form

$$(3D^3 - 2D^2 - D)y = D(3D + 1)(D - 1)y = 0.$$

While Theorem 17.3 treats the case where the coefficient of D^n is 1, the solutions of our differential equation remain the same if we divide through by 3. We see that the numbers r_i in Theorem 17.3 can be characterized as solutions of the polynomial equation

$$P(r) = 0, \tag{12}$$

whether or not the coefficient of D^n is 1. This polynomial equation (12) is the **auxiliary** or **characteristic equation** of the differential equation $P(D)y = 0$. Our characteristic equation in this example is

$$3r^3 - 2r^2 - r = r(3r + 1)(r - 1) = 0$$

and has as solutions $r_1 = 0$, $r_2 = -\frac{1}{3}$, and $r_3 = 1$. Thus our general solution is

$$y = C_1 e^{0x} + C_2 e^{-x/3} + C_3 e^x = C_1 + C_2 e^{-x/3} + C_3 e^x. \qquad \blacksquare$$

EXAMPLE 4   Solve the initial value problem

$$y''' - y'' - 6y' = 0, \qquad y(0) = 1, \qquad y'(0) = -2, \qquad y''(0) = 0.$$

Solution   The characteristic equation of the differential equation is

$$r^3 - r^2 - 6r = 0, \qquad \text{or} \qquad r(r - 3)(r + 2) = 0.$$

This characteristic equation has solutions $r = -2, 0, 3$. The general solution of the differential equation is therefore

$$y = C_1 e^{-2x} + C_2 + C_3 e^{3x}.$$

To make use of the initial conditions involving y' and y'', we differentiate this general solution and obtain

$$y' = -2C_1 e^{-2x} + 3C_3 e^{3x} \qquad \text{and} \qquad y'' = 4C_1 e^{-2x} + 9C_3 e^{3x}.$$

The initial conditions then give these equations in C_1, C_2, and C_3:

$$
\begin{aligned}
C_1 + C_2 + C_3 &= 1, & y(0) &= 1 \\
-2C_1 \phantom{{}+ C_2} + 3C_3 &= -2, & y'(0) &= -2 \\
4C_1 \phantom{{}+ C_2} + 9C_3 &= 0. & y''(0) &= 0
\end{aligned}
$$

Solving the last two equations for C_1 and C_3, we have

$$
\begin{aligned}
-4C_1 + 6C_3 &= -4 \\
4C_1 + 9C_3 &= 0 \\
\hline
15C_3 &= -4
\end{aligned}
\qquad C_3 = -\tfrac{4}{15}.
$$

Then

$$C_1 = \tfrac{1}{2}(3C_3 + 2) = \tfrac{1}{2}(-\tfrac{12}{15} + \tfrac{30}{15}) = \tfrac{9}{15}.$$

The first equation, corresponding to $y(0) = 1$, then gives

$$\tfrac{9}{15} + C_2 - \tfrac{4}{15} = 1, \qquad C_2 = 1 - \tfrac{5}{15} = \tfrac{2}{3}.$$

Our desired particular solution is therefore

$$y = \tfrac{9}{15}e^{-2x} + \tfrac{2}{3} - \tfrac{4}{15}e^{3x}.$$ ◼

Case 2: Repeated Linear Factors

We now take up the case where the polynomial $P(D)$ factors into a product of the form

$$P(D) = (D - r_1)^{n_1}(D - r_2)^{n_2} \cdots (D - r_m)^{n_m}, \tag{13}$$

where, of course, $n_1 + n_2 + \cdots + n_m = n$.

The general solution of $P(D)y = 0$ can again be obtained by repeated application of Theorem 17.1. To see what form the solution now takes, we consider a simple case where $P(D) = (D - r)^2$. Let $u = (D - r)y$, so that the equation $(D - r)^2 y = 0$ becomes $(D - r)u = 0$. We then obtain $u = C_1 e^{rx}$. Therefore

$$(D - r)y = C_1 e^{rx}$$

and

$$y = \frac{1}{e^{-rx}} \int e^{-rx} C_1 e^{rx}\, dx = \frac{1}{e^{-rx}} \int C_1\, dx$$

$$= \frac{1}{e^{-rx}}(C_1 x + C_2)$$

$$= e^{rx}(C_1 x + C_2).$$

It is straightforward to check that the general solution of $(D - r)^3 y = 0$ is

$$y = e^{rx}(C_1 x^2 + C_2 x + C_3).$$

The factors in Eq. (13) can be written in any order. For the given order, the first n_1 iterations to solve $P(D)y = 0$ show that

$$e^{r_1 x}(C_1 x^{n_1 - 1} + C_2 x^{n_1 - 2} + \cdots + C_{n_1})$$

forms a portion of the general solution. We consider that we have proved the following theorem.

THEOREM 17.4   Characteristic equation with multiple roots

Let $P(D) = (D - r_1)^{n_1} \cdots (D - r_m)^{n_m}$, where $n_1 + \cdots + n_m = n$. Then the general solution of $P(D)y = 0$ is

$$y = e^{r_1 x}(C_1 x^{n_1 - 1} + C_2 x^{n_1 - 2} + \cdots + C_{n_1}) + \cdots$$

$$+ e^{r_m x}(C_{n - n_m + 1} x^{n_m - 1} + C_{n - n_m + 2} x^{n_m - 2} + \cdots + C_n). \qquad \square$$

EXAMPLE 5   Solve $y''' + 2y'' + y' = 0$.

Solution   The characteristic equation of $y''' + 2y'' + y' = 0$ is

$$r^3 + 2r^2 + r = r(r + 1)^2 = 0,$$

and we see that the general solution of the equation is

$$y = C_1 e^{0x} + e^{-x}(C_2 x + C_3) = C_1 + e^{-x}(C_2 x + C_3).$$ ◼

HISTORICAL NOTE

THE PROCEDURE FOR SOLVING linear differential equations with constant coefficients was discovered by Leonhard Euler in 1739. It arose out of Euler's work on various problems in dynamics that led to differential equations of this type. Euler first solved the equation $a^3 y''' - y = 0$ in the form

$$y = c_1 e^{x/a}$$
$$+ c_2 e^{-x/a} \sin \frac{(c_3 + x)\sqrt{3}}{2a},$$

probably by using the obvious solution $y = e^{x/a}$ to reduce the equation to one of second order (as in the examples of this section) and then manipulating the latter equation into the form $y'' + \alpha y = 0$, for which a sine function is a solution. These manipulations, however, ultimately convinced him that he needed to deal only with the characteristic equation and its roots to find solutions. Euler announced his discovery in a letter to Johann Bernoulli on September 15, 1739, and over the next two years he convinced Bernoulli that solutions involving complex roots of the characteristic polynomial are equivalent to real solutions involving sines and cosines.

Case 3: $P(D)$ Contains Quadratic Factors

It is a theorem of algebra that every polynomial with real coefficients can be factored into a product of linear and quadratic factors with real coefficients. Namely, it can be shown that a polynomial can be factored into a product of linear factors if one allows complex coefficients. (This is the *fundamental theorem of algebra*.) If $D - (a + bi)$ is a factor, then $D - (a - bi)$ can be shown to be a factor also, and the product $D^2 - 2aD + (a^2 + b^2)$ is therefore a quadratic factor of $P(D)$ with real coefficients. We are interested in discovering what such a quadratic factor contributes to the general solution of $P(D)y = 0$.

Proceeding purely formally with complex numbers, we might expect the general solution of

$$[D^2 - 2aD + (a^2 + b^2)]y = [D - (a + bi)][D - (a - bi)]y = 0 \quad \textbf{(14)}$$

to be

$$y = C_1 e^{(a+bi)x} + C_2 e^{(a-bi)x}$$
$$= C_1 e^{ax} e^{bix} + C_2 e^{ax} e^{-bix} = e^{ax}(C_1 e^{i(bx)} + C_2 e^{-i(bx)}). \quad \textbf{(15)}$$

In Exercise 43 of Section 10.7, we asked you to verify formally Euler's formula

$$e^{ix} = \cos x + i \sin x.$$

Using this formula and proceeding formally from Eq. (15), we obtain

$$y = e^{ax}[C_1(\cos bx + i \sin bx) + C_2(\cos(-bx) + i \sin(-bx))]$$
$$= e^{ax}[(C_1 + C_2)\cos bx + (C_1 i - C_2 i)\sin bx]. \quad \textbf{(16)}$$

Replacing the arbitrary constant $C_1 + C_2$ by C_1 and replacing $C_1 i - C_2 i$ by C_2, we obtain from Eq. (16)

$$y = e^{ax}(C_1 \cos bx + C_2 \sin bx). \quad \textbf{(17)}$$

We conjecture that Eq. (17) is the general solution of Eq. (14). The preceding use of complex numbers can be justified, and our conjecture is indeed correct. We could compute directly that

$$[D^2 - 2aD + (a^2 + b^2)][e^{ax}(C_1 \cos bx + C_2 \sin bx)] = 0$$

(see Exercise 1). From our work with repeated linear factors, we would guess that a repeated quadratic factor $[D^2 - 2aD + (a^2 + b^2)]^2$ would give rise to a repetition of Eq. (17) with an additional factor x in the general solution. This can also be verified.

We can now solve any homogeneous linear differential $P(D)y = 0$ with constant coefficients, provided we are able to factor the polynomial $P(D)$ into quadratic and linear factors.

EXAMPLE 6 Solve the differential equation

$$D(D - 1)^2(D^2 + 2D + 4)y = 0.$$

Solution The characteristic equation is $r(r - 1)^2(r^2 + 2r + 4) = 0$, which has a root $r_1 = 0$, a double root $r_2 = 1$, and complex roots $-1 \pm i\sqrt{3}$ that are obtained by solving $r^2 + 2r + 4 = 0$ by the quadratic formula. The general solution of our equation is therefore

$$y = C_1 + e^x(C_2 x + C_3) + e^{-x}(C_4 \cos \sqrt{3}x + C_5 \sin \sqrt{3}x). \quad ∎$$

EXAMPLE 7   Solve the differential equation $(D^6 + 4D^4 + 4D^2)y = 0$.

Solution   We turn to the characteristic equation

$$r^6 + 4r^4 + 4r^2 = r^2(r^2 + 2)^2 = 0.$$

Here $r_1 = 0$, $r_2 = \sqrt{2}i$, and $r_3 = -\sqrt{2}i$ are all double roots. The general solution of our differential equation is

$$y = C_1 x + C_2 + (\cos \sqrt{2}x)(C_3 x + C_4) + (\sin \sqrt{2}x)(C_5 x + C_6). \qquad \blacksquare$$

EXAMPLE 8   Solve the initial value problem $y'' + y = 0$, $y(\pi/2) = 0$, $y'(\pi/2) = 1$.

Solution   The characteristic equation is

$$r^2 + 1 = 0,$$

which has complex roots $r = 0 \pm i$. The general solution of the differential equation is thus

$$y = e^{0x}(C_1\cos x + C_2\sin x), \qquad \text{or} \qquad y = C_1\cos x + C_2\sin x.$$

To make use of the initial condition $y'(\pi/2) = 1$, we differentiate and obtain

$$y' = -C_1\sin x + C_2\cos x.$$

The initial conditions then give rise to these equations:

$$C_1 \cdot 0 + C_2 \cdot 1 = 0, \qquad y(\pi/2) = 0$$
$$-C_1 \cdot 1 + C_2 \cdot 0 = 1. \qquad y'(\pi/2) = 1$$

Thus $C_2 = 0$ and $C_1 = -1$, so our desired particular solution is

$$y = -\cos x. \qquad \blacksquare$$

SUMMARY

1. The differential equation

$$y^{(n)} + b_1 y^{(n-1)} + b_2 y^{(n-2)} + \cdots + b_{n-1}y' + b_n y = 0$$

is a linear homogeneous differential equation with constant coefficients b_i. We let

$$P(D) = D^n + b_1 D^{n-1} + b_2 D^{n-2} + \cdots + b_{n-1}D + b_n$$

be the polynomial in the differential operator D (standing for differentiation). The equation then is symbolically written

$$P(D)y = 0.$$

2. If $P(D) = (D - r_1)(D - r_2) \cdots (D - r_n)$, where $r_i \neq r_j$ for $i \neq j$, then the general solution of $P(D)y = 0$ is

$$y = C_1 e^{r_1 x} + C_2 e^{r_2 x} + \cdots + C_n e^{r_n x}.$$

3. Let $P(D) = (D - r_1)^{n_1} \cdots (D - r_m)^{n_m}$, where $n_1 + \cdots + n_m = n$.

Then the general solution of $P(D)y = 0$ is

$$y = e^{r_1 x}(C_1 x^{n_1-1} + C_2 x^{n_1-2} + \cdots + C_{n_1}) + \cdots$$
$$+ e^{r_m x}(C_{n-n_m+1} x^{n_m-1} + C_{n-n_m+2} x^{n_m-2} + \cdots + C_n).$$

4. An irreducible quadratic factor in $P(D)$ having as complex roots $a + bi$ and $a - bi$ gives rise to a summand of the general solution of $P(D)y = 0$ of the form

$$e^{ax}(C_1 \cos bx + C_2 \sin bx).$$

EXERCISES 17.2

1. Convince yourself by direct computation that

$$[D^2 - 2aD + (a^2 + b^2)][e^{ax}(C_1 \cos bx + C_2 \sin bx)] = 0.$$

In Exercises 2–20, find the general solution of the given differential equation.

2. $y' + 3y = 0$

3. $2y' + 4y = 0$

4. $y'' + 4y' + 3y = 0$

5. $4y'' + 12y' + 5y = 0$

6. $y'' - 6y' + 9y = 0$

7. $4y''' + 4y'' + y' = 0$

8. $y''' - 3y'' = 0$

9. $y'' + 3y = 0$

10. $y'' + 2y' + 6y = 0$

11. $y''' - y = 0$

12. $(D^2 + 4)y = 0$

13. $(D^3 + 9D)y = 0$

14. $(D^4 - 16)y = 0$

15. $(D^4 + 2D^2 + 1)y = 0$

16. $D(D - 3)^2(D^2 + 1)y = 0$

17. $D^2(D + 2)(D^2 + 2)y = 0$

18. $D^3(D^2 + 1)^2 y = 0$

19. $(D + 1)^2(D^2 + D + 2)y = 0$

20. $D^2(D + 5)(D^2 + 3D + 5)^2 y = 0$

In Exercises 21–30, find the solution of the initial value problem.

21. $y'' - 5y' + 6y = 0$, $y(0) = 1$, $y'(0) = -1$

22. $y'' - y = 0$, $y(0) = 0$, $y'(0) = 1$

23. $y''' - y'' = 0$, $y(0) = 1$, $y'(0) = 0$, $y''(0) = 3$

24. $y''' - 4y' = 0$, $y(0) = -1$, $y'(0) = 2$, $y''(0) = 0$

25. $y^{(4)} - 9y'' = 0$, $y(0) = 0$, $y'(0) = 2$, $y''(0) = 0$, $y'''(0) = -1$

26. $y'' + y = 0$, $y(\pi/2) = 3$, $y'(\pi/2) = -2$

27. $y'' + 4y = 0$, $y(\pi) = 3$, $y'(\pi) = -4$

28. $y''' + y' = 0$, $y(0) = -1$, $y'(0) = 2$, $y''(0) = -3$

29. $y''' - 8y = 0$, $y(0) = 2$, $y'(0) = 0$, $y''(0) = 4$

30. $y''' - 2y'' + 2y' = 0$, $y(0) = 1$, $y'(0) = -1$, $y''(0) = 2$

17.3 THE NONHOMOGENEOUS CASE; APPLICATIONS

In the first two parts of this section, we consider the problem of finding the solutions of a linear differential equation with constant coefficients of the form

$$y^{(n)} + b_1 y^{(n-1)} + \cdots + b_{n-1} y' + b_n y = g(x)$$

for a continuous function $g(x)$. The existence theorem was stated in the preceding section. If $g(x) = 0$, the equation is called **homogeneous.**

The Form of the Solution

Writing the equation in the form

$$P(D)y = g(x), \tag{1}$$

suppose that $C_1 y_1(x) + \cdots + C_n y_n(x)$ is the general solution of the *homogeneous* equation obtained from Eq. (1) by replacing $g(x)$ by 0. Suppose also that $f(x)$ is any particular solution of $P(D)y = g(x)$. Then

$$y = C_1 y_1(x) + \cdots + C_n y_n(x) + f(x) \qquad (2)$$

gives solutions of Eq. (1) because

$$P(D)y = P(D)(C_1 y_1(x) + \cdots + C_n y_n(x)) + P(D)f(x)$$
$$= 0 + g(x) = g(x).$$

Also, if $h(x)$ is any solution of Eq. (1), then

$$P(D)(h(x) - f(x)) = P(D)h(x) - P(D)f(x) = g(x) - g(x) = 0,$$

so $h(x) - f(x)$ is a solution of the homogeneous equation obtained by setting $g(x) = 0$. Thus

$$h(x) = [\text{A solution of } P(D)y = 0] + f(x),$$

so any solution $h(x)$ is of the form of Eq. (2). Thus Eq. (2) is the general solution of Eq. (1). We have proved the following theorem.

THEOREM 17.5   Linearity property

Let $P(D)y = g(x)$ be a linear differential equation with constant coefficients. Let $y = H(x)$ be the general solution of the homogeneous equation $P(D)y = 0$ and let $y = f(x)$ be any solution of $P(D)y = g(x)$. Then the general solution of $P(D)y = g(x)$ is $y = H(x) + f(x)$. $\square$

In Section 17.2, we showed how to find solutions of the homogeneous equation. We therefore focus our attention on finding a particular solution $y = f(x)$ of Eq. (1). Two methods are presented.

Successive Reduction to First-Order Equations

Let $P(D)$ be a polynomial in D that factors into (not necessarily distinct) linear factors, so that

$$P(D) = (D - r_1)(D - r_2) \cdots (D - r_n).$$

Consider an equation of the form

$$P(D)y = (D - r_1)(D - r_2) \cdots (D - r_n)y = g(x). \qquad (3)$$

We can solve Eq. (3) by the method illustrated in Example 2 of Section 17.2. This technique consists of setting

$$(D - r_2) \cdots (D - r_n)y = u, \qquad (4)$$

so that Eq. (3) becomes

$$(D - r_1)u = g(x). \qquad (5)$$

Equation (5) is linear in u and can be solved using the method of Section 17.1. We have thus reduced our problem to the solution of Eq. (4), which is a differential equation of order $n - 1$. Repetition of this process n times enables us to find y.

Actually, the procedure just outlined yields the general solution of Eq. (3). Since it is easy for us to write down "most" of this general solution, namely the part in Eq. (2) involving the arbitrary constants, it is more efficient when using this technique to take all arbitrary constants equal to zero and obtain just a particular solution of Eq. (3).

EXAMPLE 1 Solve the differential equation

$$y'' - 3y' + 2y = x + 1.$$

Solution Our equation can be written in the form

$$(D - 1)(D - 2)y = x + 1.$$

We let $(D - 2)y = u$ and find a particular solution of $(D - 1)u = x + 1$. Integrating by parts or using tables, we obtain as particular solution, taking zero for all constants of integration,

$$u = \frac{1}{e^{-x}} \int e^{-x}(x + 1) \, dx = \frac{1}{e^{-x}}(-xe^{-x} - 2e^{-x})$$

$$= -x - 2.$$

Solving $(D - 2)y = u = -x - 2$ by the same method, we obtain as a particular solution of our original equation

$$y = \frac{1}{e^{-2x}} \int -e^{-2x}(x + 2) \, dx = \frac{1}{e^{-2x}}\left(\frac{1}{2} xe^{-2x} + \frac{5}{4}e^{-2x}\right)$$

$$= \frac{1}{2}x + \frac{5}{4}.$$

Our general solution is therefore

$$y = C_1 e^x + C_2 e^{2x} + \frac{1}{2}x + \frac{5}{4}. \qquad \blacksquare$$

The Method of Undetermined Coefficients

Sometimes a particular solution of $P(D)y = g(x)$ can be determined by inspection.

EXAMPLE 2 Solve the equation

$$y'' + 4y' + 4y = 12.$$

Solution Obviously a particular solution of $y'' + 4y' + 4y = 12$ is $y = 3$; just check that $y = 3$ satisfies the equation. Since the characteristic equation is $r^2 + 4r + 4 = (r + 2)^2 = 0$, we see that the general solution is

$$y = e^{-2x}(C_1 x + C_2) + 3. \qquad \blacksquare$$

EXAMPLE 3 Find by inspection a particular solution of the equation $y'' - 3y' + 2y = x + 1$, which we solved in Example 1.

Solution We try to find $y = f(x)$ such that we obtain $x + 1$ upon computing $y'' - 3y' + 2y$. Clearly $y = x/2$ will give the desired amount of x; namely for $y = x/2$, we have

$$y'' - 3y' + 2y = D^2\left(\frac{x}{2}\right) - 3D\left(\frac{x}{2}\right) + 2\left(\frac{x}{2}\right) = 0 - \frac{3}{2} + x.$$

To get the desired constant 1 rather than $-\frac{3}{2}$, we need to add to our $x/2$ a constant that when *doubled* yields $1 + \frac{3}{2} = \frac{5}{2}$ since our equation contains $2y$. Thus the needed constant is $\frac{5}{4}$, and a particular solution is $y = (x/2) + \frac{5}{4}$, as obtained in Example 1. ∎

A more systematic attack suggested by Example 3 leads to the method of undetermined coefficients.

EXAMPLE 4 Solve the differential equation

$$y'' + 2y' - 3y = 2x - 17.$$

Solution From Example 3, we would guess that $y = ax + b$ should be a solution for some a and b. For $y = ax + b$, we have

$$y = ax + b, \qquad y' = a, \qquad y'' = 0.$$

Substituting these expressions into the given differential equation, we see that we must have

$$2a - 3ax - 3b = 2x - 17.$$

Equating the coefficients of x and the constant terms, we obtain

$$-3a = 2 \qquad \text{and} \qquad 2a - 3b = -17.$$

Thus $a = -\frac{2}{3}$ and $b = \frac{47}{9}$. A particular solution is therefore

$$y = -\frac{2}{3}x + \frac{47}{9}.$$

The general solution is then

$$y = C_1 e^{-3x} + C_2 e^x - \frac{2}{3}x + \frac{47}{9}.$$ ∎

The *method of undetermined coefficients* illustrated in Example 4 works well when the function $g(x)$ on the right-hand side of Eq. (1) and derivatives of all orders of $g(x)$ involve sums of only a *finite number* of different functions, except for constant factors. For example, a polynomial of degree n and its derivatives of all orders involve only sums of constant multiples of the finite number of functions $1, x, x^2, \ldots, x^n$. Also, derivatives of $\sin ax$ or $\cos ax$ are just constant multiples of these same trigonometric functions. Derivatives of the exponential function e^{ax} involve only constant multiples of e^{ax}. If $g(x)$ contains only sums and products of these functions, then the method of undetermined coefficients is often useful. Suppose now that $g(x)$ is of this type.

Particular Solutions by Undetermined Coefficients

1. Suppose none of the summands in $g(x)$ or any of their derivatives is a solution of the homogeneous equation where $g(x)$ is replaced by zero. Then one takes as a trial particular solution a sum of all the summands in $g(x)$ and derivatives of all orders of these summands, with coefficients a, b, c, and so on, to be determined.

2. Suppose some summand of $g(x)$ or a derivative of some order of a summand is a solution of the homogeneous equation corresponding to an s-fold root of the characteristic equation. Then one should start with a trial particular solution in which that summand of $g(x)$ is multiplied by x^s before taking derivatives of it.

This seemingly complicated procedure is quite simple (although frequently tedious) in practice and is best understood by further examples.

EXAMPLE 5 Solve the differential equation

$$y'' - 3y' + 2y = 2 \cos 4x.$$

Solution Now $\cos 4x$ is not part of the solution of the homogeneous equation $y'' - 3y' + 2y = 0$, so we try as particular solution

$$y = a \cos 4x + b \sin 4x,$$

which is a sum with coefficients a and b of $\cos 4x$ and all different types of derivatives of $\cos 4x$. We then obtain

$$y = a \cos 4x + b \sin 4x,$$
$$y' = 4b \cos 4x - 4a \sin 4x,$$
$$y'' = -16a \cos 4x - 16b \sin 4x.$$

Multiplying the first equation by 2, the second by -3, and the last by 1 and adding, we see from our original differential equation that we must have

$$2 \cos 4x = (2a - 12b - 16a)\cos 4x + (2b + 12a - 16b)\sin 4x$$
$$= (-14a - 12b)\cos 4x + (12a - 14b)\sin 4x.$$

We therefore must have

$$-14a - 12b = 2,$$
$$12a - 14b = 0,$$

which yields upon solution $a = -\frac{7}{85}$ and $b = -\frac{6}{85}$. Our general solution is therefore

$$y = C_1 e^x + C_2 e^{2x} - \frac{7}{85} \cos 4x - \frac{6}{85} \sin 4x. \qquad \blacksquare$$

EXAMPLE 6 Solve the differential equation

$$y'' - 2y' + y = e^x.$$

Solution In this case, e^x is part of the solution of the homogeneous equation corresponding to the double root $r = 1$ of the characteristic equation

$$r^2 - 2r + 1 = (r - 1)^2 = 0.$$

We therefore multiply by x^2 and start with $x^2 e^x$ and functions obtained by differentiating it; that is, we first take as a trial particular solution

$$y = ax^2 e^x + bxe^x + ce^x.$$

But since e^x and xe^x are solutions of the homogeneous equation, the bxe^x and ce^x portions will contribute zero when we compute $y'' - 2y + y$. We thus simply take $y = ax^2 e^x$ as trial solution, and we find that

$$y = ax^2 e^x,$$
$$y' = ax^2 e^x + 2axe^x,$$
$$y'' = ax^2 e^x + 4axe^x + 2ae^x.$$

Multiplying the first equation by 1, the second by -2, and the last by 1 and adding, we obtain

$$e^x = 0(x^2 e^x) + 0(xe^x) + 2ae^x = 2ae^x,$$

so we must have $2a = 1$ or $a = \frac{1}{2}$. Our general solution is therefore

$$y = e^x(C_1 x + C_2) + \frac{1}{2}x^2 e^x. \qquad \blacksquare$$

Applications

Let us consider motion of a body of mass m along a straight line, which we consider to be an s-axis. By Newton's second law of motion, we have

$$F(t) = m\,\frac{d^2 s}{dt^2},$$

where $F(t)$ is the force at time t acting on the body and directed along the line. Frequently, mathematicians denote a derivative with respect to time t by a dot over a variable rather than by a prime. Thus we let $\dot{s} = ds/dt$, $\ddot{s} = d^2 s/dt^2$, and so on.

There are many physical situations in which the force on the body has a certain basic component $g(t)$ at time t together with components due to the velocity and position of the body. If these additional components due to velocity and position are constant multiples of the velocity and position, then by Newton's law we have

$$\ddot{s} = k_1 \dot{s} + k_2 s + \frac{1}{m}g(t) \qquad (6)$$

for constants k_1 and k_2. The differential equation (6) is linear with constant coefficients and might be solved by the methods we have presented. We illustrate several particular cases.

EXAMPLE 7 *Free fall in a vacuum* We consider a body falling freely near the surface of a planet without atmosphere and with gravitational acceleration g, which is essentially constant near the surface. The motion of the body is then governed by the differential equation

$$\ddot{s} = -g, \tag{7}$$

where s is measured upward from the surface of the planet. The general solution of Eq. (7) is found to be

$$s = C_1 t + C_2 - \tfrac{1}{2}gt^2,$$

and the constants C_1 and C_2 can be determined if the position s and velocity $\dot{s}$ of the body are known at a particular time t_0. ∎

EXAMPLE 8   *Free fall in a medium*   Let a body be falling freely through a medium (atmosphere) near the surface of a planet with gravitational acceleration g near its surface. Suppose that, due to the medium, the motion of the body is retarded by a force proportional to its velocity. The motion of the body is then governed by the differential equation

$$\ddot{s} = -k\dot{s} - g, \tag{8}$$

where s is measured upward from the planet and $k > 0$. The characteristic equation of (8) is $r^2 + kr = 0$, and a particular solution is found to be $s = -(g/k)t$. The general solution is therefore

$$s = C_1 + C_2 e^{-kt} - \frac{g}{k}t. \tag{9}$$

Again, C_1 and C_2 can be determined by the position s and velocity $\dot{s}$ at a particular time t_0.

Differentiating Eq. (9), we obtain for the velocity

$$\dot{s} = -kC_2 e^{-kt} - \frac{g}{k}.$$

As t becomes large, the term $-kC_2 e^{-kt}$ becomes very small; the *terminal velocity* of the body is thus $-g/k$. Note that the body approaches this terminal velocity exponentially (quite rapidly). ∎

EXAMPLE 9   *Undamped vibrating spring*   Consider a body of mass m hanging on a spring, shown in the natural position of rest at $s = 0$ on the vertical s-axis in Fig. 17.1. If the spring is stretched (or compressed) from this natural position, then the spring exerts a restoring force proportional to the displacement of the body, assuming that the elastic limit of the spring is not exceeded. If the body is set in vertical motion and this restoring force is the only force on the system, then the motion of the body must satisfy the differential equation

$$m\ddot{s} = -ks \tag{10}$$

for a spring constant $k > 0$. The characteristic equation for (10) is $mr^2 + k = 0$, and we see that the general solution is

$$s = C_1 \cos\!\left(\sqrt{\frac{k}{m}}\,t\right) + C_2 \sin\!\left(\sqrt{\frac{k}{m}}\,t\right), \tag{11}$$

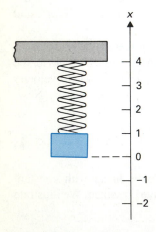

FIGURE 17.1   Body suspended by a spring in rest position at $s = 0$.

where C_1 and C_2 can be determined by the position and velocity of the body at a particular time t_0.

If we set $A = \sqrt{C_1{}^2 + C_2{}^2}$, then Eq. (11) can be written

$$s = A\left[\frac{C_1}{A}\cos\left(\sqrt{\frac{k}{m}}t\right) + \frac{C_2}{A}\sin\left(\sqrt{\frac{k}{m}}t\right)\right].$$

Since $(C_1/A)^2 + (C_2/A)^2 = 1$, there exists θ where $0 \le \theta < 2\pi$ and where $\sin\theta = C_1/A$ and $\cos\theta = C_2/A$. Thus

$$s = A\left(\sin\theta\cos\sqrt{\frac{k}{m}}t + \cos\theta\sin\sqrt{\frac{k}{m}}t\right)$$

$$= A\sin\left(\sqrt{\frac{k}{m}}t + \theta\right). \tag{12}$$

Equation (12) shows that our vibratory motion is sinusoidal, with amplitude A. Such sinusoidal undamped vibratory motion is frequently referred to as *simple harmonic motion*. ■

EXAMPLE 10 *Damped vibrating spring* Suppose the spring in Fig. 17.1 has a damping mechanism attached that exerts a force against the direction of motion and proportional to the velocity of the body. The differential equation then becomes

$$m\ddot{s} = -ks - c\dot{s} \tag{13}$$

for $k > 0$ and $c > 0$. The character of the general solution of this equation $m\ddot{s} + c\dot{s} + ks = 0$ depends on the relative sizes of m, c, and k and on the initial conditions. In the exercises, we ask you to describe the nature of particular motions in the *overdamped case* where $c^2 > 4km$, the *critically damped case* where $c^2 = 4km$, and the *underdamped case* where $c^2 < 4km$ (see Exercises 27, 28, and 29). ■

SUMMARY

1. The general solution of a linear equation with constant coefficients

$$P(D)y = g(x)$$

is of the form

$$y = [\text{General solution of } P(D)y = 0] + f(x),$$

where $f(x)$ is any particular solution of $P(D)y = g(x)$.

2. If $P(D) = (D - r_1)(D - r_2)\cdots(D - r_n)$, then $P(D)y = g(x)$ can be solved by setting $u = (D - r_2)\cdots(D - r_n)y$, so $P(D)y = g(x)$ becomes $(D - r_1)u = g(x)$. Solve for u and apply the same technique again to $(D - r_2)\cdots(D - r_n)y = u$, and so on, to find the general solution.

3. The method of undetermined coefficients to find a particular solution $f(x)$ of $P(D)y = g(x)$ is too complicated to describe in a summary. See Examples 5 and 6 and the displayed explanation preceding them.

EXERCISES 17.3

In Exercises 1–8, find the general solution of the differential equation by finding a particular solution by the method of successive reduction to first-order equations, as in Example 1.

1. $y'' - 2y' - 3y = 4x$ **2.** $y'' - 4y = e^x$

3. $y'' - y = e^x$ **4.** $y'' - y = \cos x$

5. $y'' - y' = 1$ **6.** $y'' + y' = x$

7. $y'' + 2y' + y = e^{-x}$ **8.** $y''' - y'' = x^2$

In Exercises 9–24, find the general solution of the differential equation by the method of undetermined coefficients.

9. $y'' - 3y' + 2y = x - 3$ **10.** $y'' - y = x + e^{2x}$

11. $y'' + 4y = \sin x$ **12.** $y'' + y = \sin 2x$

13. $y'' - 2y' = 2 \sin x + 3e^x$ **14.** $y'' + 2y' + y = x^2$

15. $y'' + y = xe^x$ **16.** $y'' - 4y = x \sin 2x$

17. $y'' - y = x \sin x + \sin x$

18. $y'' + 4y' + 4y = e^{2x}\cos x$

19. $y'' + 4y' = x^2$

20. $y'' - 4y' - 5y = x^2 + 2e^{-x}$

21. $y''' + 3y'' = x + e^{3x}$ **22.** $y''' - y'' = x^2 + e^x$

23. $(D^4 - 2D^3)y = x^3 + 3x^2$ **24.** $(D - 1)^3y = 4 - 3e^x$

25. A body of mass 2 slugs is dropped from an altitude of 3000 ft above the Atlantic Ocean. Suppose the motion of the body is retarded by a force due to air resistance and of magnitude $v/2$ lb, where v is the velocity of the body measured in ft/sec. Find the height s of the body above the ocean as a function of time t, and find the terminal velocity of the body. (Take $g = 32$ ft/sec^2.)

26. A body of mass 1 slug (weighing 32 lb) is attached to a vertical spring and stretches the spring 2 ft. The body is then raised 1 ft and released.

    (a) Find the displacement s of the body from its natural position at rest at the end of the spring as a function of the time after it was released. (For example, $s = 1$ when $t = 0$.)

    (b) What is the amplitude of this oscillatory motion?

    (c) What is the period of this oscillatory motion? (The period is the time required for one complete oscillation.)

27. For a damped vibrating spring with spring constant k and with a weight of mass m attached, suppose the weight is raised a height h from its position at rest and then released.

Show that if $c^2 > 4km$, the weight eases back to its original position of rest without crossing this position. See Eq. (13).

28. With reference to Exercise 27, suppose now that $c^2 = 4km$ and that the weight is given an initial velocity $v_0 < -ch/2m$ toward the position of rest. Show that the weight crosses its original position of rest and then eases back to this position.

29. With reference to Exercise 27, show that if $c^2 < 4km$, then the motion of the weight about its position of rest is oscillatory but has amplitude decreasing exponentially to zero as time increases.

30. It can be shown that if an electric circuit has a (constant) resistance R, a (constant) inductance L, (constant) capacitance C, and variable impressed electromotive force $E(t)$, then the charge $Q(t)$ on the capacitor at time t is governed by the differential equation

$$L\ddot{Q} + R\dot{Q} + \frac{1}{C}Q = E(t).$$

Suppose the electromotive force $E(t)$ is abruptly cut off, so that $E(t) = 0$ for $t > t_0$. Use Exercises 27, 28, and 29 to discuss the possibilities for the behavior of the charge Q after time t_0.

31. *Resonance.* Some possibilities for the behavior of the solutions of a differential equation $a\ddot{s} + b\dot{s} + cs = 0$, where $a, b, c > 0$, are indicated in Exercises 27–29. This exercise exhibits a phenomenon that may appear in the solutions of the differential equation $a\ddot{s} + b\dot{s} + cs = \sin kt$. A physical model for such an equation is given by the electric circuit described in Exercise 30, where the impressed electromotive force $E(t)$ is sinusoidal.

    Suppose the "damping factor" b in our equation is zero. Show that if $k = 0$, the motion described by the resulting (homogeneous) equation is oscillatory with period $2\pi\sqrt{a}/\sqrt{c}$. Then show that if the impressed sinusoidal force $\sin kt$ has the same period so that $k = \sqrt{c/a}$, the amplitude of the motion increases without bound as time increases. (If the damping factor b is nonzero but quite small, the amplitude can still get quite large if $k = \sqrt{4ac - b^2}/2a$. This phenomenon is known as "resonance" and can be very destructive. A group of men marching in step should be instructed to break step when crossing a bridge, on the outside chance that a frequency in their step might be the same as the natural frequency of vibration of the bridge.)

SUPPLEMENTARY EXERCISES FOR CHAPTER 17

1. Find the general solution of $y' - 2xy = 3x$.

2. Find the general solution of $y' + (\sin x)y = \sin x$.

3. Find the general solution of $y' + (\cot x)y = x$.

4. Solve the initial value problem $y' - e^x y = e^x$, $y(0) = 2$.

5. Find the general solution of $y''' - 6y'' + 9y' = 0$.

6. Find the general solution of the differential equation $y''' - 2y'' + y' = 0$.

7. Find the general solution of $y''' - 2y'' = 0$.

8. Find the general solution of $(D^2 - 4)(D^2 + 3)y = 0$.

9. Find the general solution of $(D^3 + 4D)y = 0$.

10. Solve the initial value problem $(D^2 - 4D + 5)y = 0$, $y(0) = 1$, $y'(0) = -1$.

11. Find the general solution of $y'' - 3y' + 2y = \sin x$.

12. Find the general solution of $y'' - 2y' + 5y = e^{3x}$.

13. Find the general solution of the differential equation $(D^3 - D^2)y = x^2$.

14. Solve the initial value problem $y'' - 3y' + 2y = x$, $y(0) = 1$, $y'(0) = -1$.

APPENDIXES
ANSWERS
INDEX

SUMMARY OF ALGEBRA AND GEOMETRY

APPENDIX 1

A. ALGEBRA

1. Inequalities

Notation	Read
$a < b$	a is less than b
$a > b$	a is greater than b
$a \leq b$	a is less than or equal to b
$a \geq b$	a is greater than or equal to b

Laws

$a \leq a$

If $a \leq b$ and $b \leq a$, then $a = b$.

If $a \leq b$ and $b \leq c$, then $a \leq c$.

If $a \leq b$, then $a + c \leq b + c$.

If $a \leq b$ and $c \leq d$, then $a + c \leq b + d$.

If $a \leq b$ and $c > 0$, then $ac \leq bc$.

If $a \leq b$ and $c < 0$, then $bc \leq ac$.

2. Absolute Value

Notation	Read
$|a|$	The absolute value of a

Properties

$$|a| = \begin{cases} a, & \text{if } a \geq 0 \\ -a, & \text{if } a < 0 \end{cases}$$

$|a| \geq 0$, and $|a| = 0$ if and only if $a = 0$

$|ab| = |a| \cdot |b|$

$|a + b| \leq |a| + |b|$

$|a - b| \geq |a| - |b|$

$|a - b|$ = distance from a to b on the number line

3. Arithmetic of Rational Numbers

$$\frac{a}{b} + \frac{c}{d} = \frac{ad + bc}{bd}, \quad \frac{a}{b} \cdot \frac{c}{d} = \frac{ac}{bd}, \quad \frac{a/b}{c/d} = \frac{ad}{bc}$$

4. Laws of Signs

$$(-a)(b) = a(-b) = -(ab), \quad (-a)(-b) = ab$$

5. Distributive Laws

$$a(b + c) = ab + ac, \quad (a + b)c = ac + bc$$

6. Laws of Exponents

$$a^m a^n = a^{m+n}, \quad (ab)^m = a^m b^m, \quad (a^m)^n = a^{mn}$$

$$a^{m/n} = \sqrt[n]{a^m} = (\sqrt[n]{a})^m, \quad a^{-n} = \frac{1}{a^n}, \quad \frac{a^m}{a^n} = a^{m-n}$$

7. Arithmetic Involving 0

$$a \cdot 0 = 0 \cdot a = 0 \qquad \text{for any number } a$$

$$a + 0 = 0 + a = a \qquad \text{for any number } a$$

$$\frac{0}{a} = 0$$

$$a^0 = 1 \text{ and } 0^a = 0 \qquad \text{if } a > 0$$

8. Arithmetic Involving 1

$$1 \cdot a = a \cdot 1 = a \qquad \text{for any number } a$$

$$\frac{a}{1} = a^1 = a \qquad \text{for any number } a$$

$$1^n = 1 \qquad \text{for any integer } n$$

9. Binomial Theorem

$$(a + b)^n = a^n + na^{n-1}b + \frac{n(n-1)}{1 \cdot 2}a^{n-2}b^2$$

$$+ \frac{n(n-1)(n-2)}{1 \cdot 2 \cdot 3}a^{n-3}b^3 + \cdots + nab^{n-1} + b^n,$$

where n is a positive integer

10. Quadratic Formula

If $a \neq 0$, then the solutions of the quadratic equation

$$ax^2 + bx + c = 0$$

are given by the formula

$$x = \frac{-b \pm \sqrt{b^2 - 4ac}}{2a}.$$

B. GEOMETRY

In the formulas that follow,

A = Area b = Length of base h = Altitude
s = Slant height r = Radius V = Volume
C = Circumference S = Surface area or lateral area

1. Triangle

$$A = \frac{1}{2}bh$$

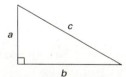

2. Similar triangles

$$\frac{a'}{a} = \frac{b'}{b} = \frac{c'}{c}$$

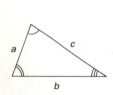
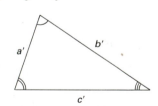

3. Pythagorean theorem

$$c^2 = a^2 + b^2$$

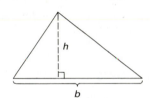

4. Parallelogram

$$A = bh$$

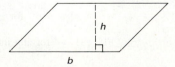

5. Trapezoid

$$A = \frac{1}{2}(b_1 + b_2)h$$

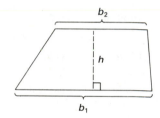

6. Circle

$$A = \pi r^2, \qquad C = 2\pi r$$

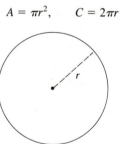

7. Right circular cylinder

$$V = \pi r^2 h, \qquad S = 2\pi rh$$

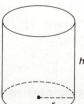

8. Right circular cone

$$V = \frac{1}{3}\pi r^2 h, \qquad S = \pi r s$$

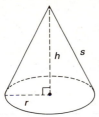

9. Sphere

$$V = \frac{4}{3}\pi r^3, \qquad S = 4\pi r^2$$

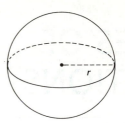

TABLES OF FUNCTIONS

APPENDIX 2

TABLE 1   Natural trigonometric functions

Angle					Angle				
Degree	Radian	Sine	Cosine	Tangent	Degree	Radian	Sine	Cosine	Tangent
0°	0.000	0.000	1.000	0.000					
1°	0.017	0.017	1.000	0.017	31°	0.541	0.515	0.857	0.601
2°	0.035	0.035	0.999	0.035	32°	0.559	0.530	0.848	0.625
3°	0.052	0.052	0.999	0.052	33°	0.576	0.545	0.839	0.649
4°	0.070	0.070	0.998	0.070	34°	0.593	0.559	0.829	0.675
5°	0.087	0.087	0.996	0.087	35°	0.611	0.574	0.819	0.700
6°	0.105	0.105	0.995	0.105	36°	0.628	0.588	0.809	0.727
7°	0.122	0.122	0.993	0.123	37°	0.646	0.602	0.799	0.754
8°	0.140	0.139	0.990	0.141	38°	0.663	0.616	0.788	0.781
9°	0.157	0.156	0.988	0.158	39°	0.681	0.629	0.777	0.810
10°	0.175	0.174	0.985	0.176	40°	0.698	0.643	0.766	0.839
11°	0.192	0.191	0.982	0.194	41°	0.716	0.656	0.755	0.869
12°	0.209	0.208	0.978	0.213	42°	0.733	0.669	0.743	0.900
13°	0.227	0.225	0.974	0.231	43°	0.750	0.682	0.731	0.933
14°	0.244	0.242	0.970	0.249	44°	0.768	0.695	0.719	0.966
15°	0.262	0.259	0.966	0.268	45°	0.785	0.707	0.707	1.000
16°	0.279	0.276	0.961	0.287	46°	0.803	0.719	0.695	1.036
17°	0.297	0.292	0.956	0.306	47°	0.820	0.731	0.682	1.072
18°	0.314	0.309	0.951	0.325	48°	0.838	0.743	0.669	1.111
19°	0.332	0.326	0.946	0.344	49°	0.855	0.755	0.656	1.150
20°	0.349	0.342	0.940	0.364	50°	0.873	0.766	0.643	1.192
21°	0.367	0.358	0.934	0.384	51°	0.890	0.777	0.629	1.235
22°	0.384	0.375	0.927	0.404	52°	0.908	0.788	0.616	1.280
23°	0.401	0.391	0.921	0.424	53°	0.925	0.799	0.602	1.327
24°	0.419	0.407	0.914	0.445	54°	0.942	0.809	0.588	1.376
25°	0.436	0.423	0.906	0.466	55°	0.960	0.819	0.574	1.428
26°	0.454	0.438	0.899	0.488	56°	0.977	0.829	0.559	1.483
27°	0.471	0.454	0.891	0.510	57°	0.995	0.839	0.545	1.540
28°	0.489	0.469	0.883	0.532	58°	1.012	0.848	0.530	1.600
29°	0.506	0.485	0.875	0.554	59°	1.030	0.857	0.515	1.664
30°	0.524	0.500	0.866	0.577	60°	1.047	0.866	0.500	1.732

TABLE 1 Natural trigonometric functions (continued)

Angle Degree	Radian	Sine	Cosine	Tangent	Angle Degree	Radian	Sine	Cosine	Tangent
61°	1.065	0.875	0.485	1.804	76°	1.326	0.970	0.242	4.011
62°	1.082	0.883	0.469	1.881	77°	1.344	0.974	0.225	4.332
63°	1.100	0.891	0.454	1.963	78°	1.361	0.978	0.208	4.705
64°	1.117	0.899	0.438	2.050	79°	1.379	0.982	0.191	5.145
65°	1.134	0.906	0.423	2.145	80°	1.396	0.985	0.174	5.671
66°	1.152	0.914	0.407	2.246	81°	1.414	0.988	0.156	6.314
67°	1.169	0.921	0.391	2.356	82°	1.431	0.990	0.139	7.115
68°	1.187	0.927	0.375	2.475	83°	1.449	0.993	0.122	8.144
69°	1.204	0.934	0.358	2.605	84°	1.466	0.995	0.105	9.514
70°	1.222	0.940	0.342	2.748	85°	1.484	0.996	0.087	11.43
71°	1.239	0.946	0.326	2.904	86°	1.501	0.998	0.070	14.30
72°	1.257	0.951	0.309	3.078	87°	1.518	0.999	0.052	19.08
73°	1.274	0.956	0.292	3.271	88°	1.536	0.999	0.035	28.64
74°	1.292	0.961	0.276	3.487	89°	1.553	1.000	0.017	57.29
75°	1.309	0.966	0.259	3.732	90°	1.571	1.000	0.000	

TABLE 2 Exponential functions

x	e^x	e^{-x}	x	e^x	e^{-x}	x	e^x	e^{-x}
0.00	1.0000	1.0000	1.5	4.4817	0.2231	4.0	54.598	0.0183
0.05	1.0513	0.9512	1.6	4.9530	0.2019	4.1	60.340	0.0166
0.10	1.1052	0.9048	1.7	5.4739	0.1827	4.2	66.686	0.0150
0.15	1.1618	0.8607	1.8	6.0496	0.1653	4.3	73.700	0.0136
0.20	1.2214	0.8187	1.9	6.6859	0.1496	4.4	81.451	0.0123
0.25	1.2840	0.7788	2.0	7.3891	0.1353	4.5	90.017	0.0111
0.30	1.3499	0.7408	2.1	8.1662	0.1225	4.6	99.484	0.0101
0.35	1.4191	0.7047	2.2	9.0250	0.1108	4.7	109.95	0.0091
0.40	1.4918	0.6703	2.3	9.9742	0.1003	4.8	121.51	0.0082
0.45	1.5683	0.6376	2.4	11.023	0.0907	4.9	134.29	0.0074
0.50	1.6487	0.6065	2.5	12.182	0.0821	5	148.41	0.0067
0.55	1.7333	0.5769	2.6	13.464	0.0743	6	403.43	0.0025
0.60	1.8221	0.5488	2.7	14.880	0.0672	7	1096.6	0.0009
0.65	1.9155	0.5220	2.8	16.445	0.0608	8	2981.0	0.0003
0.70	2.0138	0.4966	2.9	18.174	0.0550	9	8103.1	0.0001
0.75	2.1170	0.4724	3.0	20.086	0.0498	10	22026	0.00005
0.80	2.2255	0.4493	3.1	22.198	0.0450			
0.85	2.3396	0.4274	3.2	24.533	0.0408			
0.90	2.4596	0.4066	3.3	27.113	0.0369			
0.95	2.5857	0.3867	3.4	29.964	0.0334			
1.0	2.7183	0.3679	3.5	33.115	0.0302			
1.1	3.0042	0.3329	3.6	36.598	0.0273			
1.2	3.3201	0.3012	3.7	40.447	0.0247			
1.3	3.6693	0.2725	3.8	44.701	0.0224			
1.4	4.0552	0.2466	3.9	49.402	0.0202			

TABLE 3   Natural logarithms

n	$\log_e n$	n	$\log_e n$	n	$\log_e n$
0.0	*	4.5	1.5041	9.0	2.1972
0.1	7.6974	4.6	1.5261	9.1	2.2083
0.2	8.3906	4.7	1.5476	9.2	2.2192
0.3	8.7960	4.8	1.5686	9.3	2.2300
0.4	9.0837	4.9	1.5892	9.4	2.2407
0.5	9.3069	5.0	1.6094	9.5	2.2513
0.6	9.4892	5.1	1.6292	9.6	2.2618
0.7	9.6433	5.2	1.6487	9.7	2.2721
0.8	9.7769	5.3	1.6677	9.8	2.2824
0.9	9.8946	5.4	1.6864	9.9	2.2925
1.0	0.0000	5.5	1.7047	10	2.3026
1.1	0.0953	5.6	1.7228	11	2.3979
1.2	0.1823	5.7	1.7405	12	2.4849
1.3	0.2624	5.8	1.7579	13	2.5649
1.4	0.3365	5.9	1.7750	14	2.6391
1.5	0.4055	6.0	1.7918	15	2.7081
1.6	0.4700	6.1	1.8083	16	2.7726
1.7	0.5306	6.2	1.8245	17	2.8332
1.8	0.5878	6.3	1.8405	18	2.8904
1.9	0.6419	6.4	1.8563	19	2.9444
2.0	0.6931	6.5	1.8718	20	2.9957
2.1	0.7419	6.6	1.8871	25	3.2189
2.2	0.7885	6.7	1.9021	30	3.4012
2.3	0.8329	6.8	1.9169	35	3.5553
2.4	0.8755	6.9	1.9315	40	3.6889
2.5	0.9163	7.0	1.9459	45	3.8067
2.6	0.9555	7.1	1.9601	50	3.9120
2.7	0.9933	7.2	1.9741	55	4.0073
2.8	1.0296	7.3	1.9879	60	4.0943
2.9	1.0647	7.4	2.0015	65	4.1744
3.0	1.0986	7.5	2.0149	70	4.2485
3.1	1.1314	7.6	2.0281	75	4.3175
3.2	1.1632	7.7	2.0412	80	4.3820
3.3	1.1939	7.8	2.0541	85	4.4427
3.4	1.2238	7.9	2.0669	90	4.4998
3.5	1.2528	8.0	2.0794	95	4.5539
3.6	1.2809	8.1	2.0919	100	4.6052
3.7	1.3083	8.2	2.1041		
3.8	1.3350	8.3	2.1163		
3.9	1.3610	8.4	2.1282		
4.0	1.3863	8.5	2.1401		
4.1	1.4110	8.6	2.1518		
4.2	1.4351	8.7	2.1633		
4.3	1.4586	8.8	2.1748		
4.4	1.4816	8.9	2.1861		

*Subtract 10 from these entries.

ANSWERS TO ODD-NUMBERED EXERCISES

CHAPTER 1

Section 1.1

1. (a) (No solution set)

(b)

(c)

3. (a) 3 (b) 5 (c) 3 **5.** (a) 2 (b) -2 **7.** $x > -2$

9. The distance from $(a + b)/2$ to a is $\left| a - \dfrac{a + b}{2} \right| = \left| \dfrac{a}{2} - \dfrac{b}{2} \right| = \left| \dfrac{a - b}{2} \right|$. Similarly, the distance from $(a + b)/2$ to b is $\left| b - \dfrac{a + b}{2} \right| = \left| \dfrac{b}{2} - \dfrac{a}{2} \right| = \left| \dfrac{b - a}{2} \right| = \left| \dfrac{a - b}{2} \right|$. Thus $(a + b)/2$ is equidistant from a and b.

11. (a) -4.5 (b) $-\sqrt{2}/2$ (c) $(\sqrt{2} + \pi)/2$ **13.** (a) -10 (b) -8 **15.** $[2, 4]$ **17.** $[-5, 2]$ **19.** $[\frac{7}{2}, 9]$ **21.** $[-4, 4]$

23. $[-5, 3]$ **25.** $[1, \frac{13}{3}]$

27. (a)

(b)

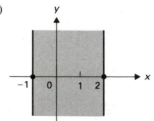

(c)

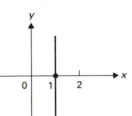

(d)

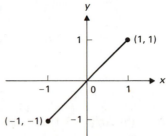

29. (a) $(1, -3)$ (b) $(2, 6)$ **31.** (a) $\sqrt{89}$ (b) $\sqrt{171}$

33. The distance from $(-1, 3)$ to $(4, 1)$ is $\sqrt{25 + 4} = \sqrt{29}$. The distance from $(-1, 3)$ to $(1, 8)$ is $\sqrt{4 + 25} = \sqrt{29}$. The distance from $(4, 1)$ to $(1, 8)$ is $\sqrt{9 + 49} = \sqrt{58}$. Since $(\sqrt{58})^2 = (\sqrt{29})^2 + (\sqrt{29})^2$, the Pythagorean relation holds and we have a right triangle. We have Area $= \frac{1}{2}$(Product of lengths of legs) $= \frac{1}{2}\sqrt{29}\sqrt{29} = \frac{29}{2}$.

35. $\sqrt{145}$ mi **37.** $\sqrt{1553}/4 \approx 9.85$ mi **39.** $(x + 1)^2 + (y - 2)^2 = 9$ **41.** $(x + 2)^2 + (y + 3)^2 = 5$

43. Center $(-3, 0)$, radius 7   **45.** Center $(2, -3)$, radius 4   **47.** Center $(\frac{3}{2}, 3)$, radius $9/(2\sqrt{2})$ **49.** $(x - 2)^2 + (y + 2)^2 = 25$
51. $(-3, -4)$ **53.** (a) $(7, 4)$ (b) $(2, 1)$ (c) $(3, -2)$ **55.** ≈ 4.88233 **57.** ≈ -0.10642 **59.** ≈ 4.47214 **61.** ≈ 29.84445
63. Center $\approx (-1.5788, 0.6177)$, radius ≈ 2.49256

Section 1.2

1. $-3/5$ **3.** Vertical line; slope undefined   **5.** 0   **7.** -8 **9.** $\frac{13}{2}$ **11.** $-2/3$ **13.** (a) A parallelogram   (b) A rectangle
15. (a) Not a parallelogram   (b) Not a rectangle   **17.** (a) 0   (b) $\frac{5}{4}$ **19.** Not collinear   **21.** Collinear   **23.** $(-13, 20)$
25. $(3, 0), (-5, 4)$
27. The midpoint of the side from $(0, 0)$ to (b, c) is $(b/2, c/2)$. The midpoint of the side from $(a, 0)$ to (b, c) is

$((a + b)/2, c/2)$. The slope of the line joining these midpoints is $\dfrac{c/2 - c/2}{a/2} = 0$, which is the slope of the line joining

$(0, 0)$ and $(a, 0)$.
29. The slope is 3000, which represents (in dollars) the average increase in cost per year.   **31.** $y - 4 = 5(x + 1)$, or $y = 5x + 9$
33. $y - 2 = 0(x - 4)$, or $y = 2$ **35.** $y + 5 = -\frac{6}{5}(x - 4)$, or $5y + 6x = -1$ **37.** $x = -3$ **39.** $2x + 3y = -1$
41. $4x - y = -11$ **43.** $y = 5x + 15$ **45.** $3y + 2x = 26$ **47.** $4y - 3x = 25$ **49.** $2x + y = 1$ **51.** Perpendicular   **53.** Neither
55. Slope 1, x-intercept 7, y-intercept -7 **57.** Slope undefined, x-intercept 4, no y-intercept
59.
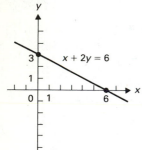
61.
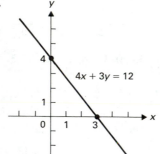
63.
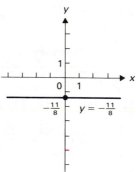

65. $(-52, 37)$
67. 2
69.

Side	Equation of the perpendicular bisector
$(-a, 0)$ to $(a, 0)$	$x = 0$ (the y-axis)
$(-a, 0)$ to (b, c)	$y - \dfrac{c}{2} = -\dfrac{a + b}{c}\left(x - \dfrac{b - a}{2}\right)$
$(a, 0)$ to (b, c)	$y - \dfrac{c}{2} = \dfrac{a - b}{c}\left(x - \dfrac{a + b}{2}\right)$

The first two meet at $\left(0, \dfrac{b^2 - a^2}{2c} + \dfrac{c}{2}\right)$

The equation of the third side is also satisfied by the point $\left(0, \dfrac{b^2 - a^2}{2c} + \dfrac{c}{2}\right)$, so all three perpendicular bisectors meet at
this point.
71. $x^2 + y^2 + 6x - 4y = -3$ **73.** $d = 13 + \frac{3}{2}(t - 3)$ in., $3 \le t \le 11$ **75.** $W = \frac{1}{6}d$ **77.** (a) $d = \frac{1}{20}t$ (b) 4:20 P.M.

Section 1.3

1. $V = x^3, x > 0$ **3.** $A = s^2/(4\pi), s > 0$ **5.** $V = d^3/(3\sqrt{3}), d > 0$ **7.** $s = (\sqrt{34})t, t \ge 0$ **9.** (a) 1   (b) 6   (c) 6
11. (a) $-\frac{5}{3}$ (b) -1 (c) 5   **13.** (a) 1   (b) 11   (c) $\frac{1}{2}$ (d) $\frac{63}{4}$ **15.** (a) $4 + 4(\Delta x) + (\Delta x)^2$ (b) $4(\Delta x) + (\Delta x)^2$
(c) $4 + \Delta x, \Delta x \ne 0$ **17.** (a) $1/(-3 + \Delta t)$ (b) $\Delta t/[3(-3 + \Delta t)]$ (c) $1/[3(-3 + \Delta t)], \Delta t \ne 0$ **19.** $x \ne 0$ **21.** $x \ne 1, 2$
23. $u \le -1$ or $u \ge 1$ **25.** $x \ge 4$ **27.** $x \ne \pm 1$ **29.** $2, x \ne 0$ **31.** $t - 2, t \ne -2$ **33.** $\begin{cases} -1, x < 0 \\ 1, x > 0 \end{cases}$ **35.** $4 + \Delta x, \Delta x \ne 0$
37. (a) 1   (b) $\frac{1}{2}$ (c) $-\frac{3}{2}$

39.

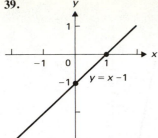

$y = x - 1$

41.

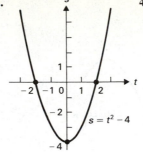

$s = t^2 - 4$

43.

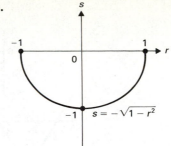

$s = -\sqrt{1 - r^2}$

45.

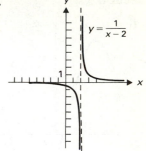

$y = \dfrac{1}{x - 2}$

47.

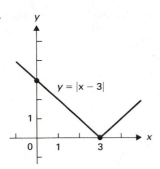

$y = |x - 3|$

49.

x	y
-1	0
$-\frac{1}{2}$	$-\frac{1}{3}$
0	-1
$\frac{1}{2}$	-3
$\frac{3}{4}$	-7
$\frac{7}{8}$	-15
$\frac{9}{8}$	17
$\frac{5}{4}$	9
$\frac{3}{2}$	5
2	3
$\frac{5}{2}$	$\frac{7}{3}$
3	2

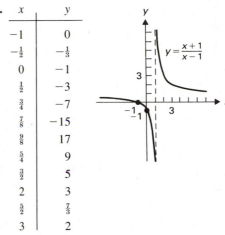

$y = \dfrac{x + 1}{x - 1}$

51. (a) ≈ 0.089692 (b) ≈ -755.06 (c) ≈ -317.77 **53.** (a) ≈ 0.56618 (b) ≈ 0.15578 (c) ≈ 8.9092 **55.** (a) ≈ 1.2354
(b) ≈ 0.91433 (c) ≈ -1.5444

57.

x	$(x + 1)/\sqrt{x^3 + 1}$
0	1.0
1	1.4142
2	1.0
3	0.7559
4	0.6202
5	0.5345
6	0.4752
7	0.4313
8	0.3974
9	0.3701
10	0.3477

$y = \dfrac{(x + 1)}{\sqrt{x^3 + 1}}$

Section 1.4

1.
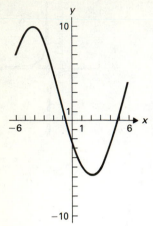

3.
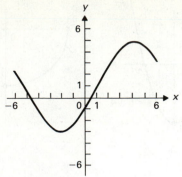

5.
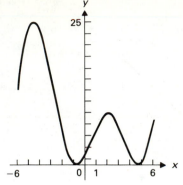

7.
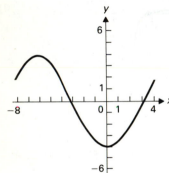

9.
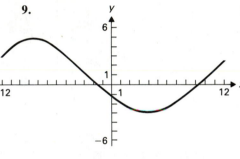

11.
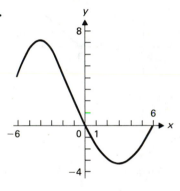

13.
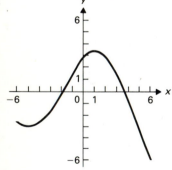

15.
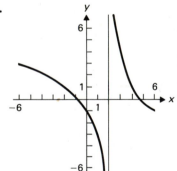

17.
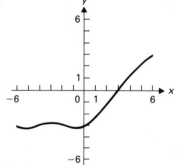

19.

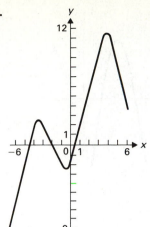

21. $g(x) = 3 + |x + 1|$

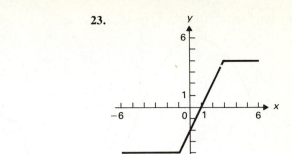

23.

25.

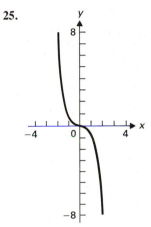

27.

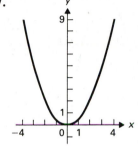

29.

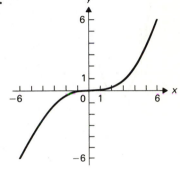

31.

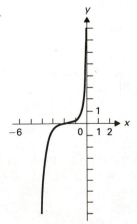

33.

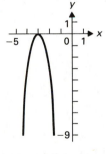

35.

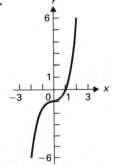

37.

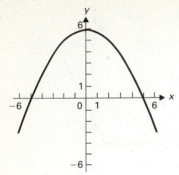

39.

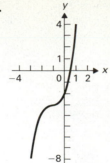

41.

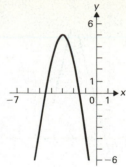

43.

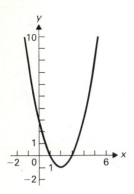

45.

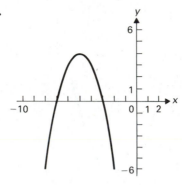

47. $\frac{1}{7}$ **49.** 4   **51.** 4   **53.** $\sqrt{2}$ **55.** 4   **57.** 1   **59.** 100

61.

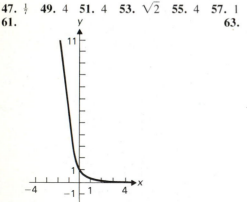

63.

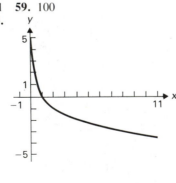

65.

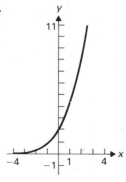

Section 1.5

1. $\sqrt{3}/2$ **3.** $-1/\sqrt{3}$ **5.** $-\sqrt{2}$ **7.** $2/\sqrt{3}$ **9.** Undefined   **11.** Undefined   **13.** $-\sqrt{2}$ **15.** $-\sqrt{2}$ **17.** -2 **19.** 0   **21.** 1
23. 0   **25.** -1 **27.** Undefined   **29.** $\sqrt{2}$ **31.** $2\sqrt{2}/3$ **33.** $-1/\sqrt{24}$ **35.** $-\sqrt{15}$ **37.** $-4\sqrt{5}/9$ **39.** $-\frac{7}{8}$ **41.** (a) $(u, -v)$
(b) $\sin(-x) = -v = -\sin x; \cos(-x) = u = \cos x$

43. (a) $(v, -u)$ (b) $\sin\left(x - \frac{\pi}{2}\right) = -u = -\cos x; \cos\left(x - \frac{\pi}{2}\right) = v = \sin x$ **45.** $\sec(-x) = \dfrac{1}{\cos(-x)} = \dfrac{1}{\cos x} = \sec x$

47. $\sin\left(x - \frac{\pi}{2}\right) = \sin x \cos\left(-\frac{\pi}{2}\right) + \cos x \sin\left(-\frac{\pi}{2}\right) = (\sin x)(0) + (\cos x)(-1) = -\cos x$

49. $\sec\left(x - \frac{\pi}{2}\right) = \dfrac{1}{\cos[x - (\pi/2)]} = \dfrac{1}{(\cos x)\cos(-\pi/2) - (\sin x)\sin(-\pi/2)}$

$$= \dfrac{1}{(\cos x)(0) - (\sin x)(-1)} = \dfrac{1}{\sin x} = \csc x$$

51. $\cos 2x = \cos^2 x - \sin^2 x = \cos^2 x - (1 - \cos^2 x) = 2\cos^2 x - 1, \ \cos 2x = \cos^2 x - \sin^2 x = (1 - \sin^2 x) - \sin^2 x = 1 - 2\sin^2 x$

53. $\frac{3}{10}$ **55.** $\sqrt{74 - 35\sqrt{2}} \approx 4.95$ **57.** $\sqrt{1 - \left(\frac{37}{40}\right)^2} \approx 0.38$

59. Amplitude 3, period 2π

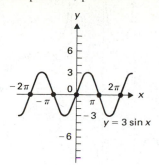

61. Amplitude $\frac{1}{2}$, period 2π

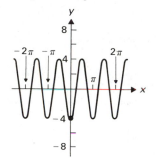

63. Amplitude 1, period 2π

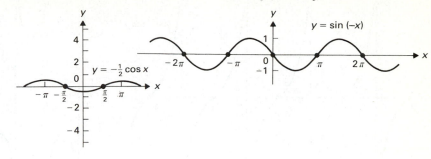

65. Amplitude 3, period $2\pi/3$

67. Amplitude 4, period π

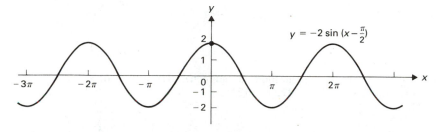

69. Amplitude 2, period 2π

71. Amplitude 5, period 4π

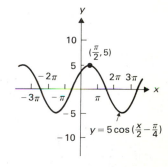

73. Amplitude 5, period 8π

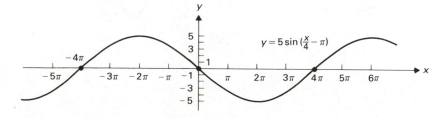

75. Period π

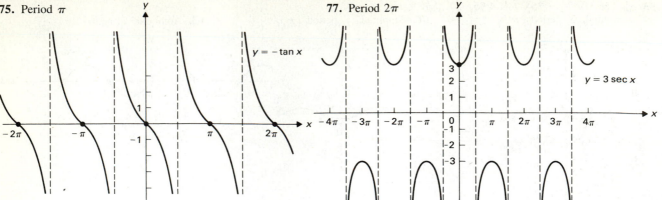

$y = -\tan x$

77. Period 2π

$y = 3 \sec x$

79. Period π

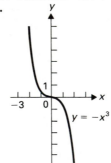

$y = \sin^2 x$

81. Period π

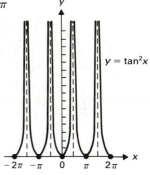

$y = \tan^2 x$

Section 1.6

1. $x \geq 3$ **3.** $t \geq -4$, $t \neq \pm 1$ **5.** $u < 4$ **7.** $g(t) \geq 0$ **9.** No, the functions are not equal. The domain of g includes -1, while -1 is not in the domain of f.   **11.** 3   **13.** Undefined   **15.** 7   **17.** $8 - 12\sqrt{2}$ **19.** $\sqrt{3}$

21.

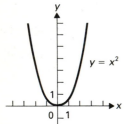

$y = x^2$

23.

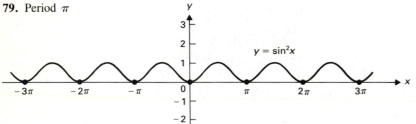

$y = -x^3$

25.

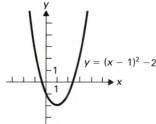

$y = (x - 1)^2 - 2$

27.

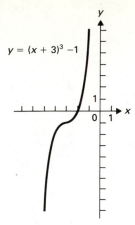

$y = (x + 3)^3 - 1$

29.

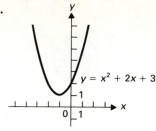

$y = x^2 + 2x + 3$

31. 3 **33.** 6 **35.** 3 **37.** 1 **39.**

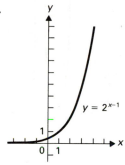

$y = 2^{x-1}$

41. $5\pi/9$ **43.** $\sqrt{3}/2$ **45.** $-\sqrt{3}/2$

47.

49.

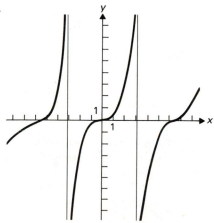

Supplementary Exercises for Chapter 1

1. (a) 7   (b)

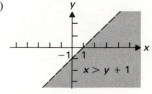

$x > y + 1$

3. (a) 10   (b) (1, 6)

5. (a) $x^2 + y^2 - 4x + 2y = 48$ (b)

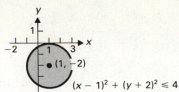

$(x - 1)^2 + (y + 2)^2 \leq 4$

7. (a) $\frac{3}{4}$ (b) -9 **9.** (a) $x = -4$ (b) $x - 3y = -7$ **11.** (a) 1 (b) $-1/2 \leq x \leq 9/2$ **13.** $4 \pm 4\sqrt{3}$
15. $x^2 + y^2 - 14x - 12y = -49$ **17.** $-3/2$ **19.** $4x + 3y = 4$ **21.** (a) 1 (b) $-7/9$ (c) $\frac{7}{4}$
23. (a) $-5 \leq x \leq 5$ (b) 4 (c)

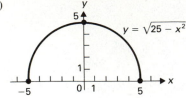

$y = \sqrt{25 - x^2}$

25. $t > -3$, $t \neq 0$ **27.**

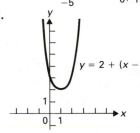

$y = 2 + (x - 1)^4$

29.

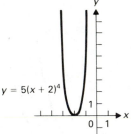

$y = 5(x + 2)^4$

31. 1 **33.** 9 **35.** 5 **37.** (a) 4 (b) 2 (c) $\frac{19}{4}$ **39.**

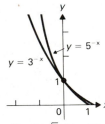

$y = 5^{-x}$

$y = 3^{-x}$

41. (a) $-1/\sqrt{3}$ (b) $-1/\sqrt{2}$ **43.** (a) -1 (b) $\sqrt{2}$ **45.** $(2\sqrt{6})/5$ **47.** $-\sqrt{3}/2$ **49.** Period 6π, amplitude $\frac{1}{2}$
51. (a) Adding:

$$-|a| \leq a \leq |a|$$
$$-|b| \leq b \leq |b|$$
$$\overline{-(|a| + |b|) \leq a + b \leq |a| + |b|}$$

so $|a + b| \leq |a| + |b|$.
(b) From part (a), we have $|a| = |(a - b) + b| \leq |a - b| + |b|$, so $|a - b| \geq |a| - |b|$.
53. From Exercise 52, $2(a_1 a_2 + b_1 b_2) \leq 2\sqrt{a_1^2 + b_1^2} \cdot \sqrt{a_2^2 + b_2^2}$. Adding $a_1^2 + a_2^2 + b_1^2 + b_2^2$ to both sides, we obtain
$(a_2 + a_1)^2 + (b_2 + b_1)^2 \leq (\sqrt{a_1^2 + b_1^2} + \sqrt{a_2^2 + b_2^2})^2$.
55. (a) Do not intersect (b) $10x + 2y = 9$ **57.** (a) 4 (b) 5 **59.** $2 - \sqrt{5} < x < 2 + \sqrt{5}$ **61.** $18\sqrt{5}/5$
63.

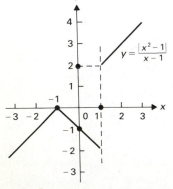

$y = \dfrac{|x^2 - 1|}{x - 1}$

65. $g(y) = (3y + 7)/(2 - y)$

CHAPTER 2

Section 2.1

1. $m_{sec} = 8.01$, $m_{tan} = 8$ **3.** $m_{sec} = -4.2$, $m_{tan} = -4$ **5.** $m_{sec} = 0.31$, $m_{tan} = 0$ **7.** $m_{sec} \approx 0.001$, $m_{tan} = 0$
9. $m_{sec} = 2a + 3.2$, $m_{tan} = 2a + 3$ **11.** $y = 8x - 16$ **13.** $x = 2$ **15.** $z = -v - 4$ **17.** -2 **19.** -2 **21.** -2 **23.** -2
25. 0 **27.** 0 **29.** $\frac{2}{5}$ **31.** 0 **33.** 2 **35.** 0 **37.** $-1/9$ **39.** (a) -4 (b) 0 (c) 9 **41.** (a) 2 (b) Does not exist (c) 2
43. (a) $\frac{1}{8}$ (b) Does not exist (c) $\frac{1}{3}$ **45.** (a) $\frac{1}{2}$ (b) Does not exist (c) Does not exist **47.** No points near -3 except -3
itself are in the domain of $\sqrt{-(x+3)^2}$. **49.** No points near 0 are in the domain of $\sqrt{x^2 - 9}$. **51.** Does not exist **53.** $\frac{1}{2}$
55. -2

Section 2.2

1. We have $\displaystyle\lim_{x \to a} [f(x) - g(x)] = \lim_{x \to a} [f(x) + (-1)g(x)]$

$$= \lim_{x \to a} f(x) + \lim_{x \to a} [(-1)g(x)] \qquad \textit{Sum property}$$

$$= \lim_{x \to a} f(x) + [\lim_{x \to a} (-1)][\lim_{x \to a} g(x)] \qquad \textit{Product property}$$

$$= L + (-1)M = L - M.$$

3. Let $f(x) = \begin{cases} 0 & \text{for } x \le 1, \\ 2 & \text{for } x > 1, \end{cases}$ and $g(x) = \begin{cases} 2 & \text{for } x \le 1, \\ 0 & \text{for } x > 1. \end{cases}$ Then $\lim_{x \to 1} f(x)$ and $\lim_{x \to 1} g(x)$ do not exist, but $f(x) + g(x) = 2$
for all x, so $\lim_{x \to 1} [f(x) + g(x)] = 2$.

5. Let $f(x) = 0$ for all x and let $g(x) = \begin{cases} 1 & \text{for } x \ge 2, \\ -1 & \text{for } x < 2. \end{cases}$ Then $\lim_{x \to 2} g(x)$ does not exist, but $f(x)g(x) = 0$ for all x, so
$\lim_{x \to 2} [f(x)g(x)] = 0$.

7. Let $f(x) = \begin{cases} 1 & \text{for } x \ge 0, \\ -1 & \text{for } x < 0, \end{cases}$ and let $g(x) = -f(x)$. Then $f(x)g(x) = -1$ for all x, so $\lim_{x \to 0} [f(x)g(x)] = -1$, although
neither $\lim_{x \to 0} f(x)$ nor $\lim_{x \to 0} g(x)$ exists.

9. Such an example is impossible by the quotient property in Theorem 2.1. **11.** Does not exist **13.** $\frac{1}{3}$ **15.** 1 **17.** 0 **19.** 1
21. 1 **23.** $\frac{1}{4}$ **25.** -1
27.

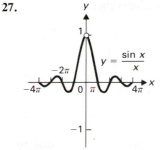

29. -0.0185 **31.** 2.718 **33.** -1 **35.** 3.42552 **37.** 1.34164

39. 9.88751 **41.** 2.582

Section 2.3

1. Does not exist **3.** ∞ **5.** ∞ **7.** $-\infty$ **9.** $-\infty$ **11.** ∞ **13.** ∞ **15.** ∞ **17.** $-\infty$ **19.** 1 **21.** -1 **23.** 0 **25.** $-\infty$
27. -2 **29.** $-\infty$ **31.** 0 **33.** $-\infty$ **35.** ∞ **37.** -1 **39.** $-\infty$ **41.** ∞ **43.** ∞ **45.** 0 **47.** Does not exist **49.** 12 ft **51.** 1
53. 1 **55.** 2.71828

Section 2.4

1.   **3.**   **5.**

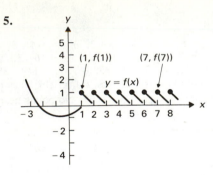

7. No; $\lim_{x \to 2} f(x) = -8 \neq 8$.

9. (a) Yes; $\lim_{x \to 3+} f(x) = 0 = \lim_{x \to 3-} f(x) = f(3)$.   (b) Yes; $\lim_{x \to 1+} f(x) = -2 = \lim_{x \to 1-} f(x) = f(1)$.   (c) Yes; continuous where $x \neq 1, 3$ as part of a polynomial function and continuous at 1 and 3 by parts (a) and (b).

11. (i) If your birth weight was 7 lb and you now weigh 138 lb, at some time you weighed exactly 71.583 lb.   (ii) If you are driving on a trip and going 47 mph at 1:00 P.M. and 63 mph at 2:00 P.M., then at some instant between 1:00 and 2:00 P.M. you were going exactly 53.4 mph.   (iii) If a tire with pressure 15 lb/in^2 is inflated to a pressure of 30 lb/in^2, then at some instant during inflation the pressure was 23 lb/in^2.

13. Since $f(a)$ and $f(b)$ have opposite sign, 0 is between $f(a)$ and $f(b)$. By Theorem 2.3, we have $f(x_0) = 0$ for some x_0 in $[a, b]$.

15. Let $f(x) = x^4 + 3x^3 + x + 4$.   (a) $f(-2) = -6$, $f(0) = 4$; by the corollary, $f(x) = 0$ has a solution in $[-2, 0]$.   (b) No; $f(x) > 0$ for $x \geq 0$ as a sum of positive quantities.

17. No; $x^2 - 4x + 5 = (x - 2)^2 + 1 > 0$ for all x.   **19.** F F T F T

21. If $f(S/2) = 96$, we are done. Suppose $f(S/2) \neq 96$. Then $f(0) = 96 -$ (Speed at $S/2$) and $f(S/2) =$ (Speed at $S/2$) $- 96 = -f(0)$. Then $f(0)$ and $f(S/2)$ have opposite sign, so for some c, where $0 < c < S/2$, we have $f(c) =$ (Speed at c) $-$ (Speed at $c + S/2$) $= 0$. Thus (Speed at c) $=$ (Speed at $c + S/2$).

23. (a) $\lim_{x \to \infty} f(x) = \lim_{x \to -\infty} f(x) = \infty$ (b) Find $K_1 > 0$ such that $f(x) > f(0)$ for $x > K_1$, and find $K_2 > 0$ such that $f(x) > f(0)$ for $x < -K_2$. Let C be the maximum of K_1 and K_2.   (c) By Theorem 2.4, f attains a minimum over $[-C, C]$. Since $f(x) > f(0)$ for $|x| > C$, this minimum value must also be the minimum attained for all x.

Section 2.5

1. $0 < \delta \leq \varepsilon$ **3.** $0 < \delta \leq \varepsilon/2$ **5.** $0 < \delta \leq \varepsilon/5$ **7.** (a) For some $\varepsilon > 0$, there does not exist $\delta > 0$.   (b) For some apple blossom, there exists no apple.   (d) Find one $\varepsilon > 0$ such that for every $\delta > 0$, there is an x_δ such that $0 < |x_\delta - a| < \delta$, but $|f(x_\delta) - c| \geq \varepsilon$.

9. Suppose the domain of f contains points x arbitrarily close to a but greater than a. Then $\lim_{x \to a+} f(x) = L$ if for each $\varepsilon > 0$ there exists $\delta > 0$ such that when $a < x < a + \delta$ and x is in the domain of f, we have $|f(x) - L| < \varepsilon$.

11. Suppose the domain of f contains points x arbitrarily close to a but greater than a. Then $\lim_{x \to a+} f(x) = -\infty$ if for each real number γ there exists $\delta > 0$ such that when $a < x < a + \delta$ and x is in the domain of f, we have $f(x) < \gamma$.

13. Suppose the domain of f contains points x arbitrarily close to a but less than a. Then $\lim_{x \to a-} f(x) = \infty$ if for each real number γ there exists $\delta > 0$ such that when $a - \delta < x < a$ and x is in the domain of f, we have $f(x) > \gamma$.

17. Let the domain of f contain arbitrarily large numbers. Then $\lim_{x \to \infty} f(x) = -\infty$ if for each real number μ there exists a real number γ such that when $x > \gamma$ and x is in the domain of f, we have $f(x) < \mu$.

19. (a) Incorrect; replace "$|f(x) - a|$" by "$|f(x) - f(a)|$."   (b) Incorrect; delete "$0 < .$"   (c) Correct   (d) Incorrect; replace "some" by "each."   (e) Incorrect; replace "$\leq$" by "$<$."   (f) Correct

21. Let $\varepsilon > 0$ be given and let δ be any positive number. The domain of f consists of all real numbers. For each real number a, we have $|f(x) - c| = |c - c| = 0 < \varepsilon$ for all x, and in particular for all x such that $0 < |x - a| < \delta$. Thus $\lim_{x \to a} f(x) = c$. Since $f(a) = c$, we see that f is continuous at $x = a$. Since a can be any point in the domain of f, we see that f is a continuous function.

23. *Product property:* Let f and g be functions such that $\lim_{x \to a} f(x) = L$ and $\lim_{x \to a} g(x) = M$, and suppose the domains of f and g contain points in common arbitrarily close to a but different from a. Then the limit of h, where $h(x) = f(x)g(x)$, exists at a, and $\lim_{x \to a} h(x) = LM$. *Proof:* Let $\varepsilon > 0$ be given. Since $\lim_{x \to a} g(x) = M$, we can find $\delta_1 > 0$ such that $|g(x) - M| < 1$ for all x in the domain of g such that $0 < |x - a| < \delta_1$. Thus for all such x, we have $M - 1 < g(x) < M + 1$. Since $|M + 1| \leq |M| + 1$ and $|M - 1| \leq |M| + 1$, we see that

$$|g(x)| < |M| + 1 \quad \text{for} \quad 0 < |x - a| < \delta_1$$

and x in the domain of g. Again using the fact that $\lim_{x \to a} g(x) = M$, we can find $\delta_2 > 0$ such that

$$|g(x) - M| < \frac{\varepsilon}{2(|L| + 1)} \quad \text{for} \quad 0 < |x - a| < \delta_2$$

and x in the domain of g. Since $\lim_{x \to a} f(x) = L$, we can find $\delta_3 > 0$ such that

$$|f(x) - L| < \frac{\varepsilon}{2(|M| + 1)} \quad \text{for} \quad 0 < |x - a| < \delta_3$$

and x in the domain of f. Now let δ be the minimum of the three numbers δ_1, δ_2, and δ_3, so that all the relations displayed above hold simultaneously when $0 < |x - a| < \delta$ and x is in the domains of both f and g, so that x is in the domain of h. For all such x, we obtain

$$
\begin{aligned}
|h(x) - LM| &= |f(x)g(x) - LM| \\
&= |f(x)g(x) - Lg(x) + Lg(x) - LM| \\
&= |g(x)[f(x) - L] + L[g(x) - M]| \\
&\leq |g(x)[f(x) - L]| + |L[g(x) - M]| \\
&= |g(x)||f(x) - L| + |L||g(x) - M| \\
&< (|M| + 1)\frac{\varepsilon}{2(|M| + 1)} + |L|\frac{\varepsilon}{2(|L| + 1)} \\
&\leq \varepsilon/2 + \varepsilon/2 = \varepsilon.
\end{aligned}
$$

Thus $\lim_{x \to a} h(x) = LM$.

25. *Root Property:* If $\lim_{x \to a} f(x) = L$ and $L > 0$, then $\lim_{x \to a} \sqrt{f(x)} = \sqrt{L}$. *Proof:* Let $\varepsilon > 0$ be given. Since $\lim_{x \to a} f(x) = L$ and $L > 0$, there exists $\delta_1 > 0$ such that $|f(x) - L| < L/2$ for all x in the domain of f such that $0 < |x - a| < \delta_1$. Thus for such x, we have $-L/2 < f(x) - L < L/2$, so in particular we have

$$f(x) > L/2 \quad \text{for} \quad 0 < |x - a| < \delta_1$$

and x in the domain of f. Using again the fact that $\lim_{x \to a} f(x) = L$, we see that there exists δ_2 such that

$$|f(x) - L| < \frac{(\sqrt{2} + 1)\sqrt{L}}{\sqrt{2}}\varepsilon \quad \text{for} \quad 0 < |x - a| < \delta_2$$

and x in the domain of f. Let δ be the minimum of δ_1 and δ_2, so that for x in the domain of f and such that $0 < |x - a| < \delta$, both the displayed relations above hold. For all such x, we have

$$
\begin{aligned}
|\sqrt{f(x)} - \sqrt{L}| &= \left| \frac{\sqrt{f(x)} - \sqrt{L}}{1} \cdot \frac{\sqrt{f(x)} + \sqrt{L}}{\sqrt{f(x)} + \sqrt{L}} \right| \\
&= \frac{1}{\sqrt{f(x)} + \sqrt{L}} \cdot |f(x) - L| \\
&\leq \frac{1}{\sqrt{L/2} + \sqrt{L}} \cdot \frac{(\sqrt{2} + 1)\sqrt{L}}{\sqrt{2}}\varepsilon \\
&= \frac{(\sqrt{2} + 1)\sqrt{L}}{\sqrt{L} + \sqrt{L}\sqrt{2}}\varepsilon = \varepsilon.
\end{aligned}
$$

Thus $\lim_{x \to a} \sqrt{f(x)} = \sqrt{L}$.

Supplementary Exercises for Chapter 2

1. $-2/3$ **3.** $1/\sqrt{2a - 3}$ **5.** $m_{\tan} = 1$, $y = x - 4$ **7.** $-\infty$ **9.** 0 **11.** $\frac{1}{4}$ **13.** Does not exist **15.** $\frac{3}{4}$ **17.** $-4/5$ **19.** ∞
21. $-1/10$ **23.** $\frac{7}{4}$ **25.** $-\infty$ **27.** $-\infty$ **29.** ∞ **31.** 0 **33.** 0 **35.** See Theorem 2.3. **37.** Let $f(x) = x^4 - 5x + 1$. Then $f(0) = 1$ and $f(1) = -3$. By the intermediate value theorem, $f(c) = 0$ for some c in $[0, 1]$. **39.** See Theorem 2.4. **41.** (a) Yes, $f(1) = 1^2 = 1$ is maximum, and 1 is in $(0, 1]$. (b) No, a minimum would have to be attained at "the first point to the right of 0," and there is no such point.

43. Let $f(x) = \begin{cases} 5 & \text{for } x \leq 2, \\ 3 & \text{for } x > 2, \end{cases}$ and $g(x) = \begin{cases} 7 & \text{for } x \leq 2, \\ 9 & \text{for } x > 2. \end{cases}$ Then $f(x) + g(x) = 12$ for all x, and this constant function is continuous.

45. Let $f(x) = \begin{cases} 1 & \text{for } x \text{ rational, that is, for } x \text{ a fraction,} \\ 2 & \text{for all other } x. \end{cases}$

CHAPTER 3

Section 3.1

1. $2x - 3$ **3.** $3u^2 + 1$ **5.** $-2/(2x + 3)^2$ **7.** $2t - (1/t^2)$ **9.** $1/\sqrt{2s - 1}$ **11.** $f'(x) = \begin{cases} 1 & \text{for } x > 0, \\ -1 & \text{for } x < 0 \end{cases}$ **13.** (a) -1

(b) -1 (c) 0 **15.** (a) $-2/15$ (b) $-1/8$ (c) $\frac{1}{120}$ **17.** (a) $\frac{1}{2}$ (b) $1/\sqrt{5}$ (c) $\dfrac{\sqrt{5} - 2}{2\sqrt{5}}$

19. 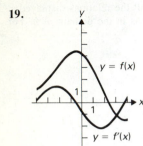   **21.** 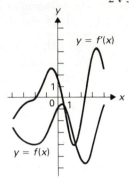   **23.**   **25.**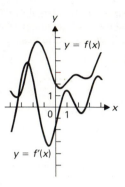

27. (a) 26 mph   (b) 23 mph   (c) 21.5 mph   (d) About 20 mph   **29.** (a) $0 \le t \le \sqrt{2}$ (b) $16\sqrt{2}$ ft/sec
(c) Speed $= 32t$ ft/sec   (d) $t = \sqrt{2}/2$ sec   **31.** (a) -0.06 dynes/sec   (b) -0.02 dynes/sec   **33.** $g(t) = |t^2 + 2t - 8|$. Other
answers are possible.

35.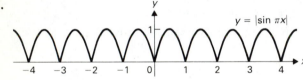

37. -0.108577 **39.** -12.51929 **41.** 45.545 **43.** $f(x) = |x|$, $a = 0$

Section 3.2

1. 3 **3.** $14u^6 + 8u$ **5.** $x - \frac{3}{2}$ **7.** $324t^3 - 160t^4$ **9.** $4s^3 + 12s^2 + 8s$ **11.** $36z^2 + 40z$ **13.** $v^2 - \frac{1}{3}$ **15.** (a) 2 (b) 0
17. (a) 28 (b) -8 **19.** (a) -1 (b) $-3/2$ **21.** (a) $y = -14x - 11$ (b) $14y = x + 240$ **23.** (a) $y = 21x - 32$
(b) $x + 21y = 212$ **25.** (a) $y = -4x - 1$ (b) $4y = x + 13$

27. (a) $1/(2\sqrt{x})$ (b) $\displaystyle\lim_{\Delta x \to 0} \frac{\sqrt{x + \Delta x} - \sqrt{x}}{\Delta x} = \lim_{\Delta x \to 0} \frac{(\sqrt{x + \Delta x} - \sqrt{x})(\sqrt{x + \Delta x} + \sqrt{x})}{\Delta x(\sqrt{x + \Delta x} + \sqrt{x})}$

$$= \lim_{\Delta x \to 0} \frac{x + \Delta x - x}{\Delta x(\sqrt{x + \Delta x} + \sqrt{x})}$$

$$= \lim_{\Delta x \to 0} \frac{1}{\sqrt{x + \Delta x} + \sqrt{x}} = \frac{1}{2\sqrt{x}}$$

(c) $\dfrac{3}{2\sqrt{x}} - 4x$ (d) $\dfrac{\sqrt{5}}{2\sqrt{x}} - \dfrac{\sqrt{7}}{2\sqrt{x}}$
29. (a) 12 in^3/sec   (b) 75 in^3/sec   **31.** (a) 256π in^2 (b) 256π in^2/sec   **33.** \$488

Section 3.3

1. $6s + 17$ **3.** $2x/3$ **5.** $-3/y^2$ **7.** $12u^2 + (4/u^3)$ **9.** $(z^2 - 1)(2z + 1) + (z^2 + z + 2)(2z)$
11. $(x^2 + 1)[(x - 1)3x^2 + (x^3 + 3)] + [(x - 1)(x^3 + 3)](2x)$ **13.** $\left(\dfrac{1}{t^2} - \dfrac{4}{t^3}\right)2 + (2t + 3)\left(\dfrac{-2}{t^3} + \dfrac{12}{t^4}\right)$
15. $4 + (3/r^2)$ **17.** $[(x + 3)2x - (x^2 - 2)]/(x + 3)^2$ **19.** $[(t^2 + 2)((t^2 + 9) + (t - 3)2t) - (t^2 + 9)(t - 3)2t]/(t^2 + 2)^2$
21. $\dfrac{(u - 1)(4u^2 + 5)[(2u + 3)2u + (u^2 - 4)2] - (2u + 3)(u^2 - 4)[(u - 1)8u + (4u^2 + 5)]}{(u - 1)^2(4u^2 + 5)^2}$

23. $\dfrac{x+1}{2x+3}\left(\dfrac{-1}{x^2}+\dfrac{2}{x^3}\right)+\left(\dfrac{1}{x}-\dfrac{1}{x^2}\right)\dfrac{1}{(2x+3)^2}$ **25.** $x\cos x+\sin x$ **27.** $2\sin u\cos u$ **29.** $\dfrac{(\cos x)^2+(\sin x)^2}{(\cos x)^2}=\sec^2 x$

31. $\dfrac{3v^2\sin v-v^3\cos v}{(\sin v)^2}$ **33.** $\dfrac{(t^2-4t)\cos t-(2t-4)\sin t}{(t^2-4t)^2}$ **35.** -3 **37.** 2 **39.** $-1380/169$ **41.** $\frac{41}{6}$

43. $f(4)=\frac{28}{17},\ f'(4)=-\frac{1}{17}$ **45.** $f'(3)=-5,\ g'(3)=2$ **47.** Tangent line: $x+y=2$; normal line: $y-x=0$

49. Tangent line: $y+5x=-3$; normal line: $5y-x=-15$

51. (a) Slope ≈ 0 (b) 0

53. (a) Slope ≈ -1.7 (See the figure in the answer to Exercise 51.)   (b) -1.8

55. (a) Slope ≈ -2 (b) -1.8

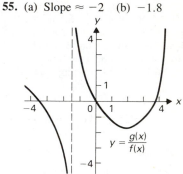

57. The marginal propensity to save (consume) when the income is I is the proportion of the next dollar earned that would be saved (spent). That is, 100 times this marginal propensity to save (consume) is the number of cents saved (spent) from the next dollar earned.

59. (a) $\$\frac{2}{9}$ per dollar income   (b) $\$\frac{7}{9}$ per dollar income   **61.** (a) 0.373

Section 3.4

1. $-x\sin x+\cos x$ **3.** $(t^2+3t)(\sec t\tan t)+(2t+3)(\sec t)$ **5.** $-2\cot v\csc^2 v$ **7.** $\sin r\sec^2 r+\cos r\tan r$

9. $\dfrac{(1-\cos x)(\cos x)-\sin^2 x}{(1-\cos x)^2}=\dfrac{-1}{1-\cos x}$ **11.** $\dfrac{-(3+2\cos s)(\csc^2 s)+(\cot s\sin s)}{(3+2\cos s)^2}$ **13.** $\cos x$

15. $-2\tan y\csc^2 y\cot y+\sec^2 y\csc^2 y=(\csc^2 y)(\sec^2 y-2)$ **17.** $2(\cos u)(1+\sin u)$ **19.** $\dfrac{(v^2-4v)(\sec^2 v)-(2v-4)(\tan v)}{(v^2-4v)^2}$

21. (a) $\cos x$ (b) $-\sin x$ (c) $-\cos x$ (d) $\sin x$ (e) $\cos x$ (f) $-\sin x$ **23.** (a) $x\cos x+\sin x$ (b) $-x\sin x+2\cos x$

(c) $-x\cos x-3\sin x$ (d) $x\sin x-4\cos x$ (e) $x\cos x+13\sin x$ (f) $-x\sin x+1034\cos x$ **25.** $y=2x+1-\dfrac{\pi}{2}$

27. $3x-6y=3\sqrt{3}+5\pi$ **29.** $y=-2x+\dfrac{3\pi}{2}-1$ **31.** $-3\pi/2,\ -\pi/2,\ \pi/2,\ 3\pi/2$ **33.** $-5\pi/6,\ -\pi/6,\ 7\pi/6,\ 11\pi/6$

35. $y=-3x,\ f'(0)=-3$ **37.** $p(x)=x^2-1$, tangent line $y=2x-2,\ f'(1)=2$

Section 3.5

1. (a) 3 (b) 3 **3.** (a) 1404 (b) 1404 **5.** $12(3x+2)^3$ **7.** $9s^2(s^2+3s)^2(s^3-1)^2+2(s^3-1)^3(s^2+3s)(2s+3)$

9. $\dfrac{16u(4u^2+1)^2-16u(8u^2-2)(4u^2+1)}{(4u^2+1)^4}$ **11.** $8x-2+5x^{2/3}$ **13.** $-\frac{1}{2}w^{-3/2}$ **15.** $\frac{2}{3}t^{-1/3}+\frac{1}{5}t^{-4/5}$ **17.** $(2x+1)^{-1/2}$

19. $-(5x^2+10x)^{-3/2}(5x+5)$ **21.** $t(t^2+1)^{-1/2}$ **23.** $\dfrac{\frac{1}{2}(y+1)y^{-1/2}-\sqrt{y}}{(y+1)^2}$

25. $\frac{8}{3}\sqrt{3t+4}\,(4t+2)^{-1/3}+\frac{3}{2}(4t+2)^{2/3}(3t+4)^{-1/2}$

27. $\sqrt{2x+1}\left[\dfrac{2(2x+5)(4x^2-3x)(8x-3)-2(4x^2-3x)^2}{(2x+5)^2}\right]+\dfrac{(4x^2-3x)^2}{2x+5}(2x+1)^{-1/2}$ **29.** $2\cos 2x$ **31.** $-(\sin\sqrt{v}/2\sqrt{v})$

33. $6\cos(2-3w)\sin(2-3w)$ **35.** $-2\sin^3 z\cos z+2\sin z\cos^3 z$ **37.** $\frac{1}{2}(1+2\cot^2 s)^{-1/2}(-4\cot s\csc^2 s)$

39. $3[\cos(\tan 3x)]\sec^2 3x$ **41.** -2 **43.** 40 **45.** $4y+3x=25$ **47.** $3y-x=5$

Section 3.6

1. $2e^{2x}$ **3.** $4e^{4t+1}$ **5.** $2xe^{x^2+1}$ **7.** $e^t(t+1)$ **9.** $\dfrac{e^r(r-1)-1}{r^2}$ **11.** $e^{3x}(2\sec^2 2x+3\tan 2x)$ **13.** $2e^{2x}\sec(e^{2x})\tan(e^{2x})$

15. $\dfrac{e^{2x}}{\sqrt{e^{2x}+1}}$ **17.** $-2e^t\cos(e^t)\sin(e^t)$ **19.** $(\ln 3)3^{\sin x}\cos x$ **21.** $(\ln 5)\,5^{t^2+t}(2t+1)$ **23.** $(\ln 10)(10^5)\cos(10^5)$

25. $e^{3x}\,10^{x+1}(3+\ln 10)$ **27.** $x+ey=e^2+1$ **29.** $(1,7)$ **31.** $\approx \$44{,}816.89$ **33.** ≈ 3.85 yr

Section 3.7

1. $y'=5x^4-12x^3$, $y''=20x^3-36x^2$, $y'''=60x^2-72x$ **3.** $y'=(1/\sqrt5)(-\tfrac12)t^{-3/2}$, $y''=(1/\sqrt5)(\tfrac34)t^{-5/2}$, $y'''=(1/\sqrt5)(-\tfrac{15}{8})t^{-7/2}$ **5.** $y'=x(x^2+1)^{-1/2}$, $y''=-x^2(x^2+1)^{-3/2}+(x^2+1)^{-1/2}$, $y'''=3x^3(x^2+1)^{-5/2}-3x(x^2+1)^{-3/2}$

7. $y'=(s+1)^{-2}$, $y''=-2(s+1)^{-3}$, $y'''=6(s+1)^{-4}$ **9.** $-1/(10\sqrt{10})$ **11.** $\tfrac34$ **13.** $e^{2t}(4t^2+8t+2)$ **15.** $-2e^s\sin s$

17. $4(\sec^3 2x+\sec 2x\tan^2 2x)$ **19.** $e^x(x+n)$ **21.** $v=\tfrac34$, $a=\tfrac{7}{64}$ **23.** (a) $v=-10/(t+1)^2$ (b) The body moves in the direction of decreasing x as time increases.   (c) $a=20/(t+1)^3$ (d) The velocity increases as time increases.   (e) Decreasing

25. Upward   **27.** Decreasing   **29.** Decreasing, downward   **31.** (a) -2.8 m/sec   (b) 2.8 m/sec   (c) 4 m   **33.** (a) 5 cm

(b) $-\dfrac{5\pi}{2}$ cm/sec   (c) $\dfrac{5\pi}{2}$ cm/sec   (d) $\dfrac{5\sqrt3\pi^2}{2}$ cm/sec^2 **35.** $a=0$ cm, $b=3/\pi$ cm   **37.** $5000(0.6)^{t/200}$ ft/sec   **39.** 1000

41. $f''(1)\approx 16.25$; $f'''(1)\approx 16.875$ **43.** $f''(3)\approx 3.84362$; $f'''(3)\approx 2.6642$ **45.** $f''(1)\approx 14.46699$; $f'''(1)\approx 25.72$

Section 3.8

1. $\tfrac34$ **3.** 1   **5.** $-\tfrac53$ **7.** -3 **9.** $\tfrac75$ **11.** $\tfrac{19}{14}$ **13.** $-\tfrac34$ **15.** $\tfrac12$ **17.** -6 **19.** $-1/5$ **21.** $3/(3x+2)$, $x>-\tfrac23$ **23.** $3/x$, $x>0$

25. $-2\tan s$ **27.** $2(\ln x)\cdot\dfrac1x$ **29.** $(9x^2-4)/(6x^3-8x)$, $3x^3-4x>0$ **31.** $\tan u+\dfrac{\sec^2 u}{\tan u}$, $\sec u\tan u>0$

33. $-2\tan x+6\cot 2x$, $\sin 2x>0$ **35.** $1/[(\ln 10)x]$, $x>0$ **37.** $(\cot s)/(\ln 2)$, $\sin s>0$

39. $\dfrac{1}{\ln 5}\left(\dfrac{2}{2x+1}-\dfrac1x\right)$, $\dfrac{2x+1}{x}>0$ **41.** $\dfrac{-7}{27}$ **43.** 4   **45.** $x^x(1+\ln x)$ **47.** $t^{\sin t}\left[\dfrac{\sin t}{t}+(\ln t)(\cos t)\right]$

49. $(x^2+1)^{x+1}\left[\dfrac{2x(x+1)}{x^2+1}+\ln(x^2+1)\right]$ **51.** $u^{(e^u)}\left[\dfrac{e^u}{u}+e^u(\ln u)\right]$ **53.** $(\sin t)^{\cos t}\,[\cos t\cot t-(\sin t)\ln(\sin t)]$

55. Computation shows that the slope of the first curve is -1 at $(2,4)$, while the second curve has slope 1 there. The curves are orthogonal.

57. Let (x_0,y_0) be a point of intersection. Since both c and k are nonzero, neither x_0 nor y_0 is zero at a point of intersection. By implicit differentiation, the slope of $y^2-x^2=c$ at (x_0,y_0) is x_0/y_0, while the slope of $xy=k$ is $-y_0/x_0$. The curves are orthogonal.

Section 3.9

1. $dy=\dfrac{1}{(x+1)^2}\,dx$ **3.** $dA=2\pi r\,dr$ **5.** $3e^{\tan 3x}\sec^2 3x\,dx$ **7.** $12\tan 2t\sec^2 2t\,dt$ **9.** 0.99   **11.** $\dfrac{47.68}{9}\approx 5.298$

13. $(\sqrt3/2)+\pi/180$ **15.** 10.05   **17.** 3.975   **19.** $\tfrac{23}{12}$ **21.** $\tfrac{93}{46}$ **23.** (a) $3/\pi$ ft   (b) The estimate is exact, because the circumference of the earth is a *linear* function of the radius.   **25.** $\tfrac{7}{24}$ ft^2 **27.** $25/(68\pi)$ ft   **29.** 6%   **31.** 0.5%

33. $\dfrac{E(\Delta x)}{\Delta x}=\Delta x$ for $\Delta x\neq 0$, so $\lim_{\Delta x\to 0}\dfrac{E(\Delta x)}{\Delta x}=\lim_{\Delta x\to 0}\Delta x=0$ **35.** $1-\dfrac{e-3}{e+1}\approx 1.08$ **37.** $4+\dfrac{1}{4(\ln 4)-1}\approx 4.22$

39. $2-\dfrac{0.05}{32(1+\ln 2)}\approx 1.99908$

Supplementary Exercises for Chapter 3

1. $f'(a)=\lim\limits_{\Delta x\to 0}\dfrac{f(a+\Delta x)-f(a)}{\Delta x}$ **3.** $f'(x)=\lim\limits_{\Delta x\to 0}\dfrac{[1/(2(x+\Delta x)+1)]-[1/(2x+1)]}{\Delta x}$

$$=\lim_{\Delta x\to 0}\dfrac{(2x+1-2x-2\cdot\Delta x-1)/[(2(x+\Delta x)+1)(2x+1)]}{\Delta x}$$

$$=\lim_{\Delta x\to 0}\dfrac{-2}{[2(x+\Delta x)+1](2x+1)}=\dfrac{-2}{(2x+1)^2}$$

5. $11y-x=100$ **7.** $8y+x=-22$ **9.** $\dfrac{(4t^3+3)(16t-2)-(8t^2-2t)(12t^2)}{(4t^3+3)^2}$ **11.** 0

13. $\dfrac{(2x^3+7)(2x-3)-(x^2-3x)(6x^2)}{(2x^3+7)^2}$ **15.** $\tfrac12(x^2-17x)^{-1/2}(2x-17)$ **17.** $-15/4$ **19.** $\tfrac{36}{7}$ **21.** $-2x\csc^2 x^2$

23. $e^{\sin t}(1 + t\cos t)$ **25.** $e^{-u^2}(-2u^4 + 3u^2)$ **27.** $5000e^{1.08} \approx \$14{,}723.40$ **29.** $\frac{81}{8}(3x + 4)^{-5/2}$ **31.** (i) 15   (ii) 14

33. $\dfrac{8x - 2xy}{3y^2 + x^2}$ **35.** $(-3, -6)$ and $(3, 6)$ **37.** $y' = \frac{2}{3}$, $y'' = \frac{4}{27}$ **39.** $\dfrac{1}{x} + \dfrac{x}{x^2 + 1} + \dfrac{2}{2x + 1}$ **41.** $4 + \log_2 3$

43. $(t + 1)^{2t}\left[\dfrac{2t}{t + 1} + 2\ln(t + 1)\right]$ **45.** $(\cos\theta)^{\tan\theta}[-\tan^2\theta + \sec^2\theta \,\ln(\cos\theta)]$ **47.** 4%   **49.** $\frac{1}{20}$ ft^2 **51.** $\frac{1}{32}$ **53.** $\frac{53}{3}$ **55.** -25

57. $\dfrac{d[f_1(x)f_2(x)\cdots f_n(x)]}{dx} = f_1'(x)f_2(x)\cdots f_n(x) + f_1(x)f_2'(x)\cdots f_n(x) + \cdots + f_1(x)f_2(x)\cdots f_n'(x)$

59. If $p(x)$ is a polynomial function with $(x - a)^m$ as a factor, then $p(a) = p'(a) = \cdots = p^{(m-1)}(a) = 0$.

61. (a) $A = \dfrac{f(a - \Delta x) + f(a + \Delta x) - 2f(a)}{2(\Delta x)^2}$, $B = \dfrac{f(a + \Delta x) - f(a - \Delta x)}{2(\Delta x)} - 2aA$, $C = f(a) - a^2A - aB$

 (b) $p'(a) = \dfrac{f(a + \Delta x) - f(a - \Delta x)}{2(\Delta x)} = m_{\text{sym}}$

63. $\dfrac{d(f(g(h(t))))}{dt}\bigg|_{t=t_1} = f'(g(h(t_1))) \cdot g'(h(t_1)) \cdot h'(t_1)$

CHAPTER 4

Section 4.1

1. $1/(4\pi)$ ft/sec   **3.** $10\sqrt{3}$ in^2/min   **5.** (a) 0.1 mi/sec   (b) 0.5 mi/sec   **7.** $125\pi/12$ cm^2/min   **9.** 4 volts/min
11. $2/\sqrt{65}$ ft/sec   **13.** $\frac{100}{13}$ ft/sec   **15.** $\frac{5}{6}$ units/sec   **17.** $\frac{15}{7}$ ft/sec   **19.** $-\sqrt{2}/14$ rad/min   **21.** $\frac{8}{63}$ rad/sec
23. Show: $dr/dt = $ Constant. Given: $dV/dt = kS = k(4\pi r^2)$. Then

$$V = \frac{4}{3}\pi r^3, \qquad \frac{dV}{dt} = 4\pi r^2 \frac{dr}{dt}, \qquad k \cdot 4\pi r^2 = 4\pi r^2 \frac{dr}{dt}, \qquad \frac{dr}{dt} = k.$$

25. $3/(4\pi)$ in./sec   **27.** $-\pi L^3/(60\sqrt{2})$ cm^3/sec   **29.** $365/\sqrt{89}$ ft/sec

Section 4.2

1. $f(-3) = f(2) = 10$, $c = -\frac{1}{2}$ **3.** $f(0) = f(\pi) = 0$, $c = \pi/2$ **5.** $f(1) = f(5) = 3$, $c = 2 + \sqrt{\frac{7}{3}}$
7. Let $f(x) = x^2 - 3x + 1$. Then $f(2) = -1$ and $f(6) = 19$ have opposite sign, so there is at least one solution. But $f'(x) = 2x - 3 = 0$ when $x = \frac{3}{2}$, which is not in $[2, 6]$, so there is at most one solution.
9. Let $f(x) = \sin x - x/4$. Then $f(\pi/2) = 1 - \pi/8 > 0$ and $f(\pi) = 0 - \pi/4 < 0$, so there is at least one solution. But $f'(x) = (\cos x) - \frac{1}{4} < 0$ in $[\pi/2, \pi]$, so there is at most one solution.
11. Let $f(x) = 0$ at points $x_1, x_2, \ldots, x_r$, listed in increasing order, in $[a, b]$. Then $f(x_i) = f(x_{i+1}) = 0$ for $i = 1, 2, \ldots, r - 1$. By Rolle's theorem, there exists c_i where $x_i < c_i < x_{i+1}$ such that $f'(c_i) = 0$. This holds for $i = 1, 2, \ldots, r - 1$.
13. All c, where $1 < c < 4$ **15.** $\sqrt{3}$ **17.** $-\frac{5}{4}$
19. (a) It is the average rate of change of $f(x)$ over $[a, b]$.   (b) It is the instantaneous rate of change of $f(x)$ at c.   (c) If f is continuous on $[a, b]$ and differentiable for $a < x < b$, then there exists c, where $a < c < b$, such that the instantaneous rate of change of $f(x)$ at c is the same as the average rate of change of $f(x)$ over $[a, b]$.
21. We compute that

$$\frac{f(x_2) - f(x_1)}{x_2 - x_1} = \frac{a(x_2^2 - x_1^2) + b(x_2 - x_1)}{x_2 - x_1} = a(x_2 + x_1) + b \quad \text{and} \quad f'\left(\frac{x_2 + x_1}{2}\right) = 2a\left(\frac{x_2 + x_1}{2}\right) + b = a(x_2 + x_1) + b$$

also. Example 3 illustrates this.
23. By the mean value theorem applied to $[a, x]$ for $a < x \le b$, $(f(x) - f(a))/(x - a) = f'(c)$ for some c, where $a < c < x$. Then $m \le (f(x) - f(a))/(x - a) \le M$, so $m(x - a) \le f(x) - f(a) \le M(x - a)$ for $a < x \le b$. Clearly this also holds at $x = a$.
25. $-9 - 5x \le f(x) \le 11 + 5x$ for x in $[-2, 4]$ **27.** $6 + 2x \le f(x) \le 6 + 5x$ for x in $[0, 6]$
29. $-6 + 5x \le f(x)$ for x in $[0, 2]$ **31.** $-4 - x \le f(x) \le -4 + 4x$ for $x \ge 0$
33. $1 + 3x \le f(x) \le 1 + 7x$ for $x \ge 0$, $1 + 7x \le f(x) \le 1 + 3x$ for $x < 0$
35. (a) $h(0) = g(0 + a) = g(a) \ne 0$ (b) $h'(t) = g'(t + a)\dfrac{d(t + a)}{dt} = g'(t + a) \cdot 1 = g'(t + a) = k \cdot g(t + a) = k \cdot h(t)$ (c) By

Theorem 4.6, $h(t) = h(0)e^{kt} = g(a)e^{kt}$, so $g(0) = g(-a + a) = h(-a) = g(a)e^{-ak} \ne 0$, contradicting our hypothesis that $g(0) = 0$. Thus our assumption that there exists a number a such that $g(a) \ne 0$ must be false, so $g(t) = 0 = 0 \cdot e^{kt} = g(0)e^{kt}$ for all t, which completes the proof.

Section 4.3

1. F T T T F   **3.** $x_0 = 0$ is not in the domain of f.   **5.** Local maximum at critical point 1   **7.** Local minimum at critical point 1   **9.** Critical point 0, no local extremum   **11.** Local minimum at critical point 0   **13.** Local maximum at critical point 0   **15.** Local maximum at critical point -1; local minimum at critical point 1   **17.** Local minima at critical points $n\pi$ for all integers n; local maxima at critical points $n\pi + (\pi/2)$ for all integers n **19.** Local maxima at critical points $n\pi$ for all integers n; local minima at critical points $(\pi/2) + n\pi$ for all integers n **21.** Maximum 16; minimum 0   **23.** Maximum 1; minimum $\frac{1}{5}$ **25.** (a) Maximum 2; minimum -6 (b) Maximum -3; minimum -7 (c) Maximum -3; minimum -7 **27.** (a) Maximum $\frac{7}{5}$; minimum $\frac{13}{10}$ (b) Maximum $\frac{3}{2}$; minimum 1   (c) Maximum 1; minimum $\frac{1}{2}$ (d) Maximum $\frac{3}{2}$; minimum $\frac{1}{2}$ **29.** (a) Maximum $\sqrt{2}$; minimum 1   (b) Maximum $\sqrt{2}$; minimum -1 (c) Maximum $\sqrt{2}$; minimum -1 (d) Maximum 1; minimum $-\sqrt{2}$ **31.** Maximum $e-1$; minimum 1   **33.** (a) Maximum 1600 ft; minimum 1200 ft   (b) Maximum 0 ft/sec; minimum -160 ft/sec   (c) Maximum 160 ft/sec; minimum 0 ft/sec   **35.** (a) Maximum 1400 ft; minimum 1336 ft   (b) Maximum 64 ft/sec; minimum -32 ft/sec   (c) Maximum 64 ft/sec; minimum 0 ft/sec

37. We may write $f(x) = x^n\left(a_n + \dfrac{a_{n-1}}{x} + \cdots + \dfrac{a_1}{x^{n-1}} + \dfrac{a_0}{x^n}\right)$, $x \ne 0$. Then $\lim_{x\to\infty} f(x) = \lim_{x\to-\infty} f(x) = \infty$. Consequently, there exists $c > 0$ such that $f(x) > f(1)$ if $|x| > c$. Then the minimum value attained on $[-c, c]$, which exists by Theorem 4.1, is also the minimum value attained on the whole x-axis.

39. -14 **41.** -27

Section 4.4

1. Vertical asymptote $x = 2$, horizontal asymptote $y = 0$, y-intercept $-\frac{1}{2}$

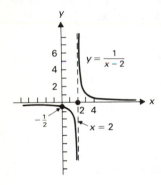

3. Vertical asymptote $t = 0$, horizontal asymptote $s = 1$, t-intercept 1

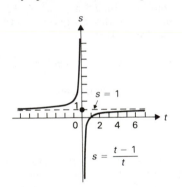

5. Vertical asymptote $x = 1$, horizontal asymptote $y = 2$, x-intercept -2, y-intercept -4

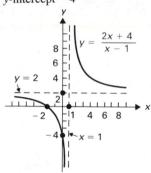

7. Polynomial function: $x + 1$

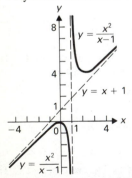

9. Polynomial function: $s^2 + s + 1$

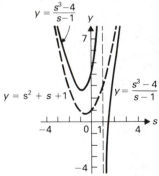

11. Increasing $x > 3$; decreasing $x < 3$ **13.** Increasing $t < -3$, $t > 1$; decreasing $-3 < t < 1$ **15.** Increasing $-2 < s \le 2$; decreasing $s < -2$, $s > 2$ **17.** Increasing $-1 < x < 1$; decreasing $x < -1$, $x > 1$ **19.** Increasing $x < -2\sqrt{3}$, $x > 2\sqrt{3}$; decreasing $-2\sqrt{3} < x < -2$, $-2 < x < 2$, $2 < x < 2\sqrt{3}$ **21.** Increasing $0 < x < 2\sqrt{2}$, $x > 2\sqrt{2}$ (actually for $x > 0$) decreasing $x < -2\sqrt{2}$, $-2\sqrt{2} < x < 0$ (actually for $x < 0$)   **23.** Increasing $-1 < x < 0$, $x > 1$; decreasing $x < -1$, $0 < x < 1$ **25.** Increasing $0 < x < 1$; decreasing $x > 1$ **27.** Increasing $-2\pi < t < -7\pi/4$, $-3\pi/4 < t < \pi/4$, $5\pi/4 < t < 2\pi$; decreasing $-7\pi/4 < t < -3\pi/4$, $\pi/4 < t < 5\pi/4$ **29.** -8 **31.** (a) $b/a = -4$ (b) $a = 3$, $b = -12$ (c) No   **33.** No   **35.** (a) $-1 < x < 3$, $x > 3$ (b) $x < -1$ (c) None   (d) -1 **37.** (a) $x < -2$, $x > 1$ (b) $-2 < x < 0$, $0 < x < 1$ (c) None   (d) None   **39.** (a) $-2 < x < 0$, $0 < x < 1$, $x > 4$ (b) $x < -2$, $1 < x < 4$ (c) None   (d) 4   **41.** $n - 1$ **43.** $n/2$, $n/2 - 1$

45. F F T T F **47.** T F T F F

49.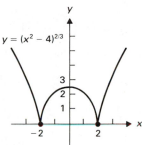

$y = x^2 - 6x + 2$

$(3, -7)$

51.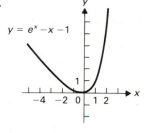

$y = \frac{1}{8}(t^3 + 3t^2 - 9t + 5)$

$(-3, 4)$

53.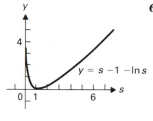

$y = x^3 + 3x^2 + 3x - 5$

$(-1, -6)$

55.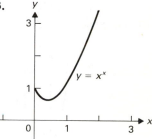

$y = \frac{x}{x^2 - 1}$

57.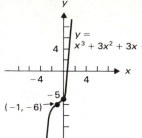

$y = (x^2 - 4)^{2/3}$

59.

$y = e^x - x - 1$

61.

$y = s - 1 - \ln s$

63.

$y = x^x$

65. Let $A(x) = \dfrac{C(x)}{x}$. Then $A'(x) = \dfrac{x \cdot C'(x) - C(x)}{x^2} = 0$ when $x \cdot C'(x) - C(x) = 0$, or when $C'(x) = \dfrac{C(x)}{x}$. This is true when the marginal cost $C'(x)$ is equal to the average cost $A(x) = \dfrac{C(x)}{x}$.

Section 4.5

1. T F F T T F T F T T

3.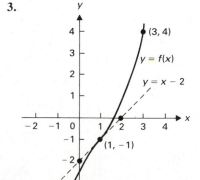

$(3, 4)$

$y = f(x)$

$y = x - 2$

$(1, -1)$

5. No. The tangent line to the graph of f at $(0, 0)$ passes through $(1, 1)$ and is below the graph of f for $x > 0$ since $f''(x) > 0$ for $x > 0$. Thus the graph of f must pass above $(1, 1)$, so $f(1) > 1$. Alternatively, $f''(x) > 0$ means $f'(x)$ is increasing, and thus $f'(x) > 1$ if $x > 0$. If $f(1) = 1$, then by the mean value theorem applied to $[0, 1]$, we have $f'(c) = 1$ for some c, where $0 < c < 1$, which is a contradiction.

7. (a) $-\infty < t < \infty$ (b) None **9.** (a) $x > -1$ (b) $x < -1$ **11.** (a) $u > 1$ (b) $u < 1$ **13.** (a) $x > 0$ (b) $x < 0$
15. (a) $x < -1/\sqrt{3},\ x > 1/\sqrt{3}$ (b) $-1/\sqrt{3} < x < 1/\sqrt{3}$ **17.** (a) $(2n - \frac{1}{2})\pi < s < (2n + \frac{1}{2})\pi$
(b) $(2n + \frac{1}{2})\pi < s < (2n + \frac{3}{2})\pi$ **19.** (a) $x > 2$ (b) $0 < x < 2$ **21.** (a) $t > \ln 2$ (b) $t < \ln 2$
23. (a) $x < -1,\ -1 < x < 0,\ x > 1$ (b) $0 < x < 1$ **25.** (a) $-3 < x < 0,\ x > 1$ (b) $x < -3,\ 0 < x < 1$
27. (a) $-3 < x < 0,\ x > 1$ (b) $x < -3,\ 0 < x < 1$ **29.** (a) $x > 2$ (b) $x < -1,\ -1 < x < 0,\ 0 < x < 2$
31. F T F T F **33.** T F T T T

35. (a) 4 at $x = 0$ (b) None (c) None
(d)

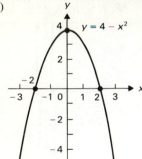

$y = 4 - x^2$

37. (a) None (b) None (c) None
(d)

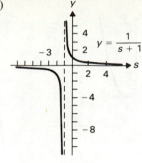

$y = \dfrac{1}{s + 1}$

39. (a) None (b) None
(c) $(-1, -\frac{19}{3})$
(d)

$y = \dfrac{x^3}{3} + x^2 + x - 6$

$(-1, -\frac{19}{3})$

41. (a) None (b) -2 at $x = 1$ (c) None
(d)

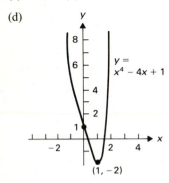

$y = x^4 - 4x + 1$

$(1, -2)$

43. (a) None (b) -1 at $t = 1$ (c) $(0, 0)$, $\left(\dfrac{2}{3}, \dfrac{-16}{27}\right)$
(d)

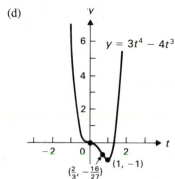

$y = 3t^4 - 4t^3$

$(\frac{2}{3}, -\frac{16}{27})$ $(1, -1)$

45. (a) $2n\pi + \frac{2}{3}\pi + \sqrt{3}$ at $x = 2n\pi + \frac{2}{3}\pi$ (b) $2n\pi + \frac{4}{3}\pi - \sqrt{3}$ at $x = 2n\pi + \frac{4}{3}\pi$
(c) $(n\pi, n\pi)$ (d)

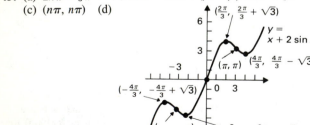

$\left(\frac{2\pi}{3}, \frac{2\pi}{3} + \sqrt{3}\right)$

$y = x + 2\sin x$

(π, π) $\left(\frac{4\pi}{3}, \frac{4\pi}{3} - \sqrt{3}\right)$

$\left(-\frac{4\pi}{3}, -\frac{4\pi}{3} + \sqrt{3}\right)$

$(-\pi, -\pi)$ $\left(-\frac{2\pi}{3}, -\frac{2\pi}{3} - \sqrt{3}\right)$

47. (a) None (b) $4 - 8\ln 2$ at $x = 2$
(c) None
(d)

$y = x^a - 8\ln x$

49. (a) None (b) $-1/e$ at $x = -1$
(c) $(-2, -2/e^2)$
(d)

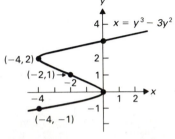

$y = xe^x$

51.

$x = y^3 - 3y^2$

$(-4, 2)$
$(-2, 1)$
$(-4, -1)$

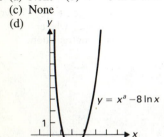

53.

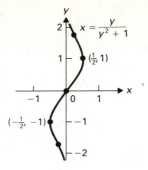

Section 4.6

1. 25 ft   **3.** $x = 2$, $y = 4$ **5.** 6, 6   **7.** $8\sqrt{2}$ units^2 **9.** 42 units^2 **11.** 108 in^2 **13.** (a) 8192 in^3 (b) 32,768/π in^3
15. 60,000 ft^2 **17.** 4 in. wide by 2 in. high   **19.** $\sqrt{2}a$ by $a/\sqrt{2}$ **21.** $4a/3$ **23.** $a^2\sqrt{3}/8$ units^2
25. 16 in. wide by 24 in. high   **27.** $2/\sqrt{5}$ mi   **29.** $a/(\sqrt[3]{c} + 1)$ ft   **31.** (a) Cut into two 50-in. pieces.   (b) Bend it all into
one triangle.   **33.** (a) $(2 + \pi)/4$ (b) $(4 + \pi)/8$ (c) 4 ft   **35.** 10:36 A.M.   **37.** Squares of sides $(a + b - \sqrt{a^2 + b^2 - ab})/6$
39. $(a^{2/3} + b^{2/3})^{3/2}$ ft   **41.** $24\sqrt{3}$ ft   **43.** (a) $90/stove   (b) 1000 stoves for $40,000 profit   **45.** (a) $78/cord   (b) $79/cord
(c) 500 cords   **47.** (a) 0 cords   (b) 1 cord. The average revenue approaches a maximum of $80/cord as $x \to 0$, but 0 cords
produces 0 revenue. The average revenue function is not defined when $x = 0$.   **49.** (a) $x = (A + B - b - t)/(2B + 2c)$
(b) $t = (A + B - b)/2$ **51.** Order 2000 records 50 times a year.   **53.** (a) Populations of 0 and of 60   (b) A population of 30
rabbits will yield a maximum sustainable harvest of 75 rabbits per year.

Section 4.7

1. $\frac{97}{56}$ **3.** $\frac{49}{20}$ **5.** $\dfrac{32{,}257}{12{,}192} \approx 2.6458$ **7.** $\dfrac{3958}{2178} \approx 1.8173$ **9.** $\frac{59}{86}$ **11.** $\frac{333}{440}$

13. (a)

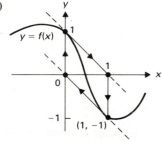

(b) (i) $d = 1$ (ii) $a + b + c + d = -1$ (iii) $c = -1$ (iv) $3a + 2b + c = -1$
(v) $f(x) = 2x^3 - 3x^2 - x + 1$

15. -2.387687 **17.** 2.924017738

19. 0.7390851332

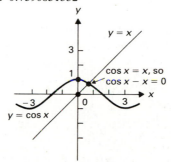

21. 2.66365 **23.** 2.129372 **25.** 3.35282

Section 4.8

1. $2x + C$ **3.** $\dfrac{t^3}{3} - \dfrac{3t^2}{2} + 2t + C$ **5.** $\frac{2}{3}x^{3/2} + C$ **7.** $2u^2 + \frac{2}{3}u^{3/2} + C$ **9.** $-\dfrac{1}{r} + \dfrac{3}{2}r^2 + r + C$ **11.** $-\dfrac{1}{x} - \dfrac{1}{2x^2} + C$

13. $\frac{6}{5}y^{5/2} - \frac{4}{3}y^{3/2} + 2\sqrt{y} + C$ **15.** $\dfrac{t^4}{4} + \dfrac{t^2}{2} + C$ **17.** $\frac{2}{3}(x - 1)^{3/2} + C$ **19.** $2\sqrt{u - 1} + C$ **21.** $-\cos x + C$

23. $-\frac{5}{8}\cos 8\theta + C$ **25.** $-\frac{1}{4}\sin(2 - 4x) + C$ **27.** $-2\cot 2\phi + C$ **29.** $-\frac{1}{4}\csc 4t + C$ **31.** $\frac{1}{2}e^{2x} + \frac{1}{3}x^3 + C$

33. $2\ln|t| + \frac{1}{3}\sin 3t + C$ **35.** $8x - 19$ **37.** $\dfrac{x^2}{2} - \cos x + 4$ **39.** $\frac{1}{2}x^2 - \frac{2}{3}x^{3/2} + \frac{1}{6} + \pi$ **41.** $y = \frac{1}{3}x^3 - 4\ln|x| - \frac{10}{3}$

43. $y = F(x) = x^2 - 2x + 1$ **45.** $y = F(x) = \frac{1}{4}\sin 2x + \frac{3}{2}x - 1$ **47.** $y = \frac{1}{9}e^{-3x} - \frac{2}{3}x + \frac{53}{9}$

49. The general solution is $y = F(x) + C_1 x + C_2$. The boundary conditions give the simultaneous equations

$$C_1 a + C_2 = \alpha - F(a),$$
$$C_1 b + C_2 = \beta - F(b),$$

which always have unique solutions C_1 and C_2 if $a \neq b$.

51. $y = G(x) = -\sin x + \dfrac{2}{\pi}x$ **53.** (a) $v = -32t + v_0$ (b) $y = -16t^2 + v_0 t + y_0$ **55.** $\dfrac{d^2 s}{dt^2} = -\dfrac{k}{m} \cdot \dfrac{ds}{dt}$; when $t = 0$, $s = 0$

and $ds/dt = 80$ **57.** $\dfrac{dP}{dt} = kP$

Supplementary Exercises for Chapter 4

1. $2/(10\pi)$ ft/min **3.** $\frac{1}{16}$ amp/sec **5.** 27 **7.** $-x^2 - x + 6 \leq f(x) \leq x^2 + x + 2$ for $1 \leq x \leq 3$ **9.** (a) $x < -2$, $-2 < x < 0$
(b) $0 < x < 2$, $x > 2$ **11.** (a) None (b) $-1, 4$ **13.** Maximum 19 at $x = 3$, minimum -1 at $x = 1$
15. Maximum 2π at $x = 2\pi$, minimum 0 at $x = 0$ **17.** (a) $-3 < x < 0$, $x > 2$ (b) $x < -3$, $0 < x < 2$ **19.** $\frac{11}{7}$, 5
21. No local extremum. **23.**

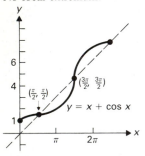

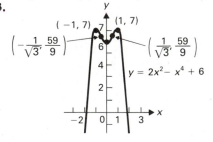

25. 2, 4 **27.** $\sqrt{3}a^2/8$ ft^2 **29.** 80 pairs **31.** $a_2 = \frac{5}{3}$, $a_3 = \frac{223}{144}$ **33.** 1.02987 **35.** $s = t^3 - 4t^2 + 2t + 5$
37. $y = \frac{1}{6}x^3 - \frac{3}{2}x^2 + \frac{1}{2}x + \frac{35}{6}$
39. Use 400 ft of fence for a square enclosure of 10,000 ft^2 and the rest for the circular enclosure. Total enclosed: 38,648 ft^2.
41. If f is n times differentiable for $a \leq x \leq b$ and $f(x)$ attains the same value at $n + 1$ distinct points in $[a, b]$, then $f^{(n)}(c) = 0$ for some c, where $a < c < b$.
43. Jog all the way.
45. This sequence will occur if $f(0) = -1$, $f'(0) = 1$, $f(1) = -1$, $f'(1) = 1$, $f(2) = -2$, $f'(2) = -1$. These six conditions should be able to be satisfied by a polynomial function $f(x) = ax^5 + bx^4 + cx^3 + dx^2 + ex + f$ with six coefficients.

CHAPTER 5

Section 5.1

1. $a_0 + a_1 + a_2 + a_3$ **3.** $a_2 + a_4 + a_6 + a_8$ **5.** $c + c^2 + c^3 + c^4 + c^5$ **7.** 30 **9.** 35 **11.** 44 **13.** $\displaystyle\sum_{i=1}^{3} a_i b_{i+1}$
15. $\displaystyle\sum_{i=1}^{3} a_i^{i+1}$ **17.** $\displaystyle\sum_{i=1}^{3} a_i^{b_{3i}}$

19. $\displaystyle\sum_{i=1}^{n}(a_i + b_i)^2 = (a_1 + b_1)^2 + \cdots + (a_n + b_n)^2 = a_1^2 + 2a_1b_1 + b_1^2 + \cdots + a_n^2 + 2a_nb_n + b_n^2$

$$= a_1^2 + \cdots + a_n^2 + 2(a_1b_1 + \cdots + a_nb_n) + b_1^2 + \cdots + b_n^2 = \sum_{i=1}^{n}a_i^2 + 2\sum_{i=1}^{n}a_ib_i + \sum_{i=1}^{n}b_i^2$$

21. $\frac{111}{70} \approx 1.586$ **23.** $R_{max} = 5$, $R_{min} = 1$ **25.** $R_{max} \approx 0.76$, $R_{min} \approx 0.63$ **27.** 153   **29.** π **31.** 40   **33.** $24 + \dfrac{9\pi}{2}$ **35.** 1.89
37. 2.006   **39.** 3.142   **41.** 19.703   **43.** ≈ 2.547

Section 5.2

1. (a)
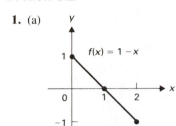
(b) 1   (c) -1 (d) 0   **3.** 2
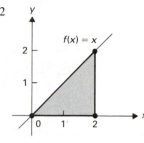
5. 14
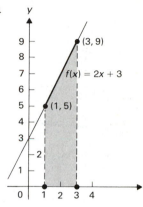

7. 2
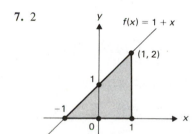
9. $9\pi/2$
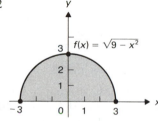
11. $12 + 4\pi$
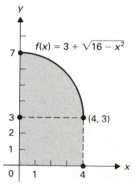

13.
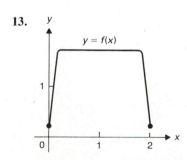
15. 1   **17.** 1   **19.** 0   **21.** -6 **23.** 2   **25.** 4   **27.** 4   **29.** $(\pi^2/4) - 3$ **31.** 6
33. 3   **35.** -1 **37.** -10 **39.** 5   **41.** $-\frac{3}{2}$ **43.** -2 **45.** $\frac{1}{4}$ **47.** $-\frac{9}{4}$
49. 2.046696   **51.** 0.459481

Section 5.3

1. Refer to Theorem 5.3 to check your answer.   **3.** $\frac{1}{4}$ **5.** $\frac{20}{3}$ **7.** $\frac{45}{4}$ **9.** $-\frac{3}{8}$ **11.** $\frac{14}{9}$ **13.** 2   **15.** 3   **17.** $3\sqrt{2}$ **19.** $2/\sqrt{3}$
21. 1   **23.** -20 **25.** $\frac{3}{2}$ **27.** $-\pi/2$ **29.** $2 + (\pi/2)$ **31.** π **33.** $\pi + (3\pi^3/4)$ **35.** $(3\pi^2/2) - 2\pi$ **37.** $e^2 - 1$ **39.** 4
41. $\ln 5$ **43.** $\frac{1}{2}e^4 - \ln 2$ **45.** 8   **47.** $1/\sqrt{2}$ **49.** $\frac{88}{15}$

51. 36

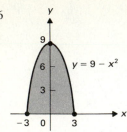

53. $\frac{256}{5}$

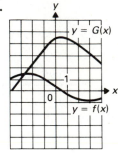

55. $\sqrt{t^2 + 1}$ **57.** $-1/(1 + t^2)$ **59.** $2\sqrt{3 + 4t^2}$
61. $-\sqrt{t^2 + 1}$ **63.** $2t\sqrt{t^4 - 6t^2 + 10}$

65.

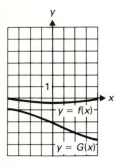

67.

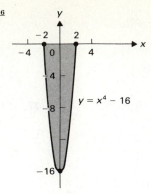

69.

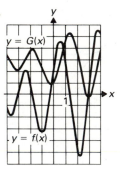

Section 5.4

1. $\frac{1}{4}x^4 + \frac{4}{3}x^3 + C$ **3.** $\frac{1}{6}(u + 1)^6 + C$ **5.** $\dfrac{-1}{4(4x + 1)} + C$ **7.** $\dfrac{-1}{18t^2 + 6} + C$ **9.** $\dfrac{1}{12(4 - 3x^2)^2} + C$ **11.** $4\sqrt{s^2 + s} + C$

13. $\dfrac{-2}{3(\sqrt{x} + 1)^3} + C$ **15.** $\frac{1}{12}(x^3 + 4)^4 + C$ **17.** $\frac{1}{3}\sin 3v + C$ **19.** $\frac{1}{2}\sin(x^2 + 1) + C$ **21.** $\frac{1}{2}\sin^2 s + C$ **23.** $-\frac{1}{16}\cos^4 4x + C$

25. $\frac{1}{2}\tan x^2 + C$ **27.** $\frac{1}{2}\tan^2 t + C$ **29.** $\frac{1}{8}\tan^4 x^2 + C$ **31.** $(1 + \cos s)^{-1} + C$ **33.** $\frac{1}{4}\sin 4x + C$ **35.** $\frac{1}{12}\sec^4 3v + C$
37. $\frac{2}{9}(1 + \sec 3t)^{3/2} + C$, $\sec 3x > 0$ **39.** $-\frac{1}{10}\csc^5 2x + C$ **41.** $-1/2 \ln|\cos 2x| + C$ **43.** $\frac{1}{2}\ln(x^2 + 2) + C$ **45.** $\frac{1}{3}e^{t^3} + C$

47. $\dfrac{1}{\ln 3} 3^{\sin y} + C$ **49.** $\frac{1}{9}(1 + 3e^{2x})^{3/2} + C$ **51.** $-3\dfrac{\sqrt{4 + t^2}}{4t} + C$ **53.** $\pi/2$ **55.** $-\dfrac{\cos 5t}{10} + \dfrac{\cos t}{2} + C$

Section 5.5

1. $\dfrac{1}{y} = -\dfrac{x^3}{3} + C$ **3.** $-\dfrac{1}{y} = \dfrac{1}{2}x^2 + C$ **5.** $\tan y = -\cos x + C$ **7.** $8\sqrt{u} = x^4 + C$ **9.** $x^2 + e^{-2y} = C$

11. $\frac{1}{3}\tan 3y = x + \frac{2}{3}x^3 + \frac{1}{5}x^5 + C$ **13.** $-\cot y = \frac{1}{2}x^2 - 2$ **15.** $\ln|y| = \frac{1}{3}x^3 + x^2 - \frac{20}{3}$ **17.** $\dfrac{-1}{y^2} = x^2 - 1$

19. $y = -4/3x$, $x \neq 0$ **21.** $-\cos 2y = 2 \sin x + 1$

23. $y + x = C$

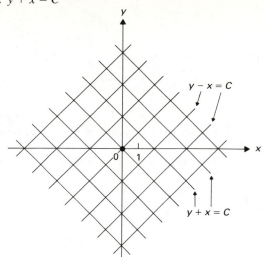

25. $y = Ce^{2x}$

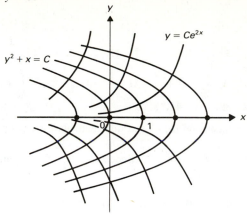

27. $(x + 1)^2 + y^2 = C$

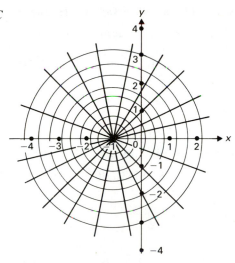

Section 5.6

1. 1.3524   **3.** 1.425   **5.** (a) 1.737   (b) 1.7321   **7.** 3.1416   **9.** $\dfrac{73\pi}{180} \approx 1.2741$ **11.** 1.1167   **13.** 0.5917   **15.** 3.6219

17. $\int_{-1}^{1} |x|\, dx$ for $n = 2$ is found exactly by the rectangular rule and the trapezoidal rule, but not by Simpson's rule.
19. 3.057355539   **21.** 3.1412554158   **23.** 33.02739148   **25.** 0.89508189   **27.** 0.8820997338   **29.** 6.043173883

Supplementary Exercises for Chapter 5

1. $\frac{15}{2}$ **3.** $\dfrac{\pi}{2}\left(\sin\dfrac{\pi}{8} + \sin\dfrac{3\pi}{8} + \sin\dfrac{5\pi}{8} + \sin\dfrac{7\pi}{8}\right) \approx 4.1047$ **5.** $6 + \dfrac{9\pi}{4}$ **7.** $9\sqrt{6}\pi/4$ **9.** $\frac{3}{4}$ **11.** -2 **13.** $\dfrac{\sqrt{2t + t^2}}{2\sqrt{t}}$

15. $12 + \dfrac{9\pi}{2}$ **17.** 0   **19.** $\frac{3}{2}$ **21.** 1   **23.** $\dfrac{\sqrt{3}}{2}$ **25.** $\frac{4}{9}$ **27.** 1   **29.** $\frac{32}{3}$ **31.** $\dfrac{-1}{9(x^3 + 1)^3} + C$ **33.** $(-1/9)\csc^3 3x + C$

35. $-2\sqrt{\csc x} + C$ **37.** $(-1/3)\ln|\cos 3x| + C$ **39.** $\frac{1}{2}e^{x^2} + C$ **41.** $-4x \sin x^2 + 2\cos x^2 + C$ **43.** $\tan 2y = x^2 + C$

45. $-\dfrac{1}{y} = \dfrac{1}{2}\sin 2x - \dfrac{1}{6}$ **47.** $\dfrac{1}{y} = -\dfrac{x^3}{3} + \dfrac{5}{6}$ **49.** ≈ 0.693 **51.** ≈ 0.8047 **53.** 2   **55.** 0   **57.** ∞

CHAPTER 6

Section 6.1

1. $\frac{32}{3}$ **3.** $\frac{1}{6}$ **5.** $\frac{4}{15}$ **7.** $\frac{44}{15}$ **9.** $\frac{9}{2}$ **11.** $\frac{9}{2}$ **13.** $\frac{7}{15}$ **15.** $\frac{4}{3} + \frac{\pi}{2}$ **17.** $\frac{8}{3}$ **19.** 4   **21.** $\frac{8\pi}{3} - 2\sqrt{3}$ **23.** $\frac{\pi - 2}{4}$ **25.** $\frac{2}{3}\pi^{3/2} - 2$

27. $\frac{15}{8} - \ln 4$ **29.** $\frac{e^2}{2} - e + \frac{3}{2}$ **31.** $\frac{\pi}{4} - \ln 2$ **33.** $2\int_0^2 (\sqrt{68 - y^2} - 2y^2)\,dy$ **35.** $25\pi - \int_{-4}^3 \left(\sqrt{25 - x^2} - \frac{x + 25}{7}\right) dx$

37. $8 - 4\sqrt{2}$ **39.** $-\frac{1}{3}$ **41.** $5^{1/4}$

43. $F(t)$ is continuous on $[a, b]$ and differentiable for $a < t < b$. Thus $\dfrac{F(b) - F(a)}{b - a} = F'(c)$ for some c, where $a < c < b$. Now

$F'(t) = f(t)$ by Theorem 5.3. Since $F(b) = \displaystyle\int_a^b f(x)\,dx$ and $F(a) = \displaystyle\int_a^a f(x)\,dx = 0$, we have $\dfrac{1}{b - a}\displaystyle\int_a^b f(x)\,dx = f(c)$ for some

c, where $a < c < b$.

45. 21.9919   **47.** 71.40633459 ($n = 10$)   **49.** 0.1356975072 ($n = 10$)   **51.** $\dfrac{3 + (4 \ln 4)}{20}$ slug   **53.** $\frac{32}{15}$ slug

Section 6.2

1. $16\pi/15$ **3.** $4\sqrt{3}\,\pi$ **5.** $128\pi/5$ **7.** $\pi/6$ **9.** 2π **11.** $\pi^2/2$ **13.** $(\pi/2)(8 + 3\pi)$ **15.** $(\pi/2)(e^2 - 1)$ **17.** $\pi(\ln 4)$

19. $\pi/2$ **21.** $8\pi/3$ **23.** $\displaystyle\int_0^h \pi\left(\frac{r}{h}x\right)^2 dx = \frac{\pi}{3}r^2h$ **25.** $\dfrac{4a^3}{\sqrt{3}}$ **27.** $\dfrac{\pi}{3}(2a^3 + b^3 - 3a^2b)$ **29.** The area of the base, where

$x = 0$, is ch^2. The volume is $\displaystyle\int_0^h c(h - x)^2\,dx = \frac{ch^3}{3} = \frac{h}{3}\cdot ch^2$.   **31.** ≈ 2.0191 ($n = 10$)   **33.** ≈ 33.994 ($n = 10$)

35. ≈ 0.27936 ($n = 20$)   **37.** ≈ 0.74878 ($n = 20$)

Section 6.3

1. $16\pi/15$ **3.** $\pi/6$ **5.** $128\pi/5$ **7.** $3\pi/10$ **9.** $27\pi/2$ **11.** $4\pi^2$ **13.** $6\pi^2$ **15.** 2π **17.** $18\pi(\ln 3)$

19. $\displaystyle\int_0^r 2\pi y\left(h - \frac{h}{r}y\right) dy = \frac{\pi}{3}r^2h$ **21.** $2\pi^2a^2b$ **23.** ≈ 1.1673 ($n = 20$)   **25.** ≈ 152.77 ($n = 40$)   **27.** ≈ 189.46 ($n = 10$)

29. ≈ 0.37779

Section 6.4

1. (a) -6 (b) 6   **3.** (a) $-\frac{16}{3}$ (b) 8   **5.** (a) $\frac{5}{6}$ (b) $\frac{5}{6}$ **7.** (a) $\frac{3}{2}$ (b) $\frac{11}{6}$ **9.** (a) 0   (b) 10   **11.** (a) $4/\pi$ (b) $4\sqrt{2}/\pi$

13. (a) $\frac{25}{2}$ (b) $\frac{25}{2}$ **15.** (a) $3(1 - e^{-t})$ (b) 3 units   **17.** (a) $t^2 - 4t + 3$ (b) $\frac{8}{3}$ **19.** (a) $4 - 2\sqrt{t + 1}$ (b) $\dfrac{20\sqrt{5} - 36}{3}$

21. (a) $\dfrac{-6}{t + 1} + 3$ (b) $6 \ln(\frac{4}{3})$

23. By Exercise 22, the time required for the body to travel a distance h is

$\displaystyle\int_0^h \frac{ds}{Ae^{ks}} = \frac{1}{A}\int_0^h e^{-ks}\,ds = \frac{-1}{kA} e^{-ks}\Big|_0^h = \frac{-1}{kA}(e^{-kh} - 1) = \frac{-1}{kA}\left(\frac{1}{e^{kh}} - 1\right)$. Taking the limit as $h \to \infty$, we find that the time

required to travel an infinite distance is $\displaystyle\lim_{h\to\infty} \frac{-1}{kA}\left(\frac{1}{e^{kh}} - 1\right) = -\frac{1}{kA}(0 - 1) = \frac{1}{kA}$, which is finite.

25. $200(e^{0.1} - 1)$ million $\approx 21{,}034{,}000$ people   **27.** \$792.80   **29.** $\dfrac{50}{0.03}(e^6 - e^3) \approx 638{,}905$ units

Section 6.5

1. $\frac{8}{27}(10^{3/2} - 1)$ **3.** $\frac{1}{6}(125 - 13^{3/2})$ **5.** $\frac{1}{27}(40^{3/2} - 13^{3/2})$ **7.** $\frac{53}{6}$ **9.** $\frac{221}{120}$ **11.** $3\sqrt{2}/100$ **13.** $\sqrt{37}/100$ **15.** $\sqrt{2}/20$

17. ≈ 2.2643 ($n = 20$)   **19.** ≈ 55.089 ($n = 20$)

Section 6.6

1. $\sqrt{17}\pi$ **3.** $5\sqrt{2}\pi$ **5.** $\dfrac{\pi}{27}(145\sqrt{145} + 10\sqrt{10} - 2)$ **7.** $2\pi\dfrac{24\sqrt{3} + 4}{15}$ **9.** 112π **11.** $\dfrac{12289\pi}{192}$
13. $4000^2\pi(2 - \sqrt{2})$ mi^2 **15.** $\pi\sqrt{(R - r)^2 + h^2}(R + r)$ **17.** ≈ 37.704 ($n = 20$)   **19.** ≈ 170.01 ($n = 20$)
21. ≈ 1353.0 ($n = 20$)   **23.** ≈ 81.221 ($n = 20$)

Section 6.7

1. 64 ft·lb   **3.** 8   **5.** (a) 9 ft·lb   (b) 36 ft·lb   **7.** 12480π ft·lb   **9.** 3 ft   **11.** 3225 ft·lb   **13.** $17k/15$ units
15. $W = \displaystyle\int_{x_{t_1}}^{x_{t_2}} F(x)\,dx = \int_{x_{t_1}}^{x_{t_2}} \left(m\dfrac{dv}{dx}v\right)dx = \int_{v_{t_1}}^{v_{t_2}} mv\,dv = \dfrac{1}{2}mv^2\Big]_{v_{t_1}}^{v_{t_2}} = \dfrac{1}{2}mv_{t_2}{}^2 - \dfrac{1}{2}mv_{t_1}{}^2 =$ Change in kinetic energy
17. $(1497.6)\pi$ lb   **19.** 55,328 lb   **21.** 550,368 lb   **23.** $(998.4)\pi$ lb   **25.** 5148π lb   **27.** 13.76 kwh   **29.** 2.95 kwh

Supplementary Exercises for Chapter 6

1. $\frac{4}{3}$ **3.** $4(\sqrt{2} - 1)$ **5.** $\frac{122}{3}$ **7.** $8\pi(\sqrt{2}\pi - 4)$ **9.** $832\pi/15$ **11.** $a^3/\sqrt{3}$ **13.** $10\pi(\ln\frac{13}{5})$ **15.** $t = 4$ **17.** $\frac{38}{3}$ ft   **19.** $\frac{67}{10}$
21. $\dfrac{7\pi}{9\sqrt{3}}$ **23.** (a) $\displaystyle\int_0^\pi (1 + 2\sin x)\sqrt{1 + 4\cos^2 x}\,dx$ (b) ≈ 11.1861 (Simpson's rule, $n = 40$)   **25.** $\dfrac{k}{20}$ ft·lb   **27.** 208,000 lb
29. $256\sqrt{2}/5$ **31.** $\dfrac{200}{3}$ hr

CHAPTER 7

Section 7.1

1. Invertible; $f^{-1}(x) = x + 1$ **3.** Invertible; $f^{-1}(x) = 3 - x$ **5.** Not invertible   **7.** Not invertible

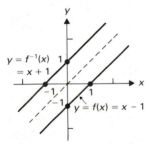

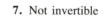

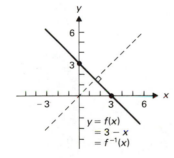

9. Invertible; $f^{-1}(x) = \sqrt[3]{x - 1}$ **11.** Invertible; $f^{-1}(x) = x^2$, $x \geq 0$ **13.** Not invertible

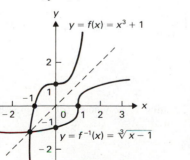

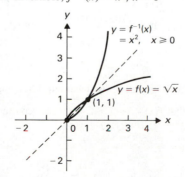

15. Invertible; $f^{-1}(x) = \dfrac{x+1}{1-x}$ **17.** Not invertible   **19.** Not invertible

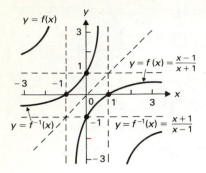

21. Invertible   **23.** Not invertible   **25.** Invertible

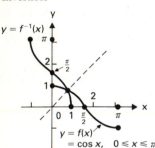

27. T T T F F T T F T T   **29.** -4 **31.** $\frac{1}{2}$ **33.** 1   **35.** $\frac{1}{2}$ **37.** 3   **39.** (a) $f(x)$ is decreasing   (b) $[-1, 1]$ (c) $-1\sqrt{1-x^2}$
(d) $-1/\sqrt{-x^2-x}$ **41.** (a) $g(x)$ is decreasing   (b) All x (c) $-1/(1+x^2)$ (d) $-2x/(1+x^4)$ **43.** (a) $f(x)$ is one to one
(b) $x \geq 1,\ x \leq -1$ (c) $-1/(x\sqrt{x^2-1})$ (d) $-1/(2x\sqrt{x-1})$

Section 7.2

1. By Theorem 5.3, $\dfrac{d}{dx}\displaystyle\int_a^x f(t)\,dt = f(x)$, so $\dfrac{d(\ln x)}{dx} = \dfrac{d}{dx}\displaystyle\int_1^x \dfrac{1}{t}\,dt = \dfrac{1}{x}$ for $x > 0$.   **3.** $\cot x$

5. If $f(x) > 0$, then by Exercise 2, $\dfrac{d}{dx}[\ln f(x)] = \dfrac{1}{f(x)} \cdot f'(x)$. If $f(x) < 0$, then by Exercise 2,

$\dfrac{d}{dx}[\ln(-f(x))] = \dfrac{1}{-f(x)}(-f'(x)) = \dfrac{1}{f(x)} \cdot f'(x)$. Thus $\dfrac{d[\ln|f(x)|]}{dx} = \dfrac{1}{f(x)} \cdot f'(x)$, so $\displaystyle\int \dfrac{f'(x)}{f(x)}\,dx = \ln|f(x)| + C$, that is,

$\displaystyle\int \dfrac{du}{u} = \ln|u| + C$.

7. We have $\dfrac{d(\ln x)}{dx} = \dfrac{1}{x} > 0$ for $x > 0$, so $\ln x$ is increasing by Theorem 4.7.

9. Since $f'(x) = g'(x)$, Theorem 4.10 tells us that $f(x) = g(x) + C$ for some constant C. Setting $x = c$ where $f(c) = g(c)$, we
find that $f(c) = g(c) + C$, so $C = f(c) - g(c) = 0$. Thus $f(x) = g(x)$.

11. $f'(x) = \dfrac{2x}{x^2+1} + \dfrac{3}{x-1} + 4\tan x$

13. $\dfrac{d(\ln x)}{dx} = \dfrac{1}{x}$ and $\dfrac{1}{x}\Big|_1 = 1$, so the graph of $\ln x$ has slope 1 at $x = 1$. Also, $\dfrac{d^2(\ln x)}{dx} = \dfrac{d(1/x)}{dx} = \dfrac{-1}{x^2} < 0$, so the graph is
concave down. See Fig. 3.24.

15. The domain of $\exp(x)$ is the range of $x = \ln y$, namely $-\infty < x < \infty$. The range of $\exp(x)$ is the domain of $x = \ln y$, namely
$y > 0$. See the graph in Fig. 3.23.

17. (a) and (b) These relations follow at once from the fact that each of $\ln x$ and e^x is the inverse of the other. See Eq. (2) of Section 7.1.

19. Let $y = e^x$. Then $x = \ln y$. Differentiating implicitly, we have $1 = \dfrac{1}{y} \cdot \dfrac{dy}{dx}$, so $\dfrac{dy}{dx} = y = e^x$. Thus $\dfrac{d(e^x)}{dx} = e^x$.

21. $3(\cos 3x)e^{\sin 3x}$ **23.** $\frac{1}{2}e^{x^2} + C$

25. We use the fact that $\ln x$ is a one-to-one function and need only check that the expressions on the sides of each equation have the same natural logarithm.   (a) $\ln(a^b a^c) = \ln(a^b)\ln(a^c) = b(\ln a) + c(\ln a) = (b + c)(\ln a)$, $\ln(a^{b+c}) = (b + c)(\ln a)$

(b) $\ln\left(\dfrac{a^b}{a^c}\right) = \ln(a^b) - \ln(a^c) = b(\ln a) - c(\ln a) = (b - c)(\ln a)$, $\ln(a^{b-c}) = (b - c)(\ln a)$ (c) $\ln((a^b)^c) = c\,\ln(a^b) = cb(\ln a)$,

$\ln(a^{bc}) = bc(\ln a) = cb(\ln a)$

27. $3^{\cos 2x}[2x - 2(\ln 3)x^2\sin 2x]$ **29.** $\dfrac{-1}{2(\ln 10)}10^{\cos 2x} + C$

31. Let $y = \log_a u$ where u is a differentiable function of x. Then $u = a^y$. Differentiating implicitly with respect to x, we obtain $\dfrac{du}{dx} = (\ln a)a^y \cdot \dfrac{dy}{dx}$, so $\dfrac{dy}{dx} = \dfrac{1}{\ln a} \cdot \dfrac{1}{a^y} \cdot \dfrac{du}{dx} = \dfrac{1}{\ln a} \cdot \dfrac{1}{u} \cdot \dfrac{du}{dx}$.

33. The domain of h includes all x in the intersection of the domains of f and of g such that $f(x) > 0$. For such x, we define $h(x) = e^{[\ln f(x)]g(x)}$. For x in the intersection of the domains of f and of g such that $f(x) = 0$ and $g(x) > 0$, we define $f(x)^{g(x)} = 0$.

Section 7.3

1. $3(\ln 2)2^{3x}$ **3.** $3^{2x-1}[2(\ln 3)x^2 + 2x]$ **5.** $(\ln 18)2^x \cdot 3^{2x}$ **7.** $x^{2x}(2 + 2\ln x)$ **9.** $x^{\tan x}\left[\dfrac{\tan x}{x} + (\sec^2 x)(\ln x)\right]$

11. $(x - 1)^{x^2}\left[\dfrac{x^2}{x - 1} + 2x\ln(x - 1)\right]$ **13.** $7^x 8^{-x^2} 100^x[(\ln 700) - 2x(\ln 8)]$

15. $\dfrac{5^{\sin x}}{2^{1/x}}\left[(\ln 5)(\cos x) + \dfrac{\ln 2}{x^2}\right]$ **17.** $\dfrac{x^2 3^x}{\tan x}\left[\dfrac{2}{x} + (\ln 3) - \dfrac{\sec^2 x}{\tan x}\right]$ **19.** $\dfrac{x^{\sin x}}{(\cos x)^x}\left[\dfrac{\sin x}{x} + (\cos x)(\ln x) + x\tan x - \ln(\cos x)\right]$

21. $y = x$ **23.** $4 + (0.04)(1 + \ln 2) \approx 4.0677$ **25.** 2.231829 **27.** 2.484709796 **29.** 3.0457

Section 7.4

1. $\dfrac{1600\ln(\frac{3}{2})}{\ln 2}$ yr   **3.** $\dfrac{\ln 2000}{\ln 2.4} \approx 8.68$ days   **5.** (a) 80 yr (b) 80 yr   **7.** (a) $\sqrt[4]{3}$ (b) $\sqrt[4]{3}$ **9.** 13,235 yr   **11.** \$27,182.82

13. \$79,048.94   **15.** (a) $50\ln(\frac{6}{5}) \approx 9.12\%$ (b) \$6,339.38   **17.** (a) $1985 + 20(\ln 50) \approx 2063$
(b) $1985 + 20(\ln 50{,}000) \approx 2201$ (c) Have a "decimal point day" when all wages and prices are divided by 10 every time the average yearly wage reaches \$100,000.   **19.** (a) $(2000)e^{2.25} \approx \$18{,}975$ (b) 3.9   **21.** $\dfrac{200}{e}$ lb   **23.** $200 \cdot e^{-1.2} \approx 60.24$ lb

25. $\dfrac{2560}{9\ln(\frac{4}{3})}$ ft   **27.** $40 + 50(\frac{3}{5})^{8/3}$ degrees   **29.** Year 4029

Section 7.5

1. $\pi/2$ **3.** $-\pi/3$ **5.** $7\pi/6$ **7.** $\pi/4$ **9.** $3\pi/2$ **11.** π **13.** 0   **15.** $5\pi/4$ **17.** $-\pi/4$ **19.** $\pi/2$ **21.** $\pi/2$ **23.** 0
25. $2/\sqrt{1 - 4x^2}$ **27.** $1/[2\sqrt{x(1 + x)}]$ **29.** $1/[x\sqrt{(1/x)^2 - 1}]$ **31.** $6(\tan^{-1}2x)^2/(1 + 4x^2)$ **33.** $-1/[(1 + x^2)(\tan^{-1}x)^2]$
35. $2(x + \sin^{-1}3x)[1 + (3/\sqrt{1 - 9x^2})]$ **37.** $\pi/2$ **39.** π **41.** $\frac{1}{3}\sec^{-1}2x + C$ **43.** 0.4889   **45.** 0.2145   **47.** 0.9463

49. 0.472   **51.** $\pi/2$ **53.** $\dfrac{2\sqrt{3}}{3} - \dfrac{\pi}{6}$ **55.** $\pi/6$ **57.** $\dfrac{2\pi}{3} + \dfrac{8}{3}$

59. Let $y = \tan^{-1}x$. Then $x = \tan y$ and $1 = (\sec^2 y)(dy/dx)$, so $\dfrac{dy}{dx} = \dfrac{1}{\sec^2 y} = \dfrac{1}{1 + \tan^2 y} = \dfrac{1}{1 + x^2}$.

61. Let $y = \sec^{-1}x$. Then $x = \sec y$ and $1 = (\sec y\tan y)(dy/dx)$, so $\dfrac{dy}{dx} = \dfrac{1}{\sec y\tan y} = \dfrac{1}{\sec y\sqrt{\sec^2 y - 1}} = \dfrac{1}{x\sqrt{x^2 - 1}}$.

(It is appropriate to substitute the *positive* quantity $\sqrt{\sec^2 y - 1}$ for $\tan y$ since $0 \le y < \pi/2$, or $\pi \le y < 3\pi/2$.)

63. (a) $\sec^{-1}(-\sqrt{2}) = \dfrac{5\pi}{4}$, while $\cos^{-1}(-1/\sqrt{2}) = 3\pi/4$.   (b) For the "new" *inverse secant*, let $y = \sec^{-1}x$, so $x = \sec y$

where $0 \leq y < \pi/2$, or $\pi/2 < y \leq \pi$. Now $\dfrac{dy}{dx} = \dfrac{1}{dx/dy} = \dfrac{1}{\sec y \tan y}$. For $x > 1$ so that $0 \leq y < \pi/2$, we have $\tan y > 0$, so $\tan y = \sqrt{\sec^2 y - 1} = \sqrt{x^2 - 1}$. For $x < -1$ so that $\pi/2 < y \leq \pi$, we have $\tan y < 0$, so $\tan y = -\sqrt{\sec^2 y - 1} = -\sqrt{x^2 - 1}$. Thus

$$\frac{dy}{dx} = \begin{cases} \dfrac{1}{x\sqrt{x^2 - 1}} & \text{for } x > 1, \\[2ex] \dfrac{1}{-x\sqrt{x^2 - 1}} & \text{for } x < -1, \end{cases} = \frac{1}{|x|\sqrt{x^2 - 1}}.$$

65. 3.14159 $(n = 20)$; 3.141592653589793 $(n = 200)$

Section 7.6

1. Divide the relation $\cosh^2 x - \sinh^2 x = 1$ by $\cosh^2 x$. **3.** $\sinh(-x) = \dfrac{e^{-x} - e^{-(-x)}}{2} = \dfrac{-e^x + e^{-x}}{2} = -\sinh x$

5. $\sinh x \cosh y + \cosh x \sinh y = \dfrac{e^x - e^{-x}}{2} \cdot \dfrac{e^y + e^{-y}}{2} + \dfrac{e^x + e^{-x}}{2} \cdot \dfrac{e^y - e^{-y}}{2}$

$$= \frac{2e^{x+y} - 2e^{-x-y}}{4} = \frac{e^{x+y} - e^{-x-y}}{2} = \sinh(x + y)$$

7. $\sinh 2x = 2 \sinh x \cosh x$; $\cosh 2x = \cosh^2 x + \sinh^2 x$

9. $D(\cosh x) = D\left(\dfrac{e^x + e^{-x}}{2}\right) = \dfrac{1}{2}D(e^x + e^{-x}) = \dfrac{1}{2}(e^x - e^{-x}) = \sinh x$

11. $D(\operatorname{sech} x) = D\left(\dfrac{1}{\cosh x}\right) = \dfrac{-\sinh x}{\cosh^2 x} = -\tanh x \operatorname{sech} x$

13. Let $y = \cosh^{-1} x$, so $x = \cosh y$. Then

$$\frac{dy}{dx} = \frac{1}{dx/dy} = \frac{1}{\sinh y} = \frac{1}{\sqrt{\cosh^2 y - 1}} = \frac{1}{\sqrt{x^2 - 1}}, \quad x > 1.$$

(Since $y = \cosh^{-1} x \geq 0$, $\sinh y \geq 0$, so the *positive* square root was appropriate.)

15. Let $y = \coth^{-1} x$, so $x = \coth y$. Then

$$\frac{dy}{dx} = \frac{1}{dx/dy} = \frac{-1}{\operatorname{csch}^2 y} = \frac{-1}{\coth^2 y - 1} = \frac{-1}{x^2 - 1} = \frac{1}{1 - x^2}, \quad |x| > 1.$$

17. Let $y = \operatorname{csch}^{-1} x$, so $x = \operatorname{csch} y$. Then

$$\frac{dy}{dx} = \frac{1}{dx/dy} = \frac{-1}{\operatorname{csch} y \coth y} = \frac{-1}{(\operatorname{csch} y)(\pm\sqrt{1 + \operatorname{csch}^2 y})} = \frac{-1}{(x)(\pm\sqrt{1 + x^2})}.$$

If $x > 0$, then $y = \operatorname{csch}^{-1} x > 0$, so $\coth y > 0$ and the *plus sign* is appropriate. If $x < 0$, then $y = \operatorname{csch}^{-1} x < 0$ so $\coth y < 0$ and the *minus sign* is appropriate. These two cases are both covered by the formula

$$D(\operatorname{csch}^{-1} x) = \frac{-1}{(|x|\sqrt{1 + x^2})}.$$

19. $2x \sinh(x^2)$ **21.** $-\dfrac{\operatorname{sech}\sqrt{x}\tanh\sqrt{x}}{2\sqrt{x}}$ **23.** $-\dfrac{\operatorname{csch}(\ln x)\coth(\ln x)}{x}$ **25.** $2\sinh^3 x \cosh x + 2\cosh^3 x \sinh x$ **27.** $\dfrac{2}{\sqrt{1 + 4x^2}}$

29. $2 \sec 2x$ **31.** $\dfrac{-2}{x\sqrt{1 - x^4}}$ **33.** $\dfrac{e^{2x}}{\sqrt{1 + e^{2x}}} + e^x \sinh^{-1}(e^x)$ **35.** $\dfrac{-3 \operatorname{csch}^2 3x}{1 - \cot^2 3x}$, $|\cot 3x| < 1$ **37.** $\ln|\sinh x| + C$

39. $\dfrac{1}{3}\cosh(3x + 2) + C$ **41.** $\dfrac{1}{3}\tanh 3x + C$ **43.** $\ln|\sinh x| + C$ **45.** $\dfrac{1}{2}(\ln \frac{4}{3})$ **47.** $\dfrac{1}{4}\sqrt{1 + 4x^2} + C$

49. $-\operatorname{sech}^{-1}(e^x) + C$ **51.** $\dfrac{\sqrt{5}}{2} + 2\sinh^{-1}\left(\dfrac{1}{2}\right) = \dfrac{\sqrt{5}}{2} + 2\ln\left(\dfrac{1 + \sqrt{5}}{2}\right)$

53. $\sqrt{16 + x^2} - 4\sinh^{-1}\left|\dfrac{4}{x}\right| + C = \sqrt{16 + x^2} - 4\ln\left(\dfrac{4 + \sqrt{16 + x^2}}{x}\right) + C$

55. $-\dfrac{9}{2}\sinh^{-1}\left(\dfrac{\sin x}{3}\right) + \dfrac{\sin x\sqrt{9 + \sin^2 x}}{2} + C = \dfrac{(\sin x)}{2}\sqrt{9 + \sin^2 x} - \dfrac{9}{2}\ln(\sin x + \sqrt{9 + \sin^2 x}) + C$

57. $8\cosh^{-1}\left(\dfrac{e^x}{4}\right) + \dfrac{e^x}{2}\sqrt{e^{2x} - 16} + C = \dfrac{e^x}{2}\sqrt{e^{2x} - 16} + 8\ln|e^x + \sqrt{e^{2x} - 16}| + C$ **59.** $\dfrac{\cosh^2 4x \sinh 4x}{12} + \dfrac{1}{6}\sinh 4x + C$

61. Let $x = \tanh y = \dfrac{e^y - e^{-y}}{e^y + e^{-y}}$. Then $xe^y + xe^{-y} = e^y - e^{-y}$ and $xe^{2y} + x = e^{2y} - 1$, so $(x-1)e^{2y} = -x - 1$ and $e^{2y} = \dfrac{1+x}{1-x}$.

Then $2y = 2 \tanh^{-1}x = \ln\left(\dfrac{1+x}{1-x}\right)$, and $\tanh^{-1}x = \dfrac{1}{2}\ln\left(\dfrac{1+x}{1-x}\right)$.

63. Let $x = \text{sech } y = \dfrac{2}{e^y + e^{-y}}$. Then $xe^y + xe^{-y} = 2$, so $xe^{2y} - 2e^y + x = 0$ and $e^y = \dfrac{2 \pm \sqrt{4-4x^2}}{2x} = \dfrac{1 + \sqrt{1-x^2}}{x}$. (The plus

sign is appropriate since $y \geq 0$ if $y = \text{sech}^{-1}x$.) Then $y = \text{sech}^{-1}x = \ln\left(\dfrac{1 + \sqrt{1-x^2}}{x}\right)$.

Supplementary Exercises for Chapter 7

1. $\dfrac{2x+1}{1-x}$ **3.** (a) No. It is not one to one.   (b) Yes. It is one to one.   **5.** No horizontal line can cut the graph of f in more

than one point.   **7.** A function f is one to one if $f(a) = f(b)$ implies that $a = b$.

9. (a) $\ln x = \displaystyle\int_1^x \dfrac{1}{t}\, dt, \quad x > 0$

(b)

11. (a) The number e is the unique real number such that $\displaystyle\int_1^e \dfrac{1}{t}\, dt = 1$.

(b)

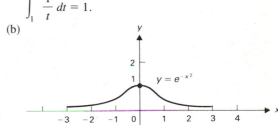

13. $-(\ln 3)/(\ln 24)$ **15.** (a) $10^{-1/5}$ (b) 49 **17.** (a) $\dfrac{e^{\sin^{-1}x}}{\sqrt{1-x^2}}$ (b) $\frac{1}{3}e^{\tan 3x} + C$ **19.** (a) $2^{3x+4}[3(\ln 2)]$ (b) $x^{5x}[5 + 5(\ln x)]$

21. (a) $(\ln 3)3^{\sin x}(\cos x)$ (b) $\dfrac{3^x \cdot 4^{x^2}}{5x}[(\ln 3) + (\ln 4)2x - (\ln 5)]$ **23.** $e^{x \sin x}(\cos x)^x[x \cos x + \sin x - x \tan x + \ln(\cos x)]$

25. $2y = -x + 2$ **27.** (a) $100(\frac{1}{2})^{t/10}$ ft/sec   (b) $\dfrac{1000}{\ln 2}[1 - (\frac{1}{2})^{t/10}]$ ft   **29.** (a) $-\pi/6$ (b) $-\pi/3$ **31.** (a) $-\pi/4$ (b) $-\pi/2$

33. $\dfrac{4}{1+16x^2}$ **35.** $\dfrac{1}{\sqrt{1-x}} \cdot \dfrac{1}{2\sqrt{x}}$ **37.** $\tan^{-1}(1/\sqrt{2})$ **39.** $\dfrac{7\pi}{12}$ **41.** (a) $\sinh x = \dfrac{e^x - e^{-x}}{2}$ (b) $-6 \text{ sech}^3 2x \tanh 2x$

43. (a)

(b) $-6 \coth^2(2x + 1)\csc h^2(2x + 1)$

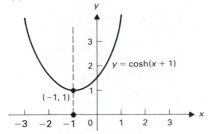

$y = \cosh(x + 1)$

$(-1, 1)$

CHAPTER 8

Section 8.1

1. $\cos x + x \sin x + C$ **3.** $\frac{1}{3}xe^{3x} - \frac{1}{9}e^{3x} + C$ **5.** $-\frac{1}{4}x^2 \cos 2x^2 + \frac{1}{8}\sin 2x^2 + C$ **7.** $-x^2\cos x + 2x \sin x + 2 \cos x + C$

9. $\frac{1}{2}x^2e^{x^2} - \frac{1}{2}e^{x^2} + C$ **11.** $x \tan x - \dfrac{x^2}{2} + \ln|\cos x| + C$ **13.** $\dfrac{x}{2}\tan 2x + \frac{1}{4}\ln|\cos 2x| + C$

15. $\dfrac{x^2}{2}\sqrt{1 + 2x^2} - \frac{1}{6}(1 + 2x^2)^{3/2} + C$ **17.** $\dfrac{x^4}{3}(1 + x^2)^{3/2} - \dfrac{4x^2}{15}(1 + x^2)^{5/2} + \dfrac{8}{105}(1 + x^2)^{7/2} + C$

19. $\dfrac{-x^2}{2 + 2x^2} + \frac{1}{2}\ln(1 + x^2) + C$ **21.** $\dfrac{x^2}{2}(\ln x) - \dfrac{x^2}{4} + C$ **23.** $x \ln(1 + x) - x + \ln(1 + x) + C$

25. $x \ln(1 + x^2) - 2x + 2 \tan^{-1}x + C$ **27.** $x \cos^{-1}x - \sqrt{1 - x^2} + C$ **29.** $x \tan^{-1}3x - \frac{1}{6} \ln(1 + 9x^2) + C$

31. $\dfrac{x^2}{2} \tan^{-1}x - \frac{1}{2}x + \frac{1}{2} \tan^{-1}x + C$ **33.** $\dfrac{x^3}{3} \sin^{-1}x + \dfrac{x^2}{3}\sqrt{1 - x^2} + \frac{2}{9}(1 - x^2)^{3/2} + C$

35. $\frac{1}{2}[x \cos(\ln x) + x \sin(\ln x)] + C$ **37.** $\dfrac{e^{ax}}{a^2 + b^2}(a \sin bx - b \cos bx) + C$ **39.** See formula 16 in the integral table.

41. See formula 86 in the integral table. **43.** See formula 67 in the integral table. **45.** See formula 92 in the integral table.

47. $\dfrac{\pi^2}{8} - 1$ **49.** $4 + \pi^2$ **51.** $2\pi(e - 2)$ **53.** $x = 1$ **55.** $k(\pi^3 + \pi^2 - 6\pi - 4)$

Section 8.2

1. $\ln|x| + 1/x + C$ **3.** $\dfrac{x^3}{3} - \dfrac{x^2}{2} - x + \ln|x + 1| + C$ **5.** $x^2/2 + \frac{1}{2} \ln|x^2 + 1| + \tan^{-1}x + C$ **7.** $\dfrac{1}{2} \ln\left|\dfrac{x - 1}{x + 1}\right| + C$

9. $\frac{1}{2}x^2 + \ln|(x + 2)(x - 2)^3| + C$ **11.** $\dfrac{x^3}{3} + \dfrac{3x^2}{2} + 5x + 14 \ln|x - 2| - 5 \ln|x - 1| + C$

13. $-\frac{4}{3} \ln|x| + \frac{13}{12} \ln|x - 3| + \frac{1}{4} \ln|x + 1| + C$ **15.** $x + \frac{5}{2} \ln|x + 3| - \frac{1}{2} \ln|x - 1| - \ln|x + 1| + C$

17. $\ln\left|\dfrac{x + 1}{x}\right| - \dfrac{2}{x + 1} + C$ **19.** $\dfrac{-1}{3(x + 1)^2} - \dfrac{8}{9(x + 1)} + \dfrac{1}{27} \ln\left|\dfrac{x - 2}{x + 1}\right| + C$

21. $\dfrac{-3}{x} - \ln|x| + 2 \ln|x - 2| + 4 \ln|x + 2| + C$ **23.** $4 \tan^{-1}x + \frac{1}{2} \ln(x^2 + 1) + 5 \ln|x - 1| + C$

25. $\ln\left|\dfrac{(x - 2)^2}{x + 2}\right| + \dfrac{1}{2} \ln(x^2 + 2) - \dfrac{1}{\sqrt{2}} \tan^{-1}\left(\dfrac{x}{\sqrt{2}}\right) + C$ **27.** $\frac{5}{6} \ln|3x^2 - 4x + 2| - \dfrac{\sqrt{2}}{3} \tan^{-1}\left(\dfrac{3x - 2}{\sqrt{2}}\right) + C$

29. $\dfrac{2x - 5}{2(x^2 - 2x + 2)} + \tan^{-1}(x - 1) + C$ **31.** $\dfrac{x - 1}{x^2 - x + 1} + \dfrac{1}{2} \ln|x^2 - x + 1| + \sqrt{3} \tan^{-1}\left(\dfrac{2x - 1}{\sqrt{3}}\right) + C$

33. $-\dfrac{3}{x} + \dfrac{2 - x}{2(x^2 + 1)} - \dfrac{3x}{4x^2 + 4} - \dfrac{3}{4} \tan^{-1}x + C$ **35.** $2\pi[8 - 6(\ln 3)]$ **37.** $\ln\left|\dfrac{y - 1}{y + 1}\right| = x^2 - \ln 2$

Section 8.3

1. $u = x$ $\quad$ $dv = (x + 1)^{1/2}\, dx$ $\quad$ $\displaystyle\int x\sqrt{x + 1}\, dx = \frac{2}{3}x(x + 1)^{3/2} - \int \frac{2}{3}(x + 1)^{3/2}\, dx$

$\quad du = dx$ $\quad$ $v = \frac{2}{3}(x + 1)^{3/2}$ $\qquad\qquad = \frac{2}{3}x(x + 1)^{3/2} - \frac{4}{15}(x + 1)^{5/2} + C$

3. $\frac{2}{3}(1 + x)^{3/2} - 2\sqrt{1 + x} + C$ **5.** $\frac{2}{7}(4 + x)^{7/2} - \frac{16}{5}(4 + x)^{5/2} + \frac{32}{3}(4 + x)^{3/2} + C$ **7.** $x - 2\sqrt{x} + 2 \ln|\sqrt{x} + 1| + C$

9. $x + 2 - 4\sqrt{x + 2} + 4 \ln|\sqrt{x + 2} + 1| + C$ **11.** $2(x + 1)^{3/2} - 2\sqrt{x + 1} + C$ **13.** $-\frac{116}{15}$ **15.** $\sqrt{1 + x^2} + C$

17. $\frac{1}{3}(1 + x^2)^{3/2} - \sqrt{1 + x^2} + C$ **19.** $\frac{1}{7}(x^2 + 1)^{7/2} - \frac{1}{5}(x^2 + 1)^{5/2} + C$ **21.** $\frac{1}{3}(x^2 + 8)^{3/4} - 9\sqrt{x^2 + 8} + C$

23. $\dfrac{-1}{\sqrt{x^2 - 1}} - \dfrac{1}{3(x^2 - 1)^{3/2}} + C$ **25.** $\frac{64}{3}$ **27.** $\frac{13}{3}$

29. $\frac{4}{7}x^{7/4} + \frac{2}{3}x^{3/2} + \frac{8}{5}x^{5/4} + 2x + \frac{8}{3}x^{3/4} + 4x^{1/2} + 8x^{1/4} + 8 \ln|x^{1/4} - 1| + C$

31. $\frac{6}{7}x^{7/6} - \frac{3}{2}x^{2/3} + 6x^{1/6} - 2 \ln|1 + x^{1/6}| + \ln|x^{1/3} - x^{1/6} + 1| - 2\sqrt{3} \tan^{-1}\left(\dfrac{2x^{1/6} - 1}{\sqrt{3}}\right) + C$ **33.** $\dfrac{e^{\cos^{-1}x}}{2}(x - \sqrt{1 - x^2}) + C$

35. $\dfrac{e^{\sin^{-1}x}}{10}[2x\sqrt{1 - x^2} - 2(1 - 2x^2)] + C$ **37.** $\dfrac{e^{\pi/2} + 1}{2}$ **39.** $\dfrac{128\pi\sqrt{5}}{105}$ **41.** $(y - 2)\sqrt{y + 1} = (x - 4)\sqrt{x + 2} + 22$

Section 8.4

1. Using $\sin^2x + \cos^2x = 1$, we have $\cos 2x = \cos^2x - \sin^2x = 2 \cos^2x - 1 = 1 - 2 \sin^2x$. Then

$$\cos^2x = \frac{1 + \cos 2x}{2} \quad \text{and} \quad \sin^2x = \frac{1 - \cos 2x}{2}$$

from which formulas (1) follow at once.

3. $\dfrac{\sin^3x}{3} - \dfrac{\sin^5x}{5} + C$ **5.** $\frac{1}{5} \sin^5x - \frac{2}{7} \sin^7x + \frac{1}{9} \sin^9x + C$ **7.** $-\frac{1}{3}(\cos 2x)^{3/2} + \frac{1}{7}(\cos 2x)^{7/2} + C$

9. $\dfrac{-2}{\sqrt{\sin x}} - \dfrac{2}{3}(\sin x)^{3/2} + C$ **11.** $\dfrac{x}{2} + \dfrac{1}{20}\sin 10x + C$ **13.** $\frac{3}{8}x + \frac{1}{8} \sin 4x + \frac{1}{64} \sin 8x + C$

15. $\frac{5}{16}x - \frac{1}{4}\sin 2x + \frac{3}{64}\sin 4x + \frac{1}{48}\sin^3 2x + C$ **17.** $-\cot x + C$ **19.** $\tan x + \frac{1}{3}\tan^3 x + C$ **21.** $\frac{\tan^3 x}{3} + C$ **23.** $\frac{\tan^5 x}{5} + C$

25. $\frac{1}{6}\tan^2 3x + C$ **27.** $-\tan x + \frac{\tan^3 x}{3} + x + C$ **29.** $\frac{1}{12}\tan^6 2x - \frac{1}{8}\tan^4 2x + \frac{1}{4}\tan^2 2x + \frac{1}{2}\ln|\cos 2x| + C$ **31.** $-\frac{1}{4}\csc^4 x + C$

33. $-\frac{1}{2}(\csc x \cot x + \ln|\csc x + \cot x|) + C$ **35.** $\frac{1}{4}\sec^3 x \tan x + \frac{3}{8}(\sec x \tan x + \ln|\sec x + \tan x|) + C$

37. $\displaystyle\int \cot^n ax \, dx = \int (\cot^{n-2} ax)(\csc^2 ax - 1) \, dx = \int \cot^{n-2} ax \csc^2 ax \, dx - \int \cot^{n-2} ax \, dx = -\frac{\cot^{n-1} ax}{a(n-1)} - \int \cot^{n-2} ax \, dx$

39. $\displaystyle\int \csc^n ax \, dx = \int \csc^{n-2} ax(1 + \cot^2 ax) \, dx = \int \csc^{n-2} ax \, dx + \int \csc^{n-2} ax \cot^2 ax \, dx$

Let

$$u = \cot ax \qquad\qquad dv = \csc^{n-2} ax \cot ax \, dx$$
$$du = -a \csc^2 ax \, dx \qquad v = -\frac{\csc^{n-2} ax}{a(n-2)}$$

for integration by parts. We obtain

$$\int \csc^{n-2} ax \cot^2 ax \, dx = -\frac{\csc^{n-2} ax \cot ax}{a(n-2)} - \frac{1}{n-2}\int \csc^n ax \, dx.$$

Thus

$$\int \csc^n ax \, dx = \int \csc^{n-2} ax \, dx - \frac{\csc^{n-2} ax \cot ax}{a(n-2)} - \frac{1}{n-2}\int \csc^n ax \, dx.$$

Solving this equation for $\int \csc^n ax \, dx$ gives the formula.

41. For integration by parts, we let

$$u = \cos^{n-1} ax \qquad\qquad dv = \cos ax \, dx$$
$$du = -(n-1)a \cos^{n-2} ax \sin ax \, dx \qquad v = \frac{1}{a}\sin ax.$$

Then

$$\int \cos^n ax \, dx = \frac{1}{a}\cos^{n-1} ax \sin ax + (n-1)\int \cos^{n-2} ax \sin^2 ax \, dx$$
$$= \frac{1}{a}\cos^{n-1} ax \sin ax - (n-1)\int \cos^n ax \, dx + (n-1)\int \cos^{n-2} ax \, dx.$$

Solving this equation for $\int \cos^n ax \, dx$ gives the formula.

43. $1 - \frac{\pi}{4}$ **45.** (a) $\frac{4}{3}$ (b) 4 **47.** (a) $\sin x = \frac{2t}{1+t^2}$ (b) $\cos x = \frac{1-t^2}{1+t^2}$ (c) $dx = \frac{2}{1+t^2}\,dt$ (d) $\ln|1 + \tan(x/2)| + C$

49. $\frac{1}{5}\ln\left(\frac{2\tan(x/2) + 1}{\tan(x/2) - 2}\right) + C$

Section 8.5

1. $x = a \sin t$, $\sqrt{a^2 - x^2} = \sqrt{a^2 - a^2\sin^2 t} = \sqrt{a^2\cos^2 t}$. Now $-\pi/2 \le t \le \pi/2$, so $\cos t \ge 0$. Thus $\sqrt{a^2\cos^2 t} = a\cos t$.
3. $x = 3\sin t$, $\int 9\cos^2 t \, dt$ **5.** $x = 4\sec t$, $\int \sec t \, dt$ **7.** $x = \sec t$, $\int (\sec^2 t + 2\sec t) \, dt$ **9.** $x = \frac{1}{2}\tan t$, $\int \frac{1}{2}\sec^3 t \, dt$

11. $x = \frac{1}{4}\sin t$, $\int \frac{1}{1024}\sin^4 t \, dt$ **13.** $x - \frac{3}{\sqrt{5}}\tan t$, $\int \frac{\sqrt{5}}{45}\cos t \, dt$ **15.** $x = \sqrt{\frac{2}{3}}\tan t$, $\int \frac{2}{\sqrt{3}}\sec^3 t \, dt$

17. $x - 2 = \sec t$, $\int \sec t \tan^2 t \, dt$ **19.** $x + 1 = \sin t$, $\int (-1 + \sin t)^2 \, dt$ **21.** $\sqrt{3}\,x + \frac{2}{\sqrt{3}} = \frac{2\sqrt{7}}{3}\sin t$, $\int \left(\frac{1}{\sqrt{3}}\right) dt$

23. $u = e^x$, $\int \frac{du}{\sqrt{15 + 2u - u^2}}$, $u - 1 = 4\sin t$, $\int 1 \, dt$ **25.** $\sqrt{1 + x^2} + C$ **27.** $\frac{4}{3}$ **29.** $\frac{136}{15}$ **31.** $\sqrt{5 + x^2} + C$

33. $\sqrt{x^2 - 16} - \ln|x + \sqrt{x^2 - 16}| + C$ **35.** $\frac{5\sqrt{6}}{2} + \frac{1}{4}\ln(5 + 2\sqrt{6})$ **37.** $\frac{-1}{\sqrt{9 + x^2}} + \frac{3}{(9 + x^2)^{3/2}} + C$

39. $-\frac{1}{3}(1 - e^{2x})^{3/2} + \frac{1}{5}(1 - e^{2x})^{5/2} + C$ **41.** $\frac{1}{2}[\sqrt{2} + \ln(1 + 2)]$ **43.** $\pi/6$
45. $2\ln|\sqrt{x^2 - 4x + 5} + x - 2| + \sqrt{x^2 - 4x + 5} + C$

47. (a) $A = 4\displaystyle\int_0^a \sqrt{a^2 - x^2}\,dx$, which yields $4\left(\frac{\pi a^2}{4}\right) = \pi a^2$ upon integration using the substitution $x = \sin t$, $dx = \cos t \, dt$.
(b) We used the formula for the area of a circle in Section 2.2 to demonstrate that $\lim_{x \to 0} [(\sin x)/x] = 1$, which was in turn

used in Section 3.4 to show that $[d(\sin x)]/dx = \cos x$, which was in turn used in the derivation of the area formula in part (a) to compute dx for the substitution $x = \sin t$.

49. $\frac{1}{2}[\sqrt{5} + \frac{1}{2}\ln(2 + \sqrt{5})]$ **51.** $\frac{\pi}{4}\left[\frac{9\sqrt{5}}{4} - \frac{1}{8}\ln(2 + \sqrt{5})\right]$

53. $\sqrt{y^2 + 1} - \ln\left(\frac{1 + \sqrt{y^2 + 1}}{y}\right) = \frac{x}{2}\sqrt{1 - x^2} + \frac{1}{2}\sin^{-1}x + C$

Section 8.6

1. $\frac{56}{3}$ **3.** $[-3/\sqrt{4 + x^2}] + C$ **5.** $-\sqrt{(10 - x)/x} + C$ **7.** $\frac{4(7x + 6)}{147}\sqrt{3 - 7x} + C$ **9.** $\frac{6x^2 - 1}{120}\sqrt{(4x^2 + 1)^3} + C$

11. $\frac{(4x^2 - 1)^8}{32}\left(\frac{4x^2 - 1}{9} + \frac{1}{8}\right) + C$ **13.** $-\frac{\sqrt{16 - x^4}}{32x^2} + C$ **15.** $-\frac{1}{4}\sqrt{\frac{4 - x^2}{x^2}} + C$ **17.** $\frac{\pi}{2}$

19. $\frac{-\sin^3x \cos x}{4} + \frac{3}{4}\left(\frac{x}{2} - \frac{\sin 2x}{4}\right) + C$ **21.** $-\frac{\cos 5x}{10} + \frac{\cos x}{2} + C$ **23.** $\frac{1}{4} \cdot \frac{\sin^2 4x}{2} + C$

25. $\frac{\sin^5 x \cos x}{6} - \frac{\sin^3 x \cos x}{24} + \frac{1}{8}\left(\frac{x}{2} - \frac{\sin 2x}{4}\right) + C$ **27.** $-\frac{1}{2} \cdot \cot\frac{x^2}{2} + C$ **29.** $\frac{1}{16} \cdot \cos 4x + \frac{x}{4} \cdot \sin 4x + C$

31. $x^3\sin x + 3x^2\cos x - 6(\cos x + x \sin x) + C$ **33.** $\frac{1}{3} \cdot \tan 3x - x + C$ **35.** $\frac{\tan^3 2x}{6} - \left(\frac{1}{2} \cdot \tan 2x - x\right) + C$

37. $\frac{1}{3}\left(\frac{x^3}{2} + \frac{\sin 2x^3}{4}\right) + C$ **39.** $\frac{(2 \sin x + 4)^9}{4}\left(\frac{\sin x + 2}{5} - \frac{4}{9}\right) + C$ **41.** $-\frac{\sqrt{4 - \sin^2x}}{4 \sin x} + C$ **43.** $-\frac{\sqrt{\sin^2 2x + 9}}{18 \sin 2x} + C$

45. $-\frac{6 \sec 2x + 1}{120}\sqrt{(1 - 4 \sec 2x)^3} + C$ **47.** $-\frac{\sqrt{25 + 4 \tan^2x}}{25 \tan x} + C$ **49.** $-\frac{1}{6}\sqrt{\frac{4 - \sec 3x}{\sec 3x}} + C$

Supplementary Exercises for Chapter 8

1. $\frac{x^2}{3}\sin 3x - \frac{2}{27}\sin 3x + \frac{2}{9}x \cos 3x + C$ **3.** $\frac{x^2}{2}\ln x - \frac{x^2}{4} + C$ **5.** $\frac{x^2}{4} - \frac{1}{4}x \sin 2x - \frac{1}{8}\cos 2x + C$

7. $x \sin x - \frac{x}{3}\sin^3x + \frac{2}{3}\cos x + \frac{1}{9}\cos^3x + C$ **9.** $\frac{x^2}{2} + \ln\left|\frac{x}{x^2 + 1}\right| + C$ **11.** $-5/2 \ln|2x + 3| + 5 \ln|x - 5| + C$

13. $x - 1 - 2\sqrt{x - 1} + 2 \ln|1 + \sqrt{x - 1}| + C$ **15.** $\frac{1}{3}(x^2 + 4)^{3/2} - 4\sqrt{x^2 + 4} + C$ **17.** $\frac{1}{6}\ln\left|\frac{\sqrt{9 - x^2} - 3}{\sqrt{9 - x^2} + 3}\right| + C$

19. $-\frac{4}{5}(x - 1)^{5/4} - 3(x - 1) - \frac{8}{3}(x - 1)^{3/4} - 12(x - 1)^{1/2} - 72(x - 1)^{1/4} - 216 \ln|(x - 1)^{1/4} - 3| + C$ **21.** $\frac{x}{2} - \frac{1}{16}\sin 8x + C$

23. $\frac{1}{4}(\sec 2x \tan 2x + \ln|\sec 2x + \tan 2x|) + C$ **25.** $\frac{1}{3}\sin^3x - \frac{1}{5}\sin^5x + C$ **27.** $-\frac{1}{3}\cot^3x - \frac{1}{5}\cot^5x + C$

29. $-\frac{1}{2}\cos 2x + \frac{1}{3}\cos^3 2x - \frac{1}{10}\cos^5 2x + C$ **31.** $\frac{1}{9}\tan^3 3x + \frac{1}{15}\tan^5 3x + C$ **33.** $\frac{1}{4}\sec^4x + C$ **35.** $\frac{1}{2}\sin^{-1}\left(\frac{2x}{3}\right) + C$

37. $2 \sin^{-1}\left(\frac{x}{2}\right) - \frac{1}{2}x\sqrt{4 - x^2} + C$ **39.** $\sin^{-1}\left(\frac{x + 1}{3}\right) + C$ **41.** $-\frac{2\sqrt{9 + x^3}}{27x^{3/2}} + C$ **43.** $(8 + 3\pi)/64$

CHAPTER 9

Section 9.1

1. The approximation of $f(x)$ by $T_1(x)$ near x_0 can be written $f(x) \approx f(x_0) + f'(x_0)(x - x_0)$ for x near x_0. Replacing x_0 by a and $x - x_0$ by Δx so that $x = x_0 + (x - x_0) = a + \Delta x$, this formula becomes $f(a + \Delta x) \approx f(a) + f'(a)\Delta x$, which is the approximation formula (7) in Section 3.9.
3. $2 + 2(x + 1)$ **5.** $12 - 5(x + 2)$ **7.** $9 - 5(x + 2) + \frac{1}{2}(x + 2)^2$ **9.** $-47 + 27(x + 3) - 4(x + 3)^2$
11. $4 - 7(x + 1) + 6(x + 1)^2 - 4(x + 1)^3 + (x + 1)^4$ **13.** $-113 + 198(x + 2) - 135(x + 2)^2 + 42(x + 2)^3 - 5(x + 2)^4$
15. $T_6(x) = 1 + x + \frac{x^2}{2!} + \frac{x^3}{3!} + \frac{x^4}{4!} + \frac{x^5}{5!} + \frac{x^6}{6!}$, so $e \approx T_6(1) \approx 2.7180556$. Using a calculator, we have $e \approx 2.7182818$, so the error is < 0.00023.
17. $T_4(x) = 1 - \frac{1}{2!}\left(x - \frac{\pi}{2}\right)^2 + \frac{1}{4!}\left(x - \frac{\pi}{2}\right)^4$, so $\sin 88° \approx T_4\left(\frac{22\pi}{45}\right) \approx 0.9993908$. Using a calculator, we have $\sin 88° \approx 0.9993908$, so the error is $< 1/10^7$.

19. $T_6(x) = 1 - (x - 1) + (x - 1)^2 - (x - 1)^3 + (x - 1)^4 - (x - 1)^5 + (x - 1)^6$, so $1/0.8 \approx T_6(0.8) \approx 1.249984$. Since $1/0.8 = 5/4 = 1.25$, we have error < 0.00002.   **21.** $-1 + x + \dfrac{3}{2}x^2 - \dfrac{1}{24}x^4$

23. $1 + x + \dfrac{x^2}{2!} + \dfrac{7}{6}x^3 + \dfrac{x^4}{4!} + \dfrac{x^5}{5!} + \dfrac{x^6}{6!} + \dfrac{x^7}{7!} + \dfrac{x^8}{8!}$ **25.** $1 + 2x + \dfrac{1}{2}x^3 - \dfrac{5}{24}x^4$

27. (a) $2x - \dfrac{8x^3}{3!} + \dfrac{32x^5}{5!} - \dfrac{128x^7}{7!}$ (b) The polynomial may be obtained by replacing x by $2x$ in the answer to Example 3.

29. (a) $1 + x^2$ (b) The polynomial is the portion of degree less than or equal to 3 of the polynomial obtained by replacing x by x^2 in the answer to Example 8.   (c) $1 + x^2 + x^4 + x^6 + x^8$

31. Differentiating $g(x) = f(rx)$ repeatedly using the chain rule, we find that $g^{(i)}(x) = r^i f^{(i)}(rx)$. The term of degree $i \leq n$ in the nth Taylor polynomial for $g(x)$ at $x_0 = 0$ is therefore

$$\frac{g^{(i)}(0)}{i!}x^i = \frac{r^i f^{(i)}(r \cdot 0)}{i!}x^i = \frac{f^{(i)}(0)}{i!}(rx)^i,$$

which is the term of degree i in $T_n(rx)$. Thus $T_n(rx)$ is the nth Taylor polynomial for $g(x)$ at $x_0 = 0$.

Section 9.2

1. If $f'(x)$ exists for $x_0 - h < x < x_0 + k$, then for each such x there exists c between x and x_0 such that $f(x) = f(x_0) + f'(c)(x - x_0)$. This is the mean value theorem.

3. (a) $\frac{82}{27}$ (b) $\frac{1}{2187}$ **5.** (a) $4\pi/3$ ft^3 (b) $\dfrac{100\pi}{12^3}$ ft^3 **7.** (a) $\dfrac{\pi^9}{2 \cdot 9! \cdot 60^9}$ (b) $\dfrac{\pi^{10}}{10! \cdot 60^{10}}$. Since $T_8(x) = T_9(x)$, we have $E_8(\pi/60) = E_9(\pi/60)$, so $E_9(\pi/60)$ can be used to estimate the error.

9. (a) 1.9094875   (b) 0.00012   **11.** (a) 0.2588190618   (b) 0.0000000084   **13.** (a) 1.175198413   (b) 0.0000048225

15. $n = 7$. The same value $n = 7$ is obtained using Taylor's formula as described.   **17.** $|E_n(x)| \leq \left| \dfrac{x^{n+1}}{} \right|$, so $\lim |E_n(x)| = 0$

19. $|E_n(x)| \leq \left| \dfrac{2e^{|x|}}{2} \cdot \dfrac{x^{n+1}}{(n+1)!} \right|$ and $\lim\limits_{n \to \infty} \dfrac{x^{n+1}}{(n+1)!} = 0$, so $\lim\limits_{n \to \infty} |E_n(x)| = 0$

21. $|E_n(x)| \leq \left| \dfrac{e^{|x|} + e^{|3x|}}{(n+1)!}x^{n+1} \right|$ and $\lim\limits_{n \to \infty} \left| \dfrac{x^{n+1}}{(n+1)!} \right| = 0$. Thus $\lim\limits_{n \to \infty} |E_n(x)| = 0$

23. $|E_n(x)| = \left| \dfrac{3 \cdot 1 \cdot 1 \cdot 3 \cdot 5 \cdots (2n - 3)}{2^{n+1}c^{(2n-1)/2}} \cdot \dfrac{(x - 1)^{n+1}}{(n+1)!} \right| \leq \left| \dfrac{3}{2} \cdot \dfrac{1}{n+1} \cdot \dfrac{1^{n+1}}{1^{(2n-1)/2}} \right| = \dfrac{3}{2n+2}$, so $\lim\limits_{n \to \infty} |E_n(x)| = 0$

Section 9.3

1. $\frac{4}{5}$ **3.** $\frac{1}{5}$ **5.** -1 **7.** -1 **9.** ∞ **11.** $\frac{1}{2}$ **13.** 0   **15.** $-\infty$ **17.** $-\infty$ **19.** $\frac{1}{2}$ **21.** $-\infty$
23. $-\infty$ **25.** 3   **27.** 6   **29.** 0   **31.** 2

33. (a) $\lim\limits_{x \to a+} \dfrac{f(x)}{g(x)} = \lim\limits_{x \to a-} \dfrac{f(x)}{g(x)} = 0$ (b) $\lim\limits_{x \to a+} \dfrac{f(x)}{g(x)} = \lim\limits_{x \to a-} \dfrac{f(x)}{g(x)} = \dfrac{f^{(n)}(a)}{g^{(n)}(a)}$

(c) $\lim\limits_{x \to a+} \dfrac{f(x)}{g(x)} = \begin{cases} \infty & \text{if } f^{(n)}(a)g^{(m)}(a) > 0, \\ -\infty & \text{if } f^{(n)}(a)g^{(m)}(a) < 0, \end{cases}$ (d) $\lim\limits_{x \to a+} \dfrac{f(x)}{g(x)} = \begin{cases} \infty & \text{if } f^{(n)}(a)g^{(m)}(a) > 0, \\ -\infty & \text{if } f^{(n)}(a)g^{(m)}(a) < 0, \end{cases}$

$\lim\limits_{x \to a-} \dfrac{f(x)}{g(x)} = \begin{cases} -\infty & \text{if } f^{(n)}(a)g^{(m)}(a) > 0, \\ \infty & \text{if } f^{(n)}(a)g^{(m)}(a) < 0 \end{cases}$ $\lim\limits_{x \to a-} \dfrac{f(x)}{g(x)} = \lim\limits_{x \to a+} \dfrac{f(x)}{g(x)}$

35. (a) By the Cauchy mean value theorem and the hypotheses, $\dfrac{f(x) - f(a)}{g(x) - g(a)} = \dfrac{f(x)}{g(x)} = \dfrac{f'(c)}{g'(c)}$ for some c where $a < c < x$. As $x \to a$, we must have $c \to a$. Thus $\lim\limits_{x \to a+} \dfrac{f(x)}{g(x)} = \lim\limits_{c \to a+} \dfrac{f'(c)}{g'(c)} = \lim\limits_{x \to a+} \dfrac{f'(x)}{g'(x)}$.   (b) We note that $\dfrac{f(a) - f(x)}{g(a) - g(x)} = \dfrac{-f(x)}{-g(x)} = \dfrac{f(x)}{g(x)}$, and then proceed as in part (a).   (c) This follows since from parts (a) and (b), the two one-sided limits of $f(x)/g(x)$ at a both equal the limit of $f'(x)/g'(x)$ at a.

37. 0.2   **39.** 0   **41.** 3

Section 9.4

1. 0   **3.** 0   **5.** 0   **7.** $e^2/3$ **9.** 0   **11.** $-1/6$ **13.** ∞ **15.** $-\infty$ **17.** -1 **19.** 1   **21.** 1   **23.** 0   **25.** ∞ **27.** 0   **29.** ∞

31. 1 **33.** 1 **35.** e^2 **37.** 1 **39.** 0 **41.** ∞ **43.** ∞ **45.** e^3 **47.** ∞

49. Since $\lim_{x \to a} f(x) = \lim_{x \to a} g(x) = \infty$, we see that $1/f(x)$ and $1/g(x)$ are defined and differentiable for x near a but $x \neq a$. Of course $\dfrac{f(x)}{g(x)} = \dfrac{1/g(x)}{1/f(x)}$. Applying l'Hôpital's rule to the 0/0-type form $\dfrac{1/g(x)}{1/f(x)}$, we find that

$$\lim_{x \to a} \frac{f(x)}{g(x)} = \lim_{x \to a} \frac{1/g(x)}{1/f(x)} = \lim_{x \to a} \frac{-g'(x)/[g(x)]^2}{-f'(x)/[f(x)]^2} = \left[\lim_{x \to a} \frac{g'(x)}{f'(x)}\right]\left[\lim_{x \to a} \left(\frac{f(x)}{g(x)}\right)^2\right].$$

Thus $L = \left[\lim_{x \to a} \dfrac{g'(x)}{f'(x)}\right] \cdot L^2$, so $\lim_{x \to a} \dfrac{f'(x)}{g'(x)} = L = \lim_{x \to a} \dfrac{f(x)}{g(x)}$.

51. -0.1668

53. This limit is difficult to establish with some calculators, since they will not compute $(\ln x)^x$ for a large enough x to show that this quantity is very small compared with x as $x \to \infty$. The correct answer for the limit is 0. Using a calculator for values of x up to 140, one would conjecture that the limit is ∞, which is as far wrong as one can get. If we suspect a problem, we can examine values of $\ln\left(\dfrac{(\ln x)^{100}}{x}\right) = 100 \ln(\ln x) - \ln x$ for large x, which is best done by letting $t = \ln x$ and evaluating $100(\ln t) - t$ for large t. A calculator indicates that this expression approaches $-\infty$ as $t \to 0$, so the given function approaches $e^{-\infty} = 0$ as $x \to \infty$.

Section 9.5

1. $\displaystyle \lim_{h \to 1^-} \int_0^h \frac{dx}{x^2 - 1}$ **3.** $\displaystyle \lim_{h \to 1^+} \int_h^2 \frac{dx}{x - 1} + \lim_{h \to \infty} \int_2^h \frac{dx}{x - 1}$ **5.** $\displaystyle \lim_{h \to -\infty} \int_h^0 e^{-x^2} dx + \lim_{h \to \infty} \int_0^h e^{-x^2} dx$

7. Not improper **9.** $\displaystyle \lim_{h \to 0^+} \int_h^1 \frac{dx}{\sqrt{x} + x^2} + \lim_{h \to \infty} \int_1^h \frac{dx}{\sqrt{x} + x^2}$

11. $\displaystyle \lim_{h \to (\pi/2)^-} \int_0^h \sec x \, dx + \lim_{h \to (\pi/2)^+} \int_h^\pi \sec x \, dx + \lim_{h \to (3\pi/2)^-} \int_\pi^h \sec x \, dx + \lim_{h \to (3\pi/2)^+} \int_h^{2\pi} \sec x \, dx$

13. $\frac{1}{2}$ **15.** $\pi/2$ **17.** 1 **19.** $2\sqrt{2}$ **21.** $\frac{10}{3}$ **23.** 1 **25.** Diverges **27.** Diverges

29. For $p \neq 1$, $\displaystyle \int_1^\infty x^{-p} \, dx = \lim_{h \to \infty} \left. \frac{x^{-p+1}}{1 - p} \right]_1^h = \lim_{h \to \infty} \left(\frac{h^{1-p}}{1 - p} - \frac{1}{1 - p} \right)$. Now $\displaystyle \lim_{h \to \infty} \frac{h^{1-p}}{1 - p} = \begin{cases} 0 & \text{if } p > 1 \\ \infty & \text{if } p < 1. \end{cases}$

Thus the given integral converges for $p > 1$ and diverges for $p < 1$. For $p = 1$, $\displaystyle \int_1^\infty \frac{1}{x} \, dx = \lim_{h \to \infty} \left. (\ln x) \right]_1^h = \lim_{h \to \infty} (\ln h) = \infty$, so the integral diverges for $p \leq 1$.

31. (a) No (b) $V = \displaystyle \int_1^\infty \frac{\pi}{x^2} \, dx = \lim_{h \to \infty} \left. -\frac{\pi}{x} \right]_1^h = \pi \lim_{h \to \infty} \left(-\frac{1}{h} + 1 \right) = \pi$. (c) No **33.** $\sqrt{2}$ **35.** $\dfrac{3k}{2}$ units

37. (a) $\displaystyle \int_0^\infty \frac{ds}{v} = \int_0^\infty \frac{ds}{Ae^{ks}} = \int_0^\infty \frac{1}{A} e^{-ks} \, ds = \lim_{h \to \infty} \left. \frac{-1}{kA} e^{-ks} \right]_0^h = \frac{1}{kA}$, which is finite. (b) $\dfrac{dv}{dt} = \dfrac{dv}{ds} \cdot \dfrac{ds}{dt} = Ake^{ks} \cdot v = kv^2$

(c) If $v = a(1 + s^2)$, which increases more slowly than Ae^{ks} for $k > 0$, the integral $\displaystyle \int_0^\infty (ds/v)$ still converges.

Section 9.6

1. Diverges **3.** Converges **5.** Diverges **7.** Diverges **9.** Converges **11.** Diverges **13.** Converges **15.** Diverges **17.** Diverges **19.** Diverges **21.** Converges **23.** Diverges **25.** Diverges **27.** Converges **29.** Converges **31.** Diverges **33.** Converges **35.** Diverges **37.** Converges **39.** Diverges **41.** Absolutely convergent **43.** Not absolutely convergent **45.** Absolutely convergent **47.** Not absolutely convergent **49.** Absolutely convergent

Supplementary Exercises for Chapter 9

1. $(x + 1)^2 - 6(x + 1) + 10$ **3.** 3 **5.** $\frac{1}{2} - \frac{1}{4}(x - 1) + \frac{1}{8}(x - 1)^2 - \frac{1}{16}(x - 1)^3$ **7.** $-2\left(x - \dfrac{\pi}{4}\right) + \dfrac{4}{3}\left(x - \dfrac{\pi}{2}\right)^3$

9. ≈ 1.6484 with error $\leq \frac{1}{1920}$ **11.** ≈ 0.03922133 with error ≤ 0.00000064

13. $E_n(x) = \dfrac{f^{(n+1)}(c)}{(n + 1)!}(x - x_0)^{n+1}$ where for each x, c is some number between x_0 and x.

15. Let $f(x) = \sin x$. Then at $x_0 = 0$, we have $E_n(x) = \dfrac{f^{(n+1)}(c)}{(n + 1)!} x^{n+1}$ for some c between x and x_0. Now $f^{(n+1)}(x)$ is either $\sin x$,

$-\sin x$, $\cos x$, or $-\cos x$, so $|f^{(n+1)}(c)| \le 1$. Thus $|E_n(x)| \le \left|\dfrac{x^{n+1}}{(n+1)!}\right|$. Since $\lim\limits_{n\to\infty} |x^n/n!| = 0$ for any value of x, we see that $\lim\limits_{n\to\infty} |E_n(x)| = 0$ for any value of x. Thus for any x, $T_n(x) \approx \sin x$ if n is large enough.

17. $\frac{2}{25}$ **19.** 0 **21.** $-1/2$ **23.** 6 **25.** $-1/3$ **27.** 0 **29.** $e^{3/2}$ **31.** e^2 **33.** $\dfrac{8\sqrt{2}}{3}$ **35.** π **37.** Diverges **39.** Diverges
41. Converges **43.** Converges **45.** Converges **47.** Converges

CHAPTER 10

Section 10.1

1. We have $\lim_{n\to\infty} a_n = \infty$ if for each real number γ there exists an integer N such that $a_n > \gamma$ provided that $n > N$.
3. Converges to 0 **5.** Converges to 1 **7.** Converges to 2 **9.** Converges to 0 **11.** Converges to 0 **13.** Diverges
15. Converges to 0 **17.** Converges to 0 **19.** Diverges to ∞ **21.** Converges to 0 **23.** Diverges to ∞ **25.** Diverges
27. Converges to 1 **29.** Converges to 1 **31.** Converges to 1 **33.** Diverges to ∞ **35.** Converges to 0 **37.** 120 ft
39. Let $\varepsilon > 0$ be given. Let $N = 5$. Then $|a_n - 1| = |1 - 1| = 0 < \varepsilon$ if $n > 5$, so the sequence converges to 1. (Any positive integer will do for N.)
41. Let $\varepsilon > 0$ be given, and find a positive integer N such that $N > 1/\varepsilon^2$. Then $\varepsilon^2 > 1/N$, so $1/\sqrt{N} < \varepsilon$. If $n > N$, then $\left|\dfrac{1}{\sqrt{n+1}} - 0\right| = \dfrac{1}{\sqrt{n+1}} < \dfrac{1}{\sqrt{n}} < \dfrac{1}{\sqrt{N}} < \varepsilon$, so the sequence converges to zero.
43. Suppose the sequence has limit c. Then, with $\varepsilon = \frac{1}{4}$, there exists a positive integer N such that $|a_n - c| < \frac{1}{4}$ for $n > N$. In particular, $|a_{N+1} - c| < \frac{1}{4}$ and $|a_{N+2} - c| < \frac{1}{4}$, which implies $|a_{N+1} - a_{N+2}| < \frac{1}{2}$. However, $|a_{N+1} - a_{N+2}| = 1$, so our assumption that the sequence has limit c must be false.
45. Let $\varepsilon > 0$ be given. We may assume $\varepsilon < 1$. Then $|a|^n < \varepsilon$ if $n \cdot \ln|a| < \ln \varepsilon$; that is, if $n > (\ln \varepsilon)/(\ln|a|)$. Thus we can let N be the largest integer less than $(\ln \varepsilon)/(\ln|a|)$.
47. 0.36787944 (which equals $1/e$) **49.** 1 **51.** $\frac{1}{2}$

Section 10.2

1. (a) 1 (b) 1 (c) 0 (d) 0 (e) 0 (f) 0 (g) 1 **3.** $a_1 = \frac{1}{2}$, $a_2 = -\frac{1}{6}$, $a_3 = -\frac{1}{12}$, $a_4 = -\frac{1}{20}$, $a_5 = -\frac{1}{30}$ **5.** Diverges to ∞
7. Converges to 1 **9.** Diverges to $-\infty$ **11.** Diverges to ∞ **13.** Converges to $-\frac{4}{3}$ **15.** Converges to $e^2/(e^2 - 1)$
17. Diverges **19.** Converges to $\frac{2}{21}$ **21.** Converges to $4/(1 + \ln 2)$ **23.** Converges to $\frac{7}{2}$ **25.** Converges to 6
27. Converges to 13 **29.** 60 ft
31. If a is the time required to travel the first mile, then the time required to travel the first mile plus the time required to travel the second plus $\ldots$ is $a + ka + k^2a + k^3a + \cdots = a/(1 - k)$ since $|k| < 1$.
33. (a) Since $\ln[(n+1)/n] = \ln(n+1) - \ln(n)$, we see that $s_n = \ln(n+1) - \ln(1) = \ln(n+1)$.
(b) $\lim_{n\to\infty} s_n = \lim_{n\to\infty} \ln(n+1) = \infty$, so the series diverges.
35. $\frac{11}{6}$ **37.** (a) $\dfrac{1}{n^2 - 1} = \dfrac{\frac{1}{2}}{n-1} - \dfrac{\frac{1}{2}}{n+1}$ (b) $\frac{7}{24}$ **39.** $1, 2, 3, 4, \ldots, n, \ldots$ and $-1, -2, -3, -4, \ldots, -n, \ldots$
41. It diverges. **43.** $\frac{247}{30}$ **45.** $\frac{4}{33}$ **47.** $\frac{27287339}{99900}$
49. (a) Impossible for us to answer. (b) Any number with a nonzero digit in some decimal place when added to
$4.12999999999 \ldots$ will yield a number greater than $4.13000000. \ldots$ (c) $\dfrac{412}{100} + \dfrac{\frac{9}{1000}}{\frac{9}{10}} = \dfrac{412}{100} + \dfrac{1}{100} = \dfrac{413}{100}$

Section 10.3

1. Converges to $\sqrt{5}$ **3.** Converges to $\frac{3}{4} + \sqrt{2} - \sqrt{3}$ **5.** T F F T F T T F T T **7.** Converges; SCT with $\sum_{n=1}^{\infty} (\frac{1}{2})^n$
9. Converges; SCT with $\sum_{n=2}^{\infty} (\frac{1}{2})^n$ **11.** Converges; SCT with $\sum_{n=1}^{\infty} (\frac{1}{2})^n$ **13.** Diverges; $\lim_{n\to\infty} a_n = \frac{1}{3} \ne 0$
15. Converges; SCT with $\sum_{n=1}^{\infty} (\frac{1}{5})^n$ **17.** Converges; sum of two convergent geometric series
19. Diverges; LCT with the harmonic series **21.** Converges; LCT with a geometric series with ratio $1/e^2$
23. Diverges; $a_n \ge 1$ for all n **25.** Diverges; $a_n \ge 1$ for all n **27.** Diverges; LCT with geometric series with ratio $1/(\ln 2) > 1$
29. Diverges; LCT with $\sum_{n=1}^{\infty} \cos^2 n$; $\lim_{n\to\infty} \cos^2 n \ne 0$ **31.** Diverges; LCT with $\sum_{n=1}^{\infty} (1/n)$
33. Converges; root test **35.** Converges; root test **37.** Diverges; $\lim_{n\to\infty} a_n \ne 0$
39. Diverges; $\lim_{n\to\infty} a_n = 1 \ne 0$ **41.** Converges; root test

Section 10.4

1. Diverges **3.** Diverges **5.** Converges **7.** $\displaystyle\int_2^{\infty} \dfrac{dx}{x(\ln x)} = \lim_{t\to\infty} \ln(\ln x)\Big]_2^t = \lim_{t\to\infty} \ln(\ln t) - \ln(\ln 2) = \infty$ **9.** Converges

11. Diverges **13.** Converges **15.** Converges **17.** Converges **19.** Diverges **21.** Diverges **23.** Diverges **25.** Converges
27. Converges **29.** Diverges **31.** Diverges **33.** Converges **35.** Converges **37.** Converges **39.** Converges **41.** Converges
43. Diverges **45.** Diverges **47.** Converges **49.** e^{1000} **51.** 2.0762 **53.** 2.607 **55.** 10.544

Section 10.5

1. $1 - 1 + 1 - 1 + 1 - 1 + \cdots$ **3.** $1 + \frac{1}{2} + \frac{1}{3} + \frac{1}{4} + \frac{1}{5} + \cdots$ **5.** F T T F F F **7.** Converges absolutely **9.** Divergent
11. Conditionally convergent **13.** Divergent **15.** Absolutely convergent **17.** Absolutely convergent **19.** Divergent
21. Conditionally convergent **23.** Conditionally convergent **25.** Conditionally convergent **27.** Divergent
29. Absolutely convergent **31.** Conditionally convergent **33.** Absolutely convergent **35.** Divergent
37. Conditionally convergent **39.** Conditionally convergent
41. (a) Since $a_n = u_n + v_n$, this follows from Corollary 1 of Theorem 10.2 (Section 10.2). (b) If, say, $\sum_{n=1}^{\infty} u_n$ converges, then convergence of $\sum_{n=1}^{\infty} a_n$ would imply convergence of $\sum_{n=1}^{\infty} (a_n - u_n) = \sum_{n=1}^{\infty} v_n$ by Corollary 1 of Theorem 10.2 (Section 10.2). (c) Let $\sum_{n=1}^{\infty} a_n = 1 - \frac{1}{2} + \frac{1}{3} - \frac{1}{4} + \frac{1}{5} - \frac{1}{6} + \cdots$.
43. 99 **45.** 4 **47.** 6 **49.** -0.823 **51.** -3.645 **53.** -0.63214

Section 10.6

1. $R = 1; -1 \le x < 1$ **3.** $R = 1; -1 < x \le 1$ **5.** $R = 1; 0 \le x \le 2$ **7.** $R = 5; -6 \le x < 4$ **9.** $R = 3; -3 < x < 3$
11. $R = 4; -4 \le x \le 4$ **13.** $R = 1; 0 \le x < 2$ **15.** $R = \infty; -\infty < x < \infty$ **17.** $R = 0; x = -5$ **19.** $R = \frac{1}{2}; -\frac{7}{2} < x \le -\frac{5}{2}$
21. $R = \frac{3}{2}; \frac{1}{2} \le x \le \frac{7}{2}$ **23.** $R = \frac{3}{2}; -\frac{9}{2} \le x < -\frac{3}{2}$ **25.** $R = \frac{1}{3}; -\frac{7}{3} \le x \le -\frac{5}{3}$ **27.** $R = \frac{1}{2}; -\frac{5}{2} \le x \le -\frac{3}{2}$

29. $\displaystyle\sum_{n=0}^{\infty} \frac{(x - \frac{5}{2})^n}{(\frac{3}{2})^n} = \sum_{n=0}^{\infty} \frac{(2x - 5)^n}{3^n}$ **31.** $\displaystyle\sum_{n=1}^{\infty} \frac{(x + 3)^n}{n \cdot 2^n}$ **33.** $1 + x^2 + x^4 + \cdots + x^{2n} + \cdots$

35. $-5 - 5 \cdot 3x - 5 \cdot 3^2 x^2 - \cdots - 5 \cdot 3^n x^n - \cdots$ **37.** $\dfrac{1}{2} + \dfrac{1}{2 \cdot 4} x + \dfrac{1}{2 \cdot 4^2} x^2 + \cdots + \dfrac{1}{2 \cdot 4^n} x^n + \cdots$

39. $2x^2 - 3x^3 + \dfrac{3^2}{2} x^4 - \dfrac{3^3}{2^2} x^5 + \cdots + \dfrac{(-1)^n 3^n}{2^{n-1}} x^{n+2} + \cdots$ **41.** $\dfrac{1}{2} - \dfrac{(x - 1)}{2^2} + \dfrac{(x - 1)^2}{2^3} - \cdots + (-1)^n \dfrac{(x - 1)^n}{2^{n+1}} + \cdots$

43. $\dfrac{1}{4} - \dfrac{1}{4 \cdot 8}(x - 2) + \dfrac{1}{4 \cdot 8^2}(x - 2)^2 + \cdots + \dfrac{(-1)^n}{4 \cdot 8^n}(x - 2)^n + \cdots$

45. $-\dfrac{(x - 3)^2}{2} + \dfrac{(x - 3)^3}{2} - \dfrac{(x - 3)^4}{2} + \cdots + (-1)^{n+1} \dfrac{(x - 3)^{n+2}}{2} + \cdots$

47. Letting $u = 1/x$, we see that $\sum_{n=0}^{\infty} a_n u^n$ converges for $u = 1/c$. By Theorem 10.11, it must converge for all u such that $|u| < |1/c|$. The given series thus converges for all x such that $|1/x| < |1/c|$, that is, for $|x| > |c|$.
49. $x < -2, x > 2$ **51.** No x **53.** $x \le -2, x \ge 6$ **55.** $x < \frac{8}{3}, x > \frac{16}{3}$

Section 10.7

1. F T T T F T **3.** $1 + x + \dfrac{3}{2} x^2 + \dfrac{x^3}{3!} + \cdots + \dfrac{x^n}{n!} + \cdots$ for $n \ge 3$ **5.** $x^2 - \dfrac{x^4}{3!} + \cdots + (-1)^n \dfrac{x^{2n+2}}{(2n+1)!} + \cdots$ for $n \ge 1$

7. $x + x^2 + x^3 + \cdots + x^{n+1} + \cdots$ **9.** $1 - \dfrac{x^6}{2!} + \cdots + (-1)^n \dfrac{x^{6n}}{(2n)!} + \cdots$ **11.** $1 + x - \dfrac{x^3}{3} - \dfrac{x^4}{6} - \dfrac{x^5}{30} - \cdots$

13. $1 - 2x + 3x^2 - \cdots + (-1)^n (n + 1)x^n + \cdots$ **15.** $1 + \dfrac{x^2}{2!} + \dfrac{5x^4}{4!} + \cdots$ **17.** $1 + 2x + \frac{5}{2}x^2 + \frac{8}{3}x^3 + \cdots$

19. $x + \dfrac{x^3}{3!} + \dfrac{x^5}{5!} + \cdots + \dfrac{x^{2n+1}}{(2n+1)!} + \cdots$ **21.** $1 - x + \dfrac{x^2}{2!} - \dfrac{x^3}{3!} + \cdots + (-1)^n \dfrac{x^n}{n!} + \cdots$ **23.** $x + \dfrac{5x^3}{6} + \cdots$

25. -24 **27.** $\frac{3}{4}$ **29.** $\dfrac{10!}{5 \cdot 4^5}$ **31.** 240 **33.** (a) $x - \frac{2}{3}x^3 + \frac{2}{15}x^5 + \cdots$ (b) $x - \frac{2}{3}x^3 + \frac{2}{15}x^5 - \cdots + (-1)^n (2^{2n}) \dfrac{x^{2n+1}}{(2n+1)!} + \cdots$
35. $1/x = -1/[1 - (x + 1)]; -1 - (x + 1) - (x + 1)^2 - \cdots - (x + 1)^n - \cdots$
37. (a) Replace x by $-x$ in Eq. (4). (b) The alternating series test is satisfied.
(c) $|E_n(1)| = \left| \dfrac{n!}{(1 + c)^{n+1}} \cdot \dfrac{1^{n+1}}{(n + 1)!} \right| = \left| \dfrac{1}{(1 + c)^{n+1}(n + 1)} \right| \le \left| \dfrac{1}{n + 1} \right|$, so $\lim_{n \to \infty} E_n(1) = 0$. It now follows from part (a) that the alternating harmonic series converges to $\ln 2$.

39. $C + x + \dfrac{x^3}{3} + \dfrac{x^5}{5 \cdot 2!} + \dfrac{x^7}{7 \cdot 3!} + \cdots + \dfrac{x^{2n+1}}{(2n + 1)n!} + \cdots$ **41.** $\pi + x - \dfrac{x^5}{5 \cdot 2!} + \dfrac{x^9}{9 \cdot 4!} - \cdots + (-1)^n \dfrac{x^{4n+1}}{(4n + 1)(2n)!} + \cdots$

43. (a) e^{ix}: $1 + ix - \dfrac{x^2}{2!} - i\dfrac{x^3}{3!} + \dfrac{x^4}{4!} + i\dfrac{x^5}{5!} - \dfrac{x^6}{6!} - \cdots + (i)^n \dfrac{x^n}{n!} + \cdots$

e^{-ix}: $1 - ix - \dfrac{x^2}{2!} + i\dfrac{x^3}{3!} + \dfrac{x^4}{4!} - i\dfrac{x^5}{5!} - \dfrac{x^6}{6!} + \cdots + (-i)^n\dfrac{x^n}{n!} + \cdots$

(b) and (c) This is obvious from part (a) and the series (2) and (3) for sin x and cos x.   (d) $\cos x = (e^{ix} + e^{-ix})/2$; $\sin x = (e^{ix} - e^{-ix})/(2i)$

(e) They are the same except for the presence of i in certain places.

45. $2x + [1/(1 + x)]$ **47.** $3x + \cos x$ **49.** $\sin x + \cos x$ **51.** $(e^x - 1) + [1/(1 - x)]$ **53.** $2/(1 - x)^3$

55. $[1/(1 + 2x^2)] + 2x^2 - 1 = 4x^4/(1 + 2x^2)$

Section 10.8

1. 6   **3.** 8   **5.** 1   **7.** $\frac{35}{16}$ **9.** 1   **11.** $\frac{5}{81}$ **13.** $1 + 0x + 0x^2 + \cdots + 0x^n + \cdots$; $\dbinom{0}{k} = 1$ for $k = 0$ and 0 for $k > 0$

15. $1 - 2x + 3x^2 - 4x^3 + 5x^4 - \cdots$; $R = 1$ **17.** $1 - \frac{1}{3}x^2 - \frac{1}{9}x^4 - \frac{5}{81}x^6 - \frac{10}{243}x^8 - \cdots$; $R = 1$

19. $1 + 5x + 5x^2 - \frac{5}{3}x^3 + \frac{5}{3}x^4 - \cdots$; $R = \frac{1}{3}$ **21.** $\dfrac{1}{5} + \dfrac{1}{10}\left(\dfrac{x}{5}\right)^2 + \dfrac{3}{40}\left(\dfrac{x}{5}\right)^4 + \dfrac{1}{16}\left(\dfrac{x}{5}\right)^6 + \dfrac{7}{128}\left(\dfrac{x}{5}\right)^8 + \cdots$; $R = 5$

23. $16 + \dfrac{8}{3}x^3 + \dfrac{1}{18}x^6 - \dfrac{1}{648}x^9 + \dfrac{80}{243}\left(\dfrac{x}{2}\right)^{12} - \cdots$; $R = 2$ **25.** Find a partial sum of $(1 - x)^{1/2}$ with $x = \frac{1}{3}$.

27. Find a partial sum of $(1 - x)^{-1/3}$ with $x = \frac{1}{2}$.   **29.** $x - \dfrac{1}{6}x^3 + \dfrac{3}{40}x^5 - \dfrac{5}{112}x^7 + \cdots$ **31.** $x + \dfrac{1}{2}x^3 + \dfrac{3}{8}x^5 + \dfrac{5}{16}x^7 + \cdots$

33. $1 + \dfrac{3}{2}x + \dfrac{11}{8}x^2 + \dfrac{23}{16}x^3 + \dfrac{179}{128}x^4 + \cdots$ **35.** $1 + \dfrac{3}{2}x - \dfrac{1}{8}x^2 + \dfrac{3}{16}x^3 - \dfrac{37}{128}x^4 + \cdots$

37. $\dfrac{1}{2}x^3 + \dfrac{1}{16}x^5 + \dfrac{3}{2^8}x^7 + \dfrac{5}{2^{11}}x^9 + \cdots$ **39.** $\dfrac{1}{3} - \dfrac{1}{7\cdot3!} + \dfrac{1}{11\cdot5!} - \dfrac{1}{15\cdot7!}$ with error less than $\dfrac{1}{19\cdot9!}$

41. $\dfrac{1}{4} - \dfrac{1}{8\cdot2!} + \dfrac{1}{12\cdot4!} - \dfrac{1}{16\cdot6!} + \dfrac{1}{20\cdot8!}$ with error less than $\dfrac{1}{24\cdot10!}$

43. $\dfrac{1}{10} - \dfrac{1}{3\cdot10^3} + \dfrac{1}{5\cdot2!\cdot10^5} - \dfrac{1}{7\cdot3!\cdot10^7}$ with error less than $\dfrac{1}{9\cdot4!\cdot10^9}$

45. $\dfrac{1}{2} - \dfrac{1}{7\cdot2^8} \approx 0.49944$ with error less than $\dfrac{1}{39\cdot2^{16}}$

47. $2\left[\dfrac{1}{2} - \dfrac{1}{10}\left(\dfrac{1}{2}\right)^5 + \dfrac{1}{24\cdot9}\left(\dfrac{1}{2}\right)^9\right] \approx 0.99376808$ with error less than $\dfrac{2}{720\cdot13}\left(\dfrac{1}{2}\right)^{13} \approx 0.000000026$

49. Checking coefficients of x^k, we need only show that $p\dbinom{p}{k} = \left[(k + 1)\dbinom{p}{k + 1}\right] + \left[k\dbinom{p}{k}\right]$. We have

$$(k + 1)\dbinom{p}{k + 1} + k\dbinom{p}{k} = (k + 1)\dfrac{p(p - 1)\cdots(p - k)}{(k + 1)(k)\cdots(2)(1)} + k\dfrac{p(p - 1)\cdots(p - k + 1)}{k!}$$

$$= \dfrac{[p(p - 1)\cdots(p - k)] + k[p(p - 1)\cdots(p - k + 1)]}{k!}$$

$$= \dfrac{p(p - 1)\cdots(p - k + 1)}{k!}(p - k + k) = \dbinom{p}{k}p.$$

51. Series 1.462650; Simpson's rule 1.462654; difference 0.000004

Supplementary Exercises for Chapter 10

1. See Definition 10.2 of Section 10.1.

3. Note that $(n - 1)/n = 1 - (1/n)$. Let $\varepsilon > 0$ be given. Find $N > 0$ such that $1/N < \varepsilon$. Then if $n > N$, we have $1/n < \varepsilon$, so $|(n - 1)/n - 1| = |[1 - (1/n)] - 1| = |1/n| < \varepsilon$. Thus $\{(n - 1)/n\}$ converges to 1.

5. $\text{Lim}_{n\to\infty} a_n = \infty$ if for each real number $K > 0$, there exists a positive integer N such that $a_n > K$ for all $n > N$.

7. Let $\{s_n\} = 2, 3, 4, 5, 6, 7, \ldots$ and $\{t_n\} = 1, 0, -1, -2, -3, -4, \ldots$.   **9.** 48   **11.** 30 ft   **13.** $-25/6$ **15.** $\frac{3}{5}$

17. Divergent; $\lim_{n\to\infty} a_n \ne 0$ **19.** Divergent; LCT with $\displaystyle\sum_{n=1}^{\infty} (1/n^{1/2})$ **21.** Convergent; ratio test gives a ratio of $\frac{1}{3} < 1$.

23. Convergent; SCT with $\displaystyle\sum_{n=1}^{\infty} (1/n^2)$ **25.** Convergent; LCT with $\displaystyle\sum_{n=1}^{\infty} (1/n^{3/2})$ **27.** Divergent; LCT with $\displaystyle\sum_{n=1}^{\infty} (1/n)$

29. Divergent; $\lim_{n\to\infty} a_n = \infty \ne 0$ **31.** Convergent; ratio test with ratio $1/\sqrt{2} < 1$ **33.** See Theorem 10.7.

35. The hypotheses are satisfied. $\displaystyle\int_2^\infty \frac{x+1}{x^2+1}\,dx = \lim_{h\to\infty} \left(\frac{1}{2}\ln|x^2+1| + \tan^{-1}x\right)\Big]_2^h$

$$= \lim_{h\to\infty} \left[\frac{1}{2}\ln|h^2+1| + \tan^{-1}h - \frac{1}{2}\ln(5) - \tan^{-1}2\right] = \infty.$$

The series is divergent.

37. See Theorem 10.8. **39.** We have $\displaystyle\lim_{n\to\infty}\frac{a_{n+1}}{a_n} = \lim_{n\to\infty}\left[\frac{(n+1)!}{100^{n+1}}\cdot\frac{100^n}{n!}\right] = \lim_{n\to\infty}\frac{n+1}{100} = \infty$, so the series diverges.

41. Conditionally convergent; satisfies the alternating series test, but diverges absolutely by the LCT with $\sum_{n=1}^\infty (1/\sqrt{n})$

43. Absolutely convergent; $|a_n| = \dfrac{\cos(1/n)}{n^{3/2}} \le 1/n^{3/2}$. Use the SCT.

45. Conditionally convergent; satisfies the alternating series test, but diverges absolutely by the integral test

47. Divergent; $\lim_{n\to\infty} a_n = \infty \ne 0$ **49.** Absolutely convergent; LCT with $\displaystyle\sum_{n=1}^\infty (1/n^{3/2})$ **51.** Absolutely convergent; root test

53. Divergent; $\lim_{n\to\infty} |a_n| = \infty \ne 0$ **55.** Divergent; $\lim_{n\to\infty} a_n \ne 0$ **57.** $-8 \le x < -2$ **59.** $\dfrac{-\sqrt{3}-1}{2} \le x \le \dfrac{\sqrt{3}-1}{2}$

61. $-\infty < x < \infty$ **63.** $x = 3$

65. (a) A function is analytic if it can be represented by a power series in a neighborhood of each point in its domain.
(b) $x + x^3 + x^5 + x^7 + \cdots + x^{2n+1} + \cdots$

67. $x^5 - \dfrac{x^7}{2} + \dfrac{x^9}{3} - \dfrac{x^{11}}{4} + \cdots + (-1)^n\dfrac{x^{2n+5}}{n+1} + \cdots$ **69.** $8 \cdot 10!$

71. $-(x-1) + \frac{1}{2}(x-1)^2 - \frac{1}{4}(x-1)^4 + \frac{1}{4}(x-1)^5 - \frac{1}{8}(x-1)^6 + \cdots$ **73.** $1 + \frac{1}{2}x^2 - \frac{1}{8}x^4 + \frac{1}{16}x^6 - \frac{5}{128}x^8$

75. $1 + \dfrac{x}{4} + \frac{3}{32}x^2 + \frac{5}{128}x^3$ **77.** $4 + \frac{1}{3}x - \frac{1}{144}x^2 + \frac{1}{2592}x^3 + \cdots$ **79.** $\frac{5}{243}$

CHAPTER 11

Section 11.1

1.
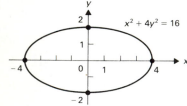

3.
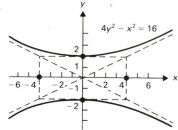

5.
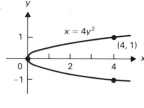

7.
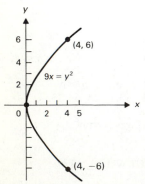

9.
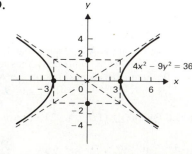

11.
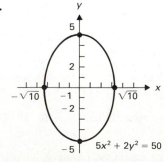

13.

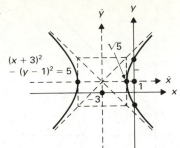

$(x + 3)^2 - (y - 1)^2 = 5$

$\sqrt{5}$

15.

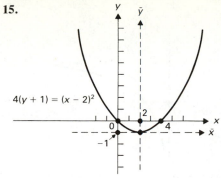

$4(y + 1) = (x - 2)^2$

17.

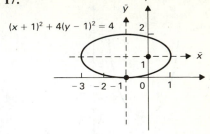

$(x + 1)^2 + 4(y - 1)^2 = 4$

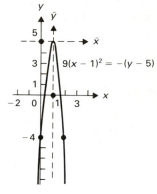

19.

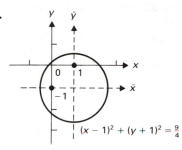

$(x - 1)^2 + (y + 1)^2 = \frac{9}{4}$

21. No points satisfy this equation.

23.

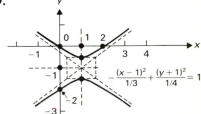

$9(x - 1)^2 = -(y - 5)$

25.

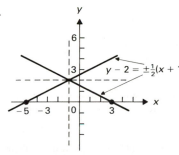

$y - 2 = \pm \frac{1}{2}(x + 1)$

27.

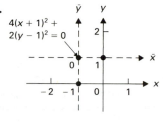

$4(x + 1)^2 + 2(y - 1)^2 = 0$

29.

$-\dfrac{(x - 1)^2}{1/3} + \dfrac{(y + 1)^2}{1/4} = 1$

31. The point $(\cosh t, \sinh t)$ lies on the hyperbola $x^2 - y^2 = 1$ for each real number t.

Section 11.2

1.

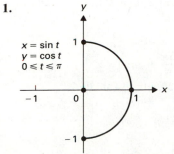

$x = \sin t$
$y = \cos t$
$0 \leqslant t \leqslant \pi$

3.

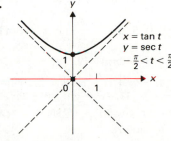

$x = \tan t$
$y = \sec t$
$-\dfrac{\pi}{2} < t < \dfrac{\pi}{2}$

5.

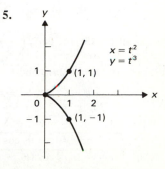

$x = t^2$
$y = t^3$

$(1, 1)$
$(1, -1)$

7.

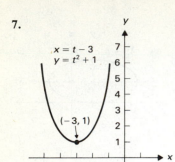

9.

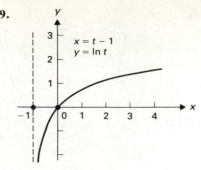

11.

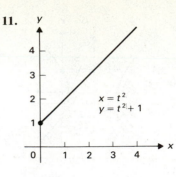

13. $x = \ln m$, $y = m$ for $0 < m < \infty$ **15.** $x = s/6$, $y = s^5/216$ for all s **17.** $x = 3 \cot \theta$, $y = 3$ for $0 < \theta \le \pi/2$

19. (a) Yes (b) No (c) Yes (d) No.   In parts (a) and (c), different points correspond to different values of the derivative, while in parts (b) and (d), (x, y) and $(-x, y)$ correspond to the same values of the derivative.

21. $x = ma/\sqrt{1 + m^2}$, $y = a/\sqrt{1 + m^2}$ for $-1 \le m \le 1$ **23.** $x = a\theta - b \sin \theta$, $y = a - b \cos \theta$

Section 11.3

1. $\frac{2}{3}t$ **3.** 0   **5.** $\frac{1}{18}$ **7.** $\dfrac{1}{\sqrt{2} - 1}$ **9.** 0   **11.** 4   **13.** $y + 3x = 3$ **15.** $y = -1$ **17.** $3y + x = 12$ **19.** (3, 0)   **21.** (1, 0)

23. $2/\sqrt{3}$ **25.** $\frac{3}{4}$ **27.** $\frac{1}{9}$ **29.** $\dfrac{1}{18t^5}$ **31.** $\frac{243}{23}e^{3t}$ **33.** 24   **35.** $6a$ **37.** $\frac{116}{15}$ **39.** 24   **41.** $3a/2$ **43.** 0.25   **45.** $8a$

47. $41\pi/3$ **49.** $\dfrac{\pi}{6}(37^{3/2} - 1)$ **51.** $\dfrac{\pi}{3}(29^{3/2} - 13^{3/2})$ **53.** ≈ 4148 **55.** ≈ 13.1

Section 11.4

1. 1   **3.** $1/(2\sqrt{2})$ **5.** $1/\sqrt{2}$ **7.** $12/(5\sqrt{5})$ **9.** $\frac{1}{5}$ **11.** $\frac{4}{25}$ **13.** $\frac{5}{16}$ **15.** 2

17. Where $d\phi/ds$ is positive, ϕ increases as s increases, and increasing ϕ corresponds to counterclockwise rotation of the tangent line, so that the curve must bend to the left. Where $d\phi/ds$ is negative, ϕ decreases as s increases, and the curve bends to the right.

19. Zero

21. If the curvature is zero at each point, then by formula (7) of the text, we have $d^2y/dx^2 = 0$ at each point. It follows from this differential equation that $y = Ax + B$ for some constants A and B.

23. $x^2 + y^2 - 4x + 2y = 0$ **25.** 0   **27.** (a) $(\pm 1/(45)^{1/4}, \pm 1/(45)^{3/4})$ (b) (0, 0)   **29.** (1, 0)   **31.** $(x - 2)^2 + (y - 2)^2 = 2$

33. $(x + \frac{11}{2})^2 + (y + \frac{16}{3})^2 = \frac{2197}{36}$ **35.** $x = -4t^3$, $y = \frac{1}{2} + 3t^2$

Section 11.5

1. $(2\sqrt{2}, 2\sqrt{2})$ **3.** $(0, -6)$ **5.** $(-\sqrt{2}, -\sqrt{2})$ **7.** $(4, 0)$ **9.** $(0, 0)$ **11.** $\left(2\sqrt{2}, \dfrac{\pi}{4} + 2n\pi\right)$ and $\left(-2\sqrt{2}, \dfrac{5\pi}{4} + 2n\pi\right)$

13. $(0, \theta)$ for all θ **15.** $\left(5, \tan^{-1}\left(\dfrac{-4}{3}\right) + 2n\pi\right), \left(-5, \tan^{-1}\left(\dfrac{-4}{3}\right) + \pi + 2n\pi\right)$

17. $(\sqrt{13}, \tan^{-1}(\frac{3}{2}) + \pi + 2n\pi)$ and $(-\sqrt{13}, \tan^{-1}(\frac{3}{2}) + 2n\pi)$ **19.** $\left(4, -\dfrac{\pi}{6} + 2n\pi\right), \left(-4, \dfrac{5\pi}{6} + 2n\pi\right)$

21. $2r \cos \theta + 3r \sin \theta = 5$ **23.** $r^2 = \sec 2\theta$ **25.** $r^2 = 4/(1 + \cos^2\theta)$ **27.** $r^3 = -4 \csc \theta$ **29.** $r = -4 \sin \theta$

31. $x^2 + y^2 = 64$ **33.** $x^2 + y^2 = -3ay$ **35.** $x = -3$ **37.** $(x^2 + y^2)^{3/2} = -2ax$ **39.** $x^2y + y^3 = 4$

41. $d = \sqrt{(x_2 - x_1)^2 + (y_2 - y_1)^2} = \sqrt{(r_2 \cos \theta_2 - r_1 \cos \theta_1)^2 + (r_2 \sin \theta_2 - r_1 \sin \theta_1)^2}$

$\quad = \sqrt{r_1^2 + r_2^2 - 2r_1r_2(\cos \theta_1 \cos \theta_2 + \sin \theta_1 \sin \theta_2)}$

$\quad = \sqrt{r_1^2 + r_2^2 - 2r_1r_2 \cos (\theta_1 - \theta_2)}$

43.

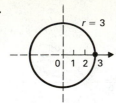

$r = 3$

45.

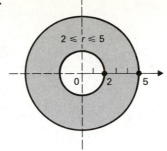

$2 \leqslant r \leqslant 5$

47.

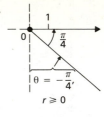

$\theta = -\dfrac{\pi}{4},$

$r \geqslant 0$

49.

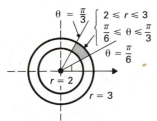

$\theta = \dfrac{\pi}{3}$ $\begin{cases} 2 \leqslant r \leqslant 3 \\ \dfrac{\pi}{6} \leqslant \theta \leqslant \dfrac{\pi}{3} \end{cases}$

$\theta = \dfrac{\pi}{6}$

$r = 2$

$r = 3$

Section 11.6

1.

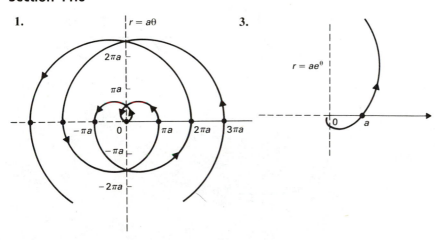

$r = a\theta$

3.

$r = ae^{\theta}$

5.

$r = a(1 - \cos \theta)$

7.

$r = 3 + 2 \cos \theta$

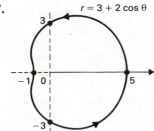

9.

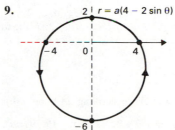

$r = a(4 - 2 \sin \theta)$

11.

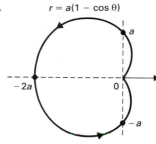

$r = 2a \sin \theta$

13.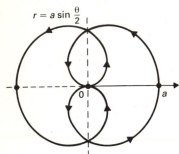

$r = -2 \sec \theta$

15.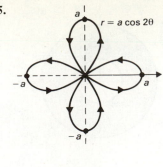

$r = a \cos 2\theta$

17.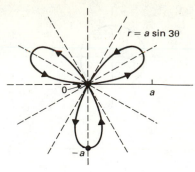

$r = a \sin 3\theta$

19.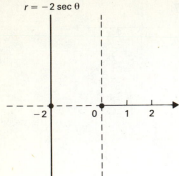

$r = a \sin \frac{\theta}{2}$

21.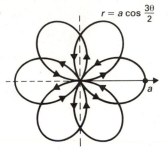

$r = a \cos \frac{3\theta}{2}$

23.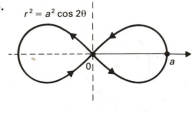

$r^2 = a^2 \cos 2\theta$

25.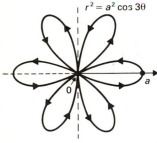

$r^2 = a^2 \cos 3\theta$

27.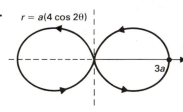

$r = a(4 \cos 2\theta)$

29.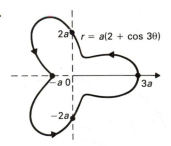

$r = a(2 + \cos 3\theta)$

31. (a) $r = 2 - \csc \theta$, $r = 2 - \dfrac{1}{\sin \theta}$, $r \sin \theta = 2 \sin \theta - 1$, $r \sin \theta = \dfrac{2}{\csc \theta} - 1$, $y = \dfrac{2}{2-r} - 1$

(b) $\lim_{r \to \infty} y = \lim_{r \to \infty} \left(\dfrac{2}{2-r} - 1 \right) = -1$ (c)

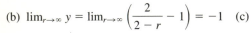

$r = 2 - \csc \theta$

33. (a) $x = a(\cos \theta + \cos^2 \theta)$, $y = a(\sin \theta + \sin \theta \cos \theta)$ (b) -1 **35.** $(a/2, 2\pi/3)$, $(a/2, 4\pi/3)$, $(2a, 0)$ **37.** $(\frac{1}{4}, \pi/4)$, $(-\frac{1}{4}, -\pi/4)$ **39.** $(a/\sqrt{2}, \pi/4)$, $(0, 0)$ **41.** $(a, \pi/12)$, $(a, 5\pi/12)$, $(a, 13\pi/12)$, $(a, 17\pi/12)$ **43.** $(\pm a, \pi/6)$, $(\pm a, 5\pi/6)$, $(\pm a, \pi/3)$, $(\pm a, 2\pi/3)$ **45.** $(\sqrt{3}/2, \pm\pi/3)$, $(\sqrt{3}/2, \pm 4\pi/3)$

Section 11.7

1. πa^2 (No. We used the formula for the area of a circle in deriving $dA = \frac{1}{2} r^2 \, d\theta$.)   **3.** $\dfrac{a^2}{4}(e^{2\pi} - 1)$ **5.** $\frac{3}{2} \pi a^2$ **7.** $\frac{1}{8} \pi a^2$

9. $\pi a^2/6$ **11.** $2a^2$ **13.** $a^2(2\pi/3 + \sqrt{3})$ **15.** $a^2(\pi + 3\sqrt{3})$ **17.** a^2 **19.** $99\pi a^2/8$ **21.** $a^2(19\pi/2 + 12)$ **23.** 4.80956

25. 53.4396

Section 11.8

1. $-\sqrt{3}$ **3.** $-2/\pi$ **5.** $\left(a\cos\dfrac{3\pi}{8},\dfrac{3\pi}{8}\right),\left(a\cos\dfrac{7\pi}{8},\dfrac{7\pi}{8}\right)$

7. $\left(\dfrac{3+\sqrt{33}}{4},\sin^{-1}\!\left(\dfrac{\sqrt{33}-1}{8}\right)\right),\left(\dfrac{3+\sqrt{33}}{4},\pi-\sin^{-1}\!\left(\dfrac{\sqrt{33}-1}{8}\right)\right),\left(\dfrac{3-\sqrt{33}}{4},\sin^{-1}\!\left(\dfrac{\sqrt{33}+1}{-8}\right)\right),$

$\left(\dfrac{3-\sqrt{33}}{4},\pi-\sin^{-1}\!\left(\dfrac{\sqrt{33}+1}{-8}\right)\right)$

9. $-\tan^{-1}(-\pi/2)\approx 57.5°$ **11.** $\tan^{-1}2\approx 63.4°$ **13.** $\tan\psi=\dfrac{a\cdot e^{\theta}}{a\cdot e^{\theta}}=1$ for all θ; $\psi=45°$ for all θ

15. $\dfrac{\pi}{2}-\tan^{-1}\!\left(\dfrac{-\sqrt{3}}{6}\right)\approx 106.1°$ **17.** $\tan^{-1}(\tfrac{4}{3})\approx 53.1°$ **19.** $8a$ **21.** $6a\displaystyle\int_0^{\pi/6}\sqrt{1+8\cos^2 3\theta}\,d\theta$ **23.** $128\pi/5$

25. If a dot over a variable denotes differentiation with respect to a parameter, then $\kappa=\left|\dfrac{\dot{x}\ddot{y}-\dot{y}\ddot{x}}{(\dot{x}^2+\dot{y}^2)^{3/2}}\right|$. Now $x=f(\theta)\cos\theta$ and

$y=f(\theta)\sin\theta$. Thus, taking θ as parameter, we have

$$\dot{x}=-f(\theta)\sin\theta+f'(\theta)\cos\theta$$
$$\ddot{x}=-f(\theta)\cos\theta-2f'(\theta)\sin\theta+f''(\theta)\cos\theta$$
$$\dot{y}=f(\theta)\cos\theta+f'(\theta)\sin\theta$$
$$\ddot{y}=-f(\theta)\sin\theta+2f'(\theta)\cos\theta+f''(\theta)\sin\theta.$$

Substituting these expressions in the formula for κ and simplifying yields the desired result.

27. $a/3$ **29.** $6.68245a$ **31.** 501.668 **33.** (a) $(0,0)$ (b) 2 units time (c) $r=\sqrt{2}e^{\pi/4}e^{-\theta}$, $\theta\geq\pi/4$ (d) 2 units
(e) It will get dizzy.

Supplementary Exercises for Chapter 11

1.

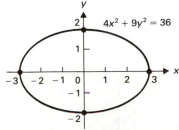

$x^2-4y^2=16$

3.

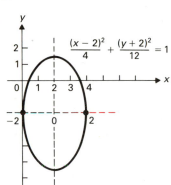

$\dfrac{(x-2)^2}{4}+\dfrac{(y+2)^2}{12}=1$

5.

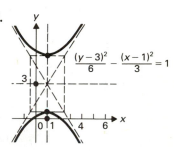

$\dfrac{(y-3)^2}{6}-\dfrac{(x-1)^2}{3}=1$

7.

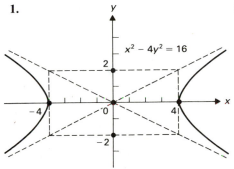

$4x^2+9y^2=36$

9.

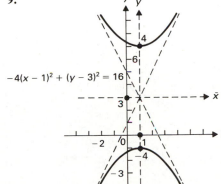

$-4(x-1)^2+(y-3)^2=16$

11.

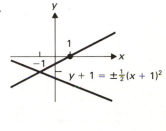

$y+1=\pm\tfrac{1}{2}(x+1)^2$

13.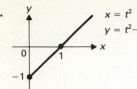
$x = t^2$
$y = t^2 - 1$

15.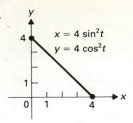
$x = 4\sin^2 t$
$y = 4\cos^2 t$

17. $\left.\dfrac{dy}{dx}\right|_{t=1} = \dfrac{1}{2}, \left.\dfrac{d^2y}{dx^2}\right|_{t=1} = \dfrac{5}{4}$ **19.** $27/(16t^4)$ **21.** $x = \sqrt[3]{h}, y = h$ **23.** $x = \dfrac{m}{2} - 2, y = \dfrac{m^2}{4} - 4, m \geq 2$

25. $\displaystyle\int_0^3 \sqrt{4t^2 + \cos^2 t}\,dt$ **27.** $\frac{1}{27}(85\sqrt{85} - 8)$ **29.** $\dfrac{2}{17\sqrt{17}}$ **31.** $x^2 + y^2 = 1$ **33.** $\dfrac{2}{29\sqrt{29}}$ **35.** 2 **37.** $\dfrac{785^{3/2}}{144}$

39. $(x + 2)^2 + (y - 4)^2 = 1$ **41.** $\left(2, \dfrac{5\pi}{6} + 2n\pi\right), \left(-2, \dfrac{11\pi}{6} + 2n\pi\right)$ **43.** $\left(8, \dfrac{4\pi}{3} + 2n\pi\right), \left(-8, \dfrac{\pi}{3} + 2n\pi\right)$

45. $\left(-\dfrac{5}{2}, -\dfrac{5\sqrt{3}}{2}\right)$ **47.** $(0, 0)$ **49.** $r^2\cos 2\theta + 4r\cos\theta = 9$ **51.** $x^2 + y^2 = x + y$ **53.** $(x^2 + y^2)^2 = x^2 - y^2$

55.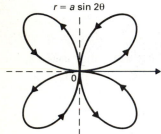
$r = a\sin 2\theta$

57.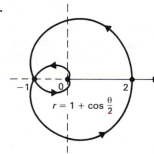
$r = 1 + \cos\dfrac{\theta}{2}$

59.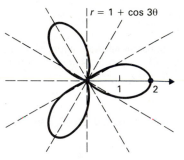
$r = 1 + \cos 3\theta$

61. $(a/\sqrt{2}, \pi/6), (a/\sqrt{2}, 5\pi/6), (a/\sqrt{2}, 7\pi/6), (a/\sqrt{2}, 11\pi/6)$ **63.** $a^2\left(1 + \dfrac{\pi}{16}\right)$ **65.** π **67.** $\sqrt{3}$ **69.** $-\sqrt{3}/7$

71. $\dfrac{\pi}{2} - \tan^{-1}\left(\dfrac{2}{\sqrt{3}}\right)$ **73.** $\pi/3$ **75.** $\dfrac{\sqrt{5}}{2}\left(e^{4\pi} - 1\right)$ **77.** ≈ 15.4038 **79.** Since $1 + f'(x)^2 \geq 1$, we have

$|\kappa(x)| = \left|-\dfrac{f''(x)}{(1 + f'(x)^2)^{3/2}}\right| \leq |f''(x)|.$ **81.** $15a/8$

CHAPTER 12

Section 12.1

1.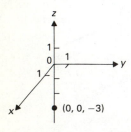
$(0, 0, -3)$

3.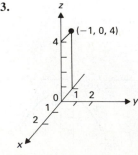
$(-1, 0, 4)$

5.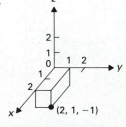
$(2, 1, -1)$

7.

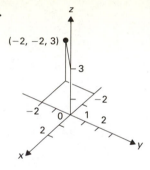

$(-2, -2, 3)$

9.

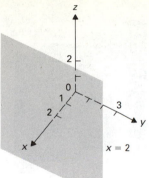

$x = 2$

11.

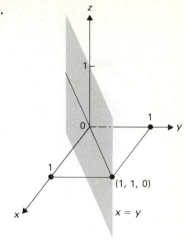

$(1, 1, 0)$

$x = y$

13.

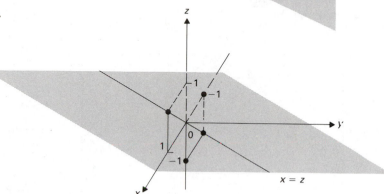

$x = z$

15.

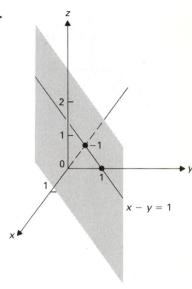

$x - y = 1$

17.

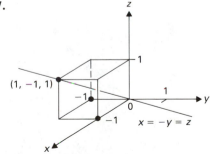

$(1, -1, 1)$

$x = -y = z$

19. $(1, -\pi, -\sqrt{10})$ **21.** $(-1, 2, 2)$ **23.** $(1, 1, 4)$ **25.** 3 **27.** 7 **29.** 5 **31.** $(x + 1)^2 + (y - 2)^2 + (z - 4)^2 = 19$

33. Center $(1, -1, 0)$, radius $\sqrt{2}$ **35.** Empty

37.

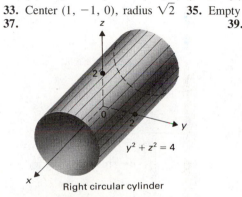

$y^2 + z^2 = 4$

Right circular cylinder

39.

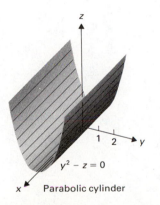

$y^2 - z = 0$

Parabolic cylinder

41.

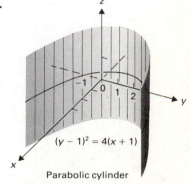

$(y - 1)^2 = 4(x + 1)$

Parabolic cylinder

43.

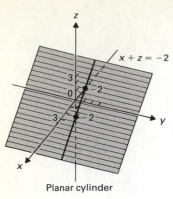

$x + z = -2$

Planar cylinder

45.

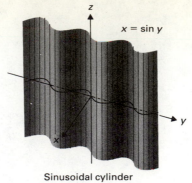

$x = \sin y$

Sinusoidal cylinder

47. 70 ft

Section 12.2

1.

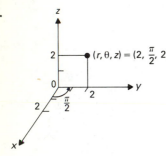

$(r, \theta, z) = (2, \frac{\pi}{2}, 2)$

3.

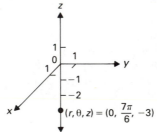

$(r, \theta, z) = (0, \frac{7\pi}{6}, -3)$

5.

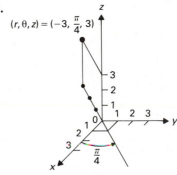

$(r, \theta, z) = (-3, \frac{\pi}{4}, 3)$

7.

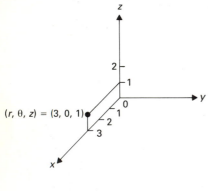

$(r, \theta, z) = (3, 0, 1)$

9.

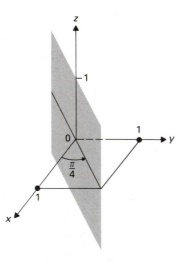

11.

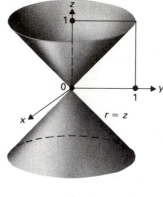

$r = z$

13.

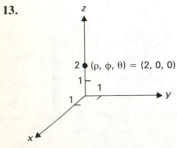

$(\rho, \phi, \theta) = (2, 0, 0)$

15.

$(\rho, \phi, \theta) = (0, \frac{\pi}{4}, \frac{\pi}{2})$

17.

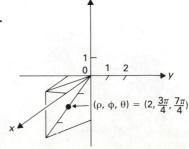

$(\rho, \phi, \theta) = (2, \frac{3\pi}{4}, \frac{7\pi}{4})$

19.
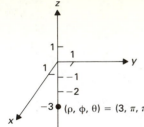
$(\rho, \phi, \theta) = (3, \pi, \pi)$

21.
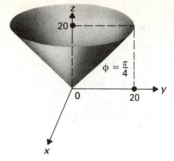
$\phi = \dfrac{\pi}{4}$

23.
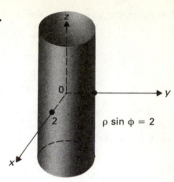
$\rho \sin \phi = 2$

25. $\left(2, -\dfrac{\pi}{3} + 2n\pi, 5\right), \left(-2, \dfrac{2\pi}{3} + 2n\pi, 5\right)$ **27.** $(-2, 2\sqrt{3}, -2)$ **29.** $\left(1, \dfrac{\pi}{2}, 1\right)$ **31.** $\left(2, \dfrac{11\pi}{6}, 2\right)$

33. $(4, \pi, \theta)$ for any θ **35.** $\left(5\sqrt{2}, \dfrac{\pi}{4}, \tan^{-1}\left(\dfrac{4}{3}\right)\right)$ **37.** $(0, 0, 0)$ **39.** $(-3, 0, -3\sqrt{3})$ **41.** $\rho = \sqrt{r^2 + z^2}$,

$\phi = \cos^{-1}\left(\dfrac{z}{\sqrt{r^2 + z^2}}\right)$, $\theta = \theta$ **43.** $\left(5, \dfrac{\pi}{2}, \tan^{-1}\left(\dfrac{4}{3}\right)\right)$ **45.** $\left(\sqrt{2}, \dfrac{7\pi}{4}, 3\right)$ **47.** $(-\sqrt{2}/2, \sqrt{3/2}, \sqrt{2})$

49. $\left(-\dfrac{3\sqrt{3}}{2}, -\dfrac{3}{2}, -5\right)$ **51.** 78π **53.** $\rho = \dfrac{\csc \phi}{\cos \theta + \sin \theta}$

Section 12.3

1.
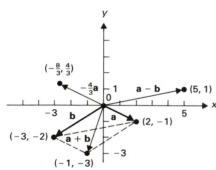
$\left(-\dfrac{8}{3}, \dfrac{4}{3}\right)$
$-\dfrac{4}{3}\mathbf{a}$ $\mathbf{a} - \mathbf{b}$ $(5, 1)$
$\mathbf{b}$ $\mathbf{a}$ $(2, -1)$
$(-3, -2)$ $\mathbf{a} + \mathbf{b}$
$(-1, -3)$

3. $-3\mathbf{i} + 9\mathbf{j} - 6\mathbf{k}$ **5.** $3\mathbf{i} + 3\mathbf{j} - 3\mathbf{k}$ **7.** $25\mathbf{i} + 9\mathbf{j} - 16\mathbf{k}$
9. $68\mathbf{i} + 36\mathbf{j} - 44\mathbf{k}$ **11.** $\sqrt{17}$ **13.** $\sqrt{17} + 3\sqrt{2}$
15. $3\sqrt{2} - \sqrt{17}$ **17.** $5\mathbf{i} - \mathbf{j}$ **19.** $7\mathbf{i} - 4\mathbf{j} - 11\mathbf{k}$ **21.** Perpendicular
23. Neither   **25.** Parallel, same direction   **27.** 3   **29.** 0   **31.** $-\dfrac{3}{2}$ **33.** None

35. -3 **37.** $\dfrac{1}{\sqrt{11}}(\mathbf{i} - \mathbf{j} + 3\mathbf{k})$ **39.** $\dfrac{1}{\sqrt{5}}(\mathbf{i} + 2\mathbf{j})$ and $\dfrac{1}{\sqrt{2}}(\mathbf{i} + \mathbf{k})$ (Many

other answers are possible.)   **41.** $\dfrac{4}{\sqrt{17}}\mathbf{i} + \dfrac{1}{\sqrt{17}}\mathbf{j}$ **43.** $x + y = -4$

45. It stays at $(-1, 4)$.
47. $(3, -2, 4) - (1, -1, 4) = (2, -1, 0) = (-2, 1, 6) - (-4, 2, 6)$; also,
$(3, -2, 4) - (-2, 1, 6) = (5, -3, -2) = (1, -1, 4) - (-4, 2, 6)$

49. $\mathbf{a} + \mathbf{b} = (a_1 + b_1)\mathbf{i} + (a_2 + b_2)\mathbf{j} + (a_3 + b_3)\mathbf{k} = (b_1 + a_1)\mathbf{i} + (b_2 + a_2)\mathbf{j} + (b_3 + a_3)\mathbf{k} = \mathbf{b} + \mathbf{a}$
51. $(r + s)\mathbf{a} = [(r + s)a_1]\mathbf{i} + [(r + s)a_2]\mathbf{j} + [(r + s)a_3]\mathbf{k} = (ra_1 + sa_1)\mathbf{i} + (ra_2 + sa_2)\mathbf{j} + (ra_3 + sa_3)\mathbf{k} = r\mathbf{a} + s\mathbf{a}$

Section 12.4

1. $\pi/2 = 90°$ **3.** $3\pi/4 = 135°$ **5.** $7\pi/12 = 105°$ **7.** $11\pi/30 = 66°$ **9.** $\pi = 180°$ **11.** $\cos^{-1}(-\tfrac{1}{3}) \approx 109.47°$
13. $\cos^{-1}(4/\sqrt{65}) \approx 60.26°$ **15.** $\cos^{-1}(\tfrac{3}{5}) \approx 53.13°$ **17.** $4\mathbf{i} - 8\mathbf{j} + 8\mathbf{k}; 12$ **19.** $\tfrac{3}{26}(\mathbf{i} + 3\mathbf{j} + 4\mathbf{k}), 3/\sqrt{26}$ **21.** $0, 0$
23. $\mathbf{i} + \mathbf{j}; \sqrt{2}$ **25.** $(\mathbf{a} - \mathbf{c}) \cdot \mathbf{b} = \left(\mathbf{a} - \dfrac{\mathbf{a} \cdot \mathbf{b}}{|\mathbf{b}|^2}\mathbf{b}\right) \cdot \mathbf{b} = \mathbf{a} \cdot \mathbf{b} - \dfrac{\mathbf{a} \cdot \mathbf{b}}{|\mathbf{b}|^2}(\mathbf{b} \cdot \mathbf{b}) = \mathbf{a} \cdot \mathbf{b} - \dfrac{\mathbf{a} \cdot \mathbf{b}}{|\mathbf{b}|^2}|\mathbf{b}|^2 = \mathbf{a} \cdot \mathbf{b} - \mathbf{a} \cdot \mathbf{b} = 0$
27. $300/\sqrt{37}$ **29.** $100/\sqrt{3}$ lb
31. (a) $(\sin \alpha)\mathbf{i} + (\cos \alpha)\mathbf{j}$ is a unit vector along the right-hand rope. Thus $\mathbf{F}_1 = T_1[(\sin \alpha)\mathbf{i} + (\cos \alpha)\mathbf{j}]$
  (b) $-T_2(\sin \beta)\mathbf{i} + T_2(\cos \beta)\mathbf{j}$
  (c) $T_1(\sin \alpha) - T_2(\sin \beta) = 0$, $T_1(\cos \alpha) + T_2(\cos \beta) = 100$
  (d) $T_1 = \dfrac{100\sqrt{2}}{\sqrt{3} + 1}$, $T_2 = \dfrac{200}{\sqrt{3} + 1}$

33. Using the hint, we show $\left|\dfrac{\mathbf{a} + \mathbf{b}}{2}\right|^2 = \left|\dfrac{\mathbf{a} - \mathbf{b}}{2}\right|^2$. $\left|\dfrac{\mathbf{a} + \mathbf{b}}{2}\right|^2 = \left(\dfrac{\mathbf{a} + \mathbf{b}}{2}\right) \cdot \left(\dfrac{\mathbf{a} + \mathbf{b}}{2}\right) = \dfrac{1}{4}(\mathbf{a} \cdot \mathbf{a} + 2\mathbf{a} \cdot \mathbf{b} + \mathbf{b} \cdot \mathbf{b})$;

$\left|\dfrac{\mathbf{a} - \mathbf{b}}{2}\right|^2 = \left(\dfrac{\mathbf{a} - \mathbf{b}}{2}\right) \cdot \left(\dfrac{\mathbf{a} - \mathbf{b}}{2}\right) = \dfrac{1}{4}(\mathbf{a} \cdot \mathbf{a} - 2\mathbf{a} \cdot \mathbf{b} + \mathbf{b} \cdot \mathbf{b})$. Now $\mathbf{a} \cdot \mathbf{b} = 0$ since the triangle is a right triangle.

Thus $\left|\dfrac{\mathbf{a} + \mathbf{b}}{2}\right|^2 = \left|\dfrac{\mathbf{a} - \mathbf{b}}{2}\right|^2 = \dfrac{1}{4}(\mathbf{a} \cdot \mathbf{a} + \mathbf{b} \cdot \mathbf{b})$.

35. Let θ_1 be the angle from $\mathbf{a}$ to $\mathbf{v} = |\mathbf{a}|\mathbf{b} + |\mathbf{b}|\mathbf{a}$ and θ_2 the angle from $\mathbf{b}$ to $\mathbf{v}$. Then

$$\cos\theta_1 = \frac{\mathbf{a}\cdot(|\mathbf{a}|\mathbf{b} + |\mathbf{b}|\mathbf{a})}{|\mathbf{a}|\,|\,|\mathbf{a}|\mathbf{b} + |\mathbf{b}|\mathbf{a}\,|} = \frac{|\mathbf{a}|(\mathbf{a}\cdot\mathbf{b}) + |\mathbf{b}|\,|\mathbf{a}|^2}{|\mathbf{a}|\,|\,|\mathbf{a}|\mathbf{b} + |\mathbf{b}|\mathbf{a}\,|} = \frac{\mathbf{a}\cdot\mathbf{b} + |\mathbf{a}|\,|\mathbf{b}|}{|\,|\mathbf{a}|\mathbf{b} + |\mathbf{b}|\mathbf{a}\,|}, \text{ and}$$

$$\cos\theta_2 = \frac{\mathbf{b}\cdot(|\mathbf{a}|\mathbf{b} + |\mathbf{b}|\mathbf{a})}{|\mathbf{b}|\,|\,|\mathbf{a}|\mathbf{b} + |\mathbf{b}|\mathbf{a}\,|} = \frac{|\mathbf{a}|\,|\mathbf{b}|^2 + |\mathbf{b}|(\mathbf{a}\cdot\mathbf{b})}{|\mathbf{b}|\,|\,|\mathbf{a}|\mathbf{b} + |\mathbf{b}|\mathbf{a}\,|} = \frac{|\mathbf{a}|\,|\mathbf{b}| + \mathbf{a}\cdot\mathbf{b}}{|\,|\mathbf{a}|\mathbf{b} + |\mathbf{b}|\mathbf{a}\,|} = \cos\theta_1. \text{ Thus } \theta_1 = \theta_2.$$

37. Let $\mathbf{a} = a_1\mathbf{i} + a_2\mathbf{j} + a_3\mathbf{k}$ and $\mathbf{b} = b_1\mathbf{i} + b_2\mathbf{j} + b_3\mathbf{k}$. Then $\mathbf{a}\cdot\mathbf{b} = a_1 b_1 + a_2 b_2 + a_3 b_3 = b_1 a_1 + b_2 a_2 + b_3 a_3 = \mathbf{b}\cdot\mathbf{a}$.
39. Let $\mathbf{a} = a_1\mathbf{i} + a_2\mathbf{j} + a_3\mathbf{k}$ and $\mathbf{b} = b_1\mathbf{i} + b_2\mathbf{j} + b_3\mathbf{k}$. Then $(r\mathbf{a})\cdot\mathbf{b} = (ra_1)b_1 + (ra_2)b_2 + (ra_3)b_3 = a_1(rb_1) + a_2(rb_2) + a_3(rb_3) = \mathbf{a}\cdot(r\mathbf{b}) = r(a_1 b_1 + a_2 b_2 + a_3 b_3) = r(\mathbf{a}\cdot\mathbf{b})$.

Section 12.5

1. -15 **3.** 15 **5.** 61 **7.** -12

9. (a)
$$\begin{vmatrix} a_1 & a_2 & a_3 \\ a_1 & a_2 & a_3 \\ c_1 & c_2 & c_3 \end{vmatrix} = a_1\begin{vmatrix} a_2 & a_3 \\ c_2 & c_3 \end{vmatrix} - a_2\begin{vmatrix} a_1 & a_3 \\ c_1 & c_3 \end{vmatrix} + a_3\begin{vmatrix} a_1 & a_2 \\ c_1 & c_2 \end{vmatrix} = a_1 a_2 c_3 - a_1 a_3 c_2 - a_1 a_2 c_3 + a_2 a_3 c_1 + a_1 a_3 c_2 - a_2 a_3 c_1 = 0$$

(b)
$$\begin{vmatrix} a_1 & a_2 & a_3 \\ b_1 & b_2 & b_3 \\ a_1 & a_2 & a_3 \end{vmatrix} = a_1\begin{vmatrix} b_2 & b_3 \\ a_2 & a_3 \end{vmatrix} - a_2\begin{vmatrix} b_1 & b_3 \\ a_1 & a_3 \end{vmatrix} + a_3\begin{vmatrix} b_1 & b_2 \\ a_1 & a_2 \end{vmatrix} = a_1 a_3 b_2 - a_1 a_2 b_3 - a_2 a_3 b_1 + a_1 a_2 b_3 + a_2 a_3 b_1 - a_1 a_3 b_2 = 0$$

11. $-6\mathbf{i} + 3\mathbf{j} + 5\mathbf{k}$ **13.** 0 **15.** $22\mathbf{i} + 18\mathbf{j} + 2\mathbf{k}$ **17.** 11 **19.** $\sqrt{374}$ **21.** $9\sqrt{2}$ **23.** 16 **25.** $\sqrt{166}/2$ **27.** 24 **29.** $\frac{20}{7}$
31. $\mathbf{a}\cdot(\mathbf{b}\times\mathbf{c}) = -8$, $\mathbf{a}\times(\mathbf{b}\times\mathbf{c}) = 2\mathbf{i} + 2\mathbf{j}$ **33.** $\mathbf{a}\cdot(\mathbf{b}\times\mathbf{c}) = -27$, $\mathbf{a}\times(\mathbf{b}\times\mathbf{c}) = 24\mathbf{i} + 54\mathbf{j} - 21\mathbf{k}$ **35.** 0 **37.** 6 **39.** $\frac{175}{6}$
41. $\frac{71}{3}$ **43.** $(\mathbf{a}\times\mathbf{b})\times\mathbf{c} = (\mathbf{a}\cdot\mathbf{c})\mathbf{b} - (\mathbf{b}\cdot\mathbf{c})\mathbf{a}$
45. Computation gives the determinant of a matrix with the first two rows the same, which is thus zero.
47. (a) The vector $\mathbf{a}\times(\mathbf{b}\times\mathbf{c})$ is perpendicular to $\mathbf{b}\times\mathbf{c}$, which is in turn a vector perpendicular to the plane containing $\mathbf{b}$ and $\mathbf{c}$. Thus $\mathbf{a}\times(\mathbf{b}\times\mathbf{c})$ lies in this plane.   **(b)** The argument is just like that in part (a).   **(c)** From parts (a) and (b), equal products $\mathbf{a}\times(\mathbf{b}\times\mathbf{c})$ and $(\mathbf{a}\times\mathbf{b})\times\mathbf{c}$ would have to be parallel to $\mathbf{b}$. A quick sketch shows that $\mathbf{a}\times(\mathbf{b}\times\mathbf{c})$ is not, in general, parallel to $\mathbf{b}$.

Section 12.6

1. $x = 3 - 8t$, $y = -2 + 4t$ **3.** $x = 4 + t$, $y = -1 - 3t$, $z = t$ **5.** $x = 1 + t$, $y = -1$, $z = 1 + 4t$ **7.** $x = 1 - 2t$, $y = -4 + 5t$ **9.** $x = 3 - 2t$, $y = -1 + 3t$, $z = 4 - 8t$ **11.** $x = 3 + 2t$, $y = -1 + t$ **13.** $x = -1 + 3t$, $y = 5 + t$
15. $x = -1 + 3t$, $y = 4 + 2t$ **17.** $x = -1 + 9t$, $y = 4 - 3t$, $z = -4t$ **19.** $x = 4 - 5t$, $y = -1 + 5t$, $z = 3 + 3t$ **21.** $x = 2$, $y = 1 + t$, $z = 4$ **23.** $x = -1 + 3t$, $y = 2 - 2t$, $z = 3 + 9t$ **25.** $(-7, 3)$ **27.** $(5, -2, 0)$ **29.** $(-1, 4)$ **31.** $(0, -2)$
33. No intersection   **35.** $\cos^{-1}(2/\sqrt{5})$ **37.** $\cos^{-1}(2/\sqrt{210})$ **39.** $\cos^{-1}(1/\sqrt{3})$
41. $\cos\alpha = d_1/\sqrt{d_1^2 + d_2^2 + d_3^2}$, $\cos\beta = d_2/\sqrt{d_1^2 + d_2^2 + d_3^2}$, $\cos\gamma = d_3/\sqrt{d_1^2 + d_2^2 + d_3^2}$ **43.** $(\frac{95}{17}, \frac{91}{17})$
45. $\left(\frac{43}{11}, \frac{58}{11}, \frac{-1}{11}\right)$ **47.** $(\frac{5}{2}, 3, 1)$ **49.** $(\frac{7}{3}, 3, 6)$
51. By the work in the text, the line segment is given by $x_1 = a_1 + (b_1 - a_1)t$, $x_2 = a_2 + (b_2 - a_2)t$, $x_3 = a_3 + (b_3 - a_3)t$ for $0 \le t \le 1$. These equations can be rewritten $x_1 = (1 - t)a_1 + tb_1$, $x_2 = (1 - t)a_2 + tb_2$, $x_3 = (1 - t)a_3 + tb_3$ for $0 \le t \le 1$.
53. $(\frac{2}{3}, \frac{5}{3}, \frac{4}{3})$ **55.** $\left(-1 - \dfrac{12}{\sqrt{10}}, 5 + \dfrac{4}{\sqrt{10}}, 6\right)$

Section 12.7

1. $x - 2y + z = -7$ **3.** $x = 2$ **5.** $4x - 2y + z = 22$ **7.** $4x - 3y + 4z = -3$ **9.** $7x - y - 2z = -5$ **11.** $2x - 3y + z = -2$
13. $x - 2y + 7z = 10$ **15.** $x + y + z = 1$ **17.** $5x + 7y - 9z = -28$ **19.** $5x + 3y - 3z = -2$ **21.** $7x - 23y - z = -101$
23. $8x + 4y + z = 6$ **25.** $35x + 14y + 5z = 141$ **27.** $5x - 12y + 22z = 13$ **29.** $x + y + z = 6$ **31.** The lines don't intersect.
33. $\pi/2$ **35.** $\pi/2 - \cos^{-1}(\frac{16}{21}) \approx 49.63°$ **37.** $(3, 6, -10)$ **39.** Empty intersection   **41.** $x = -2 + t$, $y = 1 - 2t$, $z = 4t$
43. $x = 1 + 4t$, $y = -1 - 3t$, $z = 3 + t$ **45.** $x = 3 + 4t$, $y = \frac{11}{2} + 7t$, $z = 2 - 2t$ **47.** 4 **49.** $7/\sqrt{17}$ **51.** 2
53. If $(x, y) = (a_1, a_2)$, the determinant has identical first and second rows and is thus zero. If $(x, y) = (b_1, b_2)$, the determinant has identical first and third rows and is thus zero. The linear equation gives a line and is satisfied by $(x, y) = (a_1, a_2)$ and $(x, y) = (b_1, b_2)$, so it must be the line through these two points.
55. (a) Parametric equations are $x = a_1 + d_1 t$, $y = a_2 + d_2 t$, $z = a_3 + d_3 t$. Solving for t, $t = \dfrac{x - a_1}{d_1} = \dfrac{y - a_2}{d_2} = \dfrac{z - a_3}{d_3}$.

(b) $\dfrac{x - a_1}{d_1} = \dfrac{y - a_2}{d_2}$, $\dfrac{x - a_1}{d_1} = \dfrac{z - a_3}{d_3}$, $\dfrac{y - a_2}{d_2} = \dfrac{z - a_3}{d_3}$

Section 12.8

1. A plane $z = z_0$ intersects the surface in an ellipse if $z_0 > c$. Planes $x = x_0$ and $y = y_0$ intersect the surface in hyperbolas.

3.

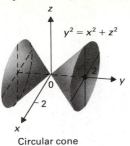

$y^2 = x^2 + z^2$

Circular cone

5.

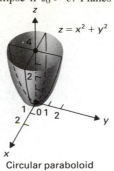

$z = x^2 + y^2$

Circular paraboloid

7.

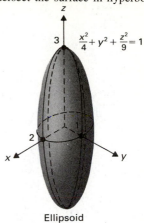

$\dfrac{x^2}{4} + y^2 + \dfrac{z^2}{9} = 1$

Ellipsoid

9.

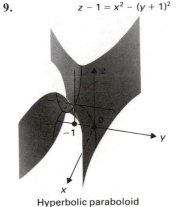

$z - 1 = x^2 - (y + 1)^2$

Hyperbolic paraboloid

11.

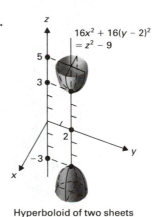

$16x^2 + 16(y - 2)^2 = z^2 - 9$

Hyperboloid of two sheets

13.

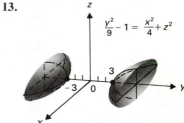

$\dfrac{y^2}{9} - 1 = \dfrac{x^2}{4} + z^2$

Hyperboloid of two sheets

15.

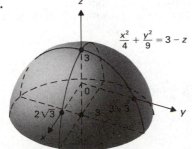

$\dfrac{x^2}{4} + \dfrac{y^2}{9} = 3 - z$

Elliptic paraboloid

17.

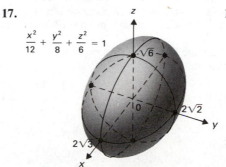

$\dfrac{x^2}{12} + \dfrac{y^2}{8} + \dfrac{z^2}{6} = 1$

Ellipsoid

19.

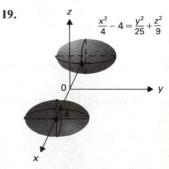

$\dfrac{x^2}{4} - 4 = \dfrac{y^2}{25} + \dfrac{z^2}{9}$

Hyperboloid of two sheets

21. Elliptic paraboloid   **23.** Hyperboloid of two sheets   **25.** Hyperbolic paraboloid   **27.** Point ellipsoid
29. Hyperbolic cylinder   **31.** Elliptic cylinder   **33.** Parabolic cylinder   **35.** Hyperboloid of two sheets

Supplementary Exercises for Chapter 12

1. $2\sqrt{22}$ **3.** $-1 \pm \sqrt{3}$ **5.** $(x + 1)^2 + (y - 1)^2 + (z - 6)^2 = 21$ **7.**

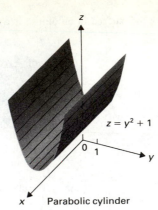

Parabolic cylinder

9. $(2, -\pi/3, -2)$ **11.** $\left(-\dfrac{3}{2\sqrt{2}}, \dfrac{2}{2\sqrt{2}}, -\dfrac{3\sqrt{3}}{2}\right)$ **13.** $(2\sqrt{2}, 3\pi/4, 2\pi/3)$ **15.** $\sqrt{14}$ **17.** $\frac{2}{3}\mathbf{i} + \frac{1}{3}\mathbf{j} - \frac{2}{3}\mathbf{k}$ **19.** $\frac{7}{2}$

21. $\theta = \cos^{-1}\left(\dfrac{-5}{14}\right)$ **23.** $\sqrt{6}$ **25.** $\frac{13}{6}(\mathbf{i} + 2\mathbf{j} + \mathbf{k})$ **27.** $11\sqrt{101}$ **29.** $-\mathbf{i} - 13\mathbf{j} - 9\mathbf{k}$ **31.** $\frac{17}{2}$ **33.** $7\mathbf{i} - 34\mathbf{j} + 14\mathbf{k}$

35. (a) 0 (b) $\mathbf{b}$ (c) $\mathbf{0}$ **37.** 52 **39.** $\frac{32}{3}$ **41.** $(11/7, 13/7, 16/7), (-5/3, 25/3, 2/3)$ **43.** $x = -1 + 4t, \; y = 5 - 6t,$
$z = 2 + 2t$ **45.** $(0, -11, -5)$ **47.** $x = t, \; y = -2t, \; z = 3t$ **49.** $2x - 3y + z = -11$ **51.** $x - 4y + 2z = 3$ **53.** $y + z = 2$
55. $14x + 13y + 10z = 45$ **57.** $(\frac{37}{16}, \frac{59}{16}, \frac{31}{16})$ **59.** $(2, 5, -4)$ **61.** $\frac{11}{3}$
63.

$\dfrac{x^2}{4} + \dfrac{y^2}{9} = 1 + z^2$

$\dfrac{x^2}{4} + \dfrac{y^2}{9} = 1 + z^2$

Hyperboloid of one sheet

65.

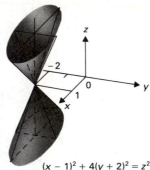

$(x - 1)^2 + 4(y + 2)^2 = z^2$

Elliptic cone

67. $1/\sqrt{299}$
69. Each equation represents a plane in space. The possibilities are as follows:
1. No solutions: (a) three parallel planes, (b) the line of intersection of two of them is parallel to the third plane.
2. Unique solution (true for "randomly" chosen equations): the line of intersection of two planes goes through the third in a unique point.
3. A line of solutions: one plane contains the line of intersection of the other two planes.
4. A plane of solutions: the three planes are the same.

71. The area is given by $|(a_1\mathbf{i} + a_2\mathbf{j} + 0\mathbf{k}) \times (b_1\mathbf{i} + b_2\mathbf{j} + 0\mathbf{k})| = \left| \begin{vmatrix} \mathbf{i} & \mathbf{j} & \mathbf{k} \\ a_1 & a_2 & 0 \\ b_1 & b_2 & 0 \end{vmatrix} \right| = |0\mathbf{i} + 0\mathbf{j} + (a_1 b_2 - a_2 b_1)\mathbf{k}| = |a_1 b_2 - a_2 b_1|,$

which is the absolute value of $\begin{vmatrix} a_1 & a_2 \\ b_1 & b_2 \end{vmatrix}$.

CHAPTER 13

Section 13.1

1. By the chain rule and Eq. (8), we have $\dfrac{d\mathbf{r}}{ds} = \dfrac{dt}{ds} \cdot \dfrac{d\mathbf{r}}{dt} = \dfrac{dt}{ds}\mathbf{v} = \dfrac{dt}{ds}\left(\dfrac{ds}{dt}\mathbf{t}\right) = \mathbf{t}$.

3. (a) $2\mathbf{i} + 3\mathbf{j}$ (b) $\sqrt{13}$ (c) $0\mathbf{i} + 0\mathbf{j} = \mathbf{0}$ **5.** (a) $-\mathbf{i}$ (b) 1 (c) $-4\mathbf{j}$ **7.** (a) $\mathbf{i}$ (b) 1 (c) $-\mathbf{i} + \mathbf{j}$ **9.** (a) $-\mathbf{i} - 2\mathbf{j}$
(b) $\sqrt{5}$ (c) $2\mathbf{i} + 6\mathbf{j}$ **11.** (a) $4\mathbf{i} + 12\mathbf{j} - \mathbf{k}$ (b) $\sqrt{161}$ (c) $2\mathbf{i} + 12\mathbf{j}$ **13.** (a) $-\mathbf{i} + \mathbf{j} + 2\mathbf{k}$ (b) $\sqrt{6}$ (c) $2\mathbf{i} - 2\mathbf{j} + 2\mathbf{k}$
15. $\mathbf{a}_{\text{tan}} = \frac{70}{17}\mathbf{i} - \frac{42}{17}\mathbf{j}$, $\mathbf{a}_{\text{nor}} = -\frac{36}{17}\mathbf{i} - \frac{60}{17}\mathbf{j}$; speed is decreasing. **17.** $\mathbf{a}_{\text{tan}} = \frac{3}{2}\mathbf{i} - \frac{3}{2}\mathbf{j}$, $\mathbf{a}_{\text{nor}} = \frac{1}{2}\mathbf{i} + \frac{1}{2}\mathbf{j}$; speed is decreasing.
19. $\mathbf{a}_{\text{tan}} = \dfrac{-4\pi}{1 + 4\pi^2}\mathbf{i} - \dfrac{8\pi^2}{1 + 4\pi^2}\mathbf{k}$, $\mathbf{a}_{\text{nor}} = \dfrac{4\pi}{1 + 4\pi^2}\mathbf{i} - 4\mathbf{j} - \dfrac{2}{1 + 4\pi^2}\mathbf{k}$; speed is decreasing. **21.** $x = 1 + t$, $y = t$, $z = 0$
23. $x = \frac{1}{2} - \frac{1}{2}t$, $y = 8 - 8t$, $z = -1 + 3t$ **25.** $x + y = 1$ **27.** $2x + 32y - 6z = 135$ **29.** $2\pi\sqrt{4a^2 + 1}$ **31.** 140 **33.** 8.23730
35. $\mathbf{a} \cdot \mathbf{b} = 2t^3 + 3t^4 + t^3 = 3t^3 + 3t^4$; $\dfrac{d(\mathbf{a} \cdot \mathbf{b})}{dt} = 9t^2 + 12t^3$. Also

$$\frac{d(\mathbf{a} \cdot \mathbf{b})}{dt} = \mathbf{a} \cdot \frac{d\mathbf{b}}{dt} + \frac{d\mathbf{a}}{dt} \cdot \mathbf{b} = [t^2\mathbf{i} - (3t + 1)\mathbf{j}] \cdot (2\mathbf{i} - 3t^2\mathbf{j}) + (2t\mathbf{i} - 3\mathbf{j}) \cdot (2t\mathbf{i} - t^3\mathbf{j})$$
$$= (2t^2 + 9t^3 + 3t^2) + (4t^2 + 3t^3) = 9t^2 + 12t^3.$$

37. By Example 8, $x = (v_0 \cos \theta)t$, $y = (v_0 \sin \theta)t - 16t^2$. Eliminating t, we have $y = (\tan \theta)x - (16/v_0^2)(\sec^2 \theta)x^2$, which is the equation of a parabola.

39. Substituting $t = (v_0 \sin \theta)/16$ from Exercise 38 into $x = (v_0 \cos \theta)t$, we have $x = \dfrac{v_0^2 \sin \theta \cos \theta}{16} = \dfrac{v_0^2 \sin 2\theta}{32}$ as range.

41. 128 ft/sec **43.** $v_0 = 80\sqrt{10}$ ft/sec; $\theta = \tan^{-1}(\frac{1}{2}) \approx 26.57°$ **45.** $\mathbf{r} = a(\cos 2\pi\beta t)\mathbf{i} + a(\sin 2\pi\beta t)\mathbf{j}$;
$\mathbf{F} = m\mathbf{a} = m\dfrac{d^2\mathbf{r}}{dt^2} = 4m\pi^2\beta^2(-\mathbf{r})$ is directed toward the center of the circle; $|\mathbf{F}| = 4m\pi^2 a\beta^2$.

Section 13.2

1. $\mathbf{i}$ **3.** 0 **5.** $9b/(4a^2)$ **7.** $-\frac{3}{5}\mathbf{i} + \frac{4}{5}\mathbf{j}$ **9.** $\frac{2}{5}$ **11.** $(2/\sqrt{5})\mathbf{i} + (1/\sqrt{5})\mathbf{j}$ **13.** $7/\sqrt{5}$ **15.** $6/(5\sqrt{5})$ **17.** 3 **19.** $2x - z = 0$
21. $2\sqrt{5}/3$ **23.** $(1/\sqrt{6})\mathbf{i} + (1/\sqrt{6})\mathbf{j} + (2/\sqrt{6})\mathbf{k}$ **25.** $(1/\sqrt{30})\mathbf{i} - (5/\sqrt{30})\mathbf{j} + (2/\sqrt{30})\mathbf{k}$ **27.** $5\sqrt{6}/3$ **29.** $\sqrt{30}/18$
31. $\sqrt{2}$ **33.** $x + y - z = -1 - \ln 2$ **35.** $2\sqrt{6}$ **37.** If $\kappa = |d\mathbf{t}/ds| = 0$, then $\mathbf{n} = \dfrac{d\mathbf{t}/ds}{|d\mathbf{t}/ds|}$ is not defined. **39.** (a) $-\sqrt{26}/13$
(b) $5/(13\sqrt{26})$

Section 13.3

1. $\mathbf{u}_r = -\mathbf{i}$, $\mathbf{u}_\theta = -\mathbf{j}$ **3.** $\sqrt{2}$ **5.** $9/(2\sqrt{2})$ **7.** (a) $\mathbf{v} = -\sqrt{3}a\mathbf{u}_r + a\mathbf{u}_\theta$, $\mathbf{a} = -2a\mathbf{u}_r - 2\sqrt{3}a\mathbf{u}_\theta$ (b) $\mathbf{v} = -\sqrt{3}a\mathbf{i} - a\mathbf{j}$,
$\mathbf{a} = 2a\mathbf{i} - 2\sqrt{3}a\mathbf{j}$ **9.** 0 **11.** $\mathbf{u}_r = -\dfrac{\sqrt{3}}{2}\mathbf{i} + \dfrac{1}{2}\mathbf{j}$, $\mathbf{u}_\theta = -\dfrac{1}{2}\mathbf{i} - \dfrac{\sqrt{3}}{2}\mathbf{j}$ **13.** $\sqrt{3}a$ **15.** $\dfrac{\sqrt{3}}{2a}$ **17.** From Eq. (31), we have
$T = 2\pi a^{3/2}/\sqrt{GM} = (\pi/\sqrt{2GM})(2a)^{3/2}$, so T can be computed if the length $2a$ of the major axis is known. **19.** No. By Eq.
(31), $a^3/T^2 = GM/(4\pi^2)$, and the mass "M" of the sun is different from the mass "M" of Jupiter. **21.** As in Eq. (11) of the
text, we have $dA/dt = \frac{1}{2}r^2\dot\theta$, so by Exercise 20, $\frac{1}{2}r^2\dot\theta = \beta$ and $r^2\dot\theta = 2\beta$.
23. Since $\mathbf{a} = (\ddot r - r\dot\theta^2)\mathbf{u}_r + (r\ddot\theta + 2\dot r\dot\theta)\mathbf{u}_\theta$, we see from Exercise 22 that we must have $\mathbf{a} = (\ddot r - r\dot\theta^2)\mathbf{u}_r$, which is directed along
the ray joining the sun and the planet.
25. From Exercise 24 we obtain, upon multiplication by r, $\dot r[r(1 - e \cos \theta)] + r^2\dot\theta e \sin \theta = 0$, or $\dot r B + 2\beta e \sin \theta = 0$.
27. By Exercise 21, we have $r\dot\theta^2 = \dfrac{(r^2\dot\theta)^2}{r^3} = \dfrac{(2\beta)^2}{r^3} = \dfrac{4\beta^2}{r^3}$. From Exercise 26, we then have

$$\ddot r - r\dot\theta^2 = \frac{4\beta^2}{r^3} - \frac{4\beta^2}{r^2 B} - \frac{4\beta^2}{r^3} = -\frac{4\beta^2}{r^2 B} = -\frac{4\beta^2}{B} \cdot \frac{1}{r^2}.$$

29. Referring to Exercise 24, we have $2a = r|_{\theta=0} + r|_{\theta=\pi} = \dfrac{B}{1 - e} + \dfrac{B}{1 + e} = \dfrac{2B}{1 - e^2}$. Thus $B = a(1 - e^2)$. Also,

$$c = a - r|_{\theta=\pi} = \frac{B}{1 - e^2} - \frac{B}{1 + e} = \frac{B}{1 - e^2} - \frac{B(1 - e)}{1 - e^2} = \frac{Be}{1 - e^2}.$$

Then

$$b^2 = a^2 - c^2 = \frac{B^2}{(1 - e^2)^2} - \frac{B^2 e^2}{(1 - e^2)^2} = \frac{B^2}{1 - e^2} = \frac{a^2(1 - e^2)^2}{1 - e^2} = a^2(1 - e^2).$$

Thus $b = a\sqrt{1 - e^2}$.
31. Since the area of an ellipse of major axis $2a$ and minor axis $2b$ is πab, we have $\pi ab = \beta T$, so $\beta = \pi ab/T$.
33. Since by Kepler's third law we have $a^3/T^2 = k$, a constant, we have, by Exercise 32, $|\mathbf{F}| = \dfrac{4\pi^2 k}{r^2}m$, so the force per unit
mass is $4\pi^2 k/r^2$.

Section 13.4

1. 0   **3.** $\sqrt{2}/[2(1 + 2t^2)^{3/2}]$ **5.** 0   **7.** $\frac{1}{3}(2\mathbf{i} - 2\mathbf{j} + \mathbf{k})$ **9.** $\frac{2}{27}$ **11.** $(1/\sqrt{5})(2\mathbf{i} - \mathbf{k})$ **13.** $2\sqrt{5}/27$ **15.** $(1/\sqrt{5})(2\mathbf{i} - \mathbf{k})$
17. $\sqrt{5}(3\sqrt{6})$ **19.** $(1/\sqrt{3})(\mathbf{i} + \mathbf{j} - \mathbf{k})$ **21.** $\sqrt{6}$ **23.** $(1/\sqrt{13})(-3\mathbf{j} + 2\mathbf{k})$ **25.** $\frac{1}{13}$ **27.** From $\dot{\mathbf{r}} = \dot{s}\mathbf{t}$, we obtain

$$\ddot{\mathbf{r}} = \ddot{s}\mathbf{t} + \dot{s}\dot{\mathbf{t}} = \ddot{s}\mathbf{t} + \dot{s}\left(\frac{ds}{dt} \cdot \frac{d\mathbf{t}}{ds}\right) = \ddot{s}\mathbf{t} - \dot{s}^2\kappa\mathbf{n}.$$

29. (a) Since $\mathbf{t}$, $\mathbf{n}$, $\mathbf{b}$ is an orthogonal unit right-hand triple, we have from Exercise 27

$$\dot{\mathbf{r}} \times \ddot{\mathbf{r}} = \begin{vmatrix} \mathbf{t} & \mathbf{n} & \mathbf{b} \\ \dot{s} & 0 & 0 \\ \ddot{s} & \dot{s}^2\kappa & 0 \end{vmatrix} = 0\mathbf{t} + 0\mathbf{n} + \dot{s}^3\kappa\mathbf{b} = \dot{s}^3\kappa\mathbf{b}.$$

    (b) From the remarks and result in part (a) and the result in Exercise 28, we have $(\dot{\mathbf{r}} \times \ddot{\mathbf{r}}) \cdot \dddot{\mathbf{r}} =$
    $0(\dddot{s} - \dot{s}^3\kappa^2) + 0(3\dot{s}\ddot{s} + \dot{s}^2\dot{\kappa}) + (\dot{s}^3\kappa)(\dot{s}^3\kappa\tau) = \dot{s}^6\kappa^2\tau.$
31. $\boldsymbol{\delta} \times \mathbf{t} = (\tau\mathbf{t} + \kappa\mathbf{b}) \times \mathbf{t} = \tau(\mathbf{t} \times \mathbf{t}) + \kappa(\mathbf{b} \times \mathbf{t}) = \tau\mathbf{0} + \kappa\mathbf{n} = \kappa\mathbf{n}$
    $\boldsymbol{\delta} \times \mathbf{n} = (\tau\mathbf{t} + \kappa\mathbf{b}) \times \mathbf{n} = \tau(\mathbf{t} \times \mathbf{n}) + \kappa(\mathbf{b} \times \mathbf{n}) = \tau\mathbf{b} + \kappa(-\mathbf{t}) = -\kappa\mathbf{t} + \tau\mathbf{b}$
    $\boldsymbol{\delta} \times \mathbf{b} = (\tau\mathbf{t} + \kappa\mathbf{b}) \times \mathbf{b} = \tau(\mathbf{t} \times \mathbf{b}) + \kappa(\mathbf{b} \times \mathbf{b}) = \tau(-\mathbf{n}) + \kappa\mathbf{0} = -\tau\mathbf{n}$
33. The Frenet formula $d\mathbf{t}/ds = \kappa\mathbf{n}$ yields $d\mathbf{t}/ds = \mathbf{0}$ if $\kappa = 0$. But then $\mathbf{t}$ is constant, so $\mathbf{t} = c_1\mathbf{i} + c_2\mathbf{j} + c_3\mathbf{k}$. Thus $d\mathbf{r}/ds = c_1\mathbf{i} + c_2\mathbf{j} + c_3\mathbf{k}$, so in terms of the parameter s, we obtain $\mathbf{r} = (c_1s + d_1)\mathbf{i} + (c_2s + d_2)\mathbf{j} + (c_3s + d_3)\mathbf{k}$, and the curve has parametric equations $x = c_1s + d_1$, $y = c_2s + d_2$, $z = c_3s + d_3$ and is a line.
35. Following the hint, we have

$$\frac{dw}{ds} = \mathbf{t} \cdot \frac{d\bar{\mathbf{t}}}{ds} + \frac{d\mathbf{t}}{ds} \cdot \bar{\mathbf{t}} + \mathbf{n} \cdot \frac{d\bar{\mathbf{n}}}{ds} + \frac{d\mathbf{n}}{ds} \cdot \bar{\mathbf{n}} + \mathbf{b} \cdot \frac{d\bar{\mathbf{b}}}{ds} + \frac{d\mathbf{b}}{ds} \cdot \bar{\mathbf{b}}$$

$$= \mathbf{t} \cdot (\kappa\bar{\mathbf{n}}) + \kappa\mathbf{n} \cdot \bar{\mathbf{t}} + \mathbf{n} \cdot (-\kappa\bar{\mathbf{t}} + \tau\bar{\mathbf{b}}) + (-\kappa\mathbf{t} + \tau\mathbf{b}) \cdot \bar{\mathbf{n}} + \mathbf{b} \cdot (-\tau\bar{\mathbf{n}}) + (-\tau\mathbf{n}) \cdot \bar{\mathbf{b}} = 0.$$

    Thus w is a constant, and since $\mathbf{t}$, $\mathbf{n}$, and $\mathbf{b}$ are equal to their respective barred counterparts at zero, we have $w(0) = 1 + 1 + 1 = 3$. Since $\mathbf{t}$, $\bar{\mathbf{t}}$, $\mathbf{n}$, $\bar{\mathbf{n}}$, $\mathbf{b}$, $\bar{\mathbf{b}}$ are unit vectors, we have $\mathbf{t} \cdot \bar{\mathbf{t}} \leq 1$ and $\mathbf{t} \cdot \bar{\mathbf{t}} = 1$ if and only if $\mathbf{t} = \bar{\mathbf{t}}$, with similar results holding for $\mathbf{n}$ and $\bar{\mathbf{n}}$ and for $\mathbf{b}$ and $\bar{\mathbf{b}}$. Thus $w = 3$ implies $\mathbf{t} = \bar{\mathbf{t}}$, $\mathbf{n} = \bar{\mathbf{n}}$, and $\mathbf{b} = \bar{\mathbf{b}}$ for all values of the parameter s. From $\mathbf{t} = \bar{\mathbf{t}}$, we have $d\mathbf{r}/ds = d\bar{\mathbf{r}}/ds$, so $\mathbf{r}(s) = \bar{\mathbf{r}}(s) + \mathbf{c}$, and from $\mathbf{r}(0) = \bar{\mathbf{r}}(0)$, we have $\mathbf{c} = \mathbf{0}$. Thus $\mathbf{r}(s) = \bar{\mathbf{r}}(s)$, so our curves are the same in terms of the arc length parameter.

Supplementary Exercises for Chapter 13

1. $\mathbf{i} + (1/\sqrt{2})\mathbf{j}$ **3.** $1/\sqrt{2}$ **5.** $-(1/\sqrt{2})\mathbf{j}$ **7.** $-4\mathbf{i}$ **9.** $2\sqrt{2}$ **11.** $\dfrac{1}{\sqrt{2}}(-\mathbf{i} - \mathbf{j})$ **13.** $8\sqrt{2}$

15. Since $\dfrac{d^2s/dt^2}{k(ds/dt)^2} = 1$ at all times, it follows that the acceleration vector $\mathbf{a} = \dfrac{d^2s}{dt^2}\mathbf{t} + \kappa\left(\dfrac{ds}{dt}\right)^2\mathbf{n}$ has the form $c\mathbf{t} + c\mathbf{n}$ at any
time, for some number $c \neq 0$. The direction of motion is given by the vector $\mathbf{t}$, and for the angle θ between $\mathbf{t}$ and $\mathbf{a}$, we then
have $\cos\theta = \dfrac{\mathbf{a} \cdot \mathbf{t}}{|\mathbf{a}||\mathbf{t}|} = \dfrac{c}{\sqrt{2c^2}} = \dfrac{c}{\sqrt{2}|c|} = \pm\dfrac{1}{\sqrt{2}}$. Thus $\theta = 45°$ or $\theta = 135°$. Since $\mathbf{F} = m\mathbf{a}$, the force vector makes the same
angle as the acceleration vector with the direction of motion.
17. $\mathbf{i} + (3/\sqrt{2})\mathbf{j}$ **19.** $3\sqrt{2}x - y - z = \dfrac{3\sqrt{2}}{4}$ **21.** $\dfrac{1}{\sqrt{2}}(-\mathbf{j} + \mathbf{k})$ **23.** $\dfrac{1}{\sqrt{17}}(3\mathbf{i} + 2\mathbf{j} - 2\mathbf{k})$

25. $-22/\sqrt{17}$ **27.** $\frac{3}{10}$ **29.** $\frac{1}{10}$ **31.** $\dfrac{4}{17\sqrt{17}}$
33. A (possibly degenerate) ellipse, parabola, or hyperbola with focus at the center of the force field; that is, a second-degree plane curve.
35. (See Section 13.3.)   **37.** (See Section 13.1.)   **39.** The torsion τ is defined by the equation $\dfrac{d\mathbf{b}}{ds} = -\tau\mathbf{n}$.

CHAPTER 14

Section 14.1

1. Does not exist.   **3.** 0   **5.** $\frac{1}{2}$ **7.** $f_x = 3$, $f_y = 4$ **9.** $f_x = 2x$, $f_y = 2y$ **11.** $f_x = \dfrac{e^{x/y}}{y}$, $f_y = \dfrac{-xe^{x/y}}{y^2}$ **13.** $f_x = y^2 + \dfrac{6x}{y^3}$,

$f_y = 2xy - \dfrac{9x^2}{y^4}$ **15.** $f_x = (x^2 + 2xy)(2x) + (y^3 + x^2)(2x + 2y)$, $f_y = (x^2 + 2xy)(3y^2) + (y^3 + x^2)(2x)$ **17.** $f_x = \dfrac{2xy - 2xy^2}{(x^2 + y)^2}$,

$f_y = \dfrac{2y(x^2 + y) - (x^2 + y^2)}{(x^2 + y)^2}$ **19.** $f_x = 2x \sec^2(x^2 + y^2)$, $f_y = 2y \sec^2(x^2 + y^2)$ **21.** $f_x = 2xye^{xy^2} \sec(x^2y) \tan(x^2y) + y^2 e^{xy^2} \sec(x^2y)$,

$f_y = x^2 e^{xy^2} \sec(x^2y) \tan(x^2y) + 2xye^{xy^2} \sec(x^2y)$ **23.** $f_x = \dfrac{2 \cot y^2}{2x + y}$, $f_y = -2y \ln(2x + y) \csc^2 y^2 + \dfrac{\cot y^2}{2x + y}$

25. $f_x = 3y \sec^3 x \tan x + y^2$, $f_y = \sec^3 x + 2xy$ **27.** $f_x = \dfrac{y^2}{1 + x^2y^4}$, $f_y = \dfrac{2xy}{1 + x^2y^4}$ **29.** 1 **31.** 18 **33.** -48 **35.** $-\frac{1}{2}$ **37.** 0

39. $-1/z^2$ **41.** 0 **43.** 0 **45.** $f_{xy} = f_{yx} = 2x$ **47.** $f_{xy} = f_{yx} = 0$ **49.** $f_x = y \cos xy$, $f_{xx} = -y^2 \sin xy$, $f_y = x \cos xy$,

$f_{yy} = -x^2 \sin xy$, $x^2 f_{xx} = -x^2 y^2 \sin xy = y^2 f_{yy}$ **51.** 3π ft^3/(unit increase in altitude) **53.** 372π ft^3/(unit increase in r)

55. Increase the y-coordinate since f_y is greater than either f_x or f_z at $(1, 2, -2)$.

Section 14.2

1. $3x + 16y - z = 21$ **3.** $3x - y - 4z = 0$ **5.** $8x_1 - 28x_2 + x_3 = -32$ **7.** $x = 1 + 2t$, $y = -1 + t$, $z = t$ **9.** $x = 2 + 9t$,
$y = -3 + 4t$, $z = 6 - t$ **11.** $x = -3 + 3t$, $y = 2 - 8t$, $z = 5 + 5t$ **13.** $(-\frac{3}{2}, 1, \frac{13}{4})$ **15.** $(-\frac{5}{3}, \frac{2}{3}, -\frac{10}{9})$ **17.** $(2, \frac{1}{2}, 2)$,
$(-2, -\frac{1}{2}, -2)$ **19.** 1.96 **21.** $1 - (\pi/90)$ **23.** 103 in^3 **25.** 85/12 ft/sec^2 **27.** 13/32 ft^3 **29.** 2.67% **31.** $3a\%$

33. $\left(\dfrac{108}{17}, \dfrac{-28}{17} \right)$ **35.** $(\frac{4}{3}, \frac{5}{6})$ **37.** $(-\frac{1}{4}, \frac{3}{4})$

39. Multiply the first equation in (10) in the text by $g_y(a_1, b_1)$ and the second by $-f_y(a_1, b_1)$ and add. We obtain

$$(f_x g_y - f_y g_x)|_{(a_1, b_1)} \cdot \Delta x = (-f \cdot g_y + g \cdot f_y)|_{(a_1, b_1)}, \text{ so } a_2 = a_1 + \Delta x = a_1 - \left(\dfrac{f \cdot g_y - g \cdot f_y}{f_x g_y - f_y g_x} \right)\bigg|_{(a_1, b_1)}. \text{ The second formula}$$

follows similarly.

41. $(10.04950857, -0.0495085669)$ **43.** $(-0.5321332411, -1.62529471)$ **45.** $(-0.0781627166, 1.817072385)$

Section 14.3

1. 0 **3.** $z = 0$ **5.** For $\Delta y = \Delta x$, Eq. (17) at $(x_0, y_0) = (0, 0)$ becomes $\sqrt{2}|\Delta x| \, |\Delta x| = \varepsilon_1 \Delta x + \varepsilon_2 \Delta x$ or
$\sqrt{2}|\Delta x| = \Delta x(\varepsilon_1 + \varepsilon_2)$. Thus $|\varepsilon_1 + \varepsilon_2| = \sqrt{2}$, so we do not have $\varepsilon_1 \to 0$ and $\varepsilon_2 \to 0$ as $\Delta x \to 0$ and $\Delta y \to 0$. Therefore f is not
differentiable at $(0, 0)$.
7. $\varepsilon_1 = \Delta x$, $\varepsilon_2 = \Delta y$, which both approach zero as $\Delta x \to 0$ and $\Delta y \to 0$.
9. $\varepsilon_1 = \dfrac{\Delta x - \Delta y + (\Delta x)(\Delta y)}{(-1 + \Delta x)(1 + \Delta y)}$, $\varepsilon_2 = -\dfrac{\Delta y}{1 + \Delta y}$, which both approach zero as $\Delta x \to 0$ and $\Delta y \to 0$.
11. $132 \, dx + 56 \, dy$ **13.** $dx - 3 \, dy$ **15.** $(\frac{1}{2} + \pi) \, dx + (\frac{1}{4} + \pi e^{4\pi}) \, dy + (4e^{4\pi} + 2) \, dz$ **17.** 2.04 **19.** -0.05 **21.** $-6\mathbf{j}$
23. $\frac{1}{3}\mathbf{i} - \frac{1}{3}\mathbf{j} - \frac{2}{9}\mathbf{k}$ **25.** $-\frac{5}{2}\mathbf{i}$ **27.** Points on a circle with center $(-1, 0)$ and radius $\frac{1}{2}$ **29.** Points on the line $3x - 4y = -3$
31. Points on the ellipsoid $\dfrac{x^2}{4} + \dfrac{y^2}{4} + z^2 = 1$ **33.** Points in the plane $3x - y - 4z = 0$

Section 14.4

1. (a) $x = 1$, $y = 2$, $z = \frac{5}{4}$ (b) $\frac{15}{4}$ (c) $z = t^4 + (t + 1)^{-2}$ (d) $\frac{15}{4}$ **3.** (a) $x = -\sqrt{2}$, $y = \sqrt{2}$, $z = -2\sqrt{2}$
(b) $\partial z/\partial r = -3\sqrt{2}$, $\partial z/\partial \theta = 2\sqrt{2}$ (c) $z = r^3 \sin^2\theta \cos\theta$ (d) $\partial z/\partial r = -3\sqrt{2}$, $\partial z/\partial \theta = 2\sqrt{2}$ **5.** 7 **7.** -4 **9.** -31 **11.** 20
13. -1 **15.** -6 **17.** $f_x = \frac{7}{10}$, $f_y = \frac{1}{5}$ **19.** 4800π in^3/min **21.** $-300m$ units/unit time **23.** $\frac{1}{5}$ (lb/ft^2)/min **25.** $-17.008G$
27. $bc(\partial w/\partial x) = ac(\partial w/\partial y) = ab(\partial w/\partial z)$

29. We have $\dfrac{\partial w}{\partial x} = f'(u) \cdot y^2$ and $\dfrac{\partial w}{\partial y} = f'(u) \cdot 2xy$, from which the desired result follows at once.

31. We have $\dfrac{\partial w}{\partial x} = f'(u)\left(\dfrac{1}{y} \right)$ and $\dfrac{\partial w}{\partial y} = f'(u)\left(\dfrac{-x}{y^2} \right)$, from which we at once obtain the desired result.

33. $\dfrac{\partial^2 z}{\partial s^2} = \dfrac{\partial z}{\partial x} \cdot \dfrac{\partial^2 x}{\partial s^2} + \dfrac{\partial z}{\partial y} \cdot \dfrac{\partial^2 y}{\partial s^2} + \dfrac{\partial^2 z}{\partial x^2}\left(\dfrac{\partial x}{\partial s} \right)^2 + 2 \dfrac{\partial^2 z}{\partial x \, \partial y} \cdot \dfrac{\partial x}{\partial s} \cdot \dfrac{\partial y}{\partial s} + \dfrac{\partial^2 z}{\partial y^2}\left(\dfrac{\partial y}{\partial s} \right)^2$

35. Following the hint, we obtain $f_x(tx_0, ty_0) \cdot x_0 + f_y(tx_0, ty_0) \cdot y_0 = 2t \cdot f(x_0, y_0)$, so, letting $t = 1$, we obtain
$x_0 f_x(x_0, y_0) + y_0 f_y(x_0, y_0) = 2 \cdot f(x_0, y_0)$. Since x_0 and y_0 could be arbitrary, we are done.

37. Let $w = g(x, y) = f(x/y)$. Then $g(tx, ty) = t^0 g(x, y)$, so by Exercise 36, $x \cdot \dfrac{\partial w}{\partial x} + y \cdot \dfrac{\partial w}{\partial y} = 0 \cdot g(x, y) = 0$.

Section 14.5

1. $\frac{36}{5}$ 3. $\frac{5}{3}$ 5. $-32/\sqrt{5}$ 7. -9 9. $-5/(13\sqrt{10})$ 11. $9/\sqrt{10}$ 13. $\frac{28}{3}$ 15. $1/\sqrt{6}$ 17. $\frac{7}{6}$ 19. $-\sqrt{2}$ 21. $2.4/\sqrt{58}$
23. (a) $2\mathbf{i} + \mathbf{j}$ (b) $2\sqrt{5}$ (c) $-2\mathbf{i} - \mathbf{j}$ (d) $-2\sqrt{5}$ 25. (a) $-7\mathbf{i} + 2\mathbf{j} + \mathbf{k}$ (b) $\sqrt{54}/2$ (c) $7\mathbf{i} - 2\mathbf{j} - \mathbf{k}$ (d) $-\sqrt{54}/2$
27. Directional derivatives in all directions will be zero.
29. (a) Along the surface, $f(x, y, z)$ has constant value c, so it has zero rate of change in a direction tangent to the surface.
(b) For a point $\mathbf{a}$ on the surface, $\nabla f(\mathbf{a}) \cdot \mathbf{u} = 0$ for any vector $\mathbf{u}$ tangent to the surface at $\mathbf{a}$, so $\nabla f(\mathbf{a})$ is normal to the tangent plane to the surface at $\mathbf{a}$.
31. $\mathbf{i}$ 33. $\mathbf{i}$ and $-\mathbf{i}$ 35. $\mathbf{j}$ and $-\mathbf{j}$

Section 14.6

1. $2/(\pi - 2)$ 3. $-\frac{4}{7}$ 5. 2 7. $(4u^3v + 3)/(2uw^3 - 6u^2v^2)$ 9. 1 11. 0 13. $-\frac{11}{2}$ 15. $-\dfrac{x^2v\,\cos(vw)}{2x\,\sin(vw) - ue^{xu}}$ 17. 2
19. 0 21. $(xu - x^2u^2)/(2yu^3 + xv)$ 23. Tangent line: $3x - 5y = -2$; normal line: $x = 1 + 3t$, $y = 1 - 5t$ 25. Tangent line: $2x + 3y = 2$; normal line: $x = 1 + 2t$, $y = 3t$ 27. Tangent plane: $13x - 3y + 10z = 53$; normal line: $x = 2 + 13t$, $y = 1 - 3t$, $z = 3 + 10t$ 29. -5 31. $-39/\sqrt{30}$ 33. (a) 8 (b) $x^2 + y^2 = 8$ (c) $-\mathbf{i} + \mathbf{j}$ (d) $16\sqrt{2}/5$
35. (a) 21 (b) $\dfrac{27}{1 + x^2 + y^2} + \dfrac{36}{1 + y^2 + z^2} = 21$ (c) $-3\mathbf{i} + 7\mathbf{j} - 4\mathbf{k}$ (d) $2\sqrt{74}$
37. Differentiating the relation $dy/dx = -G_x/G_y$ using the quotient rule and the chain rule, we have

$$\frac{d^2y}{dx^2} = -\frac{G_y\left(G_{xx} + G_{xy} \cdot \dfrac{dy}{dx}\right) - G_x\left(G_{yx} + G_{yy} \cdot \dfrac{dy}{dx}\right)}{G_y^2}$$

$$= -\frac{G_y\left(G_{xx} + G_{xy}\dfrac{-G_x}{G_y}\right) - G_x\left(G_{yx} + G_{yy}\dfrac{-G_x}{G_y}\right)}{G_y^2} = -\frac{G_y^2G_{xx} - 2G_xG_yG_{xy} + G_x^2G_{yy}}{G_y^3}.$$

39. $-\dfrac{G_z^2G_{yy} - 2G_yG_zG_{yz} + G_y^2G_{zz}}{G_z^3}$
41. At a point (x, y) of intersection, a vector normal to the first curve is $(20x^3y - 20xy^3)\mathbf{i} + (5x^4 - 30x^2y^2 + 5y^4)\mathbf{j}$ while a vector normal to the second curve is $(5x^4 - 30x^2y^2 + 5y^4)\mathbf{i} + (-20x^3y + 20xy^3)\mathbf{j}$. These vectors are perpendicular, for their dot product is zero.
43. At a point (x, y, z) of intersection, a vector normal to the first surface is $\mathbf{i} + 2y\mathbf{j} + 6z^2\mathbf{k}$, while a vector normal to the second surface is $12\mathbf{i} - (3/y)\mathbf{j} - (1/z^2)\mathbf{k}$. These vectors are perpendicular, for their dot product is zero.

Section 14.7

1. Local minimum of -4 at $(0, 0)$; no local maximum 3. Local minimum of -2 at $(-2, 1)$; no local maximum 5. No local minimum or maximum 7. Local maximum of 48 at $(0, -2)$; local minimum of -20 at $(2, 2)$ 9. Local minimum of -16 at $(\pm 2, 4)$; no local maximum 11. Local maximum of 42 at $(-8, -2)$; no local minimum 13. Local minimum of $\ln 10$ at $(-2, 3)$; no local maximum 15. Local maximum of 1 where $xy = 1 + 2n\pi$; local minimum of -1 where $xy = 1 + (2n + 1)\pi$
17. Local maximum of 3 at $(1, -1)$ 19. No local maximum; local minimum of -12 at $(\pm 1, -1, 2)$ 21. Local maximum of 11 at $(0, 0, \pm 1)$; local minimum of -7 at $(\pm 1, -2, 0)$ and $(\pm 1, 2, 0)$ 23. Local maximum of -11 at $(0, 0, 0)$ 25. Local minimum of 4 at $(x, y, z, w) = (\pm 2, 1, -2, 2)$ 27. (a) $-x^2 - y^4$ (b) $x^2 + y^4$ (c) $x^2 + y^3$ (d) If $AC - B^2 = 0$, further examination of f is needed to determine its behavior at (x_0, y_0). 29. Maximum assumed at $(1, 1)$ and $(-1, -1)$; minimum assumed at $(1, -1)$ and $(-1, 1)$ 31. Maximum assumed at $(1, -1)$ and $(-1, 1)$; minimum assumed at $(0, 0)$. [Seen from $x^2 + y^2 - xy = (\frac{1}{2}x - y)^2 + \frac{3}{4}x^2$.] 33. $(5, 2, -4)$ 35. Base 6 ft by 6 ft, height 3 ft 37. 3, 3, and 3 39. $3\sqrt{2}$
41. $8abc/(3\sqrt{3})$ 43. $y = \frac{1}{4}x + 2$ 45. $y = -x + \frac{9}{5}$

Section 14.8

1. Local maximum $2\sqrt{2}$ at $(\sqrt{2}, \sqrt{2})$; local minimum $-2\sqrt{2}$ at $(-\sqrt{2}, -\sqrt{2})$ 3. Local maximum 6 at $(4, -2)$; local minimum -6 at $(-4, 2)$ 5. Local maximum 36 at $(\pm 6, 0)$; local minimum -27 at $(0, \pm 3)$ 7. Local maximum 5 at $(3, -1)$; local minimum $\frac{7}{27}$ at $(\frac{1}{3}, \frac{1}{3})$ 9. Local maximum 128 at $(12, 2, 2)$ 11. Local maximum 40 at $(6, -9, 1)$; local minimum -40 at $(-6, 9, -1)$ 13. Local maximum 4 at $(3, -1, 0)$; local minimum 0 at $(1, 1, 0)$
15. Local maximum $\dfrac{2}{3 - 2\sqrt{2}}$ at $\left(\dfrac{1}{2 - \sqrt{2}}, \dfrac{-1}{2 - \sqrt{2}}, \dfrac{-\sqrt{2}}{2 - \sqrt{2}}\right)$; local minimum $\dfrac{2}{3 + 2\sqrt{2}}$ at $\left(\dfrac{1}{2 + \sqrt{2}}, \dfrac{-1}{2 + \sqrt{2}}, \dfrac{-\sqrt{2}}{2 + \sqrt{2}}\right)$

17. $\left(\dfrac{10}{49}, \dfrac{-15}{49}, \dfrac{30}{49}\right)$ **19.** Base 6 ft by 6 ft; height 3 ft   **21.** $\frac{32}{81}\pi a^3$ **23.** Base of rectangle $(2 - \sqrt{3})P$; sides of rectangle $\dfrac{3 - \sqrt{3}}{6}P$; sides of the triangle $\dfrac{2\sqrt{3} - 3}{3}P$ **25.** $\left(\dfrac{58}{19}, \dfrac{-18}{19}, \dfrac{44}{19}\right)$

Supplementary Exercises for Chapter 14

1. (a) $z^2 + y^2ze^{xz}$ (b) $y^2z^2e^{xz}$ (c) $2yze^{xz}$ **3.** $e^{\sin xy}[x^2y\cos(xy) + 2x]$ **5.** $2x - 4y + z = 3$ **7.** $x = 2 + t,\ y = 1 - \frac{2}{3}t,$
$z = 1 - t$ **9.** $4\,dz$ **11.** **i**   **13.** ≈ 1.84 **15.** $\approx \dfrac{2\pi + 1}{24\pi}$ ft   **17.** -5 **19.** $\frac{57}{10}$ **21.** 12   **23.** 10

25. $\dfrac{dz}{dt} = \dfrac{\partial z}{\partial x} \cdot \dfrac{d^2x}{dt^2} + \dfrac{\partial z}{\partial y} \cdot \dfrac{d^2y}{dt^2} + \dfrac{\partial^2 z}{\partial x^2}\left(\dfrac{dx}{dt}\right)^2 + 2\dfrac{\partial^2 z}{\partial x\,\partial y} \cdot \dfrac{dx}{dt} \cdot \dfrac{dy}{dt} + \dfrac{\partial^2 z}{\partial y^2}\left(\dfrac{dy}{dt}\right)^2$

27. $-2784/5$ **29.** Direction **i**, magnitude 9   **31.** Direction $\mathbf{i} + 3\mathbf{j}$, magnitude $6\sqrt{10}$ **33.** $-\dfrac{x^2 + 2yz^3}{3xz^2 + 2y^2z}$ **35.** $\frac{3}{8}$

37. $x - y + 2z = -2$ **39.** (a) $x^3y - xyz^2 + z^3 = 11$ (b) $x - 3y - 16z = -36$ **41.** Relative maximum of -1 at $(1, -1)$
43. No local extrema   **45.** 4   **47.** $\frac{17}{4}$ **49.** 9   **51.** 2   **53.** Local maxima of 13 at $(-2, 0, 3)$ and of 5 at $(2, 0, -1)$; local
minimum of 4 at $(1, \pm\sqrt{3}, 0)$
55. $f(x + \Delta x,\ y + \Delta y,\ z + \Delta z) - f(x, y, z) = [f(x + \Delta x,\ y + \Delta y,\ z + \Delta z) - f(x,\ y + \Delta y,\ z + \Delta z)]$
$$+ [f(x,\ y + \Delta y,\ z + \Delta z) - f(x, y,\ z + \Delta z)] + [f(x, y,\ z + \Delta z) - f(x, y, z)]$$
57. $(x + y) - [(x + y)^3/3!]$ **59.** $1 + xy + \frac{1}{2}x^2y^2$ **61.** -0.03 with error ≤ 0.0037

63. The binomial theorem of algebra states that $(a + b)^n = \displaystyle\sum_{i=1}^{n} \binom{n}{i} a^i b^{n-i}$, where $\binom{n}{i} = \dfrac{n!}{i!(n - i)!}$. Thus the coefficient of

$(x - x_0)^h(y - y_0)^k$ in the given expression for $T_n(x, y)$ is $\dfrac{1}{(h + k)!}\binom{h + k}{h}\dfrac{\partial^{h+k}f}{\partial x^h\,\partial y^k}\bigg|_{(x_0,\ y_0)}$. Since

$$\dfrac{1}{(h + k)!}\binom{h + k}{h} = \dfrac{1}{(h + k)!} \cdot \dfrac{(h + k)!}{h!k!} = \dfrac{1}{h!k!},$$

the result follows immediately.
65. Following the hint, $A(\Delta x)^2 + 2B(\Delta x)(\Delta y) + C(\Delta y)^2 = (\Delta x)^2(A + 2Bt + Ct^2) > 0$. Then we must have $(A + 2Bt + Ct^2) > 0$.
Since $f(x, y)$ has *continuous* second partial derivatives, we have $(f_{xx} + 2f_{xy}t + f_{yy}t^2) > 0$ for all (x, y) in some small disk
with center (x_0, y_0). Therefore $(\Delta x)^2(f_{xx} + 2f_{xy}t + f_{yy}t^2) \geq 0$ at $x = x_0 + c(\Delta x),\ y = y_0 + c(t \cdot \Delta x)$ for all Δx sufficiently
small, and $0 < c < 1$. That is, $E_1(x, y) \geq 0$ for all (x, y) sufficiently close to (x_0, y_0). If
$A(\Delta x)^2 + 2B(\Delta x)(\Delta y) + C(\Delta y)^2 < 0$, a similar argument gives $E_1(x, y) \leq 0$, and if both signs are assumed by
$A(\Delta x)^2 + 2B(\Delta x)(\Delta y) + C(\Delta y)^2$, then both signs are assumed by $E_1(x, y)$.
67. (a) Let i be as large as possible such that $\dfrac{\partial^n f}{\partial x^i\,\partial y^{n-i}}\bigg|_{(x_0,\ y_0)} \neq 0$; suppose this derivative is greater than zero. Then for

sufficiently small Δx and Δy, $\dfrac{\partial^n f}{\partial x^i\,\partial y^{n-i}}\bigg|_{(x_0+c\cdot\Delta x,\ y_0+c\cdot\Delta y)} > 0$ for $0 < c < 1$. If $|\Delta x/\Delta y|$ is sufficiently large, then

$\dfrac{\partial^n f}{\partial x^i\,\partial y^{n-i}}\bigg|_{(x_0+c\cdot\Delta x,\ y_0+c\cdot\Delta y)} (\Delta x)^i(\Delta y)^{n-i}$ dominates the other terms of $E_{n-1}(x, y)$, due to our choice of i. Now
$f(x, y) = f(x_0, y_0) + E_{n-1}(x, y)$ since partial derivatives of order less than n are zero. Taking Δx and Δy negative and
$|\Delta x/\Delta y|$ large, our dominating term of $E_{n-1}(x, y)$ is negative, since n is odd. Taking Δx and Δy positive, the dominating

term is positive. Of course, there is a similar sign change if $\dfrac{\partial^n f}{\partial x^i\,\partial y^{n-i}}\bigg|_{(x_0,\ y_0)} < 0$. Thus there is no local maximum or

minimum of $f(x, y)$ at (x_0, y_0).   (b) Of course Exercise 64(b) holds with $E_{n-1}(x, y)$ in place of $E_1(x, y)$ whether n is even
or odd, since $f(x, y) = f(x_0, y_0) + E_{n-1}(x, y)$. It is just that $E_{n-1}(x, y) \geq 0$ for all (x, y) near (x_0, y_0) is impossible if n is
odd, as is $E_{n-1}(x, y) \leq 0$ for all (x, y) near (x_0, y_0), by part (a).

CHAPTER 15

Section 15.1

1. We have integrated only over a *rectangular* region.

3. $\displaystyle\int_c^d\int_a^b f(x, y)\,dx\,dy = \int_d^c\int_a^a f(x, y)\,dx\,dy = \int_a^b\int_c^d f(x, y)\,dy\,dx = \int_b^a\int_d^c f(x, y)\,dy\,dx$

$$= -\int_c^d\int_b^a f(x, y)\,dx\,dy = -\int_d^c\int_a^b f(x, y)\,dx\,dy = -\int_a^b\int_d^c f(x, y)\,dy\,dx = -\int_b^a\int_c^d f(x, y)\,dy\,dx$$

5. ≈ 0 **7.** ≈ 5.366 **9.** ≈ 1600 **11.** 0 **13.** 1600 **15.** 13 **17.** 2 **19.** $e^3 + e - 2$ **21.** $16[1 - (1/e)]$
23. $\sqrt{2}[(\pi/4) - 1] - 2$ **25.** ≈ 0.894832 $(n = 100)$ **27.** ≈ 4.01114 $(n = 100)$ **29.** ≈ 2.1778 **31.** ≈ 23.670

Section 15.2

1. $\frac{1}{2}$ **3.** $\frac{1}{6}$ **5.** 8 **7.** $\frac{8}{15}$ **9.** $\frac{9}{4}$ **11.** 1 **13.** $\frac{1}{4}$ **15.** $\int_0^4 \int_{\sqrt{y}}^2 (\sin xy)\, dx\, dy$ **17.** $\int_0^1 \int_y^{\sqrt{y}} y \cos x\, dx\, dy$

19. $\int_{-2}^2 \int_0^{(x/2)+1} x \cos y\, dy\, dx + \int_0^1 \int_0^{1-x} x \cos^2 y\, dy\, dx$ **21.** $\int_{-4}^0 \int_{-y}^4 x^2 y^2\, dx\, dy + \int_0^4 \int_y^4 x^2 y^2\, dx\, dy$ **23.** $\int_0^1 \int_{\sin^{-1}y}^{\pi-\sin^{-1}y} x^2 y\, dx\, dy$

25. $(\pi^2/8) - 1$ **27.** $\frac{3}{2}$ **29.** 6 **31.** $\frac{40}{21}$ **33.** ≈ 0.3110367 $(n = 60)$ **35.** ≈ 1.80623 $(n = 60)$ **37.** ≈ 0.081327
39. ≈ 0.25571

Section 15.3

1.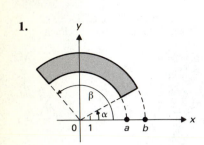

3. $\pi(b^2 - a^2)$ **5.** $2a^2 + \dfrac{a^2\pi}{4}$ **7.** $a^2\left(\dfrac{\pi}{6} - \dfrac{\sqrt{3}}{8}\right)$ **9.** $a^2\left(\sqrt{3} - \dfrac{\pi}{3}\right)$ **11.** $\dfrac{\pi a^3}{3}$

13. $a^4/8$ **15.** $\pi\left(1 - \dfrac{1}{e}\right)$ **17.** $9\left(1 - \dfrac{1}{\sqrt{2}}\right)$ **19.** $\frac{256}{3}$

Section 15.4

1. 6π **3.** $\sqrt{2}/6$ **5.** $\frac{518}{27}$ **7.** $\dfrac{1}{2}\sqrt{14} + \dfrac{5}{6}\ln\left(\dfrac{3 + \sqrt{14}}{\sqrt{5}}\right)$ **9.** $\dfrac{\pi}{6}[(1 + 4a^2)^{3/2} - 1]$ **11.** 10π **13.** $\dfrac{\pi a^2}{6}(5\sqrt{5} - 1)$

15. $\dfrac{\pi}{3}(17^{3/2} - 1)$ **17.** 37.2448 **19.** 4.05488

Section 15.5

1. (a) The first integration can be performed with respect to any one of three variables, the next with respect to any one of the two remaining variables, and the last with respect to the remaining variable to give $3 \cdot 2 \cdot 1 = 6$ orders in all. (b) $2^3 3! = 48$

3. ≈ 1.214 **5.** $\frac{95}{4}$ **7.** 32 **9.** $(\pi^2/2) - 2$ **11.** $\frac{1}{15}$ **13.** 4π **15.** 0 **17.** $\frac{1}{3}$ **19.** $\frac{2}{3}$ **21.** $\int_0^1 \int_0^{\cos^{-1}x} \int_{-2}^1 x \sin yz\, dy\, dz\, dx$

23. $\int_{-1}^0 \int_{1+x}^0 \int_{z-x-1}^0 2xy\, dy\, dz\, dx$ **25.** $\int_0^{16} \int_2^5 \int_{-\sqrt{16-z}}^0 (x + y)\, dx\, dy\, dz$ **27.** 8π **29.** 4π **31.** $(\pi/3)(2a^3 - 3a^2b + b^3)$ **33.** $3\pi/2$
35. $4096\pi/9$ **37.** $6\pi + \frac{128}{15}$ **39.** $1053\pi/4$ **41.** π

Section 15.6

1. $\dfrac{128\pi}{3}\left(1 - \dfrac{1}{\sqrt{2}}\right)$ **3.** $18\pi\left(\dfrac{2}{\sqrt{5}} + \dfrac{1}{\sqrt{2}}\right)$ **5.** $4\pi/9$ **7.** $\dfrac{\pi}{3}(2a^3 - 3a^2b + b^3)$ **9.** $4\pi a^5/5$

11. $\int_0^{2\pi} \int_0^{\pi/4} \int_0^{2\sqrt{2}} \rho^3 \sin\phi \cos\phi\, d\rho\, d\phi\, d\theta$ **13.** $\int_0^{\pi/2} \int_{\pi/6}^{\pi/3} \int_0^{3\csc\phi} \rho^4 \sin^3\phi \sin\theta \cos\theta\, d\rho\, d\phi\, d\theta$

15. $\int_{\pi/2}^\pi \int_0^{\tan^{-1}(1/3)} \int_0^{\sqrt{10}} \rho^3(\sin^2\phi)(\sin\theta - \cos\theta)\, d\rho\, d\phi\, d\theta$ **17.** $\int_0^\pi \int_{\pi/6}^{\pi/2} \int_0^{3\csc\phi} \rho^4 \sin^2\phi \cos\phi\, d\rho\, d\phi\, d\theta$

19. $\int_0^{2\pi} \int_{\pi/4}^{\pi/2} \int_{2\csc\phi}^{4\csc\phi} \rho\, d\rho\, d\phi\, d\theta$

Section 15.7

1. 17 **3.** 4 **5.** -19 **7.** 48 **9.** 8 **11.** $\frac{640}{3}$ **13.** $\frac{64}{3}$ **15.** $(4 + 3\pi)/3$ **17.** $\frac{13}{4}$ **19.** (a) $ka^2/2$ (b) $ka^3/3$

21. $\frac{5}{54}(37^{3/2} - 1)$ **23.** $k\pi a^4/2$ **25.** $M = \int_{l_1}^{l_2} \rho \cdot (x + a)\, dA = \int_{l_1}^{l_2} \rho x\, dA + a \int_{l_1}^{l_2} \rho\, dA = M_y + a \cdot m$ **27.** $\frac{2}{3}k\pi a^4$ **29.** $\frac{1}{6}k\pi a^4$

31. $\dfrac{3k\pi a^5}{10}$ **33.** $\frac{8}{15}ka^5 b$ **35.** $\dfrac{28k\pi a^5}{15}$

Section 15.8

1. A plane annular region (a disk with a hole in it) **3.** $(\frac{5}{9}, \frac{1}{2})$ **5.** $(0, \frac{3}{5})$ **7.** $\left(\dfrac{a}{3}, \dfrac{b}{3}\right)$ **9.** $(0, \frac{244}{91})$ **11.** $(\frac{12}{5}, 0)$

13. $(0, 4.7943)$ **15.** $\left(\dfrac{\pi}{2}, 0.6009\right)$ **17.** $\dfrac{3\pi - 1}{3}\sqrt{2}\rho$ **19.** (a) $\left(\dfrac{2a\sin\alpha}{3\alpha}, \dfrac{2a(1 - \cos\alpha)}{3\alpha}\right)$ (b) $\left(\dfrac{2a}{3}, 0\right)$

21. (a) $\frac{1}{6}$ (b) $(\frac{2}{3}, \frac{3}{4})$ **23.** $\left(0, \dfrac{4a}{3\pi}\right)$ **25.** $\left(0, \dfrac{3a}{2\pi}\right)$ **27.** $\left(0, 0, \dfrac{3a}{8}\right)$ **29.** $(\frac{4}{3}, 0, 0)$ **31.** (a) $\left(0, 0, \dfrac{a(n + 3)\sin^2\alpha}{2(n + 4)(1 - \cos\alpha)}\right)$

(b) $\left(0, 0, \dfrac{a\sin^2\alpha}{2(1 - \cos\alpha)}\right)$ **33.** The first moment of a body in space about the plane $z = -a$ is $M_{xy} + ma$. **35.** (a) $\dfrac{1}{\sqrt{3}}$

(b) $\dfrac{1}{\sqrt{3}}\sqrt{(a + 1)^2 - a^2}$

Section 15.9

1.
$$\begin{vmatrix} a & b & c \\ d & e & f \\ g & h & i \end{vmatrix} = a\begin{vmatrix} e & f \\ h & i \end{vmatrix} - b\begin{vmatrix} d & f \\ g & i \end{vmatrix} + c\begin{vmatrix} d & e \\ g & h \end{vmatrix}$$
$$= aei - afh - bdi + bfg + cdh - ceg$$
$$= aei - afh - dbi + dch + gbf - gce$$
$$= a\begin{vmatrix} e & h \\ f & i \end{vmatrix} - d\begin{vmatrix} b & h \\ c & i \end{vmatrix} + g\begin{vmatrix} b & e \\ c & f \end{vmatrix} = \begin{vmatrix} a & d & g \\ b & e & h \\ c & f & i \end{vmatrix}.$$

Taking $a = f_u = \partial x/\partial u$, $b = g_u = \partial y/\partial u$, and so on, we see that the determinants in Eqs. (11) and (12) are equal.
3. Jacobian 20; area change factor 20 **5.** Jacobian -8; volume change factor 8 **7.** Jacobian -48; volume change factor 48
9. (a) Recall from Section 3.9 that for small Δx, we have $\Delta y \approx \Delta y_{\text{tan}} = f'(a)\,\Delta x$, and by definition of the derivative, $\lim_{\Delta x \to 0} (\Delta y/\Delta x) = f'(a)$. Each of these relations indicates that $f'(a)$ is the local length change factor at $x = a$.
(b) If $f'(a) > 0$, then a sufficiently short positively directed line segment on the x-axis containing a is carried into a *positively* directed line segment on the y-axis containing $f(a)$. If $f'(a) < 0$, then such a line segment on the x-axis is carried into a *negatively* directed line segment on the y-axis.

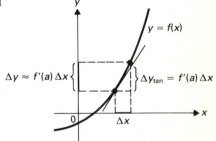

11. $\int_0^5 \int_1^3 u^4 \cdot \dfrac{1}{5}\, du\, dv = \dfrac{242}{5}$ **13.** $\int_\pi^{3\pi} \int_{-\pi}^{\pi} (u\cos^2 v)\left|-\dfrac{1}{2}\right| du\, dv = 0$ **15.** $\int_0^4 \int_3^6 u^2 e^v \cdot \dfrac{1}{5}\, du\, dv = \dfrac{63}{5}(e^4 - 1)$ **17.** Let
$x = r\cos\theta$ and $y = r\sin\theta$; $\pi a^5/10$ **19.** Let $x = u$, $y = r\sin\theta$, and $z = 2r\cos\theta$; $128\pi/5$ **21.** 12π **23.** $3(\ln 2)$ **25.** $\frac{1}{12}$

Supplementary Exercises for Chapter 15

1. 32 **3.** ≈ 3.55495 **5.** $\frac{1}{4}$ **7.** 4 **9.** $\frac{117}{2}$ **11.** $\frac{1}{6}$ **13.** πa^3 **15.** $\int_{\pi/4}^{\pi/2} \int_0^{\csc\theta} r^4\, dr\, d\theta$ **17.** $2\pi(a^2 - ab)$ **19.** 16π

21. $\int_0^4 \int_{-\sqrt{4-z}}^{\sqrt{4-z}} \int_0^{\sqrt{4-z-y^2}} yz\, dx\, dy\, dz$ **23.** $\frac{8}{3}$ **25.** $\pi/6$ **27.** $\int_{-1}^0 \int_0^{\sqrt{1-x^2}} \int_0^{x^2+y^2} \sqrt{x^2 + y^2}\, dz\, dy\, dx$ **29.** 16π

31. $\int_0^{2\pi} \int_{\pi/4}^{\pi/2} \int_0^{\cos\phi\csc^2\phi} \rho^2\sin\phi\, d\rho\, d\phi\, d\theta$ **33.** $\pi a^4/2$ **35.** $\dfrac{5k\pi a^3}{3}$ **37.** $\pi a^6/6$ **39.** $(0, 2, \frac{1}{5})$ **41.** $\frac{2541}{25}$ **43.** π **45.** $\pi^2 a^4/2$

47. $2\pi^2 a^3$ **49.** $\sum_{i,\,j=1}^{n} \left[\left(5+\dfrac{5i}{n}\right)^2 + 3\left(5+\dfrac{5i}{n}\right)\left(-2+\dfrac{4j}{n}\right)\right]\cdot\dfrac{20}{n^2}$ **51.** $\frac{2}{3}$

53. (a) $dx = \cos\theta\, dr - r\sin\theta\, d\theta$, $dy = \sin\theta\, dr + r\cos\theta\, d\theta$ (b) $rdr \wedge d\theta$

CHAPTER 16

Section 16.1

1. Exact; $\dfrac{x^3}{3} - \dfrac{y^2}{2} + C$ **3.** Not exact **5.** Exact; $\tan xy + y + C$ **7.** Not exact **9.** Exact; $x^2yz - 3xy^2 + 2xz^3 + 4z^2 + C$

11. Not exact **13.** $x\cos y + y + C$ **15.** $xe^y - \sin xy + C$ **17.** $\cos 2wz - \dfrac{1}{1+w} + \ln|z| + C$ **19.** Not exact

21. Let the given form be $dF(\mathbf{x})$. Then $\dfrac{\partial f_i}{\partial x_j} = \dfrac{\partial^2 F}{\partial x_j\, \partial x_i} = \dfrac{\partial^2 F}{\partial x_i\, \partial x_j} = \dfrac{\partial f_j}{\partial x_i}$.

23. $xy + y^2 = 6$ **25.** $x^2 + \cos xy = 1$ **27.** $x\sin y + x^2 - y^3 = 1$ **29.** $(x/y) + x^2 = 2$

Section 16.2

1. $\frac{1}{12}(5^{3/2} - 1)$ **3.** $\frac{7}{2}$ **5.** $-\frac{5}{3}$ **7.** $\dfrac{2\sqrt{5}}{3}$ **9.** 0 **11.** $\frac{486}{5}$ **13.** $\dfrac{\pi}{8} + \dfrac{5}{6} - \dfrac{2\sqrt{2}}{3}$ **15.** $58\sqrt{5}/3$ **17.** $\frac{16}{3}$ **19.** 40 **21.** $\frac{3}{2}$ **23.** $\frac{3}{2}$

25. $\dfrac{2\pi + 21}{2}$ **27.** 8 **29.** π **31.** $\frac{3}{2}$ **33.** $-9/2$

35. $\gamma_1: x = \cos t, y = \sin t, 0 \le t \le 2\pi$
$\gamma_2: x = \cos t, y = \sin t, 0 \le t \le 3\pi$

Section 16.3

1. $\frac{2}{3}$ **3.** $\frac{3}{4}$ **5.** $\frac{53}{12}$ **7.** $x = 4 - 5t, y = 3 - t$ for $0 \le t \le 1$ **9.** Initial point $(0, 0)$, terminal point $(2, 7)$; not a loop

11. Initial and terminal point $(0, 0)$; a loop **13.** Initial point $(2, 7)$, terminal point $(0, 0)$; not a loop **15.** 9 **17.** 162

19. $\mathbf{n} = \dfrac{1}{\sqrt{5}}(2\mathbf{i} - \mathbf{j})$, $ds = \sqrt{1^2 + 2^2}\, dt = \sqrt{5}\, dt$, $\displaystyle\int_\gamma (\mathbf{F}\cdot\mathbf{n})\, ds = \int_\gamma \left[\dfrac{2}{\sqrt{5}}x^2y - \dfrac{1}{\sqrt{5}}(-2x)\right] dx = \int_0^1 (4t^3 + 2t)\, dt = 2$

21. $\frac{1}{3}$ **23.** $\frac{121}{10}$ **25.** $-23/2$ **27.** 4 **29.** $-4/3$

Section 16.4

1. 18 **3.** $\frac{5}{3}$ **5.** $-\pi$ **7.** -82 **9.** 0 **11.** 2π **13.** $-\frac{1}{2}$ **15.** 0 **17.** 8 **19.** (a) $P(x, y) = 3y + k(x)$

(b) $P(x, y) = -2xy + h(y)$ **21.** (a) $P(x, y) = -(y/x)\sin xy - (1/x^2)\cos xy + k(x)$

(b) $P(x, y) = (x/y)\sin xy + (1/y^2)\cos xy + h(y)$ **23.** (a) $Q(x, y) = x^3y^2 + \frac{3}{2}x^2 + h(y)$ (b) $Q(x, y) = -\frac{1}{2}xy^4 - \frac{3}{2}y^2 + k(x)$

25. (a) $Q(x, y) = -2x + h(y)$ (b) $Q(x, y) = -3y + k(x)$

Section 16.5

1. $\sqrt{3}/6$ **3.** $\sqrt{3}/2$ **5.** 16π **7.** 0

9. Let θ be the angle between $\mathbf{F}$ and $\mathbf{n}$. We have $|\mathbf{F}\cdot\mathbf{n}| = \|\mathbf{F}\|\,\|\mathbf{n}\|\,|\cos\theta\| \le (1)(1)(1)$. Thus

$\left|\iint_\sigma (\mathbf{F}\cdot\mathbf{n})\, dS\right| \le \iint_\sigma |\mathbf{F}\cdot\mathbf{n}|\, dS \le \iint_\sigma 1\, dS = (\text{Area of } \sigma)$.

11. $\frac{19}{8}$ **13.** $\frac{4}{3}$ **15.** $128\pi/3$ **17.** $32\pi/3$

Section 16.6

1. 4π units mass/unit time **3.** -198 **5.** Both integrals equal 12. **7.** Both integrals equal 128π. **9.** Both integrals equal $\frac{1}{12}$.

11. $e^2 - \frac{5}{3}$. **13.** -96π units mass/unit time (The plates are 3 units apart.) **15.** -120 **17.** $\frac{1}{3}$ **19.** $\frac{3}{5}$ **21.** 1 **23.** $2\pi ab$

25. By the two-dimensional divergence theorem, $\dfrac{1}{2}\oint_{\partial G}(x\,dy - y\,dx) = \dfrac{1}{2}\iint_G (1+1)\,dx\,dy = \iint_G 1\cdot dx\,dy =$ Area of G.

27. 10   **29.** $\left(\dfrac{-4}{3},\dfrac{29}{8}\right)$ **31.** $\oint_{\partial G}\left(\dfrac{\partial f}{\partial x}\,dy - \dfrac{\partial f}{\partial y}\,dx\right)$ **33.** Both integrals are equal to $4\pi a^3$.   **35.** Both integrals are equal to $-64\pi/3$.   **37.** $4\pi a^3$ **39.** 32π

41. Let G be the region $a^2 \le x^2 + y^2 + z^2 \le b^2$. Then ∂G is the sphere S_b where $x^2 + y^2 + z^2 = b^2$ with outward normal, away from $(0,0,0)$, together with the sphere $-S_a$ where $x^2 + y^2 + z^2 = a^2$ with normal toward the origin. Then $\iint_{\partial G} = \iint_{S_b} + \iint_{-S_a} = \iint_{S_b} - \iint_{S_a}$. Now

$$\iint_{\partial G}(\nabla f\cdot\mathbf{n})\,dS = \iiint_G (\nabla\cdot\nabla f)\,dx\,dy\,dz\cdot\nabla f = \dfrac{-x}{(x^2+y^2+z^2)^{3/2}}\mathbf{i} - \dfrac{y}{(x^2+y^2+z^2)^{5/2}}\mathbf{j} - \dfrac{z}{(x^2+y^2+z^2)^{3/2}}\mathbf{k},$$

where $f(x,y,z) = (x^2+y^2+z^2)^{-1/2}$. Now

$$\dfrac{\partial}{\partial x}\left(\dfrac{-x}{(x^2+y^2+z^2)^{3/2}}\right) = \dfrac{-(x^2+y^2+z^2)^{3/2} + x(3/2)(x^2+y^2+z^2)^{1/2}2x}{(x^2+y^2+z^2)^3} = \dfrac{2x^2 - y^2 - z^2}{(x^2+y^2+z^2)^{5/2}}.$$

Using symmetry, $\nabla\cdot\nabla f = \dfrac{1}{(x^2+y^2+z^2)^{5/2}}[(2x^2 - y^2 - z^2) + (2y^2 - x^2 - z^2) + (2z^2 - x^2 - y^2)] = 0$. Thus $\iiint_G(\nabla\cdot\nabla f)\,dx\,dy\,dz = 0 = \iint_{\partial G}(\nabla f\cdot\mathbf{n})\,dS$, so $\iint_{S_a}(\nabla f\cdot\mathbf{n})\,dS = \iint_{S_b}(\nabla f\cdot\mathbf{n})\,dS$, as was to be shown.

43. This is immediate from the divergence theorem: $\iiint_G(\nabla^2 f)\,dx\,dy\,dz = \iiint_G(\nabla\cdot\nabla f)\,dx\,dy\,dz = \iint_{\partial G}(\nabla f\cdot\mathbf{n})\,dS$.

Section 16.7

1. F F T F T T F T F T   **3.** Both integrals equal -120.   **5.** -17

7. (a) By Green's theorem, $\iint_G(\nabla\cdot\mathbf{E})\,dx\,dy = \oint_{\partial G}(\mathbf{E}\cdot\mathbf{n})\,ds = \oint_{\partial G}(\mathbf{F}\cdot\mathbf{n})\,ds = \iint_G(\nabla\cdot\mathbf{F})\,dx\,dy$.
(b) By Green's theorem, $\iint_G(\text{curl }\mathbf{E})\,dx\,dy = \oint_{\partial G}(\mathbf{E}\cdot\mathbf{t})\,ds = \oint_{\partial G}(\mathbf{F}\cdot\mathbf{t})\,ds = \iint_G(\text{curl }\mathbf{F})\,dx\,dy$.

9. Let $P(x,y) = -y$ and $Q(x,y) = x$. By Green's theorem,
$\oint_\lambda(-y\,dx + x\,dy) = \iint_G(\partial x/\partial x - \partial(-y)/\partial y)\,dx\,dy = \iint_G 2\,dx\,dy = 2\iint_G 1\,dx\,dy = 2(\text{Area of }G)$.

11. -30π **13.** 18π **15.** 4π **17.** -2π **19.** 2π **21.** -2 **23.** -10 **25.** 2   **27.** 0

29. $\nabla f = f_x\mathbf{i} + f_y\mathbf{j}$. If ∇f is an incompressible flow, then $\nabla\cdot(\nabla f) = \dfrac{\partial(f_x)}{\partial x} + \dfrac{\partial(f_y)}{\partial y} = 0$, so $f_{xx} + f_{yy} = 0$.

31. $g(x,y) = 3x^2 y + h(y)$ **33.** 0

35. If the force field $\mathbf{F}(x,y) = P(x,y)\mathbf{i} + Q(x,y)\mathbf{j} = a\mathbf{i} + b\mathbf{j}$, then $\partial P/\partial x = \partial P/\partial y = 0$ and $\partial Q/\partial x = \partial Q/\partial y = 0$. Thus $\partial Q/\partial x - \partial P/\partial y = 0$, so the flow is irrotational, and $\partial P/\partial x + \partial Q/\partial y = 0$, so the flow is incompressible. With such a constant flow, mass is neither rotating nor collecting within any region.

37. (a) Curl $\mathbf{F} = \partial(x^2)/\partial x - \partial(2xy)/\partial y = 2x - 2x = 0$, so $\mathbf{F}$ is conservative.   (b) $u(x,y) = -x^2 y$ (c) $u(x,y) = -x^2 y + 5$
(d) 5

39. If $W(A,B)$ is the work done by the force in moving a body from A to B, then using Exercise 38, we have $u(A) - u(B) = W(A,B) = k(B) - k(A)$. Then $u(A) + k(A) = u(B) + k(B)$.

41. (a) Since $\mathbf{F} = \nabla(u) = \nabla(-H) = -\nabla H$, we see that $\mathbf{F}$ is perpendicular to level curves of $-H$, which are the same as the level curves of H, that is, the equipotential curves.   (b) The work done is $W(P_1,P_2) = H(P_2) - H(P_1)$. If P_1' is on γ_1 and P_2' is on γ_2, then $H(P_1') = H(P_1)$ and $H(P_2') = H(P_2)$ by the meaning of equipotential (level) curves. Thus $W(P_1',P_2') = H(P_2') - H(P_1') = W(P_1,P_2)$.

Section 16.8

1. $-2\mathbf{k}$ **3.** $2y^2\mathbf{i} + xy\mathbf{j} - xz\mathbf{k}$ **5.** $-\left(\dfrac{x}{y^2} + \dfrac{1}{x}\right)\mathbf{i} - \left(\dfrac{y}{z^2} + \dfrac{1}{y}\right)\mathbf{j} - \left(\dfrac{z}{x^2} + \dfrac{1}{z}\right)\mathbf{k}$ **7.** $\dfrac{1}{y}\mathbf{i} + (e^y - x\cos xy)\mathbf{k}$ **9.** Both integrals equal -512π.   **11.** Both integrals equal 0.   **13.** 12π **15.** -72π **17.** $8\pi a^3$ **19.** -405π **21.** -66π **23.** 360π

25. Curl $\mathbf{F} = 0$, so $\iint_G[(\text{curl }\mathbf{F})\cdot\mathbf{n}]\,dS = 0$, so $\oint_{\partial G}(\mathbf{F}\cdot\mathbf{t})\,ds = 0$

27. Remove a very small "disk" from G and let the remaining surface be H. Then

$$\iint_G[(\nabla\times\mathbf{F})\cdot\mathbf{n}]\,dS \approx \iint_H[(\nabla\times\mathbf{F})\cdot\mathbf{n}]\,dS,$$

since only a very small part of G was removed. In fact, the integrals can be made as nearly equal as we please by removing a sufficiently small part of G. But

$$\iint_H [(\nabla \times \mathbf{F}) \cdot \mathbf{n}]\, dS = \oint_{\partial H} (\mathbf{F} \cdot \mathbf{t})\, ds \approx 0,$$

since ∂H is a very short curve. By removing a sufficiently small part of G, we can make ∂H as short as we please, and $\oint_{\partial H} (\mathbf{F} \cdot \mathbf{t})\, ds$ as close to zero as we please. Thus, taking the limit as smaller and smaller parts of G are removed, we see that $\iint_G [(\nabla \times \mathbf{F}) \cdot \mathbf{n}]\, dS$ must be zero.

29. $\iiint_G [\nabla \cdot (\mathbf{curl\ F})]\, dx\, dy\, dz = \iint_{\partial G} [(\mathbf{curl\ F}) \cdot \mathbf{n}]\, dS$ by Exercise 28. Now ∂G is a surface that is the entire boundary of the three-dimensional region G. By Exercise 27, $\iint_{\partial G} [(\mathbf{curl\ F}) \cdot \mathbf{n}]\, dS = \iint_{\partial G} [(\nabla \times \mathbf{F}) \cdot \mathbf{n}]\, dS = 0$.

Supplementary Exercises for Chapter 16

1. $\partial P/\partial y = \partial Q/\partial x$, if the domain of definition has no holes in it.   **3.** $x^2 \sin y - x^3 y + \frac{4}{3}y^3 + C$ **5.** Impossible
7. $x^3 y + y^2 - 2x = 14$ **9.** $2\int_1^2 (t^2 - 1)^2 \sqrt{9 + 4t^2}\, dt$ **11.** $\frac{976}{27}$ **13.** $\dfrac{-2\sqrt{5}}{3} - 2$ **15.** $\ln(\sqrt{2} + 1) + \dfrac{2 - 2\sqrt{2}}{3}$ **17.** -8
19. $\pi a b$ **21.** $x = -2 \cos \pi t$, $y = 3 \sin \pi t$ for $0 \le t \le 1$ **23.** $u(x, y) = xe^y - x^2 y^3 + 12$ **25.** $\frac{11}{3}$ **27.** $243\pi/2$ **29.** $4a^5 \pi/5$
31. $224\pi/3$ **33.** 4   **35.** $\frac{128}{5}$ **37.** $\frac{27}{4}$ **39.** (a) $\partial Q/\partial x = \partial P/\partial y$ (b) $\partial P/\partial x + \partial Q/\partial y = 0$ **41.** 10   **43.** P and Q continuously differentiable and   (a) $\partial P/\partial x + \partial P/\partial y = 0$ (b) $\partial Q/\partial x - \partial P/\partial y = 0$ **45.** 0   **47.** $(xz - y)\mathbf{i} - yz\mathbf{j}$
49. (a) $2y\mathbf{i} + (3x^2 + 2xyz)\mathbf{j} - xz^2\mathbf{k}$ (b) $yz^2 + 2z$ **51.** 36π
53. Both G and G' have as boundary the circle $x^2 + y^2 = 9$ in the plane $z = 4$, but with ∂G traveled counterclockwise and $\partial G'$ clockwise as you look down on them. That is, $\partial G = -\partial G'$. Thus $\iint_G ((\mathbf{curl\ F}) \cdot \mathbf{n})\, dS = \oint_{\partial G}(\mathbf{F} \cdot \mathbf{t})\, ds = -\int_{\partial G'}(\mathbf{F} \cdot \mathbf{t})\, ds = -\iint_{G'}((\mathbf{curl\ F}) \cdot \mathbf{n})\, dS$.
55. $\partial P/\partial x = \partial Q/\partial w$, $\partial P/\partial y = \partial R/\partial w$, $\partial P/\partial z = \partial S/\partial w$, $\partial Q/\partial y = \partial R/\partial x$, $\partial Q/\partial z = \partial S/\partial x$, $\partial R/\partial z = \partial S/\partial y$
57. (a) $\dfrac{\partial P}{\partial y} = \dfrac{\partial Q}{\partial x} = \dfrac{x^2 - y^2}{(x^2 + y^2)^2}$ (b) $\displaystyle\int \dfrac{y}{x^2 + y^2}\, dx = \tan^{-1}\left(\dfrac{x}{y}\right) + A$ and $\displaystyle\int \dfrac{-x}{x^2 + y^2}\, dy = \tan^{-1}\left(\dfrac{x}{y}\right) + B$ show that $H(x, y)$ has
the desired form for $y \ne 0$. Since $\lim_{y \to 0+} \tan^{-1}(1/y) + A = (\pi/2) + A$ and $\lim_{y \to 0-} \tan^{-1}(1/y) + B = -(\pi/2) + B$, we see we must have $(\pi/2) + A = -(\pi/2) + B$, so $B = A + \pi$. We then see that we must define $H(x, 0) = A + (\pi/2)$ for $x > 0$, in order to have $H(x, y)$ continuous at $(x, 0)$ for $x > 0$. But then

$$\lim_{y \to 0+} H(-1, y) = A - \frac{\pi}{2} \quad \text{while} \quad \lim_{y \to 0-} H(-1, y) = A + \frac{3\pi}{2},$$

so it is impossible to define $H(x, 0)$ for $x < 0$ to make $H(x, y)$ continuous there.
59. $d\omega = dP \wedge dx \wedge dy \wedge dz = \left(\dfrac{\partial P}{\partial x} dx + \dfrac{\partial P}{\partial y} dy + \dfrac{\partial P}{\partial z} dz\right) \wedge dx \wedge dy \wedge dz$

$$= \frac{\partial P}{\partial x} dx \wedge dx \wedge dy \wedge dz + \frac{\partial P}{\partial y} dy \wedge dx \wedge dy \wedge dz + \frac{\partial P}{\partial z} dz \wedge dx \wedge dy \wedge dz = 0$$

since each summand contains a repeated differential, dx, dy, or dz.
61. $\int_{\partial G} \omega = \int_{\partial G} f(x) = f(b) - f(a)$ by the definition given in the exercise. Also, $\int_G d\omega = \int_{[a, b]} f(x)\, dx = \int_a^b f(x)\, dx$. Thus $f(b) - f(a) = \int_a^b f(x)\, dx$, which is the fundamental theorem of calculus.
63. $\quad d\omega = \left(\dfrac{\partial P}{\partial x} dx + \dfrac{\partial P}{\partial y} dy\right) \wedge dx + \left(\dfrac{\partial Q}{\partial x} dx + \dfrac{\partial Q}{\partial y} dy\right) \wedge dy = \left(\dfrac{\partial Q}{\partial x} - \dfrac{\partial P}{\partial y}\right) dx \wedge dy,$

$$\int_{\partial G} \omega = \oint_{\partial G} (P\, dx + Q\, dy) = \int_G d\omega = \iint_G \left(\frac{\partial Q}{\partial x} - \frac{\partial P}{\partial y}\right) dx\, dy,$$

so Eq. (1) becomes Green's theorem.
65. (a) Equation (1) becomes $\displaystyle\int_{\partial G} \omega = \iint_{\partial G} (P\, dy\, dz + Q\, dz\, dx + R\, dx\, dy) = \int_G d\omega = \iiint_G \left(\dfrac{\partial P}{\partial x} + \dfrac{\partial Q}{\partial y} + \dfrac{\partial R}{\partial z}\right) dx\, dy\, dz.$
(b) The normal unit vector $\mathbf{n}$ at (x, y, z) on G can be written as $\mathbf{n} = (\cos \alpha)\mathbf{i} + (\cos \beta)\mathbf{j} + (\cos \gamma)\mathbf{k}$, where α, β, and γ are the angles that $\mathbf{n}$ makes with the x-, y-, and z-axes. Since α is the angle between $\mathbf{n}$ and $\mathbf{i}$, it is also the angle between the tangent plane at (x, y, z) and the y,z-plane. Thus $(\cos \alpha)\, dS$ is the area of the projection of a surface element of area dS onto the y,z-plane. But this is the meaning of $dy\, dz$ in $\displaystyle\iint_{\partial G} P\, dy\, dz$. Thus $\displaystyle\iint_{\partial G} P\, dy\, dz = \iint_{\partial G} P(\cos \alpha)\, dS$, which is the contribution of P to $\displaystyle\iint_{\partial G} [(P\mathbf{i} + Q\mathbf{j} + R\mathbf{k}) \cdot \mathbf{n}]\, dS$. With similar arguments for Q and R, we see that $\displaystyle\int_{\partial G} \omega = \iint_{\partial G} [(P\mathbf{i} + Q\mathbf{j} + R\mathbf{k}) \cdot \mathbf{n}]\, dS$, so the divergence theorem is obtained.

CHAPTER 17

Section 17.1

1. $y = Ce^{x^2/2}$ **3.** $y = Ce^{-x} + \frac{3}{2}e^x$ **5.** $y = Ce^{-2x} + \frac{x^2}{2}e^{-2x} + \frac{3}{2}$ **7.** $y = C \csc x - 3x \cot x - \cot x + 3, \ 0 < x < \pi$

9. $y = Ce^{-x^2} + \frac{1}{2}$ **11.** $y = -\frac{2}{9}e^{3x} - \frac{x}{3} - \frac{7}{9}$ **13.** $y = -e^{-\tan^{-1}x} + 3$ **15.** $y = -2e^{x^2/2} - 1$

17. $y = \dfrac{1}{1 + \sin x}\left(\dfrac{x}{2} - \dfrac{\sin 2x}{4} + 6 - \dfrac{\pi}{4}\right)$ **19.** $y = e^{x^2/2}\left(C + \dfrac{x^3}{3} - \dfrac{x^5}{5 \cdot 2!} + \dfrac{x^7}{7 \cdot 2^2 \cdot 2!} - \dfrac{x^9}{9 \cdot 2^3 \cdot 3!} + \dfrac{x^{11}}{11 \cdot 2^4 \cdot 4!} - \cdots\right)$

21. $y = 2x(\ln x) + \dfrac{1 - 4(\ln 2)}{2}x$ **23.** 35 lb

25. (a) $i = e^{-Rt/L}\left(i_0 + \dfrac{1}{L}\displaystyle\int_0^t e^{Ru/L}E(u)\,du\right)$ (b) From part (a), we obtain

$$i = e^{-Rt/L}\left(i_0 + \frac{E}{R}e^{Rt/L} - \frac{E}{R}\right) = \left(i_0 - \frac{E}{R}\right)e^{-Rt/L} + \frac{E}{R}.$$

Thus $\lim_{t\to\infty} i = E/R$. (c) i approaches 0 (d) It decays exponentially from its value at $t = t_0$.

27. (a) Differentiation of $v = y^{1-n}$ yields $v' = (1 - n)y'/y^n$. Multiplying the given differential equation by $(1 - n)/y^n$, we obtain $v' + (1 - n)p(x)v = (1 - n)g(x)$, which is linear. (b) $y^2(Ce^{-2x^2} - \frac{5}{2}) = 1$

29. 0.8870206 **31.** 3.8392805 **33.** 40.77954219

Section 17.2

1. (This is a slightly tedious but routine exercise in differentiation.) **3.** $y = Ce^{-2x}$ **5.** $y = C_1e^{-5x/2} + C_2e^{-x/2}$

7. $y = C_1 + e^{-x/2}(C_2 + C_3x)$ **9.** $y = C_1 \cos \sqrt{3}x + C_2 \sin \sqrt{3}x$ **11.** $y = C_1e^x + e^{-x/2}\left(C_2 \cos \dfrac{\sqrt{3}}{2}x + C_3 \sin \dfrac{\sqrt{3}}{2}x\right)$

13. $y = C_1 + C_2 \cos 3x + C_3 \sin 3x$ **15.** $y = (C_1x + C_2)(\cos x) + (C_3x + C_4)(\sin x)$

17. $y = C_1 + C_2x + C_3e^{-2x} + C_4 \cos \sqrt{2}x + C_5 \sin \sqrt{2}x$ **19.** $y = e^{-x}(C_1 + C_2x) + e^{-x/2}\left(C_3 \cos \dfrac{\sqrt{7}}{2}x + C_4 \sin \dfrac{\sqrt{7}}{2}x\right)$

21. $y = -3e^{3x} + 4e^{2x}$ **23.** $y = -2 - 3x + 3e^x$ **25.** $y = \frac{19}{4}x - \frac{1}{54}e^{3x} + \frac{1}{54}e^{-3x}$ **27.** $y = 3 \cos 2x - 2 \sin 2x$

29. $y = e^{2x} + e^{-x}\left(\cos \sqrt{3}x - \dfrac{1}{\sqrt{3}} \sin \sqrt{3}x\right)$

Section 17.3

1. $y = C_1e^{-x} + C_2e^{3x} - \frac{4}{3}x + \frac{8}{9}$ **3.** $y = C_1e^x + C_2e^{-x} + \frac{1}{2}xe^x$ **5.** $y = C_1 + C_2e^x - x$ **7.** $y = e^{-x}(C_1 + C_2x) + \frac{1}{2}x^2e^{-x}$

9. $y = C_1e^x + C_2e^{2x} + \frac{x}{2} - \frac{3}{4}$ **11.** $y = C_1 \sin 2x + C_2 \cos 2x + \frac{1}{3} \sin x$

13. $y = C_1 + C_2e^{2x} - 3e^x - \frac{2}{5} \sin x + \frac{4}{5} \cos x$ **15.** $y = C_1 \sin x + C_2 \cos x + \frac{1}{2}xe^x - \frac{1}{2}e^x$

17. $y = C_1e^x + C_2e^{-x} - \frac{1}{2}(x \sin x + \sin x + \cos x)$ **19.** $y = C_1 + C_2e^{-4x} + \frac{1}{12}x^3 - \frac{1}{16}x^2 + \frac{1}{32}x$

21. $y = C_1 + C_2x + C_3e^{-3x} + \frac{1}{18}x^3 - \frac{1}{18}x^2 + \frac{1}{54}e^{3x}$ **23.** $y = C_1 + C_2x + C_3x^2 + C_4e^{2x} - \frac{1}{240}x^6 - \frac{3}{80}x^5 - \frac{3}{32}x^4 - \frac{3}{16}x^3$

25. Height: $s = 3512 - 512e^{-t/4} - 128t$; terminal velocity: -128 ft/sec

27. The solutions of the characteristic equation $mr^2 + cr + k = 0$ are

$$r_1 = \frac{-c + \sqrt{c^2 - 4km}}{2m} \quad \text{and} \quad r_2 = \frac{-c - \sqrt{c^2 - 4km}}{2m}.$$

Since $c^2 > 4km$, we have $r_2 < r_1 < 0$. The general solution of $m\ddot{s} + c\dot{s} + ks = 0$ is then $s = C_1e^{r_1t} + C_2e^{r_2t}$. From the initial conditions $s(0) = h$ and $\dot{s}(0) = 0$, we obtain as solution for the given problem

$$s = \frac{-r_2h}{r_1 - r_2}e^{r_1t} + \frac{r_1h}{r_1 - r_2}e^{r_2t}.$$

Since $r_1 < 0$ and $r_2 < 0$, we have $\lim_{t\to\infty} s(t) = 0$, so the body approaches its position of rest as $t \to \infty$. For s to be zero, our solution shows we would need to have $e^{(r_1-r_2)t} = r_1/r_2$. Since $r_1 - r_2 > 0$ and $0 < r_1/r_2 < 1$, we see that there is no $t > 0$ satisfying this condition, so the body never crosses its position of rest.

29. The solutions of the characteristic equation $mr^2 + cr + k = 0$ are complex, and if $a = \sqrt{4km - c^2}/2m$, the solutions of the equation have the form $s = e^{(-c/2m)t}[C_1 \cos at + C_2 \sin at]$ for constants C_1 and C_2. The desired conclusion now follows at once.

31. If $k = 0$, our equation may be written as $\ddot{s} + (c/a)s = (1/a) \sin kt = 0$. We see that any solution of this equation is of the form $s = C_1 \cos(\sqrt{c/a}\,t) + C_2 \sin(\sqrt{c/a}\,t)$ for suitable constants C_1 and C_2. The period of this oscillatory motion is $2\pi\sqrt{a}/\sqrt{c}$. If $k = \sqrt{c/a}$, then any solution of our equation is of the form $s = (C_1 + At) \cos(\sqrt{c/a}\,t) + C_2 \sin(\sqrt{c/a}\,t)$ for suitable constants C_1, C_2, and $A \neq 0$. The amplitude $\sqrt{(C_1 + At)^2 + C_2^2}$ then increases without bound as $t \to \infty$.

Supplementary Exercises for Chapter 17

1. $y = Ce^{x^2} - \frac{3}{2}$ **3.** $y \sin x = -x \cos x + \sin x + C$ **5.** $y = C_1 + e^{3x}(C_2 + C_3 x)$ **7.** $y = C_1 + C_2 x + C_3 e^{2x}$
9. $y = C_1 + C_2 \cos 2x + C_3 \sin 2x$ **11.** $y = C_1 e^x + C_2 e^{2x} + \frac{1}{10} \sin x + \frac{3}{10} \cos x$ **13.** $y = C_1 e^x + C_2 + C_3 x - \frac{1}{12}x^4 - \frac{1}{3}x^3 - x^2$

INDEX

64. $\int \sin^2 ax \, dx = \dfrac{x}{2} - \dfrac{\sin 2ax}{4a} + C$

65. $\int \cos^2 ax \, dx = \dfrac{x}{2} + \dfrac{\sin 2ax}{4a} + C$

66. $\int \sin^n ax \, dx = \dfrac{-\sin^{n-1}ax \cos ax}{na} + \dfrac{n-1}{n} \int \sin^{n-2}ax \, dx$

67. $\int \cos^n ax \, dx = \dfrac{\cos^{n-1}ax \sin ax}{na} + \dfrac{n-1}{n} \int \cos^{n-2}ax \, dx$

68. (a) $\int \sin ax \cos bx \, dx = -\dfrac{\cos(a+b)x}{2(a+b)} - \dfrac{\cos(a-b)x}{2(a-b)} + C, \quad a^2 \neq b^2$

(b) $\int \sin ax \sin bx \, dx = \dfrac{\sin(a-b)x}{2(a-b)} - \dfrac{\sin(a+b)x}{2(a+b)} + C, \quad a^2 \neq b^2$

(c) $\int \cos ax \cos bx \, dx = \dfrac{\sin(a-b)x}{2(a-b)} + \dfrac{\sin(a+b)x}{2(a+b)} + C, \quad a^2 \neq b^2$

69. $\int \sin ax \cos ax \, dx = -\dfrac{\cos 2ax}{4a} + C$

70. $\int \sin^n ax \cos ax \, dx = \dfrac{\sin^{n+1}ax}{(n+1)a} + C, \quad n \neq -1$

71. $\int \dfrac{\cos ax}{\sin ax} \, dx = \dfrac{1}{a} \ln |\sin ax| + C$

72. $\int \cos^n ax \sin ax \, dx = -\dfrac{\cos^{n+1}ax}{(n+1)a} + C, \quad n \neq -1$

73. $\int \dfrac{\sin ax}{\cos ax} \, dx = -\dfrac{1}{a} \ln |\cos ax| + C$

74. $\int \sin^n ax \cos^m ax \, dx = -\dfrac{\sin^{n-1}ax \cos^{m+1}ax}{a(m+n)} + \dfrac{n-1}{m+n} \int \sin^{n-2}ax \cos^m ax \, dx, \quad n \neq -m$

(If $n = -m$, use No. 92.)

75. $\int \sin^n ax \cos^m ax \, dx = \dfrac{\sin^{n+1}ax \cos^{m-1}ax}{a(m+n)} + \dfrac{m-1}{m+n} \int \sin^n ax \cos^{m-2}ax \, dx, \quad m \neq -n$

(If $m = -n$, use No. 93.)

76. $\int \dfrac{dx}{b + c \sin ax} = \dfrac{-2}{a\sqrt{b^2 - c^2}} \tan^{-1}\left[\sqrt{\dfrac{b-c}{b+c}} \tan\left(\dfrac{\pi}{4} - \dfrac{ax}{2}\right)\right] + C, \quad b^2 > c^2$

77. $\int \dfrac{dx}{b + c \sin ax} = \dfrac{-1}{a\sqrt{c^2 - b^2}} \ln\left|\dfrac{c + b \sin ax + \sqrt{c^2 - b^2}\cos ax}{b + c \sin ax}\right| + C, \quad b^2 < c^2$

78. $\int \dfrac{dx}{1 + \sin ax} = -\dfrac{1}{a} \tan\left(\dfrac{\pi}{4} - \dfrac{ax}{2}\right) + C$

79. $\int \dfrac{dx}{1 - \sin ax} = \dfrac{1}{a} \tan\left(\dfrac{\pi}{4} + \dfrac{ax}{2}\right) + C$

80. $\int \dfrac{dx}{b + c \cos ax} = \dfrac{2}{a\sqrt{b^2 - c^2}} \tan^{-1}\left[\sqrt{\dfrac{b-c}{b+c}} \tan\dfrac{ax}{2}\right] + C, \quad b^2 > c^2$

81. $\int \dfrac{dx}{b + c \cos ax} = \dfrac{1}{a\sqrt{c^2 - b^2}} \ln\left|\dfrac{c + b \cos ax + \sqrt{c^2 - b^2}\sin ax}{b + c \cos ax}\right| + C, \quad b^2 < c^2$

82. $\int \dfrac{dx}{1 + \cos ax} = \dfrac{1}{a} \tan\dfrac{ax}{2} + C$

83. $\int \dfrac{dx}{1 - \cos ax} = -\dfrac{1}{a} \cot\dfrac{ax}{2} + C$

84. $\int x \sin ax \, dx = \dfrac{1}{a^2} \sin ax - \dfrac{x}{a} \cos ax + C$

85. $\int x \cos ax \, dx = \dfrac{1}{a^2} \cos ax + \dfrac{x}{a} \sin ax + C$

86. $\int x^n \sin ax \, dx = -\dfrac{x^n}{a} \cos ax + \dfrac{n}{a} \int x^{n-1} \cos ax \, dx$

87. $\int x^n \cos ax \, dx = \dfrac{x^n}{a} \sin ax - \dfrac{n}{a} \int x^{n-1} \sin ax \, dx$

88. $\displaystyle\int \tan ax\, dx = -\frac{1}{a}\ln|\cos ax| + C$

89. $\displaystyle\int \cot ax\, dx = \frac{1}{a}\ln|\sin ax| + C$

90. $\displaystyle\int \tan^2 ax\, dx = \frac{1}{a}\tan ax - x + C$

91. $\displaystyle\int \cot^2 ax\, dx = -\frac{1}{a}\cot ax - x + C$

92. $\displaystyle\int \tan^n ax\, dx = \frac{\tan^{n-1}ax}{a(n-1)} - \int \tan^{n-2}ax\, dx, \qquad n \ne 1$

93. $\displaystyle\int \cot^n ax\, dx = -\frac{\cot^{n-1}ax}{a(n-1)} - \int \cot^{n-2}ax\, dx, \qquad n \ne 1$

94. $\displaystyle\int \sec ax\, dx = \frac{1}{a}\ln|\sec ax + \tan ax| + C$

95. $\displaystyle\int \csc ax\, dx = -\frac{1}{a}\ln|\csc ax + \cot ax| + C$

96. $\displaystyle\int \sec^2 ax\, dx = \frac{1}{a}\tan ax + C$

97. $\displaystyle\int \csc^2 ax\, dx = -\frac{1}{a}\cot ax + C$

98. $\displaystyle\int \sec^n ax\, dx = \frac{\sec^{n-2}ax\, \tan ax}{a(n-1)} + \frac{n-2}{n-1}\int \sec^{n-2}ax\, dx, \qquad n \ne 1$

99. $\displaystyle\int \csc^n ax\, dx = -\frac{\csc^{n-2}ax\, \cot ax}{a(n-1)} + \frac{n-2}{n-1}\int \csc^{n-2}ax\, dx, \qquad n \ne 1$

100. $\displaystyle\int \sec^n ax\, \tan ax\, dx = \frac{\sec^n ax}{na} + C, \qquad n \ne 0$

101. $\displaystyle\int \csc^n ax\, \cot ax\, dx = -\frac{\csc^n ax}{na} + C, \qquad n \ne 0$

102. $\displaystyle\int \sin^{-1}ax\, dx = x\sin^{-1}ax + \frac{1}{a}\sqrt{1 - a^2 x^2} + C$

103. $\displaystyle\int \cos^{-1}ax\, dx = x\cos^{-1}ax - \frac{1}{a}\sqrt{1 - a^2 x^2} + C$

104. $\displaystyle\int \tan^{-1}ax\, dx = x\tan^{-1}ax - \frac{1}{2a}\ln(1 + a^2 x^2) + C$

105. $\displaystyle\int x^n \sin^{-1}ax\, dx = \frac{x^{n+1}}{n+1}\sin^{-1}ax - \frac{a}{n+1}\int \frac{x^{n+1}dx}{\sqrt{1 - a^2 x^2}}, \qquad n \ne -1$

106. $\displaystyle\int x^n \cos^{-1}ax\, dx = \frac{x^{n+1}}{n+1}\cos^{-1}ax + \frac{a}{n+1}\int \frac{x^{n+1}dx}{\sqrt{1 - a^2 x^2}}, \qquad n \ne -1$

107. $\displaystyle\int x^n \tan^{-1}ax\, dx = \frac{x^{n+1}}{n+1}\tan^{-1}ax - \frac{a}{n+1}\int \frac{x^{n+1}dx}{1 + a^2 x^2}, \qquad n \ne -1$

INTEGRALS INVOLVING EXPONENTIAL AND LOGARITHM FUNCTIONS

108. $\displaystyle\int e^{ax}dx = \frac{1}{a}e^{ax} + C$

109. $\displaystyle\int b^{ax}dx = \frac{1}{a}\frac{b^{ax}}{\ln b} + C, \qquad b > 0, \; b \ne 1$

110. $\displaystyle\int xe^{ax}dx = \frac{e^{ax}}{a^2}(ax - 1) + C$

111. $\displaystyle\int x^n e^{ax}dx = \frac{1}{a}x^n e^{ax} - \frac{n}{a}\int x^{n-1}e^{ax}\, dx$

112. $\displaystyle\int x^n b^{ax}\, dx = \frac{x^n b^{ax}}{a\ln b} - \frac{n}{a\ln b}\int x^{n-1}b^{ax}\, dx, \qquad b > 0, \; b \ne 1$

113. $\displaystyle\int e^{ax}\sin bx\, dx = \frac{e^{ax}}{a^2 + b^2}(a\sin bx - b\cos bx) + C$

114. $\displaystyle\int e^{ax}\cos bx\, dx = \frac{e^{ax}}{a^2 + b^2}(a\cos bx + b\sin bx) + C$